Monograph of the Little Slit Shells

VOLUME 1
Introduction
Scissurellidae

VOLUME 2
Anatomidae
Larocheidae
Depressizonidae
Sutilizonidae
Temnocinclidae

Daniel L. Geiger

Monograph of the Little Slit Shells

Daniel L. Geiger

Santa Barbara Museum of Natural History Monographs Number 7
Studies in Biodiversity Number 5

2559 Puesta del Sol Road
Santa Barbara, CA 93105-2936 USA
http://www.sbnature.org

Library of Congress Cataloging-in-Publication Data

Geiger, Daniel L.
Monograph of the little slit shells / Daniel L. Geiger. -- 1st ed.
p. cm. -- (Studies in biodiversity ; no. 5)
Includes bibliographical references and index.
ISBN 978-0-936494-45-6 (alk. paper)
1. Scissurellidae. 2. Archaeogastropoda. I. Title.
QL430.5.S25G45 2012
594'.32--dc23
2012015342

ISBN (13) 978-0-936494-45-6

First Edition

Entire layout and design by Daniel L. Geiger. Images by Daniel L. Geiger, unless otherwise noted.
Typeset in Optima 18 point (sans serif headers) and Arno Pro 11 point (most text) 10 point (specimen records), and 9 point (synonymies).
Printed and bound in China by Everbest Printing Company through FCI Print Group.

About the author

Daniel L. Geiger grew up in Basel, Switzerland, where he started shell collecting in *Gymnasium* (high school). He soon specialized in abalone (Haliotidae) as an active member of the Swiss *Société internationale de Conchyliologie* and the German *Club Conchylia*. He received his *Diplom* in organismal biology with a *Nebenfach* (minor) in paleontology from the University of Basel, Switzerland; thesis work was carried out as a visiting post-graduate student at the University College of Swansea, Wales, Great Britain, co-advised by David G. Senn and John Ryland.

Subsequently he earned his Ph.D. in marine biology from the University of Southern California, Los Angles (USC), co-advised by Russel L. Zimmer (USC) and James H. McLean (Los Angeles County Museum of Natural History: LACM). His dissertation on systematics and evolution of abalone received the *2000 Annual Award* by the *Malacological Society of London*. After a *W. M. Keck* post-doctoral fellowship in molecular systematics at LACM, and teaching appointments as an adjunct assistant professor at USC, he took a position at the Santa Barbara Museum of Natural History in 2005, where he now is curator of malacology.

Geiger is interested in global systematic assessments of basal marine snails, with special emphasis on small-sized groups. He has published two monographs on abalone (Geiger & Poppe, 2000; Geiger & Owen, 2012) and two key works on micromollusks (Geiger *et al.*, 2007; Geiger & Ruthensteiner, 2008), amongst many other contributions. Field work has taken him from boreal St. Kilda to tropical Papua New Guinea and anywhere in between.

He is senior subject editor of Mollusca with *Zootaxa*, associate editor of *Molluscan Research*, member of the editorial board of *Malacologia*, and member of the *IUCN Mollusc Specialist Group*.

When not spending time on snails, he enjoys photography, gardening with California native plants, growing orchids, playing the viola d'amore, and spending quality time in the kitchen and above all with LACM ichthyologist Christine E. Thacker.

Fore-edge quick index

The fore-edge of the volume shows edge markings indicating the position of the various parts. It permits the reader to quickly find the beginning of the desired section.

Table of contents: Volume 1

Volume 2

Acknowledgements

This work would have been impossible to complete without the help and support from many people.

My mentors and friends kindly provided access to laboratories and gave much constructive advice on various topics: Jim McLean (LACM), David Senn (University of Basel, Switzerland), and Russel Zimmer (USC). Early encouragement was given by Bo Holmstedt † (Karolinska Institute, Stockholm, Sweden), Lukas Hottinger † (University of Basel, Switzerland), and Yehuda Mazur † (Weizmann Institute, Rehovot, Israel).

Field assistance above and below the water was given by Christine Thacker, Andrew Thompson, and the crew of NOAA *RV Shearwater*.

Personnel in SEM laboratories were most accommodating: Alicia Thompson (USC), Sue Lindsey (AMS), Jennifer Murphy (LACM). Literature assistance was provided by Dick Petit, Leslie Allen, Petar Nikolov, Rachelle Smith (University of Southern California: USC), Melinda Hayes (USC), Ricardo Pena (USC), Sue Henderson (USC), Leslie Harris (LACM: Polychaeta), Jane Myhill (Canterbury Museum), Arnold Müller (University Leipzig). Some translations were carried out by Vyacheslav Palchevskiy (USC. Russian), Stylianos Chatzimanolis (University of Tennessee at Chattanooga, formerly SBMNH. Greek), Urs-Peter Geiger (Vienna. Italian), Takenori Sasaki (UMUT. Japanese), and Karen Thacker (Palo Alto. Japanese).

Specimens were received on loan from various institutions, who generously allowed preparation of shells for SEM, radular extraction and destructive use for histology, and helped with various collection related matters:
AMNH (Paula Mikkelsen, Julia Sigwart, Marla Coppolino); AIM (Tom Trnski, Wilma Blom, Peter Quin); AMS (Ian Loch, Alison Miller, Winston Ponder, Mandy Reid, Jane Waterhouse); ANSP (Gary Rosenberg, Paul Callomon, Robert Robertson); BMSM (José Leal); BPBM (Regina Kawamoto); CASIZ (Elisabeth Kools); CCS (Noe Barrera); DMNH (Tim Pearce, Kevin Roe, Leslie Slibinski); FLMNH (Fred Thompson, John Slapcinsky); FMNH (Rüdiger Bieler, Janet Voight, Jochen Gerber); GNM (Ted von Proschwitz); GNS (John Simes, Alan Beu); HUJ (Henk Mienis); LACM (Lindsey Groves, Jim McLean); LMD (Silke Stoll); MCZ (Andreas Baldinger); MNHN (Philippe Bouchet, Virginie Héros, Philippe Maestrati, Pierre Lozouet); MMG (Ulf Linnemann, Markus Wilmsen); MV (Chris Rowley); NHMUK (Kathie Way, Amelia Campbell, Andreia Salvador); NHMW (Anita Eschner, Karl Edlinger); NMNZ (Bruce Marshall); NMR (Frans Slieker); NMSA (Dai Herbert); NMSZ (Sankurie Pye); NMW (Graham Oliver, Harriet Wood); NSMT (Kazunori Hasegawa, H. Saito, Takeshi Kubodera); MHNG (Yves Finet); MNHB (M. Landreaux); MNKD (Katrin Schniebs); MV (Chris Rowley); OMU; OSU (Thomas Watters); RM (Anthony Howell, Anthony Ricciardi); RMNH (Jeroen Goud, W. J. M. Maassen); ROM (Claire Healy, Maureen Zubowski); SAM; SBMNH (Henry Chaney, Eric Hochberg, Patricia Sadeghian, Paul Valentich-Scott); SIO (Greg Rouse, Kent Trego); SMNH (Karin Sindemark, Anders Warén); TMAG (Liz Tyler); University of Leipzig (Arnold Müller); UCMP (Carole Hickman); UMML (Nancy Voss); UMUT (Takenori Sasaki); UMZC (Richard Preece); USNM (Paul Greenhall, Jerry Harasewych, Tyjuana Nickens, Ellen Strong, Linda Ward); WAM (Shirley Slack-Smith, Corey Whisson); ZMA (Robert Moolenbeek);

ZMB (Matthias Glaubrecht, Christine Zorn); ZMBN (Christoffer Schander †); ZMUC (Ole Tendal, Tom Schiøtte); ZMUH (Angelika Brandt); ZSM (Winfried Engl, Michael Schrödl, Enrico Schwabe).

Several private collectors made specimens available for study: Emilio García (Lafayette, Louisiana, USA), Kirstie Kaiser (Puerto Vallarta, Mexico), David Kirsh (Durham, North Carolina, USA), Mathew Kosnick (Macquarie, Australia), Harry Lee (Jacksonville, Florida, USA), Felix Lorenz, Jr. (Buseck, Germany), Alwyn & Johan Marais (Center of Molluscan Studies, Pretoria, South Africa), Max Marrow (Melbourne, Australia), Gloria Pearson (Naples, Florida, USA), Bret Raines (Alamogordo, New Mexico, USA), Colin Redfern (Boca Raton, Florida, USA), Walter Renda (Italy), Stefano Rufini (Anguillara, Italy), Solly Singer (Rehovot, Israel), Carole Skoglund (Phoenix, Arizona, USA); Lynton Stephens (Sydney, Australia); John Wolff (Lancaster, Pennsylvania, USA). SEM preparation, radula extraction, and destructive use for histology was generously permitted, and voucher specimens were deposited in select institutions.

Select images were kindly made available by D. Brabant (through Virginie Héros), Peter Batson and J. Voight, A.-L. Fleddum (through Virginie Héros), Pierre Lozouet (through Virginie Héros), Philippe Maestrati (through Virginie Héros), Bruce Marshall, Jaya Nolt, Simon van Noort and Liz Hoenson (through Alwyn and Johan Marais), Jody Pesapane (Liquid Blue Media LLC), Peter Quin, Emilio Rolán, Michael Schrödl, Anders Warén, Takenori Sasaki, and Diego Zelaya.

Additional support was received from Jane Myhill (CM), Simon van Noort and Liz Hoenson (SAMCT), Dick Kilburn (NMSA), Henk Dekker, Arie Janssen, Ronald Janssen (SMF), Alan Kabat, and Michael Hollmann.

Several people read parts of the manuscript in various stages and made valuable comments: Robert Dees, Emilio García, Lindsey Groves, Carole Hickman, Harry Lee, James McLean, Bret Raines, Janet Voight, and Diego Zelaya. Colin Redfern and Christine Thacker read the entire manuscript.

Financial support was provided with visiting curatorships from AMS, BMSM, and MNHN. The production of the volumes was covered by an anonymous donor.

All the above assistance is most gratefully acknowledged. All errors herein are, however, mine.

Introduction

The present volumes are the culmination of 12 years of research. They provide a summary of knowledge on scissurellids, as well as a presentation of a large amount of novel data. The emphasis is on systematics and taxonomy, while all other aspects—from history to anatomy/histology and phylogeny—are summarized.

The first part, the introduction, is broad. It not only covers the historical developments and relevant details for using the descriptive portion, but also discusses more theoretical issues such as taxon concepts, provides a concise guide for working with these micromollusks, summarizes known biological parameters, and explores larger topics such as biodiversity patterns on a global scale. The major tables and indices (lists of species-level and genus-level taxa, distribution checklist, appendices with taxa removed from Scissurellidae and unavailable names) are placed at the end of this introduction. It keeps the systematics portion as a single, large, uninterrupted block.

The descriptive portion comprises the bulk of the two volumes. The taxa are arranged first by family, within families starting with the type genus, and within the genera alphabetically by species, the extinct species grouped at the end. For details of conventions used, see *Organization of the descriptions* below.

The literature cited section and the general index are at the end of Volume 2.

The Thumbnail Identification Cards inserted at the end of the volumes are designed as quick references to all the species (see also *Organization of descriptions: Comparisons*).

Why little slit shells?

The selection of a research topic takes careful consideration. In general, taking one's strength into account is a profitable approach, particularly when one enjoys the task (see Senn, 2010). Within molluscan systematics and taxonomy, the remaining choice is which group to treat. The little slit shells, or scissurellids, was a natural progression from monographic work on abalone (Haliotidae: Geiger & Poppe, 2000; Geiger & Owen, 2012), and broader phylogenetic work on Vetigastropoda (Geiger & Thacker, 2005; Geiger *et al.*, 2008). Scissurellids were the least known group among Vetigastropoda, and were last monographed by Pilsbry (1890) and Thiele (1912), long before the invention of the electron microscope in the 1960s (Goldstein *et al.*, 1992). Due to their small size and their scarcity in collections, scissurellids were perceived as being a difficult group to work on; hence, there was relatively little need to make arrangements with other systematic malacologists.

On the other hand, scissurellids are of no further biological consequence, except for contributing to biodiversity. They constitute just a few more species amongst the millions on the planet. Hence, it was considered a low priority project that was not going to provide any broader insights into biology. However, as one immerses oneself into a subject matter, and does not only explore it superficially, but thoroughly/holistically, many findings avail themselves, such as:

- Novel morphological structure in gastropod shells, the umbilical trough, of *Sasakiconcha elegantissima* Geiger, 2006.
- Novel radular structures, demonstrating extreme radular plasticity (Geiger & Sasaki, 2008; this study).
- New cases of sexual dimorphic shell morphologies (*Sinezona subantarctica*: this study).
- Argument for a lower size limit in gastropods based on gonad structure.
- Consequences for r/K strategy of size, and the evolution of reproductive strategies.
- Alternative patterns of bathymetric biodiversity gradients (Geiger & Sasaki, 2009).

The above brief list demonstrates that important unexpected discoveries can be made in any group if the investigator is prepared to engage in careful examination of the subject matter. Research is into the unknown, and the various biological groups are a vast source of novel discoveries. Accordingly, not the a priori assumed importance of the biological group, but rather the demonstrated ability of the investigator to carry out the task at hand is the key to making major new discoveries in any group of organisms.

I further want to stress, that by and large systematic biology is not a social undertaking, but one of determined individuals. The larger works in systematic biology are still authored by one or two individuals. There is no need for wikis, web-forums, coordination meetings or the like to complete monographs. The work is done quietly, with the least amount of unnecessary organizational overhead. I had the good fortune of collaborating with a select few individuals, such as Patty Jansen, Diego Zelaya, Takenori Sasaki, Jim McLean, Tore Høisaeter, and Bruce Marshall, who share the above approach. The Santa Barbara Museum of Natural History supported the long-term project, and recognizes the value of larger bodies of work.

History

A selection of key steps in the study and understanding of scissurellids is provided in Table 1. The ensuing narrative discusses the meandering path of history. There have been few champions of scissurellids. Most contributions have the character of occasional notes, opposed to comprehensive works. Even large expeditions result in the description of a just few more new species.

Discovery, composition. The first indications of the family are traced to Solandi (1789), a non-binominal work (Fig. 1). Scissurellids were described as "Cochleae parvae ac microscopiae," shells small and microscopic. The two illustrated specimens (vol. 1, pl. 15, fig. e–f) represent *Scissurella costata* d'Orbigny, 1824; one has the typical axial cords, the other is the smooth variant.

Linnaeus did not know scissurellids; the first species were properly described in Linnean binominal manner by d'Orbigny (1824) who also introduced the genus *Scissurella*. The species were revised three times by Philippi (1853), Pilsbry, (1890) and Thiele (1912). In the last monograph, 60 species were recognized, and few additional taxa were expected to be added. The advent of the scanning electron microscope as well as renewed deep-water sampling provided a second boost from the latter half of the 20th century onwards (e.g., Yaron, 1983; Herbert, 1986; McLean, 1989; Bandel, 1998; this study).

The group was recognized as a distinct family by Gray (1847), though his family was not considered valid by many malacologists for an extended period of time (Fischer, 1885). A key question was whether Scissurellidae should be treated separately from Pleurotomariidae, as some considered the former to be small members of the latter family (Cuvier, 1834). Some even questioned the validity of the genus *Scissurella* as distinct from *Pleurotomaria*. Deshayes (1864–1866) discussed this early debate in detail. Others placed scissurellids in different families, such as Stomatellinae (Canefri, 1869) or Haliotidae (Camerano & Lessona, 1895). The first illustrations of the animal [*Anatoma crispata* by Barrett (1856)], provided some clues and renewed interest; Clark (1856) noted the affinity with trochoideans based on the shared presence of papillated cephalic and epipodial tentacles (his

Author	**Contribution**
Solandi (1789)	First illustration of a scissurellid.
d'Orbigny (1824)	First four scissurellid taxa.
Philippi (1853)	Scissurellids in *Systematisches Conchylien Kabinet* (11 species).
Woodward (1859)	Introduction of subgenus *Anatoma*.
A. Adams (1862)	Nine new species from Japan.
Watson (1886)	*Challenger* report, with description of eight scissurellid species.
Pilsbry (1890)	Scissurellids in *Manual of Conchology* (41 species).
Schepman (1908)	Three new species from *Siboga* Expedition.
Bourne (1910)	First histological study on scissurellid (*Incisura*).
Thiele (1912)	Scissurellids in *Systematisches Conchylien Kabinet*, second edition (60 species).
Bartsch (1946)	First slit-less scissurellid (*Coronadoa*).
Batten (1975)	First SEM image of a scissurellid.
MNHN 1976–present	Many expeditions (e.g., MUSORSTOM) collecting many lots.
Yaron (1983)	Revision of Red Sea scissurellids.
Herbert (1986)	Revision of South African scissurellids.
McLean (1989)	Hydrothermal vent scissurellids (Sutilizonidae, Temnocinclidae).
Haszprunar (1989)	Anatomy and histology of vent scissurellids.
Strasoldo (1991)	Dissertation on histology and ontogeny of scissurellids.
Marshall (1993)	Recognized *Larochea* as a scissurellid, introduced *Larocheopsis*.
Bandel (1998)	Many new fossil and Recent genera and species described.
Kase & Kano (2002)	Cave dwelling *Trogloconcha* described.
Geiger (2003)	Summary of scissurellid systematics with description of new taxa.
Geiger & Thacker (2005)	Non-monophyly of *Scissurella* and *Anatoma* using molecular markers.
Nützel & Geiger (2006)	Triassic genus *Triassurella* described.
Baborka (2007)	Diploma thesis using 3D reconstruction for scissurellid anatomy.
Geiger *et al.* (2004–2012)	Series of papers describing many new genera and species.

Table 1. Brief chronological summary of key contributions in the study and understanding of scissurellids.

tentacles and lateral vibracula), and the multispiral, corneous although thin operculum. However, he doubted that *Scissurella* needed to be distinguished from *Trochus*: (pp. 270–271) "But surely this scission cannot be considered of sufficient value to entitle the animal to represent a distinct trochoidean genus, and to nullify its indisputable generic essentials with the type, *Trochus*, in which we are of the opinion it ought to be deposited, instead of burdening science with the new genus *Scissurella*."

Genera included. The second question was the genera contained in Scissurellidae. The core genera include *Scissurella, Sinezona*, and *Anatoma*, small 'archaeogastropods' with a slit or foramen. Locard (1898) also included *Seguenzia* in Scissurellidae due to the presence of the sinus, however, the genus is placed in its own family Seguenziidae in modern classifications (e.g., Beesley *et al.*, 1998), while the relationships of seguenziids and trochoidean lineages are still unsettled (e.g., Kano, 2008). Haller (1894) included *Emarginula* (today Fissurellidae) in Scissurellidae. The discovery of other microscopic archaeo/vetigastropods lead to an expansion of the family concept over time. *Incisura*, described by Hedley (1904) as a fissurellid genus, was later transferred to Scissurellidae (Thiele, 1912a, 1931). Bartsch (1946) included the slit-less *Coronadoa*. Habe & Kosuge (1964) recognized *Sukashitrochus* as a distinct genus. The newly discovered hydrothermal vents brought novel forms to light,

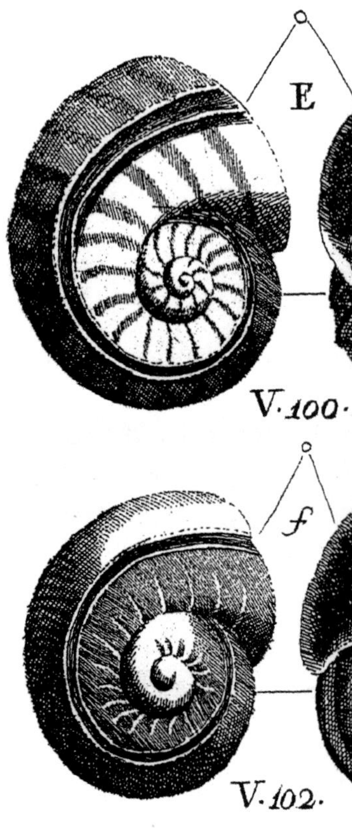

Vas C
Tab 15E
Cochleae parvae ac mycrospicae, albidae, depressiusculae, ex una parte umbilicatae, & orbe striato; ex altera spira prominula striata in gyrum, spira in plana posita, vel vix supra planum elevata, ore patulo, labio graciliore, pro indeque raro integro: melius icone E quam descriptione adumbratur. Sunt precendentibus omnibus graciliores leviorésque, ita ut in hoc Vasculo contentae n. 108 unius grani quadrantem non superent. Non videntur crescere, saltem in mediterraneo, nisi ad eum molem quam in preaicta figura exprimit status naturalis, nempe cirecellus Schemati appositus: quad si in alliis mundi pelagis grandiorus ac crassiores evaderent, quamdam cochleae scalaris specimen conficerent. In fundo maris & in concretionibus zoophyticis.

Vas CI
Cochlease, ut preacendentes, sed minimae ac prorusus mycroscopicae. Ibi.

Vas CII
Tab. 15f
Cochleae minimae ac mycroscopicae spira depressa, turbine in gyrum leviter canaliculato ac striato, ut preacedentes, a quibus differunt quod aliam partem, quae levis est & magis umbilicate, ut ostendit f. Ibi.

Figure 1. Left: Solandi's (1789) illustrations of *Scissurella costata*. Right: Transcription of descriptive part given by Solandi (1789: 21–22).

which McLean (1989) described in Temnocinclinae (*Temnocinclis, Temnozaga*) and Sutilizoninae (*Sutilizona*). Radular investigation let Marshall (1993) recognize the slit-less *Larochea* as belonging in the vicinity of Scissurellidae: Larocheinae, and not Merriidae/Lamellariidae and he added his *Larocheopsis*. Examination of shells collected in marine caves by Kase & Kano (2002) added the slit-less *Trogloconcha* to Larocheinae. Bandel (1998) erected several new genera, of which *Satondella* is considered valid. Geiger (2003) placed his monotypic *Depressizona* in Scissurellidae: Depressizoninae, although Bouchet & Rocroi (2005) synonymized the two family names without discussion, while Geiger (2009b) provided further details to justify the family level taxon. Geiger (2006b) added the most recent genus, *Sasakiconcha*, located in expedition material collected by MNHN. Figure 2 provides a graphic representation of the development of generic concepts over time.

The result was a broad family concept consisting of small to minute snails with a rhipidoglossate radula showing a serrated rachidian tooth and non-nacreous shells with a broad array of morphologies. The placement of the various lineages was largely based on radular similarities, specifically the shape of lateral tooth 5 (Scissurellidae vs. Anatomidae), and the distinctness of the separation of lateral and marginal teeth (Temnocinclidae, Sutilizonidae, Larocheidae). Other characters, such as structure of the gill or presence of a selenizone, slit, or foramen, were considered more variable. Scissurellidae was a heterogeneous assemblage, but the radular similarities were considered indicative of shared family membership. Such a view is traced to Thiele's (1931) approach to gastropod classification, which was largely based on radular features.

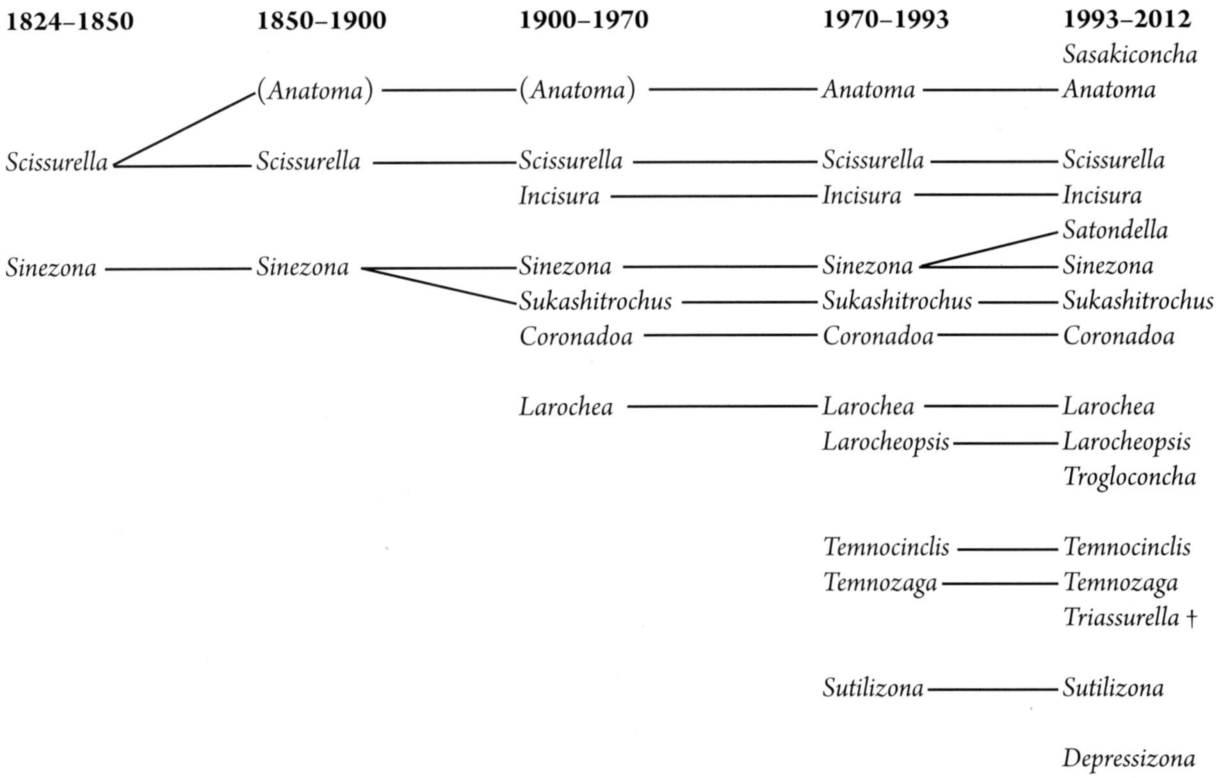

Figure 2. Graphic representation of the generic classification of scissurellids. Divergent lines show the placement of already described species into new genera. New rows indicate the introduction of new genera with new species. *Anatoma* was long considered a subgenus of *Scissurella*, which is indicated by parentheses.

Anatomical characters. Anatomical characters were also used to justify a larger scissurellid clade, often at the superfamily level: Scissurelloidea (Ponder & Lindberg, 1997; Haszprunar, 1989). Specifically, the presence of the right hypobranchial gland only was taken as a synapomorphy, while Bourne (1910) had noted a reduced left hypobranchial gland in *Incisura*, and Haszprunar (1989) in Temnocinclidae (see also Baborka, 2007 for discussion). The absence of a stomach caecum is typical of all scissurellids. The main challenge with using anatomical data to document systematic affinity is that the comparative data from related taxa are only available for much larger taxa. Hence, it is difficult to separate phylogenetic signal from effects of small size. Additional data from small members of other vetigastropod lineages would be highly desirable.

Phylogeny, classification. The monophyly of the family is questionable, as pointed out by Warén & Bouchet (2001), who consider the vent subfamilies Temnocinclidae and Sutilizonidae more closely related to Lepetodrilidae. Schwarzpaul (2002), on the other hand, found Temnocinclidae and Sutilizonidae more closely related to Fissurelloidea (Fissurellidae and Clypeosectidae) than to Lepetodrilidae using morphological data on a larger array of hydrothermal vent "archaeogastropods." Recent molecular data (Geiger & Thacker, 2005; Kano, 2008: Figs 3, 5) also suggest that Anatomidae and Scissurellidae are not monophyletic. Scissurellidae is sister group to Lepetodrilidae, whereas Anatomidae is basal to Fissurellidae. The position of Larocheidae is unresolved. There are radular differences between them and the remainder of the scissurellids, and the absence of slit/foramen further supports their separate status. *Coronadoa* on the other hand is a scissurellid despite the missing slit/foramen.

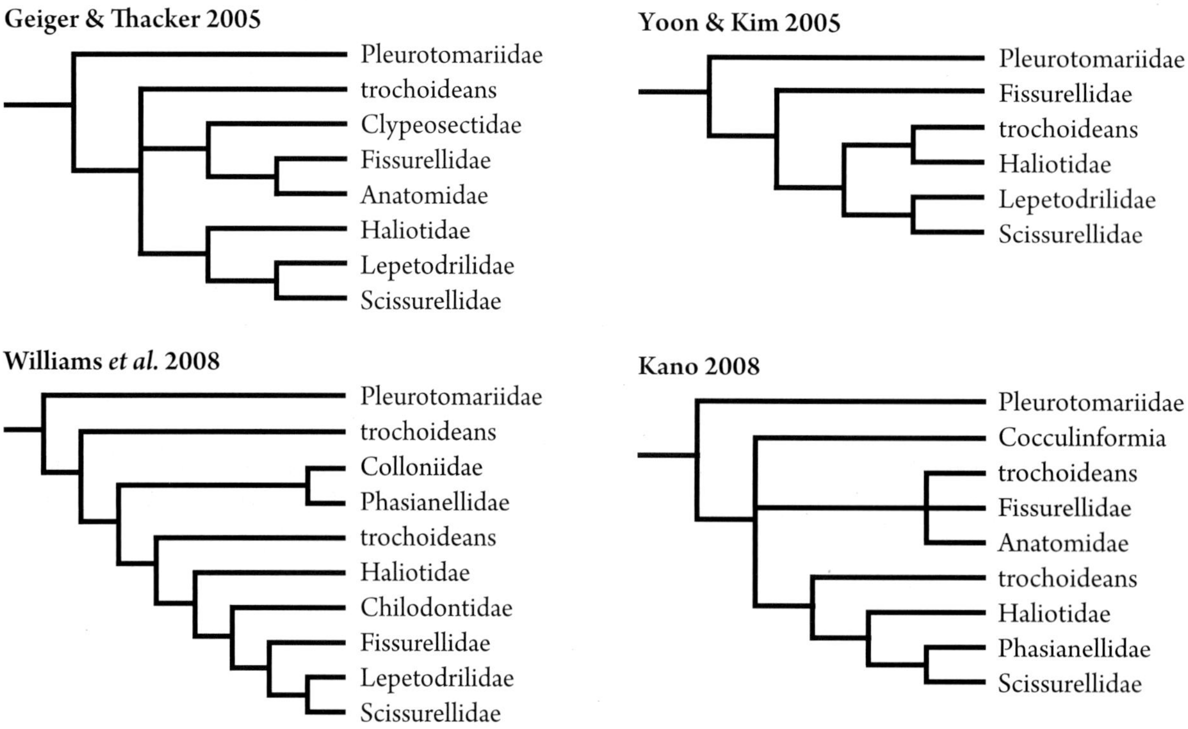

Figure 3. Some recent phylogenetic analyses including Scissurellidae s.l. All are simplified and highlight relationships of scissurellids among Vetigastropoda. Three of the four include a Scissurellidae-Lepetodrilidae clade, and both that include Anatomidae show the family close to Fissurellidae. All phylogenies agree that Scissurellidae and Pleurotomariidae are only distantly related within Vetigastropoda, and Anatomidae and Scissurellidae do not form a clade.

The ranking of the family-level taxa has been variable, with earlier accounts using sub-family rank for the groups within Scissurellidae s. l. Given the mounting evidence that the former Scissurellidae s. l. is not monophyletic, with several discrete lineages diverging independently early on in vetigastropod evolution, those higher taxa are here all treated as full families. The earlier cited radular similarities can be explained by constraints due to small size, through paedomorphosis (retention of larval characters as an adult), or as a symplesiomorphy (ancestral trait for all Vetigastropoda) (see Geiger & Thacker, 2005 for discussion). In Anatomidae, novel radular features have come to light that demonstrate a much greater diversity in the feeding structure (Geiger & Sasaki, 2008; Sasaki *et al.*, 2010; this study), and weaken the previously cited commonalities that were adduced to classify all of those lineages in a single family Scissurellidae or superfamily Scissurelloidea (Haszprunar, 1989). Some authors (Kano, 2008) expanded the concept of Scissurelloidea to include Lepetodrilidae, while Bouchet & Rocroi (2005) used Scissurelloidea for Scissurellidae (including Depressizonidae), Anatomidae and Larocheidae only, while Sutilizonidae (and Temnocinclidae as synonym) were placed in Lepetodriloidea. However, the relationships of Scissurellidae s. s., Lepetodrilidae, and minor groups such as *Bathyxylophila, Gorgoleptis, Pendroma, Clypeosectus,* Temnocinclidae, *Sutilizona,* and Larocheidae is far from satisfactorily resolved, and various lineages within "skeneimorphs" introduce further complications. Scissurellidae s. l. have variously been placed in their own superfamily (see above), or have been included with Pleurotomariidae, and various additional families, in the superfamily Pleurotomarioidea (Brookes Knight *et al.*, 1960; Götting, 1974; Beesley *et al.*, 1998).

Given the above uncertainties, it is best to abstain from placing families in Vetigastropoda into existing higher taxa or naming additional ones until more complete phylogenies are available. Before such classification schemes are proposed, multiple phylogenies with different character- and taxon-sets should converge on the same topological elements. One should restrain oneself from overstating conclusions reached particularly on data matrices with limited taxon sampling.

In the present volume, the family name Scissurellidae is used *sensu stricto* (*Scissurella, Sinezona, Sukashitrochus, Satondella, Incisura, Coronadoa*), whereas scissurellids (Scissurellidae s. l.) denotes all the small, slit-bearing Vetigastropoda covered in this contribution. The latter highlights the non-monophyly of this morphotype or grade.

Figure 4. Crossed platy shell structure in *Anatoma kopua* (AMS C.453747). Notice the absence of nacre on the inside (lower margin). Scale bar = 20 µm.

Although the membership of scissurellids and its nearest sistergroup relationships have been debated extensively, with no consensus in sight, the placement of all those lineages in Vetigastropoda is unanimously agreed upon (see Geiger *et al.*, 2008 for review; Aktipis & Giribet, 2012). The presence of bursicles on the gill is one of the most important synapomorphies that unites Vetigastropoda (including Neomphalidae and Peltospiridae). The rhipidoglossate radula, the slit or foramen in the shell, the two shell retractor muscles, epipodial tentacles with epipodial sense organs are other important characters (Haszprunar, 1987; Salvini-Plawen & Haszprunar, 1987; Geiger *et al.*, 2008). Some typical vetigastropod characters such as nacre are absent in scissurellids (Fig. 4), but are also not found in several other vetigastropod families such as Lepetodrilidae, Fissurellidae, Neomphalidae, Peltospiridae, and Pendromidae (Hedegaard, 1990; Sasaki, 1998; Geiger *et al.*, 2008). The absence of nacre in scissurellids is not related to size as it is clearly present in Seguenziidae (Hedegaard, 1990: figs 59, 60; Geiger *et al.*, 2008:

fig. 12.2 J; Geiger, unpubl. data). The shell is composed entirely of aragonite and is either of homogeneous mineralogy or has crossed platy structures (Hedegaard, 1990; Fig. 4).

Small size. The small size of the shells and their intricate structure took some time to be appreciated. Early light microscopical illustrations often missed the slit or foramen (Semper, 1865: fig. 4: *Scissurella hoernesi*), although d'Orbigny (1824) showed it. Woodward (1859) was the first to note the lack of selenizone on the early teleoconch (teleoconch I). Overly broad species concepts encompassing taxa now placed in multiple genera are found in early discussions. For instance, Jeffreys (1856e) lumped all three European members into a single species, and falsely thought that early workers had missed the closed foramen in *Anatoma crispata*. Protoconch sculpture was largely unknown, although some can be seen under the light microscope. Only with the advent of the SEM, the true diversity of scissurellid shell morphology received full attention. The first SEM images (Batten, 1975) were of single species. The first broad assessments were carried out by Yaron (1983), Herbert (1986), and McLean (1989). More comprehensive recent contributions include Bandel (1998), Marshall (2002), Geiger (2003, 2006a), Geiger & Jansen (2004a, b), Geiger & McLean (2010), Høisaeter & Geiger (2011), Zelaya & Geiger (2007), Luque *et al.* (2011), and Geiger & Marshall (2012). Only some of the most recent contributions go as far as to illustrate multiple specimens per species to document intraspecific variability.

While some original illustrations—specifically line drawings from the 18th and 19th century—were made with excessive artistic license, many are astoundingly accurate. Examples of the latter include the type species of *Anatoma*, *An. crispata*, which clearly illustrates the rounded overall shell, not to be confused with the more stepped shell shape of *An. aspera* or the lenticular shape of *An. tenuis*; *An. eximia*, which captures the key attributes of this species; and *Satondella tabulata*, which shows the more depressed shell shape, and also illustrates the unique internal rim in the aperture, which distinguishes it from *Sat. brasiliensis*, with which it had been confused for a long time. Some of the sculptural detail shown can only be seen with quality microscopes and by using appropriate imaging techniques; the presumption that historical microscopes produced low quality images is false (Ford, 2011).

Anatomical studies are even fewer. Some illustrations on the external anatomy of *An. crispata* were published early on (Barrett, 1856), while more comprehensive external anatomical descriptions were first published by Vayssière (1894) and Pelseneer (1899). The first work on the internal anatomy was carried out by Bourne (1910) on *Incisura lytteltonensis*. The next work was on the hydrothermal vent groups (Haszprunar, 1989), in the unpublished dissertation of Strassoldo (1991) on one *Scissurella* and one *Anatoma* species, and the unpublished diploma thesis by Baborka (2007) on *Scissurella* and *Larochea*. Accounts involving the external anatomy include Montouchet (1972), Sasaki (1998), and Kase & Kano (2002), Zelaya & Geiger (2007), Sasaki *et al.* (2010), and Geiger & McLean (2010).

The fragility of the specimens and the difficulty in working with them has been noted by many workers. Wood (1848: 163) remarked: "Since the above was written I regret to say that the specimen while in the hands of the engraver, was lost. The figure above referred to is, I believe, a faithful representation of what it was." The type material of *Sutilizona theca* is mostly broken and the only known specimen of *Sut. tunnicliffae* was destroyed during the preparation of the original description. Geiger (2003) and Geiger & Jansen (2004b) noted that hydroxides used for radular extraction have a destabilizing effects on the shells. Geiger *et al.* (2007) provided a summary of working techniques for micromollusks.

The biology of scissurellids still is a wide open field. Targeted field collecting and *in situ* observations are impossible. Philippi (1853: 32 [translated to English]) noted that "The scissurellids seem to live in the oceans of all zones, and that one only knows so few species so far, the cause is unquestionably, that one has given no attention to these microscopically small shells." Some observations on the swimming behavior were given by Haszprunar (1988b), Hasegawa (2004), and Porter & Hickman (2007). Reproductive behavior and sexual dimorphism was noted by Hickman (1999) and Porter & Hickman (2007).

Geiger & Thacker (unpublished data) used continuous character state reconstruction to assess inferred sizes of ancestral nodes based on their molecular phylogeny of Vetigastropoda (Fig. 5). Inferred size at the ancestral node is moderately small (6.5 mm). Core Vetigastropoda (Trochoidea, Fissurellidae, Haliotidae, Lepetodrilidae, Scissurellidae, Anatomidae) show an increased ancestral size of 25 mm. Within those core Vetigastropoda, several independent lineages have grown larger (Fissurellidae, Haliotidae, Turbinidae), while Scissurellidae and Anatomidae are secondarily reduced in size. Overall, size changes several times in Vetigastropoda back and forth, and no directional trend is detected.

Preservation. Some material examined and named is of poor preservation. Some systematists would prefer not to describe new taxa based on poorly or incompletely preserved material, few available specimens or even a combination of those two factors. For Recent taxa, anatomical data are considered by some to be crucial for the description of new taxa. While one approach does not fit all, in some cases the description of new taxa under those unfavorable conditions is justified (see also section Taxon concepts below).

Specimens of poor preservation are regularly described in paleontology, and attempts at classifying those specimens are executed as carefully as possible. Anatomical data such as radular patterns are held to be indicative of higher systematic placement. In scissurellids, this presumption is at best dubious. For one, the most common radular pattern is most likely the plesiomorphic condition of Vetigastropoda (Warén, 1990; Geiger & Thacker, 2005), and symplesiomorphies cannot be used to classify organisms. Second, highly divergent radulae are found in conchological congeners (Geiger & Sasaki, 2008; this study). Although anatomical data may provide insights, it is not guaranteed that they do. As is true in systematics in general, the arguments presented here are subject to revision when more data become available. In addition, a description may well bring to light additional specimens of recently described species that are currently lying dormant in someone's collection. The description of *Orthotheres haliotidis* Geiger & Martin, 1999 (Crustacea: Pinnotheridae) in a tropical abalone was met early on with much skepticism, but this commensal is now known to occur at infestation rates of up to 50% (E. Capinpin, pers. comm. 2000).

Additional problems were encountered with type material of already described taxa. Because specimens were sometimes stored under unsuitable conditions or in inappropriate containers, Byne's disease and glass disease have covered the surface of some specimens with recrystalized carbonates, making it impossible to assess fine sculpture and protoconch characteristics. In a few cases, the entire specimens were reduced to dust (See Geiger *et al.*, 2007 for case descriptions and proper storage solutions).

Bouchet & Strong (2010) discussed various issues with molluscan type specimens, including poor preservation with missing key identifying characters, absence of soft parts and problems with shell-morphologically cryptic species and the difficulty of setting aside existing name-bearing type material in poor condition, which they identified as an important impediment to naming the large number of still undescribed species. The above problems are not insurmountable with existing

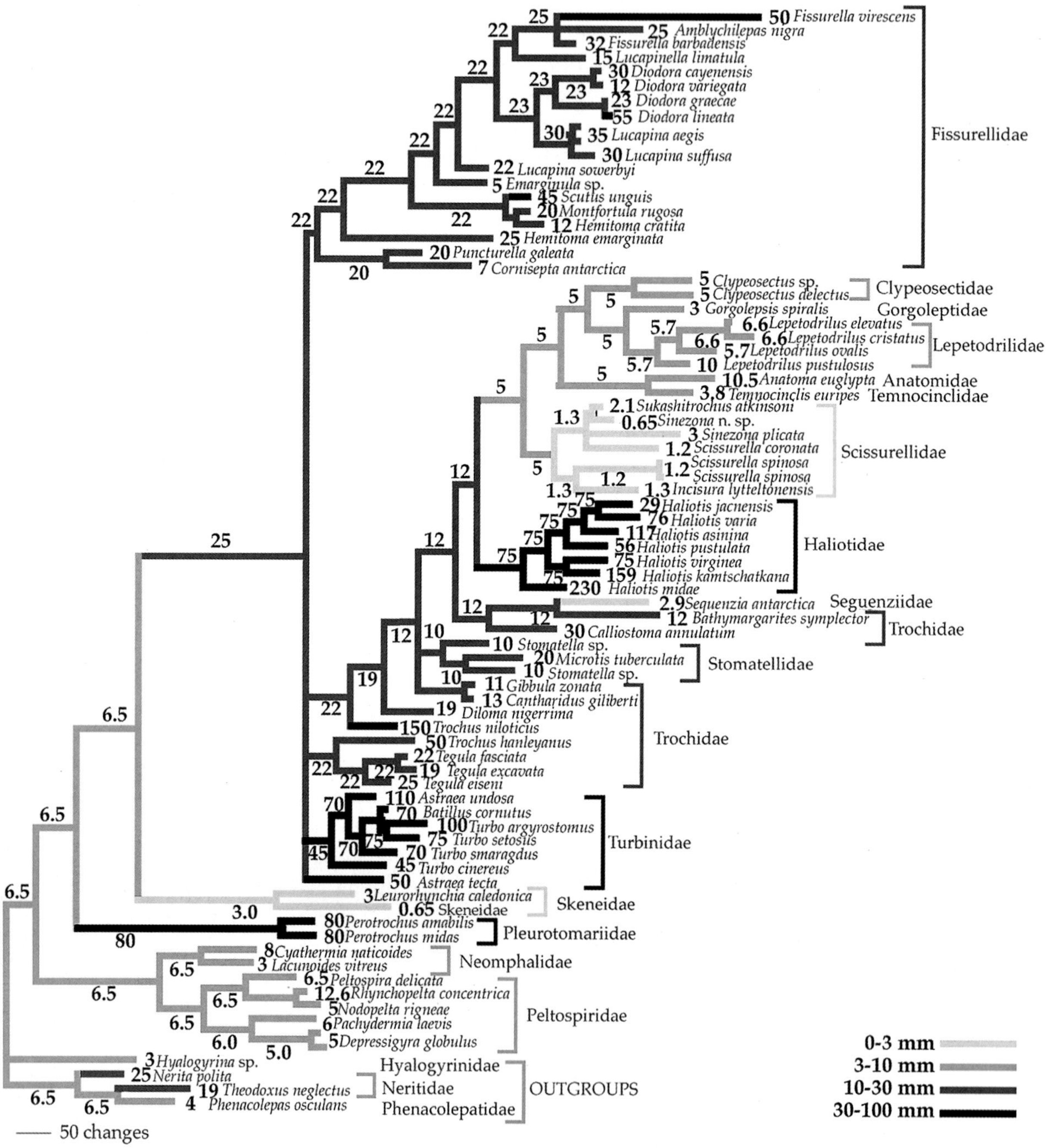

Figure 5. Molecular phylogeny of Geiger & Thacker (unpublished data: 3000 bp of 18S, COI, Histone 3, Bayesian reconstruction), with mapping of size based on continuous character reconstruction in *MacClade*. Size indications are in millimeters. Shading of branches is according to four size classes. For discussion, see text.

nomenclatorial tools, specifically applying first reviser's principle, but require a thorough understanding of the group before action is taken (see also Anon., 2002). Such built-in speed bumps ensure against poorly thought out actions. The option to designate epitypes as known in botany may be the most sensible addition to ICZN, while a wholesale clearing of the taxonomic slate similar to bacteriology is not needed. A major step forward would be the on-line availability of information on type specimens (preferentially with high resolution images) in repositories. Such projects have been carried out for instance by NMNZ, but are still wanting at many key institutions (e.g., NHMUK, MNHN, USNM). In the context of this monograph, despite some problem cases, all taxonomic-nomenclatorial issues could be addressed using existing tools.

Taxon concepts

Species. At the species level, the morphospecies concept is most frequently employed in systematic malacology: a species is composed of morphologically similar specimens. The criteria for recognizing species here are as follows. Apparent geographic separation is generally considered insufficient reason to recognize a species as distinct, even if genetic connectivity between populations must be called into question. Examples include species that are known from the Caribbean as well as the Indo-Pacific. A conservative approach to recognizing species is taken. Only specimens that can be diagnosed without ambiguity and show distinct differences are treated here as valid species; in most cases three discrete co-differentiating character states are cited. In problem cases, no new taxa are introduced, and in the remarks section of the most similar taxon the problem is explained. However, in some cases single specimens that are distinct have been named (*Depressizona axiosculpta*). Here, principles borrowed from statistics are applied: small differences can only be shown to be significant if they are consistent and found in a large number of specimens; for a single specimen to be significantly different, the amount of the difference must be large. If the material is distinct and diagnosable, and better material is not expected to become available any time soon, there is no advantage of holding back with a description. On the contrary, hiding known biodiversity in one's personal collection drawers is counterproductive in light of efforts to compile the global biodiversity inventory.

With some fossil taxa the material available is not well-preserved, and shows similarities with Recent species. If it is known that the missing portion of the fossil specimens can show species-specific character states in congeneric Recent species, and if the time span covered exceeded 10 million years, taken as a common evolutionary life span of a species, then both taxa are retained. In those rare cases the chronospecies concept is cautiously applied.

No subspecies are recognized. Those that were proposed as subspecies are either synonymized under a second species, or elevated to full species.

Genera. The genus is difficult to define, and a morpho-taxon approach was invoked. Here it is taken as a container for species that are more similar to one another than to those of any other genus. It can be conceptualized as an intuitive application of a weighed cluster analysis, based on shell morphological similarities. Ideally, a genus is recognized by the following attributes: 1) the diagnosing character state applies to all species in the genus; 2) the diagnosing character state is discrete; genera are not separated at an arbitrarily chosen character state value along the range of a continuously variable character, but by discrete character states.

Families. Family-level taxa are applied considering whether the lineage is distinct among other lineages of equal rank. It is comparative in nature, assessing the standing in relation to the remainder of the taxa in Vetigastropoda. In the absence of good phylogenetic information, shell morphological (and anatomical) distinctness is taken as a proxy for distinctness of the lineage. The higher classification of scissurellids within Vetigastropoda is expected to change.

Species groups. There have been no formal phylogenetic analyses examining the relationships amongst scissurellid species. For molecular analyses, too few taxa have ever been included to provide any meaningful insight, and the number of discernible shell morphological and radular characters are so low as to lead to underdetermined matrices (Geiger, 2003; Fig. 21). Nevertheless, some species groups, or Formenkreise, can be identified based on overall morphology, or Gestalt. Some species can be assigned to more than one group; those are marked with an asterisk (*) Those groupings include:

- *Scissurella azorensis, Sci. kaiserae, Sci. lobini, Sci. maraisorum, Sci. prendrevillei, Sci redferni, Sci. regalis, Sin. semicostata, Sin. wanganellica*: shell small (rarely over 1 mm), flat spire, underside of protoconch visible through umbilicus; includes species in traditional genera *Scissurella* and *Sinezona.*
- *Scissurella costata, Sci. mirifica, Sci. rota, Sci. staminea, Sci. sudanica, Sci. xandaros*: similar overall shape with wide umbilicus lacking a periumbilical cord, differing only in protoconch sculpture and strength and density of clathrate sculpture.
- *Scissurella lorenzi, Sci. quadrata*: similar to above, but with narrow umbilicus and distinct periumbilical cord.
- *Sinezona insignis, Sin. iota, Sin. haliotimorpha*†, *Sin. kayae, Sin. rimuloides*: distinctly elevated spire.
- *Sukashitrochus atkinsoni, Suk. lyallensis, Suk. ngatutura*†, *Suk. pulcher*: temperate Australia-New Zealand species with same protoconch (flocculent sculpture restricted to outer portion of whorl).
- *Satondella azonata, Sat. bicristata, Sat. dantarti, Sat. minuta*: small *Satondella* with sunken apex.
- *Satondella brasiliensis, Sat. cachoi, Sat. danieli, Sat. goudi, Sat. senni, Sat. tabulata*: larger *Satondella* with more or less elevated spire.
- *Anatoma aupouria, An. parageia, An. pseudoequatoria*: shells wider than tall, very short teleoconch I, and a subquadrate profile of the base.
- *Anatoma australis, An. austrolissa**, *An. flexidentata*: flexoglossate radula, but widely different shells.
- *Anatoma amydra, An. atlantica, An. biconica, An. disciformis, An. euglypta, An. japonica, An. maxima, An. rainesi**, *An. rapaensis, An. regia, An. rolani, An. tenuis, An. tenuisculpta, An. yaroni*: discoidal shell with, sharp periphery and rather narrow sutsel.
- *Anatoma aedonia, An. aspera, An. austrolissa**, *An. emilioi, An. equatoria, An. eximia, An. hyposculpta, An. micalii, An. munieri, An. plicatazona**, *An. quadraxialis, An. sinuosa**: scalar whorls.
- *Anatoma alternatisculpta, An. aspera, An. finlayi, An. indonesica, An. janusa*: strong axial lamellae on shoulder, usually finer sculpture on base.

- *Anatoma alta, An. gephyra†, An. sinuosa*, An. porcellana*: selenizone distinctly above middle of whorl.
- *Anatoma agulhasensis, An. alternatisculpta, An. aspera, An. fujikurai, An. finlayi, An. gunteri, An. japonica, An. kelseyi*, An. lamellata, An. peruviana*, An. plicatazona*, An. rainesi*, An. sagamiana*, An. schanderi**: inflated whorls with rounded periphery, shell taller than wide.
- *Anatoma kelseyi*, An. peruviana*, An. sagamiana*, An. schanderi**: reticulate protoconch, inflated whorls, distinct sutsel.
- *Anatoma conica, An. crispata, An. lyra, An. tobeyoides, An. weddelliana*: globular overall shell shape, but wider than tall.

While those observations are striking, they do not form a cohesive picture, because too many species fall either between boundaries of those groups (e.g., *An. australis, An. rhynchodentata*), or cannot be assigned to any of those groupings (e.g., *An. megascutula, An. shiraseae*). It is not possible to derive from those selective observations a classification scheme beyond what is given here. None of the groups above correspond even vaguely to any proposed generic concepts. Some species can be assigned to more than one group based on different character sets, a typical case of character conflict, which needs to be evaluated with proper phylogenetic tools. Most *Sinezona* species cannot be grouped in this fashion.

Applied taxonomy

When monographing a group of organisms on a global scale, some taxonomic changes will have to be executed. Such revisions offer the opportunity to advance our understanding not only of the diversity, but also offer windows into emergent observations relating to distribution and patterns of variable and invariable characters. On occasion, even one's own treasured discoveries need to be reassessed, and some of my own species had to be synonymized here under previously described names. Below I describe a couple of cases detailing both the rationale for introducing the taxon previously, as well for entering it into synonymy. All those examples have in common that additional material has led to the reappraisal, which is fully in keeping with scientific best practices.

Anatoma finlayi. *Anatoma finlayi* was described from New Zealand, has until now only been known from there, and is easily recognized by the strong axial lamellae on shoulder and base. New Zealand is well-known for its very diverse temperate to boreal fauna. In material from MNHN some specimens from tropical Reunion Island with strong lamellae on shoulder and base were discovered and named *An. herberti*. It was compared in the original description to *An. finlayi*, noting differences in axial sculpture on the base, as observed on the holotype of *An. finlayi*. The geographic separation and difference in temperature regime further supported the introduction of the new taxon. Examining additional material from New Zealand demonstrated the holotype of *An. finlayi* to be a specimen with rather uncharacteristic sculpture. Furthermore, the type locality of *An. finlayi* is from Three Kings Islands in northernmost New Zealand. Those islands have been the source of several other southernmost records of mostly Indo-Pacific species. Additional records from Madagascar, Indonesia, Philippines, Western Australia, New Caledonia, Fiji,

Tonga, and even French Polynesia have made it evident that there is not a tropical-temperate species pair, but a single tropical species of wide distribution and easily recognizable shell morphology.

Anatoma pseudoequatoria. This species was described from Hawaii. It has a rather low spire for the genus, a very short teleoconch I of 0.2–0.3 whorl, and a subquadrate profile of the whorls. Many specimens from Hawaii show a highly refined arrangement of internestling axial ribs on the shoulder, highly unusual among *Anatoma* species, from a place known for its high rate of endemism. The same beautiful arrangement is clearly shown in *Hainella pulchella* described from Indonesia. A species with somewhat taller whorls, and without the pattern of internestling axial ribs was described as *An. funiculata* from Queensland. After examining 60 specimens by SEM from throughout the Indo-Malayan Archipelago and Central Pacific, it has become clear that the sculpture on this species is extraordinarily variable. The various teleoconch ornamentations are found both in Hawaii, as well as in the Indo-Malayan Archipelago (demonstrated by the identical type locality of *H. pulchella* and *H. paucispiralia*), but the very short teleoconch I and the protoconch with very fine flocculent sculpture is found on all those specimens.

The alternative option of multiple species with discrete teleoconch sculpture, but shared protoconch and unusually short teleoconch I is unlikely, as intermediates are known from throughout the range of the species. In this case, synonyms were introduced, because the morphological variability of the species is by far greater than what was known from previous examinations of other species.

On the other hand, *An. parageia* from Sagami Bay, Japan, also belonging into that Formenkreis, seems to be distinct. In terms of overall shell morphology, it can easily be mistaken for *An. pseudoequatoria*. All known specimens have a rather short teleoconch I of ~0.4 whorl; the usual intraspecific variability is ±0.125 whorl. Hence, his minor distinction is of questionable importance to separate the two taxa. The teleoconch sculpture consists of very fine and dense axial cords, finer than on most *An. pseudoequatoria*, but identical to one of the paratypes of *An. pseudoequatoria*. However, the protoconch sculpture of all six specimens of *An parageia* examined by SEM consists of coarse flocculent sculpture, while all 60 specimens of *An. pseudoequatoria* examined by SEM show fine flocculent sculpture. In this case, subtle but highly consistent differences in protoconch sculpture can separate the two taxa. Unfortunately, it means that specimens from localities between the known boundaries of those two species (Okinawa – Sagami Bay), will require examination by SEM for positive identification.

Taxonomic actions taken here

Below taxonomic actions are listed, except for new species described and new synonymies. Those include designations of lectotype and paralectotype(s), neotype, *nomen oblitum/protectum*, and restriction of type locality that are otherwise difficult to find in the overall text:

- *Scissurella striatula* Philippi, 1844 [= *Scissurella costata* d'Orbigny, 1824]: here explicitly designated as type species of *Schismope* Jeffreys, 1856, becoming synonym of *Scissurella* (type species misidentified by Jeffreys, 1856c).
- *Scissurella elegans* d'Orbigny, 1824 [= *Scissurella costata* d'Orbigny, 1824]: here explicitly designated as type species of *Woodwardia* Crosse & Fischer, 1861, becoming synonym of *Scissurella* (type species misidentified by Crosse & Fischer, 1861).
- *Scissurella costata* d'Orbigny, 1824: designated as senior synonym of *Sci. elegans* d'Orbigny, 1824 by first reviser's principle. Additional synonyms include *Sci. laevigata* d'Orbigny, 1824, and *Sci. decussata* d'Orbigny, 1824.
- *Scissurella crispata* Fleming, 1828: here explicitly designated as type species of *Anatoma* Woodward, 1859 (type species misidentified by Woodward, 1859).
- *Scissurella concinna* Sowerby, 1831: here designated as *nomen oblitum* and *Sin. cingulata* designated as *nomen protectum*.
- *Scissurella affinis* O. G. Costa, 1861 (*Anatoma*): neotype here designated, becoming objective synonym of *Anatoma aspera*: see under synonymy of *Anatoma aspera*.
- *Anatomus mirificus* A. Adams, 1862 (*Scissurella mirifica*): lectotype here designated.
- *Anatomus concinnus* A. Adams, 1862 (*Anatoma concinna*): lectotype here designated.
- *Anatomus japonicus* A. Adams, 1862 (*Anatoma japonica*): lectotype here designated.
- *Anatomus lamellatus* A. Adams, 1862 (*Anatoma lamellata*): lectotype here designated.
- *Anatomus stamineus* A. Adams, 1862 (*Scissurella staminea*): lectotype here designated.
- *Anatomus turbinatus* A. Adams, 1862 [= *Anatoma munieri*]: lectotype here designated.
- *Scissurella carinata* A. Adams, 1862 (*Sukashitrochus carinatus*): lectotype here designated.
- *Schismope ferriezi* Crosse, 1867 (*Sinezona*): lectotype here designated.
- *Scissurella tenuis* Jeffreys, 1877 (*Anatoma*): lectotype here designated.
- *Scissurella umbilicata* Jeffreys, 1883 (*Anatoma*): lectotype here designated, type locality restricted.
- *Scissurella aedonia* Watson, 1886 (*Anatoma*): lectotype here designated.
- *Scissurella declinans* Watson, 1886 [= *Scissurella mirifica*]: lectotype here designated.
- *Scissurella alta* Watson, 1886 (*Anatoma*): lectotype here designated.
- *Schismope insignis* Smith, 1890 (*Sinezona*): lectotype here designated.
- *Schismope depressa* Watson, 1897 [= *Sinezona cingulata*]: lectotype here designated.
- *Scissurella jacksoni* Melvill, 1904 [= *Anatoma munieri*]: lectotype here designated.
- *Scissurella maxima* Schepman, 1908 (*Anatoma*): lectotype here designated.
- *Scissurella jucunda* Smith, 1910: lectotype here designated.
- *Schismope padangensis* Thiele, 1912 [= *Sinezona ferriezi*]: lectotype designated.

Organization of the descriptions

All supraspecific names are described and diagnosed, both for shell as well as animal characters as much as data are available. The original citation, type taxon, mode of designation (OD: original designation; M: monotypy; SD: subsequent designation) is given.

For all species level names, the following sections will provide a circumscription of the content of each of the descriptive sections and some conventions used therein.

Species name. The full name of the species is given with author in parentheses only if the species is currently classified in a different genus than the one which it had been proposed originally.

A dagger (†) at the end of the line indicates extinct species. Living species with a fossil record are not marked with a dagger, including species originally described as fossils and only subsequently discovered also in the Recent.

Chresonymy. The chresonymy, i.e., the application of a particular name or combination, is as complete as it has been able to be traced during the study. The original description is indicated by the species name followed by the authors name:

Genus species Author, date: pagination, plate, figures.

Subsequent indications of that species under that species epithet are given in chronological order; the subsequent author is separated from the species epithet by a colon:

Genus species: Subsequent-Author, date: pagination, plate, figures [notes].

Chresonymy entries that could not be verified during this study are credited to the source: [*fide* Author, date]. Occasionally, abbreviations used by these authors may have been ambiguous. For instance the generic abbreviation *S.* may have both been used for *Scissurella* as well as *Schizotrochus*. The most likely usage has been chosen here.

The certainty of the identifications by the different authors varies extensively. Mentions of names without illustrations are less certain than those with illustrations; line drawings are more difficult to positively identify than electron micrographs. I appeal to the user's intelligent reading of these synonymies. Particularly problematic listings are annotated at the end in square brackets. Compilations of names without any perceptible attempt of scholarly exactitude, evidenced for instance by pervasive spelling mistakes, as are found in certain shell lists and web-portals, have been excluded from the synonymy (e.g., Galindo, 1977; GBIF).

Separate sections are introduced for **Synonyms**, i.e., species that are considered to be conspecific with the treated species, as well as **Misidentifications**, i.e., species names of either valid species or non-synonymous species applied to the present species. Synonyms are flagged with an antecedent + sign, and subsequent uses of that synonym are listed below. Full type information for the synonyms (type material, type locality, etymology) is provided. Any additional short remarks are given in [brackets], such as local interpolations, coordinates not given in OD, notes on copies of figures.

+ *Genus synonym-species* Author, date: pagination, plate, figures. ***Type material.*** Type (collection), dimensions. ***Type locality.*** Local name (OD/SD). ***Etymology.*** Derivation of name. ***Remarks.*** Additional information.

Genus synonym-species: Subsequent-Author, date: pagination, plate, figures [notes].

The chresonymy for genera is similar to that of the species. Citations of subsequent usages of the genus are restricted to those that provide some discussion of the genus-level taxon. Simple mentions of the genus have not been included.

Type material. All indications as to nomenclatorially relevant type material are given, i.e., syntype(s), holotype, paratype(s), lectotype, paralectotype(s), and neotype, but not topotype(s), hypotype(s) or ideotype(s); see Fernald (1939) for types of types. The nature of these indications is variable: information from the original description and original dimensions, original collection in which specimens were deposited, documented location of type specimens (Author, date: figs), current whereabouts of the specimens and subsequent illustrations of type specimens (Collection registration-number: Author, date: figs.). Multiple indications are given in chronological order, such as listing of syntype material and subsequent selection of a lectotype. From most syntype series for which no lectotype had been designated, such a designation is carried out here for taxon stabilization: see *Taxonomic actions taken here* above for list.

Original measurements in imperial units have been converted to the nearest 0.05 millimeters using 1 English line = 2.12 mm, 1 French line = 1.75 mm, and 1 inch = 25.4 mm.

Most original illustrations are given; exception are some species in which the name-bearing type(s) have been investigated by SEM. Most figures have been rearranged to fit the format and to show each species in an as consistent format as possible. The quality of the reproduced original figures is variable due to several factors. Species have been illustrated at various sizes, sometimes only photocopies of uncommon works were available, and half-tone screens of published figures do interfere with reproduction. Various digital imaging techniques (cut-out, levels, highlight-shadows, curves, Gaussian blur, unsharp-mask) were applied to produce the highest quality image that is as true to the original and shows all the salient features as well as possible.

Type species. For genera, the type species, and mode of designation is provided (OD, M, SD: author, date: pagination). Type genera are self evident from the formation of family-level names.

Other genera/species. For family-level taxa, the genera assigned to that taxon are given. For genera, a listing of all species considered valid are given. Extinct species are marked with a dagger (†). Living species with a fossil record are not marked with a dagger, including species originally described as fossils and only subsequently discovered also in the Recent.

Species removed from *Genus*. For genera, any species originally described in that genus but currently classified elsewhere are given.

Type locality. The type locality from the original description (OD) is given. Type localities inferred from the title of the publication or some other non-traditional source are indicated as such: (OD: inferred from title). Precision not found explicitly in the original description such as country and coordinates are given in brackets.

Subsequent designations or restrictions of type localities are given (SD: author, date: pagination). For fossil species the type horizon is indicated as well.

Original depth indications in imperial units are converted to the nearest meter using as conversion 1 fathom = 1.8288 m and 1 foot = 0.305 m.

Coordinates of the type locality are given as in the original, either as degrees, arc-minutes, and arc-seconds, or as degrees with decimals. Interpolated coordinates from named locations are given as degrees with three decimals (±~100 m). Interpolated coordinates for vague locality names are only given, if the star symbol on the map covers the range of geographic uncertainty (Ø ~50–100 km, ~0.5–1°). For type localities such as "Japan," no coordinates are provided, and no type locality is marked on the distribution map.

Etymology. The derivation of the species name is detailed. When the derivation of the name is taken from the original description it is indicated by "(OD)."

Description. The description of the shell is given in telegraphic style in the following sequence: overall size and shape; protoconch, teleoconch I, teleoconch II: shoulder, base, umbilicus, selenizone, aperture. Operculum, radula, animal. See below for definition of terms and illustrations of traits.

Comparisons. Distinguishing characters are given for all similar species world-wide, and somewhat similar, sympatric species. The discriminating character states usually given as a comparison (e.g., stronger, weaker, taller, wider), and are only sometimes quantified (e.g., 0.1 vs. 0.5 whorl). The objective is to highlight character states when comparing the respective illustrations. Accurate differentiation based on verbal descriptions alone would take too much space and would often remain ambiguous. Although some species are clearly distinct based on geography or climate, comparisons are given for similar species to help to identify mislocalized specimens.

The Thumbnail Identification Cards inserted at the end of the volumes can be used in conjunction with the written comparisons, as some obvious aspect of the particular shell's morphology may either not be mentioned, or is difficult to express in words. Those thumbnail identifications are additionally printed in the bound portion of the book towards the end of the Introduction, so that they cannot be misplaced or lost.

The comparisons are based on shell morphological features only, despite the fact that some species can easily be separated by their distinct radulas (e.g., *An. australis* and *An. flexidentata* as the only known species with flexoglossate radula).

Distribution. The *known* geographic and bathymetric distribution is given. Note that our knowledge is still limited for most species. Habitat indications are provided if they are available.

Geologic Range. For families and genera, the time range of geologic occurrence is given.

Fossil. Fossil occurrences are listed and referenced.

Specimen records. Material examined is listed in the following format: depth, specific locality to region, coordinates (Collection registration-number, number-of-specimens). [For some collections (e.g., LACM), station numbers are cited, which do not permit identification of specific lots, because with changing identification the reference to the lot is lost. Proper registration numbers of lots is by far superior and should be implemented for all collections.]. Some collections (e.g., MNHN) do not use registration numbers; only the collection abbreviation and number of specimens can be given. The records are ordered by geographic region and in a North-South, East-West direction as well as feasible. Depth, if data is available, is given to the nearest meter; greater precision is meaningless as it is within the range of tidal fluctuations; 'intertidally' collected specimens are given the numeric depth value of 0 m. Specimens with wet-preserved animals are listed as "complete," which is considered better than 'live taken' as used by others, because live taken specimens may never have had the animal preserved. The collection-abbreviations are given below.

All non-type locality coordinates are given as degrees with three decimals (±~100 m), including those interpolated from named locations. Interpolated coordinates for vague locality names are only given if the symbol on the map covers the range of geographic uncertainty (Ø ~50–100 km, ~0.5–1°). For localities such as "Japan," no coordinates are provided.

Locality names are given by their English equivalents as much as possible. Slight adjustments from the original labels are not specifically marked. Tentative identifications of locations with distinct alternate spellings are noted in square brackets.

Literature records. Distributional records for Recent material from the literature are listed and referenced. Numbers in parentheses at end of entries refer to number of specimens collected but without indication of repository. A few personal communications of easily identifiable species are included in this section.

Remarks. Any information that did not fit into any of the above categories is given in the remarks section.

Distributional maps. The dot maps provided give type locality (star symbol ☆), specimen records (round dots ●), and literature records (squares ■). Vague type localities, e.g., "Japan," are not plotted; hence, some maps do not show a star symbol.

Collecting and curation

Collecting. Micromollusks require different collecting techniques compared to larger mollusks. The general methods have been discussed in detail elsewhere (Geiger *et al.*, 2007; also chapters in Geiger & Ruthensteiner, 2008). Here only the most relevant and scissurellid-specific details are provided. I concentrate on those methods that I have performed and found to be advantageous; select additional or alternate methods and approaches are mentioned if considered appropriate.

Scissurellids can be collected regularly in suitable habitats. Hence, habitat selection is an important initial consideration. They usually live on hard substrates, such as rocks, corals/reefs, and larger stones; shallow water sand/mud samples usually only provide empty shells. In the intertidal and subtidal, rock washings and algal and coral shaking techniques are well-suited (Fig. 6). With algal shakings, in general, green algae yield better results than either red or brown.

Sampling in deeper waters can be achieved with remote devices. Small dredges with screens of appropriately small sizes may work occasionally, but given the small mesh size required for the collection of scissurellids, a dredge will fill up with sediment in the briefest of times. Grab samplers may be more efficient and some models (e.g., *Wildco* Petite Ponar, 14 kg: see Geiger, 2006d; Figs 7, 8) permit deployment by a single person from a small boat. Sampling down to 200 m depth

Figure 6. Collecting micromollusks in the intertidal. DLG with Takenori Sasaki (right) at Cape Manazuru, Sagami Bay, Japan, eventually yielding the type specimens for *Sinezona costulata, Coronadoa hasegawai* and *Anatoma parageia*. DLG washing rocks with a dish-washing brush in a plastic tub. Note sieves on right. Photograph Christine E. Thacker, March 2008.

have been accomplished, both by hand (B. Raines, pers. comm.) as well as by using a standard capstan available on many boats (Geiger, pers. obs.).

Scissurellids comprise some of the smallest known gastropods down to just over 0.5 mm. To avoid loss of specimens, sieves with mesh-size of no greater than 0.25 mm/250 µm is best used to remove silt. For sorting, the >0.25 mm fraction is best gently sieved into several sub-fractions to facilitate picking the sediment. This even applies to live examination/sorting procedures.

Fluid preservation needs to be tailored to the intended use of the specimens. By default, specimens were preserved in 90–95% ethanol, with 2–3 initial changes of the preservative, as water bound to small particles will dilute the preservative. The reason is, that ethanol preserved specimens can be used for molecular work as well as radula extraction, and even for examination of external anatomy by SEM. Specimens/samples designated for histology were fixed in 5% formalin for a few days, then transferred to >80% ethanol. The ethanol concentration chosen is higher

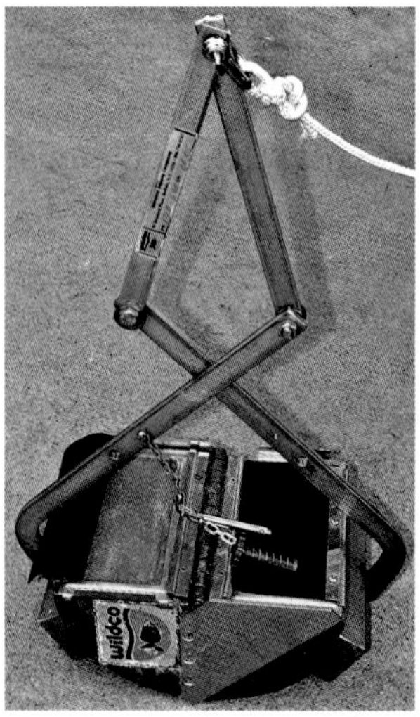

Figure 7. Petit Ponar grab sampler.

Figure 8 (facing page). The use of the Petit Ponar grab sampler in the field, during an abalone survey with California Department of Fish and Game at San Miguel Island, off Santa Barbara, southern California, September 2009. **A.** Preparing the grab sampler and inserting the spring loaded release pin. The rope is stored in the bin to the left. Note that gloves are worn to prevent rope burn. **B.** Lowering the grab sampler over the side of the boat. **C.** The grab sampler approaching the water surface. **D.** The closed sampler being pulled out of the water. **E.** Grab sampler placed into a plastic container. **F.** The top screen is removed. **G.** Inside view of the grab sampler with sediment inside. **H.** Grab sampler is emptied into container. **I.** Happy scissurellid researcher after successful haul. Photographs by Jody Pesapane / Liquid Blue Media LLC.

A
B
C
D
E
F
G
I
wildco

Figure 9. "Spearing" a scissurellid with a brush.

than what is recommended for histology (70%); small shells tend to dissolve in <70% ethanol in as little as six months (Geiger, pers. obs.); hence, a 10% safety margin is advisable.

Samples for dry material must be washed in fresh-water to remove salts, then air dried. Air-drying is optionally accelerated with the last 1–2 washes in high-percentage ethanol.

Many scissurellids are fragile and manipulation of the specimens with forceps is problematic. Manipulation with fine brushes is by far better. I use brushes in sizes 0–0000 depending on size of the specimens. Dry specimens can be rolled in a dish with a dry brush to obtain the desired view, and can be picked up with a moistened (not wet) brush; moistening with water is better than with ethanol because of water's greater surface tension and lower evaporation rate, particularly in places with low humidity. Specimens in fluid can also be manipulated with brushes, and moved by "spearing" them (Fig. 9).

Curation. Store specimens in area with <60% relative humidity in gelatin capsules (gelcaps) that are placed in a secondary container with the label. Alternatively, use high sodium carbonate glass vials (avoid high silica glasses). Labels should be of archival quality and not release any acid. Wet specimens are best stored in high percentage ethanol; even short times (6 months) in dilute ethanol can lead to complete dissolution of the shell. Samples have also been cold stored in Nunc tubes in a -80°C ultracold freezer, with no ill effects to the shell. Individual glass sample vials are stored in larger one liter jars filled with 80% ethanol. If multiple specimens are available, dry at least one good shell specimen and store it separately in the dry collection to verify the identity of the species. Specimens stored in buffered formalin show evidence of acid corrosion (Fig. 10).

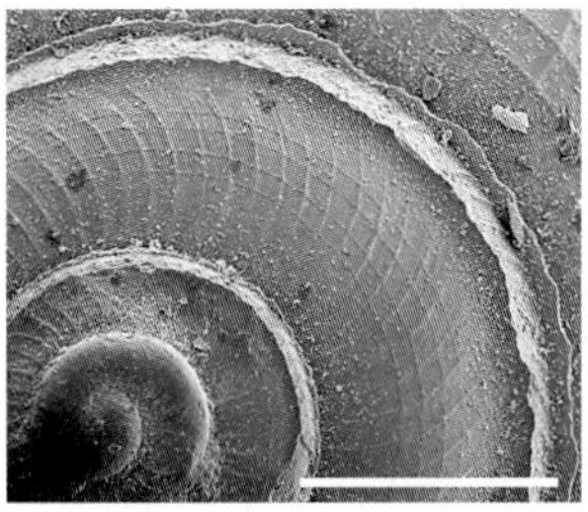

Figure 10. Shell of specimen stored in buffered formalin. Note efflorescence similar to that seen in early stages of Byne's/glass disease. *Anatoma australis* 1500 m, Lord Howe Rise, Australia, 27.567°S, 161.573°E (AMS C.166770, 2: complete). Scale bar = 0.5 mm.

Sampling effort. The present analysis is based on approximately 9,800 locality records of Scissurellidae s. l. representing approximately 73,000 specimens, an expansion of the account given by Geiger (2008b). The vast majority of locality records is accounted for by lots in approximately 50 museum and some private collections, with some select literature records added. My personal collection comprises 521 lots/3511 specimens, of which I personally collected 135 lots. From those lots approximately a dozen new taxa and neotypes were derived. Accordingly, over 95% of all lots and specimens examined are housed in other collections. Only records that were reasonably georeferenced (to 0.1° latitude and longitude) were considered. A visual impression of the sampling effort is gleaned from the density of dots plotted on the world map shown in Figure 11.

The least amount of material is available from the temperate southeastern Pacific and the Arctic. The most densely sampled areas are the Indo-Malayan Archipelago with approximately half of all lots and specimens, and Southern Australia plus New Zealand. It is somewhat surprising that Europe only shows average sampling densities, despite a long history of collecting. It becomes apparent that contemporary sampling efforts of investigators, who focus on micromollusks, determine sampling densities: Philippe Bouchet (MNHN) with the New Caledonia, Vanuatu and Philippine workshops; Winston Ponder (AMS) focusing on Australia and adjacent territories; and Bruce Marshall (NMNZ) with emphasis on New Zealand and surrounding seas. This observation is confirmed with the number of specimens and lots available from the top collections (See Geiger, 2008: table 3 for details).

Government regulations The increasing bureaucratic hurdles put in the way of systematists, while in some instances understandable, are excessive on occasion and are counterproductive to scientific advances. While the concerns about bioprospecting and the associated commercial exploitation are justified, biosystematics and revisions do not fall into that category; if anything, such studies are a drain on financial resources. Given the limited number of specialists, it is in the interest of governments to facilitate pure academic biosystematics. Two examples given below illustrate problematic regulations.

The Brazilian government requires recipients of institutional loans to sign a *Material Transfer Agreement* (MTA). The language of the MTA threatens the receiving institution with being sued in international court for any reason, and prohibits any commercial exploitation of anything associated with the specimens. Does the publication of results constitute a commercial use? What if it is in a book that is sold? How about someone being paid to examine the specimens? The language of the document is sufficiently vague, and not being trained as a lawyer, I am not sure about all the possible consequences of signing such an agreement. Add to the above the potential of overzealous government lawyers, and it results in a situation where the uncalculatable risks are too great to be acceptable. As a consequence, material in Brazilian institutional collections was not examined and this country remains a *mare fere incognitum* with respect to scissurellid biodiversity.

The Australian government requires for any material exported legally after 1991, that subsequently designated types be sent back to Australia. The best specimens of *Satondella cachoi*, and the only live collected specimen from which the radula was prepared, was not designated as the holotype of that species. It was extracted from

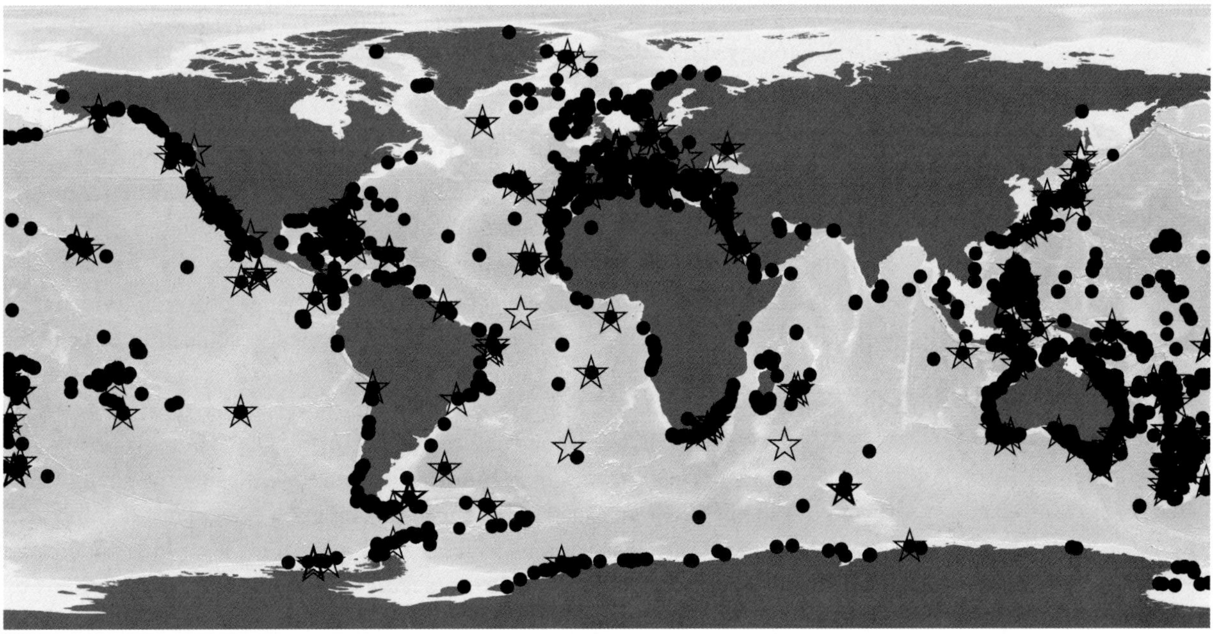

Figure 11. Global distribution of samples. ☆ type locality. ● specimen record. ■ literature record.

a bulk sediment sample, that was collected under permit and exported legally in 1994 from Western Australia to SBMNH. The repatriation demand is only effective after the sample was sorted, identified, shell and radula imaged by SEM, and the particular specimen was recognized as being new in the context of a global revision. As far as is known, the Australian government is not compensating for all this value added, which is easily assessed in the tens of thousands of dollars. Accordingly, this particular specimen is only listed as additional material examined, for which the repatriation requirement does not apply. The Australian government's regulations are in fact counterproductive to its own goals.

I would be remiss not to emphasize, that those are taxonomic impediments imposed by governments, while colleagues at the various institutions have always been helpful, accommodating, professional, and have tried to act in the best scientific interest.

Specimen preparation

Specimen preparation techniques in general have been detailed by Geiger *et al.* (2007). A rudimentary account specific to scissurellids is provided here.

Type specimens. All type specimens of previously described species were not cleaned except for light brushing with a fine artists brush (sable #0000). A few types that were glued to paper were separated from it by immersing the piece of paper into water to soften the glue; in cases where writing was present on the paper, the glue spots were repeatedly brushed with a water moistened brush until the specimen was released. Those removals of specimens were only carried after permission had been received from the person in charge at the holding institution; for an example of glued specimens, see Color Plate 2A. Types were mounted on double sided carbon tabs (Ted Pella), and imaged un-coated in variable pressure mode (30–80 Pa: see also Stokes, 2008) in an SEM using the VPSE detector. Variable pressure permitted imaging challenging specimens such as the holotype of *Sinezona rimuloides* uncoated and attached to a glass histology slide (Fig. 418). Specimens sputter coated with gold gave slightly better images than un-coated specimens (e.g., *Sci. maraisorum* Fig. 131: A coated, B–C un-coated): fewer hotspots from local charge build-up arose. These could be mitigated by higher pressure in the SEM chamber or by alternate noise reduction techniques such as frame integration or line averaging, but both produced somewhat less crisp images. The loss of resolution is of no consequence at the much reduced size the final images are shown (factor of ~5).

Shells. Shells selected for SEM were assessed for stability (presence of cracks, evidence of erosion, thickness of shell), and sturdy shells were cleaned by sonication. Specimens were first wetted with 100% ethanol to break surface tension, then immersed in distilled water and sonicated in an embryo bowl. The lower surface tension of ethanol permits fluid to enter the small apertures of the specimens. Direct immersion in water usually results in relatively large air bubbles being trapped inside the shell, which cause the shell to float on the surface. Those specimens are not sonicated properly. Some heavily encrusted specimens were cleaned manually with a tungsten needle and/or a short-bristled brush prior to sonication. Cleaning delicate specimens by immersing them alternatingly in ethanol and water was occasionally successful. After sonication, specimens were washed several times in distilled water, with a final rinse in ethanol to speed up complete drying. Specimens were mounted by using a brush to spear the specimens on double sided carbon tabs. The most effective placement is a single row ⅕ above the midline of the stub, the apex pointing upwards at an angle of 20–30° relative to the surface of the stub, the slit and the umbilicus parallel to the surface of the stub. This permits an apertural view with the stage tilted at 20–30°, and an apical view including a detail of the protoconch by rotating the stage by 180° and tilting it to 60–70°. To obtain the umbilical view, the specimens had to be re-mounted. The carbon tab was cleared of the gold coating below the spire by rubbing the sides of a pair of forceps until the black carbon reappeared. Most images were taken in variable pressure mode at 30 Pa chamber pressure using the VPSE detector. This results in more even illumination of the specimen compared to the more standard secondary electron detector. The slight loss in resolution is irrelevant; only a quarter of the native image resolution is used for print production. The specimens were partially or fully released from the carbon tab with a 100% ethanol moistened brush, and pushed over so that the spire would adhere to the cleared carbon tab. Remounting is best done after the specimens have been in atmospheric conditions for at least 1–2 hours. Removing specimens immediately after taking them out of the SEM tends to leave some carbon tab material stuck to the shell. Such residues can be removed by brushing the spots with 100% ethanol moistened brush, but it is best to avoid this extra step. Remounted specimens were re-coated with gold.

Radula. Radula preparations were mostly taken from specimens with dried animals within the shell. In a few cases, the dried animal was pulled out of the shell with a hooked tungsten needle. In most instances, the animal in the shell was soaked in water. Often, the specimen had to be placed in a partial vacuum (half atmospheric pressure) to remove air bubbles within the shell and to permit water to enter the space. The softened specimen was pulled out with a hooked tungsten needle while the shell was held between two fingers. In a few instances, limited observations on the external anatomy were possible at this stage. The specimen was then placed in an embryo bowl, one pellet of KOH was added and the specimen was macerated at ~40°C for a few hours. The presence of eyes was often detected after some maceration in KOH. The radula often has to be separated from the softened tissue using a couple of tungsten needles. The isolated radula is washed in water, and then briefly (5–10 s) dipped in commercial bleach, then washed again in water. That bleach step results in much cleaner radulae. Radulae were mounted in water onto glass cover slips attached with double sided carbon tabs to a SEM stub, sputter coated with gold, and viewed in high vacuum mode in the SEM (15 kV, probe current 20–100 pA) using the secondary electron detector.

External anatomy. Either the shell of the specimen was removed with dilute HCl, or the animal was pulled out of the shell with hooked tungsten needles. The body was put through several changes of 100% ethanol, then through several changes of 100% hexamethyldisilizane (Ted Pella, HMDS), then air-dried, mounted onto double sided carbon tabs, and sputter coated. As well-preserved animals are rare, the maximum amount of information has to be retrieved; hence, the dried animal was then further dissected in several stages. Specifically, the roof of the mantle cavity with the attached gills was removed with very fine iris scissors (FST 15000-03, 2 mm blade) and remounted to examine the pallial complex. In some instances, the radula was dissected out of the dried animal through a dorsal incision of the snout.

Conditions for SEM imaging were variable and dictated by the behavior of the specimen under the electron beam.

Photography and digital imaging

SEM. Scanning electron micrographs (SEM) were taken on several instruments: *Cambridge* Stereoscan 360 at the University of Southern California, Los Angeles, California, USA; *LEO* 435VP at the Australian Museum, Sydney, New South Wales, Australia; *Hitachi* 3000N at the Natural History Museum of Los Angeles County, Los Angeles, California, USA; the vast majority of images was taken on the *Zeiss* EVO 40XVP at the Santa Barbara Museum of Natural History, Santa Barbara, California, USA. All images were digitally captured and processed in *Photoshop*. All images were cut out from the background, tonality was adjusted, and some hot-spots were dodged out, or alleviated with highlights/shadows control. All *Photoshop* manipulations only serve the purpose of producing the most pleasing image; while the scientific integrity of all specimens was retained. Particularly, no composite specimens were created, and no defects were mended. Some authors (e.g., B. Marshall, pers. comm.) digitally remove small pieces of dirt to provide a less distracting image. As the line between justified image improvement and unacceptable data generation *ex nihilo* is too vague, the use of cloning tools and healing brush in *Photoshop* was judged entirely unacceptable.

In some instances, the specimens were too large to be imaged in a single frame even at maximum working distance. Multiple images were stitched together in *Photoshop* and due to residual lens distortion, images had to be somewhat un-distorted to provide the best, most representative image. These photographs are marked in the figure caption as 'composite image' (*e.g., An. euglypta*).

Light microscopy. Histological photomicrographs were taken on a *Zeiss* Axioskop 2 plus with a set of *Zeiss* plan-apochromatic objectives, optovars 1.2× and 1.6×, and digitally captured with an *Zeiss* Axiocam HRc (12 megapixels) using *Zeiss Axiovision* software (Carl Zeiss, Oberkochern).

Some light optical images were generated on a *Zeiss* Discovery V12 stereomicroscope with planapochromatic 0.63× and 1.5× objective lenses, objective slider, and captured with an *Zeiss* Axiocam HRc. Z-stack integration was performed with *HeliconFocus* from 37 MB 24 bit RGB .tif-files, and optimal software parameters (radius, smoothing) chosen experimentally for each specimen (e.g., Color Plates 1, 2). Z-stacking with *Photoshop* CS5 produced unusable images, because of autoalign and blending errors from microscope images. A few images were manually z-stacked by using layer masks in *Photoshop*.

Three-dimensional reconstruction from histological sections was accomplished with *Amira* 3.1 (TGS, San Diego); see Figure 29. For methodological review, see Ruthensteiner (2008).

SLR photography and digital imaging. SLR photography was performed with a *Contax* RTS III and *Zeiss* MakroPlanar 100 mm, with extension rings, bellows and a stacked *Zeiss* Tessar 50 mm f/1.4, as conditions required (e.g., Color Plate 2A; see Geiger, 2006e for review of photographic methods). *Fuji* Provia 100F was scanned on a *Nikon* Coolscan 4000ED at 4000 dpi in 14 bit Adobe 98 RGB using Nikonscan 3.1 software. Some reproduction photographs from books were taken with a *Nikon* Coolpix 8700 (8 megapixels RAW files). All files were processed in *Photoshop*. Layered master files were down-sampled to 400 dpi at 100% production files and saved as flattened uncompressed .tif files. All images were output-sharpened with *NikSharpener* 3.007 for a 175 lpi line screen.

Some original illustrations were scanned from original publications or photocopies on an *Epson* 4990 and V700 flatbed scanner and further processed in *Photoshop*. Various digital imaging techniques were applied to produce the best reproduction possible, including threshold transformation, noise reduction filters, burn and dodge tools, and dust and scratches filter to remove paper irregularities from line drawings, minimal Gaussian blur to remove print screen patterns and to avoid subsequent moirée patterns; curve manipulation to produce better grey scale, eraser with variable hardness and diameters to remove specimens from uneven background and to remove extraneous elements, rotation and scaling to show specimen in an as uniform fashion and to make various image elements proportional to each other.

Distribution maps. A global map download from http://www.ngdc.noaa.gov/mgg/topo/pictures/GLOBALeb3colshade.jpg was custom grey scale transformed in *Photoshop* and used as the base map for plotting data points with *iMap* 3.5 (Biovolutions, Leuven).

Characters

Shell shape. Shell shape varies to a great degree in the microscopic Vetigastropoda. An overall description is given with size at maturity (Fig. 12). Common expressions include:

- Trochiform with elevated spire (Fig. 12A).
- Trochiform with flat spire (Fig. 12B).
- Limpet-shaped due to increased expansion rate of the whorl (Fig. 12C).
- Limpet shape due to compression of shell axis (calyptraeiform: Fig. 12D).

Protoconch. The protoconch is separated into two growth phases: the embryonic cap that is formed as a sheath (Crofts, 1937 for *Haliotis*), and the main protoconch that is formed by continuous colabral accretion, the usual growth pattern in mollusks. Sometimes the sculptural pattern on the embryonic cap and the remainder of the protoconch are different, which is noted.

The number of protoconch whorls varies little within species. Protoconch whorls are counted here in apical view by extending the imaginary spiral (*S*) starting in the middle of the apertural margin of the protoconch towards the tip

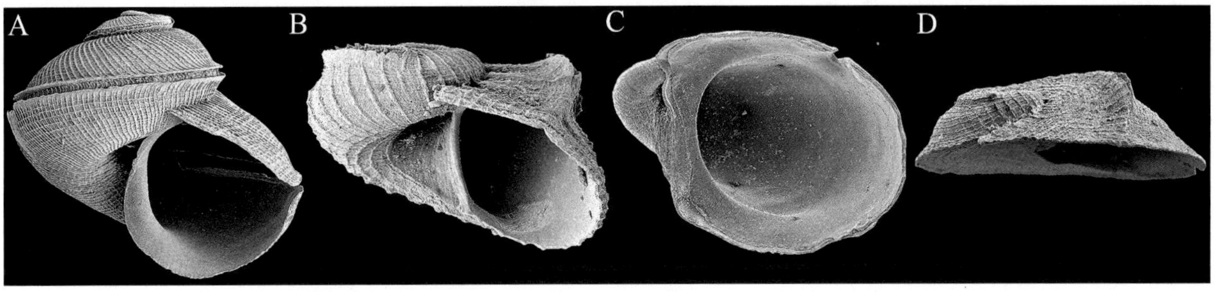

Figure 12. Shell shapes. All specimens are oriented identically with the apex at the top and the umbilicus at the bottom. **A.** *Anatoma tobeyoides*: trochiform with elevated spire. **B.** *Satondella senni*: trochiform with flat spire. **C.** *Incisura lytteltonensis*: limpet-shaped due to increased expansion rate of the whorl. **D.** *Depressizona exorum*: calyptraeiform.

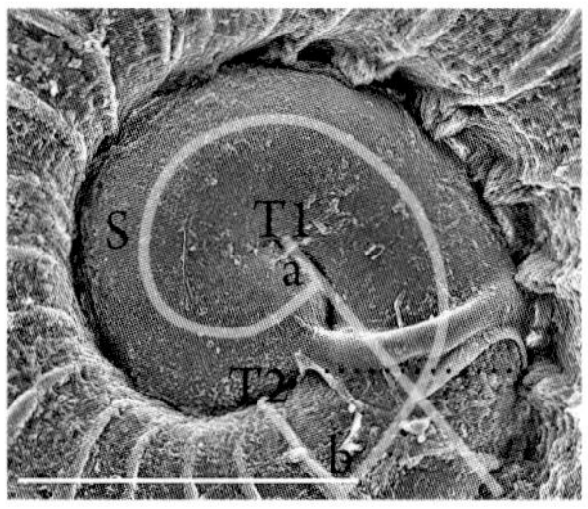

Figure 13. Counting protoconch whorls. See text for details. Scale bar = 100 µm.

of the embryonic cap. At the intersection between this imaginary spiral and the border of the embryonic cap (*a*) an imaginary line (tangent *T1*) is placed at right angles to the spiral *S*. One whorl is counted when spiral *S* intersects at right angle with a line (*T2*) parallel with *T1* at point B. Portions of whorls are determined by angular measurements between *T1* and the line going through the feature of interest (e.g., apertural margin of protoconch: *dotted line*) parallel to the growth margin (see Fig. 13).

Whorl counts are given in decimal system for ease of comparison. The conversion of fractions to decimal values generates values with up to three decimal places: ⅛ = 0.125. These values are not an indication of precision, but a consequence of numerical conversion between fractions to decimal indication. Accuracy in the determination of whorls is approximately 1/20 whorl = 0.05 whorl. Decimal values are preferred over fractions because of greater ease of comparison. Protoconch whorl count is a continuously variable character without discrete states.

Protoconch sculpture. The protoconch structures and sculpture types have been defined by Geiger (2003) and those character state definitions are adopted and expanded (Fig. 12). Common character states include:

- Smooth (Fig. 14A).
- Reticulate (Fig. 14B).
- Spirals (Fig. 14C).
- Fine irregular axials (Fig. 14D).
- Strong axials (Fig. 14E).
- Pitted.
- Flocculent (Fig. 14F).
- Microhexagonal.

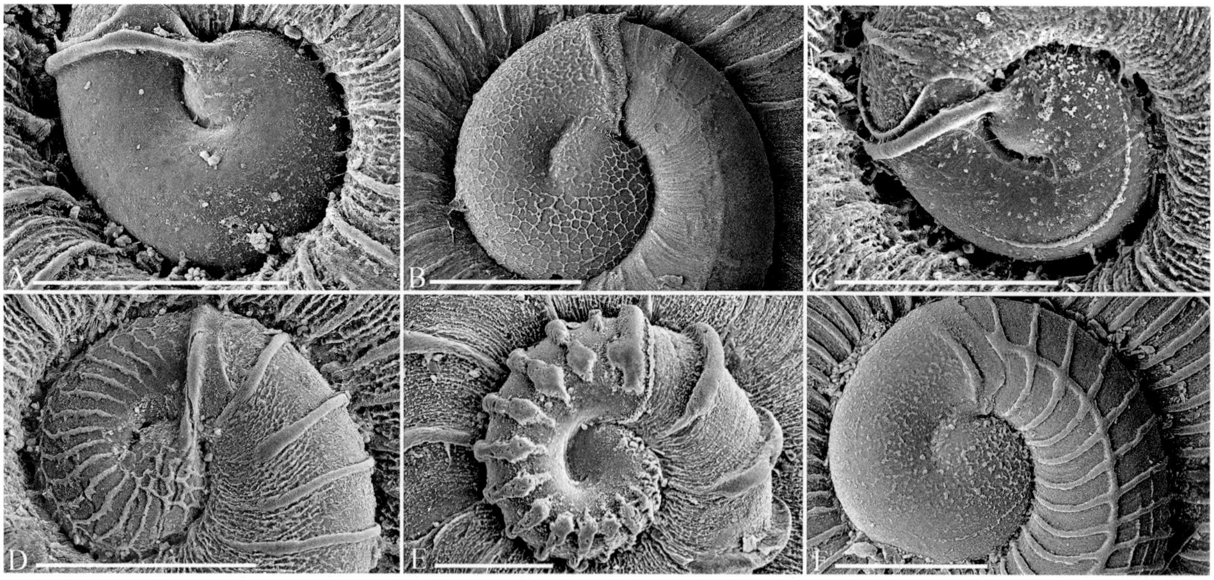

Figure 14. Protoconch character states. **A.** *Sinezona ferriezi.* **B.** *Anatoma gunteri.* **C.** *Sinezona plicata.* **D.** *Scissurella quadrata.* **E.** *Sinezona iota.* **F.** *Anatoma australis.* Protoconch sculpture. Smooth: A.—Reticulate: B.—Spiral: C.—Irregular axials: D.—Strong axials: E.—Flocculent: F.—Protoconch varix. Absent: E.—Present, not connecting to embryonic cap: B, F.—Present, connecting with embryonic cap: A, C, D.—Protoconch aperture shape. Sinusoid: A – D, F.—Rounded: E. Scale bars = 100 µm. Modified from Geiger (2003).

Protoconch varix. Some protoconchs show a distinct pre-terminal thickening, known as a varix. Common character states include:

- Absent (Fig. 14E).
- Present, not connected to apex (Fig. 14B, F).
- Present, connected to apex and deforming area of embryonic cap (Fig. 14A, C, D).

Shape of PC margin. The shape of the apertural margin of the protoconch shows a few variations. Common character states include:

- Sinusoid (Fig. 14A–D).
- Convex (Fig. 14E).
- Prosocline.

Teleoconch I/II. Two portions of postembryonic shell are distinguished. Teleoconch I starts after the protoconch and ends with the onset of the selenizone. Woodward (1859) first noted the lack of selenizone on teleoconch I in Scissurellidae. Teleoconch II is the selenizone/slit/foramen bearing portion of the shell. Some genera (*Coronadoa, Larochea, Larocheopsis, Trogloconcha*) only have teleoconch I as no slit is formed in their members.

Teleoconch I size is indicated in terms of fractional whorl. Although some authors have suggested that the teleoconch I always has approximately 0.66 whorl (Sohl, 1992: 431), there is considerable variation from 0.25 (*Anatoma tobeyoides*) to >1 (e.g., *Incisura remota, Scissurella mirifica*). Intraspecific variability of teleoconch I whorl number usually is 0.25 or ±0.125. In most instances, the variation is indicated in the description, or the modal value is given. However, some species are still known only from only a few specimens, and the intraspecific variability is insufficiently determined. In these cases, it is suggested that the user of this manual use the above guideline for intraspecific variation.

Whorl count is a continuously variable character without discrete states.

Expansion rate is only rarely indicated and measured as defined in the Glossary.

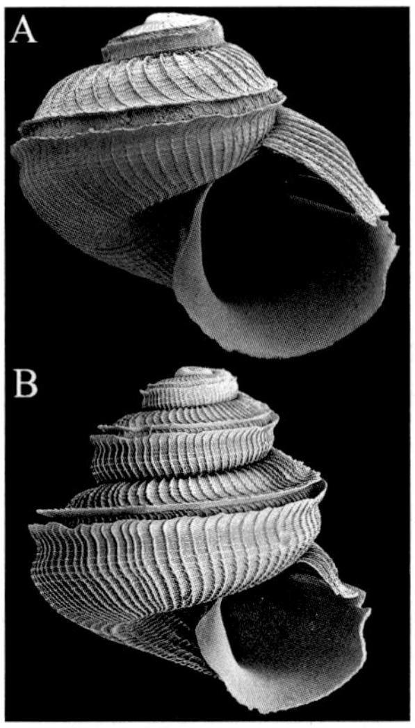

Figure 15. Shape of shoulder. **A.** *Anatoma proxima*: rounded. **B.** *Anatoma munieri*: angulate.

Shoulder. The shoulder has some variable characters, including the following character states:

- Shape: rounded (Fig. 15A) or angular (Fig. 15B).
- Sculpture: see shell sculpture below for details.

Base. The base has its full complement of sculptural elements with the following character states:

- Constriction below selenizone absent, weak (Fig. 15A), moderate, strong (Fig. 15B).
- Sculpture similar to shoulder or different.
- Number and strength of of axial and spiral elements.
- Presence of umbilical trough (*Sasakiconcha elegantissima*).

Shell sculpture. Shell sculpture is variable (Fig. 16). It usually includes both axial and spiral elements. The sculptural elements are described in the order of prominence (keels, ribs, cords, lines, threads). In most instances, counts of spirals and axials are given at well-defined growth points as appropriate and applicable: on the first whorl, on the last teleoconch whorl, at the onset of the selenizone, at the apertural margin. The onset and the distributional regularity or irregularity of spirals or axials is indicated as appropriate, as well as any interaction between spirals and axials (e.g., the formation of points or spines).

The description of shell sculpture is partitioned into three parts: teleoconch I, teleoconch II shoulder between apical suture and selenizone, teleoconch II base between selenizone and umbilicus. Characters and states include:

- Axials: onset, number at position, strength at position [keels, ribs (Fig. 16D), cords (Fig. 16C), lines, threads (Fig. 16E)], change of number/strength with growth.
- Spirals: onset, number at position, strength at position [keels (Fig. 16E), ribs, cords, lines (Fig. 16D), threads (Fig. 16C)], change of number/strength from suture to umbilicus.
- Crossing of axials and spirals: one running over other, forming point or spine (Fig. 16F).
- Occasionally, oblique elements (Fig. 16A).

Umbilicus. The umbilicus is open or closed, and often has a prominent funiculus. Common character states include:

- Absent.
- Closed with callus (Fig. 17A)
- Present without marginal carina and without funiculus (Fig. 17B).
- Present with marginal carina and without funiculus (Fig. 17C).
- Present without marginal carina and with funiculus (Fig. 17D).
- Present with marginal carina and with funiculus.

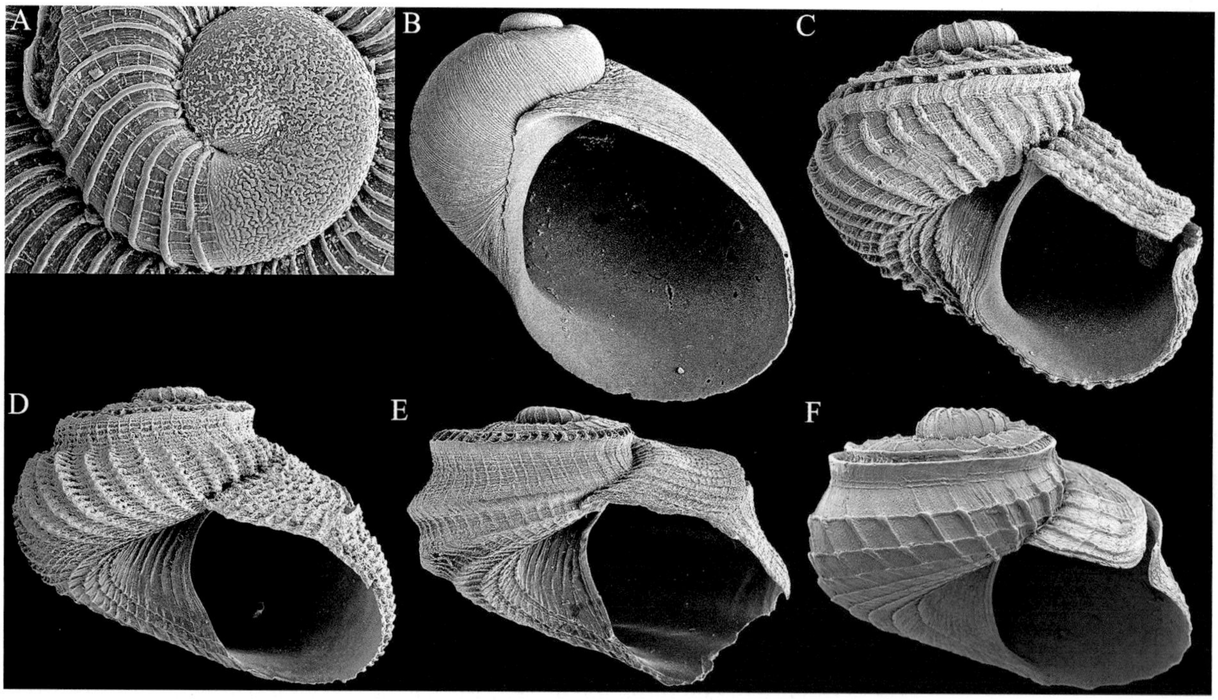

Figure 16. Sculpture elements. **A.** *Anatoma tobeyoides*: centrifugal threads on teleoconch I. **B.** *Trogloconcha christinae*: axial lines. **C.** *Scissurella alto*: axial cords intersected by spiral threads. **D.** *Sinezona plicata*: axial ribs intersected by spiral lines. Note that the ribs do not leave an impression in the internal wall. **E.** *Sukashitrochus morleti*: spiral keels. Note the keels are visible in the internal wall. **F.** *Scissurella rota*: raised points formed at intersection between spirals and axials forming cancellate pattern with raised points.

- Underside of protoconch visible.
- Walls straight.
- Cavity made by rounded portion of previous whorls (e.g., *Sinezona platyspira*).

Umbilical brood pouch. A brood pouch is occasionally encountered in some Larocheidae. Character states:
- Absent.
- Present.

Aperture. The overall shape and attributes of the aperture include the following character states:
- Round.
- Oval. (Fig. 16B)
- D-shaped (Fig. 16C).
- Subquadrate.
- Flared (Fig. 15B).
- With internal rim (*Satondella tabulata*).
- Lower adumbilical corner with small shelf (Fig. 16C).

Selenizone and slit. A few parameters are identified in relation to the slit and selenizone.
- Slit/foramen absent (Fig. 16B, C).
- Selenizone present/absent.
- Position of selenizone at periphery (Fig. 15A) or above periphery (Fig. 16F).
- Slit open, margins parallel (Fig. 16C) or converging (Fig. 16F).
- Closed anteriorly forming foramen (16D).
- Strength of keels of selenizone: weak/fragile, moderate (Fig. 16F), strong (Fig. 16C).
- Elevation of keels of selenizone: low (less then 25% the width of the slit. Fig. 16C), moderate (25–100% the width of slit at its widest point Fig. 16E, F), high (>100% of the width of the slit at its widest point).
- Sculpture on keels of selenizone.
- Strength and density of lunules.
- Cross section of keels: straight, turned inwards, curved upwards.

Start of selenizone. The start of the selenizone is indicated after the number of teleoconch I whorls, usually to a precision of 0.1 whorl. It is a continuously variable character without discrete states (Fig. 16A).

Suture. The suture is either connected flush with the previous whorl (Fig. 16B, C), or is impressed by various degrees (Fig. 16E). Most common is a moderately impressed

Figure 17. Umbilicus characters. **A.** *Sinezona subantarctica*: closed umbilicus covered with callus. **B.** *Anatoma australis*: open umbilicus without marginal carina **C.** *Sukashitrochus atkinsoni*: open umbilicus with marginal carina. **D.** *Coronadoa simonsae*: open umbilicus with funiculus/periumbilical cord. Scale bars = 200 µm. Modified from Geiger (2003).

suture. Discrete character states are difficult to define, but the character is helpful in differentiating between some taxa.

Sutsel. This new term first introduce by Geiger & Marshall (2012) refers to the portion of the base between the suture and the lower keel of the selenizone above it: *sut*ure to *sel*enizone = sutsel (Fig. 18). In species in which the suture is located immediately below the selenizone, there is no sutsel (Fig. 15A). The width of the sutsel compared to the width of the selenizone: 0× = no sutsel; 0.5× = sutsel half as wide as selenizone; 1× = sutsel as wide as selenizone; 2× = sutsel twice as wide as selenizone. The sculpture visible on the sutsel is specified. This morphological character is particularly useful for the differentiation of Anatomidae. As the aperture generally descends along the shell's axis, the sutsel becomes wider with growth. Accordingly, stage specific indications usually accompany this term, e.g., after one whorl = flush with onset of selenizone, at apertural margin of large specimen. It is a continuously variable character without discrete states.

Figure 18. The sutsel is the area between the two arrows: lower arrow at suture, upper arrow below selenizone.

Operculum. The operculum is the corneous (partial) cover of the shell's aperture. In some groups the operculum is entirely missing. The proportion of the aperture covered is derived from observations of animals in the shell with the operculum in place. The placement of the nucleus and the approximate number of whorls is given.

- Operculum: present, absent.
- Size: covering aperture, covering portion of aperture.

Radula. The radula is now known for the majority of species. It is described for each species, ideally based on at least two specimens. The description starts at the rachidian or central tooth, and proceeds towards the periphery. The degree of radular interlock terminates the description.

Rachidian tooth. The main character states include:

- Cusps arranged in row parallel with row of teeth.
- Cusps arranged in V-shaped formation.
- Relative size of central and lateral cusps.
- Number of cusps on each side of central cusp.
- Level of development: rigid teeth due to extensive tanning of chitin or floppy teeth (flexiglossate: some *Anatoma* species).

Lateral teeth. The radula has five lateral teeth. Character states include:

- Relative size of rachidian tooth to lateral teeth, and relative size among lateral teeth.
- Lateral teeth 1–3 similar or dissimilar to lateral tooth 4.
- Lateral teeth 1–3 with cusps on outer edge of tip only, or also some along inner edge (*Satondella*).
- Lateral tooth 5 enlarged by broadening (most Scissurellidae), enlarged by elongation (many *Anatoma* species), scythe-shaped (some *Anatoma* species), not distinctly different from lateral teeth 1–4 or inner marginal teeth (Larocheidae, Temnocinclidae).

Marginal teeth. Only few character states are found on the marginal teeth:

- Food-groove present or absent.

- Size of central cusp: about the same size as others, much larger.

External anatomy. The external anatomy of several species is given. Depending on the preservational state of the material, the following characters are given:

- Cephalic tentacle present or absent, surface condition.
- Eyes present or absent.
- Suboptic tentacle present or absent, relative size to cephalic tentacle.
- Accessory cephalic tentacle present or absent, surface condition.
- Epipodial tentacles: number, arrangement, and relative size.
- Epipodial sense organ: location.
- Attachment site of operculum.
- Bursicles on gills confirmed.
- Presence of mantle tentacles.

Discussion of characters

Protoconch. The sculpture of the protoconch is diverse and shows a rich complement of structural details. This fact has long been recognized (Herbert, 1986; Bandel, 1998). However, despite examination of structures by SEM, some misinterpretations have crept into the literature.

Smooth protoconch sculpture should not be confused with eroded protoconchs; the two conditions are difficult to tell apart on occasions. At higher magnifications (~2000–4000×) the surface of a smooth protoconch still appears smooth, while an eroded protoconch shows strong granularity (Fig. 19). Most eroded protoconchs are unequally eroded and remnants of the true surface sculpture may be traced near the suture. In specimens with badly eroded protoconchs, the apertural margin of the protoconch is difficult to identify, and part of teleoconch I is also devoid of sculpture. Spirals are the first sculptural elements to be lost.

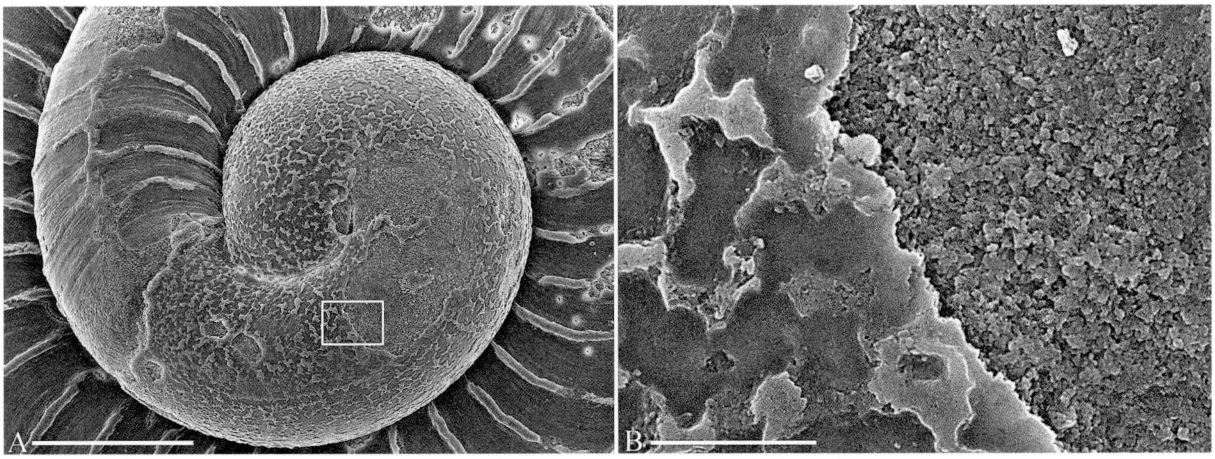

Figure 19. Distinction between true protoconch sculpture and eroded surface. *Anatoma kopua* (AMS C.453747). **A.** Entire protoconch with evidence of partial erosion. Box shows location of enlargement shown in B. Scale bar = 100 μm. **B.** Enlargement of protoconch area with original sculpture on left and eroded surface on right. Note microgranular appearance of eroded surface, while smooth areas with original sculpture is smooth even at high magnification. Scale bar = 10 μm.

Close to the apertural margin of the protoconch, a subterminal varix is often found, which in some species forms a connecting bridge to the early embryonic cap of the protoconch. The connection of the varix to the embryonic cap is variable in some species, e.g., *Sin. zimmeri*. It does not have value for supraspecific classification, but is marginally helpful when identifying species. A well-developed varix and strong axial sculpture can be seen under a good stereomicroscope and high-contrast illumination (Fig. 20). However, the inability to visualize a protoconch varix under the stereomicroscope cannot be taken as good evidence of its absence.

The shape of the protoconch aperture, sinusoidal to curved, has not been used for generic differentiations. The much more common sinusoidal condition varies from a strong curve with an amplitude of up to half the width of the whorl, to a barely perceptible undulation. The amplitude of the sinusoidal aperture shape is variable within species; the rounded aperture has a sinusoidal line with zero amplitude. Species identification based on the shape of the protoconch aperture is at best tentative.

Protoconch in systematics. Protoconch sculpture has been used to classify scissurellids for a long time. Herbert (1986) cited the smooth protoconch as diagnostic for *Sukashitrochus* based on the only examined species, *Sukashitrochus dorbignyi* (as *Suk. maraisi*). Other diagnostic character states cited include fine axials and reticulation in *Praescissurella* by Lozouet (1998), strong axials for *Scissurella* and *Sinezona*, pitted in *Sutilizona* (McLean, 1989), and reticulate in *Thieleella* (Bandel, 1998). Marshall (1993) cited specific protoconch types as diagnostic for the subfamilies discussed.

The assignment of particular species to a genus has also been based on protoconch sculpture. Burnay & Rolán (1990) based their generic assignment of "*Sinezona*" *lobini* on protoconch sculpture although no specimen with a slit closed into a foramen has been found. Given the non-diagnostic nature of protoconch sculpture and the consistently open slit, the species was reassigned to *Scissurella* by Geiger (2003). Linse (2002) justified the placement of *clathrata* in *Anatoma* instead of *Scissurella* on its putatively smooth protoconch, although her illustration showed evidence of erosion in the apical portion of the shell.

Protoconch sculpture, as with any other character, shows some degree of intraspecific variability. The protoconch types illustrated above (Fig. 14) are

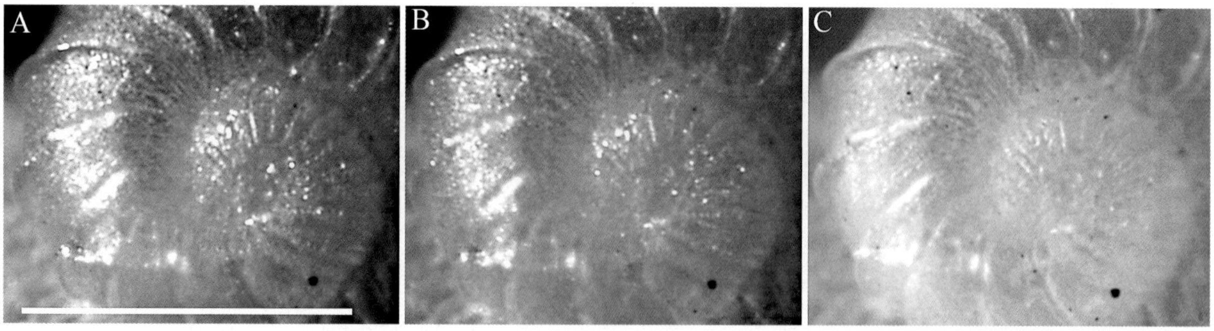

Figure 20. Stereomicroscope image of the protoconch of *Sinezona cingulata*. *Zeiss* Discovery V12 with *Zeiss* planApo S 1.5× objective lens, at maximum magnification (150×) and smallest aperture to increase depth of field. Approximately ⅓ of image area shown. All three images with autoexposure and simultaneously processed in Photoshop to provide accurate visual impression of native image. Scale bar = 0.25 mm. **A.** Unidirectional, low incident angle light from left. **B.** Dual fiber-optics light sources. **C.** Shadow-free illumination. Note decreasing level of contrast and visible detail. 3 m, Posillipo, Naples, Italy, 40.817°N, 14.217°E (DLG 317, 1).

distinct. There are some cases of gradations (e.g., between flocculent and reticulate in *Anatoma* species), but pairs of species showing such different types are distinct. Much caution has to be applied when putative species only differ by minute details in protoconch sculpture. Marshall (2002) cited unspecified minor differences in protoconch sculpture between *Sci. prendrevillei* and *Sci. marshalli*. Examination of multiple specimens from both nominal taxa revealed it to be well within intraspecific variability. In the absence of additional distinguishing characters, those two taxa were synonymized.

Bouchet (1990) noted variation in lineages of non-planktotrophic turrids (Neogastropoda: Toxoglossa) and questioned the utility of any protoconch characters in this group. Scissurellids also show non-planktotrophic development; hence, sculptural details is of a priori questionable phylogenetic utility. Geiger (2003) showed the wide intrageneric variability of scissurellid protoconch sculpture types, and demonstrated with phylogenetic techniques that most protoconch sculptures have arisen multiple times and are homoplastic (Fig. 21).

Teleoconch. The majority of species in Scissurellidae are trochiform to naticiform with spire of variable elevation; a few genera show limpet-shaped shells (*Incisura, Temnocinclis, Temnozaga, Sutilizona, Depressizona*). Most species tend to develop a more elevated spire during ontogeny; the whorls are positioned lower on the shell with increasing size. Thus, the application of overall shape parameters proves difficult due to slight differences between species, and a strong ontogenetic trend to taller shell shape. Furthermore, measurements of shell parameters requires that the image is taken in a standardized fashion, a consideration which is often neglected. Hence, despite SEM imaging, shell morphometric parameters can often not be extracted.

This ontogenetic change in shell morphology, however, also permits distinction in many cases between juvenile and mature shells. Specifically, juvenile *Sinezona* species and mature *Scissurella* species share the open slit; hence, the condition of the open slit may be misleading with respect to generic identification of a specimen. Noting whether or not the last quarter of a whorl is descending markedly along the coiling axis can help to clarify the problem, with mature specimens showing that descending portion. In Anatomidae, this phenomenon can help to decide whether a juvenile specimen with a much larger adult form is at hand, or whether a fully grown specimen of a small species is being examined.

The shape of the shoulder is often indicated in generic diagnoses, although objective criteria are difficult to apply. Bandel (1998) used overall whorl shape, scalar and rounded, for generic distinctions in *Anatoma*, but it is difficult to apply such terms unambiguously because of extensive interspecific gradations.

Numerous sculptural elements are distinguished. Most species have axial as well as spiral cords, producing a cancellate pattern. In addition, ribs and folds are encountered, among other modifications. Smooth shells are rare, but those species have been assigned to particular genera (*Incisura, Scissurona*). Quantitative differences in spiral and axial elements, and differences in sculptural elements on the shoulder and the base of the shell have been used to distinguish species, but have not been applied in generic classification of scissurellids. In some species, the apertural region of the body whorl shows slightly different sculpture than the major portion of the last whorl.

Keels on the base of the teleoconch have been used to diagnose *Sukashitrochus* (Geiger, 2003), though the strength of the keels is variable within the genus. They are more prominent at the apertural margin of juveniles with an open slit, while

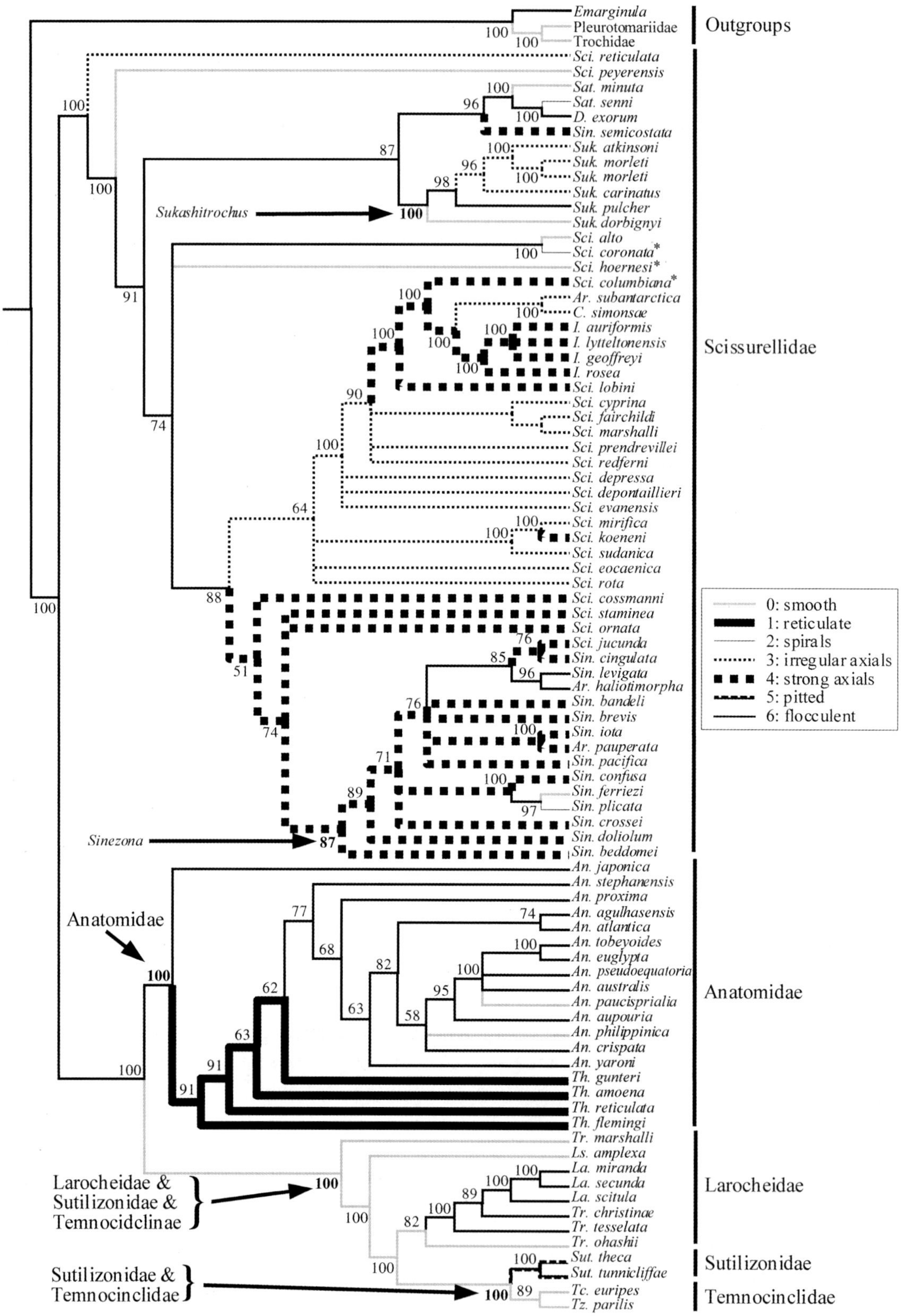

Figure 21. 50% majority rule consensus tree based on matrix for shell morphological characters. Character states for protoconch sculpture are mapped and show non-diagnostic nature of protoconch sculpture. From Geiger (2003). Species names have been updated; * indicates mixed taxon concepts.

the spiral keels become more obtuse at the aperture in mature specimens with a closed foramen.

Some authors have used different terms for equivalent structure depending on axial vs. spiral orientation. For instance, axial cords are occasionally called 'riblets,' while spiral cords are 'cords.' Although such a differentiation is encountered in the literature, it has not been adopted here because it makes it more difficult to judge relative strength of axial and spiral sculpture. While it is not clear that a shell with 'riblets' and 'cords' shows clathrate sculpture of equal strength, it is obvious when the sculpture is described with axial cords and spiral cords. Additionally, sculptural elements at a distinct angle are identified in the unified system, whereas the alternate approach leads to ambiguities.

Umbilicus. The presence or absence of an umbilicus is correlated with overall shell shape. The umbilical sculpture and associated modification of the internal lip of the aperture has been used for species diagnoses, particularly the presence of a keel. The significance of those keels is currently unknown.

The funiculus is a spiral cord in the umbilical wall and has been used as a diagnostic character for the generic diagnosis of *Praescissurella* by Lozouet (1998), but was shown to be homoplastic by Geiger (2003).

Umbilical modifications are known to be involved in brooding by females in *Larochea* and in Liotiidae (Marshall, 1993; J. McLean, pers. comm.), and sexual dimorphism has been documented in scissurellids in *Larochea* and *Larocheopsis* by Marshall (1993), as well as *Incisura auriformis* by Hickman (1999: as *Sinezona* sp.). Gender specific shell morphologies or their absence have otherwise not been mentioned in the literature, with the exception of Kase & Kano (2002) in their discussion of *Trogloconcha*; hence, these potential factors should be kept in mind, because one species may have different morphs, that are identified as two different species. The functional significance of the umbilical trough unique to *Sasakiconcha elegantissima* is unclear.

Slit and selenizone. The configuration of the slit has been considered a key character in scissurellid classification. It ranges from an open slit in which the margins of the slit remain parallel (*Anatoma, Scissurella*), to an open slit in which the margins converge towards the aperture (*Scissurella*), to an elongated slit in which the anterior margins are fused, forming a hole (*Depressizona, Sinezona, Sukashitrochus, Temnocinclis, Temnozaga, Sasakiconcha*), to a round hole without a selenizone (some *Sinezona* species), and finally to the complete absence of any trace of a slit or hole (*Coronadoa, Larochea, Larocheopsis, Trogloconcha*). The last condition is taken as the diagnostic character for Larocheidae.

The selenizone or slitband and slit or foramen is one of the family level characteristics, though it is absent in some genera (*Coronadoa, Larochea, Larocheopsis, Trogloconcha*). Some populations of single species also differ in their tendency to produce a selenizone (Marshall, 2002). The length of the selenizone has been used for generic diagnoses, but Geiger (2003) demonstrated that this character varies continuously among species and has no value for higher classification.

The position of the selenizone varies slightly from peripheral to a shoulder location. This minor positional difference has been used by Bandel (1998) to distinguish between genera in *Anatoma*, but it is difficult to apply these characters unambiguously.

The selenizone and the slit or foramen are laterally bordered by a pair of spiral keels of variable development. The development of the keel is correlated with the degree of sculpturing of the shell; smooth shells usually have a hardly perceptible keel (*Incisura, Scissurona*). The axial growth marks (= lunules) of the selenizone are often coordinated with the axials of the whorl, and have limited use at the specific level rather than the generic level.

Coloration. Most scissurellid specimens are off-white in color. One exception is *Scissurella lorenzi* with its typical axial bands, which are more prominent under the light microscope (Color Plate 2C). In some species, select specimens show some coloration: *An. philippinica* (Color Plate 1E), *An. pseudoequatoria, An. tangaroa* (Color Plate 1A), *An. japonica* (Fig. 756), and *Sci. spinosa* (Color Plate 2B) with brown axial bands; *An. indonesica* with yellow apex (Color Plate 1C). However, examination of those specimens by SEM did not reveal any additional, structural differences, which could be taken as an indication of taxonomic distinctness. The reason for such occasional appearance of color patterns is unknown. It may be related to food sources (color bands are known to be produced by alternating diets in Haliotidae: Olson, 1968a&b), mutations, or effect of the physico-chemical environment those individuals encountered.

Scissurella rota from the Indian Ocean is the only species known that has a strong periostracum. It is dark tan in color and obscures some of the sculpture in that species. A comparison of a typical white shell devoid of periostracum and one with it dark periostracum is shown in Color Plate 2F. The two specimens are from the same lot and were photographed together.

Radula. A special effort was made to document the radulae of as many species as possible. Previously, the radula of scissurellids was considered highly conserved at the family level, and examining multiple species was most likely considered without merit. Sasaki (1998) summarized differences in radular structure among the families (as subfamilies): Larocheidae, Scissurellidae, Anatomidae (R - 5 - n); Temnocinclidae (R - 3 - n); Sutilizonidae (R - 2[–4?] - n). The last two, hydrothermal vent families share a poor distinction of lateral and marginal teeth. In Temnocinclidae, the lateral teeth have a strong bend in the shaft near the cusp, while in Sutilizonidae no such distinction is possible (McLean, 1989; Warén & Bouchet, 2001). For this reason, the number of lateral teeth in Sutilizonidae is difficult to assess; Sasaki (1998) quoted a range of 2–4 lateral teeth. Nevertheless, the radulae of the vent families are more

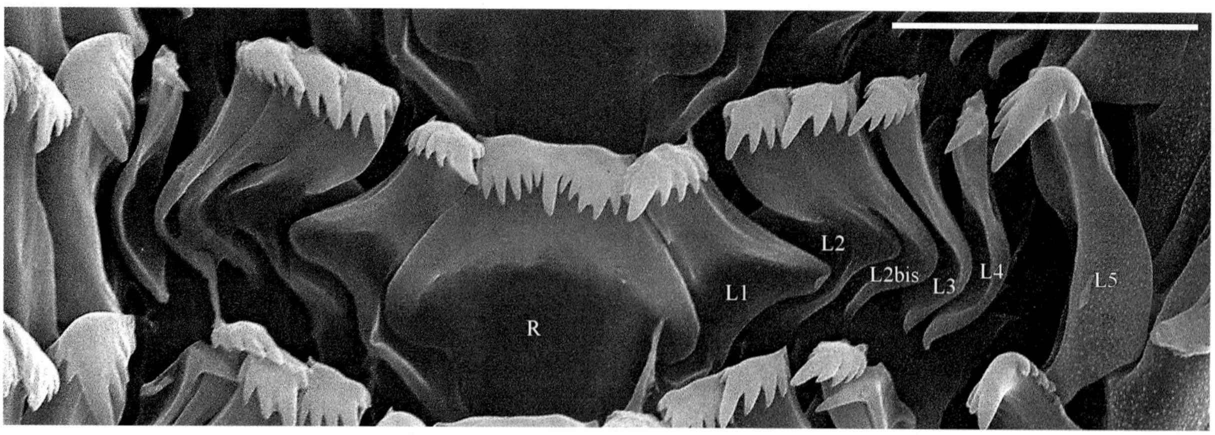

Figure 22. Aberrant radula with duplicated lateral tooth 2. *Anatoma schioettei*. 197–198 m, Greenland, Denmark, 75.015°N, 13.800°W (ZMUC-GAS 427, 428, 1, 1: complete). R: rachidian tooth. L1–5: lateral tooth 1–5. Scale bar = 20 µm.

similar to one another than either is to the remainder of scissurellid subfamilies with five laterals and an enlarged lateral tooth 5.

All indications in the literature of Scissurellidae with only four lateral teeth, or an enlarged lateral tooth 4 instead of 5 are mistaken, as can in some instances be confirmed by the accompanying illustrations (Bartsch, 1946; McLean, 1989: 5, caption fig. 6F). Ponder (1998) also cited an enlarged lateral tooth 4 as characteristic for Vetigastropoda, although the vast majority of Vetigastropoda have five lateral teeth, with the lateral tooth 5 enlarged (Geiger *et al.*, 2008), although in abalone (Haliotidae) it is lateral tooth 3.

A couple of aberrant radulae deserve mention. A single specimens of *An. schioettei* with four lateral teeth 1–3, with a duplicated lateral tooth 2, is shown in Figure 22. Other radula preparations from the same species showed the usual radular pattern (Fig. 888). The common *An. munieri* is surprisingly rarely collected alive. The only successful radula preparation (Fig. 806) suggests that one of the lateral teeth 1–3 to be missing. Whether an aberrant specimen was at hand, or whether this peculiarity is an autapomorphy of the species remains to be resolved.

As pointed out by Geiger *et al.* (2007), histological sections are not suitable for detailed examination of radular structure. Color Plate 5L (*Sin. rimuloides*) and Color Plate 5K (*An. euglypta*) show examples of histological sections. In the case of *Sin. rimuloides*, the structure is further obscured by the remaining anatomical features of the buccal mass. In the case of *An. euglypta*, after 10 years of dark storage, the tissue staining (Richardson's stain) had completely faded, while the radula retained the typical blue tones. In either case, it is only possible to ascertain the rhipidoglossate radular type, but details of tooth morphology are indiscernible.

A certain parallelism in radular variability between Haliotidae (see Geiger & Owen, 2012 for radulae of Haliotidae) and Scissurellidae is observed. The relative width of the lateral teeth is variable, while their height is tied to the height of the rachidian tooth. In Haliotidae as well as in Scissurellidae and Anatomidae, there is a predominant ratio of the width of the rachidian tooth to the five lateral teeth. In select cases, one or several lateral teeth are either much broadened (*Haliotis elegans, An. megascutula, An. rhynchodentata*) or much compressed (*H. dalli, Sin. macleani*). This phenomenon is either viewed as a developmental constraint, or as a functional constraint because the cutting edges of the cusps in a row may need to be aligned to function properly. It further demonstrates parallel diversification patterns in distinct lineages.

Lateral tooth 5. A further question arises with the identity of the fifth tooth from the rachidian tooth, specifically, whether it is lateral tooth 5 (Geiger, 2003) or the first marginal tooth (Marshall, 2002). As discussed by Geiger (2003: 59), the cusps on the tip of lateral tooth 5 (= Marshall's marginal tooth 1) are directed inward, whereas the inner marginal teeth (Marshall's marginal teeth 2–15) have the majority of cusps oriented towards the outer edge of the tip. Some of the ambiguity in the establishment of the boundary between lateral and marginal teeth stems from the ontogenetic changes documented by Warén (1990), Roberts *et al.* (1999), Kawamura *et al.* (2001), and Onitsuka *et al.* (2004). However, the dissimilarity of what most authors have called lateral tooth 5 (Marshall's marginal tooth 1) with the marginal teeth is unquestionable, and the membership of said tooth to the central field—hence, the lateral teeth—is evident.

A further point of discussion is whether this tooth is called lateral tooth 5, pluricuspid tooth, or dominant tooth. The latter two terms are often applied to the hypertrophied tooth in Fissurellidae, but not in other vetigastropod lineages such as Trochidae or Haliotidae. The homology of those enlarged teeth with each other among Vetigastropoda is not obvious, though their membership to the central field, and accordingly to the group of lateral teeth is clear. Specifically, the hypertrophied lateral tooth in Haliotidae is lateral tooth 3, whereas in Fissurellidae, Scissurellidae, and Anatomidae it is lateral tooth 5. In order to facilitate comparative examination of radular structures within Vetigastropoda, and also among other Mollusca, it is advisable not to encumber the field with obfuscatory terminology. Accordingly, this tooth is referred to here as lateral tooth 5.

Radular variability. Some early comparative examination of radular data attempted to identify commonalities in certain genera. Geiger (2003) noted that Anatomidae have the cusps of the rachidian arranged in line parallel to the tooth rows, whereas Scissurellidae have the cusps arranged on a pointed V. Furthermore Geiger (2003) noted that species of Anatomidae generally show a lateral tooth 5 with a long pointed tip that is more or less symmetrically serrated, whereas the vast majority of Scissurellidae show a broad lateral tooth 5, which is asymmetrically serrated. While the observation on lateral tooth 5 still holds, examination of additional material demonstrated much more intrageneric variability in the formation of the rachidian tooth (Zelaya & Geiger, 2007).

The radula of *Satondella* has distinctive lateral teeth 1–3 (data only for *Sat. cachoi* and *Sat. tabulata* available: Luque *et al.*, 2011), in which there is one cusp along the inner cutting edge, whereas in all other scissurellids there are no cusps along the inner edge, only along the outer.

While the radular structure of Scissurellidae has shown little variation among species, Anatomidae are highly variable (Geiger, 2006c; Geiger & Sasaki, 2008; Sasaki *et al.*, 2010; this study). The floppy teeth of *Anatoma flexidentata* and *An. australis* stand out, with other variations being found in further species. Some species in Anatomidae also display a food groove on the shafts of the marginal teeth, formed by a posterior projection on the upper portion of the shaft. Whether any of those modifications have any phylogenetic significance remains to be demonstrated. The food groove is comparable to that of *Tegula funebralis* (A. Adams, 1855), *Clanculus* species and Liotiidae (see Morris & Hickman, 1981; Hickman & McLean, 1990; Singer *et al.*, 2000).

Some alternative explanations for the occurrence of unusual radular morphologies have been explored, but on the basis of available data have to be rejected.

i) The morphological changes are not correlated with depth, because some deep-water species, e.g., *An. argentinae* from the abyssal plain of the South Atlantic, have the typical anatomid radula pattern. Additionally, a pair of sympatric anatomid species contains both a species with typical anatomid radula, and one with a highly modified radula (Geiger & Sasaki, 2008). Some shell-morphologically similar species can easily be distinguished based on their radulae (*An. flexidentata* - *An. zancliformis*).

ii) The hydrothermal vent environment is also unlikely as an explanatory factor as shown by the only moderately modified radula of *An. janetae* versus the highly modified radula in *An. fujikurai*, both collected from sites in the vicinity of hydrothermal vents.

iii) The two species with the most aberrant radulae are from the Tethys area: Japan (*An. fujikurai*) and the Indian Ocean (*An. flexidentata*). The significance of this observation is uncertain because of general poor sampling of the deep sea due to lack of suitable material, and that relatively few Atlantic species have been examined (Geiger, 2008b).

Comparison of tooth morphologies between taxa is complicated by both intraspecific as well as ontogenetic variability (Fig. 24). Intraspecific variability includes the number of minor cusps on the rachidian tooth as well as lateral teeth 1–3 and 5 (±1 of nominal count normal), and distinctness of minor points on lateral tooth 4. The ontogenetic trajectory, from forming radula at the posterior end to worn teeth at front, may also be the source of some confusion. The formation process from the posterior end (Fig. 24E) to the anterior third (Fig. 24B) is not uniform across all teeth in a row. While the marginal teeth are often already fully formed, the central field may still show limp tips. But when the rachidian tooth and lateral teeth 1–3 are at their peak, usually lateral tooth 5 already shows some wear (Fig. 24C–B). In the anterior third, the wear patterns affect mainly the central field, while the marginal teeth are usually in fairly good condition to close to the tip of the radula (Fig. 24B–A).

In some cases there are major documented differences in the tooth morphology among different rows. The development of the wing on base and shaft of lateral tooth 4 of *Sin. macleani* (Fig. 377) is restricted to the more basal part of the tooth, or extend almost over the entire length.

Given that the preparation of radula from micromollusks is not easy (see Geiger *et al.*, 2007 for methods), the images taken and shown represent an overall best-case: well-spread teeth, little dirt, fully formed but not worn teeth. However, it is impossible to fully standardize the imaging process. In some cases, the preparations have major problems (torn radula, upside down radula remounted, dirty preparations), and the investigator has to cope with those limitations as well as possible. Accordingly, minor differences in counts or shapes must be viewed with much caution, even more so if gleaned from published accounts as opposed to one's own examinations.

Operculum. The operculum is almost constant within all genera. When present, it is thin, corneous, and multispiral with central nucleus. In *Scissurella, Sinezona, Sukashitrochus,* and *Anatoma* it covers most of the aperture, while in *Incisura* it is much reduced. *Temnocinclis, Temnozaga, Sutilizona,* and presumably *Depressizona* lack an operculum due to the modified patelliform to calyptraeiform shell shape. The cross section shows multiple thin layers that are continuous across successive volutions (Fig. 23). *Incisura* sp. have more or less distinct thickenings at the junction between successive whorls of the operculum (Figs 562, 565A).

Quinn (1983) speculated that the operculum of seguenziids is paucispiral as a result of a constraint due to their small size. Given that scissurellids are by far smaller than seguenziids and have a multispiral operculum, the operculum condition in seguenziids is interpreted as an autapomorphy within vetigastropod families.

In anatomids (*Anatoma fujikurai:* Fig. 724A; *An. kelseyi:* Figs 772, 773; *An. peruviana:* Fig. 816A) the operculum is not attached to the main portion of the metapodium, but to a small tissue flap. The attachment area only occupies a wedge of approximately 60°, or ⅙ of the total area of the operculum. No other gastropod has the operculum attached in such a fashion.

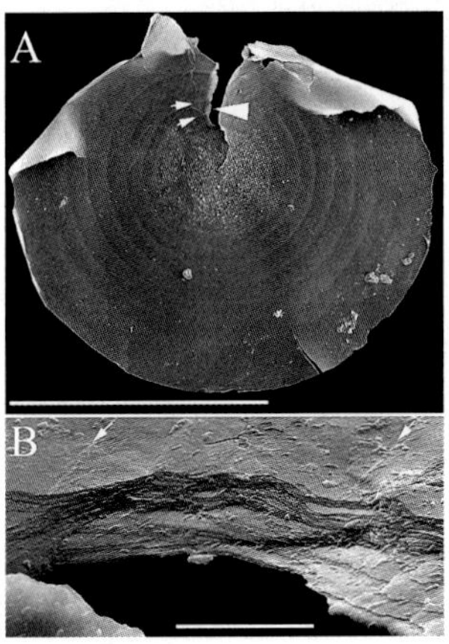

Figure 23. Operculum of *Anatoma japonica.* 370–400 m, Coral Sea, Chesterfield, 19.883°S, 158.667°E (MNHN, 1). **A.** Entire operculum. Scale bar = 1 mm. **B.** Enlargement of fracture. Scale bar = 20 µm. Large arrow: direction of view of B. Small arrows: sutures on operculum between successive volutions.

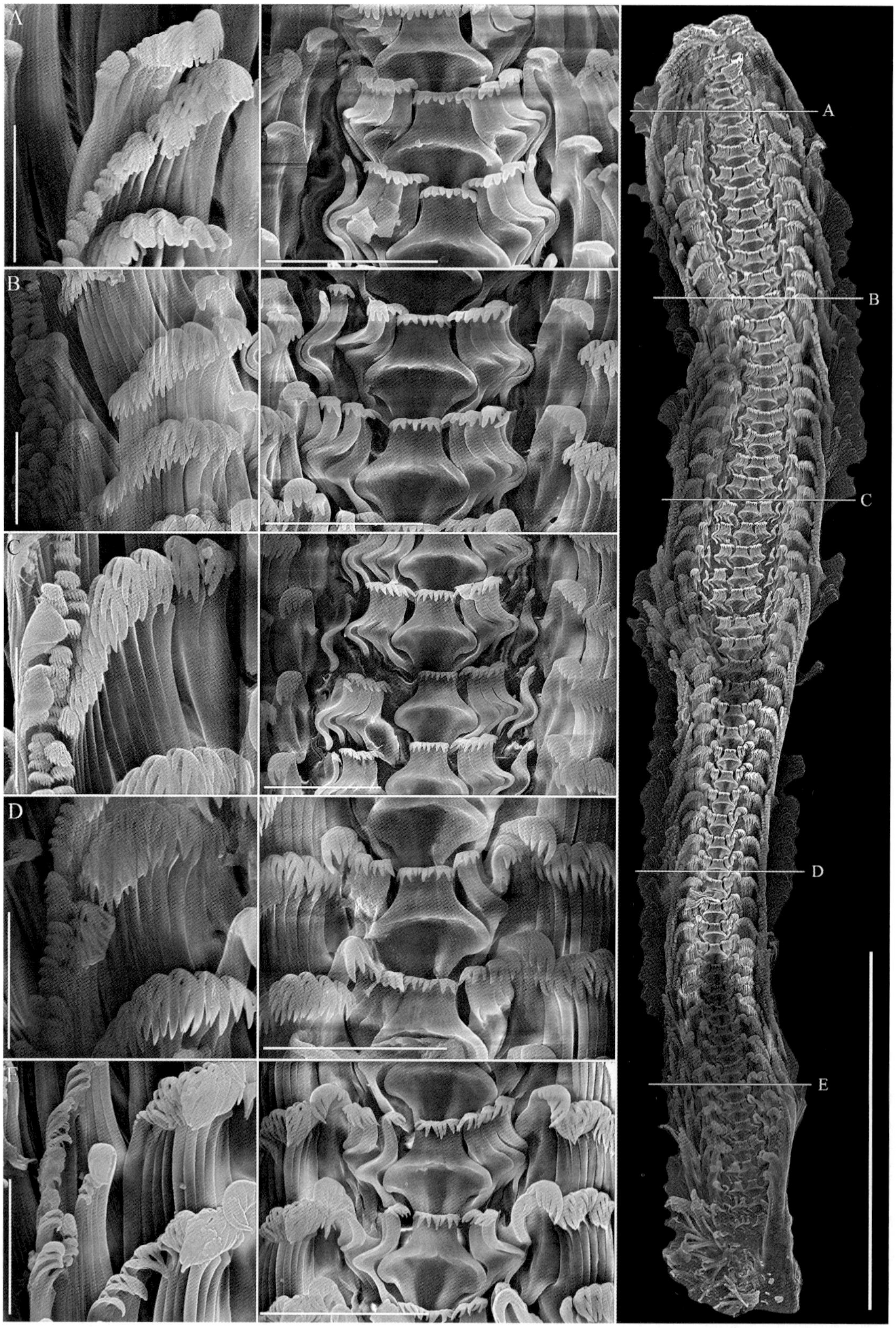

Figure 24. Radular variability due to ontogeny and wear; for discussion see text. *Sukashitrochus morleti*. 35–38 m, base of cliff, Horseshoe, Okinawa, Japan, 26.504°N, 127.858°E (DLG 972, 2). Scale bar radula = 200 µm. Scale bars central fields = 20 µm. Scale bars marginal teeth = 10 µm.

External anatomy. The head-foot complex harbors a number of tentacles (Color Plates 3–4). Starting at the front, two large, usually papillate cephalic tentacles are found. Posterio-lateral usually an eyestalk with a pigmented eye is encountered; in some species the eye and/or eyestalk is missing, particularly in deep-water species. Geiger (2003) discussed whether the lack of eyes may support the anatomid genus *Thieleella*. However, the bathyal and abyssal species *An. janetae* also lacks eyes, suggesting that their absence correlates with depth, and is not useful for classification. An additional suboptic tentacle is found in some taxa; it is more closely associated with the eye, and usually much smaller (Color Plates 4B, 4F, 5B; Fig. 773B–C). The two smooth tentacles located between the cephalic tentacle and the first epipodial tentacles are most likely homologous to what Quinn (1983), Sasaki (1998) and Kano (2009) called the accessory cephalic tentacle (See Geiger & McLean, 2010 for discussion). As these accessory cephalic tentacles (Fig. 773B–C) are paired, and as fertilization is most likely external and no ciliary tracks are associated with them, it is unlikely that they have any penial function.

Epipodial tentacles are located in various clusters in the meso- and metapodial region. An epipodial sense organ (ESO) has been detected in Scissurellidae (Baborka, 2007; Color Plate 5E) and in Anatomidae (Fig. 773B–E). The non-papillate stubby projection associated with the first two epipodial tentacles is interpreted as the ESO. Its non-papillate, blunt nature agrees with typical ESO, which usually, however, are located ventrally to epipodial tentacles and are only about has high as wide (here 3–4 times as high as wide). For further discussion, see Geiger & McLean (2010) and Sasaki *et al.* (2010).

The nomenclature of these processes is still unsettled. Specifically, the term tentacle implies some muscular development, whereas non-muscular processes are called peduncles. Given the minute size of those processes, it is difficult to assess with a high degree of certainty the presence or absence of muscle fibers with regular histology; antibody stains (myosin) may be the better approach. The terminology has further implications with respect to similarity/homology statements. For instance, either the suboptic process has states muscular and non-muscular, and is considered to be putatively homologous. Alternatively, there are two potential processes, suboptic tentacle and suboptic peduncle, each with states present and absent. Then the two structures are not considered putative homologies. In the absence of strong evidence either way, I have elected to use the most common terminology to facilitate communication about those structures.

The presumably chemosensory bursicles (see Geiger *et al.*, 2008 for review) were found by Geiger (2004) in Scissurellidae and by Sasaki (1998), Geiger & McLean (2010) and Geiger (this study) in Anatomidae (Color Plate 5J; Fig. 773G–H).

Internal anatomy. Representative histological sections through *Sinezona rimuloides* are provided in Color Plate 5. As anatomical data are only available for a few select species in the literature, the significance of those finds is difficult to assess. Accordingly, there are no comments on the internal anatomy with the species descriptions.

Aberrant specimens

Aberrant specimens are rare in Scissurellidae s.l. Two underlying reasons have been distinguished: those alterations due to traumatic events, and those that are the result of a genetic-developmental error (programming errors of Hickman, 1980).

The rare occurrence of aberrant shells due to trauma can be attributed to the most likely short life span of the animals, and the fragile nature of their shells. Both make survival of a traumatic event unlikely; hence, drastic shell repair is rarely seen. Occasionally, some stronger growth lines and some evidence of breakage of the aperture is noted, but they usually do not leave any macroscopic traces on the shell.

Given their general rarity, the few exceptions are worthy of note. Figure 25A–B shows a *Scissurella mirifica* with evidence of breakage and atypical growth in the vicinity of the affected area. In-fill of the older part of the selenizone is apparent, and the irregular width of the slit/selenizone as well as deformation of the overall shell outline is noted.

Two instances of teratological specimens are known. One is a specimen of *Scissurella morretesi* with repeated closure of the selenizone (Geiger, 2006b; Fig. 141C). This is not a case of shell repair as there is no sign of injury. It is a developmental irregularity caused by some unspecified genetic defect.

Figure 25C shows the only known specimen of a scissurellid (*Sinezona rimuloides*) with a detached whorl. Otherwise, the specimen does not show any evidence of injury that may explain the irregular growth habit. In some specimens the peristome does not fully attached to the previous whorl, but the separation of whorls usually does not extend over an extended portion of the shell.

As noted for abalone (Geiger, 1998; Geiger & Poppe, 2000; Geiger & Owen, 2012), there are no known sinistral specimens in Scissurellidae s.l. of normally dextral species, nor are there species that are usually sinistral (see also Robertson, 1993). Over 73,000 specimens were examined and not a single wrong-handed specimen was encountered.

A single aberrant radula has been found. A specimen of *Anatoma schioettei* from Norway had a central field with an extra lateral tooth 2 (Fig. 22). Lateral teeth 1–3

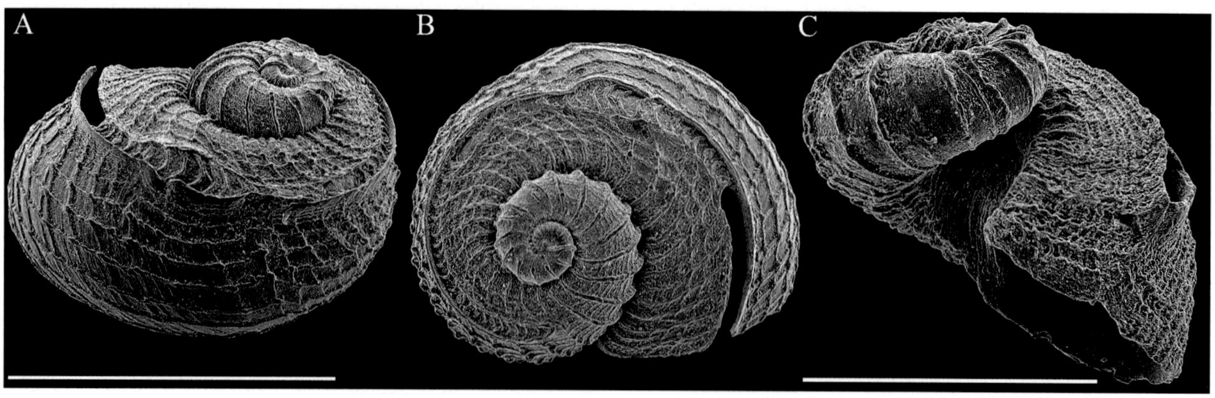

Figure 25. **A–B.** Specimen of *Scissurella mirifica* showing rare shell repair. DLG 1056. 38 m, base of cliff, Horseshoe Bay, Okinawa, 26.504°N, 127.858°E. Leg. Bret Raines. Scale bar = 1 mm. **C.** *Sinezona rimuloides* with detached whorl. 6–12 m, off Ildefonso Island, Gulf of California, Mexico, 26.632°N, 111.430°W (SBMNH 423738, 1). Scale bar = 0.5 mm.

are similar; hence, the specific duplication is difficult to establish with certainty. The identification of the duplication as applicable to lateral tooth 2 is based on the best agreement among adjacent teeth.

Occasional confusions

In general, identifying slit-bearing scissurellids to family poses no problems. The most common family-level confusions are with the slit-bearing worm-snails (Siliquariidae), followed by other small trochiform snails such as Fossariidae and Vanikoridae. How much the slit-less scissurellids (*Coronadoa*, Larocheidae) are misidentified as members of any of the many trochiform groups is open to speculation.

Distribution and diversity patterns

Bathymetric distribution. Scissurellids occur in all fully marine oceans of the globe, from the Arctic to the Antarctic at depths from the lower intertidal to at least 6,000 m depth, including hydrothermal vents.

The relative abundance of scissurellids compared to the overall malacofauna is lower than expected in the shallow water (0–30 m). It gradually increases until it reaches its maximum on the continental slope (300–1000 m), where scissurellids are much more common than expected. In the deep-sea (>1000 m) scissurellids have the same relative abundance as the rest of Mollusca (Fig. 26).

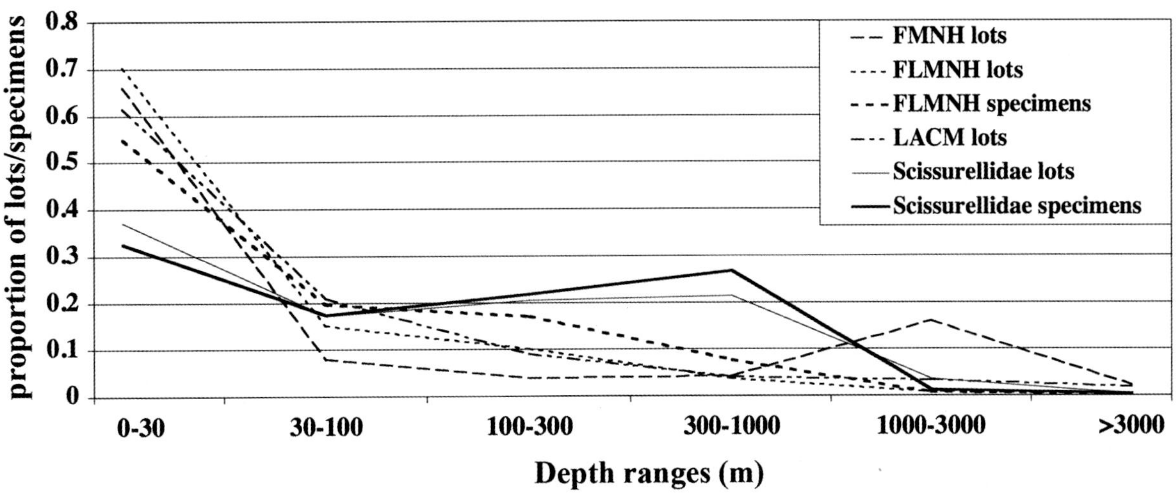

Figure 26. Available material of scissurellids compared to overall sampling effort as a function of depth. Overall sampling effort was gauged by the holdings in three major museum collections: FLMNH 32,413 lots, 134,558 specimens; FMNH 9,890 lots; LACM 78,966 lots. Data kindly provided by John Slapcinsky (FLMNH), Jochen Gerber (FMNH) and Lindsey Groves (LACM). From Geiger (2008b).

A taxonomic replacement of the Scissurellidae in shallow water with Anatomidae in greater depths is notable. Figure 27 shows the number of lots in given depth intervals segregated by genera in Scissurellidae (first three bars with black outline) and Anatomidae (last two bars with grey outline); the remaining family level taxa are too rare to permit inclusion in this analysis. In shallow water (0–25 m) Scissurellidae are about 14–15 times more abundant than Anatomidae. Equal abundance is reached in the 75–100 m depth interval. At greater depths Anatomidae remain the dominant scissurellid group. These patterns also hold if data are analyzed separately for temperate and tropical zones (Geiger & Sasaki, 2009). While this general pattern is not without exceptions, such as the shallow-water *An. parageia*, the overall trend is apparent. No typical shelf break nor the tropical 300 m break noted by Bouchet *et al.* (2008) was detected; the transition is gradual throughout (Fig. 27). The species collected from the deepest sites all belong to Anatomidae. In many instances samples with live-collected specimens are available, confirming that these species live on the abyssal plain, and are not transported from shallower environments. It is worth noting, that a few of those samples were collected below the calcium compensation depth (~4200–5000 m for various parts of the world's oceans), at which carbonates dissolve due to a combination of factors including hydrostatic pressure, temperature, and dissolved salts.

Geographic distribution. Scissurellids follow the distributional patterns of most marine invertebrates. Most species are bound to broad temperature limitations, usually circumscribed as tropical, temperate and polar. There are no species with antitropical distribution, species that occur in temperate waters of both hemispheres to the exclusion of the tropics. Species vary in the extent of their distribution. Single island endemites (e.g., some species of Easter Island) are contrasted with widely-distributed species in the Indo-Pacific such as *Sinezona plicata* and *Scissurella mirifica*. A few deep-water species have more extensive ranges and transcend traditional faunal barriers. One example is *Anatoma austrolissa* known from deep-water of the Mascarene Islands in

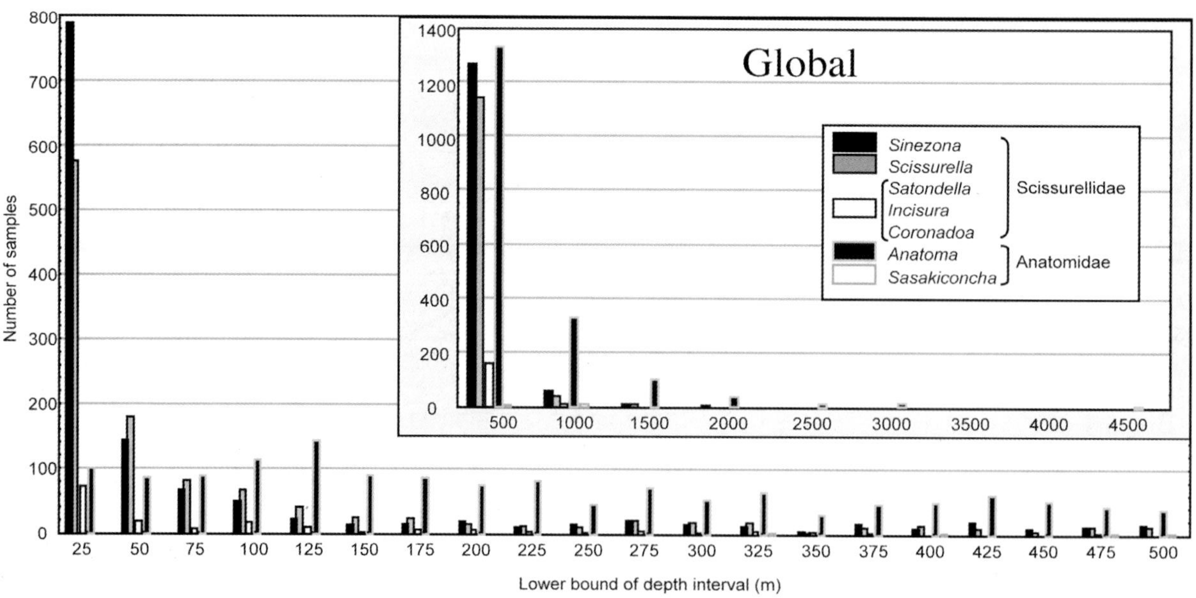

Figure 27. Metadata analysis of relative abundance of genera over depth. Note prevalence of Scissurellidae in shallow water, while Anatomidae are the dominant component in deep water. From Geiger & Sasaki (2009, modified).

the Indian Ocean and from off New South Wales, Australia. *Anatoma argentinae* is known from the abyssal plain off Argentina in the southern Atlantic as well as in the Panamic of the tropical eastern Pacific.

A few species seem to be more eurythermal. This includes *Sinezona rimuloides* from the temperate northeastern Pacific, but also extending to the more tropical Panamic, as well as *Sukashitrochus atkinsoni* from temperate southern Australia but also extending its range to more tropical Queensland.

In the Indo-Malayan Archipelago, Wallace's line is crossed as a rule. The genus *Anatoma* is an example of this pattern. Most species with records in the Indo-Malayan Archipelago cross the line (*Anatoma amydra, An. biconica, An. equatoria, An. finlayi, An. indonesica, An. japonica, An. maxima, An. munieri, An. philippinica, An. porcellana, An. pseudoequatoria, An. rapaensis, An. rhynchodentata*); species restricted to either side of the line are usually represented by few samples only (*An. flexidentata*), so are most likely a sampling artifact. Some species are exclusively found either in the Indian Ocean (e.g., *An. yaroni*) or the Western Pacific (e.g., *An. megascutula*), but those species do not approach Wallace's line by at least 4–5,000 km; other forces have to explain those patterns.

The Caribbean holds some surprises. A few specimens of Indo-Pacific species were discovered with collection data from the Caribbean. Whether those are genuine finds or mislabeled specimens remains to be determined. Those species are the extremely rare *Sinezona globosa* and the not much more common *Anatoma finlayi*. Both of those species are distinct, and the commonalities were documented at the SEM level.

If those locality data are correct, then the question of taxonomic treatment arises. They are either considered conspecific under the application of the morphological species concept, or are deemed distinct, shell-morphologically cryptic species under the biological species concept. Unfortunately, the scissurellid taxa in question all have little material available, and almost all specimens are shells only. Accordingly, molecular techniques cannot be applied here due to lack of suitably preserved material.

In some gobiid fishes, such molecular techniques have been applied (Thacker, 2004). In that study, the Indo-Pacific and Caribbean species cannot be reliably identified based on morphological characters, they are recovered as reciprocally monophyletic groups in a molecular phylogeny, though the split is about half as deep as between other species.

Other reports of scissurellid species from multiple ocean basins are based on misidentification of material and overly broad species concepts. The chief example is *Anatoma crispata*, a northeastern Atlantic species, which has been reported from virtually the entire globe. None of the records with extant voucher material outside of Europe was assigned to that species.

Diversity patterns. Table 2 shows the number of species in each genus and family, segregated by some major biogeographic areas (non-polar Atlantic vs. Pacific; tropical – temperate – polar). The patterns emerging are as follows: Unsurprisingly, the Atlantic Ocean is less diverse than the Pacific Ocean with respect to number of species (35 vs. 154: factor 4.4) and genera (5 vs. 15: factor 3) found in each ocean. Several genera and families are restricted to the Pacific (*Sukashitrochus, Sasakiconcha*, Larocheidae, Depressizonidae, Temnocinclidae). Although generic concepts are subjective, there is no doubt that the Pacific harbors greater diversity of lineages and morphotypes than the Atlantic. The tropical-subtropical regions have 1.15× the number of species compared to the temperate regions (101 vs. 88); for this analysis, each species was assigned to one or the other region based on its center of distribution. The polar regions have a much reduced number of species (7) and genera (2).

Earlier indications on the diversity in the group (Geiger, 2008b) overestimated the true diversity of the group by 14% (229 vs. 196 species). There are two main reasons. 1) Intraspecific variability is difficult to assess during preliminary examination of material under the stereomicroscope. Recognizing small specimens

Genus	Family	Atlantic non-polar	Pacific non-polar	Tropical 0–30°N/S	Temperate 30–60°N/S	Polar 60–90°N/S	Total
Scissurella	Scissurellidae	8	26	19	15	1	35
Sinezona		6	37	20	23	0	43
Incisura		0	4	0	4	0	4
Satondella		2	8	10	0	0	10
Coronadoa		0	3	0	3	0	3
Sukashitrochus		0	7	3	4	0	7
Total Scissurellidae		16	85	52	49	1	102
Anatoma	Anatomidae	18	51	41	28	6	75
Sasakiconcha		0	1	1	0	0	1
Total Anatomidae		18	52	42	28	6	76
Larochea	Larocheidae	0	4	0	4	0	4
Larocheopsis		0	2	0	2	0	2
Trogloconcha		0	5	2	3	0	5
Total Larocheidae		0	11	2	9	0	11
Depressizona	Depressizonidae	0	2	2	0	0	2
Temnocinclis	Temnocinclidae	0	1	0	1	0	1
Temnozaga		0	1	1	0	0	1
Sutilizona	Sutilizonidae	1	2	2	1	0	3
Total Minor Groups		1	6	5	2	0	7
Grand Total		**35**	**154**	**101**	**88**	**7**	**196**

Table 2. Species-level diversity of worldwide Recent Scissurellidae s.l. The number of Atlantic + Pacific + Polar, and the number of Tropical + Temperate + Polar adds up to the total number of species in the right-most column. Note that the Pacific Ocean is more diverse both in species as well as in genera. The tropical and temperate regions hold approximately the same diversity, while the polar regions are more depauperate.

as juveniles of larger specimens, and to assign damaged/eroded specimens to the correct morphospecies is difficult. 2) Initial examination of material in multiple collections posed challenges with keeping track of conspecific morphospecies in various collections, resulting in duplication, which inflated the preliminary taxon count. As a consequence, faunal lists containing many numbered morphospecies (sp. 1, sp. 2, ... sp. 297) should not necessarily be taken at face value. The problem may get compounded in global biodiversity estimates, when numbered morphospecies counts from different regional assessments are added up, while the overlap in biological species is close to 100%. This is particularly the case with two leading biodiversity inventories, namely the New Zealand data set (Spencer *et al.*, 2009), and data based on the MNHN collecting efforts. The best way of removing redundant manuscript species is preparing figures and writing the comparisons of the formal descriptions. Actual time spent working on the group and its in-depth understanding cannot be replaced by any mathematical models.

Fossils

Table 3 provides an overview of the fossil taxa described, their known temporal range and geographic occurrence. The age indications have been converted to the international system based on the most recent available literature; for details, see the full treatment of the particular species. Geographic occurrences are given only to major regions with details of records given with each taxon in question.

The earliest members are known from the Triassic (*Triassurella* species) in a group most closely resembling modern Temnocinclidae. Scissurellidae are known from the Jurassic through the Recent, while the earliest Anatomidae are known only from the uppermost Cretaceous.

The synonymy ratio among fossil taxa is notably higher than for Recent species, particularly when considering the relatively low number of specimens available. Not every microfossil unearthed is a new species. Some illustrations in the literature cannot be identified to family or genus (Traub, 1984: pl. 1, fig. 1a–c: *Scissurella*/*Sukashitrochus* sp.).

Biology

The available data on the biology of scissurellids is scant. Whether the select observations on any given species applies to others, and if so how widespread those biological parameters are, is a wide open field for study. An area of particular interest is the reproductive system, as some species are considered protrandrous hermaphrodites, whereas others may be simultaneous hermaphrodites, both in contrast to the more common gonochoristic systems in Vetigastropoda. Age determinations and generation times are also unknown. Niche partitioning in sympatric species has not been explored at all. There are no known parasites, although given their general ubiquity, some parasites should affect scissurellids.

Table 3. Overview of the occurrence of scissurellids in the fossil record. Pleistocene includes Quaternary and "subfossil" records. * Species also known from the Recent. For source of data, see individual species.

Taxon	**Cenozoic**						**Mesozoic**			**Distribution**
		Neogene		**Paleogene**						
	Pleistocene	**Pliocene**	**Miocene**	**Oligocene**	**Eocene**	**Paleocene**	**Cretaceous**	**Jurassic**	**Triassic**	
Scissurella alicae						◆				Europe
Scissurella annulata						◆				Europe
Scissurella apudornata			◆	◆						New Zealand
Scissurella bituminata				◆						Indonesia
Scissurella condita			◆					◆		New Zealand
Scissurella cossmanni				◆	◆					Europe
**Scissurella costata*	◆	◆								Europe
Scissurella depontaillieri			◆	◆	◆	◆				Europe
Scissurella eocaenica					◆					Europe
Scissurella gruendeli								◆		Europe
Scissurella hedegaardi						◆				Europe
Scissurella marchmontensis							◆			Caribbean
Scissurella peyrerensis				◆			◆			Europe
**Scissurella prendrevillei*	◆	◆								New Zealand
Scissurella ravni						◆				Europe
**Scissurella reticulata*	◆									Red Sea
**Scissurella staminea*	◆	◆								Japan
Scissurella szaboi								◆		Europe
Scissurella tinaeformis					◆					Europe
Scissurella transsylvanica			◆	◆						Europe
**Sinezona brevis*	◆									New Zealand
**Sinezona cingulata*		◆								Europe
Sinezona confusa	◆									Caribbean
Scissurella cupelliformis				◆	◆					Europe
Sinezona deshayesi					◆					Europe
**Sinezona finlayi*			◆							New Zealand
Sinezona haliotimorpha			◆	◆						Europe

Taxon	Cenozoic						Mesozoic			Distribution
	Neogene			Paleogene						
	Pleistocene	Pliocene	Miocene	Oligocene	Eocene	Paleocene	Cretaceous	Jurassic	Triassic	
Sinezona koruahina		◆								New Zealand
**Sinezona laqueus*	◆	◆								New Zealand
**Sinezona levigata*	◆	◆								New Zealand
Sinezona malloryi					◆					NW America
**Sinezona mechanica*	◆									New Zealand
**Sinezona plicata*	◆			◆						Indo-Pacific
Sinezona praecedens			◆	◆						Europe
**Sinezona rimuloides*	◆									NW America
Sinezona terquemi			◆							Europe
Sinezona tertia		◆								New Zealand
Satondella kondoi					◆					Tonga
**Sukashitrochus estotiensis*				◆						Europe/IP
Sukashitrochus kaiparaensis			◆							New Zealand
Sukashitrochus ngatutura	◆	◆								New Zealand
Incisura geoffreyi	◆	◆								New Zealand
**Anatoma aspera*	◆	◆					◆?			Europe
**Anatoma australis*		◆								Australia
**Anatoma crispata*	◆	◆								Europe
**Anatoma eximia*	◆									Europe
Anatoma gephyra					◆					New Zealand
**Anatoma japonica*	◆									Japan
Anatoma miocenica			◆							New Zealand
Anatoma muelleri				◆						Eurasia
**Anatoma lyra*	◆									California
Anatoma philippiana				◆	◆	◆				Eurasia
**Anatoma tenuisculpta*	◆	◆	◆	◆						Europe
Anatoma tongaensis					◆					Tonga
Trogloconcha marshalli				◆						Europe
Triassurella carnica									◆	Europe
Triassurella goederti							◆			NW America

Food and Feeding. Little information is available on feeding preferences of scissurellids. Limited gut content analysis have most frequently revealed amorphous material and sand grains with some diatom frustules (Geiger, 2006c; Zelaya & Geiger, 2007) and coccolithophore plates (Fig. 28). This indicates that scissurellids are micrograzers, and possibly also utilize biomass from bacterial biofilms. That latter mode of feeding is also likely found in the hydrothermal vent groups. Without reference background sediment samples it is impossible to determine whether particle ingestion is selective and to which degree (see Hickman, 1981 for discussion).

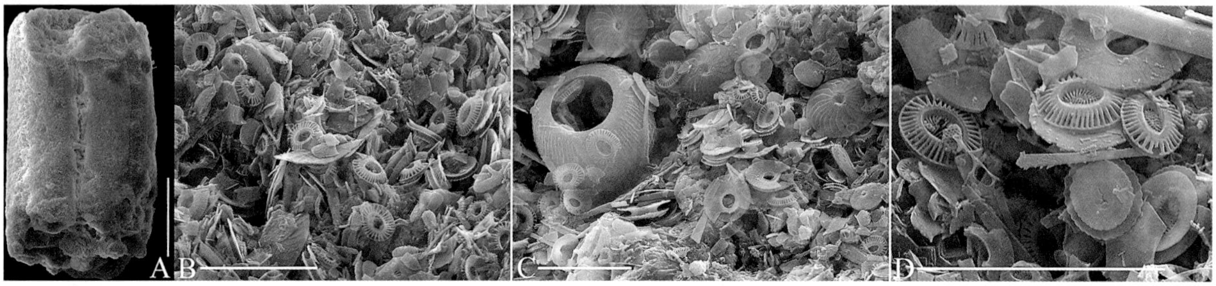

Figure 28. Gut content of *Anatoma rhynchodentata*. 657–660 m, Loyalty Ridge, New Caledonia, 23.583°S, 169.617°E (MNHN, 1). **A.** Piece of intestinal content after dissolution of animal in KOH. Note mid-ventral gut furrow preserved in gut content. Scale bar = 200 µm. **B–D.** Close-ups of gut content. Notice abundant coccolithophore plates (disks with spokes) and other amorphous contributions. Scale bars = 10 µm.

Habitat. In the intertidal and shallow subtidal scissurellids have only been found on hard bottom habitats; although shells can frequently be collected in sediment samples, live specimens are exceptionally rare. Specimens can be live-collected on macroscopically bare rock and biocarbonates as well as on algae. Limited experience indicates that scissurellids prefer green and red algae, and are less frequently encountered on brown sea weeds.

For specimens occurring on the shelf and the continental slope, detailed habitat information is wanting. Given that the prevalent shelf and slope habitat is soft-bottom substrate, this is also the presumed habitat of scissurellids (specifically Anatomidae). The relative abundance of Scissurellidae and Anatomidae is reversed with increasing depth: Scissurellidae are the dominant family in the shallow water (<75 m), whereas Anatomidae dominate depths greater than 75 m (Fig. 26). This supports the idea that Anatomidae inhabit soft substrates, while Scissurellidae prefer hard substrates.

The hydrothermal vent habitat is inhabited by two specialized lineages: Temnocinclidae and Sutilizonidae, where they are found on hard substrate associated with hydrothermal vents and most likely feed on microbial mats. Some anatomid species have been reported from environment with chemosynthetic influence (*An. janetae, An. fujikurai*). Whether these species are truly specialized or are widely found throughout the deep sea but have only been collected at comparatively well-studied vent and seep sites remains to be determined.

Reproduction, life history. The gender distribution in scissurellids is variable. Some scissurellids have separate sexes, including *Incisura* and *Sinezona* among Scissurellidae, and *Larochea* sp. (Larocheidae: Baborka, 2007). Hickman (1999) suggested there might be sexual dimorphism in *Incisura auriformis* (Scissurellidae), but this has not been confirmed in other species or lineages. Baborka (2007) reported hermaphroditic condition

Figure 29. Gonad of *Sinezona rimuloides*. 3D reconstruction from histological sections. Notice two rows of eggs.

in *Scissurella jucunda*. The female gonad is filled with a limited number of eggs; specimens of 1.2 mm size (*Sinezona rimuloides*) have two rows of eggs in the ovary (Fig. 29). One can infer that specimens of half that size (0.6 mm) only have a single row of eggs, which imposes a lower size limit for Scissurellidae. The smallest scissurellids are in fact 0.6 mm in size (*Sinezona hawaiiensis, Sin. wileyi, Coronadoa simonsae, C. demisispira*). A similar constraint is found in the smallest teleost fishes, where *Schindleria* (Teleostei: Gobiidae) also has single rows of eggs in each of the paired ovaries (Thacker & Grier, 2005).

Given the low number of eggs produced, scissurellids must be considered to be at the K-end of the reproductive strategy spectrum unlike most other Vetigastropoda (Haliotidae, Trochidae, Fissurellidae), which can produce millions of eggs (Poore, 1973; Ault, 1985; Bevelander, 1988; Rogers-Bennet *et al.*, 2004) and are text-book examples of broadcast spawning r-strategists. One report suggests that some scissurellids agglutinate eggs to the substrate (Hickman, 1999). Two reports (Hasegawa, 2004; Porter & Hickman, 2007) have observed nocturnal mass swimming in response to artificial light sources with ensuing mass release of gametes near the water surface. Convergence of individuals near the water/air interface provides efficient means to increase local gamete concentration and to increase fertilization success. Porter & Hickman (2007) also speculated on the possibility of long distance dispersal of swimming scissurellids in ocean currents.

There are no data on growth rates or longevity of any species. Such small species are presumed to only live for a short period of time, on the order of 1–6 months, but that indication is entirely conjectural.

Predators. There is only a single study that has reported scissurellids as gut contents: Schliewen & Kovačić (2008) identified a *Sinezona* sp. (as *Sinozona* [sic]) in radiographs of some gobiid fishes (Teleostei: Gobiidae) from São Tomé and Cape Verde Islands, most likely *Sinezona cingulata*. Given the fragile nature of scissurellid shells, such findings are not likely. A few specimens in collections are labeled "ex pisce" = out of a gut content of a fish [e.g., *Anatoma aspera*: Northern, fish market of Tanger, Ex. *Lepidotrigla* (Teleostei: Triglidae), Morocco, 35.785°N, 5.813°W (MNHN, 20)]. Most likely, those were bottom feeding fish, which accidentally ingested a scissurellid. Occasionally drill holes from Caenogastropoda: Muricidae are encountered on shells, but they are infrequent. From limited seasonal collecting efforts, some of the temperate species (*Sinezona rimuloides*: Geiger pers. obs.) may show cyclical abundance patterns with the summer months yielding richer samples than those collected in winter. Hence, scissurellids may engage in some type of temporal escape from predators.

Symbionts. The vast majority of specimens do not show any evidence of symbionts of any sort. Shells can serve as hard substrate for encrusting organisms. On a few rare occasions, Bryozoa (Fig. 30B), small serpulid worms (Annelida: cf. *Spirorbis* sp.), or rotaliid Foraminifera (Fig. 30A) have been located on the exterior of the shell. A few shells of the Indo-Pacific *Sinezona plicata* (USNM 374438) and *An. rhynchodentata* (Fig. 867) the southeastern Australian *An. australis* (Color Plate 2G), and the New Zealand *An. regia* as well as some of the larger Indo-Pacific species have been occupied by minute hermit crabs of uncertain identities. No parasites are known.

Conservation. Scissurellids are of too little human interest for direct pressures to affect populations or species. Given their minute size, scissurellids are possibly more likely to be adversely affected by elevated temperatures and potentially also by ocean acidification.

Supplementary information.

The state of knowledge on scissurellids is far from complete. Some problem taxa remain to be carefully evaluated, some new species will be discovered, and the distributional pattern for most species will benefit from additional records. However, the present contribution makes a major advance in our knowledge of the group.

With the understanding of the present shortcomings and based on experiences from previous work on abalone, additional information is going to be presented either as supplements or potentially in a revised and expanded edition of the present volumes in the next 10–20 years. I am looking forward to additional material, new records and any kind of information on this elusive yet fascinating group.

In the intervening time, supplemental information, such as range extensions, new references, and errata will be made public through the web under:

http://www.vetigastropoda.com/scissurellidae

Figure 30. Shells with epizoans. **A.** *Anatoma equatoria* with rotaliid foraminiferan on shell. 500–614 m, Bight Water, Fiji, 16.015°S, 2.962°E (MNHN, 1). **B.** *Incisura lytteltonensis* with a bryozoan. 0 m, Great Expedition Bay, S end of Rarawa Beach, North Island, New Zealand, 34.717°S, 173.083°E (NMNZ M.134318, 60). Scale bars = 1 mm.

Glossary of terms

Accessory cephalic tentacle. A non-papillate tentacle located between the complex of cephalic tentacle, optic stalk plus suboptic tentacle and the first epipodial tentacle (Fig. 773B–C).

Adumbilical. Close to, towards the umbilicus.

Angular measurement/count. Measurement expressed in terms of numbers or fraction of whorls, not in terms of absolute size.

Angulation. The ridge formed by adjacent portions of shell suddenly changing orientation to one another. Less pronounced than a keel.

Anomphalous. Without an umbilicus. Opposite term to umbilicate.

Anterior. Body axis to the front or head of the animal. At right angle to dorsal-ventral axis. Opposite term to posterior.

Aperture. The opening of the shell.

Apertural margin. The opening of the shell as an outline. It can be identified for any life stage of the specimen, including the juvenile shell (protoconch).

Apex. The top part of the shell.

Apical. Towards the apex, the top of the shell.

Axial. In the direction of the shell axis, i.e., from protoconch to umbilicus (see also spiral).

Axials. Axial sculptural elements. In order of increasing strength: thread/striae, lines, cords, ribs/lamellae. Axial keels unknown in group (see also spirals, pseudospirals).

Base. The portion of the shell from the slit/selenizone to the umbilicus.—The bottom portion of a radular tooth connected to the radular membrane; see also shaft and tip.

Bead. Sculptural element consisting of a small, round section of a sphere.

Bursicle. Chemosensory organ on gill, consisting of a bulbous chamber on each leaflet that opens to the outside (Color Plate 5J, Figs 773G–H, 918D–E).

Carina. See Keel 1).

Central field. Rachidian tooth plus lateral teeth on the radula.

Centrifugal. Away from the center. Opposite term to centripetal.

Centripetal. Towards the center. Opposite term to centrifugal.

Cephalic tentacle. Paired tentacles in the head region; papillate or non-papillate (Figs 724A–C, 773A–B). Compare to accessory cephalic tentacle.

Cords. Distinct sculptural elements, easily seen under the light microscope, wider than lines, lower than ribs.

Cusp. Fine points on the tip of a radular tooth.

Dorsal. Body axis to the back or visceral mass. At right angle to anterior-posterior body axis. Opposite term to ventral.

Embryonic cap. Earliest portion of the protoconch, which is secreted as a sheet, usually approximately 50 μm wide.

Epipodium. Tissue ridge running on each side of the animal from posterior to the neck to the end of the foot, between pedal sole and mantle. Bears often epipodial tentacles and the Epipodial Sense Organ (ESO: Fig. 724D; 773 B–E).

Epipodial Sense Organ (ESO). Sensory structure associated with the epipodium, typically found in Vetigastropoda. Typically without papillae, blunt (Fig. 724D; 773 B–E).

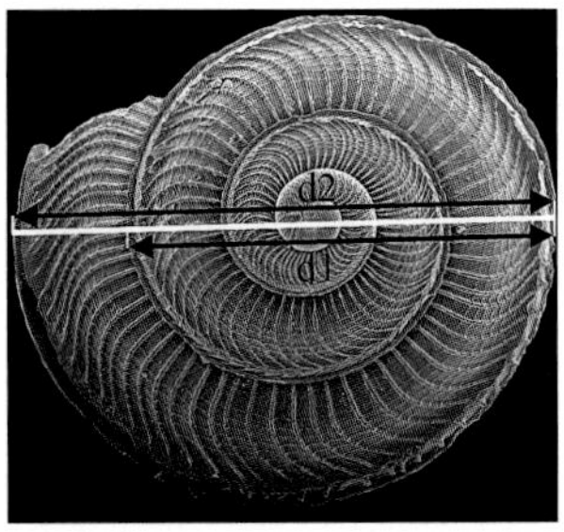

Figure 31. Expansion rate d1/d2.

Epipodial tentacle(s). Sensory structures associated with epipodium, usually found in Vetigastropoda. Typically with papillae, tapering towards tip (Fig. 724D; 773 B–E).

Expansion rate (Fig. 31). Here as ratio of maximum diameter of shell minus last whorl (d1) divided by maximum diameter of entire shell (d2). Note that in mathematical shell growth modelling, expansion rate (W) is given as quotient of radius of one whorl over radius at same angle of subsequent whorl, which requires the placement of the center or origin of the spire. Given the relatively large size of the protoconch, the precise identification of the center is difficult. The method applied here is more practical as all points are easy to identify.

Eye. Light sensory organ on head, lateral and posterior to cephalic tentacles. Situated at tip of optic stalk (Color Plate 5B, I).

Eye stalk. See optic stalk.

Flocculent. Adjective describing irregular sculpture on the protoconch, resembling sprinkles of flaky material (Fig. 14I).

Food groove. On the marginal teeth of radula, a space below the cusp and delimited by a posterior projection on the upper shaft of each marginal tooth, forming a canal (Fig. 652 E, F).

Foramen. An open portion of the slit, which is sealed again towards the apertural margin. Some authors have used the term 'hole' interchangeably but they are distinguished here (see under Hole).

Funiculus. A spiral cord in the wall of the umbilicus.

Growth lines. Finest axial marks due to the intermittent secretion process of new shell material onto the margin of the shell.

Hole. An artifact arising from damage of the specimen.—Some authors have used 'hole' interchangeably with 'foramen', but they are distinguished here.

Interlock. Describes how strongly teeth in the central field are connected to one another at the base.

Jaw. Paired chitinous plates located near the opening of the mouth (Fig. 868E).

Keel. 1) A strong protrusion from the shell, e.g., on the base of *Sukashitrochus*. The internal surface of the shell shows a depression in the position of the external keel, while ribs do not leave a trace on the internal surface of the shell.
2) The lateral borders of the selenizone, slit/foramen.

Lamella. See rib.

Lateral tooth. Usually five teeth on each side of the rachidian tooth on the radula. The teeth are numbered with 1 referring to the tooth adjacent to the rachidian tooth.

Lines. Sculptural elements that may barely be seen under the light microscope and are approximately as wide as high. Lines are wider than threads/striae, narrower than cords, and less elevated than riblets.

Lunule. The crescent-shaped growth marks on the closed portion of the slit or selenizone.

Marginal tooth. Many teeth outside of the laterals on the radula.

Mesopodium. Middle portion of the foot; see also propodium, metapodium.

Metapodium. Rear portion of foot; see also propodium, mesopodium.

Nucleus. Center of the operculum.

Operculum. Corneous cover of the aperture, attached to the metapodium of the animal.

Optic stalk. Blunt tentacle posterior to cephalic tentacle, usually with eye at tip (Color Plate 5B).

Parietal wall. The portion of the aperture making up the umbilicus.

Peristome. Rim of the aperture.

Periumbilical cord. Distinct cord at the edge of the umbilicus towards the base.

Pit, pitted. With small discrete depressions.

Posterior. Body axis to the rear or metapodium of the animal. At right angle to dorsal-ventral axis. Opposite term to anterior.

Protoconch. Earliest portion of shell, usually around 200 μm wide. It shows many species specific characteristics (embryonic cap, sculpture, varix, apertural margin).

Propodium. Anterior portion of foot; see also mesopodium, metapodium.

Pseudospirals. Nodular thickenings on axial cords arranged like spirals, but not connected to one another, producing macroscopic visual impression of spiral sculpture, although no true, continuous spirals are present (Fig. 812. See also spirals, axials).

Rachidian tooth. Middle tooth on a radula = central tooth.

Radula. The rasping organ in the foregut of snails.

Radular interlock. See interlock.

Raphe. A suture line formed by the fusion of two separate pieces of tissue or shell. In scissurellids, found in foramen-bearing species anterior to the closure of the foramen.

Rib. A shell cord that is higher than wide. Ribs do not leave a trace on the internal surface of the shell, while keels show an internal depression in their position.

Riblet. A shell line that is higher than wide.

Roof. Apertural portion of shoulder region, which often overhangs the remainder of the apertural margin.

Selenizone. The closed portion of the slit. Has lateral keels and growth markings (= lunules). The onset of the selenizone with growth defines the boundary between teleoconch I and II.

Shaft. The middle part of a radular tooth, joining the base with the tip.

Shoulder. On teleoconch II, the portion of the shell between the apical suture and selenizone.

Slit. The anteriormost portion of the selenizone, which is open at the apertural margin (cf. foramen). During shell growth, species with a foramen as adults all have an open slit when immature.

Spiral. In direction of the growth, continuous sculptural elements; approximately at right angle to growth lines (see also axial).

Spirals. Spiral sculptural elements. In order of increasing strength: thread/striae, lines, cords, ribs/lamellae, keels (see also pseudospirals, axials).

Step. Superficially looking like a spiral cord, but raises only on one side steeply and slopes gently on the other side.

Striae. Thin sculptural element, most likely not visible with the light microscope, = threads, weaker than lines.

Suboptic tentacle. Tentacle attached to the base of the eye stalk (Figs 772, 773B, C).

Sutsel. Portion of base between suture and selenizone above (Fig. 18). The width of the sutsel is compared to the width of the selenizone: 0× = no sutsel; 0.5× =

sutsel half as wide as selenizone; 1× = sutsel as wide as selenizone; 2× = sutsel twice as wide as selenizone.

Suture. 1) The line at which two whorls touch. 2) The spiral fusion line of the margins of the slit anterior to the foramen.

Teleoconch. Postembryonic part of the shell.

Teleoconch I. Postembryonic shell to the start of the selenizone.

Teleoconch II. Postembryonic shell from the start of the selenizone to the apertural margin.

Thread. Thin sculptural element, most likely not visible with the light microscope, = striae; weaker than lines.

Tip. The top part of a radular tooth (cf. shaft, base); bears cusps.

Trough. See umbilical trough.

Umbilical trough. Spiral cavity formed in base of shell by shoulder of previous whorls. Unique to *Sasakiconcha elegantissima*.

Umbilicate. With an umbilicus. Opposite term to anomphalous.

Umbilicus. Cavity in the base of the shell.

Varix. An axial thickening of the shell. In little slit shells, it can only be found on the protoconch, but not on the teleoconch.

Ventral. Body axis towards the pedal sole. At right angle to anterior-posterior body axis. Opposite term to dorsal.

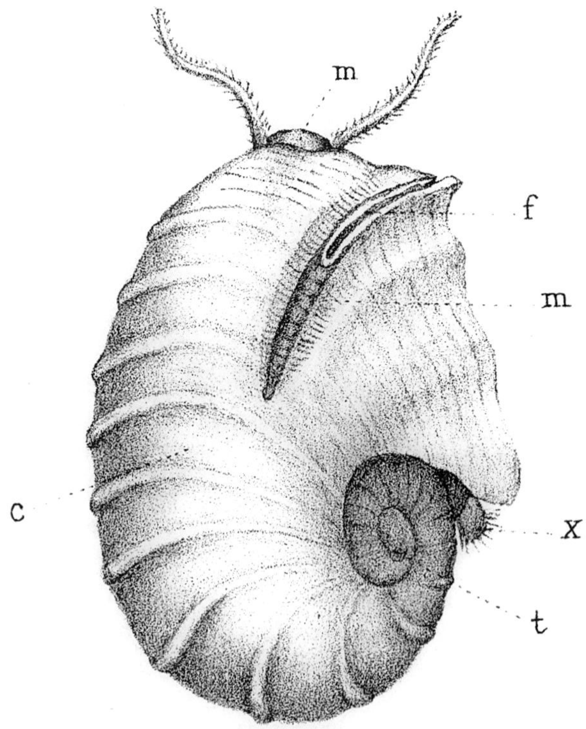

Boutan's (1886: pl. 42, Fig. 4) illustration of a *Scissurella costata* mistakenly identified as a juvenile *Fissurella*.

Collection abbreviations

ex COLL####	Separated from specified lot in specified collection; new registration number not known at this time.
AIM	Auckland Institute and Museum, Auckland, New Zealand.
AMNH	American Museum of Natural History, New York, New York, USA.
AMS	Australian Museum Sydney, New South Wales, Australia.
ANSP	Academy of Natural Sciences, Philadelphia, Pennsylvania, USA.
ARF	Alice Rasmussen Collection, Faske, Denmark.
BMSM	Bailey-Matthews Shell Museum, Sanibel, Florida, USA.
BPBM	Bernice P. Bishop Museum, Honolulu, Hawaii, USA.
CASIZ	California Academy of Sciences, Invertebrate Zoology, San Francisco, California, USA.
CCS	Center for Coastal Studies, Texas A&M University, Corpus Christi, Texas, USA.
EEGH-UNC	Centro de Investigaciones Paleobiológicas, Universidad Nacional de Córdoba, Argentina.
CM	Canterbury Museum, Christchurch, New Zealand.
CPC	Cory Pittman Collection, Farifield, Washington, USA.
CRC	Colin Redfern Collection, Boca Raton, Florida, USA.
CSC	Carol Skoglund Collection, Phoenix, Arizona, USA. Now in SBMNH.
CZTL	Centro de Zoología del Instituto de Investigasción Científica Tropical, Lisbon, Portugal.
DMNH	Delaware Museum of Natural History, Wilmington, Delaware, USA.
DKC	David Kirsh Collection, Durham, North Carolina, USA.
DLG	Daniel L. Geiger Collection, Los Angeles, California, USA. To be deposited in SBMNH.
DLG#	Internal reference number for this project: see Voucher Repository below for explanation.
EGC	Emilio García Collection, Lafayette, Louisiana, USA.
ERC	Emilio Rolán Collection, Vigo, Spain.
ESC	Enrico Schwabe Collection, Munich, Germany.
FLMNH	Florida Museum of Natural History, Gainsville, Florida, USA.
FMNH	Field Museum of Natural History, Chicago, Illinois, USA.
FSC	Frank Swinnen Collection, Lommel, Belgium.
GMH	Geological Museum of Hungary, Geological Institute of Hungary, Budapest, Hungary.
GNM	Natural History Museum, Göteborg, Sweden.
GNS	Institute of Geological and Nuclear Sciences, Lower Hutt, Wellington, New Zealand.
GPC	Gloria Pearson Collection, Naples, Florida, USA.
HLC	Harry Lee Collection, Jacksonville, Florida, USA.
HUJ	Hebrew University, Jerusalem, Israel.
IKC	Ingo Kurtz Collection, Zornheim, Germany.
ISL	Ingemann Schnetler Collection, Langå, Denmark.
IY	Isaac Yaron Collection. Now in Tel Aviv Museum, Israel.
IZSH	Instituto de Zoología y Sistemática, Havana, Cuba.
JAC	Johan van Aartsen Collection, The Netherlands.
JKC	Joan Koven Collection, Silver Spring, Maryland, USA.

JWC	John Wolff Collection, Lancaster, Pennsylvania, USA.
KBIN	Koninklij Belgisch Instituut voor Natuurwetenschappen, Brussels, Belgium [also as I.R.Sc.N.B. = Institut Royal des Sciences Naturelles de Belgique].
KLK	Kirstie Kaiser Collection, Porto Vallarta, Mexico.
LACM	Natural History Museum of Los Angeles County-Malacology, Los Angeles, California, USA.
LACMIP	Natural History Museum of Los Angeles County-Invertebrate Paleontology, Los Angeles, California, USA.
LMD	Löbbecke-Museum Düsseldorf, Germany.
LSC	Lynton Stephens Collection, Malvern, Victoria, Australia.
MAC	Maria Alessandra Conti Collection [location unknown].
Macn	Museo Argentino de Ciencias Naturales "Bernardino Rivadavia," Buenos Aires, Argentina.
MBL	Museu Bocage, Lisbon, Portugal.
MCZ	Museum of Comparative Zoology, Harvard University, Cambridge, Massachusetts, USA.
MGUH	Geological Museum of the University of Copenhagen, Denmark.
MHNG	Muséum d'Histoire Naturelle de Genève, Geneva, Switzerland.
MHNS	Museo de Historia Natural, Universidad de Santiago de Compostela, Spain.
MLP	Museo de La Plata, Argentina.
MMC	Max Marrow Collection, Melbourne, Victoria, Australia. Now in MV.
MMF	Museu Municipal do Funchal (História Natural), Madeira, Portugal.
MMG	Senckenberg Naturhistorische Sammlungen Dresden, Museum für Mineralogie und Geologie, Germany.
MNCN	Museo Nacional de Ciencias Naturales, Madrid, Spain.
MNHN	Muséum Nationale d'Histoire Naturelle, Paris, France.
MNHN-LP	Muséum Nationale d'Histoire Naturelle, Laboratoire de Paléontologie, Paris, France.
MNRJ	Museu Nacional da Universidade Federal do Rio de Janeiro, Brazil.
MOGURG	Museu Oceanográfico da Fundação Universidade de Rio Grande, Brazil.
MPIUFRGS	Museu de Paleontologia de Invertebrados da Universidade Federal Rio Grande do Sul, Porto Alegre/RS, Brazil.
MV	Museum of Victoria, Melbourne, Victoria. Australia.
MZSP	Museu de Zoologia da Universidade de São Paulo, Brazil.
NHMUK	The Natural History Museum, London, England, Great Britain.
NHMW	Naturhistorisches Museum, Wien, Vienna, Austria.
NMNZ	National Museum New Zealand, Wellington, New Zealand.
NMR	Natural History Museum Rotterdam, The Netherlands.
NMSA	Natal Museum, Pietermaritzburg, Natal, South Africa.
NMSZ	National Museum of Scotland, Zoology, Edinburgh, Scotland, Great Britain.
NMW	National Museum Wales, Cardiff, Great Britain.
NSMT	National Science Museum, Tokyo, Japan.
OMU	Oxford University Museum, Oxford, Great Britain.
OSU	Ohio State University, Columbus, Ohio, USA.
PIN	Paleontological Institute of Russian Academy of Sciences, Moscow, Russia.
PRI	Paleontological Research Institution, Ithaca, New York, USA.
RFGC	Raúl Fernández-Garcés Collection, Cienfuegos, Cuba.
RM	Redpath Museum, Montreal, Ottawa, Canada.
RMC	Robert Moore Collection, Los Angeles, California, USA.

RMNH	National Museum of Natural History, Naturalis, Leiden, The Netherlands.
ROM	Royal Ontario Museum, Toronto, Ontario, Canada.
SAM	South Australian Museum, Adelaide, South Australia, Australia.
SAMCT	South African Museum Cape Town, South Africa.
SBMNH	Santa Barbara Museum of Natural History, Santa Barbara, California, USA.
SDNHM	San Diego Natural History Museum, San Diego, California, USA.
SGPIH	Collection [Sammlung] Geological-Paleontological Institute, University of Hamburg, Germany.
SIO	Scripps Institution of Oceanography, Benthic Invertebrate Collection, La Jolla, California, USA.
SMF	Senckenberg Museum, Frankfurt am Main, Germany.
SMNH	Swedish Museum of Natural History, Stockholm, Sweden.
SSB	S. Stillman Berry Collection, California, USA. Now in CASIZ and SBMNH.
SRC	Stefano Rufini Collection, Anguillara, Italy.
SUPTC	Stanford University Paleontological Type collection, Stanford, California, USA. Now in CASIZ.
TBS	Trondheim Biologiske Stasjon, Norway.
TMAG	Tasmania Museum and Art Gallery, Hobart, Tasmania, Australia.
TMGLC	Tisselli Morena and Giunchi Luigi Collection, Ravenna, Italy.
UMML	University of Miami Marine Laboratory, Rosenstiel School of Marine and Atmospheric Sciences, Florida, USA.
UMMZ	University of Michigan, Museum of Zoology, Ann Arbor, Michigan, USA.
UMUT	University Museum University of Tokyo, Japan.
UMZC	University Museum of Zoology, Cambridge, Great Britain.
USNM	United States National Museum, Smithsonian Institution, Washington, District of Columbia, USA.
WAM	Western Australian Museum, Perth, Western Australia, Australia.
WSC	Willi Segers Collection, Aartselaar, Belgium.
ZIN	Zoological Institute of Russian Academy of Sciences, St. Petersburg, Russia.
ZMA	Zoological Museum Amsterdam, The Netherlands.
ZMB	Zoological Museum Berlin, Germany.
ZMBN	Zoological Museum in Bergen, Norway.
ZMH	Zoological Museum Hamburg, Germany.
ZMMU	Zoological Museum Moscow University, Moscow, Russia.
ZMON	Zoological Museum in Oslo, Norway.
ZMTøN	Zoological Museum in Tromsø, Norway.
ZMUC	Zoological Museum, University of Copenhagen, Denmark.
ZMUH	Zoological Museum, University of Hamburg, Germany.
ZSM	Zoological State Collection [Zoologische Staatssammlung], Munich, Germany.

Abbreviations

n.d.	No date.
M	Monotypy.
OD	Original designation; original description.
SD	Subsequent designation.
SEM	Scanning Electron Microscope, —Micrograph, —Microscopy.
VPSE	Variable Pressure Secondary Electron detector.
†	Fossil taxon.

Multiletter abbreviations for genera have been used to avoid confusion, following ICZN Recommendation 25A.

An.	*Anatoma*
Ar.	*Ariella*
C.	*Coronadoa*
D.	*Depressizona*
I.	*Incisura*
La.	*Larochea*
Ls.	*Larocheopsis*
M.	*Maxwellella*
Sas.	*Sasakiconcha*
Sat.	*Satondella*
Sci.	*Scissurella*
Sin.	*Sinezona*
Suk.	*Sukashitrochus*
Sut.	*Sutilizona*
Tc.	*Temnocinclis*
Th.	*Thieleella*
Tri.	*Triassurella*
Tro.	*Trogloconcha*
Tz.	*Temnozaga*
Z.	*Zardinitrochus*

Voucher repository

All new material obtained while preparing the present volume will be deposited in a public repository. Accessioned material obtained on loan from other institutions and collections is in their respective holdings.

All specimens were data-based separately for this project. Each single-species lot was given a unique project-number in the format DLG#1234. This internal reference is only given in case where no institutional registration or station numbers are available to unambiguously refer to specific lots.

Thumbnail identification

To assist the user with the identification of specimens, and to narrow down the number of possible species, plates with thumbnails of the apertural view from each species are provided. The first plate provides an overview over the genera. Some characters to distinguish various species may not be apparent in the apertural view. Please refer to the discussion of the respective species for details. For the smaller genera only a single plate is provided, arranged by family and genus. With the species name, the approximate *known* distribution is given.

The Thumbnail Identification figures for all species are additionally provided as separate cards located in inserts at the end of the volumes. This duplication is intentional, for the greater ease of use of the cards, and providing greater structural integrity of the figures in the bound book. The size indications for the genera are minimum–**modal**–maximum.

Geographic abbreviations used. Med.: Mediterranean. IO: Indian Ocean. IMA: Indo-Malayan Archipelago. WP: Western Pacific. CP: Central Pacific. NESW: cardinal directions.

SCISSURELLIDAE

Scissurella Recent (0.5–**1.2**–1.75 mm)

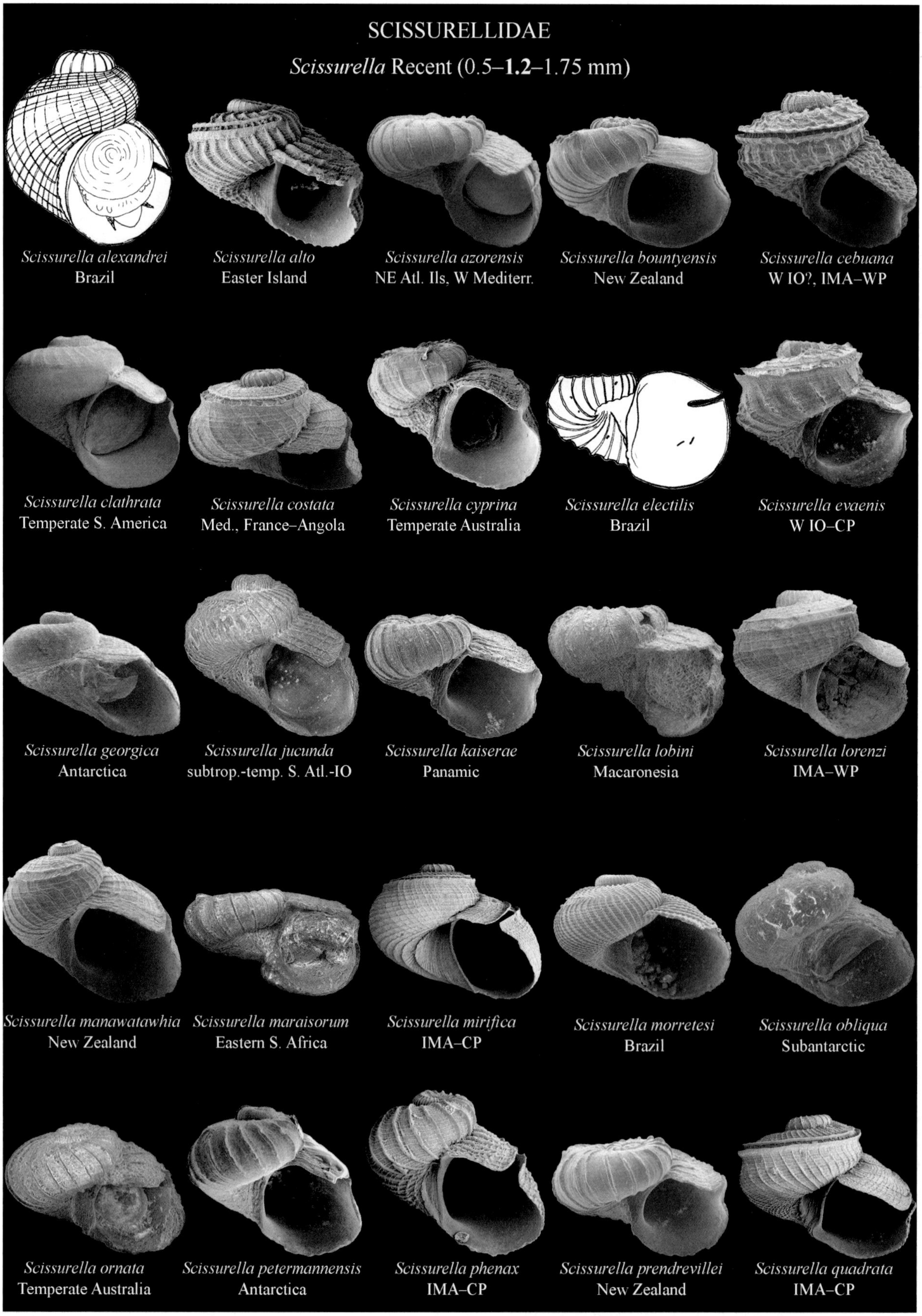

Scissurella alexandrei Brazil

Scissurella alto Easter Island

Scissurella azorensis NE Atl. Ils, W Mediterr.

Scissurella bountyensis New Zealand

Scissurella cebuana W IO?, IMA–WP

Scissurella clathrata Temperate S. America

Scissurella costata Med., France–Angola

Scissurella cyprina Temperate Australia

Scissurella electilis Brazil

Scissurella evaenis W IO–CP

Scissurella georgica Antarctica

Scissurella jucunda subtrop.-temp. S. Atl.-IO

Scissurella kaiserae Panamic

Scissurella lobini Macaronesia

Scissurella lorenzi IMA–WP

Scissurella manawatawhia New Zealand

Scissurella maraisorum Eastern S. Africa

Scissurella mirifica IMA–CP

Scissurella morretesi Brazil

Scissurella obliqua Subantarctic

Scissurella ornata Temperate Australia

Scissurella petermannensis Antarctica

Scissurella phenax IMA–CP

Scissurella prendrevillei New Zealand

Scissurella quadrata IMA–CP

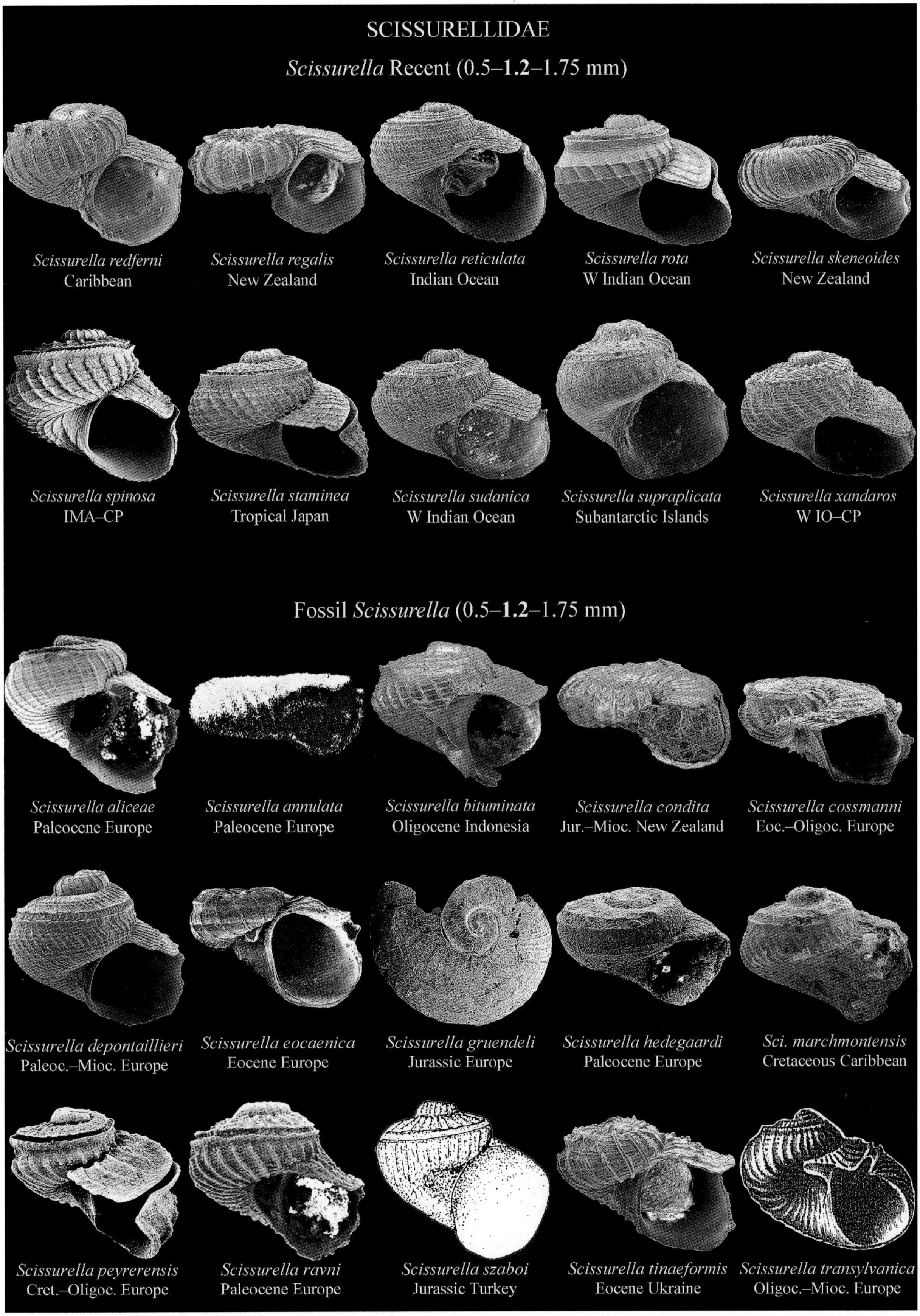

SCISSURELLIDAE

Scissurella Recent (0.5–**1.2**–1.75 mm)

Scissurella redferni Caribbean — *Scissurella regalis* New Zealand — *Scissurella reticulata* Indian Ocean — *Scissurella rota* W Indian Ocean — *Scissurella skeneoides* New Zealand

Scissurella spinosa IMA–CP — *Scissurella staminea* Tropical Japan — *Scissurella sudanica* W Indian Ocean — *Scissurella supraplicata* Subantarctic Islands — *Scissurella xandaros* W IO–CP

Fossil *Scissurella* (0.5–**1.2**–1.75 mm)

Scissurella aliceae Paleocene Europe — *Scissurella annulata* Paleocene Europe — *Scissurella bituminata* Oligocene Indonesia — *Scissurella condita* Jur.–Mioc. New Zealand — *Scissurella cossmanni* Eoc.–Oligoc. Europe

Scissurella depontaillieri Paleoc.–Mioc. Europe — *Scissurella eocaenica* Eocene Europe — *Scissurella gruendeli* Jurassic Europe — *Scissurella hedegaardi* Paleocene Europe — *Sci. marchmontensis* Cretaceous Caribbean

Scissurella peyrerensis Cret.–Oligoc. Europe — *Scissurella ravni* Paleocene Europe — *Scissurella szaboi* Jurassic Turkey — *Scissurella tinaeformis* Eocene Ukraine — *Scissurella transylvanica* Oligoc.–Mioc. Europe

SCISSURELLIDAE

Recent *Sinezona* (0.6–**1.2**–3.8 mm)

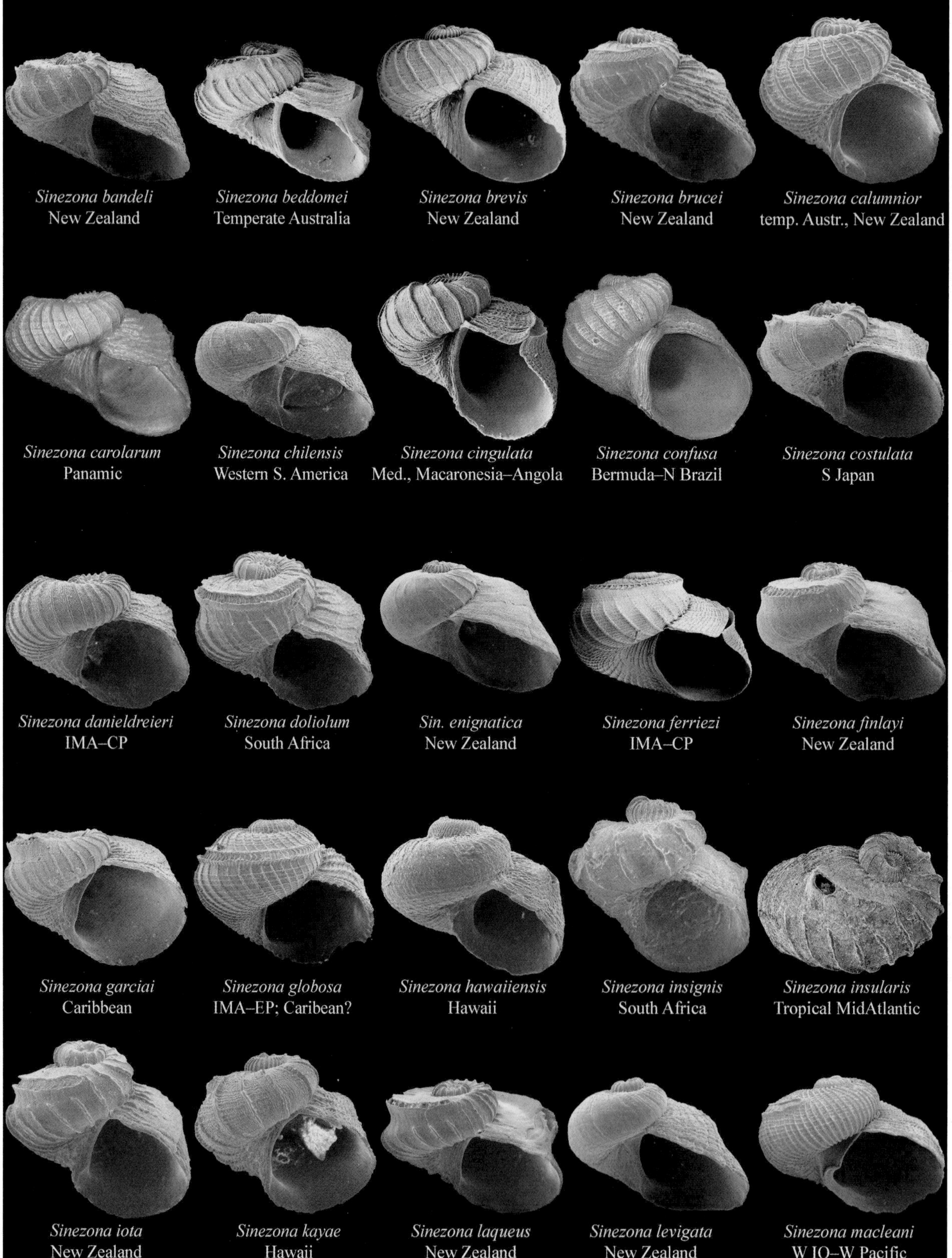

Sinezona bandeli New Zealand; *Sinezona beddomei* Temperate Australia; *Sinezona brevis* New Zealand; *Sinezona brucei* New Zealand; *Sinezona calumnior* temp. Austr., New Zealand

Sinezona carolarum Panamic; *Sinezona chilensis* Western S. America; *Sinezona cingulata* Med., Macaronesia–Angola; *Sinezona confusa* Bermuda–N Brazil; *Sinezona costulata* S Japan

Sinezona danieldreieri IMA–CP; *Sinezona doliolum* South Africa; *Sin. enignatica* New Zealand; *Sinezona ferriezi* IMA–CP; *Sinezona finlayi* New Zealand

Sinezona garciai Caribbean; *Sinezona globosa* IMA–EP; Caribean?; *Sinezona hawaiiensis* Hawaii; *Sinezona insignis* South Africa; *Sinezona insularis* Tropical MidAtlantic

Sinezona iota New Zealand; *Sinezona kayae* Hawaii; *Sinezona laqueus* New Zealand; *Sinezona levigata* New Zealand; *Sinezona macleani* W IO–W Pacific

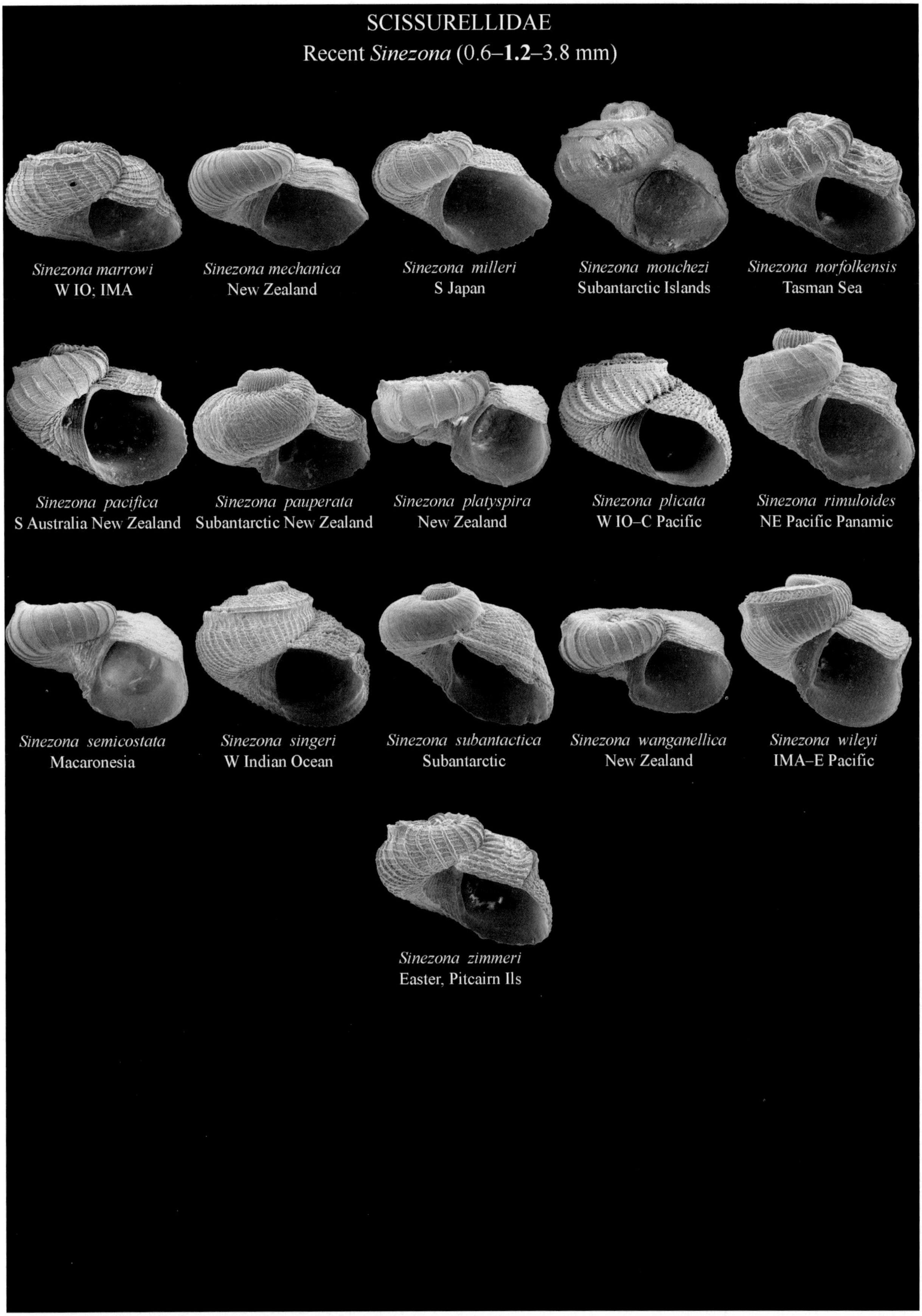

SCISSURELLIDAE
Recent *Sinezona* (0.6–**1.2**–3.8 mm)

Sinezona marrowi W IO; IMA
Sinezona mechanica New Zealand
Sinezona milleri S Japan
Sinezona mouchezi Subantarctic Islands
Sinezona norfolkensis Tasman Sea
Sinezona pacifica S Australia New Zealand
Sinezona pauperata Subantarctic New Zealand
Sinezona platyspira New Zealand
Sinezona plicata W IO–C Pacific
Sinezona rimuloides NE Pacific Panamic
Sinezona semicostata Macaronesia
Sinezona singeri W Indian Ocean
Sinezona subantactica Subantarctic
Sinezona wanganellica New Zealand
Sinezona wileyi IMA–E Pacific
Sinezona zimmeri Easter, Pitcairn Ils

SCISSURELLIDAE

Fossil *Sinezona* 0.5–**1.5**–3 mm)

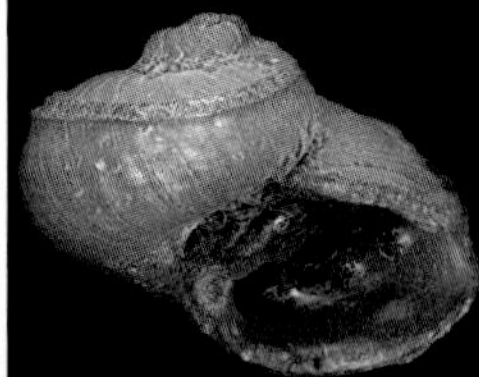

Sinezona apudornata
Miocene New Zealand

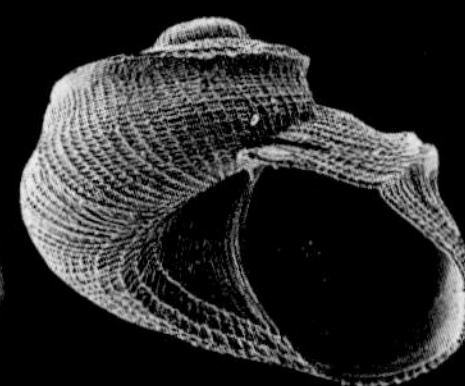

Sinezona cupelliformis
Eoc.–Oligocene Europe

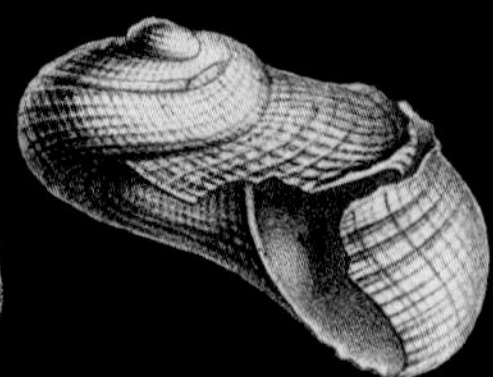

Sinezona deshayesi
Eocene Europe

Sinezona finlayi
Miocene New Zealand

Sinezona haliotimorpha
Oligo.–Mioc. Europe

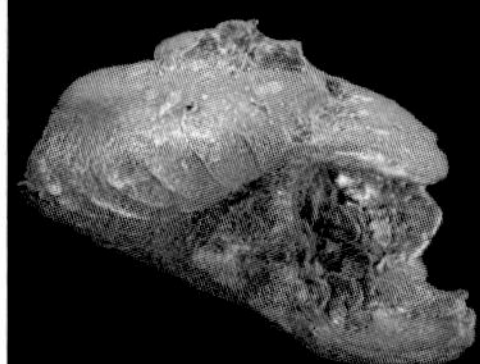

Sinezona koruahina
Late Tertiary New Zealand

Sinezona malloryi
Eocene N. America

Sinezona praecedens
Oligo.–Miocene Europe

Sinezona terquemi
Miocene Europe

Sinezona tertia
Pliocene New Zealand

Satondella Recent (0.6–**1.5**–2.1 mm)

Sat. azonata
New Zealand

Sat. bicristata
New Zealand

Sat. brasiliensis
Caribbean – Brazil

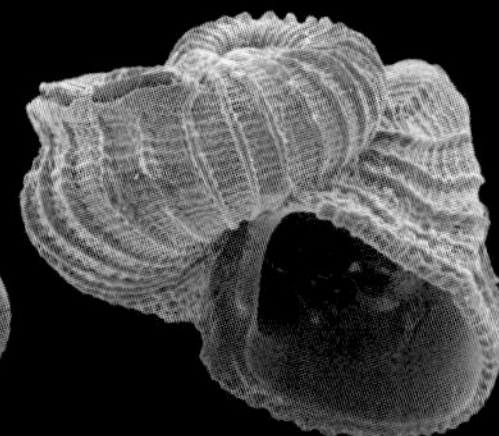

Sat. cachoi
W IO–CP

Sat. danieli
tropical N Atlantic

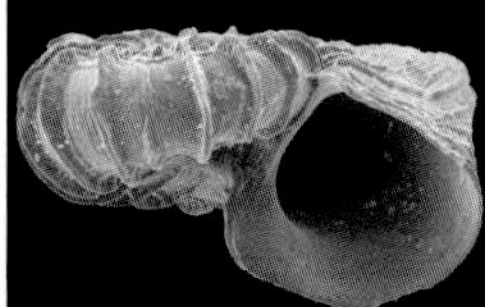

Sat. dantarti
IMA–CP

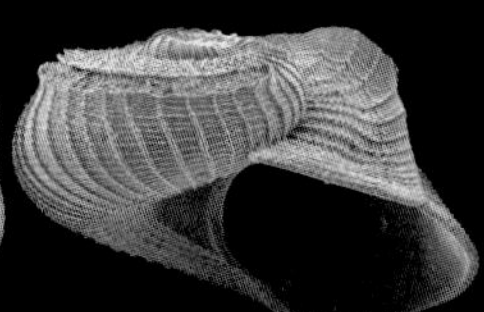

Sat. goudi
W IO; IMA

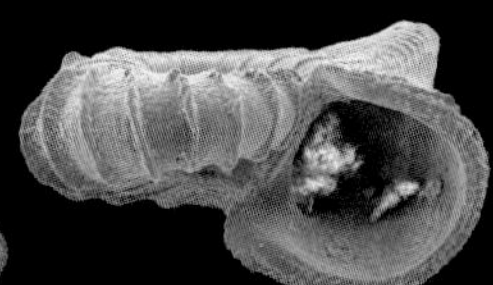

Sat. minuta
IMA–WP

Sat. senni
Easter Island

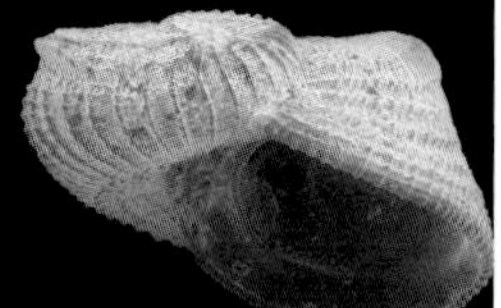

Sat. tabulata
W IO; W Pacific

Satondella fossil (1.5 mm)

Sat. kondoi
Eocene Tonga

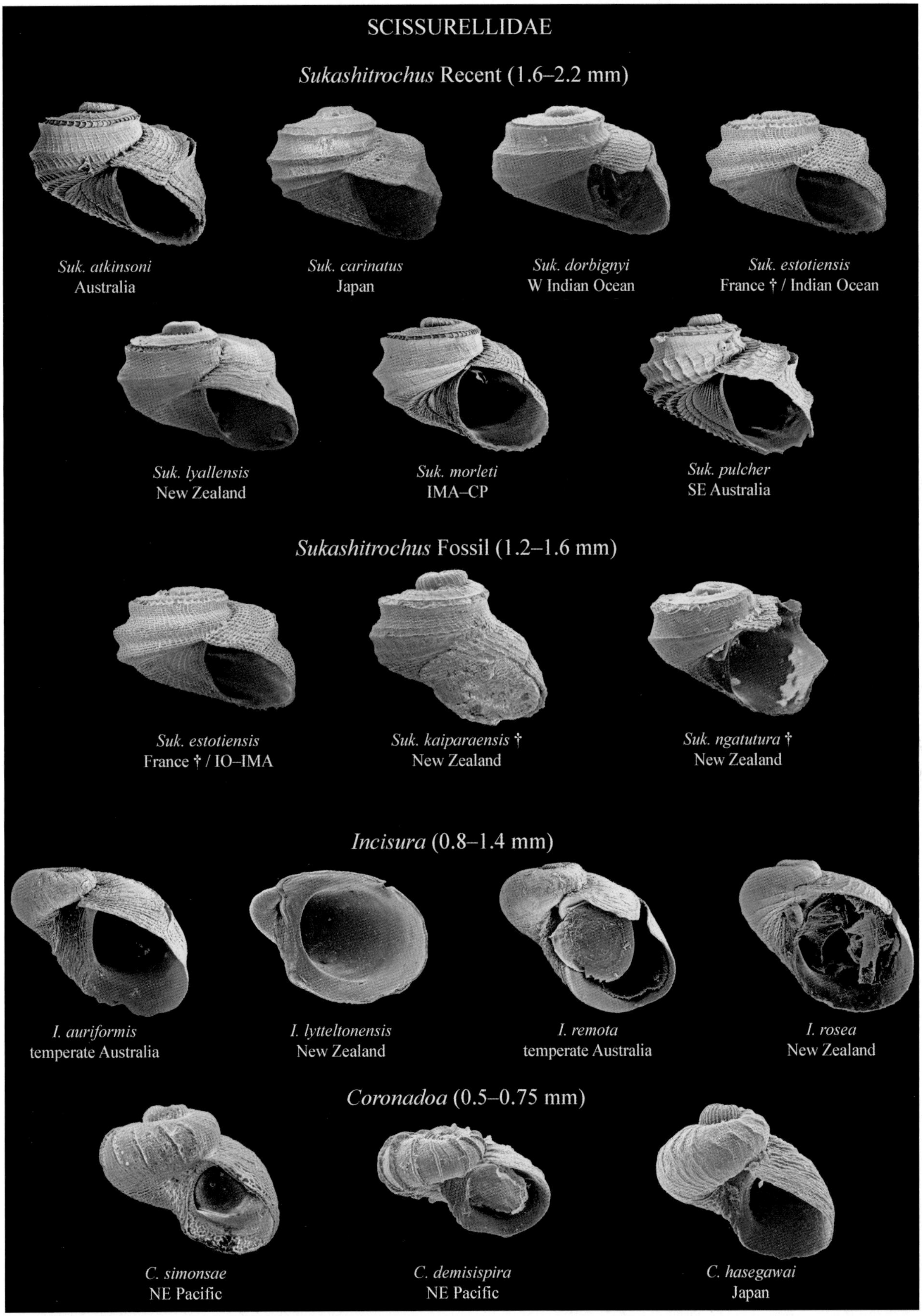
SCISSURELLIDAE
Sukashitrochus Recent (1.6–2.2 mm)
Suk. atkinsoni
Australia
Suk. carinatus
Japan
Suk. dorbignyi
W Indian Ocean
Suk. estotiensis
France † / Indian Ocean
Suk. lyallensis
New Zealand
Suk. morleti
IMA–CP
Suk. pulcher
SE Australia
Sukashitrochus Fossil (1.2–1.6 mm)
Suk. estotiensis
France † / IO–IMA
Suk. kaiparaensis †
New Zealand
Suk. ngatutura †
New Zealand
Incisura (0.8–1.4 mm)
I. auriformis
temperate Australia
I. lytteltonensis
New Zealand
I. remota
temperate Australia
I. rosea
New Zealand
Coronadoa (0.5–0.75 mm)
C. simonsae
NE Pacific
C. demisispira
NE Pacific
C. hasegawai
Japan

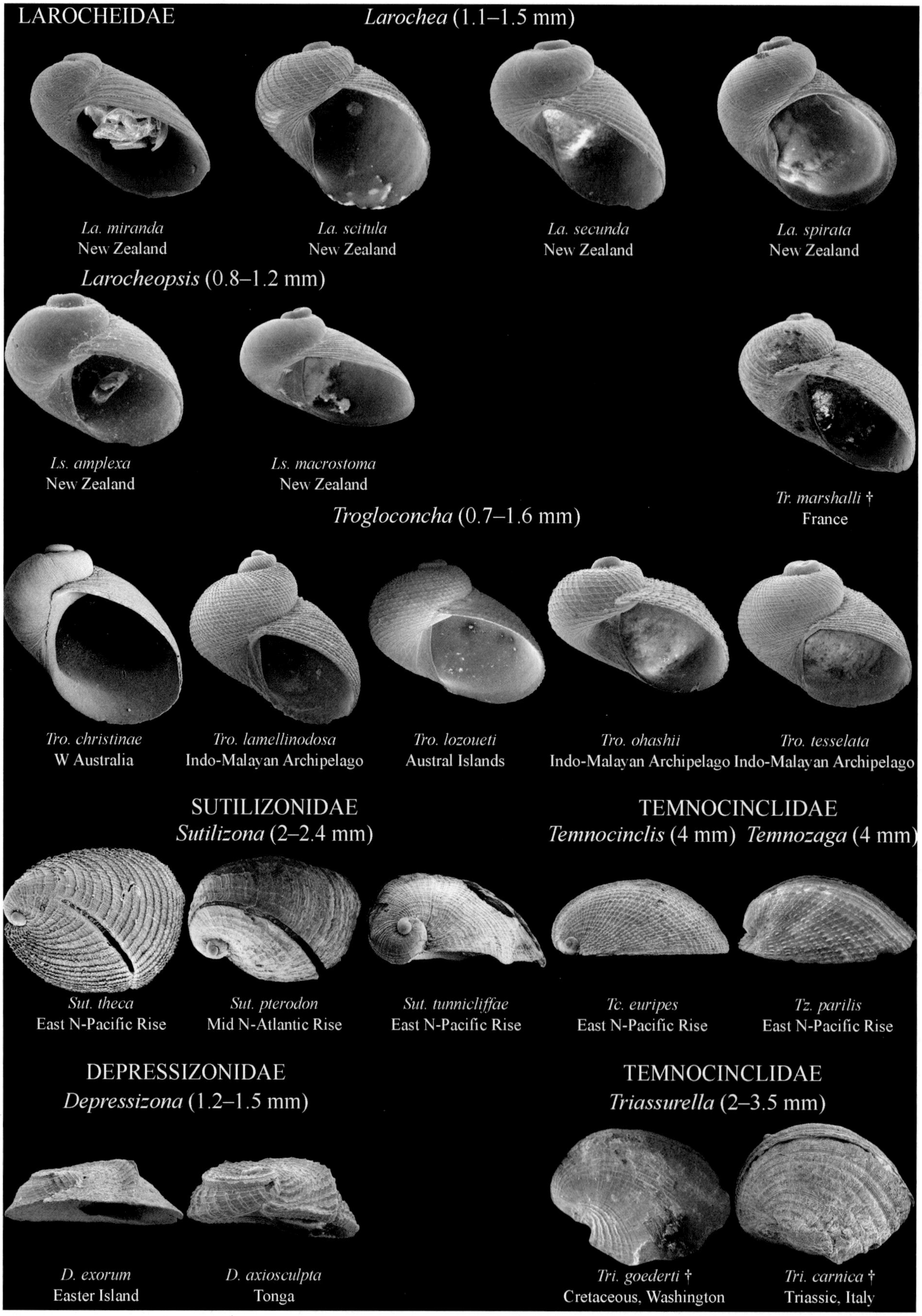
LAROCHEIDAE
Larochea (1.1–1.5 mm)
La. miranda
New Zealand
La. scitula
New Zealand
La. secunda
New Zealand
La. spirata
New Zealand
Larocheopsis (0.8–1.2 mm)
Ls. amplexa
New Zealand
Ls. macrostoma
New Zealand
Tr. marshalli †
France
Trogloconcha (0.7–1.6 mm)
Tro. christinae
W Australia
Tro. lamellinodosa
Indo-Malayan Archipelago
Tro. lozoueti
Austral Islands
Tro. ohashii
Indo-Malayan Archipelago
Tro. tesselata
Indo-Malayan Archipelago
SUTILIZONIDAE
Sutilizona (2–2.4 mm)
TEMNOCINCLIDAE
Temnocinclis (4 mm) Temnozaga (4 mm)
Sut. theca
East N-Pacific Rise
Sut. pterodon
Mid N-Atlantic Rise
Sut. tunnicliffae
East N-Pacific Rise
Tc. euripes
East N-Pacific Rise
Tz. parilis
East N-Pacific Rise
DEPRESSIZONIDAE
Depressizona (1.2–1.5 mm)
TEMNOCINCLIDAE
Triassurella (2–3.5 mm)
D. exorum
Easter Island
D. axiosculpta
Tonga
Tri. goederti †
Cretaceous, Washington
Tri. carnica †
Triassic, Italy

ANATOMIDAE

Anatoma (1–**3**–10.5 mm)

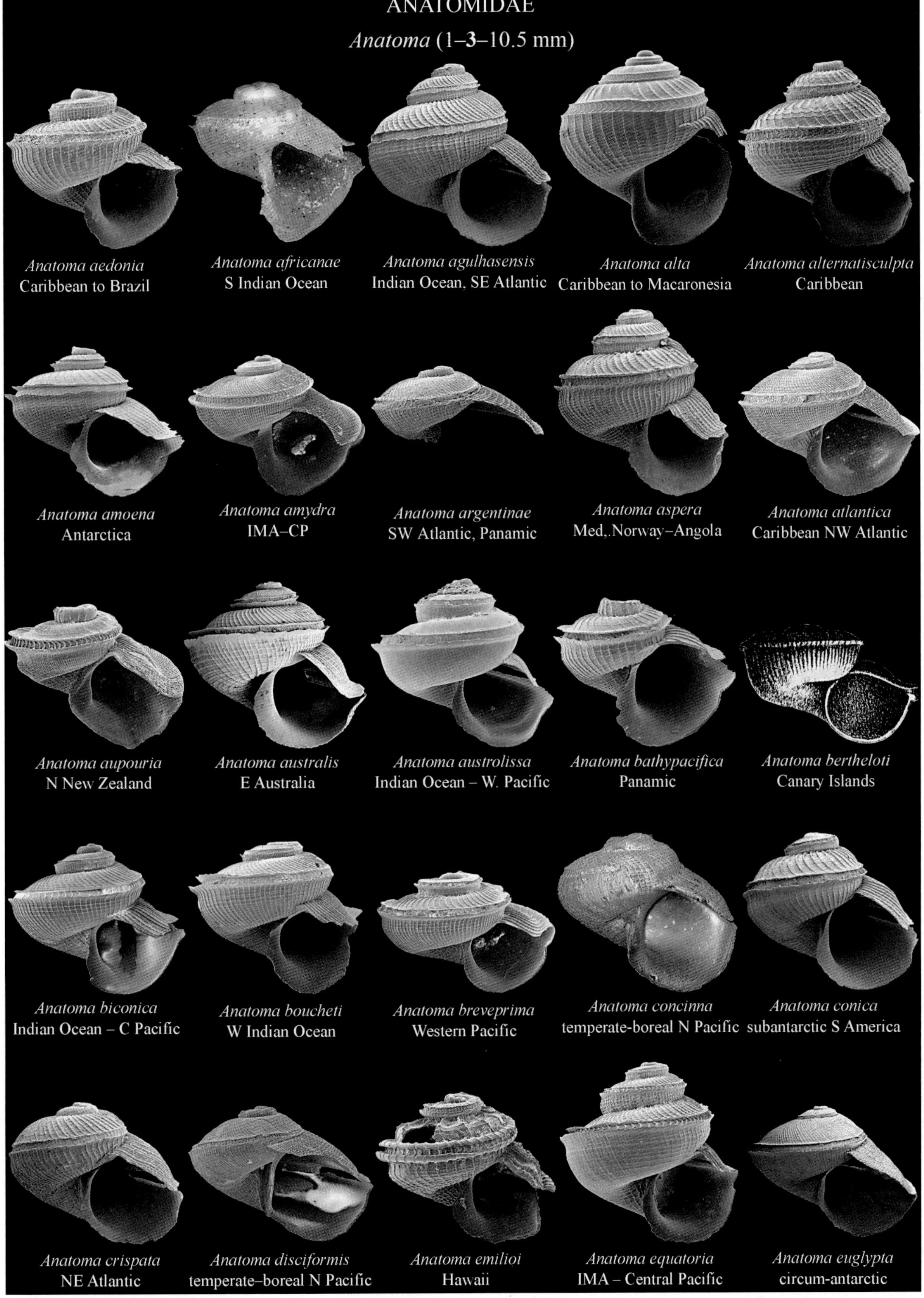

Anatoma aedonia
Caribbean to Brazil

Anatoma africanae
S Indian Ocean

Anatoma agulhasensis
Indian Ocean, SE Atlantic

Anatoma alta
Caribbean to Macaronesia

Anatoma alternatisculpta
Caribbean

Anatoma amoena
Antarctica

Anatoma amydra
IMA–CP

Anatoma argentinae
SW Atlantic, Panamic

Anatoma aspera
Med,.Norway–Angola

Anatoma atlantica
Caribbean NW Atlantic

Anatoma aupouria
N New Zealand

Anatoma australis
E Australia

Anatoma austrolissa
Indian Ocean – W. Pacific

Anatoma bathypacifica
Panamic

Anatoma bertheloti
Canary Islands

Anatoma biconica
Indian Ocean – C Pacific

Anatoma boucheti
W Indian Ocean

Anatoma breveprima
Western Pacific

Anatoma concinna
temperate-boreal N Pacific

Anatoma conica
subantarctic S America

Anatoma crispata
NE Atlantic

Anatoma disciformis
temperate–boreal N Pacific

Anatoma emilioi
Hawaii

Anatoma equatoria
IMA – Central Pacific

Anatoma euglypta
circum-antarctic

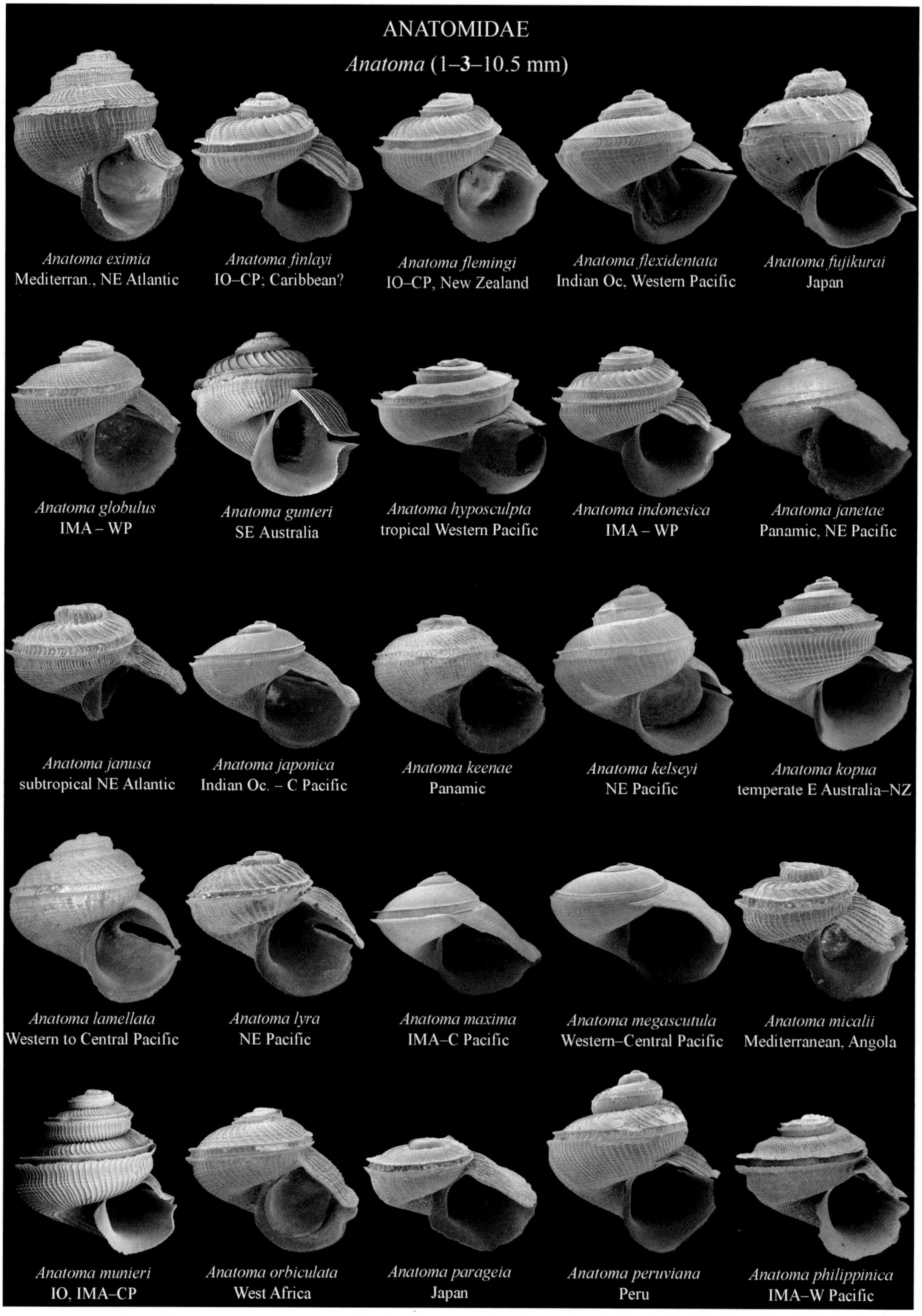

ANATOMIDAE

Anatoma (1–**3**–10.5 mm)

Anatoma eximia Mediterran., NE Atlantic

Anatoma finlayi IO–CP; Caribbean?

Anatoma flemingi IO–CP, New Zealand

Anatoma flexidentata Indian Oc, Western Pacific

Anatoma fujikurai Japan

Anatoma globulus IMA – WP

Anatoma gunteri SE Australia

Anatoma hyposculpta tropical Western Pacific

Anatoma indonesica IMA – WP

Anatoma janetae Panamic, NE Pacific

Anatoma janusa subtropical NE Atlantic

Anatoma japonica Indian Oc. – C Pacific

Anatoma keenae Panamic

Anatoma kelseyi NE Pacific

Anatoma kopua temperate E Australia–NZ

Anatoma lamellata Western to Central Pacific

Anatoma lyra NE Pacific

Anatoma maxima IMA–C Pacific

Anatoma megascutula Western–Central Pacific

Anatoma micalii Mediterranean, Angola

Anatoma munieri IO, IMA–CP

Anatoma orbiculata West Africa

Anatoma parageia Japan

Anatoma peruviana Peru

Anatoma philippinica IMA–W Pacific

ANATOMIDAE
Anatoma (1–**3**–10.5 mm)

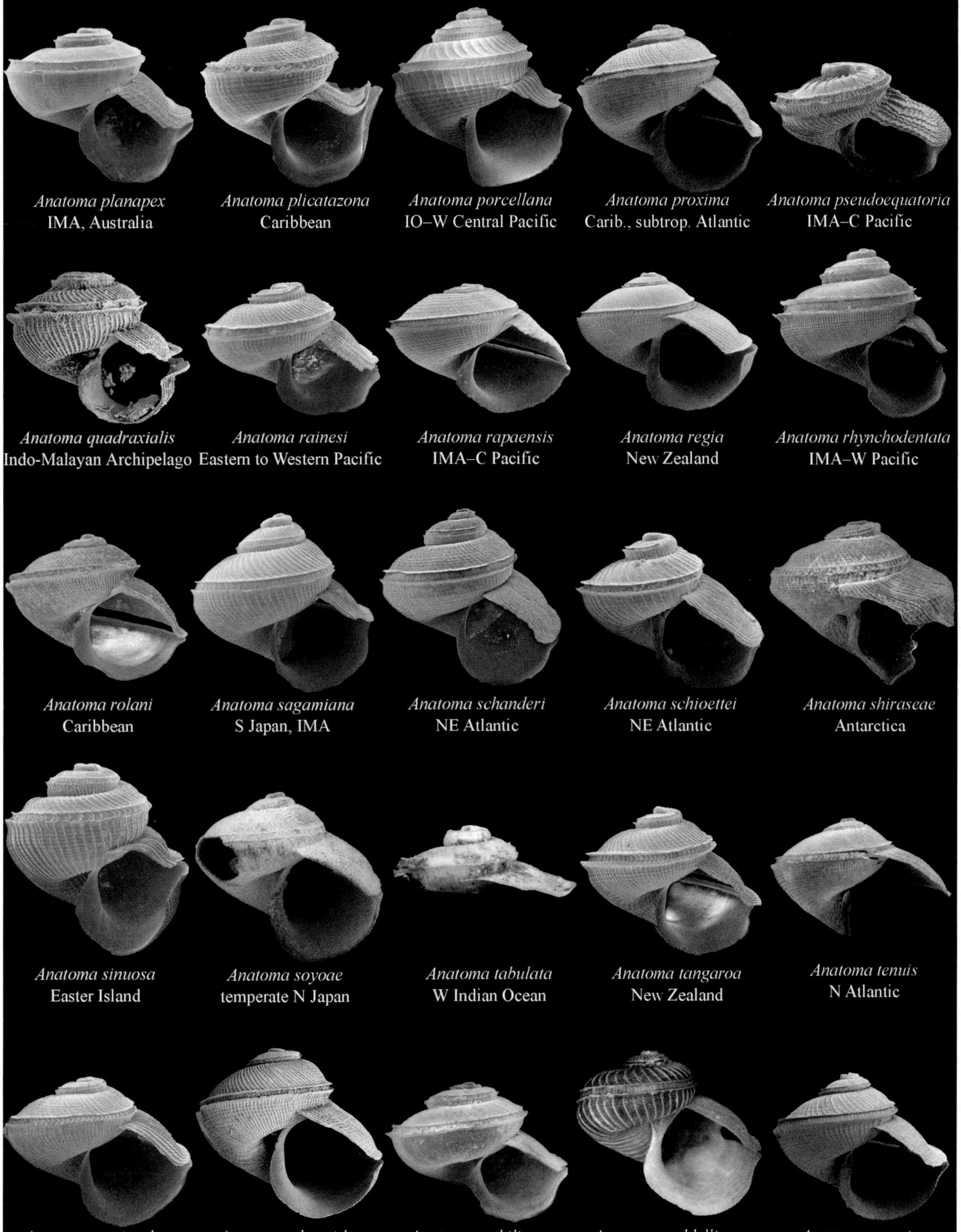

Anatoma planapex IMA, Australia
Anatoma plicatazona Caribbean
Anatoma porcellana IO–W Central Pacific
Anatoma proxima Carib., subtrop. Atlantic
Anatoma pseudoequatoria IMA–C Pacific

Anatoma quadraxialis Indo-Malayan Archipelago
Anatoma rainesi Eastern to Western Pacific
Anatoma rapaensis IMA–C Pacific
Anatoma regia New Zealand
Anatoma rhynchodentata IMA–W Pacific

Anatoma rolani Caribbean
Anatoma sagamiana S Japan, IMA
Anatoma schanderi NE Atlantic
Anatoma schioettei NE Atlantic
Anatoma shiraseae Antarctica

Anatoma sinuosa Easter Island
Anatoma soyoae temperate N Japan
Anatoma tabulata W Indian Ocean
Anatoma tangaroa New Zealand
Anatoma tenuis N Atlantic

Anatoma tenuisculpta Med., Norway–Angola
Anatoma tobeyoides SE Australia
Anatoma umbilicata Med., NE Atlantic
Anatoma weddelliana Antarctic
Anatoma yaroni W Indian Ocean

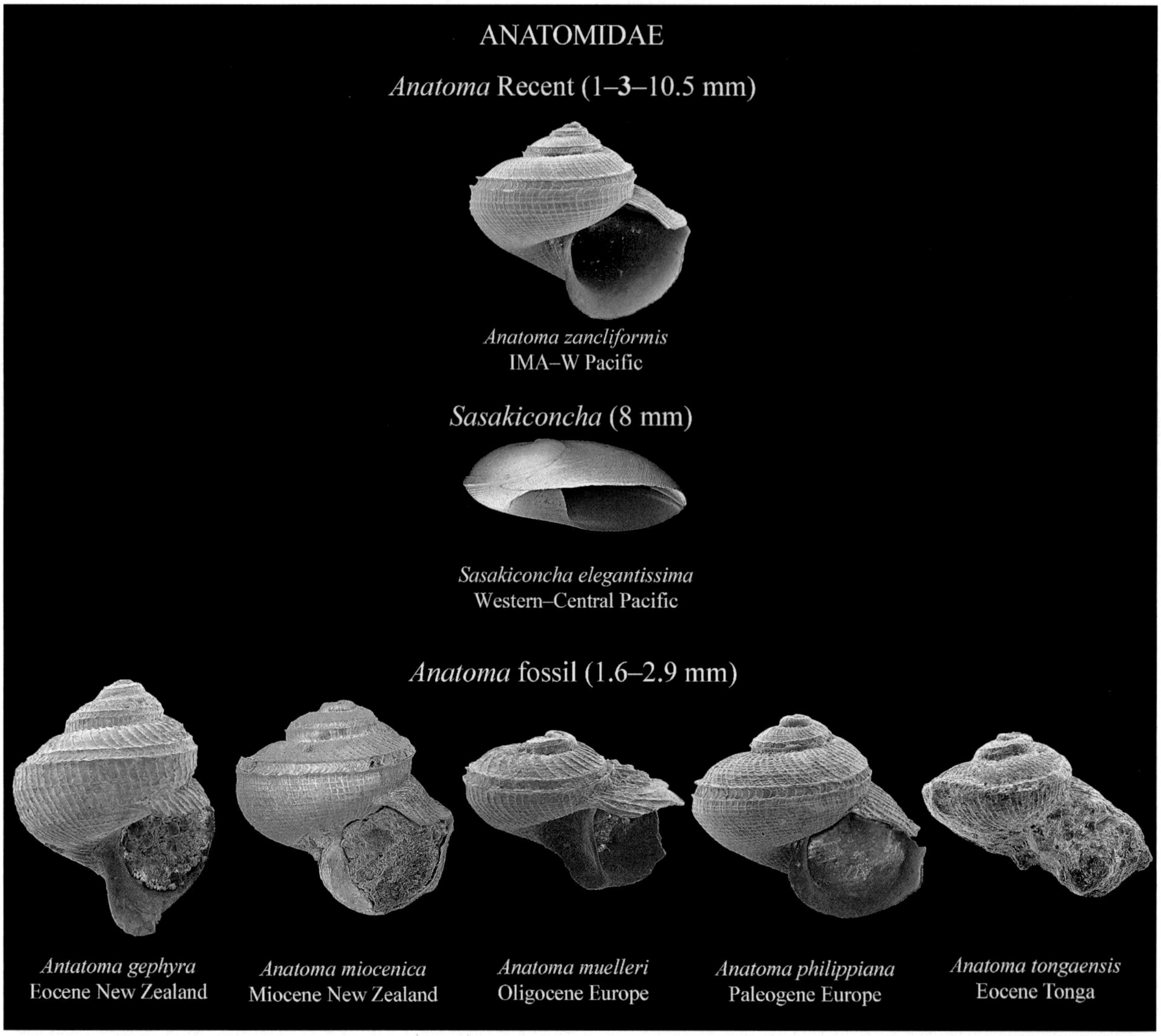
ANATOMIDAE
Anatoma Recent (1–**3**–10.5 mm)
Anatoma zancliformis
IMA–W Pacific
Sasakiconcha (8 mm)
Sasakiconcha elegantissima
Western–Central Pacific
Anatoma fossil (1.6–2.9 mm)
Antatoma gephyra
Eocene New Zealand
Anatoma miocenica
Miocene New Zealand
Anatoma muelleri
Oligocene Europe
Anatoma philippiana
Paleogene Europe
Anatoma tongaensis
Eocene Tonga

List of species by global regions.

Definitions of regions. Figure 32 shows the limits of the global areas as used here on a world map. These regions are based on zoogeographic provinces, which are thereafter modified to permit simple outlines, and contain similar number of species. The areas are defined by landmarks and latitudinal and longitudinal boundaries in the listing below.

1: Mediterranean: Mediterranean Sea and Black Sea.

2: Northeastern Atlantic: Northeastern Atlantic from Arctic circle (60°N), to western Morocco (20°N) including Canary Islands, to 40°W [~ mid-Atlantic ridge].

3: West African: 20°N including Cape Verde Islands to Namibia-South Africa border (28.6°S), to 40°W [~ mid-Atlantic ridge].

4 : South African: Namibia-South Africa border (28.6°S), to Kei River Mouth (32.7°S), to 40–20°W [~ mid-Atlantic ridge] and 89°E.

5: West Indian Ocean: Kei River Mouth (32.7°S), South Africa, to eastern limit of India (89°W), including Red Sea and Persian Gulf.

6: Indo-Malayan Archipelago: east of India (89°W), to Shark Bay, Western Australia, Port Jackson, New South Wales, to 180°E/W, to 30°N, south of Kyushu, Japan.

7: Temperate Australia: from Shark Bay, Western Australia, to Port Jackson, New South Wales, and to Antarctic; from 89°E to 160°W.

8: New Zealand: from Kermadec Islands (30°S) to Antarctic (60°S) and 160°W to 180°E/W.

9: Northwestern Pacific: from Kyushu, Japan, to 180°E/W and 60°N.

10: Tropical Central Pacific: 180°E/W 140–100°W, from 32°N to 60°S.

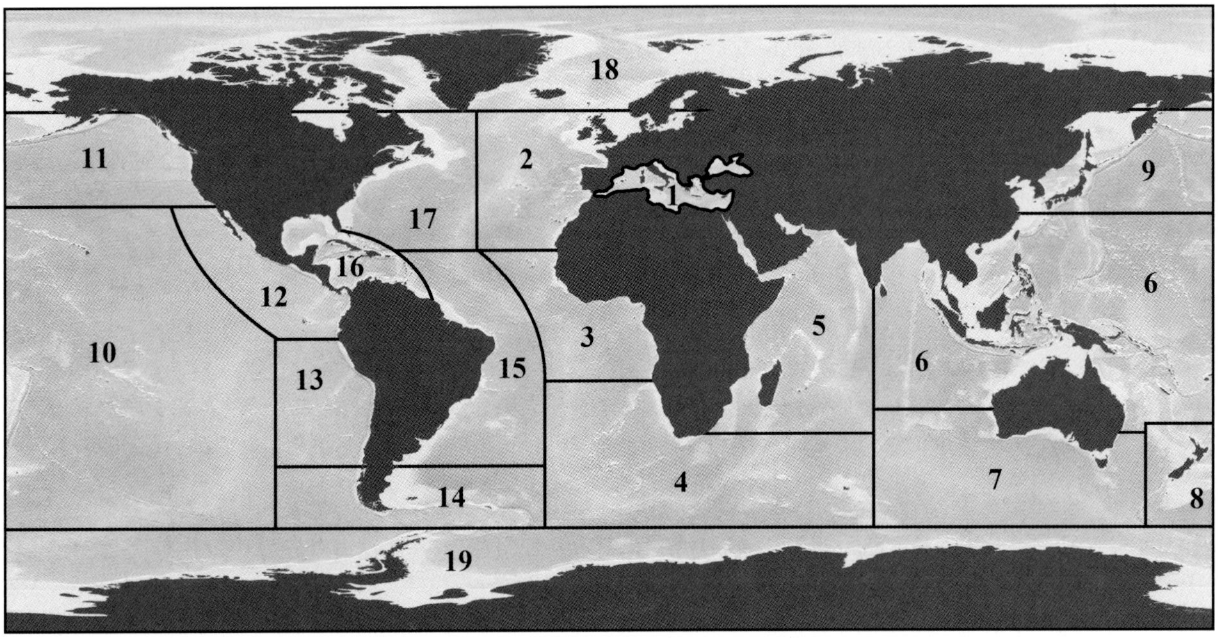

Figure 32. World map showing boundaries of zoogeographic regions used here. For definitions, see text.

11: Northeastern Pacific: 60°N–180°E/W, to US/Mexico border (~32°N).

12: Panamic: from US/Mexico Border (~32°N) to northern border of Peru (3°S).

13: Peruvian: from northern border of Peru (3°S) to Magellanic (42°S).

14: Magellanic: South of 42°S to Antarctic (60°S).

15: Argentinian: from San Matias Gulf (42°S) to the northern border of Brazil (4.1°N).

16: Caribbean: from the northern border of Brazil (4.1°N) to tip of Florida Peninsula.

17: Northwestern Atlantic: from tip of Florida Peninsula south to 20°N, to 40°W and north to 60°N.

18: Arctic: north of 60°N.

19: Antarctic: south of 60°S.

1: Mediterranean. *Scissurella azorensis*
Scissurela costata
Sinezona cingulata
Anatoma aspera
Anatoma eximia
Anatoma micalii
Anatoma tenuisculpta
Anatoma umbilicata

2: NE Atlantic. *Scissurella azorensis*
Scissurella costata
Scissurella lobini
Sinezona cingulata
Sinezona semicostata
Satondella danieli
Anatoma alta
Anatoma aspera
Anatoma bertheloti [*nomen dubium*]
Anatoma crispata
Anatoma eximia
Anatoma janusa
Anatoma micalii
Anatoma schanderi
Anatoma schioettei
Anatoma tenuis
Anatoma tenuisculpta
Anatoma umbilicata

3: West Africa. *Scissurella costata*
Sinezona cingulata
Satondella danieli
Anatoma aspera
Anatoma micalii
Anatoma orbiculata
Anatoma tenuis

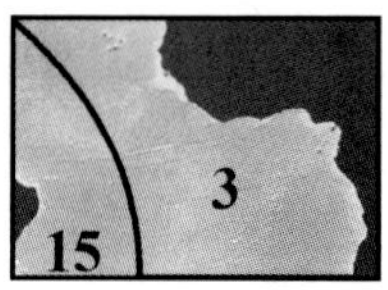

4: South Africa. *Scissurella jucunda*

Scissurella rota
Scissurella obliqua (subantarctic)
Scissurella petermannensis (subantarctic)
Scissurella reticulata
Scissurella supraplicata (subantarctic)
Sinezona doliolum
Sinezona insignis
Sinezona subantarctica (subantarctic)
Sinezona mouchezi (subantarctic)
Sukashitrochus dorbignyi
Anatoma agulhasensis
Anatoma tabulata
Anatoma yaroni

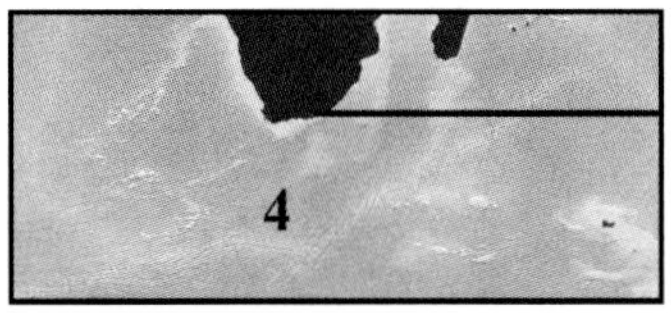

5: Indian Ocean. *Scissurella cebuana*
Scissurella evaensis
Scissurella jucunda
Scissurella maraisorum
Scissurella quadrata
Scissurella reticulata
Scissurella rota
Scissurella sudanica
Scissurella xandaros
Sinezona doliolum
Sinezona insignis
Sinezona macleani
Sinezona marrowi
Sinezona plicata
Sinezona singeri
Sukashitrochus dorbignyi
Sukashitrochus estotiensis
Satondella cachoi
Satondella goudi
Satondella tabulata
Anatoma africanae
Anatoma agulhasensis
Anatoma austrolissa
Anatoma biconica
Anatoma boucheti
Anatoma finlayi
Anatoma flexidentata
Anatoma japonica
Anatoma munieri
Anatoma porcellana
Anatoma yaroni
Trogloconcha tesselata

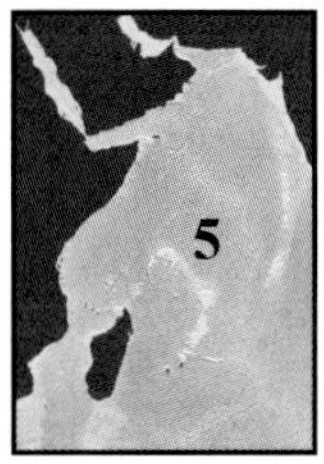

6: Indo-Malayan Archipelago. *Scissurella cebuana*
Scissurella evaensis
Scissurella lorenzi
Scissurella mirifica

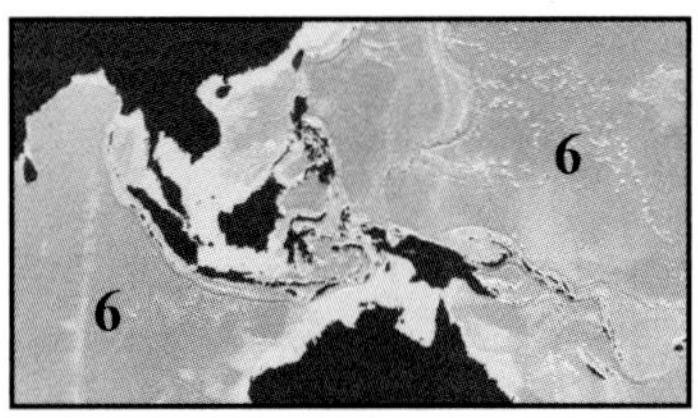

Scissurella phenax
Scissurella quadrata
Scissurella spinosa
Scissurella staminea
Scissurella xandaros
Sinezona danieldrieri
Sinezona ferriezi
Sinezona globosa
Sinezona macleani
Sinezona marrowi
Sinezona milleri
Sinezona norfolkensis
Sinezona plicata
Sinezona wileyi
Sukashitrochus atkinsoni
Sukashitrochus carinatus
Sukashitrochus morleti
Satondella cachoi
Satondella dantarti
Satondella goudi
Satondella minuta
Satondella tabulata
Anatoma amydra
Anatoma aupouria
Anatoma australis
Anatoma austrolissa
Anatoma breveprima
Anatoma biconica
Anatoma equatoria
Anatoma finlayi
Anatoma flexidentata
Anatoma globulus
Anatoma hyposculpta
Anatoma indonesica
Anatoma japonica
Anatoma lamellata
Anatoma maxima
Anatoma megascutula
Anatoma munieri
Anatoma philippinica
Anatoma planapex
Anatoma porcellana
Anatoma pseudoequatoria
Anatoma quadraxialis
Anatoma rainesi
Anatoma rapaensis
Anatoma quadraxialis
Anatoma rapaensis
Anatoma rhynchodentata

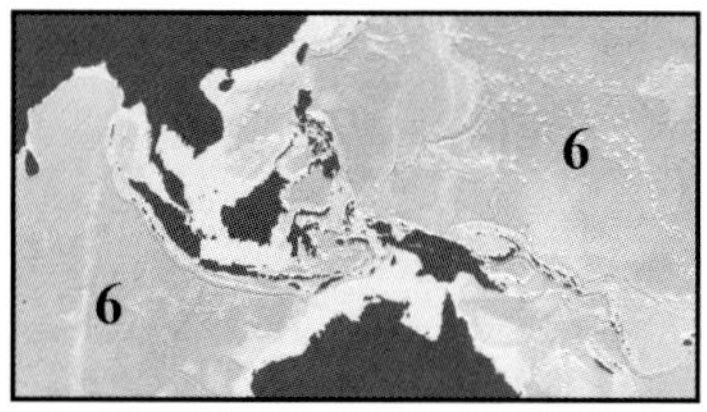

Anatoma sagamiana
Anatoma tobeyoides
Anatoma zancliformis
Saskiconcha elegantissima
Trogloconcha lamellinodosa
Trogloconcha lozoueti
Trogloconcha ohashii
Trogloconcha tesselata
Depressizona axiosculpta

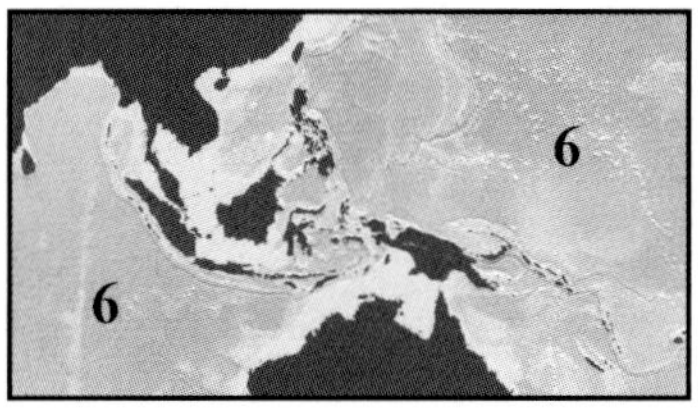

7: Temperate Australia. *Scissurella cyprina*
Scissurella ornata
Scissurella petermannensis (subantarctic)
Scissurella phenax
Sinezona beddomei
Sinezona calumnior
Sinezona pacifica
Sinezona subantarctica (subantarctic)
Sukashitrochus atkinsoni
Sukashitrochus pulcher
Incisura auriformis
Incisura remota
Anatoma australis
Anatoma austrolissa
Anatoma biconica
Anatoma gunteri
Anatoma japonica
Anatoma kopua
Anatoma munieri
Anatoma planapex
Anatoma tobeyoides
Trogloconcha christinae

8: New Zealand. *Scissurella bountyensis*
Scissurella manawatawhia
Scissurella petermannensis (subantarctic)
Scissurella prendrevillei
Scissurella regalis
Scissurella skeneoides
Sinezona bandeli
Sinezona brevis
Sinezona brucei
Sinezona calumnior
Sinezona enigmatica
Sinezona finlayi
Sinezona iota
Sinezona laqueus
Sinezona levigata
Sinezona mechanica
Sinezona norfolkensis
Sinezona pacifica

Sinezona pauperata
Sinezona platyspira
Sinezona wanganellica
Sukashitrochus lyallensis
Incisura lytteltonensis
Incisura rosea
Satondella azonata
Satondella bicristata
Satondella dantarti
Anatoma aupouria
Anatoma equatoria
Anatoma finlayi
Anatoma flemingi
Anatoma globulus
Anatoma japonica
Anatoma kopua
Anatoma pseudoequatoria
Anatoma regia
Anatoma tangaroa
Anatoma zancliformis
Larochea miranda
Larochea secunda
Larochea scitula
Larochea spirata
Larocheopsis amplexa
Larocheopsis macrostoma
Trogloconcha tesselata

9: NW Pacific. *Scissurella staminea*
Sinezona costulata
Coronadoa hasegawai
Sukashitrochus carinatus
Anatoma concinna
Anatoma disciformis
Anatoma equatoria
Anatoma fujikurai
Anatoma japonica
Anatoma lamellata
Anatoma parageia
Anatoma sagamiana
Anatoma soyoae

10: Central Pacific. *Scissurella alto*
Scissurella cebuana
Scissurella evaensis
Scissurella mirifica
Scissurella phenax
Scissurella quadrata
Scissurella spinosa
Scissurella xandaros
Sinezona danieldrieri

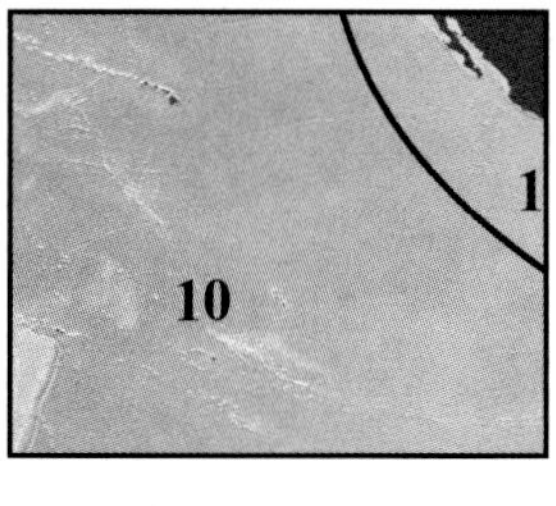

Sinezona ferriezi
Sinezona globosa
Sinezona hawaiiensis
Sinezona kayae
Sinezona plicata
Sinezona wileyi
Sinezona zimmeri
Satondella cachoi
Satondella dantarti
Satondella senni
Sukashitrochus morleti
Anatoma amydra
Anatoma biconica
Anatoma emilioi
Anatoma equatoria
Anatoma finlayi
Anatoma indonesica
Anatoma japonica

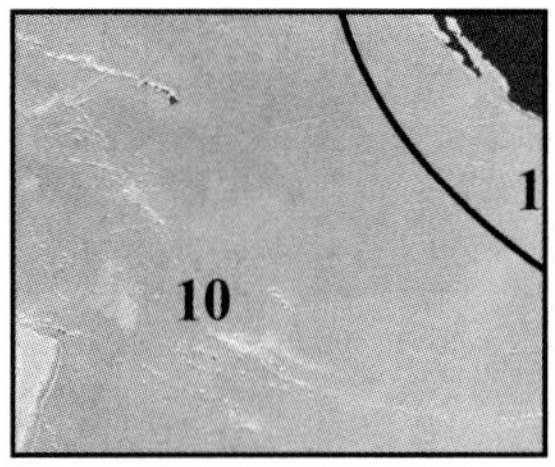

Anatoma lamellata
Anatoma maxima
Anatoma megascutula
Anatoma munieri
Anatoma porcellana
Anatoma pseudoequatoria
Anatoma rainesi
Anatoma rapaensis
Anatoma sinuosa
Sasakiconcha elegantissima
Trogloconcha tesselata
Trogloconcha ohashii
Trogloconcha lozoueti
Depressizona exorum

11: NE Pacific. *Sinezona rimuloides*
Coronadoa simonsae
Coronadoa demisispira
Anatoma concinna
Anatoma disciformis
Anatoma janetae

Anatoma kelseyi
Anatoma lyra
Sutilizona tunnicliffae
Temnocinclis euripes

12: Panamic. *Scissurella kaiserae*
Sinezona carolarum
Sinezona rimuloides
Coronadoa simonsae
Anatoma argentinae
Anatoma bathypacifica
Anatoma janetae

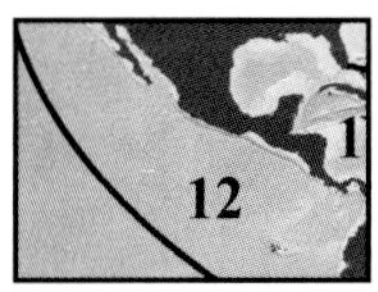

Anatoma keenae
Temnocinclis euripes
Temnozaga parilis
Sutilizona theca

13: Peruvian. *Scissurella clathrata*
Sinezona chilensis
Anatoma peruviana

14: Magellanic. *Scissurella clathrata*
Scissurella petermannensis
Scissurella supraplicata
Sinezona chilensis
Sinezona mouchezi
Sinezona subantarctica
Anatoma argentinae
Anatoma conica

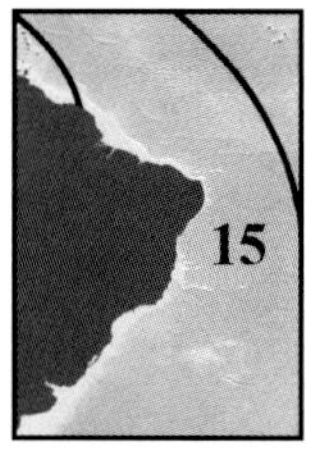

15: Argentinian. *Scissurella alexandrei* [= juv. *Sinezona confusa*?]
Scissurella electilis
Scissurella morretesi
Sinezona confusa
Sinezona insularis
Satondella brasiliensis
Anatoma aedonia
Anatoma proxima

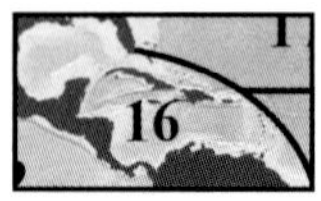

16: Caribbean. *Scissurella redferni*
Sinezona confusa
Sinezona garciai
Sinezona globosa [?]
Satondella brasiliensis
Satondella danieli
Satondella tabulata [?]
Anatoma aedonia
Anatoma alta
Anatoma alternatisculpta
Anatoma atlantica
Anatoma finlayi [?]
Anatoma plicatazona
Anatoma proxima
Anatoma rolani

17: NW Atlantic. *Satondella brasiliensis*
Sinezona confusa
Anatoma alta
Anatoma atlantica
Anatoma crispata
Anatoma tenuis
Anatoma proxima
Anatoma schanderi
Anatoma tenuis
Sutilizona pterodon

18: Arctic. *Anatoma aspera*
Anatoma crispata
Anatoma eximia
Anatoma lyra
Anatoma schanderi
Anatoma schioettei
Anatoma soyoae
Anatoma tenuis
Anatoma tenuisculpta

19: Antarctic. *Scissurella georgica*
Scissurella petermannensis
Anatoma amoena
Anatoma euglypta
Anatoma shiraseae
Anatoma weddelliana

The dredge arrives on the deck of the *Talisman*. From Marshall (1888: fig. 37).

Indices of species and genus-level taxa

Two alphabetical indices are provided. One for species-level taxa, the other for genus-level taxa.

The species names index lists all species names encountered with authority and date, synonymy status of the taxon (valid, synonym of, *nomen dubium, nomen nudum*, unavailable), original genus in which the species was described, and current placement of the species. Some species described in Scissurellidae s.l. have been transferred to other families, which is noted. The listing includes manuscript names encountered; listing of manuscript names here should not be misconstrued as a validation of the names; they are given here for the sake of completeness and details are given in the remarks section in the full treatment of the species. Some frequently encountered or otherwise meaningful misspellings are listed separately, but most misspellings are only given in the synonymies of the respective species with the adjacent "[*sic*]." Alternative inflections of adjectivic species epithets are provided if the species has been placed in genera of different genders (*Sinezona*, feminine; *Sukashitrochus*, masculine).

Some authors have occasionally referenced misidentifications by changing the authority of the taxon in question. For instance, *Sci. striatula* Philippi, 1844 sensu Jeffreys (1856a) was quoted by Weinkauff (1873) as "*striatula* Jeffreys," although Jeffreys never introduced such a species name. Such alternative taxon-authority combinations have not been included in the species-name index.

In the genus-name index, all genus-level names ever associated with Scissurellidae s.l. are given with authority, status (valid, synonym of, placed in), type species, and mode of designation (OD, M, SD).

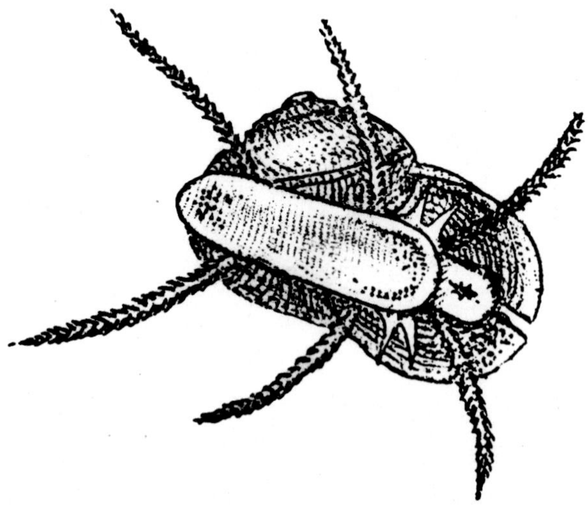

Animal of *Anatoma crispata* from Jeffreys (1865).

Index of species-level taxa

Species, Authority, Date	Status	Original Genus	Current Genus
aedonia Watson, 1886	valid	*Scissurella*	*Anatoma*
aequatoria Hedley, 1899	misspelling of *equatoria*	*Scissurella*	*Anatoma*
aetheria Melvill & Standen, 1903	*japonica*	*Scissurella*	*Anatoma*
aetherica auct.	*japonica*, misspelling of *aetheria*		
affinis O. G. Costa, 1861	*aspera*	*Scissurella*	*Anatoma*
africanae Barnard, 1963	valid	*Scissurella*	*Anatoma*
agulhasensis Thiele, 1925	valid	*Scissurella*	*Anatoma*
alba Bandel, 1998	*nomen nudum*	*Anatoma*	?
alexandrei Montouchet, 1972	*nomen dubium, confusa*?	*Scissurella*	*Scissurella*
aliceae Schnetler, Lozouet & Pacaud, 2000†	valid	*Scissurella*	*Scissurella/Sinezona*?
alta Watson, 1886	valid	*Scissurella*	*Anatoma*
alternatisculpta Geiger & McLean, 2010	valid	*Anatoma*	*Anatoma*
alto Geiger, 2003	valid	*Scissurella*	*Scissurella*
americana Bandel, 1998	*proxima*	*Anatoma*	*Anatoma*
amoena Thiele, 1912	valid	*Scissurella*	*Anatoma*
amplexa Marshall, 1993	valid	*Larocheopsis*	*Larocheopsis*
amydra new species	valid	*Anatoma*	*Anatoma*
angulata Lovén, 1846	*crispata*	*Anatomus*	*Anatoma*
angulata "Sowerby, <1853"	mis-attribution of Lovén's name.		
annulata Ravn, 1933	valid	*Scissurella*	*Scissurella*
antarctica auct.	misspelling of *subantarctica*		
apudornata Laws, 1935 †	valid	*Scissurella*	*Scissurella*
argentina Zelaya & Geiger, 2007	valid	*Thieleella*	*Anatoma*
argutecostata Seguenza, 1877 †	*tenuisculpta*	*Scissurella*	*Anatoma*
armillata Yaron, 1983	*dorbignyi*	*Sinezona*	*Sukashitrochus*
armillatus Yaron, 1983	*dorbignyi*	*Sinezona*	*Sukashitrochus*
asper Philippi, 1884	*aspera*		
aspera Philippi, 1844	valid	*Scissurella*	*Anatoma*
athleenae Pilsbry & McGinty, ms	unavailable, unpublished, *nomen nudum*	*Scissurella*	*Satondella*
atkinsoni Tenison Wood, 1876	valid	*Scissurella*	*Sukashitrochus*
atlantica Bandel, 1998	valid	*Hainella*	*Anatoma*
argentinae Zelaya & Geiger, 2007	valid	*Thieleella*	*Anatoma*

Species, Authority, Date	Status	Original Genus	Current Genus
aupouria Powell, 1937	valid	*Schizotrochus*	*Anatoma*
auriformis Geiger & Jansen, 2004	valid	*Incisura*	*Incisura*
australis Hedley, 1903	valid	*Scissurella*	*Anatoma*
austrolissa Geiger & Sasaki, 2008	valid	*Anatoma*	*Anatoma*
atlatntica Bandel, 1998	valid	*Hainella*	*Anatoma*
axiosculpta Geiger, 2009	valid	*Depressizona*	*Depressizona*
azonata Geiger & Marshall, 2012	valid	*Satondella*	*Satondella*
azorensis Nolt, 2008	valid	*Scissurella*	*Scissurella*
bandeli Marshall, 2002	valid	*Sinezona*	*Sinezona*
bathypacifica Geiger & McLean, 2010	valid	*Thieleella*	*Anatoma*
baxteri McLean, 1984	*lyra*	*Anatoma*	*Anatoma*
beddomei Petterd, 1884	valid	*Schismope*	*Sinezona*
bertheloti d'Orbigny, 1839	*nomen dubium*	*Scissurella*	*Anatoma*
beyrichi Semper, 1865 †	*sismondai*	*Scissurella*	*Pleurotomaria*
biconica new species	valid	*Anatoma*	*Anatoma*
bicristata Geiger & Marshall, 2012	valid	*Satondella*	*Satondella*
bituminata Beets, 1943 †	valid	*Scissurella*	*Scissurella*
borealis Sowerby	*nomen nudum*	*Scissurella*	
boucheti Geiger & Sasaki, 2008	valid	*Anatoma*	*Anatoma*
bountyensis Powell, 1933	valid	*Scissurella*	*Scissurella*
brasiliensis Mattar, 1987	valid	*Sinezona*	*Satondella*
breveprima new species	valid	*Anatoma*	*Anatoma*
brevis Hedley, 1904	valid	*Schismope*	*Sinezona*
brucei new species	valid	*Sinezona*	*Sinezona*
cachoi Luque, Rolán & Geiger, 2011	valid	*Satondella*	*Satondella*
caliana Dall, 1919	valid	*Schismope*	*Pseudorbis*: Skeneidae
californica "Bartsch"	*nomen nudum*: *rimuloides*	*Schismope*	*Sinezona*
campbelli Bandel, 1998	*levigata*	*Ariella*	*Sinezona*
cancellata Jeffreys, 1865	*costata*	*Scissurella*	*Scissurella*
calumnior new species	valid	*Sinezona*	*Sinezona*
carinata Watson, 1886	*atkinsoni*, secondary homomym	*Schismope*	*Sukashitrochus*
carinata A. Adams, 1862	*carinatus*	*Scissurella*	*Sukashitrochus*
carinatus A. Adams, 1862	valid	*Scissurella*	*Sukashitrochus*
carnica Nützel & Geiger, 2006 †	valid	*Triassurella*	*Triassurella*

Species, Authority, Date	Status	Original Genus	Current Genus
carolarum Geiger & McLean 2010	valid	*Sinezona*	*Sinezona*
cebuana Bandel, 1998	valid	*Anatoma*	*Scissurella*
chilensis new species	valid	*Sinezona*	*Sinezona*
chiricova Dall, 1919	*kelseyi*	*Scissurella*	*Anatoma*
christinae Geiger, 2003	valid	*Trogloconcha*	*Trogloconcha*
cingulata O. G. Costa, 1861	valid	*Scissurella*	*Sinezona*
clathrata Strebel, 1908	valid	*Scissurella*	*Scissurella*
columbiana Bandel, 1998	*confusa*	*Schismope*	*Sinezona*
condita Laws, 1939 †	valid	*Scissurella*	*Scissurella*
conica d'Orbigny, 1841	valid	*Scissurella*	*Anatoma*
concinna Sowerby I, 1831	*nomen dubium, nomen oblitum* [*cingulata*?]	*Scissurella*	*Sinezona*
concinna A. Adams, 1862	valid	*Anatomus*	*Anatoma*
concinnus A. Adams, 1862	*concinna* A. Adams		
confusa Rolán & Luque, 1994	valid	*Sinezona*	*Sinezona*
corneti Cossmann, 1914 †	*depontaillieri*	*Scissurella*	*Scissurella*
coronata Watson, 1886	*nomen dubium*	*Scissurella*	?
cossmanni Depontaillier, 1881 †	valid	*Scissurella*	*Scissurella*
costamagnaensis Marquet, 1984 †	misattributed	*Anatoma*	*Anatoma*
costamagnaensis Bandel, 1998 †	*aspera*	*Anatoma*	*Anatoma*
costata d'Orbigny, 1824	valid	*Scissurella*	*Scissurella*
costatus d'Orbigny, 1824	*costata* d'Orbigny, 1824		
costulata Geiger & Sasaki, 2009	valid	*Sinezona*	*Sinezona*
crispata Fleming, 1828	valid	*Scissurella*	*Anatoma*
crossei Folin, 1869	*cingulata*	*Trochotoma*	*Sinezona*
cupelliformis Amitrov, 1996 †	valid	*Scissurella*	*Sinezona*
cyprina Cotton & Godfrey, 1938	valid	*Scissurella*	*Scissurella*
dalli Bartsch, 1903	*nomen dubium*	*Scissurella*	*Scissurella*
danica Bandel, 1998 †	*niloticiformis*	*Anatoma*	*Leptomaria:* Pleurotomariidae
danieldreieri Geiger, 2008	valid	*Sinezona*	*Sinezona*
danieli Segers, Swinnen & Abreu, 2009	valid	*Satondella*	*Satondella*
dantarti Luque, Rolán & Geiger, 2011	valid	*Satondella*	*Satondella*
decipiens O. G. Costa, 1861	*costata*	*Scissurella*	*Scissurella*
declinans Watson, 1886	*mirifica*	*Scissurella*	*Scissurella*
decussata d'Orbigny, 1824	*costata*	*Scissurella*	*Scissurella*

Species, Authority, Date	Status	Original Genus	Current Genus
demisispira Geiger & McLean, 2010	valid	*Coronadoa*	*Coronadoa*
depontaillieri Cossman, 1879	valid	*Scissurella*	*Scissurella*
depressa Reuss, 1860 †	*transylvanica*	*Scissurella*	*Scissurella*
depressa Watson, 1897	*cingulata*	*Schismope*	*Sinezona*
depressa Locard, 1898 as var. of *umbilicata*	*tenuis*	*Scissurella* (*Schizotrochus*)	*Anatoma*
deshayesi Munier-Chalmas, 1862 †	valid	*Scissurella*	*Sinezona*
disciformis Golikov & Sirenko, 1980	valid	*Scissurella* (*Schizotrochus*)	*Anatoma*
divaricatus Monterosato, 1890	*aspera*	*Schizotrochus*	*Anatoma*
dohrniana Dunker, 1861	*nomen dubium*	*Anatomus*	*Scissurella*
dolioliformis Amitrov, 1996†	*cossmanni*	*Scissurella*	*Scissurella*
doliolum Herbert, 1986	valid	*Sinezona*	*Sinezona*
dorbignyi Audouin, 1826	valid	*Scissurella*	*Sukashitrochus*
dorbignyi Scacchi, 1836	unavailable, *costata*	*Scissurella*	*Scissurella*
elatior Sowerby I, 1820–1834	*crispata*	*Scissurella*	*Anatoma*
electilis Montouchet, 1972	valid	*Scissurella*	*Scissurella* [*Sinezona* ?]
elegans d'Orbigny, 1824 †	*costata*	*Scissurella*	*Scissurella*
elegantissima Geiger, 2006	valid	*Sasakiconcha*	*Sasakiconcha*
emilioi Geiger, 2011	valid	*Anatoma*	*Anatoma*
enigmatica Geiger & Marshall, 2012	valid	*Sinezona*	*Sinezona*
eocaenica Bandel, 1998 †	valid	*Scissurella*	*Scissurella*
epicharis McLean, 1970	*keenae*	*Scissurella* (*Anatoma*)	*Anatoma*
equatoria Hedley, 1899	valid	*Scissurella*	*Anatoma*
estotiensis Lozouet, 1999 †	valid	*Sukashitrochus*	*Sukashitrochus*
eucharista Melvill & Standen, 1912	*clathrata*	*Scissurella*	*Scissurella*
euglypta Pelseneer, 1903	valid	*Scissurella*	*Anatoma*
euripes McLean, 1989	valid	*Temnocinclis*	*Temnocinclis*
evaensis Bandel, 1998	valid	*Scissurella*	*Scissurella*
eximia Seguenza, 1877 †	valid	*Scissurella*	*Anatoma*
exorum Geiger, 2003	valid	*Depressizona*	*Depressizona*
exquisita Schepman, 1908	*japonica*	*Scissurella*	*Anatoma*
fairchildi Powell, 1933	*prendrevillei*	*Scissurella*	*Scissurella*
falunica Morgan, 1915 †	*transylvanica*	*Scissurella*	*Scissurella*
fayalensis Dautzenberg, 1889	*cingulata*	*Schismope*	*Sinezona*
ferriezi Crosse, 1867	valid	*Schismope*	*Sinezona*

Species, Authority, Date	Status	Original Genus	Current Genus
finlayi Powell, 1937	valid	*Schizotrochus*	*Anatoma*
finlayi Laws, 1948 †	valid	*Schismope*	*Sinezona*
flemingi Marshall, 2002	valid	*Thieleella*	*Anatoma*
flexidentata Geiger & Sasaki, 2008	valid	*Anatoma*	*Anatoma*
fossilis Laws, 1940 †	*geoffreyi*	*Scissurona*	*Incisura*
funnazzensis deGregorio, 1899	*tenuisculpta*	*Scissurella* (*Schizotrochus*)	*Anatoma*
funiculata Geiger & Jansen, 2004	*pseudoequatoria*	*Anatoma*	*Anatoma*
fujikurai Sasaki, Geiger & Okutani, 2009	valid	*Anatoma*	*Anatoma*
garciai Geiger, 2006	valid	*Sinezona*	*Sinezona*
gephyra Maxwell, 1992 †	valid	*Anatoma*	*Anatoma*
geoffreyi Laws 1940 †	valid	*Scissurella*	*Incisura*
georgica Davolos & Moolenbeek, 2006	valid	*Scissurella*	*Scissurella*
globosa Geiger, 2006	valid	*Sinezona*	*Sinezona*
globulus new species	valid	*Anatoma*	*Anatoma*
goederti Kiel, Campbell & Gaillard, 2010	valid	*Triassurella*	*Triassurella*
goudi new species	valid	*Satondella*	*Satondella*
gruendeli Bandel, 1998 †	valid, emendation of *gründeli*	*Maxwellella*	*Scissurella*
gründeli Bandel, 1998	*gruendeli*		
gunteri Cotton & Godfrey, 1933	valid	*Schizotrochus*	*Anatoma*
guntheri auct.	misspelling of *gunteri*		
haliotimorpha Bandel, 1998 †	valid	*Ariella*	*Ariella*
hasegawai Geiger & Sasaki, 2009	valid	*Coronadoa*	*Coronadoa*
hawaiiensis Geiger & McLean, 2010	valid	*Sinezona*	*Sinezona*
hedegaardi Bandel, 1998 †	valid	*Scissurella*	*Scissurella*
herberti Geiger & Sasaki, 2008	*finlayi*	*Anatoma*	*Anatoma*
hispanica Kiel & Bandel, 1999 †	*peyrerensis*	*Scissurella*	*Scissurella*
hoernesi Semper, 1865	*nomen dubium*	*Scissurella*	
hungarica Szöts, 1953	*depontaillieri*	*Scissurella*	*Scissurella*
hyposculpta new species	valid	*Anatoma*	*Anatoma*
indica Montfort, 1810	*indicus*		
indicus Montfort, 1810	*nomen dubium*	*Anatomus*	Annelida: Serpulidae
indonesica Bandel, 1998	valid	*Anatoma*	*Anatoma*
indonesicus Bandel, 1998	*morleti*	*Sukashitrochus*	*Sukashitrochus*

Species, Authority, Date	Status	Original Genus	Current Genus
indopacifica Lozouet, 1986	unavailable, unpublished, *nomen nudum*	*Praescissurella*	?
insignis Smith, 1890	valid	*Schismope*	*Sinezona*
insularis Simone, 2009	valid	*Sinzona*	*Sinezona*
iota Finlay, 1926	valid	*Schismope*	*Sinezona*
jacksoni Melvill, 1904	*munieri*	*Scissurella*	*Anatoma*
janetae Geiger, 2006	valid	*Anatoma*	*Anatoma*
jansenae Geiger, 2006	*japonica*	*Anatoma*	*Anatoma*
janusa new species	valid	*Anatoma*	*Anatoma*
japonica A. Adams, 1862	valid	*Anatomus*	*Anatoma*
japonicus A. Adams, 1862	*japonica*		
josephinae Odhner, 1960	*tenuis*	*Scissurella*	*Anatoma*
jucunda Smith, 1890	valid	*Scissurella*	*Scissurella*
jucunda Smith, 1910	*jucunda* Smith, 1890, primary homonym	*Scissurella*	*Scissurella*
kaiparaensis Laws, 1941 †	valid	*Schismope*	*Sukashitrochus*
kaiserae Geiger, 2006	valid	*Scissurella*	*Scissurella*
kayae Geiger & McLean, 2010	valid	*Sinezona*	*Sinezona*
kelseyi Dall, 1905	valid	*Scissurella*	*Anatoma*
keenae McLean, 1970	valid	*Scissurella* (*Anatoma*)	*Anatoma*
koeneni Semper, 1865	*nomen dubium*	*Scissurella*	?
koeneniana Janssen, 1978 †	*aspera*	*Scissurella* (*Anatoma*)	*Anatoma*
kondoi Ladd, 1970 †	valid	*Sinezona*	*Satondella*
kopua Geiger & Marshall, 2012	valid	*Anatoma*	*Anatoma*
koruahina Laws, 1936 †	valid	*Schismope*	*Sinezona*
lacuniformis Watson, 1986	?	*Schismope*	*Macromphalina*: Vanikoridae?
laevigata d'Orbigny, 1824	*costata* d'Orbigny	*Scissurella*	*Scissurella*
"*laevigata* Iredale" auct.	*levigata*	*Schismope*	*Sinezona*
lamellata A. Adams, 1862	valid	*Anatomus*	*Anatoma*
lamellatus A. Adams, 1862	*lamellata*	*Anatomus*	*Anatoma*
lamellinodosa new species	valid	*Trogloconcha*	*Trogloconcha*
lamellosa Benoist, 1875 †	valid	*Scissurella*	*Scissurella*
laquea Auct.	*laqueus* [unjustified "gender agreement" on noun]		
laqueus Finlay, 1926	valid	*Schismope*	*Sinezona*
levigata Iredale, 1908	valid	*Schismope*	*Sinezona*
lleidania Kiel & Bandel, 2000 †	*marchmontensis*	*Scissurella*	*Scissurella*

Species, Authority, Date	Status	Original Genus	Current Genus
lobini Burnay & Rolán, 1990	valid	*Sinezona*	*Scissurella*
lorenzi Geiger, 2006	valid	*Scissurella*	*Scissurella*
lozoueti Geiger, 2008	valid	*Trogloconcha*	*Trogloconcha*
lyallensis Finlay, 1926	valid	*Schismope*	*Sukashitrochus*
lyra Berry, 1947	valid	*Scissurella* (*Anatoma*)	*Anatoma*
lytteltonensis Smith, 1894	valid	*Scissurella*	*Incisura*
macleani Geiger, 2006	valid	*Sinezona*	*Sinezona*
macrostoma Geiger & Marshall, 2012	valid	*Larocheopsis*	*Larocheopsis*
malloryi Squires & Goedert, 1996 †	valid	*Scissurella*	*Sinezona*
malvinarum Zelaya & Geiger, 2007	*clathrata*	*Scissurella*	*Scissurella*
manawatawhia Powell, 1937	valid	*Scissurella*	*Scissurella*
mantelli Woodward, 1859	*nomen dubium*	*Scissurella*	*Scissurella* ?
maraisi Herbert, 1986	*dorbignyi*	*Sukashitrochus*	*Sukashitrochus*
maraisorum Geiger, 2006	valid	*Scissurella*	*Scissurella*
marchmontensis Sohl, 1992 †	valid	*Scissurella*	*Scissurella*
marrowi new species	valid	*Sinezona*	*Sinezona*
marshalli Bandel, 1998	*prendrevillei*	*Scissurella*	*Scissurella*
marshalli Lozouet, 1998 †	valid	*Larocheopsis*	*Trogloconcha*
maxima Schepman, 1908	valid	*Scissurella*	*Anatoma*
mechanica Geiger & Marshall, 2012	valid	*Sinezona*	*Sinezona*
medioplicata Thiele, 1925	*petermannensis*	*Scissurella*	*Scissurella*
mediterranea auct./Monterosato	unpublished	*Scissurella*	*Anatoma*
megascutula Geiger & Marshall, 2012	valid	*Anatoma*	*Anatoma*
micalii new species	valid	*Anatoma*	*Anatoma*
milleri Geiger & Sasaki, 2009	valid	*Sinezona*	*Sinezona*
minuta Turton, 1932	*nomen dubium*	*Scissurella*	? [not Scissurellidae]
minuta Bandel, 1998	valid	*Satondella*	*Satondella*
miocenica Laws, 1939 †	valid	*Schizotrochus*	*Anatoma*
miranda Finlay, 1927	valid	*Larochea*	*Larochea*
miranda A. Adams, 1862	*nomen dubium*	*Scissurella*	*Sinezona*
mirifica A. Adams, 1862	valid	*Anatomus*	*Scissurella*
mirificus A. Adams, 1862	*mirifica*	*Anatomus*	*Scissurella*
modesta A. Adams, 1862	*nomen dubium*	*Scissurella*	*Sinezona*

Species, Authority, Date	Status	Original Genus	Current Genus
montfortii d'Orbigny, 1824	unavailable hypothetical	*Scissurella*	
morleti Crosse, 1880	valid	*Schismope*	*Sukashitrochus*
morretesi Montouchet, 1972	valid	*Scissurella*	*Scissurella*
mouchezi Vélain, 1877	valid	*Schismope*	*Sinezona*
muelleri new species †	valid	*Anatoma*	*Anatoma*
munieri Fischer, 1862	valid	*Scissurella*	*Anatoma*
munieriana auct.	misspelling of *munieri*		
nanshaensis Feng, 1996 as ssp. of *lamellata*	*equatoria*	*Anatoma*	*Anatoma*
ngatutura Laws, 1936 †	valid	*Schismope*	*Sukashitrochus*
norfolkensis new species	valid	*Sinezona*	*Sinezona*
novozeelandica Bandel, 1998 †	*condita*	*Maxwellella*	*Scissurella*
obliqua Watson, 1886	valid	*Scissurella*	*Scissurella/Incisura* (?)
obtusata Golikov & Gublin, 1978	*concinna* (A. Adams, 1862)	*Scissurella*	*Anatoma*
ohashii Kase & Kano, 2002	valid	*Trogloconcha*	*Trogloconcha*
oligomiocaenica Lozouet, 1998 †	*depontaillieri*	*Praescissurella*	*Scissurella*
orbignyi "Audouin, 1826" auct.	*dorbignyi*	*Scissurella*	*Sukashitrochus* (?)
orbignyi Scacchi, 1836	unavailable, *costata*	*Padollus*	*Scissurella*
ornata May, 1908	valid	*Scissurella*	*Scissurella*
pacifica Oliver, 1915	valid	*Schismope*	*Sinezona*
pacifica Bandel, 1998	*pacifica* Oliver, 1915, secondary homonym	*Daizona*	*Sinezona*
padangensis Thiele, 1912	*ferriezi*	*Schismope*	*Sinezona*
palaeomphaloides Nordsieck, 1973	*peronii*	*Schizotrochus*	*Atlanta* (Atlantidae)
parageia Geiger & Sasaki, 2009	valid	*Anatoma*	*Anatoma*
parilis McLean, 1989	valid	*Temnozaga*	*Temnozaga*
parisiensis Deshayes, 1864 †	*deshayesi*	*Scissurella*	*Sinezona*
paucicostata Jeffreys, 1865	*aspera*	*Scissurella*	*Anatoma*
paucispiralia Bandel, 1998	*pseudoequatoria*	*Hainella*	*Anatoma*
paumotuensis Garrett, 1872	*nomen dubium*	*Scissurella*	*Sinezona*
pauperata Powell, 1933	valid	*Sinezona*	*Sinezona*
peilei Winckworth, ms.	unavailable. *dorbighyi*	*Sinezona*	*Sukashitrochus*
peruviana Geiger & McLean, 2010	valid	*Anatoma*	*Anatoma*
petermannensis Lamy, 1910	valid	*Scissurella*	*Scissurella*
peyrerensis Bandel, 1998 †	valid	*Scissurella*	*Scissurella*
phenax new species	valid	*Scissurella*	*Scissurella*
philippiana Semper, 1865 †	valid	*Scissurella*	*Anatoma*

Species, Authority, Date	Status	Original Genus	Current Genus
philippinica Bandel, 1998	valid	*Hainella*	*Anatoma*
plicata Hedley, 1899	valid	*Schismope*	*Sinezona*
plicata Philippi, 1836	*costata*	*Scissurella*	*Scissurella*
plicatazona Geiger & McLean, 2010	valid	*Anatoma*	*Anatoma*
porcellana new species	valid	*Anatoma*	*Anatoma*
praecedens Lozouet, 1998 †	valid	*Sinezona*	*Sinezona*
praecrispata Gougerot & LeRenard, 1977 †	*depontaillieri*	*Scissurella*	*Scissurella*
prendrevillei Powell, 1933	valid	*Scissurella*	*Scissurella*
provecta Beets, 1943 †	*plicata*	*Woodwardia*	*Sinezona*
proxima Dall, 1927	valid	*Scissurella*	*Anatoma*
pseudoequatoria Kay, 1979	valid	*Scissurella*	*Anatoma*
pterodon Warén & Bouchet, 2001	valid	*Sutilizona*	*Sutilizona*
pulchella Bandel, 1998	*pseudoequatoria*	*Hainella*	*Anatoma*
pulcher Petterd, 1884	valid	*Schismope*	*Sukashitrochus*
pulchra Petterd, 1884	*pulcher*	*Schismope*	*Sukashitrochus*
quadrata Geiger & Jansen, 2004	valid	*Scissurella*	*Scissurella*
quadraxialis new species	valid	*Thieleella*	*Anatoma*
rainesi Geiger, 2003	valid	*Anatoma*	*Anatoma*
rapaensis Geiger, 2008	valid	*Anatoma*	*Anatoma*
ravni Schnetler, Lozouet & Pacaud, 2000 †	valid	*Praeascissurella*	*Scissurella*
regia Mestayer, 1916	valid	*Scissurella*	*Anatoma*
remota Iredale, 1924	valid	*Scissurona*	*Incisura*
redferni Rolán, 1996	valid	*Sinezona*	*Scissurella*
regalis Geiger & Marshall, 2012	valid	*Scissurella*	*Scissurella*
reticulata Philippi, 1853	valid	*Scissurella*	*Scissurella*
reticulata Bandel, 1998	*sagamiana*	*Thieleella*	*Anatoma*
reunionensis auct.	*nomen nudum, singeri*	*Scissurella*	*Sinezona*
rhynchodentata new species	valid	*Anatoma*	*Anatoma*
richardi Dautzenberg & Fischer, 1896	*tenuis*	*Scissurella*	*Anatoma*
rimuloides Carpenter, 1865	valid	*Scissurella*	*Sinezona*
rolani Geiger & Fernández-Garcés, 2010	valid	*Anatoma*	*Anatoma*
rosea Hedley, 1904	valid	*Scissurella*	*Incisura*
rosenkrantzi Schnetler, Lozouet & Pacaud, 2000 †	*aspera*	*Anatoma*	*Anatoma*
rota Yaron, 1983	valid	*Scissurella*	*Scissurella*

Species, Authority, Date	Status	Original Genus	Current Genus
rubiginosa Hutton, <1910	*nomen nudum*	*Scissurella*	?
sagamiana Okutani, 1964	valid	*Scissurella*	*Anatoma*
saubadae Lozouet, 1998 †	*cupelliformis*	*Sukashitrochus*	*Sinezona*
scitula Marshall, 1993	valid	*Larochea*	*Larochea*
schanderi Høisaeter & Geiger, 2011	valid	*Anatoma*	*Anatoma*
schioettei Høisaeter & Geiger, 2011	valid	*Anatoma*	*Anatoma*
secunda Powell, 1937	valid	*Larochea*	*Larochea*
semicostata auct.	unpublished, *costata*	*Scissurella*	*Scissurella*
semicostata Burnay & Rolán, 1990	valid	*Sinezona*	*Sinezona*
senni Geiger, 2003	valid	*Satondella*	*Satondella*
shiraseae Numanami & Okutani, 1990	valid	*Anatoma*	*Anatoma*
simonsae Bartsch, 1946	valid	*Coronadoa*	*Coronadoa*
simplex Bandel, 1998	*indonesicus*	*Sukashitrochus*	*Sukashitrochus*
sinuosa new species	valid	*Anatoma*	*Anatoma*
singeri Geiger, 2006	valid	*Sinezona*	*Sinezona*
skeneoides new species	valid	*Scissurella*	*Scissurella*
smithi Thiele, 1912	*jucunda*	*Scissurella*	*Scissurella*
soyoae Habe, 1951	valid	*Schizotrochus*	*Anatoma*
spinosa Geiger & Jansen, 2004	valid	*Scissurella*	*Scissurella*
spirata Geiger & Marshall, 2012	valid	*Larochea*	*Larochea*
staminea A. Adams, 1862	valid	*Anatomus*	*Scissurella*
stamineus A. Adams, 1862	*staminea*		
stellae Fleming, 1948	*prendrevillei*	*Scissurella*	*Scissurella*
stephanensis Bandel, 1998 †	*aspera*	*Anatoma*	*Anatoma*
striatula Philippi, 1844	*costata*	*Scissurella*	*Scissurella*
subangulata Monterosato, 1878	unavailable	*Scissurella* (*Schizotrochus*)	*Anatoma*
subantarctica Hedley, 1916	valid	*Schismope*	*Sinezona*
subaspera Boettger, 1907 †	*tenuisculpta*	*Scissurella*	*Anatoma*
subitana Lozouet, 1986 †	unavailable, unpublished, *estotiensis*	*Scissurella* (*Sukashitrochus*)	*Sukashitrochus*
sudanica Bandel, 1998	valid	*Scissurella*	*Scissurella*
suessi Klipstein in Kittl, 1891 †	valid	*Temnotropis*	*Zardinitrochus*
supraplicata Smith, 1875	valid	*Scissurella*	*Scissurella*
szaboi Conti & Monari, 1991 †	valid	*Scissurella* ?	*Scissurella* ?
tabulata Watson, 1886	valid	*Schismope*	*Satondella*

Species, Authority, Date	Status	Original Genus	Current Genus
tabulata Barnard, 1964	valid	*Scissurella*	*Anatoma*
tangaroa Geiger & Marshall, 2012	valid	*Anatoma*	*Anatoma*
tasmanica Petterd, 1879	*atkinsoni*	*Schismope*	*Sukashitrochus*
tasmanicus Petterd, 1879	*atkinsoni*	*Schismope*	*Sukashitrochus*
tenera Monterosato, 1878	unavailable	*Scissurella*	?
tenuis Jeffreys, 1877	*crispata*	*Scissurella*	*Anatoma*
tenuisculpta Seguenza, 1877 †	valid	*Scissurella*	*Anatoma*
terquemi Deshayes, 1865 †	valid	*Trochotoma*	*Sinezona*
tertia Laws, 1940 †	valid	*Schismope*	*Sinezona*
tesselata Kase & Kano, 2002	valid	*Trogloconcha*	*Trogloconcha*
theca McLean, 1989	valid	*Sutilizona*	*Sutilizona*
timora Melvill & Standen, 1912	*petermannensis*	*Scissurella*	*Scissurella*
tinaeformis Amitrov, 1996 †	valid	*Scissurella*	*Scissurella*
tobeyoides Geiger & Jansen, 2004	valid	*Anatoma*	*Anatoma*
tongaensis Ladd, 1970 †	valid	*Scissurella*	*Anatoma*
transylvanica Reuss, 1860 †	valid	*Scissurella*	*Scissurella*
tricarinata Yaron, 1983	*dorbignyi*	*Sinezona*	*Sukashitrochus*
tricarinatus Yaron, 1983	*dorbignyi*	*Sinezona*	*Sukashitrochus*
trivialis auct.	unavailable. *beddomei*	*Schismope*	*Sinezona*
tshapliensis Amitrov, 1996 †	*depontaillieri*	*Scissurella*	*Scissurella*
tunnicliffae Warén & Bouchet, 2001	valid	*Sutilizona*	*Sutilizona*
turbinata A. Adams, 1862	*munieri*	*Anatomus*	*Anatoma*
turbinatus A. Adams, 1862	*munieri*	*Anatomus*	*Anatoma*
umbilicata Jeffreys, 1883	valid	*Scissurella*	*Anatoma*
unispirata Bandel, 1998	*evaensis*	*Maxwellella*	*Scissurella*
uniumbilicatus O.G. Costa, 1828	*costata*	*Argonauta*	*Scissurella*
vincentiana Cotton, 1945	valid	*Scissurona*	*Incisura*
wanganellica Geiger & Marshall, 2012	valid	*Sinezona*	*Sinezona*
weddelliana Zelaya & Geiger, 2007	valid	*Thieleella*	*Anatoma*
wileyi Geiger, 2008	valid	*Sinezona*	*Sinezona*
xandaros new species	valid	*Scissurella*	*Scissurella*
yaroni Herbert, 1986	valid	*Anatoma*	*Anatoma*
zancliformis Geiger & Marshall, 2012	valid	*Anatoma*	*Anatoma*
zimmeri Geiger, 2003	valid	*Sinezona*	*Sinezona*

Species, Authority, Date	Status	Original Genus	Current Genus
Informal Names	**Scientific Name**	**Original Genus**	**Current Genus**
East Coast	*tobeyoides*	*Anatoma*	*Anatoma*
SWA	*japonica*	*Anatoma*	*Anatoma*
square	*quadrata*	*Scissurella*	*Scissurella*
spiny	*spinosa*	*Scissurella*	*Scissurella*
auriform	*auriformis*	*Incisura*	*Incisura*

Marquis de Folin and his companions examine specimens from Cape Breton. Note the hand lenses. From Marshall (1888: fig. 38).

Index of genus-level taxa

Name, Authority, Date	Synonym of	Type Species (M, OD, SD)
Anatoma Woodward, 1859	valid	*Scissurella crispata* (M: misidentified; SD: this volume)
Ariella Bandel, 1998	*Sinezona*	*Ariella haliotimorpha* † (OD)
Coronadoa Bartsch, 1946	valid	*Coronadoa simonsae* (OD)
Daizona Bandel, 1998	*Sinezona*	*Sinezona doliolum* (OD)
Depressizona Geiger, 2003	valid	*Depressizona exorum* (OD)
Hainella Bandel 1998	*Anatoma*	*Scissurella euglypta* (OD)
Incisura Hedley, 1904	valid	*Scissurella lytteltonensis* (M)
Larochea Finlay, 1927	valid	*Larochea miranda* (M)
Larocheopsis Marshall, 1993	valid	*Larocheopsis amplexa* (OD)
Maxwellella Bandel, 1998	?	*Scissurella annulata* † (OD)
Pagodella Bandel, 1998 [*non* Swainson, 1840; *non* H. Adams, 1867]	*nomen nudum*, homonym	—
Praescissurella Lozouet, 1998	*Scissurella*	*Scissurella depontaillieri* † (OD)
Satondella Bandel, 1998	valid	*Satondella minuta* (OD)
Reussella Bandel, 1998 [*non* Galloway, 1933]	preoccupied, *Scissurella*	*Scissurella depressa* † (OD)
Sasakiconcha Geiger, 2006	valid	*Sasakiconcha elegantissima* (OD)
Schismope Jeffreys, 1856	*Scissurella*	*Scissurella striatula* sensu Jeffreys, *non* Philippi (M) *Scissurella striatula* (SD: this volume)
Schizotrochus Monterosato, 1877	*Anatoma* (objective)	*Scissurella crispata* (M)
Scissurella d'Orbigny, 1841	valid	*Scissurella laevigata* (SD: Gray, 1847)
Scissurona Iredale, 1924	*Incisura*	*Scissurella rosea* (OD)
Sinezona Finlay, 1926	valid	*Schismope brevis* (OD)
Sukashitrochus Habe & Kosuge, 1964	valid	*Scissurella carinata* (OD)
Sutilizona McLean, 1989	valid	*Sutilizona theca* (OD)
Temnocinclis McLean, 1989	valid	*Temnocinclis euripes* (OD)
Temnozaga McLean, 1989	*Anatoma*	*Temnozaga parilis* (OD)
Thieleella Bandel, 1998	valid	*Scissurella amoena* (OD)
Triassurella Nützel & Geiger, 2006 †	valid	*Triassurella carnica* † (OD)
Trogloconcha Kase & Kano, 2002	valid	*Trogloconcha ohashii* (OD)
Woodwardia Crosse & Fischer, 1861	*Scissurella*	*Scissurella elegans* d'Orbigny, 1824 sensu Woodward, 1859; *Scissurella elegans* d'Orbigny, 1824 (SD: this volume)
Zardinitrochus Bandel, 1998 †	?	*Temnotropis suessi* † (OD)

Unavailable names

A few unavailable names per ICZN criteria have been introduced. Each is discussed, with explicit reasoning for dismissal given.

Scissurella monfortii Orbigny, 1824 [unavailable hypothetical name]

Chresonymy.

Scissurella monfortii Orbigny, 1824: 343.

Type material. Iconotype Montfort (1810: text-fig. p. 278).

Remarks. Orbigny (1824: 343) gave a hypothetical name to Monfort's *Anatomus indicus* if it should turn out to belong to his genus *Scissurella*:

> "... cependant il serait possible qu l'Anatome, mieux observé et dessin plus correctement, se rapportât au genre Scissurelle; alors il ***formerait*** une cinquième espèce sous le nom de *Scissurella Montfortii*, et la courte description que cet auteur donne de l'animal, si elle est exacte, deviendrait le complément de ce Mémoire; ." [anyway it may still be possible that the Anatome, better observed and drawn more correctly, could be related to the genus *Scissurella*; therefore it ***would*** constitute the fifth species under the name of *Scissurella Monfortii*, and the short description that this author gives of the animal, if it is exact, would become the complement to this memoir]. (Orbigny, 1824: 342–343: emphasis added)

Hypothetical names are generally not recognized by ICZN (Art. 1.3.1), and this case does not fall under the exemptions of Articles 11.5.1, 15, and 11.9.6.3. This name is given here solely for the sake of completeness, and it should not be misconstrued as a taxonomic validation.

Scissurella tenera Monterosato, 1878 [unavailable name, *nomen dubium*]

Chresonymy.

Scissurella tenera Monterosato, 1878: 79, not illustrated.

Scissurella tenera: Giannuzzi-Savelli, 1990: 196 [list of names].

Type material. (wherabouts unknown).

Type locality. Palermo [38.133°N, 13.383°E] and Trapani [38.183°N, 12.717°E], [Sicily, Italy] (OD).

Etymology. Tenera: Latin for soft, delicate.

Remarks. This name was introduced as a single line entry reading in full "S. tenera, Monts. n. sp. Pal. e Trapani!." The entry does not contain any descriptive indication or reference to specimens or illustrations in previously published material. It fails the requirements of ICZN article 12.1. The mention of the type locality falls under the specific exclusions of ICZN article 12.3. Accordingly, *Sci. tenera* is unavailable. The epithet has not been validated by any subsequent author (see also Giannuzzi-Savelli & Piani, 1990). It is given here solely for the sake of completeness, and it should not be misconstrued as a taxonomic validation.

The species is also unidentifiable and therefore a *nomen dubium*. It could either be *Sci. costata* or *Sin. cingulata.*

Scissurella crispata subangulata Monterosato, 1878 [unavailable name]

Chresonymy.

Scissurella (Schizotrochus) crispata var. *subangulata* Monterosato, 1878: 80 [not illustrated, is *An.* sp.].

Type material. (whereabouts unknown).

Type locality. Sciacca [Sicily, Italy, 37.509°N, 13.089°E] (OD).

Etymology. Less angulated than *An. angulata*.

Remarks. *Scissurella crispata subangulata* was introduced as a single line entry reading in full "Var. *subangulata* (*S. angulata, Lovén,* prox., Sciacca!)" by Monterosato (1878). The entry does not contain any descriptive indication or reference to specimens or illustrations in previously published material. It fails the requirements of ICZN article 12.1. The mention of a locality falls under the specific exclusions of ICZN article 12.3. Accordingly, the epithet *subangulata* is unavailable. The epithet has not been validated by any subsequent author, and was not given in Giannuzzi-Savelli & Piani (1990). It is given here solely for the sake of completeness, and it should not be misconstrued as a taxonomic validation. Most likely it refers to *An. tenuisculpta*.

Scissurella dorbignyi Scacchi, 1836 [unavailable junior primary homonym]

Chresonymy.

Scissurella d'orbignyi Scacchi, 1836: 16, not illustrated.

Padollus obigny [sic]: Costa, 1836: 8–9, pl. 3, figs. 1A–C.

Padollus d'Orbignyi [sic]: O.G. Costa, 1844: 51.

Padollus d'orbigny [sic]: Costa, 1861: 60, pl. 10, Fig. 2 [as senior synonym of *Sci. plicata* Philippi].

Padollus orbignyi: Monterosato, 1872: 30 [as synonym of *Sci. costata*].

Padollus orbignyi: Monterosato, 1878: 79 [as synonym of *Sci. costata*].

Type material. (whereabouts unknown).

Type locality. Naples, Italy [40.833°N, 14.250°E] (inferred from title).

Etymology. Named to honor Alcide Dessalines d'Orbigny (1802–1857).

Remarks. The name is an unavailable primary homonym of *Sci. dorbignyi* Auduin, 1826. Subsequent placement of Auduin's species in *Sukashitrochus* does not affect the status of Scacchi's name as a primary homonym. Costa placed the species in *Padollus*, a synonym for *Haliotis* (Vetigastropoda: Haliotidae).

Scissurella borealis "Sowerby" [*nomen nudum*]

Chresonymy.

Scissurella borealis "Sowerby": Paetel, 1883: 47.

Scissurella borealis "Sowerby": Paetel, 1888: 288.

Scissurella borealis "Sow.": Petit, 2009: 90 [as *nomen nudum*].

Remarks. The species name attributed to Sowerby was given in Paetel (1883, 1888) but the listing lacks any description, illustration or reference, hence is a *nomen nudum*. The mention of the name here is not to be construed as a validation, and is solely given for the sake of completeness.

Given the name of the species referring to higher latitudes, and Paetel's species mainly coming from Europe, it is likely that *Scissurella borealis* refers to *Anatoma* sp. The evidence is entirely circumstantial.

Scissurella rubiginosa "Hutton" [*nomen nudum*]

Chresonymy.

Scissurella rubiginosa Hutton: Suter, 1910: 9, not illustrated.

Remarks. The species was mentioned by Suter (1910) as a Hutton species from the Pliocene of New Zealand. The name cannot be traced to any Hutton publication. Suter's listing lacks any description, illustration or reference, hence is a *nomen nudum*. The mention of the name here is not be misconstrued as a validation, and is solely given for the sake of completeness.

Schismope californica "Bartsch" [*nomen nudum*]

Chresonymy.

Schismope californica "Bartsch" Dall, 1923: 4 [not illustrated, *nomen nudum*].

Schismope californica: Keen, 1937: 45.

Remarks. Dall (1923) mentioned the name and attributed it to Bartsch without giving a specific reference. The name was not found in any of Bartsch's writing. Dall did not introduce a new species as there is no mention of specimens. The name is a *nomen nudum*. It is given here solely for the sake of completeness, and it should not be misconstrued as a taxonomic validation.

Burch (1948b: 23) quoted the following personal communication from A. Myra Keen: "In Emendations to Bull. 112, 1923, Dall cites *Schismope californica* Bartsch as a substitute for *S. coronata* Watson, 1886, as used in Bull. 112. However, as Watson's shell referred to in Bull. 112 was from Tahiti and there is no word of diagnosis or description nor figure of the California *Schismope*, it seems to me that *S. californica* Bartsch *in* Dall, 1923 must be regarded as a *nomen nudum*."

Wilson & Kennedy (1967: 261) listed "paratypes" [in quotes] of "S.[*chismope*] *californica* Dall" in SDMNH 674, 3381–3383. The quotes around paratypes indicates that those authors questioned the validity of such a designation.

Some specimens collected in 1922 in Pacific Grove, La Playa [Los Angeles], Guadalupe Island, AMB [?], TBK Reef [?], and PH [?] were identified as that species. The specimens were made available by John Wolff, who obtained them from Phyllis Diegel, who in turn purchased them from Rev. Wilkins a member of the Broward shell club. Those specimens are identifiable as *Sinezona rimuloides*.

Schismope trivialis auct. [*Sinezona beddomei*]

Remarks. The name was encountered on a label with *Sin. beddomei* (MCZ 249561), written after 1902 inferred from a reference to Tate & May (1902). It has not been published anywhere, hence, has no standing. It is given here solely for the sake of completeness, and it should not be misconstrued as a taxonomic validation. Even if it should have been described somewhere, it is unlikely to threaten the name of Petterd (1884).

Scissurella mediteranea auct./Monterosato

Remarks. A lot with three specimens of *Anatoma* sp. from the Mediterranean Sea with a label reading *Scissurella mediterrana* [*sic*] is in LMD/LÖB 112957a–c from the Löbbecke collection. The lot was purchased in 1876 from the van Lennep collection. The specimen was originally identified as *Scissurella mediterranea* without authority, but subsequently reidentified as *Solarium mediterrana* (Monteros[ato]). The genus *Solarium* is still retained in Architectonicidae, the species epithet *mediterraneum* Monterosato, 1872 (*non* Philippi, 1853) has been renamed *lepida* Bayer, 1942 and is the type species of *Basisulcata* Melone & Taviani, 1985 (Sabelli *et al.*, 1990; Bieler & Petit, 2005). Additionally, the *Scissurella mediterranea* lot was annotated with "2" in the Löbbecke sales catalog, suggesting two specimens. After Löbbecke's death, the new card catalog lists "3" specimens, which are found today. All of the above opens the possibility of a mix-up of specimens. The name has not been published anywhere (see Giannuzzi-Savelli & Piani, 1990), hence, has no standing. It is given here solely for the sake of completeness, and it should not be misconstrued as a taxonomic validation.

Scissurella athleenae Pilsbry & McGinty, ms. [unavailable, unpublished, *nomen nudum*]

Remarks. The name has appeared on a webpage (www.discoverlive.org/mp/20q?search=Scissurella+athleenae: accessed I/2008), based on a specimen label in FLMNH 150107 from Palm Beach, Florida, USA, which is a *Satondella brasiliensis*. The name has not been published anywhere, hence, has no standing. It is given here solely for the sake of completeness, and it should not be misconstrued as a taxonomic validation. John Slapcinsky (FLMNH) kindly provided information on this taxon and sent the specimen on loan.

Scissurella costata var. *semicostata auct.* (*non* Burnay & Rolán, 1990) [unpublished]

Remarks. The name *Scissurella costata* var. *semicostata* was found on a lot in KBIN (DLG#5408), attached to an old specimen form the Dautzenberg collection from Princess Alice trawl number 344. It has not been published anywhere, hence, has no standing. It is given here solely for the sake of completeness, and it should not be misconstrued as a taxonomic validation.

Praescissurella indopacifica Lozouet, 1986 [unavailable, unpublished, *nomen nudum*]

Chresonymy.

Praescissurella indopacifica Lozouet, 1986: 110, 127.

Remarks. The name was introduced in the diploma thesis of Lozouet (1986), which is unpublished by ICZN criteria. Although some discriminating characters were given, there is no explicit mention of any specimens, nor are the required types and type repositories indicated. Accordingly, the name is a *nomen nudum*.

Scissurella (*Sukashitrochus*) *subitana* Lozouet, 1986 [unavailable, unpublished]

Chresonymy.

Scissurella (*Sukashitrochus*) *subitana/estotensis* [*sic*]: Lozouet, 1986: 110, 124–125, text-fig. 35A.

Remarks. The name was introduced diploma thesis of Lozouet (1986), which is unpublished by ICZN criteria. In the text the name *subitana* was used, but the corresponding figure caption gave *estotensis* as the species' epithet. The species was validly described as *Sukashitrochus estotiensis* by Lozouet (1999).

Anatoma alba Bandel, 1998 [*nomen nudum*]

Chresonymy.

Anatoma alba Bandel, 1998: 28.

Remarks. The name was used in the comparison of *An. costamagnaensis*. It is described as being more rounded and having an earlier onset of the selenizone. No specimens are mentioned or illustrated, hence, the name is not specimen-based and, therefore, a *nomen nudum*. Most likely, *An. alba* was an earlier manuscript name for one of the other species described by Bandel (1998) that was not modified throughout the entire manuscript.

Pagodella Bandel, 1998 [*nomen nudum,* junior homonym]

Chresonymy.

Pagodella Bandel, 1998: 2 (*non* Swainson, 1840: Caenogastropoda: Littorinidae; *non* H. Adams, 1867: Pulmonata: Pupillidae). ***Type species.*** Not given (OD).

Remarks. Bandel (1998: 2) mentioned both in the German as well as the English abstract "*Pagodella* n. gen." among *Anatoma* and *Hainella* as species with large shells known from the Oligocene onwards. The abstract does not contain an indication as to the type species of *Pagodella*, and the genus cannot be found elsewhere in Bandel (1998); hence, *Pagodella* is a *nomen nudum*. *Pagodella* was most likely an earlier manuscript name, that was later exchanged for *Thieleella*. *Thieleella* is not found in the main portion of the abstract, unlike all other generic names discussed. Conversely, in the listing of new species, *Pagodella* does not appear, but *Thieleella* is used.

Pagodella Bandel, 1998, is also a twice preoccupied junior homonym.

Scissurella reunionensis auct. [*nomen nudum*]

Chresonymy.

Scissurella Reunionensis [sic] auct. http://vieoceane.free.fr/mollusques/Scissurellidae.htm [accessed 1/2007].

Remarks. The species name is mentioned on a web page authored by M. Jay as "in press." The available information falls short of the ICZN publication requirements. Maurice Jay (pers. comm. 1/2007) indicated the name to be a manuscript name never intended for publication. It is given here solely for the sake of completeness, and it should not be misconstrued as a taxonomic validation. The specimens were examined by SEM and are identified as *Sin. singeri*.

Species erroneously described/classified in Scissurellidae.

A few species have erroneously been described in Scissurellidae s. l., but have been shown not to belong in this family. Reasons for removal and (tentative) reassignments are detailed for each of the nominal taxa.

Anatomus indicus Montfort, 1810 [Annelida: Polychaeta: Serpulidae: *Spirorbis* sp.]

Figures 33–34

Chresonymy.

Vermiculi cochleati: Solandi, 1789: 33, pl. 30, vas. 143, c, c? [see Remarks].
Anatomus indicus Montfort, 1810: 279–280, text-Fig. p. 278.
"l'Anatome indien": Orbigny, 1824: 341 [as an annelid].
Anatomus indicus: Adams & Adams, 1853–1858 (1854): vol. 1, 439.
Anatomus indicus: Woodward, 1859: 203 [as synonym of "*Serpula spirorbis*].
Anatomus indicus: Mörch, 1863: 1, 428 [as synonym of *Spirorbis*].
Scissurella indica: Paetel, 1888: 289.
Anatomus indicus: Hartman, 1959: 568 [as questionable synonym of *Spirorbis* Daudin, 1800: Serpulidae].

Figure 33. Figure of Solandi (1789: pl. 30, vas 143 cc).

Misidentifications.

Scissurella indica: Philippi, 1853: 32, pl. 6, fig. 4 [is *An.* cf. *aspera*].

Type material. Iconotype Montfort (1810: text-fig. p. 278).

Type locality. 14° towards tropics of cancer, in sargassum (OD). India (Hartman, 1959).

Remarks. The species illustrated by Montfort shows a trochiform depressed shell with an apparent slit. It is described as being cemented to *Sargassum* sea weed, which rules out it being a scissurellid. Serpulid characters that are described are the operculum, two cephalic tentacles, and the absence of eyes. The eyes of scissurellids are prominent (except for deep-water and hydrothermal vent taxa) and would not be missed if the cephalic tentacles were observed. Eyes are minute in serpulids, if at all present, and may not have been seen by Montfort. The shell is described by Montfort as being iridescent green with golden reflections. Although many Vetigastropoda have a nacreous layer, Scissurellidae lack it and do not show iridescence.

Hartman (1959) listed *Anatomus* in synonymy of *Spirorbis*, though she designated the species as a *nomen dubium*. A footnote in d'Orbigny (1824) noted

the likely identity of Monfort's species as an annelid worm. Dehayes (1864–1866: 3) discussed Monfort's species in detail and also identified it as a spirorbid worm. Ten Hoven & Kupriyanova (2009: 105) erroneously re-assigned *Anatomus* to Scissurellidae.

Montfort gave in his synonymy a reference to one of Solandi's figures. Plate 30 does have several figures of number 143, but the numbering of 143 started with K–Q on plate 29, continued with R–cc on plate 30 and concluded with dd–ff on plate 31. Hence, Montfort's "c, c?" should most likely be read as cf. cc. Given that Montfort annotated this citation with a question mark, it cannot be considered an iconotype.

Philippi (1853: pl. 6, fig. 4) illustrated an anatomid shell as *Scissurella indica* Montfort. The species is not conspecific with Montfort's animal, but the identity of Philippi's specimen is difficult to assess. The pronounced axial sculpture on shoulder and base is noticeable, and all of Philippi's species were from the European-Mediterranean region. The most likely candidate, therefore, is *An. aspera*.

Figure 34. Original illustration of *Anatomus indicus*.

Pleurotomaria sismondai Goldfuss, 1844 † [Pleurotomariidae]

Figure 35

Chresonymy.

Pleurotomaria sismondai Goldfuss, 1833–1844 [1844]: 77, pl. 188, fig. 1 [*fide* Koenen, 1892].

Pleurotomaria sismondai: Koenen, 1892: 887–890, pl. 54, figs. 3–6 [Synonymy of *Sci. beyrichi*: p. 890].

Synonymy.

+ *Scisurella beyrichi* Semper, 1865: 284, 285: pl. 12, fig. 1. ***Type material.*** (Semper collection: whereabouts unknown), 2 × 1.5 mm. ***Type locality.*** Latdorf [Saxony, Germany. 51°48.5'N, 11°48'E]. Lower Oligocene (OD). Upper Eocene (Lozouet, 1999). ***Etymology.*** Named after August Heinrich Ernst Beyrich (1815–1896).

Misidentifications.

Scissurella ? (*Sukashitrochus*) *beyrichi*: Lozouet, 1986: 109.

Sukashitrochus beyrichi: Lozouet, 1999: 3 [undescribed species of *Sukashitrochus*].

Figure 35. Original illustration of *Scissurella beyrichi*.

Remarks. Koenen (1892: 890) considered Semper's specimens to be juveniles of *P. sismondai* based on an ontogenetic series. Koenen (1892) did recognize *Sci. philippiana*, so did not assume that all small slit-bearing shells must be juvenile pleurotomariids.

Lozouet's (1999) specimens belong to a yet unnamed species of *Sukashitrochus*.

Leptomaria niloticiformis (von Schlotheim, 1820) † [Pleurotomariidae]

Trochilites niloticiformis von Schlotheim, 1820: 156, not illustrated

+ *Anatoma danica* Bandel, 1998: 31–32, pl. 9, fig. 8, pl. 10, figs. 1–2. ***Type material.*** Holotype (SGPIH 3875), 0.6 mm. ***Type locality.*** Quarry of Faxe, Denmark [55.250°N, 12.130°E]. Danian, Paleocene (OD). ***Etymology.*** Named after its type locality in Denmark (OD).

Anatoma danica: Geiger, 2003: 73.

Type material. Holotype, whereabouts unknown.

Type locality. Younger chalk of Zealand Island, Denmark (OD), [55.5°N, 11.7°E].

Remarks. Schnetler *et al.* (2000: 88) synonymized the juvenile specimen illustrated by Bandel under the pleurotomariid.

Schismope lacuniformis Watson, 1886 [*Macromphalina*: Vanikoridae?]

Figures 36–37

Chresonymy.

Schismope lacuniformis Watson, 1886: 118–119, pl. 8, fig. 8.

Schismope lacuniformis: Kobelt, 1887: 428.

Schismope lacuniformis: Pilsbry, 1890: 63–64, pl. 68, figs 9–10.

Schismope lacuniformis: Pelseneer, 1906: 80, text-fig. 62 [copy Watson].

Schismope lacuniformis: Thiele, 1912a: 23–24, pl. 3, figs 14–15 [copy figures Watson].

Sinezona lacuniformis: Finlay, 1926: 341.

Scissurella (Incisura) lacuniformis: Abbott, 1974: 17.

Scissurella (Incisura) lacuniformis: Ortiz-Corps, 1985: 7.

Scissurella s.l. *lacuniformis*: Lozouet, 1986: 112.

Ariella ? *haliotimorpha*: Geiger, 2003: 75.

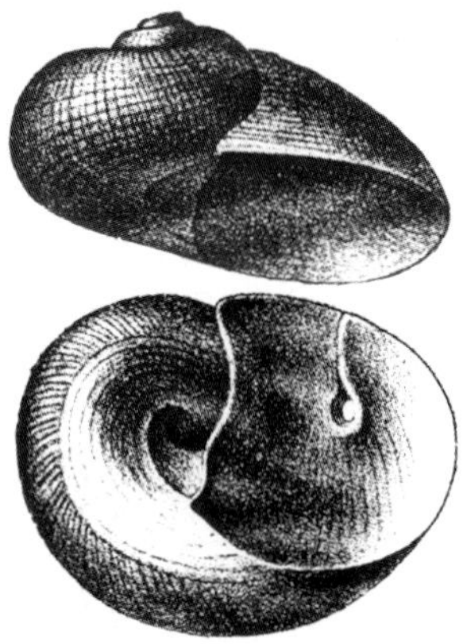

Figure 36. Original illustration of *Schismope lacuniformis*.

Type material. Five syntypes (NHMUK 1887.2.9.410–414. Fig. 37), 0.053 × 0.078" [= 1.33 × 1.95 mm].

Type locality. 390 fms. [= 713 m], station 24, off Culebra Island, [Puerto Rico,] West Indies, 18°38'30"N 65°5'3"W, pteropod ooze (OD: see Remarks).

Etymology. Named for its large lacuniform umbilicus (OD).

Remarks. The species is not a scissurellid. Two of the specimens have a tiny, almost terminal foramen, which is much too small, proportionally, for any scissurellid. Therefore, the foramen could also simply be due to shell repair. The foramen shown by Watson (1886) is stylized shell damage. The specimen may be placed in *Macromphalina* (Vanikoridae).

The known *Macromphalina* species from the Caribbean mostly have a much more protruding apex, and/or a protoconch with more than one whorl (Rolán & Rubio, 1998; Redfern, 2001). *lacuniformis* most closely resembles *Gyrodes duplinensis* Dall, 1895 [*Macromphalina* (*Gyrodisca*)] from the Miocene of North Carolina, USA, in terms of overall shell morphology and protoconch of 0.75 whorl, but the protoconch of *duplinensis* has distinct spirals (Rolán & Rubio, 1998: figs 117–119) not evident in *lancuniformis*. The genus is best known from the Atlantic, with only a few records from the Eastern Pacific (Keen, 1971). There seem to be no known species from the broad Indo-Pacific.

Figure 37. *Schismope lacuniformis*. Best syntype. 390 fms. [= 713 m], station 24, off Culebra Island, [Puerto Rico,] West Indies, 18°38'30"N 65°5'3"W, pteropod ooze (NHMUK 1887.2.9.410–414). See Remarks for possibility of mislocalization. Scale bar shell = 1 mm. Scale bar protoconch = 100 µm.

Pseudorbis caliana (Dall, 1919) [Skeneidae]

Figure 38

Chresonymy.

Schismope caliana Dall, 1919: 370, not illustrated.
? *Schismope caliana*: Dall, 1921: 183.
Adeorbis caliana: Dall, 1923: 4.
Schismope caliana: Oldroyd, 1927: 230.
Schismope caliana: Keen, 1937: 45.
Tornus calianus: Burch, 1946a: 27 [Scissurellidae].
Scissurella caliana: Abbott, 1974: 17.

Figure 38. *Schismope caliana*. Holotype. San Diego, California [37.7°N, 117.2°W] (USNM 1987609). Uncoated specimen, 3 kV, high vacuum, secondary electron detector. Note hole in shell visible in apertural view. Scale bar = 0.5 mm.

Type material. Holotype (USNM 198609: Boss *et al.*, 1968: 55. Fig. 38), 1 × 1.28 mm.

Type locality. San Diego, California [37.7°N, 117.2°W] (OD).

Etymology. Caliandrum, Latin, for hairdress (?).

Remarks. Dall (1923: 4) noted that Bartsch "refers *S. caliana* to *Adeorbis*." (Tornidae). McLean (1967: 407) noted that the "foramen" of Dall's species represents shell damage, similar to *Sin. lacuniformis* discussed above. An annotation by W. H. Dall on the type label reads "or is the hole a fracture? W.H.D." The hole in the shell has a smooth margin, which points to a potential growth anomaly or drill hole by a predator, rather than a post mortem damage to the shell.

USNM classified the species under Tornidae; Ponder & Keyzer (1998) questioned whether Tornidae is distinguishable from Vitrinellidae, and Bouchet & Rocroi (2005) use Vitrinellinae under Tornidae. McLean (pers. comm. 2/2004) places the species in the skeneid genus *Pseudorbis*.

Zardinitrochus Bandel, 1998 †

Zardinitrochus Bandel, 1998: 24–25.

Type species. *Temnotropis suessi* Klipstein *in* Kittl, 1891 (OD).

Etymology. Named after Italian paleontologist Rinaldo Zardini (1902–1988) (OD). Gender masculine.

Remarks. The type species of *Zardinitrochus* may well be misidentified and constitute a potentially undescribed species. The nomenclatural ramifications are apparent, but it is left to specialists of fossil gastropods to address these issues. The genus is currently monotypic.

Temnotropis Laube, 1869 †

Temnotropis Laube, 1869: 42.

Type species. *Sigaretus carinatus* Münster, 1841 (OD).

"*Temnotropis*" *suessi* Klipstein *in* Kittl, 1891 † [incertae sedis: Fossariidae?]

Figure 39

Chresonymy.

Temnotropis suessi Klipstein *in* Kittl, 1891: 180, pl. 1, fig. 28.
Temnotropis suessi: Delhaes, 1909: 32, pl. 3, fig. 21.
Temnotropis suessi: Zardini, 1978: 17, pl. 2, figs. 11a–d.
? *Temnotropis suessi*: Zardini, 1980: 3: pl. 1, figs. 6a–b [juvenile *fide* Zardini, 1980].
? *Zardinitrochus suessi*: Bandel, 1998: 25, pl. 7, figs. 3–4.
? *Zardinitrochus suessi*: Bandel, 2009: 20, pl. 7, figs. 101–102.

Type material. Holotype (NHMW 53: Tichy, 1970; missing from shelf, Geiger, pers. obs.), 10 mm.

Type locality. St. Cassian, Stuoresmarl [Italy, 46.6°N, 11.9°E] (OD).

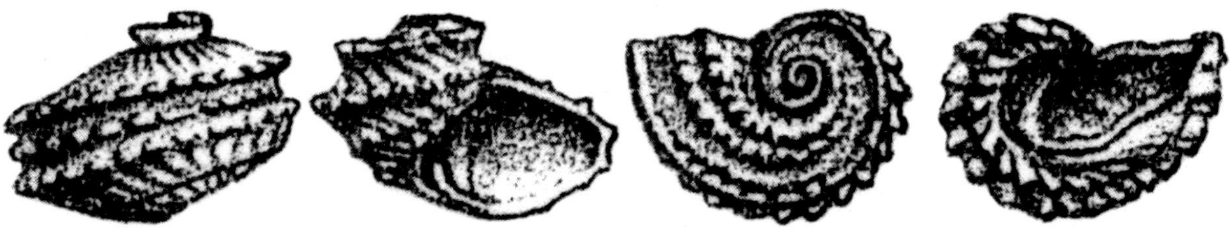

Figure 39. Original illustration of *Temnotropis suessi*.

Remarks. The identity of *Temnotropis suessi* is uncertain. The holotype was missing from the shelf at NHMW in 2001. It is further questionable whether the specimens illustrated by Bandel are conspecific with the original material. Accordingly, it is not certain that *suessi* belongs in the new genus *Zardinitrochus*; see ICZN Art. 70.3 for actions regarding misidentified type species. The original illustrations do not show a selenizone, which is apparent in Bandel's figures. However, Kittl's description is explicit in describing the shell as follows (excerpt): "[whorls] are separated by two thorny keels, whose upper one represents the slit-keel [= selenizone?] and allows to recognize cap-like lunules." Delhaes (1909) also noted the distinct "band" at the periphery of the shell.

Additionally, Kittl's illustration shows a shell of approximately 10 mm width. Slit-less scissurellids are known (Larocheidae: *Larochea, Larocheopsis, Trogloconcha.* Scissurellidae: *Coronadoa*) but all are small, not exceeding 2 mm in size, which is also the size of the specimen illustrated by Bandel (1998, 2009).

Kittl placed his *suessi* only provisionally in *Temnotropis* because it was incompletely known (Kittl, 1891: 179), further casting doubt about the classification of the species. Kittl did not mention nacre as a character, but noticed that *Sigaretus* (today Caenogastropoda: Naticidae: without nacre) had been considered closely related to *Temnotropis*, hence, the presence of nacre was most likely not considered important.

Scissurella minuta Turton, 1932 [*nomen dubium*]

Figure 40

Chresonymy.

Scissurella minuta Turton, 1932: 204, pl. 53, fig. 1415.
Scissurella minuta: Herbert, 1986: 601, 620 [as *nomen dubium*].

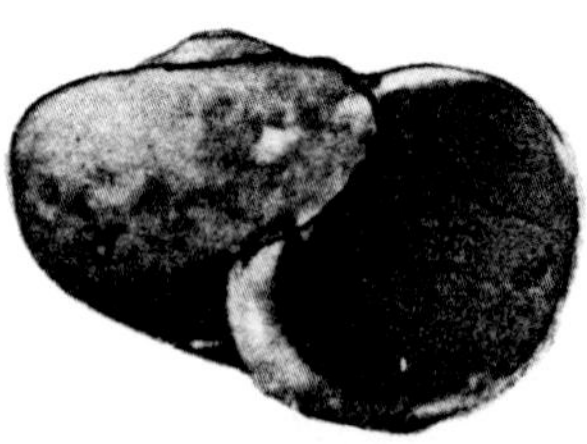

Figure 40. Original illustration of *Scissurella minuta* Turton.

Type material. Holotype (OMU: *fide* Herbert, 1986: 620), 0.7 × 1 mm, fragmented, unrecognizable.

Type locality. Port Alfred [South Africa, 33.6°S, 26.9°E] (OD).

Etymology. Most likely referring to small size.

Remarks. Herbert (1986: 620) noted that the holotype "is now an unidentifiable collection of shell fragments embedded in glue." The description and the illustration note the smooth shell surface, a characteristic not known from any scissurellid. Turton (1932: 204) indicated that "The slit is not very clear, but I think it is natural (not a damage) and this fixes the genus."

The round aperture and the roof raised above the suture are characteristics not known from other Scissurellidae s.s. The species is here considered to belong to a family other than Scissurellidae s.l., but the true affinity remains unknown.

Scissurella minuta does not threaten *Satondella minuta*. The two species clearly belong into different genera.

Atlanta peronii Lesueur, 1817 [Heteropoda: Atlantidae]

Figure 41

Atlanta peronii Lesueur, 1817: 390–391, pl. 2, fig. I.1-2.

Synonym.

+ *Schizotrochus palaeomphaloides* Nordsieck, 1973: 4, pl. 1, fig. 1. ***Type material.*** Holotype (Nordsieck collection, 02.28, now SMF 336131), 0.8 × 0.47 mm, one paratype (SMF 336132). ***Type locality***. *Meteor* station B2, 36°14'N 18°9.4'E, 3910 m, mud (OD). ***Etymology***. Not given (OD). Paleo-, Latin prefix for old or fossil; -omphaloides, Latin for umbilicus-bearing.

Scissurella (Schizotrochus) palaeomphaloides: Nordsieck, 1982: 2, pl. 1, fig. 17.

Figure 41. Original illustration of *Schizotrochus palaeomphaloides*.

Remarks. The species was originally described in *Schizotrochus*, which currently is treated as a synonym of *Anatoma*. The crude original drawing (Fig. 41) shows characters that are not associated with Scissurellidae or Anatomidae: angular profile of whorls, position of slit below the periphery, absence of selenizone, number of whorls for size, overall proportions of shell. Arie Janssen (pers. comm.) has examined the types and identified them as a juveniles of the heteropod *Atlanta peronii*.

The species has also been synonymized under *Protatlanta souleyeti* (E. A. Smith, 1888) by Ramazzotti *et al.* (2007).

Color Plate 1.
Colorations of Anatomidae shells and animals.

A–B. *Anatoma tangaroa* with axial dark bands. 120 m, 28 km S of Great Island, Three Kings Islands, New Zealand, 34.400°S, 174.280°E (NMNZ M.160285, 15). Scale bars = 1 mm.

C. *Anatoma indonesica* with yellow apex. 211–212 m, off Kerama Island, Okinawa, Japan, 26.047°N, 127.375°E (SBMNH, 9). Scale bar = 1 mm.

D. *Anatoma kelseyi* animal. LACM 23129 BFI. Stereomicroscope images, z-stacked with *HeliconFocus*. Scale bar = 1 mm. ct: cephalic tentacle; es: eye stalk; et: epipodial tentacle; j: jaws; m: mantle; mp: metapodium; op: operculum; pp: propodium; sn: snout; st: suboptic tentacle.

E. *Anatoma philippinica* with animal in shell marked with maroon spotted axial bands. 60 m, Pamilican Island, Philippines, 9.502°N, 123.918°E (MNHN barcode samples).

F. *Anatoma euglypta* live animal. Photograph by and with kind permission of Michael Schrödl and Diego Zelaya.

G. *Anatoma janetae* holotype. Lightoptical image by Peter Batson of deepseaphotography.com, courtesy J. Voight.

A
B
C
D
E
F
G
m
m
op
es
sn
mp
pp
et
st
ct
op
et
st
ct
sn
j
mp
pp
et

Color Plate 2.
Colorations of scissurellid shells and animals.

A. *Sinezona cingulata* collected in situ on old card-board mount. Notice piece of coralline red algae in middle with attached specimen. NHMUK 1911.10.26.34137–42. Image width approximately 9 mm.

B. *Scissurella cebuana* with axial color bands. 85–90 m, [10 km N of Ambil Island], Philippines, 13.917°N, 120.333°E (MNHN, 4).

C. *Scissurella lorenzi* in apical view showing typical axial hyaline bands. 35–38 m, base of cliff, Horseshoe, Okinawa, Japan, 26.504°N, 127.858°E (DLG 978, 2). Scale bar = 1 mm.

D. Live collected *Trogloconcha tesselata*. Head with cephalic tentacles and eyes on eyestalks, operculum not covering aperture, and three epipodial tentacles visible. 90–200 m, Maribohoc Bay, Bohol Island, Philippines, 6.602°N, 123.75°E (MNHN, 1: complete).

E. *Sinezona confusa*. Live on intertidal *Batophora* (Chlorophyta: Dasycladaceae), from Abaco, Bahamas. Image by and with kind permission of Colin Redfern.

F. *Scissurella rota* without and with periostracum. 0 m, Mission Rock, Natal, South Africa, 28.267°S, 32.495°E (DLG 1380, 13). Scale bar = 1 mm.

G. *Anatoma australis* with hermit crab and encrusting bryozoans as well as a serpulid annelid worm (AMS C.400809). Scale bar = 2 mm.

H. Original illustration of *Scissurella conica*.

A
B
C
D
E
F
G
H

Color Plate 3.
Live Scissurellidae from Madagascar and the Philippines.

A. *Scissurella rota*. 14–18 m, Lavanono sector, ATIMO VATAE BB03, Madagascar, 25.440°S, 44.935°E (MNHN, 1).
B. *Scissurella jucunda*. Dorsal view manually z-stacked. 0 m, False Cape, ATIMO VATAE BM 12, Madagascar, 25.571°S, 45.528°E (MNHN, 1).
C. *Sukashitrochus dorbignyi*. Notice tan colored periostracum of this not fully grown specimen. 12–14 m, NW Albatross Rock, ATIMO VATAE BS01, Madagascar, 25.470°S, 44.940°E (MNHN, 1).
D. *Scissurella mirifica*. 4 m, Biking, Panglao Island, PANGLAO 2004 B5, Philippines, 9.586°N, 123.840°E (MNHN, 9).
E. *Sinezona plicata*. Notice extensive motility of propodium. 2–4 m, Pamilacan Island, PANGLAO 2004 B11, Philippines, 9.490°N, 123.933°E (MNHN, 1).

For approximate sizes of animals, refer to particular species treatment. All images courtesy MNHN, Figs. A–C (P. Maestrati MNHN), Figs D–E (P. Lozouet MNHN).

ct: cephalic tentacle.
e: eye.
et: epipodial tentacle.
g: gill.
mp: metapodium.
op: operculum.
pp: propodium.
sn: snout.

A
ct
ct
e
B
ct
ct
ct
e
C
op
D
ct
ct
g
e
ct
et
ct
sn
g
et
e
E
et
et
e
ct
op
et
op
g
ct
pp
mp
sn
pp
ct
op

Color Plate 4.
Live scissurellids from Vanuatu.

A. *Scissurella quadrata*. Note pink hepatopancreas. 32–40 m, S Tutuba Island, SANTO 2006 DB16, Vanuatu, 15.591°S, 167.263°E (MNHN, 6).
B. *Sinezona plicata*. SANTO 2006 DB83, Vanuatu (MNHN).
C. *Scissurella cebuana*. 80–94 m, NW coast of Malo Island, SANTO 2006 EP19, Vanuatu, 15.629°S, 167.089°E (MNHN, 20).
D. *Sukashitrochus morleti*. 0 m, Baldwin Bay, SANTO 2006 VM30, Vanuatu, 15.580°S, 167.033°E (MNHN, 1).
E. *Sukashitrochus morleti*. 0 m, Aoré Island, SANTO 2006 VM46, Vanuatu, 15.575°S, 167.205°E (MNHN, 1).
F. *Anatoma rapaensis*. 80 m, W Tutuba Island, SANTO 2006 DS104, Vanuatu, 15.568°S, 167.267°E (MNHN, 11).
G. *Anatoma amydra*. 45–101 m, E Aoré Island, SANTO 2006 EP10, Vanuatu, 15.604°S, 167.150°E (MNHN, 1).
H. *Trogloconcha lozoueti*. 99 m, NW coast of Malo Island, SANTO 2006 EP21, Vanuatu, 15.628°S, 167.087°E (MNHN, 3).

For approximate sizes of animals, refer to particular species treatment. All images courtesy MNHN, Figs A–E, G–H (D. Brabant, MNHN), Fig. F (A.-L. Fleddum, Oslo University).

ct: cephalic tentacle.
e: eye.
et: epipodial tentacle.
fg: foregut.
g: gill.
m: mantle.
op: operculum.
pp: propodium.
sn: snout.
st: suboptic tentacle.

A
hp
ct
et
e
fg
B
et
st
op
C
pp
D
E
g
F
G
H
sn
r
m

Color Plate 5. Histology.

A–K: Representative 2 μm semithin resin sections of *Sinezona rimuloides.* Multiple stain (Polysciences). Scale bar A–G = 250 μm (only shown in A).
A. Snout and cephalic tentacles.
B. Eyes and propodial gland.
C. Epipodial tentacle and pedal ganglion.
D. Oesophagus.
E. Statocysts, epipodial sense organ and gills.
F. Hypobranchial gland, metapodium with operculum.
G. Intestinal loops.
H. Gonad.
I. Enlargement of the eye and suboptic tentacle. Scale bar = 50 μm.
J. Gill filaments with bursicles. Scale bar = 100 μm.
K. Radula. Scale bar = 10 μm.
L. Radula of *Anatoma euglypta*. 2 μm resin embedded semithin sections stained with Richardson's stain. Scale bar = 100 μm.

b: bursicle.
c: conchiolin (proteinaceous remnant of decalcified shell).
ct: cephalic tentacle.
e: eye.
ESO: epipodial sense organ.
gl: gill leaflet.
hp: hepatopancreas.
hpg: hypobranchial gland.
i: intestine.
l: lens of eye.
lu: lumen of bursicle.
m: mantle.
mp: metapodium.
n: nucleus.
no: nucleolus.
o: opening of bursicle.
oc: oocyte.
od: odontophore.
oe: oesophagus.
op: operculum.
pg. pedal ganglion.
pp: propodium.
ppg: propodial gland.
r: radula.
rc: radular caecum.
re: retina.
rt: rectum.
sm: shell muscle.
sot: suboptic tentacle.
R: rachidian tooth.
L1–5: lateral teeth 1–5.
M: marginal teeth.

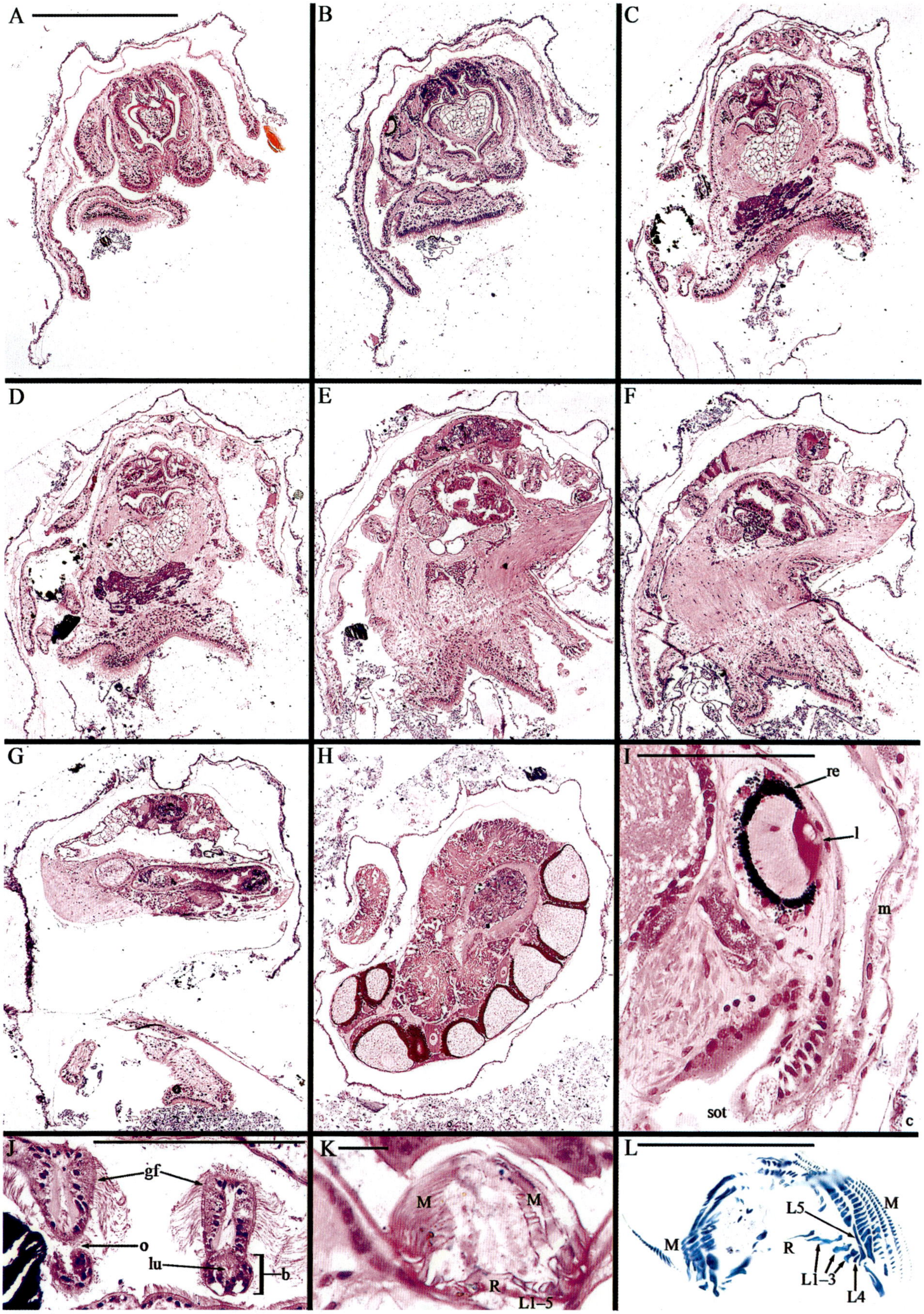
A
B
C
D
E
F
G
H
I
re
l
m
sot
c
J
gf
o
lu
b
K
M
M
R
L1–5
L
M
L5
M
R
L1–3
L4

Scissurellidae Gray, 1847

Chresonymy.

Scissurellidae Gray 1847: 146 [as Scissurellina, emended Gray, 1857: 160].
Scissurellidae: Carpenter, 1861: 216.
Scissurellidae: Kobelt, 1864–1881: 163.
Scissurellidae: Deshayes, 1866: 1–2.
Scissurellidae: Pilsbry, 1890: 49.
Scissurellidae: Thiele, 1931: 30.
Scissurellidae: Brookes Knight *et al.*, 1960: I221.
Scissurellidae: Wenz, 1961: 172.
Scissurellidae: McLean, 1989: 3–5.
Scissurellidae: Marshall, 1993: 286.
Scissurellidae: Hickman, 1998: 666–667.
Scissurellidae: Bandel, 1998: 7–8.
Scissurellidae: Marshall, 2002: 166.
Scissurellidae, Scissurellinae: Geiger, 2003: 25.
Scissurellidae: Zelaya & Geiger, 2007: 417.
Scissurellidae: Geiger & Sasaki, 2009: 2.
Scissurellidae: Geiger & McLean, 2010: 3.

Type genus. *Scissurella* d'Orbigny, 1824.

Other genera. *Coronadoa* Bartsch, 1946.
Incisura Hedley, 1904.
Satondella Bandel, 1998.
Sinezona Finlay, 1926.
Sukashitrochus Habe and Kosuge, 1964.

Description. Shell small (<3 mm), trochiform (*Coronadoa, Scissurella*), naticiform (*Satondella, Scissurella, Sinezona*), neritiform and haliotiform (*Incisura*); no nacre. Protoconch 0.5–1 whorl, sculpture variable: smooth, spirals, axials, flocculent, microhexagonal; apertural varix present or absent; apertural margin convex or sinusoidal. Teleoconch sculpture variable: smooth, axials, spirals, folds. No brood pouch. Usually with selenizone (no selenizone in some *Sinezona* and *Satondella* species), selenizone on shoulder, with slit or single foramen (absent in *Coronadoa*). Usually umbilicate, few with umbilical callus. Operculum thin, corneous, multispiral, with central nucleus.

Radula usually (except *Sin. macleani*) with rachidian tooth serrated, five serrated lateral teeth, lateral teeth 1–3 similar, lateral tooth 4 reduced, lateral tooth 5 enlarged, broadened, asymmetrically serrated. Two shell muscles or fused horseshoe-shaped muscle. Epipodial and cephalic tentacles with papillae. Eyes present. Gills with bursicles. Hypobranchial gland present.

Geologic Range. Jurassic through Recent.

Comparisons. Anatomidae have the slit (or foramen: *Sasakiconcha*) and selenizone at the periphery.

Temnocinclidae have a planispiral shell with extremely increased expansion rate of the whorl, a radula with three lateral teeth, and are found exclusively at hydrothermal vents.

Sutilizonidae have a pitted protoconch, a radula with 3–4 lateral teeth, a combination of highly sculptured and depressed shell, and are exclusively found at hydrothermal vents.

Larocheidae lack any sign of selenizone, slit, or foramen, and, in contrast to the scissurellid *Coronadoa*, do not have protoconch sculpture with axials.

Depressizonidae have a calyptraeiform shell.

Remarks. Linse (1997) erroneously included presence of nacre in the family diagnosis.

Scissurella d'Orbigny, 1824

Chresonymy.

Scissurella d'Orbigny, 1824: 343.
Scissurella: Deshayes, 1866: 2–4.
Scissurella: Tryon, 1883: 321.
Scissurella: Pilsbry, 1890: 49.
Scissurella: Brookes Knight *et al.*, 1960: I221.
Scissurella: Wenz, 1961: 173.
Scissurella: McLean, 1967: 404.
Scissurella: Montouchet, 1972: 2.
Scissurella: Herbert, 1986: 619–620.
Scissurella: Bandel, 1998: 9.
Scissurella: Marshall, 2002: 166–167.
Scissurella: Geiger, 2003: 26–27.
Scissurella: Zelaya & Geiger, 2007: 417.

Synonyms.

+ *Schismope* Jeffreys, 1856c: 321. ***Type species.*** *Scissurella striatula* Philippi, 1844 *sensu* Jeffreys, [= *Sin. cingulata*] *non* Philippi, 1844 [= *Sci. costata*] (M). *Scissurella striatula* Philippi, 1844 (here explicitly designated: see Remarks). Gender masculine. ***Etymology.*** Schisme: Greek for slit and ope: Greek for foramen. ***Remarks.*** See under Remarks of *Scissurella* for justification of synonymy.

+ *Woodwardia* Crosse & Fischer, 1861: 160. ***Type species.*** *Scissurella elegans* d'Orbigny, 1824, *sensu* Woodward (1859) [= *Sin. cingulata*] *non* d'Orbigny, 1824 [= *Sci. costata*] (M). *Scissurella elegans* d'Orbigny, 1824 (here explicitly designated). ***Etymology.*** Named after Samuel Pickworth Woodward (1821–1865: OD). Gender feminine. ***Remarks.*** See under Remarks of *Scissurella* for justification of synonymy

+ *Maxwellella* Bandel, 1998: 19. ***Type species.*** *Scissurella annulata* Ravn, 1933 (OD, †). ***Etymology.*** Named after Bill Maxwell of GNS. Gender feminine. ***Remarks.*** See under Remarks of *Scissurella* for justification of synonymy.

+ *Reussella* Bandel, 1998: 44 [preoccupied]. ***Type species.*** *Scissurella depressa* Reuss, 1860 (OD, †), not *Schismope depressa* Watson, 1897. Not *Reussella* Galloway, 1933 (Foraminifera: Buliminida: Buliminoidea: Reussellidae). ***Etymology.*** Named for A. E. Reuss (1811–1873) for his contribution to micropaleontology; for biography, see Vávra (2002). Gender feminine. ***Remarks.*** See under Remarks of *Scissurella* for justification of synonymy.

+ *Praescissurella* Lozouet, 1998: 66. ***Type species.*** *Scissurella depontaillieri* (OD, †). ***Etymology.*** Prae-, Latin for before; an early form of *Scissurella*. Gender feminine. ***Remarks.*** See under Remarks of *Scissurella* for justification of synonymy.

Type species. *Scissurella laevigata* (SD: Gray, 1847) [= *Scissurella costata* d'Orbigny, 1824: cf. Marshall, 2002].

Scissurellidae *Scissurella*

Other species. *Scissurella alexandrei* Montouchet, 1972.
Scissurella aliceae Schnetler, Lozouet & Pacaud, 2000 †.
Scissurella alto Geiger, 2003.
Scissurella annulata Ravn, 1933 †.
Scissurella apudornata Laws, 1935 †.
Scissurella azorensis Nolt, 2008.
Scissurella bituminata Beets, 1943 †.
Scissurella bountyensis Powell, 1933.
Scissurella cebuana (Bandel, 1998).
Scissurella clathrata Strebel, 1908.
Scissurella condita Laws, 1939 †.
Scissurella coronata Watson, 1886 [*nomen dubium*].
Scissurella costata d'Orbigny, 1824.
Scissurella cossmanni Depontaillier, 1881 †.
Scissurella cyprina Cotton & Godfrey, 1938.
Scissurella dalli (Bartsch, 1903) [*nomen dubium*].
Scissurella depontaillieri Cossmann, 1879 †.
Scissurella dohrniana (Dunker, 1861) [*nomen dubium*].
Scissurella electilis Montouchet, 1972.
Scissurella eocaenica Bandel, 1998 †.
Scissurella evaensis Bandel, 1998.
Scissurella georgica Davolos & Moolenbeek, 2006.
Scissurella gruendeli Bandel, 1998 †.
Scissurella hedegaardi Bandel, 1998 †.
Scissurella hoernesi Semper, 1865 [*nomen dubium*].
Scissurella jucunda Smith, 1890.
Scissurella kaiserae Geiger, 2006.
Scissurella koeneni Semper, 1865 [*nomen dubium*].
Scissurella lobini (Burnay & Rolán, 1990).
Scissurella lorenzi Geiger, 2006.
Scissurella manawatawhia Powell, 1937.
Scissurella mantelli Woodward, 1859 [*nomen dubium*].
Scissurella maraisorum Geiger, 2006.
Scissurella marchmontensis Sohl, 1992 †.
Scissurella mirifica (A. Adams, 1862).
Scissurella morretesi Montouchet, 1972.
Scissurella obliqua Watson, 1886.
Scissurella ornata May, 1908.
Scissurella petermannensis Lamy, 1910.
Scissurella peyrerensis Bandel, 1998 †.
Scissurella phenax new species.
Scissurella prendrevillei Powell, 1933.
Scissurella quadrata Geiger & Jansen, 2004.
Scissurella ravni (Schnetler, Lozouet & Pacaud, 2000) †.
Scissurella redferni (Rolán, 1996).
Scissurella regalis Geiger & Marshall, 2012.
Scissurella reticulata Philippi, 1853.
Scissurella rota Yaron, 1983.
Scissurella skeneoides new species.

Scissurella spinosa Geiger & Jansen, 2004.
Scissurella staminea (A. Adams, 1862).
Scissurella sudanica Bandel, 1998.
Scissurella supraplicata Smith, 1875.
Scissurella szaboi Conti & Monari, 1991 †.
Scissurella tinaeformis Amitrov, 1996 †.
Scissurella transylvanica Reuss, 1860 †.
Scissurella xandaros new species.

Species removed from *Scissurella*.
Anatoma alta (Watson, 1886).
Anatoma amoena (Thiele, 1912).
Anatoma australis (Hedley, 1903).
Anatoma aspera (Philippi, 1844).
Sukashitrochus atkinsoni (Tenison-Woods, 1876).
Anatoma bertheloti (d'Orbigny, 1840).
Sukashitrochus carinatus (A. Adams, 1862).
Sinezona cingulata (O. G. Costa, 1861).
Anatoma conica (d'Orbigny, 1841).
Anatoma crispata (Fleming, 1828).
Sinezona cupelliformis (Amitrov, 1996) †.
Sinezona deshayesi (Munier-Chalmas, 1862) †.
Anatoma disciformis (Golikov & Sirenko, 1980).
Sukashitrochus dorbignyi (Audouin, 1826).
Anatoma equatoria (Hedley, 1899).
Anatoma euglypta (Pelseneer, 1903).
Anatoma exquisita (Schepman, 1908).
Anatoma jacksoni (Melvill, 1904).
Anatoma kelseyi (Dall, 1905).
Incisura lytteltonensis (Smith, 1894).
Sinezona malloryi (Squires & Goeddert, 1996) †.
Anatoma maxima (Schepman, 1908).
Sukashitrochus mirandus (A. Adams, 1862).
Sinezona modesta (A. Adams, 1862).
Anatoma tenuis (Jeffreys, 1877).
Anatoma tenuisculpta (Seguenza, 1877).

Species removed from Scissurellidae. *Scissurella minuta* Turton, 1932: *nomen dubium* [not Scissurellidae: see Introduction].

Etymology. Scissura: Latin for slit; -ella: Latin diminutiv suffix. Gender feminine.

Description. Shell trochiform, shoulder rounded or angular. Sculpture usually reticulate, but also with predominating axials, or axials only; no spiral keel(s) on base. Slit open, selenizone on shoulder, starting at less than 0.75 teleoconch I whorl, with moderately elevated keels. Protoconch with variable sculpture: smooth, spirals, fine axials, strong axials; apertural varix absent or present, if present restricted to aperture, or connecting to embryonic cap; apertural margin sinusoidal or simple convex curve. Umbilicus usually open, with or without carina, some with funiculus; rarely anomphalous. Operculate. Radula R - 5 - n, lateral tooth 5 broadly enlarged, asymmetrically serrated. Anatomy essentially unknown.

Distribution. Northern temperate waters to Antarctica in all oceans (Fig. 42). For specific localities, see individual species.

Geologic Range. Jurassic through Recent.

Comparisons. In *Sinezona* the slit is always closed anteriorly to form a foramen at maturity; juvenile *Sinezona* with an open slit may be difficult to distinguish from *Scissurella*. Half the *Scissurella* species can be distinguished by the presence of a carina or a funiculus in the umbilical region. Fully grown specimens can often be identified by the marked drop of the final quarter whorl along the axis of the shell.

Sukashitrochus has prominent spiral keel(s) on the base, the umbilicus is always bordered by a carina, and the slit is closed anteriorly to form a foramen (see above).

Coronadoa lacks a selenizone, slit, or foramen.

Incisura usually has a very smooth shell and is rather neritiform than trochiform.

Differentiation of species. Species are differentiated by overall shell shape, spire height, length of teleoconch I, details of sculpture (absolute and relative strength of axials and spirals), width of umbilicus, as well as protoconch sculpture.

Anatomical characters have not been found to help in species identification or in the establishment of species boundaries. Under a very good light microscope, only prominent protoconch sculpture can be resolved sufficiently.

Intraspecific variation extends to ontogenetic change of shell/spire height, condition of the slit (notice, juvenile *Sinezona* have an open slit, hence, can be confused with *Scissurella*), and strength and density of sculptural elements (variable by approximately 25–50%).

Remarks. Marshall (2002) discussed the nomenclatorial problems associated with *Schismope* and *Woodwardia*, and his suggestion of maintaining prevalent usage of the taxa and, hence, synonymizing both generic names under *Scissurella* is followed here.

It is clear that the type species for *Schismope* was misidentified by Jeffreys (1856c) as remarked by Marshall (2002). *Scissurella striatula* can be identified as the true Mediterranean member of *Scissurella*, *Sci. costata*, and not a juvenile of the foramen-

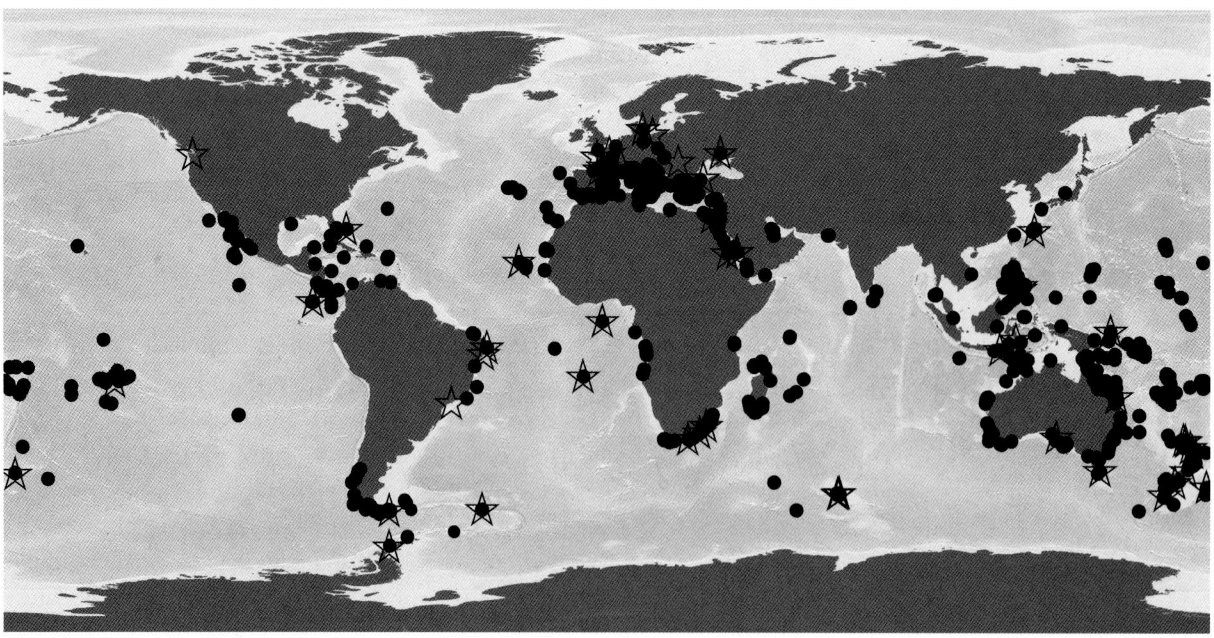

Figure 42. Global distribution of *Scissurella* species. ☆ type locality. ● specimen record. ■ literature record.

bearing species, usually known as *Sin. cingulata* (see also Woodward, 1856b for an early discussion). Jeffreys (1856c) introduced his genus *Schismope* for species with a closed foramen, hence, the type species he chose is misidentified. ICZN Art. 70.3 permits under Art. 70.3.1 as one option to declare the nominal species as the type species in order to serve in the interest of taxonomic stability. By choosing that option here, *Schismope* becomes a junior synonym of *Scissurella*. The alternative of choosing *Scissurella cingulata* as type species, would make the well-established *Sinezona* a junior synonym, which would cause highly undesirable instability.

Crosse & Fischer (1861) clearly misidentified the type species as pointed out by Marshall (2002). They referred to Woodward's (1859) illustrations where specimens with an open slit and a closed foramen were identified as growth stages. The true *Sci. elegans* illustrated by d'Orbigny (1824) shows a true *Scissurella*; particularly telling is the length of the selenizone of over one whorl, whereas all European scissurellids with a foramen have a selenizone of less than half a whorl. *Scissurella elegans* is a fossil species, which is here synonymized under *Sci. costata*. ICZN Art. 70.3 permits under Art. 70.3.1 as one option to declare the nominal species as the type species in order to serve in the interest of taxonomic stability. By choosing that option here, *Woodwardia* becomes a junior synonym of *Scissurella*. The alternative of choosing *Sinezona cingulata* as type species would make the well-established *Sinezona* a junior synonym, and would cause highly undesirable instability.

Among the trochospiral scissurellids with an open slit, Bandel (1998) diagnosed *Maxwellella* as containing those species with a depressed apex, as opposed to the low trochospiral apex in *Scissurella*. Although the species assigned by Bandel to *Maxwellella* all show a more or less depressed apex, the same condition is found in several species retained by Bandel in *Scissurella* (*Sci. marshalli*, *Sci. rota*, *Sci. eocaenica*). Furthermore, some species assigned to *Maxwellella* show a hardly depressed apex with an overall trochiform shell ("*M.*" *unispirata* [= *Sci. evaensis*]) that are difficult to separate from species placed in other genera (*Sci. evaensis*). Note that the same species cannot be assigned to two different genera. The elevation or depression of the apex in scissurellids with an open slit exhibits a continuous gradation, with no apparent separation of conditions. Accordingly, *Maxwellella* is synonymized under *Scissurella*.

Some of the fossil members of *Maxwellella*, including the type species, are unique in that the top margins of all whorls are higher than the tip of the protoconch. It may be possible to retain the genus *Maxwellella* for those species using a much restricted generic concept (e.g., Schnetler *et al.*, 2001), but too little material is available to make an informed decision. There are no covariant characters that could support a separations of those species. No Recent scissurellid species are known that show this condition.

Reussella was diagnosed by the rounded shoulder, that is flat in *Scissurella* and *Maxwellella* (Bandel, 1998). The type species of the genera show such a distinction, yet the scissurellid species cannot be assigned unambiguously to either of these two morphs due to extensive intergradation among species. The assignment of species to the genera in question by Bandel (1998) according to his own criteria is also inconsistent. *Scissurella cossmanni* and *Sci. reticulata* are close to identical in apertural profile to "*R.*" *depressa* (Reuss, 1860) [*non* Watson, 1897] or "*R.*" *plicata*. Furthermore, Bandel (1998) diagnosed *Reussella* with an open slit to separate it from those genera with a foramen, yet, "*R.*" *plicata* (Bandel, 1998: pl. 15, fig. 8, pl. 16, fig. 1) has a closed slit and a foramen. As the generic concept of *Reussella* lacks discrete diagnostic characters and cannot be made more precise, the genus was synonymized under *Scissurella* by Geiger (2003). The same conclusion was reached by Lozouet *et al.* (2001).

Praescissurella was introduced for the fossil *Sci. depontaillieri* based on the presence of a funiculus in the umbilicus. This character does not seem to diagnose a particular group, but is homoplastic. Schnetler *et al.* (2001: 84) found it difficult to assign their "*Praescissurella* ? *ravni*" to this genus with sufficient certainty, despite relatively well-preserved material, further calling into question the validity of this genus. *Praescissurella* is here synonymized under *Scissurella*.

Scissurella alexandrei Montouchet, 1972 [*nomen dubium*]

Figures 43–44

Chresonymy.

Scissurella alexandrei Montouchet, 1972: 4–8, pl. 1, figs 1–7.
Scissurella alexandrei: Rios, 1985, 10, pl. 5, fig. 31 [copy figure Montouchet, 1972].
Scissurella alexandrei: Rios, 1994, 21, pl. 5, fig. 33 [copy figure Montouchet, 1972].
Scissurella alexandrei: Lozouet, 1986: 107.
Scissurella alexandrei: Geiger, 2003: 76.
Scissurella alexandrei: Rios, 2009, 23, text-fig. [copy figure Montouchet, 1972].
Scissurella alexandrei: Dornellas & Simone, 2011: 3, figs 18–19 [holotype].

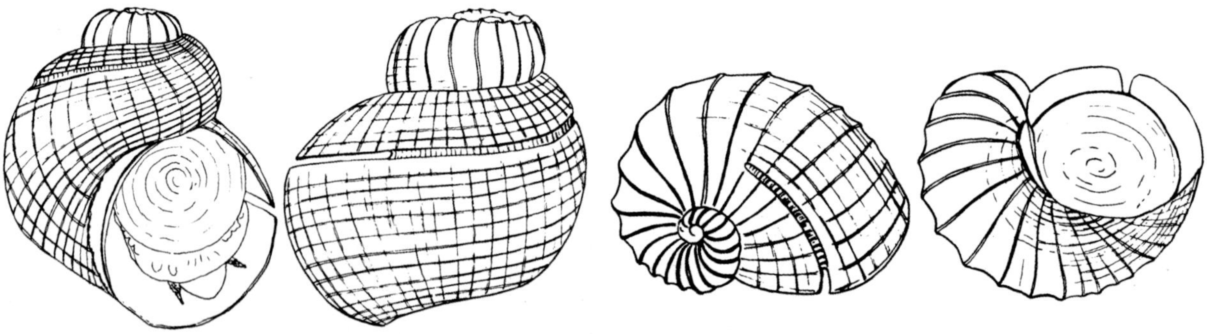

Figure 43. Original illustration of *Scissurella alexandrei*. Left: holotype.

Type material. Holotype (MZSP 18487. Dornellas & Simone, 2011: figs 18–19), 0.9 × 0.8 mm. Paratypes from type locality (MZSP 18491, 26. Only one surviving Dornellas & Simone, 2011; MZSP 18492, 3. Only one surviving: Dornellas & Simone, 2011). Off Paracuru, Ceará State (MZSP 18488, 2. Only one surviving: Dornellas & Simone, 2011). Meireles Beach, Fortaleza, Ceará State (MZSP 18489, 1). Off Itamaracá, Pernambuco State (MZSP 18490, 1). Salvador off Rio Vermelho Beach, Bahia State (MZSP 18493, 1).

Type locality. Praia dos Sete Coqueiros, Maceió, Alagoas, Brazil [9.667°S, 35.733°W]. Living on seaweeds in the lowest meso-littoral (OD).

Etymology. In memory of Alexandre Valente Boffi (OD).

Description. [Based on original figure: see Remarks]. Shell trochiform, taller than wide. Protoconch unknown. Teleoconch I of possibly >1 whorl, approximately 22 axial cords. Teleoconch II of approximately 0.8 whorl. Shoulder with approximately same density of axial cords as on teleoconch I, approximately six spiral lines. Base with same density of axials as shoulder, approximately 22 spiral lines. Anomphalous. Operculum corneous, covering aperture, central nucleus, multispiral. External anatomy and rachidian tooth plus lateral teeth 1–4 given in Montouchet (1972: figs 7–8): typical for scissurellids.

Comparisons. *Scissurella* ? *electilis* from Brazil has an almost planispiral shell with a narrow umbilicus and predominant axial sculpture.

Scissurella morretesi from Brazil has an umbilicate trochiform shell that is wider than tall and distinct reticulate sculpture over the entire shell.

Sinezona insularis from off shore Brazil has a more depressed shell, stronger axial cords, and a closed foramen when fully grown.

Distribution. Brazil (Fig. 44).

Literature records.

Brazil. Paracuru, Ceará, 3.450°S, 39.117°W. Praia Meireles, Fortaleza, Ceará, 3.750°S, 38.583°W Itamaracá, Pernambuco, 7.750°S, 34.750°W. Praia do Rio Vermelho, Slavador Bay, 13.017°S, 38.500°W (Montouchet, 1972).

Remarks. It is virtually impossible to legally obtain scientific loans from Brazil due to bureaucratic hurdles imposed by the Brazilian Government; hence, no specimens of this species in Brazilian collections could be examined, including type material. The species is not known from any specimens in collections elsewhere.

The type specimens are fully decalcified (Dornellas & Simone, 2011).

It is possible that *Sci. alexandrei* is a juvenile of *Sin. confusa*. The overall shell shape and the teleoconch sculpture as indicated on the original drawing are congruent with such an identification. However, the selenizone is too long; at the whorl count of the holotype of *Sci. electilis* (0.6. Fig. 43), *Sin. confusa* has a closed foramen (Figs 309–311). *Sinezona confusa* is mainly known from the Caribbean, but a record from Brazil at 3°S in the same geographic area as the type locality of *Sci. alexandrei* supports that conclusion. Given that the original drawing of *Sci. alexandrei* is rather crude, the type specimen is completely decalcified, and that *Sinezona confusa* is a fairly well-established name, I rather treat *Sci. alexandrei* as a *nomen dubium*.

Figure 44. Distribution of *Scissurella alexandrei*. ☆ type locality. ■ literature record.

Scissurella alto Geiger, 2003

Figures 45–47

Chresonymy.

Scissurella sp.: Raines, 2002: 12–13, fig. 2.
Scissurella alto Geiger, 2003: 27–29, 76, figs 1–2.
Scissurella alto: Geiger, 2004: text-fig. p. 6.
Scissurella alto: Geiger, 2006c: text-fig. 3.
Scissurella alto: Geiger *et al.*, 2008: fig. 12.1J.

Type material. Holotype (LACM 2914. Fig. 45). Paratypes from type locality (LACM 2915, 1: gold-coated on SEM stub. Fig. 46A; AMS C.205034, 1; ANSP 410309, 1; UMUT RM28240, 1; UMUT RM28241, 1; NHMUK 20020267, 2).

Type locality. 10–20 m, off Rosalia Point, E of Anakena, Easter Island, in sand and rubble at base of cliff, Dec. 2000. Leg. B. Raines [27.067°S, 109.333°E] (OD).

Figure 45. *Scissurella alto*. Holotype. 10–20 m, off Rosalia Point, E of Anakena, Easter Island (LACM 2914). Scale bar shell = 1 mm. Scale bar protoconch = 100 μm.

Etymology. The species honors the developers of the first personal computer, the *Alto* (Hiltzik, 1999). This invention made modern phylogenetic analysis possible. Noun in apposition (OD). The developers of the *Alto* were awarded the 2004 *Draper Prize* by the *National Academy of Engineering* of the United States of America, and the *ACM Turing Award* was bestowed upon most of the team members (1992, 2003, 2009).

Figure 46. *Scissurella alto*. **A.** Paratype. 10–20 m, off Rosalia Point, E of Anakena, Easter Island (LACM 2915, 1). **B.** 100 m, Tahai, Easter Island, Chile, 27.140°S, 109.429°W (DLG 1323, 20). **C.** Juvenile shell. 70–100 m, Hanga Roa, Easter Island, Chile, 27.145°S, 109.440°W (DLG 1332, 11). A–B. Scale bars shell = 1 mm. C. Scale bar shell = 0.5 mm. Scale bars protoconch = 100 µm.

Description. Shell medium-sized (to 1.48 mm), globular, 1.08× (holotype) and 1.25× (paratype) wider than high. Protoconch 134 μm (holotype), 140 μm (paratype), slightly sunken, smooth, with strong subterminal varix connecting to embryonic cap, aperture sinusoidal. Teleoconch I of 1.25 whorls, sculpture early on without spirals but with flocculations, after one whorl spiral lines appearing. Selenizone with low, sturdy, blunt keel, lunules coincide exactly with axials on teleoconch II. Teleoconch II of up to 1.25 whorls, shoulder almost flat, suture impressed. Base with distinct constriction below selenizone, base rounded. Strong axial cords predominate, 19 on last whorl, 3–5 irregular fine spiral lines on shoulder, approximately 15 on base, running threadlike over axials. Umbilicus narrow, open, with strong carina at margin, inner wall smooth. Aperture rotund, adumbilical portion thickened at suture and at intersection with umbilical keel, giving square appearance of adumbilical portion of aperture. Selenizone above widest part of shell; slit open, margins parallel at aperture.

Animal unknown.

Specimen records.

Chile, Easter Island. 150–200 m, Tahai, 27.140°S, 109.429°W (DLG 1319, 5). 100 m, Tahai, 27.140°S, 109.429°W (DLG 1323, 20). 30–50 m, Tahai, 27.140°S, 109.429°W (DLG 1324, 20). 30–50 m, Anakena, 27.069°S, 109.324°W (DLG 1331, 1). 70–100 m, Hanga Roa, 27.145°S, 109.440°W (DLG 1332, 11). 65 m, Motu Iti, 27.196°S, 109.450°W (DLG 1347, 3). 150–200 m, SW corner, 27.201°S, 109.463°W (DLG 1350, 2). 100–150 m, La Perouse, 27.077°S, 109.295°W (DLG 1360, 4).

Distribution. Easter Island (Fig. 47).

Comparisons. *Scissurella cebuana* from the Indo-Malayan Archipelago to the Western Pacific has strong knobs on the axial cords and stronger spiral cords on shoulder and base, and a narrower umbilicus.

Scissurella quadrata, from the Indo-Malayan Archipelago to the Central Pacific, has weaker and more numerous axials (47 vs. 19), has a slightly shorter teleoconch I (1–1.125 vs. 1.25 whorls), the keel of the selenizone is fine, has a sharp edge, the lunules do not align with the axials of teleoconch II, and the protoconch has strong axials.

Scissurella spinosa from the Indo-Malayan Archipelago to the Central Pacific has a protoconch with fine axial sculpture and has half as many but much stronger spiral cords forming distinctly elevated points at the intersection with the axial cords.

Juvenile *Sin. ferriezi*, from the Indo-Malayan Archipelago to the Central Pacific, has weaker and more numerous axials (25 in *Sin. ferriezi*, 19 in *Sci. alto*), and more numerous spirals on the shoulder (11 in *Sin. ferriezi*, 3–5 in *Sci. alto*), has a shorter teleoconch I (1.125 vs. 1.25 whorls), and the keel of the selenizone is fine, has a sharp edge, and the lunules do not align with the axials of teleoconch II.

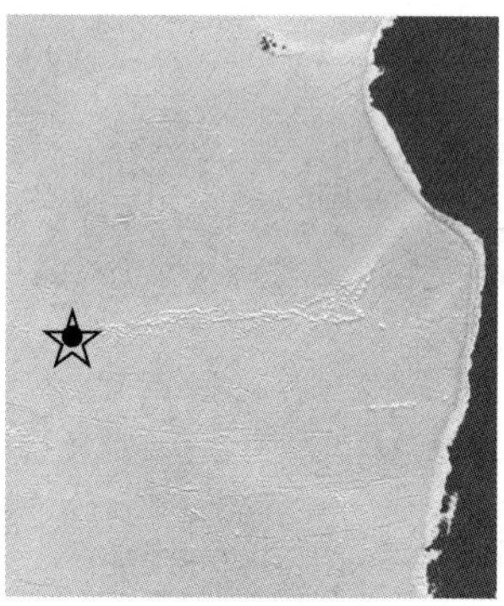

Figure 47. Distribution of *Scissurella alto*. ☆ type locality. ● specimen record.

Scissurella azorensis Nolt, 2008

Figures 48–50

Chresonymy.

Scissurella azorensis Nolt, 2008: 54–57, figs 1, 2A.

Misidentification.

not *Sinezona cingulata*: Martins et al., 2009: 23, fig. 6 [is *Sci. azorensis*].

Type material. Holotype (MNHN Moll 9640. Fig. 48). Paratypes from type locality (MNHN Moll 9641, 2. Fig. 49A). Pedra Furada, Baia do Alcaide, Silveira, Terceira Island, Azores, Portugal, 38.650°N, 27.133°W (MCZ 356976, 1. Fig. 49B).

Type locality. 0–1 m, Santa Maria Island, off Ponta do Marvão, Azores, Portugal. In rocks. June 1990. Leg. S. and C. Gofas. 36°56'N, 25°08'W [36.967°N, 25.117°W](OD).

Etymology. Named for the type locality on the Azores (OD).

Description. Shell small (to 0.75 mm. Figs 48, 49), trochiform, with flat spire, one fourth wider than high, whitish. Protoconch 120 µm long, 1.16–1.2 whorls, 24–28 strong to

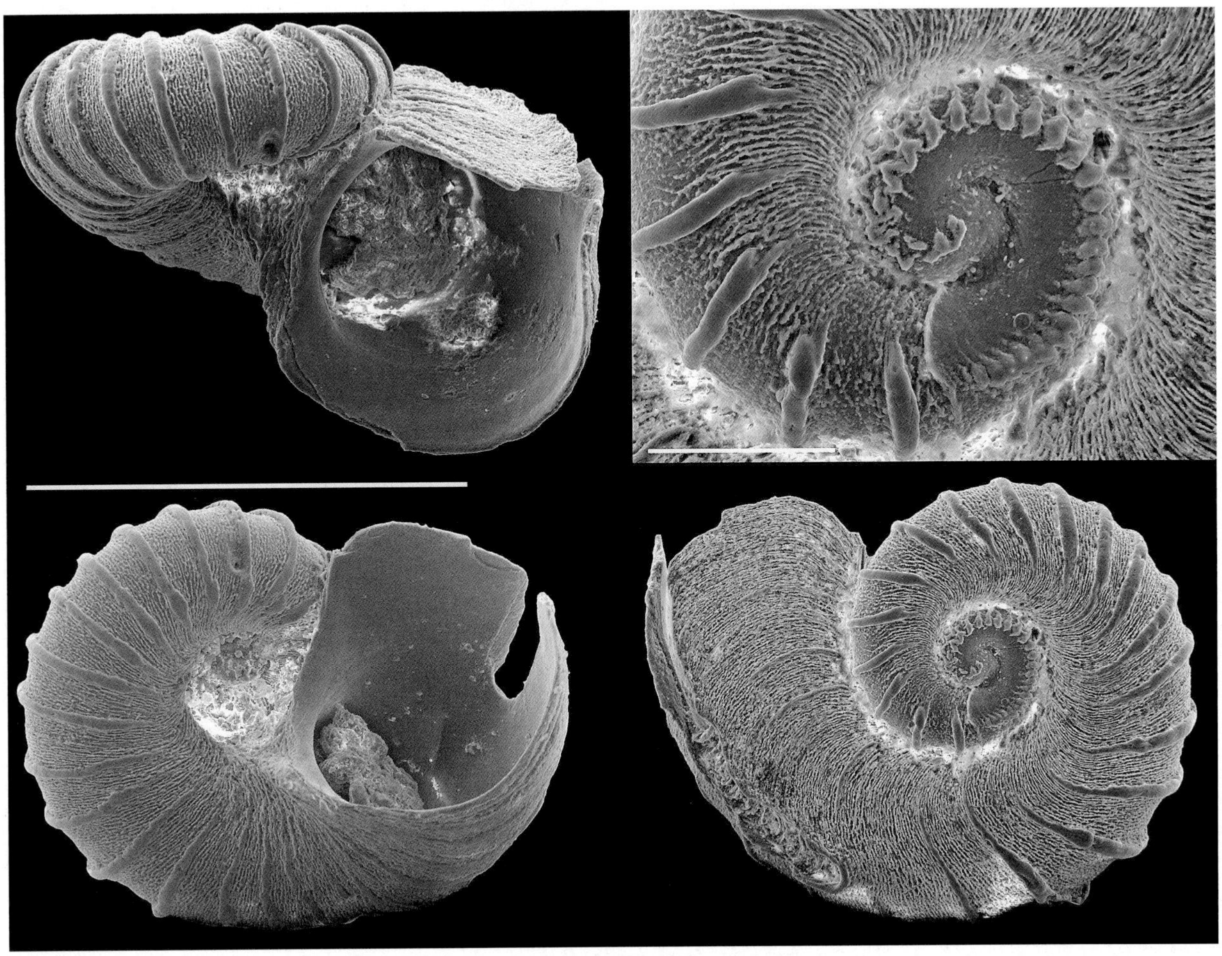

Figure 48. *Scissurella azorensis.* Holotype. 0–1 m, Santa Maria Island, Ponta do Marvão, Azores, 36.967°N, 25.117°W (MNHN 9640). Scale bar shell = 0.5 mm. Scale bar protoconch = 100 µm.

irregular axial cords, axials on outer margin of protoconch not extending full width, no apertural varix, apertural margin rounded. Teleoconch I slightly greater than one whorl, approximately 18–20 strong axial cords, fine irregular lamellae between axials, no spiral sculpture. Teleoconch II 0.75 whorl, with nine weaker axial cords, fine irregular lamellae, no spiral sculpture. Shoulder angled down from coiling axis with fine irregular lamellae. Base rounded, with 18–25 weak axial cords, fine irregular lamellae between axials. Umbilicus open with marginal carina, weak axial

Figure 49. *Scissurella azorensis*. **A.** Paratype: 0–1 m, Santa Maria Island, Ponta do Marvão, Azores, Portugal, 36.967°N, 25.117°W (MNHN 9641, 2). **B.** Paratype: Pedra Furada, Baia do Alcaide, Silveire, Terceira, Azores, Portugal, 38.650°N, 27.133°W (MCZ 356976, 1). **C.** 30 m, Secca del Cammello, Marettimo Island, Egadi Islands, Trapani, Sicily, Italy, 37.989°N, 37.989°E (DLG 1812, 3) Scale bars shell = 200 μm. Scale bars protoconch = 100 μm.

lines on base. Selenizone, slit above periphery, moderately keeled, lunules distinct. Aperture subquadrate, edge slightly projecting.

Animal unknown.

Distribution. Northeastern Atlantic off-shore islands: Azores, Madeira, to Western Mediterranean (Fig. 50).

Specimen records.

Portugal. St. 5.151 [station not listed in van der Land (1987)], Portugal (RMNH, 1). 0 m, Prainha, Pico, Azores, 38.472°N, 28.199°W (NMR 993000070938, 5).

Italy. 30 m, Secca del Cammello, Marettimo Island, Egadi Islands, Trapani, Sicily, 37.989°N, 37.989°E (DLG 1812, 3).

Literature records.

Portugal. 117–145 m, St. 28 off Ribeira das Tainhas, San Miguel, Azores, 37.700°N, 25.417°W (Martin *et al.*, 2009).

Comparisons. The holotype is a mature *Scissurella* and not a juvenile *Sinezona*, because it shows an open slit with the long angled shoulder and descending last quarter whorl.

Scissurella lobini from the Cape Verde Islands and *Scissurella redferni* from the Caribbean have protoconchs with fewer axial cords (24–28 in *Sci. azorensis*, 14–18 in *Sci. lobini* and *Sci. redferni*), the axials are more uniform in shape and the bases of *Sci. redferni* and *Sci. lobini* have spiral threads.

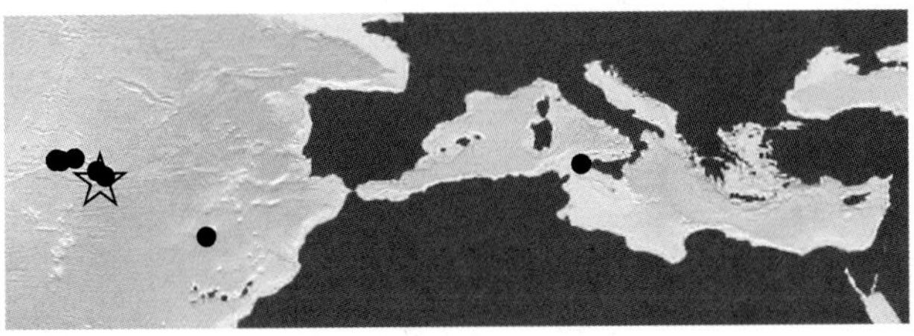

Figure 50. Distribution of *Scissurella azorensis*. ☆ type locality. ● specimen record.

Scissurella costata from the Mediterranean and the adjacent Atlantic has a flattened spire, the apertural margin of the protoconch is sinusoidal, the teleoconch sculpture contains thin irregular axial lines and spiral threads on shoulder and base, and the teleoconch II has approximately 1.3 whorls.

Scissurella maraisorum from the Indian Ocean of South Africa has axial cords that increase in strength with growth (decreasing in *Sci. azorensis*). Otherwise, these two species are exceedingly similar; however, as the subtropical Atlantic *Sci. azorensis* is separated from *Sci. maraisorum* in eastern South Africa by a distinct temperate zone, they thus are very unlikely to be conspecific.

Sinezona cingulata from the Mediterranean and the adjacent Atlantic has a protoconch sculpture with approximately 15 strong axial cords that cover more than half the width, the teleoconch I has fewer then one whorl, with 14 axial cords and the umbilicus is very narrow.

Sinezona semicostata from the Cape Verde Islands has approximately 12 protoconch axial cords over more than half the width and teleoconch I has fewer than one whorl and the foramen is closed in fully grown specimens.

Sinezona confusa from the Caribbean to Brazil has protoconch sculpture with approximately 14 axial cords that cover more than half the width, an apertural varix and spiral threads on teleoconch I and II.

Remarks. Here the second *Scissurella* species is reported from the Mediterranean Sea (Fig. 49C), which is much smaller than the predominant *Sci. costata*, hence, *Sci. azorensis* could be mistaken as a juvenile *Sci. costata*. *Scissurella azorensis* is more discoidal compared to *Sci. costata* of the same size; with a high quality light microscope and careful lighting, the protoconch sculpture can be seen, which readily distinguishes the two species.

Scissurella bountyensis Powell, 1933

Figures 51–54

Chresonymy.

Scissurella bountyensis Powell, 1933b: 34–35, pl. 6, figs 1–2.
Scissurella bountyensis: Powell, 1955: 45.
Scissurella bountyensis: Dell, 1956: 168.
Scissurella bountyensis: Powell, 1979: 34, pl. 13, figs 12–13.
Scissurella bountyensis: Lozouet, 1986: 107.
Scissurella bountyensis: Spencer & Willan, 1995: 13.
Scissurella bountyensis: Freeman *et al.*, 1997: 21.
Scissurella bountyensis: Geiger, 2003: 76.
Scissurella bountyensis: Spencer *et al.*, 2009: 202.

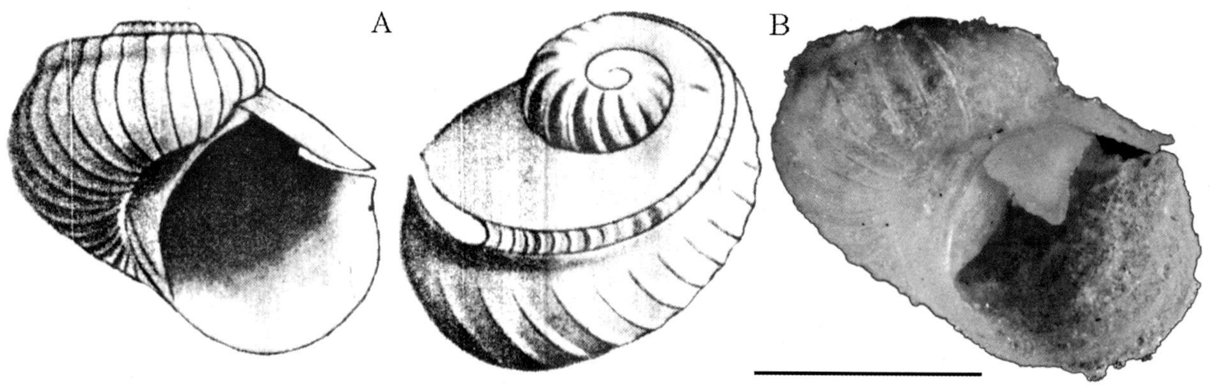

Figure 51. **A.** Original illustration of *Scissurella bountyensis.* **B.** Paratype of *Scissurella bountyensis.* 170 fathoms [= 311 m], off Bounty Islands, [New Zealand, 47.750°S, 179.000°E] (AIM AK 72370). Scale bar = 0.5 mm. Image by and courtesy of Peter Quin/AIM.

Type material. Holotype (CM M5318: Freeman *et al.*, 1997), 0.88 × 0.62 mm (W × H). Paratype from type locality (AIM AK 72370, 1).

Type locality. 170 fathoms [= 311 m], off Bounty Islands, [New Zealand, 47.750°S, 179.000°E] (OD).

Etymology. Named for its type locality on Bounty Islands, New Zealand (OD).

Description. Shell small (to 0.87 mm), trochiform, subquadrate. Protoconch one whorl, strong axial cords, apertural varix not connected to embryonic cap, apertural margin slightly sinusoidal. Teleoconch I of 0.75 whorl, approximately 10–13 strong axial cords, interstices with very fine irregular growth marks, no spirals. Teleoconch II of up to one whorl, suture hardly impressed. Shoulder shallow convex, usually only with fine irregular growth marks, some specimens with few irregular axial cords starting at suture; no spirals. Base with weak constriction below selenizone, axial cords of same strength and density as on teleoconch I, reaching from below constriction below selenizone to carina bordering umbilicus; occasionally very indistinct spiral cords in adumbilical ⅕ of base. Umbilicus narrow, bordered by more or less distinct carina, walls straight, smooth. Aperture subquadrate, roof overhanging. Selenizone above periphery; keels low, moderately strong; lunules distinct; slit open, margins converging, almost touching at apertural margin.

Animal unknown.

Comparisons. *Scissurella prendrevillei* from New Zealand is overall rounder, has a small spur in the basal adumbilical corner of the aperture, the umbilical walls are rounded, the margins of the open slit remain parallel even in mature specimens, and the protoconch has fine axial sculpture. Immature *Sci. bountyensis* also have parallel margins of the slit, however, the basal adumbilical corner of the apertural margin is not drawn out in a spur.

Scissurella skeneoides from New Zealand has a more depressed shell, a longer teleoconch I (>1 vs. 0.75 whorl), and more axial cords per whorl.

Distribution. New Zealand, South Island and Southern off-shore islands (Bounty Islands, Auckland Islands. Fig. 54).

Specimen records.

New Zealand. 37 m, off 30 Fathom Point, inside Five Fingers Peninsula, Fiordland, South Island, 45.720°S, 166.492°E (NMNZ M.150293, 15). 13 m, off outer Gilbert Island, Breaksea Sound, Fiordland, South Island, 45.600°S, 166.667°E (NMNZ M.139032, 9).

Figure 52. *Scissurella bountyensis.* 39 m, Proclamation Island, Bounty Islands, New Zealand, 47.750°S, 179.000°E (NMNZ M.119752, 50). Scale bar shell = 0.5 mm. Scale bar protoconch = 100 µm.

20 m, bay on N side of W end of Bauza Island, Doubtful Sound entrance, South Island, 45.283°S, 166.907°E (NMNZ M.141747, 50). 39 m, Proclamation Island, Bounty Islands, 47.750°S, 179.000°E (NMNZ M.119752, 50). Derry Castle Reef, Enderby Island, Auckland Islands, 50.483°S, 166.303°E (NMNZ M.049661, 2).

Literature records.

New Zealand. 311 m, off Bounty Island, 47.702°S, 179.057°E (Dell, 1956).

Figure 53. *Scissurella bountyensis.* **A–B.** 39 m, Proclamation Island, Bounty Islands, New Zealand, 47.750°S, 179.000°E (NMNZ M.119752, 50). **C.** 20 m, bay on N side of W end of Bauza Island, Doubtful Sound entrance, South Island, New Zealand, 45.283°S, 166.907°E (NMNZ M.141747, 50). Scale bars shell = 0.5 mm. Scale bars protoconch = 100 μm.

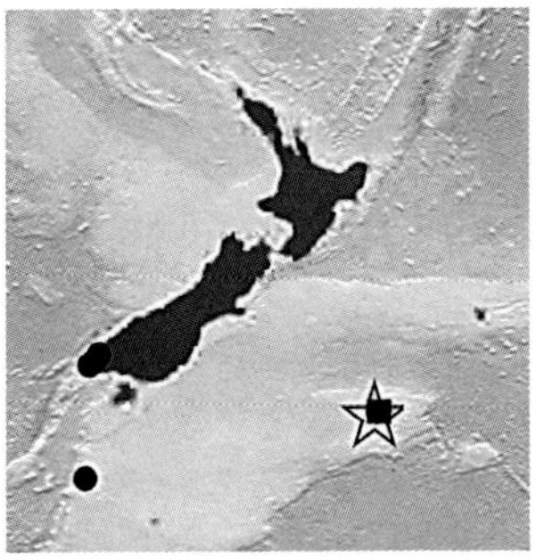

Figure 54. Distribution of *Scissurella bountyensis*. ☆ type locality. ● specimen record. ■ literature record.

Scissurella cebuana (Bandel, 1998)

Figures 55–59, Color Plates 2B, 4C

Chresonymy.

Anatoma cebuana Bandel, 1998: 32–33, pl. 10, figs 3–5.
Scissurella cebuana: Geiger, 2003: 76.
Scissurella cebuana: Geiger, 2008b: fig. 1.

Figure 55. Original illustrations of *Anatoma cebuana*.

Type material. Holotype (SGPIH 3886), 1.6 × 1.6 mm.

Type locality. Ocean side reef slope off Cebu City, Cebu, Philippines [10.283°N, 124.017°E] (OD).

Etymology. Named for its type locality on Cebu, Philippines (OD).

Description. Shell medium-sized (to 1.6 mm), globular, apex depressed; some specimens with approximately four broad brown axial bands on last whorl (Color Plate 2B). Protoconch of 0.75 whorl, smooth, apertural varix connected to embryonic cap, apertural margin deeply sinusoidal. Teleoconch I of 1.25 whorls, 25 axial cords becoming stronger with growth, with three faint spirals, one in position of selenizone, forming strong elevated knobs at intersection with axials. Teleoconch II of up to 1.3 whorls. Shoulder with 21 strong axial cords, much weaker spirals increasing from one to four from onset of selenizone to apertural margin; intersection of axials and spirals forming strong pointed knobs. Base with coordinated axials, first spiral below selenizone forming elevated pointed knobs, approximately six fine spiral threads. Selenizone slightly above periphery, keels slightly slanted towards apex. Umbilicus narrow, bordered by strong carina.

Operculum covering aperture, thin, corneous, multispiral, with central nucleus.

Radula (Fig. 58). Rachidian tooth trapezoidal, tip with 7–9 cusps, central one largest, 3–4 on each side. Lateral teeth 1–3 similar, tip with 3–4 cusps on outer edge of cusp. Lateral tooth 4 reduced, hook-shaped. Lateral tooth enlarged, tip with four distinct cusps on inner edge of tip. Inner marginal teeth, tip with axial cusp largest, 4–5 smaller cusps on each side; outer marginal teeth elongated, tip spoon-shaped with many minute cusps, outermost marginal teeth paddle-shaped, tip with many small cusps, more numerous on inner edge. Radular interlock of central field moderate.

Figure 56. *Scissurella cebuana*. 82 m, Tanguingui Island, Philippines, 11.613°N, 123.765°E (USNM 812112, 1). Radula shown in Figure 58. Scale bar shell = 1 mm. Scale bar protoconch = 100 µm.

Comparisons. *Scissurella evaensis* from the Indo-Malayan Archipelago has a single spiral row of strong spines on the base, but lacks such spines on the shoulder.

Scissurella lorenzi from the Indo-Malayan Archipelago has much less pronounced sculpture, often has axial hyaline and white color bands, and a protoconch with fine axial sculpture.

Figure 57. *Scissurella cebuana*. **A–B.** 105–110 m, Poindemie sector, New Caledonia, 20.817°S, 165.317°E (MNHN, 12). **C.** 92 m, off North West Cape, Western Australia, Australia, 22.553°S, 113.620°E (AMS C.376862, 1). Scale bars shell = 1 mm. Scale bars protoconch = 100 µm.

Scissurella mirifica from the Indo-Malayan Archipelago to the Central Pacific has an overall more globular shell, less distinct, but finer sculpture, and a protoconch with fine axial sculpture.

Scissurella spinosa from the Indo-Malayan Archipelago to the Central Pacific has a shell that is wider than high, has less distinct spines on shoulder and base, and has a protoconch with fine axials.

Scissurella quadrata from the Indo-Malayan Archipelago to the Central Pacific has much weaker, but denser sculpture, and a protoconch with fine axial sculpture.

Scissurella alto from Easter Island has a less elevated spire, much finer spiral threads, and no knobs at the intersection of axials and spirals.

Specimen records.

Oman. 285 m, Gulf of Oman (NMSZ 1961.61, 1).

Japan, Okinawa. 60 m, Onna Village, 26.493°N, 127.842°E (LACM 79-76, 1). 71–85 m, off Kerama Island, 26.159°N, 127.448°E (SBMNH 149469, 3). 211–212 m, off Kerama Island, 26.047°N, 127.375°E (SBMNH 149461, 3).

Philippines. Ocean side reef slope off Cebu City, Cebu, 10.283°N, 124.017°E (SGPIH 3886, 1). 82 m, Tanguingui Island, 11.613°N, 123.765°E (USNM 812112, 1). 60 m, Pamilican Island, 9.502°N, 123.918°E (MNHN, 1: complete). 90 m, off Momo Beach, Panglao Island, 9.608°N, 123.755°E (MNHN, 2: complete). 90–110 m, Balicasag Island, 9.515°N, 123.687°E (MNHN, 1: complete). 88 m, St. 5477, off Tacbuc Point, E Leyte, 10.746°N, 125.208°E (USNM 282905, 1). 83–86 m, off Bolod, Panglao Island, PANGLAO 2004 T44, 9.406°N, 123.815°E (MNHN, 1). 110–112 m, Cervera Shoal, W of Pamilacan Island, PANGLAO 2004 T41, 9.495°N, 123.836°E (MNHN, 1). 80–98 m, Cervera Shoal, W Pamilacan Island, PANGLAO 2004 T40, 9.495°N, 123.846°E (MNHN, 4). 55–81 m, Pamilacan Island, PANGLAO 2004 L65-68, 9.498°N, 123.918°E (MNHN, 2). 90–98 m, Balicasag Island, PANGLAO 2004 L69-73, 9.511°N, 123.682°E (MNHN, 2). 120–139 m, Balicasag Island, PANGLAO 2004 L74–75, 9.511°N, 123.682°E (MNHN, 2). 90–110 m, Balicasag Island, PANGLAO 2004 L46, 9.515°N, 123.686°E (MNHN, 5). 80–90 m, Balicasag Island, PANGLAO 2004 L42, 9.520°N, 123.678°E (MNHN, 3). 80–140 m, Balicasag, PANGLAO 2004 T28, 9.538°N, 123.705°E (MNHN, 8). 83–102 m, Bolod, Panglao Island, PANGLAO 2004 T1, 9.540°N, 123.788°E (MNHN, 7). 152 m, Bolod, Panglao Island, PANGLAO 2004 T2, 9.540°N, 123.796°E (MNHN, 3). 17–25 m, Pntod Lagoon 1, Panglao Island,

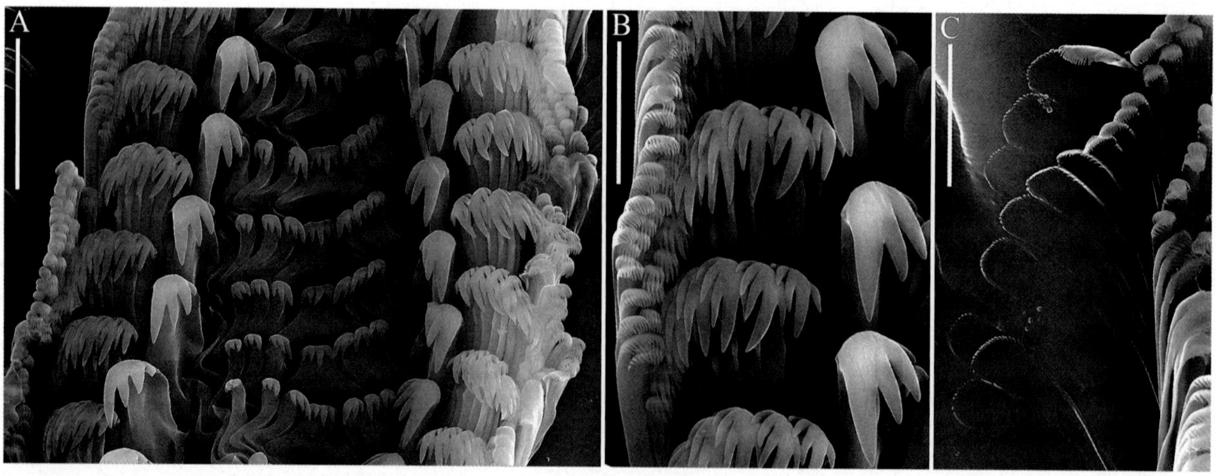

Figure 58. *Scissurella cebuana*. Radula. 82 m, Tanguingui Island, Philippines, 11.613°N, 123.765°E (USNM 812112, 1). Shell shown in Figure 56. **A.** Whole width of radula. Scale bar = 20 μm. **B.** Lateral tooth 5 and marginal teeth. **C.** Outer marginal teeth. Scale bars B, C = 10 μm.

PANGLAO 2004 B39, 9.546°N, 123.701°E (MNHN, 1). 82 m, Bolod, Panglao Island, PANGLAO 2004 T4, 9.550°N, 123.808°E (MNHN, 3). 34–82 m, W of Baclayon, Bohol Island, PANGLAO 2004 T6, 9.585°N, 123.853°E (MNHN, 4). 4–30 m, Tatarman, Panglao Island, PANGLAO 2004 B7, 9.598°N, 123.863°E (MNHN, 1). 90–200 m, Maribohoc Bay, Bohol Island, PANGLAO 2004 P1, 9.600°N, 123.750°E (MNHN, 9). 120 m, off Momo Beach, Panglao Island, PANGLAO 2004 L50, 9.615°N, 123.763°E (MNHN, 5). 43–62 m, Bingag/Tabalong, Panglao Island, PANGLAO 2004 L51-60, 9.628°N, 123.800°E (MNHN, 1). 78–95 m, Maribohoc Bay, Bohol Island, PANGLAO 2004 T11, 9.681°N, 123.833°E (MNHN, 2).

Indonesia. 100 m, E of Melolo, NE coast of Sumba, 9.883°S, 120.717°E (RMNH, 1). 365 m, S of Kaledupa Reef, Banda Sea, Tukang Besi Islands, 5.963°S, 123.775°E (RMNH, 1). 65 m, Teluk Slawi, E of Komodo, 8.613°S, 119.512°E (RMNH, 1). 175 m, SW Salayer, 6.373°S, 120.437°E (RMNH, 2). 250–290 m, NW of Baningko, Banda Sea, Tukang Besi Islands, 5.875°S, 123.975°E (RMNH, 1).

Australia. 92 m, off North West Cape, Western Australia, 22.553°S, 113.620°E (AMS C.376862, 1).

Papua New Guinea. 65 m, off Susan's Reef, Kimbe Bay, New Britain, 5.291°S, 150.128°E (DLG 734, 5).

Solomon Islands. 35 m, midway between Karumaneke Island and Cheke Village, Marovo Lagoon, Vangunu Island, 8.517°S, 158.133°E (NMNZ M.230054, 1). 35 m, midway between Karumaneke Island and Cheke Village, Marovo Lagoon, Vangunu Island, 8.522°S, 158.128°E (NMNZ M.230054, 1).

New Caledonia. 455 m, southern, 22.267°S, 166.183°E (MNHN, 1). 381–469 m, Norfolk Ridge, 23.367°S, 168.017°E (MNHN, 1). 45–56 m, Touho sector, 20.775°S, 165.263°E (MNHN, 2). 80–140 m, Touho sector, 20.817°S, 165.317°E (MNHN, 1). 105–110 m, Poindemie sector, 20.817°S, 165.317°E (MNHN, 12). 140–150 m, E coast, BATHUS 1 DE692, 20.583°S, 164.983°E (MNHN, 2). 250–350 m, St. 40, 22.500°S, 166.400°E (MNHN, 2).

Vanuatu. 85–100 m, off Aneityum Island, 20.267°S, 169.850°E (AMS C.404505, 9: complete). 360–419 m, MUSORSTOM 8 DW1065, 16.266°S, 167.350°E (MNHN, 2). 90–93 m, SW Maéa Island, SANTO 2006 EP38, 15.394°S, 167.221°E (MNHN, 2). 102–120 m, W Mavéa Island, Palikulo Bay, SANTO 2006 AT14, 15.397°S, 167.220°E (MNHN, 2). 45–50 m, Aoré Island, Aimbuei Bay, SANTO 2006 FP46–49, 15.540°S, 167.211°E (MNHN, 2). 100–105 m, NW Tutuba Island, SANTO 2006 DS99, 15.541°S, 167.281°E (MNHN, 2). 114–196 m, NE Aoré Island, SANTO 2006 AT85, 15.543°S, 167.261°E (MNHN, 1). 20–60 m, E Aoré Island, Aimbué Bay, 15.553°S, 167.209°E (MNHN, 1). 125–156 m, W Tutuba Island, SANTO 2006 EP40, 15.555°S, 167.274°E (MNHN, 1). 70–80 m, W Tutuba Island, SANTO 2006 DS103, 15.568°S, 167.266°E (MNHN, 1). 10–51 m, E Aoré Island, off Aimbué Bay, SANTO 2006 EP35, 15.583°S, 167.233°E (MNHN, 1). 45–101 m, E Aoré Island, SANTO 2006

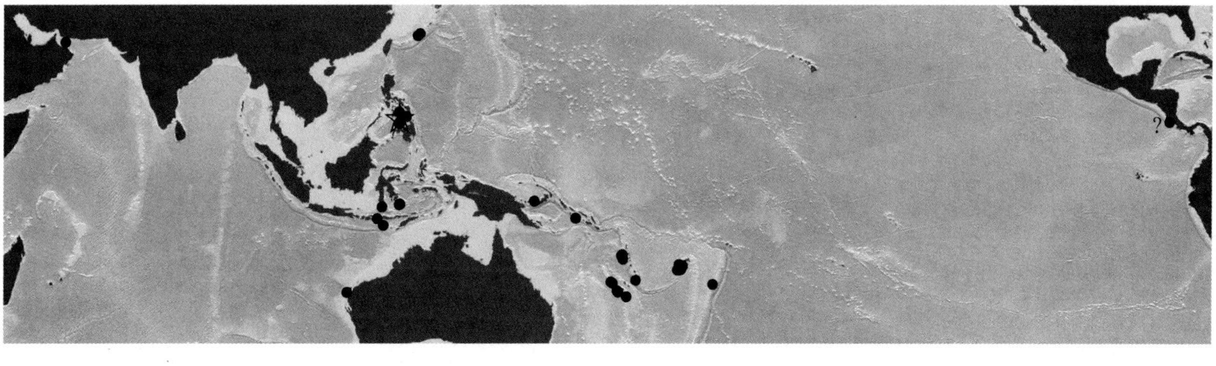

Figure 59. Distribution of *Scissurella cebuana*. ☆ type locality. ● specimen record.

EP10, 15.604°S, 167.150°E (MNHN, 1). 108–121 m, W Urélapa Island, SANTO 2006 EP24, 15.608°S, 167.021°E (MNHN, 9). 25 m, Bruat Channel, N coast of Malo Island, Palikulo Bay, SANTO 2006 DB77FP50, 15.613°S, 167.145°E (MNHN, 10). 78–91 m, NW coast of Malo Island, SANTO 2006 EP22, 15.622°S, 167.098°E (MNHN, 16). 112 m, NW coast of Malo Island, SANTO 2006 EP41, 15.628°S, 167.085°E (MNHN, 1). 99 m, NW coast of Malo Island, SANTO 2006 EP21, 15.628°S, 167.086°E (MNHN, 11). 80–94 m, NW coast of Malo Island, SANTO 2006 EP19, 15.629°S, 167.089°E (MNHN, 22). 90–110 m, NW coast of Malo Island, Palikulo Bay, SANTO 2006 DB77 EP28, 15.633°S, 167.083°E (MNHN, 7). 85–90 m, 13.917°N, 120.333°E (MNHN, 4).

Fiji. 149–168 m, S of Viti Levu, 18.207°S, 178.552°E (MNHN, 20). 656–660 m, Bligh Water, 17.268°S, 178.247°E (MNHN, 1). 144–150 m, S of Viti Levu, 18.207°S, 178.550°E (MNHN, 14). 260–305 m, S of Viti Levu, 18.308°S, 178.097°E (MNHN, 11). 275–430 m, S of Viti Levu, 18.297°S, 177.907°E (MNHN,10). 392–433 m, S of Viti Levu, 18.185°S, 178.390°E (MNHN, 15). 879–897 m, SE of Viti Levu, 17.515°S, 178.888°E (MNHN, 2).

Tonga. 433 m, NW Tongatapu, BORDAU 2 DW1569, 21.033°S, 175.316°W (MNHN, 1).

Costa Rica. 15–21 m, Ballena Bay, W side Nicolay Gulf, Puntarenas Province, 9.733°N, 84.995°W (LACM 84-152, 1) [problematic record].

Distribution. Indo-Malayan Archipelago. One specimen from Oman and one from Costa Rica (?) (Fig. 59).

Remarks. Bandel (1998) classified the species in *Anatoma.* The depressed apex and overall shape of the shell rather seem to be indicating a placement in *Scissurella*. The radula with the broadened lateral tooth 5 confirms a placement in *Scissurella*.

Scissurella clathrata Strebel, 1908

Figures 60–66

Chresonymy.

Scissurella clathrata Strebel, 1908: 77–78, pl. 6, figs 84a–d.
Scissurella clathrata: Thiele, 1912a: 9, pl. 1, figs 18–19 [copy figures Strebel, 1908].
Schizotrochus clathrata [*sic*]: Caracelles & Williamson, 1951: 251.
Scissurella clathrata: Powell, 1960: 126.
Anatoma clathrata: Lozouet, 1986: 112.
Scissurella clathrata: Castellanos & Landoni, 1988: 7, pl. 1, fig. 3.

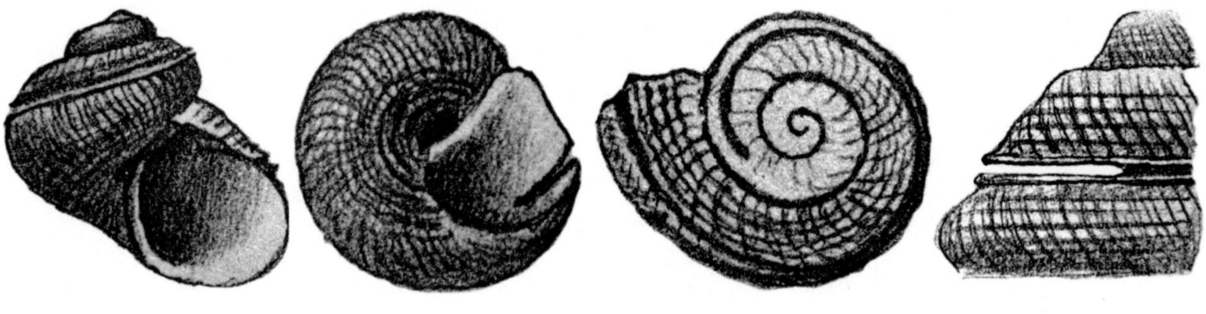

Figure 60. Original illustrations of *Scissurella clathrata* in Strebel (1908).

Scissurella clathrata: Linse, 1997: 26.

Scissurella clathrata: Linse & Brandt, 1998: 883.

Scissurella clathrata: Linse, 1999: 400.

Scissurella clathrata: Forcelli, 2000: 44, text-fig. [line drawing].

Anatoma clathrata: Linse, 2002: 64–67, fig. 4.2-2 [map], pl. 1, figs 9.1.1.5–9.1.1.8 [shell, radula].

Scissurella/Anatoma ? clathrata: Geiger, 2003: 76.

Scissurella clathrata: Zelaya & Geiger, 2007: 417–422, figs 91 [map], 93–115 [shell SEM, copy figures Strebel (1908)], 116–118 [external anatomy], 120–123 [radula].

Scissurella clathrata: Zelaya & Geiger, 2008: figs 1–6.

Scissurella clathrata: Zelaya, 2010: 466, 475, text-figs.

Scissurella clathrata: Zelaya & Gordillo, 2011: 4, fig. 2E [Holocene].

Synonymy.

+ *Scissurella eucharista* Melvill & Standen, 1912: 344, pl., figs 1–1a. ***Type material.*** Holotype (NMSZ 1921.143.595. Fig. 63B), 1 × 0.7 mm. ***Type locality.*** 56 fms. [= 102 m], Burdwood Bank, Station 346, Antarctica, 54.417°S, 57.533°E (OD). ***Etymology.*** Eucharistos, Greek for elegant, agreeable (OD).

Schizotrochus eucharista [*sic*]: Caracelles & Williamson, 1951: 251.

Scissurella eucharista: Powell, 1960: 126.

Scissurella eucharista: Castellanos & Landoni, 1988: 6, pl. 1, fig. 2.

Scissurella eucharista: Forcelli, 2000: 44, text-fig. [line drawing].

Scissurella eucharista: Geiger, 2003: 77.

+ *Scissurella malivinarum* Zelaya & Geiger, 2007: 428–430, figs 163–174. ***Type material.*** Holotype (NMSZ 2006110. Fig. 63A). Paratypes from type locality (NMSZ 1992167.60101, 15). ***Type locality.*** 52°25'S 59°03'W, Sea Lion Island, Malvinas [Falkland] Islands (OD). ***Etymology.*** Named after the type locality, the Malvinas [Falkland] Islands (OD).

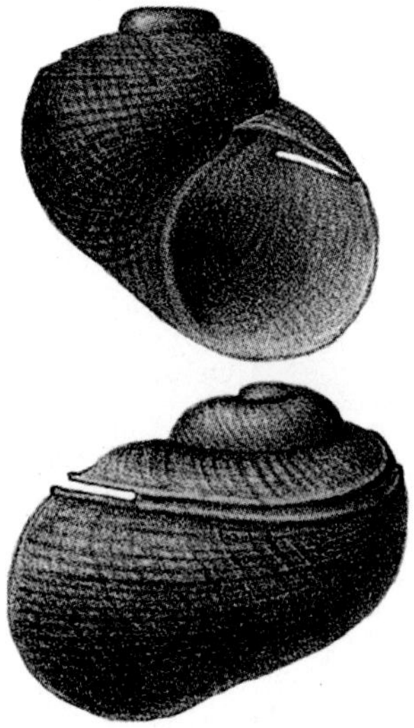

Figure 61. Original illustration of *Scissurella eucharista.*

Misidentification.

Scissurella obliqua: Geiger, 2003: fig. 16e [shell, radula. As *Sin. obliqua* in caption. Is *Sci. clathrata*].

Type material. Holotype (missing from SMNH), 1.7 × 1.6 mm (W × H).

Type locality. 36 m, Station 3, Feuerland [= Tierra del Fuego, Argentina] (54°43'S 64°8'W), boulders and gravel (OD).

Etymology. Referring to cancellate sculpture.

Description. Shell moderate size (to 1.4 mm), trochiform, inflated. Protoconch of one whorl, sculpture flocculent, no apertural varix, apertural margin convex. Teleoconch I of 1–1.125 whorls, approximately 24 low, indistinct axial cords, crossed by 0–6 obscure spiral lines; interstices with fine axial growth marks. Teleoconch II of slightly more than one whorl in fully grown specimens (1.1 mm). Shoulder somewhat convex, suture impressed, approximately same density and strength of axials on teleoconch I, crossed by approximately 4–6 irregularly spaced, somewhat smaller, indistinct cords; interstices with fine irregular growth marks. Base with slight constriction below selenizone, same density and strength of axials as on shoulder, approximately 18 spiral cords, adumbilical 3–4 becoming more distinct, forming strong spiral cords bordering umbilicus. Umbilicus narrow, walls straight, with fine axial growth marks;

no funiculus. Selenizone above periphery, keels moderately strong, low; lunules indistinct; slit open, with parallel margins. Aperture rounded obliquely D-shaped, roof overhanging.

Operculum (Fig. 65C) round, corneous, thin, multispiral, with central nucleus.

Anatomy. Color pale cream, uniform. Pallial cavity with two large bipectinate ctenidia, left one of about 10% larger. Cephalic tentacles short, flat, foliated, papillate. Eyestalks short, fused at base of cephalic tentacles; eyes black, lens half of eye in diameter. Three pairs of conical, relatively long, papillate epipodial tentacles; first epipodial tentacle divided basally forming two tentacles, with posterior branch bifurcated distally; second, third tentacles simple. Foot long, narrow; foot sole completely ciliated. Columellar muscle thick, triangular in outline. Single small tentacle located at posterio-dorsal part of mantle cavity. Salivary glands elongate, larger than odontophore. Intestine very long, with hindgut forming wide inverted "S."

Radula (Fig. 65A–B). Rachidian tooth triangular, tip with seven cusps, central one largest. Lateral teeth 1–3 similar, three cusps on outer margin of tip. Lateral tooth 4 reduced, central cusps on tip largest, one small cusp on each side. Lateral tooth

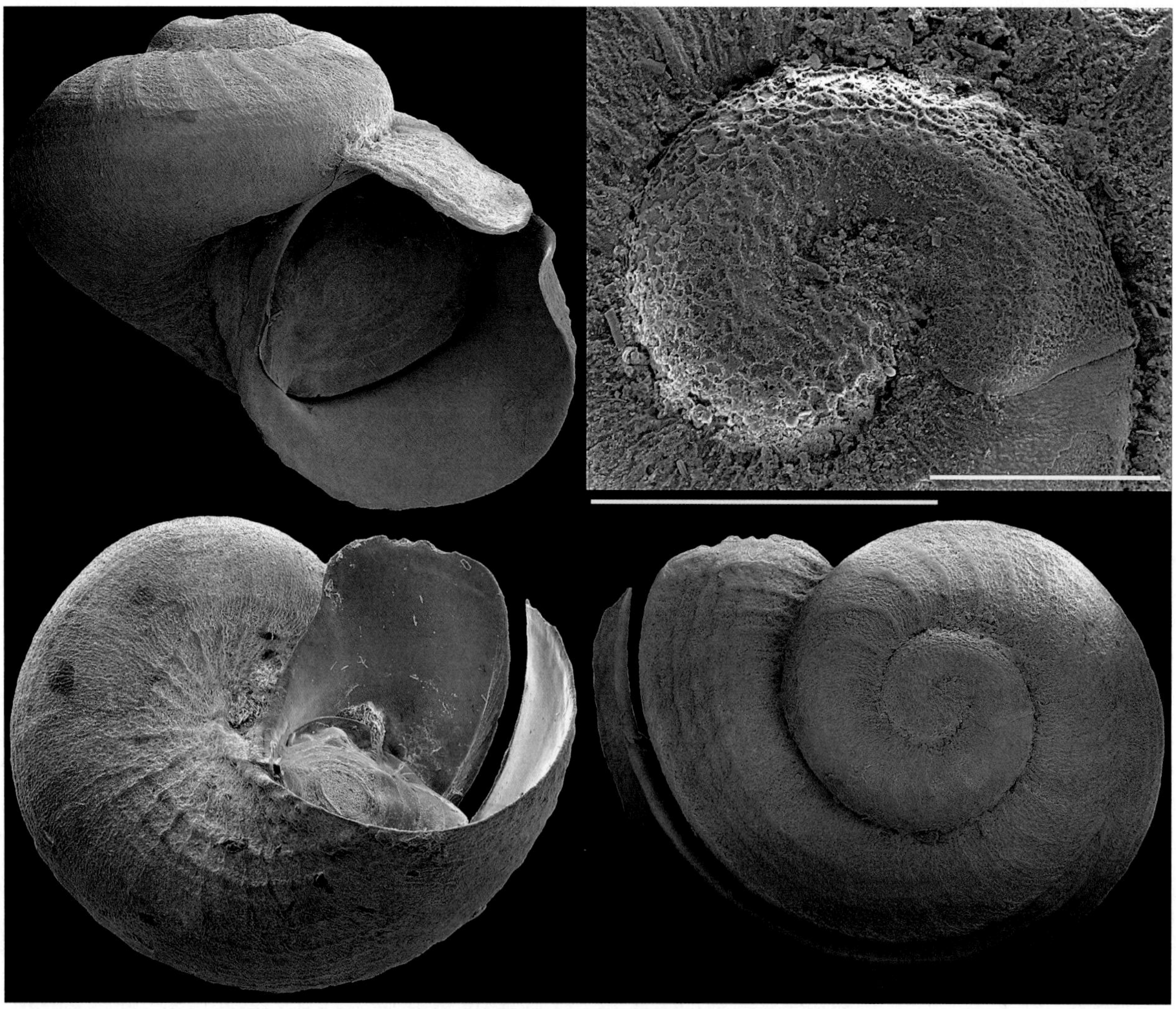

Figure 62. *Scissurella clathrata*. 18 m, Valparaiso Point, Canal Cockburn, Chile, 54.370°S, 71.362°W (LACM 73-69, 13). Scale bars shell = 0.5 mm. Scale bars protoconch = 100 μm.

5 broadened, seven cusps on inner margin of tip. Radular interlock of central field moderate. Inner marginal teeth with three cusps on inner margin of tip, terminal cusp largest, four cusps on outer margin. Outer marginal teeth spoon-shaped.

Figure 63. *Scissurella clathrata.* **A.** Holotype of *Scissurella malvinarum.* 52°25'S 59°03'W, Sea Lion Island, Malvinas [Falkland] Islands (NMSZ 2006110). **B.** Holotype of *Scissurella eucharista.* 56 fms. [= 102 m], Burdwood Bank, Station 346, Antarctica, 54.417°S, 57.533°E (NMSZ 1921.143.595), 1.2 mm wide. Uncoated specimen, 3 kV, frame integration **C.** 73 m, 5 mi N Observatorio Island, Staten Island, Tierra del Fuego, Argentina, 54.573°S, 64.167°W (LACM 71-316, 4). Scale bars shell = 0.5 mm. Scale bars protoconch = 100 µm.

Comparisons. *Scissurella obliqua* from the subantarctic region has a deeper umbilicus, the selenizone is more displaced towards periphery, lacks spiral cords surrounding the umbilicus, and the protoconch ornamentation has anastomosing zig-zag axials.

Figure 64. *Scissurella clathrata*. **A.** 9 m, Canal Darwin, Chile, 45.463°S, 74.413°W (LACM 73-74, 29). Juvenile. **B.** 22 m, Westhoff Island, Chile, 43.900°S, 73.725°W (LACM 73-75, 40). Radula shown in Figure 65. **C.** 3–6 m, E side Laitec Island, off SE end of Chiloe Island, Chiloe Province, Chile, 43.233°S, 73.600°W (LACM 75-46, 20). Scale bars shell = 0.5 mm. Scale bars protoconch = 100 µm.

Scissurella supraplicata from the subantarctic islands has a protoconch with spiral sculpture, has strong axial cords on teleoconch I and lacks the spiral cords surrounding the more narrow umbilicus.

Sinezona subantarctica from the subantarctic has a closed foramen, the umbilicus is covered by a callus, and has more developed shell sculpture.

Sinezona chilensis from southwestern South America has a protoconch with fine axial sculpture, fewer but more distinct axial cords on the teleoconch (~12), a much broader umbilicus, and in fully grown specimens, the slit closes to a foramen.

Scissurella georgica from Antarctica has a much flatter shell, a wider umbilicus, and a protoconch with axial sculpture.

Distribution. Magellanic Province, 0–110 m, living over entire bathymetric range (Fig. 66).

Specimen records.

Great Britain. Sea Lion Island, Malvinas [Falkland] Islands, 52.417°S, 59.050°W (NMSZ 2006110, 1; NMSZ 1992167.60101, 15: type material of *Sci. malvinarum*).

Figure 65. *Scissurella clathrata*. Radula and operculum. **A.** Central field of radula enlarged. 15 m, Tom Bay, Chile, 50.188°S, 74.798°W (LACM 73-72, 80). Scale bar = 10 µm. **B–C.** 22 m, Westhoff Island, Chile, 43.900°S, 73.725°W (LACM 73-75, 40). Shell shown in Figure 64B. **B.** Marginal teeth of radula. Scale bar = 20 µm. **C.** Operculum. Scale bar = 0.5 mm.

Argentina. 303–358 m, 5 mi SE Cabo Kempe, Staten Island, Tierra del Fuego, 54.920°S, 64.340°W (USNM 845411, 1). 144 m, 8 mi E Ensenada Patagones, SE Tierra del Fuego, 54.867°S, 65.083°W (USNM 845399, 1). 31 m, Puerto Vancouver, Staten Island, Tierra del Fuego, 54.780°S, 64.067°W (LACM 71-259, 3). 31 m, Staten Island, Puerto Vancouver, Tierra del Fuego, 54.780°S, 64.067°W (USNM 845407, 1). 135–137 m, Staten Island, Tierra del Fuego, 54.652°S, 63.835°W (USNM 845401, 1). 73 m, 5 mi N Observatorio Island, Staten Island, Tierra del Fuego, 54.573°S, 64.167°W (LACM 71-316, 4). 73 m, 5 mi N Overvatorio Island, Staten Island, Tierra del Fuego, 54.567°S, 64.167°W (USNM 845412, 2). 118 m, 9 mi N Cabo San Juan, E end Staten Island, Tierra del Fuego, 54.567°S, 63.833°W (LACM 71-327, 4). 116–120 m, 18 mi N Crossley Bay, Staten Island, Tierra del Fuego, 54.488°S, 64.673°W (USNM 845413, 1). 303–358 m, 5 mi SE Cabo Kempe, Staten Island, Tierra del Fuego, 54.920°S, 64.340°W (LACM 71-340, 3). 342–353 m, 2.5 mi S Fallows Point, Staten Island, Tierra del Fuego, 54.833°S, 63.842°W (LACM 71-331, 1). 0 m, W side Buen Suceso Bay, SE Tierra del Fuego, 54.797°S, 65.267°W (LACM 71-270, 3). 0 m, W side Buen Suceso Bay, SE Tierra del Fuego, 54.797°S, 65.267°W (LACM 71-270, 3, 8: complete). 73–76 m, 9 mi N Cabo San Antonio, Staten Island, Tierra del Fuego, 54.567°S, 64.500°W (LACM 71-352, 16). 116–120 m, 18 mi N Crossley Bay, Staten Island, Tierra del Fuego, 54.488°S, 64.673°W (LACM 71-347, 1).

Chile. 40 m, Wollaston Island, 55.642°S, 67.214°W (ZMH 19591, 127, 39: complete). 33 m, SE Picton Island, Beagle Channel, 55.127°S, 66.743°W (ZMH 19593 3: complete). 25 m, Picton Point, Beagle Channel, 55.122°S, 66.880°W (ZMH 19589, 10, 26: complete). 63 m, SE Picton Island, Beagle Channel, 55.115°S, 66.666°W (ZMH 19592, 204, 218: complete). 110 m, Picton Island, Beagle Channel, 55.114°S, 66.926°W (ZMH 19590, 1). 65–80 m, Beagle Channel, 55.100°S, 66.483°W (MLP 12080, 3, 6: complete). 30–35 m, Beagle Channel, 55.050°S, 66.617°W (MLP 12079, 28, 22: complete). 15–20 m, Beagle Channel, 55.033°S, 66.700°W (MLP 12078, 1, 1: complete). 103 m, Gardiner Island, Beagle Channel, 55.009°S, 66.886°W (ZMH 19594 1: complete). 30–40 m, Beagle Channel, 54.950°S, 67.017°W (MLP 12077 15: complete). 15–20 m, Beagle Channel, 54.917°S, 68.350°W (MLP 12076, 6, 6: complete). 7–13 m, Beagle Channel, 54.867°S, 68.233°W (MLP 12072 3: complete). 12–20 m, Beagle Channel, 54.867°S, 68.217°W (MLP 12068, 1). 1 m, Beagle Channel, 54.867°S, 68.117°W (MLP 12074 1: complete). 5–10 m, Beagle Channel, 54.850°S, 68.317°W (MLP 12071, 2). 5–10 m, Beagle Channel, 54.850°S, 68.317°W (MLP 12073, 1). 27 m, Beagle Channel, 54.850°S, 68.033°W (MLP 12075 2: complete). 5 m, Beagle Channel, 54.833°S, 68.333°W (MLP 12070, 4). 18–21 m, Beagle Channel, 54.833°S, 68.250°W (MLP 12069 1: complete). 17 m, Conett Island, 54.724°S, 64.238°W (LACM 73-66, 1). 18 m, Valparaiso Point, Cockburn Canal, 54.370°S, 71.362°W (LACM 73-69, 13). 0 m, Santa Ana Point, Fuerte Bulnes, Peninsula Brunswick, Magellan Strait, 53.633°S, 70.908°W (LACM 75-45, 2). 0 m, Puerto el Hambre, cove S of Askew Point, Peninsula Brunswick, Magellan Strait, 53.617°S, 70.933°W (LACM 75-49, 3). 90 m, Laredo Bay, Magellan Strait, 52.958°S, 70.683°W (ZMH 19587 1: complete). 92–101 m, Eltanin Station 958, 52.933°S, 75.000°W (NMNZ M.242830, 3). 64 m, Eltanin Station 960, 52.667°S, 74.967°W (NMNZ M.242829, 2). 13 m, Dashwood Point, Canal Smyth, 52.400°S, 73.662°W (LACM 73-71, 11). 15 m, Tom Bay, 50.188°S, 74.798°W (LACM 73-72, 80). 1 m, San Andres Bay, N Gulf de Peñas, 46.588°S, 75.510°W (LACM 73-73, 13). 9 m, Darwin Canal, 45.463°S, 74.413°W (LACM 73-74, 29). 22 m, Westhoff Island, 43.900°S, 73.725°W (LACM 73-75, 40). 3–6 m, E side Laitec Island, off SE end of Chiloe Island, Chiloe Province, 43.233°S, 73.600°W (LACM 75-46, 20). 3–15 m, Nihuel Islet, Gulf of Corcovado, Chiloe Province, 42.633°S, 72.950°W (LACM 75-43, 4). 50–70 m, Inutil Bay, Tierra del Fuego, 53.517°S, 69.811°W (DLG 2018, 4).

Figure 66. Distribution of *Scissurella clatrata*. ☆ type locality. ● specimen record. ■ literature record.

Literature records.

Chile. 40 m, St. 1200, Wollaston Island, 55.642°S, 67.214°W. 33 m, St. 1221, SE Picton Island, 55.127°S, 66.743°W. 25 m, St. 1176, Rico Point, 55.122°S, 66.883°W. 25 m, St. 1178, Rico Point, 55.122°S, 66.880°W. 67 m, St. 1216, SE Picton Island, 55.120°S, 66.670°W. 63 m, St. 1213, SE Picton Island, 55.115°S, 66.666°W. 110 m, St. 1184, Picton Island, 55.114°S, 66.926°W. 103 m, St. 1237, Gardiner Island, 55.009°S, 66.919°W. 92 m, St. 972, Estrecho, Ancho Point, St. 16, 53.480°S, 70.365°W. 90 m, VH St. 813, Estrecho, Laredo, 52.958°S, 70.683°W (Linse, 2002).

Argentina. Rio Ovado, Tierra del Fuego, 54.850°S, 68.583°W (Zelaya & Gordillo, 2011: CEGH-UNC 22447, 1)

Remarks. The holotype of *Scissurella clathrata* could not be located within SMNH, although SMNH records place it there (K. Sindemark, 3/2006, pers. comm.), nor ZSM (D. Zelaya, 9/2006, pers. obs.), where other material of Strebel is housed. However, the species is clearly recognizable from the original description and figures; hence, no neotype designation is necessary.

Scissurella eucharista is a synonym of *Sci. clathrata*. The original illustration shows an anomphalous species, though the holotype clearly has an open umbilicus. The sculpture is somewhat weaker than in typical *Sci. clathrata*, but that is due to the eroded condition of the type specimen. The original description indicated a specimen of 1.7 mm. Specimens of that size have not been recorded since, the largest being 1.1 mm. It is likely that the original dimensions were in error.

Scissurella malvinarum is here synonymized under *Sci. clathrata*. The shared attributes include all aspects of protoconch morphology, number of teleoconch I whorls, and overall shape. The sculpture of *Sci. malvinarum* has a superficially smooth appearance, but that is due to erosion/abrasion. Remnants of the original sculpture are preserved near the suture and around the umbilicus. Those sculptural remnants also agree with the sculpture of *Sci. clathrata*. The type localities of the two species are only some 500 km apart. The record of the types of *Sci. malvinarum* constitute the eastern-most occurrence of the species. Zelaya & Geiger (2007) compared *Sci. malvinarum* only to *Sci. obliqua* and *Sin. subantarctica*, but not *Sci. clathrata*.

Linse (2002) noted a smooth protoconch, though her figure 9.1.1–7 shows evidence of erosion. The smooth protoconch was used to place *clathrata* in *Anatoma*. Geiger (2003) showed that protoconch sculpture is very variable in Scissurellidae s.l. and is not suitable for generic diagnoses in Scissurellidae and is problematic in Anatomidae. The radula illustration of Linse (2002: pl. 1, fig. 9.1.1–8) shows the broad lateral tooth 5 typical for Scissurellidae, and not the elongated lateral tooth 5 typically found in Anatomidae.

Linse (2002) reported *Sci. clathrata* from 53°28.8'S 70°21.9'W, Estrecho, P. Ancho, 92 m. Material from this locality identified as *Sci. clathrata* by Linse (ZMH 19588) is here re-identified as *An. conica*.

A specimen of *Sci. clathrata* was found at raised beaches from Rio Ovando, Southern Tierra del Fuego. This area was dated in 4,100 years before present, and is located 3 m over the present sea level (Zelaya & Gordillo, 2011).

Scissurella coronata Watson, 1886 [*nomen dubium*]

Figures 67–68

Chresonymy.

Scissurella coronata Watson, 1886: 114–115, pl. 8, fig. 4.
Scissurella coronata: Kobelt, 1887: 428.
Scissurella coronata: Pilsbry, 1890: 56, pl. 68 [erroneously pl. 58 in text], figs 11–13.
Scissurella coronata: Thiele, 1912a: 19–20, pl. 3, figs 2–3 [copy figures Watson, 1886].
? *Scissurella* cf. *coronata*: Robertson, 1969: 2.
Anatoma coronata: Higo, 1973: 13.
Scissurella coronata: Richard, 1985: 414.
Scissurella coronata: Lozouet, 1986: 107.
Scissurella coronata: Tröndlé, 1986: 69.
Scissurella coronata: Kay & Johnson, 1987: 106 [Holocene of Enewetak].
Scissurella coronata: Higo & Goto, 1993: 14.
Scissurella coronata: Habe & Tuchyia, 1998: 17 [could refer to *Sin. plicata*].
Scissurella coronata: Jansen, 1999: 51, figs 22–24.
Scissurella coronata: Geiger, 2003: 76.
Scissurella coronata: Williams & Ozawa, 2006: 36 et seq. [molecular phylogeny]: Japan specimen.
? *Scissurella coronata*: Tröndlé & Boutet, 2009: 12.

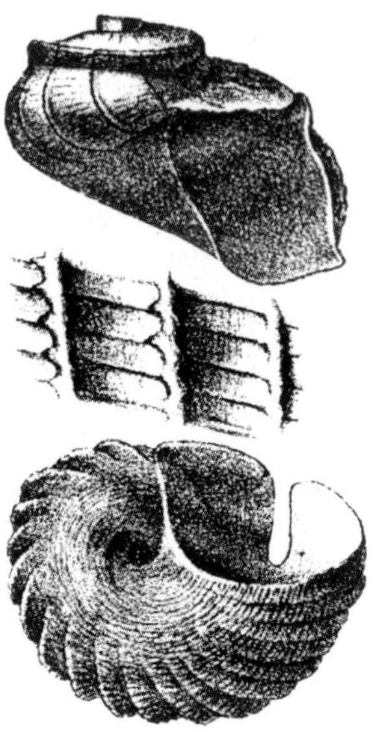

Figure 67. Original illustration of *Scissurella coronata*.

Misidentifications.

not *Schismope coronata*: Dall, 1921: 183 [is *Sin. rimuloides*].
not *Scissurella coronata*: Habe, 1963: 229 [Japanese], 233 [English], text-figs 3–4 [is *Sin. plicata*; slightly juvenile with open slit].
not *Scissurella coronata*: Okutani & Hasegawa, 2000: 37, pl. 18, fig. 10 [is *Sin. plicata*; slightly juvenile with open slit].
not *Scissurella coronata*: Ladd, 1966: 27, pl. 2, figs 7–8 [USNM specimen examined 1/2007 is *Sin. ferriezi*].
not *Scissurella "coronata"*: Geiger & Thacker, 2005: 49, fig. 1 [phylogeny. Is *Sci. spinosa*].

Type material. Syntype (NHMUK 1877.2.9.403, 1: destroyed by Byne's disease), 0.048" × 0.065" [= 1.9 × 2.6 mm]. Syntype (NMW 1955.158.601, 1. Fig. 68: see Remarks).

Type locality. 20 fathoms [= 36 m], Tahity harbour [=? Papeete, Tahiti, French Polynesia, 17.533°S, 146.567°W], near the reefs (OD).

Etymology. Corona, Latin for crown.

Remarks. The NMW syntype is certainly not conspecific with Watson's (1886) species, as already noted by A. Janssen in 1995. It appears to be a *Satondella* sp., although the apertural margin with the diagnostic shape of the foramen is severely broken. Given that the NHMUK specimen is destroyed, the NMW species is certainly not conspecific with Watson's (1886) figure, which does not permit unambiguous identification of Watson's species, I elect to treat *Sci. coronata* as a *nomen dubium*. His figure shows in apertural view noticeably fewer axial cords than in the umbilical view (cf. Fig. 67). Hence, the species may have been based on a mixture of species,

Figure 68. NMW 1955.158.601 syntype of *Scissurella coronata*, see Remarks for discussion. Scale bar shell = 1 mm. Scale bar protoconch = 100 μm/

and with the last remaining syntype not matching either of the species illustrated, it is unidentifiable.

Scissurella costata d'Orbigny, 1824

Figures 69–82

Chresonymy.

—[C-CII]: Solandi, 1789: 21–22, pl. 15, figs e–f (see Fig. 1).
Scissurella costata d'Orbigny, 1824: 344, pl. 23, fig. 2.
Scissurella costata: Sowerby, 1824: 257.
—: Cantraine, 1840: pl. 6, figs 26, 26a.
Scissurella costata: Philippi, 1853: 34, pl. 6, fig. 2 [copy figure d'Orbigny, 1824].
Anatomus costatus: Adams & Adams, 1853–1854: 439.
Scissurella costata: Chenu, 1859: 364, fig. 2716.

Scissurella costata: Munier-Chalmas, 1862: 394.
Scissurella costata: Weinkauff, 1868: 385.
Scissurella costata: Petit, 1869: 110 [*plicata, dorbignyi, cancellata* as synonyms].
Scissurella costata: Appelius, 1870: 225.
Scissurella costata: Aradas & Benoit, 1870: 156.
Scissurella costata: De Folin, 1871: 266.
Scissurella costata: Monterosato, 1872a: 30.
Scissurella costata var. *laevigata*: Monterosato, 1872a: 30.
Scissurella costata: Monterosato, 1872b: 28.
Scissurella costata: Weinkauff, 1873: 40.
Scissurella costata: Seguenza, 1876: 2 [*fide* Bucquoy *et al.*, 1886].
Scissurella costata: Monterosato, 1875a: 23.
Scissurella costata var. *laevigata*: Monterosato, 1875a: 23.
Scissurella costata: Monterosato, 1877b: 34.
Scissurella costata: Monterosato, 1877a: 416.
Scissurella costata: Monterosato, 1878b: 79.
Scissurella costata: Fischer, 1882: 1201.
Scissurella costata: Marion, 1883a: 38, 46 [*fide* Bucquoy *et al.*, 1886].
Scissurella costata: Marion, 1883b: 85, 90 [*fide* Bucquoy *et al.*, 1886].
Scissurella costata: Jeffreys, 1883a: 89.
Scissurella costata: Paetel, 1883: 47.
Scissurella costata: Grasset, 1884: 136.
Scissurella costata: Monterosato, 1884: 39.
Scissurella costata: Granger: 1885: 154.
Scissurella costata: Bucquoy *et al.*, 1882–1886: 430–432, pl. 51, figs 8–10, fig. 7 [*Sci. costata* var. *cancellata* Jeffreys], fig. 11 [*Sci. costata* var *laevigata* d'Orbigny].
Scissurella costata: Locard, 1886: 332.
Scissurella costata: Paetel, 1888: 289.
Scissurella costata: Kobelt, 1888: 258.
Scissurella costata: Norman, 1888: 21.
Scissurella costata: Monterosato, 1888: 168.
Scissurella costata: Carus, 1889–1893: vol. 2, 238.
Scissurella costata: Pilsbry, 1890: 50, pl. 50, fig. 1.
Scissurella costata: Monterosato, 1890: 143.
Scissurella costata: Vayssière, 1894: 19–29, pl. 2.
Scissurella costata: Sturnay, 1896: 4, 13.
Scissurella costata: Locard, 1899: 106.

Figure 69. Original illustration of *Scissurella costata* in Orbigny (1824).

Scissurella costata: Locard & Caziot, 1900: 174.

Scissurella costata: Pallary, 1900: 358.

Scissurella costata var. *laevigata*: Pallary, 1900: 359.

Scissurella costata: Verrill & Bush, 1900: 525 [*Sin. confusa* ? See also Jensen & Pearce, 2009: 33].

Scissurella costata: Bellini, 1901: 111.

Scissurella costata: Claudon, 1902: 197.

Scissurella costata: Sowerby & Fulton, 1902: 28.

Scissurella costata: Sowerby & Fulton, 1903: 53.

Scissurella (*Schizotrochus*) *costata*: Franceschini, 1906: 7–8.

Scissurella costata var. *laevigata*: Horst & Schepman, 1908: 486.

Scissurella costata: Dautzenberg, 1910: 105, 168.

Scissurella costata: Thiele, 1912a: 3.

Scissurella costata: Pallary, 1912: 70.

Scissurella costata: Dautzenberg, 1913: 79.

Scissurella costata: Cerulli-Irelli, 1916: 208–209, pl. 23, fig. 18.

Scissurella costata: Hildago, 1916: 632.

Scissurella costata: Peile, 1926: 75 [*fide* Jensen & Pearce, 2009: 33; based on Verrill & Bush, 1900; *Sin. confusa* ?].

Scissurella costata: Coen, 1933: 12–13.

Scissurella costata: Nobre, 1936: 180.

Scissurella costata: Coen, 1937: 6–7.

Scissurella costata: Nobre, 1938–1940: 470.

Scissurella costata: Priolo, 1948: 21–22

Scissurella costata: Nicklès, 1950: 36, fig. 5.

Scissurella costata: Wenz, 1961: 173, fig. 269 [copy d'Orbigny, 1824].

Scissurella costata: Parezan, 1961: 68.

Scissurella costata: Smith, 1961: 11.

Scissurella costata: Pasteur-Humbert, 1962: 123.

Scissurella costata: Nordsieck, 1968: 9, pl. 1, fig. 2.00.

Scissurella costata: Parenzan, 1970: 30.

Scissurella costata: Batten, 1975: 6, figs 4a, 12 [shell], figs 7, 11 [protoconch].

Scissurella costata: Fekih, 1975: 88, pl. 28, fig. 17.

Scissurella costata: Carrozza, 1976: 164, fig. 1.

Scissurella costata: Piani, 1980: 117.

Scissurella costata: Terreni, 1981: 13.

Scissurella costata: Nordsieck, 1982: 1, pl. 1, fig. 2.

Scissurella costata: Coppa & Crovato, 1983: 202, pl. 4, fig. 7.

Scissurella costata: Salvini-Plawen, 1983: 265: pl. 87.

Scissurella costata: Gofas *et al.*, n.d.: 33, figs a–b.

Scissurella costata: Rolán, 1983: 63, text-fig.

Scissurella costata: Grasso *et al.*, 1983: 36.

Scissurella costata: Bertozzi *et al.*, 1984: 75.

Scissurella costata: Ghisotti *et al.*, 1984: 104.

Scissurella costata: Miari & Micali, 1984a: 108.

Scissurella costata: Bertozzi *et al.*, 1984: 125.

Scissurella costata: Lazzari *et al.*, 1984: 167.

Scissurella costata: Miari & Micali, 1984b: 175.

Scissurella costata: Coraggio & Micali, 1985: 161.

Scissurella costata: Coraggio, 1985: 162.

Scissurella costata: Perretti, 1986: 126.

Scissurella costata: Lozouet, 1986: 107.

Scissurella costata: Tenekides, 1989: pl. 2, fig. 1.

Scissurella costata: Sabelli *et al.*, 1990: 12.

Scissurella costata: van Aartsen & Kinzelbach, 1990: 104, 108.

Scissurella sp.: Bandel, 1991: 443, text-fig. 2

Scissurella costata: Poppe & Goto, 1991: 32, 64, pl. 2, figs 4–5.

Scissurella costata form *plicata*: Poppe & Goto, 1991: 64.

Scissurella costata: Cachia *et al.*, 1991: 25, pl. 1, fig. 3.

Scissurella costata: Rolán, 1991: 124.

Scissurella costata: Barash & Danin, 1992: 17.

Scissurella costata: Doneddu & Manunza, 1992: 17.

Scissurella costata: Cossignani *et al.*, 1992: 18, fig. 026.

Scissurella costata: Cachia *et al.*, 1993: 10.

Scissurella costata: Bedulli *et al.*, 1993a: 71.

Scissurella costata: Bedulli *et al.*, 1993b: 107.

Scissurella costata: Bedulli, 1994: 77.

Scissurella costata: Bedulli *et al.*, 1994: 81.

Scissurella costata: Giannuzzi-Savelli *et al.*, 1994: 13, figs 90–91.

Scissurella costata: Bedulli & Magistrali, 1995: 45.

Scissurella costata: Arduino *et al.*, 1995: 17, text-figs.

Scissurella costata: Bandel, 1998: 9–10, pl. 1, figs 1–2.

Scissurella costata: Kuntz, 1998: 19, text-fig.

Scissurella costata: Rolán & Ryall, 1999: 21.

Scissurella costata: Macedo *et al.*, 1999: 266.

Scissurella costata: Öztürk & Ergen, 2000: 61, 62, 65, fig. 2d.

Scissurella costata: Borja & Muxika, 2001: 71.

Scissurella costata: daSilva, 2001: 85, pl. 2, figs 1–2 [*fide* Landau *et al.*, 2003].

Scissurella costata: Landau *et al.*, 2003: 14–15, pl. 1, figs 1–2.

Scissurella costata: Geiger, 2003: 76.

Scissurella costata: Demir, 2003: 103 [also as var. ex forma *laevigata*].

Scissurella costata: Geiger, 2004: 6, text-fig.

Scissurella costata: Ardovini & Cossignani, 2004: 19.

Scissurella costata: Mienis & Ben-David-Zaslow, 2004: 14.

Scissurella costata: Kabasakal & Kabasakal, 2005: 53.

Scissurella costata: Geiger, 2006c: text-fig. 3.

Scissurella costata: Peñas *et al.*, 2006: 25.

Scissurella costata: Nolt, 2008: 56, fig. 3C.

Scissurella costata: Mazziotti *et al.*, 2008: 74.

Scissurella costata: Jensen & Pearce, 2009: 33 [Bermuda; based on literature compilation; juvenile *Sin. confusa*?].

Scissurella costata: Segers *et al.*, 2009a: 3, pl. 1, figs 3a–c.

Scissurella costata: Vazzana, 2010: 69.

Scissurella costata: Cossignani & Ardovini, 2011: 96, text-fig.

Scissurella costata: Wesselingh & Pouwer, 2011: 140–141, fig. 33 [fossil].

Scissurella costata: Moreno, 2011: 95, text-figs.

Synonyms.

+ *Scissurella laevigata* d'Orbigny, 1824: 343–344, pl. 23, fig. 1. ***Type material.*** (whereabouts unknown), 2 mm. ***Type locality.*** Mediterranean, among *Fucus* and *Ceramium* (OD). ***Remarks.*** Munier-Chalmas (1865) and Buquoy *et al.* (1882–1886) noted as publication date of d'Orbigny "17 janvier, 1823."

Figure 70. Original illustration of *Scissurella laevigata* in d'Orbigny (1824).

Scissurella laevigata: Sowerby, 1824: 256–257.

Scissurella laevigata: Philippi, 1853: 33–34, pl. 6, fig. 1 [copy figure d'Orbigny, 1824].

Anatomus laevigatus: Adams & Adams, 1853–1858 (1854): vol. 1, 439.

Scissurella laevigata: Munier-Chalmas, 1865: 394.

Scissurella laevigata: Weinkauff, 1868: 385.

Scissurella laevigata: Petit, 1869: 110 [*calcaroides* as synonym].

Scissurella laevigata: Aradas & Benoit, 1870: 156.

Scissurella laevigata: Weinkauff, 1873: 40.

Scissurella laevigata: Monterosato, 1878b: 79 [as synonym of *Sci. costata*].

Scissurella laevigata: Dautzenberg, 1883: 16.

Scissurella laevigata: Paetel, 1883: 47.

Scissurella laevigata: Lorcard, 1892: 220.

Scissurella laevigata: Grossu, 1886: 130–131, fig. 25.

Scissurella laevigata: Granger, 1885: 154.

Scissurella costata var. *laevigata*: Bucquoy *et al.*, 1882–1886: pl. 51, fig. 11.

Scissurella cancellata var. *laevigata*: Pilsbry, 1890: 50, pl. 50, fig. 2.

Scissurella laevigata: Locard & Caziot, 1900: 174.

Scissurella laevigata: Sowerby & Fulton, 1903: 53.

Scissurella (*Schizotrochus*) *laevigata*: Franceschini, 1906: 8.

Scissurella laevigata form *costata*: Abuayo & Jaume, 1949: 560 [*fide* Jensen & Pearce, 2009: 33].

Scissurella laevigata: Brookes Knight *et al.*, 1960: I221, fig. 136, 1a–1b.

Scissurella costata form *laevigata*: Poppe & Goto, 1991: 64.

Scissurella laevigata: Geiger, 2003: 77 [as possible synonym of *Sci. costata*].

+ *Scissurella decussata* d'Orbigny, 1824: 344–345, pl. 23, fig. 3. ***Type material.*** (whereabouts unknown), 1 mm. ***Type locality.*** Fossil sands of Castell'Arquato [Italy, 44.850°N 9.866°E], in the Plainsantin [= late Pliocene = Piacenzian] (OD). ***Etymology.*** Decussatus, Latin, referring to the roman number 10 (X).

Scissurella decussata: Sowerby, 1824: 257.

Scissurella decussata: Chenu, 1859: 363, fig. 2715.

Scissurella decussata: Munier-Chalmas, 1865: 394.

Scissurella decussata: Petit, 1869: 110.

Scissurella decussata: Munier-Chalmas, *non* d'Orbigny: Monterosato, 1872a: 30 [as synonym of *Sci. costata*].

Scissurella decussata: Paetel, 1888: 289.

+ *Scissurella elegans* d'Orbigny, 1824: 345, pl. 23, fig. 4. ***Type material.*** (whereabouts unknown), 2 mm. ***Type locality.*** Same as number 3 (= *Sci. decussata*: fossil sands of Castell'Arquato [Italy, 44.850°N 9.866°E]), in the Plainsantin [= late Pliocene = Piacenzian] (OD). ***Etymology.*** Elegans, Latin for elegant.

Figure 71. Original illustration of *Scissurella decussata* in d'Orbigny (1824).

Scissurella elegans: Sowerby, 1824: 258.

Scissurella elegans: Munier-Chalmas, 1865: 395.

Schismope elegans: Kobelt, 1878: 163, pl. 56, fig. 29.

Scissurella elegans: Geiger, 2003: 77.

Figure 72. Original illustration of *Scissurella elegans* in d'Orbigny (1824).

+? *Argonauta uniumbilicatus* O. G. Costa, 1828: pl. 10, fig. 1 [*fide* Bucquoy *et al.* 1886. See Remarks].

Argonauta uniumbilicatus: O. G. Costa, 1829: 61–63, not illustrated. ***Type material.*** (whereabouts unknown): iconotype Costa, 1828: pl. 10, fig. 1. ***Type locality.*** Gulf of Taranto, Sicily, Italy [40.454°N, 17.200°E]. ***Etymology.*** Possibly referring to the single umbilicus of this "argonautid" shell.

Argonauta uniumbilcatus Costa, 1829: d'Orbigny, 1855: 233 [assigned to *Solarium*. See Remarks].

+ *Scissurella plicata* Philippi, 1836: 187, not illustrated. ***Type material.*** (whereabouts unknown. Philippi, 1844: pl. 25, fig. 18). ***Type locality***. Thaps Peninsula [= Magnisi Peninsula], Sicily, Italy [37.150°N, 15.233°E] (OD). ***Etymology.*** Plicatus, Latin for folded.

Scissurella plicata: Philippi, 1844: 159, pl. 25, fig. 18.

Scissurella plicata: Forbes, 1844: 138, 166, 167, 169.

Scissurella plicata: Requien, 1848: 71.

Scissurella plicata: Philippi, 1853: 34 [as synonym of *Sci. costata*].

Scissurella plicata: de Rayneval *et al.*, 1854: 11.

Scissurella plicata: Petit de la Saussaye, 1860: 251.

Scissurella plicata: Conti, 1864: 32.

Scissurella plicata: Brusina, 1864: 28.

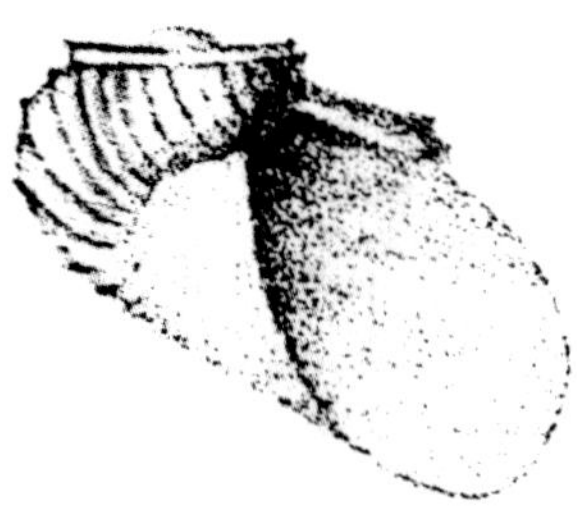

Figure 73. *Scissurella plicata* from Philippi (1844: ideotype).

Anatomus plicatus: Brusina, 1866: 80 [in Stomatellinae].

Scissurella plicata: Paetel, 1869: 62.

Scissurella plicata: Conti, 1871: 38 [*fide* Cerulli-Irelli, 1916].

Scissurella plicata: Monterosato, 1872a: 30 [as synonym of *Sci. costata*].

Scissurella plicata: Ponzi, 1875: 26 [*fide* Cerulli-Irelli, 1916].

Scissurella plicata: Monterosato, 1877b: 34 [as synonym of *Sci. costata*].

Scissurella plicata: Monterosato, 1877a: 416 [as synonym of *Sci. costata*].

Scissurella plicata: Monterosato, 1878b: 79 [as synonym of *Sci. costata*].

Scissurella plicata: Pallary, 1909: 144.

Scissurella plicata: Coen, 1933: 12–13 [as synonym of *Sci. costata*].

Scissurella plicata: Coen, 1937: 6–7 [as synonym of *Sci. costata*].

+ *Scissurella d'orbignyi* Scacchi, 1836: 16, not illustrated. Non *Sci. dorbignyi* Auduin, 1826. Is unavailable primary homonym. ***Type material.*** (whereabouts unknown). ***Type locality.*** Naples, Italy, 40.833°N, 14.250°E (inferred from title). ***Etymology.*** Named Alcide Dessalines d'Orbigny (1802–1857).

Padollus Orbigny [sic]: Costa, 1838: 8–9, pl. 3, figs 1A–C. Costa placed the species in *Padollus*, a synonym for *Haliotis* (Vetigastropoda: Haliotidae).

Padollus d'orbigny [sic]: Costa, 1861: 60, pl. 10, fig. 2 [as senior synonym of *Sci. plicata* Philippi].

Padollus orbignyi: Monterosato, 1872a: 30 [as synonym of *Sci. costata*].

Padollus orbignyi: Monterosato, 1878b: 79 [as synonym of *Sci. costata*].

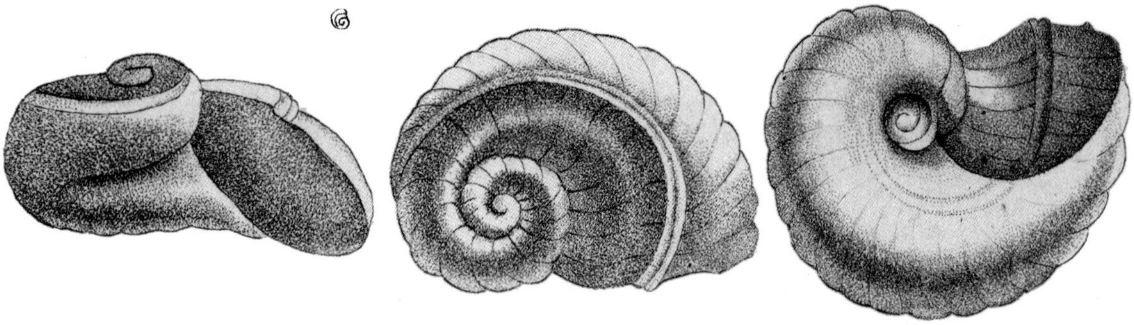

Figure 74. Costa's (1838) figure of *Padollus Orbigny* [sic], most likely *Sci. costata*.

+ *Delphinula calcaroides* Cantraine, 1842 : 341, not illustrated. ***Type material.*** (whereabouts unknown). ***Type locality.*** Gulf of Cagliari [Sardinia, Italy, ~39.2°N, 9.1°E] (OD). ***Etymology.*** Not given.

Delphinula calcaroides: Monterosato, 1872a: 30 [as synonym of *Sci. costata*].

+ *Scissurella striatula* Philippi, 1844: 160, pl. 25, fig. 33. ***Type material.*** Holotype/Syntypes? (whereabouts unknown), 1''' [= 1.75 mm]. ***Type locality.*** "Cum priore occurrit" [Occurs with the prior = *Sci. plicata*]: Magnisi Peninsula [= Thaps Peninsula, Sicily, Italy, 37.150°N, 15.233°E] (OD). ***Etymology.*** Striata, Latin for channelled. Referring to the axial sculpture of the species.

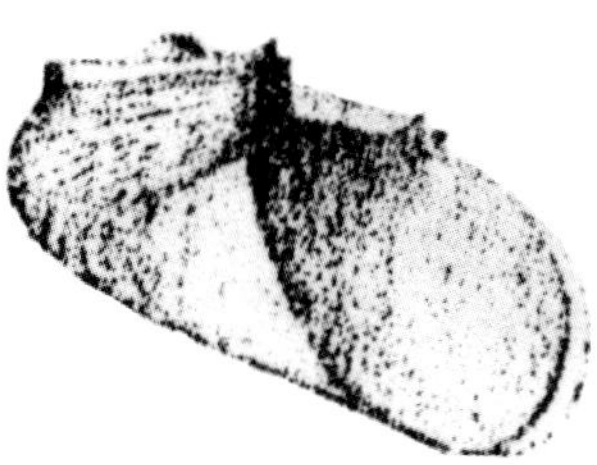

Figure 75. Original illustration of *Scissurella striatula*.

Anatomus striatulus: Adams & Adams, 1853–1858 (1854): vol. 1, 439.

Scissurella/Schismope striatula: Capellini *in* Jeffreys, 1860: 71 [lumped *Scissurella* and *Schismope* into same genus].

Scissurella striatula: Munier-Chalmas, 1865: 394.

Schismope striatula: Petit, 1869: 111.

Scissurella striatula: Canefri, 1869: 329.

Scissurella sriatula: Monterosato, 1872a: 30–31 [as synonym of *Sci. crispata* = *An. crispata*].

Figure 76. Original illustration of *Scissurella cancellata*.

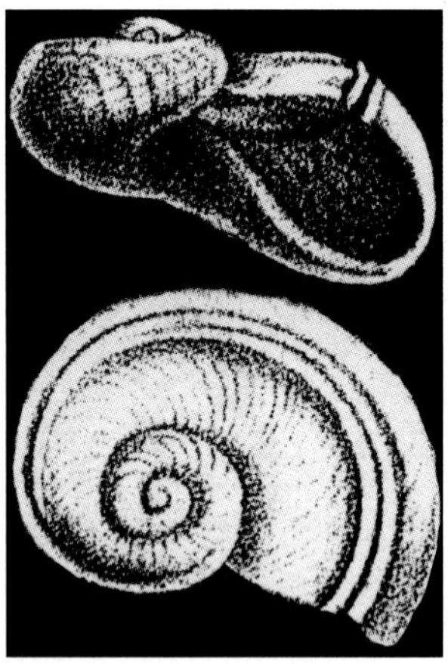

Figure 77. Original illustration of *Scissurella decipiens*.

Schismope striatula: Jeffreys, 1874: 113.

Schismope striatula: Kobelt, 1874: 341 [translation of Jeffreys, 1874 = *An. crispata*].

Scissurella striatula: Seguenza, 1877: 272 [as synonym of *Sci. crispata* = *An. crispata*].

Scissurella sriatula: Monterosato, 1878b: 79 [as synonym of *Sci. crispata* = *An. crispata*].

Scissurella striatula: Paetel, 1888: 289.

Schismope striatula: Lorcard, 1892: 220, fig. 197.

Schismope striatula: Sacco, 1986: 89.

Scissurella/Schismope striatula: Bofill i Poch & Chia, 1913: 13 [genera entered into synonymy].

Scissurella sriatula: Coen, 1933: 12–13 [as synonym of *Schizotrochus crispatus* = *An. crispata*].

Scissurella striatula: Coen, 1937: 6–7 [as synonym of *Schizotrochus crispatus* = *An. crispata*].

Scissurella striatula: Glibert, 1962: 4.

Scissurella striatula: Geiger, 2003: 78.

+ *Scissurella cancellata* Jeffreys, 1856a: 181, pl. 2 fig. 1. ***Type material.*** Holotype (NHMUK 1856.2.18.16. Fig. 78), 1.4 mm. ***Type locality.*** Sestri di Levante, Piedmont [Italy, 44.267°N, 9.383°E] (OD). ***Etymology.*** Noting the cancellate sculpture (OD). ***Remarks.*** Warén (1980) labeled the specimen as a "syntype" although the original description refers clearly to a single specimen found, as also noted by K. Way in 1981 on a label annotation. Warén annotated USNM 181648 (54 specimens remaining, some empty glue spots) "syntype" specimens of *Sci. cancellata*: "These specimens are syntypes of *Scissurella cancellata* Jeffreys 1856. Woodward identified them with *S. elegans* d'Orb. but never believed that it was a new species." Original label identifies them as *Schismope cingulata* (correct ID) from Spezzia, Greece: "These are the types of *Sc. elegans* Woodward, 1859 not of Orbigny."

Scissurella cancellata: Jeffreys, 1860: 35–36, fig. 1 [copy Jeffreys, 1856a].

Scissurella cancellata: Paetel, 1869: 62.

Scissurella cancellata: Munier-Chalmas, 1865: 395.

Scissurella cancellata: Weinkauff, 1868: 386.

Scissurella cancellata: Canefri, 1869: 329.

Scissurella cancellata: Appelius, 1869: 201.

Scissurella cancellata: Monterosato, 1872a: 30 [as synonym of *Sci. costata*].

Scissurella cancellata: Monterosato, 1878b: 79 [as synonym of *Sci. costata*].

Scissurella costata var. *cancellata*: Bucquoy *et al.*, 1886: pl. 51, fig. 7.

Scissurella cancellata: Pilsbry, 1890: listed as synonym of *Sci. costata.*

+ *Scissurella decipiens* O. G. Costa, 1861: 61, pl. 10, fig. 1. ***Type material.*** Holotype (whereabouts unknown), 1 mm. ***Type locality.*** Sardegna Island, Italy (OD). ***Etymology.*** Decipere: Latin for to deceive, to elude. ***Remarks.*** Costa (1861) noted as figure "fig. 1, A, D," However, there is no figure D on that plate. The type locality is listed for the next species (*Scissurella cingulata*) where he noted that it was found at the same place as the previous one.

Scissurella decipiens: Petit, 1869: 110.

Scissurella decipiens: Monterosato, 1872a: 30 [as synonym of *Sci. costata*].

Misidentifications

not *Scissurella decussata*: Audouin, 1826, 42, pl. 5, fig. 29 [is *Sci. reticulata*].

not *Scissurella decussata*: Audouin, 1828: 183 [is *Sci. reticulata*, *fide* Bouchet & Danrigal, 1982].

not *Scissurella striatula*: Philippi, 1853: 37, pl. 6, fig. 9 [is *Anatoma* sp.].

not *Scissurella elegans*: Woodward, 1859: 202, pl. 46, figs 1–7, 9? [is *Sin. cingulata*].

not *Scissurella elegans*: Chenu, 1859: 363, fig. 2717 [juvenile *Sin. cingulata*].

not *Scissurella elegans*: Jeffreys, 1860: 35 [is *Sin. cingulata*].

not *Schismope elegans*: Weinkauff, 1868: 386 [is *Sin. cingulata*].

not *Schismope elegans*: Aradas & Benoit, 1870: 156 [most likely is *Sin. cingulata*. Synonymized with *striatula* Philippi, *elegans, sensu* Weinkauff].

not *Schismope elegans*: Monterosato, 1872a: 30 [is *Sin. cingulata*].

not *Schismope striatula*: Tryon, 1883: 321 [is *Sin. cingulata*].

not *Schismope striatula*: Kobelt, 1888: 259 [is *Sin. cingulata*].

not *Schismope striatula*: Carus, 1889–1893: v. 2, 238 [is *Sin. cingulata*].

not *Scissurella costata*: Lorcard, 1892: 220, fig. 196 [figure shows *An. crispata*].

not *Scissurella costata*: Óskarsson, 1982: 164, fig. 4 [as synonym of *Scissurella crispata*. Figure shows *An. aspera*].

Type material. (whereabouts unknown), 1 mm.

Type locality. Same as preceding species [Mediterranean, among *Fucus* and *Ceramium*] (OD).

Etymology. Costata, Latin for ribbed.

Description. Shell medium-sized (to 1.7 mm) trochiform, squat. Protoconch of 0.75 whorl, fine axial sculpture, apertural varix connected to embryonic cap, apertural margin sinusoidal. Teleoconch I of 0.9–1.0 whorl, approximately 18 low axial lamellae to tall axial cords, diminishing in strength with growth, no spirals, interstices with very fine growth lines. Teleoconch II of up to 1.125 whorls, suture impressed. Shoulder shallow sinusoidal in profile, forming a table, approximately 30 axials, as distinct cords close to suture, soon fading out to indistinct low cords; first spiral lines after approximately 0.5 whorl, approximately 10 at apertural margin; irregularly spaced. Base with distinct constriction below selenizone, bulging; axial sculpture variable

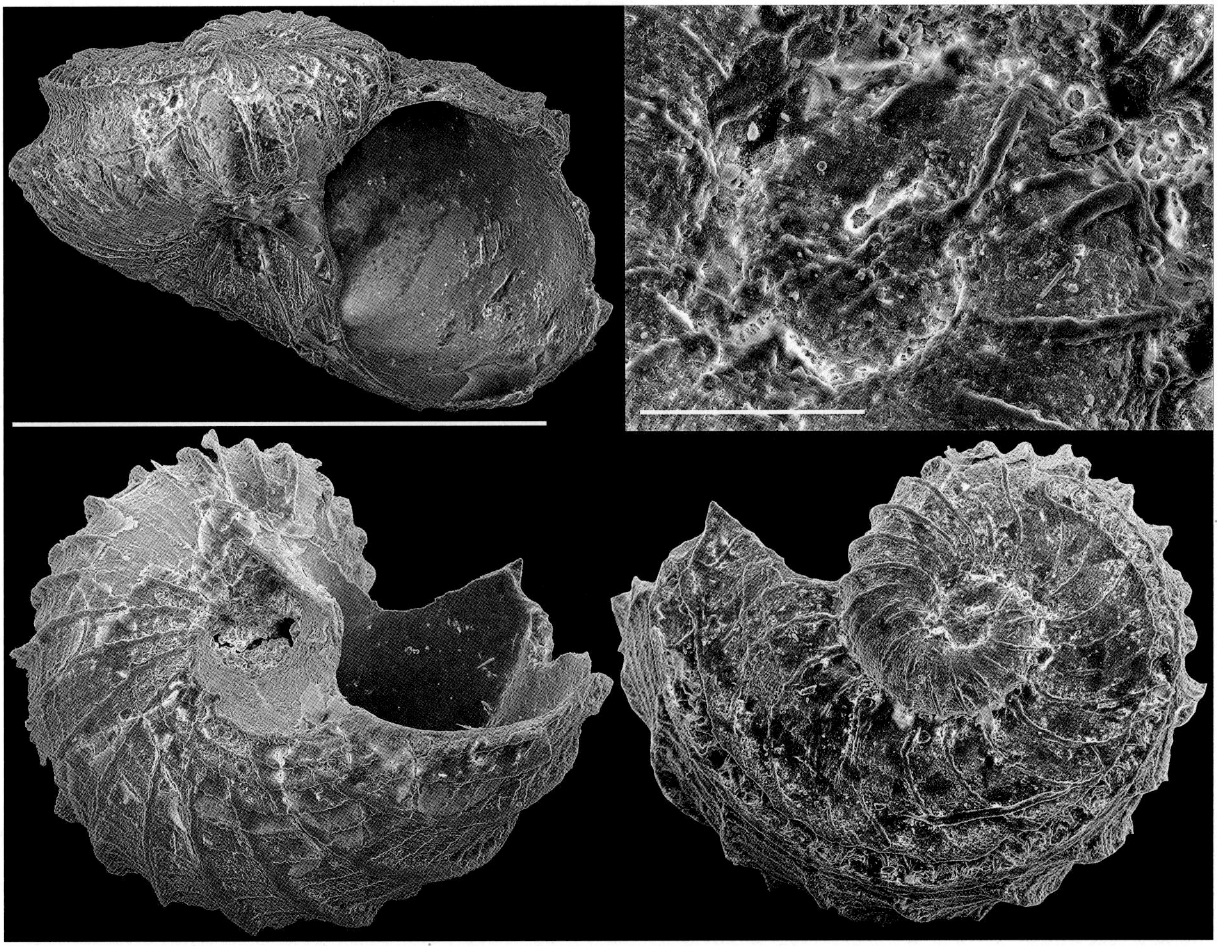

Figure 78. *Scissurella cancellata*. Holotype. Sestri di Levante, Piedmont [Italy, 44.267°N, 9.383°E] (NHMUK 1856.2.18.16). Scale bar shell = 1 mm. Scale bar protoconch = 100 µm.

from hardly perceptible to discrete low axial cords; approximately 16 irregularly spaced spiral lines, in adumbilical third turning into spiral steps. Umbilicus open, wide, bordered by fine spiral cord, walls at angle to base, straight, very fine growth lines. Aperture D-shaped, roof overhanging. Selenizone above periphery; keels quite high, moderately strong; lunules distinct, slit open, margins almost parallel.

Radula (Fig. 81). Rachidian tooth triangular, with V-shaped tip bearing nine cusps, central largest, lateral ones fanning out. Lateral teeth 1–3 similar, lateral teeth 1–2 with three cusps, lateral tooth 3 with four cusps on tip. Lateral tooth 4 reduced, hook-shaped. Lateral tooth 5 broadly enlarged, with five cusps on tip, outermost largest. Innermost marginal teeth with largest cusps on tip towards middle of radula, outer marginal teeth spoon-shaped, many fine bristles along edge. Radular interlock moderate.

Comparisons. The fossil *Sci. transylvanica* and *Sci. terquemi* from the Tertiary of Europe have a smooth protoconch. Early ontogenetic stages of *Sin. terquemi* are difficult to distinguish from *Sci. costata*.

Scissurella azorensis from the Mediterranean and the Macaronesian Islands and *Sci. lobini* from the Macaronesian Islands have a more depressed spire, a protoconch

Figure 79. *Scissurella costata*. 20 m, Cala Rossa, Capraia Island, Italy, 43.004°N, 9.812°E (AMS C.461527, 8). Scale bar shell = 1 mm. Scale bar protoconch = 100 µm.

with spiral rib and very fine axial cords on the centrifugal half, and the wide umbilicus shows the underside of the protoconch.

Scissurella jucunda from South Africa and the southern Atlantic Islands has a much taller and more rounded shell, and a narrower umbilicus.

Scissurella redferni from the Caribbean has a more depressed spire, a narrower umbilicus, and stronger axial cords aligned with the coiling axis.

Scissurella mirifica from the Indo-Malayan Archipelago to the Central Pacific has more but much finer axial cords, and distinct spiral steps surrounding the umbilicus.

Figure 80. *Scissurella costata*. **A.** 7–19 m, Sardinia, Testa Cape, bay E of lighthouse, Italy, 41.233°N, 9.133°E (AMS C.406387, 1). **B.** 20 m, Cala Rossa, Capraia Island, Italy, 43.004°N, 9.812°E (AMS C.461527, 8). **C.** 42 m, Scilla, Italy, 38.250°N, 15.717°E (DLG 329, 5). Scale bar shell = 1 mm. Scale bar protoconch = 100 μm.

Sinezona cingulata from Europe has a teleoconch I of less than one whorl, has a very narrow umbilicus, a protoconch with straight fine axial cords, and the slit closes to a foramen once fully grown.

Distribution. Mediterranean; Madeira; Portugal, Senegal (Nicklès, 1950), to Angola (Fig. 82). Found on rocks, the Phaeophyceae *Padina pavonica* (Linnaeus) Thivy and *Cystoseira* species, as well as the seagrass *Posidonia oceanica* (Linnaeus) Delile according to Öztürk & Ergen (2000).

Figure 81. *Scissurella costata*. Radula. 6–8 m, Sicily, Terrasini, Cala Rossa, Italy, 38.150°N, 13.083°E (AMS C.380710, 2). **A.** Width of radular ribbon, **B.** Central field enlarged. Scale bars = 10 µm.

Specimen records.

Portugal. Lagos, 37.102°N, 8.673°W (RMNH, 1).

Spain. Tarifa, Cadiz Province, 36.013°N, 5.603°W (RMNH, 4). 0 m, Torre de la Peña (10 km N of Tarifa), 36.056°N, 5.659°W (RMNH, 1). 0 m, Sotogrande (Torre Guadiaro), Cadix, 36.534°N, 6.299°W (MNHN, 1). 0 m, Barate, Cadix, 36.534°N, 6.299°W (MNHN, 4). 20–40 m, Rincón de la Victoria, Malaga Province, 36.717°N, 4.300°W (MNHN, 1). 24 m, de la Mona Point, Granada, 36.719°N, 3.728°W (NMR 993000033756, 2). 24 m, de la Mona Point, Granada, 36.733°N, 3.717°W (NMR 993000024210, 1). Near Valencia, 39.442°N, 0.299°W (USNM 780323, 1). 25 m, Benidorm, 38.533°N, 0.131°W (DLG 1313, 1). Barcaras, Mallorca, Balearic Islands (RMNH, 35). 0 m, El Arenal, Mallorca, Balearic Islands, 39.504°N, 2.751°E (NMR 993000033759, 11). 0 m, Cala Morlanda, Mallorca, Balearic Islands, 39.566°N, 3.373°E (NMR 993000033758, 2). Playa de Muro, Mallorca, Balearic Islands, 39.792°N, 3.130°E (RMNH, 4). NE Mallorca, Mallorca, Balearic Islands, 39.800°N, 3.100°E (RMNH, 24). 30–36 m, Islas Medas, Gerona, 42.045°N, 3.225°E (NMR 993000024713, 13). 16–20 m, Cala de Mongo, La Escala, Gerona, 42.106°N, 3.172°E (NMR 993000024245, 7).

France. Corsica (KBIN, 6; KBIN, 8). Rhone Delta (RMNH, 1). Bonifacio, Corsica, 41.386°N, 9.159°E (RMNH, 1). 10 m, Tizzano, Corsica, 41.540°N, 8.851°E (NMR 993000033725, 6). Calvi, Corsica, 42.567°N, 8.757°E (RMNH, 2). Algajola, Corsica, 42.617°N, 8.867°E (MNHN, 4). St. Florent, Corsica, 42.686°N, 9.300°E (RMNH, 1). 25 m, Cap Béar, Banyuls sur Mer, Pyrénées-Orientales, 42.517°N, 3.137°E (NMR 993000023824, 9). 112–113 m, Grande Passe, Hyères Island, 42.967°N, 6.333°E (MNHN, 11). 0–1 m, Notre Dame beach, Porquerolles, Provence, 43.010°N, 6.230°E (MNHN, 4). 7 m, Bagaud Island, Hyères Island, 43.018°N, 6.362°E (MNHN, 30). Port du Niel, 1 km SW of Giens, Presqu'île de Giens, Var, 43.036°N, 6.128°E (RMNH, 1). 135–146 m, Magaud Bank, Hyères Island, 43.050°N, 6.683°E (MNHN, 1). La Capte, 6 km SSE of Hyères, Var, 43.065°N, 6.151°E (RMNH, 21). La Capte, 43.065°N, 6.152°E (RMNH, 1). Le Brusc, near Toulon, 43.076°N, 5.801°E (RMNH, 24). 15 m, Le Brusc, Var, 43.077°N, 5.780°E (NMR 993000024469, 4). 0–1 m, Les Embiez, N coast and Petit Rouveau, 43.083°N, 5.783°E (MNHN, 7). 0–3 m, N coast and Petit Rouveau, Les Embiez, Provence, 43.083°N, 5.783°E (MNHN, 1: complete). 0–1 m, Les Embiez, beach of Salines, Provences, 43.083°N, 5.787°E (MNHN, 6). 3–5 m, Petit Rouveau, Embiez Islands, Provence, 43.085°N, 5.780°E (MNHN, 6). 0–1 m, Le Brusc beach of Cros, Provence, 43.085°N, 5.805°E (MNHN, 1). Cabasson, 43.100°N, 6.323°E (RMNH, 1; 1). 150–250 m, NE canyon de la Cassidaigne, 43.112°N, 5.550°E (MNHN, 4). Sanary Var, 43.117°N, 5.800°E (MNHN, 1). 22 m, Ile Rousse, Bandols, Var, 43.133°N, 5.733°E (NMR 993000024765, 4). 0–1 m, St. Clair, Provence, 43.137°N, 6.387°E (MNHN, 6). 33 m, Grand Congloue, Marseille, 43.177°N, 5.403°E (MNHN, 50). 22 m, Calanque de la Triperie, Cap Morgiou, Marseille, 43.203°N, 5.448°E (MNHN, 25). 8–25 m, 3 km SW of Château d'If, Marseille, 43.239°N, 5.302°E (DLG 947, 1). 0 m, Endoume, Marseille, 43.282°N, 5.350°E (MNHN, 1). Near Marseille, 43.294°N, 5.349°E (KBIN, 2). Marseille, Bouches-du-Rhône, 43.300°N, 5.367°E (LACM 147406, 4). 7 m, Prado Beach, Marseille, 43.300°N, 5.400°E (MNHN, 1). Marseille, 43.300°N, 5.400°E (MNHN, 2). Around Marseille, 43.302°N, 5.351°E (KBIN, 1). Cap Couronne, St. Croix, Bouches-du-Rhône, 43.324°N, 5.053°E (RMNH, 100). Plage St. Croix (10 km S of Martigues), Bouches-du-Rhône, 43.329°N, 5.077°E (RMNH, 7). Cette, 43.392°N, 3.703°E (KBIN, 1). St. Raphael, 43.407°N, 6.789°E (KBIN,

Figure 82. Distribution of *Scissurella costata*. ● specimen record. ■ literature record.

1; KBIN, 100; KBIN, 1). 22–30 m, Le Dramont pied de tombant, Provence, 43.412°N, 6.862°E (MNHN, 50). St. Raphael Var, 43.417°N, 6.767°E (MNHN, 6). Iles de Lerins, 43.540°N, 7.017°E (MNHN, 4). Pointe de Tire Poil, Antibes, Alpes-Maritimes, 43.544°N, 7.133°E (RMNH, 18). Port de l'Olivette, Antibes, Alpes-Maritime, 43.552°N, 7.121°E (NMR 993000025065, 2). Antibes, 43.578°N, 7.129°E (RMNH, 1). 100–110 m, off Saint-Jean-Cap-Ferrat, 43.668°N, 7.351°E (NMR 993000033727, 2). 123 m, Cap d'Aglio, near Monaco, 43.717°N, 7.400°E (MNHN, 2). 4 m, Rade d'Agay, 43.427°N, 6.863°E (KBIN 1230, 1).

Monaco. 123 m, near Monaco, 43.737°N, 7.429°E (KBIN, 3).

Italy. 40 m, Lampedusa, 35.500°N, 12.600°E (DLG 827, 11). 15 m, Baia di Guitgia, Lampedusa Island, 35.517°N, 12.583°E (AMS C.461541, 1). 5 m, St George's Bay, in small harbour, 35.933°N, 14.500°E (AMS C.404933, 1) 100 m, Moirina oli Ragusa, 36.450°N, 14.733°E (JWC 6199, 2). 2 m, Porto Palo, Siracusa, Sicily, 36.683°N, 15.133°E (AMS C.461542, 5). 60 m, off Siracusa, towards Murro di Porco Cape, Sicily, 37.044°N, 15.327°E (AMS C.461538, 2). Siracuse, Sicily, 37.049°N, 15.285°E (RMNH, 150). 5 m, Syracuse, Sicily, 37.060°N, 15.290°E (DLG 1420, 1). 5 m, Siracusa, Sicily, 37.067°N, 15.284°E (DLG 1438, 15). 168 m, Adventure Bank, 37.333°N, 12.167°E (RMNH, 2). Sicily, Magnisi Peninsula, 37.500°N, 14.000°E (MHNG 14364, 1). Sicily, 37.500°N, 14.000°E (MHNG 14361, 1; MHNG 14365, 2; NMSZ 1961.61, 5). Sicily Island, 37.500°N, 14.200°E (FLMNH 00132409, 2). Sciacca, Agrigento Province, Sicily, 37.508°N, 13.080°E (RMNH, 1). 35 m, Cannizzaro, Catania, Sicily, 37.521°N, 15.153°E (DLG 1748, 13). 5–30 m, Porto Palo, Sicily, 37.567°N, 12.900°E (DLG 327, 36). 36 m, Outside rocks of Ciclopi, Acitrezza, Sicily, 37.567°N, 15.167°E (AMS C.461519, 3). Palermo, Sicily, 37.817°N, 13.583°E (MHNG 14363, 5; NMSZ 1961.61, 1; ZMUC, 6; ZMUC, 8). 4 m, Cala Galera, Favignana Island, Sicily, 37.917°N, 12.300°E (AMS C.461540, 9). 30 m, Secca del Cammello, Marettimo Island, Egadi Islands, Trapani, Sicily, 37.989°N, 12.065°E (DLG 1811, 10). 7 m, Faraglione, Levanzo Island, W Sicily, 38.000°N, 12.333°E (AMS C.461530, 4). Trapani, Sicily, 38.017°N, 12.483°E (MNHN, 3). Trapani Beach, Trapani, 38.017°N, 12.508°E (NMR 993000024163, 1). 47 m, Calabria, Pellaro, 38.020°N, 15.631°E (DLG 1410, 18). Balestrate, Palermo, Sicily, 38.052°N, 13.008°E (RMNH, 2). 233 m, Gallina West, 38.086°N, 15.675°E (DLG 1411, 2). Palermo, Sicily Island, 38.117°N, 13.376°E (FLMNH 00132416, 2). Palermo, 38.133°N, 13.383°E (SDMNH, 1; LACM 58268, 1). 80–90 m, Palermo, 38.133°N, 13.383°E (MNHN, 7). Palermo, 38.133°N, 13.383°E (MNHN, 5; MNHN, 3; NMSZ 1961.61, 1; KBIN, 2; KBIN, 1). 50 m, Catalana Point, Terrasini, Sicily, 38.150°N, 13.083°E (AMS C.461521, 6). 48 m, Catalana Point, Terrasini, Sicily, 38.150°N, 13.083°E (AMS C.461531, 1). Messina, Sicily, 38.183°N, 15.550°E (NHMW 22023, 6). Mondello, Sicily, 38.200°N, 13.333°E (USNM 181643, 5; MNHN, 4). Mondello, Sicily, 38.201°N, 13.329°E (KBIN, 1; KBIN, 1). 42 m, Scilla, 38.250°N, 15.717°E (DLG 329, 5). 35 m, Scilla, Calabria, 38.257°N, 15.713°E (DLG 1409, 20). 50 m, Scilla, Calabria, 38.257°N, 15.715°E (DLG 1209, 2). 35 m, Costa Viola, Reggio Calabria, 38.378°N, 15.858°E (DLG 1436, 5). Secca = Banko di Amenolara, Cosenza, 39.874°N, 16.724°E (DLG 1423, 3). Rizzuto Cape Island, Crotone, 38.904°N, 17.101°E (DLG 1422, 1). 22 m, Paluniro Cape, Salerno, 40.029°N, 15.268°E (NMR 993000033739, 28). 6 m, Palinuro Cape, Salerno, 40.029°N, 15.267°E (DLG 1439, 9). 30 m, Porto Cesareo, 40.250°N, 17.900°E (AMS C.461534, 1). Gandoli, (10 km S of Teranto), 40.379°N, 17.298°E (RMNH, 2). 80–110 m, Capri, 40.550°N, 14.233°E (DLG 328, 33). 20 m, Grotto, Capri, Naples, 40.561°N, 14.206°E (DLG 1440, 12). 45 m, Gulf of Naples, La Gaiola, 40.700°N, 14.300°E (RMNH, 1). 59 m, Ischia Island, Naples, 40.719°N, 13.905°E (NMR 993000024839, 6). 3 m, Cennone Rock at Point Piopetta, Marina Granade, Procida Island, 40.750°N, 14.017°E (AMS C.461536, 5). 30 m, Ventotene Island, Latina, 40.800°N, 13.400°E (NMR 993000033737, 23). 4–8 m, Posilligo, 40.817°N, 14.217°E (AMS, 1: complete). Naples, 40.833°N, 14.250°E (MNHN,

3). Gulf of Naples, 40.833°N, 14.267°E (USNM 181644, 3). 21 m, Le Scoglietelle, Ponza Island, Latina, 40.938°N, 13.012°E (DLG 1792, 30). 7–19 m, Sardinia, Testa Cape, bay E of lighthouse, 41.233°N, 9.133°E (AMS C.406387, 1). 35 m, NW ca. 1 mi from Castel Sonnino, 41.417°N, 13.250°E (AMS C.461544, 1). 33 m, Secche di Tor Paterno, Rome, 41.600°N, 12.333°E (DLG 1824, 2). 33 m, Tor Paterno (Rome), 41.675°N, 12.371°E (DLG 852, 8). 47 m, Secche di Tor Patero, Rome, 41.681°N, 12.380°E (NMR 993000021234, 12). 30–35 m, Giannutri Island, Tuscany, 42.250°N, 11.100°E (DLG 926, 60). 28 m, Secca delle Murelle, Viterbo, 42.310°N, 11.602°E (NMR 993000033740, 1). 80 m, Ortona, 42.400°N, 14.400°E (DLG 835, 1). 17 m, Argentarola Island, Groseto, 42.419°N, 11.081°E (NMR 993000024935, 1). Calamita Cape, Elba Island, 42.700°N, 10.383°E (AMS C.461517, 1). 67 m, Della Stella Cape, Elba Island, 42.733°N, 10.317°E (AMS C.461518, 27; AMS C.461533, 23). 32 m, Capoliveri, Elba, 42.750°N, 10.367°E (NMR 993000024533, 25). 5 m, Verde Bay, Ala Point, 42.817°N, 10.733°E (AMS C.461535, 1). 200 m, off Capraia Island, Livorno, Toscana, 43.000°N, 9.800°E (DLG 825, 2). 160 m, Capraia Island, Tuscany, 43.000°N, 9.800°E (DLG 956, 50). 120 m, Carpaia Island, 43.000°N, 9.800°E (DLG 1441, 3). 20 m, Rossa Bay, Capraia Island, 43.004°N, 9.812°E (AMS C.461527, 8). 120 m, W of Capraia Island, 43.033°N, 9.817°E (DLG 318, 1). 100 m, Capraia Island, 43.033°N, 9.817°E (DLG 870, 1). 5 m, Outside Harbor, Capraia Island, 43.050°N, 9.840°E (AMS C.461528, 2). 80 m, Capraia Island, 43.050°N, 9.840°E (AMS C.461526, 1). 80 m, Capraia Island, 43.050°N, 9.840°E (AMS C.461523, 1; AMS C.461524, 18). 5 m, Antignano, Livorno, 43.496°N, 10.323°E (DLG 1765, 5). Livorno, Bagni, Fiume, 43.550°N, 10.317°E (AMS C.461543, 1). 50 m, Calambrone, Pisa, Tuscany, 43.596°N, 10.048°E (DLG 1981, 1). 30 m, Riomaggiore, 44.100°N, 9.733°E (AMS C.461532, 2). 30 m, Portofino, Cala Aurora, 44.300°N, 9.200°E (AMS C.461545, 2).

Malta. Birzebbuga, Pretty Bay, 35.825°N, 14.530°E (NMR 993000024357, 12). 3 m, St. Thomas Bay, 35.852°N, 14.566°E (NMR 993000033742, 4; NMR 993000023982, 1). 6 m, Marsaskala Bay, 35.865°N, 14.565°E (NMR 993000024031, 8). 100 m, 35.917°N, 14.417°E (DLG 326, 12). 3 m, Gnejna Bay, 35.921°N, 14.342°E (NMR 993000024103, 18). 4 m, Qalet Marku Bay, 35.946°N, 14.451°E (NMR 993000024405, 1). 40 m, Quawra Point, St. Paul's Bay, 35.948°N, 14.401°E (DLG 314, 2). 6 m, Xemxija, Mistra Bay, 35.958°N, 14.391°E (NMR 993000024668, 2). 10 m, Anchor Bay, 35.965°N, 14.341°E (NMR 993000024648, 6). 40–60 m, Qammieh, 35.974°N, 14.329°E (NMNZ M.262690, 4). 17–30 m, Cirkewwa Bay, 35.984°N, 14.330°E (NMR 993000024574, 22). 8 m, Blue Lagoon, Camino Island, 36.014°N, 14.323°E (NMR 993000024624, 1). 35 m, Wied iz Zurrieq, 38.819°N, 14.459°E (DLG 871, 2).

Croatia. 8 m, Krk Island, Bay Ogruli S of Vrbnik, 45.033°N, 14.567°E (NHMW 22026, 1). Pzokljan near Sebenico, Dalmatia, 43.732°N, 15.889°E (KBIN, 10). 45 m, Privic Island, 43.730°N, 15.800°E (DLG 1729, 30). Biograd, Hrvatska, 43.939°N, 15.440°E (RMNH, 15). 13 m, Krk Island, 45.033°N, 14.567°E (NHMW 22020, 7).

Greece. 1010 m, near Cerigo [= Kythira] Island, 36.156°N, (GNS WM 1552, 4). Spezzia, 37.267°N, 23.150°E (USNM 181645, 1). 15 m, Delos, 37.400°N, 25.267°E (MNHN, 3). NE of Skiathos, Skiathos Island, Voriai Sporadhes, 39.183°N, 23.511°E (RMNH, 1). 4 m, Orkos Beach, Paxos, Kerkyra, 39.234°N, 20.150°E (NMR 993000033729, 2). 33 m, Sithonia, Ormos Panagias, N Dhiaporos Island, 40.733°N, 24.733°E (AMS C.461516, 2). 0 m, Kato-Galatas, W Creete, 35.513°N, 23.962°E (DLG 1314, 2).

Turkey. 10–45 m, Ada Bank, Kusadasi, Izmir, 37.833°N, 27.267°E (DLG 316, 8). Ada Bank, Kusadasi, Izmir, 37.858°N, 27.231°E (EGC 29739, 5). 50 m, Bozcaada Island, 39.804°N, 25.981°E (DLG 1197, 9). 40 m, Eskina Kayaligi, Buyukada, Mamora, 40.867°N, 29.133°E (MCZ 346607, 1).

Libya. Bengasi, Cyrenaica, 32.126°N, 20.059°E (RMNH, 1).

Tunisia. 0 m, Sidi Youssef, Gharbi, Kerkenah Islands, Sfax, 34.660°N, 10.978°E (NMR 993000033768, 10).

Algeria. Oran, 35.691°N, 0.642°W (MNHN, 5; MNHN, 5; MNHN, 3). Kabylia, 36.250°N, 5.000°E (FMNH 84107, 3: complete). Algiers, 36.833°N, 3.000°E (NHMW 22023, 1). Algiers, 36.788°N, 3.066°E (KBIN, 1).

Morocco. 0 m, Asilah, Qued el Helou, 35.466°N, 6.040°W (MNHN, 25). 35–45 m, Anse Almadraba, Ceuta South, Strait of Gibraltar, 35.875°N, 5.317°W (MNHN, 1: complete). Anse Sarchal, Ceuta South, Strait of Gibraltar, 35.890°N, 5.297°W (MNHN, 1: complete). 25–40 m, Almina Point, Ceuta North, Strait of Gibraltar, 35.902°N, 5.275°W (MNHN, 2: complete; MNHN, 1: complete).

Mauritania. 14 m, 20.333°N, 17.033°E (RMNH, 1). 31 m, 20.350°N, 17.233°W (RMNH, 3). 6 m, Port Etienne, 20.900°N, 17.067°W (MNHN, 1). 20 m, Port Etienne, 20.333°N, 16.367°W (MNHN).

Senegal. Dakar region, 14.671°N, 17.438°W (MNHN, 1). 33–36 m, Dakar region, 14.671°N, 17.438°W (MNHN, 1).

Gabon. 92 m, Station C13C, Gulf of Guinea, 3.172°S, 9.820°E (ZMBN, 2: complete).

Angola. 470–60 m, Moçamedes (Praia Amelia), Moçâmedes Province, 15.196°S, 12.152°E (MNHN, 3). 40 m, Lucira (Santa Marta), Moçâmedes Province, 13.867°S, 12.533°E (MNHN, 1). 40 m, Cabo Ledo, Luanda Province, 9.653°S, 13.261°E (MNHN, 4). 90–100 m, Around Mussulo, Luanda Province, 8.891°S, 13.131°E (MNHN, 11). 40–60 m, Ilha de Luanda, Luanda Province, 8.838°S, 13.234°E (MNHN, 1). 10–20 m, Corimba, Luanda Province, 8.838°S, 13.234°E (MNHN, 1). 120 m, Ilha de Luanda, Luanda Province, 8.838°S, 13.234°E (MNHN, 1). 40–60 m, Luanda Island, Luanda Province, 8.838°S, 13.234°E (MNHN, 1: complete). 0 m, Ambrizete Region, Bango, 7.339°S, 12.919°E (MNHN, 4). 45 m, Ambrizete, 7.233°S, 12.867°E (MNHN, 8). 0 m, S of Ambrizete, Luanda Province, 7.233°S, 12.867°E (MNHN, 3). 80 m, Ambrizete, 7.233°S, 12.867°E (MNHN, 1).

Literature records.

Spain. Banco de Galicia, 42.750°N, 12.833°W (Rolán, 1983).—100 m, Matxitxako, Basque, 43.448°N, 0.901°E (Borja & Muxika. 2001).—Cala Torrent, Ibizza (RMNH AD 4614). Playa es Cavalet, Ibizza, 38.897°N, 1.447°E (RMNH AD 4748). Calpe, Alicante, Valencia, 38.627°N, 0.046°E (RMNH AD 1869). Es Pujol, Formentera, 38.718°N, 1.473°E (RMNH AD 4411). Portinatx, Ibizza, 39.113°N, 1.517°E (RMNH AD 4552) (J. van Aarsten, pers. comm. 5/2011).—30–400 m, Alboran Island, 35.951°N, 3.033°W. 30–250 m, Alboran Island, 35.951°N, 3.033°W (Peñas *et al.*, 2006).

France. Banyuls sur mer, 42.483°N, 3.133°E (Bandel, 1998).—Anse de Pampelonne, St. Tropez, 43.267°N, 6.650°E (Kuntz, 1988).—Bonifacio, Corsica, 41.386°N, 9.159°E (RMNH AD 13383, 1; RMNH AD 13422, 1). Porto Vecchio, Corsica, 41.591°N, 9.287°E (RMNH AD 13321). Pinarello, Corsica, 41.672°N, 9.395°E (RMNH AD 13661). Propriano, Corsica, 41.675°N, 8.904°E (RMNH AD 13383). 150–200 m, in front of Solenzara, Corsica, 41.859°N, 9.450°E (RMNH AD 12233). La Capte, 43.065°N, 6.152°E (RMNH AD 11542). Cabasson, 43.103°N, 6.230°E (RMNH AD 11866). Plage de l'Estragnol, Var, 43.110°N, 6.297°E (RMNH AD 16286). Les Leques, Provence, 43.180°N, 5.686°E (RMNH AD 12056). Ramatuelle, Plage de, St. Tropez, 43.242°N, 6.665°E (RMNH AD 15207). St. Tropez, 43.269°N, 6.631°E (RMNH AD 14228). St. Tropez, Plage des, 43.269°N, 6.631°E (RMNH AD 14346). Between Sète and Agde, Herault, 43.359°N, 3.609°E (RMNH AD 1534) (J. van Aarsten, pers. comm. 5/2011).

Italy. 6 m, below the lighthouse, San Domino, Tremiti Islands, 42.006°N, 15.498°E. 53 m, "Zona B," San Domino, Tremiti Islands, 42.104°N, 15.473°E. 21 m, Matano Bay, San Domino, Tremiti Islands, 42.113°N, 15.495°E. 10 m, Spido Point, San Domino, Tremiti Islands, 42.116°N, 15.497°E. 48 m, towards SE of San Nicola, Tremiti Islands, 42.116°N, 15.503°E. 7 m, Spido Point, San Domino, Tremiti Islands, 42.117°N, 15.497°E. 36 m, Cala dei Turchi, Carprara, Tremiti Islands, 42.136°N, 15.507°E. 30 m, Turchi Bay, Carprara, Tremiti Islands, 42.136°N, 15.508°E. 27 m, Secca Point, Carprara, Tremiti Islands, 42.136°N, 15.523°E. 6 m, Vecchio Faro Bay, Carprara, Tremiti Islands,

42.137°N, 15.519°E. 41 m, Secca Point, Carprara, Tremiti Islands, 42.137°N, 15.527°E. 21 m, Architiello, Carprara, Tremiti Islands, 42.138°N, 15.522°E. 57 m, Secca Point, Carprara, Tremiti Islands, 42.140°N, 15.522°E. 15 m, Cala Zio Cesare, San Domino, Tremiti Islands, 42.103°N, 15.484°E (Mazziotti *et al.*, 2008).—12–22 m, Secca di Flavia (RMNH AD 15514). 30 m, Secca delle Murelle (RMNH AD 18845). 25–27 m, delle Correnti Island, 36.645°N, 15.077°E (RMNH AD 15482). 32 m, off Vendicari, 36.793°N, 15.156°E (RMNH AD 20960). 30 m, off Catania, 37.466°N, 15.127°E (RMNH AD 26020). 80 m, off Catania, Sicily, 37.489°N, 15.142°E (RMNH AD 20140). Catania, Sicily, 37.502°N, 15.088°E (RMNH AD 16383). Acitrezza, Sicily, 37.566°N, 15.161°E (RMNH AD 15715). 90 m, Acitrezza, Sicily, 37.566°N, 15.161°E (RMNH AD 17920). San Guiliano, Trapani, Sicily, 38.026°N, 12.528°E (RMNH AD 21854). Terrassini, 38.152°N, 13.074°E (RMNH AD 21552). San Lorenzo, Sicily, 38.153°N, 13.074°E (RMNH AD 21747). Messina, Sicily, 38.183°N, 15.550°E (RMNH AD 17248). Mondello, Sicily, 38.200°N, 13.326°E (RMNH AD 4215). 44 m, Scilla, Calabria, 38.256°N, 15.713°E (RMNH AD 26183). Ferro Cape, Sardinia, 41.152°N, 9.516°E (RMNH AD 15964). Sperlonga, 41.259°N, 13.434°E (RMNH AD 27197). 30 m, Giannutri Island, Tuscany, 42.250°N, 11.100°E (RMNH AD 18788). 10 mi S of Capraia Island, 42.852°N, 9.810°E (RMNH AD 15255). Gulf of Baratti, 42.999°N, 10.506°E (RMNH AD 18310). 40 m, Gorgona Island, 43.433°N, 9.900°E (RMNH AD 20615) (J. van Aarsten, pers. comm. 5/2011).—80–100 m, Corso Cape, 43.509°N, 10.303°E (Carrozza, 1976).

Croatia. 0 m, Sebenico Islands (= Sibenik), 43.734°N, 15.886°E. 0 m, Mika Point, near Zara. 0 m, Melada Island, 44.210°N, 14.884°E (Brusina, 1886).—Istria region, 45.250°N, 14.000°E (Bandel, 1998).—Porec, Istra, 51.483°N, 9.467°E (Veldhuis, 2004).—Murter, 43.821°N, 15.592°E (RMNH AD 17154). Biograd, 43.938°N, 15.444°E (RMNH AD 2332). Porec, Istra, 45.227°N, 13.598°E (RMNH AD 1358) (J. van Aarsten, pers. comm. 5/2011).

Greece. 160 m, Station 194, between Crigo and Cerigotto, 36.050°N, 23.100°E (Sturnay, 1906).—Nitika (RMNH AD 17864). Githio/Methoni (RMNH AD 28733). 40–50 m, Kastos Island, Ionian Sea, 38.580°N, 20.920°E (RMNH AD 14563). Elounda, 35.262°N, 25.722°E (RMNH AD 24879). Rethymno, Crete, 35.367°N, 24.475°E (RMNH AD 24824). Prassonissi, Rhodos, 35.882°N, 27.761°E (RMNH AD 29046). Milos, 36.700°N, 24.400°E (RMNH AD 22283). Gulf of Saronicos, 37.900°N, 23.600°E (RMNH AD 20890). Thessaloniki, 40.635°N, 22.926°E (RMNH AD 15776) (J. van Aarsten, pers. comm. 5/2011).

Cyprus. Kato Paphos, 34.865°N, 32.341°E (RMNH AD 20047). Famagusta, 35.274°N, 33.922°E (RMNH AD 24184). Salamis/Bogasz, Famagusta, 35.312°N, 33.949°E (RMNH AD 24347) (J. van Aarsten, pers. comm. 5/2011).

Turkey. Iztuzu Beach, Dalyan, 36.805°N, 28.606°E (Aartsen & Konzelbach, 1990).

Israel. 0 m, Carmel Bay, 32.800°N, 34.983°E. 60 m, Haifa Bay, 32.867°N, 35.050°E (Barash & Danin, 1992).—10 m, Shiqmona, 32.822°N, 34.997°E (RMNH AD 19676). Haifa, 32.830°N, 34.974°E (RMNH AD 772). 44–85 m, Haifa Bay, 32.891°N, 35.009°E (RMNH AD 13069) (J. van Aarsten, pers. comm. 5/2011).

Egypt. Ramleh, 31.244°N, 29.974°E (Pallary, 1909).

Tunisia. Porto-Farina (Fekih, 1975).—Sidi Youssouf, Kerkennah Island, 34.657°N, 10.971°E (RMNH AD 25594). Nabeul, 36.441°N, 10.736°E (RMNH AD 9518). Sidi Daoud, Cape Bon, 37.088°N, 11.036°E (RMNH AD 9316) (J. van Aarsten, pers. comm. 5/2011).

Algeria. La Macta, Port-aux-Poules, Oran, 35.691°N, 0.165°W (Pallary, 1900).—Sidi Ferruch, Palm Bech, 36.761°N, 2.844°E (RMNH AD 10832). Alger, 36.762°N, 3.068°E. Sidi Ferruch, Port, 36.764°N, 2.848°E (RMNH AD 11023). La Madrague = El Dje, 36.805°N, 2.898°E (RMNH AD 10661) (J. van Aarsten, pers. comm. 5/2011).

Morocco. 64 m, Tangier, 35.705°N, 5.750°W (Pasteur-Humbert, 1962).

Angola. N'zeto, 7.233°S, 12.867°E (Gofas *et al.*, nd).

Scissurellidae
Scissurella

Specimen fossil.

Italy. Terreti - Cellantoni Bridge, 567 m above sea level Pleistocene epibathyal facies, muddy clay, 38.120°N, 15.716°E (DLG 1425, 1).

Literature fossil.

Great Britain. Pliocene, Zanclean of Raydon Hall, Suffolk (Long & Zalasievicz, 2011).

Netherlands. Westerschelde, Ritthem, de Kaloot, Zudsloe, Ouddorp (Wesselingh & Pouwer, 2011).

Italy. Pliocene (d'Orbigny, 1824: as *Sci. elegans*).—Monte Mario, Farnesina. Monte Pellegrino (Cerulli-Irelli, 1916).— Pleistocene of Campo Island, Gallipoli (Coppa & Crovato).

Tunisia. Quarternary and Pliocene (Gougerot & LeRenard, 1977). Early Pliocene (Landau *et al.*, 2003).

Spain. Early Pliocene of Estepona (Landau *et al.*, 2003).

Portugal. Middle Pliocene of Portugal, Pleistocene of Italy (Landau *et al.*, 2003 and references therein).

Remarks. Munier-Chalmas (1865) noted a publication date for d'Orbigny of "17 janvier 1823." It is unclear how he obtained that data.

d'Orbigny (1824) described the species under no less then four names in the same publication. *Scissurella costata* is customarily taken as the valid species name, whereas *Sci. laevigata, Sci. decussata,* and *Sci. elegans* have been considered synonyms. Specific indications as to the synonymy of the various taxa with *Sci. costata* are available (Monterosato, 1878: *Sci. laevigata*; Monterosato, 1872: *Sci. decussata*), which are here interpreted as an application of first reviser's principle. No such formal synonymy can be found for *Sci. elegans*, which is here designated formally as synonym of *Sci. costata* by first reviser's principle.

The synonymy of *Argonauta uniumbilicatus* Costa, 1828/1829 is problematic. Costa (1829) referred to an other publication, which was dated by Bucquoy *et al.* (1882–1886) to 1828. This 1828 publication is untraceable. Bucquoy *et al.* (1882–1886) listed *uniumbilicatus* as a synonym of *Sci. costata.* d'Orbigny (1855) dated the taxon to Costa (1829) and considered it a synonym of *Solarium*. As d'Orbigny had already described several scissurellids, one would presume that he would recognize such a distinctive genus. As the potential original description is currently untraceable, and that such divergent opinions on its synonym have been expressed previously, the synonymy with *Sci. costata* has to viewed with caution.

Padollus orbignyi/dorbignyi Costa should not be confused with *Sukashitrochus dorbignyi* from the Indian Ocean.

Philippi (1853) considered his *Sci. plicata* Philippi a synonym of *Sci. costata* d'Orbigny.

Synonymy of *Sci. striatula*. The identity of this species has been thoroughly confused. It has been synonymized with all three European species, which are classified in three distinct genera: *Sci. costata, Sin. cingulata, An. crispata.*

- Jeffreys (1856a) specified that *striatula* has a closed foramen as opposed to the open slit of other species, hence, would be classified in *Sinezona*. He must have considered Philippi's illustration a juvenile specimen with an open slit. Weinkauff (1868) listed *Sci. striatula* as a synonym of *Sin. cingulata* [as *Schismope elegans*].
- Weinkauff (1873: 40) listed "*striatula* Jeffr." under the genus *Schismope* = *Sinezona*, which suggests he was aware that Jeffreys' *striatula* was not conspecific with Philippi's species: *striatula, sensu* Jeffreys, *non* Philippi.

- Monterosato (1878) listed *Sci. striatula* as a synonym of *An. crispata*. Anistratenko & Starobogatov (1997) noted that *striatula* described an *Anatoma* species, but has been erroneously synonymized with *Sin. crossei*. Anistratenko & Starobogatov's (1997) textual analysis is mystifying. They extracted from the descriptive text of *striatula* characters that are typical for *Anatoma*. The three European genera can be separated into two groups based on overall shell shape: the more trochiform depressed *Scissurella* and *Sinezona*, and the more globular-inflated *Anatoma*. Philippi (1844) used for *Sci. plicata* [= *Sci. costata*] and *Sci. striatula* the term "testa ovata" [shell oval], whereas *Sci. aspera* [= *An. aspera]* was described with "testa orbiculato-conoidea" [shell round-conical]. This initial description rules out that *Sci. plicata* is an *Anatoma* species.
- Jeffreys (1883a) synonymized *Sci. striatula* under *Sci. costata*, without further explanation.

Philippi's (1844) illustration shows a specimen with an open slit, which applies to juvenile and adult *Sci. costata*, as well as juvenile *Sin. cingulata*. The illustration as well as the description lacks strong axial elements evident in all *Sin. cingulata*. *Scissurella costata* shows a gradation of axial sculpture from almost smooth to very well developed. *Scissurella striatula* is hence based on less sculptured specimens of *Sci. costata*.

Fossil. Fekih (1975) noted the variability of the species and that there are no differences between the Recent and the Pliocene specimens. Wesselingh & Pouwder reported the species as extinct in the Recent of the Netherlands. The occurrence in Great Britain is also restricted to fossil (Pliocene) specimens (Long & Zalasiweicz, 2011).

Scissurella cyprina Cotton & Godfrey, 1938

Figures 83–87

Chresonymy.

Scissurella cyprina Cotton & Godfrey, 1938a: 199, not illustrated.
Scissurella cyprina: Cotton & Godfrey, 1938b: 2.
Scissurella cyprina: Cotton, 1959: 45–46, fig. 16.
Scissurella cyprina: Lozouet, 1986: 107.
Scissurella cyprina: Wilson, 1993: 46, text-fig.
Scissurella cyprina: Jansen, 1999: 50–51, figs 19–21.
Scissurella cyprina: Geiger, 2002: 1, text-fig. [shell, radula].
Scissurella cyprina: Geiger, 2003: 76.
Scissurella cyprina: Geiger & Jansen, 2004b: 14–20, figs 6–8 [shell], 9A–C [radula, map].
Scissurella cyprina: Geiger, 2004: text-fig. p. 6.

Misidentifications.

Scissurella ornata: Cotton & Godfrey, 1933: 21, pl. 1, fig. 8 [is *Sci. cyprina*].
Scissurella ornata: May & Macpherson, 1958: 21, pl. 17, fig. 2 [is *Sci. cyprina*].

Type material. Holotype (SAM D9674), 2 × 1.5 mm.

Type locality. Venus Bay, South Australia, Australia [33.233°S, 134.667°E] (OD).

Etymology. Cypris, Latin title of Venus; a shell from Venus Bay.

Description. Shell small (to 2 mm), usually slightly less than 1 mm, depressed trochiform. Protoconch 1.25 whorls, with dense (approximately 30) strong axials on outer and top portion of whorl, inner portion smooth, tops of axials thickened, touching one another, forming spiral cord; no varix, apertural margin convex. Teleoconch I of 0.75–0.875 whorl, 8–13 (usually 11–12) strong axial cords to low lamellae, interstices with irregularly spaced fine growth marks, usually faint spiral in position of selenizone (may easily erode), additional spiral developing on shoulder towards end of teleoconch I. Teleoconch II of >0.5–0.75 whorl, suture strongly impressed. Shoulder with strong axial cords of approximately equal spacing as on teleoconch I, up to 10 narrow but distinct spiral cords at aperture of mature specimens. Base with constriction below selenizone; sculpture variable: axials usually as distinct as on shoulder, sometimes becoming fainter towards umbilicus. Spirals from zero to a few towards the umbilical region to 12 equally spaced ones. Umbilicus deep, smooth transition to base, no carina. Selenizone on outer shoulder, moderately

Figure 83. *Scissurella cyprina*. 0–4 m, Cape Peron, 30 mi S of Perth, Western Australia, Australia, 32.270°S, 115.685°E (AMS C.377559, 2). Scale bar shell = 0.5 mm. Scale bar protoconch = 100 μm.

elevated keels often abraded, slit open, margins somewhat converging. Aperture subquadrate, often detached from previous whorl. Operculum (Figs 84A, 85A–B) thin, corneous, multispiral, nucleus central.

Radula (Fig. 86). Rachidian tooth with broadly triangular tip, central cusp strongest, three smaller ones on each side. Lateral teeth 1–3 similar each with three cusps on tip. Lateral tooth 4 hook-shaped. Lateral tooth 5 broad, points of

Figure 84. *Scissurella cyprina*. **A.** Port Lincoln, Billy Lights Point, South Australia, Australia, 34.750°S, 135.883°E (AMS C.377571, 6). Radula shown in Figure 86. **B.** 57 m, Albany, Western Australia, Australia, 35.167°S, 117.878°E (AMS C.377547, 1). **C.** 3 m, Point Turton, W Yorke Peninsula, South Australia, Australia, 34.950°S, 137.350°E (AMS C.377569, 3). Scale bars shell = 0.5 mm. Scale bars protoconch = 100 μm.

cusps almost parallel with base; inner marginal teeth broad, outer marginal teeth spoon-shaped.

Comparisons. *Scissurella cyprina* is characterized by the distinct spirals on the shoulder that markedly increase in strength on teleoconch II towards the apertural margin.

Figure 85. *Scissurella cyprina*. **A.** 5 m, off Narooma, Montague Island, New South Wales, Australia, 36.253°S, 150.225°E (AMS C.377575, 3). **B.** 104 m, S of D'Entrecasteaux Channel, Tasmania, Australia, 43.673°S, 146.840°E (AMS C.406361). **C.** 3 m, Point Turton, W Yorke Peninsula, South Australia, Australia, 34.950°S, 137.350°E (AMS C.377569, 3). Scale bars shell = 0.5 mm. Scale bars protoconch = 100 µm.

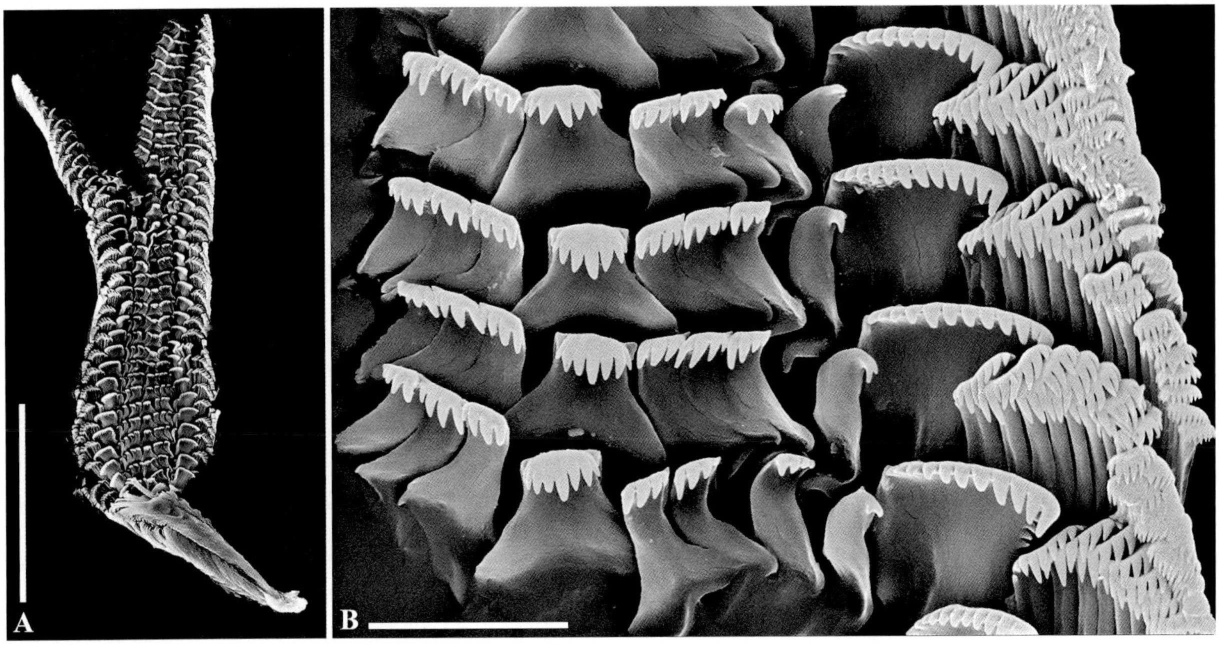

Figure 86. *Scissurella cyprina*. Radula. Port Lincoln, Billy Lights Point, South Australia, Australia, 34.750°S, 135.883°E (AMS C.377571, 6). Shell shown in Figure 84A. **A.** Entire radula. Scale bar = 100 µm. **B.** Central field enlarged with more prominent hook-shaped lateral tooth 4. Scale bar = 10 µm.

Scissurella ornata from temperate Australia has about half the number of axial lamellae (NMW syntype of *Sci. ornata*). Whether *ornata* and *cyprina* are only varieties of one species, or whether they warrant separate recognition is an open question. There is very little material from Tasmania, which could shed light onto intraspecific variability.

Scissurella prendrevillei from New Zealand has a less flared aperture, and has a longer teleoconch I (1.25 vs. 1 whorl).

Scissurella mirifica from the Indo-Malayan Archipelago to the Central Pacific is more inflated and proportionally wider than *Sci. cyprina*.

Scissurella evaensis from the Indo-Malayan Archipelago to the Central Pacific has a characteristic spiral row of elevated lamellae on the mid-base.

Scissurella quadrata from the Indo-Malayan Archipelago is overall more square in outline and has fine axial and spiral sculpture that is evenly distributed over the entire shell.

Scissurella spinosa from the Indo-Malayan Archipelago has spirals and axials of equal strength and evenly distributed over the entire shell; where spirals and axials intersect, its characteristic strong points are formed, giving it an overall spiny appearance.

Distribution. Southern Australian: from New South Wales south of 34°S to Western Australia, south of North West Cape (22.5°S. 113.5–151.5°E. Fig. 87).

Specimen records.

Australia, New South Wales. Little Coogee Bay, S of Sydney, 33.921°S, 151.262°E (AMS C.346336, 1). Collaroy Beach, N of Sydney, 33.728°S, 151.300°E (AMS C.461564, 1). Off Eden, 37.083°S, 150.000°E (AMS C.377578, 1). 5 m, off Narooma, Montague Island, 36.253°S, 150.225°E (AMS C.377575, 3). Bondi, 33.893°S, 151.264°E (MV F 161672, 2). 9–27 m, Pittwater, Hawkesbury, 33.630°S, 151.298°E (NMNZ M.273213, 1). 10 m,

N of Batesman Bay, W side of Wasp Island, 35.667°S, 150.308°E (AMS C.461547, 1). 0 m, E of Grafton, Minnie Water, 29.777°S, 153.300°E (AMS C.461563, 1). Bondi, 33.893°S, 151.264°E (MV F 161672, 2). Port Stephens, 32.394°S, 152.067°E (RMNH AD ex 30458, 3). Narrabeen, 33.714°S, 151.302°E (RMNH AD ex 30459, 4).

Australia, South Australia. 7 m, Edithburgh Jetty, Yorke Peninsula, 35.083°S, 137.750°E (AMS C.404233, 5: complete). Fisherman Point, Cape Donnington, 34.758°S, 135.983°E (AMS C.377572, 1; AMS C.377574, 2). 7 m, Kangaroo Island, N coast, Stokes Bay, 35.617°S, 137.200°E (AMS C.377598, 1, 2: complete). 18 m, Pearson Island, 33.967°S, 134.283°E (AMS C.404988). Point Souttar, Yorke Peninsula, 34.900°S, 137.267°E (AMS C.377595, 1). 3 m, Point Turton, W Yorke Peninsula, 34.950°S, 137.350°E (AMS C.377569, 3, 2: complete). 1–4 m, Point Turton, W Yorke Peninsula, 34.950°S, 137.350°E (AMS C.377573, 2). Port Lincoln, Billy Lights Point, 34.750°S, 135.883°E (AMS C.377571, 6). Venus Bay, 33.233°S, 134.683°E (SAM, 1). Bluff Beach, Yorke Peninsula, 34.733°S, 137.483°E (AMS C.402703, 1).

Australia, Tasmania. 10–30 m, Little Squally Cove, Deal Island, Bass Strait, 39.500°S, 147.333°E (AMS C.404901, 1: complete). 104 m, S of D'Entrecasteaux Channel, 43.673°S, 146.840°E (AMS C.377577, 1; AMS C.406361, 1).

Australia, Victoria. 24 m, SSE end of Gabo Island, 37.567°S, 149.910°E (AMS C.377570, 1). 28 m, SSE end of Gabo Island, 37.567°S, 149.917°E (AMS C.404987). Lorne, 38.550°S, 143.967°E (WAM, 1).

Australia, Western Australia. 91 m, 50 mi N of Boolbarly, 23.467°S, 113.467°E (AMS C.377576, 1). Carnac Island, W side, off Freemantle, 32.123°S, 115.658°E (AMS C.377548, 1). 4–8 m, Carnac Island, W side, off Freemantle, 32.117°S, 115.667°E (AMS C.377579, 1; AMS C.379972, 2; AMS C.377278, 3). 120 m, Direction Bank, off Rottnest Island, 31.768°S, 115.203°E (AMS C.377543, 1). 130 m, Direction Bank, off Rottnest Island, 31.703°S, 115.170°E (AMS C.377563, 1). 120 m, Direction Bank, off Rottnest Island, 31.723°S, 115.230°E (AMS C.377568, 1). 145 m, Direction Bank, off Rottnest Island, 31.777°S, 155.168°E (AMS C.461474, 2). 158 m, Direction Bank, off Rottnest Island, 31.730°S, 115.163°E (AMS C.461475, 3). 62 m, off Rottnest Island, 31.633°S, 115.226°E (AMS C.461556, 4). 143 m, off Rottnest Island, 31.708°S, 115.186°E (AMS C.461552, 2). 124 m, off Rottnest Island, 31.718°S, 115.208°E (AMS C.461472, 1). Ellensbrook, S Cowaramup, near Margaret River mouth, 33.883°S, 114.983°E (AMS C.379975, 1). 161–165 m, NW of Bunbury, 33.250°S, 114.600°E (AMS C.377593, 2). 71 m, Albany, 35.245°S, 118.173°E (AMS C.377592, 6). 112 m, off Albany, 35.333°S, 118.333°E (AMS C.461473, 7; AMS C.461553, 1). 170 m, Bunbury, 33°S, 114.683°E (AMS C.377565, 2). 70–90 m, off Esperance, 34.360°S, 121.160°E (AMS C.377282, 1). 150 m, Rottnest Island, 31.778°S, 115.170°E (AMS C.377283, 1). 140 m, Rottnest Island, 31.633°S, 115.133°E (AMS C.377285, 2). 124 m, Rottnest Island, 31.650°S, 115.167°E (AMS C.377286, 3). 155 m Rottnest Island, 31.732°S, 115.158°E (AMS C.377287, 1). 148 m, Rottnest Island, 31.688°S, 115.143°E (AMS C.377566, 1). 124 m, Rottnest Island, 31.940°S, 115.285°E (AMS C.377596, 1). 150 m, Rottnest Island, 31.703°S, 115.137°E (AMS C.377597, 2). 138 m, Rottnest Island, 31.737°S, 115.195°E (AMS C.377599, 1). 150 m, Rottnest Island, 31.663°S, 115.118°E (AMS C.377602, 2). 134 m, Rottnest Island, 31.777°S, 115.213°E (AMS C.377603, 2). 148 m, Rottnest Island, 31.688°S, 115.143°E (AMS C.403739, 1). 0–4 m, Cape Peron, 30 mi S of Perth, 32.270°S, 115.685°E (AMS C.377559, 2; AMS C.404907, 1: complete). 146–150 m, SW of Mandurah, 32.683°S, 114.850°E (AMS C.377601, 2). 274–219 m, W of Cape Hamelin, 34.250°S, 114.500°E (AMS C.377280, 1). 146 m, W of Dongara, 29.350°S, 114.117°E (AMS C.377281, 3). 183 m, W of Dongara, 29.142°S, 113.913°E

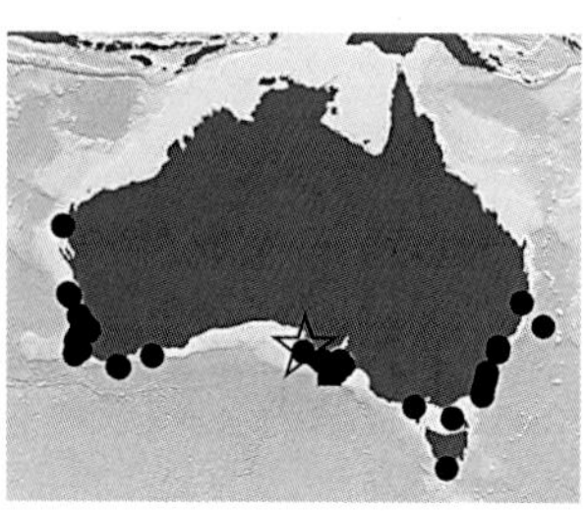

Figure 87. Distribution of *Scissurella cyprina*. ☆ type locality. ● specimen record. ■ literature record.

(AMS C.377545, 4). 201–214 m, W of Rottnest Island, 31.992°S, 115.217°E (AMS C.377279, 3). 146 m, W of Rottnest Island, 31.000°S, 114.850°E (AMS C.377567, 1). Perkin Point, Garden Island, 32.233°S, 115.683°E (WAM, 1).

Literature records.

Australia, South Australia. Cape Borda [35.750°N, 136.567°E], Minlacowie (Cotton, 1959).

Remarks. *Scissurella cyprina* was described by Cotton & Godfrey (1938) from material that was identified by them in their 1933 paper as *Sci. ornata*.

Scissurella ornata may actually be conspecific with *Sci. cyprina*. The only distinguishing character is the greater number of axial cords. Protoconch sculpture, teleoconch I whorl, flared lip, and umbilicus are all identical.

Scissurella dalli Bartsch, 1903 [*nomen dubium*]

Chresonymy.

Scissurella dalli Bartsch, 1903: 90–91, not illustrated.
Scissurella dalli: Thiele, 1912a: 8.
Scissurella dalli: Carcelles, 1950: 50.
Schizotrochus dalli: Caracelles & Williamson, 1951: 251.
Scissurella dalli: Castellanos & Landoni, 1988: 9.
Scissurella dalli: Forcelli, 2000: 44.
Scissurella dalli: Geiger, 2003: 76.
Scissurella dalli: Zelaya & Geiger, 2007: 431–432 [as *nomen dubium*].
Scissurella dalli: Zelaya & Geiger, 2008: 265–268 [as *nomen dubium*].

Type material. Holotype (USNM 171400. Missing: Zelaya & Geiger, 2007), 1.4 × 0.8 mm.

Type locality. Gulf of St. George, Patagonia [Argentina, ~ 46°S, 66°W] (OD). Leg. Ihering (USNM label).

Etymology. Named after William Healy Dall (1845–1927).

Remarks. The species was not illustrated by Bartsch (1903), and has never been figured elsewhere. Bartsch's description lacks any discernible detail necessary to evaluate the species. The type material is missing from USNM (Geiger, pers. obs. 2/2004) where it was originally deposited (*fide* Bartsch, 1903). The generic placement is tentative; Bartsch noted the selenizone "near periphery," which applies to *Scissurella*, while in *Anatoma* the selenizone is at the periphery; *Anatoma* and *Scissurella* were not used as distinct genera in the early 20th century. Whether Bartsch's species was based on an immature specimen of a *Sinezona* sp. remains an open question. Zelaya & Geiger (2008) discussed the taxon in detail.

Scissurella dohrniana (Dunker, 1861) [*nomen dubium*]

Chresonymy.

Anatomus dohrnianus Dunker, 1861: 35, not illustrated.
Scissurella dohrniana: Munier-Chalmas, 1865: 396.
Scissurella dohrniana: Issel, 1869: 228.
Scissurella dohrniana: Paetel, 1888: 289.
Anatomus dohrniana: Pilsbry, 1890: 60.
Anatomus dohrnianus: Sturnay, 1905: 146 [in part: *fide* Yaron, 1983].
Scissurella dohrniana: Thiele, 1912a: 11.
Sukashitrochus dohrnianus: Habe & Kosuge, 1964: 3.
Scissurella dohrniana: Yaron, 1983, 267–168 [*nomen dubium*].
Scissurella dohrniana: Dekker & van Capelle, 1994: 125.
Scissurella dohrniana: Dekker & Orlin, 2000: 17.
Scissurella dohrniana: Geiger, 2003: 76 [as *nomen dubium*].

Misidentification.

Scissurella dohrniana: Sturnay, 1905: 146 [is in part *Suk. dorbignyi* (as *Suk. armillatus*), *fide* Yaron, 1983].

Type material. (whereabouts unknown), 2.5 mm.

Type locality. Red Sea (OD).

Etymology. Named for Heinrich Dohrn (1838–1913).

Description. The original description reads [translated from Latin]: "Globose shell, very small, almost transparent, concentrically and very finely striated, narrowly umbilicated; three carinated whorl, spire small and stepped; fissure not deep, embracing the keel of the whorl, aperture finely sharp; aperture oblique, almost round, almost elliptical."

The German section reads: "The very thin, milky white, somewhat translucent, almost spherical, concentrically finely and densely girdled shell consists of three quite rapidly increasing whorls, which appears angulated and therefore step-like separated by a carina found on top, in which anteriorly there is a small insertion. The aperture has an almost elliptical outline; the umbilicus is rather narrow. At large magnification one notes additionally to the concentric or rather the spiral girdlets that follow the direction of the whorls, in addition also quite fine growth lines.

Compared to the in Philippi and d'Orbigny popularized *Anatomus*- (*Scissurella*-) species, none of them agrees with the ours."

Remarks. Habe & Kosuge (1964) placed the species in *Sukashitrochus* based on the indication of spiral cords mentioned in the original description. Most likely they erroneously translated the "anfractus 3 carinati" as 'whorl with three keels', instead of 'three keeled whorls'. The carina refers to the selenizone as it is described as being open anteriorly with an insertion (= slit).

Yaron (1983) discussed the identity of the species in detail and concluded it may be a synonym of *Sci. reticulata*, although there is some doubt remaining. Following Yaron (1983), *Sci. dohrniana* is treated as a *nomen dubium*. Thiele (1912) could not identify the species either, nor could he locate a specimen in Dunker's collection.

? *Scissurella electilis* Montouchet, 1972

Figures 88–89

Chresonymy.

Scissurella electilis Montouchet, 1972: 8–9, pl. 1, figs 8–10.
Scissurella electilis: Rios, 1985: 10, pl. 5, fig. 32 [copy figure Montouchet, 1972].
Sinezona electilis: Lozouet, 1986: 111.
Scissurella electilis: Absalão, 1989: 2.
Scissurella electilis: Rios, 1994: 21, pl. 5, fig. 35 [copy figure Montouchet, 1972].
Scissurella electilis: Geiger, 2003: 76.
Scissurella electilis: Rios, 2009: 24, text-fig. [copy figure Montouchet, 1972].
Scissurella electilis: Dornellas & Simone, 2011: 14, figs 20–21 [holotype].

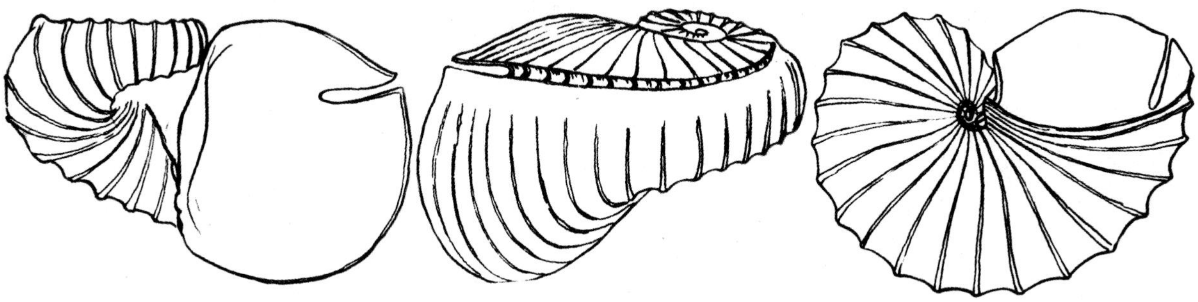

Figure 88. Original illustration of *Scissurella electilis*.

Type material. Holotype (MZSP 18494. Dornellas & Simone, 2011: figs 20–21), 0.5 × 0.3 mm. Paratypes off Ponta de Pedra, Pernambuco State (MZSP 18501, 3: one specimen surviving: Dornellas & Simone, 2011).

Type locality. Itamaracá, Pernambuco, Brazil [7.750°S, 34.750°W]. Dead shells in sea-weed samples (OD).

Etymology. Electilis, Latin for choice, dainty selected (OD).

Description. [Based on original description]. Shell small [to 0.6 mm], trochiform depressed. Protoconch unknown. Teleoconch I, number of whorls unknown, teleoconch II approximately 0.3 whorl. Shoulder with approximately 22 axial cords on last whorl, no spiral sculpture, spiral band devoid of axials above [?] selenizone. Base with same density of axial cords, no spirals, continuously sloping with narrow umbilicus. Aperture rounded, flared.

Animal unknown.

Comparisons. *Scissurella alexandrei* [= juvenile *Sin. confusa*?] from Brazil is high-spired and has axial as well as spiral sculpture.

Scissurella morretesi from Brazil has a somewhat elevated spire, has denser axial sculpture, and is anomphalous.

Sinezona insularis from off-shore Brazil has fewer axial cords (10 vs. 15 on last half whorl) and a shorter selenizone (0.15 vs. 0.3 whorl).

Scissurella redferni from the Caribbean has a more elevated spire and a distinct projection of the adumbilical corner of the aperture.

Sinezona confusa from the Caribbean has a taller spire, axial cords of increasing density towards the aperture as well as a closed foramen of a fully grown specimen.

Literature records.

Brazil. Off Ponta de Pedra, Pernambuco, 7.633°S, 34.800°W (Montouchet, 1972).—Off Esipirito Santo, ~22.3°S, 40.8°W (Absalão, 1989).

Distribution. Brazil (Fig. 89).

Remarks. It is virtually impossible to obtain scientific loans from Brazil due to bureaucratic hurdles imposed by the Brazilian Government, hence, no specimens of this species in Brazilian collections could be examined, including type material. The species is not known from any specimens in collections elsewhere.

The type specimens are fully decalcified (Dornellas & Simone, 2011).

Montouchet noted the spiral band "above" the selenizone devoid of sculpture. The illustration shows such a spiral band *below* the selenizone. It is not clear whether there was a translation error, or whether there are two such spiral bands.

Montouchet noted the lenticular shape of the slit, and speculated that it may close in the adult to form a foramen. If this is the case, then *Sci. electilis* should be properly classified in *Sinezona*. It seems to be distinct from *Sin. confusa* as well as *Sat. brasiliensis,* because the whorls of the latter two descend more along the axis of the shell at the teleoconch II whorl count of *Sci. electilis.*

Figure 89. Distribution of *Scissurella electilis.* ☆ type locality. ■ literature record.

Scissurella evaensis Bandel, 1998

Figures 90–95

Chresonymy.

Scissurella evaensis Bandel, 1998: 15–16, pl. 3, fig. 8, pl. 4, figs 1–2.
Scissurella evaensis: Geiger, 2003: 77 [see Remarks].
Scissurella evaensis: Geiger & Jansen, 2004b: 26–28, figs 13 [shell], 17c [map].

Figure 90. Original illustration of *Scissurella evaensis.*

Scissurella staminea from tropical Japan has a wider overall shell shape with wider umbilicus, lacks the distinct tubercles on the mid-base, and has reticulate sculpture on the base.

Scissurella sundanica from the Indian Ocean has a wider overall shell shape with wider umbilicus, lacks the distinct tubercles on the mid-base, and has fine reticulate sculpture on the base.

Figure 93. *Scissurella evaensis*. **A–C.** 7–10 m, Hibernia Reef, in lagoon east side, dead *Acropora* rubble, Northern Territories, Australia, 11.967°S, 123.349°E (SBMNH 348010, 3). Scale bars shell = 0.5 mm. Scale bars protoconch = 100 μm.

Scissurella xandaros from the Indo-Malayan Archipelago has a wider overall shell shape with wider umbilicus, has strong axial ribs lacking projections on the mid-base, and has a smooth protoconch.

Scissurella jucunda from South Africa and the Indian Ocean has a taller spire and lacks the projecting points on the mid-base,

Scissurella maraisorum from eastern South Africa has a more depressed spire, the protoconch has a strong mid-rib and lacks an apertural varix, and the protoconch can be seen through the wide umbilicus.

Sinezona ferriezi from the Indo-Malayan Archipelago to the Central Pacific has a wider overall shell shape with wider umbilicus, distinctly prosocline axial cords, and a closed foramen.

Distribution. Indian Ocean through Central Pacific. 0–60 m depth, usually in coral reef assemblages; shells to 460 m (Fig. 95).

Specimen records.

Madagascar. 296–307 m, St. Luce sector, ATIMO VATAE DW3534, 24.716°S, 47.533°E (MNHN, 1). 80–83 m, between Lokaro and St. Luce, ATIMO VATAE DW3519, 24.865°S, 47.466°E (MNHN, 4; MNHN, 1). 184–203 m, between Lokaro and St. Luce, ATIMO VATAE DW3515, 24.883°S, 47.466°E (MNHN, 5). 56–57 m, SW Cape St. Marie, ATIMO VATAE DW3805, 24.908°S, 44.850°E (MNHN, 1). 18 m, Flacourt Lighthouse, ATIMO VATAE TB02-03, 25.021°S, 47.008°E (MNHN, 1). 65–66 m, S False Cape, ATIMO VATAE CP3579, 25.908°S, 45.553°E (MNHN, 1). 455–458 m, S of False Cape, ATIMO VATAE DW3555, 26.150°S, 45.666°E (MNHN, 2). 284–286 m, S Cape St. Marie, ATIMO VATAE CP3615, 26.233°S, 45.150°E (MNHN, 1). 86–87 m, N of St. Luce, DW3532, 24.656°S, 47.528°E (MNHN, 1).

Japan, Okinawa. 20 m, 5 km W Tsuken-Jima (= Admiral's Island), 26.268°N, 127.920°E (LACM 78-21, 2). 50 m, 0.5 km ESE Zampa-Misaki (Bolo Point), 26.435°N, 127.708°E (LACM 78-25, 5). 49 m, 0.5 km ESE of Zampa-Misaki, 26.435°N, 127.875°E (LACM 78-100, 5). 58 m, 1 km W Onna Village, 26.493°N, 127.842°E (LACM 78-101, 17). 60 m, 1 km W Onna Village, 26.493°N, 127.842°E (LACM 79-76, 15). 50 m, 1 km NWN of Onna Village, 26.493°N, 127.842°E (LACM 79-75, 13). 33 m, 1 km W Onna Village, 26.495°N, 127.843°E (LACM 78-20, 5). 50 m, 1 km W Onna Village, 26.495°N, 127.843°E (LACM 78-99, 2). 46 m, 1 km WNW Onna Village, 26.495°N,

Figure 94. *Scissurella evaensis*. Radula. 33 m, 1 km W Onna Village, Okinawa, Japan, 26.495°N, 127.843°E (LACM 78-20, 5). **A.** Whole radula width. Scale bar = 20 µm. **B.** Lateral tooth 5 and marginal teeth. Scale bar = 10 µm. **C.** Operculum. Scale bar = 200 µm.

127.843°E (LACM 78-26, 1). 60 m, 1 km NWN of Onna Village, 26.497°N, 127.843°E (LACM 78-29, 1; LACM 78-29, 1). 50 m, 1 km WNW of Onna Village, 26.497°N, 127.843°E (LACM 78-99, 3). 33 m, Nakijin, Motobu Peninsula, 26.703°N, 127.942°E (LACM 78-22, 2). 16 m, 1 km NNW Oku, 26.847°N, 128.287°E (LACM 77-64, 4). 71–85 m, off Kerama Island, 26.159°N, 127.448°E (SBMNH 149470, 4). 211–212 m, off Kerama Island, 26.047°N, 127.375°E (SBMNH 149462, 10). 28 m, Maeda Point, 26.450°N, 127.759°E (DLG 1072, 10). 17–35 m, Maeda Point, 26.435°N, 127.759°E (DLG 965, 8). 38 m, base of cliff, Horeshoe Bay, 26.504°N, 127.858°E (DLG 1053, 24). 35–38 m, base of cliff, Horseshoe, 26.504°N, 127.858°E (DLG 973, 5). 20–30 m, Maeda Point, 26.445°N, 127.773°E (DLG 1526, 1). 43 m, Horseshoe Beach, 26.505°N, 127.853°E (DLG 1543, 6). 37 m, Big Rock, Horseshoe, 26.508°N, 127.853°E (DLG 1461, 1). 13–33 m, Flab, Horseshoe, 26.509°N, 127.856°E (DLG 1488, 3). 33 m, Horseshoe, 26.509°N, 127.856°E (DLG 1516, 1).

Philippines. 13 m, E end Santa Cruz Island, Zamboanga, 6.865°N, 122.073°E (LACM 81-7, 4). 7 m, off Tambuli resort, Mactan, 10.283°N, 124.017°E (LACM 81-2, 1). 20 m, Looc, Panglao Island, PANGLAO 2004 B32, 9.430°N, 123.743°E (MNHN, 37). 6–14 m, Pamilacan Island, PANGLAO 2004 S10, 9.490°N, 123.933°E (MNHN, 12). 90–98 m, Balicasag Island, PANGLAO 2004 L69-73, 9.511°N, 123.682°E (MNHN, 8). 120–139 m, Balicasag Island, PANGLAO 2004 L74–75, 9.511°N, 123.682°E (MNHN, 3). 85–100 m, Balicasag Island, PANGLAO 2004 L44, 9.513°N, 123.683°E (MNHN, 1). 90–110 m, Balicasag Island, PANGLAO 2004 L46, 9.515°N, 123.686°E (MNHN, 6). 80–90 m, Balicasag Island, PANGLAO 2004 L42, 9.520°N, 123.678°E (MNHN, 1). 90–100 m, front of PTA Compound, Panglao Island, PANGLAO 2004 L41, 9.521°N, 123.686°E (MNHN, 1). 24 m, BBC Point, Panglao Island, PANGLAO 2004 B4, 9.553°N, 123.805°E (MNHN, 1). 5 m, Biking, Panglao Island, PANGLAO 2004 S1, 9.588°N, 123.841°E (MNHN, 1). 24–27 m, Doljo Point, Panglao Island, PANGLAO 2004 B12, 9.593°N, 123.720°E (MNHN, 9). 24 m, N of Doljo, Panglao Island, PANGLAO 2004 B36, 9.598°N, 123.741°E (MNHN, 2). 31 m, N of Doljo, Panglao Island, PANGLAO 2004 B35, 9.598°N, 123.741°E (MNHN, 1). 4–30 m, Tatarman, Panglao Island, PANGLAO 2004 B7, 9.598°N, 123.863°E (MNHN, 1). 90–200 m, Maribohoc Bay, Bohol Island, PANGLAO 2004 P1, 9.600°N, 123.750°E (MNHN, 2). 0–3 m, Momo Beach, Panglao Island, PANGLAO 2004 M7, 9.601°N, 123.753°E (MNHN, 1). 3–14 m, Momo Beach, Banglao Island, PANGLAO 2004 B10, 9.608°N, 123.760°E (MNHN, 20). 28–32 m, Momo Beach, Panglao Island, PANGLAO

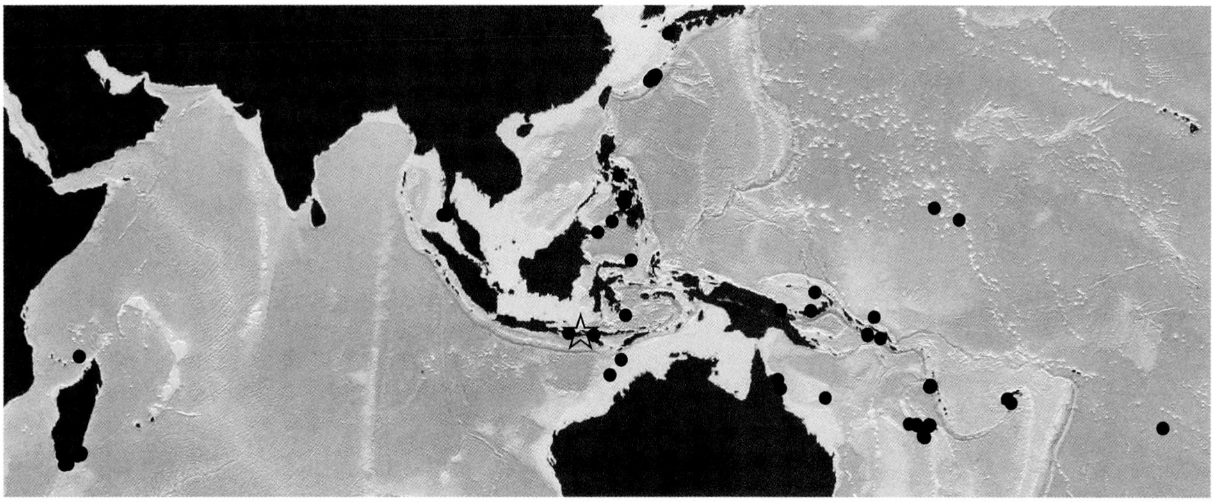

Figure 95. Distribution of *Scissurella evaensis*. ☆ type locality. ● specimen record.

2004 S8, 9.608°N, 123.760°E (MNHN, 3). 120 m, off Momo Beach, Panglao Island, PANGLAO 2004 L50, 9.615°N, 123.763°E (MNHN, 1). 30–33 m, between Momo and Napaling, Panglao Island, PANGLAO 2004 B42, 9.616°N, 123.766°E (MNHN, 4). Napaling, Panglao Island, PANGLAO 2004 B8, 9.618°N, 123.769°E (MNHN, 1). 28–32 m, Napaling, Panglao Island, PANGLAO 2004, 9.620°N, 123.773°E (MNHN, 1). 20–21 m, Napaling, Panglao Island, PANGLAO 2004 B21, 9.620°N, 123.773°E (MNHN, 9). 3–21 m, Bingag, Panglao Island, PANGLAO 2004 B17, 9.625°N, 123.781°E (MNHN, 3). 43–62 m, Bingag/Tabalong, Panglao Island, PANGLAO 2004 L51-60, 9.628°N, 123.800°E (MNHN, 42). 3–5 m, Sungcolan Bay, Panglao Island PANGLAO 2004 B18, 9.641°N, 123.828°E (MNHN, 1). 2–4 m, Sungcolan, Panglao Island, PANGLAO 2004 B15, 9.646°N, 123.818°E (MNHN, 9). 4–6 m, Cortes Takot, Bohol Island, PANGLAO 2004 S15, 9.688°N, 123.825°E (MNHN, 2). 21 m, Ubajan, Bohol Island, PANGLAO 2004 S25, 9.691°N, 123.850°E (MNHN, 1). 4–12 m, Manga, Bohol Island, PANGLAO 2004 S21, 9.695°N, 123.848°E (MNHN, 5).

Thailand. 0 m, S tip of Phromthep Cape, Phuket, 7.758°N, 98.322°E (LACM 85-6, 1). 3 m, Mai Thon Island, off Phuket, 7.762°N, 98.488°E (LACM 85-8, 6). 15 m, Hin Hmusang, rock near Pee-Pee Island, Phuket, 7.792°N, 98.550°E (LACM 85-14, 3).

Indonesia. 15 m, off Ajer and Meno Islets, Lombok, 8.367°S, 116.067°E (LACM 88-63, 2). 520 m, Sulawesi, Bunaken and Siladen Islets, 1.603°N, 124.767°E (LACM 88-55, 9). 65 m, Teluk Slawi, E of Komodo, 8.613°S, 119.512°E (RMNH 42349, 10).

New Guinea. 9 m, Yassi II reef, offshore from Nagada, Astrolabe Bay, 5.158°S, 145.837°E (LACM 80-13, 1). 30 m, channel between Wongat Island and barrier reef at Astrolabe Bay, Madang Province, 5.135°S, 145.845°E (LACM 80-26, 1). 15–35 m, Baudisson Bay, Kavieng, New Ireland, 2.743°S, 150.658°E (DLG 297, 5).

Australia. 20 m, E side, Sandy Islet, Scott Reef, Western Australia, 14.044°S, 121.780°E (SBMNH 369508, 5). 25 m, Hibernia Reef, Northern Territories, 11.967°S, 123.349°E (SBMNH 348009, 1). 7–10 m, Hibernia Reef, in lagoon E side, dead *Acropora* rubble, Northern Territories, 11.967°S, 123.349°E (SBMNH 348010, 3). 15 m, 0.4 km off Anchor Bay, W Lizard Island, Queensland, 14.670°S, 145.433°E (LACM 79-56, 1). 38 m, Outer Slope, S part Escape Reef, Great Barrier Reef, Queensland, 15.883°S, 145.817°E (AMS ex C.461466, 1). 10–15 m, SE from Observatory Cay, Lihou Reef, Coral Sea Territory, 17.157°S, 152.141°E (SBMNH 369622, 4).

Solomon Islands. 22 m, Matui Island, 8.495°S, 158.200°E (LACM, 4). 10–25 m, Sakulufat Point Mbanika Island, Russell Island, Central Province, 6.120°S, 159.158°E (SBMNH 85143, 1). 10–25 m, Tanavula Point, Nggela Sule, Florida Island, Central Province, 9.043°S, 160.065°E (SBMNH 367143, 1).

Marshall Islands. 23 m, Ocean Side, Kwajalein, 8.739°N, 167.746°E (DLG 1002, 3).

New Caledonia. 8 m, Touho sector, 20.741°S, 165.265°E (MNHN, 2). 577–780 m, south, BATHUS 2 DW754, 22.383°S, 166.216°E (MNHN, 1). 250–350 m, St. 40, 22.500°S, 166.400°E (MNHN, 1).

Vanuatu. 6–43 m, Aésé Island, SANTO 2006 DB58, 15.410°S, 167.238°E (MNHN, 1). 21 m, SE Aésé Island, SANTO 2006 DB63, 15.448°S, 167.263°E (MNHN, 2). 100–105 m, NW Tutuba Island, SANTO 2006 DS99, 15.541°S, 167.281°E (MNHN, 1). 92 m, NW Tutuba Island, SANTO 2006 DS105, 15.550°S, 167.278°E (MNHN, 1). 80 m, W Tutuba Island, SANTO 2006 DS104, 15.568°S, 167.266°E (MNHN, 1). 90–110 m, NW coast of Malo Island, Palikulo Bay, SANTO 2006 DB77 EP28, 15.633°S, 167.083°E (MNHN, 1).

Fiji. 80–120 m, SE Viti Levu, MUSORSTOM 10 CP1358, 17.808°S, 178.778°E (MNHN, 1).

Cook Islands. 22 m, Rarotonga, N of Matavera, 21.212°S, 159.725°W, (LACM 87-81, 1).

Remarks. The registration number of the holotype in the body of text is given as "3871" (Bandel, 1998: 16). According to the text, the illustrated specimen is the holotype. However, the caption to the plate 3, figure 8, page 78, gives the specimen number of the illustrated specimen as "3872."

The etymology mentions explicitly that the species is named in honor of Mrs. Eva Vinx. The construction of the species epithet with suffix -ensis instead of feminine genitive evae suggest a place name. The original spelling was maintained as an intentional error of the original author, and treated as a noun in apposition applying first reviser's principle (Geiger, 2003).

Maxwellella unispirata has been synonymized under *Sci. evaensis* by Geiger (2003). Both taxa were described from Satonda, Indonesia. The two taxa share a protoconch with fine axials and an apertural varix, the shell sculpture dominated by strong axials in between which finer irregular axials are found, and a row of raised tubercles on the strong axials on the shoulder and the base. The only difference is that in *M. unispirata* the apparent position of the last whorl with respect to the remainder of the spire is somewhat lower. However, the angle at which the SEMs were taken also differ and can account for this apparent difference. Both types are rather small specimen with only approximately 0.5 teleoconch II whorls, while fully grown specimens have more than one teleoconch II whorls.

Scissurella georgica Davolos & Moolenbeek, 2006

Figures 96–97

Chresonymy.

Scissurella georgica Davolos & Moolenbeek, 2006: 134–135, figs 8a–d.

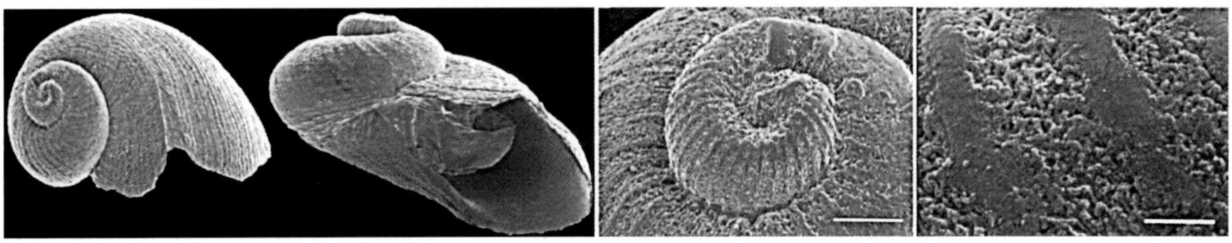

Figure 96. Original illustration of *Scissurella georgica*.

Type material. Holotype (ZMA), 1 × 0.9 mm.

Type locality. 0.25–1 m, Penguin Bay, Barff Peninsula, South Georgia [British Territory], 54°20'S 36°14'W (OD).

Etymology. Named for its type locality in South Georgia.

Description. Shell medium-sized (to 1 mm), trochiform depressed. Protoconch of one whorl, approximately 25 fine axial cords covering more than half of protoconch width, apertural varix not connected to embryonic cap, apertural margin convex. Teleoconch I of 1.25 whorls, approximately 34 indistinct axial cords, interstices with fine granular sculpture. Teleoconch II of 0.25 whorl, suture moderately impressed. Shoulder flat, with same density of indistinct axial cords as on teleoconch I, approximately eight even less distinct spiral cords. Base with slight depression below selenizone, same density of axials as on shoulder, approximately 14 spiral cords, very indistinct in region below selenizone, becoming quite prominent towards umbilicus. Umbilicus open, wide, continuously sloping with base. Selenizone above periphery, keels low, moderately strong; slit open (?). Aperture oblong D-shaped, roof overhanging.

Animal unknown.

Comparisons. *Scissurella petermannensis* from the Antarctic has strong axial lamellae on the teleoconch.

Scissurella clathrata from temperate-subantarctic South America is overall taller, has a protoconch with flocculent sculpture, shows overall reticulate teleoconch sculpture, and has a very narrow umbilicus.

Scissurella obliqua from the Antarctic has a protoconch with anastomosing network sculpture and has an overall taller shell.

Scissurella supraplicata from subantarctic islands has strong axial cords on teleoconch I, has a taller spire, and a very narrow umbilicus.

Sinezona subantarctica from the subantarctic Islands has a protoconch with irregular and anastomosing axial ribs, lacks a selenizone, and has the umbilicus covered with a callus.

Sinezona mouchezi from subantarctic islands has a taller overall shell with narrower umbilicus and a protoconch with fewer but stronger axial cords.

Figure 97. Distribution of *Scissurella georgica*. ☆ type locality.

Sinezona chilensis from southwestern South America has taller shell with narrower umbilicus, the selenizone is closer to the periphery, and the aperture is rounder.

Sinezona levigata from New Zealand has a protoconch with strong axial cords, a very short selenizone (occasionally absent) and a closed foramen, and the umbilicus is covered by a callus.

Sinezona mechanica from New Zealand has a shorter teleoconch I (<1 vs. 1.25 whorls), has more prominent axial cords, and lacks spirals.

Distribution. Known only from type locality (Fig. 97).

Remarks. The species is known only from the holotype, which is an immature specimen as the last quarter whorl does not descend. Its generic affinity (*Scissurella*/*Sinezona*) is, therefore, not resolved.

Scissurella hoernesi Semper, 1865 [*nomen dubium*]

Figures 98–99

Chresonymy.

Scissurella hoernesi Semper, 1865: 284, 288–289, pl. 12, fig. 4.
Scissurella hoernesi: Paetel, 1888: 289.
Scissurella hoernesi: Pilsbry, 1890: 51, pl. 57, figs 15–16 [copy figs Semper, 1865].
Scissurella hoernesi: Hildago, 1904: 261.
? *Scissurella hoernesi*: Thiele, 1912a: 14, pl. 2, fig. 6 [= Fig. 98B).
? *Scissurella hoernesi*: Thiele, 1925: 42–43.
Scissurella hoernesi: Faustino, 1928: 126.
Scissurella koeneni: Lozouet, 1986: 107.
? *Scissurella hoernesi*: Bandel, 1998: 10–11, pl. 1, figs 6–8 [juvenile of *Sin. ferriezi*].
Scissurella hoernesi: Geiger, 2003: 77.

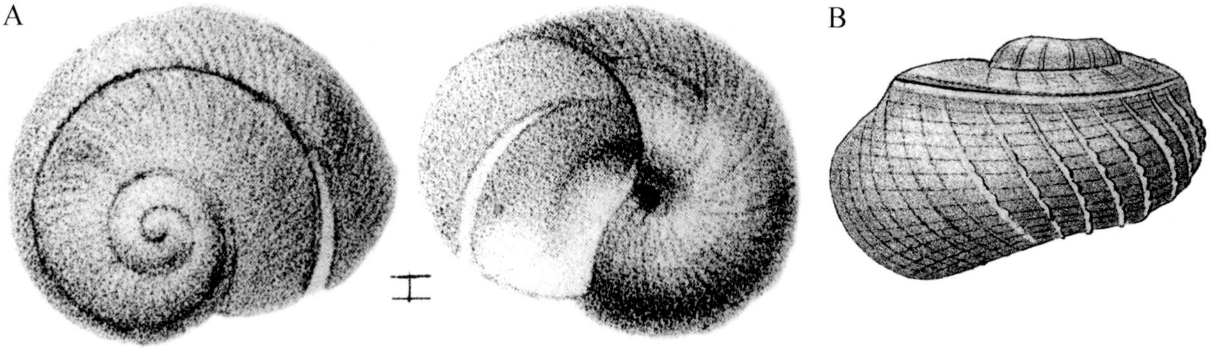

Figure 98. **A.** Original illustrations of *Scissurella hoernesi.* **B.** Illustration of *Scissurella hoernersi* from Thiele (1912); "after a specimen in the Hamburg Museum" = holotype?

Type material. Holotype (Hamburg Museum: ? Thiele, 1912a: pl. 2. fig. 6. Fig. 98B), 1.25 × 1 mm (W × H). Lost during World War II (B. Hausmann, pers. comm. 2/2007).

Type locality. Manhan, Luzon Island, Philippines [~ 16.700°N, 121.300°E](OD. Fig. 99).

Etymology. Named after the Vienna paleontologist Hörnes (OD).

Remarks. The species is here treated as a *nomen dubium* for the following reasons. The type specimens that may have been located in the Hamburg Museum (c.f., *Sci. koeneni*), were destroyed in World War II (B. Hausmann, pers. comm. 2/2007). The original illustrations and the description are very crude, at best allowing family-level identification; the specimen illustrated by Thiele (1912), which may have been the holotype, cannot be identified with certainty to species (e.g., *Sci. mirifica, Sci. lorenzi, Sin. ferriezi, Sin. marrowi*). Bandel (1998) illustrated a specimen under that name, but it is not clear how the identification was achieved.

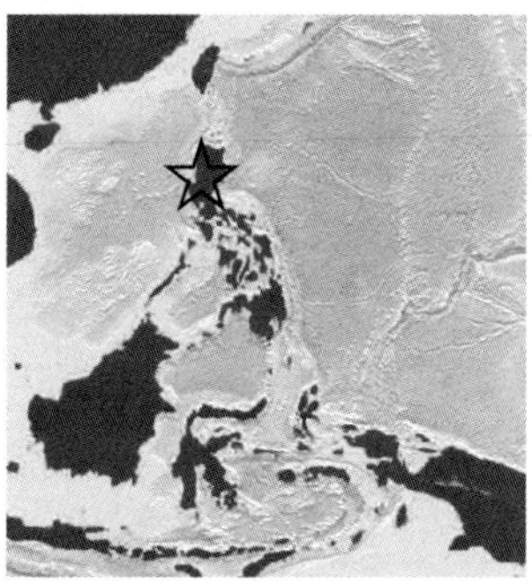

Figure 99. Distribution of *Scissurella hoernesi.* ☆ type locality.

Scissurella jucunda Smith, 1890

Figures 100–107, Color Plate 3B

Chresonymy.

Scissurella jucunda Smith, 1890: 311–312, pl. 24, figs 22, 22a.

Scissurella jucunda Smith, 1890: Kobelt, 1891: 358.

Scissurella jucunda Smith, 1890: Thiele, 1912a: 7, pl. 1, figs 12–13 [copy figures Smith, 1890].

Scissurella jucunda Smith, 1890: Thiele, 1925: 42.

Scissurella jucunda Smith, 1890: Rosewater, 1975: 9.

Scissurella jucunda Smith, 1890: Herbert, 1986: 620–622, figs 8, 21–24.

Scissurella jucunda Smith, 1890: Lozouet, 1986: 107.

Scissurella jucunda Smith, 1890: Herbert, 1986: 1, figs e–f.

Scissurella jucunda Smith, 1890: Bandel, 1998: 15.

Scissurella jucunda Smith, 1890: Geiger, 2003: 77.

Scissurella jucunda Smith, 1890: Geiger, 2004: text-fig. p. 5 [radula], text-fig. p. 6 [shell].

Scissurella jucunda Smith, 1890: Geiger, 2006c: text-figs 3 [shell], 8 [radula].

Scissurella jucunda Smith, 1890: Baborka, 2007: 16–37 [histology, 3D reconstruction].

Synonyms.

+ *Scissurella jucunda* Smith, 1910: 207–208, pl. 8, figs 2, 2a. ***Type material.*** Syntypes (NHMUK 1911.8.30.43–4, 2), 1 × 0.75 mm. The here illustrated specimen (Fig. 102) is designated as the lectotype (NHMUK 1911.8.30.43) for the express purpose of taxon stabilization in case other non-conspecific syntypes should be found. Paralectotype (NHMUK 1911.8.30.44, 1). ***Type locality.*** Four miles south of Port Elizabeth [South Africa, 33.967°S, 25.583°E] (OD). ***Etymology.*** Jucundus, Latin for agreeable, pleasant, merry.

Scissurella jucunda Smith, 1910: Bartsch, 1915: 175–176.

Scissurella jucunda Smith, 1910: Thiele, 1925: 42.

+ *Scissurella smithi* Thiele, 1912a. 7, pl. 1, figs 14–15 [copy figures Smith, 1910]. Replacement name for *Sci. jucunda* Smith, 1910 (*non* Smith, 1890). Type data as for *Sci. jucunda* Smith, 1910 (not 1890). ***Etymology.*** Named after Edgar Alan Smith (1847–1916).

Scissurella smithi: Thiele, 1925: 42.

Scissurella smithi: Tomlin, 1926: 296.

Scissurella smithi: Turton, 1932: 204.

Scissurella smithi: Barnard, 1963: 301.

Scissurella smithi: Lozouet, 1986: 108.

Scissurella smithi: Geiger, 2003: 77.

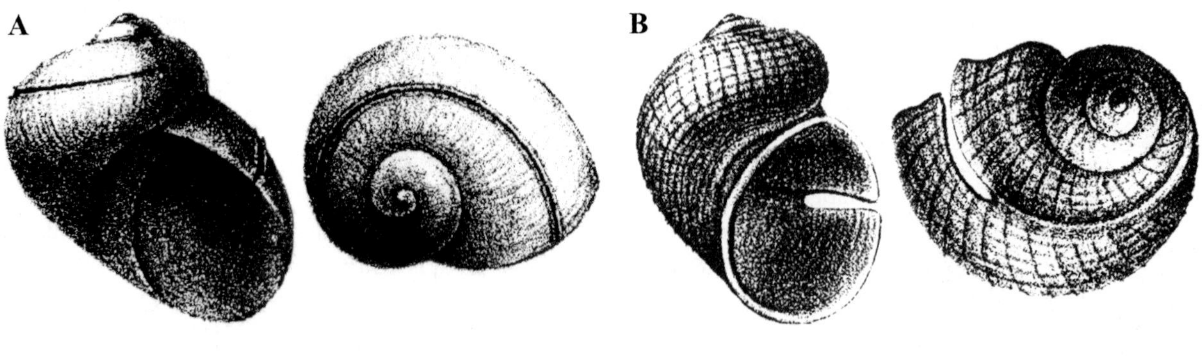

Figure 100. **A.** Original illustration of *Scissurella jucunda* Smith, 1890. **B.** Original illustration of *Scissurella jucunda* Smith, 1910.

Misidentifications.

Scissurella smithi: Kensley, 1973: 28, fig. 28 [is juvenile *Suk. dorbignyi*].

Type material. Lectotype (NHMUK 1889.10.1.1575: designated by Herbert, 1986. Fig. 101), 0.7 × 0.7 mm. Paralectotype (NHMUK 1889.10.1.1576, 1).

Type locality. St. Helena [Great Britain, 15.967°S, 5.717°W] (OD: inferred from title).

Etymology. Jucundus, Latin for agreeable, pleasant, merry.

Description. Shell medium-sized (to 1 mm), trochiform, inflated. Protoconch of 0.75 whorl, with strong axials, apertural varix present, not connected to embryonic cap, apertural margin straight. Teleoconch I of 0.875 whorl, approximately 28 moderate axial cords, no spirals. Teleoconch II of 0.66 whorl, suture hardly impressed. Shoulder slightly convex, approximately 19 axial cords on first half whorl, becoming indistinct on last quarter whorl, up to five spiral lines, increasing in strength towards aperture, decreasing in strength from suture towards selenizone. Base with slight constriction below selenizone, axials of same strength and density as on shoulder, approximately 10 spiral lines, increasing in strength somewhat towards umbilicus. Selenizone well above periphery, slightly outside middle of shoulder width, lunules dense, distinct,

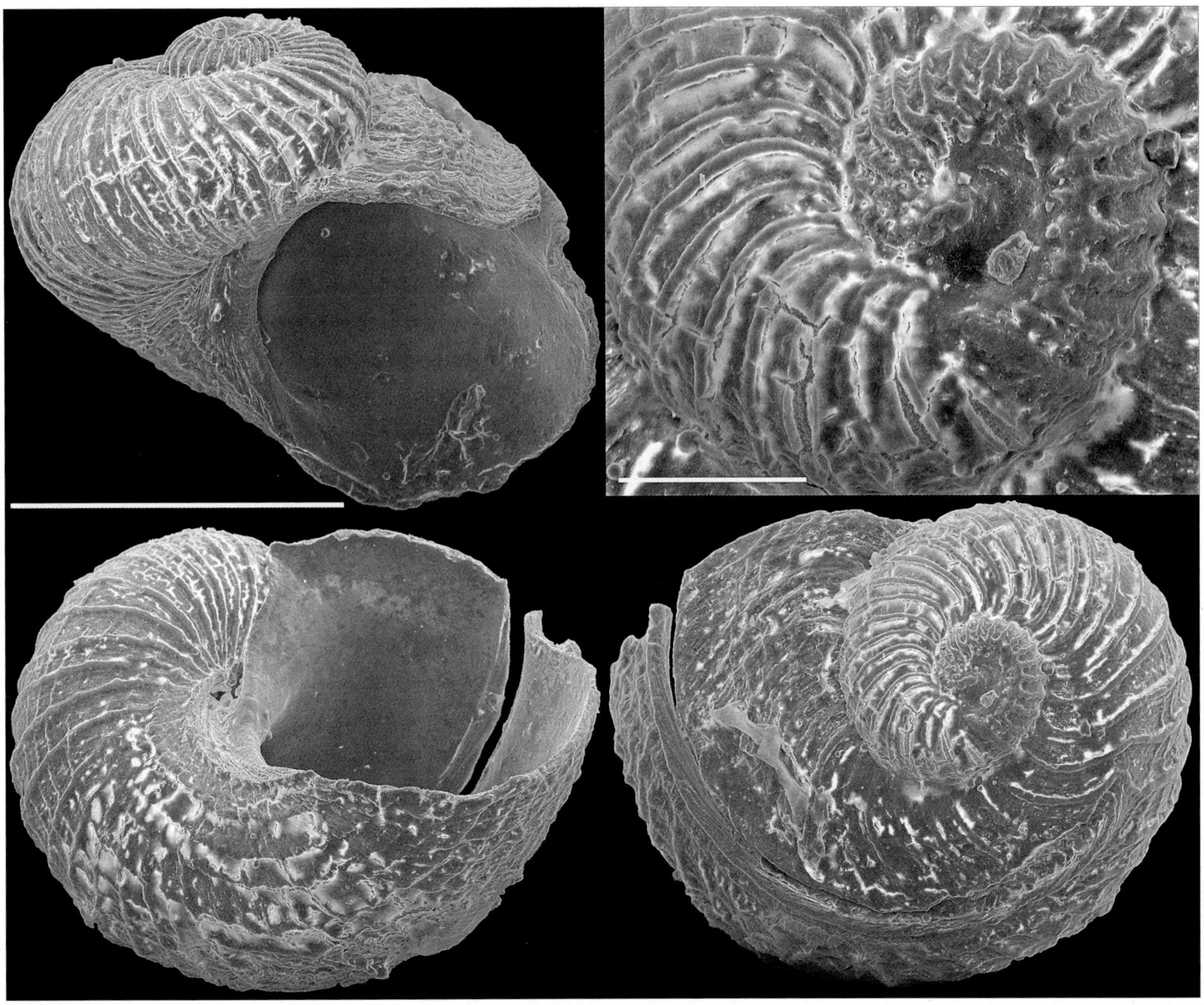

Figure 101. *Scissurella jucunda* Smith, 1890. Lectotype. St. Helena [Great Britain, 15.967°S, 5.717°W] (NHMUK 1889.10.1.1575). Scale bar shells = 0.5 mm. Scale bar protoconch = 100 μm.

low, keels low, strong; slit open with parallel margins in smaller specimens, in larger specimens converging, almost touching. Umbilicus narrow, bordered by distinct spiral cord, walls straight, with very fine growth lines. Aperture rounded, roof overhanging.

Operculum (Fig. 103, 105A) round, thin, multispiral, with central nucleus, covering most of aperture.

Radula (Fig. 106). Rachidian tooth triangular, with slightly V-shaped tip bearing five cusps on tip, central one slightly larger than peripheral ones. Lateral teeth 1–3 similar, with three (or four on lateral tooth 3) cusps on tip. Lateral tooth 4 reduced with one major cusp and occasionally with one on either side of tip. Lateral tooth 5 broadened, enlarged, with approximately seven cusps on tip, outermost largest. Inner marginal teeth with innermost cusps largest, outer marginal teeth spoon-shaped, symmetrical. Radular interlock moderate.

Animal. The anatomy of the species has been examined in detail in two theses: Strassoldo (1991) and Baborka (2007). The cephalic tentacles are papillate, has

Figure 102. *Scissurella jucunda* Smith, 1910. Lectotype. Four miles south of Port Elizabeth [South Africa, 33.967°S, 25.583°E] (NHMUK 1911.8.30.43). Scale bar shells = 1 mm. Scale bar protoconch = 100 µm.

distinct eyes, and a right suboptic tentacle. The propodium has pedal glands. Two epipodial tentacles below the operculum, the anterior distally blunt with ESO. Allegedly a second ESO at the tip of the suboptic tentacle. The posterior end of the mantle cavity bears a papillate mantle tentacle; roof with extensive hypobranchial gland. Right gill monopectinate, left gill partially bipectinate, with distal osphradia, leaflets with rods and bursicles. Diotocardian heart, traversed by rectum. Right kidney significantly larger. Hermaphroditic, most of gonad filled with approximately 30 eggs, up to three eggs parallel, posterior-most portion with spermatocytes in position of junction with gonoduct. Digestive system with two jaws, one pair of odontophore cartilages, paired salivary gland in posterior pharynx, stomach filled with detritus. Nervous system streptoneurous hypoathroid, with statocysts.

Comparisons. *Sinezona doliolum* from South Africa is taller than wide, has a distinct constriction below the selenizone on the base, and a strong carina surrounding the extremely narrow umbilicus.

Sinezona insignis from South Africa has a taller shell with very narrow umbilicus, has few axial lamellae, and lacks spirals on the shoulder.

Figure 103. *Scissurella jucunda.* Georgetown, Ascension Island, 7.933°S, 14.417°W (USNM 735075, 4). Scale bar shells = 0.5 mm. Scale bar protoconch = 100 µm.

Scissurella maraisorum from South Africa has a much flatter spire, a protoconch with finer axial sculpture and a distinct apertural varix, and a much wider umbilicus, through which the underside of the protoconch is visible.

Figure 104. *Scissurella jucunda*. **A–B.** False Bay, Dalebrook, South Africa, 34.250°S, 18.600°E (AMS C.380715, 3). **C.** False Bay, Strandfontein, South Africa, 34.250°S, 18.600°E (AMS C.380716, 1). Scale bars shells = 0.5 mm. Scale bars protoconch = 100 μm.

Scissurella reticulata from the Indian Ocean has an overall wider shell, neat reticulate sculpture, and a protoconch with an apertural varix connected to the embryonic cap.

Figure 105. *Scissurella jucunda*. **A.** False Bay, South Africa, 34.250°S, 18.600°E (AMS C.380717, 2). Radula from specimen of same lot shown in Figure 106. **B–C.** 0 m, Cape Town, South Africa, 33.917°S, 18.417°E (LACM 57-4, 2). Scale bars shell = 0.5 mm. Scale bars protoconch = 100 μm.

Scissurella rota from the Indian Ocean has a wider shell with wider umbilicus, broad reticulate sculpture, and a protoconch with an apertural varix connected to the embryonic cap.

Scissurella sudanica from the Indian Ocean has a more globular shell, an umbilicus with distinct periumbilical cord, and a protoconch with an apertural varix connected to the embryonic cap.

Scissurella mirifica from the Indo-Malayan Archipelago to the Central Pacific has a wider shell with wider umbilicus, a protoconch with wider spacing of the fine axial cords, and reticulate sculpture on the base.

Scissurella xandaros from the Indo-Malayan Archipelago to the Western Pacific has a wider shell with less elevated spire, and a smooth protoconch with an apertural varix connected to embryonic cap.

Scissurella manawatawhia from New Zealand has a protoconch with distinct spiral cord, stronger axial cords, and a shorter teleoconch I (0.75 vs. 1 whorl).

Sinezona macleani from the Indian Ocean to the Western Pacific has a wider shell with wider umbilicus, a straight parietal wall, and a distinct projection in the lower adumbilical corner of the aperture.

Sinezona semicostata from the subtropical northeastern Atlantic Islands has a flatter spire, a very strong periumbilical cord, and lacks spiral sculpture.

Sinezona cingulata from the Mediterranean to Angola, has twice as many but half as strong axial cords, a more narrow umbilicus with strong periumbilical cord, and a shorter selenizone (0.25 vs. 0.4 whorl).

Distribution. St. Helena, Ascension, South Africa (Fig. 107).

Biology. Found live intertidally among *Laurencia* and *Caulerpa* algae, and in holdfasts of *Ecklonia* kelp (Herbert, 1986).

Specimen records.

Tanzania. 0–1 m, Mbudya Island, ca. 16 km N of Dar-es-Salaam, 6.650°S, 39.250°E (AMS ex C.380205, 5).

Madagascar. 0 m, St. Luce, ATIMO VATAE TM19, 24.785°S, 47.198°E (MNHN, 4). 0 m, Crique at the NE lighthouse of Evatra, 24.978°S, 47.098°E (MNHN, 4). 2–7 m, E Point Evatra, ATIMO VATAE TB08, 24.986°S, 47.090°E (MNHN, 1). 0 m, Monseigneur Beach, ATIMO VATAE TM01, 25.035°S, 46.998°E (MNHN, 12). 7 m, Port d'Ehoala, ATIMO VATAE TP28, 25.061°S, 46.963°E (MNHN, 1). 10 m, Entrance to East Bay

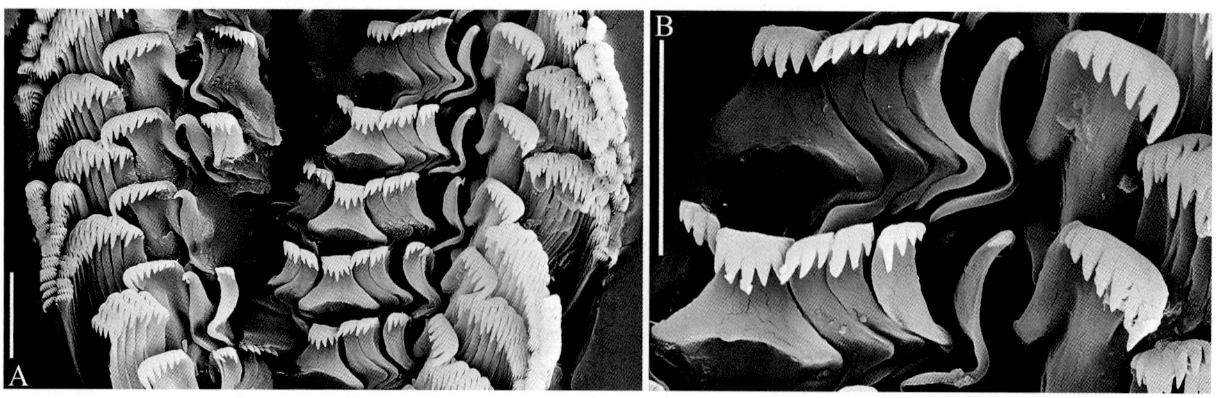

Figure 106. *Scissurella jucunda*. Radula. From same lot as specimen shown in Figure 105A. False Bay, South Africa, 34.250°S, 18.600°E (AMS C.380717, 2). **A.** Full width of radula. **B.** Central field enlarged. Scale bars = 10 µm.

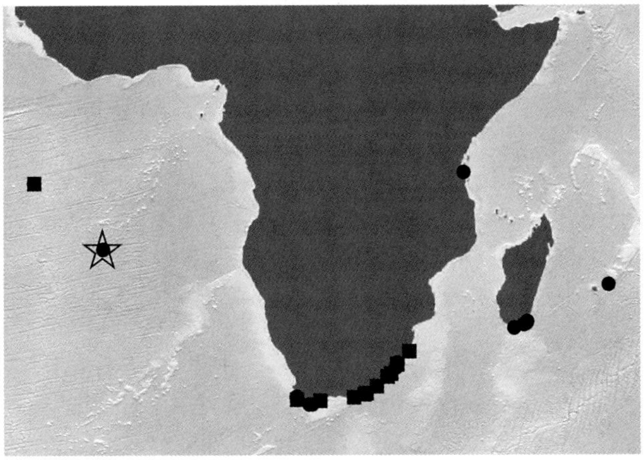

Figure 107. Distribution of *Scissurella jucunda*. ☆ type locality. ● specimen record. ■ literature record.

of Galions, ATIMO VATAE TB10, 25.155°S, 46.755°E (MNHN, 1). 0 m, False Cape, ATIMO VATAE BM12, 25.571°S, 45.528°E (MNHN, 3).

Great Britain, Ascension Island. Georgetown, 7.933°S, 14.417°W (USNM 735075, 4).

Great Britain, St. Helena. 15.967°S, 5.717°W (NHMUK 1889.10.1.1575, 1; NHMUK 1889.10.1.1576, 1; USNM 124080, 2)

Mauritius. Ficq-en Flacq, 20.300°S, 57.350°E (USNM 716533, 22).

South Africa. False Bay, Dalebrook, 34.250°S, 18.600°E (AMS C.380715, 3). False Bay, Strandfontein, 34.250°S, 18.600°E (AMS C.380716, 1). False Bay, 34.250°S, 18.600°E (AMS C.380717, 2). 0 m, Cape Agulhas, 34.825°S, 20.027°E (DLG 2323, 3). 0 m, Cape Town, South Africa, 33.917°S, 18.417°E (LACM 57-4, 2). Kellys Rocks, NW side Salt Vlei Point, Port Alfred, 33.611°S, 26.892°E (AMS C.404939, 1). Agulhas, 34.817°S, 20.017°E (AMS C.404940, 1). Table Bay, Kom, 33.883°S, 18.450°E (AMS C.404941, 1). Millers Point, 34.233°S, 18.467°E (AMS C.406350, 2). Cape Colony, Port Alfred, 33.600°S, 26.900°E (USNM 250524, 1). North Transkei (HLC, 1). Port Alfred, 33.600°S, 26.900°E (NHMW, 2). Port Alfred, Cape Colony, near Grahamstown, 33.600°S, 26.900°E (USNM 187097, 3). Port Alfred, Cape Colony, 33.600°S, 26.900°E (USNM ex 249762, 1; USNM 249761, 3). Port Alfred, Bathurst District, Cape Province, 33.600°S, 26.900°E (MCZ 101504, 2).

Literature records.

South Africa. Mapalane (Maphelana), Umfolozi River mouth, 28.400°S, 32.417°E. Reunion Rock, Natal, 29.967°S, 30.933°E. Mzamba, 31.100°S, 30.183°E (NMSA A6289). 50 m, off Mbotyi, Transkei, 31.450°S, 29.733°E (NMSA C8014). 0 m, Qolora River mouth, Transkei, 32.633°S, 28.417°E (NMSA C6583). Great Fish River mouth, Eastern Cape, 33.483°S, 27.133°E. Port Alfred, Eastern Cape, 33.600°S, 26.900°E (NMSA D245). 285 m, 4 mi S of Port Elizabeth, 33.967°S, 25.583°E (NMSA 2947). Port Elizabeth, Eastern Cape, 33.967°S, 25.583°E (NMSA 8535). Port Elizabeth, Eastern Cape, 33.967°S, 25.583°E (NMSA B6123). Still Bay, Tsitsikamma Coast, 34.367°S, 21.417°E (SAMCT A31576). 0 m, Miller's Point, False Bay, Western Cape, 34.233°S, 18.467°E (NMSA D1671) (Herbert, 1986).

Remarks. The specimen illustrated by Kensley (1973) seems to be a juvenile *Suk. dorbignyi*, based on the notable spiral keel on the base of the shell, the fairly strong angulation between shoulder and base at selenizone, and the open slit.

In some specimens the edges of the slit are almost touching. In all other respects, they are typical *Sci. jucunda*. Specimens with open slit have the last quarter of the last whorl markedly descending, indicating mature specimens. Accordingly *Sci. jucunda* is not a juvenile of a *Sinezona* sp.

Scissurella kaiserae Geiger, 2006

Figures 108–112

Chresonymy.

Scissurella kaiserae Geiger, 2006a: 2–7, figs 1–4.
Scissurella sp.: Skoglund, 2002: 3.
Scissurella kaiserae: Kaiser, 2009: 221, 228, fig. 237L–N.

Type material. Holotype (SBMNH 348769. Fig. 108). Paratypes from type locality (SBMNH 359438, 14. Fig. 109). 10–21 m, Roca Próspera Islet, Chiriqui, Gulf, Coiba National Park, Panama, 7.776°N, 81.759°W (KKC, 1. Fig. 110A). 12–18 m, E side Lava Bay, San Benedicto Island, Revillagigedo Islands, Mexico, 19.300°N, 110.790°W (KKC, 1. Fig. 110B). 16–23 m, Maria Cleofa Island, Tres Marias Islands, Mexico, 21.300°N, 106.276°W (KKC, 5. Fig. 110C). 16–23 m, Maria Cleofa Island, Tres Marias Islands, Mexico, 21.300°N, 106.276°W, ex KKC (AMNH 314981, 2; USNM 1081938, 3; ZMA Moll. 4.05.019, 2).

Type locality. 30–35 m, Baja Alcyone, Cocos Island, Costa Rica, 5.533°N, 86.983°W (OD).

Figure 108. *Scissurella kaiserae*. Holotype. Baja Alcyone, Cocos Island, Costa Rica (SBMNH 348769). Scale bar shell = 0.5 mm. Scale bar protoconch = 100 μm.

Etymology. The name honors the discoverer of the species and longtime student of the Panamic micromolluscan fauna, Kirstie Kaiser, Puerto Vallarta, Mexico (OD).

Description. Shell small (0.4–0.6 mm), trochiform depressed, off-white. Protoconch of one whorl, with fine axial sculpture from apical-most portion to periphery, adapical surface smooth, no apertural varix, apertural margin sinusoidal. Teleoconch I of 0.825 whorl, 17–20 distinct axial cords, interstices filled with fine irregular growth marks, no spiral sculpture. Teleoconch II of 0.33–0.4 whorl. Shoulder flat, axial cords of same spacing as on teleoconch I, interstices filled with fine irregular growth lines, no spiral sculpture, suture distinct. Base with distinct constriction

Figure 109. *Scissurella kaiserae*. Paratypes. **A–C.** Alcyone Bay, Cocos Island, Costa Rica (SBMNH 359438, 14). Scale bars shell = 0.25 mm. Scale bars protoconch = 100 μm.

below selenizone, same sculpture as on shoulder. Umbilicus open, moderately wide, bordered by strong spiral cord, walls straight, with axial growth marks, underside of protoconch visible through umbilicus. Selenizone above periphery, at 45° angle to coiling axis, keels moderately strong, moderately elevated, lunules distinct; slit

Figure 110. *Scissurella kaiserae*. Paratypes. **A.** 10–21 m, Roca Prospera Islet, Chiriqui Gulf, Coiba National Park, Panama, 7.776°N, 81.759°W (KKC). **B.** 12–18 m, E side Lava Bay, San Benedicto Island, Revillagigedo Islands, Mexico, 19.300°N, 110.790°W (KKC). **C.** 16–23 m, Maria Cleofa Island, Tres Marias Islands, Mexico, 21.300°N, 106.276°W (KKC). Scale bar shells = 0.5 mm. Scale bars protoconch = 100 μm.

open, with parallel margins. Aperture subquadrate, roof overhanging. Operculum (Fig. 111C) round, thin, covering aperture, multispiral with central nucleus.

Radula (Fig. 111A–B). Rachidian tooth triangular, tip with seven cusps, central one largest, arranged in V. Lateral teeth 1–3 similar, three cusps on outer margin of tip. Lateral tooth 4 reduced, with two minute cusps on tip. Lateral tooth 5 broadened, enlarged, with approximately six cusps on inner margin of tip. Radular interlock of central field moderate. Inner marginal teeth spoon-shaped, with 1–2 cusps on inner margin, 6–7 cusps on outer margin; outer marginal teeth spoon-shaped, with many fine cusps on each side of tip.

Comparisons. The species shows little intraspecific variation, both within as well as between populations. Figure 109 shows three specimens from the same population, and Figure 110 shows three specimens from three distinct islands.

Sinezona rimuloides from the Panamic and Southern and Central California, has a more elevated spire, strong axial sculpture on the protoconch, and closes the slit to a foramen.

Sinezona carolarum from the Panamic has a more globular shell with more elevated spire, the protoconch has an apertural varix and the umbilicus is wider and funnel-shaped.

Sinezona chilensis from western South America has a taller shell, a shorter selenizone, and a protoconch with apertural varix.

Scissurella redferni from the Caribbean has a similar overall shape, but has fewer axials (14 vs. 17–20 on teleoconch I), and has a protoconch with strong axial sculpture.

Scissurella maraisorum from eastern South Africa has a similar overall shape, but has somewhat fewer axials (14–18 vs. 17–20 on teleoconch I) and on the shoulder of teleoconch II the axials are becoming indistinct, while they remain well-developed on the base.

Scissurella prendrevillei from New Zealand has a protoconch with more and finer axial cords, the axial cords are usually elevated to lamellae, and has a longer teleoconch I (1.25 vs. 0.825 whorls).

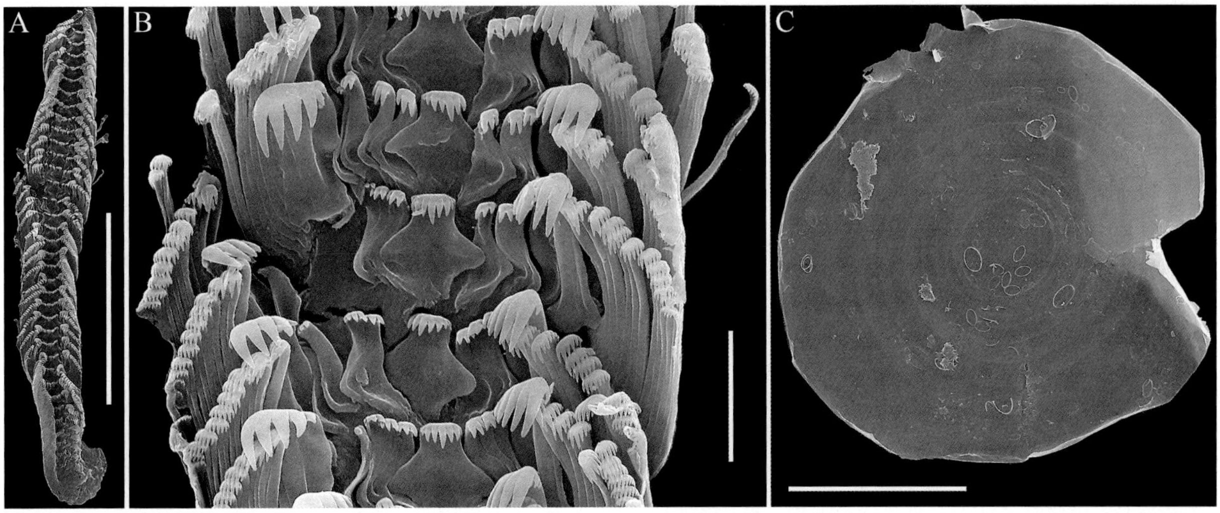

Figure 111. *Scissurella kaiserae*. Radula and operculum. 16–23 m, Maria Cleofa Island, Tres Marias Islands, Mexico, 21.300°N, 106.276°W (KKC). **A.** Entire radula. Scale bar = 100 µm. **B.** Section of radula enlarged. Scale bar = 10 µm. **C.** Operculum. Scale bar = 100 µm.

Scissurellidae *Scissurella*

Distribution. Pacific coast of Panamic Province and its offshore islands from 4°N to 28.5°N (Fig. 112).

Specimen records.

Mexico. Socorro Island, Revillagigedo Islands, 19.300°N, 111.650°W (SBMNH). 15–19 m, Cabo Middleton, Socorro Island, Revillagigedo Islands, 18.780°N, 110.990°W (KKC, 2). 12–18 m, E side Lava Bay, San Benedicto Island, Revillagigedo Islands, 19.300°N, 110.790°W (KKC, 50). 15–27 m, San Benedicto Island, Revillagigedo Islands, 19.300°N, 110.790°W (SBMNH). 16–23 m, Maria Cleofa Island, Tres Marias Islands, 21.300°N, 106.276°W (KKC, 20). 8–11 m, Roca Blanca, San Juanito Island, Tres Marias Islands, 21.738°N, 106.700°W (KKC, 3). 6–12 m, off Ildefonso Island, Gulf of California, 26.632°N, 111.430°W (SBMNH, 1). 13–30 m, SW side Guadalupe Island (Afeura Island, Adentro, 5 fm Bank, and basalt arches), Baja California, 28.900°N, 118.292°W (LACM 72-120, 1). 10–25 m, San Pedro Nolasco Island, Sonora, 28.426°N, 111.367°W (LACM 67-4). 27 m, Los Angeles Bay, Baja California, Mexico, 28.917°N, 113.500°W (LACM 75-9, 13). 3–17 m, Baja California Sur, Magdalena Bay, 24.540°N, 112.067°W (LACM 71-14). 8 m, Baja California Sur, Cabo Pulmo, 23.442°N, 109.425°W (SBMNH B-5016). 10–17 m, Baja California Sur, Ceralvo Island, 24.167°N, 109.867°W (LACM 71-24). 18–37 m, Baja California Sur, Rancho Palmilla, 22.967°N, 109.800°W (LACM 66-17). 18–37 m, Baja California Sur, Gorda Point, 23.133°N, 109.583°W (LACM 66-18). 5–10 m, Jalisco, Marietas Islands, 20.750°N, 105.500°W (LACM 65-14).

France, Clipperton Island. 12–15 m, 10.305°N, 109.198°W (KKC, 3).

Costa Rica. 9–13 m, N side del Caño Island, Puntarenas, 8.721°N, 83.885°W (LACM 72-63, 1). 13–17 m, Guanacaste, Santa Elena Point, 10.893°N, 85.964°W (LACM 72-30).

Panama. 10–21 m, Roca Próspera Islet, Chiriqui Gulf, Coiba National Park, 7.776°N, 81.759°W (KKC, 20). 0 m, Canal Zone, Farfan Beach, 8.933°N, 79.567°W (LACM 75-56).

Costa Rica, Cocos Island. 10–20 m, Rafael Island, Meule, Iglesias Bay, 5.507°N, 87.507°W (SBMNH, 1). 10–16 m, Rodrigues Point, 5.510°N, 87.090°W (KKC, 8). 26–35 m, SSW Rodrigues Point, 5.510°N, 87.090°W (KKC, 17). 20–33 m, North Islet, Dos Amigos Islands, 5.515°N, 87.102°W (KKC, 9). Alcyone Bay, 5.533°N, 86.983°W (SBMNH 348769, 16). 23–30 m, Sharkfin Rock, SSW Rodriguez Point, 5.533°N, 86.983°W (KKC, 8). 10–13 m, Westion Bay, Pajara Island, 5.547°N, (KKC, 2).

Colombia, Malpelo Island. 38 m, 4.002°N, 81.611°W (KKC, 1). 26–50 m, Hammerwall, 4.007°N, 81.604°W (KKC, 3). 13–30 m, North Rock, Little Bob Islet, 4.003°N, 81.602°W (KKC, 1). 33–62 m, Tiger Mount, 4.007°N, 81.602°W (KKC, 1). 10–43 m, NE Coral Gardens, 4.987°N, 81.604°W (KKC, 2).

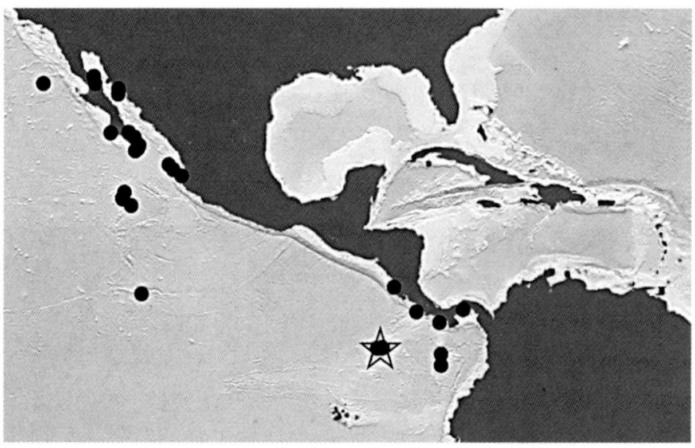

Figure 112. Distribution of *Scissurella kaiserae*. ☆ type locality. ● specimen record.

Scissurella koeneni Semper, 1865 [*nomen dubium*]

Figures 113–114

Chresonymy.

Scissurella koeneni Semper, 1865: 284, 286–288, pl. 12, fig. 3.
Scissurella koeneni: Paetel, 1888: 289.
Scissurella koeneni: Pilsbry, 1890: 51, pl. 57, figs 17–18 [copy figs Semper, 1865].
Scissurella koeneni: Hildago, 1904: 261.
Scissurella koeneni: Thiele, 1912a: 13–14, pl. 2, fig. 5 [= Fig. 113B).
Scissurella koeneni: Faustino, 1928: 126.
Scissurella koeneni: Lozouet, 1986: 107.
Scissurella koeneni: Bandel, 1998: 10, pl. 1, figs 3–5 [is *Sci. quadrata*].
Scissurella koeneni: Geiger, 2003: 77.

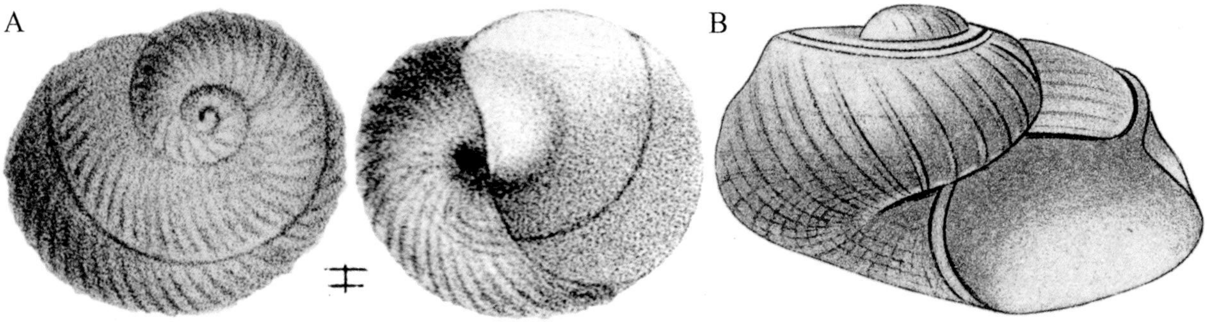

Figure 113. **A.** Original illustrations of *Scissurella koeneni*. **B.** Illustration of *Scissurella koeneni* from Thiele (1912), which may be an illustration of a member of the type series.

Type material. Holotype (Museum Hamburg: ? Thiele, 1912a: pl. 2, fig. 5), 1.75 × 1 mm. Most likely destroyed during World War II (B. Hausdorf, pers. comm. 2/2007).

Type locality. Bohol, Philippines [9.600°N, 123.750°E] (OD).

Etymology. Named for Adolf von Koenen [1837–1915], author of several works on Tertiary fauna of northern Germany (OD).

Remarks. The shell is described as greenish in color, and the illustrations do not show a slit. The "scissure" is placed very high, and is followed by angled spiral sculpture towards the base. All of this points towards Stomatellidae, not Scissurellidae, with the possible exception of *Sukashitrochus*, which is always higher in aspect.

Thiele (1912) noted, that the original illustrations are un-typical, and his figure is based on the typical [= type ?] specimen in the "Hamburg Museum." The line drawing is significantly better than the original illustration, but still does not permit distinction between several species occurring in the Indo-Pacific (*Sci. quadrata, Sci. lorenzi, Sci. mirifica, Sin. ferriezi* juv., *Sin. marrowi* juv.). The Semper material in Hamburg was destroyed in World War II (B. Hausdorf, pers. comm. 2/2007).

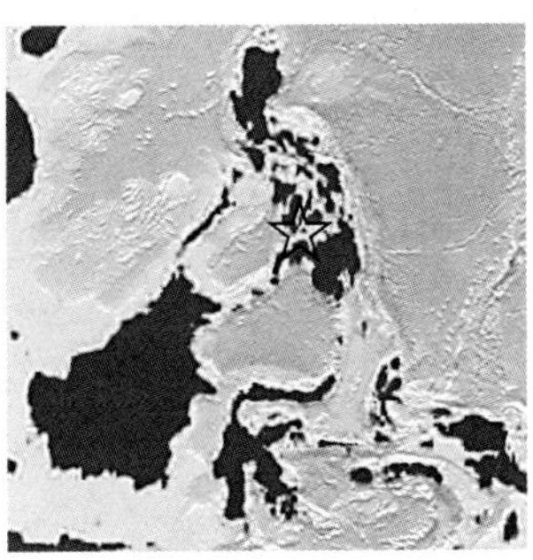

Figure 114. Distribution of *Scissurella koeneni*. ☆ type locality.

Bandel (1998) illustrated a specimen under that name, though it is unclear how such an identification was achieved.

Given that the original illustrations are of no utility, the subsequent illustration is not verifiably based on the type and cannot be unambiguously identified, and that the name was infrequently used, I here designate the species a *nomen dubium*.

Scissurella lobini (Burnay & Rolán, 1990)

Figures 115–118

Chresonymy.

Sinezona lobini Burnay & Rolán, 1990: 36–38, pl. 2, figs 6–8, 10, pl. 3, figs 15, 17, 19.

Scissurella lobini: Geiger, 2003: 77.

Sinezona lobini: Rolán, 2005: 41, pl. 6, fig. 73.

Scissurella lobini: Nolt, 2008: 56, fig. 3A.

Type material. Holotype (CZTL 1555). Paratypes from type locality (MNCN 15.05/3999, 2; MNHN, 2; SMF 308398/2, 2; NHMUK 1991122, 2; RHML, 2; MBL, 2; CER; 2; CZTL, 7).

Type locality. 61 m, Porto Novo, Sto Antão Island, Cape Verde Archipelago [17.067°N, 25.083°W] (OD).

Etymology. Named for botanist Wolfram Lobin, Bonn, Germany (OD).

Figure 115. Original illustrations of *Scissurella lobini*. Apical and umbilical view of holotype, remainder from paratypes.

Description. Shell small (to 0.65 mm. Figs 115–117), trochiform, apex depressed, whorls very loose. Protoconch of one whorl, strong axial sculpture, no apertural varix, apertural margin convex. Teleoconch I of one whorl, approximately 13 strong axial cords, interstices with very fine irregular axial lamellae, no spirals. Teleoconch II of 0.3–0.5 whorl. Shoulder convex, approximately 13 axial cords, axials more crowded towards aperture, interstices with very fine irregular axial lamellae, no spiral sculpture, suture very deep. Base with constriction below selenizone, same sculpture as shoulder. Umbilicus open, deep, walls continuous with base. Selenizone somewhat above periphery, relatively wide, keels distinct, low, lunules distinct, open slit short, with parallel margins. Aperture rounded.

Animal unknown.

Comparisons. Juvenile *Sin. semicostata* from Cape Verde Islands has more axials on the base of mature specimens, and the underside of the protoconch viewed through the umbilicus shows fine reticulations. Others characters cited by Burnay & Rolán (1990) as distinguishing, such as the extent and strength of the axial ribs on the shoulder, are known to be quite variable within species of *Scissurella* (e.g., *Sci. costata*).

Figure 116. *Scissurella lobini*. Paratype. 61 m, Porto Novo, Santo Antão Island, Cape Verde Archipelago [17.067°N, 25.083°W] (NHMUK 1991122, 2). Scale bar shell = 0.5 mm. Scale bar protoconch = 100 µm.

Scissurella azorensis from the Azores and the Western Mediterranean has a protoconch with more axial cords (24–28 vs. 14–18), has axial cords on the teleoconch that are less uniform in shape, and lacks spiral threads on the base.

Juvenile *Sin. cingulata* from the Mediterranean and northeastern Atlantic has a tall spire, a narrow umbilicus, and a protoconch with an apertural varix.

Scissurella maraisorum from eastern South Africa has a protoconch with dense fine axial cords, and the shoulder of usually lacks axial cords.

Scissurella redferni from the Caribbean has a protoconch with fewer but stronger axial cords, about 50% fewer and lower axial cords on the teleoconch, and tends to have a longer teleoconch I (>1 vs. <1 whorl).

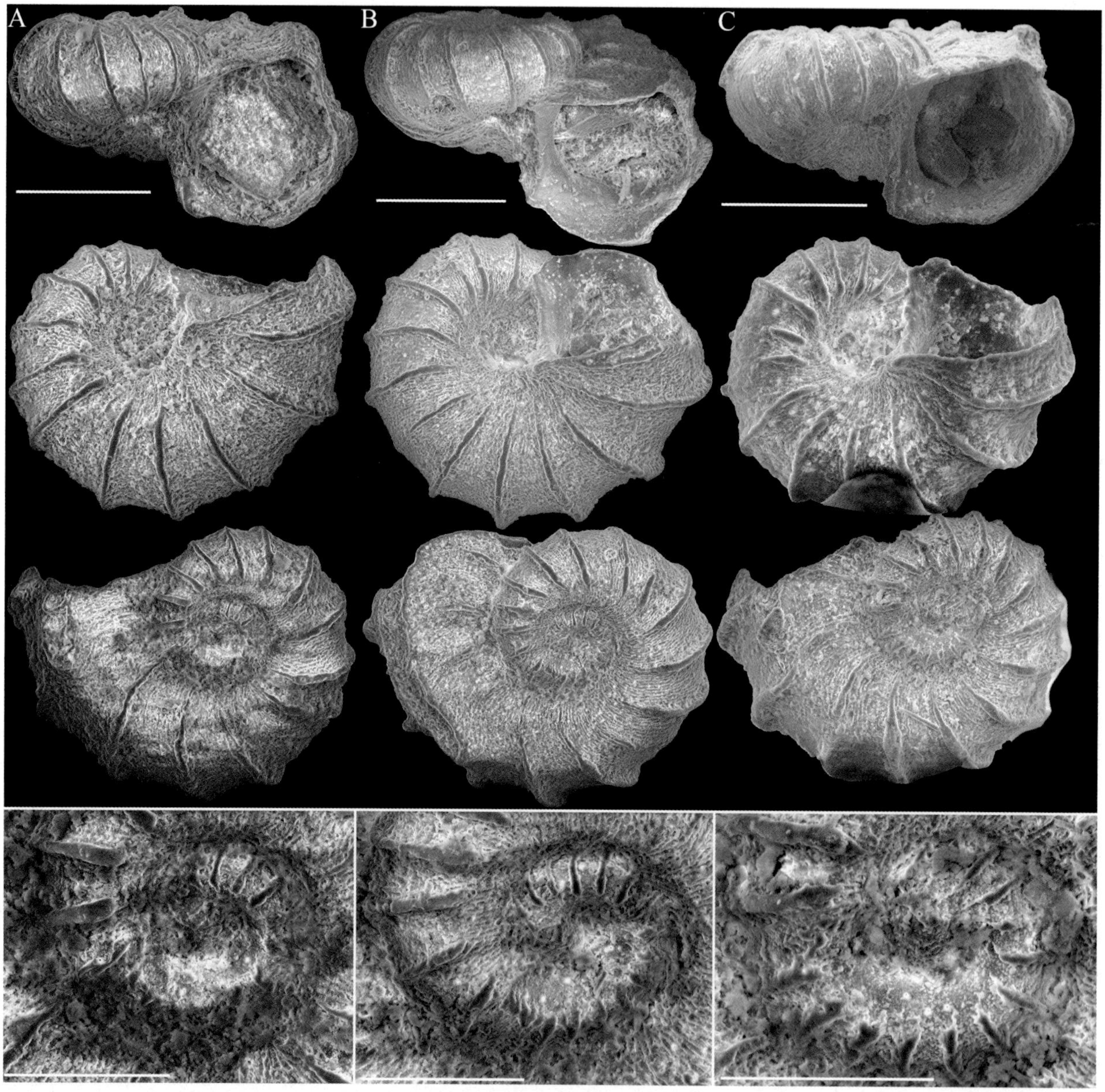

Figure 117. *Scissurella lobini*. Paratypes. 61 m, Porto Novo, Sto Antão Island, Cape Verde Archipelago [17.067°N, 25.083°W]. **A–B.** (MNHN, 2). **C.** (NHMUK 1991122, 2). Scale bar shell = 0.2 mm. Scale bar protoconch = 100 µm.

Scissurella kaiserae from the Panamic has about 50% more axial cords on the teleoconch and the axial cords on the base reach into the constriction below the selenizone.

Scissurella prendrevillei from New Zealand has a protoconch with very fine axial cords, has a longer teleoconch I (1.25 vs. 1 whorl), and usually some spiral sculpture on the base, particularly surrounding the umbilicus.

Sinezona platyspira from New Zealand has a sunken protoconch with more but finer axial cords and much more elevated keels of the selenizone

Distribution. Tropical and subtropical eastern Atlantic islands off Africa: Cape Verde Archipelago, Selvagens Archipelago, Azores (Fig. 118).

Specimen records.

Cape Verde Archipelago. 61 m, Porto Novo, Santo Antão Island, 17.067°N, 25.083°W (CZTL 1555, 1: complete; NHMUK, 2). 80 m, S of Boa Vista, 15.917°N, 22.917°W (NMR 993000033723, 5). Porto Novo, Santo Antão, 17.067°N, 25.083°W (CZTL, 1). 400 m, San Jorge Bay, S of São Nicolau, 16.550°N, 24.283°W (RMNH, 2). 80 m, S of Razo, 16.600°N, 24.600°W (RMNH, 1). 35 m, S of Branco, 16.633°N, 24.683°W (RMNH, 2). 35 m, S of Branco, 16.633°N, 24.683°W (RMNH, 2).

Portugal, Selvagens Archipelago. 524 m, 30.017°N, 16.017°W (RMNH, 1).

Portugal, Azores. 620 m, S of Santa Maria, 36.917°N, 25.117°W (RMNH, 11). 122 m, E of Santa Maria, 36.983°N, 25.033°W (RMNH, 1). 300 m, S of Faial, 38.500°N, 28.633°W (RMNH, 3). 188 m, E of Faial, 38.517°N, 28.583°W (RMNH, 8). 45 m, W of Pico, 38.533°N, 28.550°W (RMNH, 1). 92 m, E of Faial, 38.567°N, 28.567°W (RMNH, 9).

Spain, Canary Islands. 215–325 m, S of Lanzarote, 28.800°N, 13.767°W (RMNH, 1). 82 m, S of Lanzarote, 28.800°N, 13.767°W (RMNH, 30). 109–116 m, SE of Lanzarote, 28.917°N, 13.533°W (RMNH, 1). 47–50 m, SE of Lanzarote, 28.933°N, 13.567°W (RMNH, 15).

Figure 118. Distribution of *Scissurella lobini*. ☆ type locality. ● specimen record. ■ literature record.

Literature records.

Cape Verde Archipelago. 269 m, S coast, Santo Antão Island, 15.267°N, 23.133°W. 785 m, Santo Antão Island, 15.267°N, 23.133°W. 60 m, SE Santo Antão Island, 15.267°N, 23.133°W. 32 m, Porto Novo, Santo Antão Island, 17.033°N, 25.067°W. 43 m, Porto Novo, Santo Antão Island, 17.033°N, 25.067°W (Burnay & Rolán, 1990).

Remarks. The species was originally described in *Sinezona* because of the strong axial sculpture on the protoconch although no specimen with a slit closed into a foramen could be found. Geiger (2003) showed the intrageneric variability of protoconch sculpture in *Scissurella* and *Sinezona* and re-assigned the species to *Scissurella*.

Scissurella lorenzi Geiger, 2006

Figures 119–123, Color Plate 2C

Chresonymy.

Scissurella lorenzi Geiger, 2006a: 7–9, figs 5–6.

Type material. Holotype (SBMNH 350876). Paratypes from type locality (SBMNH 350879, 6).

Type locality. 15–35 m, Baudisson Bay, Kavieng, New Ireland, Papua New Guinea, 2.743°S, 150.658°E (OD).

Etymology. Named after Felix Lorenz Jr. (1967–) who kindly donated some of the material (OD).

Description. Shell trochiform, globular, medium-sized (to 1.22 mm: holotype. Fig. 120). Protoconch of 0.75 whorl, fine axial sculpture, broadly anastomosing on embryonic cap, with faint spiral zigzag line; apertural varix well developed, connected to embryonic cap; apertural margin sinusoidal. Teleoconch I of 1.125–1.3 whorls (1.25 in holotype), approximately 29–35 fine axial cords (30 in holotype), spirals absent to indistinct at beginning of selenizone. Teleoconch II of 1.0–1.25 whorls (1.25 in holotype), suture little impressed. Shoulder with axials becoming weaker and/or more widely spaced with distance from start of selenizone, 2–4 irregular spiral lines, 2–6 additional very weak spiral lines gaining prominence towards apertural margin. Under light microscope, on shoulder of teleoconch II darker and lighter radial bands, approximately eight pairs on last half of body whorl (Fig. 119). Base with more distinct axial cords compared to shoulder of teleoconch II, less distinct compared to teleoconch I, axial cords on shoulder and base mostly coordinated; approximately 15 strong spiral lines of unequal strength, on average similar to axials, producing reticulate pattern. Umbilicus narrow, open, with very thin funiculus from broadened parietal wall of aperture. Selenizone above widest part of shell, slit open, margins remaining parallel, keels moderate. Aperture round with overhanging roof.

Radula (Fig. 122). Rachidian tooth triangular, central cusps largest, three cusps on each side of tip. Lateral teeth 1–3 similar, tip with 2–3 cusps on outer margin, inner margin smooth. Lateral tooth 4 reduced, lacking tip [abnormal specimen? Preparation artifact?]. Lateral tooth 5 enlarged by broadening, middle of eight cusps on inner edge of tip largest. Marginal teeth with spoon-shaped tip, many fine cusps around fringe. Radular interlock of central field moderate.

Figure 119. Light optical z-stacked stereomicroscope image of apical view of *Scissurella lorenzi*, showing typical axial markings on shoulder. 35–38 m, base of cliff, Horseshoe, Okinawa, Japan, 26.504°N, 127.858°E (DLG 978, 2). Scale bar = 1 mm.

Comparisons. *Scissurella staminea* from Japan has a protoconch with strong axials, the axial sculpture on teleoconch II remains the same strength, there are fewer axials per whorl, the transition of the base to the umbilicus is bordered by a distinct carina, and there is no banding pattern on the shoulder of teleoconch II.

Scissurella mirifica from the Indo-Malayan Archipelago to the Central Pacific, has a protoconch with strong axial cords, the transition of the base to the umbilicus is bordered by a distinct carina, and there is no banding pattern on the shoulder of teleoconch II.

Scissurella quadrata from the Indo-Malayan Archipelago lacks the axial color bands, has an umbilicus with strong periumbilical cord, and the aperture has a rather straight parietal wall with projecting lower adumbilical corner.

Scissurella rota from the Indian Ocean lacks the axial color bands, has widely spaced but distinct reticulate sculpture on the base, and much wider, funnel-shaped umbilicus.

Scissurella sudanica from the Indian Ocean has fewer axials on teleoconch I (22–25 vs. 29–35), the spacing and strength of axial cords on the shoulder of teleoconch II does not decrease with growth, the umbilicus is slightly wider, and there is no banding pattern on the shoulder of teleoconch II.

Scissurella xandaros from the Indo-Malayan Archipelago to the Western Pacific has a smooth protoconch, strong points formed at the intersection of axials and

Figure 120. *Scissurella lorenzi.* Holotype. 15–35 m, Baudisson Bay, Kavieng, New Ireland, Papua New Guinea, 2.743°S, 150.658°E (SBMNH 350876). Uncoated specimen, low kV SEM (1 kV). Scale bar shell = 1 mm. Scale bar protoconch = 100 µm.

spirals, has a wider umbilicus with a weaker bordering carina, and lacks the spiral color bands.

Juvenile *Sin. ferriezi* from the Indo-Malayan Archipelago to the Central Pacific has a smooth protoconch, a wider, funnel-shaped umbilicus, and lacks the axial color bands.

Figure 121. *Scissurella lorenzi*. **A.** 46 m, 1 km WNW of Onna Village (Horseshoe Cliffs), Okinawa, Japan, 26.493°N, 127.842°E (LACM 79-75, 2). **B–C.** Paratypes. 15–35 m, Baudisson Bay, Kavieng, New Ireland, Papua New Guinea, 2.743°S, 150.658°E (SBMNH 350879, 6). Scale bars shell = 1 mm. Scale bars protoconch = 100 µm.

Juvenile *Sin. macleani* from the Indian Ocean to the Western Pacific has a protoconch with flocculent sculpture and no apertural varix, has a wider, funnel-shaped umbilicus, and a longer teleoconch I (1.5 vs. 1.25–1.33 whorls).

Juvenile *Sin. marrowi* from the Indo-Malayan Archipelago lacks the axial color bands, has a protoconch with strong axial cords, and a wider funnel-shaped umbilicus.

Juvenile *Sin. singeri* from the western Indian Ocean lacks the axial color bands, has a smooth protoconch, and a wider, funnel-shaped umbilicus.

Distribution. Indo-Malayan Archipelago to Western Pacific (Fig. 123).

Specimen records.

Japan, Okinawa. 1 km WNW of Onna Village, moon rock entry point, 26.497°N, 127.843°E (LACM 78-29, 1). 1 km WNW of Onna Village, Horseshoe South, 26.495°N, 127.843°E (LACM 78-99, 1). 46–55 m, 1 km WNW of Onna Village, Horseshoe Cliffs, 26.493°N, 127.842°E (LACM 79-75, 2). 33 m, 1 km W Onna Village, 127.843°N, 127.843°E (LACM 78-20, 1). 92 m, Onna Village, 26.493°N, 127.493°E (LACM 78-101, 2). 38 m, base of cliff, Horeshoe Bay, 26.504°N, 127.858°E (DLG 1060, 3). 35–38 m, base of cliff, Horseshoe, 26.504°N, 127.858°E (DLG 978, 2). 26.500°N, 127.900°E (FLMNH 399224, 1). 43 m, Horseshoe Beach, 26.505°N, 127.853°E (DLG 1544, 1). 37 m, Big Rock, Horseshoe Bay, 26.508°N, 127.853°E (DLG 1459, 2).

Wallis and Futuna. 415–420 m, Wallis Island, 13.350°S, 178.133°W (MNHN, 1).

Tonga. Naupapu Island, Vava'a Group, 18.700°S, 174.100°E (LACM 85-89, 1).

New Caledonia. 10–21 m, Santal Bay, Lifou, 20.788°S, 167.113°E (MNHN, 1). 5–20 m, Santal Bay, Lifou, 20.875°S, 167.135°E (MNHN, 3). 25–30 m, Santal Bay, Lifou, 20.822°S, 167.173°E (MNHN, 1).

Papua New Guinea. 15–35 m, Baudisson Bay, Kavieng, New Ireland, 2.743°S, 150.658°E (SBMNH 350876, 1; SBMNH 350879, 1). 28 m, off Big Malu Malu Island, Kimbe Bay, New Britain, 5.231°S, 150.098°E (DLG 399, 1). 47 m, off Schaumann Island, Kimbe Bay, New Britain, 5.304°S, 150.092°E (DLG 772, 1). 65 m, off Susan's Reef, Kimbe Bay, New Britain, 5.291°S, 150.128°E (DLG 733, 14). 25 m, Port Moresby, Horseshoe Reef, 9.467°S, 147.167°E (AMS ex C.348633, 2). 13–22 m, Lolorua Island, SW of Port Moresby, 9.500°S, 147.117°E (AMS C.348624, 1).

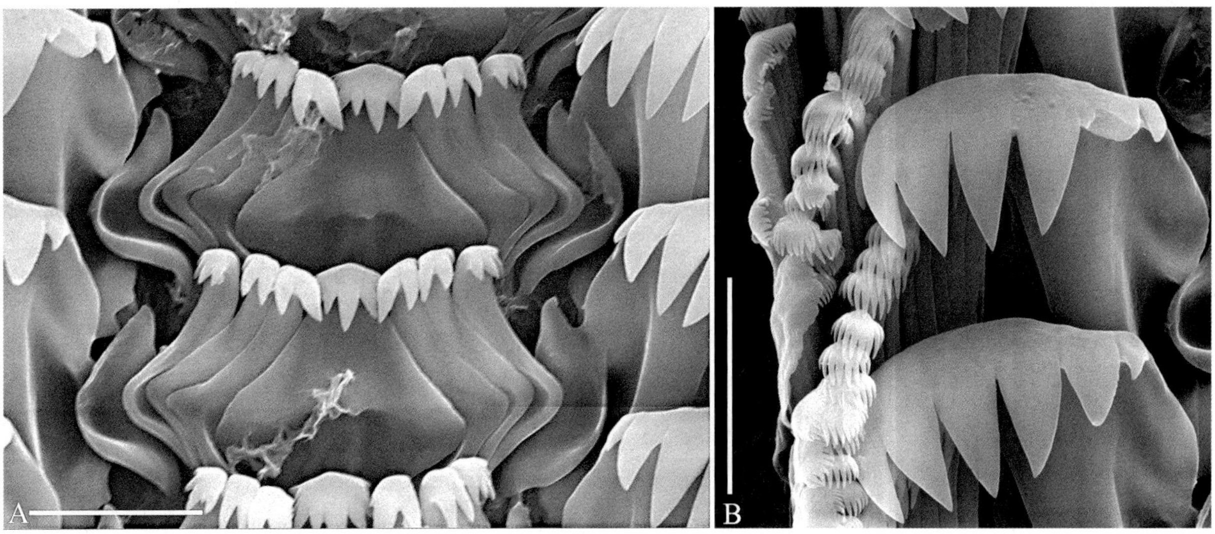

Figure 122. *Scissurella lorenzi.* Radula. 11 m, Scott Reef, Western Australia, Australia, 14.083°S, 121.833°E (AMS C.380209, 3) **A.** Central field. **B.** Lateral tooth 5 and marginal teeth. Scale bars = 10 µm.

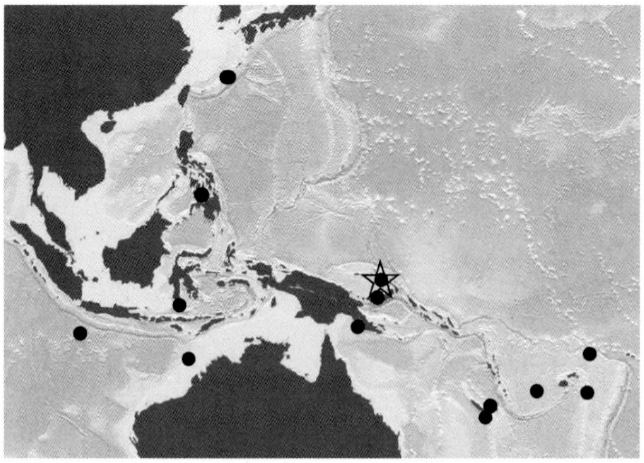

Figure 123. Distribution of *Scissurella lorenzi*. ☆ type locality. ● specimen record.

Philippines. 90–110 m, Balicasag Island, 9.515°N, 123.687°E (MNHN, 1: complete). 60 m, Pamilican Island, 9.502°N, 123.918°E (MNHN, 1: complete). 5 m, Biking, Panglao Island, PANGLAO 2004 S1, 9.588°N, 123.841°E (MNHN, 10). 24–27 m, Doljo Point, Panglao Island, PANGLAO 2004 B12, 9.593°N, 123.720°E (MNHN, 1). 31 m, N of Doljo, Panglao Island, PANGLAO 2004 B35, 9.598°N, 123.741°E (MNHN, 1). 4–30 m, Tatarman, Panglao Island, PANGLAO 2004 B7, 9.598°N, 123.863°E (MNHN, 5). 61–62 m, W of Baclayon, Bohol Island, PANGLAO 2004 T7, 9.601°N, 123.888°E (MNHN, 1). 120 m, off Momo Beach, Panglao Island, PANGLAO 2004 L50, 9.615°N, 123.763°E (MNHN, 1). 43–62 m, Bingag/Tabalong, Panglao Island, PANGLAO 2004 L51-60, 9.628°N, 123.800°E (MNHN, 9).

Indonesia. 175 m, SW Salayer, 6.373°S, 120.437°E (RMNH, 1).

Christmas Island. 183 m, off NE point, 10.417°S, 105.700°E (AMS C.406365, 1).

Australia. 11 m, Scott Reef, Western Australia, 14.083°S, 121.833°E (AMS C.380209, 3).

New Caledonia. 250–350 m, St. 40, 22.500°S, 166.400°E (MNHN, 1).

Fiji. 240–319 m, BORDAU 1 DW1494, 18.916°S, 178.483°W (MNHN, 1).

Remarks. The species is easier to recognize under the light microscope due to the unique pattern of dark and light radial bands on teleoconch II (Fig. 119). This banding pattern is not as easily discernible in the SEM (Figs 120, 121). All specimens have been found in sand and coral rubble.

Scissurella manawatawhia Powell, 1937

Figures 124–128

Chresonymy.

Scissurella manawatawhia Powell, 1937: 175–176, pl. 49, fig. 1.
Scissurella manawatawhia: Powell, 1940: 208.
Scissurella manawatawhia: Dell, 1956: 168.
Scissurella manawatawhia: Powell, 1979: 34, pl. 13, fig. 15.
Scissurella manawatawhia: Lozouet, 1986: 107.
Scissurella manawatawhia: Spencer & Willan, 1995: 13.
Scissurella manawatawhia: Geiger, 2003: 77.
Scissurella manawatawhia: Spencer *et al.*, 2009: 202.

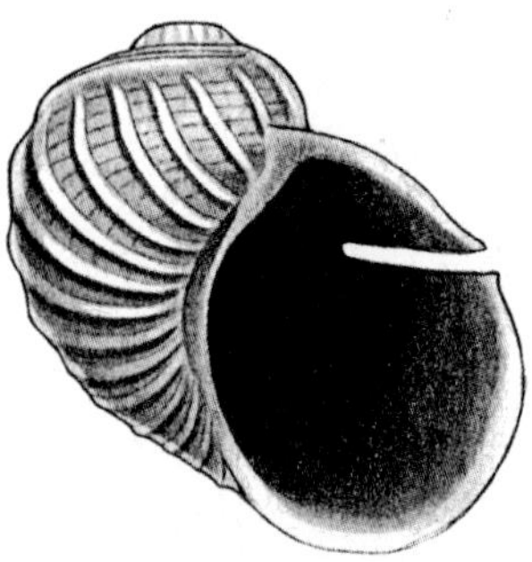

Figure 124. Original illustration of *Scissurella manawatawhia*.

Type material. Holotype (NHMUK 1962951. Fig. 125), 1.0 × 1.05 mm (H × W). Paratype from type locality (AIM AK 72217, 1).

Type locality. 260 m, off Three Kings Islands, [New Zealand, 34.090°S, 172.080°E] *Discovery* station 933 (OD).

Etymology. Manawatawhia: Maori name for Three Kings Islands (B. Marshall, pers. comm. 1/2012).

Description. Shell medium-sized (to 1.05 mm), trochiform, inflated. Protoconch of one whorl, with fine axial cords with occasional short side branches, strong, raised spiral cord on apical surface, strong, raised, broken spiral cord just above suture; no apertural varix; apertural margin convex. Teleoconch I of 0.75 whorl, approximately 12 strong axial cords, intersected by three fine spiral threads; interstices with fine growth lines. Teleoconch II of 0.75 whorl, suture little impressed. Shoulder somewhat convex, approximately 17 strong axial cords, mostly at same spacing as on teleoconch I, near apertural margin more crowded, weaker; 2–4 fine spiral threads starting after 0.33 teleoconch II whorl. Base with same axials as on shoulder, approximately 13 spiral threads not running over axials, spacing denser towards umbilicus. Umbilicus very

Figure 125. *Scissurella manawatawhia*. Holotype. 260 m, off Three Kings Islands, [New Zealand, 34.090°S, 172.080°E] *Discovery* station 933 (NHMUK 1962951). Scale bar shell = 1 mm. Scale bar protoconch = 100 µm.

narrow, with periumbilical cord, walls smooth. Selenizone above periphery, lunules indistinct, keels moderately elevated, moderately strong, margins of slit remaining parallel. Aperture obliquely rounded.

Operculum thin, round, multispiral, nucleus central, covering only ¼ of aperture (Fig. 126B).

Figure 126. *Scissurella manawatawhia*. **A.** 310 m, 11 km NW of Great Island, Three Kings Islands, New Zealand, 36.108°S, 172.078°E (NMNZ M.093858, 15). **B.** 91 m, off Three Kings Islands, New Zealand, 34.183°S, 172.167°E (NMNZ M.033654, 12). Radula shown in Figure 127. **C.** 805 m, off Three Kings Islands, New Zealand, 34.000°S, 171.917°E (NMNZ M.20894, 40). Scale bars shell = 0.5 mm. Scale bars protoconch = 100 μm.

Radula (Fig. 127). Rachidian tooth with central cusps on tip largest, three smaller cusps on each side. Lateral teeth 1–3 similar, apical cusp largest, four cusps of decreasing size on outer edge of tip. Lateral tooth 4 reduced, hook-shaped, minute points below apical cusp. Lateral tooth 5 enlarged by broadening, apical cusp on tip largest, approximately six cusps of decreasing size along inner edge of tip, three cusps at junction between tip and shaft along outer edge. Inner marginal teeth with broad triangular tip, apical cusp largest approximately six cusps on each side; no food groove. Outer marginal teeth spoon-shaped, with many fine bristles. Radular interlock moderate.

Animal with eyes. Further anatomical details unknown.

Comparisons. *Scissurella bountyensis* from New Zealand has a protoconch with strong axial cords and a distinct apertural varix, lacks spirals on shoulder and base, and has a wider umbilicus.

Scissurella prendrevillei from New Zealand has flat spire, a protoconch lacking the distinct spiral cord, and stronger axial lamellae, but weaker spiral threads.

Scissurella skeneoides from New Zealand has a flatter spire, lacks spiral sculpture, and has axial cords that are twice as dense.

Scissurella ornata from Australia has a protoconch with an apertural varix, a less elevated spire with more rapidly expanding whorls, and a strong periumbilical cord.

Juvenile *Sin. bandeli* from New Zealand has a more depressed spire, a wider umbilicus, a protoconch with strong axial cords, and more numerous spiral threads on shoulder and base.

Juvenile *Sin. brevis* from New Zealand has a shorter (or no) selenizone, a protoconch lacking the strong spiral cord, and denser spiral threads on shoulder and base.

Juvenile *Sin. iota* from New Zealand has a protoconch with strong axial cords, has a more elevated spire, stronger but fewer raised axial cords, and lacks spiral threads on shoulder and base.

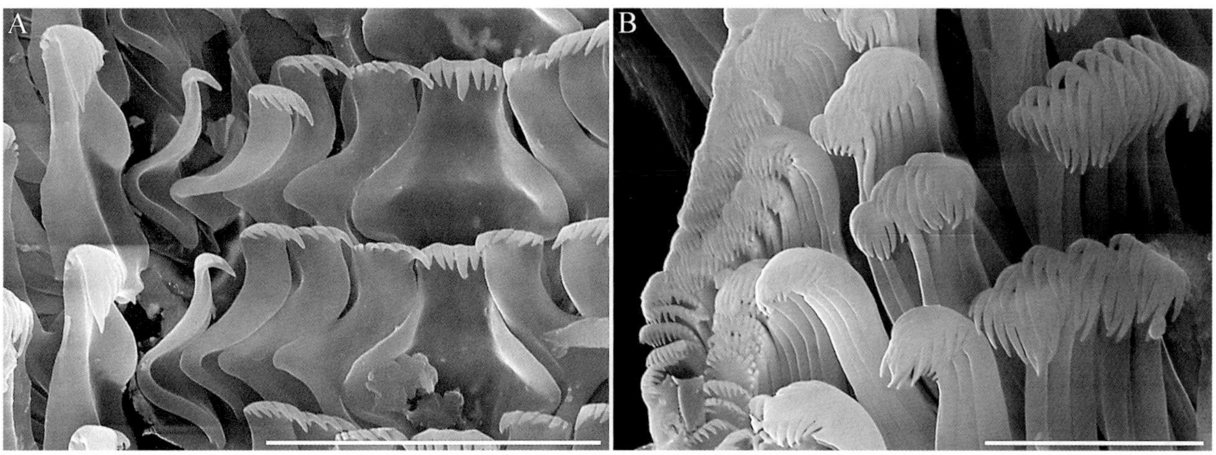

Figure 127. *Scissurella manawatawhia*. Radula. 91 m, off Three Kings Islands, New Zealand, 34.183°S, 172.167°E (NMNZ M.033654, 12). Shell shown in Figure 126B. **A.** Central field. Scale bar = 20 µm. **B.** Marginal teeth. Scale bar = 10 µm.

Juvenile *Sin. laqueus* from New Zealand has a strong constriction below the selenizone, very strong keels of the selenizone and prominent lunules, and has an overall more depressed shell shape with wider umbilicus.

Juvenile *Sin. levigata* from New Zealand has a more rounded shell shape with lower spire, a distinct callus covering the umbilicus, and a very short (or no) selenizone.

Juvenile *Sin. macleani* from the Indian Ocean to the Western Pacific including northernmost New Zealand has a protoconch with flocculent sculpture, a wider, funnel-like umbilicus, and fine reticulate sculpture on shoulder and base.

Juvenile *Sin. mechanica* from New Zealand lacks spiral sculpture on shoulder and base, has a more depressed spire, and a wider, funnel-like umbilicus.

Juvenile *Sin. pacifica* has a protoconch with sharp axial cords and no spiral cord, a less pointed spire, and lacks spiral thread on early teleoconch I.

Juvenile *Sin. platyspira* from New Zealand has a sunken protoconch without a spiral cord, a wide umbilicus through which the underside of the protoconch is visible, and lacks spiral sculpture on shoulder and base.

Distribution. Northernmost New Zealand (Fig. 128).

Specimen records.

New Zealand. 805 m, N of Three Kings Islands, 34.000°S, 171.917°E (AMS C.376863, 1). 92 m, off Three Kings Island, 34.193°S, 172.182°E (AMS C.376866, 2). 92 m, off Three Kings Islands, *Discovery* St. 934, 34.167°S, 172.083°E (NHMUK 1961658, 2). 310 m, 11 km NW of Great Island, Three Kings Islands, 36.108°S, 172.078°E (NMNZ M.093858, 15). 102 m, off Northeast Island, Three Kings Islands, 34.142°S, 172.183°E (NMNZ M.034411, 11). 710 m, 24 km NW of Great Island, Three Kings Islands, 35.098°S, 171.918°E (NMNZ M.94440, 7). 91 m, off Three Kings Islands, 34.183°S, 172.167°E (NMNZ M.033654, 12). 805 m, off Three Kings Islands, 34.000°S, 171.917°E (NMNZ M.20894, 40). 123–128 m, King Bank, N of Three Kings Islands, 33.957°S, 172.323°E (NMNZ M.160312, 40). 246–291 m, Middlesex Bank, NW of Three Kings Islands, 34.033°S, 171.733°E (NMNZ M.177154, 5). 98–103 m, Middlesex Bank, NW of Three Kings Islands, 33.950°S, 171.757°E (NMNZ M.177150, 13).

Figure 128. Distribution of *Scissurella manawatawhia*. ☆ type locality. ● specimen record.

Remarks. A note by Kathie Way (NHMUK) from 1983 indicates that no paratypes were found.

Scissurella mantelli Woodward, 1859 [*nomen dubium*]

Figure 129

Chresonymy.

Scissurella mantelli Woodward, 1859: 202, pl. 46, fig. 9 [caption referring to fig. 8 is an obvious error].

Scissurella mantelli: Hutton, 1885: 368.

Scissurella mantellii [*sic*]: Paetel, 1888: 289 [?].

Scissurella mantelli: Hutton, 1904: 83.
Scissurella mantelli: Suter, 1910: 9 [?: Pliocene of New Zealand].
Scissurella mantelli: Lozouet, 1986: 108.

Misidentifications.

not *Scissurella mantellii* [sic]: Munier-Chalmas, 1865: 396 [is *An. flemingi*].
not *Scissurella mantelli*: Pilsbry, 1890: pl. 57, fig. 12 [is *An. flemingi*].
? *Scissurella mantelli*: Hutton, 1893: 71 [referenced both Woodward and Pilsbry].
not *Scissurella mantelli*: Thiele, 1912a: 20, pl. 3, fig. 5 [is *An. flemingi*].
not *Scissurella mantelli*: Suter, 1902: 217 [is *An. flemingi*].
not *Scissurella mantelli*: Suter, 1913: 88–89, pl. 6, fig. 10 [is *An. flemingi*].
not *Scissurella mantelli*: Suter, 1915: 19 [is *An. flemingi*].
not *Schizotrochus mantelli*: Finlay, 1926: 339 [is *An. regia*: *fide* Fleming, 1948].
not *Schizotrochus mantelli*: Powell, 1940: 208 [is *An. flemingi*].
not *Schizotrochus mantelli*: Fleming, 1948: 84, pl. 8, fig. 3 [is *An. flemingi*].
not *Schizotrochus mantelli*: Fleming, 1950: 25 [is *An. flemingi*].
not *Schizotrochus mantelli*: Fleming, 1951: 136 [is *An. flemingi*].
not *Schizotrochus mantelli*: Dell, 1956: 168, 182 [is *An. flemingi*].
not *Anatoma mantelli*: Powell, 1979: 36, pl. 14, fig. 1 [is *An. flemingi*].
not *Anatoma mantelli*: Spencer & Willan, 1995: 13 [is *An. flemingi*].
Anatoma s.l. *mantelli*: Geiger, 2003: 75 [as *nomen dubium*].

Figure 129. Original illustration of *Scissurella mantelli.*

Type material. Syntype(s) (presumed lost: Fleming, 1951). Fleming's (1951: 136) "neotype" (GNS TM 426: Fleming, 1948: pl. 8, fig. 3), 3 × 2.6 mm. Neotype designation invalid because an obvious transposition in the figure caption was not recognized (ICZN Art. 75.3.5).

Type locality. New Zealand in ironsand (OD). Fiordland, Dagg's Sound, inside a rock bar, 58 fathoms [= 106 m] (Fleming's "neotype").

Etymology. Named in honor of Walter Mantell [1820–1895] (OD).

Remarks. Woodward's (1859) plate 44 showed nine specimens, one of which is clearly an *Anatoma* (fig. 8), whereas the others are referable to *Scissurella* and/or *Sinezona* (See Lozouet, 1986; Marshall, 2002). A single specimen in the caption (fig. 9) was assigned to Woodward's new subgenus *Anatoma*, with type species *Sci. crispata*; see Remarks of genus *Anatoma* for misidentified type species.

As discussed by Marshall (2002), the identity of the true *Sci. mantelli* is uncertain. Most authors had accepted the figure caption at face value, hence, identifying the most prominent member of Anatomidae from the New Zealand fauna as "*Anatoma mantelli.*"

The neotype designated by Fleming (1951) is obviously flawed, as the transposed figure caption was not recognized. Marshall (2002) disregarded the neotype designation of Fleming (1951), in accordance with ICZN Art. 75.3.5, and described the known biological species as *Th. flemingi*.

The identity of the true *Sci. mantelli* remains unclear. No New Zealand species resembles it (Marshall, 2002). Suter's (1910) single line listing is unconfirmed. The closest species is the tropical Indo-Pacific *Sin. plicata*, which does not occur in New Zealand.

Scissurella maraisorum Geiger, 2006

Figures 130–132

Chresonymy.

Scissurella maraisorum Geiger, 2006a: 10–12, figs 7–8.

Type material. Holotype (NMSA W3498/T1616. Fig. 130). Paratypes from type locality (NMSA W3499/T1617, 2; Coll. A. P. & J. P. Marais, 3. Geiger, 2006: fig. 8A; SBMNH 359327, 2. Fig. 131B. SBMNH 359324. Fig. 131C; AMS C.447877, 1; NHMUK 20050274, 1; MNHN, 1; UMUT RM29043, 1; USNM 1081939, 1).

Type locality. 20 m, Aliwal Shoal, Kwazulu-Natal south coast, South Africa, 30.250°S, 30.817°E (OD).

Etymology. Named after Dr. Johan P. and Alwyn P. Marais of South Africa, who collected the specimens and generously made them available for study, and for their continuing

Figure 130. *Scissurella maraisorum.* Holotype. 20 m, Aliwal Shoal, Kwazulu-Natal S coast, South Africa (NMSA W3489/T1616). Uncoated specimens, variable pressure, VPSE detector. Scale bar shell = 0.5 mm. Scale bar protoconch = 100 μm.

accomplishments to further the understanding of the South African malacofauna. Genitive masculine plural of surname.

Description. Shell small (to 0.72 mm: holotype), trochiform depressed. Protoconch of one whorl, fine axial sculpture with median ridge, no apertural varix, apertural margin convex. Teleoconch I of 1–1.125 whorls, approximately 14–18 strong axial cords, no spirals, interstices with fine irregular axials. Teleoconch II of 0.33 whorl, suture deeply impressed. Shoulder flat, suture deep, near suture indistinct axial cords of same spacing as on teleoconch I, mainly fine irregular axials. Base with strong

Figure 131. *Scissurella maraisorum*. **A.** 100 m, Park Rynie, Natal, South Africa, 30.357°S, 30.183°E (DLG 1399, 6). **B–C.** 20 m, Aliwal, Kwazulu-Natal south coast, South Africa. **B.** Paratype (SBMNH 359327, 1). **C.** Paratypes (SBMNH 359324, 1). Uncoated specimens, variable pressure, VPSE detector. Scale bars shell = 0.2 mm. Scale bars protoconch = 100 μm.

axial cords of same spacing as on teleoconch I from selenizone to lowermost base, fading towards umbilicus; no spirals; interstices with fine irregular axials; base continuously sloping with umbilical wall. Umbilicus wide, open, protoconch visible through umbilicus. Selenizone above periphery, lunules distinct, keels moderately strongly elevated, strong, inclined approximately 45° towards coiling axis. Aperture subquadrate, adumbilical wall straight, roof almost straight, base curved with adumbilical projection.

Animal unknown.

Comparisons. *Scissurella maraisorum* is characterized among the African species by its flat spire, the absence of spiral sculpture, and the widely open umbilicus, which shows the lower surface of the protoconch.

Scissurella jucunda from South Africa has a distinctly elevated spire, a closed umbilicus, and some spiral sculpture on shoulder as well as base.

Sinezona doliolum from South Africa has a much taller shell, distinct spiral threads on shoulder and base and a much narrower umbilicus.

Scissurella rota from East Africa has spiral lines on shoulder and base and the umbilicus is bordered by an edge towards base.

Scissurella sudanica from East Africa has an elevated spire, spirals on shoulder and base, and a narrower umbilicus that is bordered by a strong cord towards the base.

Scissurella azorensis from the Macaronesian Islands and the Western Mediterranean has axial cords that increase in strength with growth (decreasing in *Sci. maraisorum*). Otherwise, these two species are exceedingly similar; however, as the subtropical Atlantic *Sci. azorensis* is separated from *Sci. maraisorum* in eastern South Africa by a distinct temperate zone, they are very unlikely to belong to the same biological species.

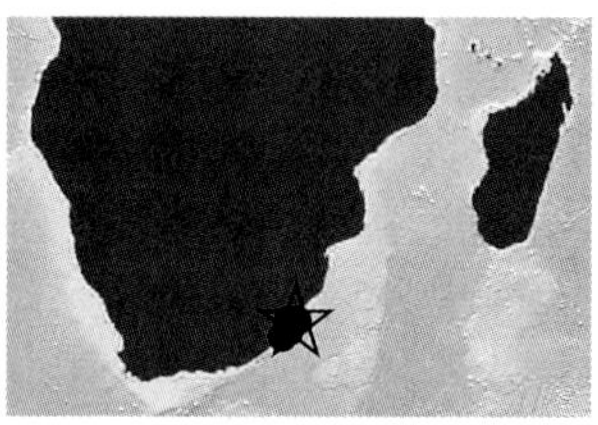

Figure 132. Distribution of *Scissurella maraisorum*. ☆ type locality. ● specimen record.

Scissurella lobini from the Cape Verde Islands has a protoconch with strong axial cords and has axial cords on the shoulder of teleoconch II.

Sinezona insignis from South Africa has a closed foramen in adults, fewer but stronger axial cords to low lamellae, has strong axials on the protoconch, and has a much more narrow umbilicus bordered by a cord towards base.

Scissurella reticulata from the Indian Ocean has reticulate sculpture on shoulder and base, a taller spire, and an overall more rounded shell shape.

Sukashitrochus dorbignyi from the Indian Ocean has distinct spiral keels on the base, a closed foramen in mature specimens, and the umbilicus is bordered by a keel.

Scissurella kaiserae from the Panamic has a protoconch with fewer axial cords, has about 50% more axial cords on the teleoconch, and the suture of the aperture is below the periphery on the previous whorl.

Scissurella prendrevillei from New Zealand has a longer teleoconch I (1.25 vs. 1 whorl), the suture of the aperture is below the periphery of the previous whorl, and the axial cords are raised into more pronounced ribs.

Sinezona platyspira from New Zealand is indistinguishable except for the closed foramen in fully grown specimens. Unlike the New Zealand *An. finlayi* and its synonym *An. herberti* described from the Indian Ocean, no specimens with the *maraisorum-platyspira* morphology are known from the Indo-Malayan Archipelago.

Distribution. Eastern South Africa (Fig. 132).

Specimen records.

South Africa. 20 m, Aliwal Shoal, Kwazulu-Natal S coast, 30.250°S, 30.817°E (NMSA W3489/T1616, 1). Mzamba, Transkei, 31.100°N, 30.183°E (DLG 1384, 1). 45 m, Landers Reef, Rocky Bay, Natal, 30.357°S, 30.796°E (DLG 1388, 2). 90 m, Mbotyi,

Transkei, 31.463°S, 29.737°E (DLG 1394, 2). 100 m, Park Rynie, Natal, 30.357°S, 30.183°E (DLG 1399, 6).

Scissurella mirifica (A. Adams, 1862)

Figures 133–139, Color Plate 3D

Chresonymy.

Anatomus mirificus A. Adams, 1862: 348, not illustrated.
Scissurella mirifica: Crosse, 1863: 110.
Anatomus mirificus: Troschel, 1863: 702.
Scissurella mirifica: Paetel, 1888: 289.
Anatomus mirificus: Pilsbry, 1890: 60.
Scissurella mirifica: Thiele, 1912a: 15, pl. 2, figs 7–8 [after a specimen in the British Museum: type? Fig. 133A].
Anatoma mirifica: Habe & Kosuge, 1964: 4.
Scissurella mirifica: Lozouet, 1986: 108.
Anatoma mirifica: Higo & Goto, 1993: 15.
Scissurella mirifica: Geiger, 2003: 77.
Scissurella mirifica: Geiger *et al.*, 2007: fig. 12D–F [SEM of radula].
Scissurella mirifica: Geiger, 2008a: 158, pl. 24, fig. 6a–b.
Scissurella mirifica: Tröndlé & Boutet, 2009: 12.
Scissurella sp.: Warén, 2011: fig. 500G [is *Sci. mirifica*].

Synonym.

+ *Scissurella declinans* Watson, 1886: 115–116, pl. 8, fig. 2. Fig. 133B. ***Type material.*** Syntypes (NHMUK 1887.2.9.404–405, 2), 0.044" × 0.063" [1.8 × 2.5 mm], 1.4 mm (broken). One here designated as lectotype (NHMUK 1887.2.9.404. Fig. 134) for the express purpose of taxon stabilization in case other non-conspecific syntypes should be found. Paralectotype (NHMUK 1887.2.9.405, 1). ***Type locality.*** Station 185b, Raine Island, Cape York, North Eastern Australia [Queensland, Australia]. 11°38'15"S 143°59'38"E, 55 fathoms [= 101 m], coral sand (OD). ***Etymology.*** Named for the descending aperture of the shell (OD).
Scissurella declinans: Kobelt, 1887: 428.
Scissurella declinans: Pilsbry, 1890: 57, pl. 68 [erroneously pl. 58 in text], figs 6–8.
Scissurella declinans: Thiele, 1912a: 18, pl. 2, figs 17–19 [copy figures Watson, 1886].
Scissurella declinans: Ladd, 1966: 12, 26–27, pl. 2 fig. 6 [USNM specimen examined].
Scissurella sp.: Maes, 1967: 103, pl. 3, fig. B. [is *Scissurella* cf. *mirifica*].

A B

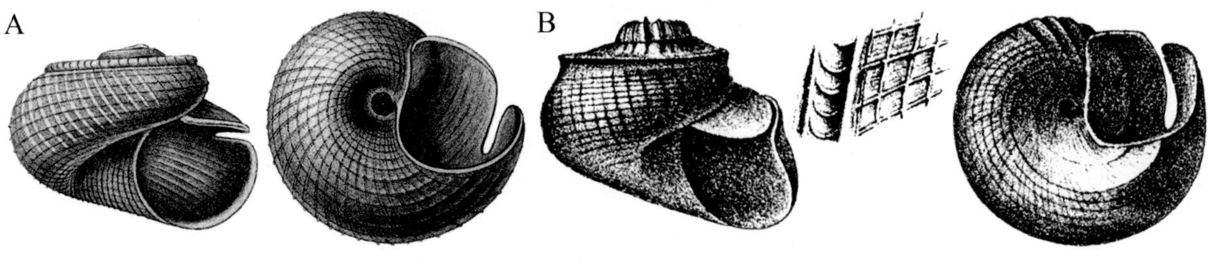

Figure 133. **A.** Illustration of *Scissurella mirifica* by Thiele (1912) "after a specimen in the British Museum," which may have been Adams' type. **B.** Original illustration of *Scissirurella declinans*.

Scissurella declinans: Batten 1975: fig. 13.
Scissurella declinans: Lozouet, 1986: 107.
Scissurella declinans: Kay & Johnson, 1987: 106 [Holocene of Enewetak].
Scissurella declinans: Jansen, 1999: 51, figs 25–27.
Scissurella declinans: Geiger, 2003: 76.
Scissurella declinans: Smith, 2003: 249.
Scissurella declinans: Geiger & Jansen, 2004b: 20–26, figs 9D [map], 10–12 [shell, radula].
Scissurella declinans: Geiger, 2004: text-fig. p. 6.
Scissurella declinans: Geiger, 2006c: text-fig. 3.

Misidentification.

not *Sukashitrochus declinans*: Bandel, 1998: 54–55, pl. 19, figs 1–3 [is *Suk. atkinsoni*].

Type material. Syntype (NHMUK 1874.5.19.63, 1. Fig. 135), 2.05 mm, here designated as lectotype with the express purpose of taxon stabilization in case other non-conspecific syntypes should be located.

Type locality. Lo-shan-kow [=? Lau-shan bay], Shantung [= Shandong, China, ~36°N, 120°E] (OD).

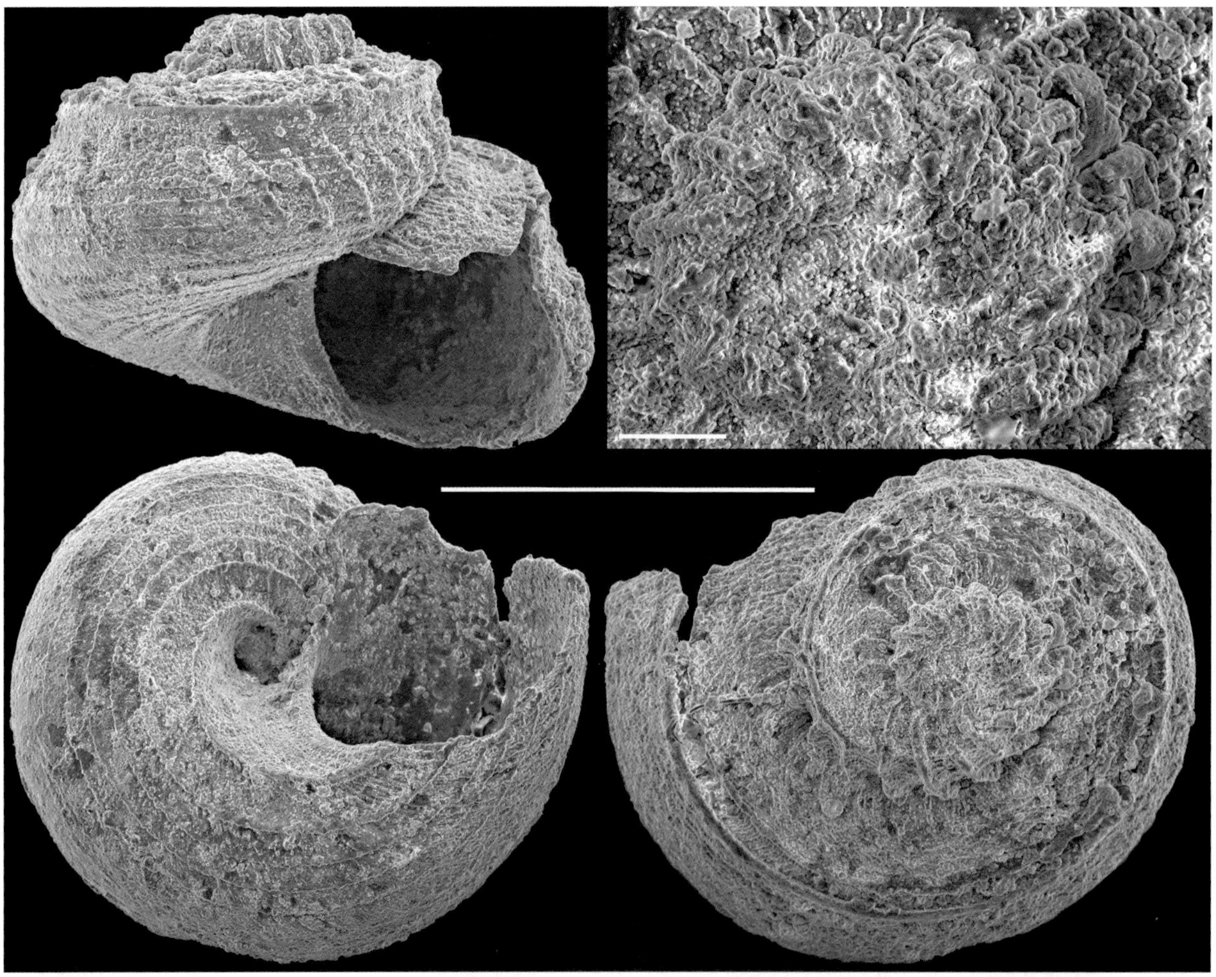

Figure 134. *Scissurella declinans.* Here designated lectotype. Station 185b, Raine Island, Cape York, North Eastern Australia [Queensland, Australia], 11°38'15"S 143°59'38"E, 55 fathoms [= 101 m] (NHMUK1887.2.9.404). Scale bar shell = 1 mm. Scale bar protoconch = 100 µm.

Etymology. Mirificus, late Latin superlative of mirabilis: the most wonderful.

Description. Shell medium-sized (to 1.75 mm), trochiform. Protoconch 0.75 whorl, with fine axial sculpture, apertural varix connected to embryonic cap, apertural margin sinusoidal. Teleoconch I of 1.125–1.25 whorls, 15–19 axial cords, approximately a dozen weaker, evenly spaced spiral lines. Teleoconch II up to 1.33 whorls, suture moderately impressed. Shoulder slightly convex, axial cords weaker than on teleoconch I, slightly stronger than spiral lines; spiral lines of similar strength over entire teleoconch. Base with weak constriction below selenizone, proximal to selenizone with axial cords and spiral lines; axials not coordinated with those on shoulder; 14–18 spiral lines from selenizone to umbilicus forming tiny knobs at intersection with axials portion proximal to selenizone; adumbilical portion of base without axials. Base and wide umbilicus at sharp angle; carina at most weak. Umbilical wall smooth, straight. Slit open, in fully mature specimens, moderately elevated keels of slit almost touching at aperture; lunules distinct, not coordinated with axials.

Operculum (Fig. 137A) round, thin, nucleus central, with few irregular concentric growth rings.

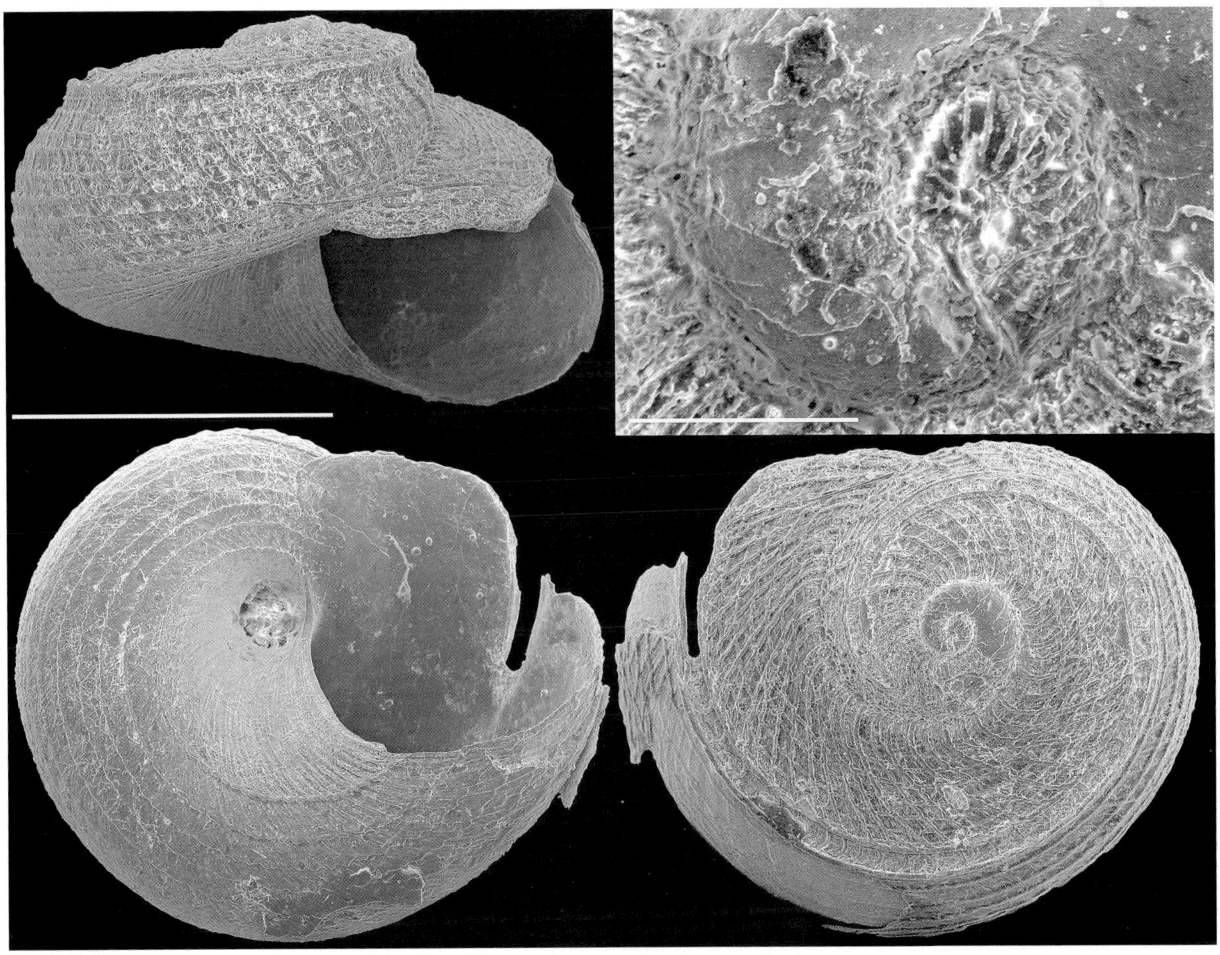

Figure 135. *Anatomus mirificus*. Here designated lectotype. Lo-shan-kow [=? Lau-shan bay], Shantung [= Shandong, China, ~36°N, 120°E] (NHMUK 1874.5.19.63). Scale bar shell = 1 mm. Scale bar protoconch = 100 μm.

Radula (Fig. 138) with pentacuspid rachidian tooth. Lateral teeth 1–3 bi- or tricuspid. Lateral tooth 4 hook-shaped. Lateral tooth 5 hypertrophied. Marginal teeth with fine bristles, finer with increasing distance from rachidian tooth. Radular interlock of central field moderate.

Comparisons. *Scissurella staminea* from Japan has stronger and less dense sculpture, with elevated axial cords, and finer spiral threads, forming small but distinct points at their intersections.

Scissurella quadrata from the Indo-Malayan Archipelago to the Central Pacific has a more subquadrate overall shell shape, a straight parietal wall in the aperture with a distinctly pointed lower adumbilical corner, and a more narrow umbilicus.

Scissurella spinosa from the Indo-Malayan Archipelago to the Central Pacific has half as many but twice as strong axial cords, which show raised points at the intersection with the spiral threads, and is overall less wide.

Scissurella reticulata from the Indian Ocean has a more rounded overall shell shape with notable clathrate sculpture on shoulder and base and the intersections of the axials and spirals are drawn out into points.

Figure 136. *Scissurella mirifica*. Sovi Bay Souvi, SW coast Viti Levu, Fiji, 18.183°S, 177.600°E (AMS C.406355, 15). Scale bar shell = 1 mm. Scale bar protoconch = 100 µm.

Scissurella rota from the Indian Ocean has very coarse clathrate sculpture on the base, and a shorter teleoconch I (1 vs. 1.25 whorls).

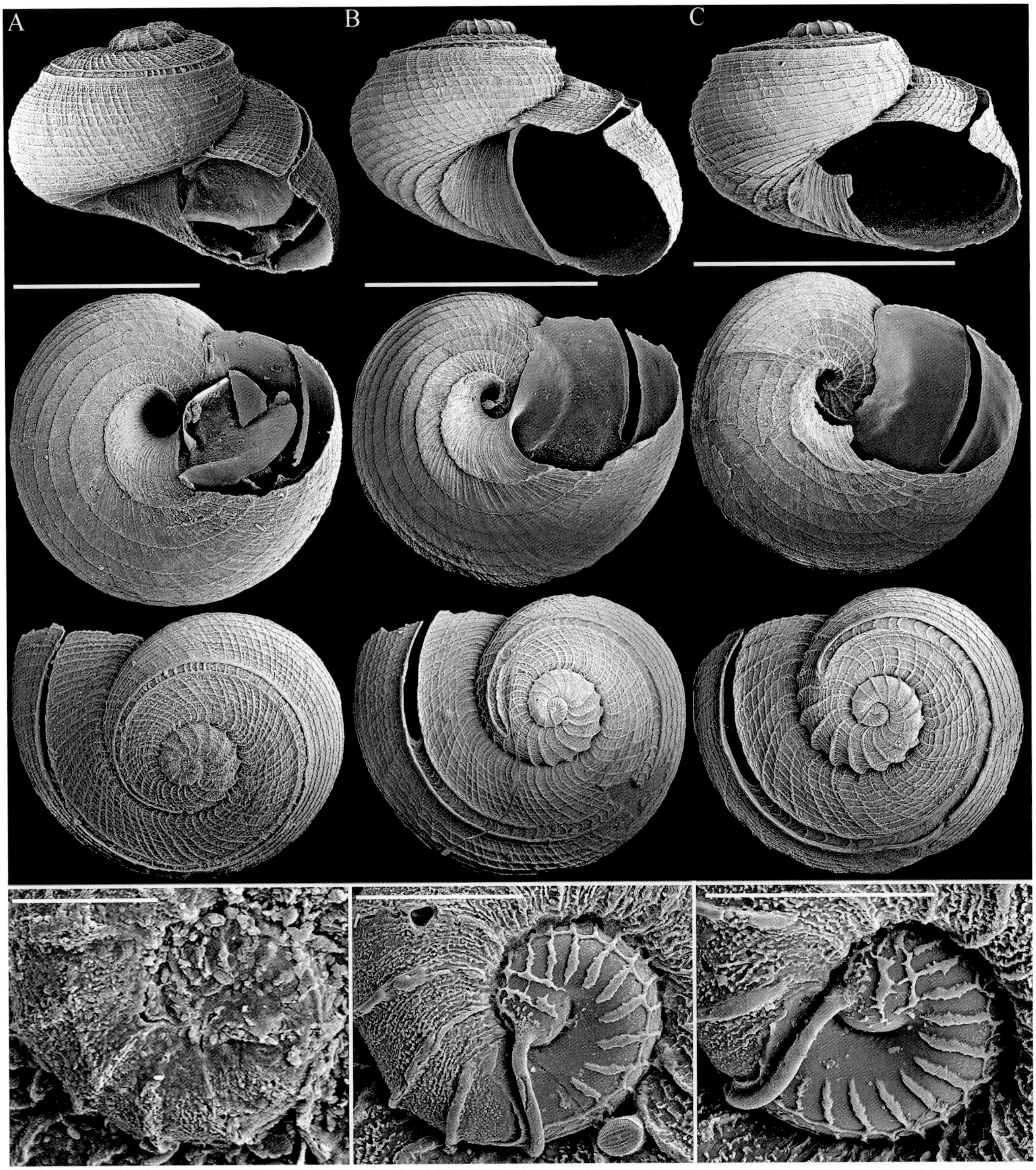

Figure 137. *Scissurella mirifica* **A.** Loyalty Islands, SE side Lifou Island, Luengoni, New Caledonia, 21.033°S, 167.417°E (AMS C.348865, 1). **B.** Wassagne, Lifou Island, Loyalty Islands, New Caledonia, 21.000°S, 167.383°E (AMS C.348642, 22). **C.** Viti Levu S coast, Korotongo, Fiji, 18.183°S, 177.533°E (AMS C.348872, 4). Scale bars shell = 1 mm. Scale bars protoconch = 100 µm.

Scissurella sudanica from the Indian Ocean has an overall more globular shell outline with more narrow umbilicus, very fine reticulate pattern on the base, and stronger axial cords on the shoulder.

Scissurella xandaros from the Indo-Malayan Archipelago to the Western Pacific has a smooth protoconch and has ewer but stronger axial cords that are drawn out into points at the intersection with the spiral threads.

Scissurella morretesi from Brazil has a more inflated shell lacking the constriction of the base below the selenizone, has a protoconch with flocculent sculpture and lacking an apertural varix, and has a longer teleoconch I (1.75 vs. 1.25 whorls).

Juvenile *Sin. ferriezi* with an open slit is angulated at the selenizone: shoulder and base does not form a continuous arc as in *Sci. mirifica*. The outline of the whorl is more triangular and less round in *Sin. ferriezi* compared to *Sci. mirifica*.

Juvenile *Sin. marrowi* from the Indo-Malayan Archipelago has stronger axial cords that are drawn out into points at the intersection with the spiral threads, the protoconch has strong axial cords, and the slit always closes to form a foramen after at most 0.6 whorl.

Juvenile *Sin. macleani* from the Indian Ocean to the Western Pacific have a wider shell with straight parietal wall and drawn out lower adumbilical corner of the aperture, and the protoconch has flocculent sculpture and lacks an apertural varix.

Juvenile *Sin. singeri* from the Indian Ocean have a stronger constriction on the base below the selenizone, a smooth protoconch, and stronger axial cords.

Distribution. Indo-Malayan Archipelago to Central Pacific (Fig. 139). 0–6 m depth live, in algae and coral rubble of reefs.

Fossil record. Ladd (1966) reported the species from Upper Miocene of the Bikini Atoll.

Specimen records.

Thailand. 4–6 m, Krabi, coral reefs outside, 8.046°N, 98.912°E (DLG 1126, 25). 0 m, Rawai Beach, Puket, 7.770°N, 98.328°E (LACM 85-2, 2). 4 m, Mai Thon island, off Phuket, 7.762°N, 98.488°E (LACM 85-8, 3).

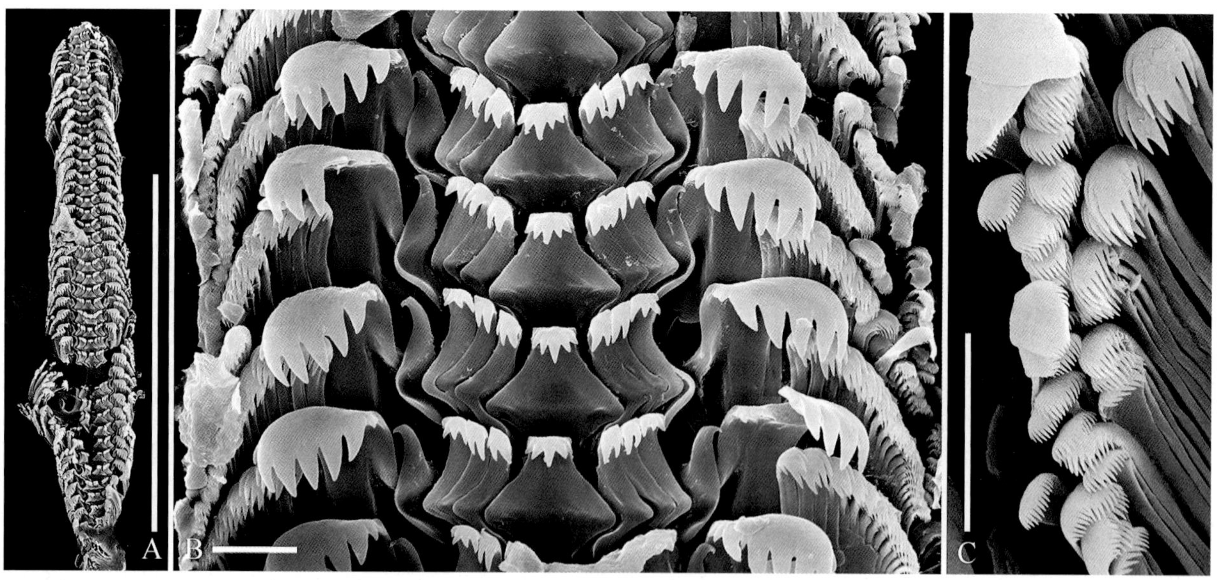

Figure 138. *Scissurella mirifica.* Radula. Wassagne, Lifou Island, Loyalty Islands, New Caledonia, 21.000°S, 167.383°E. (AMS C.348642, 22). **A.** Entire radula. Scale bar = 500 µm. **B.** Full rows of radula. Scale bar = 20 µm. **C.** Tips of marginal teeth. Scale bar = 10 µm.

Singapore. 1.200°N, 103.900°E (KBIN, 1).

Indonesia. Kalimantan, Sembuga, ca. 14 km by road E of Balikpapan, 1.233°S, 116.950°E (AMS C.349125, 1). 175 m, SW Salayer, 6.373°S, 120.437°E (RMNH, 1; RMNH, 5). SW Salayer, 6.355°S, 120.435°E (RMNH, 5). 233–274 m, SW Salayer, 6.372°S, 120.433°E (RMNH, 3). 0–10 m, bay N of Tanjung Batuangus, Selat Lembeh, N Sulawesi, 1.517°N, 125.250°E (RMNH, 1). 4–5 m, W side of Laha, outer bay, Ambon Bay, Hitu, Ambon, Moluccas, 3.715°S, 128.098°E (RMNH, 1). 3 m, Latuhalata, S coast, Leitimur, Ambon, Moluccas, 3.778°S, 128.126°E (RMNH, 1). 175–185 m, Bay of Sanggar, N of Sumbawa Bay, 8.322°S, 118.260°E (RMNH, 3). 74–83 m, E of Melolo, NE coast of Sumba, 9.892°S, 120.712°E (RMNH, 1). 280 m, NE of Binongko, Banda Sea, Tukang Besi Islands, 5.875°S, 123.975°E (RMNH, 1). 250–290 m, NW of Binongko, Banda Sea, Tukang Besi Islands, 5.875°S, 123.975°E (RMNH, 4). 385 m, NW of Binongko, Banda Sea, Tukang Besi Islands, 5.873°S, 123.975°E (RMNH, 1). 285–305 m, S of Kaledupa Reef, Banda Sea, Tukang Besi Islands, 5.958°S, 123.775°E (RMNH, 1). 465 m, S of Kaledupa Reef, Banda Sea, Tukang Besi Islands, 5.958°S, 123.775°E (RMNH, 4). 365 m, S of Kaledupa Reef, Banda Sea, Tukang Besi Islands, 5.963°S, 123.775°E (RMNH, 3). 6 m, Tg. Benteng (Galghoek), outer bay, Ambon Bay, Leitimur, Ambon, Moluccas, 3.702°S, 128.163°E (RMNH 42322, 1). 300–340 m, NW Pulau Bahuluang, SW Salayer, 6.467°S, 120.400°E (RMNH, 1). 233–274 m, SW Salayer, 6.372°S, 120.433°E (RMNH, 50). 300 m, NE of Pulay Bahuluang, SW Salayer, 6.437°S, 120.442°E (RMNH, 1; RMNH, 2). 233–274 m, SW Salayer, 6.372°S, 120.433°E (RMNH, 1). 558 m, St. 5592, off Silungan Island, Borneo, 4.212°N, 118.462°E (USNM ex 279043, 1). 870 m, St. 5585, Sibuko Bay, off Sipadan Island, Borneo, 4.117°N, 118.832°E (USNM ex 258561, 2). 5–20 m, Sulawesi, Bunaken and Siladen Islets, 1.603°N, 124.767°E (LACM 88-55, 13). 20 m, off Waibailon Rock, E coast of Flores (CPC, 6).

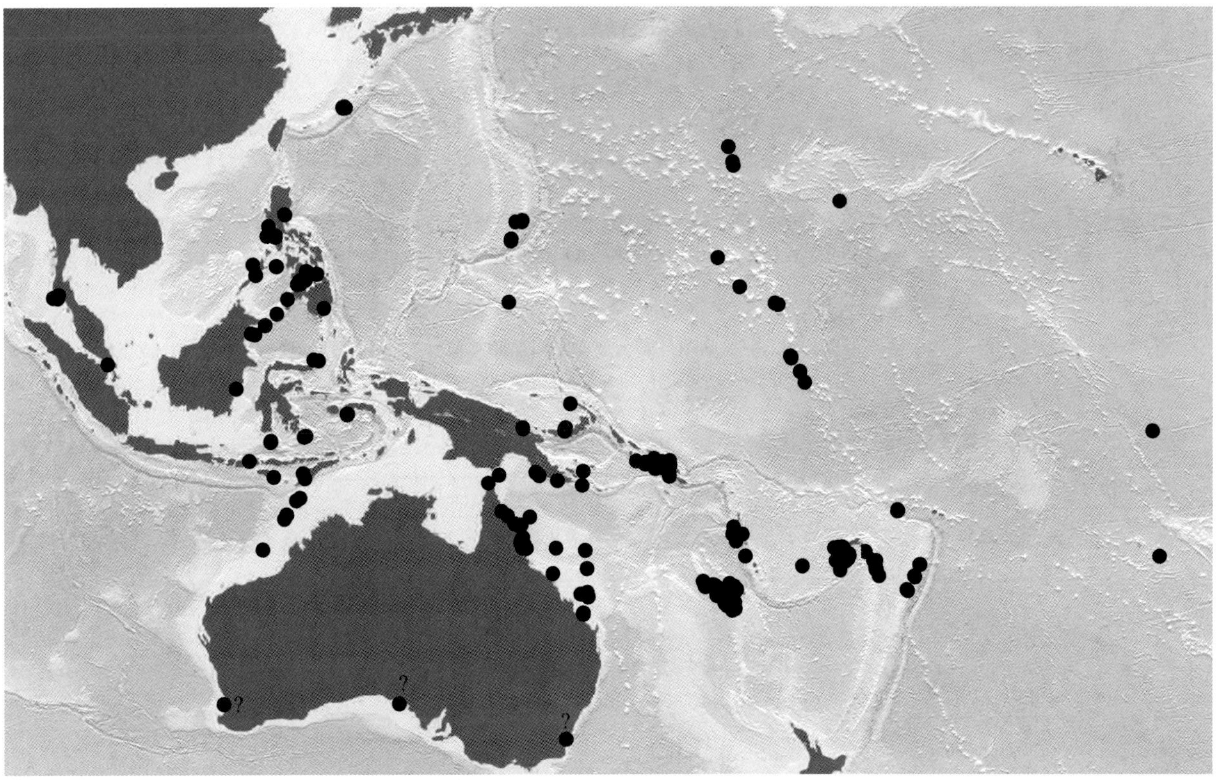

Figure 139. Distribution of *Scissurella mirifica*. ● specimen record. ? questionable record.

Malaysia. Leper Island, NE of Madang, 5.087°S, 145.821°E (MMC, 6; MMC, 1).

Japan, Okinawa. 52 m, Zampa Misaki, 26.435°N, 127.708°E (LACM 78-100, 1). 50 m, Onna Village, 26.495°N, 127.843°E (LACM 78-99, 4). 28 m, Maeda Point, 26.450°N, 127.759°E (DLG 1068, 2). 17–35 m, Maeda Point, 26.435°N, 127.759°E (DLG 968, 5). 38 m, base of cliff, Horeshoe Bay, 26.504°N, 127.858°E (DLG 1056, 16). 35–38 m, base of cliff, Horsseshoe, 26.504°N, 127.858°E (DLG 974, 5). 17 m, N side, Ike Jima, 26.384°N, 127.990°E (DLG 864, 7). 13–20 m, Kadena Steps, 26.360°N, 127.739°E (DLG 1448, 1). 20–30 m, Maeda Point, 26.445°N, 127.773°E (DLG 1530, 3). 2–3 m, Horseshoe Beach, 26.504°N, 127.858°E (DLG 1575, 9). 43 m, Horseshoe Beach, 26.505°N, 127.853°E (DLG 1542, 10). 37 m, Big Rock, Horseshoe, 26.508°N, 127.853°E (DLG 1460, 6). 13–33 m, Flab, Horseshoe Bay, 26.509°N, 127.856°E (DLG 1489, 4). 33 m, Horseshoe Bay, 26.509°N, 127.856°E (DLG 1518, 2).

Philippines. 20 m, off Bohol Island, Pamilacan Island, 9.500°N, 123.917°E (AMS C.349128, 1). Panglao Island, Cebucao Beach, 9.633°N, 123.817°E (AMS C.349131, 5). Palawan Island, Hondo Bay, N of Fondeado Island, 9.950°N, 118.917°E (AMS C.349129, 13). 2 m, Palawan Island, N of Fondeado Island, 9.950°N, 118.917°E (AMS C.349130, 2). Palawan Island, N Hondo Bay, Tadio Island, 9.950°N, 118.933°E (AMS C.349132, 1). 0 m, Buyong Beach, Mactan Island, Cebu, 10.083°N, 124.000°E (AMS C.349127, 1). NW Bohol Island, Danajon Bank, Banacon Island, 10.200°N, 124.167°E (AMS C.348644, 3, 4: complete). 200 m, Gulf of Davao, 6.700°N, 125.800°E (DLG 1219, 30). 10–12 m, off Mactan Island, Cebu, 10.322°N, 124.014°E (FLMNH 392481, 1). 10 m, Balicasag Island, Bohol, 9.518°N, 123.680°E (DLG 944, 1). 3–6 m, San Juan Reef, Sal Juan, Siquijor Island, Visayan Islands, 9.157°N, 123.485°E (USNM ex 808962, 1). 37 m, St. 5140, off Jolo, 6.146°N, 121.050°E (USNM 275329, 1). 46 m, St. 5105, off Sueste Point, W Luzon, 14.732°N, 120.214°E (USNM 262555, 4). 49 m, St. 5131, N 20°E 0.4 mi off Panabutan Point, W Mindanao, 7.586°N, 122.121°E (USNM ex 262915, 7; USNM ex 262931, 1). 48 m, St. 5132, N 15°W 0.3 mi off Panabutan Point, W Mindanao, 7.586°N, 122.121°E (USNM 293357, 1). 33 m, St. 5164, off SE Tawi Tawi, 5.028°N, 119.872°E (USNM 283706, 2). 1339 m, St. 5487, Sogod Bay, Leyte, 10.046°N, 125.093°E (USNM ex 283798, 1). 146 m, St. 5277, off Malavatuan Island, W Luzon, 13.949°N, 120.229°E (USNM ex 278089, 1). 686 m, St. 5348, Palawan Pass, off Pt. Tabonan, 10.963°N, 118.638°E (USNM ex 285852, 1). 1–4 m, Bonbonon, Antolong Bay, S Negros Oriental, 9.048°N, 2.052°E (USNM 812011, 2). 3–17 m, Bisucay Island, Cuyo Islands, Palawan, 10.810°N, 120.973°E (USNM 808557, 1). 1–2 m, Sungcolan Bay, Panglao, 9.638°N, 123.838°E (MNHN, 4: complete). 2–3 m, Abatan River, Bohol Island, 9.753°N, 123.933°E (MNHN, 1: complete). 1–2 m, Sungcolan inlet, Panglao Island, 9.638°N, 123.838°E (MNHN, 7: complete). 90–110 m, Balicasag Island, 9.515°N, 123.687°E (MNHN, 1: complete). 842 m, 13.783°N, 120.050°E (MNHN, 2). 25 m, Luzon, Ligpo Point, 13.825°N, 120.898°E (LACM 84-161, 1). 20 m, Luzon, Manig Point, 13.683°N, 120.892°E (LACM 84-163, 8). 2–6 m, N of Fodeado Island, Palawan, 9.950°N, 118.917°E (AMS, 1: complete). 1–2 m, Mactan Channel, Visayas, 10.033°S, 123.867°E (AMS, 1: complete). 2 m, Panglao Island, 9.600°N, 123.750°E (AMS, 1: complete). Cuyo Island, San Carlos, 10.800°N, 121.043°E (USNM 808074, 1). Dumaguete City, Bantayan Beach, Negros Oriental, 9.328°N, 123.314°E (USNM 812251, 1). 2 m, Bohol, Panglao Island, SW side, 9.600°N, 123.750°E (AMS C.348632, 2, –: complete). 26 m, Bohol Island, off Balicasag Island, 9.517°N, 123.683°E (AMS ex C.348621, 4). 538 m, AURORA 2007 CP2750, 15.890°N, 121.893°E (MNHN, 1). 20 m, Looc, Panglao Island, PANGLAO 2004 B32, 9.430°N, 123.743°E (MNHN, 42). 4–30 m, Catarman, Panglao Island, PANGLAO 2004 S4, 9.431°N, 123.863°E (MNHN, 3). 134–190 m, Cervera Shoal, W Pamilacan Island, PANGLAO 2004 T34, 9.470°N, 123.845°E (MNHN, 1). 470–566 m, off Balicasag Island, Bohol Sea, CP2394, 9.476°N, 123.666°E (MNHN, 1). 95–128 m, Cervera Shoal, W Palmilacan Island, PANGLAO 2004 T36, 9.488°N, 123.858°E (MNHN, 1). 5–12 m, Pamilacan Island, PANGLAO 2004 S14, 9.488°N, 123.918°E (MNHN, 1). 6–14 m, Pamilacan Island, PANGLAO

2004 S10, 9.490°N, 123.933°E (MNHN, 14). 15–20 m, Pamilacan Island, PANGLAO 2004 S22, 9.490°N, 123.933°E (MNHN, 1). 6–8 m, Pamilacan Island, PANGLAO 2004 S12, 9.490°N, 123.933°E (MNHN, 2). 356–396 m, off Balicasag Island, Bohol Sea, CP2393, 9.501°N, 123.693°E (MNHN, 5). 19–20 m, Balicasag Island, PANGLAO 2004 B37, 9.515°N, 123.680°E (MNHN, 1). 90–110 m, Balicasag Island, PANGLAO 2004 L46, 9.515°N, 123.686°E (MNHN, 2). 6 m, Black Forest, Balicasag Island, PANGLAO 2004 S3, 9.518°N, 123.688°E (MNHN, 5). 12–14 m, Black Forest, Balicasag Island, PANGLAO 2004 B6, 9.518°N, 123.688°E (MNHN, 6). 309–342 m, Balicasag Island, Bohol Sea, CP2399, 9.528°N, 123.698°E (MNHN, 3; MNHN, 5). 164–176 m, off Balicasag Island, Bohol Sea, DW2339, 9.531°N, 123.728°E (MNHN, 1; MNHN, 15). 152 m, Bolod, Panglao Island, PANGLAO 2004 T2, 9.540°N, 123.796°E (MNHN, 6). 713–731 m, off Balicasag Island, Bohol Sea, PANGLAO 2005 CP2398, 9.543°S, 123.675°E (MNHN, 3). 17–25 m, Pntod Lagoon 1, Panglao Island, PANGLAO 2004 B39, 9.546°N, 123.701°E (MNHN, 15). 100–140 m, between Panglao and Balicasag islands, PANGLAO 2004 T31, 9.550°N, 123.700°E (MNHN, 1). 8–14 m, Alona Reef, Panglao Island, PANGLAO 2004 B1, 9.550°N, 123.775°E (MNHN, 6). 5 m, Alona reef, Panglao Island, PANGLAO 2004 B2, 9.550°N, 123.775°E (MNHN, 12). 82 m, Bolod, Panglao Island, PANGLAO 2004 T4, 9.550°N, 123.808°E (MNHN, 4; MNHN, 2). 0–2 m, Pontod Islet, Panglao Island, PANGLAO 2004 D4, 9.551°N, 123.733°E (MNHN, 1). 8–10 m, Napaling, Panglao Island, PANGLAO 2004 B9, 9.551°N, 123.733°E (MNHN, 2). 24 m, BBC Point, Panglao Island, PANGLAO 2004 B4, 9.553°N, 123.805°E (MNHN, 2). 106–137 m, between Panglao and Pamilacan Islands, PANGLAO 2004 T27, 9.556°N, 123.850°E (MNHN, 1). 8 m, Arco Point, Panglao Island, PANGLAO 2004 B3, 9.558°N, 123.810°E (MNHN, 13). 0–3 m, Pontod Islet, Panglao Island, PANGLAO 2004 D5, 9.560°N, 123.725°E (MNHN, 1). 2 m, lagoon off Poblacion, Panglao Island, PANGLAO 2004 S11, 9.560°N, 123.726°E (MNHN, 1). 97–120 m, off San Isidro, Panglao Island, PANGLAO 2004 T9, 9.561°N, 123.833°E (MNHN, 4). 6 m, San Isidro, Panglao Island, PANGLAO 2004 S17, 9.576°N, 123.831°E (MNHN, 16). 642–669 m, Maribojoc Bay, Bohol Sea, CP2397, 9.581°N, 123.695°E (MNHN, 4). 80 m, Biking-Catarman, Panglao Island, PANGLAO 2004 T28, 9.583°N, 123.856°E (MNHN, 2). 34–82 m, W of Baclayon, Bohol Island, PANGLAO 2004 T6, 9.585°N, 123.853°E (MNHN, 4). 0–1 m, inside lagoon near Doljo Point Panglao Island, PANGLAO 2004 M9, 9.585°N, 123.726°E (MNHN, 1). 4 m, Biking, Panglao Island, PANGLAO 2004 B5, 9.586°N, 123.840°E (MNHN, 9). 5 m, Biking, Panglao Island, PANGLAO 2004 S1, 9.588°N, 123.841°E (MNHN, 74). 84–87 m, W of Baclayon, Bohol Island, PANGLAO 2004 T5, 9.588°N, 123.870°E (MNHN, 1). 24–27 m, Doljo Point, Panglao Island, PANGLAO 2004 B12, 9.593°N, 123.720°E (MNHN, 12). 24 m, N of Doljo, Panglao Island, PANGLAO 2004 B36, 9.598°N, 123.741°E (MNHN, 5). 31 m, N of Doljo, Panglao Island, PANGLAO 2004 B35, 9.598°N, 123.741°E (MNHN, 2). 4–30 m, Tatarman, Panglao Island, PANGLAO 2004 B7, 9.598°N, 123.863°E (MNHN, 37). 2–4 m, Momo Beach, Panglao Island, PANGLAO 2004 S24, 9.601°N, 123.750°E (MNHN, 2). 0–3 m, Momo Beach, Panglao Island, PANGLAO 2004 M7, 9.601°N, 123.753°E (MNHN, 1). 60–62 m, Baclayon, Bohol Island, PANGLAO 2004 T32, 9.606°N, 123.896°E (MNHN, 1). 3–14 m, Momo Beach, Banglao Island, PANGLAO 2004 B10, 9.608°N, 123.760°E (MNHN, 56). 28–32 m, Momo Beach, Panglao Island, PANGLAO 2004 S8, 9.608°N, 123.760°E (MNHN, 31). 0 m, Mayacabac, Panglao Island, PANGLAO 2004 M51, 9.613°N, 123.870°E (MNHN, 1). 120 m, off Momo Beach, Panglao Island, PANGLAO 2004 L50, 9.615°N, 123.763°E (MNHN, 1). 30–33 m, between Momo and Napaling, Panglao Island, PANGLAO 2004 B42, 9.616°N, 123.766°E (MNHN, 3). 2–4 m, Napaling, Panglao Island, PANGLAO 2004 S5, 9.618°N, 123.768°E (MNHN, 8). Napaling, Panglao Island, PANGLAO 2004 B8, 9.618°N, 123.769°E (MNHN, 55). 3–5 m, Baclayon Takot, Bohol Island, PANGLAO 2004 B13, 9.618°N, 123.876°E (MNHN, 30). 8–15 m, Baclayon Takot, Bohol Island, PANGLAO 2004 S13, 9.618°N,

Scissurellidae
Scissurella

123.876°E (MNHN, 50). 3–5 m, Baclayon Takot, Bohol Island, PANGLAO 2004 B13, 9.618°N, 123.876°E (MNHN, 28). 28–32 m, Napaling, Panglao Island, PANGLAO 2004, 9.620°N, 123.773°E (MNHN, 9). 20–21 m, Napaling, Panglao Island, PANGLAO 2004 B21, 9.620°N, 123.773°E (MNHN, 2). 2 m, SE of Dauis, Panglao Island, PANGLAO 2004 D1, 9.620°N, 123.870°E (MNHN, 1). 4–5 m, Baclayon, Bohol Island, PANGLAO 2004 S2, 9.623°N, 123.908°E (MNHN, 4; MNHN, 2). 3–21 m, Bingag, Panglao Island, PANGLAO 2004 B17, 9.625°N, 123.781°E (MNHN, 4). 20 m, Bingag, PANGLAO 2004 B16, 9.626°N, 123.788°E (MNHN, 6). 62 m, Bigag/Tabalong, Panglao Island, PANGLAO 2004 L51–60, 9.628°N, 123.800°E (MNHN, 1; MNHN, 2). 2 m, Sungcolan Inlet, Panglao Island, PANGLAO 2004 S34, 9.638°N, 123.838°E (MNHN, 29). 1–4 m, Sungcolan Bay, Panglao Island, PANGLAO 2004 S7, 9.641°N, 123.820°E (MNHN, 7). 2–4 m, Sungcolan Bay, Panglao Island, PANGLAO 2004 B14, 9.641°N, 123.820°E (MNHN, 3). 1–2 m, Sungcolan Inlet, Panglao Island, PANGLAO 2004 B34, 9.641°N, 123.838°E (MNHN, 3). 1–2 m, Dauis, Panglao Island, PANGLAO 2004 B31, 9.645°N, 123.840°E (MNHN, 9). 2–4 m, Sungcolan, Panglao Island, PANGLAO 2004 B15, 9.646°N, 123.818°E (MNHN, 7). 78–95 m, Maribohoc Bay, Bohol Island, PANGLAO 2004 T11, 9.681°N, 123.833°E (MNHN, 4). 12 m, Ubajan, Bohol Island, PANGLAO 2004 S27, 9.691°N, 123.850°E (MNHN, 2). 2–8 m, Ubajan, Bohol Island, PANGLAO 2004 B20, 9.691°N, 123.850°E (MNHN, 11). 21 m, Ubajan, Bohol Island, PANGLAO 2004 S25, 9.691°N, 123.850°E (MNHN, 2; MNHN, 7). 4–12 m, Manga, Bohol Island, PANGLAO 2004, 9.695°N, 123.848°E (MNHN, 14). 10 m, Manga, Bohol Island, PANGLAO 2004 S20, 9.696°N, 123.851°E (MNHN, 1). 100 m, Cortes, Bohol Island, PANGLAO 2004 B1, 9.698°N, 123.825°E (MNHN, 2). 3–4 m, Manga, Bohol Island, PANGLAO 2004 S19, 9.701°N, 123.856°E (MNHN, 5). 35–45 m, Cortes, Bohol Island, PANGLAO 2004 T23, 9.703°N, 123.843°E (MNHN, 1). 123–135 m, Cortes, Bohol Island, PANGLAO 2004 T26, 9.721°N, 123.813°E (MNHN, 1; MNHN, 1). 473 m, AURORA 2007 CP2749, 15.943°N, 121.836°E (MNHN, 1).

Papua New Guinea. 13 m, Milne Bay District, Amazon Bay, 10.317°S, 149.350°E (AMS C.406395, 2). Milne Bay District, Amazon Bay, 10.300°S, 149.333°E (AMS C.348923, 6). Kapa Kapa, ca. 45 mi SE of Port Moresby, 9.800°S, 147.500°E (AMS C.348641, 2; AMS C.348875, 7, 4: complete). Taurama, W side Pyramid Point, 9.542°S, 147.242°E (AMS C.349122, 1). 18 m, off Port Moresby, Manubada Local Island, off W end, 9.517°S, 147.167°E (AMS C.404944, 6). 23 m, off Port Moresby, Manubada Local Island, off NW side, 9.517°S, 147.167°E (AMS C.404974, 4). 1 m, Milne Bay, Egum Island, 9.367°S, 151.933°E (AMS C.402847, 2). 1 m, Madang, Kranket Island, N end, 5.200°S, 145.850°E (AMS C.348876, 1; AMS C.349123, 3). Vakunabada, Fly River, 9.467°S, 147.133°E (MV F 133035, 1). 67 m, off Schauman Island, Kimbe Bay, New Britain, 5.282°S, 150.093°E (DLG 791, 3). 15–35 m, Baudisson Bay, Kavieng, New Ireland, 2.743°S, 150.658°E (DLG 771, 12). 65 m, off Susan's Reef, Kimbe Bay, New Britain, 5.291°S, 150.128°E (DLG 764, 10; DLG 607, 8). 0–5 m, Big Malu Malu, Kimbe Bay, New Britain, 5.230°S, 150.090°E (DLG 730, 1). 47 m, off Schaumann Island, Kimbe Bay, New Britain, 5.301°S, 150.092°E (DLG 672, 2). 0–10 m, Restorf Island, Kimbe Bay, New Britain, 5.292°S, 150.104°E (DLG 661, 4). 0–5 m, Lumu Reef, Kimbe Bay, New Britain, 5.283°S, 150.131°E (DLG 647, 20). 4 m, off Anne Sophie's Reef, Kimbe Bay, 5.089°S, 150.186°E (DLG 578, 2). 70 m, off S Ema Reef, Kimbe Bay, New Britain, 5.202°S, 150.152°E (DLG 559, 1). 0–3 m, First Reef, Kimbe Bay, New Britain, 5.432°S, 150.090°E (DLG 532, 2: complete). 1 m, Hanging Garden Reef, Kimbe Bay, New Britain, 5.429°S, 150.097°W (DLG 529, 10: complete). 40 m, off Anne Sophie's Reef, Kimbe Bay, New Britain, 5.091°S, 150.186°E (DLG 445, 2). 54 m, off Schauman Island, Kimbe Bay, New Britain, 5.307°S, 150.092°E (DLG 441, 10). 28 m, off Big Malu Malu Island, Kimbe Bay, 5.231°S, 150.098°E (DLG 400, 22). 40 m, off Restorf Island, Kimbe Bay, 5.292°S, 150.101°E (DLG 393, 11). Madang Province, Astrolabe Bay, off Nagda, 5.158°S, 145.837°E (LACM 80-13, 2). 0 m, Madang, Christensen Research

Institute, 5.117°S, 145.817°E (AMS, 3: complete). Vakunabada, Fly River, 9.467°S, 147.133°E (MV F 133035, 1). Christensen Research Institute, Madang, 5.117°S, 145.817°E (AMS C.348630, 2, –: complete). Louisiade Archipelago, Conflict Group, 10.767°S, 151.833°E (AMS C.348634, 1).

Solomon Islands. 3 m, Aoki Auki Harbour, W coast Malaita Island, 8.767°S, 160.700°E (AMS C.404914, 1: complete). 23 m, New Georgia Islands, Marovo Rukiana Island, 8.367°S, 157.283°E (AMS C.402846, 1). 2 m, Santa Isabel Island, Furona Island, 8.117°S, 159.100°E (AMS C.402843, 1). 10–25 m, Tanavula Point, Nggela Sule, Florida Island, Central Province, 9.043°S, 160.065°E (SBMNH 85117, 2). 20 m, Lark Reef, Rua Sura Islands, Guadalcanal Province, 9.501°S, 160.603°E (SBMNH 369625, 1). 20–25 m, Matui Island, 8.495°S, 158.200°E (LACM, 7). 20 m, Vanguni Island, Minjanga Island, 8.667°S, 158.233°E (LACM 89-74, 6). 2–3 m, SE end Santa Isabel Island, W of Ngalignagho Island, Central District, Solomon Islands, 8.423°S, 159.908°E (LACM 78-54, 1). 24 m, off Mbatuna, Marovo Lagoon, Vangunu Island, Solomon Islands, 8.550°S, 158.117°E (NMNZ M.229116, 14). 9 m, between Telina Island and Mbareki Peninsula, Marovo Lagoon, Vangunu Island, 9.900°S, 160.650°E (NMNZ M.230485, 3). 2 m, off Northern Vangunu Island, Kolo Lagoon, Solomon Islands, 8.617°S, 158.117°E (NMNZ M228923, 8). 35 m, midway between Karumaneke Island and Cheke Villag, Marovo Lagoon, Vangunu Island, 8.517°S, 158.133°E (NMNZ M230050, 7). Yandina, Bankia Island, Russel Islands, 9.117°S, 159.217°E (NMNZ M.230680, 4; NMNZ M.273217, 1). 6 m, off Mbatuna, Marovo Lagoon, Vanungu Island, 8.550°S, 158.117°E (NMNZ M.230160, 1). 3–7 m, Aoki Auki Harbour, W coast Malaita Island, 8.767°S, 160.700°E (AMS C.404915, 1: complete).98–200 m, SALOMON 1 DW1767, 8.323°S, 160.665°E (MNHN, 5). 1327–1598 m, SALOMON 1 CP1764, 8.610°S, 160.123°E (MNHN, 3).

Australia New South Wales. Batemans Bay, Surfside Beach, 35.708°S, 150.200°E (AMS C.346286, 2) [questionable record].

Australia, Northern Territories. 8 m, Hibernia Reef, in lagoon, E side, 11.979°S, 123.359°E (SBMNH, 1).

Australia, Queensland. 15 m, Great Barrier Reef, Capricorn Group, Heron Island, 23.433°S, 151.950°E (AMS C.348698, 1). 6 m, Great Barrier Reef, Capricorn Group, Heron Island, 23.433°S, 151.950°E (AMS C.348855, 1). Great Barrier Reef, Capricorn Group, Wilson Island, N side, 23.300°S, 151.950°E (AMS C.348647, 1). Great Barrier Reef, Swain Reefs, Bylund Gillett Cay, 21.717°S, 152.417°E (AMS C.348692, 4). 64 m, Great Barrier Reef, Swain Reefs, 3 km NE of W side of Bylund Gillett Cay, 21.700°S, 152.433°E (AMS C.348635, 4). 3 m, Great Barrier Reef, Swain Reefs, Reef 21-189, 21.450°S, 151.683°E (AMS C.348700, 4, 17: complete). 10 m, Great Barrier Reef, Swain Reefs, Centennary Reef, 21.283°S, 152.333°E (AMS C.404208, 1: complete). Great Barrier Reef, Gould Reef, NE of Bowen, 19.433°S, 148.867°E (AMS C.348696, 1, 1: complete). 20 m, Marion Reef, Coral Sea, off Great Barrier Reef, 18.950°S, 152.308°E (LACM 77-120, 1). Great Barrier Reef, Briggs Reef, E of Cairns, 16.950°S, 146.183°E (AMS C.404211, 2: complete). Great Barrier Reef, 200 m SE of Michaelmas Cay, 16.600°S, 145.983°E (AMS C.404952, 4: complete; AMS C.348695, 3). Great Barrier Reef, Michaelmas Cay, 16.600°S, 145.983°E (AMS C.348640, 16). 16 m, Great Barrier Reef, Low Isles, 16.383°S, 145.567°E (AMS C.348701, 1). Great Barrier Reef, Undine Reef, 16.133°S, 145.667°E (AMS C.404201, 4: complete). Great Barrier Reef, Lizard Island, Eagle Island, 14.700°S, 145.383°E (AMS C.403661, 1). Great Barrier Reef, Lizard Island, W side Eagle Island, 14.700°S, 145.383°E (AMS C.348646, 1). Great Barrier Reef, E side of Eagle Island, 14.700°S, 145.383°E (AMS C.348693, 2). 1 m, between Lizard and Palfrey Island, Lizard Island, Great Barrier Reef, 14.683°S, 145.450°E (AMS C.404953, 1: complete). Great Barrier Reef, Lizard Island, Casuarina Research Station Beach, 14.683°S, 145.450°E (AMS C.406399, 1). 1 m, Great Barrier Reef, Lizard Island, Casuarina Beach, 14.683°S, 145.450°E (AMS C.348854, 1). Great Barrier Reef, Lizard Island, N side of Palfrey Island, 14.683°S, 145.450°E (AMS

C.348856, 2). 615 m, near lagoon entrance, SE Lizard Island, 14.683°S, 145.458°E (LACM 79-55, 1). 20 m, Great Barrier Reef, Lizard Island, off SE side Bird Islet, 14.683°S, 145.467°E (AMS C.348699, 1). 15 m, 0.4 km off Anchor Bay, W Lizard Island, 14.670°S, 145.433°E (LACM 79-56, 1). 0 m, SW side Lizard Island, 14.670°S, 145.442°E (LACM 79-53, 9). 6 m, Great Barrier Reef, Lizard Island, off Osprey Island, 14.667°S, 145.433°E (AMS C.404203, 20: complete). 5 m, Great Barrier Reef, Lizard Island, off Rocky Point, 14.667°S, 145.433°E (AMS C.404416, 1). Great Barrier Reef, Lizard Island, Bird Islet, 14.667°S, 145.467°E (AMS C.404206, 4: complete). 19 m, Great Barrier Reef, No Name Reef, SW end, 14.667°S, 145.650°E (AMS C.348639, 1). 8 m, Great Barrier Reef, Lizard Island, North Point, 14.650°S, 145.450°E (AMS C.348857, 1). 15 m, Great Barrier Reef, Howick Group, Coquet No.1 Howick Island, 14.542°S, 144.992°E (AMS C.41226, 1). 8 m, Great Barrier Reef, No. 5 Sandbank Reef 13-120, SW end, 13.750°S, 144.267°E (AMS C.348636, 1). 19 m, Great Barrier Reef, off Bow Reef, near Cape Sidmouth, 13.333°S, 143.667°E (AMS C.2661, 1). Torres Strait, Wednesday Island, 10.533°S, 142.317°E (AMS C.348702, 1). 9 m, Torres Strait, off Murray Island, 9.933°S, 144.067°E (AMS C.48356, 7). Torres Strait, Yorke Island, S side, 9.733°S, 143.417°E (AMS C.404418, 1). 17 m, Osprey Reef, 13.883°S, 146.535°E (AMS C.461454, 2). 38 m, Escape Reef, S part, Great Barrier Reef, 15.883°S, 145.817°E (AMS C.461468, 1). 0 m, between Lizard Head, and Coconut Beach, SE side Lizard Island, 14.682°S, 145.467°E (LACM 79-54, 7). 0 m, SW side Lizard Island, 14.670°S, 145.442°E (LACM 79-53, 17). 11–18 m, Cairns Reef, 16.905°S, 145.778°E (MV 161662, 3; MV F 161668, 18). 20 m, Great Barrier Reef, Lizard Island, off SE side Bird Islet, 14.683°S, 145.467°E (AMS C.404414, 1). Great Barrier Reef, Low Isles, 16.383°S, 145.567°E (AMS ex C.348616, 1). Great Barrier Reef, Gould Reef, NE of Bowen, 19.433°S, 148.867°E (AMS C.404485, 1: complete).

Australia, South Australia. 25 m, Petrel Bay, N of St. Francis Island, 32.200°S, 133.292°E (LACM 73-157, 1) [questionable record].

Australia, Western Australia. 18–20 m, N end Mermaid Reef, Rowley Shoals, 17.050°S, 119.600°E (LACM 86-250, 5). 10 m, Seringapatam Reef, inside lagoon, 13.642°S, 122.019°E (SBMNH 85131, 1). 27 m, Seringapatam Reef, 13.628°S, 122.016°E (SBMNH 85126, 2). 21 m, Seringapatam Reef, N side, 13.624°S, 122.005°E (SBMNH 85130, 4). 15–25 m, N side, Serigapatam Reef, 13.624°S, 122.005°E (SBMNH 85126, 2). 27 m, Seringapatam Reef, 13.628°S, 122.016°E (SBMNH 85128, 3). 20 m, Sandy Islet, Scott Reef, 14.044°S, 121.780°E (SBMNH 369696, 2). 20 m, E side, Sandy Islet, Scott Reef, 14.044°S, 121.780°E (SBMNH 369505, 26). Garden Island, 32.233°S, 115.683°E (WAM, 1) [questionable record].

Coral Sea. N East Herald Cay, 16.933°S, 149.183°E (AMS C.406345, 1; AMS C.348637, 2). 10–15 m, SE from Observatory Cay, Lihou Reef, 17.157°S, 152.141°E (SBMNH 369616, 4).

New Caledonia. 13 m, ESE of Ile de Crouy, SW of Noumea, 22.413°S, 21.300°E (LACM 86-361, 1). 0 m, Croissant Reef, off Noumea, 22.317°S, 166.367°E (AMS C.404954, 2: complete; AMS C.349118, 2). Noumea, Ile St. Marie, SW point, 22.300°S, 166.483°E (AMS C.348643, 3). Noumea, 22.267°S, 166.450°E (AMS C.348860, 1). 0 m, Magenta Bay, Noumea, 22.267°S, 166.483°E (AMS C.404950, 1: complete). 400 m, off S coast, 21.250°S, 167.250°E (AMS C.348859, 2). Loyalty Islands, SE side Lifou Island, Luengoni, 21.033°S, 167.417°E (AMS C.348864, 7, 8: complete). Loyalty Islands, SE side Lifou Island, Luengoni, 21.033°S, 167.417°E (AMS C.348865, 1). Wassagne, Lifou Island, Loyalty Islands, 21.000°S, 167.383°E (AMS C.348642, 22, 20: complete; AMS C.349134, 7). 0 m, Loyalty Islands, Lifou, Doueolou, 20.933°S, 167.083°E (AMS C.402848, 1). Loyalty Islands, Lifou, Doueolou, 20.933°S, 167.083°E (AMS C.348862, 8). We, Lifou Island, Loyalty Islands, 20.900°S, 167.267°E (AMS C.348867, 10). Doking, Lifou Island, Loyalty Islands, 20.700°S, 167.167°E (AMS C.348645, 7). 0 m, Doking, Lifou Island, Loyalty Islands, 20.700°S, 167.167°E (AMS C.348861, 2). Lifou, 20.900°S, 167.200°E (KBIN, 8). Redika Island, 22.520°N, 166.610°E (USNM 724143,

18). 7–10 m, Magnin Point, Noumea Lagoon, 22.316°S, 166.443°E (MNHN, 4: complete). 1–5 m, Santal Bay, Lifou, 20.782°S, 167.132°E (MNHN, 2: complete). 5–20 m, Santal Bay, Lifou, 20.875°S, 167.135°E (MNHN, 1: complete). 11 m, Touho sector, 20.717°S, 165.133°E (MNHN, 1: complete). 10 m, Touho sector, 20.825°S, 165.327°E (MNHN, 2). 11 m, Touho sector, 20.717°S, 165.133°E (MNHN, 2). 49–59 m, Touho sector, 20.733°S, 165.233°E (MNHN, 30). 0 m, Touho sector, 20.753°S, 165.272°E (MNHN, 40). 3–6 m, Touho sector, 20.778°S, 165.228°E (MNHN, 7). 8 m, Touho sector, 20.742°S, 165.265°E (MNHN, 50). 45–56 m, Touho sector, 20.775°S, 165.263°E (MNHN, 12). 6–15 m, Touho sector, 20.771°S, 165.229°E (MNHN, 1). 80–140 m, Touho sector, 20.817°S, 165.317°E (MNHN, 9). 5–25 m, Touho sector, 20.878°S, 165.325°E (MNHN, 12). 0–2 m, Touho sector, 20.775°S, 165.242°E (MNHN, 3). 2–3 m, Touho sector, 20.777°S, 165.235°E (MNHN, 1). 15–20 m, Touho sector, 20.585°S, 165.135°E (MNHN, 1). 0–1 m, Touho sector, 20.782°S, 165.230°E (MNHN, 2). 10–15 m, Touho sector, 20.660°S, 165.245°E (MNHN, 1). 50–62 m, Touho sector, 20.817°S, 165.283°E (MNHN, 1). 0 m, Touho sector, 20.770°S, 165.242°E (MNHN, 17). 15 m, Koumac sector, 20.662°S, 164.248°E (MNHN, 1). 53–71 m, Koumac sector, 20.753°S, 164.253°E (MNHN, 2). 26–40 m, Koumac sector, 20.673°S, 164.248°E (MNHN, 6). 12 m, Koumac sector, 20.667°S, 164.187°E (MNHN, 3, 1: complete). 11–13 m, Koumac sector, 20.652°S, 164.207°E (MNHN, 2). 55–57 m, Koumac sector, 20.672°S, 164.195°E (MNHN, 1). 12 m, Koumac sector, 20.562°S, 164.172°E (MNHN, 3). 0 m, Koumac sector, 20.562°S, 164.183°E (MNHN, 100). 0–2 m, Koumac sector, 20.567°S, 164.267°E (MNHN, 1). 30–63 m, Koumac sector, 20.663°S, 164.255°E (MNHN, 14). 0 m, Koumac sector, 20.487°S, 164.170°E (MNHN, 5). 0 m, Koumac sector, 20.373°S, 164.113°E (MNHN, 4). 90–115 m, Koumac sector, 20.745°S, 164.248°E (MNHN, 3). 10–11 m, Koumac sector, 20.593°S, 164.253°E (MNHN, 2). Bourail sector, 2136.917°S, (MNHN, 1). 15–20 m, Koumac sector, 20.745°S, 164.258°E (MNHN, 5). 0 m, Koumac sector, 20.583°S, 164.258°E (MNHN, 21). 12–14 m, Koumac sector, 20.573°S, 164.217°E (MNHN, 3). 0–8 m, Koumac sector, 20.629°S, 164.275°E (MNHN, 2). 2–4 m, Koumac sector, 20.587°S, 164.277°E (MNHN, 6). 33–38 m, Koumac sector, 20.647°S, 164.260°E (MNHN, 25). 20–30 m, Koumac sector, 20.690°S, 164.247°E (MNHN, 1). 3–7 m, Santal Bay, Lifou, 20.785°S, 167.152°E (MNHN, 1). 5–10 m, Santal Bay, Lifou, 20.780°S, 167.046°E (MNHN, 6). 27–31 m, Santal Bay, Lifou, 20.763°S, 167.028°E (MNHN, 1). 12 m, Santal Bay, Lifou, 20.900°S, 197.122°E (MNHN, 3). 2–4 m, Santal Bay, Lifou, 20.945°S, 167.052°E (MNHN, 1). 8–18 m, Santal Bay, Lifou, 20.792°S, 167.118°E (MNHN, 2). 5–20 m, Santal Bay, Lifou, 20.875°S, 167.135°E (MNHN, 9). 4 m, Santal Bay, Lifou, 20.878°S, 167.142°E (MNHN, 2). 4–5 m, Santal Bay, Lifou, 20.795°S, 167.156°E (MNHN, 2). 4–7 m, Santal Bay, Lifou, 20.765°S, 167.103°E (MNHN, 3). 1–5 m, Santal Bay, Lifou, 20.782°S, 167.132°E (MNHN, 10). 0 m, Santal Bay, Lifou, 20.781°S, 167.129°E (MNHN, 1). 2–5 m, Santal Bay, Lifou, 20.903°S, 167.123°E (MNHN, 2). 12–32 m, Santal Bay, Lifou, 20.892°S, 167.045°E (MNHN, 2). 48–52 m, Santal Bay, Lifou, 20.897°S, 167.122°E (MNHN, 1). 4 m, Santal Bay, Lifou, 20.785°S, 167.123°E (MNHN, 1). 170 m, E coast, 20.800°S, 165.317°E (MNHN, 6). 7–10 m, Noumea, Magnin Point, 22.315°S, 166.443°E (MNHN, 4). 0 m, Noumea, Platier Ilot Maitre, 22.340°S, 166.428°E (MNHN, 3). 0 m, Noumea, Platier Ilot Maitre, 22.340°S, 166.428°E (MNHN, 20). 425–430 m, 22.283°S, 167.233°E (MNHN, 10). 250–350 m, Vauban, 22.500°S, 166.400°E (MNHN, 100). 390 m, Vauban, 22.283°S, 167.200°E (MNHN, 1). 350–400 m, E coast, 21.033°S, 165.633°E (MNHN, 50). 160–222 m, E coast, 20.950°S, 165.583°E (MNHN, 15). 200 m, Loyalty Ridge, 20.900°S, 167.283°E (MNHN, 1). 105–110 m, E coast, 20.950°S, 165.567°E (MNHN, 2). Poindimie sector, 20.014°S, 2.755°E (MNHN, 8). 390 m, Loyalty Ridge, 20.783°S, 167.100°E (MNHN, 1). 270–282 m, E coast, 20.550°S, 165.000°E (MNHN, 3). 247–252 m, E coast, 21.700°S, 166.567°E (MNHN, 7). 1508 m, Loyalty Basin, 21.033°N, 167.033°E (MNHN, 1). 500 m, Loyalty Basin, 21.550°S, 166.400°E (MNHN, 1). 408–

440 m, E coast, 20.583°S, 165.117°E (MNHN, 100). 302–335 m, E coast, 20.967°S, 165.600°E (MNHN, 1). 105–110 m, E coast, 20.950°S, 165.567°E (MNHN, 20). 465–525 m, 22.583°S, 166.450°E (MNHN, 1). 380–400 m, E coast, 20.583°S, 165.117°E (MNHN, 75). 250 m, E coast, 21.750°S, 166.617°E (MNHN, 1). 300 m, Loyalty Ridge, 21°S, 167.417°E (MNHN, 1). 497–520 m, E coast, 20.567°S, 164.950°E (MNHN, 20). 310–315 m, Loyalty Basin, 21.533°S, 166.483°E (MNHN, 1). 450–470 m, E coast, 20.850°S, 165.467°E (MNHN, 6). 410–430 m, E coast, 20.583°S, 164.967°E (MNHN, 5). 630 m, 21.483°S, 166.333°E (MNHN, 1). 1080–1180 m, E coast, 21.700°S, 166.667°E (MNHN, 1). E coast, 20.817°S, 2.755°E (MNHN, 30). 375 m, E coast, 21.283°S, 165.950°E (MNHN, 3). 570 m, Loyalty Basin, 22.650°S, 166.567°E (MNHN, 4). 280 m, Loyalty Ridge, 20.433°S, 166.683°E (MNHN, 1). 237–298 m, E coast, 21.283°S, 165.950°E (MNHN, 50). 373 m, Loyalty Ridge, 20.683°N, 167.117°E (MNHN, 3). 455 m, southern, 22.267°S, 166.183°E (MNHN, 50). 330 m, Loyalty Ridge, 20.983°S, 167.417°E (MNHN, 1). 283 m, Loyalty Ridge, 20.700°S, 167.067°E (MNHN, 1). 237–298 m, E coast, 21.283°S, 165.950°E (MNHN, 1). 500 m, Loyalty Basin, 21.550°S, 166.400°E (MNHN, 16). 96–109 m, E coast, 20.817°S, 165.317°E (MNHN, 3). 490–500 m, southern, 22.317°S, 166.183°E (MNHN, 12). 282 m, Loyalty Ridge, 20.700°S, 167.000°E (MNHN, 1). 150–200 m, Pointe Lefèvre, Santal Bay, Lifou, 20.903°S, 167.018°E (MNHN, 2). 1–2 m, Ile Mouac, SE of Neba Island, Poum Bay, 20.217°S, 164.017°E (LACM 86-363, 2). Beach at E side of Mt. Panie, 20.600°S, 164.833°E (AMS ex C.404431, 1). 308–324 m, E coast, BATHUS 1 DE693, 20.583°S, 164.983°E (MNHN, 18). 140–150 m, E coast, BATHUS 1 DE692, 20.583°S, 164.983°E (MNHN, 1). 400–500 m, E coast, BATHUS 1 DE694, 20.600°S, 164.966°E (MNHN, 9). 491–500 m, E coast, BATHUS 1 DE682, 20.783°S, 165.350°E (MNHN, 1). 0–1 m, Poé Lagoon, Bourail sector, Montrouzier St. 1325, 21.615°S, 165.378°E (MNHN, 1). 15 m, off Nouville, Noumea sector, St. 1345, 22.268°S, 166.395°E (MNHN, 1). 577–780 m, south, BATHUS 2 DW754, 22.383°S, 166.216°E (MNHN, 100). 577–780 m, south, BATHUS 2 DW754, 22.383°S, 166.216°E (MNHN, 1). 10 m, Goeland Islet, Noumea sector, St. 1367, 22.405°S, 166.345°E (MNHN, 30). 250–350 m, St. 40, 22.500°S, 166.400°E (MNHN, 200). 465–525 m, south, BATHUS 2 DW739, 22.583°S, 166.450°E (MNHN, 100). 713–950 m, south, BATHUS 2 CP743, 22.600°S, 166.433°E (MNHN, 300). 525–547 m, south, BATHUS 2 DW721, 22.900°S, 167.283°E (MNHN, 1). 295–306 m, Norfolk Ridge, BATHUS 3 DW836, 23.033°S, 166.983°E (MNHN, 3). 397–400 m, south, BATUHUS 2 DW730, 23.050°S, 166.966°E (MNHN, 4).

Vanuatu. Port Vila, Efate Island, 17.733°S, 168.300°E (AMS C.405021). Efate Island, Port Vila, Point Ardel Malapoa, 17.733°S, 168.300°E (AMS C.349120, 1, 7: complete; AMS C.349121, 1). Opposite Matte Valu Island, Peterson's Bay, Espiritu Santo, 15.365°S, 167.191°E (USNM 542268, 1). 622–625 m, 15.667°S, 167.333°E (MNHN, 3). 154–179 m, 15.050°S, 167.117°E (MNHN, 3). 700–736 m, 14.883°S, 167.100°E (MNHN, 14). 154–179 m, MUSORSTOM 8 DW1105, 15.050°S, 167.116°E (MNHN, 6). 205–210 m, MUSORSTOM 8 DW1101, 15.066°S, 167.133°E (MNHN, 3). 7 m, S Turtle Island, Palikulo Bay, SANTO 2006 DB77DB71, 15.360°S, 167.208°E (MNHN, 4). 9 m, reef near Oyster Island, SANTO 2006 FB40, 15.381°S, 167.195°E (MNHN, 20). 20 m, SE Matewulu, SANTO 2006 DB75, 15.381°S, 167.198°E (MNHN, 12). 7 m, W Mavéa Island, SANTO 2006 DB67, 15.381°S, 167.218°E (MNHN, 4). 102–120 m, W Mavéa Island, Palikulo Bay, SANTO 2006 AT14, 15.397°S, 167.220°E (MNHN, 3). 38 m, Palikulo Bay, SANTO 2006 DB69, 15.406°S, 167.216°E (MNHN, 2). 6–43 m, Aésé Island, SANTO 2006 DB58, 15.410°S, 167.238°E (MNHN, 7). 13 m, Palikulo Bay, SANTO 2006 DB65, 15.430°S, 167.216°E (MNHN, 1). 21 m, SE Aésé Island, SANTO 2006 DB63, 15.448°S, 167.263°E (MNHN, 3). 22 m, Palikulo Bay, SANTO 2006 DS43, 15.465°S, 167.238°E (MNHN, 10). 0 m, Palikulo Peninsula, SANTO 2006 VM58, 15.476°S, 167.255°E (MNHN, 4). 5 m, Palikulo Bay, SANTO 2006 DS54, 15.480°S, 167.253°E (MNHN, 2). 2–3 m, Palikulo Bay, SANTO 2006 DB46,

15.480°S, 167.253°E (MNHN, 1). 5 m, Palikulo Bay, SANTO 2006 DB77DB53, 15.480°S, 167.253°E (MNHN, 3). 0 m, E Palikulo Peninsula, Palikulo Bay, SANTO 2006 DB77VM54, 15.481°S, 167.258°E (MNHN, 5). 5 m, Palikulo Bay, SANTO 2006 DB40, 15.496°S, 167.251°E (MNHN, 2). 22–25 m, NW Urélapa Island, SANTO 2006 DB20, 15.508°S, 167.023°E (MNHN, 4). 10–14 m, Segond Channel, vicinity of Luganville, SANTO 2006 DB14, 15.515°S, 167.183°E (MNHN, 2). 8–15 m, Segond Channel, vicinity of Luganville, SANTO 2006 DS06, 15.515°S, 167.185°E (MNHN, 1; MNHN, 3). 7 m, Segond Channel, E Sarakata river mouth, SANTO 2006 LS17, 15.518°S, 167.175°E (MNHN, 11). 5 m, Segond Channel, Caverot Point, SANTO 2006 DS31, 15.523°S, 167.161°E (MNHN, 4). 2–3 m, Segond Channel, vicinity of Maritime College, SANTO 2006 NS37, 15.523°S, 167.163°E (MNHN, 3). 7–8 m, Segond Channel, vicinity of Martitime College, SANTO 2006 LD30, 15.523°S, 167.166°E (MNHN, 1). 15–30 m, Segond Channel, *Coolidge* wreck, SANTO 2006 ZS19, 15.523°S, 167.235°E (MNHN, 1). 25 m, Segond Channel, *Coolidge* wreck, SANTO 2006 DS04, 15.523°S, 167.235°E (MNHN, 3). 2–3 m, Segond Channel, vicinity of Maritime College, SANTO 2006 NS36, 15.528°S, 167.158°E (MNHN, 6). 18–21 m, Segond Channel, vicinity of Maritime College, SANTO 2006 ED02, 15.528°S, 167.161°E (MNHN, 1). 62–71 m, Segond Channel, SANTO 2006 AT53, 15.530°S, 167.226°E (MNHN, 1). 97 m, Scorff Passage, SANTO 2006 EP12, 15.531°S, 167.251°E (MNHN, 1). 17–19 m, N Tutuba Island, SANTO 2006 DS101, 15.538°S, 167.286°E (MNHN, 1). 5 m, Aoré Island, Aimbué Bay, SANTO 2006 ZM16, 15.540°S, 167.201°E (MNHN, 2). 45–50 m, Aoré Island, Aimbuei Bay, SANTO 2006 FP46–49, 15.540°S, 167.211°E (MNHN, 1). 46–47 m, Segond Channel, SANTO 2006 EP01, 15.541°S, 167.150°E (MNHN, 1). 100–105 m, NW Tutuba Island, SANTO 2006 DS99, 15.541°S, 167.281°E (MNHN, 3). 8–20 m, Tutuba Island, SANTO 2006 FB83, 15.543°S, 167.290°E (MNHN, 1). MUSORSTOM 8 DW1069, 15.550°S, 167.233°E (MNHN, 15). 92 m, NW Tutuba Island, SANTO 2006 DS105, 15.550°S, 167.278°E (MNHN, 3). 2 m, Segond Channel, NW coast of Aoré Island, SANTO 2006 FB80, 15.551°S, 167.160°E (MNHN, 1). 20 m, Segond Channel, N Aoré Island, SANTO 2006 NB12, 15.551°S, 167.160°E (MNHN, 1). 15–25 m, NE Tutuba Island, SANTO 2006 DB01, 15.551°S, 167.296°E (MNHN, 13). 15–25 m, NE Tutuba Island, Palikulo Bay, SANTO 2006 DB77DB01, 15.551°S, 167.296°E (MNHN, 2). 20–60 m, E Aoré Island, Aimbué Bay, SANTO 2006 EP36, 15.553°S, 167.209°E (MNHN, 3). 20–60 m, E Aoré Island, Aimbué Bay, 15.553°S, 167.209°E (MNHN, 1). 100 m, NW Tutuba Island, SANTO 2006 DS108, 15.553°S, 167.276°E (MNHN, 1). 0–1 m, W Tutuba Island, SANTO 2006 VM69, 15.556°S, 167.278°E (MNHN, 3). 100 m, W Tutuba Island, Aoré Island, SANTO 2006 DS93, 15.560°S, 167.275°E (MNHN, 1). 2–4 m, Tutuba Island, SANTO 2006 FB92, 15.560°S, 167.276°E (MNHN, 1). 8–9 m, Tutuba Island, SANTO 2006 FS84, 15.560°S, 167.276°E (MNHN, 5). 75–80 m, W Tutuba Island, SANTO 2006 EP39, 15.560°S, 167.273°E (MNHN, 2). 7 m, Segond Channel, Wambu river mouth, SANTO 2006 DS91, 15.561°S, 167.140°E (MNHN, 16). 70–80 m, W Tutuba Island, SANTO 2006 DS103, 15.568°S, 167.266°E (MNHN, 2). 98–100 m, W Tutuba Island, SANTO 2006 DS102, 15.568°S, 167.266°E (MNHN, 5). 80 m, W Tutuba Island, SANTO 2006 DS104, 15.568°S, 167.266°E (MNHN, 7). 0 m, E Aoré Island, Aisari Bay, SANTO 2006 ZM36, 15.571°S, 167.206°E (MNHN, 2). 2–3 m, Aoré Island, Port Benier, SANTO 2006 FB61, 15.573°S, 167.210°E (MNHN, 4). 0 m, Aoré Island, Palikulo Bay, SANTO 2006 DB77VM46, 15.575°S, 167.205°E (MNHN, 4). 12 m, E Aoré Island, SANTO 2006 DB08, 15.576°S, 167.230°E (MNHN, 3). 3–18 m, NE Tangisi Island, Aoré Island, SANTO 2006 FB56, 15.586°S, 167.035°E (MNHN, 1). 37 m, Tangoa Island, SANTO 2006 FS70, 15.590°S, 166.995°E (MNHN, 1). Tangua Island, 15.590°S, 167.986°E (MNHN, 1). 6–30 m, S Tutuba Island, SANTO 2006 NB43, 15.593°S, 167.266°E (MNHN, 2). 12 m, Strait N Tangoa Island, SANTO 2006 FS74, 15.595°S, 166.988°E (MNHN, 3). 10–18 m, S Aoré Island, SANTO 2006 DB12, 15.610°S, 167.168°E (MNHN, 14). 6–24 m, S Aoré Island, SANTO 2006

DS10, 15.610°S, 167.168°E (MNHN, 11). 35 m, Bruat Channel, N coast of Malo Island, SANTO 2006 DS110, 15.613°S, 167.146°E (MNHN, 1). 78–91 m, NW coast of Malo Island, SANTO 2006 EP22, 15.622°S, 167.098°E (MNHN, 1). 10 m, Bruat Channel, N Coast of Malo Island, SANTO 2006 DB25, 15.628°S, 167.188°E (MNHN, 1). 80–94 m, NW coast of Malo Island, SANTO 2006 EP19, 15.629°S, 167.089°E (MNHN, 1). 15 m, W Malo Island, Palikulo Bay, SANTO 2006 DB77DB29, 15.648°S, 167.085°E (MNHN, 2). 622–625 m, MUSORSTOM 8 DW1072, 15.666°S, 167.333°E (MNHN, 1). 6 m, E Malo Island, Palikulo Bay, SANTO 2006 DB77DB83, 15.723°S, 167.250°E (MNHN, 2). 360–419 m, MUSORSTOM 8 DW1065, 16.266°S, 167.350°E (MNHN, 27).

Fiji. 1 m, S coast Viti Levu, Deuba, 18.250°S, 178.083°E (AMS C.348868, 7; AMS C.348869, 4). Komave, S coast Viti Levu, 18.217°S, 177.767°E (AMS C.402842, 2; AMS C.402849, 3, 6: complete; AMS C.348871, 3). Viti Levu S coast, Korotongo, 18.183°S, 177.533°E (AMS C.348872, 4; AMS C.349135, 18, 17: complete). Sovi Bay Souvi, SW coast Viti Levu, 18.183°S, 177.600°E (AMS C.406355, 15). 0 m, Yasawa-i-rara, Tavewa Island, Yasawa Group, 16.933°S, 177.367°E (AMS C.461463, 4). 15 m, Yaqara Bay, Viti Levu, 17.310°S, 177.994°E (LACM, 1: complete). 10 m, near Nacilau Point, Yaqara Bay, Viti Levu, 17.364°S, 178.018°E (LACM, 1). 0–1 m, Nukumbotho channel W side of Nukumbutho Island, Laucala Bay, Viti Levu, 18.182°S, 178.466°W (ZSM 20070110, 7: complete). 0–1 m, Nukumbutho Island, Laucala Bay, Viti Levu, 18.175°S, 178.472°E (ZSM 20070138, 2: complete). 0 m, Nukumbotho channel W side of Nukumbutho Island, Laucala Bay, Viti Levu, 18.182°S, 178.466°W (ZSM 20070161, 2). 0–1 m, Nukumbotho channel W side of Nukumbutho Island, Laucala Bay, Viti Levu, 18.182°S, 178.466°W (ZSM 20070153, 3). 0–1 m, Nukumbutho Island, Laucala Bay, Viti Levu, 18.175°S, 178.472°E (ZSM 20062107, 1: complete). 24 m, Vitu Levu S lagoon, 18.168°S, 178.477°E (MNHN, 1). 314–377 m, 19.667°S, 178.167°W (MNHN, 1). 300–400 m, 18.167°S, 178.733°W (MNHN, 1). 200–215 m, Bligh Water, 16.840°S, 178.208°E (MNHN, 15). 183–188 m, SE Viti Levu, 17.828°S, 178.797°E (MNHN, 20). 660–663 m, Bligh Water, 17.248°S, 178.492°E (MNHN, 30). 251–257 m, Bligh Water, 16.857°S, 178.232°E (MNHN, 1). 879–897 m, SE of Viti Levu, 17.515°S, 178.888°E (MNHN, 300). 392–433 m, S of Viti Levu, 18.185°S, 178.390°E (MNHN, 30). 959–963 m, SE of Viti Levu, 17.710°S, 178.917°E (MNHN, 20). 441–443 m, S of Viti Levu, 18.320°S, 177.862°E (MNHN, 70). 567–699 m, Bligh Water, 17.158°N, 177.938°E (MNHN, 1). 275–430 m, S of Viti Levu, 18.297°S, 177.907°E (MNHN, 300). 234–361 m, 18.310°S, 178.085°E (MNHN, 11). 260–305 m, S of Viti Levu, 18.308°S, 178.097°E (MNHN, 50). 295–302 m, S of Viti Levu, 18.212°S, 178.540°E (MNHN, 50). 102–106 m, Bligh Water, 147.322°S, 177.790°E (MNHN, 5). 497–504 m, S of Viti Levu, 18.312°S, 178.152°E (MNHN, 20). 144–150 m, S of Viti Levu, 18.207°S, 178.550°E (MNHN, 30). 420–450 m, 18.200°S, 178.617°W (MNHN, 20). 350–357 m, 18.667°S, 178.500°W (MNHN, 4). 251–257 m, Bligh Water, 16.857°S, 178.232°E (MNHN, 25). 656–660 m, Bligh Water, 17.268°S, 178.247°E (MNHN, 20). 656–660 m, Bligh Water, 17.268°S, 178.247°E (MNHN, 4). 149–168 m, S of Viti Levu, 18.207°S, 178.552°E (MNHN, 30). 0 m, Suva Point, Viti Levu, 18.170°S, 178.420°E (LACM 79-40, 16). 18 m, Kandavu, Cape Washington, 19.108°S, 177.967°E (LACM 85-138, 1). 10–20 m, off Francis Island in Beqa Lagoon, just off of Beqa Island (S of Viti Levu), 18.300°S, 178.067°E (DLG 281, 10). 500–614 m, Bligh Water, MUSORSTOM 10 CP1341, 16.875°S, 177.728°E (MNHN, 2). 820–863 m, BORDAU 1 DW1459, 17.300°S, 179.550°W (MNHN, 200). 102–106 m, Bligh Water MUSORSTOM 10 DW1329, 17.321°S, 177.790°E (MNHN, 35). 80–120 m, SE Viti Levu, MUSORSTOM 10 CP1358, 17.808°S, 178.778°E (MNHN, 2). 402–444 m, SE of Viti Levu, MUSORSTOM 10 CP1360, 17.993°S, 178.803°E (MNHN, 4). 785–820 m, BORDAU 1 CP1490, 18.850°S, 178.533°W (MNHN, 11). 240–319 m, BORDAU 1 DW1494, 18.916°S, 178.483°W (MNHN, 100). 395–540 m, BORDAU 1 DW1486, 19.016°S, 178.433°W (MNHN, 1). 395–540 m, BORDAU 1 DW1486,

19.016°S, 178.433°W (MNHN, 60). 500–516 m, BORDAU 1 DW1488, 19.016°S, 178.416°W (MNHN, 40). 700–707 m, BORDAU 1 DW1485, 19.050°S, 178.500°W (MNHN, 16).

Tonga. Kipukea, Haapai, Zunga Grow Island [spelling very uncertain] (KBIN, 1). 389–418 m, N Ha'apai group, 19.700°S, 174.533°W (MNHN, 1). 0 m, Ha'apai, 19.783°S, 174.450°W (AMS C.349133, 1). Ha'apai, 19.783°S, 174.450°W (AMS C.348874, 1). Naupapu Island, Vava'a Group, 18.700°S, 174.100°E (LACM 85-89, 3). 18 m, Langitau island, Vaka'eitu Island, Vava'a Group, 18.717°S, 174.083°E (LACM 85-90, 5). 0 m, Tola Beach, Vava'u Island, 18.600°S, 174.000°W (AMS C.402841, 5). 0 m, Stowaway Village, Nuku'Alofa, Tongatapu, 21.133°S, 175.200°W (AMS C.402850, 1). 440–487 m, Vava'u group, BORDAU 2 DW1586, 18.566°S, 173.916°W (MNHN, 25). 433 m, NW Tongatapu, BORDAU 2 DW1569, 21.033°S, 175.316°W (MNHN, 300).

Kiribati. 11 m, Gilbert Islands, Nonouti Atoll, 0.667°S, 174.350°E (AMS C.402845, 1). 8 m, Apamana Lagoon Abemama, Gilbert Islands, 0.400°N, 173.867°E (AMS C.402844, 1). 10 m, N Gilbert Islands, Abaiang, Point Bolton, 1.717°N, 172.983°E (AMS C.348638, 1). 10 m, N Gilbert Island, Abaiang Atoll, 0.25 mi N of Tedio Island, 1.850°N, 172.967°E (AMS C.402852, 2). 2 m, N Gilbert Islands, Abaiang Atoll, NW end of lagoon, 1.917°N, 172.867°E (AMS C.402851, 1).

Mariana Islands. 1 m, 100 m N marina dock, Merizo, Cocos Lagoon, Guam, 13.265°N, 144.665°E (LACM 78-17, 1). 1 m, S end Piti Bay, Guam, 13.456°N, 144.693°E (LACM 77-14, 2). 1 m, Western Shoal, Guam, 13.450°N, 144.670°E (LACM 77-20, 1). 6 m, Piti Bay, E of Apra Harbor Guam, 13.465°N, 144.690°E (LACM 77-11, 7). 20–33 m, Marpi, Saipan, 15.286°N, 145.816°E (DLG 1159, 2). 5–20 m, LaoLao Bay, Saipan, 15.137°N, 145.759°E (DLG 1075, 16). Surge channel, LaoLao Bay, Saipan, 15.137°N, 145.137°E (DLG 1024, 17). Cliffing outside caver, Marpi, Saipan, 15.286°N, 145.816°E (DLG 1015, 2). Between NW end of Falarik and Elanglap Island, Efaluk Atoll, 7.255°N, 144.432°E (USNM 616655, 1).

Marshall Islands. Arno Atoll, Ine, lagoon side, 7.000°N, 171.667°E (SBMNH 85129, 1). 20–30 m, Majuro, Anemwanot, 7.133°N, 171.317°E (SBMNH 85135, 2; SBMNH 85125, 1). 23 m, ocean side, Kwajalein, 8.739°N, 167.746°E (DLG 1001, 4). 46 m, 3 mi SW by S Bikini Island, 11.592°N, 165.521°E (USNM 586679, 1). 22 m, Majuro, Anemwanot, 7.133°N, 171.317°E (SBMNH 85120, 1).

Wallis and Futuna. Wallis Island, 13.300°S, 176.200°W (KBIN, 11). 350 m, Wallis Island, 13.317°S, 176.283°W (MNHN, 10). 455–515 m, Wallis Island, 13.200°S, 176.267°W (MNHN, 30).

French Polynesia. 20–60 m, Sediments, Afaahiti, Tahiti, 17.760°S, 149.384°W (JTC, 30).

Remarks. *Scissurella mirifica* is clearly conspecific with *Sci. declinans*. The type material of both species is in the NHMUK and agrees well with what has been considered *Sci. declinans* (*e.g.*, Geiger & Jansen, 2004b). Common characteristics include overall shell shape and sculpture, all details of protoconch sculpture, as well as umbilical characters.

The specimens from New South Wales, South Australia, and temperate Western Australia are suspicious for their geographic data and are marked with a question mark on the distribution map (Fig. 139).

Scissurella morretesi Montouchet, 1972

Figures 140–142

Chresonymy.

Scissurella morretesi Montouchet, 1972: 9–10, figs 11–12.
Scissurella morretesi: Rios, 1985: 10, pl. 5, fig. 33 [copy figure Montouchet, 1972].
Scissurella (?) *morretesi*: Lozouet, 1986: 108.
Scissurella morretesi: Rios, 1994: 21, pl. 5, fig. 36 [copy figure Montouchet, 1972].
Scissurella morretesi: Geiger, 2006b: 49–50, fig. 3.
Scissurella morretesi: Rios, 2009: 24, text-fig. [copy figure Montouchet, 1972].
Scissurella morretesi: Dornellas & Simone, 2011: 28.

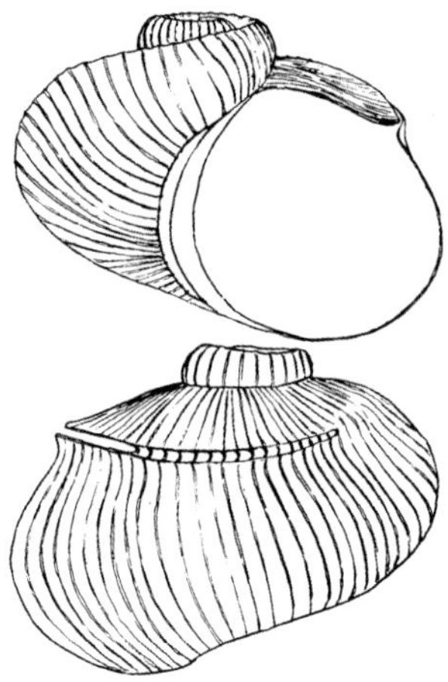

Figure 140. Original illustration of *Scissurella morretesi.*

Type material. Holotype (MZSP 14386), 0.65 × 0.5 mm.

Type locality. Baía dos Castelhanos, Ilha Bela, São Paulo, Brazil [23.783°S, 45.333°W] (OD).

Etymology. Dedicated to Frederico Lange de Morretes [1892–1954: Coan *et al.*, 2007] (OD).

Description. Shell of moderate size (to 1.7 mm), trochiform, inflated. Protoconch of 0.8 whorl, sculpture flocculent with spiral tendency, weak apertural varix not connected to embryonic cap, apertural margin sinusoidal. Teleoconch I of 1.875 whorls, approximately 30 axial cords on first whorl, first spiral cordlet after 0.875 whorl, seven in shoulder portion at onset of selenizone, additional ones on base; interstices with fine, straight growth lines. Teleoconch II of 0.66 whorl, suture impressed. Shoulder convex, same density of axials, axials as on teleoconch I, becoming somewhat stronger with growth (isometric with overall shell size); approximately 12 equispaced spiral lines at apertural margin. Base inflated, without constriction below selenizone, same density and strength of axials as on shoulder, approximately 23 equispaced spiral lines. Umbilicus open, wide, walls straight, with axial growth lines, separated from base by distinct carina, no funiculus. Selenizone above periphery, keels low, of moderate strength, lunules distinct; slit open with parallel margins. Aperture oblong, D-shaped, roof overhanging.

Animal unknown.

Comparisons. *Scissurella* ? *electilis* from Brazil has an almost planispiral shell with a narrow umbilicus and predominant axial sculpture.

Scissurella alexandrei [= juvenile *Sin. confusa*?] from Brazil has an anomphalous shell that is taller than wide.

Scissurella reticulata from the Indian Ocean has a protoconch with fine axial cords and an apertural varix and has a teleoconch I of <1.25 whorls.

Juvenile *Sin. macleani* from the Indian Ocean to the Western Pacific has an overall broader shell, a straight parietal wall with drawn out lower adumbilical corner of the aperture, and the spirals become weaker towards the umbilicus.

Distribution. Brazil (Fig. 142).

Specimen records.

Brazil. 790–940 m, N/O *Marion Dufresne* MD55, st. CB77, SE, 19.683°S, 37.800°W (MNHN, 2). 295 m, N/O *Marion Dufresne* MD55, st. DC75, SE, 18.983°S, 37.833°W (MNHN, 1).

Figure 141. *Scissurella morretesi*. **A–B.** 790–940 m, SE, Brazil, 19.683°S, 37.800°W (MNHN). **C.** Aberrant specimen with evidence of three-fold successive closure of slit to foramen (bold arrows). Square in apical view shows portion of shell shown enlarged below. Inset: Enlarged view of second closure; connected arrows show onset and end of closure. 295 m, SE, Brazil, 18.983°S, 37.833°W (MNHN). Scale bars shell = 1 mm. Scale bars protoconch = 100 µm. Scale bar inset = 0.5 mm.

Remarks. It is virtually impossible to obtain scientific loans from Brazil due to bureaucratic hurdles imposed by the Brazilian Government, hence, the type specimens of this species could not be examined. The specimens illustrated here are identified from the line drawings in the original description. The drawings do not show the umbilicus clearly referred to in the original description. The description and drawings correspond reasonably well with the additional specimens illustrated here, particularly considering that the holotype is a smaller and somewhat juvenile specimen of only 0.65 mm, whereas the species grows to at least 1.7 mm.

The holotype is fully decalcified (L. Simone, pers. comm. 6/2007).

An aberrant specimen, which shows evidence of at least three-time successive opening and closure of the selenizone, is known (Fig. 141C). The closure of the slit is not a product of injury, as the remainder of the shell shows no scarring. The growth pattern is very similar to that seen in normal specimens of *Haliotis* species, in which the older foramina are filled in with the typical lunule seen in the normal selenizone. Despite the broken nature of the specimen, it can be unambiguously identified as a member of *Sci. morretesi.*

Figure 142. Distribution of *Scissurella morretesi.* ☆ type locality. ● specimen record.

Scissurella obliqua Watson, 1886

Figures 143–146

Chresonymy.

Scissurella obliqua Watson, 1886: 116, pl. 8, fig. 5.
Scissurellla obliqua: Kobelt, 1887: 428.
Scissurellla obliqua: Pilsbry, 1890: 58, pl. 68 [erroneously pl. 58 in text], figs 20–21.
Scissurella obliqua: Thiele, 1912a: 10–11, pl. 1, figs 22–23 [copy figs Watson, 1886].
Scissurella obliqua: Powell, 1960: 126.
Scissurella obliqua: Cantera & Arnaud, 1985: 31.
Scissurona ? obliqua: Lozouet, 1986: 114.
Incisura ? obliqua: Geiger, 2003: 74.
Scissurella obliqua: Zelaya & Geiger, 2007: 426–428, figs 123 [map], 147–162 [shell SEM].

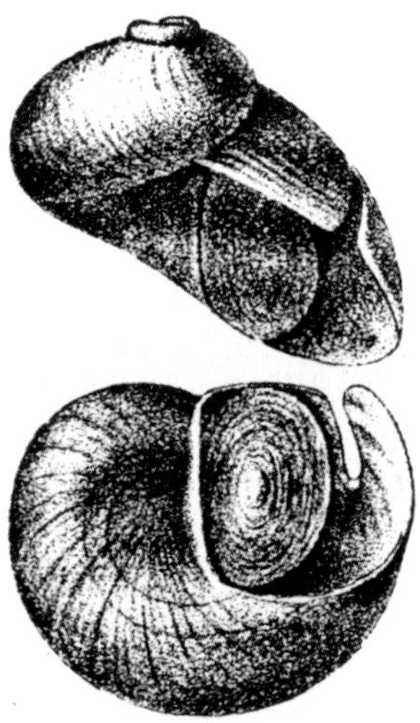

Figure 143. Original illustration of *Scissurella obliqua.*

Misidentification

not *Scissurella obliqua*: Geiger, 2003: fig. 16e [shell, radula. As *Sin. obliqua* in caption. Is *Sci. clathrata*].

Type material. Lectotype (NHMUK 187.2.9.406: designated by Zelaya & Geiger, 2007: 428. Fig. 144). Paralectotypes from type locality (NHMUK 1887.2.9.408, 6. Fig. 145), 0.037 × 0.041" [= 0.94 × 1.04 mm].

Type locality. Royal Sound, Kerguelen Islands, [France, 49.467°S, 70.300°E] shore (OD).

Etymology. Shell described as "obliquely globose."

Description. Shell medium-sized (up to 1.1 mm), trochiform, depressed, globose; spire small, low. Protoconch of one whorl, with strong, wide, flat zig-zag axials, somewhat anastomosing; apertural margin convex, without apertural varix. Teleoconch I of 1.16 –1.25 nearly smooth whorls, only sculptured with low growth marks. Teleoconch II of up to 0.5 whorl, suture little impressed. Shoulder somewhat convex, smooth, with minor growth marks; base continuous with umbilical cavity, mostly smooth with minor growth marks. Aperture rounded oblong, roof overhanging. Selenizone above periphery, keels very low, moderately strong, with irregular lunules; slit with low keels, open, with parallel margins. Umbilicus shallow, not covered by callus. Operculum (Figs 144, 145B–C) round, multispiral with central nucleus, covering approximately 80% of aperture.

Animal unknown.

Comparisons. *Scissurella clathrata* from the southern South America has a shallower umbilicus, the selenizone is less displaced towards periphery, and has spiral cords surrounding

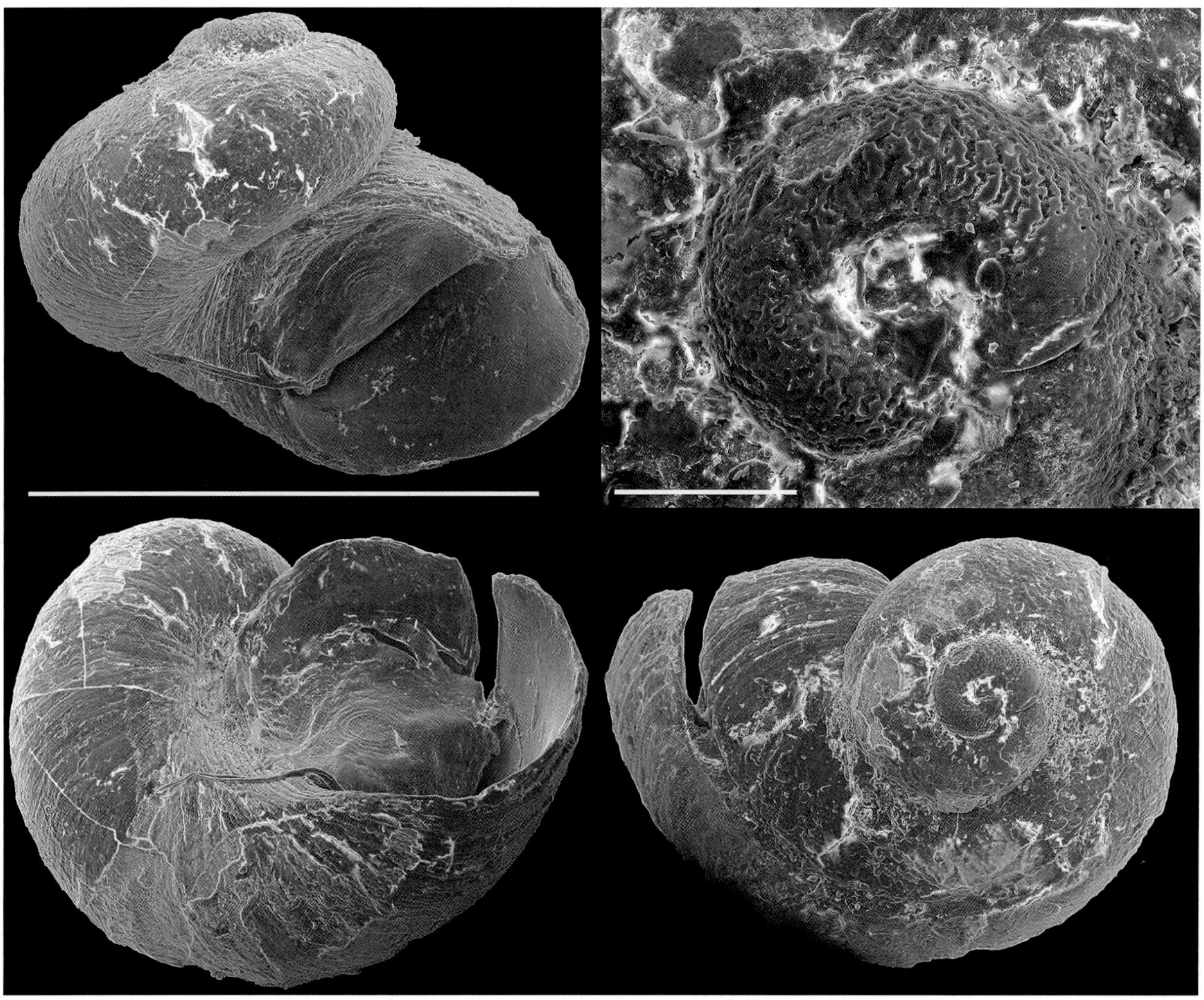

Figure 144. *Scissurella obliqua*. Lectotype. Royal Sound, Kerguelen Islands [France, 49.467°S, 70.300°E] (NHMUK 1887.2.9.406). Scale bars shell = 1 mm. Scale bar protoconch = 100 µm.

the umbilicus. The protoconch ornamentation of *Sci. clathrata* shows a mesh-like pattern, while anastomosing zig-zag axials are present in *Sci. obliqua*.

Scissurella georgica from Antarctica has a protoconch with axial cords, and has regular, low axial cords crossed by fine spiral threads on shoulder and base.

Scissurella petermannensis from Antarctica has strong axial folds on teleoconch I and at the periphery.

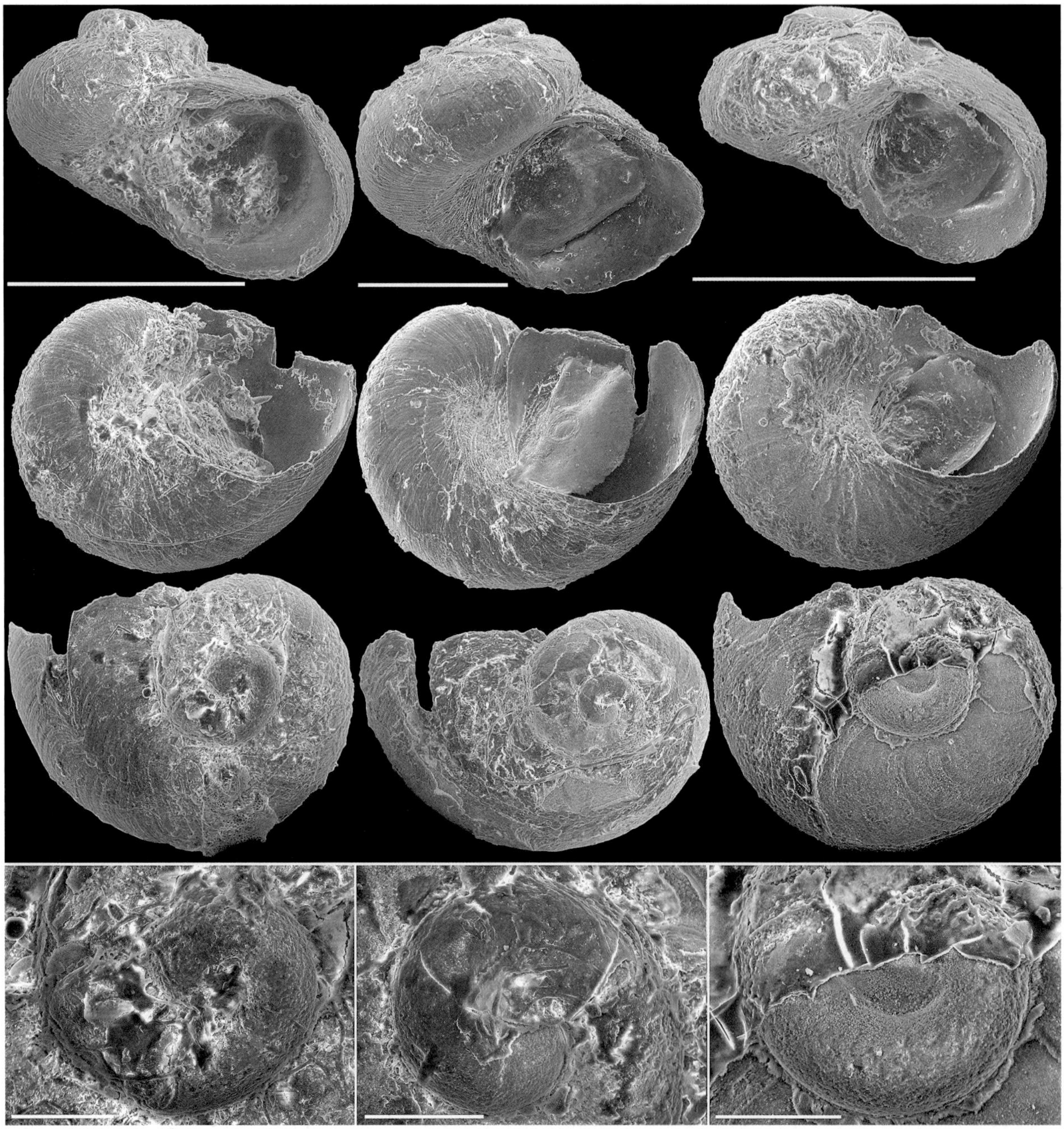

Figure 145. *Scissurella obliqua*. Paralectotypes. Royal Sound, Kerguelen Islands [France, 49.467°S, 70.300°E] (NHMUK 1887.2.9.408, 6). Scale bars shells = 0.5 mm. Scale bars protoconchs = 100 µm.

Figure 146. Distribution of *Scissurella obliqua*. ☆ type locality. ■ literature record.

Scissurella supraplicata from subantarctic islands has a taller shell, a protoconch with spiral sculpture, teleoconch I with strong axial cords, and a narrow umbilicus with distinct periumbilical cord.

Juvenile *Sin. levigata* from New Zealand has a wider shell, a distinct callus covering the umbilicus, and has reticulate sculpture of variable strength on the teleoconch.

Juvenile *Sin. pauperata* from New Zealand lacks a selenizone, has low axial and spiral cords, and has a very narrow umbilicus with distinct periumbilical cord.

Juvenile *Sin. subantarctica* from the subantarctic region has an umbilicus closed by a callus and has numerous narrow spiral bands on the last whorl.

Distribution. Known only from type locality (Fig. 146).

Literature records.

Antarctic. MD 30 Crozet, Ile de la Possession, Baie du Marin, 46.416°S, 51.000°E (Cantera & Arnaud, 1985).

Remarks. There were seven syntype specimens in the NHMUK although the range for the collection numbers on the original label only suggests three specimens. This discrepancy is best explained as an error during cataloguing and not some alteration of the type lot (J. Pickering, pers. comm. 12/2005).

The distinctly descending apertural margin indicates that the specimens are mature, and do not represent juveniles of a *Sinezona* species.

Scissurella ornata May, 1908

Figures 147–149

Chresonymy.

Scissurella ornata May, 1908: 57–58, pl. 6, figs 4–5.
Scissurella ornata: Iredale, 1924: 215.
Scissurella ornata: May, 1923: pl. 17, fig. 2.
Scissurella ornata: Cotton & Godfrey, 1933: 21, pl. 1, fig. 8.
Scissurella ornata: Iredale & McMichael, 1962: 30.
Scissurella ornata: Wilson, 1993: 46, text-fig.
Scissurella ornata: Jansen, 1998: 4, figs 7–9.
Scissurella ornata: Jansen, 1999: 51–52, figs 28–30.
Scissurella ornata: Geiger, 2003: 77.

Figure 147. Original illustration of *Scissurella ornata*.

Misidentifications.

not *Scissurella ornata*: Cotton & Godfrey, 1933: 21, pl. 1, fig. 8 [is *Sci. cyprina*].
not *Scissurella ornata*: May & Macpherson, 1958: 21, pl. 17, fig. 2 [is *Sci. cyprina*].
not *Scissurella ornata*: Jansen, 1995: 11, fig. 14 [is juvenile of *Sin. pacifica*].

Figure 148. *Scissurella ornata* **A.** Paratype. Frederick Henry Bay, [Tasmania: NMW label; Australia, 42.920°S, 147.580°E] (NMW 1955.158.604). **B.** Paratype. Frederick Henry Bay, [Tasmania, Australia, 42.920°S, 147.580°E] (MV F 30456). **C.** *Scissurella* cf. *ornata*. New South Wales, Australia (MMC). See Remarks. Scale bars shell = 0.5 mm. Scale bars protoconch = 100 μm.

Type material. Holotype (TMAG E219, damaged: pers. comm. Liz Tyler, TMAG), 1.5 × 1 mm. Paratypes from type locality (NMW 1955.158.604, 1. Fig. 148A; MV F 30456, 1. Fig. 148B). See Remarks.

Type locality. Frederick Henry Bay, [Tasmania, Australia: NMW label; 42.920°S, 147.580°E]: a few specimens in kelp roots (OD).

Etymology. Ornatus, Latin for decorated.

Description. Shell medium-sized (1.5 mm. OD; 1.18 mm. NMW paratype), trochiform, depressed. Protoconch of one whorl, strong axials, no apertural varix, apertural margin straight. Teleoconch I of 0.9 whorl, 13 raised axial cords, no spirals. Teleoconch II of 0.6 whorl, suture deeply impressed. Shoulder convex, 16 axial lamellae, 3–4 weak spiral lines near apertural margin in vicinity of selenizone. Base convex, weakly constricted below selenizone, same density and strength of axial lamellae as on shoulder. Umbilicus bordered by strong carina, narrow, walls straight, axial growth lines. Selenizone above periphery, keels low, moderately strong; slit open, margins parallel. Aperture D-shaped, flaring.

Comparisons. The comparison for this species are based on the two paratypes in NMW and MV. The MV specimen agrees rather well with the original illustration.

Scissurella cyprina from southeastern Australia has fewer and somewhat more elevated axial lamellae, and a sunken protoconch; see also Remarks.

Juvenile *Sin. beddomei* from temperate Australia has a strong constriction of the base below the selenizone, stronger raised axial cords on the base that do not reach the selenizone, a protoconch with strong axial cords, and lacks spiral sculpture on shoulder and upper base.

Juvenile *Sin. pacifica* from southern Australia and New Zealand has a protoconch with strong axial cords, the umbilicus lacks a periumbilical cord, and has half as many but stronger axial cords on teleoconch I.

Juvenile *Suk. pulcher* and *Suk. atkinsoni* from southeastern Australia both have distinct spiral keels on the base, and broad reticulate protoconch sculpture.

Figure 149. Distribution of *Scissurella ornata*. ☆ type locality. ● specimen record.

Distribution. Temperate Australia (Fig. 149).

Specimen records.

Australia. 4–8 m, Carnac Island, W side, off Freemantle, Western Australia, 32.117°S, 115.667°E (AMS C.402982, 2). Frederick Henry Bay, Tasmania, 42.883°S, 147.567°E (AMS C.29423, 1). Frederick Henry Bay, Tasmania, 42.914°S, 147.541°E (MV F 30456, 1). Off Wilson's Promontory, Victoria, 38.933°S, 146.367°E (MV F 161664, 1).

Remarks. The original description specified "Type in Tasmanian Museum," which makes the TMAG specimen the holotype. The indication in May (1980: 58) "a few specimens in kelp roots," indicates that he considered additional specimens to be conspecific with his *Sci. ornata*. Some specimens from the type locality collected by May in NMW and MV are considered paratypes. There may be additional paratypes in other collections.

As the holotype is damaged, and the two paratypes are rather distinct, it may be advisable to designate a neotype for the species, preferentially based on freshly collected material from the type locality.

The specimen from New South Wales (Fig. 148C) is tentatively identified as *Sci.* cf. *ornata*. It does not match any other species in the general area. The closest are *Sci. skeneoides* and *Sin. mechanica*, both from New Zealand.

Scissurella petermannensis Lamy, 1910

Figures 150–155

Chresonymy.

Scissurella petermannensis Lamy, 1910: 323–324, pl. 1, figs 14–16.

Scissurella petermannensis: Lamy, 1912: 16, pl. 1, figs 14–16.

Scissurella petermannensis: Thiele, 1912a: 9–10, pl. 4, figs 11–13 [copy figures Lamy, 1910].

Schizotrochus petermanensis [sic]: Carcelles, 1953: 162, pl. 1, fig. 7 [copy figure Lamy, 1910].

Scissurella petermannensis: Arnaud, 1972a: 430 [listing only].

Scissurella petermannensis: Powell, 1960: 126.

Scissurella petermannensis: Lozouet, 1986: 108.

Scissurella petermannensis: Geiger, 2003: 77.

Scissurella cf. *petermannensis*: Davolos & Moolenbeek, 2006: 135–136, figs 9a–d.

Scissurella petermannensis: Zelaya & Geiger, 2007: 422–427, figs 123 [map], 124–141 [shell SEM], 142–146 [radula].

Figure 150. Original illustration of *Scissurella petermannensis* Lamy, 1910.

Synonyms.

+ *Scissurella timora* Melvill & Standen, 1912: 345, pl. 48, figs 2, 2a. ***Type material.*** Lectotype (NMSZ 2006109: designated by Zelay & Geiger, 2007: 423. Fig. 152B), 1.75 mm. Paralectotype (NMSZ 1921.143.598, 1. Fig. 152C; NMW 1955.158.091, 1: radula extracted from dried animal: Zelaya & Geiger, 2007: figs 142–146. Fig. 154). ***Type locality.*** Station 325, Scotia Bay, South Orkneys, 9–10 fathoms [= 16–18 m], on *Macrocystis pyrifera* and other large fuci (OD). ***Etymology.*** Timoros, Greek for honoured (OD).

Scissurella timora: Powell, 1951: 61.

Scissurella timora: Powell, 1960: 126.

Schizotrochus timora [sic]: Carcelles, 1953: 161, pl.1, fig. 6 [copy figure Melvill & Standen, 1912].

Scissurella timora: Castellanos & Landoni, 1988: 8–9, pl. 1, fig. 1.

Scissurella timora: Linse, 1999: 400.

Scissurella timora: Forcelli, 2000: 44, text–fig. [line drawing].

Scissurella timora: Geiger, 2003: 78.

Scissurella cf. *timora*: Gordillo *et al.*, 2005: 342.

+ *Scissurella medioplicata* Thiele, 1925: 42, pl. 13, figs 3–4. ***Type material.*** Lectotype (ZMB-Moll 109.047: designated by Zelay & Geiger, 2007: 423. Fig. 153A). Paralectotype from type locality (ZMB-Moll 109.046, 1), 1.25 × 1 mm (W × H). ***Type locality.*** Station 160, Gazelle-Bay, Kerguele Islands [France, 49.500°S, 69.500°E] (OD). ***Etymology.*** Medio-, Latin for middle. -plicata: Latin for folded, referring to only half of the shell bearing axials.

Scissurella medioplicata: Powell, 1960: 126.

Scissurella medioplicata: Cantera & Arnaud, 1985: 32 [as synonym of *Sci. supraplicata*].

Scissurella medioplicata: Geiger, 2003: 77.

Scissurella cf. *medioplicata*: Davolos & Moolenbeek, 2006: 134, figs 7a–d.

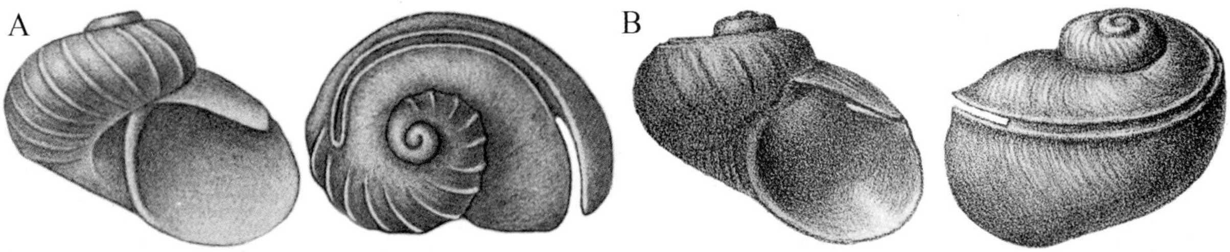

Figure 151. **A.** Original illustration of *Scissurella medioplicata* from Thiele (1925). **B.** Original illustration of *Scissurella timora* Melvill & Standen, 1912.

Type material. Holotype (MNHN. Zelaya & Geiger, 2007: figs 124, 127, 130 [shell SEM]. Fig. 152A), 1 × 0.5 mm.

Type locality. Petermann Island [65.183°S, 64.167°W] Antarctica (OD).

Etymology. Named for its type locality on Petermann Island.

Description. Shell medium-sized (to 1.3 mm), trochiform, depressed. Protoconch of one whorl, 184–193 μm in diameter, with strong, wide, flat zig-zag axial ribs, somewhat anastomosing; apertural margin convex, without apertural varix. Teleoconch I of 1–1.125 whorls, with 10–15 strong, rounded to slightly lamellate axial ribs; axial ornamentation starting sometimes at suture, sometimes at level of selenizone or slightly below periphery. Teleoconch II of up to 0.66 whorl, suture little impressed. Shoulder slightly convex, overall smooth, only sculptured with irregular growth lines, diminishing in prominence with growth. Base with axial ribs and growth lines; occasionally three spiral cords surrounding umbilicus. Umbilicus narrow, deep, with sharp edge at transition to base, walls smooth. Selenizone above periphery, keels moderately elevated, lunules indistinct; slit open, margins parallel.

Operculum (Figs 152B, 153A, C) circular with central nucleus.

Radula (Fig. 154). Rachidian tooth triangular at base; cusp nearly straight, tip with central cusp rounded at the apex; four more narrow, sharper cusps on each side. Lateral teeth 1–3 similar; tip with 4–5 cusps on outer margin. Lateral tooth 4 reduced, with central cusp on tip, two smaller cusps on outer side. Lateral tooth 5 enlarged, with eight strong cusps on inner margin of tip, rounded at apex; outer edge smooth. Inner marginal teeth with main terminal cusp, and 3–4 cusps on both inner and outer margins of tip; outer marginal teeth spoon-shaped, with narrow but long cusps. Radular interlock of central field moderate.

Comparisons. The widely spaced and strong axial ridges that extend only over the peripheral portion of the shell distinguish this species from any other known *Scissurella*.

Scissurella clathrata from subantarctic South America has a taller shell, lacks the axial lamellae, and usually has spiral threads on the shoulder.

Scissurella georgica from Antarctica has a wider shell with wider umbilicus, has much denser but lower axial cords, and has spiral cords on shoulder and base.

Scissurella obliqua from subantarctic islands lacks the strong axial lamellae and lacks a periumbilical cord.

Scissurella supraplicata from subantarctic islands lacks the strong axial lamellae, has a very narrow umbilicus, and has a protoconch with spiral sculpture.

Sinezona chilensis from western South America has twice as many axial cords, spiral threads on shoulder and base, and a protoconch with strong axial cords.

Sinezona levigata from New Zealand has an umbilicus covered by a callus, a very short (or no) selenizone, and has reticulate sculpture of variable strength on shoulder and base.

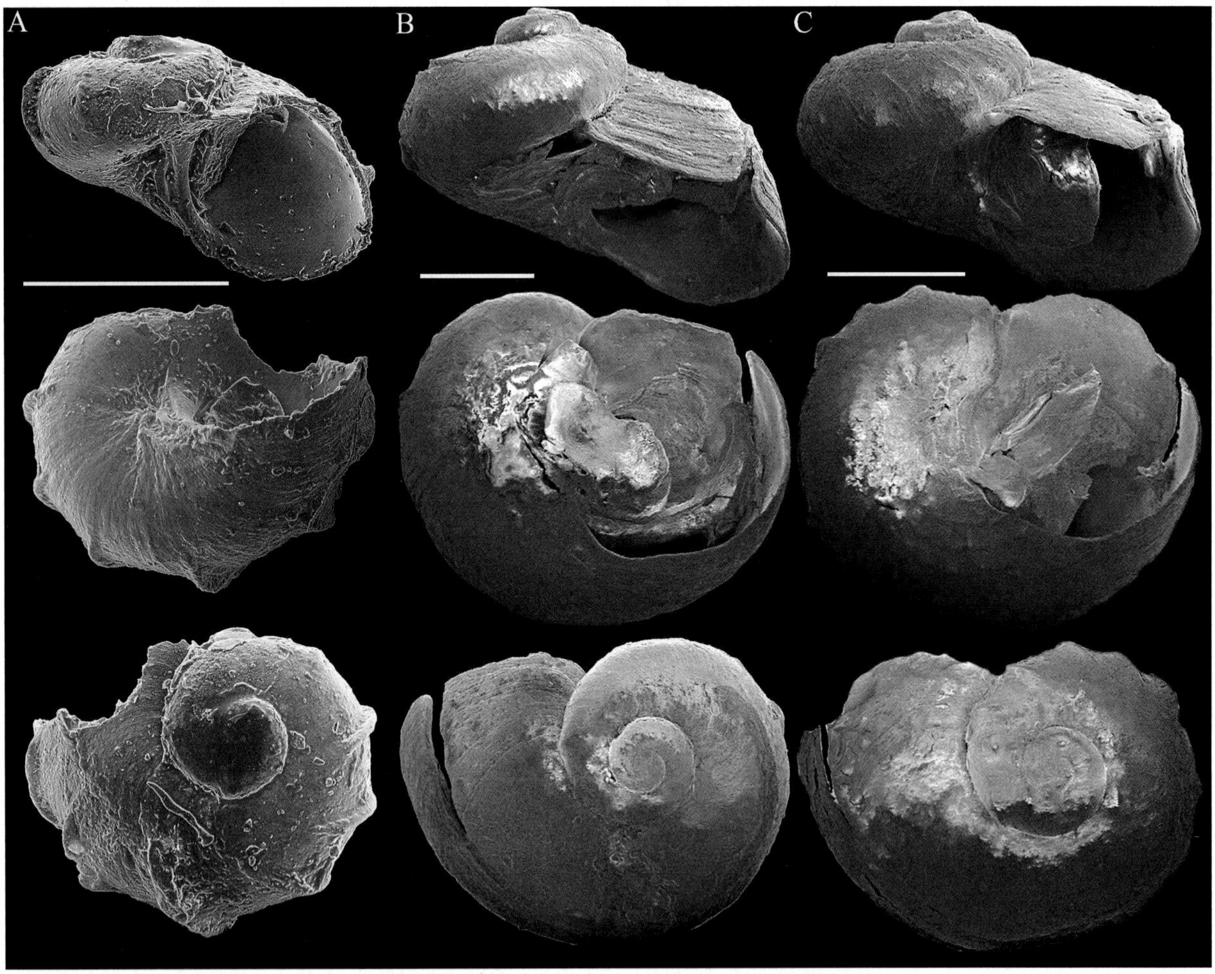

Figure 152. *Scissurella petermannensis*. **A.** *Scissurella petermannensis*. Holotype. Petermann Island [65.183°S, 64.167°W] Antarctica (MNHN). **B.** *Scissurella timora*. Lectotype. Station 325, Scotia Bay, South Orkneys, 9–10 fathoms [= 16–18 m] (NMSZ 2006109). **C.** *Scissurella timora*. Paralectotype. Station 325, Scotia Bay, South Orkneys, 9–10 fathoms [= 16–18 m] (NMSZ 1921.143.598, 1). Scale bars = 0.5 mm.

Sinezona pauperata from subantarctic New Zealand has a very narrow umbilicus, lacks the axial lamellae, lacks a selenizone, and has spiral lines on the teleoconch.

Figure 153. *Scissurella petermannensis*. **A.** *Scissurella medioplicata*. Lectotype. Station 160, Gazelle-Bay, Kerguelen Islands [France, 49.500°S, 69.500°E] (ZMBMoll 109.047). **B.** 5–10 m, Beagle Channel, Antarctica, 54.850°S, 68.317°W (MLP 12081, 2). **C.** 31 m, Puerto Vancouver, Staten Island, Tierra del Fuego, Argentina, 54.780°S, 64.067°W (LACM 71-259, 3). **D.** Green Gorge, 500 m S of hut, Macquarie Island, 54.633°S, 158.917°E (AMS C.376623, 1). Scale bars shell = 0.5 mm. Scale bars protoconch = 100 µm.

Sinezona subantarctica from subantarctic islands has the umbilicus usually covered with a callus, lacks the strong axial lamellae, has a very short selenizone, and spiral sculpture on shoulder and base

Distribution. Probably circum-(sub-)antarctic, reaching the southern tip of South America (Fig. 155), 0.25–18 m.

Specimen records.

Subantarctic/antarctic. 5–10 m, Beagle Channel, Antarctica, 54.850°S, 68.317°W (MLP 12081, 2). 31 m, Puerto Vancouver, Staten Island, Tierra del Fuego, Argentina, 54.780°S, 64.067°W (LACM 71-259, 3). King George Island, S Shetland Islands, 62.230°S, 58.670°W (ZSM-Moll 37995, 1). 7 m, Palmer Station, Antarctic Peninsula, 64.770°S, 64.050°W (NMNZ M.21630, 1). Green Gorge, 500 m S of hut, Macquarie Island, 54.633°S, 158.917°E (AMS C.376623, 1). 0 m, Cumberland East Bay, near Sorling Valley, South Georgia [British Territory], Great Britain, 54.367°S, 36.333°W

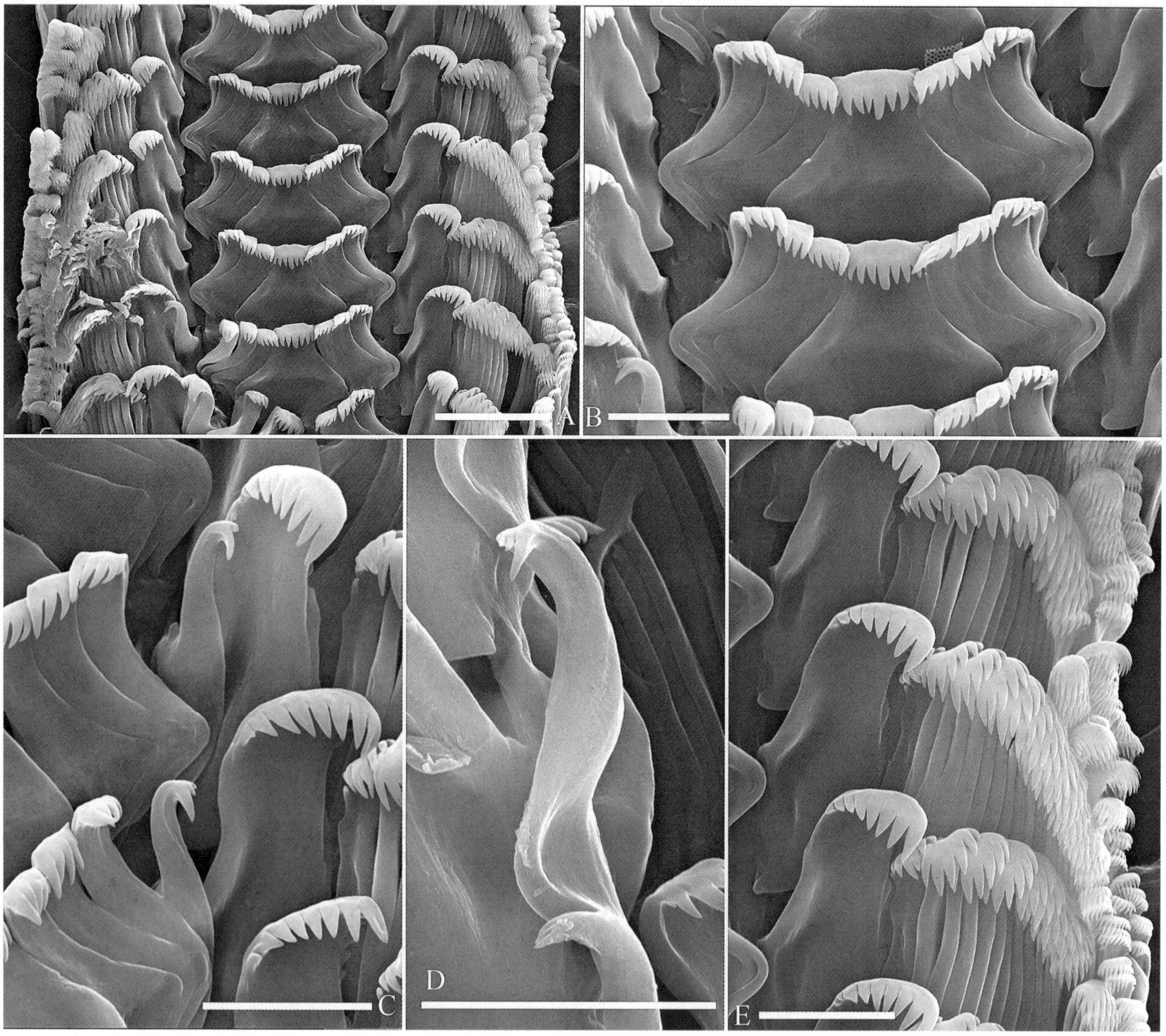

Figure 154. *Scissurella petermannensis*. Radula of one paralectotype of *Scissurella timora*. Station 325, Scotia Bay, South Orkneys, 9–10 fathoms [= 16–18 m] (NMW 1955.158.091). **A.** Full width of ribbon. **B.** Detail of rachidian tooth and lateral teeth 1–3. Detail of lateral teeth 4 and 5. **C.** Lateral teeth 1–5. **D.** Detail of lateral teeth 4. **E.** Detail of marginal teeth. Scale bar A = 20 μm. Scale bars B–E = 10 μm.

(ZMA, 1). 0.25–1 m, Penguin Bay, Barff Peninsula, South Georgia [British Territory], Great Britain, 54.333°S, 36.233°W (ZMA, 1).

Literature records.

Subantarctic/antarctic. Falkland [= Malvinas] Island, 51.800°S, 59.500°W. Strait of Magellan (Castellanos, 1980).—Quarternary raised beach, Rio Ovando, Tierra del Fuego, Argentina, 54.850°S, 68.583°W (Gordillo *et al.*, 2005).—Cumberland East Bay, 54.367°S, 36.333°W. 0.25–1 m, Penguin Bay, South Georgia, 54.333°S, 36.233°W (Davolos & Moolenbeek, 2005).

Figure 155. Distribution of *Scissurella petermannensis*. ☆ type locality. ● specimen record. ■ literature record.

Remarks. *Scissurella timora* and *Sci. medioplicata* were recognized as synonyms of *Sci. petermannensis* by Zelaya & Geiger (2007). The holotype of *Sci. petermannensis* is a juvenile specimen with only the very beginning of the open slit. The typical descending apertural margin on the last 0.25 whorl are only seen in the type specimens of *Sci. timora*. Although the lectotype and paralectotype of *Sci. timora* are very eroded, the general shell morphology and ornamentation of these specimens do not differ from that present in the holotype of *Sci. petermannensis*. The holotype of *Sci. medioplicata* also agrees in general shell outline with the type specimens of *Sci. petermannensis* and *Sci. timora*; due to this fact, this species name is also considered synonymous. However, it should be noted that the holotype of *Sci. medioplicata* has more prominent axial ribs, but this difference seems to be due to the better state of preservation of the material.

Cantera & Arnaud (1985) indicated *Sci. medioplicata* as a synonym of *Sci. supraplicata*. However, the two species are distinct. In *Sci. supraplicata* the shell is as wide as height, whereas *Sci. medioplicata* (a synonymy of *Sci. petermannensis*) has a depressed, laterally expanded shell.

Castellanos (1980) reported *Sci. petermannensis* (as *Sci. timora*) from Falkland Islands and Strait of Magellan. These specimens could not be found in the collections of the MLP or MACN, where other material studied by that author is housed (D. Zelaya, pers. obs.).

The radula of *Sci. petermannensis* has lateral teeth 1–3 with 4–5 cusps, whereas scissurellids usually have three cusps. Whether this character may be of phylogenetic significance remains to be investigated.

Scissurella phenax new species

Figures 156–159

Type material. Holotype (AMS C.377897. Fig. 156). Paratypes 1 m, Unawatuna, Sri Lanka, 6.095°N, 80.248°E (ZSM 20061090, 1, 1: complete. Fig. 157A). Near Tangalle, Sri Lanka, 6.022°N, 80.802°E (ZSM 20061094, 1: complete. Fig. 157B). Pointe Tahara, near Arue, Tahiti, Society Islands, French Polynesia, 17.500°S, 149.217°W (AMS C.377910, 1. Fig. 157C).

Type locality. 6 m, New Year Island, Northern Territories, Australia, 10.900°S, 133.033°E.

Etymology. Phenax: Greek for cheat, impostor; noun in apposition. Referring to the difficulty of separating it from *Sin. calumnior*.

Figure 156. *Scissurella phenax* new species. Holotype. 6 m, New Year Island, Northern Territories, Australia, 10.900°S, 133.033°E (AMS C.377897). Scale bars shell = 0.5 mm. Scale bars protoconch = 100 µm.

Description. Shell small (to 0.7 mm: holotype. Figs 156, 157), trochiform, rounded. Protoconch of 0.9 whorl, dense axial cords in outer portion of whorl, median thickenings not entirely connected to form spiral ridge; indistinct apertural varix not connected to embryonic cap; apertural margin straight. Teleoconch I of 0.9–1 whorl, 15–18

Figure 157. *Scissurella phenax* new species. Paratypes. **A.** 1 m, Unawatuna, Sri Lanka, 6.095°N, 80.248°E (ZSM 20061090, 1, 1: complete). **B.** Near Tangalle, Sri Lanka, 6.022°N, 80.802°E (ZSM 20061094, 1: complete). Radula shown in Figure 158. **C.** Pointe Tahara, near Arue, Tahiti, Society Islands, French Polynesia, 17.500°S, 149.217°W (AMS C.377910, 1). Scale bars shell = 0.5 mm. Scale bars protoconch = 100 µm.

strong axial cords to low lamellae, interstices with very fine growth marks; first spiral lines after 0.2 whorl, 3–4 in shoulder region at onset of selenizone. Teleoconch II of 0.6 whorl, suture moderately impressed. Shoulder slightly convex; axial cords weaker than on teleoconch I, of same density; 6–12 axial threads of unequal strength. Base with constriction below selenizone, strong axial cords to low lamellae more widely spaced as on shoulder; approximately 20 fine spiral lines, 3–4 closest to umbilicus thicker. Umbilicus narrow, bordered by periumbilical cord, wall straight, with finest growth lines. Selenizone above periphery, lunules strong, irregularly spaced; keels low, moderately strong; slit open. Aperture rounded, roof overhanging.

Operculum corneous, thin, multispiral, nucleus central, covering approximately 50% of aperture (Fig. 157C).

Radula (Fig. 158). Rachidian tooth trapezoid, tip triangular, central cusp largest, 3–4 on each side of decreasing size. Lateral teeth 1–3 similar, apical cusp largest, 2–3 along outer edge of tip. Lateral tooth 4 reduced, hook-shaped. Lateral tooth 5 enlarged by broadening, apical cusp largest, seven cusps along inner edge of tip. Inner marginal teeth with broadly triangular tips, apical cusp largest, 1–2 cusps along inner edge, 5–6 cusps along outer edge; shafts without food groove. Outer marginal teeth with broad spoon-shaped cusp, many fine bristles along edge. Radular interlock of central field moderate.

Comparisons. *Sinezona brucei* from northern-most New Zealand has a wider shell, a protoconch with more widely spaces axial lamellae without median thickenings, and the axials on the base are absent on the constriction below the selenizone on the base.

Sinezona norfolkensis from the Tasman Sea has a wider shell, a protoconch with axials lacking the median thickenings, a wider umbilicus, and the axials on the shoulder of teleoconch II are as strong as on the base.

Sinezona calumnior from temperate Australia and New Zealand has a protoconch that does not form median thickenings of the axials, a slightly more narrow

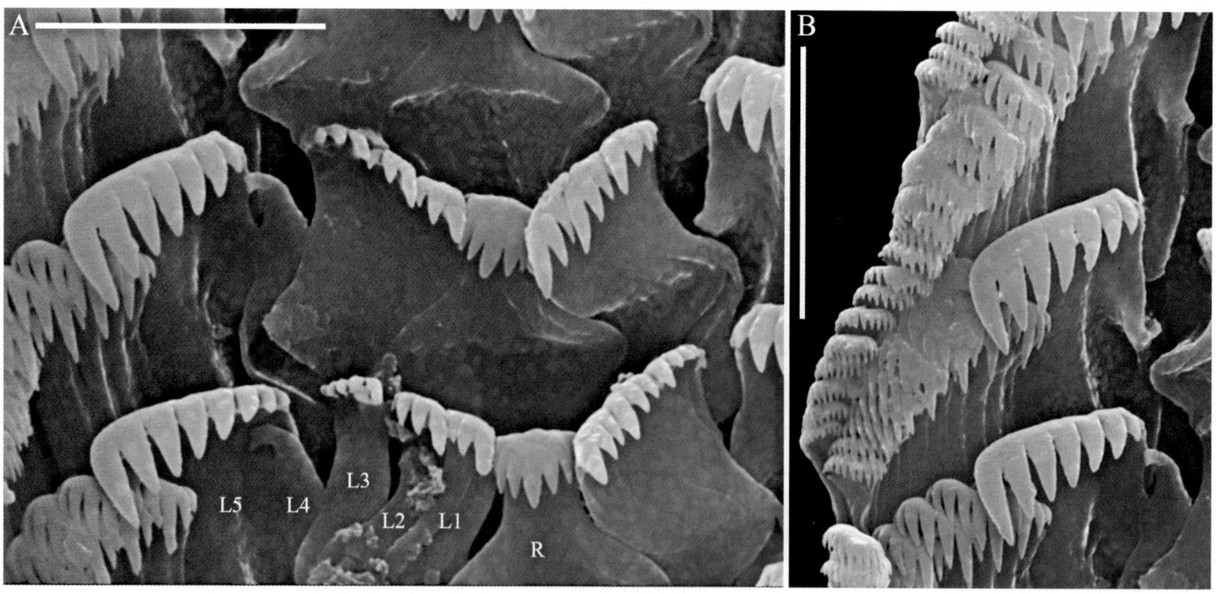

Figure 158. *Scissurella phenax* new species. Radula. Near Tangalle, Sri Lanka, 6.022°N, 80.802°E (ZSM 20061094, 1: complete). Shell shown in Figure 284B. **A.** Central field. **B.** Lateral tooth 5 and marginal teeth. R: rachidian tooth. L1–5: lateral teeth 1–5. Scale bars = 100 μm.

umbilicus, and irregularly spaced spirals on the shoulder with few at the onset of the selenizone.

Sinezona danieldreieri from the Indo-Malayan Archipelago to the Central Pacific has a lower spire, a wider umbilicus, and the axial cords on the protoconch do not form median thickenings.

Sinezona milleri from Southern Japan has a wider shell, a more funnel-like umbilicus, and denser but weaker axial cords.

Sinezona pacifica from temperate Australia and New Zealand has a wider shell, a wider umbilicus, and a protoconch with axials lacking the median thickenings.

Distribution. Indo-Malayan Archipelago to Central Pacific (Fig. 159).

Specimen records.

Sri Lanka. 1 m, Unawatuna, 6.095°N, 80.248°E (ZSM 20061090, 1, 1: complete). Near Tangalle, 6.022°N, 80.802°E (ZSM 20061094, 1: complete).

Australia, Western Australia. Narrow Neck North, Bathhurst Point Area, Rottnest Island, 32.000°S, 115.550°E (WAM ex S13378, 1). Turtle Bay, North-West Cape, 20.766°S, 115.350°E (WAM S13379, 2). Kendrew Island, Dampier Archipelago, 20.475°S, 116.533°E (WAM S13383, 1). Ningaloo Marine Park, ~22.116°S, 113.867°E (WAM S13384, 2).

Japan. 43 m, Horseshoe Beach, Okinawa, 26.505°N, 127.853°E (DLG 1546, 1).

American Samoa. 25 m, Pago-Pago Harbour, Tutuila Island, 14.279°S, 170.701°W (ESC, 1).

French Polynesia. Moorea Island, Baie de Papetoai, Passe Tareu, W side, Society Islands, 17.508°S, 149.850°W (AMS C.377909, 4). Pointe Tahara, near Arue, Tahiti, Society Islands, 17.500°S, 149.217°W (AMS C.377910, 1).

Remarks. The species was separated from *Sin. calumnior* only late in manuscript preparation; re-assessment of all previously examined material was not possible. The listed specimen records are from specimens of which SEM images were available. Additional specimens of this species may be among those listed as *Sin. calumnior* s.l. at the end of *Sin. calumnior*.

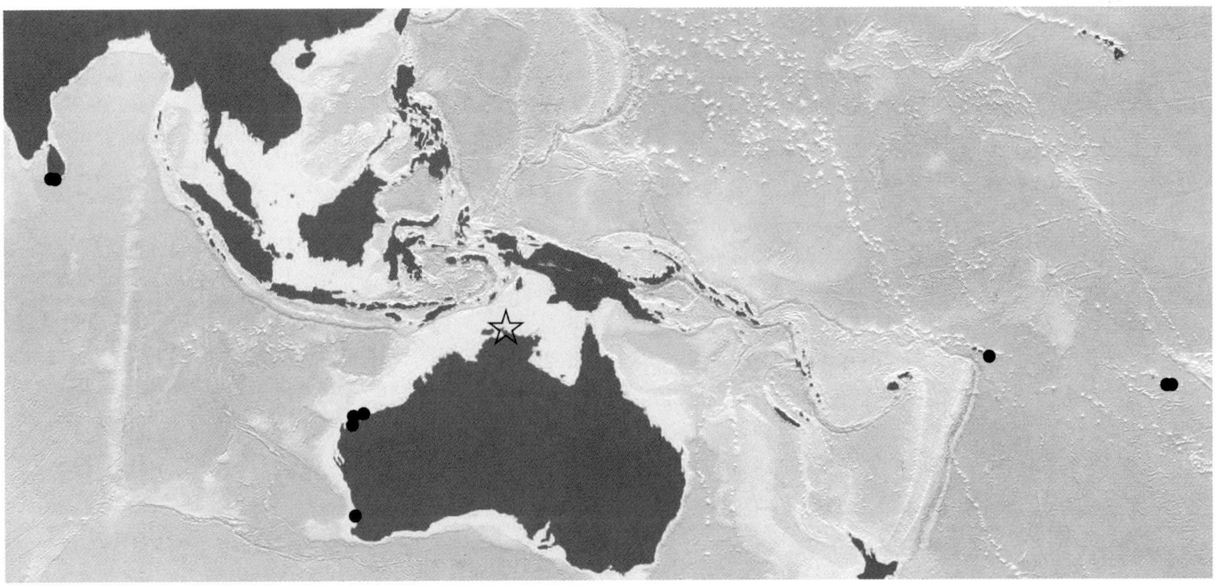

Figure 159. Distribution of *Scissurella phenax* new species. ☆ type locality. ● specimen record.

Scissurella prendrevillei Powell, 1933

Figures 160–168

Chresonymy.

Scissurella sp.: Finlay, 1928: 234.

Scissurella prendrevillei Powell, 1933a [Sept. 25]: 193, pl. 33, fig. 6.

Scissurella prendrevillei: Dell, 1960: 145, 155.

Scissurella prendrevillei: Powell, 1979: 34, pl. 13, fig. 14.

Scissurella prendrevillei: Lozouet, 1986: 108.

Scissurella prendrevillei: Spencer & Willan, 1995: 13.

Scissurella prendrevillei: Marshall, 2002: 168, figs 4–6.

Scissurella prendrevillei: Geiger, 2003: 77.

Scissurella prendrevillei: Spencer *et al.*, 2009: 202.

Figure 160. **A.** Original illustration of *Scissurella prendrevillei*. **B.** Holotype of *Scissurella prendrevillei*. Image courtesy Peter Quin/AIM. Scale bar = 1 mm. **C.** Original illustration of *Scissurella stellae*.

Synonyms.

+ *Scissurella fairchildi* Powell, 1933b [Oct. 6]: 34, pl. 6, figs 3–4. ***Type material.*** Holotype (CM M5316), 0.65 × 1 mm (H × W). Paratypes from type locality (AIM AK 72369, 12. Fig. 161B). ***Type locality.*** 170 fathoms [= 311 m], off Bounty Island, [New Zealand, 47.750°S, 179.000°E](OD). ***Etymology.*** Named for the collector Captain [D.?: see Coan *et al.*, 2007] Fairchild.

Scissurella fairchildi: Powell, 1955: 46.

Scissurella fairchildi: Dell, 1956: 168.

Scissurella fairchildi: Powell, 1979: 34, pl. 13, figs 16–17.

Scissurella fairchildi: Lozouet, 1986: 107.

Scissurella fairchildi: Spencer & Willan, 1995: 13.

Scissurella fairchildi: Freeman *et al.*, 1997: 21.

Scissurella fairchildi: Marshall, 2002: 168, figs 1–3.

Scissurella fairchildi: Geiger, 2003: 77.

Scissurella fairchildi: Spencer *et al.*, 2009: 202.

+ *Scisssurella stellae* Fleming, 1948: 84–85, pl. 6, fig. 3, pl. 8, fig. 4. ***Type material.*** Holotype (GNS TM421. Fig. 163), 1.2 × 0.7 mm. Paratypes from type locality (GNS TM 422, 1: other paratype apparently lost). Paratype 4–9 m, New Golden Hind Station 53, boat passage E of Pigeon Island, Fiordland, SW South Island [New Zealand, 45.710°S,

166.572°E] (GNS TM 424). Paratype 7–15 m, New Golden Hind Station 34, 400 m NE of anchorage, Northport, Chalky Sound, New Zealand, 49.983°S, 166.567°E (GNS TM 425). ***Type locality.*** New Golden Hind Bottom Station 41, Northport, Chalky Sound in 7 fms [= 13 m] (OD) [New Zealand, 49.983°S, 166.567°E]. ***Etymology.*** [not given].

Scissurella stellae: Fleming, 1950: 25.

Scissurella stellae: Powell, 1979: 24.

Scissurella stellae: Lozouet, 1986: 108.

Scissurella stellae: Beu & Maxwell, 1990: 376.

Scissurella stellae: Spencer & Willan, 1995: 13.

Scissurella stellae: Marshall, 2002: 168 [as synonym of *Sin. prendrevillei*].

Figure 161. **A.** Original illustration of *Scissurella fairchildi*. **B.** Paratype of *Scissurella fairchildi*. Scale bar = 1 mm. Image by and courtesy of Peter Quin/AIM. Scale bar = 0.5 mm.

+ *Scissurella marshalli* Bandel, 1998: 12–13, pl. 2, figs 4–6. ***Type material.*** Holotype (NMNZ M.93992), 0.9 × 0.75 mm. ***Type locality.*** Three Kings Islands, New Zealand (OD). 33 m, reef between Great Island and Farmer Rocks, Three Kings Islands, New Zealand, 34°09'S, 172°10'E (Marshall, 2002). ***Etymology.*** Named after Bruce Marshall of NMNZ (OD).

Scissurella marshalli: Marshall, 2002: 169, figs 7–9.

Scissurella marshalli: Geiger, 2003: 77.

Scissurella marshalli: Spencer *et al.*, 2009: 202.

Figure 162. Original illustration of *Scissurella marshalli*.

Type material. Holotype (AIM AK 707011. Fig. 160B), 0.8 × 1.07 mm (H × W).

Type locality. 10 fathoms [= 18 m], off Owenga Beach [Chatham Islands, New Zealand, 44.050°S, 176.650°W] (OD).

Etymology. Not given (OD).

Scissurellidae *Scissurella*

Description. Shell small (to 0.92 mm. Figs 163–166), trochiform subquadrate. Protoconch sunken, one whorl, sculpture of fine axials, no apertural varix, apertural margin sinusoidal. Teleoconch I of 1.25–1.33 whorls, approximately 18–20 distinct axial cords, interstices with fine irregular growth marks, no spirals. Teleoconch II of up to one whorl, suture moderately impressed. Shoulder convex, approximately 8–12 axial cords, becoming progressively more indistinct to almost invisible, lateral extent variable from restricted to adsutural third to covering entire width of shoulder; approximately eight faint irregularly spaced spiral lines at apertural margin, first appearing after 0.33 teleoconch II whorl. Base convex inflated, axials continuous from shoulder; approximately 12 indistinct spiral cordlets, strongest near umbilicus, usually increasing in strength with growth. Umbilicus open, deep, moderately narrow, bordered by carina. Selenizone above periphery; keels low to moderately elevated, moderately strong; lunules distinct; slit open, with parallel margins. Aperture round, roof overhanging.

Operculum thin, round, multispiral, central nucleus (Fig. 165A).

Figure 163. *Scissurella stellae*. Holotype (GNS TM 421). 7 fms [= 13 m], New Golden Hind Bottom Station 41, Northport, Chalky Sound [New Zealand, 49.983°S, 166.567°E] (OD). Scale bar shell = 1 mm. Scale bar protoconch = 100 μm.

Radula (Fig. 167). Rachidian tooth triangular, apical cusp largest, 2–3 cusps on each side of triangular tip. Lateral teeth 1–3 similar, apical cusp largest, 2–3 cusps on outer margin. Lateral tooth 4 reduced, hook-shaped. Lateral tooth 5 enlarged by broadening, apical cusp largest, 5–6 cusps along inner edge of tip. Inner marginal teeth with triangular tip, apical cusp largest, 3–4 cusps along each edge. Outer marginal teeth spoon-shaped, with many fine bristles. Radular interlock of central field moderate.

Comparisons. *Scissurella azorensis* from the subtropical northeastern Atlantic and Mediterranean has weaker (or faint) axial cords.

Scissurella kaiserae from the Panamic and *Sci. maraisorum* from eastern South Africa have less than one teleoconch I whorl and more and weaker axial cords.

Scissurella lobini from the subtropical northeastern Atlantic has strong axial cords on the protoconch and lacks spiral cords on shoulder and base.

Scissurella bountyensis from New Zealand's South Island and the off-shore southern Islands has a protoconch with strong axial cords, usually lacks axial cords on the shoulder, lacks the basal adumbilical spur of the aperture, and the margins of the slit are strongly converging in fully grown specimens.

Figure 164. *Scissurella prendrevillei*. 55 m, off Whero Island, Carter Passage, Stewart Island, New Zealand, 46.908°S, 168.208°E (NMNZ M.019465, 18). Scale bar shell = 0.5 mm. Scale bar protoconch = 100 µm.

Scissurella skeneoides from New Zealand has more numerous but weaker axial cords, lacks spiral sculpture, has an overall more depressed shell shape, and has a protoconch with strong axial cords.

Scissurella cyprina from temperate Australia has less strongly elevated axial cords, and a more flaring aperture.

Distribution. Late Pleistocene to Recent. New Zealand (Fig. 168). 39–311 m.

Figure 165. *Scissurella prendrevillei*. **A.** 55 m, off Whero Island, Carter Passage, Stewart Island, New Zealand, 46.908°S, 168.208°E (NMNZ M.019465, 18) **B.** 103–106 m, Ranfurly Bank, East Cape, North Island, New Zealand, 37.557°S, 178.805°E (NMNZ M.071824, 25). **C.** 28 m, Stewart Island, off Halfmoon Bay, New Zealand, 46.883°S, 168.133°E (AMS C.406360, 1). Scale bars shell = 0.5 mm. Scale bars protoconch = 100 μm.

Specimen records.

New Zealand. 55 m, off Poutama Island, South Cape, Stewart Island, 47.267°S, 167.383°E (NMNZ M.150264, 60). 7 m, W end of Iona Island, Paterson Inlet, Stewart Island, 46.908°S, 168.117°E (NMNZ M.019462, 25). 55 m, off Whero Island, Carter Passage, Stewart Island, 46.908°S, 168.208°E (NMNZ M.019465, 18). 28 m, off Halfmoon Bay, Stewart Island, 46.883°S, 168.133°E (AMS C.406360, 1). 16 m, off Point of Bay,

Figure 166. *Scissurella prendrevillei.* **A–B.** *"marshalli"*-variant. 246–291 m, Middlesex Bank, NW of Three Kings Islands, New Zealand, 34.033°S, 171.733°E (NMNZ M.150351, 40). **C.** 53 m, Wet Jacket Arm, middle of Basin, Ole Island, New Zealand, 45.581°S, 166.734°W (USNM 780324, 8). Scale bars shell A–B = 0.5 mm. Scale bar shell C = 1 mm. Scale bars protoconch = 100 μm.

N side of Edwardson Sound, Fiordland, South Island, 45.900°S, 166.668°E (NMNZ M.143843, 25). 90 m, off Otago Peninsula, South Island, 45.817°S, 170.900°E (NMNZ M.130544, 1). 21 m, off Otago, 45.800°S, 170.800°E (AMS C.379056, 1). 15 m, S side of Wet Jacket Arm entrance, Fiordland, South Island, 45.667°S, 166.750°E (NMNZ M.141215, 40). 20 m, bay N side of W end of Bauza Island, Doubtful Sound entrance, Fiordland, South Island, 45.283°S, 166.908°E (NMNZ M.141663, 80). 48 m, Deas Cove, Thompson Sound, Fiordland, South Island, 45.197°S, 166.972°E (NMNZ M.142059, 100). 18 m, off Owenga Beach, Chatham Islands, 44.033°S, 176.367°W (AIM 70701, 1). 40 m, S of Patea, Taranaki, North Island, 39.933°S, 174.433°E (NMNZ M.053566, 13). 103–106 m, Ranfurly Bank, East Cape, North Island, 37.557°S, 178.805°E (NMNZ M.071824, 25). 71–76 m, Ranfurly Bank, East Cape, North Island, 37.553°S, 178.838°E (NMNZ M.072679, 50). 89–94 m, Ranfurly Bank, East Cape, North Island, 37.552°S, 178.825°E (NMNZ M.074757, 50). 94 m, Ranfurly Bank, East Cape, North Island, 37.547°S, 178.812°E (NMNZ M.150303, 1). 38 m, off E side of Ruamahua-nui Island, Aldermen Islands, North Island, 36.953°S, 176.097°E (NMNZ M.150232, 12). 79 m, off Hen and Chicken Islands, Gulf of Hauraki, 35.917°S, 174.750°E (AMS C.461500, 1). 24 m, Whangaroa Harbour entrance, 35.100°S, 173.867°E (AMS ex C.461502, 13). Taupo Bay, 34.983°S, 173.717°E (AMS C.461509, 12). 33 m, off Taupo Bay, Whangaroa Bay, 34.983°S, 173.717°E (AMS ex C.461514, 2). 63 m, Great Exhibition Bay, North Island, 34.557°S, 173.080°E (NMNZ M.113257, 4). 113 m, Wanganella Bank, Norfolk Ridge, 32.573°S, 167.517°E (NMNZ M.174850, 18). 53 m, Wet Jacket Arm, middle of Basin, Ole Island, 45.581°S, 166.734°W (USNM 780324, 8). Doubtless Bay, 34.900°S, 173.400°E (MV ex F 96917, 10). 47 m, Spirits Bay, 34.366°S, 172.744°E (NMNZ M.174852, 1). 33 m, reef between Great Island and Farmer Rocks, Three Kings Islands, 34.150°S, 172.167°E (NMNZ M.93992, 1). 394–400 m, Reinga Ridge, NW of Cape Reinga, 34.112°S, 171.662°E (NMNZ M.171447, 1). 246–291 m, Middlesex Bank, NW of Three Kings Islands, 34.033°S, 171.733°E (NMNZ M.150351, 40). 188 m, 37 km NW of Great Island, Three Kings Islands, 34.033°S, 171.807°E (NMNZ M.150320, 9). Waitangi, Chatham Island, 43.950°S, 176.533°W (USNM ex 680550, 4). Waitangi, Chatham Island, 43.950°S, 176.533°W (USNM ex 680550, 4). 79 m,

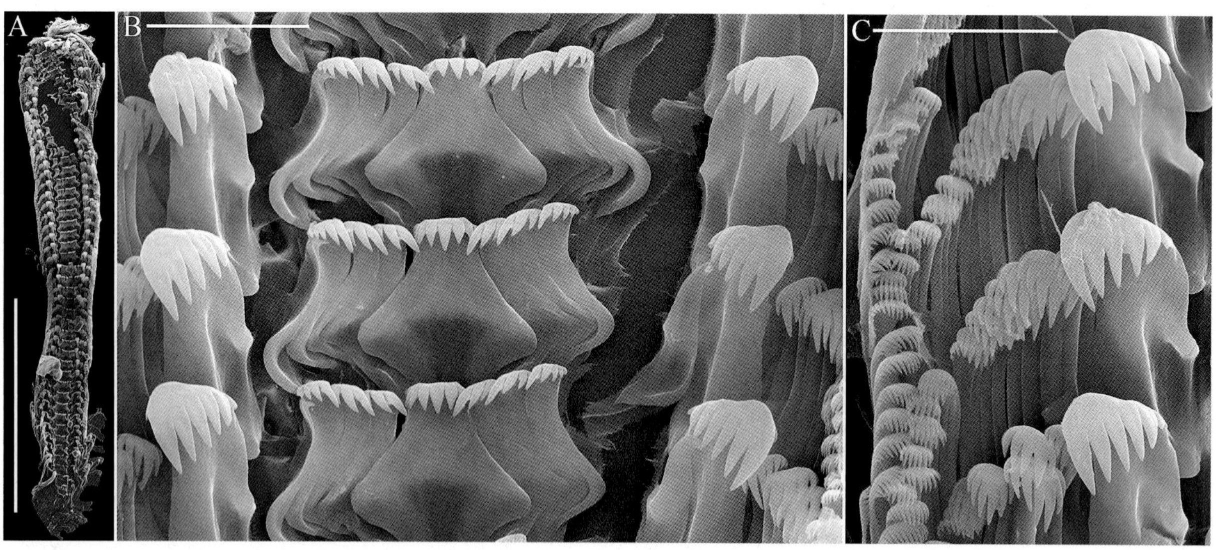

Figure 167. *Scissurella prendrevillei*. Radula. 55 m, off Whero Island, Carter Passage, Stewart Island, New Zealand, 46.908°S, 168.208°E (NMNZ M.019465, 18). **A.** Entire radula. Scale bar = 200 µm. **B.** Central field enlarged. **C.** Lateral tooth 5 and marginal teeth. Scale bars B–C = 10 µm.

off Hen and Chicken Island, Hauraki Gulf, 5.917°S, 174.750°E (AMS C.461499, 2). Doubtless Bay, 34.900°S, 173.400°E (MV ex F 96917, 10).

Literature records.

New Zealand. Northport, 4–8 fms [= 7–25 m]. Boat passage E of Pigeon Island, Dusky Sound, 2–5 fms. [= 4–9 m]. (Fleming, 1948: paratype localities).

Fossil.

New Zealand, specimens. GS12161, cliffs near The Point homestead, W of Flowerpot Bay, N coast of Pitt Island, Chatham Islands, Waipipian-Mangapanian (middle to late Pliocene), 43.710°S, 176.560°W (GNS, 6). GS15119, R28/f62, Paua Kill site on 1855 intertidal slope, c 400 m E of Turakirae Head, S Wellington, Holocene, co-seismic uplift of 1855, 41.438°S, 174.917°E (GNS, 8). GS4120, *Mactra tristis* layer, Kupe Formation, (Castlecliffian, oxygen isotope stage 17, middle Pleistocene) coast E of Omapu Stream Mouth, between Castlecliff and Kai-Iwi beaches, Wanganui, 39.907°S, 174.941°E (GNS, 1). GS4057, Kaikokopu Shellbed (Castlecliffian, oxygen isotope stage 19, middle Pleistocene, coast E of Omapu Stream Mouth, between Castlecliff and Kai-Iwi beaches, Wanganui, 39.907°S, 174.941°E (GNS, 1).

New Zealand, literature. Castlecliffian = middle Pleistocene (Beu & Maxwell, 1990).— Late Pleistocene, Ohope, between Whangaparaoa and Te Areroa. Holocene (Marshall, 2002).

Remarks. Marshall (2002) cited several characters to distinguish *Sci. prendrevillei* and *Sci. marshalli*, none of which are discriminating: *Scissurella prendrevillei* as smaller (= juvenile), with later appearing selenizone (1.75 vs. 1.66 whorls: insignificant considering known intraspecific variation of ±0.125 whorl); weaker axials closer to suture (known to be variable in many species), and unspecified differences in protoconch sculpture (well within intraspecific variability). Examination of several specimens of each nominal taxon by SEM did not reveal any consistent differences. Apparent differences in the elevation of the spire can be accounted for by size differences of the specimens as well as intraspecific variability. The smallest illustrated specimen identified in NMNZ as *marshalli* with 0.25 teleoconch II whorls has also the least elevated spire (Fig. 166B); the intermediate NMNZ-*marshalli* specimen with 0.66 teleoconch II whorls has a spire of intermediate elevation; the holotype with 0.8 teleoconch II whorls (Fig. 162) has a spire that is as elevated as the typical *Sci. prendrevillei* in Figures 164–165.

The holotype of *Sci. fairchildi* from New Zealand has fewer and stronger axial costae on the teleoconch compared to the holotype of *Sci. prendrevillei*, but *Sci. fairchilidi* is indistinguishable in this respect from *Sci. stellae*, which had been synonymized by Marshall (2002) under *Sci. prendrevillei*. Examining 28 specimens of the species by SEM demonstrates that strength and density of the axial costae is continuously variable. Such variability is known from other species such as *Sci. costata*, *Suk. atkinsoni*, and *C. simonsae*. The protoconch differences cited by Marshall (2002) are within the intraspecific variation of *Scissurella* species. Accordingly, all three nominal taxa refer to the same biological species.

Scissurella prendrevillei and *Sci. fairchildi* were both published by Powell in 1933. The publication date of *Sci. prendrevillei* is printed on the frontispiece of the article: September, 25. For *Sci. fairchildi*, Jane Myhill (pers. comm. 12/2010) of the Canterbury Museum provided the publication date as October 6; hence, *prendrevillei* has priority by two weeks.

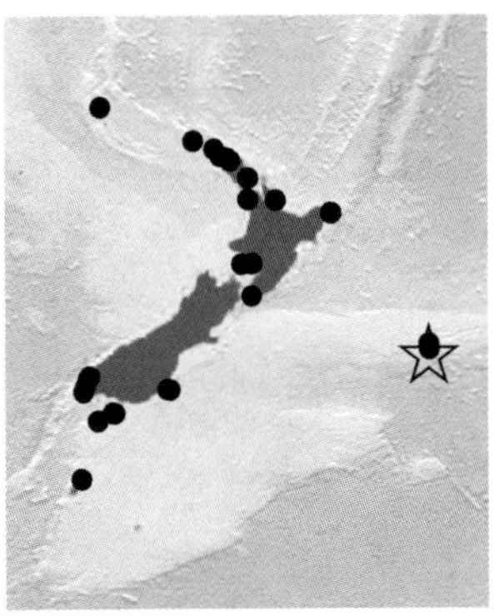

Figure 168. Distribution of *Scissurella prendrevillei*. ☆ type locality. ● specimen record.

Scissurella quadrata Geiger & Jansen, 2004

Figures 169–172, Color Plate 4A

Chresonymy.

Scissurella "square": Jansen, 1999: 53, figs 40–41.

Scissurella quadrata Geiger & Jansen, 2004b: 29–33, figs 14–16 [shell], 17d [map].

Scissurella quadrata: Tröndlé & Boutet, 2009: 12.

Type material. Holotype (AMS C.402642. Fig. 169). Paratypes from type locality (AMS C.403659, 10. Fig. 170A–B). 12 m, Great Barrier Reef, Heron Island, Pams Point, Queensland, Australia, 23.433°S, 151.950°E (AMS C.377699, 13. Fig. 171A). 12 m, North West Cape, Western Australia, Australia, 21.742°S, 114.327°E (AMS C.377645, 8. Fig. 171B).

Type locality. 21 m, Great Barrier Reef, Euston Reef, SW side, Queensland, Australia, 16.667°S, 146.217°E (OD).

Etymology. Quadratus, Latin for square, referring to the squared aperture of the species (OD).

Figure 169. *Scissurella quadrata*. Holotype. 21 m, Great Barrier Reef, Euston Reef, SW side, Queensland, Australia, 16.667°S, 146.217°E (AMS C.402642). Scale bar shell = 1 mm. Scale bar protoconch = 100 μm.

Description. Shell small (to 1.1 mm), squat. Protoconch 0.75 whorl, sculpture of fine axials, apertural varix connecting to embryonic cap, apertural margin strongly sinusoidal. Teleoconch I >1 –1.33 whorls, 13–21 axial ribs, interstices with fine irregular axials, no spiral sculpture. Teleoconch II of up to 1.33 whorls, suture lightly impressed.

Figure 170. *Scissurella quadrata*. **A–B.** Paratypes. 21 m, Great Barrier Reef, Euston Reef, SW side, Queensland, Australia, 16.667°S, 146.217°E (AMS C.403659, 10). **C.** 6 m, Great Barrier Reef, Lizard Island, off Chinamans Point, Queensland, Australia, 14.667°S, 145.450°E (AMS C.377702, 3). Scale bars shell = 1 mm. Scale bars protoconch = 100 µm.

Shoulder with approximately 45–50 axial cords becoming fainter, more crowded on last 0.25 whorl; spiral lines commencing shortly after onset of selenizone, increasing in number with growth, approximately 10–12 at apertural margin of mature specimen; 15–20% of shoulder closest to suture devoid of spirals on first teleoconch II whorl. Base rounded to slightly flared, axial cords stronger than spiral lines, number of axials similar to those on shoulder. Umbilicus with carina bordering

Figure 171. *Scissurella quadrata*. **A.** Paratype. 12 m, Great Barrier Reef, Heron Island, Pams Point, Queensland, Australia, 23.433°S, 151.950°E (AMS C.377699, 13). **B.** Paratype. 12 m, North West Cape, Western Australia, Australia, 21.742°S, 114.327°E (AMS C.377645, 8). **C.** 57 m, off Albany, Western Australia, Australia, 35.167°S, 117.878°E (AMS C.377547, 1). Scale bar shell A = 1 mm. Scale bars shell B–C = 500 μm. Scale bars protoconch = 100 μm.

base, steep, smooth, flat walls. Selenizone with distinct keels, lunules weak, not coordinated with axials, slit open. Aperture subquadrate, roof overhanging.

Animal unknown.

Comparisons. *Scissurella quadrata* is characterized by the strongly angular aperture, particularly in the lower adumbilical corner, and rather delicate axials and spirals.

Scissurella cyprina from temperate southern Australia is more turbinate in profile, and has a different sculpture consisting of spirals on the shoulder that markedly increase in strength on teleoconch II towards the apertural margin.

Scissurella lorenzi from the Indo-Malayan Archipelago to the Western Pacific typically has light colored axial bands, has a more narrow umbilicus, and has a rounder aperture.

Scissurella mirifica of broad Indo-Pacific distribution has an overall more rounded shell with broader umbilicus lacking a periumbilical cord and the aperture is rounder.

Scissurella rota from the Indian Ocean has at most half as many axial and spiral cords on shoulder and base and has a rounder aperture.

Scissurella spinosa of broad Indo-Malayan Archipelago to Central Pacific distribution shares the overall shell shape, but has fewer and more prominent spirals and axials that form its characteristic points at the intersections.

Scissurella staminea from Japan has stronger and fewer axials and spirals, as well as a protoconch with strong axials.

Scissurella sudanica from the Indian Ocean has an overall more globular shape, has finer axial cords, particularly on the base, and has a rounder aperture.

Scissurella xandaros from the Indo-Malayan Archipelago has a smooth protoconch, fewer and stronger axial cords that bear minute points at the intersection with the spirals, and lacks a periumbilical cord on the wider umbilicus.

Juvenile *Sin. ferriezi* with broad Indo-Malayan Archipelago to Central Pacific distribution has a smooth protoconch, has a wider umbilicus without periumbilical cord, and the base projects distinctly below the constriction below the selenizone.

Juvenile *Sin. marrowi* from the Indo-Malayan Archipelago has a protoconch with strong axial cords, a broader umbilicus without periumbilical cord, and the intersection of axial cords and spiral lines produce fine points.

Juvenile *Sin. singeri* from the Indian Ocean has a smooth protoconch and has a wider umbilicus without periumbilical cord.

Distribution. Indo-Malayan Archipelago and Central Pacific (Fig. 172). 0–37 m, live, in coral rubble and algae.

Specimen records.

Maldives. 2 m, W side Rannalhi Islet, S Male Atoll, 3.917°N, 73.333°E (LACM 84-4, 1).

Thailand. 3.5 m, Mai Thon island, off Phuket, 7.762°N, 98.488°E (LACM 85-8, 1).

Vietnam. Qui Nhon, Binh Dinh, 13.764°N, 109.227°E (DLG 1781, 1).

Philippines. 2 m, Bohol, Panglao Island, SW side, 9.600°N, 123.750°E (AMS C.403630, 1). 200 m, Gulf of Davao, 6.700°N, 125.800°E (DLG 1218, 4). 10–12 m, off Mactan Island, Cebu, 10.322°N, 124.014°E (FLMNH ex 392481, 2). Off Engaño Point, Mactan Island, Cebu, 10.317°N, 124.017°E (DLG 869, 1). 13 m, E end Santa Cruz Island, Zamboanga, 6.865°N, 122.073°E (LACM 81-7, 3). 20 m, Luzon, Manig Point, 13.683°N, 120.892°E (LACM 84-163, 1). 90 m, off Momo Beach, Panglao Island, 9.608°N, 123.755°E (MNHN, 1: complete). 60 m, Pamilican Island, 9.502°N, 123.918°E (MNHN, 1: complete). 83–86 m, off Bolod, Panglao Island, PANGLAO 2004 T44, 9.406°N, 123.815°E (MNHN, 1). 20 m, Looc, Panglao Island, PANGLAO 2004 B32, 9.430°N, 123.743°E (MNHN, 1). 6–8 m, Pamilacan Island, PANGLAO

2004 S12, 9.490°N, 123.933°E (MNHN, 1). 16 m, Pamilacan Island, PANGLAO 2004 B25, 9.490°N, 123.935°E (MNHN, 1). 110–112 m, Cervera Shoal, W of Pamilacan Island, PANGLAO 2004 T41, 9.495°N, 123.836°E (MNHN, 4). 55–81 m, Pamiacan Island, PANGLAO 2004 L65-68, 9.498°N, 123.918°E (MNHN, 5). 90–98 m, Balicasag Island, PANGLAO 2004 L69-73, 9.511°N, 123.682°E (MNHN, 4). 120–139 m, Balicasag Island, PANGLAO 2004 L74–75, 9.511°N, 123.682°E (MNHN, 3). 85–100 m, Balicasag Island, PANGLAO 2004 L44, 9.513°N, 123.683°E (MNHN, 1). 90–110 m, Balicasag Island, PANGLAO 2004 L46, 9.515°N, 123.686°E (MNHN, 5; MNHN, 1; MNHN, 2). 12–14 m, Black Forest, Balicasag Island, PANGLAO 2004 B6, 9.518°N, 123.688°E (MNHN, 1). 152 m, Bolod, Panglao Island, PANGLAO 2004 T2, 9.540°N, 123.796°E (MNHN, 2). 17–25 m, Pntod Lagoon 1, Panglao Island, PANGLAO 2004 B39, 9.546°N, 123.701°E (MNHN, 1). 24 m, BBC Point, Panglao Island, PANGLAO 2004 B4, 9.553°N, 123.805°E (MNHN, 1). 24 m, N of Doljo, Panglao Island, PANGLAO 2004 B36, 9.593°N, 123.741°E (MNHN, 1; MNHN, 1). 90–200 m, Maribohoc Bay, Bohol Island, PANGLAO 2004 P1, 9.600°N, 123.750°E (MNHN, 1). 3–14 m, Momo Beach, Banglao Island, PANGLAO 2004 B10, 9.608°N, 123.760°E (MNHN, 2). 20–21 m, Napaling, Panglao Island, PANGLAO 2004 B21, 9.620°N, 123.773°E (MNHN, 2). 3–5 m, Sungcolan Bay, Panglao Island PANGLAO 2004 B18, 9.641°N, 123.828°E (MNHN, 3). 35–45 m, Cortes, Bohol Island, PANGLAO 2004 T23, 9.703°N, 123.843°E (MNHN, 1). 100 m, Manga, Bohol Island, PANGLAO 2004 L48, 9.705°N, 123.831°E (MNHN, 1).

Indonesia. 3 m, Komodo Island, E arm of Slawi Bay, 8.583°S, 119.500°E (AMS C.377701, 2). New Guinea. 23 m, off Port Moresby, Manubada Local Island, off NW side, 9.517°S, 147.167°E (AMS C.404976, 3). 20 m, Hutumuri, S coast, Leitimur, Ambon, Moluccas, 3.731°S, 128.259°E (RMNH, 7). 300 m, NE of Pualau Bahuluang, SW Salayer, 6.437°S, 120.442°E (RMNH, 2). 465 m, S of Karang Kaledupa, Banda Sea, Tukang Besi Islands, 5.958°S, 123.775°E (RMNH, 1). 65 m, Teluk Slawi, E of Komodo, 8.613°S, 119.512°E (RMNH, 7). SW Salayer, 6.355°S, 120.435°E (RMNH, 1).

Papua New Guinea. 25 m, Port Moresby, Horseshoe Reef, 9.467°S, 147.167°E (AMS C.380655, 1). 23 m, off Port Moresby, Manubada Local Island, off NW side, 9.517°S, 147.167°E (AMS C.404976, 3). 13 m, Papua, Milne Bay District, Amazon Bay, 10.317°S, 149.350°E (AMS C.406396, 1). 2 m, Port Moresby, near Basilisk Passage, 9.533°S, 147.117°E (AMS C.404973, 1: complete).

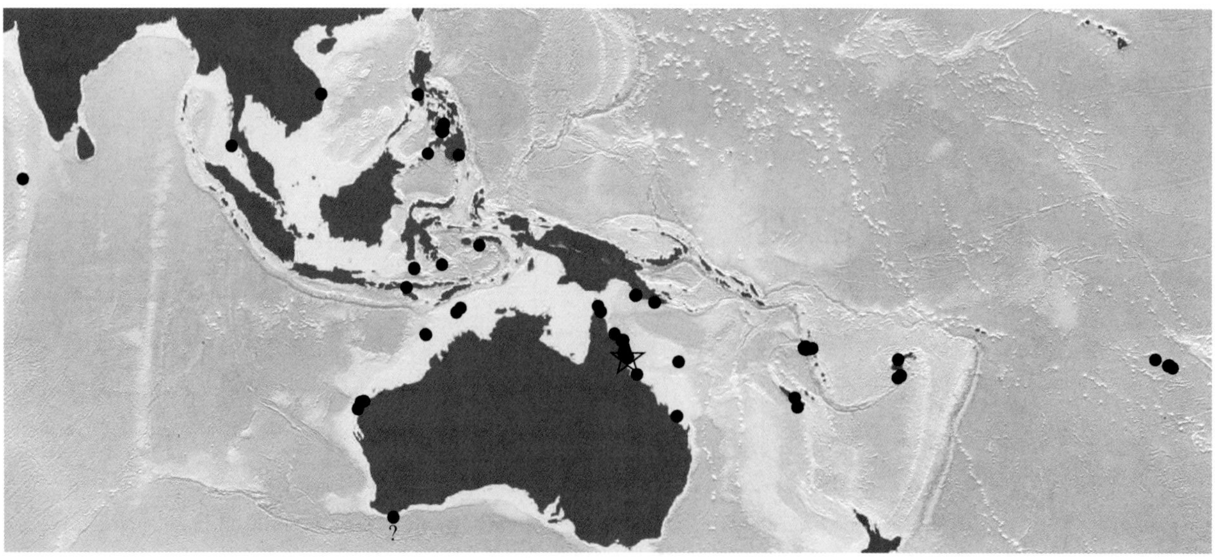

Figure 172. Distribution of *Scissurella quadrata*. ☆ type locality. ● specimen record.

Australia, Western Australia. 57 m, off Albany, 35.167°S, 117.878°E (AMS C.377547, 1) [problematic record]. 80 m, North West Cape, 22.650°S, 113.580°E (AMS C.377636, 3). 92 m, off North West Cape, 22.553°S, 113.620°E (AMS C.377638, 1). 97 m, off North West Cape, 21.783°S, 114.167°E (AMS C.377648, 1). 12 m, North West Cape, 21.742°S, 114.327°E (AMS C.377645, 8). 11 m, Scott Reef, 14.083°S, 121.833°E (AMS C.404491, 2: complete). 27 m, Sahul Bank, Timor Sea, 11.500°S, 125.500°E (AMS C.377647, 1). 20 m, E side, Sandy Islet, Scott Reef, 14.044°S, 121.780°E (SBMNH 369513, 5).

Australia, Queensland. 12 m, Great Barrier Reef, Heron Island, Pams Point, 23.433°S, 151.950°E (AMS C.377699, 13). 15 m, Great Barrier Reef, Capricorn Group, Heron Island, 23.433°S, 151.950°E (AMS C.404427, 1). Great Barrier Reef, Capricorn Group, Heron Island, S side, 23.433°S, 151.950°E (AMS C.404979, 1: complete). 14 m, Great Barrier Reef, off Cairns, Thetford Reef, 16.750°S, 146.167°E (AMS C.377652, 1). 21 m, Great Barrier Reef, Euston Reef, SW side, 16.667°S, 146.217°E (AMS C.402642, 2; AMS C.403659, 10). 37 m, Great Barrier Reef, Spur Reef, off Cairns, 16.400°S, 146.050°E (AMS C.377653, 3, 1: complete). 2 m, Great Barrier Reef, Lizard Island, between Lizard and Palfrey Island, 14.683°S, 145.450°E (AMS C.377696, 1). 5 m, Great Barrier Reef, Lizard Island, between South and Bird Island, 14.683°S, 145.467°E (AMS C.377697, 1). 3 m, Great Barrier Reef, Lizard Island, off Rocky Point, 14.667°S, 145.433°E (AMS C.377637, 4). 10 m, SE side Lizard Island, 14.683°S, 145.458°E (LACM 79-55, 8). 19 m, Great Barrier Reef, No Name Reef, SW end, 14.667°S, 145.650°E (AMS C.377640, 3). 8 m, Great Barrier Reef, Lizard Island, Rocky Point, 14.667°S, 145.433°E (AMS C.377650, 1). 6 m, Great Barrier Reef, Lizard Island, off Chinamans Point, 14.667°S, 145.450°E (AMS C.377702, 3). 3 m, Great Barrier Reef, Reef 14-034, 13.933°S, 144.600°E (AMS C.377639, 2). 4 m, Halfway Island, Cape York Peninsula, 11.383°S, 142.950°E (AMS C.377643, 1). 7 m, Cape York Peninsula, Albany Passage, 10.750°S, 142.617°E (AMS C.377700, 1). Helix Reef, 18.617°S, 147.283°E (LACM 83-41, 2; LACM 83-41, 2). 38 m, Escape Reef, S part, Great Barrier Reef, 15.883°S, 145.817°E (AMS ex C.461468, 1; AMS ex C.461466, 3). 10 m, SE side Lizard Island, 14.683°S, 145.458°E (LACM 79-55, 8). Great Barrier Reef, Capricorn Group, Heron Island, S side, 23.433°S, 151.950°E (AMS C.404979, 1: complete). 3–4 m, Rudder Reef, 30 mi NE of Mossman, 16.192°S, 145.702°E (MV F178694, 1).

Australia, Coral Sea. 10–15 m, SE from Observatory Cay, Lihou Reef, 17.157°S, 152.141°E (SBMNH 369617, 70).

New Caledonia. 12–37 m, Aboré Grand Reef, Mouméa Lagoon, 22.371°S, 166.265°E (MNHN, 4: complete). 375 m, E coast, 21.283°S, 165.950°E (MNHN, 1).

Vanuatu. 140–182 m, 15.633°S, 167.058°E (MNHN, 1). 9 m, reef near Oyster Island, SANTO 2006 FB40, 15.381°S, 167.195°E (MNHN, 2). 20 m, SE Matewulu, SANTO 2006 DB75, 15.381°S, 167.198°E (MNHN, 2). 7 m, W Mavéa Island, SANTO 2006 DB67, 15.381°S, 167.218°E (MNHN, 2). 50–61 m, SW Mavéa Island, SANTO 2006 EP37, 15.391°S, 167.220°E (MNHN, 3). 38 m, Palikulo Bay, SANTO 2006 DB69, 15.406°S, 167.216°E (MNHN, 10). 6–43 m, Aésé Island, SANTO 2006 DB58, 15.410°S, 167.238°E (MNHN, 11). 13 m, Palikulo Bay, SANTO 2006 DB65, 15.430°S, 167.216°E (MNHN, 5). 21 m, SE Aésé Island, SANTO 2006 DB63, 15.448°S, 167.263°E (MNHN, 5). 22 m, Palikulo Bay, SANTO 2006 DS43, 15.465°S, 167.238°E (MNHN, 1). 19 m, Palikulo Bay, Aoré Island, SANTO 2006 FB43, 15.473°S, 167.248°E (MNHN, 3). 5 m, Palikulo Bay, Palikulo Bay, SANTO 2006 DB77DB53, 15.480°S, 167.253°E (MNHN, 5). 5 m, Palikulo Bay, SANTO 2006 DB40, 15.496°S, 167.251°E (MNHN, 2). 22–25 m, NW Urélapa Island, SANTO 2006 DB20, 15.508°S, 167.023°E (MNHN, 14). 46 m, Segond Channel, SANTO 2006 EP03, 15.536°S, 167.160°E (MNHN, 1). 41 m, N Tutuba Island, SANTO 2006 DB61, 15.538°S, 167.281°E (MNHN, 4). 5 m, Aoré Island, Aimbué Bay, SANTO 2006 ZM16, 15.540°S, 167.201°E (MNHN, 1). 45–50 m, Aoré Island, Aimbuei Bay, SANTO 2006 FP46–49, 15.540°S,

167.211°E (MNHN, 4). 46–47 m, Segond Channel, SANTO 2006 EP01, 15.541°S, 167.150°E (MNHN, 5). 114–196 m, NE Aoré Island, SANTO 2006 AT85, 15.543°S, 167.261°E (MNHN, 1). 92 m, NW Tutuba Island, SANTO 2006 DS105, 15.550°S, 167.278°E (MNHN, 2). 35 m, Segond Channel, NW Aoré Island, SANTO 2006 FS96, 15.551°S, 167.160°E (MNHN, 1). 20 m, Segond Channel, N Aoré Island, SANTO 2006 NB12, 15.551°S, 167.160°E (MNHN, 2). 15–25 m, NE Tutuba Island, SANTO 2006 DB01, 15.551°S, 167.296°E (MNHN, 3). 15–25 m, NE Tutuba Island, Palikulo Bay, SANTO 2006 DB77DB01, 15.551°S, 167.296°E (MNHN, 1). 15–25 m, NE Tutuba Island, SANTO 2006 DB01, 15.551°S, 167.300°E (MNHN, 1). 20–60 m, E Aoré Island, Aimbué Bay, 15.553°S, 167.209°E (MNHN, 1). 100 m, W Tutuba Island, Aoré Island, SANTO 2006 DS93, 15.560°S, 167.275°E (MNHN, 2). 114 m, W Tutuba Island, SANTO 2006 DS96, 15.560°S, 167.275°E (MNHN, 2). 75–80 m, W Tutuba Island, SANTO 2006 EP39, 15.560°S, 167.273°E (MNHN, 4). 20–28 m, Segond Channel, Wambu river mouth, SANTO 2006 ED07, 15.567°S, 167.136°E (MNHN, 1). 98–100 m, W Tutuba Island, SANTO 2006 DS102, 15.568°S, 167.266°E (MNHN, 2). 80 m, W Tutuba Island, SANTO 2006 DS104, 15.568°S, 167.266°E (MNHN, 13). 12 m, E Aoré Island, SANTO 2006 DB08, 15.576°S, 167.230°E (MNHN, 4). 36–39 m, Segond Channel, SW coast of Aoré Island, SANTO 2006 FB90, 15.583°S, 167.128°E (MNHN, 2). Tangua Island, 15.590°S, 167.986°E (MNHN, 1). 32–40 m, S Tutuba Island, SANTO 2006 DB16, 15.591°S, 167.263°E (MNHN, 6). 6–30 m, S Tutuba Island, SANTO 2006 NB43, 15.593°S, 167.266°E (MNHN, 2). 16 m, Elia Island, SANTO 2006 FB72, 15.601°S, 167.975°E (MNHN, 1). 10–18 m, S Aoré Island, SANTO 2006 DB12, 15.610°S, 167.168°E (MNHN, 10). 18–30 m, NE Urélapa Island, Palikulo Bay, SANTO 2006 DB77ZB13, 15.611°S, 167.033°E (MNHN, 2). 18 m, Bruat Channel, SANTO 2006 DB80, 15.618°S, 167.125°E (MNHN, 1). 78–91 m, NW coast of Malo Island, SANTO 2006 EP22, 15.622°S, 167.098°E (MNHN, 2). 103–120 m, NW coast of Malo Island, Palikulo Bay, SANTO 2006 DB77EP30, 15.626°S, 167.089°E (MNHN, 2). 99 m, NW coast of Malo Island, SANTO 2006 EP21, 15.628°S, 167.086°E (MNHN, 1). 90–110 m, NW coast of Malo Island, Palikulo Bay, SANTO 2006 DB77 EP28, 15.633°S, 167.083°E (MNHN, 1). 5–7 m, W coast of Malo Island, SANTO 2006 ZB09, 15.676°S, 167.085°E (MNHN, 3). 7 m, Malokilikili, Aoré Island, SANTO 2006 FB52, 15.711°S, 167.235°E (MNHN, 1).

Fiji. 20 m, Herald Pass, W of Ndravuni Island, NW end of Great Astrolabe Reef, 18.767°S, 178.467°E (LACM 85-135, 5). 200–215 m, Bligh Water, 16.840°S, 178.208°E (MNHN, 5). 14 m, easternmost end of Tombani Namalata Reef, Malatta Reefs, N side Kandavu, 18.967°S, 178.167°E (LACM 85-137, 1). 1–2 m, S coast Viti Levu, Deuba, 18.250°S, 178.083°E (AMS ex C.404972, 1: complete). N Astrolabe Reef, 18.679°S, 178.526°E (JKC, 1).

French Polynesia. 4 m, Paevaeva, W side Moorea Island, 17.533°S, 149.883°W, (AMS C.404433, 8). Moorea Island, Papetoai Bay, Passe Tareu, W side, 17.508°S, 149.850°W, (AMS C.404437, 2). 85 m, Raiatea Island, 16.820°S, 151.420°W (DLG 962, 2). 20–60 m, Sediments, Afaahiti, Tahiti, 17.760°S, 149.384°W (JTC, 1).

Remarks. All known records of this species are from tropical waters, with the exception of a single specimen from Albany, on the southern, temperate, coast of Western Australia (Fig. 171C). The shell is in excellent condition and is indistinguishable from tropical specimens. Although there is no evidence that the specimen is mislocalized, I remain suspicious of the locality until further material from temperate waters is found.

The radula figured by Geiger & Jansen (2004b: fig. 17a–b) is from a juvenile *Sci. spinosa*. The shell of the specimen has at most half as many spirals compared to *Sci. quadrata*.

Scissurella redferni (Rolán, 1996)

Figures 173–177

Chresonymy.

Sinezona redferni Rolán, 1996: 109–110, figs 1–4.
Sinezona redferni: Redfern, 2001: 2, pl. 1, fig. 6.
Scissurella redferni: Geiger, 2003: 77.
Scissurella redferni: Nolt, 2008: 56, fig. 3B.
Sinezona redferni: Tunnell *et al*., 2010: 114, text-fig.

Type material. Holotype (AMNH 226480), 0.65 mm. Paratypes from type locality, 10–56 m (NHMUK 1996100, 1; MNCN 15.05/17218, 1; ZMA, 1; MNHN, 1; ERC, 1; CRC, 11).

Type locality. 21 m, Chub Rock, Abaco Island, Bahamas [26.467°N, 77.083°W] (OD).

Etymology. Named after Colin Redfern of Boca Raton, Florida, USA (OD).

Figure 173. Original illustration of *Scissurella redferni*.

Description. Shell small (to 0.65 mm), trochiform, inflated. Protoconch of one whorl, strong axial sculpture, no apertural varix, apertural margin convex. Teleoconch I of 0.875 whorls, approximately 14 axial cords, interstices with very fine irregular axial sculpture, no spiral sculpture. Teleoconch II of 0.66 whorls, suture moderately impressed. Shoulder slightly convex, approximately 10 equally spaced axial cords, becoming less distinct with growth, interstices with very fine irregular axial sculpture. Base with slightest constriction below selenizone, with more distinct axials of equal frequency, interstices with very fine irregular axial growth lines, approximately 14 very fine, indistinct, irregularly spaced spiral lines. Selenizone slightly above periphery, lunules distinct, more frequent than axial cords, keels rather low, moderately strong; slit open, with parallel margins. Umbilicus open, continuously sloping with base, walls with same sculpture as base, with funiculus. Aperture subquadrate, with flared margins.

Operculum (Fig. 176C) corneous, thin, round, multispiral, nucleus central.

Radula (Fig. 176A–B). Rachidian tooth triangular, central cusp largest, three on each side. Lateral teeth 1–3 similar, apical cusp largest, 2–3 along outer edge. Lateral tooth 4 reduced, hook-shaped. Lateral tooth 5 enlarged by broadening. Inner marginal teeth triangular tip, apical cusp largest 3–4 along inner edge, 4–6 along outer edge. Outer marginal teeth spoon-shaped, with many fine bristles along edge of tip. Radular interlock of central field moderate.

Comparisons. *Scissurella azorensis* from the subtropical northeastern Atlantic and the Mediterranean and *Sci. maraisorum* from the Indian Ocean have a protoconch with fine, but dense axial cords, have an overall flatter shell, and a proportionally wider umbilicus.

Scissurella kaiserae from the Panamic tends to have more axial cords on the protoconch, tends to have more distinct spiral cords on the base, and has fewer axials (14 vs. 17–20) on teleoconch I.

Scissurella prendrevillei from New Zealand has a protoconch with finer axial sculpture and the shell is usually proportionally wider.

Juvenile *Sin. confusa* with an open slit from the Caribbean has a distinctly elevated spire, more axial cords, and lacks the flared apertural margin.

Juvenile *Sin. garciai* with an open slit from the Caribbean shares the depressed spire, but has a more pronounced constriction below the selenizone, more numerous and more sharply elevated axial cords, and lacks the flared apertural margin.

Juvenile *Sin. platyspira* from New Zealand has fewer axials that are elevated to lamellae and a protoconch with strong axial cords.

Figure 174. *Scissurella redferni*. 6 m, Sandy Cay, Abaco, Bahamas, 26.398°N, 76.988°W (CRC 12974, 1). Scale bars shell = 0.5 mm. Scale bars protoconch = 100 μm.

Juvenile *Sin. semicostata* from subtropical northeastern Atlantic Islands has a more depressed and proportionally wider shell and a protoconch with strong axial cords.

Distribution. Caribbean, Western Atlantic from Bermuda to Venezuela (Fig. 177).

Specimen records.

Bermuda. North Rock, 32.304°N, 64.752°W (ZSM 20021661, 4: complete). 0 m, beach at Shelly Bay, 32.333°N, 64.741°W (HLC, 1).

Figure 175. *Scissurella redferni*. **A.** East Flower Garden Bank, Texas, USA, 27.908°N, 93.600°W (CCSC). **B–C.** 16 m, St. Croix, 17.800°N, 64.800°W (UMML ex30.12051). Scale bars shell = 0.5 mm. Scale bars protoconch = 100 μm.

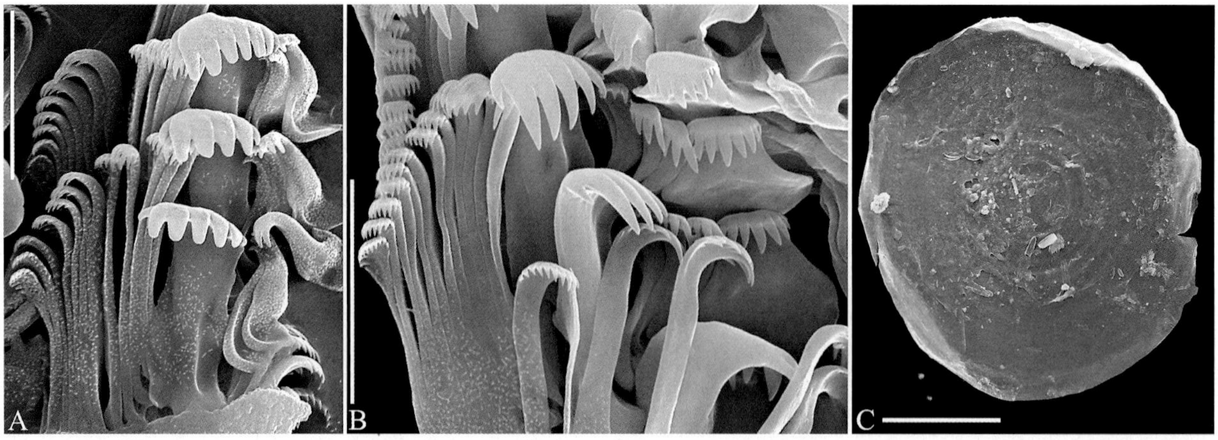

Figure 176. *Scissurella redferni*. Radula and operculum. Puerto Rico, Puerto Rico, (FLMNH ex 163127, 50). **A.** Lateral tooth 5 and marginal teeth. Scale bar = 10 µm. **B.** Central field. Scale bar = 10 µm. **C.** Operculum. Scale bar = 100 µm.

USA, Florida. 2 m, Sand Key, outer reef, W of Key West, 24.450°N, 81.875°W (LACM 70-30, 1). 2 m, NE end Sumerland Key, Monroe County, 24.663°N, 81.440°W (LACM 87-103, 1). 1 m, Carysfort Reef, Key Largo, 25.217°N, 80.200°W (UMML 30.12038, 5). 10 m, Ajax Reef, 25.396°N, 80.137°W (USNM ex 780320, 1). 21–24 m, off Hillsboro Inlet, Broward County, 26.258°N, 80.077°W (LACM 81-9, 1). 5 m, off Palm Beach, 26.705°N, 80.037°W (HLC, 1).

USA, Texas. E Flower Garden Bank, 27.908°N, 93.600°W (CCSC, 2).

Bahamas. 21 m, Abaco Island, Chub Rocks, 26.467°N, 77.083°W (AMNH 226480, 1). 10–60 m, Abaco, 26.500°N, 77.750°W (CRC, 1). 6 m, Sandy Cay, Abaco, 26.398°N, 76.988°W (CRC 12974, 1).

Turks and Caicos. 20 m, Amphitheatre, Grand Turk Islands, 21.467°N, 71.133°W (HLC, 2).

Mexico. Cancun, Quintana Roo, 21.176°N, 86.791°W (FLMNH ex 236681, 8). Mujeres Island, Quintana Roo, 21.232°N, 86.734°W (USNM 662708, 1).

Puerto Rico. (FLMNH ex 163127, 50).

St. Croix. 16 m, 17.800°N, 64.800°W (UMML ex30.12051, 2).

Trinidad & Tobago. 8–9 m, Bucco, Tobago, 18.449°N, 64.745°W (FLMNH ex 236687, 6).

British Virgin Islands. 1–2 m, Savanna Bay, Virgin Gorda, 18.469°N, 64.414°W (DLG 1317, 1).

Nicaragua. Serrana Bank, 14.400°N, 81.550°W (UMML 30.12049, 1).

Venezuela. 20 m, off Los Frailes, near Margarita Island,11.200°N, 63.750°W (DKC 3173, 1). Los Roques, Los Roques Island, 11.588°N, 66.754°W (DLG 900, 1).

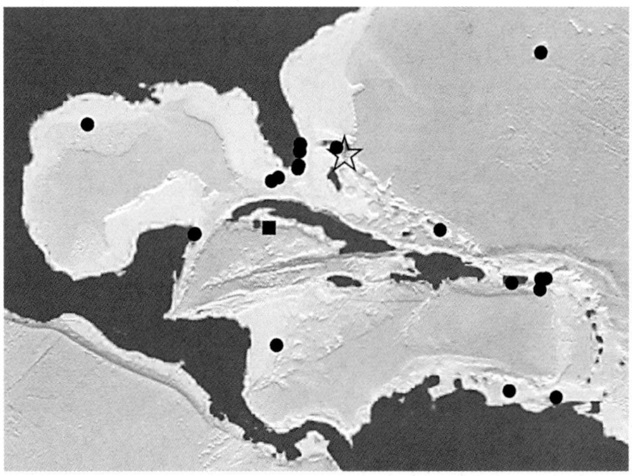

Figure 177. Distribution of *Scissurella redferni*. ☆ type locality. ● specimen record. ■ literature record.

Literature records.

Cuba. 15 m, Cayo Cantiles, 21.600°N, 82.033°W (Rolán, 1996).

Remarks. One live specimen was collected by Redfern (2001) on an algal covered rock in 14 m depth. Empty shells are found in 10–60 m (Redfern, 2001).

The species was originally classified in *Sinezona* due to the protoconch sculpture. Geiger (2003) showed that protoconch sculpture is highly variable within scissurellid genera. *Scissurella redferni* also shows an umbilicus with the same sculpture as on the base, as opposed to the smooth walls seen in *Sinezona*. It is clearly not a juvenile *Sinezona*.

Scissurella regalis Geiger & Marshall, 2012

Figures 178–179

Chresonymy.

Scissurella regalis Geiger & Marshall, 2012: 2–4, fig. 1.

Type material. Holotype (NMNZ M.173745. Fig. 179A). Paratypes from type locality (NMNZ M.303322, 12. Fig. 179B).

Type locality. 780–790 m, Three Kings Rise, New Zealand, 31.866°S, 172.433°E, RV *Rapuhia*, comminuted bryozoans and shell (OD).

Etymology. Derived for the type locality; regalis, Latin adjective for royal (OD).

Description. Shell small (to 1.16 mm: holotype. Fig. 179), trochiform depressed. Protoconch of one whorl, fine axials with strong median thickenings, no apertural varix, apertural margin straight. Teleoconch I of 1.125 whorls, 23–27 axial lamellae, no spiral sculpture. Teleoconch II of 0.66 whorl, suture strongly impressed. Shoulder with axial lamellae of increasing density (holotype), interstices with very fine growth lines, no spiral sculpture. Base with slight constriction below selenizone, same strength and density of axial lamellae as on shoulder, indistinct spiral threads on last quarter whorl from below selenizone, growing fainter towards umbilicus. Umbilicus wide, rounded whorls, underside of protoconch visible. Selenizone above periphery, keels moderately strong, moderately elevated; slit open, margins parallel.

Animal unknown.

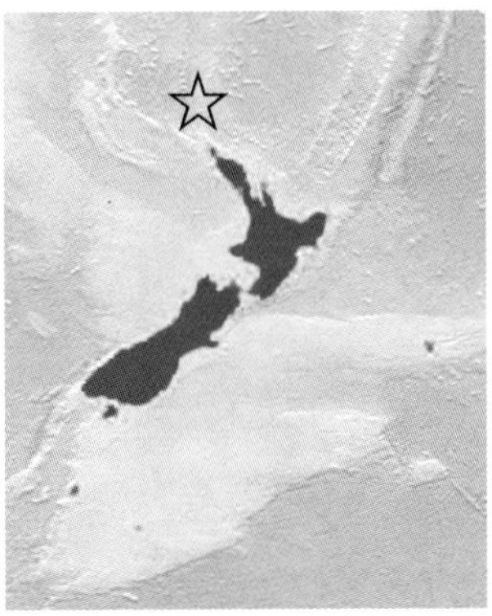

Figure 178. Distribution of *Scissurella regalis*. ☆ type locality.

Distribution. Known only from type material (Fig. 178).

Comparisons. *Scissurella prendrevillei* from New Zealand has a protoconch with fine axials but lacks median thickenings, has fewer axial lamellae on teleoconch I (17–20 vs. 23–27), and usually has broad but low spiral cords surrounding the umbilicus.

Scissurella azorensis from the Azores and the Western Mediterranean and *Sci. maraisorum* from eastern South Africa have broader and lower axial cords, which are not reaching the suture, and have no spiral sculpture on shoulder or base.

Scissurella kaiserae from the Panamic has a shorter teleoconch I (<1 vs. >1 whorl), lacks spiral sculpture on the shoulder, and the axial cords are at most as tall as wide.

Scissurella lobini from the northeastern Atlantic has a protoconch with more widely spaced axial cords devoid of a median thickening, lacks spiral sculpture on the teleoconch, and has at least 50% fewer axial cords.

Scissurella skeneoides from New Zealand has a more rounded whorl, its protoconch is not sunken, and has about 50% more but weaker axial cords.

Figure 179. *Scissurella regalis*. 780–790 m, Three Kings Rise, New Zealand, 31.866°S, 172.433°E. **A.** Holotype (NMNZ M.173745). **B.** Paratype (NMNZ M.303322). Scale bars shell = 1 mm. Scale bars protoconch = 100 μm.

Remarks. The species is known only from the type lots. It is strikingly similar to *Sin. platyspira*, but the slit has parallel rather than convergent margins and the axial lamellae extend almost to the selenizone rim on the last half whorl. The crowding of the axial lamellae towards the apertural margin in the holotype indicates that it is a fully-grown specimen and not a juvenile of a *Sinezona* species, in which the foramen has yet to close. A comparable *Scissurella-Sinezona* pair is *Sci. azorensis* and *Sin. semicostata* from the Macaronesian Islands. Small specimens of *Sci. regalis* and *Sin. platyspira* may be indistinguishable based on shell morphology.

Scissurella reticulata Philippi, 1853

Figures 180–184

Chresonymy.

[no name] Savigny, 1817: pl. 5, figs 29.1–29.3.
Scissurella reticulata Philippi, 1853: 38, pl. 6, fig. 11.
Anatomus reticulatus: Adams & Adams, 1853–1858 (1854): vol. 1, 439.
Scissurella reticulata: Munier-Chalmas, 1865: 395.
Scissurella reticulata: Issel, 1869: 227, 346.
Scissurella reticulata: Paetel, 1888: 289.
Scissurella reticulata: Pilsbry, 1890, 51, figs 49–51.
Scissurella reticulata: Sturnay, 1905: 146 [*fide* Yaron, 1983].
Scissurella reticulata: Pallary, 1926: 82 [*fide* Yaron, 1983].
Scissurella reticulata: Franc, 1956: 22.
Scissurella reticulata: Mastaller, 1979: 29, 242, table 13 [*fide* Yaron, 1983].
Scissurella reticulata: Bouchet & Danrigal, 1982: 14, 22, fig. 62.
Scissurella reticulata: Yaron, 1983: 264–265, pl. 1.
Scissurella reticulata: Lozouet, 1986: 108.
Scissurella reticulata: Vine, 1986: 172.
Scissurella reticulata: Hedegaard, 1990: 74, 145.
Scissurella reticulata: Dekker & van Capelle, 1994: 125.
Scissurella reticulata: Dekker & Orlin, 2000: 17.
Scissurella reticulata: Geiger, 2003: 77.

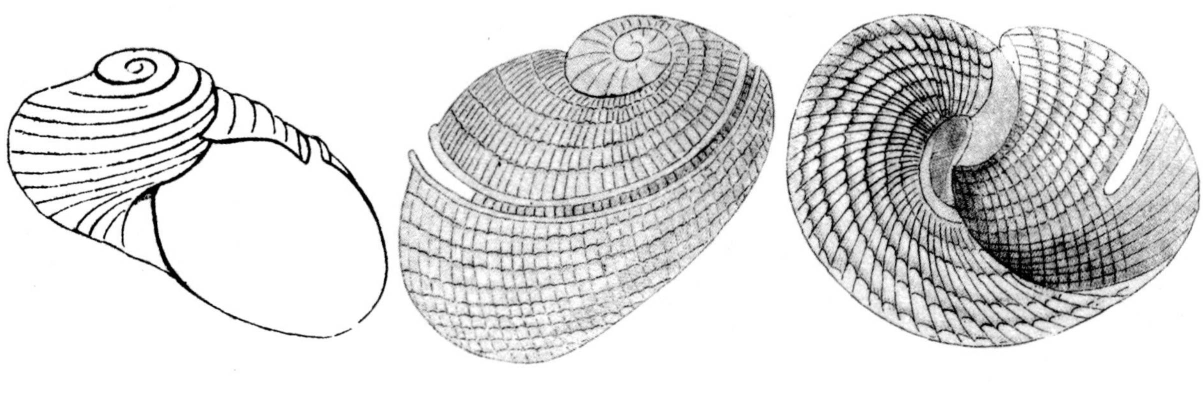

Figure 180. Original Illustration of *Scissurella reticulata*.

Misidentification.

Scissurella decussata: Audouin, 1826: 42, pl. 5, fig. 29 [is *Sci. reticulata*].

Scissurella decussata: Audouin, 1828: 183 [is *Sci. reticulata*].not *Scissurella reticulata*: Bandel, 1998 (*partim*): pl. 2, figs 7–8 [is *Sci. rota*: mislocalized: see Remarks].

not *Scissurella reticulata*: Bandel, 1998 (*partim*): pl. 3, fig. 1 [is *Sci. sudanica*].

Type material. Syntypes (MNHN, 4: Bouchet & Danrigal, 1982), 1 line [1.75 mm] (OD). Lectotype (Specimen of fig. 62 in Bouchet & Danrigal, 1982: selected by Yaron, 1983: 265. Fig. 181). Paralectotypes from type locality (MNHN, 3).

Type locality. Red Sea near Maksur and other localities (OD). Restricted to Suez, at the head of Gulf of Suez [Egypt, 29.967°N, 32.550°E] (Yaron, 1983).

Etymology. Reticulatus, Latin for with net, referring to the netlike surface sculpture.

Description. Shell medium-sized (to 1.2 mm), trochiform, inflated. Protoconch of 0.75 whorl, embryonic cap with reticulate sculpture, remainder with fine axials, finer spiral on outer portion, apertural varix connected to embryonic cap, apertural margin sinusoidal. Teleoconch I of 1.125 whorls, convex, with approximately 17–25 axial cords, interstices with very fine irregular axials, early portion without spirals, later portion with approximately five spiral lines between suture and position of

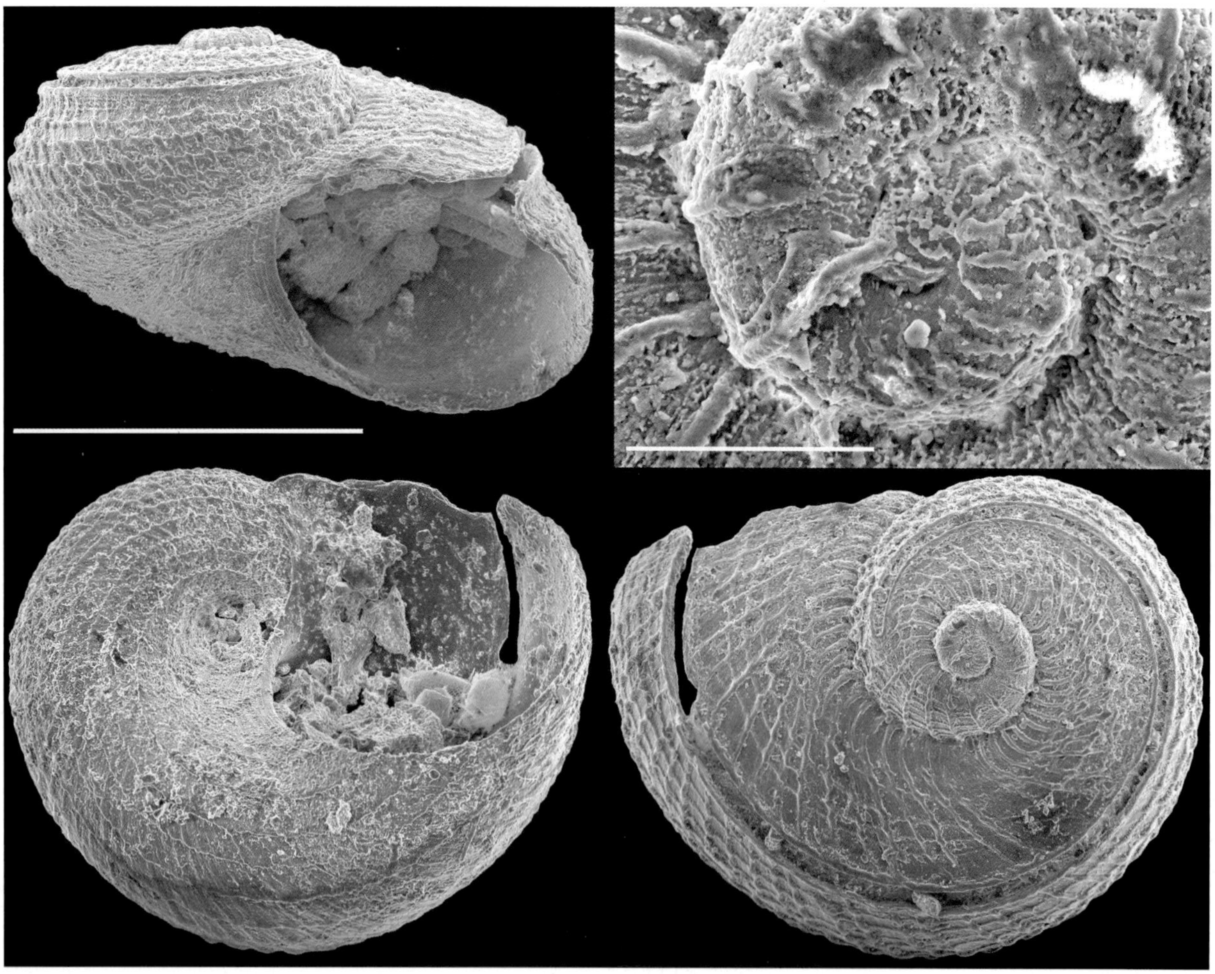

Figure 181. *Scissurella reticulata.* Lectotype. Red Sea near Maksur and other localities (MNHN). Scale bar shell = 1 mm. Scale bar protoconch = 100 µm.

selenizone; intersection of axials and spirals raised to sharp points. Teleoconch II of 1.125 whorls, suture moderately impressed. Shoulder convex, with approximately 38 axial cords on body whorl, with 3–4 more distinct spiral cords in central half of shoulder, being as strong as axials, intersections of spirals and axials raised to sharp points. Base with same density of axials, approximately eight spirals, decreasing in strength from cords below selenizone to periumbilical lines, elevated points at intersection between axials and spirals decreasing in strength from selenizone to umbilicus. Umbilicus open, at angle to base, bordered by spiral cord, walls straight,

Figure 182. *Scissurella reticulata*. **A–C.** 10–18 m, Hurghada, Egypt, 27.283°N, 33.783°E (USNM 719212, 9). Scale bars shell = 1 mm. Scale bars protoconch = 100 µm

smooth. Selenizone above periphery, with low keels of moderate strength, slit open, margins parallel. Aperture wide, oblique, roof overhanging.

Operculum (Fig. 182A–B) corneous, thin, not covering entire aperture, multispiral, central nucleus.

Radula (Fig. 183). Rachidian tooth triangular, shaft with convex margins, tip obtuse triangular, with five cusps on tip, central one slightly larger. Lateral teeth 1–3 similar, base triangular, shaft as high as rachidian tooth, tip with 2–3 cusps. Lateral tooth 3 with narrower base, tip pointed. Lateral tooth 5, broadened, enlarged, tip with 4–5 cusps, apical one largest. Marginal teeth with elongated shafts, tips with approximately five cusps. Outer marginal teeth with spoon-shaped tip with many bristles along edge. Radular interlock of central field moderate.

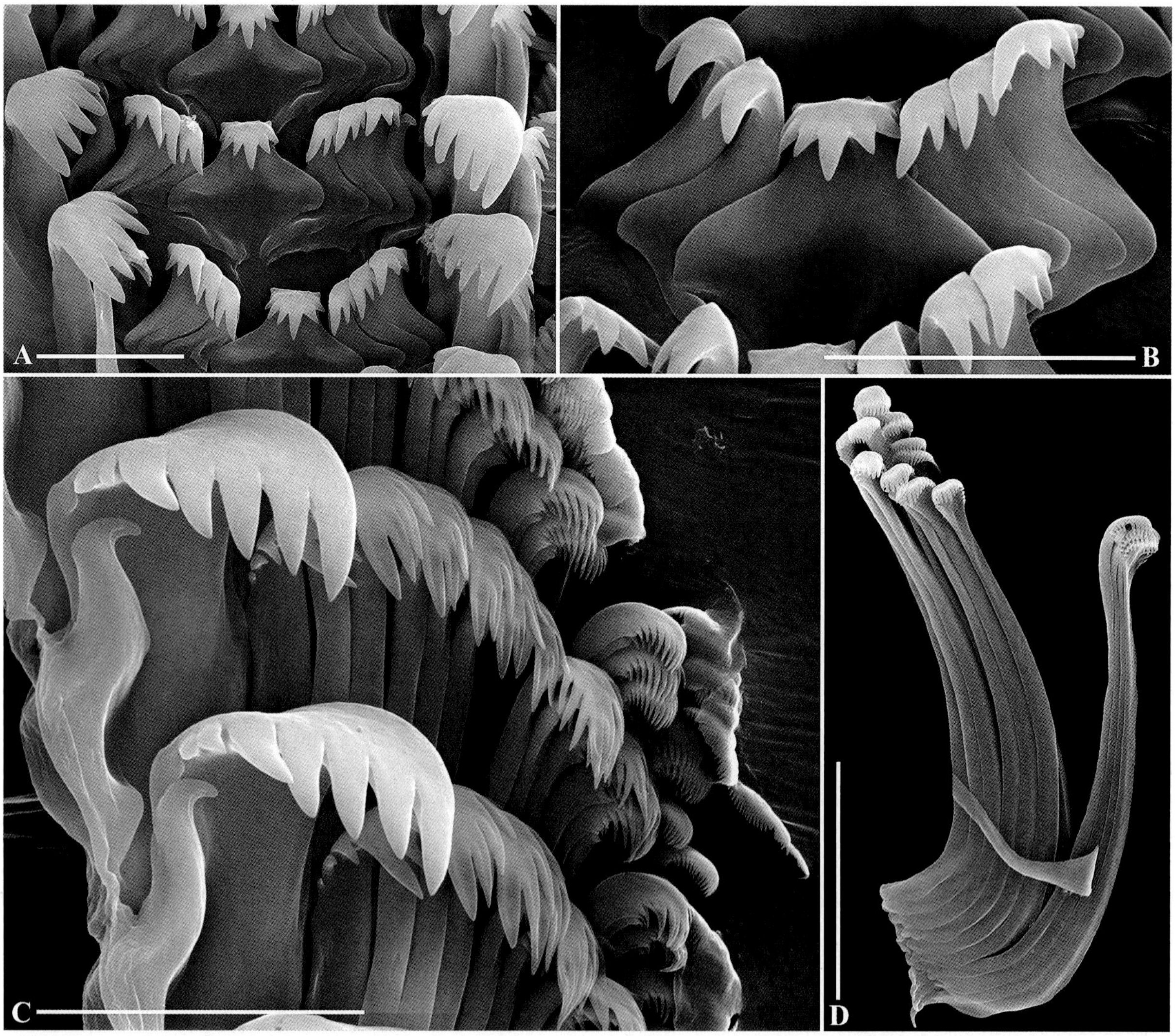

Figure 183. *Scissurella reticulata*. Radula. 10–18 m, Hurghada, Egypt, 27.283°N, 33.783°E (USNM 719212, 9). **A.** Central field with lateral teeth 5 showing smooth outer cutting edge. **B.** Rachidian tooth and lateral teeth 1–3. **C.** Hook-shaped lateral tooth 4, lateral tooth 5 showing serrated inner cutting edge, and marginal teeth. **D.** Full length of outer marginal teeth. Scale bars A–C = 20 μm. Scale bar D = 200 μm

Comparisons. *Scissurella rota* from the Indian Ocean has more sturdy elevated spines on the base of the last half whorl of mature specimens, fewer axials and spirals, has more elevated keels of the selenizone, and usually lacks spirals on teleoconch I.

Scissurella morretesi from Brazil has a protoconch with spirally arranged flecks, lacks an apertural varix on the protoconch, and has a teleoconch I of >1.5 whorls.

Scissurella sudanica from the Indian Ocean has more axials on the body whorl (40 vs. 17–25), has a narrower umbilicus, and lacks the elevated points at the intersections of axials and spirals.

Juvenile *Sin. macleani* from the Indian Ocean to the Western Pacific has a smooth protoconch without apertural varix, lacks the distinct points at the intersection of axial and spiral cords, and has a distinct shelf in the lower adumbilical corner of the aperture.

Distribution. Indian Ocean to western Indo-Malayan Archipelago (Fig. 184).

Biology. Lives under boulders on reefs in shallow water, population densities approximately 0.01/m^2 (Yaron, 1983).

Fossil. Suez, raised coral beaches (Yaron, 1983).

Specimen records.

Red Sea (KBIN, 1).

Egypt. 25 m, Dunraven Reef, Strait of Tiran, S Sinai Peninsula, N Red Sea, 27.200°N, 34.100°E (LACM 85-112, 2). 10–18 m, Hurghada, 27.283°N, 33.783°E (USNM 719212, 9). 0 m, 16 km S of Hurghada, Red Sea, 27.083°N, 33.883°E (RMNH, 14). 0 m, shipyard, Safaga, 26.731°N, 33.943°E (DLG 1643, 1). 0 m, near hotels, Safaga, 26.795°N, 33.942°E (DLG 1646, 2; DLG 1647, 3). 0 m, Makadi Bay, 26.992°N, 33.810°E (DLG 1645, 1).

Madagascar. 0–3 m, near Ampangorina, N side Nosy Komba S of Nosy Bé, Antseranana Province, 13.417°S, 48.350°E (LACM 89-56, 1).

Sri Lanka. 0–2 m, Pigeon Island, off Nilaveli, 8.725°N, 81.203°E (LACM 84-7, 2). 1 m, Unawatuna, 6.095°N, 80.248°E (ZSM ex 20061092, 1: complete).

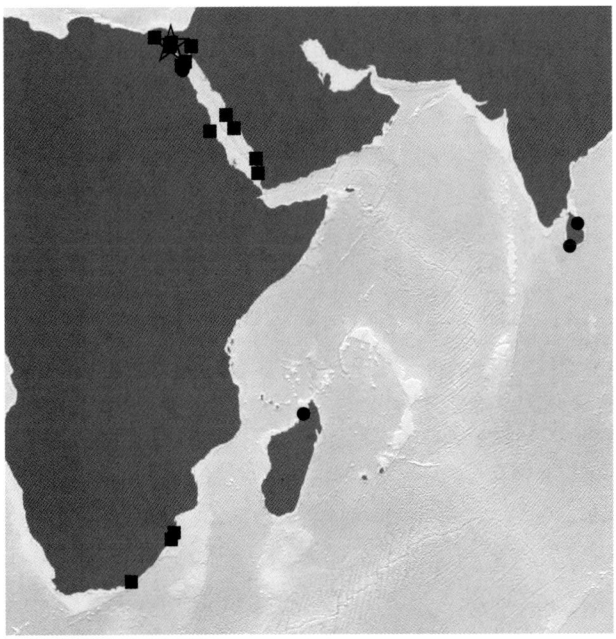

Figure 184. Distribution of *Scissurella reticulata*. ☆ type locality. ● specimen record. ■ literature record.

Literature records.

South Africa. Sodwana Bay, Zululand, 27.538°S, 32.679°E; Mapelane (Maphelana), St. Lucia, 28.383°S, 32.430°E; Port Alfred, 33.613°S, 26.891°E (NMSA D243) (Herbert, 1986: as *Sci. rota*).

Sudan. Port Sudan, 19.617°N, 37.233°E (Bandel, 1998). Port Sudan, 19.617°N, 37.233°E (NHMUK 1838) (Yaron, 1983).

Jordan. Aqaba, 29.517°N, 35.000°E (Bandel, 1998).

Indonesia. Satonda, 8.100°S, 117.750°E (Bandel, 1998).

Israel. Elat, 29.550°N, 34.950°E (HUJ 20.316, 1) (Yaron, 1983).

Egypt. Bir es Suweir, beach, 30.567°N, 30.567°E (IY 2306, 1; IY 4433, 9). Suez, 29.967°N, 32.550°E (MNHN, 4). Sinai, Ras Muhammad, 27.733°N, 34.250°E (IY 4298, 10).

Ghardaqa [= Hurghada], 27.283°N, 33.783°E (NHMW 62.379, 3; NHMW 62.528, 3). Ain Sukhna, 29.533°N, 32.400°E (NHMW 59.002, 19) (Yaron, 1983).

Saudi Arabia. Jeddah, 21.500°N, 39.167°E (NHMW 39.954, 1). Abulat Island, 19.967°N, 40.133°E (MNHN) (Yaron, 1983).

Yemen. Oreste Point, 16.367°N, 42.767°E. Al-Durayhimi, 14.667°N, 42.983°E (Dekker & Capelle, 1994).

Remarks. The description was based on an illustration in Audouin (1826), which had been misidentified by Audouin as *Sci. decussata*. Bouchet & Danrigal (1982: 14) noted that *Sci. reticulata* was a "new name" for *Sci. decussata* sensu Audouin, not d'Orbigny. The term new name has to be understood as giving a different name to a figured species identified with an existing name, and not as giving a *nomen novum* to a homonym. Yaron (1983) made a similarly confusing statement also referring to "*S. decussata* Audouin 1826 (*non* d'Orbigny, 1824)." There is no *Sci. decussata* Audouin, 1826, only a species misidentified by Audouin (1826) as *Sci. decussata* d'Orbigny, 1824.

The date of publication has been listed by Smith & England (1937).

The illustrations by Bandel (1998) for *Sci. reticulata* comprise two species. The specimen shown in pl. 2, figs 7–8 is *Sci. rota* from the Indian Ocean. The locality data for that specimen is given as Satonda, Indonesia, from where *Sci. rota* is not known. Similar species occurring in Indonesia (*Sci. mirifica*, *Sci. xandaros*, and *Sin. marrowi*) all are distinct in shoulder ornamentation and/or protoconch sculpture.

Scissurella rota Yaron, 1983

Figures 185–190, Color Plates 2F, 3A

Chresonymy.

Scissurella (*Scissurella*) *rota* Yaron, 1983: 268–270, pl. 3.
Scissurella rota: Bouchet, 1983: 5, text-fig. [copy figure Yaron, 1983].
Scissurella rota: Herbert, 1986: 622–623, figs 7 [protoconch], 25–28.
Scissurella rota: Lozouet, 1986: 108.
Scissurella rota: Vine, 1986: 172.
Scissurella rota: Herbert, 1987: 1–2.
Scissurella rota: Dekker & van Capelle, 1994: 125.
Scissurella rota: Bosch *et al.*, 1995: 28, text-fig. 4.
Scissurella rota: Bandel, 1998: 15, pl. 3, figs 5–7.
Scissurella rota: Geiger, 2003: 77.

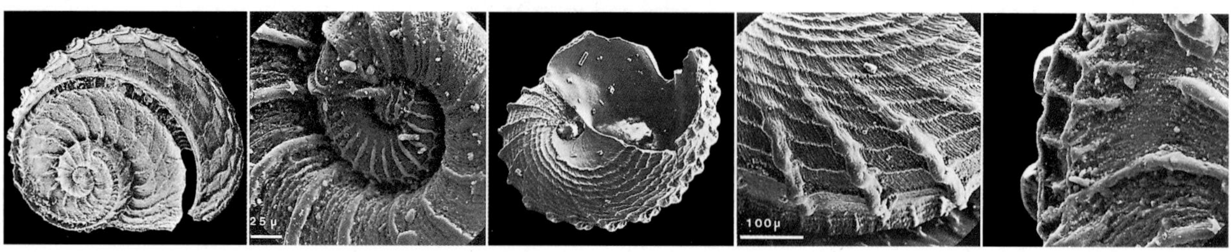

Figure 185. Original Illustration of *Scissurella rota*.

Scissurella rota: Bruyne, 2003: 48, text-fig.
Scissurella rota: Geiger, 2004: text-fig. p. 6.
Scissurella rota: Geiger, 2006c: text-fig. 3.
Scissurella rota: Zuschin *et al.*, 2009: 87, pl. 2, fig. 3.

Misidentifications.

Schismope padangensis: Franc, 1956: 21 [is *Sci. rota*: holotype].
Schismope padangensis: Mastaller, 1979: 29 [is *Sci. rota*: *fide* Yaron, 1983].
Scissurella reticulata: Bandel, 1998 (*partim*): pl. 2, figs 7–8 [is *Sci. rota*].
not *Scissurella rota*: Harasewych & Moretzsohn, 2010: 196, text-fig. [SEM is juvenile *Sin. singeri*].

Type material. Holotype (MNHN), 2.3 mm. Paratypes Ghardaqa [= Hurghada], Egypt, 27.283°N, 33.783°E (NHMW 62.594, 1). Ain Sukhna, Egypt, 28.750°N, 34.583°E (NHMW 58.902, 5).

Type locality. Abulat Island, southern Red Sea [Saudi Arabia, 19.967°N, 40.133°E] (OD). Shell sand of the north cape of Abulat Island (label by Franc).

Etymology. Rota, Latin for wheel.

Figure 186. *Scissurella rota*. 5–9 m, Reef Marker 13 North, between Bhanga Neck and Kosi Bay, Zululand, South Africa, 26.900°S, 32.867°E (NMSA S2862, 3). Scale bar shell = 0.5 mm. Scale bar protoconch = 100 µm.

Description. Shell medium-sized (to 1.6 mm), trochiform, depressed, globular. Protoconch of 0.75 whorl, fine axial sculpture, apertural varix connected to embryonic cap, apertural margin sinusoidal. Teleoconch I of 1–1.125 whorls, approximately 12 axial cords, half a dozen indistinct spiral lines. Teleoconch II of 1.25 whorls, suture impressed. Shoulder somewhat convex, approximately 19 strong, elevated axial cords, a few irregularly spaced, weaker spiral lines; intersection of strongest spiral cords with axial lines forming elevated points. Base with similar number of axial cords, axials cords forming elevated lamellae at periphery, approximately 13 spiral lines; on first 0.75 whorls, intersection of axial cords and spiral lines without modifications; on last 0.5 whorl, intersection of axial cords and spiral lines forming elevated, hollow spines; rarely axial lamellae with shingle-like projections at intersection with spirals. Base with carina bordering umbilicus with smooth walls. Selenizone near top margin of shell, more central than periphery, with strong lunules, keels strong, moderately to highly elevated; slit open, margins converging.

Operculum thin, round covering most of aperture, nucleus central, multispiral.

Radula. Rachidian tooth triangular-trapezoidal, base broad, tip slightly convex, five cusps, central cusp approximately 50% larger than lateral ones. Lateral teeth 1–3 similar, base projecting towards periphery, shaft slender, tip with 3–4 cusps in outer margin, cusp closest to tip largest. Lateral tooth 4 reduced, S-shaped, tip pointed. Lateral tooth 5 hypertrophied, broadened, 6–7 cusps on inner margin. Inner marginal teeth similar, elongated shafts, approximately four cusps on outer edge. Outer marginal teeth more symmetrical with up to a dozen teeth on either side of the tip. Radular interlock in central field moderate.

Comparisons. *Scissurella reticulata* from the Indian Ocean lacks the hollow spines, and has minute elevated points at intersection between axials and spirals, and has more axials and spirals.

Scissurella sudanica from Indian Ocean has more axials (~ 40), lacks the hollow spines on the base of mature specimens, and has a slightly sunken early whorls.

Scissurella spinosa from the Indo-Malayan Archipelago to the Central Pacific usually has a longer teleoconch I (1.125 vs. 1) and has more axials on teleoconch I (14–24 vs. 12).

Scissurella cebuana from the Indo-Malayan Archipelago to the Western Pacific has a smooth protoconch, a more narrow umbilicus with a distinct periumbilical cord, and spiral cords with distinct points on the shoulder.

Scissurella mirifica from the Indo-Malayan Archipelago to the Central Pacific has denser but finer axial and spiral cords, and a less distinct constriction of the base below the selenizone.

Scissurella quadrata from the Indo-Malayan Archipelago to the Central Pacific has a distinctly square aperture, has more numerous but finer axial lines on the shoulder (12 vs. 2–3), and a narrower umbilicus with distinct periumbilical cord.

Scissurella staminea from Japan has a teleoconch I of 1.25 whorls (vs. 1–1.125), and has a protoconch with more axial cords (9 vs. 14).

Scissurella xandaros from the Indo-Malayan Archipelago has a smooth protoconch, has more numerous spiral lines on the shoulder (>10 vs. 2–3), and has about twice as many spiral lines on the base.

Juvenile *Sin. marrowi* from the Indo-Malayan Archipelago has a protoconch with strong axials, has about twice as many spiral lines on the base, and has a broader and more inflated overall shell shape.

Juvenile *Sin. singeri* from the Indian Ocean has more than twice as many spiral lines on the base, denser axial cords, and a closed foramen when fully grown.

Distribution. Indian Ocean (Fig. 190).

Specimen records.

Egypt. Shurat el Gharqana, 28.117°N, 34.350°E (IY 4266, 2; IY 4912, 2). 23 m, Jackson Reef, Strait of Tiran, S Sinai Peninsula, N Red Sea, 27.017°N, 34.467°E (LACM

Figure 187. *Scissurella rota*. **A–B.** 20 m, Blue Hole, Dahab, Sinai, Egypt, 28.483°N, 34.533°E (DLG 100, 2). **C.** Mahé, Seychelles, 4.667°S, 55.500°E (NHMUK, 1). Scale bars shell = 0.5 mm. Scale bars protoconch = 100 μm.

85-111, 1; LACM 85-111, 1). 25 m, Dunraven Reef, Strait of Tiran, S Sinai Peninsula, N Red Sea, 27.200°N, 34.100°E (LACM 85-112, 1). Ghardaqa [= Hurghada], 27.283°N, 33.783°E (NHMW 62.594, 1). 18 m, Amphoras dive site, off Ras Um Sid, SE side of Sinai Peninsula, 27.867°N, 34.333°E (LACM 88-119, 7). 10–15 m, Coral Gardens, Strait of Tiran, S Sinai Peninsula, 27.950°N, 34.533°E (LACM 85-114, 1). 10 m, E side of Jackson Reef, Straits of Tiran, Gulf of Aquaba, SE side of Sinai

Figure 188. *Scissurella rota.* **A.** 8–14 m, 1–2 km N of Leven Point, Raggie Reef, Leadsman Shoal, Zululand, South Africa, 34.417°S, 21.367°E (NMSA E2716, 1: complete). Radula shown in Figure 189. **B–C.** 5–9 m, Reef marker 13 North, between Bhanga Neck and Kosi Bay, Zululand, South Africa, 26.900°S, 32.867°E (NMSA S2862, 3). Scale bars shell = 0.5 mm. Scale bars protoconch = 100 μm.

Peninsula, 28.000°N, 34.467°E (LACM 88-118, 1). 20 m, Blue Hole, Dahab, Sinai, 28.483°N, 34.533°E (DLG 100, 2). 10–15 m, The Alternatives, Strait of Tiran, S Sinai Peninsula, N Red Sea, 28.733°N, 34.217°E (LACM 85-110, 1). Ain Sukhna, 28.750°N, 34.583°E (NHMW 58.902, 5). Sinai, Gulf of Aqaba, Nuweiba, 28.967°N, 34.617°E (AMS C.380207, 4). Ras Burqa, 29.214°N, 34.739°E (IY 4891, 5). Bir es Suweir, 30.567°N, 32.117°E (IY 2797, 1). 10 m, 8 km N of Marsa Alam, Coral Cove, Al Bahr al Ahmar, 25.143°N, 34.861°E (NMR 993000060065, 1). 0 m, 5 km S of Râs Adabîya, Gulf of Suez, 29.867°N, 32.500°E (RMNH, 1).

Saudi Arabia. Abulat Island, 19.967°N, 40.133°E (MNHN, 1).

Seychelles. Mahé, 4.667°S, 55.500°E (NHMUK, 1). 38 m, NE of Mahé, 4.217°S, 55.567°E (RMNH, 5). 45 m, W of Aride Island, 4.217°S, 55.567°E (RMNH, 1).

Tanzania. 1 km S of Kunduchi Beach, 11 km N of Dar-es-Salaam, 6.683°S, 39.250°E (AMS C.380212, 1). 0–1 m, Mbudya Island, ca. 16 km N of Dar-es-Salaam, 6.650°S, 39.250°E (AMS C.380205, 5). Mbudya Island, 16 km N of Dar-es-Salaam, 6.650°S, 39.250°E (AMS C.404525, 1, 2: complete).

France. 400–520 m, Mayotte, S reef, 13.084°S, 45.119°E (MNHN, 1). 1300–1480 m, E Bandele Reef, Mayotte, 12.933°S, 45.303°E (MNHN, 1). 500 m, E Bandele Reef, Mayotte, 12.895°S, 45.268°E (MNHN, 5). 25–35 m, NE of the Monzambique channel, Mayotte, N Sada Passage, 12.892°S, 44.952°E (MNHN, 2). 700 m, Mayotte, 12.792°S, 44.943°E (MNHN, 1). 705 m, NE Pamanzi Island, Mayotte, 12.788°S, 45.198°E (MNHN, 1). 15–20 m, N Pamanzi Island, Mayotte, 12.752°S, 45.298°E (MNHN, 1). 15–20 m, N Pamanzi Island, Mayotte, 12.752°S, 45.298°E (MNHN, 1). 10–20 m, Northern Reef, 12.577°S, 45.087°E (MNHN, 4).

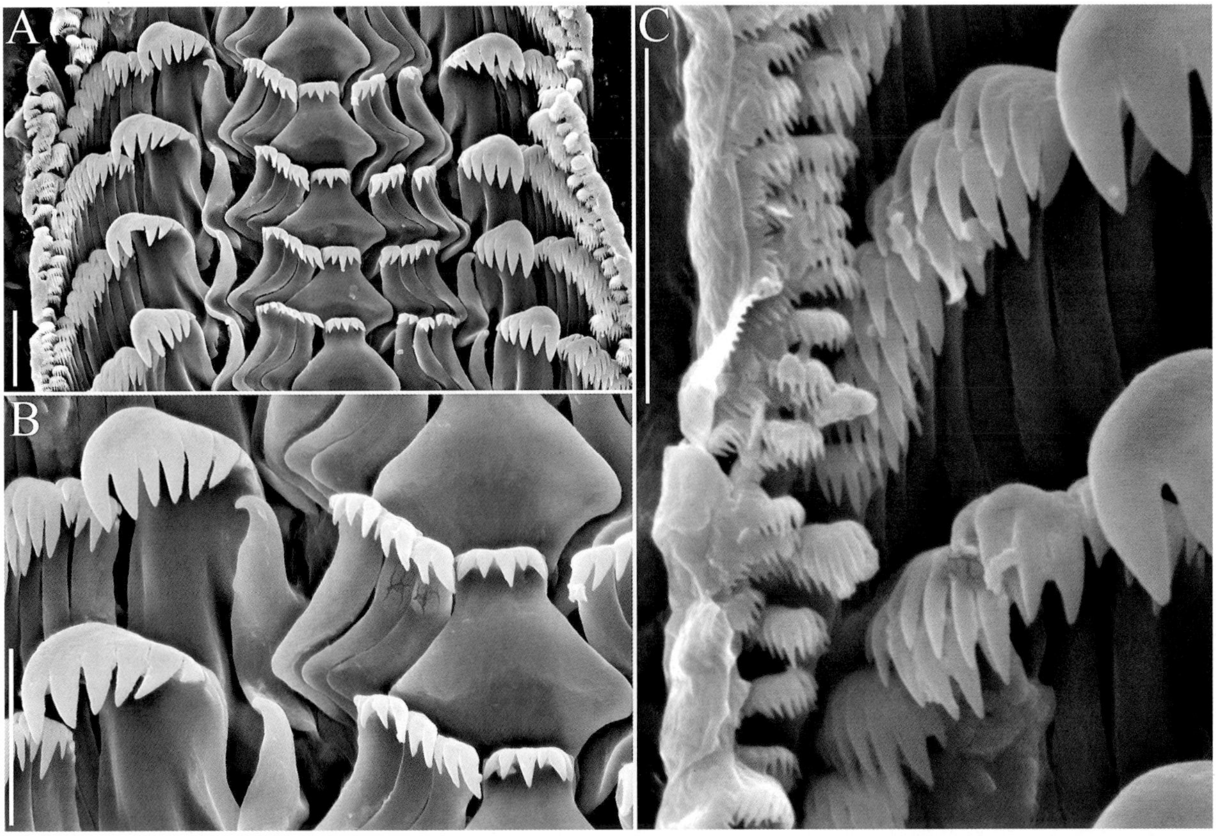

Figure 189. *Scissurella rota*. Radula of shell shown in 188A. 8–14 m, 1–2 km N of Leven Point, Raggie Reef, Leadsman Shoal, Zululand, South Africa, 34.417°S, 21.367°E (NMSA E2716, 1: complete). **A.** Whole width of radula. **B.** Central field enlarged. **C.** Marginal teeth. Scale bars = 10 µm.

Mascarene Islands. St. Brandon, Ile Paul, 16.633°S, 59.633°E (AMS C.461465, 2).
Mozambique. 0 m, da Barra Point, Inhambane, 23.792°S, 35.539°E (DLG 2325, 3).
Madagascar. 25 mi SSW of Tulear, N and NW coasts of Nosy Satrana Island, 23.726°S, 43.629°E (MCZ 315233, 1). 1 m, Nosi-Tanga, Nossi-Be, 23.600°S, 47.600°E (USNM 719733, 2). From offshore reef, Tulear, 23.417°S, 43.700°E (USNM 718424, 4). 51 m, 40 mi off Tulear, 23.417°S, 43.700°E (USNM 761256, 1). 33 m, 40 mi off Tulear, 23.417°S, 43.700°E (USNM 761213, 1). Tulear, 23.350°S, 43.667°E (MNHN, 1; MNHN, 1). 82 m, ca. 40 mi WSW off Tulear, 23.317°S, 43.600°E (USNM 717504, 14; USNM 717564, 3). 33 m, ca. 40 mi WSW off Tulear, 23.300°S, 43.600°E (USNM 717397, 4). 0–3 m, Baye de Ranobe, near Mora Mora, N of Tulear, Toliara Province, 23.083°S, 43.500°E (LACM 89-50, 1). Ile St. Marie, Ambodifototra, from harbor breakwater, S to Ilot Madame, 16.983°S, 49.850°E (MCZ 262691, 10). Nossi-Be, Point, Fievre, 13.400°S, 48.283°E (USNM 719459, 1). Nossi-Be, Point Tafondro, Ambataloaka Beach, 13.333°S, 48.250°E (USNM 719628, 5; USNM 719665, 1). 0 m, Nossi-Be, Ambariobe, 13.333°S, 48.250°E (USNM 761852, 2). 1–3 m, at Andilana, NW side Nosy Bé, Antseranana Province, 13.217°S, 48.233°E (LACM 89-53, 8). 0 m, Point Fiever, Nossi-Be, 13.400°N, 48.283°E (USNM 719488, 12). 14 m, out from Hellville, S side Nosy Bé, Anseranana Province, 13.450°N, 48.250°E (LACM 89-55, 9). 0–3 m, S and E sides Tany Kely, off S side Nosy Bé, Antseranana Province, 13.500°N, 48.233°E (LACM 89-54, 8). 4–6 m, St. Luce, SW Soiillac Islet, ATIMO VATAE TA34, 24.753°S, 47.200°E (MNHN, 1). 3 m, entrance to Lake Mananivo, ATIMO VATAE TS20, 24.766°S, 47.111°E (MNHN, 2). 0 m, St. Luce, ATIMO VATAE TM19, 24.785°S, 47.198°E (MNHN, 14). 12–17 m, NE of Lokaro, ATIMO VATAE TP07, 24.935°S, 47.125°E (MNHN, 5). 7 m, Lokaro Bay, ATIMO VATAE TS10, 24.946°S, 47.110°E (MNHN, 1). 5–6 m, S Lokaro Bay, ATIMO VATAE TS09, 24.950°S, 47.106°E (MNHN, 6). 0 m, Crique at the NE lighthouse of Evatra, 24.978°S, 47.098°E (MNHN, 1). 2–7 m, E Point Evatra, ATIMO VATAE TB08, 24.986°S, 47.090°E (MNHN, 10). 22 m, Point Evatra, ATIMO VATAE TB01, 24.996°S, 47.095°E (MNHN, 13). 12–16 m, in front of Flacourt Lighthouse, ATIMO VATAE TV07, 25.021°S, 47.003°E (MNHN, 8). 18 m, Flacourt Lighthouse, ATIMO VATAE TB02-03, 25.021°S, 47.008°E (MNHN, 21). 18 m, in front of Flacourt lighthouse, ATIMO VATAE TS02, 25.021°S, 47.008°E (MNHN, 2). 4–5 m, Flacourt Point, ATIMO VATAE TB12, 25.025°S, 47.000°E (MNHN, 2). 450–455 m, SE Barrow Point, ATIMO VATAE CP3592, 25.033°S, 43.966°E (MNHN, 1). 11–12 m, in front of Monseigneur Bay, ATIMO VATAE TB04, 25.035°S, 47.001°E (MNHN, 1). 0 m, Monseigneur Beach, ATIMO VATAE TM01, 25.035°S, 46.998°E (MNHN, 6). 22–24 m, E of Cape Antsirabe, ATIMO VATAE TS04, 25.038°S, 47.005°E (MNHN, 1). 4–5 m, Libanona Beach, ATIMO VATAE TB07, 25.041°S, 46.995°E (MNHN, 3). 17–21 m, False Bay of the Ballons, ATIMO VATAE TP03, 25.049°S, 46.986°E (MNHN, 1). 113–135 m, SE Barrow Point, ATIMO VATAE DW3588, 25.050°S, 44.000°E (MNHN, 1). 7 m, Port d'Ehoala, ATIMO VATAE TP28, 25.061°S, 46.963°E (MNHN, 10). 2–7 m, Port d'Ehoala, ATIMO VATAE TP24, 25.062°S, 46.960°E (MNHN, 1). 4–5 m, Cape Ranavalona, Tebava, ATIMO VATAE TB06, 25.073°S, 46.951°E (MNHN, 32). 0 m, Cape Ranavaiona, ATIMO VATAE TM03, 25.075°S, 46.961°E (MNHN, 1). 0 m, Gallions Bay, ATIMO VATAE TM21, 25.148°S, 46.756°E (MNHN, 9). 5–6 m, Galions Bay, ATIMO VATAE TB11, 25.153°S, 46.756°E (MNHN, 1). 5–6 m, Galions Bay, East Point, ATIMO VATAE

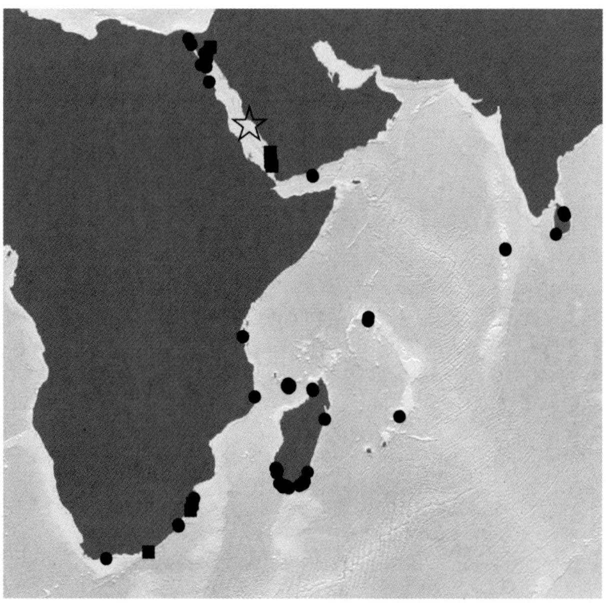

Figure 190. Distribution of *Scissurella rota*. ☆ type locality. ● specimen record. ■ literature record.

TS17, 25.153°S, 46.756°E (MNHN, 1). 10 m, Entrance to East Bay of Galions, ATIMO VATAE TB10, 25.155°S, 46.755°E (MNHN, 19). 13 m, Gallions Bay, East Point, ATIMO VATAE TS15, 25.156°S, 46.755°E (MNHN, 2). 53–54 m, SE Cape Andavaka, ATIMO VATAE CP3549, 25.281°S, 46.521°E (MNHN, 2). 12–14 m, E Cape Malaimpolka, 25.308°S, 44.621°E (MNHN, 1). 0 m, Cape Malaimploka, ATIMO VATAE BM 15, 25.343°S, 44.805°E (MNHN, 1). 14–18 m, Lavanono sector, ATIMO VATAE BB03, 25.440°S, 44.935°E (MNHN, 1). 6–9 m, Lavanono sector, ATIMO VATAE BP 23, 25.443°S, 44.936°E (MNHN, 1). 203–210 m, SE Barrow Point, ATIMO VATAE CP3584, 25.466°S, 44.416°E (MNHN, 1). 12 m, Albatross Rock, ATIMO VATAE BS 12, 25.476°S, 44.946°E (MNHN, 1). 0 m, Andramara, ATIMO VATAE BM 10, 25.480°S, 44.971°E (MNHN, 1). 0 m, Pasput, Ampasipotsy, ATIMO VATAE BM 11, 25.516°S, 45.066°E (MNHN, 1). 18–20 m, Cape St. Marie, ATIMO VATAE TA26, 25.565°S, 45.115°E (MNHN, 2). 0 m, NW St. Marie Cape, ATIMO VATAE BM 17, 25.596.000°S, 45.139.000°E (MNHN, 1; MNHN, 1). 1 5 m, Libanona Beach, TB07, 25.041°S, 46.995°E (MNHN, 2). 10 m, Lavanono, 25.429°S, 44.939°E (DLG 1749, 12).

South Africa. 8–14 m, 1–2 km N of Leven Point, Raggie Reef, Leadsman Shoal, Zululand, 34.417°S, 21.367°E (NMSA E2716, 1: complete). Port Alfred, Eastern Cape, 33.600°S, 26.900°E (NMSA D243). Mission Rock, Santa Lucia, 28.383°S, 32.417°E (HLC, 1). 5–9 m, Reef marker 13 North, between Bhanga Neck and Kosi Bay, Zululand, 26.900°S, 32.867°E (NMSA S2862, 3).

Maldives. 17–40 m, Villingili Island, N Male Atoll, 4.200°N, 73.450°E (LACM 83-8, 1). 25–35 m, 2 km N of Baros Island, N Male Atoll, 4.317°N, 73.417°E (LACM 83-9, 1).

Sri Lanka. 1 m, Hikkaduawa, 6.137°N, 80.095°E (LACM 84-9, 2). 29–91 m, Koddiyar Bay, E coast, 8.500°N, 81.242°E (AMS C.404919, 1). 0–2 m, Red Rock Beach, N of Nilaveli, 8.825°N, 81.100°E (LACM 84-6, 1).

Literature records.

Jordan. Aqaba, 29.517°N, 35.000°E (Bandel, 1998).

Yemen. Al-Durayhimi, 14.667°N, 42.983°E; As-Salif Peninsula, S coast, 15.183°N, 42.817°E; Oreste Point, 16.367°N, 42.767°E (Dekker & de Ceuninck van Capelle, 1994).

South Africa. Mapelane (Maphelana), St. Lucia, 28.400°S, 32.417°E; Sodwana Bay, Zululand, 27.550°S, 32.667°E (Herbert, 1986).

Remarks. The holotype had been identified initially as *Schismope padangensis*, and was the specimen mentioned by Franc (1956).

Very fresh specimens from South Africa have a thick, brown periostracum (Color Plate 2F).

Scissurella skeneoides new species

Figures 191–194

Chresonymy.

Scissurella sp. 2: Spencer *et al.*, 2009: 202.

Type material. Holotype (NMNZ M.174853). Paratypes from type locality (NMNZ M.301402, 6). Paratypes 33 m, off Spirits Bay, New Zealand, 34.398°S, 172.760°E (NMNZ M.174830, 3).

Type locality. 29 m, Spirits Bay, New Zealand, 34.405°S, 172.833°E.

Etymology. Reminiscent of a skeneid.

Description. Shell small (0.72 mm), trochiform, depressed. Protoconch of one whorl, strong axial sculpture, no apertural varix, apertural margin shallow sinusoidal. Teleoconch I of 1.0–1.1 whorls, 27–32 undulating axial lamellae, no spirals, interstices with very fine axial growth marks. Teleoconch II of 0.5 whorl, suture impressed. Shoulder with same density of undulating axial lamellae, no spirals. Base with moderate constriction below selenizone same sculpture as on shoulder. Umbilicus open, bordered by strong carina towards base, no distinct walls. Aperture rounded, roof

Figure 191. *Scissurella skeneoides* new species. Holotype. 29 m, Spirits Bay, New Zealand, 34.405°S, 172.833°E (NMNZ M.174853). Scale bar shell = 0.5 mm. Scale bar protoconch = 100 μm.

overhanging, Selenizone above periphery; keels of moderate strength, low; lunules distinct; slit open margins parallel.

Operculum (Fig. 193E) thin, round, multispiral with central nucleus.

Radula (Fig. 193A–D). Rachidian tooth with central cusp on tip largest, three on each side in semicircle. Lateral teeth 1–3 similar, apical cusp largest, 2–3 cusps of decreasing size on outer edge of tip. Lateral tooth 4 reduced, hook-shaped, with minute cusp below apical cusp. Lateral tooth 5 enlarged by broadening, apical cusp largest, seven cusps along inner edge of tip. Inner marginal teeth with broad triangular tip, apical cusp largest, approximately 5–6 cusps on each side, no food

Figure 192. *Scissurella skeneoides* new species. Paratypes. **A–B.** 29 m, Spirits Bay, New Zealand, 34.405°S, 172.833°E (NMNZ ex174853, 7). **C.** 30 m, off Spirits Bay, New Zealand, 34.398°S, 172.760°E (NMNZ M.174830, 3). Scale bars shell = 0.5 mm. Scale bars protoconch = 100 μm.

groove. Outer marginal teeth with spoon-shaped tip, many fine cusps along edge. Radular interlock moderate.

Animal with eyes. Further details unknown.

Comparisons. *Scissurella condita* from the Miocene of New Zealand has a protoconch that is distinctly below the shoulder of the teleoconch and has about half as many axial cords.

Scissurella prendrevillei from New Zealand has half as many but stronger axial cords to lamellae, in larger specimens the whorls attach lower on the previous whorl, and the protoconch has many fine axial cords.

Scissurella azorensis from the subtropical northeastern Atlantic and the Mediterranean and *Sci. maraisorum* from the Indian Ocean have a protoconch with many dense axials and have stronger but half as many axial cords.

Juvenile *Sin. mechanica* from New Zealand has a protoconch with strong axials that form an apical spiral ridge on the protoconch, the protoconch has a terminal varix, and the protoconch is elevated from the teleoconch whorls.

Juvenile *Sin. semicostata* from the subtropical northeastern Atlantic islands has about half as many axial cords on the teleoconch and a teleoconch I of <1 whorl.

Satondella minuta from the Indo-Malayan Archipelago has a sunken protoconch, and fewer more elevated and prominent axial lamellae, and a closed foramen with prominently elevated keels.

Distribution. Northern New Zealand (Fig. 194).

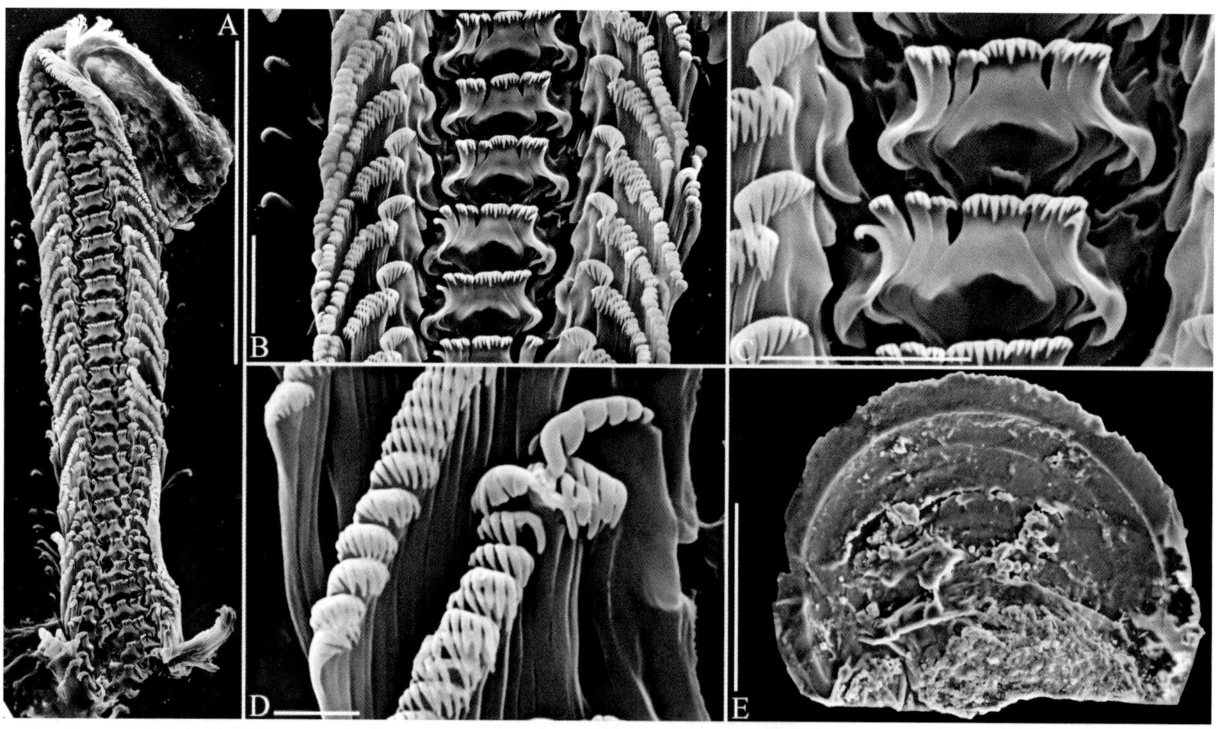

Figure 193. *Scissurella skeneoides* new species. Radula and operculum. **A.** Overview of radula. Scale bar = 100 μm. **B.** Entire width of radula. Scale bar = 10 μm. **C.** Central field enlarged. Scale bar = 10 μm. **D.** Marginal teeth. Scale bar = 2 μm. **E.** Operculum. Scale bar = 100 μm. **A–C.** 23 m, off Spirits Bay, Northland, North Island, New Zealand, 34.427°S, 172.803°E (NMNZ M.148569, 5). **D–E.** 246–291 m, Middlesex Bank, NW of Three Kings Islands, New Zealand, 34.033°S, 171.733°E (NMNZ M.150310, 2).

Specimen records.

New Zealand. 622 m, N of Three Kings Islands, 34.016°S, 172.117°E (NMNZ M.150309, 1). 246–291 m, Middlesex Bank, NW of Three Kings Islands, 34.033°S, 171.733°E (NMNZ M.150310, 2). 33 m, between Great Rock and Farmer Rocks, Three Kings Islands, 34.150°S, 172.167°E (NMNZ M.177151, 13). 33 m, reef between Great Island and Farmer Rock, Three Kings Islands, 34.150°S, 172.167°E (NMNZ M.160311, 100). 23 m, off N face of Hinemoa Island, Three Kings Islands, 34.180°S, 172.043°E (NMNZ M.150312, 1). 91 m, off Three Kings Islands, 34.183°S, 172.167°E (NMNZ M.150307, 1). 192–202 m, SE of Great Island, Three Kings Islands, 34.235°S, 172.150°E (NMNZ M.150306, 1). 63 m, E of Spirits Bay, 34.333°S, 172.763°E (NMNZ M.174824, 1). 30 m, N of Kerr Point, Tom Bowling Bay, 34.390°S, 172.986°E (NMNZ M.174843, 3; NMNZ M.174827, 5). 30 m, off Spirits Bay, 34.397°S, 172.760°E (NMNZ M.174830, 3). 29 m, Spirits Bay, 34.405°S, 172.833°E (NMNZ M.174853, holotype; NMNZ M.301402, 6 paratypes). 23 m, off Spirits Bay, Northland, North Island, 34.426°S, 172.803°E (NMNZ M.148569, 5). 4 m, Waewaetorea Passage, Bay of Islands, North Island, 35.206°S, 174.222°E (NMNZ M.049366, 3).

Figure 194. Distribution of *Scissurella skeneoides* new species. ☆ type locality. ● specimen record.

Scissurella spinosa Geiger & Jansen, 2004

Figures 195–198

Chresonymy.

Scissurella "spikey": Jansen, 1999: 53, figs 37–39.

Scissurella spinosa Geiger & Jansen, 2004b: 33–38, figs 17e [map], 18–20 [shell, radula].

Scissurella spinosa: Hickman & Porter, 2007: 10–17, figs 2, 3a–c.

Scissurella spinosa: Geiger & Thacker, 2008: fig. 2.

Scissurella spinosa: Tröndlé & Boutet, 2009: 12.

Type material. Holotype (AMS C.348694. Fig. 195). Paratypes. 5 m, Christensen Research Institute, Madang, New Guinea, 5.117°S, 145.817°E (AMS C.376853, 2. Fig. 196B).

Type locality. 27 m, Great Barrier Reef, Swain Reefs, Bylund Gillett Cay, Queensland, Australia, 21.717°S, 152.417°E (OD).

Etymology. Spinosus, Latin for thorny, referring to the pronounced spines at the intersection of axials and spirals both on the shoulder and the base (OD).

Description. Shell medium-sized (1.2 mm), trochiform to depressed trochiform. Protoconch 0.75 whorl, sculpture of fine axials, varix present, connected to embryonic cap, apertural margin sinusoidal. Teleoconch I of >1 to 1.25 whorls, approximately 14–24 axials, interstices with fine irregular axial lamellae, on outer shoulder from position of selenizone towards base on last 0.25 whorl up to five fine spirals. Teleoconch II of 1.25 whorls, suture impressed. Shoulder with approximately 19 distinct axial cords growing fainter on last 0.25 whorl, intersected by 2–3 fainter spiral lines in the middle third of shoulder, at intersection of axials and spirals strong knobs; additional fine spirals adjacent to selenizone not forming knobs. Base rounded to somewhat inflated, same number of axials as shoulder, intersected by approximately 8–9 spirals, top 2–3 spiral lines forming distinct knobs at intersection with axials, second spiral approximately at level of suture. Umbilical wall at angle with base, carina faint, wall of umbilicus smooth, flat, umbilicus of medium width. Selenizone with distinct keels; walls of keels in fresh specimens finely undulating, lunules distinct, not coordinated with axials, slit open. Aperture subquadrate roof strongly overhanging.

Operculum (Fig. 197A–B) thin, corneous, multispiral, nucleus central.

Radula (Fig. 197C–F). Rachidian tooth triangular, central cusp largest, three smaller cusps on both sides. Lateral teeth 1–3 similar with three cusps. Lateral tooth 4 hook-shaped. Lateral tooth 5 enlarged, with five cusps, triangular, edge of cusps at angle to base. Inner marginal teeth with triangular tip, apical cusp largest, 2–4 on

Figure 195. *Scissurella spinosa*. Holotype. 27 m, Great Barrier Reef, Swain Reefs, Bylund Gillett Cay, Queensland, Australia, 21.717°S, 152.417°E (AMS C.348694). Scale bar shell = 1 mm. Scale bar protoconch = 100 µm.

each side; outer marginal teeth spoon-shaped, with many fine bristles along tip; no food groove on shaft. Radular interlock of central field moderate.

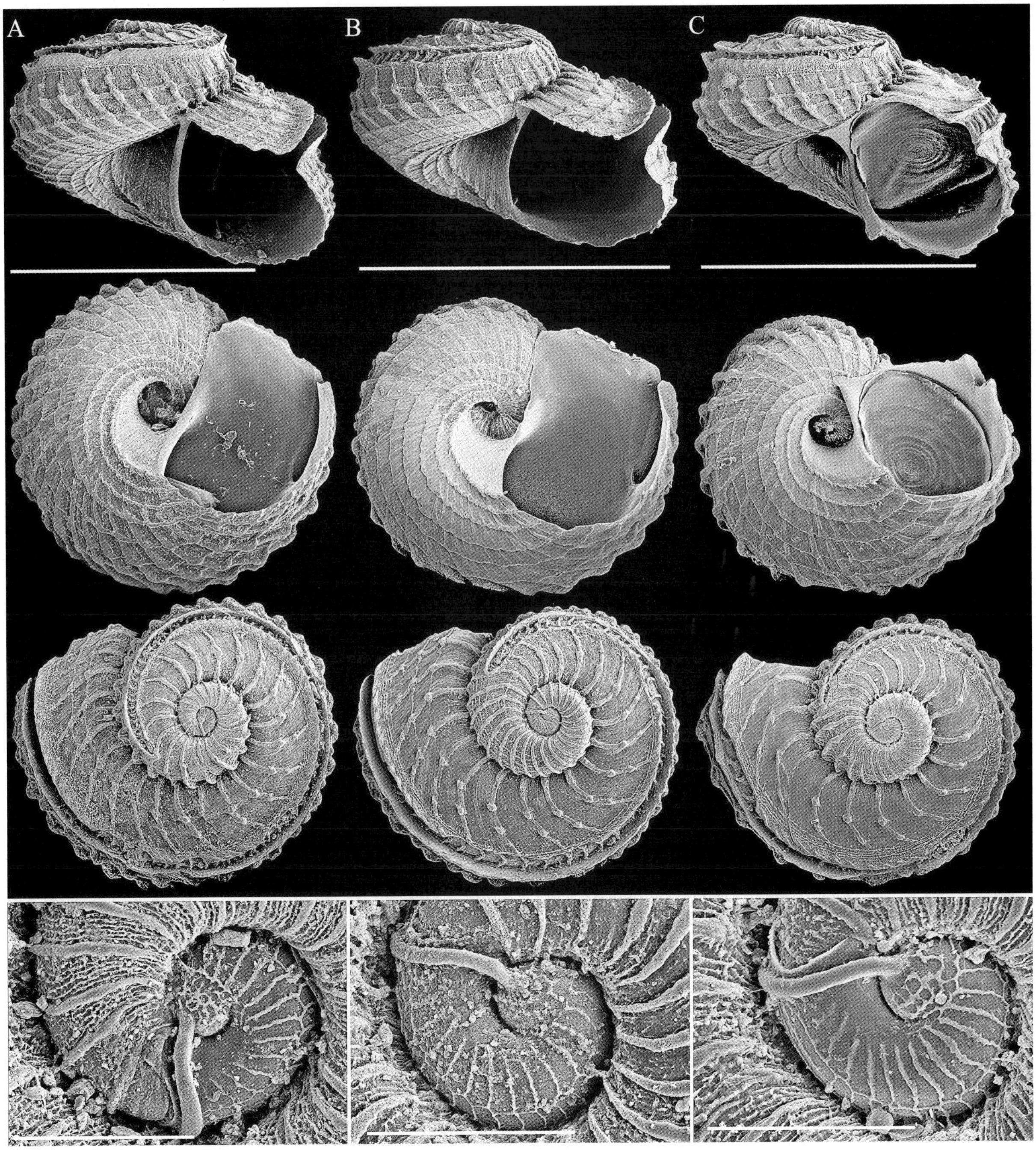

Figure 196. *Scissurella spinosa*. **A.** 64 m, Great Barrier Reef, Swain Reefs, 3 km NE of W side of Bylund Gillett Cay, Queensland, Australia, 21.700°S, 152.433°E (AMS C.379073, 20+). **B.** 5 m, Christensen Research Institute, Madang, New Guinea, 5.117°S, 145.817°E (AMS C.376853, 2). **C.** 2 m, Bohol, Panglao Island, SW side, Philippines, 9.600°N, 123.750°E (AMS C.376854, 1). Radula shown in Figure 197E, F. Scale bars shell A, B = 1 mm. Scale bar shell C = 0.5 mm. Scale bars protoconch = 100 μm.

Comparisons. *Scissurella spinosa* is characterized by the rather coarse sculpture of strong axials intersected by weaker spirals that form characteristic points at their intersections. These spiny points are also found on the shoulder, in one or usually two spiral rows.

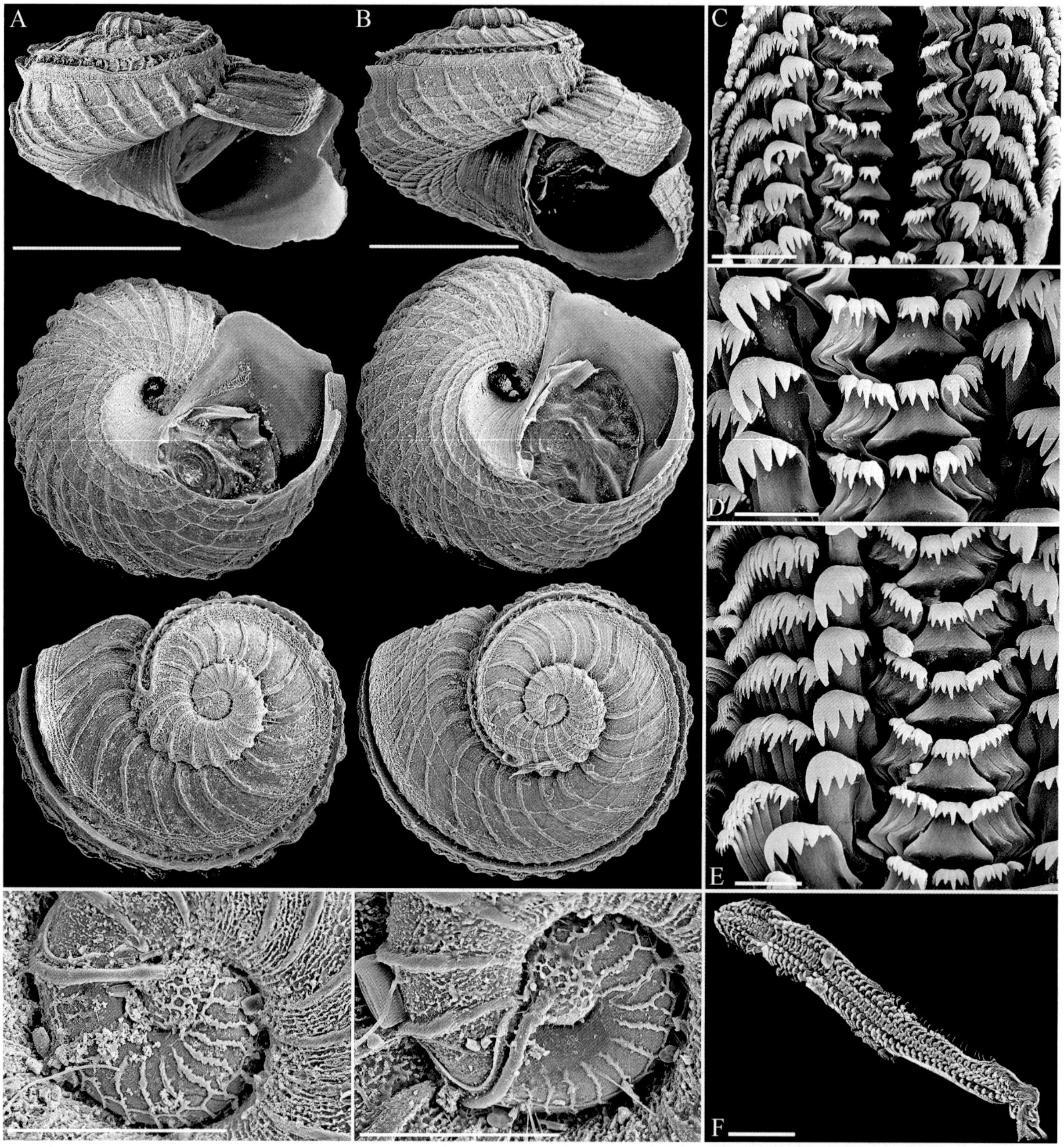

Figure 197. *Scissurella spinosa*. **A–B.** Shells. Scale bars shell = 0.5 mm. Scale bars protoconch = 100 µm. **A.** 8 m, Tiahura Island, NW Moorea Island, French Polynesia, 17.483°S, 149.917°W (AMS C.402983, 12). **B.** Point Aroa ou de Tiaia, NE corner Moorea Island, French Polynesia, 17.467°S, 149.767°W (AMS C.376855, 4). Radula shown in Figure 197CD. **C–F.** Radulae. **C–D.** Point Aroa ou de Tiaia, NE corner Moorea Island, French Polynesia, 17.467°S, 149.767°W (AMS C.376855, 4). Shell shown in Figure 197B. **E–F.** 2 m, Bohol, Panglao Island, SW side, Philippines, 9.600°N, 123.750°E (AMS C.376854, 1). Shell shown in Figure 196C. **C–E.** Radulae. Scale bars = 20 µm. **F.** Scale bar = 100 µm.

Scissurella cebuana from the Indo-Malayan Archipelago to the Western Pacific has a smooth protoconch, has more pronounced points the intersection of axials and spirals (particularly on shoulder), and has a more narrow umbilicus.

Scissurella evaensis from the Indo-Malayan Archipelago to the Central Pacific has its typical spiral row of elevated axial lamellae on the mid-base and a much more narrow umbilicus.

Scissurella lorenzi from the Indo-Malayan Archipelago to the Western Pacific typically has distinct light colored axial bands on the shoulder and has a more narrow umbilicus with distinct periumbilical cord.

Scissurella mirifica from the Indo-Malayan Archipelago to the Central Pacific has less distinct sculpture, without the raised points at the intersection of axials and the more numerous but finer spiral lines, and has a weaker constriction of the base below the selenizone.

Scissurella quadrata from the Indo-Malayan Archipelago to the Central Pacific has much weaker and denser sculpture, has a distinct periumbilical cord around the more narrow umbilicus, and has a somewhat more square aperture.

Scissurella rota from the Indian Ocean has a shorter teleoconch I (1.125 vs. 1 whorls) and has fewer axials on teleoconch I (12 vs. 14–24).

Scissurella staminea from Japan is broader in overall outline, lacks the spiny protrusions on the shoulder, and has a protoconch with strong axials as opposed to the weak axials in *Sci. spinosa*.

Scissurella xandaros from the Indo-Malayan Archipelago and the Austral Islands has a smooth protoconch, an overall more globular shell, a smooth protoconch, and fewer points formed at the intersection of axials and spirals on the shoulder (2–3 vs. >6).

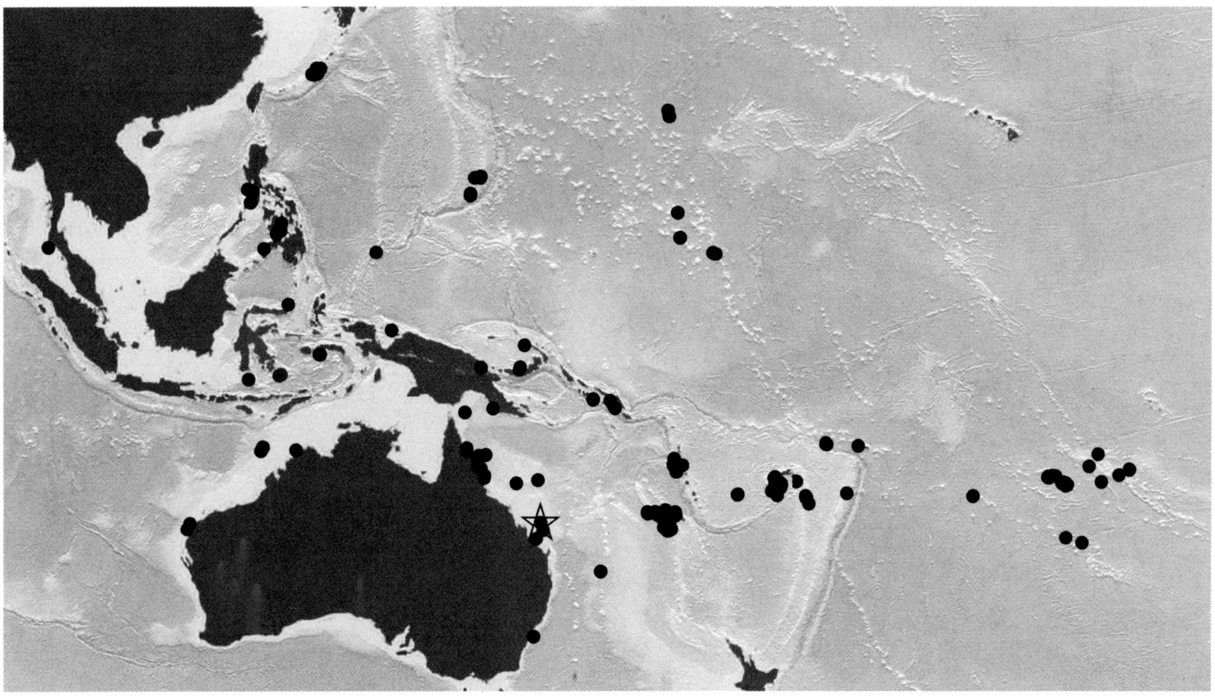

Figure 198. Distribution of *Scissurella spinosa*. ☆ type locality. ● specimen record.

Juvenile *Sin. ferriezi* of broad Indo-Malayan Archipelago to Central Pacific distribution has a smooth protoconch, and has about twice as many spiral lines on shoulder and base.

Juvenile *Sin. marrowi* from the Indo-Malayan Archipelago has a protoconch with strong axials and has about twice as many spiral lines on shoulder and base.

Juvenile *Sin. singeri* from the Indian Ocean has a smooth protoconch, at least twice as many but finer axial cords on shoulder and base, and as juveniles often a much more prominent constriction of the base below the selenizone.

Juvenile *Sin. zimmeri* from Easter and Pitcairn Islands has a protoconch with strong axial cords, has a shorter teleoconch I (0.75 vs. 1 whorl), and usually has stronger axial cords without points at the intersection of axials and spirals.

Distribution. Indo-Malayan Archipelago to Central Pacific (Fig. 198). 0–73 m, live, in coral rubble and algae.

Specimen records.

Thailand. 0 m, S tip of Phromthep Cape, Phuket, 7.758°N, 98.322°E (LACM 85-6, 1).

Indonesia. 0 m, Biak, Irian Jaya, 1.150°S, 136.050°E (LACM 88-43, 1). 5–20 m, Sulawesi, Bunaken and Siladen Islets, 1.603°N, 124.767°E (LACM 88-55, 2). 3 m, Latuhalat, S Coast, Leitimur, Ambon, Moluccas, 3.775°S, 128.102°E (RMNH 42314, 1)

Philippines. 49 m, St. 5131, S, off Panabutan Point, W Mindanao, 7.586°N, 122.121°E (USNM ex 262931, 1). 3–6 m, San Juan Reef, Sal Juan, Siquijor Island, Visayan Islands, 9.157°N, 123.485°E (USNM 808962, 1). 15–20 m, Pamilacan Island, 9.490°N, 123.933°E (MNHN, 1: complete). 20 m, off Bohol Island, Pamilacan Island, 9.500°N, 123.917°E (AMS C.376858, 1). 90–110 m, Balicasag Island, 9.515°N, 123.687°E (MNHN, 2: complete; MNHN, 1: complete). 80–140 m, Balicasag Island, 9.538°N, 123.705°E (MNHN, 1: complete). 4 m, Panglao Island, 9.587°N, 123.840°E (MNHN, 1: complete). 2 m, Bohol, Panglao Island, SW side, 9.600°N, 123.750°E (AMS C.376854, 1). 1–4 m, Sungcolan Bay, Panglao, 9.642°N, 123.820°E (MNHN, 2: complete). 92–97 m, 12.517°N, 120.650°E (MNHN, 1). 0 m, Maricaban Island, 13.649°N, 120.890°E (USNM 244610, 1). 26 m, Bohol Island, off Balicasag Island, 9.517°N, 123.683°E (AMS ex C.348621, 1). 20 m, Looc, Panglao Island, PANGLAO 2004 B32, 9.430°N, 123.743°E (MNHN, 7). 95–128 m, Cervera Shoal, Pamilacan Island, PANGLAO 2004 T36, 9.488°N, 123.858°E (MNHN, 1). 90–98 m, Balicasag Island, PANGLAO 2004 L69-73, 9.511°N, 123.682°E (MNHN, 1). 19–20 m, Balicasag Island, PANGLAO 2004 B37, 9.515°N, 123.680°E (MNHN, 1). 90–100 m, front of PTA Compound, Panglao Island, PANGLAO 2004 L41, 9.521°N, 123.686°E (MNHN, 1). 309–342 m, off Balicasag Island, Bohol Sea, CP2399, 9.528°N, 123.698°E (MNHN, 1). 164–176 m, off Balicasag Island, Bohol Sea, PANGLAO 2005 DW2339, 9.531°N, 123.728°E (MNHN, 2). 17–25 m, Pntod Lagoon 1, Panglao Island, PANGLAO 2004 B39, 9.546°N, 123.701°E (MNHN, 3). 8–14 m, Alona Reef, Panglao Island, PANGLAO 2004 B1, 9.550°N, 123.775°E (MNHN, 2). 5 m, Alona Reef, Panglao Island, PANGLAO 2004 B2, 9.550°N, 123.775°E (MNHN, 3). 8–10 m, Napaling, Panglao Island, PANGLAO 2004 B9, 9.551°N, 123.733°E (MNHN, 5). 24 m, BBC Point, Panglao Island, PANGLAO 2004 B4, 9.553°N, 123.805°E (MNHN, 1). 8 m, Arco Point, Panglao Island, PANGLAO 2004 B3, 9.558°N, 123.810°E (MNHN, 7). 0–3 m, Pontod Islet, Panglao Island, PANGLAO 2004 D5, 9.560°N, 123.725°E (MNHN, 1). 6 m, San Isidro, Panglao Island, PANGLAO 2004 S17, 9.576°N, 123.831°E (MNHN, 1). 642–669 m, Maribojoc Bay, Bohol Sea, CP2397, 9.581°N, 123.695°E (MNHN, 1). 34–82 m, W of Baclayon, Bohol Island, PANGLAO 2004 T6, 9.585°N, 123.853°E (MNHN, 2). 4 m, Biking, Panglao Island, PANGLAO 2004 B5, 9.586°N, 123.840°E (MNHN, 1). 5 m, Biking, Panglao Island, PANGLAO 2004 S1, 9.588°N, 123.841°E (MNHN, 90). 24–27 m, Doljo Point, Panglao Island, PANGLAO 2004 B12, 9.593°N, 123.720°E (MNHN, 3). 31 m, N of Doljo, Panglao Island, PANGLAO 2004 B35, 9.598°N, 123.741°E (MNHN, 5). 90–200 m, Maribohoc Bay, Bohol Island,

PANGLAO 2004 P1, 9.600°N, 123.750°E (MNHN, 1). 2–4 m, Momo Beach, Panglao Island, PANGLAO 2004 S24, 9.601°N, 123.750°E (MNHN, 4). 3–14 m, Momo Beach, Panglao Island, PANGLAO 2004 B10, 9.608°N, 123.760°E (MNHN, 15). 28–32 m, Momo Beach, Panglao Island, PANGLAO 2004 S8, 9.608°N, 123.760°E (MNHN, 12). 30–33 m, between Momo and Napaling, Panglao Island, PANGLAO 2004 B42, 9.616°N, 123.766°E (MNHN, 4). 2–4 m, Napaling, Panglao Island, PANGLAO 2004 S5, 9.618°N, 123.768°E (MNHN, 4). Napaling, Panglao Island, PANGLAO 2004 B8, 9.618°N, 123.769°E (MNHN, 13). 3–5 m, Baclayon Takot, Bohol Island, PANGLAO 2004 B13, 9.618°N, 123.876°E (MNHN, 3). 8–15 m, Baclayon Takot, Bohol Island, PANGLAO 2004 S13, 9.618°N, 123.876°E (MNHN, 7). 28–32 m, Napaling, Panglao Island, PANGLAO 2004, 9.620°N, 123.773°E (MNHN, 1). 20–21 m, Napaling, Panglao Island, PANGLAO 2004 B21, 9.620°N, 123.773°E (MNHN, 1). 62 m, Bingag/Tabalong, Panglao Island, PANGLAO 2004 L51–60, 9.628°N, 123.800°E (MNHN, 1). 1–4 m, Sungcolan Bay, Panglao Island, PANGLAO 2004 S7, 9.641°N, 123.820°E (MNHN, 3). 2–4 m, Sungcolan, Panglao Island, PANGLAO 2004 B15, 9.646°N, 123.818°E (MNHN, 2). 2–8 m, Ubajan, Bohol Island, PANGLAO 2004 B20, 9.691°N, 123.850°E (MNHN, 3). 4–12 m, Manga, Bohol Island, PANGLAO 2004, 9.695°N, 123.848°E (MNHN, 9). 3–4 m, Manga, Bohol Island, PANGLAO 2004 S19, 9.701°N, 123.856°E (MNHN, 1).

Japan, Okinawa. 0 m, 2 km SSW Fitzwoody Beach, Naha Air Force Base, 26.185°N, 127.633°E (LACM 77-59, 2). SW side Gishifu Shima, Kerama Retto, 26.223°N, 127.360°E (LACM 77-72, 1). 20 m, 5 km W Tsuken-Jima (= Admiral's Island), 26.268°N, 127.920°E (LACM 78-21, 4). 17 m, Noth side, Ike Jima, 26.384°N, 127.990°E (DLG 863, 4). 53 m, 0.5 km ESE Zampa-Misaki (Bolo Point), 26.435°N, 127.708°E (LACM 78-100, 6). 50 m, 0.5 km ESE Zampa-Misaki (Bolo Point), 26.435°N, 127.708°E (LACM 78-25, 16). 8 m, Maeda Point, 26.435°N, 127.759°E (DLG 858, 2). 50 m, 1 km WNW Onna Village, 26.493°N, 127.842°E (LACM 79-75, 1; LACM 79-75, 25). 63 m, 1 km WNW Onna Village, 26.493°N, 127.842°E (LACM 78-101, 20). 60 m, 1 km WNW Onna Village, 26.493°N, 127.842°E (LACM 79-76, 20). 50 m, 1 km W of Onna Village, 26.495°N, 127.843°E (LACM 78-99, 8). 60 m, 1 km WNW of Onna Village, 26.497°N, 127.843°E (LACM 78-29, 3; LACM 78-29, 4). 50 m, 1 km WNW of Onna village, 26.497°N, 127.843°E (LACM 78-99, 12). 77 m, Horseshoe Cliffs, Onna Village, 26.500°N, 127.848°E (USNM 822094, 5). 8 m, Nago, 26.588°N, 127.976°E (DLG 856, 1). 6–8 m, S side, Sesoko Jima, 26.641°N, 127.860°E (DLG 861, 1). 0 m, 1 km N of Nakijin village, 26.703°N, 127.942°E (LACM 78-22, 5). 33 m, 1 km W Onna Village, 26.847°N, 127.843°E (LACM 78-20, 5). 16 m, 1 km NNW Oku, 26.847°N, 128.287°E (LACM 77-64, 6). 1 m, 1 km NNW Oku, 26.847°N, 128.287°E (LACM 77-61, 1). 20–30 m, Maeda Point, 26.445°N, 127.773°E (DLG 1531, 1). 2–3 m, Horseshoe Beach, 26.504°N, 127.858°E (DLG 1574, 3).43 m, Horseshoe Beach, 26.505°N, 127.853°E (DLG 1545, 14). 37 m, Big Rock, Horseshoe Beach, 26.508°N, 127.853°E (DLG 1462, 15). 33 m, Horseshoe Bay, 26.509°N, 127.856°E (DLG 1517, 2).

Papua New Guinea. 23 m, off Port Moresby, Manubada Local Island, off NW side, 9.517°S, 147.167°E (AMS C.404977, 2). 5 m, Christensen Research Institute, Madang, 5.117°S, 145.817°E (AMS C.376853, 2). 65 m, off Susan's Reef, Kimbe Bay, New Britain, 5.291°S, 150.128°E (DLG 608, 10). 20 m, Wongat Island, Madang Province, 5.137°S, 145.845°E (LACM 80-20, 1). 29 m, Wongat Island, Madang Province, 5.135°S, 145.845°E (LACM 80-26, 2). 4 m, off Anne Sophie's Reef, Kimbe Bay, 5.089°S, 150.186°E (DLG 579, 2). 15–35 m, Baudisson Bay, Kavieng, New Ireland, 2.743°S, 150.658°E (DLG 299, 5).

Solomon Islands. 20 m, Lark Reef, Rua Sura Islands, Guadalcanal Province, 9.501°S, 160.603°E (SBMNH 36926, 1). 22 m, Vangunu Island, S end Porepore Island, 8.583°S, 158.200°E (LACM 89-75, 4). 22 m, Matui Island, Vangunu Island, New Georgia Group, 8.495°S, 158.200°E (LACM, 7). 396–411 m, SALOMON 1 DW1762, 8.665°S, 160.065°E (MNHN, 1).

Australia, New South Wales. 200 m, off Sydney, 33.872°S, 151.667°E (AMS C.402611, 1).

Australia, Queensland. 6–7 m, Great Barrier Reef, Capricorn Group, Heron Island, S side opposite marine station, 23.433°S, 151.950°E (AMS C.379454, 1). 78 m, SE of Swain Reefs, 22.518°S, 152.710°E (AMS C.402609, 1). 27–37 m, Great Barrier Reef, Swain Reefs, Bylund Gillett Cay, 21.717°S, 152.417°E (AMS C.348694, 20+). 64–73 m, Great Barrier Reef, Swain Reefs, 3 km NE of W side of Bylund Gillett Cay, 21.700°S, 152.433°E (AMS C.379073, 1). 7 m, Great Barrier Reef, off Cairns, Moore Reef, 16.900°S, 146.200°E (AMS C.376861, 1). 21 m, Great Barrier Reef, Euston Reef, SW side, 16.667°S, 146.217°E (AMS C.376856, 6). 9–18 m, Great Barrier Reef, Cairns Reef lagoon and between Cairns Reef and Hope Island, 15.700°S, 145.500°E (AMS C.404964, 2, 1: complete). 10 m, Lizard Island, 14.683°S, 145.458°E (LACM 79-55, 2). 27–30 m, Great Barrier Reef, Carter Reef, N end, 14.550°S, 145.600°E (AMS C.376860, 2). 8–10 m, Great Barrier Reef, No. 5 Sandbank Reef 13-120, SW end, 13.750°S, 144.267°E (AMS C.404486, 1: complete). 9–15 m, Torres Strait, off Murray Island, 9.933°S, 144.067°E (AMS C.402704, 1).

Australia, Western Australia. 190 m, off North West Cape, 22.387°S, 113.677°E (AMS C.376857, 2). 80 m, off North West Cape, 21.783°S, 113.990°E (AMS C.376859, 11). 20 m, E side, Sandy Islet, Scott Reef, 14.044°S, 121.780°E (SBMNH 369506, 32). 10–20 m, Cassini Island, W side, 13.952°S, 125.623°E (SBMNH 348012, 1). 18–25 m, Seringapatam reef, N side, 13.624°S, 122.005°E (SBMNH 348011, 1).

Australia, Coral Sea. 10–15 m, SE from Observatory Cay, Lihou Reef, 17.157°S, 152.141°E (SBMNH 369615, 17). 1 m, Elisabeth Reef, 26.937°S, 159.040°E (AMS, 1: complete).

New Caledonia. 250–350 m, Vauban, 22.500°S, 166.400°E (MNHN, 20). 45–50 m, Woodin Canal, 22.392°S, 166.793°E (MNHN, 1). 495 m, southern, 22.367°S, 166.233°E (MNHN, 30). 490–500 m, southern, 22.317°S, 166.183°E (MNHN, 2). 455 m, southern, 22.267°S, 166.183°E (MNHN, 20). 250 m, E coast, 21.750°S, 166.617°E (MNHN, 1). 250 m, E coast, 21.750°S, 166.617°E (MNHN, 2). 500 m, Loyalty Basin, 21.550°S, 166.400°E (MNHN, 4). 310–315 m, Loyalty Basin, 21.533°S, 166.483°E (MNHN, 1). 310–315 m, Loyalty Basin, 21.533°S, 166.483°E (MNHN, 1). 237–298 m, E coast, 21.283°S, 165.950°E (MNHN, 50). 350–400 m, E coast, 21.033°S, 165.633°E (MNHN, 30). 350–400 m, E coast, 21.033°S, 165.633°E (MNHN, 2). 1380 m, Loyalty Basin, 21.033°S, 167.033°E (MNHN, 1). 105–110 m, E coast, 20.950°S, 165.567°E (MNHN, 12). 160–222 m, E coast, 20.950°S, 165.583°E (MNHN, 30). 15–20 m, Santal Bay, Lifou, 20.947°S, 167.045°E (MNHN, 1). 2–4 m, Santal Bay, Lifou, 20.945°S, 167.052°E (MNHN, 26). 2–4 m, Santal Bay, Lifou, 20.945°S, 167.052°E (MNHN, 2: complete). 15–18 m, Santal Bay, Lifou, 20.944°S, 167.033°E (MNHN, 24). 15–18 m, Santal Bay, Lifou, 20.944°S, 167.033°E (MNHN, 1: complete). 10–20 m, Santal Bay, Lifou, 20.925°S, 167.070°E (MNHN, 4). 3–10 m, Santal Bay, Lifou, 20.922°S, 167.083°E (MNHN, 1). 5–30 m, Santal Bay, Lifou, 20.920°S, 167.012°E (MNHN, 10). 9–20 m, Santal Bay, Lifou, 20.917°S, 167.087°E (MNHN, 1). 150–200 m, Point Lefèvre, Santal Bay, Lifou, 20.903°S, 167.018°E (MNHN, 1). 12 m, Santal Bay, Lifou, 20.900°S, 167.122°E (MNHN, 5). 200 m, Loyalty Ridge, 20.900°S, 167.283°E (MNHN, 1). 12–32 m, Santal Bay, Lifou, 20.892°S, 167.045°E (MNHN, 10). 5–25 m, Touho sector, 20.878°S, 165.325°E (MNHN, 10). 5–20 m, Santal Bay, Lifou, 20.875°S, 167.135°E (MNHN, 29). 40–60 m, Santal Bay, Lifou, 20.873°S, 167.133°E (MNHN, 2). 4 m, Santal Bay, Lifou, 20.873°S, 167.142°E (MNHN, 8; MNHN, 3). 20 m, Touho sector, 20.840°S, 165.380°E (MNHN, 13). 10 m, Touho sector, 20.825°S, 165.327°E (MNHN, 10). 25–30 m, Santal Bay, Lifou, 20.822°S, 167.173°E (MNHN, 1). 25–30 m, Santal Bay, Lifou, 20.822°S, 167.173°E (MNHN, 12). 96–109 m, E coast, 20.817°S, 165.317°E (MNHN, 2). 105–110 m, Poindeimie sector, 20.817°S, 165.317°E (MNHN, 8). 105–110 m, E coast, 20.817°S, 165.317°E (MNHN, 15). 8–18 m, Santal Bay, Lifou, 20.792°S, 167.118°E (MNHN, 7). 10–21 m, Santal Bay, Lifou, 20.788°S, 167.113°E (MNHN, 22). 4 m, Santal Bay, Lifou, 20.785°S, 167.123°E (MNHN, 2). 0–1 m, Touho sector, 20.782°S, 165.230°E (MNHN,

2). 1–5 m, Santal Bay, Lifou, 20.782°S, 167.132°E (MNHN, 1). 0 m, Santal Bay, Lifou, 20.781°S, 167.129°E (MNHN, 1: complete). 5–10 m, Santal Bay, Lifou, 20.780°S, 167.046°E (MNHN, 49). 5–10 m, Santal Bay, Lifou, 20.780°S, 167.046°E (MNHN, 14: complete). 0–2 m, Touho sector, 20.775°S, 165.250°E (MNHN, 1). 45–56 m, Touho sector, 20.775°S, 165.263°E (MNHN, 2). 20 m, Santal Bay, Lifou, 20.773°S, 167.033°E (MNHN, 9). 47 m, Santal Bay, Lifou, 20.773°S, 167.033°E (MNHN, 11). 27–31 m, Santal Bay, Lifou, 20.763°S, 167.028°E (MNHN, 7). 17 m, Santal Bay, Lifou, 20.763°S, 167.028°E (MNHN, 4). 20 m, Santal Bay, Lifou, 20.763°S, 167.028°E (MNHN, 1). 0 m, Touho sector, 20.753°S, 165.272°E (MNHN, 6). 10–30 m, Touho sector, 20.750°S, 165.275°E (MNHN, 8). 15–20 m, Koumac sector, 20.745°S, 164.258°E (MNHN, 1). 15–20 m, Koumac sector, 20.745°S, 164.258°E (MNHN, 5). 15–35 m, Touho sector, 20.743°S, 165.228°E (MNHN, 5). 8 m, Touho sector, 20.742°S, 165.265°E (MNHN, 1; MNHN, 15). 549–59 m, Touho sector, 20.733°S, 165.233°E (MNHN, 3). 20–30 m, Koumac sector, 20.690°S, 164.247°E (MNHN, 11). 26–40 m, Koumac sector, 20.673°S, 164.248°E (MNHN, 5; MNHN, 1: complete). 55–57 m, Koumac sector, 20.672°S, 164.195°E (MNHN, 40). 12 m, Koumac sector, 20.667°S, 164.187°E (MNHN, 22; MNHN, 3: complete). 15–20 m, Koumac sector, 20.667°S, 164.253°E (MNHN, 2). 15 m, Koumac sector, 20.662°S, 164.248°E (MNHN, 1). 10–15 m, Touho sector, 20.660°S, 165.245°E (MNHN, 4). 33–38 m, Koumac sector, 20.647°S, 164.260°E (MNHN, 2). 15–20 m, Touho sector, 20.585°S, 165.135°E (MNHN, 24). 380–400 m, E coast, 20.583°S, 165.117°E (MNHN, 24). 408–440 m, E coast, 20.583°S, 165.117°E (MNHN, 3). 2827 m, 20.583°S, 167.217°E (MNHN, 1). 497–520 m, E coast, 20.567°S, 164.950°E (MNHN, 6). 0 m, Koumac sector, 20.562°S, 164.183°E (MNHN, 2). 270–282 m, E coast, 20.550°S, 165.000°E (MNHN, 1). 21 m, Loyalty Ridge, 20.367°S, 166.167°E (MNHN, 1). 21 m, Loyalty Ridge, 20.367°S, 166.167°E (MNHN, 2: complete). 247–252 m, E coast, 21.700°N, 166.567°E (MNHN, 4). 0 m, Loyalty Ridge, 22.340°N, 166.428°E (MNHN, 1). 400–500 m, E coast, BATHUS 1 DE694, 20.600°S, 164.966°E (MNHN, 4). 1060–1450 m, south, BATHUS 2 CP767, 22.183°S, 165.983°E (MNHN, 2). 577–780 m, south, BATHUS 2 DW754, 22.383°S, 166.216°E (MNHN, 80). 10 m, Goeland Islet, Noumea sector, St. 1367, 22.405°S, 166.345°E (MNHN, 4). 250–350 m, St. 40, 22.500°S, 166.400°E (MNHN, 180). 465–525 m, south, BATHUS 2 DW739, 22.583°S, 166.450°E (MNHN, 70). 713–950 m, south, BATHUS 2 CP743, 22.600°S, 166.433°E (MNHN, 100).

Vanuatu. 140–182 m, 15.633°S, 167.058°E (MNHN, 4). 700–736 m, 14.883°S, 167.100°E (MNHN, 3). 154–179 m, MUSORSTOM 8 DW1105, 15.050°S, 167.116°E (MNHN, 1). 20 m, SE Matewulu, SANTO 2006 DB75, 15.381°S, 167.198°E (MNHN, 1). 7 m, W Mavéa Island, SANTO 2006 DB67, 15.381°S, 167.218°E (MNHN, 1). 50–61 m, SW Mavéa Island, SANTO 2006 EP37, 15.391°S, 167.220°E (MNHN, 1). 90–93 m, SW Maéa Island, SANTO 2006 EP38, 15.394°S, 167.221°E (MNHN, 2). 102–120 m, W Mavéa Island, Palikulo Bay, SANTO 2006 AT14, 15.397°S, 167.220°E (MNHN, 1). 6–43 m, Aésé Island, SANTO 2006 DB58, 15.410°S, 167.238°E (MNHN, 1). 21 m, SE Aésé Island, SANTO 2006 DB63, 15.448°S, 167.263°E (MNHN, 4). 22 m, Palikulo Bay, SANTO 2006 DS43, 15.465°S, 167.238°E (MNHN, 1). 22–25 m, NW Urélapa Island, SANTO 2006 DB20, 15.508°S, 167.023°E (MNHN, 6). 7 m, Segond Channel, E Sarakata river mouth, SANTO 2006 LS17, 15.518°S, 167.175°E (MNHN, 1). 25 m, Segond Channel, *Collidge* wreck, SANTO 2006 DS04, 15.523°S, 167.235°E (MNHN, 1). 17–19 m, N Tutuba Island, SANTO 2006 DS101, 15.538°S, 167.286°E (MNHN, 1). 46–47 m, Segond Channel, SANTO 2006 EP01, 15.541°S, 167.150°E (MNHN, 1). MUSORSTOM 8 DW1069, 15.550°S, 167.233°E (MNHN, 2). 15–25 m, NE Tutuba Island, SANTO 2006 DB01, 15.551°S, 167.296°E (MNHN, 4). 98–100 m, W Tutuba Island, SANTO 2006 DS102, 15.568°S, 167.266°E (MNHN, 7). Tangua Island, 15.590°S, 167.986°E (MNHN, 1). 622–625 m, MUSORSTOM 8 DW1072, 15.666°S, 167.333°E (MNHN, 3). 360–419 m, MUSORSTOM 8 DW1065, 16.266°S, 167.350°E (MNHN, 3).

Fiji. 314–377 m, 19.667°S, 178.167°W (MNHN, 1). 20 m, Herald Pass, W of Ndravuni Island, NW end of Great Astrolabe Reef, 18.767°S, 178.467°E (LACM 85-135, 2). 441–443 m, S of Viti Levu, 18.320°S, 177.862°E (MNHN, 5). 497–504 m, S of Viti Levu, 18.312°S, 178.152°E (MNHN, 5). 234–361 m, 18.310°S, 178.085°E (MNHN, 1). 260–305 m, S of Viti Levu, 18.308°S, 178.097°E (MNHN, 20). 275–430 m, S of Viti Levu, 18.297°S, 177.907°E (MNHN, 30). 295–302 m, S of Viti Levu, 18.212°S, 178.540°E (MNHN, 4). 144–150 m, S of Viti Levu, 18.207°S, 178.550°E (MNHN, 5). 149–168 m, S of Viti Levu, 18.207°S, 178.552°E (MNHN, 1). 392–433 m, S of Viti Levu, 18.185°S, 178.390°E (MNHN, 3). 1 m, Suva Point, Viti Levu, 18.170°S, 178.420°E (LACM 79-40, 1). 183–188 m, SE Viti Levu, 17.828°S, 178.797°E (MNHN, 3). 959–963 m, SE of Viti Levu, 17.710°S, 178.917°E (MNHN, 4). 879–897 m, SE of Viti Levu, 17.515°S, 178.888°E (MNHN, 50). 820–863 m, 17.300°S, 179.550°W (MNHN, 1). 656–660 m, Bligh Water, 17.268°S, 178.247°E (MNHN, 6). 660–663 m, Bligh Water, 17.248°S, 178.492°E (MNHN, 9). 640–687 m, Bligh Water, 16.937°S, 178.132°E (MNHN, 1). 200–215 m, Bligh Water, 16.840°S, 178.208°E (MNHN, 7). 820–863 m, BORDAU 1 DW1459, 17.300°S, 179.550°W (MNHN, 19). 203–208 m, SE Viti Levu, MUSORSTOM 10 DW1356, 17.838°S, 178.800°E (MNHN, 1). 402–444 m, SE of Viti Levu, MUSORSTOM 10 CP1360, 17.993°S, 178.803°E (MNHN, 1). 785–820 m, BORDAU 1 CP1490, 18.850°S, 178.533°W (MNHN, 1). 240–319 m, BORDAU 1 DW1494, 18.916°S, 178.483°W (MNHN, 1). 395–540 m, BORDAU 1 DW1486, 19.016°S, 178.433°W (MNHN, 7). 500–516 m, BORDAU 1 DW1488, 19.016°S, 178.416°W (MNHN, 5). 700–707 m, BORDAU 1 DW1485, 19.050°S, 178.500°W (MNHN, 5).

Tonga. 18 m, Langitau island, Vaka'eitu Island, Vava'a Group, 18.717°S, 174.083°E (LACM 85-90, 1). Naupapu Island, Vava'a Group, 18.700°S, 174.100°E (LACM 85-89, 3). 27 m, Hunga Island, Vava'a Goup, 18.692°S, 174.133°E (LACM 85-91, 2). 440–487 m, Vava'u group, BORDAU 2 DW1586, 18.566°S, 173.916°W (MNHN, 10).

Belau. 25 m, Nghelobll, 7.183°N, 134.367°E (LACM 78-39, 1).

Mariana Islands. 1 m, Merizo Cocos lagoon, Guam, 13.253°N, 144.665°E (LACM 77-18, 1). 6 m, Piti Bay, E of Apra Harbor Guam, 13.465°N, 144.690°E (LACM 77-11, 1). 25 m, Anemwanot lagoon, Majuro, 7.133°N, 171.317°E (SBMNH 85114, 4). Bock Island, N end Rongerik, 11.405°N, 167.482°E (USNM 594643, 5).

Wallis and Futuna. 415–420 m, Wallis Island, 13.350°S, 176.133°W (MNHN, 2). 350 m, Wallis Island, 13.317°S, 176.283°W (MNHN, 2). 455–515 m, Wallis Island, 13.200°S, 176.267°W (MNHN, 30).

Cook Islands. 22 m, Aitutaki, off Arutanga, 18.872°S, 159.792°W (LACM 87-79, 17). N of Atua Tane, Aitutaki, 18.850°S, 159.770°W (DLG 1409, 5).

French Polynesia. Under blocks, reef, Papeete, Tahiti, 17.526°S, 149.553°W (JTC, 7). 29 m, Tahiti, Anchor Bay, 17.533°S, 149.567°W (AMS C.406353, 1). 0 m, Tahiti, 17.695°S, 149.584°W (DLG 477, 1: complete). Under rocks, reef, Faaone, Tahiti, 17.662°S, 149.311°W (JTC, 1). 485–560 m, Moorea, TARASOC DW3459, 17.466°S, 149.800°W (MNHN, 2). Point Aroa ou de Tiaia, NE corner Moorea, 17.467°S, 149.767°W (AMS C.376855, 4). 0 m, Gump Station, Moorea, 17.533°S, 149.833°W (DLG 239, 1: complete). 0 m, Gump Station, Moorea, 17.533°S, 149.833°W (DLG 242, 4). 8–10 m, Tiahura Island, NW Moorea Island, 17.483°S, 149.917°W (AMS C.402983, 12). E side Ilot Tiahura, NW corner Moorea, 17.483°S, 149.917°W (AMS C.380659, 5). Point Aroa ou de Tiaia, NE corner Moorea, 17.467°S, 149.767°W (AMS C.376855, 4). Moorea Island, Papetoai Bay, Passe Tareu, W side, 17.508°S, 149.850°W (AMS C.348626, 1). 2 m, Moorea, 17.500°S, 149.767°E (LACM 74-36, 14). Under blocks, reef Point Venus, Mahina, Tahiti, 17.493°S, 149.492°W (JTC, 3). Under blocks, reef Point Venus, Mahina, Tahiti, 17.492°S, 149.493°W (JTC, 11). 0 m, Pl Aroa ou de Tiara, NE corner of Moorea, 17.467°S, 149.767°W (AMS, 3: complete). 0 m, Cook Bay, Vaiare, Moorea, 17.386°S, 149.774°W (DLG 225, 1: complete; DLG 233, 1, 3: complete). Laisse de plage exterieure, Anaa, Tuamotus,

17.345°S, 145.527°W (JTC, 7). 85 m, Raiatea Island, 16.833°S, 151.500°W (JWC 11309, 1). Mavana Island, Raroia, 16.056°S, 142.418°W (USNM 720631, 1). 0 m, S end of Opakea Island, Raroia, 16.037°S, 142.403°W (USNM ex 722419, 2). 0 m, Ngarumaea Island, Raroia, 16.037°S, 142.403°W (USNM 697679, 1). 0 m, Ngarumaoa Island, Raroia, 16.020°S, 142.430°W (USNM ex 711798, 3). 3 m, N of Oneroa Island, Raroia, 16.017°S, 142.433°W (USNM ex 721376, 5). Oneroa Island, Raroia, 16.017°S, 142.433°W (USNM ex 720859, 2). 1 m, Ahe Atoll, Tuamotus, 14.467°S, 146.367°E (LACM 73-94, 8). 507–607 m, Kaukura, Tuamotus, TARASOC DW3373, 15.683°S, 146.900°W (MNHN, 1). 493–540 m, Huahine, TARASOC DW3429, 16.716°S, 150.633°W (MNHN, 1). 520–572 m, Raiatea, TARASOC DW3457, 16.750°S, 151.400°W (MNHN, 1). 440–490 m, Raiatea, TARASOC DW3451, 16.883°S, 151.350°W (MNHN, 1). 558–1000 m, Tubuai, Austral Islands, BENTHAUS DW1957, 23.313°S, 149.488°W (MNHN, 1). 100 m, N of Raivavae, Austral Islands, 23.828°S, 147.693°W (MNHN, 11).

Remarks. Hasegawa (2004) reported mass swimming in *Sci. spinosa* (as *Scissurella* sp.).

Scissurella staminea (A. Adams, 1862)

Figures 199–201

Chresonymy.

Anatomus stamineus A. Adams, 1862: 348–349, not illustrated.
Scissurella staminea: Crosse, 1863: 110.
Anatomus stamineus: Troschel, 1863: 702.
Anatomus stramineus [*sic*]: Dunker, 1882: 147.
Scissurella staminea: Paetel, 1883: 47.
Scissurella staminea: Paetel, 1888: 289.
Anatomus stamineus: Pilsbry, 1890: 60.
Scissurella staminea: Thiele, 1912a: 16, pl. 2, figs 11–12.
Scissurella staminea: Habe, 1951: 65, pl. 11, figs 1–2.
Scissurella staminea: Kuroda & Habe, 1952: 85.
Scissurella staminea: Taki & Oyama, 1954: pl. 42, fig. 21.
Scissurella staminea: Habe & Kikuchi, 1960: 60.
Scissurella staminea: Habe, 1961: 2, pl. 1, fig. 8.
Scissurella straminea [*sic*]: Habe, 1963: 229 [Japanese], 233 [English].
Scissurella staminea: Habe, 1964: 2, pl. 1, fig. 8.
Scissurella staminea: Anon., 1969: 26, pl. 1, fig. 8.
Anatoma staminea: Higo, 1973: 13.
Scissurella staminea: Itoigawa & Ogawa, 1973: 72, pl. 5, fig. 18.
Anatoma staminea: Matsumoto, 1979: 4.
Scissurella staminea: Lozouet, 1986: 108.
Scissurella staminea: Habe *et al.*, 1986: 4.
Scissurella staminea: Tsuchida *et al.*, 1991: 7–8, pl. 1, fig. 4.
Scissurella staminea: Fukuda *et al.*, 1992: 49, pl. 41, fig. 17.
Scissurella staminea: Higo & Goto, 1993: 14.
Scissurella staminea: Okutani & Hasegawa, 2000: 37, pl. 18, fig. 9.
Scissurella staminea: Higo *et al.*, 2001: 13, G75 [one of two syntypes].

Scissurella staminea: Kwon *et al.*, 2001: 16, 47, figs 23-1–23-3.

Scissurella staminea: Geiger, 2003: 77.

Scissurella staminea: Min, 2004: 85, figs 55, 55-1, 55-2.

Scissurella staminea: Geiger, 2004: text-fig. p. 6.

Scissurella staminea: Geiger, 2006c: text-fig. 3.

Scissurella staminea: Sasaki, 2008: fig. 2E.

Scissurella staminea: Koyama, 2008: 253, pl. 1, fig. 8.

Misidentifications.

not *Scissurella turbinata*: Yokoyama (1924: 35–36, pl. 5, fig. 21) [is *Sci. staminea*. Plate reprinted in Taki & Oyama, 1954: pl. 42).

not *Scissurella staminea*: Hu & Tao, 2000: 45, fig. 1 [Trochoidea?].

Type material. Syntypes (NHMUK 1878.1.28.309, 3), 1.7 mm. Lectotype here designated (NHMUK 1878.1.28.309/1. Fig. 199) for the express purpose of taxon stabilization in case non-conspecific syntypes should be discovered. Paralectotypes from type locality (NHMUK 1878.1.28.309/2–3, 2).

Type locality. 25 fms [= 46 m], Tsu-Sima, [=?= Okinawa, Tsuken Jima, Japan, 26.268°N, 127.920°E] (OD).

Figure 199. *Anatomus stamineus*. Lectotype. 25 fms [= 46 m], Tsu-Sima, [=?= Okinawa, Tsuken Jima, Japan, 26.268°N, 127.920°E] (NHMUK 1878.1.28.309/1). Scale bar shell = 1 mm. Scale bar protoconch = 100 µm.

Etymology. Stamineus, Latin for threaded, referring to the sculpture.

Description. Shell medium-sized (to 1.7 mm), trochiform, depressed. Protoconch of 0.75 whorl, strong axial cords, apertural varix connected to embryonic cap, apertural margin strongly sinusoidal. Teleoconch I of 1.125–1.25 whorls, with 16–20 strong axial cords, intersected by increasing number of spiral threads, approximately 10 at end of teleoconch I. Teleoconch II of 1.125 whorls, suture moderately impressed. Shoulder

Figure 200. *Scissurella staminea*. **A.** 50 m, 1 km NWN of Onna Village, Okinawa, Japan, 26.493°N, 127.842°E (LACM 79-75, 23). **B.** Okinawa, Onna Flats, 1.5 km WNW Onna, Japan, 26.488°N, 127.837°E (AMS C.405179, 2). **C.** Numa, Japan, Pleistocene (UMUT CT21891, 1); specimen cited by Yokoyama (1924). Scale bars shell = 1 mm. Scale bars protoconch = 100 µm.

slightly convex, axials diminishing in strength from strong cords to indistinct cords, intersected by approximately six spirals at apertural margin ranging from threads to cordlets of irregular spacing, strength. Base with same density of axial cords; approximately 11 spirals changing from lines under selenizone to cords at periphery to steps towards umbilicus. Umbilicus wide, bordered by carina, walls smooth with faint growth marks. Aperture rounded, D-shaped, roof overhanging. Selenizone with keels of moderate strength, moderate elevation; lunules distinct; slit open, converging towards apertural margin.

Animal unknown.

Comparisons. *Scissurella mirifica* of wide Indo-Malayan Archipelago to Central Pacific distribution has a protoconch with fine axials and has at least twice as many axial and spiral cordlets on shoulder and base.

Scissurella costata from the Mediterranean and adjacent Atlantic Ocean has a protoconch with fine axials and about twice as many finer spiral lines on the base.

Scissurella lorenzi from the Indo-Malayan Archipelago to the Western Pacific has a protoconch with fine axials, usually has distinct light-colored axial bands on the shoulder, and has a more narrow umbilicus bordered by a distinct periumbilical cord.

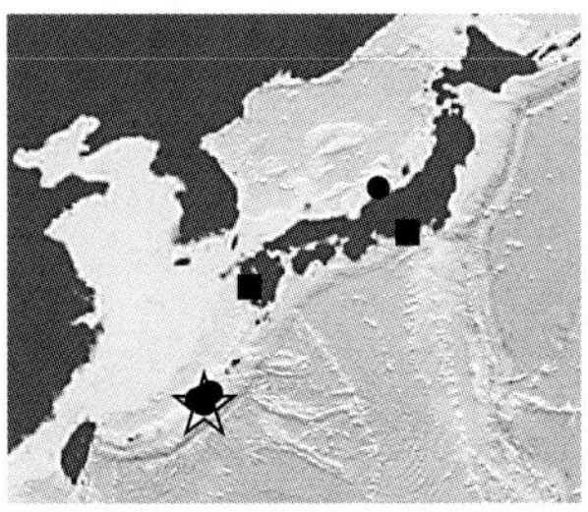

Figure 201. Distribution of *Scissurella staminea.* ☆ type locality. ● specimen record. ■ literature record.

Scissurella quadrata from the Indo-Malayan Archipelago to the Central Pacific has a protoconch with fine axial cords, a much more narrow umbilicus with distinct periumbilical cord, and has about 50% more but finer axial cords.

Scissurella rota from the Indian Ocean has a shorter teleoconch I (1 vs. 1.25 whorls), usually more elevated axial cords, particularly on the base, and the intersections of axials and spirals are elevated to very distinct points.

Scissurella sudanica from the Indian Ocean has a protoconch with fine axials, a more narrow umbilicus with distinct periumbilical cord, and about twice as many but finer axial cords on shoulder and base.

Juvenile *Sin. ferriezi* from the Indo-Malayan Archipelago to the Central Pacific has a smooth protoconch, has at least twice as many fine spiral lines on the base, and has a more pronounced constriction of the base below the selenizone.

Juvenile *Sin. marrowi* from the Indo-Malayan Archipelago closes the foramen after <0.5 teleoconch II whorls and has more spiral lines on the base (16–20 vs. 11).

Juvenile *Sin. singeri* from the Indian Ocean has a smooth protoconch, at least twice as many fine spiral lines on the base, and the axial cords are usually more elevated and denser.

Juvenile *Sin. zimmeri* from Easter Island and Pitcairn Island has a shorter teleoconch I (<1 vs. 1.25 whorls), an overall broader shell, and has more pronounced axial cords.

Distribution. Southern Japan, Ryukyu Islands (Fig. 201).

Fossil. Tertiary (youngest Pleistocene) of Numa, Japan (Yokoyama, 1924: UMUT CM21891-5-21, CM21892. Fig. 200C). Pleistocene, Hasama, Shima Peninsula, central Japan [34.369°N, 136.804°E] (Ioigawa & Ogawa, 1973).

Specimen records.

Japan. [no additional locality data] (KBIN, 1; RM 3150, 4; RM 1443, 2). Numa (UMUT CT21891, 1). 2 m, Tanoura, Noto Peninsula, Japan Sea, Ishikawa Prefecture, Honshu, 37.300°N, 137.160°E (LACM 82-9, 2).

Japan, Okinawa. 20 m, 5 km W Tsuken-Jima (= Admiral's Island), 26.268°N, 127.920°E (LACM 78-21, 1). 0 m, Kadena, 26.337°N, 127.747°E (LACM 77-55, 1). 42 m, Zampa-Misaki, 26.427°N, 127.718°E (LACM 78-24, 3). 50 m, 0.5 km ESE Zampa-Misaki

(Bolo Point), 26.435°N, 127.708°E (LACM 78-25, 14). 0 m, Onna Flats, 1.5 km WSW Onna, 26.488°N, 127.837°E (LACM 77-57, 3; LACM 77-56, 3). Onna Flats, 1.5 km WNW Onna, 26.488°N, 127.837°E (AMS C.405179, 2). 60 m, Okinawa, Onna Village, 26.493°N, 127.842°E (LACM 79-76, 21). 50 m, 1 km NWN of Onna Village, 26.493°N, 127.842°E (LACM 79-75, 23). 92 m, Okinawa, Onna Village, 26.493°N, 127.493°E (LACM 78-101, 13). 46 m, 1 km WNW Onna Village, 26.495°N, 127.843°E (LACM 78-26, 4). 33 m, 1 km W Onna Village, 26.495°N, 127.843°E (LACM 78-20, 2). 50 m, 1 km W Onna Village, 26.495°N, 127.843°E (LACM 78-99, 3). 0 m, 1 km W Onna Village, 26.495°N, 127.843°E (LACM 78-19, 1). 45 m, 1 km WNW of Onna, 26.497°N, 127.845°E (AMS C.404980, 1). 60 m, Onna Village, 26.497°N, 127.843°E (LACM 78-29, 2). 55 m, Onna Village, 26.497°N, 127.843°E (LACM 78-27, 11). 77 m, Horseshoe Cliffs, Onna Village, 26.500°N, 127.848°E (USNM 822094, 2). 33 m, Motobu Peninsula, Nakijin, 26.703°N, 127.943°E (LACM 78-22, 1). 1 m, 1 km NNW Oku, 26.847°N, 128.287°E (LACM 77-61, 2). 17 m, Hedo Point, 26.863°N, 128.250°E (LACM 78-23, 1). 46 m, Tsu-Sima, [=?= Okinawa, Tsuken Jima], 26.268°N, 127.920°E (NHMUK 1878.1.28.309, 3). 401–404 m, Nago Bay, 26.525°N, 127.760°E (SBMNH 149472, 2).

Literature records.

Japan. Strait of Korea near Tsushima, 45 m (Thiele, 1912a). Amakusa Marine Biological Laboratory [Tomoe Cove, Kumamoto, 32.529°N, 130.035°E] (Habe & Kikuchi, 1960).—Suruga Bay [Shizuoka, 34.800°N, 138.600°E] (Habe *et al.*, 1986).

Remarks. Hasegawa (2004) reported mass swimming in *Sci. staminea*. The Japanese common name is *Kuchikireebisugai*.

Scissurella sudanica Bandel, 1998

Figures 202–207

Chresonymy.

Scissurella sp.: Bandel, 1982: 14–15, pl. 10, figs 7–8.
Scissurella sp.: Yaron, 1983: 271, pl. 5.
Scissurella sudanica Bandel, 1998: 14–15, pl. 3, figs 2–4.
Scissurella sudanica: Geiger, 2003: 78.
Scissurella sudanica: Geiger, 2004: 6, text-fig.

Figure 202. Original Illustration of *Scissurella sudanica*.

Scissurella sudanica: Geiger, 2006c: text-fig. 3.
Scissurella sp.: Zuschin *et al.*, 2009: 86, pl. 2, fig. 1.

Misidentifications.

not *Scissurella reticulata*: Bandel, 1998 (*partim*): pl. 3, fig. 1 [is *Sci. sudanica*].
not *Scissurella* aff. *reticulata*: Zuschin *et al.*, 2009: 86–87, pl. 2, fig. 2 [is *Sci. sudanica*].

Type material. Holotype (SGPIH 3870), 1.2 mm.

Type locality. Port Sudan, Sudan [19.617°N, 37.233°E]. Algal mats and *Halimeda* (OD).

Etymology. Named after its type locality in Sudan (OD).

Description. Shell medium-sized (to 1.2 mm) trochiform, inflated. Protoconch of 0.75 whorl, embryonic cap with reticulate sculpture, remainder with fine axials and even finer spiral thread, apertural varix connected to embryonic cap, apertural margin sinusoidal. Teleoconch I of 1.125 whorls, approximately 23 axial cords, approximately 6–11 very fine, regularly spaced, spiral threads. Teleoconch II of 1.125 whorls, suture little impressed. Shoulder somewhat convex, approximately 40 axial cords or cordlets, eight finer spiral lines, density of spirals increasing from suture to selenizone. Base with similar density of axials, approximately 17 much weaker spiral lines. Umbilicus narrow, bordered by distinct spiral cord, walls straight, at angle to base, with very fine

Figure 203. *Scissurella sudanica*. 23 m, Jackson Reef, Strait of Tiran, S Sinai Peninsula, N Red Sea, Egypt, 27.017°N, 34.467°E (LACM 85-111, 12). Scale bar shell = 1 mm. Scale bar protoconch = 100 μm.

growth marks, no funiculus. Selenizone above periphery with distinct, moderately elevated keels, lunules somewhat weaker than axials, slit open with parallel margins. Aperture rounded, D-shaped, oblique, roof overhanging.

Figure 204. *Scissurella sudanica.* **A.** Mbudya Island, ca. 16 km N of Dar-es-Salaam, Tanzania, 6.650°S, 39.250°E (AMS C.380206, 1). For radula, see Figure 206. **B.** Near Ras Tanura, Saudi Arabia, 26.700°N, 50.100°E (AMS C.406389, 2). **C.** S of Manama, Bahrain, 26.183°N, 50.633°E (AMS C.406390). Scale bars shells = 0.5 mm. Scale bars protoconch = 100 μm.

Radula (Fig. 206). Rachidian tooth triangular with somewhat convex margins, tips triangular with five cusps, central cusp largest. Lateral teeth 1–3 similar, triangular, tips with 2–3 cusps. Lateral tooth 4 reduced, with pointed tip. Lateral

Figure 205. *Scissurella sudanica*. **A–C.** Karachi, Pakistan, 24.867°N, 67.050°E (NMSZ 1961.61, 5). Scale bars shell = 0.5 mm. Scale bars protoconch = 100 μm.

tooth 5 broadly enlarged with six cusps. Marginal teeth with long shafts, tips with increasing number of cusps. Radular interlock moderate.

Comparisons. *Scissurella costata* from the Mediterranean and adjacent Atlantic has a wider umbilicus without periumbilical cord and usually has stronger axial cords.

Scissurella lorenzi from the Indo-Malayan Archipelago to the Western Pacific usually has distinct light-colored axial bands on the shoulder, about 50% fewer axial cords, and fewer spiral lines on the shoulder (2–4 vs. 6–11).

Scissurella mirifica from the Indo-Malayan Archipelago to the Central Pacific has a wider umbilicus, no pronounced periumbilical cord, and has 50% fewer axial cords on shoulder and base.

Scissurella morretesi from Brazil has a protoconch with fine flocculations in spiral orientation, has no apertural varix on the protoconch, and has a longer teleoconch I (1.75 vs. 1.25 whorls).

Figure 206. *Scissurella sudanica.* Radula. Mbudya Island, ca. 16 km N of Dar-es-Salaam, Tanzania, 6.650°S, 39.250°E (AMS C.380206, 1). For shell, see Figure 204A. **A.** Width of radula. **B.** Central field enlarged. Scale bars = 10 μm.

Scissurella quadrata from the Indo-Malayan Archipelago to the Central Pacific has a more pronounced constriction on the base below the selenizone, the axial cords are at least twice as strong as the spiral lines, and the aperture has a subquadrate cross section.

Scissurella reticulata from the Indian Ocean has fewer but somewhat stronger sculpture on the shoulder (17–25 vs. 40 axials, 3–4 vs. 6–11 spirals), markedly elevated points at the intersection of axials and spirals, and a wider umbilicus.

Scissurella rota from the Indian Ocean has pronounced hollow spines on the base at the intersection of axials and spirals, fewer axials and spirals, and a more pronounced spiral constriction below the selenizone.

Scissurella staminea from Japan have a protoconch with strong axials, at most half as many axial and spiral cords, and has a wider umbilicus without periumbilical cord.

Juvenile *Sin. ferriezi* from the Indo-Malayan Archipelago to the Central Pacific have a smooth protoconch, has a more pronounced constriction of the base below the selenizone, and has stronger axial cords.

Juvenile *Sin. marrowi* from the Indo-Malayan Archipelago have a protoconch with strong axials, and has a wider umbilicus without periumbilical cord.

Juvenile *Sin. singeri* from the Indian Ocean has a smooth protoconch, a wide umbilicus without periumbilical cord, and as juveniles usually a more pronounced constriction below the selenizone.

Distribution. Indian Ocean (Fig. 207).

Specimen records.

Egypt. 23 m, Jackson Reef, Strait of Tiran, S Sinai Peninsula, N Red Sea, 27.017°N, 34.467°E (LACM 85-111, 12). 0 m, 16 km S of Hurghada, Red Sea, 27.083°N, 33.883°E (RMNH, 14). Hurghada, between Giftun-el-Saghir and Giftun-el-Kabir, 27.200°N, 33.917°E (AMS C.404922, 3). 25 m, Dunraven Reef, Strait of Tiran, S Sinai Peninsula, N Red Sea, 27.200°N, 34.100°E (LACM 85-112, 10). 18 m, Amphoras dive site, off Ras Um Sid, SE side of Sinai Peninsula, 27.867°N, 34.333°E (LACM 88-119, 10). 7 m, Jackson Reef, W of Tiran Island, Straits of Tiran, Red Sea, Sinai Peninsula, 27.917°N, 34.467°E (LACM 78-81, 2). Blue Hole, Dahab, Sinai, 28.483°N, 34.533°E (DLG 102, 11). 10–15 m, The Alternatives, Strait of Tiran, S Sinai Peninsula N Red Sea, 28.733°N, 34.217°E (LACM 85-110, 4). 0 m, 5 km S of Râs Adabîya, Gulf of Suez, 29.867°N, 32.500°E (RMNH, 2). 0 m, shipyard, Safaga, 26.731°N, 33.943°E (DLG 1644, 9). 0 m, near hotels, Safaga, 26.795°N, 33.942°E (DLG 1647, 3). 0 m, 3-4 km N of hotels, Safaga, 26.829°N, 33.944°E (DLG 1752, 2).

Bahrain. S of Manama, 26.183°N, 50.633°E (AMS C.406390, 2).

Sudan. 25 m, Sha'ab Rumi, 19.900°N, 37.400°E (DLG 898, 1). 28 m, Sha'ab Rumi, 19.900°N, 37.400°E (DLG 897, 1). 95 m, Sanganeb, 19.718°N, 37.425°E (DLG 894, 5).

Tanzania. Mbudya Island, ca. 16 km N of Dar-es-Salaam, 6.650°S, 39.250°E (AMS C.380206, 1).

Madagascar. 0–3 m, Baye de Ranobe, near Mora Mora, N of Tulear, Toliara Province, 23.083°S, 43.500°E (LACM 89-50, 2). 225–282 m, S Cape St. Marie, ATIMO VATAE CP3613, 26.216°S, 45.133°E (MNHN, 1).

Reunion Island. 205–215 m, 21.383°S, 55.600°E (MNHN, 1). 205–215 m, 21.383°S, 55.617°E (MNHN, 1).

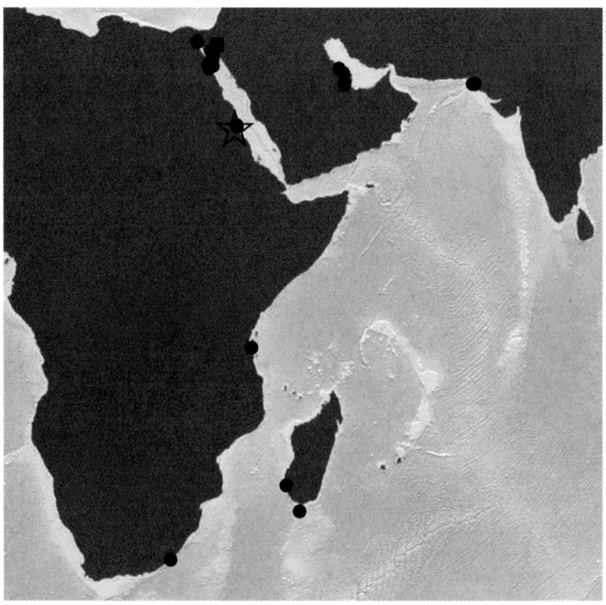

Figure 207. Distribution of *Scissurella sudanica*. ☆ type locality. ● specimen record. ■ literature record.

South Africa. Transkei, 50 m SSE of Bomvu river mouth, 31.989°S, 29.149°E (AMS C.404527, 4: complete).

Saudi Arabia. 20 m, Gulf of Salwah, Persian Gulf, 24.746°N, 50.764°E (RMNH, 100). 20 m, Gulf of Salwah, Persian Gulf, 24.819°N, 50.794°E (RMNH, 1). Near Ras Tanura, 26.700°N, 50.100°E (AMS C.406389, 2).

Pakistan. 0–3 m, Oyster Rocks, Karachi Harbor, Sind Province, 24.800°N, 67.000°E (LACM 79-9, 1). 4 m, small cove 4.8 km W atomic power plant and 7 km WNW Bulegi Point, Sind Province, 24.833°N, 66.725°E (LACM 79-1, 16). Karachi, 24.850°N, 67.033°E (NHMUK, 14). Karachi, 24.867°N, 67.050°E (NMSZ 1961.61, 5).

Literature records.

Israel. Elat, 29.550°N, 34.950°E (Yaron, 1983).

Remarks. The species is variable. The strength of the axial cords may be quite strong as shown in the specimen from Tanzania in Figure 204A, and the shell from Karachi, Pakistan, are somewhat more inflated than the typical form from the Red Sea (Fig. 205). Juveniles (Tanzania. Fig. 204A) and smaller specimens (Fig. 205A vs. Fig. 205B) are more inflated than the more fully developed ones. Shared characters of all these specimens indicating conspecificity include all protoconch characters, length of teleoconch I, number of axials and spirals, stronger axials than spirals, and umbilical characters. There is no distinction based on meristic characters, only in degree of development. Such differences in degree are found also in other species, including *Sci. rota*, *Sci. costata*, *Suk. atkinsoni*, and *Suk. morleti*.

Scissurella supraplicata Smith, 1875

Figures 208–210

Chresonymy.

Scissurella supraplicata Smith, 1875: 72, not illustrated.
Scissurella supraplicata: Kidder, 1876: 48.
Scissurella supraplicata: Crosse, 1877: 10.
Scissurella supraplicata: Smith, 1879: 176, pl. 9, figs 5, 5a.
Scissurella supraplicata: Paetel, 1888: 289.
Scissurella supraplicata: Pilsbry 1890: 54–55, pl. 68 [erroneously pl. 58 in text], figs 27–28.
Scissurella supraplicata: Thiele, 1912a: 10, pl. 1, figs 20–21.
Scissurella supraplicata: Thiele, 1925: 42.
Scissurella supraplicata: Powell, 1957: 124.
Scissurella supraplicata: Powell, 1960: 126.
Scissurella supraplicata: Cantera & Arnaud, 1985: 32.
Scissurella (?) *supraplicata*: Lozouet, 1986: 108.
Scissurella supraplicata: Forcelli, 2000: 45.
Scissurella supraplicata: Geiger, 2003: 78.
Scissurella supraplicata: Zelaya & Geiger, 2007: 430–431, figs 123 [map], 175–178 [shell SEM].

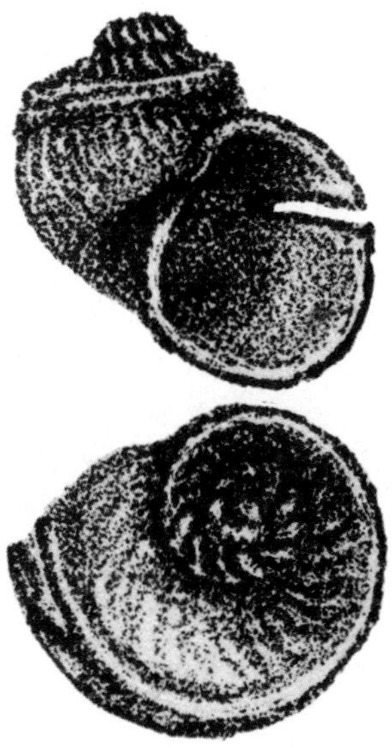

Figure 208. Illustration of *Scissurella supraplicata* Smith, 1878, from Smith (1879). Although not necessarily the illustration of type specimens, it certainly shows an ideotype, and quite likely a specimen from the type series.

Misidentifications:

Scissurella supraplicata: Melvill & Standen, 1912: 345 [is *An. conica*].
Schizotrochus supraplicata: Carcelles & Williamson, 1951: 251 [is *An. conica*].
Schizotrochus supraplicatus: Carcelles, 1953: 162 [is *An. conica*].
Scissurella supraplicata: Castellanos & Landoni, 1988: 8 [is *An. conica*].

Type material. Holotype (NHMUK. 1876.8.28.33. Zelaya & Geiger, 2007: figs 175–178. Fig. 209), 1.33 × 1.1 mm.

Type locality. Swain's Bay, Kerguelen Island [France, 49.705°S, 69.948°E] (OD).

Etymology. Supra–: Latin for above, on top. Plicata: Latin for folded.

Description. Shell medium in size (1.25 mm in holotype), trochiform, inflated-bulbous. Protoconch of one whorl; apertural margin convex, without varix; protoconch ornamentation unknown (very eroded in only known specimen). Teleoconch I of one whorl, sculptured with 17 strong axial cords. Teleoconch II of about one whorl, suture little impressed. Shoulder convex, base inflated, sculptured with axial growth

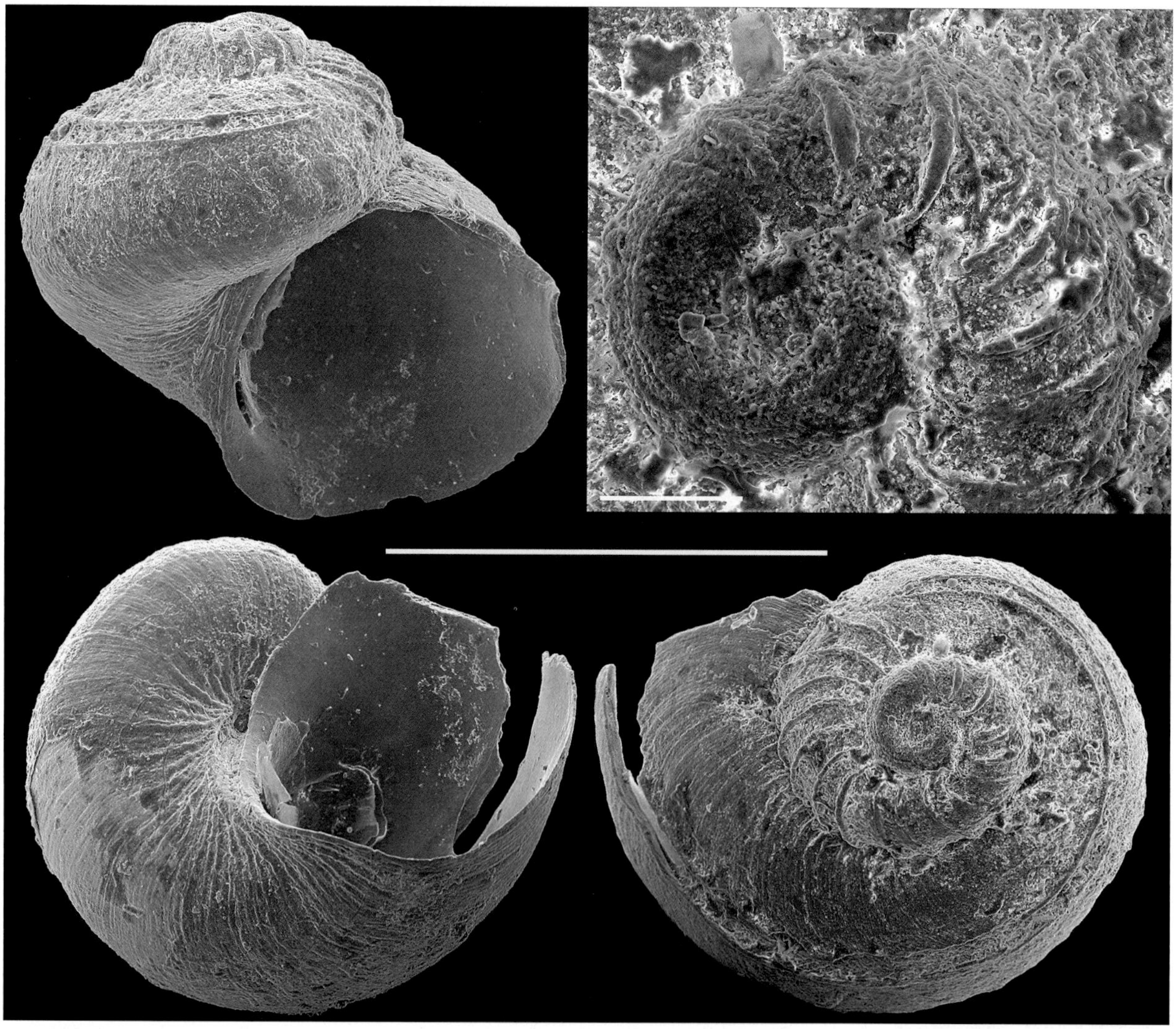

Figure 209. *Scissurella supraplicata*. Holotype. Swain's Bay, Kerguelen Island [France, 49.705°S, 69.948°E] (NHMUK 1876.8.28.33). Scale bar shell = 1 mm. Scale bar protoconch = 100 μm.

folds; axials making more prominent toward umbilicus. Selenizone above periphery, surrounded by low keels, low lunules; slit open, with parallel margins. Umbilicus narrow. Aperture rounded, roof overhanging.

Animal unknown.

Comparisons. *Scissurella clathrata* from subantarctic waters of South America has a delicate cancellate shell sculpture and a wider umbilicus.

Scissurella georgica from Antarctica has a protoconch with dense axial cords and is wider than tall with a distinct wide umbilicus.

Scissurella obliqua from the subantarctic Islands has no axial cords on teleoconch I, and lacks a periumbilical cord.

Scissurella petermannensis from Antarctica has a protoconch with extremely fine axial cords and has strong axial cords to lamellae on teleoconch I and the base of teleoconch II.

Juvenile Sin. *beddomei* from temperate Australia has a wide umbilicus, a strong constriction of the base below the selenizone, and very strong axial cords on the base.

Juvenile *Sin. chilensis* from temperate Pacific South America has a wider shell with a distinct umbilicus, distinct axial cords on the base, and a protoconch with many fine axials.

Sinezona mouchezi from antarctic waters of the Indian Ocean has a closed foramen, axial cords on the shoulder of teleoconch II, a wider umbilicus, and a protoconch with axial cords.

Juvenile *Sin. subantarctica* has a protoconch with dense axials, has a shell that is wider than tall, and the umbilicus is (occasionally partially in juveniles) covered by an umbilical callus.

Figure 210. Distribution of *Scissurella supraplicata*. ☆ type locality. ● specimen record. ■ literature record.

Distribution. Antarctica around Kerguelen Islands (Fig. 210).

Literature records.

Antarctica. Gazellen-Bucht, Kerguelen, 49.500°S, 69.500°E (Thiele, 1925).— 150 m, BANZARE Station #47, just S of Kerguelen, 49.833°S, 69.550°E (Powell, 1957).— 168 m, MD 04, DC272, 49.908°S, 70.407°E. MD 30, Crozet, île de l'Est, Baie de l'Aventure, 46.417°S, 51.000°E. 15 m, KER, D 54, S of Morbihan Bay, between Suhm and Cementery Island, Kerguelen Islands, 49.503°S, 70.188°E. 158 m, MO 03 CP59, 49.987°S, 70.032°E (Cantera & Arnaud, 1985).

Remarks. Cantera & Arnaud (1985) indicated *Sci. medioplicata* as a synonym of *Sci. supraplicata*. They are certainly mistaken; see under Remarks of *Sci. petermannensis*.

Carcelles & Williamson (1951) and Ageitos de Castellanos & Landoni (1988) included *Sci. supraplicata* in the list of species from the Magellan Region, based on the previous record of Melvill & Standen (1912), which does not correspond to this species but to *An. conica*. The same source of error seems to be the origin of the inclusion of *Sci. supraplicata* in the list of species by Carcelles (1953).

Scissurella xandaros new species

Figures 211–214

Type material. Holotype (SBMNH 149865). Paratypes from type locality (SBMNH 149866, 9).

Type locality. 15–35 m, Baudisson Bay, New Ireland, Papua New Guinea, 2.743°S, 150.658°E.

Etymology. Xandaros, Greek fabulous sea monster, noun in apposition.

Description. Shell medium-sized (to 1.4 mm), trochiform. Protoconch of 0.75 whorl, smooth, apertural varix barely connected to embryonic cap, apertural margin sinusoidal. Teleoconch I of 1.25–1.33 whorls, 28–32 distinct axial cords, first spiral lines after 0.33–0.66 whorl, approximately 9 at onset of selenizone. Teleoconch II of 1.1 whorls, suture moderately impressed. Shoulder straight to slightly convex, axial cords initially at same density and strength as on teleoconch I, increasing in strength, becoming crowded towards apertural margin; very fine irregular growth marks in interstices; approximately 6–12 strong spiral lines to weak spiral cords of equal strength or decreasing in strength from suture to selenizone; spirals and

Figure 211. *Scissurella xandaros* new species. Holotype. 15–35 m, Baudisson Bay, New Ireland, Papua New Guinea, 2.743°S, 150.658°E (SBMNH 149865). Scale bar shell = 1 mm. Scale bar protoconch = 100 μm.

axials forming distinct points at intersection. Base with weak constriction below selenizone; same strength and density of axial cords as on shoulder; 2–3 weak spiral lines in constriction below selenizone approximately 12–14 spirals, changing from distinct spiral lines to faint spiral steps from selenizone towards umbilicus. Umbilicus moderately wide, bordered by weak carina, walls straight, at distinct angle to base, with very fine growth lines. Selenizone above periphery, lunules distinct, keels

Figure 212. *Scissurella xandaros* new species. **A–B.** Paratypes. 15–35 m, Baudisson Bay, New Ireland, Papua New Guinea, 2.743°S, 150.658°E (SBMNH 149865, 9). **C.** 100 m, North of Raivavea, Austral Islands, French Polynesia, 23.828°S, 147.693°W (MNHN, 14). Scale bars shell = 1 mm. Scale bars protoconch = 100 μm.

moderately strong, moderately elevated; slit open, with parallel margins. Aperture rounded D-shaped, roof overhanging.

Radula (Fig. 213). Rachidian tooth broadly triangular-trapezoidal, central cusp largest, three on each side, of decreasing size. Lateral teeth 1–3 similar, apical cusp largest, 2–3 on outer edge of tip. Lateral tooth 5 enlarged by broadening, apical cusp largest, three cusps along inner edge. Inner marginal teeth with elongated-triangular tip, apical cusp largest, 3–4 along outer edge, 1–2 along inner edge; shaft with food groove. Outer marginal teeth with spoon-shaped tip, many fine bristles along edge. Radular interlock of central field moderate.

Comparisons. *Scissurella cebuana* from the Indo-Malayan Archipelago to the Central Pacific has at most half as many spiral cords on shoulder and base, which form more distinct knobby points, and the more narrow umbilicus is bordered by a distinct periumbilical cord.

Scissurella lorenzi from the Indo-Malayan Archipelago to the Western Pacific has a protoconch with fine axial sculpture, is overall more globular in shape, has a narrower umbilicus with a stronger bordering carina, and has the typical radiating light color rays.

Scissurella mirifica from the Indo-Malayan Archipelago to the Central Pacific has a protoconch with fine axials and weaker axial cords, which do not form points at the intersection with spiral lines on shoulder and base.

Scissurella rota from the Indian Ocean has a protoconch with fine axials and has half as many spiral lines on the base.

Scissurella spinosa from the Indo-Malayan Archipelago to the Central Pacific has a protoconch with fine axial sculpture, stronger overall teleoconch sculpture with a proportionally wider umbilicus, and has fewer spiral points at the intersection of spirals and axials on the shoulder (2–3 vs. >6).

Scissurella staminea from southern Japan has a protoconch with strong axial cords and has at most half as many spiral lines on the base.

Scissurella sudanica from the Indian Ocean has a protoconch with fine axials, a more narrow umbilicus with distinct periumbilical cord, and weaker axial cords on the base.

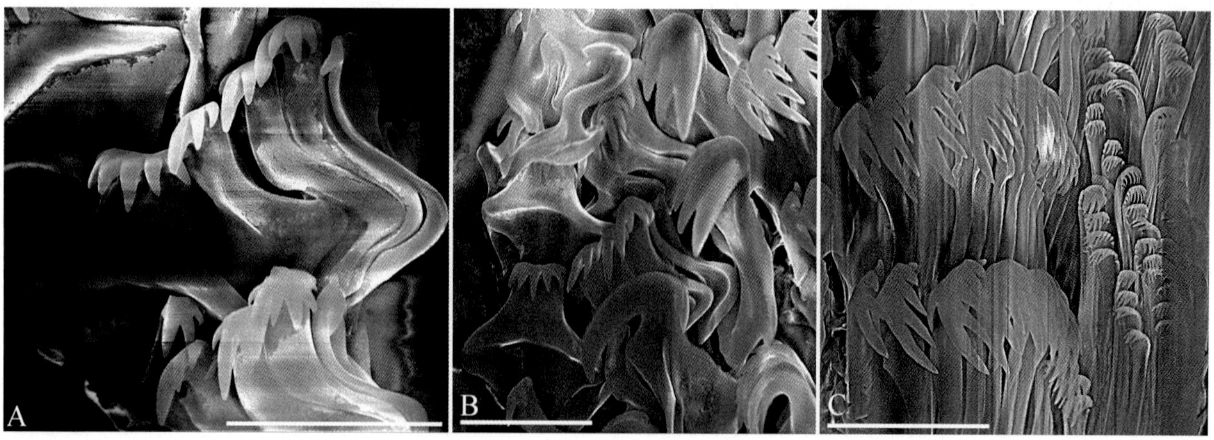

Figure 213. *Scissurella xandaros* new species. Radula. 114 m, W Tutuba Island, SANTO 2006 DS96, Vanuatu, 15.560°S, 167.275°E (MNHN, 3). **A.** Central field. **B.** Central field with lateral tooth 5. **C.** Marginal teeth. Scale bars shell = 1 mm. Scale bars = 10 μm.

Sinezona marrowi from the Indo-Malayan Archipelago has an overall wider shell shape with wider umbilicus, a protoconch with strong axials on the outer half of the whorl, lacks the strong elevated points at the intersection of axial and spirals, and in fully grown specimens has a closed foramen.

Distribution. Indo-Malayan Archipelago to Central Pacific (Fig. 214).

Specimen records.

Reunion Island. 205–215 m, 21.383°S, 55.600°E (MNHN, 1; MNHN, 1).
Coral Sea. NE Herald Cay, 16.933°S, 149.183°E (AMS C.402650, 1).
Philippines. 1–2 m, Visayas, Mactan Channel, 10.033°N, 123.867°E (AMS C.348620, 2). 26 m, Bohol Island, off Balicasag Island, 9.517°N, 123.683°E (AMS exC.348621, 1). 609–673 m, Maribojoc Bay, Bohol Sea, CP2396, 9.605°N, 123.700°E (MNHN, 2).
Papua New Guinea. 25 m, Port Moresby, Horseshoe Reef, 9.467°S, 147.167°E (AMS C.348633, 3).
New Caledonia. 150–200 m, SE Shelter Reef, Santal Bay, Lifou, Loyalty Islands, LIFOU DW1648, 20.900°S, 167.050°E (MNHN, 2).
Vanuatu. 7 m, W Mavéa Island, SANTO 2006 DB67, 15.381°S, 167.218°E (MNHN, 1). 100–105 m, NW Tutuba Island, SANTO 2006 DS99, 15.541°S, 167.281°E (MNHN, 46). 92 m, NW Tutuba Island, SANTO 2006 DS105, 15.550°S, 167.278°E (MNHN, 6). 100 m, NW Tutuba Island, SANTO 2006 DS108, 15.553°S, 167.276°E (MNHN, 1). 114 m, W Tutuba Island, SANTO 2006 DS96, 15.560°S, 167.275°E (MNHN, 3). 70–80 m, W Tutuba Island, SANTO 2006 DS103, 15.568°S, 167.266°E (MNHN, 10). 98–100 m, W Tutuba Island, SANTO 2006 DS102, 15.568°S, 167.266°E (MNHN, 4). 80 m, W Tutuba Island, SANTO 2006 DS104, 15.568°S, 167.266°E (MNHN, 3).
French Polynesia. 100 m, N of Raivavea, Austral Islands, 23.828°S, 147.693°W (MNHN, 14).

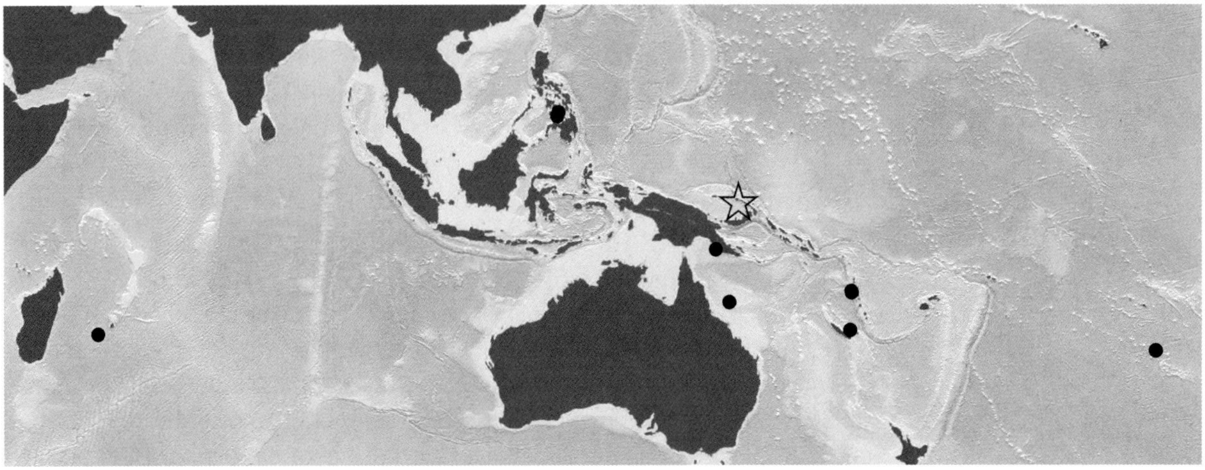

Figure 214. Distribution of *Scissurella xandaros* new species. ☆ type locality. ● specimen record.

Scissurella aliceae Schnetler, Lozouet & Pacaud, 2000 †

Figures 215–216

Chresonymy.

Scissurella aliceae Schnetler, Lozouet & Pacaud, 2000: 82–83, pl. 1, figs 1–3, pl. 3, figs 1–2.

Figure 215. Original illustration of *Scissurella aliceae*.

Type material. Holotype (MGUH 25748), 1.6 × 1.9 mm. Paratypes from type locality (MGUH 25749, 1; MGUH 25750, 1; MGUH 25758, 1; MGUH 25759, 1; MGUH unnumbered, 1; MNHN-LP R63044, 1: not found in 4/2005, DLG pers. obs.; SMF 321220, 1; ISL unnumbered, 7; ARF unnumbered, 49).

Type locality. Faske, Denmark [55.250°N, 12.130°E]. Coral Limestone, middle Danian, Paleocene (OD).

Etymology. Named for Alice Rasmussen, Faske, Denmark (OD).

Description. Shell medium-sized (to 1.9 mm), squat trochiform. Protoconch 0.75 whorl, sculptural details unknown. Teleoconch I approximately 1.5 whorls, 14 strong axials on first whorl, several equally-spaced, weaker spirals. Teleoconch II of one whorl, suture moderately impressed. Shoulder with up to 24 axial cords on body whorl, shoulder with 3–4 equally-spaced, weaker spiral cordlets. Base sculpture similar to shoulder, approximately 35 axial cords, 15 spiral cordlets. Selenizone on shoulder, keels well developed, lunules indistinct, slit open, margins parallel. Umbilicus open, wide, deep, walls smooth, with distinct carina separating base from umbilicus, lamella almost entirely obscuring umbilical opening.

Comparisons. The wide umbilicus almost completely obscured by a lamella is characteristic for this species.

Scissurella cossmanni from the Paleogene of Europe has strong axial lamellae on the base.

Scissurella bituminata from the Oligocene of Indonesia has a more narrow umbilicus and stepped spiral cords towards the umbilicus.

Scissurella depontaillieri from the Paleocene through Miocene of Europe has a very narrow umbilicus with a strong funiculus and is overall more turreted.

Figure 216. Distribution of *Scissurella aliceae*. ☆ type locality.

Scissurella eocaenica from the Eocene of Europe has a shorter teleoconch I (0.75 vs. 1.5 whorls) and lacks spirals on the first teleoconch whorl, with only a single spiral on the shoulder of the last 0.5 teleoconch whorl.

Scissurella marchmontensis from the Late Cretaceous of the Caribbean and Spain has about twice as many spiral lines and distinctly opisthocline axial cords on the base.

Scissurella peyrerensis from the Oligocene of Europe has strong axial folds on the adsutural half of the shoulder and a very strong constriction below the selenizone at the periphery of the base.

Scissurella ravni from the Paleocene of Europe has strong axial cords to low lamellae and a much more narrow umbilicus.

Scissurella szaboi from the Early Jurassic of Turkey looks very similar based on the rather crude line drawing available, but the temporal gap makes conspecificity rather unlikely.

Scissurella tinaeformis from the Eocene of Europe has strong axial lamellae on the base.

Scissurella transylvanica from the Miocene of Europe has an overall more globular shell outline and a more narrow umbilicus.

Distribution. Known only from type locality (Fig. 216).

Remarks. Schnetler *et al.* (2000) speculated on the possible function of the umbilicus as a brood pouch as has been demonstrated in *Larochea* by Marshall (1993).

Scissurella annulata Ravn, 1933 †

Figures 217–218

Chresonymy.

Scissurella annulata Ravn, 1933: 4–5, pl. 1, figs 14a–c.

Maxwellella annulata: Bandel, 1998: 19–20, pl. 5, figs 4–6.

Maxwellella annulata: Schnetler, Lozouet & Pacaud, 2000: 86, pl. 1, figs 7–9.

Type material. Nine syntypes (OD). "Holotype" [see Remarks] (MGUH 3123: Schnetler *et al.*, 2000). Figured specimen of Ravn (1933. Fig. 217): 0.6 × 1.4 mm.

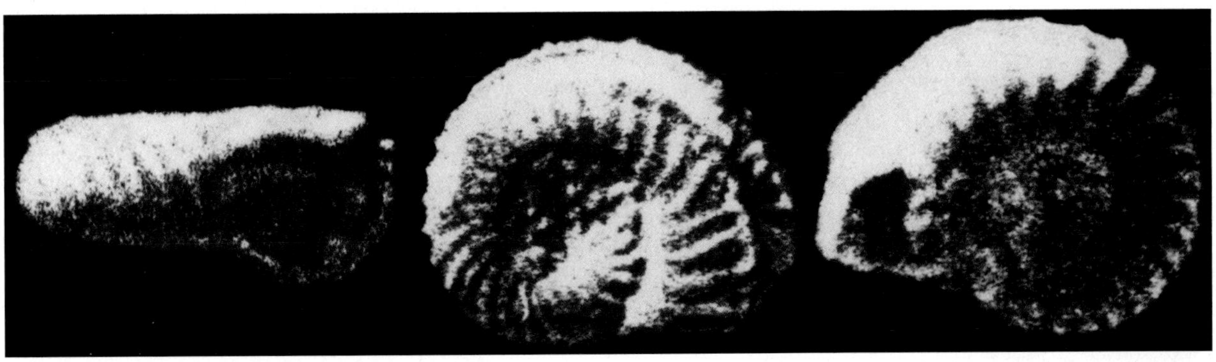

Figure 217. Original illustration of *Scissurella annulata*.

Type locality. Le Nez, Faxe [= Fakse], Denmark [55.250°N, 12.130°E]. Danian, Cretaceous [see Remarks] (OD). Danian = earliest Paleocene (Bandel, 1998; Dam *et al.*, 1998).

Etymology. Annulus: Latin for ring; possibly referring to the strong axial cords on the shell.

Description. Shell medium-sized (1.4 × 0.6 mm), planispiral. Protoconch 0.75 whorl, further details unknown. Teleoconch I of 2.33 whorls, with 28–35 distinct axial cords, very faint spiral lines, no spiral in position of selenizone. Teleoconch II of at least 0.66 whorl, suture impressed. Shoulder with axial and spiral sculpture of similar density as on teleoconch I. Selenizone slightly in above periphery, including periphery, at approximately 45° angle to shell axis. Umbilicus open, wide, base continuously sloping with umbilical wall. Aperture subquadrate.

Comparisons. The planispiral shell distinguishes the species from any other Recent or fossil species.

Scissurella gruendeli has the selenizone on the upper portion of the whorl well above the periphery and at only a slight angle towards the shell axis.

Scissurella condita from the Jurassic through Miocene of New Zealand has spiral lines on the shoulder and base of the shell, not evident in the original illustration of *Sci. annulata*.

Scissurella eocaenica from the Eocene of Europe has a much shorter teleoconch I (0.75 vs. 2.33 whorls) and many fewer axial cords.

Figure 218. Distribution of *Scissurella annulata*. ☆ type locality.

Distribution. Known only from type locality (Fig. 218).

Remarks. Ravn (1933: 9), citing also Schilder on cowries from the Danian, considered the Danian to belong to the Cretaceous. Currently, the Danian is classified as earliest Paleocene (belonging to the Tertiary, not the Cretaceous).

Schnetler *et al.* (2000) called their illustrated specimen "holotype," although the original description did not make such an explicit statement and all nine specimens constitute a syntype series. According to ICZN Art. 74.7, Schnetler *et al.*'s indication does not constitute a lectotype designation.

Scissurella apudornata (Laws, 1935) †

Figures 219–221

Chresonymy.

Scissurella apudornata Laws, 1935: 30, fig. 1.
Scissurella apudornata: Olson *in* Gage, 1957: 123.
Sinezona ? apudornata: Lozouet, 1986: 111.
Scissurella apudornata: Beu & Maxwell, 1990: 402.

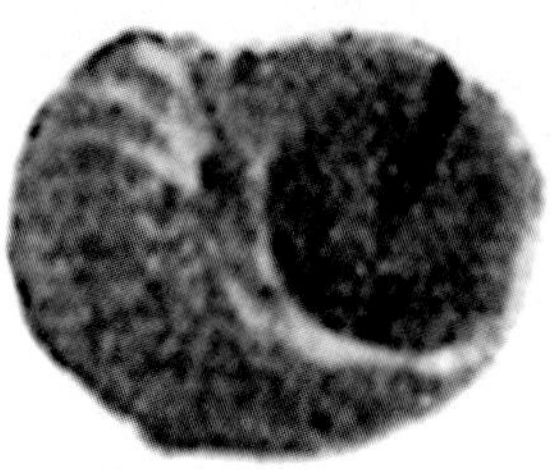

Figure 219. Original illustration of *Scissurella apudornata*.

Type material. Holotype (AIM AK 70700: destroyed by Byne's/glass disease: W. Blom pers. comm. 4/2009), 1.1 × 1.0 mm. Paratype from type locality (GNS TM 1712, 1. Fig. 220).

Type locality. Shell bed, Target Gully, Oamaru [South Island, New Zealand, 46.367°S, 169.683°E]. Awamoan [late Oligocene/early Miocene] (OD).

Etymology. Apud: Latin for close, similar referring to the similarity to *Sci. ornata* (OD).

Description. Shell medium-sized (to 1.4 mm), trochiform. Protoconch of one whorl, strong axial cords, apertural varix not connected to embryonic cap, apertural margin straight. Teleoconch I of one whorl, approximately a dozen indistinct axial cords, on last 0.125 whorl, indistinct spiral in position of selenizone. Teleoconch II of 1.1 whorls, suture lightly impressed. Shoulder convex, approximately 20 indistinct, low or distinctly raised axial cords, increasing in density towards apertural margin; no spirals, or very fine spiral lines starting after 0.1 whorl, increasing to approximately a dozen at apertural margin, forming small points at intersection with axials. Base with moderate constriction below selenizone, similar sculpture as on shoulder, if strong axial cords on shoulder then axials on base starting below constriction under selenizone; faint spiral sculpture on adumbilical quarter or spirals increasing, from

Figure 220. *Scissurella apudornata*. Paratype. Shell bed, Target Gully, Oamaru [South Island, New Zealand, 46.367°S, 169.683°E]. Awamoan (GNS TM 1712). Scale bar shell = 1 mm. Scale bar protoconch = 100 μm.

fine lines at upper border of axial cords to distinct spiral cords near umbilicus, spirals at most half as strong as axials. Umbilicus bordered by strong carina, walls straight, smooth. Selenizone with low keels of moderate strength, lunules indistinct, slit open. Aperture oblong, D-shaped, roof overhanging.

Comparisons. *Scissurella bountyensis* from the Recent of New Zealand has a shorter teleoconch I (0.66 vs. 1 whorl), at least 50% fewer axial cords, and a less protruding apex.

Scissurella costata from the Mediterranean and adjacent Atlantic has a protoconch with fine anostomizing axials, and has distinct spiral threads on shoulder and base.

Scissurella obliqua from the subantarctic has more rounded whorls, lacks axial sculpture, and has a more narrow umbilicus.

Scissurella supraplicata from the subantarctic Islands has a taller shell, a more narrow umbilicus, and lacks axial sculpture.

Sinezona cupelliformis from the Eocene of Europe has fine clathrate sculpture on the base and a wider umbilicus.

Sinezona deshayesi from the Eocene of Europe is overall more depressed and has a wider umbilicus.

Sinezona finlayi from the Miocene (possibly to Recent) of New Zealand has an overall more depressed shell shape and distinct spiral cords on the base, particularly surrounding the umbilicus.

Sinezona terquemi from the Miocene of Europe seems to have more axial cords on shoulder and base.

Sinezona tertia from the Pliocene of New Zealand has stronger axial cords that on the base do not reach the selenizone and strong spiral lines to weak cords on the base.

Sinezona malloryi from the Eocene of North America has an overall broader shell with correspondingly wider umbilicus and distinct spiral lines on shoulder and base.

Sinezona mechanica from the Recent of New Zealand has similar protoconch and teleoconch sculpture, but the spire is significantly lower, has fewer teleoconch II whorls (0.6 vs. 1.1), and is about half the size.

Figure 221. Distribution of *Scissurella apudornata*. ☆ type locality. ● specimen record.

Distribution. Oligocene/Miocene of New Zealand (Fig. 221).

Remarks. The strength of the sculpture is quite variable in this species, while the relative arrangement of the various sculptural elements, as well as the protoconch morphology, is uniform. The paratype available for examination by SEM shows some convergence of the margins of the slit.

Beu & Maxwell (1990) gave the age of the species as Altonian = early Miocene.

Scissurella bituminata Beets, 1943 †

Figures 222–223

Chresonymy.

Scissurella bituminata Beets, 1943: 259, pl. 27, figs 1–3.

Type material. Holotype (RMNH 42473. Fig. 222A), 1 × 1.3 mm (H × W). Paratype from type locality (RMNH unnumbered, 1. Fig. 222B), 1.5 mm wide.

Type locality. Waisiu [Bouton = Pulau Butung], Indonesia, [5.615°S, 122.705°E; late Oligocene: inferred from title] (OD).

Etymology. Most likely referring to the type locality in tar deposits.

Description. Shell trochiform, depressed, medium-sized (to 1.5 mm). Protoconch eroded. Teleoconch I whorl count uncertain, at least one whorl; with approximately 12 raised axial cords, approximately 10 spiral lines between position of selenizone and suture. Teleoconch II of 1.5 whorls (paratype), suture impressed. Shoulder convex, suture moderately impressed. Approximately 30 axials decreasing in strength from raised cords to cordlets; first axials at onset of selenizone, increasing to seven spiral cords with some additional fine lines after one whorl; at onset of selenizone axials stronger than spirals, both of equal strength after approximately 0.5 whorl, spirals dominating subsequently. Base with distinct constriction below selenizone. Similar change of axials and spirals with growth as on shoulder, on lower ⅖ of whorl with steps mostly without axial sculpture; step bordering umbilicus with dense,

Figure 222. *Scissurella bituminata*. Waisiu [Bouton = Pulau Butung], Indonesia, 5.615°S, 122.705°E. **A.** Holotype (RMNH 42473). **B.** Paratype (RMNH unnumbered). Scale bars = 1 mm.

rough axial cords; border towards umbilicus forming sharp upturned edge. Walls of umbilicus straight, smooth except for very fine growth lines. Selenizone above periphery; keels moderately strong, moderately elevated; lunules quite strong. Slit open, with parallel margins (original illustrations). Aperture D-shaped.

Comparisons. The strong washboard-like axials on the step surrounding the umbilicus are not known from any other species. The structure can be seen in the cavity of the umbilicus and should not be mistaken for ventral sculpture of the protoconch occasionally visible in some species, for example *Sci. lobini*.

Scissurella aliceae from the Paleocene of Europe has a very wide umbilicus and has non-stepped spiral cords towards the umbilicus.

Scissurella depontaillieri from the Miocene of Europe has an overall more turreted shell shape and a more narrow umbilicus.

Scissurella marchmontensis from the latest Cretaceous of the Caribbean and Spain has pronounced opisthocline axial cords and about twice as many spiral lines on the base.

Scissurella peyrerensis from the Oligocene of Europe has strong axial folds on the adsutural half of the shoulder and on the periphery of the base and a strong constriction of the base below the selenizone.

Scissurella ravni from the Paleocene of Europe has an umbilicus bordered by a strong carina and the axials are elevated cords to low lamellae.

Scissurella transylvanica from the Miocene of Europe is overall more globular in shape, and has a more narrow umbilicus.

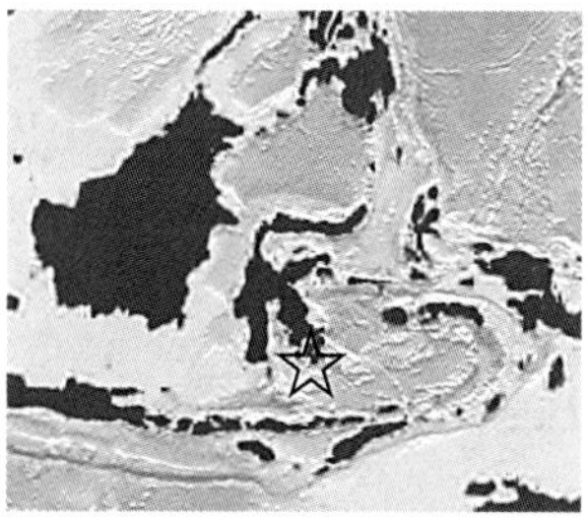

Figure 223. Distribution of *Scissurella bituminata*. ☆ type locality.

Distribution. Known only from type specimens (Fig. 223).

Remarks. The holotype has had its apertural lip broken and the lip fragment is no longer in the type lot. Despite this broken aperture, the original illustration clearly shows an open slit with parallel margins, hence, the generic assignment is well-founded.

Scissurella condita Laws, 1939 †

Figures 224–227

Chresonymy.

Scissurella condita Laws, 1939: 475, fig. 74.
Scissurella condita: Lozouet, 1986: 108.
Scissurella condita: Beu & Maxwell, 1990: 376, pl. 53, figs a–c [topotype].
Maxwellella condita: Bandel, 1998: 22.

Figure 224. Original illustration of *Scissurella condita*.

Synonym.

+ *Maxwellella novozeelandica* Bandel, 1998: 20, pl. 5, figs 7–8, pl. 6, figs 1–2. ***Type material.*** Holotype (GNS TM7253), 1.5 × 1.2 mm. ***Type locality.*** Kaiwara Valley, central [South Island], New Zealand. Early Jurassic [42.842°S, 173.135°E] (OD). ***Etymology.*** Named after source country of New Zealand (OD).

Maxwellella novozeelandica: Bandel *et al.*, 2000: pl. 1, figs 4 [holotype], 5 [paratype].

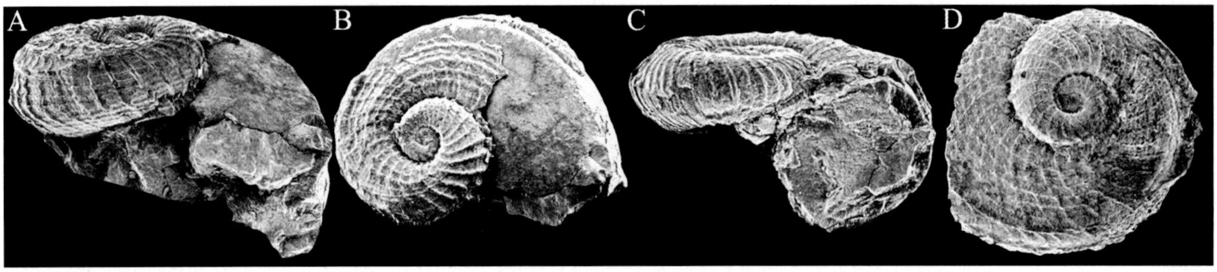

Figure 225. Original illustration of *Maxwellella novozeelandica*. **A–B.** Holotype. **C–D.** Paratypes.

Type material. Holotype (GNS TM 1348, ex coll. Laws), 0.9 × 0.4 mm. Paratypes from type locality (coll. Laws, 2; GNS TM 8632–8636, 7).

Type locality. Pakaurangi Point, Kaipara, New Zealand [36.239°S, 174.293°E]. Tertiary beds (OD: inferred from title). Holland's Point, Otaian = early Miocene (Beu & Maxwell, 1990: 376).

Etymology. Conditus: Latin has many meanings, most likely here for preserved.

Description. Shell medium-sized (to 1.5 mm), depressed turbiniform, spire depressed. Protoconch 0.75 whorl, strong axial sculpture, apertural varix present, condition of aperture unknown. Teleoconch I of 0.875–1 whorl, 19 strong axial cords, approximately 10 weaker spirals towards end. Teleoconch II of 0.875 whorl, suture deeply impressed. Shoulder with strong axial cords and approximately 12 fine spirals. Base of body whorl without constriction below selenizone, with 24 axial cords, spiral lines increasing from selenizone towards umbilicus. Umbilicus open, wide. Selenizone slightly above periphery, with low (eroded) keels, slit open. Aperture rounded, roof overhanging.

Comparisons. *Scissurella annulata* from the Paleocene of Europe seems to lack spiral lines on shoulder and base, judging from the original illustration.

Scissurella eocaenica from the Eocene of Europe has many fewer axial cords on shoulder and base and lacks any spirals.

Scissurella gruendeli from the Middle Jurassic of Europe has the selenizone higher on the shoulder with the selenizone being almost in the coiling axis of the shell.

In the Recent *Scissurella skeneoides* from northern New Zealand, the protoconch protrudes above the shoulder of the teleoconch and the number of axial lamellae is about half.

The Recent *Satondella bicristata* from New Zealand shares the overall shell shape, but lacks a selenizone, and the keels of the foramen are drawn out into a distinct chimney.

Distribution. Early Jurassic through Miocene of New Zealand (Fig. 227).

Figure 226. *Scissurella condita*. Holotype. Pakaurangi Point, Kaipara, New Zealand [36.239°S, 174.293°E] (GNS TM1348). Scale bar shell = 0.5 mm. Scale bar protoconch = 100 µm.

Specimen records.

New Zealand. GS12163, Moutapu Point, NE Pitt Island, Chatham Islands, Whenuataru Tuff, Waipipian-Mangapanian (Middle-late Pliocene), New Zealand, 36.239°S, 174.293°W (GNS TM, 17).

Remarks. Bandel (1998) did not compare his *novozeelandica* with *condita*. The latter is evidently based on a juvenile specimen, as the comparison with the juvenile specimen as well as the early portion of the larger specimens illustrated by Bandel (1998) clearly demonstrate: overall shape with sunken protoconch, identical number of axial cords on teleoconch I (19 vs. 20), onset of selenizone, deeply incised suture (Fig. 225C). The time span covered by the specimens is quite wide (Early Jurassic to Miocene), but in the absence of any morphological characters, the taxa cannot be maintained as distinct, unless the chronospecies concept is invoked.

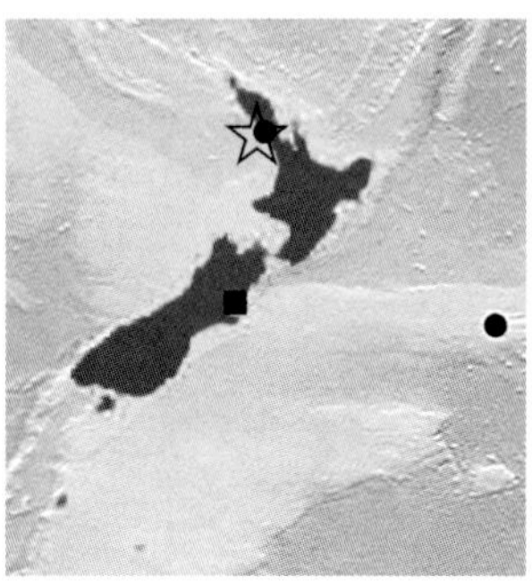

Figure 227. Distribution of *Scissurella condita*. ☆ type locality. ● specimen record. ■ literature record.

Scissurella cossmanni Depontaillier, 1881 †

Figures 228–231

Chresonymy.

Scissurella cossmanni Depontaillier, 1881: 176–178, pl. 7, fig. 2.

Scissurella cossmanni: Geiger, 2003: 76.

Figure 228. Original Illustration of *Scissurella cossmanni.*

+ *Scissurella dolioliformis* Amitrov, 1998: 98, figs 1E–G. ***Type material.*** Holotype (PIN 3595/707), 1.6 × 1.25 mm. ***Type locality.*** Rybalsk (Tschapli) quarry, Dnepropetrovsk, Ukraine [48°29'N, 35°0'E]. Mandrikovka Layers, late Eocene (OD). ***Etymology.*** Doliolum, Latin for keg.

Scissurella dolioliformis: Amitrov & Zhegallo, 1998: 7, 9, figs 2A–C [protoconch].

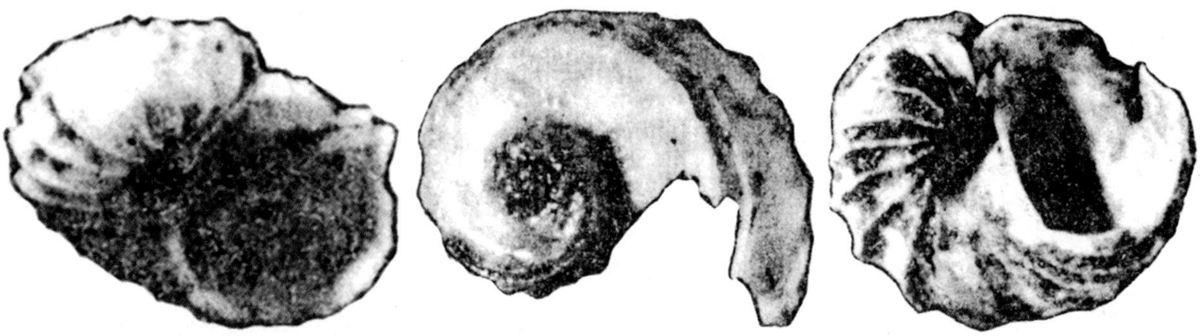

Figure 229. Original illustration of *Scissurella dolioliformis.*

Misidentifications.

not *Scissurella cossmanni*: Lozouet, 1986: 108, fig. 29D [is *Sci. tinaeformis*].

not *Scissurella cossmanni*: Lozouet, 1998: figs 4c, 4e [is *Sci. tinaeformis*].

not *Scissurella cossmanni*: Bandel, 1998: 11–12, pl. 2, figs 1–3 [is *Sci. transylvanica*].

Type material. Syntypes (whereabouts unknown, not MNHN), 1.5 × 1 mm (W × H).

Type locality. Gaas, Landes [France], 43.614°N, 1.034°W, late Tongrien [early Oligocene] (OD).

Etymology. In honor of French paleontologist Alexandre Édouard Maurice Cossmann (1850–1924).

Description. Shell medium-sized (to 1.6 mm. Fig. 230), trochiform, subquadrate. Protoconch of 0.75 whorl, with 13–18 fine axials, apertural varix present, connected to embryonic cap, apertural margin sinusoidal. Teleoconch I of 1–1.25 whorls, 11–18 axial cords

(Amitrov, 1996; Amitrov & Zhegallo, 1998), approximately eight moderately distinct spiral lines at onset of selenizone. Teleoconch II of up to 1.25whorls, suture impressed. Shoulder flat, early teleoconch II whorl with hardly any sculpture; axials increasing with growth from cords to low lamellae, initially restricted to adsutural portion of whorl, subsequently extending towards selenizone decreasing in strength from suture towards selenizone; interstices with finest growth lines; spiral lines increasing from approximately six to a dozen. Base with 8–20 strong raised axial cords from umbilicus to below selenizone, approximately one dozen weak spiral lines. Umbilicus open, without carina, walls straight, smooth. Aperture rounded. Selenizone above periphery, moderately wide; keels moderately strong, quite elevated; lunules moderately distinct, dense; slit open.

Comparisons. *Scissurella transylvanica* from the Miocene of Europe and *Sci. costata* from the Recent of Europe lack the smooth area just below the selenizone.

Scissurella tinaeformis from the Eocene of Europe has an umbilicus bordered by a carina, stronger axial sculpture on the shoulder of early teleoconch II, and the intersections of axials and spirals form strongly elevated points.

Distribution. Eocene and Oligocene of Europe (Fig. 231).

Figure 230. *Scissurella cossmanni*. Horizon U1, granite quarry Rybalskij near Dnipropetrovsk, Ukraine, 48.450°N, 35.142°E. Late Eocene (Priabonian: NP19 zone: see Müller & Rosenberg, 2003, for details. MMG UkT 14). Scale bar shell = 0.5 mm. Scale bar protoconch = 100 μm

Specimen records.

Ukraine. Granite quarry Rybalskij near Dnipropetrovsk, Ukraine, 48.450°N, 35.142°E. Late Eocene (Priabonian: NP19 zone: see Müller & Rosenberg, 2003, for details. MMG UkT 14).

Literature records.

France. Oligocene (Rupelian) of France (Armitrov, 1996).

Remarks. The number of syntypes was not mentioned in the original description, where only "a single lot" is noted.

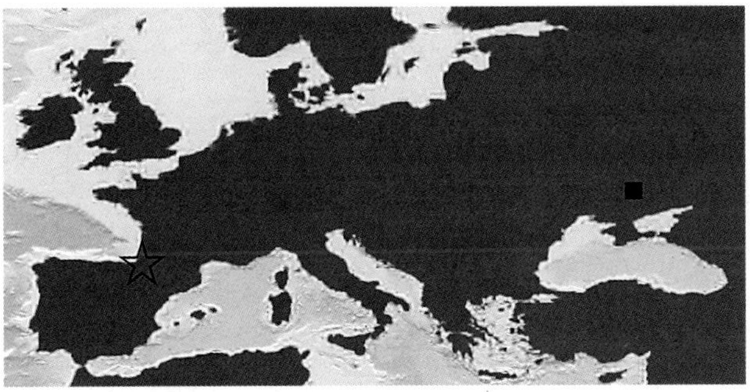

Figure 231. Distribution of *Scissurella cossmanni.* ☆ type locality. ■ literature record.

Scissurella dolioliformis was said to differ from *Sci. cossmanni* by the less pronounced spiral sculpture. The other common characters such as the flat shoulder, the strong axial ribs on the base and the carina surrounding the umbilicus, however, make the conspecificity of those specimens clear. Their age (Eocene-Oligocene) and distribution (Europe) is also compatible with them belonging to a single species.

Scissurella depontaillieri Cossmann, 1879 †

Figures 232–238

Chresonymy.

Scissurella depontaillieri Cossmann, 1879: 346–347, pl. 13, figs 8–9.
Scissurella depontaillieri: Cossmann & Lambert, 1884: 131–132, pl. 4, fig. 9.
Scissurella depontaillieri: Gougerot & LeRenard, 1977: 5.
Scissurella (*Anatoma*) sp.: Neuffer, 1978: 94–95, figs 5–7.
Praescissurella depontaillieri: Lozouet, 1986: 110, pl. 1, fig. 5, pl. 2, figs 4, 7.
Reussella depontaillieri: Bandel, 1998: 45–46, pl. 15, figs 4–7.
Praescissurella depontaillieri: Lozouet, 1998: figs 1e–f.
Scissurella depontaillieri: Geiger, 2003: 76.

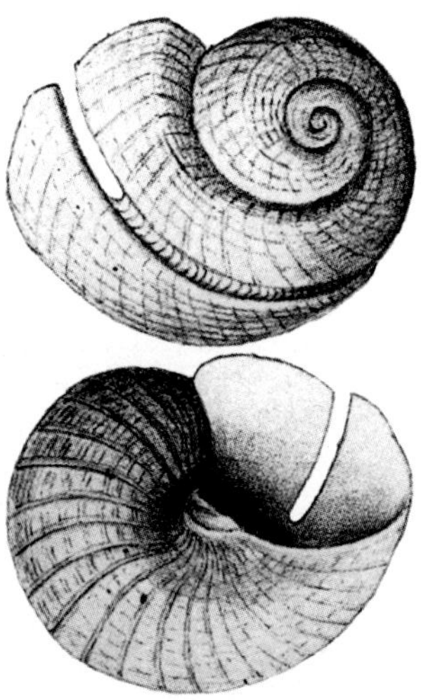

Figure 232. Original illustration of *Scissurella depontaillieri.*

Synonyms.

Scissurella corneti Rutot *in* Cossmann, 1914: 17–18, pl. 1, figs 53–55. ***Type material.*** Holotype (KBIN 3076. Glibert, 1973: fig. 1). Approximately 1 × 0.5 mm. Paratypes from type locality

(KBIN, coll. Houzeau, 2: *fide* Pacaud, 2004). ***Type locality.*** Puits Goffint [Mons. Montien of Belgium: inferred from title] (OD). Mine shaft of Goffint, Mons (Glibert, 1973). Mine shaft Coppée à Mons, Belgium (Pacaud, 2004). Montien of Belgium (Lozouet, 1986) = latest [early: *fide* Pacaud, 2004] Paleocene. ***Etymology.*** Named after François Léopold Cornet (1834–1887: see Dewalque, 1888, for a contemporary biography, Coan *et al.*, 2007). ***Remarks.*** The caption for pl. 1 indicates *Sci. corneti* to be shown in figs 63–65, an obvious error, and figs 53–55 are correct.

Scissurella corneti: Glibert, 1973: 7, pl. 1, fig. 1.

Praescissurella corneti: Lozouet, 1986: 110.

Praescissurella corneti: Pacaud, 2004: 585, 613–615, fig. 18E.

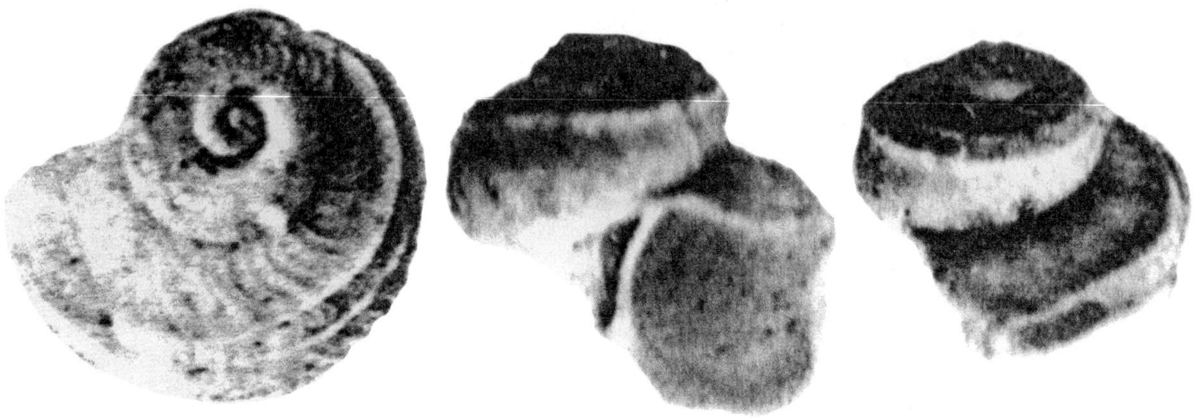

Figure 233. Original illustration of *Scissurella corneti.*

+ *Scissurella hungarica* Szöts, 1953: 22–23, pl. 1, figs 5–7. ***Type material.*** 12 lots of syntypes (GMH E2, 1; E.5581, 1: L. Kordos, pers. comm. 9/2006); 1.20 × 1.31 mm (H × W); whereabouts of other syntypes unknown. ***Type locality.*** Új-feltárás, Hosszú-harasztos, Hungary, middle Miocene (OD). = Csákvár, Transdaubia, Hungary [47.400°N, 18.450°E] (L. Kordos, pers. comm. 9/2006). ***Etymology.*** Named after its type locality in Hungary. ***Remarks.*** The type material is not illustrated because GMH does not send type material on loan. One specimen (GMH E2) was labelled as holotype; however, Szöts (1953) did not indicate a holotype, but only noted 12 lots without specifying the number of specimens in each lot or in total.

Praescissurella hungarica: Lozouet, 1986: 110, fig. 29F.

Maxwellella cf. *hungarica*: Bandel, 1998: 21–22, pl. 6, figs 6–7.

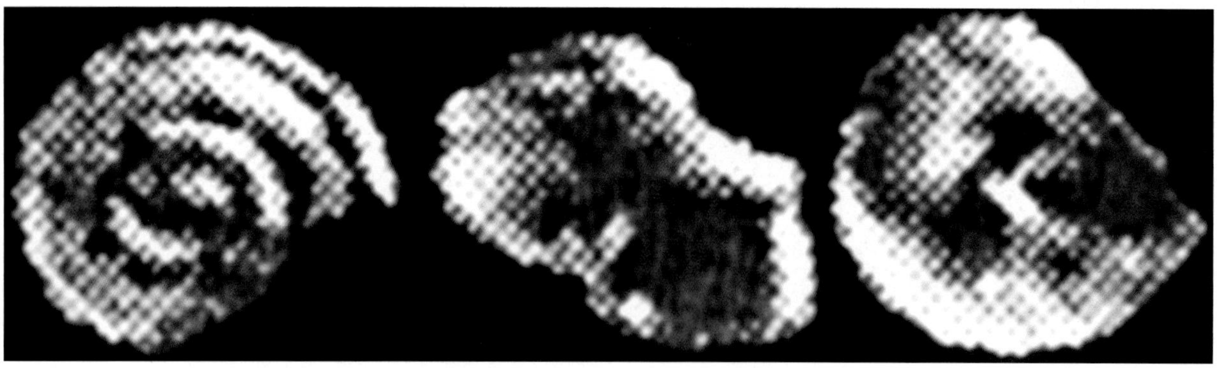

Figure 234. Original illustration of *Scissurella hungarica*.

Figure 235. Original illustration of *Scissurella praecrispata.*

+ *Scissurella* (*Anatoma*) *praecrispata* Gougerot & LeRenard, 1977: 4–6, 31, fig. 1. ***Type material.*** "Une trentaine" [some 30] syntypes (coll. Gougerot and coll. LeRenard), 1 × 1 mm. ***Type locality.*** Chaussy [ambiguous locality name], Fercourt, and Gourbesville [49.420°N, 1.420°W] (Oise) [France], middle Lutétien (zone IV of Abrand) [= middle Eocene] (OD).

+ *Scissurella tshapliensis* Amitrov, 1996: 99–100, figs 1L–N. ***Type material.*** Holotype (PIN 3595/714), 1.5 × 1.5 mm. ***Type locality.*** Rybalsk (Tschapli) quarry, Dnepropetrovsk, Ukraine [48°29'N, 35°0'E]. Mandrikovka Layers, late Eocene (OD). ***Etymology.*** Named after the type locality (OD).

Scissurella tshapliensis: Amitrov & Zhegallo, 1998: 7, 9, figs 2G–K.

+ *Praescissurella oligomiocaenica*: Lozouet, 1986: 110, 127–128, text-figs 35D–E, pl. 2, figs 5–8 [unavailable, unpublished].

Praescissurella oligomiocaenica Lozouet, 1998: 66, fig. 3d. ***Type material.*** Holotype (MNHN. Fig. 237A), 1 × 1.15 mm (H × W). Paratypes from type locality (MNHN, 25). ***Type locality.*** Villandraut (Gamachot), Gironde [France, 44°28'N, 0°22'W]. Early Miocene (Aquitanian), marnocalcaire of Porites (OD). ***Etymology.*** Known from the Oligocene to the middle Miocene in Europe (OD). ***Remarks.*** Lozouet (1998) did not indicate how his species differs from any of the ones here considered synonymous.

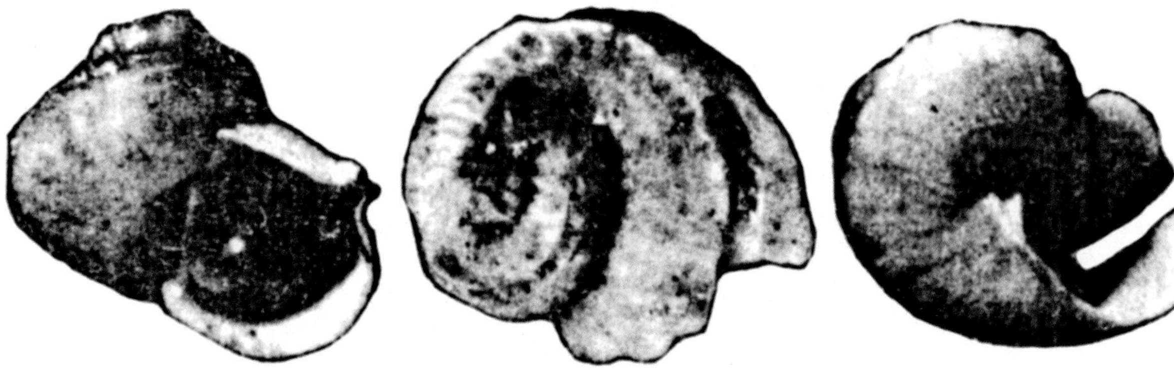

Figure 236. Original Illustration of *Scissurella tshapliensis.*

Misidentifications.

? *Scissurella* (*Anatoma*) *philippiana*: Kuster-Wendenburg, 1982: 84 [confused identification as both Semper (1865) and Neuffer (1979) are cited, the latter showing *Sci. depontaillieri*].

not *Anatoma philippiana*: Gürs, 1995: 28–29, pl. 3, figs 2a–c [is *Sci. depontaillieri*].

Type material. (whereabouts unknown), 1.5 × 1 mm.

Type locality. Jeures near Etampes, Sands of Fontainebleau [France, 48.433°N, 2.167°E] (OD). Oligocene (Gougerot & LeRenard, 1977).

Etymology. Named after Jules Alexandre Depontaillier (1850–1882: see Coan *et al.*, 2007).

Description. Based on illustration in Lozouet (1998). Shell medium-sized (to 1.1 mm. Fig. 237), trochiform, inflated. Protoconch of 0.75 whorl, embryonic cap with reticulate sculpture, remainder with fine axials, apertural varix connected to embryonic cap, shape of apertural margin sinusoidal. Teleoconch I of 1.25 whorls, approximately 24 low axial cords, single spiral of equal strength in position of selenizone. Teleoconch II of 1.3 whorls, suture hardly impressed. Shoulder with approximately 30 low, indistinct, irregularly spaced axial cords, spiral lines increasing in number from 2 to 3–12. Base with slightest constriction below selenizone, inflated, axial cords as on shoulder approximately 19 very fine, irregularly spaced spirals. Selenizone

Figure 237. *Scissurella depontaillieri*. **A.** *Scissurella oligomiocaenica*. Holotype. Villandraut (Gamachot), Gironde [France, 44°28'N, 0°22'W] (MNHN). **B.** Site M3, Mammendorf, Germany, 52.177°N, 11.437°E, early Oligocene (Rupelian: NP22 zone: see Müller, 2011, for details. MMG AnT 2). Scale bars shell = 1 mm. Scale bars protoconch = 100 μm.

above and slightly inside periphery; keels low, weak (possibly broken); lunules low, indistinct; slit open, with parallel margins. Umbilicus open, with distinct funiculus. Aperture subquadrate.

Comparisons. *Scissurella aliceae* from the Paleocene of Europe has a less turreted overall shell shape, a much wider umbilicus, and fewer spiral threads on the shoulder.

Scissurella bituminata from the Oligocene of Indonesia has stepped spiral cords near the umbilicus and the unique washboard sculpture immediately adjacent to the umbilicus.

Scissurella marchmontensis from the Late Cretaceous of the Caribbean and Spain has distinctly opisthocline axial cords on the base.

Scissurella peyrerensis from the Oligocene of Europe has strong axial lamellae on the adsutural half of the shoulder as well on the periphery of the base, while the base is strongly constricted below the selenizone.

Scissurella ravni from the Paleocene of Europe has much more distinctly elevated axial cords to low lamellae on shoulder and base, which form distinct knobs at the intersection with the spiral lines.

Scissurella transylvanica from the Miocene of France has strong axial cords on shoulder and base, a wider shell, a more rounded profile of the whorls, and lack the funiculus in the umbilicus.

Sinezona deshayesi from the Eocene of France has a wider overall shell with more rounded whorl profile and lacks a funiculus in the umbilicus.

Distribution. Latest Paleocene through middle Miocene of Europe (Fig. 238).

Specimen records.

Germany. Mammendorf, Germany, 52.177°N, 11.437°E, early Oligocene (Rupelian: NP22 zone: see Müller, 2011, for details. MMG AnT 2).

Literature records.

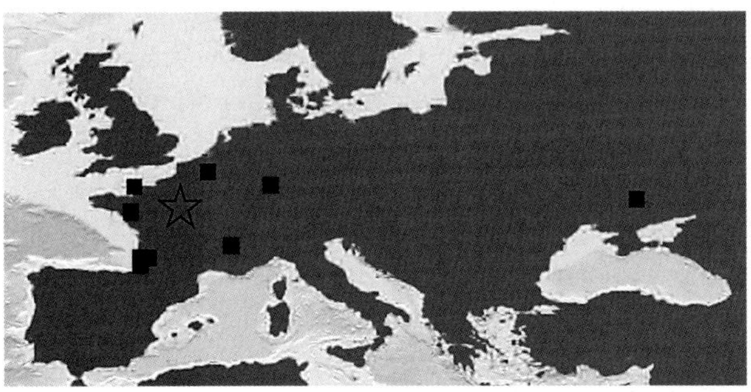

Figure 238. Distribution of *Scissurella depontaillieri*. ☆ type locality. ■ literature record.

France. Puits Coppée (Cossmann, 1914: as *Sci. corneti*).— Peyrère, SW France [44.670°N, 5.620°E], late Oligocene. Saffrè, Paris Basin [47.502°N, 1.579°W], middle Miocene (Bandel, 1998: as *Sci. hungarica*).— Gaas (Espibos, Lagourade) [43.614°N, 1.034°W], early Oligocene. St.-Paul-lès-Dax (Estoti, Abesse, Bezoye B) [44.333°N, 1.050°W], Peyrehorade (Peyrère) [44.670°N, 5.620°E], St.-Etienne-d'Orthe (Ruisseau de l'Église B) [43.611°N, 1.174°W], late Oligocene. St.-Paul-lès-Dax [44.333°N, 1.050°W], early Miocene (Aquitanian). St.-Paul-lès-Dax (Cabanes), early Miocene (early Burgidalian). St.-Martin-de-Hinx [43.578°N, 1.275°W], late Burdigalian (Lozouet, 1998: as *Sci. oligomiocaenica*).

Belgium. Paleocene (Montian) of Belgium (Armitrov, 1996: as *Sci. corneti*).

Hungary. Eocene (Lutetien) of Hungary (Armitrov, 1996: as *Sci. hungarica*).

Germany. Lattorfium–Rupelium (Oligocene) of Mainz Basin (Würzmühle, Eckelsheim, Heimberg, Welschberg, Langenlonsheim (Gürs, 1995).—Sand pit Gaul, Steigerberg near Eckelsheim/Rhh., Mainz Basin (Neuffer, 1978).

Remarks. The date of publication is uncertain. The frontispiece and text are dated 1913, while the plates are dated 1914. The latter is taken as the correct publication date.

Scissurella corneti cannot be differentiated from *Sci. depontaillieri* from the Oligocene of France. The overall shell-shape, the umbilicus with distinct funiculus, and the moderate axial sculpture on shoulder and base all are identical. Gougerot &

LeRenard (1977) noted in their differentiation from *Sci. praecrispata* a more narrow umbilicus, which is not evident from the original illustrations, and a different curve, a statement that cannot be further evaluated. Neither Rutot *in* Cossmann (1914) nor Szöts (1953) compared their *Sci. corneti* and *Sci. hungarica*, respectively, to *Sci. depontaillieri*.

Scissurella praecrispata was established mainly to segregate *An. crispata* - like shells from the Tertiary, applying the chronospecies concept. The original description does not cite any discrete characters that could separate *An. crispata* from *An. praecrispata*. The line drawings do not furnish any additional information not conveyed in the written description. The species is not a member of *Anatoma*, but rather of *Scissurella*, due to the position of the selenizone above the periphery. Lozouet (1986) synonymized it under *hungarica*.

There are no perceptible, significant differences between the six taxa discussed here. They all are small scissurellids (~1.5 mm) of a squat, stepped overall outline, with a distinct funiculus and the most common fine reticulate sculpture. All species have been described from the Paleocene through Miocene of Europe. Hence, they are here synonymized under the oldest name.

The most dissimilar taxon is *Sci. oligomiocaenica*, which has slightly elevated low axial lamellae on the middle of the base, similar to *Sci. evaensis*. Based on the name, Lozouet (1998) considered the holotype from the Miocene of France conspecific with the similarly shaped specimens from the Oligocene, for which reason *Sci. oligomiocaenica* is here synonymized under the Oligocene species. There are several possibilities of addressing this issue:

- It is a single species ranging from at least the Eocene/Oligocene to the Miocene with overall consistent teleoconch morphology. A few specimens have the short axial lamellae on the base. This is the view taken here.
- There are two species, both ranging from Eocene/Oligocene through Miocene, with and without short axial lamellae on the mid base. With spiral lamellae: *Sci. oligomiocaenica* from Oligocene through Miocene. Without spiral lamellae: *Sci. depontaillieri* = *Sci. corneti* = *Sci. praecrispata* = *Sci. tshapliensis* from the Eocene/Oligocene; = *Sci. hungarica* from the Miocene.
- There are two species, one restricted to Eocene/Oligocene, one from Miocene only; the temporal range indication for *Sci. oligomiocaenica* is based on a mixed species concept. Eocene/Oligocene: *Sci. depontaillieri* = *Sci. corneti* = *Sci. praecrispata* = *Sci. tshapliensis*. Miocene: *Sci. hungarica* = *Sci. oligomiocaenica*; the very coarse grained illustrations of *Sci. hungarica* (Fig. 234) may well hide the spiral lamellae on the middle of the base.

Only examination of larger series and all type material can resolve this question.

Scissurella eocaenica Bandel, 1998 †

Figures 239–240

Chresonymy.

Scissurella eocaenica: Lozouet, 1986: 109 [unpublished].
Scissurella eocaenica Bandel, 1998: 17–18, pl. 4, figs 5–7.
Scissurella eocaenica: Geiger, 2003: 77.

Figure 239. Original illustration of *Scisssurella eocaenica*.

Type material. Holotype (RMNH).

Type locality. Saffrè, Paris Basin [France, 47.502°N, 1.579°W], Eocene (Lutet V) (OD).

Etymology. Named after the stratigraphic position.

Description. [Based on original illustration] Shell trochiform, depressed, small (to 0.8 mm). Protoconch 0.75 whorl, 140 µm, fine axials, apertural varix connected to embryonic cap, apertural margin sinusoidal. Teleoconch I of 0.75 whorl, approximately 12 axial cords, no spiral sculpture, interstices with fine irregular growth lines. Teleoconch II of at least 0.8 whorl, suture deeply impressed. Shoulder convex, approximately 10 strong raised axial cords, one indistinct spiral cord. Base barely constricted below selenizone, same density and strength of axial cords as on shoulder, approximately 10 fine spiral lines. Umbilicus bordered by strong carina; Umbilicus with straight, smooth walls. Selenizone above periphery; keels moderately strong, moderately elevated [mostly eroded]; distinct, dense lunules; only beginning of slit visible, remainder of aperture broken. Aperture round, roof overhanging.

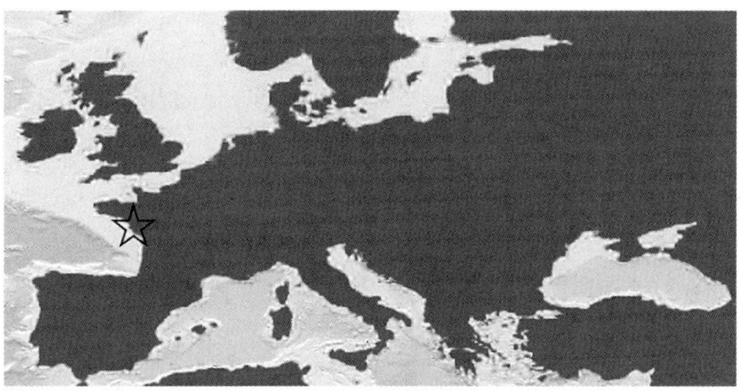

Figure 240. Distribution of *Scissurella eocaenica*. ☆ type locality.

Distribution. Known only from type locality (Fig. 240).

Comparisons. *Scissurella annulata* from the Paleocene of Europe has at least twice as many axial cords and has a lower overall shell profile.

Scissurella condita from the Late Cretaceous to Miocene of New Zealand has twice as many axial cords on the shoulder and base.

Scissurella gruendeli from the Middle Jurassic of Europe has twice as many axial cords on shoulder and base and the selenizone is higher on the shoulder.

Scissurella marchmontensis from the Late Cretaceous through Oligocene of Europe has a similar protoconch, but a much longer teleoconch I (1.3 vs. 0.75 whorls), and quite similar sculpture on the available portions of the whorls.

Remarks. *Scissurella eocaenica* is evidently based on a immature specimen.

Scissurella gruendeli (Bandel, 1998) †

Figures 241–242

Chresonymy.

Maxwellella gründeli [sic] Bandel, 1998: 20–21, pl. 6, figs 3–5.
Maxwellella gruendeli: Geiger, 2003: 24 [emendation].

Figure 241. Original illustration of *Maxwellella gruendeli.*

Type material. Holotype (Collection Gründel), 1.01 × 0.74 mm (W × H).

Type locality. Drilling at Klemmen (Kleby), Poland [53.800°N, 14.967°E], early middle Callovian (Middle Jurassic) (OD).

Etymology. Named after J. Gründel (OD).

Description. Shell medium-sized (to 1 mm) planispiral, apex slightly sunken. Protoconch of 0.75 whorl, sculpture smooth (possibly eroded), apertural varix present, apertural margin convex. Teleoconch I of 1.25 whorls, approximately 33 low-profile, broad, axial cords. Approximately 12 finer spiral cordlets sometimes eroded. Teleoconch II of at least 0.5 whorl, suture impressed. Shoulder convex, density of axials and spirals similar as on teleoconch I. Umbilicus open, wide. Selenizone above periphery, orientation almost parallel to axis, keels very low (possibly eroded).

Distribution. Known only from type locality (Fig. 242).

Comparisons. *Scissurella annulata* has the selenizone at approximately 45° angle to shell axis and the selenizone reaches the periphery.

Scissurella condita from the Late Cretaceous to Miocene of New Zealand has the selenizone at an angle of 45° to the shell axis.

Juvenile *Sci. peyrerensis* from the Oligocene of Europe lacks spiral sculpture.

Scissurella eocaenica from the Eocene of Europe has only half as many axial cords on shoulder and base.

Remarks. The spelling of the species name in the original description was not formed in accordance with ICZN Art. 32.5.2.1; Geiger (2003) corrected the spelling to *gruendeli*.

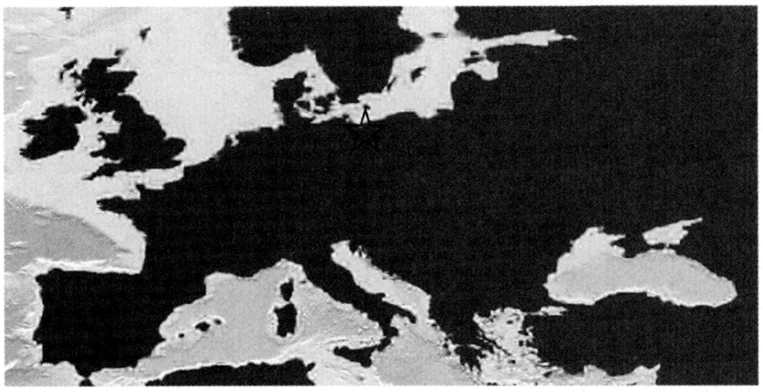

Figure 242. Distribution of *Scissurella gruendeli*. ☆ type locality.

Scissurella hedegaardi Bandel, 1998 †

Figures 243–244

Chresonymy.

Scissurella hedegaardi Bandel, 1998: 16–17, pl. 4, figs 3–4.

Misidentifications.

Anatoma hedegaardi: Schnetler *et al.*, 2000: 86–88, pl. 2, figs 1–3 [is *Anatoma* sp.].

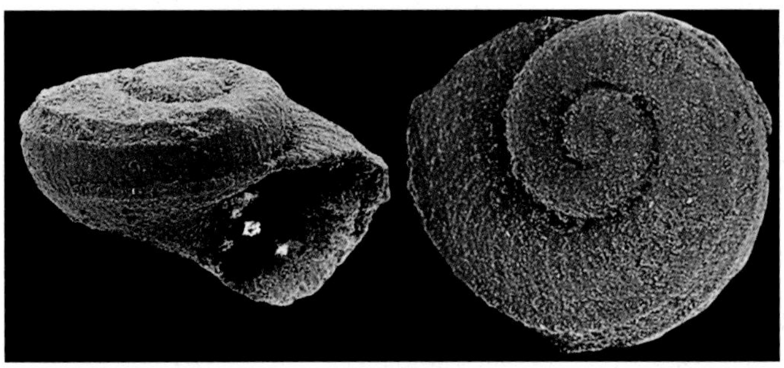

Figure 243. Original illustration of *Scissurella hedegaardi*.

Type material. Holotype (SGPIH 3872), 0.5 mm.

Type locality. Faxe, Denmark [55.250°N, 12.130°E]. [Danian, early] Paleocene (OD).

Etymology. Named after Claus Hedegaard [1963–2009], who collected the material (OD).

Description. Shell small (0.5 mm), trochiform. Protoconch 0.13 mm [Bandel, 1998], details not visible due to poor preservation. Teleoconch I of 0.75 whorl, sculpture not discernible. Teleoconch II with almost one whorl, approximately 40 axial cords. Base with faint axial lines. Other descriptive attributes not discernible.

Comparisons. *Scissurella aliceae* from the Paleocene of Europe has a longer teleoconch I (1.5 vs. 0.75 whorls) and has a very wide umbilicus.

Figure 244. Distribution of *Scisssurella hedegaardi*. ☆ type locality.

Scissurella bituminata from the Oligocene of Indonesia is not distinguishable based on currently available data.

Scissurella depontaillieri from the Oligocene to Miocene of Europe has an overall more turreted shell shape.

Scissurella marchmontensis from the Late Cretaceous of the Caribbean and Europe has distinctly opisthocline axial cords on the base.

Scissurella peyrerensis from the Oligocene of Europe has strongly elevated axial lamellae in the adumbilical half of the shoulder and the periphery of the base and a distinct constriction below the selenizone on the base.

Scissurella ravni from the Paleocene of Europe is overall more turreted and has more distinct elevated axial cords to low lamellae on shoulder and base.

Scissurella transylvanica from the Oligocene to Miocene of Europe is overall more globular in shell outline and has more distinct axial cords on shoulder and base.

Distribution. Known only from type material (Fig. 244).

Remarks. The species is rather globular but carries the selenizone clearly on the shoulder rather than at the periphery; hence, it is classified in *Scissurella*. The specimen illustrated by Bandel is of poor quality, yet, Schnetler *et al.*'s (2000) specimen is certainly not conspecific with Bandel's; Schnetler *et al.*'s specimen is an *Anatoma* species.

Scissurella marchmontensis Sohl, 1992 †

Figures 245–248

Chresonymy.

Scissurella marchmontensis Sohl, 1992: 428–431, figs 9.5–9.9, 10.1–10.6.

Scissurella marchmontensis: Kiel & Bandel, 2000: 272.

Synonym.

+ *Scissurella lleidania* Kiel & Bandel, 2000: 272–273, pl. 1, figs 10–11. ***Type material.*** Holotype (SGPIH 3966), 1 × 1.5 mm (H × W). Paratype from type locality (SGPIH 3967, 1). ***Type locality.*** Valley system around Torallola, Toralla and Sensui near Pobla de Segur in the Tremp basin of the Spanish Pyrenees [Spain, 42.250°N, 0.967°E]. Puimanyons Olisthostrom (Campanian [Late Cretaceous]) of the Valcarga Formation (OD). ***Etymology.*** Named after the province of Lleida [, Spain], to which the locality of Rorallola belongs (OD).

Scissurella lleidania "n. sp.": Kiel, 2001: 14–15, pl. 1, figs 10–11.

Figure 245. Original illustration of *Scissurella marchontensis*. Apertural view of holotype, remainder of paratypes.

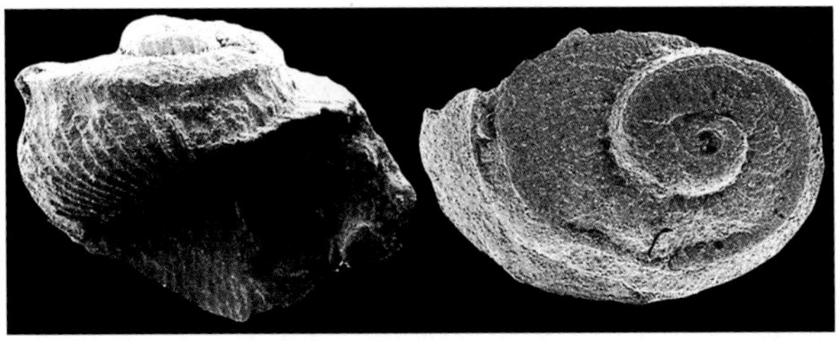

Figure 246. Original illustration of *Scissurella lleidania.*

Type material. Holotype (USNM 452684). Paratypes from type locality (USNM 452683, USNM 45285–88, 4). Marchmont, Jamaica (USNM 452689–90, 2). Near Sabana Grande, Puerto Rico (USNM 652701, 1).

Type locality. Locality 16, *Titanosarcolites* limestone. From a calcarenite interbedded in a 3 m shale section exposed in cuts of the Marchmont to Cambridge road 1.1 km NW of Marchmont, Marchmont Inlier, Westmoreland Parish, Jamaica, 18°16'9"N, 77°53'19"S . Late Maastrichian (= Late Cretaceous) (OD).

Etymology. Named for the type locality (OD).

Description. Shell medium-sized (to 1.5 mm), trochiform inflated. Protoconch of 0.75 whorl, with strong axial sculpture, apertural varix present (possibly connected to embryonic cap), apertural margin unknown. Teleoconch I of approximately 0.66 whorl, approximately 15 strong opisthocline axial cords, no spiral sculpture. Teleoconch II of at least 1.25 convex whorls, suture little impressed. Shoulder with approximately 43 strong axial cords, no spirals. Base with constriction below selenizone, followed by faint rounded ridge; density and strength of axials as on shoulder, approximately 12–20 spirals, spirals somewhat to much weaker than axial cords. Umbilicus open, wide, carina at transition to base, walls smooth. Selenizone above periphery, keels moderately strong, moderately elevated; lunules irregular, distinct; slit open.

Comparisons. *Scissurella aliceae* from the Paleocene of Europe has straight axial cords and a very pronounced umbilicus.

Scissurella bituminata from the Oligocene of Indonesia has a very narrow umbilicus bordered by the unique washboard sculpture.

Scissurella depontaillieri from the Oligocene to Miocene of Europe has an overall more turreted shell, a very narrow umbilicus with distinct funiculus, and straight axials cords.

Scissurella hedegaardi from the Paleocene of Europe has the periphery at mid-base.

Scissurella peyrerensis from the Oligocene of Europe has strongly elevated axial lamellae at the adsutural half of the shoulder and at the periphery of the base and a strong constriction of the base below the selenizone.

Scissurella ravni from the Paleocene of Europe has strongly elevated axial cords to low lamellae on shoulder and base, which are straight.

Scissurella transylvanica from the Miocene of Europe has approximately half as many but twice as strong orthocline axial cords and a more narrow umbilicus.

Distribution. Puerto Rico, Sabana Grande quadrangle in limestones with the El Rayo Formation at locality 4, Jamaica Marchmont Inlier in *Titanosarcolites* limestones at localities 13 and 16 (Sohl, 1992). Spain (Kiel & Bandel, 2000. Fig. 248). Late Cretaceous.

Ecology. Found in back reef lagoon environments, and from rudist assemblages on a carbonate platform (Sohl, 1992).

Remarks. Kiel & Bandel (2000: 272) noted their *Sci. lleidania* is "distinct only by its convex columellar lip" from *Sci. marchmontensis*. The differences in columellar lip shape, however, are only a product of differential breakage of the aperture, hence, have no taxonomic significance. Although the Caribbean and the European coastline are separated today by the Atlantic Ocean, in the uppermost Cretaceous (~70 MYA),

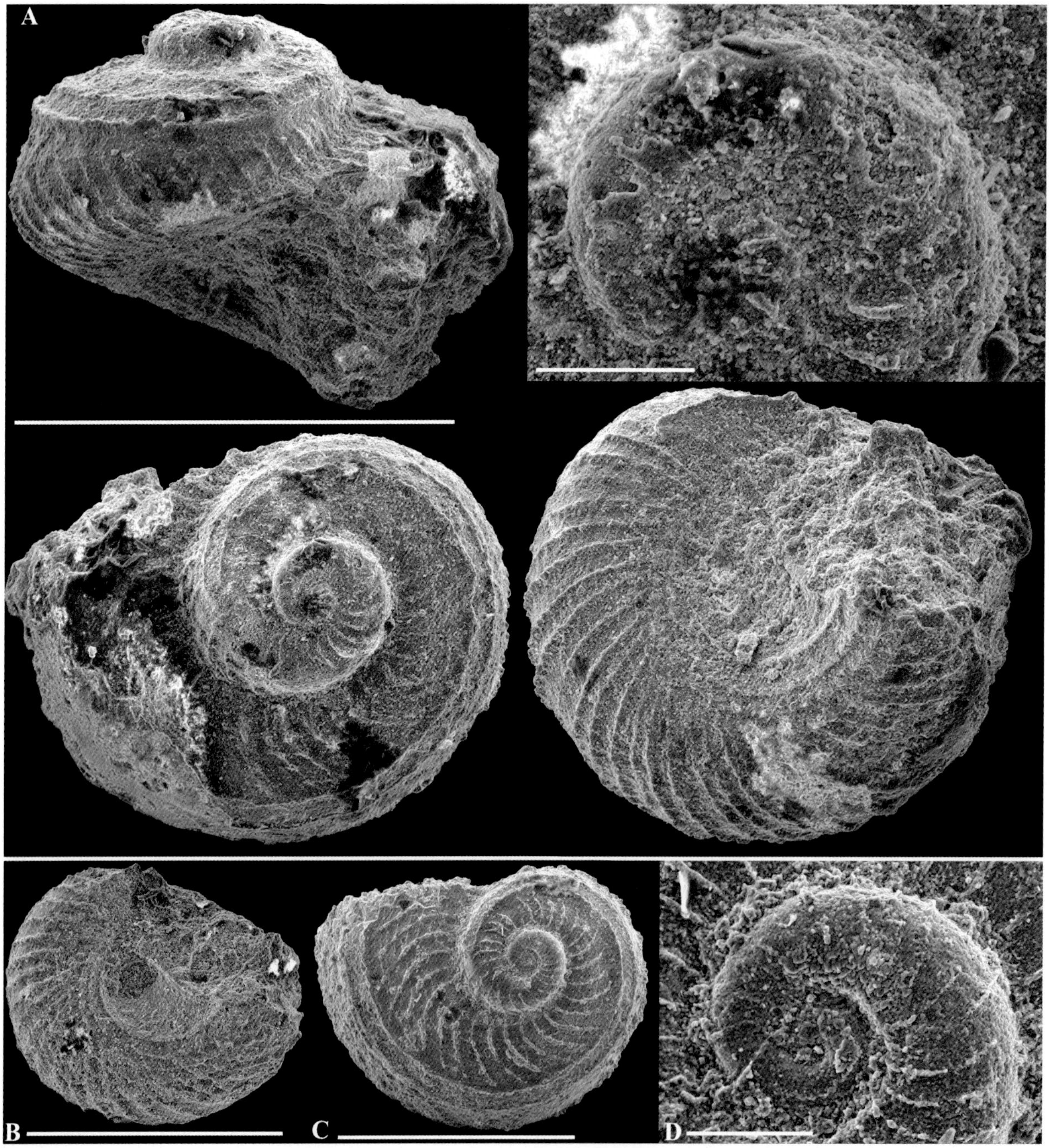

Figure 247. *Scissurella marchontensis*. Marchmont, Jamaica. **A.** Holotype (USNM 452684). **B.** Paratype (USNM 452688), umbilical view. **C–D.** Apical view and protoconch of paratype (USNM 452689). Scale bars shell = 1 mm. Scale bars protoconch = 100 µm.

Figure 248. Distribution of *Scissurella marchmontensis*. ☆ type locality. ■ literature record.

the ocean basin was much more narrow; the Atlantic only began to open around 110 MYA.

Scissurella lleidania was properly described by Kiel & Bandel (2000). Kiel's (2001) dissertation also contains the description, most likely from the manuscript of Kiel & Bandel (2000). As the dissertation is not published according to ICZN criteria, no junior homonym was introduced.

Kiel & Bandel (2000) referred to axial ribbing and two spiral cords in the umbilicus. That portion of the shell, in fact, is the lower part of the base, with the innermost spiral bordering the umbilicus proper, which has smooth walls, and no spirals.

Scissurella peyrerensis Bandel, 1998 †

Figures 249–251

Chresonymy.

Scissurella peyrerensis Bandel, 1998: 18–19, pl. 4, fig. 8, pl. 5, figs 1–3.

Praescissurella peyrerensis: Lozouet, 1986: 110, 128–129, text-fig. 29C., pl. 2, fig. 10 [unavailable, unpublished].

Synonym.

+ *Scissurella hispanica* Kiel & Bandel, 2000: 272, pl. 1, figs 8–9. ***Type material.*** Holotype (SGPIH 3965), 0.1 × 1 mm (H × W). ***Type locality.*** Valley system around Torallola, Toralla and Sensui near Pobla de Segur in the Tremp Basin of the Spanish Pyrenees [Spain, 42.150°N, 0.580°E]. Puimanyons Olisthostrom (Campanian [Upper Cretaceous]) of the Valcarga Formation (OD). ***Etymology.*** Named after Spain, the country of its occurrence (OD).

Scissurella hispanica "n. sp.": Kiel, 2001: 14, pl. 1, figs 8–9 [unpublished dissertation].

Figure 249. Original illustration of *Scissurella peyrerensis*.

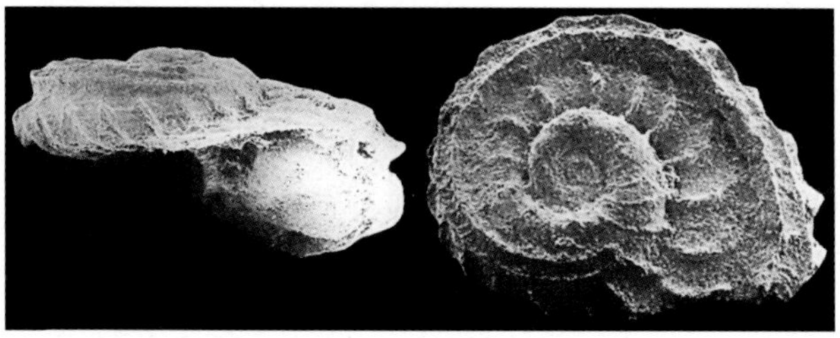

Figure 250. Original illustration of *Scissurella hispanica*.

Type material. Holotype (RMNH), 1.2 × 1.2 mm.

Type locality. Peyrère, Aquitanian, France [44.670°N, 5.620°E], late Oligocene (OD).

Etymology. Named after its type locality.

Description. Shell medium-sized (to 1.2 mm), trochiform, subquadrate. Protoconch of 0.75 whorl, smooth, apertural varix not connected to embryonic cap, apertural margin slightly sinusoidal. Teleoconch I of 1.33 whorls, approximately 26 axial cords, some incompletely formed, interstices with fine irregular axial growth lines, no spiral sculpture. Teleoconch II of 1.25 whorls, suture deeply impressed. Shoulder overall flat, with slightly sinusoidal profile. Axial sculpture becoming restricted to adsutural half of shoulder, increasing in strength with growth to form distinctly curved, low lamellae. Very faint, irregular spiral threads sometimes visible. Base constricted below selenizone, shallow spiral keel below constriction. Approximately 17 axials on body whorl, axials most elevated in position of spiral keel, making keel more prominent. Fine spiral threads present. Base with distinct keel at transition to open umbilicus with smooth walls; funiculus very weak. Selenizone above periphery; lunules indistinct; keels strong, of medium elevation; slit open, margins parallel. Aperture subquadrate.

Comparisons. *Scissurella aliceae* from the Paleocene of Europe has a much wider umbilicus and lacks the strong axial lamellae.

Scissurella bituminata from the Oligocene of Indonesia lacks the strong axial lamellae and has the characteristic washboard sculpture adjacent to the umbilicus.

Scissurella depontaillieri from the Miocene of Europe has a more narrow umbilicus with a very distinct funiculus, while any axial sculpture is at best indistinct.

Scissurella eocaenica from the Eocene of Europe has a shorter teleoconch I (0.75 vs. 1.3 whorls) and the sculpture on the available portions of the whorls are quite similar. *Scissurella eocaenica* is evidently based on an immature specimen.

Scissurella ravni from the Paleocene of Europe has a shorter teleoconch I (1 vs. 1.3 whorls) with a spiral cord in position of the selenizone and more axial cords on teleoconch II.

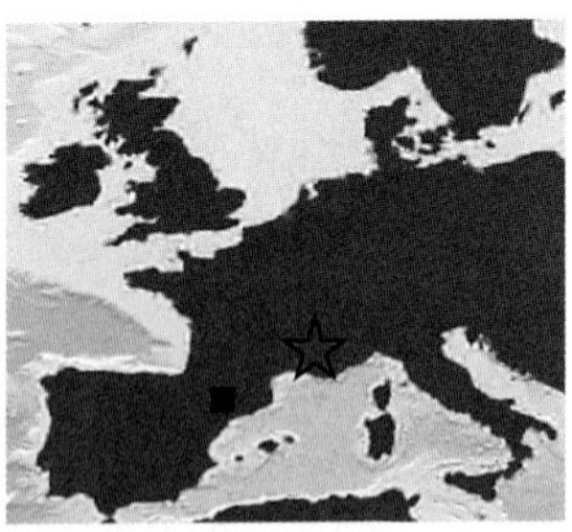

Figure 251. Distribution of *Scissurella peyrerensis*. ☆ type locality. ■ literature record.

Scissurella transylvanica from the Miocene of Europe has an overall wider and more globular rounded shell, axial cords extending all across the shoulder, and axial cords on shoulder and base that are crossed by half as many, twice as strong spiral lines.

Scissurella marchmontensis from the Late Cretaceous of the Caribbean and Europe has distinctly opisthocline axial cords.

Scissurella tinaeformis from the Eocene of Europe (a potential filler of the stratigraphic gap) shares the 1.3 teleoconch I whorls, the density and strength of the axial cords, but the axial cords on the shoulder may extend further towards the selenizone compared to *peyrerensis*. Additionally, the nature of the spiral sculpture cannot be assessed based on the original illustration. The only difference is the protoconch sculpture, which, however, usually is a good indicator for the validity of a taxon.

Distribution. Late Cretaceous to Oligocene of Europe (Fig. 251).

Remarks. Bandel (1986) used "*Scissurella peyrerensis* (Lozouet, 1986) n. sp." in the caption for the species description. The species was described by Bandel (1998), the description being taken from the dissertation of Lozouet (1986).

Scissurella hispanica was properly described by Kiel & Bandel (2000). Kiel's (2001) dissertation also contains the description, most likely from the manuscript of Kiel & Bandel (2000). As the dissertation is not published according to ICZN criteria, no junior homonym was introduced.

Scissurella peyrerensis and *Sci. hispanica* agree in all available morphological details: axial sculpture on shoulder as strong lamellae extending from near suture to two thirds across shoulder, strong constriction below selenizone, strong axial lamellae at same density and strength as on shoulder crossed by very fine spiral lines. The main difference between the nominal taxa is the height of the shell, which can be accounted for by the smaller size (1 vs. 1.2 mm) and diagenetic compression in the case of the older *hispanica*. The range of ages covered (uppermost Cretaceous to Late Oligocene) is well within the realm of possibilities. Kiel & Bandel (2000) did not compare their *hispanica* to *peyrerensis*.

Scissurella ravni (Schnetler, Lozouet & Pacaud, 2000) †

Figures 252–253

Chresonymy.

Praescissurella ? *ravni* Schnetler, Lozouet & Pacaud, 2000: 84–86, pl. 1, figs 4–6, pl. 3, figs 5a–c.

Type material. Holotype (MGUH 25752), 1.2 × 1.6 mm. Paratypes from type locality (MGUH 25751, 1; MGUH 25761, 1; MNHN-LP R63045, 1: not found in 4/2005, D. L. Geiger, pers. obs.; SMF 321221, 1; ISL, 3; ARF, 1).

Figure 252. Original illustration of *Scisssurella ravni*.

Type locality. Faske, Denmark [55.250°N, 12.130°E]. Coral Limestone, middle Danian, Paleocene (OD).

Etymology. Named for Jesper Peter Johannes Ravn (1866–1951: OD).

Description. Shell medium-sized (1.6 mm), trochiform, subquadrate. Protoconch of 1.5 whorls, sculptural details unknown. Teleoconch I of 0.625 whorl with approximately 11 axial cords, shoulder angulated in anticipation of selenizone. Teleoconch II of 1.75 whorls, suture little impressed. Shoulder with strong, raised axial cords, approximately 27 on body whorl, approximately five spiral lines running over axials, forming small pustules at intersection. Base with similar sculpture, axial cords protruding over keels of selenizone on first teleoconch II whorl in apical view, approximately 12 spiral lines. Umbilicus open wide, walls smooth, with strong keel separating umbilicus from base, no funiculus. Selenizone on margin of shoulder; keels strong, distinct; slit open, margins parallel. Aperture subquadrate, with somewhat concave umbilical region.

Distribution. Known only from type material (Fig. 253).

Comparisons. *Scissurella aliceae* from the Paleocene of Europe has much weaker axial cords and a wider umbilical cavity.

Scissurella bituminata from the Oligocene of Indonesia has weaker axial cords and the unique washboard sculpture surrounding the umbilicus.

Scissurella depontaillieri from the Oligocene to Miocene of Europe has a more narrow umbilicus and much weaker axial sculpture.

Scissurella marchmontensis from the Late Cretaceous of the Caribbean and Europe has distinctly opisthocline and much weaker axial cords and a less distinct spiral cord at the edge of the umbilicus.

Scissurella peyrerensis from the Oligocene of Europe has a more elevated spire and fewer axials (14 on body whorl) and lacks the spiral in position of the selenizone on shorter teleoconch I (1 vs. 1.3 whorls). Additionally, the axial folds only extend on the adsutural half of the shoulder and the periphery of the base.

Scissurella transylvanica from the Oligocene to Miocene of Europe has an overall more globular shell shape and a much more narrow umbilicus.

Figure 253. Distribution of *Scissurella ravni*. ☆ type locality.

Scissurella szaboi Conti & Monari, 1991 †

Figures 254–255

Chresonymy.

Scissurella? (*Scissurella*?) *szaboi* Conti & Monari, 1991: 265–266, text-figs 13–16, pl. 7, figs 1–6.

Type material. Holotype MAC. BGA, 1 × 1 mm. Paratypes from type locality (MAC 196, 1; MAC 197, 1; MAC, 194, 1; MAC, 198, 1; MAC 193, 1; BGA, 1; MAC 220 BGA 1bis, 1; MAC 224, 2).

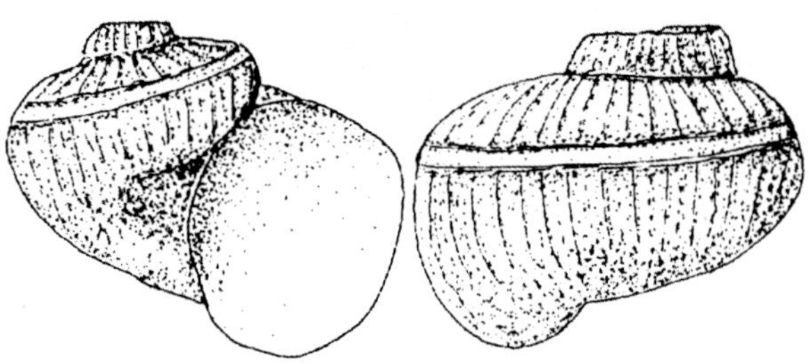

Figure 254. Original Illustration of *Scissurella szaboi*.

Type locality. Günören (BGA section), Bilecik (Western Pontides, Turkey) [40.166°N, 29.983°E], late Sinemurian/early Pliensbachian (Raricostatum to Ibex zones) [Early Jurassic] (OD).

Etymology. Named after Janos Szabó (OD).

Description. Shell medium-sized (to 1.8 × 1.6 mm), trochiform, globular. Protoconch unknown. Teleoconch I whorls unknown. Approximately one teleoconch II whorl. Shoulder with axial cords, apparently no spiral sculpture. Base with similar sculpture as on shoulder. Umbilicus open.

Comparisons. Given the very crude drawing provided as the original illustration, it is impossible to clearly distinguish this nominal taxon from other species.

Distribution. Known only from type locality (Fig. 255).

Remarks. The description here is based on the rather crude drawings from the original description. It is impossible to evaluate the status of this taxon, and it is here given for the sake of completeness.

The MAC material is stored in the Dipartimento di Scienze della Terra dall'Università di Roma "La Sapienza" collection number NS8. Some material was intended to be transferred to the Middle East Technical University, Anakara, Turkey (Conti & Monari, 1991: 247).

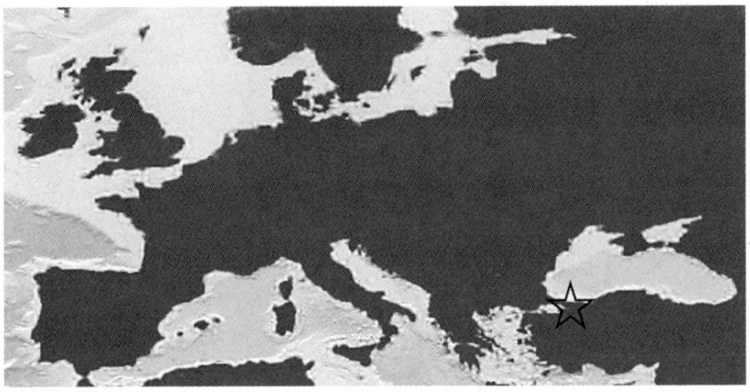

Figure 255. Distribution of *Scissurella szaboi*. ☆ type locality.

Scissurella tinaeformis Amitrov, 1996 †

Figures 256–258

Chresonymy.

Scissurella tinaeformis Amitrov, 1996: 98–99, fig. 1H–K.
Scissurella tinaeformis: Amitrov & Zhegallo, 1998: 7, 9, fig. 2D–E.

Misidentifications.

Scissurella cossmanni: Lozouet, 1986: 108, fig. 29D [is *Sci. tinaeformis*].

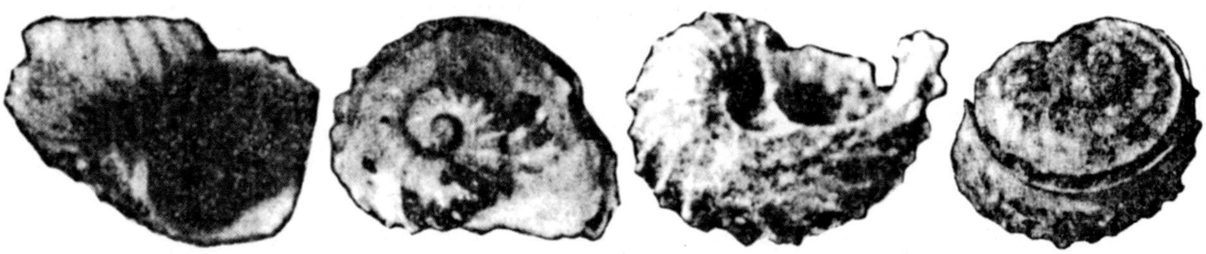

Figure 256. Original Illustration of *Scissurella tinaeformis*.

Scissurella cossmanni: Lozouet, 1998: figs 4c, 4e [is *Sci. tinaeformis*].

Type material. Holotype (PIN 3595/709), 1.65 × 1.2 mm.

Type locality. Rybalsk (Tschapli) quarry, Dnepropetrovsk, Ukraine [48°29'N, 35°0'E]. Mandrikovka Layers, late Eocene (OD).

Etymology. Tina, Latin for keg (OD).

Description. Shell medium-sized (to 1.65 mm. Fig. 257), trochiform, subquadrate. Protoconch of 0.75 whorl, with 9–14 axials, apertural varix connected to embryonic cap, apertural margin sinusoidal. Teleoconch I of 1–1.2 whorls, with 16–19 axial cords; first spiral

Figure 257. *Scissurella tinaeformis*. Granite quarry Rybalskij near Dnipropetrovsk, Ukraine, 48.450°N, 35.142°E. Late Eocene (Priabonian: NP19 zone: see Müller & Rosenberg, 2003, for details). **A–B.** Horizon U1 (UtK 15). **C.** Horizon U5 (UtK 19. Scale bars shells = 0.5 mm. Scale bars protoconch = 100 µm.

line after 0.25 whorl, 6–8 at onset of selenizone. Teleoconch II of up to one whorl, suture deeply impressed. Shoulder with approximately 15 broad axial cords; 1–3 strong spiral lines, forming strongly elevated points at intersection with axials; 2–3 very fine spiral lines in vicinity of keels of selenizone. Base with stronger, raised axial cords of same density as on shoulder; 7–9 spiral lines forming sharp points at intersection with axials, points strongest at height of widest shell, spirals in vicinity of umbilicus increasing in strength to cordlets. Umbilicus open, moderately wide, with carina; walls smooth, straight. Selenizone above periphery, moderately wide; keels moderately strong, moderately elevated; lunules moderately distinct, average density; slit open. Aperture subquadrate.

Comparisons. *Scissurella cossmanni* from the Paleogene of Europe lacks distinct axial sculpture on the shoulder, has more spiral lines on the shoulder, and a flatter spire.

Distribution. Known only from type locality (Fig. 256).

Specimen records.

Ukraine. Granite quarry Rybalskij near Dnipropetrovsk, Ukraine, 48.450°N, 35.142°E. Late Eocene (Priabonian: NP19 zone: see Müller & Rosenberg, 2003, for details. MMG UkT 15; MMG UkT 19).

Remarks. *Scissurella tinaeformis* may fill the stratigraphic gap for *Sci. peyrerensis* (Late Cretaceous and Oligocene of Europe), and may be its senior synonym. Bandel (1998) did not compare his species to *Sci. tinaeformis*. It shares the 1.3 teleoconch I whorls, the density and strength of the axial cords, but the axial cords on the shoulder may extend further towards the selenizone compared to *Sci. peyrerensis*. Additionally, the nature of the spiral sculpture cannot be assessed based on the original illustration of *Sci. tinaeformis*. The only difference is the protoconch sculpture, which, however, usually is a good indicator for the validity of a taxon.

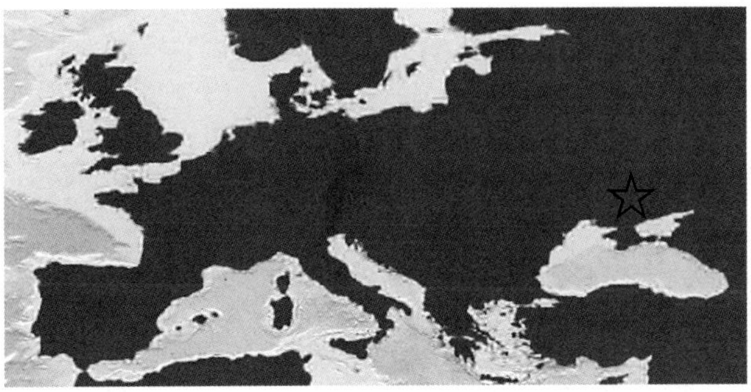

Figure 258. Distribution of *Scissurella tinaeformis*. ☆ type locality.

Scissurella transylvanica Reuss, 1860 †

Figures 259–263

Chresonymy.

Scissurella transylvanica Reuss, 1860: 267, pl. 7, fig. 6.
Scissurella transylvanica: Semper, 1865: 284.
Scissurella transsylvanica [*sic*]: Boettger, 1902: 172.
Scissurella transsylvanica [*sic*]: Boettger, 1906: 197.
Scissurella transylvanica: Friedberg, 1938: 45–46, fig. 8 [*fide* Strausz, 1966].
Scissurella plicata transylvanica: Strausz, 1955: 200.
Scissurella costata transylvanica: Strausz, 1966, 27, text-fig. 17.
Scissurella transylvanica: Baluk, 1975: 21–22, pl. 1, figs 1–2.

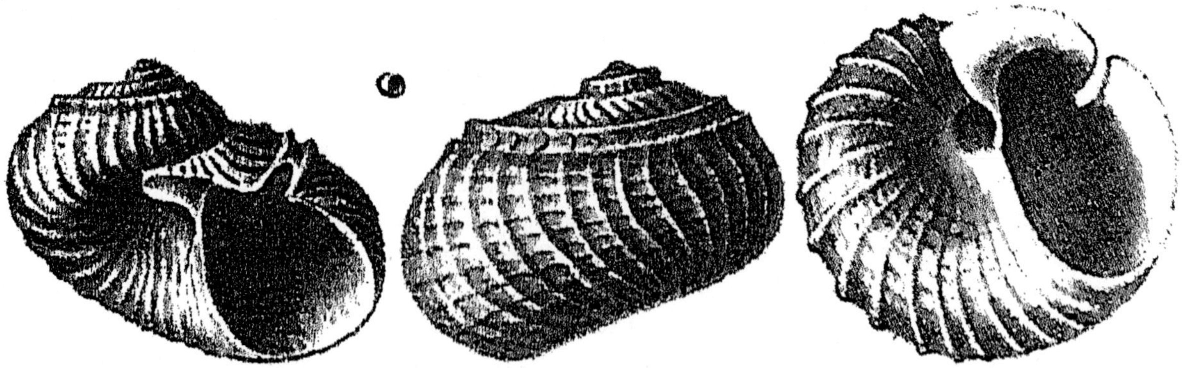

Figure 259. Original Illustration of *Scissurella transylvanica.*

Scissurella transylvanica: Lozouet, 1986: 108.

Synonyms.

+ *Scissurella depressa* Reuss, 1860: 266, pl. 7, fig. 7. ***Type material.*** Syntypes (NHMW, 2). ***Type locality.*** Rudoltice [Czech Republic, 49.898°N, 16.570°E]. ***Etymology.*** From Latin depressa, for depressed, most likely referring to the overall shell shape.

Scissurella depressa: Semper, 1865: 284.

Scissurella depressa: Lozouet, 1986 [as synonym of *Sci. transylvanica*].

Figure 260. Original Illustration of *Scissurella depressa.*

Reussella depressa: Bandel, 1998: 44–45, pl. 14, fig. 8, pl. 15, figs 1–3.

Scissurella depressa: Geiger, 2003: 76.

+ *Scissurella lamellosa* Benoist, 1875: 67, pl. 1, fig. 1a–b. ***Type material.*** Syntype(s) (MNHB), 2 × 1.5 mm. ***Type locality.*** Lariey, Saucats, [Gironde, France, 44.650°N 0.600°W] in the *Pyrula lainei* marls of the Aquitanian, lower Miocene (OD). ***Etymology.*** Most likely noting the lamellar sculpture.

Scissurella lamellosa: Lozouet, 1986: 109.

Scissurella cf. *lamellosa*: Lozouet, 1986: 123.

Scissurella lamellosa: Lozouet *et al.*, 2001: 16, pl. 1, fig. 2, pl. 2, fig. 2a–b.

+ *Scissurella falunica* Morgan, 1915: 232–233, fig. 15. ***Type material.*** Syntypes (coll. Morgan, 5: whereabouts unknown), 1 × 1 mm. ***Type locality.*** Charenton near Pont-Levoy [Touraine, France, 47.267°N, 2.983°E], Falunien (OD). Middle Miocene (Lozouet, 1986). ***Etymology.*** Named after the stratigraphic position.

Figure 261. Original illustrations of *Scissurella lamellosa.* From scan provided by M. Landreaux (MNHB).

Figure 262. Original illustration of *Scissurella falunica*.

Scissurella falunica: Lozouet, 1986: 108.

Misidentifications.

? *Trochotoma terquiemi* [sic]: du Boucher, 1884: 181 [is *Sci. transylvanica*].

Scissurella (*Sinezona*) *terquemi*: Cossmann & Peyrot, 1917: 70–71, pl. 3, figs 10–14 [is *Sci. transylvanica*].

Scissurella cossmanni: Bandel, 1998: 11–12, pl. 2, figs 1–3 [is *Sci. transylvanica*].

Type material. Unspecified number of syntypes (whereabouts unknown).

Type locality. Lapugy [= Lăpugiu de Jos, Romania: 45.888°N, 22.482°E]. Tertiary (OD), middle Miocene (Lozouet, 1986).

Etymology. Referring to its type locality in Transylvania, today Romania.

Description. [Based on Lozouet *et al.*, 2001]. Shell medium-sized (to 2 mm. OD of *Sci. lamellosa*), trochiform, depressed. Protoconch of 0.75 whorl with 9–10 moderately strong axial ribs, apertural varix connected to embryonic cap, apertural margin sinusoidal. Teleoconch I of one whorl, approximately 13 raised axial cords, spiral line in position of selenizone, interstices with fine axial growth marks. Teleoconch II of more than one whorl, suture little impressed. Shoulder with same density of axial cords as on teleoconch I, some fine spiral lines. Base with slight constriction below selenizone, with approximately 15 raised axial cords, approximately nine spiral lines running over axial cords; last spiral bordering umbilicus. Umbilicus open, relatively wide, walls smooth, at distinct angle to base. Aperture oval D-shaped, roof overhanging. Selenizone above periphery; keels moderately strong, moderately elevated; slit open, margins parallel.

Comparisons. *Scissurella aliceae* from the Paleocene of Europe is overall more angulated in shell shape and has a much wider umbilicus.

Scissurella bituminata from the Oligocene of Indonesia has a more angulated shell profile and the unique washboard sculpture surrounding the umbilicus.

Scissurella depontaillieri from the Paleocene through Miocene of Europe has a more square profile of the whorls and a distinct funiculus in the umbilicus.

Scissurella marchmontensis from the Late Cretaceous of Europe and the Caribbean has about twice as many, much weaker, and opisthocline (vs. orthocline) axial cords.

Scissurella peyrerensis from the Oligocene of Europe has strong axial lamellae in the adsutural half of the shoulder and the periphery of the base and an overall more angulated shell profile.

Scissurella ravni from the Paleocene of Europe has a more angulated shell profile and much stronger axial cords on shoulder and base.

Sinezona deshayesi from the Eocene of France lacks the strong axial cords on shoulder and base.

Distribution. Late Oligocene to Miocene of Europe (Fig. 263).

Literature records.

France. Peyrère, Aquitanian, late Oligocene, [44.670°N, 5.620°E] (Bandel, 1998).

Romania. Kostej [= Coşteiu de Sus], Părău ungurului [= B on map in Boettger (1902: 3): 45.906°N, 22.355°E] (Boettger, 1902). Lapugy: [= Lăpugiu de Jos: 45.888°N, 22.482°E], Părău muntanului (Boettger, 1902). Kostej [Coşteiu de Sus], Valea popi [= E on map in Boettger (1902: 3): 45.911°N, 22.372°E] (Boettger, 1906).

Hungary. Várpalota, 47.197°N, 18.369°E, Miocene (Strausz, 1966).

Remarks. *Scissurella depressa* is a juvenile of *Sci. transylvanica* (Lozouet *et al.*, 2001). Other specimens in the type lot of *Sci. depressa* Reuss have disintegrated (Bandel, 1998).

Four nominal taxa are here united under one name. *Scissurella transylvanica*, *Sci. depressa*, and *Sci. falunica* are clearly conspecific due to identical overall shell morphology, ornamentation, umbilicus, geographic and stratigraphic occurrence. The original illustrations of *transylvanica* and *lamellosa* show as common characters the overall shape of the shell, the density and strength of the axial cords, the size and lack of further ornamentation of the umbilicus, as well as their common geologic occurrence in the Miocene of Europe. The type of *Sci. lamellosa* could not be examined due to MHNB institutional policy of not sending types out on loan. No better images of the specimens were available.

Strausz (1955) indicated *Sci. depressa* to be a synonym of *Sci. transylvanica*; the latter species he classified as a subspecies of *Sci. plicata* Philippi (*non* Hedley), which is itself a synonym of the Recent European-Mediterranean *Sci. costata*, as recognized by Strausz (1966). Accordingly, Strausz considered *Sci. depressa* to be a fossil occurrence of the Recent *Sci. costata*.

Baluk (1975) remarked on the extreme variability of the species, both in terms of spire elevation as well as axial sculpture from strong ribs to almost smooth specimens. Baluk (1975) also synonymized *Sin. terquemi* under *Sci. transylvanica*; hence, he used an excessively wide species concept.

Lozouet (2001) illustrated specimens of the nominal taxa *transylvanica* and *lamellosa*. They are identical in all aspects of protoconch and teleoconch I morphology, and also agree as much as possible in teleoconch II sculpture (density of the axial cords) considering the size differences (1.5 vs. 1.7 mm).

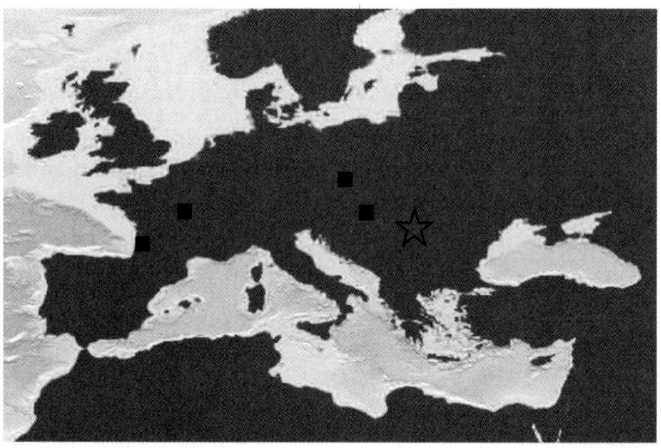

Figure 263. Distribution of *Scissurella transylvanica*. ☆ type locality. ■ literature record.

Sinezona Finlay, 1926

Chresonymy.

Sinezona Finlay, 1926: 341.
Sinezona: Brookes Knight *et al.*, 1960: I221.
Sinezona: McLean, 1967: 407.
Sinezona: Montouchet, 1972: 2–3.
Sinezona: Herbert, 1986: 623–624.
Sinezona: Bandel, 1998: 61.
Sinezona: Marshall, 2002: 169–171.
Sinezona: Geiger, 2003: 32–33.
Sinezona: Zelaya & Geiger, 2007: 435.
Sinezona: Geiger & Sasaki, 2009: 2.
Sinezona: Geiger & McLean, 2010: 4.
Schismope auct. (*non* Jeffreys, 1856): Tate, 1868: 38.
Schismope auct. (*non* Jeffreys, 1856): Tryon, 1883: 320–321.

Synonyms.

+ *Daizona* Bandel, 1998: 57. ***Type species.*** *Sinezona doliolum* Herbert, 1986 (OD). Gender feminine. ***Etymology.*** Named after Dai Herbert of the NMSA (OD).
+ *Ariella* Bandel, 1998: 63. ***Type species.*** *Ariella haliotimorpha* Bandel, 1998 (OD). Gender feminine. ***Etymology.*** Not given. Zelaya & Geiger, 2007: 432.

[For synonymy of *Schismope* and *Woodwardia*, see Remarks].

Type species. *Schismope brevis* Hedley, 1904 (OD).

Other species.
Sinezona bandeli Marshall, 2002.
Sinezona beddomei (Petterd, 1884).
Sinezona brevis (Hedley, 1904).
Sinezona brucei new species.
Sinezona calumnior new species.
Sinezona carolarum Geiger & McLean, 2010.
Sinezona chilensis new species.
Sinezona cingulata (Costa, 1861).
Sinezona concinna (Sowerby, 1831) [*nomen dubium*].
Sinezona confusa Rolán & Luque, 1994.
Sinezona costulata Geiger & Sasaki, 2009.
Sinezona cupelliformis (Amitrov, 1996) †.
Sinezona danieldreieri Geiger, 2008.
Sinezona deshayesi (Munier-Chalmas, 1862) †.
Sinezona doliolum Herbert, 1986.
Sinezona enigmatica Geiger & Marshall, 2012.
Sinezona ferriezi (Crosse, 1867).
Sinezona finlayi (Laws, 1948).
Sinezona garciai Geiger, 2006.
Sinezona globosa Geiger, 2006.
Sinezona hawaiiensis Geiger & McLean, 2010.
Sinezona insignis (Smith, 1890).
Sinezona insularis Simone, 2009.
Sinezona iota (Finlay, 1926).
Sinezona kayae Geiger & McLean, 2010.
Sinezona koruahina (Laws, 1936) †.

Sinezona laqueus (Finlay, 1926).
Sinezona levigata (Iredale, 1908).
Sinezona macleani Geiger, 2006.
Sinezona malloryi (Squires & Goeddert, 1996) †.
Sinezona marrowi new species.
Sinezona mechanica Geiger & Marshall, 2012.
Sinezona milleri Geiger & Sasaki, 2009.
Sinezona miranda (A. Adams, 1862) [*nomen dubium*].
Sinezona modesta (A. Adams, 1862) [*nomen dubium*].
Sinezona mouchezi (Vélain, 1877).
Sinezona pacifica (Oliver, 1915).
Sinezona paumotuensis (Garrett, 1872) [*nomen dubium*].
Sinezona pauperata Powell, 1933.
Sinezona platyspira Geiger & Marshall, 2012.
Sinezona plicata (Hedley, 1899).
Sinezona praecedens Lozouet, 1998 †.
Sinezona rimuloides (Carpenter, 1865).
Sinezona singeri Geiger, 2006.
Sinezona subantarctica (Hedley, 1916).
Sinezona terquemi (Deshayes, 1865) †.
Sinezona tertia (Laws, 1940) †.
Sinezona wanganellica Geiger & Marshall, 2012.
Sinezona wileyi Geiger, 2008.
Sinezona zimmeri Geiger, 2003.

Species removed from *Sinezona*. *Satondella brasiliensis* (Mattar, 1987).
Schismope caliana Dall, 1919 [Tornidae].
Schismope lacuniformis Watson, 1885 [Vanikoridae?].
Satondella kondoi (Ladd, 1970) †.
Scissurella lobini (Burnay & Rolán, 1994).
Sukashitrochus ngatutura (Laws, 1936) †.
Sukashitrochus pulcher (Petterd, 1884).

Etymology. Referring mistakenly to the lack of a selenizone. Gender feminine.

Description. Shell medium-sized (0.5–3.8 mm, usually 1–2 mm), trochiform. Protoconch usually with strong axials; spirally sculptured and smooth ones also known; apertural varix present with or without contact to embryonic cap. Anomphalous or umbilical wall forming an angle with base of shell, without carina or funiculus; rarely covered by callus. Sculpture usually with axials predominating over spirals, no spiral keel(s) on base. Selenizone on shoulder, slit closed anteriorly forming foramen, elevation of keels of selenizone and foramen of more than 0.25 up to the full width of the selenizone.

Distribution. Temperate through tropical waters of all oceans, some in subantarctic (Fig. 264). For specific localities, see individual species.

Geologic range. Eocene through Recent.

Comparisons. *Sukashitrochus* shares the anteriorly closed foramen, but has prominent spiral keel(s) on the base, and the umbilicus is always bordered by a carina.

Satondella has an elevated foramen and usually has a distinct bend in the coiling at the position of the foramen.

Scissurella never closes the slit anteriorly; juvenile *Sinezona* may be difficult to distinguish from *Scissurella*. Half the *Scissurella* species can be distinguished by the presence of a carina or a funiculus in the umbilical region. Fully grown specimens can often be identified by the marked drop of the final quarter whorl along the axis of the shell.

Coronadoa lacks a selenizone, slit or foramen.

Incisura usually has a smooth sculpture and has either a higher spire in the naticiform *I. rosea*, or is neritiform as in *I. lytteltonensis*.

Differentiation of species. Species are differentiated by overall shell shape and spire height, details of sculpture (absolute and relative strength of axials and spirals), width of umbilicus, as well as protoconch sculpture.

Occasionally, the presence of an umbilical callus, the presence/absence and length of the selenizone, the shape of the foramen, and the size at which the foramen forms can be of additional utility. Anatomical characters have not been found to help in species identification or in the establishment of species boundaries. Under a very good light microscope, only very prominent protoconch sculpture can be resolved sufficiently.

Intraspecific variation extends to ontogenetic change of shell/spire height, length of teleoconch I (± 0.125 whorl), condition of the slit (notice, juvenile *Sinezona* have an open slit, hence, can be confused with *Scissurella*), and strength and density of sculptural elements (variable by approximately 25–50%).

Remarks. The publication date for the genus has also been taken to be 1927. However, the byline under the title notes that the paper had been "issued separately, December 23rd 1926."

Millard (2003) noted a possible homonymy between *Ariella* Bandel, 1998, and *Ariella* Shasky, 1961 [Neogastropoda: Turridae]. Shasky (1961) introduced the genus *Arielia*, which was subsequently misspelled by Shuto as *Ariella*. Shuto's misspelling cannot be construed as an introduction of a new name as he consistently

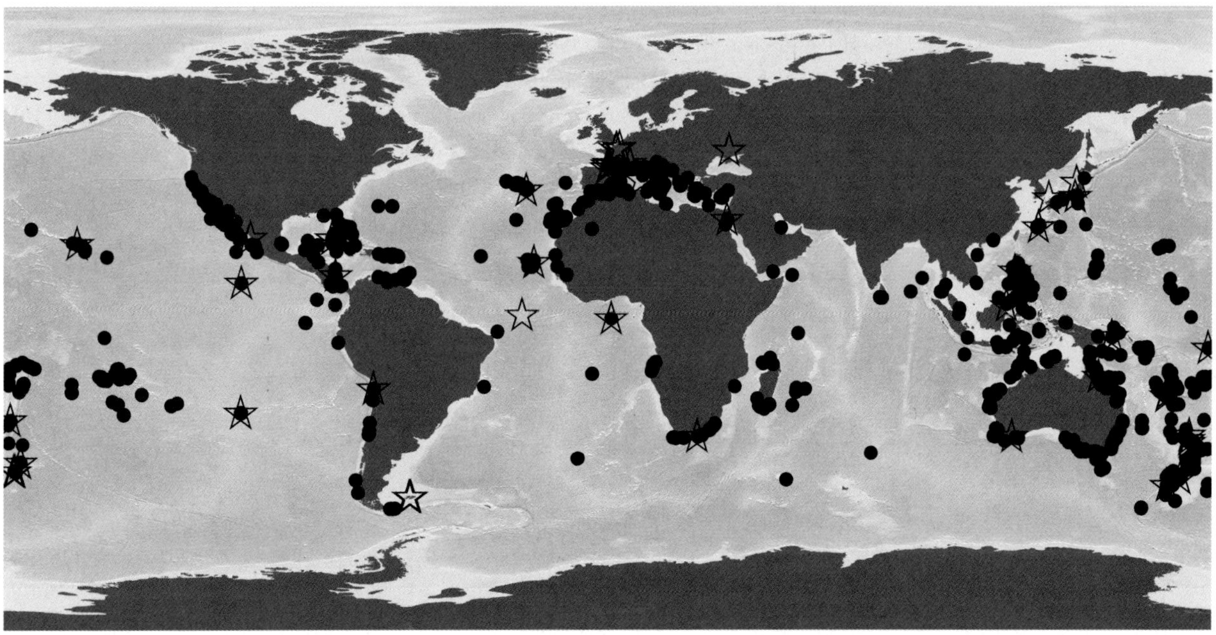

Figure 264. Global distribution of *Sinezona* species. ☆ type locality. ● specimen record. ■ literature record.

credited the genus to Shasky (1961). Tucker (2004: 171) noted Shuto's misspelling. Accordingly, there is no homonymy.

Two other genus names are often associated with *Sinezona*. *Schismope* has long been considered a synonym of *Sinezona*, but the type species had been misidentified as noted e.g., by Marshall (2002). In the interest of taxonomic stability, the nominal type species (*Sci. striatula*) is fixed here as the type species of *Schismope*, properly identified as a synonym of *Sci. costata* in *Scissurella*. See under *Scissurella* for details.

Woodwardia Crosse & Fischer, 1861 has long been considered a synonym of *Sinezona*, but the type species had been misidentified as noted e.g., by Marshall (2002). In the interest of taxonomic stability, the nominal type species (*Sci. elegans*) is fixed here as the type species of *Woodwardia*, properly identified as member of *Scissurella*. See under *Scissurella* for details.

The only diagnostic character is the slit closing anteriorly to form a foramen that has non-elevated keels. The protoconch sculpture usually consists of strong axials, but this condition is not restricted to *Sinezona*, but also applies to half the species of *Scissurella*. Accordingly, protoconch sculpture is not a diagnostic character for *Sinezona*.

Daizona was diagnosed by Bandel (1998) based on the length of the selenizone. Marshall (2002) and Geiger (2003) synonymized *Daizona* under *Sinezona*. A histogram of angles of protoconch to selenizone, and angle of selenizone for all the 14 species are shown in Figure 265. Measurements were taken to the nearest 5°. There are no discernible groupings in the plot, showing that any separation of taxa based on this character is arbitrary and indefensible. Statistical testing for deviation from normal distribution (alternate hypothesis: bimodal distribution) and rectangular distribution (null hypothesis: all values with equal frequency) were carried out with Kruskal-Wallis test for categorized data, Lilliefors test for continuous data, and chi-square test. All tests for both variables and either distribution were insignificant. *Daizona* is synonymized under *Sinezona*.

Interesting patterns emerged from the correlation coefficients between the selenizone and foramen (Table 4). Besides the two metrics introduced above, the angular measurement of the length of the foramen were also obtained. As these measurements were taken to the nearest 5°, both the parametric Pearson product-moment correlation coefficient, as well as the non-parametric Spearman rank correlation coefficient for ordinal data were calculated. There is no correlation between the onset of the selenizone and the length of the selenizone, and the correlation between onset of the selenizone and the foramen is marginally significant.

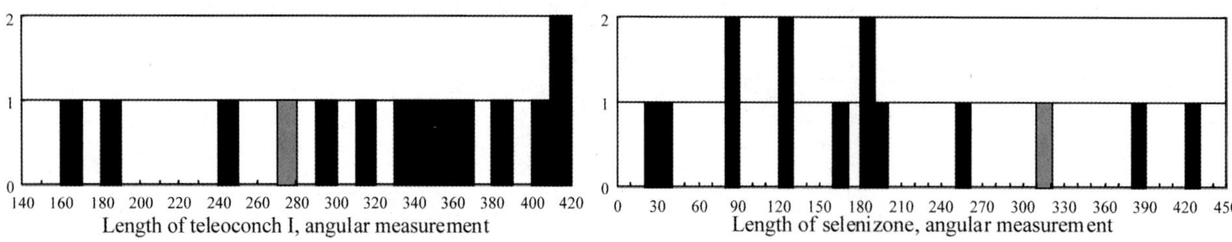

Figure 265. Histogram of two shell morphometric parameters in *Sinezona: Sin. beddomei, Sin. brevis, Sin. cingulata, Sin. confusa, Sin. doliolum, Sin. ferriezi, Sin. insignis, Sin. iota, Sin. levigata, Sin. pacifica, Sin. plicata, Sin. semicostata, Sin. zimmeri*. The type species of *Daizona, Sin. doliolum*, is highlighted in grey. Left: length of teleoconch I, as angular measurement. Right: angular measurement of length of selenizone. All measurements were taken to the nearest 5°. Distribution are not significantly different from either normal or rectangular distribution (from Geiger, 2003).

	Teleoconch I	Selenizone	Foramen
Teleoconch I	—	0.31	0.048
Selenizone	0.38	—	0.0011
Foramen	0.072	0.00046	—

Table 4. Correlation coefficients for three selenizone and foramen related metrics for 14 *Sinezona* species. Top right quadrant: Alpha error probability of parametric correlation coefficient. Bottom left quadrant: Alpha-error probability of Spearman rank correlation coefficient. Teleoconch I: angular measurement from aperture of protoconch to beginning of selenizone. Selenizone: Angular measurement of selenizone length. Foramen: Angular measurement of foramen length.

However, there is a significant correlation between length of the selenizone and the foramen. The results remain significant even if the usual significance level of $\alpha = 0.05$ is adjusted with the Bonferroni correction for multiple comparisons to $\alpha' = 0.0125$. It is unclear whether this correlation has any underlying biological significance, but it is not correlated to size ($0.36 < p < 0.77$).

Gougerot & LeRenard (1977: 6) expressed the opinion that "sans doubte" [= without a doubt] the closed foramen in *Sinezona* is only a gerontic condition of the open slit in *Scissurella*. Hence, the generic distinction between *Scissurella* and *Sinezona* was considered to be unjustified. These authors cited without illustrations incidences of *Sci. parisiensis* [= *Sin. deshayesi*] with a foramen, and an anteriorly closed slit with a soldered raphe. It is difficult to evaluate that claim, which may also be explained by two discrete species in two genera with similar shell morphology in the stage with an open slit, adult *Scissurella*, and subadult *Sinezona*. In examining >2,600 lots and over 19,000 specimens I have not seen a single specimen of *Scissurella* showing gerontic closure of the slit. There is only one specimen of *Scissurella morretesi*, which shows multiple successive closures of the slit (Fig. 141C), a clear developmental anomaly as discussed by Geiger (2006b).

Sinezona bandeli Marshall, 2002

Figures 266–270

Chresonymy.

Daizona pacifica (partim): Bandel, 1998: 58.
Sinezona bandeli Marshall, 2002: 174–176, figs 19–21 [shell], 25–26 [radula].
Sinezona bandeli: Geiger, 2003: 78.
Sinezona bandeli: Spencer *et al.*, 2009: 202 [cited as sp. 5 by Marshall, 2002].

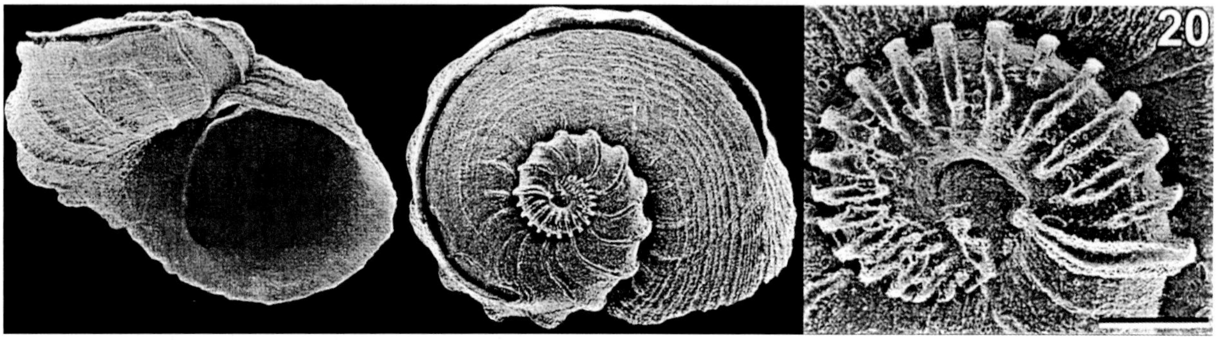

Figure 266. Original illustration of *Daizona bandeli.*

Type material. Holotype (NMNZ M.155919. Fig. 266), 1.3 mm. Paratypes from type locality (M.150283, many). Paratypes 23 m, off Rangaunu Bay, New Zealand, 34°49.6'S, 173°15.0'E (NMNZ M.100505, 42).

Type locality. 25 m, off Rangaunu Bay, New Zealand, 34°48.4'S, 173°15.0'E (OD).

Etymology. Named after Klaus Bandel, University of Hamburg, Germany (OD).

Description. Shell medium-sized (to 1.3 mm. Figs 266–268), depressed trochiform. Protoconch of 1.25 whorls, strong axial sculpture, apertural varix not connected to embryonic cap, apertural margin slightly sinusoidal. Teleoconch I of 0.75 whorl, approximately 12 strong raised axial cords, no spiral in position of selenizone. Teleoconch II of 0.875 whorl, suture little impressed. Shoulder 16 weak axial cords, stronger close to suture, spiral lines increasing from zero to approximately 20 irregularly spaced and of unequal strength, finer than axials. Base with constriction below selenizone devoid of sculpture, approximately twice as wide as selenizone; axial cords starting well below selenizone, over ⅔ of base, spiral threads approximately 12, first approximately at upper end of axials. Selenizone just below top margin of shell, slit closed to form triangular-oblong foramen, keels moderately elevated. Umbilicus open, sloping gradually with base, no funiculus. Aperture rounded obliquely oblong.

Radula (Fig. 269). Rachidian tooth triangular, shaft slightly convex, tip very broad V-shaped, central tip largest, three tips on either side. Lateral teeth 1–3 similar, overall shape sinusoidal, pointing towards the midline of the ribbon, apical cusp on tip also innermost, 2–3 additional smaller cusps along outer edge. Lateral tooth 4 reduced, hook-shaped. Lateral tooth 5 somewhat enlarged, apical cusp outermost, with approximately three to four progressively smaller cusps on inner margin of tip.

Marginal teeth similar, progressively smaller towards periphery, shaft long, slender; with 5–6 cusps on tip.

Comparisons. *Sinezona beddomei* from temperate Australia has stronger raised axial cords on the base, lacks spiral lines on the shoulder and upper base of teleoconch II, and has stronger lunules.

Sinezona brevis from New Zealand has a shorter selenizone (0.1 vs. 0.6 whorl), is overall more globular, and has a less distinct constriction on the base below the selenizone.

Sinezona brucei from New Zealand has axial cords on the shoulder, has coarser spiral cords on the shoulder, and a more rounded overall shell shape.

Sinezona calumnior from temperate Australia and New Zealand has an overall more globular shell, a less distinct constriction on the base below the selenizone, and a more narrow umbilicus.

Sinezona iota from New Zealand has a much taller shell with protruding protoconch, lacks spiral lines on shoulder and base, and has a more narrow umbilicus.

Sinezona laqueus from New Zealand has stronger axial ribs, a longer selenizone, and much stronger lunules.

Figure 267. *Sinezona bandeli*. 10–30 m, Sunday Island, Kermadec Islands, New Zealand, 29.264°S, 177.926°W (FLMNH 00049068, 2). Scale bar shell = 0.5 mm. Scale bar protoconch = 100 µm.

Sinezona levigata from New Zealand has a more globular shell, a much shorter selenizone (0.1 vs. 0.5 whorl), and the umbilicus is covered by a callus.

Sinezona mechanica from New Zealand has a more globular shell, a weaker constriction of the base below the selenizone, and lacks spiral lines on shoulder and base.

Figure 268. *Sinezona bandeli*. **A.** 40 m, New Zealand, 34.393°S, 172.862°E (NMNZ M.152479, 7). **B–C.** 32 m, New Zealand, 34.398°S, 172.793°E (NMNZ M.152482, 28). Scale bars shell = 0.5 mm. Scale bars protoconch = 100 μm.

Sinezona norfolkensis from the Tasman Sea has few spiral lines on the shoulder, a protoconch with fewer and lower axial lamellae, and a longer teleoconch I (0.8–0.9 vs. 0.75 whorl).

Sinezona pacifica from temperate Australia and New Zealand has fewer and stronger axials on the teleoconch, and closes the foramen later (>0.9 mm versus <0.8 mm).

Sinezona pauperata from New Zealand has a more globular shell, lacks a selenizone and the protoconch has fine axial cords and lacks an apertural varix.

Sinezona platyspira from New Zealand has a protoconch with very fine irregular axial cords and lacks an apertural varix, fewer strongly raised axial lamellae, and an umbilicus made by the rounded portions of the previous whorls.

Sinezona subantarctica with broad subantarctic distribution has a wider and more globular shell, a callus covering the umbilicus, and at most a very short selenizone.

Sinezona costulata from southern Japan has a narrower umbilicus, has more strong axial cords on the protoconch (17 vs. 12), and the axial cords on teleoconch I do not reach the suture.

Distribution. New Zealand, 0–805 m, live 0–102 m, locally common (Fig. 270).

Specimen records.

New Zealand. 10–30 m, Sunday Island, Kermadec Islands, 29.264°S, 177.926°W (FLMNH 00049068, 2). 103 m, off Three Kings Islands, Middlesex Bank, 33.957°S, 172.323°E (NMNZ M.150274, 3). 20 m, Awanui Bay, Northland, 35.048°S, 173.256°E (GNS RM 1036, 1). Taupo Bay, 34.983°S, 173.717°E (AMS ex C.461509, 5). 33 m, off Taupo Bay, Whangaroa Bay, 34.983°S, 173.717°E (AMS C.461514, 6). 40 m, Northland, 0.5 mi off W side Stephensons Island, 34.967°S, 173.783°E (AMS C.403743, 7). 33 m, NE side Sephenson Island, 34.967°S, 173.783°E (AMS C.461495, 1). 23 m, off Ranagaunu Bay, 34.827°S, 173.250°E (NMNZ M.100505, 50). 25 m, off Rangaunu Bay, 34.807°S, 173.325°E (NMNZ M.155919, 1). 25 m, off Rangaunu Bay, 34.807°S, 173.325°E

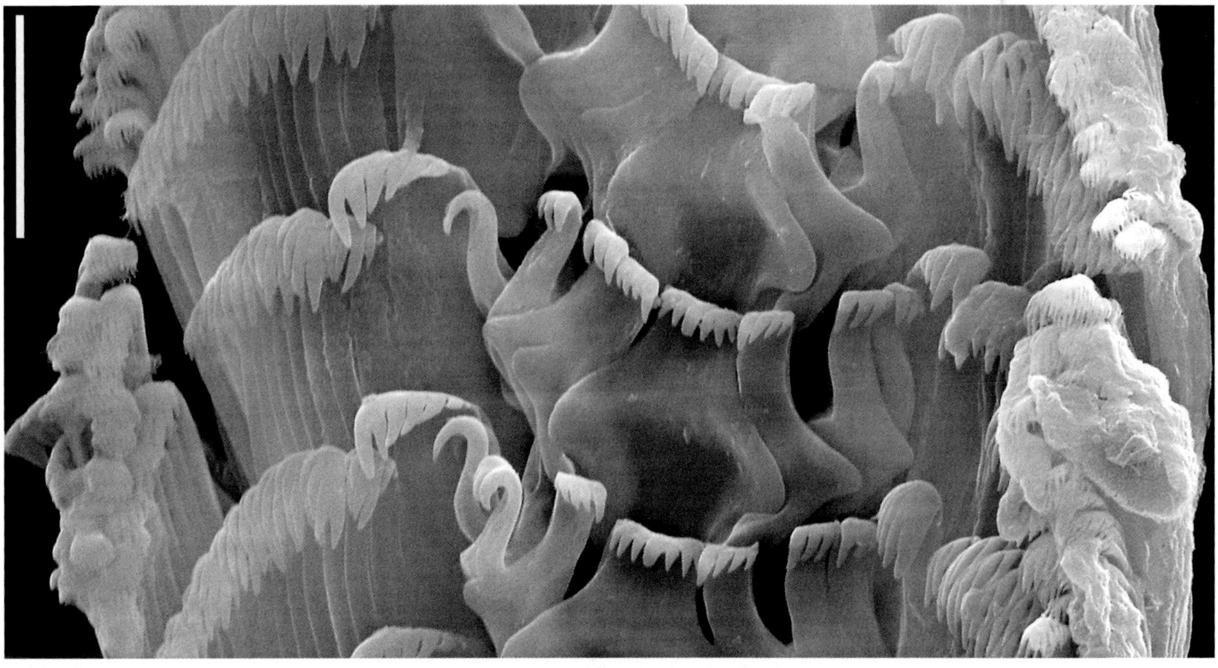

Figure 269. *Sinezona bandeli*. Radula. 25 m, off Rangaunu Bay, New Zealand, 34.807°S, 173.325°E (NMNZ M.150283, 200). Scale bar = 10 µm.

(NMNZ M.150283, 200). 34.398°S, 172.760°E (NMNZ M.152483, 49). 32 m, 34.398°S, 172.793°E (NMNZ M.152482, 28). 40 m, 34.393°S, 172.862°E (NMNZ M.152479, 7). Hahei, North Island, 36.838°S, 175.807°E (RMNH AD 29662, 1). Island Bay, Cook Strait, 41.333°S, 174.783°E (CASIZ ex 119116, 1). Bathing Beach, Stewart Island, 46.900°S, 168.117°E (AMS C.379062, 20+).

Literature records.

New Zealand. 47 m, off Waikuki Beach, 43.283°S, 172.717°E. 188–228 m, Rungapapa Knoll, WNW of White Island, 37.563°S, 176.983°E. 103–106 m, Ranfurly Bank, East Cape, 37.557°S, 178.805°E. 59–74 m, off E side on Major Island, 37.315°S, 176.270°E. 38 m, off E side of Ruamahua-nui Island, Aldermen Islands, 36.953°S, 176.097°E. 18 m, North Reef, of NW tip of Goat Island, Leigh, 36.270°S, 174.795°E. 20–25 m, between Marsden Point and Calliope Bank, Whangerei Harbour entrance, 35.837°S, 174.508°E. Off Poor Knights Islands, 35.500°S, 174.750°E. 0 m, Bartle's Bay, Aorangi Island, Poor Knights Islands, 35.475°S, 174.738°E. 7 m, 35.263°S, 174.173°E. 3–5 m, Oke Bay, 35.223°S, 174.268°E. 20–32 m, 35.213°S, 174.260°E. 4 m, Bay of Islands, Waewaetorea Passage, 35.207°S, 174.222°E. 20 m, Whangaroa Harbour entrance, main channel, 35.033°S, 173.125°E 25 m, immediately outside Whangaroa Harbour entrance, 35.006°S, 173.762°E. 63 m, Great Exhibition Bay, 34.557°S, 173.080°E. 23 m, 34.427°S, 172.803°E. 29 m, 34.417°S, 172.777°E. 29 m, 34.405°S, 172.833°E. 30 m, N of Kerr Point, Tom Bowling Bay, 34.390°S, 172.986°E. 49 m, off Spirits Bay, 34.376°S, 172.821°E. 427 m, 39 km SW of Great Island, 34.293°S, 171.755°E. 192–202 m, S of Great Island, 34.235°S, 172.150°E., 34.193°S, 172.182°E. 37 m, off West Island, 34.183°S, 172.050°E. 91 m, 34.183°S, 172.167°E (NMNZ M.150281, 6). 23 m, off N face of Hinemoa Island, 34.180°S, 172.043°E. Arch Pinnacle, Princes Rocks, 34.180°S, 172.050°E. 38 m, Arch Pinnacle, Princes Rocks, 34.175°S, 172.005°E. 14 m, off Princes Rocks, 34.167°S, 172.133°E. 33 m, off South-West Island, 34.167°S, 172.133°E. 13–15 m, South-East Bay, Great Island, 34.1580°S, 172.147°E. 23 m, North-West Bay, Great Island, 34.152°S, 172.140°E. 33 m, reef between Great Island and Farmer Rocks, 34.150°S, 172.167°E. 102 m, off North-East Island, 34.142°S, 172.183°E. 310 m, 11 km NW of Great Island, 34.108°S, 172.078°E. 508 m, 13 km N of Great Island, 34.030°S, 172.200°E. 622 m, 34.017°S, 172.117°E. 805 m, Three Kings Trough, 34.000°S, 171.917°E. 123–128 m, King Bank, 33.957°S, 172.323°E (Marshall, 2002).

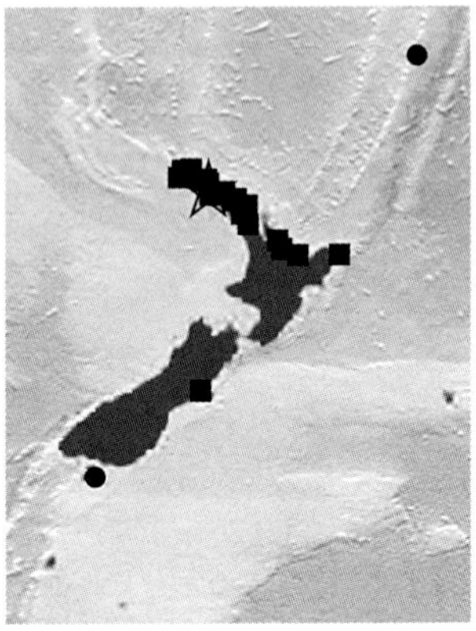

Figure 270. Distribution of *Sinezona bandeli*. ☆ type locality. ● specimen record. ■ literature record.

Remarks. Some specimens identified by Bandel (1998) as *D. pacifica* are in fact *Sin. bandeli.*

Sinezona beddomei (Petterd, 1884)

Figures 271–274

Chresonymy.

Schismope beddomei Petterd, 1884: 139, not illustrated.

Schismope beddomei: Paetel, 1888: 289.

Schismope beddomei: Pilsbry, 1890: 67–68.

Schismope beddomei: Adcock, 1893: 9 [*fide* Verco, 1910].

Schismope beddomei: Tate & May, 1902: 407, pl. 24, fig. 24 [holotype].

Schismope beddomei: Pritchard & Gatliff, 1902: 181.

Schismope beddomei: Sowerby & Fulton, 1903: 53.

Schismope beddomei: Hedley, 1904: 89.

Schismope beddomei: Pritchard & Gatliff, 1902: 65.

Schismope beddomei: Anonymous, 1906: 367.

Schismope beddomei: Suter, 1909: 7.

Schismope beddomei: Iredale & Mestayer, 1908: 412.

Schismope beddomei: Verco, 1910: 116–117.

Schismope beddomei: Thiele, 1912a: 30–31, pl. 4, fig. 1.

Schismope beddomei: Hedley, 1918: M39.

Schismope beddomei: May, 1921: 34, pl. 17, fig. 5.

Schismope beddomei: May, 1923: pl. 17, fig. 5.

Schismope beddomei: Cotton & Godfrey, 1938b: 2.

Schismope beddomei: May & Macpherson, 1958: 21, pl. 17, fig. 5.

Schismope beddomei: Cotton, 1959: 43, fig. 13.

Schismope beddomei: Macpherson & Gabriel, 1962: 34.

Sinezona beddomei: Lozouet, 1986: 110.

Sinezona beddomei: Wilson, 1993: 46.

Sinezona beddomei: Jansen, 1995: 11.

Sinezona beddomei: Jansen, 1999: 54, figs 46–48.

Sinezona beddomei: Geiger, 2003: 78.

Sinezona beddomei: Geiger *et al.*, 2008: fig. 12.2N [protoconch].

Figure 271. *Sinezona beddomei*. Port Fairy, Victoria, Australia, 38.383°S, 142.233°E (AMS C.406354). Scale bar shell = 1 mm. Scale bar protoconch = 100 µm.

Misidentifications.

not *Schismope beddomei*: Suter, 1909: 7 [is *Sin. laqueus*: Powell, 1979].

not *Schismope beddomei*: Suter, 1913: 90, pl. 6, fig. 13 [1917: is *Sin. laqueus*].

Type material. Holotype (TMAG E224: Tate & May, 1902: pl. 24, fig. 24), 1 × 0.75 mm. Shell completely broken in early 1900s (pers. comm. L. Turner).

Type locality. North-west coast [Tasmania, Australia] (OD).

Etymology. Named for C. E. Beddome (1839–1898: Roberts, 1898).

Figure 272. *Sinezona beddomei.* **A.** Kangaroo Island, N coast, Stokes Bay, South Australia, Australia, 35.617°S, 137.200°E (AMS C.379039). **B.** Bittangabee, N of Green Cape, New South Wales, Australia, 37.217°S, 150.017°E (AMS C.346368). Slightly juvenile. Radula shown in Figure 273. **C.** Port Fairy, Victoria, Australia, 38.383°S, 142.233°E (AMS C.406354). Scale bars shell = 1 mm. Scale bars protoconchs = 100 µm.

Description. Shell moderate size (to 1 mm. Figs 271, 272), trochiform depressed. Protoconch of 1.25 whorls, strong axial sculpture, no apertural varix, apertural margin convex. Teleoconch I of 0.8–0.9 whorl, with approximately 13 strong axial elevated cords not reaching suture, interstices with fine irregular growth marks, no spiral sculpture. Teleoconch II of 0.66 whorl, suture moderately impressed. Shoulder flat, with indistinct broad axials most notable near suture, interstices with fine irregular growth marks, no spirals. Base with distinct constriction below selenizone, strong axial ridges starting below constriction under selenizone, approximately 12–14 visible in ventral view; 4–5 spiral cordlets on lower base towards umbilicus. Umbilicus open, moderately wide, bordered with distinct raised cord, walls smooth. Selenizone slightly inside periphery, keels strong, moderately elevated, lunules strong, foramen elongated triangular. Aperture rounded D-shaped.

Radula (Fig. 273). Rachidian tooth triangular, tip with 5–7 cusps, arranged in V-shape, central one largest. Lateral teeth 1–3 similar, narrow, with three cusps on outer margin. Lateral tooth 4 reduced, pointed. Lateral tooth 5 enlarged broadened, with five cusps on inner margin, terminal one largest. Radular interlock of central field moderate. Inner marginal teeth with tip bearing approximately five cusps on outer edge, outer marginal teeth spoon-shaped with many cusps on either side.

Comparisons. *Sinezona bandeli* from New Zealand has distinct spiral lines on shoulder and base, less distinct axial cords, and less distinct lunules.

Sinezona brevis from New Zealand has a much shorter selenizone (0.1 vs. 0.5 whorl), is overall more globular, and has a less distinct constriction of the base below the selenizone.

Sinezona calumnior from temperate Australia and New Zealand has a more narrow umbilicus, a longer teleoconch I (0.75–1 vs. 0.6 whorl), and spiral sculpture on the shoulder and upper base.

Sinezona iota from New Zealand has a much more elevated shell with protruding protoconch, axial cords that maintain their strength to the border of the umbilicus, and a more narrow umbilicus.

Sinezona laqueus from New Zealand has a longer selenizone (0.5 vs. 0.75 whorl), has nodular thickenings of the spiral cords near the umbilicus and tends to have fewer but slightly stronger axial lamellae.

Sinezona levigata from New Zealand is more globular, has a very short selenizone (0.1 vs. 0.5 whorl), and the umbilicus is covered by a callus.

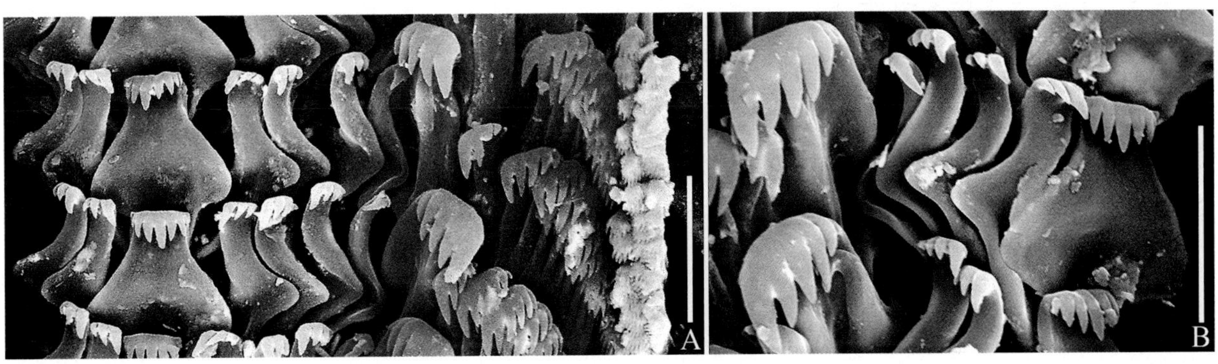

Figure 273. *Sinezona beddomei*. Radula from specimen shown in Figure 272B. Bittangabee, N of Green Cape, New South Wales, Australia, 37.217°S, 150.017°E (AMS C.346368). **A.** Half width of radula. **B.** Central field enlarged. Scale bars = 10 μm.

Sinezona mechanica from New Zealand has a more globular shell, the protoconch has a top spiral cord, and the lunules are much weaker.

Sinezona norfolkensis from the Tasman Sea has irregular spiral cordlets on the shoulder of teleoconch II, a wider umbilicus, and a less distinct constriction of the base below the selenizone.

Sinezona pacifica from New Zealand has less distinct lunules, spiral lines on shoulder and upper base, and a more narrow umbilicus.

Sinezona pauperata from New Zealand has a more globular shell, the protoconch lacks an apertural varix, and lacks a selenizone.

Sinezona platyspira from New Zealand has a more depressed spire, the umbilicus is made by the rounded portion of the previous whorls, and the protoconch has fine irregular axials and lacks an apertural varix.

Sinezona subantarctica with broad subantarctic distribution has a more globular shell, the umbilicus is covered by a callus, and has almost no selenizone.

Sinezona costulata from southern Japan has less distinct lunules, spiral lines on shoulder and base, and a less distinct constriction of the base below the selenizone.

Distribution. Temperate Australia (Fig. 274).

Specimen records.

Australia, New South Wales. Pambula, 36.933°S, 149.917°E (AMS C.11785, 1). Bittangabee, N of Green Cape, 37.217°S, 150.017°E (AMS C.346368, 1). Sydney, Port Jackson, 33.855°S, 151.247°E (AMS C.404961, 20+). Sydney, Manly Beach, 33.803°S, 151.283°E (AMS C.404957, 1).

Australia, South Australia. Kangaroo Island, N coast, Stokes Bay, 35.617°S, 137.200°E (AMS C.379039, 1). 0 m, Nora Creina Bay, Robe, 37.343°S, 139.870°E (MV F178690, 1).

Australia, Tasmania. Tinderbox Bay, S of Hobart, 43.050°S, 147.333°E (AMS C.376852, 2). Blow Hole, Tasman Peninsula, 43.150°S, 147.850°E (WAM, 2). Eaglehawk Bay, Eaglehawk Neck, 43.017°S, 147.917°E (AMS C.17727, 1). 6–20 m, Pirates Bay area, Eaglehawk Neck, 43.018°S, 147.928°E (MV F165442 [decalcified], 1: complete).

Australia. Victoria. Port Fairy, 38.383°S, 142.233°E (AMS C.406354, 8). Port Fairy, 38.383°S, 142.217°E (MV F 161686, 9). 11 m, Western Port, 38.367°S, 145.533°E (AMS C.16006, 1). Point Leo (N. edge of Bobbanaring Point), W side of Western Port Bay, 38.367°S, 145.533°E (LSC, 1). Mallacoota, 37.567°S, 149.933°E (AMS C.379044, 2). Portland, 38.342°S, 141.609°E (MV 161687, 1). 7–10 m, Southern Port Phillip Bay, 38.273°S, 144.697°E (MV F 161691, 1). 0 m, Reef Point, Port Fairy, 38.364°S, 140.663°E (MV F178689, 1). 58 m, Bass Strait, 38.895°S, 147.108°E (MV F83096 [decalcified], 1: complete). 36 m, Bass Strait, 40.633°S, 145.383°E (MV F83093 [decalcified], 1: complete). 58 m, Bass Strait, 38.971°S, 147.223°E (MV F83095, 2, 3: complete).

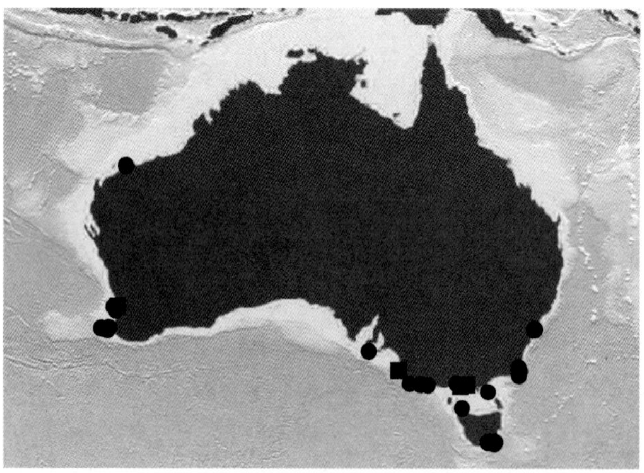

Figure 274. Distribution of *Sinezona beddomei*. ● specimen record. ■ literature record. ? questionable record.

Australia, Western Australia. Cape Naturaliste, 33.742°S, 114.435°E (WAM, 1). Ellensbrook, 33.883°S, 114.983°E (WAM, 1). Moses Rock [14 km S of Yallingup], 33.700°S, 115.100°E (WAM, 1). Rottnest Island, 32.000°S, 115.500°E (WAM, 2). The Basin, Rottnest Island, 32.000°S, 115.500°E (WAM, 1; WAM, 1). 0 m, The Basin, NE Rottnest Island, 32.000°S, 115.500°E (OSU 11920a, 2). Bothurst Point area, Rottnest Island, 31.983°S, 115.533°E (WAM, 2). Garden Island, 32.233°S, 115.683°E (WAM, 2). Penguin Island, 32.300°S, 115.683°E (WAM, 1). Kendrew Island, 20.475°S, 116.533°E (WAM, 1). Hopetoun, 33.867°S, 120.133°E (WAM S13382, 1).

Literature records.

Australia. Flinders, Western Port, Victoria (Pritchard & Gatliff, 1902).—Cottesloe, S Western Australia. Guichen Bay, Gulf of St. Vincent, Spencer Gulf, South Australia. NW coast, S, Tasmania. to 100 fms. [= 183 m] (Cotton, 1959).—Flinders, Victoria (Macpherson & Gabriel, 1962).

Remarks. *Schismope trivialis* [auct, ms.] was encountered on a label with *Sin. beddomei* (MCZ 249561). It has not been published anywhere, hence, has no standing. It is given here solely for the sake of completeness.

Sinezona brevis (Hedley, 1904)

Figures 275–279

Chresonymy.

Schismope brevis Hedley, 1904: 90, fig. 16.
Schismope brevis: Iredale & Mestayer, 1908: 412.
Schismope brevis: Iredale, 1908a: 381.
Schismope brevis: Iredale, 1908b: 391.
Schismope brevis: Iredale, 1908c: 406.
Schismope brevis: Suter, 1909: 8.
Schismope brevis: Suter, 1913: 91, pl. 6, fig. 14 [1917].
Schismope brevis: Thiele, 1912a: 31–32, pl. 4, figs 4–5 [copy figures Hedley, 1904].
Sinezona brevis: Dell, 1960: 145.
Sinezona brevis: Brookes Knight *et al.*, 1960: I221, fig. 136, 2 [unidentifiable].
Woodwardia (*Sinezona*) *brevis*: Wenz, 1961: 173, fig. 272 [copy Thiele, 1912a].
Sinezona brevis: Powell, 1979: 34, pl. 13, figs 10–11.
Sinezona brevis: Lozouet, 1986: 111.
Sinezona brevis: Beu & Maxwell, 1990: 376.
Sinezona brevis: Hedegaard, 1990: 74, 145.
Sinezona brevis: Spencer & Willan, 1995: 13.
Sinezona brevis: Bandel, 1998: 61–62, pl. 21, figs 2–5.
Sinezona brevis: Geiger, 2003: 78, fig. 16c [shell, radula].
Sinezona brevis: Geiger & McLean, 2010: 8.
Sinezona brevis: Spencer *et al.*, 2009: 202.

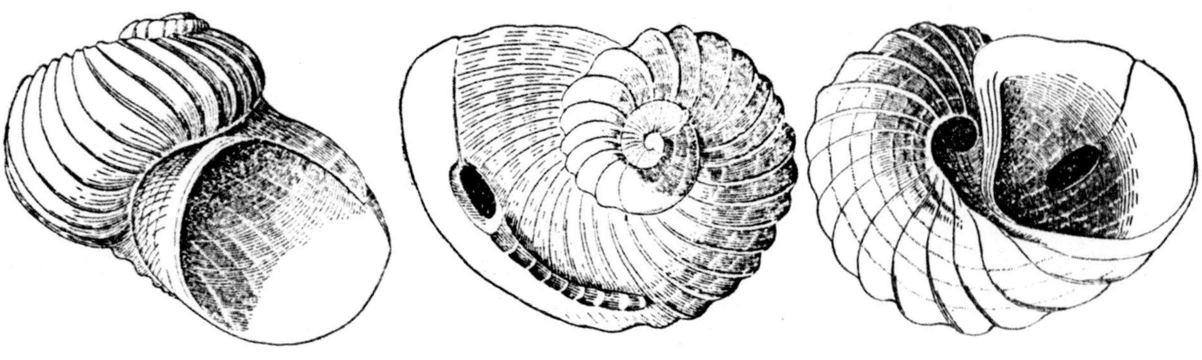

Figure 275. Original illustration of *Schismope brevis.*

Sinezona brevis: Raven & Bracegirdle, 2010: 10, text-fig.

Misidentifications.

not *Sinezona brevis*: Finlay, 1926: 341 [is *Sin. levigata*].

not *Sinezona brevis*: Bandel, 1998: 61, pl. 21, figs 2–5 [is *Sin. levigata*].

Sinezona iota: Bandel, 1998: 62–63, pl. 21, figs 6–8 [is *Sin. brevis*].

Type material. Holotype (AMS C.13591), 1.14 × 0.9 × 0.94 mm. Paratypes from type locality (AMS C.170413, 16).

Type locality. Lyall Bay, near Wellington, New Zealand [41.333°S, 174.800°W] (OD).

Etymology. Brevis, Latin for short, referring to the short selenizone.

Description. Shell medium-sized (to 1.17 mm. Figs 276, 277), trochiform depressed as juvenile, trochiform globular fully mature. Protoconch of one whorl, fine axials with median thickening, apertural varix not connected to embryonic cap, apertural margin slightly convex. Teleoconch I of 1.1–1.25 whorls, 14–18 strong axial cords, thinning towards suture, interstices with fine irregular growth lines, first spiral lines after 0.75 whorl. Teleoconch II of approximately 0.17 whorl, convex, suture lightly impressed. Shoulder with axials of same density, but reduced strength, 6–12 low spiral cords at

Figure 276. *Sinezona brevis*. Houghton Bay, Wellington, New Zealand, 41.345°S, 174.785°E (AMS C.380180). Scale bar shell = 0.5 mm. Scale bar protoconch = 100 μm.

apertural margin. Base with same density of axials of decreasing strength, 6–12 spirals increasing in strength from strong lines at periphery to low cords at umbilicus. Umbilicus open moderately narrow, narrower in senescent specimens, bordered by distinct rough cord, walls straight, with irregular growth marks. Selenizone above

Figure 277. *Sinezona brevis*. **A, C.** E. side of Island Bay, Wellington, New Zealand, 41.347°S, 174.775°E (AMS C.380172). **B.** Great Barrier Island, Rocky Bay, New Zealand, 36.167°S, 175.417°E (AMS C.380181). Scale bars shell = 0.5 mm. Scale bars protoconch = 100 μm.

periphery, oriented at 45° to coiling axis; keels low, strong; foramen lenticular, with higher keels. Aperture rounded D-shaped, roof overhanging, in fully mature specimens with point at intersection of cord bordering umbilicus and peristome.

Operculum filling aperture, nucleus central, multispiral.

Radula (Fig. 278). Rachidian tooth triangular, tip with central cusp largest, five cusps on either side arranged in quarter arch. Lateral teeth 1–3 similar, four cusps on outer edge of tip. Lateral tooth 4 reduced, pointed, with small subterminal projection on either side of main cusp. Lateral tooth 5 broadened, 7–9 cusps on inner margin. Central field with moderate interlock. Inner marginal teeth with 1–2 cusps on inner

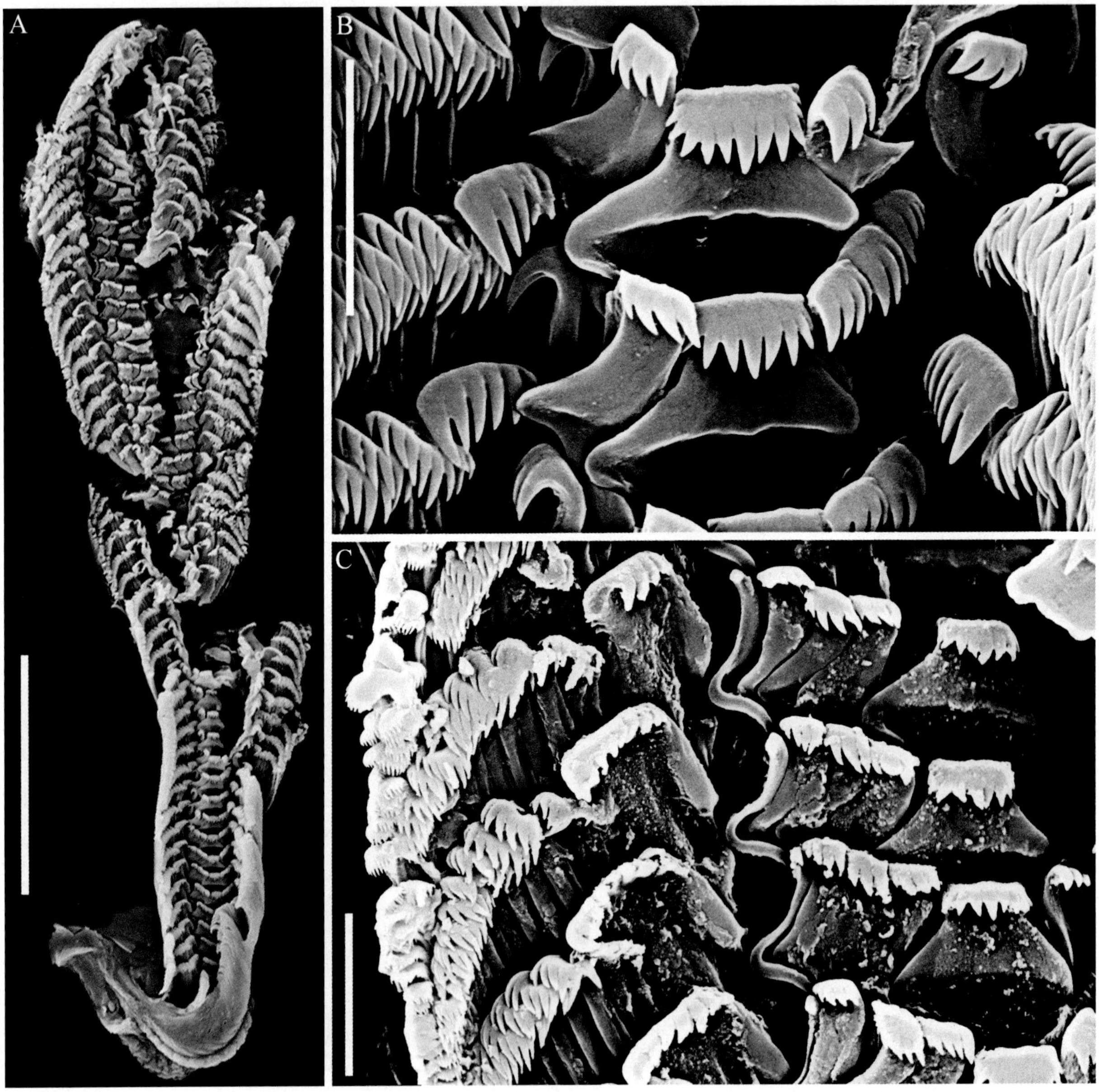

Figure 278. *Sinezona brevis*. Radula. E side of Island Bay, Wellington, New Zealand, 41.347°S, 174.775°E (AMS C.380172). **A.** Entire radula. Scale bar = 100 µm. **B.** Central field. Scale bar = 10 µm. **C.** Lateral tooth 5 and marginal teeth. Scale bar = 10 µm.

edge, terminal cusp largest, approximately five cusps on outer edge. Outer marginal teeth spoon-shaped, many fine bristles.

Comparisons. *Sinezona bandeli* from New Zealand has a more subquadrate shell, a strong constriction on the base below the selenizone, and a much longer selenizone (0.4 vs. 0.1 whorl).

Sinezona brucei from New Zealand has a longer selenizone (0.5 vs. 0–0.125 whorl), at most half as many axial lamellae on the protoconch, and a shorter teleoconch I (0.6 vs. 1 whorl).

Sinezona iota from New Zealand has a much taller shell with protruding protoconch, lacks spiral sculpture on the shoulder and base, and the axial cords maintain their strength to the border of the umbilicus.

Sinezona laqueus from New Zealand has a more subquadrate shell, lacks spiral sculpture on shoulder and base, and has much stronger lunules.

Sinezona levigata from New Zealand has a callus covering the umbilicus, an overall broader shell, and fewer but stronger axial cords on the protoconch.

Sinezona mechanica from New Zealand lacks spiral sculpture on shoulder and base, has about twice as many axial cords, and a protoconch with a top spiral cord.

Sinezona pacifica from temperate Australia and New Zealand has a more subquadrate shell, a stronger constriction of the base below the selenizone, and a longer selenizone (0.5 vs. 0.1 whorl).

Sinezona pauperata from New Zealand has a protoconch with many fine axial cords, but no apertural varix, has no selenizone, generally has weaker axial cords, and the umbilicus is very narrow.

Sinezona platyspira from New Zealand has a sunken apex, the umbilicus is produced by the rounded portion of the previous whorls, and the protoconch lacks an apertural varix.

Sinezona subantarctica from the circum-antarctic Islands has a callus over the umbilicus, a protoconch with very fine irregular axial cords and lacking an apertural varix, and the overall shell shape is wider.

Distribution. New Zealand (Fig. 279)

Fossil. Haweran = late Pleistocene of New Zealand (Beu & Maxwell, 1990).

Specimen records.

New Zealand. 50 m, Northern Arch, Te Araara Point, Poor Knights Islands, 35.450°S, 174.733°E (NMNZ M.174551, 1). Island Bay, Wellington, 41.347°S, 174.773°E (AMS C.380170, 15). 3–5 m, Oke Bay, Bay of Islands, 35.223°S, 174.268°E (NMNZ M.174840, 1). 55 m, Stewart Island, Port Adventure, 47.083°S, 168.183°E (AMS C.380171, 5). E side of Island Bay, Wellington, 41.347°S, 174.775°E (AMS C.380172, 20). 28 m, Stewart Island, off Halfmoon Bay, 46.883°S, 168.133°E (AMS C.380173, 1). Goat Island Beach, Leigh, Hauraki Gulf, 36.267°S, 174.800°E (AMS C.380174, 16). Tahunanui Beach, Nelson, 41.283°S, 173.250°E (AMS C.380176, 9). 13 m, Waterfall Inlet, Auckland Islands, 50.817°S, 166.217°E (AMS C.380178, 1). Houghton Bay, Wellington, 41.345°S, 174.785°E (AMS C.380180, 14). Great Barrier Island, Rocky Bay, 36.167°S, 175.417°E (AMS C.380181, 1). Lyall Bay, near Wellington, 41.333°S, 174.800°W (AMS C.13591, 1). Lyttleton, 43.617°S, 172.733°E (AMS C.27125, 5). Lyall Bay, near Wellington, 41.333°S, 174.800°W (AMS C.170413, 16). 9 m, between Complex Point and Sandy Beach, Northwest Bay, Campbell Island, 52.550°S, 169.050°E (AMS C.376636, 1). 57 m, near Okahari Point, 1 km W of Leigh Marine Laboratory, Leigh, North Island, 36.263°S, 174.797°E (LACM 87-83, 1). 0 m, E side Island Bay, Wellington, North Island, 41.347°S, 174.775°E (LACM 87-87, 8). Island Bay, Wellington, North Island, 41.283°S, 174.783°E (LACM 151629, 4).

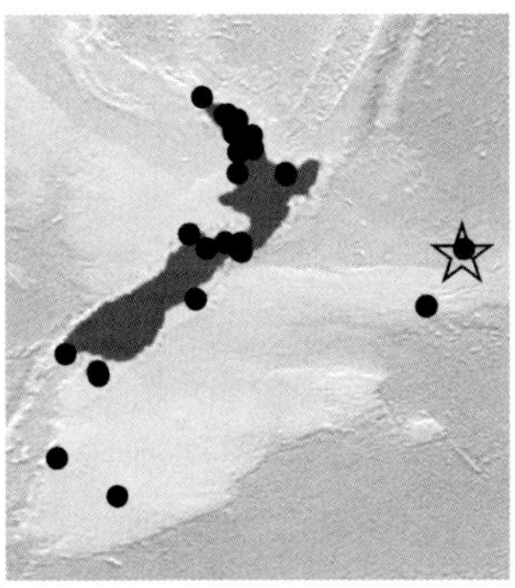

Figure 279. Distribution of *Sinezona brevis*. ☆ type locality. ● specimen record.

Island Bay, Cook Strait, 41.333°S, 174.783°E (CASIZ 119116, 5). 0 m, E side of Island Bay, Wellington, 41.350°S, 174.783°E (AMS, 20: complete). Island Bay, Wellington, 41.347°S, 174.773°E (LACM A.2777, 1). Bay of Islands, North Island, 41.350°S, 174.783°E (FLMNH 00049065, 2). Waitangi, Chatham Island, 43.950°S, 176.533°W (USNM 780329, 1). Langs Beach, N of Mangawhai, 36.050°S, 174.533°E (AMS C.461506, 3). 79 m, off Hen and Chicken Island, Hauraki Gulf, 36.917°S, 174.750°E (AMS C.461497, 1). Tapeka Point, Russel, Bay of Islands, 35.233°S, 174.117°E (AMS C.461513, 3). McGregor Bay, 36.750°S, 175.467°E (AMS C.461510, 10). 79 m, off Hen and Chicken Island, Hauraki Gulf, 35.917°S, 174.750°E (AMS C.461501, 1). 35–44 m, New Golden Hind Station B2, Narrows, Long Sound, Preservation Inlet, Fiordland, 46.090°S, 166.620°E (GNS RM 1042, 2). Lyall Bay, Wellington, North Island, 41.338°S, 174.794°E (GNS RM 5960, 3). Seaweek washings, Mount Maunganui, Tauranga, Bay of Plenty [multiple localities] (GNS RM 1041, 4). Lyall Bay, Wellington, 41.337°S, 174.792°E (GNS RM 5961, 5; GNS RM 1044, 1). Lyall Bay, W side near Wellington, 41.333°S, 174.792°E (AMS C.405008, 3: complete). Waitangi, Chatham Islands, 43.950°S, 176.560°W (NMNZ M.091957, 12). Poriua, Titahi Bay, North Island, 41.108°S, 174.825°E (NMNZ M.150130, 13). 0 m, Ohope, Otaraweirere Bay, North Island, 37.950°S, 177.017°E (NMNZ M.044088, 40). 0 m, Pukerua Bay, Wellington W coast, 41.033°S, 174.883°E (NMNZ M.130798, 35). Naim River mouth, Chatham Islands, 43.950°S, 176.558°W (NMNZ M.134355, 100). 0 m, E side of North Cape, North Island, 34.417°S, 173.050°E (NMNZ M.134352, 10). 18–20 m, off Raglan, Jackson Reef, North Island, 37.872°S, 174.725°E (NMNZ M.137816, 8). 0 m, Wellington, Bay of Islands, North Island, 41.345°S, 174.773°E (NMNZ M.150342, 25). 9 m, Bay of Plenty, off SE corner of Motuhora (Whale) Island, North Island, 37.872°S, 176.988°E (NMNZ M.160331, 11). 0 m, Climo, NW Nelson, Paturau, South Island, 40.633°S, 172.433°E (NMNZ M.150132, 16). 0 m, head of Titirangi Bay, Marlborough Sound, South Island, 41.017°S, 174.133°E (NMNZ M.040709, 13). 55 m, Stewart Island, Port Adventure, 47.083°S, 168.183°E (AMS C.379060, 1).

Fossil specimens.

New Zealand. GS15119, R28/f62, Paua Kill site on 1855 intertidal slope, c 400 m E of Turakirae Head, S Wellington, Holocene, co-seismic uplift of 1855, 41.438°S, 174.917°E (GNS, 5; GNS, 8).

Remarks. A single specimen from Western Australia that is indistinguishable from New Zealand specimens is known (138 m, Direction Bank, off Rottnest Island, Western Australia, Australia, 31.740°S, 115.203°E (AMS ex C.378641, 1). Whether it is a mislocalized specimen or not remains to be determined.

Sinezona brucei new species

Figures 280–282

Type material. Holotype (NMNZ M.160295. Fig. 280). Paratypes from type locality (NMNZ M.148571, 5). 23 m, off N face of Hinemoa Island, Three Kings Islands, New Zealand, 34.180°S, 172.043°E (NMNZ M.148571, 5. Fig. 281).

Type locality. 123–128 m, King Bank, N of Three Kings Islands, New Zealand, 33.956°S, 172.323°E.

Etymology. Named for Bruce Marshall of NMNZ for his comprehensive contribution to the malacofauna of New Zealand.

Description. Shell small (to 0.96 mm: holotype. Figs 280, 281), trochiform, inflated. Protoconch of one whorl, raised axial lamellae, apertural varix not connected to embryonic cap, apertural margin straight. Teleoconch I of 0.6 whorl, 8–11 strong axial cords to lamellae, interstices with very fine growth lines, no spiral sculpture. Teleoconch II of one whorl, suture impressed. Shoulder of sinusoidal profile, axial lamellae from suture, not reaching selenizone, first indistinct low spiral cord after 0.25 teleoconch II whorl, 3–5 at apertural margin, irregularly spaced. Base with distinct constriction below selenizone, strong axial lamellae of same density as on shoulder, from below constriction fading towards umbilicus; approximately 10 spirals, increasing in strength from fine line below constriction to strong cord bordering umbilicus. Umbilicus broad funnel-shaped, walls straight, very fine growth marks. Selenizone above periphery, keels strong, moderately elevated, lunules dense, little distinct; foramen elongated triangular. Aperture D-shaped, roof overhanging.

Figure 280. *Sinezona brucei* new species. Holotype. 123–128 m, King Bank, N of Three Kings Islands, New Zealand, 33.956°S, 172.323°E (NMNZ M.160295). Scale bar shell = 1 mm. Scale bar protoconch = 100 μm.

Animal unknown.

Distribution. Three Kings Islands, New Zealand (Fig. 282).

Comparisons. *Sinezona bandeli* from New Zealand lacks axial cords on the shoulder, has many fine spiral lines on the shoulder, and a more depressed, subquadrate overall shell shape.

Figure 281. *Sinezona brucei* new species. Paratypes. **A–C.** 23 m, off N face of Hinemoa Island, Three Kings Islands, New Zealand, 34.180°S, 172.043°E (NMNZ M.148571, 5, 5). Scale bars shell = 1 mm. Scale bars protoconch = 100 μm.

Figure 282. Distribution of *Sinezona brucei* new species. ☆ type locality. ● specimen record.

Sinezona calumnior from temperate Australia and New Zealand has a much more narrow umbilicus, a longer teleoconch I (0.75–1 vs. 0.6 whorl), and a single, strong periumbilical cord.

Sinezona finlayi from New Zealand has weaker or no axial sculpture on the shoulder, lacks a strong periumbilical cord, and has much stronger lunules.

Sinezona iota from New Zealand has a taller overall shell shape with protruding apex, a more narrow umbilicus, the axial cords/lamellae extend into the umbilicus, and lacks spiral sculpture on shoulder and base.

Sinezona laqueus from New Zealand has a flatter shell, lacks spiral sculpture on shoulder and base and much more prominent lunules.

Sinezona norfolkensis from the Tasman Sea has a longer teleoconch I (0.8–0.9 vs. 0.6 whorl), a protoconch with axials reaching well past the midline of the whorl, and is overall taller.

Sinezona pacifica from temperate Australia and New Zealand has a longer teleoconch I (0.75–1 vs. 0.6 whorl), less elevated axial cords, and a less elevated apex.

Scissurella phenax from the Indo-Malayan Archipelago to the Central Pacific has a taller shell, a protoconch with denser axials bearing median thickenings, and has axials in the constriction below the selenizone on the base.

Sinezona calumnior new species

Figures 283–286

Chresonymy.

Sinezona sp. 2: Spencer *et al.*, 2009: 202.
Sinezona sp. 3: Spencer *et al.*, 2009: 202.
Sinezona sp. 4: Spencer *et al.*, 2009: 202.

Type material. Holotype (WAM S13382. Fig. 283). Paratypes S of Cowaramup Bay, near mouth of Margaret River, Western Australia, Australia, 33.883°S, 115.017°E (AMS C.475964, 1. Fig. 284A). S. of Museum Bay, Kendrew Island, Dampier Archipelago, Western Australia, Australia, 20.475°S, 116.533°E (WAM S13388, 1. Fig. 284B). 38 m, Arch Pinnacle, Princes Rocks, Three Kings Islands, New Zealand, 34.175°S, 172.050°E (NMNZ 174856, 8. Fig. 284C).

Type locality. Hopetoun, Western Australia, Australia, 33.867°S, 120.133°E.

Etymology. From Latin, calumnior, misrepresenter, practicer of trickery, with reference to the difficulty of recognizing this species; see also *Sci. phenax*; noun in apposition (OD).

Description. Shell small (to 0.75 × 0.67 mm: holotype. W × H. Figs 283, 284), trochiform, globular. Protoconch of 1 whorl, strong axial sculpture without median thickenings, apertural varix not connected to embryonic cap, apertural margin straight. Teleoconch I of 0.75–1 whorl, approximately 18 low axial lamellae; first spiral thread after 0.5 whorl. Teleoconch II of 0.5 whorl, suture deeply impressed. Shoulder convex, same density of axial lamellae; approximately six indistinct spiral threads, irregularly distributed, increasing in number from onset of selenizone to apertural margin; axial lamellae running over spiral threads. Base with moderate constriction below selenizone, same density and strength of axials as on shoulder, approximately 23 spiral lines running under axial lamellae. Umbilicus very narrow, bordered by strong cord, walls straight, with growth marks. Aperture round, roof overhanging. Selenizone above periphery; lunules moderately distinct, irregularly spaced; keels low; foramen closed, continued anteriorly by raphe.

Operculum (Fig. 285A) covering aperture, round, thin, multispiral, with central nucleus.

Radula (Fig. 285B–D). Rachidian tooth trapezoidal, central cusp strongest, three cusps on each side. Lateral teeth 1–3 similar, apical-most cusp strongest, 3–4 cusps on outer cutting edge. Lateral tooth 4 reduced, hook-shaped. Lateral tooth 5 enlarged by broadening, apical cusp strongest, six cusps on inner cutting edge. Inner marginal teeth with broadly triangular tip, apical cusp strongest, three cusps along inner edge, 4–5 cusps along outer edge; outer marginal teeth with spoon-shaped tip, many bristles along edges; shafts without food groove. Radular interlock of central field moderate.

Figure 283. *Sinezona calumnior* new species. Holotype. Hopetoun, Western Australia, Australia, 33.867°S, 120.133°E (WAM S13382). Scale bar shell = 0.5 mm. Scale bar protoconch = 100 µm.

Figure 284. *Sinezona calumnior* new species. Paratypes **A.** S of Cowaramup Bay, near mouth of Margaret River, Western Australia, Australia, 33.883°S, 115.017°E (AMS C.475964, 1). **B.** S of Museum Bay, Kendrew Island, Dampier Archipelago, Western Australia, Australia, 20.475°S, 116.533°E (WAM S13388, 1). Radula shown in Figure 285B. **C.** 38 m, Arch Pinnacle, Princes Rocks, Three Kings Islands, New Zealand, 34.175°S, 172.050°E (NMNZ M.174856, 8). Scale bars shell = 0.5 mm. Scale bars protoconch = 100 µm.

Comparisons. *Sinezona bandeli* from New Zealand has a much wider umbilicus, an overall broader shell shape, and at least twice as many spiral threads on the shoulder.

Sinezona beddomei from temperate Australia has a wider umbilicus, an overall broader shell, and no spiral sculpture on the shoulder and upper base.

Sinezona brucei from New Zealand has a wider, funnel-like umbilicus, a shorter teleoconch I (0.6 vs. 0.75–1 whorl), and multiple, weaker periumbilical cords.

Sinezona confusa from the Caribbean to Brazil has a wider umbilicus and a stronger periumbilical cord forming a shelf in the adumbilical lower corner of the aperture.

Sinezona danieldreieri from the Indo-Malayan Archipelago to the Central Pacific has a wider umbilicus and the apex is much more sunken.

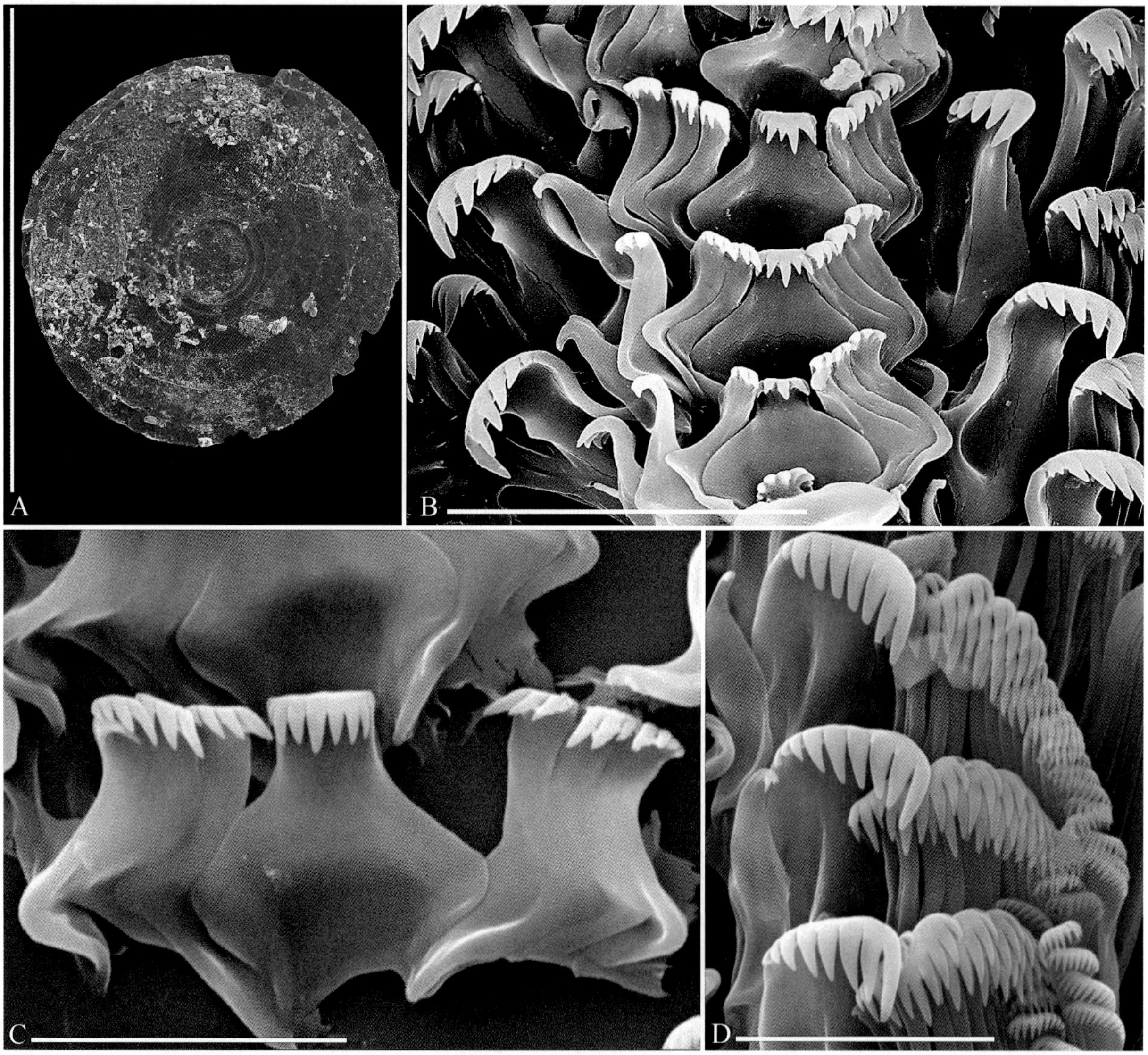

Figure 285. *Sinezona calumnior* new species. **A.** Operculum. S of Museum Bay, Kendrew Island, Dampier Archipelago, Western Australia, Australia, 20.475°S, 116.533°E (WAM S13388, 1). Shell shown in Figure 284B. Scale bars = 0.5 mm. **B–D.** Radula. **B.** 57 m, off Albany, Western Australia, Australia, 35.167°S, 117.878°E (AMS C.406403, 1). Scale bars = 20 µm. **C–D.** 38–43 m, between Pandora Bank and Cape Maria van Diemen, North Island, New Zealand, 34.490°S, 172.587°E (NMNZ M.160297, 1). **C.** Central field. **D.** Lateral teeth 4–5 and marginal teeth. Scale bars = 10 µm.

Sinezona doliolum from the western Indian Ocean has finer spiral threads on the shoulder, has a longer teleoconch II (1 vs. 0.6 whorl), and longer foramen.

Sinezona macleani from the Indian Ocean to the Western Pacific has a protoconch with flocculent sculpture and no apertural varix, a straight parietal wall with shelf at the lower adumbilical corner, and a much longer teleoconch I (1.75 vs. 1 whorls).

Sinezona norfolkensis from the Tasman Sea has a wider, funnel-like umbilicus, an overall broader shell, and strongly folded axial lamellae on the shoulder.

Sinezona praecedens from the Oligocene of Europe has a teleoconch of 0.8 whorl, and the axial cords are transformed into scabrous scales towards the umbilicus.

Sinezona rimuloides from the northeastern Pacific and the Panamic has less pronounced spiral threads on shoulder and base and the shell is overall taller.

Scissurella phenax from the Indo-Malayan Archipelago to the Central Pacific has a protoconch with axials bearing median thickenings, a slightly wider umbilicus, and regularly spaced spiral cordlets on the shoulder from the onset of the selenizone.

Distribution. Tropical and subtropical waters from Sri Lanka to French Polynesia (Fig. 286).

Specimen records.

Australia, Western Australia. S of Museum Bay, Kendrew Island, Dampier Archipelago, 20.475°S, 116.533°E (WAM S13388, 1). 0 m, The Basin, NE Rottnest Island, 32.000°S, 115.500°E (OSU 11920a, 2). Hopetoun, 33.866°S, 120.133°E (WAM S13382, 1). S of Cowaramup Bay, near mouth of Margaret River, 33.883°S, 115.017°E (AMS C.378295-2, 1). Recherche Archipelago, 34.458°S, 122.005°E (AMS C.406405, 1). 57 m, off Albany, 35.167°S, 117.878°E (AMS C.406403, 1).

Australia, South Australia. Point Souttar, Yorke Peninsula, 34.900°S, 137.267°E (AMS C.378562, 3). Kangaroo Island N coast, Stokes Bay, 35.617°S, 137.200°E (AMS C.378563, 2).

Tasman Sea. 9 m, Middleton Reef, 29.453°S, 159.113°E (AMS C.348612, 1).

New Zealand. 310 m, 11 km NW of Great Island, Three Kings Islands, 34.108°S, 172.078°E (NMNZ M.093875, 3). 38 m, Arch Pinnacle, Princes Rocks, Three Kings Islands, 34.175°S, 172.050°E (NMNZ 174856, 8). 33 m, Arch Pinnacle, Princes Rocks, Three Kings Islands, 34.175°S, 172.050°E (NMNZ M.150318, 3). 23 m, off N face of Hinemoa Island, Three Kings Islands, 34.180°S, 172.043°E (NMNZ M.160292, 12). 38–43 m, between Pandora Bank and Cape Maria van Diemen, North Island, 34.490°S, 172.587°E (NMNZ M.160297, 1). 38 m, off E side of Ruamahua-nui Island, Aldermen Islands, North Island, 36.953°S, 176.097°E (NMNZ M.112678, 6). 37–91 m, off SE side of Mayor Island, North Island, 37.266°S, 176.283°E (NMNZ M.092136, 1). 59–74 m, off E side Mayor Island, 37.315°S, 176.270°E (NMNZ M.150286, 2).

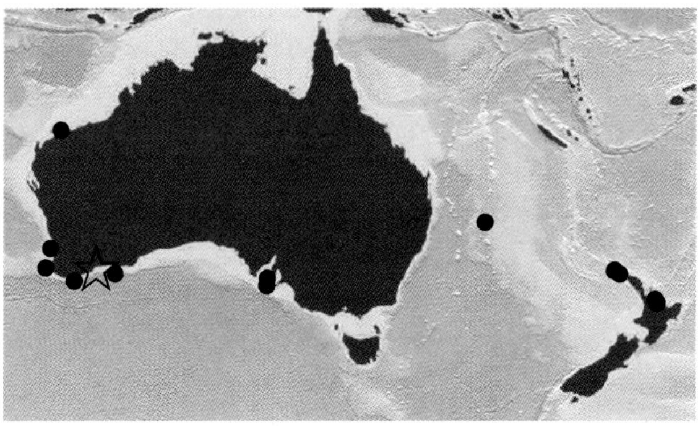

Figure 286. Distribution of *Sinezona calumnior* new species. ☆ type locality. ● specimen record. ■ literature record.

Sinezona calumnior **s.l. [see Remarks].**

Australia. 15 m, Middleton Reef, NW side, Queensland, 29.456°S, 159.062°E (AMS C.445049, 2: complete). Point Quobba, Western Australia, 24.483°S, 113.417°E (AMS C.404157, 1).

New Caledonia. 0 m, Koumac Sector, 20.373°S, 164.113°E (MNHN, 31). 0–1 m, Bourail Sector, 21.615°S, 165.378°E (MNHN, 12).

New Zealand. 219–274 m, SE of D'Arcy Point, Raoul Island, Kermadec Islands, 29.313°S, 177.903°W (NMNZ M.226873, 3). 805 m, Three Kings Trough, Three Kings Islands, 34.000°S, 171.917°E (NMNZ M.049028, 7). 310 m, 11 km NW of Great Island, Three

Kings Islands, 34.108°S, 172.078°E (NMNZ M.160293, 1). 20–22 m, Southeast Bay, Great Island, Three Kings Islands, 34.158°S, 172.147°E (NMNZ M.134758, 1). 14 m, off Princes Rocks, Three Kings Islands, 34.166°S, 172.133°E (NMNZ M.049826, 2). 110 m, W of Poor Knight Island, North Island, 35.483°S, 174.725°E (NMNZ M.035252, 1). 50 m, Cave, Poor Knight Island, 35.510°S, 174.730°E (NMNZ M.112030, 14).

French Polynesia. Moorea Island, Baie de Papetoai, Passe Tareu, W side, Society Islands, 17.508°S, 149.850°W (AMS C.377911, 14, 28: complete).

Sri Lanka [most likely* Sci. phenax*]. Ahangama, 5.972°N, 80.363°E (ZSM 20061091, 1: complete). 1–2 m, Ahangama,5.972°N, 80.363°E (ZSM 20061127, 1: complete). 1 m, Unawatuna, 6.095°N, 80.248°E (ZSM 20061092, 6: complete).

Remarks. *Sinezona calumnior* was a difficult species to treat. For a long time, I considered it a rather polymorphic species, until sufficient material had accumulated to see some patterns in shell morphology. This led to the separation of *Sin. brucei*, *Sin. norfolkensis*, and *Sci. phenax* from the initial pool of *Sin. calumnior* s.l. rather late in manuscript preparation. Although many SEM images were taken, not every single specimen was examined by SEM. Some specimens identified early on as *Sin. calumnior* (s.l.) may, therefore, represent any of the four species s.s. Those specimens of problematic s.l. identification are listed here under *Sinezona calumnior* s.l. All specimens listed as any of the four s.s. species have been documented by SEM.

The specimens from Western Australia and New Zealand are indistinguishable. The holotype from Western Australia (Fig. 283) and the specimen from New Zealand shown in Figure 284 is a case in point. There is some intraspecific variability, which is well within the normal, expected range of any given species: teleoconch I whorl count varies by the customary 0.25 whorl; the density of the axial sculpture varies by approximately 30%, with populations, within regions, and between regions (Western Australia vs. New Zealand); the exact pattern of spiral threads on the shoulder. Commonalities include all aspects of the protoconch, the very narrow umbilicus, and the general pattern of low axial lamellae and fine spiral threads. Radulae from specimens from Western Australia (Fig. 285B) and New Zealand (Fig. 285C–D) did not reveal any differences either.

Sinezona carolarum Geiger & McLean, 2010

Figures 287–291

Chresonymy.

Sinezona n. sp. Geiger, ms: Kaiser, 2009: 221, 228, fig. 237 I–K [paratype Fig. 289B].

Sinezona carolarum Geiger & McLean, 2010: 8–12, figs 5–8.

Type material. Holotype (SBMNH 83703. Fig. 287). Paratypes: from type locality (SBMNH 83704, 2. Fig. 289A–B);16–23 m, Maria Cleofa Island, Tres Marias Islands, Mexico, 21.300°N, 106.276°W (SBMNH 83705, 4, mounted radula. Figs 288C, 290); 23–30 m, Sharkfin Rock, SSW Rodriguez Point, Cocos Island, Costa Rica, 5.533°N, 86.983°W (SBMNH 83706, 3. Fig. 289C); 12–15 m, Clipperton Island, France, 10.305°N, 109.198°W (KKC 210001, 8. Fig. 288A); 12–15 m, 10.305°N, 109.198°W (LACM 3116, 5; MNHN 21265, 5; SDMNH, 5; WAM S41348, 5; USNM 1117767, 5; all ex KKC 200518). 10–38 m, Clipperton Island, France, 10.320°N, 109.223°W (SBMNH 83707, ex KKC 210111, 2. Fig. 288B).

Figure 287. *Sinezona carolarum*. Holotype. 15–30 m, East side, Clipperton Island, France, 10.295°N, 109.201°W (SBMNH 83703). Scale bar shell = 0.5 mm. Scale bar protoconch = 100 µm.

Type locality. 15–30 m, East side, Clipperton Island, France, 10.295°N, 109.201°W, leg. Kirstie Kaiser (OD).

Etymology. Named after Carole Hertz, editor of *The Festivus*, and Carol Skoglund, long time collector and student of the Panamic molluscan fauna, for their significant body of work and often collaborative efforts to better understand the taxonomy of the Panamic fauna. Genitive feminine plural of their common first name (OD).

Figure 288. *Sinezona carolarum*. Paratypes. **A.** 12–15 m, Clipperton Island, France, 10.254°N, 109.198°W (KKC 2100001, 8). **B.** Juvenile with still open slit. 10–38 m, Clipperton Island, France, 10.320°N, 109.223°W (SBMNH 83707). **C.** 16–23 m, Maria Cleofa Island, Tres Marias Islands, Mexico, 21.300°N, 106.276°W (SBMNH 83705). Scale bars shell = 0.5 mm. Scale bars protoconch = 100 μm.

Description. Shell small (to 0.54 mm. Figs 287–289), trochiform. Protoconch of 0.875 whorl, fine axials on outer portion of whorl, weak apertural varix not connected to embryonic cap, apertural margin straight. Teleoconch I of 0.8–1.1 whorls, with 15–38 axial cords, interstices with fine irregular growth marks, first spiral thread no earlier than after 0.66 whorl, increasing to 0–2 on shoulder, approximately six between position of selenizone and periphery. Teleoconch II of 0.33 whorl, suture little impressed. Shoulder almost flat, axial cords of same density as on teleoconch I, 2–5 indistinct spiral lines. Base with moderate constriction below selenizone, axials as on shoulder,

Figure 289. *Sinezona carolarum*. Paratypes. **A–B.** 15–30 m, East side, Clipperton Island, France, 10.295°N, 109.201°W (SBMNH 83704, 2). **C.** 23–30 m, Sharkfin Rock, SSW Rodriguez Point, Cocos Island, Costa Rica, 5.533°N, 86.983°W (SBMNH 83706, 3). Scale bars shell = 0.5 mm. Scale bars protoconch = 100 µm.

approximately 12 spirals, increasing in strength from lines at periphery to low cords near umbilicus, sometimes forming diffuse nodules at intersection with axials. Umbilicus open, bordered by moderately strong spiral cord, walls straight, with fine axial growth marks. Selenizone slightly above periphery, keels low, strong, lunules distinct, selenizone approximately twice as long as teardrop-shaped foramen. Aperture subquadrate, roof overhanging.

Radula (Fig. 290). Rachidian tooth triangular with five cusps on tip, central one largest, cusps arranged in V. Lateral teeth 1–3 similar, 3–4 cusps on outer edge of tip. Lateral tooth 4 reduced, hook-shaped. Lateral tooth 5 broadened, enlarged, six cusps on inner margin of tip. Radular interlock of central field moderate. Inner marginal teeth with terminal cusp largest, one on inner edge, four on outer edge. Outer marginal spoon-shaped with many cusps on either side of tip.

Comparisons. *Sinezona rimuloides* from the Panamic to central California is more elevated in overall shape, has a narrower umbilicus, and the selenizone is approximately three times as long as the foramen.

Sinezona chilensis from southwestern South America has about half as many axial cords (15–38 vs. 16–18) and a distinct periumbilical cord.

Distribution. Mexico (offshore Maria Cleofa Island), Clipperton and Cocos Islands (Fig. 291).

Specimens records.

France, Clipperton Island. 12–15 m, 10.305°N, 109.198°W (KKC 200518, 75). 11–22 m, 10.316°N, 109.214°W (KKC 210048, 5). 17 m, NW side, 10.322°N, 109.223°W (KKC

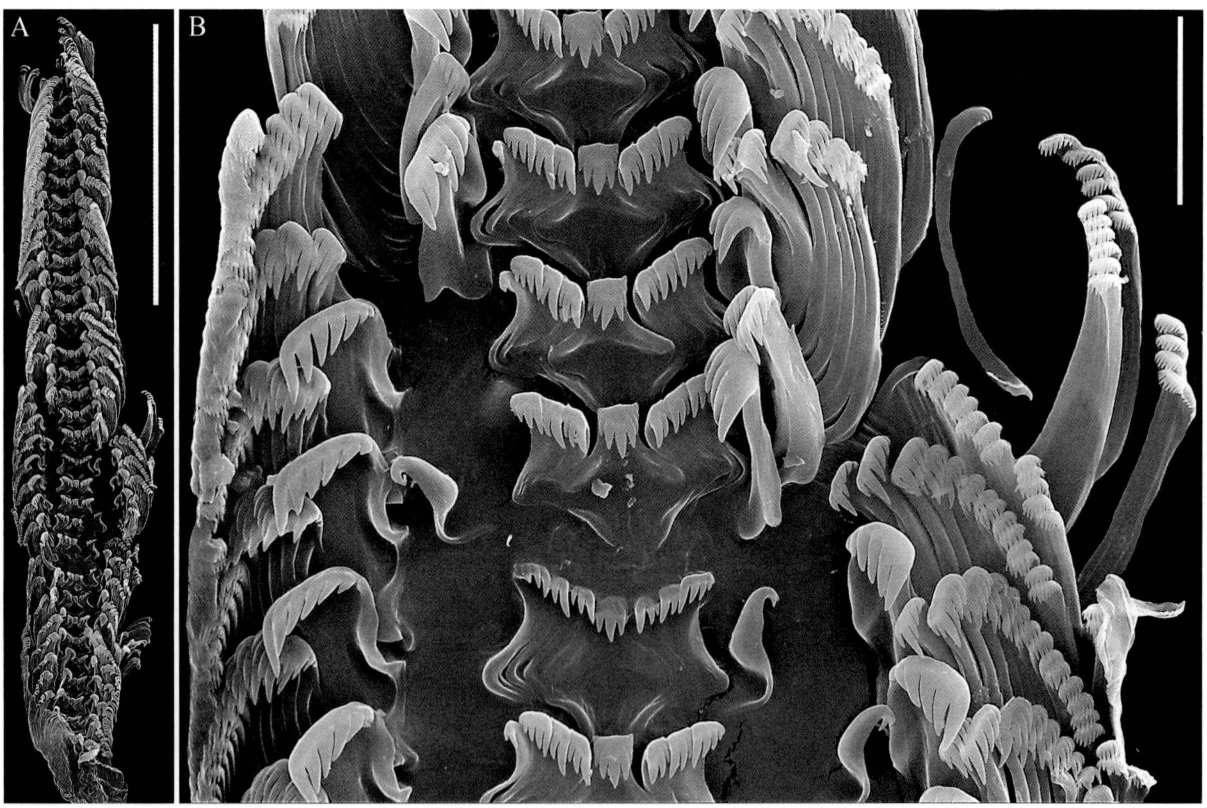

Figure 290. *Sinezona carolarum*. Radula. 16–23 m, Maria Cleofa Island, Tres Marias Islands, Mexico, 21.300°N, 106.276°W (KKC). **A.** Entire radula. Scale bar = 100 μm. **B.** Full width of radula enlarged with portions separated from one another. Scale bar = 10 μm.

210108, 1). 15–30 m, E side, 10.295°N, 109.201°W (KKC 200516, 51). 12–43 m, 10.291°N, 109.225°W (KKC 200517, 80). 10–38 m, 10.320°N, 109.223°W (KKC 210047, 8). 15–20 m, 10.314°N, 109.238°W (KKC 210000, 12). 69 m, 10.320°N, 109.235°W (KKC 200515, 47). 15 m, 10.288°N, 109.350°W (KKC 200519, 40). 10–20 m, SE, 10.291°N, 109.201°W (SBMNH 211380, 1: complete).

Costa Rica. 23–30 m, Sharkfin Rock, SSW Rodriguez Point, Cocos Island, 5.533°N, 86.983°W (KKC, 5).

Mexico. 16–23 m, Maria Cleofa Island, Tres Marias Islands, 21.300°N, 106.276°W (KKC, 7).

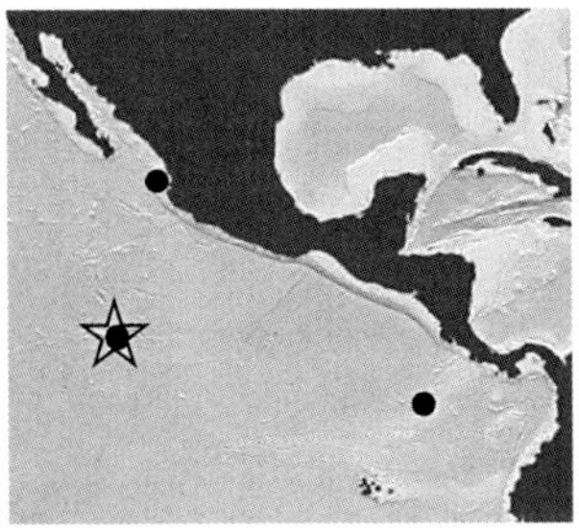

Figure 291. Distribution of *Sinezona carolarum*. ☆ type locality. ● specimen record.

Sinezona chilensis new species

Figures 292–296

Misidentifications.

Sinezona rimuloides [partim]: Abbott, 1974: 17 [Chile records most likely referring to *Sin. chilensis*].

not *Sinezona rimuloides*: Marincovich, 1973: 4, 16, fig. 26 [is *Sin. chilensis*].

Type material. Holotype (LACM 3139. Fig. 292). Paratypes 6–17 m, S Herrandura Bay, S of Coquimbo, Chile, 29.983°S, 71.367°W. (LACM 3140, 2. Fig. 293B–C).

Type locality. 0 m, Iquique, near end of Ave. Baquedana, Tarapacá Province, Chile, 20.217°S, 70.167°W.

Etymology. Referring to the country of the type locality.

Description. Shell small (to 0.91 mm. Figs 292–294), trochiform globular. Protoconch of one whorl, fine, dense axials, apertural varix not connected to embryonic cap, apertural margin convex. Teleoconch I of 1–1.25 whorls, approximately 16–18 axial cords, interstices with fine irregular growth marks, no spiral sculpture. Teleoconch II of 0.25 whorl, suture little impressed. Shoulder convex, fainter axials at same density as on teleoconch I, interstices with fine irregular growth marks, with approximately five indistinct low spiral cords at apertural margin. Base with similar axial sculpture as on shoulder, approximately 12 spirals, increasing in strength from fine lines below selenizone, to strong cords adjacent to umbilicus. Umbilicus narrowly open, bordered by strong cord, walls straight, smooth to irregular. Selenizone short (0.1–0.2 whorl), between less than length of foramen to twice length of foramen, strong keels very low, foramen broad lenticular, anteriorly continued as a raphe. Aperture rounded D-shaped, with pointed corner where cord bordering umbilicus meets peristome.

Radula (Fig. 295). Rachidian with central cusp largest, three cusps on each side. Lateral teeth 1–3 similar, four cusps on outer edge of tip, apical-most largest. Lateral tooth 4 reduced, hook-shaped, one minute point on each side. Lateral tooth 5 enlarged by broadening, approximately seven cusps on inner edge of tip, apical-most largest. Inner marginal teeth with unequilateral triangular tip, apical cusp largest, two cusps on inner margin, 3–4 on outer margin. Outer marginal teeth spoon-shaped, many very fine cusps. Radular interlock of central field moderate.

Operculum (Fig. 292) thin, round, covering entire aperture, multispiral, with central nucleus.

Comparisons. *Sinezona rimuloides* from the northeastern Pacific to the Panamic is overall taller, with the suture at the protoconch varix being below the periphery of the shell, whereas it is at or above the periphery in *Sin. chilensis*. The foramen of *Sin. chilensis* is generally shorter, only 2–3 times as long as wide, whereas the foramen of *Sin. rimuloides* generally is approximately 4–5 times as long as wide.

Sinezona carolarum from the Panamic has about twice as many axial cords (15–38 vs. 16–18), the selenizone is about twice as long as in *Sin. chilensis*, and it lacks a distinct periumbilical cord.

The circumantarctic *Sin. subantarctica* occurring in southern Chile (>45°S) shows a distinct callus covering the umbilicus and the protoconch lacks axial cords as well as an apertural varix.

Sinezona bandeli from New Zealand has a much longer selenizone (0.4 vs. 0.1 whorl), a distinct constriction below the selenizone, and a wider umbilicus.

Sinezona brevis from New Zealand has a much lower keel surrounding the foramen and a stronger periumbilical cord.

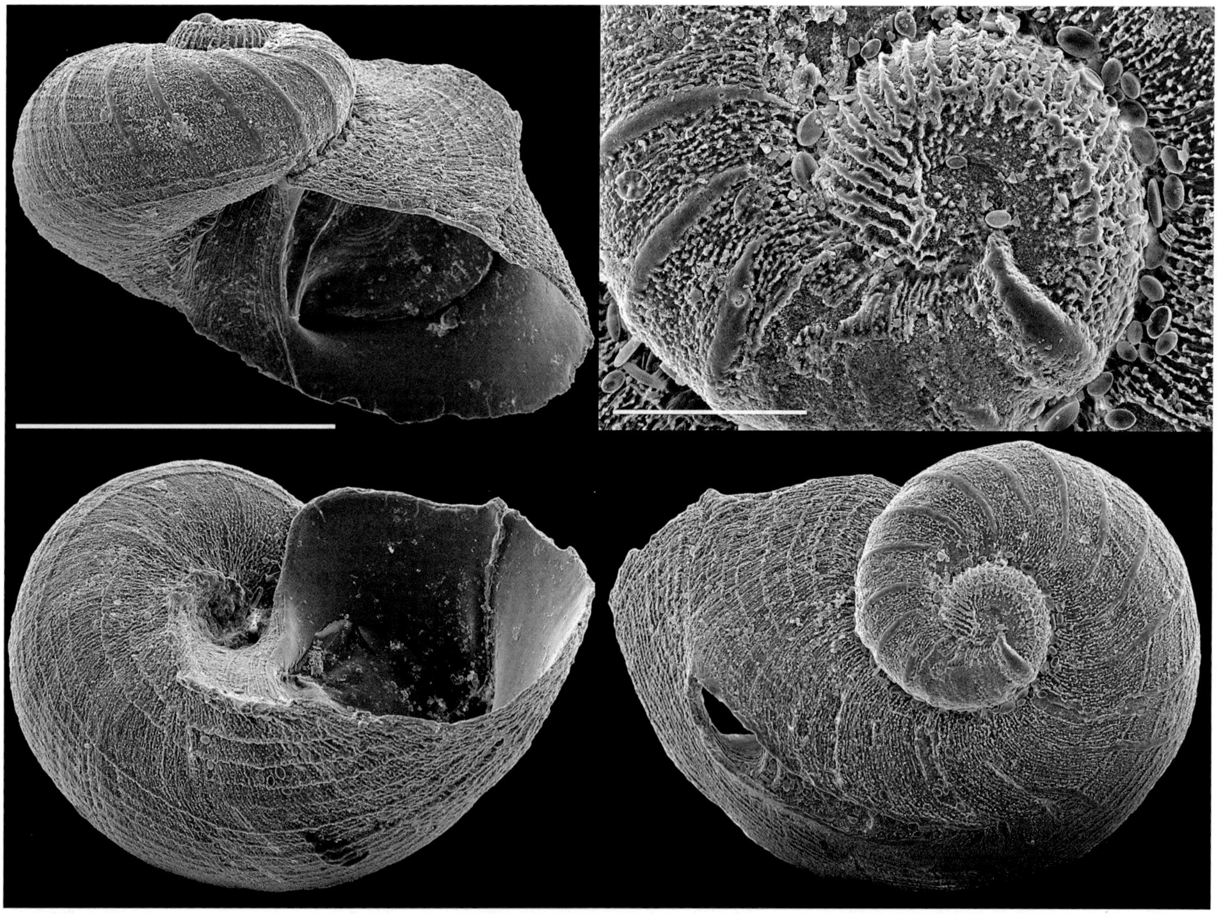

Figure 292. *Sinezona chilensis* new species. Holotype. 0 m, Iquique, near end of Ave. Baquedana, Tarapacá Province, Chile, 20.217°S, 70.167°W (LACM 3139). Scale bar shell = 0.5 mm. Scale bar protoconch = 100 µm.

Sinezona costulata from southern Japan has a protoconch with fewer but stronger raised axial cords and a much longer selenizone (0.1 vs. 0.4 whorl).

Sinezona globosa from the Indo-Malayan Archipelago to tropical Eastern Pacific (and possibly Caribbean) grows much larger (3.8 mm), has a much longer selenizone (0.1 vs. 0.8 whorl), an overall more globular profile, and the base slopes gradually into the umbilicus.

Sinezona insularis from equatorial mid-Atlantic islands has a protoconch without apertural varix and a shorter teleoconch I (<1 vs. >1–1.25 whorls).

Figure 293. *Sinezona chilensis* new species. **A.** 0 m, Iquique, near end of Ave. Baquedana, Tarapacá Province, Chile, 20.217°S, 70.167°W (LACM 64-16, 39). **B–C.** Paratypes. 6–17 m, S Herrandura Bay, S of Coquimbo, Chile, 29.983°S, 71.367°W (LACM 3140). Scale bars shell = 0.5 mm. Scale bars protoconch = 100 μm.

Sinezona pacifica from temperate Australia and New Zealand has a much longer selenizone (0.1 vs. 0.5 whorl), fewer but stronger axial cords on the protoconch, and lacks a periumbilical cord.

Sinezona zimmeri from Easter and Pitcairn Islands has a longer selenizone (0.1 vs. 0.3 whorl), maintains strong axial cords on the shoulder of teleoconch II, and has a wider umbilicus.

Figure 294. *Sinezona chilensis* new species. **A.** 0–6 m, Antofagasta Province, Los Colorados, Chile, 23.483°S, 70.367°W (AMS C.406351). **B–C.** 6–17 m, Coquimbo Province, Herradura Bay, S side, Chile, 29.983°S, 71.367°W (AMS C.406352). Scale bars shell = 0.5 mm. Scale bars protoconch = 100 μm.

Figure 295. *Sinezona chilensis* new species. Radula. 0 m, Iquique, near end of Ave. Baquedana, Tarapacá Province, Chile, 20.217°S, 70.167°W (LACM 64-16). **A.** Entire radula. Scale bar = 100 µm. **B.** Central field enlarged. **C.** Half width of radula. Scale bars B–C = 10 µm.

Juveniles can be distinguished from the southern South American *Sci. clathrata* by the protoconch sculpture (flocculent vs. axial cords), the lack of an apertural varix on the protoconch, the much more narrow umbilicus, and the more similar strength of axial and spiral sculpture, giving the shell an overall reticulate appearance.

Distribution. Peru to Chile (Fig. 296).

Specimen records.

Peru. Lobos de Afaea Island, NW and NE of Isthmus, 6.952°S, 80.705°W (LACM 74-6, 6).

Chile. 0 m, Iquique, near end of Ave. Baquedana, Tarapacá Province, 20.217°S, 70.167°W (LACM 64-16, 40). Inquique marine lab, Universidad del Norte, Taapaca Province, 20.258°S, 70.133°W (LACM 75-12, 2). 0 m, Iquique, 20.217°S, 70.167°W (UMML 30.12037, 2). Tarapaca Province, Pozo Toyo, 20.417°S, 70.175°W (LACM 75-10, 14; AMS C.380697, 18). Antofagasta Province, E of Santa Maria Island, 23.417°S, 70.600°W (LACM 75-17, 3; AMS C.380684, 1). 0–6 m, Antofagasta Province, Los Colorados, 23.483°S, 70.367°W (LACM 75-19, 14; AMS C.406351, 1). 6–17 m, S side, Herradura Bay, Coquimbo Province, 29.983°S, 71.367°W (LACM 3140, 2 [paratypes]; AMS C.406352, 10). Aconcagua Province, Los Molles, 32.233°S, 71.533°W (AMS C.380690, 5; LACM 75-28, 22). 3–7 m, Valparaiso Province, Algarrobo, 33.367°S, 71.700°W (AMS C.406332, 1). Marine Biology Station, Montemar, Valparaiso, 32.950°S, 71.533°W (LACM 75-30, 1).

Figure 296. Distribution of *Sinezona chilensis* new species. ☆ type locality. ● specimen record.

Sinezona cingulata (O. G. Costa, 1861)

Figures 297–305, Color Plate 2A

Chresonymy.

Scissurella cingulata O. G. Costa, 1861: 61–62, pl. 12, figs 9a–b.
Scissurella cingulata: Petit, 1869: 111.
Schismope cingulata: Monterosato, 1872a: 30.
Schismope cingulata: Monterosato, 1875a: 22.
Scissurella (*Schismope*) *cingulata*: Monterosato, 1877a: 416 .
Schismope cingulata: Monterosato, 1878a: 79.
Schismope cingulata: Paetel, 1883: 47.
Schismope cingulata: Monterosato, 1884: 39.
Schismope cingulata: Fischer, 1880–1887 [1885]: 846.
Schismope cingulata: Locard, 1886: 331.
Schismope cingulata: Brusina, 1886: 198–199.
Schismope cingulata: Normann, 1888: 21.
Scissurella cingulata: Paetel, 1888: 289.
Schismope cingulata: Pilsbry, 1890: 61, pl. 57, figs 1–7.
Scissurella (*Schismope*) *cingulata*: Pallary, 1900: 359.
Schismope cingulata: Thiele, 1912a: 21, pl. 3, figs 6–7.
Schismope cingulata: Dautzenberg, 1927: 215.

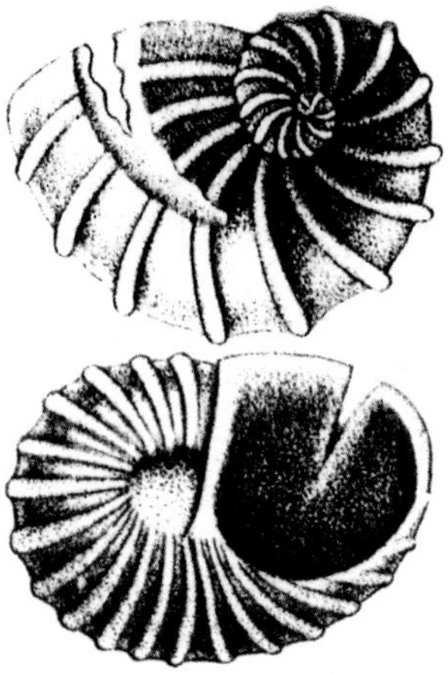

Figure 297. Original illustration of *Scissurella cingulata.*

Schismope cingulata: Coen, 1933: 12–13 .
Schismope cingulata: Coen, 1937: 6–7 .
Schismope cingulata: Priolo, 1948: 23–24.
Woodwardia (Woodwardia) cingulata: Wenz, 1961: 173, fig. 271 [copy figure Thiele, 1912a].
Woodwardia cingulata: Nordsieck, 1968: 10, pl. 1, fig. 2.30.
Woodwardia cingulata: Parenzan, 1970: 30.
Sinezona cingulata: Piani, 1980.
Scissurella (Schismope) cinculata: Nordsieck, 1982: 2, pl. 1, fig. 2.30.
Sinezona cingulata: Cuomo & Fasculo, 1983: 40.
Sinezona cingulata: Grasso *et al.*, 1983: 36.
Sinezona cingulata [partim]: García-Talavera Casañas, 1983: 38, text-fig. [including Caribbean distribution, i.e., *Sin. confusa*].
Sinezona cingulata: Ghisotti *et al.*, 1984: 104.
Sinezona cingulata: Chemello *et al.*, 1984: 169.
Schismope cingulata: Aartsen *et al*, 1984: 6–7, pl. 1, fig. 1.
Schismope cingulata: Perretti, 1986: 126.
Sinezona cingulata: Lozouet, 1986: 111.
Schismope cingulata: Luque, 1986: 40–42, fig. 7 [radula].
Schismope cingulata: Otero-Schmitt & Trigo-Trigo, 1989: 80.
Woodwardia cingulata: Tenekides, 1989: pl. 2, fig. 3.
Sinezona cingulata: Burnay & Rolán, 1990: 34–36, fig. 1 [copy figure Costa, 1861].
Sinezona cingulata: Sabelli *et al.*, 1990: 12.
Schismope cingulata: Poppe & Goto, 1991: 32, 64.
Schismope cingulata: Barash & Danin, 1992: 17–18.
Sinezona cinculata: Bedulli *et al.*, 1993a: 71.
Sinezona cinculata: Cachia *et al.*, 1993: 10.
Sinezona cinculata: Giannuzzi-Savelli *et al.*, 1994: 13, fig. 89.
Schismope cingulata: Geuze, 1994: 81.
Sinezona cingulata: Arduino *et al.*, 1995: 17, text-figs.
Schismope cingulata: Bandel, 1998: 59–60, pl. 20, figs 2–6.
Sinezona cingulata: Ávila *et al.*, 1998: 491.
Scissurella (Schismope) cingulata: Macedo *et al.*, 1999: 299.
Sinezona cingulata: Öztürk & Ergen, 2000: 61, 62, 65, fig. 2e.
Sinezona cingulata: Kurtz, 2001: 81.
Sinezona cingulata: Ávila *et al.*, 2002: 348 [fossil].
Sinezona cinculata: Landau *et al.*, 2003: 16, pl. 1, fig. 5.
Sinezona cingulata: Geiger, 2003: 78, fig. 16a, b [shell, radula].
Sinezona cingulata: Ávila *et al.*, 2004: 77.
Sinezona cingulata: Geiger, 2004: text-fig. p. 6.
Sinezona cingulata: Ardovini & Cossignani, 2004: 19.
Schismope cingulata: Mienis & Ben-David-Zaslow, 2004: 14.
Sinezona cingulata: Kabasakal & Kabasakal, 2005: 53.
Sinezona cingulata: Cardigos *et al.*, 2005: 165.
Sinezona cingulata: Ávila *et al.*, 2006: 49.
Sinezona cingulata: Geiger, 2006c: text-fig. 4.
Sinezona cingulata: Peñas *et al.*, 2006: 25.
Sinezona cingulata: Cosentino & Giacobbe, 2007: 191–193.
Sinezona cingulata: Nolt, 2008: 56–60, figs 2B, 4C, 5A [copy figure Costa, 1861], 5B [copy figure Thiele, 1912a], 6 [radula].

Sinezona cingulata: Manzziotti *et al.*, 2008: 74.

Sinezona cingulata: Albuquerque *et al.*, 2009: 68–69, text-figs.

Sinezona cingulata: Segers *et al.*, 2009a: 3.

Sinezona cingulata: Segers *et al.*, 2009b: 5.

Sinezona cingulata: Segers *et al.*, 2009c: 65, pl. 5, figs 5–5b.

Sinezona cingulata: Cossignani & Ardovini, 2011: 96, text-fig.

Sinezona cingulata: Hernández *et al.*, 2011: 61, fig. 11A–C

Sinezona cingulata: Moreno, 2011: 95, text-figs.

Synonyms.

+ *Trochotoma crossei* Folin, 1869: 144–145, pl. 22 figs 6–7 (Fig. 298). ***Type material.*** Holotype [see Remarks] 0.0005 × 0.0009 [units not given]. Original type lost (Kisch, 1960: 153). Neotype (MNHN: designated by Burnay & Rolán, 1990: 36). ***Type locality.*** São Vicente, Cape Verde Islands [16.900°N, 25.000°W] (OD & neotype). ***Etymology.*** Named after French malacologist Joseph Charles Hippolyte Crosse (1826–1898). ***Remarks.*** Neotype not found in 4/2005 (Geiger pers. obs.). The original description is explicit about the species being based on a single specimen, so should be considered the holotype.

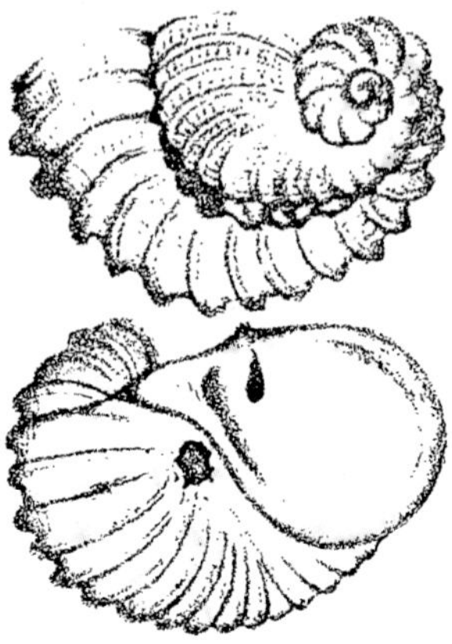

Figure 298. Original illustration of *Trochotoma crossei.*

Schismope crossei: Pilsbry, 1890: 66, pl. 57, fig. 51.

Schismope crossei: Thiele, 1912a: 23, pl. 3, figs 12–13 [copy figures Folin, 1869].

Woodwardia crossei: Nordsieck, 1968: 10, pl. 1, fig. 2.31.

Scissurella (*Schismope*) *crossei*: Nordsieck, 1982: 2, pl. 1, fig. 2.31.

Sinezona crossei: Lozouet, 1986: 111 [as possible synonym of *Sin. cingulata*].

Sinezona crossei: Burnay & Rolán, 1990: 34–36, pl. 1, figs 3–5

Sinezona crossei: Rolán & Luque, 1994: 4, figs 3–6.

Scissurella crossei: Ardovini & Cossignani, 2004: 19.

Sinezona crossei: Ardovini & Cossignani, 2004: 19.

Sinezona crossei: Rolán, 2005: 41, pl. 6, figs 70–71.

+ *Schismope fayalensis* Dautzenberg, 1889: 65, pl. 4, fig. 10. ***Type material.*** Three syntypes (KBIN IG 10591, MT756, Dautzenberg types: 525115), 0.75 × 0.75 mm. ***Type locality.*** Station 103, [38°32'5"N 30°57'4"E: from page 6], 15–20 m (OD). ***Remarks.*** The three syntype specimens are strongly affected by Byne's disease and are close to unrecognizable [Geiger, pers. obs. 7/2007].

Schismope fayalensis: Thiele, 1912a: 22, pl. 3, figs 8–9 [copy figure Dautzenberg, 1889].

Scissurella (*Schismope*) *fayalensis*: Nordsieck, 1982: 2, pl. 1, fig. 32.

Scissurella (*Sinezona*) *fayalensis*: Nordsieck & García-Talavera, 1979: 35, pl. 1, fig. 2.

Sinezona fayalensis: Lozouet, 1986: 111 [most likely a synonym of *Sin. cingulata*].

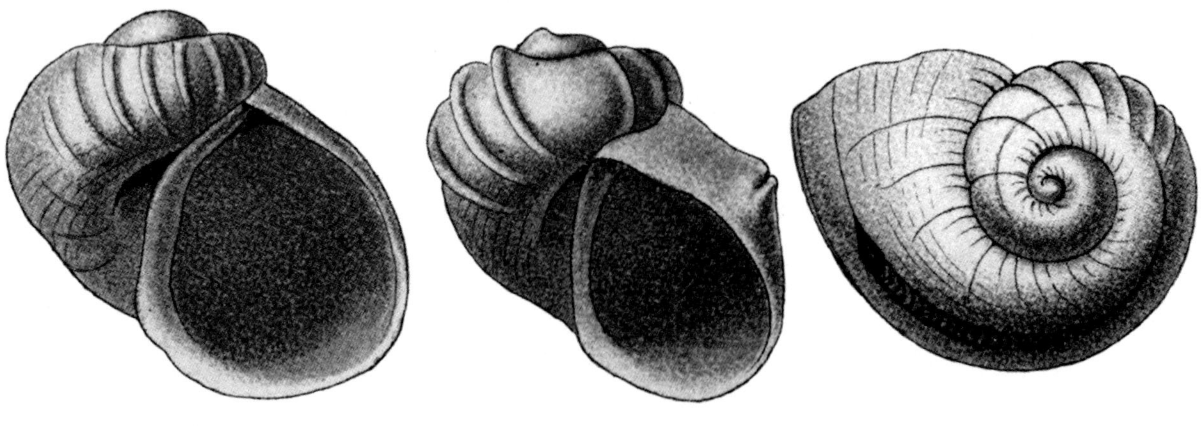

Figure 299. Original illustration of *Schismope fayalensis.*

Schismope fayalensis: Ávila & Azevedo, 1997: 324.

Sinezona fayalensis: Geiger, 2003: 78.

Scisssurella fayalensis: Ardovini & Cossignani, 2004: 19.

Sinezona fayalensis: Ardovini & Cossignani, 2004: 19.

+ *Schismope depressa* Watson, 1897 (*non* Reuss, 1860): 263, pl. 20, fig. 32a–c. ***Type material.*** Syntypes (NHMUK 1911.7.17.21–23, 3). The here illustrated specimen (Fig. 301) is designated as the lectotype (NHMUK 1911.7.17.21) with the express purpose of taxon stabilization in case non-conspecific syntypes should be found. Paralectotypes from type locality (NHMUK 1911.7.17.22–23, 2). ***Type locality.*** Funchal eastward and at Porto Santo, deep water, Madeira [Portugal] (OD). ***Etymology.*** Shell described as "obliquely depressed."

Schismope depressa: Thiele, 1912a, 22–23, pl. 3, figs 10–11 [copy figures Watson, 1897].

Schismope depressa: Nobre, 1937: 66.

Sinezona depressa: Ardovini & Cossignani, 2004: 19.

Misidentifications.

Scissurella elegans: Woodward, 1859: 202, pl. 46, figs 1–7, 9? [is *Sin. cingulata*].

Scissurella elegans: Jeffreys, 1860: 35 [is *Sin. cingulata*].

Schismope elegans: Weinkauff, 1868: 386 [is *Sin. cingulata*].

Schismope elegans: Brusina, 1870: 186 [*fide* Brusina, 1886. Is *Sin. cingulata*].

Schismope elegans: Monterosato, 1872a: 30 [is *Sin. cingulata*].

Schismope elegans: Paetel, 1888: 289 [is *Sin. cingulata*].

Schismope striatula: Tate, 1868: 38 [is *Sin. cingulata*].

Schismope striatula: Kobelt, 1888: 259 [is *Sin. cingulata*].

Schismope striatula: Carus, 1889–1893: v. 2, 238 [is *Sin. cingulata*].

not ? *Schismope cingulata*: Verrill & Bush, 1900: 525 [from Bermuda. Most likely *Sin. confusa*].

not ? *Schismope cingulata*: Peile, 1926: 75 [based on Verrill & Bush, 1900. From Bermuda. Most likely *Sin. confusa*].

not ? *Schismope cingulata*: Aguayo, 1935 [*fide* Jensen & Pearce, 2009. Based on Verrill & Bush, 1900. Bermuda: most likely *Sin. confusa*].

not *Woodwardia cingulata*: Olsson & McGinty, 1958: 10 [Caribbean: most likely *Sin. confusa*].

not *Schismope cingulata*: Warmke & Abbott, 1961: 35, fig. 10 [is *Sin. confusa*].

not *Schismope cingulata*: Warmke & Almodóvar, 1963: 165, 166, 170 [is *Sin. confusa*].

not *Scissurella cingulata*: Abbott, 1974: 17 [is *Sin. confusa*].

not *Scissurella cingulata*: Schweimanns, 1988: 25 [*fide* Jensen & Pearce, 2009. Bermuda: most likely *Sin. confusa*].

not *Scissurella cingulata*: Bandel, 1982: pl. 10 fig. 9 [is *Sin. confusa*].

not *Sinezona cingulata*: Ladd, 1982: 20–21, pl. 40, figs 1–3 [is *Sin. plicata*].

not *Scissurella cingulata*: Vokes & Vokes, 1983: 12, pl. 23, fig. 1 [is *Sin. confusa*].

not *Scissurella cingulata*: Ortiz-Corps, 1985: 6 [is *Sin. confusa*].

not *Scissurella cingulata*: DeJong & Coomans, 1988: 9, pl. 8, fig. 1 [is *Sin. confusa*].

not *Sinezona cingulata*: Odé, 1988: 12 [is *Sin. confusa*].

not *Schismope cingulata*: Cachia *et al.*, 1991: 25, pl. 1, fig. 1 [is *Sin.* cf. *confusa*].

not *Scissurella cingulata*: Rios, 1994: 21, pl. 5, fig. 34 [based on Abbott, 1974. Is *Sin. confusa*].

? *Scissurella cingulata*: Merlano & Hegedus, 1994: 108, pl. 33, fig. 316 [confused citation indicating as range Mediterranean, Bermuda, and Caribbean. Illustration shows a species with open slit, most likely juvenile *Sin. confusa*].

not *Scissurella cingulata*: Camp *et al.*, 1998: 15 [most likely *Sin. confusa*].

not *Scissurella cingulata*: Barrera, 2001: 12 [is most likely *Sin. confusa*].

not *Sinezona cingulata*: Rios, 2009: 25, text-fig. [is *Sin. confusa*].

Figure 300. Original illustration of *Schismope depressa*.

Figure 301. *Sinezona depressa*. Lectotype. Funchal eastward and at Porto Santo, deep water, Madeira [Portugal] (NHMUK 1911.7.17.21). Scale bar shell = 0.5 mm. Scale bar protoconch = 100 µm.

not *Sinezona cingulata*: Martins *et al.*, 2009: 23, fig. 6 [is *Sci. azorensis*].

Scissurella (*Schizotrochus*) *bertheloti*: Nordsieck, 1968: 10, pl. 1, fig. 15 [is juv. *Sin. cingulata*].

Type material. Holotype (whereabouts unknown), 0.7 mm.

Type locality. Sardegna Island, Italy [~40°N, 9°E] (OD).

Etymology. Cingulata, Latin for girdled.

Description. Shell small (to 0.6 mm. Figs 301–303), trochiform, slightly depressed. Protoconch of 0.875 whorl, strong axial sculpture, no apertural varix, apertural margin convex. Teleoconch I of 0.875–1.0 whorl, approximately 14 elevated axial cords, approximately up to 10 finer spiral cords, first appearing on last 0.125 teleoconch I whorl. Teleoconch II of slightly more than 0.25 whorl, suture impressed. Shoulder slightly convex, approximately 11–16 axial cords, irregularly spaced, increasing in density towards aperture, approximately 7–13 spirals at apertural margin, irregularly spaced, of irregular strength, always less strong than axials. Base with same number but stronger and elevated axial cords than on shoulder, many finer spirals, smooth to approximately 21 fine spiral threads. Selenizone above periphery, three times as long as foramen, lunules distinct, keels strong, low to moderately elevated; slit closed to oblong foramen. Umbilicus narrow, bordered by strong cord, walls straight, walls with axial growth marks, no funiculus. Aperture obliquely D-shaped, roof overhanging.

Operculum thin, corneous, multispiral, nucleus central.

Radula (Fig. 304). Rachidian tooth triangular, tip V-shaped, five cusps, central one largest. Lateral teeth 1–3 similar, tip with three cusps on outer margin, innermost largest. Lateral tooth 4 reduced, hook-shaped. Lateral tooth 5 broadened, approximately 10 cusps on inner margin, terminal one largest. Radular interlock of central field moderate. Inner marginal teeth with approximately 4–5 cusps on either side of tip; outer marginal teeth spoon-shaped, many cusps on frilled tip.

Comparisons. *Sinezona semicostata* from the tropical and subtropical northeastern Atlantic islands, has a more depressed spire, a longer selenizone (0.3 vs. 0.1 whorl), the umbilicus is made up by the rounded portion of the previous whorls, the underside of the protoconch is visible through the umbilicus, and the protoconch lacks an apertural varix.

Sinezona confusa from the Caribbean has a protoconch with an apertural varix, starts the spiral sculpture on teleoconch I earlier, has a longer teleoconch II, the foramen is triangular, the selenizone is at least four times as long as the foramen (0.4 vs. 0.1 whorl), and has about twice as many axial cords and more distinct spiral lines.

Sinezone garciai from the Caribbean has an overall wider shell with much greater expansion rate of the whorls, a flat spire, a protoconch devoid of axial cords as well as an apertural varix, and a straight parietal wall.

Sinezona insignis from South Africa has more turreted shell with fewer but stronger axial cords and a longer selenizone (0.5 vs. 0.1 whorl).

Sinezona insularis from the equatorial mid-Atlantic Islands has a more depressed spire, a much wider umbilicus, and the protoconch lacks an apertural varix.

Sinezona brevis from New Zealand has a shorter foramen (0.1 vs. 0.2 whorl), generally a wider umbilicus, and a larger teleoconch I (1.1 vs. <1 whorls).

Sinezona iota from New Zealand has an overall taller shell, lacks spiral sculpture on shoulder an base and a protoconch with fewer but stronger axial cords.

Sinezona kayae from Hawaii has a longer selenizone (0.4 vs. 0.2 whorl) and tends to have fewer axial cords.

Sinezona rimuloides from the northeastern Pacific and the Panamic has a longer selenizone (0.3 vs. 0.1 whorl), a taller shell, less distinct axial cords, and the protoconch distinctly protrudes from the spire.

In the fossil record, *Sci. praecedens* is distinguished from *Sin. cingulata* by the open slit according to Landau *et al.* (2003). *Sukashitrochus terquemi* has a more elevated spire.

Distribution. Mediterranean, Atlantic from Portugal to Angola; off shore Islands in tropical to subtropical eastern Atlantic.

Ecology. Found on rocks, the Phaeophyceae *Padina pavonia* (Linnaeus) Gaill. and *Cystoseira* species, as well as the seagrass *Posidonia oceanica* (Linnaeus) Del. according to Öztürk & Ergen (2000).

Specimen records.

Algeria. Port Gueydon, Kabrilia, 36.893°N, 4.420°E (FMNH 166879, 1; KBIN, 1; KBIN, 2).

Angola. 40–60 m, Moçâmedes, near Amelia, Moçâmedes Province, 15.200°S, 12.100°E (MNHN, 2). 0 m, Praia das Conchas, Moçâmedes Province, 15.120°S, 12.116°E

Figure 302. *Sinezona cingulata*. Cala Rossa, Terasini, Sicily, Italy, 38.150°N, 13.083°E (AMS C.380214). Scale bar shell = 0.5 mm. Scale bar protoconch = 100 µm.

(MNHN, 4). 0 m, Sao Nicolau, Moçâmedes Province, 14.267°S, 12.383°E (MNHN, 3). 0 m, Sao Nicolas, Moçâmedes Province, 14.267°S, 12.383°E (MNHN, 100). 1 m, São Nicolau, Moçâmedes Province, 14.267°S, 12.383°E (MNHN 1: complete). 10–20 m, Lucira Bay (Cesar), Moçâmedes Province, 13.867°S, 12.533°E (MNHN,

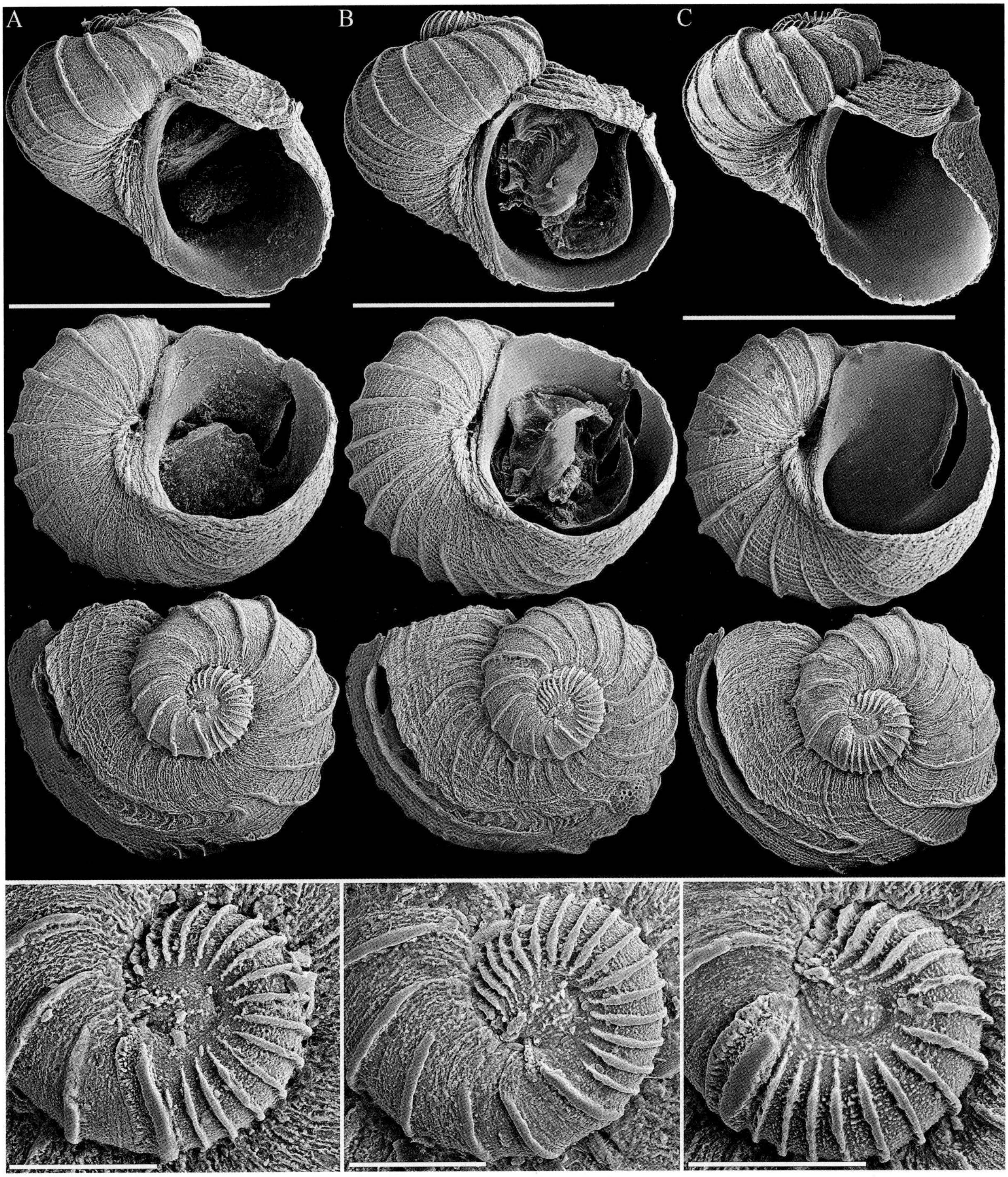

Figure 303. *Sinezona cingulata*. **A–B.** Ceuta, Benzu, Spain, 35.917°N, 5.400°W (AMS C.380708, 6). **C.** 0.5–2 m, W of lighthouse Cala San Antonio, Milazzo Capo, Italy, 38.267°N, 15.233°W (AMS C.380215, 2). Scale bars shell = 0.5 mm. Scale bars protoconch = 100 μm.

2). 8–10 m, Santa Maria Bay, Benguela Province, 13.417°S, 12.533°E (MNHN, 30). 0–2 m, Santa Maria Bay, Benguela Province, 13.417°S, 12.533°E (MNHN, 6). 0–2 m, Limagem Bay, Benguela Province, 13.268°S, 12.654°E (MNHN, 2). 1 m, Caotinha, Benguela Province, 12.590°S, 13.269°E (MNHN, 4; MNHN, 14).

Croatia. Mika Point, 2 km NW of Zadar, Ostri Rat, Hrvatska, 44.133°N, 15.202°E (RMNH, 1). Biograd, Hrvatska, 43.938°N, 15.444°E (RMNH, 30).

France. Corsica, Galeria, Stollo Point, 42.700°N, 8.650°E (AMS C.404929, 1). 0–1 m, Plage Notre Dame, Porquerolles, Provence, 43.010°N, 6.230°E (MNHN, 6). 0–3 m, Passe du Gaou Les Emblez, Provence, 43.072°N, 5.790°E (MNHN, 35). 15 m, Le Brusc, Var, 43.077°N, 5.780°E (NMR 993000024465, 4). Le Brusc (near Toulon), 43.077°N, 5.801°E (RMNH, 15). 1 m, N coast and Petit Rouveau, Les Emblez, Provence, 43.083°N, 5.783°E (MNHN, 6). 0–1 m, N coast and Petit Rouveau, Les Emblez, Provence, 43.083°N, 5.783°E (MNHN, 50). 0–1 m, Plage des Salins, Les Embiez, Provence, 43.083°N, 5.787°E (MNHN, 7). 0–1 m, St. Clari, Provence, 43.137°N, 6.387°E (MNHN, 10). 33 m, Grand Congloue, Marseille, 43.177°N, 5.403°E (MNHN, 30).

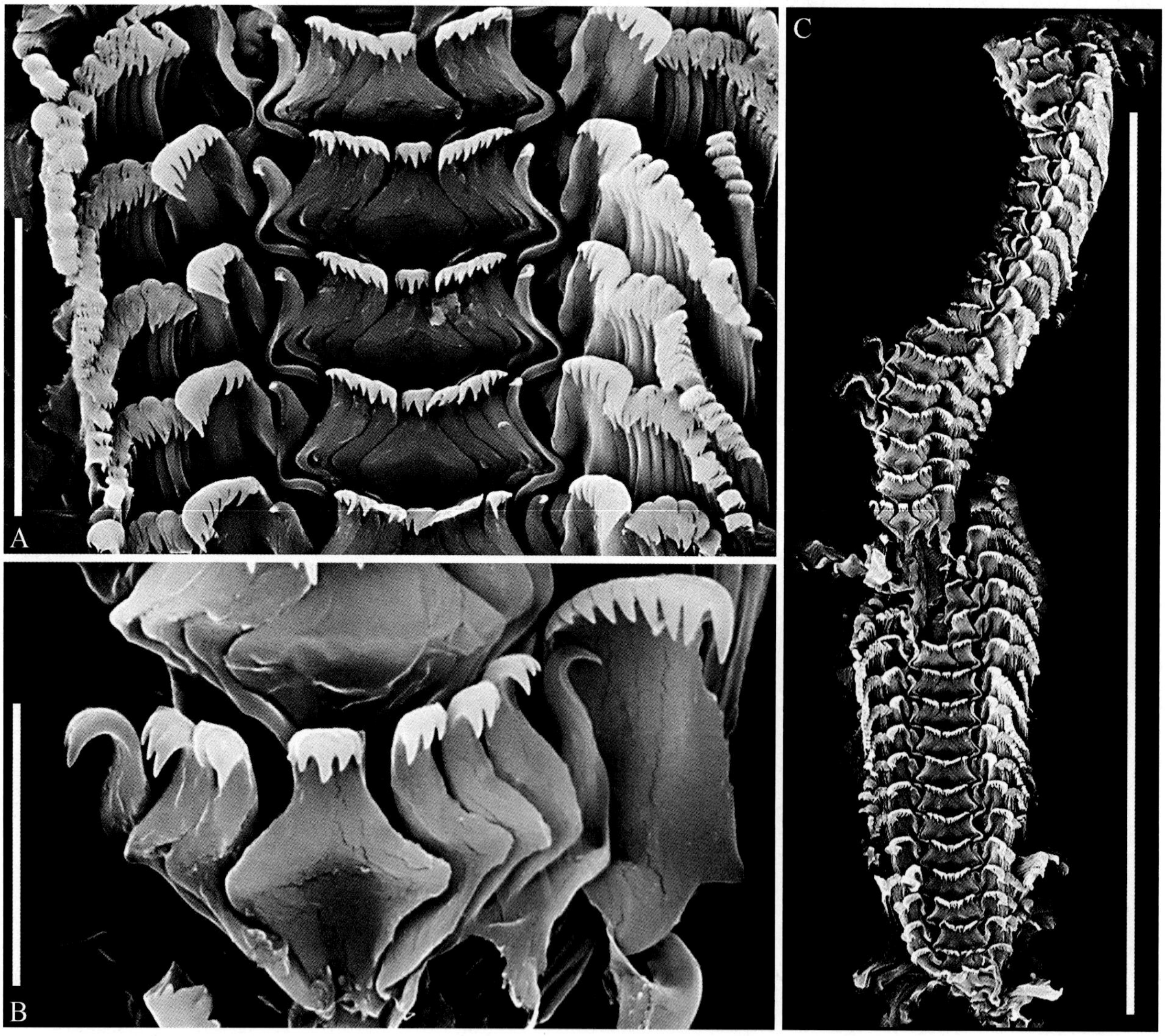

Figure 304. *Sinezona cingulata.* Radula. Ceuta, Anse Sachel, Caserne de la Reine, Spain, 35.883°N, 5.300°W (AMS C.380705, 2, 7: complete). **A.** Whole width of radula. Scale bar = 20 µm. **B.** Central field enlarged. Scale bar = 10 µm. **C.** Whole radular ribbon. Scale bar = 250 µm.

22 m, Calanque de la Triperie, Cap Morgiou, Marseille, 43.203°N, 5.448°E (MNHN, 14). 0–3 m, Cassis, Provence, 43.213°N, 5.536°E (DLG 739 1: complete; DLG 740 2: complete). St. Tropez, Departement de Var, France, 43.274°N, 6.640°E (RMNH, 1). Plage St. Croix (10 km S of Martiques), Rhone river mouth, 43.328°N, 5.077°E (RMNH, 1). 22–30 m, Le Dramont, Provence, 43.412°N, 6.862°E (MNHN, 11). 0 m, Pointe de Rire Poil, Antibes Peninsula, Alpes Maritimes, 43.545°N, 7.139°E (RMNH, 30). Alpes-Maritime, Antibes, Pointe de Tire Poil, 43.556°N, 7.131°E (RMNH, 80). Villfrance sur Mer, 43.704°N, 7.314°E (RMNH, 5).

Gibraltar. 0–4 m, Benzu, Ceuta north, Strait of Gibraltar, 35.917°N, 5.375°W (MNHN 23: complete). 0 m, Ceuta south, Anse Sarchatl, 35.890°N, 5.297°W (MNHN, 1).

Greece. Naxos, Kikladhes, 37.108°N, 25.372°E (RMNH, 3). 18–22 m, off Delos, 37.393°N, 25.268°E (USNM 107564, 7). 22 m, Delos, 37.394°N, 25.273°E (KBIN, 10). 0 m, Argostoli, Kefalonia Island, Kefallinia, 38.187°N, 20.482°E (NMR 993000033730, 3). 4 m, Orkos Beach, Paxos, Kerkyra, 39.234°N, 20.150°E (NMR 993000033728, 1).

Greece, Cyprus. 0–2 m, Pafos, 34.750°N, 32.417°E (AMS C.404926, 4). 200 m, off Cape Drepano, Paphos, 34.897°N, 32.297°E (DLG 1212, 1).

Italy. 5–20 m, Villaggio Pace, ME (DLG 1274, 1). 0–1 m, Brucoli, Sicily, 37.283°N, 15.183°E (AMS, 3: complete). 12 m, Aci Trezze, Catania, Sicily, 37.500°N, 15.100°E (DLG 324, 1). 12 m, Aci Trezza, 37.564°N, 15.164°E (DLG 828, 5). Palermo, 37.817°N, 13.367°E (NMSZ 1961.61, 1; NMSZ 1961.61, 3). 25 m, Taromina, Messina, 37.849°N, 15.302°E (DLG 1278, 1). 4 m, Cala Galera, Favignana Island, Sicily, 37.917°N, 12.300°E (AMS ex C.461540, 6). Cala Fredda, Levanzo Island, W Sicily, 38.000°N, 12.333°E (AMS C.461529, 1). 0 m, Trapani, Sicily, 38.017°N, 12.483°E (MNHN, 2). 18 m, Sicily, Aspra, Bagheria, 38.083°N, 13.500°E (AMS C.404927, 1). Sicily, Palermo, 38.117°N, 13.367°E (MHNG 14359, 2; MHNG 14358, 2). Palermo, Sicily, 38.133°N, 13.383°E (FLMNH 49066, 1). Sicily, Terrasini, Cala Rossa, 38.150°N, 13.083°E (AMS C.380214, 4). 0–1 m, Sicily, Terrasini, Cala Rossa, 38.150°N, 13.083°E (AMS C.380701, 2). Terrasini, Cala Rossa, Sicily, 38.150°N, 13.083°E (AMS C.404960 2: complete). Sicily, Trapani Province, Tonnara near San Vito Lo Capo, 38.183°N, 12.717°E (AMS C.380713, 5, 3: complete). Mondello, Sicily, 38.200°N, 13.333°E (USNM 181647, 2; Mondello, Sicily, Italy, 38.201°N, 13.329°E (RMNH, 3). 2–5 m, Pace, Messina, 38.239°N, 15.577°E (DLG 1297, 5). 42 m, Scilla, Reggio Calabria, 38.255°N, 15.714°E (DLG 1298, 5). 0–2 m, W of lighthouse Cala San Antonio, Milazzo Cape, 38.267°N, 15.233°E (AMS C.380215, 2). 0.33 m, Salina Island, 38.567°N, 14.867°E (DLG 325, 3). 6 m, Tropea, Calabria, 38.681°N, 15.896°E (DLG 958, 6). 30 m, Ventotene Island, Latina, 40.800°N, 13.400°E (NMR 993000033736, 3). 3 m, Posillipo. Naples, 40.817°N, 14.217°E (DLG 317, 1). 30–35 m, Giannutri Island, Tuscany, 42.250°N, 11.100°E (DLG 932, 1). 32 m, Capoliveri, Elba, Livorno, 42.737°N, 10.367°E (NMR 993000024522, 1). Capraia Island, 43.040°N, 9.820°E (DLG 1295, 1). 0–1 m, d'Orso Cape, Sardinia, 41.167°N, 9.417°E (AMS C.404928, 1). 35 m, Cannizzaro, Catania, Sicily, 37.521°N, 15.153°E (DLG 1747, 1). 1 m, Elba, Livorno, 42.750°N, 10.900°E (DLG 1421, 1).

Malta. 6 m, 35.900°N, 14.300°E (DLG 1296, 1). Birzebbuga, Pretty Bay, 35.825°N, 14.530°E (NMR 993000024342, 1). 3 m, St. Thomas Bay, 35.852°N, 14.566°E (NMR 99300003341, 5). 3 m, St. Thomas Bay, 35.852°N, 14.566°E (NMR ex 993000023982, 1). 6 m, Marsaskala Bay, 35.865°N, 14.565°E (NMR 993000024022, 1). Marsaskala Bay, 35.870°N, 14.568°E (AMS C.380712, 1). 3 m, Gnejna Bay, 35.924°N, 14.341°E (NMR 993000024100, 50). 200 m, off Gnejna Bay, Mgarr, 35.928°N, 14.328°E (DLG 1196, 1). 8 m, Salina Bay, 35.951°N, 14.406°E (NMR 993000024690, 1). 2–6 m, Salina Bay, 35.951°N, 14.406°E (NMR 993000024689, 3). 4 m, Qalet Marku Bay, 35.956°N, 14.451°E (NMR 993000024401, 2). 17–30 m, Cirkewwa Bay, 35.984°N, 14.333°E (NMR 993000024565, 6).

Morocco. 2 m, Ceuta, Anse Sachal, 35.885°N, 5.267°E (AMS C.380681, 17). 0 m, Essaouira, ex Mogador, Atlantic, 31.517°N, 9.783°W (MNHN, 1). 0 m, El Jadida, Grande Plage,

Atlantic, 33.267°N, 8.483°W (MNHN, 2). 0–2 m, Temara, Plage Sables d'Or, Atlantic, 33.917°N, 7.000°W (MNHN, 14). 0 m, Lahlou, Rabat, Atlantic, 34.025°N, 6.836°W (MNHN, 1). 0 m, Asilah, Oued el Helou, 35.466°N, 6.040°W (MNHN, 9). 0 m, Tanger, Strait of Gibraltar, Grande Plage, 35.785°N, 5.813°W (MNHN, 2; MNHN, 30). 0 m, Oued er Rmel, Strait of Gibraltar, 35.888°N, 5.503°W (MNHN, 12). 0 m, Cires Point, Straits of Gibraltar, 35.908°N, 5.482°W (MNHN, 3).

Portugal, Azores. 0 m, harbor of Vila Nova do Corvo, Negra Point, Corvo, 39.672°N, 31.120°W (NMR 993000062320, 30). 0 m, Porto da Barra, Santa Cruz da Graciosa, Graciosa, 39.084°N, 27.997°W (NMR 993000070937, 1). 20 m, Fores Island, Santa Cruz de Flores, 39.450°N, 31.117°W (MNHN, 30). 1 m, Flores Island, Santa Cruz de Flores, 39.450°N, 31.117°W (MNHN, 1). 0 m, Port das Lajes, SE coast, Flores, 38.400°N, 28.250°W (RMNH, 1). 0 m, Port das Lajes, SE coast, Flores, 39.383°N, 31.167°W (RMNH, 1). 0 m, Santa Cruz de Flores, Flores Island, 39.450°N, 31.117°W (MNHN 50: complete). 0 m, N of Harbor, E coast Santa Cruz, Flores, 39.450°N, 31.117°W (RMNH, 1). 0 m, Porto, Faja Grande, Flores, 39.457°N, 31.265°W (NMR 993000036871, 4). 39–40 m, Terceira, 38.650°N, 27.217°W (MNHN, 2). 35–37 m, Terceira, Ponta de Sao Diogo, W of M Brasil, 38.650°N, 27.217°W (MNHN, 1). Pedra Furada, Baia do Alcaide, Silveire, Terceira (MCZ 272409, 29; MCZ 272494, 1). 0 m, W of Anra do Heroismo, S coast, Terceita, 38.650°N, 27.250°W (RMNH, 9). Terceira, St. 21, 38.700°N, 27.200°W (RMNH, 10). 0 m, S part of Northern Beach, Praia da Vitoria, E coast, Terceita, 38.717°N, 27.067°W (RMNH, 1). 1 m, da Biscoitos Point, Terceira, 38.802°N, 27.253°W (DLG 1214, 2). 0 m, Fajã das Almas, San Jorge, 38.627°N, 28.077°W (NMR 993000062380, 35). 0 m, Velas, San Jorge, 38.677°N, 28.208°W (NMR 993000062321, 4). 1 m, Caldeira Inferno, Faial, 37.733°N, 25.200°W (MNHN, 1). 7 m, Horta, Faial, 38.533°N, 28.633°W (MNHN, 1). 56–64 m, E of Faial, 38.517°N, 28.600°W (RMNH, 3). 188 m, E of Faial, 38.517°N, 28.583°E (RMNH, 30). 110 m, E of Faial, 38.517°N, 28.600°E (RMNH, 1). 77 m, E of Faial, 38.517°N, 28.600°E (RMNH, 3). 10 m, Caldeira Inferno, SE coast of Faial, 38.517°N, 28.633°E (RMNH, 1). 0 m, Porto Pim, Horta, Faial, 38.525°N, 28.630°W (DLG 1213, 13). 140 m, E of Faial, 38.533°N, 28.583°W (RMNH, 1). 0 m, Caminho do Meio, Faial, 38.552°N, 28.607°W (NMR 993000033749, 80). 70 m, E of Faial, 38.567°N, 28.583°W (RMNH, 10). 92 m, E of Faial, 38.567°N, 28.567°W (RMNH, 3). 33 m, E of Faial, 38.850°N, 28.600°W (RMNH, 9). 16 m, W coast of Formigas, 37.267°N, 25.783°W (MNHN, 2). 88 m, E of Formigas, 37.267°N, 24.767°W (RMNH, 1). 0–1 m, Sao Miguel Island, Ilheu de Vila Franca do Campo, 36.917°N, 25.167°W (MNHN, 25; MNHN, 1; MNHN 7: complete). 0 m, San Miguel Island, Agua d'Alto, 37.717°N, 25.450°W (MNHN, 1). 0–5 m, San Miguel Island, Vila Franca do Campo, 37.717°N, 25.433°W (MNHN, 2). 13–18 m, Sao Miguel Island, Ponta da Galera, 37.730°N, 25.667°W (MNHN, 7). 13 m, Sao Miguel Island, Ponta da Piramida, 37.730°N, 25.667°W (MNHN, 4). 0 m, Sao Miguel Island, Ponta Delgada (Calheta), 37.733°N, 25.667°W (MNHN, 50; MNHN, 1). 75 m, S of São Miguel, 37.700°N, 25.500°W (RMNH, 11). 35 m, S of São Miguel, 37.700°N, 25.500°W (RMNH, 2). 4 m, Ilheu de Vila Franca, S coast, Sao Miguel, 37.700°N, 25.450°W (RMNH, 5). 0 m, Ilheu de Vila Franca, Sao Miguel, 37.706°N, 25.444°W (NMR 993000033750, 20). 0–3 m, Rosto do Cac/Cao. Point, S coast, Sao Miguel, 37.733°N, 25.633°W (RMNH, 1). 9 m, Santa Maria Island, Vila do Porto, 36.933°N, 25.150°W (MNHN, 10). 0 m, Santa Maria Island, Vila do Porto, 36.933°N, 25.150°W (MNHN, 5). 6 m, Santa Maria

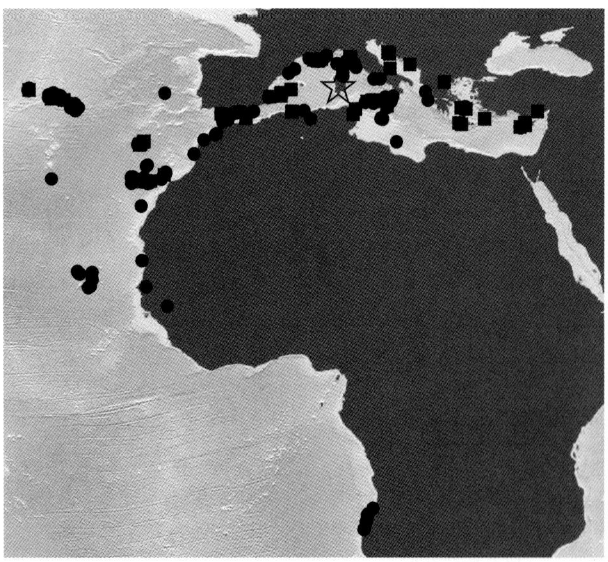

Figure 305. Distribution of *Sinezona cingulata*. ☆ type locality. ● specimen record. ■ literature record.

Island, Vila do Porto, 36.933°N, 25.150°W (MNHN, 20). 12–25 m, Santa Maria, Ponta Malbusca, 36.967°N, 25.117°W (MNHN, 1). 0–1 m, Santa Maria Island, Ponta do Marvao, 36.967°N, 25.117°W (MNHN, 100). 620 m, S of Santa Maria, 36.917°N, 25.117°W (RMNH, 10). 0 m, Anjos, Santa Maria, 37.007°N, 25.156°W (NMR 993000033747, 6). 0 m, Lages harbor, S coast, Pico, 28.400°N, 28.250°W (RMNH, 3). 85 m, W of Pico, 38.533°N, 28.567°W (RMNH, 10). 45 m, W of Pico, 38.533°N, 28.550°W (RMNH, 2). 70 m, W of Pico, 38.567°N, 28.550°W (RMNH, 11). 0 m, Prainha, Pico, 38.472°N, 28.199°W (NMR 993000070939, 3; NMR 993000070941, 100; NMR 993000070942, 20).

Cape Verde Islands. 70–88 m, S of Sao Tiago, 14.900°N, 23.500°W (RMNH, 1). 166 m, Ponta Inglez/Ponta Preta, SW of Maio, 15.100°N, 23.233°W (RMNH, 1). 53 m, SW of Boa Vista, 15.883°N, 23.000°W (RMNH, 2). 208 m, S of Razo, 16.600°N, 24.617°W (RMNH, 1). 140–160 m, S of Razo, 16.600°N, 24.600°W (RMNH, 1). 35 m, S of Branco, 16.633°N, 24.683°W (RMNH, 1). 204 m, SSE of Santa Lucia, 16.717°N, 24.783°W (RMNH, 2). 102 m, SW of Santa Luzia, 16.717°N, 24.767°W (RMNH, 1). 354 m, off Palmeira, W of Sal, 16.767°N, 23.017°W (RMNH, 1). 95 m, Baia das Gatas, NE of Sao Vincente, 16.900°N, 24.900°W (RMNH, 3).

Portugal, Madeira. 32.667°N, 16.750°W (NMNZ M.14987, 1). 32.700°N, 16.900°W (KBIN, 1; KBIN, 1). 33.017°N, 16.367°W (AMS C.034462, 6). 16 m, Ponta dos Reis Magos, 32.656°N, 16.824°W (NMR 993000033751, 16). 0 m, Ponta de Calheta, Porto Santo, 33.024°N, 16.379°W (NMR 993000033753, 19). Ponta da Calheta, Ilha de Porto Santo, 33.017°N, 16.367°E (AMS C.404931, 5).

Portugal, Selvages Archipelago. 524 m, 30.017°N, 16.017°W (RMNH, 1). 0 m, Petite Salvage, plage de l'anse extrème W, 15°N, 16.041°W (MNHN, 1). 0 m, Salvages Islands, Grand Salvage, 100 m S of the ornithology base, 30.127°N, 15.865°W (MNHN, 1). 0 m, Salvages Islands, Grande Salvage, Moines Point, 30.127°N, 15.865°W (MNHN, 50).

Spain, Canary Islands. 0 m, La Tajita, Tenerife, 18.250°N, 16.583°W (MNHN, 3). 0 m, Palm Mar, Tenerife, 25.010°N, 16.680°W (MNHN, 10). 6–8 m, Palm mar, Tenerife, 25.010°N, 16.680°W (MNHN, 12; MNHN, 2). 0 m, Las Galletas, Tenerife, 28.000°N, 16.650°W (MNHN, 15). 0 m, Los Cristianos, Tenerife, 28.050°N, 16.717°W (MNHN, 1; MNHN, 10). 0 m, Playa San Domingo, Tenerife, 28.090°N, 16.650°W (MNHN, 40). 0 m, Teno Point, Tenerife, 28.333°N, 16.917°W (MNHN, 16). 0 m, Hildago Point, Tenerife, 28.583°N, 16.317°W (MNHN, 1; MNHN, 20). Puerto Orotava, Tenerife, (RMNH, 10). 50 m, Las Galletas, Tenerife, 28.006°N, 16.656°W (NMR 993000033764, 10). Tenerife, 28.300°N, 16.600°W (KBIN, 3). 0 m, Palm-Mar, Tenerife, 25.010°N, 16.680°W (MNHN, 20). 12 m, Playa de Las Americas, Tenerife, 28.064°N, 16.732°W (DLG 1444, 1). 12 m, Garachico, Tenerife, 28.374°N, 16.767°W (DLG 1443, 7). Playa Paraiso, Tenerife, 28.120°N, 16.779°W (DLG 2347, 8). 0 m, La Isleta, Lanzarote, 28.173°N, 15.417°W (MNHN, 5). 0 m, Las Coloradas, Lanzarote, 28.910°N, 13.600°W (MNHN, 40; MNHN, 40). 0 m, Arrecife, Lanzarote, 28.950°N, 13.533°W (MNHN, 50). 0 m, Orzola, Lanzarote, 29.217°N, 13.450°W (MNHN, 30). 82 m, S of Lanzarote, 28.800°N, 13.767°W (RMNH, 1). 46–50 m, Puerto del Carmen, Lanzarote, 28.917°N, 13.633°W (DLG 321, 2). 0 m, La Caleta, Famara Beach, Lanzarote, 29.119°N, 13.566°W (NMR 993000033766, 50). 0 m, Puerto de Mogan, S coast of Gran Canaria, 27.817°N, 15.833°W (RMNH, 1). 15 m, Sardina, Gran Canaria, 27.850°N, 15.700°W (HLC, 5). Canteras, Gran Canaria, 28.137°N, 15.438°W (RMNH, 2). 141–245 m, S of Palma, 28.450°N, 17.850°W (RMNH, 1). 75 m, SW of Palma, 28.650°N, 17.967°W (RMNH, 10). 0 m, Los Cancajos, La Palma, 28.646°N, 17.756°W (NMR 993000033760, 100). Los Cancajos, La Palma, 28.647°N, 17.756°W (NMR 993000021238, 14). 0 m, Oliva Beach, Fuerteventura, 28.709°N, 13.838°W (NMR 993000033763, 17). 0 m, Charco Manso, Hierro, 27.849°N, 17.923°W (NMR 993000033761, 20).

Spain. Ceuta, Anse Sachel, Caserne de la Reine, 35.883°N, 5.300°W (AMS C.380705, 2, 7: complete). Ceuta, Piedra del Cigarro, 35.883°N, 5.300°W (AMS C.404253

3: complete). Ceuta, Anse Sachel, Caserne de la Reine, 35.883°N, 5.300°W (AMS C.404930, 1). 0 m, Strait of Gibraltar, Oued er Rmel, 35.888°N, 5.503°W (MNHN, 1). 0 m, Rarajal, Ceuta south, Strait of Gibraltar, 35.890°N, 5.308°W (MNHN, 1). 2 m, Ceuta, Club Nautica, 35.892°N, 5.325°W (AMS C.404932, 5). Ceuta, Benzu, 35.917°N, 5.400°W (AMS C.380708, 6; AMS C.380711, 1). 0 m, Benzu, Ceuta, 35.917°N, 5.400°W (AMS 8: complete). Tarifa, Cadiz Province, 36.011°N, 5.600°W (RMNH, 11). 0 m, Tarifa, Cadiz, 36.011°N, 5.600°W (RMNH, 30). 0 m, Torre de la Peña, Tarifa, Cadiz, 36.013°N, 5.606°W (MNHN, 25). 0 m, Torre de la Peña (10 km N of Tarifa), 36.056°N, 5.659°W (RMNH, 14). Algeciras, Getares, Cadiz, 36.070°N, 5.432°W (RMNH, 13). 0 m, Getares, Cadiz, 36.070°N, 5.432°W (RMNH, 2). Playa de Getares, 4 km S of Algeciras, Cadiz, 36.097°N, 5.443°W (RMNH, 1). 0 m, Cadiz, Barbate, 36.182°N, 5.948°W (MNHN, 20). 0 m, Barbate, Cadix, 36.190°N, 5.922°W (MNHN, 6). 5 m, Cabopino, Marbella, 36.488°N, 4.766°W (DLG 821, 2). 0 m, Getares, Carnero Point, Cadix, 36.534°N, 6.299°W (MNHN, 20). 250 m, Estepona, Malaga, 36.600°N, 4.300°W (DLG 822, 1). 0 m, Calahonda, Malaga, 36.700°N, 3.417°W (MNHN, 2). Calahonda, Malaga, 36.700°N, 3.417°W (MNHN, 2). 5 m, Malaga, 36.717°N, 4.417°W (DLG 309, 10). 34 m, de la Mona Point, Granada, 36.719°N, 3.728°W (NMR 993000033755, 1). 3–10 m, La Herradura, Grenada, 36.733°N, 3.733°W (MNHN, 4). de la Mona Point, Granada, 36.733°N, 3.717°W (NMR 993000024205, 2). 24 m, de la Mona Point, Granada, 36.733°N, 3.717°W (NMR ex 993000024210, 1). 0 m, La Caleta (Cadix), Cadix, 36.750°N, 4.067°W (MNHN, 1). 5 m, Playa de Los Genoveses (San José), Almeira, 36.767°N, 2.117°W (MNHN, 8). 1 m, Los Escullos (Embarcadero), Almeria, 36.800°N, 2.050°W (MNHN, 1; MNHN, 7). 25 m, Benidorm, 38.533°N, 0.131°W (DLG 1312, 1). Calpe, Alicante, 38.640°N, 0.057°E (NMR 993000027050, 1; NMR 993000027049, 1). Calpe, Alicante, 38.641°N, 0.052°E (RMNH, 3). Calpe, Alicante Province, 38.642°N, 0.050°E (RMNH, 5). 35 m, Mataró, Barcelona, 41.533°N, 2.450°E (JWC 9327A, 4). 10 m, Islas Medas, Gerona, 42.047°N, 3.223°E (NMR 993000033754, 2).

Literature records.

Algeria. Sidi Ferruch, Palm Be, 36.761°N, 2.844°E (RMNH AD 10833). Sidi Ferruch, Port, 36.764°N, 2.848°E (RMNH AD 11024). La Madrague = El Dje, 36.805°N, 2.898°E (RMNH AD 10660) (J. van Aartsen, pers. comm. 5/2011).

Croatia. 0 m, Lacroma Island [= Lokrum], 42.628°N, 18.121°E. 0 m, Ulbo Island, near Zara, 44.131°N, 15.203°E (Brusina, 1886). Biograd, 43.938°N, 15.444°E (RMNH AD 1986; RMNH AD 10348) (J. van Aartsen, pers. comm. 5/2011).

Cyprus. Paphos, Cyprus, 34.865°N, 32.341°E (RMNH AD 24003). Kormkiti, Cyprus, 35.403°N, 32.920°E (RMNH AD 23717) (J. van Aartsen, pers. comm. 5/2011).

France. St. Tropez, Var, 43.267°N, 6.633°E (Aartsen *et al.,* 1984).—Pinarello, Corsica, 41.672°N, 9.395°E (RMNH AD 13690). La Capte, 43.065°N, 6.152°E (RMNH AD 11541). Cabasson West, Var, 43.102°N, 6.320°E (RMNH AD 6694). Cabasson, 43.103°N, 6.230°E (RMNH AD 11866). Les Leques, Provence, 43.180°N, 5.686°E (RMNH AD 12057). St. Tropez, 43.269°N, 6.631°E (RMNH AD 14345) (J. van Aartsen, pers. comm. 5/2011).

Greece. Crete, 35.200°N, 24.800°E (RMNH AD 19887). Rethimnon, Crete, 35.367°N, 24.475°E (RMNH AD 24823). Prassonissi, Rhodos, 35.883°N, 27.761°E (RMNH AD 29045). Naxos, 37.107°N, 25.376°E (RMNH AD 19496). Methoni, 40.446°N, 22.588°E (RMNH AD 28826) (J. van Aartsen, pers. comm. 5/2011).

Italy. Tremiti Islands, 42.120°N, 15.500°E (Mazziotti *et al.,* 2008).—80 m, off Catania, Sicily, 37.489°N, 15.142°E (RMNH AD 20139). Catania, Sicily, 37.502°N, 15.088°E (RMNH AD 16381). Acitrezza, Sicily, 37.566°N, 15.161°E (RMNH AD 15734). 90 m, Acitrezza, Sicily, 37.566°N, 15.161°E (RMNH AD 17919). Lido San Guliano, Trapani, Sicily, 38.027°N, 12.526°E (RMNH AD 21853). Mondello, Sicily, 38.200°N, 13.326°E (RMNH AD 4217). Castiglioncello, Livorno, 43.402°N, 10.411°E (RMNH

AD 18408). Gulf of Baratti, Livorno, 43.552°N, 10.308°E (RMNH AD 18311) (J. van Aartsen, pers. comm. 5/2011).

Israel. 0 m, near Akhziv, 35.067°N, 33.033°E (Barash & Danin, 1992).

Morocco. Tangier, 35.789°N, 5.801°W (RMNH AD 12425) (J. van Aartsen, pers. comm. 5/2011).

Cape Verde Islands. Sal Island (Rolán & Luque, 1994).

Portugal, Azores. 0–45 m, Formigas islet, Stations DBUA 332, 336, 338, 340, 343, 345, 346, 350, 353, 355, 37.258°N, 24.758°E (Ávila & Azevedo, 1997). 13–50 m, D. João de Castro Bank, 38.222°N, 26.603°W (Ávila *et al.*, 2004). —2 m, DBUA 662, Baixa-de-fora, Lajes do Pico, 38.389°N, 28.256°W. 0 m, DBUA 446, 465-468, 470-472, 475, 481, 496, Lajes do Pico, Pico, 38.390°N, 28.253°W. 2–4 m, DBUA 492, 500, Foça de Pano, Lajes do Pico, Pico, 38.390°N, 28.253°W. 1–2 m, DBUA 493, 499, Lagoa, Lajes do Pico, Pico, 38.390°N, 28.253°W. DBUA 204, 268, 276, 277, Piscinas, Santa Cruz, Flores, 39.459°N, 31.126°W. 20 m, DBUA 274, Piscinas, Santa Cruz, Flores, 39.459°N, 31.126°W. 4 m, DBUA 669, Baixa-de-fora, Lajes do Pico, 38.446°N, 28.074°W. DBUA 545, Potinhas, Flores. 6 m, DBUA 564, Porto de Baleia, Flores. 0 m, DBUA 663, between Poça de Pano and Poça de Baleia, Lajes do Pico (Ávila *et al.*, 1998).—DBUA-F 139, Praínha, Santa Maria (Ávila *et al.*, 2002).

Spain. 30–250 m, Alborán Island, 35.951°N, 3.033°W (Rolán & Luque, 1994; Peñas *et al.*, 2006). — 1450 m, near Tenerife, 28.067°N, 16.825°W (Dautzenberg, 1927).—Las Canteras, Gran Canaria, Canary Islands, 28.144°N, 15.435°W (RMNH AD 4875). 40 m, del Carmen Point, Lanzarote, Canary Islands, 28.921°N, 13.671°W (RMNH AD 24271). Tarifa, 36.013°N, 5.603°W (RMNH AD 20265). Getares, Bay of Algeciras, 36.073°N, 5.433°W (RMNH AD 9867). Cadiz, 36.481°N, 6.266°W (RMNH AD 25106). Calpe, Alicante, Valencia, 38.627°N, 0.046°E (RMNH AD 1838). Es Pujol, Formentera, 38.725°N, 1.458°E (RMNH AD 4441). Playa es Cavalet, Ibizza, 38.897°N, 1.447°E (RMNH AD 4758). Portinax, Ibizza, Balearic Islands, 39.102°N, 1.493°E (RMNH AD 2252). El Arenal, Mallorca, 39.504°N, 2.752°E (RMNH AD 10030) (J. van Aartsen, pers. comm. 5/2011).

Portugal, Madeira. 2–50 m, Funchal Bay, 32.644°N, 16.908°W. 0 m, Caniçal, 32.735°N, 16.740°W. 0 m, Calheta, Porto Santo, 33.024°N, 16.380°W. 0 m, Serra di Dentro, Porto Santo, 33.084°N, 16.299°W (Segers *et al.*, 2009c).

Tunisia. Nabeul, 36.441°N, 10.743°E (RMNH AD 9517). Sidi Daoud, Cape Bon, N Side, 37.088°N, 11.036°E (RMNH AD 9317) (J. van Aartsen, pers. comm. 5/2011).

Turkey. Akkum, Mersin, 36.794°N, 34.631°E (RMNH AD 22055) (J. van Aartsen, pers. comm. 5/2011).

Fossil. Early Pliocene of Estepona, Rio del Pardón, Spain (Landau *et al.*, 2003). Pliocene of Azores (Ávila *et al.*, 2002).

Ecology. Found in fine sediments in 2–3 m depth retained by *Posidonia* rhizomes (Luque, 1986). It also lives on *Pinna* shells (Consentino & Giacobbe, 2007).

Remarks. The drawing of the radula in Luque (1986: fig. 7) shows the inner four lateral teeth all similarly formed; lateral tooth 4 is reduced and pin-like as revealed by SEM.

The specimens from the Azores, described as *Sin. fayalensis*, tend to have smoother shells with a much lower frequency of axial ribs than in Mediterranean populations (coll. MNHN). There are, however, specimens with axial ribs comparable to those seen in Mediterranean specimens. There are no discrete characters that would separate the specimens from the Azores, and hence *fayalensis* is here also regarded as a synonym of *cingulata*.

Populations of Angola also have a higher proportion of smoother specimens than those in the Mediterranean, but there are some highly costate specimens as well. There seems insufficient reason to separate those specimens as distinct species.

Burnay & Rolán (1990) argued that the da Costa's description of *cingulata* does not match the cited figures, and neither matches the Mediterranean species to which *cingulata* traditionally has been applied. In their discussion, they also noted that Thiele (1912) cited an other figure in Costa referring to *cingulata* (fig. 8A–B), though a key attribute of *cingulata*, the strong axials, are not shown in that figure. They conclude that *cingulata* is a *nomen dubium*, and the first available name referring to what is usually taken to be called *Sin. cingulata* is *Sin. crossei*.

Unfortunately, Burnay & Rolán's argument is flawed in all points. Costa (1861) only referred to his fig. 9 in the description of *cingulata*; Thiele's error also attributing Costa's figure 8 to *cingulata* cannot be used against Costa's species concept. The description of the shell matches the illustration quite well: 11–12 strong axial cords are mentioned, and the same number is shown on the last whorl of the illustration. The open umbilicus is noted and shown in the figure, and the semi-oval aperture mentioned is clearly shown. There is an inconsistency with the length of the selenizone. It is noted to be one third of a whorl long, while it is one third of a half-whorl, or one-sixth whorl long. I would attribute this to an excusable *lapsus calami*. The total count of three whorls is also quite high, yet whorl counts on these small shells is difficult. The illustrated specimen does correspond with a species well-known from the Mediterranean Sea under that name. The number of axial cords observed (11–16) agrees with the original description (11–12), the species has a narrow, open umbilicus, and has an obliquely D-shaped aperture. Teleoconch II is up to 0.25 whorl long, slightly longer than 0.16 indicated in the original description, and the slit is closed to form a foramen. The slightly shorter teleoconch II in conjunction with the open slit indicates that a slightly juvenile specimen served for the basis of the description. However, the species described and illustrated by Costa (1861) can unambiguously be assigned to what has traditionally been known as *Sin. cingulata* from the Mediterranean [see also Nolt (2008) for discussion leading to the same conclusion].

Three taxa are here considered synonymous based on their available type material and original illustrations. Three syntypes of *Sin. depressa* (NHMUK 1911.17.21–23: See Nolt 2008: fig. 2C), type locality Madeira, and seven specimens of *Sin. fayalensis* (MNHN: See Nolt 2008: fig. 4C) from Santa Maria Island, Azores where examined. The neotype of *Sin. crossei* could not be located in the MNHN collections (Geiger pers. obs. 4/2005), but was adequately illustrated by Burnay & Rolán (1990: pl. 1, figs 3–5). One type specimen of *Sin. fayalensis* is in KBIN (T. Backeljau pers. comm. 10/2006), but was not available for examination. Near topotypical specimens of *Sin. fayalensis* from the Azores tended to display fewer axial ribs and have a smoother sculpture than specimens from the Mediterranean. One syntype of *Sin. depressa* was smoother due to erosion. Sculpture and shape are the same. All the specimens have no apertural varix and an open umbilicus. The nominal taxa differ in the number of axial cords (protoconch 16–17 and teleoconch I 12–17) and axial lines (teleoconch II 15–16 and base 12–16). These differences are not sufficient to justify recognition as separate species and can be accounted for by intraspecific variations.

The distributional pattern of Azores-Madeira-Mediterranean is shared with *Scissurella azorensis*, further strengthening the synonymy of the three *Sinezona* taxa. *Sinezona cingulata* has also been applied to a Caribbean species now known as *Sin. confusa*. The species is most likely the one reported in gut content of a gobiid fish (Teleostei: Gobiidae) as *Sinozona* [*sic*] sp. by Schoiewen & Kovačić (2008).

Sinezona concinna (Sowerby, 1831) [*nomen dubium, nomen oblitum*]

Figures 306–307

Chresonymy.

Scissurella concinna Sowerby, 1821–1834 (1831): pl. 55, fig. 1 (Fig. 306).
? *Scissurella concinna*: Brown, 1843: 52, pl. 6, fig. 17 (Fig. 307).
Scissurella concinna: Woodward, 1856b: 402.
Scissurella concinna: Jeffreys, 1856d: 471.

Type material. (whereabouts unknown).

Type locality. Not given (OD).

Etymology. Concinna, Latin adjective for well-arranged, skillfully joined, beautiful, striking.

Remarks. Sowerby I (1820–1834) introduced *Sci. concinna* as *nomen et figura*, i.e., without any descriptive text going along with the name. The species was certainly described after 1828, because Sowerby (1821–1834: unnumbered page to pl. 55) stated: "Rev. Dr. Fleming has described a recent species under the appellation of *Sc. crispa*, which he found in shell sand from Noss, Zetland, after a storm." Sykes (1906) gave publication dates for Sowerby (1821–1834) although it is not clear which genera were published in which part, hence, was of no utility. Petit (2006) could date the *Scissurella* part April, 30, 1831, based on a set in original wrappers. Petit (2009) did not list this taxon.

Figure 306. Original illustration of *Scissurella concinna*.

The illustration of *Sci. concinna* shows clearly a closed foramen and no spiral keels on the base, hence, the species belongs in *Sinezona*. Given that the earliest described scissurellids all are from Europe [*Sci. costata* d'Orbigny, 1824; *An. crispata* (Fleming, 1828)], it is very likely that *Sin. concinna* is also a European species. The rather well executed drawing agrees well with the European species known as *Sin. cingulata* (O. G. Costa, 1861). Particularly the number, strength, and spacing of the axials, decreasing in strength close to the aperture, and the very fine spirals are clearly shown in Sowerby's illustration. Given the rather uncertain status of this synonymy due to the lack of a type locality and any descriptive text, I prefer to treat *Sin. concinna* as a *nomen dubium*.

Sinezona concinna was last mentioned as a valid species before 1899 and is eligible for *nomen oblitum* (ICZN Art. 23.9.1). *Sinezona cingulata* has been used as a valid species at least 25 times by at least 10 authors in the preceding 10–50 years (ICZN Art. 23.9.1.2: see chresonymy of *Sin. cingulata* for list of citations). These 25 citations do not include previous identifications of the Caribbean *Sin. confusa* as the European-Mediterranean *Sin. cingulata*. In accordance with ICZN Art. 23.9.2, *Sin. concinna* is designated as a *nomen oblitum* and if it should indeed be conspecific with *Sin. cingulata* then the latter can be designated as a *nomen protectum*.

Figure 307. Illustration of *Scissurella concinna* from Brown (1843).

The indication of the taxon in Brown (1843: 52) is perplexing. He noted explicitly that *Scissurella* has a foramen, which would clearly place it in *Sinezona*. [The slit of *Pleurotomaria* and *Emarginula* were both termed "notch"] However, the illustration provided shows a species with open slit, and has the overall shape of *Anatoma crispata*, the other known species of the time.

Sinezona concinna (Sowerby, 1831) does not pose a threat to *Anatomus concinnus* A. Adams, 1862, because Adams' species is an *Anatoma*.

Jeffreys (1856d) cited Sowerby that *cocinna* was found in Calcaire grossier, a deposit from the Paris region of Lutetien (Eocene) age, which suggests a fossil species. As this indication is not available in the original description, it has to be viewed with caution.

Sinezona confusa Rolán & Luque, 1994

Figures 308–313, Color Plate 2E

Scissurellidae
Sinezona

Chresonymy.

Scissurellidae sp.: Palazzi, 1985: 40, figs 15–16.
Sinezona sp.: Leal, 1991: 43, pl. 2, figs J–K.
Sinezona confusa Rolán & Luque, 1994: 2–4, figs 1, 2, 7, 8.
Sinezona confusa: Sevilla-Hernández, 1998 [*fide* Cruz & Gánadara, 2006].
Sinezona confusa: Redfern, 2001: 5, pl. 1, fig. 5; pl. 107, fig. 5 [live animal. Color Plate 2E].
Sinezona confusa: Geiger, 2004: text-fig. p. 6.
Sinezona confusa: Yoon & Kim, 2005: 284 [molecular phylogeny].
Sinezona confusa: Geiger, 2006c: text-fig. 4.
Sinezona confusa: Williams & Ozawa, 2006: 36 et seq. [molecular phylogeny].
Sinezona confusa: Cruz & Gánadara, 2006: 130.
Sinezona confusa: Nolt, 2008: 57, fig. 4B.
Sinezona confusa: Garica-Rios *et al.*, 2008: 1668, 1669, 1972.
Sinezona confusa: Tunnell *et al.*, 2010: 115, text-fig. [juvenile with open slit].
Sinezona confusa: Daccarett & Bossio, 2011: 51, 211, fig. 51.

Synonyms.

+ *Schismope columbiana* Bandel, 1998: 60–61, pl. 20, figs 7–8, pl. 21, fig. 1. ***Type material.*** Holotype (SGPIH 3883. Fig. 308), 0.6 mm. ***Type locality.*** Near Cartagena, in subfossil coral reef of Galerazamba, Columbia [10.800°N, 75.267°W] (OD). ***Etymology.*** Named for its type locality in Columbia (OD).

Scissurella columbiana: Geiger, 2003: 76.

Misidentifications.

Schismope cingulata: Verrill & Bush, 1900: 525 [Bermuda: most likely *Sin. confusa*].

Figure 308. Original illustration of *Schismope columbiana.*

Schismope cingulata: Peile, 1926: 75 [based on Verrill & Bush, 1900. Bermuda: most likely *Sin. confusa*].

Schismope cingulata: Aguayo, 1935 [*fide* Jensen & Pearce, 2009. Based on Verrill & Bush, 1900. Bermuda: most likely *Sin. confusa*].

Woodwardia cingulata: Olsson & McGinty, 1958: 10 [Caribbean: most likely *Sin. confusa*].

Schismope cingulata: Warmke & Abbott, 1961: 35, fig. 10 [is *Sin. confusa*].

Schismope cingulata: Warmke & Almodóvar, 1963: 165, 166, 170 [is *Sin. confusa*].

Scissurella cingulata: Abbott, 1974: 17 [is *Sin. confusa*].

Scissurella cingulata: Schweimanns, 1988: 25 [*fide* Jensen & Pearce, 2009. Bermuda: most likely *Sin. confusa*].

Scissurella cingulata: Bandel, 1982: pl. 10, fig. 9 [is *Sin. confusa*].

Scissurella cingulata: Vokes & Vokes, 1983: 12, pl. 23, fig. 1 [is *Sin. confusa*].

Scissurella cingulata: Ortiz-Corps, 1985: 6 [is *Sin. confusa*].

Scissurella cingulata: DeJong & Coomans, 1988: 9, pl. 8, fig. 1 [is *Sin. confusa*].

Sinezona cingulata: Odé, 1988: 12 [is *Sin. confusa*].

Scissurella cingulata: Rios, 1994: 21, pl. 5, fig. 34 [based on Abbott, 1974. Is *Sin. confusa*].

? *Scissurella cingulata*: Merlano & Hegedus, 1994: 108, pl. 33, fig. 316 [confused citation indicating as range Mediterranean, Bermuda, and Caribbean. Illustration shows a species with open slit, most likely juvenile *Sin. confusa*].

Scissurella cingulata: Camp *et al.*, 1998: 15 [Caribbean: most likely *Sin. confusa*].

Sinezona cingulata: Rios, 2009: 25, text-fig. [is *Sin. confusa*].

Sinezona brasiliensis: Rios, 1994: 21, pl. 5, fig. 38 [copy figure Leal, 1991: is *Sin. confusa*].

Scissurella cingulata: Barrera, 2001: 12 [is most likely *Sin. confusa*].

Sinezona brasiliense [*sic*]: Rios, 2009: 25, text-fig. [copy Leal, 1991: is *Sin. confusa*].

Type material. Holotype (MNCN 15.05/17230), 0.78 mm. Paratypes from type locality (MNCN, 1; NHMUK 1994094, 2; AMNH. 2; USNM 890852, 2; IZSH, 2; ZMA, 2; ERC, 22).

Type locality. Playa Baracoa, near Havana, Cuba, 23°03'N, 82°34'W (OD).

Etymology. Named for its past confusion with *Sin. crossei* [= *Sin. cingulata*](OD).

Description. Shell small (to 0.8 mm. Figs 308–311), trochiform, slightly depressed. Protoconch of one whorl, strong axial sculpture, apertural varix barely connected to embryonic cap, apertural margin convex. Teleoconch I of 0.875–1.125 whorls, approximately 18 strong, elevated axial cords, up to 16 fine spiral lines, first after 0.5 whorl. Teleoconch II of 0.66 whorl, suture deeply impressed. Shoulder somewhat convex, approximately 25–29 axial cords to low lamellae, increasing in density towards aperture, approximately eight spiral lines of unequal strength and spacing, usually weaker than axials, sometimes as strong as axials. Base with constriction below selenizone, same density of axials and spirals as on shoulder. Approximately 16 spiral lines. Umbilicus open, moderately wide, margin separated with strong cord from base. Selenizone with strong lunules denser than axials, keel low, strong. Slit closed to triangular foramen. Aperture subquadrate, adumbilical corners slightly distended towards umbilical cavity, roof overhanging.

Operculum (Fig. 310A) round, smaller than aperture, multispiral.

Radula (Fig. 312). Rachidian tooth triangular, tip triangular, central cusp largest, three cusps on either side decreasing in size with distance from central one. Lateral teeth 1–3 similar, S-shaped, tips with three cusps each in terminal position. Lateral tooth 4 hook-shaped, reduced, hook-shaped. Lateral tooth 5 enlarged, six distinct plus one indistinct cusps pointing towards midline of radula. Inner marginal teeth with asymmetric tip, largest cusp closer to midline of radula, inner margin with

1–2 small cusps, outer margin with five stronger cusps. Outer marginal teeth with spoon-shaped tip bearing a dozen very small cusps on either side. Radular interlock moderate of central field.

Comparisons. *Sinezona cingulata* from the eastern Atlantic and the Mediterranean has a shorter selenizone (0.1 vs. 0.4 whorl), a teleoconch I with fewer spiral lines that have a later onset, the foramen is more compressed oval rather than triangular, and the selenizone is only three times as long as the foramen.

Sinezona garciai from the Caribbean has an overall more depressed and wider shell, with a distinct umbilicus, a protoconch lacking axial cords as well as an apertural varix, has more sharply raised axial cords, and a subquadrate aperture.

Sinezona calumnior from temperate Australia and New Zealand has a more narrow umbilicus and a weaker periumbilical cord not forming a shelf in the adumbilical lower corner of the aperture.

Sinezona globosa from the Indo-Malayan Archipelago to the Eastern Pacific (and possibly Caribbean) grows much larger (3.8 mm), has an overall more globular

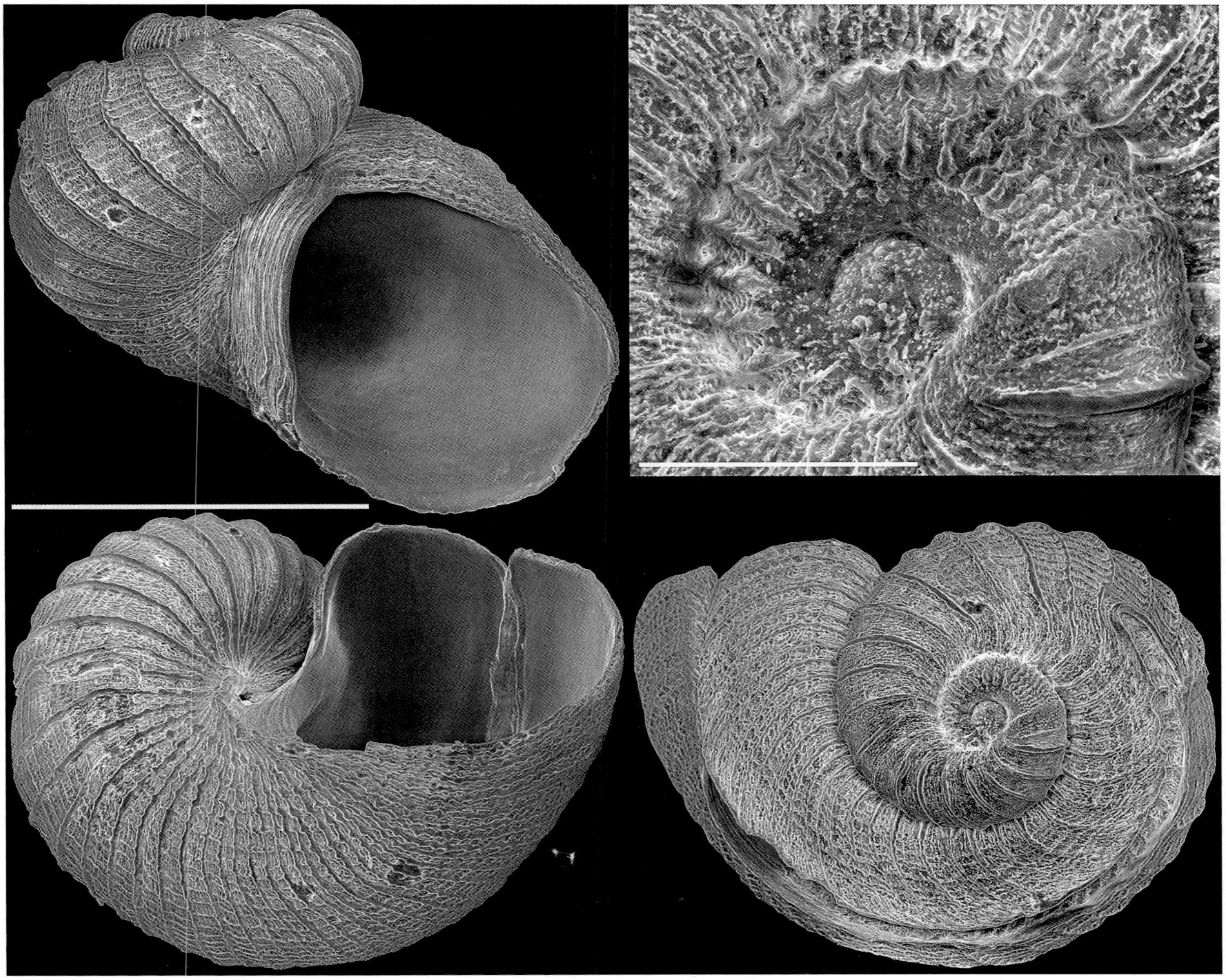

Figure 309. *Sinezona confusa*. 0 m, just W of Boca Chica, Florida Keys, USA, 24.567°N, 81.683°W (JWC 9327, 4). Uncoated specimen, VPSE detector, 49 Pa. Scale bar shell = 0.5 mm. Scale bar protoconch = 100 μm.

shell, a much longer selenizone (0.8 vs. 0.4 whorl), a protoconch with spiral lines, and the base is continuously sloping into the umbilicus.

Sinezona insularis from the equatorial mid-Atlantic islands has an overall more depressed shell, lacks an apertural varix on the protoconch, and has a wider umbilicus.

Figure 310. *Sinezona confusa*. **A.** 10–52 m, Abaco, Bahamas, 26.500°N, 77.750°W (CRC, 1). **B.** W end Grand Bahama Island, Bahamas, 26.683°N, 78.967°W (BMSM 26038, 2). **C.** Juvenile specimen with open slit. 1.5–9 m, Inner Reef near Transect, Cahuita Reef Point, off Puerto Vargas, Limon Province, Costa Rica, 9.736°N, 82.812°W (LACM 86-136). Scale bars shell = 0.5 mm. Scale bars protoconch = 100 µm.

Sinezona semicostata from the subtropical northeastern Atlantic islands has a more depressed overall shell profile, the umbilicus is formed by the rounded portion

Figure 311. *Sinezona confusa*. **A.** La Parguera, Puerto Rico, 17.972°N, 67.047°W (DMNH 23178). **B.** 0 m, just E of Racoon and just W of Boca Chica, Monroe County, Florida, USA, 24.564°N, 81.717°W (HLC). **C.** 10 m, S coast Mujeres Island, Cancun, Mexico, 21.200°N, 86.717°W (HLC). Scale bars shell = 0.5 mm. Scale bars protoconch = 100 μm.

of the previous whorl through which the underside of the protoconch is visible, and the protoconch lacks an apertural varix.

Sinezona rimuloides from the northeastern Pacific and Panamic has protoconch that is distinctly protruding from the spire and teleoconch I is larger (>1 vs. 0.8–<1 whorls).

Scissurella redferni from the Caribbean can be distinguished from juvenile *Sin. confusa* with an open slit by the much more depressed overall shape and the more flared apertural margin, and having fewer, more widely spaced axials.

Distribution. Caribbean to northern Brazil (Fig. 313).

Ecology. Live specimens occur intertidally on *Batophora* (Chlorphyta: Dasycladaceae) or *Cladophora* (Chlorphyta: Cladophoraceae) and are found sometimes on rocks covered with algae in less then 1 m depth. Empty shells are found from the intertidal to 60 m depth (Redfern, 2001). The green calacareous alga *Halimeda opuntia* (Linnaeus) J. V. Lamouroux, 1816 (Chlorphyta: Udotaceae) bears large numbers of the species (García-Rios *et al.*, 2008)

Specimen records.

Bermuda. 32.400°N, 68.800°W (DLG 1318, 1), Compass Cay, Exuma Island, 24.265°N, 76.510°W (FLMNH 236688, 4). 6–7 m, 32.283°N, 64.783°W (UMML 30.12044, 2). Shelly Bay, 32.333°N, 64.742°W (FLMNH 236676, 1). Howard Beach, Tucker Town Bay, 32.334°N, 64.692°W (FLMNH 236685, 20). 1–3 m, S side of Nonsuch Island, 32.347°N, 64.663°W (MCZ 248281, 5). Off St. George, near St. Catherine Fort, 32.382°N, 64.678°W (USNM 659007, 2).

USA, Florida. Ragged Rock (USNM 754722, 3). Sta. FMNH 177411, Key West, Monroe County (FMNH 133189, 5). 6 m, Western Dry Rocks, about 8 nm SW off Key West, Lower Florida Keys, 24.444°N, 81.927°W (FMNH 278738, 1). Ohio Key, 24.473°N, 81.244°W (USNM 754721, 3). Middle Sambo Shoals, Key West, 24.485°N, 81.678°W (FLMNH 150238, 8). Mid Sambo Reef, Monroe County, 24.500°N, 81.667°W (FLMNH 249550, 4). Middle Sambo Shoal, near Key West, 24.513°N, 81.617°W (MCZ 226830, 1). Monroe County, Key West, 24.550°N, 81.767°W (CASIZ 119130, 3). Key West, 24.561°N, 81.779°W (USNM 754726, 12; USNM 754720, 4; USNM 754717, 9). 0 m, Boca Chica Channel, Boca Chica Key, Monroe County, 24.564°N, 81.717°W (HLC). 0 m, just E of Racoon and just W of Boca Chica, Monroe County, 24.564°N, 81.717°W (HLC, 1). 0 m, just W of Boca Chica Channel, just SE Racoon Key, Monroe County, 24.564°N, 81.717°W (HLC, 61). 0 m, Florida Keys, just W of Boca Chica,

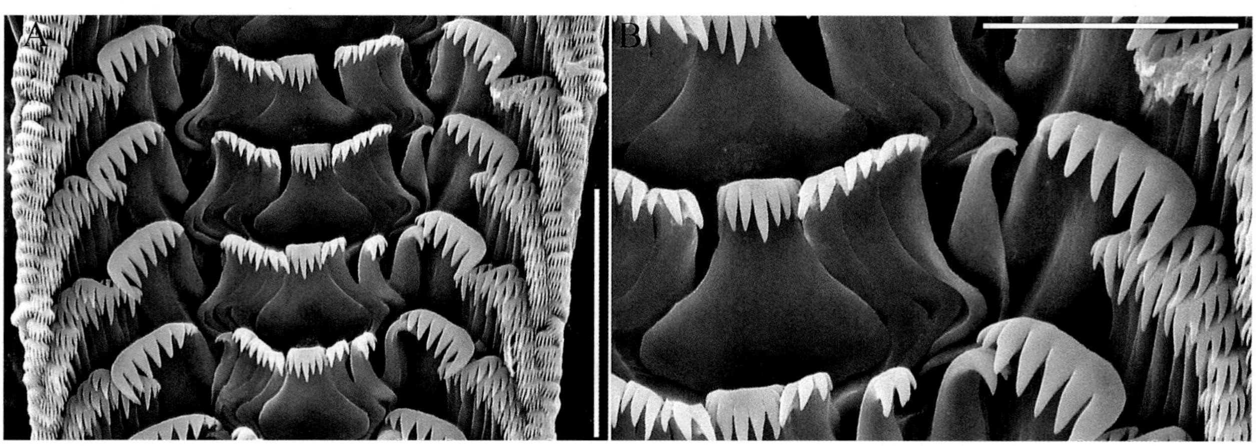

Figure 312. *Sinezona confusa*. Radula. La Parguera, Puerto Rico, 17.972°N, 67.047°W (DMNH 23178). **A.** Full width of radula. Scale bar = 20 µm. **B.** Central field enlarged. Scale bar = 10 µm.

24.567°N, 81.683°W (JWC 9327, 4). E side of Summerland Key, 24.658°N, 81.433°W (LACM 70-34, 1). W Summerland Key, 24.659°N, 81.449°W (USNM 754723, 2). Little Torch Key, Monroe County, 24.660°N, 81.470°W (FLMNH 236680, 1). Little Torch Key, Monroe County, 24.666°N, 81.386°W (FLMNH 349256, 1). 0 m, Little Torch Key, Monroe County, 24.667°N, 81.386°W (FLMNH 236678, 1). Missuori Key, 24.677°N, 81.238°W (USNM 663308, 3). 1 m, Knight's Key, Gulf, S end of Marathon, Monroe County, 24.709°N, 81.121°W (FLMNH 122363, 1). Bonefish Key, 24.734°N, 81.016°W (USNM 754718, 4). 0 m, W end Crawl Key, Monroe County, 24.755°N, 81.359°W (HLC, 1). 0 m, Gulf Side, Long Key, Monroe County, 24.820°N, 80.813°W (FLMNH 399137, 2). E of Tavernier, Key Largo, 25.014°N, 80.492°W (USNM 780322, 6). Key Largo, 25.119°N, 80.418°W (USNM 754727, 2). 1 m, Carysfort Reef, Key Largo, 25.217°N, 80.200°W (UMML 30.12038, 3). 1 m, Back side of Caksfort [=Carysfort?] Reef, 25.222°N, 80.212°W (UMML 30.12052, 2). 2 m, Patach Reef, 3 mi off Old Rhodes Key, 25.350°N, 80.233°W (UMML 30.12039, 8). 10 m, Ajax Reef, E, 25.396°N, 80.137°W (USNM 780320, 9). Soldier Key, 25.583°N, 80.150°W (UMML 30.9043, 4). 21–24 m, off Hillsboro Inlet, Broward County, 26.258°N, 80.077°W (LACM 81-9, 1). Delray Beach, Palm Beach County, 26.458°N, 80.057°W (FLMNH 236682, 1). Singer Island, N Inlet in Lake Worth, Palm Beach County, 26.770°N, 80.040°W (FLMNH 155181, 2). 0 m, Ft. Pierce, St. Lucie County, 27.439°N, 80.276°W (FLMNH 236686, 2). 0–2 m, Caesar's Creek, Monroe County, 29.706°N, 84.993°W (FLMNH 158067, 5). Off St. Augustine, 29.902°N, 81.262°W (FLMNH 285750, 25).

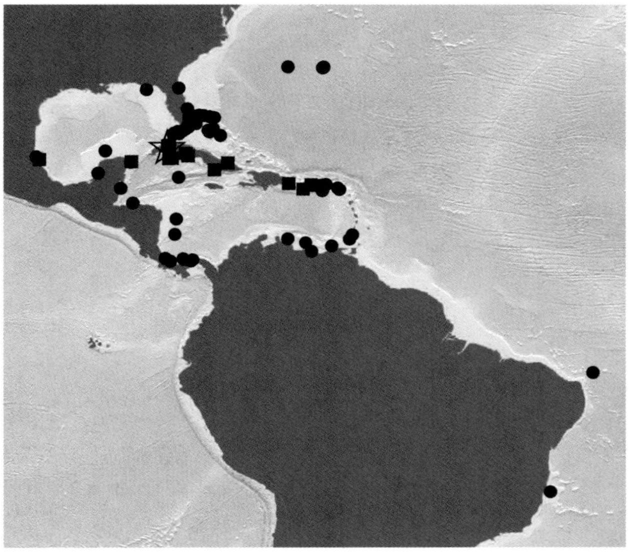

Figure 313. Distribution of *Sinezona confusa*. ☆ type locality. ● specimen record. ■ literature record.

Bahamas. Pigeon Cays, Andros Island (FLMNH 263475, 100; FLMNH 150236, 100). 4–4 m, Andros, 24.700°N, 77.767°W (UMML 30.12052, 2). Nassau, 25.081°N, 77.336°W (USNM 754719, 2). Turtle Rock, Bimini Island, 25.665°N, 79.309°W (FLMNH 150237, 1). 7 m, Abaco, 26.400°N, 77.200°W (CRC 12933, 4). 10–52 m, Abaco, 26.500°N, 77.750°W (CRC, 1). 0 m, W end Grand Bahama Island, 26.633°N, 78.417°W (HLC, 30). West End, Grand Bahama Island, 26.683°N, 78.967°W (BMSM 26038, 2).

American Virgin Islands. 0 m, Magens Bay, N coast St. Thomas, 18.344°N, 64.933°W (HLC, 1).

British Virgin Islands. 0 m, Carambala, St. Croix, 17.766°N, 64.829°W (FLMNH 236684, 2). 0 m, Virgin Gorda, 18.500°N, 64.400°W (HLC, 2). Virgin Gorda Island, 18.448°N, 64.429°W (FLMNH 236679, 3).

Cayman Islands. 0 m, Rum Point, Cayman Kai, Grand Cayman, 19.350°N, 81.283°W (HLC, 1).

Cuba. Playa Baracoa, La Habana, 23.050°N, 82.567°W (MNCN 15.05/17230, 1). Baracoa, 23.050°N, 82.567°W (NHMUK, 2). De al Habana, Playa Baracoa, 23.050°N, 82.567°W (USNM 890852, 2).

Guadeloupe. Saint Martin, 18.097°N, 63.073°W (DLG 892, 1).

Grenada. 8 m, W side of Point at at Fort Jeudy, SW side Grenada, 11.997°N, 61.710°W (LACM 86-193, 1).

Lesser Antilles. 0 m, E side, Petit St. Vincent, Windward Islands, 12.533°N, 61.383°W (HLC, 2). 0 m, near Gustavia, St. Barts, 17.900°N, 62.850°W (USNM 780318, 4).

Netherland Antilles. 20 m, Newport Reef, Curacao, 12.050°N, 68.817°W (HLC, 1).

Nicaragua. 2 m, N San Andres Island, 12.583°N, 81.717°W (LACM 70-25, 1).

Puerto Rico. (FLMNH 163127, 50). La Parguera, 17.972°N, 67.047°W (DMNH 23178, 4). La Parguera, 17.973°N, 67.048°W (MCZ 247103, 50).

Trinidad & Tobago. 8–9 m, Bucco, Tobago Island, 18.449°N, 64.745°W (FLMNH 236687, 5). 0 m, Tobago Island, 18.449°N, 64.745°W (FLMNH 236683, 1).

Mexico. Serrana Bank, 14.400°N, 81.550°W (UMML 30.12049, 1). Alacran Reef, Campeche Bank, Yucatan, 19.850°N, 90.533°W (CCSC, 6). Fiesta Americana Hotel, Cancun Island, Quintana Roo, 21.137°N, 86.754°W (FLMNH 349244, 6; FLMNH 350459, 1). Cancun, Quintana Roo, 21.176°N, 86.791°W (FLMNH 236681, 11). 10 m, S coast Islad Mujeres, Cancun, 21.200°N, 86.717°W (HLC, 2). Blanquilla Reef, Cabo Roio, 21.783°N, 97.583°W (UMML 30.12036, 10). 140 km N of Progreso, Areciffe Alacran, Yucatan, 22.467°N, 89.692°W (FLMNH 264298, 1; FLMNH 264297, 1).

Panama. Bocas del Toro, Bocas del Toro Province, 9.340°N, 82.240°W (FLMNH 231575, 4). E Colon Island, Bocas del Toro, 9.400°N, 82.283°W (MCZ 247251, 1). 0 m, Galeta Island, Caribbean side of canal zone, 9.400°N, 79.867°W (LACM 70-96, 3; USNM 743245, 3). Galeta Island, 9.583°N, 80.667°W (USNM 732941, 4). E Colon Island, Bocas del Toro Province, 9.412°N, 82.253°W (FLMNH 171265, 20). 0–2 m, 2 mi W of Portobel, 9.550°N, 79.683°W (LACM 70-19, 1).

Honduras. 2 m (USNM 780325, 9). Roatan Island, 16.300°N, 86.550°W (DLG 340, 11).

Costa Rica. 14 m, off Puerto Viejo, Limon Province, 9.663°N, 82.759°W (LACM 86-155, 3; LACM 86-157, 3). 13 m, N of Poerto Viejo, Limon Province, 9.680°N, 82.759°W (LACM 86-125, 1). 13 m, Inner Reef near Transect, Cahuita Reef Point, off Puerto Vargas, Limon Province, 9.736°N, 82.812°W (LACM 86-108, 1). 2–9 m, Inner Reef near Transect, Cahuita Reef Point, off Puerto Vargas, Limon Province, 9.736°N, 82.812°W (LACM 86-136, 2). 2–6 m, Inner Reef near Channel, Cahuita Reef Point, off Puerto Vargas, Limon Province, 9.736°N, 82.812°W (LACM 86-144, 14). 1 m, Cahuita Reef Point, off Puerto Vargas, Limon Province, 9.742°N, 82.811°W (LACM 86-130, 5). 12 m, SW of Sloth River Mouth, Cahuita Point, Limon Province, 9.745°N, 82.823°W (LACM 86-151, 11). 1–3 m, N of Puerto Viejo, Limon Province, 9.680°N, 80.759°W (LACM 86-122, 7).

Belize. (UMML 30.12048, 2). 1 m, Tres Cocos, Ambergris Caye, 18.050°N, 87.933°W (OSU 11071A).

Venezuela. 20 m, off Los Frailes, near Margarita Island, 8.812°N, 70.779°W (DKC 3175, 1). 0–2 m, Los Totumos, S side Cabo Codero, Moranda State, 10.553°N, 66.058°W (LACM 90-19, 1). Los Roques, Los Roques Island, 11.588°N, 66.754°W (DLG 901, 1).

Brazil. 82–105 m, Lagoon, Atol das Rocas, 3.833°S, 33.833°W (BMSM 26037, 3). 23 m, Bahia, Pardel dos Abrolhos, 3 km SE of Santa Barbara Island, 17.950°S, 38.732°W (BMSM 26036, 60).

Literature records.

Bahamas. Abaco (Redfern, 2001).

Cuba. 4 m, Jibacoa, Cuba, 20.250°N, 77.200°W. 3 m, El Salado, 21.054°N, 75.656°W. 20 m, Cayo Matias, Cuba, 21.567°N, 82.433°W. 20 m, Cayo Cantiles, 21.600°N, 82.033°W. 2 m, Yaguanabo, Cienfuegos, 21.892°N, 80.195°W. 56 m, Faro de los Colorados, Cienfuegos, Cuba, 22.167°N, 80.450°W. 40 m, Tamarindo Point, Cienfuegos, 22.167°N, 80.450°W. 5 m, Playa Baracoa, La Habana, 23.050°N, 82.567°W. 2 m, Comodoro, La Habana, 23.132°N, 82.364°W (Rolán & Luque, 1994).

Dominican Republic. Higüey, 18.617°N, 68.700°W (Rolán & Luque, 1994).

Mexico. 0 m, Mujeres Island, Yucatan, 21.200°N, 86.717°W (Odé, 1988).—Veracruz, Lobos Reef, 21.458°N, 97.229°W (Cruz & Gándara, 2006).

Puerto Rico. San Juan, 18.483°N, 66.133°W (Rolán & Luque, 1994).—1–2 m, Cayo Icaros, La Cordillera, 18.391°N, 65.597°W. 1 m, La Paguera, 17.957°N, 67.043°W (García-Rios *et al.*, 2008).

Remarks. The illustrated specimen in Bandel (1998) has clearly an open slit, though in the text he noted that in fully mature specimens the anterior margins of the slit touch one another. In all other aspects of the shell's morphology *columbiana* and *confusa* are identical: protoconch sculpture, length of teleoconch I, sculpture on

teleoconch I and II. Therefore, *columbiana* is a juvenile of *confusa*. Bandel (1998) did not compare his species to *Sin. confusa*. Robert Robertson (pers. comm.) first suggested the synonymy, which has also been adopted by Tunnell *et al.* (2010).

It is possible that *Sci. alexandrei* is a juvenile of *Sin. confusa*. The overall shell shape and the teleoconch sculpture as indicated on the original drawing of *Sci. alexandrei* are congruent with such an identification. *Sinezona confusa* is mainly known from the Caribbean, but a record from Brazil at 3°S in the same geographic area as the type locality of *Sci. alexandrei* supports that conclusion. Given that the original drawing of *Sci. alexandrei* is rather crude, the type specimen is completely decalcified, and that *Sin. confusa* is a fairly well-established name, I rather treat *Sci. alexandrei* as a *nomen dubium*.

Sinezona costulata Geiger & Sasaki, 2009

Figures 314–317

Chresonymy.

Sinezona costulata Geiger & Sasaki, 2009: 2–4, figs 1–3.
Sinezona sp.: Fukuda, 1993: 17–18, pl. 4, fig. 17.
Sinezona sp.: Sasaki, 1998: fig. 45a–c.
Sinezona sp. 2: Hasegawa & Okutani, 2000: 37, pl. 18, figs 7a,b [see Remarks].
Sinezona costulata: Geiger & McLean, 2010: 6.

Type material. Holotype (SBMNH 83544. Fig. 314). Paratypes from type locality (SBMNH 83545, 5. Fig. 315; SBMNH 83546, 11: complete specimens in ethanol, operculum and radula mounted; DLG 1037, 100; DLG 1038, 9: complete in ethanol). Goshikinohama, Usa, Tosa, Kochi Prefecture, Japan, 33.424°N, 133.457°E (UMUT RM27645: Sasaki, 1998: fig. 45a–c). Misaki Field Station, W coast of Miura Peninsula, 50 m to right of field station, Japan, 35.250°N, 139.667°E (AMS C.380660, 4).

Type locality. 0 m, peninsula towards shrine, Cape Manazuru, Kanagawa Prefecture, Japan, 35.141°N, 139.161°E. Rock-washings and algal shakings from moderately exposed rock and boulder shore (OD).

Etymology. Costulata, Latin adjective ribbed, referring to the many fine axial ribs on the teleoconch (OD).

Description. Shell small for genus (to 0.9 mm. Figs 314, 315), trochiform depressed. Protoconch of 0.75 whorl, strong axial cords, apertural varix not connected to embryonic cap, apertural margin straight. Teleoconch I of 0.66–0.8 whorl, 8–10 axial lamellae, in majority of specimens not reaching suture, starting out as axial cords, from mid-shoulder to mid-base as lamellae, fading towards umbilicus. No spiral sculpture. Interstices with fine irregular growth marks. Teleoconch II of 0.66 whorl, suture little impressed. Shoulder with approximately a dozen indistinct axial cords from suture to selenizone, fine spiral lines starting 0.1–0.2 whorl after onset of selenizone, increasing to approximately a dozen at apertural margin, irregularly spaced between suture and selenizone. Base with distinct constriction below selenizone, axial lamellae starting abruptly below constriction, fading toward umbilicus. Approximately

11 spirals, below suture as fine spiral lines, towards umbilicus changing to spiral steps. Umbilicus bordered by finely granulated carina, umbilical walls straight with very fine axial growth lines, no funiculus. Selenizone above periphery, keels shorter than width of selenizone (usually eroded), slit closed to foramen. Aperture rounded D-shaped, roof overhanging.

Animal. The animal has distinct eyes and papillate epipodial tentacles. The retracted state of the preserved animal did not permit additional insights.

Radula (Fig. 316). Rachidian tooth triangular, all cusps of tip approximately the same size, central cusp, three cusps on each side. Lateral teeth 1–3 similar, 3–4 cusps on outer edge of tip. Lateral tooth 4 reduced, hook-shaped, with minute point on outer edge of tip. Lateral tooth 5 enlarged by broadening, apical cusp largest, approximately seven cusps along inner edge of tip, decreasing in size from tip, outer edge smooth. Inner marginal teeth with triangular tip, apical cusp largest approximately three cusps along inner margin, approximately five cusps on outer margin. Outer marginal teeth with spoon-shaped tip, many fine cusps on each side. Radular interlock of central field moderate.

Figure 314. *Sinezona costulata*. Holotype. 0 m, peninsula towards shrine, Cape Manazuru, Kanagawa Prefecture, Japan, 35.141°N, 139.161°E (SBMNH 83544). Scale bar shell = 0.5 mm. Scale bar protoconch = 100 µm.

Comparisons. *Sinezona costulata* is the only species in the genus in its known range and easily identified by the slit closed to a foramen and the lack of spiral keels on the teleoconch.

Sinezona danieldreieri from the Indo-Malayan Archipelago to the Central Pacific has a shorter foramen (twice vs. >three times as long as wide), maintains the strength of the axial cords on the shoulder, and has 50% fewer axial cords.

Figure 315. *Sinezona costulata*. Paratypes. **A–C.** 0 m, peninsula towards shrine, Cape Manazuru, Kanagawa Prefecture, Japan, 35.141°N, 139.161°E (SBMNH 83545, 5). Scale bars shell = 0.5 mm. Scale bars protoconch = 100 µm.

Sinezona ferriezi from the Indo-Malayan Archipelago to the Central Pacific is much larger (to 2.1 m), has a smooth protoconch, many densely arranged but weaker axial cords, and a much wider umbilicus.

Sinezona milleri from southern Japan has more distinct spiral cordlets on shoulder and base, and a longer teleoconch I (1 vs. 0.8 whorl).

Sinezona plicata from the Indo-Malayan Archipelago to the Central Pacific grows much larger (2.3 mm), has a smooth protoconch occasionally with a spiral line, the apertural varix on the protoconch is connected to the embryonic cap, and has strongly raised prosocline axial lamellae crossed by distinct spiral cordlets.

Sukashitrochus carinatus from Southern Japan and the Indo-Malayan Archipelago has distinct spiral keels on the teleoconch.

Scissurella staminea from southern Japan has an open slit, is overall more globular and has distinct axials and spiral on the teleoconch forming a wide reticulate pattern, while in *Sin. costulata* the axial lamellae predominate.

Sinezona bandeli from New Zealand has more strong axial cords on the protoconch (17 vs. 12), and the axial cords on teleoconch I reach the suture while *Sin. costulata* shows a distinct gap.

Sinezona carolarum from the Panamic has about twice as many axial cords, and the umbilicus is wider.

Sinezona chilensis from western South America has a protoconch with twice as many but finer axial cords, a wider umbilicus, and a much shorter selenizone (0.1 vs. 0.4 whorl).

Sinezona cingulata from the Mediterranean and the northeastern Atlantic Ocean has a longer teleoconch I (~1 vs. 0.8 whorl), a shorter selenizone (0.1 vs. 0.4 whorl, and generally a more narrow umbilicus.

Sinezona garciai from the Caribbean has a protoconch with fine axials, lacks an apertural varix, and the teleoconch has more axial cords (25 vs. 9), which are weaker.

Distribution. Central to southern Japan (Fig. 317).

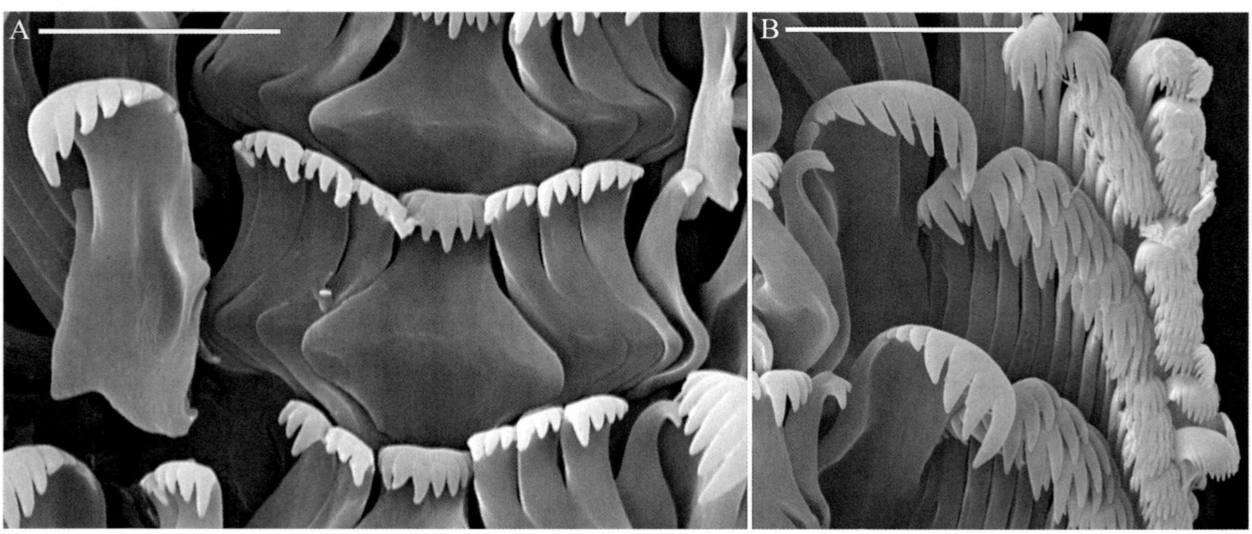

Figure 316. *Sinezona costulata.* Radula. 0 m, peninsula towards shrine, Cape Manazuru, Kanagawa Prefecture, Japan, 35.141°N, 139.161°E (Paratype SBMNH 83546, 5). **A.** Central field. **B.** Lateral tooth 5 and marginal teeth. Scale bars = 10 µm.

Specimen records.

Japan. 0 m, near Kominato, 13 mi SE of Mutsu-Minato, Aomori Prefecture, Honshu, 40.450°N, 141.683°E (LACM 87-11, 2). 1 m, Cape Banzho-zaki, near Seto Marine Laboratory, Wakayama Preference, Honshu, 34.183°N, 135.150°E (LACM 82-20, 1).

Literature records.

Japan. Miyanohama, Chichijima, Ogasawara Islands, 27.105°N, 142.194°E (Fukuda 1993). Hachijô Island, Izu Islands, 33.112°N, 139.785°E (Y. Shikano, pers. comm. cited by Fukuda 1993).

Remarks. Okutani and Hasegawa (2000) switched the descriptions of their two unnamed *Sinezona* species. *Sinezona* sp. 1 of Okutani & Hasegawa (2000) = *Sin. milleri* is placed at Honshu and Izu Islands, whereas it in fact occurs in more tropical waters. Conversely, *Sinezona* sp. 2 of Okutani & Hasegawa (2000) = *Sin. costulata* is reported by Okutani & Hasegawa (2000: 37) from "South of Amami Islands" [~28.5°N], i.e., the tropics, it is in fact a temperate species we collected on Honshu in Sagami Bay (35.2°N) and around Tosa (33.5°N) in large numbers and alive.

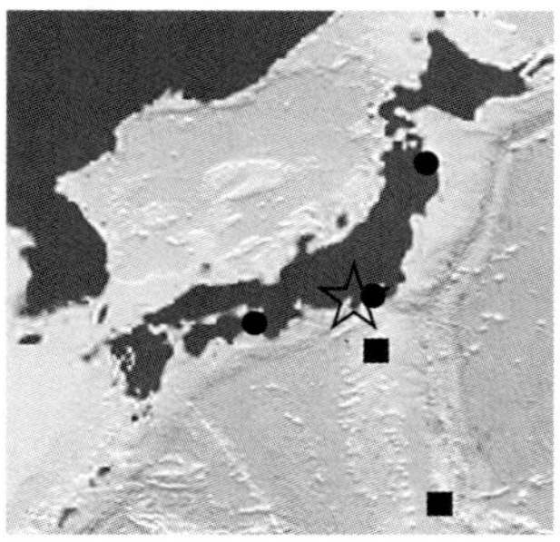

Figure 317. Distribution of *Sinezona costulata*. ☆ type locality. ● specimen record. ■ literature record.

Sinezona danieldreieri Geiger, 2008

Figures 318–322

Chresonymy.

Sinezona danieldreieri Geiger, 2008c: 186–190, figs 1–9.
Sinezona danieldreieri: Tröndlé & Boutet, 2009: 12.
Sinezona danieldreieri: Geiger & McLean, 2010: 6.

Type material. Holotype (SBMNH 83540. Fig. 318). Paratypes 3–4 m, Little Santa Cruz Island, Mindanao, 6.884°N, 122.040°E (USNM 812453, 1. Fig. 319A). 1 m, Okinawa, Oku, Japan, 26.847°N, 128.287°E (LACM 77-61, 3. Fig. 319B). 20 m, off Bohol Island, Pamilacan Island, 9.500°N, 123.917°E (AMS C.406342, 1. Fig. 319C).

Type locality. 0–5 m, Lumu Reef, Kimbe Bay, New Britain, Papua New Guinea, 5.283°S, 150.131°E. Leg. D. L. Geiger (OD).

Etymology. The name honors Daniel Dreier, son of Douglas and Hanna Dreier of Santa Barbara, who have made generous contributions to the Santa Barbara Museum of Natural History (OD).

Description. Shell small (to 0.77 mm. Figs 318–320), trochiform. Protoconch of one whorl, with strong axial sculpture not reaching apical suture, apertural varix not connected to smooth embryonic cap, apertural margin straight. Teleoconch I of 0.875 whorl,

suture at periphery, approximately 9–15 strong, raised axial cords; first spirals after 0.5 whorl. Teleoconch II of 0.6 whorl, suture deep; shell descending noticeably on last 0.25 whorl. Shoulder slightly concave, approximately 13–20 axial cords, raised near suture, as high as wide towards selenizone; approximately 10 spiral lines, distinct cords near suture, diminishing to very fine cords near selenizone; interstices with fine irregular lunules Base with constriction below selenizone, approximately 15 elevated spiral cords, run over by approximately 16 fine spiral lines. Umbilicus cone-shaped, moderately wide, wall smooth, at distinct angle to base. Aperture D-shaped, roof overhanging. Selenizone above periphery, keels quite strong, moderately elevated, distinct lunules, elongated foramen closed anteriorly.

Operculum (Fig. 319A) thin, flexible, multispiral, with central nucleus.

Radula (Fig. 321). Rachidian tooth triangular tip with approximately five equal sized cusps. Lateral teeth 1–3 similar; outer edge of tip with 4–5 cusps. Lateral tooth 4 reduced in size, hook-shaped. Lateral tooth 5 enlarged by broadening, approximately six large cusps on inner edge, 1–2 small ones on outer edge. Central cusp of inner marginal teeth large, approximately three cusps on inner edge, four on outer edge. Outer marginal teeth spoon-shaped, bilaterally symmetrical with approximately a dozen fine cusps on each side. Radular interlock moderate.

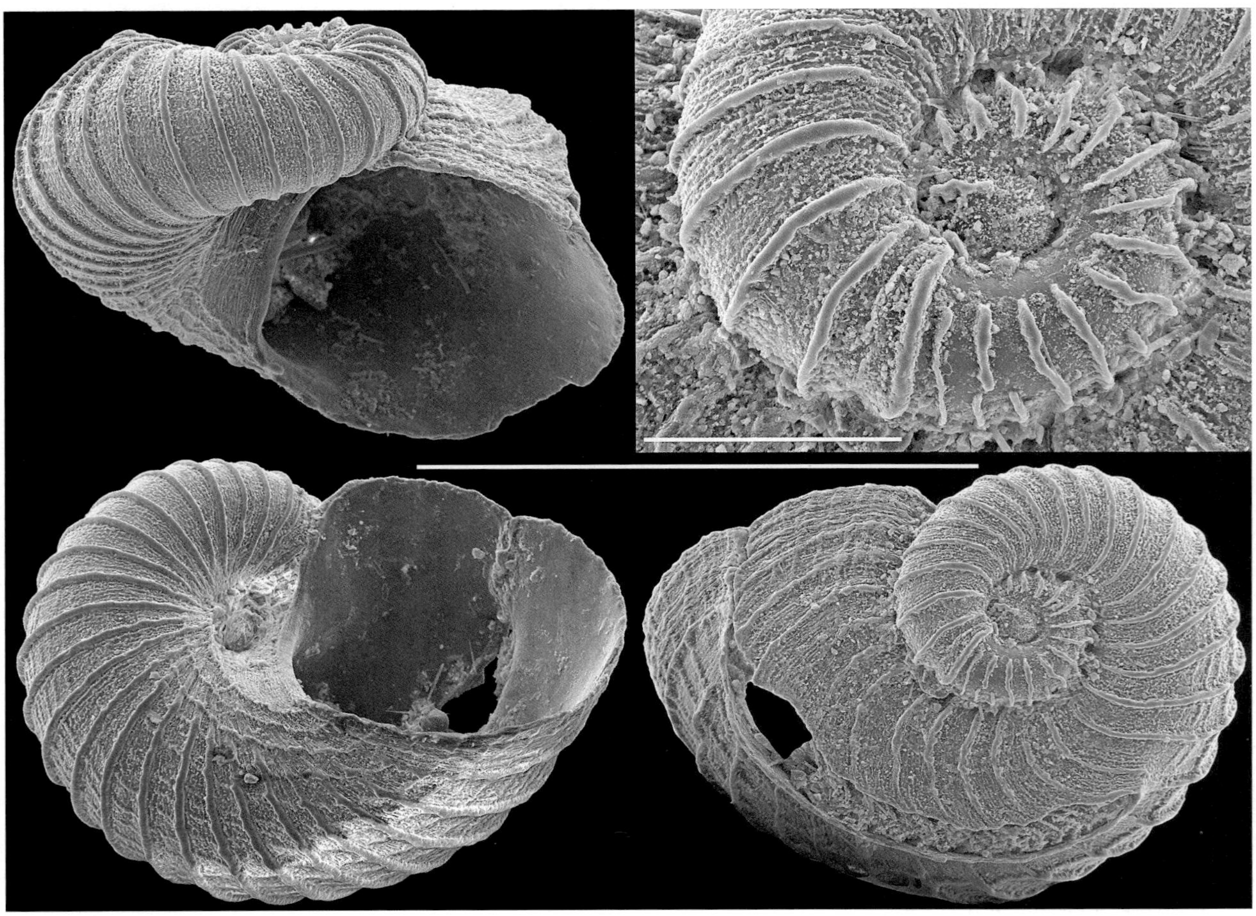

Figure 318. *Sinezona danieldreieri*. Holotype. 0–5 m, Lumu Reef, Kimbe Bay, New Britain, Papua New Guinea, 5.283°S, 150.131°E (SBMNH 83540). Scale bar shell = 0.5 mm. Scale bar protoconch = 100 μm.

Comparisons. *Sinezona calumnior* from temperate Australia and New Zealand has a very narrow umbilicus and the protoconch protrudes the shell.

Sinezona costulata from southern Japan has half as many but stronger axial cords and the axial cords strongly diminish in strength on the shoulder of teleoconch II.

Sinezona ferriezi from the Indo-Malayan Archipelago to the Central Pacific is about twice as large, has a smooth protoconch with a distinct apertural varix, and has a teleoconch I of >1 whorl.

Figure 319. *Sinezona danieldreieri*. Paratypes. **A.** 3–4 m, Little Santa Cruz Island, Mindanao, Philippines, 6.884°N, 122.040°E (USNM 812453, 1). **B.** 1 m, Okinawa, Oku, Japan, 26.847°N, 128.287°E (LACM 77-61, 3). **C.** 20 m, off Bohol Island, Pamilacan Island, Philippines, 9.500°N, 123.917°E (AMS C.406342, 1). Scale bars shell = 0.5 mm. Scale bars protoconch = 100 μm.

Sinezona globosa from the Indo-Malayan Archipelago to the Eastern Pacific is more rounded in overall appearance, lacks the elevated spiral cords, grows to a much larger size (3.8 mm), and has spiral sculpture on the protoconch with the apertural varix connecting to the embryonic cap.

Sinezona macleani from the Indian Ocean to the Western Pacific is more globular in overall outline, grows to a much larger size (2.3 mm), has a protoconch with flocculent sculpture, lacks the raised axials, and has a wider umbilicus.

Figure 320. *Sinezona danieldreieri*. **A–B.** 2 m, North of Pukitarava, Rapa, Austral Islands, French Polynesia, 27.597°S, 144.227°W (MNHN, 4). **C.** 5 m, Anarua Bay, Rapa, Austral Islands, French Polynesia, 27.605°S, 144.378°W (MNHN, 4). Radula of one specimen shown in Figure 321. Scale bars shell = 0.5 mm. Scale bars protoconch = 100 μm.

All the above species grow to much larger size and would be immature with open slit at the maximum size of *Sin. danieldreieri.*

Sinezona marrowi from the Indo-Malayan Archipelago has an overall broader shell with wide umbilicus, the apertural varix of the protoconch connects to the embryonic cap, and has a longer selenizone (>0.5 vs. 0.3 whorl).

Sinezona milleri from the Ryukyu Islands has a longer teleoconch I (1 vs. 0.75–0.8 whorl) and the protoconch has distinct thickenings along the apical ridge.

Sinezona plicata from the Indo-Malayan Archipelago to the Central Pacific is has an overall wider shell, more pronounced but fewer raised axial cords, grows to much larger size (2.3 mm), and has a protoconch with spirals.

Sinezona wileyi from the Indo-Malayan Archipelago to the tropical Central Pacific has a subquadrate overall shell shape and a protoconch with fine irregular axials but without apertural varix.

Sinezona zimmeri from Easter and Pitcairn Islands has a wider overall shell, a wider umbilicus, and stronger axial cords.

Scissurella phenax from the Indo-Malayan Archipelago to the Central Pacific has a taller spire, a more narrow umbilicus, and a protoconch with median thickenings on the axials.

Distribution. Indo-Malayan Archipelago to southern Polynesia; shallow reef environments (Fig. 322).

Specimen records.

French Polynesia, Austral Islands. 0 m, Rapa Iti Island, Rapa, 27.622°S, 144.302°W (MNHN, 1). 0 m, Kotuaie Point, Tupuaki Bay, Rapa, 27.577°S, 144.343°W (MNHN, 1). 2 m, N of Pukitarava, Rapa, 27.597°S, 144.227°W (MNHN, 4: paratypes). 2–4 m, N of Aturapa Island, Rapa, 27.572°S, 144.350°W (MNHN, 4). 2 m, N of Rapa Iti Island, Rapa, 27.620°S, 144.303°W (MNHN, 1). 2 m, N of Anatakuri Bay, Rapa, 27.623°S, 144.307°W (MNHN, 15). 3–24 m, Hiri Bay, Rapa, 27.622°S, 144.370°W (MNHN, 4). 5 m, Anarua Bay, Rapa, 27.605°S, 144.378°W (MNHN, 7: paratypes). 6 m, Mei Point, Rapa, 27.637°S, 144.303°W (MNHN, 1; MNHN, 1). 6 m, SW of Gotenaonao Point, Rapa, 27.645°S, 144.320°W (MNHN, 3). 15–20 m, Vavai, Rapa, 27.590°S, 144.381°W (MNHN, 1; MNHN, 1). 16–20 m, Mei Point, Rapa, 27.637°S, 144.303°W (MNHN, 2). 18 m, Rarapai Island, Rapa, 27.572°S, 144.368°W (MNHN, 1). 27 m, Kauira Point,

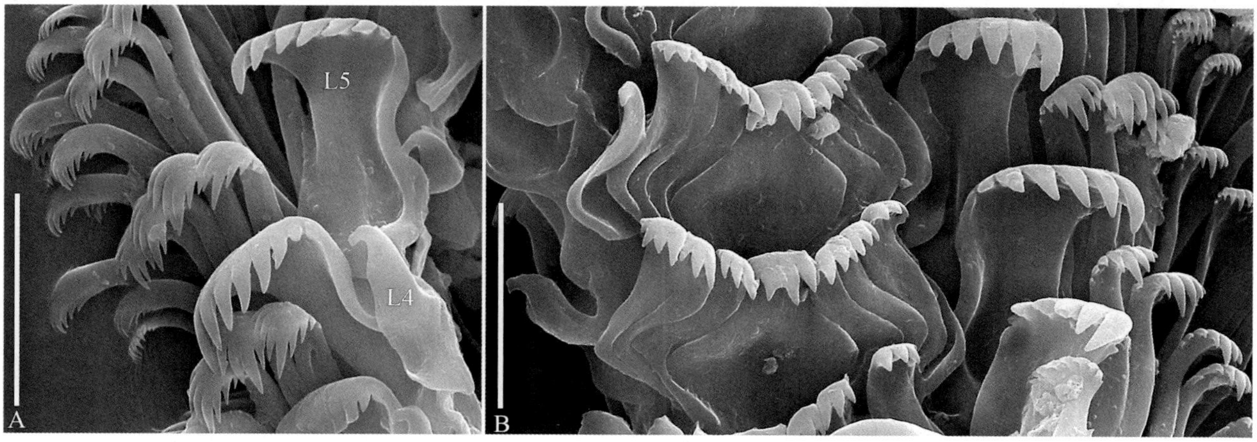

Figure 321. *Sinezona danieldreieri.* Radula. Shell from lot shown in Figure 320C. 5 m, Anarua Bay, Rapa, Austral Islands, 27.605°S, 144.378°W (MNHN, 4). **A.** Lateral teeth 4 and 5 plus marginal teeth. **B.** Half width of radula with central field. Scale bars = 10 µm.

Rapa, 27.592°S, 144.347°W (MNHN, 1). 30 m, NW of Tauna Island, Rapa, 27.605°S, 144.303°W (MNHN, 17). 36 m, around Rukuaga Point, Rapa, 27.568°S, 144.368°W (MNHN, 8).

French Polynesia, Society Islands. 0–1 m, Moorea, 17.591°S, 149.834°W (DLG 756, 9: complete). 0–1 m, Moorea, 17.591°S, 149.834°W (DLG 758, 1: complete). 0 m, Cook Bay, Vaiare, Moorea, 17.386°S, 149.774°W (DLG 224, 40: complete). 0 m, Cook Bay, Vaiare, Moorea, 17.536°S, 149.674°W (DLG 140, 15). 20–60 m, sediments, Afaahiti, Tahiti, 17.760°S, 149.384°W (JTC, 4). Laisse de plage ext., Mururoa, Tuamotu, 21.835°S, 138.897°W (JTC, 4).

Cook Islands. 0–2 m, off Titikeveka, S side Rarotonga, 21.263°S, 159.757°W (LACM 87-73, 2). 0 m, Tuoro (Black Rock), Rarotonga Island, 21.200°S, 159.825°W (AMS C.436786, 1). 4–5 m, Rutaki, Rarotonga, 278.767°S, 159.810°W (DLG 238, 25: complete). 1–2 m, Muri Beach, Rarotonga, 21.259°S, 159.732°W (DLG 235, 1: complete).

Coral Sea. 12 m, Elisabeth Reef, 29.953°S, 159.020°E (AMS C.445050, 3: complete).

Australia, Norfolk Islands. Cemetery Beach, Point Hunter Reserve, 29.067°S, 167.967°E (AMS C.404254, 15: complete). Emily Bay, Kingston, 29.067°S, 167.967°E (AMS C.404985, 5: complete).

Papua New Guinea. 0–5 m, Lumu Reef, Kimbe Bay, New Britain, 5.283°S, 150.131°E (SBMNH 83544, 1: holotype).

New Caledonia. 0 m, Noumea, Platier Ilot Maitre, 22.340°S, 166.428°E (MNHN, 2).

Philippines. 20 m, off Bohol Island, Pamilacan Island, 9.500°N, 123.917°E (AMS C.406342, 1: paratype). Canipo Island, Cuyo Islands, Palawan Province, 11.000°N, 120.948°E (USNM 808195, 1). 3–4 m, Little Santa Cruz Island, Mindanao, 6.884°N, 122.040°E (USNM 812453, 1: paratype).

Guam. 1 m, Ajayan Bay, S Guam, 13.245°N, 144.717°E (LACM 77-19, 2).

Japan, Okinawa. 2 m, Oku, 26.847°N, 128.287°E (LACM 77-61, 3: paratypes). 2–3 m, Horseshoe Beach, 26.504°N, 127.858°E (DLG 1576, 3).

USA. Hawaii. 8 m, Midway Atoll, 25.772°N, 171.736°W (CPC, 3).

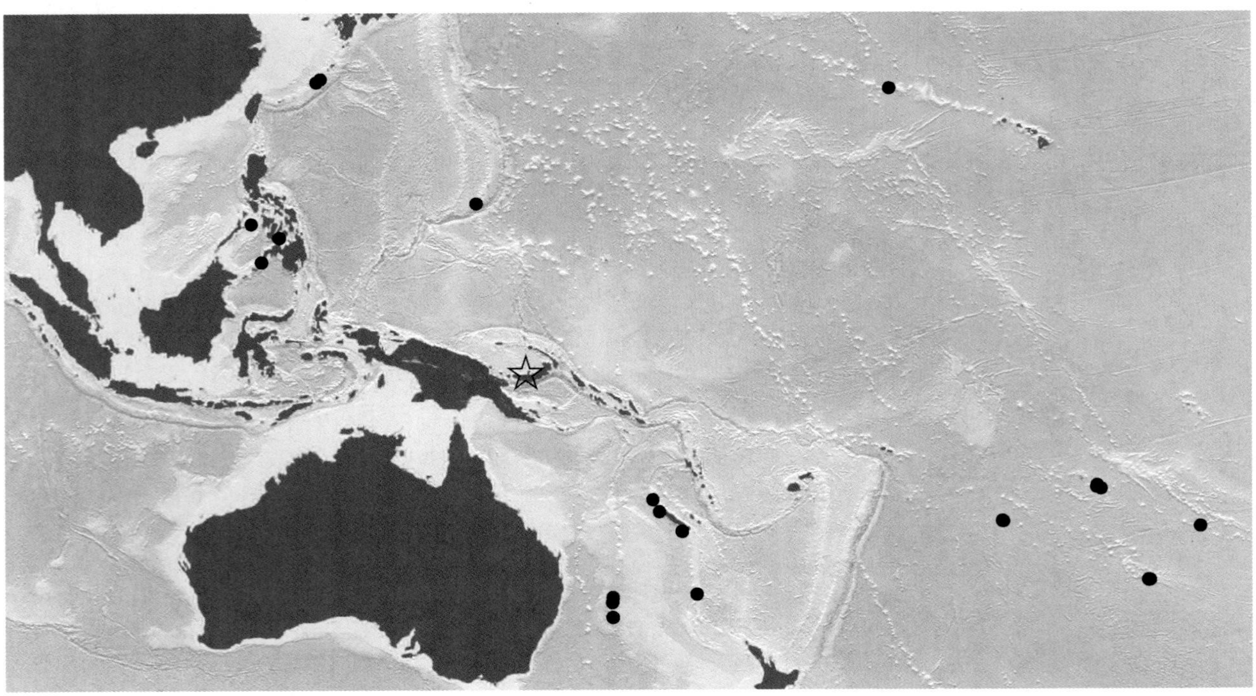

Figure 322. Distribution of *Sinezona danieldreieri*. ☆ type locality. ● specimen record. ■ literature record.

Remarks. The protoconch sculpture shows some intraspecific variability in the material examined, as the strong axial cords are sometimes discontinuous. This character is not restricted to the present species, hence, it does not seem to be of any systematic value. Those axial cords are also somewhat variable in width, the thickenings occasionally seeming to form some faint spiral elements. Given the intra- and interspecific variability of protoconch sculpture at this level of detail, the presence/absence of faint spiral elements is considered irrelevant for systematic purposes. Discrete differences can be observed between strong and weak axial cords as utilized previously in a systematic context (Geiger, 2003).

Sinezona doliolum Herbert, 1986

Figures 323–326

Chresonymy.

Sinezona doliolum Herbert, 1986: 625–626, figs 9, 28–31.
Sinezona doliolum: Herbert, 1987: 1–2.
Sinezona doliolum: Geiger, 2003: 78.
Sinezona doliolum: Kurtz, 2011: 34.

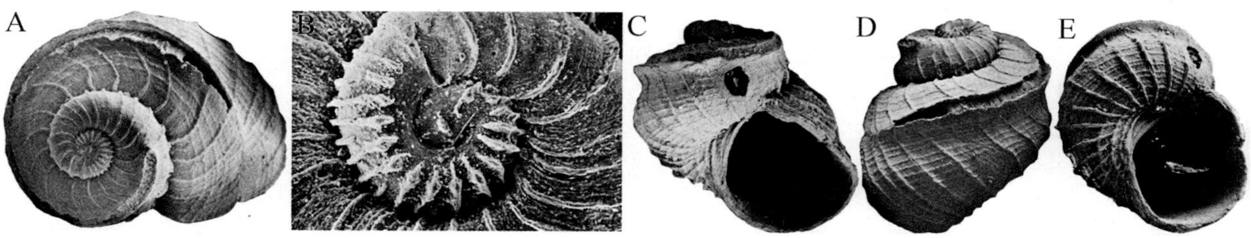

Figure 323. Original illustration of *Sinezona doliolum*. **A, B, D**. Holotype, **C, E**. Paratype (NM C7987/T3275).

Type material. Holotype (NMSA C8471/T3274. Fig. 323A, B, D), 0.95 mm. Paratypes from type locality (NMSA C8470/T3276, 1). 96 m, off Ubombo, Transkei, South Africa, 27.583°S, 32.083°E (NMSA C7987/T3275, 1. Fig. 323C. E). 106 m, off Mtamvuna River, Transkei, South Africa, 31.100°S, 30.200°E (NMSA C8020/T3277, 6).

Type locality. Off Mncwasa Point, Transkei, [South Africa] 32°6.2'S, 29°6.5'E, 68 m, sand (OD).

Etymology. Doliolum, Latin diminutive of dolium, a cask (OD).

Description. Shell small (to 1.0 mm. Figs 324, 325), trochiform, elevated higher than wide. Protoconch of one whorl, strong axial sculpture, no apertural varix, apertural margin somewhat sinusoidal. Teleoconch I of 0.75 whorl, approximately 13 axial cords, no spiral sculpture. Teleoconch II of 1.125 whorls, suture below periphery of previous whorl little impressed. Shoulder flat, with 17 axial cords, becoming weaker near lip of mature shell, approximately six spirals. Base with strong constriction below selenizone, axial density as on shoulder, approximately 15 spirals becoming stronger towards umbilicus. Selenizone above periphery, almost at periphery of

shell, lunules irregular, little distinct, keels strong, moderately elevated. Slit closed to elongated triangular foramen. Umbilicus very narrowly open, bordered by strong cord towards base.

Animal unknown.

Comparisons. *Sinezona insignis* from South Africa lacks sculpture on the shoulder, has stronger axial folds on the base, a much shorter selenizone (0.3 vs. 0.8 whorl), and a distinctly open umbilicus.

Scissurella jucunda from South Africa is wider than high, lacks the constriction below the selenizone on the base, and has a barely closed umbilicus without a cord bordering the base.

Sinezona calumnior from temperate Australia and New Zealand has stronger spiral threads on the shoulder, has a shorter teleoconch II (0.6 vs. 1 whorl), and a shorter foramen.

Figure 324. *Sinezona doliolum*. 100 m, Qora River mouth, Transkei, South Africa, 32.547°N, 28.803°E (DLG 1390, 2). Scale bar shell = 0.5 mm. Scale bar protoconch = 100 µm.

Sinezona iota from New Zealand has a more prominent protoconch, lacks spiral sculpture, has a much shorter selenizone (0.4 vs. 0.8 whorl), and has a more distinct umbilicus with more distinct periumbilical cord.

Sinezona mouchezi from subantarctic Indian Ocean has a less elevated spire, a shorter selenizone (0.6 vs. 0.8 whorl) and a longer teleoconch I (1.2 vs. 0.8 whorls).

Figure 325. *Sinezona doliolum*. **A.** 100 m, Xora, Transkei, South Africa, 32.128°S, 29.000°E (DLG 1404, 2). **B–C.** 100 m, Park Rynie, Natal, South Africa, 30.357°S, 30.851°E (DLG 1393, 11). Scale bars shell = 0.5 mm. Scale bars protoconch = 100 µm.

Sinezona singeri from the Indian Ocean has a much wider shell with distinct wide umbilicus, a smooth protoconch, and a much longer selenizone (1.6 vs. 0.8 whorls).

Sinezona rimuloides from the northeastern Pacific and the Panamic has a protoconch with distinct apertural varix, a much shorter selenizone (0.3 vs. 0.8 whorl), and a wider umbilicus.

Distribution. Eastern South Africa, Mascarene Islands, 0–106 m (Fig. 326).

Specimen records.

South Africa. 100 m, southern Natal (DLG 1370, 1). 100 m, Park Rynie, Natal, 30.357°S, 30.851°E (DLG 1393, 11). 150–165 m, Whale Rock, Transkei, 32.010°S, 29.308°E (DLG 1379, 5). 100 m, Qora River mouth, Transkei, 32.547°S, 28.803°E (DLG 1390, 2). 90 m, Mbotyi, Transkei, 31.463°S, 29.737°E (DLG 1403, 12). 100 m, Xora, Transkei, 32.128°S, 29.000°E (DLG 1404, 2).

Mauritius. 1 m, 200 m S from channel between Ile aux Cerfs and Ile de l'Est, 20.266°S, 57.805°E (IKC 10980, 1).

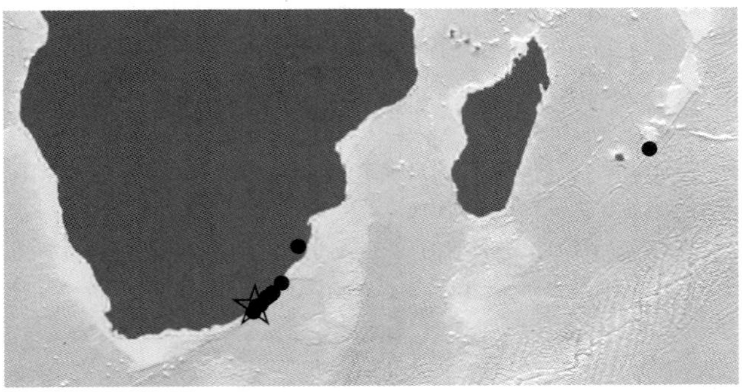

Figure 326. Distribution of *Sinezona doliolum*. ☆ type locality. ● specimen record. ■ literature record.

Literature records.

South Africa. Off Ubombo, Transkei, 96 m (paratype 1). Off Mtamvuna River, Transkei, 106 m (paratypes 3–8) (Herbert, 1986).

Remarks. Type species of *Daizona* Bandel, 1998.

Sinezona enigmatica Geiger & Marshall, 2012

Figures 327–329

Chresonymy.

Sinezona enigmatica Geiger & Marshall, 2012: 4–6, figs 2–3.

Type material. Holotype (NMNZ M.303302. Fig. 327). Paratype from type locality (AMS C.461493, 13. Fig. 328).

Type locality. McGregor Bay, Whangarei Harbour, New Zealand, 36.750°S, 175.467°E (OD).

Etymology. Enigma, Latin for something inexplicable, referring to the surprising finding of a new species in a well-collected area, adjective; see also Remarks (OD).

Description. Shell medium-sized (to 1.24 mm. Figs 327, 328), trochiform depressed. Protoconch of one whorl, strong axial sculpture with median thickenings, apertural varix not connected to embryonic cap, apertural margin straight. Teleoconch I of 0.9–1.1 whorls, 12–13 strong axial cords, diminishing in strength with growth, hardly perceptible at onset of selenizone, weaker towards suture; interstices with fine irregular growth marks. Teleoconch with moderately impressed suture, 0.25–0.4 whorl. Shoulder slightly convex, weak irregular growth marks, faint spiral lines or low indistinct cords, irregularly spaced, approximately eight at apertural margin. Base

with weak constriction below selenizone, macroscopically smooth, fine indistinct spiral lines, most distinct at apertural margin; very fine and dense lines below selenizone, fewer, more prominent cordlets near umbilicus. Umbilicus at angle to base, surrounded by weak cords; walls with very fine growth lines, straight. Aperture ovoid, roof overhanging. Selenizone above periphery; keels low, moderately strong; lunules indistinct; foramen with raphe extending anteriorly.

Animal unknown.

Comparisons. *Sinezona brevis* from New Zealand has a proportionally taller shell, its protoconch lacks the strong medial thickenings and has a strong apertural varix, has axial cords also on teleoconch II, and has distinct spiral threads on shoulder and base.

Sinezona levigata from New Zealand usually has distinct spiral and axial cords forming a reticulate pattern; little sculptured specimens have an umbilical callus and a very short or even no selenizone, at most as long as the foramen.

Sinezona finlayi from the Tertiary of New Zealand has stronger spiral sculpture particularly on the base and has axial cords also on teleoconch II.

As the smooth surface of the last whorl and the relatively fast expanding whorls are shared with *Incisura*, the species is also compared to similar ones in *Incisura*. *Incisura lyttletonensis* and *I. rosea* from the Recent fauna of New Zealand have an open slit.

Figure 327. *Sinezona enigmatica*. Holotype. McGregor Bay, Whangarei Harbour, 36.750°S, 175.467°E (NMNZ M.303302). Scale bar shell = 1 mm. Scale bar protoconch = 100 µm.

Incisura auriformis from Australia shares the closed foramen, but has more rapidly expanding whorls, and a very strong bulging cord surrounding the umbilicus.

Incisura rosea from New Zealand has an open slit and an umbilical callus.

Distribution. Known only from type lot (Fig. 329).

Remarks. The finding of the single lot in shallow water from McGregors Bay raises some questions. The bay is rather well-collected, but not a single lot of the species is available in the more comprehensive collection of NMNZ. The location harbors an important port for international shipping to and from New Zealand. Various options to account for the specimens were explored. Specimens could have been mislabeled,

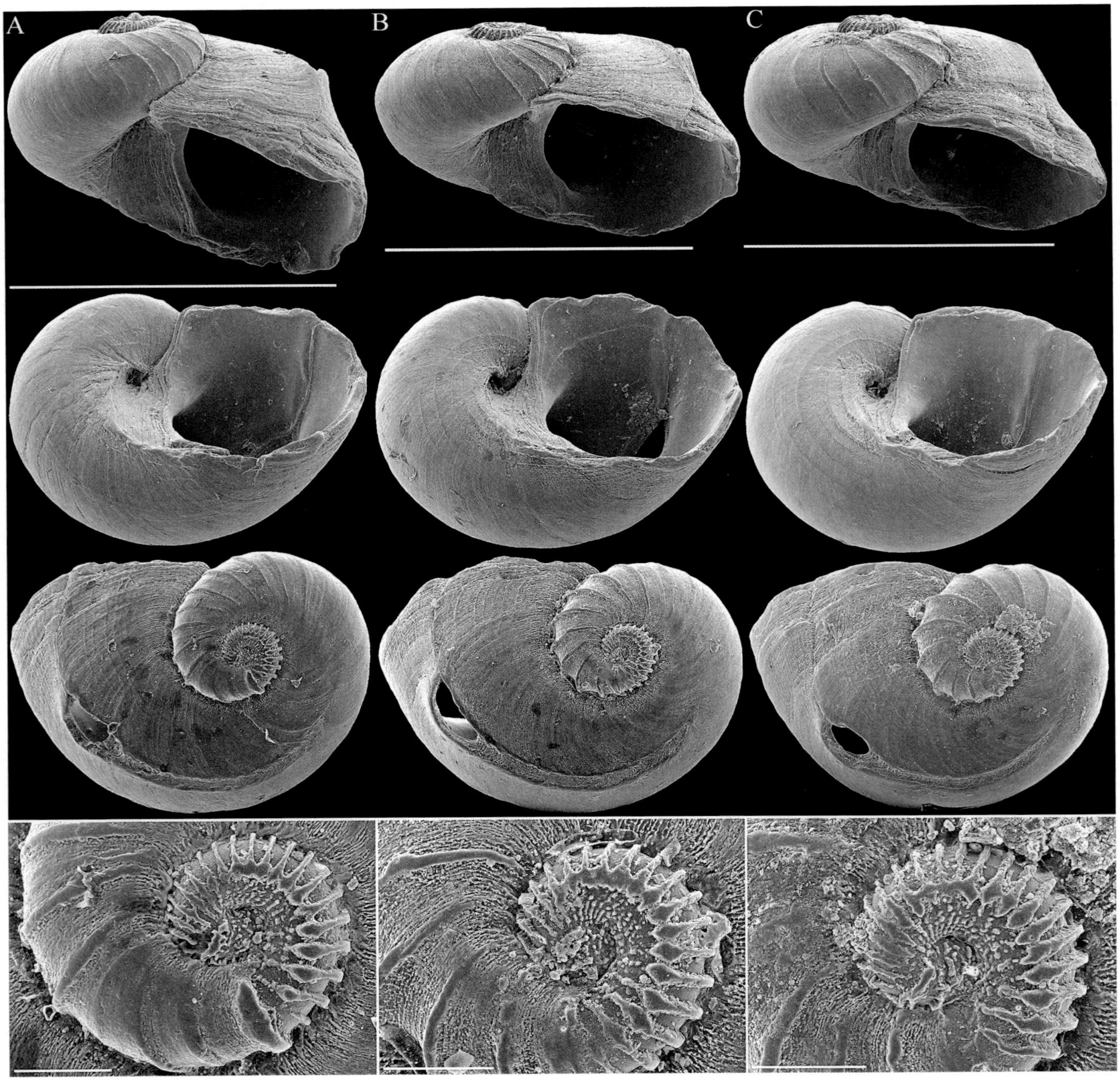

Figure 328. *Sinezona enigmatica.* Paratypes. **A–C.** McGregor Bay, Whangarei Harbour, 36.750°S, 175.467°E (AMS C.461493, 13). Scale bars shell = 1 mm. Scale bars protoconch = 100 µm.

which is more likely as no further information on collector or date of collection are available, but the adjacent catalog numbers contain typical New Zealand scissurellid taxa, suggesting a single collection event. The specimens could have been brought to New Zealand in commercial ship ballast, though the preservation of the specimens is quite good. If the specimens are originally from elsewhere, the morphospecies still is distinct. There are other examples of single lot, shallow water species from other well-collected areas, such as *Sci. azorensis* from the Western Mediterranean. Confirmation at the stated type locality, or discovery of its true native location will be desirable.

Figure 329. Distribution of *Sinezona enigmatica*. ☆ type locality.

Sinezona ferriezi (Crosse, 1867)

Figures 330–335

Chresonymy.

Schismope ferriezi Crosse, 1867: 318, pl. 11, fig. 7 (Fig. 330A).
Schismope ferriezi: Paetel, 1888: 289.
Schismope ferriezi: Pilsbry, 1890: 61, pl. 22, figs 43–44.
Schismope ferriezi: Thiele, 1912a: 29–30, pl. 3, figs 29–30 [copy fig. Crosse, 1867].
Scissurella (*Sukashitrochus*) *ferriezi*: Lozouet, 1986: 109.
Sinezona ferriezi: Jansen, 1999: 54–55, figs 52–54.

Synonym.

+ *Scissurella padangensis* Thiele, 1912a: 25–26, pl. 3, figs 23–24 (Fig. 330B). ***Type material.*** 1.5 × 1.2 mm (original dimensions). Lectotype (ZMB/MOLL 109.044) here designated with the express purpose of taxon stabilization [see Remarks of *Sin. ferriezi*]. Paralectotype from type locality (ZMB/MOLL 109.045, 1. Is *Sin. plicata* Hedley). ***Type locality.*** Padang, Sumatra [Indonesia, 1°S, 100°21'E] (OD). ***Etymology.*** Named after type locality.
Scissurella (*Sukashitrochus*) *padangensis*: Lozouet, 1986: 109.
Sinezona padangensis: Geiger, 2003: 79.

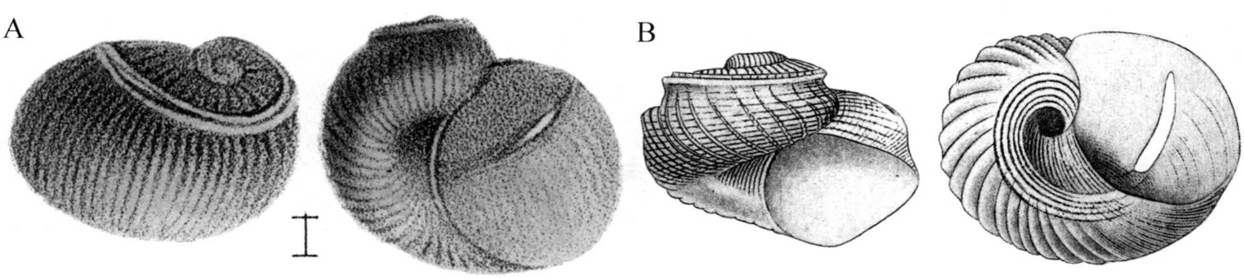

Figure 330. **A.** Original illustration of *Schismope ferriezi*. **B.** Original illustration of *Scissurella padangensis*.

Misidentifications.

not *Schismope padangensis*: Franc, 1956: 21 [is *Sci. rota*]

not *Schismope padangensis*: Mastaller, 1979: 29 [is *Sci. rota. fide* Yaron, 1983].

not *Sinezona padangensis*: Yu & Feng, 1996: pl. 1, figs 5–7 [is *Sin. plicata*].

Scissurella coronata: Ladd, 1966: 27, pl. 2, figs 7–8 [is *Sin. ferriezi*. USNM specimen examined 1/2007]

Scissurella hoernesi: Bandel, 1998: 10–11, pl. 1, figs 6–8 [is juvenile of *Sin. ferriezi*].

Type material. Syntypes (Coll. Crosse, 4), 2.25 × 1.5 mm. Syntypes from type locality (MNHN, 2: Geiger, pers. obs. 4/2005: see Remarks). The specimen illustrated in Figure 333A is here designated as the lectotype for the express purpose of taxon stabilization, in case other non-conspecific syntypes should be located. One surviving paralectotype (Fig. 333B).

Type locality. Noumea, [New Caledonia] under stones [22.267°S, 166.450°E] (OD).

Etymology. Named for the collector of the specimens Mr. E. Ferriez (OD).

Description. Shell medium-sized (to 2.1 mm. Figs 331–333), trochiform depressed. Protoconch with smooth sculpture (occasionally with hint of weakest spirals), apertural varix

Figure 331. *Sinezona ferriezi*. Sovi Bay Souvi, SW coast Viti Levu, Fiji, 18.183°S, 177.600°E (AMS C.406356). Scale bar shell = 0.5 mm. Scale bar protoconch = 100 µm.

present, connected to embryonic cap, apertural margin sinusoidal. Teleoconch I of 1–1.125 whorls, approximately 20–25 indistinct, irregularly spaced axial cords, interstices with very fine irregular axials, no spirals. Teleoconch II of up to 1.5 whorls, suture moderately impressed. Shoulder profile slightly sinusoidal, with approximately 33 fine slightly prosocline axial cords, approximately 13 very fine spiral threads at apertural margin. Base with constriction below selenizone, axials at

Figure 332. *Sinezona ferriezi*. **A.** Topotype. Noumea, New Caledonia, 22.267°S, 166.450°E (AMS C.4287, 1). Radula shown in Figure 334D. **B.** Magnetic Island, Geoffrey Bay, Queensland, Australia, 19.153°S, 146.863°E (AMS C.379951, 3). **C.** 31–37 m, Great Barrier Reef, Capricorn Group, Masthead Island, Queensland, Australia, 23.533°S, 151.750°E (AMS C.019326, 20+). Scale bars shell = 0.5 mm. Scale bars protoconch = 100 μm.

Figure 333. *Schismope ferriezi*. Noumea, [New Caledonia] under stones [22.267°S, 166.450°E]. **A.** Lectotype (MNHN). **B.** Paralectotype (MNHN). Scale bars shell = 1 mm. Scale bars protoconch = 100 μm.

same intervals as on shoulder but stronger, approximately 30 fine, regularly spaced spiral lines. Umbilicus wide open, walls straight, with sharp axial lamellae, distinctly angulated at transition to base, but without keel or ridge on border between base and umbilicus. Selenizone above periphery, keels strong moderately elevated, lunules strong at approximately the same intervals as axial sculpture elements; slit closed to very elongated foramen. Aperture rounded, roof overhanging.

Operculum (Fig. 332B) corneous, very thin, more or less covering aperture, nucleus central, multispiral.

Radula (Fig. 334). Rachidian tooth trapezoidal, tip rounded, central cusp of same size or larger than four cusps on side. Lateral teeth 1–3 similar with three long

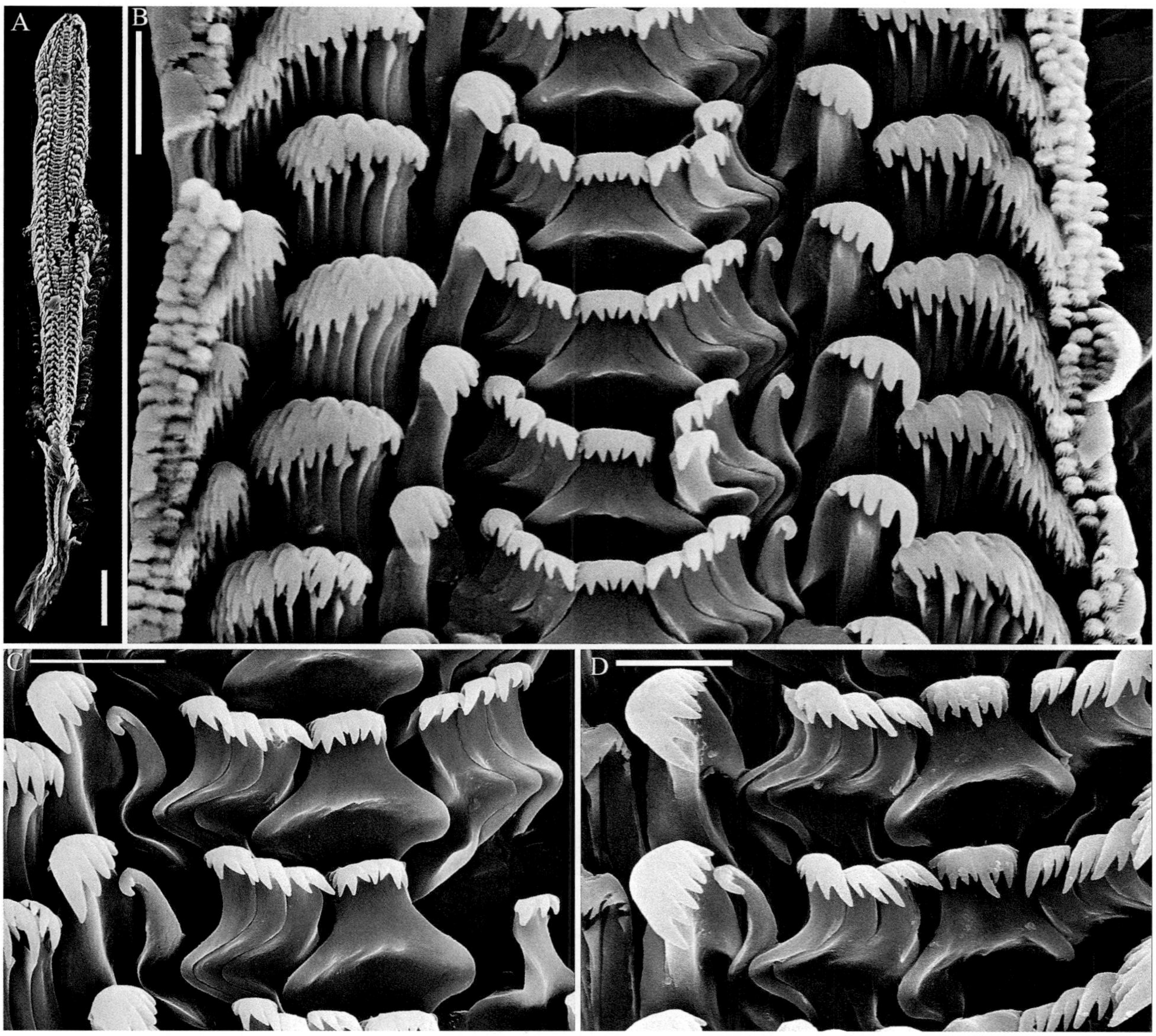

Figure 334. *Sinezona ferriezi.* Radula. **A.–C.** 3–7 m, Aoki Auki Harbour, W coast Malaita Island, Solomon Islands, 8.767°S, 160.700°E (AMS C.379954). **A.** Entire radula. **B.** Full width of radula. **C.** Central field Enlarged. **D.** Central field. Noumea, 22.267°S, 166.450°E (AMS C.4287, 1). Topotype of *Sin. ferriezi.* Shell shown in Figure 332A. Scale bar A = 100 µm. Scale bars B–D = 10 µm.

cusps on outer margin of tip. Lateral tooth 4 reduced, hook-shaped, with minute cusps on either side of tip. Lateral tooth 5 broadened, enlarged, tip with 6–7 cusps, decreasing in size from tip towards base. Inner marginal teeth with cusps on tip stronger developed on outer margin. Outer marginal teeth with small tip with many minute cusps. Radular interlock of central field moderate.

Comparisons. *Sinezona costulata* from southern Japan has fewer axial cords on the teleoconch, axial sculpture on the protoconch, a much shorter selenizone (0.4 vs. 1.6 whorls), and a more narrow umbilicus.

Sinezona danieldreieri from the Indo-Malayan Archipelago to the Central Pacific has a much shorter selenizone (0.3 vs. 1.6 whorls), a weaker constriction of the base below the selenizone, and a protoconch with axial sculpture.

Sinezona globosa from the Indo-Malayan Archipelago to tropical Eastern Pacific (and possibly Caribbean) is overall more globular, has spiral lines on the protoconch, and the base gradually slopes into the umbilicus.

Sinezona plicata from the Indo-Malayan Archipelago to the Central Pacific has a protoconch with a single spiral ridge, although it is often eroded, and has much stronger, ridge-like to folded axial sculpture.

Sinezona macleani from Indian Ocean to the Western Pacific has a more delicate shell, is overall more globular-depressed, has a shorter teleoconch I (0.6 vs. 1.6 whorls), and a protoconch with flocculent sculpture and lacking an apertural varix.

Sinezona marrowi from the Indo-Malayan Archipelago has a shorter selenizone (0.6 vs. 1.6 whorls), more distinct axial cords with fewer spiral lines on shoulder and base, and a protoconch with axial sculpture.

Sinezona milleri from southern Japan has a much shorter selenizone (0.4 vs. 1.6 whorls), a more narrow umbilicus, and strong axial sculpture on the protoconch.

Sinezona singeri from the Indian Ocean has about half as many spiral lines on shoulder and base, stronger axial cords, and stronger spiral cords in the vicinity of the umbilicus.

Sinezona wileyi from the Indo-Malayan Archipelago to the Central Pacific is much smaller (1.1 vs. 2.1 mm), has a subquadrate profile, and the protoconch has fine irregular axials and lacks an apertural varix.

Compared to *Sci. mirifica*, juvenile *Sin. ferriezi* with an open slit are angulated at the selenizone, *i.e.*, shoulder and base do not form a continuous arc as in *Sci. mirifica*. Also, the outline of the whorl is more triangular and less round in *Sin. ferriezi* compared to *Sci. mirifica*.

Sukashitrochus species have a similar overall shape and share the closed foramen with *Sin. ferriezi*, but all have a more or less developed spiral keel on the base.

Distribution. Indo-Malayan Archipelago to Central Pacific (Fig. 335).

Specimen records.

Singapore. (KBIN, 1).

Japan, Okinawa. 46–55 m, 1 km WNW of Onna Village (Horseshoe Cliffs), 26.493°N, 127.842°E (LACM 79-75, 2). 35–38 m, base of cliff, Horseshoe, 26.504°N, 127.858°E (DLG 976, 2). 38 m, base of cliff, Horeshoe Bay, 26.504°N, 127.858°E (DLG 1057, 4). 1 m, 1 km NNW Oku, 26.847°N, 128.287°E (LACM 77-61, 1). 13–20 m, Kadena Steps, 26.360°N, 127.739°E (DLG 1447, 1). 43 m, Horseshoe Beach, 26.505°N, 127.853°E (DLG 1549, 1). 37 m, Big Rock, Horseshoe Bay, 26.508°N, 127.853°E (DLG 1458, 4). 13–33 m, Flab, Horseshoe Bay, 26.509°N, 127.856°E (DLG 1490, 1).

Indonesia. 175–185 m, Bay of Sanggar, N of Sumbawa, 8.322°S, 118.260°E (RMNH, 1). 300–340 m, SW Salayer, 6.467°S, 120.400°E (RMNH, 1). 233–274 m, SW Salayer, 6.372°S, 120.433°E (RMNH, 2). 1 m, Java, Pelangi and Putri Islets, 5.600°S, 106.583°E

(LACM 86-162, 1). Ceram, Piru Bay, 3.167°S, 128.133°E (AMS C.379952, 1). 0 m, E side Auki Islet, SE of Biak Island, Irian Jaya, 1.230°S, 136.337°E (LACM 88-48, 1). 0 m, Biak, Irian Jaya, 1.150°S, 136.050°E (LACM 88-43, 1). 200 m, Makassar, Sulawesi, 1.950°S, 119.267°E (MNHN, 1).

Malaysia. W coat of Pulau Tioman, MAS-16, 2.809°N, 104.141°E (ZSM Mol-20060145, 3: complete). Sabah, Kota Kinabalu, 6.000°N, 116.000°E (AMS C.348631, 1).

Philippines. 49 m, St. 5131, gn. M. co. S, off Panabutan Point, W Mindanao, 7.586°N, 122.121°E (USNM 262912, 1). 90–110 m, Balicasag Island, 9.515°N, 123.687°E (MNHN, 1: complete). 259–280 m, Dipolog Bay, Bohol-Sulu sill, PANGLAO 2005 CP2381, 8.721°N, 123.316°E (MNHN, 1). 259–280 m, Dipolog Bay, Bohol-Sulu Sea sill, PANGLAO 2005 CP2381, 8.721°N, 123.316°E (MNHN, 1). 20 m, Looc, Panglao Island, PANGLAO 2004 B32, 9.430°N, 123.743°E (MNHN, 3). 6–14 m, Pamilacan Island, PANGLAO 2004 S10, 9.490°N, 123.933°E (MNHN, 2). 90–98 m, Balicasag Island, PANGLAO 2004 L69–73, 9.511°N, 123.682°E (MNHN, 1). 120–139 m, Balicasag Island, PANGLAO 2004 L74–75, 9.511°N, 123.682°E (MNHN, 2). 40 m, Balicasag Island, PANGLAO 2004 S44, 9.515°N, 123.680°E (MNHN, 1). 17–19 m, Balicasag Island, PANGLAO 2004 B41, 9.515°N, 123.680°E (MNHN, 1). 14 m, Balicasag Island, PANGLAO 2004 S46, 9.515°N, 123.680°E (MNHN, 1). 17 m, Balicasag Island, PANGLAO 2004 S45, 9.515°N, 123.680°E (MNHN, 1). 90–110 m, Balicasag Island, PANGLAO 2004 L46, 9.515°N, 123.686°E (MNHN, 1). 6 m, Black Forest, Balicasag Island, PANGLAO 2004 S3, 9.518°N, 123.688°E (MNHN, 1). 164–176 m, off Balicasag Island, Bohol Sea, PANGLAO 2005 DW2339, 9.531°N, 123.728°E (MNHN, 1). 152 m, Bolod, Panglao Island, PANGLAO 2004 T2, 9.540°N, 123.796°E (MNHN, 2). 713–731 m, off Balicasag Island, Bohol Sea, PANGLAO 2005 CP2398, 9.543°S, 123.675°E (MNHN, 1). 17–25 m, Pntod Lagoon 1, Panglao Island, PANGLAO 2004 B39, 9.546°N, 123.701°E (MNHN, 1). 82 m, Bolod, Panglao Island,

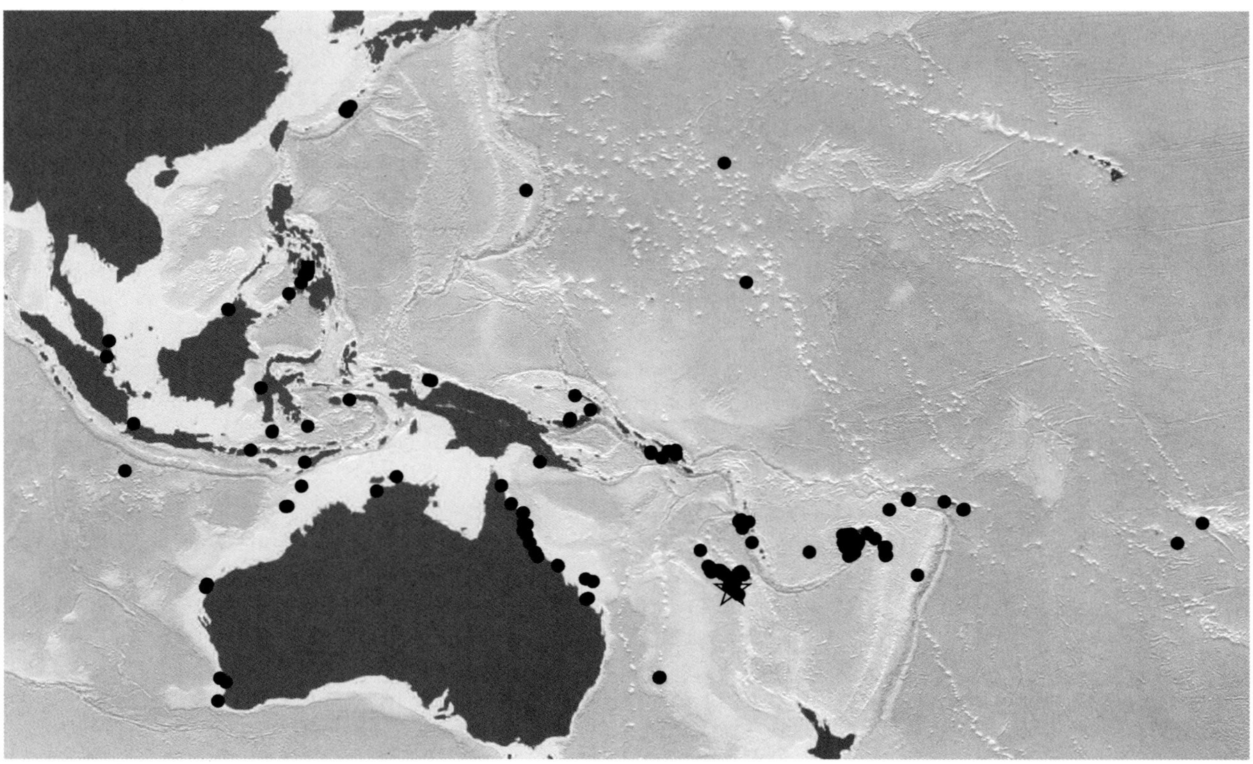

Figure 335. Distribution of *Sinezona ferriezi*. ☆ type locality. ● specimen record. ■ literature record.

PANGLAO 2004 T4, 9.550°N, 123.808°E (MNHN, 2). 8–10 m, Napaling, Panglao Island, PANGLAO 2004 B9, 9.551°N, 123.733°E (MNHN, 2). 4 m, Biking, Panglao Island, PANGLAO 2004 B5, 9.586°N, 123.840°E (MNHN, 1). 5 m, Biking, Panglao Island, PANGLAO 2004 S1, 9.588°N, 123.841°E (MNHN, 2; MNHN, 23). 31 m, N of Doljo, Panglao Island, PANGLAO 2004 B35, 9.598°N, 123.741°E (MNHN, 1). 4–30 m, Tatarman, Panglao Island, PANGLAO 2004 B7, 9.598°N, 123.863°E (MNHN, 5). 90–200 m, Maribohoc Bay, Bohol Island, PANGLAO 2004 P1, 9.600°N, 123.750°E (MNHN, 1). 3–14 m, Momo Beach, Banglao Island, PANGLAO 2004 B10, 9.608°N, 123.760°E (MNHN, 6). 28–32 m, Momo Beach, Panglao Island, PANGLAO 2004 S8, 9.608°N, 123.760°E (MNHN, 16). 30–33 m, between Momo and Napaling, Panglao Island, PANGLAO 2004 B42, 9.616°N, 123.766°E (MNHN, 1). 2–4 m, Napaling, Panglao Island, PANGLAO 2004 S5, 9.618°N, 123.768°E (MNHN, 2). Napaling, Panglao Island, PANGLAO 2004 B8, 9.618°N, 123.769°E (MNHN, 2). 8–15 m, Baclayon Takot, Bohol Island, PANGLAO 2004 S13, 9.618°N, 123.876°E (MNHN, 16). 3–5 m, Baclayon Takot, Bohol Island, PANGLAO 2004 B13, 9.618°N, 123.876°E (MNHN, 1). 28–32 m, Napaling, Panglao Island, PANGLAO 2004, 9.620°N, 123.773°E (MNHN, 4). 20–21 m, Napaling, Panglao Island, PANGLAO 2004 B21, 9.620°N, 123.773°E (MNHN, 1). 4–5 m, Baclayon, Bohol Island, PANGLAO 2004 S2, 9.623°N, 123.908°E (MNHN, 1). 3–21 m, Bingag, Panglao Island, PANGLAO 2004 B17, 9.625°N, 123.781°E (MNHN, 1). 62 m, Bingag/Tabalong, Panglao Island, PANGLAO 2004 L51–60, 9.628°N, 123.800°E (MNHN, 2). 2–4 m, Sungcolan, Panglao Island, PANGLAO 2004 B15, 9.646°N, 123.818°E (MNHN, 1). 78–95 m, Maribohoc Bay, Bohol Island, PANGLAO 2004 T11, 9.681°N, 123.833°E (MNHN, 2). 12 m, Ubajan, Bohol Island, PANGLAO 2004 S27, 9.691°N, 123.850°E (MNHN, 1). 21 m, Ubajan, Bohol Island, PANGLAO 2004 S25, 9.691°N, 123.850°E (MNHN, 1). 4–12 m, Manga, Bohol Island, PANGLAO 2004, 9.695°N, 123.848°E (MNHN, 6). 10–26 m, Cortes, Bohol Island, PANGLAO 2004 T19, 9.703°N, 123.846°E (MNHN, 2).

Papua New Guinea. 13–18 m, SW of Port Moresby, off W side of Lolorua Island, 9.500°S, 147.108°E (AMS C.379953, 1). New Britain, 4.200°S, 152.183°E (AMS C.45556, 1). 54 m, off Schauman Island, Kimbe Bay, New Britain, 5.307°S, 150.092°E (DLG 443, 1). 40 m, off Restorf Island, Kimbe Bay, 5.292°S, 150.101°E (DLG 396, 2). 65 m, off Susan's Reef, Kimbe Bay, New Britain, 5.291°S, 150.128°E (DLG 609, 8). 40 m, off Anne Sophie's Reef, Kimbe Bay, New Britain, 5.091°S, 150.186°E (DLG 444, 1). 15–35 m, Baudisson Bay, Kavieng, New Ireland, 2.743°S, 150.658°E (DLG 302, 5).

Solomon Islands. 3–7 m, Aoki Auki Harbour, W coast Malaita Island, 8.767°S, 160.700°E (AMS C.379954, 2). Vangunu Island, Karunjou Island, 8.650°S, 158.233°E (LACM 89-76, 1). 20–25 m, Matui Island, 8.495°S, 158.200°E (LACM, 1). 98–200 m, SALOMON 1 DW1767, 8.323°S, 160.665°E (MNHN, 1).

Australia, Western Australia. Kilcarnup, N side of Margaret River, 33.950°S, 114.983°E (AMS C.406358, 2). Minim Cove, Mosman Park, Swan River., 32.017°S, 115.767°E (AMS C.379070, 1). 110 m, off Rottnest Island, 31.630°S, 115.178°E (AMS C.378549, 3). 2 m, Ningaloo Reef, off Ned's Camp, 21.992°S, 113.908°E (AMS C.377284, 2). 20 m, E side, Sandy Islet, Scott Reef, 14.044°S, 121.780°E (SBMNH 369527, 1). 40 m, off North West Cape, 22.313°S, 113.765°E (AMS C.348610, 1).

Australia, Northern Territories. Oxley Island, W side, 11.000°S, 132.817°E (AMS C.404501, 1: complete). Darwin, 12.446°S, 130.829°E (RMNH AD ex 32453, 11).

Australia, Queensland. 31–37 m, Great Barrier Reef, Capricorn Group, Masthead Island, 23.533°S, 151.750°E (AMS C.019326, 20+). 4–5 m, N side of Wistari Reef, Capricorn Group, 23.467°S, 151.883°E (AMS, 1: complete). 6 m, Great Barrier Reef, Capricorn Group, Heron Island, 23.433°S, 151.950°E (AMS C.379458, 1). 36 m, Great Barrier Reef, Capricorn Group, Heron Island, 23.433°S, 151.950°E (AMS C.379460, 2). Great Barrier Reef, Capricorn Group, Heron Island, 23.433°S, 151.950°E (AMS C.379463, 4). 15 m, Great Barrier Reef, Capricorn Group, Heron Island, 23.433°S, 151.950°E

(AMS C.379468, 1). 3 m, Great Barrier Reef, Capricorn Group, SE of Heron Island, 23.433°S, 151.950°E (AMS C.379948, 4). 6 m, Great Barrier Reef, Capricorn Group, Heron Island, 23.433°S, 151.950°E (AMS C.348618, 1). 64–73 m, Great Barrier Reef, Swain Reefs, 3 km NE of W side of Bylund Gillett Cay, 21.700°S, 152.433°E (AMS C.379457, 1). 11–13 m, Great Barrier Reef, Swain's Reef, Reef 21-468, NW side, 21.500°S, 152.417°E (AMS C.377698, 5). 3 m, Great Barrier Reef, Swain Reefs, Reef 21-189, 21.450°S, 151.683°E (AMS C.348609, 1). 30 m, Whisundays Islands, Hook Island, 20.114°S, 148.956°E (SBMNH, 2). Magnetic Island, Geoffrey Bay, 19.153°S, 146.863°E (AMS C.379951, 3). 6 m, North East Bay, Great Palm Island, 18.733°S, 146.650°E (AMS C.379472, 1). Kurramine Beach, 17.767°S, 146.100°E (WAM ex S13377, 4). Yorkeys Knob, S end Halfmoon Bay, N of Cairns, 16.800°S, 145.717°E (AMS C.379476, 1). Alexander Reef, Port Douglas, 16.533°S, 145.483°E (WAM S13376, 7). 38 m, Outer Slope, S part Escape Reef, Great Barrier Reef, 15.883°S, 145.817°E (AMS ex C.461466, 1). 9–18 m, Great Barrier Reef, Cairns Reef lagoon and between Cairns Reef and Hope Island, 15.700°S, 145.500°E (AMS C.27189, 2). 9–12 m, Great Barrier Reef, Lizard Island, South Island, SW side, 14.700°S, 145.450°E (AMS C.379459, 4). 10 m, Great Barrier Reef, Lizard Island, N end South Island, 14.700°S, 145.450°E (AMS C.379462, 2). 15 m, Great Barrier Reef, Lizard Island, S end of South Island, 14.700°S, 145.450°E (AMS C.379947, 2). Great Barrier Reef, Lizard Island, N side of Palfrey Island, 14.683°S, 145.450°E (AMS C.379465, 4). 10 m, SE side Lizard Island, 14.683°S, 145.458°E (LACM 79-55, 8). 0 m, Hoalfrey Island, Lizard Island, 14.683°S, 145.450°E (AMS, 3: complete). 0–3 m, Great Barrier Reef, Lizard Island, off Osprey Island, 14.667°S, 145.433°E (AMS C.404495, 3: complete). Kurrimine Beach, 17.778°S, 146.108°E (MV F 160820, 1). 11–18 m, Cairns reef, 16.905°S, 145.778°E (MV F 161673, 1). 11–18 m, off Cairns Reef, 16.905°S, 145.778°E (MV ex F 161668, 1; MV ex F 161668, 3). Great Barrier Reef, Lizard Island, N side of Palfrey Island, 14.683°S, 145.450°E (AMS C.379465, 4, 3: complete). 1–6 m, NE corner, of South Island, Sir Charles Hardy Islands, 11.917°S, 143.283°E (AMS C.468102, 1). Great Barrier Reef, Low Isles, 16.383°S, 145.567°E (AMS ex C.348616, 1). Great Barrier Reef, Whitsunday Islands, Hayman Island, 20.060°S, 148.892°E (AMS C.348617, 2). Great Barrier Reef, Capricorn Group, Heron Island, 23.433°S, 151.950°E (AMS C.406379, 3). 8–10 m, Great Barrier Reef, No. 5 Sandbank reef 13-120, SW end, 13.750°S, 144.267°E (AMS C.402652, 1). Great Barrier Reef, Swain Reefs, Bylund Gillett Cay, 21.717°S, 152.417°E (AMS C.348613, 1). Great Barrier Reef, Lizard Island, 14.667°S, 145.467°E (AMS C.348614, 1). Great Barrier Reef, Low Isles, 16.383°S, 145.567°E (AMS C.404971, 1: complete). 0 m, 4 mi Beach, Port Douglas, 16.492°S, 145.468°E (MV F178692, 1).

Australia, Lord Howe. 34 m, Lord Howe Shelf, 31.564°S, 159.012°E (AMS C.461433, 1). Signal Point, 31.525°S, 159.065°E (AMS C.404255, 3: complete; AMS C.404256, 1: complete).

Australia, Norfolk Islands. Cemetery Beach, Point Hunter Reserve, 29.067°S, 167.967°E (AMS C.404254, 15: complete).

New Caledonia. Noumea, 22.267°S, 166.450°E (MNHN, 2). 390–400 m, North, CONCALIS DW3001, 18.536°S, 163.148°E (MNHN, 1). 3–9 m, NW end Ile Neba, off Poum, 20.150°S, 163.933°E (LACM 86-366, 1). 1–4 m, Reef Bay, Poum, 20.217°S, 164.050°E (AMS C.379950, 1). 0 m, Koumac sector, 20.373°S, 164.113°E (MNHN, 13; MNHN, 3). 0 m, Koumac sector, 20.486°S, 164.170°E (MNHN, 1). 0 m, Koumac sector, 20.486°S, 164.170°E (MNHN, 45). 12 m, Koumac sector, 20.561°S, 164.172°E (MNHN, 1). 0–2 m, Koumac sector, 20.566°S, 164.267°E (MNHN, 10, 4: complete). 0–2 m, Koumac sector, 20.566°S, 164.317°E (MNHN, 2). 491–533 m, E coast, 20.566°S, 164.950°E (MNHN, 4). 497–520 m, E coast, 20.566°S, 164.950°E (MNHN, 25). 570–650 m, E coast, 20.566°S, 164.967°E (MNHN, 7). 12–14 m, Koumac sector, 20.573°S, 164.217°E (MNHN, 5). 3–7 m, Koumac sector, 20.575°S, 164.258°E (MNHN, 1). 0 m, Koumac sector, 20.583°S, 164.258°E (MNHN, 22). 410–430 m, E

coast, 20.583°S, 164.967°E (MNHN, 2). 308–324 m, E coast, BATHUS 1 DE693, 20.583°S, 164.983°E (MNHN, 12). 408–440 m, E coast, 20.583°S, 165.117°E (MNHN, 12). 15–20 m, Touho sector, 20.585°S, 165.135°E (MNHN, 20). 2–4 m, Koumac sector, 20.586°S, 164.277°E (MNHN, 21, 4: complete). 10–11 m, Koumac sector, 20.593°S, 164.253°E (MNHN, 2). Beach at E side of Mount Panie, 20.600°S, 164.833°E (AMS ex C.404431, 1). 400–500 m, E coast, BATHUS 1 DE694, 20.600°S, 164.967°E (MNHN, 8). 0–8 m, Koumac sector, 20.629°S, 164.275°E (MNHN, 2, 2: complete). 0–8 m, Koumac sector, 20.637°S, 164.275°E (MNHN, 8). 33–38 m, Koumac sector, 20.646°S, 164.260°E (MNHN, 6). 11–13 m, Koumac sector, 20.651°S, 164.207°E (MNHN, 2). 0 m, Koumac sector, 20.653°S, 164.170°E (MNHN, 1). 10–15 m, Touho sector, 20.660°S, 165.245°E (MNHN, 1). 10–15 m, Touho sector, 20.660°S, 165.245°E (MNHN, 8). 10–15 m, Touho sector, 20.660°S, 165.245°E (MNHN, 2: complete). 15 m, Koumac sector, 20.661°S, 164.248°E (MNHN, 9; MNHN, 14). 15–20 m, Koumac sector, 20.666°S, 164.253°E (MNHN, 5). 55–57 m, Koumac sector, 20.671°S, 164.195°E (MNHN, 2). 55–57 m, Koumac sector, 20.671°S, 164.195°E (MNHN, 19). 55–57 m, Koumac sector, 20.671°S, 164.195°E (MNHN, 2). 10–60 m, Koumac sector, 20.673°S, 164.248°E (MNHN, 1). 26–40 m, Koumac sector, 20.673°S, 164.248°E (MNHN, 17). 20–30 m, Koumac sector, 20.690°S, 164.247°E (MNHN, 7). 0–1 m, Doking, Lifou Island, Loyalty Islands, 20.700°S, 167.167°E (AMS C.402654, 3). 0 m, Touho sector, 20.713°S, 165.145°E (MNHN, 5). 11 m, Touho sector, 20.716°S, 165.133°E (MNHN, 1). 11 m, Touho sector, 20.716°S, 165.133°E (MNHN, 2: complete). 25–28 m, Touho sector, 20.733°S, 165.183°E (MNHN, 1). 49–59 m, Touho sector, 20.733°S, 165.233°E (MNHN, 25). 8 m, Touho sector, 20.741°S, 165.265°E (MNHN, 50). 15–35 m, Touho sector, 20.743°S, 165.228°E (MNHN, 20). 90–115 m, Koumac sector, 20.745°S, 164.248°E (MNHN, 1). 15–20 m, Koumac sector, 20.745°S, 164.258°E (MNHN, 12). 15–20 m, Touho sector, 20.750°S, 165.163°E (MNHN, 1). 10–35 m, Touho sector, 20.750°S, 165.275°E (MNHN, 8). 53–71 m, Koumac sector, 20.753°S, 164.253°E (MNHN, 2). 0 m, Touho sector, 20.753°S, 165.272°E (MNHN, 12). 27–31 m, Santal Bay, Lifou, 20.763°S, 167.028°E (MNHN, 6). 17 m, Santal Bay, Lifou, 20.763°S, 167.028°E (MNHN, 4). 20 m, Santal Bay, Lifou, 20.763°S, 167.028°E (MNHN, 2). 17 m, Santal Bay, Lifou, 20.763°S, 167.028°E (MNHN, 2). 27–31 m, Santal Bay, Lifou, 20.763°S, 167.028°E (MNHN, 1). 4–7 m, Santal Bay, Lifou, 20.765°S, 167.103°E (MNHN, 1). 7 m, Touho sector, 20.768°S, 165.243°E (MNHN, 2). 0 m, Touho sector, 20.770°S, 165.242°E (MNHN, 20; MNHN, 4: complete). 6–15 m, Touho sector, 20.770°S, 165.229°E (MNHN, 17). 47 m, Santal Bay, Lifou, 20.773°S, 167.033°E (MNHN, 8). 20 m, Santal Bay, Lifou, 20.773°S, 167.033°E (MNHN, 1). 0–2 m, Touho sector, 20.775°S, 165.242°E (MNHN, 2). 0–2 m, Touho sector, 20.775°S, 165.250°E (MNHN, 39). 45–56 m, Touho sector, 20.775°S, 165.263°E (MNHN, 11). 2–3 m, Touho sector, 20.776°S, 165.235°E (MNHN, 5). 3–6 m, Touho sector, 20.778°S, 165.228°E (MNHN, 20). 2–4 m, Santal Bay, Lifou, 20.778°S, 167.052°E (MNHN, 1). 5–10 m, Santal Bay, Lifou, 20.780°S, 167.046°E (MNHN, 3). 5–10 m, Santal Bay, Lifou, 20.780°S, 167.046°E (MNHN, 1: complete). 0 m, Santal Bay, Lifou, 20.780°S, 167.129°E (MNHN, 4). 0–1 m, Touho sector, 20.781°S, 165.230°E (MNHN, 3). 1–5 m, Santal Bay, Lifou, 20.781°S, 167.132°E (MNHN, 2). 55–80 m, Santal Bay, Lifou, 20.783°S, 167.050°E (MNHN, 2). 4 m, Santal Bay, Lifou, 20.785°S, 167.123°E (MNHN, 2). 3–7 m, Santal Bay, Lifou, 20.785°S, 167.152°E (MNHN, 1). 10–21 m, Santal Bay, Lifou, 20.788°S, 167.113°E (MNHN, 17). 20–25 m, Santal Bay, Lifou, 20.791°S, 167.118°E (MNHN, 11). 8–18 m, Santal Bay, Lifou, 20.791°S, 167.118°E (MNHN, 1). 20–25 m, Santal Bay, Lifou, 20.791°S, 167.118°E (MNHN, 3). 10 m, Santal Bay, Lifou, 20.793°S, 167.170°E (MNHN, 1). 4–5 m, Santal Bay, Lifou, 20.795°S, 167.156°E (MNHN, 2). 4–5 m, Santal Bay, Lifou, 20.795°S, 167.156°E (MNHN, 1). 170 m, E coast, 20.800°S, 165.317°E (MNHN, 3). 105–110 m, E coast, 20.816°S, 165.317°E (MNHN, 1). 105–110 m, Poindimie sector, 20.816°S, 165.317°E (MNHN, 2). 96–109 m, E coast,

20.816°S, 165.317°E (MNHN, 1). 80–140 m, Touho sector, 20.816°S, 165.317°E (MNHN, 3). 25–30 m, Santal Bay, Lifou, 20.821°S, 167.173°E (MNHN, 11). 25–30 m, Santal Bay, Lifou, 20.821°S, 167.173°E (MNHN, 7). 10 m, Touho sector, 20.825°S, 165.327°E (MNHN, 7). 20 m, Touho sector, 20.840°S, 165.380°E (MNHN, 9). 450–470 m, E coast, 20.850°S, 165.467°E (MNHN, 1). 4 m, Santal Bay, Lifou, 20.873°S, 167.142°E (MNHN, 2; MNHN, 3; MNHN, 35). 5–20 m, Santal Bay, Lifou, 20.875°S, 167.135°E (MNHN, 2; MNHN, 1: complete). 5–25 m, Touho sector, 20.878°S, 165.325°E (MNHN, 30). 12–32 m, Santal Bay, Lifou, 20.891°S, 167.045°E (MNHN, 30). 48–52 m, Santal Bay, Lifou, 20.896°S, 167.122°E (MNHN, 13). 150–200 m, Shelter Reef, Santal Bay, Lifou, Loyalty Islands, LIFOU DW1648, 20.900°S, 167.050°E (MNHN, 3). 12 m, Santal Bay, Lifou, 20.900°S, 167.122°E (MNHN, 13). Lifou, 20.900°S, 167.200°E (KBIN, 23). 150–200 m, Pointe Lefèvre, Baie du Santal, Lifou, 20.903°S, 167.018°E (MNHN, 2). 2–5 m, Santal Bay, Lifou, 20.903°S, 167.123°E (MNHN, 1). 21–30 m, Santal Bay, Lifou, 20.910°S, 167.035°E (MNHN, 2). 9–20 m, Santal Bay, Lifou, 20.916°S, 167.087°E (MNHN, 11). 5–30 m, Santal Bay, Lifou, 20.920°S, 167.012°E (MNHN, 4). 5 m, Santal Bay, Lifou, 20.926°S, 167.075°E (MNHN, 1). 15–18 m, Santal Bay, Lifou, 20.944°S, 167.033°E (MNHN, 1; MNHN, 5; MNHN, 1). 2–4 m, Santal Bay, Lifou, 20.945°S, 167.052°E (MNHN, 5). 105–110 m, E coast, 20.950°N, 165.567°E (MNHN, 6). 160–222 m, E coast, 20.950°S, 165.583°E (MNHN, 15). Loyalty Islands, SE side Lifou Island, Luengoni, 21.033°S, 167.417°E (AMS C.348627, 1). 350–400 m, E coast, 21.033°S, 165.633°E (MNHN, 20). 237–298 m, E coast, 21.283°S, 165.950°E (MNHN, 20; MNHN, 1; MNHN, 1). 21.483°S, 166.333°E (MNHN, 2). 310–315 m, Loyalty Basin, 21.533°S, 166.483°E (MNHN, 3). 500 m, Loyalty Basin, 21.550°S, 166.400°E (MNHN, 8; MNHN, 2). 250 m, E coast, 21.750°S, 166.617°E (MNHN, 1). 1060–1450 m, South, BATHUS 2 CP767, 22.183°S, 165.983°E (MNHN, 2). 455 m, Southern, 22.266°S, 166.183°E (MNHN, 20; MNHN, 2). 490–500 m, Southern, 22.266°S, 166.183°E (MNHN, 10). Noumea, 22.266°S, 166.450°E (NMSZ 1961.61, 1; AMS C.402649, 1; AMS C.3773, 1; AMS C.4287, 1). 15 m, off Nouville, Noumea sector, St. 1345, 22.268°S, 166.395°E (MNHN, 1). Noumea, 22.296°S, 166.434°E (KBIN, 4). 0 m, Noumea, Platier Ilot Maitre, 22.340°S, 166.428°E (MNHN, 20). 6–8 m, Ilôt Maître, Noumea sector, St. 1362, 22.348°S, 166.418°E (MNHN, 1). 12–37 m, Aboré Grand Reef, Mouméa Lagoon, 22.371°S, 166.265°E (MNHN, 6: complete). 577–780 m, south, BATHUS 2 DW754, 22.383°S, 166.217°E (MNHN, 60). 10 m, Goeland Islet, Noumea sector, St. 1367, 22.405°S, 166.345°E (MNHN, 5). Ile Amedee, 22.483°S, 166.467°E (AMS C.28771, 1). 250–350 m, southern, 22.500°S, 166.400°E (MNHN, 4; MNHN, 100). 465–525 m, South, BATHUS 2 DW739, 22.583°S, 166.450°E (MNHN, 45). 713–950 m, south, BATHUS 2 CP743, 22.600°S, 166.433°E (MNHN, 200). 570 m, Loyalty Basin, 22.650°S, 166.567°E (MNHN, 3). 400–402 m, Norfolk Ridge, 23.016°S, 166.933°E (MNHN, 2). 397–400 m, south, BATUHUS 2 DW730, 23.050°S, 166.967°E (MNHN, 1).

Vanuatu. 7 m, W Mavéa Island, SANTO 2006 DB67, 15.381°S, 167.218°E (MNHN, 1). 102–120 m, W Mavéa Island, Palikulo Bay, SANTO 2006 AT14, 15.397°S, 167.220°E (MNHN, 3). 6–43 m, Aésé Island, SANTO 2006 DB58, 15.410°S, 167.238°E (MNHN, 3). 21 m, SE Aésé Island, SANTO 2006 DB63, 15.448°S, 167.263°E (MNHN, 2). 22–25 m, NW Urélapa Island, SANTO 2006 DB20, 15.508°S, 167.023°E (MNHN, 3). 10–14 m, Segond Channel, vicinity of Luganville, SANTO 2006 DB14, 15.515°S, 167.183°E (MNHN, 1). 7 m, Segond Channel, E Sarakata river mouth, SANTO 2006 LS17, 15.518°S, 167.175°E (MNHN, 1). 2–3 m, Segond Channel, vicinity of Maritime College, SANTO 2006 NS37, 15.523°S, 167.163°E (MNHN, 1). 25 m, Segond Channel, *Collidge* wreck, SANTO 2006 DS04, 15.523°S, 167.235°E (MNHN, 3). 17–19 m, N Tutuba Island, SANTO 2006 DS101, 15.538°S, 167.287°E (MNHN, 7). 100–105 m, NW Tutuba Island, SANTO 2006 DS99, 15.541°S, 167.282°E (MNHN, 8). 8–20 m, Tutuba Island, SANTO 2006 FB83, 15.543°S, 167.290°E (MNHN,

1). 97–101 m, E Aoré island, SANTO 2006 AT04, 15.550°S, 167.225°E (MNHN, 6). MUSORSTOM 8 DW1069, 15.550°S, 167.233°E (MNHN, 1). 155 m, Tutuba Island, SANTO 2006 EP27, 15.550°S, 167.267°E (MNHN, 1). 92 m, NW Tutuba Island, SANTO 2006 DS105, 15.550°S, 167.278°E (MNHN, 3). 100 m, NW Tutuba Island, SANTO 2006 DS108, 15.553°S, 167.277°E (MNHN, 1). 100 m, W Tutuba Island, Aoré Island, SANTO 2006 DS93, 15.560°S, 167.275°E (MNHN, 2). 8–9 m, Tutuba Island, SANTO 2006 FS84, 15.560°S, 167.277°E (MNHN, 2). 7 m, Segond Channel, Wambu river mouth, SANTO 2006 DS91, 15.561°S, 167.140°E (MNHN, 1). 70–80 m, W Tutuba Island, SANTO 2006 DS103, 15.568°S, 167.267°E (MNHN, 5). 98–100 m, W Tutuba Island, SANTO 2006 DS102, 15.568°S, 167.267°E (MNHN, 3). 80 m, W Tutuba Island, SANTO 2006 DS104, 15.568°S, 167.267°E (MNHN, 10). 37 m, Tangoa Island, SANTO 2006 FS70, 15.590°S, 166.995°E (MNHN, 2). 32–40 m, S Tutuba Island, SANTO 2006 DB16, 15.591°S, 167.263°E (MNHN, 1). 16 m, Elia Island, SANTO 2006 FB72, 15.601°S, 167.975°E (MNHN, 1). 10–18 m, S Aoré Island, SANTO 2006 DB12, 15.610°S, 167.168°E (MNHN, 2). 6–24 m, S Aoré Island, SANTO 2006 DS10, 15.610°S, 167.168°E (MNHN, 4). 78–91 m, NW coast of Malo Island, SANTO 2006 EP22, 15.622°S, 167.098°E (MNHN, 1). 10 m, Bruat Channel, N coast of Malo Island, SANTO 2006 DB25, 15.628°S, 167.188°E (MNHN, 1). 5–7 m, W coast of Malo Island, SANTO 2006 ZB09, 15.676°S, 167.085°E (MNHN, 3). 7 m, Malokilikili, Aoré Island, SANTO 2006 FB52, 15.711°S, 167.235°E (MNHN, 1). 360–419 m, MUSORSTOM 8 DW1065, 16.266°S, 167.350°E (MNHN, 2). Efate Island, Port Vila, Point Ardel Malapoa, 17.733°S, 168.300°E (AMS C.379955, 2, -: complete).

Fiji. 102–106 m, Bligh Water, 147.321°S, 177.790°E (MNHN, 7). 600–660 m, 16.816°S, 179.900°E (MNHN, 1). 200–215 m, Bligh Water, 16.840°S, 178.208°E (MNHN, 5). 251–257 m, Bligh Water, 16.856°S, 178.232°E (MNHN, 2). 500–614 m, Bligh Water, MUSORSTOM 10 CP1341, 16.875°S, 177.728°E (MNHN, 1). 0 m, Yasawa-i-rara, Tavewa Island, Yasawa Group, 16.933°S, 177.367°E (AMS C.461461, 1). 660–663 m, Bligh Water, 17.248°S, 178.492°E (MNHN, 2). 656–660 m, Bligh Water, 17.268°S, 178.247°E (MNHN, 12). 282–322 m, Bligh Water, 17.273°S, 177.830°E (MNHN, 5). 820–863 m, BORDAU 1 DW1459, 17.300°S, 179.550°W (MNHN, 30). 102–106 m, Bligh Water MUSORSTOM 10 DW1329, 17.321°S, 177.790°E (MNHN, 11). 879–897 m, SE of Viti Levu, 17.515°S, 178.888°E (MNHN, 50). 959–963 m, SE of Viti Levu, 17.710°S, 178.917°E (MNHN, 8). 9–35 m, Viti Levu, Nadi Bay Tomba Ko Nandi, FIJI, 17.733°S, 177.417°E (AMS C.348625, 4). 80–120 m, MUSORSTOM 10 CP1358, 17.808°S, 178.778°E (MNHN, 3). 183–188 m, SE Viti Levu, 17.828°S, 178.797°E (MNHN, 5). 402–444 m, SE of Viti Levu, MUSORSTOM 10 CP1360, 17.993°S, 178.803°E (MNHN, 3). 0–1 m, Nukumbotho channel W side of Nukumbutho Island, Laucala Bay, Viti Levu, 18.181°S, 178.466°W (ZSM 20070153, 1). Sovi Bay Souvi, SW coast Viti Levu, FIJI, 18.183°S, 177.600°E (AMS C.406356, 8). 392–433 m, S of Viti Levu, 18.185°S, 178.390°E (MNHN, 15). 420–450 m, 18.200°S, 178.617°W (MNHN, 2). 144–150 m, S of Viti Levu, 18.206°S, 178.550°E (MNHN, 12). 149–168 m, S of Viti Levu, 18.206°S, 178.552°E (MNHN, 20). 295–302 m, S of Viti Levu, 18.211°S, 178.540°E (MNHN, 30). 1 m, Viti Levu, Korolevu, 18.220°S, 177.718°E (LACM 79-42, 2). S coast Viti Levu, Deuba, 18.250°S, 178.083°E (AMS C.348629, 1). 1–2 m, S coast Viti Levu, Deuba, 18.250°S, 178.083°E (AMS exC.404972, 1: complete). 0–1 m, S coast Viti Levu, Deuba, 2 km offshore, 18.267°S, 178.083°E (AMS C.348628, 3). 275–430 m, S of Viti Levu, 18.296°S, 177.907°E (MNHN, 100). 10–20 m, off Francis Island in Beqa Lagoon, just off of Beqa Island (S of Viti Levu), 18.300°S, 178.067°E (DLG 278, 10). 260–305 m, S of Viti Levu, 18.308°S, 178.097°E (MNHN, 50). 234–361 m, 18.310°S, 178.085°E (MNHN, 5). 497–504 m, S of Viti Levu, 18.311°S, 178.152°E (MNHN, 10). 321 m, S of Viti Levu, 18.318°S, 177.987°E (MNHN, 1). 441–443 m, S of Viti Levu, 18.320°S, 177.862°E (MNHN, 30). Great Astrolabe Reef, 18.746°S, 178.300°E (JKC, 1). Drop-off at old

fallen beacon on NW tip of Dravuni, 18.754°S, 178.524°E (JKC, 1). Astrolabe Reef, Kadavu Group, 18.767°S, 178.453°E (LACM 91-189, 8). 20 m, Herald Pass, W of Ndravuni Island, NW end of Great Astrolabe Reef, 18.767°S, 178.467°E (LACM 85-135, 3). 20 m, Herald Pass, W of Ndravuni Island, NW end of Great Astrolabe Reef, 18.767°S, 178.467°E (LACM 85-135, 11). 785–820 m, BORDAU 1 CP1490, 18.850°S, 178.533°W (MNHN, 3). 240–319 m, BORDAU 1 DW1494, 18.916°S, 178.483°W (MNHN, 18). 15 m, off Ham Reef, Ngasele Bay, N side Kandavu, 18.933°S, 178.300°E (LACM 85-136, 1). 14 m, Easternmost end of Tombani Namalata Reef, Malatta Reefs, N side Kandavu, 18.966°S, 178.167°E (LACM 85-137, 1). 395–540 m, BORDAU 1 DW1486, 19.016°S, 178.433°W (MNHN, 10). 500–516 m, BORDAU 1 DW1488, 19.016°S, 178.417°W (MNHN, 11). 700–707 m, BORDAU 1 DW1485, 19.050°S, 178.500°W (MNHN, 10). 18 m, Kandavu, Cape Washington, 19.108°S, 177.967°E (LACM 85-138, 3).

Tonga. 24 m, S side Maungaui Peninsula, Pangaimotu Island, Vava'a Group, 18.700°S, 174.000°E (LACM 85-88, 7). 433 m, NW Tongatapu, BORDAU 2 DW1569, 21.033°S, 175.316°W (MNHN, 300).

Mariana Islands. 9 m, Pagan Island, 18.117°N, 145.767°E (AMS C.379956, 2).

Marshall Islands. 23 m, Ocean Side, Kwajalein, 8.739°N, 167.746°E (DLG 997, 4).

Wallis and Futuna. 419–425 m, Futuna Island, 14.333°S, 178.117°W (MNHN, 1). 415–420 m, Wallis Island, 13.350°S, 176.133°W (MNHN, 1; MNHN, 8; MNHN, 5). 350 m, Wallis Island, 13.317°S, 176.283°W (MNHN, 1). 455–515 m, Wallis Island, 13.200°S, 176.267°W (MNHN, 100).

American Samoa. 25 m, Pago-Pago Harbour, Tutuila Island, 14.280°S, 170.701°W (ESC, 20). 15 m, Leone Bay, Tutuila Island, 14.338°S, 170.793°W (LACM 99-89, 1). 0–2 m, by Hotel Rainmaker, Pago-Pago Harbor, Tutuila Island, 14.283°S, 170.667°W (LACM 76-33, 1).

Western Samoa. 4 m, Middle Vaisala Lagoon, Vaisala, 13.512°S, 172.644°W (ESC, 5).

French Polynesia. 507–607 m, Kaukura, Tuamotus, TARASOC DW3373, 15.683°S, 146.900°W (MNHN, 3). 300–650 m, Tahiti, Society Islands, TARASOC DW3484, 17.750°S, 149.383°W (MNHN, 1).

Literature records.

Philippines. Cebu, 10.283°N, 124.017°E (Bandel, 1998).

Remarks. Fischer-Piette (1950: 69) indicated there to be a holotype and one paratype. The holotype was not specifically designated in the original description; hence, the two specimens constitute syntypes. The subsequent holotype indication cannot be construed as a lectotype designation (ICZN Art. 74.5). The here designated lectotype fits the stylized original illustration better than the paralectotype.

The strong concentric sculpture shown in Thiele's ventral drawing is an artefact of flat lighting; the shells have the same sculpture over the entire base.

A lectotype is here selected for *Sci. padangensis* with the express purpose for taxon stabilization; the paralectotype is identified as *Sin. plicata*.

Sinezona finlayi (Laws, 1948)

Figures 336–338

Chresonymy.

Schismope finlayi Laws, 1948: 144–145, pl. 11, fig. 21.
Sinezona finlayi: Lozouet, 1986: 111.
Sinezona finlayi: Beu & Maxwell, 1990: 402.

Figure 336. Original illustration of *Schismope finlayi.*

Type material. Holotype (GNS TM 1488, ex coll. Laws. Fig. 337A), 1.0 × 0.6 mm. Three paratypes (GNS TM 8656–8658). Other paratypes?: "numerous specimens collected."

Type locality. Hokianga South Head. A low outcrop at top of beach below the house of the keeper of the signal station.—Low cliffs along bed of creek draining from the north in to Waimamaku River, one mile NW of Waimamaki Post Office [New Zealand, 33.550°S, 173.383°E]. Tertiary (OD).

Etymology. Named for New Zealand malacologist H. J. Finlay (1901–1951: see Coan *et al.*, 2007).

Description. Shell medium-sized (to 1 mm), trochiform. Protoconch eroded, except for two indistinct strong axial cords. Teleoconch I approximately 0.875 whorl, approximately 12 axial cords approximately six fine spiral lines between position of selenizone and suture with next whorl. Teleoconch II 0.66 whorl, suture not impressed. Shoulder slightly concave, approximately 20 weak axial cords, most distinct near suture, first spiral line soon after onset of selenizone, approximately seven at apertural margin, forming minute points at intersection with axial cords. Base with distinct constriction below selenizone, density of axials as on shoulder, strength diminishing to unrecognizable towards apertural margin; approximately 12 spirals, lines below selenizone, strong steps towards umbilicus. Umbilicus open, margin at angle to base, details obscured by in-fill. Selenizone above periphery; keels low, strong; slit closed forming elongate foramen. Aperture obscured by in-fill (holotype).

Comparisons. *Scissurella apudornata* from the Miocene of New Zealand has fewer but stronger axial cords and a wider umbilicus.

Sinezona brucei from New Zealand has stronger axial sculpture on the shoulder, a strong periumbilical cord, and much weaker lunules.

Sinezona norfolkensis from the Tasman Sea has a longer teleoconch I (0.8–0.9 vs. 0.75 whorl), axials on the shoulder reaching from the suture to the selenizone, and weaker lunules.

Sinezona cupelliformis from the Eocene of Europe has fine clathrate sculpture on the base.

Sinezona deshayesi from the Eocene of Europe and *Sin. terquemi* from the Miocene of Europe have more but finer axial cords on the base.

Sinezona koruahina from the Pliocene of New Zealand has strong axial lamellae on the lower base and a very distinct constriction below the selenizone devoid of those axial lamellae.

Figure 337. **A**. *Sinezona finlayi*. Holotype. Hokianga South Head. A low outcrop at top of beach below the house of the keeper of the signal station.—Low cliffs along bed of creek draining from the north in to Waimamaku River, 1 mi NW of Waimamaki Post Office [New Zealand, 33.550°S, 173.383°E]. Tertiary (GNS TM 1488). Scale bar shell = 0.5 mm. Scale bar protoconch = 100 μm. **B–C.** *Sinezona* cf. *finlayi*. **B.** 12 m, inside Entrance Island, Nancy Sound, Fiordland, New Zealand, 45.103°S, 167.020°E (NMNZ M.140629, 1). **C.** 11–20 m, SE of Dale Point, S side, outer Milford Sound, Fiordland, New Zealand, 44.608°S, 167.833°E (NMNZ M.142611, 1). Scale bars shell = 1 mm. Scale bars protoconch = 100 μm.

Sinezona malloryi from the Eocene of North America is indistinguishable based on the available data. However, the geographic separation makes it likely that they are distinct, and details of protoconch and early teleoconch may show significant differences. Accordingly, the two species are retained.

Sinezona tertia from the Miocene of New Zealand is overall less broad, has stronger axial cords/lamellae reaching from the suture to the umbilicus, and retains their strength over entire shell.

Distribution. Miocene [? to Recent] of New Zealand (Fig. 338).

Specimen records.

New Zealand. 12 m, inside Entrance Island, Nancy Sound, Fiordland, 45.103°S, 167.020°E (NMNZ M.140629, 1). 11–20 m, SE of Dale Point, S Side, outer Milford Sound, Fiordland, 44.608°S, 167.833°E (NMNZ M.142611, 1).

Fossil. Otaian = early Miocene of New Zealand (Beu & Maxwell, 1990).

Remarks. Two close localities are given in the original description; it is not clear from which of the two localities the holotype was collected.

Two Recent specimens best agree with *Sinezona finlayi*, though this identification is not entirely satisfactory. The specimens agree in overall shape, the distinct axial cords on teleoconch I, show some signs of fine spiral cords on the shoulder at the apertural margin of teleoconch II, shows the same pattern of spiral cords on the base. However, the spirals on the shoulder are much less distinct. The two Recent specimens show axial cords of strongly dissimilar strength on teleoconch II. However, all protoconch, teleoconch I, and shoulder characters are shared, indicating that they are conspecific. Similar differences in strength of the axial sculpture are well-known in other species such as *Sci. costata* from the Mediterranean. The implied geological time span for this species would be rather extensive from the Miocene to the Recent.

Figure 338. Distribution of *Sinezona finlayi*. ☆ type locality. ● specimen record.

Alternative identifications are equally unsatisfactory. The Recent specimens are not *Sin. laqueus* with little ornamentation, despite the overall similarity of the shell shape. They are not *Sukashitrochus* sp. with very weak keel on the base—such specimens are known in *Suk. atkinsoni*—as the protoconch sculpture does not agree with any potential species in that genus. The only further alternative would be an undescribed species based on two specimens.

Sinezona garciai Geiger, 2006

Figures 341–340

Chresonymy.

Sinezona garciai Geiger, 2006a: 12–14, figs 9–10.

Scissurella garciai: Daccarett & Bossio, 2011: 51.

Type material. Holotype (USNM 1081940: ex UMML 30.12043. Fig. 339). Paratypes from type locality (SBMNH 359317, 2: ex UMML 30.12043. Fig. 340A, B; UMML 30.12043, 1). Paratype: 16 m, St. Croix, US Virgin Islands, 17.800°N, 64.800°W (UMML 30.12051, 1. Fig. 340C).

Type locality. 1 m, behind outer reef, Courtown Cays, San Andres y Providencia, Colombia, 12.383°N, 81.433°W (OD).

Etymology. Named for Emilio García of Lafayette, Louisiana, USA, for his continuing contribution to the knowledge of the Caribbean malacofauna, and his generous support of ongoing research (OD).

Description. Shell small (to 0.76 mm. Figs 339, 340), trochiform depressed, last 0.125 whorl distinctly descending. Protoconch of one whorl, fine, sharp axials in outer half of whorl, inner half smooth, no apertural varix, apertural margin convex. Teleoconch I of 0.66 whorl. Shoulder slightly convex, suture distinct; approximately 10–12 axial cords, no spirals, interstices with fine irregular growth marks. Teleoconch II of 0.66 whorl, same density and strength of axials as on teleoconch I, onset of broad, low axial cords slightly after onset of selenizone, increasing in number to approximately 10–12, somewhat unevenly spaced. Base with distinct constriction below selenizone, same density of more distinct axial cords as on shoulder, crossed by approximately 16–18 weaker spiral cords, cords mirror increase in strength and number with growth to those on shoulder. Selenizone above periphery, keels strong, moderately elevated, lunules very distinct; foramen elongated. Umbilicus

Figure 339. *Sinezona garciai*. Holotype. 1.5 mi behind outer reef, Courtown Cays, Colombia, 12.383°N, 81.433°W (USNM 1081940: ex. UMML 30.12043). Scale bar shell = 0.5 mm. Scale bar protoconch = 100 μm.

wide, bordered by strong spiral cord, walls straight, with fine axial growth marks. Aperture obliquely D-shaped, roof overhanging.

Animal unknown.

Comparisons. *Sinezona confusa* from the Caribbean to northern Brazil has a more elevated overall shape, a very narrow, barely open umbilicus, and a protoconch with fewer and broader axials as well as an apertural varix.

Figure 340. *Sinezona garciai*. Paratypes **A–B.** 1 m, Behind outer reef, Courtown Cays, Colombia, 12.383°N, 81.433°W (SBMNH 359317, 2). **C.** 16 m, St. Croix, 17.800°N, 64.800°W (UMML 30.12051, 1). Scale bars shell = 0.5 mm. Scale bars protoconch = 100 μm.

Sinezona insularis from the equatorial mid-Atlantic islands has a much shorter selenizone (0.1 vs. 0.4 whorl), a wider umbilicus, and the umbilicus is bordered by a moderately distinct periumbilical cord.

Sinezona carolarum from the Panamic has a protoconch with stronger axial cords and an apertural varix, a more elevated spire, and fewer spiral lines on the shoulder.

Sinezona costulata from southern Japan has a protoconch with strong axial cords and an apertural varix, the spire is more elevated, and the axial cords diminish in strength on the shoulder of teleoconch II.

Scissurella electilis from Brazil has a much more narrow umbilicus, lacks the strong cord separating the umbilicus from the base, and retains an open slit at maturity.

Satondella brasiliensis from Bermuda to Brazil has a more strongly descending aperture on the last quarter whorl and the foramen is drawn out into a strongly elevated chimney.

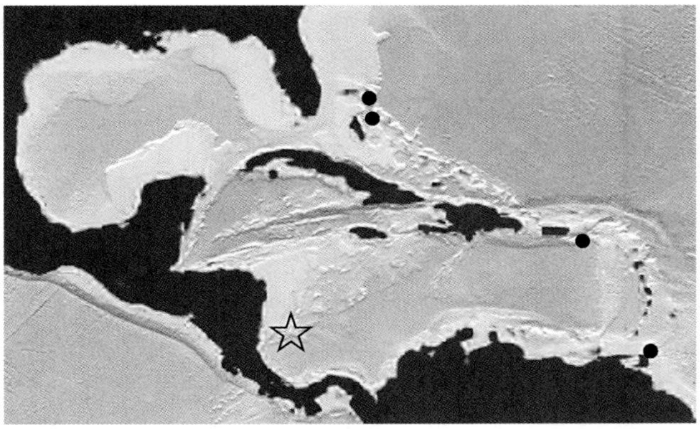

Figure 341. Distribution of *Sinezona garciai*. ☆ type locality. ● specimen record.

Distribution. Colombia to Bahamas (Fig. 341).

Specimen records.

Bahamas. 18 m, Samphire Cay, NW Nassau, 25.183°N, 77.017°W (HLC, 1).

Remarks. *Sinezona garciai* is distinct from *Sin. confusa*, and not simply a form with more depressed shells. All five specimens of *Sin. garciai* available show highly congruent features with very little intraspecific variability. The same applies to 68 lots of *Sin. confusa* inspected ranging from Bermuda to Venezuela.

Sinezona globosa Geiger, 2006

Figures 342–344

Chresonymy.

Sinezona globosa Geiger, 2006a: 15–16, fig. 11.

Sinezona globosa: Bouchet *et al.*, 2008: 26.

Type material. Holotype (MNHN. Fig. 343A). Paratype 415–420 m, Wallis Island, Wallis and Futuna, 13.350°S, 176.133°W (MNHN, 1. Fig. 343B).

Type locality. 310–315 m, Loyalty Basin, New Caledonia, 21.533°S, 166.483°E (OD).

Etymology. Globosa, Latin adjective, noting the overall globular shape of the shell (OD).

Description. Shell large (to 3.77 mm. Figs 342, 343), trochiform globular. Protoconch often sunken, of 0.875 whorl, embryonic cap with flocculent sculpture, remainder smooth with 1–4 fine spiral threads, apertural varix connected to embryonic cap, apertural margin sinusoidal. Teleoconch I of 1.66 whorls, approximately 46–50 distinct axial cords, run over equally spaced spiral lines, first spiral after 0.33 whorl, 14 at beginning of selenizone, interstices with broad growth marks. Teleoconch II of up to

1.6 whorls, suture moderately impressed. Shoulder quite convex, axial cords of same density as on teleoconch I, forming small points at intersection with spiral lines. Base slightly constricted below selenizone, with same density and strength of axial cords, approximately 20 spiral lines, becoming stronger and more closely spaced towards umbilicus. Umbilicus narrow in small specimens (paratype), moderately wide in large specimens (holotype), becoming partially covered by callus (Fig. 342), bordered by last spiral line of base, walls straight, smooth. Selenizone above periphery, keels elevated but often eroded due to their fragility, lunules distinct, slit closed to elongated foramen. Aperture subrotund, lip inflated, roof overhanging.

Animal unknown.

Comparisons. *Sinezona confusa* from the Caribbean to northern Brazil has a protoconch with axial sculpture, shorter teleoconch I (1 vs. 1.5 whorls), and the umbilicus is bordered by a distinct periumbilical cord.

Sinezona danieldreieri from the Indo-Malayan Archipelago to the Central Pacific has a protoconch with axial sculpture, the apertural varix not connected to the embryonic cap, and has a much shorter teleoconch I (0.8 vs. 1.5 whorls).

Figure 342. *Sinezona globosa*. 28–32 m, Napaling, Panglao Island, PANGLAO 2004 S28, Philippines, 9.620°N, 123.773°E (MNHN, 6). Scale bar shell = 1 mm. Scale bar protoconch = 100 μm.

Sinezona ferriezi from the Indo-Malayan Archipelago to the Central Pacific has a completely smooth protoconch, a shorter teleoconch I (1 vs. 1.5 whorls) and a distinct constriction of the base below the selenizone.

Figure 343. *Sinezona globosa*. **A.** Holotype. 310–315 m, Loyalty Basin, New Caledonia, 21.533°S, 166.483°E (MNHN). **B.** Paratype 415–420 m, Wallis Island, Wallis and Futuna, 13.350°S, 176.133°W (MNHN). **C.** 25–30 m, Santal Bay, Lifou, New Caledonia, 20.820°S, 167.173°E (MNHN, 1). Scale bars shell = 1 mm. Scale bars protoconch = 100 μm.

Sinezona garciai from the Caribbean has a more depressed spire, a protoconch with fine axial sculpture and no apertural varix, and a distinct constriction of the base below the selenizone.

Sinezona macleani from the Indian Ocean to the Western Pacific is more depressed, less globular, has a protoconch with flocculent sculpture, lacks an apertural varix, has a sharp elevated keel bordering the umbilicus, and usually a projecting shelf in the basal adumbilical corner of the aperture.

Sinezona marrowi from the Indo-Malayan Archipelago has a protoconch with strong axial sculpture, a shorter teleoconch I (1 vs. 1.5 whorls), and a wider, funnel-shaped umbilicus.

Sinezona milleri from southern Japan has a protoconch with strong axial sculpture, a shorter teleoconch I (1 vs. 1.5 whorls), and a shorter selenizone (0.4 vs. 1 whorl).

Sinezona plicata from the Indo-Malayan Archipelago to the Central Pacific is overall more depressed, has a smooth protoconch occasionally with a fine spiral line, and has strong, elevated lamellae on the teleoconch, and a strong constriction of the base below the selenizone.

Sinezona wileyi from the Indo-Malayan Archipelago to the Central Pacific has an overall subquadrate shell outline, a distinct constriction of the base below the selenizone, a protoconch with fine axial sculpture and no apertural varix, and a shorter teleoconch I (0.8 vs. 1.5 whorls).

Sinezona zimmeri from Easter Island has a protoconch with strong axial sculpture, a distinct constriction of the base below the selenizone, a wider funnel-like umbilicus, and a shorter teleoconch I (0.8 vs. 1.5 whorls).

Distribution. Tropical Indo-Malayan Archipelago to tropical Eastern Pacific; Caribbean (?) (Fig. 344).

Specimen records.

Philippines. 903 m, [494 fms.], off Magabao Island, E Mindanao, 8.983°N, 126.283°E (USNM 1022139, 3). 20 m, Looc, Panglao Island, PANGLAO 2004 B32, 9.430°N, 123.743°E (MNHN, 2). 28–32 m, Napaling, Panglao Island, PANGLAO 2004, 9.620°N, 123.773°E (MNHN, 6).

New Caledonia. 25–30 m, Santal Bay, Lifou, 20.820°S, 167.173°E (MNHN, 1).

Vanuatu. 100–105 m, NW Tutuba Island, SANTO 2006 DS99, 15.541°S, 167.281°E (MNHN, 1). 70–80 m, W Tutuba Island, SANTO 2006 DS103, 15.568°S, 167.266°E (MNHN, 1). 98–100 m, W Tutuba Island, SANTO 2006 DS102, 15.568°S, 167.266°E (MNHN, 3).

Chile. 30–50 m, Tahai, Easter Island, 27.140°S, 109.429°W (DLG 1326, 1).

Barbados. 172 m, off Lazaretto, 13.167°N, 59.567°W (USNM 500687, 1: see Remarks).

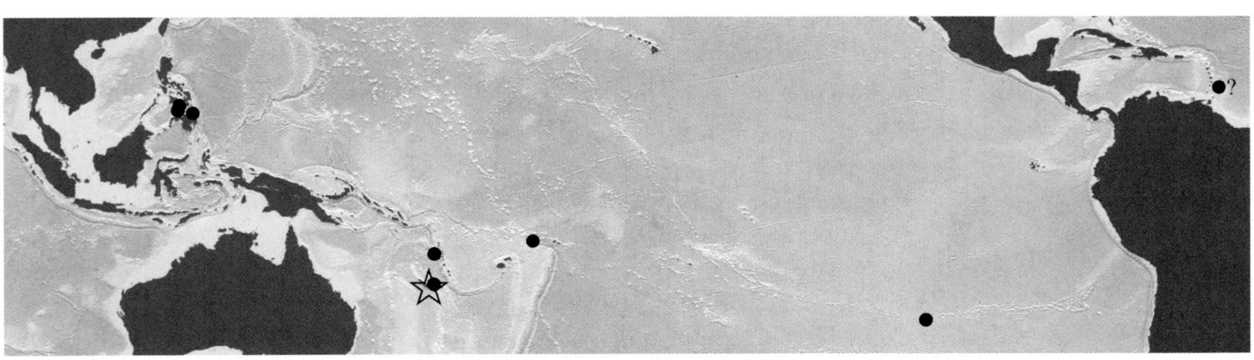

Figure 344. Distribution of *Sinezona globosa*. ☆ type locality. ● specimen record. ? questionable locality.

Remarks. The paratype is slightly immature evidenced by the hardly descending aperture on the last quarter whorl, while the holotype is a fully grown specimen. It appears that the closure of the foramen occurs early on.

The Caribbean specimen is conspecific with *Sin. globosa*, but the locality data in the Caribbean are very suspicious. The record from Easter Island makes the Caribbean locality data slightly more credible.

Sinezona hawaiiensis Geiger & McLean, 2010

Figures 345–346

Chresonymy.

Sinezona hawaiiensis Geiger & McLean, 2010: 8, fig. 5.
Sinezona hawaiiensis: Geiger, 2011: 38, pl. 3, fig. 6.

Type material. Holotype (LACM 2466. Fig. 345).

Type locality. 30 m, off Waikiki, Oahu, Hawaii, USA, 21.270°N, 157.840°W (OD).

Etymology. Named for its type locality in Hawaii, USA (OD).

Description. Shell small (0.5 mm. Fig. 345), trochiform, globular. Protoconch of one whorl, with fine axial cords, no apertural varix, apertural margin prosocline. Teleoconch I of 1.25 whorls. Shoulder broad indistinct cords, approximately 30 on first whorl, without discernible spiral sculpture. Base with axials becoming more distinct towards umbilicus, approximately 13 low spiral cords between periphery and umbilicus, barely noticeable at periphery, becoming as strong as axials towards umbilicus. Teleoconch II of 0.125 whorl, suture little impressed, same sculpture as on teleoconch I. Umbilicus very narrow, bordered by strong cord, walls with distinct growth marks, no funiculus. No selenizone; foramen oval, keels low, strong. Aperture rounded, roof overhanging.

Animal unknown.

Comparisons. *Sinezona kayae* from Hawaii has a protoconch with distinct axial cords and an apertural varix, a distinct selenizone, and a shorter teleoconch I (0.8 vs. 1.25 whorls).

Sinezona pauperata from New Zealand is indistinguishable based on shell morphology (see Remarks).

Sinezona haliotimorpha from the Oligocene of France has a protoconch with fewer and stronger axials, a teleoconch with stronger axials and consistently finer spiral threads on base, and a foramen with more elevated keels.

Distribution. Known only from type locality, Hawaii (Fig. 346).

Remarks. The shell appears mature due to the sloping apertural margin of the last portion of the whorl.

It is intriguing that *Sin. hawaiiensis* from Hawaii is indistinguishable from *Sin. pauperata* from New Zealand. Given the different ecological settings (temperate vs. tropical) on either side of the equator, it is unlikely that the Hawaiian specimen is conspecific with the New Zealand species. The precise locality information for the Hawaiian specimen makes an error in location equally unlikely. Additionally, *Sin. pauperata* is far from common in collections, further doubting a problem with collection management. Specimens can be transported in ballast of boats, but

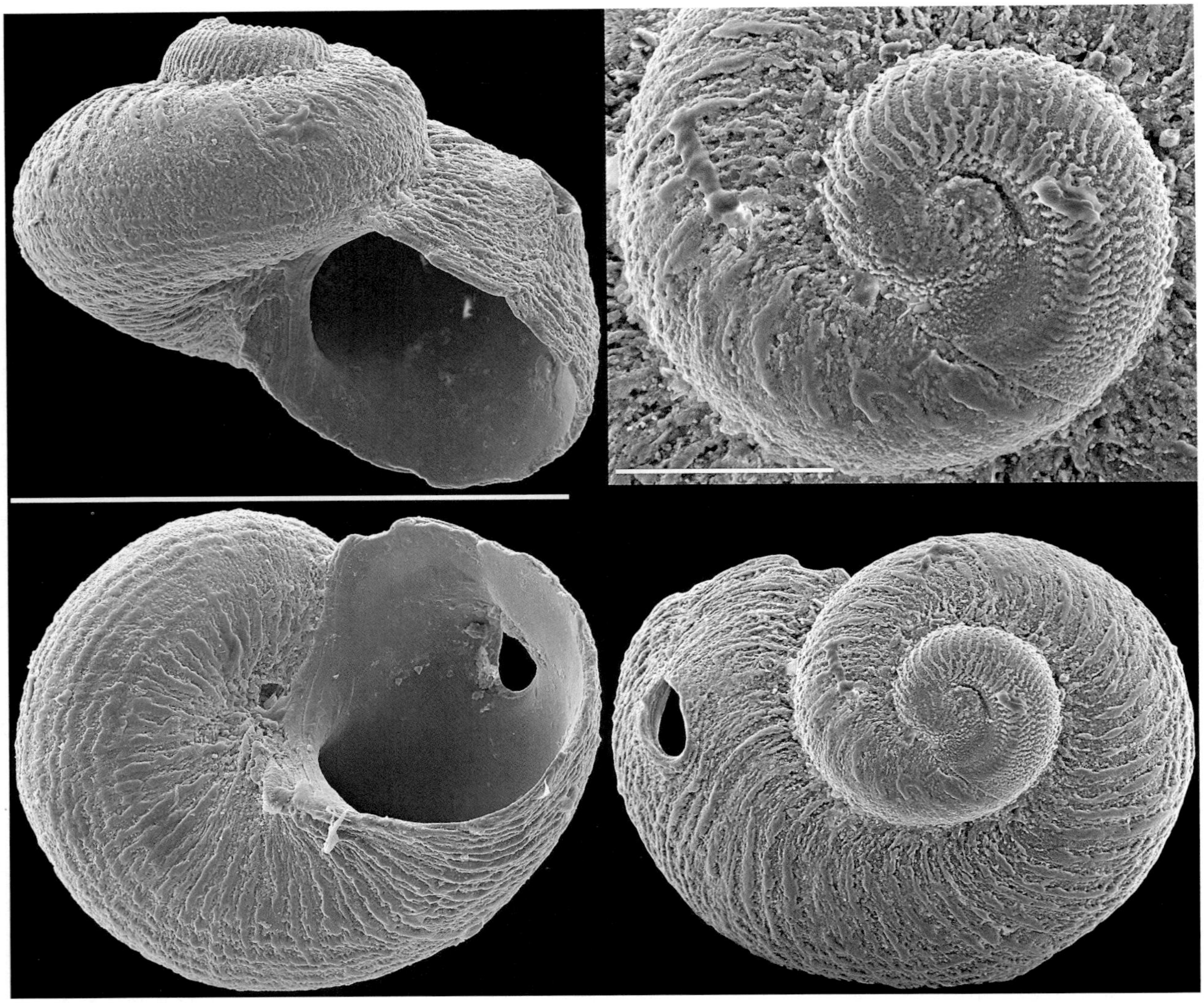

Figure 345. *Sinezona hawaiiensis*. Holotype. 30 m, off Waikiki, Oahu, Hawaii, USA, 21.270°N, 157.840°W (LACM 2466). Scale bar shell = 0.5 mm. Scale bar protoconch = 100 μm.

the exquisite preservation of *Sin. hawaiiensis* makes this possibility also unlikely. Additional specimens from Hawaii would be desirable to confirm its existence and to remove any lingering doubt.

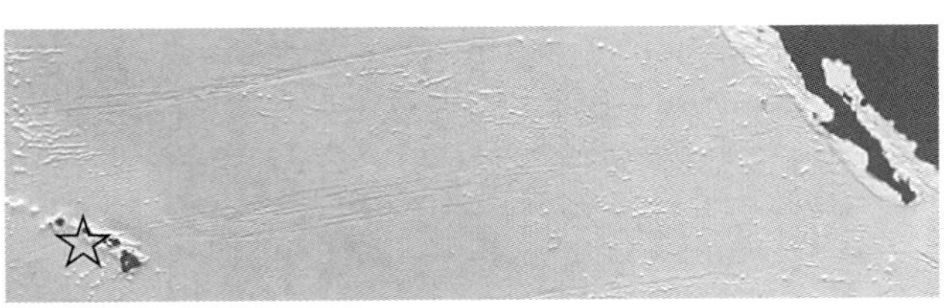

Figure 346. Distribution of *Sinezona hawaiiensis*. ☆ type locality.

Sinezona insignis (Smith, 1890)

Figures 347–351

Chresonymy.

Schismope insignis Smith, 1890: 208, pl. 8, fig. 4.
Schismope insignis: Thiele, 1912a: 24–25, pl. 3, figs 19–20 [copy figures Smith, 1890].
Schismope insignis: Bartsch, 1915: 176.
Sinezona insignis: Kensley, 1973: 28, figs 29 [two figs].
Sinezona sp.: Gofas *et al.*, n.d.: 33, fig. c.
Sinezona insignis: Herbert, 1986: 626–628, figs 10 [protoconch], 32–35.
Sinezona insignis: Lozouet, 1986: 111.
Sinezona insignis: Herbert, 1987: 1, figs g–h.
? *Sinezona* cf. *insignis*: Hedegaard, 1990: 74, 145 [identification uncertain as neither locality nor repository of voucher was given].
Sinezona insignis: Geiger, 2003: 78.
Sinezona insignis: Geiger & McLean, 2010: 6, 33.

Figure 347. Original illustration of *Schismope insignis*.

Misidentification.

not *Sinezona insignis*: Kay, 1979: 39, figs 7C, 9D–F [is *Sin. kayae*].

Type material. Syntypes (NHMUK 1911.8.30.27–28, 2), 0.8 × 0.5 mm. Lectotype here designated (NHMUK 1911.8.30.27. Fig. 348) for the express purpose of taxon stabilization in case other non-conspecific syntypes should be found. Paralectotype from type locality (NHMUK 1911.8.30.28, 1).

Type locality. Four miles south of Port Elizabeth [South Africa, 33.967°S, 25.583°E] (J. Farquhar) (OD).

Etymology. Insignis, Latin for remarkable, notable.

Description. Shell medium-sized (to 1.0 mm. Figs 348–349), turbiniform, as wide as high, or slightly higher. Protoconch of 0.75 whorl, strong axial sculpture, apertural varix present, apertural margin slightly sinusoidal. Teleoconch I of one whorl, with nine strong axial cords, becoming peripherally dislocated with growth, no spirals. Teleoconch II of 0.625 whorl, suture deep. Shoulder slightly convex, very fine growth lines, approximately seven spiral lines arising near aperture in mature specimens. Base with constriction below selenizone, approximately seven strong axial lamellae, approximately 11 fine, indistinct spiral lines. Selenizone above periphery, slightly inside, lunules indistinct, keels strong, moderately elevated. Slit closed to foramen, foramen elongated triangular, rather short for genus. Umbilicus bordered by spiral cord, wall smooth, no funiculus. Aperture round, roof overhanging.

Operculum (Fig. 349C) corneous, thin, round, covering aperture, nucleus central, multispiral.

Radula (Fig. 350). Rachidian tooth triangular with broad base, tip with 5–7 cusps, central one largest, in V-shape. Lateral teeth 1–3 similar, three cusps on outer margin of tip. Lateral tooth 4 reduced, pointed, with one minute protrusion on either side of central cusp. Lateral tooth 5 broadened, approximately seven cusps, terminal cusp largest. Radular interlock of central field moderate. Inner marginal teeth with two cusps on inner margin of tip, four cusps on outer margin, median one largest; shafts without food groove. Outer marginal teeth spoon-shaped, many bristles.

Comparisons. *Sinezona doliolum* from South Africa has axial sculpture on the shoulder, has finer axial cords on the base, is taller, a longer selenizone (0.8 vs. 0.5 whorl), and an extremely narrow umbilicus.

Sinezona cingulata from the Mediterranean and adjacent Atlantic ocean has finer axial cords on the teleoconch, a protoconch with finer axial cords, and distinct spiral lines on the shoulder and base.

Sinezona insularis from equatorial mid-Atlantic Islands has a much flatter spire, the protoconch lacks an apertural varix, has much weaker axial cords, and the umbilicus is wider.

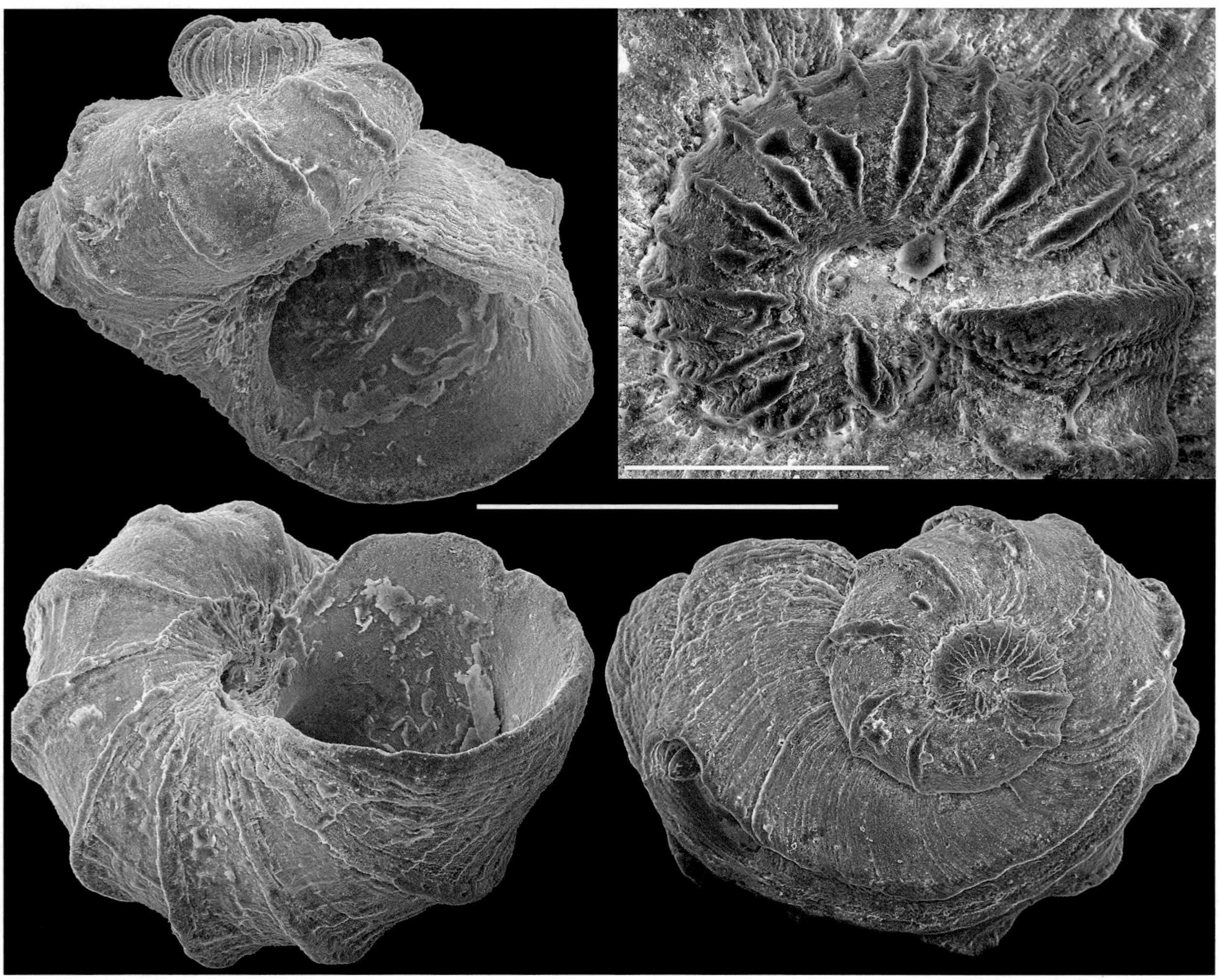

Figure 348. *Sinezona insignis*. Lectotype. Four miles south of Port Elizabeth [South Africa, 33.967°S, 25.583°E] (NHMUK 1911.8.30.27). Scale bar shell = 0.5 mm. Scale bar protoconch = 100 µm.

Sinezona iota from New Zealand has axial cords on the shoulder near the suture and the axial cords on teleoconch I reach the suture.

Figure 349. *Sinezona insignis*. **A–B.** Port Elizabeth, South Africa, 33.967°S, 25.583°E (NMSZ 1961.61). **C.** 9–15 m, Aliwal Shoal, Natal, South Africa, 30.250°S, 30.817°E (NMSA E6195). Radula shown in Figure 350. Scale bars shell = 0.5 mm. Scale bars protoconch = 100 µm.

Sinezona kayae from Hawaii has a protoconch with finer axial cords, the axial cords on teleoconch I reach the suture, and spiral lines on shoulder and base are distinct.

Sinezona mouchezi from subantarctic Indian Ocean has distinct spiral sculpture on shoulder and base, the axial cords on teleoconch I reach the suture, and has axial cords on the shoulder of teleoconch II in the vicinity of the suture.

Sinezona rimuloides from the northeastern Pacific and the Panamic has axial sculpture on teleoconch I reaching the suture, axial cords on teleoconch I, and shows distinct spiral lines on shoulder and base.

Sinezona semicostata from the subtropical eastern Atlantic off-shore islands has a much flatter shell, a protoconch without an apertural varix visible through the umbilicus, and much finer axial cords.

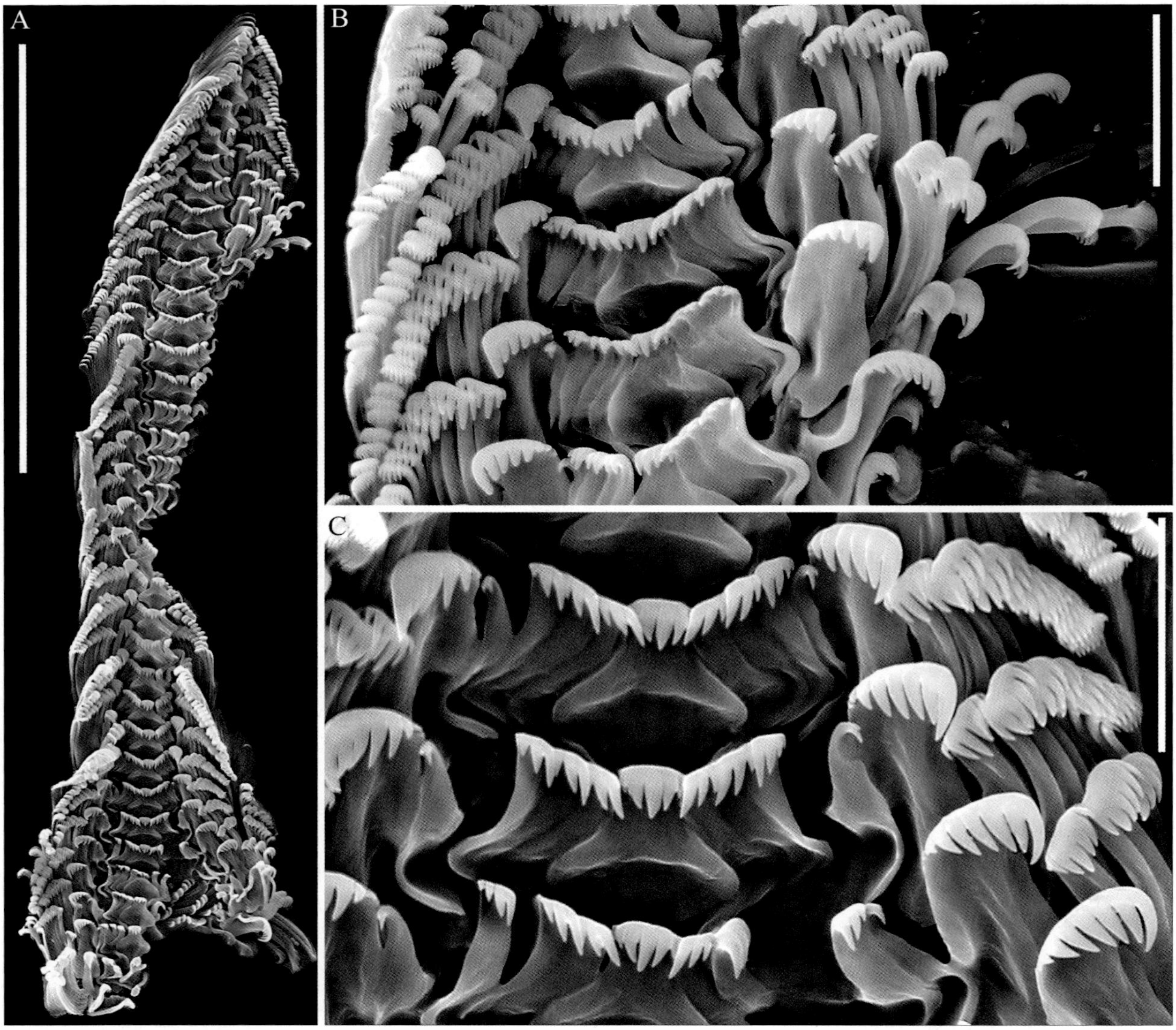

Figure 350. *Sinezona insignis*. Radula. Port Elizabeth, South Africa, 33.967°S, 25.583°E (NMSZ 1961.61). Shell shown in Figure 349C. **A.** Entire radula. Scale bar = 100 μm. **B.** Whole width of radular ribbon. **C.** Central field enlarged. Scale bars B–C = 10 μm.

Sinezona singeri from the Indian Ocean has a shell that is wider than tall, a smooth protoconch with the apertural varix connecting to the embryonic cap, and many more but much finer axial cords.

Sinezona subantarctica with broad subantarctic distribution has a flatter shell, a protoconch with fine irregular axial sculpture and no apertural varix, and the umbilicus is closed by a callus.

Distribution. Angola to South Africa, off-shore temperate Atlantic islands (Fig. 351). Intertidal to 50 m.

Ecology. Found live intertidally among *Laurencia* and *Caulerpa* algae, and in holdfasts of *Ecklonia* kelp (Herbert, 1987).

Specimen records.

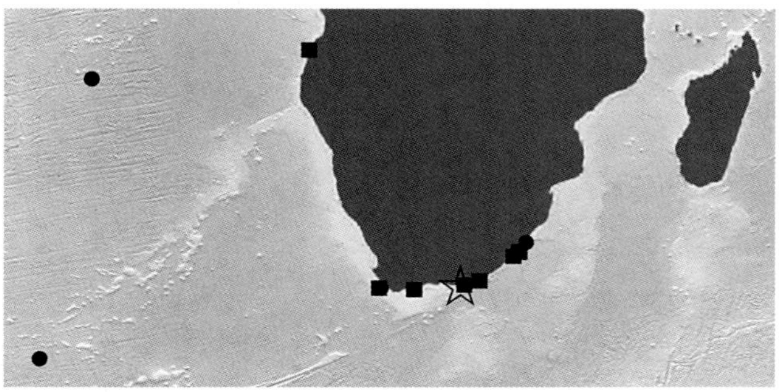

Figure 351. Distribution of *Sinezona insignis*. ☆ type locality. ● specimen record. ■ literature record.

South Africa. Algoa Bay, 38.833°S, (HLC, 1). 0 m, Miller's Point, False Bay, Western Cape, 34.233°S, 18.467°E (NMSA A1768). 10 m, False Bay, 34.139°S, 18.634°E (DLG 2324, 3). False Bay, Dalebrook, 34.250°S, 18.600°E (AMS C.406349, 1). Still Bay, Tsitsikamma coast, 34.366°S, 21.416°E (SAMCT A31577). 4 mi S of Port Elizabeth, 33.967°S, 25.583°E (NHMUK, 2; NMSZ 1961.61, 3). 4 mi S of Port Elizabeth, 33.967°S, 25.600°E (NMSA B2866, –; NMSA 2948, –). Port Alfred, 33.600°S, 26.900°E (NHMW 22045, 1; NMSA D244, –; NMSA 8536, –; USNM 249763, 1; USNM 249762, 3). 50 m, off Mbotyi, 31.450°S, 29.733°E (NMSA C7956). 9–15 m, Aliwal Shoal, Natal, 30.250°S, 30.817°E (NMSA E6195, 3).

Ascencsion Island. Georgetown, 7.933°S, 14.4170°W (USNM 735075, 4).

Great Britain [Overseas Territories]. St. Helena, 15.950°S, 5.720°W (USNM 123993, 1). 115 m, *Discovery* station 399, 1 mi SE of SW point of Gough Island, 40.370°S, 10.050°W (NHMUK, 26).

Literature records.

Angola. Santa Maria, Benguela, 13.417°S, 12.533°E (Gofas *et al.*, n.d.).

South Africa. Still Bay, Tsitsikamma coast, 34.366°S, 21.416°E. 0 m, Miller's Point, False Bay, Western Cape, 34.233°S, 18.467°E. 4 mi S of Port Elizabeth, 33.967°S, 25.600°E. Port Alfred, Eastern Cape, 33.600°S, 26.900°E. 50 m, off Mbotyi, 31.450°S, 29.733°E. Mzamba, 31.100°S, 30.183°E. Port Edward, Natal, 31.050°S, 30.233°E (Herbert, 1986).

Remarks. *Sinezona insiginis* had been misidentified from Hawaii; see Remarks under *Sin. kayae*.

Sinezona insularis Simone, 2009

Figures 352– 353

Chresonymy.

Sinezona insularis Simone, 2009: 12–14, figs 2–9.
Sinezona sp.: Oliveira *et al.*, 2009: 180, pl. A, fig. 8.
Sinezona insularis: Dornellas & Simone, 2011: 19.

Figure 352. *Sinezona insularis*. São Pedro e São Paulo Archipelago, Brazil, 00°55'01"N, 26°20'44"W. Shell of holotype MZSP 87357, protoconch from paratype (MZSP 87355, 1). Scale bar shell = 0.5 mm. Scale bar protoconch = 100 µm. Images taken and courtesy by L. Simone.

Type material. Holotype (MZSP 87357. Fig. 352). Paratypes from type locality (MZSP 87355, 1; MZSP 87356, 2; MZSP 88191, 7; MNRJ 12868, 2; USNM 1121781, 2; MZSP 86665, 1).

Type locality. São Pedro e São Paulo Archipelago, Brazil, 00°55'01"N, 26°20'44"W (OD).

Etymology. Named for the insular locality, restricted to the Archipelago (OD).

Description. Shell small (to 0.7 mm: holotype. Fig. 352), trochiform depressed. Protoconch of 0.875 whorl, many strong axial cords, no apertural varix, apertural margin straight. Teleoconch I of 0.9 whorl, 11–13 raised axial cords, no spirals. Teleoconch II of 0.33 whorl, suture moderately impressed. Shoulder flat; axial cords close to suture, more closely spaced than on teleoconch I or on base; approximately six irregularly spaced, very fine spiral lines. Base with distinct constriction below selenizone, low axial lamellae at same density as on teleoconch I, close to apertural margin 14 spiral lines, increasing in strength from lines just above periphery to strong cords surrounding umbilicus. Umbilicus wide, walls straight, with very fine growth marks. Selenizone above periphery, keels low (abraded?), rather strong, lunules distinct, closed foramen, anteriorly closed forming raphe. Aperture rounded D-shaped, roof overhanging.

Animal unknown.

Comparisons. *Sinezona cingulata* from Europe and the eastern Atlantic has a very narrower umbilicus, has a distinct apertural varix on the protoconch, and usually attains a size of greater 1 mm.

Sinezona confusa from the Caribbean to northern Brazil has a more narrow umbilicus, tends to have more but weaker axial cords, has an apertural varix on the protoconch, and usually attains a size of greater 1 mm.

Sinezona garciai from the Caribbean has a flatter spire, a protoconch with finer axial cords, a straight parietal wall, and twice as many axial cords.

Sinezona insignis from South Africa and temperate south Atlantic Islands has a much taller shell with protruding protoconch, much stronger axial lamellae, and a very narrow umbilicus.

Sinezona levigata from New Zealand has an apertural varix on the protoconch and the umbilicus is closed by a callus.

Sinezona semicostata from the subtropical eastern Atlantic off-shore islands has an umbilicus made by the rounded portions of the inner walls of the whorls and the underside of the protoconch is visible through the umbilicus.

Sinezona chilensis from southwestern South America has a protoconch with distinct apertural varix, weaker axial cords, and stronger spiral cords particularly surrounding the umbilicus.

Figure 353. Distribution of *Sinezona insularis*. ☆ type locality.

Distribution. Known only from type locality (Fig. 353).

Sinezona iota (Finlay, 1926)

Figures 354–357

Chresonymy.

Schismope iota Finlay, 1926: 340, not illustrated.
Schismope iota: Fleming, 1950: 25.
Schismope iota: Powell, 1941: 254.
Schismope iota: Powell, 1955: 46.
Schismope iota: Dell, 1956: 168, 182.
Schismope iota: Cernohorsky, 1972: 233, fig. 3.
Sinezona iota: Powell, 1979: pl. 13, fig. 9.
Sinezona iota: Lozouet, 1986: 111.
Sinezona iota: Spencer & Willan, 1995: 13.
Sinezona iota: Geiger, 2003: 78, fig. 15e [protoconch].
Sinezona iota: Geiger, 2008b: fig. 1.
Sinezona iota: Spencer *et al.*, 2009: 202.

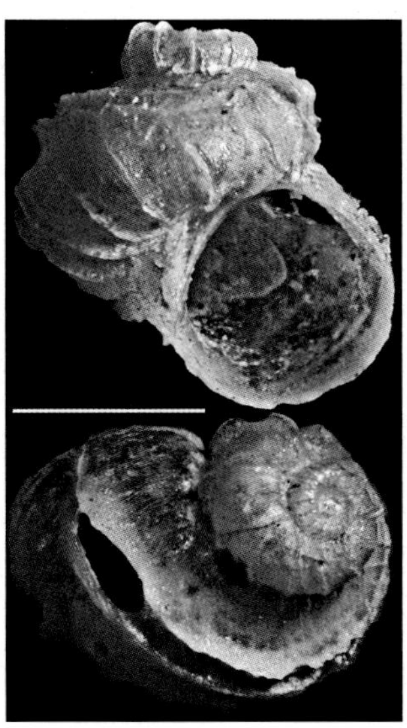

Figure 354. Holotype of *Schismope iota*. Image courtesy Peter Quin/AIM. Scale bar = 0.5 mm.

Figure 355. *Sinezona iota*. **A.** Bathing Beach, Stewart Island, New Zealand, 46.900°S, 168.117°E (AMS C.379062). **B.** 55 m, Stewart Island, Port Adventure, New Zealand, 47.083°S, 168.183°E (AMS C.380677). Radula shown in Figure 356. **C.** 28 m, Stewart Island, off Halfmoon Bay, New Zealand, 46.883°S, 168.133°E (AMS C.406380). Scale bars shell = 0.5 mm. Scale bars protoconch = 100 μm.

Misidentification.

not *Sinezona iota*: Bandel, 1998: 62–63, pl. 21, figs 6–8 [is *Sin. brevis*].

Type material. Holotype (AIM TM-693: Cernohorsky, 1972: 233, fig. 3. AIM AK 70693. Fig. 354), 0.85 × 0.82 mm. Paratypes from type locality (AIM AK 72519, 7; AMS C.51591, 1).

Type locality. Snares Island, [New Zealand, 48.017°S, 166.583°E] in 50 fms [= 92 m] (OD).

Etymology. Iota, Greek letter, associated with small size.

Description. Shell small (to 0.9 mm. Figs 354–355), trochiform turreted. Protoconch of one whorl, strong axial sculpture (sometimes even raised lamellae: see Remarks), covering outer half of whorl, beginning of axials at top broadened, no apertural varix, apertural margin straight to slightly convex. Teleoconch I of one whorl, approximately 11–13 strong axial cords, somewhat reduced near suture, interstices with fine irregular growth marks, no spirals. Teleoconch II of 0.66 whorl, suture deeply impressed. Shoulder without axial cords or cords restricted to near suture, remainder with fine irregular growth marks, no spiral sculpture. Base with strong axial cords to broad lamellae at same density as on teleoconch I, interstices with fine irregular growth marks, no spirals. Umbilicus very narrow, bordered by spiral cord, walls smooth. Selenizone slightly inside periphery, oriented 30°towards coiling axis of shell, keels strong, moderately elevated, lunules indistinct, foramen elongated triangular. Aperture rounded D-shaped.

Operculum, thin, round, multispiral, with central nucleus.

Radula (Fig. 356). Rachidian tooth triangular, tip with nine cusps, central cusp largest. Lateral teeth 1–3 similar, tip with four cusps on outer edge. Lateral tooth 4 reduced, pointed with one indistinct subterminal projection. Lateral tooth 5 broadened, 7–8 cusps on inner edge of tip. Radular interlock of central field

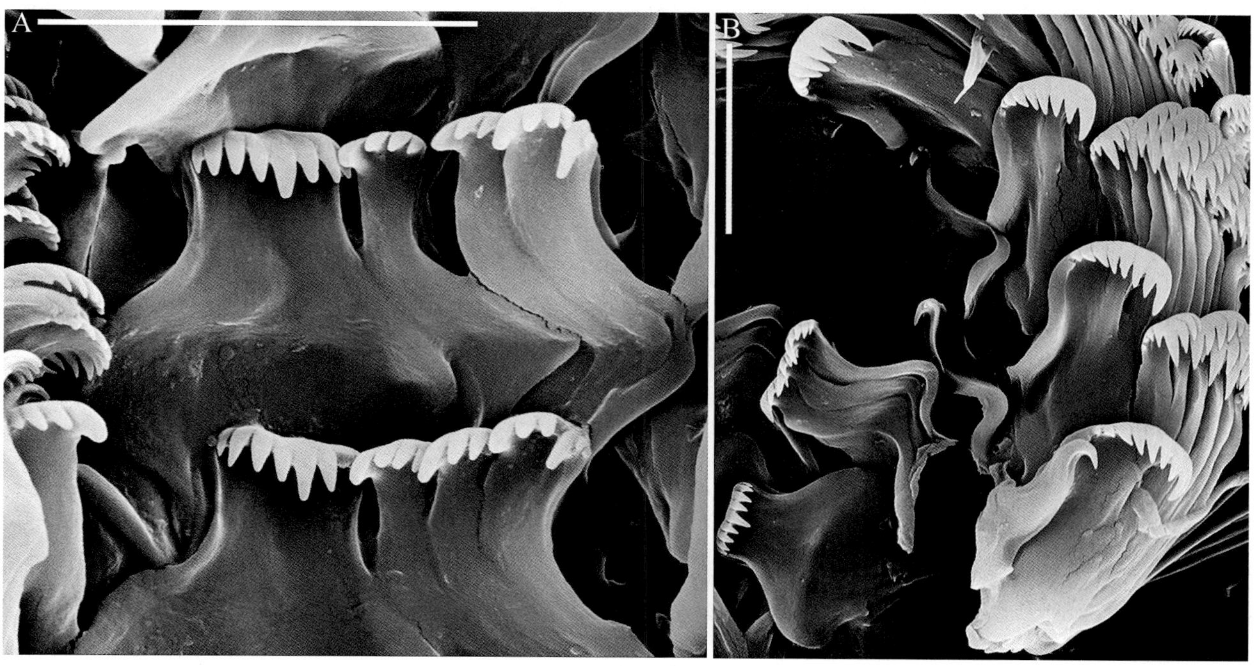

Figure 356. *Sinezona iota*. Radula. 55 m, Stewart Island, Port Adventure, New Zealand, 47.083°S, 168.183°E (AMS C.380677). Shell shown in Figure 355B. **A.** Central field. **B.** Outer lateral teeth and marginal teeth. Scale bars = 10 µm.

moderate. Inner marginal teeth with two small cusps on inner margin, terminal cusp largest, approximately five stronger cusps on outer edge. Outer marginal teeth spoon-shaped, with many fine bristles; shafts without food groove.

Comparisons. *Sinezona bandeli* from New Zealand has a much more depressed spire and spiral sculpture on shoulder and base.

Sinezona beddomei from temperate Australia has a much flatter spire and distinct spiral sculpture on the base.

Sinezona brevis from New Zealand has a much flatter spire and spiral sculpture on shoulder and base.

Sinezona brucei from New Zealand has a wider overall shell shape with flat apex, a wider umbilicus, the axial cords/lamellae do not extend into the umbilicus, and has spiral sculpture on shoulder and base.

Sinezona laqueus from New Zealand has a much flatter spire and a distinct umbilicus.

Sinezona mechanica from New Zealand has a much flatter spire, a distinct umbilicus, a protoconch without apertural varix, and at least twice as many weaker axial cords.

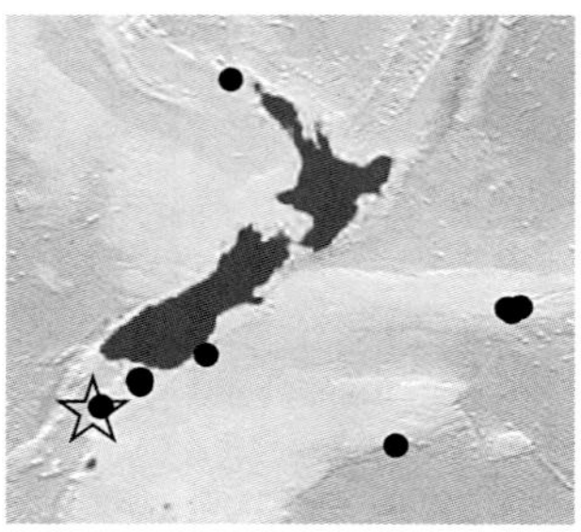

Figure 357. Distribution of *Sinezona iota*. ☆ type locality. ● specimen record.

Sinezona levigata from New Zealand has a much flatter spire, spiral sculpture, and the umbilicus is covered by a callus.

Sinezona norfolkensis from the Tasman Sea has a wider shell, the axials on the protoconch only reach to the midline of the whorl, and has spirals on shoulder and base.

Sinezona pacifica from temperate Australia and New Zealand has a much flatter spire and spiral lines on shoulder and base.

Sinezona pauperata from New Zealand has a much flatter spire, the protoconch lacks an apertural varix, lacks a selenizone, and has a round foramen.

Sinezona platyspira from New Zealand has a much flatter spire, more raised axial lamellae, and the umbilicus is produced by the rounded portions of the previous whorls.

Sinezona subantarctica of broad subantarctic distribution has a much flatter shell, the protoconch lacks an apertural varix, and the umbilicus is covered by a callus.

Sinezona doliolum from South Africa has a much longer selenizone (0.8 vs. 0.5 whorl), a protoconch without apertural varix, and spiral lines on shoulder and base.

Sinezona insignis from South Africa has axial lamellae on teleoconch I not reaching the suture and diminishing in strength towards the umbilicus.

Sinezona mouchezi from the subantarctic islands of the Indian Ocean has spiral sculpture on shoulder and base and much weaker axial cords best developed mid-base.

Distribution. New Zealand (Fig. 357).

Specimen records.

New Zealand. 113 m, off Antipodes Islands, 49.670°S, 178.738°E (NMNZ M.150200, 25). 91 m, Snares Islands, 48.037°S, 166.550°E (USNM 680546, 2). 90 m, Snares Islands, 48.017°S, 166.567°E (GNS RM 5966, 2). 91 m, off The Snares Islands, S of South Island, 48.017°S, 166.583°E (AMS C.406339, 2; AMS C.51591, 1). 55 m, Stewart Island, Port Adventure, 47.083°S, 168.183°E (AMS C.404153, 1). Ocean Beach, Stewart Island, 46.968°S, 168.177°E (RMNH, 1). 55 m, off Whero Island, Carter Passage, Stewart Island, 46.908°S, 168.208°E (NMNZ ex M.019465, 1). Bathing Beach, Stewart Island, 46.900°S, 168.117°E (AMS C.379062, 20+). 28 m, Stewart Island, off Halfmoon Bay, 46.883°S, 168.133°E (AMS C.406380, 1). 46 m, off Fast Rock, Stewart Island, 46.883°S, 168.200°E (NMNZ M.063856, 2). 26 m, off Temarama (Herekopare) Island, Foveaux Strait, 46.867°S, 168.217°E (NMNZ M.019464, 11). 90 m, off Otago Peninsula,

45.817°S, 170.900°E (NMNZ M.130569, 2). 18 m, Heaphe Shoal, SW of Waitangi, Chatham Island, 43.967°S, 176.600°W (NMNZ M.160342, 3). 12–15 m, Ocean Bay, Chatham Island, 43.833°S, 176.783°W (NMNZ M.160345, 13). 0 m, Watson, Okawa Point, Chatham Island, 43.775°S, 176.242°W (NMNZ M.150186, 11). 805 m, off Three Kings Islands, 34.000°S, 171.917°E (NMNZ ex M.20894, 1).

Remarks. The publication date for the species has usually been taken to be 1927. However, the byline under the title notes that the paper had been "issued separately, December 23rd 1926."

One specimen with a protoconch showing raised lamellae and indiscernible border between protoconch and teleoconch I from off Three Kings Islands is known (NMNZ ex M.20894 DLG#7431). In all other aspects it agrees perfectly with all other specimens of *Sin. iota*, hence, is considered a slightly unusual specimen of this species.

Sinezona kayae Geiger & McLean, 2010

Figures 358–361

Chresonymy.

Sinezona sp.: Lozouet, 1986: 111 [noted difference from true *insignis*].

Sinezona kayae Geiger & McLean, 2010: 4–7, figs 1–3.

Sinezona kayae: Geiger, 2011: 38, pl. 3, fig. 5.

Misidentifications.

not *Sinezona insignis* (Smith, 1910): Kay, 1979: 39, figs 7C, 9D–F [is *Sin. kayae*].

Type material. Holotype (SBMNH 83542. Fig. 358). Paratypes from type locality (SBMNH 83543, 5. Fig. 359A–B). 0 m, NE coast, Oahu, Hawaii, USA, 21.620°N, 157.910°W (UF 350462, 1). 0 m, off Kala Place, Diamond Head, Oahu, Hawaii, USA, 21.250°N, 157.800°W (SBMNH 351023, 5: formalin fixed and ethanol preserved; SBMNH 351024, 4: formalin fixed and dried; SBMNH 351025, 1: ethanol fixed and preserved). Hakipu'u Reef, Kaneohe Bay, Oahu, Hawaii, USA, 21.450°N, 157.800°W (BPBM 261475, 1: complete). Moku Manu Island, Kaneohe Bay, Oahu, Hawaii, USA, 21.466°N, 157.717°W (BPBM 263630, 1: complete).

Type locality. 0 m, Haleiwa, Oahu, Hawaii, USA, 21.590°N, 158.113°W (OD).

Etymology. Named in honor of E. Alison Kay of the University of Hawaii for her many accomplishments with respect to the Hawaiian malacofauna (OD).

Description. Shell small (to 0.65 mm. Figs 358, 359), depressed ovoid. Protoconch of 0.75 whorl, with 19–20 strong axials, no apertural varix, apertural margin slightly sinusoidal. Teleoconch I of 0.75 whorl, only axial sculpture, approximately 13 fine, distinct ribs, interstices with fine irregular growth lines. Teleoconch II slightly >0.5 whorl, suture impressed. Shoulder with axial sculpture slightly diminished in strength towards selenizone/foramen, approximately five indistinct axial cords appear towards aperture; base with fine, distinct raised axial cords only, base forming angle with walls of umbilicus. Selenizone approximately 0.25 whorl, slit closed to form foramen. Aperture ovoid, roof strongly protracted, adumbilical margin straight. Umbilicus open, quite wide, border to base with funiculus.

Operculum (Fig. 360C) thin, round, multispiral, with central nucleus.

Radula (Fig. 360A–B). Rachidian triangular, tip with five cusps, central one largest. Lateral teeth 3–5 similar, 3–4 cusps on outer margin. Lateral tooth 4 reduced, pointed. Lateral tooth 5 broadened with approximately 6–8 cusps on inner margin. Radular interlock of central field moderate. Inner marginal teeth with two small cusps on inner margin of tip, terminal cusp largest, approximately five cusps on outer margin. Outer marginal teeth spoon-shaped, many fine bristles on each side of tip.

Comparisons. *Sinezona hawaiiensis* from Hawaii has a more globular shell, lacks a selenizone, has a closed umbilicus, and the protoconch lacks an apertural varix.

Sinezona insignis from South Africa has stronger and fewer axial lamellae and lacks spiral sculpture on shoulder and base.

Sinezona cingulata from the Mediterranean and the adjacent Atlantic ocean has a shorter selenizone (0.2 vs. 0.3 whorl), less distinct axial cords, and a longer teleoconch I (1 vs. 0.8 whorl).

Figure 358. *Sinezona kayae*. Holotype. 0 m, Haleiwa, Oahu, Hawaii, USA, 21.590°N, 158.113°W (SBMNH 83542). Scale bar shell = 0.5 mm. Scale bar protoconch = 100 µm.

Sinezona confusa from the Caribbean to northern Brazil has a longer selenizone (0.5 vs. 0.3 whorl), more but weaker axial cords, and a weaker constriction of the base below the selenizone.

Figure 359. *Sinezona kayae*. **A–B.** Paratypes. 0 m, Haleiwa, Oahu, Hawaii, USA, 21.590°N, 158.113°W (SBMNH 83543, 5). **C.** 0 m, NE coast, Oahu, Hawaii, USA, 21.620°N, 157.910°W (FLMNH 350462, 1). Scale bars shell = 0.5 mm. Scale bars protoconch = 100 μm.

Sinezona costulata from southern Japan is overall wider, lacks the strong spiral cord bordering the umbilicus, and the axial lamellae on the base do not reach the selenizone.

Sinezona danieldreieri from the Indo-Malayan Archipelago to the Central Pacific has an overall broader shell, more and finer axial cords on the teleoconch, axial cords on the shoulder reaching from the suture to the selenizone, and a protoconch with fewer (9–15 vs. 19–20) broader strong axial cords.

Sinezona iota from New Zealand lacks spiral sculpture on shoulder and base and the axial cords retain their strength to the border of the umbilicus.

Sinezona rimuloides from the northeastern Pacific and the Panamic generally has weaker axial cords, a wider umbilicus, and half as many spiral lines on the base.

Distribution. Hawaii, USA (Fig. 361).

Specimen records.

USA, Hawaii. 5–21 m, Honaunau Bay, Big Island, 19.433°N, 155.930°W (LACM 86-421, 1). 0 m, Kona Village Resort, Big Island, 19.831°N, 155.988°W (DLG 485 3: complete; DLG 751, 6). 23 m, Maume Beach (Mauna Kea Beach), N of Hapuna Bay, Big Island,

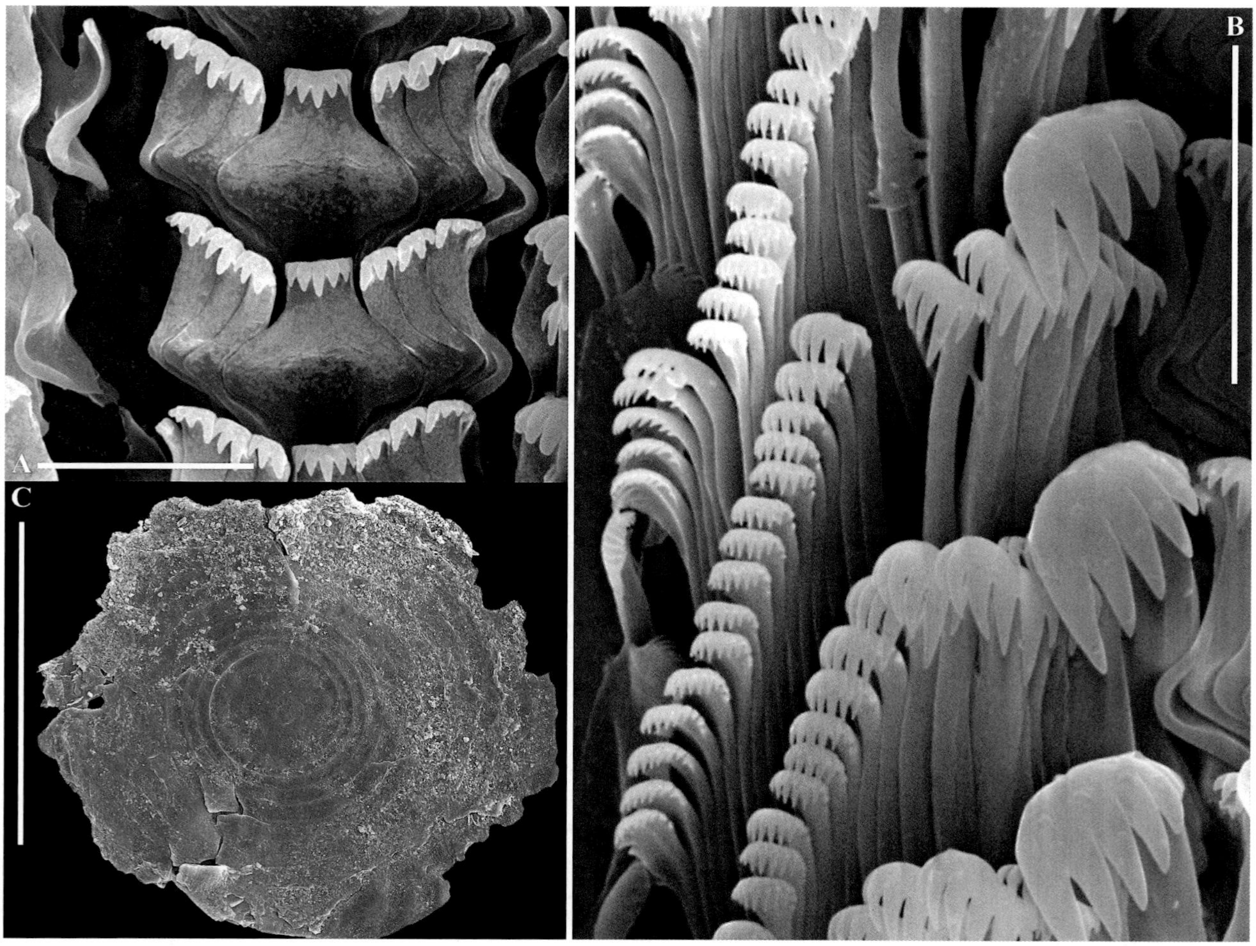

Figure 360. *Sinezona kayae*. Radula and operculum. 0 m, Kona Village Resort, Big Island, Hawaii, USA, 19.831°N, 155.988°W (DLG 485, 3: complete). **A–B.** Radula. **A.** Central field. **B.** Lateral tooth 5 and marginal teeth. **C.** Operculum. Scale bars radula = 10 µm. Scale bar operculum = 200 µm.

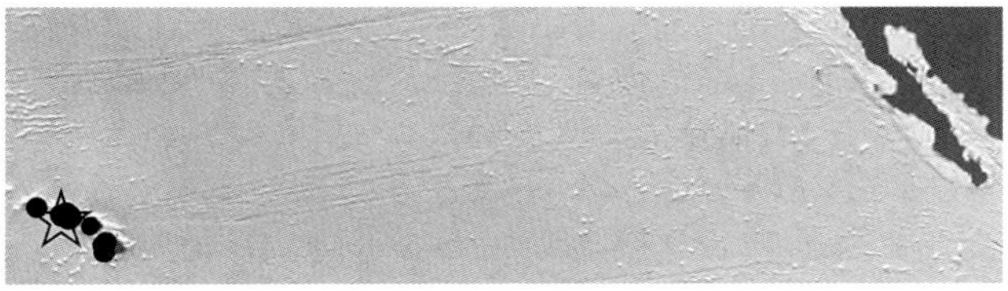

Figure 361. Distribution of *Sinezona kayae*. ☆ type locality. ● specimen record.

20.008°N, 155.828°W (LACM 72-220, 1). 0 m, Makuleia Bay, Maui, 21.012°N, 156.643°W (CPC, 8). 0 m, off Kala Place, Diamond Head, Oahu, 21.250°N, 157.800°W (DLG 223, 12; DLG 231, 11: complete). 0 m, Kala, Diamond Head, Oahu, 21.261°N, 157.787°W (DLG 495, 7: complete; DLG 752, 1; DLG 753, 4). 30 m, off Waikiki, Oahu, 21.270°N, 157.840°W (LACM 75-96, 1). 0 m, Makai Pier, Makapu Beach, Oahu, 21.319°N, 157.670°W (FLMNH 350463, 1). 10 m, Maile, Oahu, 21.418°N, 158.183°W (FLMNH 350460, 1; FLMNH 350476, 1). Hakipu'u Reef, Kaneohe Bay, Oahu, 21.450°N, 157.800°W (BPBM 261475 1: complete). Moku Manu Island, Kaneohe Bay, Oahu, 21.466°N, 157.717°W (BPBM 263630 1: complete). 60 m, Kepuhi Point, Oahu, 21.483°N, 158.228°W (FLMNH 350461, 2). 0 m, Ka'awa, Oahu, 21.555°N, 157.851°W (DLG 473, 2). 0 m, Haleiwa, Oahu, 21.590°N, 158.113°W (SBMNH 83542, 1: holotype; SBMNH 83543, 5: paratypes). 0 m, NE coast, Oahu, 21.620°N, 157.910°W (FLMNH 350462, 1). 30 m, Koloa, Kauai, 21.883°N, 159.475°W (LACM 75-97, 1).

Remarks. Herbert (1986) included Kay's (1979) listing and illustrations of this new species in the synonymy of the South African species, although noting that (p. 628) "the identity of her specimens requires confirmation as the occurrence of *S. insignis*, an essentially warm temperate species, in the tropical Pacific is surprising."

Sinezona laqueus (Finlay, 1926)

Figures 362–368

Chresonymy.

Schismope laqueus Finlay, 1926: 341, pl. 13, figs 30–31.
Schismope laqueus: Powell, 1940: 208.
Schismope laqueus: Powell, 1941: 254.
Schismope laqueus: Fleming, 1950: 25.
Schismope laqueus: Powell, 1955: 46.
Schismope laqueus: Dell, 1956: 168, 182.
Schismope laqueus: Dell, 1960: 145.
Schismope laqueus: Cernohorsky, 1972, 232, fig. 2.
Sinezona laqueus: Powell, 1979: 34–35, pl. 13, fig. 8.
Sinezona laqueus: Lozouet, 1986: 111.
Sinezona laquea [sic]: Beu & Maxwell, 1990: 376.
Sinezona laqueus: Spencer & Willan, 1995: 13.
Sinezona laqueus: Geiger, 2003: 78.
Sinezona laqueus: Spencer *et al.*, 2009: 202.

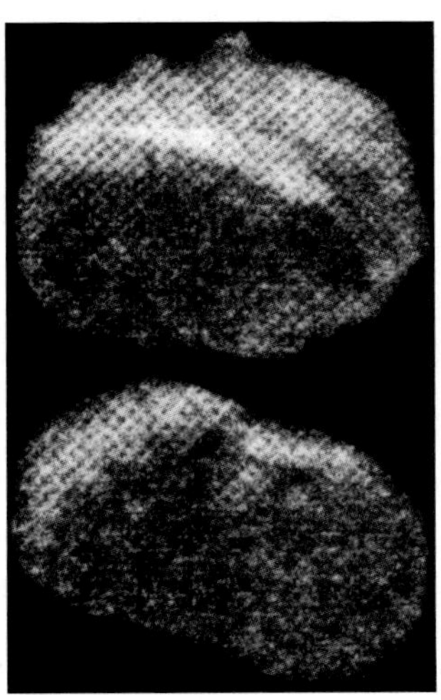

Figure 362. Original illustration of *Schismope laqueus*.

Misidentifications.

Schismope beddomei: Suter, 1909: 7 [is *Sin. laqueus*: Powell, 1979].

Schismope beddomei: Suter, 1913: 90, pl. 6, fig. 13 [1917: is *Sin. laqueus*].

Type material. Holotype (AIM TM-695: Cernohorsky, 1972: 232, fig. 2. AIM AK 70695. Fig. 363), 1.25 × 0.95 mm. Paratype from type locality (AMS C.51590, 1).

Type locality. Snares Island, [New Zealand, 48.017°S, 166.583°E] in 50 fathoms [= 92 m] (OD).

Etymology. Laqueus: Latin for snare, noose (noun, masculine), referring playfully to the type locality.

Description. Shell medium-sized (to 1.5 mm. Figs 363–366), trochiform depressed, Protoconch of 0.75 whorl, strong axials, apertural varix not connected to embryonic cap, apertural margin straight. Teleoconch I of 0.75 whorl, approximately 12 broad axial lamellae not quite reaching suture, interstices with fine irregular growth marks, no spiral sculpture. Teleoconch II of over one whorl, suture little impressed. Shoulder concave, usually no macroscopic sculpture, only fine irregular growth lines; some with axial cords in adsutural quarter. Base with distinct constriction below selenizone and subsequent distension, strong axial lamellae starting below constriction, highest at periphery, gradually becoming indistinct towards umbilicus; approximately five spirals, increasing in strength from fine lines below periphery to low cords at entrance to umbilicus. Umbilicus open wide, bordered by indistinct cord, walls almost straight, with fine axial growth lines. Selenizone above periphery, oriented at approximately 60°towards coiling axis, keels strong, high, lunules very distinct, approximately three times as dense as axial lamellae on shell; foramen elongated tear-drop-shaped, closed anteriorly by raphe. Aperture oblique, roof overhanging, D-shaped.

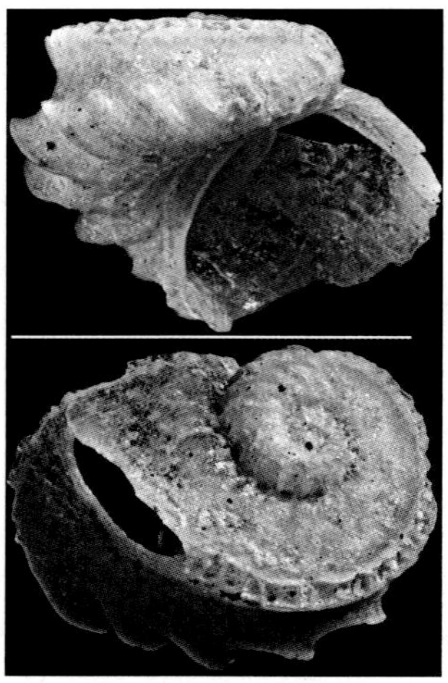

Figure 363. Holotype of *Schismope laqueus*. Image courtesy Peter Quin/AIM. Scale bar = 1 mm.

Operculum (Fig. 367E) covering more or less aperture, corneous, thin, nucleus central, multispiral.

Radula (Fig. 367). Rachidian tooth broad triangular, apical cusp not enlarged, three cusps on each side. Lateral teeth 1–3 similar, apical cusp largest, 2–4 cusp along outer edge of tip. Lateral tooth 4 reduced, hook-shaped. Lateral tooth 5 enlarged by broadening, apical cusp largest, 5–6 cusps along inner edge of tip. Inner marginal teeth with palmate tip, apical cusp largest, 1–2 cusp along inner edge, 4–5 cusps on outer. Outer marginal teeth with spoon-shaped tip, many fine bristles along edge. Outermost marginal teeth paddle-shaped with fine bristles at tip. Radular interlock of central field moderate.

Comparisons. *Sinezona bandeli* from New Zealand has spiral lines on the shoulder, a shorter selenizone (0.5 vs. 0.75 whorl), and weaker lunules in the selenizone.

Sinezona beddomei from temperate Australia has a shorter selenizone (0.5 vs. 0.75 whorl), lacks the nodular thickenings of the spiral cords near the umbilicus, and tends to have more but slightly weaker axial lamellae.

Sinezona brevis from New Zealand has an overall more globular shell, weaker axial lamellae, and distinct spiral lines on shoulder and base.

Sinezona brucei from New Zealand has a taller overall shell shape, spiral sculpture on shoulder and base, and much weaker lunules.

Sinezona iota from New Zealand has taller shell, much weaker lunules, and a much more narrow umbilicus.

Sinezona koruahina from the Pliocene of New Zealand has strong axial sculpture on the shoulder.

Sinezona levigata from New Zealand has a globular shell, spiral sculpture on shoulder and base, and the umbilicus is covered by a callus.

Sinezona mechanica from New Zealand has a more globular shell, much weaker and numerous axial cords, and much weaker keels of the selenizone and lunules.

Sinezona norfolkensis from the Tasman Sea has a weaker constriction on the base below the selenizone, spiral sculpture on the shoulder of teleoconch II, and weaker lunules.

Sinezona pacifica from temperate Australia and New Zealand has spiral sculpture on shoulder and base, a less distinct constriction of the base below the selenizone, and a more narrow umbilicus.

Sinezona pauperata from New Zealand has a globular shell, lacks a selenizone and a very narrow umbilicus.

Sinezona platyspira from New Zealand has a flat spire, a protoconch with very fine axial cords and lacking an apertural varix, and the umbilicus is constructed by the rounded portions of the previous whorls.

Figure 364. *Sinezona laqueus.* 55 m, Carter Passage, off Whero Island, Stewart Island, New Zealand, 46.908°S, 168.208°E (NMNZ M.019459, 30). Radula shown in Figure 367. Scale bar shell = 1 mm. Scale bar protoconch = 100 µm.

Sinezona subantarctica with broad subantarctic distribution has an overall rounded shell, lacks a selenizone and the umbilicus is covered by a callus.

Sinezona costulata from southern Japan has distinct spiral lines on shoulder and base, much less distinct lunules, and a shorter selenizone (0.4 vs. 0.8 whorl).

Distribution. New Zealand endemic, Pliocene to Recent (Fig. 368).

Figure 365. *Sinezona laqueus*. **A.** 55 m, off Fishing Rock, Foveaux Strait, New Zealand, 46.567°S, 168.000°E. (USNM 609795). **B–C.** 201–216 m, Middlesex Bank, NW of Three Kings Island, New Zealand, 34.015°S, 171.745°E (NMNZ M.093303, 6). Scale bars shell = 1 mm. Scale bars protoconch = 100 µm.

Specimen records.

New Zealand. 201–216 m, Middlesex Bank, NW of Three Kings Island, 34.015°S, 171.745°E (NMNZ M.093303, 6). 46 m, North Cape, North Island, 34.417°S, 173.033°E (AMS C.461505, 5). Doubtless Bay, 34.900°S, 173.400°E (MV F 96917, 1). 46 m, off Cavalli Islands, 34.967°S, 173.967°E (AMS C.461515, 6). Taupo Bay, 34.983°S, 173.717°E (AMS ex C.461509, 4). Shell sand, Spirits Bay, N Northland, 34419.000°S, 172.856°E (GNS RM 1037, 1). 11–18 m, Mangonui, 35.000°S, 173.567°E (MCZ 084665, 2).

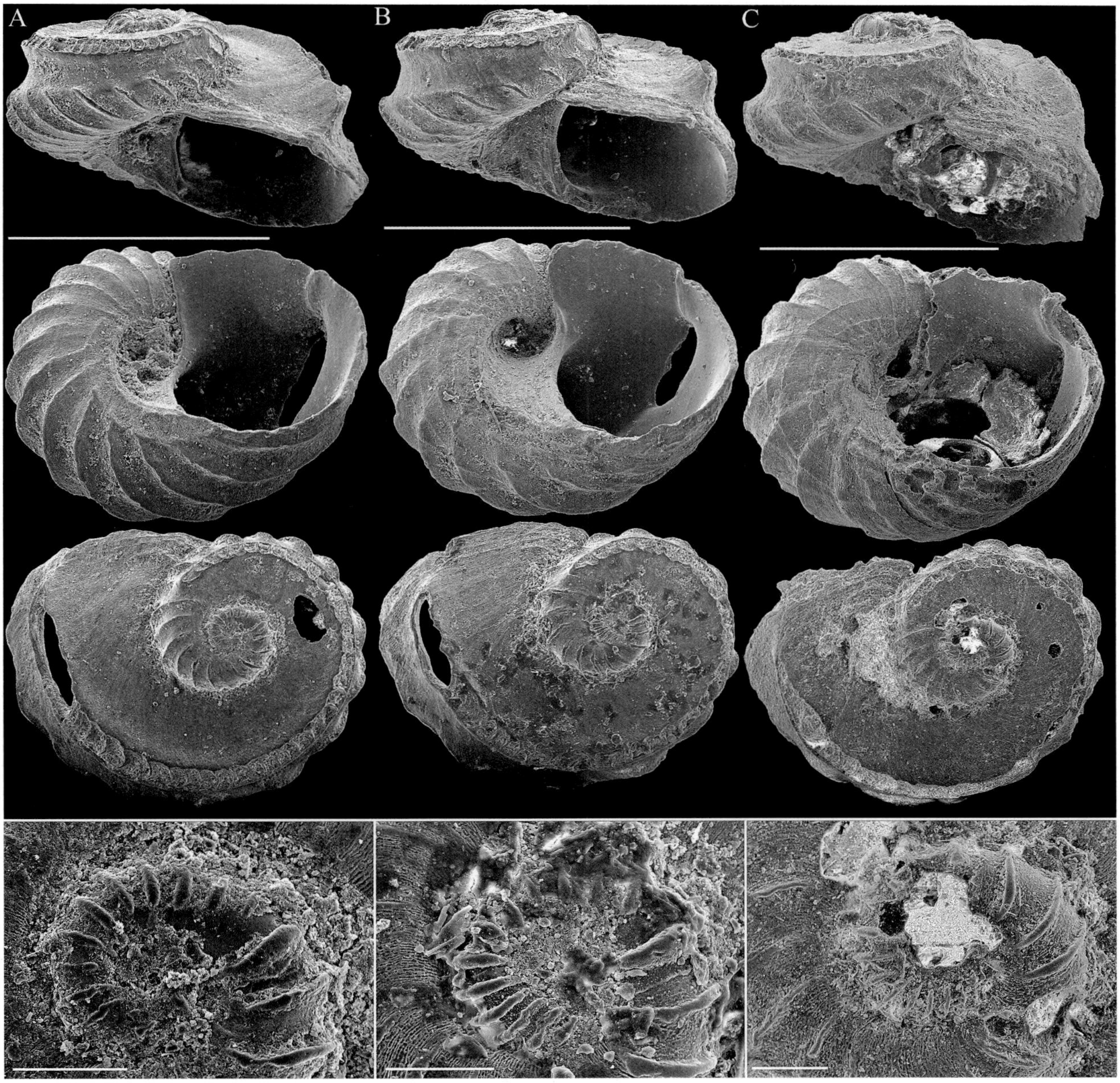

Figure 366. *Sinezona laqueus*. Pleistocene fossils. **A–B.** GS4174, Oturi Shellbed, Hauriri Terrace cover beds (oxygen isotope stage 5a, late middle Pleistocene), Waipipi, Waverly Beach, W of Wanganui, New Zealand, 39.830°S, 174.627°E (GNS, 2). **C.** GS14697, high terrace remnant (Hawaran, oxygen isotope stage 5c?, late middle Pleistocene) on NE corner of Haumuri Bluff, Marlborough, NE South Island, New Zealand, 42.552°S, 173.519°E (GNS, 7). Scale bars shell = 1 mm. Scale bars protoconch = 100 µm.

Awanui Bay, Northland, 35.047°S, 173.256°E (GNS RM 1038, 1). 24 m, Whangaroa Harbour entrance, 35.100°S, 173.867°E (AMS C.461502, 6). 41–43 m, Poor Knight Island, N end of North Island, 35.442°S, 174.733°E (LACM 65-165, 12). Poor Knights, 35.500°S, 174.750°E (WAM, 10). Langs Beach, N of Mangawhai, 36.050°S, 174.533°E (AMS ex C.461506, 1). 18 m, W side Goat Island, off Leigh Marine Laboratory, Leigh, North Island, 36.267°S, 174.792°E (LACM 87-84, 2). 0 m, near Leigh Marine Laboratory, Leigh, North Island, 36.270°S, 174.775°E (LACM 87-82, 2). McGregor Bay, 36.750°S, 175.467°E (AMS C.461492, 3). Leask's Bay, Stewart Island, 40.900°S, 173.900°E (HLC, 1). Waitangi, Chatham Island, 43.950°S, 176.533°W (USNM 680550, 11). 10–14 m, St. Anne Bay, Milford Sound entrance, 44.575°S, 167.783°E (NMNZ M.142933, 1). 62 m, New Golden Hind, entrance to George Sound, Fiordland, 44.839°S, 167.348°E (GNS RM 1039, 3). 12 m, inside Entrance Island, Nancy Sound, Fiordland, 45.103°S, 167.020°E (NMNZ exM.140629, 1). 106 m, inside rock bar, Daggs Sound, Fiordland, SW South Island, 45.312°S, 166.978°E (GNS RM 1046, 5). 75 m, off Otago, 45.800°S, 170.883°E (AMS C.379058, 1). 73–91 m, Cape Saunders, Otago Head, 45.866°S, 170.750°E (USNM 680549, 3). Curio Bay, South Island, 46.455°S, 169.104°E (RMNH AD 29461, 1). 55 m, off Fishing Rock, Foveaux Strait, 46.567°S, 168.000°E (USNM 609795, 6). 55 m, Fast, Foveaux Strait, 46.567°S, 168.000°E (USNM 609855, 2). 9 m, Zife Rock, Foveaux Strait, South Island, 46.567°S, 168.000°E (USNM 609777, 1). 37 m, Middle Grounds, Foveaux Strait, 46.567°S, 168.000°E (USNM 609889, 3). 37 m, 0.5 mi NW of Fish Rock, Foveaux

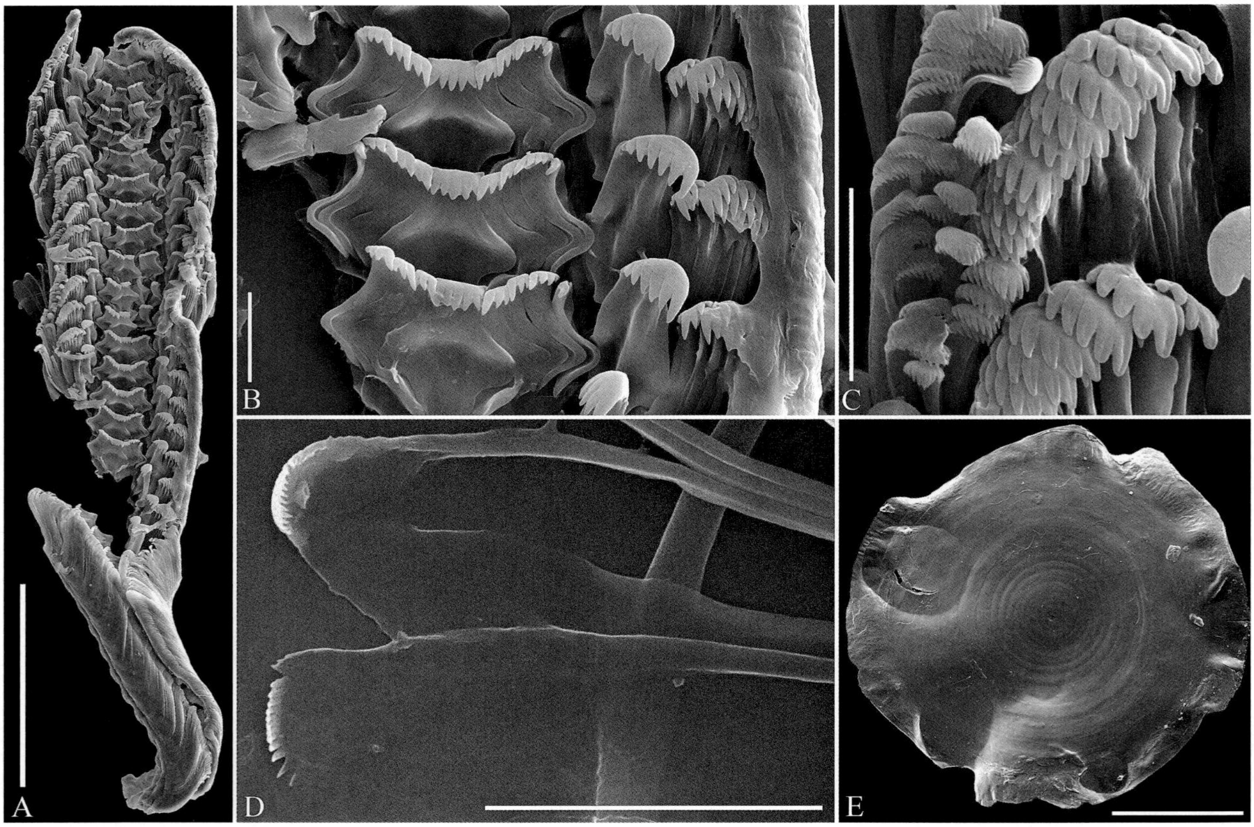

Figure 367. *Sinezona laqueus*. Radula and operculum. 55 m, Carter Passage, off Whero Island, Stewart Island, New Zealand, 46.908°S, 168.208°E (NMNZ M.019459, 30). Shell shown in Figure 364. **A–D.** Radula. **A.** Entire radula. Scale bar = 100 µm. **B.** Half width of radular rows. **C.** Marginal teeth. **D.** Outermost marginal teeth. Scale bars B–D = 10 µm. **E.** Operculum. Scale bar = 200 µm.

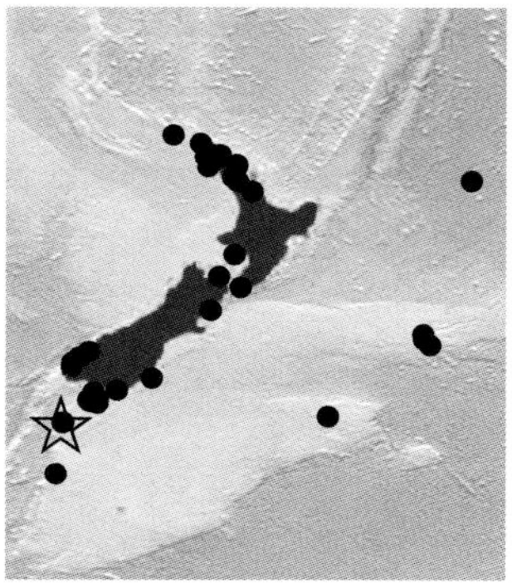

Figure 368. Distribution of *Sinezona laqueus*. ☆ type locality. ● specimen record.

Strait, 46.567°S, 168.000°E (USNM 609919, 1). 46 m, Middle Ground near Ruapuke Island, Foveaux Strait, 46.567°S, 168.000°E (USNM 609803, 1). 55 m, off Fish Rock, Foveaux Strait, South Island, 46.567°S, 168.000°E (USNM 610110, 1). 28 m, Stewart Island, off Halfmoon Bay, 46.883°S, 168.133°E (AMS C.379045, 1; AMS C.379057, 12). 4 m, off Fast Rock, Stewart Island, 46.883°S, 168.200°E (NMNZ M.019461, 3). Sand, bathing beach (i.e., Halfmoon Bay), Stewart Island, 46.899°S, 168.130°E (GNS RM 2393, 8). Bathing Beach, Stewart Island, 46.900°S, 168.117°E (AMS C.379062, 20+). Stewart Island, 46.900°S, 168.167°E (SBMNH 85175, 4). 55 m, Carter Passage, off Whero Island, Stewart Island, 46.908°S, 168.208°E (NMNZ M.019459, 30). 55 m, Jacky Lee, off Paterson Inlet, Stewart Island, 46.916°S, 168.050°E (USNM 610436, 1). Mason Bay, Stewart Island, 46.926°S, 167.757°E (NMNZ M.091940, 13). Ocean Beach, Stewart Island, 46.968°S, 168.177°E (USNM 617420, 1; MCZ 204561, 3; RMNH, 10). Bathing Beach, Stewart Island, 47.000°S, 167.833°E (USNM 610511, 2). 55 m, Stewart Island, Port Adventure, 47.083°S, 168.183°E (AMS ex C.379060, 1). 457 m, Bounty Islands, 47.700°S, 179.067°E (AMS C.17091, 1). 90 m, Captain Bollons, Bounty Islands, 47.718°S, 179.079°E (GNS RM 5965, 3). 90 m, Snares Islands, 48.016°S, 166.567°E (GNS RM ex5966, 11). 91 m, Snares Islands, S of South Island, 48.017°S, 166.583°E (AMS C.51590, 1). Snares Island, 48.017°S, 166.583°E (AIM TM 695, 1). Snares Islands, 48.036°S, 166.550°E (MHNG 14356, 2). 91 m, Snares Islands, 48.036°S, 166.550°E (USNM 200974, 2). 4–18 m, Awanui, N Auckland Island, 50.533°S, 166.233°E (USNM 680551, 1).

New Zealand, fossil. GS12161, cliffs near "The Point" homestead, W of Flowerpot Bay, N cost of Pitt Island, Chatham Islands, Waipipian-Mangapanian (mid-late Pliocene), 43.710°S, 176.560°W (GNS, 3). GS12163, Moutapu Point, NE Pitt Island, Chatham Islands, Whenuataru Tuff, Waipipian-Mangapanian (Mid-late Pliocene), New Zealand, 36.239°S, 174.293°W (GNS TM, 17). GS14697, high terrace remnant (Hawaran, oxygen isotope stage 5c?, late middle Pleistocene) on NE corner of Haumuri Bluff, Marlborough, NE South Island, 42.552°S, 173.519°E (GNS, 7). GS4174, Oturi Shellbed, Hauriri Terrace cover beds (oxygen isotope stage 5a, late middle Pleistocene), Waipipi, Waverly Beach, W of Wanganui, 39.830°S, 174.627°E (GNS, 2). Kaawa Creek, coast S of Waikato Heads, SW Auckland; Opoitan, early Pliocene, 35.550°S, 173.383°E (GNS TM 8661, 1).

Literature records.

New Zealand. 50 fathoms [= 91 m], Snares Islands. 15 fathoms [= 27 m], off Leeward Island, Antipodes Islands, New Zealand (Powell, 1955).

New Zealand, fossil. Haweran = late Pleistocene of New Zealand (Beu & Mawell, 1990).

Remarks. Beu & Maxwell (1990) noted the feminine gender of *Sinezona* and produced a gender agreement for the masculine species epithet. However, the species epithet is derived from a Latin noun, while gender agreement only applies to adjectives. Hence, the inclination introduced by these authors is inappropriate.

The publication date for the species has also been taken to be 1927. However, the byline under the title notes that the paper had been "issued separately, December 23rd 1926."

Sinezona levigata (Iredale, 1908)

Figures 369–374

Chresonymy.

Schismope brevis levigata Iredale, 1908a: 381, not illustrated.
Schismope brevis levigata: Suter, 1913: 91–92.
Schismope brevis var. *laevigata* [sic]: Finlay, 1924: 517.
Sinezona laevigata [sic]: Fleming, 1926: 341.
Sinezona laevigata [sic]: Fleming, 1950: 25.
Sinezona laevigata [sic]: Dell, 1951: 16.
Sinezona laevigata [sic]: Powell, 1955: 46.
Sinezona laevigata [sic]: Dell, 1960: 145.
Sinezona levigata: Powell, 1979: 35.
Sinezona levigata: Lozouet, 1986: 111.
Schismope brevis levigata: Freeman *et al.*, 1997: 36.
Sinezona levigata: Spencer & Willan, 1995: 13.
Sinezona levigata: Marshall, 2002: 171–172, figs 10–12.
Sinezona levigata: Geiger, 2003: 78.
Sinezona levigata: Spencer *et al.*, 2009: 202.

Synonyms.

+ *Ariella campbelli* Bandel, 1998: 64, pl. 22, figs 5–7. ***Type material.*** Holotype (AMS C.162251. Fig. 369), 1 mm. ***Type locality.*** Coastal region of Campbell Islands [New Zealand, 52.533°S, 169.200°E] (OD). ***Etymology.*** Named for its type locality in the Campbell Islands (OD).

Figure 369. Original illustration of *Ariella campbelli.*

Misidentifications.

Sinezona brevis: Fleming, 1926: 341 [is *Sin. levigata*].
Sinezona brevis: Bandel, 1998: 61, pl. 21, figs 2–5 [is *Sin. levigata*].

Type material. Probable holotype (AMS C.27126). Type (CM: Suter, 1913). Holotype (CM?: Powell, 1955). Holotype probably lost; not in CM (Freeman *et al.*, 1997; Marshall, 2002).

Type locality. Sandfly Bay, Otago Peininsula, [New Zealand, 45.900°S, 170.650°E] (OD).

Etymology. Laevigata, levis, Latin for smooth.

Description. Shell medium-sized (to 1.2 mm. Figs 370–372), trochiform depressed. Protoconch of one whorl, strong axial cords, apertural varix barely connected to embryonic cap, apertural margin straight. Teleoconch I of 1.125–1.33 whorls; 20–35 axials, low number of low lamellae, high number of low cords; spiral lines starting as early as 0.25 teleoconch I whorl, increasing to approximately five at onset of selenizone. Teleoconch II of 0.25–0.33 whorl, suture impressed. Shoulder flat, same strength and density of axials, 5–7 spiral lines to broad low spiral cords at apertural margin, spacing irregular. Base rounded, same axial sculpture as on shoulder, 12–18 spiral lines regularly or irregularly spaced with tendency to be more distinct in vicinity of umbilicus. Umbilicus partially closed by distinct callus from parietal margin, wall straight, at blunt angle to base. Selenizone above periphery, short (as long as foramen), foramen closed, with anterior raphe. Aperture D-shaped, roof overhanging.

Radula (Fig. 373). Rachidian, central cusp as large as adjacent ones, 3–4 cusps on each side. Lateral teeth 1–3 similar, 4–5 cusps on outer edge of tip, apical-most cusp largest. Lateral tooth 4 reduced, hook-shaped, without minute points. Lateral tooth 5 enlarged by broadening, approximately eight cusps along inner edge of tip, apical-most cusp largest. Inner marginal teeth with tip forming inequilateral triangle, apical-most cusp largest, inner edge with four cusps, outer edge with 4–5 cusps.

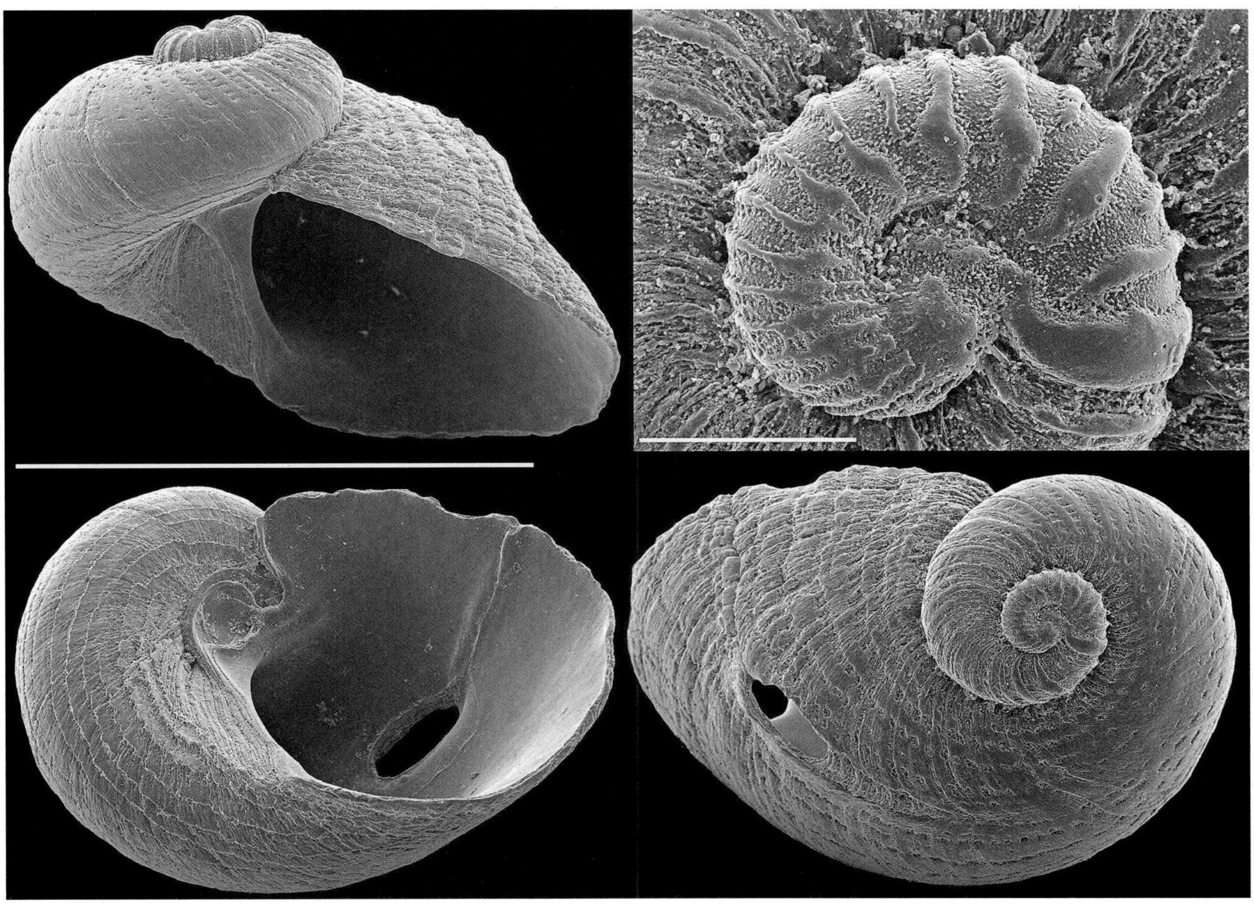

Figure 370. *Sinezona levigata*. 113 m, off Antipodes Island, New Zealand, 49.670°S, 178.738°E (NMNZ M.150138, 5). Scale bar shell = 1 mm. Scale bar protoconch = 100 μm.

Outer marginal teeth spoon-shaped, with many fine bristles along edge. No food groove on shafts of marginal teeth. Radular interlock of central field moderate.

Comparisons. *Sinezona bandeli* from New Zealand has a more subquadrate shell shape, lacks the umbilical callus, and has a much longer selenizone (0.4 vs. 0.1 whorl).

Figure 371. *Sinezona levigata*. **A–B.** 91 m, off Big South Cape Island, Stewart Island, New Zealand, 47.250°S, 167.400°E (NMNZ M.019455, 100). **C.** Juvenile with open slit. 32 m, outer George Sound, Fiordland, South Island, New Zealand, 44.867°S, 167.367°E (NMNZ M.150204, 15). Scale bars shell = 0.5 mm. Scale bars protoconch = 100 μm.

Sinezona brevis from New Zealand has an open umbilicus bordered by a strong periumbilical cord.

Sinezona iota from New Zealand has a much more elevated spire, a longer selenizone, and the umbilicus not covered by a callus.

Sinezona laqueus from New Zealand has a subquadrate overall shell shape, strong axial lamellae, a wide and open umbilicus, and a much longer selenizone.

Figure 372. *Sinezona levigata*. **A.** Bathing Beach, Stewart Island, New Zealand, 46.900°S, 168.117°E (AMS C.380179). **B.** Auckland Islands, Lindley Point, New Zealand, 50.533°S, 166.233°E (AMS C.380177). **C.** Cliffs near "The Point" homestead, W of Flowerpot Bay, N coast of Pitt Island, Chatham Islands, Whenuataru Tuff, Waipipian-Mangapanian (middle to late Pliocene) (GS 12161). Scale bars shell = 0.5 mm. Scale bars protoconch = 100 µm.

Sinezona mechanica from New Zealand has a protoconch without apertural varix, a longer selenizone (0.3 vs. 0.1 whorl), and an open umbilicus bordered by a periumbilical cord.

Sinezona pacifica from temperate Australia to New Zealand has a much longer selenizone (0.6 vs. 0.1 whorl), a more depressed spire, and the umbilicus is not covered by a callus.

Sinezona pauperata from New Zealand has a protoconch without apertural varix, lacks a selenizone, has a round foramen, and the umbilicus is not covered by a callus.

Sinezona platyspira from New Zealand has a flat spire, the protoconch lacks an apertural varix, and the wide umbilicus is not covered by a callus.

Sinezona subantarctica with broad subantarctic distribution has a protoconch with fine irregular axial cords and no apertural varix and a proportionally wider shell.

Distribution. Southern North Island, Stewart, Snares, Auckland, Campbell, Antipodes, Bounty, Chatham Islands, New Zealand, 0–113 m, live 0–10 m from algae (Marshall, 2002). Macquarie Island, Crozet Islands, Magellan Region. 0–15 m (living specimens: 1–5 m).

Specimen records.

New Zealand. 0 m, Wellington, Island Bay, North Island, 41.345°S, 174.773°E (NMNZ 010293, 3). 8–12 m, Inner NE side of Jackson Head, 43.960°S, 168.624°E (NMNZ 160397, 8). 62 m, Entrance to George Sound, Fiordland, 44.839°S, 167.348°E (GNS RM 1049, 2). 32 m, outer George Sound, Fiordland, South Island, 44.866°S, 167.367°E (NMNZ 150204, 15). 12 m, inside Entrance Island, Nancy Sound, Fiordland, South Island, 45.103°S, 167.020°E (NMNZ 140662, 13). 106 m, New Golden Hind Station B80, inside rock bar, Daggs Sound, Fiordland, South Island, 45.400°S, 166.810°E (GNS RM 1043, 1). Curio Bay, South Island, 46.455°S, 169.104°E (RMNH AD 29462, 1). 0 m, Butterfield Beach, Stewart Island, 46.896°S, 168.132°E (NMNZ 010295, 21). Bathing Beach, Stewart Island, 46.900°S, 168.117°E (AMS C.380179, 20+). Paterson Inlet, The Neck, Stewart Island, 46.940°S, 168.090°E (BPBM 226029, 1). Ocean Beach, Stewart Island, 46.968°S, 168.177°E (RMNH, 6; GNS RM 3081, 30). 91 m, off Big South Cape Island, Stewart Island, 47.250°S, 167.400°E (NMNZ 019455, 100). 55 m, off Putama Island, South Cape, Stewart Island, 47.266°S, 167.383°E (NMNZ 028376, 100). 37 m, off The Breaks, Snares Islands, 48.015°S, 168.446°E (NMNZ 143235, 50). 90 m, Snares Islands, 48.016°S, 166.567°E (GNS

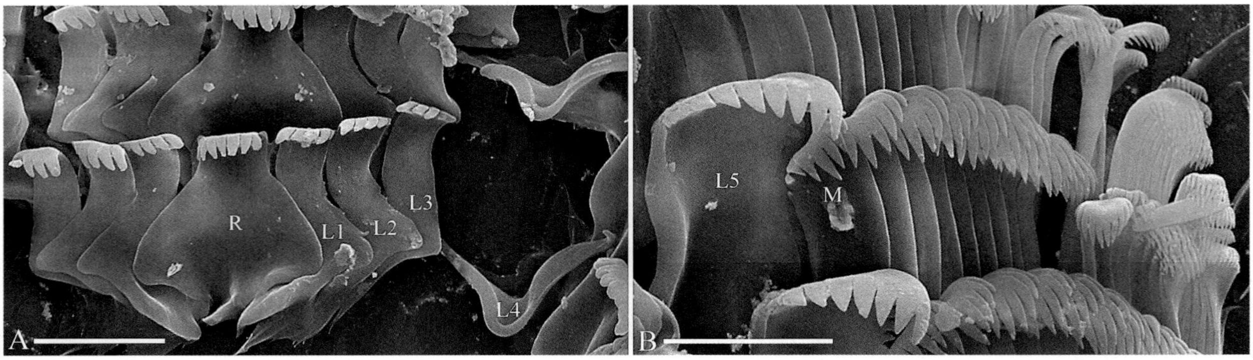

Figure 373. *Sinezona levigata*. Radula. 16 m, E side of Perpendicular Head, Antipodes Islands, New Zealand, 49.650°S, 178.883°E (NMNZ M.150210, 50). **A.** Central field. **B.** Lateral tooth 5 and marginal teeth. R: Rachidian tooth. L1–5: Lateral teeth 1–5. M: Marginal teeth. Scale bars = 10 μm.

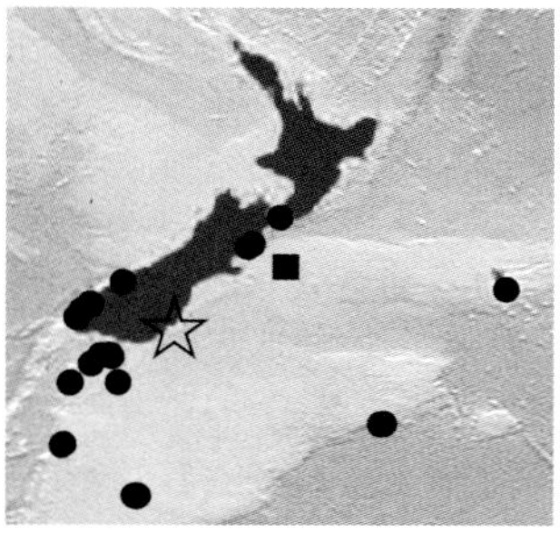

Figure 374. Distribution of *Sinezona levigata*. ☆ type locality. ● specimen record.

RM 5956, 2; GNS RM 5967, 5). 91 m, off The Snares Islands, S of South Island, 48.017°S, 166.583°E (AMS C.380175, 13). Snares Islands, 48.036°S, 166.550°E (MHNG 14354, 2). 91 m, Snares Island, 48.036°S, 166.550°E (USNM 680544, 1). 15 m, Clio Bay, Antipodes Islands, 49.650°S, 178.883°E (NMNZ 100499, 11). 16 m, E side of Perpendicular Head, Antipodes Islands, 49.650°S, 178.883°E (NMNZ 150210, 50). 113 m, off Antipodes Island, 49.669°S, 178.738°E (NMNZ 150138, 5). 0 m, Derry Caslte Reef, Enderby Island, Auckland Islands, 50.483°S, 166.303°E (NMNZ 049658, 16). Auckland Islands, Lindley Point, 50.533°S, 166.233°E (AMS C.380177, 2). 10 m, NW side first bay NW of Boulder Beach, Smoothwater Bay, Campbell Island, 52.533°S, 169.200°E (AMS C.376624, 14). 4 m, NW side of Boulder Beach, Smoothwater Bay, Campbell Island, 52.533°S, 169.200°E (AMS C.376638, 2). 8 m, NW side first bay NW of Boulder Beach, Smoothwater Bay, Campbell Island, 52.533°S, 169.200°E (AMS C.376645, 11, –: complete). 8 m, Smoothwater Bay, Campbell Island, 52.533°S, 169.200°E (AMS C.376646, 7, –: complete). E side Windlass Bay, Campbell Island, 52.550°S, 169.067°E (AMS C.376627, 1). 3–6 m, E side Windlass Bay, Campbell Island, 52.550°S, 169.067°E (AMS C.376629, 3). 8 m, Mouth of small cove E of Limestone Pt, Northwest Bay, Campbell Island, 52.550°S, 169.067°E (AMS C.376633, 5). 8 m, mouth of Windlass Bay, Campbell Island, 52.550°S, 169.067°E (AMS C.376639, 7). 6 m, E side Windlass Bay, Campbell Island, 52.550°S, 169.067°E (AMS C.376640, 3, –: complete). 10 m, E side Vire Point, Perseverance Harbour, Campbell Island, 52.550°S, 169.167°E (AMS C.376625, 3). 3 m, E side Vire Point, Perseverance Harbour, Campbell Island, 52.550°S, 169.167°E (AMS C.376630, 2; AMS C.376632, 4: complete). 10 m, W of East Cape, Smoothwater Bay, Campbell Island, 52.550°S, 169.217°E (AMS C.376631, 3). 3 m, Caves E of Divers Point, Perseverance Harbour, Campbell Island, 52.567°S, 169.183°E (AMS C.376641, 7). 16 m, W side Davis Point, Perseverance Harbour, Campbell Island, 52.567°S, 169.217°E (AMS C.376626, 5; AMS C.376628, 1). W side Davis Point, Perseverance Harbour, Campbell Island, 52.567°S, 169.217°E (AMS C.376643, 1). 8 m, W side Davis Point, Perseverance Harbour, Campbell Island, 52.567°S, 169.217°E (AMS C.376644, 1). 2 m, W side Southeast Harbour, Campbell Island, 52.600°S, 169.150°E (AMS C.376634, 20+, –: complete). W side Southeast Harbour, Campbell Island, 52.600°S, 169.150°E (AMS C.376637, 2).

Fossil.

New Zealand. GS 12161, cliffs near "The Point" homestead, W of Flowerpot Bay, N coast of Pitt Island, Chatham Islands, Whenuataru Tuff, Waipipian-Mangapanian (middle to late Pliocene), 44.240°S, 176.239°W (GNS, 5). GS14697, high terrace remnant (Hawaran, oxygen isotope stage 5c?, late middle Pleistocene) on NE corner of Haumuri Bluff, Marlborough, NE South Island, 42.552°S, 173.519°E (GNS, 50). GS17989, Kaikoura peninsula, high terrace (Oxygen isotype stage 5c?, late middle Pleistocene, hand-augured hole near high point of Kaikoura Peninsula, Marlborough, NE South Island, 42.422°S, 173.693°E (GNS, 50).

Literature records.

New Zealand. 95 m, Mernoo Bank, 43.350°S, 175.000°E (Dell, 1951).

Remarks. The shell of the species is quite variable. The strength of the sculpture ranges from indistinct reticulate to prominent low lamellae. Shapewise, compressed shells as well as squat specimens are equally common. There is every conceivable transitional specimen even in the same lot, hence, segregation of the species into several subspecies or forms is not defensible. The protoconch structure is remarkably similar among all specimens examined.

Sinezona macleani Geiger, 2006

Figures 375–378

Chresonymy.

Sinezona macleani Geiger, 2006a: 16–19, figs 12–13.

Sinezona macleani Bouchet *et al.* 2008: 26.

Sinezona macleani: Spencer *et al.*, 2009: 202.

Type material. Holotype (USNM 289748. Fig. 375), 2.3 mm. Paratypes from type locality (USNM 276829, 2). Paratypes 820–863 m, Fiji, 17.300°S, 179.550°W (MNHN, 1. Fig. 376A). 675–680 m, New Caledonia, 23.167°S, 167.167°E (MNHN, 2. Figs. 376B–C). 1000 m, off Curtis Island, Kermadec Islands, New Zealand, 30.467°S, 178.622°W (NMNZ M257059, 1).

Type locality. 534 m, Station 5584, Borneo, Sibuko Bay, off Si Amil Island, Indonesia [3.773°N, 118.364°E] (OD).

Etymology. Named for James H. McLean, my graduate advisor, mentor, and friend (OD).

Figure 375. *Sinezona macleani*. Holotype. 534 m, Station 5584, Sibuko Bay, off Si Amil Island, Borneo, Indonesia, 3.773°N, 118.364°E (USNM 289748). Scale bar shell = 1 mm. Scale bar protoconch = 100 μm.

Description. Shell medium-sized (holotype 2.3 mm wide. Figs 375, 376), depressed trochiform inflated. Protoconch of 0.75 whorl, smooth, no apertural varix, apertural margin straight. Teleoconch I of 1.75 whorls, 27 distinct axial cords on first whorl, spiral threads equispaced, increasing in number with growth, approximately 10 on shoulder at onset of selenizone. Teleoconch II of 0.66 whorl, suture little impressed. Shoulder irregularly convex, axials of similar density as on teleoconch I, slightly increasing in strength with growth, spiral lines increasing somewhat in strength, always weaker than axials, forming minute tubercles at intersection with axial cords, most prominent near apertural margin. Base inflated, axials of similar strength and density as on shoulder, axial cords becoming weaker towards umbilicus, some

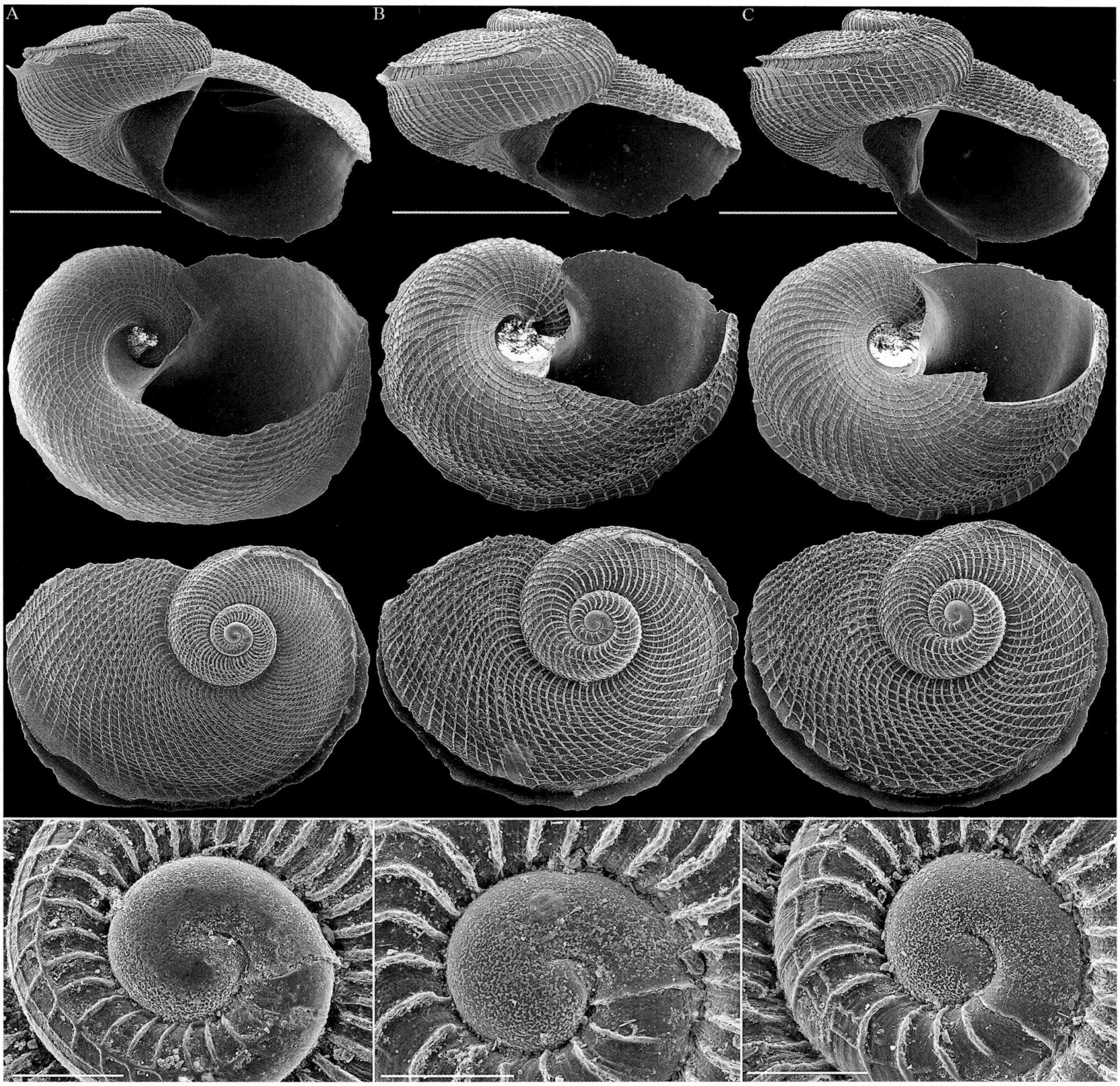

Figure 376. *Sinezona macleani*. Paratypes. **A.** 820–863 m, Fiji, 17.300°S, 179.550°W (MNHN). **B–C.** 675–680 m, New Caledonia, 23.167°S, 167.167°E (MNHN). Scale bars shell = 1 mm. Scale bars protoconch = 100 µm.

terminating before reaching umbilicus. Umbilicus open, wide, bordered by cord, walls smooth. Selenizone above periphery, lunules weak, keels of moderate height and strength, foramen elongate-teardrop-shaped. Aperture subquadrate, with indentation in lower corner of parietal wall, roof overhanging; past indentations visible inside umbilicus.

Radula (Fig. 377). Rachidian tooth trapezoidal, central cusp largest, 3–4 cusps on each side. Lateral teeth 1–4 reduced; lateral tooth 1 with 7–8 cusps of decreasing size on outer edge of tip, lateral tooth 2 with 5–6 cusps of decreasing size on outer edge of tip; lateral teeth 3–4 progressively reduced, hook-shaped; lateral tooth 4 with base and shaft variably drawn out in outward-facing wing (Fig. 377C–E); lateral tooth 5 massive, sturdy, apical cusp dominant, two progressively smaller cusps on each side. Inner marginal teeth without food groove, apical cusp largest, 2–3 cusps along inner edge, 4–6 cusps along outer edge of tip. Outer marginal teeth paddle-shaped, lateral bristles larger than apical ones. Interlock of central field weak.

Operculum (Fig. 377H) round, multispiral, very thin, with central nucleus.

Comparisons. *Sinezona bandeli* from New Zealand has a sturdier shell, a protoconch with axial sculpture and apertural varix, and the raised axial cords are at least 3–4× as strong as the spiral lines.

Sinezona concinna [not *Anatoma concinna*], with unknown type locality or distribution, has markedly decreasing strength of the axials towards the aperture, whereas in *Sin. macleani* the axials are rather increasing in strength towards the aperture. A more detailed discussion is provided under *Sin. concinna.*

Sinezona costulata from southern Japan has a less globular shell outline, a protoconch with axial cords and an apertural varix, and less than half as many but stronger axial cords on the teleoconch.

Sinezona danieldreieri from the Indo-Malayan Archipelago to the Central Pacific has a protoconch with axial cords and an apertural varix, a more narrow umbilicus, and a curved parietal wall.

Sinezona ferriezi from the Indo-Malayan Archipelago and the Western Pacific has a smooth protoconch with a distinct apertural varix, a shorter teleoconch I (1 vs. 1.5 whorls), and a distinct constriction of the base below the selenizone.

Sinezona globosa from the Indo-Malayan Archipelago to tropical Eastern Pacific (and possibly Caribbean) is overall larger, has a more globular shell, a protoconch with spiral striae and an apertural varix, and the base gradually slopes into the umbilicus.

Sinezona marrowi from the Indo-Malayan Archipelago has a protoconch with strong axial cords and an apertural varix, a more distinct constriction of the base below the selenizone, and a shorter teleoconch I (1.25 vs. 1.5 whorls).

Sinezona milleri from southern Japan has a protoconch with strong axial cords and an apertural varix, a shorter teleoconch I (1 vs. 1.5 whorls), and the axial cords are 3–4× as strong as the spiral lines on teleoconch I.

Sinezona plicata from the Indian Ocean to the Central Pacific has strong axial folds and a well-developed apertural varix on the protoconch.

Sinezona singeri from the Indian Ocean has a protoconch with a strong apertural varix, a distinct constriction on the base below the selenizone, and a much shorter teleoconch I (1 vs. 1.5 whorls).

Sinezona wileyi from the Indo-Malayan Archipelago and the Western Pacific has a subquadrate shell about as wide as high, a protoconch with fine irregular axial lines, and the umbilicus is bordered by a periumbilical cord of variable strength.

Distribution. Indo-Malayan Archipelago to Western Pacific, 25–1600 m (Fig. 378).

Specimen records.

Reunion. 1575–1600 m, MD32/REUNION st DS139, 20.783°S, 55.633°E (MNHN, 1: complete). 1150–1180 m, MD32/REUNION, ST. DC64, 21.200°S, 55.067°E (MNHN, 1).

Indonesia. 534 m, Station 5584, Borneo, Sibuko Bay, off Si Amil Island, 3.773°N, 118.364°E (USNM 289748, 1).

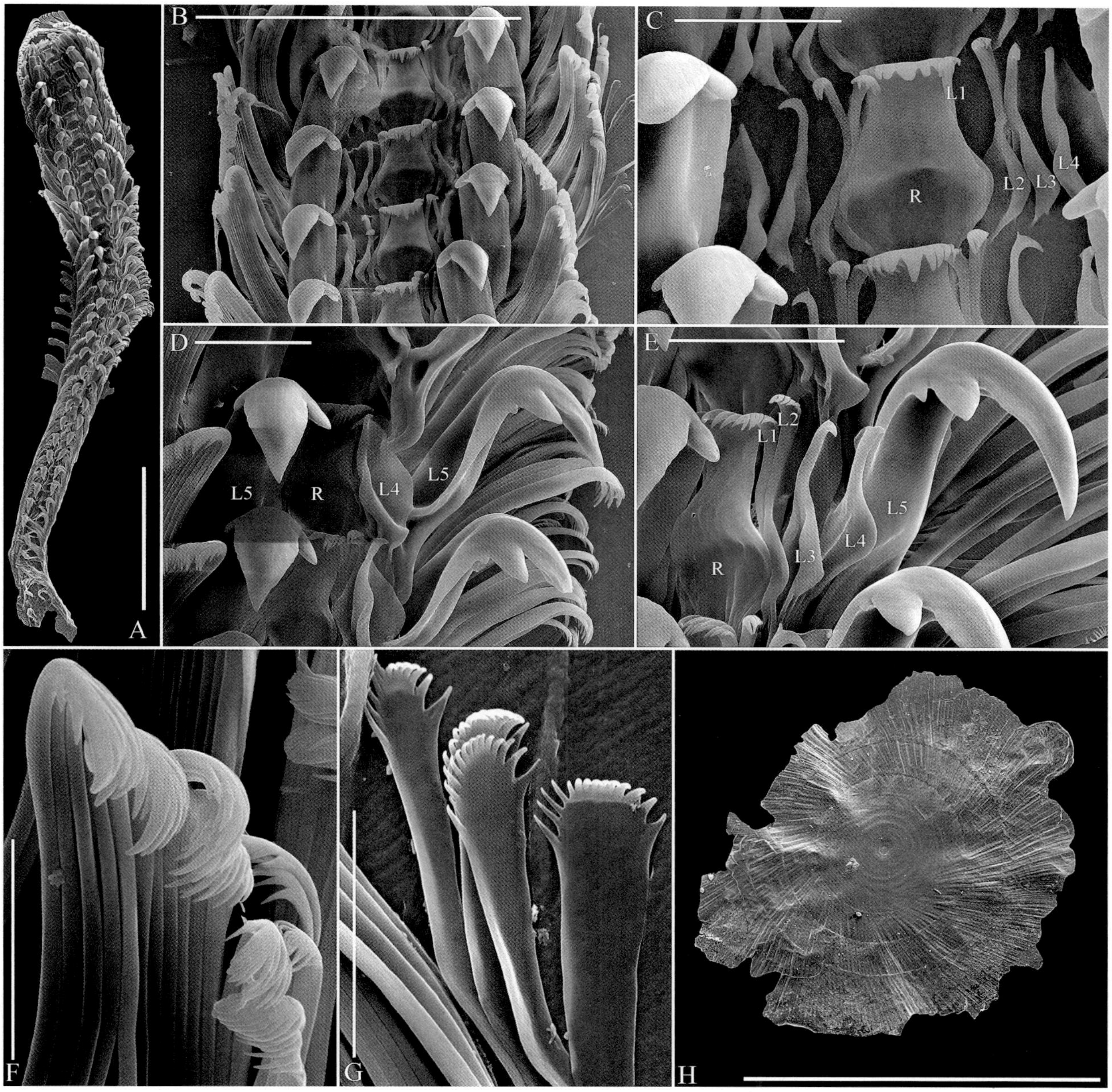

Figure 377. *Sinezona macleani*. Radula and operculum. 427–433 m, south, NORFOLK 1 DW1733, New Caledonia, 22.933°S, 167.250°E (MNHN, 10). **A.** Entire radula. Scale bar = 200 µm. **B.** Full width of radula. Scale bar = 100 µm. **C–E.** Portions of central field enlarged. Scale bars = 20 µm. **F.** Inner marginal teeth. Scale bar = 10 µm. **G.** Outer marginal teeth. Scale bar = 10 µm. **H.** Operculum. Scale bar = 1 mm.

Philippines. 538 m, AURORA 2007 CP2750, 15.883°N, 121.893°E (MNHN, 1). 903 m [494 fms.], off Magabao Island, E. Mindanao, 8.983°N, 126.264°E (USNM 276829-4, 2). 427 m, Bohol-Sulu Seas Sill, PANGLAO 2005 DW2364, 9.011°N, 123.425°E (MNHN, 2). 473 m, AURORA 2007 CP2749, 15.943°N, 121.837°E (MNHN, 1).

New Caledonia. 1600–1640 m, Loyalty Basin, BIOGEOCAL DW313, 20.983°S, 166.983°E (MNHN, 3). 1820 m, Loyalty Basin, 21.450°S, 166.433°E (MNHN, 1). 425–430 m, Vauban, 1978-79 St. 2, southern, 22.283°S, 167.233°E (MNHN, 1). 440 m, BIOCAL DW44, 22.783°S, 167.233°E (MNHN, 14). 440–450 m, BIOCAL DW44, 22.783°S, 167.233°E (MNHN, 5). 444–445 m, south, BATHUS2 DW719, 22.800°S, 167.267°E (MNHN, 44). 400 m, south, BATHUS 2 DW729, 22.866°S, 167.200°E (MNHN, 1). 530–541 m, BATHUS 2 DW720, 22.866°S, 167.267°E (MNHN, 1). 427–433 m, south, NORFOLK 1 DW1733, 22.933°S, 167.250°E (MNHN, 10). 525 m, SMIB 3 DW21, 22.983°S, 167.317°E (MNHN, 3). 775 m, 23.000°S, 167.483°E (MNHN, 1; MNHN, 1). 503 m, South, SMIB 3 DW22, 23.050°S, 167.317°E (MNHN, 5). 825–830 m, 23.050°S, 167.533°E (MNHN, 1). 1004–1009 m, Ile des Pins, NORFOLK 2 DW2144, 23.150°S, 167.450°E (MNHN, 1; MNHN, 11). 675–680 m, BIOCAL DW33, 23.166°S, 167.167°E (MNHN, 2). 965 m, BIOCAL DW70, 23.416°S, 167.883°E (MNHN, 1). 785–810 m, Jumeau East Bank, Norfolk Ridge, NORFOLK 2 DW2046, 23.733°S, 168.017°E (MNHN, 1). 770–830 m, Loyalty Ridge, 24.733°S, 170.133°E (MNHN, 2). 508–532 m, Sponge Bank, Mountain B, SMIB 8 DW146-147, 24.916°S, 168.367°E (MNHN, 13).

New Zealand. 1000 m, off Curtis Island, Kermadec Islands, 30.466°S, 178.622°W (NMNZ M257059, 1). 785–800 m, W Norfolk Ridge, W of Cape Reinga, 34.288°S, 168.430°E (NMNZ M.172123, 13).

Remarks. This species is a remarkable member of *Sinezona*. The smooth protoconch is known only from this species and *Sin. singeri*, further demonstrating the extensive plasticity of protoconch sculpture in Scissurellidae as discussed by Geiger (2003). The absence of an apertural varix is also uncommon in Scissurellidae. The species is also rather large; most *Sinezona* are 1–1.5 mm in size, whereas *Sin. macleani* is over 2 mm in size. The indentation in the lower corner of the parietal wall is a further unique feature in *Sinezona*; its function is unknown. It does not seem to be accidental as deeper in the umbilical wall traces of past such indentations are evident. Some

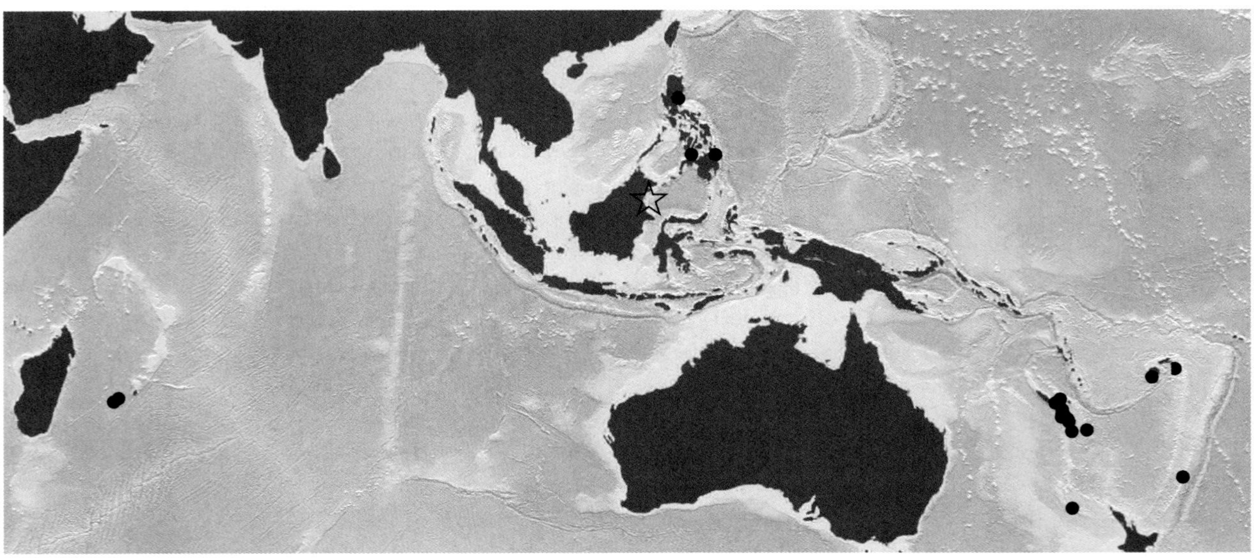

Figure 378. Distribution of *Sinezona macleani*. ☆ type locality. ● specimen record.

modifications of the parietal wall have been associated with brooding (*Larochea*: Marshall, 1993), though the indentation in *Sin. macleani* seems to be too small for such a purpose. Despite all these rather atypical characteristics, the species clearly belongs in *Sinezona* with the slit closed to form the foramen, and the cord bordering the umbilicus.

The radula of the species is also unusual. Lateral teeth 1–3 are much more slender than in all other species for which the radula is known. The winged shaft and base of lateral tooth 4 (Fig. 377C–E) is also unique, and lateral tooth 5 shows the strongest development among all Scissurellidae.

Sinezona is a shallow water representative of the minute Vetigastropoda. Most species are found much shallower than 100 m. The species ranges in depth from 25 to 1600 m; it is unlikely that it lives over this entire depth range, but it is unclear whether it is rather found in shallow, intermediate, or deep settings. A single live specimen was collected in 427–433 m.

Sinezona marrowi new species

Figures 379–381

Type material. Holotype (USNM 229503. Fig. 379A). Paratypes from type locality (USNM 1156980, 1). Great Barrier Reef, Capricorn Group, Heron Island, S side, Queensland, Australia, 23.433°S, 151.950°E (AMS C.404979, 1: complete). Great Barrier Reef, Lizard Island, W side Eagle Island, Queensland, Australia, 14.700°S, 145.383°E (AMS C.402653, 3). Barjaya Resort Area, SW Coast, Tioman Island, Malaysia, 2.810°N, 104.143°E (MMC, 1. Fig. 379B). 4–5 m, Libanona beach, TB07, Madagascar, 25.041°S, 46.995°E (MNHN IM-2010-22881, 1. Figs 379C, 380).

Type locality. 0 m, Maricaban Island, Philippines, 13.649°N, 120.890°E.

Etymology. Named for Max Marrow of Australia, the collector of first known material of the species (OD).

Description. Shell medium-sized (to 1.16 mm. Fig. 379), trochiform depressed. Protoconch of 0.75 whorl, strong axial sculpture, apertural varix connected to embryonic cap, apertural margin sinusoidal. Teleoconch I of approximately 1.1–1.2 whorls, 17–27 distinct axial cords to low lamellae, first spiral threads after 0.6–0.9 whorl, interstices with very fine irregular growth marks. Teleoconch II of up to one whorl, suture little impressed. Shoulders slightly convex, same density of axial cords as on teleoconch II, 6–13 fine spiral lines at apertural margin, at intersection with axials forming points. Base with modest constriction below selenizone, axial cords of same strength and density as on shoulder, 16–20 spiral lines, from mid-base towards umbilicus as spiral steps. Umbilicus open, wide, bordered by weak cord, forming angle to base; walls straight, with very fine growth lines. Aperture D-shaped, roof overhanging. Selenizone above periphery; keels moderately strong, moderately elevated; slit closed to foramen with anterior raphe.

Radula (Fig. 380A–C). Rachidian tooth triangular, tip with apical cusp largest, 2–3 cusps on each side, of decreasing size. Lateral teeth 1–3 similar, apical cusp largest, 2–3 cusps along outer edge of tip. Lateral tooth 4 reduced, hook-shaped.

Lateral tooth 5 enlarged by broadening, apical cusp largest, four cusps along inner edge of tip. Inner marginal teeth with triangular tip, apical cusp largest, 2–3 cusps along inner edge, 4–5 cusps along outer edge; outer marginal teeth with spoon-shaped tip, many fine bristles along edge; no food groove on shaft. Radular interlock of central field moderate.

Operculum (Fig. 380D), thin, round, corneous, multispiral, with central nucleus.

Figure 379. *Sinezona marrowi* new species. **A.** Holotype. 0 m, Maricaban Island, Philippines, 13.649°N, 120.890°E (USNM 229503). **B.** Paratype. Barjaya Resort Area, SW Coast, Tioman Island, Malaysia, 2.810°N, 104.143°E (MMC, 1). **C.** Paratype. 4–5 m, Libanona beach, TB07, Madagascar, 25.041°S, 46.995°E (MNHN IM-2010-22881, 15: radula shown in Figure 380). Scale bars shell = 1 mm. Scale bars protoconch = 100 µm.

Comparisons. *Sinezona costulata* from southern Japan has half as many but stronger axial cords, a shorter teleoconch I (0.75 vs. 1.1 whorls), and a more narrow umbilicus.

Sinezona danieldreieri from the Indo-Malayan Archipelago and Central Pacific has a shorter teleoconch I (<1 vs. 1.1 whorls), a shorter selenizone (0.3 vs. 0.6 whorl), and is smaller (0.8 vs. 1.2 mm).

Sinezona ferriezi from the Indo-Malayan Archipelago and the Western Pacific has a smooth protoconch with apertural varix connected to embryonic cap, at least twice as many fine spiral lines on shoulder and base, and grows much larger (2.1 vs. 1.2 mm).

Sinezona globosa from the Indo-Malayan Archipelago to the tropical Eastern Pacific (and possibly Caribbean) has a more globular shell, grows larger (3.8 vs. 1.2 mm) and the base gradually slopes into the umbilicus.

Sinezona macleani from the Indian Ocean to the Western Pacific has a more globular shell, a protoconch with flocculent sculpture and no apertural varix, and a straight parietal wall.

Sinezona milleri from southern Japan has a more narrow umbilicus and about twice as many but finer spiral lines on the base.

Sinezona plicata from the Indian Ocean to the Central Pacific has strong raised axial lamellae, a more pronounced constriction on the base below the selenizone,

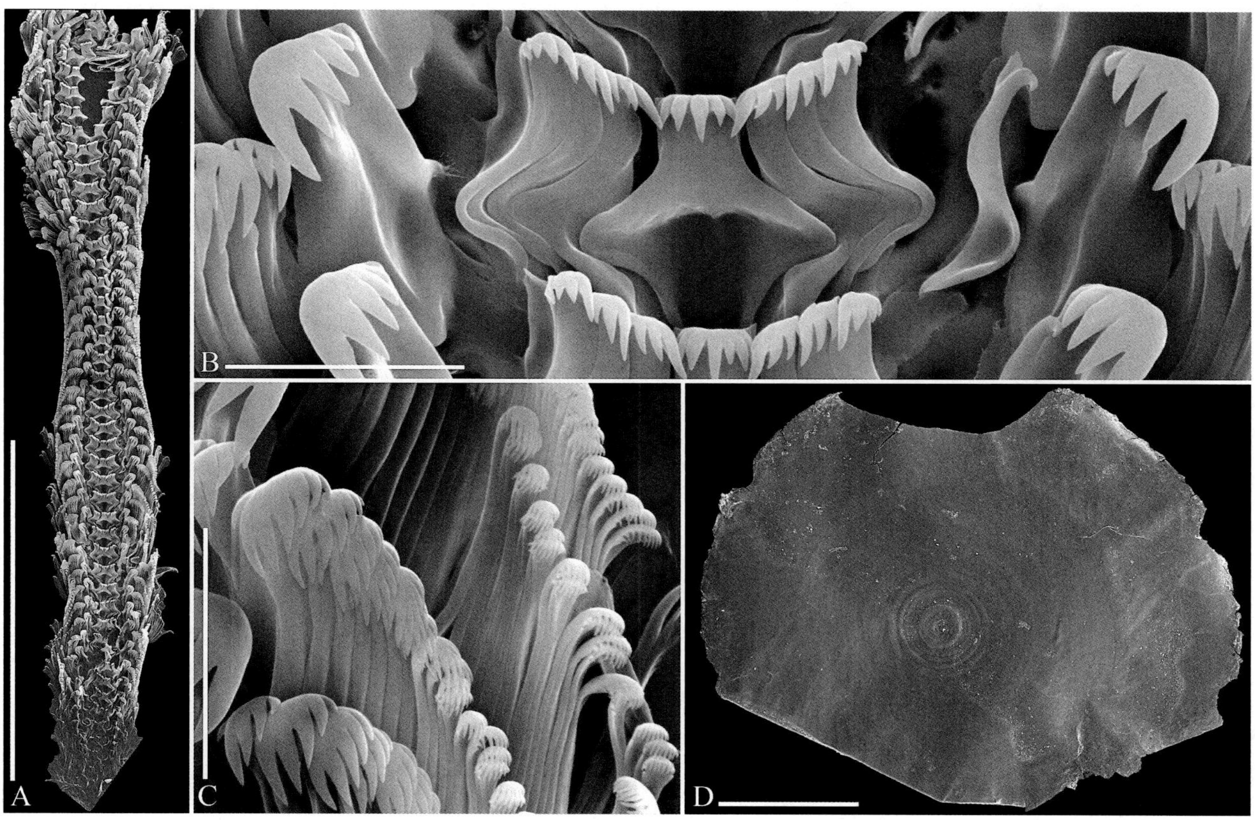

Figure 380. *Sinezona marrowi* new species. 4–5 m, Libanona beach, TB07, Madagascar, 25.041°S, 46.995°E (MNHN IM-2010-22881, 15: shell shown in Figure 379B). **A.** Entire radula. Scale bar = 200 µm. **B.** Central field enlarged. Scale bar = 10 µm. **C.** Marginal teeth. Scale bar = 10 µm. **D.** Operculum. Scale bar = 200 µm.

and a smooth protoconch with a spiral line and with the apertural varix connected to the embryonic cap.

Sinezona singeri from the Indian Ocean has a smooth protoconch with the apertural varix connected to the embryonic cap, a much larger teleoconch II (1.3 vs. 1 whorls), and a stronger constriction of the base below the selenizone.

Sinezona wileyi from the Indo-Malayan Archipelago and the Central Pacific has a protoconch with very fine axial cords and no apertural varix, is overall subquadrate, and has a more narrow umbilicus.

Sinezona zimmeri from Easter and Pitcairn Islands has much stronger axial cords and a shorter teleoconch I (0.8 vs. 1.2 whorls).

Scissurella mirifica from the Indo-Malayan Archipelago to the Central Pacific has a protoconch with fine axial sculpture and maintains the open slit in mature specimens (0.6–1.25 teleoconch I whorls).

Scissurella lorenzi from the Indo-Malayan Archipelago to the Western Pacific is overall more globular, has a protoconch with fine axial sculpture, and a more narrow umbilicus with a distinct funiculus.

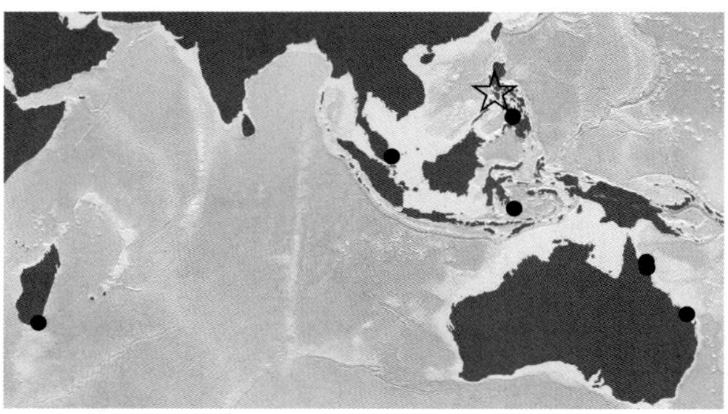

Figure 381. Distribution of *Sinezona marrowi* new species. ☆ type locality. ● specimen record.

Sinezona norfolkensis from the Tasman Sea has a stronger constriction on the base below the selenizone, lacks spiral on the shoulder of teleoconch I, and the spirals on teleoconch II are of irregular strength and spacing.

Scissurella staminea from southern Japan has fewer spiral lines on the base (11 vs. 16–20) and keeps the slit open after >0.5 whorl.

Scissurella xandaros from the Indian Ocean to the Central Pacific has a smooth protoconch, a more globular overall shell outline with correspondingly narrower umbilicus, strong elevated points at the intersections of axials and spirals, and retains an open slit in fully grown specimens with characteristically descending aperture.

Distribution. Indian Ocean, Indo-Malayan Archipelago (Fig. 381).

Specimen records.

Madagascar. 4–5 m, Libanona beach, TB07, 25.041°S, 46.995°E (MNHN, 15).

Philippines. 0 m, Maricaban Island, 13.649.000°N, 120.89.000°E (USNM 229503, 2). 382–434 m, Maribojoc Bay, Bohol Sea PANGLAO 2005 CP2395, 9.603°N, 123.730°E (MNHN, 1). 356–396 m, off Balicasag Island, Bohol Sea, CP2393, 9.501°N, 123.693°E (MNHN, 1).

Indonesia. 250–290 m, NW of Beninkgko, Banda Sea, Tukang Besi Island, 5.875°S, 123.975°E (RMNH 42315, 1).

Australia, Queensland. Great Barrier Reef, Lizard Island, W side Eagle Island, 14.700°S, 145.383°E (AMS C.402653, 3). 9–18 m, Great Barrier Reef, Cairns Reef lagoon and between Cairns Reef and Hope Island, 15.700°S, 145.500°E (AMS C.379081, 4). Great Barrier Reef, Capricorn Group, Heron Island, S side, 23.433.000°S, 151.95.000°E (AMS C.404979, 1: complete).

Sinezona mechanica Geiger & Marshall, 2012

Figures 382–385

Chresonymy.

Sinezona mechanica Geiger & Marshall, 2012: 6–9, figs 4–6.

Type material. Holotype (NMNZ M.174828. Fig. 382). Paratypes from type locality (NMNZ M.045402, 2. Fig. 383A–B). 360 m, Saunders Canyon, off Otago Peninsula, South Island, New Zealand, 45.767°S, 170.900°E (NMNZ M.160300, 2. Fig. 383C). Aerial Cove, Macquarie Island, 54.483°S, 158.950°E (AMS C.404496, 20: complete).

Type locality. 12–15 m, Ocean Bay, Chatham Island, New Zealand, 43.833°S, 176.783°W (OD).

Etymology. Latin adjective referring to the sculpture of shell and protoconch reminding of mechanical gears (OD).

Description. Shell small (to 0.89 mm. Figs 382, 383), depressed trochiform globular. Protoconch of one whorl, many dense axial cords on centrifugal portion of whorls, thickened at top forming (near) continuous connecting spiral cord. Apertural varix terminal (Fig. 383B), not connected to embryonic cap, apertural margin

Figure 382. *Sinezona mechanica*. Holotype. 12–15 m, Ocean Bay, Chatham Island, New Zealand, 43.833°S, 176.783°W (NMNZ M.174828, 1). Scale bar shell = 0.5 mm. Scale bar protoconch = 100 µm.

straight. Teleoconch I of 0.9–1 whorl, approximately 17–20 strong (occasionally barely perceptible) axial cords, no spirals. Teleoconch II of 0.6 whorl, suture little impressed. Shoulder slightly concave, same density and strength of axial cords as on teleoconch I, very fine growth marks in interstices, no spirals. Base with weak constriction below selenizone, rounded, same density and strength of axial cords as on shoulder; approximately four weak, granular spiral cordlets in adumbilical

Figure 383. *Sinezona mechanica*. Paratypes. **A–B.** 105 m, NE of Cape Saunders, Otago Peninsula, South Island, New Zealand, 45.833°S, 170.933°E (NMNZ M.045402, 1). **C.** 360 m, Saunders Canyon, off Otago Peninsula, South Island, New Zealand, 45.767°S, 170.900°E (NMNZ M.160300, 2). Scale bars shell = 0.5 mm. Scale bars protoconch = 100 μm.

⅕ of base. Umbilicus of moderate width, at sharp angle to base, bordered by weak to moderate strength spiral cord, wall straight, smooth. Aperture rounded, roof overhanging. Selenizone above periphery; keels low, moderate strength; lunules distinct; slit open, margins somewhat converging.

Radula (Fig. 384). Rachidian tooth trapezoidal central cusp largest, three cusps on each side. Lateral teeth 1–3 similar, apical cusps largest, 4–6 cusps on outer cutting edge. Lateral tooth 4 reduced, hook-shaped, with one minute point on each side of tip. Lateral tooth 5 enlarged by broadening, apical cusp largest, 5–7 cusps along inner edge. Inner marginal teeth triangular, apical cusp largest, 1–2 cusps along inner edge, 4–6 cusps along outer edge of tip; shafts without food groove. Outer marginal teeth with spoon-shaped tip bearing many fine bristles. Radular interlock of central field moderate.

Comparisons. *Scissurella apudornata* from the Miocene of New Zealand has a spire that is significantly more elevated and a longer teleoconch II (1.1 vs. 0.6 whorls).

Sinezona bandeli from New Zealand has a more subquadrate shell, much stronger constriction of the base below the selenizone, and the protoconch lacks the top spiral cord.

Sinezona brevis from New Zealand has shorter selenizone (0.1 vs. 0.4 whorl), a more distinct periumbilical cord, and fine spiral lines on the shoulder.

Sinezona iota from New Zealand is much taller, has strong axial lamellae, and lacks spiral sculpture on the base.

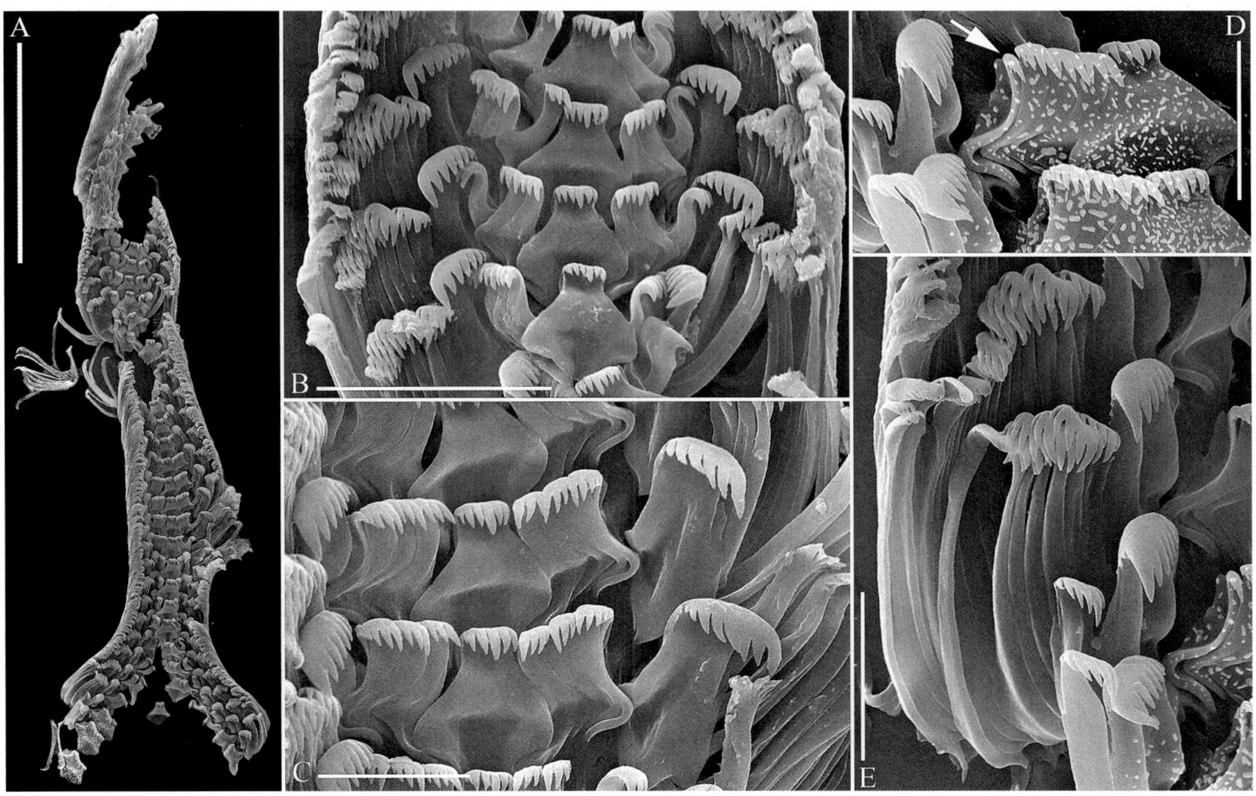

Figure 384. *Sinezona mechanica*. Radula. **A.** Entire radula. Scale bar = 100 µm. **B.** Full width of radula. Scale bar = 20 µm. **C.** Central field enlarged. Scale bar = 10 µm. **D.** Lateral tooth 4 (arrow). Scale bar = 10 µm. **E.** Marginal teeth. Scale bar = 10 µm. Aerial Cove, Maquarie Island, 54.483°S, 158.950°E (AMS C.404496, 20: complete).

Sinezona laqueus from New Zealand has a subquadrate overall shell shape, strong prosocline axial folds, and strong lunules in the selenizone.

Sinezona levigata from New Zealand has an umbilical callus, an overall wider shell, and spiral sculpture on shoulder and base.

Sinezona pacifica from temperate Australia and New Zealand has spiral sculpture on shoulder and base, a more narrow umbilicus, and lacks the top spiral cord on the protoconch.

Sinezona pauperata from New Zealand lacks a selenizone, has a round foramen, and a very narrow umbilicus.

Sinezona platyspira from New Zealand has a flatter spire, the umbilicus is produced by the rounded portions of the previous whorls, and the protoconch lacks the top spiral as well as an apertural varix.

Sinezona subantarctica of broad subantarctic distribution has a callus over the umbilicus, a very short (or absent) selenizone, and the protoconch lacks the top spiral cord and an apertural varix.

Scissurella skeneoides from New Zealand has a protoconch with fine axial cords that do not form an apical spiral cord, lacks an apertural varix on the protoconch, and the protoconch does not protrude over the teleoconch.

Distribution. Pleistocene to Recent of New Zealand (Fig. 385).

Specimen records.

New Zealand. 94 m, Ranfurly Bank, East Cape, North Island, 37.546°S, 178.812°E (NMNZ M.303294, 1). 89–94 m, Ranfurly Bank, East Cape, North Island, 37.551°S, 178.825°E (NMNZ M.303293, 15). 71–76 m, Ranfurly Bank, East Cape, North Island, 37.553°S, 178.838°E (NMNZ M.303295, 50). 103–106 m, Ranfurly Bank, East Cape, North Island, 37.556°S, 178.805°E (NMNZ M.303327, 5). 12–15 m, Ocean Bay, Chatham Island, 43.833°S, 176.783°W (NMNZ M.174828, 1). 360 m, Saunders Canyon, off Otago Peninsula, South Island, 45.766°S, 170.900°E (NMNZ M.160300, 2). 105 m, NE of Cape Saunders, Otago Peninsula, South Island, 45.833°S, 170.933°E (NMNZ M.045402, 2). 91 m, off Taieri, S of Dunedin, South Island, 46.250°S, 170.483°E (NMNZ M.066283, 2). 55 m, off Poutama Island, South Cape, Stewart Island, 47.266°S, 167.383°E (NMNZ M.303296, 60).

Macquarie Island. Aerial Cove, 54.483°S, 158.950°E (AMS C.404496, 20: complete).

New Zealand, fossil. GS4057, Kaikokopu Shellbed (Castlecliffian, oxygen isotope stage 19, middle Pleistocene), coast E of Omapu Stream Mouth, between Castlecliff and Kai-Iwi beaches, Wanganui, 39.907°S, 174.941°E (GNS, 1).

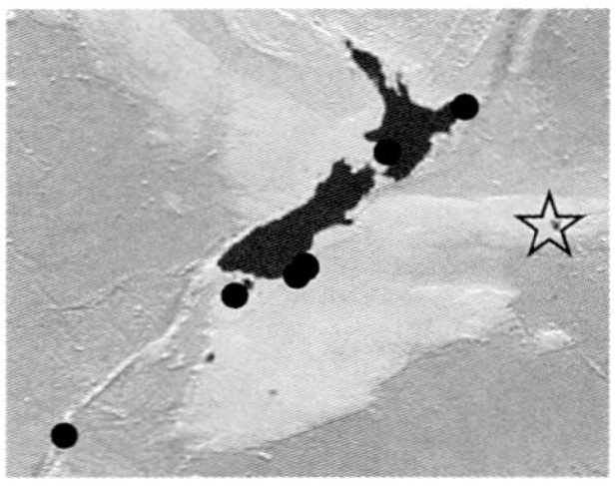

Figure 385. Distribution of *Sinezona mechanica*. ☆ type locality. ● specimen record.

Scissurellidae *Sinezona*

Sinezona milleri Geiger & Sasaki, 2009

Figures 386–387

Chresonymy.

Sinezona milleri Geiger & Sasaki, 2009: 5–6, fig. 4.

Sinezona sp. 1: Hasegawa & Okutani, 2000: 37, pl. 18, figs 6a,b [see Remarks].

Type material. Holotype (SBMNH 83548. Fig. 386). Paratypes from type locality (SBMNH 83549, 2).

Type locality. 17 m, north side, Ike Jima, Okinawa, Japan, 26.384°N, 127.990°E (OD).

Etymology. Named in honor of the collector of the type specimens, Shawn Miller of Nagahama, Okinawa, for his continued support in malacological research by providing marine sediment samples of Okinawa (OD).

Description. Shell trochiform depressed, small for genus (holotype 0.76 mm. 0.9 mm *fide* Hasegawa & Okutani, 2000: 37. Fig. 386). Protoconch of 0.75 whorl, strong axials with thickening in apical region, apertural varix not connected to embryonic cap, apertural margin straight. Teleoconch I of one whorl, approximately 19 raised axial

Figure 386. *Sinezona milleri*. Holotype. 17 m, north side, Ike Jima, Okinawa, Japan, 26.384°N, 127.990°E (SBMNH 83548). Scale bar shell = 0.5 mm. Scale bar protoconch = 100 µm.

cords, first spiral line after 0.5 whorl in position of selenizone. Teleoconch II of 0.75 whorl, suture moderately impressed. Shoulder flat, approximately 20 axials on first 0.5 whorl, decreasing in strength from distinct cords to indistinct cords, spirals increasing in strength from five lines at onset of selenizone to approximately 13 lines and cords at apertural margin of fully grown specimen; axials and spirals forming reticulate pattern towards apertural margin. Base barely constricted below selenizone; axials as on shoulder, approximately 20 spirals, increasing in strength from fine lines below selenizone to cords towards umbilicus, the latter forming small thickenings at intersection with axials. Umbilicus at angle to base, bordered by granulated spiral cord; walls straight, with fine axial growth lines, no funiculus. Selenizone above periphery, keels low, less then 20% of width of selenizone, moderately strong, slit closed anteriorly to foramen. Aperture rounded D-shaped to subquadrate, roof overhanging.

Animal unknown.

Comparisons. *Sinezona costulata* from the Japanese mainland does not show reticulate sculpture and has as stronger constriction of the base below the selenizone.

Sinezona danieldreieri from the Indo-Malayan Archipelago to the Central Pacific has a shorter teleoconch I (0.75–0.8 vs. 1 whorl) and the axial cords on the protoconch lack a medial thickening.

Sinezona ferriezi from the Indo-Malayan Archipelago to the Western Pacific has a smooth protoconch with the apertural varix connecting to the embryonic cap and about twice as many but finer axial cords and spiral lines.

Sinezona globosa from the Indo-Malayan Archipelago to the tropical Eastern Pacific (and possibly Caribbean) has a protoconch with spiral lines and the apertural varix connecting to the embryonic cap, an overall more globular shell, and the base continuously slopes into the umbilicus.

Sinezona macleani from the Indian Ocean to the Western Pacific has a protoconch with flocculent sculpture and no apertural varix, a much longer teleoconch I (1.5 vs. 1 whorls), and a distinct shelf in the lower adumbilical corner of the peristome.

Sinezona marrowi from the Indo-Malayan Archipelago has a protoconch with the apertural varix connected to the embryonic cap and about half as many spiral lines on shoulder and base.

Sinezona plicata from the Indian Ocean to the Central Pacific has strong axial keels, a much stronger constriction of the base below the selenizone, and grows to much larger size (2.3 vs. <1 mm).

Sinezona singeri from the Indian Ocean has a smooth protoconch with apertural varix connecting to embryonic cap and a much stronger constriction of the base below the selenizone.

Sinezona wileyi from the Indo-Malayan Archipelago and the Central Pacific has an overall subquadrate shell shape, a protoconch with fine irregular axials and no apertural varix, and usually as strong periumbilical cord.

Scissurella staminea from Japan has much wider spacing of the reticulate pattern, which is also developed equally over entire teleoconch.

Scissurella evaensis from the Indian Ocean to the Central Pacific bears a spiral series of distinct axial tooth-like projections on the mid-base.

Scissurella mirifica from the Indo-Malayan Archipelago to the Central Pacific has an overall more globular appearance and a fine reticulate pattern on shoulder and base over entire teleoconch.

Scissurella lorenzi from Indo-Malayan Archipelago to the Western Pacific is overall more globular in shape and shows a distinct color pattern of broad axial rays on the shoulder.

Scissurella phenax from the Indo-Malayan Archipelago to the Central Pacific has a taller spire, a more narrow umbilicus and stronger and more widely spaced axial cords to low lamellae on the teleoconch.

Scissurella spinosa from the Indo-Malayan Archipelago to the Central Pacific is overall more globular and has reticulate sculpture, in which the intersections of axials and spirals are raised to distinct points.

Sukashitrochus carinatus from the mainland of Japan to Okinawa has distinct spiral keels on the base of the teleoconch.

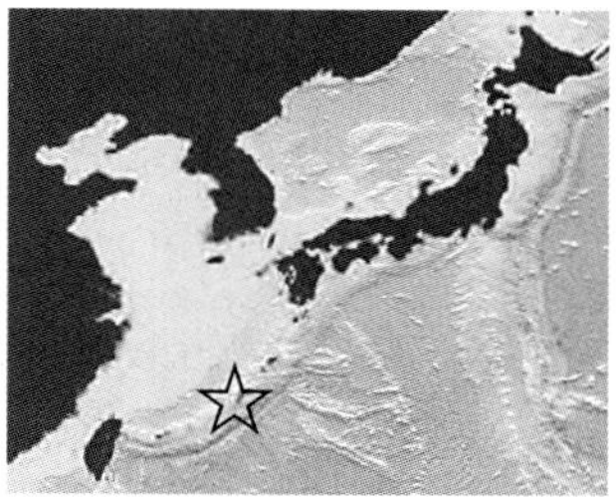

Figure 387. Distribution of *Sinezona milleri*. ☆ type locality.

Distribution. Known from type locality, Okinawa, and Amami Islands (see Remarks. Fig. 387).

Remarks. Okutani and Hasegawa (2000) switched the descriptions of their two unnamed *Sinezona* species. *Sinezona* sp. 1 of Okutani & Hasegawa (2000) = *Sin. milleri* is placed at Honshu and Izu Islands, whereas it in fact occurs in more tropical waters. Conversely, *Sinezona* sp. 2 of Okutani & Hasegawa (2000) = *Sin. costulata* is reported by Okutani & Hasegawa (2000: 37) from "South of Amami Islands" [~28.5°N], i.e., the tropics, it is in fact a temperate species we collected on Honshu in Sagami Bay (35.2°N) and around Tosa (33.5°N) in large numbers and alive.

Sinezona miranda (A. Adams, 1862) [*nomen dubium*]

Figure 388

Chresonymy.

Scissurella miranda A. Adams, 1862: 346–347, not illustrated.
Schismope miranda: Crosse, 1863: 109.
Scissurella miranda: Troschel, 1863: 702.
Scissurellam [*sic*] *mirandam* [*sic*]: Dunker, 1882: 147 [see Remarks].
Schismope miranda: Paetel, 1888: 289.
Scissurella miranda: Pilsbry, 1890: 67.
Schismope miranda: Pilsbry, 1895: 106.
Schismope miranda: Thiele, 1912a: 27.
Woodwardia miranda: Habe, 1951: 68–69.
Woodwardia miranda: Kuroda & Habe, 1952: 98.
Woodwardia miranda: Kawamoto & Tanabe, 1956: 3.
Sukashitrochus miranda [sic]: Habe & Kosuge, 1964: 3.
Sukashitrochus miranda [sic]: Higo, 1973: 14.
Sukashitrochus miranda [sic]: Higo, 1993: 15.
Sinezona ? *miranda*: Tsuchida *et al.*, 1991: 8–9.
Sukashitrochus mirandus: Geiger, 2003: 79.

Type material. (whereabouts unknown. Not in NHMUK: Thiele, 1912a). Some of A. Adams types have been found elsewhere (e.g., Kuroda & Habe, 1954).

Type locality. Mino-shima [= Mishima Island, Hagi City, Yamaguchi Prefecture, Japan (34.767°N, 131.166E): see also Kawamoto and Tanabe (1956)], 63 fms. [= 115 m] (OD).

Etymology. Mirandus, Latin adjective for wonderful, strange, singular.

Description. [Translation of OD] "*Scissurella* with ovate shell, depressed, widely umbilicate; spire somewhat flat; on the last whorl above keel [= selenizone] very finely concentrically striated, below the keel [= selenizone] ornated with oblique somewhat nodular longitudinally separated folds, on the base with concentric lines; aperture completely oblique, transversely oval.

Hab. Mino-Sima; 63 fathoms.

In this small but very pretty species the last whorl is nodosely plicate below the fissural keel."

Distribution. Only known from type locality (Fig. 388).

Remarks. Adams (1862) described the species as not having keels, unlike the previous, his *Sci. carinata* = *Suk. carinatus*. He also placed all species with a foramen in *Scissurella* (not *Sinezona*), while modern *Scissurella* species were described by him in *Anatomus* (e.g., *stamineus*). Accordingly, *mirandus/miranda* is a *Sinezona* species.

Confusingly, Adams also used the Latin term carina = keel to refer to the selenizone. In the Latin description of *carinata* he (p. 346) referred to the spiral keels on the base of *Sukashitrochus* as "cingulis elevatis," elevated rings.

Dunker (1882: 147) noted that the species should be referred to Jeffreys' genus *Schismope* [= *Sinezona*].

The species has never been illustrated, and no type material could be located in the NHMUK, where most other Adams' types are housed. The description does not offer sufficient detail to permit positive identification of his species. It might refer to *Sin. plicata*, a species known from Japan. However, as this identification is very tentative, *miranda* is here considered a *nomen dubium*. A case for a *nomen oblitum/protectum* cannot be made as the species has been used as valid after 1899.

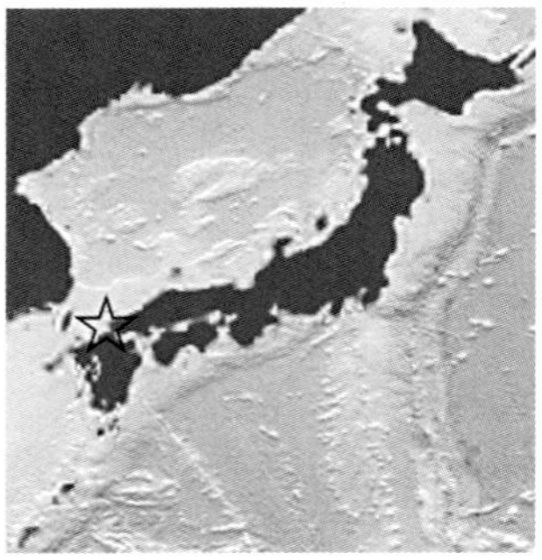

Figure 388. Distribution of *Sinezona miranda*. ☆ type locality.

Sinezona modesta (A. Adams, 1862) [*nomen dubium*]

Figure 389

Chresonymy.

Scissurella modesta A. Adams, 1862: 346, not illustrated.
Scissurella modesta: Crosse, 1863: 109.
Scissurella modesta: Troschel, 1863: 702.
Scissurellam [*sic*] *modestam* [*sic*]: Dunker, 1882: 147.
Schismope modesta: Paetel, 1888: 289.
Schismope modesta: Pilsbry, 1895: 106.

Schismope modesta: Thiele, 1912a: 27.
Woodwardia modesta: Habe, 1951: 68.
Woodwardia modesta: Kuroda & Habe, 1952: 98.
Sukashitrochus modesta [sic]: Higo, 1973: 14.
Sukashitrochus modesta [sic]: Higo & Goto, 1993: 15.
Sinezona ? *modesta*: Geiger, 2003: 78.

Type material. (whereabouts unknown. Not in NHMUK: Thiele, 1912a). Some of A. Adams types have been found elsewhere (e.g., Kuroda & Habe, 1954).

Type locality. Tabushima [= Tobishima Island, NW Honshu], Japan [39.200°N, 139.533°E] (OD).

Etymology. Not given.

Description. [Translation of OD] "*Scissurella* with oval, depressed, stomatelliform shell, narrowly umbilicate; small spire hardly elevated; on last whorl above keel [= selenizone] concentrically striated, below keel [= selenizone] longitudinally folded, on the base with spiral lyrae ornated; aperture oblique, broadly oval.

Hab. Tabu-Sima; 25 fathoms.

A small simple white species, without keels, striated above the somewhat rounded periphery, and obsoletely plicate below."

Distribution. Only known from type locality (Fig. 389).

Remarks. Adams (1862) described the species as not having keels, unlike the previous, his *Sci. carinata* = *Suk. carinatus*. He also placed all species with a foramen in *Scissurella* (not *Sinezona*), while modern *Scissurella* species were described by him in *Anatomus* (e.g., *stamineus*). Accordingly, *modesta* is a *Sinezona* species.

Confusingly, Adams also used the Latin term carina = keel to refer to the selenizone. In the Latin description of *carinata* he (p. 346) referred to the spiral keels on the base of *Sukashitrochus* as "cingulis elevatis," elevated rings.

Dunker (1882: 147) noted that the species should be referred to Jeffreys' genus *Schismope* [= *Sinezona*].

The species has never been illustrated, and no type material could be located in the NHMUK, where most other Adams' types are housed. The description does not offer sufficient detail to permit positive identification of his species. The description of the shell as both stomatelliform, i.e., much wider than tall and rapid expansion rate of the whorls, and with narrow umbilicus does not match any known scissurelid. Therefore, it is here considered a *nomen dubium*. A case for a *nomen oblitum/protectum* cannot be made as the species has been used as valid after 1899.

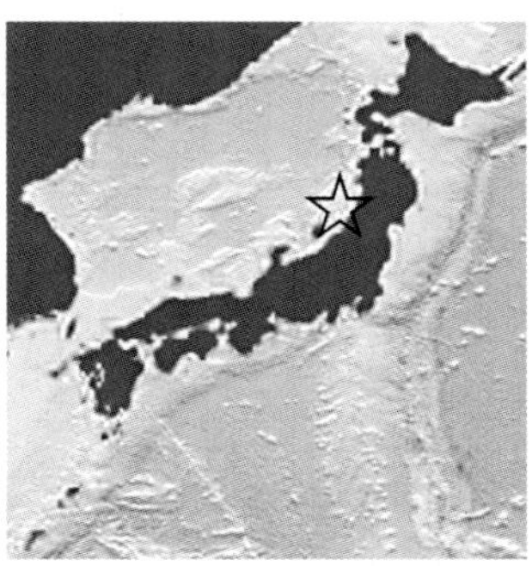

Figure 389. Distribution of *Sinezona modesta*. ☆ type locality.

Sinezona mouchezi (Vélain, 1877)

Figures 390–393

Chresonymy.

Schismope mouchezi: Vélain, 1876: 285 [*nomen nudum*].
Schismope mouchezi Vélain, 1877: 119–120, pl. 4, figs 7–8.
Schismope mouchezi: Paetel, 1888: 289.
Schismope mouchezi: Pilsbry, 1890: 62, pl. 57, figs 18–19.
Schismope mouchezi: Thiele, 1912a: 25, pl. 3, figs 21–22 [copy figures Vélain, 1877].
Sinezona mouchezi: Lozouet, 1986: 111.
Sinezona mouchezi: Geiger, 2003: 79.
Sinezona mouchezi: Zelaya & Geiger, 2007: 435–437, figs 179 [map], 197–212 [shell SEM].

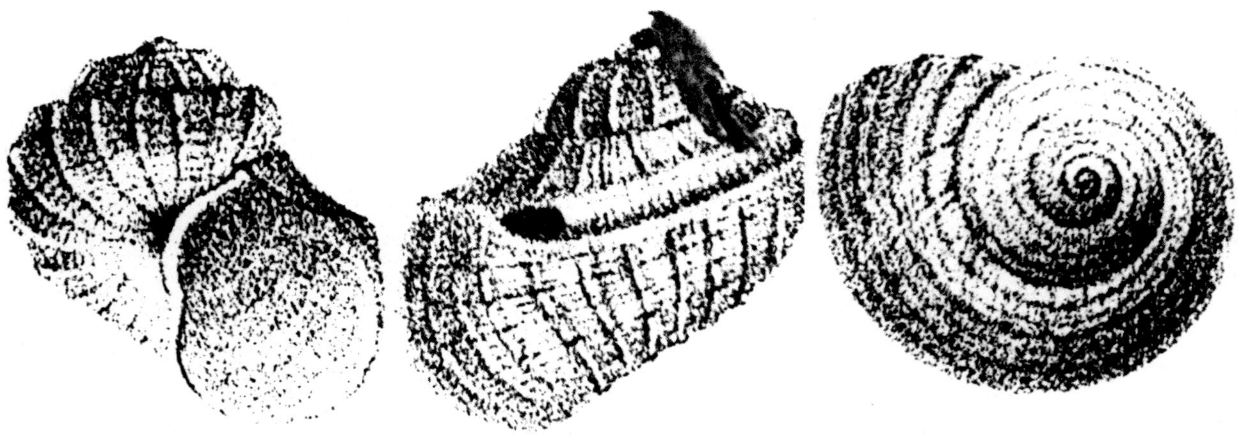

Figure 390. Original illustration of *Schismope mouchezi*.

Type material. Lectotype (MNHN: designated by Zelaya & Geiger, 2007: 435. Fig. 391). Paralectotypes from type locality (MNHN, 3. Fig. 392), 1 × 0.75 mm (H × W).

Type locality. St. Paul Island [France, 38.730°S, 77.500°E], subtidal to 30–45 m (OD).

Etymology. Named after Commandant Mouchez (OD).

Description. Shell medium in size (up to 2 mm. Figs 391, 392), trochiform, inflated, last whorl somewhat laterally expanded. Protoconch of one whorl, sculptured with 13–15 strong axial cords, distinct varix connected to embryonic cap. Teleoconch I of 1.125 whorls, sculptured with 13–14 strong axial cords, six narrower spiral threads. Teleoconch II of up to 0.875 whorl, suture moderately impressed. Shoulder with inflated base, sculptured with strong axial cords, spiral threads; 14–16 axial cords on shoulder, about 13 on base diminishing in prominence toward umbilicus; spirals regularly distributed, 6–7 on shoulder, 7–8 on base. Selenizone above periphery, surrounded by low keels, having low lunules; slit closed forming foramen. Umbilicus wide. Aperture rounded.

Animal unknown.

Comparisons. *Sinezona insignis* from South African and southern middle Atlantic islands has fewer but stronger axial cords, lacks axial cords on the shoulder of teleoconch II, and has a somewhat longer teleoconch I (1.1 vs. 1.25 whorls)

Sinezona iota from New Zealand has stronger axial lamellae that maintain their strength to the border of the umbilicus, a shorter teleoconch I (1.1 vs. 1.2 whorls), and lacks spirals on the base.

Scissurella supraplicata from the subantarctic Indian Ocean has no axial cords on the shoulder of teleoconch II, a protoconch with flocculent sculpture, an open slit and a more narrow umbilicus.

Distribution. Known only from type lot (Fig. 393).

Remarks. Among the four specimens that represent the type series of *Sin. mouchezi* two "varieties" can be recognized: three of the specimens have the above described set of characters; the other specimen has a taller spire, a less laterally expanded shell, and the axial and spiral ornamentation is almost absent on the last whorl; these characteristics are correlated with a larger overall size of teleoconch II. However, the number of protoconch and teleoconch I whorls, and their ornamentation is the same

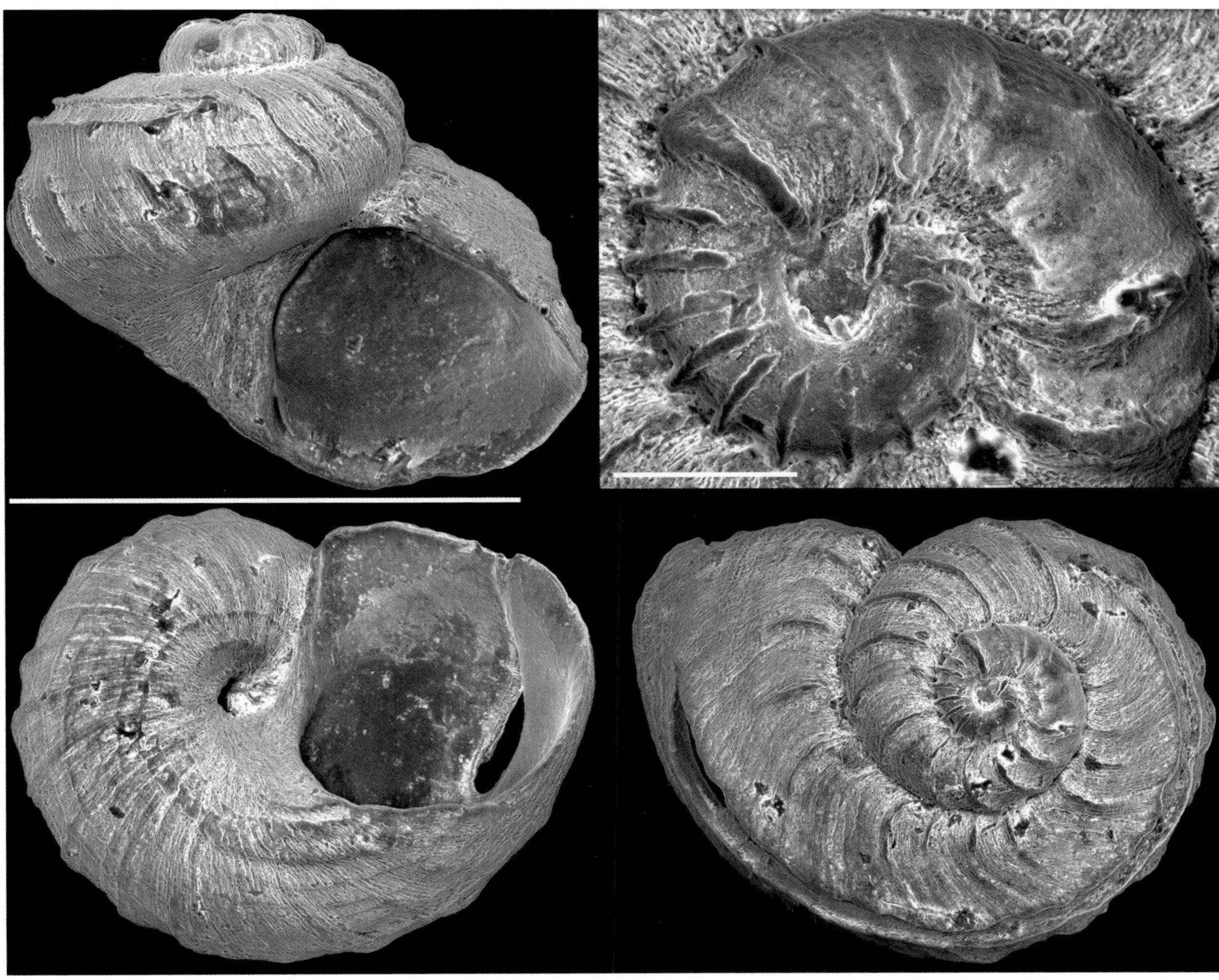

Figure 391. *Sinezona mouchezi*. Lectotype. St. Paul Island [France, 38.730°S, 77.500°E] (MNHN). Scale bar shell = 1 mm. Scale bar protoconch = 100 μm.

in all the specimens; consequently, these differences are considered infraspecific variability. None of the above-mentioned specimens correspond to the material figured by Vélain (1877), which has the teleoconch with a larger number of whorls (3.25), a taller spire, and a more prominent shell sculpture. As young specimens of

Figure 392. *Sinezona mouchezi*. Paralectotypes. St. Paul Island [France, 38.730°S, 77.500°E] (MNHN, 3). Scale bars shell = 0.5 mm. Scale bars protoconch = 100 μm.

Sin. mouchezi could be confused with *Sci. supraplicata*, a lectotype was designated of the former by Zelaya & Geiger (2007).

Figure 393. Distribution of *Sinezona mouchezi*. ☆ type locality.

Sinezona norfolkensis new species

Figures 394–396

Type material. Holotype (NMNZ M.174823. Fig. 394). Paratypes from type locality (NMNZ 226874, 2. Fig. 395A). 135–146 m, E of Dayrell Islet, Raoul Island, Kermadec Islands, New Zealand, 29.245°S, 177.839°W (NMNZ M.226874, 2. Fig. 395B). 10 m, Middleton Reef, S side, Tasman Sea, 29.485°S, 159.135°E (AMS C.349520, 3. Fig. 395C).

Type locality. 2–5 m, Kingston Lagoon, Norfolk Island, New Zealand, 29.060°S, 167.955°E.

Etymology. Named for its type locality from Norfolk Island.

Description. Shell small (to 0.9 mm: NMNZ paratype. Figs 394, 395), trochiform, depressed. Protoconch of one whorl, irregular broad axial to midline, apertural varix not connected to embryonic cap, apertural margin straight. Teleoconch I of 0.8–0.9 whorl, 15–17 axials increasing in strength from cords to low lamellae. Teleoconch II of up to one whorl, suture impressed. Shoulder slightly to strongly convex, strong axial cords, distinctly raised to low lamellae in adsutural half of shoulder, first low indistinct spiral cord after 0.25–0.5 teleoconch II whorl, 5–7 at apertural margin, irregularly spaced, absutural ones as finer lines, interstices with very fine growth lines. Base with strong constriction below selenizone, bulging, axials not coordinated with those on shoulder, as thin cord below selenizone, rapidly changing to strong lamellae, fading out in on last quarter to fine cords towards umbilicus; approximately 12 spirals, increasing in strength from lines over most of base to lumpy cords in adumbilical quarter. Umbilicus funnel-shaped, walls smooth with very fine growth lines, last spiral on base as periumbilical cord. Aperture D-shaped, roof overhanging. Selenizone above periphery, keels strong, quite elevated, foramen elongated triangular, lunules increasing in spacing with growth, rather strong.

Animal unknown.

Comparisons. *Sinezona bandeli* from New Zealand has many fine spiral lines on the shoulder, a protoconch with more numerous and taller axial lamellae, and a shorter teleoconch I (0.75 vs. 0.8–0.9 whorl).

Sinezona beddomei from temperate Australia lacks spiral sculpture on the shoulder and the upper portion of the base, has stronger and more widely spaced lunules, and the suture is below the periphery of the protoconch.

Sinezona brucei from New Zealand has an overall taller shell, the axials of the protoconch reach past the midline of the whorl, and has a shorter teleoconch I (0.6 vs. 0.8–0.9 whorl).

Sinezona calumnior from temperate Australia and New Zealand has a very narrow umbilicus, an overall taller shell, and weaker axial sculpture on the shoulder.

Sinezona finlayi from New Zealand has protoconch with the axial lamellae reaching almost to the center of the spire, less elevated axial cords on teleoconch II, and the axial cords on the shoulder (if present) do not reach the selenizone.

Sinezona iota from New Zealand is much taller, lacks spiral sculpture on shoulder and base, and has a much more narrow umbilicus.

Sinezona laqueus from New Zealand has a stronger constriction on the base below the selenizone, stronger lunules, lacks spiral sculpture on shoulder and upper base, and a more narrow umbilicus.

Sinezona marrowi from the western Indian Ocean to the Indo-Malayan Archipelago has finer and more regular spiral lines on the shoulder, spiral lines on teleoconch I, and a very weak constriction of the base below the selenizone.

Sinezona pacifica from Southern Australia and New Zealand has a shorter teleoconch I (0.75 vs. 0.8–0.9 whorl), a more narrow umbilicus, and the axials of the protoconch extend more than halfway towards the center of the spire.

Scissurella phenax from the Indo-Malayan Archipelago to the Central Pacific has a taller shell, a protoconch with median thickenings on the axials, a more

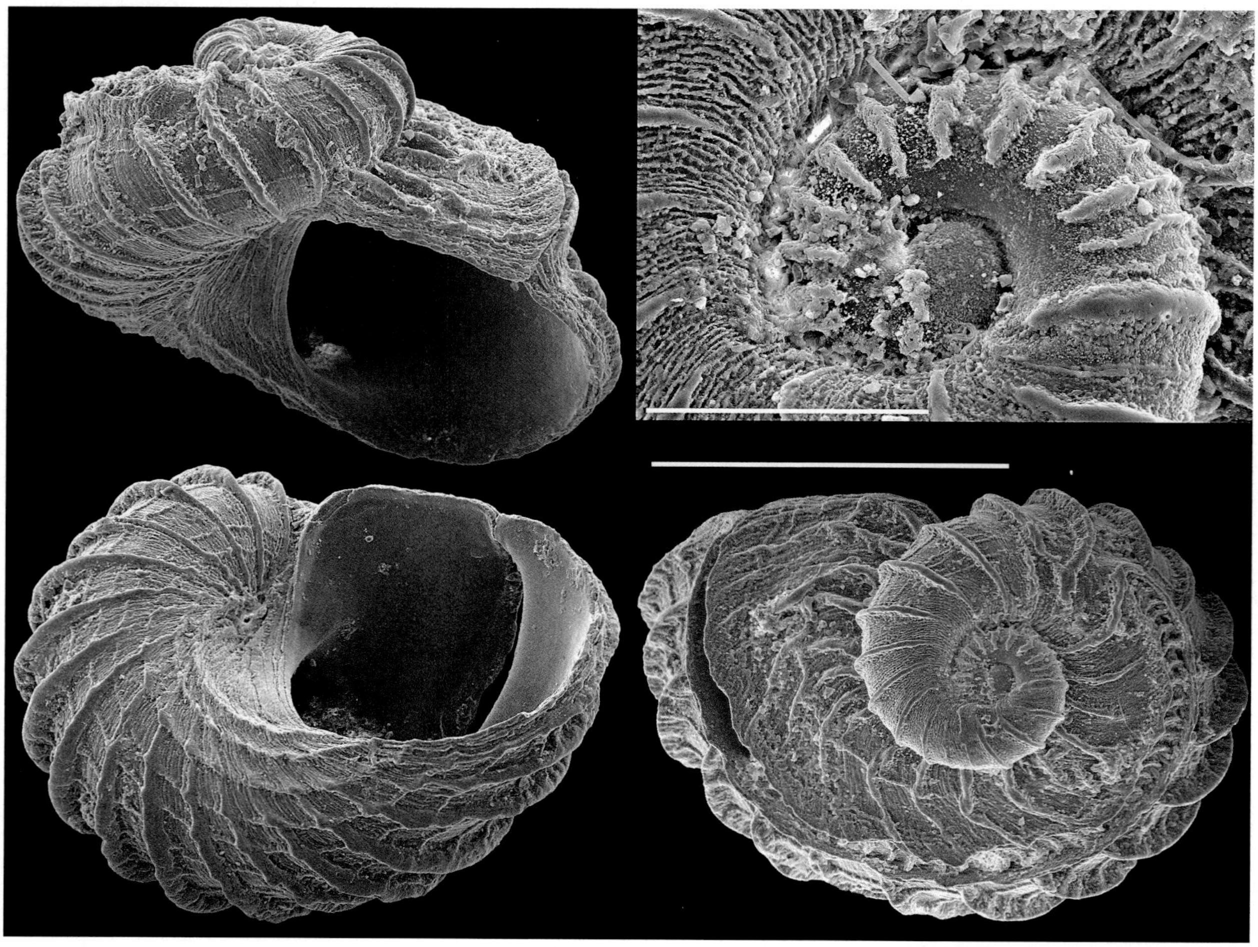

Figure 394. *Sinezona norfolkensis* new species. Holotype. 2–5 m, Kingston Lagoon, Norfolk Island, New Zealand, 29.060°S, 167.955°E (NMNZ M.174823). Scale bar shell = 0.5 mm. Scale bar protoconch = 100 μm.

narrow umbilicus, and strong axials on the base compared to the shoulder of teleoconch II.

Distribution. Tasman Sea (Fig. 396).

Specimen records.

Australia. Emily Bay, Kingston, Norfolk Island, 29.067°S, 167.967°E (AMS C.349528, 8).

Figure 395. *Sinezona norfolkensis* new species. **A.** Paratype. 2–5 m, Kingston Lagoon, Norfolk Island, New Zealand, 29.060°S, 167.955°E (NMNZ M.226874, 2). **B.** 135–146 m, E of Dayrell Islet, Raoul Island, Kermadec Islands, New Zealand, 29.245°S, 177.839°W (NMNZ M.226874, 2). **C.** Paratype. 10 m, Middleton Reef, S side, Tasman Sea, 29.485°S, 159.135°E (AMS C.349520, 3). Scale bars shells = 0.5 mm. Scale bars protoconch = 100 μm.

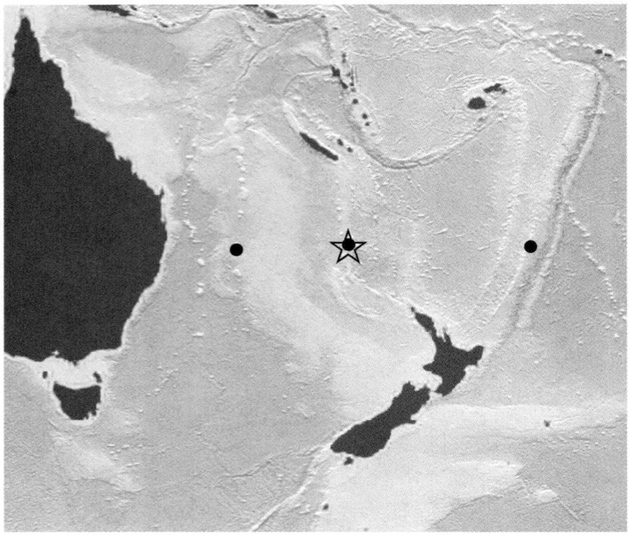

New Zealand. 135–146 m, E of Dayrell Islet, Raoul Island, Kermadec Islands, 29.245°S, 177.839°W (NMNZ M.226874, 2).

Figure 396. Distribution of *Sinezona norfolkensis* new species. ☆ type locality. ● specimen record.

Sinezona pacifica (Oliver, 1915)

Figures 397–402

Chresonymy.

Schismope pacifica Oliver, 1915: 514–515, pl. 9, figs 6, 6a.
Scissurella ornata: Jansen, 1995: 11, fig. 14 [is juvenile of *Sin. pacifica*].
Schismope pacifica: Jansen, 1998: 4, figs 4–6.
Sinezona pacificus [sic]: Brook & Marshall, 1998: 214
Sinezona pacifica: Jansen, 1999: 55, figs 55–57.
Sinezona pacifica: Marshall, 2002: 173, figs 16–18.
Sinezona pacifica: Geiger, 2003: 79.
Sinezona pacifica: Spencer *et al.*, 2009: 202.

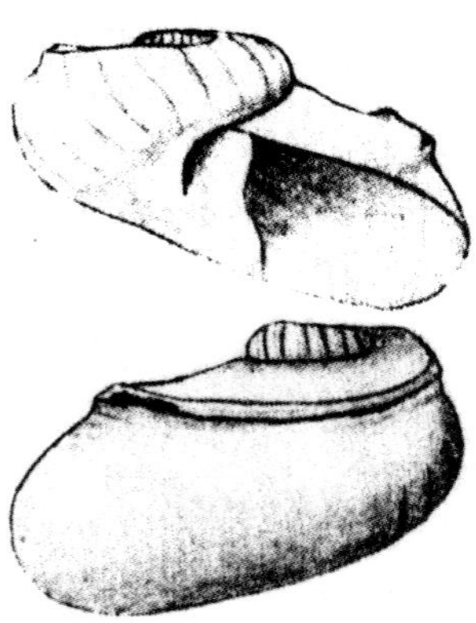

Figure 397. Original illustration of *Schismope pacifica* Oilver, 1915.

Synonyms.

+ (objective, secondary homonym) *Daizona pacifica* Bandel, 1998: 58–59, pl. 19, figs 6–7 [holotype], 8, pl. 20, fig. 1 [additional specimen]. ***Type material.*** Holotype (AMS C.162250 = neotype of *Schismope pacifica* Oliver, 1915. Fig. 398), 1 × 0.8 mm. ***Type locality.*** Kermadec [New Zealand, 29.250°S, 177.870°W] (OD). ***Etymology.*** Named for its wide distribution in the Pacific Ocean.

Figure 398. Original illustration of *Daizona pacifica* Bandel, 1998. Apertural and apical view of holotype (= neotype of *Schismope pacifica* Oliver, 1915); protoconch of different specimen.

Type material. Holotype (CM: lost), 1 mm. Paratypes from type locality (AMS C.36639, 9). Possible paratypes from type locality (AMS C.349555, 16; USNM 214806, 9). Neotype (AMS C.162250 = holotype of *Daizona pacifica* Bandel, 1998. Fig. 398), 1 × 0.8 mm: designated by Marshall (2002).

Type locality. Dead shells dredged on gravelly bottom in 10 m to 30 m near Sunday Island [= Raoul Island, 29.250°N, 177.870°W, Kermardec Island] (OD). Raoul Island, Kermadec Islands, New Zealand (neotype).

Etymology. Named for its wide distribution in the Pacific Ocean.

Description. Shell small size (to 0.9 mm. Figs 399–395), trochiform depressed. Protoconch of one whorl, strong axial sculpture, no apertural varix, apertural margin sinusoidal. Teleoconch I of 0.75–1 whorl, with approximately 10 to 17 strong axial cords, interstices with fine irregular growth marks, fine spiral lines below periphery starting after 0.5 whorl. Teleoconch II of 0.6 whorl, suture moderately impressed. Shoulder somewhat convex, axials barely visible to distinct cords, stronger near suture, at same density as on teleoconch I, interstices with fine irregular growth marks; fine spiral lines, increasing in number with growth, approximately 5–7 at apertural margin. Base with proportionally stronger axial sculpture compared to shoulder, approximately a dozen spiral lines, increasing in strength from selenizone to umbilicus; 3–4 closest to umbilicus can become knobby, last spiral marking border to umbilical wall. Umbilicus open, moderately wide, walls smooth. Selenizone above periphery, keels moderately strong, moderately elevated, at 45° to coiling axis, foramen elongated triangular. Aperture rounded D-shaped. Operculum corneous, nucleus central, multispiral.

Radula (Fig. 401). Rachidian tooth with triangular tip, central cusp largest, 3–4 cusps on each side. Lateral teeth 1–3 similar, apical cusp largest, 3–4 cusps along outer cutting edge of tip. Lateral tooth 4 reduced, hook-shaped, with one point on each side. Lateral tooth 5 enlarged by broadening, apical cusp largest, five cusps of decreasing size along inner edge of tip. Inner marginal teeth with triangular tip, apical cusp largest, 3–5 cusps on each side. Outer marginal teeth with spoon-shaped tip, many fine bristles along edge. No food groove on shaft of marginal teeth. Radular interlock of central field moderate.

Comparisons. *Sinezona bandeli* from New Zealand has a stronger constriction below the selenizone and more spiral lines on the shoulder.

Sinezona brucei from New Zealand has a shorter teleoconch I (0.75 vs. 1 whorl), more elevated axial cords, and a more elevated apex.

Sinezona brevis from New Zealand has a shorter selenizone (0.1 vs. 0.5 whorl), a less distinct constriction of the base below the selenizone, and a stronger periumbilical cord.

Sinezona iota from New Zealand has a much taller shell, lacks spiral lines on shoulder and base, and has much stronger raised axial cords.

Sinezona laqueus from New Zealand has a subquadrate shell, a much stronger constriction on the base below the selenizone, and stronger lunules.

Sinezona levigata from New Zealand has a wider shell, a callus covering the umbilicus, and a shorter selenizone.

Sinezona mechanica from New Zealand has a wider shell, lacks spiral lines on shoulder and base, and the protoconch lacks an apertural varix.

Sinezona norfolkensis from the Tasman Sea has a broader overall shell with wider umbilicus, a protoconch with the axials only reaching to the midline of the whorl, and a longer teleoconch I (0.8–0.9 vs. 0.75 whorl).

Sinezona pauperata from New Zealand lacks a selenizone, has a round foramen, and a protoconch with very fine irregular axials and no apertural varix.

Sinezona subantarctica has a wider shell, the umbilicus is covered by a callus and the selenizone is at most very short (<0.1 whorl).

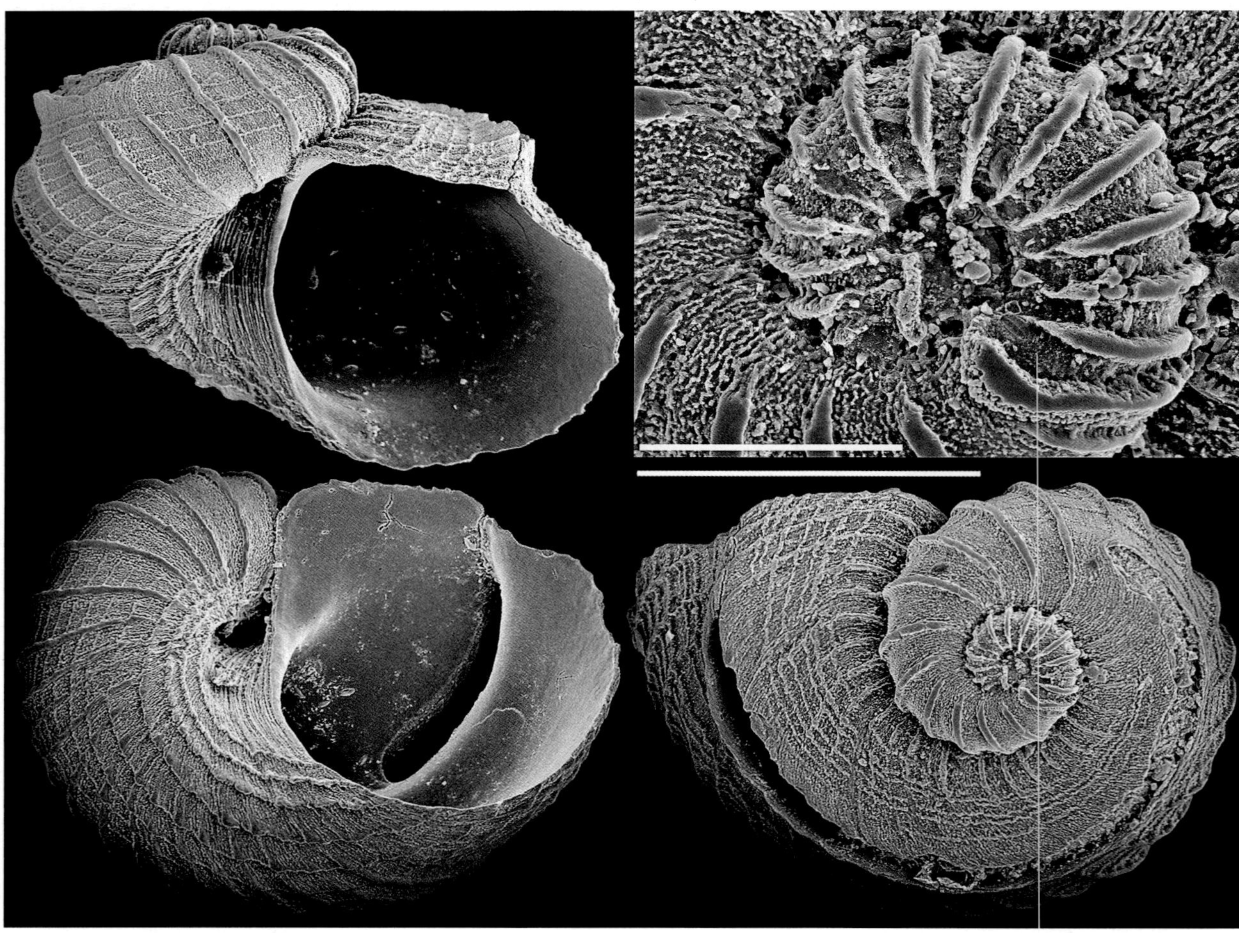

Figure 399. *Sinezona pacifica*. S end of lagoon, Lord Howe Island, 31.567°S, 159.073°E (AMS C.349523). Scale bar shell = 0.5 mm. Scale bar protoconch = 100 μm.

Sinezona platyspira from New Zealand has a flat spire, the umbilicus is wide and composed of the rounded portion of the previous whorls, and it has half as many axial lamellae.

Scissurella ornata from temperate Australia has a protoconch with median thickenings of the more numerous axials and about 50% more axial cords on the teleoconch.

Figure 400. *Sinezona pacifica*. **A.** Emily Bay, Kingston, Norfolk Island, Australia, 29.067°S, 167.967°E (AMS C.349528). Radula shown in Fig. 401B. **B.** 2 m, Sydney, Boat Harbour, N of Cronulla, New South Wales, Australia, 34.042°S, 151.200°E (AMS C.346354, 2). **C.** 9 m, SW Solitary Island, Groper Island, New South Wales, Australia, 30.162°S, 153.227°E (AMS C.346363). Scale bars shell = 0.5 mm. Scale bars protoconch = 100 μm.

Sinezona beddomei from temperate Australia has a strong constriction on the base below the selenizone, lacks any spirals on the shoulder, and spirals on the base are restricted to the adumbilical half.

Scissurella phenax from the Indo-Malayan Archipelago to the Central Pacific has a taller spire, a more narrow umbilicus, and a protoconch with median thickenings on the axials.

Distribution. Temperate Australia; Raoul Island, Kermadec Islands, Norfolk Islands, New Zealand, 0–100 m (Marshall, 2002).

Specimen records.

Australia, New South Wales. Watson Bay, Sydney Harbour, 33.845°S, 151.278°E (AMS C.461559, 2). 0 m, North Solitary Island, 29.927°S, 153.388°E (AMS C.461489, 2). Bittangabee, N of Green Cape, 37.217°S, 150.017°E (AMS C.346362, 8). 6 m, N of Green Cape, Bittangabee, 37.217°S, 150.018°E (AMS ex C.461485, 2). Twofold Bay, Eden, Nullica Beach, 37.100°S, 149.875°E (AMS C.378663, 1). Twofold Bay, 37.083°S, 149.900°E (AMS C.346357, 1). 0 m, Merimbula Beach, 36.9520°S, 149.917°E (AMS C.461482, 1). Bermagui, Shelly Beach, 36.417°S, 150.083°E (AMS C.346366, 1). Batemans Bay, 35.717°S, 150.200°E (AMS C.402948, 1). 10 m, W side Wasp Island, N of Batemans Bay, 35.667°S, 150.308°E (AMS C.461558, 1). Ulladulla, inside breakwater, S side, 35.358°S, 150.478°E (AMS C.346364, 2). S side of Ulladulla, 35.350°S, 150.483°E (AMS, 10: complete). 0 m, S Jervis Bay, 35.125°S, 150.753°E (LACM 79-43, 1). Jervis Bay, Callala Point, 35.008°S, 150.722°E (AMS C.346353, 1). Shell Harbor, 34.583°S, 150.867°E (WAM, 3; WAM, 3; AMS ex C.468098, 1). Port Hacking, 34.067°S, 151.100°E (AMS C.403741, 5). Gunnamatta Bay, 34.067°S, 151.133°E (WAM, 3). 0 m, N end of Pool enclosure, Gunnamatta Bay, Port Hacking, Sydney, 34.061°S, 151.150°E (AMS C.461562, 1). 2 m, Sydney, Boat Harbour, N of Cronulla, 34.042°S, 151.200°E (AMS C.346354, 2). 0 m, Kurnell, Botany Bay, Sydney, 34.009°S, 151.206°E (AMS C.461481, 3). Sydney, Little Coogee Bay, 33.922°S, 151.2600°E (AMS C.346278, 1). Sydney, Port Jackson, Vaucluse, 33.857°S, 151.272°E (AMS C.346355, 1). 11–16 m, Sydney, Port Jackson, Sow and Pigs Reef, 33.838°S, 151.268°E (AMS C.402951, 1). 11–14 m, Sydney, Port Jackson, off Sow and Pigs Reef, 33.838°S, 151.268°E (AMS C.402953, 1). 3 m, Sydney, Port Jackson, off Sow and Pigs Reef, 33.838°S, 151.268°E (AMS C.115475, 10). 4–8 m, Sydney, Middle Harbour, off Chinamans Beach, 33.816°S, 151.248°E (AMS C.402952, 7). Sydney, North

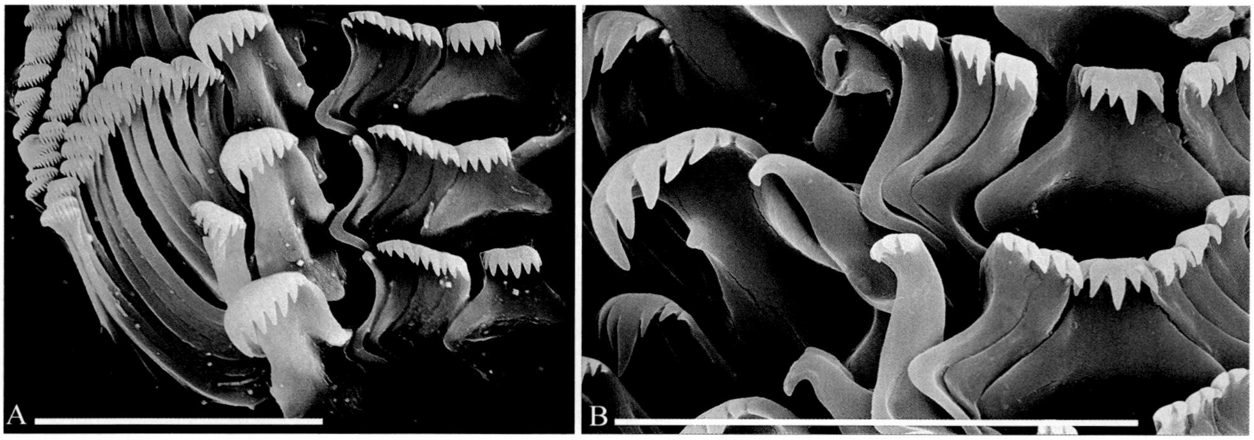

Figure 401. *Sinezona pacifica*. Radula. **A.** Half row with marginal teeth. Bittangabee, N of Green Cape, New South Wales, Australia, 37.217°S, 150.017°E (AMS C.346362, 8). **B.** Central field enlarged. Emily Bay, Kingston, Norfolk Island, 29.067°S, 167.967°E (AMS C.349528, 8). Shell shown in Figure 400A. Scale bars = 20 µm.

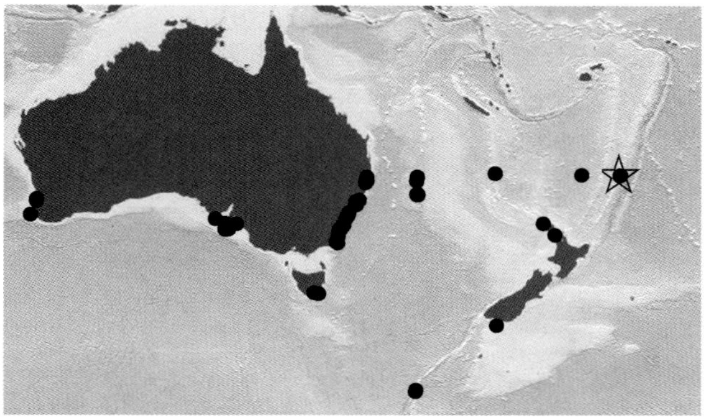

Figure 402. Distribution of *Sinezona pacifica*. ☆ type locality. ● specimen record. ■ literature record.

Harbour, SE side of Reef Bay, 33.808°S, 151.273°E (AMS C.404978, 7: complete). 4–7 m, Sydney, North Harbour, Fairlight, 33.803°S, 151.273°E (AMS C.346361, 1). Manly, Sydney, 33.803°S, 151.283°E (AMS C.469099, 5). Sydney, Manly, Shelly Beach, outer coast, 33.802°S, 151.297°E (AMS C.404238, 20: complete). 0 m, E of Shelly Beach, Manly, Sydney, 33.802°S, 151.297°E (AMS C.461549, 1). 0 m, E end rock platform, Shelly Beach, Manly, Sydney, 33.802°S, 151.297°E (AMS C.461476, 4). 0 m, E Manly Beach, Sydney, 33.802°S, 151.297°E (AMS ex C.4680097, 2). 0 m, Fairlight, N side Port Jackson, Sydney, 33.799°S, 151.269°E (LACM 82-55). Sydney N, Collaroy, Long Reef, NE side, 33.743°S, 151.310°E (AMS C.346365, 3). 0 m, Long Reef, Sydney, 33.742°S, 151.320°E (LACM 79-58, 10). Sydney N, Collaroy, Long Reef, Fishermans Beach, 33.738°S, 151.310°E (AMS C.404257, 14: complete). Sydney N, Collaroy, Long Reef, N side of platform, near Fishermans Beach, 33.738°S, 151.310°E (AMS C.404267, 1: complete). Sydney N, Collaroy, Long Reef, N side, 33.738°S, 151.310°E (AMS C.346367, 3). 0 m, N side Long Reef, Collaroy, 10 mi N of Sydney, 33.738°S, 151.317°E (LACM 75-118, 1). 0 m, Sydney N Collaroy, Long Reef, 33.733°S, 151.310°E (AMS C.445052, 30: complete). North Fingal Bay, near Port Stephens, 32.746°S, 152.175°E (AMS C.346351, 1). Port Stephens, Nelson Head, 32.713°S, 152.160°E (AMS C.346358, 8). Forster, 32.183°S, 152.517°E (AMS C.405024). 9 m, SW Solitary Island, Groper Island, 30.162°S, 153.227°E (AMS C.346363, 1). E of Grafton, Minnie Water, 29.777°S, 153.300°E (AMS C.346360, 5, 2: complete). Angourie, 29.483°S, 153.367°E (AMS C.346359, 1). Port Stephens, 32.394°S, 152.067°E (RMNH AD ex 30458, 2; RMNH AD ex 30717, 1). Avalon, 33.635°S, 151.333°E (RMNH AD 32128, 3). Narrabeen, 33.714°S, 151.302°E (RMNH AD 30459, 15). Dee Why, 33.753°S, 151.297°E (RMNH AD 32263, 2).

Australia, South Australia. Kangaroo Island, N coast, Stokes Bay, 35.617°S, 137.200°E (AMS C.378563, 2). Bay of Shoals, NE Kangaroo Island, 35.600°S, 137.617°E (AMS C.404237, 1: complete). Adelaide, Glenelg Beach, 34.983°S, 138.517°E (AMS C.378564, 1). Point Souttar, Yorke Peninsula, 34.900°S, 137.267°E (AMS C.378562, 3).

Australia, Tasmania. Blow Hole, Tasman Peninsula, 43.150°S, 147.850°E (WAM, 3). 15 m, Tinderbox Bay, S of Hobart, 43.050°S, 147.333°E (AMS C.402950, 1).

Australia, Western Australia. 4–8 m, Carnac Island, W side, off Freemantle, 32.117°S, 115.667°E (AMS C.404912 3: complete). Safety Bay, 32.300°S, 115.700°E (WAM, 1). Minim Cove, Mosman Park, Swan River, 32.017°S, 115.767°E (AMS C.404439, 1). S of Cowaramup Bay, near mouth of Margaret River, 33.883°S, 115.017°E (AMS C.404155, 1).

Australia, Lord Howe Island. 45 m, Lord Howe Shelf, 31.574°S, 159.201°E (AMS C.461430, 2). 50 m, Lord Howe Shelf, 31.551°S, 159.150°E (AMS C.461438, 2). 0 m, 31.550°S, 159.083°E (WAM, 2). 25 m, Lord Howe Shelf, 31.547°S, 159.105°E (AMS C.461439, 2). 48 m, Lord Howe Shelf, 31.510°S, 159.098°E (AMS C.461437, 1). 35 m, Lord Howe Shelf, 31.498°S, 159.031°E (AMS C.461434, 1). 35 m, Lord Howe Shelf, 31.473°S, 159.129°E (AMS C.461432, 8). 96 m, Lord Howe Shelf, 31.444°S, 159.127°E (AMS C.461436, 1). S end of lagoon, 31.567°S, 159.073°E (AMS C.349523, 1; AMS C.349525, 8). 36 m, 31.550°S, 159.083°E (AMS C.349522, 2). Off N Head Phillip Point, 31.522°S, 159.042°E (AMS C.404981, 1: complete). 8 m, N Head Phillip Point, 31.522°S, 159.042°E (AMS C.349527, 1). Neds Beach, 31.517°S, 159.067°E (AMS C.349524, 2). Neds Beach, 31.517°S, 159.067°E (AMS C.349526, 1).

Tasman Sea. 10–24 m, Elizabeth Reef, 29.962°S, 159.047°E (AMS C.349521, 1). 12 m, Elizabeth Reef, W side, 29.953°S, 159.020°E (AMS C.349519, 1). 1 m, Elizabeth Reef,

29.937°S, 159.040°E (AMS C.349517, 1, –: complete). Elizabeth Reef, 29.923°S, 159.045°E (AMS C.349512, 4). 18 m, Elizabeth Reef, NW side, 29.913°S, 159.047°E (AMS C.349514, 2). 12 m, Elizabeth Reef, NW side, 29.913°S, 159.047°E (AMS C.349518, 3). 8 m, Middleton Reef, 29.480°S, 159.125°E (AMS C.349516, 2). 6 m, Middleton Reef, N side, 29.462°S, 159.090°E (AMS C.349515, 5). 9 m, Middleton Reef, 29.453°S, 159.113°E (AMS C.349513, 1).

Australia, Norfolk Island. Emily Bay, Kingston, 29.067°S, 167.967°E (AMS C.349528, 8; AMS C.349529, 1; AMS C.349530, 1). Ball Bay, 29.055°S, 167.972°E (AMS C.349532, 5). 29.033°S, 167.950°E (AMS C.349531, 11).

New Zealand, Kermadec Islands. 29.267°S, 177.917°W (USNM 680237, 2; USNM 680237, 2; NMSZ 1961.61, 1; NMSZ 1961.61, 1; NMSZ 1961.61, 2). 10–30 m, off Sunday Island, 29.264°S, 177.926°W (GNS WM 2254, 2). Raoul, Sunday Island, 29.250°S, 177.867°W (AMS C.36639, 9; AMS C.349533, 1; AMS C.349555, 16; AMS C.349557, 1; AMS C.162250, 1). SE of Smith Bluff, 29.315°S, 177.940°W (NMNZ M.226900, 4).

New Zealand. Doubtless Bay, 34.917°S, 173.467°E (MV F 96919, 2). 28 m, Stewart Island, off Halfmoon Bay, 46.883°S, 168.133°E (AMS C.378574, 1). Goat Island Beach, Leigh, Hauraki Gulf, 36.267°S, 174.800°E (AMS C.405027, 8: complete).

Macquarie Island. 69 m, off Lusitania Bay, 54.683°S, 158.900°E (AMS C.376613, 2). 14 m, Rocks at end of Garden Bay Peninsula, 54.500°S, 158.950°E (AMS C.376619, 1). Aerial Cove, 54.483°S, 158.950°E (AMS C.376620, 1; AMS C.376622, 18, –: complete).

Literature records.

New Zealand. Raoul Island, Kermadec Islands, 29.250°S, 177.867°W. 15 m, off W side of Meyer Island, 29.233°S, 177.833°E (Marshall, 2002).

Remarks. Oliver (1915: 514) noted that he designated a single specimen ("the type") to be deposited in the Canterbury Museum (CM). As there is no mention of paratypes in the original description, the standing of the AMS specimens is questionable.

Freeman *et al.* (1997: *fide* Marshall, 2002) did not list the holotype of *Sin. pacifica* Oliver among the holdings of CM; hence, Marshall (2002) presumed it lost and designated a neotype.

Bandel (1998: 58) listed under material examined several additional lots, but did not designate those as paratypes, hence, these lots have no type status.

Sinezona paumotuensis (Garrett, 1872) [*nomen dubium*]

Chresonymy.

Scissurella paumotuensis Garrett, 1872: 878, not illustrated.
Scissurella paumotuensis: Paetel, 1888: 289.
Schismope paumotuensis: Thiele, 1912a: 32.
Sinezona paumotuensis: Geiger, 2003: 79.

Type material. (wherabouts unknown, not CASIZ), 2 × 1.5 mm.

Type locality. Paumotu Islands [= Tuamotus, French Polynesia, ~16°S, 145°E] (OD).

Etymology. Named for its type locality.

Description. [Transcription of original description]. "Shell umbilicate, depressedly suborbicular, thin, vitreous, pellucid, white; spire very slightly elevated; whorls 2½, strongly keeled, the last constricted beneath the keel, longitudinally ribbed; ribs small, rather distant, obliquely curved, smaller, more complex and flexuous above, decussated

with very fine crisp-like spiral lines; fissure linear, a little back from the margin of the lip; umbilicus large, deep and somewhat angulated on the margin; aperture subcircular; outer lip very oblique, sharp, entire, and rounded; columella thin and slightly arched."

Remarks. Garrett mentioned "2.5, strongly keeled" whorls, which suggests *Sukashitrochus*. It may also refer to the keels of the selenizone, and not the spiral keels on the base of the shell. The indication "fissure ... a little back from the margin of the lip" suggests a foramen, and hence a placement in *Sinezona* or *Sukashitrochus*. A size of 2 mm and the combination of "longitudinally ribbed" [= axial cords/ribs] and "very fine crisp-like spiral lines" [= spiral lines] is in agreement with the fairly large and common *Sin. plicata* known from French Polynesia. As such a synonymy is rather conjectural, *Sin. paumotuensis* is rather regarded here as a *nomen dubium*.

Sinezona pauperata Powell, 1933

Figures 403–407

Chresonymy.

Sinezona pauperata Powell, 1933a: 193, pl. 33, figs 4–5.
Sinezona pauperata: Dell, 1960, 1945, 155.
Sinezona pauperata: Powell, 1979: 35, pl. 13, figs 18–19.
Sinezona ? pauperata: Lozouet, 1986: 111.
Sinezona pauperata: Spencer & Willar., 1995: 13.
Ariella pauperata: Geiger, 2003: 75.
Ariella pauperata: Spencer *et al.*, 2009 202.

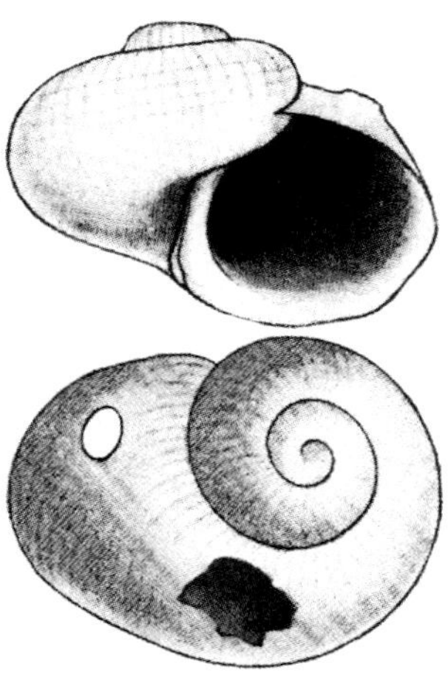

Figure 403. Original illustration of *Sinezona pauperata*.

Type material. Holotype (AIM AK 70710: destroyed by Byne's/glass disease: W. Blom, pers. comm. 4/2009), 0.4 × 0.5 mm (H × W).

Type locality. Off Owenga Beach, Chatham Islands, [New Zealand, 43.950°S, 176.550°W] 10 fathoms [= 18 m] (OD).

Etymology. Pauper, Latin for poor, penniless.

Description. Shell small (to 0.58 mm. Figs 404, 405), trochiform rounded. Protoconch of 0.9 whorl, dense fine axials, no apertural varix, apertural margin opisthocline. Teleoconch I of 1.2 whorls, approximately 45 broad, dull axial cords, approximately four slightly less distinct low, dull spiral cords on shoulder. Teleoconch II of 0.125 whorl, suture hardly impressed. Shoulder with same sculpture as on teleoconch I. Base without constriction below selenizone, same density and strength of axial

cords as on shoulder, about 10 spiral cords. Umbilicus very narrow, bordered by strong spiral cord, walls not visible. Aperture obliquely rounded. No selenizone; foramen round, anteriorly continued by raphe; keels moderately strong, moderately elevated.

Radula (Fig. 406A–B). Rachidian tooth triangular, apical cusp slightly larger than two on each side. Lateral teeth 1–3 similar, apical cusp largest, 3–5 cusps on outer margin. Lateral tooth 4 much reduced, without tip. Lateral tooth 5 enlarged by broadening, apical cusp largest, six cusps along inner edge. Inner marginal teeth with broad triangular tip, apical cusp largest, 1–2 cusps along inner margin, approximately six along outer margin. Outer marginal teeth with spoon-shaped tips, many small cusps along edge. Radular interlock moderate.

Operculum (Fig. 406C) corneous, round, thin, multispiral, with central nucleus.

Comparisons. *Sinezona bandeli* from New Zealand has a strong constriction on the base below the selenizone, a distinct umbilicus, and a distinct selenizone.

Sinezona brevis from New Zealand has a distinct umbilicus, usually a short selenizone, and a protoconch with strong axial cords and an apertural varix.

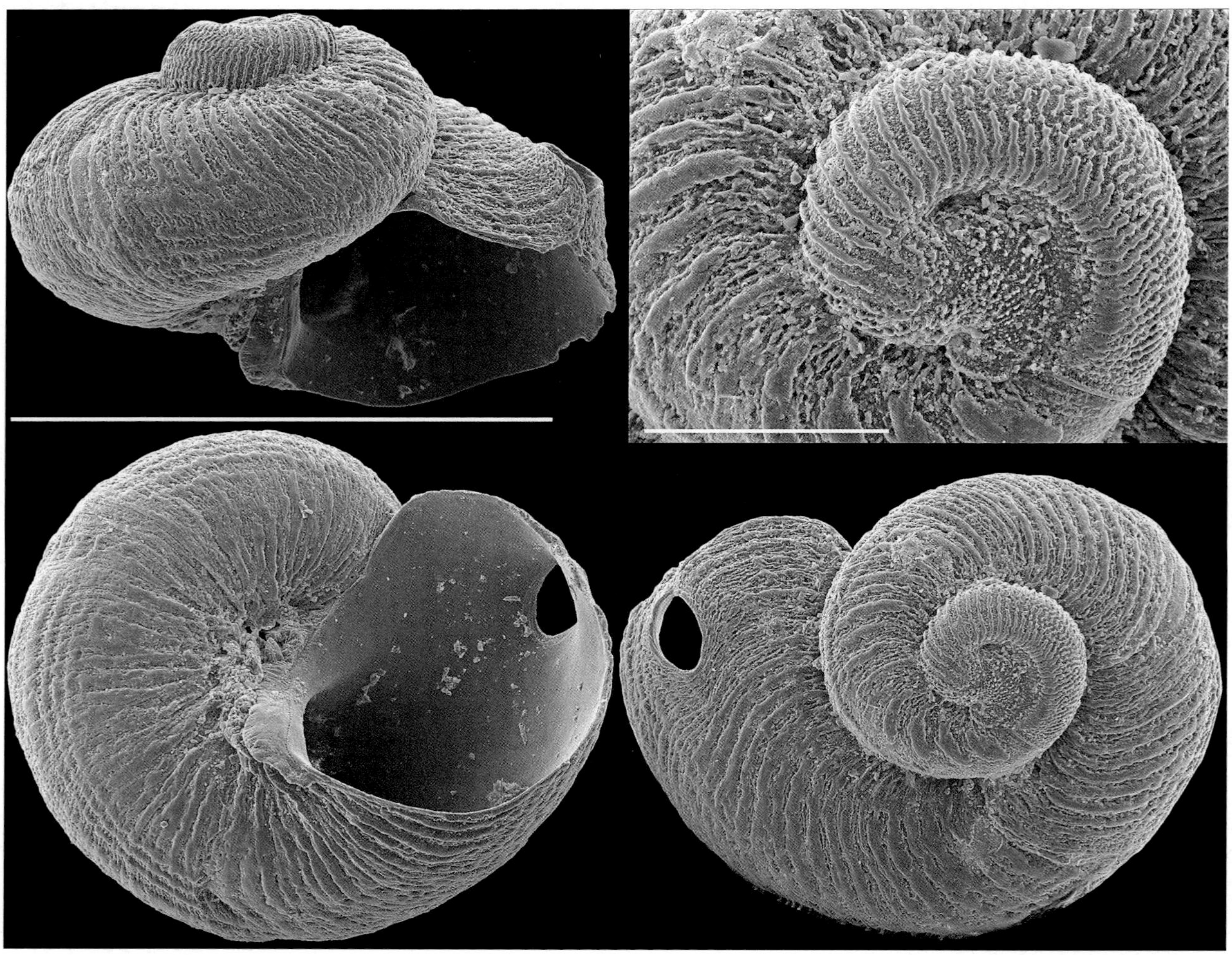

Figure 404. *Sinezona pauperata*. 33–37 m, off "The Breaks," Snares Islands, New Zealand, 48.015°S, 166.613°E (NMNZ M.143247, 7). Scale bar shell = 0.5 mm. Scale bar protoconch = 100 μm.

Sinezona iota from New Zealand has a taller shell with protruding protoconch, strong axial cords on the teleoconch, and no spiral cords on shoulder and base.

Sinezona laqueus from New Zealand has a strong constriction on the base below the selenizone, a distinct selenizone, and strong axial folds.

Sinezona levigata from New Zealand has a callus covering the umbilicus, usually a short selenizone, and the protoconch has strong axial cords and an apertural varix.

Figure 405. *Sinezona pauperata*. **A.** 33–37 m, off "The Breaks," Snares Islands, New Zealand, 48.015°S, 166.613°E (NMNZ M.143247, 7). **B.** 37 m, off "30 Fathom Point," inside Five Finger Peninsula, Fiordland, South Island, New Zealand, 45.720°S, 166.492°E (NMNZ M.147225, 30). **C.** 20 m, Boundary Rock, S Chatham Island, New Zealand, 44.062°S, 176.383°W (NMNZ M.160379, 2). Scale bars shell = 0.5 mm. Scale bars protoconch = 100 μm.

Sinezona mechanica from New Zealand has a wider shell, a wider umbilicus, and a protoconch with a top spiral cord.

Sinezona pacifica has a distinct constriction on the base below the selenizone, a distinct umbilicus, and distinct axial cords.

Sinezona platyspira from New Zealand has a flat spire, a wide umbilicus, and strong raised axial lamellae on the teleoconch.

Sinezona subantarctica of broad subantarctic distribution is overall wider, has a callus covering the umbilicus, and an oval foramen.

Sinezona haliotimorpha from the Oligocene of France has a protoconch with fewer and stronger axials, a teleoconch with stronger axials and consistently finer spiral threads on base, and a foramen with more elevated keels.

Sinezona hawaiiensis known from a single specimen from Hawaii is indistinguishable from *Sin. pauperata* (see Remarks under *Sin. hawaiiensis*).

Sinezona subantarctica from subantarctic waters has an overall more compressed shell, a distinct callus covering the umbilicus, a more elongated foramen, and a protoconch with an apertural varix.

Distribution. Subantarctic Islands of Australia and New Zealand (Fig. 407).

Specimen records.

Australia. Garden Bay, Macquarie Island, 54.500°S, 158.950°E (AMS, 20: complete). 5 m, Aerial Cove, Macquarie Island, 54.483°S, 158.950°E (AMS C.404552, 4: complete; AMS C.404556, 20: complete). Handspike Point, Macquarie Island, 54.500°S, 158.883°E (AMS C.404554, 17: complete). Aerial Cove, near cove entrance, Macquarie Island, 54.483°S, 158.950°E (AMS C.404600, 20: complete).

New Zealand. 3 m, E side Vire Point, Perseverance Harbour, Campbell Island, 52.550°S, 169.167°E (AMS C.404241, 4: complete). 4 m, NW side of Boulder Beach, Smoothwater Bay, Campbell Island, 52.533°S, 169.200°E (AMS C.404242, 6: complete). 8 m, NW side first bay NW of Boulder Beach, Smoothwater Bay, Campbell Island, 52.533°S, 169.200°E (AMS C.404250, 13: complete). 6–8 m, NW side First Bay NW of Boulder Beach, Smoothwater Bay, Campbell Island, 52.533°S, 169.200°E (AMS C.404252, 5: complete). 8 m, Smoothwater Bay, Campbell Island, 52.533°S, 169.200°E (AMS C.404546, 4: complete). 3 m, E side Vire Point, Perseverance Harbour, Campbell Island, 52.550°S, 169.167°E (AMS C.404547, 4: complete). 6 m, E side Windlass Bay, Campbell Island, 52.550°S, 169.067°E (AMS C.404548, 7: complete). 2 m, W side Southeast Harbour, Campbell Island, 52.600°S, 169.150°E (AMS C.404549, 17:

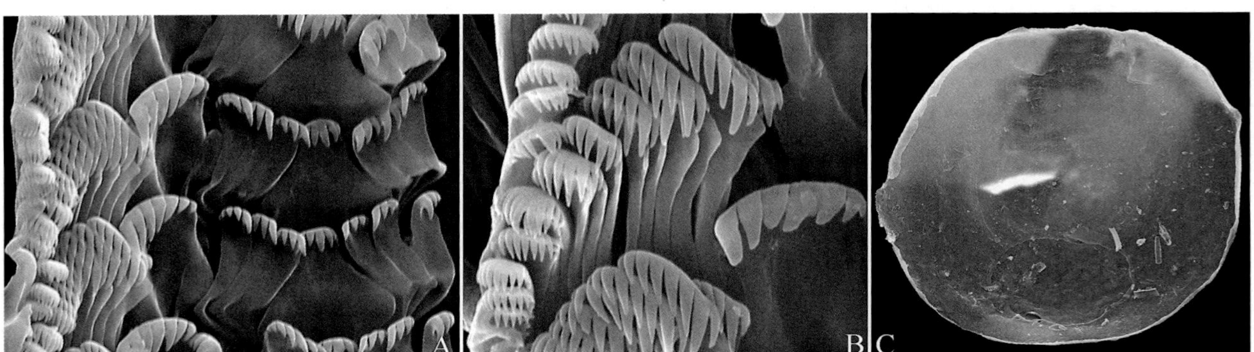

Figure 406. *Sinezona pauperata*. **A–B.** Radula. 10 m, NW point of Leeward Island, Antipodes Islands, New Zealand, 49.650°S, 178.883°E (NMNZ M.147227, 4). **A.** Central field and one side with marginal teeth. **B.** Marginal teeth. Scale bars A–B = 10 µm. **C.** Operculum. 16 m, E side of Perpendicular Head, Antipodes Islands, New Zealand, 49.650°S, 178.883°E (NMNZ M.147217, 5). Scale bar = 100 µm.

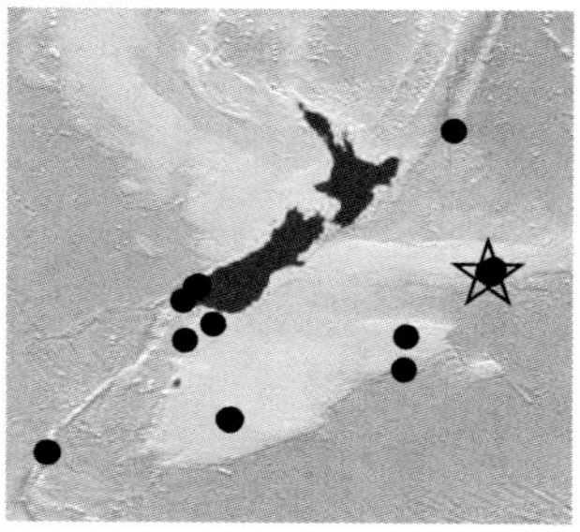

Figure 407. Distribution of *Sinezona pauperata*. ☆ type locality. ● specimen record.

complete). 1–2 m, W side Southeast Harbour, Campbell Island, 52.600°S, 169.150°E (AMS C.404551, 1: complete). 8 m, NW side of Boulder Beach, Smoothwater Bay, Campbell Island, 52.533°S, 169.200°E (AMS C.404553, 6: complete). 55 m, Port Adventure, Stewart Island, 47.083°S, 168.183°E (AMS C.406340, 1). Outer George Sound, Fjordland, 44.850°S, 167.350°E (NMNZ M.177152, 4). 220–420 m, Rumble III Volcano, S Kermadec Ridge, 35.740°S, 178.498°W (NMNZ M.174861, 1). 33–37 m, off The Breaks, Snares Islands, 48.015°S, 166.613°E (NMNZ M.143247, 7). 15 m, Two Thumb Bay, Fiordland, South Island, 44.955°S, 167.183°E (NMNZ M.174220, 2). 20 m, Boundary Rock, S Chatham Island, 44.062°S, 176.383°W (NMNZ M.160379, 2). 0 m, Okawa Point, Watson, Chatham Islands, 43.775°S, 176.242°W (NMNZ M.160377, 1). 39 m, Proclamation Island, Bounty Islands, 47.750°S, 179.000°E (NMNZ M.119753, 10). 37 m, off 30 Fathom Point, inside Five Finger Peninsula, Fiordland, South Island, 45.720°S, 166.492°E (NMNZ M.147225, 30). 0 m, Waitangi, Oliver, WRB, Chatham Islands, 43.950°S, 176.560°W (NMNZ M.013681, 9). 16 m, E side of Perpendicular Head, Antipodes Islands, 49.650°S, 178.883°E (NMNZ M.147217, 5). 10 m, NW point of Leeward Island, Antipodes Islands, 49.650°S, 178.883°E (NMNZ M.147227, 4).

Remarks. The very reduced lateral tooth 4 is unique among Scissurellidae, but known also from one species of Anatomidae, *An. equatoria*. Although the holotype is destroyed, there is no need to designate a neotype, because the species is easily recognizable and has not been confused in the literature.

Sinezona platyspira Geiger & Marshall, 2012

Figures 408–409

Chresonymy.

Sinezona platyspira Geiger & Marshall, 2012: 9–11, fig. 7.

Type material. Holotype (NMNZ M.301400. Fig. 408A). One paratype from type locality (NMNZ M.301399, 2. Fig. 408B).

Type locality. 785–800 m, W of Cape Reinga, West Norfolk Ridge, New Zealand, 34.298°S, 168.430°E (OD).

Etymology. Platy-, flat, -spira, spire: referring to the flat spire of the species. Adjective (OD).

Description. Shell small (to 0.95 mm. Fig. 408), trochiform depressed. Protoconch one whorl, strong axials, no apertural varix, apertural margin convex. Teleoconch I of 1.1 whorls, 12–26 axial lamellae, interstices with very fine growth marks, no spirals. Teleoconch II of 0.6 whorl, suture deeply impressed. Shoulder flat; approximately 12–21 axial cords, often restricted to adsutural quarter, in some specimens almost absent; no spirals. Base with distinct constriction below selenizone; on teleoconch I, axial lamellae continuous into umbilicus; on teleoconch II, fewer axial lamellae (approximately six) fading out towards umbilicus; no spirals. Umbilicus open, deep, cavity produced by rounded portion of previous whorls, bordered by strong spiral cord, underside of protoconch visible. Aperture tilted D-shaped, roof overhanging. Selenizone above periphery, keels strong, quite elevated; lunules distinct, foramen closed with short anterior raphe.

Animal unknown.

Figure 408. *Sinezona platyspira*. 785–800 m, W of Cape Reinga, West Norfolk Ridge, New Zealand, 34.298°S, 168.430°E. **A.** Holotype. (NMNZ M.301400). **B.** Paratype (NMNZ M.301399, 1). Scale bars shell = 0.5 mm. Scale bars protoconch = 100 µm.

Comparisons. *Sinezona bandeli* from New Zealand has a strong constriction on the base below the selenizone, an umbilicus with straight walls, and the protoconch has an apertural varix.

Sinezona brevis from New Zealand has an umbilicus with straight walls, a much shorter selenizone (0.1 vs. 0.4 whorl), and a more elevated spire.

Sinezona iota from New Zealand has a much taller shell with protruding protoconch, the umbilicus has straight walls, and the protoconch has an apertural varix.

Sinezona laqueus from New Zealand has a stronger constriction of the base below the selenizone, the umbilicus has straight walls, and has axial folds as opposed to lamellae.

Sinezona levigata from New Zealand has a callus covering the umbilicus, lacks a constriction of the base below the selenizone, and has no axial lamellae.

Sinezona mechanica from New Zealand has an umbilicus with straight walls, about twice as many finer axial cords, and axial cords over the entire width of the shoulder on teleoconch II.

Sinezona pacifica from temperate Australia and New Zealand has an umbilicus with straight walls, is overall taller, and has a protoconch with apertural varix.

Sinezona pauperata from New Zealand lacks a selenizone, has a very narrow umbilicus, and lacks axial lamellae.

Sinezona subantarctica with broad subantarctic distribution has a callus covering the umbilicus, a very short selenizone (<0.1 whorl), and spiral lines on the lower base.

Sinezona kayae from Hawaii has a taller shell, fewer and more distinct axials on the protoconch, and a much more narrow umbilicus.

Sinezona semicostata from the Macaronesian Islands has a protoconch with at most half as many much stronger axials, a shorter teleoconch I (0.875 vs. 1.1 whorls), and a wider aperture.

Satondella minuta from the Indo-Pacific has much more elevated keels of the foramen.

Scissurella azorensis from the Azores to the Western Mediterranean retains an open slit and has a shorter teleoconch II (0.3 vs. 0.6 whorl).

Scissurella maraisorum from East Africa has a slightly shorter teleoconch I (<1 vs. 1.1 whorls).

Scissurella kaiserae from the Panamic has a protoconch with about 50% fewer axials, retains an open slit, and has a slightly narrower umbilicus.

Scissurella lobini from the Macaronesian Islands has a shorter teleoconch II (0.3 vs. 0.6 whorl), retains an open slit, and has 50% fewer axials on the protoconch.

Figure 409. Distribution of *Sinezona platyspira*. ☆ type locality.

Distribution. Northern New Zealand (Fig. 409).

Specimen records.

New Zealand. 785–800 m, W of Cape Reinga, W Norfolk Ridge, 34.285°S, 168.430°E (NMNZ M.172128, 31).

Remarks. The most similar species is *Sci. maraisorum* from East Africa. Both species are known from only few specimens and the difference in teleoconch length of only 0.1 whorl is well below the usual intraspecific variability of 0.25 whorl. Unlike *An. finlayi* from northern New Zealand and its synonym *An. herberti* described from off East Africa, there are no specimens of the species known from intermediate localities. The two species are considered distinct.

Sinezona plicata (Hedley, 1899)

Figures 410–417, Color Plates 3E, 4B

Chresonymy.

Schisomope [*sic*] *plicata*: Anon., 1899: 520 [*nomen nudum*].
Schismope plicata Hedley, 1899: 552, fig. 62.
Schismope plicata: Thiele, 1912a: 32, pl. 4, fig. 6 [copy figure Hedley, 1889].
Scissurella (*Sukashitrochus*) *plicata*: Lozouet, 1986: 109.
scissurellid: Bandel, 1991: 444, text-fig. 4.
Reussella plicata: Bandel, 1998: 46–47, pl. 15, fig. 8, pl. 16, figs 1–2.
Sinezona plicata: Jansen, 1999: 5, figs 58–60.
Scissurella sp. 2. Hasegawa *et al.*, 2001: 4, pl. 1, fig. 2.
Sinezona cf. *plicata*: Smith, 2003: 149.
Sinezona plicata: Geiger, 2003: 79, 15c [*Sci. plicata* in caption in error].
Sinezona plicata: Geiger & Jansen, 2004b: 38–46, figs 21–24 [shell, radula, map].
Sinezona plicata: Geiger, 2004: text-fig. p. 6.
Sinezona plicata: Geiger, 2006c: text-fig. 4.
Sinezona plicata: Hickman & Porter, 2007: 10–17, fig. 3d–f.
Sinezona plicata: Geiger, 2008a: 158, pl. 24, figs 3a–b, 4a–b.
Sinezona plicata: Tröndlé & Boutet, 2009: 12.

Figure 410. Original illustration of *Schismope plicata*.

Synonyms.

+ *Woodwardia provecta* Beets, 1943: 260, pl. 27, figs 4–7. ***Type material.*** Holotype (RMNH 42474), 2 × 2.5 mm (H × W). ***Type locality.*** Kabungka [Bouton, Pulau Butung, Indonesia, 5.615°S, 122.705°E; upper Oligocene: inferred from title] (OD). ***Etymology.*** Latin for advanced in age.

Figure 411. Original illustration of *Woodwardia provecta*.

Misidentifications.

Scissurella coronata: Habe, 1963: 229 [Japanese], 233 [English], textfigs 3–4 [is *Sin. plicata*, slightly juvenile with open slit].
Sinezona cingulata: Ladd, 1982: 20–21, pl. 40, figs 1–3 [is *Sin. plicata*].

Scissurella coronata: Okutani & Hasegawa, 2000: 37, pl. 18, fig. 10 [is *Sin. plicata*, slightly juvenile with open slit].

Scissurella coronata: auct., 2005: 36, text-fig. 2 [juvenile *Sin. plicata* with open slit].

Sinezona padangensis: Yu & Feng, 1996: pl. 1, figs 5–7 [is *Sin. plicata*].

Type material. Holotype (AMS C.5640), 2.3 × 1.7 × 2 mm. Paratypes: 366 m, Funafuti Atoll, off Tutanga Tutaga Island, Tuvalu, 8.617°S, 179.083°E (AMS C.5641, 7). 91–110 m, Funafuti Atoll, off Tutanga Tutaga Island, Tuvalu, 8.617°S, 179.083°E (AMS C.5642, 5). 274 m, Funafuti Atoll, off Funamanu Beacon Island, Tuvalu, 8.567°S, 179.150°E (AMS C.170414, 1).

Type locality. 150 fms. [= 274 m], off Beacon Islet (Funamanu), Funafuti Atoll, [Tuvalu, 8.567°S, 179.150°E] (OD).

Etymology. Plicatus, Latin adjective for folded, referring to the strong axial folds on the shell.

Description. Shell large (2.3 mm), heavy, depressed trochiform. Protoconch 0.75 whorl, smooth or with single fine spiral towards periphery, varix connected to embryonic cap, aperture sinusoidal. Teleoconch I 0.9–1.25 whorls, 15 axial cords, interstices with fine densely packed irregular axial lamellae. Teleoconch II up to 1.5 whorls, suture moderately to strongly impressed. Shoulder profile S-shaped, with ridge towards

Figure 412. *Sinezona plicata*. 21 m, Great Barrier Reef, Euston Reef, SW side, Queensland, Australia, 16.667°S, 146.217°E (AMS C.379083). Scale bar shell = 1 mm. Scale bar protoconch = 100 μm.

suture, with channel towards selenizone. Shoulder sculpture with strong axial lamellae crossed by 14–16 distinct but somewhat weaker spiral cordlets. Base with constriction below selenizone; axials in constriction rather weak, abapical margin of constriction forming suture in early teleoconch; suture descends in last 0.25 whorl to base of penultimate whorl at level of aperture in fully mature shells. Whorls distinctly inflated in juveniles, less in mature shells, lateral margin always extending

Figure 413. *Sinezona plicata*. **A.** Moorea Island, Papetoai Bay, Passe Tareu, W side, Society Islands, French Polynesia, 17.508°S, 149.850°W (AMS C.379421). **B.** 3 m, Great Barrier Reef, Lizard Island, between Bird & South Islets, Queensland, Australia, 14.683°S, 145.450°E (AMS C.379078). **C.** Marutea Atoll, Tuamotu Archipelago, French Polynesia, 17.000°S, 143.167°W (AMS C.28241). Scale bars shell = 1 mm. Scale bars protoconch = 100 μm.

beyond selenizone and keels. Lower part of base with pronounced axial lamellae, crossed by distinct spiral threads. Umbilicus shallow, wide, walls smooth, with sharp edge towards base. Aperture rotund, roof projecting. Selenizone with distinct and strong keels, strong lunules not coordinated with axials, foramen elongated tear-drop-shaped.

Operculum (Fig. 413A) corneous, thin, round, multispiral, central nucleus.

Figure 414. *Sinezona plicata*. **A.** 78 m, off North West Cape, Western Australia, Australia, 22.967°S, 113.687°E (AMS C.379091). **B.** 2 m, Bohol, Panglao Island, SW side, Philippines, 9.600°N, 123.750°E (AMS C.380674). **C.** North East Herald Cay, Coral Sea, 16.933°S, 149.183°E (AMS C.334062). Scale bars shell = 1 mm. Scale bars protoconch = 100 μm.

Radula (Fig. 416). Rachidian tooth trapezoidal, central cusp largest, three cusps on each side. Lateral teeth 1–3 similar, apical cusp largest 2–3 somewhat smaller cusps along outer edge. Lateral tooth 4 reduced, apical cusp largest, one small point on each side. Lateral tooth 5 enlarged by broadening, approximately four large cusps in apical part, additional 3–4 progressively smaller cusps in transition region to shaft. Inner marginal teeth with palmate tip, apical cusp largest, 2–4 smaller cusps on each side. Outer marginal teeth with spoon-shaped tip, many fine bristles along edge. Radular interlock of central field moderate.

Comparisons. The pronounced expansion of the base below the selenizone, in combination with the strong axials folds crossed by strong and dense spirals giving the shell the appearance of a filigreed network, readily distinguish this species from any other. Juveniles, although having an open slit and not a foramen, also display these sculptural characteristics. The usual, smooth protoconch, which occasionally also shows the spiral element, is rare in Scissurellidae, although caution should be taken not to confuse an eroded protoconch with a genuinely smooth one.

Figure 415. *Woodwardia provecta*. Holotype. Kabungka [Bouton, Pulau Butung, Indonesia, 5.615°S, 122.705°E, upper Oligocene] (RMNH 42474). Scale bar shell = 1 mm. Scale bar protoconch = 100 μm.

Sinezona costulata from southern Japan has a more narrow umbilicus, lacks axial folds, and has finer spiral threads.

Sinezona danieldreieri from the Indo-Malayan Archipelago and the Central Pacific has a more narrow umbilicus, lacks axial folds, and has finer spiral threads.

Sinezona ferriezi from the Indo-Malayan Archipelago to the Central Pacific has always a smooth protoconch, much finer axial cords, and finer spiral threads.

Sinezona globosa from the Indo-Malayan Archipelago to the tropical Eastern Pacific (and possibly Caribbean) has a more globular shell, a longer teleoconch I (1.5 vs. 1 whorls), and the base gradually slopes into the umbilicus.

Sinezona macleani from the Indian Ocean to the Western Pacific lacks the axial lamellae, lacks an apertural varix on the protoconch, and has a straight parietal wall.

Sinezona marrowi from the Indo-Malayan Archipelago has a weaker constriction of the base below the selenizone, lacks strong axial folds, and has a protoconch with axial cords.

Sinezona milleri from southern Japan lacks axial folds, has a protoconch with axial cords, and a much narrower umbilicus.

Sinezona singeri from the Indian Ocean lacks axial folds, has a weaker constriction of the base below the selenizone, and finer spiral lines.

Sinezona wileyi from the Indo-Malayan Archipelago and the Western Pacific has a subquadrate overall shell shape, lacks strong axial folds, and has a more narrow umbilicus.

Satondella kondoi from the Eocene of Tonga has more axial cords that reach the selenizone on the base and the umbilicus is bordered by a more strongly angulated carina.

Distribution. Indian Ocean to Central Pacific, 0–366 m, in sand and coral rubble, as well as on coral reef algae (Fig. 417).

Fossil records.

Indonesia. Oligocene of Kabungka [Bouton, Pulau Butung, 5.615°S, 122.705°E] (RMNH 42474, 1: holotype of *Sin. provecta*).

Vanuatu. Pleistocene (Ladd, 1982).

Figure 416. *Sinezona plicata*. Radula. **A.** Tanna Island, Hotel Tanna, Vanuatu, 19.533°S, 169.267°E (AMS C.379418). Scale bar = 100 μm. **B–C.** E side Ilot Tiahura, NW corner Moorea, Society Islands, French Polynesia, 17.483°S, 149.917°W (AMS C.380658). Scale bars = 10 μm.

Specimen records.

Seychelles. 15 m, Anse Lasio, Chevalier Bay, Praslin Island, 4.283°S, 55.683°E (AMS C.404924, 1).

Andaman Islands. 3 m, Snake Island, off Corbyn's Cove, 11.633°N, 92.750°E (LACM 89-138, 3).

Sri Lanka. 1 m, Hikkaduawa, 6.137°N, 80.095°E (LACM 84-9, 5).

Thailand. 1 m, Phuket, Kaew Yai Island, 7.745°N, 98.308°E (LACM 85-5, 2). 1 m, Phuket, Hey Island, 7.747°N, 98.370°E (LACM 85-1, 1). 3 m, Phuket, Mai Thon Island, 75.000°N, 98.488°E (LACM 85-8, 2). Phuket, S tip Phromthep Cape, 7.758°N, 89.322°E (LACM 85-6, 2). 4–6 m, Krabi, coral reefs outside, 8.046°N, 98.912°E (DLG 1125, 5). 0 m, Ko Tao, Gulf of Thailand, 9.700°N, 99.950°E (DKC 3193, 1). 15 m, Koh Samah, SE Koh Phiphidon, SE of Phuket Island, 7.716°N, 98.783°E (SBMNH 85153, 2).

Indonesia. 0–10 m, bay N of Tanjung Batuangus, Selat Lembeh, N Sulawesi, 1.516°N, 125.250°E (RMNH, 11). 5–20 m, Sulawesi, Bunaken and Siladen Islets, 1.603°N, 124.767°E (LACM 88-55, 15). 10 m, Kaitetu (near Hila), N coast, Hitu, Ambon, Moluccas, 3.595°S, 128.141°E (RMNH, 1). 6 m, Suli, Baguala Bay, Hitu, Ambon, Moluccas, 3.628°S, 128.257°E (RMNH, 1). Tg. Benteng (= Galghoek), outer bay, Ambon Bay, Leitimur, Ambon, Moluccas, 3.702°S, 128.163°E (RMNH, 1). Batumerah (near Ambon City), outer bay, Ambon Bay, Leitimur, Ambon, Moluccas, 3.703°S, 128.168°E (RMNH, 1). Wside of Laha, outer Bay, Ambon Bay, Hitu, Ambon, Moluccas, 3.722°S, 128.080°E (RMNH, 4). 20 m, Hutumuri, S coast, Leitimur, Ambon, Moluccas, 3.731°S, 128.259°E (RMNH, 7). 3 m, Latuhalat, S coast, Leitimur, Ambon, Moluccas, 3.778°S, 128.126°E (RMNH, 2). 558 m, St. 5592, off Silungan Island, Borneo, 4.212°N, 118.462°E (USNM ex 279043, 2). 988 m, St. 5650, gn. M. 40.1°, Celebes, Gulf of Boni, off Lamulu Point, 4.895°S, 121.483°E (USNM 280015, 1). 20 m, Java, Pelangi and Putri Islets, 5.600°S, 106.583°E (LACM 86-163, 20). 1.25 m, Java, Pelangi and Putri Islets, 5.600°S, 106.583°E (LACM 86-162, 12). 250–290 m, NW of Beningko, Banda Sea, Tukang Besi Island, 5.875°S, 123.975°E (RMNH, 2). 280 m, NE of Beningko, Banda Sea, Tukang Besi Island, 5.875°S, 123.975°E (RMNH, 1; RMNH, 1). 465 m, S of Kaledupa reef, Banda Sea, Tukang Besi Island, 5.958°S, 123.775°E (RMNH, 1). 185–305 m, S of Kaledupa reef, Banda Sea, Tukang Besi Island, 5.958°S, 123.775°E (RMNH, 1). 365 m, S of Kaledupa Reef, Banda Sea, Tukang Besi Island, 5.963°S, 123.775°E (RMNH, 1). SW Salayer, 6.355°S, 120.435°E (RMNH, 2). 233–274 m, SW Salayer, 6.371°S, 120.433°E (RMNH, 5;

Scissurellidae *Sinezona*

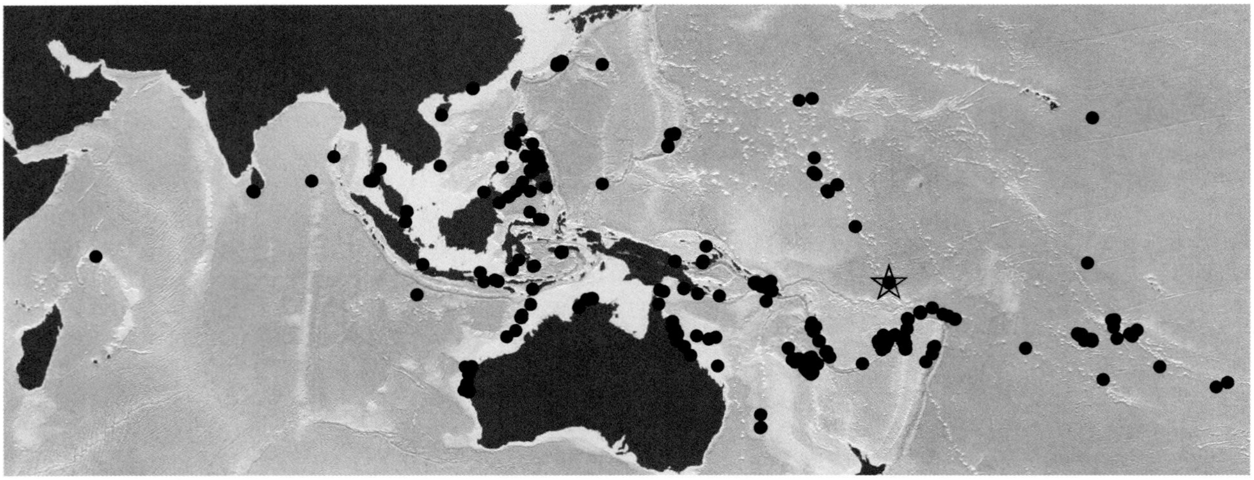

Figure 417. Distribution of *Sinezona plicata*. ☆ type locality. ● specimen record. ■ literature record.

RMNH, 20). 175 m, SW Salayer, 6.373°S, 120.437°E (RMNH, 1). 300–340 m, NW Pulau Bahuluang, SW Salayer, 6.466°S, 120.400°E (RMNH, 14). Sapeken, Anakkangean Eilanden, 6.900°S, 115.500°E (RMNH, 1). 175–185 m, Bay of Sanggar, N of Sumbawa, 8.321°S, 118.260°E (RMNH, 2). 5–20 m, Lombok, off Ajer, 8.367°S, 116.067°E (LACM 88-63, 4).

Malaysia. W coat of Pulau Tioman, MAS-16, Malaysia, 2.809°N, 104.141°E (ZSM Mol-20060145, 1: complete). 0 m, Coral Island [Pulau Tulai], off Tioman Island, N tip, 2.920°N, 104.103°E (MV F178687, 1). Pig (Tab) Island, NE of Madang, 5.130°S, 145.817°E (MMC, 6). 1–2 m, Tabad Island, off Jais Aben, NE of Madang, 5.156°S, 145.806°E (MMC, 1). 1–3 m, Sinub Island, off Jais Aben, NE of Madang, 5.156°S, 145.806°E (MMC, 1). 6 m, N Palau Sapi, off Kota Kinabalu, Sabah, 5.983°N, 116.067°E (SBMNH 85155, 1). 7 m, Police Bay, Palau Gaya, off Kota Kinabalu, Sabah, 5.983°N, 116.067°E (SBMNH 85154, 1).

Philippines. Aliguay Island, 2.746°N, 123.224°E (DLG 1588, 1). 20 m, off Bohol Island, Pamilacan Island, 9.500°N, 123.917°E (AMS C.379417, 1). 2 m, Bohol, Panglao Island, SW side, 9.600°N, 123.750°E (AMS C.380674, 4). 3 m, Palawan Island, N Hondo Bay, Tadio Island, 9.950°N, 118.933°E (AMS C.379416, 3). 2 m, Palawan Island, N of Fondeado Island, 9.950°N, 118.917°E (AMS C.379424, 2). Palawan Island, N Hondo Bay, Tadio Island, 9.950°N, 118.933°E (AMS C.379425, 3). 7 m, Cebu, Mactan Island, 10.283°N, 124.017°E (LACM 81-2, 3). 20 m, Luzon, Manig Point, 13.683°N, 120.892°E (LACM 84-163, 4). 25 m, Luzon, Ligpo Point, 13.825°N, 120.898°E (LACM 84-161, 2). 50 m, Carigara, Leyte, 11.299°N, 124.676°E (DLG 872, 2). 538 m, AURORA 2007 CP2750, 15.890°N, 121.893°E (MNHN, 5). 20 m, Looc, Panglao Island, PANGLAO 2004 B32, 9.430°N, 123.743°E (MNHN, 4). 4–30 m, Catarman, Panglao Island, PANGLAO 2004 S4, 9.431°N, 123.863°E (MNHN, 4). 242–400 m, off Balicasag Island, Bohol Sea, PANGLAO 2005 CP2392, 9.483°N, 123.685°E (MNHN, 1). 6–14 m, Pamilacan Island, PANGLAO 2004 S10, 9.490°N, 123.933°E (MNHN, 3). 2–4 m, Pamilacan Island, PANGLAO 2004 B11, 9.490°N, 123.933°E (MNHN, 1). 15–20 m, Pamilacan Island, PANGLAO 2004 S22, 9.490°N, 123.933°E (MNHN, 1). 356–396 m, off Balicasag Island, Bohol Sea, CP2393, 9.501°N, 123.693°E (MNHN, 1). 120–139 m, Balicasag Island, PANGLAO 2004 L74–75, 9.511°N, 123.682°E (MNHN, 3). 87–111 m, Balicasag Island, PANGLAO 2004 L613/30/201164, 9.515°N, 123.680°E (MNHN, 1). 17–18 m, Balicasag Island, Panglao Island, PANGLAO 2004 B38, 9.515°N, 123.680°E (MNHN, 2). 90–110 m, Balicasag Island, PANGLAO 2004 L46, 9.515°N, 123.686°E (MNHN, 1). 6 m, Black Forest, Balicasag Island, PANGLAO 2004 S3, 9.518°N, 123.688°E (MNHN, 3). 12–14 m, Black Forest, Balicasag Island, PANGLAO 2004 B6, 9.518°N, 123.688°E (MNHN, 3). 309–342 m, Balicasag Island, Bohol Sea, CP2399, 9.528°N, 123.698°E (MNHN, 5). 164–176 m, off Balicasag Island, Bohol Sea, PANGLAO 2005 DW2339, 9.531°N, 123.728°E (MNHN, 12). 83–102 m, Bolod, Panglao Island, PANGLAO 2004 T1, 9.540°N, 123.788°E (MNHN, 1). 152 m, Bolod, Panglao Island, PANGLAO 2004 T2, 9.540°N, 123.796°E (MNHN, 12). 713–731 m, off Balicasag Island, Bohol Sea, PANGLAO 2005 CP2398, 9.543°S, 123.675°E (MNHN, 2). 17–25 m, Pntod Lagoon 1, Panglao Island, PANGLAO 2004 B39, 9.546°N, 123.701°E (MNHN, 4). 0–1 m, Alona Beach, Panglao Island, PANGLAO 2004 M1, 9.548°N, 123.776°E (MNHN, 1). 100–140 m, between Panglao and Balicasag Island, PANGLAO 2004 T31, 9.550°N, 123.700°E (MNHN, 1). 5 m, Alona reef, Panglao Island, PANGLAO 2004 B2, 9.550°N, 123.775°E (MNHN, 1). 82 m, Bolod, Panglao Island, PANGLAO 2004 T4, 9.550°N, 123.808°E (MNHN, 23). 8–10 m, Napaling, Panglao Island, PANGLAO 2004 B9, 9.551°N, 123.733°E (MNHN, 8). 24 m, BBC Point, Panglao Island, PANGLAO 2004 B4, 9.553°N, 123.805°E (MNHN, 1). 106–137 m, between Panglao and Pamilacan Islands, PANGLAO 2004 T27, 9.556°N, 123.850°E (MNHN, 1). 2 m, lagoon off Poblacion, Panglao Island, PANGLAO 2004 S11, 9.560°N, 123.726°E (MNHN, 1). 97–120 m, off San Isidro, Panglao Island,

PANGLAO 2004 T9, 9.561°N, 123.833°E (MNHN, 5). 83–86 m, off Bolod, Panglao Island, PANGLAO 2004 T44, 9.573°N, 123.815°E (MNHN, 1). 6 m, San Isidro, Panglao Island, PANGLAO 2004 S17, 9.576°N, 123.831°E (MNHN, 1). 6 m, San Isidro, Panglao Island, PANGLAO 2004 S17, 9.576°N, 123.831°E (MNHN, 5). 80 m, Biking-Catarman, Panglao Island, PANGLAO 2004 T28, 9.583°N, 123.856°E (MNHN, 7). 4 m, Biking, Panglao Island, PANGLAO 2004 B5, 9.586°N, 123.840°E (MNHN, 3). 5 m, Biking, Panglao Island, PANGLAO 2004 S1, 9.588°N, 123.841°E (MNHN, 42). 24–27 m, Doljo Point, Panglao Island, PANGLAO 2004 B12, 9.593°N, 123.720°E (MNHN, 4). 2–3 m, Looc, Panglao Island, PANGLAO 2004 S32, 9.596°N, 123.743°E (MNHN, 1). 4–30 m, Tatarman, Panglao Island, PANGLAO 2004 B7, 9.598°N, 123.863°E (MNHN, 10). 2–4 m, Momo Beach, Panglao Island, PANGLAO 2004 S24, 9.601°N, 123.750°E (MNHN, 5). 0–3 m, Momo Beach, Panglao Island, PANGLAO 2004 M7, 9.601°N, 123.753°E (MNHN, 1). 382–434 m, Maribojoc Bay, Bohol Sea, PANGLAO 2005 CP2395, 9.603°N, 123.730°E (MNHN, 1). 609–673 m, Maribojoc Bay, Bohol Sea, CP2396, 9.605°N, 123.700°E (MNHN, 1). 90 m, Looc, Panglao Island, PANGLAO 2004 L35, 9.606°N, 123.758°E (MNHN, 1). 3–14 m, Momo Beach, Banglao Island, PANGLAO 2004 B10, 9.608°N, 123.760°E (MNHN, 7). 28–32 m, Momo Beach, Panglao Island, PANGLAO 2004 S8, 9.608°N, 123.760°E (MNHN, 6). 3 m, Napaling, Panglao Island, PANGLAO 2004 B8, 9.618°N, 123.768°E (MNHN, 1). 2–4 m, Napaling, Panglao Island, PANGLAO 2004 S5, 9.618°N, 123.768°E (MNHN, 6). Napaling, Panglao Island, PANGLAO 2004 B8, 9.618°N, 123.769°E (MNHN, 8). 3–5 m, Baclayon Takot, Bohol Island, PANGLAO 2004 B13, 9.618°N, 123.876°E (MNHN, 1). 8–15 m, Baclayon Takot, Bohol Island, PANGLAO 2004 S13, 9.618°N, 123.876°E (MNHN, 4). 28–32 m, Napaling, Panglao Island, PANGLAO 2004, 9.620°N, 123.773°E (MNHN, 3). 4–5 m, Baclayon, Bohol Island, PANGLAO 2004 S2, 9.623°N, 123.908°E (MNHN, 9). 3–21 m, Bingag, Panglao Island, PANGLAO 2004 B17, 9.625°N, 123.781°E (MNHN, 4). 20 m, Bingag, PANGLAO 2004 B16, 9.626°N, 123.788°E (MNHN, 2). 1–4 m, Sungcolan Bay, Panglao Island, PANGLAO 2004 S7, 9.641°N, 123.820°E (MNHN, 3). 78–95 m, Maribohoc Bay, Bohol Island, PANGLAO 2004 T11, 9.681°N, 123.833°E (MNHN, 16). 2–8 m, Ubajan, Bohol Island, PANGLAO 2004 B20, 9.691°N, 123.850°E (MNHN, 1). 4–12 m, Manga, Bohol Island, PANGLAO 2004, 9.695°N, 123.848°E (MNHN, 3). 80–100 m, Cortes, Bohol Island, PANGLAO 2004 T18, 9.696°N, 123.831°E (MNHN, 2). 100 m, Cortes, Bohol Island, PANGLAO 2004 B1, 9.698°N, 123.825°E (MNHN, 3). 123–135 m, Cortes, Bohol Island, PANGLAO 2004 T26, 9.721°N, 123.813°E (MNHN, 3). 121–137 m, Maribojoc Bay, Bohol Sea, PANGLAO 2005 CP2408, 9.725°N, 123.785°E (MNHN, 1). 473 m, AURORA 2007 CP2749, 15.943°N, 121.836°E (MNHN, 1).

Japan, Okinawa. Onna Flats, 1.5 km WNW Onna, 26.488°N, 127.837°E (AMS C.405177, 2). 20 m, Tsuken Jima, 26.268°N, 127.920°E (LACM 78-21, 3). 0 m, Kadena, 26.337°N, 134.467°E (LACM 77-55, 1). 42 m, Zampa-Misaki, 26.427°N, 127.718°E (LACM 78-24, 4). 50 m, Zampa Misaki, 26.435°N, 127.708°E (LACM 78-25, 20). 52 m, Zampa Misaki, 26.435°N, 127.708°E (LACM 78-100, 1). 60 m, Onna Village, 26.493°N, 127.842°E (LACM 79-76, 20). 50 m, Onna Village, 26.493°N, 127.842°E (LACM 79-75, 20). 92 m, Onna Village, 26.493°N, 127.493°E (LACM 78-101, 11). 46 m, Onna Village, 26.495°N, 127.843°E (LACM 78-26, 5). 33 m, Onna Village, 26.495°N, 127.843°E (LACM 78-20, 6). 50 m, Onna Village, 26.495°N, 127.843°E (LACM 78-99, 4). 55 m, Onna Village, 26.497°N, 127.843°E (LACM 78-27, 12). 60 m, Onna Village, 26.497°N, 127.843°E (LACM 78-29, 12). 33 m, Motobu Peninsula, Nakijin, 26.703°N, 127.943°E (LACM 78-22, 7). 16 m, Oku, 26.847°N, 128.287°E (LACM 77-64, 10). 1 m, Oku, 26.847°N, 128.287°E (LACM 77-61, 2). 10 m, Oku, 26.847°N, 128.287°E (LACM 77-63, 1). 17 m, Hedo Point, Japan, 26.863°N, 128.250°E (LACM 78-23, 2). 13–20 m, Kadena Steps, 26.360°N, 127.739°E (DLG 1446, 1). 20–30 m, Maeda Point, 26.445°N, 127.773°E (DLG 1533, 7). 13–30 m, Toilet Bowl, Onna

Scissurellidae *Sinezona*

Point, 26.504°N, 127.846°E (DLG 1445, 1). 2–3 m, Horseshoe Beach, 26.504°N, 127.858°E (DLG 1577, 3). 43 m, Horseshoe Beach, 26.505°N, 127.853°E (DLG 1548, 16). 37 m, Big Rock, Horseshoe, 26.508°N, 127.853°E (DLG 1457, 19). 13–33 m, Flab, Horseshoe, 26.509°N, 127.856°E (DLG 1491, 7). 33 m, Horseshoe, Okinawa, 26.509°N, 127.856°E (DLG 1519, 10).

China. 0 m, Hong Kong, E side Mirs Bay Tai Pang Wang, SE of Ping Chau Island, 22.550°N, 114.400°E (AMS C.404425, 1). 5–10 m, Hong Kong, Titau, 22.466°N, 114.333°E (AMS, 1: complete).

Paupa New Guinea. Milne Bay District, Amazon Bay, 10.300°S, 149.333°E (AMS C.379432, 1). 13 m, Milne Bay District, Amazon Bay, 10.317°S, 149.350°E (AMS C.379419, 2). Misima Island, Lousisade Archipelago, 10.650°S, 152.700°E (AMS C.379433, 1). Madang, N end Kranket Island, 5.200°S, 145.850°E (AMS C.379420, 1). 25 m, :Port Moresby, Horseshoe Reef, 9.467°S, 147.167°E (AMS C.379415, 1). 18–22 m, off Port Moresby, Manubada Local Island, off W end, 9.517°S, 147.167°E (AMS C.404943, 1). 4 m, off Anne Sophie's Reef, Kimbe Bay, 5.089°S, 150.186°E (DLG 577, 4). 20 m, Madang Province, Astrolabe Bay, Wonsat Island, 5.137°S, 145.845°E (LACM 80-20). Madang Province, Astrolabe Bay, off Nagda, 5.158°S, 145.837°E (LACM 80-13, 5). 70 m, off S Ema Reef, Kimbe Bay, New Britain, 5.202°S, 150.152°E (DLG 558, 1). 28 m, off Big Malu Malu Island, Kimbe Bay, 5.231°S, 150.098°E (DLG 401, 1). 0–5 m, Lumu Reef, Kimbe Bay, New Britain, 5.283°S, 150.131°E (DLG 648, 4). 65 m, off Susan's Reef, Kimbe Bay, New Britain, 5.291°S, 150.128°E (DLG 606, 1). 40 m, off Restorf Island, Kimbe Bay, 5.292°S, 150.101°E (DLG 395, 3). 35 m, off Walindi Plantation Resort, Kimbe Bay, 5.411°S, 150.095°E (DLG 397, 3). 1 m, Hanging Garden Reef, Kimbe Bay, New Britain, 5.429°S, 150.097°W (DLG 510, 3). 1 m, First Reef, Kimbe Bay, New Britain, 5.432°S, 150.090°E (DLG 798, 1).

Christmas Island. 27 m, 10.433°S, 105.667°E (AMS C.406385, 3). 183 m, off NE point, 10.417°S, 105.700°E (AMS C.404925, 1; AMS C.406362, 4; AMS C.406381, 2).

Coral Sea. NE Herald Cay, 16.933°S, 149.183°E (AMS C.334062, 57). 3 m, Middleton Reef, 29.460°S, 159.112°E (AMS, 1: complete).

Australia, Western Australia. Coral Bay, 11.183°S, 132.050°E (WAM, 1). 15–25 m, Seringapatam reef, N side, 13.624°S, 122.005°E (SBMNH 85139, 4). 12 m, SE side, North Reef, Scott Reef, 14.021°S, 121.977°E (SBMNH 85148, 4). 20 m, E side, Sandy Islet, Scott Reef, 14.044°S, 121.780°E (SBMNH 369507, 10). 20 m, Sandy Islet, Scott Reef, 14.044°S, 121.780°E (SBMNH 369694, 7). 19 m, N end Mermaid Reef, 17.050°S, 119.600°E (LACM 86-250, 8). North West Cape, 21.750°S, 114.167°E (WAM, 1). Turtle Beach, W side of North West Cape, 21.800°S, 114.167°E (AMS C.378277, 1). 8–9 m, Bundegi Reef, Exmouth Gulf, 21.817°S, 113.183°E (AMS C.379089, 3). 2 m, Ningaloo Reef, off Ned's Camp, 21.991°S, 113.908°E (AMS C.402657, 1). Cape Range National Park, S Ned's Camp, 22.000°S, 113.917°E (AMS C.379086, 3). 65 m, off North West Cape, 22.025°S, 113.962°E (AMS C.379088, 5; AMS C.406398, 11). 40 m, off North West Cape, 22.313°S, 113.765°E (AMS C.379090, 2). 78 m, off North West Cape, 22.967°S, 113.687°E (AMS C.379091, 6). Point Quobba, 24.483°S, 113.417°E (WAM, 1; WAM, 3). Point Quobba, North of Carnarvon, 24.483°S, 113.417°E (WAM, 2). 1 m, Point Quobba, 24.483°S, 113.417°E (WAM, 1). Turtle Bay, NW Cape, 25.483°S, 113.000°E (WAM, 3; WAM, 5). 0 m, Turtle Bay, 25.494°S, 112.990°E (MV F178688, 1) Shark Bay, Monkey Mia, E side Peron Peninsula, 25.800°S, 113.717°E (AMS C.404990, 1: complete).

Australia, Northern Territories. 5 m, Oxley Island, W end, 11.000°S, 132.817°E (AMS C.379087, 1). 25 m, Hibernia Reef, N side, 11.967°S, 123.349°E (SBMNH, 3). 8 m, McCleur Island, NW end, 11.033°S, 132.967°E (AMS C.348619, 1). Darwin, 12.466°S, 130.833°E (WAM, 3).

Australia, Queensland. 1–6 m, NE corner, of South Island, Sir Charles Hardy Islands, 11.917°S, 143.283°E (AMS C.468102, 1). 5 m, Great Barrier Reef, Lizard Island, off Rocky Point, 14.667°S, 145.433°E (AMS C.379075, 2). 0 m, SW side Lizard Island,

Scissurellidae *Sinezona*

14.670°S, 145.442°E (LACM 79-53, 1). 3–5 m, Great Barrier Reef, Lizard Island, between Bird and South Islets, 14.683°S, 145.450°E (AMS C.379078, 2). Great Barrier Reef, Lizard Island, Casuarina Research Station Beach, 14.683°S, 145.450°E (AMS C.379082, 2). 10 m, SE side Lizard Island, 14.683°S, 145.458°E (LACM 79-55, 6). 20 m, Great Barrier Reef, Lizard Island, off SE side Bird Islet, 14.683°S, 145.467°E (AMS C.379074, 1). Between Palfrey and South Island, Lizzard Islands, Great Barrier Reef, 14.700°S, 145.450°E (AMS C.461471, 2). 9–18 m, Great Barrier Reef, Cairns Reef lagoon and between Cairns Reef and Hope Island, 15.700°S, 145.500°E (AMS C.404963, 3, 1: complete). 10–8 m, Great Barrier Reef, Endeavour Reef, 15.783°S, 145.583°E (AMS C.402688, 2). 38 m, S part Escape Reef, Great Barrier Reef, 15.883°S, 145.817°E (AMS C.461467, 2). Great Barrier Reef, Outer Barrier Reef, St. Crispins Reef, 16.117°S, 145.867°E (AMS C.46044, 3). 20 m, Great Barrier Reef, Michaelmas Cay, 16.600°S, 145.983°E (AMS C.379072, 1). Great Barrier Reef, Michaelmas Cay, 16.600°S, 145.983°E (AMS C.379076, 2; AMS C.348640, 16). 21 m, Great Barrier Reef, Euston Reef, SW side, 16.667°S, 146.217°E (AMS C.379083, 14). 11–18 m, off Cairns Reef, 16.905°S, 145.778°E (MV ex F 161668, 1). 0 m, Kurrimine, King Reef, 17.732°S, 146.096°E (WAM, 1). Kurramine Beach, 17.766°S, 146.100°E (WAM S13377, 2). Helix reef, 18.617°S, 147.283°E (LACM 83-41, 7). Bowen, 20.016°S, 148.233°E (WAM, 2). 64–73 m, Great Barrier Reef, Swain Reefs, 3 km NE of W side of Bylund Gillett Cay, 21.700°S, 152.433°E (AMS C.379073, 20+). Torres Strait, Yorke Island, S side, 9.733°S, 143.417°E (AMS C.379080, 1). 9–15 m, Torres Strait, off Murray Island, 9.933°S, 144.067°E (AMS C.48355, 13).

Lord Howe Island. 31.550°S, 159.063°E (AMS C.404393, 2). 53 m, Lord Howe Shelf, 31.490°S, 159.153°E (AMS C.461440, 1).

New Caledonia. 502–516 m, 18.866°S, 163.383°E (MNHN, 1). 470–502 m, Northern, 18.866°S, 163.400°E (MNHN, 1). 344–360 m, northern, 18.916°S, 163.400°E (MNHN, 1). 5–10 m, Santal Bay, Lifou, 20.280°S, 167.046°E (MNHN, 8). 280 m, Loyality ridge, 20.433°S, 166.683°E (MNHN, 1). 270–282 m, E coast, 20.550°S, 165.000°E (MNHN, 20). 0 m, Koumac sector, 20.561°S, 164.183°E (MNHN, 6). 491–533 m, E coast, 20.566°S, 164.950°E (MNHN, 1). 497–520 m, E coast, 20.566°S, 164.950°E (MNHN, 3). 0 m, Koumac sector, 20.583°S, 164.258°E (MNHN, 1). 308–324 m, E coast, BATHUS 1 DE693, 20.583°S, 164.983°E (MNHN, 5). 380–400 m, E coast, 20.583°S, 165.117°E (MNHN, 20). 380–400 m, E coast, 20.583°S, 165.117°E (MNHN, 13). 380–400 m, E coast, 20.583°S, 165.117°E (MNHN, 12). 380–400 m, E coast, 20.583°N, 165.117°E (MNHN, 12). 408–440 m, E coast, 20.583°S, 165.117°E (MNHN, 10). 15–20 m, Touho sector, 20.585°S, 165.135°E (MNHN, 1). 15–20 m, Touho sector, 20.585°S, 165.135°E (MNHN, 5). Beach at E side of Mount Panie, 20.600°S, 164.833°E (AMS C.4107, 1). 400–500 m, E coast, BATHUS 1 DE694, 20.600°S, 164.967°E (MNHN, 1). 538 m, between Lifou and Uvea, 20.616°S, 166.967°E (MNHN, 1). 1230–1270 m, Loyality Bassin, 20.633°S, 167.167°E (MNHN, 1). 33–38 m, Koumac sector, 20.646°S, 164.260°E (MNHN, 1). 1420 m, Loyality Basin, 20.650°S, 167.200°E (MNHN, 3). 15 m, Koumac sector, 20.661°S, 164.248°E (MNHN, 1). 30–63 m, Koumac sector, 20.663°S, 164.255°E (MNHN, 7). 12 m, Koumac sector, 20.666°S, 164.187°E (MNHN, 8, 1: complete). 55–57 m, Koumac sector, 20.671°S, 164.195°E (MNHN, 5). 55–57 m, Koumac sector, 20.671°S, 164.195°E (MNHN, 4). 26–40 m, Koumac sector, 20.673°S, 164.248°E (MNHN, 4, 1: complete). 10–60 m, Koumac sector, 20.673°S, 164.248°E (MNHN, 1). 20–30 m, Koumac sector, 20.690°S, 164.004°E (MNHN, 1). 20–30 m, Koumac sector, 20.690°S, 164.247°E (MNHN, 5). 49–59 m, Touho sector, 20.733°S, 165.233°E (MNHN, 1). 8 m, Touho sector, 20.741°S, 165.265°E (MNHN, 1). 8 m, Touho sector, 20.741°S, 165.265°E (MNHN, 13). 15–35 m, Touho sector, 20.743°S, 165.228°E (MNHN, 1). 90–115 m, Koumac sector, 20.745°S, 164.248°E (MNHN, 1). 15–20 m, Koumac sector, 20.745°S, 164.258°E (MNHN, 2). 10–35 m, Touho sector, 20.750°S, 165.275°E (MNHN, 4). 53–71 m, Koumac sector, 20.753°S,

164.253°E (MNHN, 1). 20 m, Santal Bay, Lifou, 20.763°S, 167.028°E (MNHN, 1). 17 m, Santal Bay, Lifou, 20.763°S, 167.028°E (MNHN, 2). 27–31 m, Santal Bay, Lifou, 20.763°S, 167.028°E (MNHN, 2). 27–31 m, Santal Bay, Lifou, 20.763°S, 167.028°E (MNHN, 1). 17 m, Santal Bay, Lifou, 20.763°S, 167.028°E (MNHN, 2). 4–7 m, Santal Bay, Lifou, 20.765°S, 167.103°E (MNHN, 2). 0 m, Touho sector, 20.770°S, 165.242°E (MNHN, 1). 47 m, Santal Bay, Lifou, 20.773°S, 167.033°E (MNHN, 3). 20 m, Santal Bay, Lifou, 20.773°S, 167.033°E (MNHN, 3). 20 m, Santal Bay, Lifou, 20.773°S, 167.033°E (MNHN, 2). 47 m, Santal Bay, Lifou, 20.773°S, 167.033°E (MNHN, 4). 0–2 m, Touho sector, 20.775°S, 165.242°E (MNHN, 1). 45–56 m, Touho sector, 20.775°S, 165.263°E (MNHN, 2). 45–56 m, Touho sector, 20.775°S, 165.263°E (MNHN, 4). 5–10 m, Santal Bay, Lifou, 20.780°S, 167.046°E (MNHN, 2). 5–10 m, Santal bay, Lifou, 20.780°S, 167.046°E (MNHN, 1: complete). 4–5 m, Santal Bay, Lifou, 20.780°S, 167.120°E (MNHN, 3). 4–5 m, Santal Bay, Lifou, 20.780°S, 167.120°E (MNHN, 1). 0 m, Santal Bay, Lifou, 20.780°S, 167.129°E (MNHN, 2). 0–1 m, Touho sector, 20.781°S, 165.230°E (MNHN, 1). 1–5 m, Santal Bay, Lifou, 20.781°S, 167.132°E (MNHN, 2). 4 m, Santal Bay, Lifou, 20.785°S, 167.123°E (MNHN, 1). 4 m, Santal Bay, Lifou, 20.785°S, 167.123°E (MNHN, 2). 10–21 m, Santal Bay, Lifou, 20.788°S, 167.113°E (MNHN, 13). 10–21 m, Santal Bay, Lifou, 20.788°S, 167.113°E (MNHN, 3). 10–21 m, Santal bay, Lifou, 20.788°S, 167.113°E (MNHN, 1: complete). 8–18 m, Santal Bay, Lifou, 20.791°S, 167.118°E (MNHN, 4). 20–25 m, Santal Bay, Lifou, 20.791°S, 167.118°E (MNHN, 1). 8–18 m, Santal Bay, Lifou, 20.791°S, 167.118°E (MNHN, 2). 4–5 m, Santal Bay, Lifou, 20.795°S, 167.156°E (MNHN, 2). 105–110 m, E coast, 20.816°S, 165.317°E (MNHN, 2). 105–110 m, Poindimie sector, 20.816°S, 165.317°E (MNHN, 12). 96–109 m, E coast, 20.816°S, 165.317°E (MNHN, 2). 69–109 m, E coast, 20.816°S, 165.317°E (MNHN, 2). 80–140 m, Touho sector, 20.816°S, 165.317°E (MNHN, 3). 80–140 m, Touho sector, 20.816°S, 165.317°E (MNHN, 1). 25–30 m, Santal Bay, Lifou, 20.821°S, 167.173°E (MNHN, 1). 10 m, Touho sector, 20.825°S, 165.327°E (MNHN, 4). 20 m, Touho sector, 20.840°S, 165.380°E (MNHN, 11). 4 m, Santal Bay, Lifou, 20.873°S, 167.142°E (MNHN, 1). 5–20 m, Santal Bay, Lifou, 20.875°N, 167.135°E (MNHN, 3). 5–20 m, Santal Bay, Lifou, 20.875°S, 167.135°E (MNHN, 10). 5–25 m, Touho sector, 20.878°S, 165.325°E (MNHN, 5). 12–32 m, Santal Bay, Lifou, 20.891°S, 167.045°E (MNHN, 5). 12–32 m, Santal Bay, Lifou, 20.891°S, 167.045°E (MNHN, 9). 150–200 m, Shelter Reef, Santal Bay, Lifou, Loyalty Islands, LIFOU DW1648, 20.900°S, 167.050°E (MNHN, 2). Lifou, 20.900°S, 167.200°E (KBIN, 2). Lifou, 20.900°S, 167.200°E (KBIN, 2). Lifou, 20.900°S, 167.200°E (KBIN, 1). Lifou, 20.900°S, 167.200°E (KBIN, 30). 200 m, Loyality Ridge, 20.900°S, 167.283°E (MNHN, 1). 150–200 m, Point Lefèvre, Santal Bay, Loyalty Islands, Lifou, 20.901°S, 167.018°E (MNHN, 9). 21–30 m, Santal Bay, Lifou, 20.910°S, 167.035°E (MNHN, 2). 9–20 m, Santal Bay, Lifou, 20.916°S, 167.087°E (MNHN, 1). 5–30 m, Santal Bay, Lifou, 20.920°S, 167.012°E (MNHN, 7). 5–30 m, Santal Bay, Lifou, 20.920°S, 167.012°E (MNHN, 5). 10–20 m, Santal Bay, Lifou, 20.925°S, 167.070°E (MNHN, 3). 10–20 m, Santal Bay, Lifou, 20.925°S, 167.070°E (MNHN, 1). 5 m, Santal Bay, Lifou, 20.926°S, 167.075°E (MNHN, 1). Loyalty Islands, Lifou, Doueolou, 20.933°S, 167.083°E (AMS C.379410, 1). 15–18 m, Santal Bay, Lifou, 20.944°S, 167.033°E (MNHN, 7). 15–18 m, Santal Bay, Lifou, 20.944°S, 167.033°E (MNHN, 5). 2–4 m, Santal Bay, Lifou, 20.945°S, 167.052°E (MNHN, 1). 2–4 m, Santal Bay, Lifou, 20.945°S, 167.052°E (MNHN, 12). 15–20 m, Santal Bay, Lifou, 20.946°S, 167.045°E (MNHN, 1). 105–110 m, E coast, 20.950°S, 165.567°E (MNHN, 12). 160–222 m, E coast, 20.950°S, 165.583°E (MNHN, 5). 160–222 m, E coast, 20.950°S, 165.583°E (MNHN, 20). 302–335 m, E coast, 20.966°S, 165.600°E (MNHN, 1). 250–300 m, Nnorthern, 21.016°S, 164.467°E (MNHN, 2). 425 m, Loyality ridge, 21.016°S, 167.517°E (MNHN, 2). Loyalty Islands, SE side Lifou Island, Luengoni, 21.033°S, 167.417°E (AMS C.379412, 1). Loyalty Islands, SE side Lifou Island, Luengoni,

21.033°S, 167.417°E (AMS C.379426, 2). 350–400 m, E coast, 21.033°S, 165.633°E (MNHN, 7). 350–400 m, E coast, 21.033°S, 165.633°E (MNHN, 20). 320–344 m, 21.116°S, 164.467°E (MNHN, 30). 237–298 m, E coast, 21.283°S, 165.950°E (MNHN, 1). 237–298 m, E coast, 21.283°S, 165.950°E (MNHN, 20). 350 m, Loyality Ridge, 21.383°S, 167.983°E (MNHN, 1). 310–315 m, Loyality Bassin, 21.533°S, 166.483°E (MNHN, 1). 310–315 m, Loyality Basin, 21.533°S, 166.483°E (MNHN, 1). 247–252 m, E coast, 21.700°S, 166.567°E (MNHN, 20). 250–350 m, Northern, 22.033°S, 165.933°E (MNHN, 2). 250–300 m, Northern, 22.083°S, 165.967°E (MNHN, 3). 455 m, Southern, 22.266°S, 166.183°E (MNHN, 50). 490–500 m, Southern, 22.316°S, 166.183°E (MNHN, 25). 495 m, Southern, 22.366°S, 166.233°E (MNHN, 20). 495 m, Southern, 22.366°S, 166.233°E (MNHN, 150). 12–37 m, Aboré Grand Reef, Mouméa Lagoon, 22.371°S, 166.265°E (MNHN, 2: complete). 577–780 m, Southern, 22.383°S, 166.217°E (MNHN, 1). 577–780 m, south, BATHUS 2 DW754, 22.383°S, 166.217°E (MNHN, 50). 51–52 m, Cape N Doua, BATHUS 1 DW1235, 22.401°S, 166.925°E (MNHN, 1). 10 m, Goeland Islet, Noumea sector, St. 1367, 22.405°S, 166.345°E (MNHN, 3). 250–350 m, 22.500°S, 166.400°E (MNHN, 100). 250–350 m, St. 40, 22.500°S, 166.400°E (MNHN, 200). 465–525 m, 22.583°S, 166.450°E (MNHN, 20). 465–525 m, Southern, 22.583°S, 166.450°E (MNHN, 3). 465–525 m, South, BATHUS 2 DW739, 22.583°S, 166.450°E (MNHN, 65). 713–950 m, south, BATHUS 2 CP743, 22.600°S, 166.433°E (MNHN, 300). 570 m, Loyality Basin, 22.650°S, 166.567°E (MNHN, 2). 202–227 m, south, BATHUS 2 DW715, 22.650°S, 167.183°E (MNHN, 1). 595 m, Loyality Basin, 22.666°S, 166.550°E (MNHN, 1). 975 m, Bassin des Loyaute, 22.700°S, 166.550°E (MNHN, 1). 444–445 m, south, BATHUS 2 DE719, 22.800°S, 167.267°E (MNHN, 1). 400–402 m, Norfolk Ridge, 23.016°S, 166.933°E (MNHN, 2). 397–400 m, south, BATUHUS 2 DW730, 23.050°S, 166.967°E (MNHN, 8).

Vanuatu. 55 m, off Aneityum Island, 20.283°S, 169.800°E (AMS C.379414, 1). Tanna Island, Hotel Tanna, 19.533°S, 169.267°E (AMS C.379418, 2, 1: complete). 2 m, Efate, Erakor Reef, 17.780°S, 168.193°E (LACM 77-37, 1). 20 m, Efate, Mele Bay, 17.778°S, 168.257°E (LACM 77-40, 1). 154–179 m, MUSORSTOM 8 DW1105, 15.050°S, 167.116°E (MNHN, 8). 205–210 m, MUSORSTOM 8 DW1101, 15.066°S, 167.133°E (MNHN, 1). 7 m, S Turtle Island, Palikulo Bay, SANTO 2006 DB77DB71, 15.360°S, 167.208°E (MNHN, 1). 9 m, reef near Oyster Island, SANTO 2006 FB40, 15.381°S, 167.195°E (MNHN, 4). 20 m, SE Matewulu, SANTO 2006 DB75, 15.381°S, 167.198°E (MNHN, 5). 7 m, W Mavéa Island, SANTO 2006 DB67, 15.381°S, 167.218°E (MNHN, 3). 90–93 m, SW Maéa Island, SANTO 2006 EP38, 15.394°S, 167.221°E (MNHN, 1). 102–120 m, W Mavéa Island, Palikulo Bay, SANTO 2006 AT14, 15.397°S, 167.220°E (MNHN, 4). 6–43 m, Aésé Island, SANTO 2006 DS59, 15.410°S, 167.238°E (MNHN, 2). 13 m, Palikulo Bay, SANTO 2006 DB65, 15.430°S, 167.216°E (MNHN, 1). 19 m, Palikulo Bay, Aoré Island, SANTO 2006 FB43, 15.473°S, 167.248°E (MNHN, 1). 0 m, Palikulo Peninsula, SANTO 2006 VM12, 15.481°S, 167.258°E (MNHN, 1). 0 m, E Palikulo Peninsula, Palikulo Bay, SANTO 2006 DB77VM54, 15.481°S, 167.258°E (MNHN, 3). 5 m, Palikulo Bay, SANTO 2006 DB40, 15.496°S, 167.251°E (MNHN, 1). 22–25 m, NW Urélapa Island, SANTO 2006 DB20, 15.508°S, 167.023°E (MNHN, 15). 8–15 m, Segond Channel, vicinity of Luganville, SANTO 2006 DS06, 15.515°S, 167.185°E (MNHN, 3). 5–10 m, Segond Channel, Sarakata river mouth, SANTO 2006 DS18, 15.521°S, 167.173°E (MNHN, 1). 5 m, Segond Channel, Caverot Point, SANTO 2006 DS31, 15.523°S, 167.161°E (MNHN, 1). 25 m, Segond Channel, *Collidge* wreck, SANTO 2006 DS04, 15.523°S, 167.235°E (MNHN, 2). 20–31 m, Segond Channel, *Collidge* wreck, SANTO 2006 FS54, 15.523°S, 167.685°E (MNHN, 2). 62–71 m, Segond Channel, SANTO 2006 AT53, 15.530°S, 167.226°E (MNHN, 1). 97 m, Scorff Passage, SANTO 2006 EP12, 15.531°S, 167.251°E (MNHN, 1). 23–27 m, Segond Channel, vicinity of Maritime College, SANTO 2006 ED17, 15.533°S, 167.160°E (MNHN, 1).

17–19 m, N Tutuba Island, SANTO 2006 DS101, 15.538°S, 167.286°E (MNHN, 8). 8–20 m, Tutuba Island, SANTO 2006 FS82, 15.538°S, 167.290°E (MNHN, 1). 5 m, Aoré Island, Aimbué Bay, SANTO 2006 ZM16, 15.540°S, 167.201°E (MNHN, 1). 46–47 m, Segond Channel, SANTO 2006 EP01, 15.541°S, 167.150°E (MNHN, 1). 8–20 m, Tutuba Island, SANTO 2006 FB83, 15.543°S, 167.290°E (MNHN, 3). 5 m, Aoré Island, Aimbuei Bay, SANTO 2006 FS88, 15.544.000°S, 167.191.000°E (MNHN, 2). 97–101 m, E Aoré island, SANTO 2006 AT04, 15.550°S, 167.225°E (MNHN, 35). MUSORSTOM 8 DW1069, 15.550°S, 167.233°E (MNHN, 6). 92 m, NW Tutuba Island, SANTO 2006 DS105, 15.550°S, 167.278°E (MNHN, 2). 20 m, Segond Channel, N Aoré Island, SANTO 2006 NB12, 15.551°S, 167.160°E (MNHN, 1). 29 m, Segond Channel, N Aoré Island, SANTO 2006 FS77, 15.551°S, 167.160°E (MNHN, 1). 15–25 m, NE Tutuba Island, SANTO 2006 DB01, 15.551°S, 167.296°E (MNHN, 5). 15–25 m, NE Tutuba Island, Palikulo Bay, SANTO 2006 DB77DB01, 15.551°S, 167.296°E (MNHN, 1). 71–102 m, Aoré Island, Aimbuei Bay, Palikulo Bay, SANTO 2006 DB77EP18, 15.553°S, 167.200°E (MNHN, 1). 20–60 m, E Aoré Island, Aimbué Bay, SANTO 2006 EP36, 15.553°S, 167.209°E (MNHN, 1). 2–4 m, Tutuba Island, SANTO 2006 FB92, 15.560°S, 167.276°E (MNHN, 6). 8–9 m, Tutuba Island, SANTO 2006 FS84, 15.560°S, 167.276°E (MNHN, 5). 7 m, Segond Channel, Wambu river mouth, SANTO 2006 DS91, 15.561°S, 167.140°E (MNHN, 2). 98–100 m, W Tutuba Island, SANTO 2006 DS102, 15.568°S, 167.266°E (MNHN, 2). 2–3 m, Aoré Island, Port Benier, SANTO 2006 FB61, 15.573°S, 167.210°E (MNHN, 5). 12 m, E Aoré Island, SANTO 2006 DB08, 15.576°S, 167.230°E (MNHN, 6). 14–25 m, E Aoré Island, SANTO 2006 DB33, 15.578°S, 167.230°E (MNHN, 2). 36–39 m, Segond Channel, SW coast of Aoré Island, SANTO 2006 FB90, 15.583°S, 167.128°E (MNHN, 1). 36 m, Segond Channel, SW coast of Aoré Isalnd, SANTO 2006 ZS25, 15.585°S, 167.126°E (MNHN, 1). 10–12 m, Strait of Tangoa Island, SANTO 2006 LD29, 15.588°S, 166.990°E (MNHN, 2). Tangua Island, 15.590°S, 166.986°E (MNHN, 6). 32–40 m, S Tutuba Island, SANTO 2006 DB16, 15.591°S, 167.263°E (MNHN, 4). 6–30 m, S Tutuba Island, SANTO 2006 NB43, 15.593°S, 167.266°E (MNHN, 4). 15–20 m, S Santo, Tanoa Bay, SANTO 2006 ZB20, 15.601°S, 167.090°E (MNHN, 5). 84–105 m, N Urélapa Island, SANTO 2006 AT43, 15.606°S, 167.038°E (MNHN, 1). 108–121 m, W Urélapa Island, SANTO 2006 EP24, 15.608°S, 167.021°E (MNHN, 1). 86–116 m, N Urélapa Island, SANTO 2006 AT44, 15.608°S, 167.045°E (MNHN, 3). 10–18 m, S Aoré Island, SANTO 2006 DB12, 15.610°S, 167.168°E (MNHN, 2). 6–24 m, S Aoré Island, Aoré Island, SANTO 2006 DS10, 15.610°S, 167.168°E (MNHN, 1). 18–30 m, NE Urélapa Island, Palikulo Bay, SANTO 2006 DB77ZB13, 15.611°S, 167.033°E (MNHN, 1). 83–120 m, E Urélapa Island, SANTO 2006 AT30, 15.611°S, 167.043°E (MNHN, 1). 35 m, Bruat Channel, N coast of Malo Island, SANTO 2006 DS110, 15.613°S, 167.146°E (MNHN, 1). 18 m, Bruat Channel, SANTO 2006 DB80, 15.618°S, 167.125°E (MNHN, 2). 10 m, Bruat Channel, N Coast of Malo Island, SANTO 2006 DB25, 15.628°S, 167.188°E (MNHN, 1). 80–94 m, NW coast of Malo Island, SANTO 2006 EP19, 15.629°S, 167.090°E (MNHN, 1). 90–110 m, NW coast of Malo Island, Palikulo Bay, SANTO 2006 DB77 EP28, 15.633°S, 167.083°E (MNHN, 1). 13 m, NE Malo Island, SANTO 2006 DB86, 15.641°S, 167.251°E (MNHN, 1). 13 m, NE Malo Island, Palikulo Bay, SANTO 2006 DB77DB86, 15.641°S, 167.251°E (MNHN, 1). 15 m, W Malo Island, SANTO 2006 DB29, 15.648°S, 167.085°E (MNHN, 1). 15 m, W Malo Island, Palikulo Bay, SANTO 2006 DB77DB29, 15.648°S, 167.085°E (MNHN, 1). 5–7 m, W coast of Malo Island, SANTO 2006 ZB09, 15.676°S, 167.085°E (MNHN, 6). 7 m, Malokilikili, Aoré Island, SANTO 2006 FB52, 15.711°S, 167.235°E (MNHN, 1). 6 m, E Malo Island, Palikulo Bay, SANTO 2006 DB77DB83, 15.723°S, 167.250°E (MNHN, 2). 360–419 m, MUSORSTOM 8 DW1065, 16.266°S, 167.350°E (MNHN, 28).

Solomon Islands. 396–411 m, SALOMON 1 DW1762, 8.665°S, 160.065°E (MNHN, 2). 557–655 m, 11.485°S, 159.957°E (MNHN, 1). 314–586 m, N Malaita, SALOMONBOA 3 DW2790, 8.316°S, 160.617°E (MNHN, 1). 98–200 m, SALOMON 1 DW1767, 8.323°S, 160.665°E (MNHN, 36). 98–200 m, SALOMON 1 DW1767, 8.323°S, 160.665°E (MNHN, 2). 22 m, Vangunu Island, Kokoana Passage, Marovo Lagoon, Matiu Island, 8.497°S, 158.200°E (LACM 89-77, 12). 35 m, Midway betweeen Karumaneke Island and Cheke Village, Marovo Lagoon, Vangunu Island, 8.516°S, 158.133°E (NMNZ M.230051, 1). 24 m, off Mbatuna, marovo Lagoon, Vangunu Island, 8.550°S, 158.117°E (NMNZ M230150, 3). 6 m, off Mbatuna, Marovo Lagoon, Vangunu Island, 8.550°S, 158.117°E (NMNZ M230191, 3). 22 m, Vangunu Islnda, S end Porepore Island, 8.583°S, 158.200°E (LACM 89-75, 5). 1327–1598 m, SALOMON 1 CP1764, 8.610°S, 160.123°E (MNHN, 10). Vangunu Island, Karunjou Island, 8.650°S, 158.233°E (LACM 89-76, 2). 396–411 m, 8.665°S, 160.065°E (MNHN, 1). 20 m, Vangunu Island, Minjanga Island, 8.667°S, 158.233°E (LACM 89-74, 4). 17 m, Vangunu Island, Kicha Island, 8.767°S, 158.317°E (LACM 89-78, 1). 20 m, Central District, Russell Islands, Hanavisi Island, 8.988°S, 159.225°E (LACM 78-73, 1). 10–20 m, Fonagho Island, Russel Islands, Central Province, 9.092°S, 159.296°E (SBMNH 85146, 3). 5–10 m, Soghonangola Island, Florida Islands, Central Province, 9.116°S, 160.153°E (SBMNH 85144, 3). Yandina, Banika Island, Russell Islands, 9.116°S, 159.217°E (NMNZ M.230070, 1; NMNZ M230081, 1). 10–25 m, Sakulufat Point, Mbanika Island, Russell Islands, Central Province, 9.120°S, 159.158°E (SBMNH 85142, 6). 20 m, Central District, Russell Islands, Dry Sand Cay, 9.150°S, 159.192°E (LACM 78-75, 4). 15–20 m, W side, Mbungana Island, Florida Islands, Central Province, 9.179°S, 160.199°E (SBMNH 85140, 1). 3 m, Central District, Guadalcanal, Ndoma, 9.283°S, 159.817°E (LACM 78-64, 1). 10 m, Central District, Guadalcanal, Marau Sound, 9.817°S, 160.867°E (LACM 78-68.10, 1). 356–329 m, 9.871°S, 160.857°E (MNHN, 3). 9 m, between Telina Island and Mbareki Peninsula, Marovo Lagoon, Vangunu Island, 9.900°S, 160.650°E (NMNZ M.230484, 5).

Fiji. 545–651 m, 16.816°S, 179.900°E (MNHN, 1). 200–215 m, Bligh Water, 16.840°S, 178.208°E (MNHN, 20; MNHN, 20). 251–257 m, Bligh Water, 16.856°S, 178.232°E (MNHN, 20; MNHN, 3). 402–410 m, 17.133°S, 178.983°W (MNHN, 3). 190–308 m, 17.183°S, 178.717°W (MNHN, 1). 660–663 m, Bligh Water, 17.248°S, 178.492°E (MNHN, 9; MNHN, 3). 656–660 m, Bligh Water, 17.268°S, 178.247°E (MNHN, 1; MNHN, 20). 820–863 m, BORDAU 1 DW1459, 17.300°S, 179.550°W (MNHN, 1; MNHN, 50; MNHN, 1; MNHN, 19). 879–897 m, SE of Viti Levu, 17.515°S, 178.888°E (MNHN, 20; MNHN, 4; MNHN, 2). 0 m, SE side, Naigani Island, Lomaiviti Group, 17.586°S, 178.678°E (MV F178686, 1). 959–963 m, SE Viti Levu, 17.710°S, 178.917°E (MNHN, 1; MNHN, 7; MNHN, 1). 9–35 m, Viti Levu, Nadi Bay Tomba Ko Nandi, 17.733°S, 177.417°E (AMS C.379431, 1). 183–188 m, SE Viti Levu, 17.828°S, 178.797°E (MNHN, 10). 402–444 m, SE Viti Levu, 17.993°S, 178.803°E (MNHN, 3). 20 m, Vitu Levu S lagoon, 18.136°S, 178.415°E (MNHN, 1). 290–300 m, 18.150°S, 178.650°W (MNHN, 1). 300–400 m, 18.166°S, 178.733°W (MNHN, 12). 24 m, Vitu Levu S lagoon, 18.168°S, 178.477°E (MNHN, 1). 13 m, Vitu Levu S lagoon, 18.168°S, 178.510°E (MNHN, 1). 1 m, Viti Levu, 0.5 mi S of Suva Point, 18.170°S, 178.420°E (LACM 79-40, 2). 3–4 m, Nukumbotho channel W side of Nukumbutho Island, Laucala Bay, Viti Levu, 18.181°S, 178.466°W (ZSM 20061627, 1: complete). 0–1 m, Nukumbotho channel W side of Nukumbutho Island, Laucala Bay, Viti Levu, 18.181°S, 178.466°W (ZSM 20070110, 1: complete). "Viti Levu S coast, Korotongo," 18.183°S, 177.533°E (AMS C.404598, 1: complete). Sovi Bay Souvi, SW coast Viti Levu, 18.183°S, 177.600°E (AMS C.379430, 5). 392–433 m, S of Viti Levu, 18.185°S, 178.390°E (MNHN, 1; MNHN, 20). 123 m, Vitu Levu S lagoon, 18.196°S, 178.507°E (MNHN, 1). 420–450 m, 18.200°S, 178.617°W (MNHN, 1). 144–150 m, S of Viti Levu, 18.206°S, 178.550°E (MNHN, 1; MNHN, 12; MNHN, 14). 149–168 m, S of Viti Levu, MUSORSTOM 10 CP1366, 18.206°S, 178.552°E (MNHN, 50;

MNHN, 2; MNHN, 3). 295–302 m, S of Viti Levu, 18.211°S, 178.540°E (MNHN, 2; MNHN, 70; MNHN, 6). 1 m, Viti Levu, Korolevu, 18.220°S, 177.718°E (LACM 79-42, 1). 275–430 m, S of Viti Levu, 18.296°S, 177.907°E (MNHN, 150; MNHN, 1). 10–20 m, off Francis Island in Beqa Lagoon, just off of Beqa Island (S of Viti Levu), 18.300°S, 178.067°E (DLG 277, 1). 230–251 m, 18.306°S, 178.043°E (MNHN, 4). 260–305 m, S of Viti Levu, MUSORSTOM 10 DW1384, 18.308°S, 178.097°E (MNHN, 50; MNHN, 1; MNHN, 4). 234–361 m, 18.310°S, 178.085°E (MNHN, 4). 497–504 m, S of Viti Levu, 18.311°S, 178.152°E (MNHN, 5). 441–443 m, S of Viti Levu, MUSORSTOM 10 DW1382, 18.320°S, 177.862°E (MNHN, 2; MNHN, 6). 350–357 m, 18.666°S, 178.500°W (MNHN, 1). Great Astrolabe Reef, 18.746°S, 178.300°E (JKC, 2). 22 m, Kadavu Group, Astrolabe Reef, 18.763°S, 178.472°E (LACM 91-188, 2). Astrolabe Reef, Kadavu Group, 18.767°S, 178.453°E (LACM 91-189, 2). 20 m, Herald Pass, W of Ndravuni Island, NW end of Great Astrolabe Reef, 18.767°S, 178.467°E (LACM 85-135, 1). Great Astrolabe Reef, 18.769°S, 178.466°E (JKC, 1). 240–319 m, BORDAU 1 DW1494, 18.916°S, 178.483°W (MNHN, 10). 395–540 m, BORDAU 1 DW1486, 19.016°S, 178.433°W (MNHN, 12). 700–707 m, 19.050°S, 178.500°W (MNHN, 8). 18 m, Kandavu, Cape Washington, 19.108°S, 177.967°E (LACM 85-138, 4).

Tonga. Vava'a group, Nuapapu Island, 18.700°S, 174.100°W, (LACM 85-89, 8). 27 m, Vava'a Group, Hunga Island, 18.692°S, 174.133°W, (LACM 85-91, 1). 440–487 m, Vava'u group, BORDAU 2 DW1586, 18.566°S, 173.916°W (MNHN, 100). 433 m, NW Tongatapu, BORDAU 2 DW1569, 21.033°S, 175.316°W (MNHN, 300).

American Samoa. 15 m, Leone Bay, Tutuila Island, 14.338°S, 170.793°W, (LACM 99-89, 2). 25 m, Pago-Pago Harbour, Tutuila Island, 14.279°S, 170.701°W (ESC, 4).

Western Samoa. 4 m, Middle Vaisala Lagoon, Vaisala, 13.512°S, 172.644°W (ESC, 1). Bay Afria [= Apia?], 13.829°S, 171.763°W (KBIN, 2).

Marshall Islands. Jabor Island, Jaluit Atoll, 5.917°N, 169.650°E (SBMNH 124050, 1).

Cook Islands. 22 m, Aitutaki, off Arutanga, 18.872°S, 159.792°W, (LACM 87-79, 10).

Tuvalu. 366 m, Funafuti Atoll, off Tutanga Tutaga Island, 8.617°S, 179.083°E (AMS C.5641, 7). 91 m, Funafuti Atoll, off Tutanga Tutaga Island, 8.617°S, 179.083°E (AMS C.5642, 5). 274 m, Funafuti Atoll, off Funamanu Beacon Island, 8.567°S, 179.150°E (AMS C.5640, 1; AMS C.170414, 1).

Belau. 11 m, Mutomdu Reef, 7.266°N, 134.500°E (LACM 78-37, 1).

Kiribati. 0 m, windward Tibatiba Abemama Atoll, 0.407°N, 173.849°E (CPC, 3).

Guam. 1 m, NE part of Tumon Bay, 13.143°N, 144.750°E (LACM 77-1, 5). 2 m, Bile Bay, 13.275°N, 144.660°E (LACM 77-7, 1). 1 m, Western Shoal, 13.450°N, 144.670°E (LACM 77-20, 2). 1 m, S end Piti Bay, 13.456°N, 144.693°E (LACM 77-14, 2). 6 m, Piti Bay, 13.465°N, 144.690°E (LACM 77-11, 3).

French Polynesia, Society Islands. 4 m, Paevaeva, W side Moorea Island, 17.533°S, 149.883°W, (AMS C.379428, 7). 2 m, Moorea, Club Mediterranean, 17.500°S, 149.767°W, (LACM 74-36, 10). 25 m, Moorea, Pt. Hauru, 17.485°S, 149.920°W, (LACM 71-114, 1). Moorea Island, Papetoai Bay, Passe Tareu, W side, 17.508°S, 149.850°W, (AMS C.379421, 1; AMS C.380675, 2). 29 m, Tahiti, 17.533°S, 149.567°W, (AMS C.379434, 2). 15 m, Tiahura Island, NW Moorea, 17.483°S, 149.917°W, (AMS C.379422, 1). 8 m, Tiahura Island, NW Moorea Island, 17.483°S, 149.917°W, (AMS C.380657, 8). E side Ilot Tiahura, NW corner Moorea, 17.483°S, 149.917°W, (AMS C.380658, 2; C.404426, 1). 914 m, Bora Bora, TARASOC DW3416, 16.583°S, 151.733°W (MNHN, 1). 700–785 m, Huahine, TARASOC DW3434, 16.700°S, 151.050°W (MNHN, 3). 520–572 m, Raiatea, TARASOC DW3457, 16.750°S, 151.400°W (MNHN, 20). 440–490 m, Raiatea, TARASOC DW3451, 16.883°S, 151.350°W (MNHN, 30). 485–560 m, Moorea, TARASOC DW3459, 17.466°S, 149.800°W (MNHN, 6). 610 m, Moorea, TARSOC DW3481, 17.483°S, 149.750°W (MNHN, 2). 435–490 m, Moorea, TARASOC DW3476, 17.483°S, 149.750°W (MNHN, 4). 300–650 m, Tahiti, TARASOC DW3484, 17.783°S, 149.383°W (MNHN,

2). 400–440 m, Tahiti, TARASOC DW3487, 17.783°N, 149.350°W (MNHN, 2). 85 m, Raiatea Island, 16.820°S, 151.420°W (DLG 960, 5). 0 m, inner edge of barrier reef, Cooks Bay, Moorea, 17.486°S, 149.828°W (DLG 268, 1: complete). 0 m, Gump Station, Moorea, 17.533°S, 149.833°W (DLG 239-2, 1: complete).

French Polynesia, Tuamotus. Marutea Atoll, 17°S, 143.167°W, (AMS C.28241, 1). Manihi Atoll, King George Islands, 15.542°S, 146.058°W, (LACM 83-34, 1). 1 m, Ahe Atoll, 14.467°S, 146.367°W, (LACM 73-94, 6). 0 m, SW side, Manihini, 14.447°S, 146.067°W (EGC 29782, 5). Manihi Atoll, 14.400°S, 145.900°W (EGC 29783, 12).

Great Britain, Pitcairn Island. Isaac's Pool, 25.064°S, 130.100°W (DLG 1417, 1).

Literature records.

Japan. Okinawa (Bandel, 1998).

Indonesia. Satonda [8.100°S, 117.750°E] (Bandel, 1998).

China. Xiaodonghai, Hainan. Dadonghai, Hainan [18.221°N, 109.510°E] (Hasegawa *et al.*, 2001).

Remarks. *Sinezona plicata* Hedley should not be confused with the European *Sci. plicata* Philippi, 1836 (junior synonym of *Sci. costata*). The protoconch of the species has superficially two morphologies: smooth, and smooth with single spiral. The spiral is easily eroded (e.g., Bandel, 1998: pl. 16 fig. 2), and it is possible that this character is in fact not polymorphic but highly susceptible to erosion.

The description of the fossil *Sin. provecta* reads very similar to *Sin. plicata*, which is confirmed by the original illustration and the examination of the holotype by SEM (Fig. 415). Common features include the strong axial folds run over by spiral threads, the spiral band of elevated axial riblets on the shoulder, the wide umbilicus, and the long narrow foramen. *Sinezona provecta* was not compared to any species (fossil or Recent).

Hasegawa (2004) reported mass swimming in *Sin. plicata*. One lot with a shell occupied by a hermit crab is known (USNM 374438).

Sinezona rimuloides (Carpenter, 1865)

Figures 25C, 418–421, Color Plates 5A–K

Chresonymy.

Scissurella rimuloides: Carpenter, 1864: 548 [*nomen nudum*].
Scissurella rimuloides Carpenter, 1865: 271, not illustrated.
Scissurella rimuloides: Paetel, 1888: 289.
Schismope rimuloides: Pilsbry, 1890: 68–69.
Schismope rimuloides: Pilsbry, 1903: 84.
Schismope rimuloides: Thiele, 1912a: 27–28.
Schismope rimuloides: Orcutt, 1915: 166.
Schismope rimuloides: Dall, 1921: 183.
Schismope coronata: Dall, 1921: 183.
Schismope rimuloides: Oldroyd, 1927: 229.
Schismope rimuloides: Keen, 1937: 45.
Schismope rimuloides: Smith, 1944: 3.
Schismope rimuloides: Burch, 1946a: 27.

Schismope rimuloides: Burch, 1946b: 23.

Scissurella rimuloides: Palmer, 1951: 73, pl. 1, figs 20–22.

Sinezona rimuloides: Keen, 1958: 254, text-fig.

Sinezona rimuloides: McLean, 1966: 148–150, pl. 4, figs 26–27.

Scissurella rimuloides: Brann, 1966: 47, pl. 22, fig. 698.

Sinezona rimuloides: Keen, 1968: 403, text-fig. 26a–c [holotype].

Sinezona rimuloides: McLean, 1969: 13, fig. 3-1.

Sinezona rimuloides: McLean, 1971: 309, fig. 5.

Sinezona rimuloides: Draper, 1973: 8.

Sinezona rimuloides: Abbott, 1974: 17 [US records. Chile records refer most likely to *Sin. chilensis*].

Sinezona rimuloides: Batten, 1975: figs 6, 14 [shell], fig. 15 [mineralogy].

Sinezona rimuloides: McLean *in* Keen, 1978: 309, fig. 5.

Sinezona rimuloides: McLean, 1978: 13, fig. 3-1.

Sinezona rimuloides: Lozouet, 1986: 111.

Sinezona rimuloides: McLean, 1989: fig. 6c.

Sinezona rimuloides: Shasky, 1989: 8.

Sinezona rimuloides: Finet, 1991: 268 [from Galapagos].

Sinezona rimuloides: Skoglund, 1992: 4.

Sinezona rimuloides: Kaiser, 1993: 38–39, fig. 3 [from Galapagos].

Sinezona rimuloides: Finet, 1993: 38–39, figs 3–4 [from Galapagos].

Sinezona rimuloides: Finet, 1994a: 18, pl. 2, fig. 2 [from Galapagos].

Sinezona rimuloides: Finet, 1994b: 11 [from Galapagos].

Sinezona rimuloides: McLean & Coan, 1996: 307.

Sinezona rimuloides: Kaiser & Bryce, 2001: 16, pl. 11, figs 1, 1a.

Sinezona rimuloides: Skoglung, 2002: 3.

Sinezona rimuloides: Geiger, 2003: 79.

Sinezona rimuloides: Geiger, 2004: 4, unnumbered text-fig. [histological sections].

Sinezona rimuloides: Liff-Grieff, 2005: 2, text-fig. [photograph reversed by accident: see Liff-Grieff, 2006: 2, text-fig.].

Sinezona rimuloides: Hendrickx & Brusca, 2005: 227.

Sinezona rimuloides: Geiger, 2006c: text-fig. 9 [two histological sections; bursicles].

Sinezona rimuloides: McLean, 2007: 722, pl. 356, fig. F.

Sinezona rimuloides: Geiger *et al.*, 2007: fig. 13 [histology, 3D reconstruction].

Sinezona rimuloides: Geiger *et al.*, 2008: fig. 12.2A [histology of gill].

Sinezona rimuloides: Geiger & McLean, 2010: 11.

Synonym.

+ *Schismope californica* "Bartsch" Dall, 1923: 4 [*nomen nudum*].

Schismope californica: Keen, 1937: 45 [not validated].

Misidentifications.

not *Sinezona rimuloides*: Marincovich, 1973: 4, 16, fig. 26 [is *Sin. chilensis*].

Schismope coronata: Dall, 1921: 183 [is *Sin. rimuloides*].

Sinezona rimuloides: Paredes *et al.*, 1999: table 3 [from Peru: most likely *Sin. chilensis*].

Type material. Holotype ("Liverpool collection" tablet/Mazatlan collection 2532: NHMUK slide 1532: McLean, 1966. Keen, 1968: text-figs 26a–c. Finet, 1993: fig. 3: mounted holotype. Fig. 418), 0.023 × 0.03 '' [= 0.8 × 0.6 mm].

Type locality. Mazatlan [Sinaloa, Mexico, 23.180°N, 106.430°W], off *Spondylus calcifer* [(Carpenter, 1857)] (OD).

Etymology. Rimula, diminuitive of rima, Latin for cleft, fissure.

Description. Shell small (to 0.8 mm) trochiform. Protoconch of one whorl, strong axial sculpture, apertural varix not connected to embryonic cap, apertural margin slightly sinusoidal. Teleoconch I of slightly more than one whorl, approximately 13–15 axial cords, no spirals, interstices with fine irregular growth lines. Teleoconch II of 0.6 whorl, suture impressed. Shoulder slightly convex, axial cords gradually increasing in density, becoming less prominent on outer portion towards selenizone, first spiral cordlet at onset of selenizone or somewhat later, increasing in number to approximately 3–6 at apertural margin. Base with constriction below selenizone, with indistinct low axial cords, crossed by approximately nine somewhat irregularly spaced spiral cords. Umbilicus open, narrow, straight walls, keel at border to base. Selenizone above periphery, keels low, moderately strong, lunules distinct; selenizone approximately three times as long as oblong foramen. Aperture rounded, oblique to coiling axis, roof overhanging.

Operculum (Fig. 420B) corneous, thin, round, multispiral, central nucleus.

Radula (Fig. 420A). Rachidian tooth triangular, tip with seven cusps, central one largest, V-shaped. Lateral teeth 1–3 similar, 3–4 cusps on outer edge of tip. Lateral tooth 4 reduced, pointed, two small projections on either side of tip. Lateral tooth 5

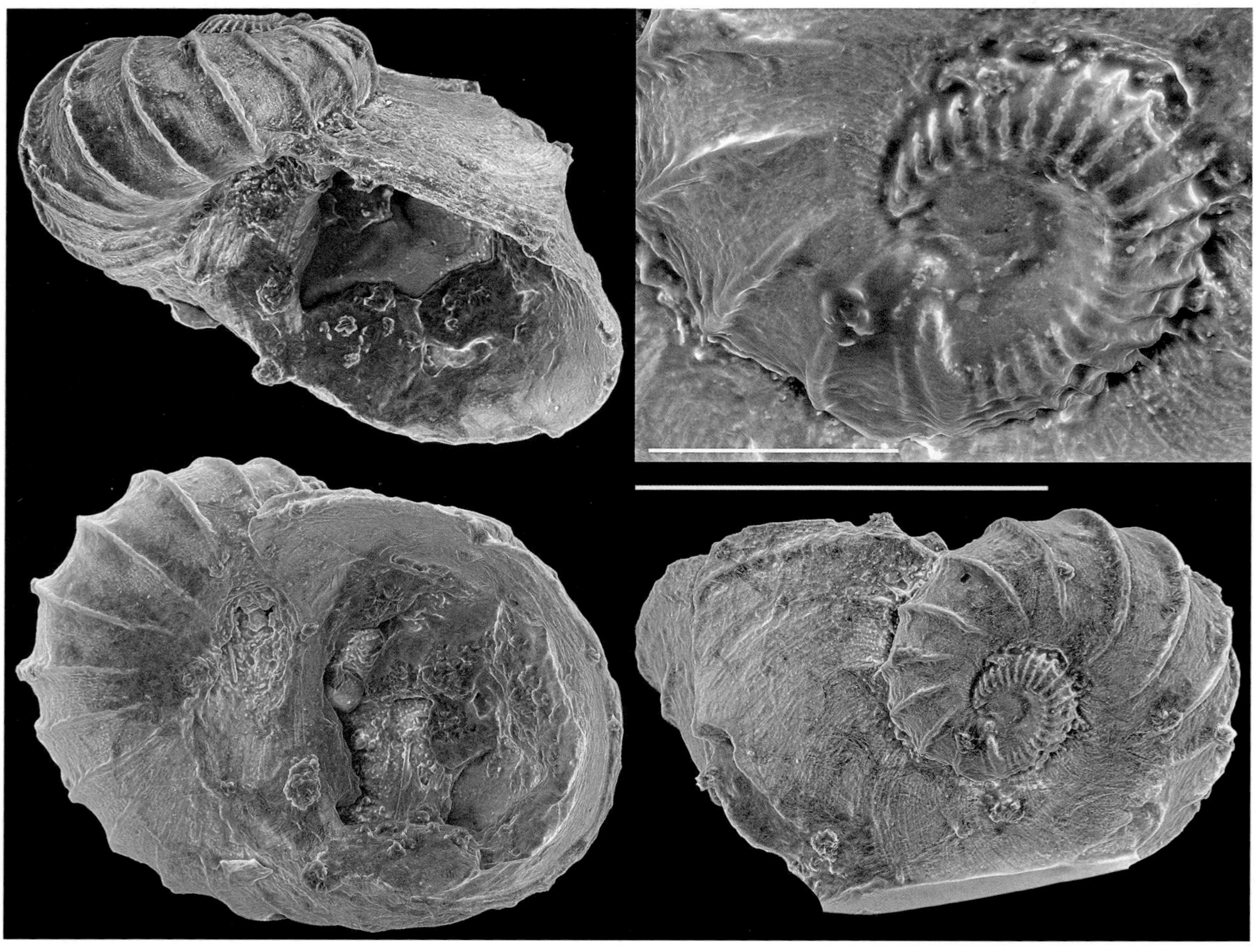

Figure 418. *Sinezona rimuloides*. Holotype. Mazatlan [Sinaloa, Mexico, 23.180°N, 106.430°W], off *Spondylus calcifer* [(Carpenter, 1857)] (NHMUK Mazatlan Coll. 2532). Specimen glued to original glass slide, imaged uncoated using variable pressure at 78 Pa. Scale bar shell = 0.5 mm. Scale bar protoconch = 100 μm.

broadened, seven cusps on inner edge, terminal one largest. Inner marginal teeth with two cusps on inner margin, terminal cusp largest, 4–5 cusps on outer margin. Outer marginal teeth spoon-shaped, many fine bristles.

Comparisons. *Sinezona carolarum* from Mexico and Clipperton Island is wider than tall, has a wider umbilicus, and the selenizone is only twice as long as the foramen.

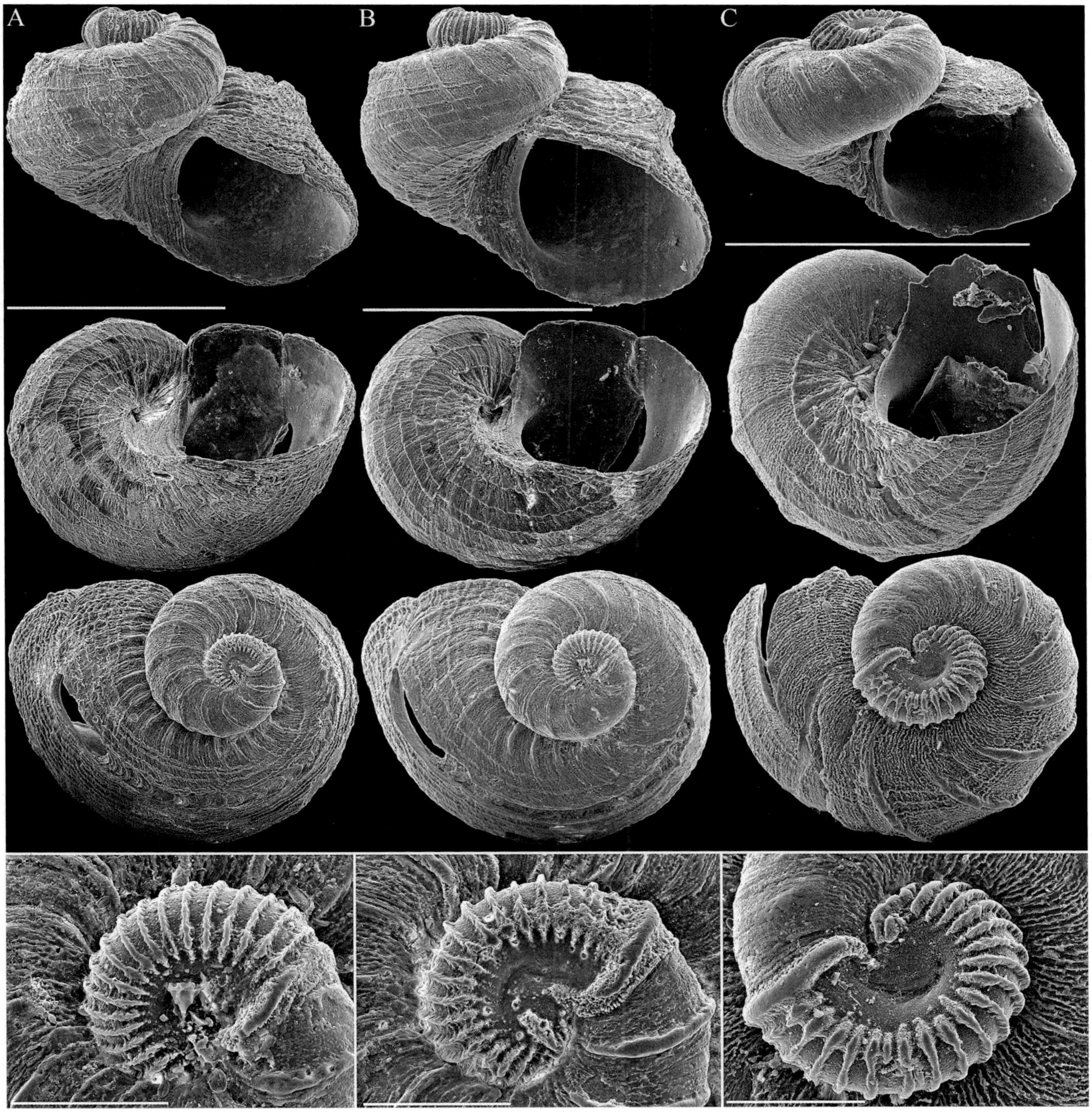

Figure 419. *Sinezona rimuloides*. **A–B.** 2–4 m, just W of Saladita Bay, Guaymas, Sonora, Mexico, 27.883°N, 110.950°W (SBMNH 211378). **C.** Juvenile with open slit. 0 m, Carmel Point, Carmel, Monterey County, California, USA, 36.667°N, 112.000°W (LACM 66-57, 51). Scale bars shell = 0.5 mm. Scale bars protoconch = 100 μm.

Sinezona chilensis from the southwestern South America is overall shorter with the suture at the protoconch varix being at or above the periphery of the shell and generally has a shorter foramen (2–3 vs. 4–5 times as long as wide).

Sinezona calumnior from temperate Australia and New Zealand has more distinct spiral lines, a narrower umbilicus, and more distinct axial cords on the shoulder of teleoconch II.

Sinezona cingulata from the Mediterranean and the adjacent Atlantic has stronger axial cords on teleoconch I, a shorter selenizone (0.2 vs. 0.4 whorl), and a narrower umbilicus.

Sinezona confusa from the Caribbean to northern Brazil has more axial cords, tends to have a longer selenizone, and has a shorter teleoconch I (0.8 vs. 1.1 whorls).

Sinezona insignis from South Africa has much stronger raised axial cords and lacks spiral lines on shoulder and base.

Sinezona iota from New Zealand has much stronger raised axial cords and lacks spiral lines on shoulder and base.

Sinezona mouchezi from the subantarctic of the Indian Ocean has a longer teleoconch I (1.25 vs. 1.1 whorls) and fewer axial cords that are often also less distinct.

Distribution. Northern California (41°N) to Ecuador (1°S), and offshore Islands (Galapagos, Clipperton).

Specimen records.

Ecuador, Galapagos. 10–25 m, Champion, Floreana Island, 1.250°S, 90.333°W (SBMNH 85166, 1). 0 m, W of Comorant Island, Floreana/Charles Island, 1.217°S, 90.417°W (SBMNH 85170, 1).

France, Clipperton Island. 11–22 m, 10.316 N, 109.214 W (CSC, 1). 10–20 m, SE, 10.291°N, 109.201°W (SBMNH 358511, 1).

Mexico. Flat Rock, S end of Barra de Navidad, Jalisco, 19.203°N, 104.684°W (SBMNH 114004, 2). 30–45 m, Caletilla Banda, San Benedicto Island, Revillagigedo Islands, 19.317°N, 110.800°W (SBMNH 85164, 1). 0 m, Tisate, Bandera Bay, Nayarit, 20.617°N, 105.250°W (CSC, 10). 0 m, Cabeza Ballena, headland 5 mi E of Cape San Lucas, Baja California, 22.900°N, 109.833°W (LACM 46-4, 8). 0 m, Cape San Lucas, Baja California, 22.867°N, 109.883°W (CSC, 1). Mazatlan, 23.180°N, 106.430°W (NHMUK Mazatlan #2532, 1). 10–25 m, Fourth Reef, Cabo Pulmo, Baja California Sur, 23.431°N, 109.422°W (SBMNH 85173, 2). 3–6 m, NE end Santa Margarita Island, Baja California Sur, 24.450°N, 111.833°W (SBMNH 85162, 1). 13 m, SW tip Partida Island, Baja California Sur, 24.533°N, 110.433°W (SBMNH 85169, 1). 12 m, Los Islotes, Pardita Island, Baja California Sur, 24.600°N, 110.433°W (SBMNH

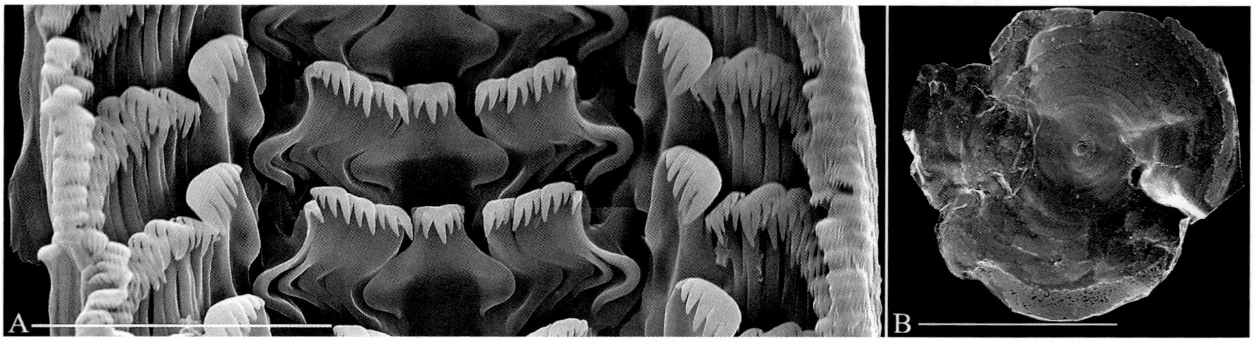

Figure 420. *Sinezona rimuloides*. **A.** Radula. Scale bar = 20 μm. **B.** Operculum. Scale bar = 200 μm. 0 m, Royal Palms, Palos Verdes, Los Angeles, California, USA, 33.720°N, 118.327°W (DLG 245, 50).

85163, 10). 2–5 m, about 1 mi from Point Huches, Santa Maria Bay, Baja California, 24.767°N, 112.267°W (LACM 67-69, 6). 1 m, Baja California Sur, Gulf of California, del Espiritu Santo Island, San Gabriel Bay, 24.500°N, 110.367°W (CASIZ 119106, 30). San Juancito, Baja California Sur, 25.383°N, 111.167°W (FLMNH 355571, 2). 60–90 m, off NE Danzante Island, Gulf of California, 25.785°N, 111.250°W (CSC, 1; SBMNH, 1). Ildefonso Island, Gulf of California, 26.600°N, 111.433°W (SBMNH, 40). 6–12 m, off Ildefonso Island, Gulf of California, 26.632°N, 111.430°W (SBMNH 423738, 1; SBMNH, 20). 10 m, Asuncion Island, Baja California Sur, 27.117°N, 114.283°W (SBMNH 116310, 1). 10 m, Ascuncion Island, Baja California Sur, 27.117°N, 114.283°W (SBMNH 116311, 8). 5 m, N side Tortugas Bay, Baja California Sur, 27.646°N, 114.856°W (SBMNH 116210, 40). 5 m, N side Tortugas Bay, Baja California Sur, 27.646°N, 114.856°W (SBMNH 116211, 17). 26 m, 12 Fathom Bank, 1 mi SW of Natividad Island, outer coast, Baja California, 27.867°N, 115.211°W (LACM 71-165, 4). 31 m, San Carlos Bay, Sonora, 27.900°N, 110.867°W (SBMNH 85177, 3). San Carlos Bay, Sonora, 27.900°N, 110.867°W (SBMNH, 1). Colorado Point, Sonora, 27.908°N, 110.972°W (SBMNH 119692, 50). 91 m, SE San Antonio Point, Sonora, 27.933°N, 111.117°W (SBMNH 85158, 1). 18 m, off San Carlos Bay, Sonora, 27.933°N, 111.067°W (SBMNH 85160, 12). Near Cerro de Tetas de Cabrta, Caleta Lalo, Sonora, 27.933°N, 111.050°W (SBMNH 132624, 2). 0–2 m, W side Bocochibampo Bay, Sonora, 27.933°N, 110.967°W (SBMNH 85165, 2). 10–25 m, Ventana Island, Bocochibampo Bay, Sonora, 27.933°N, 110.967°E (SBMNH 85168, 4). 5–10 m, Ensenada San Francisco, Sonora, 27.950°N, 111.017°W (SBMNH 32218, 18). 12 m, San Pedro Nolasco Island, Sonora, 27.964°N, 111.379°W (SBMNH, 3). 13 m, San Pedro Nolasco Island, Sonora, 27.964°N, 111.379°W (SBMNH, 50; SBMNH, 4). San Pedro Nolasco Island, Sonora, 27.964°N, 111.379°W (SBMNH, 5). 16–22 m, SE tip San Pedro Nolasco Island, Sonora, 27.967°N, 111.417°W (SBMNH 85161, 1). 12 m, E side San Pedro Nolasco Island, Sonora, 27.972°N, 111.374°W (SBMNH, 3). 13–16 m, 5 Fathom Bank about 4 mi SW of SE tip of Cedros Island, outer coast Baja California, 27.983°N, 115.212°W (LACM 71-93, 22). 1 m, Chapetona Island, Guaymas, Sonora, 27.950°N, 111.033°W (LACM 76-65, 3). 2–4 m, just W of Saladita Bay, Guaymas, Sonora, 27.883°N, 110.950°W (SBMNH 211378, 12). 1 m, Chapetona Island, Guaymas, 27.933°N, 110.900°W (SBMNH 85156, 20; SBMNH 211379, 7). 0 m, Cove between Algodones and San Carlos Bays, Sonora, 27.967°N, 111.067°W (CSC, 2). 18–22 m, S side Piedra Colorada, SW tip Cedros Island, Baja California, 28.137°N, 115.360°W (LACM 72-115, 4). 10–25 m, 2 mi N of Villa Cedros, Cedros Island, Baja California Norte, 28.217°N, 115.167°W (SBMNH 113929, 1). 13 m, W side of East Island, San Benito Islands, Baja California, 28.297°N, 115.545°W (LACM 72-119, 2). 16–26 m, 1 mi N of Cedros Village, E side Cedros Island, outer coast Baja California, 28.128°N, 115.172°W (LACM 71-92, 12). San Pedro Point S at Morto Colorado, Sonora, 28.300°N, 111.450°W (SBMNH, 7). 11 m, San Pedro Point S at Morto Colorado, Sonora, 28.300°N, 111.450°W (SBMNH, 1). 13 m, San Pedro Point S at Morto Colorado, Sonora, 28.300°N, 111.450°W (SBMNH, 1). 11 m, San Pedro Point S at Morto Colorado, Sonora, 28.300°N, 111.450°W (SBMNH, 1). 13 m, San Pedro Point S at Morto Colorado, Sonora, 28.300°N, 111.450°W (SBMNH, 2). 0 m, Rosarito Point, outer Coast Baja California, 28.567°N, 114.158°W (LACM 46-1, 1). 20–33 m, Rocas Pinaculo, 1 mi W of West San Benito Island, Baja California, 28.258°N, 115.620°W

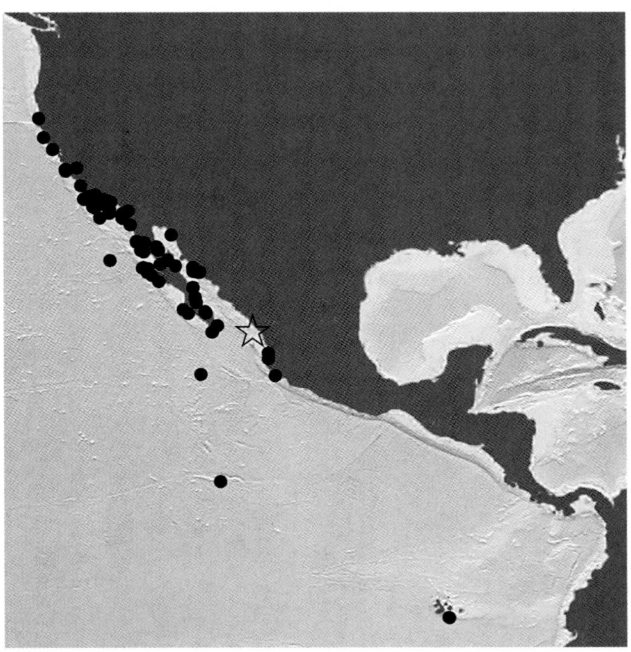

Figure 421. Distribution of *Sinezona rimuloides*. ☆ type locality. ● specimen record.

Scissurellidae *Sinezona*

(LACM 72-118, 10). 13–30 m, SW side Guadalupe Island (Afeura Island, Adentro, 5 Fathom Bank, and basalt arches), Baja California, 28.900°N, 118.292°W (LACM 72-120, 9). 0 m, San Francisquito Bay, near Guaymas, Sonora, 28.433°N, 112.883°W (LACM 46-6, 2). 7–20 m, La Cruz, Nayarit, 21.100°N, 105.246°W (FLMNH 355570, 1). 20–40 m, off La Gringa Point, Los Angeles Bay, Baja California, 28.917°N, 113.533°W (CSC, 2). 6–13 m, Sacramento Reef, just S of San Gernonimo Island, outer coast Baja California, 29.728°N, 115.760°W (LACM 71-91, 15). 5–7 m, N entrance, San Luiz Gonzaga Bay, Baja California Norte, 29.767°N, 114.300°W (SBMNH 127586, 1). 0 m, San Carlos, Baja California Sur, 29.610°N, 115.500°W (FLMNH 355613, 1). 0 m, Willard Island, San Luiz Gonzaga Bay, Baja California Norte, 29.817°N, 114.367°W (SBMNH 85159, 3). 18–30 m, Ben Rock, 3 mi S of San Martin Island, Baja California, 30.433°N, 116.117°W (LACM 72-112, 28). 2–9 m, N end of El Cholludo Island, (S of Puertocitos) Baja California, 30.025°N, 114.450°W (LACM 69-26, 8). 20 m, South Anchorage, San Martin Island, Baja California, 30.467°N, 116.117°W (LACM 67-60, 30). 8–12 m, Puerto Escondido, Baja California Sur, 30.383°N, 115.467°W (CSC, 1). 0 m, San Martin Island, Baja California, 30.484°N, 116.103°W (DLG 1311, 2). 0 m, Ensenada, Baja California, 31.867°N, 116.617°W (FLMNH 398598, 1). N Coronado Island, 32.455°N, 117.297°W (SDMNH ex 8850, 3; SDMNH, 5; SDMNH 41139, 4; SDMNH 57749, 3; SDMNH 2193, 20; LACM A.2777, 1). San Martin Island, Baja California Norte, 30.482°N, 116.089°W (SBMNH 114111, 1). 3–5 m, N side San Jorge Island, Sonora, 31.007°N, 113.243°W (SBMNH, 5).

USA, California. San Diego, 33.033°N, 116.800°W (SDMNH 21915, 1). San Diego County, San Diego, 32.700°N, 117.150°W (CASIZ 119089, 3). Mission Bay, San Diego, 32.750°N, 117.167°W (AMS C.405004). La Jolla, San Diego County, 32.800°N, 117.267°W (SBMNH 103982, 1). 0 m, Lunada Bay, Palos Verdes, Los Angeles, 33.770°N, 118.206°W (DLG 106, 7). 8 m, Italian Garden, Catalina Island, 33.378°N, 118.382°W (DLG 107, 20; DLG 222, 2: complete). Fisherman Cove, Catalina Island, Los Angeles County, 33.444°N, 118.484°W (DLG 1315, 1). 0 m, Royal Palms, Palos Verdes, Los Angeles, 33.720°N, 118.327°W (DLG 108, 4; DLG 110, 20; DLG 115, 30; DLG 117, 20; DLG 118, 4; DLG 122, 3; DLG 229, 10: complete; DLG 244, 3; DLG 245, 50; DLG 476, 30: complete; DLG 478, 20: complete). 0 m, Lunada Bay, Palos Verdes, Los Angeles, 33.770°N, 118.206°W (DLG 230, 22: complete). 0 m, N side White Cove, Santa Catalina Island, Los Angeles County, 33.394°N, 118.367°W (LACM 1406-41, 5). 0 m, S side White's Point, Los Angeles County, 33.713°N, 118.316°W (LACM 86-246, 20). 2 m, breakwater, opposite entrance to basin 2, Redondo Beach, Los Angeles County, 33.817°N, 118.383°W (LACM 55671, 1: complete). 23 m, San Nicolas Island, N side, 33.283°N, 119.517°W (LACM 72-101, 3). 10–23 m, Iron Bound Cove, S side of Santa Catalina Island, Los Angeles County, 33.450°N, 118.575°W (LACM 69-31, 9). 23–33 m, Begg Rock, NW of San Nicolas Island, 33.362°N, 119.693°W (LACM 72-99, 1). 1–2 m, Flat Rock Point, Palos Verdes Peninsula, 33.767°N, 118.417°W (AMS C.405007). 0 m, Kings, Harbor, Redondo Beach, Los Angeles, 33.843°N, 118.394°W (DLG 819, 21). 8 m, Pirate Cove, Catalina Island, 34.004°N, 118.811°W (DLG 109, 100). 20–27 m, Pyramid Point, SE tip of San Clemente Island, Los Angeles County, 32.817°N, 118.350°W (LACM 69-29, 40). 0 m, E boundary Carpenteria State Beach, Carpenteria, Santa Barbara County, 34.388°N, 119.515°W (LACM 77-30, 4). 10 m, Anacapa Island (S side of main island), Santa Barbara County, 34.067°N, 119.433°W (LACM 69-33, 3). 0 m, Forney's Point, Santa Cruz Island, Santa Barbara County, 34.000°N, 119.740°W (SBMNH 141271, 7). 0 m, E of Christi Beach, Santa Cruz Island, Channel Islands, Santa Barbara County, 34.026°N, 119.876°W (SBMNH 142247, 1). 0 m, Near Black Point, Santa Cruz Island, Channel Islands, Santa Barbara County, 34.038°N, 119.887°W (SBMNH 141270, 1). 14 m, Tyler Bight, San Miguel Island, 34.008°N, 120.405°W (DLG 883, 8). 8 m, Wyckoff Bay, San Miguel Island, 34.017°N, 120.388°W (DLG 884, 1). 18 m, San Miguel Island, 34.025°N, 120.445°W (DLG 1032, 1). 13 m, San Miguel Island, 34.027°N, 120.041°W

(DLG 881, 1). 13–20 m, Cortez Bank, ½ mi S of Bishop Rock, Los Angeles County, 32.433°N, 119.125°W (LACM 69-34, 19). 33 m, Cortez Bank, Santa Barbara County, 34.250°N, 120.050°W (SBMNH 349056, 1). 0 m, 0.4 km NW of South Point, Shell Beach, San Luis Obispo County, 35.158°N, 120.681°W (DLG 1316, 2). 15 m, Monterey County, Whaler's Cove, 36.517°N, 121.933°W (CASIZ 119121, 1). 0 m, Carmel Point, Carmel, Monterey County, 36.667°N, 112.000°W (LACM 66-57, 51). 0 m, Carmel Point, Monterey, 36.517°N, 121.950°W (LACM 81-47, 1). Hopkins Reef about 1.4 mi off Hopkins Marine Station, Pacific Grove, Monterey County, 36.667°N, 112.000°W (LACM 66-56, 1). 13–42 m, Carmel Submarine Canyon, N end of San Jose Creek Beach, Monterey County, 36.533°N, 121.933°W (LACM 60-24, 7). Mount Carmel, 36.370°N, 121.910°W (FLMNH 357179, 1). Monterey County, Carmel, Fanshell Beach, 36.550°N, 121.917°W (CASIZ 119113, 3). 0 m, Carmet (1 mi N of Salmon Creek), Sonoma County, 36.550°N, 121.917°W (LACM 1627-48, 2). Tomales Point, Marin County, 38.217°N, 122.983°W (LACM AHF 1608-48, 2). 11 m, Dark Gulch, Mendocino County, 39.233°N, 123.750°W (LACM 152127, 2: complete). 0 m, Humboldt Bay, 40.850°N, 124.147°W (DLG 820, 1: complete). 32 m, NE coast of Santa Cruz Island, 34.043°N, 119.610°W (DLG 2006, 50).

Fossil.

Mexico. S side Coronado Sur Island, Baja California Norte, 32.404°N, 117.243°W (SBMNH 85176, 1).

Remarks. Carpenter (1864: 548) listed the name in an entry reading in full "698. *Scissurella rimuloides*, n. s., 1 sp.," which does not consists of a proper species description per ICZN.

The epithet *californica* is not contained in the listing of Bartsch names by Ruhoff (1973).

McLean (2007) indicated an interstitial habitat for the species. Although most dead shells are obtained from sand samples, the species can be collected alive in the rocky intertidal from rock washings and algal shakings (Geiger, pers. obs.).

Sinezona semicostata Burnay & Rolán, 1990

Figures 422–425

Chresonymy.

Sinezona semicostata Burnay & Rolán, 1990: 38–39, pl. 2, figs 9, 11–14, pl. 3, figs 16, 18, 20.
Sinezona semicostata: Geiger, 2003: 79.
Sinezona semicostata: Rolán, 2005: 41, pl. 6, fig. 72.
Sinezona semicostata: Nolt, 2008: 60–61, figs 4A, 7.
Sinezona semicostata: Albuquerqu *et al.*, 2009: 70–71, text-figs.
Sinezona semicostata: Segers *et al.*, 2009a: 3, pl. 1, figs 4a–c.
Sinezona semicostata: Segers *et al.*, 2009b: 5.
Sinezona semicostata: Segers *et al.*, 2009c: 65–66, pl. 5, figs 3–3a.
Sinezona semicostata: Hernández *et al.*, 2011: 61–62, fig. 11D–I

Type material. Holotype (MNCN 15.05/1032: Templado *et al.*, 1993, Hernández *et al.*, 2011: fig. 11D). Paratypes from type locality (CZTL 1556, 1; MNHN, 1; SMF 308399/1, 1; RMNH, 1; MBL, 1; CZTL, 1); CER, 13: Hernández *et al.*, 2011: fig. 11E). Furna,

Brava Island, Cape Verde Islands. (NHMUK 1991123: is *Sin. cingulata*, see Remarks and Fig. 424C).

Type locality. 3–5 m, Baia Teodora, Sal Rei, Boavista Island, Cape Verde Islands [16.198°N, 22.921°W] (OD).

Etymology. Semi-, Latin for half. Costata, Latin for ribbed.

Description. Shell small (0.67 mm. Figs 422–424B), trochiform, depressed. Protoconch of one whorl, strong axial sculpture, no apertural varix, apertural margin convex. Teleoconch I of 0.875 whorl, approximately 16 axial cords, interstices with fine irregular axial lamellae, no spiral sculpture. Teleoconch II of 0.3–0.5 whorl, suture impressed. Shoulder slightly convex, approximately 11 axial cords, pronounced immediately adjacent to suture, otherwise indistinct to indiscernible, interstices with many fine irregular axial lamellae. Base with slight constriction below selenizone, sculpture similar to that of shoulder, but with more distinct axials starting at a distance as wide as the selenizone below the selenizone to entrance to umbilicus. Umbilicus wide, open, walls continuous with base. Selenizone above periphery, relatively wide, keels low, lunules distinct, slit closed anteriorly to form foramen. Aperture rounded.

Operculum round, thin, multispiral, nucleus central.

Comparisons. *Sinezona cingulata* from the Mediterranean Sea and the adjacent Atlantic ocean has a taller shell, a very narrow umbilicus not offering a view of the underside of the protoconch, and the protoconch has an apertural varix.

Sinezona confusa from the Caribbean to northern Brazil has a taller shell, a very narrow umbilicus not offering a view of the underside of the protoconch, and the protoconch has an apertural varix.

Sinezona garciai from the Caribbean has a wider shell, a protoconch with very fine irregular axial cords, and an umbilicus with straight walls.

Figure 422. Original illustration of *Sinezona semicostata*. Apical view is of holotype; remainder of paratypes.

Scissurella lobini from Cape Verde Islands has fewer axials on the base of mature specimens and the underside of the protoconch viewed through the umbilicus shows strong axials. Other characters cited by Burnay & Rolán (1990) as distinguishing, such as the extent and strength of the axial ribs on the shoulder, are known to be quite variable within species of *Scissurella* (e.g., *Sci. costata*).

Scissurella azorensis has a teleoconch I of >1 whorl and about twice as many fine axial ribs on the protoconch.

Distribution. Macaronesian Islands (Fig. 425).

Specimen records.

Cape Verde Islands. 15.020°N, 23.340°W (MNHN, 1). Furna, Brava Island, 14.883°N, 24.683°W (NHMUK 1991123, 1). 34–40 m, S of Lanzarote, Portugal, 28.850°N, 13.850°W, (RMNH, 1). 354 m, off Palmeira, W of Sal, 16.767°N, 23.017°W (RMNH, 1). 47–50 m, SE of Lanzarote, 28.933°N, 13.567°W (RMNH, 15). 82 m, S of Lanzarote, 28.800°N, 13.767°W (RMNH, 30). 37 m, S of Lanzarote, 28.817°N, 13.800°W (RMNH, 2). 34–40 m, S of Lanzarote, 28.850°N, 13.850°W (RMNH, 5).

Portugal, Madeira. 32.667°N, 16.750°W (NMSZ 1961.61, 1).

Spain, Canary Islands. 0 m, Salvage Island, Grand Salvage, Monks Point, Canary Islands, 30.150°N, 15.867°W (MNHN, 14). 40 m, del Carmen Point, Lanzarote, 28.916°N,

Figure 423. *Sinezona semicostata*. 0 m, Salvage Island, Grand Salvage, Monks Point, Canary Islands, Spain, 30.150°N, 15.867°W (MNHN, 14). Images by Jaya Nolt. Scale bar shell = 0.5 mm. Scale bar protoconch = 100 µm.

13.673°W (RMNH AD 24271, 6). Playa Famara, Lanzarote, 29.118°N, 13.550°W (RMNH AD 27607, 2).

Literature records.

Cape Verde Archipelago. Rigona, Sal Island, 16.750°N, 22.917°W (CER, 2); Palmeira, Sal Island, 16.767°N, 22.983°W (CER, 1). (Burnay & Rolán, 1990).

Figure 424. *Sinezona semicostata.* **A–B.** 0 m, Salvage Island, Grand Salvage, Monks Point, Canary Islands, Spain, 30.150°N, 15.867°W (MNHN, 14). SEM images by Jaya Nolt. **C.** Paratype. 3–5 m, Baia Teodora, Sal Rei, Boavista Island, Cape Verde Islands [16.198°N, 22.921°W] (NHMUK 1991123), is *Sin. cingulata* (see Remarks). Scale bars shell = 0.5 mm. Scale bars protoconch = 100 μm.

Scissurellidae *Sinezona*

Remarks. The name *Scissurella costata* var. *semicostata* was found on a lot in KBIN (DLG#5408), attached to an old specimen from the Dautzenberg collection from Pr. Alice trawl number 344. The name does not seem to be published.

The paratype of *Sin. semicostata* in the NHMUK (see Fig. 424C) is in fact a *Sin. cingulata*. The overall taller shell shape, the very narrow umbilicus, through which the underside of the protoconch cannot be seen, as well as the protoconch with an apertural varix all agree with *Sin. cingulata* and are at variance with the original illustrations of the holotype of *Sin. semicostata*. As the paratype is not name-bearing, it has no taxonomic consequences.

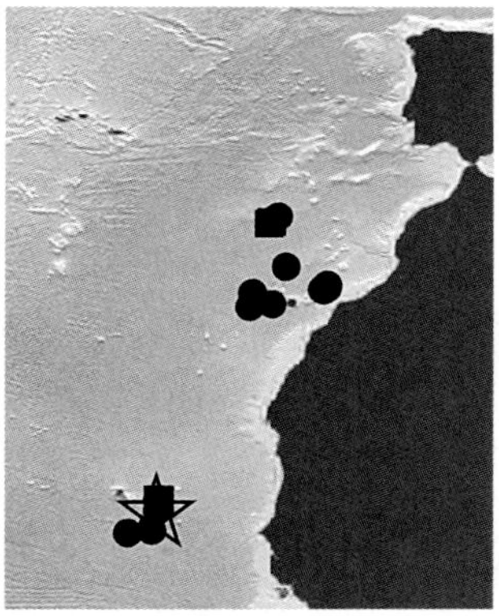

Figure 425. Distribution of *Sinezona semicostata*. ☆ type locality. ● specimen record. ■ literature record.

Sinezona singeri Geiger, 2006

Figures 426–429

Chresonymy.

Sinezona singeri Geiger, 2006a: 19–22, figs 14–16.

Sinezona singeri, Singer, 2006: 35, text-fig. [copy figure Geiger, 2006a].

Misidentification.

not *Scissurella rota*: Harasewych & Moretzsohn, 2010: 196, text-fig. [SEM is juvenile *Sin. singeri*].

Type material. Holotype (SBMNH 359439. Fig. 426B). Paratype from type locality (SBMNH 359440, 1. Fig. 426A). 170–225 m, Reunion Island, 21.083°S, 55.200°E (MNHN, 1. Fig. 427A). 1300–1480 m, Mayotte, E Bandele Reef, France, 12.933°S, 45.303°E (MNHN, 1. Fig. 427B). 200–500 m, Mayotte, E Bandele Reef, France, 12.902°S, 45.252°E (MNHN, 1. Fig. 427C).

Type locality. Intertidal, Blue Hole, Dahab, Janub Sina', Egypt, 28.483°N, 34.533°E. Leg. B. Singer Oct. 1990 (OD).

Etymology. Named for Benjamin (Solly) Singer, of Rehovot, Israel, for his continued contributions to Red Sea malacology and in recognition of his service to the malacological community at large, particularly that of Israel (OD).

Description. Shell medium-sized (1.6 mm. Figs 426, 427), trochiform, depressed. Protoconch of 0.75 whorl, smooth, apertural varix connected to embryonic cap, apertural margin deeply sinusoidal. Teleoconch I of approximately 1.125 whorls, 12–13 distinct axial cords, interstices filled with fine irregular axials, no spirals. Teleoconch II of 1.33 whorls, suture moderately impressed. Shoulder slightly convex, approximately 23–26 fine axial cords on first whorl, approximately half as strong as axial cords on teleoconch I, interstices with very fine growth lines; approximately a dozen somewhat irregularly spaced fine spirals forming barely noticeable points at intersection with axials. Base with strong constriction below selenizone, then inflated; 21–26 axials starting as thin thread on keels of selenizone, becoming strong elevated costae at

periphery, thinning out to threads towards transition to umbilicus; approximately 20 finer spirals, becoming stronger from selenizone towards umbilicus, running over axials. Umbilicus open, wide, bordered by spiral cord towards base, walls smooth. Selenizone above periphery, keels moderately elevated, moderately strong; strong lunules; foramen narrow tear-drop-shaped, extending approximately 0.16 whorl. Aperture rounded, roof strongly overhanging.

Figure 426. *Sinezona singeri.* **A.** Paratype. Intertidal, Blue Hole, Dahab, Janub Sina', Egypt, 28.483°N, 34.533°E. (SBMNH 359440, 1) **B.** Holotype. Intertidal, Blue Hole, Dahab, Janub Sina', Egypt, 28.483°N, 34.533°E (SBMNH 359439). **C.** 0–3 m, Baye de Ranobe, near Mora Mora, N of Tulear, Toliara Province, Madagascar, 23.083°S, 43.500°E (LACM 89-50, 1). Radula shown in Figure 428. Scale bars shells = 0.5 mm. Scale bars protoconch = 100 μm.

Radula (Fig. 428). Rachidian tooth triangular, seven cusps on tip, central cusp largest, parallel to radular ribbon. Lateral teeth 1–3 similar, three cusps on outer margin. Lateral tooth 4 reduced, hook-shaped, with three minute points at tip. Lateral tooth 5 broadened, enlarged, approximately five cusps along inner margin.

Figure 427. *Sinezona singeri*. Paratypes. **A.** 170–225 m, Reunion Island, 21.083°S, 55.200°E (MNHN, 1). **B.** 1300–1480 m, Mayotte, E Bandele Reef, France, 12.933°S, 45.303°E (MNHN, 1). **C.** 200–500 m, Mayotte, E Bandele Reef, France, 12.902°S, 45.252°E (MNHN, 1). Scale bars shell = 1 mm. Scale bars protoconch = 100 μm.

Radular interlock of central field moderate. Inner marginal teeth with palmate tip with 3–4 cusps on each side; outer marginal teeth spoon-shaped, with many fine cusps on tip; shafts without food groove.

Comparisons. *Sinezona doliolum* from South Africa has a taller shell, a much more narrow umbilicus, and a protoconch with axial cords.

Sinezona ferriezi from the Indo-Malayan Archipelago to the Central Pacific has about twice as many spiral lines on shoulder and base, weaker axial cords, and weaker spiral cords in the vicinity of the umbilicus.

Sinezona globosa from the Indo-Malayan Archipelago to the tropical Eastern Pacific (and possibly Caribbean) has a more globular shell, a longer teleoconch I (1.5 vs. 1 whorls), and a protoconch with spiral lines.

Sinezona insignis from South Africa is much taller, has a protoconch with strong axial sculpture, and a much narrower umbilicus.

Sinezona macleani from the Indian Ocean to the Western Pacific has a much wider shell, a straight parietal wall, and a shelf on the adumbilical lower portion of the peristome.

Sinezona marrowi from the Indo-Malayan Archipelago has a protoconch with strong axial cords, a somewhat weaker constriction of the base below the selenizone, and half as many spiral lines on shoulder and base.

Sinezona plicata from the Indian Ocean to the Central Pacific has strong axial folds, a wider umbilicus, and a much stronger constriction of the base below the selenizone.

Sukashitrochus dorbignyi from the Indian Ocean shows distinct spiral keels on the base.

Scissurella rota from the Indian Ocean has an open slit, fewer spirals on the base that form pronounced hollow points at the intersection with the axials, and a protoconch with fine axial sculpture.

Scissurella reticulata and *Sci. sudanica* from the Indian Ocean have an open slit, a less pronounced constriction below the selenizone, and a protoconch with fine axial sculpture.

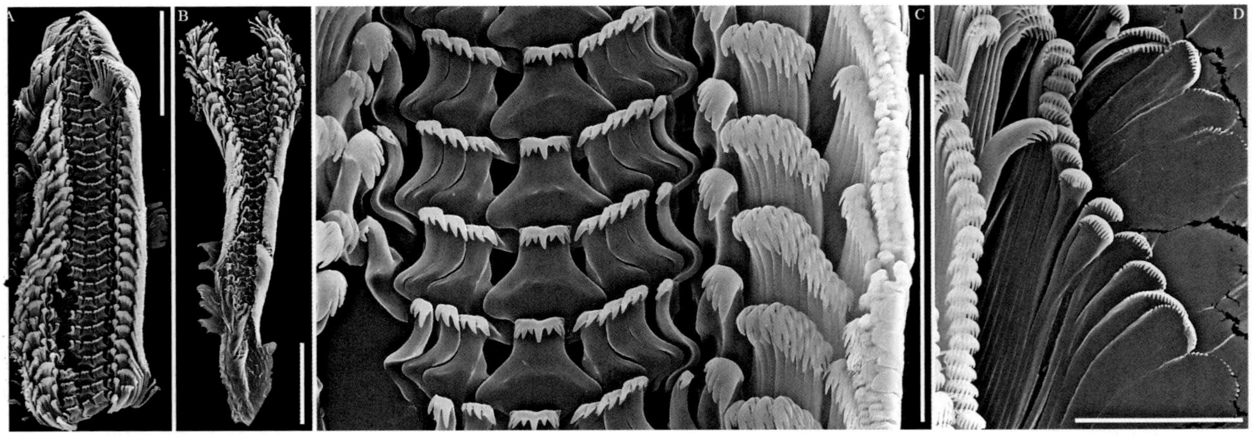

Figure 428. *Sinezona singeri*. Radula. 0–3 m, Baye de Ranobe, near Mora Mora, N of Tulear, Toliara Province, 23.083°S, 43.500°E (LACM 89-50, 1). Shell shown in Figure 426C. **A.** Anterior half of radular ribbon. Scale bar = 100 μm. **B.** Posterior half of radula ribbon. Scale bar = 100 μm. **C.** Full width of radula. Scale bar = 50 μm. **D.** Outer marginal teeth. Scale bar = 10 μm.

Distribution. Red Sea to Mascarene Islands, western Indian Ocean (Fig. 429).

Specimen records.

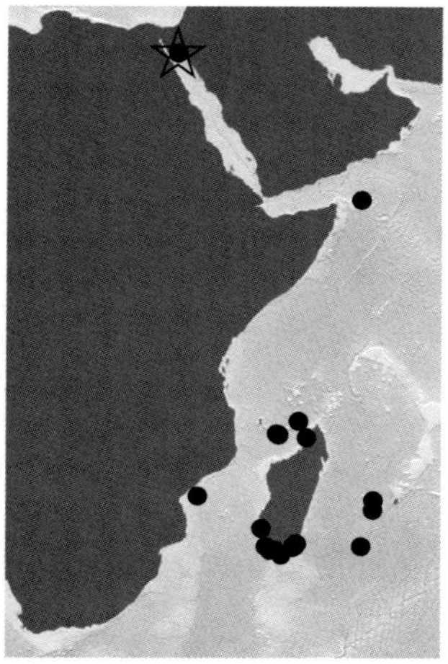

Figure 429. Distribution of *Sinezona singeri*. ☆ type locality. ● specimen record.

Madagascar. 0–3 m, Baye de Ranobe, near Mora Mora, N of Tulear, Toliara Province, 23.083°S, 43.500°E (LACM 89-50, 1). 1–3 m, at Andilana, NW side Nosy Bé, Antseranana Province, 13.217°S, 48.233°E (LACM 89-53, 3). 4–6 m, St. Luce, SW Soiillac Islet, ATIMO VATAE TA34, 24.753°S, 47.200°E (MNHN, 1). 450–455 m, SE Barrow Point, ATIMO VATAE CP3592, 25.033°S, 43.966°E (MNHN, 4). 11–12 m, in front of Monseigneur Bay, ATIMO VATAE TB04, 25.035°S, 47.001°E (MNHN, 1). 4–5 m, Libanona Beach, ATIMO VATAE TB07, 25.041°S, 46.995°E (MNHN, 3). 5–6 m, Balions Bay, ATIMO VATAE TB11, 25.153°S, 46.756°E (MNHN, 1). 203–210 m, SE Barrow Point, ATIMO VATAE CP3584, 25.466°S, 44.416°E (MNHN, 1). 88–90 m, S of False Cape, ATIMO VATAE CP3580, 25.976°S, 45.581°E (MNHN, 1). 10 m, Lavanono, 25.429°S, 44.939°E (DLG 1750, 4).

Mozambique. 352–357 m, Zambeze transect, Mozambique channel, MAINBAZA CP3151, 19.566°S, 36.750°E (MNHN, 2).

France, Mayotte. (KBIN, 10). E Bandele reef, 12.902°S, 45.252°E (MNHN, 4). 520–830 m, E Bandele reef, 12.897°S3 45.270°E (MNHN, 1). 1300–1480 m, E Bandele reef, 12.933°S, 45.300°E (MNHN, 1). 500 m, E Bandele reef, 12.895°S, 45.268°E (MNHN, 3). 1300–1480 m, E Bandele reef, 12.933°S, 45.303°E (MNHN, 1). 330–530 m, Glorieuses, N Ile du Lys, 11.440°S, 47.372°E (MNHN, 1).

Reunion Island. 170–225 m, 21.083°S, 55.200°E (MNHN, 4; MNHN, 2). 20 m, Cap La Houssaye, Saint Paul, 20.017°S, 55.234°E (MJC ex 1881, 1).

Yemen. Bay of Africa, Soccoro [= Socotra?], 12.500°N, 53.800°E (KBIN, 5).

Remarks. The MNHN paratypes were inadvertently omitted from the listing of type material in the original description, although they were identified as such in the figure caption. They should be considered as paratypes.

Sinezona subantarctica (Hedley, 1916)

Figures 430–435

Chresonymy.

Schismope subantarctica Hedley, 1916a: 36, pl. 5, figs 54–55.

Schismope subantarctica: Finlay, 1927: 341.

Sinezona subantarctica: Powell, 1955: 46.

Sinezona subantarctica: Powell, 1960: 126.

Sinezona cf. *antarctica* [*sic*]: Dell, 1960: 145, 156.

Sinezona subantarctica: Powell, 1979: 35.

Sinezona ? *subantarctica*: Lozouet, 1986: 111.

? *Sinezona* sp.: Branch *et al.*, 1991: 47, 58 (text-fig), 62.

Sinezona subantarctica: Spencer & Willan, 1995: 13.

Ariella ? *subantarctica*: Geiger, 2003: 75, fig. 13g [umbilical callus].

Ariella subantarctica: Zelaya & Geiger, 2007, 432–435, figs 179 [map], 180–181 [copy figures Hedley (1916a), 182–193 [shell SEM], 194–194 [radula].

Figure 430. Original illustration of *Schismope subantarctica*.

Misidentification.

Scissurella obliqua: Cantera and Arnaud, 1985: 31 [is *Sin. subantarctica*].

Type material. Holotype (AMS: Powell, 1955), not seen in 2001 (Geiger, pers. obs.), 1.15 × 0.75 mm.

Type locality. Aerial Cove, Macquarie Island [54.483°S, 158.950°E] (OD).

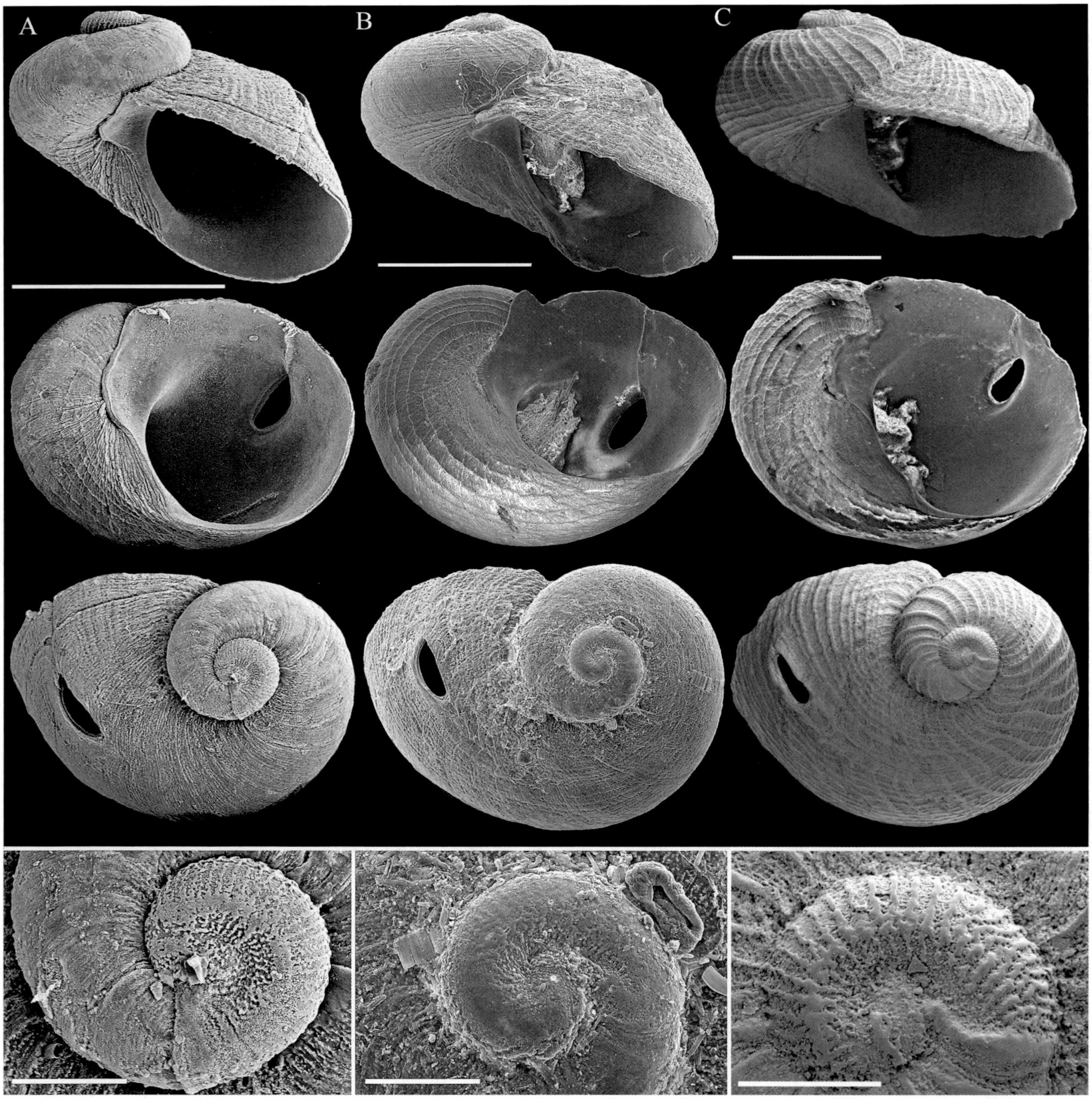

Figure 431. *Sinezona subantarctica*. **A.** Handspike Point, Macquarie Island, Australia, 54.500°S, 158.883°E (AMS C.376612, 18). **B.** 5 m, Aerial Cove entrance, Macquarie Island, Australia, 54.490°S, 158.939°E (NMNZ M152863, 2). **C.** 1 m, San Andres Bay, N Gulf de Peñas, Chile, 46.588°S, 75.510°W (LACM 73-73, 11). Scale bars shell = 0.5 mm. Scale bars protoconch = 100 μm.

Etymology. Referring to its type locality on a subantarctic island.

Description. Shell medium-sized (to 1.64 mm. Figs 431, 432), trochiform, depressed. Protoconch of 1.1–1.5 whorls, many dense, strong axials, widened apically forming backbone, some axials occasionally bifurcated, apertural varix not connected to embryonic cap, apertural margin straight. Teleoconch I of 1.16–1.625 whorls, first third whorl with approximately six strong axial cords only, interstices with fine irregular growth lines, subsequently regularly spaced spirals increasing in strength from lines at inception to broad low cords at apertural margin, giving shell reticulate appearance. Teleoconch II

Figure 432. *Sinezona subantarctica*. Notice open umbilicus or closed with callus. **A–C.** Barbecue Beach, Sea Lion Island, Falkland Islands, 51.420°S, 59.067°W (NMSZ 992167.60103, 25). Scale bars shell A–B = 1 mm. Scale bar shell C= 0.5 mm. Scale bars protoconch = 100 μm.

of 0.125–0.17 whorl, suture little impressed. Shoulder slightly convex, axial cords becoming broader, lower, less distinct, spiral cords increasing in strength with growth. Base with 8–9 spiral cords of decreasing strength vanishing one third below periphery, approximately 13 spiral cords increasing in strength from low cords at periphery to stepped ledges near umbilicus. Umbilicus open (Fig. 432B, C) or closed by callus (Figs 431, 432A) at maturity, open in juveniles, visible part of cavity bordered by indistinct cord, walls slightly concave, with irregular growth lines. Selenizone short or absent, only as long as broad teardrop-shaped foramen, keels very low, strong,

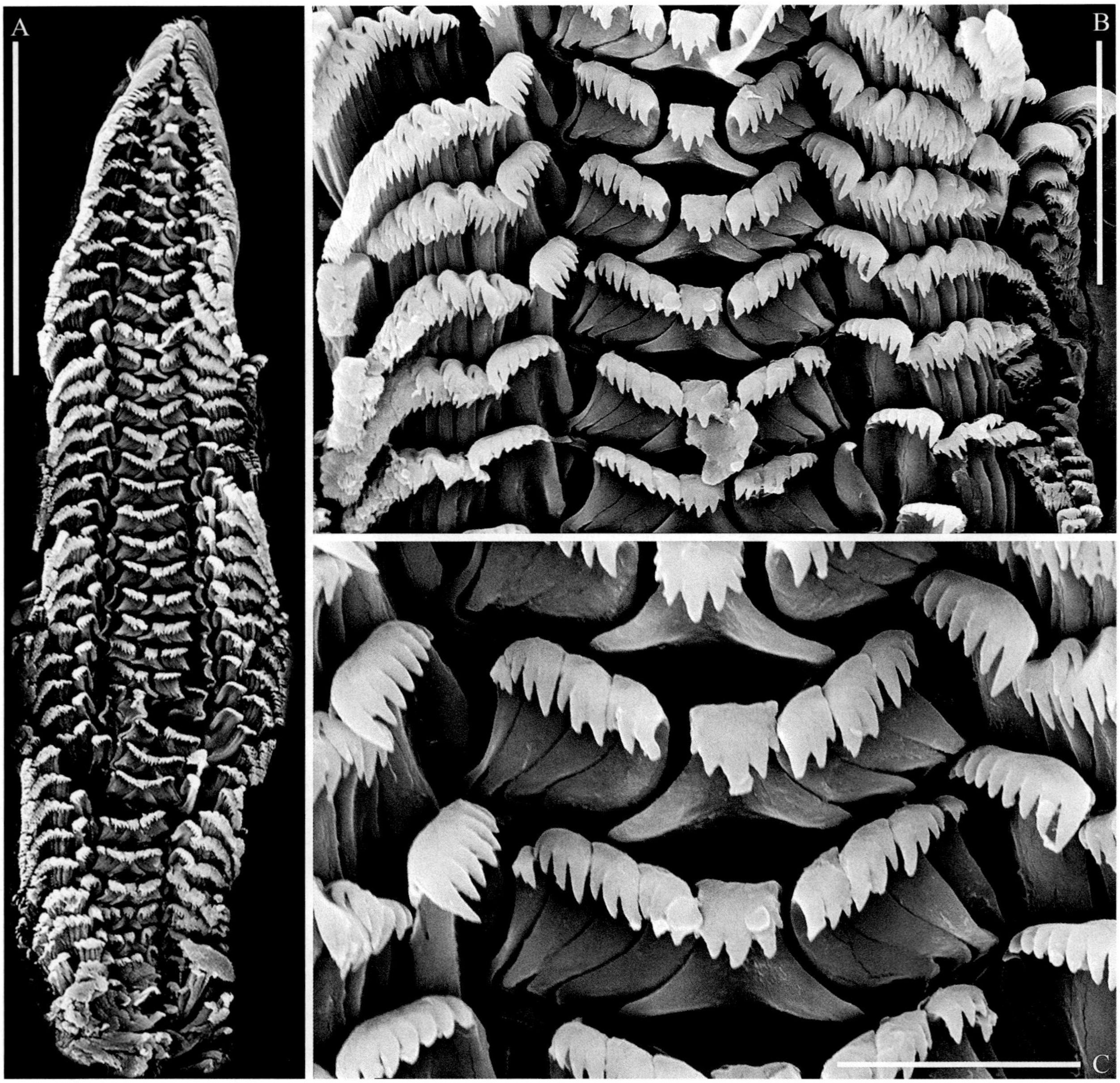

Figure 433. *Sinezona subantarctica*. Radula. From topotype. 5 m, Aerial Cove, Macquarie Island, Australia, 54.483°S, 158.950°E (AMS C.376621). **A.** Entire radula. Scale bar = 100 µm. **B.** Full width of radular ribbon. **C.** Central field enlarged. Scale bars B–C = 10 µm.

foramen anteriorly continued by raphe. Aperture oblique, oblong D-shaped, roof overhanging, adumbilical peristome extended to form umbilical callus.

Operculum very thin, multispiral, with central nucleus, covering approximately 60% of aperture width.

Radula (Figs 433, 434). Rachidian tooth with broadly triangular base, tip nearly straight at lateral edges, V-shaped distally, with one main rounded cusp and 2–3 smaller cusps on each side, reducing size towards base. Lateral teeth 1–3 similar, with three strong, sharp cusps on outer edge. Lateral tooth 4 reduced, with one small point on either side of main cusp. Lateral tooth 5 larger, wider than other lateral teeth, tip smooth on outer edge, with seven cusps on inner edge. Radular interlock of central field moderate. Inner marginal teeth with one main central cusp, two cusps on inner edge of tip, and three cusps on outer edge; shafts without food groove. Outer marginal teeth spoon-shaped.

Comparisons. *Sinezona bandeli* from New Zealand has a much longer selenizone (0.66 vs. 0.125 whorl), a more depressed spire, and a more oblique-oblong aperture.

Sinezona brevis from New Zealand has much stronger axials and hardly perceptible spirals on the shoulder and fine spiral threads on the base.

Sinezona iota from New Zealand lacks a callus over the umbilicus, has a much taller shell with protruding protoconch, and a distinct selenizone.

Sinezona laqueus from New Zealand has a pronounced constriction on the base below the selenizone, strong lunules on the selenizone, and lacks the callus covering the umbilicus.

Sinezona levigata from New Zealand generally has stronger axial sculpture, the protoconch bears half as many strong axial cords, and usually has a much stronger apertural varix.

Sinezona mechanica from New Zealand has many distinct axial cords, lacks spiral lines, and the umbilicus is not covered by a callus.

Sinezona pacifica from temperate Australia and New Zealand has a much longer selenizone (0.66 vs. 0.125 whorl), a more depressed spire, and a more oblique-oblong aperture.

Sinezona pauperata from New Zealand has a round foramen, lacks a callus covering the umbilicus, and is overall more globular.

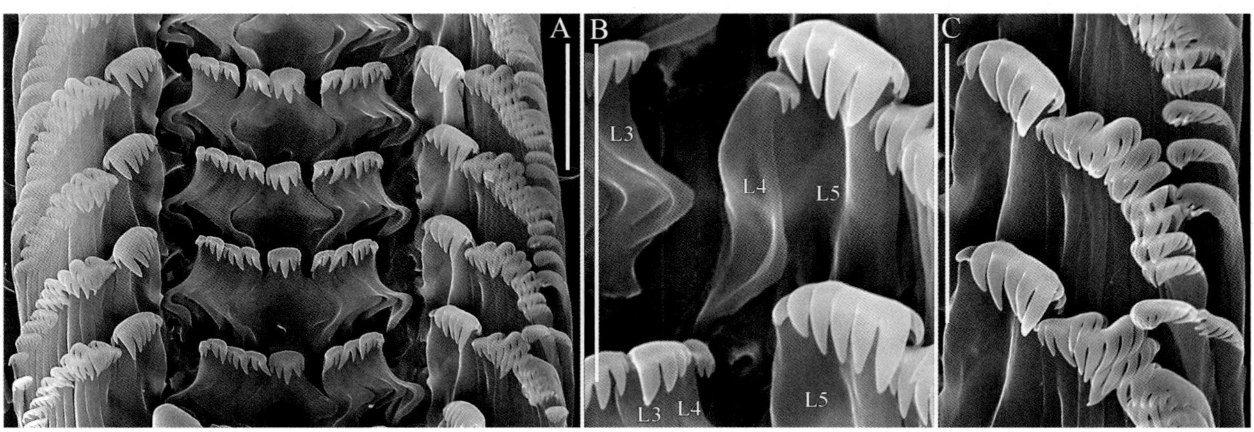

Figure 434. *Sinezona subantarctica*. Radula. Pebble Island, Falkland Islands, 51.314°N, 59.517°W (NMSZ 2009109.02-13, 11, 2: complete). **A.** Full width of radula. **B.** Lateral tooth 4. L3–5: Lateral tooth 3–5. **C.** Marginal teeth. Scale bars = 10 μm.

Sinezona platyspira from New Zealand has a flat spire, lacks a callus covering the umbilicus, and has a few strongly raised axial lamellae on the teleoconch.

Sinezona mouchezi from the subantarctic islands of the Indian Ocean has a long selenizone, a higher spire, more distinct axial cords on the teleoconch, and fewer straight axials on the protoconch.

Distribution. Subantarctic Islands. 0–15 m (living specimens 1–5 m. Fig. 435).

Specimen records. ***Crozet Islands.*** S Marin Bay (sud Baied eu Marin), 46.430°S, 52.200°E (MNHN, 2).

Argentina. 0 m, W side B. Buen Suceso, SE Tierra del Fuego, 54.796°S, 65.267°W (LACM 71-270, 1).

Falkland Islands. Pebble Island, 51.314°S, 59.517°W (NMSZ 2009109.02–13, 11, 2: complete; NMSZ 2009109.01, 1). Barbecue Beach, Sea Lion Island, 51.420°S, 59.067°W (NMSZ 992167.60102, 1; NMSZ 992167.60103, 25).

Macquarie Islands, Australia. 5 m, Aerial Cove entrance, 54.490°S, 158.939°E (NMNZ M152863, 2). Aerial Cove, 54.483°S, 158.950°E (AMS C.404556, 20: complete). Aerial Cove, near cove entrance, 54.483°S, 158.950°E (AMS C.376610, 15, –: complete). Aerial Cove, 54.483°S, 158.950°E (AMS C.376611, 15, –: complete). 5 m, Aerial Cove, 54.483°S, 158.950°E (AMS C.376621, 6). Handspike Point, 54.500°S, 158.883°E (AMS C.376612, 18). 3 m, Garden Bay, 54.500°S, 158.950°E (AMS C.376606, 1). 1 m, Garden Bay, 54.500°S, 158.950°E (AMS C.376617, 7, –: complete). Green Gorge, 500 m S of hut, 54.633°S, 158.917°E (AMS C.376607, 2).

Chile. 1 m, San Andres Bay, N Gulf de Peñas, 46.588°S, 75.510°W (LACM 73-73, 11). 15 m, Tom Bay, 50.188°S, 74.798°W (LACM 73-72, 2).

Great Britain. Gough Island, 40.317°S, 9.939°W (DLG 1442, 3).

New Zealand. Auckland Island, 50.716°S, 166.133°E (WAM, 3).

Literature records.

New Zealand. 91 m, off Snares, 48.020°S, 166.610°E. 174 m, N off Auckland Island, 50.360°S, 166.190°E. 0 m, Faith Harbour, Auckland Island, 50.522°S, 166.141°E. Rindove Bay, Antipodes Islands, 49.704°S, 178.784°E. 27 m, Leeward Island, Antipodes Islands, 49.688°S, 178.811°E. 43 m, Galathea Station 595, Campbell Island, 52.540°S, 169.180°E (Powell, 1955).

Remarks. Dell (1960: 145) misspelled the name as "*antarctica* (Hedley)." As no characters are provided in his simple listing, it cannot be construed as an introduction of a new name.

Based on the samples from the Falkland Islands, two distinct shell morphological varieties can be identified, half have an open umbilicus, half have the umbilicus closed by a distinct callus. This 50/50% split suggests sexual dimorphism as it is known from Liotiidae, in which the open umbilicus serves as a brood pouch, similarly to what is encountered in the Vetigastropoda: Larocheidae: *Larochea* (see Marshall, 1993). No specimens of *Sin. subantarctica* with early larvae in the umbilicus were encountered, though the limited available material was unsuitable for such an

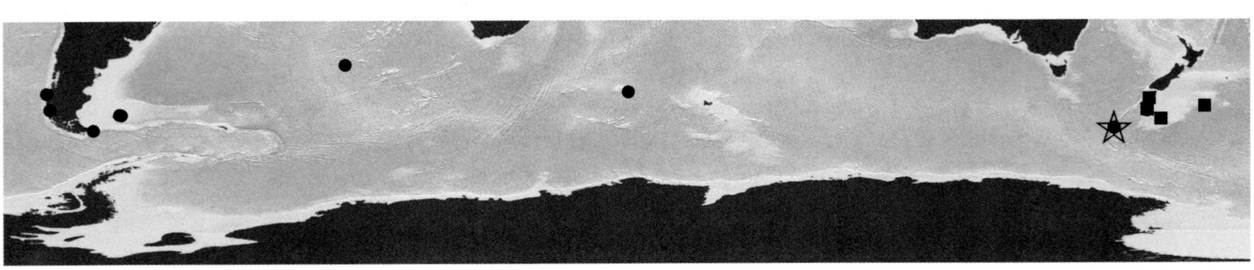

Figure 435. Distribution of *Sinezona subantarctica*. ☆ type locality. ● specimen record. ■ literature record.

assessment. The live collected material consisted mostly of very juvenile specimens without any indication of a slit, and the larger specimens are all shell only.

Many shells showed strong delamination of the outer ostracum, the layer bearing most of the finer sculpture. In the ethanol preserved material, this delamination was most prominent and most likely due to liquid storage in an aqueous solution, which adversely affects shells of micromolluscs (see Geiger *et al.*, 2007 for discussion). In the dry lots, some abrasion in the natural environment can account for the condition. The interior surface of the best shells still looked shiny, hence, the shells do not appear to have remained in the ocean's water for long periods of time. Accordingly, specimens of this species may appear quite smooth, even if freshly collected. The overall broad shell shape, the spiral steps on the base, which seems to be most resistant to the delamination process, and the umbilical callus in 50% of the specimens, will still permit easy identification of this species.

The cusps on the rachidian tooth are proportionally wider and lateral tooth 4 is comparatively better developed in *Sin. subantarctica* than in most other *Sinezona* species.

Sinezona wanganellica Geiger & Marshall, 2012

Figures 436–438

Chresonymy.

Sinezona wanganellica Geiger & Marshall, 2012: 11–13, figs 8–9.

Type material. Holotype (NMNZ M.303304. Fig. 436). Seven paratypes from type locality (NMNZ M.178433). 120–126 m, Wanganella Bank summit, West Norfolk Ridge, New Zealand, 32.606°S, 167.584°E (NMNZ M.171826, 14. Fig. 437B–C). 113 m, Wanganella Bank, Norfolk Ridge, summit, New Zealand, 32.573°S, 167.517°E (NMNZ M.174832, 14. Fig. 437A).

Type locality. 133 m, Wanganella Bank, Norfolk Ridge, summit, New Zealand, 32.543°S, 167.487°E (OD).

Etymology. Named for the type locality; adjective (OD).

Description. Shell small (to 0.7 mm. Figs 436, 437), trochiform, depressed. Protoconch of one whorl, 19–21 sharp axials from midline to outer suture, embryonic cap with some irregular flocculations, no apertural varix, apertural margin slightly convex. Teleoconch I of 0.875–0.9 whorl with 12–22 sharp to elevated axial cords with onset close to suture on early teleoconch I shifting to past position of selenizone with onset of selenizone; interstices with very fine growth marks, no spiral sculpture. Teleoconch II of 0.66 whorl, suture impressed. Shoulder overall flat, concave towards selenizone; mainly very fine growth marks, 0–12 axial cords in adsutural quarter, no spiral sculpture. Base with distinct constriction below selenizone, same density and strength of axial cords as on teleoconch I; some axial cords not reaching umbilicus. Umbilicus formed by rounded portion of previous whorls, underside of protoconch visible. Selenizone above periphery; keels moderately strong, quite elevated; lunules very strong; slit closed anteriorly to foramen. Aperture rounded, D-shaped, roof overhanging.

Animal unknown.

Distribution. Northernmost New Zealand (Fig. 438).

Specimen records.

New Zealand, Norfolk Ridge. 133 m, Wanganella Bank, 32.543°S, 167.487°E(NMNZ M.303304, 1: holotype; NMNZ M.174833, 7: paratypes). 113 m, Wanganella Bank, 32.573°S, 167.517°E (NMNZ M.174832, 14: paratypes). 120–126 m, Wanganella Bank, 32.606°S, 167.584°E (NMNZ M.171826, 14, paratypes). 133 m, Wanganella Bank, 32.653°S, 167.528°E (NMNZ M.174834, 1). 133 m, Wanganella Bank, 32.666°S, 167.560°E (NMNZ M.174841, 5). 785–800 m, W of Cape Reinga, 34.284°S, 168.430°E (NMNZ M.172128, 31).

Comparisons. *Sinezona bandeli* Marshall, 2002 from New Zealand has an overall taller shell, spiral threads on the shoulder and base of the teleoconch, and a narrower umbilicus occluding the underside of the protoconch.

Sinezona mechanica from New Zealand lacks a sunken apex, has a more rounded whorl profile, and at least 50% more axial cords.

Sinezona platyspira from New Zealand has a longer teleoconch I (1.125 vs. 0.875 whorls) and at least 50% more axials on the protoconch.

Figure 436. *Sinezona wanganellica*. Holotype. Wanganella Bank, Norfolk Ridge, summit, New Zealand, 133 m (NMNZ M.303304). Scale bar shell = 0.5 mm. Scale bar protoconch = 100 μm.

Figure 437. *Sinezona wanganellica*. Paratypes. **A.** Wanganella Bank, Norfolk Ridge, summit, New Zealand, 113 m (NMNZ M.174832). **B–C.** Wanganella Bank summit, West Norfolk Ridge, W of Cape Maria van Diemen, New Zealand, 120–126 m (NMNZ M.171826). Scale bars shell = 0.5 mm. Scale bars protoconch = 100 µm.

Scissurella regalis from New Zealand has a protoconch with at least 50% more axials, which bear strong median thickenings, a longer teleoconch (1.125 vs. 0.875 whorls), and retains an open slit.

Figure 438. Distribution of *Sinezona wanganellica*. ☆ type locality. ● specimen record.

Sinezona wileyi Geiger, 2008

Figures 439–442

Chresonymy.

Sinezona sp.: Geiger *et al.*, 2007: fig. 1B.

Sinezona wileyi Geiger 2008c: 189–193, figs 10–16.

Sinezona wileyi: Tröndlé & Boutet, 2009: 12.

Type material. Holotype (SBMNH 83541. Fig. 439). Paratypes. 0–5 m, Lumu Reef, Kimbe Bay, New Britain, Papua New Guinea, 5.283°S, 150.131°E (DLG 639, 19. Fig. 441B). 36 m, around Rukuaga Bay, Rapa, Austral Islands, 27.568°S, 144.368°W (MNHN, 20. Fig. 440A, C). Laisse de plage extérieure, Mururoa, Tuamotu, French Polynesia, 21.842°S, 138.895°W (JTC, 2. Fig. 441B). 20 m, Herald Pass, W of Ndravuni Island, NW end of Great Astrolabe Reef, Fiji, 18.767°S, 178.467°E (LACM 85-135, 2.

Figure 439. *Sinezona wileyi*. Holotype. 0–5 m, Lumu Reef, Kimbe Bay, New Britain, Papua New Guinea, 5.283°S, 150.131°E (SBMNH 83541). Scale bar shell = 0.5 mm. Scale bar protoconch = 100 μm.

Fig. 441A). 20 m, Herald Pass, W of Ndravuni Island, NW end of Great Astrolabe Reef, Fiji, 18.767°S, 178.467°E (LACM 85-135, 2). 10–20 m, off Francis Island in Beqa Lagoon, just off of Beqa Island (south of Viti Levu), 18.300°S, 178.067°E (DLG 279, 10. Fig. 441C).

Type locality. 0–5 m, Lumu Reef, Kimbe Bay, New Britain, Papua New Guinea, 5.283°S, 150.131°E (OD).

Figure 440. *Sinezona wileyi*. Paratypes. **A, C.** 36 m, around Rukuaga Bay, Rapa, Austral Islands, French Polynesia, 27.568°S, 144.368°W (MNHN, 20). **B.** Laisse de plage extérieure, Mururoa, Tuamotu, French Polynesia, 21.842°S, 138.895°W (JTC, 2). Scale bars shell = 0.5 mm. Scale bars protoconch = 100 μm.

Etymology. The name honors the collector of the first known specimens of the species Tony Wiley of Riverside, California, USA (OD).

Description. Shell medium-sized (to 1.08 mm. Figs 439– 441) trochiform turreted. Protoconch slightly sunken, of one whorl, fine irregular axials, no apertural varix, apertural margin straight. Teleoconch I of 0.6–0.7 whorl, suture below periphery, little impressed,

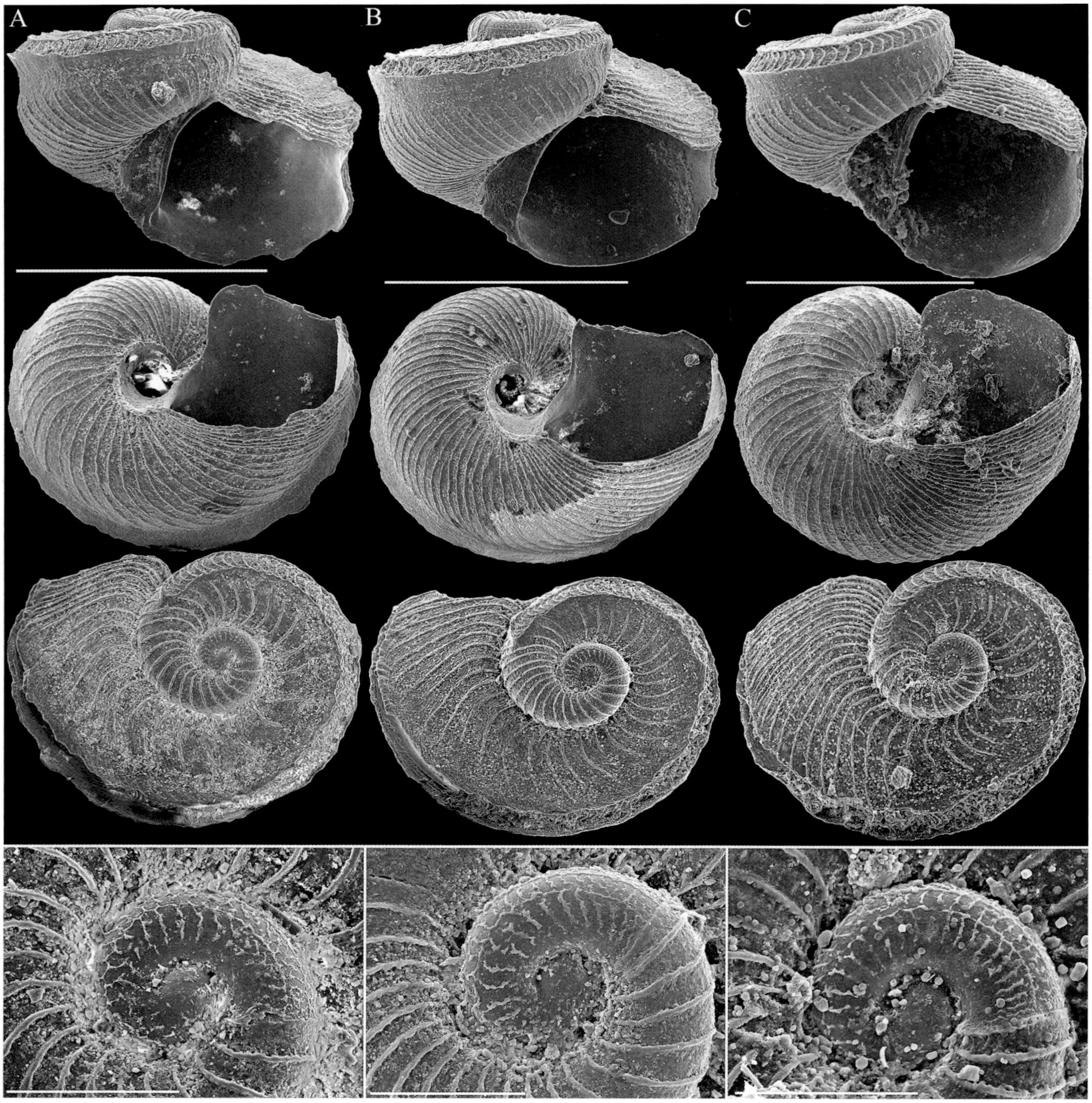

Figure 441. *Sinezona wileyi*. Paratypes. **A.** 20 m, Herald Pass, W of Ndravuni Island, NW end of Great Astrolabe Reef, Fiji, 18.767°S, 178.467°E (LACM 85-135, 2). **B.** 0–5 m, Lumu Reef, Kimbe Bay, New Britain, Papua New Guinea, 5.283°S, 150.131°E (DLG 639, 19). **C.** 10–20 m, off Francis Island in Beqa Lagoon, just off of Beqa Island (south of Viti Levu), Fiji, 18.300°S, 178.067°E (DLG 279, 10). Scale bars shell = 0.5 mm. Scale bars protoconch = 100 μm.

usually with approximately 14–22 faint to distinct fine axial cords, usually no spirals (occasionally faintest spiral threads only visible by SEM), interstices with irregular thickenings. Teleoconch II of up to 1.25 whorls, suture below periphery. Shoulder with approximately 32–44 barely perceptible to distinct fine comarginal axial cords, stronger near suture becoming less distinct towards selenizone; approximately eight barely perceptible to very indistinct spiral cords, concentrated in middle of shoulder; occasional specimens with predominant spiral sculpture. Base distinctly constricted below selenizone, axial cords of same strength and density as on shoulder, starting in lower portion of constriction below selenizone; approximately 17 fine, irregularly spaced, spiral cords, starting in lower portion of constriction with onset of axials. Umbilicus wide, walls straight, smooth, periphery at distinct angle to base, bordered by carina. Aperture subquadrate, D-shaped, roof overhanging. Selenizone above periphery, keels moderately strong, quite elevated, lunules distinct at regular interval; foramen narrow, elongated, margins converging and touching, without raphe.

Animal unknown.

Comparisons. *Sinezona costulata* from southern Japan has a wider shell, a wider umbilicus, and fewer but stronger axial cords.

Sinezona danieldreieri from Indo-Malayan Archipelago and the Central Pacific has a more globular shell, the axial cords on the base reach the selenizone, and the protoconch with strong axial cords has an apertural varix.

Sinezona ferriezi from the Indo-Malayan Archipelago and the Western Pacific is about twice the size, has a smooth protoconch with distinct apertural varix, a teleoconch I of more than whorl, is overall more rounded, and has a distinct raphe anterior to the closed foramen.

Sinezona globosa from the Indo-Malayan Archipelago to the tropical Eastern Pacific (and possibly Caribbean) has a protoconch with spiral lines, has a longer teleoconch I (1.5 vs. 0.6 whorls) and the base gradually slopes into the umbilicus.

Sinezona marrowi from the Indo-Malayan Archipelago has a wider shell, strong axial cords on the protoconch, and a wider umbilicus.

Sinezona milleri from southern Japan has a wider shell, a wider umbilicus, and a protoconch with strong axial cords.

Sinezona plicata from the Indian Ocean to the Central Pacific has a much wider shell, strong axial folds, and much stronger spiral lines.

Distribution. Tropical Indo-Malayan Archipelago to Central Pacific; shallow shelf (Fig. 442).

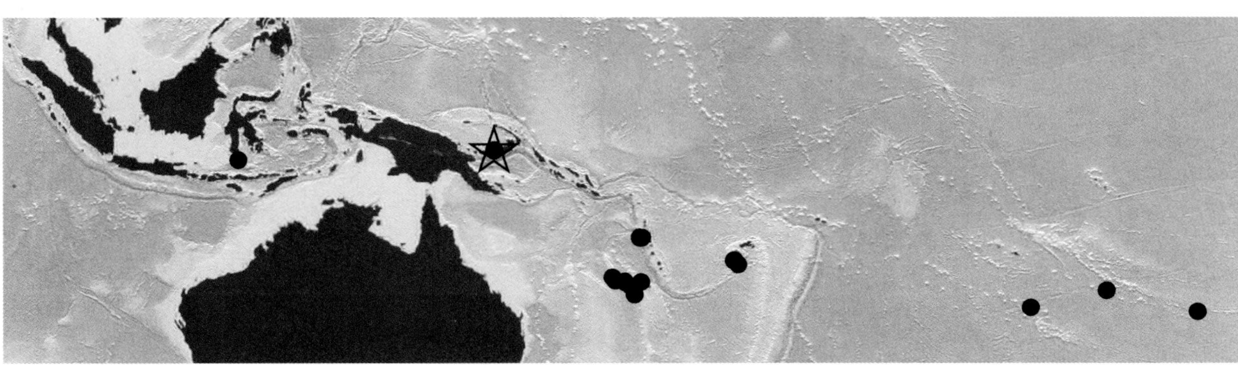

Figure 442. Distribution of *Sinezona wileyi*. ☆ type locality. ● specimen record.

Specimen records.

Indonesia. 175 m, off SW Salayer, 6.373°S, 120.437°E (RMNH 42316, 1).

Papua New Guinea. 0–5 m, Lumu Reef, Kimbe Bay, New Britain, 5.283°S, 150.131°E (SBMNH 83541: holotype; DLG 639, 19: paratypes). 70 m, off S Ema Reef, Kimbe Bay, New Britain, 5.202°S, 150.152°E (DLG 768, 3).

New Caledonia. 25–30 m, Santal Bay, Lifou, 20.822°S, 167.173°E (MNHN, 1). 55–57 m, Koumac sector, 20.672°S, 164.195°E (MNHN, 1). 5–25 m, Touho sector, 20.878°S, 165.325°E (MNHN, 1). 8 m, Touho sector, 20.742°S, 165.265°E (MNHN, 1). 245–268 m, North, BATHUS 4 DW894, 20.266°S, 163.866°E (MNHN, 1). 0 m, Maître Islet Plateau, Noumea, St. 1351, 22.340°S, 166.428°E (MNHN, 7).

Vanuatu. 22–25 m, NW Urélapa Island, SANTO 2006 DB20, 15.508°S, 167.023°E (MNHN, 1). 8–20 m, Tutuba Island, SANTO 2006 FB83, 15.543°S, 167.290°E (MNHN, 2). 15–25 m, NE Tutuba Island, SANTO 2006 DB01, 15.551°S, 167.296°E (MNHN, 6). 32–40 m, S Tutuba Island, SANTO 2006 DB16, 15.591°S, 167.263°E (MNHN, 2). 100 m, off N Urélapa Island, SANTO 2006 EP32, 15.610°S, 167.033°E (MNHN, 1).

Fiji. 10–20 m, off Francis Island in Beqa Lagoon, just off of Beqa Island (S of Viti Levu), 18.300°S, 178.067°E (DLG 279, 10). 20 m, Herald Pass, W of Ndravuni Island, NW end of Great Astrolabe Reef, 18.767°S, 178.467°E (LACM 85-135, 2: paratypes).

French Polynesia. 36 m, around Rukuaga Bay, Rapa, Austral Islands, 27.568°S, 144.368°W (MNHN, 20: paratypes). 2–4 m, N of Aturapa Island, Rapa, Austral Islands, 27.572°S, 144.350°W (MNHN, 1). 5 m, Anarua Bay, Rapa, Austral Islands, 27.605°S, 144.378°W (MNHN, 4). 45 m, Haurei Bay, Rapa, Austral Islands, 27.613°S, 144.305°W (MNHN, 1). Raivavae, Austral Islands, 23.860°S, 147.674°W (JLC, 1). Laisse de plage extérieure, Mururoa, Tuamotu, 21.842°S, 138.895°W (JTC, 2: paratypes).

Henderson Island. 12 m, East Beach, 24.352°S, 128.305°W (UMZC, 4).

Remarks. The organization of the foramen seems intermediary between the typical open slit in *Scissurella*, and the closed foramen in *Sinezona*. Although the margins of the slit are converging and are fused anteriorly, a raphe is not evident. The fact that there is a marked downward deflection of the last quarter whorl, and that the roof of the peristome attaches below the periphery of the previous whorl demonstrates that those specimen are fully grown.

The species seems to have variable strengths of sculpture. Most specimens have predominant axial sculpture (Figs 440B, 441A, C), while in occasional specimens (Fig. 440A) the axial sculpture is barely perceptible and the spiral structures, particularly on the adumbilical base, becomes most prominent. As there are intermediate specimens (Fig. 440C) with distinct adumbilical spirals and relatively weak axials, combined with identical condition of the protoconch and very small size, these sculptural difference are considered to fall within intraspecific variability.

Sinezona zimmeri Geiger, 2003

Figures 443–445

Chresonymy.

Sinezona zimmeri Geiger, 2003: 33–37, 79, figs 5–9.
Sinezona zimmeri: Geiger, 2004: text-fig. p. 6.
Sinezona zimmeri: Geiger, 2006c: text-fig. 4.

Type material. Holotype (LACM 2912). Paratypes from type locality (LACM 2913, 3; AMS C.404165, 2; AMS C.406377, 1; AMS C.205033, 1; ANSP 410308, 2; UMUT RM28239, 1; NHMUK 20020266, 1). Paratype, Yuko Haoa Avaka, Akahanga, Easter Island, Chile, 27.148°S, 109.337°W (AMNH 301625, 1).

Type locality. 10–20 m, off Rosalia Point, E of Anakena, Easter Island, Chile, in sand and rubble at base of cliff, Dec. 2000. Leg. B. Raines [27.067°S, 109.333°W] (OD).

Figure 443. *Sinezona zimmeri*. Holotype. 10–20 m, off Rosalia Point, E of Anakena, Easter Island, Chile [27.067°S, 109.333°W] (LACM 2912). Scale bar shell = 0.5 mm. Scale bar protoconch = 100 µm.

Etymology. The species honors my advisor, mentor, and friend Russel L. Zimmer, University of Southern California, Los Angeles, USA, in appreciation of his generous support as well as for his significant contributions to invertebrate zoology (OD).

Description. Shell small (holotype: 0.68 mm, to 0.83 mm. Figs 443, 444), globular (width/height 1.12 in holotype, mean = 1.38, SD = 0.38). Protoconch 150 µm (holotype): mean = 152 µm, SD = 5.0 µm, one whorl; embryonic cap smooth or irregularly

Figure 444. *Sinezona zimmeri*. **A.** Paratype. 10–20 m, off Rosalia Point, E of Anakena, Easter Island, Chile [27.067°S, 109.333°W] (LACM 2913). **B.** 30–50 m, Tahai, Easter Island, Chile, 27.140°S, 109.429°W (DLG 1325, 100). **C.** Paratype. 10–20 m, off Rosalia Point, east of Anakena, Easter Island, Chile, 10–20 m [27.067°S, 109.333°W] (LACM 2913), juvenile. Scale bars shell = 0.5 mm. Scale bars protoconch = 100 µm.

sculptured, remainder with strong, broad axials from just inside of highest point of whorl towards periphery, otherwise smooth; axials approximately 12, may bifurcate towards periphery. Subterminal varix strong, connecting to embryonic cap only in some specimens. Aperture usually round, in some specimens weakly sinusoidal. Teleoconch I of 0.75 to 0.875 whorl. Teleoconch II of approximately 0.66 whorl, suture moderately impressed. Shoulder flat to slightly convex. 17–25 axial raised axial cords (21 in holotype) on last whorl, intersected by finer spiral lines forming only weak knots at intersection. Spiral lines appearing ontogenetically progressively from base to suture. Umbilicus open, funnel like, bordered by weak edge, interior surface straight, smooth. Selenizone above widest part of shell, closed anteriorly in mature specimens; lunules distinct, more numerous than and not coordinated with axials of whorl; keel low, often eroded in earlier portion. Foramen elongate-lanceolate, rounded posteriorly, pointed anteriorly. Aperture subrectangular, obtuse angle at inner base, upper lip strongly overhanging.

Animal unknown.

Comparisons. *Sinezona pacifica* from temperate Australia and New Zealand has spiral sculpture on the base that is in the form of slightly sloping steps.

Scissurella rota from East Africa has a protoconch with fine irregular axials, the slit remains open after over a full teleoconch II whorl, and the sculpture includes pointed processes at the intersection of spirals and axials.

Scissurella staminea from Japan has spiral sculpture on the base that is in the form of slightly sloping steps and the slit remains open after over a full whorl after the start of the selenizone.

Scissurella evaensis from the western Indian Ocean to the Central Pacific lacks spiral sculpture and has a protoconch with fine irregular axials.

Sinezona danieldreieri from Indo-Malayan Archipelago to the Central Pacific has a narrower overall shell, a narrower umbilicus, and weaker axial cords.

Distribution. Easter Island, Pitcairn Island (Fig. 445).

Specimen records.

Chile, Easter Island. 30–50 m, Tahai, 27.140°S, 109.429°W (DLG 1325, 100). 70–100 m, Hanga Roa, 27.145°S, 109.440°W (DLG 1334, 3). 65 m, Motu Iti, 27.196°S, 109.450°W (DLG 1348, 4). 100–150 m, La Perouse, 27.077°S, 109.295°W (DLG 1361, 2).

Great Britain, Pitcairn Island. 1–2 m, Ted's Side, 25.063°S, 130.104°W (DLG 1410, 1). 7 m, Big Pool at St. Paul's Point, 25.076°S, 130.089°W (DLG 1415, 1). 7 m, St. Paul's, 25.075°N, 25.089°E (DLG 2320, 16). 1 m, Isaac's Pool, 25.064°S, 130.100°W (DLG 1416, 1).

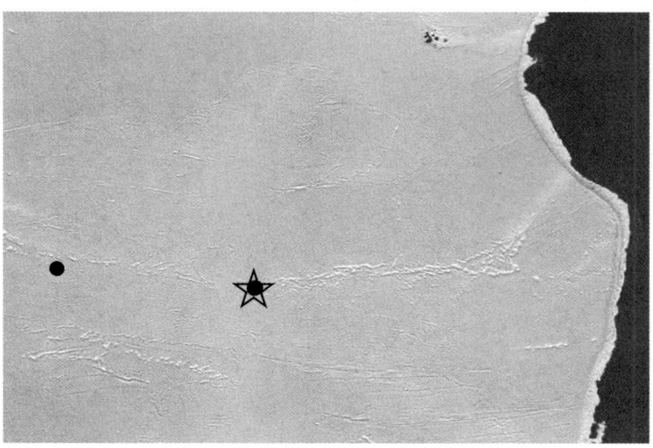

Figure 445. Distribution of *Sinezona zimmeri*. ☆ type locality. ● specimen record.

Sinezona cupelliformis (Amitrov, 1996) †

Figures 446–448

Chresonymy.

Scissurella cupelliformis Amitrov, 1996: 95, fig. 1A–D.

Scissurella cupelliformis: Amitrov & Zhegallo, 1998: 7, 9, fig. 1G–K [protoconch].

+ *Scissurella* (*Sukashitrochus*) *saubadei*: Lozouet, 1986: 110, 123–124, text-fig. 35B, pl. 3, fig. 2. [unavailable, unpublished].

Figure 446. Original illustration of *Scissurella cupelliformis*.

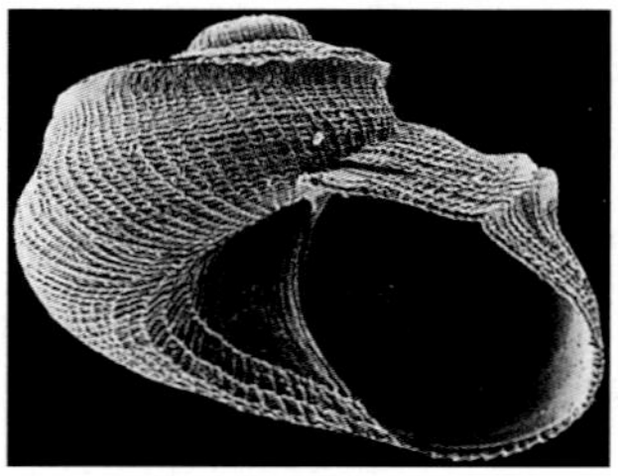

Sukashitrochus saubadae Lozouet, 1998: 68, fig. 4f. ***Type material.*** Holotype (MNHN: not found in MNHN 4/2005), 1.5 × 1.85 mm (H × W). Two paratypes (whereabouts unknown). ***Type locality.*** Gaas [43.614°N, 1.034°N] (Espibos [43.607°N, 1.054°N, France]), early Oligocene of the Adour, marl of *Oostrombus* (OD). ***Etymology.*** Named after A. M. Vergneau-Saubade, author of a malacological revision of Gaas (OD).

Figure 447. Original illustration of *Sukashitrochus saubadae*.

Misidentifications.

? *Scissurella* (*Sukashitrochus*) *terquemi*: Lozouet, 1986: 110 [is *Sin. cupelliformis*? Supported by illustrations in Lozouet *et al.* (2001)].

Sukashitrochus terquemi: Lozouet *et al.*, 2001: 17, pl. 1 fig. 3, pl. 2, figs 1a–b [juv. *Sin. cupelliformis*. See also under Remarks of *Sin. terquemi*.].

Type material. Holo/Syntypes (PIN 3595/701), 1.8 × 1.45 mm.

Type locality. Rybalsk (Tschapli) quarry, Dnepropetrovsk, Ukraine [48.483°N, 35.000°E]. Mandrikovka Layers, late Eocene (OD).

Etymology. Cupellum, Latin for keg.

Description. [Based on published illustrations]. Shell medium-sized (to 1.8 mm), trochiform, depressed. Protoconch of 0.75 whorl, smooth, apertural varix present, connected to embryonic cap, apertural margin unknown. Teleoconch I of 1.2 whorls, 28–35 axial cords, fine spiral lines. Teleoconch II >1 whorl, with lightly impressed suture, sutsel 2–3× as wide as selenizone, with similar sculpture as on teleoconch I, overall whorl-shape angulated at selenizone. Base rounded, spirals more pronounced towards umbilicus. Umbilicus open, wide, not bordered by carina, no funiculus, walls at angle to base, smooth. Aperture rounded, roof overhanging. Selenizone above periphery, with medium to strong keels, moderately elevated, slit closed to form foramen in fully grown specimens.

Comparisons. *Scissurella apudornata* from the Miocene of New Zealand lacks the clathrate sculpture on the base and has a more narrow umbilicus.

Sinezona deshayesi from the Eocene of Europe is overall more depressed, has a wider umbilicus, and more distinct axial cords on the base.

Sinezona finlayi from the Miocene (to Recent?) of New Zealand has few distinct axial cords on the base.

Sinezona malloryi from the Eocene of North America has a more depressed shell and lacks the fine clathrate sculpture on the base

Sinezona terquemi from the Miocene of Europe seems to have more fine axial cords on the base and a more narrow umbilicus.

Distribution. Eocene to Oligocene of Europe (Fig. 448).

Remarks. Most likely, *saubadae* was classified in *Sukashitrochus* because of the smooth protoconch sculpture considered diagnostic for the genus by Herbert (1986), although Lozouet (1998) did not justify the placement. The species lacks the spiral keels on the base of the shell, and also lacks the strong carina bordering the umbilicus, hence is here assigned to *Sinezona*.

cupelliformis was classified in *Scissurella* due to the open slit. However, in the apertural view of the original illustration, the margins of the slit almost touch, indicating a slightly immature *Sinezona* species. The fine reticulate sculpture over the entire shell clearly visible in both nominal taxa make their synonymy obvious.

The two nominal taxa share the very distinct reticulate pattern on the base, the strong spiral cords towards the umbilicus, and have identical overall shell shapes. They both are found in the Paleogene of Europe. Lozouet did not compare his *saubadae* with *cupelliformis*.

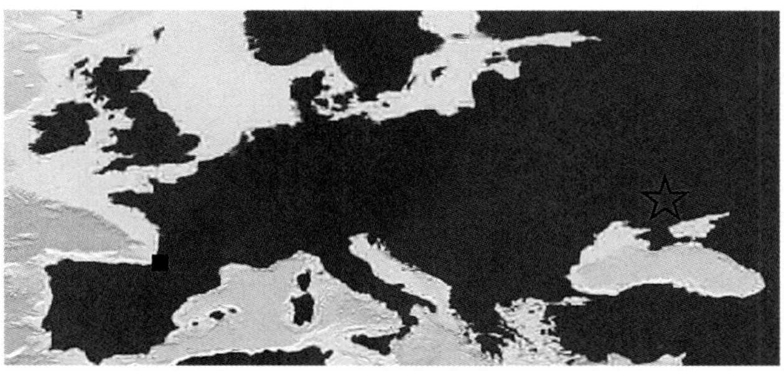

Figure 448. Distribution of *Sinezona cupelliformis*. ☆ type locality. ■ literature record.

Sinezona deshayesi (Munier-Chalmas, 1862) †

Figures 449–451

Chresonymy.

Scissurella deshayesi Munier-Chalmas, 1862: 391–392, pl. 16, figs 1–4.

Scissurella deshayesi: Munier-Chalmas, 1865: 395.

Scissurella deshayesi: Deshayes, 1864–1866 (1864): 4–5, pl. 65, figs 4–7.

Schismope deshayesi: Cossmann, 1883: 173, pl. 7, fig. 8.

Schismope deshayesi: Cossmann, 1888: 47, pl. 2, fig. 27.

Schismope deshayesi: Cossmann & Pissarro, 1906: pl. 2, fig. 12-2 [specimen with open slit: *Scissurella*?].

Scissurella deshayesi: Glibert, 1962: 4.

Scissurella ? (*Sukashitrochus*) *atkinsoni*: Lozouet, 1986: 109.

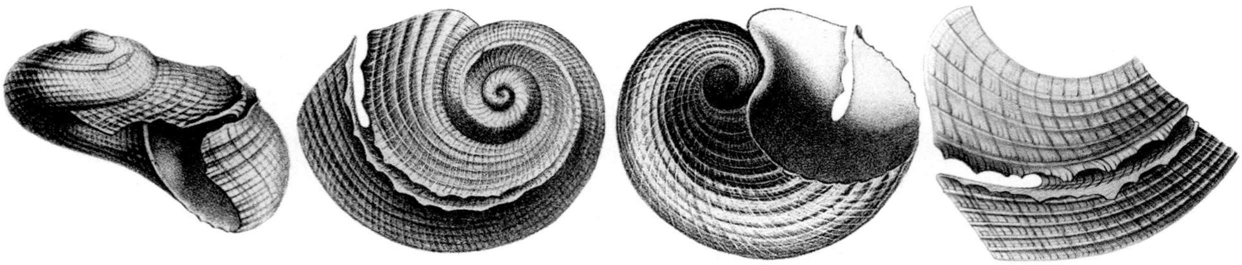

Figure 449. Original illustration of *Scissurella deshayesi.*

Synonym.

+ *Scissurella parisiensis* Deshayes, 1864–1866 (1864): 5, pl. 65, figs 8–10. ***Type material.*** Holotype (*fide* Gougerot & LeRenard, 1977: 5; École des Mines, Paris: Fischer, 1962), 1.5 × 0.75 mm (W × H). ***Type locality.*** Houdan [48.789°N, 1.602°E], Maulette [48.792°N, 1.619°E], Yvelines [48.785°N, 1.826°E, ~ 50 km ESE of Paris, France]. Middle Lutétien [Middle Eocene] (OD). ***Etymology.*** Named for its type locality in the Paris Basin.

Scissurella parisiensis: Semper, 1865: 284.

Scissurella parisiensis: Cossmann, 1888: 46 [handwritten note: voir suppl. p. 42].

Schismope parisiensis: Cossmann, 1892: 38.

Scissurella parisiensis: Cossmann & Pissarro, 1901: 281–282, pl. 29, fig. 3 [figure shows clearly a foramen].

Scissurella parisiensis: Cossmann, 1902: 24, pl 2, figs 12–13.

Scissurella parisiensis: Cossmann & Pissarro, 1906: pl. 2, fig. 12-1.

Scissurella parisiensis: Cossmann, 1923: 152, pl. 1, fig. 27 [unrecognizable].

Scissurella parisiensis: Fischer, 1962: 102.

Scissurella parisiensis: Glibert, 1962: 4.

Scissurella parisiensis: Gougerot & LeRenard, 1977: 5, 30, fig. 2.

Sukashitrochus parisiensis: Lozouet, 1998: 110, figs 4D [protoconch], 4G.

Sukashitrochus parisiensis: Lozouet, 1999: 3.

Figure 450. Original illustration of *Scissurella parisiensis.*

Type material. Two syntypes (MNHN. Fisher-Piette, 1950: 20), 1.5 × 1 mm. The syntypes were not found in the collection of MNHN in 4/2005 (Geiger pers. obs.).

Type locality. Guépelle near Senlis [France, 49°12'N, 2°35'E]. In the middle sand of the Parisien, early Eocene (OD).

Etymology. Named after French zoologist [Gérard Paul] Deshayes [1796–1875] (OD).

Description. [based on illustrations in Lozouet, 1998] Shell medium-sized (to 3 mm), trochiform, depressed. Protoconch of 0.75 whorl, smooth, apertural varix connected to embryonic cap, apertural margin sinusoidal. Teleoconch I of 1.33 whorls, approximately 43 dense, distinct axials, on last 0.33 whorl faint spiral in position of selenizone. Teleoconch II of one whorl, suture moderately impressed. Shoulder with approximately 45 fine axial cords, approximately 10 somewhat finer spiral lines at apertural margin. Base with somewhat less dense axial cords of similar strength as on shoulder, many fine spiral lines. Selenizone above periphery, lunules distinct, keels low medium strength. Foramen closed, elongate-triangular. Umbilicus wide, continuously sloping with base. Aperture rounded, roof overhanging.

Comparisons. *Scissurella apudornata* from the Miocene of New Zealand and *Sin. finlayi* from the Miocene of New Zealand have few but stronger axial cords on the base.

Sinezona cupelliformis from the Eocene of Europe has fine clathrate sculpture on the base and an overall more angulated shell outline.

Sinezona malloryi from the Eocene of North America has half as many axial cords and fewer spiral lines on the base.

Sinezona terquemi from the Miocene of Europe seems to have a more narrow umbilicus and many fine axial cords.

Scissurella depontaillieri from the Miocene through Paleocene of Europe has a more square shell profile and a more narrow umbilicus with a distinct funiculus.

Scissurella transylvanica from the Miocene of Europe has very strong axial cords on shoulder and base.

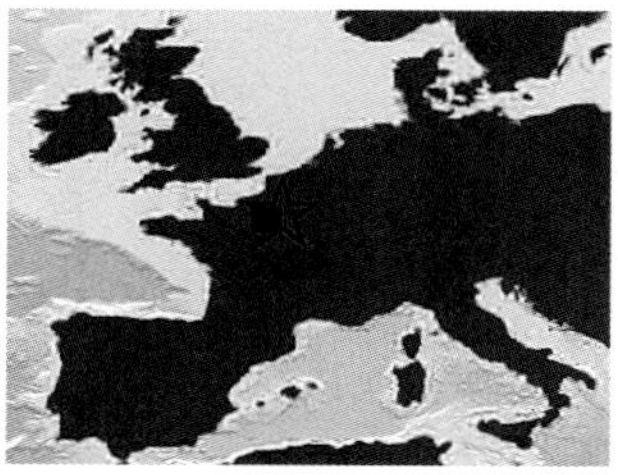

Figure 451. Distribution of *Sinezona deshayesi*. ☆ type locality. ■ literature record.

Distribution. Eocene of France (Fig. 451).

Literature records.

France. Le Bois Gouët [47.492°N, 1.608°W], single specimen, coll. Allix., from the Eocene of the Lower Loire (Cossmann, 1923).—Thionville-sur-Opton, Yvelines [48.776°N, 1.604°E] (Gougerot & LeRenard, 1977).—St.-Aignan-de-Grand-Lieu [47.123°N, 1.633°W], middle Eocene (Lozouet, 1998).

Remarks. The species has variously been classified. Most recently, Gougerot & Le Renard (1977) considered the closure of the slit to a foramen in very large specimens as a deformation due to old age. Lozouet (1998) accurately commented on the normality of such a closure, and placed the species in the genus *Sukashitrochus,* most likely due to the smooth protoconch. The species is here placed in *Sinezona* due to the closure of the foramen, the absence of spiral keels on the base of the shell otherwise diagnostic for *Sukashitrochus,* and because protoconch sculpture has been demonstrated to be highly homoplastic (Geiger, 2003).

A closed foramen is also evident in the illustration given by Cossmann & Pissarro (1901) but these authors considered the genera *Scissurella* and *Schismope* as synonymous.

The publication date indicated for *Sci. parisiensis* varies significantly. At the end of Deshayes (1864–1866) a table of publication dates for the various deliveries can be found. The publication date for *Sci. parisiensis* (pages 1–40, pls. 63–70 of vol. 3, no delivery number) is May 2, 1864.

Dehayes (1864–1866) cited specimens of *deshayesi* in Munier's and his own collection, and indicated a size of up to 3 mm.

The original illustrations of the two nominal taxa quite clearly show the same species. There is a slight variation in the strength of the axials, which may, however, also be attributed either to intraspecific variability or to different interpretation when drawing the specimens. It may have been that the two descriptions were produced independently of one another, given that only two years lie between the publications. Unfortunately, the type material of neither was available for study.

Sinezona haliotimorpha (Bandel, 1998)†

Figures 452–454

Chresonymy.

Sinezona haliotimorpha: Lozouet, 1986: 111, 129–130, text-fig. 29E, pl. 2, fig. 11 [unavailable, unpublished].

Ariella haliotimorpha Bandel, Sept. 1998: 63–64, pl. 22, figs 1–4.

Ariella haliotimorpha: Geiger, 2003: 75.

Sinezona haliotimorpha: Geiger & McLean, 2010: 8.

Synonyms.

+ *Sinezona haliotimorpha* Lozouet, Dec. 1998: 64, figs 3a–b [secondary homonym and synonym]. ***Type material.*** Holotype (MNHN. Fig. 453), 0.55 × 0.42 mm. 44 paratypes (MNHN). ***Type locality.*** Landes, St.-Paul-lès-Dax (Estoti) [France, 44.333°N, 1.050°W], late Oligocene (Chattien), *Miogysinoides* shell marl (OD). ***Etymology.*** Shaped like a young *Haliotis* (OD). ***Remarks.*** Secondary homonym and junior synonym of *Sinezona haliotimorpha* (Bandel, Sept. 1998).

Sinezona haliotimorpha: Lozouet *et al.*, 2001: 17.

Figure 452. Original illustration of *Sinezona haliotimorpha* Bandel.

Type material. Holotype (RMNH), 0.5 mm.

Type locality. Peyrère in, SW France [44.670°N, 5.620°E]. Oligocene (OD).

Etymology. *Haliotis*-like, due to the elevated hole.

Description. Shell small (to 0.5 mm) trochiform, globular. Protoconch of 0.75 whorl, with fine axial sculpture, condition of apertural varix and apertural margin unknown. Teleoconch I of 1.125 inflated whorls, approximately 19 distinct axials on first whorl, approximately 20 fine spirals increasing in strength with growth. Teleoconch II of 0.25 whorl, suture little impressed. Shoulder with increasing density of distinct axials, approximately eight very indistinct spirals. Base with irregularly spaced distinct axial cords, approximately a dozen fine irregular spiral lines. Umbilicus barely open, bordered by distinct funiculus. No selenizone; single foramen closed by straight suture towards apertural margin. Aperture round, roof overhanging.

Comparisons. *Sinezona praecedens* from the Oligocene of Europe has a distinct selenizone and lacks the elevated foramen, and the protoconch has half as many axial cords.

Sinezona hawaiiensis from the Recent of Hawaii and *Sin. pauperata* from the Recent of New Zealand have a protoconch with more and weaker axials, more numerous and less distinct axial cords on the teleoconch, and a foramen with lower keels.

Distribution. Oligocene to Miocene of France (Fig. 454).

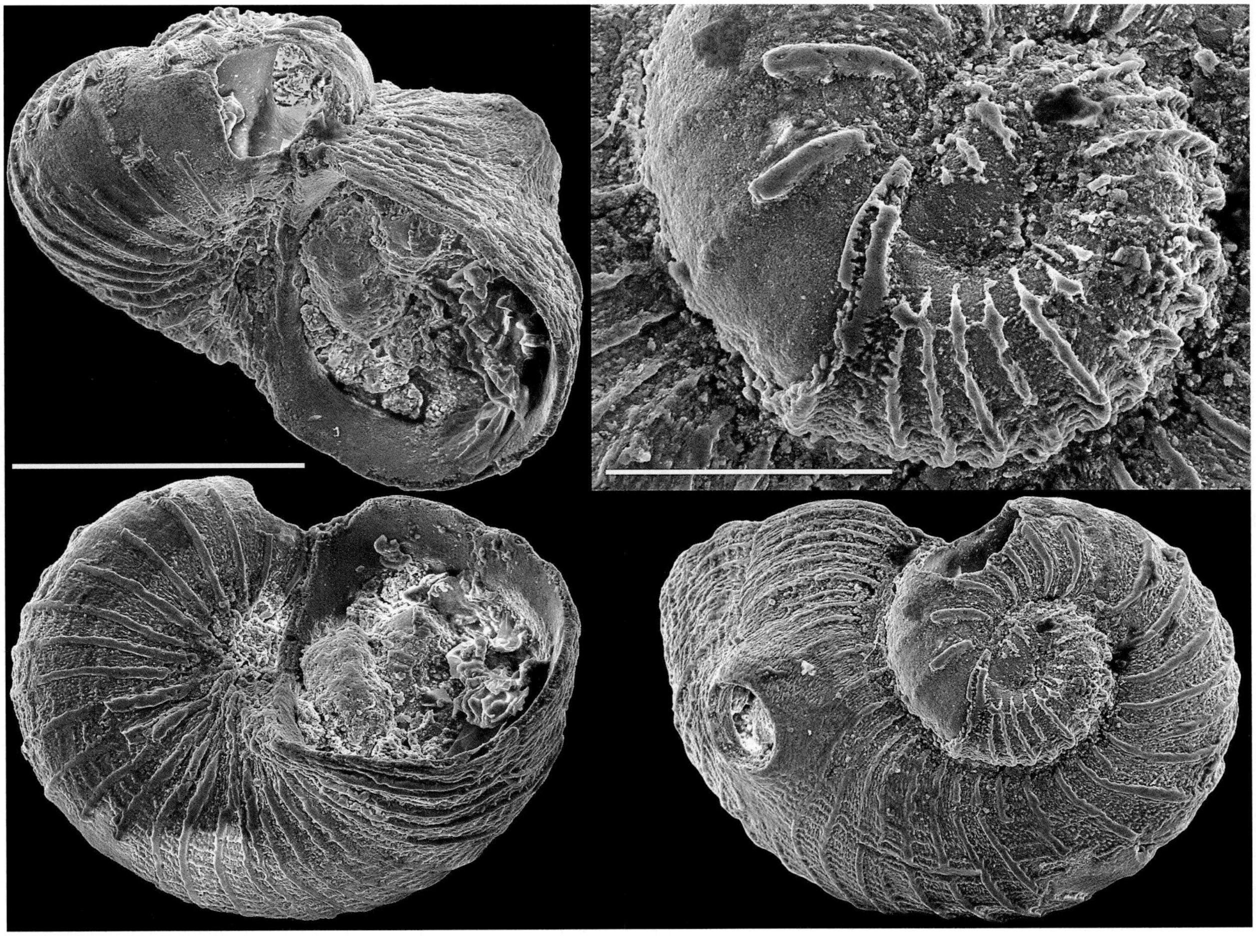

Figure 453. *Sinezona haliotimorpha* Lozouet (*non* Bandel). Holotype. Landes, St.-Paul-lès-Dax (Estoti) [France, 44.333°N, 1.050°W], late Oligocene (Chattien) *Miogysinoides* shell marl. Scale bar shell = 0.25 mm. Scale bar protoconch = 100 µm.

Literature records.

France. Pontonx, Upper Oligocene. Saucats and Villandraut, early Miocene (Lozouet, 1998).

Remarks. Bandel (1998) used "*Ariella haliotimorpha* (Lozouet, 1986) n. sp." in the caption for the species description. The species was described by Bandel (1998), the description being taken from the dissertation of Lozouet (1986). Lozouet *et al.* (2001) used "Lozouet *in* Bandel, 1998" whereas Marshall (2002) used "Bandel, 1998." The proper authority for the taxon is Bandel, 1998.

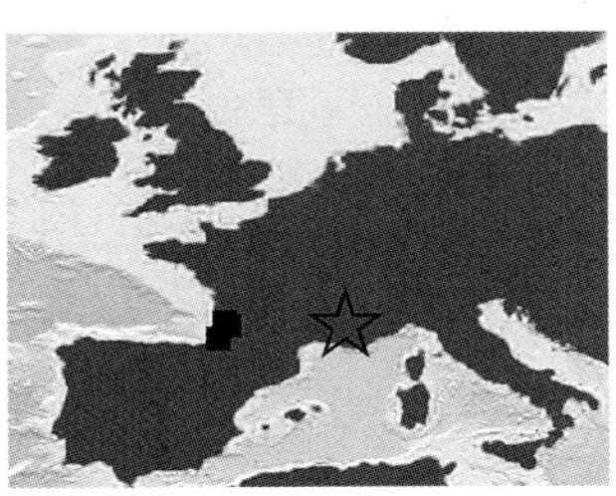

Figure 454. Distribution of *Sinezona haliotimorpha*. ☆ type locality. ■ literature record.

Sinezona koruahina (Laws, 1936) †

Figure 455–457

Chresonymy.

Schismope koruahina Laws, 1936: 100–101, pl. 15, figs 45–47.
Schismope koruahina: Powell, 1941: 254.
Sinezona koruahina: Lozouet, 1986: 112.
Sinezona koruahina: Beu & Maxwell, 1990: 376.

Figure 455. Original illustration of *Schismope koruahina*.

Type material. Holotype (AIM: Powell, 1941: 254), 1.3 × 1.0 mm (W × H). Three paratypes (GNS TM 1713, 4906–4907) from type locality.

Type locality. Kaawa Creek, New Zealand [35.550°S, 173.383°E]. Waitotaran Stage, Wanganuian (OD: inferred from title). Opotian (early Pliocene): specimen label GNS paratype.

Etymology. Seems to refer to a place, given other species names such as *Austrodrillia koruahinensis* Powell & Bartrum, 1929.

Description. Shell small (to 1.3 mm. Fig. 456), trochiform depressed. Protoconch with strong axials. Teleoconch I of approximately 1.125 whorls, some strong axial cords, eroded. Teleoconch II of 0.375 whorl, suture moderately impressed. Shoulder convex, approximately nine strong axial lamellae. Base with strong constriction below selenizone, constriction smooth; below with approximately nine strong axial lamellae. Umbilicus open, moderately wide, deep, possibly bordered by moderate keels, walls smooth. Aperture rounded. Selenizone above periphery, keels moderately elevated, quite strong, foramen triangular, most likely enlarged in type by damage.

Comparisons. *Scissurella apudornata* from the Miocene of New Zealand and *Sin. finlayi* from the Miocene (to Recent?) of New Zealand lack the strong axial lamellae on the base and the strong constriction below the selenizone.

Sinezona malloryi from the Eocene of North America lacks the distinct constriction below the selenizone and has weaker spiral cords on the base.

Sinezona tertia from the Pliocene of New Zealand lacks the strong constriction below the selenizone.

Scissurella laqueus from the Recent of New Zealand has no axial sculpture on teleoconch I and much narrower umbilicus.

Distribution. Pliocene of New Zealand (Fig. 457).

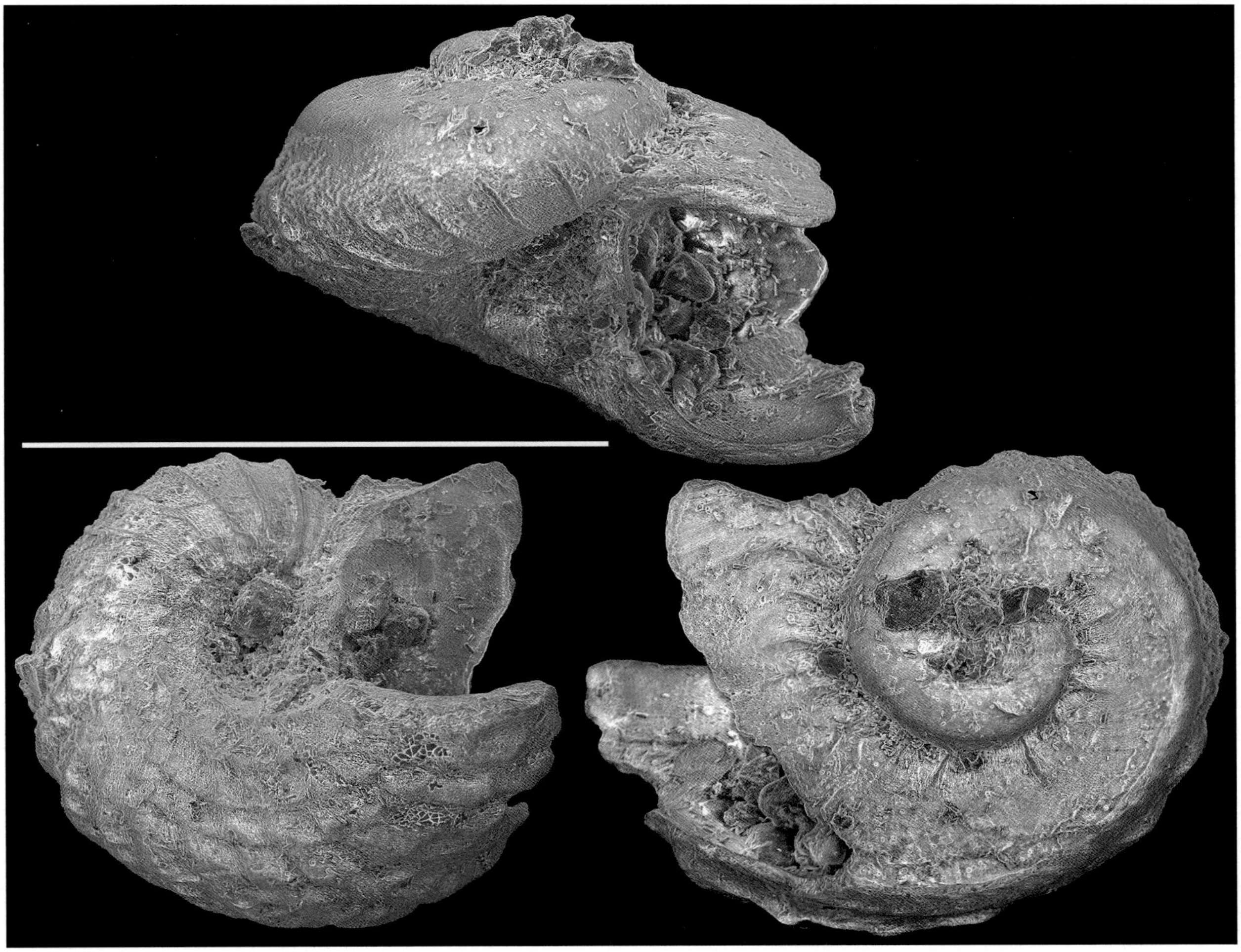

Figure 456. *Schismope koruahina*. Paratype. Kaawa Creek, SW Auckland, New Zealand, 35.550°S, 173.383°E (GNS TM 4906, 1). Scale bar = 1 mm.

Specimen records.

New Zealand. Kaawa Creek, SW Auckland, 35.550°S, 173.383°E (GNS TM 1713, 1; GNS TM 4906, 1; GNS TM 4907, 1).

Remarks. The axial sculpture stops abruptly at the periphery, where a spiral element is found. That spiral element almost looks like a keel as in *Sukashitrochus*. However, southern hemisphere *Sukashitrochus* do not show protoconch sculpture with strong axials, and the "keel" does not show an impression in the inner surface of the whorls, hence, is not comparable to the keel of *Sukashitrochus*. Accordingly, a placement of the species in *Sinezona* is warranted.

Figure 457. Distribution of *Sinezona koruahina*. ☆ type locality.

Sinezona malloryi (Squires & Goeddert, 1996) †

Figures 458–460

Chresonymy.

Scissurella (*Scissurella*) *malloryi* Squires & Goeddert, 1996: 228–230, figs 2–7.

Type material. Holotype (LACMIP 11354), 1 × 1 mm. Paratypes (LACMIP 11355–11357).

Type locality. California State University Northridge locality 1563 [LACMIP loc. 16655], Larch Mountain, Washington, USA, 47°59'03"N 123°8'2"W (OD). Early Eocene, Crescent Formation (inferred from title).

Etymology. Named for V. Standish Mallory (1919–2003) for his important study of Pacific coast of North America benthic foraminiferal biostratigraphy (OD).

Description. Shell medium-sized (to 1 mm. Figs 458, 459), trochiform depressed. Protoconch of 0.75 whorl, fine axial sculpture, apertural varix connected to embryonic cap,

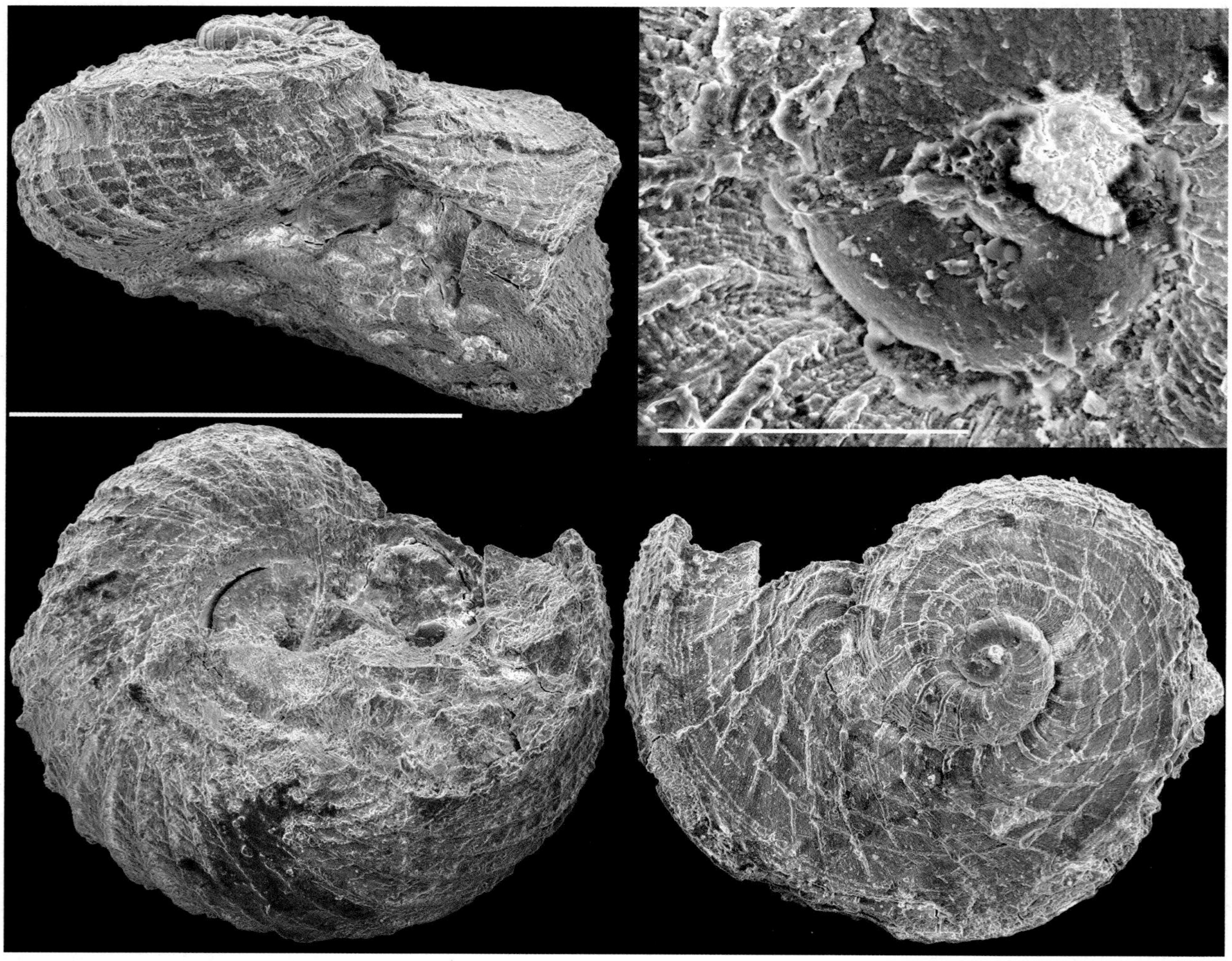

Figure 458. *Scissurella malloryi.* Holotype. California State University Northridge locality 1563 [LACMIP loc. 16655], Larch Mountain, Washington, USA, 47°59'03"N 123°8'2"W (LACMIP 11354). Scale bar shell = 1 mm. Scale bar protoconch = 100 μm.

apertural margin sinusoidal. Teleoconch I of 1.25 whorls, approximately 28 distinct axial cords, up to approximately 12 fine spiral threads. Teleoconch II of 0.66 whorl, suture moderately impressed. Shoulder somewhat convex, approximately 38 somewhat indistinct axial cords, up to approximately six spiral lines. Base with distinct constriction below selenizone, density and strength of axial cords comparable to those on shoulder, approximately 15 finer spiral lines. Selenizone slightly above periphery, growth lamellae distinct, keels moderately high, strong. Slit closed to form foramen. Umbilicus open, wide, walls smooth, continuously sloping with base. Aperture rounded.

Comparisons. *Sinezona malloryi* is the known only North American species in the genus.

Scissurella apudornata from the Miocene of New Zealand has an overall taller shell with correspondingly more narrow umbilicus and lacks distinct spiral lines on shoulder and base.

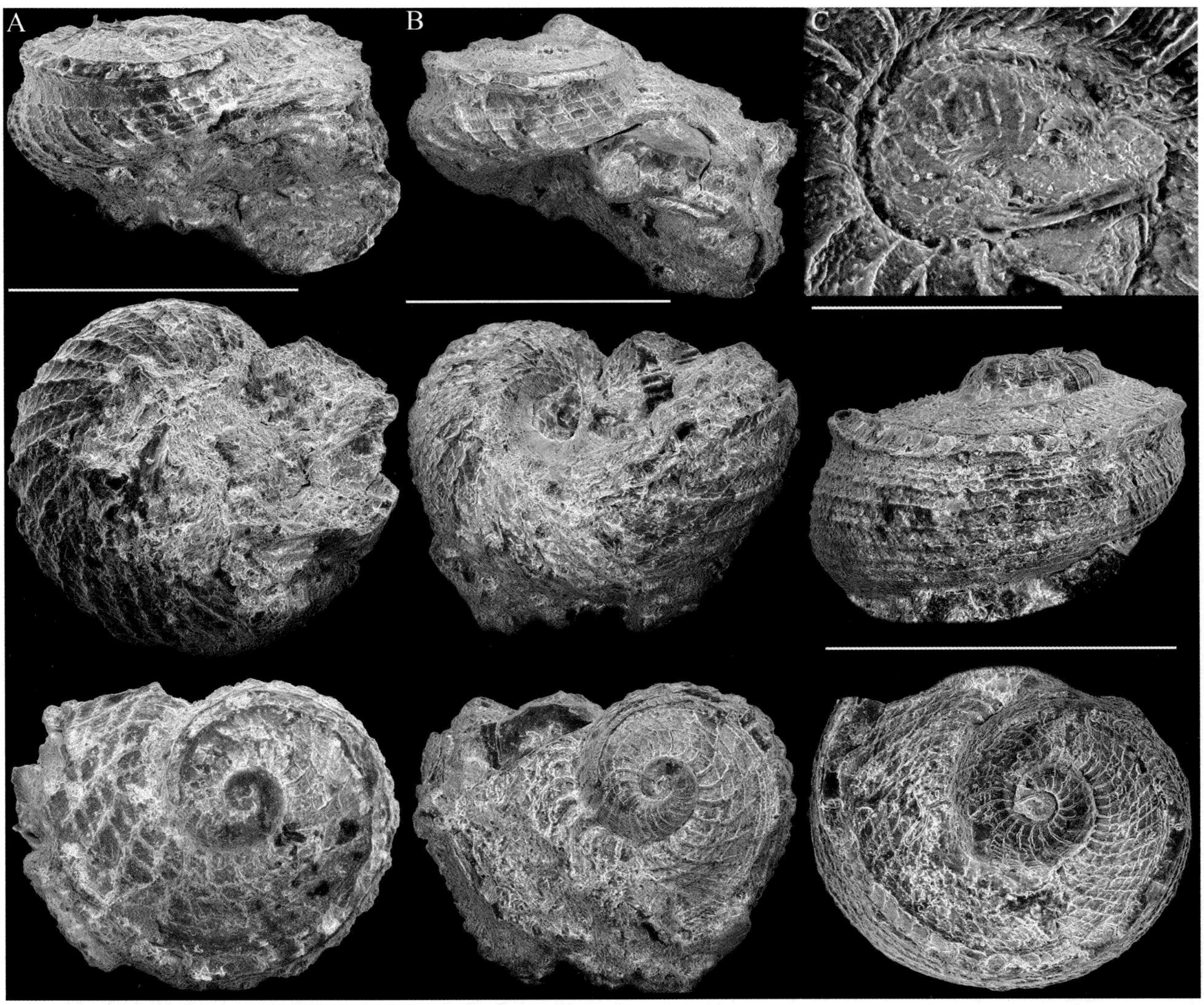

Figure 459. *Sinezona malloryi*. Paratypes. California State University Northridge locality 1563 [LACMIP loc. 16655], Larch Mountain, Washington, USA, 47°59'03"N 123°8'2"W **A.** LACMIP 11355. **B.** LACMIP 11356. **C.** LACMIP 11357. Scale bars shell = 1 mm. Scale bar protoconch = 100 μm.

Sinezona cupelliformis from the Eocene to Oligocene of Europe is less depressed shell and has distinctive fine clathrate sculpture on the base.

Sinezona deshayesi from the Eocene of Europe has twice as many axial cords and stronger spiral lines on the base.

Sinezona finlayi from the Miocene [to Recent?] of New Zealand is indistinguishable based on the available data. However, the geographic separation makes it likely that they are distinct, and details of protoconch and early teleoconch may show significant differences. Accordingly, the two species are retained.

Sinezona koruahina from the Late Tertiary of New Zealand has a distinct constriction below the selenizone and has stronger spiral cords on the base.

Sinezona terquemi from the Miocene of Europe has more but weaker axial cords, lacks distinct spiral lines on shoulder and base, and has a more narrow umbilicus.

Sinezona tertia from the Pliocene of New Zealand has strong axial cords on the protoconch and a much taller shell with more distinct axial cords.

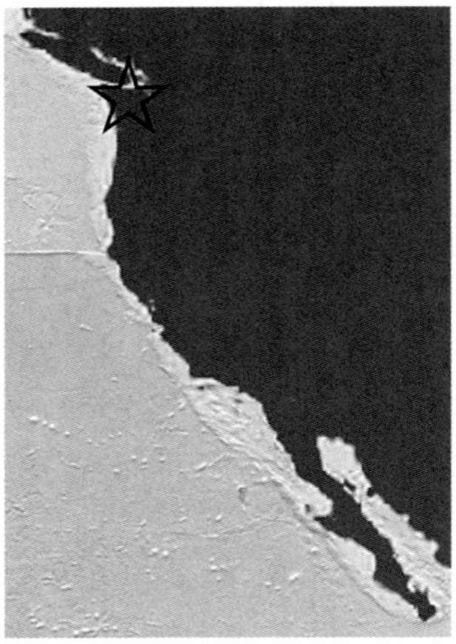

Figure 460. Distribution of *Sinezona malloryi*. ☆ type locality.

Distribution. Known only from type material (Fig. 460).

Remarks. The species is here reassigned to *Sinezona* due to the clearly closed foramen of the holotype. The other specimens are somewhat smaller with broken aperture, hence, the slit is still open.

Sinezona praecedens Lozouet, 1998 †

Figures 461–462

Chresonymy.

Sinezona praecedens: Lozouet, 1986: 112, 130, pl. 2, fig. 9 [unavailable, unpublished].
Sinezona praecedens Lozouet, 1998: 64, fig. 3c.
Sinezona praecedens: Lozouet *et al.*, 2001: 17.

Type material. Holotype (MNHN. Fig. 462), 0.8 × 0.9 mm. 87 paratypes (MNHN).

Type locality. Landes, St.-Paul-lès-Dax (Estoti) [France, 44.333°N, 1.050°W], late Oligocene (Chattien), *Miogypsinoides* shell marl (OD).

Etymology. Announces the Recent *Sinezona* (OD).

Description. Shell small (to 0.9 mm. Fig. 462) trochiform. Protoconch of one whorl, with distinct axial sculpture. Teleoconch I of 0.875 whorl, approximately 15 raised axial cords, interstices with fine irregular growth marks first spiral lines appearing after 0.25 whorl, approximately 12 fine spiral lines at onset of selenizone. Teleoconch II of 0.875 whorl, suture moderately impressed. Shoulder slightly convex, approximately 22 raised axial cords, approximately four strong spiral lines concentrated in central portion of shoulder. Base with axials of similar density and strength as on shoulder, approximately 13 spiral lines, becoming more angular towards umbilicus;

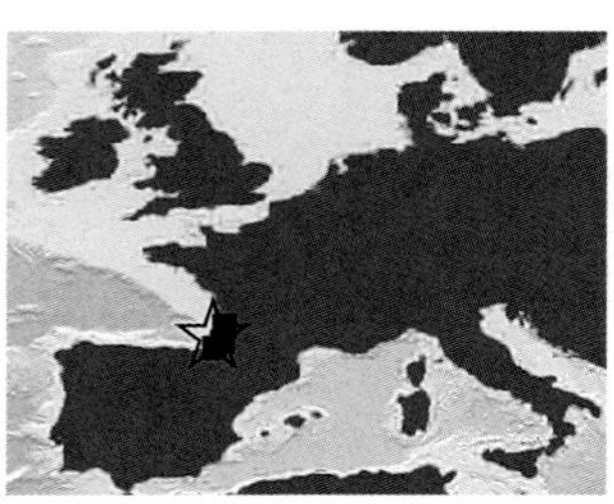

Figure 461. Distribution of *Sinezona praecedens*. ☆ type locality. ■ literature record.

intersection of axials and spirals forming small nodules, particularly on lower base towards umbilicus. Umbilicus barely open, walls smooth, last spiral of base forming border with umbilicus, no funiculus. Selenizone above periphery, keels moderately strong, moderately elevated; slit closed to foramen. Aperture obliquely D-shaped, roof overhanging.

Comparisons. *Sinezona haliotimorpha* from the Oligocene of Europe lacks a selenizone and has a very much elevated foramen.

Sinezona calumnior from the Recent Indo-Malayan Archipelago to the Central Pacific has a teleoconch I of 0.75–1 whorl and lacks the scabrous scales on the base of the shell.

Distribution. Oligocene to Miocene of France (Fig. 461).

Literature records.

France. Late Oligocene of Peyrehorade (Peyrère, 44.670°N, 5.620°E). Early Miocene (Aquitanian) of Saucats (Lariey 44652°N, 0.596°W). Early Miocene (Aquitanian) Meilhan of (Vives, 43.865°N, 0.705°W) (Lozouet, 1998).

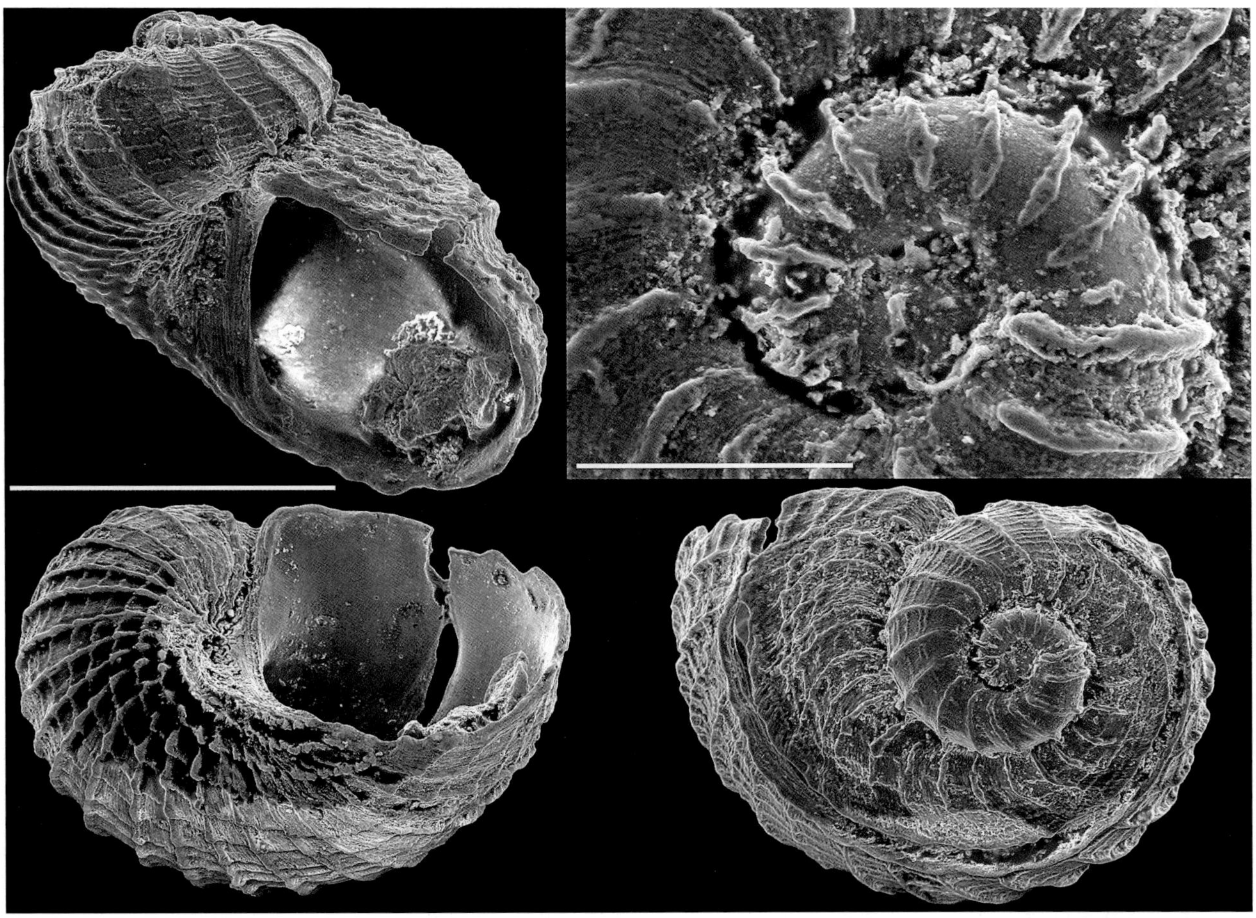

Figure 462. *Sinezona praecedens.* Holotype. Landes, St.-Paul-lès-Dax (Estoti) [France, 44.333°N, 1.050°W], late Oligocene (Chattien), *Miogypsinoides* shell marl (MNHN). Scale bar shell = 0.5 mm. Scale bar protoconch = 100 μm.

Remarks. It may appear that the slit of the holotype is open, but close inspection reveals that the raphe near the peristome was chipped; some of the intact raphe is clearly visible in dorsal view.

Sinezona terquemi (Deshayes, 1865) †

Figures 463–464

Chresonymy.

Trochotoma terquemi Deshayes, 1865: 236–237, pl. 7, fig. 1.
Ditremaria terquemi: Fischer, 1962: 101.

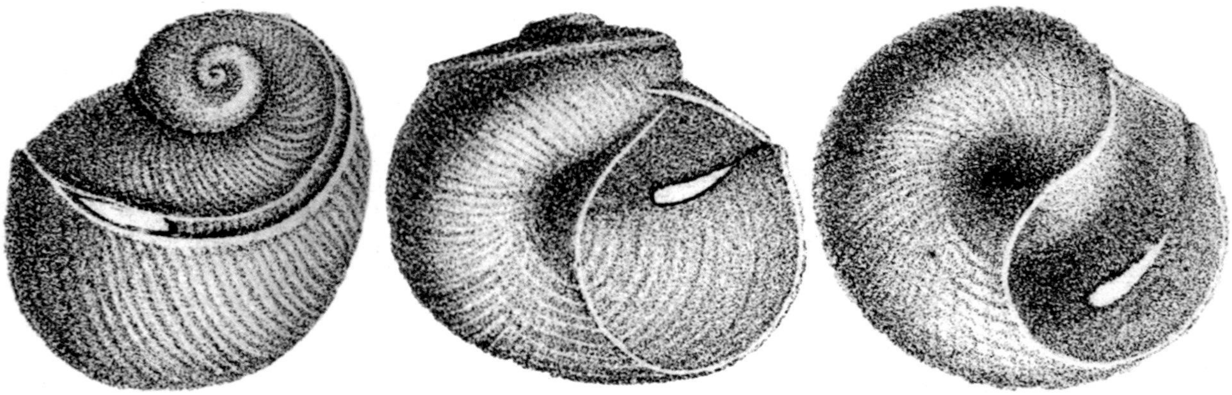

Figure 463. Original illustration of *Sinezona terquemi.*

Misidentification.

not: *Scissurella* (*Schismope*) *terquiemi* [*sic*]: Cossmann & Peyrot, 1917: 226–227, pl. 3, figs 10–14: is *Sci. transylvanica* [synonymy *fide* Lozouet *et al.*, 2001: as *Sci. lamellosa*].

not? *Scissurella* (*Sukashitrochus*) *terquemi*: Lozouet, 1986: 110 [is *Sin. cupelliformis*? Supported by illustrations in Lozouet *et al.* (2001)].

not *Sukashitrochus terquemi*: Lozouet *et al.*, 2001: 17, pl. 1 fig. 3, pl. 2, figs 1a–b [is *Sin. cupelliformis*].

Type material. Holo/Syntype(s) (Collection Deshayes. École des Mines, Paris: Fischer, 1962: probable type), 2 × 1 mm. Current repository unknown.

Type locality. Sands of Bordeaux, France [44.833°N, 0.567°W] (OD: inferred from title], early Miocene (Fischer, 1962; Lozouet, 1986).

Etymology. Named for French micropaleontologist Olry Terquem (1797–1887).

Description. [based on original illustration] Shell medium-sized (to 2 mm), trochiform, depressed. Protoconch unknown. Teleoconch II of 1.1 whorls, suture moderately impressed. dense axial cords, indistinct spiral cords restricted to adumbilical quarter of base. Umbilicus moderately wide, continuously sloping with base. Selenizone above periphery, keels moderately strong, moderately elevated; lunules fine, dense; slit closed to elongated foramen. Aperture rotund, roof overhanging.

Comparisons. *Sinezona cupelliformis* from the Eocene of Europe has fine reticulate sculpture on entire base.

Sinezona deshayesi from the Eocene of Europe has an overall more depressed shell with a much wider umbilicus.

Sinezona malloryi from the Eocene of North America has fewer but stronger axial cords, distinct spiral lines on shoulder and base, and a wider umbilicus.

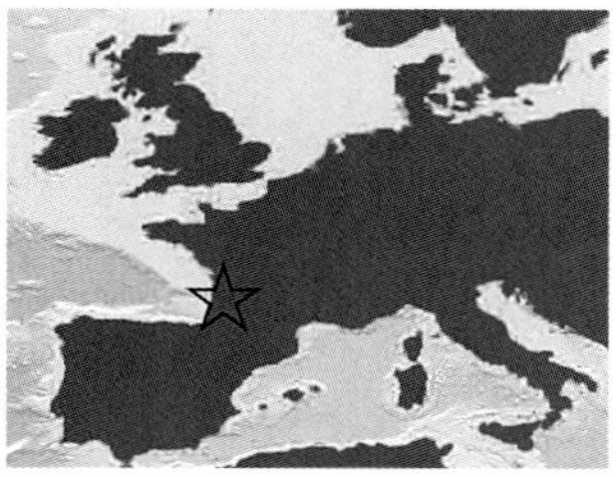

Figure 464. Distribution of *Sinezona terquemi*. ☆ type locality.

Distribution. Miocene of France (Fig. 464).

Remarks. The number of specimens examined was not detailed by Deshayes, though he indicated that the species is rare.

The placement of the new species of Deshayes was carried out after a detailed discussion highlighting the overall shape similarity between *Schismope* and *Trochotoma*, and emphasizing that size differences—though distinct—should not be sufficient to justify a distinct genus. Crosse (1865) questioned Deshayes' reasoning on the immediately following pages. He pointed to the absence of nacre in Scissurellidae and emphasized the importance of this structural distinction.

The identification of specimens as *terquemi* by Lozouet (1986) and Lozouet *et al.* (2001) poses some problems. The specimen illustrated by Lozouet et al. (2001) has distinct spiral sculpture over the entire base that is not shown at all in the original illustration. It seems unlikely that this prominent feature was not shown by Deshayes. If, however, Lozouet *et al.* (2001) correctly identified the specimen, then *Sin. cupelliformis* is a junior synonym.

Sinezona tertia (Laws, 1940) †

Figures 465–468

Chresonymy.

Schismope tertia Laws, 1940a: 434–435, pl. 60, fig. 29.
Sinezona ? *tertia*: Lozouet, 1986: 112.
Sinezona tertia: Beu & Maxwell, 1990: 376.

Figure 465. Original illustration of *Sinezona tertia*.

Type material. Holotype (GNS TM 1479, ex coll. Laws), 1.0 × 1.0 mm. Paratypes from type locality (coll. Laws, 4; GNS TM 8630–8631, 2: ex coll. Laws).

Type locality. Waitotaran Faunule, Kaawa Creek, New Zealand [35.550°S, 173.383°E]. Tertiary (OD). Opoitian = early Pliocene (Beu & Maxwell, 1990).

Etymology. Tertia, Latin for third.

Description. Shell medium-sized (to 1.0 mm. Figs 466, 467), trochiform, with low but distinct spire. Protoconch one whorl, strong axial cords, apertural varix not connected to embryonic cap, apertural margin straight. Teleoconch I 0.825 whorl, 14 strong raised axial cords/lamellae, no spiral sculpture, interstices with very fine growth lines. Teleoconch II of >0.66 whorl (aperture broken), suture moderately impressed. Shoulder convex, same density and strength of axials as on teleoconch I, first spiral line after 0.125 teleoconch II whorl, increasing to four fine spiral cords at apertural margin of holotype. Base rounded, without constriction below selenizone, very strong, sharp, prosocline lamellae, crossed by approximately 12 irregular spiral lines. Umbilicus open, bordered by strong spiral cord, walls straight, smooth. Selenizone with strong keels of moderate elevation; lunules moderately strong; slit/foramen condition unknown (broken). Aperture rounded, D-shaped, oblique, roof overhanging.

Figure 466. *Sinezona tertia*. Holotype. Waitotaran Faunule, Kaawa Creek, New Zealand [35.550°S, 173.383°E] (GNS TM 1479). Scale bar shell = 1 mm. Scale bar protoconch = 100 µm.

Comparisons. *Sinezona finlayi* from the Miocene of New Zealand has an overall broader shell and much less distinct axial cords predominant near suture and on earlier portion of teleoconch.

Scissurella apudornata from the Oligocene/Miocene of New Zealand has weaker axial cords that reach the selenizone and lacks spiral lines/cords on the base near the umbilicus.

Figure 467. *Sinezona tertia*. **A–C.** Shellbed, N[th.] Bank of Mangahao River, 1 mi south of Mangahao, north Wairarapa, New Zealand, 40.449°S, 175.784°E (NMNZ M.42686, 15). Scale bars shell = 1 mm. Scale bars protoconch = 100 μm.

Sinezona koruahina from the Pliocene of New Zealand has a strong constriction below the selenizone and is much more depressed in overall outline of the shell.

Sinezona malloryi from the Eocene of North America has fine axial sculpture on the protoconch and a much broader shell with less distinct axial cords.

Distribution. Early Pliocene of New Zealand (Fig. 468).

Specimen records.

New Zealand. Shellbed, N[th.] Bank of Mangahao River, 1 mi S of Mangahao, N Wairarapa, New Zealand, 40.449°S, 175.784°E (NMNZ M.42686, 15).

Remarks. How Laws (1940a) arrived at the generic assignment is unclear, because the generic characteristics cannot be observed in the illustration provided by Laws (1940a) or the holotype. Additional specimens (Fig. 467C) of the species clearly display a closed foramen, which confirms the generic placement.

The holotype has a rather low spire, a character that seems to be quite variable (Fig. 467C: low. Fig. 467B: high). As the holotype is an immature specimen with only 0.5 teleoconch II whorl as opposed to one teleoconch II whorl in fully grown specimens, it has a less elevated spire as is the case in early ontogenetic stages of most species.

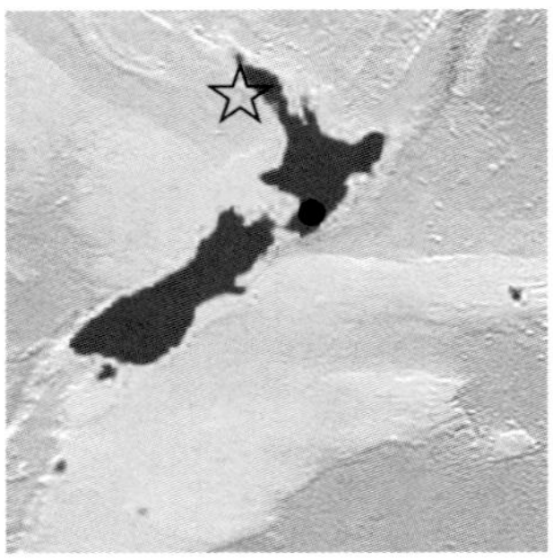

Figure 468. Distribution of *Sinezona tertia*. ☆ type locality. ● specimen record.

Sukashitrochus Habe & Kosuge, 1964

Chresonymy.

Sukashitrochus Habe & Kosuge, 1964: 3.
Sukashitrochus: Herbert, 1986: 628–629.
Sukashitrochus: Bandel: 1998: 50–51.
Sukashitrochus: Okamoto, 2001: 27 [copy original description].
Sukashitrochus: Geiger, 2003: 40–41.

Type species. *Scissurella carinata* A. Adams, 1862 (OD).

Other species. *Sukashitrochus atkinsoni* (Tennison-Woods, 1877).
Sukashitrochus dorbignyi (Audouin, 1826).
Sukashitrochus estotiensis Lozouet, 1999 †.
Sukashitrochus kaiparaensis (Laws, 1941) †.
Sukashitrochus lyallensis (Finlay, 1926).
Sukashitrochus morleti (Crosse, 1880).
Sukashitrochus ngatutura (Laws, 1936) †.
Sukashitrochus pulcher (Petterd, 1884).

Species removed from *Sukashitrochus*. *Sinezona saubadae* (Lozouet, 1998) † [see *Sin. cupelliformis*].
Sukashitrochus sp. nov. (Beck *et al.*, 2006) [is *Satondella danieli*].

Etymology. Sukashi-, gerund of Japanese verb *sukasu*, to look through; to make an opening; to leave space, most likely referring to the foramen; -trochus, shaped like a top shell (Trochidae). Gender masculine.

Description. Shell trochiform, with spiral keel(s) on base, shoulder rounded or angular. Shell with foramen, selenizone on shoulder, moderately keeled, usually starting after 0.75 whorl (except *Suk. pulcher*: <0.75 whorl). Umbilicus with carina. Shell sculpture variable: predominant axials, reticulate, or predominant spirals. Protoconch sculpture variable: smooth, fine irregular axials, flocculent; protoconch varix present, either not touching embryonic cap or forming bridge to embryonic cap, shape of protoconch aperture sinusoidal. Radula R - 5 - n with lateral tooth 5 greatly enlarged, broadened, serration asymmetrical. Operculum present.

Distribution. Indian Ocean through Central Pacific, north-subtropical to south-temperate waters. One fossil record in Europe (Fig. 469). For specific localities, see individual species.

Geologic Range. Oligocene through Recent.

Comparisons. *Sinezona* shares the foramen, but lacks the keels on the shell.

Scissurella has an open slit and lacks the keels on the shell.

Some juvenile *Turbo* species (e.g., *T. cornutus*: see Hayashi, 1983) also show strong spiral keels. However, some keels are found above the periphery, and the onset of the keels is earlier, unlike in *Sukashitrochus* species, in that they are visible after only 0.125 teleoconch whorl. The onset of the slit in all *Sukashitrochus* species is after no more than 1.5 teleoconch I whorls.

Differentiation of species. Species are differentiated by details of sculpture (absolute and relative strength of axials and spirals), as well as by protoconch sculpture. Under a very good light microscope, only very prominent protoconch sculpture can be resolved sufficiently. Anatomical characters have not been found to help in species identification or in the establishment of species boundaries.

Intraspecific variation extends to the strength and density of sculptural elements (variable by approximately 25–50%) and the distinctness of the spiral keel(s) on the base, which sometimes are only very weakly developed.

Notice, although juvenile *Sukashitrochus* have an open slit, they can be reliably distinguished from juvenile *Sinezona* and *Scissurella* by the keels on the base. These are often more prominent on early teleoconch II than in very small or fully grown specimens.

Remarks. The gender of the genus is masculine; hence, some of the specific epithets had to be adjusted (*pulchra - pulcher; carinata - carinatus*).

Sukashitrochus has been diagnosed (Herbert 1986) with a smooth protoconch, and by McLean (1989: 5) by "reticulate" [= flocculent of this contribution] protoconch sculpture. Bandel (1998) noted other protoconch sculptures among the species assigned to this genus. This character was shown by Geiger (2003) to be highly variable within the genus. Three protoconch sculptures are found in the genus; hence, none can be called diagnostic.

Herbert (1986) noted the open slit of *Suk. dorbignyi* and cautioned the use of the distinction between open slit and closed foramen as a generic character. The open slit in *Suk. dorbignyi* is due to the juvenile nature of the examined specimen (see Remarks under that species), and the closed foramen in *Sukashitrochus* remains a diagnostic feature. The conjunction of keels on the shells with a closed foramen in fully grown specimens is diagnostic for *Sukashitrochus*.

The distribution of protoconch sculpture shows an interesting biogeographic pattern. Species with flocculent protoconch sculpture are found in the northern temperate Japanese *Suk. carinatus* as well as in the southern temperate Australian *Suk. atkinsoni* and *Suk. pulcher*, as well as in the southern temperate New Zealand *Suk. lyallensis*. This is a distinct antitropical distribution.

Haszprunar (1988b) described heteropod-type swimming in a *Sukashitrochus* sp. Hasegawa (2004) reported swimming amongst others also in *Sukashitrochus*

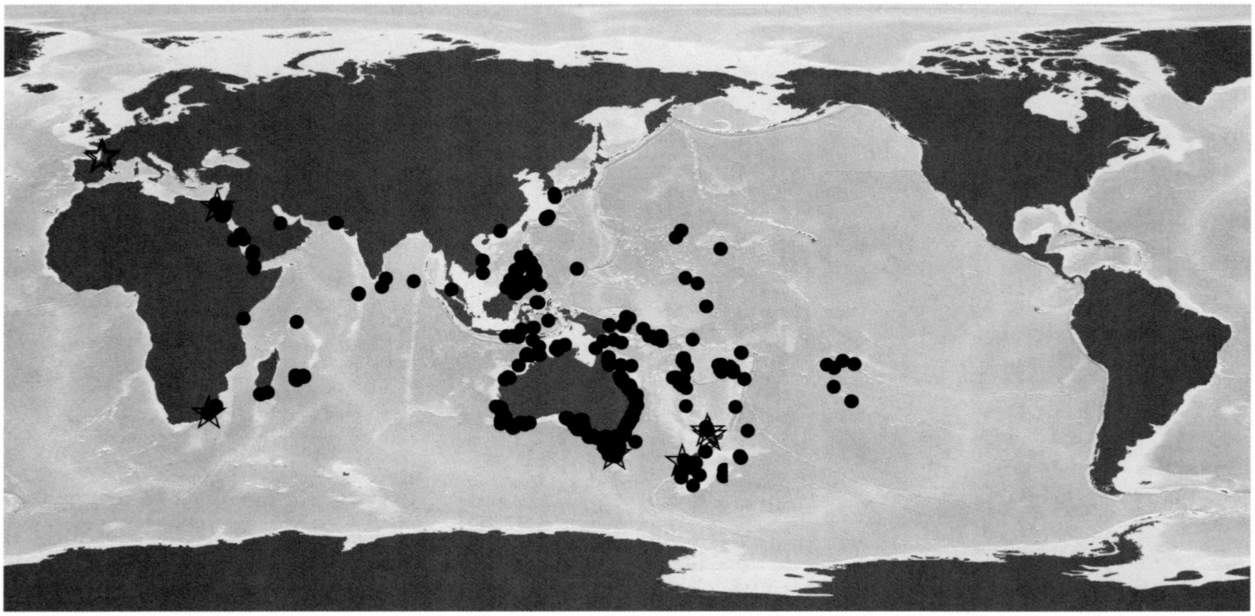

Figure 469. Global distribution of *Sukashitrochus* species. ☆ type locality. ● specimen record. ■ literature record.

sp., most likely *Suk. carinatus*. How wide-spread this behavior is in *Sukashitrochus* is unknown.

Sukashitrochus atkinsoni (Tenison-Woods, 1877)

Figures 470–474

Chresonymy.

Scissurella atkinsoni Tenison-Woods, 1877: 149–150, not illustrated.
Schismope atkinsoni: Tenison-Woods, 1878: 43.
Scissurella atkinsoni: Angas, 1878: 868.
Schismope atkinsonii [sic]: Petterd, 1884: 139.
Schismope atkinsoni: Paetel, 1888: 289.
Schismope atkinsoni: Pilsbry, 1890: 66.
Schismope atkinsoni: Tate & May, 1902: 407.
Schismope atkinsoni: Pritchard & Gatliff, 1902: 181.
Schismope atkinsoni: Sowerby & Fulton, 1902: 28.
Schismope atkinsoni: Sowerby & Fulton, 1903: 53.
Schismope atkinsoni: Hedley, 1903: 329.
Schismope atkinsoni: Hedley, 1905: 42.
Schismope atkinsoni: Anonymous, 1906: 367.
Schismope atkinsoni: Hedley, 1907a: 478.
Schismope atkinsoni: Hedley, 1907b: 288.
Schismope atkinsoni: Hedley & May, 1908: 109.
Schismope atkinsoni: Verco, 1910: 116.
Schismope atkinsoni: Hedley, 1911: 92.
Schismope atkinsoni: Verco, 1912a: 191.
Schismope atkinsoni: Verco, 1912b: 201.
Schismope atkinsoni: Thiele, 1912a: 28–29, pl. 3, fig. 26.
Schismope atkinsoni: Hardy, 1916: 63 [type list].
Schismope atkinsoni: Hedley, 1918: M39.
Schismope atkinsoni: May, 1921: 34, pl. 17, fig. 4
Schismope atkinsoni: May, 1923: pl. 17, fig. 4.
Schismope atkinsoni: Cotton & Godfrey, 1933: 23–24, pl. 1, fig. 12.
Schismope atkinsoni: Allan, 1934: 279, text-fig. p. 280.
Schismope atkinsoni: Cotton & Godfrey, 1938b: 2.
Schismope atkinsoni: May & Macpherson, 1958: 21, pl. 17, fig. 4.
Schismope atkinsoni: Cotton, 1959: 42–43, fig. 12.
Schismope atkinsoni: Allen, 1959: 51, text-fig. 9.
Schismope atkinsoni: Macpherson & Gabriel, 1962: 33–34, fig. 46.
Schismope atkinsoni: Iredale & McMichael, 1962: 30.
Scissurella (*Sukashitrochus*) *atkinsoni*: Lozouet, 1986: 109.
Sinezona atkinsoni: Wilson, 1993: 46.
Scissurella atkinsoni: Richmond, 1992: 13, fig. 1.
Sinezona atkinsoni: Jansen, 1995: 11, fig. 13.
Sinezona atkinsoni: Jansen, 1998: 4, figs 1–3.

Sukashitrochus atkinsoni: Bandel, 1998: 53–54, pl. 18, figs 5–8.

Sinezona sp.: Hickman, 1998: fig. 15.52 C–D.

Sinezona atkinsoni: Jansen, 1999: 53, figs 43–45.

Sukashitrochus atkinsoni: Geiger, 2003: 79, fig. 15I [umbilicus].

Sukashitrochus atkinsoni: Geiger & Jansen, 2004b: 46–55, figs 25–27 [shell, radula], 32a [map].

Sukashitrochus atkinsoni: Geiger, 2004: text-fig. p. 6.

Sukashitrochus atkinsoni: Geiger & Thacker, 2005: 49, fig. 1 [phylogeny].

Sukashitrochus atkinsoni: Geiger, 2006c: text-fig. 4.

Synonyms.

+ *Schismope carinata* Watson, 1886: 119, pl. 8, fig. 6 (Fig. 470) [non *Scissurella carinata* A. Adams, 1862 = *Suk. carinatus*]. ***Type material.*** Two syntypes (NHMUK 1877.2.9.414a–417, both destroyed by Byne's disease: A. Campbell, pers. com.). 0.052 × 0.062" [= 2.1 × 2.5 mm]. ***Type locality.*** Recent, April, 23, 1874, Port Jackson, Sydney, [New South Wales,] Australia [33.855°S, 151.247°E]. 6–15 fathoms [= NHMUK 1877.2.9.414a, 15–27 m]. Recent, Station 185B, August 31 1874. 11°38'15"S, 143°59'38"E. Raine Island, Cape York, [Queensland,] North-Eastern Australia. 155 fathoms [= 283 m], coral Sand [= NHMUK 1877.2.9.415–417] (OD). ***Etymology.*** Carina, Latin for rim. ***Remarks.*** As *carinata* Watson and *carinata* A. Adams belong to the same genus, they are also secondary homonyms (see Remarks of *Suk. carinatus*).

Schismope carinata: Kobelt, 1887: 428.

Schismope carinata Watson: Pilsbry, 1890: 65–66, pl. 68, figs 17–19.

Schismope carinata: Horst & Schepman, 1908: 487.

+ *Schismope tasmanica* Petterd, 1879: 104, not illustrated. ***Type material.*** Four syntypes (TMAG E222), broken (pers. com. L. Turner), 2 × 2 mm. ***Type locality.*** Blackman's Bay, Tasmania [Australia, 43.017°S, 147.317°E] (OD). ***Etymology.*** Named for its type locality in Tasmania.

Schismope tasmanica: Tate & May, 1902: 407, pl. 24, fig. 23 [one of syntypes].

Schismope tasmanica: Pilsbry, 1890: 68.

Schismope tasmanica: Thiele, 1912a, 30, pl. 3, fig. 31.

Sukashitrochus tasmanica [sic]: Bandel, 1998: 55–56, pl. 19, figs 4–5.

Sukashitrochus tasmanicus: Geiger, 2003: 79.

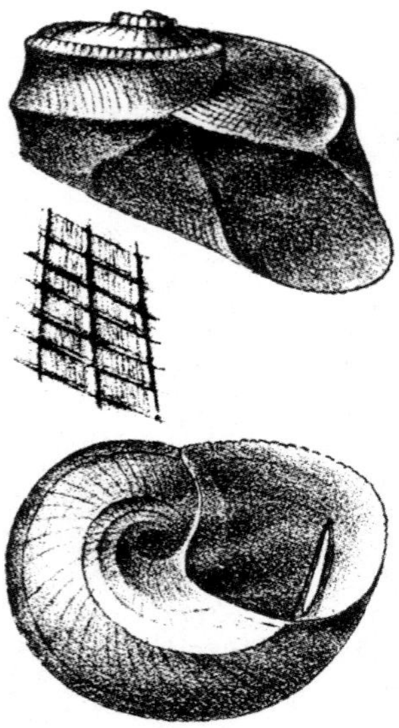

Figure 470. Original illustration of *Schismope carinata*.

Misidentifications.

Sukashitrochus declinans: Bandel, 1998: 54–55, pl. 19, figs 1–3 [is *Suk. atkinsoni*].

not *Schismope atkinsoni*: Suter, 1907: 268 [is *Suk. lyallensis*].

not *Schismope atkinsoni*: Suter, 1909: 7 [is *Suk. lyallensis*].

not *Schismope atkinsoni*: Suter, 1915: 19 [is *Suk. lyallensis*].

not *Schismope atkinsoni*: Suter, 1913: 90, pl. 6, fig. 12. [is *Suk. lyallensis*].

not *Schismope atkinsoni*: Marshall, 1915: 378 [fossil, is *Suk. lyallensis* or *Suk. ngatutura*].

Type material. "Type" (Tasmanian Museum, Hobart: Suter, 1913: see Remarks), 2 × 1 mm (OD), 2.5 × 2.25 mm (Suter, 1913). Three syntypes (TMAG E222A), one broken (pers. comm. L. Turner); listed as T.M. 5571 by Hardy (1916: 63).

Type locality. Blackman's Bay [Tasmania, Australia, 43.017°S, 147.317°E] (OD).

Etymology. Named for Rev. H. D. Atkinson (OD).

Description. Shell medium-sized to large (to 2.1 mm, usually 1–1.5 mm. Figs 471–473), trochiform. Protoconch almost one whorl, irregular reticular sculpture on outer

half, inner half smooth, varix present, connected to embryonic cap, aperture sinusoidal. Teleoconch I of 0.875–1.125 whorls, 18–31 axials, interstices with irregular crenulations with tendency to form axial pattern, no spirals. Teleoconch II to 1.3 whorls, suture impressed. Shoulder with moderately distinct to weak axial lines, axials most distinct near suture, interstices with fine crenulated axials, spirals usually present, first after 0.25 teleoconch II whorl, approximately 8–9 spirals at apertural margin. Base with usually eight distinct spiral cords; first keel at level of suture before drop of aperture in final 0.25 whorl of mature specimens, first keel of highest elevation above base but dullest, cords becoming less elevated but sharper towards umbilicus. Keel often less to hardly developed in juveniles (Fig. 472B). Some specimens, particularly from south Western Australia, devoid of keels. Umbilicus narrow, open, walls straight, carina separating at angulation to base. Umbilicus in juveniles very narrow (Fig. 472B). Selenizone with strong keels, keels rather elevated around foramen; strong lunules not coordinated with axial sculpture. Aperture rotund, roof overhanging.

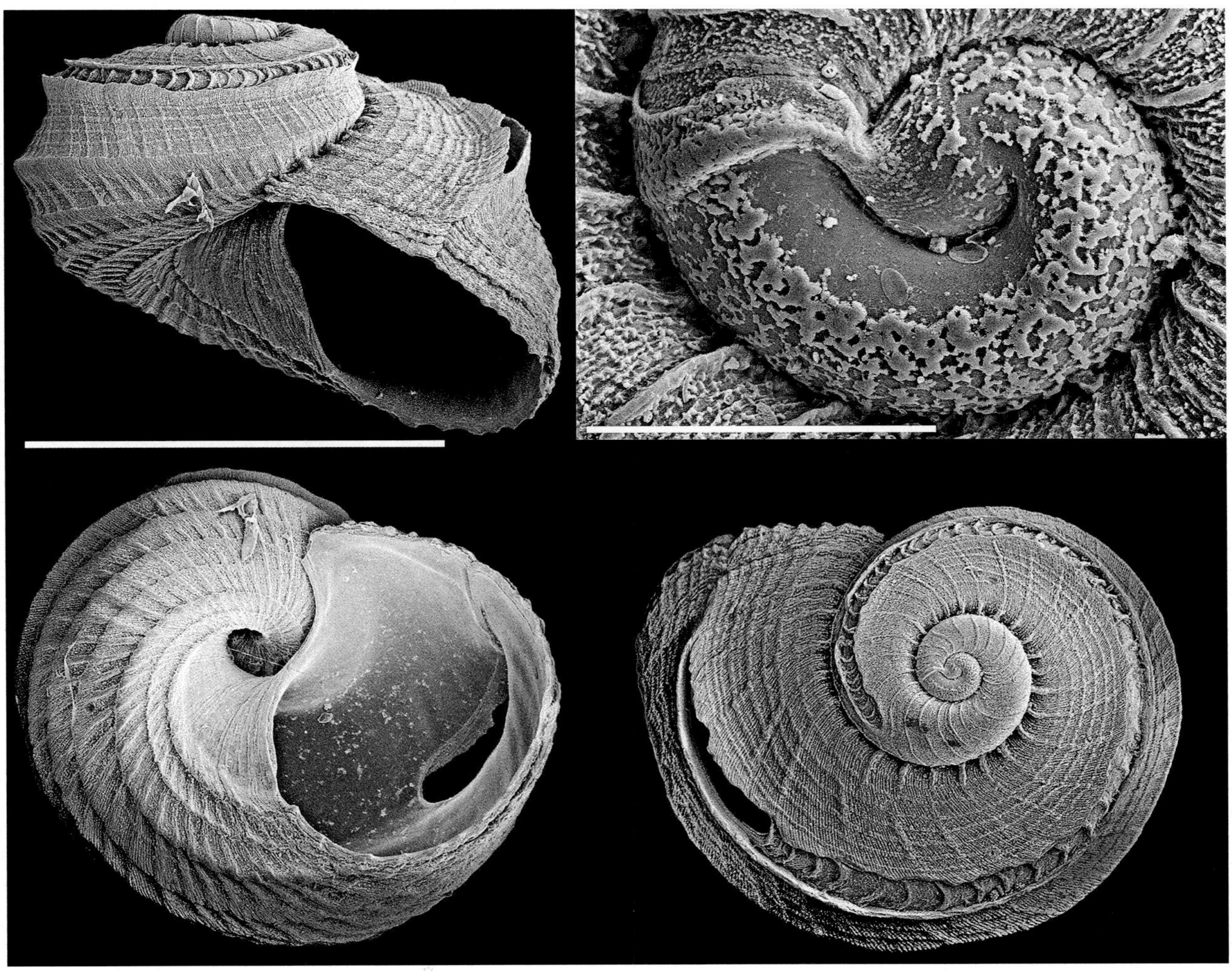

Figure 471. *Sukashitrochus atkinsoni.* 50 m, W of Green Head, Western Australia, Australia, 29.750°S, 114.713°E (AMS C.378290, 2). Scale bar shell = 1 mm. Scale bar protoconch = 100 µm.

Radula (Fig. 474). Rachidian tooth trapezoidal, tip with nine cusps of similar size, central cusp recessed. Lateral teeth 1–3 similar, base with outward-directed projection, shaft more or less straight, tip pointing inward; tips with 3–4 cusps on outer edge; inner edge smooth. Lateral tooth 4 reduced, base with slightly outward-directed projection, shaft slightly curved, tip pointed, with one indistinct cusp on either side. Lateral tooth 5 enlarged, broad base, tapering shaft, elongated-triangular

Figure 472. *Sukashitrochus atkinsoni*. **A.** Bittangabee, N of Green Cape, New South Wales, Australia, 37.217°S, 150.017°E (AMS C.346309, 11). **B.** 667 m, off Cape Martin, South Australia, Australia, 38.125°S, 140.000°E (AMS C.378547, 8). **C.** Woody Head, near Iluka, New South Wales, Australia, 29.367°S, 153.375°E (AMS C.346273, 1). Scale bar shell A = 1 mm. Scale bars shell B–C = 0.5 mm. Scale bars protoconch = 100 μm.

Scissurellidae *Sukashitrochus*

tip; tip with approximately 6–7 cusps on inner edge, outer edge smooth. Radular interlock of central field moderate. Inner marginal teeth with long shafts without food groove, tips asymmetrical, narrowly palmate, with 2–4 cusps on outer edge. Outer marginal teeth becoming more symmetrical, spoon-shaped tip, many fine bristles along margin.

Figure 473. *Sukashitrochus atkinsoni*. **A.** 8 m, Sandy Point, Western Port, just N of North Arm, Victoria, Australia, 38.405°S, 145.237°E (AMS C.377973). Radula shown in Figure 474C–E. **B.** South Cowaramup, Western Australia, Australia, 33.883°S, 114.983°E (AMS C.379071). **C.** 174 m, 40 mi S of Cape Wiles, South Australia, Australia, 35.650°S, 136.667°E (AMS C.31843). Scale bars shell = 1 mm. Scale bars protoconch = 100 µm.

Comparisons. *Sukashitrochus carinatus* from Japan has flocculent sculpture over a wider stretch of the protoconch, only leaving at most the inner 20% of the width smooth and a more narrow umbilicus.

Sukashitrochus dorbignyi from the Indian Ocean has a smooth protoconch and usually stronger keels.

Sukashtrochus estotiensis from the Oligocene of Europe and the Recent of the Indian Ocean has fine reticulate pattern over the entire shell and a smooth protoconch.

Sukashitrochus lyallensis from New Zealand has more regular reticulate protoconch sculpture and generally weaker overall sculpture on the teleoconch.

Sukashitrochus morleti from the Indo-Malayan Archipelago to the Central Pacific has fine axial protoconch sculpture.

Sukashitrochus pulcher from Ryukyu Islands, Japan, to Australia has strong cancellate sculpture, particularly with raised points on the spiral keels.

Distribution. Australia, more common in temperate waters, 0–165 m, under stones, and in algae, on semi-exposed shores (Fig. 475).

Specimen records.

Australia, New South Wales. 203 m, 12.5 mi E of Cape Byron, 28.633°S, 153.867°E (AMS C.19842, 5). 185 m, 24 km E of Ballina, 28.827°S, 153.855°E (AMS C.402612, 1). 0 m, Woody Head, 29.365°S, 153.371°E (MV F178381, 1). Woody Head, near Iluka, 29.367°S, 153.375°E (AMS C.346273, 1). Clarence River mouth, 29.425°S, 153.350°E (AMS C.346300, 1). 13 m, E of Yamba, 29.500°S, 153.367°E (AMS C.346327, 1). Nambucca Heads, 30.642°S, 153.017°E (AMS C.346307, 1). 0 m, Port Stephens, North Head, Yacaaba, 31.700°S, 152.208°E (LACM 82-51, 2). Port Stephens, 32.394°S, 152.067°E (RMNH AD 30458, 30). 35 km S of Forster, Seal Rocks, 32.439°S, 152.530°E (AMS C.346330, 1). Port Stephens, 32.694°S, 152.066°E (RMNH AD 30717, 12). 45–73 m, off Port Stephens, 32.708°S, 152.250°E (AMS C.346270, 20+).

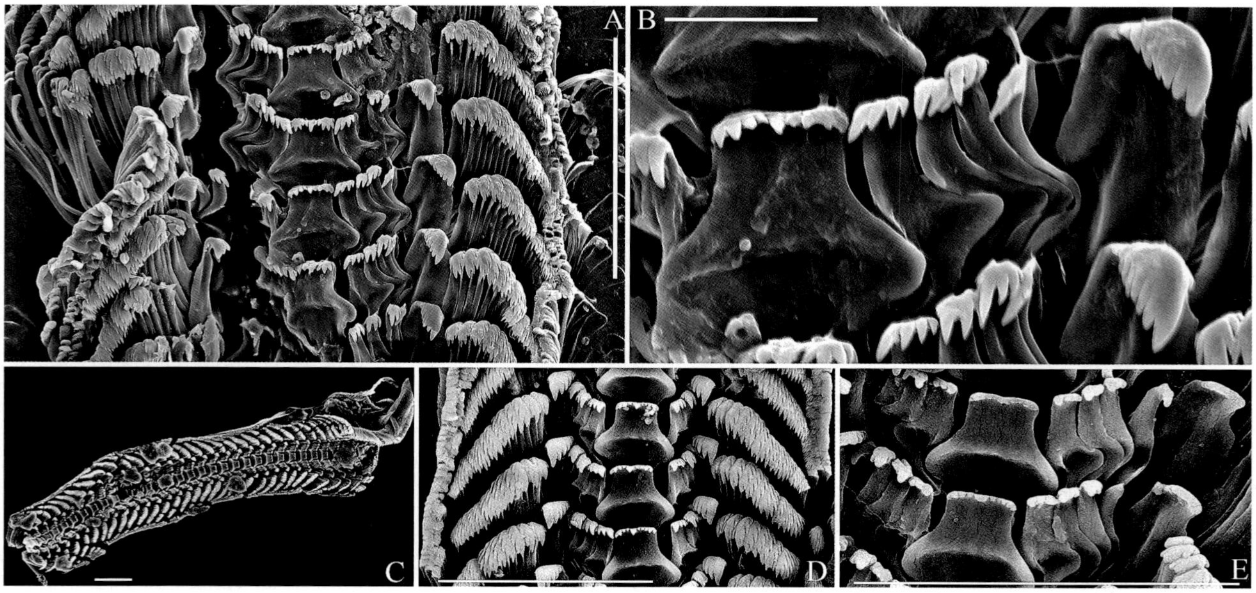

Figure 474. *Sukashitrochus atkinsoni*. Radula. **A–B.** Port Jackson, New South Wales, Australia (NMSZ unnumbered). **A.** Width of whole radula. **B.** Central field enlarged. **C–E.** 7 m, Sandy Point, Western Port, just N of North Arm, Victoria, Australia, 38.405°S, 145.237°E (AMS C.377973, 2). Shell shown in Figure 473A. **C.** Whole radula. **D.** Full width of radular row. **E.** Central field in anterior portion of used teeth, showing hook-shaped lateral tooth 4. Scale bars A, C–E = 50 µm. Scale bar B = 10 µm.

Scissurellidae *Sukashitrochus*

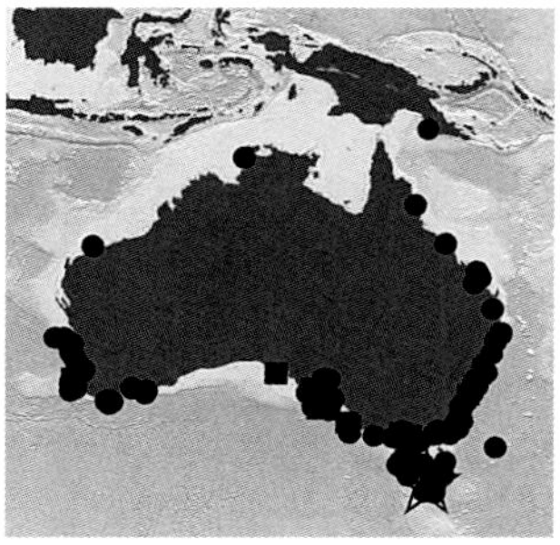

Figure 475. Distribution of *Sukashitrochus atkinsoni*. ☆ type locality. ● specimen record. ■ literature record.

North Fingal Bay, near Port Stephens, 32.746°S, 152.175°E (AMS C.346285, 2). Port Stephens, Fingal Bay, 32.750°S, 152.175°E (AMS C.346319, 6; AMS C.346343, 1). 0 m, Fingal Bay, Port Stephens, 32.750°S, 152.175°E (AMS C.461560, 1). 22 m, between Grey Cliff and Shark Island, Port Jackson, 32.750°S, 152.200°E (NMNZ M.228615, 2). Fingal Head, Port Stephens, 32.757°S, 152.175°E (AMS C.346269, 1; AMS C.404226, 2: complete; AMS C.404231, 2: complete). Balmoral, 33.050°S, 151.586°E (RMNH, 1). 0 m, The Entrance, 33.343°S, 151.504°E (MV F178685, 12). The Entrance, 33.344°S, 151.503°E (MMC). Terrigal, 33.450°S, 151.450°E (AMS C.461491, 21). Avoca Beach, 33.466°S, 151.436°E (MMC). 9–27 m, Hakesbury River, 33.570°S, 151.300°E (NMNZ M.301391, 6). Sydney, Broken Bay, Pittwater, 33.617°S, 151.300°E (AMS C.346298, 11). 146 m, 22 mi E of Narrabeen, N of Sydney, 33.700°S, 151.717°E (AMS C.25793, 1). Narrabeen, 33.714°S, 151.302°E (RMNH AD ex 30459, 2; RMNH, 20). Collaroy Beach, N of Sydney, 33.728°S, 151.300°E (AMS C.346289, 1). N of Sydney, Collaroy Beach, 33.728°S, 151.300°E (AMS C.346331, 20+). Sydney, S end of Collaroy Beach, 33.733°S, 151.300°E (AMS C.346290, 1). Collaroy Beach, N of Sydney, 33.733°S, 151.300°E (AMS C.346317, 10). Collaroy, Long Reef, Fishermans Beach, Sydney N, 33.738°S, 151.310°E (AMS C.404258, 1: complete). Collaroy, Long Reef, N side, Sydney N, 33.738°S, 151.310°E (AMS C.404969, 12: complete). Collaroy, Long Reef, N side of platform, near Fishermans Beach, Sydney N, 33.738°S, 151.310°E (AMS C.346329, 12, 3: complete). 0 m, Long Reef, Sydney, 33.742°S, 151.320°E (LACM 79-58, 2). Collaroy, Long Reef, Sydney N, 33.743°S, 151.310°E (AMS C.161107, 2: complete; AMS C.346292, 20+; AMS C.346320, 1). 0 m, Long Reef, Collaroy, 33.743°S, 151.317°E (MV F178669, 7). Collaroy, Long Reef, E end, Sydney N, 33.743°S, 151.320°E (AMS C.346303, 1). Manly, Sydney, 33.799°S, 151.286°E (MMC). 0 m, Manly, 33.799°S, 151.291°E (MV F178678, 2). Sydney, Manly Beach, 33.800°S, 151.283°E (AMS C.346312, 9). Manly, Cabbage Tree Bay, Sydney, 33.800°S, 151.295°E (AMS C.346277, 6). 0 m, Manly, Shelly Beach, Sydney, 33.801°S, 151.297°E (AMS, 4: complete). Middle Harbour, Sydney, 33.802°S, 151.233°E (AMS C.32585, 9). 5 m, Middle Harbour, Sydney, 33.802°S, 151.233°E (AMS C.346304, 1). Middle Harbour, Sydney, 33.802°S, 151.233°E (AMS C.346350, 18). Manly, Shelly Beach, outer coast, Sydney, 33.802°S, 151.297°E (AMS C.404227, 2: complete). 3 m, Manly, Shelly Beach, Sydney, 33.802°S, 151.297°E (AMS C.346310, 1). 4–5 m, Manly, Shelly Beach, Sydney, 33.802°S, 151.297°E (AMS C.346332, 1). Manly, Shelly Beach, Sydney, 33.802°S, 151.297°E (AMS C.346356, 9). 0 m, E end rock platform, Shelly Beach, Manly, Sydney, 33.802°S, 151.297°E (AMS C.461477, 2). 0 m, E of Shelly Beach, Manly, Sydney, 33.802°S, 151.297°E (AMS C.461478, 11; AMS C.4680097, 5). North Harbour, Fairlight, Sydney, 33.803°S, 151.273°E (AMS C.346294, 1; AMS C.346338, 2). 4–7 m, North Harbour, Fairlight, Sydney, 33.803°S, 151.273°E (AMS C.346306, 10). Manly, Sydney, 33.803°S, 151.283°E (AMS C.346295, 2; AMS C.346311, 3). 0 m, Shelly Beach, Manly, Sydney, 33.803°S, 151.283°E (DLG 234, 10: complete). Manly, 33.803°S, 151.283°E (WAM, 2; WAM, 3). Middle Harbour, Clontarf, Sydney, 33.808°S, 151.252°E (AMS C.346342, 1). North Harbour, SE side of Reef Bay, Sydney, 33.808°S, 151.273°E (AMS C.404970, 7: complete). North Harbour, Sydney, 33.808°S, 151.275°E (AMS C.346335, 13). 4–8 m, Middle Harbour, off Chinamans Beach, Sydney, 33.816°S, 151.248°E (AMS C.346271, 21). 11 m, off Chinamans Beach, Middle Harbour, Sydney, 33.816°S, 151.248°E (AMS C.461561, 2). Port Jackson, 33.820°S, 151.270°E (MV F 161671, 17). 12 m, Middle Harbour, N end Balmoral Beach, Sydney, 33.825°S, 151.251°E (AMS C.404967, 1: complete). 3–9 m, Middle Harbour, off Balmoral, Sydney, 33.825°S, 151.257°E (AMS C.346301, 1). Balmoral, 33.825°S, 151.257°E (WAM, 2). 3 m, off Balmoral, Middle Harbour, Sydney, 33.825°S, 151.257°E (AMS C.461565, 1). Middle Harbour, Balmoral Beach, Sydney, 33.828°S, 151.250°E (AMS C.403662, 4; AMS C.346272, 1). S end of Balmoral Beach, Middle Harbour, Sydney, 33.828°S, 151.255°E (AMS C.461486, 1). 27 m, Western Channel, Port Jackson, Sydney, 33.833°S, 151.267°E (AMS C.346302, 3). 3 m, Port Jackson, off

Sow and Pigs Reef, Sydney, 33.838°S, 151.268°E (AMS C.404154, 1). off Sow and Pigs Reef, Sydney Harbour, 33.838°S, 151.268°E (AMS C.346279, 3; AMS C.346281, 2). 11–15 m, Port Jackson, off Sow and Pigs Reef, Sydney, 33.838°S, 151.268°E (AMS C.346288, 5). Western Channel, off Sow and Pigs Reef, Sydney Harbour, 33.838°S, 151.268°E (AMS C.346308, 1). 11–16 m, Port Jackson, Sow and Pigs Reef, Sydney, 33.838°S, 151.268°E (AMS C.346347, 1). Off Georges Head, Port Jackson, Sydney, 33.842°S, 151.262°E (AMS C.346296, 1). 22 m, off Georges Head, Port Jackson, Sydney, 33.842°S, 151.262°E (AMS C.346344, 1). Green Point Port Jackson, Watsons Bay, Sydney, 33.842°S, 151.275°E (AMS C.2704, 5). 6–9 m, Sydney, Port Jackson, Watsons Bay, off Green Point, 33.842°S, 151.275°E (AMS C.346305, 3). Sydney Harbour, Nielson Park, Bottle and Glass Rocks, 33.849°S, 151.269°E (AMS C.402954, 2; AMS C.17545, 2). 9 m, Sydney Harbour, Nielson Park, off Bottle and Glass Rocks, 33.849°S, 151.269°E (AMS C.346348, 16). 33 m, Sydney, Port Jackson, off Balls Head, 33.850°S, 151.192°E (AMS C.346274, 5). Off Watson's Bay, Port Jackson, 33.850°S, 151.250°E (NMNZ M.226149, 1). 33 m, between Balls Head and Goat Island, Port Jackson, 33.852°S, 151.197°E (AMS C.346276, 3). 9 m, off Parsley Bay, Port Jackson, Sydney, 33.852°S, 151.276°E (AMS C.461488, 1). Sydney, 33.853°S, 151.212°E (MV F 161678, 2; MV F 161660, 4; MV F 161670, 2; KBIN, 1; KBIN, 4). Port Jackson, 33.855°S, 151.247°E (NMSZ 1961.61, 8). Port Jackson, 33.858°S, 151.242°E (RMNH, 1; RMNH, 2). Sydney, Little Coogee Bay, 33.922°S, 151.260°E (AMS C.346339, 3). Long Bay, Sydney, 33.941°S, 151.257°E (AMS C.468100, 9). Sydney, N of Cronulla, Boat Harbour, 34.042°S, 151.200°E (AMS C.404228, 2: complete). Cronulla, 34.050°S, 151.150°E (WAM, 1). 0 m, Gunnamatta Beach, Cronulla, 34.059°S, 151.149°E (MV F178682, 7). 5 m, Sydney, Port Hacking, Gunnamatta Bay, 34.066°S, 151.142°E (AMS C.406400, 1). Sydney, Port Hacking, Gunnamatta Bay, 34.066°S, 151.142°E (AMS C.346318, 2; AMS C.346341, 2; AMS C.461566, 2). Gunnamatta Bay,34.066°S, 151.133°E (WAM, 1). Port Hacking, 34.067°S, 151.100°E (AMS C.403740, 20+). Sydney S, Port Hacking, SW end of Gunnamatta Bay, 34.072°S, 151.145°E (AMS C.346337, 1). Port Hacking, Bundeena, Simpsons Beach, 34.083°S, 151.133°E (AMS C.346314, 1). 183 m, 16 mi E of Wollongong, 34.417°S, 151.250°E (AMS C.17989, 1). 115–137 m, 5–8 mi off Port Kembla, 34.465°S, 151.075°E (AMS C.16222, 2). Shell Harbour, 34.583°S, 150.867°E (AMS C.346280, 4; WAM, 30; AMS C.468098, 1). 0 m, Kiama, 34.670°S, 150.856°E (MV F178681, 7). Kiama, 34.671°S, 150.856°E (MMC). Werri Beach, Gerringong, 34.738°S, 150.833°E (AMS C.346297, 3). Jervis Bay, Callala Point, 35.008°S, 150.722°E (AMS C.404224, 2: complete). Jervis Bay, Huskisson, 35.042°S, 150.667°E (AMS C.346349, 1). Huskisson, Jervis Bay, 35.050°S, 150.667°E (AMS C.461487, 3). 0 m, Tapalla Point, Huskisson, Jervis Bay, 35.050°S, 150.679°E (MV F178666, 3). Honeymoon Beach, Jervis Bay, 35.058°S, 150.775°E (AMS C.346345, 1). 0 m, Plantation Point, Jervis Bay, 35.071°S, 150.697°E (MV F178667, 2). 0 m, S Jervis Bay, 35.125°S, 150.753°E (LACM 79-43, 3). Ulladulla, 35.358°S, 150.475°E (AMS C.346313, 1). Ulladulla, 35.358°S, 150.475°E (AMS C.346333, 9). 12 m, NW ide of Brush Island, N of Bateman Bay, 35.528°S, 150.416°E (AMS C.461483, 1). Murramarang Point, 35.533°S, 150.408°E (AMS C.461484, 2). Pebbly Beach, 35.613°S, 150.328°E (AMS C.346340, 6). Batemans Bay, 35.717°S, 150.200°E (AMS C.346275, 2). 24 m, S of Batemans Bay, N of Burrewarra Point, 35.830°S, 150.235°E (AMS C.448652, 4: complete). 23 m, S of Batemans Bay, N of Burrewarra Point, 35.831°S, 150.235°E (AMS C.448653, 1: complete). 0 m, Mossey Point, Batemans Bay, 35.843°S, 150.180°E (MV F178672, 10). 23 m, N of Burrewarra Point, S of Batemans Bay, 35831.000°S, 150.234°E (AMS C.461480, 2). 9–16 m, off Montague Island, Narooma, 36.250°S, 150.217°E (AMS C.346328, 1). Merimbula Jetty, Merimbula, 36.900°S, 149.925°E (AMS C.346334, 1). 36 m, 10 mi N of Eden, 36.950°S, 150.000°E (AMS C.346291, 1). Off Eden, 37.007°S, 150.327°E (AMS C.468101, 1). 0 m, Aslings Beach, Twofold Bay, 37.050°S, 149.908°E (AMS C.461479, 6). 0 m, Eden, 37.063°S, 149.910°E (MV F178668, 2). 0 m, Eden Beach, 37.063°S,

149.910°E (MV F178671, 7). Eden Beach, southern, 37.073°S, 149.897°E (MMC). Quarantine Bay, Twofold Bay, 37.077°S, 149.885°E (MV F162111, 1: complete). 0–3 m, Murrumbulga Point, Twofold Bay, 37.079°S, 149.885°E (AMS C.404225, 2: complete). 2–9 m, Murrumbulga Point, Twofold Bay, 37.079°S, 149.885°E (AMS C.404230, 2: complete). Twofold Bay, 37.083°S, 149.900°E (AMS C.402949, 2; AMS C.346316, 2). 37 m, Twofold Bay, 37.083°S, 149.950°E (AMS C.346299, 1). Off Eden, 37.083°S, 150.000°E (AMS C.402707, 2). 0 m, Nullica Beach, Eden, Twofold Bay, 37.100°S, 149.875°E (AMS C.461548, 2). Bittangabee, N of Green Cape, 37.217°S, 150.017°E (AMS C.346309, 11). 6 m, N of Green Cape, Bittangabee, 37.217°S, 150.018°E (AMS C.ex 461485, 1; AMS C.461490, 2). 10 m, Green Cape, S side, 37.267°S, 150.050°E (AMS C.346287, 3). 300 m, off Cape Howe, 37.506°S, 149.979°E (NMNZ M.226320, 1). NE of Grafton, S of Yamba, Angourie, 39.483°S, 153.358°E (AMS C.461557, 1).

Australia, Northern Territories. 27–37 m, 36 km off Point Charles, Darwin, 12.167°S, 130.367°E (AMS C.78621, 2).

Australia, Queensland. Great Barrier Reef, Michaelmas Cay, 16.600°S, 145.983°E (AMS C.348640, 16). 22 m, off Black Island, near Hayman Island, 20.082°S, 148.840°E (NMNZ M.226003, 1). 24 m, Great Barrier Reef, Whitsunday Passage, 20.225°S, 148.792°E (AMS C.378582, 2). 64 m, NE of Rockhampton, 22.833°S, 151.650°E (AMS C.378586, 5). 36 m, Great Barrier Reef, Capricorn Group, Heron Island, 23.433°S, 151.950°E (AMS C.378600, 1). 13 m, Port Curtis, 23.917°S, 151.383°E (AMS C.378580, 1). Noosa Heads, 26.383°S, 153.100°E (AMS C.378602, 1).

Australia, South Australia. 20–30 m, Nuyts Archipelago, N of St Francis Island, 32.483°S, 133.300°E (AMS C.378548, 1). 0 m, Corvisart Bay, Eyre Peninsula, 32.899°S, 134.077°E (MV F178683, 2). Corvisart Bay, Eyre Peninsula, 32.899°S, 134.077°E (MMC). Whyalla, mangrove forest near town, 33.033°S, 137.583°E (AMS C.378298, 1). The Bluff, S of Port Rickaby, Spencer Gulf, 33.100°S, 138.167°E (WAM, 2). Spencer Gulf, Eyre Peninsula, Arno Bay, 33.933°S, 136.583°E (AMS C.378303, 2; AMS C.378305, 1). Wallaroo, 33.933°S, 137.633°E (WAM, 4). Spencer Gulf, Tumby Bay, 34.367°S, 136.133°E (AMS C.378300, 5). Yorke Peninsula, Bluff Beach, 34.733°S, 137.483°E (AMS C.378314, 1; AMS C.402637, 2; AMS C.402639, 1). 0 m, Bluff Beach, Yorke Peninsula, 34.733°S, 137.483°E (AMS C.: complete). Spencers Gulf, Hardwicke Bay, 34.750°S, 137.300°E (AMS C.13409, 7; AMS C.13410, 7). 0 m, Outer Harbour, 34.785°S, 138.479°E (MV F178382, 2). Outer Harbor, 34.787°S, 138.481°E (MMC). Point Souttar, Yorke Peninsula, 34.900°S, 137.267°E (AMS C.378306, 1). Corney Point, Yorke Peninsula, 34.917°S, 137.083°E (MMC). 0 m, Corney Point, Yorke Peninsula, 34.919°S, 137.084°E (MV F178675, 1). Adelaide, Henley Beach, 34.920°S, 138.494°E (AMS C.23174, 1). Henley Beach, 34.933°S, 138.517°E (AMS C.378317, 1). 0 m, Point Turton, Magazine Bay, 34.948°S, 137.360°E (MV F178680, 5). Glenelg Beach, S of Adelaide, 34.960°S, 138.530°E (AMS C.436772, 4). 0 m, Glenelg Beach, Adelaide, 34.980°S, 138.510°E (MV F178679, 3). Glenelg Beach, Adelaide, 34.980°S, 138.513°E (MMC). West Beach, Adelaide, 34.983°S, 138.517°E (WAM, 2). Outer Harbor, Adelaide, 34.983°S, 138.517°E (WAM, 1). Holdfast Bay, 35.000°S, 138.500°E (AMS C.11803, 4). 174–183 m, 40 mi S of Cape Wiles, 35.650°S, 136.667°E (AMS C.378307, 7; AMS C.31843, 7). American Beach, Kangaroo Island, 35.765°S, 137.882°E (MV F178385, 1). American Beach, Kangaroo Island, 35.766°S, 137.552°E (MMC). 75–155 m, between Cape Jaffa and Kangaroo Island, 36.000°S, 138.167°E (AMS C.377985, 1). Gulf of St. Vincent, 38.044°S, 144.907°E (MV F 161667, 2). 0 m, Racecourse Reef, Port Macdonnell, 38.056°S, 140.752°E (MV F178684, 1). Racecourse Reef, Port Maconnell, 38.056°S, 140.752°E (MMC). 667 m, off Cape Martin, 38.125°S, 140.000°E (AMS C.378302, 3; AMS C.378547, 8).

Australia, Tasmania. 64 m, N of Hummock Island, 40.158°S, 145.425°E (AMS C.377976, 1). 40 m, Bass Strait, 40.370°S, 145.283°E (MV F 161663, 1). Cable, Bass Strait, 40.500°S, 145.600°E (MV F 15687, 6). 27 m, Bass Strait, 40.515°S, 144.933°E (MV

F 161658, 8). 44 m, Bass Strait, 40.600°S, 145.383°E (MV F 161659, 1). Station 117 S, Central Bass Strait, 20 km NNE of North Point, 40.630°S, 145.380°E (MV F 161665, 1). Stanley, 40.765°S, 145.294°E (MV F161661, 7). 399 m, off Cape Naturaliste, 40.843°S, 148.775°E (AMS C.377977, 5). West Head, Green's Beach, Tamar River mouth, 41.083°S, 146.750°E (AMS C.377917, 1). 12 m, Horseshoe Reef, off Devonport, 41.150°S, 146.417°E (AMS C.378008, 2). 88 m, S of West Point, 41.153°S, 144.403°E (AMS C.377967, 1). 113 m, off Long Point, N of Bicheno, 41.758°S, 148.517°E (AMS C.377987, 3; AMS C.406336, 15). 113 m, N of Great Oyster Bank, 41.758°S, 148.517°E (LACM 73-156, 3). 28 m, S of Cape Lodi, 42.000°S, 148.300°E (AMS C.377963, 1). 148 m, E of Cape Lodi, 42.000°S, 148.588°E (LACM 73-155, 1). 205 m, off Cape Forestier, 42.166°S, 148.578°E (AMS C.436785, 5). Schouten Island, 42.318°S, 148.291°E (MCZ 023252, 3). 45 m, Great Oyster Bay, 42.333°S, 148.217°E (AMS C.377970, 1). 82 m, Maria Island, 2.5 mi NE Beaching Bay, 42.458°S, 148.200°E (AMS C.377975, 1). Derwent Estuary, 42.917°S, 147.392°E (AMS C.377969, 3). Frederick Henry Bay, 42.917°S, 147.583°E (AMS C.10770, 1). D'Entrecasteaux Channel, 43.045°S, 147.347°E (AMS C.378009, 3). 15 m, Tinderbox Bay, S of Hobart, 43.050°S, 147.333°E (AMS C.377974, 6). Blow Hole, Tasman Peninsula, 43.150°S, 147.850°E (WAM, 3). 8 m, S Bruny Island, Great Taylor Bay, 43.450°S, 147.167°E (AMS C.404904 2: complete). 73 m, 2 mi S Tasman Head, S Bruny Island, 43.563°S, 147.323°E (AMS C.377968, 1). 104 m, S of D'Entrecasteaux Channel, 43.673°S, 146.840°E (AMS C.402706). 212 m, S of Storm Bay, 43.783°S, 147.808°E (AMS C.402635, 3).

Australia, Victoria. Station 12A1, Southern Port Philip (MV F88152, 9). Mallacoota, 37.564°S, 149.765°E (MMC; AMS C.50390, 20+). 0 m, Mallacoota, 37.569°S, 149.763°E (MV F178676, 20). 95 m, 15 km S of Gabo Island, 37.717°S, 149.957°E (AMS C.377981, 1). 0 m, Lakes Entrance, 37.880°S, 147.992°E (MV F178677, 1). Off Lakes Entrance, 37.880°S, 148.000°E (MV F 161683, 3). 75–78 m, between Cape Howe and Lakes Entrance, 37.917°S, 149.000°E (AMS C.404966, 2). 15 m, off Point Cook, 37.930°S, 144.800°E (MV F 161679, 5). 2–5 m, Port Philip Bay, 37.961°S, 144.805°E (MV F795287: complete). Lakes Entrance, 37.879°S, 147.992°E (MMC). 55 m, Eastern Bass Strait, 38.083°S, 148.142°E (MV F 161681, 9). NW of Popes Eye, Port Phillip, 38.100°S, 144.883°E (WAM 13391, 1). Portarlington, 38.120°S, 144.670°E (MV F 20669, 9). Frankston, 38.143°S, 145.117°E (MV F 161677, 17; MV F 161684, 2). Port Phillip, NW of Popes Eye, 38.150°S, 144.867°E (WAM, 1). 50 m, Bass Strait, 20 km S of Lakes Entrance, 38.250°S, 148.000°E (AMS C.377979, 1). 165–274 m, Bass Strait, ca. 27 mi SE of Cape Everard, 38.250°S, 149.200°E (AMS C.378299, 2). 47 m, Bass Strait, SW of Lakes Entrance, 38.283°S, 147.667°E (AMS C.377984, 1). 5–9 m, Western Port, 38.367°S, 145.533°E (AMS C.18667, 4). Point Leo (N edge of Bobbanaring Point), W side of Western Port Bay, 38.367°S, 145.533°E (LSC, 1). Western Port Bay, 38.367°S, 145.533°E (MV F 160840, 40; MV 161674, 2; MV 161666, 2). 10 m, between Stony Point and French Island, Western Port Bay, 38.372°S, 145.240°E (LACM 87-89, 17). Crib Point area, Western Port Bay, 38.378°S, 145.223°E (MMC). Off Crib Point, 38.381°S, 145.224°E (MMC). Port Fairy, 38.383°S, 142.233°E (AMS C.377978, 14). Tankerton Jetty Area, French Island, 38.383°S, 145.272°E (MMC). 0 m, Tankerton, French Island, 38.389°S, 145.277°E (MV F178674, 4). 0 m, Griffiths Island, Port Fairy, 38.392°S, 142.245°E (MV F178383, 2). Griffiths Island, Port Fairy, 38.392°S, 142.247°E (MMC). 7 m, Sandy Point, Western Port, just N of North Arm, 38.405°S, 145.237°E (AMS C.377973, 2). Shoreham, Western Port, 38.430°S, 145.030°E (MV F 161676, 2). 16 m, Western Port, NE of Cowes, 38.439°S, 145.253°E (AMS C.377980, 1). Western Port Bay, 38.460°S, 145.340°E (MMC). Rhyll, 38.463°S, 145.312°E (MV F 884, 100). Western Port Bay, 38.500°S, 145.300°E (MV F178384, 1). 3–5 m, San Remo, 38.519°S, 145.363°E (MV F88383, 2, 5: complete). Lorne, 38.550°S, 143.967°E (WAM, 1; WAM, 2). 0 m, Maitland Beach, Bear Gully, 38.893°S, 145.982°E (MV F178673, 1). Maitland Beach, Bear Gully, Waratah Bay, 38.893°S, 145.982°E (MMC).

Off Wilson's Promontory, 38.933°S, 146.367°E (MV F 161680, 2). 81-HK-1, Station 138, Central Bass Strait, 32 km WSW of Settlement Point, Flinders Island, 40.166°S, 147.531°E (MV F83187, 5: complete).

Australia, Western Australia. 33 m, Kendrew Island, Dampier Archipelago, 20.475°S, 116.533°E (WAM S13387, 1). 165 m, W of Dongara, 29.142°S, 113.092°E (AMS C.378666, 6). 183 m, W of Dongara, 29.142°S, 113.913°E (AMS C.378661, 1). 219 m, W of Dongara, 29.183°S, 113.900°E (AMS C.378273, 7). 152–157 m, off Dongara, 29.550°S, 114.325°E (AMS C.378651, 1). 137 m, W of Cliff Head, 29.563°S, 114.283°E (AMS C.378286, 1). 274–283 m, NW of Beagle Island, 29.717°S, 114.283°E (AMS C.378664, 2). 183 m, NW of Beagle Island, 29.725°S, 114.333°E (AMS C.378665, 1). 50 m, W of Green Head, 29.750°S, 114.713°E (AMS C.378290, 2). 223–245 m, off Jurien Bay, 30.133°S, 114.500°E (AMS C.378656, 3). 154 m, W of Cervantes, 30.550°S, 114.533°E (AMS C.378660, 5). 212–215 m, WNW of Lancelin, 30.966°S, 114.817°E (WAM, 3). 154 m, WNW of Lancelin, 30.967°S, 114.883°E (AMS C.378287, 2). 146 m, W of Rottnest Island, 31.000°S, 114.850°E (AMS C.378644, 1). 140 m, off Rottnest Island, 31.633°S, 115.133°E (AMS C.378657, 2). 62 m, off Rottnest Island, 31.633°S, 115.227°E (AMS C.378280, 1). 160 m, off Rottnest Island, 31.650°S, 115.080°E (AMS C.378643, 3). 124 m, off Rottnest Island, 31.650°S, 115.167°E (AMS C.378633, 1). Off Rottnest Island, 31.670°S, 115.155°E (AMS C.378653, 2). 116 m, off Rottnest Island, 31.673°S, 115.198°E (AMS C.378552, 8). 100 m, Direction Bank, off Rottnest Island, 31.677°S, 115.160°E (AMS C.378659, 1). 148 m, off Rottnest Island, 31.688°S, 115.143°E (AMS C.378288, 4; AMS C.378645, 1). 164 m, off Rottnest Island, 31.690°S, 115.098°E (AMS C.378632, 3). 140 m, off Rottnest Island, 31.702°S, 115.177°E (AMS C.404955, 1). 110 m, off Rottnest Island, 31.702°S, 115.222°E (AMS C.377965, 1). 150 m, off Rottnest Island, 31.703°S, 115.137°E (AMS C.378635, 2). 143 m, off Rottnest Island, 31.708°S, 115.187°E (AMS C.378276, 4). 120 m, Direction Bank, off Rottnest Island, 31.723°S, 115.230°E (AMS C.402689, 1). 158 m, Direction Bank, off Rottnest Island, 31.730°S, 115.163°E (AMS C.378274, 4). 155 m, off Rottnest Island, 31.732°S, 115.158°E (AMS C.378284, 1). 150 m, Direction Bank, off Rottnest Island, 31.735°S, 115.177°E (AMS C.378654, 2). 138 m, Direction Bank, off Rottnest Island, 31.740°S, 115.203°E (AMS C.378641, 1). 118 m, off Rottnest Island, 31.747°S, 115.230°E (AMS C.378271, 1). 105 m, Direction Bank, off Rottnest Island, 31.747°S, 115.250°E (AMS C.378649, 1). 165 m, off Rottnest Island, 31.748°S, 115.138°E (AMS C.378640, 2). 93 m, Direction Bank, off Rottnest Island, 31.750°S, 115.273°E (AMS C.378291, 3). 134 m, off Rottnest Island, 31.777°S, 115.213°E (AMS C.378658, 2). 150 m, off Rottnest Island, 31.778°S, 115.170°E (AMS C.378275, 1). 164 m, off Rottnest Island, 31.793°S, 115.157°E (AMS C.379970, 2). Bathurst Point, Basin side of Rottnest Island, 31.983°S, 115.533°E (AMS C.379069, 1). Rottnest Island, 32.000°S, 115.500°E (WAM, 2; WAM, 20; WAM, 3; WAM, 1). Bathhurst Point Area, Rottnest Island, 32.000°S, 115.550°E (WAM S13378, 15). Cockburn Sound, Jervoise Groyne, 1 mi S of Woodmans Pt, 32.150°S, 115.767°E (AMS C.378282, 1). Garden Island, 32.233°S, 115.683°E (WAM, 8). Perkin Point, Garden Island, 32.233°S, 115.683°E (WAM, 1). Garden Island, 32.233°S, 115.683°E (WAM, 4; WAM S13381, 17; WAM ex S13381, 17). 210–212 m, W of Garden Island, 32.250°S, 115.117°E (AMS C.378630, 1). 2–3 m, Cape Peron, 30 mi S of Perth, 32.267°S, 115.683°E (AMS C.402622, 1). Penguin Island, Rockingham, 32.300°S, 115.683°E (WAM S13375, 2; WAM ex S13375, 2). Rockingham, Penguin Island, 32.300°S, 115.683°E (WAM S13385, 2). Safety Bay, 32.300°S, 115.700°E (WAM, 2). 146–150 m, SW of Mandurah, 32.683°S, 114.850°E (AMS C.378647, 1). 192–179 m, SW of Mandurah, 32.717°S, 114.800°E (AMS C.378283, 1). 220 m, SW of Mandurah, 32.750°S, 114.783°E (AMS C.378292, 2). 201–228 m, NW of Bunbury, 33.250°S, 114.533°E (WAM, 1). 161–165 m, NW of Bunbury, 33.250°S, 114.600°E (AMS C.378655, 2). 155 m, NW of Bunbury, 33.250°S, 114.617°E (AMS C.378278, 1). 183–176 m, SW of Cape Naturaliste, 33.585°S, 114.523°E (AMS C.378650, 3). Gracetown,

33.670°S, 115.010°E (WAM, 3; WAM, 17). Hopetoun, 33.866°S, 120.133°E (NMNZ M.245175, 2; WAM, 1; WAM, 9; WAM, 5). S Cowaramup, 33.883°S, 114.983°E (AMS C.379071, 7). S Cowaramup, 33.883°S, 114.983°E (AMS C.379071, 7). S of Cowaramup Bay, near mouth of Margaret River, 33.883°S, 115.017°E (AMS C.378295, 4; AMS C.378295-2, 4). Hopetoun, 33.945°S, 120.126°E (MMC). Kilcarnup, N side of Margaret River, 33.950°S, 114.983°E (AMS C.406359, 4). 1–2 m, Near Hopetoun Jetty, 33.950°S, 120.117°E (AMS C.378293, 1). Hopetoun, Two Mile Beach, 33.950°S, 120.117°E (AMS C.378546, 2). 81 m, Recherche Archipelago, 34.150°S, 121.450°E (AMS C.378652, 4). 148 m, W of Cape Hamelin, 34.175°S, 114.517°E (WAM, 1). 158 m, Great Australian Bight, E of Hood Point, 34.417°S, 121.333°E (AMS C.378551, 2). Albany, 35.017°S, 117.892°E (WAM, 3). 62 m, off Albany, 35.068°S, 118.107°E (AMS C.406348, 1; AMS C.378279, 1). 75 m, off Albany, 35.240°S, 118.342°E (AMS C.378550, 7). 150 m, off Albany, 35.407°S, 118.047°E (AMS C.378294, 1).

Literature records.

Australia. Flinders [38.483°S, 145.017°E], off Rhyll [38.467°S, 145.300°E], Western Port [38.367°S, 145.533°E]; Cape Schanck [38.500°S, 144.883°E]; Puebla Coast, Victoria (Pritchard & Gatliff, 1902).—Port Kembla [34.500°S, 150.900°E], 63–75 fms. (115–137 m) (Hedley, 1903).—Bunbury [33°S, 114.683°E]. Guichen Bay [37.167°S, 139.750°E], Beachport [37.500°S, 140.017°E], St. Francis Island [32.533°S, 133.333°E], Arno Bay [33.933°S, 136.583°E] (Weeding): South Australia. Blackman's Bay, Tasmania. (Cotton, 1959).—Rhyll, Western Port, Flinders, Cape Schanck, Portarlington [38.117°S, 144.650°E], Victoria (Macpherson & Gabriel, 1962).

Remarks. The publication date for *atkinsoni* is usually given as 1876. Although the species was described in the volume for the year 1876, that volume was actually published in 1877.

Suter (1913: 90) referred to the "type" as a singular in the Tasmanian Museum, Hobart, and gave dimensions of the specimens that are different from those cited in the original description. Bruce Marshall (pers. com.) pointed out that Suter never visited Tasmania and never saw the type specimens. Suter's indication is not regarded as a lectotype designation, and the source of Suter's measurements is not known.

The epithet *carinata* Watson, 1886, is a synonym of *Suksashitrochus atkinsoni*, and because it belongs in the same genus as *Sukashitrochus carinatus*, is also a secondary homonym.

The specimen illustrated by Bandel (1998: pl. 19, figs 1–3) as "*Suk. declinans*" [= *mirifica*] from Sydney, New South Wales, is actually *Suk. atkinsoni*. *Scissurella mirifica* has an open slit even in a fully mature shell, a rounded base with no trace of a keel, an umbilicus whose slope is continuous with the rounding of the base, and has a more tropical distribution, extending south only to Queensland (Watson, 1886; Jansen, 1999).

The differences between *Suk. atkinsoni*, Bandel's (1998) "*Suk. declinans*," and *Suk. tasmanica* [sic] cited by Bandel (1998) cannot be maintained. The first two supposedly show differences in the placement of the selenizone, whereas his illustrations are nearly identical. The cited differences in protoconch sculpture between *Suk. atkinsoni* and *Suk. tasmanica* cannot be seen in his illustrations. All three of his specimens clearly belong to the same species.

Most specimens have been obtained from temperate Australia. In tropical Australia, mostly *Suk. morleti* (= *Suk. indonesicus*, = *Suk. simplex*) is found. The two species are quite similar in overall shell morphology, but the protoconch sculpture readily distinguishes them. *Sukashitrochus atkinsoni* shares with *Suk. lyallensis* flocculent sculpture, whereas *Suk. morleti* has fine axials. Some specimens from tropical Queensland show the typical *Suk. atkinsoni* protoconch (AMS C.348640

from Michaelmas Cay, Great Barrier Reef). One cannot discount mislabeling of those specimens, although there are no indications of such problems. A cryptic species could also be postulated, although nothing else beyond the distributional anomaly points to that possibility. *Sukashitrochus atkinsoni* may also be an eurythermal species.

Some specimens of *Suk. atkinsoni* are devoid of the typical spiral keels that define the genus (AMS C.378552, WAM S13387). Other characters such as all protoconch characteristics, teleoconch I whorl number, number of axials and spirals, as well as umbilical characters cannot separate those specimens from the ones with spiral keels in the base. There are also all intermediate stages, from strong keels to weak keels to faint keels, so that the absence of keels is just one further step in that series. Hence, there is no justification for separating the keel-less specimens.

The New Zealand records of *Suk. atkinsoni* most likely refer to *Suk. lyallensis* (cf. Powell, 1979; more likely) or *Sin. iota* (cf. Powell, 1955). The synonymy between *atkinsoni* and *carinata* Watson (*non* A. Adams) was first suggested by Hedley (1903).

Sukashitrochus carinatus (A. Adams, 1862)

Figures 476–478

Chresonymy.

Scissurella carinata A. Adams, 1862: 346, not illustrated [non *Schismope carinata* Watson, 1886 = *Suk. atkinsoni*].
Schismope carinata: Crosse, 1863: 109.
Scissurella carinata: Troschel, 1863: 702.
Scissurellam [*sic*] *carinatam* [*sic*]: Dunker, 1882: 147.
Scissurella carinata: Paetel, 1883: 47.
Scissurella carinata: Paetel, 1888: 289.
Scissurella carinata: Pilsbry, 1890: 67.
Schismope carinata: Pilsbry, 1895: 106.
Schismope carinata: Thiele, 1912a: 26–27, pl. 3, fig. 25.
Woodwardia carinata: Kuroda & Kinoshita, 1951: 8.
Woodwardia carinata: Habe, 1951: 68, pl. 11, figs 5–6.
Woodwardia carinata: Kuroda & Habe, 1952: 98.
Woodwardia carinata: Kawamoto & Tanabe, 1956: 3, pl. 2, fig. 14 [copy figure Habe, 1951].
Woodwardia carinata: Habe, 1961: 2, pl. 1, fig. 9.
Sukashitrochus carinata: Habe, 1964: 2, fig. 9.
Sukashitrochus carinatus: Kuroda *et al.*, 1971: 4, pl. 106, figs 2–3.
Sukashitrochus carinatus: Higo, 1973: 14.
Sukashitrochus carinatus: Honma & Kitami, 1979: 28.
Sukashitrochus carinatus: Inaba, 1982: 72
Sukashitrochus carinatus: Inaba, 1983: 72.
Sukashitrochus carinatus: Habe *et al.*, 1986: 4.
Scissurella (*Sukashitrochus*) *carinata*: Lozouet, 1986: 109.
Sukashitrochus carinatus: Tsuchida *et al.*, 1991: 6–7, pl. 1, figs 8–10.
Sukashitrochus carinatus: Fukuda *et al.*, 1992: 49, pl. 41, fig. 18.

Sukashitrochus carinatus: Higo & Goto, 1993: 15.

Sukashitrochus carinata: Okutani & Hasegawa, 2000: 37, pl. 18, fig. 8.

Sukashitrochus carinatus: Kwon *et al.*, 2001: 16, 47, figs 24-1–24-3.

Woodwardia carinata: Okamoto, 2001: 27, text-fig. [copy Habe, 1951].

Sukashitrochus carinatus: Higo *et al.*, 2001: 14: G83 [one of two syntypes].

Sukashitrochus carinatus: Geiger, 2003: 79.

Sukashitrochus carinatus: Min, 2004: 85, figs 56, 56-1, 56-2.

Sukashitrochus carinatus: Anon., 2005: 36, text-fig. 2.

Sukashitrochus carinatus: Sasaki, 2008: fig. 2F.

Sukashitrochus carinata [*sic*]: Koyama, 2008: 253, pl. 1, fig. 7.

Misidentifications.

not *Schismope carinata*: Smith, 1910: 207, pl. 8, fig. 3 [is *Suk. dorbignyi*].

not *Scissurella carinata*: Smith, 1961: 11, text-fig. [unrecognizable].

not *Sinezona carinata*: Ladd, 1982: 20, pl. 1, figs 3–4 [is likely *Suk. morleti,* Pleistocene of Vanuatu].

not *Sinezona carinata*: Jansen, 1999: 54, figs 49–51 [is *Suk. morleti*].

Figure 476. *Sukashitochus carinatus*. Lectotype (here designated) of *Scissurella carinata*. 35 fathoms [= 64 m], Okosiri (= Okujiri Island), Hokkaido [Japan, 42.170°N, 139.520°E] (NHMUK 1878.1.28.515/1). Scale bar shell = 1 mm. Scale bar protoconch = 100 µm.

Type material. Three (originally four) syntypes (NHMUK 1878.1.28.515. Higo *et al.*, 2001), 2.0 mm, 2.0 mm. The best specimen, shown here in Figure 476, is designated as the lectotype NHMUK 1878.1.28.515/1 with the express purpose of taxon stabilization in case additional non-conspecific syntypes should be located. Paralectotypes from type locality (NHMUK 1878.1.28.515/2–3, 2).

Type locality. Okosiri, 35 fathoms [= 64 m], [Hokkaido, Japan]; Seto-Uchi, 16 fathoms [= 29 m], [Kyushu, Japan]; and Goto, 71 fathoms [= 130 m], [Kyushu, Japan] (OD). Type locality explicitly restricted to Okosiri (= Okujiri Island), Hokkaido [Japan, 42.170°N, 139.520°E] 35 fathoms [= 64 m] (Kuroda *et al.*, 1971: 4).

Figure 477. *Sukashitochus carinatus*. **A–B.** Japan (RM 1475, 2). **C.** 211–212 m, off Kerama Island, Okinawa, Japan, 26.047°N, 127.375°E (SBMNH 149459, 9). Scale bars shell = 1 mm. Scale bars protoconch = 100 μm.

Scissurellidae
Sukashitrochus

Etymology. Carina, Latin for keel, referring to the spiral keels on the base of the shell.

Description. Shell medium-sized (to 2 mm. Figs 476, 477) trochiform, aperture of mature specimens distinctly deflected downwards. Protoconch of 0.75 whorl, reticulate sculpture, apertural varix connected to embryonic cap, apertural margin convex. Teleoconch I of slightly more than one whorl, approximately 25 fine axial cords, interstices with fine, sharp growth markings. Teleoconch II of 1.33 whorls, suture deeply impressed. Shoulder profile shallowly sigmoid, approximately 34 indistinct axial cordlets, axials more distinct towards suture; fine, indistinct spiral cordlets starting after 0.5 teleoconch II whorl, more distinct at apertural margin of mature specimens, approximately six spirals forming weak knots at intersection with axials; interstices with fine, sharp growth marks. Base with approximately five strong spiral keels, uppermost at level of suture strongest, same number and strength of axials as on shoulder, intersected by occasional weak spiral cordlets. Umbilicus open, bordered by weak carina, walls straight, smooth, at angle to base. Aperture irregularly, obliquely oval. Selenizone with keels of moderate elevation, moderate strength; lunules indistinct; slit closed to foramen in mature specimens.

Animal unknown.

Comparisons. *Sukashitrochus atkinsoni* from Australia has a protoconch with flocculent sculpture restricted to the outer ⅔ of the whorl.

Sukashitrochus dorbignyi from the Indian Ocean and the Indo-Malayan Archipelago has a smooth protoconch.

Sukashitrochus estotiensis from the Oligocene of Europe and the Recent of the Indian Ocean has distinctly reticulate sculpture on shoulder and base and a smooth protoconch.

Sukashitrochus lyallensis from New Zealand has regularly formed reticulate protoconch sculpture and a much weaker apertural varix on the protoconch

Sukashitrochus morleti from the Indo-Malayan Archipelago to the Central Pacific has a protoconch with fine axial sculpture.

Sukashitrochus pulcher from Southern Australia has much stronger axial cords.

Distribution. Southern Japan, Philippines, Fiji (?. Fig. 478). On gravel bottoms from intertidal to 50 m (Kuroda *et al.*, 1971; K. Hasegawa, pers. comm.).

Specimen records.

Japan, Kyushu. 30 m, Tamioka Bay, near Amakusa Marine Lab, Kumamoto Prefecture, 32.527°N, 130.050°E (LACM 93-23, 2). 20–40 m, Tamioka Bay, near Amakusa Marine Lab, Kumamoto Prefecture, 32.527°N, 130.050°E (LACM 83-23, 2).

Japan, Okinawa. (RM F3155, 6; RM 1475, 2; FLMNH 399256, 1). 20 m, Tsuken Jima, 26.268°N, 127.920°E (LACM 78-21, 1). 17 m, N side, Ike Jima, 26.384°N, 127.990°E (DLG 867, 3). 50 m, Zampa-Misaki, 26.435°N, 127.708°E (LACM 78-25, 4). 49 m, Bolo Point, 0.5 m ESE of Zampa-Misaki, 26.435°N, 127.708°E (LACM 79-100, 4). 28 m, Maeda Point, 26.450°N, 127.759°E (DLG 1070, 4). 92 m, Onna Village, 26.493°N, 127.493°E (LACM 78-101, 4). 50 m, Onna Village, 26.493°N, 127.842°E (LACM 79-75, 1). 46–55 m, 1 km WNW of Onna Village

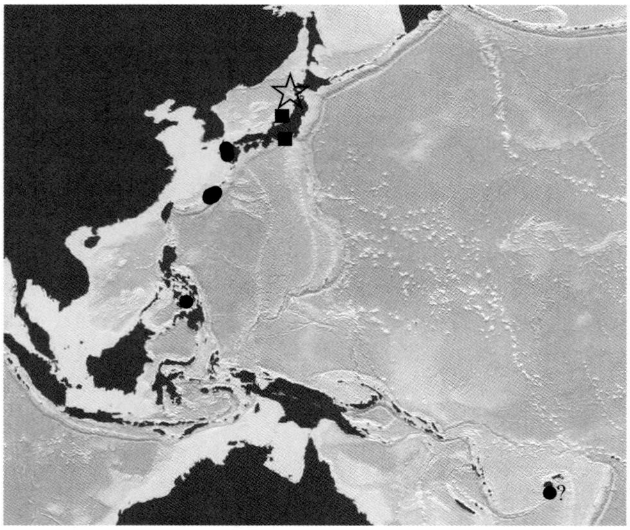

Figure 478. Distribution of *Sukashitrochus carinatus*. ● specimen record. ■ literature record. ? questionable locality.

(Horseshoe Cliffs), 26.493°N, 127.842°E (LACM 79-75, 1). 60 m, Onna Village, 26.493°N, 127.842°E (LACM 79-76, 4). 46 m, Onna Village, 26.495°N, 127.843°E (LACM 78-26, 1). 50 m, Onna Village, 26.495°N, 127.843°E (LACM 78-99, 1). 50 m, 1 km WNW of Onna Village, 90 ft cave area at Horseshoe South, 26.497°N, 127.843°E (LACM 78-27, 1). 38 m, base of cliff, Horeshoe Bay, 26.504°N, 127.858°E (DLG 1057, 13). 33 m, Motobu Peninsula, Nakijin, 26.703°N, 127.943°E (LACM 78-22, 1). 16 m, Oku, 26.847°N, 128.287°E (LACM 77-64, 1). 211–212 m, off Kerama Island, 26.047°N, 127.375°E (SBMNH 149459, 9). 71–85 m, off Kerama Island, 26.159°N, 127.448°E (SBMNH 149467, 1).

Philippines. 0–1 m, Hilutungan Islet, off Mactan Island, E side Cebu Island, Cebu, 10.208°N, 123.987°E (LACM 81-5, 1). 0 m, Cavile Reef, out of Puerto Princessa, Palawan, 9.717°N, 118.783°E (LACM 82-62, 1).

Fiji. 15 m, off Ham Reef, Ngasele Bay, N side Kandavu, 18.933°S, 178.300°E (LACM 85-136, 1).

Literature records.

Japan. Kamekisho [Sagami Bay, 34.680°N, 139.000°E] (Kuroda *et al.*, 1971). 0 m, Yobuko-Cho, Saga, 33.533°N, 129.883°E (4: Yamada, 2004).— Sado Marine Biological Station [Niigata, 38.075N, 138.246E] (Honma & Kitami, 1979).—Suruga Bay [Shizuoka, 34.800°N, 138.600°E] (Habe *et al.*, 1986).

Remarks. The epithet *carinata/carinatus* has been introduced twice for members of *Sukashitrochus*. These two epithets were introduced in two different genera, so there is no primary homonymy, but only secondary. The secondary homonym of Watson (1886) is both a junior synonym (of *Suk. atkinsoni*) and a junior secondary homonym [of *Suk. carinatus* (A. Adams, 1862)].

Dunker (1882: 147) noted that the species should be referred to Jeffreys' genus *Schismope* [= *Sinezona*].

Higo *et al.* (2001) noted two syntypes, although there are at least three in the collection, and there is a mark left from a possible fourth (missing) specimen on the mounting paper.

Hasegawa (2004) reported swimming of *Sukashitrochus* sp., most likely *Suk. carinatus*. The Japanese common name is *Sukashi–Ebisu*.

Sukashitrochus dorbignyi (Audouin, 1826)

Figures 479– 485, Color Plate 3C

Chresonymy.

[no name] Savigny, 1817: pl. 5, fig. 30.1–30.3.

scissurella [*sic*] *Dorbignii* [*sic*] Audouin, 1826: 182–183 [not *Sci. dorbignyi* Scacchi, 1836].

Scissurella dorbignii [*sic*]: Audouin, 1828: 183.

Scissurella d'orbignyi [*sic*]: Philippi, 1853: 38, pl. 6, fig. 10.

Anatomus d'orbignyi [*sic*]: Adams & Adams, 1853–1858 (1854): vol. 1, 439.

Scissurella orbignyi [sic]: Munier-Chalmas, 1865: 395.

Scissurella orbignyi [sic]: Issel, 1869: 228.

Scissurella orbignyi [sic]. Jousseaume, 1888: 194.

Scissurella dorbignyi: Pilsbry, 1890: 50–51, figs 85–87.

Scissurella d'orbignyi [*sic*]: Sturnay, 1905: 146 [*fide* Yaron, 1983].

Scissurella orbignyi [sic]: Pallary, 1926: 82 [*fide* Yaron, 1983].

Scissurella orbignyi [sic]: Franc, 1956: 21.

Scissurella orbignyi [sic]: Mastaller, 1979: 29 [*fide* Yaron, 1983].

Scissurella (Scissurella) dorbignyi: Yaron, 1983: pl. 2.

Scissurella dorbignyi: Bouchet & Danrigal, 1982: 12, 22, fig. 63.

Scissurella (Scissurella) dorbignyi: Yaron, 1983: 266–267, pl. 2.

Scissurella (Sukashitrochus) dorbignyi: Lozouet, 1986: 109.

Scissurella d'orbignyi [sic]: Vine, 1986: 172.

Scissurella d'orbignyi [sic]: Hedegaard, 1990: 74, 145.

Sukashitrochus dorbignii [sic]: Dekker & van Capelle, 1994: 125.

Sukashitrochus dorbignyi: Bosch *et al.*, 1995: 28, text-fig. 3.

Sukashitrochus dorbignyi: Dekker & Orlin, 2000: 17 [synonymized *Sin. tricarinata* Yaron, 1983 under it].

Sukashitrochus ? *dorbignyi*: Geiger, 2003: 79.

Sukashitrochus dorbignii [sic]: Zuschin *et al.*, 2009: 87, pl. 2, fig. 4.

Sukashitrochus dorbignyi: Kurtz, 2011: 34.

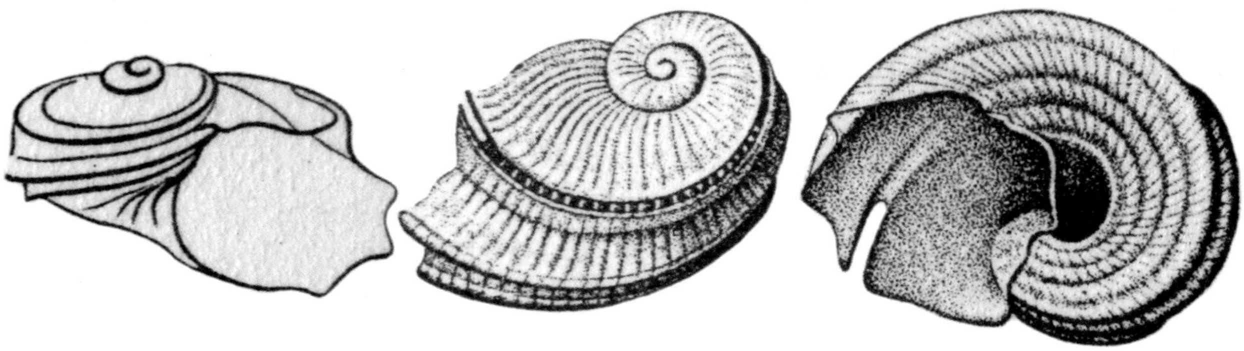

Figure 479. Original illustration of *Scissurella dorbignyi* from Savigny (1817).

Synonyms.

+ *Sinezona tricarinata* Yaron, 1983: 273–274, pl. 7, figs 1–4 (Fig. 480). ***Type material.*** Holotype (NHMW 62.639a). Paratype from type locality (NHMW 62.639b, 1; IY 4910, 1). Ain Sukhna, Egypt, 28.750°N, 34.583°E (NHMW 59.004, 6). ***Type locality***. Ghardaqa [= Hurghada], Egypt, northern Red Sea [27.283°N, 33.783°E] (OD). ***Etymology***. Denotes the three distinct spiral keels on base of shell.

Scissurella (Sukashitrochus) tricarinata: Lozouet, 1986: 109.

Sinezona tricarinata: Vine, 1986: 172.

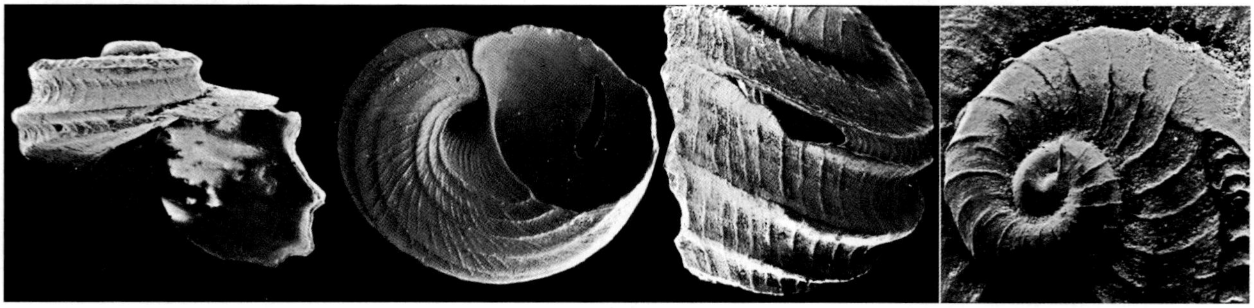

Figure 480. Original illustrations of *Sinezona tricarinata*.

Sukashitrochus tricarinatus: Geiger, 2003: 79.

+ *Sinezona armillata* Yaron, 1983: 272–273, pl. 6 (Fig. 481). ***Type material***. Holotype (NHMW 39.954), 1.7 mm. Paratype from type locality (NHMW 39.956, 1). ***Type locality***. Recent, Jeddah, Red Sea [Saudi Arabia, 21.500°N, 39.167°E] (OD). ***Etymology***. Armillatus, Latin for ornamented with a bracelet; also armed, armored.

Scissurella (Sukashitrochus) armillata: Lozouet, 1986: 109.

Sinezona armillata: Vine, 1986: 172.

Sukashitrochus armillatus: Dekker & Orlin, 2000: 17.

Sukashitrochus armillatus: Geiger, 2003: 79.

Figure 481. Original illustrations of *Sinezona armillata*.

+ *Sukashitrochus maraisi* Herbert, 1986: 629–631, figs 6, 36–39 (Fig. 482). ***Type material***. Holotype (NMSA D684/T3271), 1.45 mm. Paratypes from type locality (NMSA D686/T3272, 2; NMSA D685/T3273, 2); paratypes 5–8 (J. Marais collection); paratypes 9–18 (NMSA D3483/T3368). ***Type locality***. Mzamba, Transkei [South Africa], beach drift (OD), 31.100°S, 30.183°E. ***Etymology***. Named after Dr. Johan P. Marais, who collected much of the material currently available (OD).

Figure 482. Original illustrations of *Sukashitrochus maraisi*. Apertural and umbilical views as well as protoconch of holotype, remainders from paratypes.

Sukashitrochus maraisi Herbert, 1987: 1, figs c–d.
Sukashitrochus maraisi: Bouchet, 1987: 4, text-fig. [copy Herbert, 1986].
Sukashitrochus maraisi: Bandel, 1998: 51, pl. 17, figs 4–6.
Sukashitrochus maraisi: Geiger, 2003: 79.
+ *Schismope peilei* Winkworth: Glayzer *et al.*, 1984: 318 [*nomen nudum*, see Remarks].
Schismope peilei Winkworth, ms.: Bandel, 1998; Geiger, 2003.

Figure 483. *Sukashitrochus dorbignyi*. **A.** Lectotype. Suez, Gulf of Suez [Egypt, 29.967°N, 32.550°E] (MNHN). **B.** 25 m, Dunraven Reef, Strait of Tiran, S Sinai Peninsula, N Red Sea, Egypt, 27.200°N, 34.100°E (LACM 85-112, 3). **C.** 1 m, Hikkaduwa, Sri Lanka, 6.137°N, 80.095°E (LACM 84-9, 1). Radula shown in Figure 485. Scale bars shell = 1 mm. Scale bars protoconch = 100 μm.

Misidentifications.

Scissurella dohrniana: Sturnay, 1905: 146 [*fide* Yaron, 1983; is *Suk. dorbignyi*].
Schismope carinata: Smith, 1910, 208: pl. 8, fig. 3 [is *Suk. dorbignyi*].
Scissurella smithi: Kensley, 1973: 28, fig. 28 [is juvenile *Suk. dorbignyi*].

Type material. Five syntypes (MNHN: Bouchet & Danrigal, 1982), 1.75 mm. Lectotype (specimen fig. 63 of Bouchet & Danrigal, 1982: designated by Yaron, 1983: 267. Fig. 483A). Three paralectotypes (MNHN).

Type locality. Suez, Gulf of Suez [Egypt, 29.967°N, 32.550°E] (SD: Yaron, 1983: 267).

Etymology. Named for Alcide Dessalines d'Orbigny (1802–1857).

Figure 484. *Sukashitrochus dorbignyi* **A–C.** Karachi, Pakistan, 24.850°N, 67.033°E. Shells labeled "*peilei* Winck." (NHMUK). **B.** Specimen with deformed protoconch, also shown in Bandel (1998: pl. 17, figs 4–6). Scale bars shell = 0.5 mm. Scale bars protoconch = 100 µm.

Description. Shell medium-sized (to 1.7 mm. Figs 480–484), trochiform, depressed, turreted. Protoconch of 0.75 whorl, smooth, apertural varix connected to embryonic cap, apertural margin sinusoidal. Teleoconch I of 1.25–1.33 whorls, approximately 20 axial cordlets, no spirals. Teleoconch II of up to 1.33 whorls, suture little impressed. Shoulder flat, approximately 33 axial cords, strongest near suture, approximately 10 finer spiral lines, somewhat irregularly spaced. Base with slight constriction below selenizone, 3–4 spiral keels, top keel at periphery much stronger then remainder, axials less distinct compared to shoulder, very fine spiral threads between keels on base. Numerous fine axials, no additional spiral sculpture. Suture mostly at level of strongest keel at periphery, except for most mature portion of apertural roof dipping slightly below. Selenizone above periphery; with strong lunules; strong, moderately elevated keels; slit closed to teardrop-shaped foramen. Umbilicus open, wide, bordered by distinct cord, walls smooth, no funiculus. Aperture rounded, roof overhanging.

Radula (Fig. 485). Rachidian tooth trapezoidal, central cusp largest, five cusps on each side. Lateral teeth 1–3 similar, 3–4 cusps on outer edge, tip-most largest. Lateral tooth 4 reduced, hook-shaped with one minute cusp on each side. Lateral tooth 5 enlarged, broadened, approximately seven cusps on inner edge, apical-most largest. Inner marginal teeth with 2–4 cusps on each side, outer marginal teeth spoon-shaped with many small cusps on each side, outermost marginal teeth paddle-shaped with many minute bristles towards tip; no food groove on shaft. Radular interlock of central field moderate.

Comparisons. *Sukashitrochus atkinsoni* from Australia has flocculent protoconch sculpture.

Sukashitrochus carinatus from Japan has flocculent protoconch sculpture and a more narrow umbilicus.

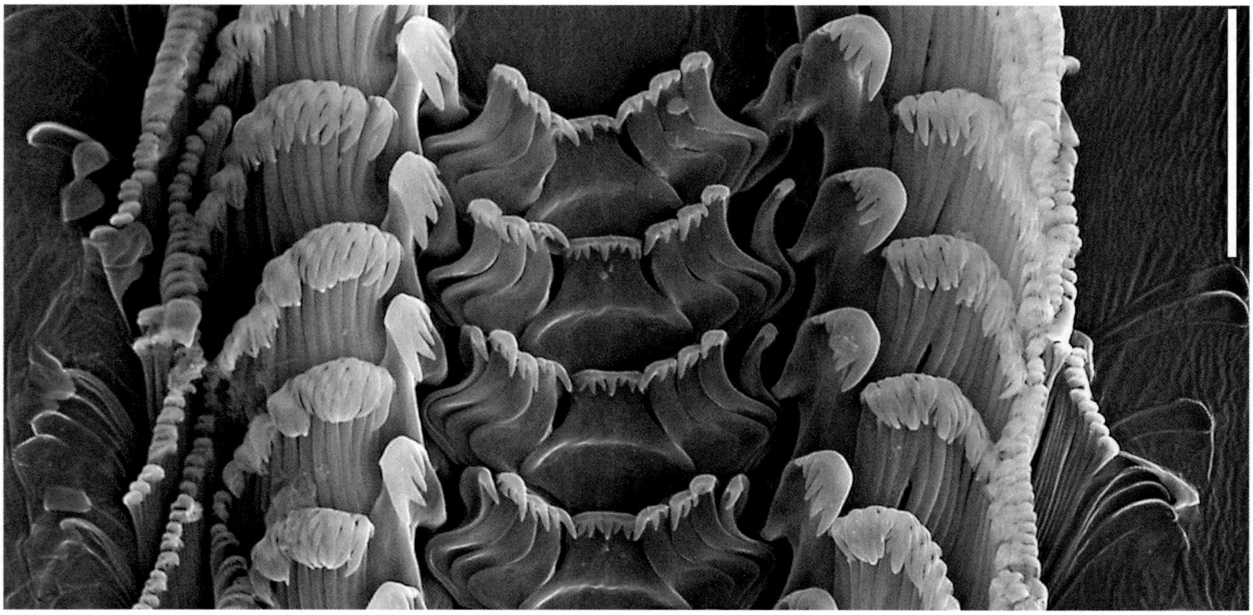

Figure 485. *Sukashitrochus dorbignyi.* Radula. 1 m, Hikkaduwa, Sri Lanka, 6.137°N, 80.095°E (LACM 84-9, 1). Shell shown in Figure 483C. Scale bar = 20 µm. Image by Jaya Nolt.

Sukashitrochus estotiensis from the Oligocene of France and the Recent of the Indian Ocean and Indo-Malayan Archipelago has rounded keels and axials and spirals of similar strength forming a reticulate pattern.

Sukashitrochus lyallensis from New Zealand has flocculent protoconch sculpture and a weak or absent apertural varix on the protoconch.

Sukashitrochus morleti from the Indo-Malayan Archipelago to the Central Pacific has a protoconch with fine axial sculpture, has more distinct keels, and axials on teleoconch I that are much stronger than those on teleoconch II.

Sukashitrochus pulcher from Southern Australia has flocculent protoconch sculpture and much more distinct axial cords on the teleoconch.

Distribution. Tropical and subtropical Indian Ocean. Rarely Indo-Malayan Archipelago (Fig. 486)

Specimen records.

Red Sea (KBIN, 2).

Israel. Elat, 29.550°N, 34.950°E (HUJ 20.314, 3).

Egypt. Ghardaqa [= Hurghada], 27.283°N, 33.783°E (NHMW 62.608, 1; NHMW 69.874, 1; NHMW 22028, 1). Dishet El Daba (25 km SSE of Hurghada), 27.044°N, 33.906°E (RMNH, 1). Ras Muhammad, 27.733°N, 34.250°E (IY 4921, 1). Shurat el Gharqana, 28.117°N, 34.350°E (IY 4269, 1). Suez, 29.967°N, 32.550°E (MNHN, 5). Bir es Suweir, 29.243°N, 34.736°E (IY 2306a, 3; IY 2884, 3); IY 2886, 3; IY 4534, 3). Ras Burqa, 29.214°N, 34.739°E (IY 4890, 5). 2–3 m, Jackson Reef, Strait of Tiran, S Sinai Peninsula, N Red Sea, 27.017°N, 34.467°E (LACM 85-111, 2). Dishet El Daba (25 km SSE of Hurghada), 27.044°N, 33.906°E (RMNH, 1). 0 m, 16 km S of Hurghada, Red Sea, 27.083°N, 33.883°E (RMNH, 5). 25 m, Dunraven Reef, Strait of Tiran, S Sinai Peninsula, N Red Sea, 27.200°N, 34.100°E (LACM 85-112, 3). Ghardaqa [= Hurghada], 27.283°N, 33.783°E (NHMW 22028, 1). Blue Hole, Dahab, Sinai, 28.483°N, 34.533°E (DLG 103, 2). 0 m, 5 km S of Râs Adabîya, Gulf of Suez, 29.867°N, 32.500°E (RMNH, 5). 0 m, Makadi Bay, 26.992°N, 33.810°E (DLG 1648, 1).

Saudi Arabia. Abulat Island, 19.967°N, 40.133°E (MNHN, 21). 20 m, Gulf of Salwah, Persian Gulf, 24.801°N, 50.798°E (RMNH, 1). 2–20 m, Jedda Jidda, 21.517°N, 39.219°E (USNM 425608, 1).

Bahrain. Sitrah Island, 26.150°N, 50.617°E (AMS C.404911, 1).

Sudan. 95 m, Sanganeb, 19.718°N, 37.425°E (DLG 893, 2).

Djibouti. 11.594°N, 43.164°E (KBIN, 9).

Kenya. Malindi, Watama, 3.233°S, 40.133°E (AMS C.404898, 1).

Seychelles. 45 m, W of Aride Island, 4.217°S, 55.567°E (RMNH, 2).

Madagascar. 22 m, Point Evatra, ATIMO VATAE TB01, 24.996°S, 47.095°E (MNHN, 1). 7 m, Port d'Ehoala, ATIMO VATAE TP28, 25.061°S, 46.963°E (MNHN, 1). 12–14 m, NW Albatross Rock, ATIMO VATAE BS01, 25.470°S, 44.940°E (MNHN, 1; MNHN, 1). 16 m, Cape St. Marie, ATIMO VATAE BS14, 25.600°S, 45.145°E (MNHN, 1).

Mauritius. 10 m, Trou aux Biches, 20.050°S, 57.533°E (MNHN, 1). Le Chaland, 20.433°S, 57.683°E, (AMS C.404913, 1).

Reunion. 20 m, Cap La Houssaye, Saint Paul, 20.951°S, 55.211°E (MJC 718, 2). 205–215 m, 21.383°S, 55.617°E (MNHN, 1).

South Africa. 100 m, Richards Bay, Natal, 28.920°S, 32.234°E (DLG 1407, 3). 20 m, Aliwal Shoal = Unkomaas, Natal, 30.208°S, 30.805°E (DLG 1408, 4). 45 m, Landers

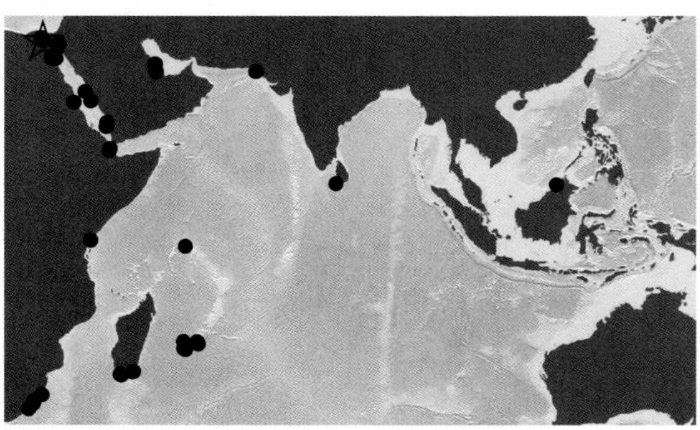

Figure 486. Distribution of *Sukashitrochus dorbignyi*. ☆ type locality. ● specimen record. ■ literature record.

Reef, Rocky Bay, Natal, 30.357°S, 30.796°E (DLG 1389, 1). 100 m, Park Rynie, Natal, 30.357°S, 30.851°E (DLG 1406, 2). 0 m, Mzamba, Transkei, 31.100°S, 30.183°E (DLG 1387, 12). NE Transkei, 31.300°S, 30.000°E (DLG 1365, 2). N Transkei (HLC, 6).

Pakistan. Karachi (NHMUK, 1; NHMUK, 2; NHMUK, 15).

Sri Lanka. (USNM 64371, 1). 1 m, Hikkaduwa, 6.137°N, 80.095°E (LACM 84-9, 1).

Malaysia. 8 m, Police Bay, Palau Gaya, off Kota Kinabalu, Sabah, 5.983°N, 116.067°E (SBMNH 85111, 3).

Literature records.

Egypt. Shurat el Ghardana, Aquaba; Ain Sukhna [29.592°N, 32.344°E] (Yaron, 1983).

Yemen. Al-Murk Island, 15.633°N, 42.633°E. Al-Luhayyah (= Loheia), 15.700°N, 42.683°E. Oreste Point, 16.367°N, 42.767°E (Dekker & Ceuninck van Capelle, 1994).

Djibouti. Obock, Surcouf, 11.963°N, 43.291°E (Jousseaume, 1888).

South Africa. Transkei (Bandel, 1998).—60 m, off Durban Bluff, 4 km, 128°N of Cooper Light, 29.883°S, 31.000°E (NMSA A6075). 0 m, Mzamba, Transkei, 31.100°S, 30.183°E (NMSA D684/T3272, NMSA D684/T3273, NMSA D3T3368483)(Herbert, 1986).—Durban (Smith, 1910).

Gumfudde, Red Sea (Philippi, 1853). Persian Gulf (Yaron, 1983).

Remarks. The spelling of the species varies considerably, though none of the spelling variations are significant according to ICZN; hence, no new taxa with minor differences in spellings were introduced. The original spelling *Dorbignii* was capitalized, which has to be emended to lower case. The apostrophe in the surname of d'Orbigny, however, cannot be used in the latinized species epithet. The question of whether the ending *-ii* or *-yi* should be used can be viewed from two perspectives. ICZN Art. 35.2.1 notes that incorrect transliterations or latinizations are not to be considered errors in the original publication, and any subsequent spellings are simple errors. ICZN Art. 33.3.1, however, gives preference to prevailing usage even in the case of an incorrect subsequent spelling. Only three out of 21 subsequent spellings use the original *-ii*; hence, prevailing use clearly is with *-yi*. I maintain prevailing use.

Munier-Chalmas (1865) dated Audouin at "1817." He also placed the species in "group ß" with last whorl biangular, which supports the placement in *Sukashitrochus*. Philippi (1853: pl. 6, fig. 10) clearly showed a *Sukashitrochus*, and SEM images of the lectotype correspond to the original illustration.

Some specimens from Karachi in the NHMUK labeled "*Schismope peilei* Wink. [worth, ms]" plus two unidentified lots from the same locality are referable to *Suk. dorbignyi*. The name does not seem to ever have been published; hence, it is treated as an unvalidated manuscript name. The mentions in Glayzer *et al.* (1984) and Bandel (1998) do not constitute a validation.

Sinezona tricarinata, *Sin. armillata* (see also Zuschin *et al.*, 2009), and *Suk. maraisi* are shell morphological variants of *Suk. dorbignyi*. *Sinezona tricarinata* is a juvenile form identical to the lectotype of *Suk. dorbignyi*. *Sinezona armillata* is a more strongly sculptured morph, particularly with better developed axials, while *Suk. maraisi* is a smoother morph. All share the smooth protoconch, the apertural varix is connected to the embryonic cap, the apertural margin is sinusoidal, and teleoconch I is of 1.125–1.33 whorls. The variation in the strength of the axials is well within the variability observed in other *Sukashitrochus* species, such as *Suk. atkinsoni*. There is no geographic co-variation for the sculpture; the smoother *maraisi* morph is found in South Africa as well as in Pakistan.

Sukashitrochus dorbignyi should not be confused with *Padollus orbigny/dorbignyi* Costa, 1838. The latter is a synonym of *Scissurella costata* from the Mediterranean and adjacent Atlantic.

Sukashitrochus estotiensis Lozouet, 1999

Figures 487–489

Chresonymy.

Scissurella (*Sukashitrochus*) *subitana/estotensis* [*sic*]: Lozouet, 1986: 110, 124–125, text-fig. 35A [unavailable, unpublished, see Remarks]

Sukashitrochus estotiensis Lozouet, 1999: 2–3, pl. 2, figs 5–7.

Sukashitrochus estotiensis: Geiger, 2008b: fig. 1.

Misidentification.

Sukashitrochus indonesicus. Geiger & Jansen, 2004b [partim]: fig. 28C [shell], D–F [radula. Is *Suk. estotiensis*].

Type material. Holotype (MNHN. Fig. 487), 0.93 × 0.74 mm. Paratypes from type locality (MNHN, 10).

Type locality. Landes, St.-Paul-lès-Dax (Estoti) [France, 44.333°N, 1.050°W], late Oligocene, *Miogypsinoides* shell marl (OD).

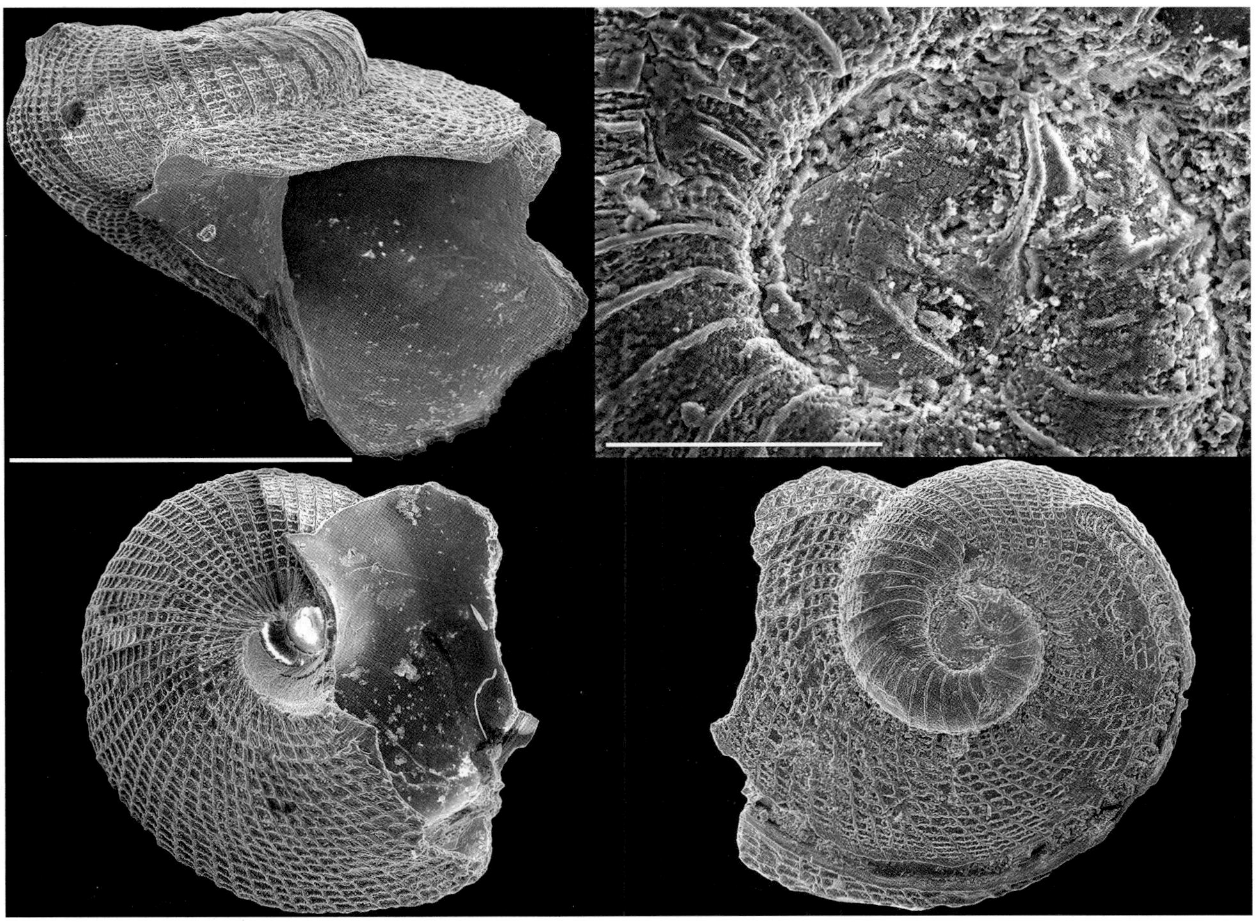

Figure 487. *Sukashitrochus estotiesnsis*. Holotype. Landes, St.-Paul-lès-Dax (Estoti) [France, 44.333°N, 1.050°W] (MNHN). Scale bar shell = 0.5 mm. Scale bar protoconch = 100 µm.

Scissurellidae *Sukashitrochus*

Etymology. Derived from the type locality (OD).

Description. Shell medium-sized (to 1.6 mm. Figs 487, 488), trochiform depressed. Protoconch of 0.75 whorl, smooth, apertural varix not connected to embryonic cap, apertural margin sinusoidal. Teleoconch I of 1.125 whorls, approximately 40 distinct axial cordlets, interstices smooth, approximately 20–25 somewhat finer spiral cordlets giving fine reticulate appearance. Teleoconch II of >0.66 whorl, suture impressed. Shoulder flat, density of sculpture comparable to teleoconch I, spirals becoming

Figure 488. *Sukashitrochus estotiensis*. The specimens are indistinguishable from the European holotype from the Oligocene. **A–B.** 20 m, Cap La Houssaye, Saint Paul, Reunion Island, 20.017°S, 55.234°E (MJC 1881, 40). Radula shown in Figure 489. **C.** 17–40 m, Villingili Island, North Male Atoll, Maldives, 4.200°N, 73.450°E (LACM 83-8, 1). Scale bars shell = 1 mm. Scale bars protoconch = 100 μm.

Scissurellidae *Sukashitrochus*

of equal to or stronger than axials. Base with broad constriction below selenizone, mid-base with distinct elevated ridge; between mid-base ridge and umbilicus a second obtuse ridge; third ridge bordering umbilicus. Reticulate sculpture similar to remainder of shell. Selenizone above periphery, lunules less frequent than axials, indistinct; keels moderately elevated, slit closed to foramen. Umbilicus large, deep, walls smooth. Aperture ovoid, roof overhanging.

Radula (Fig. 489). Rachidian tooth trapezoidal, central cusp slightly larger than 3–4 in arc on either side. Lateral teeth 1–3 similar, apical cusp largest, 3–4 along outer edge. Lateral tooth 4 reduced, hook-shaped. Lateral tooth 5 enlarged by broadening, apical cusp largest, approximately 5–6 along inner edge. Inner marginal teeth with palmate tip, shaft with food groove. Outer marginal teeth with spoon-shaped tip, many fine bristles along edge. Radular interlock of central field moderate.

Comparisons. *Sukashitrochus estotiensis* is the only species with the fine reticulate pattern on the teleoconch and easily recognized by this character alone.

Suksashitrochus dorbigny from the Indian Ocean has distinct spiral keels, regardless of the rather variable sculpture, which never consists of equally developed axials and spiral.

Sukashitrochus morleti from the Indo-Malayan Archipelago to the Central Pacific has a protoconch with fine axials and very fine growth lines between axial cords.

Sukashitrochus carinatus from Japan has flocculent to reticulate protoconch sculpture and never reticulate teleoconch sculpture.

Sukashitrochus lyallensis from New Zealand has flocculent to reticulate protoconch sculpture and never reticulate teleoconch sculpture

Sukashitrochus atkinsoni from Australia has coarse flocculent to reticulate sculpture on the protoconch, and if it has axial cords, the interstices show very fine growth lines.

Sukashitrochus pulcher from southern Australia has coarse flocculent to reticulate sculpture on the protoconch, and if it has axial cords, the interstices show very fine growth lines.

Figure 489. *Sukashitrochus estotiensis*. Radula. 20 m, Cap La Houssaye, Saint Paul, Reunion Island, 20.017°S, 55.234°E (MJC 1881, 40). Shells from lot shown in Figure 488A, B. Scale bar = 10 μm.

Distribution. Oligocene of France, Recent of Indian Ocean (Fig. 490).

Specimen records.

Mauritius. Tip of Arsenal Bay, 20.033°S, 57.567°E (USNM 761597, 1). 15 m, Trou aux Biches, 20.050°S, 57.533°E (MNHN, 1). 5 m, Trou aux Biches, 20.050°S, 57.533°E (MNHN, 1). Balaclava, 20.033°S, 57.567°E (MNHN, 1).

Reunion Island. 20 m, Cap La Houssaye, Saint Paul, 20.017°S, 55.234°E (MJC 1881, 40; MJC 1536, 1). 205–215 m, 21.383°S, 55.617°E (MNHN, 1).

Maldives. 17–40 m, Villingili Island, N Male Atoll, 4.200°N, 73.450°E (LACM 83-8, 1).

Fossil records.

France. Peyrehonrande (Tauziède A), St.-Paul-lès-Dax (Estoti), late Oligocene (MNHN, 1; MNHN, 10).

Remarks. Lozouet (1986) introduced in his unpublished diploma thesis in the text the name *subitana*, but the corresponding figure caption used *estotensis* [*sic*] as the specific epithet. The name was validated by Lozouet (1999).

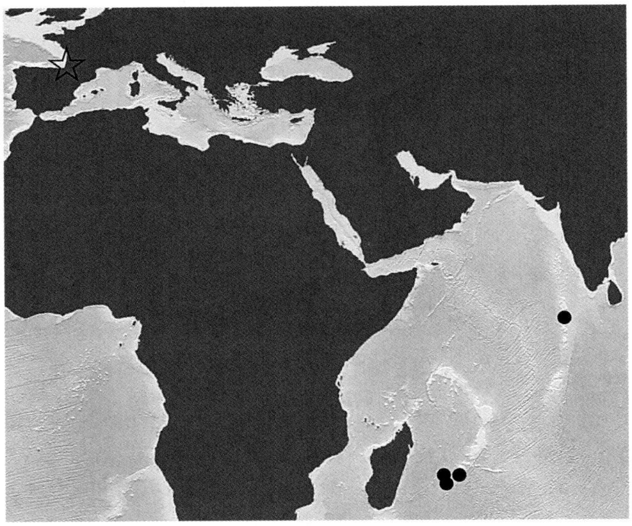

Figure 490. Distribution of *Sukashtrochus. estotiensis.* ☆ type locality. ● specimen record.

Specimens from the Recent of the Maldive and the Mascarene Islands are indistinguishable from the type from the Oligocene of France. Shared characters include the smooth protoconch with apertural varix connecting to the embryonic cap and with sinusoidal apertural margin; teleoconch I sculpture with only axials for the first half whorl, and then increasing number of spirals; teleoconch II with reticulate ornamentation; umbilicus without bordering cord and smooth walls. The only difference is the distinctness of the spiral keels in the base, which is explained by the holotype to be a more juvenile specimen at 0.97 mm than that from the Maldives of 1.15 mm. Comparing the earlier whorl of the Recent specimens with the holotype reveals that the spiral keels become stronger with growth. Invoking rather the morphospecies concept as opposed to the chronospecies concept, the Recent species should be identified as conspecific with the fossil one.

Sukashitrochus lyallensis (Finlay, 1926)

Figures 491–495

Chresonymy.

Schismope lyallensis: Finlay, 1926: 340, not illustrated.
Schismope lyallensis Finlay, 1950: 25.
Schismope lyallensis: Powell, 1937: 62.
Schismope lyallensis: Powell, 1940: 208.
Schismope lyallensis: Powell, 1941: 254.
Schismope lyallensis: Fleming, 1950: 25.
Schismope lyallensis: Dell, 1956: 168, 181.
Schismope lyallensis: Fleming, 1966: 36.
Schismope lyallensis: Cernohorsky, 1972: 232, fig. 1.
Sinezona lyallensis: Powell, 1979: 35, pl. 13, fig. 7.
Scissurella (*Sukashitrochus*) *lyallensis*: Lozouet, 1986: 109.
Sinezona lyallensis: Beu & Maxwell, 1990: 376.
Sukashitrochus lyallensis: Marshall, 1993: fig. 3.
Sukashitrochus lyallensis: Spencer & Willan, 1995: 13.

Figure 491. Holotype of *Schismope lyallensis*. Image courtesy Peter Quin/AIM. Scale bar = 1 mm.

Sukashitrochus lyallensis: Geiger, 2003: 79.

Sukashitrochus lyallensis: Spencer *et al*., 2009: 202.

Sukashitrochus lyallensis: Raven & Bracegirdle, 2010: 10, text-fig.

Misidentifications.

Schismope atkinsoni: Suter, 1907: 268 [is *Suk. lyallensis*].

Schismope atkinsoni: Suter, 1909: 7 [is *Suk. lyallensis*].

Schismope atkinsoni: Suter, 1913: 90, pl. 6, fig. 12 [is *Suk. lyallensis*].

? *Schismope* n. sp.: Marshall & Uttley, 1914: 279 [most likely same specimens as referenced in Marshall (1915)].

? *Schismope atkinsoni*: Marshall, 1915: 378 [fossil, is *Suk. lyallensis* or *Suk. ngatutura*].

Schismope atkinsoni: Suter, 1915: 19 [is *Suk. lyallensis*].

Type material. Holotype (AIM TM-696: Cernohorsky, 1972: 232, fig. 1. AIM AK 70696. Fig. 491), 2.2 × 1.8 mm. One paratype (AIM AK 73331), 1.5 mm.

Type locality. Lyall Bay [New Zealand, 45.250°S, 166.983°E], in shell sand (OD).

Etymology. Named after the type locality.

Description. Shell medium-sized (to 2.2 mm. Figs 492, 493), trochiform depressed. Protoconch of 0.75 whorl, reticulate sculpture, weak apertural varix connected to embryonic cap or missing, apertural margin somewhat sinusoidal. Teleoconch I of 1.25 whorls,

Figure 492. *Sukashitrochus lyallensis*. 55 m, South Cape, off Poutama Island, Stewart Island, New Zealand, 47.267°S, 167.383°E (NMNZ M.028375, 30). Scale bar shell = 1 mm. Scale bar protoconch = 100 μm.

with approximately 30 very indistinct low axial cords, interstices with fine irregular growth marks, no spiral sculpture. Teleoconch II of 1.3 whorls; last portion after closure of foramen distinctly descending; suture impressed. Shoulder somewhat convex, with distinct axial cords just at suture, otherwise only fine axial growth marks, no spiral sculpture. Base with one prominent spiral keel at periphery, one additional keel one third towards umbilicus. Between selenizone and periphery approximately three irregularly spaced low cords; between periphery and umbilicus

Figure 493. *Sukashitrochus lyallensis.* **A.** 55 m, South Cape, off Poutama Island, Stewart Island, New Zealand, 47.267°S, 167.383°E (NMNZ M.028375, 30). **B.** 91 m, Kane-te-to, Island off Paterson Inlet, Stewart Island, New Zealand, 46.917°S, 168.050°E (USNM 610425, 1). **C.** Bounty Islands, New Zealand, 47.7180°N, 179.079°E (GNS RM 5956, 7). Scale bars shell = 1 mm. Scale bars protoconch = 100 μm.

approximately six irregularly spaced spiral cords of unequal strength. Umbilicus open, moderately wide, bordered by spiral cord, at angle to base, walls straight, smooth. Selenizone above periphery with moderately strong keels of moderate elevation; lunules distinct; foramen elongated triangular, continued anteriorly by raphe. Aperture rounded, D-shaped, oblique, roof overhanging.

Radula (Fig. 494). Rachidian tooth trapezoidal, central cusp largest, six cusps on each side, parallel to row of teeth. Lateral teeth 1–3 similar, apical cusp largest, three cusps on outer margin of cutting edge. Lateral tooth 4 reduced, hook-shaped. Lateral tooth 5 enlarged by broadening, apical cusp largest, eight cusps on inner edge of tip, approximately three cusps at junction between tip and shaft. Marginal teeth with triangular tip, approximately six cusps on each side, no food groove. Outer marginal teeth with spoon-shaped tip, many fine bristles along edge. Radular interlock of central field moderate.

Operculum (Fig. 494E) corneous, thin, round, multispiral, with central nucleus.

Comparisons. This is the only Recent *Sukashitrochus* species from New Zealand.

Sukashitrochus atkinsoni from southern and eastern Australia has more pronounced axial sculpture, particularly on the base, and the protoconch sculpture is lacking on the centripetal side of the whorl.

Figure 494. *Sukashitrochus lyallensis*. Radula and operculum. 91 m, off Taieri, Dunedin, South Island, New Zealand, 46.250°S, 170.483°E (NMNZ M.066298, 17). **A.** Entire radula. Scale bar = 200 µm. **B.** Entire width of radula. Scale bar = 100 µm. **C.** Central field. Scale bar = 20 µm. **D.** Marginal teeth. Scale bar = 10 µm. **E.** Operculum. Scale bar = 0.5 mm.

Sukashitrochus carinatus from Japan has flocculent sculpture over most of the protoconch sculpture and a more narrow umbilicus.

Sukashitrochus dorbignyi from the Indian Ocean and the Indo-Malayan Archipelago has a smooth protoconch.

Sukashitrochus estotiensis from the Oligocene of Europe and the Recent of the Indian Ocean has distinct reticulate sculpture over the entire teleoconch and a smooth protoconch.

Sukashitrochus morleti from the Indo-Malayan Archipelago has a more pronounced axial sculpture and a protoconch with fine axial ribs.

Sukashitrochus pulcher from temperate Australia and the Indo-Malayan Archipelago has a more pronounced axial sculpture, which forms raised points at intersection with spiral cords, and the protoconch lacks sculpture on the centripetal side of the whorl.

Distribution. New Zealand (Fig. 495).

Specimen records.

New Zealand. 42–47 m, East Ancorage, Raoul Island, Kermadec Islands, 29.266°S, 177.860°W (NMNZ M.226765, 1). Whangaroa Harbour, Northland, 35.100°S, 173.867°E (GNS RM 5958, 1). 62 m, New Golden Hind, entrance to George Sound, Fiordland, 44.839°S, 167.348°E (GNS RM 1040, 1). Lyall Bay, 45.250°S, 166.983°E (AIM TM 696, 1). 360 m, Saunders Canyon, off Otago Peninsula, South Island, 45.766°S, 170.900°E (NMNZ ex 160300, 1). 183 m, off Otago Heads, 45.783°S, 170.750°E (AMS C.17018, 1). 75 m, off Otago, 45.800°S, 170.883°E (AMS C.376864, 1). 91 m, off Taieri, Dunedin, South Island, 46.250°S, 170.483°E (NMNZ M.066298, 17). Foveaux Strait, 46.567°S, 168.000°E (AMS C.13665, 2; AMS C.13666, 1). 37 m, Middle Grounds, Foveaux Strait, 46.567°S, 168.000°E (USNM 609877, 1). 55 m, off Fishing Rock, Foveaux Strait, 46.567°S, 168.000°E (USNM 609794, 4). 18 m, Fishing Rock, Foveaux Strait, South Island, 46.567°S, 168.000°E (USNM 6100075, 1). 55 m, Herekopere, Foveaux Strait, South Island, 46.567°S, 168.000°E (USNM 610055). Foveaux Strait, 46.567°S, 168.000°E (WAM, 2). 140 m, SE of Nugget Point, Otago, South Island, 46.766°S, 170.000°E (NMNZNMNZ M.065998, 2). 28 m, Stewart Island, off Halfmoon Bay, 46.883°S, 168.133°E (AMS C.378322, 2). Off Halfmoon Bay, Stewart Island, 46.883°S, 168.133°E (AMS C.22337, 1). 91 m, Kane-te-to, Island off Paterson Inlet, Stewart Island, 46.916°S, 168.050°E (USNM 610425, 1). 55 m, South Cape, off Poutama Island, Stewart Island, 47.266°S, 167.383°E (NMNZ M.028375, 30). 64 m, Stewart Island, 47.450°S, 167.800°E (NMNZ M.019451, 1). 0 m, Stewart Island, 48.036°S, 166.550°E (USNM 610683, 1). Bounty Islands, 47.718°S, 179.079°E (GNS RM 5956, 7). 90 m, Snares Islands, 48.016°S, 166.567°E (GNS RM 5957, 8). Snares Islands, 48.036°S, 166.550°E (MHNG 14355, 2). 91 m, Snares Islands, 48.036°S, 166.550°E (USNM 200975, 2). 140 m, Pukaki Rise, 49.333°S, 171.983°E (NMNZ 174842, 1). 19 m, E side of Perpendicular Head, Antipode Islands, 49.650°S, 178.883°E (NMNZ M.150190, 3). 160–244 m, Campbell Plateau, 52.383°S, 170.033°E (NMNZ M.59253, 1).

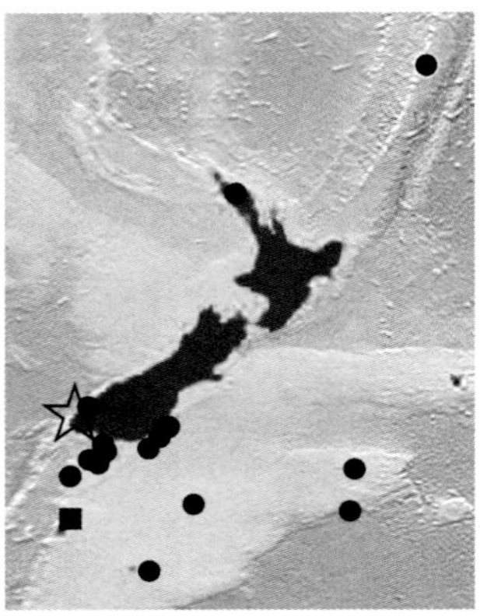

Figure 495. Distribution of *Sukashitrochus lyallensis*. ☆ type locality. ● specimen record. ■ literature record.

Literature records.

New Zealand. 116 m, Edwardson Sound, opposite entrance to Northport, 49.983°S, 166.567°E (Fleming, 1950).

Fossil. Hawaran = late Pleistocene of New Zealand (Beu & Maxwell, 1990).

Remarks. The publication date for the species has been taken to be 1927 (Powell, 1979). However, the byline under the title notes that the paper had been "issued separately, December 23rd 1926."

Sukashitrochus morleti (Crosse, 1880)

Figures 496–503, Color Plate 4D–E

Chresonymy.

Schismope morleti Crosse, 1880: 144–146, pl. 4, fig. 3 (Fig. 496).
Schismope morleti: Paetel, 1888: 289.
Schismope morleti: Pilsbry, 1890: 62, pl. 22, figs 37–39.
Schismope morleti: Thiele, 1912a: 29, pl. 3, figs 27–28 [copy figures Crosse, 1880].
Schismope morleti: Fischer, 1950: .
Scissurella ? (*Sukashitrochus*) *morleti*: Lozouet, 1986: 109.
Scissurella (*Sukashitrochus*) sp.: Lozouet, 1986: fig. 29A.
Sinezona morleti: Geiger, 2003: 78.
Sukashitrochus morleti: Geiger, 2006a: 23–24, fig. 17 [lectotype].
Sukashitrochus morleti: Geiger, 2006c: text-fig. 4.
Sukashitrochus morleti: Geiger, 2008a: 158, pl. 24, figs 1a–b, 5.
Sukashitrochus morleti: Tröndlé & Boutet, 2009: 12.
—: Bouchet *et al.*, 2011: fig. 490, row 3, second from right.

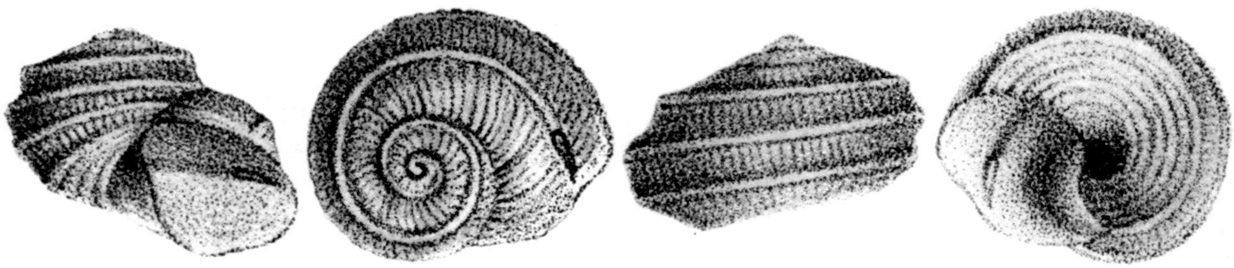

Figure 496. Original illustrations of *Schismope morleti*.

Synonyms.

+ *Sukashitrochus indonesicus* Bandel, 1998: 51–52, pl. 17, figs 7–8, pl. 18, fig. 1 (Fig. 497). ***Type material.*** Holotype (SGPIH 3881), 1.7 mm. ***Type locality.*** Near Satonda, Indonesia [8.100°S, 117.750°E] (OD). ***Etymology.*** Named for its type locality in Indonesia (OD).

Figure 497. Original illustrations of *Sukashitrochus indonesicus*.

Sukashitrochus indonesicus: Geiger & Jansen, 2004b [partim]: 55–58, figs 28 A–B [shell], 32b [map]. fig. 28C–F is *Suk. estotiensis.*

Sukashitrochus indonesicus: Geiger, 2003: 79.

Sukashitrochus indonesicus: Geiger, 2004: text-fig. p. 6.

+ *Sukashitrochus simplex* Bandel, 1998: 52–53, pl. 18, figs 2–4 (Fig. 498). ***Type material.*** Holotype (SGPIH 3882), 1.4 × 1 mm. ***Type locality.*** Near Satonda, Indonesia [8.100°S, 117.750°E] (OD). ***Etymology.*** Simplex, Latin for the simple shape of the shell (OD).

Sukashitrochus simplex: Geiger, 2003: 79.

Figure 498. Original illustrations of *Sukashitrochus simplex.*

Misidentifications.

Sinezona carinata: Ladd, 1982: 20, pl. 1, figs 3–4 [Pleistocene of Vanuatu, is *Suk. morleti*].

Sinezona carinata: Jansen, 1999: 54, figs 49–51 [is *Suk. morleti*].

not *Sukashitrochus indonesicus*: Geiger & Jansen, 2004b [partim]: fig. 28C [shell; is *Suk. morleti*], D–F [radula; is *Suk. estotiensis*].

Type material. Lectotype (MNHN: designated by Geiger, 2006a: 23. Fig. 499), 1.5 × 1.25 mm. No known paralectotypes.

Type locality. New Caledonia, intertidal sands (OD).

Etymology. Name after L. Morlet (OD).

Description. Shell medium-sized (to 1.7 mm. Figs 499–501), trochiform, depressed. Protoconch of 0.75 whorl, embryonic cap with reticular sculpture, then with fine axials, axials connected with very fine spiral thread, apertural varix connected to embryonic cap, apertural margin deeply sinusoidal. Teleoconch I of slightly more than one whorl, approximately 18 strong axial cords, interstices with fine, dense, irregular axial growth lines, no spirals. Teleoconch II of at least 1.33 whorls, suture impressed. Shoulder more or less flat, slightly undulating, almost at right angle to shell axis. Axial cordlets finer than on teleoconch I, approximately 37 on first whorl. Spiral threads increasing in number with growth, approximately 13 at apertural margin, forming faintest nodules at intersection with axials. Base with 3–4 spiral keels, uppermost strongest, just below periphery, density and strength of axials comparable to shoulder, spiral lines stronger, forming small, more distinct crenulations at intersections with axials. Umbilicus wide, walls smooth, transition to base with strong carina. Aperture subquadrate, roof overhanging. In mature specimens, apertural margin dipping below level of first, strongest keel on base. Selenizone at uppermost margin of whorl, inside of periphery, pointing at 45° angle towards axis of shell; strong lunules not coordinated with axials; keels strong, of moderate elevation. Slit closed to elongated, teardrop-shaped foramen, length of foramen approximately 0.125 whorl.

Radula (Fig. 502). Rachidian tooth trapezoidal, median cusp smaller than largest one of five cusps on each side. Lateral teeth 1–3 similar, apical cusp largest, 3–4 cusps on outer edge of tip. Lateral tooth 4 reduced, hook-shaped, with minute point on inner edge of tip. Lateral tooth 5 enlarged by broadening, apical cusp largest, 6–7 cusps of decreasing size along inner edge of tip. Inner marginal teeth with triangular tip, apical cusp largest, 2–4 cusps on each side; food groove present on upper portion of shafts. Outer marginal teeth with spoon-shaped tip, many fine bristles along edge. Outermost marginal teeth paddle-shaped, with many fine bristles along edge. Radular interlock of central field moderate.

Operculum (Fig. 502E) corneous, extremely thin, round, multispiral, nucleus central.

Comparisons. *Sukashitrochus atkinsoni* from Australia has a protoconch with flocculent sculpture towards the outer portion and more numerous spiral keels of variable strength.

Sukashitrochus carinatus from Japan has flocculent sculpture on the protoconch and a more narrow umbilicus.

Sukashitrochus estotiensis from the Oligocene of Europe and the Recent Indian Ocean has fine reticulate sculpture over the entire teleoconch.

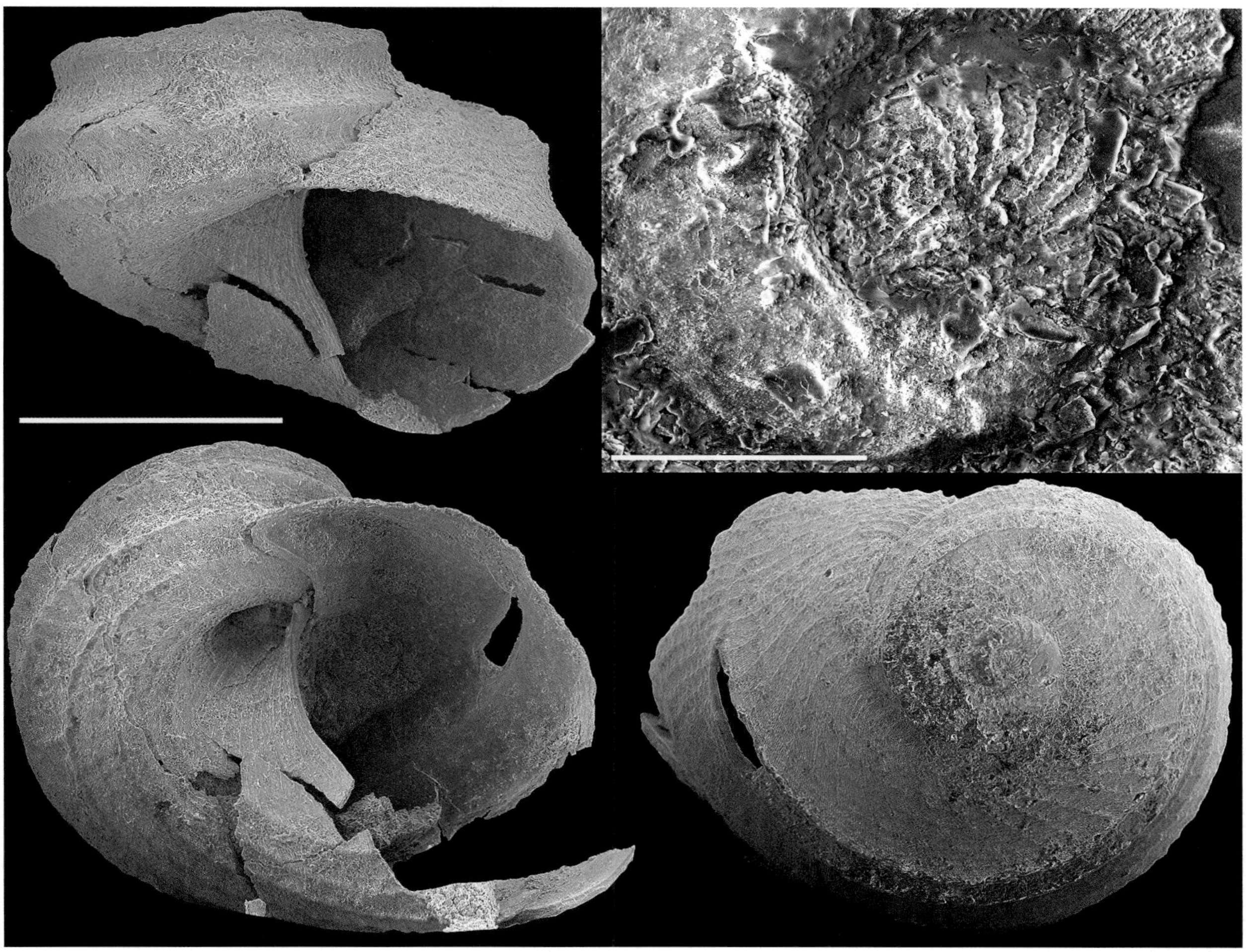

Figure 499. *Sukashitrochus morleti*. Lectotype. New Caledonia, intertidal sands (MNHN). Scale bar shell = 1 mm. Scale bar protoconch = 100 μm.

Sukashitrochus lyallensis from New Zealand has flocculent sculpture on the protoconch.

Sukashitrochus pulcher from temperate Australia has a strongly cancellate sculpture with elevated knobs at the intersection of axials and spiral cords on the base.

Sukashitrochus dorbignyi from the Indian Ocean has a smooth protoconch, a somewhat smaller teleoconch I of less than one whorl, more numerous axial cords on teleoconch I that are of similar strength to those on teleoconch II, and finer spiral sculpture on the base.

Distribution. Indo-Malayan Archipelago to Central Pacific. 0–60 m depth, live, in muddy sand (Fig. 503).

Fossil. The specimen figured by Ladd (1982) from the Pleistocene of Vanuatu (USNM 314324, 1) was examined by SEM and protoconch sculpture confirms its identification as *Suk. morleti.*

Specimen records.

Japan, Okinawa. 17–35 m, Maeda Point, 26.435°N, 127.759°E (DLG 967, 3). 46–55 m, 1 km W of Onna Village, 26.495°N, 127.843°E (DLG 1175, 1). 35–38 m, base of cliff, Horseshoe, 26.504°N, 127.858°E (DLG 972, 2). 43 m, Horseshoe Beach, 26.505°N,

Figure 500. *Sukashitrochus morleti.* 6 m, New Year Island, Northern Territories, Australia, 10.900°S, 133.033°E (AMS C.378619, 2). Scale bar shell = 0.5 mm. Scale bar protoconch = 100 μm.

127.853°E (DLG 1547, 3). 37 m, Big Rock, Horseshoe Bay, 26.508°N, 127.853°E (DLG 1463, 5). 33 m, Horseshoe Bay, 26.509°N, 127.856°E (DLG 1515, 1).

China. 5 m, Round Island, Pak Sha Chau, Mirs Bay, Hong Kong, 22.550°N, 114.333°E (AMS C.378669, 1). 0 m, Hoi Sing Wan, Tolo Channel, New Territories, Hong Kong, 22.435°N, 114.247°E (LACM 85-12, 1).

Vietnam. 2–3 m, Hon Mun Island, Nha Trang region, 10.167°N, 109.296°E (DLG 954, 2). Qui Nhon, Binh Dinh, 13.764°N, 109.227°E (DLG 1780, 1).

Figure 501. *Sukashitrochus morleti.* **A.** 15–35 m, Baudisson Bay, Kavieng, New Ireland, Papua New Guinea, 2.743°S, 150.658°E (DLG 300, 5). **B.** Norfolk Island, Australia, 29.033°S, 167.950°E (AMS C.378318, 1). **C.** 20 m, New Year Island, S side, Northern Territories, Australia, 10.900°S, 133.033°E (AMS C.378620, 2). Scale bars shell = 1 mm. Scale bars protoconch = 100 μm.

Thailand. 27 m, Koh Mersan (ZMUC, 1).

Philippines. 10–12 m, off Mactan Island, Cebu, 10.326°N, 124.005°E (FLMNH 392483, 2). 640 m, St. 5407, off Ponson Island, Camotes Islands, 10.860°N, 124.348°E (USNM 288833, 1). 686 m, St. 5348, Palawan Pass, off Point Tabonan, 10.962°N, 118.638°E (USNM 285852, 1). 271 m, St. 5194, off Chocolate Island, N Cebu, 11.258°N, 124.183°E (USNM 259608, 4). 84 m, St. 5336, Linapacan Strait off Observatory Island, 11.629°N, 119.767°E (USNM 282559, 1). 598 m, St. 5385, Ragay Gulf, off Arena Point, Luzon, 13.414°N, 123.064°E (USNM 285366, 1). 20 m, Luzon, Manig Point, 13.683°N, 120.892°E (LACM 84-163, 4). 25 m, Luzon, Ligpo Point, 13.825°N, 120.898°E (LACM 84-161, 1). Binanga Port, Subic Bay, 14.740°N, 120.254°E (USNM 229880, 1). 478–480 m, AURORA 2007 DW2706, 15.070°N, 121.715°E (MNHN, 1). 507 m, St. 5576, off Mount Dromedarie, N Tawi Tawi, 5.432°N, 120.061°E (USNM 257870, 3). 37 m, St. 5140, off Jolo, 6.145°N, 121.050°E (USNM ex 275329, 1). 200 m, Gulf of Davao, 6.700°N, 125.800°E (DLG 1220, 1). 13 m, E end Santa Cruz Island, Zamboanga, 6.865°N, 122.073°E (LACM 81-7, 14). 49 m, St. 5131, 0.4 mi N 15°W off Panabutan Point, W Mindanao, 7.585°N, 122.121°E (USNM 262976, 1; USNM 262915, 2). 124 m, St. 5357, NE off Balabac, Balabac Island, 8.100°N, 117.286°E (USNM 285978, 1). 5 m, Pipe Point, Pandanan Island, 8.200°N, 117.191°E (GPC 245235, 1). 20 m, Looc, Panglao Island, PANGLAO 2004 B32, 9.430°N, 123.743°E (MNHN, 5). 4–30 m, Catarman, Panglao Island, PANGLAO 2004 S4, 9.431°N, 123.863°E (MNHN, 1). 134–190 m, Cervera Shoal, W Pamilacan Island, PANGLAO 2004 T34, 9.470°N, 123.845°E (MNHN, 1). 95–128 m, Cervera Shoal, Pamilacan Island, PANGLAO 2004 T36, 9.488°N, 123.858°E (MNHN, 1). 5–12 m, Pamilacan

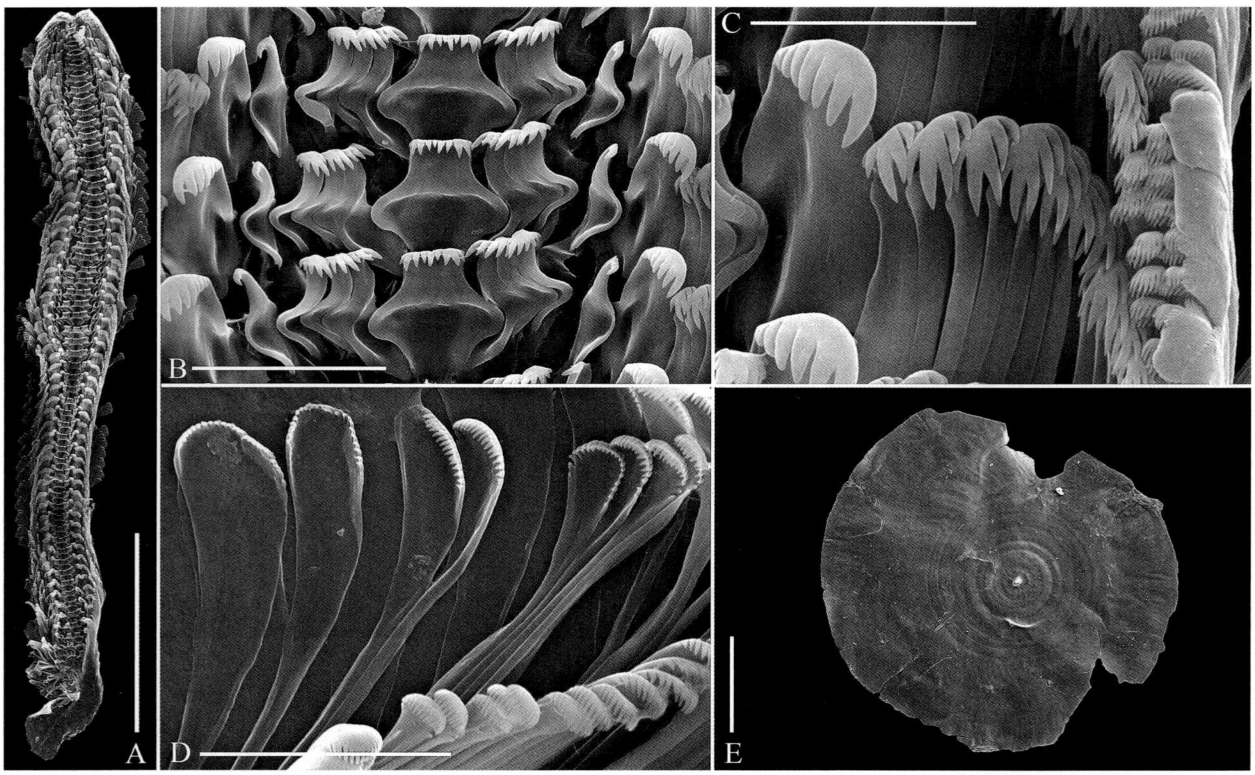

Figure 502. *Sukashitrochus morleti*. Radula and operculum. 35–38 m, base of cliff, Horseshoe, Okinawa, Japan, 26.504°N, 127.858°E (DLG 972, 2). **A.** Entire radula. Scale bar = 200 µm. **B.** Central field. Scale bar = 20 µm. **C.** Lateral tooth 5 and marginal teeth. Scale bar = 10 µm. **D.** Outermost marginal teeth. Scale bar = 10 µm. **E.** Operculum. Scale bar = 200 µm.

Island, PANGLAO 2004 S14, 9.488°N, 123.918°E (MNHN, 3). 2–4 m, Pamilacan Island, PANGLAO 2004 B11, 9.490°N, 123.933°E (MNHN, 1). 6–14 m, Pamilacan Island, PANGLAO 2004 S10, 9.490°N, 123.933°E (MNHN, 2). 38 m, Pamilacan Island, PANGLAO 2004 B24, 9.490°N, 123.933°E (MNHN, 1). 24 m, Pamilacan Island, PANGLAO 2004 S31, 9.490°N, 123.935°E (MNHN, 1). 20 m, off Bohol Island, Pamilacan Island, 9.500°N, 123.917°E (AMS C.78671, 1). 356–396 m, off Balicasag Island, Bohol Sea, CP2393, 9.501°N, 123.693°E (MNHN, 2). 60 m, Pamilican Island, 9.501°N, 123.918°E (MNHN, 1: complete). 120–139 m, Balicasag Island, PANGLAO 2004 L74–75, 9.511°N, 123.683°E (MNHN, 1). 14 m, Balicasag Island, PANGLAO 2004 S46, 9.515°N, 123.680°E (MNHN, 1). 17–18 m, Balicasag Island, Panglao Island, PANGLAO 2004 B38, 9.515°N, 123.680°E (MNHN, 1). 19–20 m, Balicasag Island, PANGLAO 2004 B37, 9.515°N, 123.680°E (MNHN, 1). 90–110 m, Balicasag Island, PANGLAO 2004 L46, 9.515°N, 123.687°E (MNHN, 1). 26 m, Bohol Island, off Balicasag Island, 9.517°N, 123.683°E (AMS C.78672, 2). 180 m, Balicasag Island, 9.518°N, 123.680°E (GPC 189140, 1). 309–342 m, Balicasag Island, Bohol Sea, CP2399, 9.528°N, 123.698°E (MNHN, 6). 164–176 m, off Balicasag Island, Bohol Sea, PANGLAO 2005 DW2339, 9.531°N, 123.728°E (MNHN, 2; MNHN, 9). 152 m, Bolod, Panglao Island, PANGLAO 2004 T2, 9.540°N, 123.797°E (MNHN, 5). 713–731 m, off Balicasag Island, Bohol Sea, PANGLAO 2005 CP2398, 9.543°S, 123.675°E (MNHN, 8). 0–2 m, Danao, Panglao Island, PANGLAO 2004 M3, 9.546°N, 123.752°E (MNHN, 1). 8–14 m, Alona Reef, Panglao Island, PANGLAO 2004 B1, 9.550°N, 123.775°E (MNHN, 1). 5 m, Alona Reef, Panglao Island, PANGLAO 2004 B2, 9.550°N, 123.775°E (MNHN, 2). 82 m, Bolod, Panglao Island, PANGLAO 2004 T4, 9.550°N, 123.808°E (MNHN, 2). 8–10 m, Napaling, Panglao Island, PANGLAO 2004 B9, 9.551°N, 123.733°E (MNHN, 4). 8 m, Arco Point, Panglao Island, PANGLAO 2004 B3, 9.558°N, 123.810°E (MNHN, 1). 97–120 m, off San Isidro, Panglao Island, PANGLAO 2004 T9, 9.561°N, 123.833°E (MNHN, 4). 6 m, San Isidro, Panglao Island, PANGLAO 2004 S17, 9.576°N, 123.832°E (MNHN, 4). 642–669 m, Maribojoc Bay, Bohol Sea, CP2397, 9.581°N, 123.695°E (MNHN, 2). 80 m, Biking-Catarman, Panglao

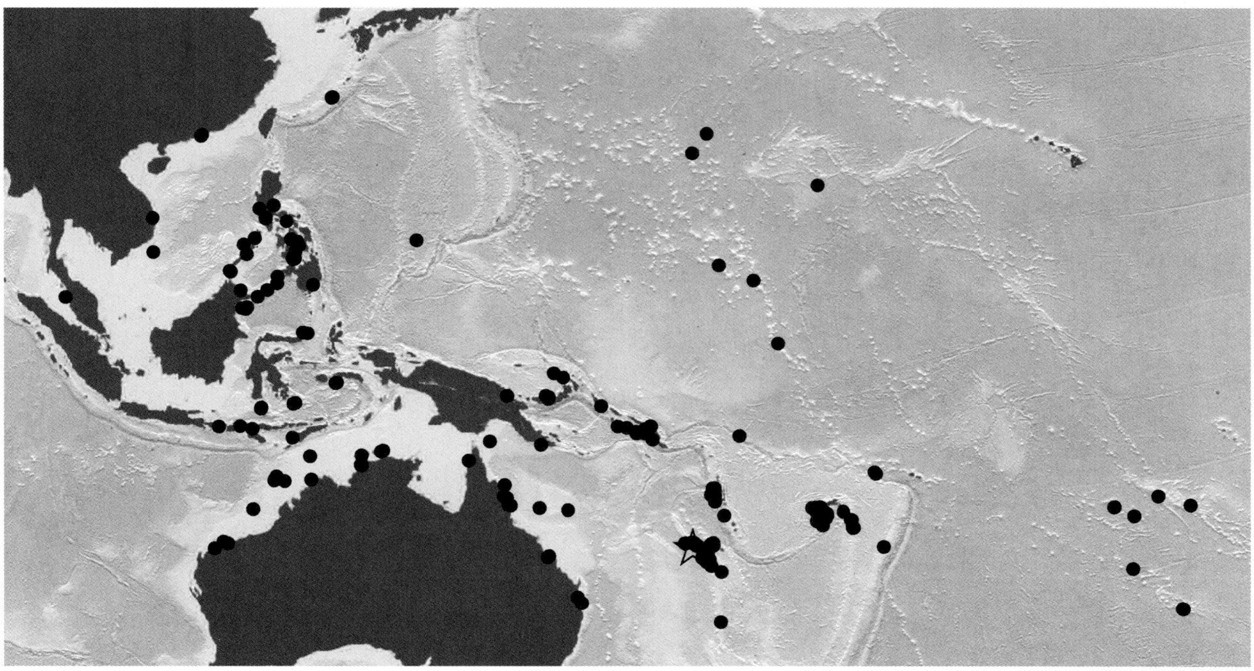

Figure 503. Distribution of *Sukashitrochus morleti*. ☆ type locality. ● specimen record. ■ literature record.

Island, PANGLAO 2004 T28, 9.583°N, 123.857°E (MNHN, 1). 4 m, Biking, Panglao Island, PANGLAO 2004 B5, 9.586°N, 123.840°E (MNHN, 2). 5 m, Biking, Panglao Island, PANGLAO 2004 S1, 9.588°N, 123.842°E (MNHN, 2). 24–27 m, Doljo Point, Panglao Island, PANGLAO 2004 B12, 9.593°N, 123.720°E (MNHN, 4). 2–3 m, Looc, Panglao Island, PANGLAO 2004 S32, 9.596°N, 123.743°E (MNHN, 1). 4–30 m, Tatarman, Panglao Island, PANGLAO 2004 B7, 9.598°N, 123.863°E (MNHN, 3). 2 m, Bohol, Panglao Island, SW side, 9.600°N, 123.750°E (AMS ex 348632, 1, –: complete). 90–200 m, Maribohoc Bay, Bohol Island, PANGLAO 2004 P1, 9.600°N, 123.750°E (MNHN, 7). 2–4 m, Momo Beach, Panglao Island, PANGLAO 2004 S24, 9.601°N, 123.750°E (MNHN, 6). 0–3 m, Momo Beach, Panglao Island, PANGLAO 2004 M7, 9.601°N, 123.753°E (MNHN, 5: complete; MNHN, 1). 382–434 m, Maribojoc Bay, Bohol Sea, PANGLAO 2005 CP2395, 9.603°N, 123.730°E (MNHN, 2). 3–14 m, Momo Beach, Panglao Island, PANGLAO 2004 B10, 9.608°N, 123.760°E (MNHN, 4). 28–32 m, Momo Beach, Panglao Island, PANGLAO 2004 S8, 9.608°N, 123.760°E (MNHN, 10). 30–33 m, between Momo and Napaling, Panglao Island, PANGLAO 2004 B42, 9.616°N, 123.767°E (MNHN, 1). 3 m, Napaling, Panglao Island, PANGLAO 2004 B8, 9.618°N, 123.768°E (MNHN, 1). 2–4 m, Napaling, Panglao Island, PANGLAO 2004 S5, 9.618°N, 123.768°E (MNHN, 7). Napaling, Panglao Island, PANGLAO 2004 B8, 9.618°N, 123.769°E (MNHN, 3). 8–15 m, Baclayon Takot, Bohol Island, PANGLAO 2004 S13, 9.618°N, 123.877°E (MNHN, 2). 3–5 m, Baclayon Takot, Bohol Island, PANGLAO 2004 B13, 9.618°N, 123.877°E (MNHN, 2). 28–32 m, Napaling, Panglao Island, PANGLAO 2004 S28, 9.620°N, 123.773°E (MNHN, 8). 0–3 m, Bingag/Tabalong, Panglao Island, PANGLAO 2004 M10, 9.630°N, 123.807°E (MNHN, 1). 1–4 m, Sungcolan Bay, Panglao Island, PANGLAO 2004 S7, 9.641°N, 123.820°E (MNHN, 5: complete; MNHN, 12). 2–4 m, Sungcolan Bay, Panglao Island, PANGLAO 2004 B14, 9.641°N, 123.820°E (MNHN, 1). 1–4 m, Sungcolan Bay, Panglao Island, PANGLAO 2004 S6, 9.641°N, 123.820°E (MNHN, 1). 3–5 m, Sungcolan Bay, Panglao Island, PANGLAO 2004 B18, 9.641°N, 123.828°E (MNHN, 1). 2–4 m, Sungcolan, Panglao Island, PANGLAO 2004 B15, 9.646°N, 123.818°E (MNHN, 1). 78–95 m, Maribohoc Bay, Bohol Island, PANGLAO 2004 T11, 9.681°N, 123.833°E (MNHN, 5). 160–210 m, Cortes, Bohol Island, PANGLAO 2004 T25, 9.685°N, 123.822°E (MNHN, 1). 256–268 m, Maribojoc Bay, Bohol Sea, PANGLAO 2005 CP2407, 9.688°N, 123.808°E (MNHN, 1). 12 m, Ubajan, Bohol Island, PANGLAO 2004 S27, 9.691°N, 123.850°E (MNHN, 5). 2–8 m, Ubajan, Bohol Island, PANGLAO 2004 B20, 9.691°N, 123.850°E (MNHN, 3). 21 m, Ubajan, Bohol Island, PANGLAO 2004 S25, 9.691°N, 123.850°E (MNHN, 6). 4–12 m, Manga, Bohol Island, PANGLAO 2004, 9.695°N, 123.848°E (MNHN, 3). 80–100 m, Cortes, Bohol Island, PANGLAO 2004 T18, 9.696°N, 123.832°E (MNHN, 3). 100 m, Cortes, Bohol Island, PANGLAO 2004 B1, 9.698°N, 123.825°E (MNHN, 1). 3–4 m, Manga, Bohol Island, PANGLAO 2004 S19, 9.701°N, 123.857°E (MNHN, 1). 35–45 m, Cortes, Bohol Island, PANGLAO 2004 T23, 9.703°N, 123.843°E (MNHN, 1). 10–26 m, Cortes, Bohol Island, PANGLAO 2004 T19, 9.703°N, 123.847°E (MNHN, 1). 70–96 m, Cortes, Bohol Island, PANGLAO 2004 T43, 9.704°N, 123.832°E (MNHN, 1). 123–135 m, Cortes, Bohol Island, PANGLAO 2004 T26, 9.721°N, 123.813°E (MNHN, 1). Palawan Island, N Hondo Bay, Tadio Island, 9.950°N, 118.933°E (AMS C.78668, 2).

Malaysia. Monkey Beach, Penang, 5.400°N, 100.233°E (WAM, 1).

Indonesia. 20 m, off Waibailon Rock, E coast of Flores (CPC, 4). 0–10 m, bay N of Tanjung Batuangus, Selat Lembeh, N Sulawesi, 1.516°N, 125.250°E (RMNH, 1). 8–40 m, Pisok Point, N of Manado, N Celebes, 1.567°N, 124.792°E (LACM 82-39, 1). 5–20 m, Sulawesi, Bunaken and Siladen Islets, 1.603°N, 124.767°E (LACM 88-55, 13). 20 m, Hutumuri, S coast, Leitimur, Ambon, Moluccas, 3.731°S, 128.259°E (RMNH, 3). Latuhalat, S coast, Leitimur, Ambon, Moluccas, 3.778°S, 128.126°E (RMNH, 1). 870 m, St. 5585, Sibuko Bay, off Sipadan Island, Borneo, 4.116°N, 118.832°E (USNM

Scissurellidae *Sukashitrochus*

258561, 9). 759 m, St. 5587, Sibuko Bay, off Sipidan Island, Borneo, 4.176°N, 118.620°E (USNM 278694, 6). 556 m, St. 5592, off Silungan Island, Borneo, 4.212°N, 118.462°E (USNM 291067, 2). 558 m, St. 5592, off Silungan Island, Borneo, 4.212°N, 118.462°E (USNM 279043, 7). 1628 m, St. 5582, off Si Amil Island, Borneo, 4.331°N, 118.977°E (USNM 289817, 1). 250–290 m, NW of Baningko, Banda Sea, Tukang Besi Islands, 5.875°S, 123.975°E (RMNH, 1). 280 m, NE of Binongki, Banda Sea, Tukang Besi Island, 5.875°S, 123.975°E (RMNH, 3). 465 m, S of Karanga Kaledupa, Banda Sea, Tukang Besi Islands, 5.958°S, 123.775°E (RMNH, 6). 365 m, S of Kaledupa Reef, Tukang Besi Islands, 5.963°S, 123.775°E (RMNH, 3). 71 m, St. 5358, off Sandakan Light, Borneo, 6.111°N, 118.304°E (USNM 285719, 2). SW Salayer, 6.355°S, 120.435°E (RMNH, 1). 233–274 m, SW Salayer, 6.371°S, 120.433°E (RMNH, 30). 175 m, SW Salayer, 6.373°S, 120.437°E (RMNH, 3). 300–340 m, NW of Pulau Bahuluang, SW Salayer, 6.466°S, 120.400°E (RMNH, 1). 175–185 m, Bay of Sanggar, N of Sumbawa, 8.321°S, 118.260°E (RMNH, 5). 5–20 m, off Ajer (Gili Air) and Meno Islet, NW side Lompok, 8.367°S, 116.067°E (LACM 88-63, 4). 65 m, Teluk Slawi, E of Komodo, 8.613°S, 119.512°E (RMNH, 8).

Australia, Northern Territories. Darwin, 12.467°S, 130.833°E (WAM, 5; RMNH AD 32453, 4). Darwin, off Emery Point, 12.450°S, 130.817°E (AMS C.378611, 12). Darwin, Sandbar, 12.433°S, 130.800°E (AMS C.378615, 25). Darwin, Sandbar No. 1, 12.433°S, 130.800°E (AMS C.378624, 2). Darwin, Sandbar No. 2, 12.400°S, 130.733°E (AMS C.378622, 1). 8–10 m, Darwin, Fannie Bay, W of East Point, 11.408°S, 130.808°E (AMS C.378614, 1). 14 m, Oxley Island, W side, 11.000°S, 132.817°E (AMS C.378616, 3). 5 m, Oxley Island, W end, 11.000°S, 132.817°E (AMS C.378625, 3). 14 m, Oxley Island, W side, 11.000°S, 132.817°E (AMS C.404490 2: complete). 5 m, Oxley Island, W end, 11.000°S, 132.817°E (AMS C.406402, 2). 6 m, New Year Island, 10.900°S, 133.033°E (AMS C.378619, 2). 20 m, New Year Island, S side, 10.900°S, 133.033°E (AMS C.378620, 2). 10 m, New Year Island, 10.900°S, 133.033°E (AMS C.378623, 1, 5: complete).

Australia, Queensland. 128–183 m, off Cape Moreton, 27.000°S, 153.583°E (AMS C.436781, 1). Noosa Heads, 26.383°S, 153.150°E (AMS C.461551, 1). 11 m, Thirsty Sound, Broad Sound, 22.122°S, 150.030°E (AMS C.461470, 1). 20 m, SW of Marble Island, Duke Group, 21.995°S, 150.162°E (AMS C.461469, 13). 21 m, Great Barrier Reef, Euston Reef, SW side, 16.667°S, 146.217°E (AMS C.405198, 2). 20 m, Great Barrier Reef, Michaelmas Cay, 16.600°S, 145.983°E (AMS C.348697, 16). 38 m, Escape Reef, S part, Great Barrier Reef, 15.883°S, 145.817°E (AMS ex C.461468, 1). 9–18 m, Great Barrier Reef, Cairns Reef lagoon and between Cairns Reef and Hope Island, 15.700°S, 145.500°E (AMS C.27172, 20+). 27–30 m, Great Barrier Reef, Carter Reef, N end, 14.550°S, 145.600°E (AMS C.402655, 1). 18 m, Great Barrier Reef, Carter Reef, off Platform Bommie, 14.533°S, 145.583°E (AMS C.378577, 1). Gulf of Carpentaria, beaches around Mapoon, 11.967°S, 141.883°E (AMS C.378588, 2). 9–15 m, Torres Strait, off Murray Island, 9.933°S, 144.067°E (AMS C.48356, 7).

Australia, Western Australia. 1 m, Beagle Island, 21.183°S, 115.633°E (WAM, 1). 8–9 m, Kendrew Island, Dampier Archipelago, 20.475°S, 116.533°E (WAM, 1). 18–20 m, N end Mermaid Reef, Rowley Shoals, 17.050°S, 119.600°E (LACM 86-250, 1). 256 m, North West Shelf, ca. 278 km N of Cape Leveque, 14.117°S, 122.867°E (AMS C.378289, 1). 20 m, E side, Sandy Islet, Scott Reef, 14.044°S, 121.780°E (SBMNH 369525, 5). 20 m, Sandy Islet, Scott Reef, 14.044°S, 121.780°E (SBMNH 369698, 2). 20 m, E side Sandy Islet, Scott Reef, 14.044°S, 121.780°E (SBMNH 369511, 5). 15 m, W side, Cassini Island, 13.952°S, 125.623°E (SBMNH 85137, 2). 20 m, N side, North Reef, Scott Reef, 13.877°S, 121.859°E (SBMNH 85138, 1). 15–25 m, Seringapatam Reef, 13.628°S, 122.016°E (SBMNH 85136, 1). 15–25 m, Seringapatam Reef, N side, 13.624°S, 122.005°E (SBMNH 85134, 4). 27 m, Sahul Banks, Timor Sea, 11.500°S, 125.500°E (AMS C.378628, 1). 0 m, Cleaverville, 20.658°S, 116.997°E (MV F178670, 22).

Australia, Coral Sea. 10–15 m, SE from Observatory Cay, Lihou Reef, 17.157°S, 152.141°E (SBMNH 369623, 1). NE Herald Cay, 16.933°S, 149.183°E (AMS C.406344, 2).

Papua New Guinea. Milne Bay District, Amazon Bay, 10.300°S, 149.333°E (AMS C.402651, 2). Lemau, W coast New Ireland, 3.200°S, 151.633°E (AMS C.378667, 1). 0 m, off Walindi Plantation Resort, Kimbe Bay, New Britain, 5.436°S, 150.087°E (DLG 500 1: complete; DLG 508, 4). 1 m, First Reef, Kimbe Bay, New Britain, 5.432°S, 150.090°E (DLG 800, 5). 54 m, off Schauman Island, Kimbe Bay, New Britain, 5.307°S, 150.092°E (DLG 442, 2). 40 m, off Restorf Island, Kimbe Bay, 5.292°S, 150.101°E (DLG 394, 1). 0–10 m, Restorf Island, Kimbe Bay, New Britain, 5.292°S, 150.104°E (DLG 635, 1). 65 m, off Susan's Reef, Kimbe Bay, New Britain, 5.291°S, 150.128°E (DLG 603, 1). 67 m, off Schauman Island, Kimbe Bay, New Britain, 5.282°S, 150.093°E (DLG 792, 1). 28 m, off Big Malu Malu Island, Kimbe Bay, 5.231°S, 150.098°E (DLG 402, 2). 0–5 m, Big Malu Malu, Kimbe Bay, New Britain, 5.230°S, 150.090°E (DLG 731, 1). 30 m, Wongat Island, Madang Province, 5.135°S, 145.845°E (LACM 80-26, 1). 1–3 m, Miliat Harbor, Astrolabe Bay, Nadang Province, 5.120°S, 145.820°E (LACM 80-17, 3). 0 m, Madang, Christensen Research Institute, 5.117°S, 149.817°E (AMS, 1: complete). 15–35 m, Baudisson Bay, Kavieng, New Ireland, 2.743°S, 150.658°E (DLG 300, 5).

Solomon Island. 611–636 m, 9.758°S, 160.875°E (MNHN, 1). 10–25 m, wreck of *Kinugawa Maru*, Bonege, Guadalcanal, Guadalcanal Province, 9.378°S, 169.871°E (SBMNH 85124, 1). Yandina, Bankia Island, Russel Islands, 9.117°S, 159.217°E (NMNZ M.230680, 2). 5–10 m, Soghonangula, Florida Islands, Central Province, 9.116°S, 160.153°E (SBMNH 85145, 1). 10–20 m, Fonagho Island, Russell Islands, Central Province, 9.093°S, 159.296°E (SBMNH 85147, 1). 18 m, E side, Njanjelakalau Island, Florida Island, 8.951°S, 159.992°E (SBMNH 85116, 1). 20–25 m, Matiu Island, Kokoana Passage, Marovo Lagoon, NE side Vangunu Island, New Georgia Group, 8.495°S, 158.200°E (LACM 89-77, 2). 23 m, New Georgia Islands, Marovo Rukiana Island, 8.367°S, 157.283°E (AMS C.378674, 1). 194–286 m, 8.357°S, 160.697°E (MNHN, 1). St. 45 R/V *Vega* cruise, E shore of Arawa Bay, Bougainville Island, 6.206°S, 155.570°E (USNM 761689, 2). 98–200 m, SALOMON 1 DW1767, 8.323°S, 160.665°E (MNHN, 42). 1327–1598 m, SALOMON 1 CP1764, 8.610°S, 160.123°E (MNHN, 1). 396–411 m, SALOMON 1 DW1762, 8.665°S, 160.065°E (MNHN, 1). 766–1100 m, SALOMON 2 DW2167, 8.913°S, 159.121°E (MNHN).

New Caledonia. 465–525 m, southern, 22.583°S, 166.450°E (MNHN, 1). 250–350 m, 22.500°S, 166.400°E (MNHN, 12). 45–50 m, Woodin Canal, 22.392°S, 166.793°E (MNHN, 1). 495 m, 22.367°S, 166.233°E (MNHN, 12). 490–500 m, southern, 22.317°S, 166.183°E (MNHN, 3). 455 m, southern, 22.267°S, 166.183°E (MNHN, 8). 250 m, E coast, 21.750°S, 166.617°E (MNHN, 1). 247–252 m, E coast, 21.700°S, 166.567°E (MNHN, 4). 500 m, Loyalty Basin, 21.550°S, 166.400°E (MNHN, 1). 237–298 m, E coast, 21.283°S, 165.950°E (MNHN, 16). 350–400 m, E coast, 21.033°S, 165.633°E (MNHN, 10). 1820–1980 m, Loyalty Basin, 21°S, 166.967°E (MNHN, 2). 1600–1640 m, Loyalty Basin, 20.983°S, 166.983°E (MNHN, 4). 1639 m, 20.983°S, 167.000°E (MNHN, 1). 105–110 m, E coast, 20.950°S, 165.567°E (MNHN, 2). 160–222 m, E coast, 20.950°S, 165.583°E (MNHN, 5). 15–20 m, Santal Bay, Lifou, 20.947°S, 167.045°E (MNHN, 1). 2–4 m, Santal Bay, Lifou, 20.945°S, 167.052°E (MNHN, 1). 15–18 m, Santal Bay, Lifou, 20.944°S, 167.033°E (MNHN, 6). 10–20 m, Santal Bay, Lifou, 20.925°S, 167.070°E (MNHN, 2). 5–30 m, Santal Bay, Lifou, 20.920°S, 167.012°E (MNHN, 2). 21–30 m, Santal Bay, Lifou, 20.910°S, 167.035°E (MNHN, 1). Lifou, 20.900°S, 167.200°E (KBIN, 2; KBIN, 4; KBIN, 4; KBIN, 15). 48–52 m, Santal Bay, Lifou, 20.897°S, 167.122°E (MNHN, 4). 12–32 m, Santal Bay, Lifou, 20.892°S, 167.045°E (MNHN, 5). 5–25 m, Touho sector, 20.878°S, 165.325°E (MNHN, 5). 5–20 m, Santal Bay, Lifou, 20.875°S, 167.135°E (MNHN, 5). 4 m, Santal Bay, Lifou, 20.873°S, 167.142°E (MNHN, 1). 4 m, Santal Bay, Lifou, 20.873°S, 167.142°E (MNHN, 1). 20 m, Touho sector, 20.840°S, 165.380°E (MNHN, 2). 10 m, Touho sector, 20.825°S, 165.327°E (MNHN, 3). 25–30 m, Santal Bay, Lifou, 20.822°S,

167.173°E (MNHN, 2). 50–62 m, Touho sector, 20.817°S, 165.283°E (MNHN, 1). 105–110 m, E coast, 20.817°S, 165.317°E (MNHN, 2). 80–140 m, Touho sector, 20.817°S, 165.317°E (MNHN, 1). 170 m, E coast, 20.800°S, 165.317°E (MNHN, 1). 1660 m, Loyalty Basin, 20.800°S, 166.883°E (MNHN, 1). 20–25 m, Santal Bay, Lifou, 20.792°S, 167.118°E (MNHN, 2). 8–18 m, Santal Bay, Lifou, 20.792°S, 167.118°E (MNHN, 1). 140–21 m, Santal Bay, Lifou, 20.788°S, 167.113°E (MNHN, 4). 5–10 m, Santal Bay, Lifou, 20.780°S, 167.046°E (MNHN, 2). 3–6 m, Touho sector, 20.778°S, 165.228°E (MNHN, 7). 2–3 m, Touho sector, 20.777°S, 165.235°E (MNHN, 1). 0–2 m, Touho sector, 20.775°S, 165.242°E (MNHN, 7). 45–56 m, Touho sector, 20.775°S, 165.263°E (MNHN, 3). 20 m, Santal Bay, Lifou, 20.773°S, 167.033°E (MNHN, 5). 47 m, Santal Bay, Lifou, 20.773°S, 167.033°E (MNHN, 5). 6–15 m, Touho sector, 20.771°S, 165.229°E (MNHN, 10). 0 m, Touho sector, 20.770°S, 165.242°E (MNHN, 3). 7 m, Touho sector, 20.768°S, 165.243°E (MNHN, 1). 20 m, Santal Bay, Lifou, 20.763°S, 167.028°E (MNHN, 1). 27–31 m, Santal Bay, Lifou, 20.763°S, 167.028°E (MNHN, 2). 17 m, Santal Bay, Lifou, 20.763°S, 167.028°E (MNHN, 1). 0 m, Touho sector, 20.753°S, 165.272°E (MNHN, 6). 15–20 m, Touho sector, 20.750°S, 165.163°E (MNHN, 1). 10–35 m, Touho sector, 20.750°S, 165.275°E (MNHN, 3). 8 m, Touho sector, 20.742°S, 165.265°E (MNHN, 12). 49–59 m, Touho sector, 20.733°S, 165.247°E (MNHN, 3). 11 m, Touho sector, 20.717°S, 165.133°E (MNHN, 1). 282 m, Loyalty Ridge, 20.700°S, 167.000°E (MNHN, 1). 20–30 m, Koumac sector, 20.690°S, 164.247°E (MNHN, 3). 26–40 m, Koumac sector, 20.673°S, 164.248°E (MNHN, 5). 55–57 m, Koumac sector, 20.672°S, 164.195°E (MNHN, 12). 15–20 m, Koumac sector, 20.667°S, 164.253°E (MNHN, 1). 30–63 m, Koumac sector, 20.663°S, 164.255°E (MNHN, 1). 15 m, Koumac sector, 20.662°S, 164.248°E (MNHN, 1). 1230–1270 m, Loyalty Basin, 20.633°S, 167.167°E (MNHN, 1). Beach at E side of Mount Panie, 20.600°S, 164.833°E (AMS C.004108, 2). 15–20 m, Touho sector, 20.585°S, 165.135°E (MNHN, 9). 380–400 m, E coast, 20.583°S, 165.117°E (MNHN, 4). 380–400 m, E coast, 20.583°S, 165.117°E (MNHN, 12). 408–440 m, E coast, 20.583°S, 165.117°E (MNHN, 20). 491–533 m, E coast, 20.567°S, 164.950°E (MNHN, 2). 497–520 m, E coast, 20.567°S, 164.950°E (MNHN, 14). 570–650 m, E coast, 20.567°S, 164.967°E (MNHN, 5). 270–282 m, E coast, 20.550°S, 165.000°E (MNHN, 2). 410–430 m, E coast, 20.583°N, 164.967°E (MNHN, 3). 1350 m, Loyalty Basin, 22.633°N, 166.483°E (MNHN, 1).

Vanuatu. Port Vila, Efate Island, 17.733°S, 168.300°E (AMS C.378675, 3). 360–419 m, 16.267°S, 167.350°E (MNHN, 4). 622–625 m, 15.667°S, 167.333°E (MNHN, 1). USGS locality 25718, Kere River, Santo, 15.567°S, 166.929°E (USNM 314324, 1). Opposite Matte Valu Island, Peterson's Bay, Espiritu Santo, 15.365°S, 167.191°E (USNM 542268, 1). 154–179 m, 15.050°S, 167.117°E (MNHN, 1). 700–736 m, 14.883°S, 167.100°E (MNHN, 1). 700–736 m, 14.883°S, 167.100°E (MNHN, 2). 1550–1620 m, 14.867°S, 167.300°E (MNHN, 2). Efate Island, Port Vila, Point Ardel Malapoa, 17.733°S, 168.300°E (AMS C.405011, 2: complete). Port Vila, Efate Island, 17.733°S, 168.300°E (AMS C.405023, 1: complete). 154–179 m, MUSORSTOM 8 DW1105, 15.050°S, 167.116°E (MNHN, 3). 7 m, S Turtle Island, Palikulo Bay, SANTO 2006 DB77DB71, 15.360°S, 167.208°E (MNHN, 1). 9 m, Reef near Oyster Island, SANTO 2006 FB40, 15.381°S, 167.195°E (MNHN, 1). 7 m, W Mavéa Island, SANTO 2006 DB67, 15.381°S, 167.218°E (MNHN, 2). 102–120 m, W Mavéa Island, Palikulo Bay, SANTO 2006 AT14, 15.397°S, 167.220°E (MNHN, 10). 38 m, Palikulo Bay, SANTO 2006 DB69, 15.406°S, 167.216°E (MNHN, 8). 6–43 m, Aésé Island, SANTO 2006 DB58, 15.410°S, 167.238°E (MNHN, 3). 13 m, Palikulo Bay, SANTO 2006 DB65, 15.430°S, 167.216°E (MNHN, 3). 21 m, SE Aésé Island, SANTO 2006 DB63, 15.448°S, 167.263°E (MNHN, 3). 22 m, Palikulo Bay, SANTO 2006 DS43, 15.465°S, 167.238°E (MNHN, 18). 19 m, Palikulo Bay, Aoré Island, SANTO 2006 FB43, 15.473°S, 167.248°E (MNHN, 2). 0 m, Palikulo Peninsula, SANTO 2006 VM58, 15.476°S, 167.255°E (MNHN, 3). 0 m, E Palikulo Peninsula, Palikulo Bay, SANTO

2006 DB77VM54, 15.481°S, 167.258°E (MNHN, 14). 5 m, Palikulo Bay, SANTO 2006 DB40, 15.496°S, 167.251°E (MNHN, 1). 22–25 m, NW Urélapa Island, SANTO 2006 DB20, 15.508°S, 167.023°E (MNHN, 13). 8–15 m, Segond Channel, vicinity of Luganville, SANTO 2006 DS06, 15.515°S, 167.185°E (MNHN, 1). 0 m, W Million Dollar Point, SANTO 2006 VM70, 15.521°S, 167.235°E (MNHN, 3). 2–3 m, Segond Channel, vicinity of Maritime College, SANTO 2006 NS37, 15.523°S, 167.163°E (MNHN, 1). 4–7 m, Segond Channel, vicinity of Maritime College, SANTO 2006 LD24, 15.523°S, 167.166°E (MNHN, 1). 15–30 m, Segond Channel, *Coolidge* wreck, SANTO 2006 ZS19, 15.523°S, 167.235°E (MNHN, 1). 25 m, Segond Channel, *Coolidge* wreck, SANTO 2006 DS04, 15.523°S, 167.235°E (MNHN, 3). 2–3 m, Segond Channel, vicinity of Maritime College, SANTO 2006 NS36, 15.528°S, 167.158°E (MNHN, 2). 25 m, Segond Channel, Caverot Point, SANTO 2006 DS22, 15.528°S, 167.161°E (MNHN, 1). 17–19 m, N Tutuba Island, SANTO 2006 DS101, 15.538°S, 167.286°E (MNHN, 2). 45–50 m, Aoré Island, Aimbuei Bay, SANTO 2006 FP46–49, 15.540°S, 167.211°E (MNHN, 1). 46–47 m, Segond Channel, SANTO 2006 EP01, 15.541°S, 167.150°E (MNHN, 11). 100–105 m, NW Tutuba Island, SANTO 2006 DS99, 15.541°S, 167.281°E (MNHN, 4). 160–175 m, NW Tutuba Island, SANTO 2006 AT02, 15.544°S, 167.271°E (MNHN, 1). 1–42 m, Segond Channel, NW coast of Aoré Island, Palikulo Bay, SANTO 2006 DB77FR19, 15.545°S, 167.166°E (MNHN, 3). 97–101 m, E Aoré island, SANTO 2006 AT04, 15.550°S, 167.225°E (MNHN, 108). MUSORSTOM 8 DW1069, 15.550°S, 167.233°E (MNHN, 15). 92 m, NW Tutuba Island, SANTO 2006 DS105, 15.550°S, 167.278°E (MNHN, 4). 2 m, Segond Channel, NW Aoré Island, SANTO 2006 FS79, 15.551°S, 167.160°E (MNHN, 2). 27–29 m, W Aoré Island, SANTO 2006 ZS22, 15.551°S, 167.160°E (MNHN, 2). 15–25 m, NE Tutuba Island, SANTO 2006 DB01, 15.551°S, 167.296°E (MNHN, 4). 71–102 m, Aoré Island, Aimbuei BayPalikulo Bay, SANTO 2006 DB77EP18, 15.553°S, 167.200°E (MNHN, 2). 89–109 m, NE Aoré Island, Aimbué Bay, SANTO 2006 EP04, 15.553°S, 167.205°E (MNHN, 3). 0–1 m, W Tutuba Island, SANTO 2006 VM69, 15.556°S, 167.278°E (MNHN, 10). 8–9 m, Tutuba Island, SANTO 2006 FS84, 15.560°S, 167.276°E (MNHN, 1). 7 m, Segond Channel, Wambu river mouth, SANTO 2006 DS91, 15.561°S, 167.140°E (MNHN, 1). 0 m, E Aoré Island, SANTO 2006 VM57, 15.563°S, 167.208°E (MNHN, 5). 20–28 m, Segond Channel, Wambu river mouth, SANTO 2006 ED07, 15.567°S, 167.136°E (MNHN, 1). 70–80 m, W Tutuba Island, SANTO 2006 DS103, 15.568°S, 167.266°E (MNHN, 7). 98–100 m, W Tutuba Island, SANTO 2006 DS102, 15.568°S, 167.266°E (MNHN, 3). 80 m, W Tutuba Island, SANTO 2006 DS104, 15.568°S, 167.266°E (MNHN, 5). 0 m, E Aoré Island, Aisari Bay, SANTO 2006 ZM36, 15.571°S, 167.206°E (MNHN, 2). 2–3 m, Aoré Island, Port Benier, SANTO 2006 FB61, 15.573°S, 167.210°E (MNHN, 7). 0 m, Aoré Island, SANTO 2006 VM46, 15.575°S, 167.205°E (MNHN, 1). 0 m, Aoré Island, Palikulo Bay, SANTO 2006 DB77VM46, 15.575°S, 167.205°E (MNHN, 15). 0 m, E Amoré Island, SANTO 2006 VM37, 15.575°S, 167.210°E (MNHN, 4). 12 m, E Aoré Island, SANTO 2006 DB08, 15.576°S, 167.230°E (MNHN, 2). 14–25 m, E Aoré Island, SANTO 2006 DB33, 15.578°S, 167.230°E (MNHN, 1). 0 m, Baldwin Bay, SANTO 2006 VM30, 15.580°S, 167.033°E (MNHN, 2). 0 m, Baldwin Bay, Palikulo Bay, SANTO 2006 DB77VM30, 15.580°S, 167.033°E (MNHN, 1). 84–105 m, N Urélapa Island, SANTO 2006 AT43, 15.606°S, 167.038°E (MNHN, 1). 108–121 m, W Urélapa Island, SANTO 2006 EP24, 15.608°S, 167.021°E (MNHN, 1). 86–116 m, N Urélapa Island, SANTO 2006 AT44, 15.608°S, 167.045°E (MNHN, 1). 10–18 m, S Aoré Island, SANTO 2006 DB12, 15.610°S, 167.168°E (MNHN, 2). 6–24 m, S Aoré Island, Aoré Island, SANTO 2006 DS10, 15.610°S, 167.168°E (MNHN, 6). 83–120 m, E Urélapa Island, SANTO 2006 AT30, 15.611°S, 167.043°E (MNHN, 1). 35 m, Bruat Channel, N coast of Malo Island, SANTO 2006 DS110, 15.613°S, 167.146°E (MNHN, 1). 78–91 m, NW coast of Malo Island, SANTO 2006 EP22, 15.622°S, 167.098°E (MNHN, 3). 99 m, NW coast of Malo Island, SANTO 2006 EP21, 15.628°S, 167.086°E

(MNHN, 3). 80–94 m, NW coast of Malo Island, SANTO 2006 EP19, 15.629°S, 167.090°E (MNHN, 1). 90–110 m, NW coast of Malo Island, Palikulo Bay, SANTO 2006 DB77 EP28, 15.633°S, 167.083°E (MNHN, 1). 0 m, NW Malo Island, SANTO 2006 ZM15, 15.635°S, 167.098°E (MNHN, 3). 13 m, NE Malo Island, Palikulo Bay, SANTO 2006 DB77DB86, 15.641°S, 167.251°E (MNHN, 1). 10–17 m, W coast of Malo Island, SANTO 2006 DS49, 15.645°S, 167.086°E (MNHN, 1). 160–167 m, NE coat of Malo Island, SANTO 2006 AT65, 15.671°S, 167.265°E (MNHN, 1). 2–3 m, W Malokilikili Island, SANTO 2006 FS51, 15.711°S, 167.251°E (MNHN, 1). 250–372 m, W Malekula, BOA 1 DW2454, 16.216°S, 167.366°E (MNHN, 1). 360–419 m, MUSORSTOM 8 DW1065, 16.266°S, 167.350°E (MNHN, 28).

Asutralia, Norfolk Islands. 29.033°S, 167.950°E (AMS C.378318, 1; AMS C.378319, 1).

Fiji. 700–707 m, 19.050°S, 178.500°W (MNHN, 15). 20 m, Herald Pass, W of Ndravuni Island, NW end of Great Astrolabe Reef, 18.767°S, 178.467°E (LACM 85-135, 1). 441–443 m, S of Viti Levu, 18.320°S, 177.862°E (MNHN, 20). 497–504 m, S of Viti Levu, 18.312°S, 178.152°E (MNHN, 1). 234–361 m, 18.310°S, 178.085°E (MNHN, 2). 260–305 m, S of Viti Levu, 18.308°S, 178.097°E (MNHN, 25). 10–20 m, off Francis Island in Beqa Lagoon, just off Beqa Island (S of Viti Levu), 18.300°S, 178.067°E (DLG 276, 1). 275–430 m, S of Viti Levu, 18.297°S, 177.907°E (MNHN, 20). 295–302 m, S of Viti Levu, 18.212°S, 178.540°E (MNHN, 20). 144–150 m, S of Viti Levu, 18.207°S, 178.550°E (MNHN, 3). 149–168 m, S of Viti Levu, 18.207°S, 178.552°E (MNHN, 12). 420–450 m, 18.200°S, 178.617°W (MNHN, 6). 392–433 m, S of Viti Levu, 18.185°S, 178.390°E (MNHN, 6). 300–400 m, 18.167°S, 178.733°W (MNHN, 1). 183–188 m, SE Viti Levu, 17.828°S, 178.797°E (MNHN, 5). 959–963 m, SE of Viti Levu, 17.710°S, 178.917°E (MNHN, 8). 879–897 m, SE of Viti Levu, 17.515°S, 178.888°E (MNHN, 40). 820–863 m, 17.300°S, 179.550°W (MNHN, 20). 567–699 m, Bligh Water, 5, 17.158°N, 177.938°E (MNHN, 4). 656–660 m, Bligh Water, 17.268°S, 178.247°E (MNHN, 12). 660–663 m, Bligh Water, 17.248°S, 178.492°E (MNHN, 15). 640–687 m, Bligh Water, 16.937°S, 178.132°E (MNHN, 1). 0 m, Yasawa-i-rara, Tavewa Island, Yasawa Group, 16.933°S, 177.367°E (AMS C.461462, 1). 251–257 m, Bligh Water, 16.857°S, 178.232°E (MNHN, 12). 200–215 m, Bligh Water, 16.840°S, 178.208°E (MNHN, 15). 80–120 m, MUSORSTOM 10 CP1358, 17.808°S, 178.778°E (MNHN, 4). 500–614 m, Bligh Water, MUSORSTOM 10 CP1341, 16.875°S, 177.728°E (MNHN, 1). 820–863 m, BORDAU 1 DW1459, 17.300°S, 179.550°W (MNHN, 8). 102–106 m, Bligh Water, MUSORSTOM 10 DW1329, 17.321°S, 177.790°E (MNHN, 2). 402–444 m, SE of Viti Levu, MUSORSTOM 10 CP1360, 17.993°S, 178.803°E (MNHN, 2). 785–820 m, BORDAU 1 CP1490, 18.850°S, 178.533°W (MNHN, 2). 240–319 m, BORDAU 1 DW1494, 18.916°S, 178.483°W (MNHN, 9). 395–540 m, BORDAU 1 DW1486, 19.016°S, 178.433°W (MNHN, 9). 500–516 m, BORDAU 1 DW1488, 19.016°S, 178.416°W (MNHN, 3). 700–707 m, BORDAU 1 DW1485, 19.050°S, 178.500°W (MNHN, 5). Great Astrolabe Reef, 18.768°S, 178.468°E (JKC, 2).

Tonga. 433 m, NW Tongatapu, BORDAU 2 DW1569, 21.033°S, 175.316°W (MNHN, 4).

Marshall Islands. 20–30 m, lagoon, Amenwanot Island, Majuro Atoll, 7.133°N, 171.317°E (SBMNH 85115, 1). 23 m, Ocean Side, Kwajalein, 8.739°N, 167.746°E (DLG 998, 2). N end Book Island, 11.405°N, 136.482°E (USNM 594641, 17).

Kiribati. 0 m, windward Tibatiba Abemama Atoll, 0.407°N, 173.849°E (CPC, 7).

French Polynesia. Afaahiti [= Ataiti?], Tahiti, 17.760°S, 149.384°W (JTC, 3). 85 m, Raiatea Island, 16.833°S, 151.500°W (JWC 11309, 1). 85 m, Raiatea Island, 16.820°S, 151.420°W (DLG 959, 2). 40–83 m, Passe Arikitamiro, Makemo, 16.620°S, 143.565°W (JLC, 2). 507–607 m, Kaukura, Tuamotus, TARASOC DW3373, 15.683°S, 146.900°W (MNHN, 1). 45 m, Haurei Bay, Rapa, Austral Islands, 27.613°S, 144.305°W (MNHN, 1). 33 m, Haurei Bay, Rapa, Austral Islands, 27.612°S, 144.318°W (MNHN, 1). 52–57 m, SE of Tauna Island, Rapa, Austral Islands, 27.608°S, 144.295°W (MNHN, 10; MNHN, 4). 30 m, NW of Tauna Island, Rapa, Austral Islands, 27.605°S, 144.303°W

(MNHN, 3). 36 m, Around Rukuaga Point, Rapa, Austral Islands, 27.568°S, 144.368°W (MNHN, 3). 80–150 m, Tubuai, Austral Islands, BENTHAUS DW1958, 23.326°S, 149.505°W (MNHN, 2).

Wallis and Futuna. 415–420 m Wallis Island, 13.350°S, 176.133°W (MNHN, 9; MNHN, 15). 355–360 m, Wallis Island, 13.217°S, 176.250°W (MNHN, 1). 455–515 m, Wallis Island, 13.200°S, 176.267°W (MNHN, 75).

Specimens fossil.

Vanuatu. USGS locality 25718, Kere River, Santo, 15.566°S, 166.929°E (USNM 314324, 1) [Specimen figured by Ladd (1982)].

Remarks. The single specimen in MNHN is a syntype because Crosse did not designate a single specimen to be the name bearer. Fischer-Piette (1950: 152) used the term "holotype" inappropriately, and this use should not be construed as a lectotype designation (ICZN Art. 74.5). The specimen was expressedly designated as the lectotype by Geiger (2006a). There are currently no known paralectotypes.

The lectotype is conspecific with Bandel's *Suk. indonesicus*; the shared similarities are a protoconch with fine reticulate sculpture (visible in lectotype of *Suk. morleti*), a teleoconch I of approximately one whorl, and an overall sculpture of axials, spirals, and keels.

Sukashitrochus simplex was synonymized under *Suk. indonesicus* [= *Suk. morleti*] by Geiger & Jansen (2004b), applying first reviser's principle. Bandel (1998) cited as the sole difference the stronger basal keels in *Suk. indonesicus* as compared to *Suk. simplex*. The degree to which keels are developed varies considerably in *Sukashitrochus*, as seen in the most abundant material available for *Suk. atkinsoni* from Australia. Shared similarities of the two nominal taxa include reticulate sculpture on embryonic cap and fine axials on protoconch, protoconch varix connected to embryonic cap, apertural margin of protoconch sinusoidal, the slightly more than one whorl of teleoconch I, number of axials on teleoconch I, absence of spiral sculpture on teleoconch I, 1.33 teleoconch II whorls, density of sculpture on shoulder and base, position of first keel that is strongest, dipping of apertural roof just below first keel on base in mature specimens. Bandel (1998) did not compare either of the species with *Suk. morleti*.

Thiele (1912) noted the similarities with *Suk. atkinsoni*; however, the protoconch of *Suk. atkinsoni* shows strong flocculent sculpture, whereas *Suk. morleti* has fine axials.

Sukashitrochus pulcher (Petterd, 1884)

Figures 504–508

Chresonymy.

Schismope pulchra Petterd, 1884: 139, not illustrated.
Schismope pulchra: Paetel, 1888: 289.
Schismope pulchra: Pilsbry, 1890: 68.
Schismope pulchra: Hedley, 1901: 726, fig. 25.
Schismope pulchra: Tate & May, 1902: 407.
Schismope pulchra: Pritchard & Gatliff, 1902: 182.
Schismope pulchra: Sowerby & Fulton, 1903: 53.
Schismope pulchra: Pritchard & Gatliff, 1906: 65.
Schismope pulchra: Anonymous, 1906: 367.
Schismope pulchra: Hedley & May, 1908: 109.
Schismope pulchra: Horst & Schepman, 1908: 486.
Schismope pulchra: Verco, 1910: 117.
Schismope pulchra: Hedley, 1911: 92.
Schismope pulchra: Thiele, 1912a: 31, pl. 4, figs 2–3 [copy figures Hedley, 1901].
Schismope pulchra: Verco, 1912a: 191.
Schismope pulchra: Verco, 1912b: 201.
Schismope pulchra: May, 1921: 35, pl. 17, fig. 6.
Schismope pulchra: May, 1923: pl. 17, fig. 6.
Schismope pulchra: Cotton & Godfrey, 1933: 24.
Schismope pulchra: Cotton & Godfrey, 1938b: 2.
Schismope pulchra: May & Macpherson, 1958: 21, pl. 17, fig. 6.
Schismope pulchra: Cotton, 1959: 43–44, fig. 14.
Schismope pulchra: Macpherson & Gabriel, 1962: 34.
Scissurella (*Sukashitrochus*) *pulchra*: Lozouet, 1986: 109.
Scissurella pulchra: Richmond, 1992: 13, fig. 2.
Sinezona pulchra: Wilson, 1993: 46.
Sinezona pulchra: Jansen, 1999: 55, figs 61–63.
Sukashitrochus pulcher: Geiger, 2003: 79.
Sukashitrochus pulcher: Geiger & Jansen, 2004b: 58–62, figs 29–31 [shell, radula], 32c [map].
Sukashitrochus pulcher: Geiger, 2004: text-fig. p. 6.
Sukashitrochus pulcher: Geiger, 2006c: text-fig. 4.

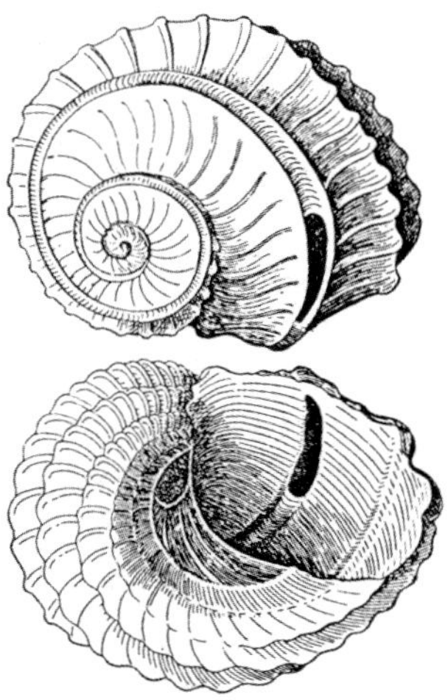

Figure 504. First illustration of *Sukashitrochus pulcher* by Hedley (1899).

Type material. Type(s) (whereabouts unknown), 2 × 1 mm.

Type locality. Northwest coast [Tasmania, Australia] (OD).

Etymology. Pulchra, Latin adjective for beautiful.

Description. Shell medium-sized (to 2.2 mm. Figs 505, 506), wider than tall. Protoconch almost one whorl, reticulate sculpture on outside with spiral apical thickening, smooth inside; varix connecting to embryonic cap; aperture sinusoidal. Teleoconch I of 0.8–1.2 whorls, with approximately 16–18 axial cords, no spirals, interaxials smooth, base without spiral keels. Teleoconch II approximately 1.5 whorls. Shoulder flat, with approximately 28 axial cords on body whorl more prominent than approximately six indistinct spiral threads. Profile of shoulder with 1–2 spiral folds; axials at intersection with fold usually with elevated projection. Base with 6–7

Scissurellidae *Sukashitrochus*

spiral keels decreasing in prominence from below selenizone to umbilical margin, intersected with approximately 28 axial cords on body whorl, axials running over spiral keels; axials on shoulder and base poorly coordinated. Umbilicus prominent, deep, bordered by last spiral keel on base, wall of umbilicus at steep angle to base. Selenizone inside periphery, keels at approximately 45° angle to coiling axis, lunules not coordinated with axials on shoulder or base. Foramen elongated lanceolate. Aperture subrotund; interior margin almost straight, roof overhanging.

Radula (Fig. 507). Rachidian tooth trapezoidal, tip with 5 cusps. Lateral teeth 1–3 similar, 3–4 cusps on outer cutting edge of tip. Lateral tooth 4 reduced, hook-shaped. Lateral tooth 5 enlarged by broadening, six cusps on inner edge of tip. Inner marginal teeth with triangular tip, 3–5 cusps on each side. Outer marginal teeth with spoon-shaped tip, many cusps on each side; shafts with food groove. Radular interlock of central field moderate.

Comparisons. The strong cancellate sculpture of *Suk. pulcher* with the elevated knobs at the intersection of axials and spiral keels on the base sets it apart from any other scissurellid species.

Figure 505. *Sukashitrochus pulcher*. 80 m, W of West Point, Tasmania, Australia, 41.020°S, 144.358°E (AMS C.379967). Scale bar shell = 1 mm. Scale bar protoconch = 100 μm.

Sukashitrochus atkinsoni from temperate Australia lacks the axial component and does not form elevated points on the spirals.

Sukashitrochus morleti from the Indo-Malayan Archipelago lacks the strong sculpture and has a protoconch with fine axial sculpture.

Figure 506. *Sukashitrochus pulcher*. **A.** 18 m, Knob Bluff, N coast Kangaroo Island, South Australia, Australia, 35.617°S, 137.233°E (AMS C.378297). Notice the smooth base of the juvenile shell. Radula shown in Figure 507. **B.** 82 m, W of Port Davey, Tasmania, Australia, 43.338°S, 145.803°E (AMS C.379964). **C.** St. Vincents Gulf, South Australia, Australia, 35°S, 138°E (AMS C.026706). Scale bars shell = 1 mm. Scale bars protoconch = 100 µm.

Figure 507. *Sukashitrochus pulcher.* Radula.18 m, Knob Bluff, N coast Kangaroo Island, South Australia, Australia, 35.617°S, 137.233°E (AMS C.378297) from specimen shown in Figure 506A. Scale bar = 10 µm.

Distribution. Ryukyu to Tasmania, Western Australia through Vanuatu, 0–667 m depth, 0–80 m depth live, from algae-covered rocks. (Fig. 508).

Specimen records.

Japan, Okinawa. 92 m, Onna Village, 26.493°N, 127.493°E (LACM 78-101, 1).

Australia, New South Wales. Bittangabee, N of Green Cape, 37.217°S, 150.017°E (AMS C.404918, 1: complete).

Australia, Queensland. 2–9 m, Great Barrier Reef, E of Lizard Island, Yonge Reef, N end, 14.583°S, 145.617°E (AMS C.404493, 1: complete). 11–18 m, off Cairns Reef, 16.905°S, 145.778°E (MV ex F 161668, 1). 35 m, Great Barrier Reef, E of Mackay, 20.867°S, 149.483°E (AMS C.378609, 1). Port Curtis, Facing Island, 23.817°S, 151.367°E (AMS C.378608, 3). 73 m, off S end Fraser Island, 25.800°S, 153.767°E (AMS C.378610, 1). 77 m, off Moreton Bay, 27.456°S, 153.650°E (AMS C.378584, 3).

Australia, South Australia. Spencer Gulf (RMNH, 1). 20–30 m, Petrel Bay, N of St. Francis Island, 32.483°S, 133.300°E (AMS C.378304, 1). 0 m, Point Brown, Eyre Peninsula, 32.533°S, 133.863°E (MV F17836, 1). 0 m, Corvisart Bay, Eyre Peninsula, 32.899°S, 134.077°E (MV F178364, 1). The Bluff, S of Point Richabys, Spencer Gulf, 33.100°S, 138.167°E (WAM, 15). 0 m, Port Lincoln, 34.720°S, 135.857°E (MV F178691, 3). Bluff Beach, Yorke Peninsula, 34.733°S, 137.483°E (AMS C.380167, 1, –: complete; AMS C.380168, 1, –: complete). 0 m, Bluff Beach, Yorke Peninsula, 34.733°S, 137.483°E (AMS, 2: complete). Hardwicke Bay, Spencers Gulf, 34.750°S, 137.300°E (AMS C.13408, 3). 0 m, Corney Point, Yorke Peninsula, 34.919°S, 137.084°E (MV F178365, 5). Glenelg Beach, Adelaide, 34.983°S, 138.517°E (AMS C.380165, 2). St. Vincents Gulf, 35.000°S, 138.000°E (AMS C.26706, 3; MV F 157449, 2). St. Vincent, 35.216°S, 138.250°E (WAM, 2). 0 m, Aldinga Beach, 35.280°S, 138.443°E (MV F178370, 4). Point Willunga, 35.283°S, 138.550°E (WAM, 6). Little Gorge, Normanville, 35.466°S, 138.317°E (WAM, 2). 18 m, Knob Bluff, N coast Kangaroo Island, 35.617°S, 137.233°E (AMS C.378297, 1). 174–183 m, 40 mi S of Cape Wiles, 35.650°S, 136.667°E (AMS C.380169, 3; AMS C.31844, 3). 0 m, Robe, 37.162°S, 139.755°E (MV F178367, 1). 0 m, Nora Creina Bay, Robe, 37.343°S, 139.870°E (MV F178363, 14). 13 m, off Middle Point, near Cape Northumberland, 38.050°S, 140.617°E (AMS C.380166, 2). 0 m, Racecourse Reef, Port Macdonnell, 38.056°S, 140.752°E (MV F178371, 2). 0 m, Cape Northumberland, Port Maconnell, 38.061°S, 140.663°E (MV F178366, 1). 667 m, off Cape Martin, 38.125°S, 140.000°E (AMS C.379961, 2).

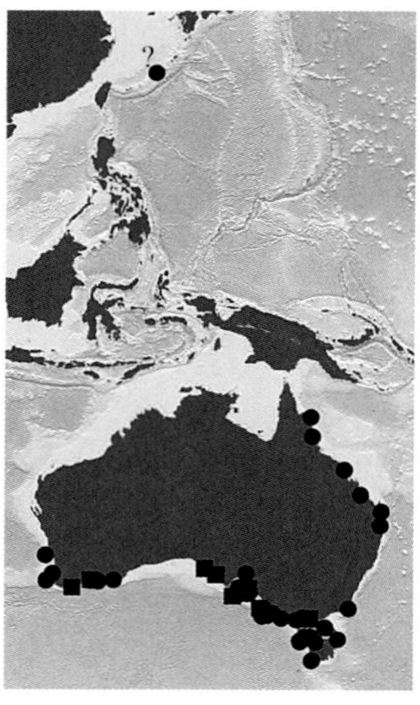

Figure 508. Distribution of *Sukashitrochus pulcher.* ● specimen record. ■ literature record. ? questionable record.

Scissurellidae *Sukashitrochus*

Australia, Tasmania. 144 m, off Port Davey, 43.375°S, 145.742°E (AMS C.379966, 1). 82 m, W of Port Davey, 43.338°S, 145.803°E (AMS C.379964, 1). Ulverstone, 41.158°S, 146.183°E (AMS C.4939, 2). 88 m, S of West Point, 41.153°S, 144.403°E (AMS C.379963, 1). 80 m, W of West Point, 41.020°S, 144.358°E (AMS C.379967, 2). 399 m, off Cape Naturaliste, 40.843°S, 148.775°E (AMS C.379962, 1). 10 m, Little Squally Cove, Deal Island, Bass Strait, 39.500°S, 147.333°E (AMS C.379969, 1, 1: complete). 6 m, East Cove, Deal Island, Bass Strait, 39.500°S, 147.333°E (AMS C.404982, 1: complete). 6 m, Erith Deal Island, Bass Strait, 39.450°S, 147.283°E (AMS C.379968, 2). 27 m, 9 km SSW of Cape Adansan, Three Hummock Island, Central Bass Strait, 40.515°S, 144.933°E (MV F 157448, 1). Cable, Bass Strait, 40.500°S, 145.600°E (MV F 15688, 5). Stanley, 40.765°S, 145.294°E (MV ex F161661, 1).

Australia, Victoria. Ocean Beach, Flinders, 38.483°S, 145.017°E (AMS C.379957, 2). Flinders, 38.483°S, 145.033°E (MV F 888, 25). Port Fairy, 38.383°S, 142.233°E (AMS C.379958, 20). 11 m, Western Port, 38.367°S, 145.533°E (AMS C.16006, 1). 7–10 m, Southern Port Phillip Bay, 38.283°S, 144.693°E (MV F 157445, 5). 0 m, Reef Point, Port Fairy, 38.364°S, 142.286°E (MV F178362, 3). 0 m, Griffiths Island, Port Fairy, 38.392°S, 142.245°E (MV F178361, 2). 0 m, Warrnambool, 38.394°S, 142.484°E (MV F178369, 1). 0 m, Maitland Beach, Bear Gully, 38.893°S, 145.982°E (MV F178368, 1). 17 m, 5 km S off Point Lonsdale towards Ocean Grove, 38.276°S, 144.559°E (MV F113284, 1: complete). 57 m, Bass Strait, 39.246°S, 147.525°E (MV F83094, 1).

Australia, Western Australia. 154 m, WNW of Lancelin, 30.967°S, 114.883°E (AMS C.378296, 1). Myalup, 33.100°S, 115.700°E (AMS C.79971, 1). Hopetoun, 33.866°S, 120.133°E (NMNZ M.245174, 2; WAM, 50; WAM, 20; WAM, 6). Ellensbrook, 33.883°S, 114.983°E (WAM, 5). S of Cowaramup Bay, near mouth of Margaret River, 33.883°S, 115.017°E (AMS C.404156, 4). Alexander Bay [around Esperance], 33.883°S, 122.769°E (WAM, 1). Kilcarnup, N side of Margaret River, 33.950°S, 114.983°E (AMS C.79973, 1). Hopetoun, Two Mile Beach, 33.950°S, 120.117°E (AMS C.379974, 8). 1–2 m, Hopetoun Jetty, 33.950°S, 120.117°E (AMS C.379977, 1). Off Esperance, 34.030°S, 121.030°E (AMS C.461554, 1).

Literature records.

Australia. St. Francis Island, South Australia, 32.533°S, 133.333°E. Venus Bay, South Australia, 33.233°S, 134.667°E. Largs, South Australia, 34.817°S, 138.483°E. Robe, South Australia, 37.150°S, 139.750°E. Beachport, South Australia, 37.500°S, 140.017°E Cotton & Godfrey. 1933).—Western Port, Vicoria, Australia, 38.367°S, 145.533°E. Off Rhyll, Victoria, 38.466°S, 145.300°E (Pritchard & Gatliff. 1902).—St. Francis Island, South Australia, 32.500°S, 133.300°E. Venus Bay, South Australia, 33.233°S, 134.683°E. St. Vincent Bay, South Australia, 35.216°S, 138.250°E. Cape Borda, South Australia, 35.753°S, 136.594°E. Robe, South Australia, Australia, 37.216°S, 139.767°E. Beachport, South Australia, 37.500°S, 140.017°E (Verco, 1910).—Hopetoun, Western Australia, 33.866°S, 120.133°E. King George Sound Beach, Western Australia, 34.738°S, 117.973°E (Verco, 1912b).

Remarks. The species has a predominantly temperate Australian distribution, but does seem to reach to tropical Australia (Queensland).

Sukashitrochus kaiparaensis (Laws, 1941) †

Figures 509–512

Chresonymy.

Schismope kaiparaensis Laws, 1941: 137, fig. 9.
Scissurella (Sukashitrochus) kaiparaensis: Lozouet, 1986: 110.
Sinezona kaiparaensis: Beu & Maxwell, 1990: 376.

Figure 509. Original illustration of *Schismope kaiparaensis*.

Type material. Holotype (GNS TM 1493, ex coll. Laws. Fig. 510), 1.3 × 1.0 mm. Paratypes from type locality (GNS TM 8637–8641, 5).

Type locality. Pakaurangi Point, Kaipara, New Zealand (OD: inferred from title) [36.239°S, 174.293°E]. Otaian [= early Miocene (Lozouet, 1986; specimen label)].

Figure 510. *Schismope kaiparaensis*. Holotype. Pakaurangi Point, Kaipara, New Zealand [36.239°S, 174.293°E] (GNS TM 1493). Scale bar shell = 1 mm. Scale bar protoconch = 100 μm.

Etymology. Named for the type locality at Kaipara.

Description. Shell medium-sized (to 1.3 mm: holotype. Figs. 510, 511), trochiform, depressed. Protoconch eroded in holotype. Teleoconch I approximately one whorl, weak axial cords, spirals uncertain. Teleoconch II >1.125 whorls, suture deeply impressed.

Figure 511. *Sukashitrochus kaiparaensis*. **A–B.** 0 m, Kaipara Harbour, small harbour c 1.6 km NW of Pakaurangi Point (Q8/26513, N28/F9828, GS9730). Otaian, early Miocene, New Zealand (NMNZ M.174839, 35). **C.** Paratype. Pakaurangi Point, Kaipara Harbour, Northland, New Zealand, 36.239°S, 174.293°E (GNS TM 8637, 1). Scale bars shell = 1 mm. Scale bars protoconch = 100 μm.

Scissurellidae
Sukashitrochus

Shoulder with sinusoidal profile, approximately 30 indistinct axial cords, most visible near suture; approximately six indistinct spiral lines; interstices with very fine growth marks. Base with strong constriction below selenizone, one distinct spiral keel at periphery, one weak one below; weak axial cords, approximately five spiral lines between selenizone and major spiral keel, four between spiral keels, four stronger spiral lines towards umbilicus, the latter forming irregular thickenings at intersection with axials. Umbilicus bordered by granulated carina, walls straight, at angle to base, with very fine growth lines. Selenizone above periphery; keels low, strong; lunules distinct; condition of slit unknown (broken in holotype), presumably closed to form foramen. Aperture rounded, D-shaped, obscured by infill.

Figure 512. Distribution of *Sukashitrochus kaiparaensis*. ☆ type locality. ● specimen record.

Comparisons. *Sukashitrochus ngatutura* from the Pliocene/Pleistocene of New Zealand has finer and denser sculpture on the base and is otherwise is identical down to all aspects of protoconch sculpture to *Suk. kaiparaensis*. Whether this distinction is sufficient to maintain them as separate species is an open question.

Sukashitrochus lyallensis from the Recent of New Zealand has a protoconch in which the reticulate sculpture covers its entire width.

Distribution. Miocene of New Zealand (Fig. 512).

Fossil. Otaian - Altonian = early Miocene of New Zealand (Beu & Maxwell, 1990).

Sukashitrochus ngatutura (Laws, 1936) †

Figures 513–515

Chresonymy.

Schismope ngatutura Laws, 1936: 100, pl. 15, figs 43–44.
Schismope ngatutura: Powell, 1941: 254.
Scissurella (Sukashitrochus) ngatutura: Lozouet, 1986: 110.
Sinezona ngatutura: Beu & Maxwell, 1990: 376.
Schismope ngatueura [sic]: Yu & Feng, 1996: pl. 1, figs 8–9.

Figure 513. Original illustration of *Schismope ngatutura*.

Misidentification.

? *Schismope* n. sp.: Marshall & Uttley, 1914: 279 [most likely same specimens as reference in Marshall (1915)].

? *Schismope atkinsoni*: Marshall, 1915: 378 [fossil, is *Suk. lyallensis* or *Suk. ngatutura*].

not *Schismope ngatutura*: Feng & Todd, 2007: 35, pl. 1 figs 5–6 [is *Sinezona* sp.].

Type material. Holotype (AIM. Powell, 1941: 254. Fig. 513), 1.2 × 0.9 mm (W × H).

Type locality. Kaawa Creek, New Zealand. Waitotaran Stage, Wanganuian (OD: inferred from title), early Pliocene (Lozouet, 1986). New Zealand Grid Reference R13 102 641. Approximate coordinates 37.500°S, 174.833°E.

Etymology. Most likely referring to nearby Ngatutura Point, 37.520°S, 174.733°E.

Description. Shell medium-sized (to 1.2 mm. Figs 514, 515), trochiform depressed. Protoconch with coarse reticulate sculpture on centrifugal 80%, apertural varix connected to embryonic cap, apertural margin sinusoidal. Teleoconch I approximately one whorl. Shoulder rounded, with broad, indistinct axial growth marks; some with indistinct low axial cords; no spiral sculpture. Teleoconch II of 0.66 whorl, suture variably impressed. Shoulder convex, with increasingly sharp axials, approximately

Figure 514. *Sukashitrochus ngatutura*. GS12163, Moutapu Point, NE Pitt Island, Chatham Islands, Whenuataru Tuff, Waipipian–Mangapanian (middle to late Pliocene), New Zealand (GNS, 30). Scale bar shell = 1 mm. Scale bar protoconch = 100 µm.

10 very fine spiral threads more closely spaced towards selenizone. Base with strong constriction below selenizone, then expanding to form strong spiral keel, additional weaker keels below; spiral sculpture prominent, axials sculpture indistinct; base sculpture overall cancellate. Selenizone in upper portion of whorl, keels low, strong;

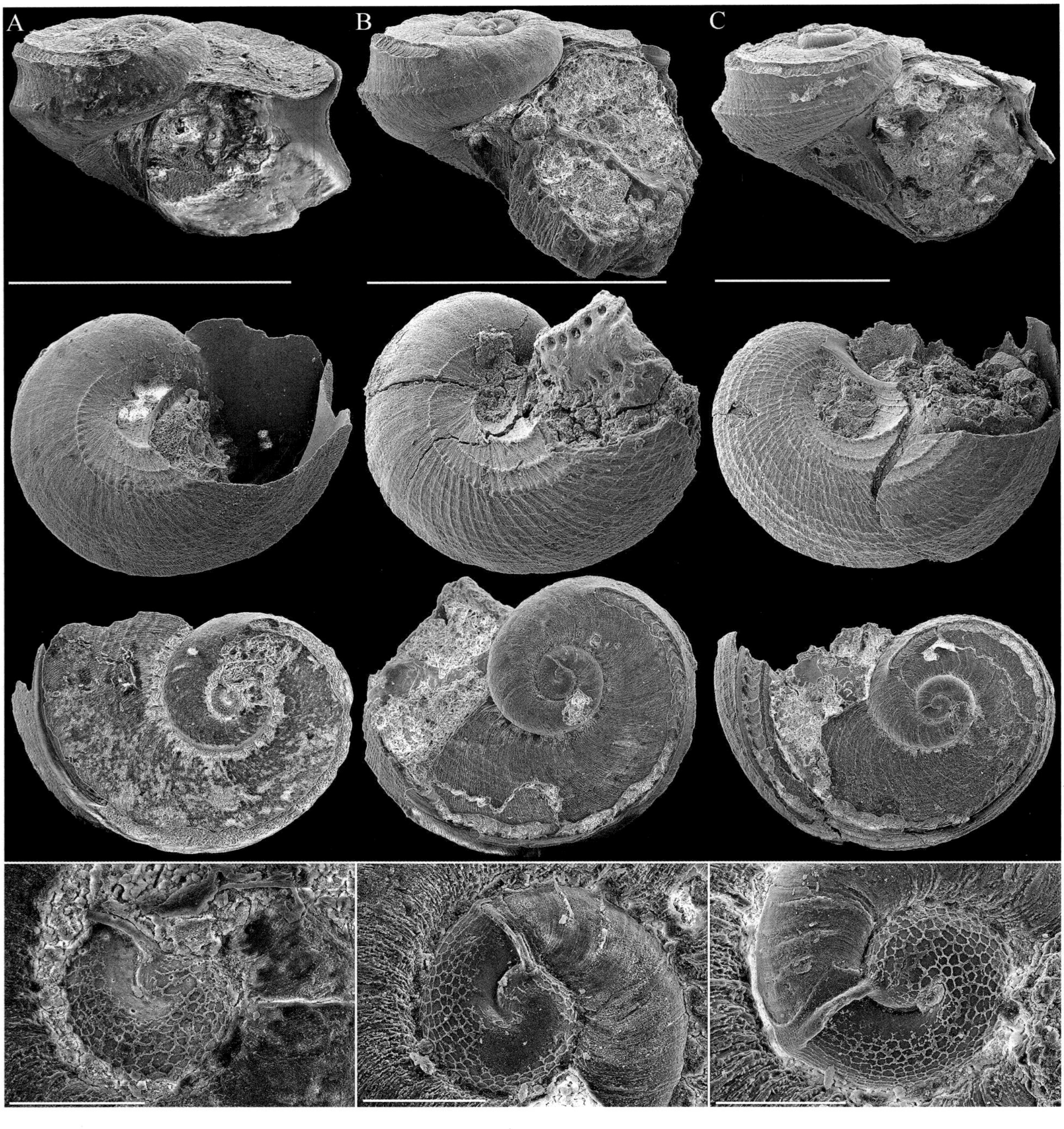

Figure 515. *Sukashitrochus ngatutura*. **A.** GS12163, Moutapu Point, NE Pitt Island, Chatham Islands, Whenuataru Tuff, Waipipian–Mangapanian (middle to late Pliocene), New Zealand (GNS, 30). **B–C.** GS12161, cliffs near "The Point" homestead, W of Flowerpot Bay, N coast of Pitt Island, Chatham Islands, Waipipian-Mangapanian (middle to late Pliocene), New Zealand, 43.710°S, 176.560°W (GNS, 5). Scale bars shell = 1 mm. Scale bars protoconch = 100 μm.

lunules strong; slit open (likely juvenile). Umbilicus open, bordered by carina, walls straight, smooth. Aperture rounded.

Comparisons. *Sukashitrochus kaiparaensis* from the Miocene of New Zealand has coarser and more widely spaced sculpture on the base, otherwise is identical down to all aspects of protoconch sculpture to *Suk. ngatutura*. Whether this distinction is sufficient to maintain them as separate species is an open question.

Sukashitrochus lyallensis from the Recent of New Zealand has a protoconch in which the reticulate sculpture covers its entire width.

Distribution. Pliocene to Pleistocene of New Zealand (Fig. 516).

Specimen records.

New Zealand. GS12161, cliffs near "The Point" homestead, W of Flowerpot Bay, N coast of Pitt Island, Chatham Islands, Waipipian-Mangapanian (middle to late Pliocene), 43.710°S, 176.560°W (GNS, 5). GS12163, Moutapu Point, NE Pitt Island, Chatham Islands, Whenuataru Tuff, Waipipian–Mangapanian (middle to late Pliocene), 36.239°S, 174.293°W (GNS, 30). GS14698, Kaikoura Peninsula, high terrace (Haweran,Oxygen isotype stage 5c?, late middle Pleistocene) hand-augured hole near high point of Kaikoura Peninsula, Marlborough, NE South Island, 42.422°S, 173.693°E (GNS, 2).

Figure 516. Distribution of *Sukashitrochus ngatutura*. ☆ type locality. ● specimen record.

Fossil. Opoitian = early Pliocene of New Zealand (Beu & Maxwell, 1990).

Remarks. The slit is still open in the type specimen illustrated, and is quite likely a juvenile specimen. It is confirmed by the non-descending last ¼ whorl.

Satondella Bandel, 1998

Chresonymy.

Satondella Bandel, 1998: 64–65.
Satondella: Geiger, 2003: 29–30.
Satondella: Luque *et al.*, 2011: 12.

Type species. *Satondella minuta* Bandel, 1998 (OD).

Other species. *Satondella azonata* Geiger & Marshall, 2012.
Satondella bicristata Geiger & Marshall, 2012.
Satondella brasiliensis (Mattar, 1987).
Satondella cachoi Luque, Geiger & Rolán, 2011.
Satondella danieli Segers, Swinnen & Abreu, 2009.
Satondella dantarti Luque, Geiger & Rolán, 2011.
Satondella goudi new species.
Satondella kondoi (Ladd, 1970) †.
Satondella tabulata (Watson, 1886).
Satondella senni Geiger, 2003.

Etymology. Named after the occurrence of the type species at Satonda Island, Indonesia (OD). Gender feminine.

Description. Shell small, early whorls more or less planispiral, last 0.25 whorl dipping markedly. Protoconch smooth, with spiral sculpture, or with fine hexagonal pattern; no apertural varix. Axials predominant, spirals much weaker. Umbilicate. Selenizone with strong keels, terminally elevated, forming chimney at anterior margin of foramen. Outline of shell with distinct angle anterior to anterior margin of foramen. Lateral teeth 1–3 of radula similar, with one cusp along inner edge of tip.

Distribution. Subtropical and tropical waters of all oceans (Fig. 517). For specific localities, see individual species.

Geologic Range. Eocene through Recent.

Comparisons. The more or less flat-topped shell with the bend anterior to the anterior margin of the foramen, in conjunction with the apex lacking a terminal varix, characterizes *Satondella*.

Scissurella maintains an open slit, is often markedly trochospiral, and the protoconch shows an apertural varix in many species, but the keels of the selenizone are less elevated. On the radula, lateral teeth 1–3 have cusps only on the outer edge of the tip. Juveniles may be difficult to distinguish.

Mature *Sinezona* shares the foramen with *Satondella*, but shows an apertural varix of the protoconch on many species, and the keels of the foramen are less elevated. On the radula, lateral teeth 1–3 have cusps only on the outer edge of the tip. Juveniles may be difficult to distinguish.

Sukashitrochus has prominent keels on the base of the shell and has a more elevated spire. On the radula, lateral teeth 1–3 have cusps only on the outer edge of the tip.

Coronadoa lacks a selenizone, slit or foramen. On the radula, lateral teeth 1–3 have cusps only on the outer edge of the tip.

Incisura usually has a smooth shell and has either a higher spire (in the naticiform *I. rosea*), or is neritiform (as in *I. lytteltonensis*). On the radula, lateral teeth 1–3 have cusps only on the outer edge of the tip.

Differentiation of species. Species are differentiated by overall shell shape and spire height, details of sculpture (absolute and relative strength of axials and spirals), width of umbilicus, as well as protoconch sculpture.

Occasionally, a modification in the adumbilical portion of the peristome can be of additional utility. Anatomical characters have not been found to help in species identification or in the establishment of species boundaries. For protoconch sculpture, only the very best light microscope may be sufficient, but SEM is usually required.

Intraspecific variation extends to ontogenetic change of shell/spire height, condition of the slit (notice, juvenile *Satondella* have an open slit; hence, they can be confused with *Scissurella* as well as juvenile *Sinezona*), and strength and density of sculptural elements (variable by approximately 25–50%).

Remarks. The bend in the shell is not unique for *Satondella* within Vetigastropoda. Szabó *et al.* (1993) discussed a similar feature in *Adeorbisina/Mariottia* (Trochoidea: Ataphridae) from the Mesozoic of Europe.

The Japanese name for the genus is *Entotsu-ebisu* (Hasegawa, 2005).

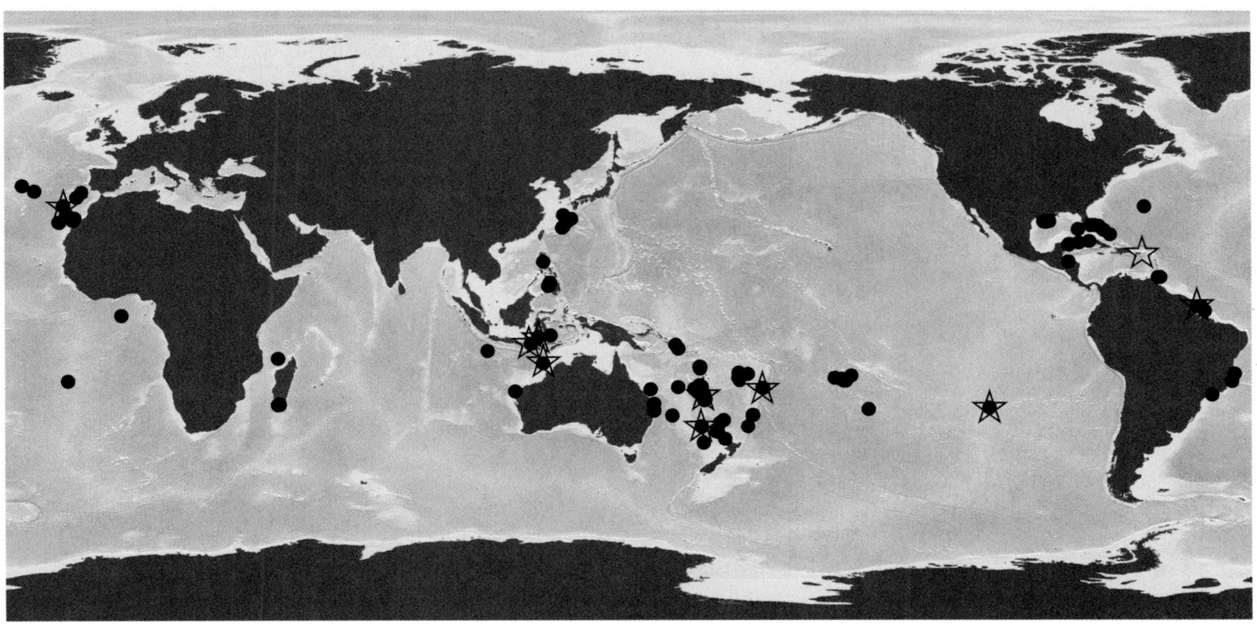

Figure 517. Global distribution of *Satondella* species. ☆ type locality. ● specimen record. ■ literature record.

Satondella azonata Geiger & Marshall, 2012

Figures 518–520

Chresonymy.

Scissurella ? sp.: Spencer *et al.*, 2009: 202.

Satondella azonata Geiger & Marshall, 2012: 13–15, figs 10–11.

Type material. Holotype (NMNZ M.301398. Fig. 518). Paratypes from type locality (NMNZ M.301398, 4. Fig. 519A–B; NMNZ M.174825, 5. Fig. 519C).

Type locality. 133 m, Wanganella Bank summit, West Norfolk Ridge, W of Cape Maria van Diemen, New Zealand, 32.543°S, 167.487°E (OD).

Etymology. A-, Latin prefix for without, -zonata referring to the selenizone. It describes the lack of a selenizone in this species. Adjective (OD).

Description. Shell small (to 0.63 mm. Figs 518, 519), trochiform, depressed, apex sunken. Protoconch of one whorl, fine irregular axial cords on outer portion, apical cap and inner portion of whorl smooth; no apertural varix, apertural margin sinusoidal. Teleoconch I of 1.125 whorls; suture impressed; 12–16 axials, low cords at onset

Figure 518. *Satondella azonata*. Holotype. 133 m, Wanganella Bank summit, West Norfolk Ridge, W of Cape Maria van Diemen, New Zealand, 32.543°S, 167.487°E (NMNZ M.301398). Scale bar shell = 0.5 mm. Scale bar protoconch = 100 µm.

Scissurellidae *Satondella*

of teleoconch I gradually rising to strong lamellae, onset of lamellae shifting from suture to position of foramen on shoulder, from suture to mid-base on underside; interstices with fine irregular growth lines. Teleoconch II of 0.125 whorl, suture deeply impressed. Shoulder flat, sculpture consisting of only fine irregular growth lines. Base with 2–3 strong axial lamellae as on teleoconch I. Umbilicus wide, distinct funiculus, underside of protoconch visible. No selenizone; foramen oblong, raised to chimney. Aperture subquadrate, peristome thickened.

Animal unknown.

Comparisons. *Satondella minuta* from the Indo-Malayan Archipelago to the Western Pacific has a smooth protoconch, a selenizone, and distinct spirals over the entire teleoconch.

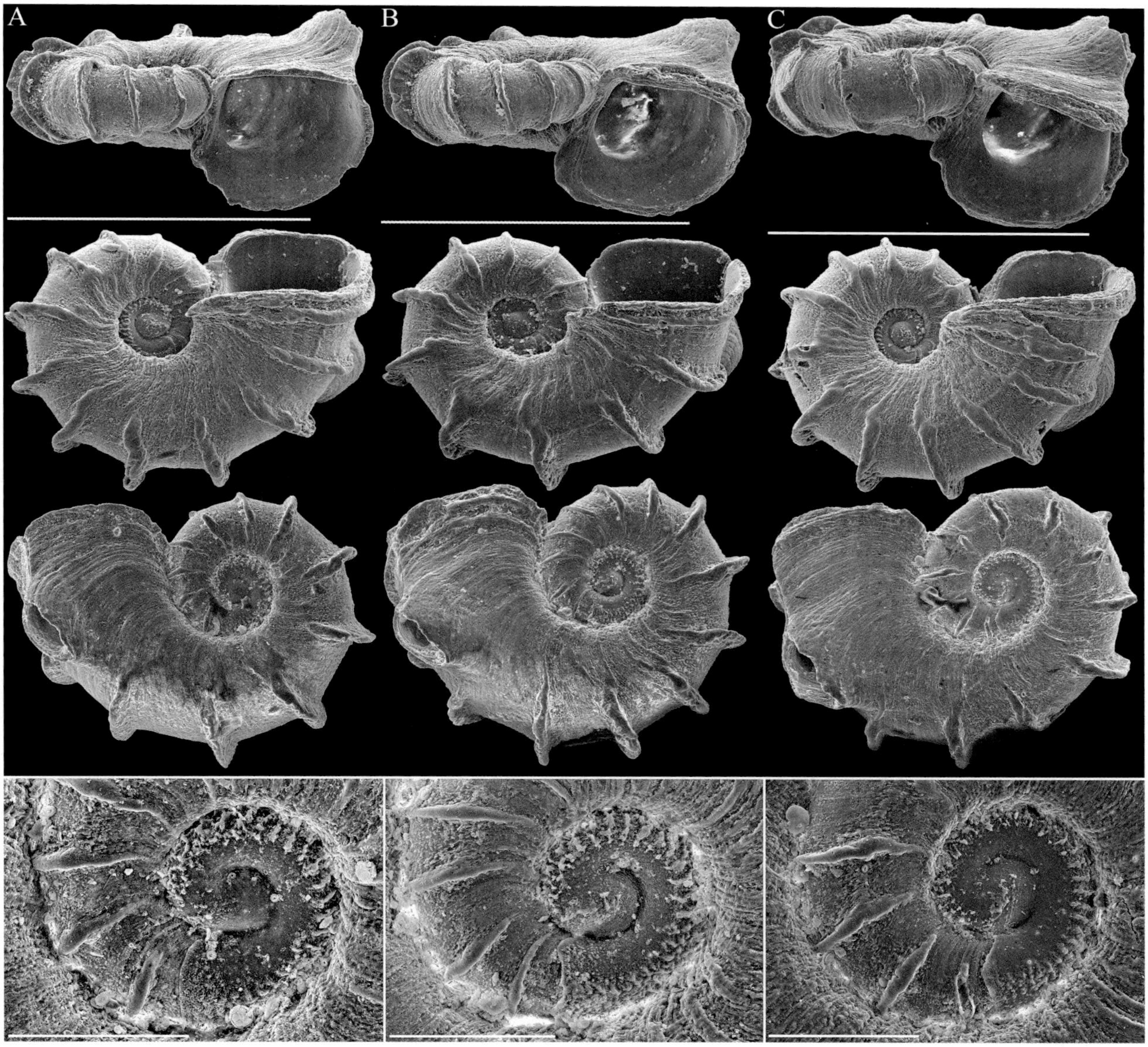

Figure 519. *Satondella azonata.* Paratypes. 133 m, Wanganella Bank summit, West Norfolk Ridge, W of Cape Maria van Diemen, New Zealand, 32.543°S, 167.487°E. **A–B.** (NMNZ M.301398, 4). **C.** (NMNZ M.174825, 5). Scale bars shell = 0.5 mm. Scale bars protoconch = 100 μm.

Satondella bicristata from New Zealand has twice as many but lower axial cords/lamellae and distinct spirals, particularly on the base.

Satondella dantarti has a selenizone, distinct spirals on the teleoconch, and axial lamellae on the shoulder.

Distribution. Northern New Zealand (Fig. 520).

Specimen records.

New Zealand. 246–291 m, Middlesex Bank, NW of Three Kings Islands, 34.033°S, 171.733°E (NMNZ M.148570, 1). 622 m, N of Three Kings Islands, 34.017°S, 172.117°E (NMNZ M.034601, 2). 805 m, NW of Three Kings Islands, Three Kings Trough, 34°N, 171.917°E (NMNZ M.160302, 1). 33 m, reef between Great Island and Farmer Rocks, Three Kings Islands, 34.150°S, 172.167°E (NMNZ M.160303, 1). 133 m, Wanganella Bank, Norfolk Ridge, summit, 32.653°S, 167.528°E (NMNZ M.174826, 1). 113 m, Wanganella Bank, Norfolk Ridge, summit, 32.537°S, 167.512°E (NMNZ M.257269, 1). 219–274 m, SE of D'Arcy Point, Raoul Island, Kermadec Islands, 29.313°S, 177.903°W (NMNZ M.026873, 3).

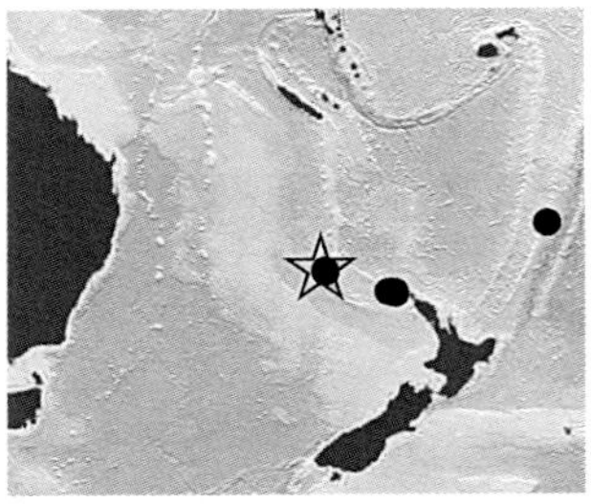

Figure 520. Distribution of *Satondella azonata*. ☆ type locality. ● specimen record.

Remarks. The thickening of the peristome is reminiscent of the condition seen in Liotiidae and is a clear sign of fully grown specimens. Accordingly, the absence of a selenizone is not an indication of juvenile specimens, but is a distinct, species-specific character.

Satondella bicristata Geiger & Marshall, 2012

Figure 521–522

Chresonymy.

Satondella bicristata: Geiger & Marshall, 2012: 15–16, fig. 12.

Type material. Holotype (NMNZ M.137457. Fig. 522C). Paratypes from type locality (NMNZ M.301393, 2. Fig. 522A–B).

Type locality. 538 m, seamount 130 km S of Esperance Rock, Kermadec Ridge, New Zealand, 32.472°S, 179.265°W (OD).

Etymology. Bi-, Latin for two; -cristata, Latin for crest-bearing, referring to the two elevated ridges formed on shoulder and base by the axial lamellae (OD).

Description. Shell small (to 0.67 mm. Fig. 522), depressed, trochiform, almost planispiral. Protoconch sunken, one whorl, fine axials from highest point centrifugally to suture, no apertural varix, apertural margin sinusoidal. Teleoconch I of 1.25–1.5 whorls; approximately 32–36 axials, near suture and peripherally as axial cords, in position of selenizone and on mid-base as raised lamellae; spiral cordlet in position of selenizone and on mid-base, crossing with tallest portion of axial lamellae; occasionally some additional spiral lines. Teleoconch II of 0.125 whorl,

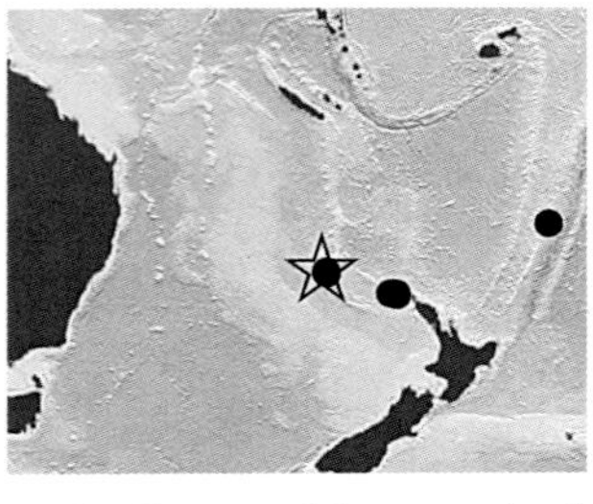

Figure 521. Distribution of *Satondella bicristata*. ☆ type locality. ● specimen record.

suture deeply impressed. Shoulder convex, same sculpture as on teleoconch I. Base with distinct constriction below foramen, sculpture as on teleoconch I. Umbilicus open, underside of protoconch visible, towards base bordered by strong bulge, no distinct wall. Aperture rounded, D-shaped. No selenizon;, foramen above periphery, anteriorly closed by raphe; keels strongly elevated, moderately strong.

Animal unknown.

Distribution. Kermadec Ridge, New Zealand. Known only from type locality (Fig. 521).

Comparisons. *Satondella minuta* from the Indo-Malayan Archipelago and Western Pacific has less depressed shell, fewer axial lamellae, and a distinct selenizone.

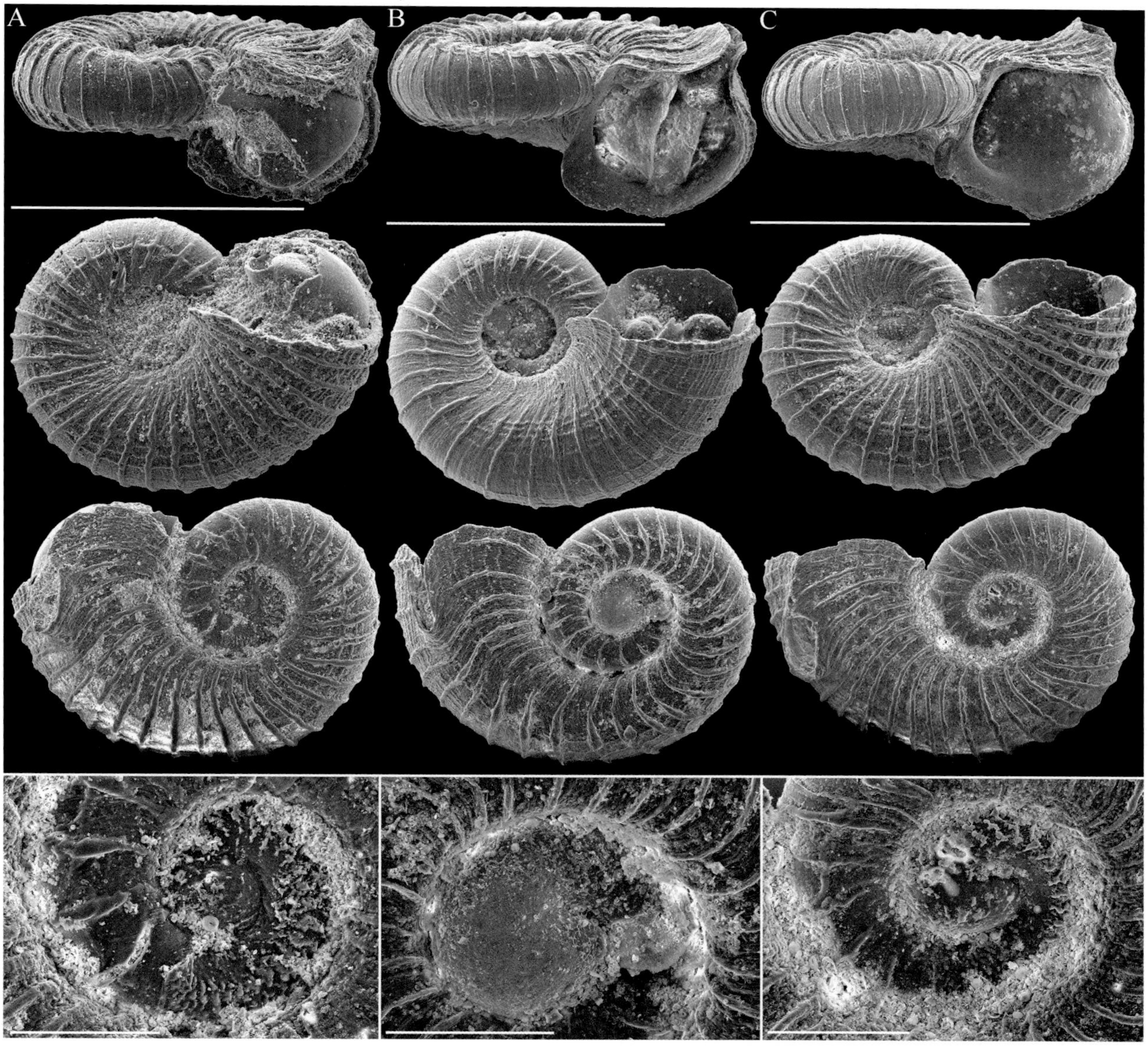

Figure 522. *Satondella bicristata*. 538 m, seamount 130 km S of Esperance Rock, Kermadec Ridge, New Zealand, 32.472°S, 179.265°W. **A–B.** Paratypes (NMNZ M.301393). **C.** Holotype (NMNZ M.137457). Scale bars shell = 0.5 mm. Scale bars protoconch = 100 µm.

Satondella azonata from New Zealand has about half as many raised axial lamellae, and lacks spirals on the base.

Scissurella condita from the Miocene of New Zealand has a similar overall shell shape, but has a distinct selenizone and an open slit.

All other *Satondella* species are markedly more elevated.

Remarks. The species is placed in *Satondella* out due to the elevation of the keels of the foramen.

Satondella brasiliensis (Mattar, 1987)

Figures 523–528

Chresonymy.

Sinezona brasiliensis Mattar, 1987: 76–78, pl. 1, figs 1–6 (Fig. 523).
Sinezona brasiliensis: Geiger, 2003: 78.
Satondella brasiliensis: Geiger & McLean, 2010: 2.
Satondella brasiliensis: Luque *et al.*, 2011: 5–8, figs 3A–L, 4A–H.

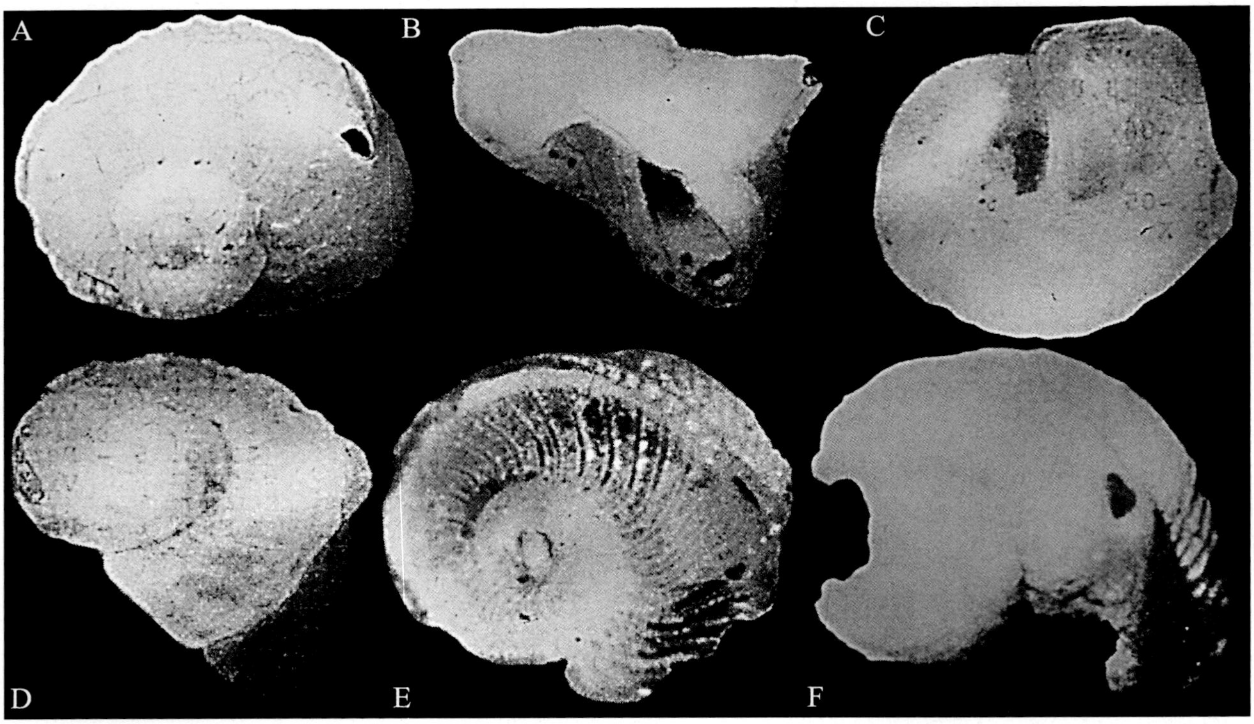

Figure 523. *Satondella brasiliensis*. Original illustrations. **A–D.** Holotype. 75 m, Site 3567, Amapá, Brazil, 3.130°N, 49.429°W. Quartz sand with argila (MPIUFRGS MP-Mol.-04). **E.** Paratype (MPIUFRGS MP-Mol.-06). **F.** Paratype (MPIUFRGS MP-Mol.-03).

Scissurellidae *Satondella*

Misidentifications.

Sinezona tabulata: Redfern, 2001: 2, pl. 1, fig. 7 [is *Sat. brasiliensis*].

Sinezona tabulate [*sic*]: Hasegawa, 2005: 142 [referred to Redfern, 2001, pl. 1, fig. 7; is *Sat. brasiliensis*].

Satondella tabulata: Geiger, 2008b: fig. 1 [is *Sat. brasiliensis*].

Satondella tabulata: Geiger *et al.*, 2008: fig. 12.2O [protoconch, is *Sat. brasiliensis*].

Satondella tabulata: Segers *et al.*, 2009a: pl. 1, figs 2a–c [is *Sat. brasiliensis*].

Scissurella cf. *tabulata*: Barrera, 2001: 12, 38 [is most likely *Sat. brasiliensis*].

Satondella tabulata: Tunnell *et al.*, 2010: figs 3.34-10–10a, 115, text-fig. [is *Sat. brasiliensis*].

not *Sinezona brasiliensis*: Rios, 1994: 21, pl. 5, fig. 38 [copy figure Leal, 1991: is *Sinezona confusa*].

not *Sinezona brasiliense* [*sic*]: Rios, 2009: 25, text-fig. [copy Leal, 1991: is *Sinezona confusa*].

Type material. Holotype (MPIUFRGS MP-Mol.-04, site 3567. Fig. 523A–D), 0.65 × 0.85 mm (H × W). Paratypes (MPIUFRGS MP-Mol.-05 [site 3318], MPIUFRGS MP-Mol.- 06 [site 3849]. Fig. 523E), 1.03 × 1.23 mm, 0.73 × 0.91 mm. Paratype(s) (MOGURG 25.450).

Figure 524. *Satondella brasiliensis*. 33–48 m, Tobago, east end, Tobago, 11.250°N, 60.667°W (SBMNH). Scale bar shell = 0.5 mm. Scale bar protoconch = 100 μm.

Type locality. 75 m, Site 3567, Amapá, Brazil, 3.130°N, 49.429°W. Quartz sand with argila (OD).

Etymology. Named for its type locality in Brazil (OD).

Description. Shell medium-sized (to 1.23 mm. Figs 523–525), trochiform, subquadrate. Color off white. Protoconch 0.75 whorl, with fine microhexagonal sculpture (Fig. 526) usually

Figure 525. *Satondella brasiliensis*. **A.** 56 m, Faro de los Colorados, Cienfuegos, Cuba, 22.010°N, 80.500°W (HLC, 2). **B–C.** 72 m, Gulf of Mexico, 27.933°N, 92.033°W (USNM 780310, 25). Scale bars shell = 1 mm. Scale bars protoconch = 100 μm.

Scissurellidae *Satondella*

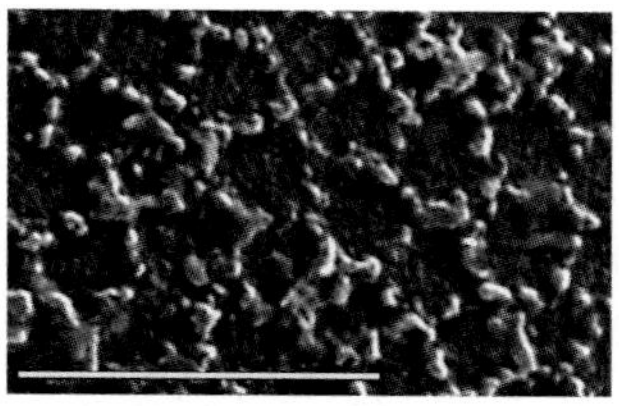

Figure 526. *Satondella brasiliensis*. Detail of microhexagonal protoconch sculpture. 72 m, Gulf of Mexico, 27.933°N, 92.033°W (USNM 780310, 25). Scale = 5 µm.

better preserved towards the suture, no apertural varix, apertural margin sinusoidal. Teleoconch I of 1.125 whorls, 17–24 axial cords, interstices with irregular fine axial patterns. Teleoconch II of 1.25 whorls, suture moderately impressed, sculpture variable. Shoulder with 18–49 axial cords, no axials in immediate vicinity of peristome in fully mature specimens, interaxials with fine irregular axial sculpture; up to seven faint and indistinct spiral lines, forming minute pustules at intersection with axials. Base sculpture similar to shoulder, or axials fainter, or axials restricted to mid-base, forming discrete short lamella; approximately six faint axials between selenizone and suture, 12 axials from suture to umbilical margin, sometimes becoming stronger towards umbilicus. Aperture subquadrate, roof overhanging, peristome thickened, outer lip crenulated. Umbilicus wide, deep, internal walls smooth, with thickened carina. Selenizone on shoulder, keels pronounced, inclined towards spire; lunules distinct; foramen with elevated keels forming chimney, terminating approximately at one teleoconch II whorl; low cord continuing anteriorly in position of foramen. Peripheral outline of shell with distinct bend at anterior end of foramen.

Radula (Fig. 527). Rachidian tooth trapezoidal, central cusp largest, four smaller cusps on each side. Lateral teeth 1–3 similar, central cusp largest, three cusps along outer edge, one cusp on inner edge. Lateral tooth 4 reduced, hook-shaped, central cusp largest, two smaller cusps on each side. Lateral tooth 5 broadened, enlarged, 4–5 cusps along inner edge of tip. Inner marginal teeth with four cusps on outer edge of tip. Outer marginal teeth with spoon-shaped tip with approximately 10 cusps on each side. Outermost marginal teeth paddle-shaped with many small bristles along tip and upper part of shaft with food groove. Radular interlock of central field moderate.

Comparisons. *Satondella tabulata* from the Indian Ocean to the Western Pacific and the Caribbean has an overall flatter shell, a distinct spiral cord in the position of the selenizone on teleoconch I, and a small shelf in the adumbilical corner of the peristome.

Satondella danieli from the tropical eastern Atlantic to the Caribbean has about half as many axial cords as well as half as many spiral lines, and a teleoconch II of 0.6 whorl.

Satondella senni from Easter Island has spiral sculpture on the protoconch and fewer axials.

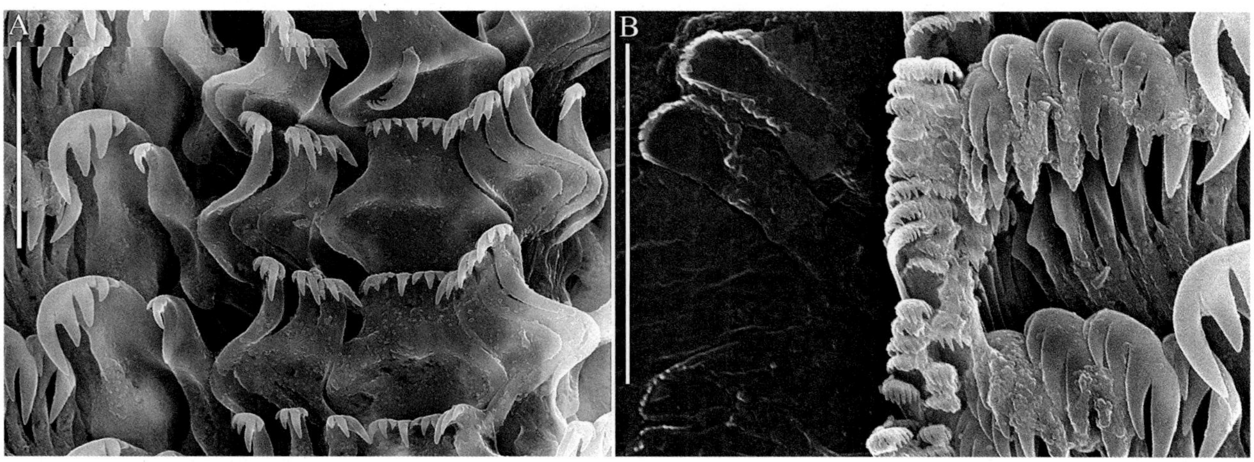

Figure 527. *Satondella brasiliensis*. Radula. 81–90 m, due S of Castle Roads, Bermuda, 32.250°N, 64.667°W (BMSM 26039, 1). **A.** Central field. **B.** Lateral tooth 5 through outer marginal teeth. Scale bars = 10 µm.

Satondella cachoi from the Indian Ocean to the Central Pacific has half as many axial cords, a teleoconch II of 0.66 whorl, a smooth protoconch, and stronger spiral cordlets on the base.

Satondella dantarti from the Indo-Malayan Archipelago has a protoconch with strong axial cords, a flat-topped shell, and much stronger elevated axial lamellae.

Satondella goudi from the Indo-Malayan Archipelago has an overall wider shell, the umbilicus lacks the funiculus, and the axial cords/lamellae are much finer, producing fine points at the intersection with the axial lines.

Biology. Lives on sea weeds (Rios, 1975).

Distribution. Bermuda, Caribbean to Brazil [Amapá to Espírito Santo ~3°N–20°S] (Fig. 528).

Specimen records.

Bermuda. 82–91 m, due S Castle Road, 32.250°N, 64.667°W (HLC, 2). 81–90 m, due S of Castle Road, 32.250°N, 64.667°W (BMSM 26039, 1).

USA. E Flower Garden Bank, Texas, 27.908°N, 93.600°W (CCSC, 8). 110 m, Dry Tortugas Islands, Florida, 25.583°N, 83.700°W (USNM 621366, 1). 91 m, 27.867°N, 93.800°W [185 km SSE of Galveston, Texas] (USNM 780321, 5). 72 m, Gulf of Mexico, 27.933°N, 92.033°W (USNM 780310, 25).

Mexico. [Yucatan Channel], 21.183°N, 86.317°W (UMML 30.12042, 1). 198 m, [off Cozumel], 20.917°N, 86.467°W (UMML 30.12041, 1). [Off Cozumel], 21.067°N, 86.417°W (UMML 30.120406, 3). [Off Cozumel], 21.067°N, 86.417°W (UMML 30.12040a, 2).

Bahamas. 35–52 m, Abaco, 26.500°N, 77.750°W (CRC, 1). 40 m, 12 mi NNW of W end of Grand Bahama Island, 26.750°N, 79.100°W (HLC, 1). 28 m, Long Cay mooring, Long Cay [= Long Bay?], 24.000°N, 74.517°W (HLC, 2). 39 m, W side of Cape Eleuthera, Eleuthera, 25.283°N, 76.317°W (HLC, 1).

Trinidad and Tobago. 30 m, E end of Tobago, 11.250°N, 60.667°W (ANSP, 8). 33–48 m, E end of Tobago, 11.250°N, 60.667°W (SBMNH, 2). 48 m, E end of Tobago, 11.250°N, 60.667°W (SBMNH, 5). 33 m, Goat Island, Tobago, 11.300°N, 60.700°W (HLC, 2). 30 m, Goat Island, Tobago Island, 11.300°N, 60.700°W (FLMNH 236689, 20). Speyside, Tobago Island, 11.300°N, 60.533°W (FLMNH 243701, 4).

Cuba. 56 m, Faro de los Colorados, Cienfuegos, 22.010°N, 80.500°W (HLC, 2; UAM, 50). 40–50 m, Cienfuegos Bay, Cuba, 22.025°N, 80.463°W (UAM, 8, 2: complete). 51 m, de Pedernales Point, de Pinos Island, 21.583°N, 83.500°W (UAM, 3).

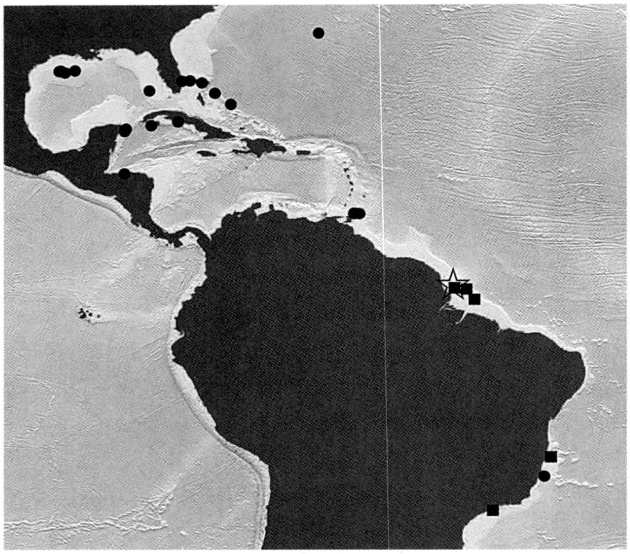

Figure 528. Distribution of *Satondella brasiliensis*. ☆ type locality. ● specimen record. ■ literature record.

Honduras. 20 m, Honduras, Bay Island, Little Hog Island, Cochino Pegueño, 15.950°N, 86.500°W (DLG 360, 1).

Literature records.

Bahamas. Abaco (Redfern, 2001).

Brazil. Ubatuba, S. P. (Rios, 1975).

Remarks. The original description and poor-quality illustration (Mattar, 1987: figs 1–6. Fig. 523) show a shell very close to the material studied here from localities other than Brazil, with a foramen with strongly elevated keels, suggesting a membership in *Satondella*. The very square overall look and the very strongly protruding foramen (Mattar, 1987: fig. 2), the umbilicus with straight walls (Mattar, 1987: fig. 3), and

Scissurellidae *Satondella*

the sculpture on the shoulder (Mattar, 1987: fig. 5) all agree with the species found throughout the Caribbean. The type material of *Sat. brasiliensis* was described as having a smooth protoconch, but the fine hexagonal protoconch sculpture (Fig. 526) is visible only at high magnification under SEM in well preserved specimens (fine hexagons are only 1.5 μm wide, and so cannot be seen under the light microscope). Regarding spiral sculpture, Mattar (1987) did not mention it in the description, but the discussion states that spiral "lirae" were not observed at the base of the last whorl; nevertheless, spiral sculpture is especially faint in this zone in *Sat. tabulata*, and in several specimens it is hardly appreciable even in SEM micrographs. Given those distinct similarities, the Caribbean *Satondella* specimens are considered conspecific with those from Brazil.

The species is quite variable in shell sculpture, with all intermediate forms being documented. All specimens with sufficiently well preserved protoconch show the unique microhexagonal pattern. Accordingly, there is not justification for separating those sculptural morphs taxonomically.

It is virtually impossible to legally obtain scientific loans from Brazil due to bureaucratic hurdles imposed by the Brazilian Government; hence, no specimens of this species in Brazilian collections could be examined, including type material.

Satondella cachoi Luque, Geiger & Rolán, 2011

Figures 529–532

Chresonymy.

Satondella cachoi Luque, Geiger & Rolán, 2011: 8–10, fig. 5A–M.

Satondella sp.: Hasegawa, 2005: 141–142, figs 3C–D.

Type material. Holotype (MNHN 22618: Luque *et al.*, 2011: fig. 5E, G–H). Paratypes Norfolk Ridge, 23°19'S 168°00'E, 601–608 m, N/O "Alis," BATHUS 3, Stn. 824 (MNHN 22617, 3. Luque *et al.*, 2010: fig. 5F, J).

Type locality. Loyalty Basin, 20°35'S 166°55'E, 470–480 m, N/O *Coriolis*, BIOGEOCAL, Stn. DW307 (OD).

Etymology. The species honours Raúl Fernández-Garcés ("Cacho"), a keen Cuban malacologist who helped us with the study of scissurellids (OD).

Description. Shell small (up to 2.1 mm. Figs 529, 530), depressed, trochiform. Protoconch 0.75 whorl, sculpture flocculent, no varix, aperture sinusoidal. Protoconch depressed and lower than shoulder of teleoconch. Teleoconch I of 1.25–1.33 whorls, 20 raised axial cords on first whorl, interstices with fine growth marks, indistinct spirals commence after 0.5 whorl. Teleoconch II of 0.66 whorl, suture impressed. Shoulder with axial cords run over with fine spiral lines, approximately 14 spirals on shoulder at aperture. Shoulder at aperture in lower half of penultimate whorl. Umbilicus open, deep, continuously sloping from base, with funiculus. Selenizone at upper periphery of shoulder, keels pronounced, elevated to form chimney at foramen. Aperture subquadrate, adumbilical portion recessed, lower adumbilical corner drawn out, forming funiculus. Apertural margin thickened, finely serrated externally by spirals.

Radula (531). Rachidian tooth trapezoidal, central cusp largest, three cusps on each side. Lateral teeth 1–3 similar, apical-most cusp largest, 2–3 cusps along outer margin, one cusp along inner margin. Lateral tooth 4 reduced, hook-shaped, with one minute point on each side of shaft. Lateral tooth 5 enlarged by broadening, apical-most cusp largest, 3–4 cusps along inner edge. Inner marginal teeth with palmate tip, apical-most cusp largest, one cusp along inner edge, 3–4 cusps along outer edge; shaft with food groove. Outer marginal teeth spoon-shaped, with many small bristles. Outermost marginal teeth paddle-shaped with many small teeth along apical edge. Radular interlock of central field moderate.

Distribution. Indian Ocean through Central Pacific (Fig. 532).

Specimen records.

Madagascar. 264–280 m, S of False Cape, ATIMO VATAE DW3552, 26.116°S, 45.650°E (MNHN, 3). 280–333 m, S of False Cape, ATIMO VATAE DW3553, 26.133°S, 45.650°E (MNHN, 1). 455–458 m, S of False Cape, ATIMO VATAE DW3555, 26.150°S, 45.666°E (MNHN, 1).

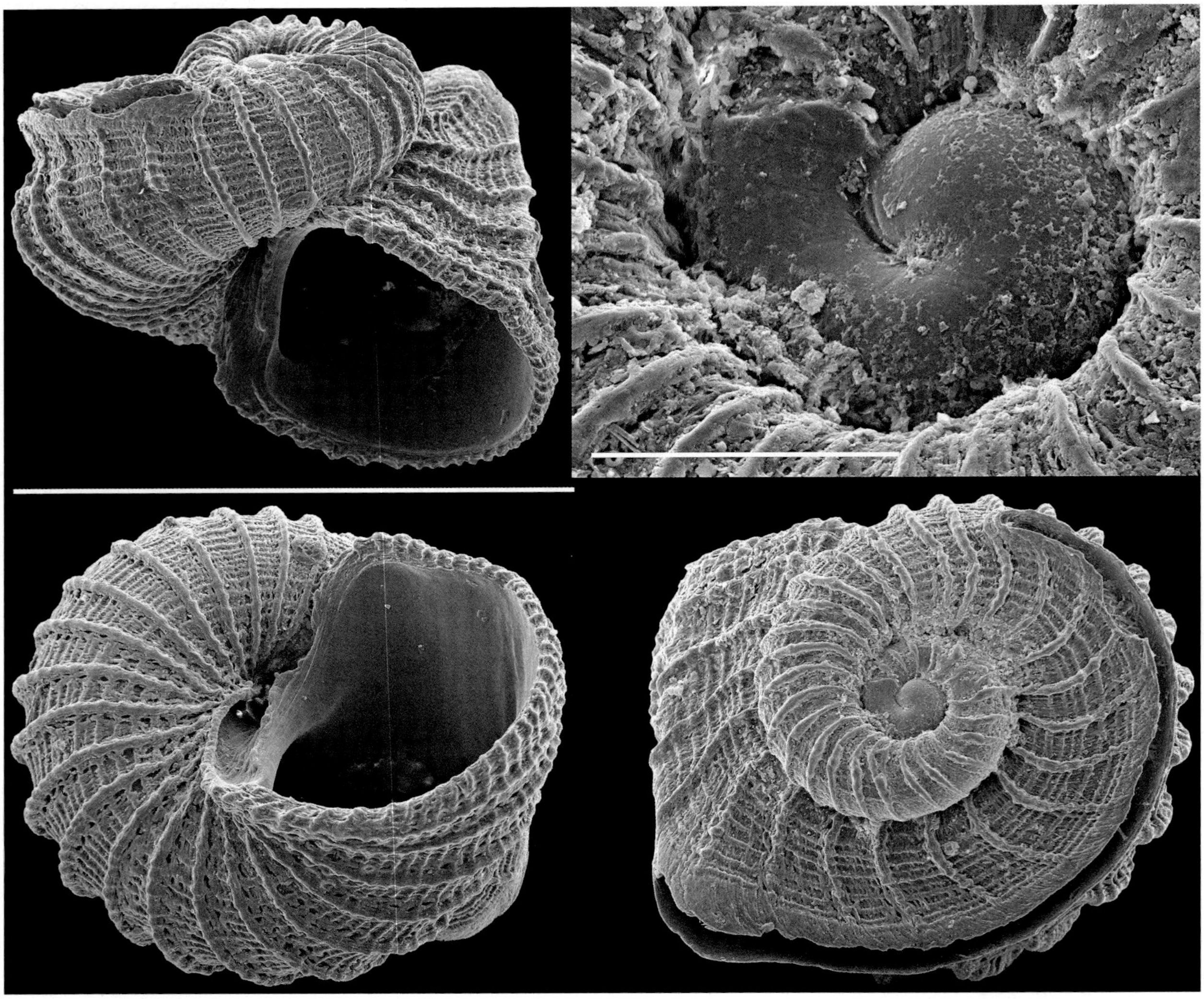

Figure 529. *Satondella cachoi.* 25–30 m, N side, Seringapatam Reef, Western Australia, Australia, 13.628°S, 122.016°E (SBMNH 422862, 1). For radula, see Figure 531. Scale bar shell = 1 mm. Scale bar protoconch = 100 μm.

Japan. 211–212 m, off Kerama Island, Okinawa, 26.047°N, 127.375°E (SBMNH 149460, 30).

Philippines. 150–163 m, Diplog Bay, Bohol-Sulu Sea Sill, PANGLAO 2005 CP2380, 8.688°N, 123.296°E (MNHN, 1). 95–128 m, Cervera Shoal, Pamilacan Island, PANGLAO 2004 T36, 9.488°N, 123.858°E (MNHN, 1). 152 m, Bolod, Panglao

Figure 530. *Satondella cachoi*. **A.** 183 m, off NE point, Christmas Island, 10.417°S, 105.700°E (AMS C.406391, 1). **B.** 260 m, SE of Tongatapu, Tonga, 21.345°S, 175.042°W (AMS C.406374, 1). **C.** 183 m, off NE point, Christmas Island, 10.417°S, 105.700°E (AMS C.406391, 1). Scale bars shell = 1 mm. Scale bars protoconch = 100 μm.

Island, PANGLAO 2004 T2, 9.540°N, 123.796°E (MNHN, 1). 12 m, Cortes, Bohol Island, PANGLAO 2004 T21, 9.713°N, 123.843°E (MNHN, 1).

Indonesia. 175 m, SW Salayer, 6.373°S, 120.437°E (RMNH 42345, 1). 175–185 m, Bay of Sanggar, N of Sumbawa, 8.322°S, 118.260°E (RMNH 42344, 2).

Australia. 120 m, off North West Cape, Western Australia, 22.293°S, 113.725°E (AMS C.377594, 1). 113 m, Great Barrier Reef, E of Swain Reef, Queensland, 21.717°S, 152.850°E (AMS C.406370, 1). 201 m, off S end Fraser Island, Queensland, 27.952°S, 153.851°E (AMS C.406371, 1). 210–219 m, off S end Fraser Island, Queensland, 25.980°S, 153.850°E (AMS C.406373, 1). 25–30 m, N side, Seringapatam Reef, Western Australia, 13.628°S, 122.016°E (SBMNH 422862, 1). 273 m, Middleton Reef, 29.401°S, 159.073°E (AMS C.461435, 1). 201 m, Norfolk Islands, 37.337°S, 168.100°E (NMNZ M.224762, 21).

Christmas Island. 183 m, off NE Point, 10.417°S, 105.700°E (AMS C.406391, 2).

New Caledonia. 250–350 m, Vauban, 22.500°S, 166.400°E (MNHN, 1). 455 m, southern, 22.267°S, 166.183°E (MNHN, 1). 430 m, Loyalty Islands, 21.033°S, 167.533°E (MNHN, 1). 470–480 m, Loyalty Basin, 20.583°S, 166.917°E (MNHN, 1). 538 m, between Lifou and Uvea, 20.617°S, 166.967°E (MNHN, 1). 282 m, Loyalty Ridge, 20.700°S, 167.000°E (MNHN, 1). 105–110 m, Poindimie sector, 20.817°S, 165.317°E (MNHN, 6). 569–583 m, S Lansdowne, 21.100°S, 160.767°E (MNHN, 1). 124 m, S of, 22.633°S, 167.167°E (MNHN, 27). 601–608 m, Norfolk Ridge, 23.317°S, 168.000°E (MNHN, 3). 447–450 m, Stylaster Bank, 23.617°S, 167.700°E (MNHN, 1). 430–452 m, southern, Stylaster Bank, 23.633°S, 167.717°E (MNHN, 1). 520 m, 24.933°S, 168.367°E (MNHN, 1). 444–445 m, south, BATHUS 2 DE719, 22.800°S, 167.266°E (MNHN, 1). 540 m, Norfolk Ridge, NORFOLK 1 DW1722, 23.300°S, 168.016°E (MNHN, 1). 433–450 m, Stylaster Bank, SMIB 8 DW166, 23.633°S, 167.716°E (MNHN, 2). 285 m, Norfolk Ridge, NORFOLK 2 DW2040, 23.683°S, 168.016°E (MNHN, 1).

New Zealand. SE of Smith Bluff, Kermadec Islands, 29.315°S, 177.940°W (NMNZ ex M.226900, 1). 146–165 m, SE of Nugent Island, Raoul Island, Kermadec Islands, 29.245°S, 177.823°W (NMNZ M.225752, 4). 538 m, 130 km S of Esperance Rock, Kermadec Ridge, 32.472°S, 179.265°W (NMNZ M.137477, 4). 219–274 m, SE

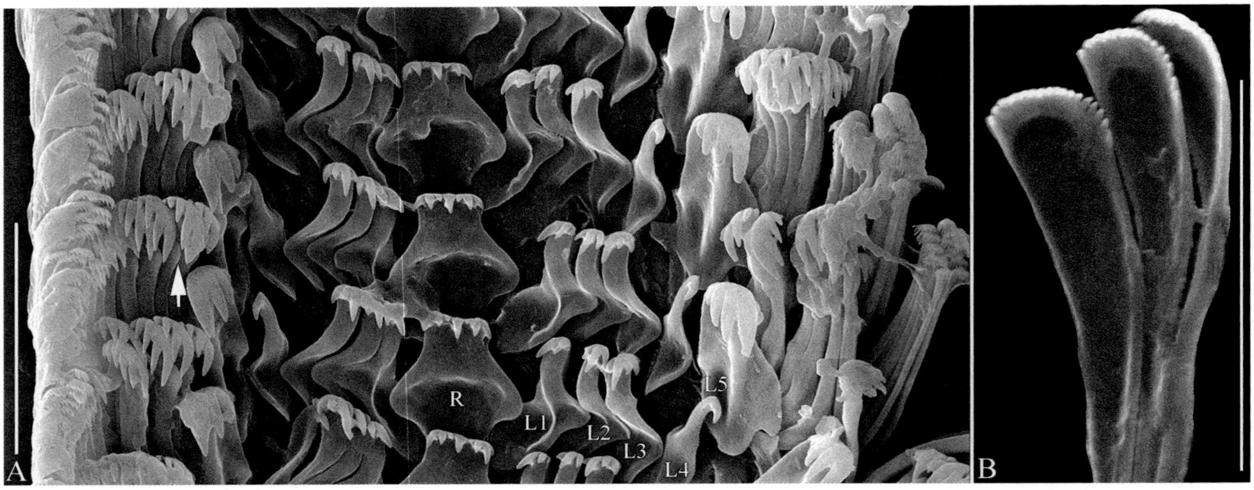

Figure 531. *Satondella cachoi.* Radula. 5–30 m, N side, Seringapatam Reef, Western Australia, Australia, 13.628°S, 122.016°E (SBMNH 422862, 1). For shell, see Figure 529. **A.** Full width of radula. Scale bar = 20 μm. **B.** Outermost marginal teeth. Scale bar = 10 μm. R: Rachidian tooth. L1–5: Lateral teeth 1–5. Arrow points to projection on shaft of marginal teeth forming food groove.

of D'Archy Point, Raoul Island, Kermadec Islands, 29.313°S, 177.903°W (NMNZ M.226874, 5).

Solomon Islands. 215–300 m, 15.105°S, 166.925°E (MNHN, 1).

Vanuatu. 154–179 m, 15.050°S, 167.117°E (MNHN, 1). 205–210 m, 15.050°S, 167.133°E (MNHN, 1). 215–300 m, 15.105°S, 166.925°E (MNHN, 1). 140–182 m, 15.633°S, 167.058°E (MNHN, 2). 205–210 m, MUSORSTOM 8 DW1101, 15.066°S, 167.133°E (MNHN, 1).

Fiji. 656–660 m, Bligh Water, 17.268°S, 178.247°E (MNHN, 1). 820–863 m, 17.300°S, 179.550°W (MNHN, 5). 700–707 m, BORDAU1, DW1485, 19.050°S, 178.500°W (MNHN, 1).

Tonga. 260 m, SE of Tongatapu, 21.345°S, 175.042°W (AMS C.406374, 1).

French Polynesia. 787–792 m, Mount Honu, Tarava Seamounts, TARASOC DW3340, 18.400°S, 154.150°W (MNHN, 5). 788–836 m, Mount Otaha, Tarava Seamounts, TARASOC DW3328, 18.766°S, 152.250°W (MNHN, 1).

Fossil.

New Zealand. Kaipara Harbour, small bay c 1.6 km NW of Pakaurangi Point. Otaian, early Miocene. GS9730, 36.217°S, 174.283°E (NMNZ M.174838, 1).

Literature records.

Japan. 520 m, DY02-5, Nansei Islands, 28.454°N, 129.337°E (1). 288–311 m, TY01-7, Nansei Islands, 28.099°N, 129.114°E (2). 199–202 m, TY03-9, Nansei Islands, 29.825°N, 127.700°E (2, 1: complete). TY04-24, Nansei Islands (3). 293–320 m, KT04-24-G3, Nansei Islands, 28.702°N, 129.973°E (1: complete) (Hasegawa, 2005).

Distribution. Tropical Indian Ocean to Central Pacific (Fig. 532).

Comparisons. *Satondella minuta* from the Indo-Malayan Archipelago to the Western Pacific is much more flattened and has fewer axials.

Satondella dantarti from the Indo-Malayan Archipelago to the Central Pacific has a protoconch with strong axial cords, a flat-topped shell, strongly elevated axial lamellae, and a wider umbilicus, which is formed by the rounded portion of the inner whorls.

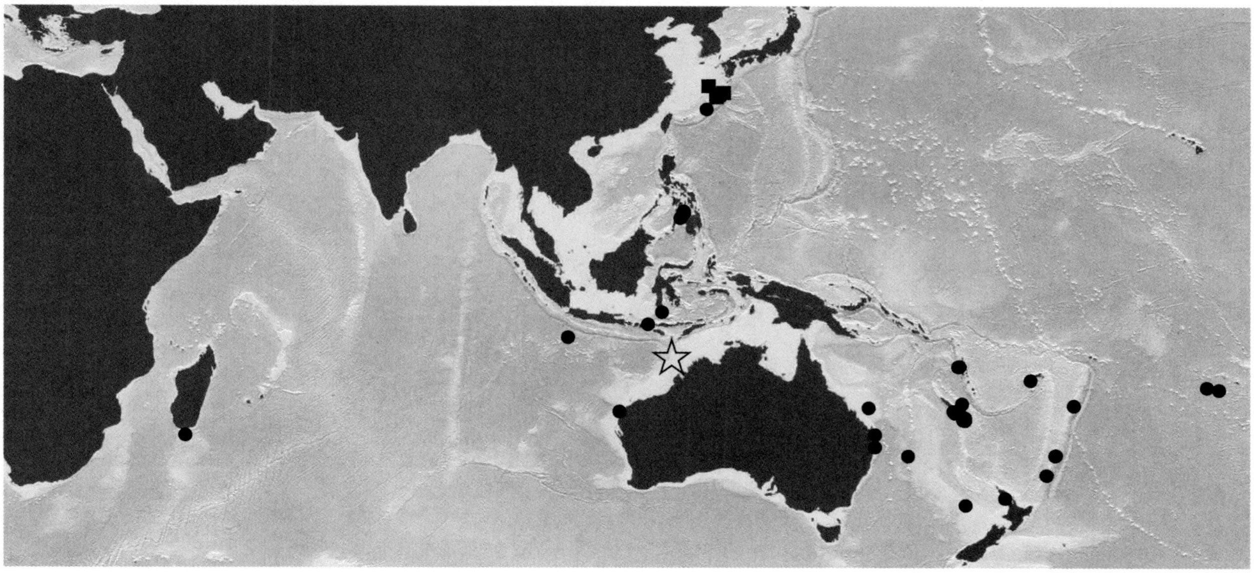

Figure 532. Distribution of *Satondella cachoi*. ☆ type locality. ● specimen record. ■ literature record.

Satondella goudi from Indonesia has an overall slightly flatter shell, a wider umbilicus lacking a funiculus, and weaker axial cords.

Satondella brasiliensis from the Caribbean to Brazil has a protoconch with fine hexagonal sculpture and finer spiral lines on the base of the teleoconch.

Satondella danieli from the tropical and subtropical North Atlantic has half as many spiral threads on shoulder and base.

Satondella tabulata from the Indian Ocean to the Central Pacific and the Caribbean has a more solid shell, a flat protoconch, and different teleoconch sculpture.

Satondella senni from Easter Island has a protoconch with spiral sculpture, has a teleoconch I of one whorl, and has fewer spirals on the base.

Remarks. One specimen from Middleton Reef, off New South Wales/Queensland border (AMS C.461435, 1) shows a protoconch with some faint spiral sculpture. Most available specimens of the species do not have exquisitely preserved protoconchs, and may not show those traces of spiral sculpture.

The Japanese name for the species is *Entotsu-ebisu* (Hasegawa, 2005).

Satondella danieli Segers, Swinnen & Abreu, 2009

Figures 533–535, 548C

Chresonymy.

Sukashitrochus sp. nov.: Beck *et al.*, 2006: 44.
Satondella danieli: Segers *et al.*, 2009a: 2–3. pl. 1, figs 1a–c.
Satondella danieli: Segers *et al.*, 2009b: 5.
Satondella danieli: Segers *et al.*, 2009c: 66, pl. 5, figs 4–4a [holotype].
Satondella danieli: Luque *et al.*, 201: 8.

Type material. Holotype (ZMA 4.09.29). Paratypes from type locality (ZMA 4.09.30–36, 7). 364 m, Porto Novo, Madeira, Portugal, 32.659°N, 16.791°W (MMF 40311-15, 5). 407 m, Ponta da Atalaia, Madeira, Portugal, 32.639°N, 16.808°W (CFS, 8; CWS 8).

Type locality. 228–240 m, W of Deserta Grande, Desertas Islands, Portugal, 32.483°N, 16.533°W (OD).

Etymology. Named after Daniel Geiger, SBMNH, for his help throughout (OD).

Description. Shell medium-sized (to 1.3 mm. Figs 533, 534), trochiform, depressed. Protoconch of 0.75 whorl, sculpture flocculent, no apertural varix, apertural margin sinusoidal. Teleoconch I of 1.3–1.5 whorls, initially only with growth marks, onset of axial cords after 0.5 whorl, approximately 30, no spirals. Teleoconch II of 0.6 whorl, suture moderately impressed. Shoulder shallow sinusoidal, axials becoming stronger from low cords to low lamellae, onset of spiral lines variable, approximately 10 spirals at apertural margin. Base constricted below selenizone, distinct axial lamellae at same density as on shoulder; approximately 10 spiral lines forming acute points at intersection with axial lamellae. Umbilicus narrow, bordered by distinct spiral cord, walls straight, with very fine growth marks. Aperture rounded, D-shaped, roof overhanging, with distinct lower adumbilical shelf; margin crenulated. Selenizone above periphery, keels moderately strong, moderately elevated; lunules distinct; foramen with elevated keels, continued anteriorly by raphe.

Animal unknown.

Distribution. Macaronesian Islands, Caribbean (see Remarks).

Specimen records.

Portugal, Azores. 620 m, S of Santa Maria, 36.917°N, 25.083°W (RMNH, 10).

Portugal, Madeira. 407 m, Station 17c, Atalaia, 32.640°N, 16.786°W (DLG 1309, 1). 360 m, S of, 32.617°N, 16.850°W (RMNH, 1).

Spain, Canary Islands. 160 m, off de la Restinga Point, S of Hierro, 27.617°N, 18.000°W (RMNH, 1). 340–480 m, Station 2.114, off de Ochilla Point, SW of Hierro, 27.683°N, 18.150°W (RMNH, 1).

Puerto Rico. 390 fathoms [= 700 m], *Challenger* Station 24, off Culebra Island, 18.641°N, 65.092°W (NMW 155.158.607, 1: paralectotype of *Schismope tabulata*).

Literature records.

NE Atlantic. 63–607 m, Gettysburg Seamount, 36.541°N, 11.504°W. 110–800 m, Ampère Seamount, 35.064°N, 12.930°W (Beck *et al.*, 2006).

Comparisons. *Satondella brasiliensis* from the Caribbean to Brazil has a protoconch with microhexagonal sculpture, a teleoconch II of more than one whorl, and at least twice as many axial cords.

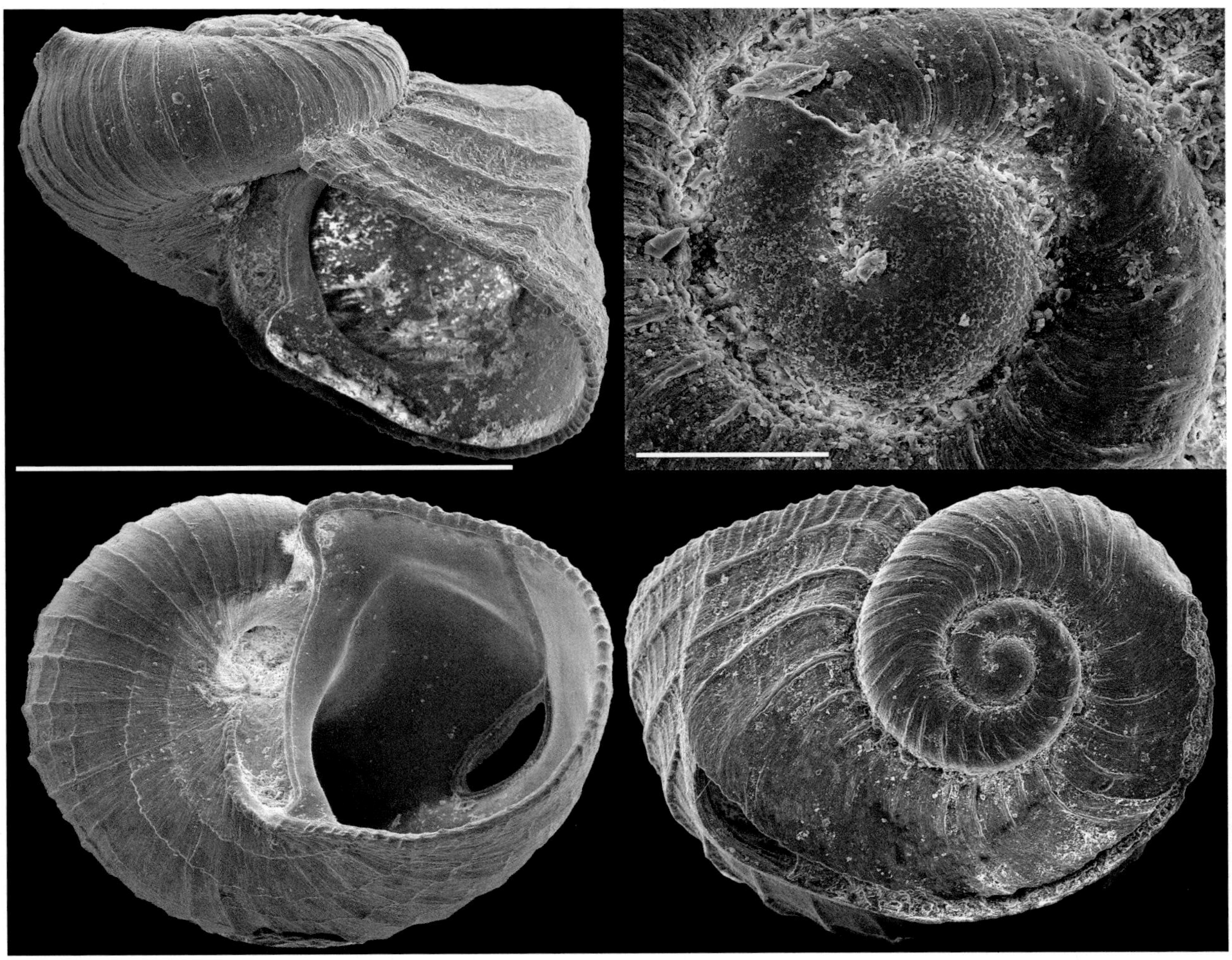

Figure 533. *Satondella danieli*. 620 m, S of Santa Maria, Azores, Portugal, 36.917°N, 25.083°W (RMNH, 10). Scale bar shell = 1 mm. Scale bar protoconch = 100 µm.

Satondella tabulata from the Caribbean and the Indo-Pacific has a proportionally wider shell and at least 50% more spiral lines on the base.

Satondella cachoi from the Indian Ocean through Central Pacific has at least twice as many spiral lines on the base and the keels of the foramen are more prominently

Figure 534. *Satondella danieli*. **A–B.** 620 m, S of Santa Maria, Azores, Portugal, 36.917°N, 25.083°W (RMNH, 10). **C.** 160 m, off de la Restinga Point, S of Hierro, Canary Islands, Spain, 27.617°N, 18.000°W (RMNH, 1). Scale bars shell = 1 mm. Scale bars protoconch = 100 μm.

Figure 535. Distribution of *Satondella danieli*. ☆ type locality. ● specimen record. ■ literature record.

elevated, with the spiral lines running into those keels and forming a distinct angle at that point.

Satondella dantarti from the Indo-Malayan Archipelago to the Central Pacific has a flat-topped shell, strong axial cords on the protoconch, fewer axial lines on the teleoconch, and the umbilicus is formed by rounded portion of the inner whorls.

Satondella goudi from the Indo-Malayan Archipelago has at least twice as many spiral lines on the base and more delicate axial lamellae.

Satondella senni from Easter Island has a protoconch with spiral sculpture and a wider umbilicus.

Remarks. The paralectotype NMW 1955.158.607 of *Schismope tabulata* (Fig. 548C) is here identified as *Sat. danieli*, which suggests that the species has an amphiatlantic distribution pattern.

Satondella dantarti Luque, Geiger & Rolán, 2011

Figures 536–538

Chresonymy.

Satondella dantarti: Luque, Geiger & Rolán, 2011: 10, fig. 6A–D .

Type material. Holotype (MNHN 22619. Fig. 536). Paratypes from type locality (MNHN 22620, 3. Fig. 537).

Type locality. 320–367 m, Aztec Bank, New Caledonia, 23.308°S, 168.083°E (OD).

Etymology. The species honours the late Luis Dantart, a keen zoologist from Barcelona University who collaborated with the first [Angel Luque] and third [Emilio Rolán] authors in a work on *Satondella tabulata* started several years ago (OD).

Description. Shell small (to 0.72 mm. Figs 536, 537), trochiform, depressed. Protoconch of one whorl, strong axial cords, no apertural varix, apertural margin sinusoidal. Teleoconch I of 1.2 whorls, 17–25 axial lamellae, first spiral line in position of selenizone after 0.2–0.3 whorl, lines running under axial lamellae Teleoconch II of 0.3–0.5 whorl, suture deeply impressed. Shoulder with same density and strength of axial lamellae, approximately six spiral lines at apertural margin. Base with distinct constriction below selenizone, axial lamellae gradually disappearing towards umbilicus, first spiral line on mid-base after 0.3–0.5 whorl, approximately a dozen spiral lines of unequal strength at apertural margin. Umbilicus wide, underside of protoconch visible, margin with distinct bulge. Aperture D-shaped, peristome flared, adumbilical corner forming bulge, outer margin not crenulated. Selenizone above periphery; lunules strong; keels strong, above periphery; slit closed to foramen with further elevated keels, closed anteriorly by raphe.

Animal unknown.

Comparisons. *Satondella cachoi* from the Indian Ocean through the Central Pacific has a more elevated spire, a smooth protoconch and more numerous but weaker axial cords.

Scissurellidae *Satondella*

Satondella goudi from the Indo-Malayan Archipelago has a more elevated spire, much less distinct axial cords, and an umbilicus with straight walls.

Satondella tabulata from the Indian Ocean to the Western Pacific and the Caribbean has a more elevated spire, an overall broader shell, and an umbilicus with straight walls.

Satondella senni from Easter Island has a protoconch with spiral sculpture, a more elevated spire, and the umbilicus is formed by the curved portion of the inner whorls.

Satondella minuta from the Indo-Malayan Archipelago to the Western Pacific has a smooth protoconch, a more elevated foramen, with the spiral threads on the shoulder forming a distinct angle on the keels of the foramen proper, and more spiral threads over the entire shell.

Satondella bicristata from New Zealand is almost planispiral and the axial lamellae are elevated in position of selenizone and on the mid-base.

Distribution. Western Pacific (Fig. 538).

Specimen records.

Solomon Islands. 53 m, SALOMON 1 CP1810, 9.795°S, 160.841°E (MNHN, 1).
Vanuatu. 99 m, NW coast of Malo Island, SANTO 2006 EP21, 15.628°S, 167.086°E (MNHN, 1).

Figure 536. *Satondella dantarti*. Holotype. 320–367 m, Aztec Bank, New Caledonia, 23.308°S, 168.083°E (MNHN 22619). Scale bar shell = 0.5 mm. Scale bar protoconch = 100 μm.

New Caledonia. 447–450 m, Stylaster Bank, 23.617°S, 167.700°E (MNHN, 1). 601–608 m, Norfolk Ridge, 23.317°S, 168.000°E (MNHN, 3). 1–3 m, Channel of Poé Lagoon, Bourail sector, MONTROUZIER St. 1326, 21.620°S, 165.378°E (MNHN, 1). 775 m, BIOCAL DW48, 23.000°S, 167.483°E (MNHN, 1). 569–616 m, Norfolk Ridge, Unfindable Bank, NORFOLK 1 DW1697, 24.650°S, 168.633°E (MNHN, 20). 565–600 m, south, BERYX 11, DW10, 24.883°S, 168.350°E (MNHN, 2). 507–967 m, Norfolk Ridge, Sponge Bank, NORFOLK St DW1692, 24.933°S, 168.350°E (MNHN, 5; MNHN, 19). 518–586 m, Norfolk Ridge, NORFOLK 2 DW2067, 24.933°S, 168.366°E (MNHN, 7).

New Zealand. 294–298 m, Kiwi Seamount, 10 mi from, New Zealand, 30.750°S, 173.850°E (NMNZ M.25341, 4). 1216–1385 m, N Three Kings Rise, North Island, 31.512°S, 172.830°E (NMNZ M.160299, 6).

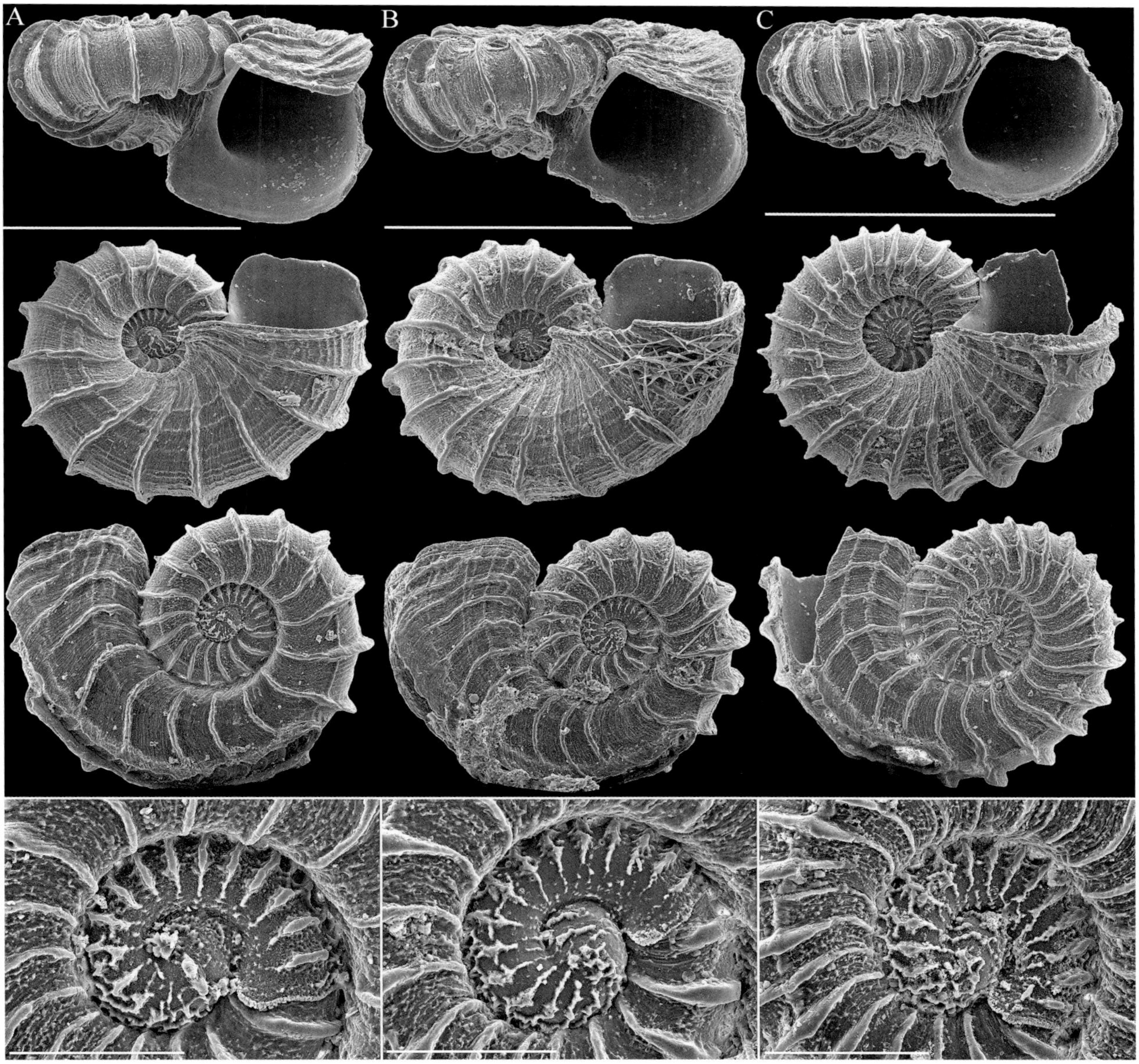

Figure 537. *Satondella dantarti*. Paratypes. 320–367 m, Aztec Bank, New Caledonia, 23.308°S, 168.083°E (MNHN 22620, 3). Scale bars shell = 0.5 mm. Scale bars protoconch = 100 µm.

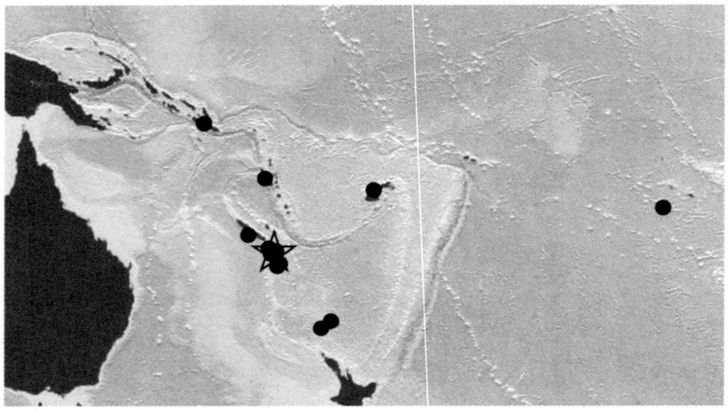

Fiji. 251–257 m, Bligh Water, MUSORSTOM 10 DW1334, 16.856°S, 178.231°E (MNHN, 1).
French Polynesia. 795–975 m, Mount Otaha, Tarava Seamounts, TARASOC DW3333, 18.750°S, 152.300°W (MNHN, 1).

Figure 538. Distribution of *Satondella dantarti*. ☆ type locality. ● specimen record.

Satondella goudi new species

Figure 539– 540

Type material. Holotype (RMNH 42364. Fig. 539A). Paratype 250–290 m, NW of Beningko, Banda Sea, Tukang Besi Island, Indonesia, 5.875°S, 123.975°E (RMNH 42363, 1. Fig. 539B).

Type locality. 233–274 m, SW Salayer, Indonesia, 6.372°S, 120.433°E.

Etymology. Named for Jeroen Goud of Naturalis, Leiden, for his support of micromollusks and kind hospitality at his institution.

Description. Shell medium-sized (to 1.65 mm. Fig. 539), trochiform, subquadrate. Protoconch 0.75 whorl, sculpture flocculent, no apertural varix, apertural margin sinusoidal. Teleoconch I of 1.5 whorls, 28–31 strong axial cords with apical thickening, diminishing in strength after approximately one whorl, first spiral lines after 0.1–0.5 whorl, approximately 15 at onset of selenizone, forming nodules at intersection with axials. Teleoconch II of one whorl, suture flush. Shoulder with same density of axial cords, 4–6 spiral lines at onset of selenizone, 15–20 at apertural margin. Base with moderate constriction below selenizone, same density of axials as on shoulder, approximately 30 spiral lines forming small nodules at intersection with axials. Umbilicus open, straight walls bordered by fine carina, smooth, at angle to base. Selenizone above periphery, keels initially low (possibly eroded), of moderate strength; lunules distinct; foramen moderately elevated, moderately strong, closed anteriorly by raphe. Aperture ovate, roof overhanging, outer lip finely crenulated; in umbilical view, adumbilical portion of base forming convex portion of peristome.

Animal unknown.

Comparisons. *Satondella cachoi* from the Indian Ocean to the Central Pacific has a similar overall shell-shape, although proportionally a bit taller, and has much stronger axial cords to low lamellae, while the umbilicus is narrower with a distinct funiculus.

Satondella dantarti from the Indo-Malayan Archipelago to the Western Pacific has a flat spire and the umbilicus is formed by the rounded portion of the inner whorls.

Satondella senni from Easter Island has a protoconch with spiral sculpture, a teleoconch of one whorl, and lacks the pointed nodules at the intersection of spirals and axials.

Satondella danieli from the tropical North Atlantic has half as many spiral lines and a distinct funiculus in the umbilicus.

Figure 539. *Satondella goudi* new species. **A.** Holotype. 233–274 m, SW Salayer, Indonesia, 6.372°S, 120.433°E (RMNH 42364, 1). **B.** Paratype. 250–290 m, NW of Beningko, Banda Sea, Tukang Besi Island, Indonesia, 5.875°S, 123.975°E (RMNH 42363, 1). **C.** 284–286 m, S Cape St. Marie, ATIMO VATAE CP3615, Madagascar, 26.233°S, 45.150°E (MNHN, 1). Scale bars shell = 1 mm. Scale bars protoconch = 100 μm.

Satondella brasiliensis from the Caribbean to Brazil has microhexagonal protoconch sculpture, a teleoconch II of more than one whorl, and a distinct funiculus in the umbilicus.

Satondella tabulata from the Indian Ocean to the Western Pacific and the Caribbean has spiral lines increasing in strength towards the umbilicus and a distinct internal rim in the aperture.

Distribution. Indonesia. Madagascar (Fig. 540).

Remarks. The bend in the shell at the foramen is not as pronounced as in other species, but still perceptible. In conjunction with the somewhat more elevated foramen, a placement in *Satondella* is indicated.

The specimen from Madagascar (Fig. 539C) is considered conspecific with the types from Indonesia. Shared characters include all details of protoconch and teleoconch sculpture, as well as the convex portion of the peristome contributed by the adumbilical portion of the base. The specimen is proportionally taller, and has an apparent lower expansion rate of the whorls.

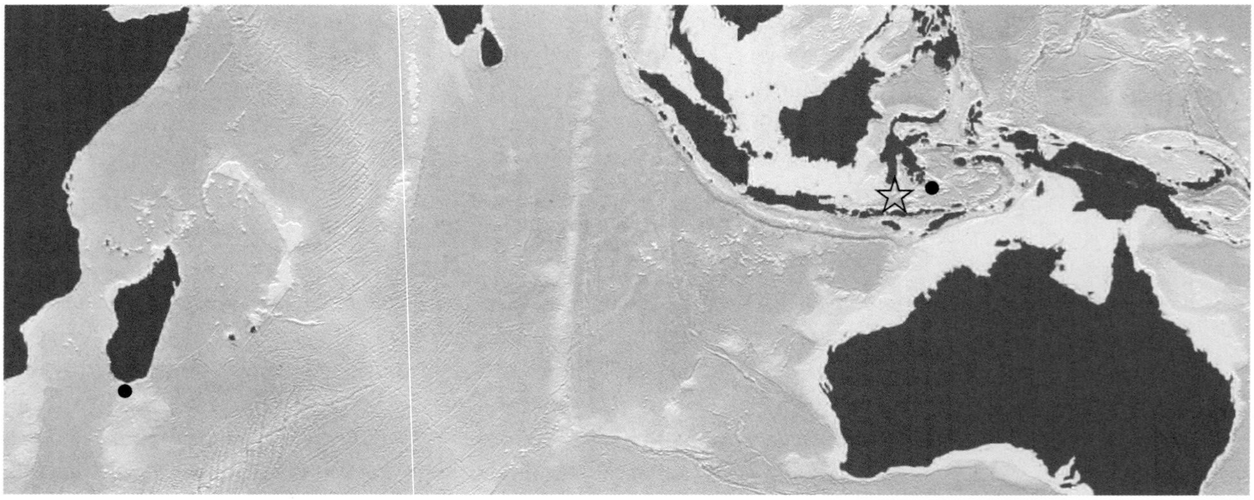

Figure 540. Distribution of *Satondella goudi* new species. ☆ type locality. ● specimen record.

Satondella minuta Bandel, 1998

Figures 541–543

Chresonymy.

Satondella minuta: Bandel, 1998: 65, pl. 22, fig. 8, pl. 23, fig. 1 [holotype: Fig. 541A, D]; pl. 23, figs 2–3 [Paratype: Fig. 541B, C].

Satondella minuta: Geiger, 2003: 76.

Satondella minuta: Luque *et al.*, 2011: 2–3, fig. 1A–I.

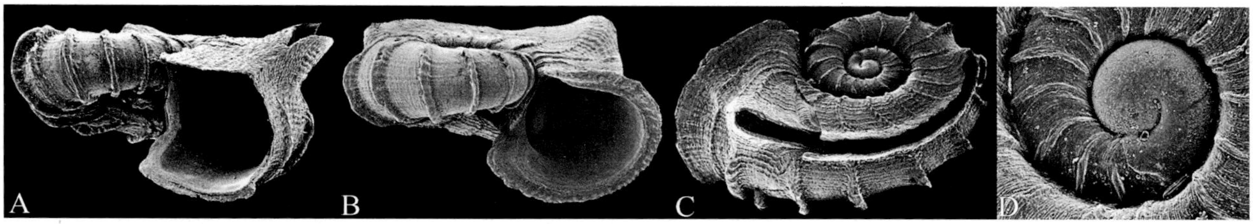

Figure 541. Original illustration of *Satondella minuta*. Sediments near the coast of Satonda, Indonesia [8.100°S, 117.750°E]. **A, D.** Holotype (SGPIH 3884). **B–C.** Paratype (SGPIH?).

Type material. Holotype (SGPIH 3884. Fig. 541A, D), 0.8 mm. Unspecified paratypes (Fig. 541BC. SGPIH ? see Remarks).

Type locality. Sediments near the coast of Satonda, Indonesia [8.1°S, 117.75°E] (OD).

Etymology. Minutus, Latin for small.

Description. Shell small (<1 mm. Figs 541, 542), depressed, trochiform. Protoconch 0.75 whorl, almost smooth, some specimens with indistinct spirals, no varix, apertural margin sinusoidal. Teleoconch I of 1.25–1.5 whorls, 15–16 strong axials on first whorl, interstices with fine growth marks, indistinct spirals appearing after one whorl. Teleoconch II approximately 0.5 whorl, suture deeply impressed. Shoulder of body whorl higher than apex to upper half of penultimate whorl, strong axial lamellae overrun by approximately 12 finer spiral lines. Base with similar sculpture as on shoulder. Umbilicus open, continuously sloping from base, funiculus present. Selenizone on shoulder at approximately 60° angle to coiling axis of shell; keels prominent, distinctly elevated in area of foramen. Aperture distinctly recessed adumbilically; lip thickened, somewhat recurved, slightly serrated; lower adumbilical corner extended, forming funiculus in umbilicus; outer lip crenulated.

Animal unknown.

Comparisons. *Satondella azonata* from the New Zealand lacks a selenizone and has fine axial sculpture on the centrifugal side of the protoconch.

Satondella bicristata from New Zealand has more axial lamellae and lacks a distinct selenizone.

Satondella cachoi from the Indian Ocean to the Central Pacific has a more elevated spire and a much more narrow umbilicus with straight walls.

Satondella dantarti from the Indo-Malayan Archipelago to the Central Pacific has a protoconch with strong axial sculpture and fewer spiral lines, which are not

angled over the less elevated keels of the foramen. These differences are discrete and consistent among specimens. One lot examined contained both species.

Satondella goudi from the western Indian Ocean and the Indo-Malayan Archipelago has a more elevated spire and a wider umbilicus with straight walls.

Satondella tabulata from the Indian Ocean to the Western Pacific and the Caribbean has much denser axials, a shelf surrounding the peristome, and a keel bordering the umbilicus.

Satondella senni from Easter Island has spiral sculpture on the protoconch and fewer axials.

Distribution. Indo-Malayan Archipelago and Western Pacific (Fig. 543).

Figure 542. *Satondella minuta*. **A–C.** 320–367 m, Aztec Bank, New Caledonia, 23.308°S, 168.083°E (MNHN, 1). Scale bars shell = 0.5 mm. Scale bars protoconch = 100 µm.

Scissurellidae *Satondella*

Specimen records.

Indonesia. Satonda [8.100°S, 117.750°E] (SGPIH 3884: holotype).
Solomon Islands. 396–411 m, 8.665°S, 160.060°E (MNHN, 1).
New Caledonia. 320–367 m, Aztec Bank, 23.308°S, 168.083°E (MNHN, 1). 282 m, Loyalty Islands, 20.700°S, 167.000°E (MNHN, 1).
Fiji. 656–660 m, Bligh Water, 17.268°S, 178.247°E (MNHN, 1). 392–433 m, S of Viti Levu, 18.185°S, 178.390°E (MNHN, 3). 144–150 m, S of Viti Levu, 18.207°S, 178.550°E (MNHN, 2). 260–305 m, S of Viti Levu, 18.308°S, 178.097°E (MNHN, 1).

Literature records.

Japan. 285–288 m, TY-01-07-1, 28.356°N, 129.227°E (2). 288–311 m, TY-01-07, 28.149°N, 129.114°E (2) (Hasegawa, 2005).

Remarks. Bandel (1998: 65) indicated "einige weitere Schalen" [a few additional shells] collected from the type locality, one of which was figured. According to ICZN (1985) Art. 72a(iii), those are to be considered paratypes. The repository of those specimens is unclear.

The Japanese name is *Hira-entotsu-ebisu* (Hasegawa, 2005).

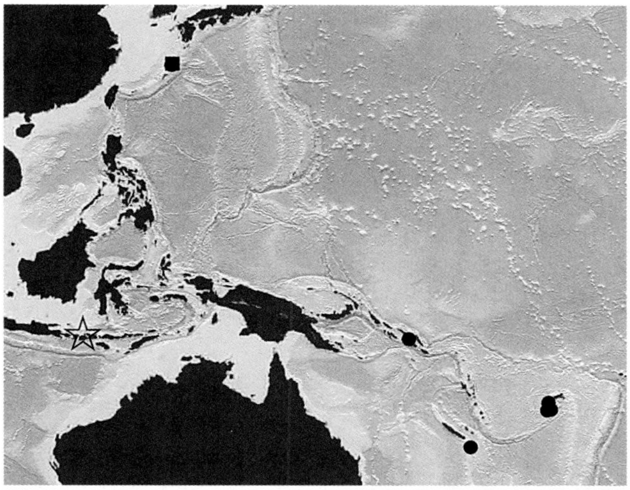

Figure 543. Distribution of *Satondella minuta*. ☆ type locality. ● specimen record. ■ literature record.

Satondella senni Geiger, 2003

Figures 544–546

Chresonymy.

Satondella sp.: Raines, 2002: 13, figs 3–4.
Satondella senni Geiger, 2003: 30–31, 76, fig. 3.
Satondella senni: Geiger, 2004: text-fig. p. 7.
Satondella senni: Geiger, 2006c: text-fig. 6.
Satondella senni: Luque *et al.*, 2011: 8.

Type material. Holotype (LACM 2911. Fig. 544). Paratypes from type locality (LACM 2923, 1: gold-coated and mounted on SEM stub; AMS C.205032, 1; ANSP 410307, 1; UMUT RM28238, 1; NHMUK 20020265, 1).

Type locality. 10–20 m, off Rosalia Point, E of Anakena, Easter Island, in sand and rubble at base of cliff, Dec. 2000. Leg. B. Raines [27.067°S, 109.333°E] (OD).

Etymology. The species honors my mentor David G. Senn, University of Basel, Switzerland, for his varied introduction to marine biology and for instilling a holistic approach to scientific inquiry (OD).

Description. Shell small (1.08 mm. Figs 544, 545), 1.5× wider than high. Color off-white. Protoconch 142 μm, 0.75 whorl, in one plane with top of shell, embryonic cap smooth, early protoconch with irregular granulation, last half whorl with three faint but distinct spiral cords, apertural margin sinusoidal, no varix. Teleoconch I of 1.1 whorls, 21–30 axials of increasing strength from low cords to distinct cords elevated to lamellae towards suture of next whorl; first spirals after 0.66 whorl, increasing in number with growth to approximately six at onset of selenizone, increasing in strength from fine threads to distinct lines from selenizone towards suture of next whorl. Teleoconch II of one whorl; suture impressed. Shoulder almost flat, with approximately six fine but distinct spiral cords, first spiral at onset of selenizone; spirals intersected by broad but less distinct curved axials cords, 13 axials on last half whorl, first axial immediately after protoconch. Base with distinct constriction below selenizone, with similar sculpture as on shoulder, strength of spirals increasing from fine lines below selenizone to strong cords towards umbilicus. Umbilicus wide, open, smooth, straight wall at distinct angle to base. Aperture subrectangular, obtuse angle at inner base, adumbilical margin thickened, upper lip strongly overhanging, outer

Figure 544. *Satondella senni.* Holotype. 10–20 m, off Rosalia Point, E of Anakena, Easter Island, Chile, in sand and rubble at base of cliff, Dec. 2000. Leg. B. Raines [27.067°S, 109.333°E] (LACM 2911). Scale bar shell = 1 mm. Scale bar protoconch = 100 μm.

margin finely crenulated. Selenizone above and inside periphery, at approximately 45° angle to shell axis, keels strong, moderately elevated; lunules strong, foramen drawn out.

Animal unknown.

Comparisons. *Satondella brasiliensis* from the Caribbean to Brazil has microhexagonal protoconch sculpture, a more squat overall shell shape, and a more elevated spire.

Satondella azonata and *Sat. bicristata*, both from New Zealand, have a more depressed shell, an umbilicus formed by the rounded portion of the previous whorls, and lack a selenizone.

Satondella minuta from the Indo-Malayan Archipelago to the Western Pacific has a depressed spire with smooth protoconch, a stronger keel of the selenizone and fewer and stronger axials on the teleoconch, and an umbilicus formed by the rounded portion of the previous whorls.

Satondella cachoi from the Indian Ocean to the Central Pacific has an almost smooth protoconch, stronger axial cords to low lamellae, and a distinct funiculus in the narrower umbilicus.

Satondella goudi from the western Indian Ocean and the Indo-Malayan Archipelago has a protoconch with flocculent sculpture, a teleoconch I of 1.5 whorls, and the intersection of axials and spirals produces fine nodules.

Figure 545. *Satondella senni.* 70–100 m, Hanga Roa, Easter Island, Chile, 27.145°S, 109.440°W (DLG 1335, 1). Scale bar shell = 1 mm. Scale bar protoconch = 100 μm.

Satondella tabulata from the western Indian Ocean to the Western Pacific, has more numerous axials, a distinct keel bordering the umbilicus, and the adumbilical part of the aperture shows a narrow shield (absent in *Sat. senni*).

Distribution. Easter Island, Chile (Fig. 546).

Specimen records.

Chile, Easter Island. 70–100 m, Hanga Roa, 27.145°S, 109.440°W (DLG 1335, 1).

Remarks. The protoconch sculpture of *Sat. senni* is unique among the known *Satondella* species. The protoconch sculpture is most similar to the single spiral found on fresh specimens of *Sin. plicata* and the multiple spirals of *Sin. globosa*.

Figure 546. Distribution of *Satondella senni*. ☆ type locality. ● specimen record.

Satondella tabulata (Watson, 1886)

Figures 547–550

Chresonymy.

Schismope tabulata Watson, 1886: 117–118, pl. 8, fig. 7 (Fig. 547).
Schismope tabulata: Kobelt, 1887: 428.
Schismope tabulata: Pilsbry, 1890: 62–63, pl. 68, figs 14–16 [copy figures Watson, 1886].
Schismope tabulata: Thiele, 1912a: 24, pl. 3, figs 16–18 [copy figures Watson, 1886].
Scissurella (*Incisura*) *tabulata*: Abbott, 1974: 17.
Scissurella (*Incisura*) *tabulata*: Ortiz-Corps, 1985: 7.
Scissurella s.l. *tabulata*: Lozouet, 1986: 112.
Schismope tabulata: Faber & Moolenbeek, 1987: 71.
Satondella tabulata: Geiger, 2003: 76.
Satondella tabulata: Luque *et al.*, 2011: 3–5, fig. 2D–I, K–L.

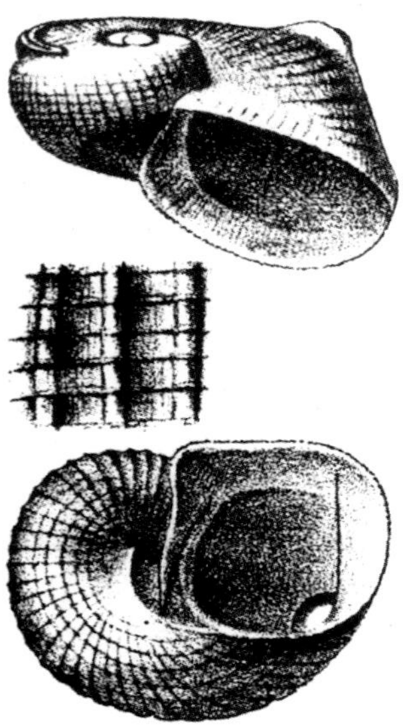

Figure 547. Original illustration of *Schismope tabulata*.

Misidentifications.

not *Sinezona tabulata*: Redfern, 2001: 2, pl. 1, fig. 7 [is *Sat. brasiliensis*].
not *Sinezona tabulate* [*sic*]: Hasegawa, 2005: 142 [referred to Redfern, 2001].
not *Satondella tabulata*: Geiger, 2008b: fig. 1.
not *Satondella tabulata*: Geiger *et al.*, 2008: fig. 12.2O [protoconch; is *Sat. brasiliensis*].
not *Satondella tabulata*: Segers *et al.*, 2009a: pl. 1, figs 2a–c [protoconch; is *Sat. brasiliensis*].
not *Scissurella* cf. *tabulata*: Barrera, 2001: 12, 38 [is most likely *Sat. brasiliensis*].
not *Satondella tabulata:* Tunnell *et al.*, 2010: 115, text-fig. [is *Sat. brasiliensis*].

Type material. One syntype (NHMUK 1887.2.9.409. Fig. 548A), subsequently designated as lectotype by Luque *et al.* (2011: 3). Possible topotypical syntypes (NHMUK 20050690, 1. Fig. 548B; NMW 155.158.607, 1. Fig. 548C: is *Sat. danieli*), 0.053 × 0.047", designated as paralectotypes by Luque *et al.* (2011).

Type locality. 390 fathoms [= 700 m], Recent, Station 24, March 25, 1873, 18°38'30"N, 65°5'30"W, off Culebra Island, West Indies [Puerto Rico], Pteropod ooze (OD).

Etymology. Most likely referring to the stepped, overall appearance of the shell.

Description. Shell medium-sized (to 2.0 mm. Figs 548, 549), trochiform, depressed. Protoconch of 0.75 whorl, pitted [erosion?], no apertural varix, apertural margin convex. Teleoconch I of 1.2–1.3 whorls, 23–30 axials increasing in strength from cords to lamellae, increasing in density with growth; spiral cord in position of selenizone distinct at onset of teleoconch I, fading out after approximately one whorl, forming strong knobs at intersection with axials; additional spiral lines mostly adumbilically from major spiral cord. Teleoconch II of 0.66 whorl, suture moderately impressed. Shoulder almost flat, 23–28 low axial lamellae, indistinct spiral lines increasing in number with growth, approximately 15 at apertural margin, forming thickenings at intersection with axials. Base moderately constricted below selenizone, sculpture similar to shoulder, approximately 15 more distinct spiral lines, increasing in strength to weak cords towards umbilicus, but weaker than axials; forming spirals running over axials, forming points close to umbilicus. Umbilicus open, bordered by distinct spiral cord, walls straight, smooth with very fine growth marks. Selenizone above periphery, keels moderately strong, moderately elevated; foramen with more elevated keels, but less pronounced compared to other *Satondella* species, closed anteriorly, continued by raphe. Aperture rounded, D-shaped, peristome flared, with distinct internal rim, roof overhanging.

Animal unknown.

Comparisons. *Satondella brasiliensis* from the Caribbean to Brazil has an overall more square-shaped shell, microhexagonal pattern on the protoconch, a narrower umbilicus, lacks a spiral cord in the position of the selenizone on teleoconch I, and lacks the internal rim of the aperture.

Satondella danieli from tropical and subtropical North Atlantic is proportionally more narrow, has a at least 50% fewer spiral lines on the base, and a smooth protoconch.

Satondella cachoi from the Indian Ocean to the Central Pacific has a taller overall shell and a much more narrow umbilicus with a distinct funiculus.

Satondella dantarti from the Indo-Malayan Archipelago to the Western Pacific has a flat spire and the umbilicus is produced by the rounded inner margins of the previous whorls.

Satondella senni from Easter Island has a protoconch with spiral sculpture, a wider umbilicus and lacks the lower adumbilical shelf on the apertural margin.

Satondella goudi from the western Indian Ocean and the Indo-Malayan Archipelago has the spiral lines on the base that remain the same strength around the umbilicus and, lacks the internal rim of the aperture.

Distribution. Western Indian Ocean to Central Pacific; Caribbean type locality in error [see Remarks] (Fig. 550).

Figure 548. *Satondella tabulata.* Recent, Station 24, March 25, 1873, 18°38'30"N, 65°5'30"W, off Culebra Island, West Indies [Puerto Rico], 390 fathoms [= 700 m]. **A.** Lectotype (NHMUK 1887.2.9.409). **B.** Paralectotype (NHMUK 20050690, 1). **C.** Paralectotype (NMW 155.158.607, 1; is *Sat. danieli*). Scale bars shell = 1 mm. Scale bars protoconch = 100 µm.

Specimen records.

Comoro Islands. 675 m, E Passe N'Zamboro, Mayotte, St. 27, 12.633°S, 45.198°E (MNHN, 1).

Philippines. 538 m, AURORA 2007 CP2750, 15.890°N, 121.893°E (MNHN, 3).

New Caledonia. 600–620 m, Norfolk Ridge, 24.900°S, 168.383°E (MNHN, 1). 320–350 m, South, BERYX 11, DW11, 24.733°S, 168.167°E (MNHN, 1). 569–583 m,

Figure 549. *Satondella tabulata.* **A.** 20–350 m, S New Caledonia, BERYX 11, DW11, 24.733°S, 168.167°E (MNHN, 1). **B.** 675 m, St. 27, E Passe N'Zamboro, Mayotte, Comoro Islands, 12.633°S, 45.198°E (MNHN, 1). Scale bars shell = 0.5 mm. Scale bars protoconch = 100 μm. Images by and with kind permission of E. Rolán.

Scissurellidae
Satondella

S Lansdowne, 21.100°S, 160.767°E (MNHN, 1). 538 m, between Lifou and Uvea, 20.617°S, 166.967°E (MNHN, 1). 470–480 m, Loyalty Basin, 20.583°S, 166.917°E (MNHN, 1). 435 m, 20.567°S, 166.900°E (MNHN, 1). 250–350 m, St. 40, 22.500°S, 166.400°E (MNHN, 1). 900–950 m, Jumeau East Bank, Norfolk Ridge, NORFOLK 2 DW2055, 23.650°S, 168.266°E (MNHN, 1). 555–565 m, Unfindable Bank, Norfolk Ridge, NORFOLK 2 DW2057, 24.666°S, 168.650°E (MNHN, 1).

Fiji. 700–707 m, 19.050°S, 178.500°E (MNHN, 1). 820–863 m, 17.300°S, 180.350°W (MNHN, 5).

French Polynesia. 300–650 m, Tahiti, Society Islands, TARASOC DW3484, 17.750°S, 149.383°W (MNHN, 1). 787–792 m, Mount Honu, Tarava Seamounts, TARASOC DW3340, 18.400°S, 154.150°W (MNHN, 1). 788–836 m, Mount Otaha, Tarava Seamounts, TARASOC DW3328, 18.766°S, 152.250°W (MNHN, 1). 614–664 m, Mount Ari'i Moana, Tarava Seamounts, TARASOC DW3309, 19.200°S, 151.583°W (MNHN, 1). 519–520 m, Mount Ari'i Moana, Tavara Seamounts, TARASOC DW3316, 19.233°S, 151.550°W (MNHN, 5). 480 m, Mount Ari'i Moana, Tarava Seamounts, TARASOC DW3319, 19.233°S, 151.533°W (MNHN, 12). 600–660 m, Mount Punu Taipu, Tarava Seamounts, TARASOC DW3302, 19.250°S, 150.950°W (MNHN, 2).

Remarks. The species has been highly confused. As noted by Luque *et al.* (2011), true *Sat. tabulata* has never been collected in the rather well-studied Caribbean. All records to that species in fact refer to *Sat. brasiliensis*. A species, which is shell-morphologically indistinguishable from the types of *Sat. tabulata*, is known from the Indo-Pacific. Shells of *Sat. tabulata* examined from Comoro Islands (Luque *et al.*, 2011: fig. 2E–I, K–L), New Caledonia, Norfolk Ridge, Loyalty Basin, and Fiji (Luque *et al.*, 2011: fig. 2D) are indistinguishable from the types of *Sat. tabulata*. They may attain a size of up to 2.4 mm, have also a solid shell with a flat, smooth protoconch of about one whorl, similar teleoconch I, teleoconch II of about 1 whorl, and similar sculpture.

The pitted protoconch sculpture is not known from any other scissurellid, but is rarely encountered in other lineages, such as Anatomidae (*An. austrolissa*), Larocheidae (*Larocheopsis* species), and Sutilizonidae (*Sutilizona* species). The presence of very similar pitting on the early teleoconch makes it more likely to be an artifact. Only additional material will shed more light on this question.

One paralectotype of *Sat. tabulata* (NMW 155.158.607. Fig. 548C) is not conspecific with the lectotype, and is here recognized as a specimen of *Sat. danieli* described from the Macaronesian Islands. It can readily be distinguished from *Sat. brasiliensis* from the Caribbean by the overall more globular shell shape and the less distinct constriction of the base below the selenizone.

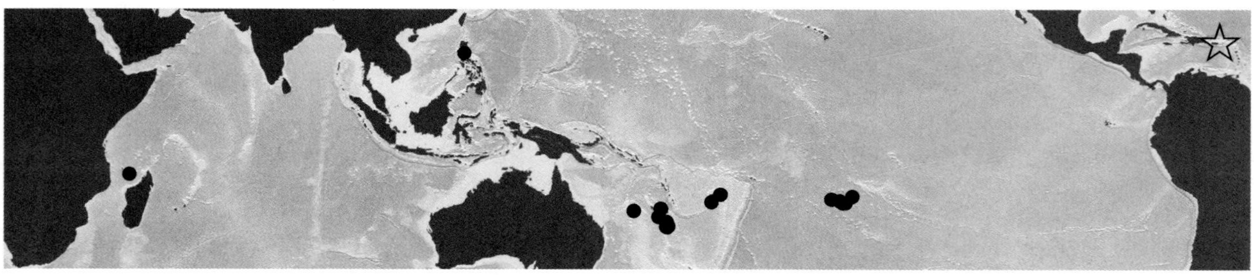

Figure 550. Distribution of *Satondella tabulata*. ☆ type locality. ● specimen record.

Satondella kondoi (Ladd, 1970) †

Figures 551–552

Chresonymy.

Sinezona kondoi Ladd, 1970: C2–3, pl. 2, figs 5–8.
Sinezona kondoi: Lozouet, 1986: 111.
Satondella kondoi: Luque *et al.*, 2011: 10–12, fig. 6E–K.

Type material. Holotype (USNM 650549. Fig. 552A), 1.2 × 1.7 mm. Paratype (USNM 650550, 1. Fig. 552B), 1.3 × 1.6 mm, from type locality.

Type locality. Eua, Tonga [21.380°S, 174.920°W], late Eocene (OD).

Etymology. Named in honor of Yoshio Kondo [1910–1990] of Bernice P. Bishop Museum (OD).

Description. Shell small (to 1.7 mm. Fig. 552), depressed, trochiform. Protoconch of 0.75 whorl, sculpture eroded. Teleoconch I of 1.1 whorls, 14 raised cords on last half of whorl; onset of fine spiral cords after 0.6 whorl, nine cords at onset of selenizone. Teleoconch II of 0.8 whorl, suture moderately impressed. Shoulder profile sinusoidal with selenizone, 17 strong raised cords crossed by nine fine cords running over axials. Base with constriction below selenizone, same number and strength of axials as on shoulder; 14 spiral cords running over axials, more or less evenly distributed over base. Umbilicus open, wide; straight, smooth walls at distinct angle to base. Aperture obscured by infill. Selenizone above periphery, keels of moderate strength, moderate elevation; keels surrounding foramen of moderate strength, strongly elevated.

Comparisons. *Satondella goudi* from the Recent western Indian Ocean and the Indo-Malayan Archipelago has an overall broader shell, weaker axials cords and the intersection of axials, and spirals does not produce the strongly elevated points.

Satondella cachoi from the Recent Indian Ocean to the Central Pacific shares the quadrate overall shape, but at the intersection of axial cords and spiral lines no elevated points are produced.

Satondella dantarti from the Recent Indo-Malayan Archipelago to the Western Pacific has a flat spire and the umbilicus is produced by the rounded portions of the inner whorls.

Satondella senni from the Recent of Easter Island has distinct spiral sculpture on the protoconch, a broader overall shell shape and weaker axial sculpture, and axial cords and spiral lines that do not form elevated points at their intersection.

Satondella tabulata from the Recent of the Indian Ocean to the Western Pacific and the Caribbean has a broader overall shell shape, the lower adumbilical corner of the aperture forms a distinct shelf, and it has about half as many axial cords (23 vs. 39) on the last whorl.

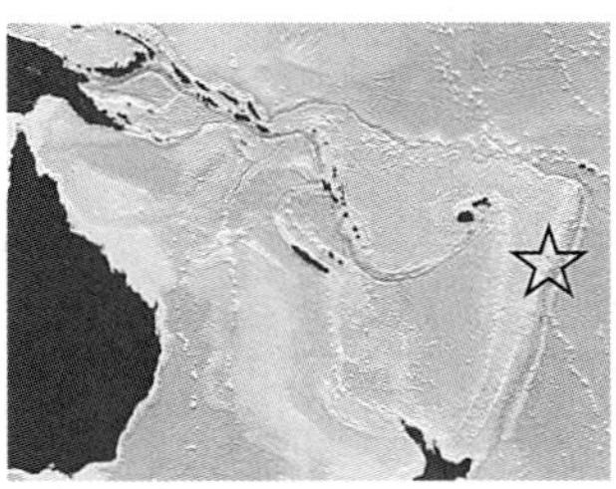

Figure 551. Distribution of *Satondella kondoi*. ☆ type locality.

Distribution. Known only from type locality (Fig. 551).

Figure 552. *Satondella kondoi*. Eua, Tonga [21.380°S, 174.920°W]. **A.** Holotype (USNM 650549). Arrow indicates suggestion of protoconch/teleoconch I boundary. **B.** Paratype (USNM 650550, 1). Scale bars shell = 1 mm. Scale bar protoconch = 100 μm.

Incisura Hedley, 1904

Chresonymy.

Incisura Hedley, 1904: 17–19.
Incisura: Thiele, 1912a: 33.
Incisura: Brookes Knight *et al.*, 1960: I221.
Incisura: Mountouchet, 1972: 2.
Incisura: Bandel, 1998: 48.
Incisura: Geiger, 2003: 37–38.

Synonym.

+ *Scissurona* Iredale, 1924: 215–216. ***Type species.*** *Scissurella rosea* Hedley, 1904 (OD). ***Etymology***. [not given].
Scissurona: Bandel, 1998: 47.

Type species. *Scissurella lytteltonensis* Smith, 1894 (M).

Other species. *Incisura auriformis* Geiger & Jansen, 2004.
Incisura geoffreyi (Laws, 1940) †.
Incisura remota (Iredale, 1924).
Incisura rosea (Hedley, 1904).

Etymology. Most likely referring to the short, open slit of the type species. Gender feminine.

Description. Shell naticiform to haliotiform, shoulder rounded, sculpture macroscopically smooth, with spirals predominating or broad reticulate sculpture, no spiral keel(s) on base. Protoconch with strong axials, no varix, aperture rounded, convex. Umbilicus with carina or funiculus. Slit open, or closed into foramen, selenizone on shoulder, starting at less than 0.75 teleoconch I whorl, keels very low. Operculate.

Radula. Lateral tooth 5 broadly enlarged, with asymmetrically serrated tip.

Distribution. Temperate Australia and New Zealand (Fig. 553). For specific localities, see individual species.

Geologic Range. Pliocene through Recent.

Comparisons. The usually macroscopically smooth sculpture and the protoconch with the strong axials distinguishes *Incisura* from any other scissurellid genus. The overall shell shape of *Incisura* and juvenile *Pseudorimula* is striking, particularly between *I. lytteltonensis* and *P. midatlantica* McLean, 1992, as illustrated by Warén & Bouchet (2001). However, the protoconchs are distinct; *Incisura* has broad axial ribs, whereas in *Pseudorimula* the protoconch shows tight spiral sculpture of undulating nature that is finely pitted.

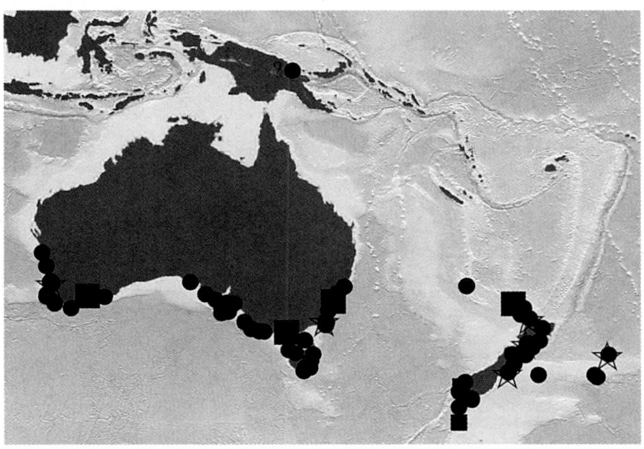

Figure 553. Global distribution of *Incisura* species. ☆ type locality. ● specimen record. ■ literature record. ? questionable record.

Differentiation of species. Species are distinguished by spire height, expansion rate of the whorls, presence or absence of spiral or axial sculpture, and foramen vs. open slit.

Remarks. *Incisura* was originally placed in Fissurellidae, a position maintained until at least 1979 (Powell, 1979), although others, such as Thiele (1912, 1931), had classified

Incisura in Scissurellidae. The fissurellid hypothesis was based on Boutan's illustrations of juveniles of *Fissurella,* which—as McLean (1984) pointed out—is a misidentification of an adult *Scissurella*. Accordingly, the placement of *Incisura* in Fissurellidae was based on a misinterpretation in Boutan's work that was accepted by Hedley (1904).

Incisura auriformis Geiger & Jansen, 2004

Figures 554–557

Chresonymy.

Sinezona auriform: Jansen, 1999: 55, figs 73–75.
Incisura 'auriform': Geiger, 2003: 75.
Incisura auriformis Geiger & Jansen, 2004b: 7–10, figs 1–2 [shell], 3a–c [radula], 3g [map].

Figure 554. *Incisura auriformis.* Holotype. 0 m, Cape Peron, 56 km S of Perth, Western Australia, Australia, 32.270°S, 115.685°E (AMS C.402714). Scale bar shell = 1 mm. Scale bar protoconch = 100 μm.

Type material. Holotype (AMS C.402714: mounted on AMS stub 4366. Fig. 554). Paratypes. 0–4 m, Cape Peron, 30 mi S of Perth, Western Australia, Australia, 32.27°S, 115.685°E (AMS C.402714, 8, 8: complete). 0–3 m, Garden Island, S of Perth, Western Australia, Australia, 32.233°S, 115.683°E (AMS C.402716, 18, 11: complete. Fig. 555B–C). Kilcarnup, N side of Margaret River, Western Australia, Australia, 33.950°S, 114.983°E (AMS C.406357, 1. Fig. 555A).

Type locality. 0 m, Point Peron, 30 mi S of Perth, Western Australia, Australia, 32.270°S, 115.685°E (OD).

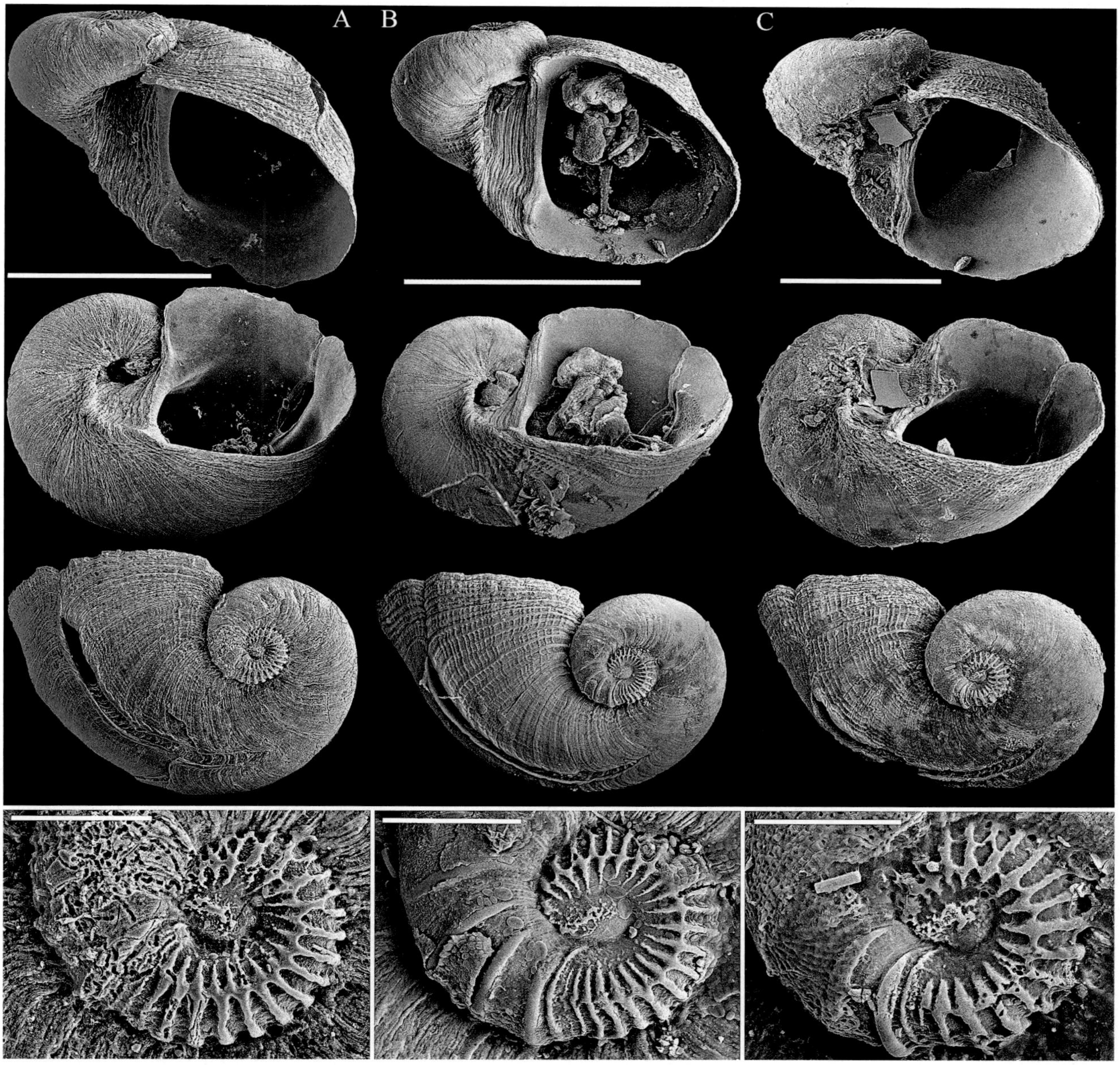

Figure 555. *Incisura auriformis.* Paratypes. **A.** Kilcarnup, N side of Margaret River, Western Australia, Australia, 33.950°S, 114.983°E (AMS C.406357, 1). **B–C.** 0 m, Garden Island, S of Perth, Western Australia, Australia, 32.233°S, 115.683°E (AMS C.402716, 11, 18: complete). Scale bars shell = 1 mm. Scale bars protoconch = 100 µm.

Etymology. Auris: Latin for ear; -formis: Latin shaped like, referring to the ear-shape of the shell (OD).

Description. Shell small to medium-sized (1.0 mm. Figs 554, 555), naticiform. Protoconch 0.75 whorl, with distinct axials, apical area of axials connecting forming spiral cord, no varix, aperture convex. Teleoconch I of 0.8–0.9 whorl, 2–3 axial cords immediately following protoconch, otherwise smooth. Teleoconch II of approximately 0.5 whorl, suture moderately to deeply impressed. Shoulder with approximately 20 fine spiral lines, irregular low axial cords. Base smooth, with spiral lines towards umbilicus. Umbilicus narrow, deep, walls almost vertical, forming shield adjacent to aperture; umbilical walls at strong angle to base, bordered by strong periumbilical cord. Selenizone above periphery; lunules irregular; keels hardly developed; foramen narrow, lanceolate; raphe anterior to foramen as long or longer than foramen in fully mature specimens. Aperture subrotund, adumbilical portion straight, parallel to coiling axis. Operculum round, smaller than outline of apertural rim.

Radula (Fig. 556). Rachidian tooth with triangular tip, apical cusp largest, 4–6 cusps in straight row on each side. Lateral teeth 1–3 similar of decreasing size, apical cusp largest, 4–2 cusps along outer edge of tip. Lateral tooth 4 reduced, hook-shaped. Lateral tooth 5 broadened, enlarged, with a dozen cusps of slightly decreasing size. Inner marginal teeth with broad tip, apical cusp largest, approximately seven cusps along outer margin, approximately 1–2 cusps along inner margin of tip; shafts without food groove. Outer marginal teeth spoon-shaped, tip with many fine bristles. Radular interlock of central field moderate.

Comparisons. *Incisura remota* from southern Australia has an open slit and an indistinct apertural shelf at the end of the periumbilical cord.

Incisura rosea from New Zealand has distinct spirals on base from the level of the suture to the umbilicus, lacks the periumbilical cord, and the protoconch has more dense axials that extend into the inner portion of the protoconch.

Incisura lyttletonensis from New Zealand has an open slit and lacks the thickened border of the almost closed umbilicus.

Distribution. Western Australia, South Australia, 0–4 m depth (Fig. 557).

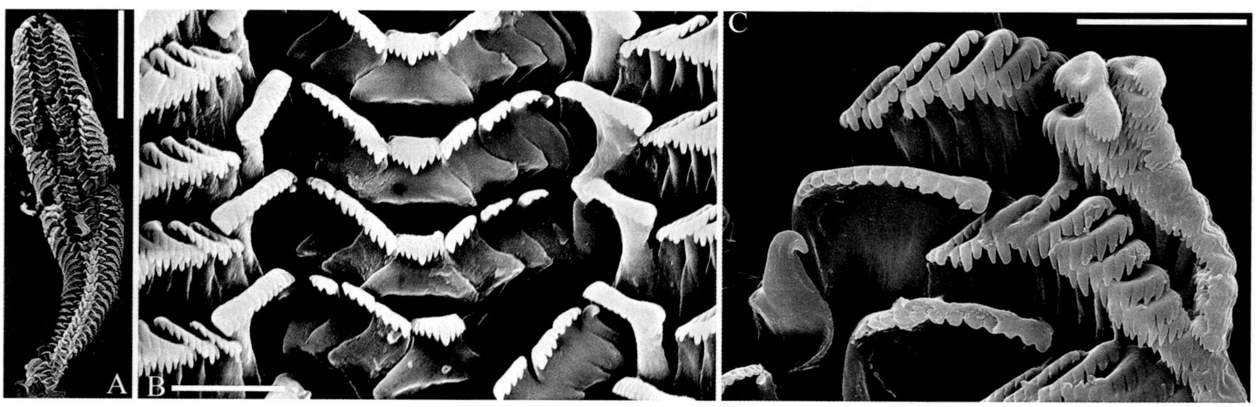

Figure 556. *Incisura auriformis.* Radula from paratype. 0 m, Garden Island, S of Perth, Western Australia, Australia, 32.233°S, 115.683°E (AMS C.402716, 11, 18: complete). **A.** Overview of entire radula. Scale bar = 100 µm. **B.** Central field enlarged, with hook-shaped lateral tooth 4. **C.** Hook-shaped lateral tooth 4, broad lateral tooth 5, and marginal teeth. Scale bars B–C = 10 µm.

Specimen records.

Australia, South Australia. Kangaroo Island, Emu Bay, 35.583°S, 137.517°E (AMS C.404523, 4: complete).

Australia, Victoria. 0 m, Western Port Bay, West Head, Flinders, 38.508°S, 145.020°E (LACM 79-47, 1). 0 m, Green Island, Port Fairy, 34.462°S, 137.404°E (MV ex F178372, 1).

Australia, Western Australia. S Cowaramup, 33.883°S, 114.983°E (WAM, 1). 0–2 m, Mississippi Bay, 30 mi E of Esperance, 33.983°S, 122.283°E (AMS C.402715, 4). Kilcarnup, N side of Margaret River, 33.950°S, 114.983°E (AMS C.406357, 1). Hopetoun, 33.867°S, 120.133°E (WAM, 3). Yallingup, 33.650°S, 115.017°E (AMS C.404989, 1: complete). Penguin Island, 32.300°S, 115.683°E (WAM, 1). 0–4 m, Cape Peron, 30 mi S of Perth, 32.270°S, 115.685°E (AMS C.402714, 8, 8: complete). 2–3 m, Cape Peron, 30 mi S of Perth, 32.267°S, 115.683°E (AMS C.402960, 1). 0–3 m, Garden Island, S of Perth, 32.233°S, 115.683°E (AMS C.402716, 18, 11: complete). Garden Island, 32.233°S, 115.683°E (WAM, 4; WAM, 3). Perkin Point, Garden Island, 32.233°S, 115.683°E (WAM, 2). 4–8 m, Carnac Island, W side, off Freemantle, 32.117°S, 115.667°E (AMS C.402961, 2, 8: complete). Fish Hook Bay, Rottnest Island, 32.017°S, 115.450°E (WAM S13389, 12: complete). Outer channel, The Basin, Rottnest Island, 32.000°S, 115.500°E (WAM, 1). The Basin, Rottnest Island, 32.000°S, 115.500°E (WAM, 25: complete). Bathurst Point, Rottnest Island, 31.983°S, 115.533°E (WAM, 2; WAM, 1). Green Head, 30.072°S, 114.967°E (WAM, 1).

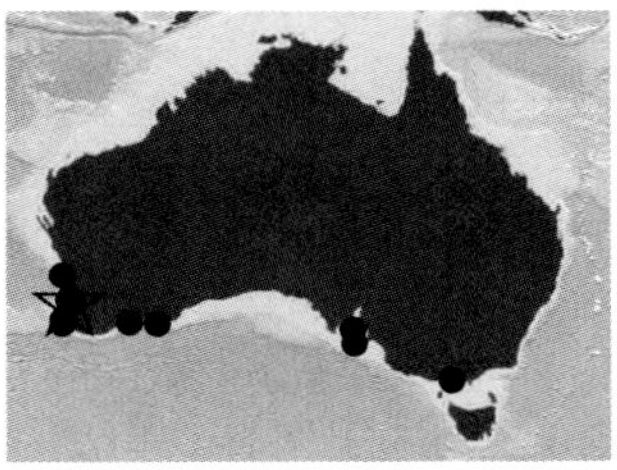

Figure 557. Distribution of *Incisura auriformis*. ☆ type locality. ● specimen record.

Incisura lytteltonensis (Smith, 1894)

Figures 558–563

Chresonymy.

Scissurella lytteltonensis Smith, 1894: 57–58, pl. 7, figs 1–2.
Scissurella lytteltonensis: Suter, 1898: 259.
Scissurella lyttletonensis [*sic*]: Sowerby & Fulton, 1903: 53.
Scissurella lytteltonensis: Hutton, 1904: 83.
Incisura lytteltonensis: Hedley, 1904: 91–92, fig. 18.
Incisura lytteltonensis: Murdoch, 1905: 231.
Incisura lytteltonensis: Iredale & Mestayer, 1908: 412.
Incisura lytteltonensis: Iredale, 1908a: 381–382.
Incisura lytteltonensis: Iredale, 1908b: 392.
Incisura lytteltonensis: Iredale, 1908c: 406.
Incisura lytteletonensis [*sic*]: Horst & Schepmann, 1908: 487.
Incisura lytteltonensis: Suter, 1909: 8.
Incisura lytteltonensis: Bourne, 1910: 1–47, pls. 1–5.
Incisura lytteltonensis: Thiele, 1912a, 33, pl. 1, fig. 3 [radula] pl. 4, fig. 8–10 [shell].
Incisura lytteltonensis: Suter, 1913: 98, pl. 6, fig. 15.
Incisura lytteltonensis: Finlay, 1924: 517.
Incisura lytteltonensis: Powell, 1940: 208.
Incisura lytteltonensis: Fleming, 1950: 25.
Incisura lytteltonensis: Powell, 1955: 48.
Incisura lytteltonensis: Dell, 1960: 145
Incisura (*Incisura*) *lytteltonensis*: Wenz, 1961: 173–174, fig. 273 [copy Smith].

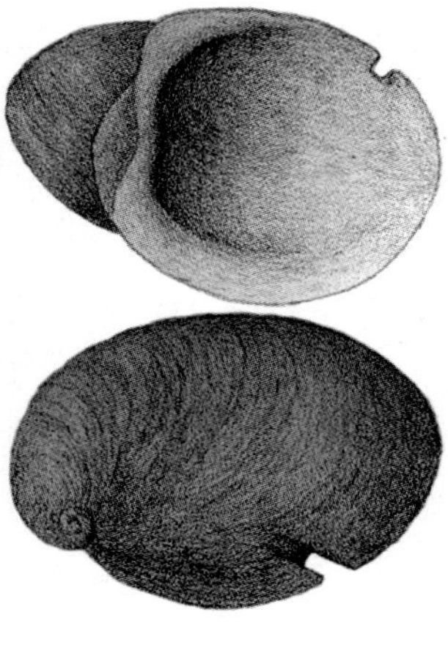

Figure 558. Original illustration of *Incisura lytteltonensis* Smith, 1894.

Scissurellidae
Incisura

Incisura (*Incisura*) *lytteltonensis*: Batten, 1975: fig. 5.
Incisura lytteltonensis: Powell, 1979: 38, text-fig. 3, fig. 9.
Incisura lytteltonensis: Lozouet, 1986: 114, fig. 31A.
Incisura lytteltonensis: McLean, 1989: figs 6B [shell], E [protoconch], G [radula].
Incisura lytteltonensis: Beu & Maxwell, 1990: 402.
Incisura lytteltonensis: Hedegaard, 1990: 74, 145.
Incisura lytteltonensis: Spencer & Willan, 1995: 13.
Incisura lytteltonensis: Geiger, 2003: 75, fig. 10 [shell, operculum, radula].
Incisura lytteltonensis: Geiger, 2004: text-fig. p. 7.
Incisura lytteltonensis: Geiger, 2006c: text-fig. 6.
Incisura lytteltonensis: Spencer *et al.*, 2009: 202.
Incisura lytteltonensis: Raven & Bracegirdle, 2010: 10, text-fig.

Misidentifications.

not *Incisura lytteltonensis*: Brookes Knight *et al.*, 1960: I221, fig. 136. 3 [is *I. rosea*]

Type material. Syntypes (NHMUK1893.5.27.42–48, 7). The specimen illustrated here is designated as the lectotype (NHMUK1893.5.27.42. Fig. 559) for the express purpose of taxon stabilization in case other non-conspecific syntypes should be found. Paralectotypes from type locality (NHMUK1893.5.27.43–48, 6).

Type locality. Lyttelton Harbour, New Zealand [43.617°S, 172.733°E] (OD).

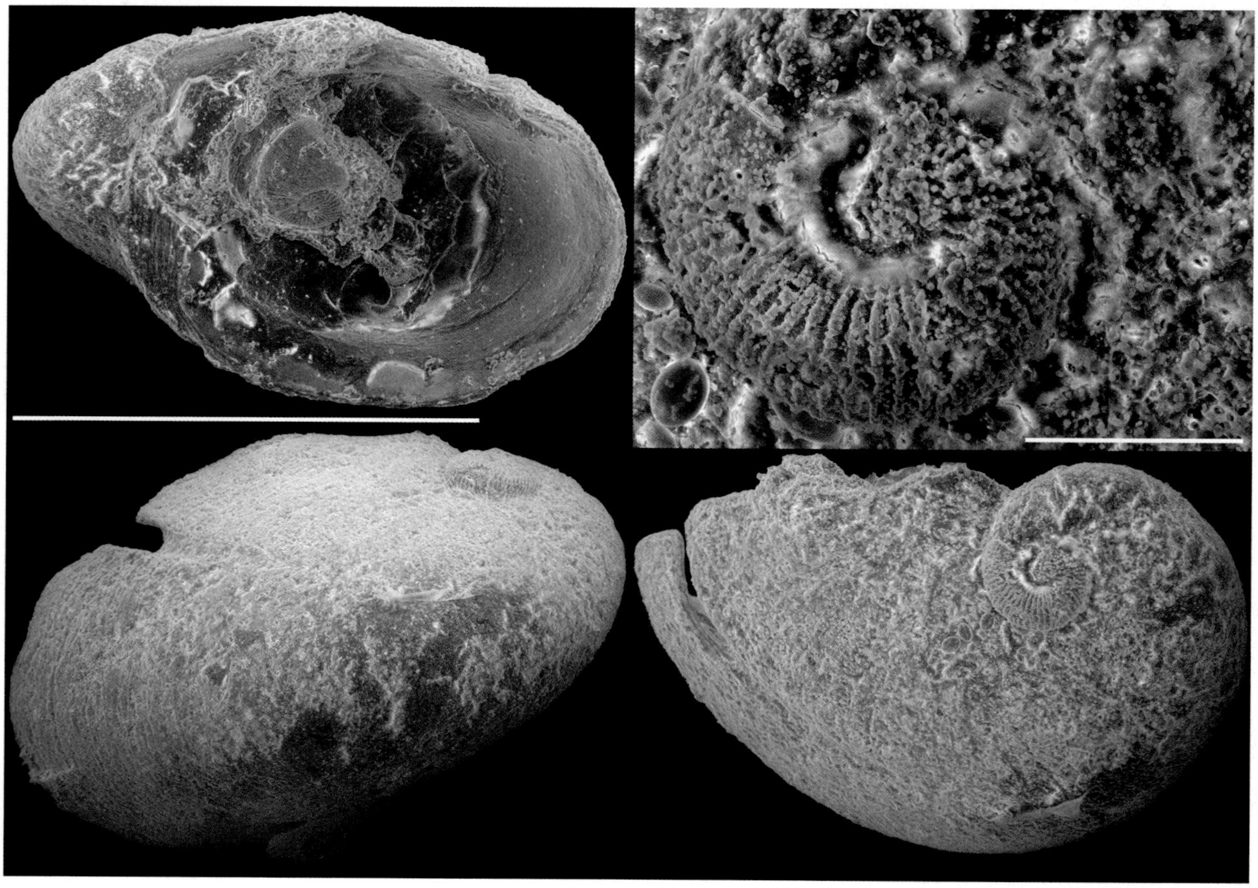

Figure 559. *Incisura lytteltonensis*. Lectotype. Lyttelton Harbour, New Zealand [43.617°S, 172.733°E] (NHMUK1893.5.27.42). Scale bar shell = 1 mm. Scale bar protoconch = 100 μm.

Etymology. Named for its type locality, Lyttelton Harbour, New Zealand.

Description. Shell medium-sized (to 1.3 mm. Figs 559, 560), neritiform. Protoconch of one whorl, fine axials with spiral connecting band, no apertural varix, apertural margin convex. Teleoconch I of 0.75 whorl, almost smooth, very fine axial growth lines. Teleoconch II of 0.25 whorl, suture deeply impressed. Sculpture as on teleoconch I. Selenizone in outer third of shoulder, very short, keels extremely low, lunules faint, irregular. Slit open, margins parallel. Aperture flared, with expanded shelf on upper and lower corner of adumbilical region. Umbilicus open, narrow, no funiculus.

Radula (Fig. 561). Ribbon almost symmetrical. Rachidian tooth triangular, tip V-shaped, nine cusps, central cusp largest. Lateral teeth 1–3 similar, triangular base, shafts slightly convex, tip with approximately four cusps. Lateral tooth 4 reduced,

Figure 560. *Incisura lytteltonensis.* **A–C.** 0 m, Great Expedition Bay, S end of Rarawa Beach, North Island, New Zealand, 34.717°S, 173.083°E (NMNZ M.134318, 60). Scale bars shell = 1 mm. Scale bars protoconch = 100 µm.

hook-shaped. Lateral tooth 5 broadened, tip with approximately 15 cusps. Marginal teeth with long shafts, no food groove, tips broadened, with approximately ten cusps. Radular interlock moderate.

Operculum (Fig. 562) round, thin, multispiral, with distinct suture between adjacent volutions.

Comparisons. *Incisura rosea* from New Zealand has a teleoconch II of up to 0.66 whorl, the selenizone is in the outer fifth of the shoulder, the body whorl is expanding less rapidly, and has an umbilical callus.

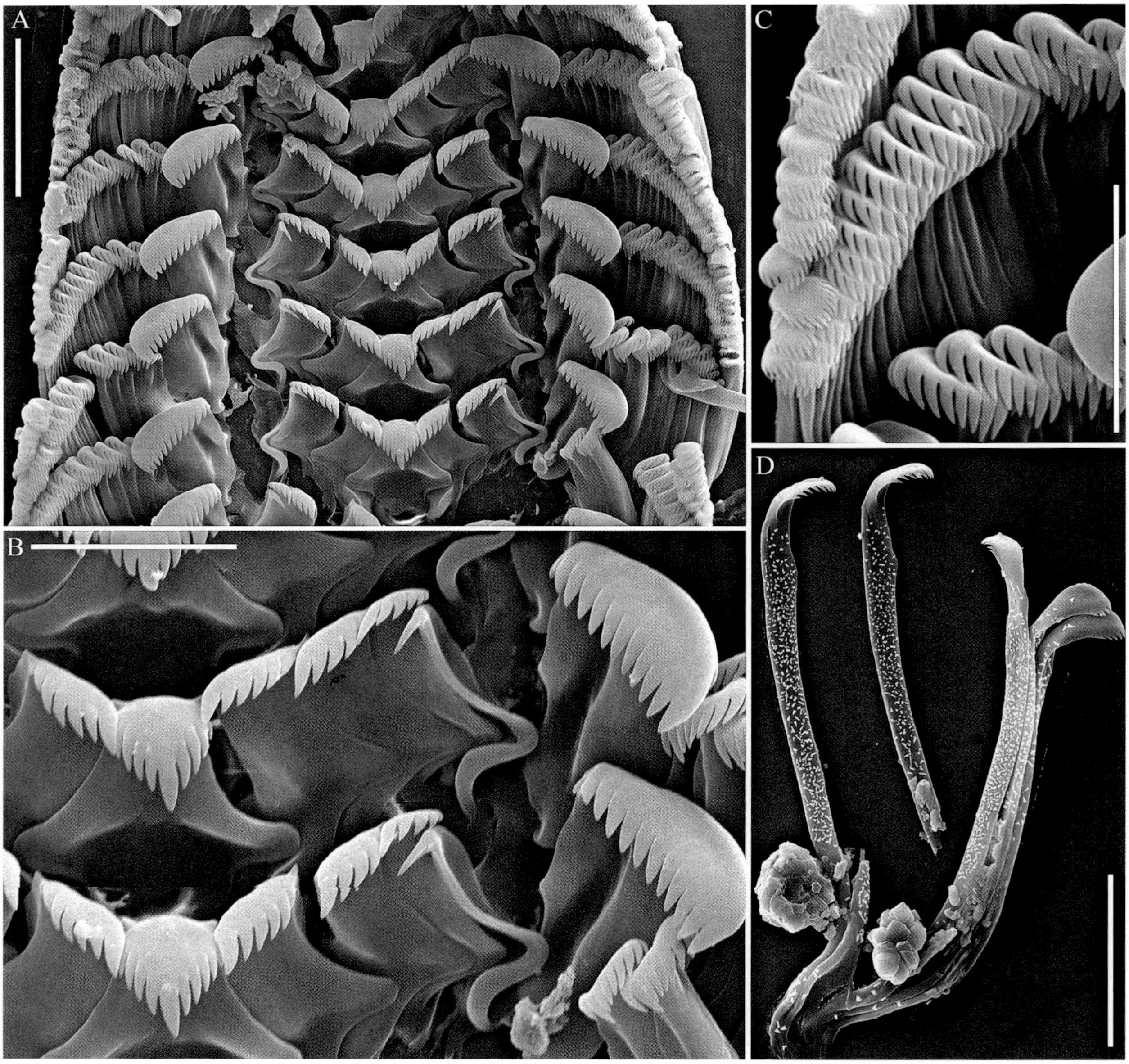

Figure 561. *Incisura lytteltonensis.* Radula. 0 m, Wellington, North Island, New Zealand, 41.333°S, 174.800°W (DLG 283, 40: complete). **A.** Full width of radula. Scale bar = 20 µm. **B.** Central field enlarged. Scale bar = 10 µm. **C.** Tips of marginal teeth. Scale bar = 10 µm. **D.** Entire marginal teeth. Scale bar = 10 µm.

Incisura fossilis from the Tertiary of New Zealand has indistinct to pronounced reticulate sculpture, the selenizone in the outer fifth of the shoulder, and a teleoconch II of up to 0.66 whorl.

Habitat. The species is found intertidally on algae, such as *Cystophora* sp. (Iredale, 1908a).

Distribution. New Zealand (Fig. 563).

Specimen records.

New Zealand. 133 m, Wanganella Bank, Norfolk Ridge, 32.653°S, 167.528°E (NMNZ M.177153, 1). 0 m, Great Expedition Bay, S end of Rarawa Beach, North Island, 34.716°S, 173.083°E (NMNZ M.134318, 60). 33 m, NE side Stephensons Island, 34.967°S, 173.783°E (AMS C.461496, 1). 12–16 m, Hole in Rock, Te Anaputa Island, Cavalli Island, North Island, 35.000°S, 173.967°E (LACM 77-125, 1). 41–43 m, Poor Knight Island, N end of North Island, 35.442°S, 174.733°E (LACM 65-165, 1). 79 m, off Hen and Chicken Islands, Gulf of Hauraki, 35.917°S, 174.750°E (AMS C.ex C.461508, 1; AMS C.461498, 1). 57 m, near Okahari Point, 1 km W of Leigh Marine Laboratory, Leigh, North Island, 36.263°S, 174.797°E (LACM 87-83, 7). 18 m, W side Goat Island, off Leigh Marine Laboratory, Leigh, North Island, 36.267°S, 174.792°E (LACM 87-84, 2). 0 m, near Leigh Marine Laboratory, Leigh, North Island, 36.270°S, 174.775°E (LACM 87-82, 8). McGregor Bay, 36.750°S, 175.467°E (AMS C.461494, 8). Ohope Beach, large bend of Wainui Stock Road, North Island, 37.975°S, 177.056°E (NMNZ M.096492, 2). Cook Strait, 41.000°S, 174.500°E (AMS C.380189, 3; FLMNH 193330, 2). Wellington, Island Bay, 41.283°S, 174.783°E (MCZ 083705, 20). Lyall Bay, Wellington, 41.329°S, 174.795°E (GNS RM 5964, 4). 0 m, Lyall Bay, Wellington, 41.331°S, 174.798°E (GNS RM 1064, 11). Lyall Bay, near Wellington, 41.333°S, 174.800°W (AMS C.406394, 35). 0 m, Wellington, North Island, 41.333°S, 174.800°W (DLG 813, 1; DLG 283, 40: complete). Lyall Bay, W side near Wellington, 41.333°S, 174.792°E (AMS C.380193, 1). Island Bay, Cook Strait, 41.333°S, 174.767°E (MV F 161689, 4). Lyall Bay, Wellington, 41.333°S, 174.800°E (USNM 680561, 1). 0 m, seaweed washings, Island Bay, Wellington, S coast, 41.345°S, 174.773°E (GNS RM 1061, 2). Houghton Bay, Wellington, 41.345°S, 174.785°E (AMS C.380197, 4; AMS C.380200, 20+). Island Bay, Wellington, 41.347°S, 174.773°E (AMS C.380192, 9, –: complete). E side of Island Bay, Wellington, 41.347°S, 174.775°E (AMS C.380188, 2; AMS C.380191, 19, –: complete). 0 m, E side Island Bay, Wellington, North Island, 41.347°S, 174.775°E (LACM 87-87, 17). 0 m, E side Island Bay, Wellington, 41.350°S,

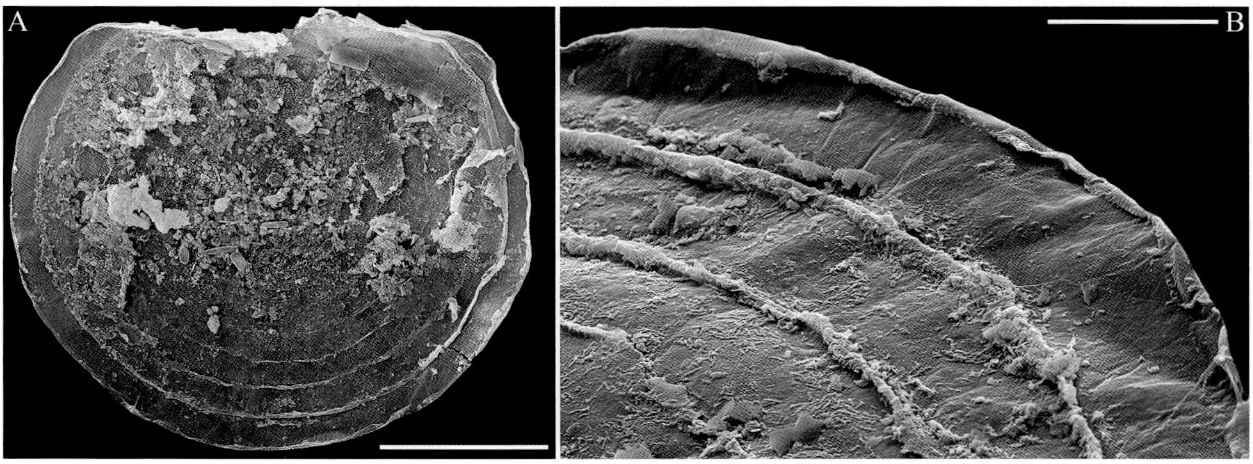

Figure 562. *Incisura lytteltonensis*. Operculum. 0 m, Wellington, North Island, New Zealand, 41.333°S, 174.800°W (DLG 283, 40: complete). **A.** Entire operculum. Scale bar = 100 µm. **B.** Oblique view of portion showing upturned outer margins of each volution. Scale bar = 20 µm.

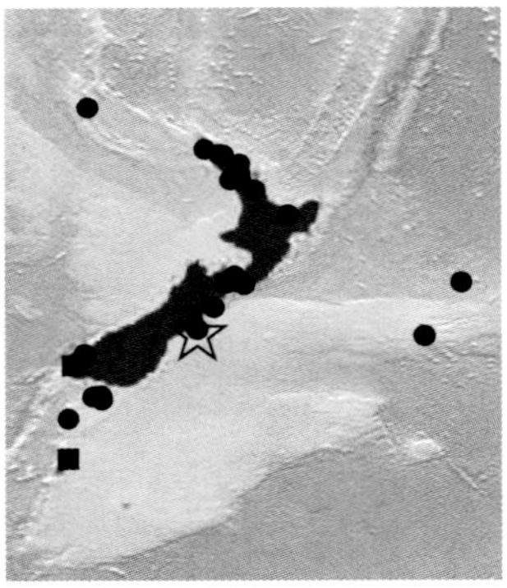

Figure 563. Distribution of *I. lytteltonensis*. ☆ type locality. ● specimen record. ■ literature record.

174.123°E (AMS, 10: complete). 0 m, E side of Island Bay, Wellington, 41.350°S, 174.783°E (AMS, 20: complete; AMS, 2: complete). Lyttelton Harbour, 43.600°S, 172.700°E (ZMB 47.034, 12; NHMW 22041, 8). Lyttelton, 43.600°S, 172.717°E (MHNG 14367, 5; USNM 109830, 22; USNM 680560, 1). Lyttelton Harbour, 43.607°S, 172.717°E (UMMZ, 7). Lyttelton Harbour, 43.616°S, 172.733°E (MV F 161690, 5; NHMUK, 7). Lyttelton Harbour, Christchurch, 43.619°S, 172.717°E (GNS RM 5962, 100). Lyttelton Harbour, 43.619°S, 172.719°E (RMNH, 7). 0 m, Waitangi, Chatham Islands, 43.950°S, 176.567°W (GNS RM 1063, 25; LACM 33-1, 13). Waitangi, Chatham Island, 43.950°S, 176.533°W (USNM 650647, 2). 62 m, New Golden Hind, entrance to George Sound, Fiordland, 44.839°S, 167.348°E (GNS RM 1062, 6). Lyall Bay, 45.250°S, 166.983°E (LACM 54536, 5). Halfmoon Bay, Stewart Island, 46.883°S, 168.133°E (AMS C.380187, 1). Bathing Beach, Stewart Island, 46.900°S, 168.117°E (AMS C.380190, 20+). 0 m, Mason's Bay, Stewart Island, 46.933°S, 167.757°E (LACM 47-3.1, 30). Ocean Beach, Stewart Island, 46.968°S, 168.177°E (RMNH, 18). 55 m, Stewart Island, Port Adventure, 47.083°S, 168.183°E (AMS C.380196, 1). 90 m, Captain Bollons, Snares Islands, 48.016°S, 166.567°E (GNS RM 5963, 2). 91 m, off Snares Islands, S of South Island, 48.017°S, 166.583°E (AMS C.380195, 3).

New Zealand, fossil. GS14697, high terrace remnant (Hawaran, oxygen isotope stage 5c?, late middle Pleistocene) on NE corner of Haumuri Bluff, Marlborough, NE South Island, 42.552°S, 173.519°E (GNS TM, 5). GS15119, R28/f62, Paua kill site on 1855 intertidal slope c. 400 m E of Turakirae Head, S Wellington, Holocene (co-seismic uplift of 1855), 41.433°S, 174.947°E (GNS TM, 16). GS12163, Moutapu Point, NE Pitt Island, Chatham Islands, Whenuataru Tuff, Waipipian-Mangapanian (middle to late Pliocene), 36.239°S, 174.293°W (GNS TM, 2).

Literature records.

New Zealand. 50 fathoms [= 91 m], near Snares Islands (Suter, 1913).—12 m, Northport, ¼ mi NE of anchorage, 49.983°S, 166.567°E. 116 m, Daggs Sound, inside a rock bar, 45.383°S, 166.767°E. 94 m, Anchorage Arm, Daggs Sound, 45.383°S, 166.767°E (Fleming, 1950).

New Zealand, fossil. Nukumaruan, early Pleistocene of New Zealand (Powell, 1979; Beu & Maxwell, 1990).

Remarks. Powell (1955) cited a single holotype in the NHMUK. According to ICZN Article 75.4, this does not constitute a valid lectotype designation as no specific specimen was selected.

The operculum with the upturned outermost margin is unique among *Incisura* species. It is neither seen in *I. auriformis* nor in *I. remota*, the only two species for which data are available.

Some specimens have epizoic serpulid worms, and some are partially covered with encrusting coralline red algae (e.g., NMNZ M.027418). Lots with the largest number of specimens are from Stewart Island.

Incisura remota (Iredale, 1924)

Figures 564–567

Chresonymy.

Scissurona rosea remota Iredale, 1924: 215, not illustrated.
Scissurona remota: Cotton & Godfrey, 1933: 21–22, pl. 1, fig. 9.
Scissurona remota: Cotton & Godfrey, 1938b: 2.
Scissurona rosea remota: Iredale & McMichael, 1962: 30.
Scissurella remota: Macpherson & Gabriel, 1962: 34.
Scissurona rosea remota: Lozouet, 1986: 114.
Scissurella rosea remota: Jansen, 1999: 52, figs 31–33.
Scissurella rosea remota: Jansen, 1995: 11.
Incisura rosea remota: Geiger, 2003: 75, fig. 10 [shell, radula].
Incisura remota: Geiger & Jansen, 2004b: 10–14, figs 3d–f [radula], 3g [map], 4, 5a–b [shell].
Incisura remota: Geiger, 2008b: fig. 1.

Synonyms.

+ *Scissurona vincentiana* Cotton, 1945: 150, pl. 12, figs 5–6 (Fig. 564). ***Type material***. Holotype (SAM D.14109), 1.2 × 1.35 mm. ***Type locality***. Glenelg, Gulf of St. Vincent, South Australia, Australia [34.983°S, 138.517°E](OD). ***Etymology***. Named for its type locality in the Gulf of St. Vincent.
Scissurona vincentiana: Cotton, 1959: 46–47, fig. 17 [copy figures Cotton, 1945].
Scissurona vincentiana: Cotton, 1963: pl. 9, fig. 165.
Scissurella (?) *vincentiana*: Lozouet, 1986: 108.
Incisura vincentiana: Wilson, 1993: 46.
Incisura vincentiana: Geiger, 2003: 75.

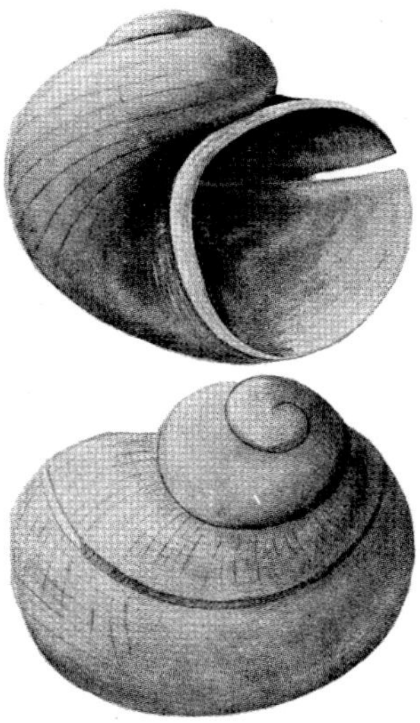

Figure 564. Illustration of *Scissurona vincentiana* from Cotton (1959).

Misidentifications.

Scissurella obliqua: Pritchard & Gatliff, 1902: 181 [is *I. remota*].
Scissurella obliqua: Verco, 1910: 115–116 [is *I. remota*].
Scissurella rosea: May, 1921: 34, pl. 17, fig. 3 [is *I. remota*].
Scissurella rosea: May, 1923: pl. 17, fig. 3 [is *I. remota*].
Scisurona rosea: May & Macpherson, 1958: 21, pl. 17, fig. 3 [is *I. remota*].

Type material. Holotype (whereabouts unknown).

Type locality. Twofold Bay [New South Wales, Australia, 37.083°S, 149.900°E] (OD).

Etymology. Remotus, Latin for distant.

Description. Shell small to medium-sized (0.8 mm. Fig. 565), broadened. Protoconch of 0.8–0.9 whorl, distinct axials, broadened apically, forming spiral band, no varix, aperture convex to weakly sinusoidal. Teleoconch I usually slightly more than one whorl,

axials if present broad, decreasing in prominence from protoconch to teleoconch II and from suture to periphery; many shells with smooth appearance. Teleoconch II of up to 0.5 whorl, suture deeply impressed. Shoulder smooth, sometimes with faint axial lines near suture. Base usually smooth, sometimes with partial axials near umbilicus. Umbilicus open, deep, walls steep, bordered by keel, walls of umbilicus and base at distinct angle. Selenizone above periphery, keels faint, lunules indistinct; slit open, margins parallel. Aperture rounded.

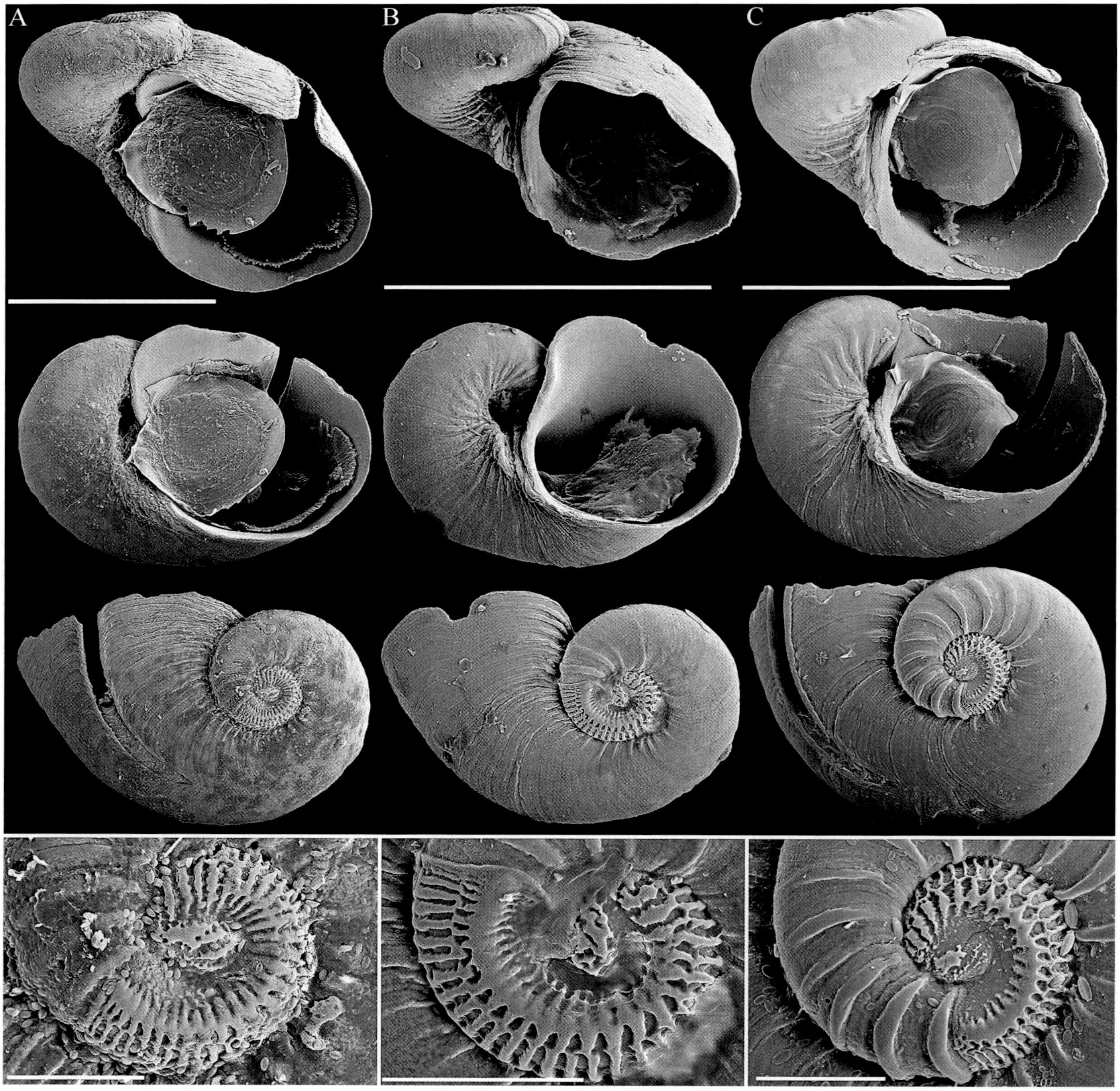

Figure 565. *Incisura remota*. **A.** Kangaroo Island, N coast, Western River Cove, South Australia, Australia, 35.683°S, 136.967°E (AMS C.402700, 1). **B.** 8 m, Gabo Island, near Monument Bay, Victoria, Australia, 37.567°S, 149.917°E (AMS C.402701, 3). **C.** 22 m, Maria Island, Shoal Halfmoon Bay, Tasmania, Australia, 42.075°S, 148.075°E (AMS C.402699). Scale bars shell = 0.5 mm. Scale bars protoconch = 100 μm.

Scissurellidae *Incisura*

Operculum (Fig. 565) round, central nucleus, smaller than aperture at apertural rim, approximately 10 indistinct growth rings visible.

Radula (Fig. 566). Rachidian tooth with triangular tip, apical cusp largest, four cusps on each side in straight line. Lateral teeth 1–3 similar with asymmetrical tips, apical cusp largest, 3–4 cusps along outer edge of tip. Lateral tooth 4 reduced, hook-shaped. Lateral tooth 5 broadened, enlarged, series of approximately 10 cusps slightly decreasing in size. Inner marginal teeth broadened, three cusps along inner margin, 5–6 cusps along outer margin; shaft without food groove. Outer marginal teeth spoon-shaped tips with many fine bristles. Radular interlock of central field moderate.

Comparisons. *Incisura auriformis* from Western Australia and South Australia has a strong umbilical shield, even as a juvenile with open slit, and closes the slit to form a foramen in adults.

Incisura rosea from New Zealand has distinct spiral bands on the base from the level of the suture towards the umbilicus and an umbilical callus.

Distribution. Temperate Australia (Fig. 567).

Specimen records.

Australia, Tasmania. 6 m, Bass Strait, Deal Island, 39.450°S, 147.283°E (AMS C.402777, 4). 30–50 m, Murray Pass, Deal Island, Bass Strait, 39.467°S, 147.300°E (AMS C.404888, 1: complete). 10–30 m, Little Squally Cove, Deal Island, Bass Strait, 39.500°S, 147.333°E (AMS C.402781, 2, 1: complete). Stanley, 40.765°S, 145.294°E (MV ex F161661, 1). Ulverstone, Picnic Point, 41.133°S, 146.167°E (AMS C.402697, 4). 5 m, St. Helens

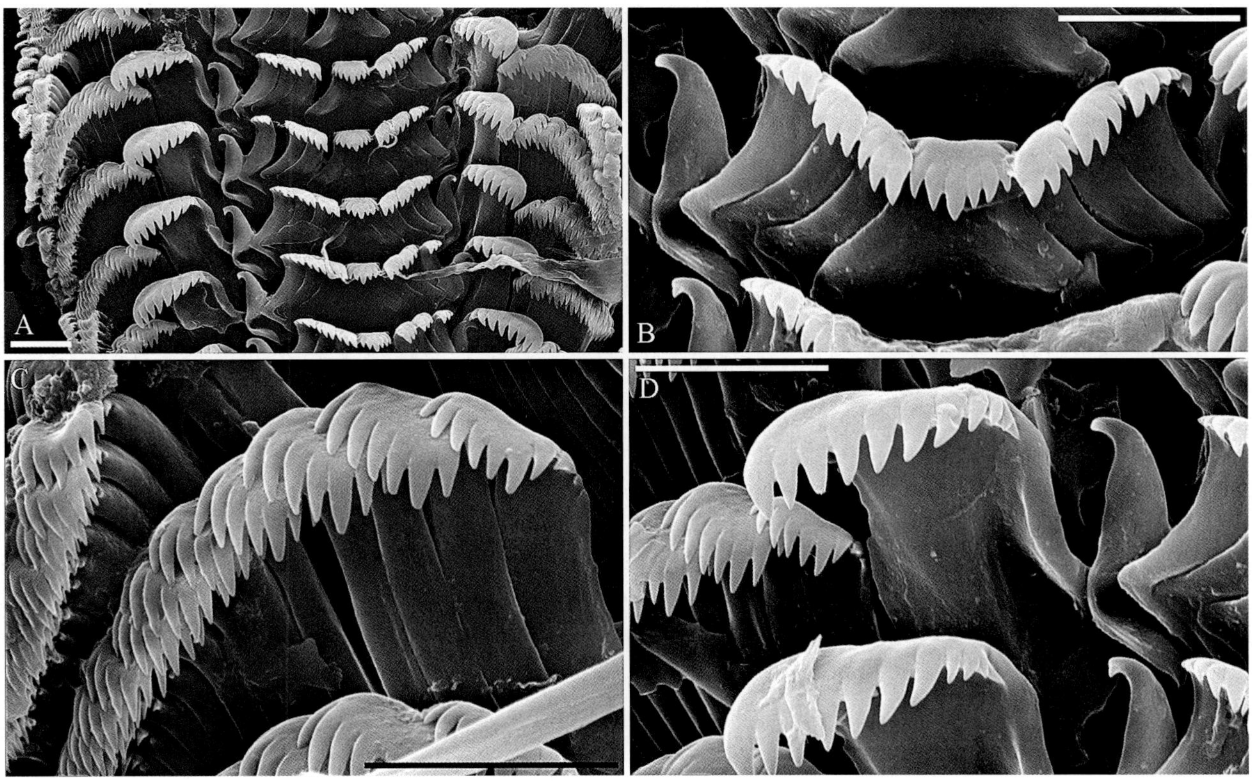

Figure 566. *Incisura remota*. Radula. Wimbie Beach, Batemans Bay, New South Wales, Australia, 35.783°S, 150.233°E (AMS C.402698, 13). **A.** Whole width of radula. **B.** Central field; notice hook-shaped lateral tooth 4. **C.** Enlarged lateral tooth 5 with hook-shaped lateral tooth 4. **D.** Marginal teeth. Scale bars = 10 µm.

Point, 41.280°S, 148.345°E (AMS C.404219, 1: complete). 22 m, Maria Island, Shoal Halfmoon Bay, 42.075°S, 148.075°E (AMS C.402699, 7, 9: complete). Roches Beach, W side of Frederick Henry Bay, 42.900°S, 147.500°E (AMS C.402780, 2). Primrose Point, E side of Frederick Henry Bay, 42.900°S, 147.683°E (AMS C.402782, 20+, 5: complete; AMS C.402784, 7, 7: complete; AMS C.402785, 12, 5: complete). Eaglehawk Bay, Eaglehawk Neck, 43.017°S, 147.917°E (AMS C.17704, 3). Pirates Bay, Eaglehawk Neck, 43.017°S, 147.933°E (AMS C.402778, 1; AMS C.404889, 15: complete; AMS C.404890, 3: complete). Eaglehawk Neck, N end Pirates Bay, 43.017°S, 147.933°E (AMS C.404892, 18: complete; AMS C.402643, 2; AMS C.402779, 1; MV F165431, 3: complete). Tinderbox Bay, S of Hobart, 43.050°S, 147.333°E (AMS C.402786, 3). 0 m, Tinderbox Bay, S of Hobart, 473.050°S, 147.333°E (AMS, 6: complete). 8 m, S Bruny Island, Great Taylor Bay, 43.450°S, 147.167°E (AMS C.404903, 1: complete).

Australia, Victoria. E of Grassy Creek (MV F 133038, 1: complete). 0 m, Green Island, Port Fairy, 34.462°S, 137.404°E (MV F178372, 2). 8–12 m, Gabo Island near Monument Bay, 37.566°S, 149.917°E (AMS, 2: complete; AMS C.402701, 3). 12–14 m, SSE side Gabo Island, 37.567°S, 149.917°E (AMS C.404218, 1: complete). Mallacoota, 37.567°S, 149.933°E (AMS C.50399, 5; AMS C.50565, 1). Near mouth of Shipwreck Creek, 37.620°S, 149.650°E (MV F171895 [decalcified], 1: complete). Port Philllip, NW of Popes Eye, 38.150°S, 144.867°E (WAM 13390, 1: complete). 3–4 m, St. Leonards Pier, 38.166°S, 144.717°E (MV F122396 [decalcified], 2: complete). 7–10 m, S Port Phillip Bay, 38.273°S, 144.697°E (MV F 160819, 1). 6–12 m, Popes Eye Bank, 38.274°S, 144.707°E (MV F162108 [decalcified], 1: complete; MV F162109, 4: complete). 6 m, S Port Philip Bay, 38.281°S, 144.633°E (MV F67854 [decalcified], 2: complete; MV F67629 [decalcified], 1: complete). Mornington Peninsula, Queenscliffe, 38.283°S, 144.667°E (AMS C.402795, 2; AMS C.402796, 3, 2: complete). 10–12 m, Popes Eye Bank, Port Phillip Bay, 38.283°S, 144.683°E (MV F159101, 4: complete). 6–10 m, Popes Eye Bank, Port Phillip Bay, 38.283°S, 144.683°E (MV F162110, 30: complete). 15 m, S Port Phillip Bay, 38.284°S, 144.668°E (MV F67156 [decalcified], 1: complete). Point Lonsdale, near Geelong, 38.300°S, 144.617°E (AMS C.402798, 1; AMS C.402799, 1; AMS C.404891, 1: complete). Portland, 38.341°S, 141.622°E (MV F. 157451, 10). 0 m, Reef Point, Port Fairy, 38.364°S, 140.663°E (MV F178376, 2). Flinders, N side of Kennon Cove, W side Western Port Bay, 38.367°S, 145.533°E (LSC, 1). Port Fairy, 38.383°S, 142.233°E (AMS C.402792, 7). Port Fairy, 0.25 mi W of town, 38.383°S, 142.233°E (AMS C.402794, 1, 1: complete; AMS C.404884, 1: complete; AMS C.404886, 2: complete). Rhyll, 38.463°S, 145.312°E (MV ex F 884, 1). Flinders, 38.483°S, 145.017°E (NMSZ 1961.61, 6). Flinders Ocean Beach, 38.483°S, 145.033°E (MV F 882, 100). N side of West Head, Flinders Ocean Beach, 38.483°S, 145.033°E (MV F 157452, 3). Western Port, West Head, near Flinders, 38.487°S, 145.025°E (AMS C.14474, 7). Loch Ard Gorge, 38.643°S, 143.056°E (MV F113334, 2: complete). Moonlight Beach, W of Cape Otway, 38.754°S, 143.211°E (MV F133028, 1: complete). 0 m, Maitland Beach, Bear Gully, 38.893°S, 145.982°E (MV F178379, 2).

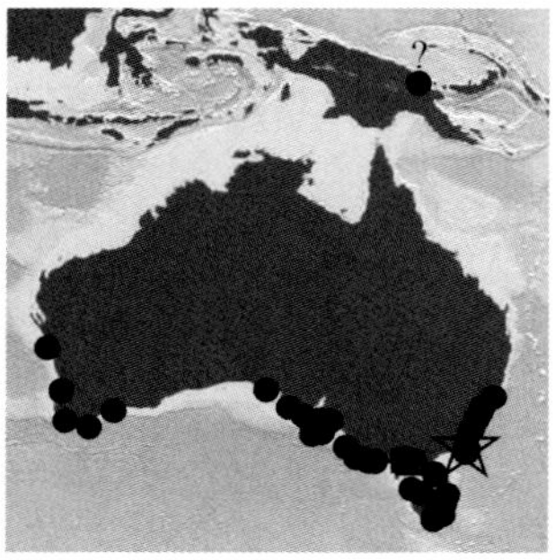

Figure 567. Distribution of *Incisura remota*. ☆ type locality. ● specimen record. ■ literature record. ? dubious record.

Australia, New South Wales. 46–73 m, off Port Stephens, 32.708°S, 152.250°E (AMS C.402956, 1). 0 m, Fingal Bay, Port Stephens, 32.750°S, 152.175°E (AMS ex C.461560, 1). 0 m, Birubi Beach, Nelson Bay Area, 32.783°S, 152.060°E (MV F178373, 1). Patonga, Broken Bay, 33.550°S, 151.267°E (AMS C.402616, 2). Sydney, Broken Bay, Pittwater, 33.617°S, 151.300°E (AMS C.402955, 1). Port Hacking, 34.067°S, 151.100°E (AMS C.403742, 1). Shell Harbour, 34.583°S, 150.867°E (WAM, 1). 0 m, S side of Ulladulla, 35.350°S, 150.483°E (AMS, 20: complete). Wimbie Beach, Batemans Bay, 35.783°S, 150.233°E (AMS C.402698, 13; AMS C.402773, 2). Twofold Bay, Eden, Aslings Beach, 37.050°S, 149.908°E (AMS C.403663, 4). 0 m, Eden Beach, 37.061°S, 149.910°E (MV F178380, 2). Twofold Bay, 37.083°S, 149.900°E (AMS C.402775, 2). Nadgee, Saltwater Creek, 37.170°S, 150.000°E (AMS C.461546, 3). Bittangabee, N of Green Cape, 37.217°S, 150.017°E (AMS C.402702, 5; AMS C.404917, 10: complete).

6 m, N of Green Cape, Bittangabee, 37.217°S, 150.018°E (AMS C.461485, 2). 2–4 m, Green Cape, S side of tip, 37.267°S, 150.050°E (AMS C.402774, 2).

Australia, South Australia. 0–1 m, Point Sinclair, 32.100°S, 132.983°E (AMS C.402822, 1; AMS C.402823, 3). Elliston Reef, 33.650°S, 134.883°E (AMS C.402801, 3, 14: complete). 0–5 m, Elliston, 33.650°S, 134.883°E (AMS C.402803, 3). Spencer Gulf, Tumby Bay, The Heap, 34.367°S, 136.133°E (AMS C.402812, 1). Yorke Peninsula, Bluff Beach, 34.733°S, 137.483°E (AMS C.402619, 2). Gulf St. Vincent, Largs Bay, 34.817°S, 138.483°E (AMS C.402813, 1). 0 m, Corney Point, Yorke Peninsula, 34.919°S, 137.084°E (MV F178378, 1). 0–3 m, Pondalowie & Marion Bays, Yorke Peninsula, 35.233°S, 136.833°E (AMS C.402810, 1). Kangaroo Island N coast, Stokes Bay, 35.617°S, 137.200°E (AMS C.402776, 2; AMS C.402804, 5). Kangaroo Island N coast, Western River Cove, 35.683°S, 136.967°E (AMS C.402700, 1; AMS C.402807, 1; AMS C.402808, 2; AMS C.402809, 1). Kangaroo Island S coast, Pennington Bay, 35.867°S, 137.750°E (AMS C.404887, 1: complete). Kangaroo Island S coast, Hanson Bay, 36.033°S, 136.850°E (AMS C.402805, 8, 10: complete). Cape Dombey, Robe, below Obelisk, 37.150°S, 139.750°E (AMS C.402820, 9). 0 m, Robe, 37.162°S, 139.755°E (MV F178374, 2). 0 m, Carpenters Rocks, Port Macdonnell, 37.912°S, 140.396°E (MV F178375, 1). 13 m, off Middle Point, near Cape Northumberland, 38.050°S, 140.617°E (AMS C.402818, 1). 6 m, Middle Point, 600 m offshore, near Cape Northumberland, 38.050°S, 140.617°E (AMS C.402819, 1). W side Cape Northumberland, Port Macdonnell, 38.050°S, 140.650°E (AMS C.402815, 1). 0 m, Cape Northumberland, Port Maconnell, 38.061°S, 140.663°E (MV F178377, 1).

Australia, Western Australia. Horrocks Beach, N of Geraldton, 28.383°S, 114.433°E (AMS C.402832, 1). 4–8 m, Carnac Island, W side, off Freemantle, 32.117°S, 115.667°E (AMS C.402829, 1). 2–3 m, Cape Peron, 30 mi S of Perth, 32.267°S, 115.683°E (AMS C.402621, 2). Hopetoun, Two Mile Beach, 33.950°S, 120.117°E (AMS C.402826, 1). 1–2 m, near Hopetoun Jetty, 33.950°S, 120.117°E (AMS C.402827, 3). Windy Harbour, 34.833°S, 116.000°E (AMS C.402831, 1). 150 m, off Albany, 35.407°S, 118.047°E (AMS C.402712, 1).

Papua New Guinea. 5 m, Christensen Research Institute, Madang, 5.117°S, 145.817°E (AMS C.380676, 1) [questionable record, see remarks].

Literature records.

Australia. Flinders [38.483°S, 145.017°E], San Remo [38.517°S, 145.367°E], Point Nepean, Victoria [38.350°S, 144.919°E] (Macpherson & Gabriel, 1962).

Remarks. Geiger & Jansen (2004b) used *I. remota* as a full species, not as a subspecies of *I. rosea*, which is followed here. There is no evidence of peripheral mixing of populations or of specimens with intermediate characters, hence, the shell morphologies (as indicators of genepools) are distinct.

Incisura vincentiana (Cotton, 1945) seems to be based on large, partially eroded specimens lacking any further distinctive characters. Geiger & Jansen (2004b) synonymized *I. vincentiana* under *I. remota*.

The record from tropical Papua New Guinea of an otherwise temperate species needs to be taken with caution. There are no tropical species in this genus.

Incisura rosea (Hedley, 1904)

Figures 568–572

Chresonymy.

Scissurella rosea Hedley, 1904: 90–91, fig. 17.
Scissurella rosea: Suter, 1907: 7.
Scissurella rosea: Iredale & Mestayer, 1908: 412.
Scissurella rosea: Iredale, 1908b: 391.
Incisura rosea: Thiele, 1912a: 33–34, pl. 4, fig. 7.
Scissurella rosea: Suter, 1913: 89, pl. 6, fig. 11.
Scissurella rosea: Iredale, 1924: 215–216.
Scissurona rosea: Powell, 1940: 208.
Scissurona rosea: Fleming, 1950: 25.
Scissurona rosea: Powell, 1955: 46.
Scissurona rosea: Dell, 1960: 145.
Incisura (*Scissurona*) *rosea*: Wenz, 1971: 174, fig. 274 [copy Hedley, 1904].
Scissurona rosea: Powell, 1979: 35, pl. 13, fig. 20.
Scissurona rosea: Lozouet, 1986: 114, fig. 31B.
Scissurona rosea: Spencer & Willan, 1995: 13.
Incisura rosea rosea: Geiger, 2003: 75.
Incisura rosea: Geiger & Jansen, 2004b: fig. 5c [shell].
Incisura rosea: Nützel & Geiger, 2006: fig. 2D–G.
Incisura rosea: Spencer *et al.*, 2009: 202.
Incisura rosea: Raven & Bracegirdle, 2010: 10, text-fig.

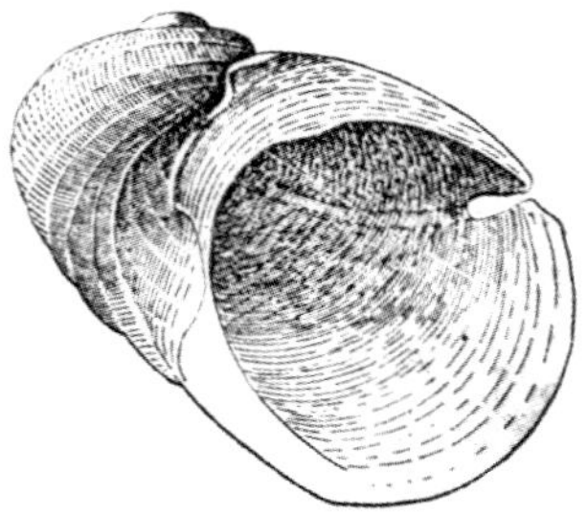

Figure 568. Original illustration of *Incisura rosea.*

Misidentifications.

Incisura lytteltonensis: Brookes Knight *et al.*, 1960: I221, fig. 136, 3 [is *I. rosea*].

Type material. Holotype (AMS C.13590).

Type locality. Lyall Bay, near Wellington, [41.333°S, 174.800°W] (OD).

Etymology. Referring to the rose-tinged protoconch.

Description. Shell medium-sized (to 1.35 mm. Figs 569, 570), neritiform. Protoconch of 0.875 whorl, fine axial sculpture with connecting spiral band, apertural varix present, apertural margin convex. Teleoconch I of one whorl, almost smooth, with fine spiral cords. Teleoconch II of 0.66 whorl, suture moderately to deeply impressed. Shoulder with similar sculpture as on teleoconch I. Base with edge at border towards umbilicus. Selenizone in outer fifth of shoulder; lunules indistinct; keels low; slit open, margins parallel. Aperture round. Umbilicus open, very narrow.

Radula (Fig. 571). Rachidian tooth trapezoidal, central cusp largest, three cusps on each side. Lateral teeth 1–3 similar, apical cusp largest, three cusps on outer edge of tip. Lateral tooth 4 reduced, hook-shaped, apical cusp largest, one minute point on each side. Lateral tooth 5 enlarged by broadening, apical cusp largest, six cusps on inner margin. Inner marginal teeth with triangular tip, apical cusp largest, three cusps on inner edge, five cusps on outer; shafts without food groove. Outer marginal teeth with spoon-shaped tip, many small cusps along margin. Radular interlock of central field moderate.

Comparisons. *Incisura lytteltonensis* from New Zealand has a more rapidly expanding body whorl, a shorter, smooth teleoconch II (0.75 vs. 1 whorl), the selenizone in the outer third of the shoulder, and lacks the umbilical callus.

Incisura remota from New Zealand has variable sculpture ranging from indistinct spirals and axials to distinct reticulate sculpture and lacks the umbilical callus.

Distribution. New Zealand (Fig. 572).

Specimen records.

New Zealand. 90 m, Snares Islands, 48.017°S, 166.567°E (GNS RM 5968, 2). 55 m, Stewart Island, Port Adventure, 47.083°S, 168.183°E (AMS C.406341, 1). Ocean Beach, Stewart Island, 46.968°S, 168.177°E (RMNH, 8). 35–44 m, New Golden Hind Station B2, Narrows, Long Sound, Preservation Inlet, Fiordland, 46.090°S, 166.620°E (GNS RM 1042, 3). 13 m, North Port, Chalky Sound, Fiordland, SW South Island, 46.023°S, 166.569°E (GNS RM 1047, 1). 0 m, Wellington, Bay of Islands, 41.345°S, 174.773°E (NMNZ M.134335, 25). Lyall Bay, near Wellington, 41.333°S, 174.800°W (AMS C.13590, 1). Lyall Bay, Wellington, 41.300°S, 174.783°E (USNM 680538, 2). Tahunanui, Nelson, 41.283°S, 173.250°E (USNM 602651, 4). Blind Bay, Nelson, 41.268°S, 173.271°E (GNS RM 5969, 1). N of Mangawhai, Langs Beach, 36.050°S, 174.533°E (AMS C.461507, 3). 79 m, off Hen and Chicken Islands, Gulf of Hauraki, 35.917°S, 174.750°E (AMS C.461508, 1). 0 m, North Island, Bay of Islands, Deep Water

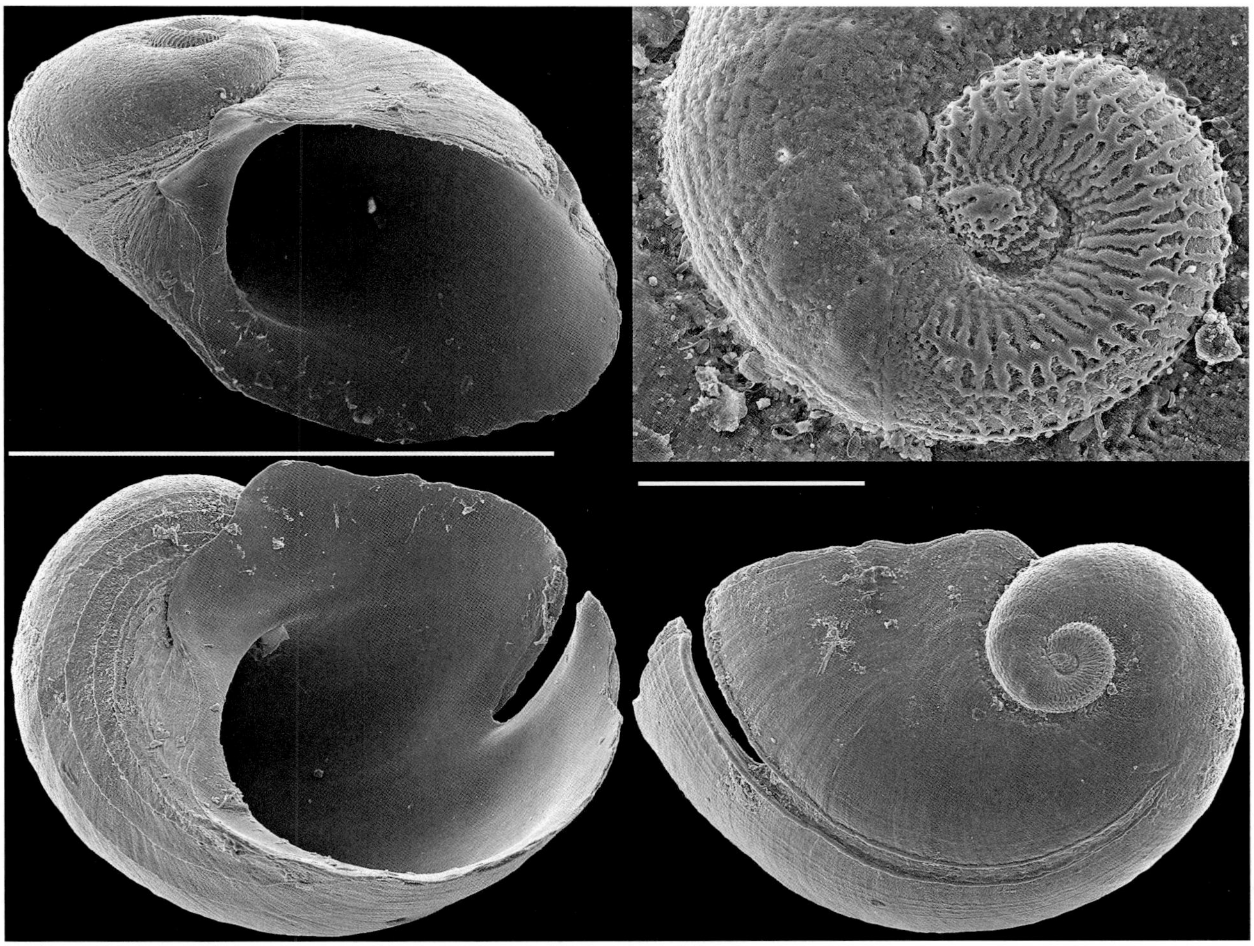

Figure 569. *Incisura rosea.* 0 m, Wellington, Bay of Islands, New Zealand, 41.345°S, 174.773°E (NMNZ M.134335, 25). Scale bar shell = 1 mm. Scale bar protoconch = 100 µm.

Scissurellidae *Incisura*

Cove, 35.200°S, 174.300°E (DMNH 203169, 1). 24 m, Whangaroa Harbour entrance, 35.100°S, 173.867°E (AMS C.461512, 1). 26 m, between Waimahana and Whekarua, 34.950°S, 173.617°E (AMS C.461504, 1). 133 m, Wanganella Bank, Norfolk Ridge summit, 32.653°S, 167.528°E (NMNZ M.174862, 4). 120–126 m, Wanganella Bank summit, W Norfolk Ridge, 32.611°S, 167.584°E (NMNZ M.171794, 1). Doubtless Bay, 34.900°S, 173.400°E (MV F 96920, 1).

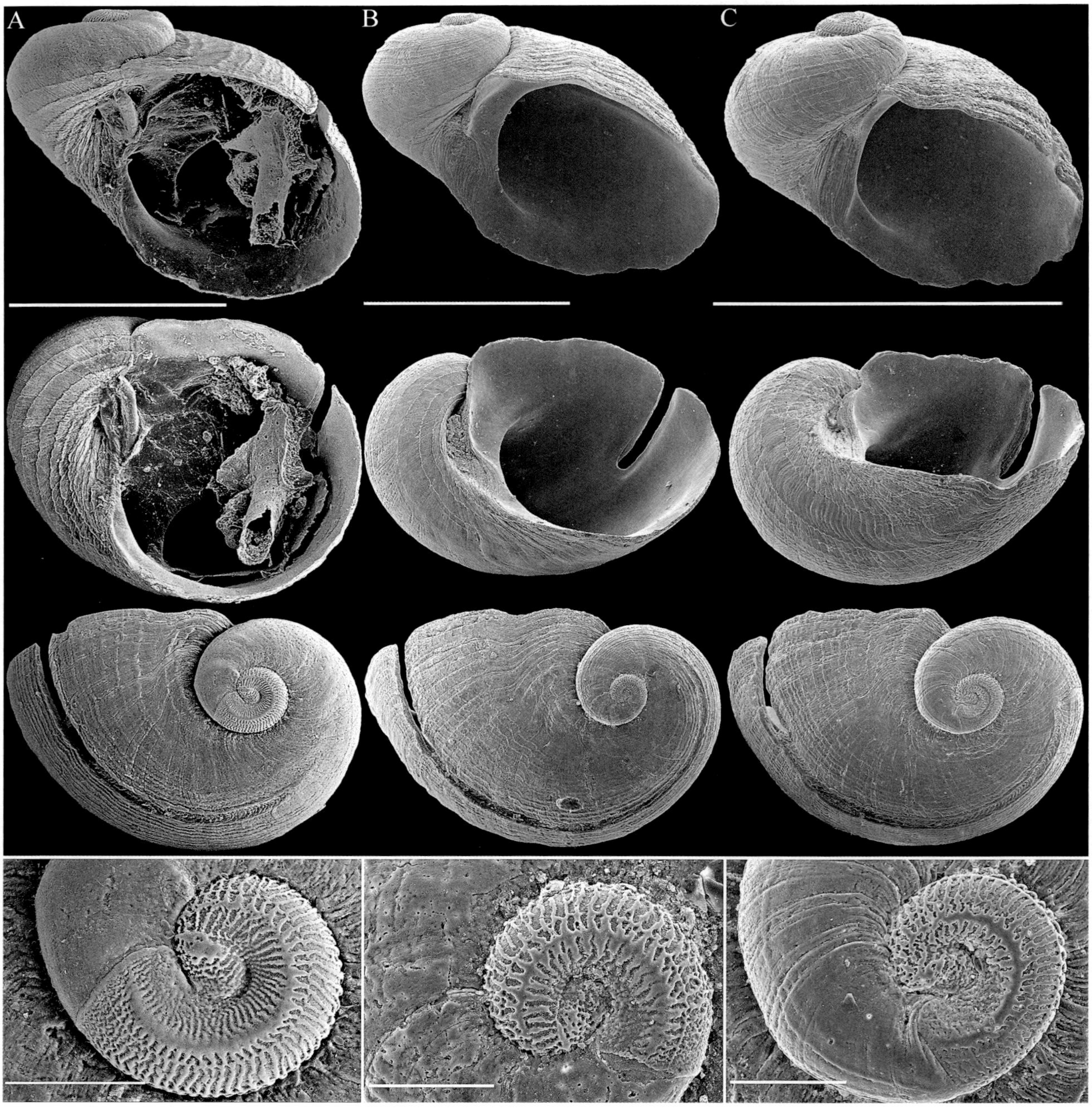

Figure 570. *Incisura rosea*. **A.** 55 m, Stewart Island, Port Adventure, New Zealand, 47.083°S, 168.183°E (AMS C.406341, 1). **B.** 13 m, North Port, Chalky Sound, Fiordland, SW South Island, New Zealand, 46.023°S, 166.569°E (GNS RM 1047, 1). **C.** 133 m, Wanganella Bank, Norfolk Ridge summit, New Zealand, 32.653°S, 167.528°E (NMNZ M.174862, 4). Scale bars shell = 1 mm. Scale bars protoconch = 100 µm.

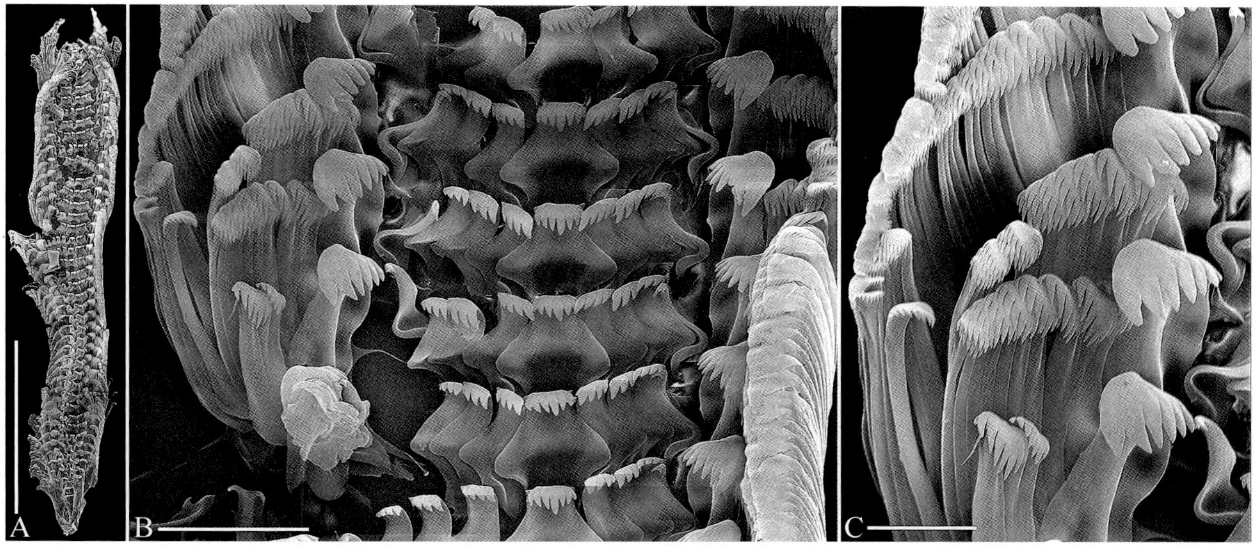

Figure 571. *Incisura rosea*. Radula. 0 m, Wellington, Bay of Islands, New Zealand, 41.345°S, 174.773°E (NMNZ M.134335, 25). **A.** Entire radula. Scale bar = 200 µm. **B.** Full width of radula. Scale bar = 20 µm. **C.** Lateral tooth 5 and marginal teeth enlarged. Scale bar = 10 µm.

New Zealand, fossil. GS 12161, cliffs near The Point homestead, W of Flowerpot Bay, N coast of Pitt Island, Chatham Islands, Whenuataru Tuff, Waipipian-Mangapanian (Mid-late Pliocene), 44.240°S, 176.239°W (GNS, 2).

Literature records.

New Zealand. Snares Islands (Powell, 1955: Finlay coll. AIM).—116 m, Edwardson Sound, opposite entrance to Northport, 49.983°S, 166.567°E. Northport Chalky Sound, 49.983°S, 166.567°E. 116 m, Daggs Sound, inside a rock bar, 45.383°S, 166.767°E. 94 m, Anchorage Arm, Daggs Sound, 45.383°S, 166.767°E (Fleming, 1950).

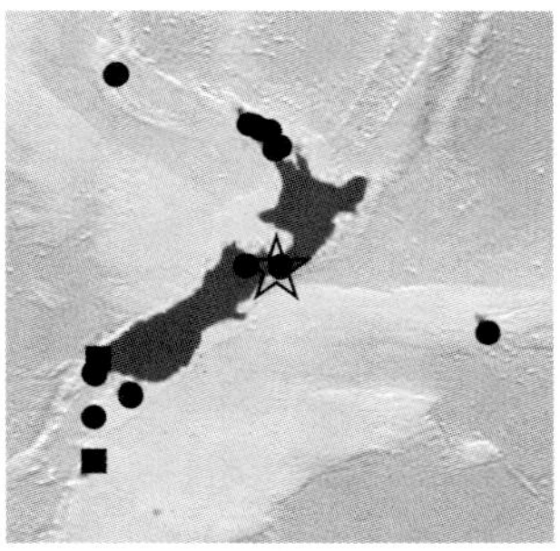

Figure 572. Distribution of *Incisura rosea*. ☆ type locality. ● specimen record. ■ literature record.

Remarks. Most specimens have a rust-colored apical region.

Incisura geoffreyi (Laws, 1940) †

Figures 573–576

Chresonymy.

Scissurella geoffreyi Laws, March 1940a: 434, pl. 60, fig. 28 (Fig. 573).

Scissurona ? *geoffreyi*: Lozouet, 1986: 114.

Scissurella geoffreyi: Beu & Maxwell, 1990: 376.

Figure 573. Original illustration of *Scissurella geoffreyi*.

Synonym.

+ *Scissurona fossilis* Laws, June 1940b: 48, fig. 20. (Fig. 574) ***Type material.*** Holotype (GNS TM 1478. Fig. 576A), 1.3 × 1 mm. Paratypes from type locality (GNS TM 8643–8644, 2). ***Type locality.*** Nukumaru Beach, Wanganui, North Island, [New Zealand, 39.930°S, 175.050°E]. Nukumaruan [= 2.6 mya, Late Pliocene] (OD, specimen label). ***Etymology.*** Fossilis, Latin for fossil, referring to the fossil nature of the species.

Scissurona fossilis: Lozouet, 1986: 114.

Scissurona fossilis: Beu & Maxwell, 1990: 402.

Scissurona fossilis: Bandel, 1998: 47, pl. 16, figs 3–6

Incisura fossilis: Geiger, 2003: 75.

Figure 574. Original illustration of *Scissurona fossilis*.

Type material. Holotype (GNS TM 1360, ex coll. Laws. Fig. 575), 1.0 × 1.0 mm. Paratypes from type locality (coll. Laws, 3: OD; GNS TM 8642, 1, ex coll. Laws; whereabouts unknown, 2).

Type locality. Waitotaran Faunule, Kaawa Creek, [North Island,] New Zealand [37.500°S, 174.833°E]. Opotian [= early Pliocene: specimen label] (OD).

Etymology. Named for Geoffrey [= Étienne Geoffroy-Saint-Hilaire (1772–1844)?].

Description. Shell medium-sized (to 1.3 mm. Figs 575, 576), trochiform, expanded-neritiform, depressed, moderately elevated to low spire. Protoconch of one whorl, fine axial sculpture with apical backbone, no apertural varix, apertural margin convex. Teleoconch I of 0.66 whorl, approximately 18 indistinct axial cords, approximately a dozen weaker spiral cords of unequal strength; some specimens with distinct reticulate sculpture. Teleoconch II of 0.5 whorl, suture impressed. Shoulder convex, with strong, equally spaced axial cords, approximately eight spiral cords of same strength as axials at apertural margin, forming wide reticulate pattern; slight thickening at intersections of axials and spirals. Base rounded without constriction below selenizone, similar sculpture to shoulder; overall appearance of cancellation. Selenizone with very low keels; lunules of same strength as axial and spiral cords; slit open, with parallel margins. Umbilicus open, narrow, bordered by strong cord,

Scissurellidae
Incisura

walls smooth, no funiculus. Aperture rounded, slightly flared, oblique, oval, roof overhanging.

Comparisons. *Incisura lytteltonensis* from the Recent of New Zealand is almost smooth, has the selenizone almost parallel with growth axis of the shell and in outer third of shoulder, and has a more rapidly expanding body whorl.

Incisura rosea from the Recent of New Zealand is almost smooth and has an apertural varix on the protoconch.

Sinezona finlayi from the Miocene [to Recent?] of New Zealand has a constriction below the selenizone, a wider umbilicus, and a closed foramen in fully grown specimens.

Distribution. Pliocene to Pleistocene of New Zealand (Fig. 577).

Specimen records.

New Zealand. GS 4360, Waipuru Shellbed (late Nukumaruan, late Pliocene), stream E of Ikirae Road, Whangaehu Valley, E of Wanganui, 39.890°S, 175.300°E (GNS, 1). GS 1063, Petane Group (Nukumaruan, latest Pliocene) Okawa Creek, 4 km W of Fernhill, Central Hawke's Bay, 39.575°S, 176.727°E (GNS, 1). Nukumaru Brown Sands, 39.930°S, 175.050°E (NMNZ M.8227, 1). Shellbed, N[th] Bank of Mangahao River, 1 mi S of

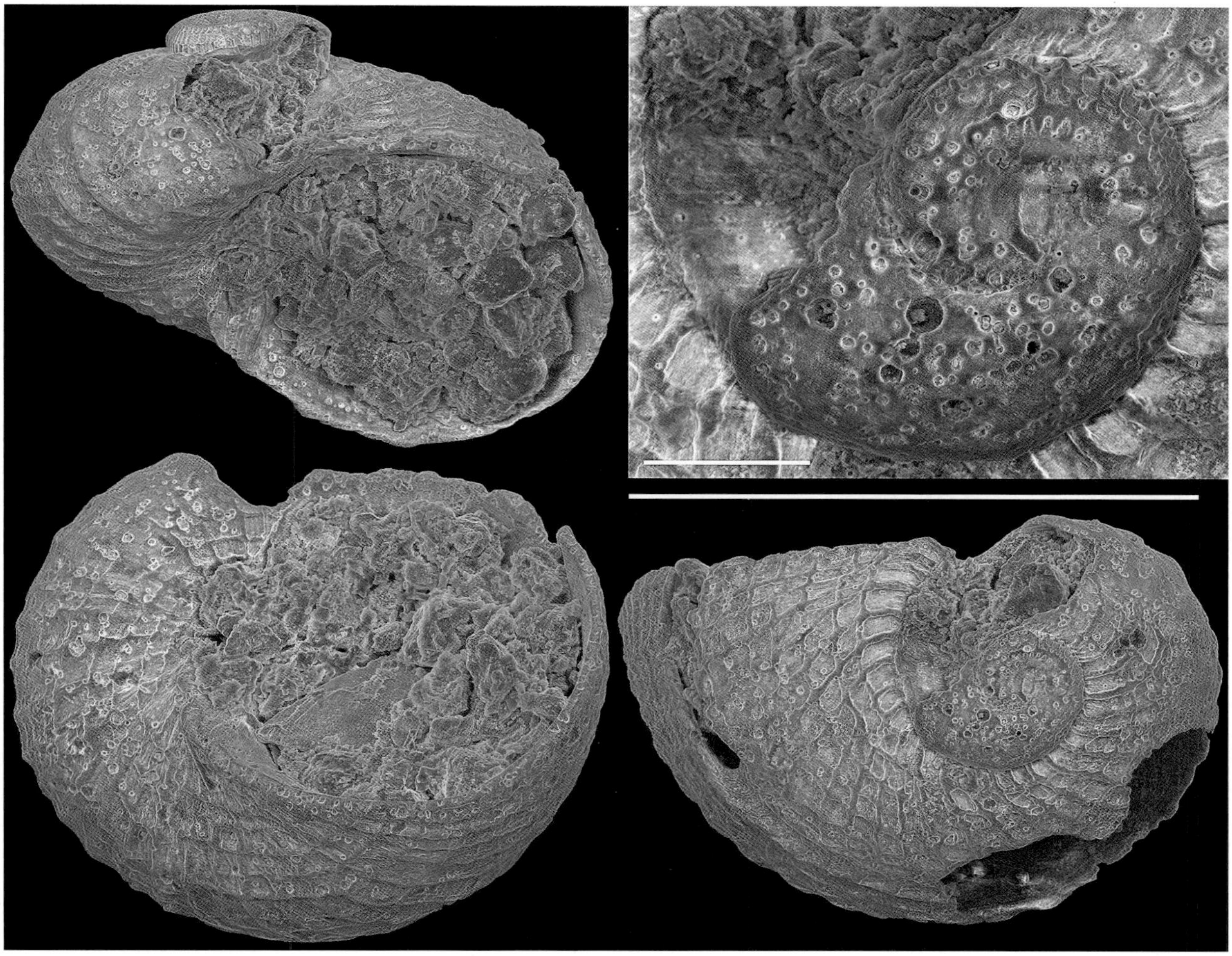

Figure 575. *Scissurella geoffreyi*. Holotype. Waitotaran Faunule, Kaawa Creek, [North Island,] New Zealand [37.500°S, 174.833°E] (GNS TM 1360). Scale bar shell = 1 mm. Scale bar protoconch = 100 μm.

Mangahao, N Wairarapa, Wn, 40.448°S, 175.789°E (NMNZ M.42674, 40). Shellbeds and blue muddy siltstone below old reservoir in stream foot of Whakatane-Ohope Road, Ohope, Wc, 37.950°S, 177.000°E (NMNZ M.43067, 23). GS12163, Moutapu Point, NE Pitt Island, Chatham Islands, Whenuataru Tuff, Waipipian-Mangapanian (middle to late Pliocene), 36.239°S, 174.293°W (GNS TM, 7).

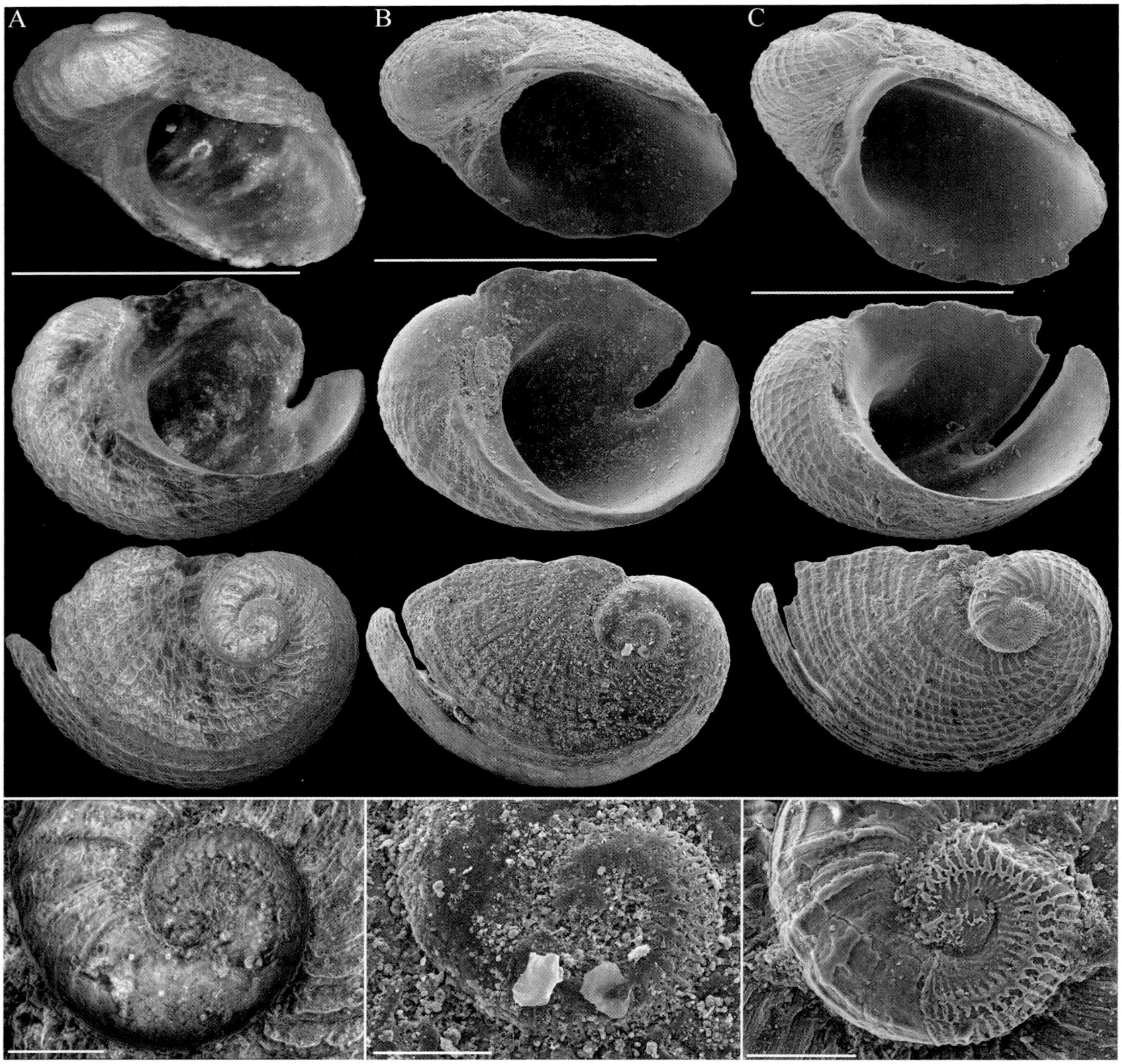

Figure 576. *Incisura geoffreyi.* **A.** Holotype of *Scissurona fossilis.* Nukumaru Beach, Wanganui, North Island, [New Zealand, 39.930°S, 175.050°E]. Nukumaruan [= 2.6 mya, Late Pliocene] (GNS TM 1478). **B.** GS 1063, Petane Group (Nukumaruan, latest Pliocene) Okawa Creek, 4 km W of Fernhill, Central Hawke's Bay, 39.575°S, 176.727°E (GNS, 1). **C.** GS 4360, Waipuru Shellbed (late Nukumaruan, late Pliocene), stream E of Ikirae Road, Whangaehu Valley, E of Wanganui, 39.890°S, 175.300°E (GNS, 1). Scale bars shell = 1 mm. Scale bars protoconch = 100 μm.

Scissurellidae *Incisura*

Literature records.

New Zealand. Pliocene, Mangahao, Ohope [as *fossilis*] (Bandel, 1998).—Waipipian to Nukumaruan = late Pliocene to early Pleistocene of New Zealand [as *fossilis*]. Opoitian = early Pliocene of New Zealand (Beu & Maxwell, 1990).

Remarks. Although both taxa (*Scissurella geoffreyi, Scissurona fossilis*) were described by Laws in 1940, they were introduced in two publications in the same journal. The former was published in volume 69 with a byline on the article noting issued separately, March, 1940, whereas the latter in volume 70, with a byline on the article noting issued separately, June, 1940. Hence, *Scissurella geoffreyi* has priority.

The elevation of the spire is quite variable in the species as is typical for *Incisura*. The holotype of *geoffreyi* (Fig. 575) is a more elevated form, whereas the holotype of *fossilis* (Fig. 576A) is slightly more depressed, although not the flattest known specimen. Details of teleoconch sculpture and preserved protoconch sculpture clearly indicates that all those specimens are conspecific.

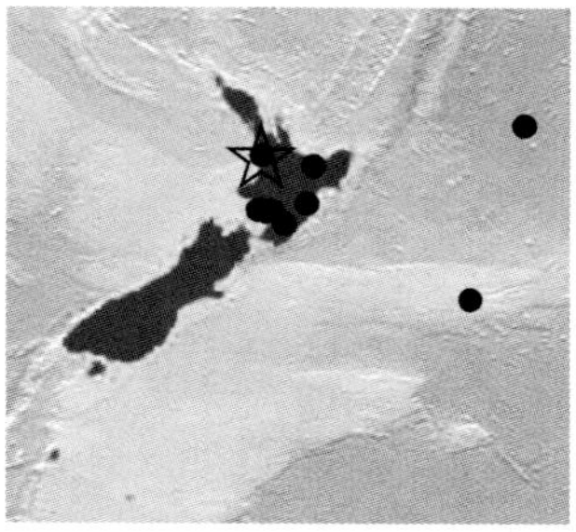

Figure 577. Distribution of *Incisura geoffreyi*. ☆ type locality. ● specimen record.

Coronadoa Bartsch, 1946

Chresonymy.

Coronadoa Bartsch, 1946: 447–448.
Coronadoa: Keek & Pearson, 1952: 15.
Coronadoa: Abbott, 1974: 55 [Cyclostrematidae].
Coronadoa: Geiger, 2003: 41.
Coronadoa: Geiger & Sasaki, 2009: 6.
Coronadoa: Geiger & McLean, 2010: 12.

Type species. *Coronadoa simonsae* Bartsch, 1946 (OD).

Other species. *Coronadoa demisispira* Geiger & McLean, 2010.
Coronadoa hasegawai Geiger & Sasaki, 2009.

Etymology. Most likely named for the type locality of the type species, Coronado Islands, Mexico.

Description. Shell trochiform, shoulder rounded. Sculpture axials only, no spiral keel. Protoconch sculpture of fine axials, no varix. Umbilicate, with funiculus. No slit, foramen, or selenizone. Operculate. Radula R - 5 - n, lateral teeth 1–3 similar, lateral tooth 4 reduced, hook-shaped, lateral tooth 5 broadened, asymmetrically serrated.

Comparisons. The absence of a slit, foramen, or selenizone distinguishes *Coronadoa* from all other scissurellid genera. The protoconch sculpture of fine axials separates this genus from all larocheid genera, which share with *Coronadoa* the absence of a slit, foramen, or selenizone, but have a smooth or flocculent protoconch sculpture. The radula of Larocheidae have lateral teeth 1–4 similar, whereas in *Coronadoa,* lateral tooth 4 is reduced.

Differentiation of species. Species are distinguished mainly by the height of the spire.

Distribution. North Pacific (Fig. 578). For specific localities, see individual species.

Remarks. The difficulty of identifying *Coronadoa* species is recognizing small trochiform shells with no slit or foramen as belonging to this scissurellid genus. Unfortunately, there is no easily recognizable, positive diagnostic character for the genus; the genus is usually identified by exclusion of other possible trochiform genera. The absence of nacre can be of help, but nacre is often not distinct in vetigastropod shells of minute size and is absent in all Caenogastropoda. The typical rhipidoglossate radula

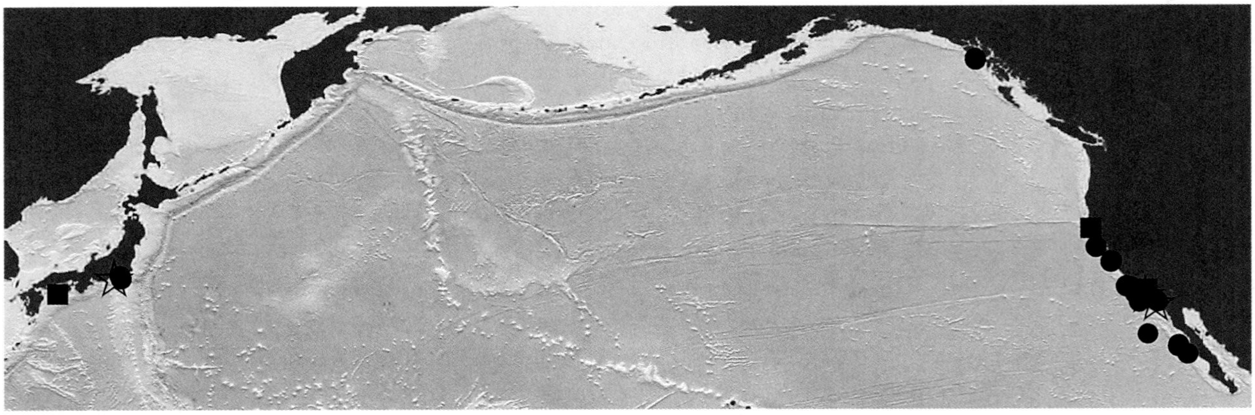

Figure 578. Global distribution of *Coronadoa* species. ☆ type locality. ● specimen record. ■ literature record.

provides positive confirmation of the placement of *Coronadoa* in Vetigastropoda, and the absence of nacre at the microscopic level can confirm the placement in *Coronadoa*.

The placement of *Coronadoa* in Scissurellidae is indicated by the protoconch sculpture and the radula. Bartsch (1946) had missed the lateral tooth 4, using a light microscope. The absence of a slit, foramen, or selenizone had been taken as a sign of being juvenile (McLean 1967, 1971); however, *C. simonsae* maintains this condition even when fully grown. Juveniles of the sympatric *Sin. rimuloides*, with which it had been synonymized, can be clearly separated from *C. simonsae* (Marshall, 2002; Geiger & McLean, 2010). Marshall (2002) tentatively synonymized *Coronadoa* under *Sinezona*. I agree with Marshall (2002) that the sole distinguishing character is the lack of a selenizone, slit, or foramen. I consider this character significant as it is most likely caused by anatomical transformation of the mantle, as is the case in Fissurellidae (McLean & Geiger, 1998).

Coronadoa demisispira Geiger & McLean, 2010

Figure 579–580

Chresonymy.

Coronadoa demisispira Geiger & McLean, 2010: 12–15, fig. 9.

Type material. Holotype (LACM 3113. Fig. 579C). Paratypes from type locality (LACM 3114, 1). 13–30 m, Cortez Bank, 0.5 mi S of Bishop Rock, California, USA, 32.433°N, 119.125°W (LACM 3115, 2. Fig. 579A–B). 26 m, San Miguel Island, California, USA, 34.020°N, 120.402°W (DLG 1131, 1). 46 m, off Santa Cruz Island, California USA, 34.056°N, 119.961°W (DLG, 1263, 3).

Type locality. 26–40 m, Farnsworth Bank, off S side of Catalina Island, California, USA, 33.350°N, 118.517°W (OD).

Etymology. demisi-: Latin for sunken; -spira: Latin for spire. Referring to the flat spire of the species (OD).

Description. Shell small (to 0.75 mm. Fig. 579), trochiform, depressed. Protoconch of 0.875 whorl, embryonic cap without sculpture, remainder with strong axial ribs, no apertural varix, apertural margin prosocline. Teleoconch of 1.5 whorls, 11–13 raised axial cords, indistinct at suture and towards umbilicus, most pronounced at periphery, interstices with irregular axial growth marks; no spiral sculpture. Umbilicus open, occasionally under-surface of protoconch visible, bordered by knobby carina towards base. Aperture subquadrate, D-shaped, lower adumbilical corner pronounced, forming knobby carina of umbilicus.

Operculum thin, round, multispiral, nucleus central.

Comparisons. The sympatric *C. simonsae* is proportionally taller (width/height: *simonsae* mean = 1.28, SD = 0.047, n = 4; *demisispira* mean = 1.79, SD = 0.22, n = 4; t-test t = -4.36, p = 0.022). As a corollary, its umbilicus is wider and tends to have fewer axial cords per unit whorl. The degree of axial sculpturing is variable in both species, from distinct cords to raised lamellae.

Coronadoa hasegawai from temperate Japan is as tall as *C. simonsae*.

Distribution. Northern California and Pacific coast of Baja California, Mexico (Fig. 580).

Specimen records.

USA, California. 37–52 m, Granite pinnacle, Cordell Bank, off Reyes Point, Marin County, 38.017°N, 123.417°W (LACM 81-50, 20). 26 m, San Miguel Island, 34.020°N, 120.402°W (DLG 1131, 1). 46 m, off Santa Cruz Island, 34.056°N, 119.961°W (DLG 1263, 3).

Mexico. 10–23 m, NW side of Guadalupe Island, Pilot Rock and 5 mi S, Baja California, 29.188°N, 118.253°W (LACM 72-121, 1). 20–26 m, SW side Natividad Island, Baja California, 27.888°N, 115.212°W (LACM 72-117, 1). 13–17 m, 5 Fathom Bank, about

Figure 579. *Coronadoa demisispira*. **A–B.** Paratypes. 13–30 m, Cortez Bank, 0.5 mi S of Bishop Rock, California, USA, 32.433°N, 119.125°W (LACM 3115, 2). **C.** Holotype. 26–40 m, Farnsworth Bank, off S side of Catalina Island, California, USA, 33.350°N, 118.517°W (LACM 3113,). Scale bars shell = 0.5 mm. Scale bars protoconch = 100 μm.

4 mi SW of SE tip of Cedros Island, outer coast Baja California, 27.983°N, 115.212°W (LACM 71-93, 2).

Remarks. *Coronadoa demisispira* is much less common than *C. simonsae*; if the two species are present in the same lot, *C. demisispira* comprises at most 10% of specimens in the genus *Coronadoa*. This skewed ratio suggests that the two are not dimorphic males and females of the same species. In general, Vetigastropoda show a male:female ratio of close to 1:1 (Poore, 1973; Bretos *et al.*, 1985; Hooker & Creese, 1995; Wells & Malay, 1995), although there are no data on sex ratios in scissurellids; some may be hermaphroditic (Baborka, 2007).

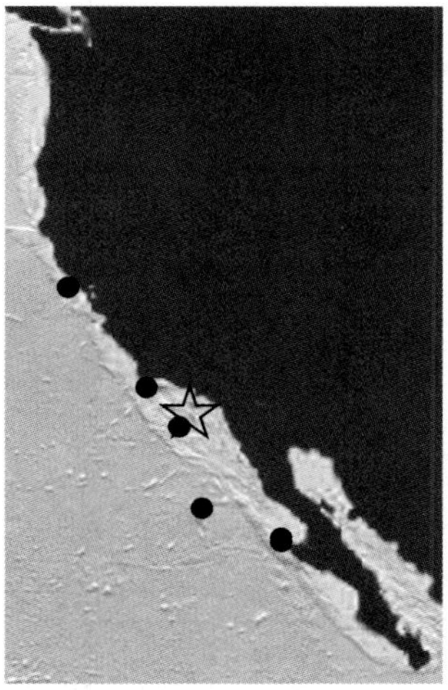

Figure 580. Distribution of *Coronadoa demisispira*. ☆ type locality. ● specimen record.

Coronadoa hasegawai Geiger & Sasaki, 2009

Figures 581–584

Chresonymy.

Scissurellidae gen. et sp.: Sasaki, 1998: fig. 45d–f.

Coronadoa hasegawai Geiger & Sasaki, 2009: 6–9, figs 5–7.

Coronadoa hasegawai: Geiger & McLean, 2010: 14.

Type material. Holotype (SBMNH 83550. Fig. 582). Paratypes from type locality (SBMNH 83551, 5. Figs 583A–C, 584A–B; DLG 1041, 15). Goshikinohama, Usa, Tosa, Kochi Prefecture, Japan, 33.424°N, 133.457°E (UMUT RM27646. Sasaki, 1998: fig. 45d–f). 0–1 m, 50 m to right of Misaki Field Station, W coast Miura Peninsula, Japan, 35.250°N, 139.667°E (AMS C.380664, 3). Misaki Field Station, W coast Miura Peninsula, Japan, 35.250°N, 139.667°E (AMS C.380665, 1. Fig. 584C). Banda, Tateyama, Chiba Prefecture, Japan, 34.975°N, 139.770°E (UMUT RM30034-30043, 10).

Type locality. Peninsula towards shrine, Cape Manazuru, Kanagawa Prefecture, Japan, 35.141°N, 139.161°E. Rock and algal washings from moderately exposed rock and boulder shore in intertidal (OD).

Etymology. The species is named for Kazunori Hasegawa of the National Museum of Science and Technology in recognition of his many contributions to Japanese malacology (OD).

Description. Shell small (to 0.61 mm. Figs 582–584), trochiform globular. Protoconch of 0.9 whorl, embryonic cap with irregular elevated sculptural elements, of same height as subsequent axial cords; approximately 25 axial cords from inner third to suture (some restricted to outer portion of whorl), no apertural varix, apertural margin straight to slightly prosocline. Teleoconch of up to 1.5 whorls, suture deeply impressed, approximately six distinct axial cords on first half whorl, density subsequently increasing approximately 2–3 fold, interstices with irregular growth lines, no spiral sculpture. Base with same sculpture as on shoulder, distinct angle at border to umbilicus, adorned with thickenings of axial cords. Umbilicus deep, walls with axials. Aperture D-shaped, lower adumbilical corner flared, giving rise to periumbilical cord.

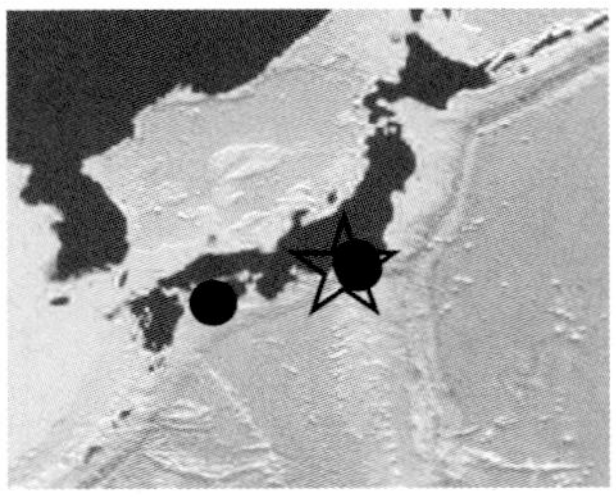

Figure 581. Distribution of *Coronadoa hasegawai*. ☆ type locality. ● specimen record.

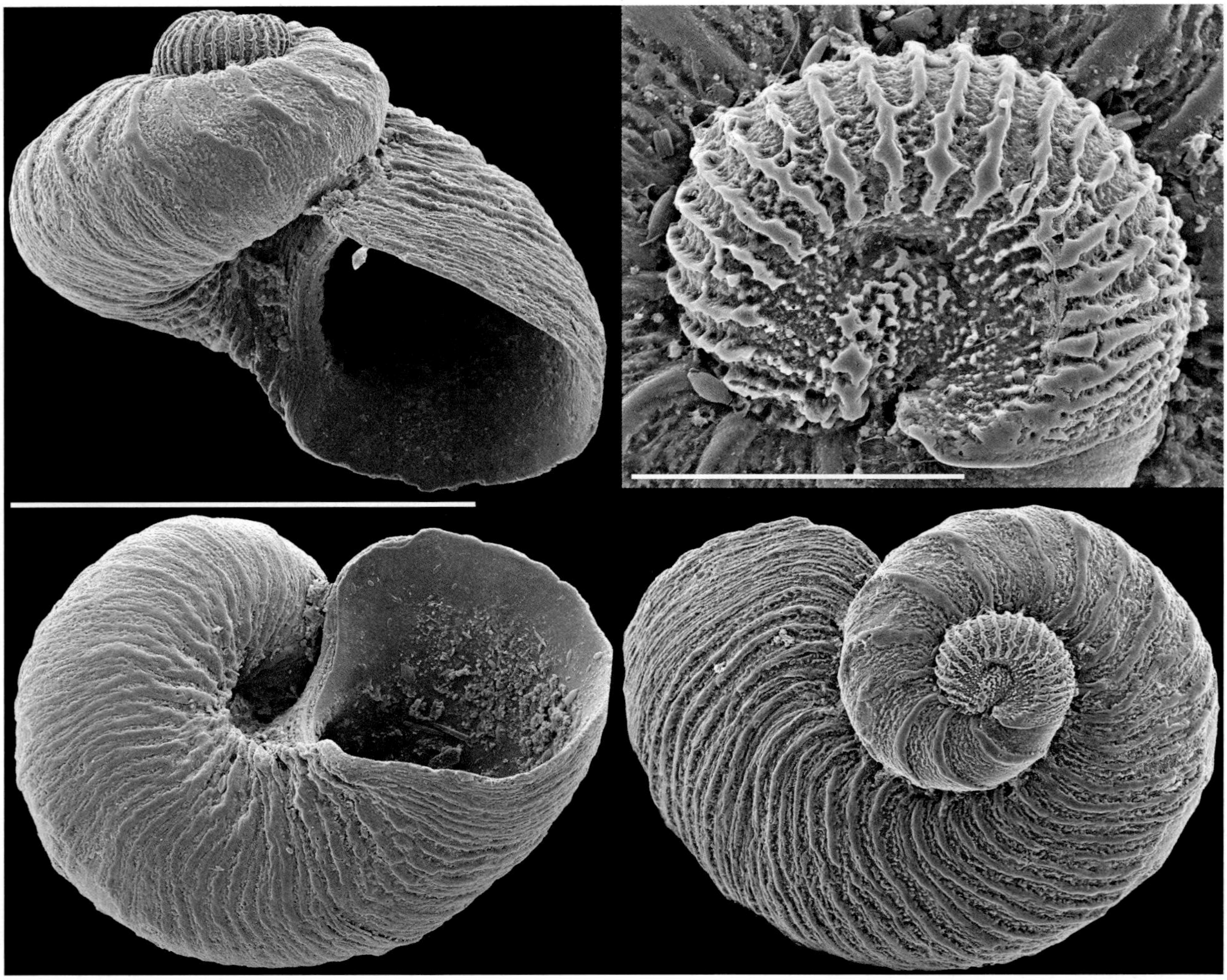

Figure 582. *Coronadoa hasegawai*. Holotype. 0 m, peninsula towards shrine, Cape Manazuru, Kanagawa Prefecture, Japan, 35.141°N, 139.161°E (SBMNH 83550). Scale bar shell = 0.5 mm. Scale bar protoconch = 100 μm.

Operculum thin, round, multispiral, nucleus central. Further data on animal not available.

Comparisons. *Coronadoa simonsae* from the Panamic and Northeast Pacific has an embryonic cap that lacks irregular but distinct granules and ridges, and the apertural margin of the protoconch is prosocline in *C. simonsae*.

Distribution. Southern Japan (Fig. 581).

Figure 583. *Coronadoa hasegawai*. Paratype. **A–C.** 0 m, peninsula towards shrine, Cape Manazuru, Kanagawa Prefecture, Japan, 35.141°N, 139.161°E (SBMNH 83551, 5). Scale bars shell = 0.5 mm. Scale bars protoconch = 100 μm.

Remarks. Although the radula could not be obtained for this species, the similarities in the shells of *C. hasegawai* and *C. simonsae* make the generic placement of *C. hasegawai* quite certain. The trochoid protoconch (see Geiger *et al.*, 2008 for discussion), and the very similar sculpture on protoconch and teleoconch make the placement straightforward. The distinct dip of the last quarter whorl and the tendency to show some separation of the peristome from the previous whorl indicate that the

Figure 584. *Coronadoa hasegawai*. Paratypes. **A–B.** Peninsula towards shrine, Cape Manazuru, Kanagawa Prefecture, Japan, 35.141°N, 139.161°E (SBMNH 83551, 5). **C.** Misaki Field Station, W coast Miura Peninsula, Japan, 35.250°N, 139.667°E (AMS C.380665, 1). Scale bars shell = 0.5 mm. Scale bars protoconch = 100 μm.

specimens are fully grown and not juveniles of another, larger species. The more than one teleoconch whorl without a slit indicate that this is not a juvenile slit-bearing scissurellid which has not yet developed the slit.

The species is commonly found in southwestern Japan (H. Fukuda, pers. comm. 1/2009).

Coronadoa simonsae Bartsch, 1946

Figures 585–589

Chresonymy.

Coronadoa simonsae Bartsch, 1946: 281, figs 1 [radula], 2 [operculum], 3 [shell] (Fig. 585).

Coronadoa simonsae: Geiger, 2003: 41, 75, fig. 15J [shell, basal view], 16D [shell, apertural view; radula].

Coronadoa simonsae: Geiger, 2008b: fig. 1.

Coronadoa simonsae: Geiger & McLean, 2010: 14–15, fig. 9.

Type material. Holotype (USNM 346654), 0.6 × 0.5 mm. Paratypes from type locality (CASIZ 36-1-14, 3: opercula; SDMNH 673, 3376–3380, 6: *fide* Wilson & Kennedy, 1967: 249).

Type locality. North Coronado Island, Baja California [Mexico, 32.455°N, 117.297°W] (OD).

Etymology. Named for Mrs. Carrie L. Simons (OD).

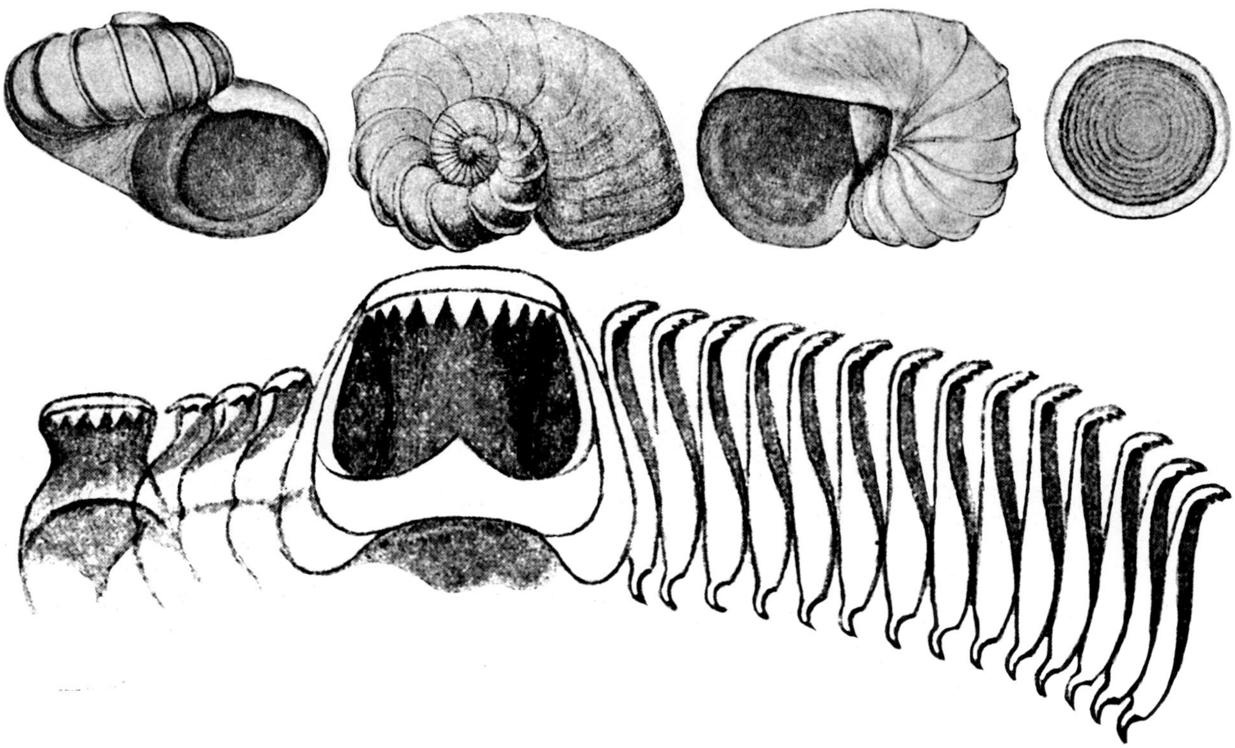

Figure 585. Original illustration of *Coronadoa simonsae*.

Description. Shell small (to 0.6 mm. Figs 586, 587), trochiform. Protoconch of one whorl, strong axials, no apertural varix, apertural margin convex. Teleoconch of 1.5 whorls, approximately 10 strong axial cords per half whorl, often eroded on early portion of whorl, becoming more crowded and less distinct towards mature apertural margin; interstices with irregular axials, no spirals. Umbilicus very narrow, bordered by strong spiral cord, walls with axial growth marks, no funiculus. No selenizone, slit, or foramen. Aperture obliquely D-shaped, roof overhanging.

Operculum (Fig. 588B) round, multispiral with central nucleus.

Radula (Fig. 588A). Rachidian tooth triangular with convex shaft, tip triangular, five cusps, central cusp largest. Lateral teeth 1–3 similar to one another, triangular, with convex shaft, tip asymmetrical, innermost tip longest of three. Lateral tooth 4 reduced to hook with 1–2 minute cusps. Lateral tooth 5 enlarged, squarish, row of tips parallel to tooth row, outermost cusp largest of six. Marginal teeth similar to one another, decreasing in size outwards, tip spoon-shaped with increasing number

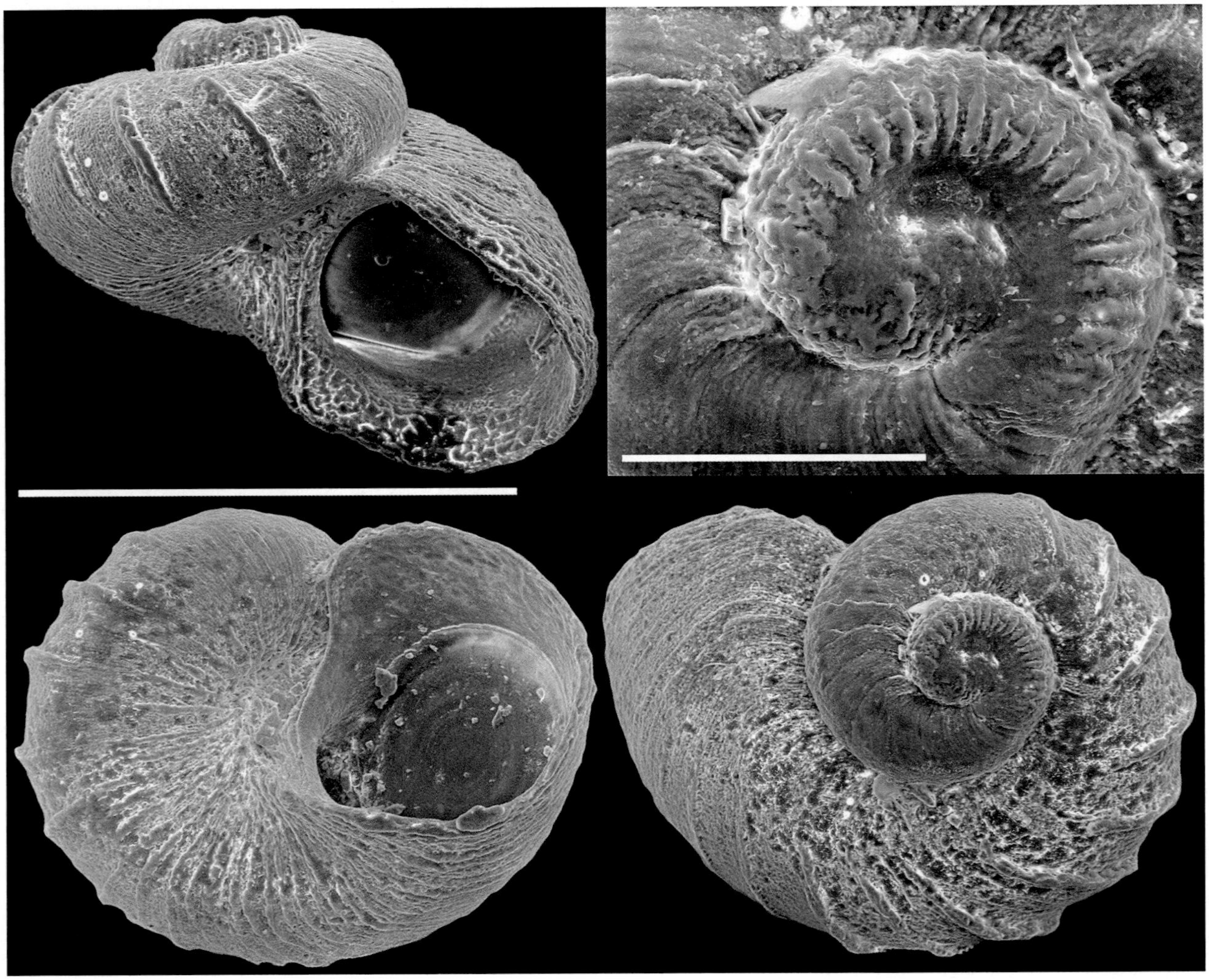

Figure 586. *Coronadoa simonsae*. North Coronado Island, Mexico, 32.455°N, 117.297°W (SDMNH 8850, 20). Scale bar shell = 0.5 mm. Scale bar protoconch = 100 μm.

of cusps of decreasing size with distance from center; shafts without food groove. Radular interlock of central field moderate.

Comparisons. *Coronadoa demisispira* from the northeastern Pacific has an overall more depressed shell.

Figure 587. *Coronadoa simonsae*. **A, C.** 13–30 m, Cortez Bank, 0.5 mi south of Bishop Rock, Los Angeles County, California, USA, 32.433°N, 119.125°W (LACM 69-34, 4). **B.** 26–40 m, Farnsworth Bank, off S side Catalina Island, California, USA, 33.350°N, 118.517°W (LACM 70-27, 2). Scale bars shell = 0.5 mm. Scale bars protoconch = 100 μm.

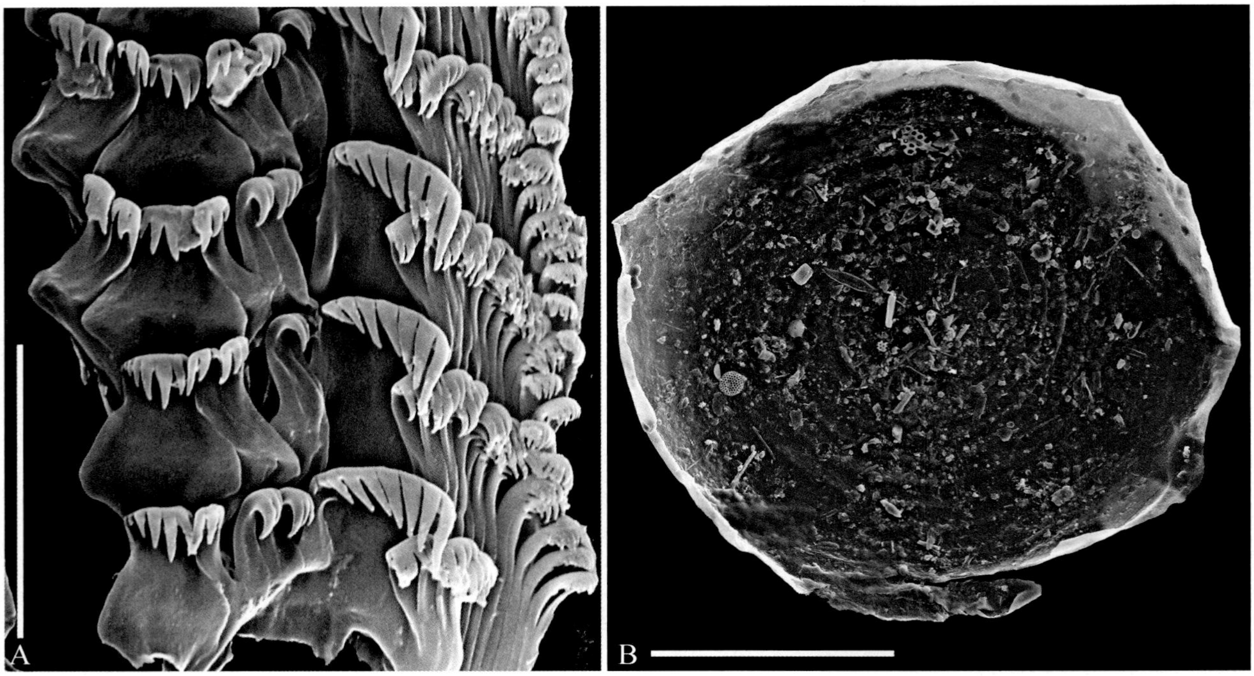

Figure 588. *Coronadoa simonsae.* 0 m, Carmel Point, Carmel, Monterey County, California, USA, 36.517°N, 112.950°W (LACM 66-57.23, 31). **A.** Radula. Scale bar = 20 µm. **B.** Operculum. Scale bar = 100 µm.

Coronadoa hasegawai from Japan has an embryonic cap that has irregular but distinct granules and ridges and the apertural margin of the protoconch is straight.

Distribution. Alaska to Pacific coast of middle Baja California, Mexico (Fig. 589).

Specimen records.

USA, Alaska. 3–12 m, outside Pirate Cove, Sitka Sound, Baranof Island, 56.987°N, 135.367°W (LACM 73-15, 2). 0 m, SE of Halibut Point, Sitka Sound, Baranof Island, 57.100°N, 135.400°W (LACM 73-16, 1).

USA, California. 13–30 m, Cortez Bank, 0.5 mi S of Bishop Rock, Los Angeles County, 32.433°N, 119.125°W (LACM 69-34, 4). 42 m, off W side San Clemente Island, 32.905°N, 118.564°W (RMC, 7). 26 m, San Nicolas Island, N side, 33.283°N, 119.517°W (LACM 72-101, 1). 27 m, Farnsworth Bank, Catalina Island, 33.343°N, 118.518°W (RMC, 16). 26–40 m, Farnsworth Bank, off S side Catalina Island, 33.350°N, 118.517°W (LACM 70-27, 2). 10–22 m, Iron Bound Cove, S side Santa Catalina Island, Los Angeles County, 33.450°N, 118.575°W (LACM 69-31, 6). Paradise Cove, Los Angeles County, 33.733°N, 118.367°W (LACM 66-58, 1). San Miguel Island, 34.000°N, 120.400°W (DLG 1145, 2). 8 m, Wyckoff Bay, San Miguel Island, 34.017°N, 120.388°W (DLG 885, 2). 26 m, San Miguel Island, 34.020°N, 120.402°W (DLG 1132, 2). 15 m, San Miguel Island, 34.025°N, 120.401°W (DLG 879, 1). 13 m, San Miguel Island, 34.027°N, 120.041°W (DLG 880, 1). 32 m, NE coast of Santa Cruz Island, 34.043°N, 119.610°W (DLG 2007, 8). 9–14 m, NW side San Miguel Island, 34.055°N, 120.425°W (LACM 82-58, 1). 23–27 m, Granite Pinacles, Wilson Rock, off San Miguel Island, 34.106°N, 120.395°W (LACM 82-59, 2). 0 m, Carmel Point, Carmel, Monterey County, 36.516°N, 121.950°W (LACM 66-57.23, 31). 13–42 m, Carmel Submarine Canyon, N end of San Jose Creek Beach, Monterey County, 36.533°N, 121.933°W (LACM 60-24, 4). 37 m, off Hopkins Marine Station, Pacific Grove, Monterey Bay, 36.625°N, 121.900°W (LACM 60-23, 1). 37–52 m, granite pinnacle, Cordell Bank, off Reyes Point, Marin County, 38.016°N, 123.417°W

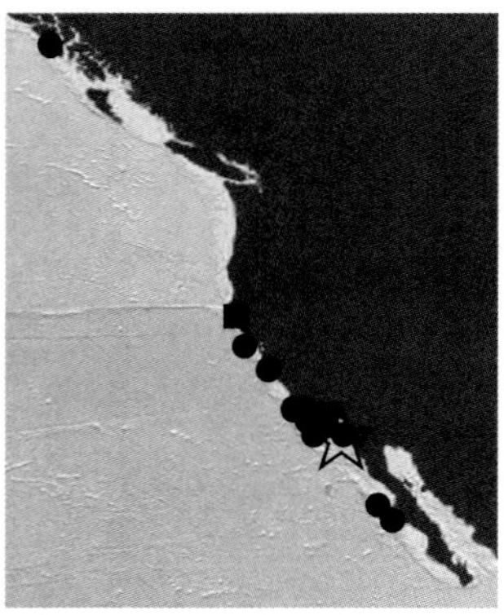

Figure 589. Distribution of *Coronadoa simonsae.* ☆ type locality. ● specimen record.

(LACM 81-50, 20). 0 m, N end Bear Landing, Bear Harbor Ranch, Mendocino County, 39.913°N, 123.937°W (LACM 76-9, 2).

Mexico. 10 m, Asuncion Island, Baja California Sur, 27.117°N, 114.283°W (SBMNH 116312, 1). 18 m, 2 mi N of Cedros Village, Cedros Island, Baja California Norte, 28.074°N, 115.182°W (SBMNH 113930, 3). N Coronado Island, 32.455°N, 117.297°W (SDMNH 8850, 20; SDMNH 21917; SDMNH 41157; LACM A.2777, 1).

Remarks. *Coronadoa simonsae* had been synonymized under *Sinezona rimuloides* by McLean (1967, 1971) and Abbott (1974), as it was considered a juvenile specimen lacking the selenizone, slit, or foramen. Marshall (2002) and Geiger & McLean (2012) recognized it as a distinct species. The strength of the axial sculpture is highly variable.

Monograph of the Little Slit Shells

VOLUME 1
Introduction
Scissurellidae

VOLUME 2
Anatomidae
Larocheidae
Depressizonidae
Sutilizonidae
Temnocinclidae

Daniel L. Geiger

Monograph of the Little Slit Shells

Daniel L. Geiger

Santa Barbara Museum of Natural History Monographs Number 7
Studies in Biodiversity Number 5

2559 Puesta del Sol Road
Santa Barbara, CA 93105-2936 USA
http://www.sbnature.org

Library of Congress Cataloging-in-Publication Data

Geiger, Daniel L.
Monograph of the little slit shells / Daniel L. Geiger. -- 1st ed.
p. cm. -- (Studies in biodiversity ; no. 5)
Includes bibliographical references and index.
ISBN 978-0-936494-45-6 (alk. paper)
1. Scissurellidae. 2. Archaeogastropoda. I. Title.
QL430.5.S25G45 2012
594'.32--dc23
2012015342

ISBN (13) 978-0-936494-45-6

First Edition

Entire layout and design by Daniel L. Geiger. Images by Daniel L. Geiger, unless otherwise noted.
Typeset in Optima 18 point (sans serif headers) and Arno Pro 11 point (most text) 10 point (specimen records), and 9 point (synonymies).
Printed and bound in China by Everbest Printing Company through FCI Print Group.

Fore-Edge Quick Index

The fore-edge of the volume shows edge markings indicating the position of the various parts. It permits the reader to quickly find the beginning of the desired section.

Anatomidae McLean, 1989

Chresonymy.

Anatominae McLean, 1989: 5.
Anatominae: Marshall, 1993: 286.
Anatominae: McLean, 1996: 16.
Anatominae: Geiger, 2003: 42–43.
Anatomidae: Geiger & Jansen, 2004a: 2–3, 5.
Anatomidae: Zelaya & Geiger, 2007: 395.
Anatomidae: Geiger & Sasaki, 2009: 9.
Anatomidae: Geiger & McLean, 2010: 15.

Synonyms.

+ Schizotrochidae Iredale & McMichael, 1962: 30 [unavailable, no diagnosis, see Bouchet & Rocroi, 2005].

Type genus. *Anatoma* Woodward, 1859.

Other genus. *Sasakiconcha* Geiger, 2006.

Description. Shell trochiform (*Anatoma*), depressed (*Sasakiconcha*); no nacre, thin, shoulder rounded or angulated. Sculpture variable: axials only, predominantly axials, reticulate; no spiral keel(s) on base. Protoconch sculpture smooth, flocculent, reticulate; varix if present not connecting to embryonic cap; apertural margin sinusoidal or convex. Umbilicus without carina or keel, in profile describing smooth curve with base of shell; with or without funiculus. *Sasakiconcha* with umbilical trough. No brood pouch. Slit open (*Anatoma*), closed to foramen (*Sasakiconcha*), selenizone at periphery, start of selenizone on teleoconch variable, keels of moderate height. Operculum coiled with central nucleus. Radula R - 5 - n, variable; if similar to condition of Scissurellidae, lateral tooth 5 little to strongly elongated, but not broadened.

Distribution. Arctic to antarctic of all oceans (Fig. 591). For specific localities, see individual species.

Geologic range. Paleocene through Recent.

Comparisons. The trochiform, brittle shells with the peripheral selenizone and slit, and an umbilicus continuously sloping with the base, distinguish Anatomidae from the other families.

Scissurellidae have the selenizone on the shoulder and a radula with the lateral tooth 5 broadened and not elongated.

Sutilizonidae and Temnocinclidae have a limpet-shaped shell, 2–4 lateral teeth, and are found exclusively at hydrothermal vents.

Larocheidae lack a slit, foramen, or selenizone and have a radula with lateral teeth 1–4 similar.

Depressizonidae have a calyptraeiform shell.

Remarks. McLean (1989) introduced the subfamily Anatominae for *Anatoma* and *Sukashitrochus*. The inclusion of *Sukashitrochus* in Anatominae/Anatomidae has been controversial (Numanami & Okutani, 1990; Marshall, 1993; Sasaki, 1998; Bandel, 1998; Lozouet, 1998; Amitov & Zhegallo, 1998). Phylogenetic analysis of both shell characters (Geiger, 2003) and molecular data (Geiger & Thacker, 2005) showed that *Sukashitrochus* is more closely related to *Scissurella* than to *Anatoma*. According to McLean (1989), *Sukashitrochus* shares radular similarities with

Anatoma. However, radular similarities between *Sukashitrochus* and other genera in Scissurellidae s. s. are greater; in particular, the much broadened and asymmetrically serrated lateral tooth 5 is shared with other scissurellid genera, whereas in the anatomid species with radulae similar to Scissurellidae, lateral tooth 5 is enlarged by elongation and is more symmetrically serrated.

McLean (1989: 5) also cited similar "reticulate" [= flocculent of this contribution] protoconch sculpture in *Sukashitrochus* and *Anatoma*. It is not clear on what observations that was based. At the time, the only published protoconch for *Sukashitrochus* was the smooth protoconch of *Suk. mariasi* by Herbert (1986), who considered it diagnostic for the genus. As discussed by Geiger (2003), *Sukashitrochus* has three protoconch types, namely smooth, flocculent, and irregular axials; hence, protoconch sculpture cannot be considered diagnostic for *Sukashitrochus*.

Anatoma Woodward, 1859

Chresonymy.

Anatoma Woodward, 1859: 204.
Scissurella (*Anatoma*): Brookes Knight *et al.*, 1960: I221.
Scissurella (*Anatoma*): McLean, 1967: 404–405.
Anatoma: Montouchet, 1972: 2.
Anatoma: Herbert, 1986: 610.
Anatoma: McLean, 1996: 17.
Anatoma: Bandel, 1998: 26.
Anatoma: Geiger, 2003: 41–42.
Anatoma: Geiger & Jansen, 2004a: 6.
Anatoma: Zelaya & Geiger, 2007: 395–396.
Anatoma: Geiger & Sasaki, 2009: 9.
Anatoma: Geiger & McLean, 2010: 15.
Anatoma: Høisaeter & Geiger, 2011: 90.

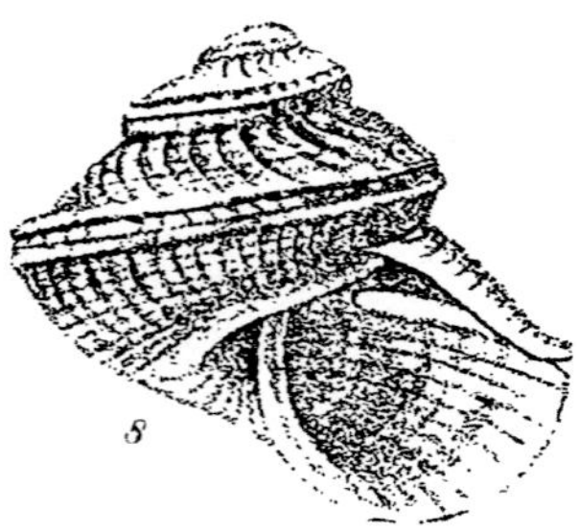

Figure 590. Woodward's (1859) misidentified illustration of *Scissurella* (*Anatoma*) *crispata*.

Synonyms.

+ (objective) *Schizotrochus* Monterosato, 1877a: 416. ***Type species.*** *Scissurella crispata* Fleming, 1828 (M).
+ *Hainella* Bandel, 1998: 36–37. ***Type species.*** *Scissurella euglypta* Pelseneer, 1903 (OD).
+ *Thieleella* Bandel, 1998: 35. ***Type species.*** *Scissurella amoena* Thiele, 1912 (OD). ***Etymology.*** Named after Johannes Thiele (1860–1935: Bieler & Boss, 1991).
+ *Pagodella* Bandel, 1998: 2 [*nomen nudum*].

Misidentification.

not *Anatomus* Montfort, 1810 (Annelida: Polychaeta: Serpulidae. See McLean, 1967; Herbert, 1986).

Type species. *Scissurella crispata* Fleming, 1828 (M: misidentified). The specimen figured as *Scissurella* (*Anatoma*) *crispata* by Woodward (1859: pl. 46, fig. 8. Fig. 590) from Norway is here identified as *An. aspera* based on the strength and density of the

sculpture on the shoulder. The overall shell shape is slightly more similar to the sympatrically occurring *An. tenuisculpta*, which, however, has much finer axial cords on the shoulder (see Høisaeter & Geiger, 2011). Woodward's figure certainly does not show *An. crispata*. Notice that the figure captions for Woodward's plate 46 are transposed for figures 8 and 9, respectively. This transposition caused complications with *Scissurella/Anatoma mantelli*; see Remarks under *Sci. mantelli* and *An. flemingi* for details. *Scissurella crispata* is here explicitly designated as the type species of *Anatoma* according to ICZN Art. 70.3.1.

Other species. *Anatoma aedonia* (Watson, 1886).
Anatoma africanae (Barnard, 1963).
Anatoma agulhasensis (Thiele, 1925).
Anatoma alta (Watson, 1886).
Anatoma alternatisculpta Geiger & McLean, 2010.
Anatoma amoena (Thiele, 1912).
Anatoma amydra Geiger & Marshall, 2012.
Anatoma argentinae (Zelaya & Geiger, 2007).
Anatoma aspera (Philippi, 1844).
Anatoma atlantica (Bandel, 1998).
Anatoma aupouria (Powell, 1937).
Anatoma australis (Hedley, 1902).
Anatoma austrolissa Geiger & Sasaki, 2008.
Anatoma bathypacifica (Geiger & McLean, 2010)
Anatoma bertheloti (d'Orbigny, 1840) [*nomen dubium*].
Anatoma biconica new species.
Anatoma boucheti Geiger & Sasaki, 2008.
Anatoma breveprima new species.
Anatoma concinna (A. Adams, 1862).
Anatoma conica (d'Orbigny, 1841).
Anatoma disciformis (Golikov & Sirenko, 1980).
Anatoma emilioi Geiger, 2011.
Anatoma equatoria (Hedley, 1899).
Anatoma euglypta (Pelseneer, 1903).
Anatoma eximia (Seguenza, 1877).
Anatoma finlayi (Powell, 1937).
Anatoma flemingi (Marshall, 2002).
Anatoma flexidentata Geiger & Sasaki, 2008.
Anatoma fujikurai Sasaki, Geiger & Okutani, 2010.
Anatoma gephyra Maxwell, 1992 †.
Anatoma globulus new species.
Anatoma gunteri (Cotton & Godfrey, 1933).
Anatoma hyposculpta new species.
Anatoma indonesica Bandel, 1998.
Anatoma janetae Geiger, 2006.
Anatoma janusa new species.
Anatoma japonica (A. Adams, 1862).
Anatoma keenae McLean, 1970.
Anatoma kelseyi (Dall, 1905).
Anatoma kopua Geiger & Marshall, 2012.
Anatoma lamellata (A. Adams, 1862).

Anatoma lyra (Berry, 1947).
Anatoma maxima (Schepman, 1908).
Anatoma megascutula Geiger & Marshall, 2012.
Anatoma micalii new species.
Anatoma miocenica (Laws, 1939) †.
Anatoma muelleri new species †.
Anatoma munieri (Fischer, 1862).
Anatoma orbiculata new species.
Anatoma parageia Geiger & Sasaki, 2009.
Anatoma peruviana (Geiger & McLean, 2010).
Anatoma philippinica (Bandel, 1998).
Anatoma planapex new species.
Anatoma plicatazona Geiger & McLean, 2010.
Anatoma porcellana new species.
Anatoma proxima (Dall, 1927).
Anatoma pseudoequatoria (Kay, 1979).
Anatoma quadraxialis new species.
Anatoma rainesi Geiger, 2003.
Anatoma rapaensis Geiger, 2008.
Anatoma regia (Mestayer, 1916).
Anatoma rhynchodentata new species.
Anatoma rolani Geiger & Fernández-Garcés, 2010.
Anatoma sagamiana (Okutani, 1964).
Anatoma schanderi Høisaeter & Geiger, 2011.
Anatoma schioettei Høisaeter & Geiger, 2011.
Anatoma shiraseae Numanami & Okutani, 1990.
Anatoma sinuosa new species.
Anatoma soyoae (Habe, 1951).
Anatoma tabulata (Barnard, 1964).
Anatoma tangaroa Geiger & Marshall, 2012.
Anatoma tenuis (Jeffreys, 1877).
Anatoma tenuisculpta (Seguenza, 1877).
Anatoma tobeyoides Geiger & Jansen, 2004.
Anatoma tongaensis Ladd, 1970 †.
Anatoma umbilicata (Jeffreys, 1883).
Anatoma weddelliana (Zelaya & Geiger, 2010).
Anatoma yaroni Herbert, 1986.
Anatoma zancliformis new species.

Species removed from *Anatoma*. *Schizotrochus palaeompaloides* Nordsieck, 1973 [*Atlanta peronii*: Atlantidae].
Anatoma danica Bandel, 1998 [*Leptomaria niloticiformis*: Pleurotomariidae].
Scissurella cebuana (Bandel, 1998).

Etymology. Ana- return, tomus: cut. A cut along the whorl. Derived from *Anatomus* Montfort "spelled properly" (Woodward, 1859: 204; OD). Gender feminine.

Woodward's spelling correction related to the Adams & Adams (1853–1858) assignment of *Scissurella crispata* to Montfort's genus *Anatomus*. Woodward noted that *Anatomus* referred to a serpulid polychaete, and that *Scissurella crispata* should be distinguished from other members of *Scissurella*: "If it should still be considered desirable to have a subgeneric name for *Scissurella crispata*, I have no objection

to the adoption of *Anatoma*, provided it be spelled properly, and not attributed to Montfort." Accordingly, Woodward did not emend the genus name, but proposed a new one with explicit designation of the type species. The two spellings are significantly different by ICZN criteria and cannot be considered alternate spellings of the same genus.

Description. Shell large, trochiform. Protoconch smooth, with flocculent or reticulate sculpture, varix absent or faint, not forming bridge to embryonic cap. Spire prominent. Slit deep, open. Slit and selenizone at periphery or slightly above; keels usually prominent, at right angle to shell axis. Umbilicus open, often very constricted, no carina, with or without funiculus. Operculate. Radula as for family.

Distribution. From Arctic to Antarctic (Fig. 591), 20–6000 m, usually 100–1000 m. The greatest species diversity is found in the Indo-Malayan Archipelago. Deep records of Anatomidae are rare because deep sampling is uncommonly executed (see Geiger, 2008b, Geiger & Sasaki, 2009; Figs 26, 27). The deepest localities are those of *An. tenuis* (5500–5987 m) and *An. argentinae* (5200 m).

Geologic range. Paleocene through Recent.

Comparisons. *Sasakiconcha* has an overall depressed shell, a closed foramen at maturity, and a very short teleoconch I of less than 0.1 whorl.

Differentiation of species. Species are distinguished by overall shell shape (changes with ontogeny to a certain degree), protoconch sculpture, length of teleoconch I, sculpture on various portions of the shell, presence or absence of a funiculus, and width of the sutsel, particularly at onset of the selenizone and when fully grown.

Characters that are occasionally useful include: details of selenizone (orientation, height, and strength of keels; strength and spacing of lunules; slit geometry), and the radula (rhipidoglossate, flexoglossate, details of individual tooth morphology).

Variable characters include length of teleoconch I (± 0.125 whorl = range of 0.25 whorl), strength and density of sculpture, generally by about 25%, and overall shell

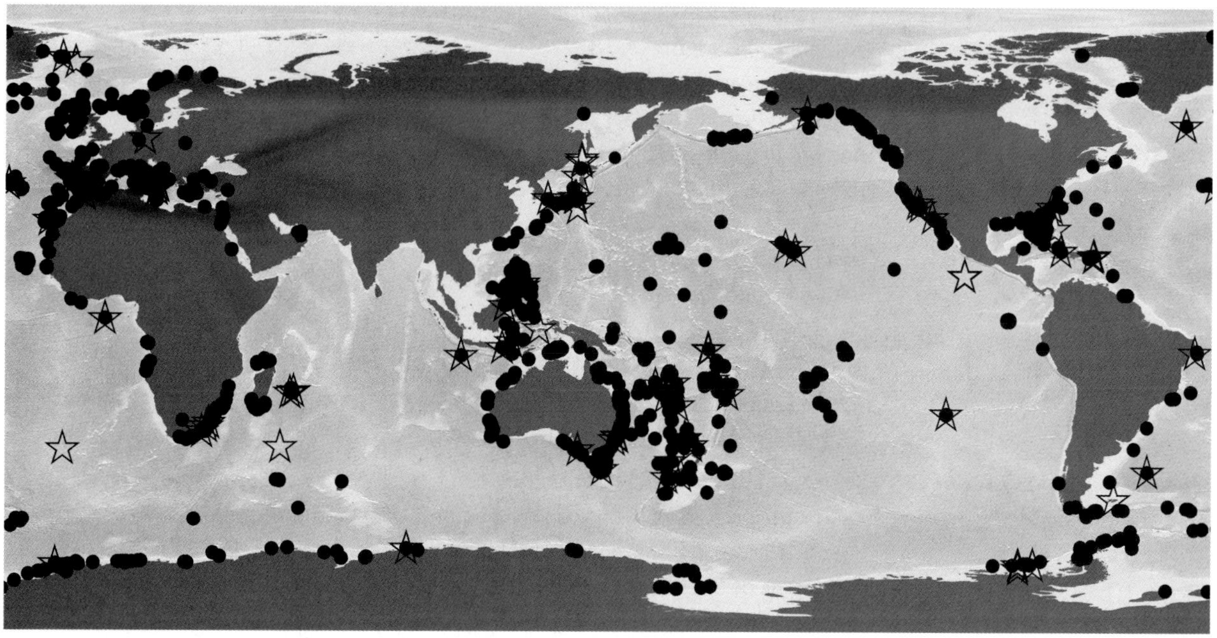

Figure 591. Global distribution of *Anatoma* species. ☆ type locality. ● specimen record. ■ literature record.

shape (more globular in small specimens, taller in large specimens). Caution needs to be exercised with respect to protoconch sculpture and the keels of the selenizone to recognize erosion and abrasion.

Remarks. Woodward (1859) introduced the name *Anatoma* and gave as type species *Sci. crispata*. He also illustrated *Sci.* (*An.*) *crispata* on plate 46, figure 8. The caption to plate 44 identifies figure 9 as *crispata*, but this is obviously in error, as discussed by Marshall (2002). Only one specimen was assigned to Woodward's new subgenus, and only one of the specimens illustrated is an *Anatoma*. The other eight specimens are clearly *Scissurella* or *Sinezona*.

Bandel (1998) introduced two new anatomid genera, recognizing a total of three: *Anatoma* Woodward, 1859 (type *Sci. crispata*), *Hainella* Bandel, 1998 (type *Sci. euglypta*), and *Thieleella* Bandel, 1998 (type *Sci. amoena*). The diagnostic characters listed by Bandel (1998) are summarized in Table 5.

The distinction between rounded and scalar whorls made by Bandel is difficult to understand given his generic assignments and illustrations. For instance, the rounded, trochiform *Anatoma* s. s. includes compatible species such as *An. americana*, *An. proxima*, and *An. alta*, but also species that have a very scalar appearance such as *An. indonesica* and *An. agulhasensis*. Furthermore, *Th. amoena* is almost identical to *An. indonesica* as illustrated by Bandel (1998), while *Th. reticulata* is almost identical to the illustrated *An. alta*. *Hainella philippinica* is intermediate with respect to overall shell shape between *An. indonesica* and *An. proxima*, and not unlike *Th. amoena*. With respect to height to width ratio, *Hainella* tends to contain shells that are somewhat wider than those in the other two generic concepts of Bandel. However, no discrete boundary can be determined. Accordingly, differences in overall shell shape cannot be recognized, and this supposedly diagnostic character is dismissed. Given the lack of differentiation between *Anatoma* s.s. and *Hainella*, Geiger (2003) synonymized the latter under the former.

The validity of *Thieleella* has been uncertain as discussed in Geiger & Jansen (2004). Given that protoconch sculpture has been shown to be of no utility in the generic classification in *Scissurella*, *Sinezona*, and *Sukashitrochus* (Geiger, 2003), the value of the character in Anatomidae is rather suspicious (Geiger & Jansen, 2004). Geiger (2003) speculated on limited data that the absence of eyes reported in some *Thieleella* species might be a synapomorphy for the genus. However, several *Anatoma* species without eyes have been found since. Reticulate protoconch has been used as one of the characters to diagnose *Retiskenea* Bouchet & Warén,

Character	***Anatoma***	***Hainella***	***Thieleella***
Whorls	rounded	scalar	scalar
Apical side	—	flattened	—
Selenizone	lateral with lamellae	lateral, peripheral with lamellae	lateral with lamellae
Shell shape	wider than high	as high as wide wider than high	higher than wide
Protoconch sculpture	smooth faint	granular smooth	reticulate

Table 5. Diagnostic characters for generic distinction of *Anatoma*, *Hainella*, and *Thieleella*. After Bandel (1998).

2001 (Vetigastropoda: Neomphalidae), though Okutani & Fujikura (2002) have suggested a placement among skeneids.

Examination of the protoconch of additional species has shown that reticulate sculpture is not a discrete state, but that there are fluid transitions from open flocculent, to clustered flocculent (*An. lyra, An. aspera, An. gunteri*), to irregular reticulate (*An. bathypacifica*), to coarse hexagonal-reticulate (*An. flemingi*), and to tight quadrate-reticulate (*An. argentinae*). Drawing any boundaries for the diagnostic character state would be arbitrary, hence, unjustifiable. As a consequence, *Thieleella* is here considered a synonym of *Anatoma*, following Høisaeter & Geiger (2011).

The genus *Pagodella* was used in the abstract of Bandel (1998) but is not found elsewhere in that publication. Although all new names are indicated in the abstract, *Thieleella* is not given. It appears that *Pagodella* was an earlier manuscript name for *Thieleella*, and that the replacement was not carried out throughout the manuscript.

Anatoma aedonia (Watson, 1886)

Figures 592–594

Chresonymy.

Scissurella aëdonia Watson, 1886: 114, pl. 8, fig. 3 (Fig. 592).
Scissurella aëdonia Kobelt, 1887: 428.
Scissurella aedonia: Pilsbry, 1890: 56, pl. 68 [erroneously pl. 58 in text], figs 1–3.
Scissurella aedonia: Thiele, 1912a: 6, pl. 1, figs 10–11.
Scissurella aedonia: Morretes, 1949: 55.
Anatoma aedonia: Rios, 1985: 10, pl. 5, fig. 34 [copy figure Watson, 1886].
Anatoma aedonia: Lozouet, 1986: 112.
Anatoma aedonia: Absalão, 1989: 2.
Anatoma aedonia: Rios, 1994: 21, pl. 5, fig. 37 [copy figure Watson, 1886].
Anatoma aedonia: Barros *et al.* 2001: 12, fig. 2B.
Anatoma s.l. *aedonia*: Geiger, 2003: 74.
Anatoma aedonia: Geiger *et al.*, 2007: fig. 5C [SEM of NHMUK syntype].
Anatoma aedonia: Rios, 2009: 24, text-fig. [copy figure Watson, 1886].
Anatoma aedonia: Geiger & McLean, 2010: 22.
Anatoma aedonia Geiger & Fernández-Garcés, 2010: 67.

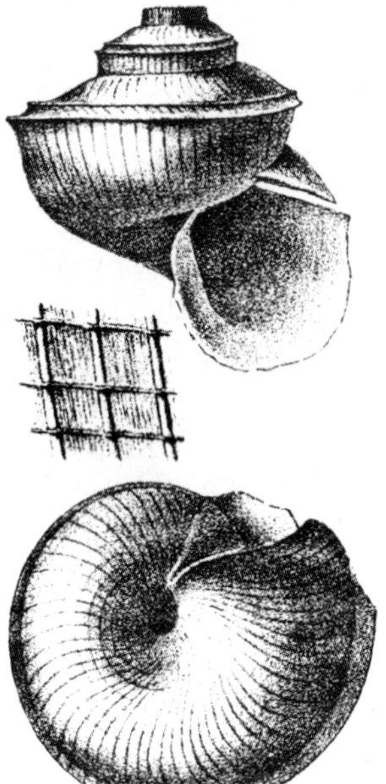

Figure 592. Original illustration of *Scissurella aedonia*.

Type material. Syntypes Pernambuco (NHMUK 1887.2.9.398–401, 4: all but one syntype NHMUK 1887.2.9.398 destroyed by Byne's/glass disease: Geiger *et al.*, 2007: fig. 5C; RMNH, 2; NMW 155.158.600, 6). Tristan da Cunha = Nightingale Island (NHMUK 1887.2.9.402, 1: 0.099" × 0.09" [= 2.52 × 2.27 mm: L × W]).

NMW 155.158.600 (Fig. 593C) designated here as lectotype with the express purpose of taxon stabilization.

Type locality. Not fixed (OD), syntypes from Station 122, 350 fms [= 640 m], off Pernambuco [Brazil], 9°5'S, 34°50'W, and station 135c, 100–155 fms. [= 183–274 m], Nightingale Island, Tristan de Cunha [St. Helena], 37°25'30"S, 12°28'30"W. Station 122 here restricted as type locality as the locality of the lectotype.

Etymology. Aedonios, Greek, belonging to a nightingale (OD), referring to one of the localities, which is here restricted.

Description. Shell moderate size (2.5 mm. Fig. 593) trochiform, biconical. Protoconch of 0.75 whorl, almost smooth, some flocculent elements, faint spiral line, apertural varix not connected to embryonic cap, apertural margin sinusoidal. Teleoconch I of 0.66 whorl, approximately 19 axial cords, interstices with hint of faintest reticulate pattern only visible under high magnification in SEM. Teleoconch II of at least 1.6 whorls (1.5 mm specimen). Shoulder almost straight with strongly impressed suture; axials proportionally increasing in strength with growth from thin to thick cords, maintaining same density with growth. First spiral line shortly after onset of selenizone, increasing in number forming reticulate pattern with axials, but always significantly weaker than axial cords; at intersection of spiral lines and

axial cords distinct, somewhat elevated points. Base with slight constriction below selenizone, sculpture similar to shoulder, axials 50% denser, spirals at same density but somewhat stronger, forming axially elongated rectangles; spirals not reaching

Figure 593. *Anatoma aedonia*. **A–B.** Puerto Rico, 18.633°N, 65.158°W (USNM 780328, 5). **C.** *Scissurella aedonia.* Lectotype. 350 fms [= 640 m], off Pernambuco [Brazil], 9°5'S, 34°50'W (NMW 155.158.600). Scale bars shell = 1 mm. Scale bars protoconch = 100 μm.

to selenizone (one spiral "missing"). Intersection of axial cords and spiral lines forming distinct, somewhat elevated points. Umbilicus very narrow, with distinct funiculus. Aperture rounded, flared. Selenizone slightly above periphery; keels of moderate strength and elevation, slightly adapically deflected; slit open with parallel margins.

Animal unknown.

Comparisons. *Anatoma alta* from the Caribbean and the northeastern Atlantic has less distinct sculpture, is usually taller than wide, and has a very short teleoconch I (0.167 vs. 0.66 whorl).

Anatoma alternatisculpta from the Caribbean has a more rounded profile of the whorls and the sutsel on the early whorls is much narrower than the width of the selenizone.

Anatoma atlantica from the Caribbean and adjacent Atlantic has a disc-shaped shell with no sutsel on the early whorls.

Anatoma plicatazona from the Caribbean has distinctly upturned keels of the selenizone.

Anatoma proxima from the Caribbean and the northeastern Atlantic is overall more globular, has denser sculpture on the teleoconch, and lacks a distinct spiral cord in the position of the selenizone.

Anatoma rolani from the Caribbean is more discoidal in overall shape and has very dense axial cords, with interstices narrower than cords.

Anatoma boucheti from East Africa has a more narrow sutsel on the early whorls and a teleoconch I of >1 whorl.

Figure 594. Distribution of *Anatoma aedonia*. ☆ type locality. ● specimen record.

Anatoma equatoria from the Indo-Malayan Archipelago to the Central Pacific has a shorter teleoconch I (0.5 vs. >0.75 whorl) and the spiral lines on the base reach up to the selenizone.

Anatoma eximia from the Mediterranean and the adjacent Atlantic has a more narrow umbilicus, usually a shorter teleoconch I (0.5 vs. >0.75 whorl), and the base bulges further outward.

Anatoma munieri from the Indian Ocean to the Western Pacific is more turreted, has a wider sutsel (>2× vs. <2× width of selenizone), and a sharp constriction below the selenizone with subsequent spiral keel on the mid-base.

Anatoma kopua from temperate eastern Australia is larger (4.5 vs. 2.5 mm), has a protoconch without apertural varix, and 30–50% denser spiral lines on the base.

Anatoma rhynchodentata from the Indo-Malayan Archipelago has much more finely reticulate sculpture, a shallow but distinct constriction below the selenizone with subsequent shallow spiral keel, and the sutsel on early whorls is more narrow (> vs. < than width of selenizone).

Distribution. Caribbean, Western Atlantic to Brazil (Fig. 594).

Specimen records.

Bahamas. 713 m, off Cay Sal Bank, 23.720°N, 80.750°W (FLMNH 261556, 1).

Barbados. 183 m, off Pelican Island, 13.167°N, 59.567°W (USNM ex 500680, 1).
Brazil. 640 m, *Challenger* expedition St. 122, Pernambuco, 9.083°S, 34.833°W (RMNH, 2). 600 m, Pernambuco, Brazil, 9.200°S, 34.100°W (DLG 1129, 2).
Puerto Rico. 18.633°N, 65.158°W (USNM 780328, 5).
USA. 201 m, off Western Dry Dock Rocks, Florida, 24.445°N, 81.928°W (USNM 449425, 1). 538 m, off Fernandina, Florida, 30.669°N, 81.463°W (USNM 108147, 15). 101 m, Gulf of Mexico, 27.950°N, 92.017°W (USNM 780312, 3).

Specimen records.

Brazil. Off Esipirito Santo, 22.300°S, 40.800°W (Absalão, 1989).

Remarks. Although the majority of the *Challenger* expedition material is found in the NHMUK, some has been dispersed as noted by Lingwood (1981).

Of the six syntypes in NMW, one was separated by F. Scarabino in 1996 to be designated as the lectotype. The formal designation has not been published; hence, that specimen has no elevated standing among the other syntypes. The specimen is here designated explicitly as the lectotype.

Anatoma africanae (Barnard, 1963)

Figures 595–596

Chresonymy.

Scissurella africanae Barnard, 1963: 13, not illustrated.
Scissurella africanae: Kensley, 1973: 28.
Anatoma africanae: Herbert, 1986: 661.
Anatoma africana [*sic*]: Lozouet, 1986: 112.
Anatoma s.l. *africana* [*sic*]: Geiger, 2003: 74.

Type material. Holotype (SAM A29933. Fig. 595), 1.75 × 1.5 mm.

Type locality. *Africana II* Station A1248, 400 m (OD); 36°48'S, 52°08'E (D. Kilburn & A. Marais, pers. comm.) [Gallieni Fracture Zone, ~2,500 km E of Port Elizabeth].

Etymology. Named after RV *Africana II* (OD).

Description. Shell small (1.75 × 1.55 mm W × H), trochiform, globular. Protoconch unknown. Teleoconch I unknown. Teleoconch II of approximately two whorls, suture moderately impressed; sutsel approximately as wide as selenizone. Shoulder moderately convex, approximately 100–120 axial cords on last whorl, approximately 25–30 very fine spiral lines at apertural margin. Base without constriction below selenizone, same density and strength of sculpture as on shoulder, aperture rounded. Selenizone slightly above periphery, quite wide; keels moderately strong, moderately elevated; lunules moderately distinct; slit open [broken].

Animal unknown.

Comparisons. *Anatoma agulhasensis* from South Africa has a distinct constriction below the selenizone on the base and has at least 50% fewer axial cords as well as spiral lines on shoulder and base.

Anatoma boucheti from the Indian Ocean has only half as many axial cords and spiral lines on the base and a depressed spire.

Figure 595. *Scissurella africanae.* Holotype. Station A1248, 400 m, [36°48'S, 52°08'E Gallieni Fracture Zone, ~2,500 km E of Port Elizabeth] (SAM A29933). Scale bar = 1 mm. Images by Simon van Noort and Liz Hoenson (SAM).

Anatoma austrolissa from the Indian Ocean and the Western Pacific has a more turreted overall shell shape, hardly any sculpture and a wider sutsel (1× vs. 1.5× width of the selenizone).

Anatoma finlayi from the Indian Ocean through the Indo-Malayan Archipelago has strong axial lamellae on shoulder and base and a more narrow selenizone.

Anatoma flexidentata from the Indian Ocean to the Western Pacific has only few spiral lines on the shoulder.

Anatoma japonica from the Indian Ocean through the Western Pacific has a more discoidal overall shell shape and a more narrow selenizone.

Anatoma yaroni from the Indian Ocean has a more discoidal overall shell shape and a more narrow selenizone.

Anatoma biconica from eastern Australia and New Zealand has a more biconical shell shape and at most one quarter the density of axial and spiral sculpture on shoulder and base.

Anatoma equatoria from the Indo-Malayan Archipelago to the Central Pacific has a more turreted overall shell shape, a wider sutsel (>2× vs. 1× width of the selenizone), and at most half the density of axial and spiral sculpture on shoulder and base.

Anatoma globulus from the Indo-Malayan Archipelago has an overall more globular shell shape, at most half the density of axial and spiral sculpture on shoulder and base, and a more narrow selenizone.

Anatoma orbiculata from West Africa has at most half the number of all sculptural elements, a more lenticular overall shell morphology, and the selenizone more towards the periphery.

Anatoma rhynchodentata from the Indo-Malayan Archipelago has half as many spiral lines on the shoulder at same whorl count (2 whorls) and at least 50% less dense reticulate sculpture on the base.

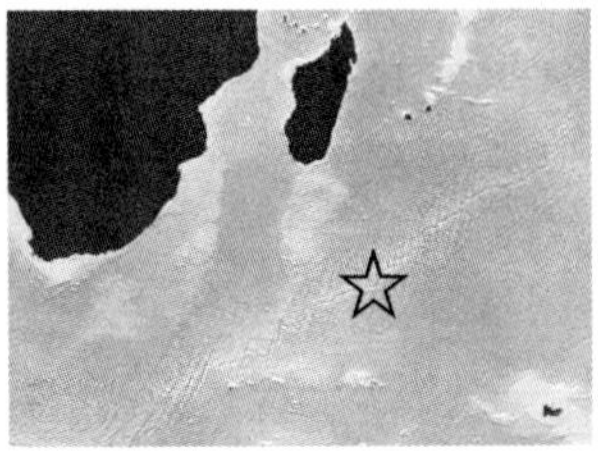

Figure 596. Distribution of *Anatoma africanae.* ☆ type locality.

Anatoma zancliformis from the Indo-Malayan Archipelago has about one quarter the axial cords and spiral lines and a more narrow sutsel on early whorls (<0.5× vs. 1× width of selenizone).

Distribution. Known only from type locality (Fig. 596).

Remarks. The description is moderately incomplete because the holotype was not available for examination by SEM. The species is most similar to *An. boucheti* from the Indian Ocean; the differences in the density of the sculpture on the base seem sufficiently distinct, though the similarities in overall shell shape and relatively wide selenizone are notable. *Anatoma boucheti* has a long teleoconch I (0.9 whorl), which could help to distinguish it from *An. africanae*, but information on the early whorl is unavailable for the latter.

Anatoma agulhasensis (Thiele, 1925)

Figures 597–603

Chresonymy.

Scissurella agulhasensis Thiele, 1925: 41(7)–42(8), pl. 13 (1), fig. 2 (Fig. 597).
Scissurella agulhasensis: Barnard, 1963: 301–302.
Scissurella agulhasensis: Kensley, 1973: 28, fig. 27.
Anatoma agulhasensis: Herbert, 1986: 611–613, figs 3, 11–14.
Anatoma agulhasensis: Lozouet, 1986: 112.
Anatoma agulhasensis: Herbert, 1987: 1, figs a–b.
Anatoma agulhasensis: Geiger, 2003: 73.
Anatoma agulhasensis: Marais & Marais, 2003: 12, text-fig.
Anatoma agulhasensis: Geiger & Jansen, 2004a: 6–7.

Figure 597. Original illustration of *Anatoma agulhasensis.*

Misidentifications.

not *Anatoma agulhasensis*: Bandel, 1998: 34–35, pl. 11, figs 4–6 [is *An. munieri*].
not *Anatoma agulhasensis*: Jansen, 1999: 48, figs 1–3 [is *An. munieri*].
Scissurella (Anatoma) jacksoni: Yaron, 1983: 270–271, pl. 4, figs 1–3 [is *An. agulhasensis*].
Anatoma jacksoni: Lozouet, 1986: 113 [is *An. agulhasensis*: reference to Yaron, 1983].
Scissurella jacksoni: Vine, 1986: 172 [is *An. agulhasensis*: Red Sea reference].
? *Anatoma jacksoni*: Dekker & Orlin, 2000: 17 [is *An. agulhasensis*: Red Sea reference].
Anatoma jacksoni: Geiger, 2003: 73 [mixed taxon concept *agulhasensis-munieri*].

Type material. Three syntypes (ZMB uncataloged, destroyed by Byne's disease: note Kilias, 1973; Geiger, pers. obs. 2002). Neotype [NMSA W181/T1916, separated from NMSA C7967, designated by Geiger & Jansen (2004: 6 = Herbert, 1986: figs 12, 14). Fig. 599A].

Type locality. (35°16'S 22°26.7'E), 155 m (OD); off Mossel Bay (Herbert, 1986; Kilburn, 1996). Neotype 120–140 m, off Mtamvuna River, Transkei [South Africa], 31.100°S, 30.200°E, live taken specimen.

Etymology. Agulhas- referring to the Agulhas Current around South Africa, -ensis indicating a locality: from the Agulhas Current region (OD).

Description. Shell of moderate size (to 2.5 mm. Fig. 598–600), high turbiniform, turreted. Protoconch of 0.75 whorl, sculpture flocculent, apertural varix present, apertural margin sinusoidal. Teleoconch I of 0.75 whorl, approximately 25 axial cords, spiral cord in position of selenizone. Teleoconch II of over three whorls. Sutsel 0.5× width of selenizone at onset of selenizone, 1.5× width of selenizone at apertural margin, showing 2–3 spirals between selenizone and shoulder on base of previous whorl. Shoulder somewhat convex to straight, approximately 70 raised axial cords, running over keel of selenizone, approximately six finer spiral lines. Base with slight constriction below selenizone, density of axials as on shoulder, approximately 20

Figure 598. *Anatoma agulhasensis*. 275–285 m, SSE of Tsitsikamma, South Cape, South Africa, 34.817°S, 24.150°E (NMSA V3623, 3). Scale bar shell = 1 mm. Scale bar protoconch = 100 μm.

Figure 599. *Anatoma agulhasensis*. **A.** Neotype. 120–140 m, off Mtamvuna River, Transkei [South Africa], 31.100°S, 30.200°E live taken specimen (NMSA W181/T1916). **B–C.** 120–140 m, off Mtamvuna River, Transkei [South Africa], 31.100°S, 30.200°E live taken specimen (NMSA C7967). Scale bars shells = 1 mm. Scale bars protoconch = 100 μm.

Figure 600. *Anatoma agulhasensis*. **A.** 275–285 m, SSE of Tsitsikamma, South Cape, South Africa, 34.817°S, 24.150°E (NMSA V3623, 3). **B.** 63–70 m, NE of Gypsey Hill, Natal, South Africa, 27.773°S, 32.657°E (NMSA S3480, 1). Radula shown in Figure 602. **C.** 110 m, 40 mi E of Quissico, Mozambique, 24.726°S, 34.766°E (USNM 761526, 1). Scale bars shell = 1 mm. Scale bars protoconch = 100 μm.

spirals of unequal strength, some as strong as axials, producing fine reticulate pattern. Selenizone at periphery; weak lunules at regular intervals; keels moderately strong, moderately elevated; slit open, margins parallel. Umbilicus open, moderately wide, no funiculus. Aperture round, inner lip somewhat flared.

Operculum (Figs 601E, 602C) thin, round, multispiral, nucleus central, covering entire aperture.

Radula (Figs 601, 602). Rachidian tooth trapezoidal, wide, central cusp largest, approximately seven cusps on each side. Lateral tooth 1 similar to rachidian tooth, apical cusp largest, five cusps on outer edge of tip. Lateral teeth 2–3 similar, apical cusp largest, 3–4 cusps on outer edge of tip. Lateral tooth 4 reduced, hook-shaped, with 2–3 minute points on inner edge of tip. Lateral tooth 5 somewhat enlarged by elongation, apical cusp largest, 4–5 cusps on each side. Inner marginal teeth with triangular tip, approximately seven cusps on each side of tip. Outer marginal teeth with spoon-shaped tip, many fine cusps on each side. Outermost marginal teeth flattened with fan of apical bristles. Radular interlock of central field moderate.

Comparisons. *Anatoma munieri* from the Indian Ocean to the Western Pacific is strongly turreted with a distinct lateral compression at the suture and has a much stronger constriction of the base below the selenizone.

Anatoma africanae from the Indian Ocean has twice as many axial cords and spiral lines on shoulder and base and lacks a constriction on the base below the selenizone.

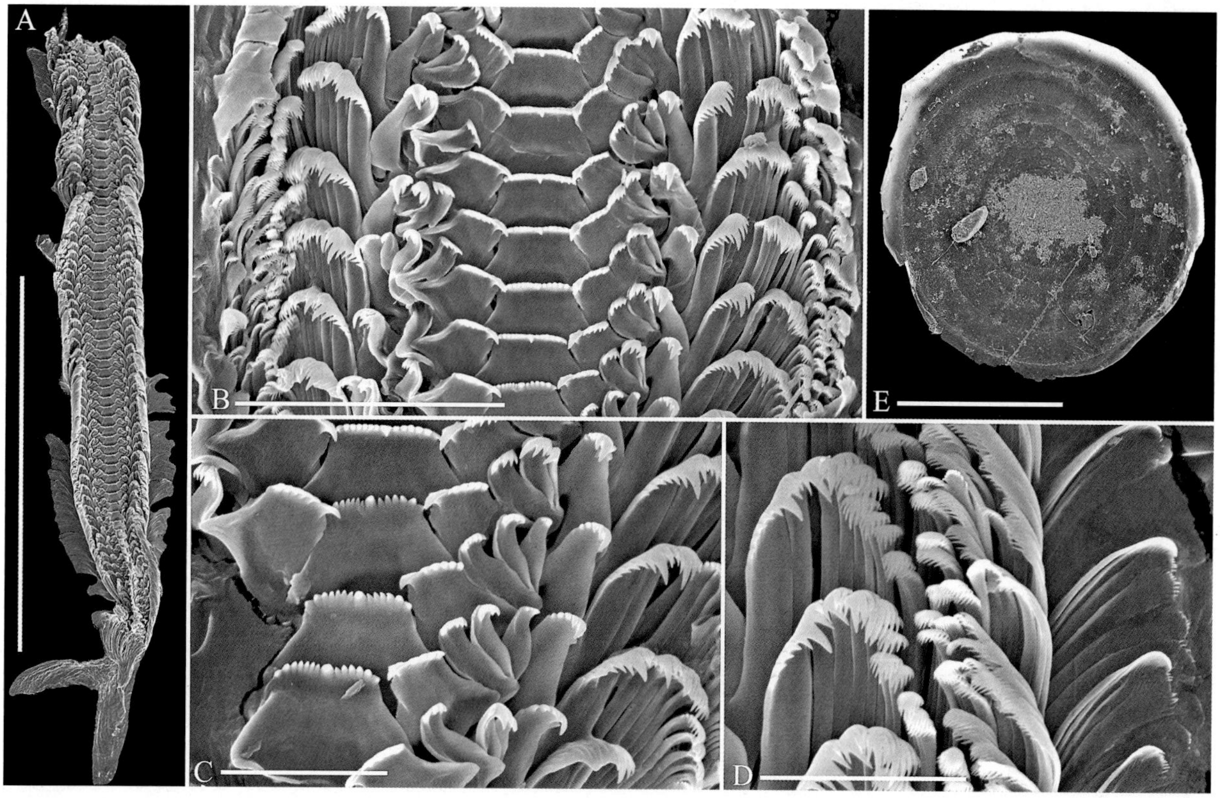

Figure 601. *Anatoma agulhasensis*. 130 m, off Mtamvuna River, Transkei, South Africa, 31.100°S, 30.200°E (NMSA C7967). **A–D.** Radula. **A.** Entire radula. Scale bar = 500 µm. **B.** Entire width of radula. Scale bar = 50 µm. **C.** Central field enlarged. Scale bar = 20 µm. **D.** Marginal teeth. Scale bar = 20 µm. **E.** Operculum. Scale bar = 500 µm.

Anatoma tabulata from eastern South Africa has a more depressed overall shell shape and fewer axial cords and spiral lines on shoulder and base.

Anatoma yaroni from the Indian Ocean has a more lenticular overall shell shape, very narrow sutsel on the early whorls, and weaker axial cords.

Anatoma japonica from the Indian Ocean to the Western Pacific has a more lenticular overall shell shape, a very narrow sutsel on the early whorls, and a proportionally narrower selenizone.

Anatoma boucheti from the Mascarene Islands has a more depressed apex, a longer teleoconch I (1 vs. 0.75 whorl), and weaker axial sculpture on the shoulder.

Anatoma flexidentata from the Indian Ocean has a shorter teleoconch I (0.25 vs. 0.75 whorl) and weaker as well as less dense axial and spiral sculpture.

Anatoma austrolissa from the Indian Ocean and the Western Pacific has a shorter teleoconch I (0.3 vs. 0.75 whorl) and almost no sculpture.

Anatoma aspera from the Mediterranean and adjacent Atlantic has a more stepped profile of the shell, a sunken apex, and usually fewer and sharper axial cords on the shoulder.

Anatoma australis from eastern Australia has a more discoidal profile of the shell, several fine spiral lines on teleoconch I, and the onset of the spiral lines on the shoulder is very close to the start of teleoconch II.

Anatoma conica from subantarctic South America lacks an apertural varix on the protoconch, has stronger and more elevated keels of the selenizone, and lacks the fine constriction below the selenizone on the base.

Anatoma eximia from the Mediterranean and adjacent Atlantic has a more turreted overall shell shape, a taller shell, and a wider sutsel on later whorls (>2× vs. <2× width of selenizone).

Anatoma orbiculata from West Africa has a wider shell, a protoconch with irregular reticulate sculpture and duller axial cords on the shoulder.

Anatoma kelseyi from the northeastern Pacific has reticulate sculpture on the protoconch and a more rounded whorl profile.

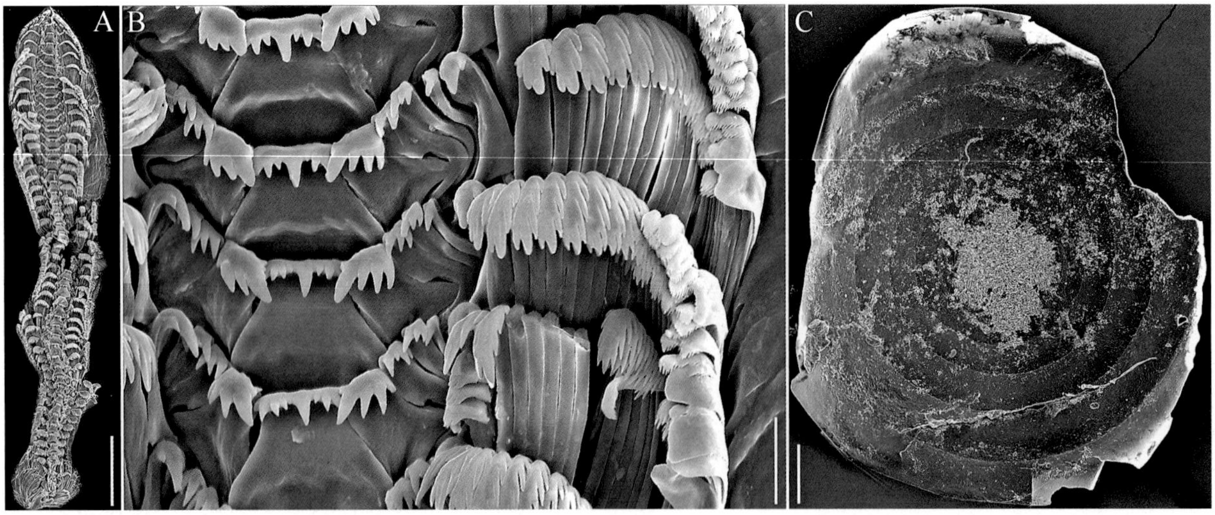

Figure 602. *Anatoma agulhasensis*. Radula and operculum. 63–70 m, NE of Gypsey Hill, Natal, South Africa, 27.773°S, 32.657°E (NMSA S3480, 1). Shell shown in Figure 600B. **A.** Whole radula. **B.** Half width of radula. **C.** Operculum. Scale bars A, C = 100 μm. Scale bar B = 10 μm.

Anatoma peruviana from Peru has reticulate protoconch sculpture and more spirals on the sutsel (6–7 vs. 3–4).

Anatoma sagamiana from Japan to the Indo-Malayan Archipelago has reticulate protoconch sculpture and a more narrow sutsel on early teleoconch II (< vs. > width of selenizone).

Anatoma schanderi from the boreal northeastern Pacific has reticulate protoconch sculpture and finer axial and spiral sculpture.

Anatoma tobeyoides from southeastern Australia lacks an apertural varix on the protoconch, has a much shorter teleoconch I (0.16 vs. 0.5–0.75 whorl), and very fine prosocline lines on teleoconch I and shoulder of early teleoconch II.

Distribution. Western Indian Ocean (Fig. 603).

Specimen records.

Israel. 49–55 m, off Mersat Abu Samra, 29.272°N, 34.745°E (AMS C.406384, 2). 137–146 m, off Nuweiba el Tarabin, 28.967°N, 34.617°E (AMS C.406386, 1).
France, Mayotte. 500 m, E Bandele Reef, 12.895°S, 45.268°E (MNHN, 1).
Mozambique.110 m, 40 mi E of Quissico, 24.726°S, 34.766°E (USNM 761526, 1).
South Africa. 275–285 m, SSE of Tsitsikamma, South Cape, 34.817°S, 24.150°E (NMSA V3623, 3). 106 m, off Mtamvuna River, Transkei, 31.100°S, 30.200°E (NMSA C7962, 2, 1: complete). 130 m, off Mtamvuna River, Transkei, 31.100°S, 30.200°E (NMSA C7967, 1: complete; NMSA W181/T1916, 1). 63–70 m, NE of Gypsey Hill, Natal, 27.773°S, 32.657°E (NMSA S3480, 1).
Senegal. Petite Côte St. L4, 14.583°N, 17.583°W (MNHN, 6).

Literature records.

South Africa. 229 m, off Cape St. Blaize, Tsitsikamma coast, 34.183°S, 22.150°E (SAM A9334). 227 m, off Cape Recife, Eastern Cape, 34.033°S, 25.750°E (SAM A9333). 100 m, off, Cape Morgan Lighthouse, Eastern Cape, 32.700°S, 28.367°E (NMSA B132). 345 m, off Qolora River, Transkei, 32.617°S, 28.433°E (NMSA C8760). 500 m, off Shixini Point, 32.383°S, 28.717°E (NMSA C9841). 75 m, off Mbashe River, 32.250°S, 28.883°E (NMSA C7912). 225 m, off Nthlonyane River, 32.050°S, 28.800°E (NMSA C7916). 68 m, off Mncwasa Point, Transkei, 32.050°S, 29.017°E (NMSA C7901). 74 m, off Mncwasa Point, Transkei, 32.050°S, 29.017°E (NMSA C7896). 71 m, off Whale Rock, Transkei, 31.917°S, 29.217°E (NMSA C7910). 90 m, off Whale Rock, Transkei, 31.917°S, 29.217°E (NMSA C7892). 410 m, off Whale Rock, Transkei, 31.917°S, 29.217°E (NMSA C7939). 155 m, off Rame Head, Transkei, 31.800°S, 29.350°E (NMSA C7891). 180 m, off Mgazi, Transkei, 31.633°S, 29.533°E (NMSA C7950; NMSA C7949). 48 m, off Mgazi, Transkei, 31.633°S, 29.533°E (NMSA C7971). 50 m, off N'tafufu River, Transkei, 31.550°S, 29.633°E (NMSA C7415). 140 m, off Mtamvuna River, Transkei, 31.100°S, 30.200°E (NMSA C284). 100 m, off Mtamvuna River, Transkei, 31.100°S, 30.200°E (NMSA C7960, 1: complete). 115 m, off Mtamvuna River, Transkei, 31.100°S, 30.200°E (NMSA C7958). 111 m, off Mtamvuna River, Transkei, 31.100°S, 30.200°E (NMSA C7980). 137 m, off Mtamvuna River, Transkei, 31.100°S, 30.200°E (NMSA C7984). 70 m, off Port Shepstone, Natal, 30.733°S, 30.467°E (NMSA B3670). 120 m, off Park Rynie, Natal, 30.333°S, 30.750°E (NMSA B3807). 130 m, off Park Rynie, Natal, 30.333°S, 30.750°E (NMSA

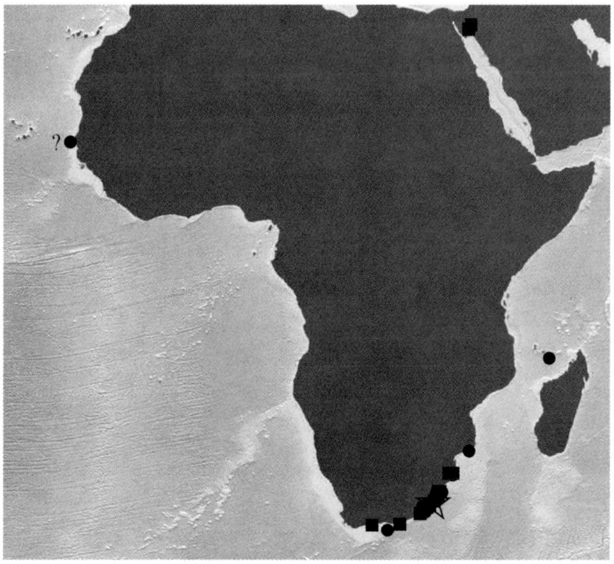

Figure 603. Distribution of *Anatoma agulhasensis*. ☆ type locality. ● specimen record. ■ literature record. ? questionable record.

D239). 150 m, off Umlass Canal, Natal, 29.967°S, 30.967°E (NMSA D1672). 82 m, off Port Grosvenor, Natal, 29.917°S, 31.000°E (NMSA C7995). 107 m, off Port Grosvenor, Natal, 29.917°S, 31.000°E (NMSA C1340). 105 m, off Port Grosvenor, Natal, 29.917°S, 31.000°E (NMSA C7975). 97 m, off Port Grosvenor, Natal, 29.917°S, 31.000°E (NMSA C7976). 210 m, off Durban, Natal, 29.883°S, 31.000°E (NMSA A2953). 80 m, off Ubombo, Transkei, 27.583°S, 32.083°E (NMSA 7954). 96 m, off Ubombo, Transkei, 27.583°S, 32.083°E (NMSA C7944; NMSA C7943). 100 m, off Sodwana Bay, Zululand, 27.550°S, 27.667°E (NMSA A5784) (Herbert, 1986).

Israel. 80 m, Elat, harbor, 29.550°N, 34.950°E. Elat, 29.550°N, 34.950°E. Elat, opposite electric power station, 29.550°N, 34.950°E (Yaron, 1983).

Egypt. 77 m, off Nuweiba el Tarabeen, Sinai, 28.967°N, 34.617°E (Yaron, 1983).

Oman. 205 m, Gulf of Oman. 285 m, Gulf of Oman (Yaron, 1983) [Questionable records, see Remarks].

Remarks. *Anatoma agulhasensis* has been reported from the Australian fauna (Bandel, 1998; Jansen, 1999). However, those specimens represent misidentified *An. munieri* and were discussed by Geiger & Jansen (2004: as *An. turbinata*). Small specimens (Fig. 600C) have a slightly more globular overall shell shape, and the small kink in the base, approximately in the position of the suture, is barely noticeable.

Literature records of *An. jacksoni* are somewhat problematic. The type is conspecific with *An. munieri*, a species occurring from the Indo-Pacific to Oman. Records of *An. jacksoni* from the Red Sea (Yaron, 1983) can be referred to *An. agulhasensis*. All specimens identified as *An. jacksoni* from Oman are in fact *An. munieri*. Specimens identified as *An. jacksoni* by Yaron (1983) from Oman may either be misidentified *An. munieri*, or the two species (*agulhasensis, munieri*) live sympatrically in Oman.

The rachidian tooth of the species is quite variable. It can either be wide with about six cusps on each side or more narrow with only approximately 3–4 cusps on each side. This type of variability has been observed in other species (e.g., *An. japonica*. Figs 757–759).

Anatoma alta (Watson, 1886)

Figures 604–607

Chresonymy.

Scissurella alta Watson, 1886: 113, pl. 8, fig. 1 (Fig. 604).
Scissurella alta Kobelt, 1887: 428.
Scissurella alta: Dautzenberg, 1889: 66.
Scissurella alta: Pilsbry, 1890: 55, pl. 68 [erroneously pl. 58 in text], figs 4–5.
Scissurella alta: Dall, 1889: 168.
Scissurella alta: Thiele, 1912a: 6, pl. 1, figs 8–9 [copy figure Watson, 1886].
Scissurella alta: Johnson, 1934: 68.
Scissurella alta: Jaume & Sarasúa, 1943: 56.
Scissurella alta: Poirier, 1954: 12.
Scissurella (*Schizotrochus*) *aspera alta*: Nordsieck, 1968: 10, pl. 1, fig. 14.
Scissurella alta: Abbott, 1974: 17.
Scissurella alta: Ortiz-Corps, 1985: 7.
Anatoma alta: Lozouet, 1986: 112.
Anatoma alta: Bandel, 1998: 33, pl. 10, figs 6–8.
Scissurella (*Schizotrochus*) *alta*: Nordsieck, 1982: 2, pl. 1, fig. 14.
Anatoma alta: Geiger, 2003: 73.
Scissurella alta: Ardovini & Cossignani, 2004: 19.
Anatoma cf. *crispata*: Segers *et al.*, 2009a: 65, pl. 5, figs 2–2a.
Anatoma cf. *crispata*: Segers *et al.*, 2009b: 5 [identification according to illustrations in Segers *et al*, 2009a].
Anatoma alta: Geiger & McLean, 2010: 15.
Anatoma alta Geiger & Fernández-Garcés, 2010: 67.

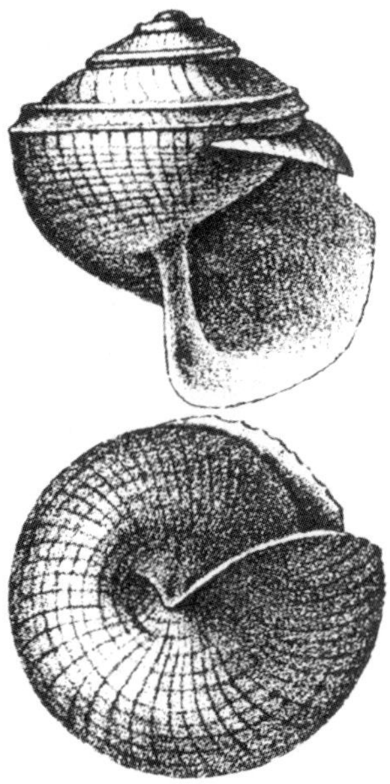

Figure 604. Original illustration of *Scissurella alta.*

Type material. Syntypes: 390 fms [= 713 m], *Challenger* Station 24, off Culebra Island [Puerto Rico, 18.333°N, 65.333°W] (NHMUK: 1887.2.9.396, 2: destroyed by Byne's/glass disease; NMW 1955.158.605, 5). 450 fms [= 823 m], *Challenger* station 75, off Fayal [= Faial, Azores, 38.583°N, 28.700°W] (NHMUK: 1887.2.9.397, 1: destroyed by Byne's/glass disease, 0.082 × 0.08" [= 2.08 × 2.03 mm. L × W]).

Lectotype (NMW 1955.158.605/1. Fig. 606C) here designated with the express taxonomic purpose of taxon stabilization. The remaining syntypes become paralectotypes.

Type locality. 390 fms [= 713 m], *Challenger* Expedition Station 24, off Culebra Island [Puerto Rico, 18.333°N, 65.333°W]. 450 fms [= 823 m], *Challenger* Expedition Station 75, off Fayal [= Faial, Azores, 38.583°N, 28.700°W] (OD).

Here explicitly restricted by selection of lectotype to locality of lectotype (*Challenger* Station 24).

Etymology. Not given (OD). Altus, Latin adjective for tall.

Description. Shell large (to 2.1 mm. Figs 605, 606), trochiform, biconical, inflated. Protoconch of 0.75 whorl; sculpture flocculent, somewhat axially arranged; no apertural varix, apertural margin slightly sinusoidal. Teleoconch I of 0.167 whorl, approximately

5–6 axial cordlets more distinct from suture to single axial in position of selenizone. Teleoconch II of >4 whorls. Shoulder convex, approximately 40 neat axial cordlets on first whorl, approximately 54 on fourth whorl of large specimen, cordlets somewhat more crowded on last third whorl; first fine spiral line after one third whorl, approximately 2–3 at apertural margin of large specimen. Base with slightly more dense straight axials, crossed by approximately 12 axial cords of similar strength forming fine reticulate pattern; first axial below suture; intersections of axials and spirals raised to minute points. Base continuously sloping with closed

Figure 605. *Anatoma alta.* [35 km NNE of Culebra Island, Puerto Rico, USA], 18.633°N, 65.158°W (USNM 614117, 22). Scale bar shell = 1 mm. Scale bar protoconch = 100 μm.

Figure 606. *Anatoma alta*. **A.** [35 km NNE of Culebra Island, Puerto Rico, USA], 18.633°N, 65.158°W (USNM 614117, 22). **B.** Between Cuba and Florida (USNM 94309, 9). **C.** Lectotype. 390 fms [= 713 m] *Challenger* Expedition Station 24, off Culebra Island [Puerto Rico, 18.333°N, 65.333°W] (NMW 1955.158.605/1). Scale bars shell = 1 mm. Scale bars protoconch = 100 μm.

umbilicus. Selenizone above periphery; keels low, of moderate strength; lunules fine and regular, more dense than axials.

Animal unknown.

Comparisons. *Anatoma alternatisculpta* from the Caribbean has the selenizone approximately in the middle of the whorl and distinct clathrate sculpture on the base.

Anatoma atlantica from the Caribbean and adjacent Atlantic has discoidal overall shell shape and finer axial and spiral cords.

Anatoma finlayi from the Indian Ocean to the Western Pacific (and possibly Caribbean) has the selenizone approximately in the middle of the whorl and much stronger axial lamellae on the shoulder.

Anatoma plicatazona from the Caribbean has the keels of the selenizone folded upwards and fine reticulate sculpture on shoulder and base.

Anatoma proxima from the Caribbean and eastern Atlantic has a longer teleoconch I (0.5 vs. 0.167 whorl), a shorter base compared to the spire, and denser sculpture.

Anatoma rolani from the Caribbean has discoidal shell shape with very dense axial cords, in which the interstices are more narrow than the cords.

Anatoma eximia from the Mediterranean and the adjacent Atlantic has a pronounced sutsel at least as wide as the selenizone, has a longer teleoconch I, the sculpture usually is more distinct, and the umbilicus is slightly wider.

Anatoma sinuosa from Easter Island has an overall more box-shaped, turreted shell profile, a flatter shoulder with sinusoidal profile, and fine clathrate sculpture on shoulder and base.

Distribution. Gulf of Mexico to Azores and Madeira in the Atlantic (Fig. 607).

Specimen records.

USA, Texas. 800–1200 m, 28.225°N, 87.067°W (MNHN, 2). 1097–1189 m, 28.333°N, 87.050°W (MNHN, 14). 123 m [222 km E of Corpus Cristi], 27.917°N, 94.817°W (USNM 780815, 2).

USA, Florida. Off Boynton, Palm Beach, 26.517°N, 80.067°W (FMNH 29982, 3). 503 m, in the Gulf Stream, off Boynton, Palm Beach County, 26.525°N, 80.067°W (FMNH

Figure 607. Distribution of *Anatoma alta*. ☆ type locality. ● specimen record. ■ literature record.

29982, 3). 201 m, off Fowey Light, 25.583°N, 80.100°W (USNM 449407, 1). Off Florida (UMMZ, 8).

USA, Puerto Rico. 713 m, off Culebra Island, 18.333°N, 65.333°W (NHMUK 1887.2.9.396, 1). 900 m, off S coast, S of Magueyes Island, 18.022°N, 66.888°W (DMNH, 6). [35 km NNE of Culebra Island], 18.633°N, 65.158°W (USNM 614117, 22).

Cuba. 23.633°N, 82.083°W (USNM 780314, 1). Between Cuba and Florida (USNM 94309, 9). 915 m, off Santa Clara Bay, 23.350°N, 80.600°W (MCZ 320437, 2).

Virgin Islands. (RMNH, 1).

Azores. 823 m, Fayal [= Faial], 38.583°N, 28.700°W (NHMUK 1887.2.9.397, 1).

Portugal, Madeira. 680 m, SE of Madeira, 32.667°N, 16.767°W (RMNH, 4). Covos 7 (FSC, 3). 587 m, [Funchal Bay], 32.633°N, 16.886°W (FSC, 1). 800 m, Fiade #5, S of Madeira, 32.600°N, 16.900°W (FSC, 1). 600 m, Fiade #7, S of Madeira, 32.600°N, 16.900°W (FSC, 1).

Literature records.

USA, Florida. 800 m, Key West. 20 m, Cape Fear (Bandel, 1998).

Cuba. 64 m, Gibara, 21.150°N, 76.183°W (1) (Jaume & Sarasúa 1943).

Azores. 165 m, Fayal, 38.583°N, 28.700°W. 823 m, Fayal, 38.583°N, 28.700°W (Dautzenberg, 1899).

Portugal, Madeira. [off Ponta del Sol], 32.654°N, 17.099°W (Segers *et al.*, 2009b).

Remarks. Both NHMUK syntypes were destroyed by Byne's disease by 1982.

Anatoma alta is typically taller than wide in fully grown specimens, but smaller specimens (Fig. 606A) are about as tall as wide. There is not a single specimen known with a complete apertural lip. Whether this is because the lip is very fragile and is commonly broken when alive, whether it is an indication of predation, or whether it is a collecting artifact is unknown.

Anatoma alternatisculpta Geiger & McLean, 2010

Figures 608–610

Chresonymy.

Anatoma alternatisculpta Geiger & McLean, 2010: 15–16, figs 11–12.

Anatoma alternatisculpta Geiger & Fernández-Garcés, 2010: 67.

Type material. Holotype (USNM 1132066. Fig. 608). Paratypes from type locality (USNM 1132067, 9). 137–146 m, off Pelican Island, Barbados, 13.167°N, 59.567°W (USNM 500678, 15. Fig. 609A–B). 165–183 m, off Lazaretto, Barbados, 13.133°N, 59.617°W (USNM 1132065, 1. Fig. 609C).

Type locality. 137 m, off Pelican Island, Barbados, 13.167°N, 59.567°W (OD).

Etymology. Alternati-, Latin for differing on two sides, -sculpta, Latin for sculptured, referring to the predominant axial sculpture on shoulder, while the base has predominantly spiral sculpture. Adjective (OD).

Description. Shell medium-sized (to 2.09 × 2.15 mm: W × H. Figs 608, 609), trochiform, globular. Protoconch of 0.875 whorl, 175 µm wide in holotype, embryonic cap with some flocculent sculpture, remainder smooth, weak apertural varix not connected to embryonic cap, apertural margin slightly sinusoidal. Teleoconch I of 0.66–0.75 whorl, 14–19 axial cords, some with spiral cord in position of selenizone. Teleoconch II of 2.66 whorls, suture moderately impressed. Shoulder strongly

convex, approximately 34 distinct equally spaced axial cords; first spiral line shortly after onset of selenizone, 3–4 thick spiral lines at apertural margin, concentrated in middle of shoulder, running over axial cords, forming minute thickenings. Base with weak constriction below selenizone, continuously sloping into umbilicus, approximately 20 axial cords of decreasing strength from selenizone towards middle of base, then increasing in strength towards umbilicus; about twice as many thick axial lines as on shoulder, weak below selenizone, increasing in strength to cords towards umbilicus. Umbilicus open, moderately wide. Selenizone at periphery; keels moderately strong, strongly elevated; lunules distinct; slit open, with parallel margins. Aperture round, flared, roof overhanging.

Figure 608. *Anatoma alternatisculpta*. Holotype. 137 m, off Pelican Island, Barbados, 13.167°N, 59.567°W (USNM 1132066). Scale bar shell = 1 mm. Scale bar protoconch = 100 μm.

Figure 609. *Anatoma alternatisculpta*. Paratypes. **A–B.** 137–146 m, off Pelican Island, Barbados, 13.167°N, 59.567°W (USNM 500678, 15). **C.** 165–183 m, off Lazaretto, Barbados, 13.133°N, 59.617°W (USNM 1132065, 1). Scale bar shell = 1 mm. Scale bar protoconch = 100 µm.

Animal unknown.

Comparisons. *Anatoma proxima* from the southeastern USA to Brazil has similar sculpture on shoulder and base, a less convex shoulder, and a protoconch with flocculent sculpture.

Anatoma atlantica from the Caribbean and the adjacent Atlantic is proportionally wider, has finer sculpture, and a much larger protoconch.

Anatoma aedonia from the Caribbean and southwestern Atlantic has a more scalar overall shell shape, a flatter shoulder, and finer axial sculpture on the shoulder.

Anatoma alta from the Caribbean to the northeastern Atlantic is more globular in shape, the selenizone is distinctly above the periphery, and has similar sculpture on shoulder and base.

Anatoma orbiculata from West Africa has a protoconch with irregular reticulate sculpture, spiral lines surrounding the umbilicus, and stronger axial sculpture on the shoulder.

Anatoma plicatazona from the Caribbean has a more turreted shell form and the keels of the selenizone are strongly folded towards the apex.

Anatoma rolani from the Caribbean has a more discoidal overall shell shape and much denser axial cords with interstices more narrow than the cords.

Anatoma finlayi from the Indian Ocean to the Western Pacific (and possibly Caribbean) has a more depressed overall shape and much stronger axial lamellae on the shoulder.

Anatoma janusa from the Azores has a wider sutsel on the early teleoconch (> vs. < width of selenizone) and more tightly spaced axial cords on the base.

Anatoma aspera from the Mediterranean and adjacent Atlantic has a spiral cord in on teleoconch I in position of the selenizone and usually about 50% more spiral lines on shoulder and base.

Anatoma philippiana from the Paleogene of Europe has an overall broader shell, more numerous spiral cordlets on the shoulder, and a protoconch with flocculent sculpture but without an apertural varix.

Anatoma gunteri from southeastern Australia has reticulate protoconch sculpture, sharper axial cords on the shoulder, and about 30% denser axials and spirals on the base.

Distribution. Known only from type locality (Fig. 610).

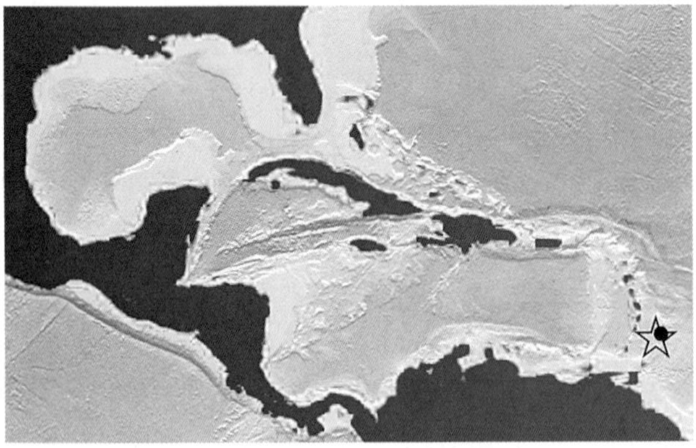

Figure 610. Distribution of *Anatoma alternatisculpta*. ☆ type locality. ● specimen record.

Anatoma amoena (Thiele, 1912)

Figures 611–615

Chresonymy.

Figure 611. Original illustration of *Scissurella amoena.*

Scissurella amoena: Thiele, 1912b: 187, pl. 11, fig. 11 (Fig. 611).
Scissurella amoena: Thiele, 1912a: 9, pl. 1, fig. 17.
Schizotrochus amoena [*sic*]: Carcelles, 1953: 161.
Schizotrochus amoenus: Powell, 1960: 126.
Schizotrochus amoenus: Egorova, 1982: 11, fig. 71 [copy figure Thiele, 1912b].
Anatoma amoena: Lozouet, 1986: 112.
Schizotrochus amoenus: Castellanos & Landoni, 1988: 10.
Anatoma amoena: Dell, 1990: 75, fig. 130.
Anatoma amoena: Numanami & Okutani, 1990: 100–102, figs 9–12 [shell], 13 [protoconch].
Anatoma amoena: Numanami & Okutani, 1992: 172.
Anatoma amoena: Branch *et al.*, 1991: 47, 58 (text-fig.), 62.
Anatoma amoena: Numanami, 1996: 18–19, figs 8A–E.
Thieleella amoena: Bandel, 1998: 35, pl. 11, figs 7–8 [shell], pl. 12, fig. 1 [protoconch].
Anatoma amoena: Gutt *et al.*, 2000: 44.
Thieleella amoena: Geiger, 2003: 74.
Thieleella amoena: Zelaya & Geiger, 2007: 408–411, figs 51 [map], 56–64 [shell SEM], 65–70 [external anatomy, radula].

Type material. Lectotype (ZMB-Moll 63047-a. Fig. 612A) (designated by Zelaya & Geiger, 2007: 409). Paralectotypes (ZMB-Moll 63047–b, 2). 3.5 mm.

Type locality. Gauss Station, Antarctica [66.033°S, 89.633°E] (OD).

Etymology. Amoena, Latin adjective for pleasant, delightful.

Description. Shell medium-sized (to 3.5 mm. Figs 612, 613), broadly trochiform, broader than high. Protoconch of 0.75 whorl, reticulate sculpture, apertural varix not connected to embryonic cap, apertural margin convex. Teleoconch I of 2.33 whorls, approximately 22 indistinct axial cords, spiral cord in position of selenizone forming strong angulation. Teleoconch II of 1.5 whorls; sutsel as wide as selenizone. Shoulder with approximately 30 axial cords, first spirals after 0.33 teleoconch II whorl, increasing in number to ten in outer two thirds of shell. Base with stronger spiral lines than on shoulder, weaker than axial cords on base. Selenizone at periphery, with elevated keels; lunules indistinct; slit open, with parallel margins. Umbilicus open, moderately wide, without funiculus.

Head-foot. Whitish, uniform. Cephalic tentacles long, circular, papillate. No eyes, eyestalks. Neck wide, with two smooth accessory cephalic tentacles on each side. Three pairs of long, sparsely micropapillate epipodial tentacles (Fig. 614B); first epipodial tentacle bifurcated at base with one additional short posterior pteriform appendix (possibly an epipodial sense organ); second, third tentacles simple. Papillae tubular, about 27–33 μm in length, arranged in whorls, with distal cilia. Pallial cavity with two large bipectinate ctenidia, left one about 10% larger. A relatively long papillate conical tentacle extending anteriorly, starting in posterior part of mantle sinus. Hindgut forming inclined S-shape.

Radula (Fig. 614A). Rachidian tooth trapezoidal at base, with nearly straight tip, showing main central cusp, five narrower cusps on each side. Lateral teeth 1–3

similar; tips elongate, tongue-like, with one prominent main cusp, 4–6 cusps on outer margin, 3–4 narrower cusps on inner margin. Lateral tooth 4 narrower than lateral teeth 1–3, with more delicate serration at tip. Lateral tooth 5 wide at base, with prominent main cusp, edges strongly serrated along elongated tip. Marginal teeth long, narrow, main cusp straight, sharp, numerous serrations on both sides; shafts without food groove. Radular interlock of central field moderately strong.

Comparisons. *Anatoma conica* from subantarctic waters of the Magellan Region has more and broader axial cords, a more narrow selenizone, and a protoconch with flocculent sculpture.

Anatoma euglypta from Antarctica has a biconical shell that is not turreted, extremely fine and dense axial and even finer spiral sculpture, and a protoconch with flocculent sculpture.

Anatoma shiraseae from Antarctica has a distinct keel on the shoulder, a distinct constriction below the selenizone, and a protoconch with flocculent sculpture.

Figure 612. *Anatoma amoena*. **A.** Lectotype. Gauss Station, Antarctica, 66.033°S, 89.633°E (ZMB-Moll 63047-b). **B–D.** 1579–1584 m, Weddell Sea, Antarctica, 62.779°S, 57.043°W (ZSM Moll 20062197, 1). Scale bars shell = 1 mm. Images by Diego Zelaya, from Zelaya & Geiger (2007: modified),

Anatoma weddelliana from Antarctica has a protoconch with very finely reticulate sculpture, a shorter teleoconch (0.75 vs. 1 whorl), and stronger, sharp axial riblets on the base.

Figure 613. *Anatoma amoena*. **A–B.** 530–549 m, Elatin Station 1997, Antarctica, 72.000°S, 172.467°E (NMNZ M242840, 2). **C.** 392 m, Antarctica, 72.083°S, 172.250°E (USNM 613010, 6). Scale bars shell = 1 mm. Scale bars protoconch = 100 μm.

Anatoma gunteri from eastern Australia has noticeably stronger sculpture on the shoulder compared to the base and more numerous and fainter spirals.

Anatoma aspera from the Mediterranean and adjacent Atlantic has a protoconch with flocculent sculpture, stronger axial cords particularly on the shoulder, and a 50% wider sutsel.

Anatoma boucheti from the Indian Ocean has a protoconch with flocculent sculpture and about twice as many axial cords per whorl.

Anatoma lamellata from Japan has a protoconch with strong flocculent sculpture, a shorter teleoconch I (0.5 vs. 1 whorl), and stronger axial and spiral cords on shoulder and base.

Anatoma peruviana from Peru has a sutsel that is twice as wide, more pronounced spiral lines on shoulder and base, and the suture is below the periphery of the previous whorl.

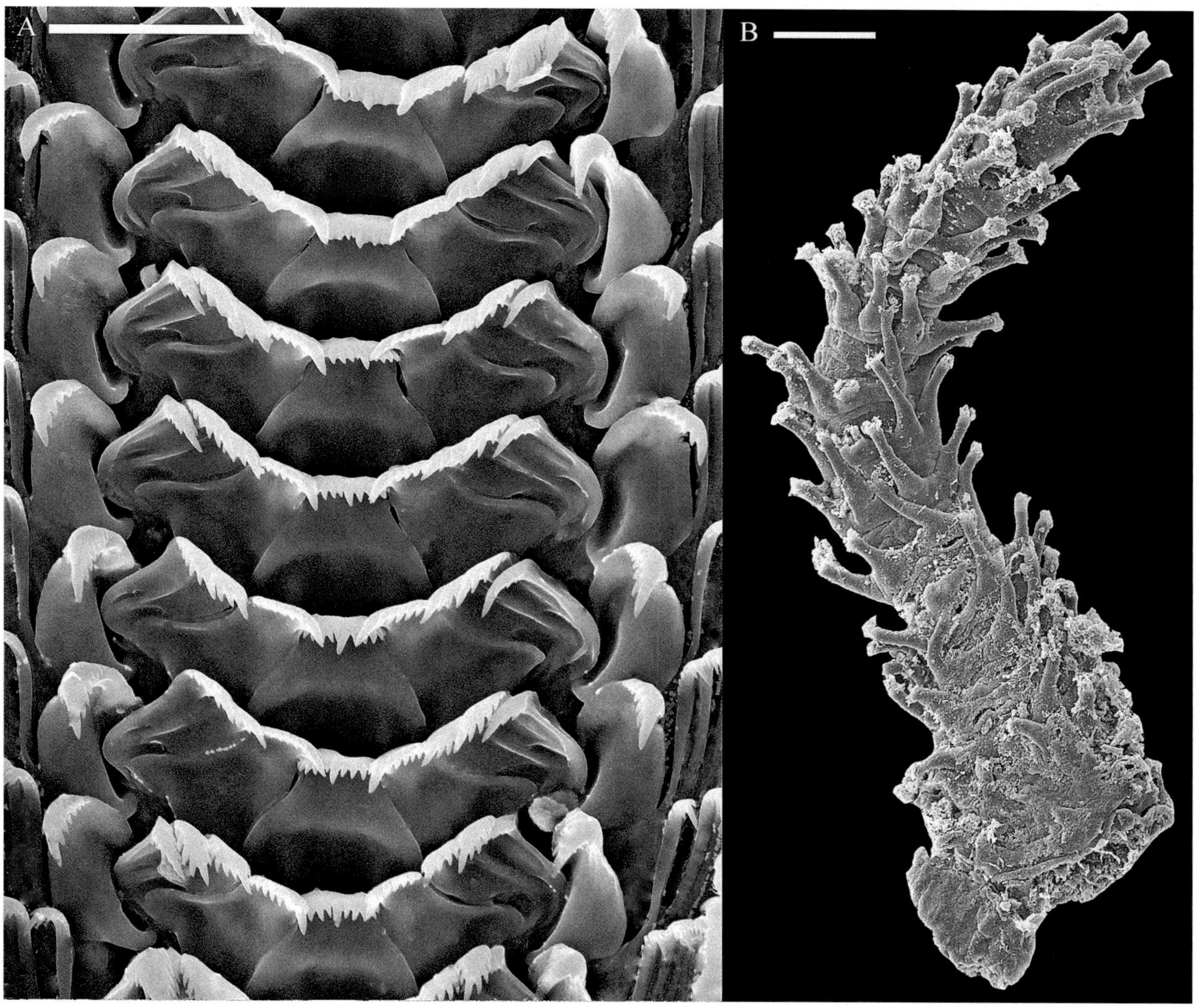

Figure 614. *Anatoma amoena*. 1579–1584 m, Weddell Sea, Antarctica, 62.779°S, 57.043°W (ZSM 20062197, 1). **A.** Radula. Scale bar = 20 µm. **B.** Epipodial tentacle. Scale bar = 50 µm. Images by Diego Zelaya, from Zelaya & Geiger (2007: modified),

Antomidae *Anatoma*

Anatoma porcellana from the Indo-Malayan Archipelago to the Western Pacific is proportionally higher, has a narrower umbilicus, and the spiral cord on teleoconch I does not form an angulation.

Anatoma quadraxialis from the Indo-Malayan Archipelago to the Western Pacific has much denser axial sculpture (85 vs. 22 on the body whorl) and lacks a spiral cord on teleoconch I.

Distribution. Circumantarctic. 161–1584 m (living specimens: 280–1584 m. Fig. 615).

Specimen records.

Antarctica. 392 m, Ross Sea, 72.083°S, 172.250°E (USNM 613010, 6; USNM 1084418, 1). 530–549 m, Elatin Station 1997, Ross Sea, 72.000°S, 172.467°E (NMNZ FM21630, 2; NMNZ M242840, 2). 1579–1584 m, Weddell Sea, 62.779°S, 57.043°W (ZSM 20062197, 1). 432 m, Lazarev Sea, 69.717°S, 10.767°E (SMNH 83287, 11). 161–185 m, Lazarev Sea, 69.983°S, 8.000°E (SMNH 83288, 6; SMNH 83289, 7; SMNH 83290, 1). 280–298 m, Lazarev Sea, 69.950°S, 6.333°E (SMNH 83291, 5: complete).

Literature records.

Antarctica. Weddell Sea (Bandel, 1998). 433 m, Ross Sea, 72.133°S, 172.167°E (Dell, 1990).—295–310 m, Station 7, Breid Bay, 709.107°S, 1.900°E (1, 1: complete) (Numanami & Okutani, 1990).—271–310 m, Breid Bay, 70.167°S, 24.000°E (2) (Numanami & Okutani, 1992).—123 m, AG 06 24, Cape Norvegia area, 71.086°S, 11.538°W (Gutt *et al.*, 2000).

Figure 615. Distribution of *Anatoma amoena*. ☆ type locality. ● specimen record. ■ literature record.

Anatoma amydra Geiger & Marshall, 2012

Figures 616–619, Color Plate 4G

Chresonymy.

Anatoma amydra Geiger & Marshall, 2012: 16–21, figs 13–15.

Type material. Holotype (MNHN 24957. Fig. 616). Paratypes 570–610 m, N/O *J. Charcot* BIOCAL DW46, New Caledonia, 22.883°S, 167.283°E (MNHN 24958, 18; NMNZ M.303303, 1. Fig. 617).

Type locality. 600–616 m, Grand Passage, N/O *Alis* BATHUS 4 DW914, New Caledonia, 18.817°S, 163.250°E (OD).

Etymology. Amydros, Greek for indistinct, dim, obscure. Refers to the lack of prominent features (OD).

Description. Shell medium-sized (to 3.4 mm) trochiform, turreted. Protoconch of 0.75 whorl, sculpture flocculent, apertural varix barely connected to embryonic cap, apertural margin shallow sinusoidal. Teleoconch I of 0.66 whorl, 13–19 axial cords, spiral cord in position of selenizone. Teleoconch II of 2.75 whorls, suture little impressed, sutsel less than width of selenizone on early shell showing 1–2 spiral lines, 2–3× width of selenizone at apertural margin, showing approximately six spiral lines. Shoulder convex, many fine axial cords extending onto keels of selenizone, interstices as wide as cords at suture, twice as wide near selenizone; first spiral line after 0.1–0.2 teleoconch II whorl, 4–10 after one whorl, 12–15 at apertural margin, evenly distributed over width of shoulder; spiral threads running over axial cords. Base barely constricted below selenizone, continuously sloping with umbilicus, dense axial cords extending onto keels of selenizone, interstices twice as wide as

Figure 616. *Anatoma amydra*. Holotype. 600–616 m, Grand Passage, N/O *Alis* BATHUS 4 DW914, New Caledonia, 18.816°S, 163.250°E (MNHN 24957). Scale bar shell = 1 mm. Scale bar protoconch = 100 µm.

cords near selenizone, less than width of cords near umbilicus; approximately 25 spiral lines from immediately below selenizone into umbilicus, spiral threads from selenizone to mid-base, then turning into steps, increasing in strength twofold towards umbilicus. Umbilicus of moderate width, no funiculus. Aperture rounded, adumbilical portion flared, roof overhanging. Selenizone at periphery; keels

Figure 617. *Anatoma amydra*. Paratypes. **A–C.** 570–610 m, N/O *J. Charcot* BIOCAL DW46, New Caledonia, 22.883°S, 167.283°E (MNHN 24958, 18). Scale bars shell = 1 mm. Scale bars protoconch = 100 µm.

moderately elevated, moderately strong; fine lunules distinct; slit open, margins converging in fully grown specimens.

Operculum (Fig. 618D) corneous, thin, multispiral, with central nucleus.

Radula (Fig. 618A–B). Rachidian tooth trapezoidal, central cusp slightly larger, 3–4 cusps on each side, arranged as fan. Lateral teeth 1–3 similar, apical cusp largest, 3–2 cusps along outer edge, one minute point on inner edge. Lateral tooth 4 reduced, hook-shaped, apical cusp largest, one minute point on each side. Lateral tooth 5 somewhat enlarged, apical cusp largest, four cusps along inner edge, one small point on outer edge. Inner marginal teeth with triangular tip, apical cusp largest, 1–2 cusps on inner margin, 3–5 along outer margin; shafts with weak food groove. Outer marginal teeth spoon-shaped, with many fine bristles along edge. Radular interlock of central field moderate.

Jaw (Fig. 618C) teardrop-shaped, composed of many small rhomboidal platelets.

Comparisons. *Anatoma australis* from Australia has a longer teleoconch I (0.5 vs. 0.75 whorl), multiple spiral lines on teleoconch I, and wider spacing of spiral lines on the base (3 vs. 7 on sutsel of 1.5× width of selenizone).

Anatoma biconica from tropical Australia and the Indo-Malayan Archipelago has less dense clathrate sculpture with axials and spirals at about half the density and the axials not fully extending across the keels of the selenizone.

Anatoma equatoria from the Indo-Malayan Archipelago to the Western Pacific has a sutsel about as wide as the selenizone at the onset of the selenizone and at most 50% less dense axial and spiral sculpture.

Anatoma flexidentata from the Indian Ocean to the Western Pacific lacks the apertural varix on the protoconch, has a shorter teleoconch I (0.25 vs. 0.5 whorl), more widely spaced axial cords particularly on the shoulder, and an at least 50% more narrow sutsel.

Anatoma pseudoequatoria from the Indo-Malayan Archipelago has a shorter teleoconch I (0.25 vs. 0.5 whorl), an at least 50% more narrow sutsel, and the intersection of the axials and spirals on the shoulder form distinct points.

Anatoma japonica from the Indian Ocean and the Central Pacific has a very narrow umbilicus, a sutsel that is more narrow after one whorl (<0.5× vs. >0.5× width of selenizone), and about 50% less dense spacing of axial cords (interstices approximately as wide or wider as cords vs. more narrow then cords).

Anatoma rainesi from the Western Pacific to the Eastern Pacific has a protoconch with distinct axial cords, axial cords on teleoconch I of markedly decreasing strength, and the axial cords on the base are at least twice as widely spaced.

Anatoma rhynochodentata from the Indo-Malayan Archipelago to the Western Pacific has weak but distinct constriction of the base below the selenizone and a wider sutsel on early whorls (<0.3× vs. >0.5× width of selenizone).

Anatoma atlantica from the Caribbean and the adjacent Atlantic has a much more narrow sutsel (<0.5× width of selenizone), a longer teleoconch I (0.5 vs. 0.75 whorl), and lacks a spiral cord in the position of the selenizone on teleoconch I.

Anatoma proxima from the Caribbean and the southwestern Atlantic has an overall more globular shape and has wider spacing of the axial cords on shoulder and base.

Anatoma tobeyoides from southeastern Australia has a protoconch with reticulate sculpture, very fine obliquely radiating lines on teleoconch I, and a more narrow selenizone.

Anatoma yaroni from the western Indian Ocean has a sutsel that is at most half as wide, an overall more discoidal shape, and more widely spaced axial cords on the base.

Distribution. Indo-Malayan Archipelago and Western Pacific, 250–1000 m.

Specimen records.

Philippines. 92–97 m, 12.517°N, 120.650°E (MNHN, 7: complete). 60 m, Pamilacan Island, 9.502°N, 123.918°E (MNHN, 1: complete). 90–110 m, Balicasag Island, 9.515°N, 123.687°E (MNHN, 2: complete). 90 m, off Momo Beach, Panglao Island, 9.608°N, 123.755°E (MNHN, 3: complete). 95–128 m, Cervera Shoal, W Palmilacan Island, PANGLAO 2004 T36, 9.488°N, 123.858°E (MNHN, 22; MNHN, 1). 110–112 m, Cervera Shoal, W of Pamilacan Island, PANGLAO 2004 T41, 9.495°N, 123.837°E (MNHN, 3). 100–138 m, Cervera, W Pamilacan Island, PANGLAO 2004

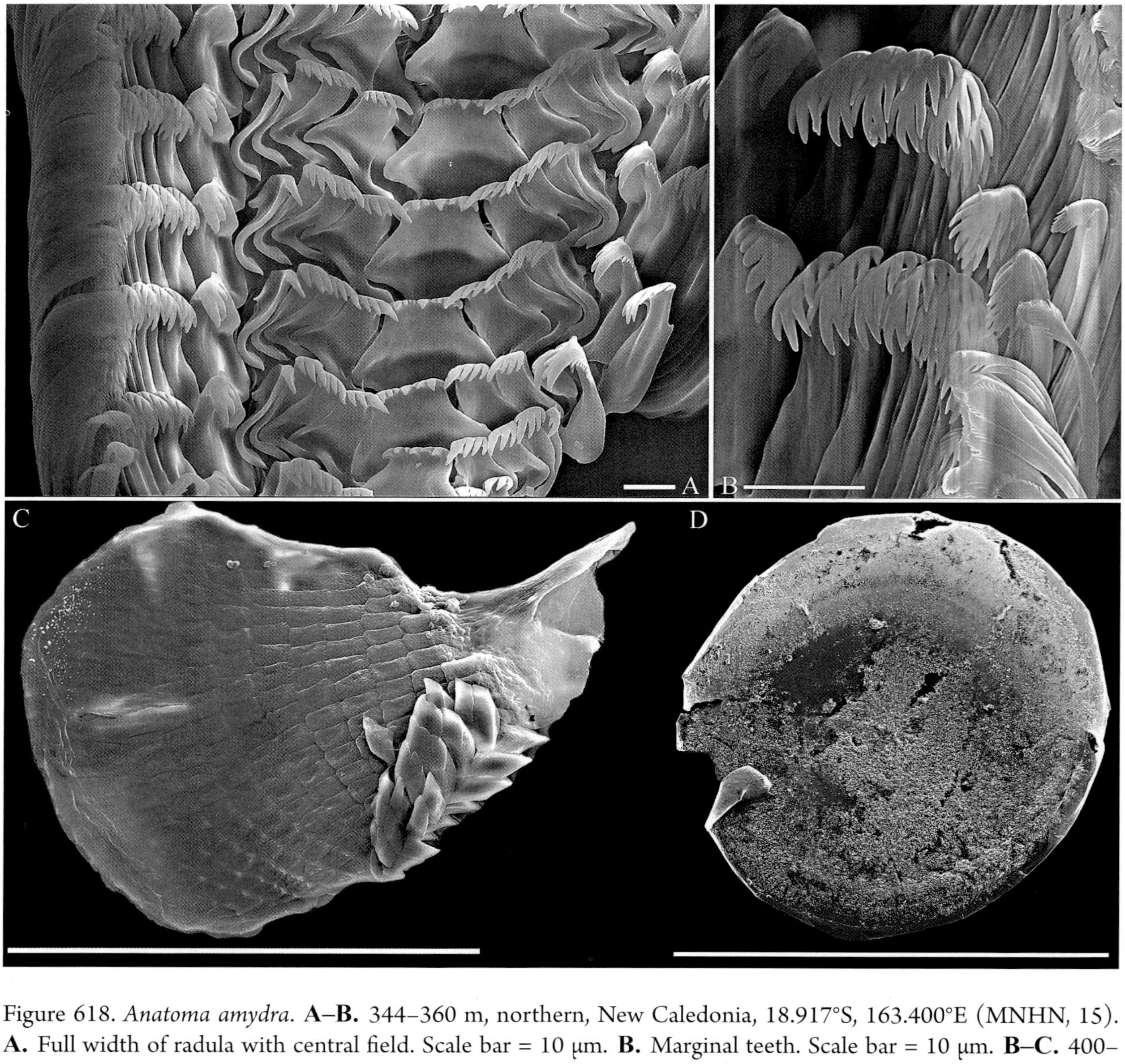

Figure 618. *Anatoma amydra.* **A–B.** 344–360 m, northern, New Caledonia, 18.917°S, 163.400°E (MNHN, 15). **A.** Full width of radula with central field. Scale bar = 10 μm. **B.** Marginal teeth. Scale bar = 10 μm. **B–C.** 400–420 m, Norfolk Ridge, Jumeau east, New Caledonia, 23.750°S, 168.267°E (MNHN, 4). **C.** Jaw. Scale bar = 100 μm. **D.** Operculum. Scale bar = 1 mm.

T39, 9.502°N, 123.840°E (MNHN, 1; MNHN, 1: complete). 87–111 m, Balicasag Island, PANGLAO 2004 L613/30/201164, 9.515°N, 123.680°E (MNHN, 2). 90–110 m, Balicasag Island, PANGLAO 2004 L46, 9.515°N, 123.687°E (MNHN, 1). 90–200 m, Maribohoc Bay, Bohol Island, PANGLAO 2004 P1, 9.600°N, 123.750°E (MNHN, 3). 62 m, Bigag/Tabalong, Panglao Island, PANGLAO 2004 L51–60, 9.628°N, 123.800°E (MNHN, 4). 473 m, AURORA 2007 CP2749, 15.943°N, 121.837°E (MNHN, 1). 624–647 m, off Aligbay Island, Bohol-Sulu Sill, PANGLAO 2005 CP2384, 8.770°N, 123.268°E (MNHN, 1).

Indonesia. 796 m, St. 5619, Molucca Pass, off March Island, 0.617°N, 127.250°E (USNM 312781, 1).

Solomon Islands. 53 m, 9.795°S, 160.842°E (MNHN, 1). 396–411 m, SALOMON 1 DW1762, 8.665°S, 160.065°E (MNHN, 3).

Fiji. 144–150 m, S of Viti Levu, 18.207°S, 178.550°E (MNHN, 1). 395–540 m, BORDAU 1 DW1486, 19.017°S, 178.433°W (MNHN, 1). 700–707 m, BORDAU 1 DW1485, 19.050°S, 178.500°W (MNHN, 1).

Wallis and Futuna. 286 m, Wallis Island, 13.350°S, 176.150°W (MNHN, 2). 415–420 m, Wallis Island, 13.350°S, 176.133°W (MNHN, 6).

Vanuatu. 90–110 m, NW coast of Malo Island, Palikulo Bay, SANTO 2006 DB77 EP28, 15.633°S, 167.083°E (MNHN, 1). 45–101 m, E Aoré Island, SANTO 2006 EP10, 15.604°S, 167.150°E (MNHN, 1). 92 m, NW Tutuba Island, SANTO 2006 DS105, 15.550°S, 167.278°E (MNHN, 2).

New Caledonia. 508–532 m, Sponge Bank, B mound, 24.917°S, 168.367°E (MNHN, 3). 430–450 m, southern, 24.7330°S, 168.117°E (MNHN, 100). 750–760 m, Loyalty Ridge, 24.717°S, 170.117°E (MNHN, 1). 943–1080 m, Norfolk Ridge, 24.667°S, 168.150°E (MNHN, 2). 569–616 m, Norfolk Ridge, Banc Introuvable, 24.650°S, 168.633°E (MNHN, 3). 400–420 m, Norfolk Ridge, Jumeau East, 23.750°S, 168.267°E (MNHN, 4). 410–415 m, Norfolk Ridge, 23.750°S, 168.283°E (MNHN, 1). 278–310 m, Norfolk Ridge, 23.683°S, 168.017°E (MNHN, 8). 433–450 m, Stylaster Bank, S of, 23.633°S, 167.717°E (MNHN, 20). 430–452 m, southern, 23.633°S, 167.717°E (MNHN, 20). 433–450 m, southern, Stylaster Bank, 23.633°S, 167.717°E (MNHN, 1). 447–450 m, southern, Stylaster Bank, 23.617°S, 167.700°E (MNHN, 1). 695–705 m, 23.583°S, 167.200°E (MNHN, 1). 655 m, Loyalty Ridge, 23.583°S, 169.617°E (MNHN, 1). 965 m, 23.417°S, 167.883°E (MNHN, 1). 381–469 m, Norfolk Ridge, 23.367°S, 168.017°E (MNHN, 20). 386–390 m, Norfolk Ridge, 23.350°S, 168.033°E (MNHN, 20). 601–608 m, Norfolk Ridge, 23.317°S, 168.000°E (MNHN, 50). 305–367 m, southern, Aztec Bank, 23.308°S, 168.083°E (MNHN, 1). 305–367 m, southern, 23.308°S, 168.083°E (MNHN, 100). 305–310 m, Aztec Bank, 23.300°S, 168.083°E (MNHN, 30). 400–402 m, Aztec Bank, 23.300°S, 168.100°E (MNHN, 7). 266–267 m, Norfolk Ridge, P Bank, 23.300°S, 168.250°E (MNHN, 5). 680–700 m, 23.083°S, 167.750°E (MNHN, 1). 680–700 m, 23.083°S, 167.750°E (MNHN, 1). 503 m, southern, 23.050°0S, 167.317°E (MNHN, 1). 360 m, 23.000°S, 167.250°E (MNHN, 1). 491–558 m, Île des Pins, SE South Reef, 22.992°S, 168.367°E (MNHN, 3: complete). 525 m, 22.983°S, 167.317°E (MNHN, 1). 570–610 m, 22.883°S, 167.283°E (MNHN, 19). 414–436 m, SE Reef, Île des Pins, 22.875°S, 168.208°E (MNHN, 3). 400 m, southern, 22.867°S, 167.200°E (MNHN, 1). 440 m, 22.783°S, 167.233°E (MNHN, 20). 440 m, 22.783°S, 167.233°E (MNHN, 1). 440–450 m, 22.783°S, 167.233°E (MNHN, 30). 390–400 m, southern, 22.767°S, 167.200°E (MNHN, 1). 400 m, 22.767°S, 167.250°E (MNHN). 410 m, SW Île des Pins, 22.767°S, 167.333°E (MNHN, 1). 310–315 m, Loyalty Basin, 21.533°S, 166.483°E (MNHN, 1; MNHN, 6). 310 m, Loyalty Ridge, 21.150°S, 167.917°E (MNHN, 1). 320–344 m, northern, 21.117°S, 164.467°E (MNHN, 1). 302–335 m, E coast, 20.967°S, 165.600°E (MNHN, 5). 40–60 m, Santal Bay, Lifou, 20.873°S, 167.133°E (MNHN, 2). 105–110 m, Poindimie sector, 20.817°S, 165.317°E (MNHN, 1). 490 m, Loyalty Ridge, 20.633°S, 167.117°E (MNHN, 1). 380–400 m, E coast, 20.583°S, 165.117°E (MNHN,

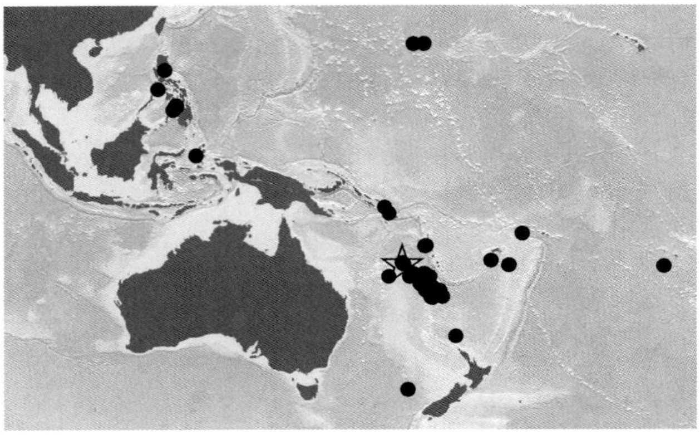

Figure 619. Distribution of *Anatoma amydra*. ☆ type locality. ● specimen record.

1). 735 m, 20.583°S, 166.883°E (MNHN, 1). 460 m, 20.583°S, 166.900°E (MNHN, 1). 470–480 m, Loyalty Basin, 20.583°S, 166.917°E (MNHN, 1). 435 m, 20.567°S, 166.900°E (MNHN, 11). 370–405 m, 18.917°S, 163.400°E (MNHN, 1). 344–360 m, northern, 18.917°S, 163.400°E (MNHN, 15). 668 m, 18.700°S, 163.217°E (MNHN, 1). 550 m, 18.650°S, 163.183°E (MNHN, 2). 538 m, between Lifou and Uvea, 20.617°N, 166.967°E (MNHN, 2). 49–59 m, Touho sector, 20.733°S, 165.233°E (MNHN, 1). 20 m, Santal Bay, Lifou, 20.763°S, 167.028°E (MNHN, 1). 27–31 m, Santal Bay, Lifou, 20.763°S, 167.028°E (MNHN, 5). 20 m, Santal Bay, Lifou, 20.773°S, 167.033°E (MNHN, 11). 45–56 m, Touho sector, 20.775°S, 165.263°E (MNHN, 1). 0–1 m, Touho sector, 20.782°S, 165.230°E (MNHN, 1). 25–30 m, Santal Bay, Lifou, 20.822°S, 167.173°E (MNHN, 4). 5–20 m, Santal Bay, Lifou, 20.875°S, 167.135°E (MNHN, 7). 5–25 m, Touho sector, 20.878°S, 165.325°E (MNHN, 1). 12–32 m, Santal Bay, Lifou, 20.892°S, 167.045°E (MNHN, 10). 5–30 m, Santal Bay, Lifou, 20.920°S, 167.012°E (MNHN, 2). 12–37 m, Aboré Grand Reef, Mouméa Lagoon, 22.371°S, 166.265°E (MNHN, 2: complete). 440 m, 22.783°S, 167.233°E (MNHN, 1: complete). 440–450 m, 22.783°S, 167.233°E (MNHN, 4: complete). 570–610 m, 22.883°S, 167.283°E (MNHN, 1: complete). 503 m, southern, 23.050°S, 167.317°E (MNHN, 1; MNHN, 10). 295–306 m, Norfolk Ridge, BATHUS 3 DW836, 23.033°S, 166.983°E (MNHN, 2). 518–586 m, Norfolk Ridge, NORFOLK 2 DW2087, 24.933°S, 168.367°E (MNHN, 4). 250–270 m, south, BERYX 11 DW18, 24.800°S, 168.150°E (MNHN, 3). 231–233 m, Kaimon-Maru Bank, Norfolk Ridge, NORFOLK 1 DW1675, 24.750°S, 168.150°E (MNHN, 6). 320–350 m, S, BERYX 11 DW11, 24.733°S, 168.167°E (MNHN, 12). 237–550 m, Norfolk Ridge, BATHUS 3 CH802, 23.683°S, 168.000°E (MNHN, 1). 285 m, Norfolk Ridge, NORFOLK 2 DW2040, 23.683°S, 168.017°E (MNHN, 27). 230–360 m, S BERYX 11 CH41, 23.650°S, 168.000°E (MNHN, 1). 571–610 m, Stylaster Bank, Norfolk Ridge, NORFOLK 2 DW2036, 23.633°S, 167.650°E (MNHN, 3). 381–469 m, Norfolk Ridge, BATHUS 3 DW827, 23.367°S, 168.017°E (MNHN, 4). 347–1063 m, Norfolk Ridge, NORFOLK 1 DW1732, 23.333°S, 168.267°E (MNHN, 3). 320–367 m, S SMIB 8 DW182–184, 23.308°S, 168.083°E (MNHN, 7). 540 m, Norfolk Ridge, NORFOLK 1 DW1722, 23.300°S, 168.017°E (MNHN, 10). 405–456 m, Norfolk Ridge, NORFOLK 2 DW2132, 23.283°S, 168.233°E (MNHN, 7). 1004–1009 m, Île des Pins, NORFOLK 2 DW2144, 23.150°S, 167.450°E (MNHN, 3). 427–433 m, south, NORFOLK 1 DW1733, 22.933°S, 167.250°E (MNHN, 16). 525–547 m, S, BATHUS 2 DW721, 22.900°S, 167.283°E (MNHN, 5). 400 m, south, BATHUS 2 DW729, 22.867°S, 167.200°E (MNHN, 1). 444–445 m, south, BATHUS2 DW719, 22.800°S, 167.267°E (MNHN, 40). 465–525 m, BATHUS 2 DW739, 22.583°S, 166.450°E (MNHN, 50). 569–583 m, S Lansdowne, EBISCO DW2629, 21.100°S, 160.767°E (MNHN, 1). 600–616 m, Grand Passage, N/O Alis BATHUS 4 DW914, 18.816°S, 163.250°E (MNHN 24957, 1: holotype). 613–647 m, north, BATHUS 4 DW918, 18.817°S, 163.267°E (MNHN, 2). 501 m, N, Grand Passage, CONCALIS CP3005, 18.600°S, 163.160°E (MNHN, 1).

New Zealand. 780–790 m, Three Kings Rise, 31.867°S, 172.433°E (NMNZ M.173749, 9). 950–1000 m, Louisville Ridge, 41.450°S, 164.133°E (NMNZ M.119213, 4).

French Polynesia. 557–569 m, Mount Ari'i Moana, Tarava Seamounts, TARASOC DW3318, 19.250°S, 151.517°W (MNHN, 1).

Remarks. Juvenile specimens, as usual, have a more lenticular overall shell shape. The details of the early teleoconch permit identification of the species: protoconch with flocculent sculpture, teleoconch I of 0.66 whorl with spiral in position of selenizone, and first spiral lines initially in outer half of shoulder.

Anatoma argentinae (Zelaya & Geiger, 2007)

Figures 620–622

Chresonymy.

Thieleella argentinae Zelaya & Geiger, 2007: 415–416, figs 51 [map], 85–88 [shell SEM], 89 [operculum], 90 [radula].

Figure 620. *Anatoma argentinae*. 4500 m, ECAOI, North Equatorial Pacific, NW of Clipperton Island, 14.666°N, 125.392°W (LACM H353, 1). Scale bar shell = 1 mm. Scale bar protoconch = 200 µm.

Type material. Holotype (MCZ 356975: operculum and radula on SEM stub, shell. Fig. 621); Paratypes from type locality (MCZ 345437, 15: broken shells with fluid preserved animals). 2707 m, Atlantis II-60 Expedition Station 245, Argentina, 36.928°S, 53.023°W (MCZ 345380, 3: small shells with fluid preserved animals).

Figure 621. *Anatoma argentinae*. Holotype. 5208–5223 m, Argentine Basin, Atlantis II-60 Expedition Station 247, 36°55.7'–43°33'S, 53°01.4'–48°58.1'W (MCZ 356975). Scale bar shell = 1 mm. Scale bar protoconch = 200 µm. Scale bar operculum = 1 mm. Scale bar radula = 20 µm.

Type locality. 5208–5223 m, Argentine Basin, Atlantis II-60 Expedition Station 247, 36°55.7'–43°33'S, 53°01.4'–48°58.1'W (OD).

Etymology. Named for its type locality in Argentina (OD).

Description. Shell large (up to ~5 mm), trochiform, broader than high. Protoconch of 0.75 whorl, globose, about 370–480 µm in diameter, with reticulate sculpture; apertural margin convex, with prominent varix. Teleoconch I of 0.3–0.5 whorl, 11 axial cords; spiral cord in position of selenizone. Teleoconch II of 1.5 whorls at 1.3 mm (broken in the aperture); narrow sutsel at onset of selenizone (Fig. 621), subsequently no sutsel (Fig. 620); shoulder with 57 axial cords on the first whorl, and narrower spiral threads then increasing in number and strength with growth. Base with prominent spiral cords surrounding wide umbilicus; same strength and density of axial cords as on shoulder. Selenizone at periphery, with elevated keels; lunules indistinct; condition of slit unknown.

Operculum (Fig. 621) corneous, multispiral, slightly smaller than aperture.

Cephalic tentacles long, papillate; no eye, eyestalks; two neck tentacles; first epipodial tentacle with three branches, second with two.

Radula (Fig. 621). Rachidian tooth pyriform at base; cutting edge straight, with one main central cusp, 4–5 smaller, sharper cusps on each side. Lateral teeth 1–3 relatively wide at bases, with triangular tips, one main apical cusp, 3–4 narrower cusps along outer margin. Lateral tooth 4 narrower, smaller. Lateral tooth 5 strong, with wide base, sharply serrated inner margin of tip. Inner marginal teeth similar to lateral tooth 5; shafts without food groove; outer marginal teeth more delicate, with finely denticulated tips. Radular interlock of central field weak.

Comparisons. The species is highly distinct despite the shell being known only from broken early whorls: the very large protoconch with uncommon reticulate sculpture, and the uncommonly short teleoconch I of 0.3–0.5 whorl, distinguish this species from all others.

Anatoma bathypacifica from the Panamic has a smaller protoconch (240 µm vs. 370–480 µm) with coarser reticulate sculpture, a strong angulation along the spiral cord in position of the selenizone, and at most half as many spiral lines on shoulder and base.

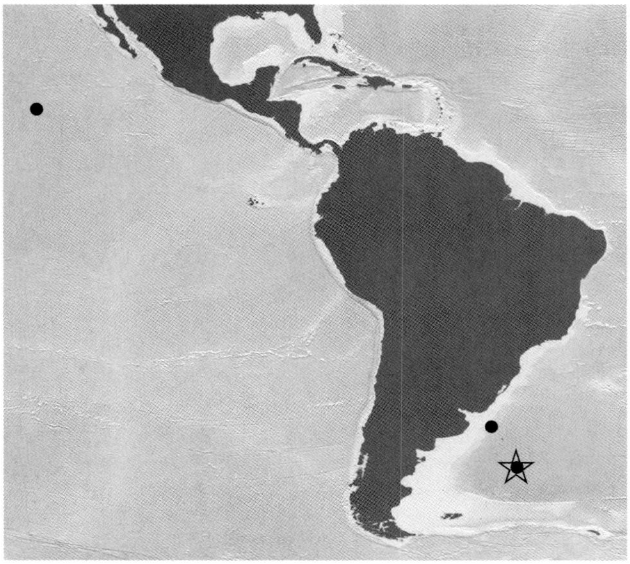

Figure 622. Distribution of *Anatoma argentinae*. ☆ type locality. ● specimen record.

Anatoma peruviana from Peru has a much taller shell with a wide sutsel even on early whorls, a lower expansion rate of the whorls, and stronger axial cords on the shoulder.

Anatoma proxima from the Caribbean and the southwestern Atlantic has a protoconch with flocculent sculpture and an apertural varix, and a longer teleoconch I (0.3–0.5 vs. 0.75 whorl).

Anatoma janusa from the Azores has a protoconch with flocculent sculpture, a teleoconch I with fewer axial cords and no spiral cord in position of the selenizone, and much stronger axial cords on the shoulder with no spiral lines on early teleoconch II.

Anatoma tenuis from the deep North Atlantic has a much smaller protoconch (300 µm vs. 370–480 µm) with the typical spiral coiling and flocculent sculpture, and a longer teleoconch I (0.6 vs. 0.25 whorl) with more spiral cords (18 vs. 11).

Distribution. Argentine Basin, north tropical Pacific Plate, 2707–5223 m (Fig. 622).

Specimen records

International Waters. 4500 m, ECAOI, North Equatorial Pacific, 1800 km WNW of Clipperton Island, 1800 km WSW off tip of Baja California, Pacific Plate, 14.666°N, 125.392°W (LACM H353, 1).

Remarks. The species is only know from two widely separated localities but from very similar bathymetry (abyssal plain). Shell morphological characters agree exceptionally well considering the rather poor preservation of the material: extremely large protoconch with very fine reticulate sculpture, same length and sculpture of teleoconch I, same density and strength of teleoconch II sculpture, as much as it can be assessed. Deep-sea species can be wide-spread; hence, the conspecificy of those specimens is quite plausible.

Anatoma aspera (Philippi, 1844)

Figures 623–632

Chresonymy.

Scissurella aspera Philippi, 1844: 160, pl. 25, fig. 17 (Fig. 623).
Scissurella aspera: Wood, 1848: 163 [as possible synonym of *Sci. crispata*].
Scissurella aspera: Philippi, 1853: 35–36, pl. 6, fig. 6.
Anatomus asper: Adams & Adams, 1853–1858 (1854): 439.
Scissurella aspera: Chenu, 1859: 363, fig. 2699.
Scissurella aspera: Weinkauff, 1862: 349.
Scissurella aspera: Munier-Chalmas, 1865: 396.
Scissurella aspera: Hildago, 1867: 144–145.
Scissurella aspera: Monterosato, 1875a: 22 [as synonym of *Sci. crispata*].
Scissurella aspera: Monterosato, 1877a: 34.
Scissurella aspera: Seguenza, 1877: 272, pl. 16, fig. 31.
Scissurella crispata var. *aspera*: Jeffreys, 1883a: 88.
Scissurella (*Schizotrochus*) *aspera*: Monterosato, 1878a: 149.
Schizotrochus asper: Monterosato, 1884: 39.
Scissurella aspera: Brusina, 1886: 198.
Scissurella aspera: Locard, 1886: 332.
Scissurella aspera: Paetel, 1888: 289.
Scissurella aspera: Norman, 1888: 21.
Scissurella aspera: Kobelt, 1888: 258.
Scissurella aspera: Carus, 1889–1893: vol. 2, 237.
Schizotrochus asper: Monterosato, 1890: 143.
Scissurella aspera: Hildago, 1890: 143.
Schizotrochus asper: Dautzenberg, 1891: 609, 617, pl. 17, figs 10–11.
Scissurella aspera: Marshall, 1895: 30, 40.
Scissurella aspera: Sturnay, 1896: 4, 13.
Scissurella aspera: Locard, 1898: 70–71.
Scissurella aspera: Locard, 1899: 107.
Schizotrochus asper: Pallary, 1900: 359.

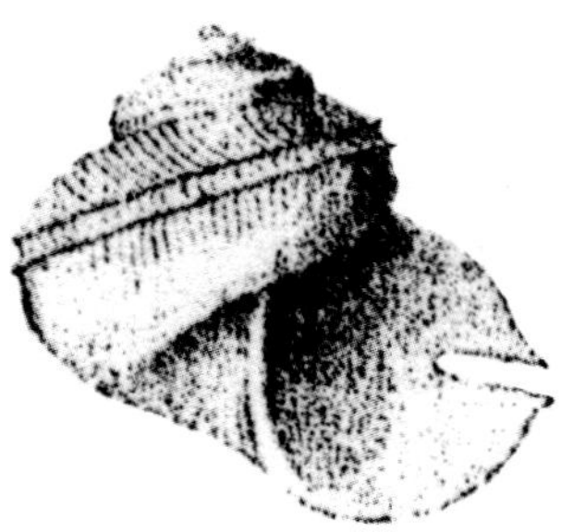

Figure 623. Original illustration of *Scissurella aspera.*

Scissurella aspera: Bellini, 1901: 111.

Schizotrochus asper: Claudon, 1902: 197.

Scissurella aspersa [*sic*]: Sowerby & Fulton, 1903: 53.

Scissurella (*Schizotrochus*) *aspera*: Franceschini, 1906: 7.

Scissurella crispata var. *aspera*: Horst & Schepman, 1908: 486.

Scissurella crispata var. *aspera*: Odhner, 1912: 13, pl. 2, fig. 25.

Scissurella aspera: Thiele, 1912a: 3 [under *Sci. crispata*], pl. 1, fig. 5.

Scissurella aspera: Thiele, 1925: 41.

Scissurella (*Schizotrochus*) *aspera*: Dautzenberg, 1927: 213–214.

Scissurella aspera: Priolo, 1948: 22–23.

Scissurella aspera: Smith, 1961: 11, text-fig.

Scissurella (*Schizotrochus*) *aspera*: Nordsieck, 1968: 10, fig. 2.13.

Scissurella aspera: Parenzan, 1970: 29, pl. 1, fig. 3.

Scissurella (*Schizotrochus*) *aspera*: Carrozza, 1976: 165, fig. 3.

Scissurella (*Anatoma*) *aspera*: Piani, 1980: 117.

Scissurella (*Schizotrochus*) *aspera*: Nordsieck, 1982: 2, pl. 1, fig. 13.

Scissurella aspera: Cuomo & Fasculo, 1983: 40.

Scissurella cf. *aspera*: Coppa & Crovato, 1983: 202.

Scissurella aspera: Bertozzi *et al.*, 1984: 75.

Scissurella aspera: Miari & Micali, 1984b: 175.

Scissurella aspera: Biagi *et al.*, 1985: 51.

Scissurella aspera: Tenekides, 1989: pl. 2, fig. 2.

Scissurella aspera: Marin & Ros, 1991: 202, 204

Anatoma aspera: Anistratenko & Starobogatov, 1997: 76, fig. 4.

Scissurella aspera: Geiger, 2003: 76.

Anatoma aspera: Ardovini & Cossignani, 2004: 19; 2 text-figs p. 68 [as *An. crispata*].

Anatoma aspera: Peñas *et al.*, 2006: 25, figs 11–13.

Anatoma aspera: Cossignani & Ardovini, 2011: 97, text-figs.

Anatoma aspera (partim): Hernández *et al.*, 2011: 62, fig. 11J, N [figs 11K–M is *An. eximia*].

Anatoma aspera: Høisaeter & Geiger, 2011: 94–98, figs. 19–33.

Anatoma aspera: Moreno, 2011: 95, text-figs.

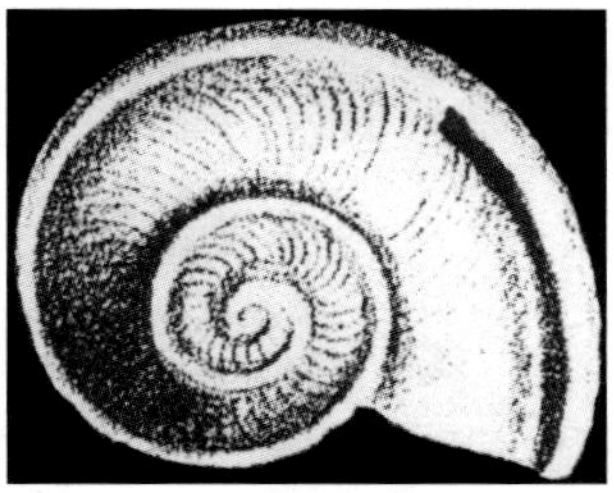

Figure 624. Original illustration of *Scissurella affinis.*

Synonyms.

+ *Scissurella affinis* O. G. Costa, 1861: 62, pl. 10, fig. 2B (Fig. 624). ***Type material.*** Syntypes (whereabouts unknown), 1.7 mm. Neotype here designated (SBMNH 149681). ***Type locality.*** Coast of Africa (OD). Neotype. Terreti - Cellantoni Bridge, 567 m above sea level Pleistocene epibathyal facies, muddy clay, Italy, 38.120°N, 15.716°E (SBMNH 149681). ***Etymology.*** Latin for Adjacent, closely related. ***Remarks***. See Remarks of *An. aspera* for details.

Scissurella affinis: Petit, 1869: 111.

Scissurella affinis: Monterosato, 1872a: 30 [as synonym of *Sin. elegans* = *Sin. cingulata*].

Scissurella affinis: Monterosato, 1878b: 79 [as synonym of *Sci. costata*].

Schizotrochus affinis: Monterosato, 1890: 143

Scissurella (*Schizotrochus*) *affinis*: Franceschini, 1906: 8.

+ *Scissurella crispata* var. *paucicostata* Jeffreys, 1865: 284, not illustrated. ***Type material.*** Holotype (USNM 181580: Warén, 1980: pl. 2, figs 5–6. Fig. 626). ***Type locality.*** Not given (OD). Zetland [= Shetland, Great Britain], on label. ***Etymology.*** Named for the fewer number of axials.

Scissurella crispata var. *paucicostata*: Jeffreys, 1887: 31.

Scissurella crispata var. *paucicostata*: Marshall, 1895: 30, 40 [as synonym of *aspera*].

Scissurella crispata var. *paucicostata*: Chaster, *et al.*, 1901: 16.

Scissurella crispata var. *paucicostata*: Odhner, 1912: 13 [as synonym of var. *aspera*].

+ *Schizotrochus divaricatus* Monterosato, 1890: 143, not illustrated. ***Type material.*** Syntypes (?: whereabouts unknown), 1–2 mm. ***Type locality.*** Deep waters off Palermo, Sicily, Italy (OD: inferred from title). ***Etymology.*** Latin for spread apart. ***Remarks.*** The species is said to be a small form of *S. asper* and *S. affinis*, and the stepped teleoconch was noted, hence, it is most likely a juvenile of *An. aspera*. *Anatoma micalii* has fewer and stronger axial cords, which are not mentioned in the description of *divaricatus*; hence, those two taxa are not synonymous. The number of specimens on which the description was based is uncertain; Monterosato mentioned that it was "found only one time" [Trovato una sola volta].

Schizotrochus divaricatus: Giannuzzi-Savelli & Piani, 1990: 161 [list of name only].

+ *Anatoma costamagnaensis*: Bandel, 1998: 27–28, pl. 8, figs 1–3. ***Type material.*** Syntype(s) (whereabouts unknown), 1.1 × 1.2 mm (Fig. 625). ***Type locality.*** Le Puyet near Nice, southern France [Le Puyet: 46.501°N 3.527°E; Nice: 43.703°N 7.266°E], Pliocene sediments of the Mediterranean (OD). ***Etymology.*** Geographic reference to "Costa Magna," of unknown location. ***Remarks.*** The species was attributed by Bandel to Marquet (1984), without reference to pagination or figure(s). Similar specimens were illustrated by Marquet (1984: pl. 3, fig. 2; 1998: 30, text-fig.) as *An. crispata*. Accordingly, Bandel (1998) introduced the name and is the authority of the species epithet. Bandel's specimen is indistinguishable from the common Mediterranean *An. aspera*. I was unable to trace a type locality of "Le Puyet" in the vicinity of Nice. The coordinates of the only traceable "Le Puyet" in France, and those of Nice proper, are given.

Anatoma costamagnaensis: Geiger, 2003: 73.

Figure 625. Original illustration of (one of the?) syntype(s?) of *Anatoma costamagnaensis.*

Misidentifications.

? *Scissurella indica*: Philippi, 1853: 32, pl. 6, fig. 4 [see Remarks].

Scissurella (Anatoma) crispata: Woodward, 1859: pl. 46, fig. 8. [is *An. aspera*, not fig. 9 as in caption, an obvious error as noted by Marshall, 2002].

Scissurella crispata: Kobelt, 1878: 163, pl. 56, fig. 27 [is *An. aspera*].

Scissurella (Schizotrochus) crispata: Nordsieck, 1968: 10, pl. 1, fig. 10 [is *An. aspera*].

Scissurella (Schizotrochus) crispata: Nordsieck, 1982: 2, pl. 1, fig. 10 [is *An. aspera*].

Scissurella crispata: Rolán, 1983: 64, text-fig. [is *An. aspera*].

Scissurella (Anatoma) crispata: Marquet, 1984: 338 [is *An. aspera*: based on illustrations on same locality by Marquet, 1995, 1998].

Scissurella crispata: Poppe & Goto, 1991: 32, 64, pl. 2, figs 7–8 [is *An. aspera*].

Anatoma crispata: Cossignani *et al.*, 1992: 18, fig. 027 [Mediterranean, is *An. aspera*].

Anatoma crispata: Nicolay & Angioy, 1993: 53, text-fig. [is *An. aspera*].

Anatoma crispata: Giannuzzi-Savelli *et al.*, 1994: 13, figs 92a–b [is *An. aspera*].

Anatoma umbilicata: Giannuzzi-Savelli *et al.*, 1994: 13, figs 93a–b [is juvenile *An. aspera*].

Anatoma crispata: Arduino *et al.*, 1995: 17, text-figs [is *An. aspera*].

Anatoma crispata: Marquet, 1995: 63, pl. 2, fig. 4 [is *An. aspera*].

Anatoma crispata: Bandel, 1998: 26–27, pl. 7, fig. 7 [is *An. aspera*].
Anatoma crispata: Marquet, 1998: 30, text-fig. [is *An. aspera*].
Anatoma crispata: Ardovini & Cossignani, 1999: 32, 2 text-figs [is *An. aspera*].
not *Anatoma affinis*: Palazzi & Villari, 2000: 9, figs 13–19 [is *An. micalii*].
Anatoma crispata: Doneddu & Trainito, 2005: 50, fig. 28 [is *An. aspera*].
not *Anatoma aspera*: Beck *et al.*, 2006, 44, text-fig. [is *An. eximia*].
Anatoma crispata: Portalatina, 2008: 155, fig. 2H, I [is *An. aspera*].
Anatoma crispata: Segers *et al.*, 2009b: 5 [most likely is *An. aspera*: *cf.* Segers *et al.*, 2009c].
Anatoma crispata: Segers *et al.*, 2009c: 64–65, pl. 5, figs 1–1a [is *An. aspera*].
not (partim) *Anatoma aspera*: Hernández *et al.*, 2011, 62, figs 11K–M [is *An. eximia*].

Type material. (whereabouts unknown: McLean, 1967), 1.33 × 1.33 ''' [~ 2.75 × 2.75 mm. Size is taken from natural size auxiliary figure in Philippi's original illustration]. Neotype (SBMNH 149681, designated by Høisaeter & Geiger 2011: 98. Fig. 627).

Figure 626. *Scissurella crispata* var. *paucicostata*. Holotype. Zetland [Shetland], Great Britain (USNM 181580). Scale bar shell = 1 mm. Scale bar protoconch = 100 µm.

Type locality. Near Rhegium, Calabria [Italy, 38.100°N, 15.650°E] (OD). Quartenary of Sicily (Locard, 1898). Neotype: Terreti - Cellantoni Bridge, 567 m above sea level Pleistocene epibathyal facies, muddy clay, Italy, 38.120°N, 15.716°E (Høisaeter & Geiger, 2011).

Etymology. Asper, Latin for rough, harsh.

Description. Shell moderate size (to 2.4 mm. Figs 627–629), trochiform, inflated, taller than wide. Protoconch of 0.75 whorl, sculpture flocculent, weak apertural varix barely connected to embryonic cap, apertural margin slightly sinusoidal. Teleoconch I of 0.75–0.825 whorl, approximately 15–27 axial cords, strong spiral in position of selenizone forming thickenings at intersection with axials. Teleoconch II of >3 whorls, suture deep, space between selenizone and suture of subsequent whorl as wide as selenizone, showing part of base. Shoulder strongly convex, with 28–50 axial lamellae per whorl, on keel of selenizone with additional fine axial lines; spiral lines starting at onset of selenizone to after 0.25 whorl, approximately 8–12 at apertural margin, running over axial lamellae. Base without or with weak constriction below selenizone, density of sharp axial cords about 10–20% higher than on shoulder, density of spiral lines similar as on shoulder; approximately 90 axial cords below suture, often every other axial disappearing at mid-base, particularly towards apertural margin in mature specimens, crossed by approximately 23 spiral lines. Umbilicus open, continuously sloping with base, weak funiculus present. Aperture rounded, flared, roof overhanging. Selenizone at periphery; lunules moderately distinct; keels strongly elevated, moderately strong; slit open, margins parallel.

Operculum (Fig. 630D) covering aperture, round, thin, multispiral, with central nucleus.

Radula (Figs 630–631). Rachidian tooth with central cusp strongest, curved arrangement of 3–4 cusps on each side. Lateral teeth 1–3 similar, apical cusp strongest, 3–4 cusps on outer edge. Lateral tooth 4 smaller, apical cusp strongest, three smaller cusps on inner margin. Lateral tooth 5 enlarged by elongation, apical cusp largest, approximately six cusps on inner edge. Inner marginal teeth inequilaterally triangular, apical cusp largest, three cusps along inner margin, 4–5 cusps along outer edge. Outer marginal teeth with spoon-shaped tip, many fine bristles. Outermost marginal teeth paddle-shaped, with many small bristles along distal edge; no food groove. Radular interlock of central field moderate.

Comparisons. *Anatoma crispata* from northeastern Atlantic has a larger protoconch (~180 μm vs. 150 μm), which protrudes from the shell, has weaker, more numerous axial ribs, the suture is immediately below the selenizone, and the overall shape is wider than in *An. aspera*. The easiest way to distinguish *An. aspera* from *An. crispata* is by the relative strength of the sculpture on shoulder and base. In *An. aspera* the shoulder sculpture is much stronger than that of the base, whereas in *An. crispata* the sculpture is equally well developed on shoulder and base.

Anatoma bertheloti from the Canary Islands has a much wider shell, at least twice as many axial cords, and the axial cords on shoulder and base are of equal strength.

Anatoma micalii from the Mediterranean and the Atlantic to Angola lacks an apertural varix on the protoconch, has a shorter teleoconch I (0.5 vs. 0.75 whorl), and less dense spiral sculpture on the upper portion of the base.

Anatoma eximia from the Western Mediterranean and the adjacent Atlantic has a shorter teleoconch I (0.5 vs. 0.75 whorl), a less convex shoulder, and a more narrow umbilicus.

Anatoma janusa from West Africa and the Macaronesian Islands has stronger axial cords on the shoulder with later onset of spiral lines and denser axial cords on the base.

Anatoma orbiculata from West Africa has a more globular overall shell shape, a more narrow sutsel on early whorls (<0.5× vs. 1× width of selenizone), stronger clathrate sculpture on the base, and a wider umbilicus.

Anatoma tenuisculpta from the Mediterranean Sea and the adjacent Atlantic has a more biconical overall shell shape, lacks a distinct funiculus, has more axial cords, and more distinct spiral lines.

Figure 627. *Anatoma aspera*. Neotype of *Scissurella aspera*. Also Neotype of *Scissurella affinis*. Terreti - Cellantoni Bridge, 567 m above sea level Pleistocene epibathyal facies, muddy clay, Italy, 38.120°N, 15.716°E (SBMNH 149681). Scale bar shell = 1 mm. Scale bar protoconch = 100 μm.

Anatoma philippiana from the Paleogene of Europe usually lacks an apertural varix on the protoconch, has a rather wide range of teleoconch I whorl (0.4–0.8), has stronger axial cords, and has a more narrow sutsel on intermediate whorls (<0.5× vs. 1× width of selenizone.

Anatoma alternatisculpta from the Caribbean has stronger but at most half as many axial cords on the shoulder, half as many spiral lines on the shoulder, and the axial and spiral cords on the base are of the same strength.

Anatoma japonica from the Indian Ocean to the Central Pacific has coarse flocculent sculpture and much less dense but coarser clathrate sculpture on the base.

Anatoma peruviana from Peru has a wider sutsel on early whorls, a protruding protoconch (not flat or sunken), and lacks the weak constriction on the base below the selenizone.

Anatoma rhynchodentata from the Indo-Malayan Archipelago to the Western Pacific has much finer and denser axial cords on shoulder and base.

Distribution. Mediterranean, tropical and subtropical northeastern Atlantic (Fig. 632).

Specimen records.

Norway. Liholmene, 60.362°N, 5.175°E (RMNH AD, 2). 50 m, Sletvik, Trondjheimfjord, 63.582°N, 9.521°E (RMNH AD ex 16807, 2).

Faroe Islands. 250–400 m, off, 64.000°N, 7.000°W (NMR 993000037760, 1).

Great Britain. 657 m, Bank Porcupine, 53.770°N, 13.950°W (NMR 993000033714, 5). 773 m, Rockall Trough, off W Scotland, 55.733°N, 16.133°W (NMR 993000033718, 30). 557 m, Rockall Trough, off W Scotland, 55.790°N, 15.800°W (NMR 993000033719, 25). 584 m, Rockall Trough, off W Scotland, 55.833°N, 15.133°W (NMR 993000033715, 200). 673 m, Rockall Trough, off W Scottland, 55.833°N, 15.132°W (NMR 993000033721, 50: complete). 679 m, Rockall Trough, off W Scotland, 55.833°N, 15.132°W (NMR 993000033722, 100). 1271 m, *Challenger* Station ES20, 56.767°N, 9.283°W (NMSZ 1994128.10103, 3: complete).

Portugal. 2500 m, *Challenger* station ES 347, 41.633°N, 11.333°W (NMSZ 1994128.60101, 3).

Spain. 706 m, Cadiz, 36.229°N, 7.124°W (NMSZ 1979025.01, 4). 250 m, Estepona, Malaga, 36.410°N, 5.129°W (DLG 1127, 2). 2192 m, off Cadiz, 36.533°N, 6.300°W (NHMW, 2). 250 m, Malaga, 36.717°N, 4.417°W (DLG 308, 10). 1870 m, [Bay of Biscay, 200 km N of Foz], 45.426°N, 7.269°W (FSC, 8). 250 m, off Estepona, Malaga, 36.344°N, 5.092°W (EGC 29730, 1). 530 m, *Challenger* station RD 304, 36.767°N, 8.183°W (NMSZ 1994128.60108, 2).

France. Rhone delta (RMNH, 4). 534–634 m, Gulf of Lion, 42.358°N, 3.380°E (MNHN, 2). 1129–1193 m, Gulf of Lion, 42.742°N, 4.490°E (MNHN, 1). 450–800 m, NE canyon of Toulon, 42.955°N, 6.022°E (MNHN, 75). 112–113 m, Grande Passe, Hyères Islands, 42.967°N, 6.333°E (MNHN, 20). 135–146 m, Magaud Bank, Hyères Island, 43.050°N, 6.683°E (MNHN, 30). 40–100 m, Le Brusc, Cap Sicié, Provence, 43.067°N, 5.800°E (MNHN, 3). 200–300 m, NE Cassidaigne Canyon, 43.110°N, 5.458°E (MNHN, 13). 150–250 m, NE Cassidaigne Canyon, 43.112°N, 5.550°E (MNHN, 30). 33 m, Grand Congloue, Marseille, 43.177°N, 5.403°E (MNHN, 2). Marseille, 43.300°N, 5.367°E (MNHN, 2). 60–75 m, Carry le Rouet, 43.320°N, 5.156°E (KBIN, 1). 60 m, St. Raphael, 43.404°N, 6.790°E (KBIN, 28; KBIN, 1). 100–110 m, off Saint-Jean-Cap-Ferrat, 43.668°N, 7.351°E (NMR 993000033726, 5; NMR 993000037870, 4). Gulf of Gascogne, 43.750°N, 1.750°W (MNHN, 1). 510 m, 47.567°N, 7.200°W (MNHN, 4). 150–200 m, Solenzara, Corse, 41.859°N, 9.450°E (RMNH AD 12223, 35). SE coast, St. Tropez, 43.271°N, 6.651°E (RMNH AD 14229, 1).

Italy. 70–100 m, Northern Tyrhenian Sea, 180.000°N, (DLG 311, 6). 100–150 m, SW Sardegna, 180.000°N, (DLG 320, 2). 32 m, Vendicari Island, Sicily, 36.792°N,

Figure 628. *Anatoma aspera*. **A.** Palermo, Italy, 38.133°N, 13.383°E, (NHMUK 1911.10.26.34124–34132, 14). **B.** 160 m, 36.050°N, 23.100°E, (NHMW, 4). **C.** Sporaden, Greece, ~36°N, 27°E, (NHMW 241, 6). Scale bars shell = 1 mm. Scale bars protoconch = 100 μm.

15.105°E (RMNH AD 19149, 2). Siracuse, Sicily, 37.051°N, 15.283°E (RMNH, 4). 22 mi E of Porto Empedocle, Sicily, 37.273°N, 13.133°E (KBIN, 80). Cape Empedoche, Sicily, 37.280°N, 13.529°E (KBIN 368, 9). 90 m, Acitrezza, Sicily, 37.564°N, 15.162°E (RMNH AD 17921, 3). Palermo, Sicily, 37.816°N, 13.583°E (USNM 889964, 1).

Figure 629. *Anatoma aspera*. **A–B.** 2500 m, *Challenger* station ES 347, 41.633°N, 11.333°W (NMSZ 1994128.60101, 3). **C.** 2192 m, off Cadiz, Spain, 36.533°N, 6.300°W (NHMW, 2). Radula shown in Figure 630. Scale bars shell = 1 mm. Scale bars protoconch = 100 μm.

Terreti - Cellantoni Bridge, 567 m above sea level Pleistocene epibathyal facies, muddy clay, 38.12°N, 15.716°E (DLG 1642, 12; SBMNH 149681, 1: neotype). 60–100 m, Palermo, 38.133°N, 13.383°E (MNHN, 4). Palermo, Sicily, 38.133°N, 13.383°E (MNHN, 2; MNHN, 7; MNHN, 5; MNHN, 5; KBIN, 3; KBIN, 3; FLMNH 00132410, 1). Archi, Calabria, 57 m above sealevel, Pleistocene, 38.148°N, 15.663°E (DLG 1641, 10). 50 m, Catalana Point, Terrasini, Sicily, 38.150°N, 13.083°E (AMS C.461522, 8). 48 m, Catalana Point, Terrasini, Sicily, 38.150°N, 13.083°E (AMS ex C.461531, 1). 274–366 m, SW Sardegna, 38.900°N, 8.200°E (DLG 1310, 2). 22 m, Paluniro Cape, Salerno, 40.029°N, 15.268°E (NMR 993000033738, 7). 80–110 m, Capri, 40.550°N, 14.233°E (DLG 330, 48). 59 m, Ischia, Napoli, 40.733°N, 13.950°E (NMR 993000024836, 1). 30 m, Ventotene Island, Latina, 40.794°N, 13.432°E (NMR 993000021243, 1). Gulf of Naples, 40.833°N, 14.267°E (USNM 181628, 1). Naples, 40.883°N, 14.417°E (USNM 181627, 5). 50–70 m, off Ferro Cape, Sardegnia, 41.152°N, 9.516°E (RMNH AD 15963, 15). 1–15 m, off Gaeta, Napoli, 41.207°N, 13.564°E (RMNH AD 14494, 10). Gulf of Naples, Liberi, 41.233°N, 14.283°E (USNM 181629, 1). 500 m, Anzio, Rome, 41.396°N, 12.564°E (NMR 993000033731, 9). 400–500 m, Fiumicino, Rome, 41.700°N, 12.000°E (DLG 904, 5). 350 m, Fiumicino, Roma, 41.700°N, 12.000°E (NMR 993000033735, 1; NMR

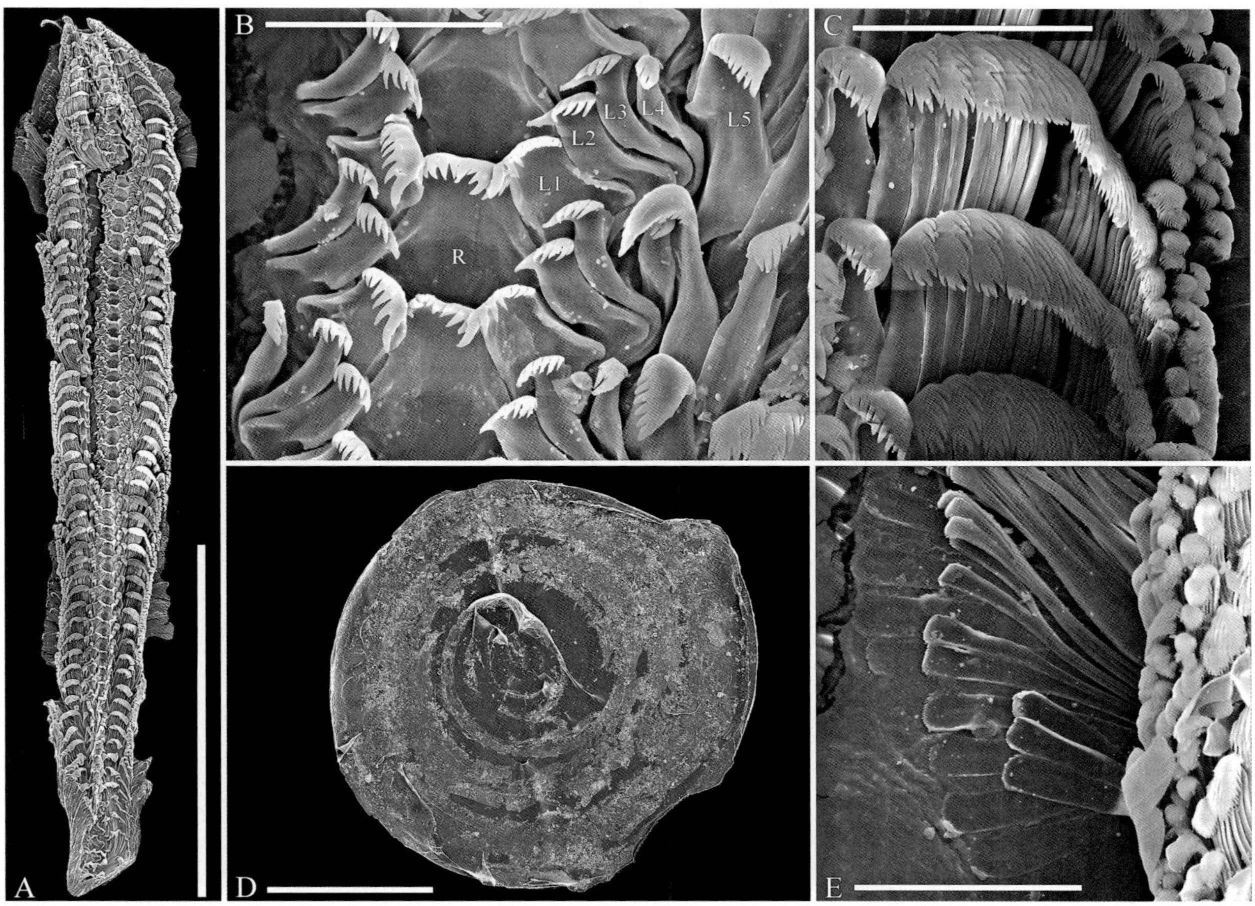

Figure 630. *Anatoma aspera*. Radula and operculum. 2500 m, *Challenger* station ES 347, 41.633°N, 11.333°W (NMSZ 1994128.60101, 3). Shell shown in Figure 629A. **A–C, E.** Radula. **A.** Entire radula. Scale bar = 500 µm. **B.** Central field enlarged. **C.** Lateral tooth 5 and marginal teeth. **E.** Outermost marginal teeth. Scale bars B, C, E = 30 µm. **D.** Operculum. Scale bar = 500 µm.

993000033734, 18). 500 m, off Fuimicino, Central Tyrrenian Sea, 41.743°N, 11.917°E (DLG 1211, 2). 500 m, Civitavecchia, 42.040°N, 11.665°E (NMR 993000033732, 20). 504 m, Civitavecchia, Rome, 42.040°N, 11.665°E (NMR 993000033733, 8). 504 m, Civitavecchia, Roma, 42.090°N, 11.758°E (NMR 993000021241, 1). 20–30 m, Tor Valdariga n Civitavecchia, Rome, 42.100°N, 11.800°E (DLG 333, 4). 85 m, Secca delle Vedove, 42.125°N, 15.507°E (RMNH AD 18043, 13). 350–400 m, 10 mi S of Capraia, 42.853°N, 9.809°E (RMNH AD 15254, 2). Corse Cape, Sardinia, 42.984°N, 9.480°E (RMNH AD 20218, 1). 160 m, Capraia Island, Tuscany, 43.000°N, 9.800°E (DLG 957, 2). Cap Corse, Capraia, 43.010°N, 9.648°E (RMNH AD 23417, 14). 70 m, between Capraia and St. Andrea, 43.021°N, 9.813°E (RMNH AD 16471, 2). 100–200 m, Capraia, Livorno, 43.033°N, 9.817°E (RMNH AD 23364, 10). 380 m, Capraia Island, 43.040°N, 9.820°E (DLG 1302, 10). Capraia, 43.050°N, 9.833°E (DLG 331, 8). 80 m, Capraia Island, 43.050°N, 9.840°E (AMS C.461525, 6). 80–100 m, Capraia, 43.050°N, 9.900°E (DLG 322, 3). W of Capraia Island, 43.050°N, 9.900°E (DLG 323, 5). 160 m, between Capraia Island, Corso Cape Point, 43.050°N, 9.900°E (AMS C.461539, 1). 150 m, between Capraia and [?], 43.050°N, 9.900°E (RMNH AD 12649, 11). 80 m, Capraia, Tuscany, 43.076°N, 9.890°E (DLG 1633, 8). 100–150 m, Gorgona Island, Tuscany, 43.429°N, 9.898°E (RMNH AD 16529, 3). 200 m, Corso Cape, Livorno, 43.509°N, 10.303°E (DLG 1128, 2). 50 m, Calambrone, Tuscany, Pisa, 43.596°N, 10.048°E (DLG 1980, 3).

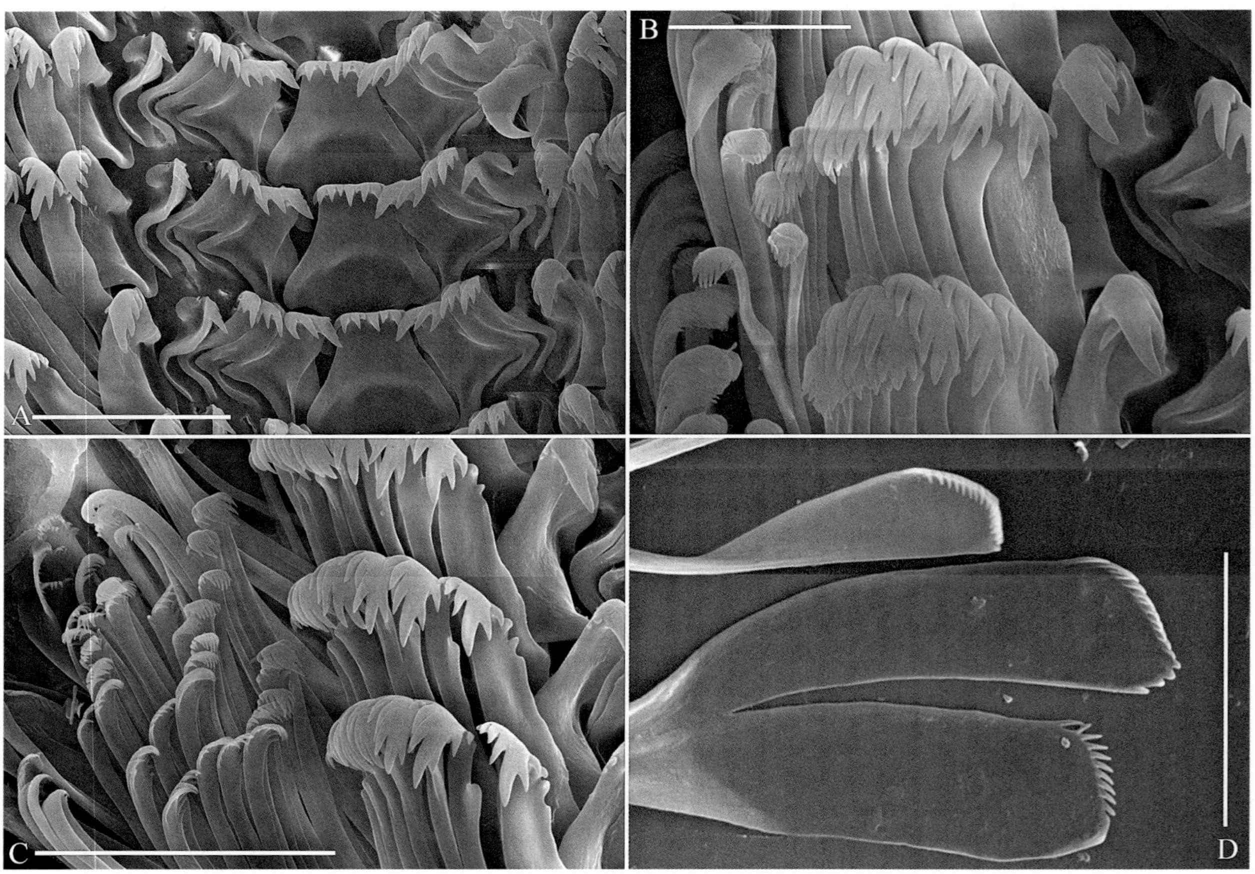

Figure 631. *Anatoma aspera*. Radula. 70–55 m, Liholmsrennen, Raunefjorden, Hordaland, Norway, 60°18'N, 5°09'E. **A.** Central field. **B–C.** Marginal teeth. **D.** Outermost marginal teeth. Scale bars A, C = 20 μm. Scale bars B, D = 10 μm.

Croatia. 25 m, Korcula Island, 42.900°N, 17.000°E (DLG 1303, 2). 32–35 m, Cape Kljuc, near Vela Luca, Korcula Island, 42.915°N, 16.662°E (DLG 1305, 6).

Greece. 140 m, Kyklades Plateau, Kyklades, 37.000°N, 25.600°E (NMR 993000040661, 50). 75–150 m, Aegean A32 (75 m) A 33 (130 m) (RMNH AD 22467, 40). Cyclades A15-A3 (RMNH AD 22104, 40). 180 m, Aegean Sea (RMNH AD 22209, 30). 130 m, Cyclades, 37.050°N, 25.500°E (RMNH AD 26090, 50). Sporaden, ~36°N, 27°E, (NHMW 241, 6).

Turkey. 120 m, Bozcaada Island, 39.804°N, 25.981°E (DLG 1198, 9). 40 m, Eskina Kayaligi, Buyukada, Mamora, 40.867°N, 29.133°E (MCZ 346608, 1).

Libya. 509 m, Atlantis II-59, St. 211, 33.950°N, 15.137°E (MCZ 346367, 7: complete).

Morocco. 125 m, off Cape Hadid, 32.017°N, 9.950°W (RMNH, 6). 100 m, off Cap Blac du Nord, 33.250°N, 8.800°W (RMNH, 1). 174 m, Langier [= Tangier?] Bay, 35.805°N, 5.783°W (USNM 181616, 4). Northern, fishmarket of Tanger, ex *Lepidotrigla*, 35.785°N, 5.813°W (MNHN, 20).

Mauretania. 200 m, 18.750°N, 16.700°W (RMNH, 2). 167 m, 46.000°N, 16.667°W (RMNH, 1).

Portugal, Madeira. 228–240 m, W of Deserta Grande, 32.483°N, 16.533°W (RMNH, 1). 144 m, W of Deserta Grande, 32.517°N, 16.533°W (RMNH, 6). 360 m, 32.617°N, 16.850°W (RMNH, 2). 386 m, Funchal Bay, 32.619°N, 16.886°W (FSC, 3). 587 m, [Funchal Bay] 32.633°N, 16.886°W (FSC, 2). 391–407 m, RV Auriga, Atalaia, 32.640°N, 16.786°W (FSC, 6). 220 m, St. Cruz, 32.678°N, 16.769°W (FSC, 20). 340 m, SE of, 32.683°N, 16.767°W (RMNH, 13). 189–190 m, Frente Aeropouerto, 32.684°N, 16.752°W (FSC, 20). 108 m, 32.700°N, 16.733°W (RMNH, 1). 280–300 m, 32.717°N, 16.717°W (RMNH, 1). 100–122 m, 32.717°N, 16.717°W (RMNH, 2).

Portugal, Azores. 622–786 m, Josephine Bank, 36.667°N, 14.250°W (USNM 181619, 1). 620 m, S of Santa Maria, 36.917°N, 25.117°W (RMNH, 20). 88 m, E of Formigas, 37.267°N, 24.767°W (RMNH, 3). 240–245 m, E of Formigas, 37.267°N, 24.733°W (RMNH, 11). 140–170 m, E of Formigas, 37.267°N, 24.733°W (RMNH, 10). 300 m, E of Faial, 38.500°N, 28.633°W (RMNH, 8). 85 m, W of Pico, 38.533°N, 28.567°W (RMNH, 14). 45 m, W of Pico, 38.533°N, 28.550°W (RMNH, 1). 4400 m, N of São Jorge, 38.650°N, 27.900°W (RMNH, 1). 2500 m, *Challenger* Station ES347, 41.633°N, 11.333°W (NMSZ 1994128.60102, 5: complete).

Figure 632. Distribution of *Anatoma aspera*. ☆ type locality. ● specimen record. ■ literature record.

Senegal. 7–15 m, Hann Bay, 14.728°N, 17.421°W (FSC, 1). 78 m, Around Gorée, Dakar Region, 14.317°N, 17.383°W (MNHN, 2). 145–170 m, Around Gorée, Dakar Region, 14.669°N, 17.399°W (MNHN, 3). 95 m, S of Gorée, Dakar region, 14.669°N, 17.399°W (MNHN, 4). 250 m, Dakar region, SW Cap Manuel, 14.671°N, 17.500°W (MNHN, 1; MNHN, 3). SW Cap Manuel, Dakar Region, 14.671°N, 17.500°W (MNHN, 1). Dakar Region, 14.671°N, 17.438°W (MNHN, 30).

Spain, Canary Islands. 1050 m, off de la Restinga Point, S of Hierro, 27.583°N, 18.000°W (RMNH, 1). 600–1000 m, off de la Restinga Point, S of Hierro, 27.583°N, 17.983°W (RMNH, 1). 336 m, off de la Restinga Point, S of Hierro, 27.616°N, 18.133°W (RMNH, 1). 160 m, off de la Restinga Point, S of Hierro, 27.616°N, 18.000°W (RMNH, 1). 330–430 m, off de Orchilla Point, SW of Hierro, 27.683°N, 18.167°W (RMNH, 17). 340–480 m, off de Orchilla Point, SW of Hierro, 27.683°N, 18.150°W (RMNH, 29). 83–97 m, de Jandia Point, S of Fuerteventura, 28.050°N,

14.500°W (RMNH, 1). 550 m, de Jandia Point, Fuerteventura, 28.033°N, 14.500°W (RMNH, 6). 500–700 m, de Jandia Point, Fuerteventura, 28.033°N, 14.500°W (RMNH, 9). 530 m, de Jandia Point, Fuerteventura, 28.033°N, 14.483°W (RMNH, 6). 225 m, de Jandia Point, Fuerteventura, 28.050°N, 14.500°W (RMNH, 7). 100–300 m, de Jandia Point, Fuerteventura, 28.050°N, 14.483°W (RMNH, 11). 125 m, de Jandia Point, Fuerteventura, 28.050°N, 14.483°W (RMNH, 1). 140–200 m, de Jandia Point, Fuerteventura, 28.050°N, 14.483°W (RMNH, 6). 90 m, de Jandia Point, Fuerteventura, 28.166°N, 14.033°W (RMNH, 6). 670 m, Punda de Gran Tarajal, off Fuerteventura, 28.183°N, 13.950°W (RMNH, 2). 100 m, Punda de Gran Tarajal, off Fuerteventura, 28.233°N, 13.850°W (RMNH, 5). 200–202 m, Fuerteventura, 28.284°N, 14.769°W (FSC, 15). 200 m, Fuerteventura, 28.156°N, 14.577°W (FSC, 20). 101 m, SW Fuerteventura, 28.176°N, 14.352°W (FSC, 118). 200–202 m, Fuerteventura, 28.285°N, 14.769°W (FSC, 15). 400 m, El Cotillo, Fuerteventura, 28.784°N, 13.985°W (FSC, 8). 100–120 m, Agaete, Gran Canaria, 28.104°N, 15.724°W (FSC, 4). 145 m, NW Gran Canaria, 28.135°N, 15.769°W (FSC, 2). 1340 m, Tenerife, 28.300°N, 16.600°W (KBIN, 8). 503 m, S of Palma, 28.433°N, 17.850°W (RMNH, 3). 400 m, SW of Palma, 28.666°N, 17.983°W (RMNH, 2). 400 m, Toston Point, Gran Canaria, 28.768°N, 13.986°W (FSC, 6). 313 m, S of Lanzarote, 28.800°N, 13.767°W (RMNH, 4). 215–325 m, S of Lanzarote, 28.800°N, 13.767°W (RMNH, 28). 150 m, S of Lanzarote, 28.800°N, 13.767°W (RMNH, 15). 82 m, S of Lanzarote, 28.800°N, 13.767°W (RMNH, 1). 509 m, SE of Lanzarote, 28.800°N, 13.750°W (RMNH, 5). 317–500 m, Playa Honda, Lanzarote, 28.915°N, 13.569°W (FSC, 11). 200–220 m, SE of Lanzarote, 28.916°N, 13.550°W (RMNH, 35). 160 m, SE of Lanzarote, 28.916°N, 13.550°W (RMNH, 19). 109–116 m, SE of Lanzarote, 28.916°N, 13.533°W (RMNH, 20). 92 m, E of Lanzarote, 29.133°N, 13.417°W (RMNH, 25). 40 m, off Puerto de Carmen, Lanzarote, 18.490°N, 66.840°W (HLC, 1).

Cape Verde Islands. 310 m, S of São Tiago, 14.883°N, 23.500°W (RMNH, 15). 920–970 m, bay on W coast of São Tiago, 15.000°N, 23.750°W (RMNH, 1). 123–142 m, off Palmeira, W of Sal, 16.750°N, 23.017°W (RMNH, 1).

Angola. 45 m, Ambrizete region, 6.950°S, 12.383°E (MNHN, 1). 120 m, Station 14, 12.350°S, 13.367°E (DLG 953, 6). 10–20 m, Lucira Bay (Cesar), Moçâmedes Province, 13.867°S, 12.533°E (MNHN, 15).

Literature records.

Algeria. Ain el Turk, Oran, 35.691°N, 0.642°W (Pallary, 1900).—Coasts of Algeria (Locard, 1898).

France. 1250 m, between Pico and S Jorge, 38.592°N, 28.096°W. 166 m, Gulf of Gascogne, France, 46.450°N, 6.500°W (Dautzenberg 1927). — 555 m, around Marseille. Gulf of Gascogne (Locard, 1898).

Western Sahara. Cape Bojador (station 28), 23.933°N, 15.817°W (Thiele, 1925).

Italy. 160 m, Station 194, between Crigo and Cerigotto, 36.050°N, 23.100°E. 597 m, Station 213, N of Stampaglia, Sporads, 36.783°N, 26.483°E (Sturnay, 1896).—Sicily. Cerigo, Sporades (Locard, 1898).—250–700 m, Santa Maria di Leuca, 39.781°N, 18.689°E. 150–200 m, Santa Maria di Leuca, 39.900°N, 18.540°E. 50–70 m, Manca Versa, 39.982°N, 17.931°E (Portalatina, 2008).—80–100 m, Corso Cape, 43.509°N, 10.303°E (Carrozza, 1976).

Spain. 1450 m, near Tenerife, 28.067°N, 16.825°W. 1229 m, E of Flores, 39.417°N, 31.367°W (Dautzenberg, 1927).—3 m, Cabo de Palmos, 37.633°N, 1.683°E (Marin, & Ros. 1987).— 322 m, E of Spain. 1037 m, 1094 m, 894 m, N of Spain. Balearic Islands (Locard, 1898).—60–400 m, Alboran Island, 35.951°N, 3.033°W (Peñas *et al.*, 2006).

Fossil. Pleistocene Campo Island, Gallipoli, Italy (Coppa & Crovato). Cenomanian [= Late Cretaceous] of Conac, France (Chenu, 1859). Madeira, Portugal (Segers *et al.*, 2009c). Pliocene (Kattendijk Formation).

Remarks. *Anatoma affinis.* The species has been variously synonymized with *Sinezona cingulata* and *Scissurella costata*, both common shallow water species of the Mediterranean, and has been treated as valid species of *Anatoma* or as a synonym of another congener. Although only the apical view of the species was provided by Costa (1861), the selenizone is notably close to the periphery of the shell. In his illustrations of the scissurellid species, the selenizone is shown as markedly on the shoulder, with the base showing to the outside of the selenizone in apical view. Hence, Costa's species is a member of Anatomidae.

The comparison of *affinis* to *dorbignyi* [= *Sci. costata*] by Costa unfortunately does not help with the determination of the generic placement of *affinis*. The early literature on European scissurellids is plagued with a bewildering array of synonymies, going as far as combining *Sin. cingulata*, *Sci. costata*, and *An. crispata* into a single pan-European species! The position of the selenizone at the periphery, the expansion rate of the whorls, and the rather large number of whorls evident in Costa's illustrations strongly suggests membership in Anatomidae.

The true identity of the species is difficult to assess, as the original type material is missing; no scissurellid types of Monterosato have been found in Rome (A. Zilli, pers. comm.); repeated inquiries to the Milan Museum were not answered. An old lot in ZMUC from Palermo identified as *Schizotrochus affinis* contained three specimens of *An. tenuisculpta*. Specimens in MNHN from Palermo identified as *Schizotrochus affinis* by Locard, one lot ex col. Monterosato, are *Anatoma* sp., with teleoconch I of 0.5–0.625 whorl.

The common Mediterranean species is *An. aspera*, while the often cited *An. crispata* has not been confirmed in the Mediterranean. *Anatoma umbilicata* is exceedingly rare and has a notably smooth surface, while *An. tenuisculpta* is generally larger, *An. eximia* has a more stepped profile, and *An. micalii* has stronger and sparser axial sculpture on the shoulder and a smaller expansion rate of the whorls. Accordingly, *An. affinis* is most likely a synonym of *An. aspera*. Palazzi & Villari (2000) and Cossignani & Ardovini (2011) have tentatively but erroneously identified what is here described as *An. micalii* as *An. affinis* (see under *An. micalii*). Given that the identity of the name is controversial and that the original syntype(s) cannot be located, a neotype is here designated for the species. I designate here the neotype of *Scissurella aspera* as the neotype of *Scissurella affinis*, by which the two taxa become objective synonyms. Given that the species epithet has been used as a valid species after 1900, it cannot be declared a *nomen oblitum*.

Original description. [Translation aided by U.-P. Geiger, Vienna] "As was mentioned just before, this *Scissurella* is not dissimilar from *Scissurella d'orbignyi* judging by its general traits. Nevertheless, other than being smaller than the other, and all other traits being the same, it is much smoother on top, the peripheral edge is more pronounced, and the entire surface is ornamented by transverse striae, most delicate, and dense: and while the spire is smooth, from the lower part grooves arise, and the aperture is narrow. And, indeed, I believe, one should under no circumstances confuse it with that one [*Sci. d'orbignyi*], whereas I do not want to deny the possibility, that the characters can vary among several individuals of the same species.

I found in fact a few specimens, in which the edge expanded in such a way to constitute a lamella, and the transverse striae stretch out on the external surrounding resulting in some small teeth."

Anatoma costamagnaensis. There are no differences in shell morphology that could justify a taxonomic distinction from *An. aspera* once size differences are taken into account. It is identical to the holotype of *An. paucicostata* from the Recent, which would take priority. *Anatoma costamagnaensis* is of Pliocene age (5.3–2.6 Mya), which is commonly accepted to contain species conspecific with Recent material. The only difference is the slightly shorter teleoconch I (0.66 vs. 0.75 whorl), which is still well within the average intraspecific range of 0.25 whorl.

Other remarks. Jeffreys (1868) noted that *An. aspera* is just the southern form of *An. crispata*. Jeffreys (1883a) listed *Sci. aspera* as synonym of *Sci.* [= *An.*] *crispata* and equated it with what he called *paucicostata*. Dautzenberg (1927) also noted the gradual changes between *crispata* and *aspera*, but elected to treat them as separate species anyway.

Brusina (1886), however, explicitly contradicted Jeffreys' observations and noted that *crispata* does not occur in the Adriatic, where only *aspera* can be found, and that both are distinct.

Weinkauff (1868: 368) noted that his earlier indication of *An. aspera* from Algier was based on a very badly preserved and very juvenile specimen of *Trochus millegranus*. I have not been able to trace that earlier citation.

Philippi (1853: pl. 6, Fig. 4) illustrated an anatomid shell as *Scissurella indica* Montfort. The species is not conspecific with Montfort's animal, but the identity of Philippi's specimen is difficult to assess. The pronounced axial sculpture on shoulder and base is noticeable, and all of Philippi's species were from the European-Mediterranean region. The most likely candidate, therefore, is *An. aspera*.

Høiseater & Geiger (2011) discussed the taxonomic confusion of the European *Anatoma* species in detail.

Anatoma atlantica (Bandel, 1998)

Figures 633–635

Chresonymy.

Hainella atlantica Bandel, 1998: 38–39, pl. 13, figs 1–3 (Fig. 633).
Anatoma atlantica: Geiger, 2003: 73.
Anatoma atlantica: Geiger & McLean, 2010: 15.
Anatoma atlantica Geiger & Fernández-Garcés, 2010: 67.
Anatoma [cf.] *crispata:* Tunnell *et al.*, 2010: figs 3.30-5–5a, 114, text-fig. [is *An. atlantica*].

Type material. Holotype (SGPIH 3877. Fig. 633), 2.4 × [2.2: estimate] mm.

Type locality. Atlantic, off southern Florida, [USA] 200 m (OD).

Etymology. Named after its type locality in the Atlantic Ocean (OD).

Description. Shell medium-sized (to 2.4 mm. Figs 633, 634), broadly elliptical; maximum width of shell to width of shoulder at periphery = 0.36 in holotype. Protoconch of 0.75 whorl, sculpture flocculent, no apertural varix, apertural margin barely sinusoidal. Teleoconch I of 0.66 whorl, approximately 23 fine axial cords, no spiral sculpture. Teleoconch II of 1.75 whorls, suture little impressed. Shoulder with numerous fine axial cords, approximately 100–120 on body whorl; approximately 15 spiral lines at aperture. Base with axials and spirals of similar strength, similar density as on shoulder. Selenizone slightly above periphery; lunules dense, little distinct; keels moderately elevated, more elevated towards aperture; slit open, margins converging towards aperture. Umbilicus open, no funiculus. Aperture round below, flattened under shoulder.

Animal unknown.

Comparisons. *Anatoma alternatisculpta* from the Caribbean is taller than wide, has fewer but coarser axial cords on the shoulder, and a smooth protoconch.

Anatoma proxima from the Caribbean and the adjacent Atlantic is more globular, has a smaller expansion rate (1.45 vs. 1.51), and less pronounced, less dense axial sculpture.

Anatoma rolani from the Caribbean has much denser axial cords with interstices more narrow than width of cords, a smooth protoconch without apertural varix, and a shorter teleoconch I (0.5 vs. 0.75 whorl).

Anatoma japonica from the Indian Ocean to the Central Pacific usually has a spiral cord in the position of the selenizone on teleoconch I, about 30% fewer but stronger axial cords on teleoconch II, and usually spiral steps on the base closer to the umbilicus.

Anatoma amydra from the Indo-Malayan Archipelago to the Central Pacific has a shorter teleoconch I (0.5 vs. 0.75 whorl) with a spiral cord in position of the selenizone and denser axial cords on the base (interstices < vs. > width of cords).

Anatoma crispata from the northeastern Atlantic has a protoconch without apertural varix, a more globular overall shell shape, and less dense axial cords on shoulder and base.

Anatoma disciformis from the North Pacific has a protoconch without apertural varix, teleoconch I with axial cords that are at most very faint, and the sculpture on larger shells is much fainter.

Anatoma euglypta from Antarctica has at most a very narrow sutsel, a protoconch without apertural varix, a spiral cord in position of the selenizone on teleoconch I, and either fainter (small specimens) or much denser (large specimens) sculpture on teleoconch II.

Anatoma rainesi from the Eastern to the Western Pacific has a shorter teleoconch I (0.5 vs. 0.75 whorl) and a teleoconch I with axial cords decreasing in strength with growth.

Figure 633. Original illustration of holotype of *Hainella atlantica*.

Anatoma rapaensis from the Indo-Malayan Archipelago to the Central Pacific lacks a sutsel on early whorls, has a shorter teleoconch I (0.3 vs. 0.75 whorl), and an overall more discoidal shell shape.

Anatoma regia from New Zealand has a shorter teleoconch I (0.5 vs. 0.75 whorl) with a spiral cord in the position of the selenizone and fainter axial and spiral sculpture on teleoconch II.

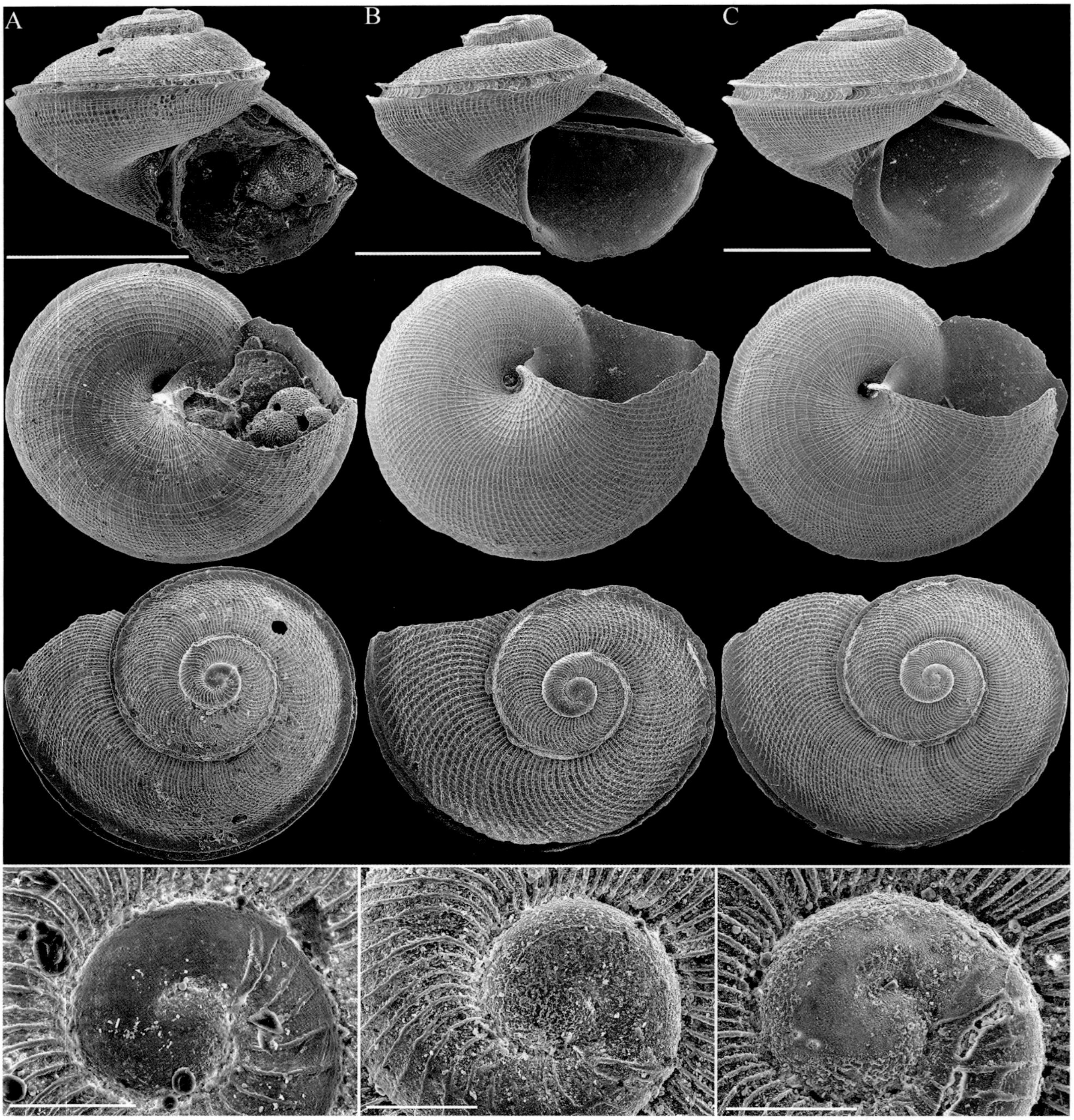

Figure 634. *Anatoma atlantica*. **A.** Off Louisiana, USA, 27.873°N, 91.848°W (CCSC, 4). **B.** 210 m, off Sambo Reef, Florida, USA, 24.500°N, 81.717°W (USNM 449412, 7). **C.** 223–233 m, off Charleston, South Carolina, USA, 32.738°N, 75.089°W (USNM 875208, 7). Scale bars shell = 1 mm. Scale bars protoconch = 100 µm.

Anatoma tenuis from the North Atlantic is much larger (to >5 mm), lacks a sutsel, has shorter teleoconch I (0.5 vs. 0.75 whorl), and has an overall shell shape that is more discoidal.

Anatoma tenuisculpta from the Mediterranean and the adjacent Atlantic has a teleoconch I with spiral cord in position of the selenizone and usually stronger but less dense axial cords.

Anatoma yaroni from the Indian Ocean has a spiral cord in position of the selenizone on teleoconch I and all sculptural elements of at most half the density.

Distribution. Caribbean and adjacent western Atlantic (Fig. 635).

Specimen records.

USA, Florida. Florida Keys, Monroe County, 24.366°N, 81.429°W (FMNH 326073, 3). 79 m, Carysfort Reef, Monroe County, 24.265°N, 80.174°W (FMNH 326057, 1). 190 m, Florida Keys, Monroe County, 24.353°N, 81.430°W (FMNH 326074, 7). 182 m, Florida Keys, Monroe County, 24.389°N, 81.430°W (FMNH 326072, 5). 175 m, Florida Keys, Monroe County, 24.401°N, 81.430°W (FMNH 326071, 5). 153 m, Ramrod, Monroe County, 24.448°N, 81.431°W (FMNH 326069, 3). 210 m, off Sambo Reef, 24.500°N, 81.717°W (USNM 449412, 7). 201 m, Key Colony Beach, Florida Keys, Gulf Stream, 24.553°N, 80.955°W (FMNH ex 326080, 24). 155 m, Florida Keys, Monroe County, 24.588°N, 80.967°W (FMNH ex 326066, 1). 128–137 m, 170° off Sombrero Light, Monroe County, 24.610°N, 81.100°W (FLMNH 249924, 1). 183 m, Windley Key, Monroe County, 24.795°N, 80.431°W (FMNH 326065, 6). 153 m, Windley Key, Monroe County, 24.814°N, 80.448°W (FMNH 326064, 1). 270 m, due W Egmont Key, Hillsborough County, 27.600°N, 79.900°W (HLC, 1).
USA, Louisiana. 27.832°N, 91.870°W (CCSC, 1). 27.873°N, 91.848°W (CCSC, 4).
USA, South Carolina. 223–233 m, off Charleston, 32.738°N, 75.089°W (USNM 875208, 7).
USA, Georgia. 805 m [~30.700°N, 79.000°W](USNM 108420, 6).
Barbados. 13.167°N, 59.567°W (USNM 500689, 1).

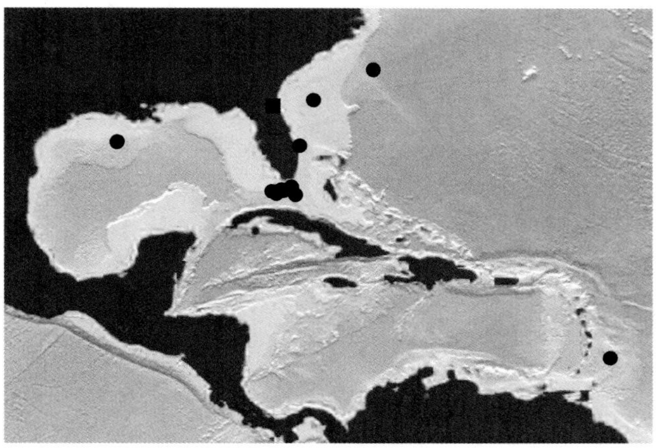

Figure 635. Distribution of *Anatoma atlantica*. ● specimen record. ■ literature record.

Literature records.

USA. 700 m, off Jacksonville, Florida, 30.317°N, 81.650°W (Bandel, 1998).

Remarks. The description of *An. atlantica* is intriguing. The cited size of the specimen is very large for the number of whorls, and the dimensions of the bulbous protoconch are also on the large size for anatomids. The species was compared to a misidentified *An. umbilicata*, but not to the sympatric *An. proxima* or Bandel's (1998) *An. americana* (= *An. proxima*). The early teleoconch I is well within the range of variability of *An. proxima*, and the wider broader shape is commonly seen in smaller specimens with low number of whorls (<2 vs. >2.5). *Anatoma atlantica* is here tentatively retained as a valid species.

Anatoma aupouria (Powell, 1937)

Figures 636–640

Chresonymy.

Schizotrochus aupouria: Powell, 1937: 176, pl. 49, fig. 3.
Schizotrochus aupouria: Powell, 1940: 208.
Schizotrochus aupouria: Dell, 1956: 168.
Anatoma aupouria: Powell, 1979: 35–36, pl. 14, fig. 2.
Anatoma aupouria: Lozouet, 1986: 112.
Anatoma aupouria: Spencer & Willan, 1995: 13.
Anatoma sp. cf. *aupouria*: Brook & Marshall, 1998: 214.
Anatoma aupouria: Geiger, 2003: 73.
Anatoma aupouria: Geiger & Jansen, 2004a: 7–9, figs 1–2.
Anatoma aupouria: Geiger, 2004: text-fig. p. 5.

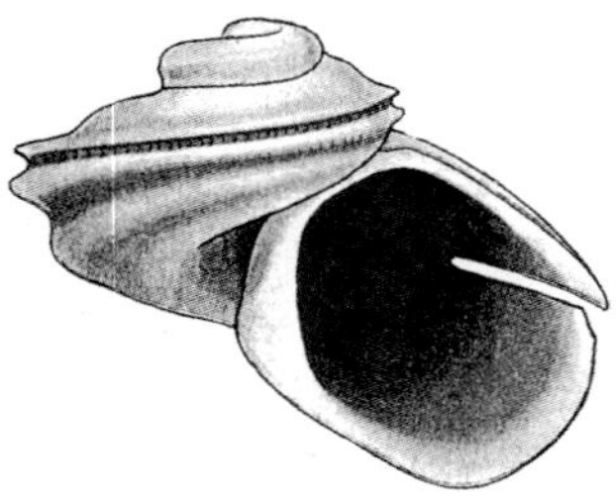

Figure 636. Original illustration of *Schizotrochus aupouria*.

Figure 637. *Anatoma aupouria*. 805 m, N of Three Kings Islands, New Zealand, 34.000°S, 171.917°E (AMS C.402696, 1). Scale bar shell = 1 mm. Scale bar protoconch = 100 μm.

Anatoma aupouria: Geiger, 2006c: text-fig. 5.
Anatoma aupouria: Spencer *et al.*, 2009: 201.

Type material. Holotype (NHMUK 1962952. Fig. 458A), 1.25 × 0.9 mm (W × H). Paratypes from type locality (NMW 1955.158.606, 1: labelled "syntype," 1.4 mm; AIM AK 72218, 1).

Figure 638. *Anatoma aupouria*. **A.** Holotype. 260 m, off Three Kings Islands, [New Zealand, 34.090°S, 172.080°E] *Discovery* station 933 (NHMUK 1962952). **B–C.** 246–291 m, Middlesex Bank, NW of Three Kings Islands, New Zealand, 34.033°S, 171.733°E (NMNZ M.160275, 17). Scale bars shell = 1 mm. Scale bars protoconch = 100 µm.

Type locality. 260 m, off Three Kings Islands, [New Zealand, 34.090°S, 172.080°E] *Discovery* station 933 (OD).

Etymology. Aupouri: a Maori tribe from Northern New Zealand (B. Marshall, pers. comm.).

Description. Shell medium-sized (to 1.2 mm. Figs 637–639), somewhat turreted. Protoconch 0.75 whorl, sculpture flocculent, no varix, apertural margin straight. Teleoconch I of

Figure 639. *Anatoma aupouria*. **A.** 252 m, off Three Kings Islands, New Zealand, 34.017°S, 172.200°E (NMNZ M.033979, 12). **B.** Sunday Island [= Raoul], Kermadec Islands, New Zealand, 29.264°S, 177.926°W (NMNZ M.174858, 1). **C.** 82–100 m, SE of Smith Bluff, Raoul Island, Kermadec Islands, New Zealand, 29.315°S, 177.940°W (NMNZ M.28895, 10). Scale bars shell = 1 mm. Scale bars protoconch = 100 μm.

0.25 whorl, approximately seven axials. Teleoconch II > 1.5 whorls, suture moderately impressed; sutsel 0.5–1× width of selenizone. Shoulder with many fine axial cords; axials variably internesting, producing complex patterns. Axials intersected with approximately 6–8 fine spiral cords in central third of shoulder. Base not constricted below selenizone, with weak angulation on mid-base, with similarly complex pattern of axials, and finer spirals. Umbilicus open, deep, continuously sloping from base; funiculus curved away from apertural margin. Selenizone at periphery; lunules distinct; keels strong, elevated, external surfaces with many fine axial lines; slit open, margins somewhat converging towards aperture. Aperture subquadrate, somewhat flared, particularly towards umbilicus where apertural margin fuses with funiculus.

Animal unknown.

Comparisons. *Anatoma quadraxialis* from the Indo-Malayan Archipelago has a protoconch with reticulate sculpture, is more globular in general outline, and lacks a funiculus.

Anatoma pseudoequatoria from the Indo-Malayan Archipelago to the Central Pacific lacks the posteriorly oriented embayment of the axials on the mid-base, the shoulder bears about 20 very distinct axial cords starting at the suture, but not reaching the selenizone, and the curvature of the funiculus is towards the apertural margin.

Anatoma munieri from the Indian Ocean to the Central Pacific is much more turreted, has a proportionally smaller aperture, and lacks a funiculus.

Anatoma parageia from Japan lacks the complex pattern of internesting axial cords on the shoulder, has dense and simple axial cords from suture to selenizone, and does not have the additional fine axial lines on the keels of the selenizone.

Anatoma philippinica from the Indo-Malayan Archipelago to the Western Pacific has an overall more globular shell, a longer teleoconch I (0.5 vs. 0.25 whorl), and axial cords extending from suture to selenizone and over the entire base.

Distribution. New Zealand, Lord Howe, Norfolk Islands (Fig. 640).

Specimen records.

Australia. 31 m, off Duncombe Bay, Norfolk Island, 29.000°S, 167.933°E (AMS C.378320, 1). 44 m, Lord Howe Rise, 31.637°S, 159.060°E (AMS C.404151, 1).

New Zealand. 805 m, N of Three Kings Islands, 34.000°S, 171.917°E (AMS C.402696, 1; NMNZ M.049031, 12). 260 m, off Three Kings Islands, St. 933, 260 m, 34.090°S, 172.080°E (NHMUK 1962952, 1). 252 m, off Three Kings Islands, 34.016°S, 172.200°E (NMNZ M.033979, 12). 38 m, off Three Kings Islands, Arch Pinnacle, Princes Rock, 34.175°S, 172.050°E (NMNZ M.160309, 1). 246–291 m, Middlesex Bank, NW of Three Kings Islands, 34.033°S, 171.733°E (NMNZ M.177155, 3). 31–45 m, between Dayrell and Chanter Islands, Herald Islets, Kermadec Islands, 29.250°S, 177.848°W (NMNZ M.227088, 3). Sunday Island [= Raoul], Kermadec Islands, 29.264°S, 177.926°W (NMNZ M.174858, 1). 82–100 m, SE of Smith Bluff, Raoul Island, Kermadec Islands, 29.315°S, 177.940°W (NMNZ M.28895, 10). 427 m, 39 km SW of Great Island, Three Kings Islands, 34.293°S, 171.755°E (NMNZ M.094335, 3). 246–291 m, Middlesex Bank, NW of Three Kings Islands, 34.033°S, 171.733°E (NMNZ M.160275, 17).

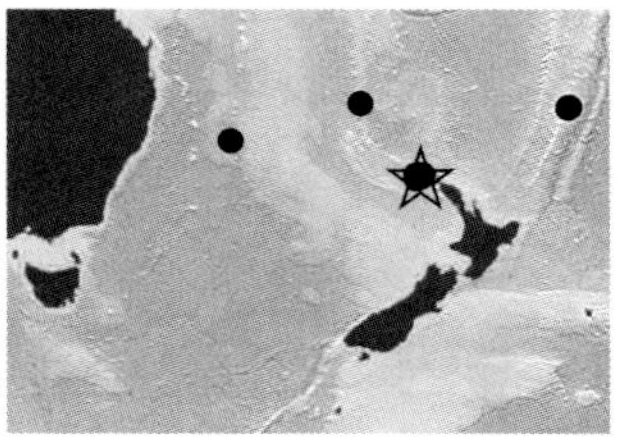

Figure 640. Distribution of *Anatoma aupouria*. ☆ type locality. ● specimen record.

Remarks. The drawings of Powell (1937, 1979) show discrete spiral keels on the base, none of which are apparent on the shell; only excessive artistic license can explain the discrepancy.

Anatoma australis (Hedley, 1903)

Figures 641–646, Color Plate 2G

Chresonymy.

Scissurella australis Hedley, 1903: 329–330, fig. 63.
Scissurella australis: Anonymous, 1906: 367.
Scissurella australis: Hedley, 1907b: 288.
Scissurella australis: Verco, 1910: 115.
Scissurella australis: Gatliff & Gabriel, 1910: 95.
Scissurella australis: Hedley, 1911: 92.
Scissurella australis: Verco, 1912a: 191.
Scissurella australis: Thiele, 1912a: 19, pl. 3, fig. 1 [copy figure Hedley, 1903].
Scissurella australis: Hedley, 1918: M39.
Scissurella australis: May, 1921: 34, pl. 17, fig. 1.
Scissurella australis: May, 1923: pl. 17, fig. 1.
Schizotrochus australis: Cotton & Godfrey, 1933: 22–23, pl. 1, fig. 10.
Schizotrochus australis: May & Macpherson, 1958: 21, pl. 17, fig. 1.
Schizotrochus australis: Iredale & McMichael, 1962: 30.
Schizotrochus australis: Macpherson & Gabriel, 1962: 34.
Anatoma australis: Lozouet, 1986: 112.
Scissurella (Anatoma) australis: Wilson, 1993: 46.
Anatoma australis: Jansen, 1995, 11.
Anatoma australis: Hickman, 1998: fig. 15.52 A–B.
Anatoma australis: Jansen, 1999: 48, figs 4–6.
Anatoma australis: Geiger, 2003: 73, figs 15f, h [protoconch, umbilicus].
Anatoma australis partim: Geiger & Jansen, 2004a: 9–14, figs 3–4, 18 [map].
Anatoma australis/crispata?: Geiger & Jansen, 2004a: 14, fig. 6 [see *An. crispata* for discussion].
Anatoma australis: Geiger, 2006a: fig. 19 [from Geiger & Jansen, 2004a].

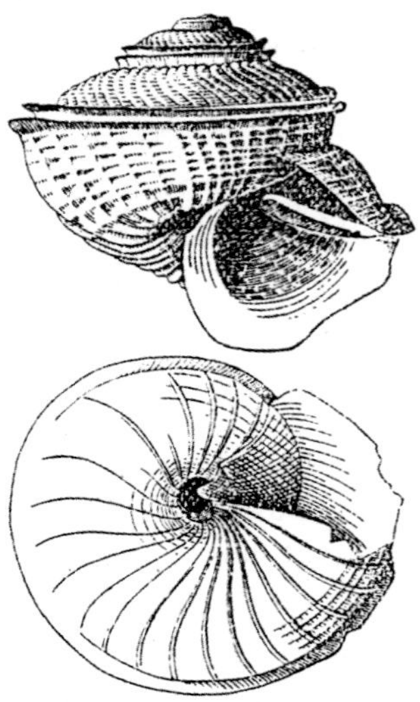

Figure 641. Original illustration of *Scissurella australis.*

Misidentifications.

? *Scissurella australis*: Hedley & May, 1908: 109 [most likely AMS C.29025 = *An. tobeyoides*].
not *Anatoma australis* partim: Geiger & Jansen, 2004a: 9–14, figs 5, 18 [map: is *An. japonica*].

Type material. Holotype (AMS C.16223), 3 × 2.5 × 2.5 mm. Paratypes from type locality (AMS C.170412, 3).

Type locality. [5–8 mi] off Port Kembla, [New South Wales, Australia, 34.333°S, 151.300°E] 63–75 fms. [= 110–140 m] (OD).

Etymology. Most likely referring to the occurrence in the southern hemisphere.

Description. Shell globular, medium-sized to large (3.6 mm), trochiform, biconical. Protoconch 0.75 whorl, sculpture flocculent, varix not connected to embryonic cap, apertural margin sinusoidal. Teleoconch I of >0.5 whorl, with spiral cord in position of selenizone, often with additional spiral cords. Teleoconch II of up to 2.5 whorls, suture little impressed. Shoulder with of 1–4 spiral lines at the start of teleoconch II, approximately 12 spiral lines at aperture of fully grown specimens (2.5 mm), spiral lines initially much weaker than axial cords, at most as strong as axials at aperture; approximately 70–75 axial cords shoulder and base of last whorl in fully grown shell. Base with approximately a dozen spiral cords. Spirals on base as strong as axials; density of axial cords below selenizone same as on shoulder, only every 3–4 axials reaching from selenizone to umbilicus; spirals running over axials forming weak nodes, spirals less dense in adumbilical third of base; adumbilical spirals as steps. Umbilicus sloping continuously from base, narrow, deep. Selenizone at periphery, keels distinct; lunules not coordinated with axials; margins of slit converging towards apertural margin of shell. Aperture rounded, somewhat flared in adumbilical corner.

Operculum round, thin, flexible, multispiral, nucleus central, covering entire aperture (Fig. 644F).

Animal without pigmented eyes, cephalic tentacles papillate.

Radula (Fig. 644A–F) rhipidoglossate, flexoglossate. Rachidian tooth floppy, triangular, apical cusp largest, mid-height distinct cusp on each side, approximately 10 small cusps on each side. Lateral teeth 1–3 similar to rachidian tooth, more

Figure 642. *Anatoma australis*. Capricorn Channel, 16.8 mi NE of North Reef, Queensland, Australia, 23.140°S, 152.205°E (AMS C.400817, 20+). Scale bar shell = 1 mm. Scale bars spire and protoconch = 100 μm.

Figure 643. *Anatoma australis*. **A.** Capricorn Channel, 16.8 mi NE of North Reef, Queensland, Australia, 23.140°S, 152.205°E (AMS C.400817, 20+). **B.** E of Caloundra, Queensland, Australia, 26.800°S, 153.583°E (AMS C.400805, 20+). **C.** Capricorn Channel, E of Lady Musgrave Island, Queensland, Australia, 23.835°S, 152.535°E (AMS C.400811, 21). Scale bars shell = 1 mm. Scale bars spire and protoconch = 100 µm.

slender. Lateral tooth 4 with strongly elongated tip, apical cusp largest, 3–4 cusps on each side. Lateral tooth 5 enlarged by elongation, apical cusp as long as height of shaft, approximately eight minute points on each side of tooth. Radula interlock between lateral teeth 4 and 5 very strong, weak among remainder of teeth. Marginal teeth elongated, with elongated shafts without food groove, serrated in upper third of tooth.

Comparisons. *Anatoma biconica* from the Indian Ocean to the Central Pacific has a more stepped overall shell profile, a wider sutsel on early whorls (> vs. < width of selenizone), lacks the spiral steps on the base of the shell, and has fewer, if any, spiral lines in addition to the one in the position of the selenizone.

Anatoma flexidentata from the Indian Ocean and the Western Pacific has a protoconch without apertural varix, a shorter teleoconch I (0.25 vs. >0.5 whorl), has a proportionally wider selenizone, and fewer and finer spiral lines on the shoulder.

Anatoma kopua from temperate eastern Australia to New Zealand has an overall more turreted shell shape, a shorter teleoconch (0.3–0.4 vs. 0.75 whorl), and all axial cords on the base extend from the suture to the selenizone.

Anatoma tobeyoides from south-eastern Australia lacks the protoconch varix, has a shorter teleoconch I (<0.5 vs. 0.75 whorl), which lacks a spiral cord in position of selenizone, and has very fine centrifugal lines, but no true spiral sculpture.

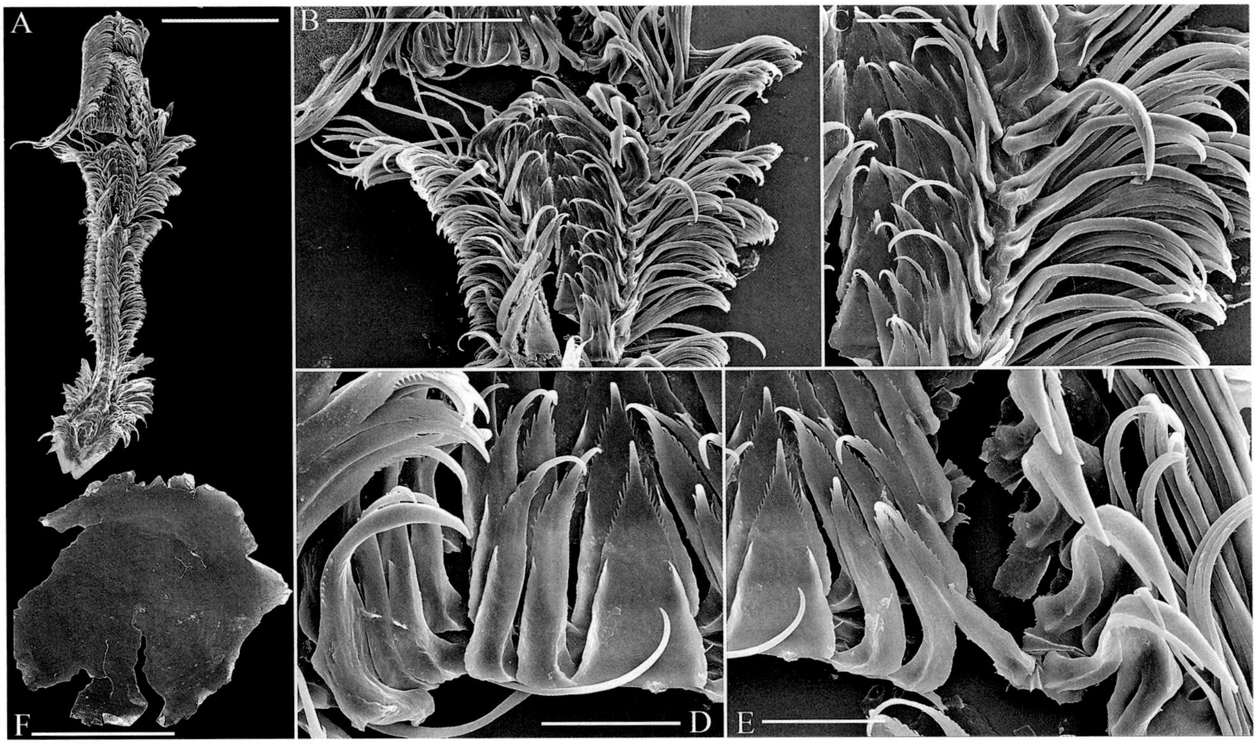

Figure 644. *Anatoma australis*. Radula and operculum. 139 m, SE Broken Bay, New South Wales, Australia, 33.717°S, 151.617°E (AMS C.461459, 2). **A.** Entire radula. **B.** Full width of radula. **C.** Radular interlock of lateral teeth 4 and 5. **D.** Central field and lateral tooth 5 showing serration at junction between tip and base. **E.** Central field and interlocked lateral teeth 4 and 5 forming a functional unit. **F.** Operculum. Scale bar A = 200 µm. Scale bar B = 100 µm. Scale bars C–E = 20 µm. Scale bar F = 1 mm.

Anatoma agulhasensis from the western Indian Ocean has an overall more globular shell with a weak constriction of the base below the selenizone, lacks additional spiral lines on the outer portion of teleoconch I, and has sharper spiral cords on the shoulder of teleoconch II.

Anatoma alternatisculpta from the Caribbean has no spiral cord in the position of the selenizone on teleoconch I, the axial cords on the shoulder are twice as widely spaced, and the spiral cords on the shoulder of early teleoconch II are not evenly distributed from suture to selenizone.

Anatoma amoena from Antarctica has a protoconch with reticulate sculpture and no apertural varix and the finer spiral lines do not form points at intersection with axial cords, neither on shoulder nor base.

Anatoma amydra from the Indo-Malayan Archipelago to the Central Pacific is overall taller, has a wider sutsel on early whorls (< vs. > width of selenizone), and finer and more numerous spiral lines on the base.

Anatoma japonica from the Indian Ocean to the Central Pacific has an overall wider shell, lacks a sutsel on early whorls, and has a more narrow sutsel on later whorls (<< vs. > width of selenizone).

Anatoma maxima from the Indo-Malayan Archipelago to the Western Pacific has no sutsel on early whorls, a proportionally wider selenizone with wider spaced lunules, and all axial cords on the base extend from the selenizone into the umbilicus.

Anatoma proxima from the Caribbean and southwestern Atlantic is overall more inflated, lacks a spiral cord in position of the selenizone on teleoconch I, and has denser axial sculpture on the base with all axials extending from selenizone to beyond mid-base.

Anatoma rainesi from the Eastern to the Western Pacific is overall broader, has fewer but strong axial cords on teleoconch I without the spiral cord in position of the selenizone, and a strong funiculus.

Anatoma rhynchodentata from the Indo-Malayan Archipelago to the Western Pacific has a more stepped profile of the shell, finer reticulate sculpture on shoulder and base, and a notable constriction of the base below the selenizone.

Anatoma sagamiana from Japan and the Indo-Malayan Archipelago has a protoconch with reticulate sculpture, fewer and weaker spiral lines on the shoulder, and finer reticulate sculpture on the base.

Anatoma schanderi from the northeastern Atlantic has an overall more globular shell shape, a protoconch with reticulate sculpture and no apertural varix, and a longer teleoconch I (1 vs. 0.75 whorl).

Anatoma zancliformis from the Indo-Malayan Archipelago to the Western Pacific has a shorter teleoconch I (0.5 vs. 0.75 whorl), the axial cords on the base extend from the selenizone to the umbilicus, and early whorls lack a sutsel.

Distribution. South Australia to Queensland, Australia, 78–1,463 m, 176 m live (Fig. 645).

Fossil. Muddy Creek, near Hamilton, NW Victoria, Australia (AMS, 2). Hamilton, Victoria, Australia, Muddy Creek, SW below Clifton Homestead, in MacDonald's Bank, Grange Burn 'Coquina" (Upper Beds), Pliocene (Kalimnan) (AMS, 2).

Specimen records.

Australia, Queensland. 201 m, off S end Fraser Island, 27.952°S, 153.851°E (AMS C.400802, 3). 1500 m, Lord Howe Rise, 27.567°S, 161.573°E (AMS C.166770 2: complete). 128–183 m, off Cape Moreton, 27.000°S, 153.583°E (AMS C.400806, 17; AMS C.400807, 17). 115–119 m, NE of Cape Moreton, 26.925°S, 153.558°E

(AMS C.400810, 3). 115–124 m, NE of Cape Moreton, 26.925°S, 153.558°E (NMNZ M228762, 20). 183 m, NNE of Cape Moreton, 26.875°S, 153.600°E (AMS C.400809, 20+). 128–183 m, E of Caloundra, 26.800°S, 153.583°E (AMS C.400805, 20+). 200 m, off Maroochydore, 26.687°S, 153.640°E (AMS C.C.400808, 9). 128 m, ESE of Noosa Heads, 26.567°S, 153.667°E (AMS C.400804, 18). 210 m, off Fraser Island, 24.965°S, 153.622°E (AMS C.400818, 6). 150 m, NE of Lady Elliot Island, 24.062°S, 152.823°E (AMS C.400813, 4). 175–200 m, E of Lady Musgrave Island, 23.853°S, 152.575°E (AMS C.400814, 20+). 296 m, E of Lady Musgrave Island, 23.871°S, 152.712°E (AMS C.448756, 200). 132 m, Capricorn Channel, E of Lady Musgrave Island, 23.835°S, 152.535°E (AMS C.400811, 21). 284 m, Capricorn Channel, E of North West Island, 23.253°S, 152.402°E (AMS C.400816, 4). 155 m, Capricorn Channel, 23.143°S, 152.277°E (AMS C.400815, 12). 115 m, Capricorn Channel, 16.8 mi NE of North Reef, 23.140°S, 152.205°E (AMS C.400817, 20+). 78 m, SE of Swain Reefs, 22.518°S, 152.710°E (AMS C.400812, 3; AMS C.402610, 1). 187 m, SE of Swain Reefs, 22.438°S, 153.285°E (AMS C.400803, 20+; AMS C.402615, 3). 1468–1477 m, Great Barrier Reef, off Cairns, 16.760°S, 146.622°E (AMS C.448755, 1).

Australia, New South Wales. 988–1015 m, E of Nowra, 34.867°S, 151.242°E (AMS C.448764, 1). 741 m, off Sydney, 34.767°S, 151.717°E (AMS C.448771, 1). Shell Harbor, 34.583°S, 150.867°E (WAM, 1). 115–137 m, 5–8 mi off Port Kembla, 34.465°S, 151.075°E (AMS C.170412, 3; AMS C.16223, 1). 1200 m, off Wollongong, 34.442°S, 151.450°E (AMS C.448760, 2). 183 m, 16 mi E of Wollongong, 34.417°S, 151.250°E (AMS C.17988, 3). 161 m, off Port Kembla, 34.333°S, 151.300°E (AMS C.400796, 1). 205 m, S of Port Hacking, 34.160°S, 151.438°E (AMS C.346346, 1). 750 m, off Sydney, 34.083°S, 151.727°E (AMS C.402694, 11). 79–295 m, off Sydney, 34.050°S, 151.617°E (AMS C.400793, 3; AMS C.402613, 1). 150 m, off Sydney, 33.967°S, 151.483°E (AMS C.400792, 14). 750 m, off Sydney, 33.917°S, 151.167°E (NHMUK 89.10.26.33, 1). 200 m, off Sydney, 33.872°S, 151.667°E (AMS C.400798, 8). 1463 m, 35 mi E of Sydney, 33.850°S, 151.967°E (AMS C.26669, 4). 109 m, off Port Jackson, 33.833°S, 151.450°E (AMS C.400801, 1). Sydney harbor, 33.803°S, 151.273°E (AMS C.456628, 2). 176 m, off Sydney, 33.767°S, 151.717°E (AMS C.400791, 20+, 2: complete). 135 m, off Sydney, 33.705°S, 151.633°E (AMS C.448766, 1). 263 m, off Broken Bay, 33.733°S, 151.800°E (AMS C.448772, 1). 322 m, off Broken Bay, 33.733°S, 151.950°E (AMS C.461458, 1). 139 m, SE Broken Bay, 33.717°S, 151.617°E (AMS C.461459, 2). 143 m, off Broken Bay, 33.717°S, 151.667°E (AMS C.400795, 2; AMS C.400797, 2). 174 m, off Sydney, 33.717°S, 151.767°E (AMS C.400794, 1). 146 m, 22 mi E of Narrabeen, N of Sydney, 33.700°S, 151.717°E (AMS C.25803, 11). 1097 m, off Sydney, 33.600°S, 152.150°E (AMS C.448763, 6). 134 m, off Sydney, 33.583°S, 151.683°E (AMS C.461460, 1). 132 m, off Norah Head, 33.333°S, 151.933°E (AMS C.400799, 1). 183 m, NE of Port Macquarie, 31.383°S, 153.200°E (AMS C.37724, 2). 185 m, 24 km E of Ballina, 28.827°S, 153.855°E (AMS C.400800, 9).

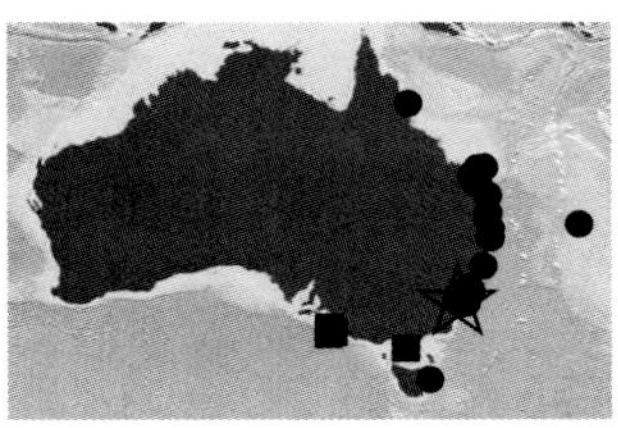

Figure 645. Distribution of *Anatoma australis*. ☆ type locality. ● specimen record. ■ literature record.

Australia, Tasmania. 113 m, N of Great Oyster Bank, 41.758°S, 148.517°E (LACM 73-156, 2).

Literature records.

Australia. Off Wilson's Promontory, Victoria, 38.933°S, 146.367°E (Macpherson & Gabriel, 1962).— 366 m, off Beachport, South Australia, 37.665°S, 139.628°E. 274 m, off Beachport, South Australia, 30.500°S, 140.017°E. 238 m, off Cape Jaffa, South Australia, 36.950°S, 139.750°E. 549 m, off Cape Jaffa, South Australia, 36.950°S, 139.667°E (Verco, 1910).

Remarks. Geiger & Jansen (2004) erroneously lumped *Anatoma* "SWA" of Jansen (1999) with *An. australis*. Those specimens were subsequently described as *An. jansenae* before they were recognized as *An. japonica*.

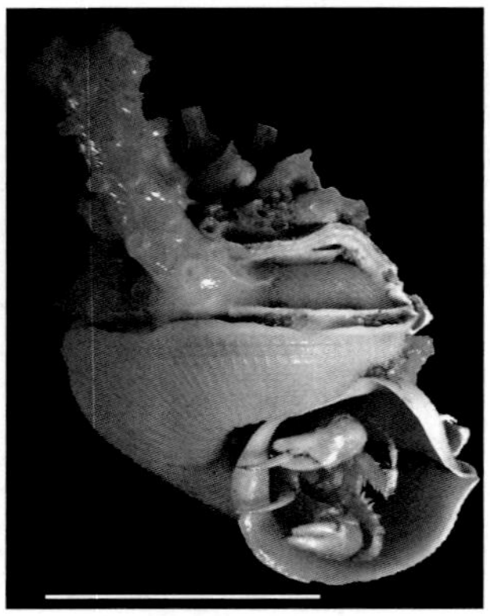

Hedley & May (1908) reported *Sci. australis* from off Cape Pillar, Tasmania in 100 fms [=183 m] depth. AMS C.29025 matches those data precisely, and is *An. tobeyoides*.

Some specimens have been found inhabited by hermit crabs (e.g., AMS C.400809: Color Plate 2G, Fig. 646).

The radula with floppy teeth is shared with *An. flexidentata* from Reunion Island and *An. austrolissa* from Indian Ocean and New South Wales. All three species occur in rather deeper water, but otherwise have typical anatomid shells.

Figure 646. *Anatoma australis* with hermit crab, serpulid worms, and encrusting bryozoans (AMS C.400809, 20+).

Anatoma austrolissa Geiger & Sasaki, 2008

Figures 647–650

Chresonymy.

Anatoma austrolissa Geiger & Sasaki, 2008: 254–257, figs 4–6.

Anatoma austrolissa: Geiger & McLean, 2010: 33.

Type material. Holotype (AMS C.212812: Geiger & Sasaki, 2008: fig. 4. Fig. 647). Paratypes 1575–1600 m, Reunion Island, MD32, DS139, 20.783°S, 55.633°E (MNHN 20631, 2 dry shells, 1 wet preserved animal: Geiger & Sasaki, 2008. Fig. 648B–C). 1200 m, off Wollongong, New South Wales, Australia, 34.442°S, 151.450°E (AMS C.448762, 1: Geiger & Sasaki, 2008. Fig. 648A). 768–713 m, off Coffs Harbour, New South Wales, Australia, 30.017°S, 153.617°E (AMS C.402711, 2).

Type locality. 1143–1106 m, off Sydney, New South Wales, Australia, 33.600°S, 152.083°E (OD).

Etymology. Austro- referring to the occurrence of the species in the southern hemisphere; -lissa, Latin adjective for smooth, referring to the overall smooth shell (OD).

Description. Shell medium-sized (to 2.2 mm. Figs 647, 648), trochiform, biconical, tabulated, thin-shelled, fragile. Protoconch of 0.75 whorl, large (262, 265, 269 µm diameter), sculpture flocculent, no apertural varix, apertural margin slightly convex. Teleoconch I of 0.2 whorl, macroscopically smooth, with minute pitting, round to twice as long as wide, 1–2 µm (Fig. 649C). Teleoconch II of 3.25 whorls, suture incised, descending on shell axis with growth, sutsel 1× width of selenizone at 1.4 mm, 3× width at 2.2 mm. Shoulder moderately sinusoidal in profile, macroscopically smooth, with very fine growth lines, minute pitting, round to twice as long as wide, 1–2 µm (Fig. 649C). Base barely constricted below selenizone, continuously sloping into rather wide umbilicus, upper half of base smooth except for minute pitting, adumbilical half of base with approximately 30 axial cords, starting very abruptly mid-base, remaining distinct for approximately width of umbilical cavity, diminishing in strength into umbilicus; no funiculus. Selenizone at periphery, onset indistinct; keels moderately elevated, moderately strong; lunules mostly indistinct, occasionally more distinct lunules; slit open, margins parallel. Aperture subquadrate, peristome flaring.

Operculum corneous, thin, multispiral, with central nucleus (Fig. 649D).

Radula with floppy teeth in central field (Fig. 649A, B). Central tooth triangular, sides concave, upper half (= tip) with approximately nine fine cusps on each side. Lateral teeth 1–3 similar to rachidian tooth, curving towards midline of radula, decreasing in size from height equal to rachidian tooth to approximately ⅔ of it, approximately nine, six, four fine cusps on each side of respective teeth. Lateral tooth 4 reduced (30% of height of rachidian tooth) without cusps. Lateral tooth 5 enlarged by elongation, approximately 2.2× length of rachidian tooth, sturdy, in lower portion of tip with approximately eight fine cusps, terminal cusp almost as strong as shaft. Marginal teeth thin, elongated, approximately nine cusps on outer

Figure 647. *Anatoma austrolissa.* Holotype. 1143–1106 m, off Sydney, New South Wales, Australia, 33.600°S, 152.083°E (AMS C.212812). Scale bar shell = 1 mm. Scale bar protoconch = 100 μm.

edge of tip, approximately three on inner edge, terminal cusp as large as others; shafts without food groove. Radular interlock minimal.

Jaw (Fig. 649E) trapezoidal, with recurved front edge, composed of rhombic elements.

Figure 648. *Anatoma austrolissa*. Paratypes. **A.** 1575–1600 m, Reunion Island, 20.783°S, 55.633°E (MNHN 20631, 2). Radula shown in Figure 649A–B. **B–C.** 1200 m, off Wollongong, New South Wales, Australia, 34.442°S, 151.450°E (AMS C.448762, 1: detail of teleoconch sculpture shown in Figure 649C). Scale bars shell = 1 mm. Scale bars protoconch = 100 μm.

Comparisons. *Anatoma austrolissa* is characterized by the large protoconch, short teleoconch I, indistinct onset of the selenizone, overall smooth sculpture with microscopic pitting, and the axial sculpture restricted to the adumbilical half of the base.

Anatoma umbilicata from the European Atlantic and Mediterranean Seas has a more globular overall shape with the shoulder at a steeper angle towards the coiling axis, a longer teleoconch I (0.5 vs. 0.3 whorl) lacking the microscopic pitting, shows very faint spiral threads on the shoulder, and more prominent spiral threads in the umbilical region, but lacks any axial sculpture on the base.

Anatoma planapex from deepwater Australia has a less turreted overall shape, a longer teleoconch I (0.8–0.9 vs. 0.3 whorl), fine axial and spiral sculpture over most of teleoconch, and a radula with sturdy teeth and with a trapezoidal rachidian tooth.

Anatoma hyposculpta from tropical Western Pacific has an apertural varix on the protoconch, a more discrete onset of the selenizone, weak, yet distinct spiral sculpture on shoulder and base, and lacks the pitting on the teleoconch.

Distribution. Indian Ocean to Indo-Malayan Archipelago and New South Wales (Fig. 650).

Specimen records.

Reunion Island. 1575–1600 m, 20.783°S, 55.633°E (MNHN 20631, 3, 1: complete).

Australia, New South Wales. 768–713 m, off Coffs Harbour, 30.017°S, 153.617°E (AMS C.402711, 2). 1143–1106 m, off Sydney, 33.600°S, 152.083°E (AMS C.212812, 1). 1200 m, off Wollongong, 34.441°S, 151.450°E (AMS C.448762, 1).

Vanuatu. 700–736 m, 14.883°S, 167.100°E (MNHN, 1).

Fiji. 640–687 m, Bligh Water, 16.937°S, 178.132°E (MNHN, 1). 700–707 m, BORDAU 1 DW1485, 19.050°S, 178.500°W (MNHN, 2).

Remarks. The microscopic pitting is seen in all specimens examined and represents a unique character on the teleoconch. In scissurellids pitting is known on the protoconch of *Sutilizona* species (Vetigastropoda: Sutilizonidae. Figs 1016, 1020, 1021, 1024)

Figure 649. *Anatoma austrolissa*. Radula and operculum. 1200 m, off Wollongong, New South Wales, Australia, 34.442°S, 151.450°E (AMS C.448762, 1: see Figure 648C for shell). **A.** Radula after bleach treatment of dirty specimen shown in Figure 649B. Scale = 20 µm. **B.** Overview of dirty radula, which was insufficiently spread to show central field. Scale bar = 100 µm. **C.** Detail of teleoconch of specimen illustrated in Figure 648B having minute pitting. Scale bar = 10 µm. **D.** Operculum in aperture of specimen (MNHN 20631). Scale bar = 500 µm. **E.** Jaw. Scale bar = 50 µm.

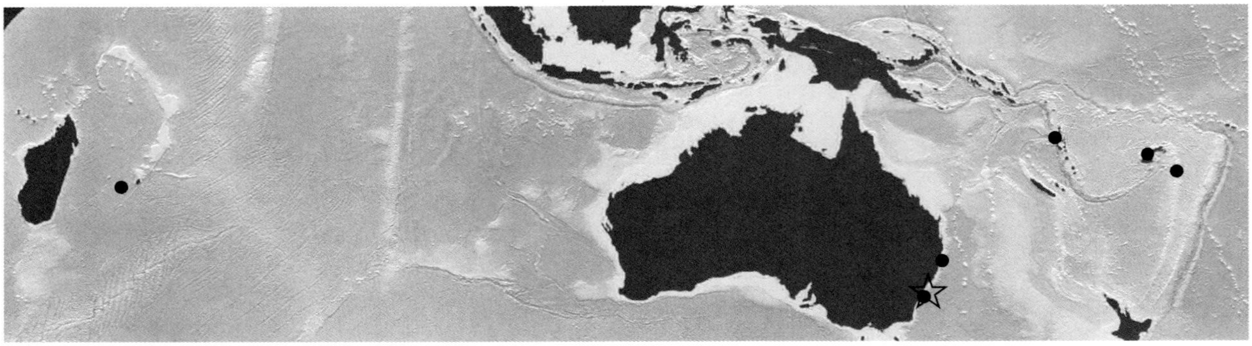

Figure 650. Distribution of *Anatoma austrolissa*. ☆ type locality. ● specimen record.

from hydrothermal vent environments (McLean, 1989; Warén & Bouchet, 2001). In *An. austrolissa*, it does not seem to be a product of either biogenic or physical-chemical deterioration. Specimens are currently known from rather widely separated localities: Reunion Island, New South Wales, Vanuatu, and Fiji (Fig. 650). The specimens agree in several characters uncommonly seen in Anatomidae: large protoconch (262–269 µm), pitting of shell, lack of macroscopic sculpture on shoulder and upper half of base, and the axial sculpture on adumbilical half of base. The onset of the selenizone is slightly more marked in the specimens from Australia, though still not as distinct as in most other *Anatoma* species. They also share a habitat on the upper slope (600–1600 m) where specimens from both ends of the distributional range were collected live. It indicates that *An. austrolissa* is a species with broad deep-water distribution in the southern Pacific.

Anatoma bathypacifica (Geiger & McLean, 2010)

Figures 651–653

Chresonymy.

Thieleella bathypacifica Geiger & McLean, 2010: 31–33, figs 23-24.

Type material. Holotype (FMNH 307884. Fig. 651), shell, operculum and radula mounted on SEM stubs.

Type locality. 2572 m, Dive 3938, *Genesis*, East Pacific Rise, 12.811°N, 103.940°W [approximately in the middle between Acapulco, Mexico, and Clipperton Island]. Sample with anemone and serpulid rocks. *Atlantis*, submersible grab, leg. J. Voight (OD).

Etymology. Bathy- referring to the deep-sea (bathyal) habitat of the species; -pacifica referring the Pacific Ocean; adjective (OD).

Description. Shell trochiform, globular (to 1.85 mm. Fig. 651). Protoconch of 0.75 whorl, reticulate sculpture, no apertural varix, apertural margin straight. Teleoconch I of 0.875 whorl, 15 indistinct axial cords, prominent spiral cord forming ridge in position of selenizone, at intersection forming thickenings. Teleoconch II of at least 1.5 whorls; sutsel slightly less wide than selenizone. Shoulder convex, 26 distinct raised axial cords, first spiral line after 0.33 teleoconch II whorl, seven spiral lines after 1.5 whorls on outer two thirds of shoulder, spiral lines running over axial cords. Base convex, barely constricted below selenizone, continuously sloping with umbilicus, same density of axial cords as on shoulder, crossed by spiral lines, 23 after 1.25 teleoconch II whorls, spiral lines running over axial cords. Selenizone at periphery, keels of moderate strength, high (higher than width of selenizone), regular lunules; slit open, with parallel margins.

Operculum (Fig. 651) round, thin, multispiral, central nucleus.

Radula. Rachidian tooth trapezoidal, central cusp often largest, approximately five cusps on each side (Fig. 652A–C: R). Lateral teeth 1–3 similar, six, four, and four cusps, respectively (Fig. 652B–C: L1–3). Lateral tooth 4 reduced, hook-shaped, apical cusp largest, two small cusps on inner cutting edge (Fig. 652C: L4). Lateral tooth 5 enlarged by broadening, inner edge with 4–5 cusps, outer edge with three cusps (Fig. 652B–C: L5). Inner marginal teeth triangular with approximately

Figure 651. *Anatoma bathypacifica.* Holotype. 2572 m, Dive 3938, *Genesis,* East Pacific Rise, 12.811°N, 103.940°W (FMNH 307884). Scale bar shell, operculum = 1 mm. Scale bar protoconch = 100 μm.

3–4 cusps on each side (Fig. 652D–E); with food groove (except marginal tooth 1. Fig. 652C: M1); outer marginal teeth spoon or paddle-shaped with many fine cusps along apical edge (Fig. 652E–F).

Comparisons. *Anatoma lyra* from the northeastern Pacific has elevated lamellar axial sculpture, flocculent protoconch sculpture, and lacks a distinct spiral cord in the position of the selenizone on teleoconch I.

Anatoma argentinae from the abyssal plain of tropical eastern Pacific and off Argentina has a much larger protoconch (370–480 vs. 240 μm), lacks the sharp angulation along the spiral cord in position of the selenizone on teleoconch I, and has at least twice as many spiral lines on shoulder and base.

Anatoma janetae from deep water of the eastern Pacific has a protoconch with flocculent sculpture in spiral orientation, the spiral cord on teleoconch I does not form a marked angulation, and has a marked switch-over of sculpture from predominant axial elements to exclusively spiral elements after 1.25 teleoconch II whorls.

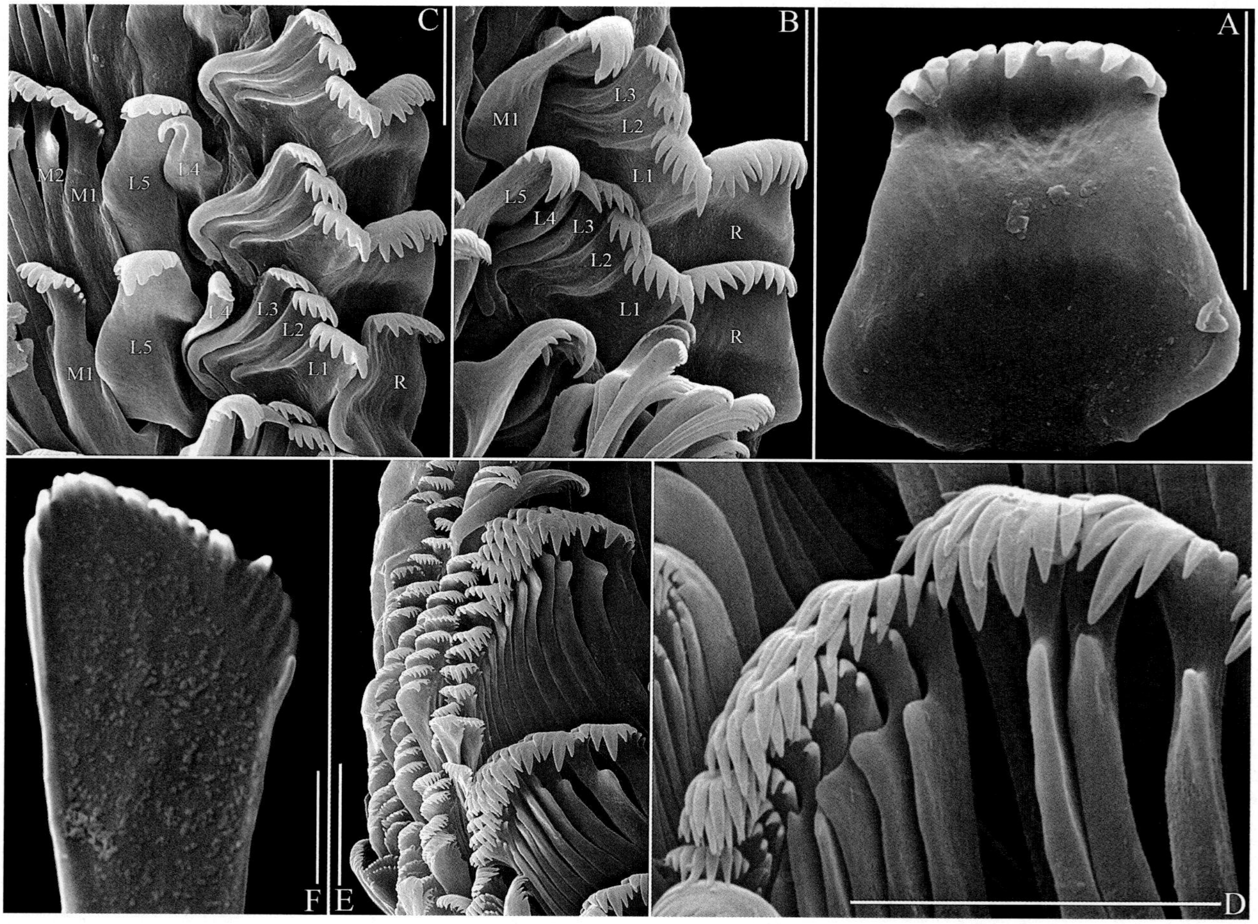

Figure 652. *Anatoma bathypacifica.* Radula from holotype. 2572 m, Dive 3938, *Genesis,* East Pacific Rise, 12.811°N, 103.940°W (FMNH 307884). Shell shown in Figure 651. **A.** Isolated rachidian tooth. **B.** Central field with well-developed teeth. **C.** Central field with somewhat worn lateral teeth and showing lateral tooth 4. **D.** Inner marginal teeth with food groove. **E.** Entire series of marginal teeth. **F.** Outermost marginal tooth. Scale bars A–E = 10 μm. Scale bar F = 2 μm. R: rachidian tooth. L1–5: lateral tooth 1–5. M1–3: marginal tooth 1–3.

Anatoma keenae from the Panamic has an overall more globular shell shape, has a protoconch with dense flocculent sculpture, and lacks any angulation along the spiral cord on teleoconch I.

Anatoma kelseyi from the northeastern Pacific is overall more globular and lacks a spiral cord in the position of the selenizone.

Anatoma peruviana from deep water off Peru has on teleoconch I only a weak spiral cord not forming an angulation on the shell's profile and has more (42 vs. 15) and weaker axial cords on teleoconch I.

Anatoma rainesi from the Eastern to the Western Pacific has flocculent sculpture on the protoconch, lacks the angulation along the spiral cord on teleoconch I, and is overall more discoidal in shape with a more narrow sutsel.

Anatoma aedonia from the Caribbean and the SW Atlantic does not form an angulation at the spiral cord on teleoconch I, has flocculent protoconch sculpture, and a wider sutsel on early teleoconch whorls.

Anatoma alternatisculpta from the Caribbean has a smooth protoconch and usually lacks a spiral cord in position of the selenizone.

Anatoma finlayi from the Indo-Malayan Archipelago (and possibly Caribbean) has flocculent protoconch sculpture, lacks the angulation along the spiral cord on teleoconch I, and has raised axial lamellae on shoulder and base.

Anatoma lamellata from Japan to the Central Pacific has a protoconch with flocculent sculpture and a middle spiral cord, lacks the angulation along the spiral cord on teleoconch I, has raised axial lamellae on shoulder and base of teleoconch II, and has stronger spiral cords on shoulder and base.

Anatoma sinuosa from Easter Island has a smooth protoconch, a shorter teleoconch I (0.4 vs. 0.75 whorl) without spiral on position of the selenizone, and at least twice as dense axial cords.

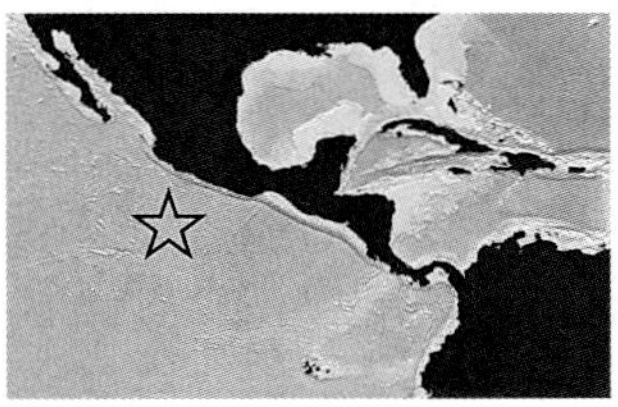

Figure 653. Distribution of *Anatoma bathypacifica*. ☆ type locality.

Distribution. Known only from type locality (Fig. 653).

Remarks. The only specimen at hand is most likely not fully mature, because the final quarter whorl does not markedly descend along the coiling axis. The specimen was preserved with the animal in the shell; hence, was collected live in its natural deep-sea habitat. The radula of the species shows a food groove on the shafts from marginal tooth 2 onwards.

Anatoma bertheloti (d'Orbigny, 1840) [*nomen dubium*]

Figures 654–655

Chresonymy.

Scissurella bertheloti d'Orbigny, 1840 (1836–1844): 96, pl. 7, figs 2–5.

Scissurella berthelotti [*sic*]: MacAndrew, 1852: 102.

Anatomus bertheloti, "Webb.": Adams & Adams, 1853–1858 (1854): vol. 1, 439.

Scissurella berthelotti [*sic*]: MacAndrew, 1854: 32.

Scissurella bertheloti: Gray, 1854b: 18, 30.

Scissurella bertheloti: MacAndrew, 1856: 119, 147.

Scissurella bertheloti: Chenu, 1859: 363, fig. 2714 [redrawn from d'Orbigny, 1836–1844].

Scissurella bertheloti: Munier-Chalmas, 1865: 394.

Scissurella bertheloti: Tryon, 1883: 321, pl. 82, fig. 92 [copy Chenu, 1859].

Scissurella bertheloti: Paetel, 1888: 289.

Scissurella bertheloti: Pilsbry, 1890: 53, pl. 68, fig. 26 [erroneously pl. 58 in text, copy figure d'Orbigny, 1836–1844].

Scissurella (Schizotrochus) bertheloti: Nordsieck, 1982: 2, pl. 1, fig. 15.

Anatoma ? bertheloti: Lozouet, 1986: 112.

Scissurella bertheloti: Geiger, 2003: 76.

Scissurella bertheloti: Ardovini & Cossignani, 2004: 19.

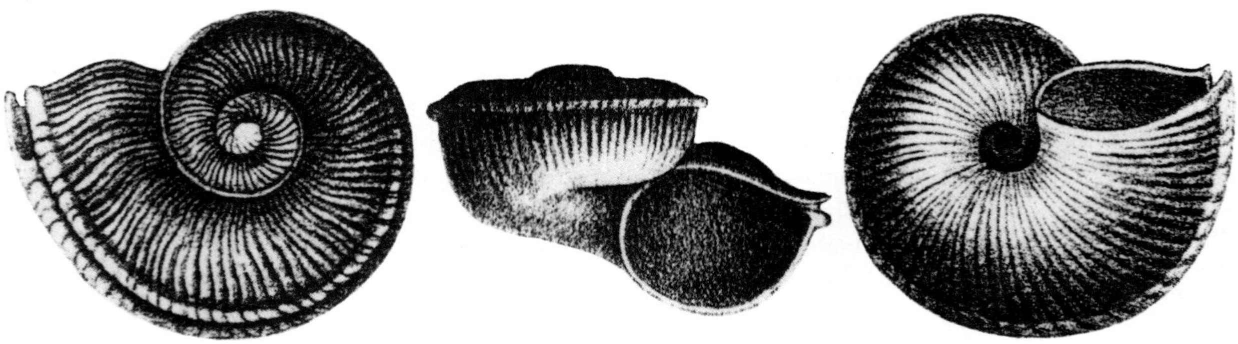

Figure 654. Original illustration of *Scissurella bertheloti* d'Orbigny, 1840.

Misidentifications.

not *Scissurella (Schizotrochus) bertheloti*: Nordsieck, 1968: 10, pl. 1, fig. 15 [is juv. *Sin. cingulata*].

not *Scissurella (Schizotrochus) bertheloti*: Nordsieck & García-Talavera, 1979: 35–36, pl. 1, fig. 3 [is *An.* cf. *orbiculata*].

Type material. Syntype(s) (whereabouts unknown. Not NHMUK, not MNHN, most likely lost), 0.75 × 1.2 mm

Type locality. Tenerife [Canary Islands, Spain, 28.250°N, 16.583°W] (OD).

Etymology. Dedicated to Mr. [Sabin] Berthelot (OD).

Description. (based on original description and illustration. Fig. 654). Shell medium-sized (to 1.2 mm), trochiform, subquadrate, spire low. Protoconch unknown. Teleoconch I whorl count estimated at 0.5–0.75, approximately 15 axial cords, equally spaced. Teleoconch II of 1.75 whorls, suture deeply impressed, sutsel wider than selenizone. Shoulder convex, with many dense axial cords, no spirals according to original illustration [see Remarks]. Base without constriction below selenizone, same sculpture as on shoulder. Umbilicus open, moderately wide, no funiculus. Selenizone slightly above periphery; keels distinct; lunules smooth (OD). Aperture round, roof slightly overhanging.

Animal unknown.

Comparisons. *Anatoma orbiculata* from West Africa has a more lenticular overall shell shape, at most half as many axial cords on shoulder and base, and the axial cords on the shoulder are much stronger than on the base.

Anatoma aspera from the Mediterranean and the adjacent Atlantic has a taller overall shell shape, at most half as many axial cords on the shoulder, and the axial cords on the shoulder are stronger than those on the base.

Anatoma eximia from the Mediterranean and the adjacent Atlantic has a much taller shell that is taller than wide and about 50% fewer axial cords.

Anatoma micalii from the Mediterranean and the adjacent Atlantic has at most half as many axial cords on the shoulder and the axial cords on the shoulder are much stronger than on the base.

Anatoma tenuisculpta from the Mediterranean and the adjacent Atlantic has a more lenticular overall shell shape and the selenizone is at the periphery.

Anatoma umbilicata from the Mediterranean and the adjacent Atlantic has hardly any sculpture and the selenizone is more towards the periphery.

Anatoma janusa from the Azores has at most half as many axial cords and the axial cords on the shoulder are much stronger than those on the base.

Distribution. Known only from type locality with certainty (Fig. 655).

Literature records.

Spain. 91 m, Orotava, Canary Islands, 28.420°N, 16.530°W (McAndrew, 1854).

Remarks. D'Orbigny (1840) did not clearly state how many specimens he had available. Although he remarked that the species is very rare, he stopped short of specifying a single specimen. Accordingly, the name-bearing type(s) must be considered one to many syntype(s).

Woodward (1856a: 37) noted that "in the British Museum (to go no further) there is the type of *S. Bertheloti*, D'Orb., in a collection we [Woodward and Jeffreys] had both lately been examining." The mention of "the type" by Woodward cannot be construed as a lectotype designation as it lacks specificity. Gray (1854a) in his list of d'Orbigny's type specimens in the NHMUK noted *Scissurella bertheloti* with "?B.M." The specific meaning of the question mark was not detailed, but some doubt either about its existence, standing, or identification is sufficiently clear. Dance (1966) indicated that the Webb & Berthelot specimens investigated by d'Orbigny are housed in the NHMUK. The specimens could not be found among the type material during the visit in 2002.

Figure 655. Distribution of *Anatoma bertheloti*. ☆ type locality. ■ literature record.

The identity of this species is uncertain. The position of the selenizone at the periphery of the shell clearly puts the species in Anatomidae. The low spire, rather boxy overall appearance, and dense and prominent axial cords extending from the suture to the selenizone, and over the entire base, should make this species recognizable, but no specimens fitting the original illustration could be located. Whether the very uncommon absence of spiral sculpture, as shown in the original figure, is accurate, is impossible to assess without the type specimen(s). Some specimens from the Azores in MNHN were identified as *bertheloti*, but the specimens have less than half as many axial cords compared to the original illustration; this species is here described as *An. janusa* (Fig. 747).

D'Orbigny (1840) indicated three rounded whorls. While the whorl count of teleoconch II is clear at 1.75 whorls from the onset of the selenizone shown in the original illustration, the remainder of the whorl count is more difficult to assess. Given the non-standard nature of the whorl counts in general, no significance is attached to count given in the original description.

Hernández *et al.* (2011) tentatively identified *bertheloti* as *Sin. cingulata*. This synonymy does not agree with the original illustration, as *Sin. cingulata* has less than half a teleoconch II whorl, while the original illustration shows about 1.75 teleoconch II whorls.

The closest species is *An. eximia*, which is known from the Canary Islands. Characters agreeing include the overall boxy shape of the whorls and the approximately 1.75 teleoconch II whorls at 1.2 mm. However, the illustration of *An. bertheloti* differs by the flatter apical portion of the whorl, the absence of spiral sculpture on the base, and the wider umbilicus. Accordingly, *An. bertheloti* is not considered a senior synonym of *An. eximia*. It cannot be matched to any species and is here treated as a *nomen dubium*.

Anatoma biconica new species

Figures 656–659

Type material. Holotype (AMS C.475693. Fig. 656). Paratypes from type locality (AMS C.453760, 7). 786–841 m, off Nowra, New South Wales, Australia, 34.833°S, 151.250°E (AMS C.448773, 6). 896–924 m, off Sydney, New South Wales, Australia, 33.633°N, 152.050°E (AMS C.448768, 2). 932 m, St. 5658, off C. Loko Loko, Burton Strait, Celebes, Indonesia, 3.544°S, 120.525°E (USNM 234423, 2. Fig. 657A). 771–776 m, Maputo transect, Mozambique Channel, MAINBAZA CC3171, Mozambique, 25.983°S, 34.700°E (MNHN IM-2010-22883, 17. Fig. 657B). 475 m, St. 5591, Sibuko Bay, off Mabul Island, Borneo, Indonesia, 4.196°N, 118.639°E (USNM 280225, 1. Fig. 657C).

Type locality. 322 m, off Broken Bay, New South Wales, Australia, 33.733°S, 151.950°E.

Etymology. Describing the overall shell morphology of the species.

Description. Shell medium-sized (to 2.8 mm; 2.5 mm: holotype. Figs 656, 657), trochiform, biconical. Protoconch of 0.75 whorl, sculpture flocculent, no apertural varix, apertural margin straight. Teleoconch I of approximately 0.66–0.9 whorl, 20–39 axial cords often not continuous across fine spiral cord in position of selenizone, additional 1–3 fine spiral lines in adumbilical portion starting as early as 0.33 teleoconch I whorl. Teleoconch II of up to three whorls, suture moderately impressed. Shoulder convex, distinct dense axials from suture reaching onto keels of selenizone, increasing in strength from distinct cords to low ribs, approximately 60 on first whorl, approximately 90 on last whorl of fully grown specimen; fine equispaced spiral lines to low cords running over axials, approximately eight at apertural margin of fully grown specimen, 20% of shoulder's width closest to suture without spirals. Base continuously sloping into umbilicus; same density of axials; axials either as cords of equal strength from below selenizone into umbilicus, or cords diminishing into low cords mid-base, turning into axial folds in umbilicus. Spirals either as fine cords to distinct lines of equal strength, or as distinct lines below selenizone turning into spiral steps from mid-base onwards. Umbilicus open, of moderate width, without funiculus. Aperture rounded, peristome flared adumbilically, roof overhanging. Selenizone at periphery; keels moderately to strongly elevated, of moderate strength; lunules distinct; slit open, margin converging.

Operculum (Fig. 658G) corneous, thin, multispiral with central nucleus.

Radula (Fig. 658). Rachidian tooth trapezoidal, central cusp largest, five cusps of at most half size of central cusp on each side. Lateral teeth 1–3 similar, apical cusp largest, 6, 3, 3 cusps on outer cutting edge of tip, respectively, one cusp on inner edge of lateral tooth 3. Lateral tooth 4 reduced, apical cusp largest, 1–2 cusps on each side. Lateral tooth 5 enlarged by elongation, apical cusp largest, approximately eight cusps along inner edge, three cusps at base of tip on outer edge. Inner marginal teeth with strong apical cusp, 3–4 cusps on each side; shafts without food groove. Outer marginal teeth spoon-shaped, with many fine bristles on tip. Outermost marginal teeth paddle-shaped, with fine fringe of bristles at tip. Radular interlock of central field moderate.

Comparisons. *Anatoma australis* from eastern Australia has an overall more globular shell shape, a shorter teleoconch I (0.75 vs. 1 whorl), and only every 3–4 axial cords on the base extend from the selenizone into the umbilicus.

Anatoma austrolissa from the Indo-Malayan Archipelago and the Western Pacific has an almost entirely smooth shell and a shorter teleoconch I (0.2 vs. 1 whorl).

Figure 656. *Anatoma biconica* new species. Holotype. 322 m, off Broken Bay, New South Wales, Australia, 33.733°S, 151.950°E (AMS C.475693). Scale bar shell = 1 mm. Scale bar protoconch = 100 µm.

Figure 657. *Anatoma biconica* new species. Paratypes. **A.** 932 m, St. 5658, off C. Loko Loko, Burton Strait, Celebes, Indonesia, 3.544°S, 120.525°E (USNM 234423, 2). From lot from which radula shown in Figure 658 was extracted. **B.** 771–776 m, Maputo transect, Mozambique Channel, MAINBAZA CC3171, Mozambique, 25.983°S, 34.700°E (MNHN IM-2010-22883, 17). **C.** 475 m, St. 5591, Sibuko Bay, off Mabul Island, Borneo, Indonesia, 4.196°N, 118.639°E (USNM 280225, 1). Scale bars shell = 1 mm. Scale bars protoconch = 100 μm.

Anatoma amydra from the Indo-Malayan Archipelago to the Central Pacific has twice as dense clathrate sculpture and the axials extend over the full width of the keels of the selenizone.

Anatoma boucheti from the western Indian Ocean has a flat apex, at least 50% denser axial cords on shoulder and base, and the selenizone is situated slightly higher on the whorl.

Anatoma flexidentata from the Indian Ocean and the Western Pacific has a shorter teleoconch I (0.25 vs. 1 whorl), an indistinct onset of the selenizone, and a flexoglossate radula.

Anatoma japonica from the Indian Ocean to the Central Pacific has a more narrow sutsel on early whorls (0.5× vs. >1× width of selenizone), a more discoidal overall shell shape, and at least 50% denser axial cords on shoulder and base.

Anatoma kopua from temperate eastern Australia and New Zealand has a taller shell, a shorter teleoconch I (0.3–0.4 vs. 1 whorl), and lacks spiral lines in the constriction of the base below the selenizone.

Anatoma tangaroa from northernmost New Zealand has a shorter teleoconch I (<0.4 vs. 1 whorl) with twice as widely spaced axial cords and a more narrow sutsel on early whorls (0.5× vs. 1× width of selenizone).

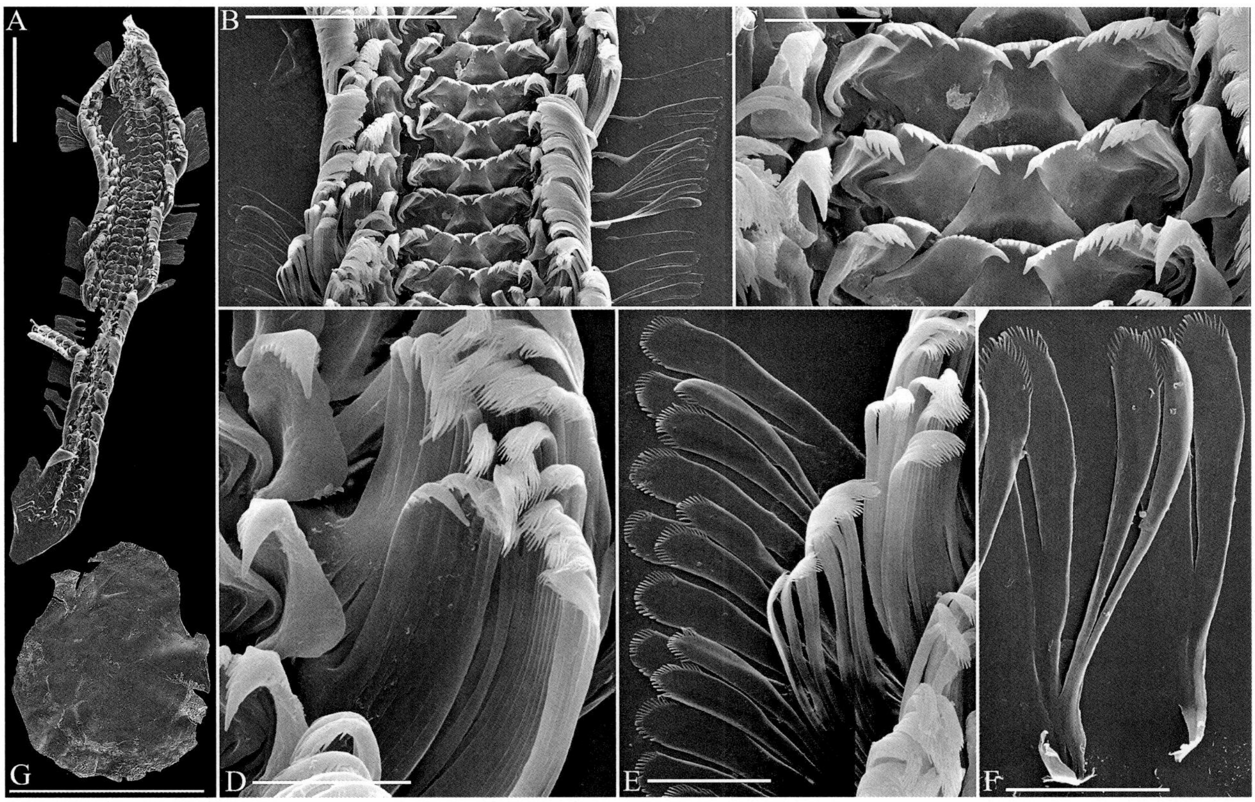

Figure 658. *Anatoma biconica* new species. Radula and operculum. Paratype 932 m, St. 5658, off C. Loko Loko, Burton Strait, Celebes, Indonesia, 3.544°S, 120.525°E (USNM 234423, 2). One shell from lot shown in Figure 657A. **A–F.** Radula. **A.** Entire radula. Scale bar = 200 μm. **B.** Full width of the radula. Scale bar = 100 μm. **C.** Central field. Scale bar = 20 μm. **D.** Lateral tooth 5 and marginal teeth. Scale bar = 20 μm. **E.** Outer marginal teeth. Scale bar = 20 μm. **F.** Outermost marginal teeth. Scale bar = 20 μm. **G.** Operculum. Scale bar = 1 mm.

Anatoma tobeyoides from southeastern Australia is more globular in overall shell shape, has a much more narrow sutsel on early whorls (<0.5× vs. >1× width of selenizone), and at least twice as many axial cords on shoulder and base.

Anatoma amoena from Antarctica has fine reticulate protoconch sculpture, a stronger axial cord in position of the selenizone, and a more globular overall shape.

Distribution. Indian Ocean to Central Pacific (Fig. 659).

Specimen records.

Madagascar. 729–732 m, Lavanono sector, ATIMO VATAE CP3597, 25.717°S, 44.400°E (MNHN, 1). 618–624 m, E of False Cape, ATIMO VATAE CP3566/DW3565, 25.608°S, 46.350°E (MNHN, 1).

Mozambique. 771–776 m, Maputo transect, Mozambique Channel, MAINBAZA CC3171, 25.983°S, 34.700°E (MNHN, 18). 949–952 m, Maputo transect, Mozambique Channel, MAINBAZA CC3170, 25.967°S, 34.783°E (MNHN, 2). 700–707 m, Maputo transect, Mozambique Channel, MAINBAZA CP3138, 25.217°S, 35.350°E (MNHN, 3).

Philippines. 903 m, [494 fms.], off Magabao Island, E Mindanao, 8.984°N, 126.264°E (USNM 276829, 13).

Indonesia. 475 m, St. 5591, Sibuko Bay, off Mabul Island, Borneo, 4.197°N, 118.639°E (USNM 280225, 1). 870 m, St. 5585, Sibuko Bay, off Sipadan Island, Borneo, 4.117°N, 118.832°E (USNM 1099254, 1). 932 m, St. 5658, off Cape Loko Loko, Burton Strait, Celebes, 3.544°S, 120.525°E (USNM 234423, 2).

Solomon Islands. 804–936 m, 9.985°S, 161.097°E (MNHN, 3).

Fiji. 640–687 m, Bligh Water, 16.937°S, 178.132°E (MNHN, 1).

Tonga. 433 m, NW Tongatapu, BORDAU 2 DW1569, 21.033°S, 175.317°W (MNHN, 6). 440–487 m, Vava'u group, BORDAU 2 DW1586, 18.567°S, 173.917°W (MNHN, 8).

Vanuatu. 154–179 m, MUSORSTOM 8 DW1105, 15.050°S, 167.117°E (MNHN, 1).

New Caledonia. 508–541 m, Sponge Bank, Norfolk Ridge, NORFOLK 1 DW1684, 24.917°S, 168.367°E (MNHN, 8).

French Polynesia. 914 m, Bora Bora, Society Islands, TARASOC DW3416, 16.583°S, 151.733°W (MNHN, 1).

Remarks. The radula from a single specimen showed some differences to that shown in Figure 658. The marginal teeth do not show the typical finely feathered cusps all along the edge of the tip, but the apical cusp is distinctly larger, similar to the condition seen in *An. janetae*. Given the rather subtle nature of the differences, the

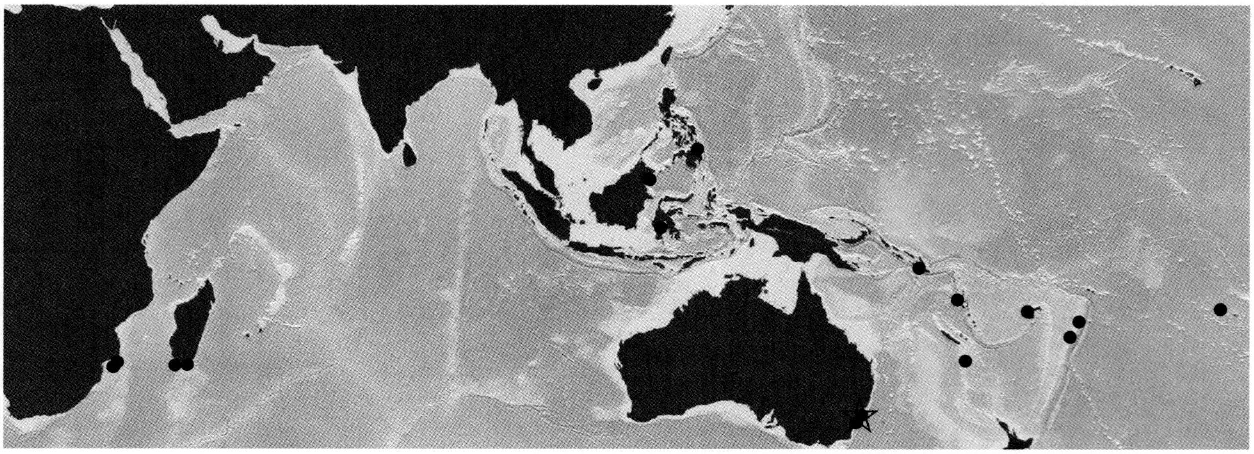

Figure 659. Distribution of *Anatoma biconica* new species. ☆ type locality. ● specimen record.

low number of available specimens, and the absence of any distinguishing characters on the shell, I treat those specimens as conspecific.

One lot from the Solomon Islands has a slightly more prominently sculptured protoconch, a stronger spiral cord in position of the selenizone resulting in a slight angling of teleoconch I, and an upturned start of the selenizone. Otherwise, the specimens are indistinguishable from the remainder examined. The Solomon Island lot is tentatively considered to be conspecific with *An. biconica*.

Anatoma boucheti Geiger & Sasaki, 2008

Figures 660–661

Chresonymy.

Anatoma boucheti Geiger & Sasaki, 2008: 252–253, fig. 3.
Anatoma boucheti: Sasaki *et al.*, 2010: 70.

Type material. Holotype (MNHN 20635: Geiger & Sasaki, 2008. Fig. 660).

Type locality. 1150–1180 m, Reunion Island, MD32, DC64, 21.200°S, 55.067°E (OD).

Etymology. Honoring Philippe Bouchet (MNHN) for his accomplishments with respect to advancing our knowledge of micromollusks (OD).

Description. Shell medium-sized (2.85 mm. Fig. 660), trochiform, tabulated. Protoconch sunken, 0.75 whorl, sculpture flocculent, weak apertural varix not connected to embryonic cap, apertural margin convex. Teleoconch I of 1.1 whorls, 39 distinct axial cords, spiral cord in position of selenizone, very fine spiral lines below periphery. Teleoconch II of two whorls; suture moderately impressed; sutsel as wide as selenizone in early whorls, 3× width of selenizone at apertural margin. Shoulder moderately convex, same density of axial cords as on teleoconch I (67 on first teleoconch II whorl), crossed by fine spiral lines forming minute thickenings at intersection with axials, first spiral line starting at onset of selenizone on outer half of shoulder, approximately 12 fine lines after one teleoconch II whorl, approximately 15 at apertural margin, spiral lines becoming weaker near keels of selenizone. Base barely constricted below selenizone, continuously sloping to moderately wide umbilicus without funiculus; sculpture similar to shoulder, approximately 35 fine spiral lines, additional finer lines towards keels of selenizone. Selenizone at periphery; keels distinctly elevated, rather weak; lunules regularly spaced; slit open, margins somewhat converging. Aperture round, roof overhanging, peristome flaring.

Animal unknown.

Comparisons. *Anatoma boucheti* can be recognized by the sunken protoconch, the >1 teleoconch I whorls, the tabulated shell shape, and the fine reticulate sculpture with predominant axial cords.

Anatoma japonica from the Indian Ocean to the Central Pacific has an overall biconical shell shape and a shorter teleoconch I (0.5–0.75 vs. 1.1 whorl).

Anatoma agulhasensis from the western Indian Ocean has a shorter teleoconch I (0.75 vs. 1.1 whorl) and an overall more globular shell shape with protruding protoconch.

Anatoma biconica from the Indian Ocean to the Central Pacific has an elevated early spire, about 50% fewer axial cords on shoulder and base, and the selenizone is situated lower on the whorl.

Anatoma flexidentata from the Indian Ocean to the Western Pacific has a more narrow sutsel over the entire shell, is more globular in overall shell outline, and has a much shorter teleoconch I (0.25 vs. >1 whorl).

Anatoma finlayi from the Indian Ocean to the Central Pacific (and possibly Caribbean) has strong axial lamellae on shoulder and base and fine axial lines on the keels of the selenizone.

Anatoma yaroni from East Africa has an overall biconical shell shape with a more narrow sutsel (< vs. > width of selenizone) and a shorter teleoconch I (0.75 vs. 1.1 whorl).

Anatoma munieri from the Indian Ocean to the Central Pacific has a shorter teleoconch (0.5–0.75 vs. 1.1 whorl), is overall more turreted, has proportionally

Figure 660. *Anatoma boucheti*. Holotype. 1150–1180 m, Reunion Island, 21.200°S, 55.067°E (MNHN 20635). Scale bar shell = 1 mm. Scale bar protoconch = 100 μm.

stronger and less dense sculpture, and a much more pronounced constriction below the selenizone.

Anatoma tabulata from the western Indian Ocean has a very elevated apex, the sutsel on the early whorl is at least as wide as the selenizone, and the base has fewer spiral cords producing a wider reticulate pattern.

Anatoma quadraxialis from the Indo-Malayan Archipelago has stronger axial cords on shoulder and base, reticulate protoconch sculpture, and a shorter teleoconch I (<0.75 vs. 1 whorl).

Anatoma rhynchodentata from the Indo-Malayan Archipelago to the Western Pacific has denser axial cords and spiral lines, a distinct constriction of the base below the selenizone, and a projecting apex.

Anatoma sinuosa from Easter Island has a more sinusoidal profile of the shoulder, a shorter teleoconch I (<0.5 vs. 1 whorl), and a smooth protoconch.

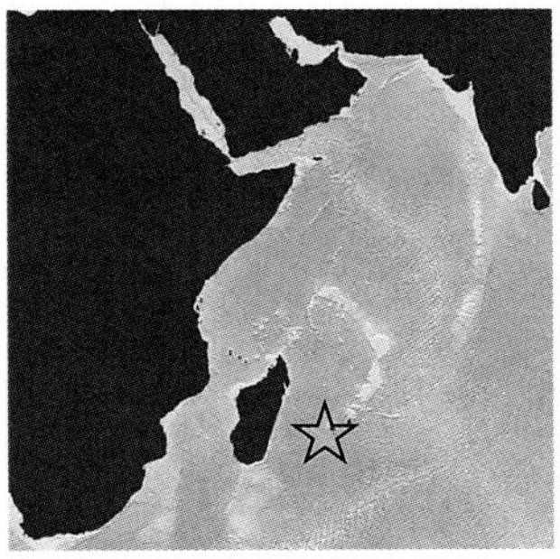

Figure 661. Distribution of *Anatoma boucheti*. ☆ type locality.

Distribution. Known only from type locality (Fig. 661).

Remarks. This species is currently known only from the holotype. It is sufficiently distinct to be recognized as a distinct species as noted in the comparisons above.

Anatoma breveprima new species

Figures 662–664

Type material. Holotype (MNHN 25100. Fig. 662). Paratypes from type locality (MNHN 25101, 5. Fig. 663).

Type locality. 495 m, Dumbéa Passage, N.O. *Alis*, BATHUS 2 CP755, New Caledonia, 22.367°S, 166.233°E.

Etymology. Breve-, Latin for short; -primus, -a, -um, Latin adjective for first. Referring to the short teleoconch I.

Description. Shell medium-sized (to 2.2 mm. Figs 662, 663) trochiform depressed, turreted. Protoconch of 0.75 whorl, sculpture flocculent, apertural varix not connected to embryonic cap, apertural margin shallow sinusoidal. Teleoconch I of 0.125 whorl, approximately four weak axial cords, no spirals. Teleoconch II of up to 3.1 whorls, suture moderately impressed, sutsel approximately 0.1× width of selenizone on early whorls, 2.5× width of selenizone at apertural margin. Shoulder slightly sinusoidal, axial cords increasing in strength from low cords to lamellae, approximately 40 on first whorl, interstices about 3–4× as wide as axials; axial cords on early teleoconch II extending from suture but not reaching selenizone; first spiral lines after 0.2 whorl, 2–4 after one whorl, 12–18 at apertural margin, regularly spaced, spirals running over axials without forming points. Base with 50% denser axial cords, axial cords alternating between reaching into umbilicus and running out two thirds down base, resulting in impression of even spacing of axial cords from selenizone into umbilicus, interstices 2–3× as wide as axial cords; approximately 35 very fine spiral lines running over axial cords. Umbilicus wide, no funiculus. Aperture rounded, peristome slightly flared. Selenizone at periphery; keels turned upwards, of moderate strength, quite elevated; lunules low, distinct, spacing similar to axial cords on base; slit open, margins converging.

Animal unknown.

Comparisons. *Anatoma amydra* from the Indo-Malayan Archipelago to the Central Pacific has a longer teleoconch I (0.5 vs. 0.125 whorl), an overall taller and more biconical shell shape, and the suture attaches below the periphery after one teleoconch II whorl.

Anatoma boucheti from the western Indian Ocean has a much longer teleoconch I (>1 vs. 0.125 whorl), the axial cords on the shoulder of early teleoconch I extend from suture to the selenizone, and the spiral lines cover the entire shoulder after one teleoconch II whorl.

Anatoma equatoria from the Indo-Malayan Archipelago to the Central Pacific has a longer teleoconch I (0.5 vs. 0.125 whorl), a wider sutsel (<2× vs. >2× width of selenizone), and usually about 50% less dense axial and spiral cords.

Figure 662. *Anatoma breveprima* new species. Holotype. 495 m, Dumbéa Passage, N.O. *Alis*, BATHUS 2 CP755, New Caledonia, 22.367°S, 166.233°E (MNHN 25100). Scale bar shell = 1 mm. Scale bar protoconch = 100 μm.

Anatoma finlayi from the Indian Ocean to the Central Pacific (and possibly Caribbean) has a longer teleoconch I (>0.6 vs. 0.125 whorl), much stronger axial lamellae on shoulder and base, and axial lines on the keels of the selenizone.

Figure 663. *Anatoma breveprima* new species. Paratypes. **A–C.** 495 m, Dumbéa Passage, N.O. *Alis*, BATHUS 2 CP755, New Caledonia, 22.367°S, 166.233°E (MNHN 25101, 5). Scale bars shell = 1 mm. Scale bars protoconch = 100 µm.

Anatoma hyposculpta from the tropical western Pacific has a longer teleoconch I (0.25 vs. 0.125 whorl), barely perceptible sculpture, and the axial cords on the base start at the level of the mid-base.

Anatoma indonesica from the Indo-Malayan Archipelago to the Western Pacific has a longer teleoconch I (0.75 vs. 0.125 whorl), strong axial lamellae on the shoulder, and an overall taller shell shape.

Anatoma munieri from the Indian Ocean to the Central Pacific has a longer teleoconch I (>0.6 vs. 0.125 whorl), a much taller overall shell shape, and a distinct constriction of the base below the selenizone.

Anatoma pseudoequatoria from the Indo-Malayan Archipelago to the Central Pacific has a more discoidal shell, the major axial cords on the shoulder of early teleoconch II extend from suture to selenizone, and the umbilicus bears a strong funiculus.

Anatoma planapex from the Indo-Malayan Archipelago and Australia has a longer teleoconch I (1 vs. 0.125 whorl), a spiral cord in position of the selenizone on teleoconch I, and weaker sculpture on shoulder and base.

Anatoma quadraxialis from the Indo-Malayan Archipelago has a protoconch with fine reticulate sculpture, a longer teleoconch I (>0.6 vs. 0.125 whorl), and at least twice as dense axial cords on teleoconch II.

Anatoma aedonia from the Caribbean to Brazil has a longer teleoconch I (>0.75 vs. 0.125 whorl), strong axial cords with a spiral cord in position of the selenizone on teleoconch I, and at least 50% denser axial cords on shoulder and base of teleoconch II.

Anatoma micalii from the Mediterranean to Angola has a longer teleoconch I (0.5 vs. 0.125 whorl) with an axial cord in position of the selenizone and axial cords on early teleoconch II extending from suture to selenizone.

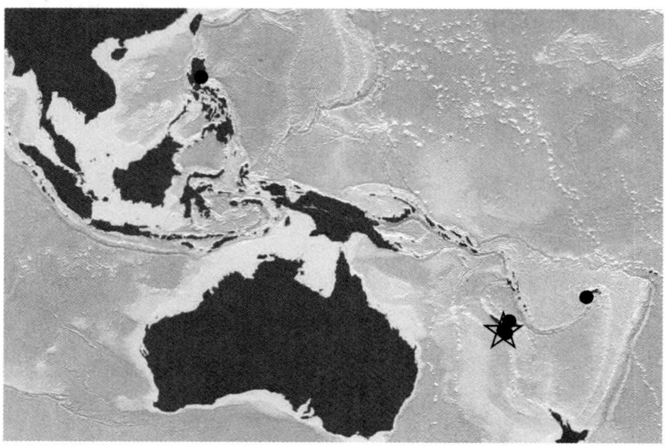

Figure 664. Distribution of *Anatoma breveprima* new species. ☆ type locality. ● specimen record.

Distribution. Indo-Malayan Archipelago, 150–650 m (Fig. 664).

Specimen records.

Philippines. 538 m, AURORA 2007 CP2750, 15.883°N, 121.893°E (MNHN, 2).

Fiji. 656–660 m, Bligh Water, 17.268°S, 178.247°E (MNHN, 12).

New Caledonia. 150–200 m, Point Lefèvre, Santal Bay, Loyalty Islands, Lifou, 20.902°S, 167.018°E (MNHN, 1). 490–500 m, southern, 22.317°S, 166.183°E (MNHN, 2). 495 m, Dumbéa Passage, N.O. *Alis*, BATHUS 2 CP755, 22.367°S, 166.233°E (MNHN 25100, 1: holotype; MNHN 25101, 5: paratypes). 570 m, Loyalty Basin, 22.650°S, 166.567°E (MNHN, 1). 250–350 m, south, St. 40, 22.500°S, 166.400°E (MNHN, 3). 577–780 m, southern, BATHUS 2 DW754, 22.383°S, 166.217°E (MNHN, 44). 495 m, southern, 22.367°S, 166.233°E (MNHN, 6).

Anatoma concinna (A. Adams, 1862)

Figures 665–669

Chresonymy.

Anatomus concinnus A. Adams, 1862: 348, not illustrated.
Scissurella concinna: Crosse, 1863: 110.
Anatomus concinnus: Troschel, 1863: 702.
Anatomus concinnus: Dunker, 1882: 147.
Scissurella concinna: Paetel, 1888: 289.
Anatomus concinnus: Pilsbry, 1890: 59–60.
Scissurella concinna: Pilsbry, 1895: 106.
Scissurella concinna: Thiele, 1912a: 17, pl. 2, figs 15–16 [after specimen in NHMUK: type?].
Scissurella (*Schitotrochus*) *concinna*: Kuroda & Kinoshita, 1951: 8.
Schizotrochus ? *concinnus*: Habe, 1951: pl. 11, figs 10–11.
Scissurella concinna: Kuroda & Habe, 1952: 84.
Anatoma concinna: Habe & Kosuge, 1964: 4.
Anatoma concinna: Higo, 1973: 13.
Anatoma concinna: Lozouet, 1986: 113.
Scissurella concinna: Kay & Johnson, 1987: 106.
Anatoma concinna: Higo & Goto, 1993: 15.
Anatoma s.l. *concinna*: Geiger, 2003: 74.

Synonym.

+ *Scissurella obtusata* Golikov & Gublin, 1978: 171–172, fig. 3 (Fig. 665). ***Type material.*** Holotype (ZIN 27738/1), 1.8 × 1.9 mm. ***Type locality.*** Near Chirpoi Island, 20 m [Russia, 46.501°N, 150.851°E] (Kantor & Sysoev, 2006). ***Etymology.*** Obtusus, Latin for blunt, dull.
Anatoma obtusata: Baxter, 1983: 24.
Anatoma obtusata: Golikov *et al.*, 2001: 155.
Anatoma s.l. *obtusata*: Geiger, 2003: 75.
Scissurella obtusata: Kantor & Sysoev, 2005: 29.
Scissurella obtusata: Kantor & Sysoev, 2006: 25, pl. 9, figs D [copy original figures], E [photograph of holotype].
Anatoma obtusata: Gulbin, 2006: 11.
Anatoma obtusata: Geiger & McLean, 2010: 20.
Anatoma obtusata: Sasaki *et al.*, 2010: 71.

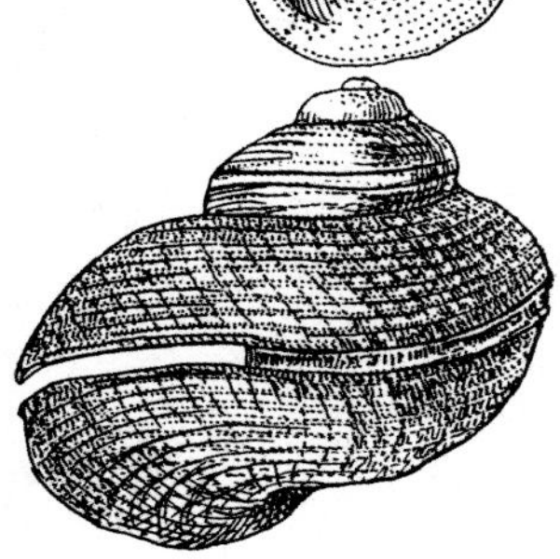

Figure 665. Original illustration of *Scissurella obtusata.*

Misidentifications.

Scissurella (*Anatoma*) *soyoae*: McLean, 1967: 407, pl. 56, fig. 7 [*fide* McLean, pers. comm.].
Anatoma lamellata: Kuroda *et al.*, 1971: 3, pl. 106, fig. 1 [is *An. concinna*].
Scissurella soyoae: Abbott, 1974: 17 [West American records most likely *An. concinna*].
Scissurella soyae [*sic*]: Foster, 1981: 13, 43 [is *An. concinna*].

Type material. Syntype (NHMUK 1874.5.19.64, 1), 1.35 mm, here designated as the lectotype with the express purpose of taxon stabilization in case other non-conspecific syntypes should be located.

Type locality. 35 fms. [= 140 m], Rifunsiri [= Rebun-tô, Japan, 45.417°N, 141.067°E] (OD).

Etymology. Concinnus: Latin for well-arranged, skillfully joined, beautiful, striking.

Description. Shell trochiform, globose, to 1.9 mm (holotype of *Sci. obtusata*. Figs 666–668). Protoconch of 0.75 whorl, sculpture flocculent, no apertural varix, apertural margin convex. Teleoconch I of 0.875–1 whorl, many dense indistinct low axial cords, a dozen spiral cords approximately 30% weaker, irregularly spaced between suture and selenizone. Teleoconch II of >1.5 whorls; suture impressed; sutsel approximately 2× width of selenizone. Shoulder concave, 80–100 low axial cords per whorl; approximately 1–6 spiral cords at onset of selenizone, 12 at apertural margin, forming reticulate pattern. Base rounded, without constriction below selenizone, continuously sloping with umbilicus, similar sculpture as on shoulder, approximately 25 spiral cords. Umbilicus narrow, deep, bordered by strong carina,

Figure 666. *Anatomus concinna*. Lectotype, here designated. 35 fms. [= 140 m], Rifunsiri [= Rebun-tô, Japan, 45.417°N, 141.067°E] (NHMUK 1874.5.19.64). Scale bar shell = 1 mm. Scale bar protoconch = 100 μm.

funiculus present. Selenizone slightly above periphery; keels very low; lunules as strong as axial cords on shoulder and base; slit open, margins parallel. Aperture rounded, lower adumbilical portion slightly thickened giving rise to funiculus.

Operculate (Kantor & Sysoev, 2006: pl. 9, fig. E: photograph of holotype).

Animal unknown.

Comparisons. *Anatoma soyoae* from Japan lacks spiral sculpture on teleoconch I, has at least twice as dense axial cords on the teleoconch and more distinct keels of the selenizone, and lacks the thickened shelf in the adumbilical corner of the aperture.

Anatoma disciformis from the North Pacific has a more depressed shell with the suture immediately below the selenizone, a shorter teleoconch I (0.75 vs. 0.875–1 whorl), and a distinct funiculus in the umbilicus.

Anatoma fujikurai from Japan has undulating keels of the selenizone, a wider selenizone, and more distinct axial cords on shoulder and base.

Figure 667. *Anatoma concinna*. 7 m, Middle Rocks, Chicagof Harbor, Bering Sea side of Attu Island, Near Islands, Aleutian Islands, County, Alaska, USA, 52.940°N, 173.265°W (LACM 79-71, 4). Scale bar shell = 1 mm. Scale bar protoconch = 100 µm.

Anatoma kelseyi from the northeastern Pacific has a more delicate shell with much finer cancellate sculpture and a protoconch with reticulate sculpture.

Figure 668. *Anatoma concinna*. **A.** 7 m, Middle Rocks, Chicagof Harbor, Bering Sea side of Attu Island, Near Islands, Aleutian Islands, County, Alaska, USA, 52.940°N, 173.265°W (LACM 79-71, 4). **B–C.** 13-37 m, Port Dick, Alaska, USA, 55.290°N, 151.133°W (LACM 504, 7). Scale bars shell = 1 mm. Scale bars protoconch = 100 µm.

Anatoma lamellata from the Western to the Central Pacific has a protoconch with a spiral band, much stronger axial lamellae on the shoulder, fewer spiral lines on the base, and much stronger keels of the selenizone.

Anatoma sagamiana from the Western Pacific to the Indo-Malayan Archipelago has reticulate protoconch sculpture, a much thinner shell, finer axial and spiral sculpture, and more distinct keels of the selenizone.

Anatoma lyra from the Northeastern Pacific has a more depressed shell particularly in smaller specimens, shows strong axial raised cords to low lamellae, and has a distinct funiculus in the umbilicus.

Anatoma conica from subantarctic South America has distinct axial lamellae crossed by much finer spiral lines on teleoconch II and usually more distinct keels on the selenizone.

Anatoma crispata from the northeastern Atlantic has a shell that is wider than tall, sharper axial cords on the shoulder of teleoconch II, and more distinct keels of the selenizone.

Anatoma keenae from the Panamic has a more narrow sutsel on early whorls (<1× vs. ~2× width of selenizone), more distinct keels of the selenizone, and 30–50% denser axial cords on shoulder and base.

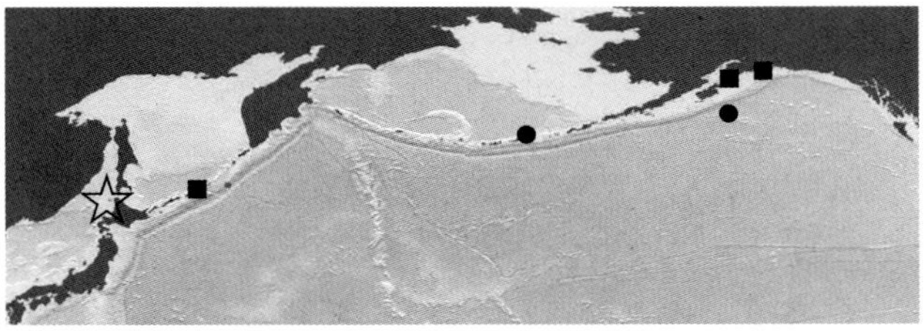

Figure 669. Distribution of *Anatoma concinna*. ☆ type locality. ● specimen record. ■ literature record.

Distribution. Temperate to arctic North Pacific (Fig. 669), 20–70 m (Golikov *et al.*, 2001).

Specimen records.

USA. 13–37 m, Port Dick, Alaska, 55.290°N, 151.133°W (LACM 504, 7). 7 m, Middle Rocks, Chicagof Harbor, Bering Sea side of Attu Island, Near Islands, Aleutian Islands, County, Alaska, 52.940°N, 173.265°W (LACM 79-71, 7).

Remarks. *Scissurella concinna* Sowerby, 1831 does not pose a threat to *An. concinna* (A. Adams, 1862) because Sowerby's species is a *Sinezona*.

Thiele (1912: pl. 2, figs 15–16; refigured by Habe, 1951) provided the only illustration of the species to date. The figure was executed "after a specimen in the British Museum" and it corresponds reasonably well with the lectotype, though the umbilicus was drawn too wide compared to the SEM images.

The lectotype seems to be slightly immature as there is no sloping of the apertural area.

The holotype of *Sci. obtusata* is severely damaged due to storage in acidic alcohol (Kantor & Sysoev, 2006).

The two nominal taxa are here synonymized based on identical protoconch sculpture, same length of teleoconch I, same teleoconch I and II sculpture, same configuration of the umbilicus, and same low keels of the selenizone. Additionally, the type localities of the two taxa are only separated by approximately 700 km. Although *An. concinna* has rarely been applied, the case for *nomen oblitum-protectum* cannot be made, because *concinna* has been used after 1899, and *obtusata* has not been used frequently enough.

Anatoma conica (d'Orbigny, 1841)

Figures 670–673, Color Plate 2H

Chresonymy.

Scissurella conica d'Orbigny, 1841: 470, pl. 78, figs 7–9.
Scissurella conica: Gray, 1854a: 50.
Anatomus conicus: Adams & Adams, 1853–1858 (1854): vol. 1, 439.
Scissurella conica: Munier-Chalmas, 1865: 396.
Scissurella conica: Paetel, 1888: 289.
Scissurella conica: Rochebrune & Mabille, 1889: H75 [not found at Cape Horn].
Scissurella conica: Pilsbry, 1890: 53, pl. 57, figs 8–9.
Scissurella conica: Carcelles, 1950: 49.
Schizotrochus conica [sic]: Carcelles & Williamson, 1951: 251.
Anatoma conica: Lozouet, 1986: 113.
Scissurella conica: Castellanos & Landoni, 1988: 6–7, pl. 1 fig. 4.
Anatoma conica: Dell, 1990: 76, fig. 128.
Anatoma conica: Linse, 1999: 400.
Scissurella conica: Forcelli, 2000: 44, text-fig. [SEM of shell copy figure Dell, 1990].
Anatoma s.l. *conica*: Geiger, 2003: 74.
Anatoma conica: Zelaya & Geiger, 2007: 396–398, figs 1 [map], 2–3 [copy figs d'Orbigny (1841)], 4–15 [shell SEM].

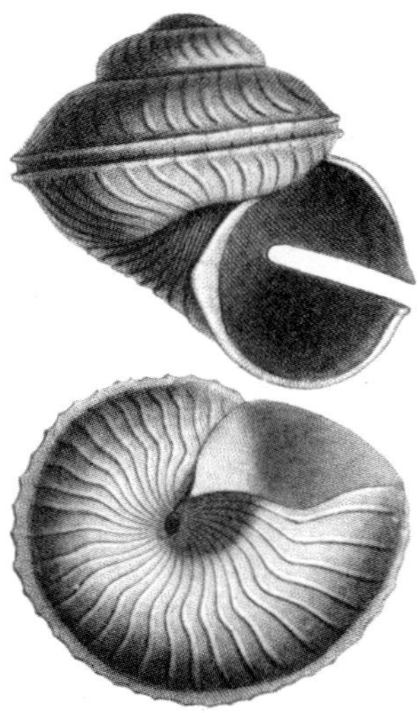

Figure 670. Original illustration of *Scissurella conica*. See also Color Plate 2H.

Misidentifications.

Scissurella supraplicata: Melvill & Standen, 1912: 345 [is *An. conica*].
Schizotrochus supraplicata: Carcelles & Williamson, 1951: 251 [is *An. conica*].
Schizotrochus supraplicatus: Carcelles, 1953: 162 [is *An. conica*].
Scissurella supraplicata: Ageitos de Castellanos & Landoni, 1988: 8 [is *An. conica*].

Type material. (NHMUK: Gray, 1854a), most likely lost.

Type locality. Falkland Islands (OD) [~ 51.5°S, 59.0°W].

Etymology. Conica, Latin for cone-shaped.

Description. Shell trochiform, globular, inflated, medium-sized (up to 1.8 mm. Figs 671, 672), sutsel approximately as wide as selenizone. Protoconch of approximately one whorl, sculpture flocculent; margin of protoconch convex, without apertural varix apertural margin straight. Teleoconch I of 0.5–0.66 whorl, with spiral cord in position of selenizone on early teleoconch I, disappearing close to onset of selenizone; 10–21, occasionally bifurcated, axial cords. Teleoconch II of at least two rounded whorls, suture moderately impressed; sutsel 0.5–1× width of selenizone. Shoulder convex, with strong axial cords, 26 cords on shoulder of first whorl, 31–34 cords on last whorl (occasionally bifurcating near selenizone); spiral threads weak, beginning at start of selenizone, gradually increasing in number with growth, approximately 18–20 at apertural margin. Base with approximately 60 axial cords, 25 spiral threads. Selenizone at periphery, relatively wide, sculptured with low axial cords, surrounded by narrow keels; lunules widely spaced, distinct. Umbilicus open, wide, deep; walls with continuous transition to base; funiculus present. Aperture rounded.

Animal unknown.

Comparisons. *Anatoma amoena* from Antarctica has a narrower shoulder, weaker axial cords, and a reticulate protoconch sculpture.

Anatoma argentinae from the southwestern Atlantic and the Panamic has a much larger protoconch (480 µm vs. 215 µm), a shorter teleoconch I (0.3–0.5 vs. 0.6 whorl), and finer axial cords on the shoulder.

Anatoma euglypta from Antarctica has a more lenticular, biconical overall shell shape and more numerous but less elevated axial cords, particularly on the shoulder.

Anatoma shiraseae from Antarctica has a distinct spiral keel on the shoulder and lacks axial lamellae on the shoulder.

Anatoma weddelliana from Antarctica has a proportionally wider shell, a wider umbilicus, and at least 50% denser spiral lines on shoulder and base.

Anatoma lyra from the boreal northeastern Pacific has about 50% fewer axial lamellae and usually has a more narrow sutsel on early whorls.

Figure 671. *Anatoma conica*. 115 m, Elatin Station 219, Antarctica, 55.783°S, 67.283°W (NMNZ M.242839, 4). Scale bar shell = 1 mm. Scale bar protoconch = 100 µm.

Anatoma crispata from the northeastern Atlantic has finer axial cords on shoulder and base and usually a shell that is a little wider than tall.

Anatoma flemingi from New Zealand has reticulate protoconch sculpture and at least twice as dense axial cords on the base.

Figure 672. *Anatoma conica*. **A–B.** 121 m, off South America [S of Tierra del Fuego], 56.328°S, 67.163°W (NHMUK 20000096, 12). **C.** 1 m, San Andres Bay, N Gulf of Peñas, Chile, 46.588°S, 75.510°W (LACM 73-73, 1). Scale bars shell = 1 mm. Scale bars protoconch = 100 μm.

Anatoma finlayi from the Indian Ocean to the Central Pacific (and possibly Caribbean) has a protoconch with an apertural varix, much stronger axial lamellae on shoulder and base, a wider sutsel on early whorls (< vs. > width of selenizone), and a more distinct constriction of the base below the selenizone.

Anatoma gunteri from southern Australia has reticulate protoconch sculpture and has at least 50% denser axial cords on the base.

Distribution. Magellanic region, 115–265 m (Fig. 673).

Specimen records.

Figure 673. Distribution of *Anatoma conica*. ☆ type locality. ● specimen record.

Antarctica. 92 m, Paso Ancho, Magellan Strait, 53.480°S, 70.365°W (ZMH 19588, 5, 3: complete). 121 m, 56.328°S, 67.163°W (NHMUK ex20000096, 12). 121 m, Drake Passage, 56.325°S, 67.163°W (NMSZ 42838 1: complete). 55 m, W Carlos III Island, Magellan Strait, 53.650°S, 72.250°W (MLP 12060, 1). 12 m, *Discovery* St. 388, off Cape Horn, 56.325°S, 67.163°W (NMNZ M.242838, 1). 115 m, Elatnin Station 219, 55.783°S, 67.283°W (NMNZ M.242839, 4). 102 m, Burwood Bank, 54.417°S, 57.533°W (NMSZ 1921.143.596, 1).

Argentina. 438–548 m, 7 mi SE Cape San Bartolome, Staten Island, Tierra del Fuego, 55.000°S, 64.812°W (LACM 71-342, 1). 771–903 m, 5 mi S Ventana Point, S side Staten Island, Tierra del Fuego, 54.908°S, 63.933°W (LACM 71-329, 2; USNM 845405, 1). 144 m, 8 mi E Ensenada Patagones, SE Tierra del Fuego, 54.867°S, 65.083°W (USNM 845403, 1; LACM 71-305, 2). 67–71 m, Capitan Canepa Bay, Staten Island, Tierra del Fuego, 54.837°S, 64.490°W (LACM 71-344, 6; USNM 845400, 1). 76 m, 2 mi W Cape Kendall, Staten Island, Tierra del Fuego, 54.833°S, 64.167°W (LACM 71-334, 1). 73–76 m, 9 mi N Cape San Antonio, Staten Island, Tierra del Fuego, 54.567°S, 64.500°W (LACM 71-352, 5; USNM 845402, 3). 116–120 m, 18 mi N Crossley Bay, Staten Island, Tierra del Fuego, 54.488°S, 64.673°W (LACM 71-347; USNM 845404, 1). 109 m, 14 ml N Cape Colnett, Staten Island, Tierra del Fuego, 54.483°S, 64.333°W (LACM 71-314, 3).

Chile. 1 m, San Andres Bay, N Gulf of Peñas, 46.588°S, 75.510°W (LACM 73-73, 1).

Great Britain. Falkland Islands, 51.500°S, 59.000°W (NHMUK, 1).

Literature records.

Argentina. 229–265 m, E of Tierra del Fuego, 54.933°S, 65.050°W. 115 m, off Cape Horn, 55.783°S, 66.283°W (Dell, 1990).

Remarks. Gray (1854a) classified the species in Haliotidae. For details on the date of publication of d'Orbigny's work, see Sherborn & Griffin (1934).

The type of *An. conica* is probably lost; it was originally deposited in the NHMUK (Gray, 1854a), but is not presently there (Geiger, pers. obs.; K. Way, pers. comm.). However, the original description and illustrations are adequate to recognize this species; in addition, topotypic specimens examined confirm the identity of this taxon (Zelaya & Geiger, 2007). Accordingly, a neotype designation is not necessary.

The specimens from Burdwood Bank, previously referred by Melvill & Standen (1912) as *Scissurella supraplicata* (NMSZ 1921.143.596), were re-identified as *An. conica* by Zelaya & Geiger (2007).

Anatoma crispata (Fleming, 1828)

Figures 674–680

Chresonymy.

Scissurella crispata Fleming, 1828: 385, pl. 6, fig. 3.

Scissurella crispata Fleming, 1828b: 366, not illustrated.

Scissurella crispa [sic]: Sowerby I (1820–1834: 1831): unnumbered page [see remarks for dating].

Scissurella crispata: Fleming, 1837: pl. 13, fig. 48 [copy Fleming, 1828].

Scissurella crispata: Wood, 1842: 530–531.

Scissurella crispa [sic]: Brown, 1843: 52.

Scissurella crispata: Thorpe, 1844: 152–153.

Scissurella crispata: Brown, 1844: 62.

Scissurella crispata: Loven, 1846: 20.

Scissurelle ventrue: Chenu, 1847: 180, text-fig. 597.

Scissurella crispata: Wood, 1848: 163, pl. 15, fig. 13.

Scissurella crispata: Forbes & Hanley, 1850 (1848–1853): 544–546, pl. 64, fig. 6.

Scissurella crispata: Forbes 1850: 222, 256, 260.

Scissurella crispata: Woodward, 1851: 174, pl. 10, fig. 23.

Scissurella crispata: Philippi, 1853: 32, pl. 6, fig. 3.

Scissurella crispata: Mac Andrew, 1854: 7, 45.

Scissurella crispata: Sowerby, 1854: 136.

Anatomus crispatus: Adams & Adams, 1853–1858 (1854): vol. 1, 439; pl. 50, fig. 4.

Scissurella crispata: Clark, 1855: 322.

Scissurella crispata: Barrett, 1856, 206, text-fig. of animal.

Scissurella crispata: Gosse, 1856: 117, text-fig. 172.

Scissurella crispata: Mac Andrew, 1856: 119, 147.

Scissurella crispata: Clark, 1856: 269–272.

Figure 674. Original illustration of *Scissurella crispata* Fleming, 1828. From volume in MNHN.

Scissurella crispata: Jeffreys, 1856b: 271–272 [includes false synonymy with *decussata* and *elegans*].

Scissurella crispata: MacAndrew, 1856: 119, 147.

Scissurella crispata: Mörch, 1857: 87.

Scissurella crispata: Mac Andrew, 1858: 9, fig. 20–21 [*fide* Dautzenberg & Fischer, 1912].

Scissurella crispata: Sars, 1858: 73 [*fide* Dautzenberg & Fischer, 1912].

Scissurella crispata: Danielssen, 1859: 143.

Scissurella crispata: Sowerby, 1859: pl. 11, fig. 26.

Scissurella crispata: MacAndrew, 1861: 8.

Scissurella crispata: Sars, 1863: 272.

Scissurella crispata: Munier-Chalmas, 1865: 396.

Scissurella crispata: Jeffreys, 1865: vol 3: 283–285, pl. 7, fig. 2 [shell and animal]; vol. 5: 201, pl. 9, fig. 3.

Scissurella crispata: Woodward, 1866: 269–270, text-fig. 116 [live animal: copy figure Barrett, 1856], pl. 10, fig. 23 [copy plate Woodward, 1851].

Scissurella crispata: Jeffreys, 1868: 301, 309.

Scissurella crispata: Sars, 1868: 258 [*fide* Dautzenberg & Fischer, 1912].

Scissurella crispata: Weinkauff, 1868: 385–386.

Scissurella crispata: Appelius, 1869: 201.

Scissurella crispata: Mörch, 1869: 24.

Scissurella crispata: Paetel, 1869: 62.

Scissurella crispata: Petit, 1869: 110 [*angulata, aspera*? as synonyms].

Scissurella crispata: Fischer, 1869: 127 [*fide* Dautzenberg & Fischer, 1912].

Scissurella crispata: Appelius, 1869: 201.

Scissurella crispata: Appelius, 1870: 201, 225, 269.

Scissurella crispata: Sars: 1870: 61.

Scissurella crispata: Aradas & Benoit, 1870: 156.

Scissurella crispata: Woodward, 1870: 269–270, text-fig. 116 [live animal: copy figure Barrett, 1856], pl. 10, fig. 23 [copy plate Woodward, 1851].

Scissurella crispata: Jeffreys, 1870a: 75.

Scissurella crispata: Jeffreys, 1870b: 444 [description of animal].

Scissurella crispata: Monterosato, 1872a: 30–31.

Scissurella crispata: Monterosato, 1872b: 28.

Scissurella crispata: Friele, 1873: 15.

Scissurella crispata: Weinkauff, 1873: 40.

Scissurella crispata: Monterosato, 1875a: 22.

Scissurella crispata: Monterosato, 1875b: 11.

Scissurella crispata var. *angulata*: Monterosato, 1875b: 11.

Scissurella crispata: Friele, 1877: 307 [should read 207: T. Høisaeter, pers. comm. 11/2006], pl. 3, figs 4a–d [radula] 5 [jaws].

Scissurella crispata: Jeffreys, 1877: 233.

Scissurella crispata: Seguenza, 1877: 272, 320.

Scissurella (*Schizotrochus*) *crispata*: Monterosato, 1877a: 416.

Scissurella crispata: Monterosato, 1877b: 34.

Scissurella crispata var. *angulata:* Monterosato, 1877b: 34.

Scissurella (*Schizotrochus*) *crispata*: Monterosato, 1878b: 79.

Scissurella crispata: Fischer, 1878: 181.

Scissurella crispata: Normann, 1879: 24.

Scissurella crispata: Cooke, 1882: 362.

Scissurella crispata: Fischer, 1882: 1201.

Scissurella crispata: Paetel, 1883: 47.

Scissurella crispata: Jefferys, 1883a: 88.

Scissurella crispata: Jeffreys, 1883b: 395.

Scissurella crispata: Tryon, 1883: 321, pl. 82, fig. 93 [copy Barrett, 1856].

Scissurella crispata: Marion, 1883a: 38, 46 [*fide* Dautzenberg & Fischer, 1912].

Scissurella crispata: Marion, 1883b: 85 [*fide* Dautzenberg & Fischer, 1912].

Schizotrochus crispatus: Monterosato, 1884: 39.

Scissurella crispata: Grasset: 1884: 136.

Scissurella crispata: Fischer, 1885: 846, text-fig. 598 [copy Barrett, 1856], pl. 10, fig. 23 [copy Woodward, 1851].

Scissurella crispata: Bush, 1885: 710.

Scissurella crispata: Granger, 1885: 154, pl. 11, fig. 16.

Scissurella crispata: Herzenstein, 1885: 669 [record from Lappland].

Scissurella crispata: Locard, 1886: 332.

Scissurella crispata: Sparre-Schneider, 1886: 102 [*fide* Dautzenberg & Fischer, 1912].

Scissurella crispata: Brusina, 1886: 198.

Scissurella crispata: Kobelt, 1887: 409.

Scissurella crispata: Marshall, 1897: 68.

Scissurella crispata: Paetel, 1888: 289.

Scissurella crispata: Norman, 1888: 21 [also var. *angulata*].

Scissurella crispata: Kobelt, 1888: 257–258.

Scissurella crispata: Bizet, 1888: 120.

Scissurella crispata: Dautzenberg, 1889: 66.

Scissurella crispata: Dall, 1889a: 396.

Scissurella crispata: Dall. 1889b: 168, pl. 48, fig. 15 [copy Barrett, 1856].

Scissurella crispata: Carus, 1889–1893: vol. 2, 237.

Scissurella crispata: Dautzenberg, 1889: 66.

Scissurella crispata: Pilsbry, 1890: 52, pl. 50, fig. 18 [radula copy Sars, 1878], pl. 68 [erroneously pl. 58 in text], figs 22–25.

Schizotrochus crispatus: Monterosato, 1890: 143.

Scissurella crispata: Thiele: 1891: 271, pl. 25. fig. 32 [radula].

Scissurella crispata: Locard, 1892: 220.

? *Scissurella crispata:* Camerano & Lessona, 1895: 183, text-fig. 342 [shell too tall for *crispata*, cf. *An. schanderi*].

Scissurella crispata: Grieg, 1896: 17.

Scissurella (*Schizotrochus*) *crispata*: Dautzenberg & Fischer, 1896: 488–489.

Scissurella (*Schizotrochus*) *crispata*: Dautzenberg & Fischer, 1897: 179.

Scissurella crispata: Locard, 1898: 69–70.

Scissurella crispata: Posselt, 1898: 120–121.

Scissurella crispata: Locard, 1899: 106.

Scissurella crispata: Nobre, 1900: 135.

Scissurella crispata: Friele & Grieg, 1901: 59.

Scissurella crispata: Chaster *et al.*, 1901: 16.

Scissurella crispata: Gordon, 1901: 138, pl. 21, fig. 334.

Scissurella crispata: Step, 1901: 197–198.

Scissurella crispata: Friele, 1902: 15.

Scissurella crispata: Hedley, 1902: 22 [mis-localized specimen].

Scissurella crispata: Sowerby & Fulton, 1903: 53.

Scissurella (*Schizotrochus*) *crispata*: Franceschini, 1906: 7.

Scissurella crispata: Hoel, 1907: 9.

Scissurella crispata: Odhner, 1912: 13, 36–37, pl. 2, figs 25–27 [also var. *angulata*].
Scissurella crispata: Dautzenberg & Fischer, 1912: 286–288.
Scissurella crispata: Pallary, 1912: 70.
Scissurella crispata: Buchner, 1913: 126, text-fig. 100.
Scissurella crispata: Grieg, 1913: 37.
Scissurella crispata: Øyer, 1915: 435.
Scissurella crispata: Hildago, 1916: 632.
Scissurella crispata: Soot-Ryen, 1924: 50.
Scissurella crispata: Dall, 1927a: 110.
Scissurella (*Schizotrochus*) *crispata*: Dautzenberg, 1927: 213.
Scissurella crispata: Thiele, 1929: 30, fig. 16.
Scissurella crispata: Nobre, 1932: 242.
Scissurella (*Schizotrochus*) *crispata:* Winckworth, 1932: 219.
Schizotrochus crispatus: Coen, 1933: 12–13.
— *crispata:* Spärck & Thorson, 1933.
Scissurella costata: Nobre, 1936: 179.
Schizotrochus crispatus: Coen, 1937: 6–7.
Scissurella crispata: Nobre, 1938–1940: 469.
— *crispata*: Thorson, 1944: [*fide* Schiøtte, 2005].
— *crispata*: Gorbunov, 1946a [*fide* Kantor & Sysoev, 2006].
Scissurella crispata: Abbott, 1954: 91 [also for NW Atlantic species: *alta, americana*?].
Scissurella crispata: Rendall, 1956: 152 [Orkney Islands].
Scissurella (*Anatoma*) *crispata*: Brookes Knight *et al.*, 1960: I221, fig. 136, 4.
Scissurella (*Schizotrochus*) *crispata*: Wenz, 1961: 173, fig. 270 [copy Forbes & Hanley, 1848–1853].
Scissurella crispata: Parezan, 1961: 68.
Scissurella crispata: Smith, 1961: 214, pl. 10, fig. 23 [copy Woodward, 1851].
Scissurella crispata: Clarke, 1962: 6.
Scissurella crispata: Pasteur-Humbert, 1962: 123.
Scissurella crispata: Fretter & Graham, 1962: 488, figs 255–256.
Scissurella crispata: Ziegelmeier, 1966: 7, pl. 1, figs 1a–c.
Scissurella crispata: Graham, 1971: 36, text-figs A–B [copy Fretter & Graham, 1962].
Scissurella crispata: Abbott, 1974 [overextended species concept, with range from Europe to West Indies, Alaska, Baja California, Japan. Illustration unidentifiable].
Scissurella crispata: Bouchet & Warén, 1979: 220.
Scissurella (*Anatoma*) *crispata*: Piani, 1980: 117.
Scissurella crispata: Terreni, 1981: 13.
Scissurella crispata: Miari, 1983: 17.
Scissurella crispata: Albesi *et al.*, 1986: 124.
Anatoma crispata: Høisaeter, 1986: 81.
Anatoma crispata: Lozouet, 1986: 113.
Scissurella crispata: Hayward *et al.*, 1990: 646, fig. 12.4.
Scissurella cf. *crispata*: Hedegaard, 1990: 74, 145.
Anatoma crispata: Sabelli *et al.*, 1990: 12.
Anatoma crispata: Smith & Heppell, 1991: 11.
Anatoma crispata: Doneddu & Manunza, 1992: 17.
Anatoma crispata: Cachia *et al.*, 1993: 10.
Anatoma crispata: Bodon *et al.*, 1995: 14.
Anatoma crispata: Marquet, 1995: 63, pl. 2, fig. 4.
Scissurella crispata: Hickman, 1998: fig. 15.52 E–F [radula].

Anatoma crispata: Ávila *et al.*, 1998: 491.
Anatoma crispata: Rolán & Ryall, 1999: 21.
Anatoma crispata: Borja & Muxika, 2001: 71.
Anatoma crispata: Golikov *et al.*, 2001: 155.
Anatoma crispata: Geiger, 2003: 73.
Anatoma crispata: Hansson, 2003: 1–2.
Anatoma crispata: Landau *et al.*, 2003: 15, pl. 1. fig. 3.
Anatoma crispata: Demir, 2003: 103 [Mediterranean Turkey, most likely misidentified].
— *crispata*: Golikov & Sirenko, 2004 [*fide* Kantor & Sysoev, 2006].
Anatoma crispata: Kantor & Sysoev, 2005: 29.
Anatoma crispata: Kantor & Sysoev, 2006: 24, pl. 9, fig. A, pl. 139, fig. D.
Scissurella crispata: Brokenshire, 2006: 6, text-fig.
Anatoma crispata:Jensen & Pearce, 2009: 33 [Bermuda, questionable identification].
Anatoma crispata: Geiger & McLean, 2010: 33.
Anatoma crispata: Høisaeter, 2010: 239–240, 242, fig. 6A.
Anatoma crispata: Høisaeter & Geiger, 2011: 90–94, figs 1–18.

Synonyms.

+ *Scissurella elatior* Sowerby I, 1821–1834 (1831): pl. 55, figs 2–3. ***Type material.*** (whereabouts unknown). ***Type locality.*** Not given (OD). Recent on the coast of Britain (Sowerby II, 1842, 1852). ***Etymology.*** ? ***Remarks***. Sowerby I (1821–1834) introduced *Scissurella elatior* as *nomen et figura*, i.e., without any descriptive text going along with the name. The introduction of the taxon is valid by ICZN criteria, but the absence of any further information about this species makes it difficult to treat it. The species was certainly described after 1828, because Sowerby I (1821–1834: unnumbered page to pl. 55) stated: "… the Rev. Dr. Fleming has described a recent species under the appellation of *Sc. crispa*, which he found in shell sand from Noss, Zetland, after a storm." Petit (2006) could date the *Scissurella* section, part 35, April, 30, 1831, based on a set in original wrappers. Petit (2009) did not list this taxon.

Sowerby II (1852: 271) spelled the species "*Scissurella elatoir*," which seems to be an obvious spelling mistake. The caption for plate 15, figure 340, on page 327, however, gives the species name as "*Scissurella elatior*" plus the note "(from Sow. Gen.)." The species name was also spelled *elatior* in the text of Sowerby II (1842).

Given the source of the shell from coast of Britain (Sowerby II, 1842, 1852), and that the species illustrated is quite clearly an *Anatoma*, *Sci. elatior* is a synonym of *An. crispata*.

Jeffreys (1856d) cited Sowerby that *elatior* was found in "Calcaire grossier," a deposit from the Paris region of Lutetien (Eocene) age, which suggests a fossil species. As this indication is not available in the original description, it has to be viewed with caution.

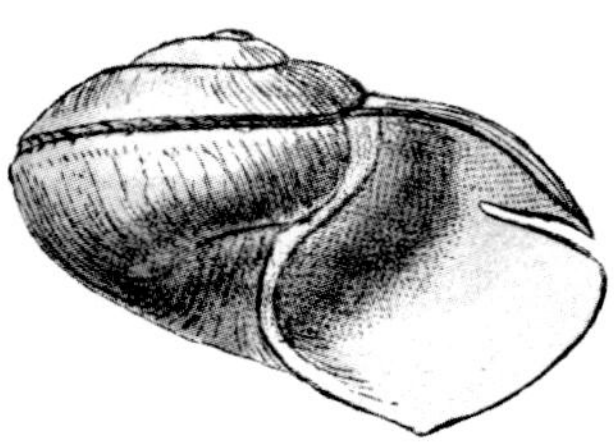

Figure 675. Original illustration of *Scissurella elatior*.

Scissurella elatior: Swainson, 1840: 344.
Scissurella elatior: Sowerby II, 1842: 254, 304 pl. 15, fig. 340.
Scissurella elatior: Sowerby II, 1852: 271, 327, pl. 15, fig. 340.
Scissurella elatior: Woodward, 1856b: 402.
Scissurella elatior: Munier-Chalmas, 1865: 396.
Scissurella elatior: Paetel, 1888: 289.

+ *Scissurella angulata* Lovén, 1846: 152, not illustrated. ***Type material.*** Lectotype (SMNH 4394: designated by Høisaeter & Geiger, 2011: 93. Fig. 676), 2.4 × 1.7 mm. ***Type locality.*** Finmark (OD). ***Etymology.*** Not given (OD). Angulatus, Latin for angled; reference unclear. ***Remarks***. The species epithet *angulata* Lovén has been attributed to Sowerby by Adams & Adams (1853–1854) and Semper (1865); the latter had not located the description. I am unable to find a homonym published by Sowerby and conclude that the Sowerby indication is a misattribution of Lovén's name.

Scissurella angulata: Sars, 1851: 181.
Scissurella angulata: Philippi, 1853: 36, pl. 6, fig. 7.
Anatomus angulatus "Sow.": Adams & Adams, 1853–1858 (1854): 439.
Scissurella angulata: Woodward, 1856a: 37.
Scissurella angulata: Mac Andrew, 1856: 119, 147.
Scissurella angulata: Mac Andrew, 1858: 20 [*fide* Dautzenberg & Fischer, 1912].
Scissurella angulata "Sowerby/Lovén": Semper, 1865: 283–284
Scissurella angulata: Munier-Chalmas, 1865: 395.
Scissurella (*Schizotrochus*) *angulata*: Monterosato, 1878b: 80.
Scissurella angulata: Kobelt, 1888: 258.
Scissurella angulata: Paetel, 1888: 288.
Scissurella crispata var. *angulata*: Pilsbry, 1890: 52.
Schizotrochus angulatus: Monterosato, 1890: 143.
Scissurella crispata var. *angulata*: Odhner, 1912: 13, pl. 2, figs 26–27.
Scissurella (*Schizotrochus*) *angulata*: Nordsieck, 1969: 254.

Figure 676. *Scissurella angulata*. Lectotype. Finmark (SMNH 4394). Scale bar shells = 1 mm. Scale bar protoconch = 100 μm.

Misidentifications.

Hainella umbilicata: Bandel, 1998: 37–38, pl. 12, fig. 8 [is *An. crispata*].

Scissurella costata: Óskarsson, 1982: 164, fig. 4 [as synonym of *Scissurella crispata*. Figure shows *An. aspera*].

not *Scissurella (Anatoma) crispata*: Woodward, 1859: pl. 46, fig. 8. [is *An. aspera*, not fig. 9 as in caption, an obvious error as noted by Marshall, 2002].

not *Scissurella crispata*: Kobelt, 1878: 163, pl. 56, fig. 27 [is *An. aspera*].

not *Scissurella crispata*: Sars, 1878: 126–127, pl. 3 fig. 4 [radula], pl. 8, fig. 7, pl. 18, fig. 1 [operculum] [is *An. tenuisculpta*].

not *Scissurella crispata*: Watson, 1886: 112 [Listing from Station 24 from West Indies could refer to *An. proxima*, *An. alta*, or *An. atlantica*. Listing from Station 145 from Prince Edward and Marion Islands, Antarctica, most likely referring to *An. euglypta*: see Gaillard, 1971: 295].

not *Scissurella crispata*: Gregorio, 1889, 14, pl. 4, fig. 10a–c [is *An. aspera*].

not *Scissurella crispata*: Poirier, 1954: 12 [NW Atlantic species].

not *Anatoma crispata*: Okutani & Hasegawa, 2000: 37, pl. 18, figs 2a–b [Japan: is *An. sagamiana*].

? *Scissurella crispata*: Verrill, 1882: 533 [off Martha's Vineyard, USA].

not *Scissurella crispata:* Norman, 1893: 360–361 [is *An. tenuisculpta*].

not *Scissurella crispata:* Brøgger, 1901: pl. 17, fig. 3 [is *An. tenuisculpta*].

not *Scissurella* cf. *crispata*: Morse, 1909: 95 [New England record].

not *Scissurella crispata*: Thiele, 1912a: 3–4, pl. 1, fig. 4 [as var. *angulata:* is *An. tenuisculpta*].

? *Scissurella crispata*: Johnson, 1915: 87 [New Hampshire].

? *Scissurella crispata*: Johnson, 1934: 68 [Greenland to New Hampshire].

not *Scissurella crispata:* Thorson, 1941: 4, 136 [is *An. tenuisculpta*: reference to Sars, 1878].

not *Scissurella crispata*: Smith, 1951: 78 [NW Atlantic species].

not *Scissurella (Anatoma) crispata*: McLean, 1967: 405, pl. 56, figs 1–4 [included *An. chiricova* and *An. japonica;* illustrated specimen is *An. kelseyi*].

not *Scissurella (Schizotrochus) crispata*: Nordsieck, 1968: 10, pl. 1, fig. 02.10 [is *An. aspera*].

not *Scissurella crispata*: Parenzan, 1970: 30, pl. 1, fig. 5 [possibly *An. micalii*].

not *Scissurella crispata*: Sneli, 1970: 64, fig. 1 [map: confused species concept, most likely *An. tenuisculpta*].

not ? *Scissurella (Schizotrochus) crispata*: Garavelli & Melone, 1971: 11 [Mediterranean record].

not *Scissurella crispata*: Macpherson, 1971: 10, pl. 2, fig. 2, map 4 [figure shows *An. tenuisculpta*].

not *Scissurella crispata*: Hubendick & Warén, 1974: 25, fig. 205 [is *An. tenuisculpta*].

not *Scissurella crispata*: Sneli, 1974: 13 [confused species concept, most likely *An. tenuisculpta*].

not *Scissurella (Anatoma) crispata*: Batten, 1975: 5, fig. 3a, fig. 8 [protoconch], figs 9, 16, 17, 20–24 [shell cross section], fig. 10 [upper whorl]. Overly extended species concept including specimens from Mariana Islands, Canary Islands, and additional specimens without locality data; identification impossible.

not *Scissurella crispata*: Fretter & Graham, 1976: 2, figs 1–2 [is *An. tenuisculpta*].

not *Scissurella (Schizotrochus) crispata*: Carrozza, 1976: 164, fig. 2 [is *An. tenuisculpta*].

not *Scissurella crispata*: Tunnell *et al.*, 1978: 31 [NW Gulf of Mexico].

not *Scissurella crispata*: Sneli, 1979: 266, fig. 3A [map: confused species concept, most likely *An. tenuisculpta*].

not *Scissurella crispata*: Foster, 1981: 13, 41–42 [most likely *An. kelseyi*].

not *Scissurella crispata*: Flor *et al.*, 1982: 63 [overextended species concept].

not *Scissurella (Schizotrochus) crispata*: Nordsieck, 1982: 2, pl. 1, fig. 10 [is *An. aspera*].

not *Scissurella crispata*: Bandel, 1982: pl. 11, figs 4, 6 [protoconch: does not match any species known from "deep water W of Hebrides," but it matches *An. proxima* from the Caribbean to Brazil].

not *Scissurella crispata*: García-Talavera Casañas, 1983: 37, text-fig. [overextended distribution based on misidentifications, not identifiable].

not *Anatoma crispata*: Baxter, 1983: 24 [Alaskan species].

Scissurella crispata (partim): Llera Gonzales *et al.*, 1983: 16–17, text-figs [map: overly extended species concept: Europe, NE and NW Pacific; live animal: copy Fretter & Graham, 1962: fig. 256: is *An. crispata*; shell is *An. tenuisculpta*].

not *Scissurella crispata*: Rolán, 1983: 64, 2 text-figs [is *An. aspera*].

not *Scissurella (Anatoma) crispata*: Marquet, 1984: 338 [is *An. aspera*: based on illustrations on same locality by Marquet, 1995, 1998].

not *Scissurella (Anatoma) crispata*: Ortiz-Corps, 1985: 7 [from Puerto Rico].

not *Scissurella crispata*: Graham, 1988: 60, figs 13A–C [C: living animal, is *An. tenuisculpta*].

? not *Anatoma crispata*: McLean, 1989: fig. 6F [radula, identification doubtful, most likely *An. kelseyi*].

? not *Scissurella crispata*: Bullock *et al.*, 1990: 43 [most likely *Sin. cingulata*: common intertidal species].

not *Anatoma crispata*: Burnay & Rolán, 1990: 33–34, pl. 1, figs 1–2 [is *An. aspera*].

not *Scissurella crispata*: Poppe & Goto, 1991: 32, 64, pl. 2, figs 7–8 [is *An. aspera*].

not *Scissurella crispata*: Cachia *et al.*, 1991: 25, pl. 1, fig. 2 [Malta].

not *Anatoma crispata*: Cavallo & Repetto, 1992: 34, fig. 16 [Italy].

not *Anatoma crispata*: Cossignani *et al.*, 1992: 18, fig. 027 [Mediterranean, is *An. aspera*].

not (?) *Scissurella crispata*: Barash & Danin, 1992: 16–17 [overextended species concept].

not *Anatoma crispata*: Nicolay & Angioy, 1993: 53, text-fig. [is *An. aspera*].

not *Anatoma crispata*: Higo & Goto, 1993: 15 [Japanese species].

not *Anatoma crispata*: Giannuzzi-Savelli *et al.*, 1994: 13, figs 92a–b [is *An. aspera*].

not *Scissurella crispata*: Paulmier, 1994: 5, fig. 4 [is *An. aspera*].

not *Anatoma crispata*: Golikov, 1995: 4, 15, fig. 1 [copy from Sars, 1878, is *An. aspera*].

not *Anatoma crispata*: Marquet, 1995: 63, pl. 2, fig. 4 [is *An. aspera*].

not *Anatoma crispata*: Arduino *et al.*, 1995: 17, text-figs [is *An. aspera*].

not *Anatoma crispata*: McLean, 1996: 17, fig. 1.2 [is *An. kelseyi*].

not *Anatoma crispata*: Bandel, 1998: 26–27, pl. 7, fig. 7 [is *An. aspera*].

not *Anatoma crispata*: Bandel, 1998: 26–27, pl. 7, figs 5–6, 8 [is *Anatoma* sp. from Florida].

not *Anatoma crispata*: Marquet, 1998: 30, text-fig. [is *An. aspera*].

not *Anatoma crispata*: Adrovini & Coossignani, 1999: 32, 2 text-figs [is *An. aspera*].

not *Scissurella (Anatoma) crispata*: Macedo *et al.*, 1999: 267 [overextended species concept].

not *Anatoma crispata*: Redfern, 2001: 2–3, pl. 1, fig. 8 [is *Anatoma* sp.].

not *Anatoma australis/crispata*?: Geiger & Jansen, 2004: 14, fig. 6 [is *An. australis*].

not *Anatoma crispata*: Ardovini & Cossignani, 2004: 19, 2 text-figs p. 68 [is *An. aspera*].

not *Anatoma crispata*: García, 2005: 27, fig. 6 [is *An. proxima*].

not *Anatoma crispata*: Rolán, 2005: 40–41, pl. 6, fig. 69 [is *An. orbiculata*].

not *Anatoma crispata*: Sneli *et al.*, 2005: 27, fig. 4, map p. 28 [is *An. tenuisculpta*].

not *Anatoma crispata*: Doneddu & Trainito, 2005: 50, fig. 28 [is *An. aspera*].

not *Anatoma crispata*: Gulbin, 2006: 10 [Sea of Japan species].

not *Anatoma crispata*: Peñas *et al.*, 2006: 25, figs 14–17 [is *An. tenuisculpta*].

not *Anatoma crispata*: Koyama, 2008: 253, pl. 1, fig. 3 [is *An. sagamiana*].

not *Anatoma crispata*: Portalatina, 2008: 155, fig. 2H, I [is *An. aspera*].

not *Anatoma crispata*: Segers *et al.*, 2009a: 64–65, pl. 5, figs 1–1a [is *An. aspera*].

not *Anatoma crispata*: Segers *et al.*, 2009b: 5 [most likely *An. aspera*: see Segers *et al.*, 2009a].

not *Anatoma crispata*: Harasewych & Moretzsohn, 2011: 199, text-fig. [is *Anatoma* sp.: too little detail visible].

not *Anatoma crispata*: Cossignani & Ardovini, 2011: 97, text-fig. [is *An. tenuisculpta*].

not *Anatoma crispata*: Hernández *et al.*, 2011: 62, fig. 11O–Q [is *An. tenuisculpta*].

not *Anatoma* cf. *crispata*: Wesselingh & Pouwer, 2011, 141, fig. 34 [is *An. tenuisculpta*].

not *Anatoma crispata*: Gofas, 2011: 30, text-fig. [is *An. tenuisculpta*].

Type material. Syntypes (whereabouts unknown: McLean, 1967: 405), 1/15" [= 1.67 mm] (OD). Neotype (NHMUK 1849.10.5.5–7: designated by Hoisaetter & Geiger, 2011: 92; see Remarks), 1.83 mm.

Type locality. In shell-sand at Noss [Island], Zetland [Shetland Islands, Scotland, Great Britain], after a storm (OD). Shetland, Great Britain, 60.500°N, 1.250°W (Neotype).

Etymology. Latin adjective for curled, wrinkled.

Figure 677. *Scissurella crispata*. Neotype. Shetland, Great Britain, 60.500°N, 1.250°W (NHMUK 1849.10.5.5–7). Scale bar shell = 1 mm. Scale bar protoconch = 100 μm.

Description. Shell moderate size (usually to 1.9 mm, exceptionally to 4.5 mm: NMR 993000033746. Figs 677, 678), trochiform, inflated, slightly wider than tall. Protoconch of 0.75 whorl, sculpture flocculent, no apertural varix, apertural margin straight to slightly convex. Teleoconch I of 0.75 whorl, approximately 22 distinct axial cords, spiral cord in position of selenizone; apical region frequently eroded in mature specimens. Teleoconch II >2 whorls, suture impressed; no sutsel,

Figure 678. *Anatoma crispata*. **A–B.** Shetland, Great Britain, 60.500°N, 1.250°W, (NHMUK 49.12.10.102-1–5/52.1.19.56-57, 6). **B.** Dröback, Norway, 59.663°N, 10.631°E (USNM 181591, 34). **C.** Shetland, Great Britain, 60.500°N, 1.250°W (NHMUK 1019.10.26.34095–34102, 9). Scale bars shells = 1 mm. Scale bars protoconch = 100 μm.

except in last 0.25 whorl of mature specimen with descending aperture. Shoulder with axial cords of similar strength as on teleoconch I, first spiral lines at start of selenizone, approximately 13 at apertural margin of mature specimen, no spirals in adsutural 15% of shoulder. Base with axial cords of same density and strength as on shoulder, approximately 18 regularly spaced spiral lines. Umbilicus moderately open, continuously sloping with base, funiculus present. Aperture rounded, roof overhanging. Selenizone at periphery; keels moderately elevated, moderately strong, usually eroded; lunules distinct; slit open with slightly converging margins.

Radula (Fig. 679). Rachidian tooth trapezoidal, central cusp isolated, on either side four cusps arranged in fan-shape. Lateral teeth 1–3 similar, innermost of 3–4 cusps largest. Lateral tooth 4 reduced, outermost of four cusps largest. Lateral tooth 5 enlarged, four cusps on inner margin, terminal cusp largest, one cusp on outer margin. Inner marginal teeth with triangular tip, terminal cusp largest, inner margin with four small cusps, outer margin with five larger cusps; shafts without food groove. Outer marginal teeth spoon-shaped, approximately eight fine cusps on either side.

Comparisons. *Anatoma aspera* from Europe has a smaller protoconch (~180 µm vs. 150 µm), fewer but stronger axial ribs, a much wider sutsel (>1× vs. <0.5× width of selenizone), and the overall shape is taller than in *An. crispata*. The easiest way to distinguish *An. aspera* from *An. crispata* is by the relative strength of the sculpture on shoulder and base; in *An. aspera*, the shoulder sculpture is much stronger than that of the base, whereas in *An. crispata* the sculpture is equally well developed on shoulder and base.

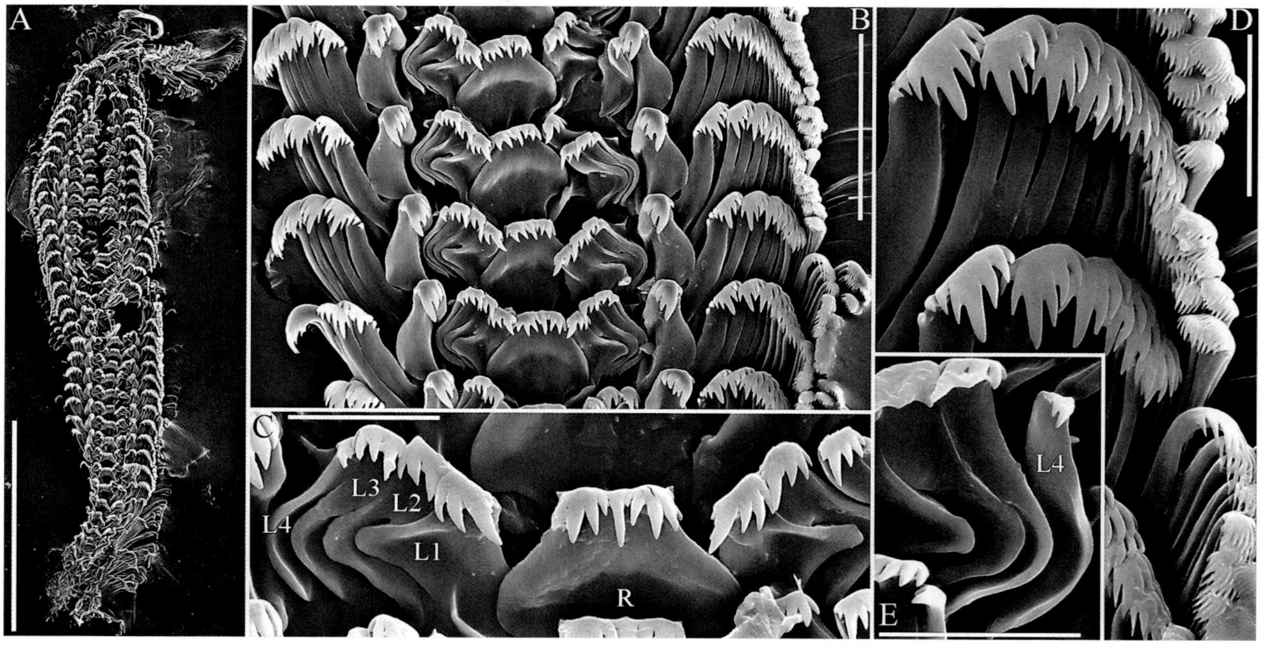

Figure 679. *Anatoma crispata*. Radula. Hordaland, Raunafjorden, Fleslnadsskjeer, Norway, 60.294°N, 5.215°E, (DMNH 23176, 1). **A.** Whole radula. Scale bar = 200 µm. **B.** Whole width of radula. Scale bar = 30 µm. **C.** Central field with rachidian tooth. Scale bar = 10 µm. **D.** Marginal teeth. Scale bar = 10 µm. **E.** Lateral tooth 4. Scale bar = 10 µm.

Anatoma eximia from the Mediterranean and adjacent Atlantic has a taller shell with stepped profile and a much wider sutsel (>2× vs. <1× width of selenizone).

Anatoma janusa from the Azores has a wider sutsel (> vs. < width of selenizone), marked axial lamellae on the shoulder, and fine reticulate sculpture on the base.

Anatoma micalii from the Mediterranean to Angola has a more stepped profile, a wider sutsel (> vs. < width of selenizone), and stronger but fewer axial lamellae on the shoulder.

Anatoma orbiculata from West Africa has a less convex profile of the shoulder, a distinct sutsel on the early whorl, and stronger axial cords on shoulder and base.

Anatoma schanderi from the northeastern Atlantic has a more turreted shell shape with wider sutsel (> vs. < width of selenizone), a thinner shell, a protoconch with reticulate sculpture, and a longer teleoconch I (1 vs. 0.75 whorl).

Anatoma schioettei from the northeastern Atlantic has a shorter teleoconch I (0.5 vs. 0.75 whorl), a more lenticular shell shape, a wider sutsel on early whorls (>0.5× vs. <0.5× width of selenizone), a thinner shell, fewer axial cords, and very fine spiral lines.

Anatoma tenuis from the North Atlantic is much larger (5 vs. 1.8 mm), more lenticular in overall shape, and has no sutsel.

Anatoma tenuisculpta from the Mediterranean and from Norway to Angola is overall more biconical, has a distinct apertural varix on the protoconch, and lacks a distinct funiculus.

Anatoma umbilicata from the Mediterranean and adjacent Atlantic has a more stepped profile and very little sculpture, with periumbilical axial cords being the most prominent.

Anatoma australis from eastern Australia has spiral lines on teleoconch I between the spiral cord in position of the selenizone and the suture, only every third of the axial cords on the base extend from the selenizone to the umbilicus, and the overall shell shape is more lenticular.

Anatoma lyra from the northeastern Pacific has at least 50% fewer axial lamellae on the shoulder and about 50% more spiral lines on the shoulder.

Anatoma conica from subantarctic South America has about 50% fewer axial lamellae on shoulder and base and about 50% more fine spiral lines on the shoulder.

Anatoma keenae from the Panamic forms fine points at the intersection of axial cords and spiral lines, has an overall more globular shape, and lacks a funiculus in the umbilicus.

Anatoma proxima from the Caribbean to Brazil is much larger, has at least twice as many axial cords on shoulder and base, and usually forms small points at the intersection of axial cords and spiral lines.

Anatoma tobeyoides from southeastern Australia has a much shorter teleoconch I (0.125 vs. 0.75 whorl) and at least twice as many axial cords on shoulder and base.

Distribution. Northeastern Atlantic (Fig. 680). 15–1320 m (Golikov *et al.*, 2001).

Specimen records.

North Atlantic. 2530 m, 64.317°N, 1.667°E (MNHN, 7: complete).

Bermuda. 1254 m, Gay Head, 35.775°N, 70.753°W (MCZ 346030 4: complete).

Canada. 100 m, Cabot Strait, Newfoundland, 46.000°N, 59.000°W (HLC, 1).

Denmark, Greenland. 104 m, 66.983°N, 55.450°W (USNM 181594, 1). 599 m, Station 32 [130 km W of SW Greenland], 66.583°N, 56.633°W (ZMUC, 3: complete). 100 m, St. Hellefiskeb, 66.683°N, 55.150°W (ZMUC, 2: complete; ZMUC, 2: complete; ZMUC, 1: complete; ZMUC, 1: complete). 74 m, St. Hellefiskebanke, 66.733°N,

57.000°W (ZMUC, 1: complete). 57 m, Hellefiskebanke, 66.800°N, 55.867°W (ZMUC, 6; ZMUC 1: complete; ZMUC, 3: complete). 64–70 m, St. Hellefiskebanke, 67.000°N, 54.333°W (ZMUC, 3: complete; ZMUC, 1: complete; ZMUC, 9). 70 m, Hellefiskebanke, Greenland, 67.050°N, 54.200°W (ZMUC, 1: complete; ZMUC, 1: complete; ZMUC, 2: complete); ZMUC, 10: complete). 80–95 m, St. Hellefiskebanke, 67.083°N, 54.117°W (ZMUC, 3; ZMUC, 3: complete; ZMUC, 5: complete; ZMUC, 1: complete; ZMUC, 6: complete; ZMUC, 1: complete; ZMUC, 3: complete; ZMUC, 4: complete). 80–95 m, St. Hellefiskebanke, 67.008°N, 54.117°W (ZMUC 6: complete). 197–198 m, off E, 75.016°N, 13.800°W (ZMUC, 1). 70 m, Bylot Sound, 76.542°N, 69.383°W (ZMUC, 3). 160–170 m, Independence Fjord, off the mouth of Jørgen Brønlund Fjord, 82.100°N, 29.917°W (ZMUC). Breidifjord, N for Sndfellsnes (ZMUC, 8). Brlidifjord/Breidifjord, N of Sindfellones, Ingolf St. 87, (ZMUC, 40).

Iceland. 1097–1280 m, shelf off, 61.610°N, 23.428°W (MNHN, 1). 165 m, S Ø. for Myrebgt, S Ø. Iceland, 64.333°N, 15.500°W (ZMUC, 1). 252 m, V fn Mündingen of Patreksfjord V Island, 65.635°N, 24.124°W (ZMUC, 1).

Faroe Islands. 150 m, Munken, 61.333°N, 6.667°W (ZMUC, 20). 247 m, [38 km W of N Suduroy], 61.667°N, 7.667°W (ZMUC, 1). 400 m, off, 64.000°N, 6.700°W (NMR 993000033724, 1).

Finland. Finmark, Sydvaranger, 69.650°N, 30.033°E (NHMUK 1911.10.26.34085–34094, 11). Vardo, 70.367°N, 31.100°E (NHMUK 88.3.10.46-47, 2).

Norway. (KBIN, 4; NHMUK, 5; USNM 83386, 3; RMNH, 1). E side of Stor Vand, near Trondheimfjorden, Sør-Trøndelag, postglacial 10–12 m above sea level (NMR 993000010970, 1). Off Smellingen, Trondheimfjorden, Sør-Trøondelag, 63.633°N, 9.517°E (NMR 993000056063, 3; NMR 993000010080, 25). 73–110 m, Christiansund, 58.167°N, 8.000°E (USNM 181664, 1). 100–200 m, Slevik, 59.193°N, 10.807°E (FSC, 3). 110–183 m, Drobak, 59.662°N, 10.616°E (NMSZ 1961-61, 2; NMSZ, 2). Hordaland, Raunafjorden, Fleslnadsskjeer, 60.294°N, 5.215°E (DMNH 23176, 1). 60–130 m, Raunefjord, 60.300°N, 5.200°E (MV F 161688, 3). Bergen, 60.391°N, 5.325°E (MNHN, 1). Bergen, 60.411°N, 5.276°E (KBIN, 2). 250–260 m,

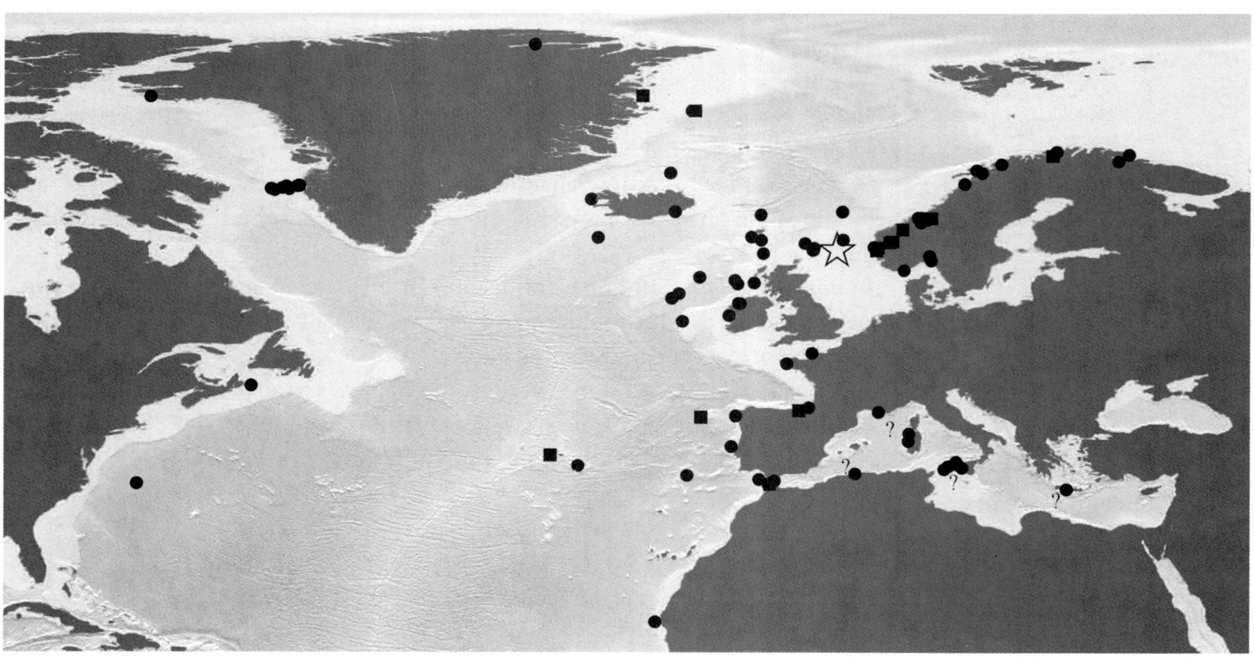

Figure 680. Distribution of *Anatoma crispata*. ☆ type locality. ● specimen record. ■ literature record. ? questionable locality.

Sadnkappneset off Blomöy, Hjelte Fjord, 60.544°N, 4.933°E (RMNH, 3). 30–50 m, Near the Mefjorboen, Hjeltefjord, 60.600°N, 4.917°E (RMNH, 50). Trondheim, 63.200°N, 9.750°E (DLG 315, 1). 200 m, Sletvik, Trondheimfjorid, Sør-Trøndelag, 63.497°N, 10.419°E (NMR 993000033746, 12). 50 m, Sletvik, Trondjheimfjord, 63.582°N, 9.521°E (RMNH AD 16807, 50). 50–300 m, off Brettingsnes, Trondheimfjorden, Søor-Trøondelag, 63.500°N, 9.817°E (NMR 993000037950, 12). 550 m, Trondheims Fjord, 63.525°N, 10.426°E (RMNH, 2). 80–130 m, Trondheims fjord, 63.525°N, 10.426°E (RMNH, 1). 7–50 m, Trondheims Fjord, 63.525°N, 10.426°E (RMNH, 1). 50–200 m, off Kivnaebnes, Trondheimsfjorden, Sør-Trøndelag, 63.533°N, 9.850°E (NMR 993000010649, 10). 60–80 m, between Selva and Agdenes Fyr, Trondheimfjorden, Sør-Trøondelag, 63.579°N, 9.515°E (NMR 993000037951, 3). 100 m, Selva, Trondheimfjord, Sør-Trøndelag, 63.625°N, 9.766°E (NMR 993000033745, 7). 10–100 m, off Garten, Trondheimsleia, Sør-Trøondelag, 63.633°N, 9.533°E (NMR 993000010539, 3). 70–80 m, between Slevia and Abdenes Fyr, Entrance Trondheimfjord, Trondheim, Sør-Trøndelag, 63.634°N, 9.634°E (NMR 993000026919, 4). 50 m, off Brettingsnes, Trondheimfjorden, Sør-Trøondelag, 63.650°N, 9.817°E (NMR 993000010412, 1). 15–20 m, off Brekstad, Trondheimsleia, Sør-Trøndelag, 63.683°N, 9.700°E (NMR 993000010608, 1). 150 m, Brekstad, Trondheimfjord, Sør-Trøndelag, 63.686°N, 9.687°E (NMR 993000033744, 16). Fevag, Trondheimsleia, Sør-Trøndelag, 63.686°N, 9.863°E (NMR 993000022727, 6). 200 m, Grande Viken, Trondheim Fjord, 63.700°N, 9.450°E (MCZ 339822, 1). 50–100 m, Grande Viken, Trondheimfjord, Trondheim, Sør-Trøndelag, 63.700°N, 9.450°E (NMR 993000037954, 5). Bodö, 67.251°N, 14.220°E (RMNH, 1). 250–280 m, Lodingen, Nordland, 68.407°N, 15.992°W (SBMNH 85480, 1). 457 m, Ledingen, 68.413°N, 16.000°E (LACM K-1593, 1). 92 m, Vesteralen, 68.696°N, 15.431°E (FSC, 30). 0 m, Lenvik, Senja Island, Troms, 69.345°N, 17.990°E (NMR 993000033743, 4). Hammerfest, 70.664°N, 23.668°E (NMSZ 1884.11.17, 4). Hammerfest, 70.667°N, 23.700°E (MCZ 279824, 1). Osterfiord [ambiguous locality data], (USNM 181592, 6). Liholmene, 60.362°N, 5.175°E (RMNH AD, 6).

Great Britain. Shetland Islands, 60.300°N, 1.300°W (NMSZ, 2). 763 m, *Challenger* Station 5, epibenthic sled, *Challenger* Cruise E73, 56.717°N, 9.083°W (NMSZ 1976004.10101, 16: complete). 1900 m, *Challenger* Station ES112, 55.200°N, 15.833°W (NMSZ 1994128.10105, 2: complete). 1443 m, Rockall Trough, off Western Scotland, 55.700.N, 15.102°W (NMR 993000033716, 1). 174 m, E of Mingulay, Outer Hebrides, Scotland, 56.830°N, 7.400°W (NMR 993000042907, 20). 992 m, *Challenger* Station SBC235 [W of S Outer Hebrides], 57.100°N, 9.367°W (NMSZ 1994128.10107, 7). 639 m, Scotia [Rockall Bank W of S Outer Hebrides], 57.432°N, 12.980°W (NMSZ 1979038.10101, 6). 400 m, *Challenger* Station SMG76 [between Outer Hebrides and Faroe], 59.917°N, 6.450°W (NMSZ 1993188.5, 4). Shetland, 60.270°N, 1.350°W (MNHN, 7). Shetland, 60.500°N, 1.250°W (NHMUK, 2; NHMUK 1849.10.5.5-7 [netotype], 1; NHMUK 1911.10.26.34095-34102, 9; NHMUK 49.12.10.102-1-5/52.1.19.56-57, 6). Shetland Islands, 61.000°N, 2.167°W (NMSZ 1906.73, 3). 150–160 m, near Murchison oil platform, North Sea between Shetland Islands and Norway, 61.367°N, 1.733°E (NMSZ 1999002.007 1: complete).

Ireland. 490 m, *Challenger* Station PB16 [Porcupine Bank, W of Limerick], 52.775°N, 14.745°W (NMSZ 1981089.10101, 1). Rathlin Island, 54.660°N, 8.830°W (NMSZ, 1). 416 m, *Challenger* Station ES209 [due W off Malin More], 54.683°N, 9.000°W (NMSZ 1994128.10106, 20). Dog's Bay, 53.381°N, 9.970°W (RMNH AD 10272, 1).

Portugal. 2003 m, Portugal (KBIN, 1). 201–219 m, Josephine Bank, Azores, 36.667°N, 14.250°W (USNM 181620, 1). 366–549 m, Villa Franca, Azores, 37.700°N, 25.450°W (USNM 181617, 1). 1097–2002 m, off Lisbon, 39.700°N, 9.717°W (AMS C.24121, 8).

Spain. 706 m, off Cadiz, 36.250°N, 6.867°W (AMS C.24122, 11). 402 m, S off C. Finistère, 42.883°N, 9.267°W (USNM 181601, 1).

Gibraltar. Meteor 21 KG 452, 1970 36.100°N, 5.253°W (ZSM 20020604, 2).

France. Gulf of Gascogne, 43.750°N, 1.750°W (MNHN, 1). 1036 m, off Finistère, 48.333°N, 4.000°W (USNM 181597, 2). 558 m, W off Finistère, 48.333°N, 4.000°W (USNM 181598, 3). 654 m, W off Finistère, 48.333°N, 4.000°W (USNM 181599, 3). 985 m, W off Finistère, 48.333°N, 4.000°W (USNM 181600, 4).

Mauretania. Station 244, 21.352°N, 17.475°W (ZSM 20021736, 1).

Morocco. 174 m, Langier [= Tangier?] Bay, 35.805°N, 5.783°W (USNM 181616, 4). Spartel Bank, 35.783°N, 5.933°W (MNHN, 1).

Dubious Mediterranean specimens.

France. 555 m, Gulf of Marseille, 43.300°N, 5.400°E (USNM 181621, 1).

Italy. 73–219 m, Adventure Bank and Sherke Bank, 37.333°N, 12.167°E (USNM 181630, 10). 168 m, Adventure Bank, 37.333°N, 12.167°E (USNM 181623, 4). Sicily, Magnisi Peninsula, 37.500°N, 14.000°E (MHNG 14364, 1). 200–220 m, Sicily, off Mazara del Vallo, towards Adventure Bank, 37.650°N, 12.583°E (AMS C.404938, 1). Palermo, 37.817°N, 13.583°E (NMSZ 1961.61, 4). Palermo, Sicily, 37.817°N, 13.583°E (USNM 126631, 1). Palermo, 38.133°N, 13.383°E (NHMW, 2; NHMUK 1911.10.26.34124-34132, 14; MNHN, 3; MNHN, 1); KBIN 10591, 3; KBIN, 1). 60–100 m, Palermo, Sicily, 38.133°N, 13.383°E (MNHN, 3). 150–250 m, Sardinia, Bosa Marina, 40.283°N, 8.483°E (AMS C.404937, 3). 400 m, Sardinia, off Asinara Gulf, 41.049°N, 8.539°E (AMS C.4835, 1).

Greece. 128–219 m, Crete Island, 35.250°N, 24.750°E (USNM 181631, 1).

Algeria. 9–93 m, off Algiers, 36.850°N, 3.000°E (AMS C.24120, 10).

Literature records.

Norway. Ligesaa (Grieg, 1896).—100–400 m, Lyktvor, 61.100°N, 6.533°E. 100–400 m, Suphammer. 100–400 m, Tveit, 60.544°N, 5.159°E. 100–400 m, Hanehaug, 61.186°N, 6.808°E (Grieg, 1913).—457–549 m, Rödberg, Trondhjem Fiord, 63.650°N, 10.817°E. 146–219 m, Leerosin, 60.235°N, 5.179°E. 91–182 m, off and a little to the W of Bernestangen, Oster Fiord, 60.533°N, 5.300°E. 686 m, off and a little to the W of Bernestangen, Oster Fiord, off Eide, 60.533°N, 5.300°E. 385 m, off Haakelsund, Kors Fiord, 62.450°N, 7.867°E. 329 m, Kors Fiord, S Side, 70.236°N, 23.270°E (Norman, 1893).

Denmark. 197 m, 75.017°N, 13.480°W. Danmarkshavn, Greenland, 76.609°N, 18.860°W (Schiøtte, T. 2005).

France. 2018 m, E of Cape Finistère (Locard, 1898).

Spain. 1022 m, near Rosles Point (St. Jorge), Spain, 38.794°N, 28.283°W (Dautzenberg, 1927).—Banco de Galicia, Spain, 42.750°N, 12.833°W (Rolán, 1983).—100 m, Matxitxako, Basque, 43.448°N, 2.754°W (Borja & Muxika. 2001).— 392 m, N of Spain (Locard, 1898).

Morocco. 64 m, Tangier, 35.750°N, 5.750°W (Pasteur-Humbert, 1962).— 1900 m, E of Morocco (Locard, 1898).

Dubious records [most likely misidentified, overextended species concept]. 37 m, between Bat Yam and Ashdod, Israel, 31.800°N, 34.633°E (Barash & Danin, 1992).— Off Iturup Island [Etorofu-tō], Japan, 49.999°N, 147.999°E (Golikov, 1995).—30 m, Marina di Camerota, Salerno, Italy, 40.667°N, 14.767°E (Garavelli & Melone. 1971).—555 m, around Marseille (Locard, 1898).

Fossil. In glacier beds of Norway (Sars, 1878). Middle to Upper Pliocene of Calabria, Italy (Seguenza, 1877). Coralline Crag, England [Pleiocene, Zanclean: Long & Zalasiewicz, 2011] (Wood, 1842, 1848; Verrill, 1882). Quaternary of Brittain, Sicily, Rodos and Norway (Locard, 1898). Pliocene of Belgium (Marquet, 1984). Early Pliocene of Estepona, Spain; Italy; Pleistocene of central Mediterranean (Landau *et al.*, 2003 and references therein). Pleistocene of Campo Island, Gallipoli, Italy (Coppa & Crovato, 1983).

Remarks. Nomenclature. The date of publication indicated for the species is either 1828 or 1832 (Bouchet & Warén, 1979). The frontispiece of volume 6 of the *Memoirs of the Wernerian Natural History Society* has a date of 1832, yet the article was read before the society on May 19, 1827. Fleming (1828b) refers to his species; hence, this publication provides the oldest credible evidence of publication of the original description. Sowerby I (1821–1834: 1831. See below for dating) also mentioned the species, providing independent evidence for a pre-1832 publication of the name.

The type material of Fleming is believed to have been lost. Given the excessively confusing history of this taxon, and its application to many anatomid species around the globe, a neotype was designated for the type species of the genus *Anatoma* by Høisaeter & Geiger (2011). Fairly well-preserved material from as near the type locality as possible was chosen. It is similar to the original material from the Shetland Islands, though without further precision. The dimension of 1.9 mm is quite close to the cited 1⁄15" [= 1.67 mm]. Of all the European species known, the neotype agrees best with the original figures. The very rounded overall shell form is notable, while the remaining details are not clearly distinguishable.

Identification. *Anatoma crispata*, described from Europe, is a species that is listed from virtually the entire globe. The usage of this species name has been overly extended. Høisaeter & Geiger (2011) reviewed the convoluted history in detail.

A possible Australian listing was given by Hedley (1902) from *Challenger* Station 164b off Sydney. Hedley (1902) discussed the problematic locality data associated with the lot and concluded that this specimen is actually from *Challenger* Station 64, a mid-North Atlantic station. The remarks by Hedley (1902) were repeated by Herbert (1986). Iredale & McMichael (1962: 4) cast further doubt unto the Australian origin of that sample. The *Challenger* specimen (NHMUK 89.10.26.33) was investigated by SEM (Geiger & Jansen 2004a. Fig. 6); it can be identified as *An. australis* due to the more acute angle of the shell at the selenizone and the wider spacing of the basal reticulation (contra Geiger & Jansen, 2004a). Hedley (1903) did not refer to this specimen in his description of *An. australis*, the holotype of which is at AMS.

Burnay & Rolán (1990) noted the variability of specimens from a single locality in the Mediterranean and the Azores, which encompass both the *crispata* and the *umbilicata-richardi* form. Burnay & Rolán (1990) accepted the synonymy of *umbilicata* with *richardi*, most likely following Locard (1898), while *An. umbilicata* is here considered a valid species, and *richardi* a synonym of *An. tenuis*. The type specimens of *umbilicata* in the NHMUK and the USNM are very smooth and porcelain-like shells of up to 2.7 mm, and are not conspecific with *An. crispata*, though Warén (1980) doubted its distinctness from *An. crispata*. Jeffreys (1868) noted that *aspera* is just the southern form of *crispata*.

As this is the first species of *Anatoma* to be described, the historical note as in the original description is of interest (Fleming, 1828: 366): “I found this shell in 1809, and specimens then transmitted to Mr Montagu, were pronounced by him as fry of a *Trochus*. It is, however, a well marked shell, and belongs to the genus *Scissurella* of M. D’Orbigny. [described in 1824]”

Biology. Bouchet & Warén (1979: 220) noted the ability of the species to swim “by twisting and beating its foot.”

Anatoma disciformis (Golikov & Sirenko, 1980)

Figures 681–683

Chresonymy.

Scissurella (*Schizotrochus*) *disciformis* Golikov & Sirenko, 1980: 107–108, fig. 2.
Anatoma disciformis: Lozouet, 1986: 113.
Anatoma disciformis: Golikov *et al.*, 2001: 155.
Anatoma s.l. *disciformis*: Geiger, 2003: 74.
Scissurella disciformis: Kantor & Sysoev, 2005: 29.
Scissurella disciformis: Kantor & Sysoev, 2006: 24–25, pl. 9, figs B [copy original figure], C [photograph of holotype].
Anatoma disciformis: Gulbin, 2006: 11.
Anatoma disciformis: Geiger & McLean, 2010: 16–20, figs 13–14.

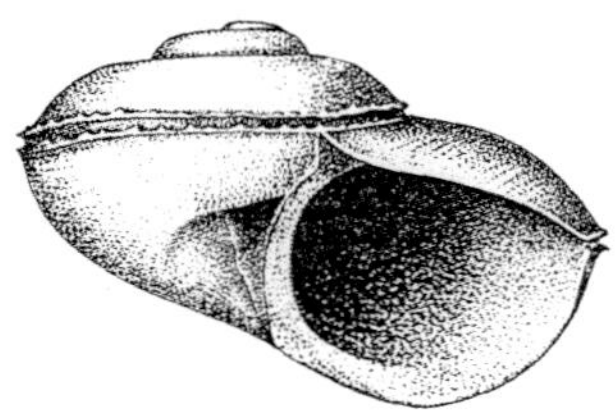

Figure 681. Original illustration of *Scisssurella* (*Schizotrochus*) *disciformis*.

Type material. Holotype (ZIN 32372), 1.32 × 1.05 mm (W × H). One to four paratypes (repository uncertain: possibly ZIN).

Type locality. Near Moneron Island, [Russia, 46.250°N, 141.250°E] 50–70 m. Leg. B. I. Sirenko (OD).

Etymology. More disk-shaped than other species (*crispata, japonica*) (OD).

Description. Translation of Golikov & Sirenko (1980: 107-108); translation of the Russian original description was kindly provided by Vyacheslav Palchevskiy (USC): "The shell is small, almost disk-like with thin walls and with 2.5 slightly convex whorls separated by a deeply impressed suture. The last whorl makes a small concave platform near the suture. The upper part of the whorl has a weakly convex, smooth slope towards the periphery, restricted by two raised keels that border the slit near the outer edge of the aperture. The base of the shell under the keels is moderately convex but forms a deep depression in the region of the wide umbilicus. The diameter of the last whorl is 1.45× as wide as high and 1.3× greater than the height of the shell.

The surface of the shell is milky-white though in one of the paratypes, the color of the shell is slightly yellow. The axial sculpture of the shell consists only of thin, non-contrasted growth lines. Only in the region of the umbilicus, there are slightly separated, barely visible (only under magnification) wrinkles. The spiral sculpture is absent except for separate, non-distinct, unshaped lines.

The aperture is round-oval, essentially shifted in respect to the axis of the shell; its height is ¾ of the shell's height. The outer apertural lip has a slight angle in the upper part near the parietal wall of the whorl and slopes smoothly towards a deep and narrow slit; the region where the keels frame the slit is slightly raised. The inner lip near the columella is strongly curved outside making the narrow fold, and during the transition towards the base of the shell, forms a smooth, circular and dull angle."

Redescription based on material from Alaska. Shell medium-sized (to 2 mm. Fig. 682), trochiform, biconical. Protoconch of 0.75 whorl, sculpture coarsely

Figure 682. *Anatoma disciformis*. **A.** 121 m, off SE side Semisopochoi Island, Rat Islands, Aleutian Islands County, Alaska, USA, 51.889°N, 179.760°W (LACM 86-322, 3). **B.** 152 m, S of Semichi Island, E of Cape Sabak, Agattu Island, Near Islands, Aleutian Islands County, Alaska, USA, 52.379°N, 174.121°W (LACM 86-356, 1). Scale bars shell = 1 mm. Scale bars protoconch = 100 μm.

flocculent, no apertural varix, apertural margin straight. Teleoconch I of 0.5 whorl, approximately 20 low, broad axial cords; no spiral sculpture. Teleoconch II up to 1.75 whorls, suture deeply impressed. Sutsel narrow on early whorls. Shoulder profile very convex, with very dense low axial cords, interstices more or less as wide as cords; first spiral lines after 0.2 teleoconch II whorl, forming fine, dull, clathrate pattern. Base without constriction below selenizone, with similar sculpture as on shoulder. Umbilicus open, distinct funiculus. Selenizone slightly above periphery; lunules moderately distinct; keels low, often broken, moderate strength; slit open, margins slightly converging. Aperture rounded, D-shaped, roof overhanging.

Animal unknown.

Comparisons. From Golikov & Sirenko (1980): "... differs from *S. crispata* and *S. japonica* by less convex turns that have almost disk-like form and an absence of axis combs."

Anatoma concinna from the boreal North Pacific is more globular in overall shell shape, has a longer teleoconch I (0.875–1 vs. 0.5 whorl), a very narrow umbilicus, and no funiculus.

Anatoma lyra from the northeastern Pacific generally has less dense but stronger axial cords, a wider sutsel particularly after more than one teleoconch I whorls (> vs. < width of selenizone), and is overall as tall as wide.

Anatoma keenae from the Panamic is overall more globular, has a wider sutsel throughout ontogeny, and at least twice as widely spaced axial cords.

Anatoma kelseyi from the northeastern Pacific has an overall more turreted shell shape, a protoconch with reticulate sculpture, and fine axial cords on shoulder and base crossed by finer spiral lines.

Anatoma soyoae from Japan has an overall more globular shell shape, most likely a longer teleoconch I (0.75 vs. 0.5 whorl), and a less impressed suture.

Anatoma crispata from the northeastern Atlantic has at least 50% less dense axial cords on shoulder and base, at least 50% fewer spiral lines on shoulder and base, which do not form fine points at their intersections, and an overall more globular shell shape.

Anatoma euglypta from Antarctica grows much larger (10.5 vs. 2 mm), has much finer sculpture, and a spiral cord in the position of the selenizone on teleoconch I.

Anatoma japonica from the Indian Ocean to the Central Pacific has an overall more lenticular shell shape, a spiral cord in the position of the selenizone, and a wider sutsel at the apertural margin of fully grown specimens (< vs. > width of selenizone).

Anatoma megascutula from the Western Pacific has a much larger shell (6.8 vs. 2 mm), a longer teleoconch I (0.75 vs. 0.5 whorl), and small specimens have a proportionally wider umbilicus without a funiculus.

Anatoma rainesi from the Western to Eastern Pacific has a sunken protoconch with fine flocculent sculpture and an apertural varix, fewer but stronger axial cords teleoconch I, and a lower expansion rate of the whorls.

Anatoma proxima from the Caribbean to Brazil has a protoconch with fine flocculent sculpture and an apertural varix, a longer teleoconch I (0.75 vs. 0.5 whorl), and lacks a distinct funiculus in the umbilicus.

Anatoma rapanensis from the Indo-Malayan Archipelago to the Central Pacific has a slightly shorter teleoconch I (0.3–0.4 vs. 0.5 whorl), finer axial cords, and the spiral lines on the base turn into spiral steps towards the umbilicus.

Anatoma regia from New Zealand has a protoconch with an apertural varix, finer axial and spiral sculpture on shoulder and base, and a less convex shoulder.

Anatoma rolani from the Caribbean has even denser axial lamellae, a smooth protoconch, and teleoconch I has more widely spaced, but more distinct axial cords.

Anatoma tenuis from the North Atlantic grows much larger (>5 vs. 2 mm), has no sutsel at 2–3 mm, a protoconch with fine flocculent sculpture, and finer and more widely spaced axial and spiral cordlets.

Distribution. North Pacific from Moneron Island to Alaska, 7–210 m (empty shells only. Fig. 683).

Specimen records.

USA, Alaska. 121 m, off SE side Semisopochoi Island, Rat Islands, Aleutian Islands County, 51.889°N, 179.760°W (LACM 86-322, 3). 152 m, S of Semichi Island, E of Cape Sabak, Agattu Island, Near Islands, Aleutian Islands County, 52.379°N, 174.121°W (LACM 86-356, 1).

Remarks. Five specimens were collected, and reference is made to at least one paratype, but it is unclear whether only one or more paratypes were designated. The holotype was illustrated by Kantor & Syosev (2006) and is strongly decalcified. Multiple attempts at obtaining the specimen on loan failed. The specimens form Alaska are considered conspecific, as they agree as well as possible with the original illustration, the subsequent photograph of the holotype, and the original description. The overall shape of the specimen, the strong funiculus, and the distribution in the boreal Pacific all support the identification of the Alaskan specimens as *An. disciformis*. The original illustration shows a specimen with a very rounded aperture, whereas some of the specimens shown here have a somewhat more angular D-shaped aperture. The latter condition is encountered in specimens with a broken lower adumbilical portion of the peristome; it is an artifact of no taxonomic significance.

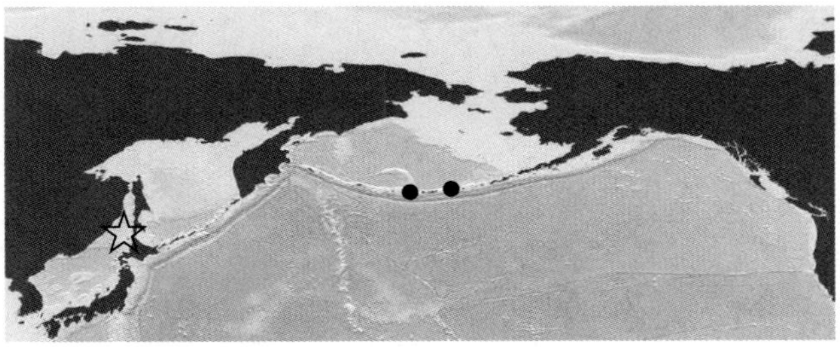

Figure 683. Distribution of *Anatoma disciformis*. ☆ type locality. ● specimen record.

Anatoma emilioi Geiger, 2011

Figures 684–685

Chresonymy.

Anatoma emilioi Geiger, 2011c: 534–535.

Anatoma emilioi: Geiger, 2011b: 40, pl. 4, fig. 3.

Type material. Holotype (SBMNH 149586). Paratypes from type locality (SBMNH 149587, 2).

Type locality. 37–183 m, Mamala Bay, Oahu, Hawaii, USA, 21.290°N, 157.950°W (OD).

Etymology. Named for Emilio García of Lafayette, Louisiana, USA, a champion of micromollusks and generous supporter of ongoing research, who kindly made the specimens available for study (see also *Sinezona garciai*. OD).

Description. Shell of moderate size (holotype 1.75 mm. Fig. 684) trochiform, turreted. Protoconch of 0.75 whorl, sculpture flocculent with very fine spiral line, apertural

Figure 684. *Anatoma emilioi*. **A.** Holotype (SBMNH 149586). **B.** Paratype (SBMNH 149587). 37–183 m, Mamala Bay, Oahu, Hawaii, USA, 21.290°N, 157.950°W. Scale bars shell = 1 mm. Scale bar protoconch = 100 µm.

varix not connected to embryonic cap, apertural margin sinusoidal. Teleoconch I of 0.66 whorl, approximately 10–11 irregularly spaced axial cords, no spirals. Teleoconch II of 2.25 whorls, suture moderately impressed, sutsel as wide as selenizone. Shoulder initially convex, strongly convex after one whorl,with S-shaped profile; strong axial cords acting like ribs supporting draping cloth, sagging between adjacent ribs; supernumerary axial cord on early teleoconch II (~0.25–1 whorl), particularly towards keels of selenizone, subsequently only on keels of selenizone; first spiral threads after 0.8 teleoconch I whorl, increasing to eight at apertural margin, forming blunt thickenings at intersection with axials. Base with distinct constriction below selenizone, thereafter continuously sloping with umbilicus; weak axial cords coordinated with cords on keels of selenizone; two weak spiral threads in constriction below selenizone, approximately 18 spiral cords always weaker than axial cords, forming blunt to knobby thickenings at intersection with axials. Umbilicus of moderate width, no funiculus. Aperture subquadrate, roof overhanging. Selenizone at periphery; keels strong, moderately elevated, distinctly inwardly curved; lunules distinct; slit open, details not observable due to damage to shells.

Animal unknown.

Comparisons. *Anatoma pseudoequatoria* from the Indo-Malayan Archipelago to the Central Pacific has a much flatter overall shell shape, a smooth protoconch, and a strong funiculus.

Anatoma finlayi from the Indian Ocean to the Central Pacific has strong axial cords to lamellae on shoulder and base, more and finer supernumerary axial threads on the keels of the selenizone, and the keels of the selenizone are finer and extend straight from the body of the shell.

Anatoma indonesica from the Indo-Malayan Archipelago to the Western Pacific has more numerous and finer supernumerary fine axial threads on the keels of the selenizone, which also extend straight from the body of the shell, teleoconch I with a spiral cord, and the axial cords on the base do not reach the lower keels of the selenizone.

Anatoma lamellata from the Western to the Central Pacific has a protoconch with dense granular sculpture and an apical spiral element, a spiral cord in position of the selenizone, and the keels of the selenizone are projecting straight from the whorl.

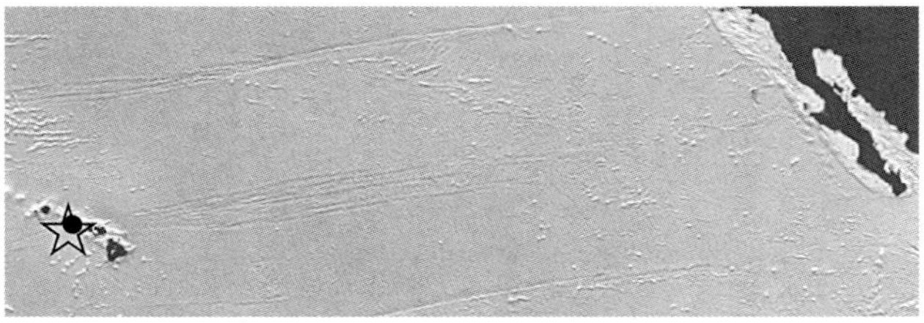

Figure 685. Distribution of *Anatoma emilioi*. ☆ type locality. ● specimen record.

Anatoma janusa from the subtropical northeastern Atlantic lacks an apertural varix on the protoconch and has at least twice as many spiral lines and weaker axial cords on the base.

Anatoma micalii from the Mediterranean to Angola lacks an apertural varix on the protoconch, has at least twice as many axial lamellae on shoulder and base, and lacks a constriction of the base below the selenizone.

Distribution. Known only from type locality (Fig. 685).

Anatoma equatoria (Hedley, 1899)

Figures 686–693

Chresonymy.

Scissurella equatoria Anon., 1899: 520 [*nomen nudum*].
Scissurella equatoria Hedley, 1899: 551–552, fig. 61.
Scissurella aequatoria [sic]: Hedley, 1903: 330.
Scissurella equatoria: Thiele, 1912a: 20, pl. 3, fig. 4 [copy figure Hedley, 1899].
Anatoma equatoria: Habe & Kosuge, 1964: 4.
Anatoma equatoria: Lozouet, 1986: 113.
Scissurella equatoria: Kay & Johnson, 1987: 106 [Holocene of Enewetak].
Anatoma equatoria: Geiger, 2003: 73.
Anatoma sp. 2: Spencer *et al.*, 2009: 201.

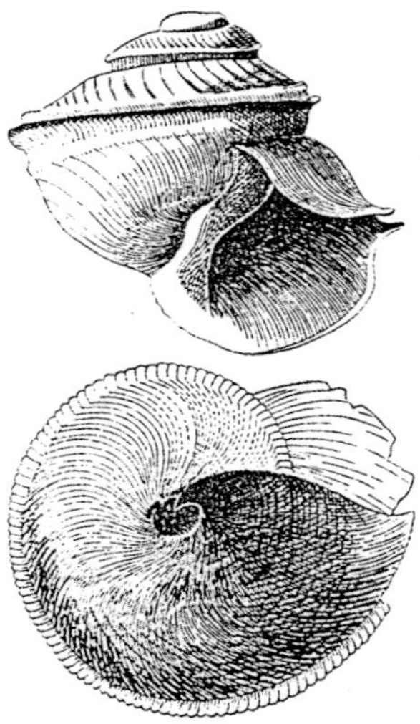

Figure 686. Original illustration of *An. equatoria.*

Synonym.

+ *Anatoma lamellata nanshaensis* Feng, 1996: 161, pl. 1, figs 1–3. ***Type material.*** Holotype (122376: repository institution unknown: Marine Biology Museum, Chinese Academy of Sciences, Qingdao?), 0.99 × 1.1 mm. ***Type locality.*** NS87-46 station, South China Sea, 100 m, sandy mud bottom (OD). ***Etymology.*** Named for its occurrence in the Nansha Sea, China (OD).

Anatoma lamellata nanshaensis: Sasaki *et al.,* 2010: 70–71.

Figure 687. Original illustration of *Anatoma lamellata nanshaensis.*

Misidentification

not *Thieleella equatoria*: Geiger & Jansen, 2004a: 26–27, figs 14, 18 [map] [is *An. quadraxialis*].

Anatoma lamellata: Numanami & Okutani, 1990: figs. 7–8 [is *An. equatoria*].

Type material. Holotype (AMS C.5639. Fig. 688), 3 × 2.5 × 2.68 mm.

Type locality. 200 fms. [= 366 m], Tutaga Islet, Funafuti Atoll, [Tuvalu, 8.500°S, 179.200°E] (OD).

Etymology. Most likely named for its occurrence near the equator.

Description. Shell trochiform, globular (Figs 688–690), medium-sized (to 2.66 mm [holotype Fig. 688]). Protoconch of 0.75 whorl, sculpture flocculent, apertural varix not connected to embryonic cap, apertural margin shallow sinusoidal or convex. Teleoconch I of 0.45–0.66 whorl, approximately 10–18 axial cords, fine spiral cord in position of selenizone. Teleoconch II of at least three whorls, suture deeply impressed. Shoulder convex; approximately 50 axial cords on first whorl, density somewhat decreasing with growth but with increasing strength; first spiral line after 0.2–0.3 whorl, approximately eight spiral lines after one whorl; approximately 12 at apertural margin; spiral lines running over axial cords, forming fine points at

Figure 688. *Scissurella equatoria*. Holotype. 200 fms. [= 366 m], Tutaga Islet, Funafuti Atoll, [Tuvalu, 8.500°S, 179.200°E] (AMS C.5639). Scale bars shell = 1 mm. Scale bar protoconch = 100 μm.

intersections. Base usually not constricted below selenizone, occasionally somewhat constricted in larger specimens; sculpture similar to that on shoulder, approximately 20–27 spiral lines, about 1.5× as strong as those in shoulder, equally spaced from selenizone into umbilicus, forming macroscopically clathrate pattern. Umbilicus open, moderately wide; funiculus barely perceptible to distinct. Selenizone slightly above periphery; lunules moderate to strong, regularly spaced; keels of moderate strength, low [= eroded]; slit open, margins somewhat converging. Aperture rounded-oval, adumbilical portion of peristome flared; roof overhanging.

Radula (Figs 691, 692). Rachidian tooth triangular, apical cusp slightly larger than four cusps on each side. Lateral teeth 1–3 similar, apical cusp largest, 4–2 cusps along outer edge of tip. Lateral tooth 4 reduced, without tip, shield-like. Lateral tooth 5 enlarged by elongation, apical cusp largest, 4–5 cusps on each side. Marginal teeth with spoon-shaped tip bearing many fine cusps; shafts without food groove. Interlock of central field moderately weak.

Figure 689. *Anatoma equatoria*. 134 m, St. 5178, off Point Origon, Tablas, Philippines, 12.7170°N, 122.104°E (USNM 281463, 5). Scale bar shell = 1 mm. Scale bar protoconch = 100 µm.

Figure 690. *Anatoma equatoria*. **A–B.** 134 m, St. 5178, off Point Origon, Tablas, Philippines, 12.7170°N, 122.104°E (USNM 281463, 5). **C.** 400 m, off S coast, New Caledonia, 21.250°S, 167.250°E (AMS C.402754, 4). Scale bars shell = 1 mm. Scale bars protoconch = 100 µm.

Comparisons. *Anatoma amydra* from the Indo-Malayan Archipelago to the Central Pacific has a sutsel of less than the half the width of the selenizone at the start of the selenizone and at least 50% denser axial and spiral sculpture compared to the most densely sculptured *An. equatoria.*

Anatoma munieri from the Indian Ocean to the Central Pacific has a much stronger spiral keel below the selenizone, is more turreted, the sculpture on the base is stronger, and the umbilicus is broader.

Anatoma kopua from temperate eastern Australia to New Zealand has no spiral lines in the constriction of the base below the selenizone, no apertural varix on the protoconch, and fewer spiral lines on the shoulder.

Anatoma philippinica from the Indo-Malayan Archipelago to the Western Pacific has a distinct constriction on the base below the selenizone (occasionally seen in *An. equatoria*), >100 dense and ruffled axial cords, while spiral lines are not discernible.

Anatoma aedonia from the Caribbean to Brazil has a longer teleoconch I (>0.75 vs. 0.5 whorl) and its base bulges slightly less below the selenizone.

Anatoma boucheti from the western Indian Ocean has a flatter apex, a longer teleoconch I (1 vs. 0.5 whorl), and a more narrow sutsel on early whorls (< vs. > width of selenizone).

Anatoma eximia from the Mediterranean and adjacent Atlantic usually has finer reticulate sculpture on shoulder and base.

Distribution. Indo-Malayan Archipelago (26°N–30°S), to mid-western Pacific (118°E–176°W), 46–700 m (Fig. 693).

Specimens records.

Japan, Okinawa. 46–55 m, Hoseshoe South, 1 km W of Onna Village, 26.495°N, 127.843°E (DLG 1176, 1). 401–404 m, Nago Bay, 26.525°N, 127.760°E (SBMNH 149455, 12).

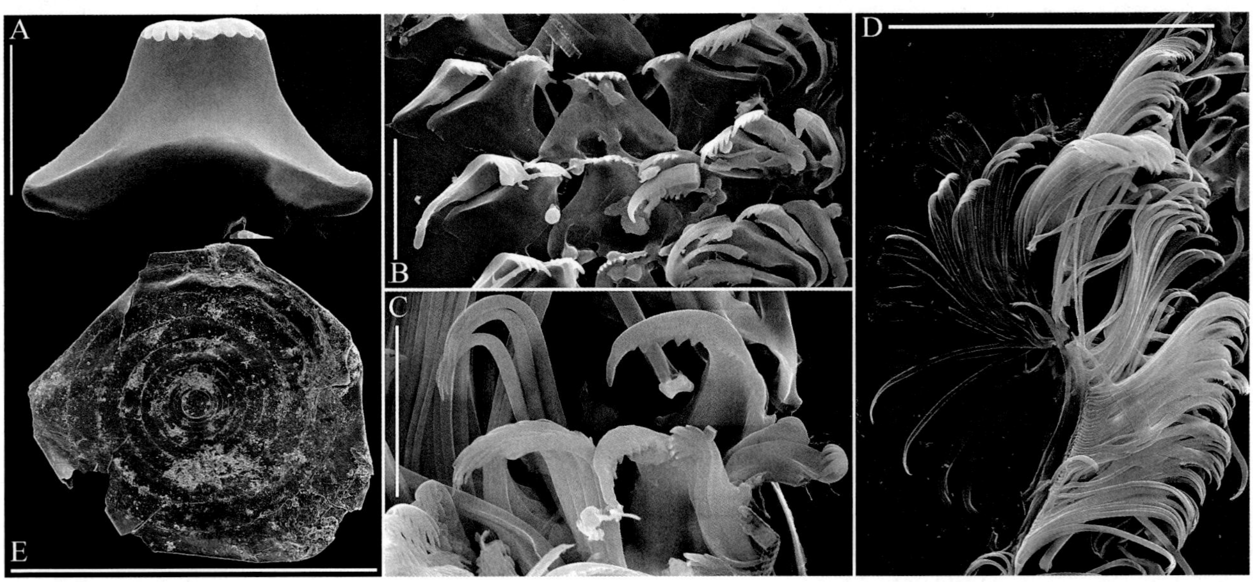

Figure 691. *Anatoma equatoria.* 391 m, St. 5290, off Matocot Point, W Luzon, Philippines, 13.669°N, 120.842°E (USNM 277902, 1). Radula and operculum. **A–D.** Radula. **A.** Rachidian tooth (central cusp damaged). Scale bar = 10 µm. **B.** Central field. Scale bar = 20 µm. **C.** Lateral tooth 5. Scale bar = 20 µm. **D.** Marginal teeth. Scale bar = 100 µm. **E.** Operculum. Scale bar = 1 mm.

Philippines. 622 m, St. 5574, off Simaluc Island, Tawi Tawi Islands, 5.513°N, 120.133°E (USNM ex 1098720, 2). 622 m, St. 5571, off Simaluc Island, Tawi Tawi Islands, 5.513°N, 132.617°E (USNM 287444, 7). 110–120 m, W Pamilacan Island, Cervera Shoal, 9.495°N, 123.837°E (MNHN 1: complete; MNHN 1: complete). 100–138 m, W Pamilacan Island, Cervera Shoal, 9.502°N, 123.840°E (MNHN 1: complete). 100–140 m, between Panglao and Balicasag Island, 9.550°N, 123.700°E (MNHN 1: complete). 93–99 m, 11.717°N, 122.567°E (MNHN, 1). 247 m, St. 5391, off Adyagan Island, E Masbate, 12.221°N, 124.084°E (USNM 281776, 1). 134 m, St. 5178, off Point Origon, Tablas, 12.717°N, 122.104°E (USNM 281463, 5). 146 m, St. 5277, off Malavatuan Island, W Luzon, 13.949°N, 120.229°E (USNM 278089, 10). 189–199 m, 14.000°N, 120.300°E (MNHN, 1). 194–196 m, 14.000°N, 120.317°E (MNHN, 1). 216 m, St. 5272, SW, off Corregidor Light, 14.000°N, 120.375°E (USNM 278324, 1). 192 m, 14.017°N, 120.300°E (MNHN, 2). 362 m, off Matoco Point, W Luzon, 13.632°N, 121.070°E (USNM 276751, 16). 90–200 m, Maribohoc Bay, Bohol Island, 9.602°N, 123.750°E (MNHN, 1: complete). 90–110 m, Balicasag Island, 9.515°N, 123.687°E (MNHN, 1: complete). 124 m, St. 5357, NE off Balabac, Balabac Island, 8.100°N, 117.286°E (USNM 1099256, 1). 402 m, St. 5269, off Matocot Point, W Luzon, 13.664°N, 120.992°E (USNM 278433, 22). 315 m, St. 5289, off Matocot Point, W Luzon, 13.697°N, 120.975°E (USNM 278030, 5). 311 m, St. 5268, off Matocot Point, W Luzon, 13.700°N, 120.954°E (USNM 278571, 4). 391 m, St. 5290, off Matocot Point, W Luzon, 13.669°N, 120.842°E (USNM 277902, 1). 247 m, St. 5110, SW off Corregidor Light, 13.989°N, 120.763°E (USNM 294342, 1). 446 m, St. 5294, off Escarceo Point, N Mindoro, 13.538°N, 121.033°E (USNM 277145, 2). 152 m, St. 5371, Tayabas Bay, off Tayabas Light, Luzon, 5.513°N, 120.133°E (USNM 285558, 1). 172–175 m, Dipolog Bay, Bohol-Sulu Sea Sill, PANGLAO 2005 DW2371, 8.580°N, 123.270°E (MNHN, 1). 679–740 m, Bohol-Sulu Sea Sill, CP2362, 8.942°N, 123.545°E (MNHN, 1). 427 m, Bohol-Sulu Seas Sill, PANGLAO 2005 DW2364, 9.012°N, 123.425°E (MNHN, 1). 273–356 m, off Pamilacan Island,

Figure 692. *Anatoma equatoria*. 408–440 m, E coast, New Caledonia, 20.583°S, 165.117°E (MNHN, 30). **A–D.** Radula. **A.** Entire radula. Scale bar = 100 µm. **B.** Central field enlarged. Scale bar = 20 µm. **C.** Central field with L4. Scale bar = 20 µm. **D.** Marginal teeth. Scale bar = 20 µm. **E.** Operculum. Scale bar = 1 mm.

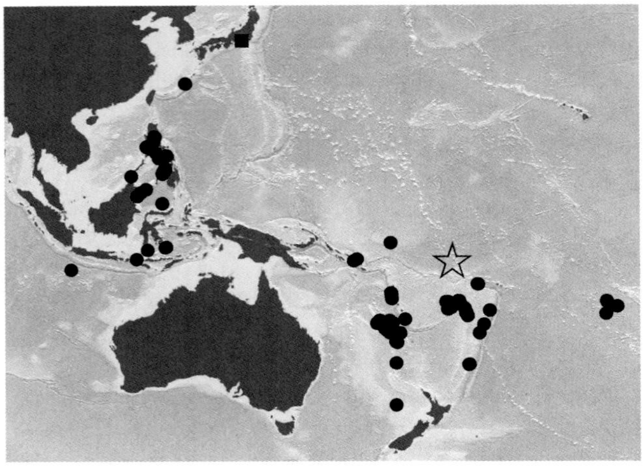

Figure 693. Distribution of *Anatoma equatoria*. ☆ type locality. ● specimen record.

Bohol Sea, PANGLAO 2005 CP2343, 9.457°N, 123.823°E (MNHN, 1). 134–190 m, Cervera Shoal, W Pamilacan Island, PANGLAO 2004 T34, 9.470°N, 123.845°E (MNHN, 6). 164–176 m, off Balicasag Island, Bohol Sea, DW2339, 9.532°N, 123.728°E (MNHN, 13). 128–142 m, off Pamilacan Island, Bohol Sea, PANGLAO 2005 CP2344, 9.473°N, 123.835°E (MNHN, 4). 470–566 m, off Balicasag Island, Bohol Sea, CP2394, 9.477°N, 123.667°E (MNHN, 5). 242–400 m, off Balicasag Island, Bohol Sea, PANGLAO 2005 CP2392, 9.483°N, 123.685°E (MNHN, 1). 95–128 m, Cervera Shoal, Pamilacan Island, PANGLAO 2004 T36, 9.488°N, 123.858°E (MNHN, 4). 110–112 m, Cervera Shoal, W of Pamilacan Island, PANGLAO 2004 T41, 9.495°N, 123.837°E (MNHN, 3). 6 m, Black Forest, Balicasag Island, PANGLAO 2004 S3, 9.518°N, 123.688°E (MNHN, 1). 145–163 m, between Libaong and Pamilacan, PANGLAO 2004 T34, 9.522°N, 123.857°E (MNHN, 1). 309–342 m, off Balicasag Island, Bohol Sea, PANGLAO 2005 CP2399, 9.528°N, 123.698°E (MNHN, 2). 164–176 m, off Balicasag Island, Bohol Sea, DW2339, 9.532°N, 123.728°E (MNHN, 1). 152 m, Bolod, Panglao Island, PANGLAO 2004 T2, 9.540°N, 123.797°E (MNHN, 3). 34–82 m, W of Baclayon, Bohol Island, PANGLAO 2004 T6, 9.585°N, 123.853°E (MNHN, 2). 90–200 m, Maribohoc Bay, Bohol Island, PANGLAO 2004 P1, 9.600°N, 123.750°E (MNHN, 1). 123–135 m, Cortes, Bohol Island, PANGLAO 2004 T26, 9.722°N, 123.813°E (MNHN, 1). 368–442 m, AURORA 2007 DW2707, 15.068°N, 121.685°E (MNHN, 1). 538 m, AURORA 2007 CP2750, 15.883°N, 121.893°E (MNHN, 5). AURORA 2007 CC2746, 15.943°N, 121.747°E (MNHN, 1). 473 m, AURORA 2007 CP2749, 15.943°N, 121.837°E (MNHN, 1). Aliguay Island, 2.746°N, 123.224°E (DLG 1590, 2).

Indonesia. 175–185 m, Bay of Sanggar, N of Sumbawa, 8.322°S, 118.260°E (RMNH 42280, 2). 175 m, SW Salayer, 6.373°S, 120.437°E (RMNH 42278, 1). 465 m, S of Karanga Kaledupa, Banda Sea, Tukang Besi Islands, 5.958°S, 123.775°E (RMNH 42274, 2). 280 m, NE of Binongko, Banda Sea, Tukang Besi Islands, 5.875°S, 123.975°E (RMNH 42276, 1). 296 m, St. 5580, Darvel Bay, off Sibutu Island, Borneo, 4.879°N, 119.113°E (USNM 1098750, 1). 558 m, St. 5592, off Silungan Island, Borneo, 4.212°N, 118.462°E (USNM ex 279050, 2). 300–340 m, NW of Pulau Bahuluang, SW Salayer, 6.467°S, 120.400°E (RMNH 42279, 2).

Christmas Islands. 183 m, off NE point, 10.417°S, 105.700°E (AMS C.404916, 2; AMS C.406383, 1).

Solomon Islands. 396–411 m, SALOMON 1 DW1762, 8.665°S, 160.065°E (MNHN, 11). 98–200 m, SALOMON 1 DW1767, 8.323°S, 160.665°E (MNHN, 17).

Vanuatu. 252–280 m, 20.317°S, 169.883°E (MNHN, 3). 360–419 m, MUSORSTOM 8 DW1065, 16.267°S, 167.350°E (MNHN, 9). 160–175 m, NW Tutuba Island, SANTO 2006 AT02, 15.544°S, 167.272°E (MNHN, 2). 281–288 m, 15.083°S, 167.183°E (MNHN, 1). 205–210 m, MUSORSTOM 8 DW1101, 5.067°S, 167.133°E (MNHN, 35).

Fiji. 660–663 m, Bligh Water, 17.248°S, 178.492°E (MNHN, 6). 500–614 m, Bligh Water, 16.875°S, 177.728°E (MNHN, 1). 300–400 m, 18.167°S, 178.733°W (MNHN, 1). 820–863 m, 17.300°S, 179.550°W (MNHN, 10). 700–707 m, BORDAU 1 DW1485, 19.050°S, 178.500°W (MNHN, 20, MNHN, 1). 395–540 m, BORDAU 1 DW1486, 19.017°S, 178.433°W (MNHN, 1). 500–516 m, BORDAU 1 DW1488, 19.017°S, 178.417°W (MNHN, 2). 402–444 m, SE Viti Levu, MUSORSTOM 10 CP1360, 17.993°S, 178.803°E (MNHN, 10). 80–120 m, SE Viti Levu, MUSORSTOM 10 CP1358, 17.808°S, 178.778°E (MNHN, 1). 314–377 m, 19.667°S, 178.167°W

(MNHN, 16; MNHN, 13). 497–504 m, S of Viti Levu, 18.312°S, 178.152°E (MNHN, 1). 234–361 m, 18.310°S, 178.085°E (MNHN, 3). 260–305 m, S of Viti Levu, 18.308°S, 178.097°E (MNHN, 30). 200–215 m, Bligh Water, 16.840°S, 178.208°E (MNHN, 3). 400–460 m, 16.750°S, 179.992°E (MNHN, 3).

Tonga. 260 m, SE of Tongatapu, 21.345°S, 175.042°W (AMS C.402753, 3). 482–504 m, seamounts, 23.050°S, 175.883°W (MNHN, 1). 440–487 m, Vava'u group, BORDAU 2 DW1586, 18.567°S, 173.917°W (MNHN, 3).

Wallis and Futuna. 440–458 m, Wallis Island, 13.367°S, 176.133°W (MNHN, 1). 415–420 m, Wallis Island, 13.350°S, 176.133°W (MNHN, 1). 350 m, Wallis Island, 13.317°S, 176.283°W (MNHN, 20).

New Caledonia. 527 m, southern, 24.917°S, 168.367°E (MNHN, 1). 250 m, 24.800°S, 168.150°E (MNHN, 2). 235–300 m, Kaimon Maru Bank, 24.767°S, 168.133°E (MNHN, 3). 400–402 m, Norfolk Ridge, 23.017°S, 166.933°E (MNHN, 4). 270–300 m, Norfolk Ridge, 23.650°S, 168.000°E (MNHN, 1). 344–358 m, 22.800°S, 167.433°E (MNHN, 4). 380–410 m, 22.750°S, 167.200°E (MNHN, 12). 350–393 m, southern, 22.733°S, 167.283°E (MNHN, 30; MNHN, 9). 465–525 m, southern, 22.583°S, 166.450°E (MNHN, 1). 250–350 m, southern, 22.500°S, 166.400°E (MNHN, 20). 490–500 m, southern, 22.317°S, 166.183°E (MNHN, 1). 390 m, southern, 22.283°S, 167.200°E (MNHN, 5). 425–430 m, 22.283°S, 167.233°E (MNHN, 10). 455 m, southern, 22.267°S, 166.183°E (MNHN, 30). 250–300 m, northern, 22.083°S, 165.967°E (MNHN, 1). 250 m, E coast, 21.750°S, 166.617°E (MNHN, 1). 310–315 m, Loyalty Basin, 21.533°S, 166.483°E (MNHN, 8). 375 m, E coast, 21.283°S, 165.950°E (MNHN, 6). 400 m, off S coast, 21.250°S, 167.250°E (AMS C.402754, 4). 320–344 m, northern, 21.117°S, 164.467°E (MNHN, 2; MNHN, 1). 350–400 m, E coast, 21.033°S, 165.633°E (MNHN, 40). 425 m, Loyalty Ridge, 21.017°S, 167.517°E (MNHN, 1). 410–430 m, E coast, 20.583°S, 164.967°E (MNHN, 1). 408–440 m, E coast, 20.583°S, 165.117°E (MNHN, 70). 380–400 m, E coast, 20.583°S, 165.117°E (MNHN, 30). 435 m, 20.567°S, 166.900°E (MNHN, 3). 105–110 m, E coast, 20.950°N, 165.567°E (MNHN, 1). 237–550 m, Norfolk Ridge, BATHUS 3 CH802, 23.683°S, 168.000°E (MNHN, 1). 290–378 m, N Bank, Norfolk Ridge, NORFOLK 1 DW1652, 23.435°S, 167.838°E (MNHN, 7). 397–400 m, S BATHUS 2 DW730, 23.050°S, 168.967°E (MNHN, 2). 215–270 m, Munida Bank, Norfolk Ridge, NORFOLK 2 DW2133, 23.017°S, 168.300°E (MNHN, 1). 400 m, S, BATHUS 2 DW729, 22.867°S, 167.200°E (MNHN, 1). 444–445 m, south, BATHUS2 DW719, 22.800°S, 167.267°E (MNHN, 1). 344–358 m, south, BATHUS 2 DW724, 22.800°S, 167.433°E (MNHN, 2). 465–525 m, south, BATHUS 2 DW739, 22.583°S, 166.450°E (MNHN, 3). 250–350 m, St. 40, 22.500°S, 166.400°E (MNHN, 31). 400–500 m, E coast, BATHUS 1 DE694, 20.600°S, 164.967°E (MNHN, 2). 408–440 m, E coast, 20.583°S, 165.117°E (MNHN, 70).

Australia, Norfolk Islands. 326 m, 40, 29.047°S, 10.000°E (NMNZ M.224841, 3). 201 m, off, 37.337°S, 168.150°E (NMNZ M.224761, 8).

New Zealand. 219–274 m, SE of D'Arcy Point, Raoul Island, Kermadec Islands, 29.313°S, 177.903°W (NMNZ M.226852, 4). 82–100 m, SE of Smith Bluff, Raoul Island, Kermadec Islands, 29.315°S, 177.940°W (NMNZ M.227177, 5). 256–348 m, E of Chanter Islets, Raoul Island, Kermadec Islands, 29.255°S, 177.822°W (NMNZ M.225392, 1).

French Polynesia. 480 m, Mount Ari'i Moana, Tarava Seamounts, TARASOC DW3319, 19.233°S, 151.533°W (MNHN, 1). 300–650 m, Tahiti, Society Islands, TARASOC DW3484, 17.783°S, 149.383°W (MNHN, 1). 440–490 m, Raiatea, Society Islands, TARASOC DW3451, 16.883°S, 151.350°W (MNHN, 2). 430 m, between Raiatea and Tahaa, Society Islands, TARASOC DW3436, 16.717°S, 151.433°W (MNHN, 2).

Literature records.

China. Station NS88-62, 84 m; Jian Zhangjiao reef (Feng, pers. comm. 2003).

Japan. 250 m, Suruga Bay, 35.065°N, 138.697°E (Numanami & Okutani. 1990.).

Remarks. The holotype of *An. nanshaensis* is strongly chipped at the aperture and certainly not a fully grown specimen at 1.1 mm.

Small specimens have an overall much broader shape, but the sculptural pattern is identical in small and fully grown specimens. There is some variability in the density of the clathrate pattern on the base, but there are no discrete groupings that would justify taxonomic recognition of those forms. In some specimens there is a weak spiral keel below the selenizone. This spiral keel is usually present in the more turreted *An. munieri*, but in some specimens is rather faint. The converse is true for *An. equatoria*.

The species is rather variable in the details of teleoconch sculpture. Although 75 specimens were documented by SEM, no discrete groupings could be identified, which leads to the conclusion that it represents one somewhat variable species, as opposed to multiple species with minute discriminating characters. Select radulae examined did not reveal any striking differences. The protoconch with flocculent sculpture, the teleoconch I of 0.5–0.75 whorl, the axial cords extending onto the upper surface of the selenizone without supernumerary ones, the sutsel being already approximately as wide as the selenizone at the onset of the selenizone and increasing in width by a factor of 2–3× unite all those specimens. Variable attributes include the distinctness of the constriction of the base below the selenizone, the number of spiral threads on the shoulder, and the fineness of the reticulate sculpture on the base. As is true for most *Anatoma* species, small specimens are proportionally wider than larger ones.

Anatoma equatoria from the Indo-Malayan Archipelago is very difficult to distinguish on shell morphology from *An. eximia* from the Mediterranean and adjacent Atlantic, and to a lesser degree from *An. aedonia* from the Caribbean to Brazil. Given the geographic isolation, those taxa are considered distinct.

Anatoma euglypta (Pelseneer, 1903)

Figures 694–700, Color Plates 1F, 5L

Chresonymy.

Scissurella euglypta Pelseneer, 1903: 17–18, 38, pl. 4, fig. 43.

Scissurella euglypta: Thiele, 1912a: 8, pl. 1, figs 1 [external anatomy: copy Pelseneer, 1903], 2 [radula], 16 [shell].

Scissurella euglypta: Thiele, 1912b: 187.

Scissurella euglypta: Melvill & Standen, 1912: 345.

Scissurella euglypta: Hedley, 1917: 85.

Schizotrochus euglyptus: Powell, 1951: 79 [distribution record].

Schizotrochus euglyptus: Carcelles, 1953: 162.

Schizotrochus euglyptus: Powell, 1958: 169, 179.

Schizotrochus euglyptus: Powell, 1960: 126.

Schizotrochus euglyptus: Arnaud, 1972b: 113 [distribution remarks only].

Schizotrochus euglyptus: Arnaud, 1974: 641.

Schizotrochus euglyptus: Egorova, 1982: 11, fig. 72 [copy figure Pelseneer].

Schizotrochus euglyptus: Cantera & Arnaud, 1984: 32.

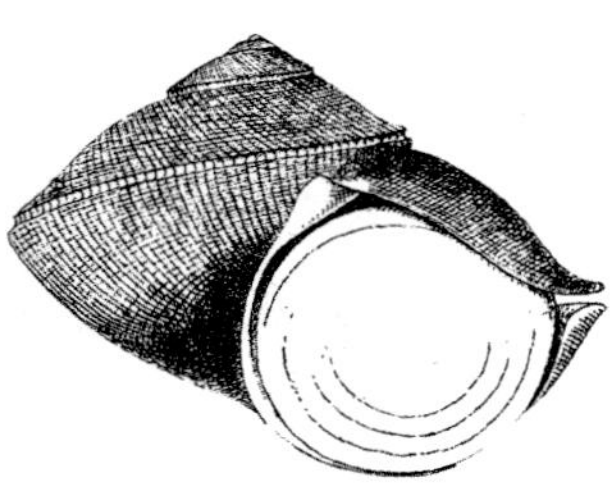

Figure 694. Original illustration of the shell of *Scissurella euglypta.*

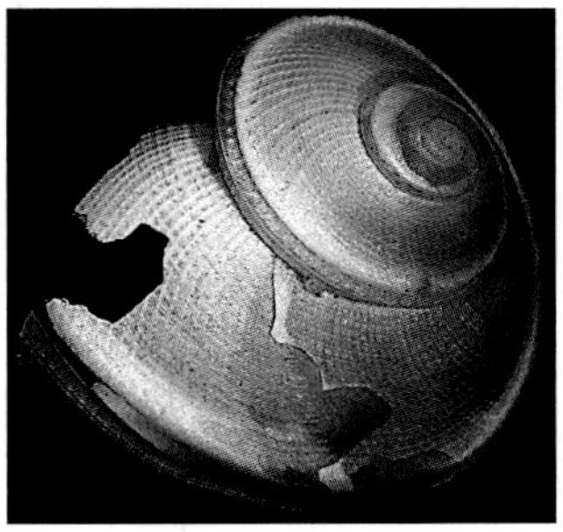

Figure 695. One of the KBIN syntypes of *Scissurella euglypta*. Image KBIN.

Anatoma euglypta: Lozouet, 1986: 113.
Schizotrochus euglyptus: Castellanos *et al.*, 1987: 58, pl. 1, fig. 1.
Schizotrochus euglyptus: Castellanos & Landoni, 1988: 9–10, pl. 1, fig. 7.
Anatoma euglyptus [sic]: Hain, 1990: 33–34, pl. 1, fig. 1; pl. 28, fig. 7 [radula].
Anatoma euglypta: Dell, 1990: 75, fig. 129.
Anatoma euglyptus [sic]: Numanami & Okutani, 1990: 94–100, figs 2–4 [shell], 5 [radula].
Anatoma euglyptus [sic]: Numanami & Okutani, 1992: 172.
Anatoma euglypta: Numanami, 1996: 12–18, figs 6A–D [radula].
Hainella euglypta: Bandel, 1998: 37, pl. 12, figs 5–7.
Anatoma euglyptus [sic]: Forcelli, 2000: 45, text-fig. [copy SEM Dell, 1990].
Anatoma euglypta: Gutt *et al.*, 2000: 44.
Anatoma euglypta: Geiger, 2003: 73.
Anatoma euglypta: Geiger & Thacker, 2005: 49, fig. 1 [phylogeny].
Anatoma euglypta: Schiaparelli *et al.*, 2006: 620.
Anatoma euglypta: Zelaya & Geiger, 2007: 399–406, figs 16 [map], 17 [copy shell Pelseneer, 1903], 18–40 [shell SEM], 41–49 [external anatomy, jaw, radula], 50 [histology].
Anatoma euglypta: Aldea & Troncoso, 2008: 50–51, fig. 5.

Misidentifications.

not *Scissurella crispata*: Watson, 1886: 113 [Station 145 record from 46°S. See Gaillard, 1971 for discussion: from Antarctica, hence most likely *An. euglypta*]

Type material. Syntypes (KBIN, 8), 5 × 3.5 mm.

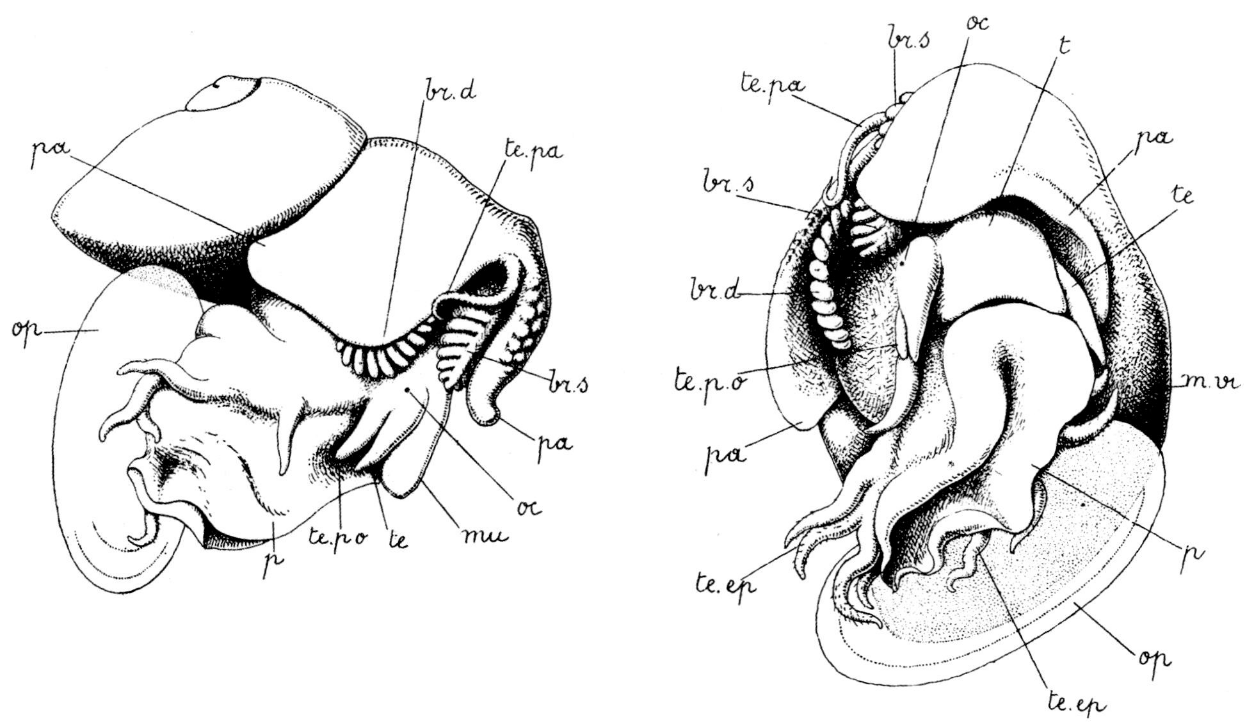

Figure 696. Anatomical illustrations of *Scissurella euglypta* by Pelseneer (1903). br. s.: left gill. br. d.: right gill. mu.: snout. oc.: eye. op.: operculum. p.: foot. pa.: mantle. t.: head. te.: [cephalic] tentacle. te. ep.: epipodial tentacle. te. pa.: pallial tentacle. te. p. o.: post-ocular tentacle.

Type locality. Not designated. Three localities: Station 780: 70°15'S, 84°06'W, 569 m. Station 998: 71°19'S, 87°37'W, 400 m. Station 1030: 70°20'S, 83°23'W, 459 m. Castellanos *et al.* (1987) only gave the first of the originally cited localities as the type locality.

Etymology. Glypho, Greek for to incise, cut in.

Description. Shell large (up to 10 mm. Figs 697, 698), trochiform, biconical, wider than high, particularly in early juveniles. Protoconch of one whorl, sculpture flocculent, apertural margin slightly sinusoidal, without apertural varix. Teleoconch I of 0.5–0.825 whorl, sculptured with 15–25 fine axial threads; interstices with fine, low growth lines; axials often increasing in strength with growth. Teleoconch II of at least three whorls (in a 5.6 mm specimen), sutsel at onset of selenizone about as wide as selenizone, becoming more narrow with growth, increasing up to 2× width of selenizone in large specimens (>5 mm). Base slightly convex, with approximately 50 fine axial threads in specimens of 1 mm, approximately 150 in specimens of 5.6 mm, crossed by approximately 15 finer spiral threads at 1 mm, 50 finer spiral threads at 5.6 mm. Selenizone at periphery; lunules distinct; keels low, thin, fragile; slit open, with margins parallel in small specimens, converging to almost touching at apertural margin in fully-grown specimens. Umbilicus open, continuously sloping with base, wide in juveniles, progressively narrowing with growth. Funiculus present in specimens up to approximately 1.5 mm, not evident in larger ones. Aperture rounded.

Operculum multispiral, with central nucleus, yellowish in color.

Head-foot. Color pale cream, uniform. Cephalic tentacles long, circular, densely papillate; papillae tubular, somewhat enlarged at base, about 25–32 μm in length, arranged as whorls in single plane (Fig. 699D–F). Eye stalks short, fused at base of cephalic tentacles; eyes black, with lens. Head separated from foot by wide neck, with one pair of smooth tentacles on each side; anterior neck tentacle foliaceous, posterior neck tentacle cylindrical, sometimes bifurcated. Three pairs of conical, densely papillate, epipodial tentacles, all with one postero-basal foliated expansion (possibly an epipodial sense organ); first epipodial tentacle divided basally forming two tentacles. Papillae as in cephalic tentacles. Foot long, narrow, with central longitudinal groove; anterior pedal mucus gland located in longitudinal groove at anterior region of foot. Columellar muscle thick, trapezoidal in outline.

Pallial cavity. Extending over 0.25 whorl, deeply sinuate in median line, with enlarged mantle border, densely papillate internally. One papillate conical tentacle, as long as depth of mantle cavity, located at end of median sinuation; smaller, smooth tentacles surrounded it at both sides. Two large bipectinate ctenidia spanning most of dorsal mantle cavity. Gill with long, inflated basal third, distally rounded filaments; left gill larger by 10%, with anterior outwards fold, provided with small indentations in margin; right gill without distal fold. Osphradium not found.

Digestive system. Snout broad, tubular, rounded terminally. Jaw plates: small, pteriform in outline, composed of rows of rhomboidal shingles, dark brownish in color. Odontophores large, elongate, enlarged posteriorly, formed by one pair of fused cartilaginous plates. Radular ribbon longer than odontophore. Esophagus as large as radula; anterior, medial esophagus walls with transversal folds; anterior, medial, posterior esophagus similar in width. Salivary glands elongate, sac-like, larger than odontophore, pinkish; salivary ducts narrow, long, opening ventrally into esophagus. Stomach half as long as esophagus. Intestine very long, twisted; hindgut forming two widely open loops, dorsally located. Anus opening into postero-central part of pallial cavity. Digestive gland extending for first 1.5 whorls.

Radula (Fig. 699A–B). Rachidian tooth trapezoidal at base, tip with cusps parallel to radula rows; central cusp largest, six narrower pointed cusps on each side. Lateral teeth 1–3 similar in outline, overlapping at their widely expanded bases; tips elongate, tongue-like, with a sharp main cusp, narrower, slightly curved

Figure 697. *Anatoma euglypta*. **A.** Station PS60.500-1, Antarctica, 57.680°S, 24.415°W (ZSM 20021094, 2, 9: complete). **B.** Station 111-19, Antarctica, 71.125°S, 11.472°W (ZSM 20012468, 10). **C.** Station 115-1, Antarctica, 70.802°S, 10.703°W (ZSM 20012473, 1). Scale bars shell = 1 mm. Scale bars protoconch = 100 μm.

cusps at both sizes: 6–7 cusps on outer side, 4–5 cusps on inner side. Lateral tooth 4 narrower than lateral teeth 1–3, with one sharp main cusp, 3–5 additional tiny cusps on outer side. Lateral tooth 5 claw-shaped, well separated at base from other lateral teeth, with one main terminal cusp, several smaller cusps on both sides.

Figure 698. *Anatoma euglypta*. **A.** Station 159-1, Antarctica, 62.917°S, 57.658°W (ZSM 20012475, 1). **B.** Station 116-1, Antarctica, 70.828°S, 10.638°W (ZSM 20012469, 1). **C.** 342-353 m, 2.5 mi S Fallows Point, Staten Island, Tierra del Fuego, Argentina, 54.833°S, 63.842°W (LACM 71-331, 1). Scale bars shell = 1 mm. Scale bars protoconch = 100 µm.

Marginal teeth numerous (~30), long, narrow, becoming more delicate towards edge of radula; first marginal tooth enlarged laterally at base; shaft, tip similar to that of remaining marginal teeth; shafts without food groove. Radular interlock of central field moderate.

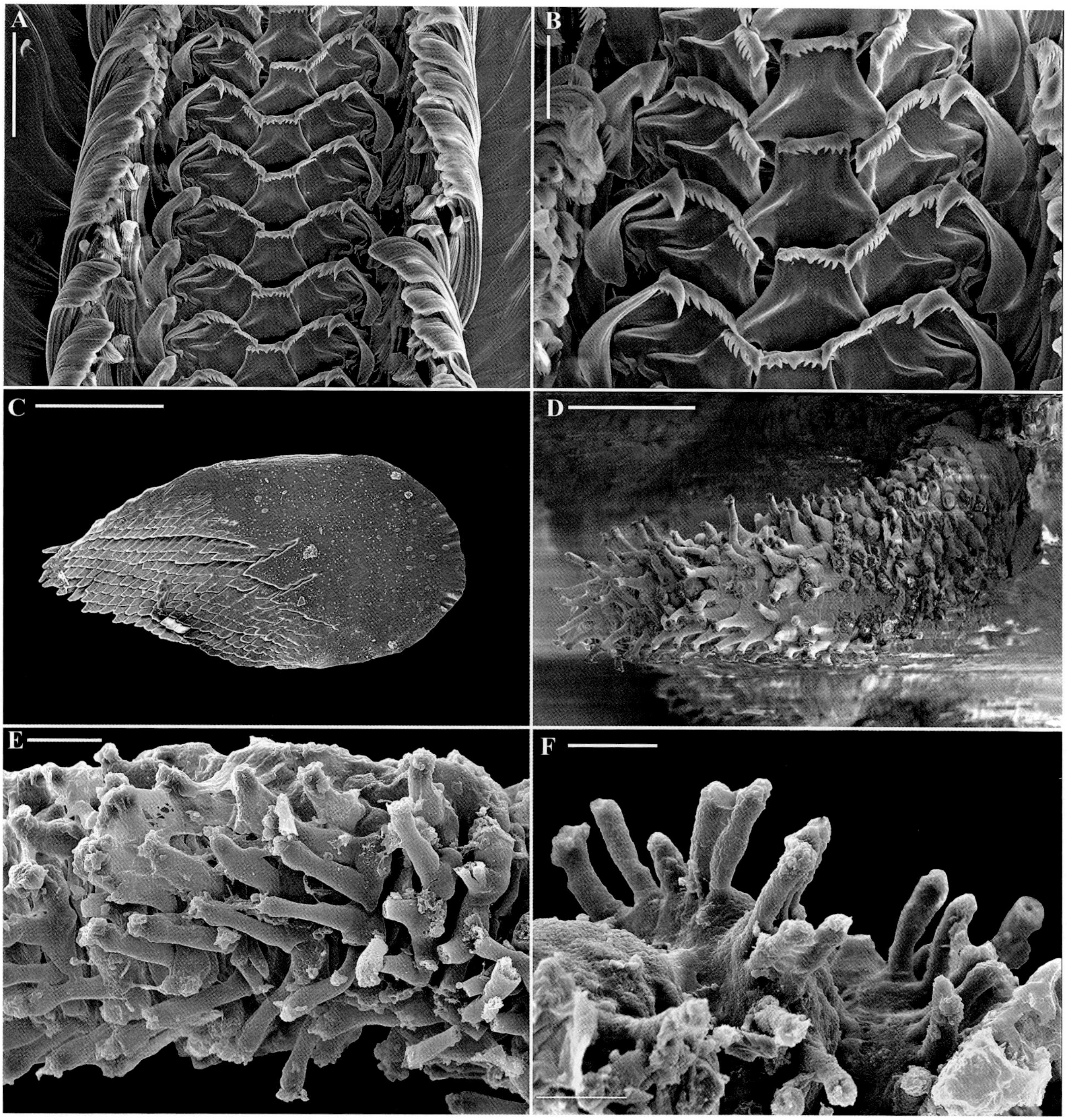

Figure 699. *Anatoma euglypta*. **A–C.** 309 m, South Sandwich Islands, Antarctica, 57.667°S, 26.417°W (MLP 12063, 15, 76: complete). **A–B.** Radula. **A.** Entire width of radula. Scale bar = 50 µm. **B.** Central field enlarged. Scale bar = 20 µm. **C.** Jaw. Scale bar = 100 µm. **D–F.** Cephalic tentacle. **D.** Entire tentacle. Scale bar = 100 µm. **E.** Section enlarged. Scale bar = 20 µm. **F.** Papillae strongly magnified. Scale bar = 20 µm. From Zelaya & Geiger (2007, modified).

Nervous system. Pair of fused cerebro-pleural ganglia located on sides of buccal bulb; pedal ganglia large, located ventral to oral tube.

Comparisons. *Anatoma amoena* from Antarctica has a more turreted shell that is as wide as it is tall, fewer and much more prominent axials on shoulder, a distinct reticulate pattern on the base, and reticulate protoconch sculpture.

Anatoma conica from Antarctica has much stronger and less dense sculpture on teleoconch II, particularly strong axial lamellae from early teleoconch II onwards on shoulder and base.

Anatoma shiraseae from Antarctica has a distinct keel on the shoulder, a distinct constriction below the selenizone, and coarser sculpture.

Anatoma weddelliana from Antarctica has much stronger but at most half as many raised axial cords, a longer teleoconch I (0.75 vs. 0.5 whorl), and a wider umbilicus.

Anatoma atlantica from the northwestern Atlantic has a protoconch with apertural varix, a longer teleoconch I (0.66–0.75 vs. 0.5 whorl), and stronger axial cords forming minute points at the intersection with the spiral lines on shoulder and base.

Anatoma australis from southeastern Australia has a wider sutsel, only every 2–3 axial cords on the base reach from the selenizone to the umbilicus, and a longer teleoconch I (0.75 vs. 0.5 whorl).

Anatoma disciformis from the North Pacific has denser and stronger axial cords with interstices only more or less the width of the cords and lacks a spiral cord in position of the selenizone on teleoconch I.

Anatoma japonica from the Indian Ocean to the Central Pacific has stronger axial cords that form points at the intersection with the spiral lines on shoulder and base and a wider sutsel after two whorls (>2× vs. 1× width of selenizone).

Anatoma maxima from the Indo-Malayan Archipelago to the Central Pacific has a wider selenizone, a longer teleoconch I (0.75 vs. 0.5 whorl), and taller keels of the selenizone.

Anatoma regia from New Zealand has a protoconch with apertural varix, small specimens have less convex whorls, and large specimens have a wider sutsel (2× vs. 1× width of selenizone).

Anatoma tenuis from the North Atlantic has a protoconch with crisp flocculent sculpture, a proportionally wider selenizone, and no sutsel to at least 4.65 mm size.

Anatoma tenuisculpta from the Mediterranean and adjacent Atlantic has a longer teleoconch I (0.75 vs. 0.5 whorl), an apertural varix on the protoconch, and stronger axial cords that form minute points at the intersection with the spiral lines.

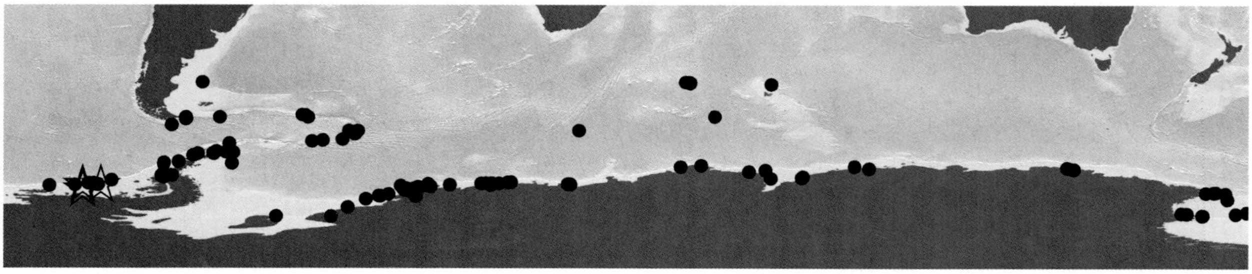

Figure 700. Distribution of *Anatoma euglypta*. ☆ type locality. ● specimen record. ■ literature record.

Anatoma yaroni from east Africa has a protoconch with indistinct apertural varix, a longer teleoconch I (0.75 vs. 0.5 whorl), and only half as many axial cords.

Distribution. Circumantarctic, 22–4420 m (living specimens: 94–4420 m). In fine sediments at several hundreds meters depth in water of <0°C (Bandel, 1998. Fig. 700).

Specimen records.

Atlantic sector of the Magellan Region. 342–353 m, 2.5 mi S Fallows Point, Staten Island, Tierra del Fuego, Argentina, 54.833°S, 63.842°W (LACM 71-331, 1). 771–903 m, 5 mi S Ventana Point, S side Staten Island, Tierra del Fuego, Argentina, 54.908°S, 63.933°W (LACM 71-329, 2; USNM 845406, 2).

Drake Passage. 121 m, off, South Africa, 56.328°S, 67.163°W (NHMUK 20000096, 12).

Burdwood Bank, Antarctica. 286–290 m, Station PS61/150-1, 54.504°S, 56.137°W (MLP 120602: 5, 8 complete, ZSM 20021528, 3: complete; ZSM-Moll 20021278: 1, 5: complete).

South Georgia, Antarctica. 94 m, 54.500°S, 35.833°W (MLP 7281, 3: complete; MACN-In 36273, 4: complete). 138 m, 53.966°S, 37.150°W (MLP 7301: 2: complete; MACN-In 36274, 1: complete).

South Sandwich Islands, Antarctica. 309 m, 57.666°S, 26.417°W (MLP 12063, 15, 76: complete). Station PS61/200-1, 57.680°S, 24.415°W (ZSM 20021094, 2, 9: complete). Station PS61/197-1, 57.681°S, 26.442°W (ZSM 20021351, 3, 20: complete). 263–270 m, 57.666°S, 26.400°W (MLP 12064: 6, 14: complete). Station PS61/194-1, 57.676°S, 26.419°W (ZSM 20021459, 4: complete). 587 m, 57.666°S, 26.450°W (MLP 12065: 3, 17 complete). Station PS61-207-1, 57.673°S, 26.465°W (ZSM 20021491, 5: complete). PS61/207-1, 57.673°S, 26.465°W (ZSM 20021328, 50: complete). 521 m, 59.900°S, 32.467°W (MLP 12066: 2, 7: complete). Station PS61/217-1, 59.916°S, 32.472°W (ZSM 20021449, 1: complete). Station PS61/214-1, 59.709°S, 27.950°W (ZSM 20021488, 1: complete). Station PS61/141-9, 58.414°S, 25.014°W (ZSM 20021126, 2: complete). Station PS61/140-7, 58.267°S, 24.896°W (ZSM 20021580, 1: complete; ZSM 20021142, 1: complete). Station PS61/141-8, 58.416°S, 25.017°W (ZSM 20021123, 2: complete). 774–756 m, 58.416°S, 25.167°W (ZSM-Moll 20041498: 5: complete. ZSM-Moll 20041496: 4: complete). 2313–2281 m, 58.417°S, 25.017°W (ZSM-Moll 20041499, 1).

Discovery Bank, Antarctica. 374 m, 60°08' S 34°56' W (MLP 12067, 2 complete). Station PS61/223-1, 60.136°S, 34.926°W (ZSM 20021076, 1: complete). 1040–1048 m, Discovery Bank, 71.300°S, 13.967°W (ZSM 20051033, 1, 1: complete; ZSM-Moll 20051016, 1, 5: complete). 4384–4420 m, 70.518°S, 14.581°W (ZSM 20051091, 1: complete). 1579–1584 m, 62.778°S, 53.042°W (ZSM 20051106, 26: complete).

South Shetland Island, Antarctica. 210–220 m, Elatin Station 1003, 62.667°S, 54.750°W (NMNZ M.242836, 3). Station PS61/046-7, 60.653°S, 53.948°W (ZSM 20021203, 1: complete; ZSM 20021215, 1: complete). 311–365 m Station, 174-1, 63.018°S, 61.152°W (ZSM 20010184, 1: complete). 119–124 m, Mero Cruise 691, Station 26, 63.433°S, 62.250°W (NMNZ M.242837, 2).

Crozet Island, Antarctica. 22 m, NE Île de la Possession, MD30 Sta 6, DC209, 46.400°S, 51.780°E (MNHN, 2). 22 m, 46.383°S, 51.833°E (MNHN, 1).

Kerguelen Islands. 1218 m, 46.800°S, 70.500°E (MNHN, 1).

Antarctic Peninsula. Station 158-1, 63.078°S, 57.527°W (ZSM 20010042, 12: complete). 214–218 m, Station 159-1, 62.917°S, 57.658°W (ZSM 20012475, 1; ZSM-Moll 20012470, 1). 94–95 m, 63.078°S, 57.527°W (ZSM-Moll 20012476, 1). 909–911 m, 62.500°S, 56.930°W (ZSM-Moll 20012889: 1 complete). 214–218 m, Station 159-1, 62.917°S, 57.658°W (ZSM 20012887, 2: complete). 94–95 m Station 158-1, 63.078°S, 57.527°W (ZSM 20012883, 1: complete). 93–97 m, Station 155-6, 63.080°S, 57.555°W (ZSM 20012884, 4: complete). 94–95 m, Station 158-1, 63.078°S, 57.527°W (ZSM 20012888, 5: complete). 64 m, Marguerite Bay, 68.117°S, 67.100°W (USNM 488965, 1). 795–830 m, N of Fram Bank, 68.813°S, 70.416°E (MV F 72734, 1). 720 m, SW of Davis Base, 68.703°S, 77.504°E (MV F 72731, 1). 179 m, NW of Davis Base, 68.431°S, 77.784°E (MV F 72730, 11). 118 m, MacRoberston Shelf, NE of Nielsen Deep, 67.274°S, 65.419°E (MV F 72732, 19). 518 m, edge of Fram Bank, 66.914°S, 69.205°E (MV F 72733, 1).

Weddell Sea, Antarctica. Station 116-1, 70.828°S, 10.638°W (ZSM 20012469, 1). 250 m, Station 133-2, 70.950°S, 10.469°W (ZSM 20012471, 1). 112 m, Station 111-19, 71.125°S, 11.472°W (ZSM 20012468, 10). 765–840 m, Station 138-1, 71.148°S, 13.213°W (ZSM 20012474, 1; ZSM 20012886, 1). 500 m, Station 115-1, 70.802°S, 10.703°W (ZSM 20012473, 1). 100 m, Station 59-4, 70.671°S, 7.668°W (ZSM 20012472, 1). 316 m, Station 116-1, 70.828°S, 10.638°W (ZSM 20012885, 1: complete). 251–271 m, Station 136-1, 70.837°S, 10.590°W (ZSM 20010077, 1). 2081–2087 m, Station PS61/132-3, 65.309°S, 53.379°W (ZSM 20021138, 1: complete). 229 m, 75.517°S, 26.717°W (USNM 612957, 1). 77.617°S, 43.250°W (USNM 612857, 1). 384 m, 77.667°S, 30.583°W (USNM 638831, 1).

Ross Sea, Antarctica. McMurdo Sound, SE of Mount Larsen, 76.917°S, 164.750°E (AMS C.31466, 1). 433 m, E of Cape Hallett, McMurdo Sound, Burton Island, 72.133°S, 172.167°E (USNM 613121, 5; USNM 878219, 2). 348–352 m, E of Eape Hallett, Moubray Bay, Victoria Land, 72.083°S, 172.133°E (USNM 904244, 2: complete). 321 m, McMurdo Sound, 77.433°S, 169.650°E (USNM 612780, 6). 392 m, 72.083°S, 172.250°E (USNM 613010, 7). 344–351 m, Elatin Station 1996, 72.083°S, 172.133°E (NMNZ M.242835, 5). 521 m, Coulman Islands, 73.667°S, 175.283°E (USNM 612841, 1). 342–360 m, Elatin Staion 1995, 72.050°S, 172.633°E (NMNZ M.242834, 3). Scott Base, Cape Armitage, McMurdo Base, 77.000°S, 166.000°E (NMNZ M.218133, 2). 465–468 m, NZOI Station A. 463, 72.333°S, 174.833°E (NMNZ M.242833, 2). 362 m, NZOI Station A.449, 77.083°S, 177.200°E (NMNZ M.242831, 4). 318–472 m, NZOI Station A450, 76.700°S, 179.733°E (NMNZ M.242832, 2).

East Antarctica. 300 m, off Enderby Land, 66.167°S, 49.683°E (AMS C.376618, 3). 80 m, Burton Island, Davis Sea, 66.558°S, 93.015°E (USNM 878215, 1).

Lazarev Sea. 161–185 m, 69.983°S, 8°E (SMNH 82903, 1). 338–343 m, 69.767°S, 10.017°E (SMNH 82902, 4). 568–644 m, 70°S, 3.933°E (SMNH 82900, 20). 432 m, 69.717°S, 10.767°E (SMNH 82897, 20). 191–204 m, 70.317°S, 3.267°W (SMNH 82895, 30). 118–126 m, 70.400°S, 6.133°E (SMNH 82896, 4). 161–185 m, 69.983°S, 8.000°E (SMNH 82904, 30). 161–185 m, 69.983°S, 8.000°E (SMNH 82901, 30; SMNH 82898, 1). 210–213 m, 69.950°S, 5.067°E (SMNH 82899, 8). 661–742 m, 69.982°S, 5.14.000°E (SMNH 82905, 20: complete). 280–298 m, 69.950°S, 6.333°E (SMNH 82906, 20: complete).

Literature records.

Antarctica. 459 m, 70.333°S, 89.383°W. 569 m, 70.250°S, 84.100°W (Carcelles, 1953). —64 m, 68.500°S, 69.000°W. 80 m, 66.549°S, 93.015°E. 384 m, 77.667°S, 30.583°W. 321 m, 77.433°S, 169.500°E. 203 m, 72.200°S, 170.333°E. 433 m, 72.133°S, 172.167°E. 5–21 m, 73.667°S, 175.283°E. (Dell, 1990).—93–130 m, Bismarck Strait, Palmer Archipelago, 64.933°S, 65.583°W. 155–178 m, off mouth of Stromness Harbour, South Georgia, 54.017°S, 36.442°W (Powell, 1951).—850 m, BT14, S of Vestcape, 73.602°S, 22.595°W. 362 m, AG06, Cape Norvegia area, 71.535°S, 13.735°W. 254 m, EB05, Cape Norvegia area, 71.675°S, 12.695°W. 560 m, AG09, Cape Norvegia area, 71.578°S, 12.443°W. 181 m, MG2902, Cape Norvegia area, 71.310°S, 12.423°W. 170 m, AG02, Cape Norvegia area, 71.312°S, 12.285°W. 123 m, GK24, Cape Norvegia area, 71.136°S, 11.538°W. 2315 m, DR30, Cape Norvegia area, 70.088°S, 8.333°W (Gutt *et al.*, 2000).—275–283 m, Station 5, Breid Bay, 70.150°S, 23.772°E. 295–310 m, Station 7, Breid Bay, 70.152°S, 24.032°E. 270 m, Station 8, Breid Bay, 70.142°S, 24.280°E. 276–289 m, Station 9, Breid Bay, 70.228°S, 24.428°E. (Numanami, 1996).—271–310 m, Breid Bay, 70.167°S, 24°E (Numanami & Okutani. 1992).—300 m, station 39 off Enderby Land, 66.167°S, 49.683°E. 220 m, station 42 off Enderby Land, 65.833°S, 54.383°E. 640 m, station 90 off Adelie Land, 66.350°S, 138.467°E (Powell, 1958).—290 m, MD 08

DC296, 46.297°S, 50.797°E. 375 m, MD 08 BB183 46.312°S, 51.492°E. 190 m, MD 08 DC186, 46.352°S, 51.565°E. 195 m, MD 30 DC209, 46.533°S, 51.845°E. 1218 m, MD 04, CP243 46.795°S, 70.503°E (Cantera & Arnaud, 1985).—102 m, Burdwood Bank, Station 346, 54.417°S, 57.533°E (Melvill & Standen, 1912).—Gauss station, 66.033°S, 89.633°E (Thiele, 1912b).— 180–200 m, Cap Géodésie, 66.667°S, 139.850°E. 230–250 m, Cap Géodésie, 66.667°S, 139.850°E. 170–180 m, Cap Géodésie, 66.667°S, 139.850°E. 135–140 m, Archipel de la Pointe Géologie, between Bernard and Curie Islands, 66.667°S, 140.017°E (Arnaud, 1972b).—570–600 m, 46.250°S, 60.150°W (Castellanos *et al.,* 1987).—347 m, 72.967°S, 11.117°W. 449 m, 70.817°S, 10.750°W. 401 m, 72.883°S, 19.217°W. 420 m, 72.850°S, 19.333°W. 300 m, 72.467°S, 17.383°W. 481 m, 72.867°S, 19.617°W 267 m, 70.500°S, 8.333°W (Hain, 1989).— 780 m, MB2, 70.487°S, 95.247°W. 1426 m, MB31, 69.949°S, 86.321°W. 438 m, MB33, 70.265°S, 84.191°W. 603 m, MB34, 70.136°S, 84.861°W. 1117 m, MB35, 69.933°S, 85.188°W. 1324 m, MB38, 69.234°S, 81.020°W. 157 m, PA39, 68.127°S, 69.603°W. 350 m, PA41, 65.471°S, 69.029°W. 1272 m, PA42, 65.166°S, 68.936°W (Aldea & Troncoso, 2008).

Remarks. The type material of *An. euglypta* could only be examined by light microscopy, and looks conspecific with what has usually been identified as this species. It can also be clearly recognized from the original description by Pelseneer (1903). Designation of a lectotype is unnecessary.

This is the largest of all Anatomidae with >10 mm shell size. A 10.5 mm specimen had a wet weight of 0.23 g (ZSM 20021142).

Anatoma euglypta resembles in general anatomy *An. yaroni, An. sagamiana* (described by Sasaki, 1998 as *Anatoma* sp.), and *Anatoma* sp. of Strasoldo (1991). *Anatoma euglypta* differs from *An. yaroni* by having the postero-basal expansion of tentacles (possibly an epipodial sense organ) in the three epipodial tentacles instead only in the two latter as in *An. yaroni,* and by having three pallial tentacles instead of only one, as reported for *An. yaroni* by Herbert (1986). Five pairs of epipodial tentacles were reported for *An. sagamiana,* while only three pairs are present in *An. euglypta*. Strasoldo (1991) noted an anterior threefold tentacle, one with an epipodial sense organ at its tip; a second threefold group is situated further posterior and, additionally, a single tentacle on the left side is encountered.

The morphology of the radula in *An. euglypta* is very similar to that of *An. lyra* (Geiger, 2003: fig. 16F: as *An. baxteri*), *An. yaroni* (Herbert, 1986: figs 1–2) and *An. sagamiana* (Sasaki, 1998: fig. 43g, h as *Anatoma* sp.), but in the former, lateral tooth 5 is narrower and claw-shaped in outline. The illustration of the radula of *An. euglypta* provided by Numanami & Okutani (1990: fig. 5; refigured in Numanami, 1996: fig. 6D) was imaged in the anterior, worn region; the more rounded tooth morphology is due to wear and is taxonomically insignificant.

Anatoma eximia (Seguenza, 1877)

Figures 701–705

Chresonymy.

Scissurella eximia Seguenza, 1877: 272–273, pl. 16, fig. 32.
Anatoma eximia: Lozouet, 1986: 114.

Figure 701. Original illustration of *Scissurella eximia* Seguenza 1877.

Misidentifications

Anatoma umbilicata: Landau *et al.*, 2003: pl. 1, fig. 4 [is *An. eximia*].
Anatoma aspera: Beck *et al.*, 2006, 44, text-fig. [is *An. eximia*].
(partim) *Anatoma aspera*: Hernández *et al.*, 2011, 62, figs 11K–M [is *An. eximia*].

Type material. Holotype (whereabouts unknown: not in Florence: Bertolosi & Palazzi, 2000), 2.9 × 3.5 mm.

Type locality. Gallina [Calaria, Italy, 38.087°N, 15.679°E] (OD). Pliocene (Lozouet, 1986).

Etymology. Latin for exceptional, uncommon.

Description. Shell medium-sized (to 3.5 mm. Figs 702, 703), trochiform, turreted. Protoconch of 0.75 whorl, sculpture flocculent, apertural varix not connected to embryonic cap, apertural margin sinusoidal. Teleoconch I of 0.5–0.75 whorl, 12–19 axial fine to distinct axial cords, spiral cord in position of selenizone; axial cords occasionally off-set on either side of spiral cord. Teleoconch II of up to three whorls; suture moderately impressed; sutsel at least as wide as selenizone, up to three times as wide in fully grown specimens; sutsel with fine reticulate pattern. Shoulder with many fine to distinct axial cords; first spiral line after approximately 0.2 teleoconch II whorl, 3–10 after one whorl, 6–9 after two whorls, spiral threads running over axial cords; spirals rarely disappearing with growth (Fig. 703C). Base without constriction below selenizone, continuously sloping into umbilicus; sculpture similar to shoulder; approximately 20–30 spirals, as lines or thin cords below selenizone, changing mid-base into spiral steps. Umbilicus moderately narrow, no funiculus. Aperture subquadrate, peristome flared in lower and adumbilical region. Selenizone slightly above periphery, keels moderately low, weak; lunules wider spaced than axial cords; slit open, margins parallel.

Radula (Fig. 704A–D). Rachidian tooth trapezoidal, tip triangular, apical cusp largest, four cusps on each side. Lateral teeth 1–3 similar, apical cusp largest, 4–5 cusps along outer edge of tip. Lateral tooth 4 unknown. Lateral tooth 5 not significantly enlarged, apical cusp largest, irregular cusps on each side (possibly worn). Marginal teeth with triangular cusp, apical cusp largest, approximately five cusps on each side, no food groove. Radular interlock of central field moderate.

Operculum corneous, round, thin, multispiral, nucleus central (Fig. 704E).

Comparisons. *Anatoma aspera* from the Mediterranean and Norway to Angola has stronger axial sculpture on the shoulder than on the base and a wider umbilicus and a much narrower sutsel, particularly on the early whorls.

Anatoma bertheloti from the Canary Islands has a lower shell profile, more numerous axial cords, and a wider umbilicus.

Anatoma micalii from the Mediterranean to Angola has stronger axial sculpture on the shoulder, a more open umbilicus, and a protoconch lacking an apertural varix.

Figure 702. *Anatoma eximia*. 250 m, Malaga, Spain, 36.717°N, 4.417°W (DLG 308, 10). Scale bar shell = 1 mm. Scale bar protoconch = 100 µm.

Figure 703. *Anatoma eximia*. **A.** 764 m, Galicia, off, Spain, 42.751°N, 11.736°W (RMNH 111811, 20). **B.** Canary Islands, Spain, 28.000°N, 16.700°W (FSC, 1). **C.** 587 m, [Funchal Bay], Madeira, Portugal, 32.633°N, 16.886°W (FSC, 1). Scale bars shell = 1 mm. Scale bars protoconch = 100 μm.

Anatoma janusa from the subtropical northeastern Atlantic has stronger axial cords on the shoulder, a fine but distinct reticulate pattern on the base, and additional fine axial lines on the keels of the selenizone between the major axial cords.

Anatoma crispata from the temperate and arctic North Atlantic has an overall lenticular and inflated shell shape and lacks a sutsel in all but the latest teleoconch whorls of fully grown specimens.

Anatoma orbiculata from West Africa has a more rounded, elliptical outline, much stronger axial cords on shoulder and base, and a more narrow sutsel on early whorls (<0.5× vs. >1× width of selenizone).

Anatoma philippiana from the Paleogene of Europe has a more rounded overall shell, a consistently more narrow sutsel, and coarser sculpture on shoulder and base.

Anatoma schanderi from the northeastern Atlantic has a more globular shell, reticulate protoconch sculpture, and finer axial sculpture.

Anatoma schioettei from the northeastern Atlantic has a more lenticular overall shell shape, a higher expansion rate of the whorls, and fewer axial cords.

Anatoma alta from the Caribbean and the subtropical eastern Atlantic Ocean has sutsel less than the width of the selenizone and a very narrow umbilicus.

Anatoma equatoria from the Indo-Malayan Archipelago has a slightly more bulging base below the selenizone and tends to have fewer axials and spirals.

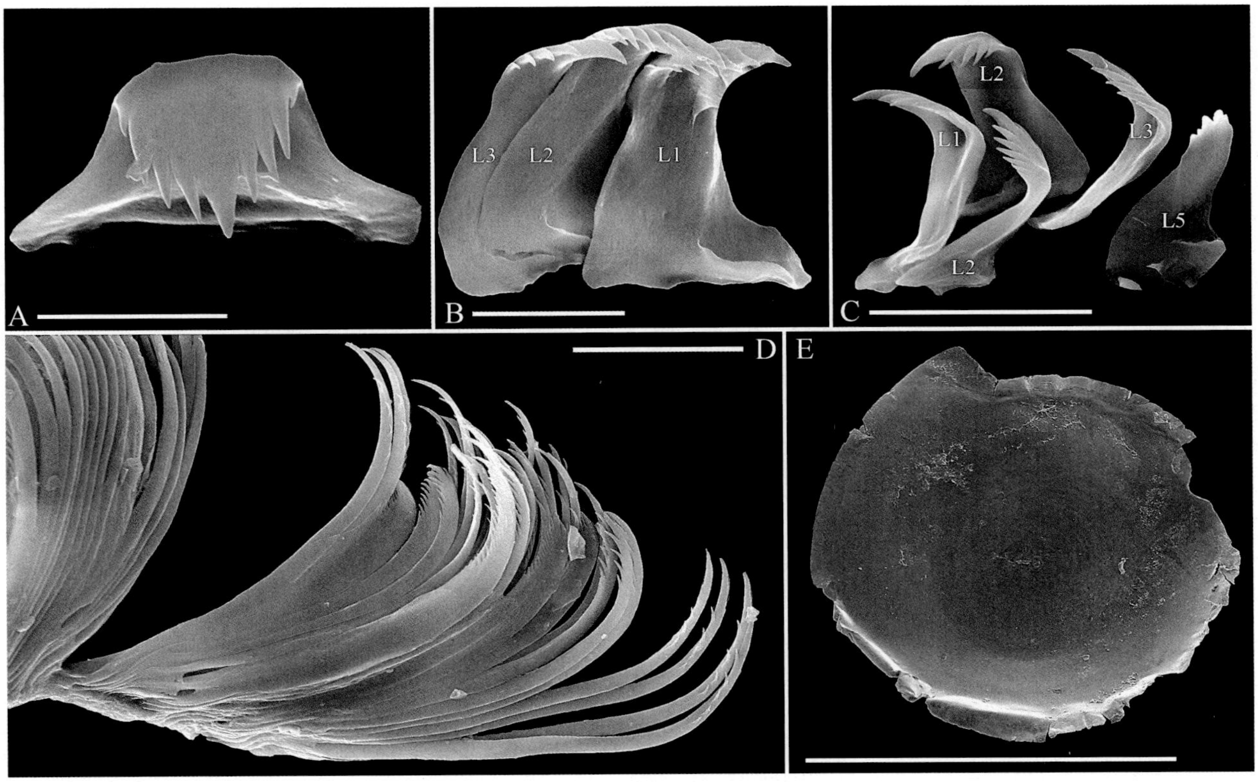

Figure 704. *Anatoma eximia*. 835 m, [155 km W of S Suduroy], Faroe Islands, 61.117°N, 9.500°W (ZMUC, 1). **A–D.** Radula. **A.** Rachidian tooth. Scale bar = 10 µm. **B.** Lateral teeth 1–3. Scale bar = 10 µm. **C.** Lateral teeth 1–3, 5. Scale bar = 20 µm. **D.** Marginal teeth. Scale bar = 20 µm. **E.** Operculum. Scale bar = 1 mm. L1–5: lateral tooth 1–5.

Anatoma aedonia from the Caribbean to Brazil has a longer teleoconch I (0.9 vs. 0.5–0.75 whorl), a slightly wider selenizone, and usually less dense axial cords.

Anatoma munieri from the Indian Ocean to the Central Pacific has a taller shell, a very distinct constriction of the base below the selenizone, and a wider sutsel on early whorls.

Anatoma kopua from temperate eastern Australia to New Zealand has no apertural varix on the protoconch and a slightly shorter teleoconch I (0.3–0.4 vs. 0.5–0.75 whorl).

Anatoma sinuosa from Easter Island has a smooth protoconch, a slightly shorter teleoconch I (<0.5 vs. 0.5–0.75 whorl), and a selenizone that is positioned higher on the whorls.

Distribution. Western Mediterranean, adjacent Atlantic Ocean to Liberia (Fig. 705).

Specimen records.

Faroe Islands. 835 m, [155 km W of S Suduroy], 61.117°N, 9.500°W (ZMUC, 1).
Spain. 250 m, Malaga, 36.717°N, 4.417°W (DLG 308, 10). 764 m, off Galicia, 42.751°N, 11.736°W (RMNH 111811, 20). Expeditie bautismal RV *Taliarte* 12/1998 Station 1, Canary Islands, 28.000°N, 16.700°W (FSC, 1). 480–1500 m, [34 km N of San Sebastian], 43.617°N, 2.133°W (ZMUC, 5).
Portugal, Madeira. 587 m, [Funchal Bay], 32.633°N, 16.886°W (FSC, 1). 587 m, [Funchal Bay], 32.633°N, 16.886°W (FSC, 2). 410–580 m, ZARCO, St. 58, 32.707°N, 16.677°W (MNHN, 2). 410–580 m, Madeira, 32.707°N, 16.677°W (MNHN, 1).
Liberia. 646 m, off, ~5.400°N, ~10.100°W (SIO, 1: complete).

Literature records.

Spain. Canary Islands, NE Grand Canaria, 28.200°N, 15.400°W (Hernández *et al.*, 2011).
France. 170 m, Seine Seamount, 33.833°N, 14.333°W. 110–800 m, Ampère Seamount, 35.064°N, 12.930°W. 750 m, Sedlo Seamount, 40.417°N, 26.917°W (Beck *et al.*, 2006).

Fossil. Pliocene of Italy (Seguenza, 1877). Pliocene of Spain (Landau *et al.*, 2003: as *An. umbilicata*).

Remarks. The type locality is given as "Gallina," of which there are five in Italy. The determination of the Gallina in Reggio Calabria is based on the title of the Sequenza (1877) volume. Type material of about 40% of Seguenza's taxa is extant in the collection of the Museum of Geology and Paleontology of Florence, Italy, but none of the scissurellid taxa are among them (Bertolosi & Palazzi, 2000).

The species described from the type locality is indistinguishable from the Recent specimens here assigned to *An. eximia*. Notable similarities are the overall boxy appearance of the spire, the sculpture of similar strength on shoulder and base, the very extensive sutsel with distinct reticulate pattern, and the very narrow umbilicus. Plio-/Pleistocene species are well-known to be found in the Recent. Seguenza's illustration captures the key characters of the species perfectly.

Figure 705. Distribution of *Anatoma eximia*. ☆ type locality. ● specimen record.

A similar situation is found with *An. tenuisculpta* and *An. aspera*, both species based on fossil material, but customarily applied also to Recent specimens; see Høisaeter & Geiger (2011).

The existence and whereabouts of Seguenza's material is unknown; most likely it is lost or *de facto* inaccessible. While it would be desirable to designate a neotype for taxon stabilization, no such action is taken as the available specimens are neither from the type horizon, nor from anywhere near the type locality.

Anatoma finlayi (Powell, 1937)

Figures 706–710

Chresonymy.

Schizotrochus finlayi Powell, 1937: 176, pl. 49, fig. 2.
Schizotrochus finlayi Powell, 1940: 208.
Anatoma finlayi: Powell, 1979: 36, pl. 14, fig. 4.
Anatoma finlayi: Lozouet, 1986: 113.
Anatoma finlayi: Spencer & Willan, 1995: 13.
Anatoma finlayi: Jansen, 1999:48–49, figs 7–9.
Anatoma finlayi: Geiger, 2003: 73.
Anatoma finlayi: Spencer *et al.*, 2009: 201.

Figure 706. Original illustration of *Anatoma finlayi*.

Synonyms.

+ *Anatoma herberti* Geiger & Sasaki, 2008: 249–252, figs 1–2. ***Type material.*** Holotype (MNHN 20632: Geiger & Sasaki, 2008: fig. 1. Fig. 707). Paratypes: 630–710 m, Reunion Island, MD32, DR62, 21.150°S, 55.200°E (MNHN 20634, 3: Geiger & Sasaki, 2008: fig. 2. Fig. 709A–B). 825–890 m, Reunion Island, MD32, DR63, 21.167°S, 55.150°E (MNHN 20633). ***Type locality.*** 280–340 m, Reunion Island, MD32, DC128, 20.850°S, 55.600°E (OD). ***Etymology.*** The name honors Dai Herbert of the Natal Museum, South Africa, for his detailed works on the South African scissurellid fauna (OD).

Anatoma herberti: Sasaki *et al.*, 2010: 70.

Type material. Holotype (NHMUK 1962953. Fig. 708C), 1.9 × 1.7 mm. One paratype (AIM AK 72219: badly affected by Byne's/glass disease: W. Blom pers. comm. 4/2009).

Type locality. 92 m, off Three Kings Islands, [New Zealand, 34.000°S, 171.917°E] *Discovery* station 934 (OD).

Etymology. Named after J. H. Finlay (1901–1951: see Coan *et al.*, 2007).

Description. Shell medium-sized (to 2 mm), trochiform, globular. Protoconch of 0.75–0.827 whorl, sculpture flocculent, apertural varix not connected to embryonic cap, apertural margin slightly sinusoidal. Teleoconch I of 0.6–0.66 whorl, faint spiral cord in position of selenizone usually indistinct, eroded; approximately 12 raised axial cords, interstices with irregular flocculent sculpture. Teleoconch II to at least 2.33 whorls; suture impressed. Shoulder strongly convex, raised axial cords changing to lamellae with growth; approximately 23–25 on first teleoconch II whorl, approximately 26–29 on last whorl, most lamellae regularly spaced; first spiral line after 0.2–0.6 whorl, increasing in number with growth, initially concentrated in mid-shoulder, approximately 16–17 at apertural margin, spread more or less evenly between suture and selenizone. Base with weak to moderate constriction below selenizone, continuously sloping into umbilicus; axial sculpture changing from cords to low lamellae with growth, always less elevated than axials on shoulder, always decreasing in strength towards umbilicus; spiral lines starting after 0.2 teleoconch II whorl, increasing to approximately 27 in fully grown specimen, equally spaced with tendency to show a gap just below selenizone. Umbilicus open, moderately narrow, funiculus indistinct. Selenizone at periphery; lunules distinct; keels strong, strongly elevated, with fine axial lines on external surfaces; slit open, margins parallel.

Animal unknown.

Comparisons. *Anatoma agulhasensis* from the western Indian Ocean has an overall more rounded shell that lacks the strong axial ribs, has less elevated keels of the selenizone, and the protoconch lacks an apertural varix.

Anatoma aupouria from New Zealand has a flatter overall shell shape, a subquadrate aperture, and at least twice as dense axial cords on the base.

Anatoma flemingi from New Zealand has a protoconch with reticulate sculpture and much weaker axial cords on the base, which are at least twice as dense.

Anatoma gunteri from southeastern Australia has a protoconch with a honeycomb pattern and much reduced sculpture on the base.

Anatoma indonesica from the Indo-Malayan Archipelago to the Western Pacific has much weaker sculpture on the base, which is smooth immediately below the selenizone.

Anatoma lamellata from the Western to the Central Pacific lacks an apertural varix on the protoconch, has less raised axial cords, and lacks the supernumerary axial lines on the outer surface of the keels of the selenizone.

Anatoma munieri from the Indian Ocean to the Central Pacific lacks an apertural varix on the protoconch, has an overall more turreted shell, and lacks the strong axial ribs as well as the supernumerary axial lines on the outer surface of the keels of the selenizone.

Anatoma kopua from temperate eastern Australia to New Zealand has no axial lamellae on shoulder or base, a flatter shoulder, and at most half as many spiral lines on shoulder or base.

Anatoma regia from New Zealand has a more discoidal overall shell shape, lacks lamellae on shoulder or base, and lacks or has a very slim sutsel.

Anatoma yaroni from the western Indian Ocean has biconical overall shell morphology, lacks the raised axial ribs, and has smooth external surfaces of the keels of the selenizone.

Anatoma alternatisculpta from the Caribbean has less raised axial lamellae on the shoulder, axial cords on the base, and fewer spiral lines on shoulder and base.

Anatoma emilioi from the Central Pacific has at most half as many axial lamellae, very strong and inturned keels of the selenizone, and fewer axial lines on the base.

Anatoma janusa from the subtropical northeastern Atlantic has weaker axial lamellae on the shoulder and axial cords on the base that are at least twice as dense.

Anatoma weddelliana from Antarctica has a greater expansion rate of the whorls, less elevated axial lamellae on shoulder and base, and a more narrow sutsel on early whorls (<0.5× vs. 1× width of selenizone).

Distribution. India Ocean to Central Pacific (and possibly Caribbean. Fig. 710).

Specimen records.

Glorioso Islands. 440 m, western, 11.475°S, 47.295°E (MNHN, 1).

Madagascar. 284–286 m, S Cape St. Marie, ATIMO VATAE CP3615, 26.233°S, 45.150°E (MNHN, 2). 225–282 m, S of Cape St. Marie, ATIMO VATAE CP3613, 26.217°S, 45.133°E (MNHN, 1). 280–333 m, S of False Cape, ATIMO VATAE DW3553, 26.133°S, 45.650°E (MNHN, 1). 264–280 m, S of False Cape, ATIMO VATAE DW3552, 26.117°S, 45.650°E (MNHN, 1). 209–229 m, S Barrow Point, ATIMO VATAE DW3581, 25.500°S, 44.267°E (MNHN, 1). 203–210 m, SE Barrow Point, ATIMO VATAE CP3584, 25.467°S, 44.417°E (MNHN, 1). 184–203 m, between Lokaro and St. Luce, ATIMO VATAE DW3515, 24.883°S, 47.467°E (MNHN, 4). 296–307 m, St. Luce sector, ATIMO VATAE DW3534, 24.717°S, 47.533°E (MNHN, 1). 307–319 m, Manantenina sector, ATIMO VATAE DW3524, 24.383°S, 47.533°E (MNHN, 2). 424–438 m, Manantenina sector, DW3528, 24.400°S, 47.550°E (MNHN, 1).

Reunion. 825–890 m, 21.167°S, 55.150°E (MNHN, 1).

Figure 707. *Anatoma finlayi*. Holotype of *Anatoma herberti* (MNHN 20632). 280–340 m, Reunion Island, 20.850°S, 55.600°E. Scale bar shell = 1 mm. Scale bar protoconch = 100 µm.

Figure 708. *Anatoma finlayi.* **A.** 805 m, Three Kings Trough, NW of Three Kings Islands, New Zealand, 37.000°S, 171.917°E (NMNZ M.049027, 30). **B.** 246–291 m, Middlesex Bank, NW of Three Kings Islands, New Zealand, 34.033°S, 171.733°E (NMNZ M.160264, 25). **C.** Holotype. 92 m, off Three Kings Islands, [New Zealand, 34.000°S, 171.917°E] *Discovery* station 934 (NHMUK 1962953). Scale bars shell = 1 mm. Scale bars protoconch = 100 μm.

Indonesia. 175 m, SW Salayer, 6.373°S, 120.437°E (RMNH, 10).

Philippines. 903 m, off Magabao Island, E Mindanao, 8.984°N, 126.264°E (USNM 1098753, 1). 427 m, Bohol-Sulu Seas Sill, PANGLAO 2005 DW2364, 9.012°N, 123.425°E (MNHN, 2).

Figure 709. *Anatoma finlayi*. **A–B.** Paratypes of *Anatoma herberti* (MNHN 20634). 630–710 m, Reunion Island, 21.150°S, 55.200°E. **C.** 175 m, SW Salayer, Indonesia, 6.373°S, 120.437°E (RMNH, 10). Scale bars shell = 1 mm. Scale bars protoconch = 100 μm.

Australia, Western Australia. 201–214 m, W of Rottnest Island, 31.992°S, 115.217°E (AMS C.402755, 1). 135 m, W of Guilderton, 31.567°S, 115.050°E (AMS C.402757, 1). 320–283 m, W of Rottnest Island, 31.000°S, 114.850°E (AMS C.402756, 8). 256–192 m, NW of Cervantes, 30.500°S, 114.633°E (AMS C.402758, 3). 223–245 m, off Jurien Bay, 30.133°S, 114.500°E (AMS C.402661, 8).

New Caledonia. 200–291 m, Norfolk Ridge, P Bank, 23.283°S, 168.233°E (MNHN, 20). 266–267 m, Norfolk Ridge, P Bank, 23.300°S, 168.250°E (MNHN, 6). 433–450 m, Stylaster Bank, 23.633°S, 167.717°E (MNHN, 1). 750–760 m, Loyalty Ridge, 24.717°S, 170.117°E (MNHN, 1). 770–830 m, Loyalty Ridge, 24.733°S, 170.133°E (MNHN, 1). 381–469 m, Norfolk Ridge, 23.367°S, 168.017°E (MNHN, 1). 569–616 m, Norfolk Ridge, Banc Introuvable, 24.650°S, 168.633°E (MNHN, 1). 250–350 m, southern, 22.500°S, 166.400°E (MNHN, 2). 1820–1980 m, Loyalty Basin, 21.000°S, 166.967°E (MNHN, 1). 435 m, 20.567°S, 166.900°E (MNHN, 4). 538 m, between Lifou and Uvea, 20.617°S, 166.967°E (MNHN, 1).900–950 m, Jumeau East Bank, Norfolk Ridge, NORFOLK 2 DW2055, 23.650°S, 168.267°E (MNHN, 2). 250–350 m, south, St. 40, 22.500°S, 166.400°E (MNHN, 1).

New Zealand. 805 m, Three Kings Trough, NW of Three Kings Islands, 37.000°S, 171.917°E (NMNZ M.049027, 30). 785–800 m, W of Cape Reinga, W Norfolk Ridge, 34.298°S, 168.430°E (NMNZ ex M.172126, 15). 92 m, off Three Kings Islands, *Discovery* station 934 (NHMUK 1962953: holotype).

Fiji. 820–863 m, BORDAU 1 DW1459, 17.300°S, 179.550°W (MNHN, 2).

Tonga. 440–487 m, Vava'u group, BORDAU 2 DW1586, 18.567°S, 173.917°W (MNHN, 1). 260 m, SE of Tongatapu, 21.345°S, 175.042°W (AMS C.402677, 1).

French Polynesia. 700–785 m, Huahine, Society Islands, TARASOC DW3434, 16.700°S, 151.050°W (MNHN, 1).

Remarks. There is one lot of this species with collection data from the Caribbean: 198 m, [off Cozumel, Mexico], 20.917°N, 86.467°W (UMML 30.12045, 2). Those specimens are indistinguishable from the Pacific specimens when examined by SEM. Whether this is a case of conchologically indistinguishable sibling species, or whether those Caribbean specimens have been mislocalized remains to be determined.

In specimens with less than 1.5 teleoconch II whorls, the base shows very fine axial cords similar to those seen in *An. indonesica*. It may not be possible to identify those small specimens from Indo-Pacific localities accurately as either *An. indonesica* or *An. finlayi*. There are no known specimens of *An. indonesica* from the Indian Ocean. Larger specimens can be differentiated consistently for which reason the similarities of young specimens is not taken to indicate synonymy.

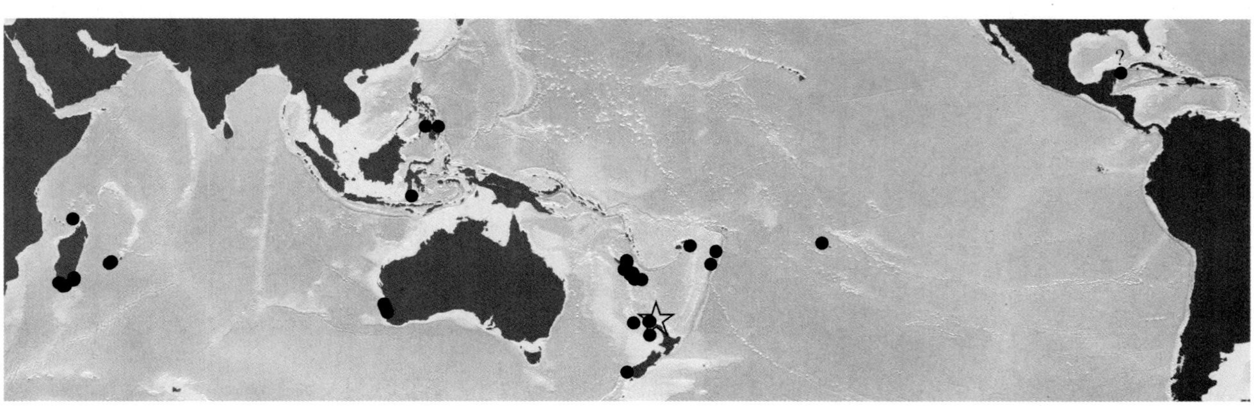

Figure 710. Distribution of *Anatoma finlayi*. ☆ type locality. ● specimen record.

Anatoma herberti is here synonymized under *An. finlayi*. The previously cited difference of relatively weaker axial cords on the base, compared to the shoulder, is no longer tenable after examining larger series of the species. The geographic separation of the type localities, Reunion Island and New Zealand, has also been bridged by additional specimens (Fig. 710). Although New Zealand is typically associated with a temperate fauna, the type locality of *An. finlayi*, Three Kings Islands off northernmost New Zealand has been the source of many species generally of tropical Indo-Pacific distribution.

Anatoma flemingi (Marshall, 2002)

Figures 711–715

Chresonymy.

Thieleella flemingi Marshall, 2002: 177–179, figs 22–24.
Thieleella flemingi: Geiger, 2003: 74.
Thieleella flemingi: Spencer *et al.*, 2009: 201.
Thieleella flemingi: Raven & Bracegirdle, 2010: 10, text-fig.

Figure 711. Original illustration of *Thieleella flemingi*.

Misidentifications.

Scissurella mantelli: Pilsbry, 1890: 54, pl. 57, fig. 12 [is *An. flemingi*].
Scissurella mantelli: Thiele, 1912a: 20, pl. 3, fig. 5 [is *An. flemingi*].
Scissurella mantelli: Suter, 1902: 217 [is *An. flemingi*].
Scissurella mantelli: Suter, 1913: 88–89, pl. 6, fig. 10 [is *An. flemingi*].
Schizotrochus mantelli: Fleming, 1948: 84, pl. 8, fig. 3 [is *An. flemingi*].
Schizotrochus mantelli: Powell, 1940: 208 [is *An. flemingi*].
Schizotrochus mantelli: Fleming, 1950: 25 [is *An. flemingi*].
Schizotrochus mantelli: Fleming, 1951: 136 [is *An. flemingi*].
Schizotrochus mantelli: Dell, 1956: 168, 182 [is *An. flemingi*].
Anatoma mantelli: Powell, 1979: 36, pl. 14, fig. 1 [is *An. flemingi*].
Anatoma mantelli: Spencer & Willan, 1995: 13 [is *An. flemingi*].

Type material. Holotype (GNS TM 426. Fig. 711) = "neotype" of *Schizotrochus mantelli* sensu Fleming, 1951 (cf. Marshall, 2002). The neotype designation for *Sci. mantelli* is based on an accidentally transposed original figure caption and is invalid according to ICZN Art. 75.3.5.

Type locality. Daggs Sound, inside a rock bar (= moraine), Fiordland, New Zealand [45.383°S, 166.767°E] 106 m (OD).

Etymology. Named after Sir Charles Fleming (OD).

Description. Shell large (to 3.7 mm. Figs 711–713), globular, inflated. Protoconch of 0.75 whorl, reticulate sculpture, apertural varix barely connected to embryonic cap, apertural varix sinusoidal. Teleoconch I of 0.75 whorl with approximately 20 axials, single spiral in position of selenizone. Teleoconch II of over two whorls, suture moderately impressed; sutsel 1–1.5× width of selenizone. Shoulder with approximately 41 colabral axials, 13 equispaced fine spirals. Base with finer axials at approximately

Figure 712. *Anatoma flemingi*. 785–800 m, W of Cape Reinga, West Norfolk Ridge, 34.298°S, 168.430°E (NMNZ M.172126, 150). Scale bar shell = 1 mm. Scale bar protoconch = 100 µm.

double density compared to shoulder, crossed by approximately 17 spiral lines of equal strength forming fine reticulate pattern. Selenizone slightly above periphery; lunules fine, regularly spaced; keels distinct, sharp, axial density on upper and lower surface equal to that on base of shell; slit open, margins parallel. Umbilicus open, deep, funiculus present. Aperture round, inner lip slightly flared.

Figure 713. *Anatoma flemingi*. **A.** Holotype. Daggs Sound, inside a rock bar (i.e., moraine), Fiordland, New Zealand [45.383°S, 166.767°E] 106 m (GNS TM 426). **B–C.** 415 m, New Zealand, 45.288°S, 167.043°E (NMNZ M.138409, 10). Scale bars shell = 1 mm. Scale bars protoconch = 100 µm.

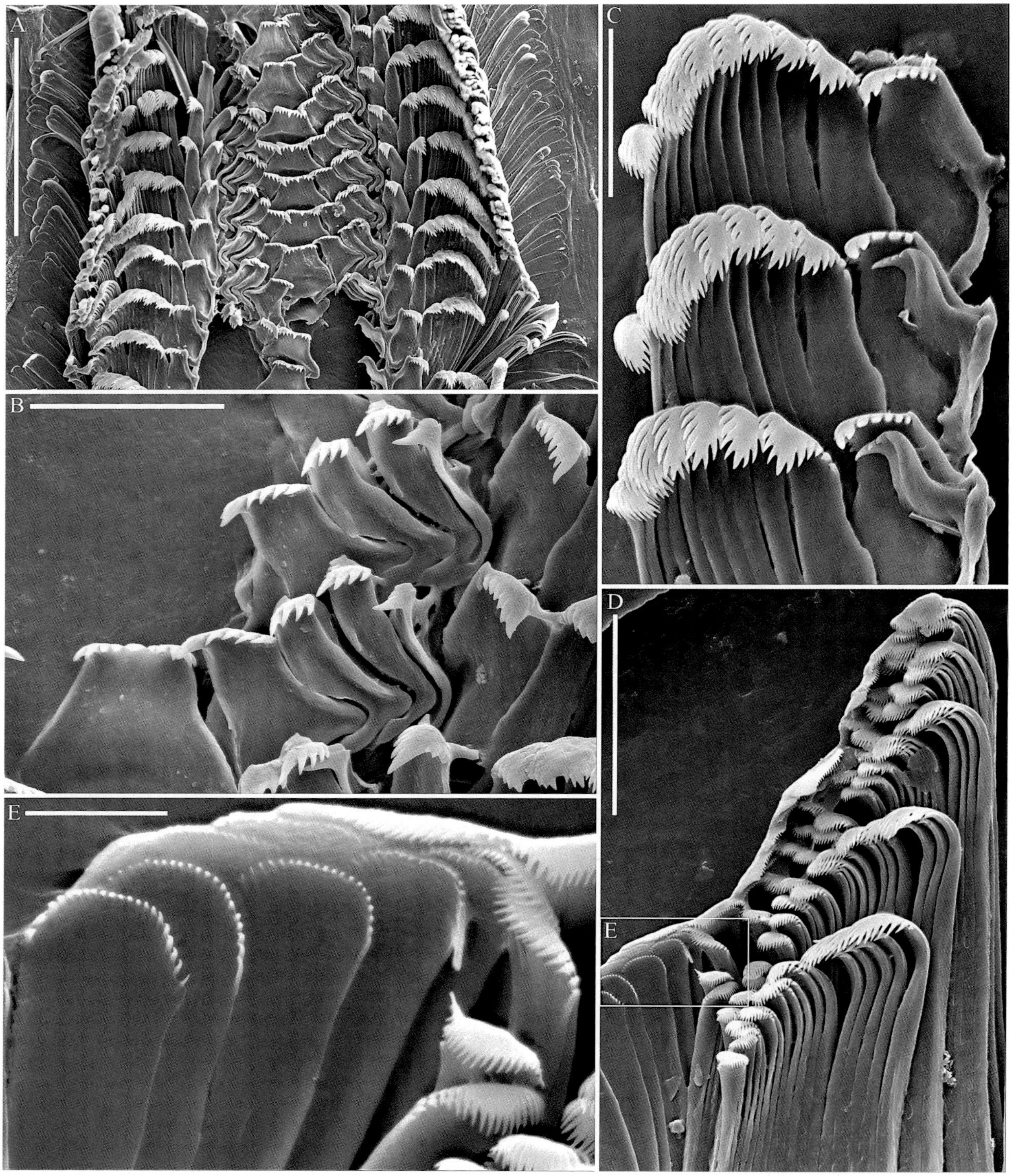

Figure 714. *Anatoma flemingi*. Radula. 340–620 m, floor and right (south) wall of upper Papanui, 20 km off Taiaroa Head, Dunedin, Otago, South Island, New Zealand, 45.861°S, 171.014°E (DLG 98, 10). **A.** Full width of radula. Scale bar = 100 µm. **B.** Central field enlarged. **C.** Lateral tooth 4 (bottom two rows) or 5 (top row) to middle marginal teeth. **D.** Middle to outer marginal teeth. Frame shows location of panel E. Scale bars B–D = 25 µm. **E.** Outer marginal teeth. Scale bar = 5 µm.

Radula (Fig. 714). Rachidian tooth trapezoidal, tip with central cusp largest, 4–5 somewhat smaller cusps on each side, parallel to radular row. Lateral teeth 1–3 similar, 4–5 cusps on outer margin of tip, terminal cusp on tip largest. Lateral tooth 4 reduced, pointed, with three cusps on inner margin, smooth outer margin. Lateral tooth 5 with enlarged base, tip with terminal cusp largest, approximately seven cusps on inner margin, 1–2 cusps on outer margin. Little radular interlock between rachidian tooth and lateral tooth 1, interlock between lateral teeth moderate. Inner marginal teeth with elongated base, triangular to spoon-shaped tip, terminal cusps somewhat larger than remainder, margins with six cusps each; shafts without food groove. Outer marginal teeth with spoon-shaped tip, many small cusps.

Comparisons. *Anatoma aupouria* from New Zealand is less inflated, has a conspicuous pattern of axials on the base that imparts the impression of spiral bands, and has a protoconch of flocculent sculpture.

Anatoma finlayi from the Indian Ocean to the Central Pacific (and possibly Caribbean) has stronger axial lamellae on shoulder and base and has 3–4 additional fine spiral lines on the keels of the selenizone between each axial cord.

Anatoma regia from New Zealand has a protoconch with flocculent sculpture and lacks the spiral cord on teleoconch I.

Anatoma alternatisculpta from the Caribbean has weaker sculpture on the base, has fewer spiral lines on the shoulder, and spiral lines on the base that fade out towards the umbilicus.

Anatoma australis from southeastern Australia has fewer axial cords, of which only every 3–4 reach from the selenizone to the umbilicus, and has additional spiral lines on teleoconch I between the spiral in position of the selenizone and the suture below.

Anatoma conica from subantarctic South America has a protoconch without apertural varix, has at least 50% fewer axial cords on the base, and has about 30–50% more but finer spiral lines on the shoulder.

Anatoma gunteri from southeastern Australia has stronger raised axial cords on the shoulder and about 50% fewer spiral lines on the shoulder.

Anatoma japonica from the Indian Ocean to the Central Pacific has twice as fine and dense axial cords on shoulder and base, and has a protoconch with flocculent sculpture.

Anatoma lamellata from the Western to the Central Pacific has at least twice as strong and widely spaced axial cords and has a protoconch with flocculent sculpture and additionally a spiral cord.

Anatoma tobeyoides from southeastern Australia has a protoconch with flocculent sculpture, very fine oblique lines on teleoconch I, and weaker axial cords.

Anatoma orbiculata from West Africa has much stronger spiral lines on shoulder and base in comparison to the axial cords and usually no axial line on the keels of the selenizone in addition to those on the shoulder.

Distribution. New Zealand, living 400–1006 m (Fig. 715).

Specimen records.

New Zealand. 376 m, Kellard Basin floor, 45.355°S, 167.056°E (NMNZ M.138598, 1). 415 m, 45.288°S, 167.043°E (NMNZ M.138409, 10). 400 m, off Kaikoura, Conway Rise, South Island, 42.570°S, 173.870°E (NMNZ M.150211, 1: complete). 415 m, Fiordland, Bradshaw Sound, Bradshaw Basin floor, 45.288°S, 167.043°E (NMNZ exM.138425, 3). 785–800 m, W of Cape Reinga, W Norfolk Ridge, 34.298°S, 168.430°E (NMNZ M.172126, 150). 840–850 m, S Cavalli Seamount, ENE of

North Cape, 34.266°S, 174.103°E (NMNZ M.174860, 9). 106 m, inside rock bar, Daggs Sound, Fiordland, SW South Island, 45.311°S, 166.976°E (GNS RM 5948, 1; GNS TM 8659, 1; GNS TM 426, 1). 35–44 m, Narrows, Long Sound, Preservation inlet, Fiordland, South Island, 46.065°S, 166.686°E (GNS RM 2212, 1). 540–490 m, Papanui Canyon, off Cape Saunders, E Otago, 45.833°S, 171.017°E (GNS RM 5949, 2). 991–1375 m, Barbara Thomas Pinnacle, Auckland Islands, 50.073°S, 165.898°E (NMNZ M.118901, 1). 0 m, Okains Bay, Banks Peninsula, South Island, 43.695°S, 173.072°E (DLG 1601, 3).

Literature records.

New Zealand. 19–22 fathoms [= 35–40 m], Narrows, Long Sound, Preservation Inlet (Fleming, 1948).—238 m, E of Forty Fours, Chatham Islands, 44.067°S, 175.392°W. 219–229 m, E of Iaingaroa, Chatham Island, 43.583°S, 176.058°W (Dell, 1956).—640 m, S of Kaikoura Peninsula, 42.583°S, 173.683°E. 400 m, wall of Pegasus Canyon NE of Banks Peninsula, 43.233°S, 173.650°E. Conway Rise, off Kaikoura, 42.417°S, 173.683°E. 350 m, Thompson Basin Floor, Thompson Sound, 45.217°S, 166.966°E. 350 m, 45.218°S, 166.967°E. 319–324 m, 45.239°S, 166.983°E. 340–362 m, 45.240°S, 166.981°E. 417 m, Bradshaw Basin Floor, Bradshaw Sound, 45.288°S, 167.041°E. 400 m, Doubtful Sound, Utah Basin Floor, 45.298°S, 166.925°E. 54–60 m, Crooked Arm entrance sill, 45.357°S, 167.030°E. 35–40 m, Long Sound Narrows, Preservation Inlet, 46.067°S, 166.750°E. 585–530 m, Head of Karitane Canyon, NE of Taiaroa Head, 45.642°S, 171.083°E. 520–600 m, off E Otago, 45.750°S, 171.033°E. 450–476 m, off Antipodes Islands, 49.667°S, 178.883°E. (Marshall, 2002).

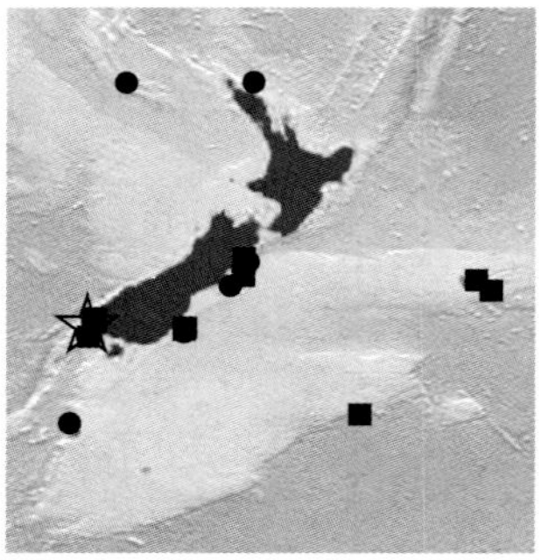

Figure 715. Distribution of *Anatoma flemingi*. ☆ type locality. ● specimen record. ■ literature record.

Remarks. The identity of *Scissurella mantelli* of Woodward has been mistaken because the captions for Woodward's (1859) figures 8 [*Scissurella mantelli*] and 9 [*Scissurella (Anatoma) crispata*] on plate 46 were transposed. Figure 8 shows *Anatoma crispata* [in fact, misidentified *An. aspera*], while figure 9 shows an unidentified *Scissurella* that cannot be matched to any species from New Zealand. As the *Anatoma* specimen illustrated by Woodward was labelled *Scissurella mantelli*, it was generally assumed that *mantelli* refers to some New Zealand *Anatoma*. Marshall (2002) uncovered this long-standing error and described the species properly for the first time. Further confusion relating to this illustration is discussed in Remarks of genus *Anatoma* in this volume.

Fleming (1951), however, designated a neotype for *Anatoma mantelli*, conforming with the generally accepted species concept of that taxon, i.e., an *Anatoma* species from New Zealand. Marshall (2002) disregarded the neotype designation by Fleming (1951) in agreement with ICZN Art. 75.3.5. because the neotype designation was clearly made in error, and introduced the new name *An. flemingi* for that species. To ensure stability of nomenclature, Marshall (2002) designated Fleming's (1951) "neotype" as the holotype of *An. flemingi*.

A few specimens (DLG 1601) have very weak axial cords, but the axial cords are still stronger than the spiral lines, and the constriction on the base below the selenizone permits identification. Small specimens are wider than tall, whereas large specimens may be taller than wide.

Anatoma flexidentata Geiger & Sasaki, 2008

Figures 716–720

Chresonymy.

Anatoma flexidentata Geiger & Sasaki, 2008: 257–261, figs 5–7.

Type material. Holotype (MNHN 20636: Geiger & Sasaki, 2008: fig. 6. Fig. 716). Paratypes: from type locality (MNHN 20637, 16: Geiger & Sasaki, 2008: fig. 7A. Fig. 717A). 280–340 m, Reunion Island, MD32, DC128, 20.850°S, 55.600°E (MNHN 20638, 2: Geiger & Sasaki, 2008: fig. 7B–C. Fig. 717B–C).

Type locality. 3300–3240 m, MD32, DS151, Reunion, 20.850°S, 56.050°E (OD: see Remarks).

Etymology. Flexi- Latin for flexible, pliable, -dentata, Latin for tooth-bearing, referring to the pliable nature of the radular ribbon and its teeth (OD).

Figure 716. *Anatoma flexidentata*. Holotype. 3300–3240 m, MD32, DS151, Reunion, 20.850°N, 56.050°E (MNHN 20636). Scale bar shell = 1 mm. Scale bar protoconch = 100 μm.

Description. Shell medium-sized (to 2.9 mm. Figs 716–718), trochiform, globular. Protoconch of 0.75 whorl, sculpture flocculent, no apertural varix, apertural margin straight. Teleoconch I of 0.25–0.33 whorl, 7–9 fine axial cords, usually with fine spiral cord

Figure 717. *Anatoma flexidentata*. Paratypes. **A.** 3300–3240 m, MD32, DS151, Reunion, 20.850°N, 56.050°E (MNHN 20637). **B–C.** 280–340 m, MD32 DC128, Reunion Island, 20.850°S, 55.600°E (MNHN 20638). Scale bars shell = 1 mm. Scale bars protoconch = 100 µm.

in position of selenizone (some may be eroded). Teleoconch II of three whorls, suture moderately impressed, sutsel on first two whorls approximately 0.5× width of selenizone, at apertural margin 1.5× width of selenizone. Shoulder moderately

Figure 718. *Anatoma flexidentata*. Indo-Pacific specimens. **A.** 1360 m, Vanuatu, 14.817°S, 167.250°E (MNHN, 3). **B.** 640–687 m, Bligh Water, Fiji, 16.937°S, 178.132°E (MNHN, 2). **C.** 1490–1620 m, New Caledonia, 23.733°S, 166.967°E (MNHN, 1). Scale bars shell = 1 mm. Scale bars protoconch = 100 μm.

convex; approximately 42 axial cords, a few not reaching suture; first fine spiral line after 0.3–1 teleoconch II whorl, increasing to 3–6 irregularly spaced spirals at apertural margin, occasionally secondarily decreasing in number. Base barely constricted below selenizone, continuously sloping into rather narrow umbilicus without funiculus (proportionally wider in small specimens. Fig. 717B), reticulate sculpture with approximately 50 axial cords of even strength from selenizone to umbilicus on last half whorl, crossed by 12–17 spiral cords, forming minute points at intersections. Selenizone at periphery, relatively wide; keels low, moderately strong; lunules distinct, evenly spaced; slit open, with parallel margins. Aperture rounded, parietal wall straight, with overhanging roof, peristome somewhat flared.

Operculum (Fig. 719E, F) thin, corneous, multispiral, with central nucleus.

Radula (Fig. 719) flexoglossate. Rachidian tooth triangular, twice as high as wide, approximately eight very small cusps on each side. Lateral teeth 1–4 similar, triangular, 3–4× as high as wide. Lateral tooth 5 enlarged by elongation, approximately 10× as long as wide, without noticeable cusps. Marginal teeth, elongated, curved, with 2–5 small barbs; shafts without food groove. Radular interlock minimal.

Jaw (Fig. 719D) triangular, composed of rhomboidal plates; hole in center artifact of handling.

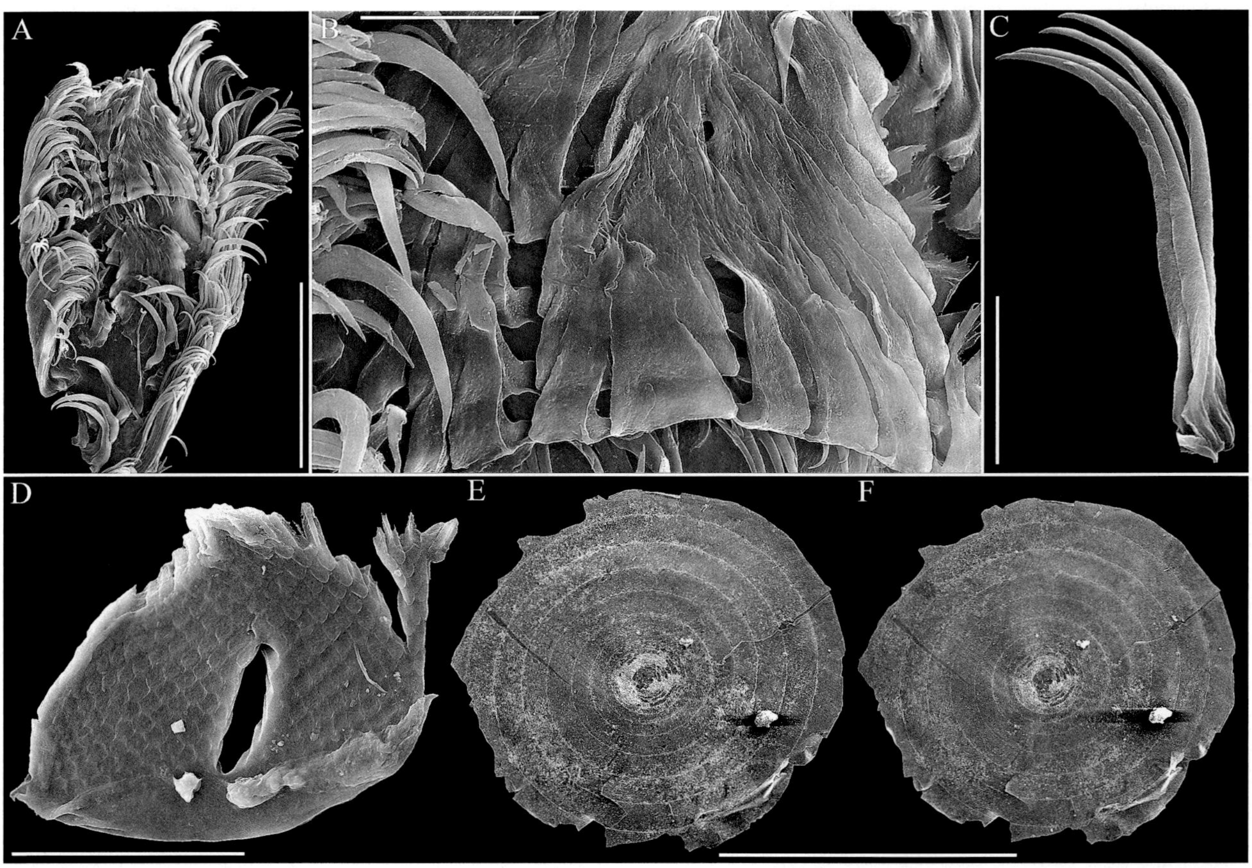

Figure 719. *Anatoma flexidentata.* Radula, jaw and operculum. 280–340 m, Reunion Island, 20.850°S, 55.600°E (MNHN 20638). **A.** Overview of radula. Scale bar = 100 µm. **B.** Central field of radula. Scale bar = 20 µm. **C.** Isolated marginal teeth. Scale bar = 20 µm. **D.** Jaw. Scale bar = 100 µm. **E–F.** Operculum. Scale bar = 1 mm. **E.** Operculum imaged at 10 kV. **D.** Operculum imaged at 30 kV.

Comparisons. *Anatoma flexidentata* is characterized in terms of shell morphology by the globular overall shape, the short teleoconch I (0.25–0.33 whorl), wide selenizone, and the fine reticulate sculpture.

Anatoma agulhasensis form the Indian Ocean has an overall more turreted shell with a distinct constriction below the selenizone, a longer teleoconch I (0.75 whorl), and the axial cords on the shoulder are at least three times as strong as the spiral lines.

Anatoma boucheti from the western Indian Ocean has a longer teleoconch (1 vs. 0.3 whorl), a depressed apex, and a much wider sutsel after one teleoconch II whorl (2× vs. 1× width of selenizone).

Anatoma biconica from the Indian Ocean to the Central Pacific has a longer teleoconch I (~1 vs. 0.3) with distinct onset of selenizone and a typical anatomid radula.

Anatoma maxima from the Indo-Malayan Archipelago to the Central Pacific has a longer teleoconch I (0.9 vs. 0.3 whorl), a smooth early teleoconch II, and more pronounced sculpture on teleoconch II.

Anatoma rhynchodentata from the Indo-Malayan Archipelago to the Western Pacific has a longer teleoconch I (0.6 vs. 0.3 whorl), at least twice as many axial and spiral lines that form minute points at their intersections, and usually a constriction on the base below the selenizone.

Anatoma zancliformis from the Indo-Malayan Archipelago to the Western Pacific has twice as many spiral lines on the shoulder and a radula with typical central field (not flexoglossate).

Anatoma tabulata from the western Indian Ocean has a longer teleoconch I (1 vs. 0.3 whorl), a sutsel as wide as the selenizone on the early whorls, and a flatter shoulder.

Anatoma munieri from the Indian Ocean to the Central Pacific is much more turreted with a very distinct constriction below the selenizone, has a protoconch with an apertural varix, a teleoconch I of >0.5 whorl, and the overall sculpture is stronger with the axial cords on the shoulder at least three times as strong as the spiral lines.

Anatoma kopua from temperate eastern Australia to New Zealand has a more biconical shell shape, a distinct constriction below the selenizone, about 50% denser spiral lines on the base, and a more narrow sutsel after 1–2 whorls (< vs. > width of selenizone).

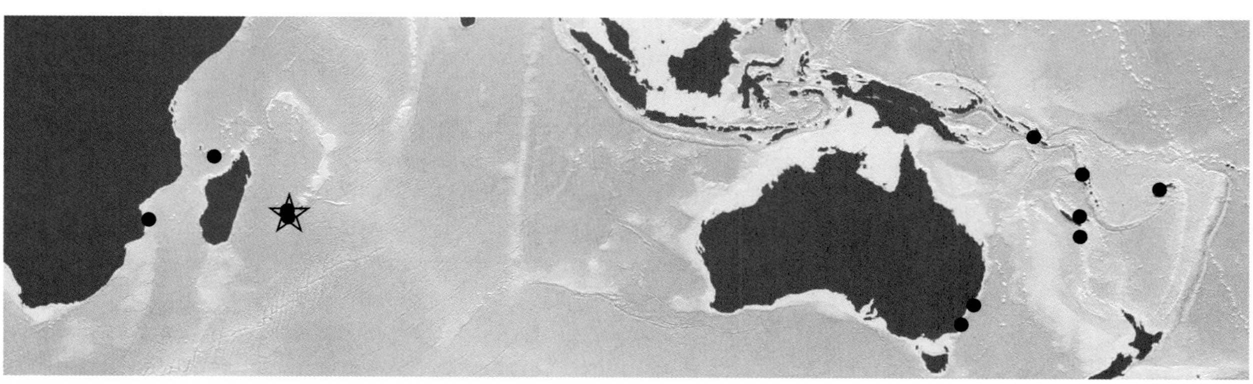

Figure 720. Distribution of *Anatoma flexidentata*. ☆ type locality. ● specimen record.

Anatoma japonica from the Indian Ocean to the Central Pacific has an overall biconical shape, a teleoconch I of >0.5 whorl, and stronger sculpture with the axial cords on the shoulder at least three times as strong as the spiral lines.

Anatoma conica from the subantarctic of South America has a longer teleoconch I (0.6 vs. 0.3 whorl), raised axial cords on the shoulder and base, and more spiral lines on the base.

Anatoma kelseyi from the northeastern Pacific has a protoconch with flocculent sculpture, a longer teleoconch I (0.75 vs. 0.3 whorl), and at least twice as many spiral lines on shoulder and base.

Anatoma proxima from the Caribbean to Brazil has a more narrow selenizone located more towards the middle of the whorl, a longer teleoconch I (0.75 vs. 0.3 whorl), and at least twice as many spiral lines on shoulder and base.

Anatoma sagamiana from Japan and the Indo-Malayan Archipelago has reticulate protoconch sculpture, a longer teleoconch I (0.75 vs. 0.3 whorl), and at least twice as many spiral lines on shoulder and base.

Distribution. Indian Ocean to Western Pacific, 280–1620 m (Fig. 720).

Specimens records.

Reunion Island. 310 m, ND32, DC26, 21.367°S, 55.783°E (MNHN, 2). 705 m, NE Pamanzi Island, Mayotte, France, Bethendi 1977, St. 28, 12.705°S, 45.332°E (MNHN, 1). 310 m, 21.367°S, 55.783°E (MNHN, 2). 3500–3510 m, MD32, DS149, 20.433°S, 55.667°E (MNHN, 52: complete).

Mozambique. 1220–1248 m, Bazaruto transect, Mozambique Channel, MAINBAZA CC3158, 21.767°S, 36.200°E (MNHN, 1).

Australia. 354–384 m, E of Bermagui, New South Wales, 36.400°S, 150.333°E (AMS C.448770, 1). 1143–1106 m, off Sydney, New South Wales, 33.600°S, 152.083°E (AMS ex C.448769, 1).

New Caledonia. 1490–1620 m, 23.733°S, 166.967°E. 1620 m, Loyalty Basin, 20.817°S, 166.900°E (MNHN, 1).

Solomon Islands. 264–273 m, 9.417°S, 160.432°E (MNHN, 1).

Vanuatu. 1360 m, 14.817°S, 167.250°E (MNHN, 3).

Fiji. 640–687 m, Bligh Water, 16.937°S, 178.132°E (MNHN, 2).

Remarks. The latitude of the type locality was erroneously given as "20.850°N," instead of 20.850°S, a clear *lapsus calami.*

The shell of *An. flexidentata* is unremarkable amongst other anatomids. The radula, however, is at significant variance to those known from most other species; its flexiglossate nature is only shared with *An. australis*. The very flexible and limp radula folded over very easily during preparation for SEM. The rachidian tooth is not trapezoidal with distinct recurved tip, but is triangular with fine bristles. Lateral teeth 1–4 are similar to one another as in Larocheidae, but unlike all other Anatomidae, where lateral teeth 1–3 are similar, and lateral tooth 4 is reduced, hook-shaped. Lateral tooth 5 shows the usual enlargement by elongation, but lacks distinct cusps on the tip, similar to *An. zancliformis* with a fully formed central field. The inner marginal teeth are more elongated, while the outer marginal teeth do not show the typical spoon-shaped tip. The possibility of radular contamination or mix up of specimens was considered; however two specimens showed identical radular morphology. The very flexible overall nature of the radula might suggest that only the still developing, posterior portion of the radula was examined, which may account for the unfamiliar morphology. However, lateral tooth 5 as well as the marginal teeth show the usual degree of sturdiness. Examinations of early portions of radulae from other anatomids never showed the morphology of *An. flexidentata*, but rather the mature tooth morphology but with limp teeth across the entire radular row.

The morphological similarity of the shell to *An. zancliformis*, particularly the rather uncommonly short teleoconch I, and shared sickle-shaped lateral tooth 5 have to be noted. The radular differences in the central field suggest that they are distinct, which is supported by a subtle shell morphological variance, the number of spiral lines on the shoulder.

Anatoma fujikurai Sasaki, Geiger & Okutani, 2010

Figures 721–725

Chresonymy.

Anatoma s.l. *fujikurai* Sasaki, Geiger & Okutani, 2010: 64–71, figs 1–4.

Type material. Holotype 3.2 × 3.1 mm (UMUT RM29549. Fig. 721). Paratype 3.1 × 2.7 mm from type locality (UMUT RM29550).

Type locality. 1224 m, Myojin Knoll, south off Izu Islands, Japan, 32°06.20'N, 139°52.17'E (OD).

Distribution. Known only from the type locality; hydrothermal vents.

Etymology. The specific name is given in honor of Dr. Katsunori Fujikura (Japan Agency for Marine-Earth Science and Technology) for his substantial contribution in biological studies of chemosynthesis-based communities (OD).

Description. Shell turbiniform, medium-sized (3.2 × 3.1 mm: holotype. Fig. 721). Protoconch eroded. Teleoconch I with nine axial cords on last quarter whorl, no spirals; earlier portion eroded or broken off. Teleoconch II of 1.75 whorls, suture impressed, sutsel as wide as selenizone. Shoulder rounded, approximately 29 comarginal fine, equally spaced axial ribs on first whorl, approximately 10 fine, equally spaced spiral threads running over axial ribs, interstices with fine irregular growth lines. Base rounded, axial ribs similar in number and strength on first 1.5 teleoconch II whorls, subsequently becoming less distinct; axial ribs slightly more blunt than on shoulder; approximately 20 fine spiral lines, more or less equally spaced between selenizone and umbilicus, spiral lines running over axials. Umbilicus open, deep, moderately wide, no funiculus. Aperture rounded, basal adumbilical portion flared, roof overhanging. Selenizone at periphery; keels quite strong, height equal to width of selenizone, undulating, mostly broken off; lunules weak; slit open, with parallel margins.

Operculum (Fig. 721D) thin, round, covering aperture, multispiral, nucleus central; margin (om) slightly thinner, undulating when dried.

Radula (Fig. 723). Rachidian tooth taller than wide, with V-shaped tip, apical cusp largest, approximately 5 cusps on each side gradually decreasing in size. Lateral teeth 1–3 similar, inner edge of tip smooth, outer edge with 3–4 strong cusps. Lateral tooth 4 of ⅔ size of lateral teeth 1–3, tip triangular, with approximately 10 cusps on inner edge, four cusps on outer edge. Lateral tooth 5 enlarged by broadening, tip triangular, apical cusp largest, approximately eight on each side, decreasing in size towards base. Inner marginal teeth with triangular tip, apical cusp largest, approximately eight on each side, decreasing in size towards base; shafts without food groove. Outer marginal teeth spoon-shaped, tip with many fine cusps. Radular interlock of central field moderate.

Animal (Figs 722, 724). Anterior portion pale in fixed condition except for dark food matter in intestine (i: Fig. 722). Head (Fig. 724) with papillate cephalic tentacles, papillae all around circumference of tentacle; no eyestalks or eyes (Figs 722C, D, 724A); suboptic tentacle short, simple; five pairs of epipodial tentacles, all but tentacles 3 (ESO: Fig. 724) papillate, papillae in two discrete, opposite rows;

Figure 721. *Anatoma fujikurai*. Holotype. 1224 m, Myojin Knoll, south off Izu Islands, Japan, 32°06.20'N, 139°52.17'E (UMUT RM29549). **A.** Apertural view of the shell. **B.** Back view of the shell. **C.** Enlarged view of the body whorl. **D.** Outer surface of the operculum. Left. Right. Paratype. Scale bars A–C= 1 mm. Scale bar D = 0.5 mm. SEM images by and with kind permission T. Sasaki; from Sasaki *et al.* (2010).

opercular attachment drop-shaped, muscle fibers of scaly appearance (Fig. 724A, F). Intestinal loops 1.5 (Fig. 722C, D). Pallial cavity deep; one pair of monopectinate ctenidia attached to pallial roof; hypobranchial gland (hg: Fig. 722) surrounded by loops of intestine in pallial cavity between ctenidia. Anus (a: Fig. 722) opening at anterior central part of pallial cavity.

Comparisons. *Anatoma concinna* from the North Pacific is overall more depressed (height/width ratio: 0.86 vs. 1.03), has reticulate sculpture with axials and spiral of similar strength, and less distinct (possibly eroded) keels of the selenizone.

Anatoma japonica from the Indian Ocean to the Central Pacific has an overall biconical shape, at least twice the density of the axials, spirals that are almost as strong as the axials (⅕ in *An. fujikurai*), and keels of the open slit that converge towards the apertural margin.

Anatoma lamellata from the Western and Central Pacific has a teleoconch I with a distinct spiral cord in the position of the selenizone, a shorter teleoconch I (<0.5 vs. >0.75 whorl), spiral cords that are approximately one third the strength of the axial lamellae (less than ⅕ in *An. fujikurai*), a wider sutsel (1.5× vs. 1× width of the selenizone), and a distinct funiculus.

Anatoma sagamiana from Japan to the Indo-Malayan Archipelago has a protoconch with reticulate sculpture, a slightly shorter teleoconch I (0.66 vs.

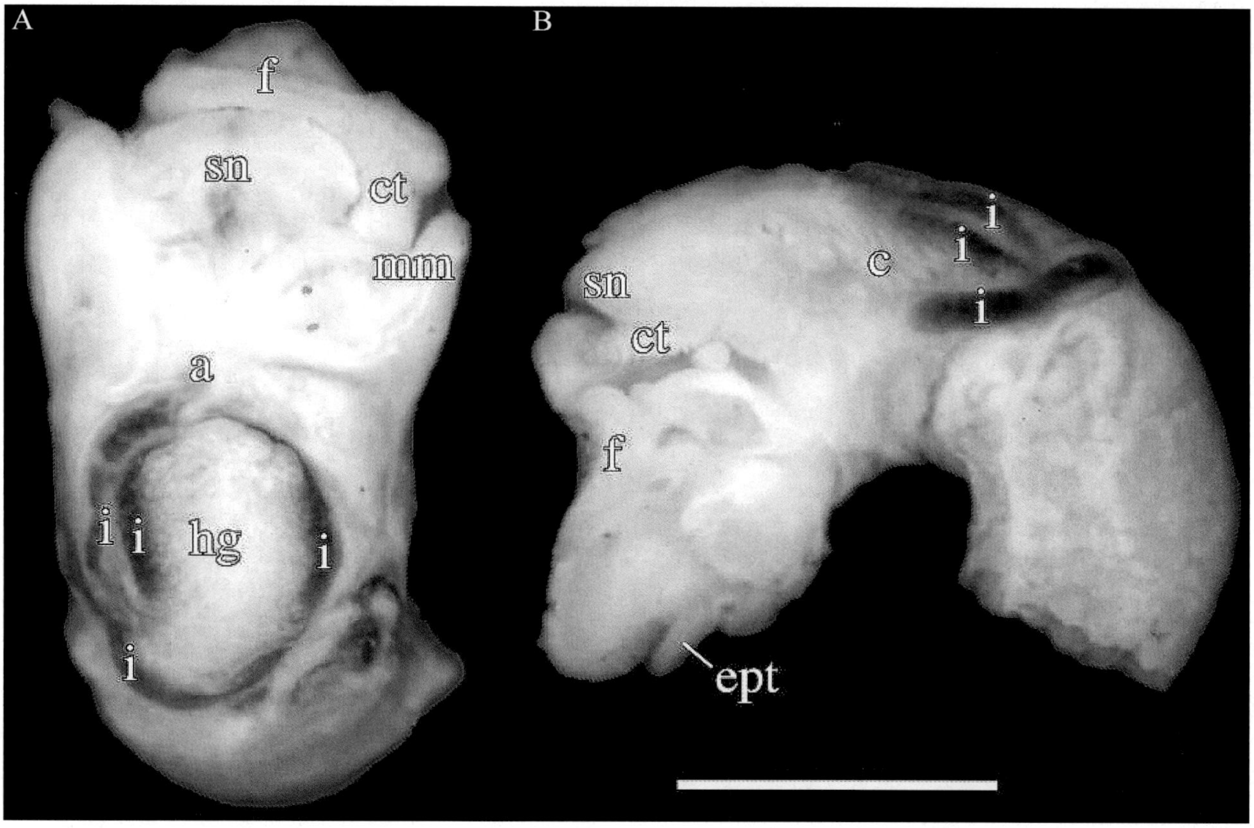

Figure 722. *Anatoma fujikurai*. Animal of paratype. 1224 m, Myojin Knoll, south off Izu Islands, Japan, 32°06.20'N, 139°52.17'E (UMUT RM29550). **A.** Dorsal view of the head-foot and pallial region. **B.** Left lateral view of the same part. c: ctenidium. ct: cephalic tentacles. f: foot. hg: hypobranchial gland. i: intestine. mm: mantle margin. sn: snout. Scale bar = 1 mm. Images by and with kind permission T. Sasaki; from Sasaki *et al.* (2010: modified).

0.75 whorl) with slightly more axial cords (11 vs. 9 on last quarter whorl), axial cords on teleoconch II that are not elevated to low lamellae, and is anomphalous.

Anatoma soyae from Japan has an overall globular shape, denser axial sculpture (17 vs. 9 on the last quarter of teleoconch I), no sutsel, and the keels of the selenizone are low [eroded?].

Figure 723. *Anatoma fujikurai*. Radula. 1224 m, Myojin Knoll, S off Izu Islands, Japan, 32°06.20'N, 139°52.17'E **A, D.** Entire width of the radular rows. **B, E.** Rachidian and lateral teeth enlarged. **C.** Bases of rachidian and lateral teeth. **F.** Enlarged view of marginal teeth. **A–D.** Holotype (UMUT RM29549). **E–F.** Paratype (UMUT RM29550). Images by and with kind permission T. Sasaki; from Sasaki *et al.* (2010).

Anatoma indonesica from the Indo-Malayan Archipelago to the Western Pacific has a spiral cord in the position of the selenizone on teleoconch I, more strongly elevated spiral lamellae on the shoulder that are crossed by approximately 4–8 fine

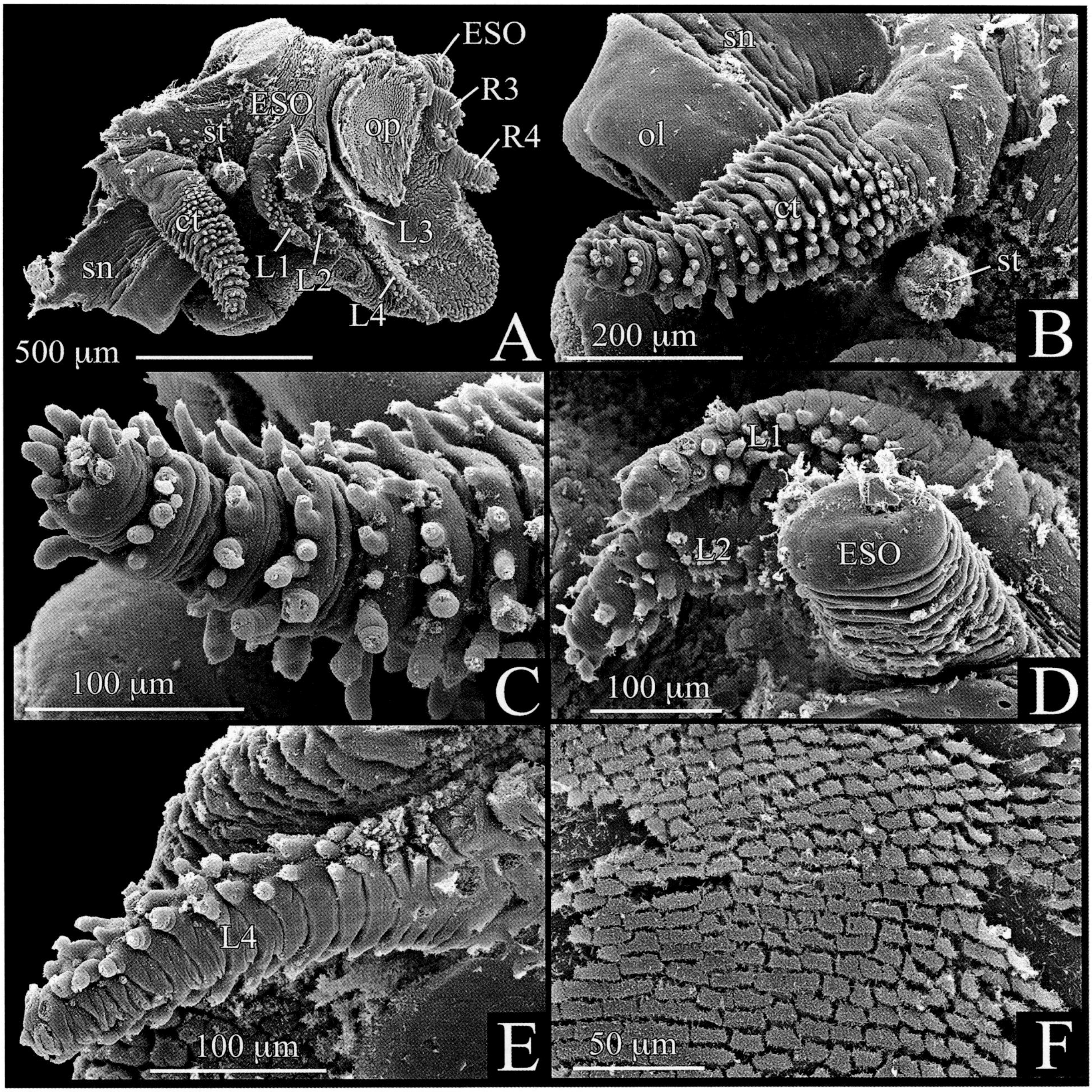

Figure 724. *Anatoma fujikurai*. Head-foot of animal of paratype. 1224 m, Myojin Knoll, S off Izu Islands, Japan, 32°06.20'N, 139°52.17'E (UMUT RM29550). **A.** Left lateral view of head-foot with buccal mass removed. **B.** Enlargement of head. **C.** Papillate tip of left cephalic tentacle. **D.** First to second left epipodial tentacles with epipodial sense organ. **E.** Fourth left epipodial tentacle. **F.** Surface of opercular attachment. ct: cephalic tentacles. ESO: epipodial sense organ. L1–L4: left epipodial tentacles 1–4. ol: outer lip of mouth. op: opercular lobe. R3–R4: right epipodial tentacles 3–4. sn: snout. st: suboptic tentacle. SEM images by and with kind permission T. Sasaki; from Sasaki *et al.* (2010).

spiral lines concentrated in the middle of the shoulder, and fine cancellate sculpture on the base.

Anatoma boucheti from the western Indian Ocean, has a teleoconch I of >1 whorl, shows fine reticulate sculpture of axials and spiral lines of approximately equal strength, and has the keels of the selenizone distinctly elevated.

Anatoma finlayi from the Indian Ocean to the Central Pacific (and possibly Caribbean) has more strongly elevated axial lamellae and a distinct constriction of the base below the selenizone that bears strongly elevated keels with 3–4 fine axial striae between each axial cord.

Anatoma equatoria from the Indo-Malayan Archipelago to the Central Pacific has a spiral cord in the position of the selenizone on teleoconch I, a shoulder that is flatter, a selenizone in the upper third of the whorl, and fine reticulate sculpture on the base.

Anatoma aspera from the Mediterranean and adjacent Atlantic has a more narrow selenizone and a depressed apex.

Anatoma bathypacifica from the Panamic has a more narrow selenizone, a depressed apex, and less convex shoulder.

Anatoma lyra from the northeastern Pacific has a more narrow selenizone, a flatter, discoidal overall shell shape, and a more narrow sutsel (<0.5× vs. 1× width of selenizone).

Anatoma conica from subantarctic South America has a more narrow sutsel (0.5× vs. 1× width of selenizone), 50% denser axial cords on shoulder and base, and fine axial lines on the keels of the selenizone.

Anatoma porcellana from the Indo-Malayan Archipelago to the Western Pacific has fewer spiral lines on the shoulder, the constriction of the base below the selenizone is devoid of spiral lines, and has a more sharply impressed suture.

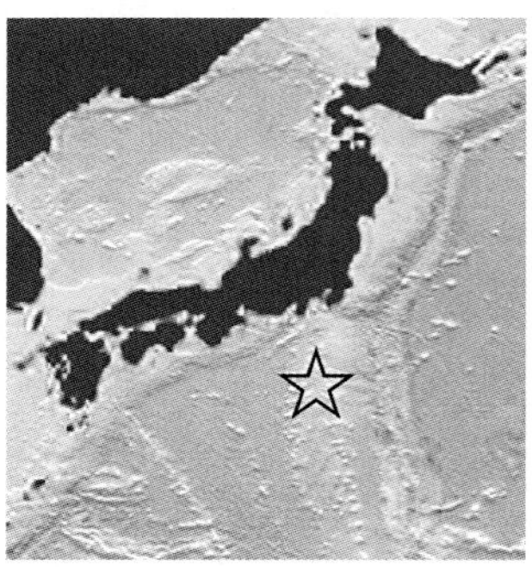

Figure 725. Distribution of *Anatoma fujikurai*. ☆ type locality.

Distribution. Known only from type locality (Fig. 725).

Remarks. The shell morphological features clearly place the species in Anatomidae, including the size, number of whorls for the size, the presence and placement of a slit and selenizone, and the common combination of axial and spiral sculptural elements. The presence of axial lamellae is unusual in Anatomidae, but is also known from the Japanese *An. lamellata*.

The radula, on the other hand, is exceptional. Most species have a trapezoidal rachidian tooth with M-shaped tip with the central cusp distinctly larger than the remainder (not V-shaped with cusps of gradually decreasing size). Lateral tooth 4 usually is strongly reduced and hook-shaped (not slightly reduced, with distinct tip bearing multiple cusps); lateral tooth 5 has strongly elongated tip, at least twice as large as those of the inner marginal teeth, with straight edges and acute angle (not broadened, only slightly larger than the inner marginal teeth, and with convex edges).

Anatoma globulus new species

Figures 726–728

Chresonymy.

Anatoma sp. 1: Spencer *et al.* 2009: 201.

Type material. Holotype (NMNZ M.303298. Fig. 726). Paratypes from type locality (NMNZ M.225753, 10). 470 m, Norfolk Ridge, N/O *Coriolis* CHACAL 2 DW76, New Caledonia, 23.666°S, 167.750°E (MNHN 24953, 2. Fig. 727A). Paratypes 320 m, St. 5579, N off Sibutu Island, Sulu Archipelago, Philippines, 4.904°N, 119.164°E (USNM 287292, 5. Fig. 727B). Paratypes 219–274 m, SE of D'Arcy Point, Raoul Island, Kermadec Islands, New Zealand, 29.313°S, 177.903°W (NMNZ M.226872, 4. Fig. 727C).

Type locality. 146–165 m, SE of Nugent Island, Raoul Island, Kermadec Islands, New Zealand, 29.245°S, 177.823°W.

Etymology. Diminutive of Latin globus, sphere, referring to the very spherical overall shape of the shell; noun in apposition.

Description. Shell medium-sized (to 2.75 mm. Fig. 727A), trochiform, globular. Protoconch of 0.75 whorl, sculpture weakly flocculent, indistinct apertural varix, apertural margin straight. Teleoconch I of 0.66–0.75 whorl, about 20 weak spiral cords, weak spiral cord in position of selenizone. Teleoconch II of up to 2.25 whorls; suture weakly impressed; sutsel 0.5–1× width of selenizone at onset of selenizone, 0.5–1.5× width of selenizone at apertural margin. Shoulder with axials often increasing in strength from fine to raised axial cords, about 60–80 per whorl; first spiral line after 0–0.2 teleoconch II whorls, 10–15 at apertural margin; intersection of axials and spirals with raised points, strength of points proportional to strength of axial cords. Base without constriction below selenizone; sculpture with axials and spirals of same strength, strength intermediate to axial cords and spiral lines on shoulder; density of axials same as on shoulder, 15–20 spiral lines. Umbilicus without carina or funiculus. Aperture rounded, with flared adumbilical portion of peristome, roof overhanging. Selenizone slightly above periphery, rather wide; lunules distinct, spaced more widely than in typical *Anatoma*; keels moderately strong, moderately elevated, axial sculpture extending onto keels. Slit open, margins slightly converging.

Animal unknown.

Comparisons. *Anatoma flemingi* from temperate New Zealand has reticulate protoconch sculpture, usually stronger raised axial cords, and a distinct constriction of the base below the selenizone.

Anatoma finlayi from the Indian Ocean to the Central Pacific has very strong, raised axial lamellae on shoulder and base and a band devoid of spiral lines immediately below the selenizone.

Anatoma regia from temperate New Zealand has a more lenticular shell shape, a more narrow sutsel at the onset of the selenizone (<0.5× vs. 0.5–1× width of selenizone), and a more narrow selenizone.

Anatoma aupouria from New Zealand has a quadrate aperture, has axial sculpture consisting of internestling cords and a much shorter teleoconch I (0.25 vs. 0.66–0.75 whorl).

Anatoma australis from eastern Australia has coarser sculpture on the base, particularly with a decrease in density of the spirals in the third closest to the umbilicus, and only every 3–4 axial cords on the base reach the umbilicus.

Anatoma amydra from the Indo-Malayan Archipelago to the Central Pacific has a wider sutsel after two teleoconch II whorls (> vs. < width of selenizone), a shorter teleoconch I (0.5 vs. 0.66–0.75 whorl), and finer axial cords and spiral lines on teleoconch II.

Anatoma biconica from the Indian-Ocean to the Central Pacific has a more biconical shell, a longer teleoconch I (0.9 vs. 0.66–0.75 whorl), and at least 50% fewer axial cords on shoulder and base.

Anatoma flexidentata from the Indian Ocean to the Western Pacific has a shorter teleoconch I (0.3 vs. 0.66–0.75 whorl), overall weaker sculpture, and a proportionally wider selenizone.

Figure 726. *Anatoma globulus* new species. Holotype. 146–165 m, SE of Nugent Island, Raoul Island, Kermadec Islands, New Zealand, 29.245°S, 177.823°W (NMNZ M.303298). Scale bar shell = 1 mm. Scale bar protoconch = 100 µm.

Anatoma pseudoequatoria from the Indo-Malayan Archipelago to the Central Pacific is less globular and more angular in overall shape, has a teleoconch I of approximately 0.5 whorl, and a distinct funiculus in the umbilicus.

Figure 727. *Anatoma globulus* new species. Paratypes. **A.** 470 m, Norfolk Ridge, N/O *Coriolis* CHACAL 2 DW76, New Caledonia, 23.666°S, 167.750°E (MNHN 24953, 2). **B.** 320 m, St. 5579, N off Sibutu Island, Sulu Archipelago, Philippines, 4.904°N, 119.164°E (USNM 287292, 5). **C.** 219–274 m, SE of D'Arcy Point, Raoul Island, Kermadec Islands, New Zealand, 29.313°S, 177.903°W (NMNZ M.226872, 4). Scale bars shell = 1 mm. Scale bars protoconch = 100 μm.

Anatoma gunteri from southeastern Australia has a more globular shell, strong axial lamellae on the shoulder, and reticulate protoconch sculpture.

Anatoma rainesi from the Eastern to the Western Pacific has a protoconch with fine axial sculpture, fewer but stronger axial cords on teleoconch I that have nodular thickenings in the position of the selenizone, and an overall more discoidal shape.

Anatoma rhynchodentata from the Indo-Malayan Archipelago to the Western Pacific has an overall taller shape with more scalar whorls, a distinct constriction on the base below the selenizone, and at least 50% denser axial and spiral cords.

Anatoma sagamiana from Japan and the Indo-Malayan Archipelago has a protoconch with reticulate sculpture, a weak constriction on the base below the selenizone, and generally weaker axial sculpture.

Anatoma tenuisculpta from the Mediterranean and adjacent Atlantic has a more discoidal overall shell shape and a less convex shoulder and base.

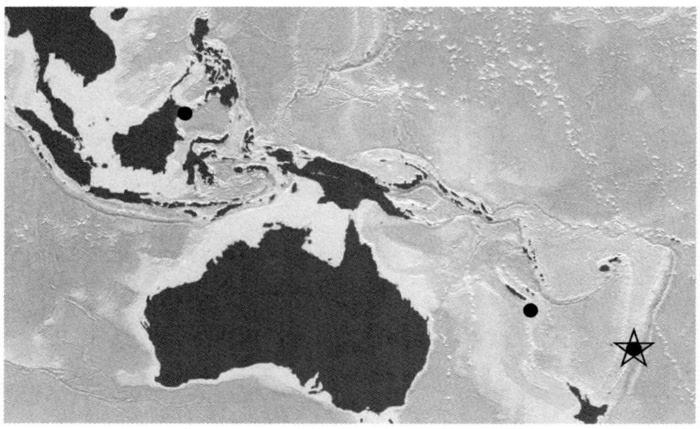

Figure728. Distribution of *Anatoma globulus* new species. ☆ type locality. ● specimen record.

Anatoma tobeyoides from southeastern Australia lacks the protoconch varix and the teleoconch I of less than 0.5 whorls lacks a spiral cord in the position of the selenizone.

Anatoma zancliformis from the Indo-Malayan Archipelago to the Western Pacific has overall weaker sculpture, at most half as many spiral lines on the shoulder and base, and no sutsel on early whorls.

Anatoma agulhasensis from the western Indian Ocean has a constriction on the base below the selenizone, a taller overall shell shape, and low axial lamellae on the shoulder.

Anatoma orbiculata from West Africa has an overall more lenticular shell shape, fewer and much stronger axial cords on shoulder and base, and different sculpture on shoulder and base.

Distribution. Indo-Malayan Archipelago (Fig. 728).

Specimen records.

Philippines. 320 m, St. 5579, N off Sibutu Island, Sulu Archipelago, 4.904°N, 119.164°E (USNM 287292, 5: paratypes).

New Caledonia. 470 m, Norfolk Ridge, N/O Coriolis Chalcal 2 DW76, 23.666°S, 167.750°E (MNHN 24953, 2: paratypes).

New Zealand. 146–165 m, SE of Nugent Island, Raoul Island, Kermadec Islands, 29.245°S, 177.823°W (NMNZ M.303298: holotype; NMNZ M.225753, 10: paratypes). 256–348 m, E of Chanter Islet, Raoul Island, Kermadec Islands, 29.255°S, 177.822°W (NMNZ M.225395, 1). 219–274 m, SE of D'Arcy Point, Raoul Island, Kermadec Islands, 29.313°S, 177.903°W (NMNZ M.226872, 4: paratypes).

Remarks. The strength of the axial sculpture, and to a lesser degree the spiral sculpture, is variable. It applies to material from the same island as the type locality (Raoul Island, Kermadec Islands) as illustrated by the strongly sculptured holotype (Fig. 726) and the weakly sculptured paratype shown in Figure 727C. The overall globular shell shape readily identifies this species.

Anatoma gunteri (Cotton & Godfrey, 1933)

Figures 729–732

Chresonymy.

Schizotrochus gunteri Cotton & Godfrey, 1933: 23: pl. 1, fig. 11 (Fig. 729).
Schizotrochus gunteri: Cotton & Godfrey, 1938b: 2.
Schizotrochus gunteri: Cotton, 1945: 150.
Schizotrochus gunteri: Cotton, 1959: 41, fig. 15.
Anatoma gunteri: Lozouet, 1986: 113.
Scissurella (*Anatoma*) *gunteri*: Wilson, 1993: 46.
Anatoma guntheri [sic]: Jansen, 1999: 49, figs 10–12.
Thieleella gunteri: Geiger, 2003: 74, fig. 15B [protoconch].
Thieleella gunteri: Geiger & Jansen, 2004a: 27–29, figs 15–17, 18 [map].

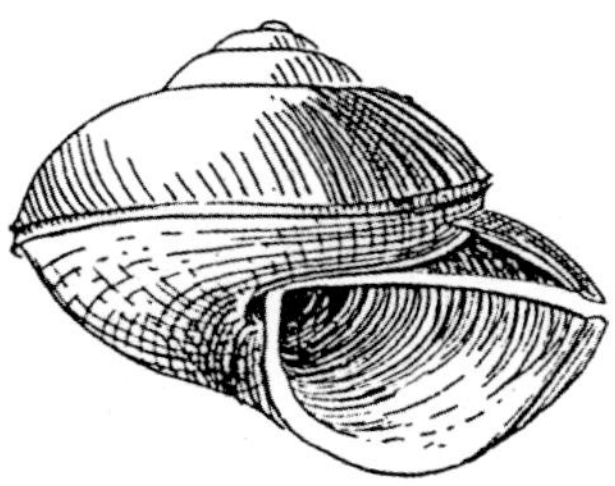

Figure 729. Original illustration of *Schizotrochus gunteri.*

Type material. Holotype (SAM D.10769: Cotton, 1959; Jansen, 1999), 1.5 × 1.1 mm.

Type locality. 150 fms. [= 274 m], Beachport, South Australia, [Australia, 37.500°S, 140.017°E] (OD).

Etymology. Named after Rev. H. Gunter from South Australia (OD).

Description. Shell globular, medium-sized to large (to 3 mm. Figs 730, 731). Protoconch 0.75 whorl, reticulate sculpture, varix faint, not connected to embryonic cap, apertural margin sinusoidal. Teleoconch I 0.6–0.85 whorl, 16–22 axial cords, with spiral cord in position of selenizone, interstices with fine growth lines. Teleoconch II of 2.66 whorls, no sutsel at onset of selenizone, wider than selenizone in fully grown specimens; suture moderately impressed; sutsel 0.5–1× width of selenizone. Shoulder with strong axial cords, approximately 45 on body whorl, approximately 20 fine spiral lines running over axials, equally spaced between suture and selenizone. Base with weak constriction below selenizone, more numerous but finer axial cordlets compared to shoulder, approximately 25 spiral lines, base sloping continuously into umbilicus. Umbilicus narrow, deep, with indistinct funiculus. Selenizone at periphery; keels elevated, fragile; lunules faint, not coordinated with axials on shoulder or base; slit open, margins parallel. Aperture rounded, flared in lower adumbilical portion.

Animal unknown.

Comparisons. *Anatoma australis* from eastern Australia has axials on shoulder of base of equal strength, only every 3–4 axial cord on the base reach from the selenizone to the umbilicus, and has a protoconch with flocculent sculpture.

Anatoma biconica from the Indian Ocean to the Central Pacific has a protoconch with coarse flocculent sculpture, a biconical overall shell shape, and at most half as many spiral lines on shoulder and base.

Anatoma japonica from the Indian Ocean to the Central Pacific has a protoconch with flocculent sculpture, same sculpture on shoulder and base, and no constriction of the base below the selenizone.

Anatoma kopua from temperate eastern Australia to New Zealand has a protoconch with flocculent sculpture, a shorter teleoconch I (0.3 vs. 0.75 whorl), and at most half as many spiral lines on shoulder and base.

Anatoma tobeyoides from southeastern Australia has axials on shoulder and base of equal strength, a protoconch with flocculent sculpture, and a shorter teleoconch I (<0.5 vs. 0.75 whorl).

Anatoma amoena from Antarctica has very faint sculpture and the spiral cord in the position of the selenizone on teleoconch I forms a strong angulation in the profile of the whorl.

Anatoma indonesica from the Indo-Malayan Archipelago to the Western Pacific is broader than high, has fewer spirals on the shoulder that are restricted to the center of the shoulder, and a smooth band just below the selenizone on the base.

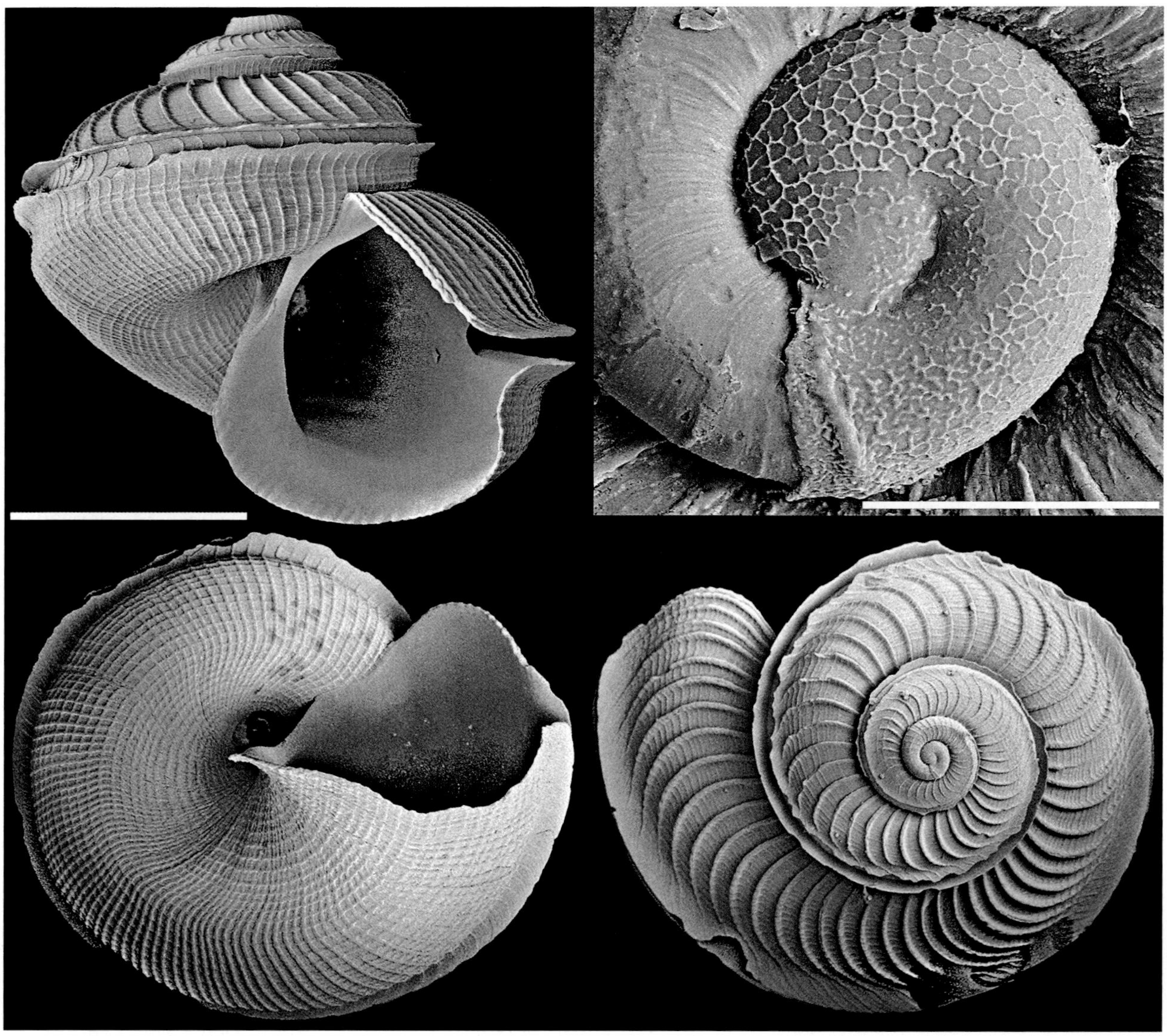

Figure 730. *Anatoma gunteri*. 174–183 m, 40 mi S of Cape Wiles, South Australia, Australia (AMS C.31845, 20). Scale bar shell = 1 mm. Scale bar protoconch = 100 µm.

Anatoma finlayi from the Indian Ocean to the Central Pacific (and possibly Caribbean) has a protoconch with flocculent sculpture, stronger, raised axial cords on the base, and distinct supernumerary axial lines on the keels of the selenizone.

Figure 731. *Anatoma gunteri*. **A.** Off Cape Naturaliste, Tasmania, Australia (AMS C.402692, 2). **B.** 40 mi S of Cape Wiles, South Australia, Australia (AMS C.402690, 20). **C.** Off Cape Martin, South Australia, Australia (AMS C.402691, 3). Scale bars shell = 1 mm. Scale bars protoconch = 100 μm.

Anatoma flemingi from New Zealand has stronger spiral lines compared to the axial cords, axials on shoulder and base that are less dissimilar, and a more narrow umbilicus.

Anatoma quadraxialis from the Indo-Malayan Archipelago has much denser axial sculpture that is of equal strength on shoulder and base and is less globular.

Anatoma porcellana from the Indo-Malayan Archipelago to the Western Pacific has weaker sculpture on the shoulder that is of equal strength with that on the base, half as many spiral lines on shoulder and base, and flocculent protoconch sculpture.

Anatoma alternatisculpta from the Caribbean has a protoconch with coarsely flocculent sculpture, half as many spiral lines on the shoulder and base, and a slightly wider sutsel.

Anatoma aspera from the Mediterranean and adjacent Atlantic has twice as wide a sutsel at the same whorl count, often a weak constriction on the base below the selenizone, and 50% fewer spiral lines on shoulder and base.

Figure 732. Distribution of *Anatoma gunteri*. ☆ type locality. ● specimen record. ■ literature record.

Distribution. Tasmania and South Australia, Australia, 219–667 m, sandy and silty bottoms (Fig. 732).

Specimen records.

Australia, South Australia. 667 m, off Cape Martin, 38.125°S, 140.000°E (AMS C.402691, 3). 174–183 m, 40 mi S of Cape Wiles, 35.650°S, 136.667°E (AMS C.402690, 20; AMS C.31845, 20). 182 m, South of Cape Wiles, 35.650°S, 136.667°E (WAM, 12).

Australia, Tasmania. 212 m, S of Storm Bay, 43.783°S, 147.808°E (AMS C.402634, 3). 570 m, 9.5 mi NE of Tasman Island, 43.208°S, 148.229°E (AMS C.402693, 6). 113 m, NE of Cape Pillar, 43.167°S, 148.112°E (AMS C.402723, 1). 399 m, off Cape Naturaliste, 40.843°S, 148.775°E (AMS C.402628, 22; AMS C.402692, 2). 176 m, E of D'Entrecasteaux Channel, 44.037°S, 146.842°E (AMS C.402722, 2).

Australia, Western Australia. Point Quobba, Carnarvon, 24.483°S, 113.417°E (WAM, 1).

Literature records.

Australia. 274 m, 366 m, Beachport, South Australia, 37.500°S, 140.017°E; 183 m, 238 m, 549 m, Cape Jaffa, South Australia, 36.967°S, 139.650°E; 183 m, Cape Wiles, South Australia, 34.950°S, 135.683°E (Cotton, 1945).

Remarks. Cotton & Godfrey (1933) also cited Cape Pillar, Tasmania, 100 fms, as locality for this species. This could indicate paratypes, but the whereabouts of these specimens is currently unknown (possibly SAM).

Anatoma hyposculpta new species

Figures 733–736

Type material. Holotype (MNHN 25102. Fig. 733). Paratypes from type locality (MNHN 25128, 2). 700–707 m, Lau Ridge (Yangasa Cluster), N.O. *Alis,* BORDAU 1 DW1485, 19.050°S, 178.500°W (MNHN 25103, 8. Fig. 734).

Type locality. 820–863 m, N.O. *Alis,* BORDAU 1 DW1459, Lau Ridge (Yacata), Fiji, 17.300°S, 179.550°W (OD).

Etymology. Hypo-, Greek prefix for weak; -sculpta, Latin adjective for sculptured, referring to the very faint yet distinct sculpture.

Figure 733. *Anatoma hyposculpta* new species. Holotype. 820–863 m, N.O. *Alis,* BORDAU 1 DW1459, Lau Ridge (Yacata), Fiji, 17.300°S, 179.550°W (MNHN 25102). Radula shown in Figure 735. Scale bar shell = 1 mm. Scale bar protoconch = 100 μm.

Description. Shell medium-sized (to 2.2 mm. Figs 733, 734), low trochiform, turreted. Protoconch of 0.75 whorl, sculpture flocculent, apertural varix not connected to embryonic cap, apertural margin straight. Teleoconch I of 0.25 whorl, approximately 0–3 low, indistinct axial cords, no spirals. Teleoconch II of up to 2.75 whorls, suture deeply impressed; sutsel 0–0.2× width of selenizone at onset of selenizone, 2× width of selenizone at apertural margin. Shoulder sinusoidal in

Figure 734. *Anatoma hyposculpta* new species. Paratypes **A.** 820–863 m, N.O. *Alis*, BORDAU 1 DW1459, Lau Ridge (Yacata), Fiji, 17.300°S, 179.550°W (MNHN 25128, 2). **B–C.** 700–707 m, Fiji, 19.050°S, 178.500°W (MNHN 25103, 8). Scale bars shell = 1 mm. Scale bars protoconch = 100 µm.

profile; early on smooth or with low, indistinct axial cords restricted to adsutural quarter, disappearing after approximately 1.5 whorls; first very fine spiral threads after approximately two whorls, initially restricted to adsutural portion of shoulder, eventually covering entire shoulder, approximately 15 at apertural margin. Base rounded, without constriction below selenizone, weak spiral angulation on middle slope, continuously sloping with umbilicus; very fine spiral lines in upper third of base, starting at least after 0.2 teleoconch II whorls, i.e., onset of spirals on base before onset on shoulder; weak, low axial cords from upper third of base into umbilicus, becoming less distinct with growth. Umbilicus open, moderately wide, no funiculus. Selenizone at periphery; keels rather weak, quite tall; lunules widely spaced, fine, sharp; slit open, margins parallel. Aperture rounded, adumbilical portion flared.

Operculum (Fig. 733) corneous, thin round, multispiral, with central nucleus, covering aperture.

Radula (Fig. 735). Rachidian tooth trapezoidal, central cusp 50% larger than lateral cusps, approximately 10 tightly spaced parallel cusps at obtuse angle to base.

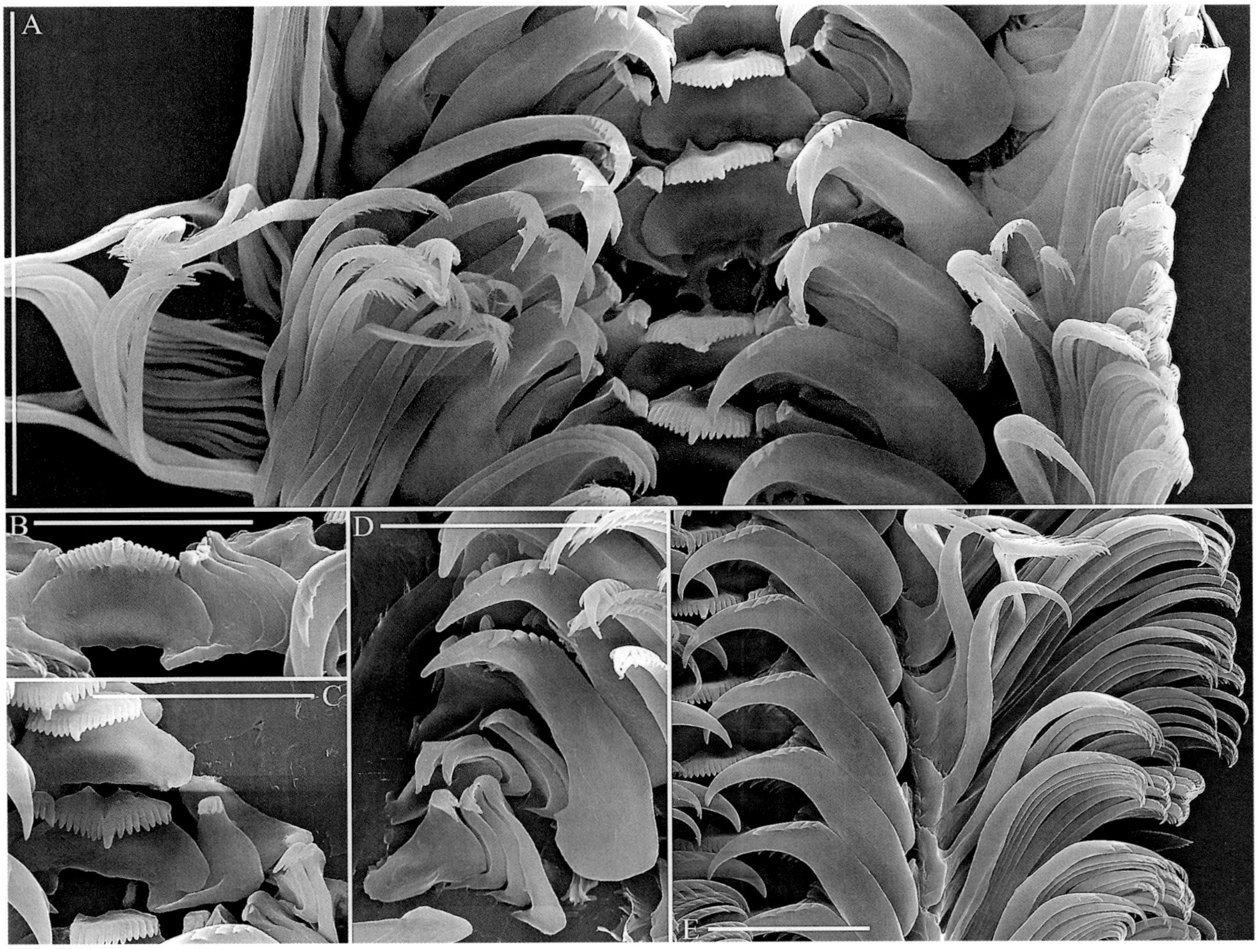

Figure 735. *Anatoma hyposculpta* new species. Radula of holotype. Shell show in Figure 733. 820–863 m, N.O. *Alis*, BORDAU 1 DW1459, Lau Ridge (Yacata), Fiji, 17.300°S, 179.550°W (MNHN 25102). **A.** Full width of the radula. Scale bar = 50 μm. **B–C.** Rachidian tooth and lateral teeth 1–4. **D.** Lateral teeth 1–5. **E.** Lateral tooth 5 and marginal teeth. Scale bars B–E = 20 μm.

Lateral teeth 1–3 similar, much smaller than rachidian tooth, 3–2 cusps on outer edge of tip. Lateral tooth 4 more reduced, tip hook-shaped, with one minute cusp on each side. Lateral tooth 5 enlarged by elongation, central cusp largest, 3–4 smaller cusps on each side. Inner marginal teeth with elongated tip, apical cusp largest, 3–5 cusps on each side; shafts without food groove. Outer marginal teeth with spoon-shaped tip, many cusps on each side. Radular interlock of central field moderate.

Comparisons. *Anatoma austrolissa* from the southern Indian Ocean and the Western Pacific lacks an apertural varix on the protoconch, has fine pitting on the shell, lacks spiral sculpture, and the radula is of the flexoglossate type.

Anatoma breveprima from the Western Pacific has a shorter teleoconch I (0.125 vs. 0.25 whorl), distinct sculpture, and axial cords extending from the selenizone into the umbilicus.

Anatoma planapex from the Indo-Malayan Archipelago and Australia has a much larger teleoconch I (1 vs. 0.125 whorl), a spiral cord in the position of the selenizone on teleoconch I, and lateral tooth 1 is less reduced.

Anatoma umbilicata from the Mediterranean and adjacent Atlantic has a longer teleoconch I (0.5 vs. 0.25 whorl) and a more narrow sutsel after one whorl (>0.5× vs. <0.5× width of the selenizone).

Distribution. Indo-Malayan Archipelago to Western Pacific, 700–1600 m (Fig. 736).

Specimen records.

New Caledonia. 731–751 m, Loyalty Ridge, 23.783°S, 169.817°E (MNHN, 1). 1490–1620 m, 23.733°S, 166.967°E (MNHN, 1).

Fiji. 820–863 m, 17.300°S, 179.550°W (MNHN, 1).

Wallis and Futuna. 1011 m, Tuscaroa Bank, 11.767°S, 178.450°W (MNHN, 1).

Remarks. Although the shell shows some similarity with *An. austrolissa*, their radulae are completely different and confirm the subtle but consistent differences in their respective shell morphologies.

The radula of *An. hyposculpta* shows several peculiarities among *Anatoma* species. The rachidian tooth with its many fine serrations is known only from the more narrow rachidian tooth of *An. fujikurai*. Lateral teeth 1–3 are more reduced compared to the remainder of the teeth on the radula, while lateral tooth 5 is comparatively speaking strongly enlarged for Anatomidae.

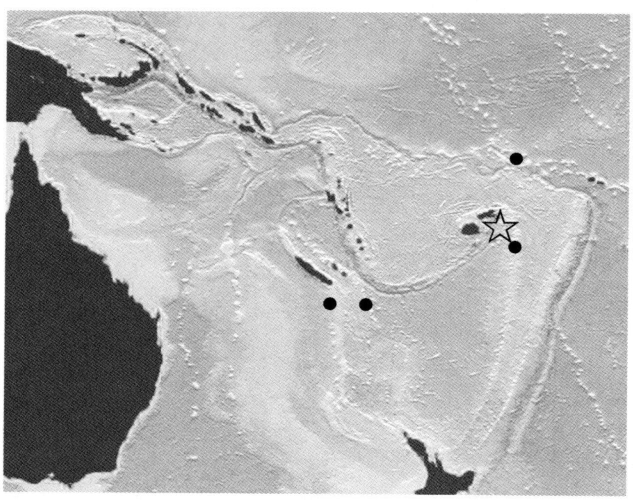

Figure 736. Distribution of *Anatoma hyposculpta* new species. ☆ type locality. ● specimen record.

Anatoma indonesica Bandel, 1998

Figures 737–740, Color Plate 1C

Chresonymy.

Anatoma indonesica Bandel, 1998: 30–31, pl. 9, figs 5–7 (Fig. 737).
Anatoma indonesica: Geiger, 2003: 73.
Anatoma indonesica: Hasegawa, 2005: 140, figs 3A–B.
Anatoma indonesica: Sasaki *et al.*, 2010: 70.
Anatoma sp. 3: Spencer *et al.*, 2009: 201.

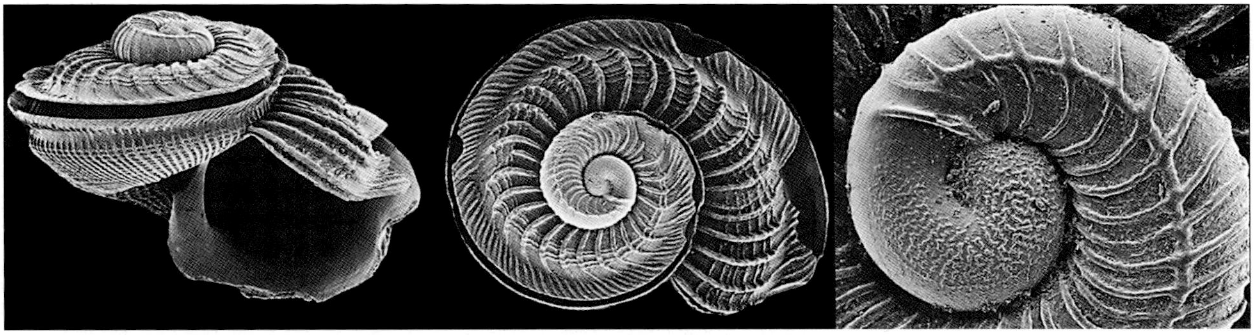

Figure 737. Original illustration of *Anatoma indonesica*.

Type material. Holotype (SGPIH 3874. Fig. 737), 1.1 × 1 mm.
Type locality. Satonda, Indonesia [8.100°S, 117.750°E] (OD).
Etymology. Named for is type locality in Indonesia (OD).
Description. Shell small (1.1 mm. Figs 738, 739), wider than high. Protoconch of 0.75 whorl, sculpture flocculent, apertural varix not connected to embryonic cap, apertural margin sinusoidal. Teleoconch I of 0.66 whorls, 18 axials, spiral cord in position of selenizone slightly stronger than axials. Teleoconch II of 1.33 whorls; sutsel 0.5–1× width of selenizone at onset of selenizone, 1–1.5× width of selenizone at apertural margin. Shoulder approximately 29 strong, elevated axial lamellae; axial lamellae overrun by up to eight fine spiral lines concentrated in central half of shoulder; onset of spiral lines within 0.25 teleoconch II whorl. Base with slight, smooth constriction below selenizone, approximately three times as many axials as on shoulder, finer than axials on shoulder, half of axials terminating at mid-base, remainder continuing to umbilicus; approximately 20 finer spiral lines. Selenizone slightly above periphery; lunules distinct, dense; keels strong and wide, upper surface with 1–3 fine axials in between each strong axial of shoulder; slit open, margins parallel. Umbilicus open, narrow. Aperture rounded.

Animal unknown.

Comparisons. *Anatoma pseudoequatoria* from the Indo-Malayan Archipelago to the Central Pacific lacks an apertural varix on the protoconch, has a shorter teleoconch I (0.25 vs. 0.66 whorl) without spiral sculpture, and the axials and spirals on the shoulder are of equal strength forming small pointed nodules at their intersection.

Anatoma finlayi from the Indian Ocean to the Central Pacific (and possibly Caribbean) has strong axial ribs on the base, no smooth band on the base below the selenizone, and all axial cords on the base reach the umbilicus.

Anatoma gunteri from southeastern Australia is globular inflated in overall shape, has the selenizone and slit at the periphery of the shell, more spirals on the shoulder that are equally spaced between suture and selenizone, and lacks the smooth band below the selenizone on the base.

Anatoma quadraxialis from the Indo-Malayan Archipelago has more axials (approximately 80 on body whorl), barely perceptible spiral, and lacks a spiral cord on teleoconch I.

Anatoma alternatisculpta from the Caribbean has weaker axial cords on the shoulder, at most half as many spiral lines on shoulder and base, and axial cords on the base that are duller.

Figure 738. *Anatoma indonesica*. 175 m, SW Salayer, Indonesia, 6.373°S, 120.437°E (RMNH, 5). Scale bar shell = 1 mm. Scale bar protoconch = 100 μm.

Anatoma aspera from the Mediterranean and adjacent Atlantic lacks the smooth band on the base below the selenizone and has a wider sutsel on later whorls (<2× vs. >2× width of selenizone).

Figure 739. *Anatoma indonesica*. **A.** 175 m, SW Salayer, Indonesia, 6.373°S, 120.437°E (RMNH, 5). **B–C.** 285–305 m, S of Karanga Kaledupa, Banda Sea, Tukang Besi Islands, Indonesia, 5.958°S, 123.775°E (RMNH 42272, 1). Scale bars shell = 1 mm. Scale bars protoconch = 100 μm.

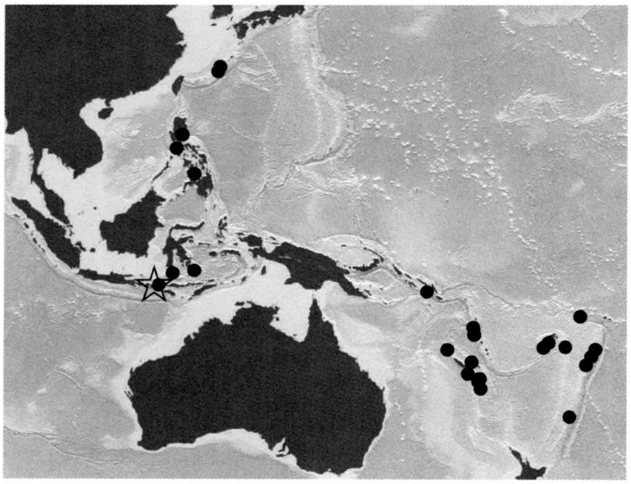

Figure 740. Distribution of *Anatoma indonesica*. ☆ type locality. ● specimen record. ■ literature record.

Anatoma orbiculata from West Africa has a protoconch with irregular reticulate sculpture, no smooth band on the base below the selenizone, and 50% fewer axial cords and spiral lines on the base.

Anatoma emilioi from Hawaii has much stronger but at most half as many axial lamellae on the shoulder, inturned keels of the selenizone, and a much stronger constriction of the base below the selenizone.

Anatoma flemingi from New Zealand has reticulate protoconch sculpture, lacks the bare spiral band on the base below the selenizone, and the 50% more numerous spiral lines on the shoulder are distributed evenly between suture and selenizone.

Anatoma munieri from the Indian Ocean to the Central Pacific has a much taller shell, a more distinct constriction of the base below the selenizone, and 1–2 more or less distinct spiral keels on the base.

Anatoma porcellana from the Indo-Malayan Archipelago to the Western Pacific has less elevated spiral cords, at most half as many spiral lines on shoulder and base, and lacks the smooth band on the base below the selenizone.

Anatoma rhynchodentata from the Indo-Malayan Archipelago to the Western Pacific lacks the smooth band on the base below the selenizone and strong axial lamellae on the shoulder, and the intersection of axials and spirals form minute but distinct points.

Anatoma janusa from the subtropical northeastern Atlantic has a later onset of spiral lines on the shoulder (after 1 vs. 0.25 teleoconch II whorl) and lacks a smooth band on the base below the selenizone.

Anatoma lamellata from the Western to the Central Pacific has a protoconch with a spiral cord, lacks the smooth band on the base below the selenizone, and has coarser axial cords.

Distribution. Indo-Malayan Archipelago to Western Pacific(Fig. 740).

Specimen records.

Philippines. 90–200 m, Maribohoc Bay, Bohol Island, PANGLAO 2004 P1, 9.600°N, 123.750°E (MNHN, 1). 538 m, AURORA 2007 CP2750, 15.883°N, 121.893°E (MNHN, 1). 169–176 m, AURORA 2007 DW2757, 15.902°N, 121.833°E (MNHN, 1). 311 m, St. 5268, off Matocot Point, W Luzon, 13.700°N, 120.954°E (USNM 1099055, 1).

Indonesia. 175–185 m, Bay of Sanggar, N of Sumbawa, 8.322°S, 118.260°E (RMNH 42273, 4). 285–305 m, S of Karanga Kaledupa, Banda Sea, Tukang Besi Islands, 5.958°S, 123.775°E (RMNH 42272, 1). 175 m, SW Salayer, 6.373°S, 120.437°E (RMNH, 5).

New Caledonia. 381–469 m, Loyalty Ridge, 23.367°S, 168.017°E (MNHN, 1). 244–278 m, Norfolk Ridge, 23.683°S, 168.000°E (MNHN, 1). 601–608 m, Norfolk Ridge, 23.317°S, 168.000°E (MNHN, 1). 230–290 m, southern, 23.683°S, 168.008°E (MNHN, 1). 538 m, Loyalty Basin, 20.617°S, 166.967°E (MNHN, 1). 230–290 m, Jumeau Bank East, southern, 23.683°S, 168.008°E (MNHN, 1). 550 m, 18.650°S, 163.183°E (MNHN, 1). 635–680 m, south, BERYX 11 DW09, 24.867°S, 168.367°E (MNHN, 1). 237–550 m, Norfolk Ridge, BATHUS 3 CH802, 23.683°S, 168.000°E (MNHN, 1). 230–360 m, south, BERYX 11 CH41, 23.650°S, 168.000°E (MNHN,

2). 187–197 m, Cryptherlia Bank, Norfolk Ridge, NORFOLK 3 DW2123, 23.300°S, 168.250°E (MNHN, 8). 190–212 m, Norfolk Ridge, NORFOLK 1 DW1727, 23.283°S, 168.233°E (MNHN, 1). 250–350 m, St. 40, 22.500°S, 166.400°E (MNHN, 4).

Fiji. 420–450 m, 18.200°S, 178.617°W (MNHN, 1). 260–305 m, S of Viti Levu, 18.308°S, 178.097°E (MNHN, 7). 879–897 m, SE of Viti Levu, 17.515°S, 178.888°E (MNHN, 3).

Tonga. 389–418 m, N Ha'apai group, 19.700°S, 174.533°W (MNHN, 1). 456–510 m, Tongatapu, 21.083°S, 175.367°W (MNHN, 2). 440–487 m, Vava'u group, BORDAU 2 DW1586, 18.567°S, 173.917°W (MNHN, 2).

Wallis and Futuna. 350 m, Wallis Island, 13.317°S, 176.283°W (MNHN, 1).

Vanuatu. 305–314 m, 15.083°S, 167.200°E (MNHN, 1). 360–419 m, MUSORSTOM 8 DW1065, 16.267°S, 167.350°E (MNHN, 1).

New Zealand. 82–100 m, SE of Smith Bluff, Raoul Island, Kermadec Islands, 29.331°S, 177.940°W (NMNZ M.226899, 1).

Literature records.

Japan. 199–202 m, TY03-9, Nansei Islands, 26.825°N, 127.700°E (Hasegawa, K. 2005).

Remarks. Specimens of *An. finlayi* with less than 1.5 teleoconch II whorls have a base with very fine axial cords similar to those seen in *An. indonesica*. It may not be possible to identify those small specimens from Indo-Pacific localities accurately as either *An. indonesica* or *An. finlayi*. There are no known specimens of *An. indonesica* from the Indian Ocean. Larger specimens can be differentiated consistently, for which reason the similarities of young specimens are not taken to indicate synonymy.

Several specimens from one lot from Okinawa showed bright yellow protoconchs (Color Plate 1C). SEM of the specimens did not reveal any structural differences to other *An. indonesica* specimens.

The Japanese common name is *Kawaribori-kuchikire-ebisu* (Hasegawa, 2005).

Anatoma janetae Geiger, 2006

Figures 742–745, Color Plate 1G

Chresonymy.

Anatoma janetae Geiger, 2006f: 108–112, figs 1 [SEM of shell], 2 [animal], 3 [SEM or radula].

Type material. Holotype FMNH 307218 (Fig. 742A–C, Color Plate 1G), broken shell on SEM stub, radula on stub; animal formalin fixed, stored in 75% ethanol. Paratype shell only. 2606 m, Baby Bare Seamount, Cascadia Basin, USA, 47.710°N, 127.786°W (FMNH 282781. Fig. 742D–F).

Type locality. 2574 m, Dive 3941 R/V *Alvin*, East Pacific Rise, North Pacific Ocean, 12°42.702' N, 103°54.452' W, 22 November 2003, manipulator arm grab, white biobox (OD).

Etymology. Named for Janet Voight (FMNH) for providing the fragmotype material along with other specimens used in ongoing work (OD).

Description. Shell medium-sized (to 3.8 mm. Figs 742, 741), trochiform, biconical, very thin, fragile, white. Protoconch of 0.75 whorl, sculpture flocculent with spiral orientation, apertural varix present [contra original description], shape of apertural margin slightly sinusoidal. Teleoconch I of 0.6–0.75 whorl, with strong spiral cord in position of selenizone, approximately 32 distinct, raised axial cords forming nodules at intersection with axial cord; interstices almost smooth with faint irregular growth lines, some indistinct crenulation. Teleoconch II of at least 2.25 whorls, no sutsel on early whorls, in large specimens sutsel less than width of selenizone; suture moderately impressed. Shoulder at start of selenizone with distinct, raised, axial cords, approximately 15 on first 0.5 teleoconch II whorl; axials becoming lower, less distinct, vanishing after 1.25 teleoconch II whorls. Spiral threads starting shortly after onset of selenizone, approximately six at one teleoconch II whorl, running over axials, not forming points or thickenings at intersections with axials; with disappearance of axials, spirals turning into low steps, somewhat irregularly spaced between selenizone and moderately impressed suture, approximately 12 at apertural margin. Interstices with very fine axial growth lines. Base of early teleoconch with axials of similar strength, spacing as on shoulder, spirals starting adumbilically, increasing in number to six at 0.5 teleoconch II whorl; with onset of selenizone additional fine spiral thread starting below selenizone, increasing in number with growth; base of last whorl with approximately 20 irregularly spaced spiral steps,

Figure 741. *Anatoma janetae*, 850–900 m, Nittnat Canyon, NW Washington State, USA, 48.167°N, 126.000°W (HLC, 1). Scale bar shell = 1 mm. Scale bar protoconch = 100 μm.

Figure 742. Shell of *Anatoma janetae*. **A–C.** Holotype 2574 m, Dive 3941 R/V *Alvin*, East Pacific Rise, North Pacific Ocean, 12°42.702' N, 103°54.452' W (FMNH 307218). Fragments aligned in *Photoshop*. The arrows highlight the end of the protoconch (C). Scale bar shell = 1 mm. Scale bar protoconch = 100 µm. **D–F.** Paratype 2606 m, Baby Bare Seamount, Cascadia Basin, USA, 47.710°N, 127.786°W. (FMNH 282781). Scale bar = 500 µm.

very fine axial growth lines. Umbilicus open, moderately wide, continuously sloping with base; weak funiculus in paratype. Selenizone at periphery; lunules faint, dense; keels moderately elevated, thin; slit open. Aperture rounded.

External anatomy (Fig. 743). Cephalic tentacles papillate, optic stalk at base, no eyes. Four papillate epipodial tentacles on either side, no cephalic lapplet.

Operculum (Fig. 743) thin, round, multispiral, with central nucleus, most likely covering aperture.

Radula (Fig. 744). Rachidian tooth trapezoidal with serrated tip, cusps parallel to row, central cusp largest, 2–3 cusps on each side. Lateral teeth 1–4 similar, decreasing in size from lateral tooth 1 to 4, number of cusps decreasing from 3–4 to 2 cusps on outer surface of tip. Lateral tooth 5 elongated, enlarged, tip with strong median cusp, three inner cusps larger than 2–3 outer cusps. Inner marginal teeth similar to lateral tooth 5, but two prominent outer cusps, two small inner cusps; shafts without food groove; outer marginal teeth spoon-shaped, with many cusps on either side; outermost marginal teeth flattened, with approximately 25 cusps mainly on inner edge; terminal marginal tooth without cusps. Radular interlock of central field moderate.

Comparisons. *Anatoma janetae* is characterized by the marked switch in teleoconch sculpture from distinct raised axial cords to fine spiral steps. Some species have markedly different sculpture on shoulder and base (*An. alternatisculpta, An. flemingi, An janusa, An. finlayi, An. gunteri*) and in many species select sculptural elements are added or fade out with growth, but a full switch is not known from any other species.

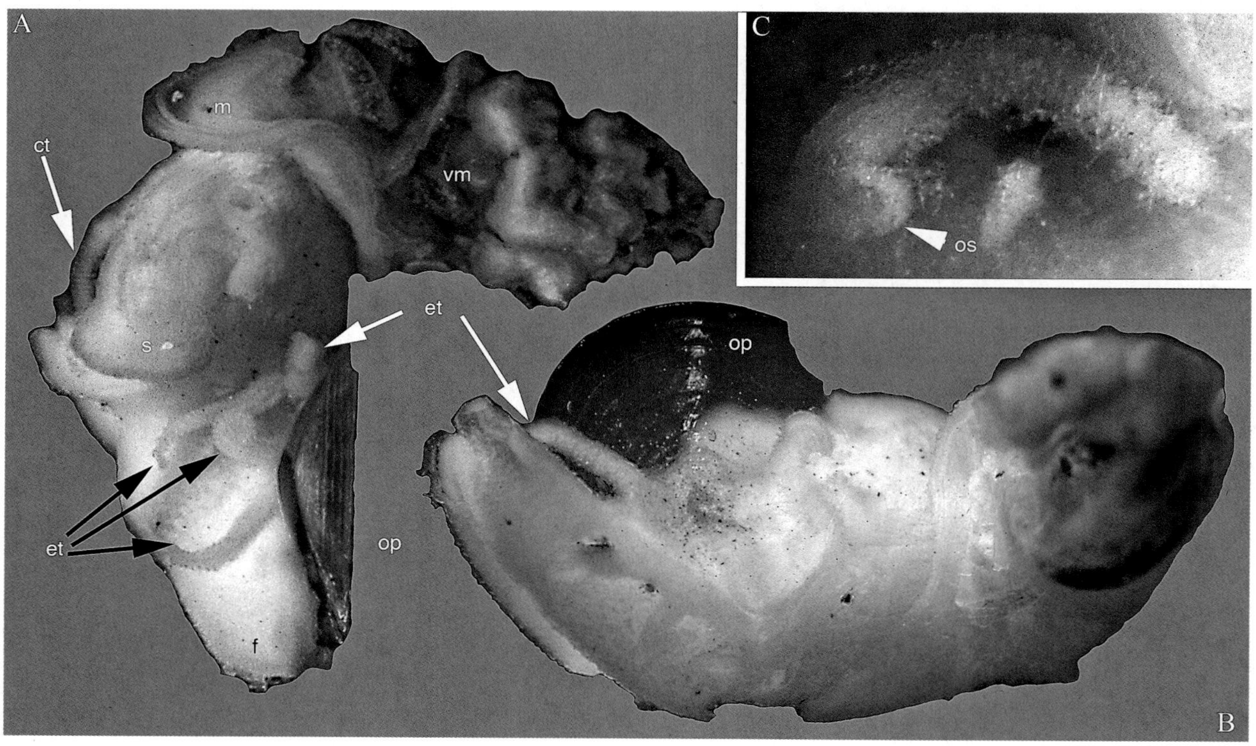

Figure 743. Animal of *Anatoma janetae*. Holotype 2574 m, Dive 3941 R/V *Alvin*, East Pacific Rise, North Pacific Ocean, 12°42.702' N, 103°54.452' W (FMNH 307218). **A.** Anterior view. **B.** Posterior view. **C.** Cephalic tentacle with optic stalk. Whole animal length = 3.2 mm. ct: cephalic tentacle. et: epipodial tentacle. f: foot. m: mantle. op: operculum. os: optic stalk. s: snout. vm: visceral mass.

Anatoma argentinae from the abyssal plain of the Panamic and off Argentina has a much larger protoconch (370–480 vs. 175 µm) with very fine reticulate sculpture, spiral lines on shoulder and base, and a shorter teleoconch I (0.3 vs. 0.7 whorl).

Anatoma bathypacifica from the Panamic has a distinct angulation along the spiral cord in the position of the selenizone on teleoconch I, wide reticulate protoconch sculpture, and spiral sculpture on shoulder and base that is lines, not steps.

Anatoma concinna from the North Pacific has prominent reticulate sculpture over the entire shell, a more globular shell, and very low keels of the selenizone.

Anatoma disciformis from the North Pacific has coarsely flocculent sculpture on the protoconch and fine reticulate sculpture on shoulder and base after one teleoconch II whorl.

Anatoma keenae from the Panamic has distinct reticulation on shoulder and base of the entire teleoconch, and spiral lines on shoulder and spiral cords on base.

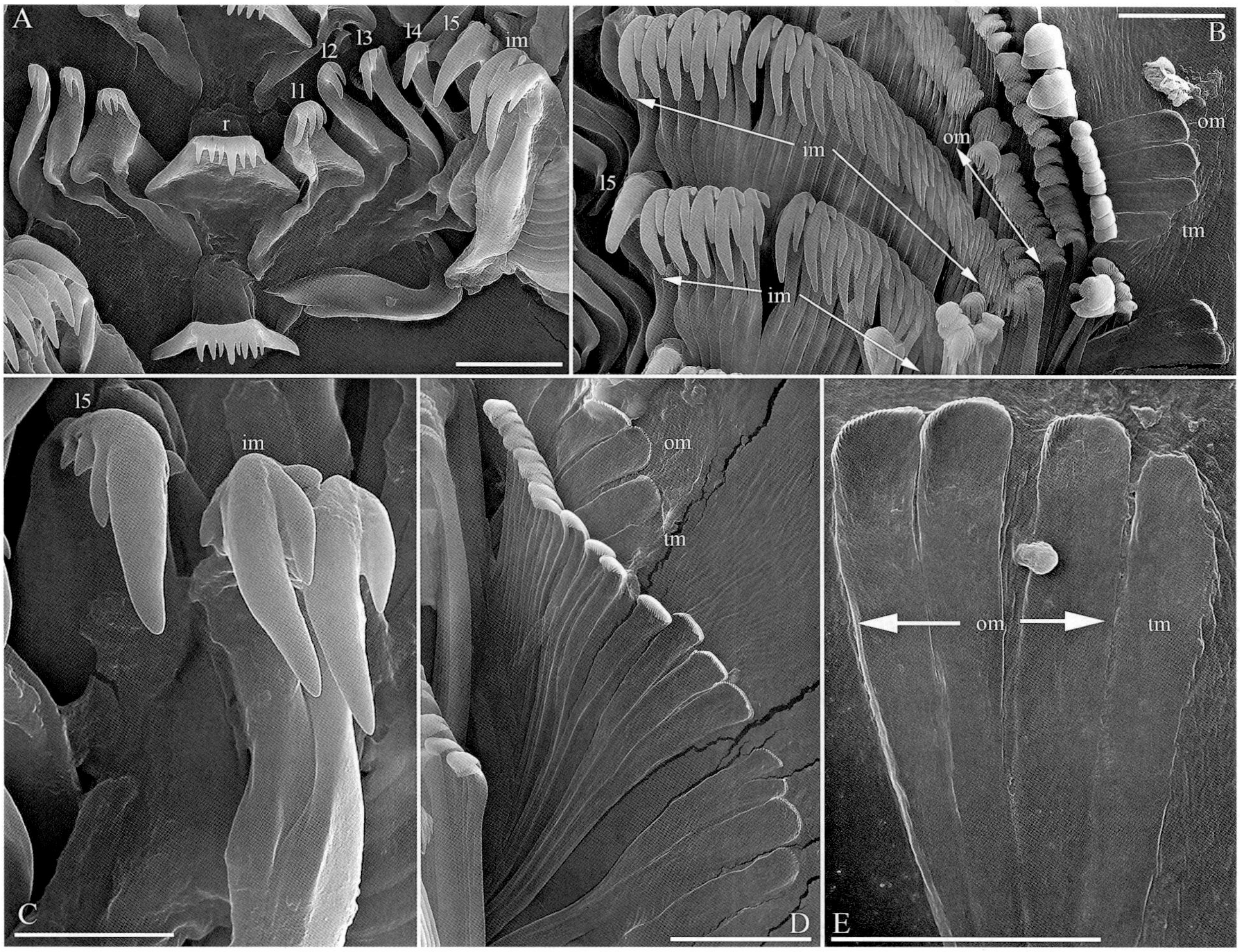

Figure 744. Radula of *Anatoma janetae*. Holotype. 2574 m, Dive 3941 R/V *Alvin*, East Pacific Rise, North Pacific Ocean, 12°42.702' N, 103°54.452' W (FMNH 307218). **A.** Central field. **B.** Marginal teeth. **C.** Lateral tooth 5 and inner marginal teeth 1. **D.** Outer marginal teeth. **E.** Outer marginal teeth and terminal marginal tooth. Scale bars = 20 µm. r: rachidian tooth. l1–5: Lateral tooth number 1–5. im: inner marginal teeth. om: Outer marginal teeth. tm: terminal marginal tooth.

Anatoma kelseyi from the northeastern Pacific has a protoconch with reticulate sculpture, a much wider sutsel, and axial cords crossed by finer spiral lines over the entire shell.

Anatoma lyra from the northeastern Pacific has distinct reticulation or strong axial cords to low lamellae on shoulder and base of entire teleoconch, and a teleoconch I of 0.33–0.5 whorl.

Anatoma soyoae from the North Pacific has a more globular shell outline and very fine spiral lines and axial cords over entire shell.

Anatoma sagamiana and *An. soyoae*, both from Japan, have fine reticulate sculpture over the entire shell.

Anatoma crispata from the northeastern Atlantic has a protoconch without apertural varix, a spiral cord in position of the selenizone on teleoconch I, and spiral lines instead of steps.

Anatoma rainesi from the Western to the Eastern Pacific, has a less convex shoulder, a sunken protoconch, and a smaller expansion rate of the whorls (1.3 vs. 1.4).

Figure 745. Distribution of *Anatoma janetae*. ☆ type locality. ● specimen record.

Distribution. Bathyal of northeastern Pacific and Panamic (Fig. 745).

Specimen records.

USA. 850–900 m, Nittnat Canyon, NW Washington State, 48.167°N, 126.000°W (HLC, 1). 2606 m, Baby Bare Seamount, Cascadia Basin, Washington State, 47.710°N, 127.786°W (FMNH 282781, 1).

Remarks. The two type specimens are broken. The holotype is close to mature, while the paratype consists of the early whorls only, missing the protoconch. The two specimens are considered conspecific due to the very similar structure of the early teleoconch. Teleoconch I of the specimens differ slightly in whorl numbers (0.6 and 0.75), but intraspecific variation in the family is generally near 0.25 whorls (Geiger, 2003); the observed difference of 0.15 whorl is unremarkable. The spacing of the fine axial lamellae, as well as the strength and number of the spiral threads are identical in both specimens, as is the superimposition of axials and spirals. The protoconch of the holotype is mostly eroded. To better show the remaining sculpture, the protoconch is illustrated not in the normal apical orientation, but at an angle, and the end of the protoconch is highlighted with arrows (Fig. 742C).

One further specimen from Washington is well-preserved, and shows the typical transition of axial cords to spiral steps on shoulder and base.

The marginal teeth of *An. janetae* show unique characteristics. In all examined anatomid species, the marginal teeth have multiple small cusps on either side, while in *An. janetae* the two enlarged cusps on the outer edge are unique among anatomids as well as scissurellids s.s.

Collection of the holotype with the animal at 2574 m depth documents that the species lives in the area, rather than having been transported from shallower depths. The absence of eyes in this species is consistent with the species living below the photic zone.

Anatoma janusa new species

Figures 746–748

Chresonymy.

Anatoma sp.: Beck *et al.*, 2006: 44, text-fig. (Fig. 746).

Type material. Holotype (NMR 993000071563). Paratype from type locality (NMR 993000071564, 1). 0 m, Prainha, Pico, 38.472°N, 28.199°W (NMR 993000071565, 1).

Type locality. 0 m, São Roque do Pico, Pico, Azores, Portugal, 38.521°N, 28.307°W.

Etymology. Derived from the Roman god Janus, usually represented with two faces, referring to the different sculpture on the shoulder and the base of the species; jaunus, -a, -um: Janusian, Janus-like; adjective.

Description. Shell small (to 1.2 mm: holotype), trochiform, globular. Protoconch of 0.75 whorl, sculpture flocculent, no apertural varix, apertural margin straight to slightly convex. Teleoconch I of 0.5 whorl, approximately 10 strong axial cords, no spirals. Teleoconch II of 1.75 whorl, suture moderately impressed, sutsel as wide as selenizone. Shoulder sinusoidal in profile, strong axial cords in adsutural quarter elevated to low lamellae; 2–3 additional fine axial cords between each major cord on keels of selenizone; first spiral lines after 1–1.5 teleoconch II whorls, more concentrated in absutural half of shoulder, up to eight at apertural margin of large specimens. Base with weak constriction below selenizone; many very dense axial cords, interstices as wide as cords, continuing onto lower keels of selenizone, select cords evanescing towards umbilicus; several fine spiral lines in distinct pattern: no lines immediately below selenizone; approximately 6–8 in quick sequence forming fine reticulate pattern with axials, towards umbilicus irregularly widely spaced. Umbilicus narrow, with funiculus. Aperture rounded, flared lower adumbilical corner. Selenizone at periphery, rather wide; lunules distinct; keels quite elevated, moderately strong; slit open, margins parallel.

Figure 746. *Anatoma janusa* new species from Beck *et al.* (2006).

Comparisons. *Anatoma aspera* from the northeastern Atlantic and the Mediterranean has a protoconch with an apertural varix, a teleoconch I of 0.75 whorl, and axial cords that are of the same density on shoulder and base.

Anatoma bertheloti from the Canary Islands is more box-shaped in overall profile, has at least twice as dense axial cords on the shoulder, and the density of the axial cords is the same on shoulder and base.

Anatoma crispata from the boreal northeastern Atlantic has a rounder overall shell shape, a teleoconch I of 0.75 whorl, and axial cords that have the same density on shoulder and base.

Anatoma eximia from the subtropical northeastern Atlantic and the Western Mediterranean has a shell with a stepped profile, a protoconch with an apertural varix, and overall weaker sculpture that is similar on shoulder and base.

Anatoma micalii from the Mediterranean to Angola has a spiral cord in the position of the selenizone on teleoconch I, more axial cords on teleoconch I (17 vs. 10), and axial cords on the base that are more widely spaced at twice the width of the cords (as wide as cords in *An. janusa*).

Anatoma orbiculata from West Africa has a more globular shell shape, irregular reticulate protoconch sculpture, and a more narrow sutsel on early whorls (<0.5× vs. >1× width of selenizone).

Figure 747. **A.** *Anatoma janusa* new species. **A.** Holotype. 0 m, São Roque do Pico, Pico, Azores, Portugal, 38.521°N, 28.307°W (NMR 993000071563). **B.** Paratype. 0 m, São Roque do Pico, Pico, Azores, Portugal, 38.521°N, 28.307°W (NMR 993000071564, 1). **C.** Paratype. 0 m, Prainha, Pico, Azores, Portugal, 38.472°N, 28.199°W (NMR 993000071565, 1). Scale bars shell = 0.5 mm. Scale bars protoconch = 100 μm.

Anatoma philippiana from the Paleogene of Europe has duller axial cords on the shoulder, lacks supernumerary axials on the keels of the selenizone, has at most half as dense axials on the base, and spiral lines over the entire base.

Anatoma rolani from the Caribbean has a smooth protoconch and much denser axial cords on the shoulder, in which the interstices are more narrow than the width of the cords.

Anatoma schioettei from the boreal northeastern Atlantic has a teleoconch I of 0.66 whorls and weaker axial cords of same density on shoulder and base.

Anatoma tenuis from the northeastern Atlantic has a much larger shell (>5 mm), no sutsel, and much finer sculpture on shoulder and base.

Anatoma tenuisculpta from the Mediterranean and Norway to Angola has a more disciform overall shape, a sutsel that is more narrow than the width of the selenizone, and sculpture on shoulder and base that is finer and the same.

Anatoma umbilicata from the Mediterranean and northeastern Atlantic has a wider umbilicus and very faint sculpture (if any) that is the same on shoulder and base.

Anatoma lyra from the northeastern Pacific has a teleoconch I of 0.66 whorl and the widely spaced axial cords are of the same density on shoulder and base.

Anatoma alternatisculpta from the Caribbean has more widely spaced axial cords on the base (interstices >2× vs. 1× width of cords) and has a wider sutsel on the early whorls (> vs. < width of selenizone).

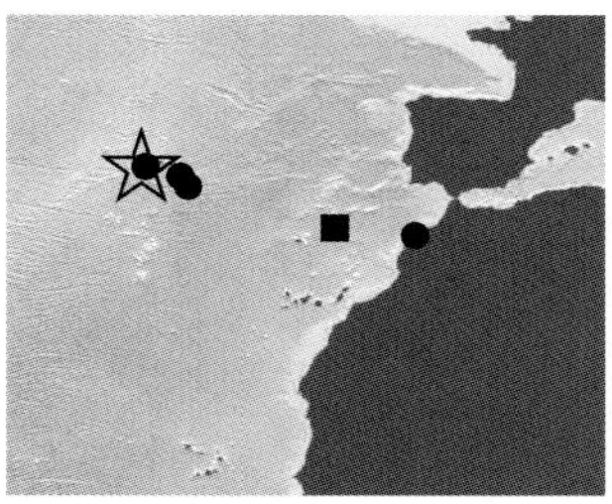

Figure 748. Distribution of *Anatoma janusa* new species. ☆ type locality. ● specimen record. ■ literature record.

Distribution. Tropical and subtropical northeastern Atlantic (Azores, Morocco. Fig. 748)

Specimen records.

Portugal, Azores. 6 m, Vila do Porto, Santa Maria Island, 36.980°N, 25.100°W (MNHN, 2). 0–1 m, Ponta do Marvão, Santa Maria Island, 36.980°N, 25.100°W (MNHN, 1). 0 m, Prainha, Pico, 38.472°N, 28.199°W (NMR 993000070940, 3). Ponta Delgada, San Miguel, 37.740°N, 25.662°W (RMNH AD 10067, 1).

Morocco. 0 m, El Jadida, 33.266°N, 8.483°W (MNHN, 2).

Literature records.

France. 170 m, Seine Seamount, 33.833°N, 14.333°W (Beck *et al.*, 2006: as *Anatoma* sp.).

Remarks. The specimen illustrated in Beck *et al.* (2006: 44, text-fig. Fig. 746) is tentatively identified as *An. janusa*. It agrees in overall shell shape, sculpture of shoulder and base and width of the sutsel. Repeated attempts at contacting the authors to locate the material were unsuccessful. Therefore, the best specimen shown in Beck *et al.* (2006) could not be designated as the holotype.

Anatoma japonica (A. Adams, 1862)

Figures 749–760

Chresonymy.

Anatomus japonicus A. Adams, 1862: 347, not illustrated.
Scissurella japonica: Crosse, 1863: 109.
Anatomus japonicus: Troschel, 1863: 702.
Anatomus japonicus: Dunker, 1882: 147.
Scissurella japonica: Paetel, 1883: 47.
Scissurella japonica: Paetel, 1888: 289.
Anatomus japonicus: Pilsbry, 1890: 59.
Scissurella japonica: Pilsbry, 1895: 106.
Scissurella japonica: Thiele, 1912a: 17, pl. 2, fig. 14.
Scissurella (*Schizotrochus*) *japonica*: Kuroda & Kinoshita, 1951: 8.
Schizotrochus japonicus: Habe, 1951: 65–66, pl. 11, fig. 9.
Scissurella japonica: Kuroda & Habe, 1952: 84.
Scissurella (*Schizotrochus*) *japonica*: Kawamoto & Tanabe, 1956: pl. 2, fig. 12 [copy figure Thiele, 1912a].
Schizotrochus japonicus: Habe, 1961: 2, pl. 1, fig. 7.
Anatoma japonicus [*sic*]: Habe & Kosuge, 1964: 4.
Anatoma japonica: Habe, 1964: 2, pl. 1, fig. 7.
Scissurella (*Anatoma*) *japonica*: McLean, 1967: 405–406, pl. 56, figs 3–4 [one of the syntypes].
Anatoma japonica: Higo, 1973: 14.
Anatoma japonica: auct, 1976: 25, pl. 4, fig. (syntype) [*fide* Tsuchida *et al.*, 1991].
? *Scissurella* (*Anatoma*) *japonica*: Ladd, 1982: 20, pl. 1, figs 1–2.
Anatoma japonica: Inaba, 1982: 72.
Anatoma japonica: Inaba, 1983: 72.
Anatoma japonica: Okutani, 1986: 31, text-fig.
Anatoma ? *japonica*: Herbert, 1986: 617–619, figs 5, 18–20.
Anatoma japonica: Habe *et al.*, 1986: 4.
Anatoma japonica: Watanaba, 1988: 28.
Anatoma japonica: Habe & Masuda, 1990: 6.
Anatoma japonica: Tsuchida *et al.*, 1991: 2–4, figs 1–2 [copy Thiele, 1912a].
Anatoma japonica: Higo & Goto, 1993: 15.
Hainella japonica: Bandel, 1998: 39–40, pl. 13, figs 4–6.
Anatoma japonica: Sasaki, 1998: figs 44c–d.
Anatoma japonica: Okutani & Hasegawa, 2000: 37, pl. 18, figs 1a–b.
Anatoma japonica: Higo *et al.*, 2001: 13, G78 [one of two syntypes].
Anatoma japonica: Geiger, 2003: 73.
Anatoma japonica: Marais & Marais, 2003: 12, text-fig.
Anatoma japonica: Koyama, 2008: 253, pl. 1, fig. 5.
Anatoma sp. 1: Spencer *et al.*, 2009: 201.
Anatoma japonica: Sasaki *et al.*, 2010: 68–69.

Synonyms.

+*Scissurella exquisita* Schepman, 1908: 83, pl. 6, fig. 10 (Fig. 749). ***Type material.*** Holotype (ZMA 3.08.101: Geiger, 2008c: fig. 24. Fig. 751B), 2.75 × 3.5 mm in OD; 2.47 × 3.5 mm from SEM (H × W). ***Type locality.*** Siboga station 95, 5°43.5'N, 119°40'E, 522 m [between

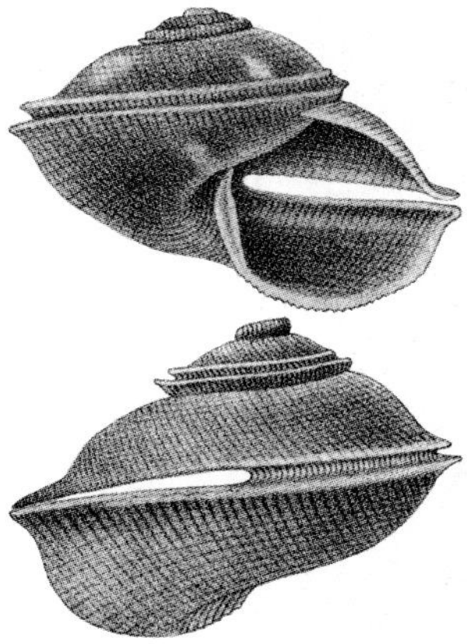

Sabah, Malaysia, and southeastern Tawitawi Group, Philippines] (OD). ***Etymology.*** Exquisitus, Latin for choice, excellent, fine.

Scissurella exquisita: Thiele, 1912a: 13, pl. 2, figs 3–4 [copy figs Schepman, 1908].

Schizotrochus exqisita [sic]: Okutani, 1964: 376–377, pl. 5, fig. 1.

Anatoma exquisita: Higo, 1973: 14.

Anatoma exquisita: Lozouet, 1986: 113.

Anatoma s.l. *exquisita*: Geiger, 2003: 75.

Anatoma s.l. *exquisita*: Geiger, 2008c: 196, fig. 24 [SEM of holotype].

Scissurella/Anatoma s.l. *exquisita*: Van der Bijl *et al.*, 2010: 55.

Anatoma exquisita: Poppe & Tagaro, 2011: 546, pl. 1264, figs 5, 6a–b.

Figure 749. Original illustration *Scissurella exquisita.*

+ *Scissurella aetheria* Melvill & Standen, 1903: 300–301, pl. 21, fig. 6 (Fig. 750A). ***Type material.*** Syntypes from type locality (NMSZ 1961.61, 1; NHMUK 1903.12.15.78–9, 2; NMW 1955.158.88, 5: see Remarks), 1.28 × 2 mm (H × W). NHMUK 1903.12.15.78 (Fig. 752A) designated here as lectotype with the express taxonomic purpose of taxon stabilization. All other syntypes become paralectotypes. ***Type locality.*** 156 fms [= 285 m], Gulf of Oman, 24°58'N, 56°54'E (OD). ***Etymology.*** Aither, Greek for the upper atmosphere.

Scissurella aetheria/oetheria n. sp. [*sic*!]: Melvill & Standen, 1904: 95, pl. B, fig. 6 (Fig. 750B) [see Remarks].

Scissurella aetheria: Thiele, 1912a: 12, pl. 1, fig. 25.

Sinezona sp.: Yaron, 1983: 274–275, pl. 8, figs 1–4 [is *An. aetheria*].

Anatoma aetheria: Bosch *et al.*, 1995: 28, text-fig. 1.

Anatoma s.l. *aetheria*: Geiger, 2003: 74.

Anatoma aetherica [*sic*]: Bruyne, 2003: 48, text-fig.

Figure 750. **A.** Original illustration of *Scissurella aetheria.* **B.** Illustrations from Melvill & Standen (1904).

+ *Anatoma jansenae* Geiger, 2006a: 24–28, fig. 18. ***Type material.*** Holotype (AMS C.402717. Fig. 752B). Paratypes from type locality (AMS C.450272, 7). 274 m, NW of Beagle Island, Western Australia, Australia, 29.717°S, 114.283°E (AMS C.402720, 7). 197 m, NW of Green Head, Western Australia, Australia, 29.967°S, 114.450°E (AMS C.402721, 1). ***Type locality.*** 238–183 m, SW of Cape Naturaliste, Western Australia, Australia, 33.742°S, 114.435°E (OD). ***Etymology.*** Named after Patty Jansen, New South Wales, Australia, who first recognized the distinctness of the species, and for her contributions to the knowledge of Australian micromollusks (OD).

Anatoma SWA: Jansen, 1999: 50, figs 16–18.

Anatoma australis partim: Geiger & Jansen, 2004a: fig. 5.

Figure 751. *Anatoma japonica*. **A.** *Scissurella japonica*. Lectotype, here designated. Approximately 113 m, Yamaguchi Prefecture, in Japan Sea (NHMUK 1878.1.28.514). **B.** *Scissurella exquisita*. Holotype. 522 m, Siboga Station 95 [between Sabah, Malaysia, and southeastern Tawitawi Group] (ZMA 3.08.101). Scale bars shell = 1 mm. Scale bars protoconch = 100 μm.

Figure 752. *Anatoma japonica*. **A.** *Scissurella aetheria*. Lectotype, here designated. 156 fms [= 285 m], Gulf of Oman, 24°58'N, 56°54'E (NHMUK 1903.12.15.78). **B.** *Anatoma jansenae*. Holotype. 238–183 m, SW of Cape Naturaliste, Western Australia, Australia, 33.742°S, 114.435°E (AMS C.402717). Scale bars shell = 1 mm. Scale bars protoconch = 100 μm.

Misidentifications.

Anatoma maxima: Okutani & Hasegawa, 2000: 37, pl. 18, figs 3a–b [is *An. japonica*].
Anatoma cf. *japonica*: Sasaki, 2006: 105, fig. 5A [is *An.* cf. *pseudoequatoria*].

Type material. Syntypes (NHMUK 1878.1.28.514, 4. McLean, 1967: pl. 56, figs 3–4. Higo *et al.*, 2001: fig. G78), 2.5, 2.7 mm. The best specimen (NHMUK 878.1.28.514/1. Fig. 751A) is here designated as the lectotype for the express purpose of taxon stabilization in case that other non-conspecific syntypes should be found. Paralectotypes (NHMUK 878.1.28.514/2–4, 3).

Type locality. Mino-Sima, 63 fms [= 115 m], Seto-Uchi, Goto, O-Sima [Japan] (OD). Mino-Sima [= Mishima Island, Hagi City, Yamaguchi Prefecture, Japan (34.767°N, 131.166E): see also Kawamoto and Tanabe (1956)], Yamaguchi Prefecture, in Japan Sea, approximately 113 m (explicitly restricted: Herbert, 1986: 617).

Etymology. Named for its type locality in Japan.

Description. Shell large (to 4.4 mm: Okutani, 1964. Figs 751–756), trochiform, biconical, somewhat inflated. Protoconch of 0.75 whorl, sculpture flocculent, apertural varix not connected to embryonic cap, apertural margin sinusoidal. Teleoconch I of 0.6–0.75 whorl, approximately 24 axial cords, spiral cord in position of selenizone. Teleoconch II of 2.5 whorls, early whorls with hardly any sutsel to sutsel of 0.5× width of selenizone; sutsel at apertural margin 1–2× width of selenizone. Shoulder slightly to strongly convex, suture hardly impressed, approximately 65–80 axial cords on first whorl, >100 on last whorl; first spiral lines after 0.75 teleoconch II whorl, approximately 20–50 at apertural margin of mature specimen, running over axials, forming weak knots to distinctly elevated points. Base with same density of axial cords as on shoulder, approximately 20 spiral lines running over axials, forming weak knots to distinctly elevated points. Umbilicus very narrow, continuously sloping with base, without funiculus. Aperture rounded to moderately elliptical, basoadumbilical portion flared, roof overhanging. Selenizone at periphery; lunules moderately distinct; keels moderately elevated, moderately strong; slit open, margins converging.

Animal with eyes, no further data on soft-parts available.

Operculum (Figs 757C, 758E, 759D), round, thin, corneous, multispiral, nucleus central.

Radula (Figs 757–759). Rachidian tooth trapezoidal, central cusp largest, approximately 4–8 smaller cusps on each side. Lateral teeth 1–3 similar, apical cusp largest, 4–2 smaller cusps on outer edge. Lateral tooth 4 reduced, hook-shaped, with one minute point below apex. Lateral tooth 5 enlarged by elongation, apical cusp largest, 6–7 cusps on inner cutting edge. Inner marginal teeth with triangular tip, apical cusp largest, approximately eight cusps on each side; shaft without food groove. Outer marginal teeth with spoon-shaped tip, with many fine bristles along edge. Outermost marginal teeth paddle-shaped with many fine bristles along edge. Radular interlock of central field moderate.

Comparisons. *Anatoma lamellata* from the Western to the Central Pacific has a protoconch with a distinct spiral, fewer but stronger axial cords to low lamellae on the teleoconch, and strong axial lamellae.

Anatoma sagamiana from Japan and the Indo-Malayan Archipelago has reticulate sculpture on the protoconch, an overall more globular shell shape with more stepped whorls, a much larger shell, and about 50% less dense axial and spiral sculpture producing a coarser reticulate pattern.

Anatoma parageia from southern Japan lacks an apertural varix on the protoconch, a spiral in the position of the selenizone on teleoconch I, and has a distinct angulation on the base of the shell.

Anatoma maxima from the Indo-Malayan Archipelago and the Central Pacific has a wider shell, lacks an apertural varix on the protoconch and a spiral cord in the position of the selenizone on teleoconch I, and has a wider selenizone with more elevated keels.

Anatoma agulhasensis from the western Indian Ocean has an overall taller shell with more stepped whorls and a sutsel of continuous width.

Anatoma tabulata from the western Indian Ocean has a more scalar shell shape, a sutsel wider than the selenizone on early whorls, and at most half as many axial cords.

Anatoma yaroni from the western Indian Ocean has finer sculpture and fewer spiral cords (at two teleoconch whorls, 4 vs. 10).

Figure 753. *Anatoma japonica*. 220 m, off Kushimoto, Kii Prefecture, Japan, 33.860°N, 136.350°E (DLG 1189, 4). Radula shown in Figure 757. Scale bar shell = 1 mm. Scale bar protoconch = 100 µm.

Anatoma australis from eastern Australia has the selenizone proportionally higher on the whorl and only every 3–4 axial cords on the base reach the umbilicus.

Anatoma biconica from the Indian Ocean to the Central Pacific has a more sharply biconical shell, is taller, and has a longer teleoconch I (1 vs. 0.5–0.75 whorl).

Figure 754. *Anatoma japonica*. **A.** 161 m, St. 5381, Ragay Gulf, off Arena Point, Luzon, Philippines, 13.238°N, 122.746°E (USNM 284043, 6). **B.** 133 m, summit, Wanganella Bank, Norfolk Ridge, 32.543°S, 167.487°E (NMNZ M.160280, 15). Radula shown in Figure 758. **C.** 350 m, Wallis Island, 13.316°S, 176.283°W (MNHN, 8). Scale bars shell = 1 mm. Scale bars protoconch = 100 μm.

Anatoma flexidentata from the Indian Ocean to the Western Pacific has a more globular overall shell shape, a wider selenizone, and a shorter teleoconch I (0.2 vs. 0.5–0.75 whorl).

Anatoma planapex from the Indo-Malayan Archipelago and Australia has weaker sculpture of axial and spiral lines, lacks sculpture immediately below the selenizone on the base, and has an overall more boxy shell shape.

Figure 755. *Anatoma japonica*. **A.** 305 m, Nuku Hiva Island, Marquesas Islands, 9.816°S, 140.283°W (MNHN, 1). **B.** 100 m, off southern Natal, South Africa (DLG 347, 6). **C.** 350 m, Wallis Island, Wallis and Futuna, 13.316°S, 176.283°W (MNHN, 2). Scale bars shell = 1 mm. Scale bars protoconch = 100 μm.

Anatoma globulus from the Indo-Malayan Archipelago has an overall more globular shell, lower keels of the selenizone, and more constant width of the selenizone.

Anatoma rainesi from the Western to Eastern Pacific has fine axial cords on the protoconch, strong axial cords on teleoconch I with nodular thickenings in the position of the selenizone, and more marked difference in the strength of axial cords to spiral lines.

Anatoma rapaensis from the Indo-Malayan Archipelago to the Central Pacific has a protoconch without apertural varix, a shorter teleoconch I (0.4 vs. 0.5–0.75 whorl) without spiral cord in position of the selenizone, and spiral steps on the shoulder, particularly close to the suture.

Anatoma regia from temperate New Zealand has a more elongated aperture, more convex shoulder, and is taller.

Anatoma proxima from the Caribbean to Brazil has a slightly more globular shell shape, usually denser and finer axial cords, particularly on the base, and less elevated keels of the selenizone.

Anatoma atlantica from the northwestern Atlantic always lacks a spiral in the position of the selenizone on teleoconch I and tends to have denser axial sculpture.

Anatoma tenuisculpta from the Mediterranean and North Atlantic tends to have denser axial cords particularly on the base.

Anatoma tobeyoides from southeastern Australia has a more globular overall shell outline, oblique very fine spiral lines on the early teleoconch, and a shorter teleoconch I (0.2 vs. 0.5–0.75 whorl).

Distribution. Recent: Indian Ocean to the Central Pacific. Pleistocene of Kyushu, Japan, and Vanuatu (Fig. 760).

Specimen records.

South Africa. 100 m, Zululand (DLG 1369, 2). 100 m, off Richards Bay, Zululand, 28.901°S, 32.226°E (DLG 1378, 11). 100 m, Park Rynie, Natal, 30.357°S, 30.851°E (DLG 1385, 11). 100 m, off S Natal, 30.600°S, 30.700°E (DLG 347, 6). 90 m, Mbotyi, Transkei, 31.463°S, 29.737°E (DLG 1377, 10).

Reunion Island. 80–83 m, 21.100°S, 55.217°E (MNHN, 1). 205–215 m, 21.383°S, 55.617°E (MNHN, 1). 20 m, Cap La Houssaye, Saint Paul, 20.017°S, 55.234°E (MJC ex 1881, 1; MJC 2816, 12).

Figure 756. *Anatoma japonica*. Z-stacked light micrograph showing dark axial color bands. 100 m, Park Rynie, Natal, South Africa, 30.357°S, 30.851°E (DLG 1385, 11). Scale bar = 1 mm.

Madagascar. 200–220 m, Manantenina sector, ATIMO VATAE DW3523, 24.383°S, 47.517°E (MNHN, 1; MNHN, 5). 307–319 m, Manantenia sector, ATIMO VATAE DW3524, 24.383°S, 47.533°E (MNHN, 22). 424–438 m, Manantenina sector, ATIMO VATAE DW3528, 24.400°S, 47.550°E (MNHN, 2). 86–87 m, N of St. Luce, ATIMO VATAE DW3532, 24.656°S, 47.528°E (MNHN, 1). 296–307 m, St. Luce sector, ATIMO VATAE DW3534, 24.716°S, 47.533°E (MNHN, 5). 80–83 m, between Lokaro and St. Luce, ATIMO VATAE DW3519, 24.865°S, 47.467°E (MNHN, 1). 184–203 m, between Lokaro and St. Luce, ATIMO VATAE DW3515, 24.883°S, 47.467°E (MNHN, 12). 113–135 m, SE Barrow Point, ATIMO VATAE DW3588, 25.050°S, 44.000°E (MNHN, 2). 140–144 m, Fort Dauphin sector, ATIMO VATAE CP3512, 25.250°S, 47.283°E (MNHN, 1). 209–229 m, S Barrow Point, ATIMO VATAE DW3581, 25.500°S, 44.267°E (MNHN, 1). 618–624 m, E of False Cape, ATIMO VATAE CP3566/DW3565, 25.608°S, 46.350°E (MNHN, 2). 122–123 m, Lavanono sector, ATIMO VATAE DW3599, 25.750°S, 44.417°E (MNHN, 1). 88–90 m, S of False Cape, ATIMO VATAE CP3580, 25.976°S, 45.582°E (MNHN, 1). 264–280 m, S of False Cape, ATIMO VATAE DW3552, 26.116°S, 45.650°E (MNHN, 2). 282–

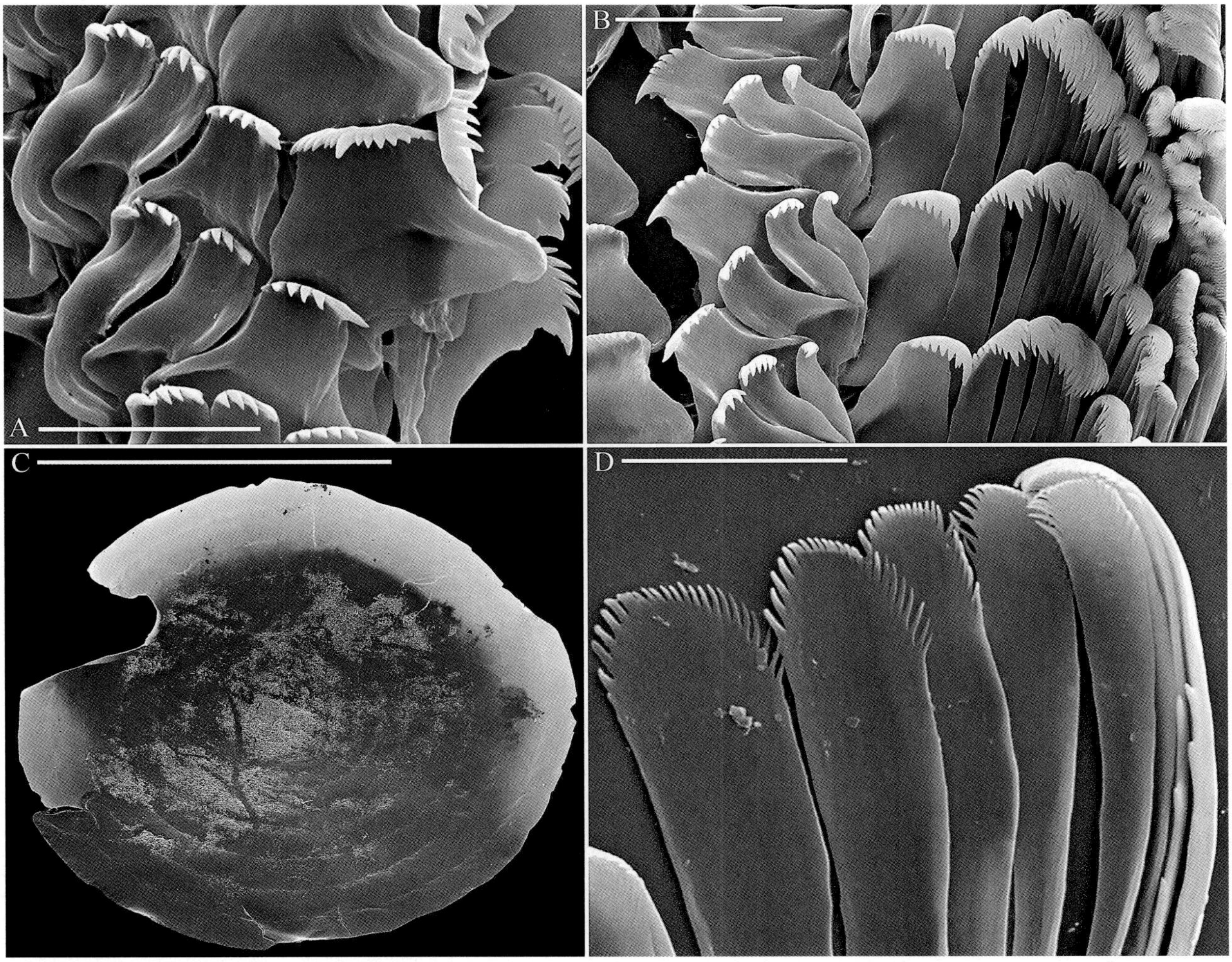

Figure 757. *Anatoma japonica*. Radula and operculum from specimen shown in Figure 753. 220 m, off Kushimoto, Kii Prefecture, Japan, 33.860°N, 136.350°E (DLG 1189, 4). **A, B, D.** Radula. **A.** Central field. Scale bar = 20 µm. **B.** Lateral teeth and marginal teeth. Scale bar = 20 µm. **D.** Outermost marginal teeth. Scale bar = 10 µm. **C.** Operculum. Scale bar = 1 mm.

333 m, S of False Cape, ATIMO VATAE DW3557, 26.133°S, 45.650°E (MNHN, 4). 280–333 m, S of False Cape, ATIMO VATAE DW3553, 26.133°S, 45.650°E (MNHN, 1). 455–458 m, S of False Cape, ATIMO VATAE DW3555, 26.150°S, 45.667°E (MNHN, 1). 225–282 m, S of Cape St. Marie, ATIMO VATAE CP3613, 26.216°S, 45.133°E (MNHN, 3; MNHN, 1). 284–286 m, S Cape St. Marie, ATIMO VATAE CP3615, 26.233°S, 45.150°E (MNHN, 7). 250–300 m, S of Cape St. Marie, ATIMO VATAE CP3614, 26.233°S, 45.150°E (MNHN, 6).

Mozambique. 352–357 m, Zambeze transect, Mozambique Channel, MAINBAZA CC3151, 19.566°S, 36.750°E (MNHN, 6). 101–102 m, Maputo transect, MAINBAZA CP3132, 25.183°S, 35.033°E (MNHN, 1).

Sudan. 95 m, Sanganeb, 19.718°N, 37.425°E (DLG 895, 5). 60 m, Sha'ab Rumi, 19.900°N, 37.400°E (DLG 896, 1; DLG 899, 1).

Egypt. 25 m, Dunraven Reef, Strait of Tiran, S Sinai Peninsula, N Red Sea, 27.200°N, 34.100°E (LACM 85-112, 4).

Oman. Gulf of Oman, (FLMNH 00132407, 1; NHMUK 1903.12.15.78–9, 2; NMSZ 1961.61, 1; NMW 1955.158.88).

Australia, Western Australia. 150 m, North West Shelf, N of Port Headland, 18.666°S, 117.917°E (AMS exC.436908, 1). 27 m, Sahul Banks, Timor Sea, 11.500°S, 125.500°E (AMS C.377600, 1). 210 m, approx. 170 mi N of Broome, 15.450°S, 121.517°E (AMS, 1). 238 m, North West Shelf, ca. 230 mi W Roebuck Bay, 18.500°S, 118.050°E (AMS C.402676, 1). 150 m, North West Shelf, N of Port Headland, 18.666°S, 117.917°E (AMS exC.436908, 1). 190 m, off North West Cape, 22.386°S, 113.677°E (AMS, 1). 108 m, off North West Cape, 22.705°S, 113.540°E

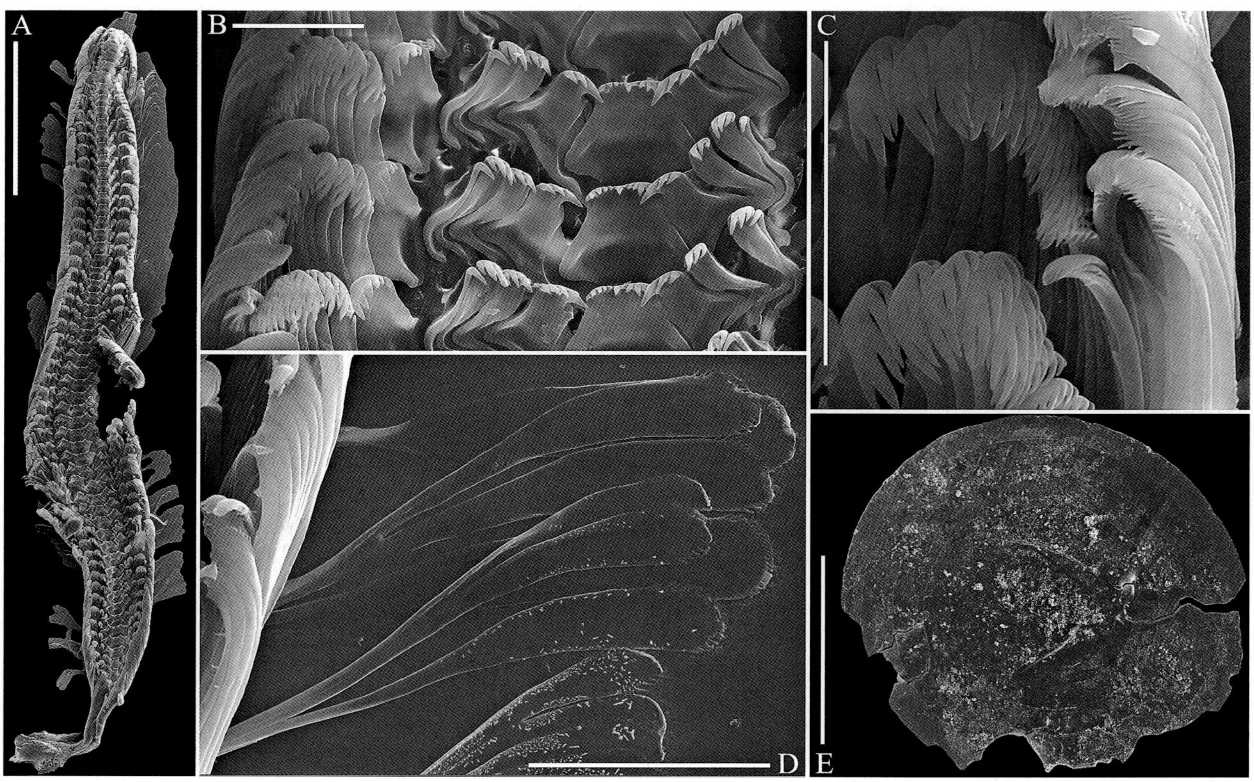

Figure 758. *Anatoma japonica*. Radula. 133 m, summit, Wanganella Bank, Norfolk Ridge, 32.543°S, 167.487°E (NMNZ M.160280, 15). Shell shown in Figure 754B. **A.** Entire radula. **B.** Half width of three rows. **C.** Marginal teeth. **D.** Outermost marginal teeth. **E.** Operculum. Scale bars A–D = 20 µm. Scale bar E = 500 µm.

(AMS C.402644, 1; AMS C.402645, 1). Point Quobba, N of Carnarvon, 24.483°S, 113.417°E (AMS C.379976, 1). 183 m, W of Dongara, 29.142°S, 113.913°E (AMS C.402743, 1). 219 m, W of Dongara, 29.183°S, 113.900°E (AMS C.402744, 1). 192–219 m, NW of Beagle Island, 29.346°S, 114.095°E (WAM, 2). 146 m, W of Dongara, 29.350°S, 114.117°E (AMS C.402742, 1). 216–256 m, NW of Beagle Island, 29.716°S, 114.333°E (WAM, 3). 274–283 m, NW of Beagle Island, 29.717°S, 114.283°E (AMS C.402720, 7). 183 m, NW of Beagle Island, 29.725°S, 114.333°E (AMS C.402746, 1). 197–219 m, NW of Green Head, 29.967°S, 114.450°E (AMS C.402721, 1). 223–245 m, off Jurien Bay, 30.133°S, 114.500°E (AMS C.402745, 1). 256–192 m, NW

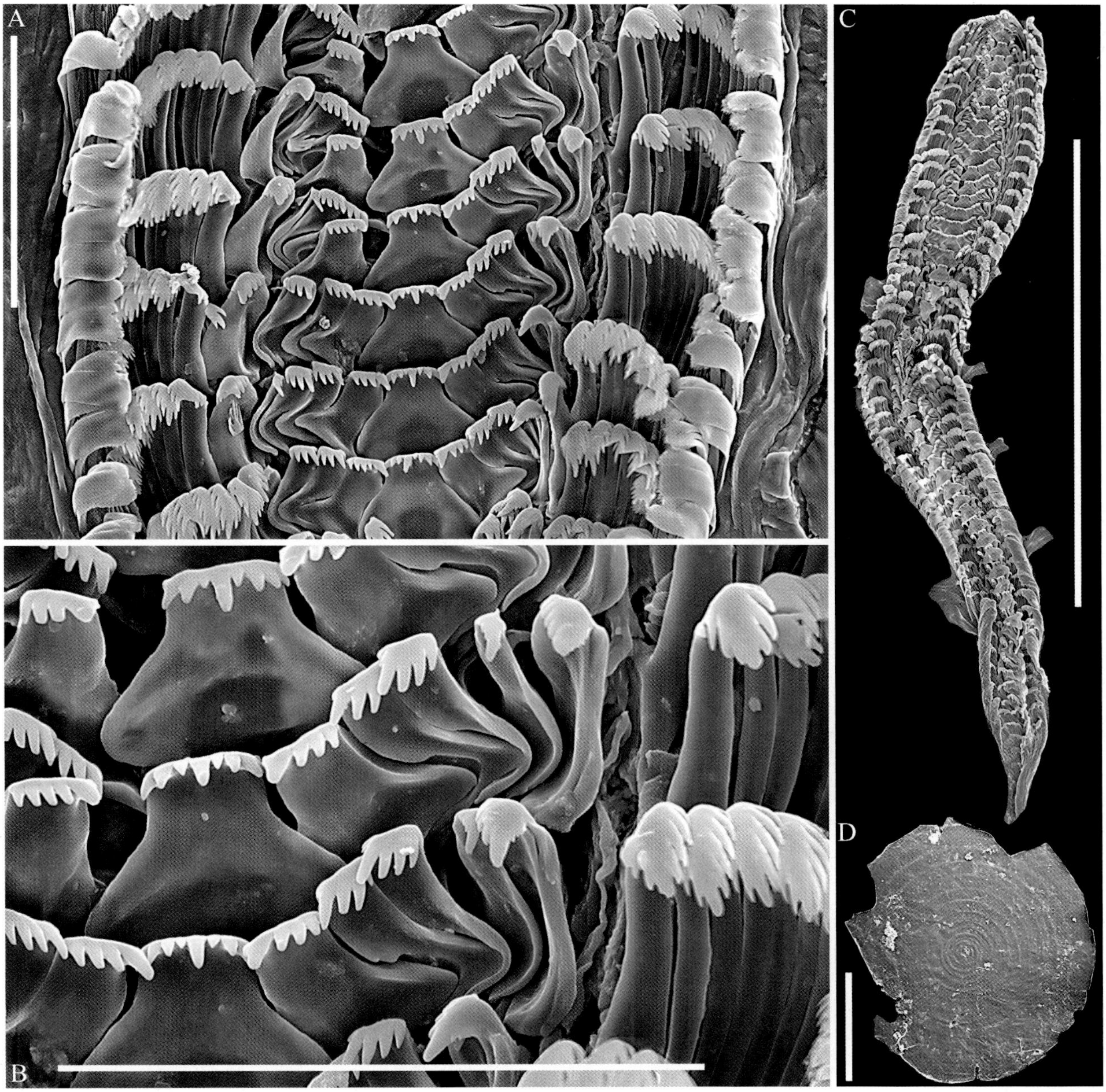

Figure 759. *Anatoma japonica*. Radula and operculum. 88 m, off Çooper Light, Natal, South Africa, 29.937°S, 31.170°E (NMSA D4422, 3). **A.** Full width of radula. Scale bar = 50 µm. **B.** Central field and inner marginal teeth enlarged. Scale bar = 50 µm. **C.** Entire radula. Scale bar = 500 µm. **D.** Operculum. Scale bar = 500 µm.

of Cervantes, 30.500°S, 114.633°E (AMS C.402718, 10). 238–247 m, off Cervantes Island, 30.533°S, 114.683°E (AMS C.402752, 2). 237–274 m, W of Green Head, 30.750°S, 114.767°E (AMS C.402748, 1). 212–215 m, WNW of Lancelin, 30.966°S, 114.817°E (WAM, 1). 183 m, WNW of Lancelin, 30.966°S, 114.850°E (WAM, 1). 732 m, W of Rottnest Island, 31.083°S, 114.767°E (AMS C.402751, 1). 160 m, off Rottnest Island, 31.650°S, 115.080°E (AMS C.402749, 1). 210–212 m, W of Garden Island, 32.250°S, 115.117°E (AMS C.402750, 2). 176–182 m, W of Garden Island, 32.262°S, 115.112°E (AMS C.402741, 1). 200–221 m, NW of Bunbury, 33.000°S, 114.617°E (AMS C.402747, 6). 183 m, NW of Bunbury, 33.000°S, 114.683°E (WAM, 5). 201–228 m, NW of Bunbury, 33.250°S, 114.533°E (WAM, 16). 155 m, NW of Bunbury, 33.250°S, 114.617°E (AMS C.402719, 1). 238–183 m, SW of Cape Naturaliste, 33.742°S, 114.435°E (AMS C.402717, 1; AMS C.450272, 7).

Indonesia. 558 m, St. 5592, off Silungan Island, Borneo, 4.212°N, 118.462°E (USNM 279050, 9).

Philippines. 100 m, off Hadsan Beach Resort, Mactan Island, 10.281°N, 123.995°E (GPC 221118, 1). 161 m, St. 5381, Ragay Gulf, off Arena Point, Luzon, 13.237°N, 122.746°E (USNM 284043, 6). 192 m, St. 5217, Ragay Gulf, off N Burias, 13.333°N, 123.238°E (USNM 280513, 1). Aliguay Island, 2.746°N, 123.224°E (DLG 1591, 2). 320 m, St. 5579, N off Sibutu Island, Sulu Archipelago, 4.904°N, 119.164°E (USNM 287292, 5). 439 m, St. 5577, N of Tawi Tawi, 5.343°N, 119.981°E (USNM 286890, 4). 622 m, off Simalac Island,Tawitawi Islands, Sulu Archipelago, 5.512°N, 120.133°E (USNM 1099224, 7). 37 m, St. 5140, off Jolo, 6.145°N, 121.050°E (USNM 275324, 1). 172–175 m, Dipolog Bay, Bohol-Sulu Sea Sill, PANGLAO 2005 DW2371, 8.580°N, 123.270°E (MNHN, 2). 259–280 m, Dipolog Bay, Bohol-Sulu Sea Sill, PANGLAO 2005 CP2381, 8.721°N, 123.317°E (MNHN, 1). 134–190 m, Cervera Shoal, W Pamilacan Island, PANGLAO 2004 T34, 9.470°N, 123.845°E (MNHN, 1). 95–128 m, Cervera Shoal, Pamilacan Island, PANGLAO 2004 T36, 9.488°N, 123.858°E (MNHN, 35). 55–81 m, Pamiacan Island, PANGLAO 2004 L65-68, 9.498°N, 123.918°E (MNHN, 2). 90–98 m, Balicasag Island, PANGLAO 2004 L69-73, 9.511°N, 123.683°E (MNHN, 2). 90–110 m, Balicasag Island, PANGLAO 2004 L46, 9.515°N, 123.687°E (MNHN, 3; MNHN, 1). 145–163 m, between Libaong and Pamilacan, PANGLAO 2004 T34, 9.521°N, 123.857°E (MNHN, 1; MNHN, 1). 309–342 m, off Balicasag Island, Bohol Sea, CP2399, 9.528°N, 123.698°E (MNHN, 1). 164–176 m, off Balicasag Island, Bohol Sea, DW2339, 9.531°N, 123.728°E (MNHN, 1). 152 m, Bolod, Panglao Island, PANGLAO 2004 T2, 9.540°N, 123.797°E (MNHN, 1; MNHN, 46). 90–200 m,

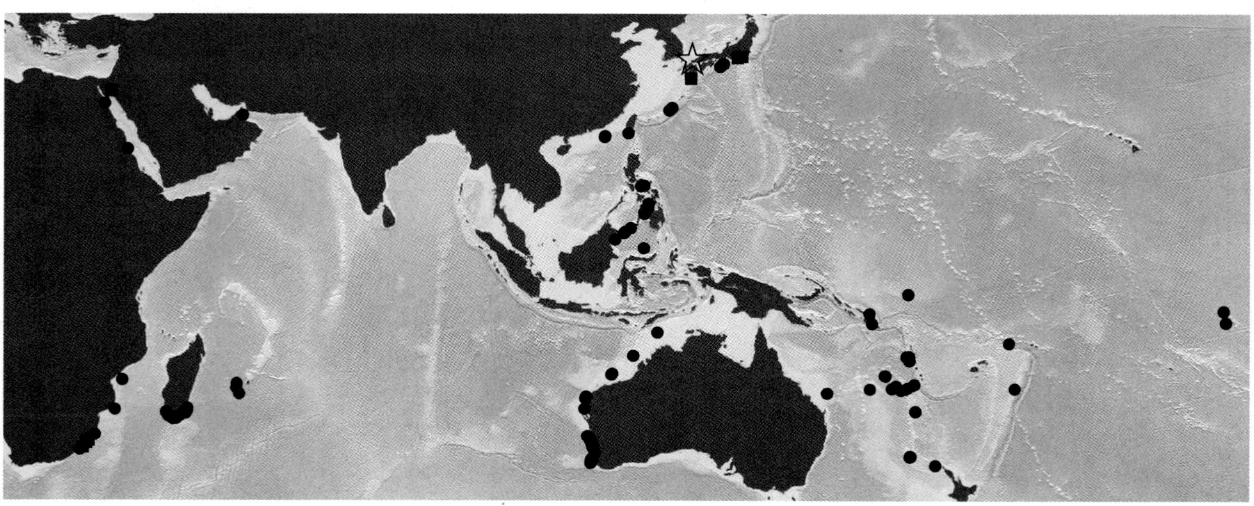

Figure 760. Distribution of *Anatoma japonica.* ☆ type locality. ● specimen record. ■ literature record.

Maribohoc Bay, Bohol Island, PANGLAO 2004 P1, 9.600°N, 123.750°E (MNHN, 4). 120 m, off Momo Beach, Panglao Island, PANGLAO 2004 L50, 9.615°N, 123.763°E (MNHN, 1). 43–62 m, Bingag/Tabalong, Panglao Island, PANGLAO 2004 L51-60, 9.628°N, 123.800°E (MNHN, 1; MNHN 3). 62 m, Bigag/Tabalong, Panglao Island, PANGLAO 2004 L51–60, 9.628°N, 123.800°E (MNHN, 2). 100 m, Manga, Bohol Island, PANGLAO 2004 L48, 9.705°N, 123.832°E (MNHN, 1). 624–647 m, off Aligbay Island, Bohol-Sulu Sill, PANGLAO 2005 CP2384, 8.770°N, 123.268°E (MNHN, 3).

China Sea. 274 m, St. 5313, off Pratas Island, 21.500°N, 116.717°E (USNM 284940, 3). 223 m, St. 5314, off Pratas Island, 21.683°N, 116.767°E (USNM 284576, 1).

Taiwan. 190–200 m, SW coast, TAIWAN 2002 CP162, 22.160°N, 120.632°E (MNHN, 4).

Japan. (RM 1442, 1). 211–212 m, off Kerama Island, Okinawa, 26.046°N, 127.375°E (SBMNH 149454, 5). 50 m, Okinawa, Onna Village, 26.493°N, 127.842°E (LACM 79-75, 1). 200 m, off Kushimoto, Kii, 33.396°N, 135.772°E (DLG 1258, 3). 220 m, off Kushimoto, Kii Prefecture, 33.860°N, 136.350°E (DLG 1189, 4). Mino-Sima, 34.767°N, 131.166°E (NHMUK 1878.1.28.514, 4).

Solomon Islands. 98–200 m, SALOMON 1 DW1767, 8.323°S, 160.665°E (MNHN, 3). 804–936 m, 9.985°S, 161.097°E (MNHN, 1).

Coral Sea. 6 m, Saumarez Reef, 21.817°S, 153.667°E (AMS C.402669, 1).

Australia, Norfolk Island. 201 m, off, 20.333°S, 168.150°E (NMNZ M.224760, 3).

Norfolk Ridge. 508–541 m, Sponge Bank, NORFOLK 1 DW1684, 24.916°S, 168.367°E (MNHN, 5). 507–967 m, Sponge Bank, NORFOLK DW1692, 24.933°S, 168.350°E (MNHN, 6). 133 m, summit, Wanganella Bank, 32.543°S, 167.487°E (NMNZ M.160280, 15). 120–126 m, Wanganella Bank summit, West, 32.610°S, 167.584°E (NMNZ M.171870, 40). 133 m, summit, Wanganella Bank, 32.653°S, 167.528°E (NMNZ M.160277, 5). 660–757 m, Wanganella Bank, 32.680°S, 167.447°E (NMNZ M.160395, 4). 394–400 m, NW of Cape Reinga, Reinga Ridge, 34.111°S, 171.662°E (NMNZ M.171448, 8).

New Caledonia. 613–647 m, northern, 18.816°S, 163.267°E (MNHN, 1). 502–516 m, northern, 18.866°S, 163.383°E (MNHN, 2). 380–400 m, E coast, 20.583°S, 165.117°E (MNHN, 2). 408–440 m, E coast, 20.583°S, 165.117°E (MNHN, 2). 150–200 m, SE Shelter Reef, Santal Bay, Lifou, Loyalty Islands, LIFOU DW1648, 20.900°S, 167.050°E (MNHN, 1). 250–300 m, northern, 21.016°S, 164.467°E (MNHN, 1). 569–583 m, South Lansdowne EBISCO DW2629, 21.100°S, 160.767°E (MNHN, 1). 237–298 m, E coast, 21.283°S, 165.950°E (MNHN, 1). 565–600 m, southern, 24.883°S, 168.350°E (MNHN, 1).

Vanuatu. 100–105 m, NW Tutuba Island, SANTO 2006 DS99, 15.541°S, 167.282°E (MNHN, 10). 100 m, W Tutuba Island, Aoré Island, SANTO 2006 DS93, 15.560°S, 167.275°E (MNHN, 2). 45–101 m, E Aoré Island, SANTO 2006 EP10, 15.595°S, 167.156°E (MNHN, 1). 360–419 m, MUSORSTOM 8 DW1065, 16.266°S, 167.350°E (MNHN, 3). 205–210 m, MUSORSTOM 8 DW1101, 5.066°S, 167.133°E (MNHN, 1).

Tonga. 433 m, NW Tongatapu, BORDAU 2 DW1569, 21.033°S, 175.317°W (MNHN, 2).

Wallis and Futuna. 350 m, Wallis Island, 13.316°S, 176.283°W (MNHN, 2; MNHN, 8).

Marquesas Islands. 1000 m, Eiao Island, 7.866°S, 140.650°W (MNHN, 1). 305 m, Nuku Hiva Island, 9.816°S, 140.283°W (MNHN, 1).

Fossil specimen.

Vanuatu. Station 43, Navaka River, Santo Island, Pleistocene, 15.600°S, 166.863°E (USNM 214325, 1).

Literature records.

Japan. 900 m, Sagami Bay, 35.103°N, 139.440°E (Okutani, 1964: as *An. equisita*).—Suruga Bay [Shizuoka, 34.800°N, 138.600°E] (Habe *et al.*, 1986).—Chochi [31.309°N, 130.579°E] (Watanabe, 1988).

Egypt. 550–700 m, S of Nuweiba, 28.885°N, 34.694°E (Yaron,1983: as *An. aetheria*).

South Africa. 88 m, off Cooper Light, Natal, 29.937°S, 31.170°E. 130 m, off Park Rynie, Natal, 30.333°S, 30.750°E. 130 m, off Park Rynie, Natal, 30.333°S, 30.750°E. 75 m, off Umlaas Canal, Natal, 29.967°S, 30.967°E. 130 m, off Mtamvuna River, Transkei, 31.100°S, 30.200°E (Herbert, 1986).

Literature fossil.

Japan. Pleistocene of Kyushu (Habe, 1951).

Remarks. *Anatoma japonica* is a difficult species, even after examining over 70 specimens by SEM, in addition to many more images of other species for comparative purposes. It seems to be a rather variable and wide-spread species, which runs counter to experience with most other species of *Anatoma*. It lacks any distinctive characters, it embodies the archetypical *Anatoma* shell morphology: more or less biconical, with regular axial cords crossed by spiral lines, forming points of variable strength at intersections.

Several characters are unusually variable:

- width of sutsel at onset of selenizone: almost none to 0.5× width of selenizone.
- width of sutsel at apertural margin: <0.5× to >1× width of selenizone.
- strength of axials: dull, indistinct cords to distinct cords.
- strength of spirals: almost none to distinct lines.
- intersection of axials and spirals: no interaction to strongly elevated points.
- shape of aperture: rounded to somewhat elliptical.
- slope of the shoulder: more or less convex (covariant of aperture shape).
- rachidian tooth: wide base with eight cusps on each side to narrower base with four cusps on each side.

One can refer to the endpoints of the spectrum as:

- *japonica*-type: low sculpture, narrow sutsel, rounded aperture.
- *exquisita*-type: high relief sculpture, wide sutsel, elliptical aperture.
- *aetheria*-type: intermediate between *japonica*- and *exquisita*-type.

The *japonica*-type is more common in Japan, Western Australia and the Western Pacific, while the *exquisita*-type is more prevalent in the Indo-Malayan Archipelago, with *aetheria*-forms in the Western Pacific and the Indian Ocean. Small specimens have proportionally more pronounced sculpture and more elevated keels. In some cases, multiple specimens from the same lot were examined by SEM. While a given population has a tendency towards one modal morphology, there is significant intra-population variability for those characters, specifically the width of the sutsel, the distinctness of the sculpture, the degree of point formation, and the shape of the aperture. The differences between the forms are less than those found in most other species pairs. This distributional pattern and shift in modal morphology is reminiscent of the abalone *Haliotis diversicolor* Reeve, 1846, with a low sculptured *diversicolor* form in Japan, an intermediate sculpture "squamicolor"-form in the Indo-Malayan Archipelago, and a high relief *squamata*-form from Western Australia. In that case, the two named forms have been treated as subspecies (Geiger & Owen, 2012).

Anatoma japonica and its forms are not treated as subspecies, as the boundaries are far less clear with intermediate specimens of every conceivable arrangement known. Taxonomies based on modal differences are best viewed cautiously. Randall & Greenfield (2001) distinguished gobiid fishes based on modal fin ray counts, but those divisions turned out to be artificial, based on discrete color pattern differences supported by molecular analyses (Thacker, 2004).

Given the extensive morphological spectrum and geographic distribution of the species, 11 specimens from throughout the geographic and morphological range are illustrated here:

Japan (Fig. 751A: lectotype of *japonica*; Fig. 753), Philippines (Fig. 754A), Indonesia (Fig. 751B: holotype of *exquisita*), Norfolk Ridge (Fig. 754B), Wallis Island (Fig. 754C), Marquesas Islands (755A), Western Australia (Fig. 752B: holotype of *jansenae*) South Africa (Fig. 755B), Mozambique (Fig. 755C), and Oman (Fig. 752A: lectotype of *aetheria*).

Typical *japonica*-type (Fig. 751A: lectotype of *japonica*; Fig. 751C: holotype of *jansenae,* Fig. 753), typical *exquisita*-type (Fig. 751B: holotype of *exquisita,* Figs 754A, 755C), intermediate *aetheria*-type (Fig. 752A: lectotype of *aetheria*; Figs 754B–C; 755A–B).

The three radulae illustrated do not show variability that may cause concern regarding excessive taxonomic lumping. The nominotypical form from Japan exhibits a slightly wider rachidian tooth with eight side cusps (shell: Fig. 753, radula: Fig. 757) compared to the radulae from specimens from the Norfolk Ridge only has six side cusps (shell: Fig. 754B, radula: 758) and the specimen from South Africa has a more narrow rachidian tooth with only four side cusps (shell: Fig. 755B, radula: Fig. 759). However, no striking modifications such as flexoglossate radula, extreme elongation of the central cusp of the rachidian (*An. rhynchodentata*: Figs 868–870), or scythe-shaped lateral teeth 5 (*An. zancliformis*: Fig. 939; *An. sagamiana*: Fig. 879) were found.

Given the extensive examination of material, three taxa are synonymized under *An. japonica*. *Scissurella exquisita,* described from Indonesia and synonymized under *An. japonica* by Habe & Kosuge (1964), is the more lenticular, highly sculptured form more common in the Indo-Malayan Archipelago. *Anatoma jansenae* from Western Australia has now been recognized to be part of the distributional range of *An. japonica,* and cannot be distinguished based on shell morphology alone. *Scissurella aetheria* shares the overall shape, protoconch attributes, teleoconch I length and sculpture, and has the same generic teleoconch II sculpture. It is the intermediate form between *japonica* and *exquisita.*

Melvill & Standen published the *Scissurella aetheria* description twice, first in the *Annals and Magazine of Natural History* (1903) and then in the *Journal of the Bombay Society of Natural History* (1904). The first was illustrated with a single apertural view of the species, while the latter also included a back view. Both descriptions were clearly marked as introductions of new names. The spelling of the second species epithet is ambiguous, as a capital letter ligature was used. The second description is considered here a *lapsus calami* with no further taxonomic consequences (neither introduction of primary homonym, nor introduction of new species).

The five NMW syntypes of *Scissurella aetheria* are not conspecific. The one sputter coated specimen does not agree with the remainder. The four un-sputter-coated specimens are conspecific with one another, as well as with the NMSZ and NHMUK syntypes and the original illustration. To avoid taxonomic confusion, the NHMUK syntype is here designated as the lectotype (Fig. 752A).

Thiele (1912: 12) misprinted the depth record of the type locality as "205 m." This is an obvious misprint.

Yaron (1983) classified the unidentified shell in *Sinezona*. While the margins of the selenizone in this species are converging, they do not fuse to form a foramen, and the selenizone is at the periphery of the shell. Both of those characters preclude a classification in *Sinezona* or *Sasakiconcha*, and make it a rather typical *Anatoma*.

Other select species support the wide geographic range of *An. japonica. Anatoma finlayi*, described from northernmost New Zealand, is a highly distinctive species with strong axial lamellae on shoulder and base (Figs 707–709). It also ranges from the Indian Ocean (described as *herberti*) to the Western Pacific, including records from Western Australia and the Philippines. *Anatoma munieri* (Figs 804–805) described from the Sea of China extends from the Western Pacific through Japan (described as *turbinata*) to the Indian Ocean (described as *jacksoni*). Accordingly, there are other, more distinctive and easily recognizable species, which show similar ranges; hence, they lend support to the wide species concept of *An. japonica*.

Anatoma globulus (Figs 726, 727) from the Indo-Malayan Archipelago exhibits similar range in the strength of the axial cords, and in the strength of point formation at the intersection of axials and spirals.

A specimen with dark axial bands is shown in Figure 756. Several species are known to produce such specimens, and its occurrence is of no taxonomic value.

The Japanese common name is *Nippon-Kuchikire-ebisu*.

Anatoma keenae McLean, 1970

Figures 761–766

Chresonymy.

Scissurella (Anatoma) keenae McLean, 1970: 117, fig. 4.
Scissurella (Anatoma) keenae: McLean, 1971: 309, fig. 4.
Scissurella (Anatoma) keenae: Sphon, 1971: 20 [typelist].
Scissurella keenae: Draper, 1973: 8.
Anatoma keenae: Lozouet, 1986: 113.
Anatoma keenae: Skoglund, 1992: 4.
Anatoma keenae: Skoglund, 2002: 3.
Anatoma s.l. *keenae*: Geiger, 2003: 75.
Anatoma keenae: Hendrickx & Brusca, 2005: 227.
Anatoma keenae: Geiger & McLean, 2010: 23, 33.

Synonym.

+ *Scissurella (Anatoma) epicharis* McLean, 1970: 117, fig. 3. ***Type material.*** Holotype (LACM 1370: Finet, 1993: figs 1–2. Fig. 761), 1.9 × 1.3 mm. Holotype lost (Fig. 761). Paratype (LACM 1371. Fig. 763B), from type locality. ***Type locality.*** 12 fathoms [= 23 m], Cartago Bay, Isabela Island, Galapagos Islands, Ecuador (0°35'S, 90°57'W) (OD). ***Etymology.*** Greek for graceful (OD).

Scissurella (Anatoma) epicharis: McLean, 1971: 309, fig. 3.
Scissurella (Anatoma) epicharis Sphon, 1971: 20 [typelist].
Scissurella epicharis: Draper, 1973: 8.
Scissurella (Anatoma) epicharis: Finet, 1985: 12.
Anatoma epicharis: Lozouet, 1986: 113.
Scissurella epicharis: Finet, 1991: 268.
Anatoma epicharis: Skoglund, 1992: 4.
Anatoma epicharis: Finet, 1993: 36, 38, figs 1–2.

Anatoma epicharis: Finet, 1994a: 17–18, pl. 2, fig. 1.
Anatoma epicharis: Finet, 1994b: 11.
Anatoma s.l. *epicharis*: Geiger, 2003: 75.

Figure 761. Holotype *Scissurella epicharis* from Finet (1993). The specimen has since been lost.

Type material. Holotype (LACM 1372. Fig. 763A), 2.0 × 2.0 mm. Paratypes from type locality (LACM 1373, 6. Fig. 763C; USNM 679566, 1).

Type locality. 46 fms. [= 84 m], off northwest side, Angel de la Guarda Island, Gulf of California, Mexico, 29°32'N, 113°46'W (OD).

Etymology. Named for McLean's graduate advisor Myra Keen (OD).

Description. Shell moderate size to large (to 3.2 mm. Figs 761–764), trochiform, inflated, as high as wide in fully grown specimens, flatter in juveniles. Protoconch of 0.75 whorl, sculpture flocculent, no apertural varix, apertural margin almost straight; sutsel 0.5× width of selenizone on early whorls, 2× width of selenizone at apertural margin of fully grown specimen. Teleoconch I of 0.66–0.75 whorl, approximately 22–28 axial cords, spiral in position of selenizone. Teleoconch II of up to two whorls, suture moderately impressed. Shoulder convex, axial cords of same strength and density as on teleoconch I, spiral threads increasing in number from 1–2 shortly after start of selenizone to approximately 12 at apertural margin, at intersection with axials forming small nodules. Base inflated, with similar sculpture to shoulder; base continuously sloping into umbilicus. Umbilicus moderately wide, without funiculus. Aperture round, somewhat flattened. Selenizone slightly above periphery; lunules moderately distinct; keels low (usually eroded) of moderate strength; slit open, with parallel margins.

Radula (Fig. 765). Rachidian tooth trapezoidal, central cusp largest, approximately six cusps in semicircle on each side. Lateral teeth 1–3 similar, apical cusp largest, 3–4 cusps on outer cutting edge, two small projections along inner cutting edge. Lateral tooth 4 reduced, apical cusp largest, two cusps on each side. Lateral tooth 5 enlarged by broadening, apical cusp largest, approximately eight cusps on inner edge, two cusps on outer edge. Inner marginal teeth with triangular tip, apical cusp largest, five smaller cusps on each side; shafts without food groove. Intermediate marginal teeth with spoon-shaped tip, approximately 12 cusps on each side. Outermost marginal teeth fan-shaped, with series of approximately 20 cusps at tip. Radular interlock of central field moderate.

Operculum (Fig. 765F) corneous, thin, round, multispiral, with central nucleus.

Comparisons. *Anatoma argentinae* from abyssal plain of Argentina and the Panamic has a more discoidal overall shell shape, a much larger protoconch (370–480 vs. 190 µm) with fine reticulate sculpture, and about 50% fewer axials and spirals.

Anatoma bathypacifica from the bathyal of the Panamic has a spiral cord in the position of the selenizone on teleoconch I forming a distinct angulation, reticulate protoconch sculpture, and an overall taller shell.

Anatoma disciformis from the North Pacific has an overall more lenticular shell shape, almost no sutsel, and finer sculpture.

Anatoma kelseyi from the northeastern Pacific has a protoconch with reticulate sculpture, an overall taller shell, and finer sculpture.

Anatoma lyra from the northeastern Pacific has a smoother early teleoconch with distinctly increasing strength of the sculpture with growth, comparatively much stronger axials than spirals, and is usually more lenticular in overall shape.

Anatoma concinna from the North Pacific has a longer teleoconch I (1 vs. 0.75 whorl), less elevated keels of the selenizone, and axials and spirals that are more equal in strength.

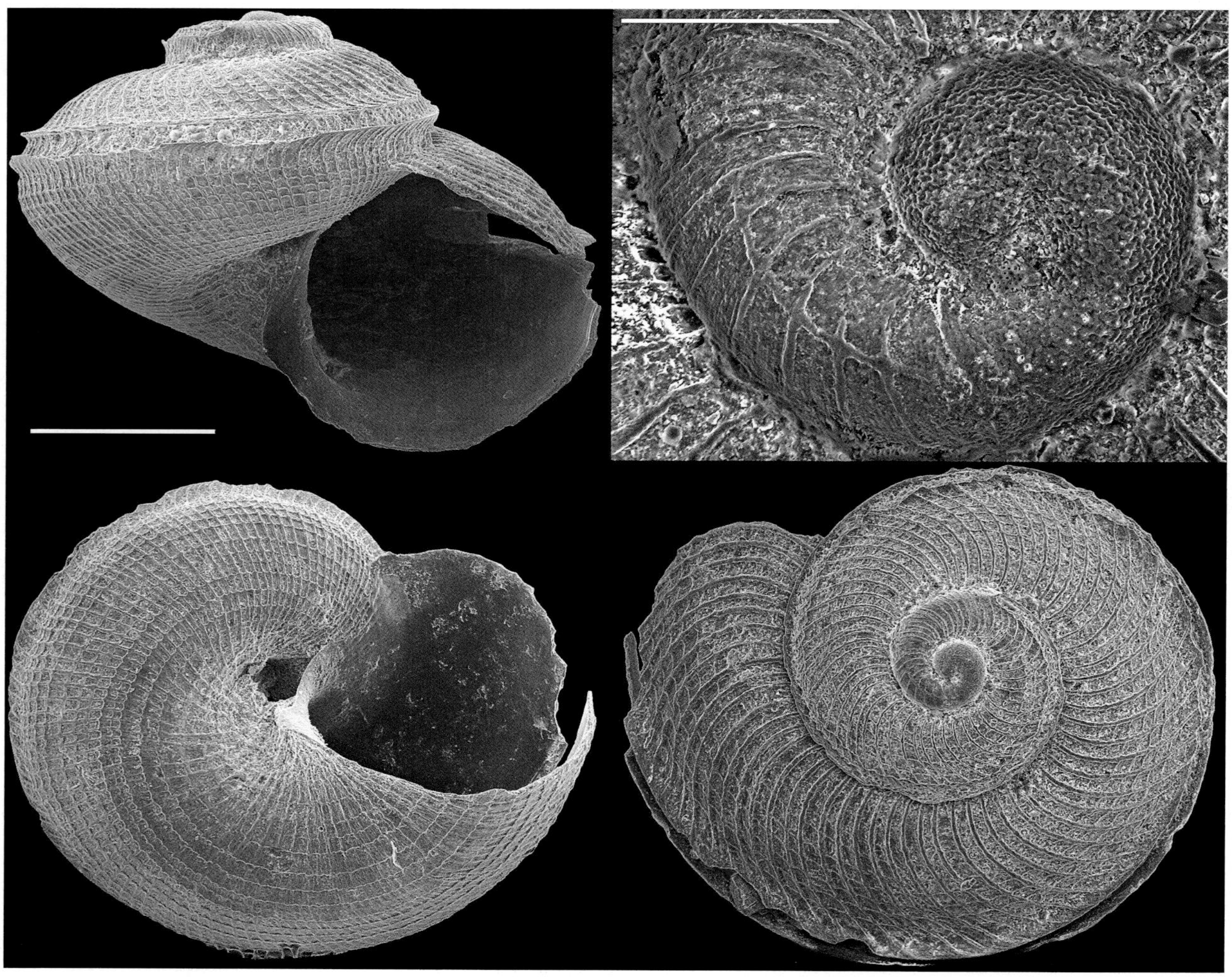

Figure 762. *Anatoma keenae*. 120–170 m, off NW Smith Island, Los Angeles Bay, Baja California, Mexico, 29.066°N, 113.533°W (CSC, 17). Scale bar shell = 0.5 mm. Scale bar protoconch = 100 µm.

Figure 763. *Anatoma keenae.* **A.** Holotype of *Scissurella* (*Anatoma*) *keenae.* 46 fms. [= 84 m], off NW side, Angel de la Guarda Island, Gulf of California, Mexico, 29.533°N, 113.633°W(LACM 1372). **B.** Paratype of *Scissurella epicharis.* 12 fathoms [= 23 m], Cartago Bay, Isabela Island, Galapagos Islands, Ecuador, 0.583°N, 90.950°W (LACM 1371). **C.** Paratype of *Scissurella* (*Anatoma*) *keenae.* 46 fms. [= 84 m], off NW side, Angel de la Guarda Island, Gulf of California, Mexico, 29.533°N, 113.633°W (LACM 1373). Scale bars shell = 1 mm. Scale bars protoconch = 100 μm.

Anatoma crispata from the northeastern Atlantic has more convex whorls, a more narrow sutsel on early whorls, and weaker overall sculpture.

Anatoma globulus from the Indo-Malayan Archipelago has a protoconch with less sculpture and a taller shell.

Figure 764. *Anatoma keenae.* Ontogenetic series. **A–C.** 60–90 m, off NE Danzante Island, Gulf of California, Mexico, 25.785°N, 111.250°W (SBMNH, 7). Scale bars shell = 0.5 mm. Scale bars protoconch = 100 µm.

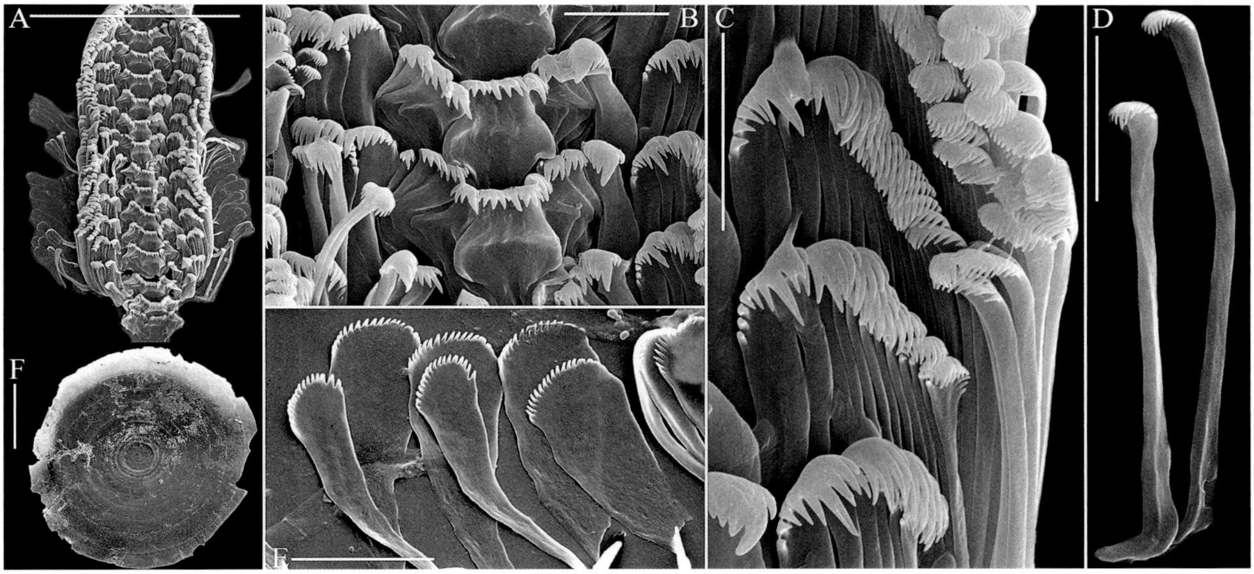

Figure 765. *Anatoma keenae*. Radula and operculum. 73 m, Raza Island, Baja California, Mexico, 28.800°N, 113.000°W (LACM AHF 275, 4). **A–D.** Radula. **A.** Entire radula. **B.** Central field enlarged. **C.** Tips of marginal teeth. **D.** Entire middle marginal teeth. **E.** Entire outermost marginal teeth. **F.** Operculum. Scale bar A = 100 μm. Scale bars B–D = 10 μm. Scale bar F = 200 μm.

Anatoma japonica from the Indian Ocean to the Western Pacific has a protoconch with finer sculpture and an apertural varix, always a less convex shoulder, and a more lenticular overall shell shape.

Anatoma lamellata from the Western to the Central Pacific has a protoconch with a spiral cord, an overall taller shell, a wider sutsel, and less dense spiral lines on shoulder and base.

Distribution. Panamic, Galapagos (Fig. 766).

Specimen records.

Mexico. 120–170 m, off NW Smith Island, Los Angeles Bay, Baja California, 29.067°N, 113.533°W (CSC, 17). 60–90 m, off NE Danzante Island, Gulf of California, 25.785°N, 111.250°W (CSC, 7). 110 m, N Angel de la Guarda Island, Gulf of California, 29.583°N, 113.583°W (LACM 37-196, 2). 150 m, San Pedro Nolasco Island, Sonora, 27.969°N, 27.000°E (LACM AHF 2051, 2). 73 m, Raza Island, Baja California, 28.800°N, 113.000°W (LACM AHF 275, 4). 110 m, N of San Pedro Nolasco Island, Gulf of California, 27.978°N, 111.403°W (LACM AHF 573-36, 1). 119 m, Puerto Refugio, Angel de la Guarda Island, Baja California, 29.567°N, 113.550°W (LACM AHF 268, 1). 101 m, off Ildefonso Island, Baja California, 26.600°N, 111.433°W (LACM AHF 2024, 1). 53 m, off N end of Partida Island, Baja California, 24.600°N, 110.408°W (LACM AHF 249, 1). 137 m, San Jaime Bank, Baja California, 22.250°N, 110.283°W (LACM AHF 2005, 1). 91 m, Outer Gorda Bank, Baja California, 23.017°N, 109.467°W (LACM AHF 2065, 1). 108–178 m, Outer Gorda Bank, Baja California, 23.015°N, 109.489°W (LACM AHF 2098, 5). 102 m, Outer Gorda Bank, Baja California, 23.067°N, 109.467°W (LACM AHF 2063, 6). 146 m, Inner Gorda Bank, Baja California, 23.033°N, 109.508°W (LACM AHF 2010, 4). 108 m, Inner Gorda Bank, Baja California, 23.033°N, 109.517°W (LACM AHF 2156, 6).

Ecuador, Galapagos Islands. 146 m, off Tagus Cove, Albemarle Island, 0.283°S, 91.383°W (LACM AHF 461, 1). 91–110 m, Sulivan Bay, James Island, 0.264°S, 90.593°W (LACM AHF 795a-38, 1).

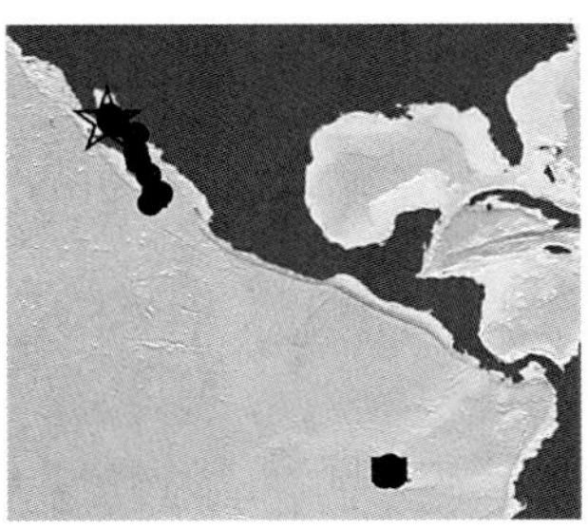

Figure 766. Distribution of *Anatoma keenae*. ☆ type locality. ● specimen record.

Remarks. *Anatoma epicharis* is a juvenile of *An. keenae*. The only difference between the two species is the size (1.3 mm vs. 2.0 mm) and the overall shape of the shell. McLean (1970) noticed the small size of *An. epicharis* and wondered whether it may be a juvenile. Growth series of *An. keenae* from Mexico have confirmed this suspicion (Fig. 764). As the two species were described in the same publication, I use first reviser's principle and choose *An. keenae* as the valid name for three reasons: it is based on a mature specimen, it is from the core distribution of the species, and it keeps the memory of Myra Keen alive.

Anatoma kelseyi (Dall, 1905)

Figures 767–774, Color Plate 1D

Chresonymy.

Scissurella (*Schizotrochus*) *kelseyi* Dall, 1905: 124–125, not illustrated.
Scissurella kelseyi: Thiele, 1912a: 18.
Scissurella kelseyi: Dall, 1921: 183.
Scissurella kelseyi: Oldroyd, 1924: 180.
Scissurella kelseyi: Oldroyd, 1927: 228.
Scissurella kelseyi: Keen, 1937: 45.
Scissurella kelseyi: Burch, 1946a: 27.
Scissurella kelseyi: Draper, 1973: 8.
Thieleella kelseyi: Geiger & McLean, 2010: 20, 22–28, 30, figs 16–19.

Synonym.

+ *Scissurella chiricova* Dall, 1919: 370, not illustrated. ***Type material.*** Holotype (USNM 206509: complete. McLean, 1967: pl. 56, fig. 2; Boss *et al.*, 1968. Fragmented most likely due to glass disease [Geiger, pers. obs. 3/2007] as evidenced by crystalline precipitations particularly on the largest fragment of the base: Fig. 768), 3 × 2.5 mm (W × H). ***Type locality.*** USFC sta. 3340, SE Chirikof Island, Alaska, USA, 695 fms [= 1270 m, ~55.4°N, 155.25°W] (McLean, 1966). ***Etymology.*** Named after its type locality.
Scissurella chiricova: Dall, 1921: 183.
Scissurella chiricova: Oldroyd, 1927: 228–229.
Scissurella chiricova: Keen, 1937: 45.
Scissurella chiricova: Burch, 1946a: 27.
Scissurella chiricova: Draper: 1973: 8

Figure 767. Holotype of *Scissurella chiricova* from McLean (1984).

Misidentifications.

not *Scissurella chiricova*: Talmandge, 1966: 83 [is *An. soyoae*: identification *fide* McLean, 1967].
not *Scissurella kelseyi*: Talmadge, 1966: 83 [identification *fide* McLean, 1967, is *An. lamellata*. McLean (1984) identified "*An. lamellata*" sensu McLean (1967) from the eastern Pacific as *An. baxteri*, but did not list the Talmadge record].
not *Scissurella crispata*: Foster, 1981: 13, 41–42 [most likely referring to *An. kelseyi*].
not *Anatoma crispata*: McLean, 1996: 17, fig. 1.2 [is *An. kelseyi*].
not *Scissurella* (*Anatoma*) *crispata*: McLean, 1967: 405, pl. 56, figs 1–4 [includes *An. chiricova* and *An. japonica* illustrated specimen is *An. kelseyi*].

Type material. Holotype (USNM 181820. Lost: McLean, 1967: 405; Boss *et al.*, 1968: 175 [without number]; Geiger pers. obs. 2/2004), 5.5 × 6 mm (H × W). Neotype (SBMNH 103215. Fig. 769: see Remarks) designated by Geiger & McLean (2010: 22; see Remarks).

Type locality. California at U.S. Fish Commission Station 4353, also off San Diego (OD), 639 fms, 15 mi from Pt. Loma. 47°W (USNM label). Neotype locality: 183 m, off La Jolla, San Diego County, California, USA [32.843°N, 117.392°W].

Etymology. Named in honor of Prof. F. W. Kelsey, of San Diego (OD).

Description. Shell large (to 6 mm: *fide* Dall, 1905. Fig. 770), trochiform, biconical, white. Protoconch of 0.75 whorl, with reticulate sculpture, apertural varix barely connected to embryonic cap, apertural margin straight. Teleoconch I of 0.75 whorl, with approximately 38 spiral cordlets, one spiral cordlet in position of selenizone, interstices with flocculent sculpture. Teleoconch II of 3.125 whorls in 3.2 mm specimen; suture little impressed; sutsel as wide as selenizone showing approximately top three spirals from base. Shoulder somewhat convex, suture lightly impressed, axials approximately of same density and strength as on teleoconch I, approximately 22 equispaced fine spiral lines running over cordlets. Base with slight constriction below selenizone, axials of same strength and density as on shoulder, crossed by approximately 32 equispaced lines, interstices with fine irregular growth lines. Umbilicus very narrow, continuously sloping with base, funiculus present. Selenizone slightly above periphery; keels of moderate strength, moderate elevation; lunules rather weak; slit open with parallel margins. Aperture rounded, adumbilical portion flared, roof overhanging.

Operculum (Fig. 773A) covering entire aperture, corneous, thin, multispiral with central nucleus.

Radula (Fig. 771). Rachidian tooth trapezoidal, central cusp largest, approximately six cusps on each side. Lateral teeth 1–3 similar, apical cusp largest; 3–4 cusps on outer cutting edge; 1–2 cusps on along inner edge, half size of outer cusps. Lateral tooth 4 reduced, hook-shaped, apical cusp largest, 3–4 minute cusps on inner

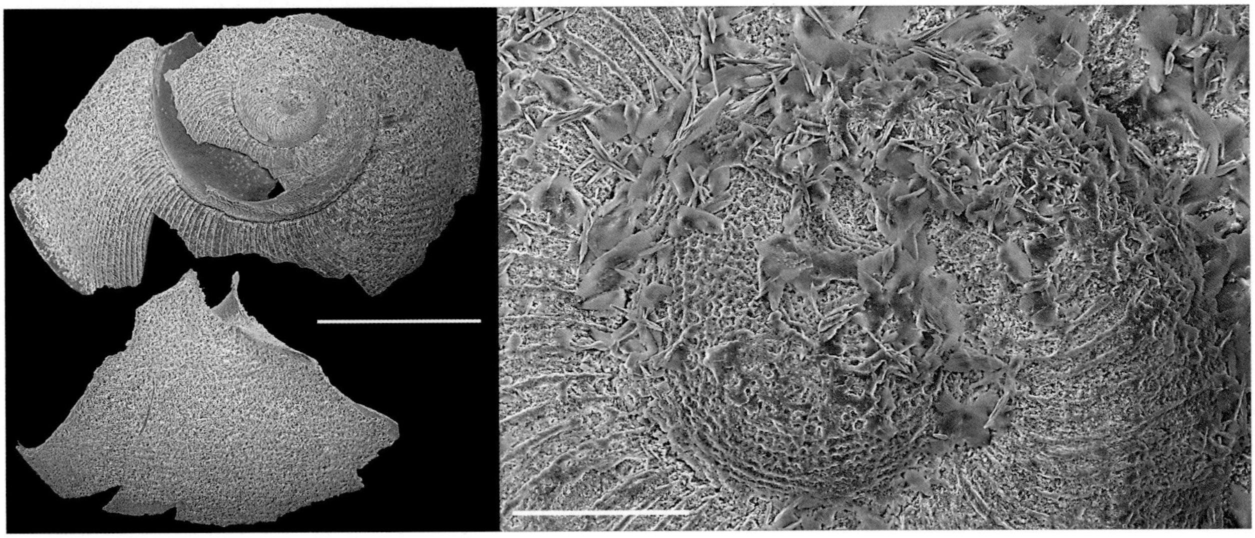

Figure 768. *Scissurella chiricova.* Holotype. 695 fms [= 1270 m], USFC sta. 3340, SE Chirikof Island, Alaska, USA, ~55.4°N, 155.25°W (USNM 206509). Scale bar shell = 1 mm. Scale bar protoconch = 100 μm.

edge. Lateral tooth 5 enlarged by elongation, apical cusp largest, approximately 12 cusps along inner edge, approximately four cusps along distal half of tip's outer edge. Inner marginal teeth with triangular tip on long shafts without food groove, apical cusp largest, approximately six cusps on each side. Outer marginal teeth thin, broadened, with many small bristles along edge of tip. Radular interlock of central field moderate.

Body (Color Plate 1D; Figs 772, 773). Tentacles in anterior-posterior order: cephalic tentacles papillate; ocular stalk without pigmented eye; suboptic tentacle, accessory cephalic tentacle; bifid papillate epipodial tentacles with single posterior blunt, stubby, non-papillate structure about half as long as second epipodial tentacle (epipodial sense organ: ESO); two single papillate epipodial tentacles. Operculum attached to discrete fine flap of metapodium.

Mantle with papillate tentacle at posterior junction of slit near anal papilla (Fig. 773F, G). Two monopectinate gills (Fig. 773G), approximately 25–30 gill

Figure 769. *Scissurella kelseyi*. Neotype. 183 m, off La Jolla, San Diego County, California, USA, 32.843°N, 117.392°W (SBMNH 103215). Scale bar shell = 1 mm. Scale bar protoconch = 100 μm.

Figure 770. *Anatoma kelseyi*. **A, C.** 346 m, S of Puffin Bay, Barinot Island, W of Crowley Point, Kulu Island, Sitka County, Alaska, USA, 56.100°N, 134.809°W (LACM 87-393, 12). Radula shown in Figure 771A–C. **B.** Between Tanner and Cortez Bank, California, USA, 32.450°N, 119.250°W (SBMNH 133350, 1). Radula shown in Figure 771D–E. Scale bars shell = 1 mm. Scale bars protoconch = 100 μm.

leaflets each, distinct bottle-shaped bursicles in distal half of each filament, opening of bursicles in middle of filament (Fig. 773G, H).

Comparisons. *Anatoma keenae* from the Panamic province has a protoconch with flocculent sculpture, lacks a spiral cord in the position of the selenizone on teleoconch I, and forms small but distinct points at the intersection of spiral and axial cords on teleoconch II.

Anatoma lyra from the northeastern Pacific has a protoconch with flocculent sculpture, a shorter teleoconch I (0.5 whorls vs. 0.75 in whorl), and axial sculpture consisting of raised lamellae.

Anatoma agulhasensis from the Indian Ocean has a constriction of the base below the selenizone, a protoconch with flocculent sculpture, and fewer but sharper axial cords on the shoulder.

Anatoma concinna from the North Pacific has a longer teleoconch I (1 vs. 0.75 whorl), an overall more globular shell shape, and coarser and less dense clathrate sculpture.

Anatoma crispata from the northeastern Atlantic has a protoconch with flocculent sculpture, an overall more discoidal shell shape, and a more narrow sutsel on early whorls.

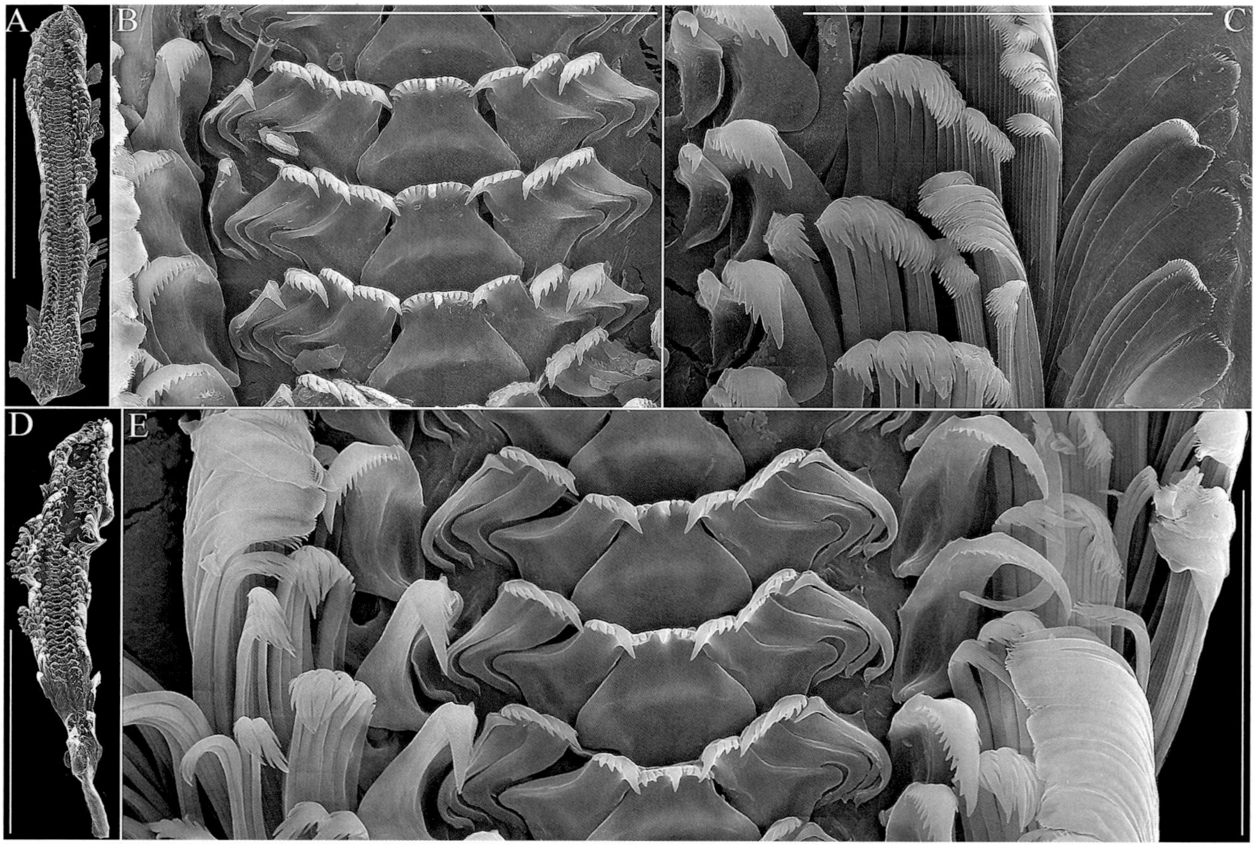

Figure 771. *Anatoma kelseyi*. Radulae. **A–C.** 346 m, S of Puffin Bay, Barinot Island, W of Crowley Point, Kulu Island, Sitka County, Alaska, USA, 56.100°N, 134.809°W (LACM 87-393, 12). For shell, see Figure 770A. **D–E.** Between Tanner and Cortez Bank, California, USA, 32.450°N, 119.250°W (SBMNH 133350, 1). For shell, see Figure 770B. Scale bars A, D = 500 µm. Scale bars B, C, E. = 50 µm.

Anatoma globulus from the Indo-Malayan Archipelago has a more globular overall shell shape and a protoconch with weak flocculent sculpture.

Anatoma peruviana from Peru has a taller shell with a wider sutsel (>2× vs. <2× width of selenizone) and stronger axial cords and spiral lines forming more distinct points at their intersections.

Anatoma sagamiana from Japan to the Indo-Malayan Archipelago, has a less convex shoulder and a very narrow or closed umbilicus.

Anatoma schanderi from the northeastern Atlantic has a spiral cord in position of the selenizone on teleoconch I, a longer teleoconch I (0.9–1 vs. 0.75 whorl), and less distinct lunules.

Anatoma soyoae from Japan has an overall more globular shell, denser sculpture with more equal axial cords and spiral lines, and a more narrow sutsel.

Distribution. Aleutian Islands to South Coronado Island (Fig. 774).

Specimen records.

USA, California. 1503 m, US Fish Commission Station 2923, off San Diego (USNM 209412, 4). 1033–1244 m, 16 mi SW off Point Loma, 32.700°N, 117.246°W (USNM 211166, 4). 32.639°N, 117.430°W (DLG 99, 1: complete).

San Diego Trough. 1171–1241 m, 32.417°N, 117.467°W (SIO M1974, 1: complete). 1166–1222 m, 32.450°N, 117.467°W (SIO M1991, 1, 5: complete). 1207–1234 m, SIO69-487, 32.450°N, 117.525°W (SIO M2070, 3: complete). 1183–1216 m, R-28, 32.583°N, 117.558°W (SIO 2171, 1: complete). 1241 m, R-42, 32.440°N, 117.483°W (SIO M1778, 1: complete). 1199–1241 m, R-17, 32.447°N, 117.467°W (SIO M1834, 2: complete).

San Clemente Island. 390 m, off Tanner Bank, 41 mi 80°T of China Point, 32.683°N, 119.234°W (LACM 76-313, 6). 1047 m, 30.0 mi 164°T of China Point, 32.585°N, 118.965°W (LACM 76-378, 6). 933 m, 30.8 mi 65°T of China Point, 32.583°N, 118.982°W (LACM 76-377, 2). 973 m, 30.2 mi 63°T of China Point, California, USA, 32.569°N, 118.984°W (LACM 76-381, 5). 292 m, 38 mi 68.5°T of China Point,

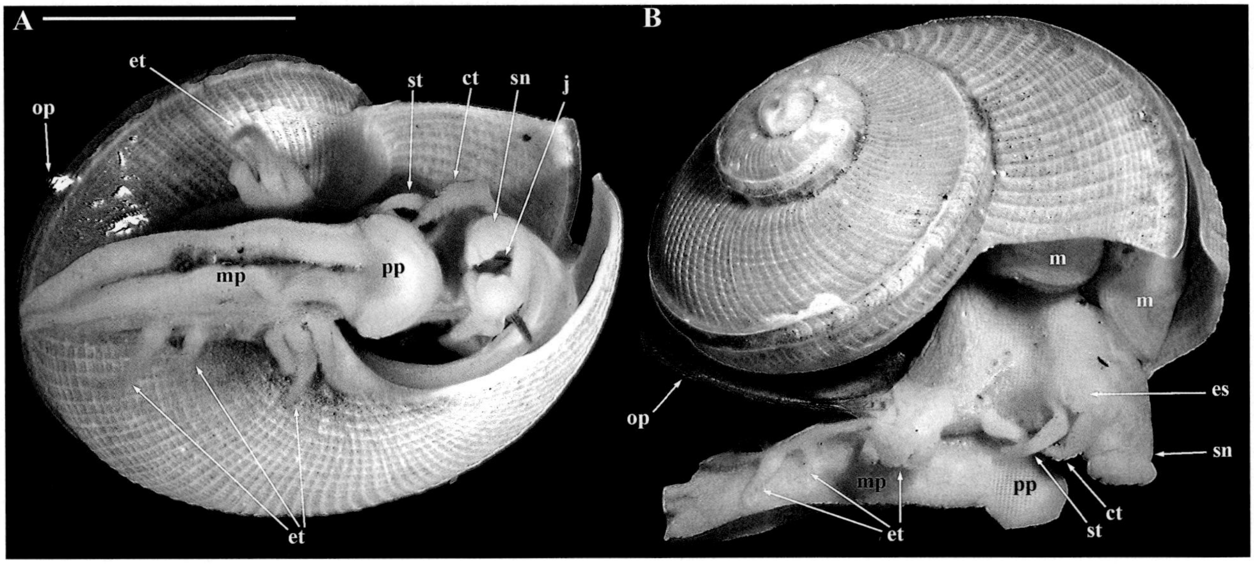

Figure 772. *Anatoma kelseyi*. Animal, stereomicroscope images, z-stacked with *HeliconFocus*. 421 m, 2.3 mi 48°T to South Point Light, Santa Rosa Island, USA, 33.85°N, 120.183°W (LACM 23129BFI, 2: complete). Scale bar = 1 mm. **A.** Ventral view. **B.** Lateral view. ct: cephalic tentacle. es: eye stalk. et: epipodial tentacle. j: jaws. m: mantle. mp: metapodium. op: operculum. pp: propodium. sn: snout. st: suboptic tentacle.

32.566°N, 119.131°W (LACM 76-387, 4). 201 m, 32.772°N, 118.379°W (LACM AHF 1082, 1).

Between Tanner Bank and Cortez Bank. 32.450°N, 119.250°W (SBMNH 1333350, 1). 1200 m, off N Coronado Island, 32.750°N, 117.500°W (USNM 208901, 3; USNM

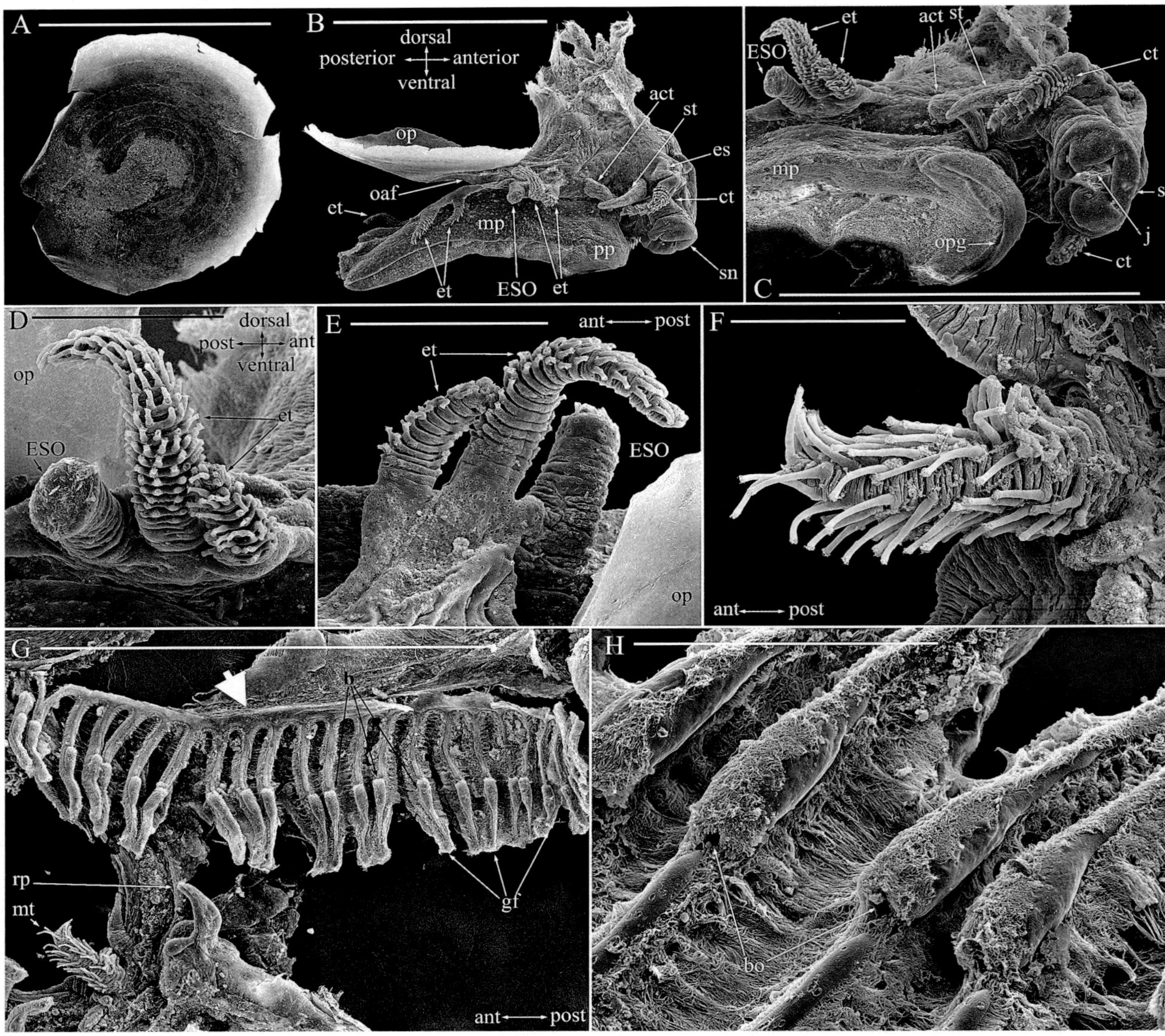

Figure 773. *Anatoma kelseyi*. Animal, SEM images. 421 m, 2.3 mi 48°T to South Point Light, Santa Rosa Island, USA, 33.85°N, 120.183°W (LACM 23129BFI, 2: complete). **A.** Operculum. Scale bar = 1 mm. **B.** Entire animal, HMDS treated. Scale bar = 1 mm. **C.** Ventral view of anterior portion. Scale bar = 1 mm. **D.** Ventral view of anterior epipodial tentacles plus ESO. Scale bar = 200 µm. **E.** Dorsal view of epipodial tentacles and ESO; note biserial papillation of epipodial tentacles. Scale bar = 200 µm. **F.** Mantle tentacle at posterior end of mantle enlarged. Scale bar = 100 µm. **G.** Overview of gill and mantle. Fat arrow: direction of view of H. Scale bar = 1 mm. **H.** Bursicles with openings. Scale bar = 100 µm. Abbreviations: act: accessory cephalic tentacle. b: bursicle. bo: bursicle opening. ct: cephalic tentacle. es: eye stalk. ESO: epipodial sense organ. et: epipodial tentacles. gf: gill filaments. j: jaw. mp: metapodium. mt: mantle tentacle. oaf: operculum attachment flap. op: operculum. opg: opening of propodial gland. pp: propodium. rp: rectal papilla. sn: snout. st. suboptic tentacle.

209054, 3). 183 m, off La Jolla, San Diego County, 32.8430°N, 117.392°W (SBMNH 3215, 1).

Santa Rosa Island. 70 m, 5 mi 7.5°T to East Point, 33.531°N, 119.982°W (LACM 77-303, 1). 115 m, 5.1 mi 331°T to East Point, 33.865°N, 119.917°W (LACM 75-300, 6). 61 m, 5.7 mi 33°T of East Point, 33.865°N, 120.030°W (LACM 75-306, 1). 105 m, 3.4 mi 279°T to East Point, 33.934°N, 119.901°W (LACM 75-480, 1). 442 m, 7.3 mi 70°T to Cluster Point, 33.884°N, 120.316°W (LACM 76-213, 8). 421 m, 2.3 mi 48°T to South Point Light, 33.850°N, 120.183°W (LACM 23129BFI, 2: complete). 492 m, off Point Sal, Santa Barbara County, California, USA, 34.970°N, 121.298°W (SBMNH 350392, 1; SBMNH 350393, 1).

USA, Oregon. 1580 m, 52.3 mi 262.6°T from Tillamook Head, Clatsop County, 45.828°N, 125.233°W (LACM 67-167, 1). 640 m, WNW of Tillamook Bay, Tillamook County, 45.583°N, 124.750°W (SBMNH 140030, 1).

USA, Washington. 146–219 m, WSW of Grays Harbor, Grays Harbor County, 46.750°N, 124.750°W (SBMNH 140037, 1). 128–146 m, off Cape Flattery, Clallam County, 48.333°N, 125.583°W (SBMNH 140024, 1).

Canada, British Columbia. 2176 m, Estevan Point Light, Hesquiat Peninsula, Noetka Sound, Vancouver Island, 48.635°N, 126.967°W (LACM 71-374. 7, 3: complete). 1602 m, off Queen Charlotte Sound, 51.500°N, 129.000°W (USNM 206690, 1). 55 m, Naden Harbor, N side of Graham Island, Queen Charlotte Island, 54.000°N, 132.667°W (LACM 66-66, 0). 260 m, Dixon Entrance, N of Graham Island, 54.483°N, 133.200°W (LACM 67-28, 2). 1602 m, off Queen Charlotte Sound, 51.500°N, 130.000°W (USNM 206690, 1).

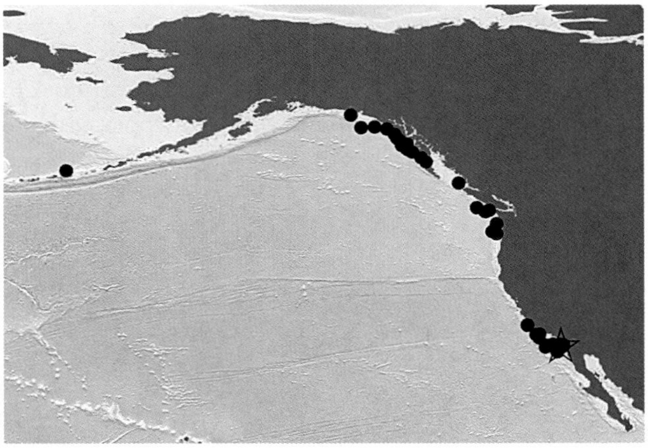

Figure 774. Distribution of *Anatoma kelseyi*. ☆ type locality. ● specimen record.

USA, Alaska. 88 m, NW of Gibson Island, off NE side Attu Island, Near Islands, Aleutian Islands County, 52.958°N, 173.237°W (LACM 86-312, 5). 179 m, S of Coronation Island, W of N end Dall Island, Prince of Wales County, 55.143°N, 134.210°W (LACM 87-377, 2). 219 m, WNW Forrester Island, 54.885°N, 134.314°W (LACM 87-381, 1). 278 m, W of Gulf of Esquibel, S of Byron Bay, 55.537°N, 134.891°W (LACM 87-369, 27). 205 m, W of N end Baker Island, S of Cape Ommaney, 55.394°N, 134.675°W (LACM 87-370, 20). 443 m, W of Nation Point, Coronation Island, SW of Cape Ommaney, 55.925°N, 135.438°W (LACM 87-365, 8). 252 m, SW of Cape Ommaney, S end, 56.000°N, 135.000°W (LACM 87-357, 2). 364 m, W of Point Howard, Kulu Island, S of Sitka, 56.037°N, 135.307°W (LACM 87-364, 13). 362 m, W of Point Crowley, Kulu Island, S of Necker Bay, 56.090°N, 135.150°W (LACM 87-363, 9). 240 m, S of Kruzof Island, W of Necker Bay, 56.598°N, 135.782°W (LACM 87-358, 100). W of Whale Bay, SSW of Biorka Island, 56.514°N, 135.661°W (LACM 87-395, 29). 390 m, SW of mouth of Sitka Sound, SSW of Cape Edgecumbe, S end of Kruzof Island, Sitka County, 56.778°N, 135.970°W (LACM 87-356, 25). 274 m, S of Cape Edward, W of Gilmer Bay, Kruzof Island, Sitka County, 57.219°N, 136.234°W (LACM 87-354, 17). 408 m, S of Icy Point, W of Point Theodore, Yakobi Island, Skagway-Yukutat County, 57.853°N, 137.111°W (LACM 87-346, 1). 214 m, SSW of Cape Fairweather, Chichagof Island, W of Cape Bingham, Skagway-Yukutat County, 58.073°N, 138.476°W (LACM 87-330, 3). 384 m, W of Yakobi Island, S of Yakutat Bay, Skagway-Yukutat County, 58.000°N, 140.000°W (LACM 87-335, 1). 178 m, SE of Icy Bay, SW of Yakutat Bay, Cordova-McCarthy County, 59.453°N, 141.150°W (LACM 87-336, 1). 353 m, S of Icy Bay, E of Dry Bay, Cordova-McCarthy County, 59.118°N, 141.495°E (LACM 87-333, 10).

Remarks. The specimen illustrated as the neotype by Geiger & McLean (2010: fig. 16A), here reproduced in Fig. 770A, accidentally showed a different specimen (LACM 87-393 from Alaska). The true neotype from a locality as close to the original type locality as possible (San Diego, California) is here shown in Figure 769. The two specimens are conspecific; hence, the species concept remains identical. As the collection, repository number, and locality are decisive regarding the identity of the neotype, the ICZN non-mandatory illustration is irrelevant in this context. The error is here rectified, having no further nomenclatorial consequences.

Scissurella chiricova is considered a synonym of *An. kelseyi*. The holotype figured by McLean (1967: pl. 56, fig. 2) has fragmented and is covered by crystalline precipitations due to Byne's/glass disease (Fig. 768). The protoconch still shows very clear reticulate pattern, and the dense axial cords of moderate strength are evident on the shoulder. Geiger & McLean (2010) doubted the synonymy of the two taxa due to the somewhat different overall shell shape. Given the known ontogenetic change of overall shell shape, and the non-standard specimen orientation of the only image of the intact type of *chiricova* in McLean (1967), those alleged differences need to be viewed with caution. On the other hand, the reticulate protoconch sculpture is clearly shown, with no other vaguely similar *Anatoma* species being known from the northeastern Pacific. Accordingly, *chiricova* is here synonymized under *kelseyi*.

Anatoma sagamiana is extraordinarily similar in terms of shell morphology, but shows some significant differences in the radula: the tip of the rachidian tooth is V-shaped, whereas in *An. kelseyi* the cusps are arranged on a ridge parallel to the rows of teeth; the lateral teeth 5 of *An. sagamiana* have cusps on both sides more or less equally distributed, whereas in *An. kelseyi* the inner edge has much better developed cusps up to the tip, and the outer edge has fewer cusps restricted to the transition between shaft and tip. The combination of slight conchological and radular differences in addition to the disjunct distribution suggests that they are separate species.

Anatoma kopua Geiger & Marshall, 2012

Figures 775–777

Chresonymy.

Anatoma kopua Geiger & Marshall, 2012: 21–23, figs 16–17.

Type material. Holotype (NMNZ M.174848). Paratypes 1143–1106 m, off Sydney, New South Wales, Australia, 33.600°S, 152.083°E (AMS C.453747, 5). 1186 m, Lord Howe Rise, SE of Lord Howe Island, New Zealand, 34.272°S, 163.102°E (NMNZ M.174864, 1). 1005–1009 m, Challenger Plateau, New Zealand, 40.835°S, 168.247°E (NMNZ M.174847, 1).

Type locality. 1029 m, Challenger Plateau, New Zealand, 40.713°S, 167.933°E (OD).

Etymology. Deep in New Zealand Maori. Adjective (OD).

Description. Shell large (4.27 × 4.58 mm W × H), trochiform, biconical, turreted. Protoconch large (275 µm), 0.75 whorl, sculpture flocculent, no apertural varix, apertural margin straight. Teleoconch I of 0.3–0.4 whorl, 12–15 axial cords, spiral cord in position of selenizone. Teleoconch II of four whorls, suture moderately impressed, sutsel less than width of selenizone on early whorls, about twice as wide as selenizone at apertural margin of large specimen. Shoulder moderately convex, regular axial cords, approximately 40 on first whorl, interstices about 4× as wide as cords; after 1–2 teleoconch II whorls, about 2–5 irregularly spaced spiral lines at apertural margin. Base with constriction below selenizone, same strength and density of axial cords as on shoulder, spiral lines regularly spaced, stronger than those on shoulder, approximately 20, noticeable gap of spirals immediately below selenizone. Umbilicus open, narrow, obscured by reflexed adumbilical portion of peristome. Selenizone

Figure 775. *Anatoma kopua*. Holotype. 1029 m, Challenger Plateau, New Zealand, 40.713°S, 167.933°E (NMNZ M.174848). Scale bar shell = 1 mm. Scale bar protoconch = 100 μm.

Figure 776. *Anatoma kopua*. **A.** 1143–1106 m, off Sydney, New South Wales, Australia, 33.600°S, 152.083°E (AMS C.453747, 5). **B.** 1186 m, Lord Howe Rise, SE of Lord Howe Island, New Zealand, 34.272°S, 163.102°E (NMNZ M.174864, 1). **C.** 1005–1009 m, Challenger Plateau, New Zealand, 40.835°S, 168.247°E (NMNZ M.174847, 1). Scale bars shell = 1 mm. Scale bars protoconch = 100 µm.

slightly above periphery, rather wide; lunules rather strong; keels moderately strong, rather elevated (Fig. 776A). Aperture subquadrate, peristome flared.

Animal unknown.

Comparisons. *Anatoma biconica* from the Indian Ocean to the Western Pacific has a longer teleoconch I (1 vs. 0.5 whorl) and the spiral lines on the base reach to the selenizone.

Anatoma equatoria from the Indo-Malayan Archipelago to the Central Pacific has a more turreted shell shape, spiral lines on the base immediately below the selenizone, and a more inflated base.

Anatoma flexidentata from the Indian Ocean to the Western Pacific has a shorter teleoconch I (0.3 vs. 0.5 whorl), a more globular whorl profile, and spiral lines on the base immediately below the selenizone.

Anatoma porcellana from the Indo-Malayan Archipelago to the Western Pacific lacks spiral sculpture around the umbilicus, has a strongly impressed suture, and has a sinusoidal profile of the shoulder.

Anatoma zancliformis from the Indo-Malayan Archipelago to the Western Pacific has a more narrow sutsel on early whorls (0× vs. >0.5× width of selenizone), spiral lines immediately below the selenizone, and a rounder aperture.

Anatoma maxima from the Indo-Malayan Archipelago to the Western Pacific has the selenizone more at the periphery, an overall more discoidal shell shape, and at least four times as many spiral lines on the shoulder.

Anatoma aedonia from the Caribbean to Brazil has a more turreted overall shell shape, a wider sutsel on early whorls (2× vs. 1× width of selenizone), a longer teleoconch I (0.75–1 vs. 0.5 whorl), and a more narrow selenizone.

Anatoma amoena from Antarctica has a protoconch with reticulate sculpture, a longer teleoconch I (1 vs. 0.5 whorl), and spiral lines immediately below the selenizone on the base.

Distribution. New South Wales, Australia, to New Zealand, approximately 1000 m (Fig. 777).

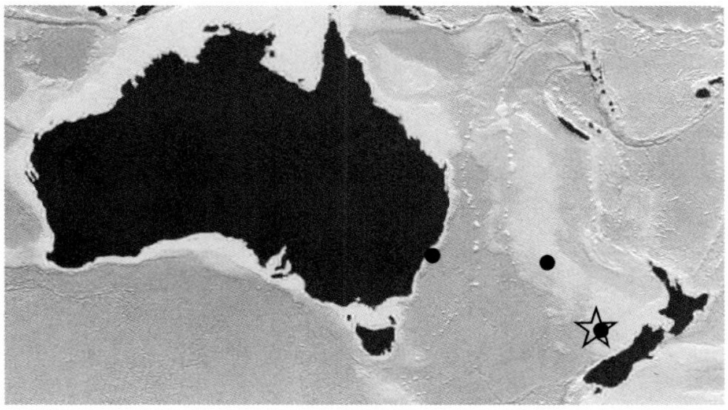

Figure 777. Distribution of *Anatoma kopua*. ☆ type locality. ● specimen record.

Anatoma lamellata (A. Adams, 1862)

Figures 778–780

Chresonymy.

Anatomus lamellatus Adams, 1862: 347, not illustrated.
Anatomus lamellatus: Troschel, 1863: 702.
Anatomus lamellatus: Dunker, 1882: 147.
Scissurella lamellata: Paetel, 1883: 47.
Scissurella lamellata: Paetel, 1888: 289.
Anatomus lamellatus: Pilsbry, 1890: 59.
Scissurella lamellata: Pilsbry, 1895: 106.
Scissurella lamellata: Thiele, 1912a: 17, pl. 2, fig. 13.
Scissurella lamellata: Kuroda & Habe, 1952: 84.
Scissurella (*Schizotrochus*) *lamellata*: Kawamoto & Tanabe, 1956: 3, pl. 2, fig. 13 [copy figure Thiele, 1912a].
Schizotrochus lamellatus: Azuma, 1960: 2.
Schizotrochus lamellatus: Habe, 1961: 2, pl. 1, fig. 6.
Anatoma lamellata: Habe, 1964: 32, pl. 1, fig. 6.
Anatoma lamellatus [*sic*]: Habe & Kosuge, 1964: 4.
Schizotrochus lamellatus: Okutani, 1964: 376.
Anatoma lamellata: Higo, 1973: 14.
Anatoma lamellata: Matsumoto, 1979: 4.
Anatoma lamellata: Inaba, 1982: 72.
Anatoma lamellata: Inaba, 1983: 72.
Anatoma lamellata: Lozouet, 1986: 113.
Anatoma lamellata: Okutani, 1986: 31, text-fig.
Anatoma lamellata: Watanabe, 1988: 28.
Anatoma lamellata: Tsuchida *et al.*, 1991: 4–5, text-fig. 3, pl. 1, figs 1, 6–7.
Anatoma lamellatus [*sic*]: Higo & Goto, 1993: 15.
Anatoma lamellata: Tsuchida & Hori, 1996: 222–223, pl. 1, fig. 1a–b.
Anatoma lamellata: Higo *et al.*, 2001: 14, G79 [one of two syntypes].
Anatoma s.l. *lamellata*: Geiger, 2003: 75.
Scissurella cf. *lamellata*: Smith, 2003 [most likely misidentified].
Anatoma lamellata: Hasegawa, 2005: 138–139.
Anatoma lamellata: Sasaki *et al.*, 2010: 65–68.
Anatoma lamellata: Geiger, 2011b: 40, pl. 4, fig. 2; fig. 4 [as *Anatoma* sp.].

Misidentifications.

Scissurella kelseyi: Talmadge, 1966: 83 [*fide* McLean, 1967].
not *Schizotrochus lamellatus*: Habe, 1951: 66, pl. 11, figs 7–8 [most likely *An. porcellana*].
not *Scissurella* (*Anatoma*) *lamellata*: McLean, 1967: 406, pl. 56, fig. 8 [is *An. lyra*].
not *Anatoma lamellata*: Kuroda *et al.*, 1971: 3, pl. 106, fig. 1 [is *An. concinna*].
not *Scissurella lamellata*: Abbott, 1974: 17 [West American records most likely *An. lyra*].
not *Scissurella lamellata*: Foster, 1981: 13, 42–43 [is *An. lyra*].
not *Anatoma lamellata*: Numanami & Okutani, 1990: 99, figs 7 [shell], 8 [radula] [is *An. equatoria*].
not *Anatoma lamellata*: Izawa & Matsuoka, 1999: 14, pl. 6, fig. 8 [most likely *An. porcellana*].
not *Anatoma lamellata*: Okutani & Hasegawa, 2000: 37, figs 1a–b. [is *An. porcellana*].
not *Anatoma lamellata*: Hasegawa, 2001: 126 [*An. equatoria* or *An. porcellana*: see Remarks].

not *Anatoma lamellata*: Koyama, 2008: 253, pl. 1, fig. 2 [is *An. sagamiana*].

Type material. Syntypes (NHMUK 1878.1.28.513, 3). One specimen here designated as the lectotype (NHMUK 1878.1.28.513/1. Fig. 778) for the express purpose of taxon stabilization in case other non-conspecific syntypes should be located. Paralectotypes (NHMUK 1878.1.28.513/2–3, 2).

Type locality. Mino-sima [= Mishima Island, Hagi City, Yamaguchi Prefecture, Japan, 34.767°N, 131.166°E: see also Kawamoto and Tanabe (1956)], 63 fathoms [= 115 m]. Goto, 71 fathoms [= 130 m]. O-Sima; 26 fathoms [= 47 m] (OD). The type locality is here explicitly restricted to Mino-sima in agreement with the specimen label of the here designated lectotype.

Etymology. Named for the strong lamellae.

Figure 778. *Anatomus lamellata*. Lectotype here designated. 63 fms [= 115 m], Mino-sima [= Mishima Island, Hagi City, Yamaguchi Prefecture, Japan, 34.767°N, 131.166°E] (NHMUK 1878.1.28.513/1). Scale bar shell = 1 mm. Scale bar protoconch = 100 μm.

Description. Shell medium-sized (to 1.5 mm. Figs 778, 779), trochiform, inflated. Protoconch of 0.75 whorl, sculpture flocculent, median spiral ridge, no apertural varix, apertural margin sinusoidal. Teleoconch I of 0.5–0.6 whorl, approximately 17 distinct axials,

Figure 779. *Anatoma lamellata*. **A.** Japan (NHMUK 1828). **B.** Japan (RM 1441). **C.** Mamala Bay, Oahu, Hawaii, USA, 21.290°N, 157.950°W (EGC 29829, 5). Scale bars shell = 1 mm. Scale bars protoconch = 100 µm.

spiral in position of selenizone. Teleoconch II of at least 1.5 whorls, suture little impressed. Shoulder with strong elevated axial cords, very fine spiral threads, first threads at 0.25 teleoconch II whorl, approximately 10 threads at apertural margin in adsutural ⅔ of shoulder. Base with slightly more closely spaced axial cords, tapering to axial threads towards umbilicus; approximately 11 equispaced spiral lines. Selenizone slightly above periphery; lunules moderately distinct; with strong keels of moderate elevation; slit open, margins parallel. Umbilicus open, moderate width, no funiculus. Aperture rounded, adumbilically flared.

Animal unknown.

Comparisons. *Anatoma concinna* from the North Pacific has a more globular shell outline, lacks distinct axial lamellae, and has a much longer teleoconch I (0.9–1 vs. 0.5–0.6 whorl).

Anatoma fujikurai from Japan has a wider sutsel, less dense axial sculpture on the teleoconch, and undulating keels of the selenizone.

Anatoma finlayi from the Indian Ocean to the Indo-Malayan Archipelago (and possibly Caribbean) has sharper axial lamellae on shoulder and base and a protoconch without a spiral cord.

Anatoma indonesica from the Indo-Malayan Archipelago to the Western Pacific has as smooth band on the base below the selenizone, 2–3 fine spiral lines in the sutsel, and usually denser axial cords on the base.

Anatoma quadraxialis from the Indo-Malayan Archipelago has denser axial cords, an overall more stepped profile of the whorls, and a protoconch with reticulate sculpture.

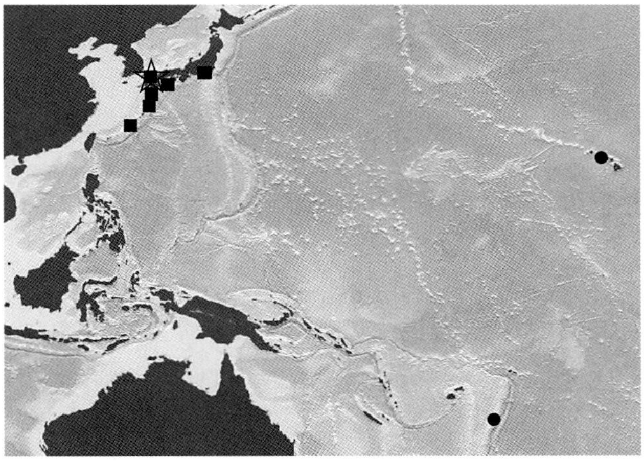

Figure 780. Distribution of *Anatoma lamellata*. ☆ type locality. ● specimen record. ■ literature record.

Anatoma emilioi from Hawaii has fewer but much stronger axial cords and a more stepped profile of the whorls.

Anatoma janusa from the subtropical northeastern Atlantic has a protoconch without spiral cord, a later onset of spiral lines on the shoulder (1 vs. 0.25 teleoconch II whorl), and weaker axial cords on the base.

Anatoma keenae from the Panamic has a protoconch without spiral cord, an overall more globular shell shape, and almost no keels of the selenizone.

Anatoma lyra from the northeastern Pacific has a more lenticular shell shape, a protoconch without spiral cord, and a more narrow sutsel on early whorls (0.5–1× vs. >1× width of selenizone).

Anatoma orbiculata from West Africa has a protoconch with irregular reticulate sculpture, a shell that is wider than tall, and a more narrow sutsel on early whorls (<0.5× vs. 1× width of selenizone).

Anatoma philippiana from the Paleogene of Europe lacks the median spiral ridge on the protoconch and has a more narrow sutsel (<0.5× vs. >0.5× width of selenizone).

Distribution. Western to Central Pacific (Fig. 780).

Specimen records.

Japan. [no specific locality] (RM 1441, 1; RM F3149, 1; NHMUK, 20). Mino-Shima, 34.767°N, 131.167°E (NHMUK 1878.1.28.513, 3).

Coral Sea. 675–680 m, Chesterfield, MUSORSTOM 5 DW340, 19.800°S, 158.683°E (MNHN, 1).
Tonga. 260 m, SE of Tongatapu, 21.345°S, 175.042°W (AMS C.402647, 2).
USA. Mamala Bay, Oahu, Hawaii, 21.290°N, 157.950°W (EGC 29829, 5).

Literature records.

Japan. 966–1004 m, KT04-24, Nansei Islands, 29.692°N, 130.663°E. 394–404 m, TY02-11, 26.536°N, 127.733°E (Hasegawa, 2005).—315 m, Tosa Bay, 33.177°N, 133.573°E. 290 m, Tosa Bay, 33.177°N, 133.573°E. 295 m, Tosa Bay, 33.193°N, 133.593°E. 300 m, Tosa Bay, 33.218°N, 133.628°E (Hasegawa, 2001).—93–95 m, St. HK-3, 34.503°N, 130.871°E (Tsuchida, & Hori. 1996).—550 m, Sagami Bay, 35.089°N, 139.311°E. 700 m, Sagami Bay, 35.150°N, 139.507°E (Okutani, 1964).—Chochi [31.309°N, 130.579°E] (Watanabe, 1988).

Remarks. The Japanese common name is *Kodemari-kuchikire-ebisu* (Hasegawa, 2001, 2005). Hasegawa (2001) remarked that two species are involved in *An. lamellata,* and that his are conspecific with those shown by Numanami & Okutani (1990: is *An. equatoria*) and Okutani & Hasegawa (2000: pl. 18, figs 1a–b: is *An. porcellana*). The identity of Hasegawa's (2001) specimens is therefore uncertain.

Habe (1951) illustrated a species with a distinct absence of spiral cords just below the selenizone, a character not found in *An. lamellata,* but quite typical for *An. porcellana.* Habe's specimen also appears to be anomphalous, whereas *An. lamellata* has a distinct umbilicus.

Anatoma lyra (Berry, 1947)

Figures 781–788

Chresonymy.

Scissurella (*Anatoma*) *lyra* Berry, 1947: 14–16, pl. 1, figs 10–11 (Fig. 781).
Scissurella lyra: Brann & Kent, 1960: 785 [type list].
Scissurella (*Anatoma*) *lyra*: McLean, 1966: 144–145, pl. 4, fig. 25.
Scissurella (*Anatoma*) *lyra*: McLean, 1967: 406–407, pl. 56, figs 5–6.
Scissurella (*Anatoma*) *lyra*: Smith, 1978: 429 [type list].
Anatoma lyra: McLean, 1989: figs 6A [shell], 6D [protoconch].
Anatoma lyra: Lozouet, 1986: 113.
Anatoma lyra: Geiger, 2003: 73.
Anatoma lyra: Geiger & McLean, 2010: 23.

Figure 781. Original illustration of *Scissurella lyra.*

Synonym.

+ *Anatoma baxteri* McLean, 1984: 233, fig. 1. ***Type material.*** Holotype (LACM 1991. Fig. 783B), 2.3 × 1.8 mm. Paratypes from type locality (LACM 1992, 25; CASIZ 033360, 2; USNM 784743, 2). ***Type locality.*** Nine miles off N side Hesketh Island, Kachemark Bay,

Kenai Peninsula, Cook Inlet, Alaska, USA (59°30.5'N, 151°31.0'W) (OD). ***Etymology.*** Named for Rae Baxter, Alaska Department of Fish and Game, Bethel, Alaska, the collector of the specimens (OD).

Anatoma sp. new: Baxter, 1983: 24.

Anatoma baxteri: Lozouet, 1986: 112.

Anatoma s.l. *baxteri*: Geiger, 2003: 74, fig. 16f [shell, radula].

Thieleella baxteri: Geiger & McLean, 2010: 20, 23, 30, 33.

Misidentifications.

Scissurella (*Anatoma*) *lamellata*: McLean, 1967: 406, pl. 56, fig. 8 [is *An. lyra*].

Scissurella lamellata: Foster, 1981: 13, 42–43 [is *An. lyra*].

Scissurella kelseyi: Talmadge, 1966: 83 [*fide* McLean, 1967, is *An. lamellata*. McLean (1984) identified "*An. lamellata*" sensu McLean (1967) from the eastern Pacific as *An. baxteri*, but did not list the Talmadge record].

Type material. Holotype (SSB 12000. Now SBMNH 34531. Fig. 783A), 1.4 × 1.0 mm. Paratypes from type locality (SSB 12009, 2. Now in PRI 20503, 2: Brann & Kent, 1960: 785; SUPTC 7869: Smith, 1978: 429, now CASIZ).

Type locality. Lower Pleistocene (Lomita formation), near 2nd and Pacific Street, San Pedro, Los Angeles, California [USA, 33.750°N, 118.317°W] (OD).

Etymology. Lyre, Greek for strings, referring to the axials sculpture (OD).

Description. Shell medium-sized (to 2.3 mm. Figs 782–784), trochiform, globular. Protoconch of 0.75 whorl, sculpture flocculent, no apertural varix, apertural margin convex. Teleoconch I of 0.5–0.6 whorl, smooth (Fig. 784A), or axial cords barely perceptible initially, becoming fine cords at onset of selenizone (Fig. 784B); no spirals. Teleoconch II of two whorls, suture strongly impressed; sutsel 0.5–1× width of selenizone at onset of selenizone, usually 1–1.5× width of selenizone at apertural margin. Shoulder slightly convex early on, distinctly convex in mature specimens; approximately 50 fine axial cords (Fig. 783C) to low axial lamellae (Fig. 783B) on last whorl, spiral lines starting slightly after onset of selenizone, approximately 10 unequally spaced indistinct spiral lines at apertural margin. Base with indistinct constriction below selenizone, same density and strength of axials as on shoulder, approximately 15 somewhat weaker equally spaced spiral cordlets. Umbilicus narrow, continuously sloping with base, funiculus present. Selenizone slightly above periphery; keels moderately strong, moderately elevated; lunules distinct; slit open, margins slightly converging towards aperture. Aperture rounded, roof overhanging.

Radula (Fig. 786). Rachidian tooth trapezoidal, central cusp largest, 3–5 cusps on each side. Lateral teeth 1–3 similar, 4–5 cusps on outer edge. Lateral tooth 4 reduced, hook-shaped. Lateral tooth 5 somewhat enlarged, broadened, five cusps on inner edge, one cusp on outer edge. Inner marginal teeth with triangular tip, three larger cusps, approximately five smaller cusps on each side; shafts without food groove; outer marginal teeth spoon-shaped, with many bristles, outermost marginal teeth paddle-shaped, with many small projections. Radular interlock of central field moderate.

Jaw composed of many small shingles.

Operculum corneous, thin, multispiral, central nucleus (Fig. 785).

Comparisons. *Anatoma bathypacifica* from the Panamic has a protoconch with reticulate sculpture, a distinct angulation along the spiral cord on teleoconch I, and an overall more scalar shell profile.

Anatoma kelseyi from the northeastern Pacific is more globular particularly in smaller specimens, has a protoconch with flocculent sculpture and more, low axial cords, and lacks a funiculus.

Anatoma disciformis from the North Pacific has an overall more depressed shell and at least twice as many less distinct axial cords that are intersected by spirals, which are about half as strong as the axials.

Anatoma keenae from the Panamic has weaker axial cords, which are about twice as dense, and relatively stronger spiral lines.

Anatoma concinna from the North Pacific has an overall more globular shell, teleoconch I of 0.875–1 whorl, blunt, coarse, reticulate sculpture, and a narrow umbilicus without a funiculus.

Figure 782. *Anatoma lyra*. 37–52 m, granite pinnacle, Cordell Bank, off Point Reyes, Marin County, California, USA, 38.017°N, 123.417°W (LACM 81-50, 30). Scale bar shell = 1 mm. Scale bar protoconch = 100 µm.

Anatoma soyoae from the temperate North Pacific has axial cords instead of lamellae, which are at least twice as dense, and lacks a spiral cord in position of the selenizone.

Figure 783. *Anatoma lyra*. **A.** Holotype of *Scissurella lyra*. Lower Pleistocene (Lomita formation), near 2nd and Pacific Street, San Pedro, Los Angeles, California, USA, 33.750°N, 118.317°W (SBMNH 34531). **B.** Holotype of *Anatoma baxteri*. 9 mi off N side Hesketh Island, Kachemark Bay, Kenai Peninsula, Cook Inlet, Alaska, USA, 59°30.5'N, 151°31.0'W (LACM 1991). **C.** 121 m, N of Carlisle Island, Islands of Four Mountains, Aleutian Islands County, Alaska, USA, 52.975°N, 170.043°W (LACM 86-333, 1). Scale bars shell = 1 mm. Scale bars protoconch = 100 µm.

Anatoma conica from subantarctic South America usually has weaker axial cords or low lamellae on the base, which have a distinct bend at mid-base, and about 50% more axial lamellae per whorl.

Figure 784. *Anatoma lyra*. Ontogentic series. **A.** 7 m, Middle Rocks, Chicagof Harbor, Bering Sea side of Attu Island, Near Islands, Aleutian Islands, County, Alaska, USA, 52.940°N, 173.265°W (LACM 79-71, 3). **B.** 37–55 m, WSW of Point Sur, Monterey County, California, USA, 36.283°N, 121.967°W (LACM 88-356, 9). **C.** 210 m, S of Attu Island, W of Agattu Island, Near Island, Aleutian Islands, Alaska, USA, 52.495°N, 172.908°W (LACM 86-310, 2). Scale bars shell = 1 mm. Scale bars protoconch = 100 μm.

Anatoma crispata from the northeastern Atlantic has less elevated axial cords on shoulder and base, about 50% denser axials, and a proportionally wider umbilicus.

Anatoma finlayi from the Indian Ocean to the Central Pacific (and possibly Caribbean) has stronger axial lamellae on shoulder and base, axial lines on the keels of the selenizone, and a distinct constriction of the base below the selenizone.

Anatoma lamellata from the Western and the Central Pacific has a wider sutsel on early teleoconch whorls (>0.5 vs. <0.5 whorl), a protoconch with a spiral ridge, and only half as many spiral lines on shoulder and base.

Anatoma orbiculata from West Africa has a protoconch with irregular reticulate sculpture and an apertural varix, a longer teleoconch I (0.75 vs. 0.5 whorl), and a more narrow umbilicus.

Anatoma weddelliana from Antarctica has more widely spaced axial lamellae on shoulder and base, a protoconch with very fine reticulate sculpture, and a wider umbilicus.

Distribution. Aleutian Islands and Alaska to Southern California (Fig. 788).

Fossil. Pleistocene of Southern California (type material).

Specimen records.

USA, Alaska, Aleutian Islands County. 210 m, S of Attu Island, W of Agattu Island, Near Island, 52.496°N, 172.908°W (LACM 86-310, 2). 121 m, N of Carlisle Island, Islands of Four Mountains, 52.976°N, 170.043°W (LACM 86-333, 1). 82 m, SE of Red Bluff, NE end Kiska Island, Rat Islands, 52.074°N, 177.696°W (LACM 86-315, 3). 7 m, Middle Rocks, Chicagof Harbor, Bering Sea side of Attu Island, Near Islands, 52.940°N, 173.265°W (LACM 79-71, 4). 388 m, NE of Semisopochnoi Island, Rat Islands, 52.250°N, 179.653°W (LACM 86-325, 2). Port Dick, 55.290°N, 151.133°W (LACM 505, 5). 27–64 m, Kachemak Bay, 59.550°N, 151.983°W (LACM 505B, 2). 148 m, SSE of Vega Point, S end Kiska Island, Rat Islands, 51.763°N, 177.356°W (LACM 86-354, 4). 121 m, off SE side Semisopochoi Island, Rat Islands, 51.889°N, 179.760°W (LACM 86-322, 4). 95 m, off SE side Semisopochol Island, Rat Islands, 51.892°N, 179.743°W (LACM 86-323, 4). 168 m, S of Sturdevant Rock, off NW side Kiska Island, Rat Islands, 52.059°N, 177.344°W (LACM 86-314, 1). 168 m, NE of Semisopochnoi Island, Rat Islands, 52.357°N, 179.821°W (LACM 86-329, 1). 219 m, NE of Semisopochnoi Island, Rat Islands, 52.402°N, 179.714°W (LACM 86-330, 2). 119 m, W of Agattu Island, Near Islands, 52.470°N, 173.171°W (LACM 86-308, 2). 263 m, SSW of Ingenstrem Rocks, E of Agattu Island, Near Islands, 52.485°N, 174.366°W (LACM 86-209, 1). 112 m, SW of Ingenstrem Rocks, E of Agattu Island, Near Islands, 52.509°N, 174.242°W (LACM 86-306, 4). 159 m, S of Attu Island, W of Agattu Island, Near Islands, 52.509°N, 172.944°W (LACM 86-309, 1). 121 m, N of Carlisle Island, Islands of Four Mountains, 52.976°N, 170.043°W (LACM 86-333, 1).

USA, Alaska. 179 m, S of Coronation Island, W of N end Dall Island, Prince of Wales County, 55.143°N, 134.210°W (LACM 87-377, 1). 179 m, S of Coronation Island, W of N end Dall Island, Prince of Wales outer Ketchikan County, 55.143°N, 134.210°W (LACM 78-377, 1). 110 m, S of Warren Island, W of Baker Island, Prince of Wales County, 55.285°N, 133.894°W (LACM 87-373, 1). 102 m, S of Point St. Albans, S end Kulu Island, W of Baker Island, Prince of Wales County, 55.365°N, 134.005°W (LACM 87-374, 1). 205 m, W of N end Noyes Island, S of Whale Bay, Baranof Island, Prince of Wales County, 55.546°N, 135.051°W (LACM 87-368, 1). 201 m, S of Cape

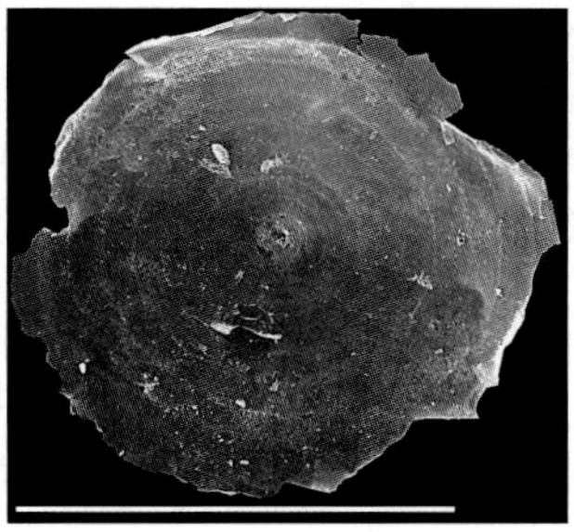

Figure 785. *Anatoma lyra*. Operculum. 102 m, S of Point St. Albans, S end Kulu Island, W of Baker Island, Prince of Wales County, Alaska, USA, 55.365°N, 134.005°W (LACM 87-374, 1). Scale bar = 1 mm.

Ommaney, Barinof Iland, W of Heceta Island, Sitka County, 55.749°N, 134.656°W (LACM 87-390, 1). 192 m, S of Puffin Bay, Barinof Island, W of Hazy Island, Sitka County, 55.860°N, 134.803°W (LACM 87-391, 1). 263 m, SSW of Biorka Island, WNW of Puffin Bay, Baranof Island, Sitka County, 56.263°N, 135.558°W (LACM 87-361, 1). 177 m, S of Biorka Island, W of Whale Bay, Baranof Island, Sitka County, 56.506°N, 135.597°W (LACM 87-359, 1). 123 m, S of Dry Bay, W of Icy Point, Skagway-Yukutat County, 58.413°N, 138.512°W (LACM 87-341, 2). 119 m, S of Dry Bay, W of Lituya Bay, Skagway-Yukutat County, 58.584°N, 138.454°W (LACM

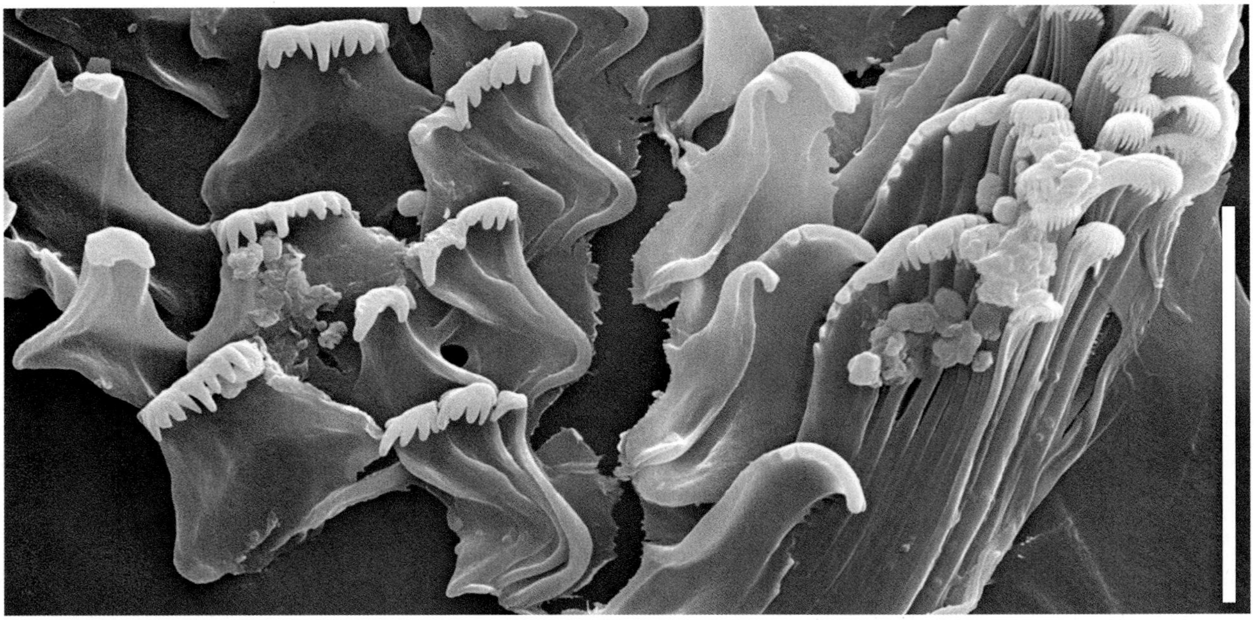

Figure 786. *Anatoma lyra*. Radula. 13–42 m, Carmel Submarine Canyon, N end of San Jose Creek Beach, Monterey County, California, USA, 36.533°N, 121.933°W (LACM 60-24, 4). Scale bar = 20 µm.

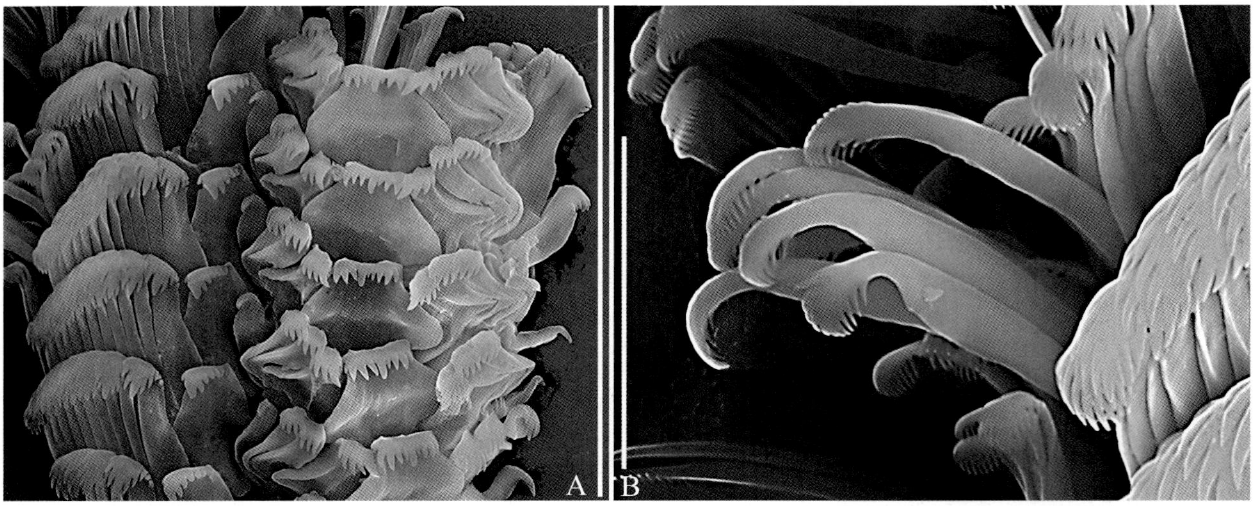

Figure 787. *Anatoma lyra*. Radula. 121 m, off SE side Semisopochoi Island, Rat Islands, Aleutian Islands County, Alaska, USA, 51.889°N, 179.760°W (LACM 86-322, 5). **A.** Half row of radula. Scale bar = 100 µm. **B.** Marginal teeth. Scale bar = 20 µm.

87-340, 3). 112 m, S of Dry Bay, W of Cape Fairweather, Skagway-Yukutat County, 58.767°N, 138.578°W (LACM 87-339, 7; USNM 869258, 3). 21–32 m, Taylor Bay Corner, Port Dick, S coast Kenai Peninsula, Kenai-Cook Inlet County, 59.275°N, 151.100°W (LACM 80-76, 4). 33–46 m, Kasitsna Bay, Kachemak Bay, Cook Inlet, 59.483°N, 151.533°W (LACM 16360, 1). 25 m, off mouth of Jakolof Bay, Kasitsna Bay, Kachemak Bay, Cook Inlet, Kenai-Cook Inlet County, 59.483°N, 151.533°W (LACM 79-117, 2). 12–16 m, Hesketh Island, 0.5 mi SW of Yukon Island, S shore Kachemak Bay, Cook Inlet, Kenai-Cook Inlet County, 59.510°N, 151.529°W (LACM 83-106, 9). 18–64 m, Kachemak Bay, Cook Inlet, Kenai-Cook Inlet County, 59.517°N, 151.567°W (LACM 57-1, 20). 79 m, NE of Wessles Reef, S of mouth of glacier River, Cordova-McCarthy County, 59.929°N, 145.652°W (LACM 87-319, 1). 46–73 m, Orca Inlet, E of Hawkins Island, Prince William Sound, Cordova-McCarthy County, 60.517°N, 145.867°W (LACM 66-262, 3). Off NE side of Stuart Island, Norton Sound, Nome County, 63.830°N, 162.300°W (LACM 76-80, 4).

USA, California. 229 m, Cortez Bank, 32.442°N, 119.125°W (SBMNH 1333331, 1). 119 m, near San Nicolas Island, 33.249°N, 119.499°W (SBMNH 133373, 1). 23–33 m, Begg Rock, NW of San Nicolas Island, 33.362°N, 119.693°W (LACM 72-99, 1). Near 2nd and Pacific Street, San Pedro, 33.742°N, 118.288°W (MCZ 352332, 1: complete). 37–55 m, WSW of Point Sur, Monterey County, 36.283°N, 121.967°W (LACM 88-356, 9). 13–42 m, Carmel Submarine Canyon, N end of San Jose Creek Beach, Monterey County, 36.533°N, 121.933°W (LACM 60-24, 4). 10–15 m, SE Farallon Island, 37.700°N, 123.000°W (LACM 62-9, 1). 37–52 m, granite pinnacle, Cordell Bank, off Point Reyes, Marin County, 38.017°N, 123.417°W (LACM 81-50, 30). 105 m, 3.4 mi 279°T to East Point, Santa Rosa Island, 33.934°N, 119.901°W (LACM 75-480, 26).

Literature records.

USA, Alaska. 140 m, Eastern Gulf, Alaska, USA, 59.027°N, 151.850°W (Foster, 1981).—9 mi off N side Hesketh Island, Kachemark Bay, Kenai Peninsula, Cook Inlet, 59.508°N, 151.517°W (McLean, 1970).

USA, California. 15 m, SE Farallon Island, 37.733°N, 123.033°W. 30 m, SE Farallon Island, 37.733°N, 123.033°W (McLean, 1967).

Remarks. *Anatoma lyra* has been described from Pleistocene material, but occurs in the Recent. The Recent populations agree very well with the fossil type: they have the same protoconch details, same teleoconch length, same overall shape, same sutsel. McLean (1967) considered that the species may be a Pacific analog to *An. aspera* (as *An. umbilicata*: see under these species for history of confusion), with less sculpture than *An. crispata* in the European fauna.

Two specimens in MCZ 352332 had been labeled as "paratypes" but are only topotypes.

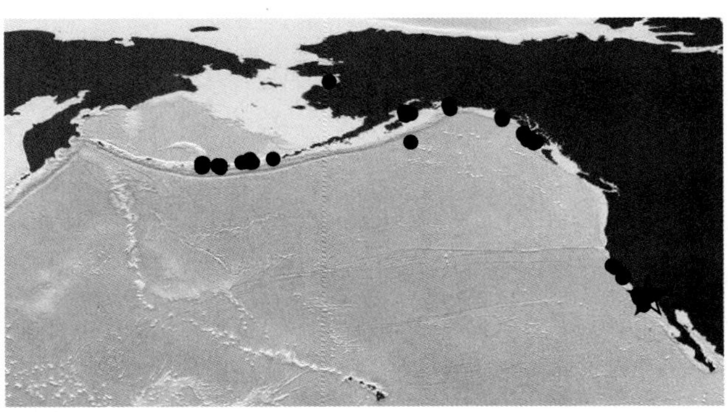

Figure 788. Distribution of *Anatoma lyra*. ☆ type locality. ● specimen record. ■ literature record.

The radula has a lateral tooth 5 that is not as elongated as in other species of *Anatoma*, but is not as wide as in Scissurellidae and also has a cusp on the outer margin not seen in Scissurellidae.

Anatoma baxteri is here synonymized under *An. lyra*. The two holotypes examined by SEM (Fig. 783A–B) are identical in every observable aspect: protoconch sculpture, teleoconch I length and sculpture, teleoconch II sculpture on shoulder and base. The Pleistocene fossil is not as well preserved as the Recent specimen, but there is no doubt that they belong to the same species. While there is some variability in sculpture in this species (see below), the two types are an exceedingly close match.

There is significant intraspecific as well as ontogenetic variability in this species, which can lead to some confusion. Fully grown specimens can have strong axial lamellae (Fig. 783B), distinct spiral cords (Fig. 784B-C), or indistinct spiral cords (Fig. 783C). Spiral lines may be indistinct and few (Fig. 784B), indistinct but numerous (Fig. 783C), or distinct and numerous (Figs 782, 784C). Teleoconch I and early teleoconch II always have less sculpture, but may range from smooth (Fig. 784A) to indistinct low axial cords (Fig. 782). Protoconch sculpture and the strong funiculus are invariant. Smaller specimens with well-preserved keels of the selenizone may appear to have a more quadrate outline of the whorl (Fig. 784B). All those variables are continuous and are found in both distribution centers: California and Alaska.

Small shells are significantly more discoidal and have generally little sculpture (Fig. 784A). Comparisons of ontogenetic series and comparisons with the early portion of larger shells show that those small specimens belong to *An. lyra*.

Anatoma maxima (Schepman, 1908)

Figures 789–793

Chresonymy.

Scissurella maxima Schepman, 1908: 83, pl. 6, fig. 9 (Fig. 789).
Scissurella maxima: Thiele, 1912a: 12–13, pl. 2, figs 1–2.
Anatoma maxima: Habe & Kosuge, 1964: 4.
Anatoma maxima: Higo, 1973: 14
Anatoma maxima: Lozouet, 1986: 113.
Anatoma maxima: Higo & Goto, 1993: 15.
Anatoma s.l. *maxima*: Geiger, 2003: 75.
Scissurella/Anatoma s.l. *maxima*: Van der Bijl *et al.*, 2010: 68.

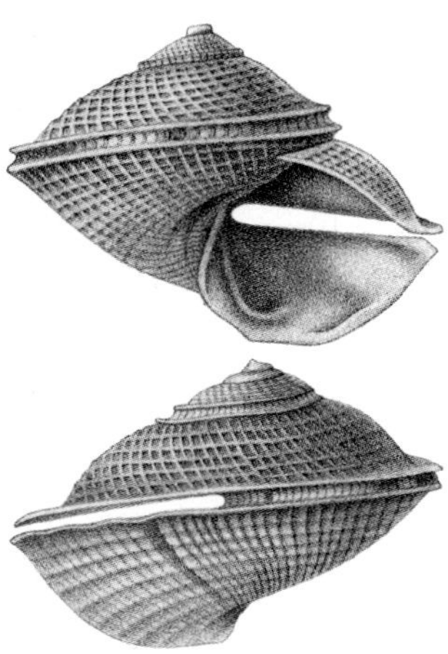

Figure 789. Original illustration of *Scissurella maxima*.

Misidentifications.

Anatoma maxima: Okutani & Hasegawa, 2000: 37, pl. 18, figs 3a–b [is *An. japonica*].

Type material. Syntypes (ZMA 3.08.099, 2. Fig. 791A), 4.5 × 6.5 mm in OD, 5.08 × 3.90 mm from SEM of figured syntype (H × W); whereabouts of other syntype unknown (possibly lost). Here designated as lectotype, in case non-conspecific syntypes should be located.

Type locality. 835 m, Siboga station 178, 2°40'S, 128°37.5'E [~25 km NW of Soabuwe, Seram, Indonesia] (OD).

Etymology. Maximus, Latin superlative for big, most likely named for its large size.

Description. Shell large (to 7.6 mm. Figs 790, 791), trochiform, biconical. Protoconch of 0.75 whorl, sculpture flocculent, very weak apertural varix, apertural margin sinusoidal. Teleoconch I of 0.8 whorl in lectotype, with estimated 35 weak axial cordlets, intersected by approximately three very faint spiral lines at onset of selenizone. Teleoconch II of at least 2.6 whorls (lectotype); suture little impressed; no sutsel on early whorls, at most 0.5× width of selenizone in large specimens. Shoulder with dense regular axial cords (approximately 55 on last half whorl of 5 mm specimen), single slightly weaker spiral cord at onset of selenizone, approximately five cords after one whorl, approximately 23 cords at apertural margin of 5 mm specimen, equally distributed between suture and selenizone, forming fine points at intersection with axials. Base with similar sculpture as on shoulder, approximately 17 spiral cords between selenizone and umbilicus. Umbilicus quite narrow, without funiculus. Selenizone at periphery; lunules rather weak; keels moderately strong, moderately elevated; slit open, margins converging towards peristome. Aperture rounded, with distinct adumbilical folding of lip.

Radula (Fig. 792). Rachidian tooth trapezoidal, central cusp largest, 5–6 cusps on each side arranged parallel to rows of teeth. Lateral teeth 1–3 similar, apical cusp largest, 8, 6, 4 cusps on outer edge of each tip, respectively. Lateral tooth 4 reduced, hook-shaped with one minute point on each side of tip. Lateral tooth 5 enlarged by elongation, approximately six cusps on inner edge of tip. Inner marginal teeth with triangular tip, few cusps; outer marginal teeth spoon-shaped, with many fine bristles on each side of tip; shafts of marginal teeth without food groove. Radular interlock of central field weak.

Jaw (Fig. 792D) trapezoidal, composed of multiple rhomboidal scales.

Operculum (Fig. 792E) thin, corneous, multispiral, with central nucleus.

No eyes detected [based on re-hydrated and partially macerated specimen from which radula was extracted]. Further anatomical details unknown.

Comparisons. *Anatoma maxima* is characterized by the very wide selenizone with very elevated keels and the large size of the shell.

Anatoma amydra from the Indo-Malayan Archipelago to the Central Pacific has a wider sutsel from early whorls on, a proportionally more narrow selenizone, and a wider umbilicus.

Anatoma australis from southeastern Australia has only every 3–4 axial cords on the base reaching the umbilicus, a taller overall shell shape, and a wider sutsel on early whorls (0.5× vs. 0× width of selenizone).

Anatoma biconica from the Indian Ocean to the Central Pacific has a wider sutsel (0.5–1× vs. 0× width of selenizone), a more turreted overall shell shape, and a more narrow selenizone.

Anatoma flexidentata from the Indian Ocean to the Western Pacific has a shorter teleoconch I (0.25 vs. 0.75 whorl) and fewer and irregularly spaced spiral lines on the shoulder.

Anatoma pseudoequatoria from the Indo-Malayan Archipelago to the Central Pacific has a shorter teleoconch I (0.25 vs. 0.75 whorl), a wider sutsel on early whorls (0.5× vs. 0 × width of selenizone), and a more narrow selenizone.

Anatoma globulus from the Indo-Malayan Archipelago has a wider sutsel on early whorls (>0.5× vs. 0× width of selenizone), a more globular overall shell shape, and a more distinct umbilicus.

Anatoma japonica from the Indian Ocean to the Central Pacific is overall more globular, has an apertural varix on the protoconch, and a spiral cord in the position of the selenizone on teleoconch I.

Anatoma megascutula from the Western Pacific is flatter and wider in overall shell shape and the spiral sculpture is stronger than the axial.

Anatoma philippinica from the Indo-Malayan Archipelago to the Western Pacific has a taller shell with distinct sutsel on early whorls and much denser axial cords on shoulder and base.

Anatoma planapex from the Indo-Malayan Archipelago and Australia has a sutsel on early whorls, a more stepped overall shell profile, and a sunken apex.

Anatoma rainesi from the Eastern to the Western Pacific has a sutsel on early whorls and few but very distinct axial cords on teleoconch I.

Anatoma rapaensis from the Indo-Malayan Archipelago to the Central Pacific has a shorter teleoconch I (0.3 vs. 0.75 whorl), spiral steps on shoulder and base, and a more narrow selenizone.

Anatoma zancliformis from the Indo-Malayan Archipelago to the Western Pacific has an overall taller shell, a shorter teleoconch I (0.4 vs. 0.75 whorl), and fewer spiral lines on the shoulder.

Figure 790. *Anatoma maxima*. 820–863 m, Fiji, 17.300°S, 179.550°W (MNHN, 6). Umbilical and apical views composite images. Scale bar shell = 2 mm. Scale bar protoconch = 100 µm.

Anatoma regia from temperate New Zealand has a more narrow selenizone, a slightly wider sutsel on the first whorl (0.25× vs. 0× width of selenizone), and less

Figure 791. *Anatoma maxima*. **A.** Surviving syntype, here designated as lectotype. 835 m, Siboga station 178 [~25 km NW of Soabuwe, Seram, Indonesia] (ZMA 3.08.099). **B.** 1175–1210 m, Vanuatu, 20.550°N, 169.600°E (MNHN, 5). Radula, jaw, and operculum from specimen shown in Figure 792. **C.** 477–493 m, Fiji, 17.333°S, 178.733°W (MNHN, 11). Scale bars shell = 1 mm. Scale bars protoconch = 100 μm. Umbilical and apical views of B and C composite images.

elevated keels of the selenizone.

Anatoma tenuis from the North Atlantic has a shorter teleoconch I (0.4 vs. 0.75 whorl) and 50% denser spiral lines on shoulder and base.

Distribution. Indo-Malayan Archipelago to Central Pacific (Fig. 793).

Specimen records.

Japan. 283 m, off Miyake Island, Izu Islands, Honshu, 34.087°N, 139.435°E (SBMNH 345565, 1).

Taiwan. 246–256 m, NE coast, off Tashi, 24.805°N, 122.140°E (MNHN, 1).

Philippines. 624–647 m, of Aligbay Island, Bohol Sulu Seas Sill PANGLAO 2005 CP2384, 8.770°N, 123.268°E (MNHN, 1). 478–480 m, AURORA 2007 DW2706, 15.070°N, 121.715°E (MNHN, 2). 490 m, Station 5567, off Dammi Island, Sulu Archipelago, 5.796°N, 120.408°E (USNM 289848-3, 5).

Malaysia. 635 m, 4.114°N, 118.789°E (USNM 1098803, 3).

Indonesia. 175–185 m, Bay of Sanggar, N of Sumbawa, 8.322°S, 118.260°E (RMNH 42263, 3). 835 m, Siboga Station 178, 2.667°S, 128.625°E (ZMA 3.08.099, 2). 1183 m, Gulf of Tomini, Celebes, 2.518°S, 120.797°E (USNM 239190, 1). 175 m, SW Salayer, 6.373°S, 120.437°E (RMNH, 4). 534 m, Station 5584, Borneo, Sibuko Bay, off Si Amil Island, 4.000°N, 118.433°E (USNM 289748-5, 3).

Solomon Islands. 611–636 m, 9.758°S, 160.875°E (MNHN, 1). 442–451 m, 9.313°S, 160.382°E (MNHN, 1). 570–756 m, 8.263°S, 160.673°E (MNHN, 1).

Australia. 1143–1106 m, off Sydney, New South Wales, 33.600°S, 152.083°E (AMS C.448769, 2). 785–800 m, W of Cape Reinga, W Norfolk Islands, 34.298°S, 168.430°E (NMNZ M.172124, 50). 750–774 m, N of Norfolk Island, Norfolk Ridge, 26.432°S, 167.181°E (NMNZ M.171920, 8)

Coral Sea. 150 m, ARGO Bank, 22.800°S, 169.400°E (MNHN, 1: complete). 1000 m, 21.333°S, 158.033°E (MNHN, 1).

Vanuatu. 1175–1210 m, 20.550°N, 169.600°E (MNHN, 5). 1058–1086 m, 18.067°S, 168.900°E (MNHN, 4).

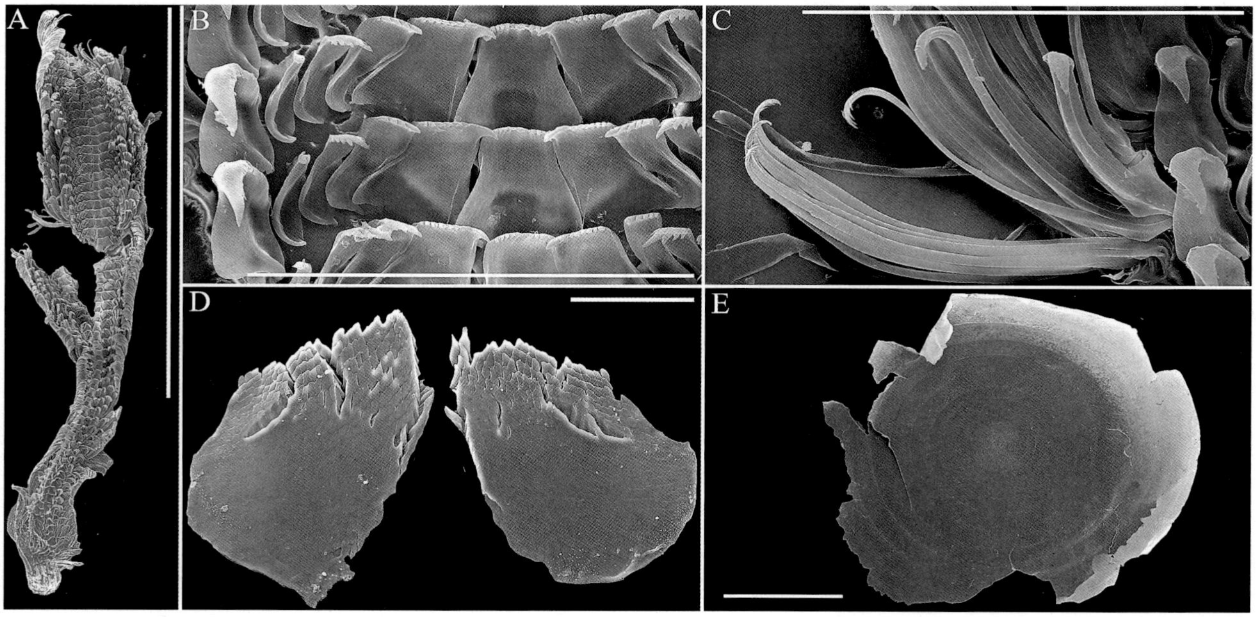

Figure 792. *Anatoma maxima*. Radula, jaw, operculum. 1175–1210 m, Vanuatu, 20.550°N, 169.600°E (MNHN, 5). For shell, see Figure 791B. **A.** Entire radula. Scale bar = 1 mm. **B.** Central field enlarged. Scale bar = 100 µm. **C.** Marginal teeth. Scale bar = 100 µm. **D.** Jaws. Scale bar = 100 µm. **E.** Operculum. Scale bar = 1 mm.

New Caledonia. 770–830 m, Loyalty Ridge, 24.733°S, 170.133°E (MNHN, 5). 298–324 m, Norfolk Ridge, 24.717°S, 168.167°E (MNHN, 1). 923 m, Norfolk Ridge, Stylaster Bank, 23.717°S, 167.717°E (MNHN, 2). 601–608 m, Norfolk Ridge, 23.317°S, 168.000°E (MNHN, 1). 540 m, Norfolk Ridge, Antigona Bank, 23.300°S, 168.017°E (MNHN, 2). 675–680 m, 23.167°S, 167.167°E (MNHN, 3). 2100–2110 m, 22.167°S, 167.550°E (MNHN, 1). 1260–1300 m, Loyalty Basin, 21.467°S, 166.383°E (MNHN, 1). 1320–1380 m, 20.667°S, 166.867°E (MNHN, 2). 735 m, 20.583°S, 166.883°E (MNHN, 4). 1140 m, 23.150°N, 166.683°E (MNHN, 1). 347–1063 m, Norfolk Ridge, NORFOLK 1 DW1732, 23.333°S, 168.267°E (MNHN, 2). 680–700 m, BIOCAL DW51, 23.083°S, 167.750°E (MNHN 4: complete). 775 m, 23.000°N, 167.650°E (MNHN, 1). 360 m, 23.000°S, 167.250°E (MNHN, 1).

New Zealand. 12–16 m, Three Kings Rise, 31.511°S, 172.830°E (NMNZ M.174863, 1; NMNZ M.174844, 7). 840–850 m, South Cavalli Seamount, ENE of North Cape, 34.265°S, 174.103°E (NMNZ M.174859, 30).

Tonga. 427–436 m, SW Tongatapu, 21.267°S, 175.300°W (MNHN, 9). 970–983 m, Eua, 21.200°S, 174.933°W (MNHN, 1). 476–478 m, S Nomuka Group, 20.633°S, 175.050°W (MNHN, 1). 511 m, S of Nomuka Group, 20.633°S, 175.017°W (MNHN, 1). 389–418 m, N Ha'apai Group, 19.700°S, 174.533°W (MNHN, 3). 630–710 m, Vava'u Group, 18.667°S, 175.867°W (MNHN, 1). 281 m, Vava'u Group, 18.650°S, 173.900°W (MNHN, 1). 578 m, Vava'u Group, 18.550°S, 173.950°W (MNHN, 5).

Fiji. 700–707 m, 19.050°S, 178.500°W (MNHN, 2). 477–493 m, 17.333°S, 178.733°W (MNHN, 11). 820–863 m, 17.300°S, 179.550°W (MNHN, 6). 820–863 m, 17.300°S, 179.550°W (MNHN, 2). 550–561 m, 16.033°S, 179.500°W (MNHN, 1). 959–963 m, S of Viti Levu, 17.71°S, 178.917°E (MNHN, 2).

Wallis and Futuna. 440–458 m, Wallis Island, 13.367°S, 178.133°W (MNHN, 1).

French Polynesia. 600–620 m, E of Rapa, 27.613°S, 144.262°W (MNHN, 1). 790–880 m, Mount Otaha, Tarava Seamounts, TARASOC DW3332, 18.750°S, 152.300°E (MNHN, 1). 782–847 m, Huahine, Society Islands, TARASOC DW3421, 16.767°S, 151.067°W (MNHN, 1). 520–572 m, Raiatea, TARASOC DW3457, 16.750°S, 151.400°W (MNHN, 2). 703–790 m, Kaukura, Tuamotus, TARASOC DW3374, 15.650°S, 146.900°W (MNHN, 1). 981–992 m, between Tikehau and Rangiroa, Tuamotus, TARASOC DW3350, 15.083°S, 148.050°W (MNHN, 1).

Marquesas. 388–406 m, Hiva Oa Island, 9.833°S, 139.167°W (MNHN, 1).

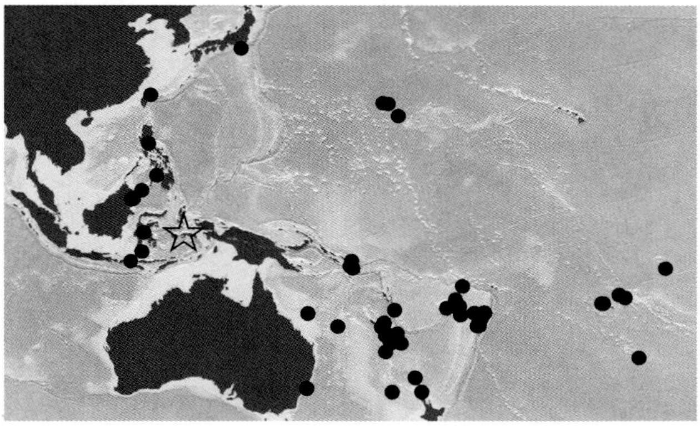

Figure 793. Distribution of *Anatoma maxima*. ☆ type locality. ● specimen record.

Remarks. The original illustration shows a specimen of similar habitus, but with less dense sculpture compared to the specimen examined by SEM. I attribute this discrepancy to limitations of sufficient detail rendition by traditional drafting.

Anatoma megascutula Geiger & Marshall, 2012

Figures 794–797

Chresonymy.

Anatoma megascutula Geiger & Marshall, 2012: 24–26, figs 18–20.

Type material. Holotype (MNHN 24954. Fig. 794). Paratypes: 820–1220 m, N/O *Tangaroa* HALIPRO 2 BT71, New Caledonia, 24.866°S, 168.183°E. (MNHN 24957, 1. Fig. 795A). 456–510 m, Tongatapu, N/O *Alis* BENTHAUS DW1509, Tonga, 21.083°S, 175.367°W (MNHN 24955, 1. Fig. 795B). 820–863 m, Lau Ridge (Yacata), N/O *Alis* BENTHAUS DW1459, Fiji, 17.300°S, 179.550°W (MNHN 24956, 1. Fig. 795C).

Type locality. 800–850 m, E of Rapa, Austral Islands, French Polynesia, N/O *Alis* BENTHAUS CP1891, 27.618°S, 144.257°W (OD).

Etymology. Mega-, Greek prefix for large; -scutula, Latin noun for flat dish, saucer, referring to the large size of this flat-shelled species. Noun in apposition (OD).

Figure 794. *Anatoma megascutula*. Holotype. 800–850 m, E of Rapa, Austral Islands, French Polynesia, N/O *Alis* BENTHAUS CP1891, 27.618°S, 144.257°W (MNHN 24954, 1). Scale bar shell = 1 mm. Scale bar protoconch = 100 µm.

Description. Shell large (to 6.8 mm. Figs 794, 795), trochiform, depressed, discoidal. Protoconch of 0.75 whorl, sculpture flocculent, apertural varix not connected to embryonic cap, apertural margin shallow sinusoidal. Teleoconch I of 0.75–0.875 whorl, approximately 20–30 axial cords, no spirals. Teleoconch II of 3.3 whorls; suture lightly impressed; no sutsel except for last quarter whorl with slight descent

Figure 795. *Anatoma megascutula*. Paratypes. **A.** 820–1220 m, Norfolk Ridge, N/O *Tangaroa* HALIPRO 2 BT71, New Caledonia, 24.866°S, 168.183°E (MNHN 24597, 1). **B.** 456–510 m, Tongatapu, N/O *Alis* BENTHAUS DW1509, Tonga, 21.083°S, 175.367°W (MNHN 24955, 1). **C.** 820–850 m, Lau Ridge (Yacata), N/O *Alis* BENTHAUS DW1459, Fiji, 17.300°S, 179.550°W (MNHN 24956, 1). Scale bars shell = 1 mm. Scale bars protoconch = 100 μm.

Antomidae
Anatoma

showing one spiral thread. Shoulder broadly convex, many barely perceptible, low axial cords becoming somewhat more noticeable with growth; first low spiral thread after 0.125–0.8 whorl, increasing in strength becoming low to distinct spiral cords, approximately a dozen cords after one teleoconch I whorl, approximately 25 cords at apertural margin, forming indistinct thickenings at intersections with axials. Base without constriction below selenizone, continuously sloping with umbilicus; axial sculpture as on shoulder, usually somewhat stronger; Approximately 22 spirals, distinct lines from selenizone onwards, changing at mid-base into steps. Umbilicus variable: open, bordered by carina, walls straight, smooth; additionally with distinct funiculus; closed by parietal callus [see Remarks]. Aperture oblong, D-shaped. Selenizone at periphery; keels moderately strong, moderately elevated (usually low due to erosion/abrasion); lunules distinct; slit open, converging margins.

Radula (Fig. 796). Rachidian tooth trapezoidal, apical cusp largest, 4–5 cusps on each side in continuous arc. Lateral tooth 1 broadened, apical cusp slightly larger than 7–8 distal cusps on cutting edge. Lateral teeth 3–4 similar, apical cusp largest, 3–5 cusps along outer margin of cutting edge. Lateral tooth 4 reduced, apical cusp largest, three small cusps on each side. Lateral tooth 5 enlarged by moderate broadening, apical cusp largest, about a dozen small cusps along inner edge, approximately eight cusps not reaching apical cusp along outer edge. Inner marginal teeth with triangular tip, apical cusp largest, approximately eight minor cusps on each side; shafts without food groove. Outer marginal teeth with indistinct separation of shaft and tip; tip with multiple fine bristles. Radular interlock of central field moderate.

Operculum corneous, thin, round, multispiral, with central nucleus, covering approximately 70% of aperture.

Comparisons. *Anatoma japonica* from the Indian Ocean to the Central Pacific has a wider sutsel on whorl 2 (0.5–1× vs. 0× width of selenizone), stronger axial than spiral sculpture, and a more narrow selenizone.

Anatoma maxima from the Indo-Malayan Archipelago to the Central Pacific has coarser sculpture (14 vs. 25 spirals on shoulder at apertural margin; distinct vs. indistinct axials), in which axials and spirals form points at intersections, a less rapid

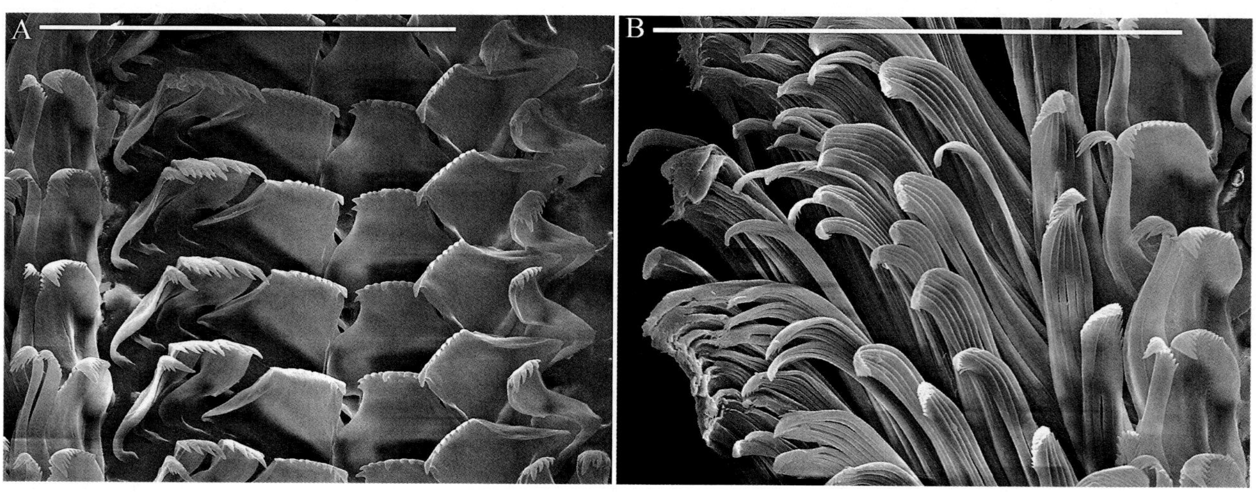

Figure 796. *Anatoma megascutula*. Radula. 680–980 m, Porthos Bank, Norfolk Ridge, NORFOLK 2 DW2068, New Caledonia, 25.333°S, 168.95°E (MNHN, 1). **A.** Central field. **B.** Marginal teeth. Scale bars = 100 μm.

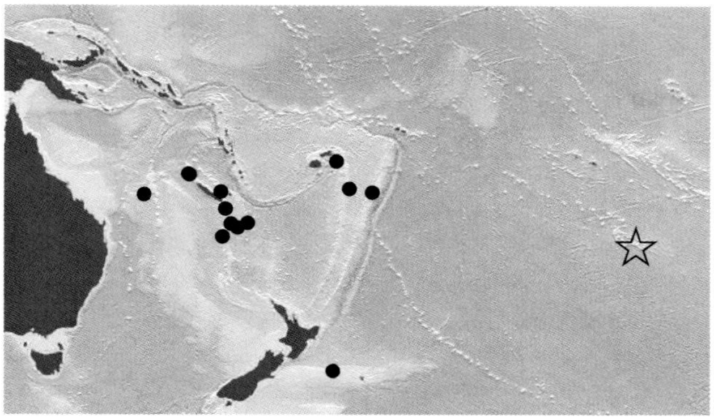

Figure 797. Distribution of *Anatoma megascutula*. ☆ type locality. ● specimen record.

expansion of the whorls (1.4 vs. 1.8), and a wider sutsel at the apertural margin showing 2–3 spiral lines.

Anatoma rainesi from the Eastern to the Western Pacific has a sutsel on the early whorls, strong axial cords on teleoconch I, and the selenizone is closer to the periphery of the shell.

Anatoma rapaensis from the Indo-Malayan Archipelago to the Central Pacific has a protoconch without an apertural varix, a shorter teleoconch I (0.2 vs. >0.75 whorl), and more pronounced axial sculpture that forms sharp points at the intersection with the spirals.

Anatoma zancliformis from the Indo-Malayan Archipelago to the Western Pacific has a lower expansion rate of the whorls, a shorter teleoconch I (0.3 vs. 0.75 whorl) and fewer and irregularly spaced spiral lines on the shoulder.

Anatoma regia from temperate New Zealand has a narrow but noticeable sutsel on early whorls, a more narrow selenizone, and axial sculpture that is stronger than the spiral.

Anatoma disciformis from the boreal North Pacific has an overall taller shell, a protoconch with strong flocculent sculpture, and much stronger axial cords.

Anatoma tenuis from the temperate North Atlantic has a shorter teleoconch I (0.4 vs. 0.75 whorl) and stronger axial than spiral sculpture.

Distribution. Western to Central Pacific. 450–1650 m (Fig. 797).

Specimen records.

Australia. 750–774 m, N of Norfolk Island, Norfolk Ridge, 26.432°S, 167.181°E (NMNZ ex M.171920, 7).

Coral Sea. 970 m, 21.317°S, 157.967°E (MNHN, 1).

New Caledonia. 525 m, 18.900°S, 163.317°E (MNHN, 1). 825–830 m, 23.050°S, 167.533°E (MNHN, 1). 850–855 m, Loyalty Ridge, 24.767°S, 170.117°E (MNHN, 1). 820–1220 m, Norfolk Ridge, N/O *Tangaroa* HALIPRO 2 BT71, 24.866°S, 168.183°E (MNHN 24597, 1). 1600–1640 m, Loyalty Basin, 20.983°S, 166.983°E (MNHN, 3). 683–669 m, Grand Passage, 18.808°S, 163.190°E (MNHN, 2). 680–980 m, Porthos Bank, Norfolk Ridge, NORFOLK 2 DW2068, 25.333°S, 168.950°E (MNHN, 1).

New Zealand. 757–875 m, Graveyard Seamount, Chatham Rise, 42.766°S, 179.989°W (NMNZ M.174849, 1).

Fiji. 820–850 m, Lau Ridge (Yacata), N/O *Alis* BENTHAUS DW1459, 17.300°S, 179.550°W (MNHN 24956, 1: paratype).

Tonga. 476–478 m, 20.633°S, 178.050°W (MNHN, 1). 456–510 m, Tongatapu, N/O *Alis* BENTHAUS DW1509, 21.083°S, 175.367°W (MNHN 24955, 1: paratype).

French Polynesia. 800–850 m, E of Rapa, Austral Islands, N/O *Alis* BENTHAUS CP1891 27.618°S, 144.257°W (MNHN 24954, 1: holotype).

Remarks. The count of axials on teleoconch I is difficult due to erosion of the early parts of the whorl. The condition of the umbilicus is quite variable in the limited number of specimens available. The more open condition is found in smaller specimens, while the callus closing the umbilicus is encountered in the largest specimens, which suggest an ontogenetic change. Other conchological characters suggest that the specimens belong to a single species.

The radula of *An. megascutula* shows an unusual arrangement in the central field. Lateral teeth 1–3 are not similar to one another. Lateral tooth 1 is much broadened.

Lateral tooth 4 is not as reduced as in most other species with the notable exception of *An. rhynchodentata*.

Anatoma micalii new species

Figures 798–800

Misidentification.

not *Anatoma affinis*: Palazzi & Villari, 2000: 9, figs 13–19 [is *An. micalii*].

not *Anatoma umbilicata*: Portalatina, 2008: 155, fig. 2L–M [is *An. micalii*].

not *Anatoma affinis* "sensu Palazzi & Villari, 2000": Cossignani & Ardovini, 2011: 96, text-fig. [is *An. micalii*].

not *Anatoma umbilicata*: Hernández *et al.*, 2011: 62, fig. 12A–B [is *An. micalii*].

Type material. Holotype (SBMNH 149938. Fig. 798). Paratypes from type locality (SBMNH 149939, 3). 38 m, Scilla, Reggio Calabria, Italy, 38.255°N, 15.714°E (DLG 1300, 10). 40–44 m, Scilla, Reggio Calabria, Italy, 38.255°N, 15.714°E (SBMNH 149940, 1. Fig. 799C). 40–42 m, Porto Infreschi, near Marina di Camerata, Italy, 40.000°N, 15.426°E (SBMNH 149941, 5. Fig. 799B). 42 m, Dente del Diavolo rock, Scilla, Reggio Calabria, at N inlet, Italy (PMC, 2). 25 m, Infreschi, near Marina dei Camerota, Sardinia, Italy, 39.998°N, 15.431°E (PMC, 3). 42 m, Scilla Reggio Calabria, Italy, 38.255°N, 15.714°E (TMGLC, 1). 40–44 m, Scilla Reggio Calabria, Italy, 38.255°N, 15.714°E (TMGLC, 4). 38 m, Scilla Reggio Calabria, Italy, 38.255°N, 15.714°E (TMGLC, 5).

Type locality. 42 m, Scilla, Italy, 38.255°N, 15.713°E.

Etymology. Named for Pasquale Micali of Italy, who brought the species to my attention. He also collected and contributed much of the type material.

Description. Shell medium-sized (to 1.5 mm. Figs 798, 799), trochiform, stepped. Protoconch of 0.75 whorl, sculpture flocculent, no apertural varix, apertural margin straight. Teleoconch I of 0.66–0.75 whorl, approximately 17 axial cords, spiral cord in position of selenizone, interstices smooth. Teleoconch II of up to 1.75 whorls; suture moderately impressed; sutsel slightly wider than selenizone. Shoulder convex, strong axial cords to low lamellae, approximately 30 per whorl, in fully grown specimens, extending over apical surface of keel of selenizone; becoming more crowded and less distinct towards apertural margin; secondary, weak axial cords starting at outermost 10% of shoulder extending over apical surface of keel of selenizone, 1–2 between adjacent strong axial cords; strong axial cords occasionally bifurcating in area of fine axial cords; first spiral threads after 0.3 teleoconch I whorl, 6–12 threads at apertural margin, spiral thread running over axial cords/lamellae. Base without constriction below selenizone, approximately 50 axial cords per half whorl, interstices about twice as wide as cords; becoming denser, less distinct towards apertural margin of fully grown specimens, axial cords extending onto lower keels of selenizone, no supernumerary axials on lower keels of selenizone; approximately 13–20 fine spiral cords more or less equally spaced; spirals running over axials forming indistinct points. Umbilicus open moderately wide, distinct funiculus. Aperture rounded, peristome flared in lower adumbilical region. Selenizone slightly above periphery;

keels moderately strong, strongly elevated; lunules regularly spaced, moderately distinct; slit open, margins parallel.

Animal unknown.

Comparisons. *Anatoma aspera* from the western Mediterranean and adjacent Atlantic Ocean has equally strong axial cords on the shoulder, an apertural varix on the protoconch, fewer additional axial cords on the keels of the selenizone, and tends to have a longer teleoconch I (>0.75 vs. 0.5–0.75 whorl). These differences are independent of specimen size.

Anatoma bertheloti from the Canary Islands has a more depressed shell profile with a wider umbilicus, more numerous axial cords on shoulder and base, and sculpture on shoulder and base that are alike.

Anatoma eximia from the western Mediterranean and adjacent Atlantic has an overall taller shell, a distinct apertural varix on the protoconch, a wider sutsel, and a much narrower umbilicus.

Figure 798. *Anatoma micalii* new species. Holotype. 42 m, Scilla, Italy, 38.255°N, 15.713°E (SBMNH 149938). Scale bar shell = 1 mm. Scale bar protoconch = 100 μm.

Anatoma janusa from the Azores has a shorter teleoconch I (0.5 vs. 0.66 whorl) with fewer axial cords (10 vs. 17) and denser axial cords on the base, in which the interstices are as wide as the cords.

Figure 799. *Anatoma micalii* new species. Paratypes. **A.** 38 m, Scilla, Reggio Calabria, Italy, 38.255°N, 15.714°E (DLG 1300, 10). **B.** 40–42 m, Porto Infreschi, near Marina di Camerata, Italy, 40.000°N, 15.426°E (SBMNH 149941, 5). **C.** 40–44 m, Scilla, Reggio Calabria, Italy, 38.255°N, 15.714°E (SBMNH 149940, 1). Scale bars shell = 1 mm. Scale bars protoconch = 100 μm.

Figure 800. Distribution of *Anatoma micalii* new species. ☆ type locality. ● specimen record.

Anatoma orbiculata from West Africa has a more lenticular overall shell profile, a protoconch with irregular reticulate sculpture and an apertural varix, and a more narrow sutsel on early whorls (<0.5× vs. 1× width of sutsel).

Anatoma tenuisculpta from the Mediterranean and the northeastern Atlantic has a more discoidal overall shell shape with no sutsel on the early whorls and grows to a much larger size.

Anatoma umbilicata from the Mediterranean and adjacent Atlantic Ocean has much less sculpture and a wider umbilicus.

Anatoma philippiana from the Paleogene of Europe has a more globular overall shell shape and lacks supernumerary axial lines on the less elevated keels of the selenizone.

Anatoma aedonia from the Caribbean to Brazil has a slightly wider sutsel, lower keels of the selenizone, and a longer teleoconch I (0.75–1 vs. 0.5–0.75 whorl).

Anatoma alternatisculpta from the Caribbean has a more rounded shoulder, no axials on the keels of the selenizone between the axials on the shoulder, an overall taller shell, and the axial cords on the base weaken significantly towards the umbilicus.

Anatoma bathypacifica from the Panamic has an angulation along the spiral cord in position of the selenizone on teleoconch I, at most half as many axial and spiral cords on the base, and reticulate protoconch sculpture.

Anatoma breveprima from the Western Pacific has a more depressed shell, a shorter teleoconch I (0.1 vs. >0.5 whorl), and a wider umbilicus.

Anatoma munieri from the Indian Ocean to the Central Pacific has a distinct constriction on the base below the selenizone, an overall much taller shell, and two distinct spiral keels on the base.

Distribution. Mediterranean Sea to Angola (Fig. 800).

Specimen records.

Angola. 10–20 m, Lucira Bay (Cesar), Moçâmedes Province, 13.867°S, 12.533°E (MNHN, 11). 8–10 m, Santa Maria Bay, Benguela Province, 13.417°S, 12.533°E (MNHN, 2).

Morocco. 60–80 m, Northern, Straits of Gibraltar, 35.850°N, 5.500°W (MNHN, 1).

France. 450 m, Cap Corse, Corsica, 43.100°N, 9.300°E (DLG 1428, 1). 160 m, Bastia, Corsica, 42.711°N, 9.522°E (DLG 1417, 1). Propriano, Corsica, 41.678°N, 8.899°E (RMNH AD 13382, 11).

Italy. 168 m, Adventure Bank, S of Sicily, 37.500°N, 12.167°E (AMS C.24118, 2). 23 m, Cannizzaro, N of Catania, Sicily, 37.535°N, 15.138°E (PMC, 1). Scilla, Reggio Calabria, 38.250°N, 15.717°E (DLG 339, 2). 44 m, Scilla, Reggio Calabria, 38.257°N, 15.715°E (RMNH AD 26182, 12). 42 m, Scilla, Reggio Calabria, 38.261°N, 15.7160°E (DLG 1277, 1). 42 m, Scilla, 38.255°N, 15.713°E (DLG 1306, 3). 38 m, Scilla, Reggio Calabria, 38.255°N, 15.714°E (DLG 1300, 10). 40–44 m, Scilla, Reggio Calabria, 38.255°N, 15.714°E (DLG 1301, 1). 50 m, Scilla, Calabria, 38.257°N, 15.715°E (DLG 1210, 2). 50 m, Scilla, Reggio Calabria, 38.261°N, 15.715°E (DLG 2346, 12). 42 m, Dente del Diavolo rock, Scilla, Reggio Calabria, at N inlet (PMC, 16). 25 m, Infreschi near Marina dei Camerota, 39.998°N, 15.431°E (PMC, 87).

Palinuro, Campania, 40.033°N, 15.283°E (DLG 338, 8). 400–500 m, off Fiumicino, Tyrrehnian Sea (DLG 337, 2). 42 m, Secca Matesi, Ragusa (DLG 1304, 3). 30–35 m, Giannutri Island, Tuscany, 42.250°N, 11.100°E (DLG 927, 9). 32 m, Secca Point, Giglio Island, 42.372°N, 10.881°E (PMC, 1). 52 m, Lazzaro, Calabria, 37.963°N, 15.669°E (DLG 1414, 32). 200 m, Corso Cape, Livorno, 43.509°N, 10.303°E (DLG 1364, 1). 20 m, Capri, 40.600°N, 14.200°E (DLG 1435, 25). 18–25 m, SE coast Sicily (RMNH AD 16998, 6). Catania, Sicily, 37.501°N, 15.098°E (RMNH AD 16382, 8). 80 m, off Catania, Sicily, 37.501°N, 15.098°E (RMNH AD 20141, 11). 30 m, off Catania, Sicily, 37.501°N, 15.098°E (RMNH AD 26019, 6). Lido San Guiliano, Trapani, Sicily, 38.026°N, 12.528°E (RMNH AD 21855, 12). 40 m, Gorgona Island, 43.433°N, 9.900°E (RMNH AD 20614, 3). 30 m, Secca del Cammello, Marettimo Island, Egadi Islands, Trapani, Sicily, 37.989°N, 12.065°E (DLG 1810, 1). 21 m, Le Scoglietelle, Ponza Island, Latina, 40.938°N, 13.012°E (DLG 1791, 3).

Malta. 26 m, 35.940°N, 14.370°E (DLG 1424, 1).

Turkey. 25–35 m, Uc Adalar, 40.800°N, 29.100°E (DLG 1308, 1; PMC, 1).

Literature records.

Italy. 80–100 m, Castro Marina, 39.983°N, 18.503°E (Portalatina, 2008).

Remarks. Palazzi & Villari (2000) illustrated *An. micalii* by SEM in their re-description of what they identified as *An. affinis*. However, their specimens cannot be regarded as belonging to *An. affinis*, because they have much fewer and stronger axial cords compared to the original illustration of *An. affinis* (see also discussion of *An. affinis* under *An. aspera*).

Much of the material of the species described herein was kindly donated by Tisselli Morena and Giunchi Luigi.

Anatoma munieri (Fischer, 1862)

Figures 801–807

Chresonymy.

Scissurella munieri Fischer, October 1, 1862: 390–391, not illustrated.

Scissurella munieri: Troschel, 1963: 702.

Scissurella munieri: Munier-Chalmas, 1865: 397.

Scissurella munieri: Fischer, 1867: 305, 468, pl,. 9, fig. 4 (Fig. 801) [the figure caption on plate 9 for figure 4 reads *Scissurella munieriana*, an error noted on the errata page 468 of the volume].

Scissurella munieri: Paetel, 1888: 289.

Scissurella munieri: Pilsbry, 1890: 54.

Scissurella munieri: Thiele, 1912a: 14–15.

scissurellid: Bandel, 1991: pl. 2, fig. 7.

Scissurella munieri: Higo & Goto, 1993: 15.

Scissurella ? *munieri*: Geiger, 2003: 77.

Anatoma munieri: Geiger, 2003: 28–30, fig. 20 [SEM of lectotype and paralectotype].

Anatoma munieri: Geiger, 2006c: text-fig. 5.

Anatoma munieri: Geiger, 2008a: 158, pl. 24, fig. 2a–b.

Anatoma munieri: Sasaki, 2008: fig. 2. D.

Figure 801. Illustration of *Scissurella munieri* from Fischer. (1867).

Synonyms.

+ *Anatomus turbinatus* A. Adams, November, 1862: 347–348, not illustrated. ***Type material.*** Syntype (NHMUK 1874.5.19.62. Higo *et al.*, 2001: G82. Fig. 803B), 2.5 mm. Here designated as the lectotype for the express purpose of taxon stabilization in case other non-conspecific syntypes should be located. ***Type locality.*** Mino-sima, [= Mishima Island, Hagi City, Yamaguchi Prefecture, Japan [34.767°N, 131.166°E]: see also Kawamoto and Tanabe (1956)], 63 fms. [= 115 m] (OD). ***Etymology.*** Turbinatus: Latin adjective: that which whirls; whirlwind, tornado; spinning top; spiral. Referring to the high spired shell shape.

Scissurella turbinata: Crosse, 1863: 109.

Anatomus turbinatus: Troschel, 1863: 702.

Anatomus turbinatus: Dunker, 1882: 147.

Scissurella turbinata: Paetel, 1888: 289.

Anatomus turbinatus: Pilsbry, 1890: 59.

Scissurella turbinata: Pilsbry, 1895: 106.

Scissurella turbinata: Thiele, 1912a: 15, pl. 2, figs 9–10 [after a specimen in the British Museum: type?].

Schizotrochus turbinatus: Habe, 1951: 68, pl. 11, figs 12–13.

Scissurella turbinata: Kuroda & Habe, 1952: 85.

Scissurella (*Schizotrochus*) *turbinata*: Kawamoto & Tanabe, 1956: 3, pl. 2, fig. 11 [copy figure Thiele, 1912a].

Anatoma turbinata: Habe & Kosuge, 1964: 4.

Anatoma turbinata: Higo, 1973: 13.

Anatoma turbinata: Tsuchida *et al.*, 1991: 5–6, pl. 1, figs 2–3, 5.

Anatoma turbinata: Higo & Goto, 1993: 15.

Anatoma turbinata: Yu & Feng, 1996: pl. 1, figs 1–4.

Anatoma turbinata: Okutani & Hasegawa, 2000: 37, fig. 5.

Anatoma turbinata: Higo *et al.*, 2001:14, G82 [unspecified type].

Anatoma turbinata: Geiger, 2003: 74.

Anatoma turbinata: Geiger & Jansen, 2004a: 18–21, figs 9–10, 18 [map].

Anatoma turbinata: Geiger, 2004: text-fig. p. 5.

Anatoma turbinata: Koyama, 2008: 253, pl. 1, fig. 6.

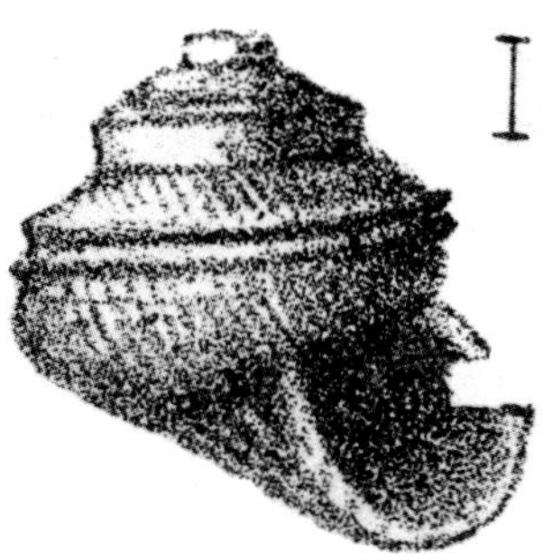

Figure 802. Original illustration of *Scissurella jacksoni.*

+ *Scissurella jacksoni* Melvill, 1904: 160–161, pl. 10, fig. 5. ***Type material.*** Syntypes from type locality (NHMUK 1905.7.14.15–17, 3: 1.75 × 1.75 mm; ANSP 164838, 1 of 2 surviving: Robertson *et al.*, 1981; NMW 155.158.89, 10: one sputter coated; NMW 155.158.90, 6; Boxton/Buxton Museum: Jackson collection, 7). The specimen here illustrated from NMW 155.158.89 (Fig. 803C) is here designated as the lectotype with the express taxonomic purpose of taxon stabilization in case non-conspecific syntypes should be located. ***Type locality.*** 156 fms [= 285 m], Gulf of Oman, 24°58'N, 56°54'E, (OD). ***Etymology.*** Named after Mr. John Wilfrid Jackson (OD).

Scissurella jacksoni: Thiele, 1912a: 11–12, pl. 1, fig. 24 [copy figure Melvill, 1904].

Scissurella jacksoni: Robertson *et al.*, 1981: 3, 16, 21.

Scissurella jacksoni: Glayzer *et al.*, 1984: 318.

? *Anatoma jacksoni*: Lozouet, 1986: 113 [could also refer to *An. agulhasensis*].

? *Scissurella jacksoni*: Vine, 1986: 172 [could also refer to *An. agulhasensis*].

Scissurella jacksoni: Hedegaard, 1990: 74, 145, pl. 58, fig. 1 [SEM of shell structure].

Anatoma jacksoni: Bosch *et al.*, 1995: 28, text-fig. 2.

Anatoma jacksoni: Bandel, 1998, 33–34, pl. 11, fig. 1.

? *Anatoma jacksoni*: Dekker & Orlin, 2000: 17 [could also refer to *An. agulhasensis*].

Misidentifications.

not *Scissurella turbinata*: Yokoyama, 1924: 5, 35–36, pl. 5, fig. 21 [is *Scissurella staminea.* Plate reprinted in Taki & Oyama, 1954: pl. 42].

Anatoma agulhasensis: Bandel, 1998: 34–35, pl. 11, figs 4–6 [is *An. munieri*].

Figure 803. *Anatoma munieri.* **A.** *Scissurella munieri.* Lectotype. Seas of China, in bottom sands (MNHN). **B.** *Anatomus turbinatus.* Lectotype here designated. Mino-sima, [= Mishima Island, Hagi City, Yamaguchi Prefecture, Japan [34.767°N, 131.166°E] (NHMUK 1874.5.19.62). **C.** *Scissurella jacksoni.* Lectotype here designated. 156 fms. [= 285 m], Gulf of Oman, 24°58'N, 56°54'E (NMW 1955.158.89). Scale bars shell = 1 mm. Scale bars protoconch = 100 μm.

Anatoma agulhasensis: Jansen, 1999: 48, figs 1–3 [is *An. munieri*].
not *Scissurella* (*Anatoma*) *jacksoni*: Yaron, 1993: 270–271, pl. 4, figs 1–3 [is *An. agulhasensis*].
not *Anatoma jacksoni*: Geiger, 2003: 73 [is *An. agulhasensis*].

Problem citation

? *Anatoma* cf. *jacksoni*: George, 2005: 358 [most likely *An. munieri*: Persian Gulf].

Type material. Lectotype (MNHN: selected by Geiger, 2006a: 30. Fig. 803A). Paralectotype (MNHN, 1), 1.5 × 1.66 mm.

Type locality. Seas of China, in bottom sands (OD).

Etymology. Named after Ernest Charles Philippe Auguste Munier-Chalmas [1843–1903] (OD).

Description. Shell turreted, medium-sized (2.4 mm), as high as wide, or wider. Some shells with radial brown bands. Protoconch of 0.75 whorl, sculpture flocculent, varix present, not connecting to embryonic cap. Teleoconch I of slightly more than 0.5 whorl, axials sculpture strong, often spiral rib in position of selenizone. Teleoconch II up to 3.25 whorls (1.7 mm shell), suture impressed. Shoulder of undulating profile, with

Figure 804. *Anatoma munieri*. Direction Bank, off Rottnest Island, Western Australia, Australia (AMS C.402664). Scale bar shell = 1 mm. Scale bar protoconch = 100 µm.

Figure 805. *Anatoma munieri*. Protoconchs of A–C eroded. **A.** Arafura Sea, ca. 330 km NW of Wessel Islands, Northern Territories, Australia (AMS C.402663, 1). **B.** Arafura Sea, ca. 210 mi NE Croker Island, Northern Territories, Australia (AMS C.402662, 2). **C.** North West Shelf, ca. 100 mi NW of Broome, Western Australia, Australia (AMS C.402665, 1). **D.** Protoconch profile of A. **E–F.** 260 m, SE of Tongatapu, Tonga, 21.345°S, 175.042°W (AMS C.402648, 1). Scale bars shell = 1 mm. Scale bar D–F = 100 µm.

axial cords stronger than spiral lines, approximately 85 axial cords on last whorl, eight spiral lines on shoulder at aperture, at intersection of axials and spirals minute raised tubercle. Base with strong constriction below selenizone, spirals increasing in strength from selenizone to umbilicus becoming as strong as axial cords; intersection of axials and spirals forming weak nodes. Umbilicus continuously sloping from base, wide, deep, with funiculus. Aperture subquadrate rounded, only flared towards umbilicus. Selenizone at periphery, above center of whorl; keels strong; lunules not coordinated with axials; slit open, margins parallel.

Radula (Fig. 806A–B). Rachidian tooth trapezoidal, apical cusp largest, five cusps on each side. Lateral teeth 1–2 similar, apical cusp largest, three cusps along outer margin, one cusp along inner margin. [Lateral tooth 3 seemingly absent: see Remarks.] Lateral tooth 4 reduced, hook-shaped, with minute point in transition to shaft. Lateral tooth 5 enlarged by elongation, apical cusp largest, four strong cusps along inner margin, five smaller cusps along outer margin. Inner marginal teeth with palmate tip, apical cusp largest, approximately six cusps on each side; shafts without food groove. Outer marginal teeth with spoon-shaped tip with many fine bristles. Radular interlock of central field moderately weak.

Operculum (Fig. 806C) corneous, round, thin, multispiral, central nucleus.

Comparisons. *Anatoma munieri* is characterized by the tall overall shell morphology, the wide sutsel, the distinct constriction on the base below the selenizone, and the two spiral keels on the base.

Anatoma equatoria from the Indo-Malayan Archipelago to the Central Pacific has a less turreted shell, a weaker spiral keel below the selenizone, and sculpture on the base that is finer and forms a reticulate pattern.

Anatoma aedonia from the Caribbean to Brazil has a less turreted shell, a proportionally much larger last whorl, weaker axial sculpture, and lacks the spiral keels on the base.

Anatoma emilioi from the Central Pacific (Hawaii) has fewer but very strong axial ridges, much stronger sculpture on the shoulder compared to the base, and a more narrow sutsel.

Anatoma eximia from the Mediterranean to Angola lacks the strong constriction on the base below the selenizone and the spiral keels on the base, and has a proportionally larger last whorl.

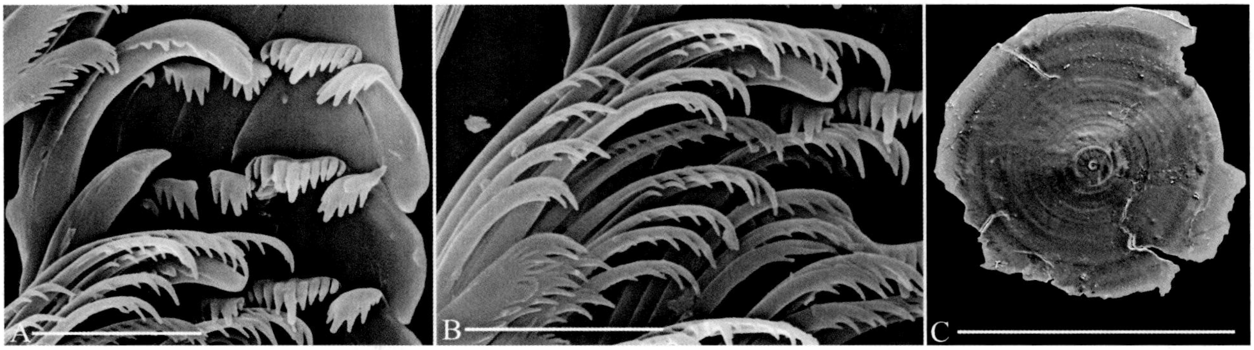

Figure 806. *Anatoma munieri.* Radula and operculum. 240–319 m, BORDAU 1 DW1494, Fiji, 18.916°S, 178.483°W (MNHN, 5). **A.** Central field. Scale bar = 10 μm. **B.** Marginal teeth. Scale bar = 10 μm. **C.** Operculum. Scale bar = 0.5 mm.

Anatoma finlayi from the Indian Ocean to the Central Pacific (and possibly Caribbean) has much stronger axial lamellae, a more narrow sutsel, and a proportionally larger last whorl.

Anatoma philippinica from the Indo-Malayan Archipelago to the Western Pacific has a more narrow sutsel, at least twice as many axial cords on shoulder and base, and a taller last whorl.

Anatoma rhynchodentata from the Indo-Malayan Archipelago to the Western Pacific has a more narrow sutsel, at least twice as many axial cords on shoulder and base, and a proportionally taller last whorl.

Anatoma agulhasensis from the western Indian Ocean has a shell that is distinctly higher than wide, stronger spirals on the shoulder, a less distinct spiral groove beneath the selenizone, and a more rounded aperture.

Anatoma aspera from the Mediterranean to Angola lacks the strong constriction on the base below the selenizone and has irregular reticulate sculpture on the protoconch.

Distribution. Indian Ocean to Western Pacific. 15–2650 m [shells only]; 53–63 m, 240–319 m [with animals] (Fig. 807).

Specimen records.

Oman. Gulf of Oman (FLMNH 00132408, 2; USNM 424849, 1; NMSZ 1961.61, 2). 280 m, Gulf of Oman (AMS C.404921, 1).

Australia, Western Australia. 156 m, W of Bunbury, 33.050°S, 114.733°E (AMS C.402763, 1). 146–150 m, W of Rottnest Island, 32.000°S, 115.250°E (AMS C.402767, 1). 150 m, Direction Bank, off Rottnest Island, 31.735°S, 115.177°E (AMS C.402762, 1). 155 m, off Rottnest Island, 31.732°S, 115.158°E (AMS C.402764, 1). 124 m, off Rottnest Island, 31.718°S, 115.208°E (AMS C.402765, 1). 150 m, off Rottnest Island, 31.703°S, 115.137°E (AMS C.402667, 1). 140 m, Direction Bank, off Rottnest Island,

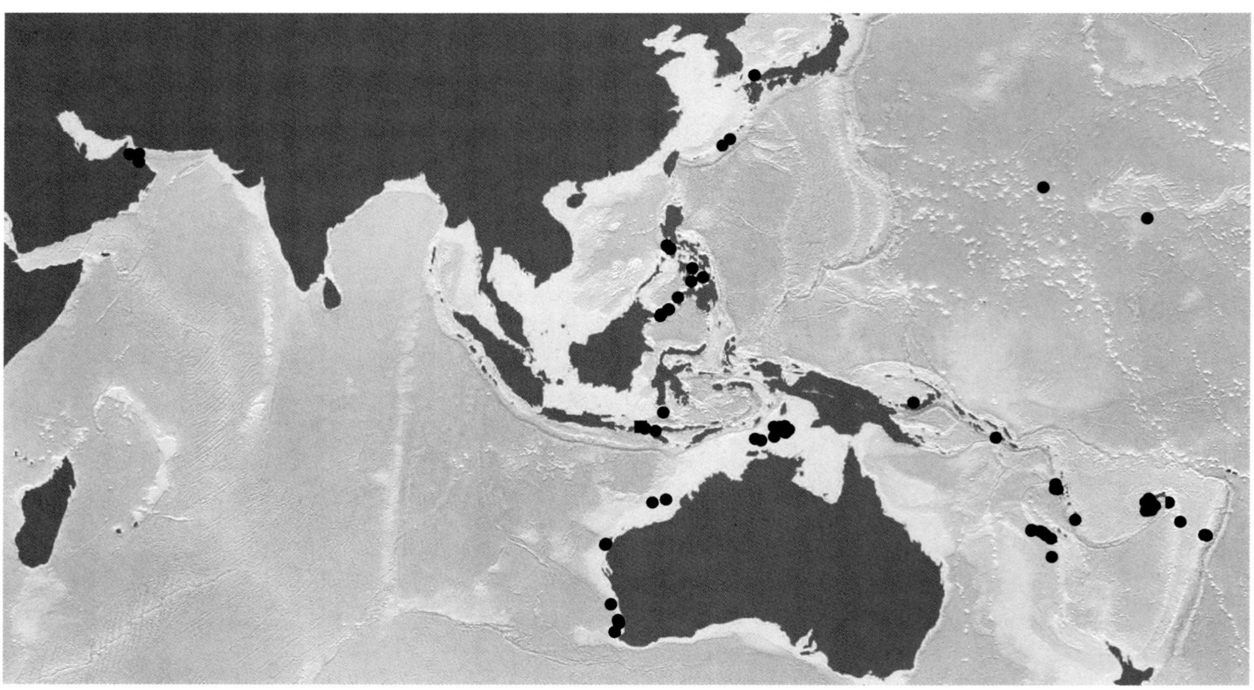

Figure 807. Distribution of *Anatoma munieri*. ● specimen record. ■ literature record.

31.627°S, 115.133°E (AMS C.402664, 1). 274–283 m, NW of Beagle Island, 29.717°S, 114.283°E (AMS C.402666, 2). 264 m, off North West Cape, 22.500°S, 113.597°E (AMS C.402766, 1). 190 m, off North West Cape, 22.387°S, 113.677°E (AMS C.402761, 2). 266 m, S of Rowley Shoals, Station UMD 6920 17.333°S, 119.167°E (WAM, 3). 194 m, North West Shelf, ca. 100 mi NW of Broome, 16.967°S, 120.783°E (AMS C.402665, 1).

Australia, Northern Territories. 108 m, Arafura Sea, ca. 95 mi N Cobourg Peninsula, 9.750°S, 132.067°E (AMS C.402770, 2). 135 m, Arafura Sea, ca. 110 mi N of Melville Island, 9.567°S, 131.367°E (AMS C.402769, 1). 135 m, Arafura Sea, 230 km N of Goulburn Island, 9.300°S, 133.633°E (AMS C.402772, 1). 100 m, Arafura Sea, 365 km N of Milingimbi Island, Arnhem Land, 8.800°S, 134.967°E (AMS C.402771, 1). 82 m, Arafura Sea, N of Wessel Islands, 8.600°S, 135.133°E (AMS C.402668, 1; AMS C.402768, 2). 75 m, Arafura Sea, ca. 330 km NW of Wessel Islands, 8.433°S, 135.367°E (AMS C.402663, 1). 132 m, Arafura Sea, ca. 210 mi NE Croker Island, 8.300°S, 133.967°E (AMS C.402662, 2). 113 m, N of Gouldbourn Island, Arafura Sea, Australia, 8.100°S, 133.600°E (AMS C.461456, 3). 110 m, N of Gouldburn Island, Arafura Sea, 8.100°S, 134.300°E (AMS C.461455, 1). 104 m, Arafura Sea, ca. 260 mi NE Croker Island, 8.017°S, 134.883°E (AMS C.461457, 1).

China. China Sea (KBIN, 100; MNHN, 2).

Japan, Okinawa. 211–212 m, off Kerama Island, 26.047°N, 127.375°E (SBMNH, 30). 16 m, Oku, 26.847°N, 128.287°E (LACM 77-64, 11).

Philippines. 90 m, St. 5153, off Tocanhi Point, Tawi Tawi, 5.303°N, 120.049°E (USNM 274608, 3). 622 m, St. 5574, off Simaluc Island, Tawi Tawi Islands, 5.513°N, 120.133°E (USNM 1098720, 7). 721 m, St. 5117, Sombrero Island, 13.873°N, 120.773°E (USNM 261938a, 1). 37 m, St. 5137, off Jolo, 6.074°N, 120.975°E (USNM 230362, 1). 37 m, St. 5140, off Jolo, 6.146°N, 121.050°E (USNM ex 275329, 5). 69 m, St. 5133, 1.5 mi off Panabutan Point, W Mindanao, 7.586°N, 122.121°E (USNM 1099053, 2). 80–140 m, Balicasag Island, 9.538°N, 123.705°E (MNHN 1: complete). 1339 m, St. 5487, Sogod Bay, Leyte, 10.046°N, 125.093°E (USNM 283798, 1). 59 m, St. 5192, SE off Bantayan Island, 11.154°N, 123.833°E (USNM 280951, 1). 155–167 m, 13.462°N, 121.202°E (MNHN, 1). 92–96 m, Dipolog Bay, Bohol-Sulu Sea, PANGLAO 2005 DW2370, 8.562°N, 123.143°E (MNHN, 7). 172–175 m, Dipolog Bay, Bohol-Sulu Sea Sill, PANGLAO 2005 DW2371, 8.580°N, 123.270°E (MNHN, 1). 427 m, Bohol-Sulu Seas Sill, PANGLAO 2005 DW2364, 9.012°N, 123.425°E (MNHN, 2). 101–110 m, Maribohoc Bay, Bohol Island, PANGLAO 2004 T14, 9.075°N, 123.822°E (MNHN, 2). 83–86 m, off Bolod, Panglao Island, PANGLAO 2004 T44, 9.407°N, 123.815°E (MNHN, 7). 134–190 m, Cervera Shoal, W Pamilacan Island, PANGLAO 2004 T34, 9.470°N, 123.845°E (MNHN, 19). 128–142 m, off Pamilacan Island, Bohol Sea, PANGLAO 2005 CP2344, 9.473°N, 123.835°E (MNHN, 1). 95–128 m, Cervera Shoal, Pamilacan Island, PANGLAO 2004 T36, 9.488°N, 123.858°E (MNHN, 98). 15–20 m, Pamilacan Island, PANGLAO 2004 S22, 9.490°N, 123.933°E (MNHN, 1). 110–112 m, Cervera Shoal, W of Pamilacan Island, PANGLAO 2004 T41, 9.495°N, 123.837°E (MNHN, 11). 100–138 m, Cervera, W Pamilacan Island, PANGLAO 2004 T39, 9.502°N, 123.840°E (MNHN, 16). 90–98 m, Balicasag Island, PANGLAO 2004 L69-73, 9.512°N, 123.683°E (MNHN, 1). 145–163 m, between Libaong and Pamilacan, PANGLAO 2004 T34, 9.522°N, 123.857°E (MNHN, 6). 164–176 m, off Balicasag Island, Bohol Sea, PANGLAO 2005 DW2339, 9.532°N, 123.728°E (MNHN, 1). 80–140 m, Balicasag, PANGLAO 2004 T28, 9.538°N, 123.705°E (MNHN, 1). 152 m, Bolod, Panglao Island, PANGLAO 2004 T2, 9.540°N, 123.797°E (MNHN, 2; MNHN, 100). 83–102 m, Bolod, Panglao Island, PANGLAO 2004 T1, 9.540°N, 123.788°E (MNHN, 2). 53–63 m, Bolod, Panglao Island, PANGLAO 2004 D15, 9.547°N, 123.802°E (MNHN, 1). 100–140 m, between Panglao and Balicasag Island, PANGLAO 2004 T31, 9.550°N, 123.700°E (MNHN, 2). 82 m, Bolod, Panglao Island, PANGLAO 2004 T4, 9.550°N,

123.808°E (MNHN, 1172). 106–137 m, between Panglao and Pamilacan Islands, PANGLAO 2004 T27, 9.557°N, 123.850°E (MNHN, 6). 97–120 m, off San Isidro, Panglao Island, PANGLAO 2004 T9, 9.562°N, 123.833°E (MNHN, 46). 83–86 m, off Bolod, Panglao Island, PANGLAO 2004 T44, 9.573°N, 123.815°E (MNHN, 4). 80 m, Biking-Catarman, Panglao Island, PANGLAO 2004 T28, 9.583°N, 123.857°E (MNHN, 1). 34–82 m, W of Baclayon, Bohol Island, PANGLAO 2004 T6, 9.585°N, 123.853°E (MNHN, 4). 90–200 m, Maribohoc Bay, Bohol Island, PANGLAO 2004 P1, 9.600°N, 123.750°E (MNHN, 1). 80–100 m, Cortes, Bohol Island, PANGLAO 2004 T18, 9.697°N, 123.832°E (MNHN, 29). 100 m, Cortes, Bohol Island, PANGLAO 2004 B1, 9.698°N, 123.825°E (MNHN, 17). 123–135 m, Cortes, Bohol Island, PANGLAO 2004 T26, 9.722°N, 123.813°E (MNHN, 1). Aliguay Island, 2.746°N, 123.224°E (DLG 1589, 3).

Indonesia. 65 m, Teluk Slawi, E of Komodo, 8.613°S, 119.512°E (RMNH, 1). 175–185 m, Bay of Sanggar, N of Sumbawa, 8.322°S, 118.260°E (RMNH, 9). 175 m, SW Salayer, 6.373°S, 120.437°E (RMNH, 14). SW Salayer, 6.355°S, 120.435°E (RMNH, 2).

Papua New Guinea. 95 m, off Anne Sophie's Reef, Kimbe Bay, New Britain, 5.100°S, 150.185°E (DLG 438, 1).

New Caledonia. 2650 m, 23.933°S, 166.683°E (MNHN, 1). 247–252 m, E coast, 21.700°S, 166.567°E (MNHN, 10). 237–398 m, E coast, 21.283°S, 165.950°E (MNHN, 100). 160–222 m, E coast, 20.950°S, 165.583°E (MNHN, 1; MNHN, 50). 105–110 m, Poindimie sector, 20.817°S, 165.317°E (MNHN, 12). 53–71 m, Deverd Passage, Koumac sector, 20.753°S, 164.253°E (MNHN, 4). 90–115 m, Deverd Passage, Koumac sector, 20.745°S, 164.248°E (MNHN, 2). 30–63 m, Chenal de la Passe de Koumac, Koumac sector, 20.663°S, 164.255°E (MNHN, 1). 105–110 m, E coast, 20.950°N, 165.567°E (MNHN, 1). 250–350 m, St. 40, 22.500°S, 166.400°E (MNHN, 2). 51–52 m, Cape N Doua, BATHUS 1 DW1235, 22.402°S, 166.925°E (MNHN, 1).

Solomon Islands. 253–356 m, 9.390°S, 159.975°E (MNHN, 13). 396–411 m, SALOMON 1 DW1762, 8.665°S, 160.065°E (MNHN, 1). 98–200 m, SALOMON 1 DW1767, 8.323°S, 160.665°E (MNHN, 2).

Fiji. 314–377 m, 19.667°S, 178.167°W (MNHN, 12). 441–443 m, S of Viti Levu, 18.320°S, 177.862°E (MNHN, 1). 234–361, 18.310°S, 178.085°E (MNHN, 7). 260–305 m, S of Viti Levu, 18.308°S, 178.097°E (MNHN, 40). 275–430 m, S of Viti Levu, 18.297°S, 177.907°E (MNHN, 26). 295–302 m, S of Viti Levu, 18.212°S, 178.540°E (MNHN, 40). 144–150 m, S of Viti Levu, 18.207°S, 178.550°E (MNHN, 40). 149–168 m, S of Viti Levu, 18.207°S, 178.552°E (MNHN, 50). 392–433 m, S of Viti Levu, 18.185°S, 178.390°E (MNHN, 20). 183–188 m, SE Viti Levu, 17.828°S, 178.797°E (MNHN, 20). 959–963 m, SE of Viti Levu, 17.710°S, 178.917°E (MNHN, 1). 820–863, 17.300°S, 179.550°W (MNHN, 2). 282–322 m, Bligh Water, 17.273°S, 177.830°E (MNHN, 3). 656–660 m, Bligh Water, 17.268°S, 178.247°E (MNHN, 2). 660–663 m, Bligh Water, 17.248°S, 178.492°E (MNHN, 1). 251–257 m, Bligh Water, 16.857°S, 178.232°E (MNHN, 4). 200–215 m, Bligh Water, 16.840°S, 178.208°E (MNHN, 10). 567–699 m, Bligh Water, 17.158°N, 177.938°E (MNHN, 3). 240–319 m, BORDAU 1 DW1494, 18.917°S, 178.483°W (MNHN, 6). 144–150 m, S of Viti Levu, MUSORTSOM 10 CP1363, 18.207°S, 178.550°E (MNHN, 1). 203–208 m, SE Viti Levu, MUSORSTOM 10 DW1356, 17.838°S, 178.800°E (MNHN, 1). 102–106 m, Bligh Water MUSORSTOM 10 DW1329, 17.322°S, 177.790°E (MNHN, 30). 143–173 m, Bligh Water, MUSORSTOM 10 CP1323, 17.268°S, 177.762°E (MNHN, 4).

Tonga. 260 m, SE of Tongatapu, 21.345°S, 175.042°W (AMS C.402648, 1). 427–436 m, SW Tongatapu, 21.267°S, 175.300°W (MNHN, 8). 433 m, NW Tongatapu, BORDAU 2 DW1569, 21.033°S, 175.317°W (MNHN, 5).

Vanuatu. 160–182 m, 19.417°S, 169.450°E (MNHN, 1). 622–625 m, 15.667°S, 167.333°E (MNHN, 2). 140–182 m, 15.633°S, 167.058°E (MNHN, 2). 154–179 m, MUSORSTOM 8 DW1105, 15.050°S, 167.117°E (MNHN, 13; MNHN, 20).

Literature records.

Indonesia. Satonda [8.100°S, 117.750°E]. South China Sea (Bandel, 1998: misidentified as *An. agulhasensis*).

Remarks. Fischer-Piette (1950: 69) indicated there to be a holotype and one paratype. The holotype was not specifically designated in the original description; hence, the two specimens constitute syntypes. Similarly, Higo *et al.* (2001) labelled the syntype of *An. turbinata* as "holotype." These subsequent holotype indications cannot be viewed as lectotype designations (ICZN Art. 74.5). Geiger (2006a. Fig. 20A) expressly designated the lectotype (Fig. 803A).

Anatoma munieri has only one month priority over *An. turbinata*. The date on the first page of the issue in which *Scissurella munieri* was described agrees with the publication dates given by Winckworth (1936). Although the latter species name is somewhat better known, the general rule of priority should be enforced in this case. It is not possible to apply the *nomen oblitum/conservandum* rules, because *munieri* has been used as a valid species after 1900. Very few specimens were ever identified as *An. munieri* (e.g., KBIN "China Sea" DLG #5424).

The type specimens of *An. jacksoni* are clearly conspecific with *An. munieri*. ANSP had once two syntypes of *An. jacksoni*; only one is now retained (Robertson *et al.*, 1981). The two nominal taxa share the stepped shell profile, the umbilical configuration, and the overall sculptural pattern. The protoconch condition cannot be compared as protoconchs are eroded in all former syntypes including the lectotype here designated (Fig. 803C). Bandel (1998) reported *An. jacksoni* from Satonda, Indonesia, and Cebu, Philippines. Bandel's specimens are more finely sculptured specimens of *An. munieri* (Geiger & Jansen, 2004a). Yaron (1983) had misidentified and illustrated *An. agulhasensis* as *An. jacksoni*.

Australian specimens of *An. munieri* have been misidentified as *An. angulhasensis* from the western Indian Ocean.

This species, regularly found in collections, is surprisingly rarely collected alive. Only two lots with three specimens in more than 130 lots with over 1300 specimens contained animals. The radula preparation from two specimens were not ideal. The outer central field was obscured by overlapping marginal teeth. It seems that this species has only two of the usual three lateral teeth 1–3. Whether this is an error due to the problems with the preparation, whether an aberrant specimen was prepared (a specimen with a fourth lateral tooth 1–3 is known: Fig. 22), or whether it is an autapomorphy of this species, remains to be demonstrated.

Anatoma orbiculata new species

Figures 808–811

Misidentification.

not *Scissurella* (*Schizotrochus*) *bertheloti*: Nordsieck & García-Talavera, 1979: 35–36, pl. 1, fig. 3 [is *An.* cf. *orbiculata*].

not *Anatoma crispata*: Rolán, 2005: 40–41, pl. 6, fig. 69 [is *An. orbiculata*].

Type material. Holotype (MNHN 25104. Fig. 808). Paratypes from type locality (MNHN 25130, 17. Fig. 809A–B). 30–40 m, Santa Maria Bay, Benguela Province, Angola, 13.416°S, 12.533°E (MNHN 25105, 9. Fig. 809C).

Type locality. 0 m, Asilha, Qued el Helou, Morocco, 35.466°N, 6.040°W.

Etymology. Adjective of diminutive of orbis, orbiculus, Latin for small globe, referring to the overall shape of a rotational ellipsoid, the strong axial and spiral cords reminding of latitudinal and longitudinal lines. Adjective.

Description. Shell medium-sized (to 1.7 × 1.4 mm W × H: holotype), trochiform, globular. Protoconch of 0.75 whorl, irregular reticulate sculpture, usually with apertural varix, apertural margin slightly sinusoidal. Teleoconch I of 0.66 whorl, 22–25 axial cords, distinct spiral cord in position of the selenizone. Teleoconch II of up to 2.2 whorls, suture moderately impressed; sutsel 0.5–1× width of selenizone at onset of selenizone, approximately 1× width of selenizone at apertural margin. Shoulder convex, strong axial cords separated by 1–2× width of cords, approximately 30 axial cords on second teleoconch II whorl; first spiral lines after 0.1 teleoconch II whorl, approximately 12 lines at apertural margin, spiral lines running over axial cords. Base with weak constriction below selenizone, at least twice as many axial cords, approximately 20 spiral cords from selenizone to umbilicus forming reticulate pattern. Umbilicus narrow, with distinct funiculus. Aperture rounded, roof overhanging. Selenizone at periphery; keels strong, elevated, occasionally with additional axial cord between those of shoulder; lunules distinct; slit open, margins parallel.

Operculum thin, corneous, round, multispiral, nucleus central (Fig. 810D).

Radula (Fig. 810A–C). Rachidian tooth trapezoidal, central cusp largest, 3–4 cusps on each side. Lateral teeth 1–3 similar, strongly decreasing in size, apical cusp largest, 2–3 cusps on outer cutting edge. Lateral tooth 4 reduced, hook-shaped, with single point below tip. Lateral tooth 5 comparatively small, apical cusp largest, two large cusps on each side. Inner marginal teeth with larger tip than lateral tooth 5, apical cusp largest, 2–3 somewhat smaller cusps on each side; shafts without food groove. Outer marginal teeth with spoon-shaped tip, many fine bristles along edge. Outermost marginal teeth paddle-shaped, with many fine bristles along edge (Fig. 810C). Radular interlock of central field moderate.

Comparisons. *Anatoma janusa* from the subtropical northeastern Atlantic has at least twice as many axial cords on the base, at most half as many spiral lines on the shoulder, and a protoconch with flocculent sculpture.

Anatoma agulhasensis from the western Indian Ocean has a taller shell, a protoconch with flocculent sculpture, and sharper axial cords on the shoulder.

Anatoma aspera from the Mediterranean to Angola has a taller and more turreted shell shape, a wider sutsel on early whorls (1× vs. <0.5× width of selenizone), weaker clathrate sculpture on the base, and a more narrow umbilicus.

Anatoma bertheloti from the Canary Islands has a more depressed shell, a subquadrate whorl profile, more axial cords on shoulder and base, and the same sculpture on shoulder and base.

Anatoma eximia from the Mediterranean to Liberia has a much more stepped profile, much weaker axial sculpture on shoulder and base, and a wider sutsel on early whorls (>1× vs. <0.5× width of selenizone).

Anatoma micalii from the Mediterranean to Angola has a more subquadrate overall shell shape, a protoconch with flocculent sculpture and without apertural varix, and a wider sutsel on early whorls (1× vs. <0.5× width of selenizone).

Figure 808. *Anatoma orbiculata* new species. Holotype. 0 m, Asilha, Qued el Helou, Morocco, 35.466°N, 6.040°W (MNHN 25104: radula shown in Fig. 810). Scale bar shell = 1 mm. Scale bar protoconch = 100 μm. Images by Jaya Nolt.

Anatoma tenuisculpta from the Mediterranean and Norway to Angola has a more discoidal overall shell shape, weaker axial cords on the shoulder, and the same sculpture on shoulder and base.

Anatoma africanae from the western Indian Ocean has at least twice the number of all sculptural elements, a rounder overall shell morphology, and the selenizone slightly above the periphery.

Figure 809. *Anatoma orbiculata* new species. Paratypes. **A–B.** 0 m, Asilha, Qued el Helou, Morocco, 35.466°N, 6.040°W (MNHN 25130, 17). **C.** 30–40 m, Santa Maria Bay, Benguela Province, Angola, 13.417°S, 12.533°E (MNHN 25105, 9). Scale bars shell = 1 mm. Scale bars protoconch = 100 µm. Images by Jaya Nolt.

Anatoma alternatisculpta from the Caribbean has a smooth protoconch, at most half as many spiral lines on the shoulder, and lacks spiral lines in the vicinity of the umbilicus.

Anatoma crispata from the temperate North Atlantic has a more convex shoulder, no sutsel on early whorls, and weaker axial sculpture on the shoulder.

Anatoma flemingi from temperate New Zealand has much finer spiral lines on shoulder and base in comparison to the axial cords and 1–2 axial lines on the keels of the selenizone between each axial cord extending over the shoulder.

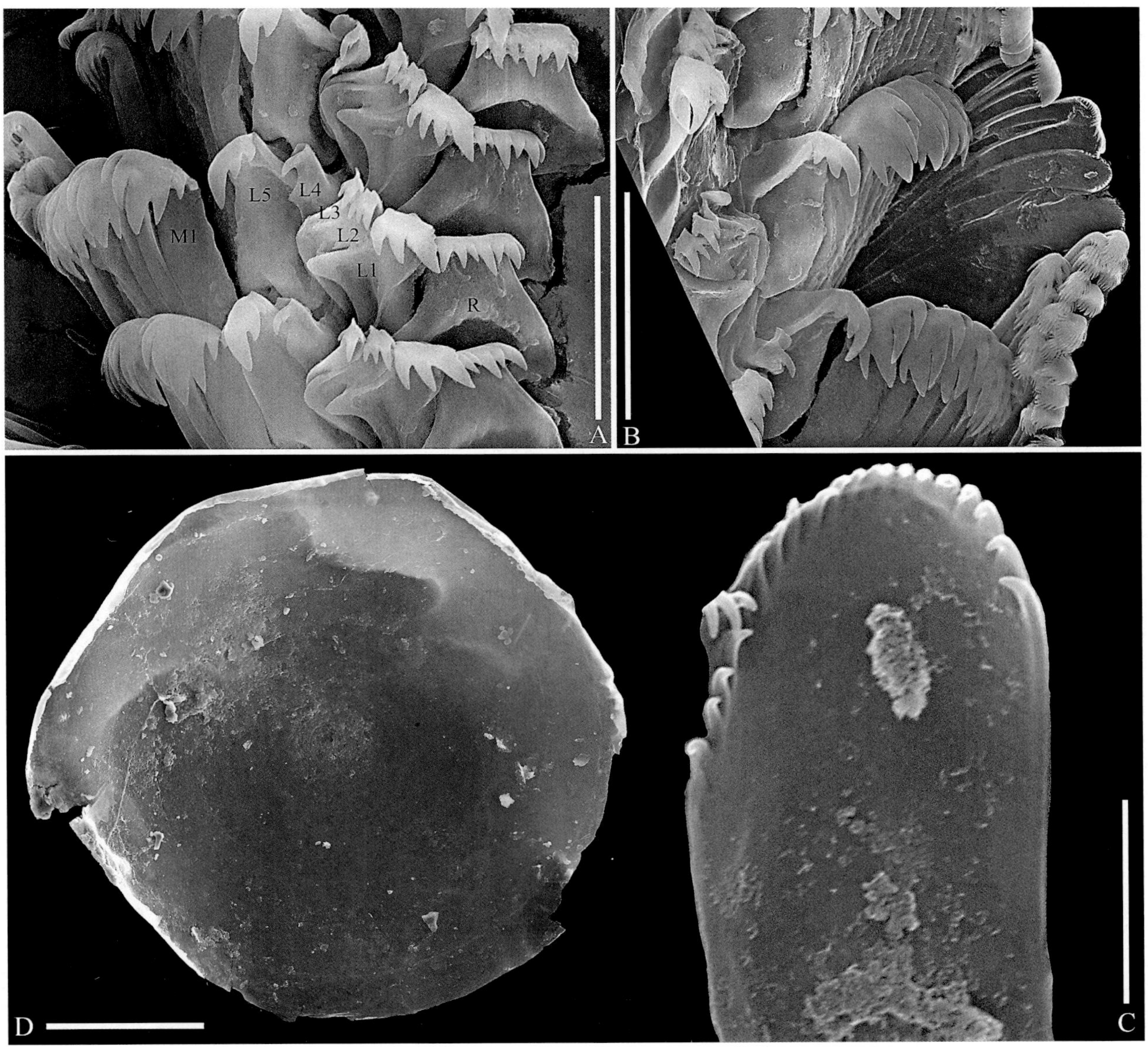

Figure 810. *Anatoma orbiculata* new species. Radula and operculum from holotype. 0 m, Asilha, Qued el Helou, Morocco, 35.466°N, 6.040°W (MNHN 25104: shell shown in Figure 808). **A.** Central field. Scale bar = 20 µm. **B.** Lateral tooth 5 and marginal teeth. Scale bar = 20 µm. **C.** High magnification image of tip of outer marginal tooth. Scale bar = 2 µm. **D.** Operculum. Scale bar = 200 µm. R: rachidian tooth. L1–5. Lateral teeth 1–5. M1 = Marginal tooth 1. Images by Jaya Nolt.

Anatoma globulus from the Indo-Malayan Archipelago has an overall more globular shell shape, finer and more numerous axial cords on shoulder and base, and the same sculpture on shoulder and base.

Anatoma indonesica from the Indo-Malayan Archipelago to the Western Pacific has a protoconch with flocculent sculpture, a smooth band on the base below the selenizone, and 50% more axial cords and spiral lines on the base.

Anatoma lamellata from the Western and Central Pacific has a protoconch with a spiral cord, a shell that is as tall as wide, and a wider sutsel on early whorls (1× vs. <0.5× width of selenizone).

Anatoma lyra from the northeastern Pacific has a protoconch with fine flocculent sculpture and no apertural varix, a shorter teleoconch I (0.5 vs. 0.75 whorl), and a wider umbilicus.

Anatoma schanderi from the northeastern Atlantic has a less convex shoulder, fewer and finer axial cords on the shoulder, and same strength and density of axials on shoulder and base.

Anatoma philippiana from the Paleogene of Europe has a protoconch without apertural varix and usually less dense spiral lines on the shoulder.

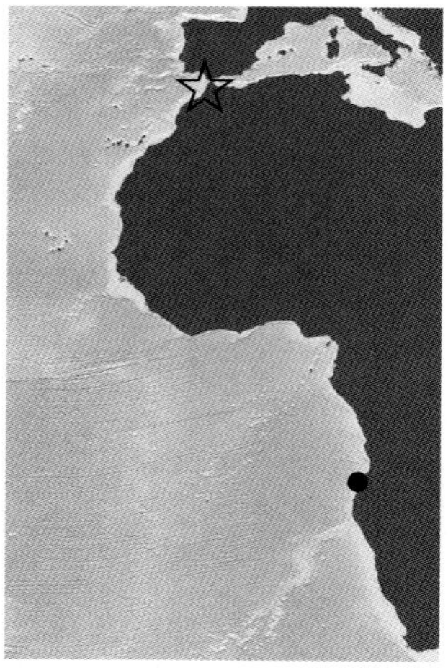

Figure 811. Distribution of *Anatoma orbiculata* new species. ☆ type locality. ● specimen record.

Distribution. West Africa, known only from type material (Fig. 811).

Remarks. The largest specimen, also containing an animal, was selected as the holotype. Its protoconch does not show an apertural varix; it is quite likely that the last portion of the protoconch was sheared off as evidenced by the slightly smaller whorl count of the protoconch proper, and the very smooth sculpture of teleoconch I immediately following the protoconch.

The radula shows some minor modifications compared to most *Anatoma* radulae. Lateral tooth 5 is remarkably small, and is smaller than marginal tooth 1. Usually, lateral tooth 5 is the largest of all teeth, with enlargement either by elongation or by broadening.

Anatoma parageia Geiger & Sasaki, 2009

Figures 812–814

Chresonymy.

Anatoma parageia Geiger & Sasaki, 2009: 10–13, figs 8–9.

Misidentifications

Not *Anatoma japonica*: Sasaki, 1998: fig. 44c–d [is *An. parageia*].

Type material. Holotype (SBMNH 83552: Geiger & Sasaki, 2009: fig. 10. Fig. 812). Paratypes from type locality (SBMNH 83553, 3: Geiger & Sasaki, 2009: fig. 8A–B. Fig. 813A–B; DLG 1036, 11; UMUT RM 29235: Geiger & Sasaki, 2009: fig. 9C. Fig. 813C). Off Shimoda, Ize Peninsula, Japan (UMUT RM 27644, 1: Sasaki, 1998: fig. 44c–d]).

Type locality. Peninsula towards shrine, Cape Manazuru, Kanagawa Prefecture, Japan, 35.141°N, 139.161°E. Rock and algal washings from moderately exposed rock and boulder shore in the intertidal (OD).

Etymology. Parageios, Greek for pertaining to shallow water, in reference to the rather unusual bathymetric records for a species in Anatomidae (OD).

Figure 812. *Anatoma parageia.* Holotype. 0 m, peninsula towards shrine, Cape Manazuru, Kanagawa Prefecture, Japan, 35.141°N, 139.161°E (SBMNH 83552). Scale bar shell = 0.5 mm. Scale bar protoconch = 100 μm.

Description. Shell trochiform, biconical, small for genus (to 1.05 mm, 0.93 × 0.66 mm in holotype. Figs 812, 813). Protoconch 182 × 122 μm in holotype, length 182–204 μm (n = 4), 0.75 whorl, sculpture coarsely flocculent, no apertural varix, apertural margin straight. Teleoconch I of 0.33 whorl, approximately 10 distinct axial cords, crowded towards onset of selenizone, no spirals. Teleoconch II of up to 1.33 whorls; sutsel <0.5× at onset of selenizone, <1× at apertural margin of mature specimen. Shoulder slightly convex, suture deeply impressed, many fine axial cordlets, regularly spaced and developed in first half teleoconch II whorl,

Figure 813. *Anatoma parageia*. Paratypes. 0 m, peninsula towards shrine, Cape Manazuru, Kanagawa Prefecture, Japan, 35.141°N, 139.161°E. **A–B.** (SBMNH 83553, 3). **C.** (UMUT RM29235, 1). Scale bars shell = 0.5 mm. Scale bars protoconch = 100 μm.

subsequently either extending from suture to keel (full axials), or starting up to ⅕ away from suture (short axials), or arising as bifurcation of full-width axial in adsutural third; most frequent pattern of alternating full axials and short axials. Spirally arranged small points on axials forming pseudospirals, starting after 0.5 teleoconch II whorl increasing variably to 6–12 rows in absutural half of shoulder, giving overall impression of rough surface, dull under the light microscope. Base not constricted below selenizone, with more or less distinct spiral keel in middle of base; axials similar to those on shoulder, slightly undulating in region of spiral keel; pseudospirals from below selenizone to umbilicus omitting area of spiral keel. Umbilicus open, with funiculus. Selenizone at periphery; lunules faint, dense; keels about as wide as selenizone, moderately strong; slit open, margins converging. Aperture rounded with basal adumbilical corner flared, forming funiculus.

Animal unknown.

Comparisons. *Anatoma pseudoequatoria* from Indo-Malayan Archipelago to the Central Pacific has a shorter teleoconch I (0.25 vs. 0.33 whorl), a protoconch with very fine flocculent sculpture, and usually less dense sculpture.

Anatoma aupouria from New Zealand has a shoulder ornamented with strong axial cords close to the suture, and distinct spiral lines in the adumbilical half of the base.

Anatoma philippinica from the Indo-Malayan Archipelago to the Western Pacific has a distinct constriction on the base below the selenizone, more elevated axial lamellae, and a wider sutsel on early whorls (>0.5× vs. <0.5× width of selenizone).

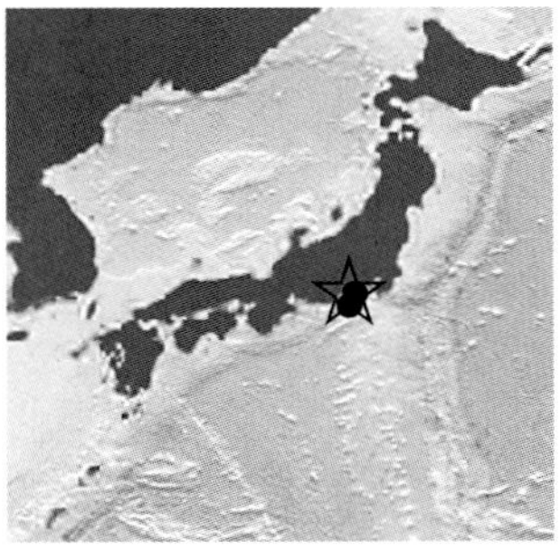

Figure 814. Distribution of *Anatoma parageia*. ☆ type locality. ● specimen record.

Distribution. Southern Japan (Fig. 814).

Remarks. Anatomids tend to occupy deeper habitats and are rarely collected in the intertidal or the shallow subtidal. The only other known exception is *Anatoma pseudoequatoria*. Although *An. parageia* was not collected alive in the intertidal, the relatively large number of empty shells, several in excellent state of preservation, suggests that it occupies a very shallow water habitat. It is unlikely that this may be due to a rare storm event because the species has been collected there repeatedly, and no other anatomids known to occur in deeper water of Sagami Bay were collected.

The most similar species, *An. pseudoequatoria*, has extraordinarily variable teleoconch sculpture. All 60 specimens of *An. pseudoequatoria* and all six specimens of *An. parageia* examined by SEM showed consistent but subtle protoconch sculpture differences: *An. pseudoequatoria* has very fine flocculent sculpture, whereas *An. parageia* has coarse flocculent sculpture. The more temperate occurrence of *An. parageia* supports it being distinct from *An. pseudoequatoria*. Critical determination of distributional boundaries between Okinawa and Honshu will require examination of shells by SEM.

Anatoma peruviana (Geiger & McLean, 2010)

Figures 815–818

Chresonymy.

Thieleella peruviana Geiger & McLean, 2010: 28–31, 33, figs 20–22.

Type material. Holotype (FLMNH 383672: shell, mounted soft parts, with radula removed, mounted radula. Figs 815–817). Paratypes from type locality (FLMNH 424420, 4: fluid preserved shells with preserved bodies).

Type locality. 991–1015 m, off N Peru [approximately 175 km WNW of Trujillo], Peru-Chile Trench, Peru, 7.983°S, 80.617°W (OD).

Etymology. Named for its type locality off Peru (OD).

Description. Shell medium-sized for genus (up to 3.0 mm), trochiform, globular. Protoconch of 0.75 whorl, reticulate sculpture (arrows), condition of apertural varix and shape of apertural margin unknown. Teleoconch I of one whorl, approximately 42 fine axial cords, spiral cord in position of selenizone. Teleoconch II of 2.5 whorls, suture impressed, below periphery of previous whorl; sutsel about 3× width of selenizone. Shoulder convex, approximately 43 axial cords on first whorl; spiral lines starting at onset of selenizone approximately 10 spiral lines after one whorl, 12 spirals at apertural margin of holotype, slightly irregularly spaced between suture and selenizone. Base with slight constriction below selenizone, 1.5× as many axials cords as on shoulder, approximately 28 spiral lines stronger than on shoulder, forming points at intersection with axials. Umbilicus narrow, weak funiculus. Aperture D-shaped, with columellar portion almost straight. Selenizone slightly above periphery; keels moderately strong, moderately elevated; lunules irregular, distinct; slit with parallel margins.

Operculum (Fig. 817D) covering aperture, corneous, thin, round, multispiral, nucleus central.

Radula (Fig. 817A–C). Rachidian tooth triangular, tip with central cusp largest, 4–5 cusps on each side, decreasing in size laterally. Lateral teeth 1–3 similar, decreasing in size two fold, 4, 4, 3 cusps on outer edge of tip, respectively. Lateral tooth 4 reduced, hook-shaped. Lateral tooth 5 enlarged by broadening, tip with penultimate cusp largest, one smaller cusp on outer edge, five cusps on inner edge, decreasing in size towards base. Marginal teeth hook-shaped with many small cusps on each side, shafts without food groove. Radular interlock of central field moderate.

Animal (Fig. 816) pale cream in color. Head with non-papillate cephalic tentacles (ct), no eye stalk or eye; suboptic tentacle (st), accessory cephalic tentacle (act) present. Four epipodial tentacles (et), biserially papillate on anterior and posterior edge, apical papillary tuft. Possible epipodial sense organ (ESO) on ventral base of second epipodial tentacle [or base of broken secondary ramus of epipodial tentacle?]. Opercular attachment (oa) with scale-like muscle fibres. Two mantle tentacles at posterior convergence of mantle slit.

Comparisons. *Anatoma argentinae* from the abyssal plain off Argentina and the tropical Panamic has a larger protoconch (370–480 vs. 230 μm), no sutsel on early whorls, and a later onset of spiral lines on the shoulder (0 vs. 0.5 teleoconch II whorl).

Anatoma lyra from the northeastern Pacific has a lenticular overall shell shape, a shorter teleoconch I (0.66 vs. 1 whorl), and raised axial lamellae.

Anatoma kelseyi from the northeastern Pacific has a longer teleoconch I (0.75 vs. 0.66 whorl), a less turreted overall shell shape, a more narrow sutsel (1× vs. 3× width of selenizone), and the axials that are about twice as dense, particularly towards the apertural margin of large specimens.

Anatoma bathypacifica from the Panamic has a strong spiral cord in position of the selenizone forming a distinct angulation and fewer (15 vs. 42) and stronger axial cords on teleoconch I.

Figure 815. *Anatoma peruviana*. Holotype. 991–1015 m, off N Peru [approximately 175 km WNW of Trujillo], Peru-Chile Trench, Peru, 7.983°S, 80.617°W (FLMNH 383672). Arrows indicate remnant of protoconch sculpture; most of apical protoconch surface eroded. Scale bar shell = 1 mm. Scale bar protoconch = 100 μm.

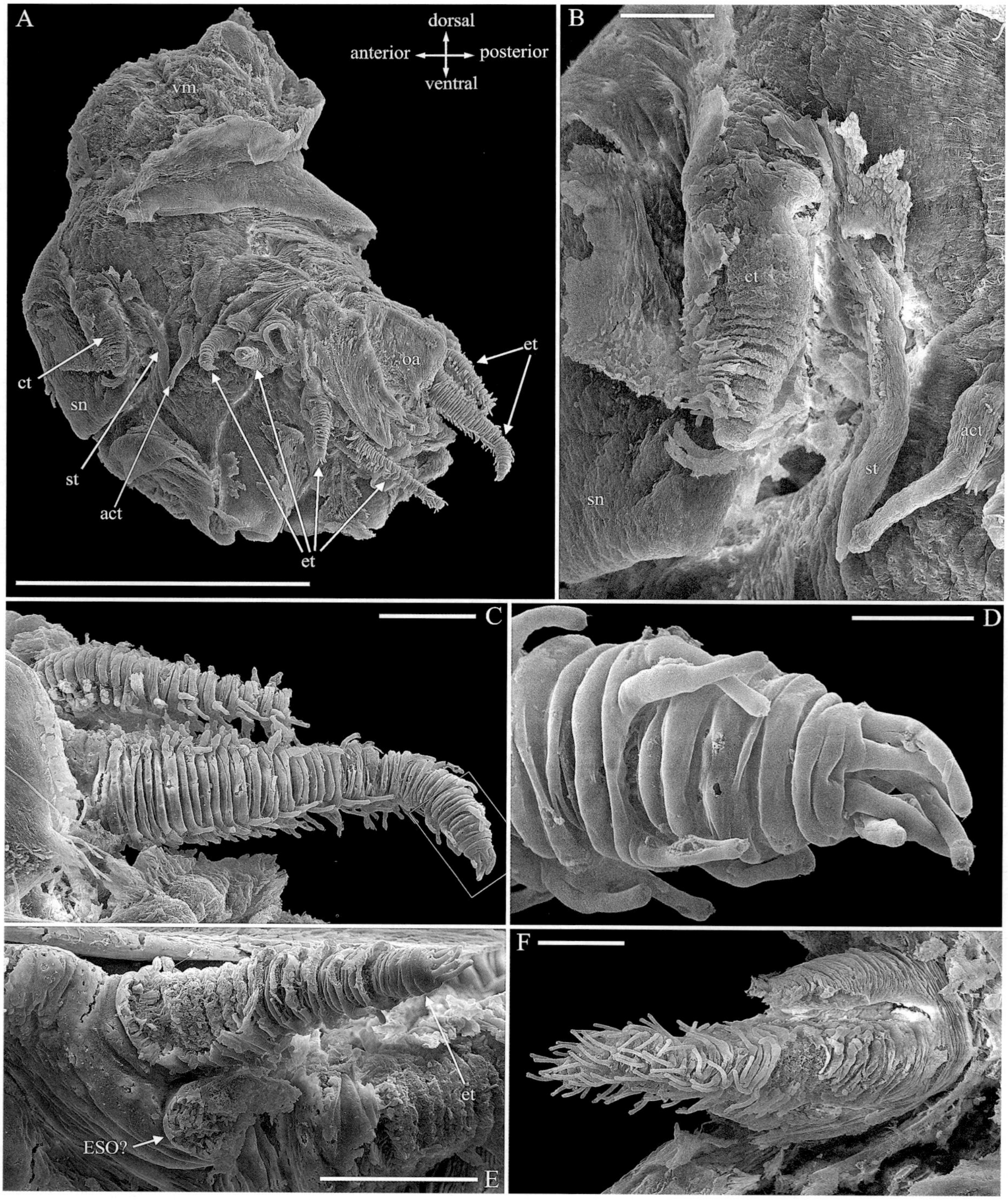

Figure 816. *Anatoma peruviana.* External anatomy of holotype. **A.** Head-foot complex from left side. Scale bar = 1 mm. **B.** Cephalic tentacle and suboptic tentacle enlarged; note absence of eye or eyestalk. Scale bar = 100 µm. **C.** Epipodial tentacles with bi-serial papillae on anterior and posterior margin. Box shows enlarged area shown in D. Scale bar = 100 µm. **D.** Tip of epipodial tentacle with papillae Scale bar = 20 µm. **E.** Ventral view of epipodial tentacle and potential epipodial sense organ. Scale bar = 100 µm. **F.** Mantle tentacles at posterior junction of slit. Scale bar = 100 µm. act = accessory cephalic tentacle. ct = cephalic tentacle. ESO = epipodial sense organ. et = epipodial tentacle. oa = opercular attachment site. sn = snout. st = suboptic tentacle. vm = visceral mass.

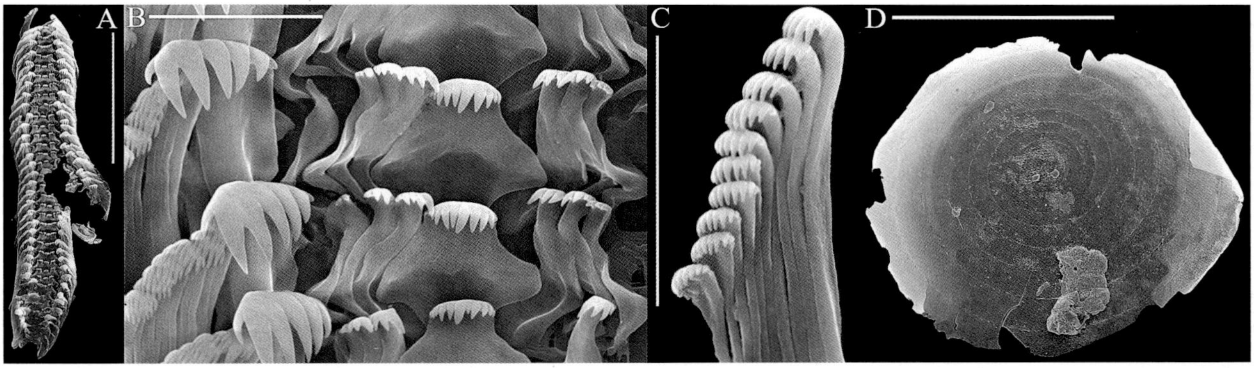

Figure 817. *Anatoma peruviana*. Radula and operculum from holotype. 991–1015 m, off N Peru [approximately 175 km WNW of Trujillo], Peru-Chile Trench, Peru, 7.983°S, 80.617°W (FLMNH 383672). **A.** Overview of entire radula. Scale bar = 100 µm. **B.** Central field enlarged. Scale bar = 10 µm. **C.** Series of marginal teeth. Scale bar = 100 µm. **D.** Operculum. Scale bar = 1 mm.

Anatoma sinuosa from the Eastern Pacific has the selenizone much higher on the whorl, a wider umbilicus, and at least twice as dense axial cords on shoulder and base.

Anatoma agulhasensis from the western Indian Ocean has a protoconch with flocculent sculpture, a more narrow sutsel at any position of the teleoconch, and a weak constriction on the base below the selenizone.

Anatoma aspera from the Mediterranean to Angola has a protoconch with irregular reticulate sculpture, a weak constriction of the base below the selenizone, and a shorter teleoconch I (0.75 vs. 1 whorl).

Anatoma sagamiana from Japan and the Indo-Malayan Archipelago has a more narrow sutsel anywhere along the teleoconch, a shorter teleoconch I (0.75 vs. 1 whorl), and a very narrow to obscured umbilicus.

Anatoma schanderi from the northeastern Atlantic has a more narrow sutsel anywhere along the teleoconch, finer axial and spiral sculpture, and a wider umbilicus.

Figure 818. Distribution of *Anatoma peruviana*. ☆ type locality.

Distribution. Known only from type locality (Fig. 818).

Anatoma philippinica (Bandel, 1998)

Figures 819–823, Color Plate 1E

Chresonymy.

Hainella philippinica Bandel, 1998: 40, pl. 13, figs 7–8, pl. 14, fig. 1 (Fig. 819).
Anatoma philippinica: Geiger, 2003: 74.

Figure 819. Original illustration of *Hainella philippinica*.

Type material. Holotype (SGPIH 3878. Fig. 819), 1.8 × 1.6 mm.

Type locality. Ocean side coral reef off Cebu City, Cebu, Philippines [10.300°N, 123.900°E] (OD).

Etymology. Named for its type locality in the Philippines (OD).

Description. Shell medium-sized (to 2.9 mm. Figs 820–821), trochiform, biconical, subquadrate. Some specimens with spiral series of brown dots (Color Plate 1E); three dots on shoulder, approximately three axials long, approximately 1.5× as wide, first row of dots adsuturally, second in middle of shoulder, third on keels of selenizone; three to four rows of dots on base, first on keels of selenizone, second and third on spiral keels on base, fourth occasionally in adumbilical region; approximately 18 dots on last whorl. Protoconch of 0.75 whorl, sculpture indistinctly flocculent, apertural varix not connected to embryonic cap, apertural margin indistinctly sinusoidal. Teleoconch I of 0.5 whorl, approximately 7–8 raised axial cords, no spirals. Teleoconch II of >2 whorls, suture moderately impressed. Shoulder with over 100 ruffled axial cords to low, forward slanted lamellae, aligned axial ruffles producing pseudospirals. Base with faint constriction below selenizone; two moderately distinct spiral keels, first keel at width of selenizone below selenizone, second keel at lowest point of base; sculpture on base similar to shoulder. Selenizone at periphery; lunules moderately distinct; keels prominent; slit open, margins parallel. Umbilicus open, narrow, funiculus inconspicuous. Aperture round, abapical keel distinct, keel below selenizone indistinct.

Radula (Fig. 822). Rachidian tooth trapezoidal, central cusp largest, four cusps in semicircle on each side. Lateral tooth 1–3 similar, apical cusp largest, two cusps on outer cutting edge of tip; lateral tooth 1 about twice as large as lateral teeth 2–3. Lateral tooth 4 reduced, hook-shaped. Lateral tooth 5 enlarged by elongation, apical cusp largest, 1–2 cusps on outer edge of tip. Inner marginal teeth with triangular tip, apical cusp largest, 3–4 cusps on outer edge, 1–2 cusps on inner edge; shafts with food groove. Outer marginal teeth with spoon-shaped tip, many fine bristles along edge. Radular interlock of central field moderate; base of lateral tooth 1 extending unusually far underneath rachidian tooth.

Operculum (not shown). Thin, round, corneous, multispiral, nucleus central.

Comparisons. *Anatoma equatoria* from the Indo-Malayan Archipelago to the Central Pacific has a weak constriction on the base below the selenizone and fewer axial ribs that are clearly separated and are crossed by distinct spiral lines, both on shoulder and base.

Anatoma quadraxialis from the Indo-Malayan Archipelago to Australia has reticulate protoconch sculpture, stronger sculpture on the shoulder that becomes abruptly more dense and fainter on the keels of the selenizone, and lacks the two spiral keels on the base.

Figure 820. *Anatoma philippinica*. 200–215 m, Bligh Water, Fiji, 16.840°S, 178.208°E (MNHN, 4). Scale bar shell = 1 m. Scale bar protoconch = 100 µm.

Anatoma rhynchodentata from the Indo-Malayan Archipelago to the Western Pacific has axial cords that are not slanted forward, true spiral lines, no spiral keel on the mid-base, and the constriction of the base below the selenizone is never as strong, if at all present.

Figure 821. *Anatoma philippinica*. **A.** 90 m, off Momo Beach, Paglao Island, PANGLAO 2004 L49, Philippines, 9.608°N, 123.755°E (MNHN, 1). **B.** 120–139 m, Balicasag Island, PANGLAO 2004 L74–75, Philippines, 9.511°N, 123.683°E (MNHN, 5). **C.** 43 m, Horseshoe Beach, Okinawa, Japan, 26.505°N, 127.853°E (DLG 1552, 1). Scale bars shell = 1 mm. Scale bars protoconch = 100 μm.

Anatoma aupouria from New Zealand has a more discoidal overall shell shape, a more subquadrate outline of the whorls, and a shorter teleoconch I (0.25 vs. 0.5 whorl).

Anatoma biconica from the Indian Ocean to the Central Pacific has a more biconical overall shell shape, at most half as many axial cords, and lacks the spiral keels on the base.

Anatoma breveprima from the Western Pacific has a more depressed overall shell shape, a much shorter teleoconch I (0.2 vs. 0.5 whorl), and lacks spiral keels on the base.

Anatoma pseudoequatoria from the Indo-Malayan Archipelago to the Central Pacific has a shorter teleoconch I (0.25 vs. 0.5 whorl), axial cords on the shoulder that are continuous from suture to selenizone, and at most half as many axial cords on the shoulder.

Anatoma rainesi from the Eastern to the Western Pacific has a longer teleoconch I (0.75 vs. 0.5 whorl), a more discoidal shell outline, at most half as many axial cords on the shoulder, and lacks spiral keels on the base.

Distribution. Indo-Malayan Archipelago (Fig. 823).

Specimen records.

Japan, Okinawa. 43 m, Horseshoe Beach, 26.505°N, 127.853°E (DLG 1552, 1; DLG 1551, 1). 35–38 m, base of cliff, Horseshoe, 26.504°N, 127.858°E (DLG 2020, 1). 20–30 m, Maeda Point, 26.445°N, 127.773°E (DLG 1534, 1). 8 m, Maeda Point, 26.435°N, 127.759°E (DLG 857, 2). 17–35 m, Maeda Point, 26.435°N, 127.759°E (DLG 964, 4). 6–8 m, S side, Sesoko Jima, 26.641°N, 127.860°E (DLG 862, 1). 1 m, Oku, 26.847°N, 128.287°E (LACM 77-61, 1). 33 m, Onna Village, 26.495°N, 127.843°E (LACM 78-20, 6). 60 m, Onna Village, 26.493°N, 127.842°E (LACM 79-76, 4). 55 m, Onna Village, 26.497°N, 127.843°E (LACM 78-27, 5). 46 m, Onna Village, 26.495°N, 127.843°E (LACM 78-26, 2). 92 m, Onna Village, 26.493°N, 127.493°E (LACM 78-101, 1). 20 m, Tsuken Jima, 26.268°N, 127.920°E (LACM 78-21, 7). 33 m, Motobu Peninsula, Nakijin, 26.703°N, 127.943°E (LACM 78-22, 1). 50 m, Zampa Misaki, 26.435°N, 127.708°E (LACM 78-25, 4). SW side Gishifu Shima, Kerama Retto, 26.223°N, 127.360°E (LACM 77-72, 1). 0 m, 2 km SSW Fitzwoody Beach, Naha Air Force Base, 26.185°N, 127.633°E (LACM 77-59, 3).

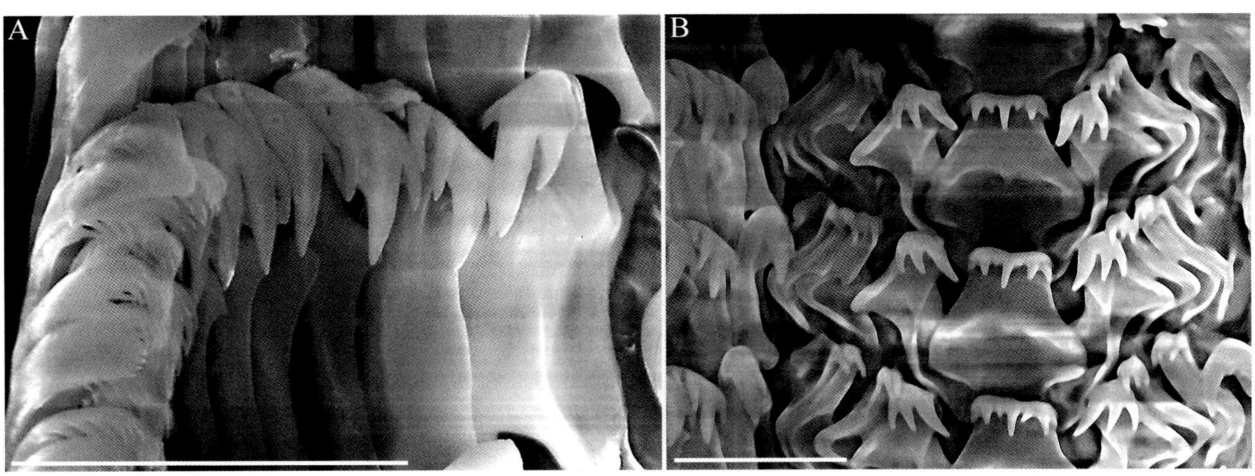

Figure 822. *Anatoma philippinica*. Radula. 55–81 m, Pamiacan Island, PANGLAO 2004 L65-68, Philippines, 9.498°N, 123.918°E (MNHN, 1). **A.** Lateral tooth 5 and marginal teeth. **B.** Central field. Scale bars = 10 µm.

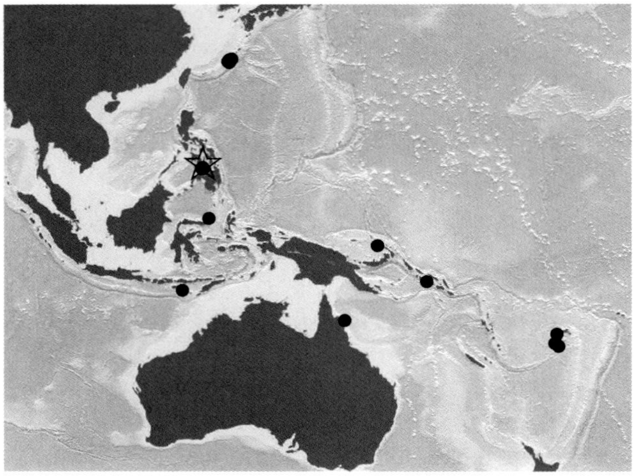

Figure 823. Distribution of *Anatoma philippinica.* ☆ type locality. ● specimen record.

Philippines. 20–25 m, Black Forest, Balicasag, 9.518°N, 123.688°E (MNHN, 1, 1: complete). 120 m, off Momo Beach, Panglao Beach, 9.615°N, 123.763°E (MNHN, 1: complete). 60 m, Pamilican Island, 9.502°N, 123.918°E (MNHN, 1: complete). 55–81 m, Pamilican Island, PANGLAO 2004 L65-68, 9.498°N, 123.918°E (MNHN, 1). 90–98 m, Balicasag Island, PANGLAO 2004 L69-73, 9.512°N, 123.683°E (MNHN, 2). 120–139 m, Balicasag Island, PANGLAO 2004 L74–75, 9.512°N, 123.683°E (MNHN, 5). 85–100 m, Balicasag Island, PANGLAO 2004 L44, 9.513°N, 123.683°E (MNHN, 1). 90–110 m, Balicasag Island, PANGLAO 2004 L46, 9.515°N, 123.687°E (MNHN, 5). 80–90 m, Balicasag Island, PANGLAO 2004 L42, 9.520°N, 123.678°E (MNHN, 1). 90–200 m, Maribohoc Bay, Bohol Island, PANGLAO 2004 P1, 9.600°N, 123.750°E (MNHN, 3). 90 m, off Momo Beach, Paglao Island, PANGLAO 2004 L49, 9.608°N, 123.755°E (MNHN, 1). 43–62 m, Bingag/Tabalong, Panglao Island, PANGLAO 2004 L51-60, 9.628°N, 123.800°E (MNHN, 2). 62 m, Bingag/Tabalong, Panglao Island, PANGLAO 2004 L51–60, 9.628°N, 123.800°E (MNHN, 2). 100 m, Manga, Bohol Island, PANGLAO 2004 L48, 9.705°N, 123.832°E (MNHN, 1).

Indonesia. 5–20 m, Sulawesi, Bunaken and Siladen Islets, 1.603°N, 124.767°E (LACM 88-55, 5). 74–83 m, E of Melolo, NE coast of Suba, 9.892°S, 120.712°E (RMNH 42268, 1).

New Guinea. 15–35 m, Baudisson Bay, Kavieng, New Ireland, 2.743°S, 150.658°E (DLG 301, 1).

Australia. 19 m, No Name Reef, Great Barrier Reef, Queensland, 14.667°S, 145.650°E (AMS ex C.461453, 3).

Solomon Islands. 22 m, Vangunu Island, Kokoana Passage, Marovo Lagoon, Matiu Island, 8.497°S, 158.2°E (LACM 89-77, 22).

Fiji. 230–251 m, 18.307°S, 178.043°E (MNHN, 10). 275–430 m, S of Viti Levu, 18.297°S, 177.907°E (MNHN, 8). 200–215 m, Bligh Water, 16.84°S, 178.208°E (MNHN, 4). Base of pinnacle, S of Herald, Great Astrolabe Reef, GAR-0118, 18.765°S, 178.471°E (JKC, 1).

Remarks. Some specimens with maroon spots are known (Color Plate 1E).

Anatoma planapex new species

Figures 824–827

Type material. Holotype (AMS C.448765. Fig. 824). Paratypes 604 m, E of Sandy Cape, Queensland, Australia, 24.728°S, 153.557°E (AMS C.448759, 2. Fig. 825A). 830 m, off Sandy Cape, Queensland, Australia, 24.413°S, 153.392°E (AMS C.448758, 5. Fig. 825B–C).

Type locality. 1000 m, off Broken Bay, New South Wales, Australia, 33.633°S, 152.100°E.

Etymology. Plan-, Latin for flat; apex, referring to the flat-topped shell. Noun in apposition.

Description. Shell medium-sized (to 2.7 mm. Figs 824, 825), biconical, somewhat turreted. Protoconch of 0.875 whorl, sculpture flocculent, weak apertural varix not connected to embryonic cap, apertural margin weakly sinusoidal. Teleoconch I of 0.8–0.9 whorl, distinct spiral cord in position of selenizone, approximately 30 axials increasing in strength from barely perceptible lines to fine cords. Teleoconch II of 2.33 whorls, suture impressed; sutsel 0.5× width of selenizone on early whorls, 2× width of selenizone at apertural margin of mature specimen. Shoulder convex, many very fine axial cords, on early whorls reaching from suture to mid-shoulder, at apertural margin reaching to selenizone. First spiral line after 0.125–1.25 whorls, approximately 12 at apertural margin. Base not constricted below selenizone, continuously sloping with umbilicus; axial lines of same strength and density as on shoulder; first spiral lines approximately 1–2× width of selenizone below selenizone, regularly spaced, increasing slightly in strength inside umbilicus; axials and spirals forming delicate clathrate pattern. Umbilicus moderately wide, no funiculus. Selenizone at periphery; keels rather weak, quite strongly elevated; lunules faint; slit open, margins slightly converging. Aperture rounded, roof overhanging.

Figure 824. *Anatoma planapex* new species. Holotype. 1000 m, off Broken Bay, New South Wales, Australia, 33.633°S, 152.100°E (AMS C.448765). Scale bar shell = 1 mm. Scale bar protoconch = 100 µm.

Operculum (Fig. 826D) round, thin, corneous, multispiral, nucleus central.

Radula (Fig. 826). Rachidian tooth trapezoidal, central cusp twice as large as 5–6 cusps on each side. Lateral teeth 1–3 similar, apical cusp largest, outer edge of tip with 4–6 cusps, inner margin with 3–4 minute points. Lateral tooth 4 moderately reduced, elongated, with 2–3 minute cusps on each side. Lateral tooth 5 enlarged by elongation, apical cusp largest, on inner cutting edge approximately 16 minute points of equal size, on outer edge approximately six larger cusps in transition area

Figure 825. *Anatoma planapex* new species. Paratypes. **A.** 604 m, E of Sandy Cape, Queensland, Australia, 24.728°S, 153.557°E (AMS C.448759, 2). For radula, see Figure 826. **B–C.** 830 m, off Sandy Cape, Queensland, Australia, 24.413°S, 153.392°E (AMS C.448758, 6). Scale bars shell = 1 mm. Scale bars protoconch = 100 μm.

between tip and shaft. Inner marginal teeth distinctly smaller than lateral teeth, tip spoon-shaped with approximately five larger cusps on each side; outer marginal teeth with spoon-shaped tip, many fine cusps on each side; outermost marginal teeth fan-shaped with many fine bristles; shafts of marginal teeth without food groove. Radular interlock of central field weak.

Comparisons. *Anatoma amydra* from the Indo-Malayan Archipelago to the Central Pacific has a more discoidal overall shell shape, has a shorter teleoconch I (<0.6 vs. >0.8 whorl), shoulder at a greater angle to the coiling axis, and sculpture immediately below the selenizone.

Anatoma austrolissa from deep water of the Indian Ocean to the Western Pacific has a more turreted overall shell shape, no sculpture, a shorter teleoconch I (0.125 vs. >0.8 whorl), and radula with floppy teeth.

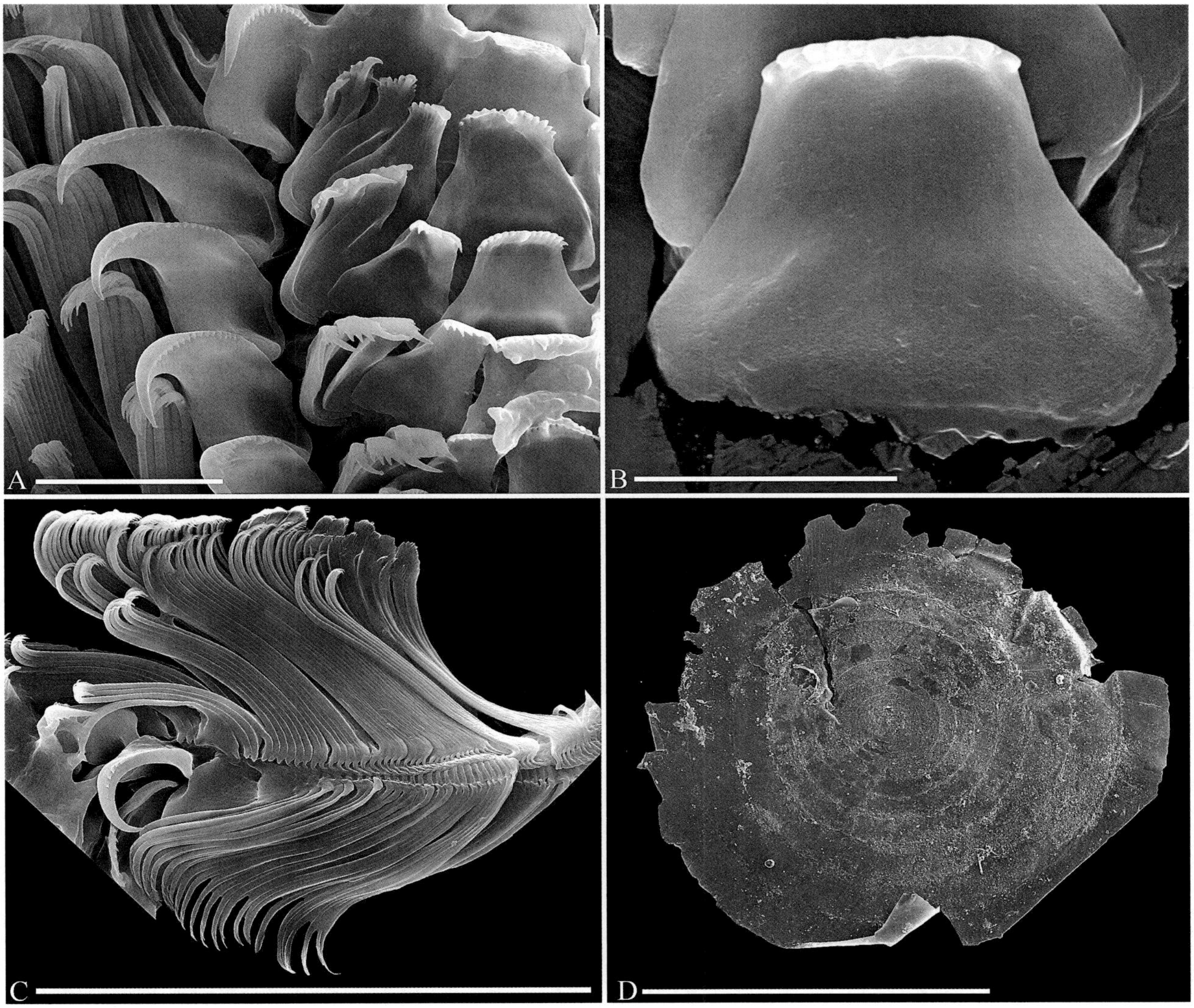

Figure 826. *Anatoma planapex* new species. Radula and operculum from specimen shown in Figure 825A. 604 m, E of Sandy Cape, Queensland, Australia, 24.728°S, 153.557°E (AMS C.448759, 2). **A.** Central field and inner marginal teeth. Scale bar = 20 µm. **B.** Rachidian tooth. Scale bar =10 µm **C.** Marginal teeth. Scale bar = 100 µm. **D.** Operculum. Scale bar = 1 mm.

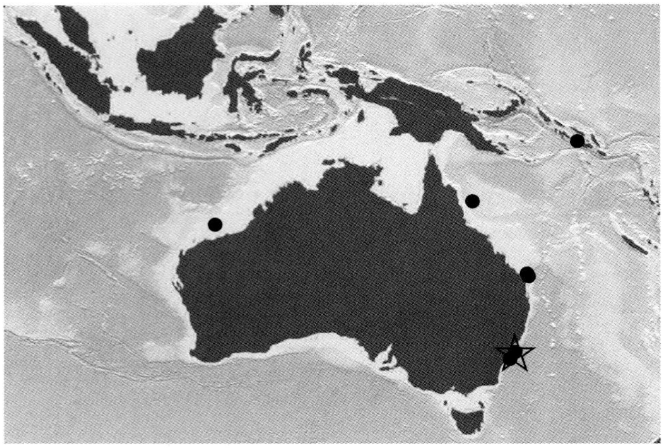

Figure 827. Distribution of *Anatoma planapex* new species. ☆ type locality. ● specimen record.

Anatoma biconica from the Indian Ocean to the Central Pacific has a more biconical overall shell shape, sculpture immediately below the selenizone, and at most half as many axial and spiral elements on shoulder and base.

Anatoma breveprima from the Western Pacific has a shorter teleoconch I (0.2 vs. >0.8 whorl), sculpture immediately below the selenizone, and a proportionally wider shell.

Anatoma equatoria from the Indo-Malayan Archipelago to the Central Pacific has a shorter teleoconch I (0.5 vs. >0.8 whorl), sculpture immediately below the selenizone showing clearly in the sutsel, and a more narrow umbilicus.

Anatoma flexidentata from the Indian Ocean to the Western Pacific has a wider selenizone, a more convex shoulder, and sculpture immediately below the selenizone.

Anatoma pseudoequatoria from the Indo-Malayan Archipelago to the Central Pacific has a shorter teleoconch I (0.25 vs. >0.8 whorl), sculpture immediately below the selenizone, and stronger axial and spiral sculpture on shoulder and base.

Anatoma hyposculpta from the Western Pacific has a shorter teleoconch I (0.125 vs. >0.8 whorl), lacks a spiral cord in position of the selenizone on teleoconch I, and lateral tooth 1 is very much reduced.

Anatoma rhynchodentata from the Indo-Malayan Archipelago to the Western Pacific has sculpture immediately below the selenizone, the shoulder at a greater angle to the coiling axis, and at least twice as many axial cords on shoulder and base.

Anatoma umbilicata from the Mediterranean and adjacent Atlantic has sculpture immediately below the selenizone, a shorter teleoconch I (0.5 vs. >0.8 whorl), and a more narrow sutsel at the onset of the selenizone (<0.25× vs. >0.5× width of selenizone).

Distribution. Tropical to moderately temperate waters of Australia and Indo-Malayan Archipelago, 150–1371 m (Fig. 827).

Specimen records.

Australia, New South Wales. 1200 m, off Wollongong, 34.442°S, 151.450°E (AMS C.448761, 1). 1000 m, off Broken Bay, 33.633°S, 152.100°E (AMS C.448765, 1: holotype). 896–924 m, off Sydney, 33.633°N, 152.050°E (AMS C.448768, 2).

Australia, Queensland. 604 m, E of Sandy Cape, 24.728°S, 153.557°E (AMS C.448759, 2: paratypes). 830 m, off Sandy Cape, 24.413°S, 153.392°E (AMS C.448758, 5: paratypes). 1371 m, NE of Cairns, 15.975°S, 147.217°E (AMS C.448757, 1).

Australia, Western Australia. 150 m, North West Shelf, N of Port Headland, 18.667°S, 117.917°E (AMS C.436908, 1).

Solomon Islands. 766–1100 m, N Russel Island, SALOMON 2 DW2167, 8.913°S, 159.122°E (MNHN, 1).

Remarks. In one specimen (AMS C.448757) the spiral sculpture on the base starts immediately below the selenizone, while in all other specimens there is a pronounced gap. In all other aspects, that specimen conforms with the remainder specimens; hence, it is tentatively identified as a member of this species.

Anatoma plicatazona Geiger & McLean, 2010

Figures 828–829

Chresonymy.

Anatoma n. sp.: Geiger, 2008b: fig. 1.

Anatoma plicatazona Geiger & McLean, 2010: 21–22, fig. 15.

Anatoma plicatazona Geiger & Fernández-Garcés, 2010: 67, fig. 1B.

Type material. Holotype (SBMNH 83518. Fig. 828A). Paratype from type locality (BMSM 17981, 1. Fig. 828B).

Type locality. 7 m, Sandy Cay, Abaco, Bahamas, 26.399°N, 76.988°W. August 13, 2005, leg. Colin Redfern (OD).

Etymology. Plicata, Latin for folded, -zona referring to the selenizone. Describing the diagnosing feature of the strongly upwardly folded selenizone (OD). Noun in apposition.

Description. Shell small (to 1.5 mm. Fig. 828), trochiform, turreted. Protoconch 0.75 whorl, sculpture flocculent, no apertural varix, apertural margin straight. Teleoconch I of 0.25–0.33 whorl, 11–12 axial cords, no spiral sculpture, interstices with irregularities. Teleoconch II >2 whorls. Shoulder convex, approximately 50 distinct axial cords on first whorl, becoming gradually less distinct to diffuse towards apertural margin; first spiral after 0.125–0.25 teleoconch II whorl, half to third strength of axials, spirals increasing to 5–7 after one teleoconch II whorl, of equal strength as axials, forming raised points at intersection with axials; raised points best developed after 1.25 teleoconch II whorls; sculpture becoming less distinct with growth, forming subdued cancellate pattern. Suture little impressed, sutsel 1.5× width of selenizone at onset of selenizone, approximately 3× width of selenizone at apertural margin. Base convex, evenly curving into umbilicus, barely constricted below selenizone; sculpture becoming weaker with growth and from selenizone towards base; reticulate sculpture with raised points at intersection below selenizone, subsequently approximately 13 low spiral steps. Aperture rounded, adumbilical portion of peristome flared. Selenizone slightly above periphery; keels of moderate strength, broken in early portion of whorl, appearing to have horizontal orientation, on last 0.5–0.66 whorl orientation of keels parallel to coiling axis; slit open, margins parallel.

Animal unknown.

Comparisons. *Anatoma plicatazona* is easily distinguished by the strongly upturned keels of the selenizone and the very wide sutsel, particularly after one teleoconch II whorl.

Anatoma aedonia from the Caribbean has a longer teleoconch I (1 vs. <0.33 whorl), a distinct spiral cord in the position of the selenizone, sculpture on teleoconch II remaining the same throughout growth, with distinct axial cords, and very fine spiral lines, and axial sculpture on the base extending to the umbilicus.

Anatoma alta from the Caribbean to the tropical eastern Atlantic has a wider selenizone without upturned keels of the selenizone, the selenizone higher on the whorl, and at most half as many axial cords on shoulder and base.

Anatoma alternatisculpta from the Caribbean lacks the upturned keels of the selenizone, has at most half as many spiral lines on the shoulder, and a more narrow sutsel on early whorls (<0.5× vs. >0.5× width of selenizone).

Figure 828. *Anatoma plicatazona*. **A.** Holotype. 7 m, Sandy Cay, Abaco, Bahamas, 26.399°N, 76.988°W, leg. Colin Redfern (SBMNH 83518). **B.** Paratype. 7 m, Sandy Cay, Abaco, Bahamas, 26.399°N, 76.988°W, leg. Colin Redfern (BMSM 17981). **C.** 40 m, Rancho Luna, Cienfuegos, Cuba, 22.017°N, 80.430°W. Images of C by and with kind permission of Emilio Rolán. Scale bars shell = 1 mm. Scale bars protoconch = 100 μm.

Anatoma proxima from the Caribbean to Brazil has a more discoidal overall shell shape, no upturned keels of the selenizone, and the selenizone more at the periphery of the shell.

Anatoma rolani from the Caribbean has a more discoidal overall shell shape, no upturned keels of the selenizone, and at least twice as dense axial lamellae on shoulder and base.

Anatoma equatoria from the Indo-Malayan Archipelago to the Central Pacific has no upturned keels of the selenizone, a spiral cord in the position of the selenizone on teleoconch I, and no elevated points at the intersection of axials and spirals.

Anatoma eximia from the Mediterranean to Liberia has no upturned keels of the selenizone, no elevated points at the intersection of axials and spirals, and the shoulder at a greater angle to the coiling axis.

Anatoma micalii from the Mediterranean to Angola has no upturned keels of the selenizone, fewer but stronger axial cords on the shoulder, and coarsely reticulate sculpture on the base.

Anatoma munieri from the Indian Ocean to the Central Pacific has a taller shell, a strong constriction on the base below the selenizone, and no upturned keels of the selenizone.

Anatoma kopua from eastern Australia and New Zealand has no upturned keels of the selenizone, no spiral lines immediately below the selenizone, and a protruding protoconch.

Anatoma rhynchodentata from the Indo-Malayan Archipelago to the Western Pacific has no upturned keels of the selenizone, a distinct constriction on the base below the selenizone, and a spiral cord in position of the selenizone on teleoconch I.

Anatoma sinuosa from the tropical Eastern Pacific has a distinct bulge on the base, a more deeply impressed suture, and no elevated points at the intersection of axials and spirals.

Distribution. Caribbean (Fig. 829).

Specimen records.

Cuba. 40 m, Rancho Luna, Cienfuegos, 22.017°N, 80.430°W (CFG, 2; MHNS, 5. Fig. 828C).

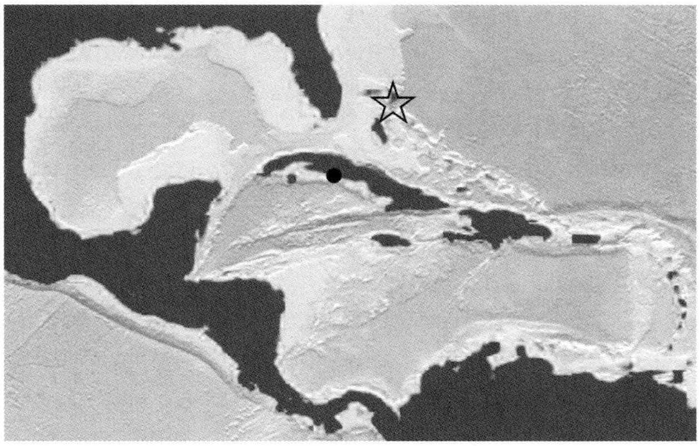

Figure 829. Distribution of *Anatoma plicatazona*. ☆ type locality. ● specimen record

Anatoma porcellana new species

Figures 830–833

Misidentifications.

not *Schizotrochus lamellatus*: Habe, 1951: 66, pl. 11, figs 7–8 [most likely *An. porcellana*].
not *Anatoma lamellata*: Izawa & Matsuoka, 1999: 14, pl. 6, fig. 8 [most likely *An. porcellana*].
not *Anatoma lamellata*: Okutani & Hasegawa, 2000: 37, figs 1a–b. [is *An. porcellana*].
not *Thieleella reticulata*: Geiger, 2004: text-fig. p. 5 [is *An. porcellana*].
not *Thieleella reticulata*: Geiger, 2006c: text-fig. 5 [is *An. porcellana*].

Type material. Holotype (MNHN 25114. Fig. 830). Paratypes from type locality (MNHN 25113, 20, one mounted radula. Figs 831A–B, 832A–D). 146–233 m, Tanimbar Island, N.O. *Baruna Jaya 1*, KARUBAR CP67 Indonesia, 8.966°S, 132.100°E (MNHN 25115, 1, mounted radula and operculum. Figs 831C, 832E).

Type locality. 273–356 m, off Pamilacan Island, Bohol Sea, PANGLAO 2005 CP2343, Philippines, 9.456°N, 123.823°E.

Etymology. Latin adjective, referring to the porcellaneous sheen of the shell under the light microscope.

Description. Shell medium-sized to large (to 3.2 mm. Fig. 830), trochiform, globular. Protoconch of 0.75 whorl, sculpture flocculent, apertural varix connected to embryonic cap, apertural margin convex. Teleoconch I of 0.6–0.7 whorl, rounded, with 17–20 distinct axial ribs between suture and spiral cord in position of selenizone; some supernumerary axials between spiral cord and base; interstices with fine growth lines only; first spiral lines after 0.125–0.5 whorl. Teleoconch II of 3.2 whorls, suture deeply impressed; sutsel about as wide as selenizone on early whorls, about 3× as wide at apertural margin of large specimen. Shoulder with shelf adjacent to suture, same density of raised axial cords as on teleoconch I; usually 1–2 fine spiral lines in middle of shoulder, occasionally disappearing with growth. Base usually with indistinct constriction below selenizone, occasionally distinct followed by spiral keel (Fig. 831C); axial cords usually at same density as on shoulder, occasionally somewhat more numerous; spiral cords somewhat weaker than axial cords, usually starting at periphery of base, extending to lower third of base; adumbilical third of base without spirals; axials distinct or barely perceptible. Umbilicus closed, occasionally narrow (Fig. 831C). Selenizone slightly above periphery; lunules moderately distinct, evenly spaced; keels moderately elevated (usually abraded), moderately strong; slit open, margins parallel (usually broken). Aperture rounded, adumbilical portion of peristome flared.

Operculum (Fig. 832E) round, corneous, thin, multispiral with central nucleus, covering entire aperture.

Radula (Fig. 832A–D). Rachidian tooth triangular with V-shaped tip, tip with nine cusps, central cusp almost twice as large as lateral ones. Lateral teeth 1–4 similar, trapezoidal, progressively smaller. Tip of lateral tooth 1 restricted to cutting edge parallel to base, with five cusps, innermost approximately 50% larger than outer four cusps of equal size. Lateral teeth 2–4 with tips progressively turning to be at right angle with base, strongest cusps at least twice as large as lateral cusps on both sides of strongest cusp. Lateral tooth 5 enlarged by elongation, central cusp much larger than seven small lateral barbs. Inner marginal teeth hook-shaped, with minute barbs at base of tip, becoming smaller towards periphery, outer marginal teeth spoon-shaped with multiple small setae; shafts without food groove. Radular interlock weak to moderate.

Comparisons. *Anatoma amoena* from Antarctica has a proportionally broader shell, a wider umbilicus, and a spiral cord on teleoconch I forming a discrete angulation.

Anatoma equatoria from the Indo-Malayan Archipelago and Central Pacific is more square, has the selenizone and slit at the periphery of the shell, much denser axials (85 vs. 30 on body whorl), and lacks a spiral cord in the position of the selenizone on teleoconch I.

Figure 830. *Anatoma porcellana* new species. Holotype. 273–356 m, off Pamilacan Island, Bohol Sea, PANGLAO 2005 CP2343, Philippines, 9.456°N, 123.823°E (MNHN 25114). Scale bar shell = 1 mm. Scale bar protoconch = 100 μm.

Anatoma flexidentata from the Indian Ocean to the Western Pacific has a shorter teleoconch I (0.3 vs. >0.6 whorl), a wider selenizone, and weaker axial cords on

Figure 831. *Anatoma porcellana* new species. Paratypes. **A–B.** 273–356 m, off Pamilacan Island, Bohol Sea, PANGLAO 2005 CP2343, Philippines, 9.456°N, 123.823°E (MNHN 25113, 20). Radula shown in Figure 832A–D. **C.** 146–233 m, Tanimbar Island, N.O. *Baruna Jaya 1*, KARUBAR CP67 Indonesia, 8.966°S, 132.100°E (MNHN 25115, 1). Operculum shown in Figure 832D. Scale bars shell = 1 mm. Scale bars protoconch = 100 µm.

shoulder and base.

Anatoma gunteri from eastern Australia has noticeably stronger sculpture on the shoulder compared to the base and more numerous and fainter spirals.

Anatoma indonesica from the Indo-Malayan Archipelago to the Western Pacific has stronger axial lamellae crossed by at least twice as many spiral lines on the shoulder and no sculpture on the base immediately below the selenizone.

Anatoma kopua from temperate eastern Australia and New Zealand has weaker axial cords, a less impressed suture, and a wider sutsel after one teleoconch II whorl (1× vs. <0.5× width of selenizone).

Anatoma sagamiana from the southern Japan and the Indo-Malayan Archipelago has a protoconch with reticulate sculpture, denser spiral sculpture on shoulder and base, and an indistinct spiral keel below the selenizone.

Anatoma zancliformis from the Indo-Malayan Archipelago to the Western Pacific has a less impressed suture, has weaker axial cords, and distinct spiral lines on the shoulder.

Anatoma alta from the Caribbean and the tropical eastern Atlantic has weaker axial cords on the shoulder and base, at least 50% more spiral lines on the base, and a much shorter teleoconch I (0.3 vs. >0.6 whorl).

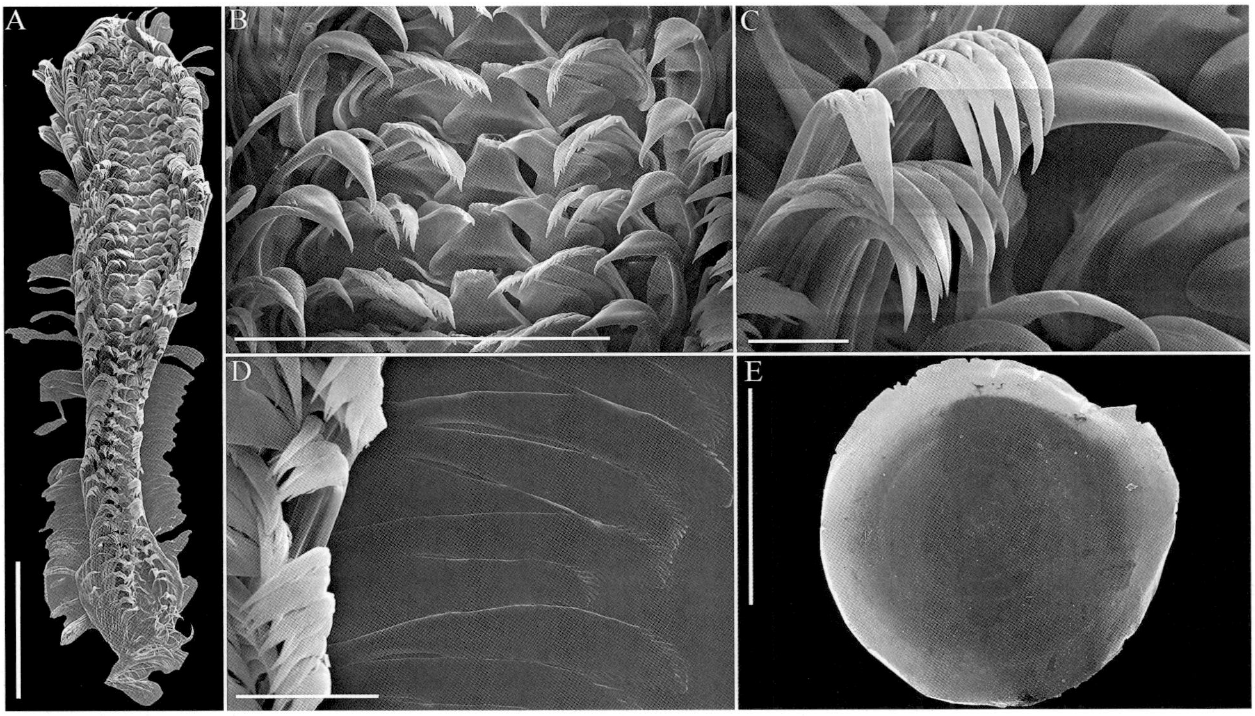

Figure 832. *Anatoma porcellana* new species. **A–D.** Radula. 273–356 m, off Pamilacan Island, Bohol Sea, PANGLAO 2005 CP2343, Philippines, 9.456°N, 123.823°E (MNHN 25113, 20); shells from lot shown in Figure 831A–B. **A.** Entire radula. Scale bar = 200 μm. **B.** Central field enlarged. Scale bar = 100 μm. **C.** Lateral tooth 5 and inner marginal teeth. Scale bar = 10 μm. **D.** Outer marginal teeth. Scale bar = 20 μm. **E.** Operculum. 146–233 m, Tanimbar Island, N.O. *Baruna Jaya 1*, KARUBAR CP67 Indonesia, 8.966°S, 132.100°E (MNHN 25115, 1); shell shown in Figure 831C. Scale bar = 1 mm.

Anatoma lamellata from the Western to the Central Pacific has a protoconch with a strong spiral, the selenizone more towards the middle of the whorl, and distinct spiral lines on the shoulder.

Anatoma sinuosa from the tropical eastern Pacific has many distinct spiral lines on shoulder and base, at least twice as many axial cords, and an open umbilicus.

Distribution. Possibly tropical Indian Ocean; Southern Japan (see Remarks), Indo-Malayan Archipelago to the Western Pacific (Fig. 833).

Specimen records.

Madagascar. 1190 m, 35 mi WSW off Tulear, 23.717°S, 43.417°E (USNM ex 717353, 1). 695 m, ca. 40 mi off Tulear, 23.333°S, 43.533°E (USNM 717614, 1). 729–732 m, Lavanono sector, ATIMO VATAE CP3597, 25.717°S, 44.400°E (MNHN, 1).

China Sea. 622 m, St. 5318, off Pratas Island, 21.533°N, 117.767°E (USNM 284370, 1).

Philippines. 2570 m, 4.633°N, 119.817°E (MNHN, 1). 507 m, St. 5576, off Mount Dromedarie, N Tawi Tawi, 5.432°N, 120.061°E (USNM 257932, 1). 622 m, St. 5574, off Simaluc Island, Tawi Tawi Islands, 5.513°N, 120.133°E (USNM 287431, 1). 604 m, St. 5570, off Simaluc Island, Tawi Tawi, 5.538°N, 120.216°E (USNM 287151, 3). 554 m, St. 5569, off Simaluc Island, Tawi Tawi Island, 5.554°N, 120.258°E (USNM 286350, 1). 490 m, Station 5567, off Dammi Island, Sulu Archipelago, 5.797°N, 120.408°E (USNM 289848, 14). 410 m, St. 5563, off Dammi Island, Sulu Archipelago, 5.803°N, 120.513°E (USNM 298990, 1). 444 m, St. 5565, off Dammi Island, Sulu Archipelago, 5.862°N, 120.508°E (USNM 286539, 5). 313 m, St. 5244, Pujada Bay, E Mindanao, 6.868°N, 126.238°E (USNM 276430, 1). 481 m, St. 5549, off Jolo Island, 7.102°N, 125.669°E (USNM 286559, 3). 49 m, St. 5426, off Mantaquin Island, E Palawan, 9.200°N, 118.467°E (USNM 288305, 12). 929 m, Station 5423, Jolo Sea, off Cagayan Island, 9.600°N, 121.233°E (USNM 288491, 9). 622 m, St. 5424, Jolo Sea, off Cagayan Island, 9.618°N, 121.210°E (USNM 288656, 1). 905 m, St. 5425, off Cagayan Island, Jolo Sea, 9.629°N, 121.183°E (USNM 288099, 3). 1401 m, St. 5429, off Fondeado Island, E Palawan, 9.692°N, 118.839°E (USNM 289236, 7). 265 m, St. 5261, off Balanja Point, SE Mindoro, 12.515°N, 121.573°E (USNM 276215, 1). 362 m, off Matoco Point, W Luzon, 13.632°N, 121.070°E (USNM 276751, 1). 410 m, St. 5563, off Dammi Island, Sulu Archipelago, 13.683°N, 123.963°E (USNM 286787, 1). 721 m, Station 5119, off Cape Santiago, W Luzon, 13.750°N, 121.367°E (USNM 23501, 30). 842 m, 13.783°N, 120.050°E (MNHN, 1; MNHN, 1). 640–668 m, 13.783°N, 120.500°E (MNHN, 50). 512 m, St. 5283, off Malavatuan Island, W Luzon, 13.808°N, 120.478°E (USNM 278056, 1).

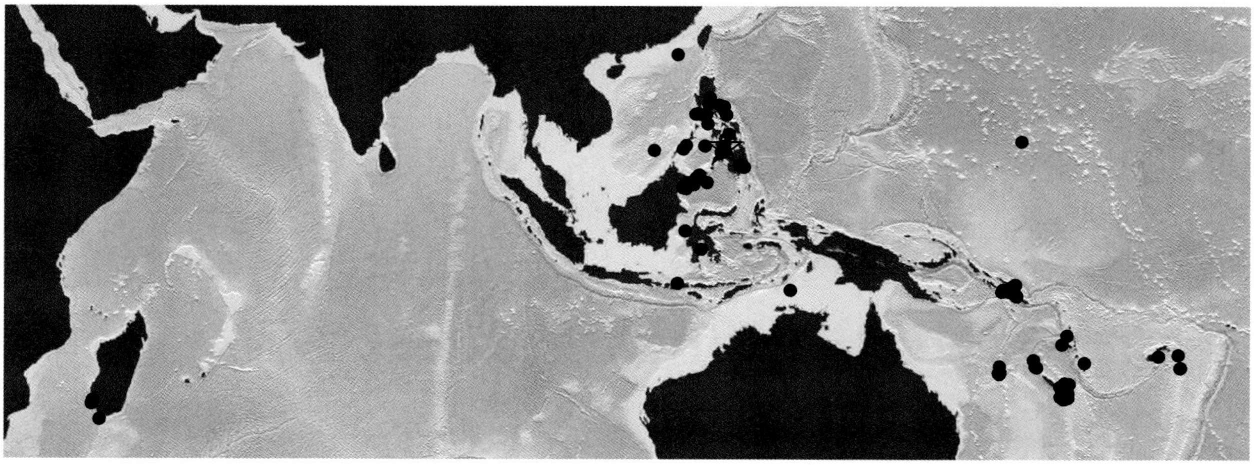

Figure 833. Distribution of *An. porcellana* new species. ☆ type locality. ● specimen record.

497 m, Station 5285, off Malavatuan Island, W Luzon, 13.861°N, 120.346°E (USNM 244948, 8). 259–280 m, Dipolog Bay, Bohol-Sulu Sea Sill, PANGLAO 2005 CP2381, 8.722°N, 123.317°E (MNHN, 7). 569–583 m, Bohol-Sulu Sea Sill, PANGLAO 2005 CP2358, 8.868°N, 123.618°E (MNHN, 1). 679–740 m, Bohol-Sulu Sea Sill, CP2362, 8.942°N, 123.545°E (MNHN, 1). 427 m, Bohol-Sulu Seas Sill, PANGLAO 2005 DW2364, 9.012°N, 123.425°E (MNHN, 15). 273–356 m, off Pamilacan Island, Bohol Sea, PANGLAO 2005 CP2343, 9.457°N, 123.823°E (MNHN, 21). 470–566 m, off Balicasag Island, Bohol Sea, CP2394, 9.477°N, 123.667°E (MNHN, 2). 356–396 m, off Balicasag Island, Bohol Sea, CP2393, 9.502°N, 123.693°E (MNHN, 27). 309–342 m, off Balicasag Island, Bohol Sea, CP2399, 9.528°N, 123.698°E (MNHN, 12). 713–731 m, off Balicasag Island, Bohol Sea, PANGLAO 2005 CP2398, 9.543°N, 123.675°E (MNHN, 1). 642–669 m, Maribojoc Bay, Bohol Sea, CP2397, 9.582°N, 123.695°E (MNHN, 1). 357–367 m, AURORA 2007 CP2695, 14.770°N, 123.668°E (MNHN, 2). 944–1004 m, AURORA 2007 CC2702, 14.922°N, 123.198°E (MNHN, 3). 996–1037 m, AURORA 2007 CP2686, 14.937°N, 123.157°E (MNHN, 9). 1071–1147 m, AURORA 2007 CP2690, 14.963°N, 123.048°E (MNHN, 1). 431–438 m, AURORA 2007 DW2675, 15.045°N, 121.717°E (MNHN, 4). 478–480 m, AURORA 2007 DW2706, 15.070°N, 121.715°E (MNHN, 6). 473 m, AURORA 2007 CP2749, 15.943°N, 121.837°E (MNHN, 2). 624–647 m, off Aligbay Island, Bohol-Sulu Sill, PANGLAO 2005 CP2384, 8.770°N, 123.268°E (MNHN, 14). 333 m, St. 5403, off Capitancillo Island, N Cebu, 11.166°N, 124.288°E (USNM 287809, 1).

Indonesia. 146–233 m, Tanimbar Island, 8.967°S, 132.100°E (MNHN, 1). 1472 m, St. 5654, Cape Tabako Gulf of Boni, 3.700°S, 120.750°E (USNM 279692, 1). 870 m, St. 5585, Sibuko Bay, off Sipadan Island, 4.117°N, 118.832°E (USNM 258529, 25; USNM 258632, 7). 558 m, St. 5592, off Silungan Island, Borneo, 4.212°N, 118.462°E (USNM ex 279050, 5). 296 m, St. 5580, Darvel Bay, off Sibutu Island, Borneo, 4.879°N, 119.113°E (USNM 287087, 1). 780 m, Galathea Station 477, S of Bali, 9.017°N, 114.800°E (ZMUC, 14). 2160 m, St. 5670, Macassar Strait, off Chenoki Point, Celebes, 1.316°S, 118.717°E (USNM 279523, 1). 988 m, St. 5650, Gulf of Boni, off Lamulu Point, Celebes, 4.896°N, 121.483°E (USNM 291011, 1).

Coral Sea. 675–680 m, Chesterfield, 19.817°S, 158.683°E (MNHN, 3). 305–310 m, Chesterfield, 19.667°S, 158.533°E (MNHN, 1). 620 m, Chesterfield, 19.633°S, 158.783°E (MNHN, 1). 580 m, Chesterfield, 19.617°S, 158.717°E (MNHN, 1). 630 m, Chesterfield, 19.617°S, 158.767°E (MNHN, 9). 675–680 m, Chesterfield, 18.817°S, 158.683°E (MNHN, 1: complete).

New Caledonia. 558–647 m, southern, 23.033°S, 166.783°E (MNHN, 1: complete). 558–647 m, southern, 23.033°S, 166.950°E (MNHN, 1). 1675 m, Loyalty Basin, 22.750°S, 166.417°E (MNHN, 1). 975 m, Loyalty Basin, 22.700°S, 166.550°E (MNHN, 7). 825–860 m, 22.317°S, 167.383°E (MNHN, 4). 1080–1180 m, E coast, 21.700°S, 166.667°E (MNHN, 16). 760–790 m, Loyalty Basin, 21.567°S, 166.450°E (MNHN, 4). 760–790 m, Loyalty Basin, 21.567°S, 166.450°E (MNHN, 1: complete). 1040 m, Loyalty Basin, 21.517°S, 166.417°E (MNHN, 2). 1260–1300 m, Loyalty Basin, 21.467°S, 166.383°E (MNHN, 3). 1810 m, 21.467°S, 166.433°E (MNHN, 1). 600 m, Loyalty Ridge, 21.100°S, 167.550°E (MNHN, 5). 1600–1640 m, Loyalty Basin, 20.983°S, 166.983°E (MNHN, 1). 325–330 m, northern, 18.950°S, 163.417°E (MNHN, 1). 525 m, 18.900°S, 163.317°E (MNHN, 1). 600–616 m, northern, 18.817°S, 163.250°E (MNHN, 4). 1160–1220 m, 15.967°S, 166.633°E (MNHN, 1). 570–590 m, Île des Pins, TERRASSES DW3119, 22.867°S, 167.383°E (MNHN, 3). 616–628 m, N Grand Passage, CONCALIS CP3012, 18.835°S, 163.237°E (MNHN, 2). 700–730 m, N Grand Passage, CONCALIS DW2993, 17.992°S, 163.025°E (MNHN, 1).

Solomon Islands. 435–461 m, 9.617°S, 160.695°E (MNHN, 1). 611–636 m, 9.758°S, 160.875°E (MNHN, 6). 621–708 m, 9.632°S, 160.828°E (MNHN, 1). 505–510 m, 9.223°S, 160.130°E (MNHN, 2). 1327–1598 m, 8.610°S, 160.123°E (MNHN, 4). 570–756 m, 8.263°S, 160.673°E (MNHN, 9). 381–383 m, 10.213°N, 161.477°E

(MNHN, 4). 720–849 m, SALOMON 1 CP1857, 9.663°S, 160.810°E (MNHN, 3). 749–799 m, SALOMON 1 CP1751, 9.173°S, 159.883°E (MNHN, 6). 600–875 m, SALOMON 2 CP2176, 9.157°S, 158.987°E (MNHN, 1). 766–1100 m, N Russel Island, SALOMON 2 DW2167, 8.913°S, 159.122°E (MNHN, 1). 1327–1598 m, SALOMON 1 CP1764, 8.610°S, 160.123°E (MNHN, 1). 98–200 m, SALOMON 1 DW1767, 8.323°S, 160.665°E (MNHN, 5). 460–550 m, N Malaita, SALOMONBOA 3 DW2792, 8.267°S, 160.717°E (MNHN, 1).

Vanuatu. 487–507 m, 18.367°S, 169.467°E (MNHN, 4). 1360 m, 14.817°S, 167.250°E (MNHN, 7).

Fiji. 700–707 m, 19.050°S, 178.500°W (MNHN, 2). 395–540 m, 19.017°S, 178.433°W (MNHN, 2). 879–897 m, SE of Viti Levu, 17.515°S, 178.888°E (MNHN, 7). 879–897 m, 17.515°S, 178.888°E (MNHN, 1). 477–493 m, 17.333°S, 178.733°W (MNHN, 3). 395–540 m, BORDAU 1 DW1486, 19.017°S, 178.433°W (MNHN, 6).

Remarks. The porcellaneous appearance of the shell is uncommon among *Anatoma* species. The only other species showing a similar shell appearance is *An. alta* from the subtropical North Atlantic. The light micrograph in Okutani & Hasegawa (2000: pl. 18, fig. 1) shows that aspect very well. Most specimens have a heavily broken peristome, which gives the shell a slightly taller appearance, but the color and sculpture are distinct, making this species easy to recognize.

The elongated lateral tooth 5 is uncommon, but it has been found in *An. sagamiana* from Japan, and in *An. zancliformis* from New Zealand.

The species is illustrated by Okutani & Hasegawa (2000: 37) from Japan under the name *An. lamellata*. Given their mixed species concept, the indicated distribution (Sagami Bay to Shikoku and Kyushu) has to be viewed with caution.

Anatoma proxima (Dall, 1927)

Figures 834–839

Chresonymy.

Scissurella proxima Dall, 1927a: 110, not illustrated.
Scissurella proxima new species [sic]: Dall, 1927b: 10 [see Remarks].
Scissurella proxima: Johnson, 1934: 68.
Scissurella proxima: Lyman, 1948: 76, text-fig. p. 77.
Scissurella proxima: Abbott, 1954: 91, fig. 20.
Scissurella proxima: Poirier, 1954: 12.
Scissurella proxima: Abbott, 1974:17.
Scissurella proxima: Odé, 1986: 72.
Anatoma sp. 1: Leal, 1991: 43–44, pl. 2, figs L–M.
Anatoma proxima: Bandel, 1998: 29, pl. 8, figs 7–8, pl. 9, fig. 1, pl. 11, figs 2–3.
Anatoma proxima: Geiger, 2003: 74.
Anatoma proxima: Geiger, 2004: text-fig. p. 5.
Anatoma proxima: Geiger, 2006c: text-fig. 5.
Anatoma proxima: Geiger *et al.*, 2007, fig. 7.
Anatoma proxima: Geiger & McLean, 2010: 15.
Anatoma proxima Geiger & Fernández-Garcés, 2010: 67.
Anatoma proxima: Tunnell *et al.*, 2010: 114, text-fig.

Synonym.

+*Anatoma americana* Bandel, 1998: 28–29, pl. 8, figs 4–6. ***Type material***. Holotype (SGPIH 3885. Fig. 834), 1.5 × 1.2 mm. ***Type locality***. 150 m, Cape Lookout, [North Carolina] USA [34.583°N, 76.533°W] (OD). ***Etymology***. Named for its type locality in America (OD).

Anatoma americana: Geiger, 2003: 73.

Figure 834. Original illustration of *Anatoma americana*.

Misidentification.

Anatoma crispata: García, 2005: 27, fig. 6 [is *An. proxima*].

Type material. Holotype (USNM 322962: Boss *et al.*, 1968: 267; missing since Feb. 2, 1996: USNM collection records; not found Geiger pers. obs. 2/2004), 1.7 × 1.3 mm (H × W). Probable paratype 434 fms [= 794 m], Station 2662, off St. Augustine Harbor [Florida, USA, ~30.063°N, 79.640°W] (USNM 93353, 1).

Type locality. Dredged off South Carolina by the United States Bureau of Fisheries Steamer *Albatross* at station 2314, in 159 fathoms [= 291 m] coarse sand, bottom temperature 47.4°F [32.717°N, 77.850°W](OD).

Etymology. Not given (OD). Proximus, Latin adjective for close, similar; reference unclear.

Description. Shell medium-sized for genus (2.5 mm. Figs 835–837), trochiform, biconical, somewhat inflated. Protoconch of 0.75 whorl, sculpture flocculent, often no apertural varix, apertural margin straight. Teleoconch I of 0.6–0.8 whorls, approximately 17 distinct axial cords, with or without single spiral in position of selenizone. Teleoconch II of at least 2.5 whorls, suture moderately impressed; usually no sutsel on first teleoconch II whorl, at apertural margin up to 3× width of selenizone showing approximately six spiral lines. Shoulder convex early in growth, almost straight in mature shells; with approximately 75 axial cords on first whorl, approximately 70 on last 0.25 whorl, three weak spiral lines starting 0.25 teleoconch II whorl, approximately 15 spiral lines at apertural margin of mature shell; spiral lines forming discrete points at intersection with axial cords. Base with same number of axial cords as on shoulder, intersected by approximately 15 stronger spiral lines, strength of spirals increasing towards umbilicus, cords in umbilicus; spacing variable, closer near selenizone and umbilicus, wider on mid-base. Umbilicus open, moderately narrow, sloping continuously with base, with weak axial and strong spiral sculpture. Selenizone at periphery, keels moderately strong to strong, distinctly elevated, lunules about as strong as axial cords; slit open, margins parallel. Aperture round, parietal peristome flared.

Operculum (Fig. 836A) thin, round, multispiral, with central nucleus.

Radula (Fig. 838). Rachidian tooth trapezoidal, middle tip strongly elongated, three smaller tips on each side. Lateral teeth 1–3 similar, four cusps on outer cutting edge. Lateral tooth 4 reduced, hook-shaped with two cusps on outer margin, one cusp on inner margin. Lateral tooth 5 enlarged, moderately elongated, approximately 8–9 cusps on inner edge up to the apex of the tip, approximately six cusps on outer edge with first cusp at inflection between tip and shaft. Inner marginal teeth with triangular tip, approximately eight cusps on each side; shafts without food groove. Middle marginal teeth elongated to spoon-shaped with increasing number of finer cusps along edge of tip. Outermost marginal teeth fan-shaped, with approximately 15–20 fine bristles on each side of tip. Radular interlock of central field moderate.

Comparisons. *Anatoma aedonia* from the Caribbean to Brazil has more scalar whorls, the selenizone higher on the whorl, and a wider umbilicus.

Figure 835. *Anatoma proxima*. 146 m, Fowey Light, USA, 25.583°N, 80.100°W (USNM 449419, 7). Scale bar shell = 1 mm. Scale bar protoconch = 100 μm.

Anatoma alta from the Caribbean and the northeastern Atlantic has a shorter teleoconch I (0.167 vs. 0.5 whorl) and much less dense sculpture on the teleoconch.

Figure 836. *Anatoma proxima*. **A.** 216 m, off Sambo Reef, Florida, USA, 24.499°N, 81.645°W (USNM 449411, 5). Radula of specimen shown in Figure 838. Specimen with operculum *in situ*. **B.** 219 m, off Sand Key, Sand Keys, Biscayne Bay, Florida, USA, 25.500°N, 80.030°W (USNM 449403, 3). **C.** 165 m, off Caesars Creek, Florida, USA, 29.822°N, 84.993°W (USNM 449423, 12). Scale bars shell = 1 mm. Scale bars protoconch = 100 μm.

Anatoma alternatisculpta from the Caribbean has a taller overall shell, stronger axial cords on the shoulder, and at most half as many spiral lines on the shoulder.

Figure 837. *Anatoma proxima*. **A.** 457 m, Yucatan Channel, Mexico (USNM 780317, 10). **B.** [off Alabama, USA], 29.583°N, 87.333°W (UMML 30.12034, 3). **C.** 110–220 m, off Macaé, Rio de Janeiro State, Brazil, 22.383°S, 41.783°W (DLG 105, 2). Scale bars shell = 1 mm. Scale bars protoconch = 100 µm.

Anatoma atlantica from the northwestern Atlantic has a more lenticular overall shell shape, a greater expansion rate (1.51 vs. 1.45), and less pronounced and less dense axial sculpture.

Anatoma plicatazona from the Caribbean has strongly upturned keels of the selenizone, a wider sutsel on early whorls (<0.5× vs. 1× width of selenizone), and the selenizone higher on the whorl.

Anatoma australis from southeastern Australia has stronger axial cords on the base, only every 3–4 axial cords on the base reaching the umbilicus, and more than one spiral cord on teleoconch I.

Anatoma crispata from the northeastern Atlantic has a more convex shoulder, fewer and/or stronger axial cords on shoulder and base, and a distinct funiculus.

Anatoma globulus from the Indo-Malayan Archipelago has an overall more globular shell, a more convex shoulder, and a spiral cord in the position of the selenizone on teleoconch I.

Anatoma japonica from the Indian Ocean to the Central Pacific has a spiral cord in the position of the selenizone on teleoconch I, a more narrow selenizone, and a more discoidal overall shell shape.

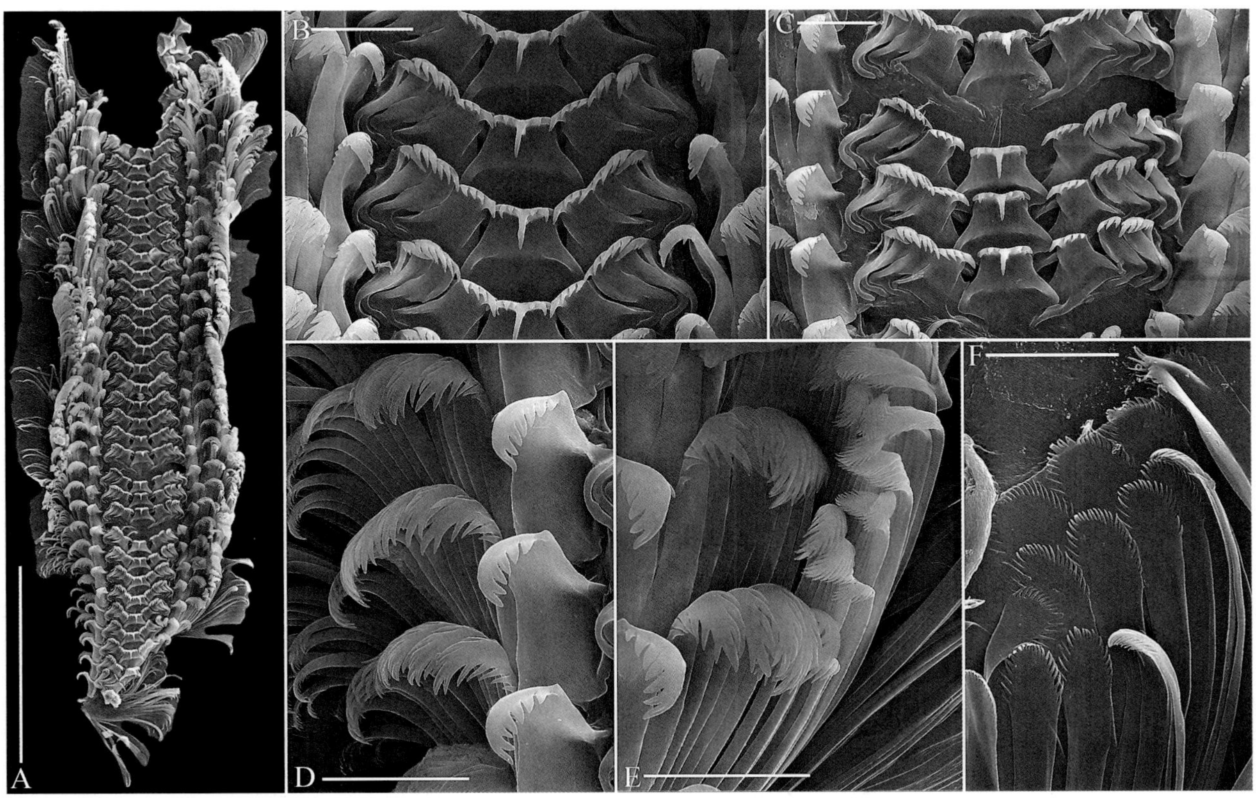

Figure 838. *Anatoma proxima*. Radula from specimen shown in Figure 836A. 216 m, off Sambo Reef, Florida, USA, 24.499°N, 81.645°W (USNM 449411, 5). **A.** Entire radula. **B.** Central field showing outer, serrated edge of lateral tooth 5. **C.** Central field with isolated lateral tooth 4. **D.** Lateral tooth 5 and inner marginal teeth. Note absence of food groove in marginal teeth. **E.** Inner and middle marginal teeth. **F.** outer marginal teeth. Scale bar A = 200 μm. Scale bars B–E = 20 μm. Scale bar F = 10 μm.

Anatoma keenae from the Panamic has a more globular overall shell shape, coarser reticulate sculpture on shoulder and base, and no apertural varix on the protoconch.

Anatoma kelseyi from the northeastern Pacific has a protoconch with reticulate sculpture, a more convex shoulder, and overall weaker sculpture.

Anatoma orbiculata from West Africa has a more depressed shell, much stronger but at most half as many axial cords on the shoulder, and markedly different sculpture on shoulder and base.

Anatoma peruviana from Peru has a protoconch with reticulate sculpture, a much wider selenizone anywhere on the shell, and the selenizone higher on the whorl.

Anatoma rainesi from the Eastern to the Western Pacific has a more discoidal shell, distinct fine irregular axial cords on the protoconch, and nodular thickenings on the axial cords on the teleoconch I in position of the selenizone.

Anatoma sagamiana from Japan to the Indo-Malayan Archipelago has reticulate protoconch sculpture, a weak constriction on the base below the selenizone, and a more narrow umbilicus.

Anatoma schanderi from the northeastern Atlantic has a protoconch with reticulate sculpture, a longer teleoconch I (>0.9 vs. <0.75 whorl), and a more convex shoulder.

Anatoma tenuis from the North Atlantic has no sutsel, particularly on early whorls, a more discoidal overall shell shape, and grows much larger (>5 vs. 2.5 mm).

Anatoma tenuisculpta from the Mediterranean and adjacent Atlantic has a more discoidal shell, a spiral cord in position of the selenizone on teleoconch I, and a more narrow selenizone.

Anatoma tobeyoides from southeastern Australia has a much shorter teleoconch I (0.25 vs. >0.5 whorl), oblique spiral lines on teleoconch I and early teleoconch II, and a more narrow selenizone.

Distribution. South Carolina to Caribbean, Brazil, 115–305 m (Fig. 839).

Specimen records.

USA, South Carolina. 32.733°N, 78.095°W (USNM 875046, 4). 213–223 m, off Charleston, 32.733°N, 78.095°W (USNM 875142, 20). 223–233 m, off Charleston, 32.738°N, 75.089°W (USNM ex 875208, 20).

USA, Virginia. 128 m, Chesapeake Bay, 37.000°N, 76.000°W (USNM 44536, 1).

USA, Georgia. 492 m, E Savannah, 31.851°N, 79.174°W (HLC, 2).

USA, Florida. 308–323 m, [~170 km W of Clearwater], 27.863°N, 84.497°W (EGC 29799, 1). 247 m, Eolis station 329, off Sambo Reef (USNM 449409, 10). 120 m, off Sambo Reef, Monroe County, 24.500°N, 81.667°W (FLMNH 155183, 1). 366–421 m, SW Egmont Key (KBIN, 3). 128 m, Key Largo (FLMNH 393633, 2). 219 m, Dry Tortugas (FLMNH 357185, 1). 366–421 m, SW Egmont Key (FLMNH 233960, 1). 165 m, off Caesars Creek, Monroe County (FLMNH 155177, 1). 400 m, ca. 150 mi W Cape Romano, Collier County (FLMNH 128260, 2). 119 m, off Delray, Palm Beach County (FLMNH 145080, 2). 146 m, off Sombrero Key Light, Monroe County (FLMNH 146524, 1). 183–201 m, 205–215° off Sombrero Light, Monroe County (FLMNH 249938, 2). 119–137 m, off Key West, Monroe County (FLMNH 257789, 5). 165 m, off Western Dry Rocks, 24.445°N, 81.928°W (USNM 449413, 1). 216 m, off Sambo Reef, 24.499°N, 81.645°W (USNM 449411, 5). 210 m, off Sambo Reef, 24.500°N, 81.717°W (USNM 449412, 7). 219 m, off Sambo Reef, 24.500°N, 81.667°W (USNM 449410, 13). 201 m, off Key West, 24.554°N, 81.792°W (USNM 449415, 6). 165 m, off Key West, 24.555°N, 81.783°W (USNM 449414, 2). 137–146 m, Key West, 24.555°N, 81.783°W (USNM 449416, 1). 128–146 m, Sombrero

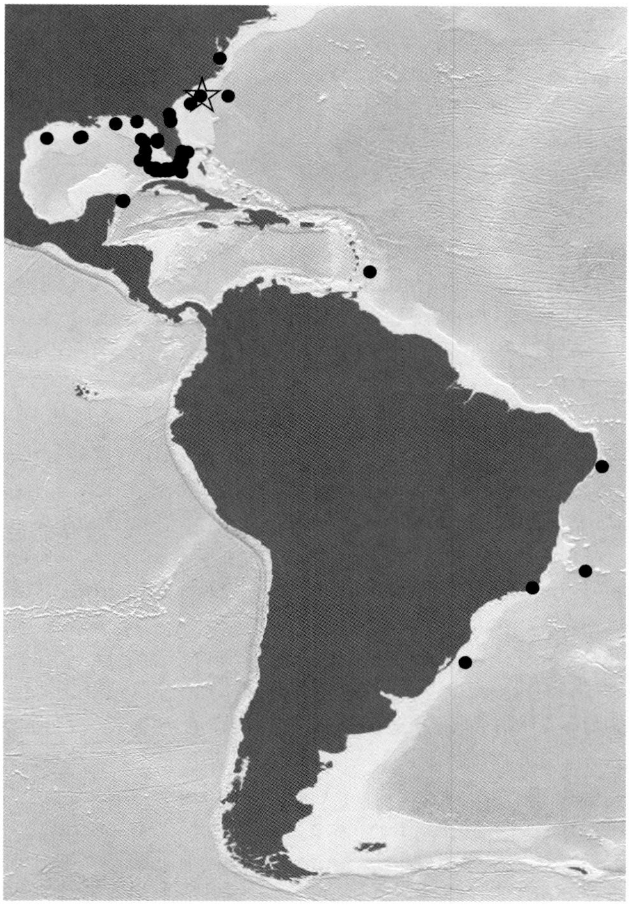

Figure 839. Distribution of *Anatoma proxima*. ☆ type locality. ● specimen record. ■ literature record.

Light, Monroe County, 24.570°N, 81.100°W (FLMNH 249671, 8). 137 m, SSE off Sombrero Key Light, Monroe County, 24.580°N, 81.090°W (FLMNH 249998, 8). 183–201 m, 205° off Sombrero Light, Monroe County, 24.600°N, 81.100°W (FLMNH 249525, 5). 128–137 m, 145–176° off Sombrero Light, Monroe County, 24.600°N, 81.100°W (FLMNH 249953, 1). 201 m, 50 mi W off Dry Tortugas, 24.650°N, 83.533°W (UMML 30.12031, 1). 128–165 m, Triumph Reef, 25.486°N, 80.115°W (USNM 450581, 3). 219 m, off Sand Key, Sand Keys, Biscayne Bay, 25.500°N, 80.030°W (USNM 449403, 3). 220 m, off Sand Key, Sand Keys, Biscayne Bay, 25.500°N, 80.030°W (MCZ 315231, 1). 128 m, off Sand Key, Sand Keys, Biscayne Bay, 25.500°N, 80.030°W (USNM 449421, 2). 155 m, Sand Key, Sand Keys, Biscayne Bay, 25.500°N, 80.030°W (USNM 449422, 1). 210 m, off Sand Key, Sand Keys, Biscayne Bay, 25.500°N, 80.030°W (USNM 449402, 1). 165 m, Sand Key Sand Keys, Biscayne Bay, 25.500°N, 80.030°W (USNM 449408, 1). 183 m, Ragged Key, 25.531°N, 80.172°W (USNM 449426, 5). 183 m, off Fowey Light, 25.583°N, 80.100°W (USNM 449417, 1). 302 m, off Fowey Light, 25.583°N, 80.100°W (USNM 449404, 1). 155 m, off Fowey Light, 25.583°N, 80.100°W (USNM 449414, 2). 165 m, Fowey Light, 25.583°N, 80.100°W (USNM 449418, 4). 274–366 m, off Fowey, 25.583°N, 80.100°W (USNM 82961, 1). 201 m, Fowey Light, 25.583°N, 80.100°W (USNM 449420, 1). 146 m, Fowey Light, 25.583°N, 80.100°W (USNM 449419, 7). 37 m, off Government Cut, Miami, 25.767°N, 80.183°W (USNM 757602, 1). 320–366 m, off Delray, 26.450°N, 84.067°W (FMNH 133207, 7). 320–364 m, off Delray, 26.461°N, 80.073°W (FMNH 133207, 6). 550 m, off Boynton, Palm Beach County, 26.517°N, 80.067°W (LACM 116022, 5). 503 m, off Boynton, 26.525°N, 80.067°W (USNM 889877, 1). W of Anna Maria, Manatee County, 27.531°N, 82.734°W (BMSM 825, 1). 62 m, SW Johns Pass, 27.820°N, 82.780°W (HLC, 1; FLMNH 233961, 4). 165 m, off Caesars Creek, 29.822°N, 84.993°W (USNM 449423, 12). 794 m, off St. Augustine Harbor, 29.894°N, 81.315°W (USNM 93353, 1). 538 m, off Fernandina, 30.669°N, 81.463°W (USNM 87360, 1; USNM ex 108147, 5; USNM 87353, 1). 128–165 m, off Fowey Light, 25.583°N, 80.100°W (USNM 449406, 2). Off Boynton, Palm Beach, 26.517°N, 80.067°W (FMNH 29982, 3).

USA, Louisiana. 71 m, 28.094°N, 91.011°W (EGC 24459, 1). 28.008°N, 91.238°W (CCSC, 1).

USA, Alabama. 29.583°N, 87.333°W (UMML 30.12034, 3). 29.583°N, 87.333°W (UMML 30.12033, 2).

Gulf of Mexico. 123 m, 27.917°N, 94.817°W (USNM 780815, 2). NW Gulf of Mexico, (USNM 780316, 1).

Mexico. 198 m, [off Cozumel], 20.917°N, 86.467°W (UMML 30.12045, 2). 102 m, [off Cozumel], 20.950°N, 86.567°W (UMML 30.12047, 1). 25 m, [off Cozumel], 20.983°N, 86.400°W (UMML 30.12046, 1). 457 m, Yucatan Channel (USNM 780317, 10).

Barbados. 229 m, off Pelican Island, 13.167°N, 59.567°W (USNM 500688, 1). 137–146 m, off Pelican Island, 13.167°N, 59.567°W (USNM 500678, 1). 13.167°N, 59.567°W (USNM 126817, 1). 268 m, off Pelican Island, 13.167°N, 59.567°W (USNM 500677, 3). 60 m, off Pelican Island, 13.167°N, 59.567°W (USNM 500679, 1). 183 m, off

Pelican Island, 13.167°N, 59.567°W (USNM 500680, 1). 137 m, off Pelican Island, 13.167°N, 59.567°W (USNM 500695, 1).

Brazil. 224 m, [off Porto Alegre], 30.717°S, 49.083°W (HLC, 1). 110–220 m, off Macaé, Rio de Janeiro State, 22.383°S, 41.783°W (DLG 105, 2). 82–105 m, Jaseur Seamount, 20.500°S, 36.000°W (BMSM 26033, 1).

Literature records.

USA. S Florida, Atlantic. 700 m, off Jacksonville, Florida (Bandel, 1998: as *An. americana*).

Remarks. One specimen (USNM 93353, 1) was annotated as syntype by F. Scarabino 7/1998. Dall (1927a) clearly listed one lot prominently, with dimensions and precise locality data, giving it elevated status. Although the lot has been missing from the shelf since 1996, USNM records note a single specimen for that lot; accordingly, it constitutes the holotype. Dall (1927a: 110) mentioned additional specimens examined from St. Augustine in 434 fathoms, and "numerous localities on the coast and off the keys of Florida, in 20 to 200 fathoms." USNM 93353 matches the former collection data, and was collected in 1888 (prior to the publication of the name); hence, it is a probable paratype. The other indications are too vague to permit identification of other paratypes.

Dall (1927a&b) described the species twice in two separate publications with identical text. The earlier publication (Dall, 1927a) in the *Proceedings of the United States National Museum* 2667 should be taken as the proper original description. The second description by Dall (1927b) was published in the next issue in the same serial (2668). It is considered a *lapsus calami* and not an introduction of a primary homonym and objective synonym.

Anatoma americana is a junior synonym of *An. proxima*. The shared characters include the flocculent protoconch sculpture, the dense axial cords on teleoconch I, and the globular whorls. Bandel (1998) listed the more extensive teleoconch I of 0.875 whorls in *An. americana* compared to 0.75 whorl in *An. proxima*. The observed difference is within the intraspecific variability of *Anatoma* species. The cited height/width difference is only 7%.

Anatoma proxima seems to be quite variable. The typical variety is shown in Figures 835 and 836. Two varieties considered conspecific are shown in Figure 837. In the Caribbean, there are specimens with more crisp sculpture—more elevated axial cords, intersection with spiral lines elevated to distinct points, umbilicus with raised tubercles—and a pattern of axial cords on the shoulder alternating between reaching the suture and stopping shortly before. This variety is what Bandel (1998) described as *An. americana*. There are some specimens that show characteristics of both morphs (e.g., USNM 449423); all details of the early shell, protoconch and teleoconch I are identical; hence, those differences are not considered sufficient to justify the introduction of an additional species name.

Whether *An. atlantica* is also a synonym of *An. proxima* remains to be examined (see Remarks under *An. atlantica*).

Specimens from Brazil (see also Leal, 1991: pl. 2, figs L, M) have a single spiral cord in the position of the selenizone, whereas a second specimen from the same lot only has a very weak, almost indiscernible spiral. Specimens from the Caribbean and the American Atlantic mostly lack that spiral cord, or have it only very weakly developed. It seems more likely that this is a slightly variable character, rather than that the Brazil specimens represent an undescribed species. The Brazil specimens are otherwise identical to the specimens from the more northern range of the species.

Anatoma pseudoequatoria (Kay, 1979)

Figures 840–848

Chresonymy.

Scissurella pseudoequatoria Kay, 1979: 39, fig. 9A–C (Fig. 840).
Scissurella pseudoequatoria: Hedegaard, 1990: 74, pl. 58, fig. 3 [SEM of shell structure].
Scissurella pseudoequatoria: Geiger, 2003: 77.
Anatoma sp. 4: Spencer *et al*., 2009: 201.
Anatoma sp. 6: Spencer *et al*., 2009: 201.
Anatoma pseudoequatoria: Geiger, 2011b: 40, pl. 4, fig. 1.

Figure 840. Original illustration of *Scissurella pseudoequatoria*.

Synonyms.

+ *Hainella pulchella* Bandel, 1998: 41–42, pl. 14, figs 5–7 (Fig. 841). ***Type material.*** Holotype (SGPIH 3880. Fig. 841), 0.8 × 0.65 mm. ***Type locality.*** Near Satonda, Indonesia [8.100°S, 117.750°E] (OD). ***Etymology.*** Diminutive of Latin adjective pulcher, meaning small, beautiful.

Anatoma pulchella: Geiger, 2003: 74.

Figure 841. Original illustration of *Hainella pulchella*.

+ *Hainella paucispiralia* Bandel, 1998: 41, pl. 14, figs 2–4. Fig. 842. ***Type material.*** Holotype (SGPIH 3879), 0.65 × 0.45 mm. ***Type locality.*** Near Satonda, Indonesia [8.100°S, 117.750°E] (OD). ***Etymology.*** Named for its low number of whorls (OD).

Figure 842. Original illustration of *Hainella paucispiralia*.

Anatoma paucispiralia: Geiger, 2003: 74.

+ *Anatoma funiculata* Geiger & Jansen, 2004a: 15–18, figs 7–8, 18 [map]. ***Type material.*** Holotype (AMS C.403660. Fig. 844C). Paratypes from type locality (AMS C.403660, >20). 27–30 m, Great Barrier Reef, Carter Reef, N end, Queensland, Australia, 14.550°S, 145.600°E (AMS C.402656, 1). ***Type locality.*** 21 m, SW side Euston Reef, Great Barrier Reef, Queensland, Australia, 16.667°S, 146.217°E (OD). ***Etymology.*** Funiculus, Latin: strand of material, referring to the shell strand in the umbilicus.

Misidentifications.

Anatoma cf. *japonica*: Sasaki, 2006: 105, fig. 5A [is *An.* cf. *pseudoequatoria*].

Type material. Holotype (BPBM 9740). Paratypes from type locality (AMS: *fide* Kay, 1979. AMS C.127718, 40: paper record, none in AMS pers. comm. I. Loch, 2002, A. Miller, 8/2010; USMN 806697, 4; NHMUK 1982253, 3).

Type locality. Hilo, Hawaii [USA, 19.717°N, 155.083°W], on algae in intertidal zone (OD).

Etymology. Superficially resembling *An. equatoria*, adjective (OD).

Description. Shell medium-sized (to 1.6 mm. Figs 843–846; holotype 1.1 mm), trochiform, depressed. Protoconch of 0.75 whorl, sculpture finely flocculent, no apertural varix, apertural margin straight. Teleoconch I of 0.2–0.33 whorls, with approximately 6–8 axial cords; approximately six faint, indistinct spiral lines. Teleoconch II of two whorls, suture strongly impressed; sutsel variable, 0–0.5× width of selenizone at onset of selenizone, 1–1.5× width of selenizone at apertural margin of fully grown specimen. Shoulder somewhat convex, with approximately 28–70 axial cords of variable pattern:

- a) straight axials from suture to selenizone (Fig. 845A);
- b) cords starting at suture, becoming less distinct towards keels of selenizone; 1× width of selenizone plus selenizone itself with two additional fine axial cords in between each complete axial cord (Fig. 846B, C);
- c) axials starting at suture, stopping mid-shoulder leaving sculpture-free band, axials starting again and running over keels of selenizone; usually adsutural axials as strong raised cords, absutural axials as fine cords or lines (Figs 843, 844B).

2–7 indistinct to weaker spiral lines; c-pattern mostly on early portion of shell, changing to a-pattern with growth. Intersection of axials and spirals often forming minute to raised points, points arranged as pseudospirals; some specimens with true spiral lines to thin cords, particularly towards apertural margin. Base with modest (juvenile) to distinct (fully grown) constriction below selenizone, major axials either at same strength and density as on shoulder or denser; on lower keels of selenizone to constriction below selenizone one additional somewhat weaker axial cord between each major cord; approximately 10–15 slightly weaker spiral cords, often forming small points at intersection, sometimes giving shells with relatively few axials reticulate appearance, some specimens with smooth band immediately below selenizone (Fig. 843). Umbilicus open, narrow, deep, bordered by relatively sharp edge forming steep funiculus. Selenizone at periphery; elevated, delicate keels at 30° angle to coiling axis; lunules moderately distinct; slit open, with parallel margins. Aperture rounded, lower, adumbilical corner slightly flared, forming funiculus with growth, funiculus curved towards apertural margin; parietal wall with small protrusion.

Operculum unknown.

Radula (Fig. 847). Rachidian tooth trapezoidal, all cusps more or less of equal size, four cusps on each side of central cusp. Lateral teeth 1–3 similar, apical cusp largest, three cusps on outer cutting edge. Lateral tooth 4 reduced, hook-shaped, one minute cusp on each side. Lateral tooth 5 somewhat enlarged, three cusps along inner edge, one small cusp on outer edge. Inner marginal teeth with broad triangular tip, apical cusp largest, two cusps on inner edge, five cusps on outer edge; shafts with food groove. Outer marginal teeth with spoon-shaped tip, many fine bristles along edge. Radular interlock of central field moderate.

Figure 843. *Anatoma pseudoequatoria*. 0 m, Kalihiwai, Kauai, Hawaii, USA (FLMNH 350479, 1). Scale bar shell = 0.5 mm. Scale bar protoconch = 100 μm.

Comparisons. *Anatoma aupouria* from New Zealand has axials on the base with a distinct posteriorly oriented indentation, a complex pattern of internesting axial cords, fine axial lines on the keels of the selenizone, and a funiculus that curves away from the apertural margin.

Anatoma parageia from temperate Japan has a protoconch with coarse flocculent sculpture, a slightly longer teleoconch I (0.4 vs. <0.3 whorl), and finer and at least 50% denser axial cords on shoulder and base.

Figure 844. *Anatoma pseudoequatoria*. **A–B.** Paratypes of *Scissurella pseudoequatoria*. Hilo, Hawaii [USA, 19.717°N, 155.083°W] (NHMUK 1982253). **C.** Holotype of *Anatoma funiculata*. 21 m, Great Barrier Reef, Euston Reef, SW side, Queensland, Australia, 16.667°S, 146.217°E (AMS C.403660). Scale bars shell = 0.5 mm. Scale bars protoconch = 100 μm.

Anatoma amydra from the Indo-Malayan Archipelago to the Central Pacific has a longer teleoconch I (0.6 vs. <0.3 whorl), fewer spiral lines on the shoulder, and less distinct sculpture on shoulder and base.

Anatoma japonica from the Indian Ocean to the Central Pacific has a longer teleoconch I (>0.5 vs. <0.3 whorl), a rounder aperture, and a less projecting apex.

Figure 845. *Anatoma pseudoequatoria*. **A.** LaoLao Bay, Saipan, Northern Mariana Islands, 15.150°N, 145.750°E (DLG 1014, 22). **B.** 15–25 m, NE Tutuba Island, SANTO 2006 DB01, Vanuatu, 15.551°S, 167.297°E (MNHN, 3). **C.** 720 m, Moorea, Society Islands, TARASOC DW3474, French Polynesia, 17.466°S, 149.833°W (MNHN, 1). Scale bars shell A–B = 0.5 mm. Scale bar shell C = 1 mm. Scale bars protoconch = 100 μm.

Anatoma equatoria from the Indo-Malayan Archipelago to the Central Pacific has a protoconch with a weak apertural varix, a longer teleoconch I (0.5 vs. <0.3 whorl) with a spiral cord in the position of the selenizone, the sculpture on shoulder and base that is uninterrupted by bare areas, and no funiculus.

Figure 846. *Anatoma pseudoequatoria*. **A.** 55–57 m, Grand Reef of Koumac, Koumac sector, New Caledonia, 20.671°S, 164.203°E (MNHN, 25). **B.** 35 m, Lord Howe Shelf, Australia, 31.473°S, 159.129°E (AMS C.461450, 10). **C.** 5–7 m, NE Herald Cay, Coral Sea, 16.933°S, 149.183°E (AMS C.436788, 1). Scale bars shell = 1 mm. Scale bars protoconch = 100 µm.

Anatoma quadraxialis from the Indo-Malayan Archipelago has a protoconch with reticulate sculpture, a longer teleoconch I (>0.5 vs. <0.3 whorl), a wider umbilicus, and more numerous and stronger axials.

Anatoma philippinica from the Indo-Malayan Archipelago to the Western Pacific has forward folded axial lamellae, a distinct constriction of the base below the selenizone, and a spiral keel on the mid-base.

Anatoma rainesi from the Eastern to the Western Pacific has an overall more discoidal shell shape, a rounder aperture, and a longer teleoconch I (0.6 vs. <0.3 whorl).

Anatoma australis from eastern Australia is more globular in overall shape, has a teleoconch I of >0.5 whorls and with spiral sculpture, and lacks a funiculus.

Anatoma tobeyoides from southeastern Australia is more globular in outline, has the irregular centripetal markings on teleoconch I, less pronounced keels on the selenizone, and lacks a funiculus.

Anatoma munieri from the Indian Ocean to the Central Pacific is much more high spired, has a teleoconch I of >0.5 whorl with spiral sculpture, and lacks a funiculus.

Anatoma atlantica from the NW Atlantic has a longer teleoconch I (0.75 vs. 0.2 whorl), no funiculus, and a rounder aperture.

Distribution. Indo-Malayan Archipelago to Central Pacific (Fig. 848).

Specimen records.

Japan, Okinawa. 28 m, Maeda Point, 26.450°N, 127.759°E (DLG 1073, 8). 38 m, base of cliff, Horeshoe Bay, 26.504°N, 127.858°E (DLG 1054, 1). 35–38 m, base of cliff, Horseshoe, 26.504°N, 127.858°E (DLG 979, 1). 2–3 m, Horseshoe Beach, 26.508°N, 127.858°E (DLG 1578, 1). 13–33 m, Flab, Horseshoe Bay, 26.509°N, 127.856°E (DLG 1492, 1).

Northern Mariana Islands. 5–20 m, LaoLao Bay, Saipan, 15.137°N, 145.759°E (DLG 1077, 52). Saipan. LaoLao Bay, 15.150°N, 145.750°E (DLG 1014, 22). Cliffing outside

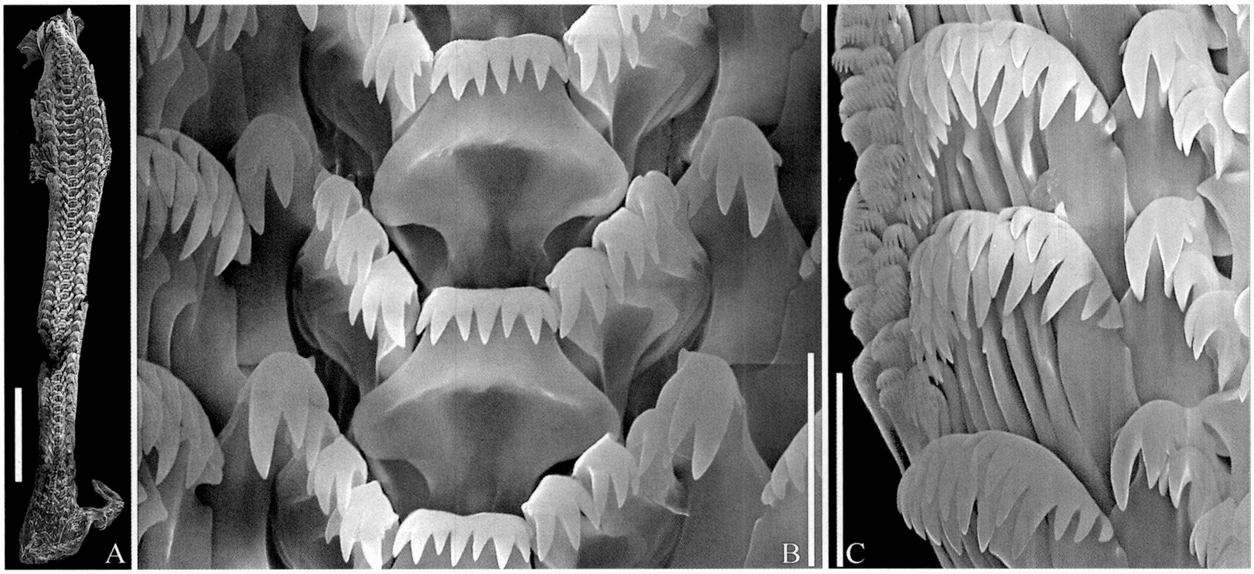

Figure 847. *Anatoma pseudoequatoria*. Radula. 12 m, Grand Récif de Koumac, Koumac sector, New Caledonia, 20.666°S, 164.187°E (MNHN, 1). **A.** Entire radula. Scale bar = 100 μm. **B.** Central field enlarged. Scale bar = 10 μm. **C.** Lateral tooth 5 and marginal teeth. Scale bar = 10 μm.

caver, Marpi, 15.286°N, 145.816°E (DLG 1017, 1). Surge channel, LaoLao Bay, 15.137°N, 145.137°E (DLG 1025, 1).

Marshall Islands. 24 m, Jy-Boh, towards Majuro, Arno Atoll, 7.100°N, 171.600°E (DLG 891, 7).

Philippines. 0 m, Cavile Reef, out of Puerto Princessa, Palawan, 9.717°N, 118.783°E (LACM 82-62, 1).

Australia. 0 m, Elizabeth Reef, 29.923°S, 159.045°E (AMS C.436909, 1).

Australia, New South Wales. 44 m, Lord Howe Rise, Lord Howe Island, 31.637°S, 159.060°E (AMS C.404151, 1).117 m, Lord Howe Shelf, 31.728°S, 159.074°E (AMS C.461445, 10). 35 m, Lord Howe Shelf, 31.473°S, 159.129°E (AMS C.461450, 10). 37 m, Balls Pyramid, Lord Howe Islands, 31.776°S, 159.223°E (AMS C.461429, 1). 45 m, Lord Howe Shelf, 31.574°S, 159.201°E (AMS C.461428, 1).

Australia. Queensland. 21 m, Great Barrier Reef, Euston Reef, SW side, 16.667°S, 146.217°E (AMS C.402640, 1; AMS C.403660, 20+). 27–30 m, Great Barrier Reef, Carter Reef, N end, 14.550°S, 145.600°E (AMS C.402656, 1). 9–15 m, Torres Strait, off Murray Island, 9.933°S, 144.067°E (AMS C.402705, 1). 19 m, No Name Reef, Great Barrier Reef, 14.667°S, 145.650°E (AMS C.461453, 1).

Coral Sea. 5–7 m, NE Herald Cay, 16.933°S, 149.183°E (AMS C.436788, 1). 27 m, Osprey Reef, Rapid Horn, 14.000°S, 146.633°E (AMS C.436787, 1).

New Zealand. 219–274 m, SE of D'Arcy Point, Raoul Island, Kermadec Islands, 29.313°S, 177.903°W (NMNZ M.226813, 7). 15 m, W side of Meyer Island, Raoul Island, Kermadec Islands, 29.242°S, 177.880°W (NMNZ M.153985, 10).

New Caledonia. 237–298 m, E coast, 21.283°S, 165.950°E (MNHN, 2). 350–400 m, E coast, 21.033°S, 165.633°E (MNHN, 1). 170 m, E coast, 20.800°S, 165.317°E (MNHN, 1). 12 m, Grand Reef of Koumac, Koumac sector, 20.667°S, 164.187°E (MNHN, 1). 20–30 m, Grand Reef of Koumac, Koumac sector, 20.690°N, 164.247°E (MNHN, 1). 55–57 m, Grand Reef of Koumac, Koumac sector, 20.672°S, 164.203°E (MNHN, 25).

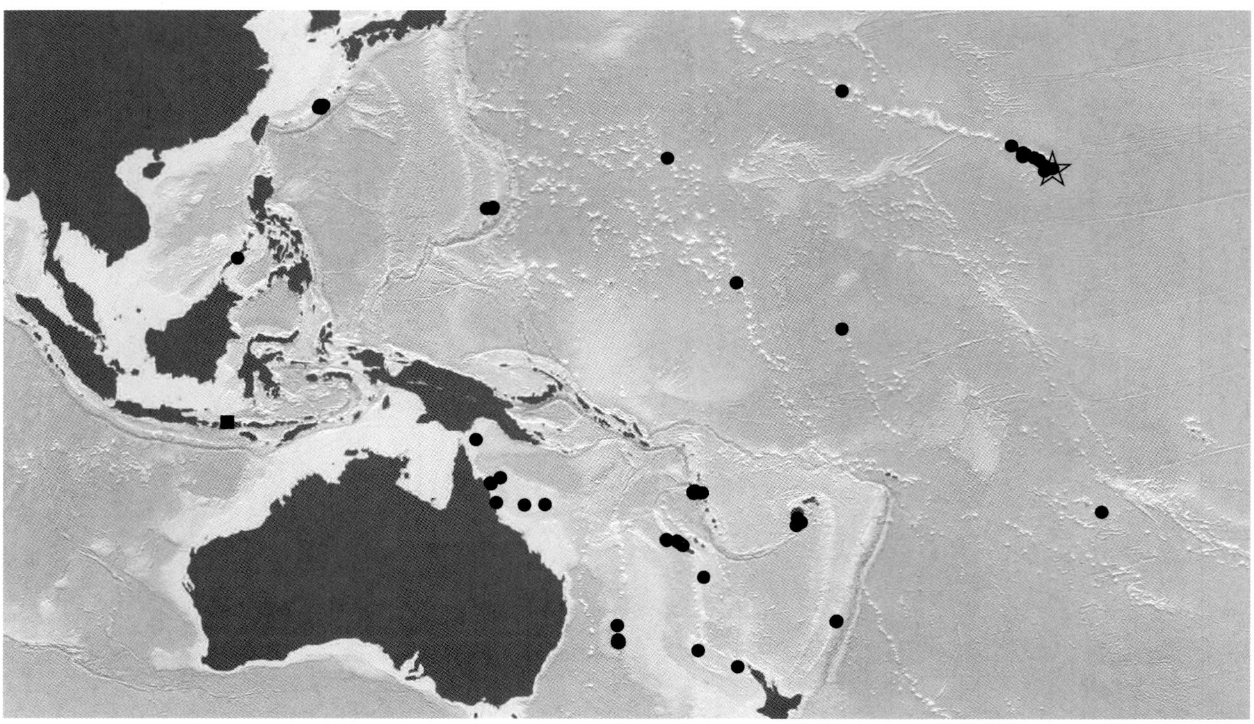

Figure 848. Distribution of *Anatoma pseudoequatoria*. ☆ type locality. ● specimen record. ■ literature record.

105–110 m, Poindimie sector, 20.817°S, 165.317°E (MNHN, 1; MNHN 3). 15–20 m, Deverd Passage, Koumac sector, 20.745°S, 164.258°E (MNHN, 3). 230 m, southern, 24.700°S, 168.167°E (MNHN, 1).

Vanuatu. 5–7 m, W coast of Malo Island, SANTO 2006 ZB09, 15.677°S, 167.085°E (MNHN, 1). 15 m, W Malo Island, Palikulo Bay, SANTO 2006 DB77DB29, 15.648°S, 167.085°E (MNHN, 1). 18–30 m, NE Urélapa Island, Palikulo Bay, SANTO 2006 DB77ZB13, 15.612°S, 167.033°E (MNHN, 1). 16 m, Elia Island, SANTO 2006 FB72, 15.602°S, 167.975°E (MNHN, 1). 12 m, E Aoré Island, SANTO 2006 DB08, 15.577°S, 167.230°E (MNHN, 1). 75–80 m, W Tutuba Island, SANTO 2006 EP39, 15.561°S, 167.273°E (MNHN, 2). 15–25 m, NE Tutuba Island, SANTO 2006 DB01, 15.552°S, 167.297°E (MNHN, 9; MNHN, 3). 92 m, NW Tutuba Island, SANTO 2006 DS105, 15.550°S, 167.278°E (MNHN, 1). 8–20 m, Tutuba Island, SANTO 2006 FB83, 15.543°S, 167.290°E (MNHN, 11). 22–25 m, NW Urélapa Island, SANTO 2006 DB20, 15.508°S, 167.023°E (MNHN, 25). 19 m, Palikulo Bay, Aoré Island, SANTO 2006 FB43, 15.473°S, 167.248°E (MNHN, 2). 21 m, SE Aésé Island, SANTO 2006 DB63, 15.448°S, 167.263°E (MNHN, 1). 6–43 m, Aésé Island, SANTO 2006 DB58, 15.410°S, 167.238°E (MNHN, 4). 7 m, W Mavéa Island, SANTO 2006 DB67, 15.382°S, 167.218°E (MNHN, 2).

Fiji. Astrolabe Reef, Kadavu Group, 18.767°S, 178.453°E (LACM 91-189, 4). 22 m, Kadavu Group, Astrolabe Reef, 18.763°S, 178.472°E (LACM 91-188, 1). 20 m, Herald Pass, W of Ndravuni Island, NW end of Great Astrolabe Reef, 18.767°S, 178.467°E (LACM 85-135, 9). 10–20 m, off Francis Island in Beqa Lagoon, just off of Beqa Island (S of Viti Levu), 18.300°S, 178.067°E (DLG 280, 10). 18 m, Kandavu, Cape Washington, 19.108°S, 177.967°E (LACM 85-138, 1).

French Polynesia. 440–490 m, Raiatea, Society Islands, TARASOC DW3451, 16.883°S, 151.350°E (MNHN, 1). 720 m, Moorea, Society Islands, TARASOC DW3474, 17.466°S, 149.833°W (MNHN, 1).

USA. Hawaii. 5–8 m, Makua Beach, W Oahu, 21.032°N, 158.237°W, (LACM 86-426, 2). 1–2 m, off Picket Pt, NW side Sand Island, Midway Island, 28.211°N, 177.390°W, (LACM 85-120, 2). 0–26 m, N end Puako Bay, NW Hawaii, 19.978°N, 155.835°W, (LACM 86-424, 1). 8 m, reef adjacent to The Hotel Reef, N Reef of Midway Island, 2.272°N, 177.363°W, (LACM 85-121, 1). 5–21 m, Hunaunau Bay, 19.433°N, 155.930°W, (LACM 86-421, 2). 20 m, NW end Lanai Island, 20.917°N, 157.050°W, (LACM 74-67, 2). 60 m, Kepuhi Point, Oahu (FLMNH 350480, 3). 10 m, Maile, Oahu (FLMNH 350477, 2). 0 m, Kalihiwai, Kauai (FLMNH 350479, 1). 66 m, Molokini, Maui, 20.634°N, 156.497°W (CPC, 17).

Literature records.

Indonesia. Near Satonda, 8.100°S, 117.750°E (holotypes of *An. pulchella, An. paucispiralia*) (Bandel, 1998).

Remarks. *Hainella pulchella* is here synonymized under *An. pseudoequatoria*. The shared characters include the protoconch with flocculent sculpture and no apertural varix, the very short teleoconch I without spiral in the position of the selenizone, the gap in axial ornamentation on the shoulder, extension of the axial cords onto the keels of the selenizone, and the gap in the ornamentation on the base below the selenizone.

Bandel (1998: 42) noted that the species was illustrated in "Bandel (1991)" from material from Cebu, Philippines, without reference to a specific figure. Although three Bandel (1991) references were cited in Bandel (1998), the "Bandel (1991)" in Bandel (1998) must have referred to Bandel (1991) of this contribution. The only anatomid illustrated on plate 2, figure 7, clearly is an *An. munieri*, which is very distinct from *An. pulchella* illustrated in Bandel (1998). The statement in Bandel (1998) seems to have been a *lapsus calami.*

Hainella paucispiralia is here synonymized under *An. pseudoequatoria*. The shared characters include a short teleoconch, overall shell shape, and teleoconch ornamentation that falls within the known spectrum. In fact, *H. pulchella* and *H. paucispiralia* share the same type locality, and show the exact same variants of teleoconch sculpture as the paratypes of *An. pseudoequatoria* (Figs 841, 842 vs. Fig. 844A–B).

Anatoma funiculata is here synonymized under *An. pseudoequatoria*. The shared characters include a protoconch with fine flocculent sculpture, a short teleoconch I, and a teleoconch sculpture that falls within the known range of the species. The figure of the holotype (Fig. 844C) appears taller because the apertural view was taken at an angle.

The high variability of teleoconch sculpture is supported by several lines of evidence:

- Geographically widely separated populations show the same divergent morphologies: dense fine axial lines from Hawaii (Fig. 844A) and Indonesia (Fig. 842); strong axial cords near suture followed by smooth band and more numerous axial lines on keels of selenizone from Hawaii (Figs 843, 844B) and Indonesia (Fig. 841).
- Sculpture varies ontogenetically. On many specimens, the early sculpture is more widely spaced, becoming denser with growth. This pattern is additionally variable in the degree of differentiation: strong difference (Fig. 843); moderate difference (Fig. 845A–B), little difference (Fig. 846A–B).
- Protoconch and teleoconch I are identical in all specimens: fine flocculent sculpture, no apertural varix, apertural margin straight, length of teleoconch I, strength and density of axial cords on teleoconch I, absence of spiral cords in position of selenizone on teleoconch I (Figs 843–846).

The alternative interpretation of 2–3 sympatric species, distinguished by partially overlapping teleoconch ornamentation and same early whorl morphology, possibly invoking hybrids to account for intermediates, is more circuitous. Other cases in which some species in a group are more variable than others have been well documented in Vetigastropoda. The Mediterranean *Haliotis tuberculata* Linnaeus, 1758, has a lamellate and a smooth form, including every shade of intermediates, and ontogenetic change in sculpture (Geiger & Owen, 2012).

The most similar species is *An. parageia* from the southern mainland of Japan. *Anatoma pseudoequatoria* has extraordinarily variable teleoconch sculpture, but all 60 specimens of *An. pseudoequatoria* and all six specimens of *An. parageia* examined by SEM, showed consistent but subtle protoconch sculpture differences: *An. pseudoequatoria* has very fine flocculent sculpture, whereas *An. parageia* has coarse flocculent sculpture. The more temperate occurrence of *An. parageia* supports it being distinct from *An. pseudoequatoria*. Critical determination of distributional boundaries between Okinawa and Honshu will require examination of shells by SEM.

Anatoma quadraxialis new species

Figures 849–850

Misidentification.

not *Thieleella equatoria*: Geiger & Jansen, 2004a: 26–27, figs 14, 18 [map] [is *An. quadraxialis*].

Type material. Holotype (AMS C.405015. Fig. 849B). Paratype 555–565 m, Norfolk Ridge, NORFOLK 2 DW2057, New Caledonia, 24.666°S, 168.650°E (MNHN IM-2010-22883, 1. Fig. 849A).

Type locality. 183 m, off NE point, Christmas Island, Australia, 10.417°S, 105.700°E.

Etymology. Quadr(a)-, referring to the square overall shell shape; -axialis, referring to the prominent and dense axial cords over the entire shell. Adjective.

Description. Shell globular, medium-sized (to 1.83 mm: holotype. Fig. 849). Protoconch 0.8 whorl, reticulate sculpture, apertural varix not connected to embryonic cap, shape of apertural margin sinusoidal. Teleoconch I of 0.6 whorl, 15 axial cords, no spirals. Teleoconch II with up to 2.25 whorls, suture moderately impressed. Shoulder convex, 75 axial cords on last whorl, very fine indistinct spiral lines. Base with faint constriction below selenizone, continuous with umbilical cavity, same strength of axials and spirals as on shoulder. Umbilicus moderately wide, deep. Aperture rounded with flared adumbilical portion of peristome. Selenizone slightly above periphery; keels quite elevated moderately strong, with about twice as many fine axial lines on external edges as on shoulder/base; lunules faint; slit open, margins parallel.

Animal unknown.

Comparisons. *Anatoma munieri* from the Indian Ocean to the Central Pacific has flocculent protoconch sculpture, a more turreted shell, a smaller aperture as compared to overall shell size, fewer axials, and more prominent spirals, and a distinct constriction on the base below the selenizone.

Anatoma porcellana from the Indo-Malayan Archipelago to the Western Pacific is more globular in overall shape, has the selenizone and slit slightly above the periphery of the shell, much fewer axials (30 vs. 75 on body whorl), and a spiral cord on teleoconch I.

Anatoma amoena from Antarctica has fewer axials (22 versus 75 on body whorl) and a spiral cord in position of the selenizone on teleoconch I, which forms a strong angulation in the whorl's profile.

Anatoma gunteri from eastern Australia has noticeably stronger sculpture on the shoulder compared to the base and more numerous and fainter spirals.

Anatoma philippinica from the Indo-Malayan Archipelago to the Western Pacific has an almost smooth protoconch, forward slanted axial lamellae, pseudospirals on the shoulder, and a constriction on the base below the selenizone.

Anatoma rhynchodentata from the Indo-Malayan Archipelago has a constriction on the base below the selenizone, a protoconch with flocculent sculpture, and a less convex shoulder.

Anatoma sinuosa from Easter Island has the selenizone higher on the whorl, a more impressed suture, and a smooth protoconch.

Figure 849. *Anatoma quadraxialis* new species. **A.** Paratype 555–565 m, Norfolk Ridge, NORFOLK 2 DW2057, New Caledonia, 24.666°S, 168.650°E (MNHN IM-2010-22883, 1). **B.** Holotype. 183 m, off NE point, Christmas Island, Australia, 10.417°S, 105.700°E (AMS C.405015). Scale bars shell = 1 mm. Scale bars protoconch = 100 µm.

Anatoma africanae from the western Indian Ocean has the selenizone higher on the whorl and about twice as many spiral lines on shoulder and base.

Anatoma aspera from the Mediterranean and adjacent Atlantic has irregular reticulate protoconch sculpture and a spiral cord in the position of the selenizone on teleoconch I.

Anatoma boucheti from the western Indian Ocean has a longer teleoconch I (1 vs. 0.75 whorl), flocculent protoconch sculpture, and the selenizone higher on the whorl.

Anatoma breveprima from the Western Pacific has a much shorter teleoconch I (0.1 vs. >0.6 whorl), a less convex shoulder, and a more narrow selenizone.

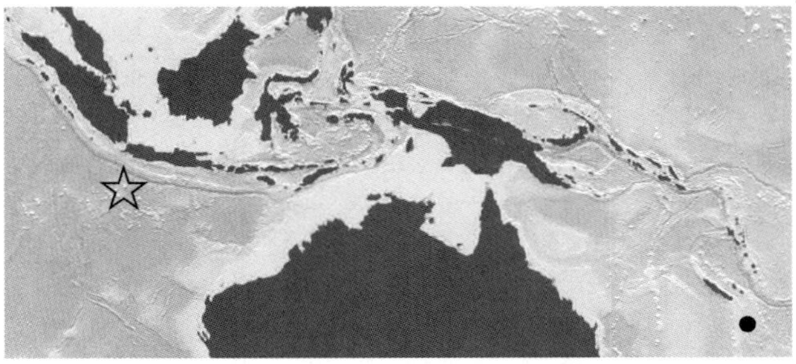

Figure 850. Distribution of *Anatoma quadraxialis* new species. ☆ type locality. ● specimen record.

Anatoma equatoria from the Indo-Malayan Archipelago to the Central Pacific has at most half as many axial cords on shoulder and base, a protoconch with flocculent sculpture, and a less convex shoulder.

Anatoma eximia from the Mediterranean to Angola has a protoconch with flocculent sculpture, spiral lines that are comparatively stronger than the axial cords on shoulder and base, and a more narrow umbilicus.

Anatoma indonesica from the Indo-Malayan Archipelago to the Western Pacific has a protoconch with flocculent sculpture, stronger axial lamellae on the shoulder, and different sculpture on shoulder and base.

Distribution. This species is known only from the type material (Fig. 850).

Anatoma rainesi Geiger, 2003

Figures 851– 855

Chresonymy.

Anatoma sp.: Raines, 2002: 12, fig. 1.

Anatoma rainesi Geiger, 2003: 44–46, 74, figs 11–12.

Anatoma rainesi: Geiger, 2004: text-fig. p. 5.

Anatoma rainesi: Geiger, 2006c: text-fig. 5.

Type material. Holotype (LACM 2909. Fig. 851), 1.5 mm. Paratypes from type locality (LACM 2910, 1; AMS C.205031, 1; ANSP 410306, 1; UMUT RM28236, 1; UMUT RM28237, 1; NHMUK 20020264, 3).

Type locality. 10–20 m, off Rosalia Point, E of Anakena, Easter Island, in sand and rubble at base of cliff, Dec. 2000. Leg. B. Raines [27.067°S, 109.333°W] (OD).

Etymology. The species honors the collector and long time student of the malacofauna of Easter Island, Bret Raines, of Victorville, California [now Alamogordo, New Mexico], USA. He collected and kindly made available the material here described (OD).

Description. Shell medium-sized (to 1.7 mm. Figs 851–853), 1.3× as wide as high. Color off white. Protoconch 0.75 whorl, 130 µm (paratype) to 150 µm (holotype), slightly sunken, smooth from suture to highest point on shoulder, outer shoulder with irregularly jagged axials, apertural varix barely connected to embryonic cap, apertural margin sinusoidal. Teleoconch I of 0.66 whorl, approximately 10 strong primary axial cords, secondary axial cords appearing between primary from adapical suture to strong spiral cord in position of selenizone forming nodes at intersection with primary axials. Teleoconch II of up to 2.125 whorls, suture moderately impressed; sutsel 0–0.1× width of selenizone at onset of selenizone, 0.5–1× width of selenizone at apertural margin of fully grown specimen [see also Remarks]. Teleoconch II whorls with convex, rounded shoulder and base. Sculpture variable, decreasing in strength with growth, switching from predominantly axial to more distinct spiral sculpture; in juveniles not always obvious (Figs 852C, 853A). Shoulder with strong to weak axial cords, decreasing in strength to weak axial cords or growth lines; first spiral threads or lines after 0–0.2 teleoconch II whorls, increasing to 3–20 spiral threads, lines, or steps. Base without constriction below the selenizone; sculpture similar to that on shoulder. Umbilicus deep, narrow, strong funiculus. Aperture round; inner lip flared in lower adumbilical portion. Selenizone slightly above periphery; keels moderately strong, moderately elevated; slit open with parallel margins. Animal with eyes.

Figure 851. *Anatoma rainesi*. Holotype. 10–20 m, off Rosalia Point, E of Anakena, Easter Island, in sand and rubble at base of cliff, Dec. 2000. Leg. B. Raines [27.067°S, 109.333°E] (LACM 2909). Scale bar shell = 1 mm. Scale bar protoconch = 100 μm.

Figure 852. *Anatoma rainesi*. **A.** 100–150 m, La Perouse, Easter Island, Chile, 27.077°S, 109.295°W (DLG 1359, 1). Radula show in Figure 854. **B–C.** 95–105 m, President Thiers Bank, French Polynesia, 24.650°S, 146.027°W (MNHN, 3). **C.** Juvenile; note that suture is immediately below selenizone, whereas in larger specimens suture is well below. Scale bars shell = 1 mm. Scale bars protoconch = 100 µm.

Figure 853. *Anatoma rainesi*. **A.** 60 m, Onna Village, Okinawa, Japan, 26.493°N, 127.842°E (LACM 79-76, 6). **B.** 71–85 m, off Kerama Island, Okinawa, Japan, 26.159°N, 127.448°E (SBMNH, 6). **C.** 720 m, Moorea, Society Islands, TARASOC DW3474, French Polynesia, 17.466°S, 149.833°W (MNHN, 3). Scale bars shell = 1 mm. Scale bars protoconch = 100 µm.

Radula (Fig. 854). Rachidian tooth trapezoidal, central cusp largest, on either side 3–5 cusps arranged in semicircle of variable sizes and arrangements. Lateral teeth 1–3 similar, 3–5 cusps on each tip, number of cusps on particular tooth variable between rows. Lateral tooth 4 reduced, hook-shaped. Lateral tooth 5 moderately enlarged by elongation, apical cusp largest, 3–4 cusps along inner edge of tip, 1–2 cusps along outer edge. Inner marginal teeth with tip showing thickened axis; apical cusp largest, 2–3 cusps along inner edge, 4–5 cusps on outer edge; shafts without food groove. Outer marginal teeth with spoon-shaped tip, many very fine bristles along edge. Outermost marginal teeth paddle-shaped with very fine serration along anterior margin of tip only. Radular interlock of central field moderate.

Comparisons. *Anatoma amydra* from the Indo-Malayan Archipelago to the Central Pacific has a more biconical overall shell shape, finer axial and spiral sculpture, and lacks a funiculus.

Anatoma atlantica from the northwestern Atlantic has a more discoidal overall shell shape, at least twice as many half as strong axial cords on teleoconch I, and at least twice as many axial and spiral cords on teleoconch II.

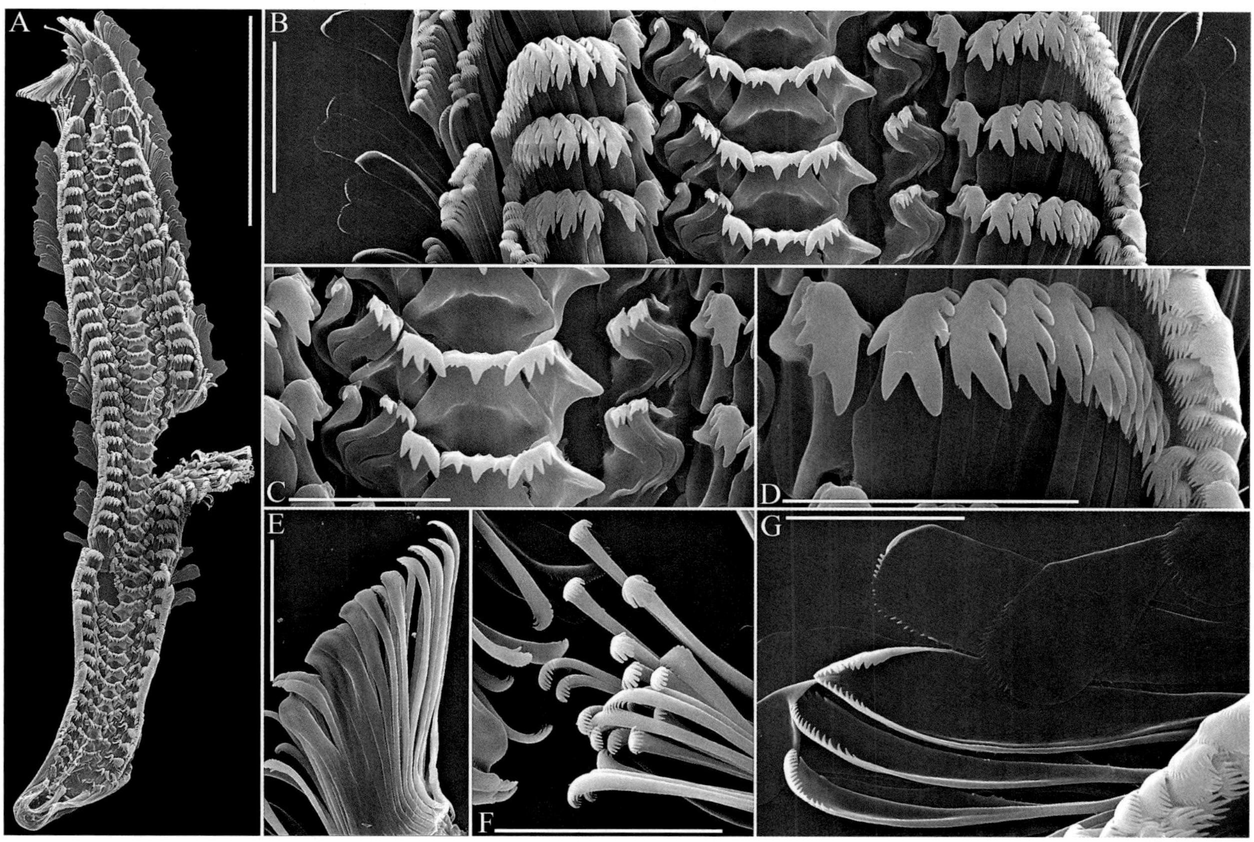

Figure 854. *Anatoma rainesi*. Radula. 100–150 m, La Perouse, Easter Island, Chile, 27.077°S, 109.295°W (DLG 1359, 1). Shell shown in Figure 852A. **A.** Entire radula. Scale bar = 200 µm. **B.** Full width of ribbon. Scale bar = 25 µm. **C.** Central field enlarged. Scale bar = 20 µm. **D.** Lateral tooth 5 and marginal teeth. Scale bar = 20 µm. **E.** Shafts of marginal teeth. Scale bar = 20 µm. **F.** Tips of intermediate marginal teeth. Scale bar = 20 µm. **G.** Outermost marginal teeth. Scale bar = 10 µm.

Anatoma australis from southeastern Australia has only every 3–4 axial cords on the base reaching the umbilicus, more than one spiral cord on teleoconch I, and no sunken protoconch.

Anatoma biconica from the Indian Ocean to the Central Pacific has an overall more biconical shell shape, at most half as many axial cords and spiral lines on shoulder and base, and no funiculus.

Anatoma concinna from the North Pacific has a protoconch with coarse flocculent sculpture and no apertural varix, a more globular overall shell shape, and lower keels of the selenizone.

Anatoma crispata from the northeastern Atlantic has a protoconch with flocculent sculpture and no apertural varix, a more globular overall shell shape, and a much finer spiral cord in position of the selenizone on teleoconch I.

Anatoma tangaroa from northernmost New Zealand has a protoconch with flocculent sculpture and no apertural varix, a shorter teleoconch I (0.3 vs. 0.66 whorl), and no funiculus.

Anatoma globulus from the Indo-Malayan Archipelago has a protoconch with fine flocculent sculpture, more globular overall shell shape, and lacks a funiculus.

Anatoma japonica from the Indian Ocean to the Central Pacific retains the strength of the sculpture on the shoulder throughout its ontogeny, or it becomes even stronger with increasing shell size, has more spiral threads (15 vs. 7–9) on the shoulder, and lacks a funiculus.

Anatoma keenae from the Panamic has an overall more globular shell shape, a more convex shoulder, and spiral sculpture on the shoulder from the selenizone to the suture.

Anatoma lyra from the northeastern Pacific has a protoconch with very fine flocculent sculpture and no apertural varix, at best faint low axial cords on teleoconch I, and does not have a sunken protoconch.

Anatoma orbiculata from West Africa has different sculpture on shoulder and base, a protoconch with irregular reticulate sculpture, and at least twice as many half as strong axial cords on teleoconch I.

Anatoma proxima from the Caribbean to Brazil has a protoconch with flocculent sculpture, lacks a spiral in position of the selenizone on teleoconch I, and lacks a funiculus.

Anatoma rapaensis from the Indo-Malayan Archipelago to the Central Pacific has a more impressed suture, is more discoidal, a shorter teleoconch I (0.3 vs. 0.66 whorl) with finer axial cords, and no spiral cord in position of the selenizone.

Anatoma yaroni from the western Indian Ocean has the protoconch elevated above all subsequent whorls, a body whorl with 10–15 spirals, and weaker lamellae along the selenizone.

Distribution. Indo-Malayan Archipelago to Eastern Pacific (Fig. 855).

Specimen records.

Indonesia. 175 m, SW Salayer, 6.373°S, 120.437°E (RMNH, 1).

Japan, Okinawa. 16 m, Oku, 26.847°N, 128.287°E (LACM 77-64, 2). 92 m, Onna Village, 26.493°N, 127.493°E (LACM 78-101, 4). 50 m, Onna Village, 26.493°N, 127.842°E (LACM 79-75, 2). 60 m, Onna Village, 26.493°N, 127.842°E (LACM 79-76, 6; FLMNH 399220, 2). 71–85 m, off Kerama Island, 26.159°N, 127.448°E (SBMNH, 6).

Fiji. 820–863 m, BORDAU 1 DW1459, 17.300°S, 179.550°W (MNHN, 2).

Tonga. 433 m, NW Tongatapu, BORDAU 2 DW1569, 21.033°S, 175.317°W (MNHN, 1).

French Polynesia. 95–105 m, President Thiers Bank, 24.650°S, 146.027°W (MNHN, 3). 85 m, Raiatea Island, 16.820°S, 151.420°W (DLG 961, 2). 50 m, Sediments, Afaahiti,

Tahiti, 17.760°S, 149.384°W (JTC, 2). 720 m, Moorea, Society Islands, TARASOC DW3474, 17.466°S, 149.833°W (MNHN, 3).

Chile, Easter Island. 30–50 m, Tahai, 27.140°S, 109.429°W (DLG 1327, 1). 70–100 m, Hanga Roa, 27.145°S, 109.440°W (DLG 1336, 4). 150 m, Mataveri, 27.154°S, 109.441°W (DLG 1341, 1). 120 m, Ovahe, 27.076°S, 109.309°W (DLG 1342, 2). 150–200 m, SW corner, 27.201°S, 109.463°W (DLG 1351, 1). 100 m, SW corner, 27.201°S, 109.463°W (DLG 1354, 2). 100–150 m, La Perouse, 27.077°S, 109.295°W (DLG 1359, 1).

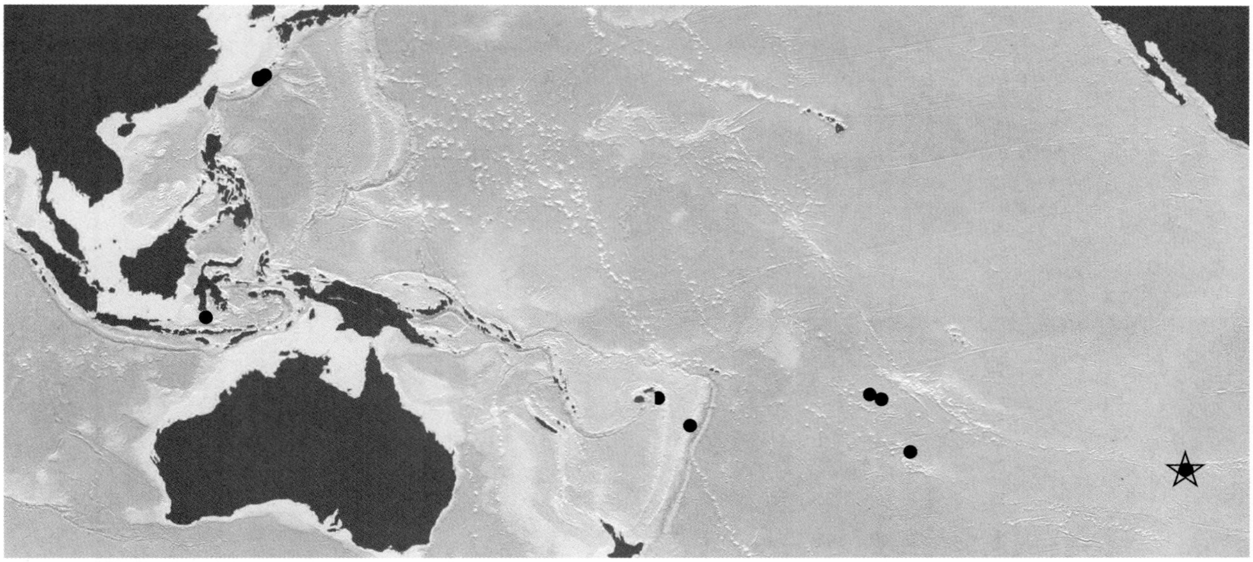

Figure 855. Distribution of *Anatoma rainesi*. ☆ type locality. ● specimen record.

Remarks. *Anatoma rainesi* is another variable species (Figs 851–853). It can be recognized by the sunken protoconch with same sculpture in all 32 specimens examined by SEM, the teleoconch I of moderate length with strong axial cords forming nodular thickenings in the position of the selenizone, usually connected by a spiral cord, and the strong funiculus. Additionally, the ontogenetic switch from predominantly axial sculpture to more prominent spiral sculpture is known only for one other species: *An. janetae* from the northeastern Pacific (Figs 742, 741).

Variable characters include the strength and type of axial and spiral sculpture on shoulder and base of teleoconch II, the width of the sutsel at the onset of the selenizone and, as a corollary, the height to width ratio of the shell, even at same whorl count. There is no apparent pattern of co-variation of those variable characters. Compare the apertural view of Figure 853B to Figure 853C or 852A, and note the different overall shell shape; the other views—umbilical, apical, protoconch—are nearly identical.

Anatoma rapaensis Geiger, 2008

Figures 856–859, Color Plate 4F

Chresonymy.

Anatoma rapaensis Geiger, 2008c: 193–196, figs 17–23.

Anatoma rapaensis: Tröndlé & Boutet, 2009: 12.

Type material. Holotype (MNHN 20823. Fig. 856). Paratypes from type locality (MNHN 20824, 29. Fig. 857A–B). 36 m, around Rukuaga Point, Rapa, Austral Islands, 27.568°S, 144.368°W (MNHN, 30. Fig. 857C). 145 m, Eiao Island, 7.972°S, 140.728°E (MNHN, 1. Fig. 858 [radula, operculum]).

Type locality. 45 m, St. 32, Haurei Bay, Rapa, Austral Islands, French Polynesia, 27.613°S, 144.305°W (OD).

Etymology. The species is named for its type locality on Rapa Island, French Polynesia (OD).

Figure 856. *Anatoma rapaensis* Holotype. 45 m, St. 32, Haurei Bay, Rapa, Austral Islands, French Polynesia, 27.613°S, 144.305°W (MNHN 20823). Scale bar shell = 1 mm. Scale bar protoconch = 100 μm.

Description. Shell medium-sized (to 1.82 mm. Figs 856, 857), trochiform, biconical. Protoconch of 0.75 whorl, sculpture flocculent, no apertural varix, apertural margin slightly sinusoidal. Teleoconch I of 0.3–0.4 whorl, approximately 9–12 axial cords, interstices with fine flocculent sculpture, occasionally (holotype) flocculent sculpture somewhat concentrated in position of selenizone. Teleoconch II of up to 2.25 whorls, suture immediately below selenizone in early growth, sutsel 1×

Figure 857. *Anatoma rapaensis*. Paratypes. **A–B.** 45 m, St. 43, Haurei Bay, Rapa, Austral Islands, French Polynesia, 27.613°S, 144.305°W (MNHN 20824). **C.** 36 m, St. 48, around Rukuaga Point, Rapa, Austral Islands, French Polynesia, 27.568°S, 144.368°W (MNHN, 30). Scale bars shell = 1 mm. Scale bars protoconch = 100 µm.

width of selenizone in fully grown specimens. Shoulder convex, approximately 53–66 (n = 4) axial cords on first teleoconch II whorl, same density on remainder of whorls; first fine spiral cords after 0.125 whorl, approximately 7–10 cords after one teleoconch II whorl, approximately 15–22 at apertural margin of fully grown specimen, cords becoming less distinct and more unevenly spaced towards apertural margin; intersections of axials and spirals forming minute points. Base continuously sloping with narrow umbilicus; axials of same density and strength as on shoulder, approximately 19 fine spiral cords below selenizone turning into low steps from mid-base onwards; intersection of spirals and axials with fine points. Aperture ovoid, D-shaped, roof overhanging, basal adumbilical portion flared. Selenizone at periphery; lunules distinct; keels moderately strong, moderately elevated; slit open, margins converging towards apertural margin.

Radula (Fig. 858A–B). Rachidian tooth trapezoidal, central cusp with 3–4 cusps on each side. Lateral teeth 1–3 similar, development of tip reduced peripherally, terminal cusp largest, 4–2 cusps along outer edge. Lateral tooth 4 reduced, hook-shaped, with one minute point on each side. Lateral tooth 5 enlarged with four strong cusps along inner margin. Inner marginal teeth elongated, terminal cusp largest, 3–4 smaller cusps on inner margin, three larger cusp on outer margin. Outer marginal teeth with spoon-shaped tip with many fine bristles. Radular interlock of central field moderate.

Operculum (Fig. 858C) covering aperture, thin, corneous, nucleus central, multispiral.

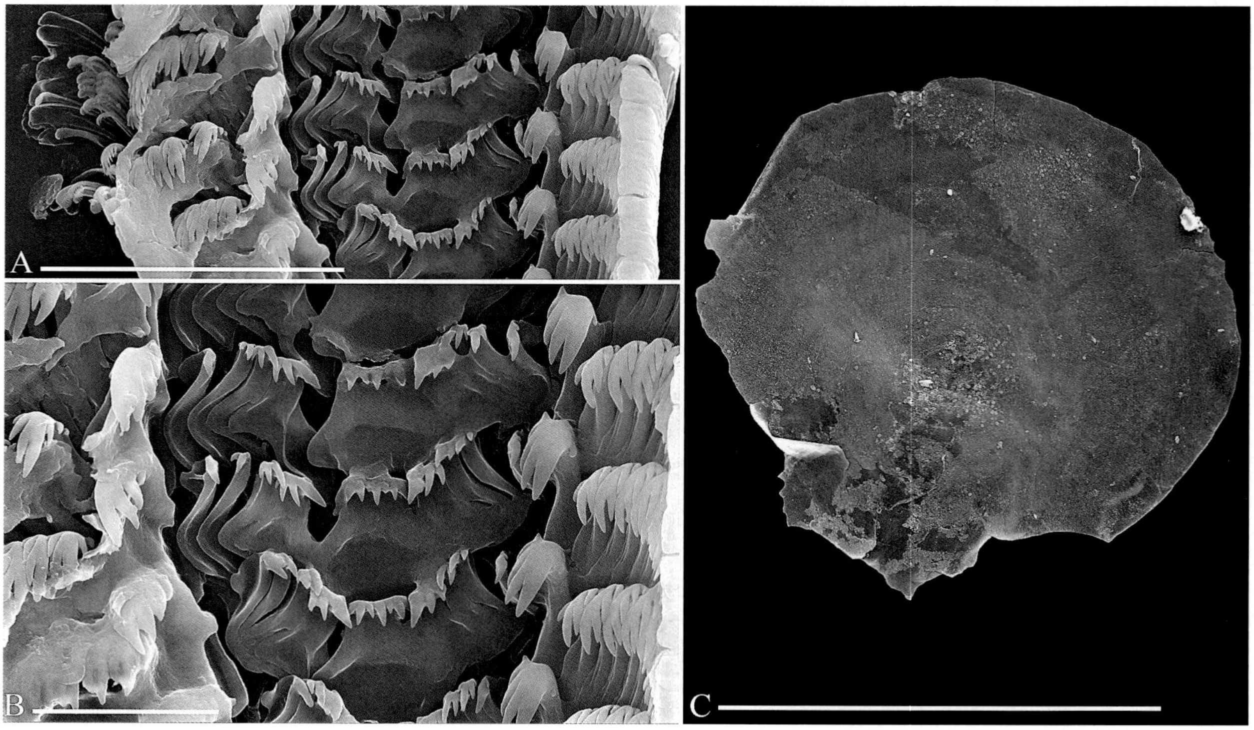

Figure 858. *Anatoma rapaensis*. 145 m, Eiao Island, Marquesa Islands, 7.972°S, 140.728°E (MNHN, 1). **A–B.** Radula. **A.** Full width of radula. Scale bar = 50 µm. **B.** Central field enlarged. Scale bar = 20 µm. **C.** Operculum. Scale bar = 0.5 mm.

Comparisons. *Anatoma rapaensis* is characterized by the rather flattened overall shape, the short teleoconch I of less then 0.5 whorl, and the fine reticular sculpture that is axial-dominated on the first half teleoconch II whorl, subsequently becoming spiral-dominated.

Anatoma argentinae from the abyssal plain of Argentina and the Panamic has a much larger protoconch (370–480 µm vs. 180 µm) with reticulate sculpture, at most half as many axial cords and spiral lines, and a teleoconch I with spiral cord in position of the selenizone.

Anatoma atlantica from the northwestern Atlantic has a less impressed suture, no funiculus, and a longer teleoconch I (0.75 vs. 0.3 whorl).

Anatoma australis from southeastern Australia has more than one spiral line on teleoconch I, only every 3–4 axial cords on the base that reach the umbilicus, and a wider sutsel at any given whorl count.

Anatoma disciformis from the North Pacific has a protoconch with coarse flocculent sculpture and maintains the same strength of axials and spirals over the entire teleoconch II on shoulder and base.

Anatoma tangaroa from northernmost New Zealand has a wider sutsel at any given whorl count, a more globular overall shell shape, and no funiculus.

Anatoma euglypta from Antarctica has a more rounded overall shell shape in small specimens, or a more biconical overall shell shape in large specimens, lacks a funiculus, and has a rounder aperture.

Anatoma japonica from the Indian Ocean to the Central Pacific has slightly taller shell, spirals and axials that are denser forming more prominent points at their intersection, and axials that are slightly stronger than the spirals even in larger specimens (3.5 mm vs. 1.8 mm).

Anatoma maxima from the Indo-Malayan Archipelago to the Western Pacific has a less convex shoulder, lacks a funiculus, and has a wider selenizone.

Anatoma megascutula from the Western to the Central Pacific has an apertural varix on the protoconch, a longer teleoconch I (>0.75 vs. 0.3 whorl), and barely noticeable axial sculpture, which does not form points at the intersection with the spirals.

Anatoma proxima from the Caribbean to Brazil has a more globular overall shell shape, a longer teleoconch I (0.75 vs. 0.3 whorl), and no funiculus.

Anatoma rainesi from the Eastern to the Western Pacific has a protoconch with fine axial sculpture and a distinct apertural varix, axial cords on teleoconch I that bear thickenings or a spiral cord in position of the selenizone, and often less regular sculpture on teleoconch II.

Anatoma regia from temperate New Zealand has a longer teleoconch I (0.6 vs. 0.3 whorl) with a spiral cord in position of the selenizone, a less convex shoulder, and a wider umbilicus.

Anatoma tenuis from the North Atlantic has a less convex shoulder, lacks a funiculus, and grows much larger (>5 vs. 1.8 mm).

Anatoma tenuisculpta from the Mediterranean and adjacent Atlantic has a longer teleoconch I (0.75 vs. 0.3 whorl) with a spiral cord in the position of the selenizone and a wider umbilicus.

Anatoma tobeyoides from southeastern Australia has very fine oblique lines on teleoconch I and early teleoconch II, a wider umbilicus, and a more globular overall shell shape.

Anatoma yaroni from the western Indian Ocean has a less convex shoulder, a longer teleoconch I (0.75 vs. 0.3 whorl) with a spiral cord in the position of the selenizone, and a narrow sutsel even on early whorls.

Distribution. Indo-Malayan Archipelago to Central Pacific; shallow shelf to upper slope (Fig. 859).

Specimens.

Japan, Okinawa. 33 m, Horseshoe Bay, 26.509°N, 127.856°E (DLG 1520, 1).

Philippines. 92–97 m, 12.516°N, 120.650°E (MNHN, 8: complete). 90–98 m, Balicasag Island, PANGLAO 2004 L69-73, 9.511°N, 123.683°E (MNHN, 1). 120–139 m, Balicasag Island, PANGLAO 2004 L74–75, 9.511°N, 123.683°E (MNHN, 1). 85–100 m, Balicasag Island, PANGLAO 2004 L44, 9.513°N, 123.683°E (MNHN, 1). 17–25 m, Pntod Lagoon 1, Panglao Island, PANGLAO 2004 B39, 9.546°N, 123.702°E (MNHN, 4). 90–200 m, Maribohoc Bay, Bohol Island, PANGLAO 2004 P1, 9.600°N, 123.750°E (MNHN, 12). 120 m, off Momo Beach, Panglao Island, PANGLAO 2004 L50, 9.615°N, 123.763°E (MNHN, 1). 43–62 m, Bingag/Tabalong, Panglao Island, PANGLAO 2004 L51-60, 9.628°N, 123.800°E (MNHN, 4). 62 m, Bingag/Tabalong, Panglao Island, PANGLAO 2004 L51–60, 9.628°N, 123.800°E (MNHN, 1; MNHN, 32). 2–4 m, Sungcolan, Panglao Island, PANGLAO 2004 B15, 9.646°N, 123.818°E (MNHN, 1).

Indonesia. 280 m, NE of Binongko, Banda Sea, Tukang Besi Islands, 5.875°S, 123.975°E (RMNH 42265, 1). SW Salayer, 6.371°S, 120.435°E (RMNH 42267, 2). 175 m, SW Salayer, 6.373°S, 120.437°E (RMNH 42266, 1). 100 m, E of Melolo, NE coast of Sumba, 9.883°S, 120.717°E (RMNH, 1).

Papua New Guinea. 70 m, off S Ema Reef, Kimbe Bay, New Britain, 5.202°S, 150.152°E (DLG 557, 1). 65 m, off Susan's Reef, Kimbe Bay, New Britain, 5.291°S, 150.128°E (DLG 604, 8).

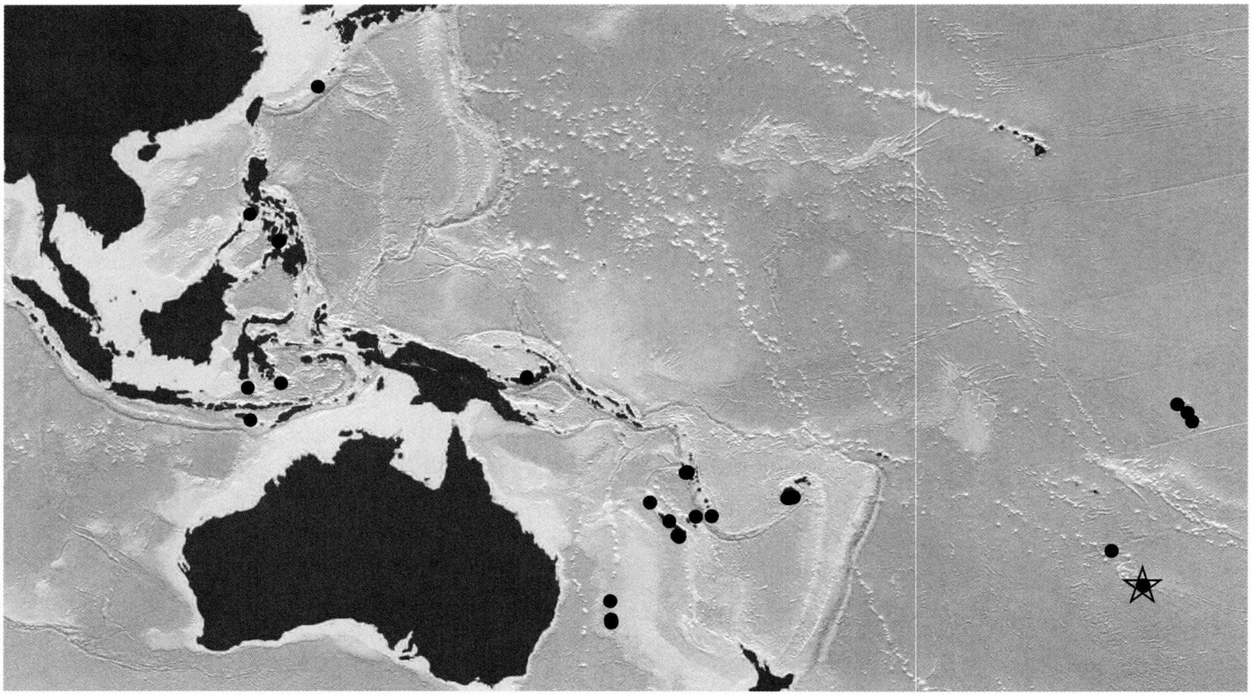

Figure 859. Distribution of *Anatoma rapaensis*. ☆ type locality. ● specimen record.

Australia. 90 m, Middleton Reef, 29.457°S, 159.028°E (AMS C.461449, 1). 96 m, Lord Howe Shelf, 31.444°S, 159.127°E (AMS C.461451, 1). 35 m, Lord Howe Shelf, 31.606°S, 159.166°E (AMS C.461447, 1). 157 m, Balls Pyramid, Lord Howe Island, New South Wales, 31.776°S, 159.161°E (AMS C.461446, 2).

New Caledonia. 600–616 m, northern, 18.816°S, 163.250°E (MNHN, 3). 105–110 m, Poindimie sector, 20.816°S, 165.317°E (MNHN, 9). 495 m, southern, 22.366°S, 166.233°E (MNHN, 3). 250–350 m, southern, 22.500°S, 166.400°E (MNHN, 1; MNHN, 1). 250–350 m, St. 40, 22.500°S, 166.400°E (MNHN, 1; MNHN, 18).

Vanuatu. 45–50 m, Aoré Island, Aimbuei Bay, SANTO 2006 FP46–49, 15.540°S, 167.212°E (MNHN, 1). 100–105 m, NW Tutuba Island, SANTO 2006 DS99, 15.541°S, 167.282°E (MNHN, 9). 92 m, NW Tutuba Island, SANTO 2006 DS105, 15.550°S, 167.278°E (MNHN, 11). 20–60 m, E Aoré Island, Aimbué Bay, 15.553°S, 167.209°E (MNHN, 1). 100 m, NW Tutuba Island, SANTO 2006 DS108, 15.553°S, 167.277°E (MNHN, 1). 125–156 m, W Tutuba Island, SANTO 2006 EP40, 15.555°S, 167.274°E (MNHN, 1). 100 m, W Tutuba Island, Aoré Island, SANTO 2006 DS93, 15.560°S, 167.275°E (MNHN, 8). 114 m, W Tutuba Island, SANTO 2006 DS96, 15.560°S, 167.275°E (MNHN, 1). 70–80 m, W Tutuba Island, SANTO 2006 DS103, 15.568°S, 167.267°E (MNHN, 14). 98–100 m, W Tutuba Island, SANTO 2006 DS102, 15.568°S, 167.267°E (MNHN, 14). 80 m, W Tutuba Island, SANTO 2006 DS104, 15.568°S, 167.267°E (MNHN, 11). 78–91 m, NW coast of Malo Island, SANTO 2006 EP22, 15.622°S, 167.098°E (MNHN, 3). 99 m, NW coast of Malo Island, SANTO 2006 EP21, 15.628°S, 167.087°E (MNHN, 1). 80–94 m, NW coast of Malo Island, SANTO 2006 EP19, 15.629°S, 167.089°E (MNHN, 1). 85–100 m, off Aneityum Island, 20.267°S, 169.850°E (AMS C.404522, 2: complete).

New Zealand. 201 m, off Norfolk Island, 20.333°S, 168.150°E (NMNZ ex M.224760, 1).

Fiji. 149–168 m, S of Viti Levu, 18.206°S, 178.552°E (MNHN, 1). 275–430 m, S of Viti Levu, 18.296°S, 177.907°E (MNHN, 1). 260–305 m, S of Viti Levu, 18.308°S, 178.097°E (MNHN, 6). 441–443 m, S of Viti Levu, 18.320°S, 177.862°E (MNHN, 1).

Marquesas Islands. 145 m, Eiao Island, 7.971°S, 140.728°W (MNHN, 1). 200–220 m, Ua Huka Island, 8.900°S, 139.633°W (MNHN, 6). 352–358 m, Hiva Oa Island, 9.850°S, 139.150°W (MNHN, 1).

French Polynesia, Austral Islands. 100 m, N of Raivavae, 23.828°S, 147.693°W (MNHN, 2). 36 m, around Rukuaga Point, Rapa, 27.568°S, 144.368°W (MNHN, 30). 30 m, NW of Tauna Island, Rapa, 27.605°S, 144.303°W (MNHN, 50). 52–57 m, SE of Tauna Island, Rapa, 27.608°S, 144.295°W (MNHN, 50; MNHN, 20). 33 m, Haurei Bay, Rapa, 27.611°S, 144.318°W (MNHN, 10). 45 m, Haurei Bay, Rapa, 27.613°S, 144.305°W (MNHN 20823, 1; MNHN, 29). 10–15 m, Pake Bay, Rapa, 27.616°S, 144.310°W (MNHN, 2).

Remarks. Only few species of anatomids have a change of sculpture on teleoconch II, namely *An. janetae* from the temperate northeastern Pacific and *An. rainesi* from the Eastern to the Western Pacific.

Anatoma regia (Mestayer, 1916)

Figures 861–865

Chresonymy.

Scissurella regia Mestayer, 1916: 128, pl. 12, figs 3, 3a–c (Fig. 861).
Schizotrochus regius: Fleming, 1948: 83, pl. 8, fig. 1.
Schizotrochus regius: Fleming, 1950: 25.
Schizotrochus regia [*sic*]: Dell, 1956: 168.
Anatoma regia: Powell, 1979: 36, pl. 14, fig. 3.
Anatoma regia: Spencer & Willan, 1995: 13.
Anatoma regia: Geiger, 2003: 74.
Anatoma regia: Spencer et al., 2009: 201.
Anatoma regia: Raven & Bracegirdle, 2010: 10, text-fig.

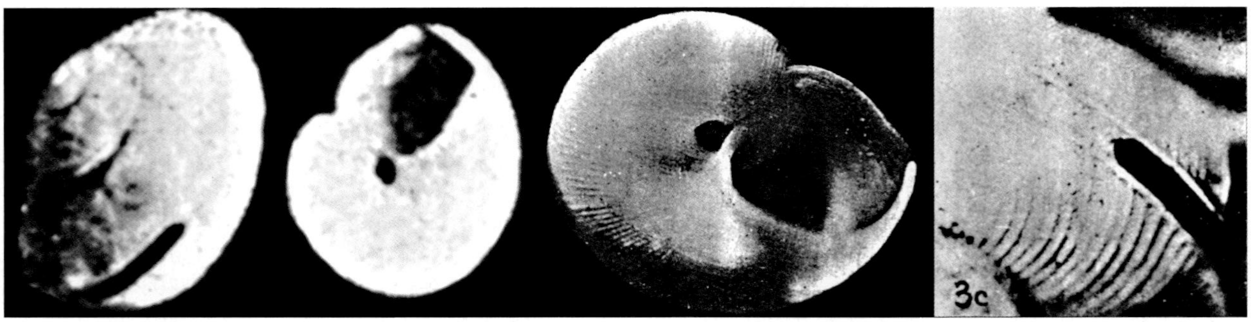

Figure 860. Original illustration of *Scissurella regia*.

Misidentification.

Schizotrochus mantelli: Finlay, 1926: 339 [is *An. regia*].

Type material. Holotype (NMNZ M.001744, ex coll. Mestayer), 2 × 2.5 mm (H × W: Mestayer, 1919: 130). Paratypes from type locality (Dominion Museum, 7; AMS, 1 *fide* Mestayer, 1916. Not found in AMS in 2001; GNS TM 8660, 1: ex coll. Suter; coll. Mestayer, 19; coll. Finlay, ?: Finlay, 1926: 339; NHMUK 1928.11.8.1, 1; USNM 217976, 1).

Type locality. 98 fathoms [= 179 m], Big King dredging, [northern coast of New Zealand: inferred from title] (OD). 15' S of Big King Island [New Zealand, 34.417°S, 172.150°E] (Fleming, 1948).

Etymology. Regius, Latin for King. Named after the latinized type locality.

Description. Shell medium-sized (to 2.6 mm. Figs 862, 863), trochiform, biconical. Protoconch 0.75 whorl, sculpture flocculent, apertural varix not connected to embryonic cap, apertural margin convex. Teleoconch I of 0.6 whorl, approximately 14 low axial cords, single spiral cord in position of selenizone. Teleoconch II of 2.66 whorls, suture impressed, no sutsel in early whorls, sutsel at apertural margin of fully grown specimen 1–1.5× width of selenizone showing approximately six spiral lines. Shoulder slightly convex, approximately 100 fine axial cords on last whorl, approximately 22 fine equispaced spiral lines running over axial cords. Base with axials of same strength and spacing, approximately 40 spirals lines below selenizone,

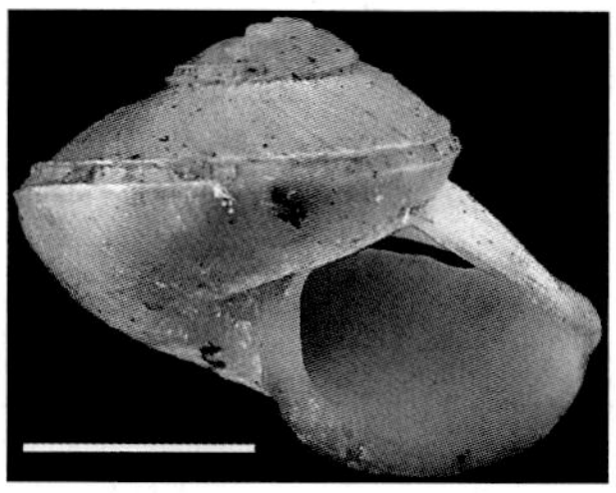

Figure 861. Photograph of holotype of *Scissurella regia*. From NMNZ collections database, photograph B. Marshall. Scale bar = 1 mm.

spiral steps towards umbilicus. Umbilicus narrow, continuously sloping with base, no funiculus. Aperture rounded, basoadumbilical region flared. Selenizone at periphery; lunules moderately distinct; keels of moderate strength, moderate elevations; slit open, margins converging.

Animal with eyes. Operculum thin, round, multispiral, nucleus central, covering aperture.

Radula (Fig. 864). Rachidian tooth triangular, central cusp largest, six cusps on either side arranged parallel to radula row. Lateral teeth 1–3 similar, tips with 5–6 cusps, terminal cusp largest, set off from remainder by noticeable space. Lateral tooth 4 reduced, pointed, central cusp largest, with 1–2 smaller cusps on each side. Lateral tooth 5 enlarged, base broadened, tip elongated, terminal cusp largest, six cusps along inner margin, three cusps along outer margin. Little radular interlock between rachidian tooth and lateral tooth 1, moderate interlock between lateral teeth. Inner marginal teeth elongated, tip with four cusps on inner edge, terminal cusp largest, five cusps on outer margin, shafts with food groove. Outer marginal teeth with spoon-shaped tip with many bristles.

Figure 862. *Anatoma regia*. 540–490 m, Papanui Canyon, off Cape Saunders, E Otago, New Zealand, 45.833°S, 171.017°E (GNS RM 5950, 100). Scale bar shell = 1 mm. Scale bar protoconch = 100 μm.

Comparisons. *Anatoma aupouria* from New Zealand has a strong constriction on the base below the selenizone, a subquadrate profile of the whorls, and a smooth band on the base below the selenizone.

Figure 863. *Anatoma regia*. **A–B.** 427 m, 39 km SW of Great Island, Three Kings Islands, New Zealand, 34.293°S, 171.755°E (NMNZ M.094267, 100). **C.** 340–620 m, S side of head of Papanui Canyon, off Otago Peninsula, New Zealand, 45.861°S, 171.014°E (NMNZ M.174836, 500). Scale bars shell = 1 mm. Scale bars protoconch = 100 μm.

Anatoma tangaroa from northernmost New Zealand has a wider sutsel at any whorl count, a shorter teleoconch I (0.3 vs. 0.6 whorl), and at most half as many spiral cords on the shoulder.

Anatoma finlayi from the Indian Ocean to the Central Pacific (and possibly Caribbean) has strongly raised axial lamellae on shoulder and base and much more convex shoulder and base.

Anatoma flemingi from temperate New Zealand has a protoconch with flocculent sculpture, at most half as many, but at least twice as strong, raised axial cords on the shoulder, stronger reticulate sculpture on the base, and a more convex profile of the shoulder.

Anatoma amydra from the Indo-Malayan Archipelago to the Central Pacific has a wider sutsel at any corresponding whorl count, a shell that is at least as tall as it is wide, and at most half as many spiral lines on the shoulder.

Anatoma argentinae from the abyssal plain of Argentina and the Panamic has a much larger protoconch (370–480 µm vs. 200 µm), at most half the density of axial cords and spiral lines, and a greater expansion rate (1.72 vs. 1.42).

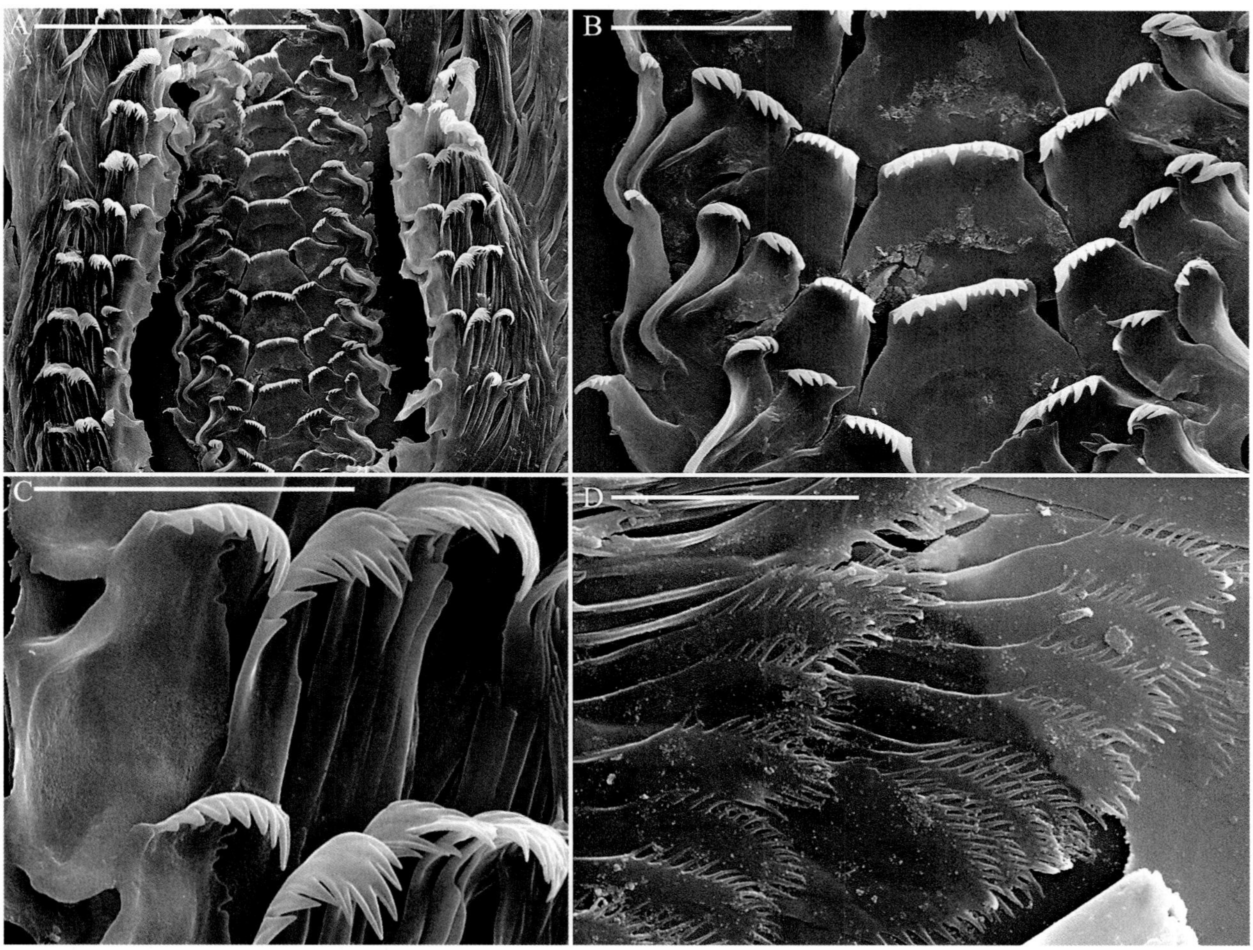

Figure 864. *Anatoma regia*. Radula. 540–490 m, Papanui Canyon, off Cape Saunders, E Otago, New Zealand, 45.833°S, 171.017°E (GNS RM 5950, 100). **A.** Full width of radula. Scale bar = 100 µm. **B.** Central field enlarged. Scale bars = 20 µm. **C.** Lateral tooth 5 and inner marginal teeth showing food groove. Scale bars = 20 µm. **D.** Outer marginal teeth. Scale bar = 10 µm.

Anatoma atlantica from the northwestern Atlantic has a longer teleoconch I (0.75 vs. 0.6 whorl) without spiral cord in position of the selenizone and has stronger axials and spirals, which form small points at their intersections.

Anatoma australis from eastern Australia has only every 3–4 axial cords on the base reaching the umbilicus, more than one spiral on teleoconch I, and a more convex shoulder.

Anatoma disciformis from the North Pacific has a protoconch with coarse flocculent sculpture and no apertural varix, lacks a spiral cord in position of the selenizone on teleoconch I, and has at least twice as many axial cords on teleoconch I with the interstices being more narrow than the cords.

Anatoma euglypta from Antarctica has a protoconch with an apertural varix and stronger and more distinct axial cords on teleoconch I.

Anatoma japonica from the Indian Ocean to the Central Pacific has stronger axial and spiral sculpture forming minute points at their intersections, and usually at least 50% less dense axial and spiral sculpture on the shoulder.

Anatoma maxima from the Indo-Malayan Archipelago to the Central Pacific has a wider selenizone, at most half as many spiral lines on the shoulder, and at least 50% fewer axial cords.

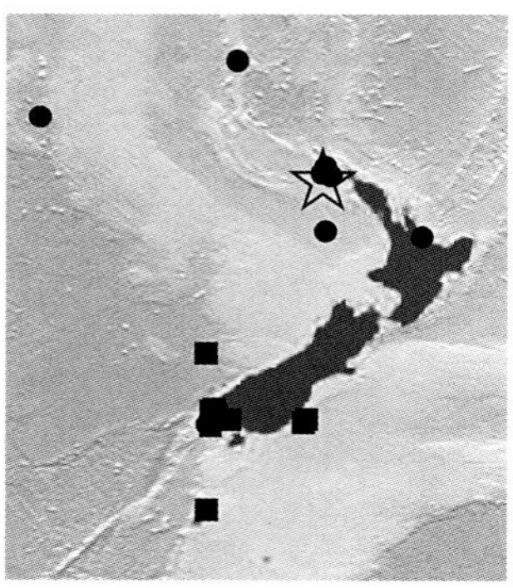

Figure 865. Distribution of *Anatoma regia*. ☆ type locality. ● specimen record. ■ literature record.

Anatoma proxima from the Caribbean to Brazil has a more convex shoulder, stronger axial and spiral sculpture forming minute points at their intersections, and lacks a spiral cord in the position of the selenizone on teleoconch I.

Anatoma rainesi from the Eastern to the Western Pacific has fewer but very strong axial cords on teleoconch I and a sunken protoconch with irregular fine axial lines.

Anatoma rapaensis from the Indo-Malayan Archipelago to the Central Pacific has a more convex shoulder with more impressed suture, a shorter teleoconch I (0.3 vs. 0.6 whorl), and lacks a spiral cord in position of the selenizone on teleoconch I.

Anatoma tenuis from the North Atlantic grows much larger (>5 vs. 2.6 mm), has no sutsel to at least 4.5 mm, and at most half as many spiral lines on the shoulder.

Anatoma tenuisculpta from the Mediterranean and adjacent Atlantic has at least 50% fewer spiral lines on the shoulder, stronger axial cords, and a slightly longer teleoconch I (0.75 vs. 0.6 whorl).

Anatoma tobeyoides from southeastern Australia has very fine oblique threads on teleoconch I and early teleoconch II, more convex shoulder and base, and stronger axial cords.

Anatoma yaroni from the western Indian Ocean has at most half as many spiral lines on the shoulder, a longer teleoconch I (0.75 vs. 0.6 whorl), and at most half as many axial cords.

Distribution. New Zealand, Lord Howe Island, Norfolk Islands (Fig. 865).

Specimen records.

Australia. 44 m, Lord Howe Rise, 31.637°S, 159.060°E (AMS C.378321, 1).

New Zealand. 260 m, off Three Kings Islands, 34.222°S, 172.200°E (NHMUK, 12). 72 m, upper end of the Narrows, Long Sound, Preservation Inlet, Fiordland, South Island, 46.078°S, 166.687°E (GNS RM 5954, 2). 35–44 m, Narrows, Long Sound, Preservation Inlet, Fiordland, South Island, 46.065°S, 166.686°E (GNS RM 5953, 1). 340–620 m, floor and right (south) wall of upper Papanui, 20 km off Taiaroa Head, Dunedin, Otago, South Island, 45.861°S, 171.014°E (DLG 98, 10; DLG 97, 10). 106 m, Edwardson Sound, opposite entrance to Northport Anchorage, Chalky Sound, Fiordland, South Island, 45.996°S, 166.593°E (GNS RM 5955, 1). 340–620 m, S side of head of Papanui Canyon, off Otago Peninsula, 45.861°S, 171.014°E (NMNZ M.174836, 500). 540–490 m, Papanui Canyon, off

Cape Saunders, E Otago, 45.833°S, 171.017°E (GNS RM 5950, 100). 550 m, off E Otago coast, Canyon Bay, 45.757°S, 171.083°E (NHMUK, 2). 58 m, off Johns Island, East Channel, Breaksea Sound, Fiordland, South Island, 45.549°S, 166.865°E (GNS RM 2223, 2). 106 m, inside rock bar, Daggs Sound, Fiordland, SW South Island, 45.320°S, 166.987°E (GNS RM 1022, 20; GNS TM 428, 1). 86 m, Anchorage Arm, Daggs Sound, Fiordland, SW South Island, 45.312°S, 166.979°E (GNS RM 5952, 1). 415 m, Fiordland, Bradshaw Sound, Bradshaw Basin floor, 45.288°S, 167.043°E (NMNZ M.138425, 30). 91 m, Southland, Doubtful Sound, 45.283°S, 166.850°E (AMS C.406392, 1). 176 m, 1.5 mi NNE of Mayor Island, Bay of Plenty, 37.267°S, 176.250°E (USNM 602636, 6). 805 m, Three Kings Trough, NW of Three Kings Islands, 37.000°S, 171.917°E (NMNZ ex M.049027, 2; NMNZ M.020895, 100). 179 m, 15'S of Big King Island, 34.417°S, 172.150°E (NHMUK 1928.11.8.1, 1; GNS TM 8660, 1; AMS C.40524, 1). 427 m, 39 km SW of Great Island, Three Kings Islands, 34.293°S, 171.755°E (NMNZ M.094267, 100). 154 m, Bay of Plenty, off Mayor Island, 37.300°S, 176.267°E (AMS C.404200, 1).

Australia. Norfolk Islands, 29.033°S, 167.950°E (AMS C.404159, 1).

Literature records.

New Zealand. 116 m, Edwardson Sound, opposite entrance to Northport, 49.983°S, 166.567°E. 42 m, Narrows, Long Sound, Preservation Inlet, 46.067°S, 166.750°E. 48 m, Edwardson Sound, opposite Little Island, 45.783°S, 167.633°E. 64 m, E channel between John's Islands, Breaksea Sound, 45.567°S, 166.800°E. 116 m, Daggs Sound, inside a rock bar, 45.383°S, 166.767°E. 94 m, Anchorage Arm, Daggs Sound, 45.383°S, 166.767°E. 76 m, center of basin opposite Groznoz Bay, Doubtful Sound, opposite stream delta, 45.300°S, 166.917°E. 70 m, NE of Parrot Island, Dusky Sound, 42.700°S, 166.533°E (Fleming, 1950).—457–512 m, Alert 54-9, 45.920°S, 170.940°E. 549 m, BS 190, 45.757°S, 171.083°E (Dell, 1956).

Remarks. Powell (1937) considered *An. regia* a synonym of *An. flemingi* [as *An. mantelli*], but later (1979) recognized both species as distinct. Some specimens with a hermit crab are known (NMNZ Sta. NZOI W426; NMNZ M.134309).

Anatoma rhynchodentata new species

Figures 866–871

Type material. Holotype (MNHN 25110. Fig. 866). Paratypes 604 m, St. 5570, off Simaluc Island, Tawi Tawi, Philippines, 5.538°N, 120.216°E (USNM 1099192, 1. Fig. 867A). 396–411 m, N.O. *Alis*, SALOMON 1 DW1762, Solomon Islands, 8.665°S, 160.065°E (MNHN 25111, 1. Fig. 867B). 560 m, Grand Passage, N.O. *Vauban*, MUSORSTOM 4 DW197, New Caledonia, 18.850°S, 163.350°E (MNHN 25112, 1. Fig. 867C).

Type locality. 770–830 m, Loyalty Ridge, Mount K, N.O. *Alis*, BATHUS 3, DW776 New Caledonia, 24.733°S, 170.133°E.

Etymology. Rhyncho-, Greek for projection, nose; -dentata, Latin for toothed, referring to the much enlarged central cusp on the rachidian tooth of the radula.

Description. Shell large (to 4.8 mm. Figs. 866, 867), trochiform, globular. Protoconch of 0.75 whorl, sculpture flocculent, weak apertural varix not connected to embryonic cap, apertural margin shallowly sinusoidal. Teleoconch I of 0.7–0.8 whorls, approximately 27–29 axial cords, spiral cord in position of selenizone forming small thickenings at intersection with axials. Teleoconch II of up to 3.1 whorls, suture little impressed, sutsel <0.5× width of selenizone on early whorl, slightly >1× width of selenizone at apertural margin. Shoulder convex early on, becoming flatter with growth, many fine axial cords extending onto keels of selenizone, interstices less than width of cord at suture, about as wide as cords near selenizone; first strong spiral lines at onset of selenizone, approximately eight after one teleoconch II whorl, approximately 30–35 at apertural margin, spirals running over axials forming small but distinct point, producing an overall rough appearance of the shell. Base with weak constriction below selenizone, continuously sloping with umbilicus; same sculpture as on shoulder. Umbilicus of moderate width, no funiculus. Aperture subquadrate, peristome flared in adumbilical lower corner, roof overhanging.

Figure 866. *Anatoma rhynchodentata* new species. Holotype. 770–830 m, Loyalty Ridge, Mount K, N.O. *Alis*, BATHUS 3, DW776, New Caledonia, 24.733°S, 170.133°E (MNHN 25110). Scale bar shell = 1 mm. Scale bar protoconch = 100 µm.

Selenizone at periphery; keels moderately strong, moderately elevated; lunules dense, weak; slit open, margins converging.

Figure 867. *Anatoma rhynchodentata* new species. Paratypes. **A.** 604 m, St. 5570, off Simaluc Island, Tawi Tawi, Philippines, 5.538°N, 120.216°E (USNM 1099192, 1). **B.** 396–411 m, N.O. *Alis*, SALOMON 1 DW1762, Solomon Islands, 8.665°S, 160.065°E (MNHN 25111, 1). **C.** 560 m, Grand Passage, N.O. *Vauban*, MUSORSTOM 4 DW197, New Caledonia, 18.850°S, 163.350°E (MNHN 25112, 1). Scale bars shell = 1 mm. Scale bars protoconch = 100 μm.

Operculum (Fig. 868F) corneous, round, thin, multispiral, with central nucleus.

Radula (Figs 868–870). Rachidian tooth trapezoidal, central cusp 3–4× as long as largest of 3–4 cusps on each side. Lateral tooth 1 much enlarged by broadening, apical cusp largest, seven cusps on outer margin, one small cusp on inner margin. Lateral tooth 2 smaller version of lateral tooth 1 with six cusps on outer margin. Lateral teeth 3 and 4 similar, with more spoon-shaped tip, 6–5 cusps on each side. Lateral tooth 5 enlarged by broadening, apical cusp largest, eight cusps along inner edge, three cusps along outer edge. Two innermost marginal teeth with pointed, spoon-shaped tip, approximately five cusps on each side. Middle and outer marginal teeth with spoon-shaped tip, many fine bristles. Outermost marginal teeth paddle-shaped with many fine bristles along apical margin. Shafts of marginal teeth without food groove. Radular interlock of central field moderate.

Jaw (Fig. 868E) triangular, composed of many rhomboidal elements.

Comparisons. *Anatoma agulhasensis* from the western Indian Ocean has at most half as many axial cords and spiral lines on shoulder and base, which do not form points at their intersections, and a wider umbilicus.

Figure 868. *Anatoma rhynchodentata* new species. Radula, jaw, and operculum. 444–452 m, northern, New Caledonia, 18.933°S, 163.367°E (MNHN, 3). **A.** Entire radula. Scale bar = 1 mm. **B.** Central field enlarged. Scale bar = 20 µm. **C.** Lateral tooth 5 and marginal teeth. Scale bar = 20 mm. **D.** Outermost marginal teeth. Scale bar = 10 µm. **E.** Jaws. Scale bar = 100 µm. **F.** Operculum. Scale bar = 1 mm.

Anatoma amydra from the Indo-Malayan Archipelago to the Central Pacific has a more biconical overall shell shape, weaker axial cords, at most half as many axial cords and spiral lines on the shoulder, and a wider umbilicus.

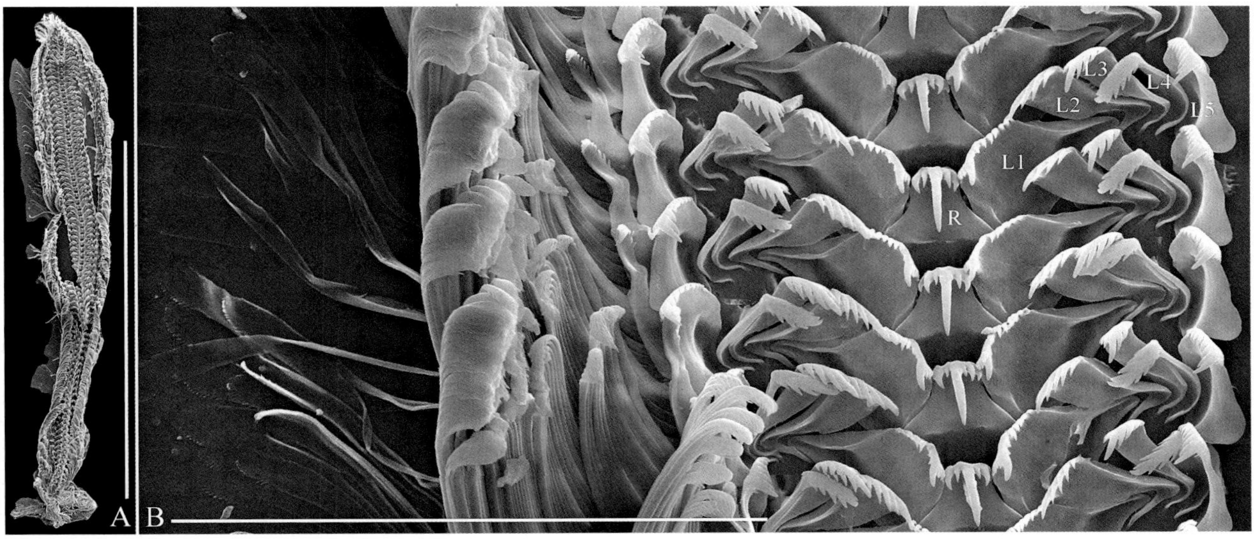

Figure 869. *Anatoma rhynchodentata* new species. Radula. 502–516 m, northern, New Caledonia, 18.867°S, 163.383°E (MNHN, 2). **A.** Entire radula. Scale bar = 1 mm. **B.** Full width of radula with rachidian tooth (R) and lateral teeth 1–5 (L1–5) identified. Scale bar = 100 µm.

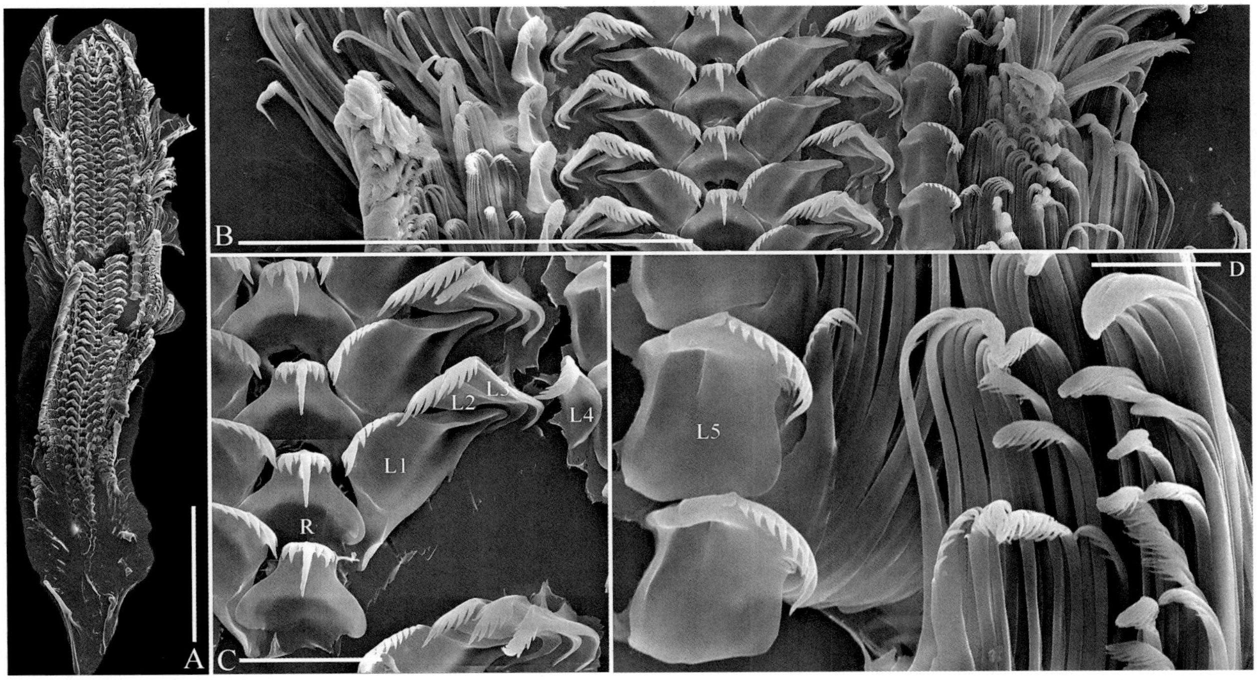

Figure 870. *Anatoma rhynchodentata* new species. Radula. 505–515 m, New Caledonia, 24.917°S, 168.367°E (MNHN, 1). **A.** Entire radula. Scale bar = 200 µm. **B.** Full width of radula. Scale bar = 100 µm. **C.** Rachidian tooth (R) and lateral teeth 1–4 (L1–4). Scale bar = 20 µm. **D.** Lateral tooth 5 (L5) and marginal teeth. Scale bar = 10 µm.

Anatoma boucheti from the western Indian Ocean has a longer teleoconch I (0.9 vs. 0.75 whorl), a sunken protoconch, and the selenizone higher on the whorl.

Anatoma tangaroa from northernmost New Zealand has a more biconical shell shape, a shorter teleoconch I (0.3 vs. 0.75 whorl), and at most half as many axial cords and spiral lines on shoulder and base.

Anatoma equatoria from the Indo-Malayan Archipelago to the Central Pacific has reticulate sculpture that is about twice as coarse, a taller overall shell shape, and a shorter teleoconch I (0.5 vs. 0.75 whorl).

Anatoma globulus from the Indo-Malayan Archipelago has a more globular overall shell shape and a more narrow sutsel at any whorl count.

Anatoma indonesica from the Indo-Malayan Archipelago to the Western Pacific has strong axial lamellae on the shoulder, different sculpture on shoulder and base, and a smooth band on the base below the selenizone.

Anatoma japonica from the Indian Ocean to the Central Pacific has a more lenticular overall shell shape, a narrower sutsel at any whorl count, and a more narrow selenizone.

Anatoma munieri from the Indian Ocean to the Central Pacific has a stronger constriction on the base below the selenizone, one additional spiral keel on the base, and a wider sutsel at any whorl count.

Anatoma philippinica from the Indo-Malayan Archipelago to the Western Pacific has axial lamellae that are slanted forward, pseudospirals on the shoulder, and two distinct spiral keels on the base.

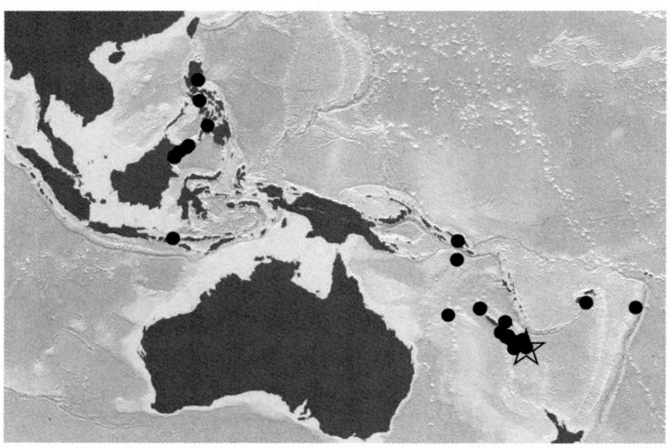

Figure 871. Distribution of *Anatoma rhynchodentata* new species. ☆ type locality. ● specimen record.

Anatoma quadraxialis from the Indo-Malayan Archipelago has a more convex shoulder, lacks the constriction on the base below the selenizone, and has reticulate protoconch sculpture.

Anatoma rainesi from the Eastern to the Western Pacific has a sunken protoconch, fewer but stronger axial cords on teleoconch I, and a distinct funiculus.

Anatoma schanderi from the North Atlantic has reticulate protoconch sculpture, lacks the constriction of the base below the selenizone, and has a more convex shoulder.

Anatoma zancliformis from the Indo-Malayan Archipelago to the Western Pacific has the wider selenizone higher on the whorl, and at most half as many axial cords and spiral lines on shoulder and base, which do not form raised points at their intersections.

Distribution. Indo-Malayan Archipelago and Western Pacific, 250–1100 m (Fig. 871).

Specimen records.

Philippines. Sulu Sea (USNM 1098723, 1). 604 m, St. 5570, off Simaluc Island, Tawi Tawi, 5.538°N, 120.216°E (USNM 1099192, 1). 444 m, St. 5565, off Dammi Island, Sulu Archipelago, 5.862°N, 120.508°E (USNM 1099181, 2). 318–322 m, off Balicasag Island, Bohol Sea, PANGLAO 2005 CP2368, 8.935°N, 123.277°E (MNHN, 1). 169–176 m, AURORA 2007 DW2757, 15.902°N, 121.833°E (MNHN, 4). 538 m, AURORA 2007 CP2750, 15.883°N, 121.893°E (MNHN, 3). 134 m, St. 5178, off Point Origon, Tablas, 12.716°N, 122.104°E (USNM 1099170, 2). 320 m, St. 5579, N off Sibutu Island, Sulu Archipelago, 4.904°N, 119.164°E (USNM 1098724, 3).

Malaysia. Sabah (USNM 1098752, 1).

Indonesia. 175–185 m, Bay of Sanggar, N of Sumbawa, 8.321°S, 118.260°E (RMNH ex 42263, 1). 534 m, Station 5584, Borneo, Sibuko Bay, off Si Amil Island, 4.000°N, 118.433°E (USNM 289748-4, 4). 558 m, St. 5592, off Silungan Island, Borneo, 4.212°N, 118.462°E (USNM ex 279050, 1).
Coral Sea. 370–400 m, Chesterfield, 19.883°S, 158.667°E (MNHN, 1).
Solomon Islands. 557–655 m, 11.485°S, 159.957°E (MNHN, 2). 396–411 m, 8.665°S, 160.065°E (MNHN, 1).
New Caledonia. 560 m, 18.850°S, 163.350°E (MNHN 25112, 1: paratype). 525 m, 18.900°S, 163.317°E (MNHN, 10). 444–452 m, northern, 18.933°S, 163.367°E (MNHN, 3). 425 m, Loyalty Ridge, 21.017°S, 167.517°E (MNHN, 1). 250–350 m, southern, 22.500°S, 166.400°E (MNHN, 1). 755 m, 23.000°S, 167.483°E (MNHN, 2). 675–680 m, 23.167°S, 167.167°E (MNHN, 40). 657–660 m, Loyalty Ridge, 23.583°S, 169.617°E (MNHN, 1). 650–730 m, Norfolk Ridge, 23.650°S, 167.983°E (MNHN, 2). 770–830 m, Loyalty Ridge, 24.733°S, 170.133°E (MNHN, 1). 850–855 m, Loyalty Ridge, 24.767°S, 170.117°E (MNHN, 1). 565–600 m, southern, 24.883°S, 168.350°E (MNHN, 1). 380–400 m, E coast, 20.583°S, 165.117°E (MNHN, 2). 408–440 m, E coast, 20.583°S, 165.117°E (MNHN, 2). 250–300 m, northern, 21.017°S, 164.467°E (MNHN, 1). 237–298 m, E coast, 21.283°S, 165.950°E (MNHN, 3). 505–515 m, 24.917°S, 168.367°E (MNHN, 1). 150–200 m, SE Shelter Reef, Santal Bay, Lifou, Loyalty Islands, LIFOU DW1648, 20.900°S, 167.050°E (MNHN, 3).
Fiji. 1058–1091 m, SE Viti Levu, 18.000°S, 178.895°E (MNHN, 4).
Tonga. 440–487 m, Vava'u group, BORDAU 2 DW1586, 18.567°S, 173.917°W (MNHN, 3).

Remarks. The radula of this species exhibits yet another configuration. It has the rachidian tooth with the much enlarged central cusp. The lateral teeth show very different size and similarity patterns compared to most anatomids; lateral tooth 4 has a large tip. The quick transition in the morphology of the marginal teeth from triangular to spoon-shaped tip is remarkable. This configuration was found in all three radulae examined; hence, it is unlikely to be an aberration, or a gender specific modification. The elongated central cusp on the rachidian tooth is shared with *An. proxima* from the Caribbean to Brazil. The relatively wide lateral tooth 1 is shared with *An. megascutula*.

Smaller specimens are distinctly more lenticular in overall shape.

Anatoma rolani Geiger & Fernández-Garcés, 2010

Figures 872–874

Chresonymy.

Anatoma rolani Geiger & Fernández-Garcés, 2010: 65–67, fig. 1.

Type material. Holotype (LACM 3120. Fig. 872). Paratypes from type locality (LACM 3121, 9; LACM 3122, 1). 40 m, Rancho Luna, Cientfuegos, Cuba, 22.017°N, 80.430°W (RFGC, 7; MHNS, 4). 10–20 m, Puerto Morelos, Yucatan, Mexico, 20.818°N, 86.817°W (MHNS, 20).

Type locality. 35 m, Kittery Beach, Guantanamo Bay, Oriente Province, Cuba, 19.883°N, 75.117°W (OD).

Etymology. The name honors Emilio Rolán for his contributions to the knowledge of the Caribbean malacofauna (OD).

Description. Shell medium-sized (to 1.7 mm: holotype. Figs. 872, 873), trochiform, biconical. Protoconch of 0.75 whorl, smooth, no apertural varix, apertural margin slightly convex. Teleoconch I of 0.6 whorl, 13 axials, initially tall cords, then low lamellae; no spirals. Teleoconch II of two whorls, suture moderately impressed, sutsel >1× width of selenizone. Shoulder with many fine axial lamellae, on first half whorl upright, then with strong forward inclination, as if pressed down; spiral sculpture very diffuse; first signs due to slight thickenings on lamellae after one whorl; approximately 12 teeth in serrated profile of lamellae forming pseudospirals, no serration close to suture. Base without constriction below selenizone, sculpture as on shoulder with approximately 12 teeth in serrated profile of lamellae. Umbilicus very narrow, continuously sloping with base, no funiculus. Aperture rounded, roof overhanging. Selenizone at periphery; keels moderately strong, quite elevated; lunules distinct, dense; slit open, margins converging.

Animal unknown.

Figure 872. *Anatoma rolani*. Holotype. 35 m, Kittery Beach, Guantanamo Bay, Oriente Province, Cuba, 19.883°N, 75.117°W (LACM 3120). Scale bar shell = 1 mm. Scale bar protoconch = 100 µm.

Comparisons. The pressed-down lamellae are a sculptural element unique among Anatomidae; only *An. philippinica* shows forward slanted lamellae.

Anatoma alternatisculpta from the Caribbean has less than half the number of axial cords, but more distinct spiral lines, and a base that shows a distinct reticulate pattern.

Figure 873. *Anatoma rolani*. Paratypes. **A–B.** 35 m, Kittery Beach, Guantanamo Bay, Oriente Province, Cuba, 19.883°N, 75.117°W (LACM 3121, 9). **C.** 40 m, Rancho Luna, Cientfuegos, Cuba, 22.017°N, 80.430°W (MHNS, 4: images by and with kind permission of E. Rolán). Scale bars shell = 1 mm. Scale bars protoconch = 100 μm.

Anatoma atlantica from the northwestern Atlantic has distinct spiral lines on shoulder and base and a more narrow sutsel after two teleoconch II whorls (< vs. > width of selenizone).

Anatoma proxima from the Caribbean to Brazil has upright lamellae on shoulder and base, more distinct spirals, and a protoconch with flocculent sculpture as well as an apertural varix.

Anatoma disciformis from the North Pacific has a teleoconch I with more dense but less distinct axial cords, comparatively stronger spiral lines on shoulder and base, and a protoconch with coarse flocculent sculpture.

Anatoma tangaroa from northernmost New Zealand lacks the forward-slanted axial ribs, has a wider sutsel on early whorls (>0.5× vs. <0.5× width of selenizone), and at most half as many axial cords.

Anatoma japonica from the Indian Ocean to the Central Pacific lacks the forward-slanted axial ribs, has a protoconch with an apertural varix and flocculent sculpture, and spiral lines on the base.

Anatoma maxima from the Indo-Malayan Archipelago to the Central Pacific has a wider selenizone, lacks forward-slanted lamellae, and has spiral lines on the base.

Anatoma philippinica from the Indo-Malayan Archipelago to the Western Pacific has a more turreted overall shell shape, a distinct constriction on the base below the selenizone, and a more convex shoulder.

Anatoma rainesi from the Eastern to the Western Pacific has a sunken protoconch, lacks forward-slanted axial lamellae, and ontogenetically switches from predominantly axial sculpture to more prominent spiral sculpture.

Anatoma rapaensis from the Indo-Malayan Archipelago to the Central Pacific has a more impressed suture, ontogenetically switches from predominantly axial sculpture to more prominent spiral sculpture, and has a more convex shoulder.

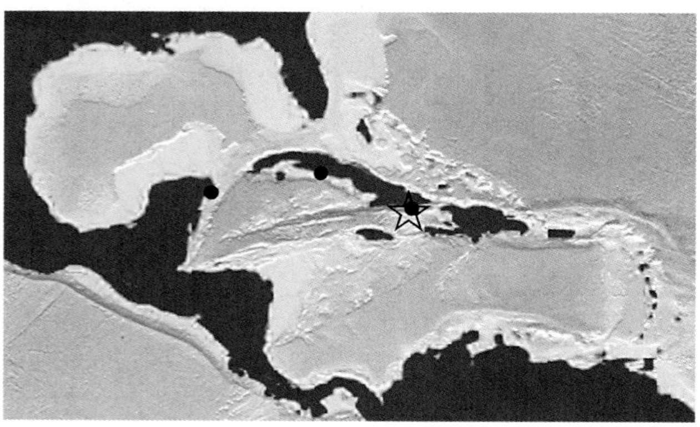

Figure 874. Distribution of *Anatoma rolani*. ☆ type locality. ● specimen record.

Anatoma regia from temperate New Zealand has no forward-slanted axial lamellae, a more narrow sutsel, and a teleoconch I with spiral in position of the selenizone.

Anatoma tenuis from the North Atlantic has no sutsel up to a size of at least 4.4 mm, no forward slanted axial lamellae, and spiral lines on the base.

Anatoma tenuisculpta from the Mediterranean and adjacent Atlantic lacks forward-slanted axial lamellae, has a longer teleoconch I (0.75 vs. 0.6 whorl) with a spiral cord in the position of the selenizone, and has spiral lines on the base.

Anatoma yaroni from the western Indian Ocean has no forward-slanted axial lamellae, spiral lines on the base, and fewer axial cords.

Distribution. Caribbean. Known only from type material (Fig. 874).

Anatoma sagamiana (Okutani, 1964)

Figures 875–880

Chresonymy.

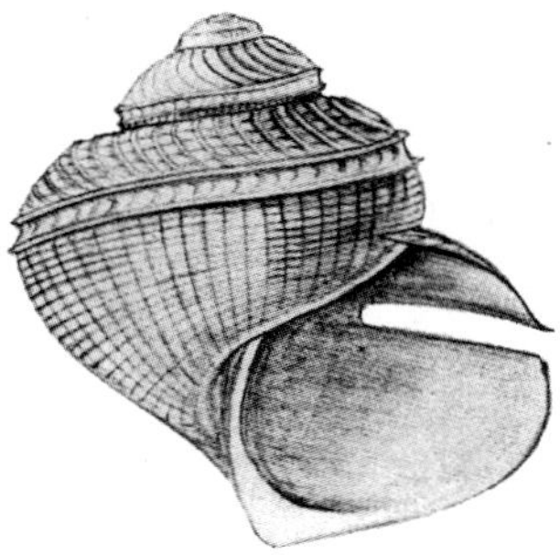

Figure 875. Original illustration of *Scissurella sagamiana*.

Scissurella sagamiana Okutani, 1964 377–379, pl. 5, fig. 2.

Schizotrochus sagamiana [sic]: Okutani, 1966: 14–15.

Anatoma sagamiana: Higo, 1973: 13.

Anatoma sagamiana: Habe *et al.*, 1986: 4.

Scissurella sagamiana: Higo & Goto, 1993: 15.

Scissurella sagamiana: Committee, 1996: pl. 1, fig. 1 [holotype].

—: Nagai & Ikebe, 1997: [*fide* Hasegawa, 2001].

Anatoma sp.: Sasaki, 1998: 67–71, figs 43 [SEM of animal, radula], 44a–b [SEM shell] [is *An. sagamiana*].

Anatoma sagamiana: Hasegawa, 2001: 126.

Anatoma sagamiana: Geiger, 2003: 74.

Anatoma sp.: Geiger *et al.*, 2007: fig. 1A.

Thieleella sp.: Geiger, 2008b: fig. 1.

Anatoma sagamiana: Koyama, 2008: 253, pl. 1, fig. 4.

Thieleella sagamiana: Sasaki *et al.*, 2010: 69-70, figs 6–9.

Synonym.

Thieleella reticulata Bandel, 1998: 36, pl. 12, figs 2–3 [holotype], 4 [protoconch of different specimen]. ***Type material.*** Holotype (SGPIH 3876. Fig. 876), 1.5 × 1.5 mm. ***Type locality.*** Satonda, Indonesia [8.1°S, 117.75°E] (OD). ***Etymology.*** Not given (OD). It may refer to the reticulate protoconch sculpture, or the reticulate sculpture of teleoconch II.

Thieleella reticulata: Geiger, 2003: 74.

Figure 876. Original illustration of *Thieleella reticulata*. Apertural and apical views of holotype; protoconch of other specimen.

Misidentifications.

Anatoma crispata: Okutani & Hasegawa *in* Okutani, 2000: pl. 18, figs 2a–b [is *An. sagamiana*].

Anatoma lamellata: Koyama, 2008: 253, pl. 1, fig. 2 [is *An. sagamiana*].

Anatoma crispata: Koyama, 2008: 253, pl. 1, fig. 3 [is *An. sagamiana*].

Figure 877. *Anatoma sagamiana.* **A.** Holotype of *Scissurella sagamiana.* 710–770 m, Station T29 (15/11/58), 35°5.5'N 139°29.3'E, Sagami Bay [Japan] (UMUT RM 8808). **B.** Paratype of *Scissurella sagamiana.* 710–770 m, Station T29 (15/11/58), 35°5.5'N 139°29.3'E, Sagami Bay [Japan] (NSMT-Mo. 69582). Scale bars shell = 1 mm. Scale bar protoconch = 100 μm.

Type material. Holotype (UMUT RM 8808. Committee, 1996: pl. 1, fig. 1. Fig. 877A), 3.75 × 3.5 mm. Paratype from type locality (NSMT-Mo. 69582. Fig. 877B), 3.9 × 4.2 mm.

Type locality. 710–770 m, Station T29 (15/11/58), 35°5.5'N 139°29.3'E, Sagami Bay [Japan] (OD).

Etymology. Referring to the type locality in Sagami Bay.

Description. Shell trochiform, rounded (to 3.86 mm: holotype. Figs 877–878). Protoconch of 0.75 whorl, reticulate sculpture, apertural varix connected to embryonic cap, shape of apertural margin straight . Teleoconch I of 0.75 whorl, approximately 25 axial cords, fine spiral line in position of selenizone. Teleoconch II of up to three whorls, sutsel half as wide as selenizone at onset of selenizone, 1.5× width at apertural margin of fully grown specimen. Shoulder moderately convex, with slightly raised axial cords, crossed by finer, equally spaced spiral lines, approximately 22 at apertural margin of fully grown specimen. Base with weak constriction below selenizone, sculpture slightly less developed, no raised axial cords; approximately 35 fine spiral cords stronger than on shoulder, weaker than axial cords on base. Anomphalous. Aperture round, roof moderately overhanging. Selenizone slightly above periphery; slit open with slightly converging margins; lunules little distinct; keels of moderate strength and elevation.

Radula (Fig. 879). Rachidian tooth trapezoidal, triangular tip, central cusp largest, strongest, approximately six cusps on each side of similar size. Lateral teeth 1–3 similar, tip pointing towards center of radula, 6–4 cusps along outer cutting edge. Lateral tooth 4 reduced, minute serration on outer cutting edge. Lateral tooth 5 enlarged by elongation, serrated on both edges with approximately six low cusps. Inner marginal teeth with elongated tip, rear of tip serrated with 4–5 long cusps; shafts without food groove. Radular interlock of central field moderate.

Operculum round, as large as aperture, central nucleus, multispiral.

Comparisons. *Anatoma concinna* from the North Pacific has an overall more globular shell, a very narrow umbilicus, a protoconch with flocculent sculpture, and no apertural varix.

Anatoma japonica from the Indian Ocean to the Central Pacific has a protoconch with flocculent sculpture, has an overall more depressed shell shape, and finer but more distinct axial and spiral cords.

Anatoma fujikurai from hydrothermal vent-associated fields in southern Japan has a distinct umbilicus and much more pronounced and undulating keels of the selenizone and fewer but more distinct axials cords on shoulder, which are about equally strong on the base and the shoulder.

Anatoma lamellata from the Western and the Central Pacific has a protoconch with flocculent sculpture and middle spiral ridge and strong axial lamellae on shoulder and base.

Anatoma parageia from southern Japan has a protoconch with flocculent sculpture and no apertural varix and an overall more depressed shell and a distinct funiculus with lower adumbilical corner of the aperture being drawn out.

Anatoma porcellana from the Indo-Malayan Archipelago to the Western Pacific has a distinct constriction below the selenizone, sharp axial cords on the shoulder, and much coarser reticulate sculpture.

Anatoma soyoae from northern Japan has an overall more globular shell with much finer reticulate sculpture.

Anatoma agulhasensis from the western Indian Ocean has finer sculpture on the base, a protoconch with flocculent sculpture, and a wider umbilicus.

Figure 878. *Anatoma sagamiana*. **A.** 2029–2045 m, KT-86-6 KN6, S of Daiozaki, Japan, 33.777°N, 136.667°E (SBMNH 83432, 1, 11: complete). **B.** 922–959 m, KT-99-6, St. 13, SE of Kamogawa, Japan, 34.812°N, 140.627°E (SBMNH 83433, 2, 16: complete). Radula and operculum shown in Figure 879. **C.** 437 m, off Cape Shionomisaki, Wakayama Prefecture, Japan, 33.450°N, 135.783°E (DLG 119, 1). Scale bars shell = 1 mm. Scale bars protoconch = 100 μm.

Anatoma alternatisculpta from the Caribbean has at most half as many spiral lines on shoulder and base, 50% fewer but stronger axial cords on the shoulder, and a wider umbilicus.

Anatoma australis from eastern Australia has more than one spiral cord on teleoconch I, a wider umbilicus, and only every 3–4 axial cords on the base reach the umbilicus.

Anatoma biconica from the Indian Ocean to the Central Pacific has a more biconical shell, at least 50% fewer axial cords and spiral lines, and a wider umbilicus.

Anatoma flexidentata from the Indian Ocean to the Western Pacific has a wider selenizone, at most half as many axial cords and spiral lines, and a wider umbilicus.

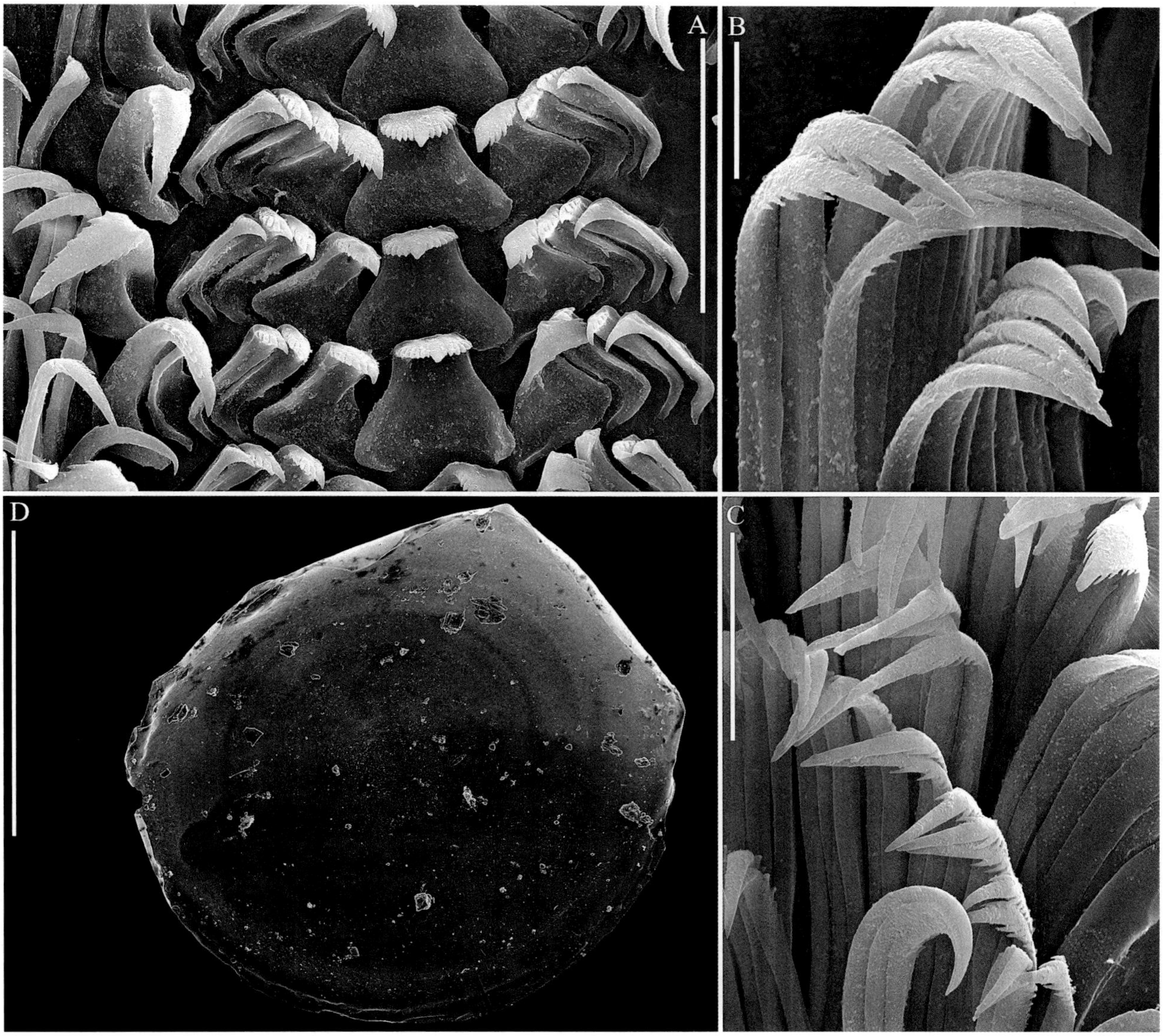

Figure 879. *Anatoma sagamiana*. Radula. 922–959 m, KT-99-6, St. 13, SE of Kamogawa, Japan, 34.813°N, 140.627°E (SBMNH 83433, 2, 16: complete). Shell shown in Figure 883B. **A.** Central field. Scale bar = 50 µm. **B.** Inner marginal teeth. Scale bar = 10 µm. **C.** Outer marginal teeth. Scale bar = 20 µm. **D.** Operculum. Scale bar = 1 mm.

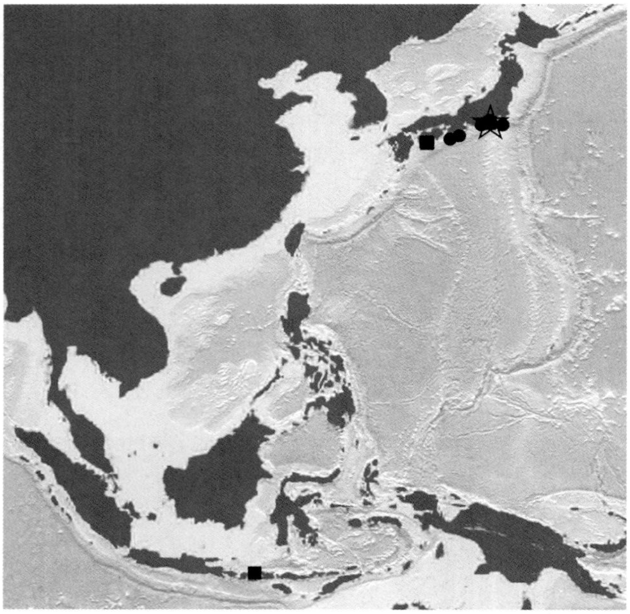

Figure 880. Distribution of *Anatoma sagamiana*. ☆ type locality. ● specimen record. ■ literature record.

Anatoma rhynchodentata from the Indo-Malayan Archipelago to the Western Pacific has a more turreted overall shell shape, more dense axial cords forming points at the intersections with the spiral lines, and a wider umbilicus.

Anatoma kelseyi from the northeastern Pacific has a wider umbilicus and a less convex shoulder.

Anatoma peruviana from Peru has a wider sutsel at any given whorl count, a wider umbilicus, and about 50% fewer spiral lines on the shoulder.

Anatoma schanderi from the northeastern Atlantic has a more convex shoulder, a less bulging base, and a wider umbilicus.

Anatoma tobeyoides from southeastern Australia has a teleoconch I with very fine oblique threads on teleoconch I and early teleoconch II, a more convex shoulder, and a wider umbilicus.

Anatoma zancliformis from the Indo-Malayan Archipelago to the Western Pacific has a more globular shell outline, no sutsel on early whorls, and at most half as many axial cords on the base.

Distribution. Honshu, Japan, Indonesia (Fig. 880)

Specimen records.

Japan. 436 m, off Cape Shionomisaki, Wakayama, 33.450°N, 135.783°E (DLG 119, 1). 200 m, off Kushimoto, Kii, 33.467°N, 135.783°E (HLC, 1). 2029–2045 m, S of Daiosaki, Kumano Basin, Mie, 33.777°N, 136.627°E (SBMNH 83432, 1). 922–959 m, SE of Kamogawa, Chiba, 34.814°N, 140.627°E (SBMNH 83433, 2, 16: complete). 920 m, St. 5053, off Hondo, 34.822°N, 138.671°E (USNM 205832, 1). 472 m, St. 5056, Suruga Gulf, 34.960°N, 138.726°E (USNM 205831, 1). 710–770 m, Sagami Bay St. T29, 35.092°N, 139.488°E (NSMT 69582, 1; UMUT 8808, 1). Sagami Bay, off Misaki, 35.250°N, 139.417°E (ZSM 20002629, 1).

Literature records.

Japan. 840 m, Tosa Bay, 33.017°N, 133.642°E. 301 m, Tosa Bay, 33.107°N, 133.718°E. 760 m, Tosa Bay, 33.117°N, 133.702°E. 301 m, Tosa Bay, 33.177°N, 133.573°E (Hasegawa, 2001).—1473–1500 m, St. T21(2) Sagami Bay, 34.947°N, 139.350°E. 1005–1020 m, St. T22(2) Sagami Bay, 35.042°N, 139.450°E (Okutani, 1966).—Suruga Bay [Shizuoka, 34.800°N, 138.600°E] (Habe *et al.*, 1986).

Indonesia. Satonda, 8.100°S, 117.750°E (Bandel, 1998).

Remarks. *Thieleella reticulata* is here synonymized under *An. sagamiana*. The two nominal taxa share the rather uncommon reticulate protoconch sculpture, have same teleoconch I whorl number, same strength and density of axial cords and spiral threads, share the slight spiral keel below the selenizone, and have the same sutsel. The illustrated holotype of *Th. reticulata* is a somewhat small specimen for the species with only 1.5 teleoconch II whorls, while large specimens have three whorls.

Anatoma sagamiana is extraordinarily similar to *An. kelseyi* from the northeastern Pacific in terms of shell morphology, but shows some significant differences in the radula: the tip of the rachidian tooth of *An. sagamiana* is V-shaped, whereas the cusps are arranged on a ridge parallel to the rows of teeth in *An. kelseyi*; the lateral teeth 5 of *An. sagamiana* have cusps on both sides more or less equally distributed, whereas in *An. kelseyi* the inner edge has much better developed cusps up to the tip, while the outer edge has fewer restricted to the transition between shaft and tip. The combination of slight conchological and radular differences with disjunct distribution suggests that they are separate species.

The Japanese common name is *Sagami-kuchikire-ebitsu* (Hagesawa, 2001).

Anatoma schanderi Høisaeter & Geiger, 2011

Figures 881–885

Chresonymy.

Anatoma schanderi Høisaeter & Geiger, 2011: 106–109, figs. 68–83.

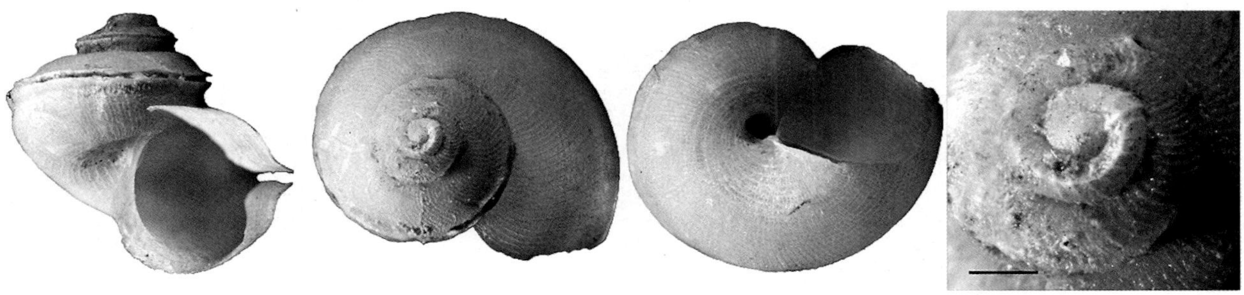

Figure 881. *Anatoma schanderi*. Holotype. 2560–2580 m, H2DEEP ROV sample 7-2008. S margin of the Greenland Basin, Norwegian Sea, 73°34' N, 7°50'–8°01' W [= 73.567°N, 8.925°W] (ZMBN 87296). Shell = 2.5 mm. Scale bar protoconch = 200 μm.

Type material. Holotype (ZMBN 87296. Fig. 881). Paratypes from type locality (ZMBN 87297–8729, 4).

Type locality. 2560–2580 m, H2DEEP ROV sample 7-2008. S margin of the Greenland Basin Norwegian Sea, 73°34'N, 7°50'–8°01'W [= 73.567°N, 8.925°W] In a sample of semi-lithified sediment (OD).

Etymology. Named after Professor Christoffer Schander [1960–2012] from the University of Bergen, who recognized the significance of what now are the type specimens during sorting of material (OD).

Description. Shell medium-sized (to 4.5 mm. Figs 881–883), trochiform, biconical. Protoconch of 0.75 whorl, reticulate sculpture, no apertural varix, apertural margin straight. Teleoconch I of 0.9 whorl, with approximately 30 axial cords, spiral cord in position of selenizone. Teleoconch II of up to 3.2 whorls, suture lightly impressed, sutsel twice as wide as selenizone. Shoulder slightly convex; numerous, fairly weak axial cords, approximately 65–70 per whorl in fully-grown specimens; first spiral thread after 0.2–0.3 whorl, 2–3 after one teleoconch II whorl, 4–7 after two teleoconch II whorls, up to 20 at apertural margin of fully grown specimen. Base with weak constriction below selenizone, axial cords as on shoulder, approximately 25 fine spiral cords regularly spaced over entire base. Umbilicus open, moderately wide, funiculus present. Aperture rounded, with somewhat flaring lip. Selenizone at periphery; keels low, weak; lunules indistinct; slit open, margins somewhat converging.

Operculum (Fig. 884F) corneous, thin round, multispiral, nucleus central.

Radula (Fig. 884). Rachidian tooth trapezoidal, central cusp largest, 7–8 cusps in straight line on each side. Lateral teeth 1–3 similar, apical cusp largest, 4–5 cusps along outer edge of tip. Lateral tooth 4 reduced, hook-shaped. Lateral tooth 5 enlarged, apical cusp largest, 5–6 cusps along inner edge, 3–4 cusps along outer edge of tip. Inner marginal teeth with oval tip, apical cusp largest, 6–10 cusps along each side. Outer marginal teeth with spoon-shaped tip, many fine bristles along edge. Outermost marginal teeth paddle-shaped, with many fine bristles along tip. No food groove on shafts of marginal teeth. Radular interlock of central field moderate.

Figure 882. *Anatoma schanderi*. 749–774 m, Greenland, Denmark, 75.009°N, 12.625°W (ZMUC, 20). Radula shown in Figure 884. Scale bar shell = 1 mm. Scale bar protoconch = 100 µm.

Comparisons. *Anatoma crispata* from the boreal North Atlantic is smaller (2.2 vs. 4.5 mm), has more pronounced axial sculpture, a distinct funiculus, and a protoconch with flocculent sculpture.

Figure 883. *Anatoma schanderi*. **A–B.** 1509–1525 m, off E Greenland, Denmark, 74.893°N, 12.500°W (ZMUC, 20). **C.** 120 m, 10–15 km NW of Digby Neck, Nova Scotia, Canada, 44.600°N, 65.750°W (JWC 7286, 1). Scale bars shell = 1 mm. Scale bars protoconch = 100 μm.

Anatoma aspera from the Mediterranean and Norway to Angola is smaller (3.3 vs. 4.5 mm), has a taller, more turreted shell, and much stronger axial sculpture on the shoulder.

Anatoma eximia from the Mediterranean to Angola has a much more stepped profile of the shell, the selenizone higher on the whorl, and flocculent protoconch sculpture.

Anatoma orbiculata from West Africa has a more convex shoulder, more and stronger axial cords, and much stronger axial cords on the shoulder compared to the base.

Anatoma tenuis from the North Atlantic has no sutsel up to at least 4.4 mm, a less convex shoulder, and stronger clathrate sculpture on shoulder and base.

Anatoma tenuisculpta from the Mediterranean and Norway to Angola grows somewhat larger (5.5 vs. 4.5 mm), has an overall more disciform shell, and the axials are stronger with a crenulated appearance due to the crossing spirals.

Anatoma schioettei from the boreal northeastern Atlantic is much smaller (2.2 vs. 4.5 mm), has much more distinct axials, and a protoconch with flocculent sculpture.

Anatoma umbilicata from the Mediterranean and adjacent Atlantic has a more depressed shell, more turreted overall shell shape, weaker sculpture, and a wider umbilicus.

Anatoma agulhasensis from the western Indian Ocean has raised axial cords on the shoulder, stronger reticulate sculpture on the base, and less inflated whorls.

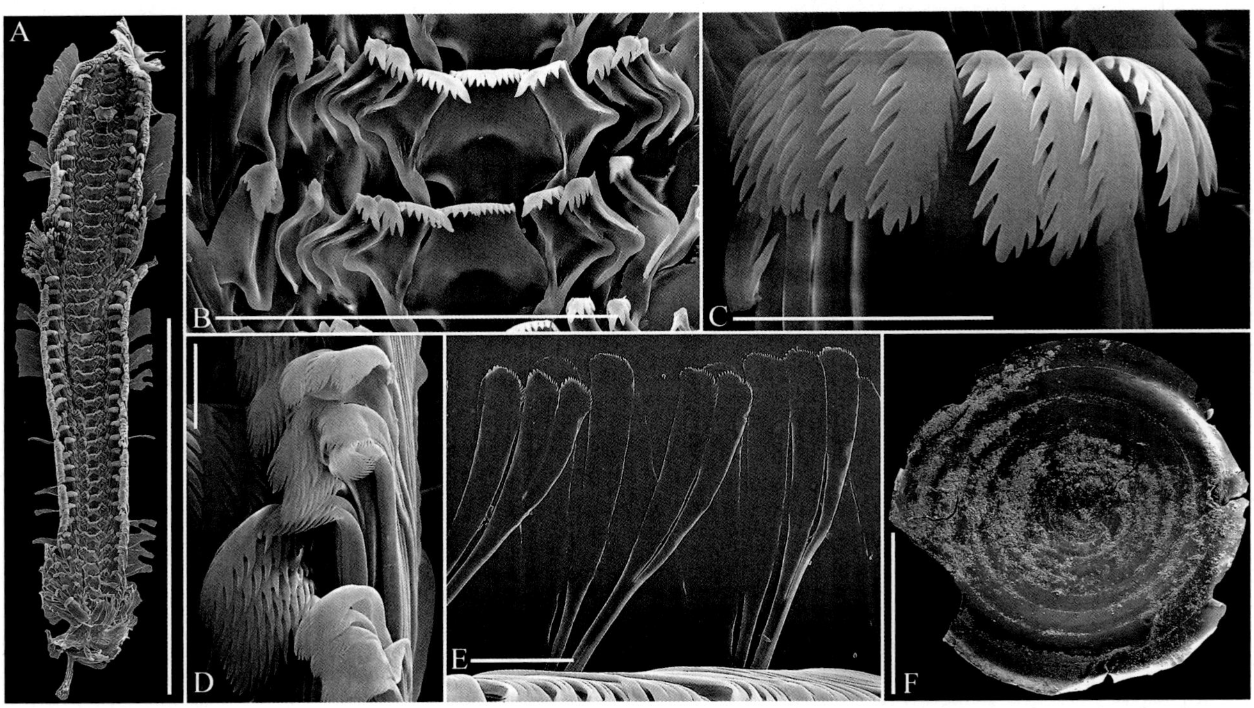

Figure 884. *Anatoma schanderi*. Radula and operculum. 749–774 m, Greenland, 75.009°N, 12.625°W (ZMUC, 20). Shell shown in Figure 882. **A.** Entire radula. Scale bar = 1 mm. **B.** Central field enlarged. Scale bar = 100 µm. **C.** Inner marginal teeth. Scale bar = 20 µm. **D.** Outer marginal teeth. Scale bar = 10 µm. **E.** Outermost marginal teeth. Scale bar = 20 µm. **F.** Operculum. Scale bar = 1 mm.

Anatoma alternatisculpta from the Caribbean has strong axial cords on the shoulder, at most half as many spiral lines on shoulder and base, and different sculpture on shoulder and base.

Anatoma concinna from the North Pacific has lower keel of the selenizone, coarser but less distinct clathrate pattern on shoulder and base, and a protoconch with flocculent sculpture.

Anatoma tangaroa from northernmost New Zealand has a shorter teleoconch I (0.3 vs. 0.9 whorl), much stronger axial and spiral sculpture, and less inflated whorls.

Anatoma flexidentata from the Indian Ocean to the Western Pacific has a wider selenizone, at least 50% fewer axial cords and spiral lines on shoulder and base, and no sutsel on early whorls.

Anatoma fujikurai from Japan has strong axial cords on the shoulder and base, undulating keels of the selenizone, and a wider sutsel on early whorls.

Anatoma globulus from the Indo-Malayan Archipelago has stronger sculpture on shoulder and base, a shorter teleoconch I (0.6 vs. 0.9 whorl), and more distinct lunules.

Anatoma kelseyi from the northeastern Pacific has a shorter teleoconch I (0.6 vs. 0.9 whorl) without a spiral cord in the position of the selenizone and an apertural varix on the protoconch.

Anatoma peruviana from Peru has a wider sutsel at any given whorl count, raised axial cords on the shoulder, and stronger sculpture on the base.

Anatoma proxima from the Caribbean to Brazil has a more depressed overall shell shape, flocculent protoconch sculpture, and stronger teleoconch sculpture.

Anatoma rhynchodentata from the Indo-Malayan Archipelago to the Western Pacific has a strong constriction on the base below the selenizone, stronger axial sculpture on shoulder and base, and flocculent protoconch sculpture.

Anatoma sagamiana from Japan and the Indo-Malayan Archipelago has a weak constriction on the base below the selenizone, a more narrow umbilicus, and stronger axial cords on the shoulder.

Anatoma soyoae from Japan has no sutsel on early whorls, fine clathrate sculpture with twice as many axials and spirals, and less elevated keels of the selenizone.

Anatoma zancliformis from the Indo-Malayan Archipelago and the Western Pacific has no sutsel on early whorls, at most half as many spiral lines on shoulder and base, and a selenizone that is higher on the whorl.

Figure 885. Distribution of *Anatoma schanderi*. ☆ type locality. ● specimen record.

Distribution. North Atlantic (Fig. 885).

Specimen records.

Canada. 120 m, 10–15 km NW of Digby Neck, Nova Scotia, 44.600°N, 65.750°W (JWC 7286, 1).

Anatoma schioettei Høisaeter & Geiger, 2011

Figures 886–889

Chresonymy.

Anatoma schioettei Høisaeter & Geiger, 2011: 103–105, figs 48–60.

Type material. Holotype (ZMH K–42704). Paratypes from type locality (ZMH K 42705–42710). 197–198 m, Greenland, Denmark, 75.016°N, 13.800°W (ZMUC GAS–427, 428).

Type locality. 749–774 m, the upper slope off E Greenland, 75°01'N, 12°38'W (75.009°N, 12.625°W) (OD).

Etymology. Named after Tom Schiøtte at ZMUC. He generously provided material that he had been working on himself, which provided the radulas for the two new species [*An. schioettei, An. schanderi*] (OD).

Figure 886. *Anatoma schioettei*. Holotype. 749–774 m, Greenland, 75.009°N, 12.625°W (ZMH K–42704). Scale bar shell = 1 mm. Scale bar protoconch = 100 µm.

Description. Shell medium-sized (to 2.15 mm. Figs 886, 887), trochiform, discoidal. Protoconch of 0.5 whorl, sculpture flocculent, no apertural varix, apertural margin slightly convex. Teleoconch I of 0.75 whorl, approximately 15 axial cords, spiral cord in position of selenizone. Teleoconch II of up to 1.5 whorls, sutsel 1–2× as wide as selenizone. Shoulder slightly convex, approximately 50–52 strong axial cords per whorl in fully grown specimens; faintest spiral threads (not visible under light

Figure 887. *Anatoma schioettei*. Paratypes. **A–C.** 749–774 m, Greenland, 75.009°N, 12.625°W (ZMH K 42705–42710). Scale bars shell = 1 mm. Scale bars protoconch = 100 μm.

microscope). Base with 30–55 axial cords, 4–8 indistinct spiral lines on inner part of base towards umbilicus, becoming gradually weaker towards periphery. Umbilicus open, wide; distinct, narrow funiculus. Aperture rounded, oblong. Selenizone at periphery; lunules indistinct; keels moderately elevated, weak; slit open, margins converging.

Operculum (Fig. 888E) corneous, thin round, multispiral, nucleus central.

Radula (Fig. 888). Rachidian tooth trapezoidal, central cusp largest, four cusps in arc on each side of tip. Lateral teeth 1–3 similar, apical cusp largest, 4–5 progressively smaller cusps on outer edge of tip. Lateral tooth 4 reduced, hook-shaped, with 1–2 cusps on outer edge. Lateral tooth 5 enlarged, apical cusp largest, 6–7 cusps along inner edge, 2–3 cusps along outer edge. Inner marginal teeth with triangular tip, apical cusp largest, 4–5 cusps on each side of tip; shafts without food groove. Outer marginal teeth with spoon-shaped tip, many small cusps along edge. Outermost marginal teeth paddle-shaped with many fine bristles along tip. Radular interlock of central field moderate.

Comparisons. *Anatoma crispata* from the boreal North Atlantic has a more narrow sutsel, at least twice as many axials, crossed by stronger spirals, and a more narrow umbilicus.

Anatoma tenuisculpta from the Mediterranean and Norway to Angola is much larger (5.5 vs. 2.2 mm), the suture is closer to the selenizone of the previous whorl, has at least twice as many axials with a crenulated appearance due to the crossing spirals, and lacks a funiculus.

Figure 888. *Anatoma schioettei*. Radula and operculum. 749–774 m, Greenland, Denmark, 75.009°N, 12.625°W (ZMH K–42705). **A.** Entire radula. Scale bar = 200 µm. **B.** Central field. Scale bar = 20 µm. **C.** Lateral tooth 5 and marginal teeth. Scale bar = 10 µm. **D.** Outermost marginal teeth. Scale bar = 10 mm. **E.** Operculum. Scale bar = 1 mm

Figure 889. Distribution of *Anatoma schioettei*. ☆ type locality. ● specimen record.

Anatoma aspera from the Mediterranean and Norway to Angola has a taller overall shell, stronger axials on the shoulder, and no funiculus.

Anatoma schanderi from the temperate North Atlantic has a taller shell with the base visible between suture and selenizone at an angle towards the spindle, is much larger (4.5 vs. 2.2 mm), has spirals and axials of approximately equal strength, and a protoconch with more regular net sculpture.

Distribution. Arctic North Atlantic (Fig. 889).

Specimen records.

Denmark, Greenland. 749–774 m, 75.009°N, 12.625°W (ZMH K-42704-42710). 100–110 m, 76.548°N, 69.003°W (ZMUC, 1: complete). 70 m, 76.542°N, 69.383°W (ZMUC, 3: complete). 197–198 m, 75.016°N, 13.800°W (ZMUC GAS-427, 428). 574 m, H2DEEP ROV dive 22-2008, the Trollveggen vent field just NE of Jan Mayen, 71.300°N, 5.783°W.

Remarks. A specimen with aberrant radula is known, in which a duplication of one of the lateral teeth 1–3 occurred (Fig. 22).

Anatoma shiraseae Numanami & Okutani, 1990

Figures 891 –890

Chresonymy.

Anatoma shiraseae Numanami & Okutani, 1990: 102–104, figs 14–16.

Anatoma shiraseae: Numanami & Okutani, 1992: 172.

Anatoma shiraseae: Numanami, 1996: 20–21, figs 9A–C [paratype].

Anatoma shiraseae: Committee, 1996: 68.

Anatoma s.l. *shiraseae*: Geiger, 2003: 75.

Anatoma shiraseae: Zelaya & Geiger, 2007: 406–408, figs 51 [map], 52–55 [holotype shell SEM].

Type material. Holotype (NSMT Mo. 69097: Zelaya & Geiger, 2007: figs 52–55. Fig. 890), 2.25 × 3.0 mm. Paratype from type locality (NSMT Mo. 69098, 1), 3.86 × 4.21 mm.

Type locality. 281–282 m, Günnerus Bank, Antarctica [67.392°S, 32.125°E] (OD).

Etymology. Named for the icebreaker *Shirase* (OD).

Description. Shell large (4.2 mm. Fig. 890), biconical, somewhat turreted. Protoconch 0.75 whorl, sculpture flocculent, no apertural varix, apertural margin convex. Shoulder with many fine axial lines, approximately 95 on body whorl, intersected by approximately 21 spiral lines of same strength, producing microclathrate pattern. Shoulder with distinct, broad, spiral keel. Base with similar spiral keel below selenizone, second spiral keel at lowest point of base. Base with approximately

115 fine axial lines, 29 somewhat stronger spiral lines. Umbilicus wide, open, sloping continuously with base. Parietal portion of aperture slightly flared. Umbilicus narrow, funnel-shaped, no funiculus. Aperture subquadrate. Selenizone at periphery with fragile keels of moderate elevation; lunules distinct; slit open, margins converging anteriorly.

Operculum slightly smaller than aperture, corneous multispiral. Position of nucleus unknown.

Comparisons. The spiral keel on the shoulder is unique to this species.

Distribution. Antarctica. Known only from type locality (Fig. 891).

Remarks. The holotype is composed of the shell with the fluid preserved animal stored separately. The animal is female, as indicated by the distinct eggs in the apical portion of the visceral mass. The paratype has been sputter coated, but was broken before it was imaged (K. Hasegawa, pers. comm. 2006); the protoconch portion is missing entirely.

Figure 890. *Anatoma shiraseae*. Holotype. 281–282 m, Günnerus Bank, Antarctica [67.392°S, 32.125°E] (NSMT Mo. 69097). Scale bar shell = 1 mm. Scale bar protoconch = 100 μm.

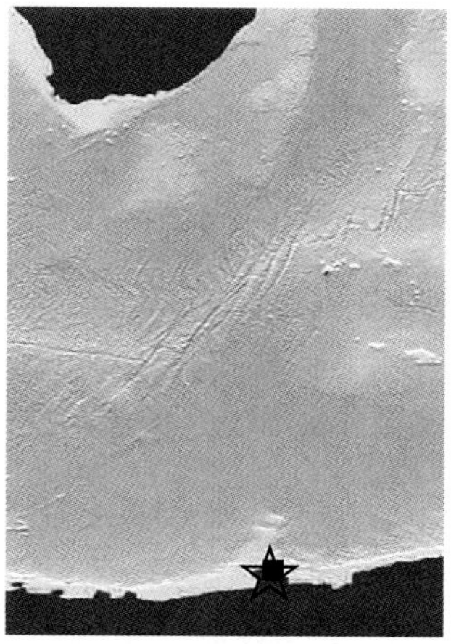

Figure 891. Distribution of *Anatoma shiraseae*. ☆ type locality. ■ literature record.

Anatoma sinuosa new species

Figures 892–894

Type material. Holotype (SBMNH 149941. Fig. 892). Paratypes from type locality (DLG 1353, 24. Fig. 893A). 150–200 m, SW corner, Easter Island, Chile, 27.201°S, 109.463°W (DLG 1352, 20. Fig. 893B–C).

Type locality. 100 m, SW corner, Easter Island, Chile, 27.201°S, 109.463°W.

Etymology. Sinuosus, Latin adjective for curved, referring to the sinus-shaped profile of the shoulder.

Description. Shell medium-sized (to 1.77 × 1.76 mm H × W. Figs 892, 893), trochiform, globular inflated. Protoconch of 0.75 whorl, embryonic cap with some flocculent sculpture, remainder smooth, apertural varix barely connected to embryonic cap, apertural margin sinusoidal. Teleoconch I of 0.4–0.5 whorl, 9–13 axial cords, no spirals. Teleoconch II of up to 2.75 whorls, suture strongly impressed. Shoulder strongly sinusoidal, dense axial cords, approximately 50–60 on first teleoconch II whorl; first spiral line after 0.1–0.2 teleoconch II whorl, approximately 15 spirals after one whorl, approximately 22 at apertural margin; no spirals in immediate vicinity of suture (10% of width of shoulder). Base with indistinct constriction below selenizone, base inflated, continuously sloping into umbilicus; same density of axial cords and spiral threads as on shoulder. Aperture rounded, flared lip in lower adumbilical corner. Umbilicus open, without funiculus. Selenizone slightly above periphery; keels low (possibly eroded), moderately strong; lunules distinct but weak; slit open, margins parallel.

Animal unknown.

Comparisons. The very strongly sinusoidal profile of the shoulder and the very high placement of the selenizone on the whorl make this species easy to recognize.

Anatoma rainesi from Easter Island has a proportionally wider shell, a sutsel that is absent or only as wide as the selenizone, a longer teleoconch I (0.75 vs. <0.5 whorl), and a narrow umbilicus with a distinct funiculus.

Anatoma aedonia from the Caribbean to Brazil has a wider sutsel on early whorls (> vs. < width of selenizone), a protoconch with flocculent sculpture, and a less convex shoulder.

Anatoma africanae from the western Indian Ocean has weaker axial cords on the base, axials and spiral on the base that are of the same strength, and a less bulging base.

Anatoma alta from the Caribbean to Macaronesia has at most half as many axial and spiral cords on shoulder and base, a base that protrudes beyond the selenizone, and a more narrow sutsel on early whorls (0× vs. >0.5× width of selenizone).

Figure 892. *Anatoma sinuosa* new species. Holotype. 100 m, SW corner, Easter Island, Chile, 27.201°S, 109.463°W (SBMNH 149941). Scale bar shell = 1 mm. Scale bar protoconch = 100 μm.

Anatoma aspera from the Mediterranean and adjacent Atlantic has stronger but fewer axial cords on the shoulder, different sculpture on shoulder and base, and a longer teleoconch I (0.75 vs. 0.4 whorl).

Figure 893. *Anatoma sinuosa* new species. Paratypes. **A.** 100 m, SW corner, Easter Island, Chile, 27.201°S, 109.463°W (DLG 1353, 24). **B–C.** 150–200 m, SW corner, Easter Island, Chile, 27.201°S, 109.463°W (DLG 1352, 20). Scale bars shell = 1 mm. Scale bars protoconch = 100 µm.

Anatoma boucheti from the western Indian Ocean has a less bulging base, a longer teleoconch I (0.9 vs. 0.4 whorl), and a flatter apex.

Anatoma equatoria from the Indo-Malayan Archipelago to the Central Pacific has a protoconch with flocculent sculpture, a shoulder that is much less convex, a sutsel on the early whorls that is about twice as wide, and generally less dense reticulate pattern, particularly on the base.

Anatoma eximia from the Mediterranean and adjacent Atlantic has a wider sutsel on early whorls (>1× vs. 0.5× width of selenizone), a less bulging base, and a more narrow umbilicus.

Anatoma kelseyi from the northeastern Pacific has a taller shell, a selenizone that is more towards the periphery, and a protoconch with reticulate sculpture.

Anatoma peruviana from Peru has a taller shell, a selenizone that is more towards the periphery, and a protoconch with reticulate sculpture.

Anatoma quadraxialis from the Indo-Malayan Archipelago has the selenizone more at the periphery, more prominent axial sculpture, and protoconch with reticulate sculpture.

Figure 894. Distribution of *Anatoma sinuosa* new species. ☆ type locality. ● specimen record.

Distribution. Known only from Easter Island (Fig. 894).

Specimen records.

Chile, Easter Island. 150–200 m, Tahai, 27.140°S, 109.429°W (DLG 1321, 6). 100 m, Tahai, 27.140°S, 109.429°W (DLG 1322, 3). 30–50 m, Tahai, 27.140°S, 109.429°W (DLG 1328, 4). 70–100 m, Hanga Roa, 27.145°S, 109.440°W (DLG 1337, 4). 150 m, Mataveri, 27.154°S, 109.441°W (DLG 1340, 7). 150–200 m, SW corner, 27.201°S, 109.463°W (DLG 1352, 20). 100 m, SW corner, 27.201°S, 109.463°W (DLG 1353, 25). 100–150 m, La Perouse, 27.077°S, 109.295°W (DLG 1358, 3).

Remarks. The material was collected and kindly made available by Bret Raines.

Anatoma soyoae (Habe, 1951)

Figures 895–896

Chresonymy.

Schizotrochus soyoae Habe, 1951: 66, pl. 11, figs 3–4 (Fig. 895).
Anatoma soyoae: Habe & Kosuge, 1964: 4.
Anatoma soyoae: Higo, 1973: 13.
Schizotrochus soyoae: Kosuge, 1979: 10.
Anatoma soyoae: Higo & Goto, 1993: 15.
Schizotrochus soyoae: Okamoto, 2001: 26 [copy original description].
Anatoma soyoae: Higo *et al.*, 2001: 14, G80 [holotype].
Anatoma s.l. *soyae* [*sic*]: Geiger, 2003: 75.
Anatoma soyoae: Sasaki *et al.*, 2010: 69, fig. 5 [SEM of holotype].

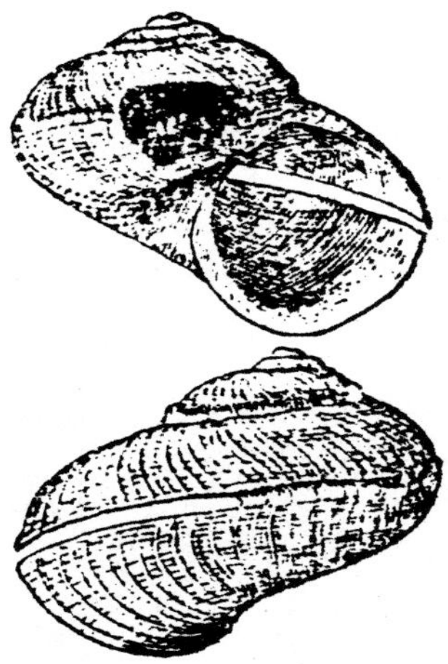

Figure 895. Original illustration of *Schizotrochus soyoae*.

Misidentifications.

Scissurella chiricova: Talmadge, 1966: 83 [*fide* McLean, 1967].
not *Scissurella* (*Anatoma*) *soyoae*: McLean, 1967: 407, pl. 56, fig. 7. [is *An. obtusata* = *An. concinna:* McLean, pers. comm.]
not *Scissurella soyoae*: Abbott, 1974: 17 [West American records most likely *An. concinna*].
not *Scissurella soyae* [*sic*]: Foster, 1981: 13, 43 [is *An. concinna*].

Type material. Holotype (NSTM Mo-38615. Fig. 896).

Type locality. 86 m, off Tsugaru Peninsula, N Honshu (*Soyo-Maru* Station 647) [Japan, 39.567°N, 141.930°E] (OD).

Etymology. Named for the research vessel *Soyo-maru.*

Description. Shell medium-sized (to 2.5 mm. Fig. 896), trochiform, inflated. Protoconch 0.75 whorl, sculpture unknown, most likely no apertural varix, apertural margin convex. Teleoconch I of 0.75 whorl, with approximately 30 axial cordlets, no spirals. Teleoconch II of at least two whorls; no sutsel at onset of selenizone, about as wide as selenizone at apertural margin. Shoulder with many axial cordlets (approximately 60 on first whorl), crossed by approximately 15 indistinct spiral threads forming minute points at intersections with axials. Base with similar sculpture as shoulder, reticulation pattern slightly more distinct. Selenizone at periphery; lunules dense, indistict; keels moderately strong, low; slit open, margins slightly converging. Umbilicus open, narrow, continuously sloping with base, no funiculus.

Animal unknown.

Comparisons. *Anatoma concinna* from the North Pacific has coarser sculpture on the teleoconch, distinct spiral sculpture on teleoconch I, and a distinct shelf on the adumbilical portion of the aperture.

Anatoma fujikurai from Japan has a taller shell, a much wider selenizone, a wider sutsel (1× vs. 0.25× width of selenizone), and strong undulating keels of the selenizone.

Anatoma lamellata from the Western to the Central Pacific has a shorter teleoconch I (0.6 vs. 0.75 whorl), at most half as many much stronger axial lamellae on the shoulder, and at most half as many spiral lines on the shoulder and base.

Anatoma sagamiana from Japan and the Indo-Malayan Archipelago has a taller overall shell shape, a more narrow umbilicus, and a slight constriction on the base below the selenizone.

Anatoma kelseyi from the northeastern Pacific is overall more globular with a proportionally taller shell, has reticulate protoconch sculpture, a shorter teleoconch I (0.75 vs. 1 whorl), and a more narrow umbilicus.

Anatoma lyra from the northeastern Pacific has a more discoidal overall shell shape, at most half as many but much stronger axial cords on the shoulder, and ontogenetically increasing strength of sculpture.

Figure 896. *Schizotrochus soyoae*. Holotype. 86 m, off Tsugaru Peninsula, Northern Honshu (*Soyo-Maru* Station 647) [Japan, 39.567°N, 141.930°E] (NSTM Mo-38615). Scale bar shell = 1 mm. Scale bar protoconch = 100 μm.

Anatoma japonica from the Indian Ocean to the Central Pacific has an overall more discoidal shell, a shorter teleoconch I (0.75 vs. 1 whorl), and more elevated keels of the selenizone.

Anatoma agulhasensis from the western Indian Ocean has a wider sutsel on early whorls (>0.5× vs. 0.25× width of selenizone), fewer but sharper axial cords on the shoulder, and different sculpture on shoulder and base.

Anatoma crispata from the northeastern Atlantic has at most half as many axial cords and spiral lines on shoulder and base and a distinct funiculus.

Anatoma disciformis from the North Pacific has coarser but duller axial cords and spiral lines, a distinct funiculus, and a more deeply impressed suture

Anatoma globulus from the Indo-Malayan Archipelago has a slightly shorter teleoconch I (0.6 vs. 0.75 whorl), at least 50% fewer spiral lines on the shoulder, and a wider sutsel on early whorls (~0.5× vs. 0.25× width of selenizone).

Anatoma proxima from the Caribbean to Brazil has a wider shell shape and a less convex shoulder.

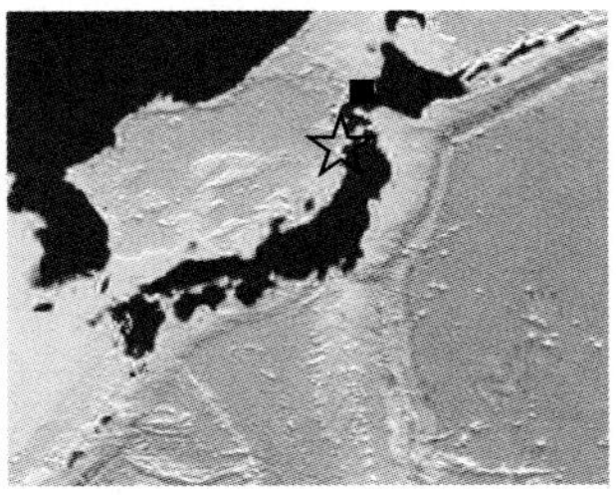

Figure 897. Distribution of *Anatoma soyoae*. ☆ type locality. ■ literature record.

Distribution. Northern Japan (Fig. 897).

Literature records.

Japan. 85 m, Ishikari Bay, 43.369°N, 140.843°E (Kosuge, 1979).

Anatoma tabulata (Barnard, 1964)

Figures 898–901

Chresonymy.

Scissurella tabulata Barnard, 1964: 21, figs 2e–f (Fig. 898).
Scissurella tabulata: Kensley, 1973: 28.
Anatoma tabulata: Herbert, 1986: 619.
Anatoma s.l. *tabulata*: Geiger, 2003: 75.

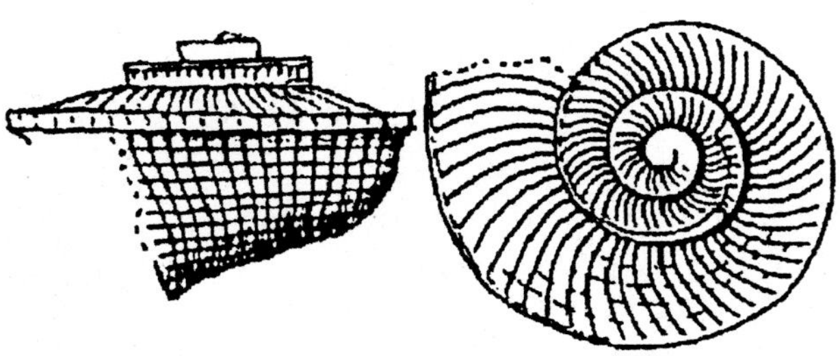

Figure 898. Original illustration of *Scissurella tabulata*.

Type material. Holotype (SAM A9345. Fig. 899), 2.3 mm.

Type locality. 440 fathoms [= 805 m], off Cape Natal (Durban) [South Africa, 29.883°S, 31.000°E] (OD).

Etymology. Most likely referring to the stepped overall shell shape.

Description. Shell medium-sized (to 2.1 mm: holotype. Fig. 899), trochiform, angulated whorls. Protoconch 0.75 whorl, sculpture flocculent, apertural varix not connected to embryonic cap. Teleoconch I of estimated 1–1.125 whorls, approximately 35 distinct axial cords, fine spiral line in position of selenizone in older portion of whorl. Teleoconch II of 1.5 whorls, suture moderately impressed, sutsel 1.5× as wide as selenizone. Shoulder with axial cords of same density and strength, approximately 48 on first teleoconch II whorl; 4–6 very fine spiral threads barely perceptible in light micrograph, forming faint points at intersections with axials, onset uncertain. Base without constriction below selenizone, continuous rounding into presumably open umbilicus; same density of slightly weaker axial cords, approximately 20 unequally spaced spiral cords, forming small points at intersection with axials; overall somewhat irregular clathrate sculpture. Aperture unknown. Selenizone at periphery, rather wide; lunules fine, distinct, crowded; keels moderately elevated (eroded), of moderate strength.

Operculum round, thin, corneous, 0.5 mm in diameter (Barnard, 1964).

Comparisons. *Anatoma agulhasensis* from the western Indian Ocean has an overall taller shell with more rounded whorls, more distinct and denser axial cords, crossed by more prominent and numerous spiral lines.

Anatoma austrolissa from the Indian Ocean and the Western Pacific has a shorter teleoconch I (0.3 vs. 1 whorl) and almost no sculpture.

Figure 899. *Scissurella tabulata.* Holotype. 440 fathoms [= 805 m], off Cape Natal (Durban) [South Africa, 29.883°S, 31.000°E] (SAM A9345). Images by Simon van Noort and Liz Hoenson. Scale bar = 0.5 mm.

Anatoma boucheti from the western Indian Ocean has a depressed apex and denser axial as well as spiral sculpture on shoulder and base.

Anatoma finlayi from the Indo-Malayan Archipelago to the Central Pacific (and possibly Caribbean) has strong axial lamellae on the shoulder and different sculpture on shoulder and base.

Anatoma flexidentata from the Indian Ocean and the Western Pacific has a more rounded shoulder, a much shorter teleoconch I (0.25 vs. 1 whorl), and generally fewer spiral lines on shoulder and base.

Anatoma janusa from the subtropical northeastern Atlantic has strong axial lamellae on the shoulder, fine but strong reticulate sculpture on the upper portion of the base, and at least twice as dense axial cords on the base.

Anatoma japonica from the Indian Ocean to the Central Pacific has a more depressed overall shell shape, a more narrow umbilicus, and at least twice as dense axial sculpture on shoulder and base.

Figure 900. *Anatoma tabulata*. 184–203 m, between Kokaro and St. Luce, Madagascar, 24.883°S, 47.467°E (MNHN, 1). Scale bar shell = 1 mm. Scale protoconch = 100 µm.

Anatoma yaroni from the western Indian Ocean has a more lenticular overall shell shape, a very narrow sutsel on the early whorls, and denser axial cords.

Anatoma aedonia from the Caribbean to Brazil has a wider sutsel on early whorls (> vs. < width of selenizone) and a more narrow umbilicus.

Anatoma bathypacifica from the Panamic has a protoconch with reticulate sculpture, a spiral cord on the teleoconch I forming a strong angulation, and the intersection of axials and spiral on the shoulder form minute points.

Anatoma equatoria from the Indo-Malayan Archipelago to the Central Pacific has a shorter teleoconch I (0.5 vs. 0.75 whorl), a more narrow umbilicus, and a more inflated base.

Anatoma eximia from the Mediterranean to Liberia has a wider sutsel on early whorls (>1.5× vs. 1× width of selenizone), a more narrow umbilicus, and spiral steps on the base.

Anatoma indonesica from the Indo-Malayan Archipelago to the Western Pacific has strong axial lamellae on the shoulder, a smooth band on the base below the selenizone, and fine reticulate sculpture on the base.

Anatoma micalii from the Mediterranean to Angola has strong axial cords on the shoulder, a shorter teleoconch I (0.5 vs. >0.75 whorl), and fine striations on the lower keel of the selenizone between the extensions of the axial cords on the base.

Anatoma munieri from the Indian Ocean to the Central Pacific has a wider sutsel (>2× vs. 1× width of selenizone), a very strong constriction on the base below the selenizone, and a second spiral keel on the base.

Anatoma rhynchodentata from the Indo-Malayan Archipelago to the Western Pacific has a constriction on the base below the selenizone, the intersections of axials and spirals forming distinct points, and at least 50% more axial cords.

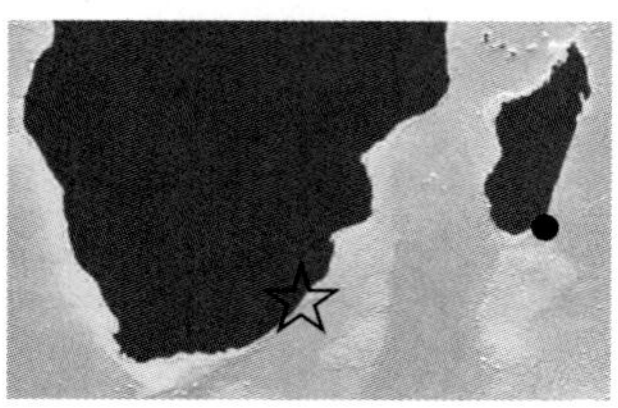

Figure 901. Distribution of *Anatoma tabulata*. ☆ type locality.

Distribution. Western Indian Ocean (Fig. 901).

Specimen record.

Madagascar. 184–203 m, between Kokaro and St. Luce, 24.883°S, 47.467°E (MNHN, 1).

Remarks. The general geographic area of the type locality (off Durban, 400–600 m) has been searched for this species, but it has not been recollected (A. Marais, pers. comm., 4/2011). Special thanks to Johan and Alwyn Marais for their instrumental help in obtaining images of the holotype.

Anatoma tangaroa Geiger & Marshall, 2012

Figures 902–904, Color Plate 1A–B

Chresonymy.

Anatoma sp. 6 Spencer *et al.*, 2009: 201.

Anatoma tangaroa Geiger & Marshall, 2012: 27–29, figs 21–22.

Type material. Holotype (NMNZ M.303297. Fig. 902). Paratypes from type locality (NMNZ M.150335, 10. Fig. 903A). 120 m, 28 km S of Great Island, Three Kings Islands, New Zealand, 34.400°S, 174.280°E (NMNZ M.160285, 15. Fig. 903B). 246–291 m, Middlesex Bank, NW of Three Kings Islands, New Zealand, 34.033°S, 171.733°E (NMNZ M.148568, 25. Fig. 903C).

Type locality. 188 m, NW of Three Kings Islands, New Zealand, 34.033°S, 171.806°E (OD).

Etymology. After R. V. *Tangaroa*, from which the type material was dredged (OD).

Figure 902. *Anatoma tangaroa*. Holotype. 188 m, NW of Three Kings Islands, New Zealand, 34.033°S, 171.806°E (NMNZ M.303297). Scale bar shell = 1 mm. Scale bar protoconch = 100 μm.

Description. Shell medium-sized (to 2.2 mm), trochiform, rounded, biconical. Protoconch of 0.75 whorl, sculpture coarsely flocculent, no apertural varix. Teleoconch I of 0.3 whorl, 5–7 axial cords, spiral cord in position of selenizone forming thickenings at intersection with axial cords. Teleoconch II of 2.2 whorls, suture moderately

Figure 903. *Anatoma tangaroa*. Paratypes. **A.** 188 m, NW of Three Kings Islands, New Zealand, 34.033°S, 171.806°E (NMNZ M.150335, 10). **B.** 120 m, 28 km S of Great Island, Three Kings Islands, New Zealand, 34.400°S, 174.280°E (NMNZ M.160285, 15). **C.** 246–291 m, Middlesex Bank, NW of Three Kings Islands, New Zealand, 34.033°S, 171.733°E (NMNZ M.148568, 25). Scale bars shell = 1 mm. Scale bars protoconch = 100 µm.

impressed; sutsel at onset of selenizone 0.5× width of selenizone, at apertural margin 1–1.5× width of selenizone. Shoulder moderately convex, approximately 40 strong axial cords on first whorl; first spiral line after 0.2–0.5 teleoconch II whorl, approximately 10–15 at apertural margin, forming small thickenings at intersection with axial cords. Base without constriction below selenizone, same density and strength of axial cords as on shoulder; approximately 20 spiral lines, regularly spaced. Umbilicus open, continuously sloping with base, no funiculus. Aperture rounded, roof overhanging. Selenizone at periphery, rather wide; keels moderately stronger, moderately elevated; lunules distinct; slit open, margins converging towards aperture.

Animal unknown.

Comparisons. *Anatoma amydra* from the Indo-Malayan Archipelago to the Central Pacific has weaker spiral lines on the base, a protoconch with an apertural varix, and a longer teleoconch I (>0.5 vs. 0.3 whorl).

Anatoma biconica from the Indian Ocean to the Central Pacific has a less rounded periphery, has a longer teleoconch I (>0.8 vs. 0.4 whorl), finer spiral lines not forming thickenings at the intersections with the axial cords, and a wider sutsel on early whorls (1× vs. 0.5× width of selenizone).

Anatoma maxima from the Indo-Malayan Archipelago to the Central Pacific has a longer teleoconch I (0.75 vs. 0.3 whorl) and no sutsel on early whorls.

Anatoma rainesi from the Eastern to the Western Pacific has a protoconch with a varix and a longer teleoconch I (0.6 vs. 0.3 whorl).

Anatoma flexidentata from the Indian Ocean to the Western Pacific has a wider selenizone, a more narrow sutsel on early whorls (<0.25× vs. 0.5× width of selenizone), and weaker axial cords.

Anatoma pseudoequatoria from the Indo-Malayan Archipelago to the Central Pacific has a more subquadrate whorl profile, no spiral cord in position of the selenizone on teleoconch I, and a distinct funiculus.

Anatoma globulus from the Indo-Malayan Archipelago has a more globular overall shell outline, a longer teleoconch I (0.6 vs. 0.3 whorl), and a more convex shoulder.

Anatoma japonica from the Indian Ocean to the Central Pacific has a longer teleoconch (>0.5 vs. 0.3 whorl), a protoconch with an apertural varix, and a protoconch that is not projecting at the apex.

Anatoma rapaensis from the Indo-Malayan Archipelago to the Central Pacific has no sutsel on early teleoconch II, a more impressed suture, and spiral steps on shoulder and base.

Figure 904. Distribution of *Anatoma tangaroa*. ☆ type locality. ● specimen record.

Distribution. Northernmost New Zealand (Fig. 904).

Specimen records.

New Zealand. 192–202 m, S of Great Island, Three Kings Islands, 34.235°S, 172.150°E (NMNZ M.160289, 10).

Remarks. Some specimens have axial color bands (Color Plate 1A–B).

Antomidae
Anatoma

Anatoma tenuis (Jeffreys, 1877)

Figures 905–910

Chresonymy.

Scissurella tenuis Jeffreys, 1877: 234, not illustrated.

Scissurella tenuis: Paetel, 1888: 289.

Scissurella tenuis: Kobelt, 1888: 258.

Scissurella tenuis: Pilsbry, 1890: 53.

Scissurella (*Schizotrochus*) *tenuis*: Nordsieck, 1968: 10, pl. 1, fig. 2.12.

Scissurella tenuis: Geiger, 2003: 78.

Synonyms.

+ *Scissurella (Schizotrochus) richardi* Dautzenberg & H. Fischer, 1896: 487, pl. 21, figs 2–3 (Fig. 905). ***Type material.*** (whereabouts unknown, not MNHN), 3.2 × 2.4 mm. ***Type locality.*** Azores. *Hirondelle* (1888), Station 69 [38°33'25"N, 30°28'54"W], 1300 m; *Princess Alice* 1895, Station 46 [37°42'40"N, 27° 27'30"W], 1385 m (OD). ***Etymology.*** [not given].

Scissurella (*Schizotrochus*) *richardi*: Dautzenberg & H. Fischer, 1897: 179.

Scissurella richardi: Locard, 1898: 69 [only variety of *An. umbilicata*].

Scissurella richardi: Thiele, 1912a: 5.

Scissurella (*Schizotrochus*) *richardi*: Dautzenberg, 1927: 214, pl. 7, figs 8–9.

Scissurella richardi: Clarke, 1962: 7.

Scissurella (*Schizotrochus*) *richardi*: Nordsieck, 1982: 2, pl. 1, fig. 16.

Scissurella richardi: Geiger, 2003: 77.

Scissurella richardi: Ardovini & Cossignani, 2004: 19.

Figure 905. Original illustration *Scissurella richardi.*

+? *Scissurella* (*Schizotrochus*) *umbilicata* var. *depressa* Locard 1898: 69, not illustrated. ***Type material.*** (whereabouts unknown). ***Type locality.*** Azores, 1165–1919 m (OD). ***Etymology.*** Describing the more depressed shape of the shell (OD).

+ *Scissurella josephinae* Odhner, 1960: 383–387, figs 12–17. ***Type material.*** Holotype (GNM Gastrop. 5339: Sandberg & Warén, 1993. Odhner, 1960: figs 15–17. Figs 906, 908) 3.7 × 2.9 mm (W × H), complete specimen. One paratype (whereabouts unknown: Odhner, 1960: figs 12–17), 5.2 × 4.5 mm: Swedish Deep-Sea Expedition. ***Type locality.*** St. 373 (sledge trawl), 28°25'N, 61°05'W–28°05'N, 60°49'W. Aug. 23–24 1948, 5500–5987 m [northeastern Nares Deep, ~ 650 km southeast of Bermuda] (OD). ***Etymology.*** Named for the Swedish Josephine Expedition, during which the holotype was collected.

Scissurella josephinae: Sandberg & Warén, 1993: 126.

Scissurella josephinae: Geiger, 2003: 77.

Type material. Syntypes (USNM, 181649, 1: Warén, 1980: pl. 2, figs 13–14, 4.8 mm, here designated as lectotype [see Remarks] 4.4 mm wide, Fig. 907; USNM 766602, 1: paralectotype), 0.25 × 0.2" [= 6.65 × 5 mm].

Type locality. 1450 fms. [= 2651 m], Station 12 [Valorous Expedition] (OD), 56°11'N 37°41'W (Warén, 1980).

Etymology. Tenuis, Latin for thin.

Description. Shell large (lectotype 4.8 mm. 5.2 mm: *josephinae*; 6.6 mm *fide* Jeffreys, 1877. Figs 907, 908), trochiform, biconical. Protoconch of 0.75 whorl, 290 µm, sculpture flocculent, no apertural varix, apertural margin straight. Teleoconch I approximately 0.5 whorl, approximately 12 axial cords, no spiral sculpture. Teleoconch II of 2.5 whorls [lectotype], suture weakly incised, immediately below lower keel of selenizone. Shoulder rounded, approximately 55 fine axial cords per half-whorl; first fine spiral cord after 0.25 whorl, seven spiral cords after one whorl, 21 spiral cords after 2.5 whorls, quite regularly spaced, differing 2× in strength, forming minute points at intersections with axials. Base continuously sloping with umbilicus, without constriction below selenizone, sculpture same as on shoulder. Umbilicus open, very narrow, no funiculus. Selenizone at periphery; keels moderately elevated, moderately strong; lunules distinct; slit most likely open. Aperture rounded.

Operculum multispiral, nucleus central (Odhner, 1960). Remainder of animal unknown.

Comparisons. *Anatoma tenuis* is characterized by its fine reticulate sculpture and the position of the suture immediately below the selenizone of the previous whorl. The species does not show the usual drop of the final quarter whorl, and at 5.2 mm size cannot be regarded as being a juvenile specimen.

Anatoma crispata from the northeastern Atlantic has a more arched shoulder and base, giving it an inflated appearance, and has a distinct funiculus.

Anatoma eximia from the Mediterranean to Liberia has a more turreted overall shell shape, a wide sutsel already on early whorls, and a narrow umbilicus.

Anatoma micalii from the Mediterranean to Angola has a more turreted shell shape, strong axial cords on the shoulder, and different sculpture on shoulder and base.

Figure 906. Original illustration of holotype (top row) and paratype (bottom row) of *Scissurella josephinae*.

Anatoma orbiculata from West Africa has strong axial cords on the shoulder, different sculpture on shoulder and base, and usually a sutsel even on the earliest whorls.

Anatoma tenuisculpta from the Mediterranean and Norway to Angola has a sutsel on early whorls, a longer teleoconch I (0.75 vs. 0.5 whorl), and a wider umbilicus.

Anatoma atlantica from the NW Atlantic has a wider sutsel (0.5–1× vs. 0× width of selenizone), a longer teleoconch I (0.75 vs. 0.5 whorl), and a smaller protoconch (190 vs. 290 μm).

Anatoma proxima from the Caribbean to Brazil has the suture placed well below the selenizone of the previous whorl, showing minimally 3–4 spirals on the upper part of the base above the suture.

Anatoma argentinae from the abyssal plain of the Argentine Basin and the tropical Eastern Pacific has a larger protoconch (370–480 vs. 290 μm) with very fine reticulate sculpture and a shorter teleoconch I (0.3 vs. 0.5 whorl) with fewer spiral cords (11 vs. 18).

Anatoma disciformis from the North Pacific has a more impressed suture, a protoconch with coarse flocculent sculpture, and denser clathrate sculpture formed by duller axials and spirals.

Figure 907. *Scissurella tenuis*. Lectotype here designated. 2651 m, Valorous Station 12, 56.183°N, 37.683°W (USNM 181649, 1). Scale bar shell = 1 mm. Scale bar protoconch = 100 μm.

Anatoma euglypta from Antarctica has a wider sutsel on the earliest whorls and about twice as many axial cords and spiral lines at the same whorl count.

Anatoma japonica from the Indian Ocean to the Central Pacific has a sutsel already at the onset of the selenizone, a wider umbilicus, and a more narrow selenizone.

Anatoma maxima from the Indo-Malayan Archipelago to the Central Pacific has a longer teleoconch I (0.75 vs. 0.5 whorl), more elevated keels of the selenizone, and a wider selenizone.

Anatoma megascutula from the Western to the Central Pacific has a longer teleoconch I (0.75 vs. 0.5 whorl), predominantly spiral sculpture on shoulder and base, and the selenizone higher on the whorl.

Anatoma regia from temperate New Zealand has a wider umbilicus, a more impressed suture, and a distinct sutsel after 1–2 teleoconch II whorls.

Anatoma yaroni from the western Indian Ocean has a distinct sutsel after 1–2 teleoconch II whorls, stronger axial cords, and a less convex shoulder.

Figure 908. *Scissurella josephinae*. Holotype. 5500–5987 m, St. 373, Bermuda, 28.25°N, 60.917°W (GNM Gastrop. 5339, 1: complete). Z-stacked light microscopy. Scale bar = 1 mm.

Distribution. North Atlantic deep sea, 310–6,000 m (Fig. 909).

Specimen records.

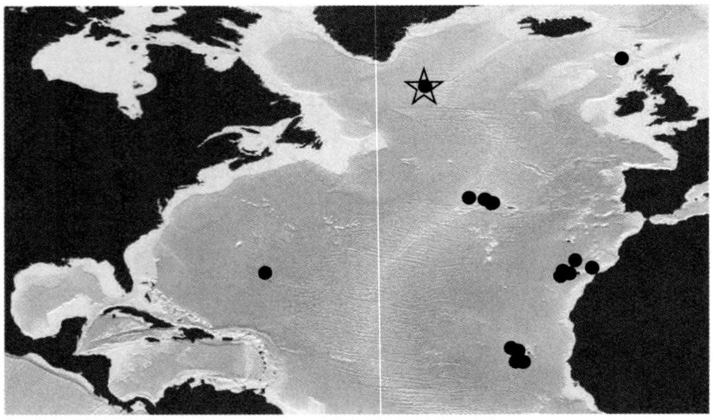

Figure 909. Distribution of *Anatoma tenuis.* ☆ type locality. ● specimen record.

Portugal, Azores. 1229 m, near Flores, (KBIN, 2). 1600 m, (KBIN, 4). 1250 m, [Travailleur and Talisman] Station 1349 [off Sao Jorge], 38.592°N, 28.096°W (KBIN, 2). 1340 m, Tenerife, [Travailleur and Talisman] St. 1713, (KBIN, 1). 400 m, N of São Jorge, 38.650°N, 27.900°W (RMNH, 3).

Cape Verde Islands. 310 m, S of São Tiago, 14.883°N, 23.500°W (RMNH, 6). 320 m, SE of São Tiago, 14.900°N, 23.633°W (RMNH, 1). 700 m, Ponta Grande da Cidade, S of São Tiago, 14.900°N, 23.633°W (RMNH, 1). 590–610 m, SE of Cima, 14.933°N, 24.633°W (RMNH, 3). 405 m, S Jorge Bay, S of São Nicolau, 16.550°N, 24.267°W (RMNH, 1). 400–430 m, SW of Razo, 16.600°N, 24.617°W (RMNH, 3). 312 m, SW of Santo Antao, 16.967°N, 25.333°W (RMNH, 1). 524 m, Selvagens Archipelago, 30.017°N, 16.017°W (RMNH, 1).

Spain, Canary Islands. 900 m, off de Orchilla Point, SW of Hierro, 27.683°N, 18.167°W (RMNH, 1). 503 m, S of Palma, 28.433°N, 17.850°W (RMNH, 1). 450–480 m, SE of Lanzarote, 28.900°N, 13.550°W (RMNH, 2).

Literature records.

Portugal, Azores. 1600 m, SE of Corvo, 39.183°N, 29.096°W. 1229 m, E of Flores, 39.417°N, 31.367°W. 1450 m, near Tenerife, 28.067°N, 16.825°W. 1300 m, between Pico and San Jorge, 38.556°N, 28.144°W. 1250 m, between Pico and San Jorge, 38.592°N, 28.096°W (Dautzenberg, 1927).

Remarks. The best syntype of *Scissurella tenuis* is here selected as lectotype with the express intent of taxon stabilization, in case other, non-conspecific syntypes should be discovered. The surviving syntype does not show the descending last portion of the last whorl; hence, it may still be an immature specimen at 4.4 mm.

Scissurella richardi is here synonymized under *An. tenuis*. The original description and illustrations show a large species (3.7 mm. Fig. 905) without a sutsel, and fine and dense axial as well as spiral cords. It was collected in deep waters (1300 m) in the North Atlantic. All those indications agree with *An. tenuis*.

Scissurella umbilicata var. *depressa* is here tentatively synonymized under *An. tenuis*, despite being described as a variety of *umbilicata*. The description barely complies with ICZN criteria: it is specimen-based, discriminating characters are provided, and minimal locality information is given. Locard (1898: 69) considered *umbilicata* and *richardi* [= *tenuis*] to be synonymous, and noted that his var. *depressa* is closer to the *richardi* [= *tenuis*] form than to *umbilicata*. It was distinguished by the overall more depressed shape, and a more flattened rounding ("galbe plus surbaissé"), and it was described as being of the same size as *umbilicata-richardi*. Given the extensive size range in those two taxa (to 2.9 mm: *An. umbilicata*; 5–6 mm: *An. tenuis*) it is impossible to consider ontogenetic shape changes that could account for the more depressed shell shape. There is no *Anatoma* species in the North Atlantic that has a more depressed spire than *An. tenuis*.

Scissurella josephinae is here synonymized under *An. tenuis*. Both nominal taxa occur in deep water of the North Atlantic, are particularly large for the genus, have no or a very narrow sutsel, the same protoconch ornamentation (Fig. 910), same teleoconch I whorl count and sculpture, same sculpture on shoulder and base, and the same umbilical configuration.

Figure 910. *Scissurella josephinae*. Protoconch of holotype. Z-stacked light microscopy. Scale bar = 0.5 mm.

The holotype of *Sci. josephinae* is stored in ethanol with an extraordinarily brittle shell. The animal is contained in the shell, but had dried out previously (as is evident from the shrunken dark colored tissue), and was subsequently rehydrated (Fig. 908).

Anatoma tenuisculpta (Seguenza, 1877)

Figures 911–919

Chresonymy.

Scissurella tenuisculpta Seguenza, 1877: 273, pl. 16, fig. 29 (Fig. 911A).
Anatoma tenuisculpta: Lozouet, 1986: 114.
Anatoma tenuisculpta: Palazzi & Villari, 1994: 76, fig. 5.
Anatoma tenuisculpta: Palazzi & Villari, 1996: 243.
Scissurella tenuisculpta: Vazzana, 1996: 150, fig. 13.
Anatoma tenuisculpta: Høisaeter & Geiger, 2011: 98–103, figs 34–47.

Synonyms.

+ *Scissurella argutecostata* Seguenza, 1877: 273, pl. 16, fig. 30 (Fig. 911B). ***Type material.*** Syntypes (whereabouts unknown), 3.1 × 4.4 mm. ***Type locality.*** Siderno [Italy, 38.270°N, 16.297°E] (OD). Pliocene (Lozouet, 1986). ***Etymology.*** Argutus, Latin for sharp; with sharp ribs. ***Remarks.*** The figured specimen shows nicely that the suture is adjacent to the selenizone of the previous whorls and the overall rounded profile of the whorl. The name

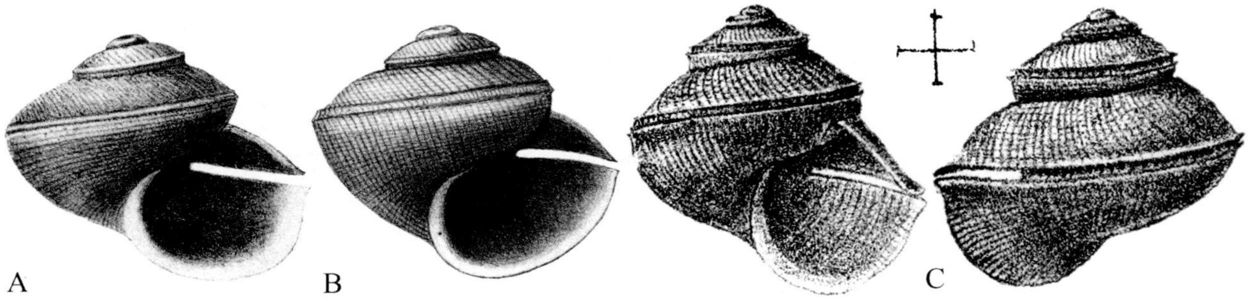

Figure 911. **A.** Original illustration of *Scissurella tenuisculpta* Seguenza (1877). **B.** Original illustration of *Scissurella argutecostata* Seguenza (1877). **C.** Illustration of *Anatoma crispata* Fleming, 1828, from Sars (1878).

was synonymized under *Sci. tenuisculpta* by Høisaeter & Geiger (2011: 102) applying first reviser's principle.

Anatoma argutaecostata [*sic*]: Lozouet, 1986: 114.

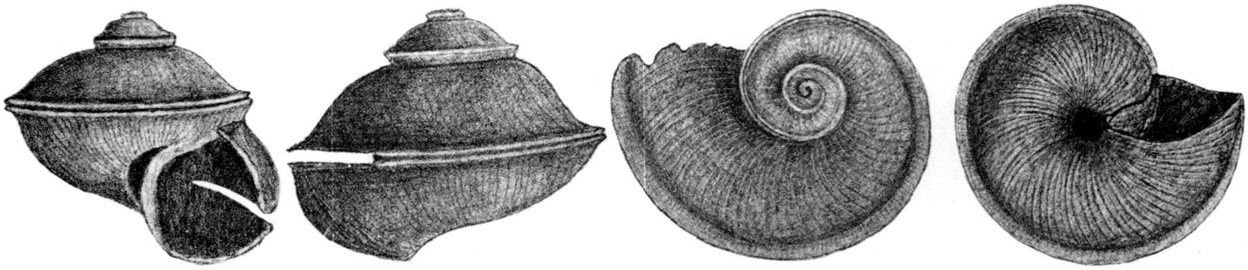

Figure 912. Original illustration of *Scissurella funnazzensis* from Gregorio (1889).

+ *Scissurella* (*Schizotrochus*) *funnazzensis* Gregorio, 1889: 13–14, pl. 1, figs 9a–d (Fig. 912). ***Type material.*** (whereabouts unknown), 2 mm [from scale bar on plate]. ***Type locality.*** Funnazzi [near Palermo, Sicily, Italy, ~38.133°N, 13.383°E, caption to pl. 1(4)], 75–100 fms [137–183 m: introduction] (OD). ***Etymology.*** Derived from *funnu*, a Sicilian variant of the Italian word for seabed, *fondali* (Luigi Giunchi, pers. comm. 1/2010).

Schizotrochus funnazzensis: Monterosato, 1890: 143 [as synonym of *Schizotrochus affinis*].

Scissurella funnazzensis: Pilsbry, 1890: 291[var. of *crispata*].

+ *Scissurella* (*Anatoma*) *subaspera* Boettger, 1907: 197, Nr. 665, not illustrated. ***Type material.*** Lectotype (SMF 2376A: subsequently designated by Zilch, 1934: 200, pl. 2, figs 23a–b). Paralectotype from type locality (SMF 2376B–C, 2). Lapugy [= Lăpugiu de Jos, Romania: 45.888°N, 22.482°E]: Valea coşului (whereabouts unknown, 1). ***Type locality.*** Miocene of Kostej [= Coşteiu de Sus, Romania], Parau ungurului [= B on map in Boettger (1902: 3): 45.906°N, 22.355°E] (Gougerot & LeRenard, 1977). Middle and upper Miocene (Lozouet, 1986), early Badenian (Langhian: 16.0–13.7 MYA). ***Etymology.*** Sub-, Latin for below or less. -aspera: Latin for rough, also a species of Anatoma.

Scissurella subaspera: Zilch, 1934: 200, pl. 2, fig. 23a–b (Fig. 913).

Scissurella subaspera: Musial & Jakubowski, 1975: 337, 338.

Scissurella subaspera: Gougerot & LeRenard, 1977: 5.

Anatoma cf. *subaspera*: Lozouet, 1986: 114, 131, pl. 2, fig. 3.

Anatoma cf. *subaspera*: Lozouet, 1998: fig. 4a [protoconch].

Figure 913. Lectotype of *Scissurella subaspera* from Zilch (1934: pl. 2, fig. 23a–b).

Misidentifications.

Scissurella crispata: Sars, 1878: 126–127, pl. 3 fig. 4 [radula], pl. 8, fig. 7, pl. 18, fig. 1 [operculum] [is *An. tenuisculpta*].

Scissurella crispata: Norman, 1893: 360–361 [is *An. tenuisculpta*].

Scissurella crispata: Brøgger, 1901: pl. 17, fig. 3 [is *An. tenuisculpta*].

Scissurella crispata: Thiele, 1912a: 3–4, pl. 1, fig. 4 [as var. *angulata:* is *An. tenuisculpta*].

Scissurella crispata: Thorson, 1941: 4, 136 [is *An. tenuisculpta*: reference to Sars, 1878].

Scissurella crispata: Macpherson, 1971: 10, pl. 2, fig. 2, map 4 [figure shows *An. tenuisculpta*].

Scissurella crispata: Hubendick & Warén, 1974: 25, fig. 205 [is *An. tenuisculpta*].

Scissurella crispata: Fretter & Graham, 1976: 2, figs 1–2 [is *An. tenuisculpta*].

Scissurella (*Schizotrochus*) *crispata*: Carrozza, 1976: 164, fig. 2 [is *An. tenuisculpta*].

Scissurella crispata: Graham, 1988: 60, figs 13A–C [C: living animal. Is *An. tenuisculpta*].

Anatoma umbilicata: Adrovini & Coossignani, 1999: 32, 2 text-figs [is *An. tenuisculpta*].

Anatoma umbilicata: Ardovini & Cossignani, 2004: 19, text-fig. p. 65 [is *An. tenuisculpta*].
Anatoma crispata: Sneli *et al.*, 2005: 27, fig. 4, map p. 28 [is *An. tenuisculpta*].
Anatoma umbilicata: Beck *et al.*, 2006: 43, text-fig. [is *An. tenuisculpta*].
Anatoma crispata: Peñas *et al.*, 2006: 25, figs 14–17 [is *An. tenuisculpta*].
Anatoma crispata: Cossignani & Ardovini, 2011, 97, text-fig. [is *An. tenuisculpta*].
Anatoma umbilicata: Cossignani & Ardovini, 2011, 97, text-figs [is *An. tenuisculpta*].
Anatoma crispata: Hernández *et al.*, 2011, 62, fig. 11O–Q [is *An. tenuisculpta*].
Anatoma cf. *crispata*: Wesselingh & Pouwer, 2011, 141, fig. 34 [is *An. tenuisculpta*].
Anatoma crispata: Gofas, 2011: 30, text-fig. [is *An. tenuisculpta*].

Figure 914. *Anatoma tenuisculpta*. Neotype. Terreti-Cellantoni Bridge, 567 m above sea level Pleistocene, epibathyal facies, muddy clay, Italy, 38.120°N, 15.716°E (SBMNH 149680). Scale bar shell = 1 mm. Scale bar protoconch = 100 μm.

Type material. Syntypes (whereabouts unknown: not in Florence Bertolosi & Palazzi, 2000), 3.5 × 3.8 mm. Neotype (SBMNH 149680. Fig. 914), designated by Høisaeter & Geiger (2011: 102).

Type locality. Gallina [Calabria, Italy, 38.087°N, 15.679°E] (OD). Pliocene (Lozouet, 1986). Terreti-Cellantoni Bridge, 567 m above sea level Pleistocene, epibathyal facies, muddy clay, Italy, 38.120°N, 15.716°E (Neotype).

Etymology. Latin, tenuis = fine, thin; sculpta = sculptured: finely sculptured

Description. Shell medium-sized (to 5.0 mm. Figs 914–916), trochiform, biconical. Protoconch of 0.75 whorl, apertural varix not connected to embryonic cap, apertural margin sinusoidal. Teleoconch II of 0.5–0.85 whorl, 17–27 axial cords, spiral cord in position of selenizone. Teleoconch II of up to 2.8 whorls, suture moderately impressed, sutsel 1× width of selenizone. Shoulder convex, same density of axial cords as on teleoconch I, first spiral thread 0.1–0.2 whorl after onset of selenizone, 7–9 spirals after one teleoconch II whorl, approximately 14–16 spirals at apertural margin, at intersection with axials forming distinct points; spiral threads often offset between adjacent axial cords. Base barely constricted below selenizone, continuously sloping

Figure 915. *Anatoma tenuisculpta*. Dröback, Norway, 59.663°N, 10.631°E (USNM 181591, 34). Scale bar shell = 1 mm. Scale bar protoconch = 100 µm.

with umbilicus, sculpture as on shoulder. Umbilicus open, without funiculus. Selenizone at periphery; keels moderately elevated, moderately strong; lunules distinct; slit open, margin parallel. Aperture rounded, adumbilical portion flared, roof overhanging.

Radula (Fig. 917A–C). Rachidian tooth with central cusp largest, three cusps on each side. Lateral teeth 1–3 similar, apical cusp largest, three cusps on outer cutting edge of tip. Lateral tooth 4 somewhat reduced, apical cusp largest, one small cusp

Figure 916. *Anatoma tenuisculpta*. **A–B.** Dröback, Norway, 59.663°N, 10.631°E (USNM 181591, 34). **C.** 400 m, Capraia, Italy, 43.050°N, 9.900°E (SRC, 1). Scale bars shell = 1 mm. Scale bars protoconch = 100 μm.

on each side. Lateral tooth 5 enlarged by elongation, apical cusp largest, 10 cusps along inner edge of tip. Inner marginal teeth unequilaterally triangular, apical cusp largest, three cusps along inner edge, 4–5 cusps on outer edge. Outer marginal teeth spoon-shaped, with many bristles. Outermost marginal teeth paddle-shaped, with fine terminal bristles. No food groove on shaft of marginal teeth. Radular interlock of central field moderate.

Operculum (Fig. 917D) round, covering aperture, thin, multispiral, with central nucleus.

Animal (Fig. 918). Head with papillate cephalic tentacles, basal eye on eyestalk, large smooth suboptic tentacle. Accessory cephalic tentacle between suboptic tentacle and first trifurcate cluster of epipodial tentacles; anterior two tentacles of first cluster papillated epipodial tentacles, third stubby tentacle ESO; Arrangement of further posterior papillate epipodial tentacles uncertain due to preservational state of available material. Metapodium with distinct triangular opercular attachment point. Gill leaflets with bursicles.

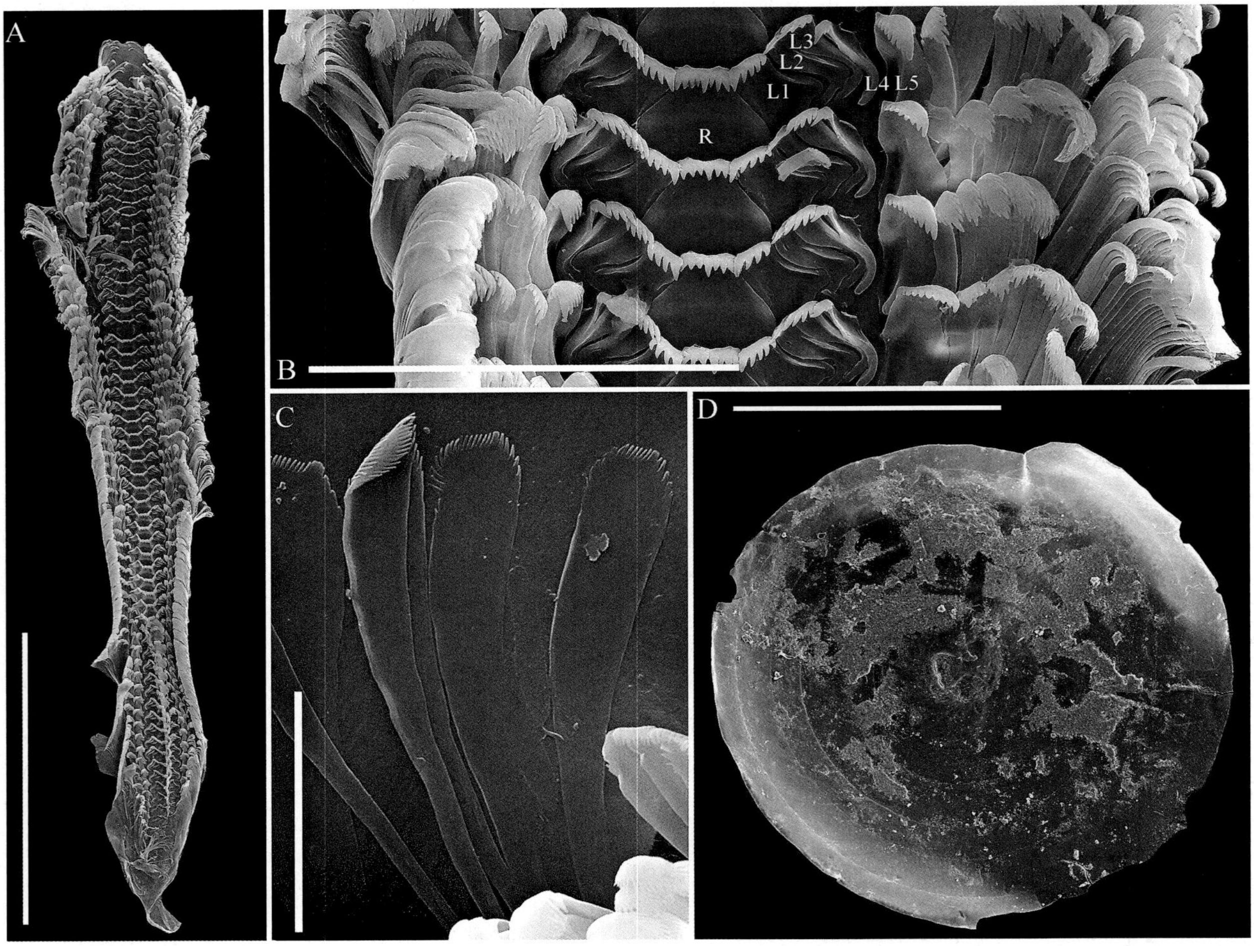

Figure 917. *Anatoma tenuisculpta*. 60–130 m, Raunefjord, Norway, 60.271°N, 5.144°E (NMSZ 1973.59.1, 5). **A–C.** Radula. **A.** Entire radula, scale bar = 500 µm. **B.** Full width of ribbon. R: Rachidian tooth. L1–5: lateral tooth 1–5. Scale bar = 100 µm. **C.** Outermost marginal teeth. Scale bar = 20 µm. **D.** Operculum. Scale bar = 1 mm.

Comparisons. *Anatoma crispata* from the North Atlantic is overall more globular with a more rounded shoulder and base, lacks an apertural varix on the protoconch, and has a distinct funiculus in the umbilicus.

Anatoma bertheloti from the Canary Islands has a more subquadrate whorl profile, a wider umbilicus, denser axial sculpture on shoulder and base, and lacks spiral sculpture.

Anatoma eximia from the Mediterranean to Liberia has a more turreted overall shell shape, a much wider sutsel (>2× vs. <0.5× width of selenizone), and weaker axial sculpture.

Anatoma micalii from the Mediterranean to Angola has a wider sutsel (>1× vs. <0.5× width of selenizone), a more turreted overall shell shape, and fewer but stronger axial cords on the shoulder.

Anatoma orbiculata from West Africa has a more elevated and globular overall shell shape, stronger axial cords on the shoulder, and much stronger axial cords on the shoulder than the base.

Anatoma schanderi from the North Atlantic has a wider sutsel at any given whorl count, a more globular overall shell shape, and a protoconch with flocculent sculpture.

Anatoma schioettei from the northeastern Atlantic has stronger but at most half as many axial cords on shoulder and base, a more impressed suture, and at most half as many spiral lines on shoulder and base.

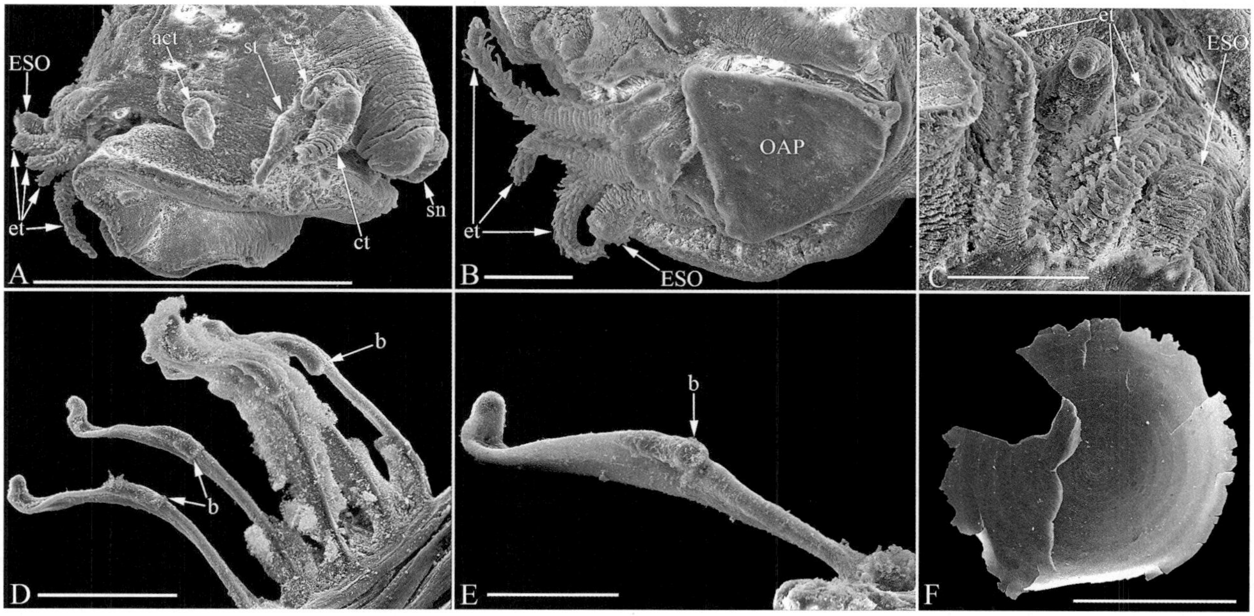

Figure 918. *Anatoma tenuisculpta*. **A–E.** 2500 m, *Challenger* Station ES347, Portugal, 41.633°N, 11.333°W (NMSZ 1994128.60110, 18: complete). **A.** Right side of head-foot of animal. Scale bar = 1 mm. **B.** Ventral view of epipodial tentacles of first trifurcate cluster with two epipodial tentacles (et) and epipodial sense organ (ESO). Scale bar = 200 µm. **C.** Dorsal view of first epipodial cluster. Scale bar = 200 µm. **D.** Bundle of gill leaflets with bursicles (b). Scale bar = 200 µm. **E.** Single gill leaflet with bursicle. Scale bar = 100 µm. **F.** 1920 m, *Challenger* Station SE342, Portugal, 39.017°N, 10.667°W (NMSZ 1994128.60109, 8: complete). Operculum. Scale bar = 1 mm. act: accessory cephalic tentacle. b: bursicle. ct: cephalic tentacle. e: eye. ESO: epipodial sense organ. et: epipodial tentacle. OAP: operculum attachment point. sn: snout.

Anatoma tenuis from the North Atlantic has no sutsel to at least 4.4 mm, a shorter teleoconch I (0.5 vs. 0.75 whorl), and a more narrow umbilicus.

Anatoma philippiana from the Paleogene of Europe has a protoconch without apertural varix, a more globular overall shell shape, and at least 50% less dense axial sculpture on shoulder and base.

Anatoma amydra from the Indo-Malayan Archipelago to the Central Pacific has a wider sutsel on early whorls (>1× vs. <0.5× width of selenizone), a more turreted overall shell shape, and at most half as many spiral lines on the shoulder.

Anatoma atlantica from the northwestern Atlantic has a teleoconch I without a spiral cord in position of the selenizone and usually weaker but denser axial cords.

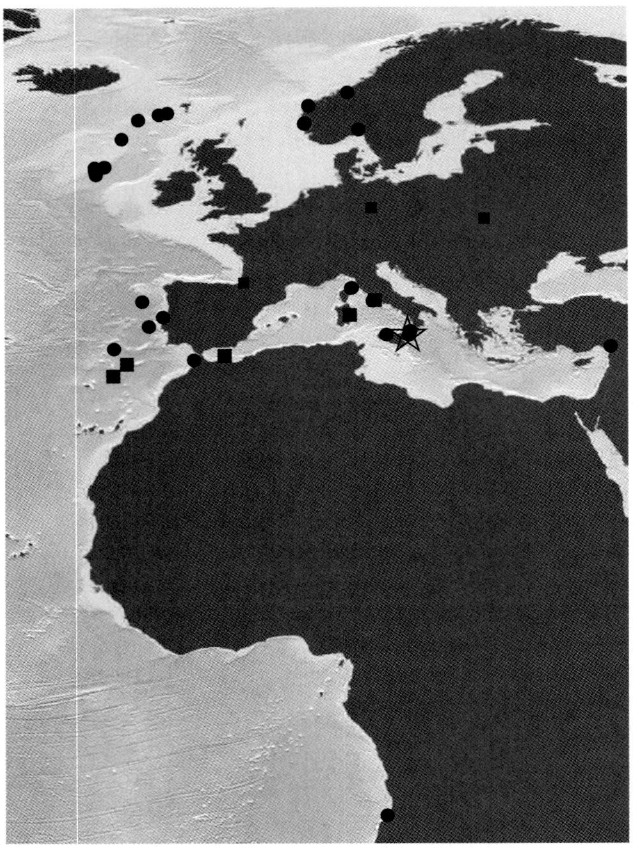

Figure 919. Distribution of *Anatoma tenuisculpta*. ☆ type locality. ● specimen record. ■ literature record.

Anatoma disciformis from the North Pacific has no sutsel on early whorls, wider but duller axial cords and spiral lines, and a shorter teleoconch I (0.4 vs. 0.75 whorl).

Anatoma tangaroa from northernmost New Zealand has about 50% fewer axial cords and spiral lines on shoulder and base, a shorter teleoconch I (0.4 vs. 0.75 whorl), and a wider selenizone.

Anatoma globulus from the Indo-Malayan Archipelago has a more convex shoulder and base and a more globular overall shell shape.

Anatoma japonica from the Indian Ocean to the Central Pacific tends to have less dense axial cords particularly on the base.

Anatoma rainesi from the Eastern to the Western Pacific has a sunken protoconch, a teleoconch I with at most half as many very strong axial cords, and an ontogenetic transition of predominantly axial to more distinct spiral sculpture.

Anatoma regia from temperate New Zealand has at least 50% more spiral lines on the shoulder, weaker axial cords, and a slightly shorter teleoconch I (0.6 vs. 0.75 whorl).

Anatoma yaroni from the western Indian Ocean has a more narrow umbilicus, a slightly less convex shoulder and base, and a more narrow sutsel after two whorls (0.5–1× vs. >1× width of selenizone).

Distribution. Mediterranean, northeastern Atlantic ocean (Fig. 919).

Specimen records.

Faroe Islands. 700 m, [286 km SW of], 60.696°N, 11.675°W (ZMUC, 9). 900 m, [156 km W of S Suduroy, 61.250°N, 9.583°W (ZMUC, 1). 900 m, 170 km SW of, 61.250°N, 9.583°W (ZMUC, 4). 352 m, [125 km SW of], 61.411°N, 8.733°W (ZMUC, 5).

Norway. (NHMUK, 1). Dröback, 59.663°N, 10.631°E (USNM 181590, 119). Dröback, 59.663°N, 10.631°E (USNM 181591, 34; USNM 181589, 8). 60–130 m, Raunefjord, 60.270°N, 5.144°E (NMSZ 1973.59.1, 5). Liholmene, 60.362°N, 5.175°E (RMNH AD, 2). 150–200 m, Møre og Romsdal, Rovdefjorden, NE of Kopperevet, 62.195°N, 5.583°E (T 70048, 4). 50 m, Sletvik, Trondjheimfjord, 63.582°N, 9.521°E (RMNH AD ex 16807, 3).

Great Britain. 765 m, Rockall Trough, off W Scotland, 55.733°N, 16.117°W (NMR 993000033717, 32). 560 m, Rockall Trough, off W Scotland, 55.816°N, 15.133°W (NMR 993000033720, 6). 1430 m, George Bligh Bank, 58.716°N, 13.386°W (NMSZ 2008.131.390, 1: complete).

Ireland. 650–750 m, W of, 55.000°N, 16.000°W (DLG 1216, 1). 784 m, Rockall Trough, 55.400°N, 16.100°W (DLG 1215, 2). 587 m, W of Rockall Trough, 55.500°N, 15.800°W (DLG 1217, 1).

Portugal. 621–786 m, Josephine Bank, Azores, 36.666°N, 14.250°W (USNM 181618, 1). 1920 m, *Challenger* Station SE342, 39.016°N, 10.667°W (NMSZ 1994128.60109, 8: complete). 1818 m, W of, 40.000°N, 9.250°W (USNM 181602, 6). 2500 m, *Challenger* Station ES347, 41.633°N, 11.333°W (NMSZ 1994128.60111, 21; NMSZ 1994128.60110, 18: complete).

Spain. 543 m, Cadiz Bay, 35.500°N, 6.132°W (NMR 993000033757, 12).

Italy. 70–100 m, N Thyrrhenian Sea (SRC, 1). Palermo, 38.133°N, 13.383°E (ZMUC, 3). 400 m, off Gioia Tauro. Calabria, 38.503°N, 15.820°E (DLG 1640, 10). 350 m, Fiumicino, Rome, 41.700°N, 12.000°E (NMR ex 993000033735, 1). 440 m, Capraia Island, Livorno, 43.040°N, 9.820°E (DLG 1281, 1). 400 m, Capraia, 43.050°N, 9.900°E (SRC, 1).

Turkey. Dortyol, Iskendrum, 36.852°N, 36.129°E (RMNH AD 29542, 1).

Angola. 120 m, Station 14, 12.350°S, 13.367°E (DLG 2019, 1).

Literature records.

France. 170 m, Seine Seamount, 33.833°N, 14.333°W. 110–800 m, Ampère Seamount, 35.064°N, 12.930°W (Beck *et al.* 2006).

Italy. 450 m, Fiumicino, Italy, 41.767°N, 12.233°E (Ardovini & Cossignani, 1999).—Vallone Catrica, Italy, Pleistocene (Vazzana, 1996).—100–200 m, E Sardinia, 40.200°N, 9.700°E (Carrozza, 1976: as *An. crispata*).

Spain. 60–400 m, Alboran Island, 35.951°N, 3.033°W (Peñas *et al.*, 2006).

Fossil. The species was described from the Plio-/Pleistocene of Reggio Calabria, Italy, and has been reported from there subsequently (Vazzana, 1996). The neotype for the species was designated from specimens from that region by Høisaeter & Geiger (2011).

Literature records fossils.

Poland. Roztocze, 50.273°N, 23.444°E (Musial & Jakubowski, 1975: as *An. subaspera*).

France. Peyrehorade (Peyrère), 43.545°N, 1.104°W (Lozouet, 1998: as *An. subaspera*).

Remarks. The type locality is given as "Gallina," of which there are five in Italy. The restriction to the Gallina in Reggio Calabria is based on the title of the Seguenza volume. Type material of about 40% of Seguenza's taxa is extant in the collection of the Museum of Geology and Paleontology of Florence, Italy, but none of the scissurellid taxa are among them (Bertolosi & Palazzi, 2000).

The publication date for *Scissurella subaspera* has also been cited as 1906 (Lozouet, 1986). Boettger's second work on the Miocene Fauna of Kostej was printed in two volumes in 1906 and 1907. *Scissurella subaspera* was published in the 1907 volume. One should not be lead astray by the continuous pagination across two volumes, which was done intentionally as explained in the meeting notes of April 3, 1906, point 4, on page XLIX of volume 55 from 1907.

Boettger (1907: 197) noted that "Bemerkung. Sicher der Vorfahre der im Mittelmeer lebenden *S. aspera* Phil., von der sie sich nur bei ganz genauer Vergleichung unterscheiden lässt und von der sie vielleicht nur eine Varietät darstellt." [Note. Certainly the ancestor of *S. aspera* Phil. living in the Mediterranean, from which it can only be differentiated through very close comparison and of which it possibly is only a variety]. The comparison of *An. subaspera* to *An. aspera* by Boettger (1907) is founded on the wide-spread confusion about the identities of *An. aspera*, *An. tenuisculpta*, and *An. crispata* (see Høisaeter & Geiger, 2011 for detailed discussion). *Anatoma subaspera* is a synonym of *An. tenuisculpta*, not of *An. aspera*, based on the same overall shell shape, same sutsel, and same sculpture on shoulder and base of the teleoconch; the condition of the protoconch of the lectotype of *subaspera* is unknown. Additionally, *subaspera* occurs within the geographic and geologic range of *tenuisculpta*.

Anatoma tobeyoides Geiger & Jansen, 2004

Figures 920–922

Chresonymy.

Anatoma "East coast": Jansen, 1999: 49, figs 13–15.
Anatoma tobeyoides Geiger & Jansen, 2004a: 21–25, figs 10–12, 18 [map].
Anatoma tobeyoides: Geiger, 2006c: text-fig. 5.

Misidentification.

Scissurella australis: Hedley & May, 1908: 109 [AMS C.29025: is *An. tobeyoides*].

Type material. Holotype (AMS C.431072. 1.47 mm. Fig. 920). Paratypes from type locality (AMS C.402673, 7. Fig. 921B). 183 m, off Cape Pillar, Tasmania, Australia (AMS C.29025, 8. Fig. 921A). 27 m, Great Barrier Reef, Swain Reefs, Bylund Gillett Cay, Queensland, Australia (AMS C.402671, 1. Geiger & Jansen, 200a. Fig. 12B). 219 m, E of Brush Island, New South Wales, Australia (AMS C.402672, 1. Geiger & Jansen, 2004a: fig. 12C). Off Moreton Bay, Queensland, Australia (AMS C.402674, 1. Geiger & Jansen, 2004a: fig. 13B). Off Long Point, N of Bicheno, Tasmania, Australia (AMS C.406337, 5. Fig. 921C).

Type locality. West of D'Entrecasteaux Channel, SE Tasmania, Australia, 43.045°S, 147.347°E (OD).

Etymology. The sculpture of fine, irregularly intersecting lines reminds of the all-over paintings of Mark Tobey (1890–1976): tobey-: referring to the painters surname; -oides: Latin, like, similar to (OD).

Description. Shell globular, medium-sized to large (4.2 mm. Figs 920, 921). Protoconch 0.75 whorl, sculpture flocculent, no apertural varix, apertural margin straight. Teleoconch I <0.5 whorl, no spiral cord in position of selenizone, fine centrifugal markings, strong axials present. Teleoconch II up to 2.25 whorls (4.2 mm shell). Sculpture with axial cords predominant, approximately 85 on body whorl on fully grown shell, approximately 15 spiral threads on shoulder at apertural margin. Approximately 25 spiral threads on base. Umbilicus continuously sloping from base, narrow, deep, no funiculus. Selenizone at periphery; keels distinct, moderately elevated; lunules poorly coordinated with axials; slit open, margins parallel. Aperture rounded, somewhat flared.

Animal unknown.

Comparisons. *Anatoma australis* from eastern Australia has a protoconch with an apertural varix, a longer teleoconch I (>0.5 vs. <0.5 whorl), a spiral cord in position of the selenizone on teleoconch I, and only every 3–4 axial cords on the base reach the umbilicus.

Anatoma biconica from the Indian Ocean to the Central Pacific has a more turreted overall shell shape, a longer teleoconch I (0.8 vs. 0.2 whorl), and a wider sutsel at any whorl count.

Anatoma planapex from the Indo-Malayan Archipelago and Australia has more turreted whorls, a wider sutsel at any whorl count, and a wider umbilicus.

Anatoma atlantica from the northwestern Atlantic has a more lenticular overall shell shape, a longer teleoconch I (0.75 vs. 0.2 whorl), and a less convex shoulder and base.

Anatoma crispata from the North Atlantic has a longer teleoconch I (0.7 vs. 0.2 whorl), a prominent funiculus, and >50% fewer axial cords on shoulder and base.

Anatoma tangaroa from northernmost New Zealand has a wider sutsel after one teleoconch II whorl (1× vs. <0.5× width of selenizone) and at most half as many axial cords and spiral lines on shoulder and base.

Figure 920. *Anatoma tobeyoides*. Holotype. W of D'Entrecasteaux Channel, Tasmania, Australia (AMS C.431072). 1.47 mm. Scale bar shell = 1 mm. Scale bar protoconch = 100 μm.

Anatoma globulus from the Indo-Malayan Archipelago has a longer teleoconch I (0.6 vs. 0.2 whorl) with a spiral cord in position of the selenizone, a protoconch with apertural varix), and a more narrow umbilicus.

Figure 921. *Anatoma tobeyoides.* Paratypes. **A.** 183 m, off Cape Pillar, Tasmania, Australia (AMS C.29025, 8). **B.** W of D'Entrecasteaux Channel, Tasmania, Australia (AMS C.402673, 7). **C.** Off Long Point, N of Bicheno, Tasmania, Australia (AMS C.406337, 5). Scale bars shell = 1 mm. Scale bars protoconch = 100 μm.

Anatoma japonica from the Indian Ocean to the Central Pacific has a longer teleoconch I (>0.5 vs. 0.2 whorl), a more lenticular overall shell shape, and intersections of axials ad spirals forming minute points.

Anatoma keenae from the Panamic has a more globular overall shell shape, a longer teleoconch I (>0.6 vs. 0.2 whorl), and at most half as many spiral lines on shoulder and base.

Anatoma lamellata from the Western and the Central Pacific has a strong spiral cord on the protoconch, a longer teleoconch I (>0.6 vs. 0.2 whorl), strong axial lamellae on shoulder and base, and has at most half as many spiral lines on shoulder and base.

Anatoma proxima from the Caribbean to Brazil has a longer teleoconch I (0.75 vs. 0.2 whorl), a more biconical overall shell shape, and straight axial cords on the base.

Anatoma rainesi from the Eastern to the Western Pacific has a sunken protoconch, very strong axial cords on teleoconch I with thickenings or a spiral cord in position of the selenizone and an ontogenetic shift of predominantly axial sculpture to more prominent spiral sculpture.

Anatoma soyoae from Japan has a more globular overall shell shape, a longer teleoconch I (0.75 vs. 0.2 whorl), and less elevated keels of the selenizone.

Anatoma tenuisculpta from the Mediterranean and the adjacent Atlantic has a longer teleoconch I (>0.5 vs. 0.2 whorl), a less convex shoulder, and a more narrow umbilicus.

Anatoma yaroni from the western Indian Ocean has a more lenticular overall shell shape, a longer teleoconch I (>0.5 vs. 0.2 whorl), and a more narrow umbilicus.

Anatoma zancliformis from the Indo-Malayan Archipelago to the Western Pacific has the selenizone higher on the whorl, at most half as many axial and spiral cords, and a distinct constriction on the base below the selenizone.

Distribution. Victoria, Tasmania to central Queensland, 27–399 m (Fig. 922).

Specimen records.

Australia, Queensland. 75 m, off Moreton Bay, 27.517°S, 153.667°E (AMS C.402674, 1). 77 m, off Moreton Bay, 27.456°S, 153.650°E (AMS C.402724, 18, 5: complete). Amity Point, Stradbroke Island, Moreton Bay, 27.400°S, 153.433°E (AMS C.402726, 1). 27 m, Great Barrier Reef, Swain Reefs, Bylund Gillett Cay, 21.717°S, 152.417°E (AMS C.402671, 1). 64 m, Great Barrier Reef, Swain Reefs, 3 km NE of W side of Bylund Gillett Cay, 21.700°S, 152.433°E (AMS C.402725, 7).

Australia, New South Wales. 219 m, E of Brush Island, 35.433°S, 150.633°E (AMS C.402672, 1). 40 m, N of Sydney, 33.757°S, 151.360°E (AMS C.402675, 1). 45 m, off Port Stephens, 32.708°S, 152.250°E (AMS C.402727, 1). 73 m, 9 mi NE of Coffs Harbour, 30.250°S, 153.317°E (AMS C.402730, 1). 13 m, E of Yamba, 29.500°S, 153.367°E (AMS C.402731, 1). 86 m, off Ballina, 29.170°S, 153.728°E (AMS C.402728, 1). 103 m, off Ballina, 29.167°S, 153.783°E (AMS C.402729, 1). Narrabeen, 33.714°S, 151.302°E (RMNH AD 30460, 1).

Australia, Tasmania. 212 m, S of Storm Bay, 43.783°S, 147.808°E (AMS C.402636, 6). 82 m, W of Port Davey, 43.338°S, 145.803°E (AMS C.402734, 1). 183 m, off Cape Pillar, 43.217°S, 148.083°E (AMS C.29025, 8). 95 m, N of Cape Pillar, 43.167°S, 148.023°E (AMS C.402732, 1). 82 m, Maria Island, 2.5 mi NE Beaching Bay, 42.458°S, 148.200°E (AMS C.402733, 2). 205 m, off Cape Forestier, 42.167°S, 148.578°E (AMS C.402737, 6). 113 m, off Long Point, N of Bicheno, 41.758°S, 148.517°E (AMS C.402735, 1; AMS C.406337, 4). 88 m, S of West Point, 41.153°S, 144.403°E (AMS C.402736, 1). 399 m, off Cape Naturaliste, 40.843°S, 148.775°E (AMS C.402626, 11). 154 m, Taupo Guyot, off Newcastle, Tasman Sea, 33.103°S,

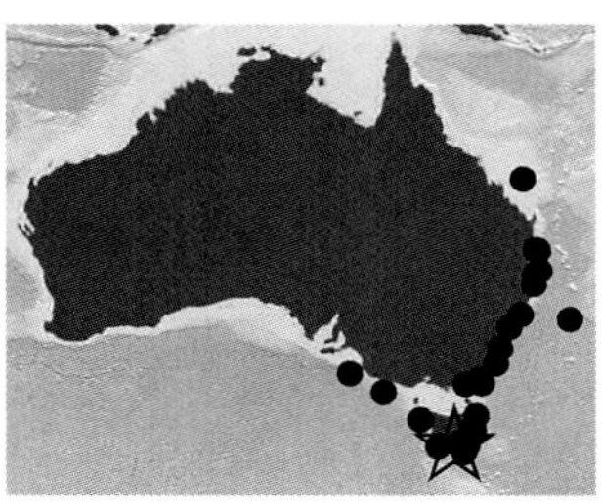

Figure 922. Distribution of *Anatoma tobeyoides*. ☆ type locality. ● specimen record.

156.155°E (AMS C.404497, 1: complete). W of D'Entrecasteaux Channel, 43.045°S, 147.347°E (paratype).

Australia, Victoria. 2000 m, Bass Strait, 30 mi S of Cape Nelson, 38.958°S, 141.542°E (AMS C.402739, 1). 60 m, Bass Strait, ca. 40 km S of Lakes Entrance, 38.317°S, 147.917°E (AMS C.402740, 1). 165 m, Bass Strait, ca. 27 mi SE of Cape Everard, 38.250°S, 149.200°E (AMS C.402738, 8). 75 m, between Cape Howe & Lakes Entrance, 37.917°S, 149.000°E (AMS C.404965, 1).

Remarks. Hedley & May (1908) reported *Sci. australis* from off Cape Pillar, Tasmania in 100 fms [=183 m] depth. AMS C.29025 matches those data precisely and is *An. tobeyoides*.

Anatoma umbilicata (Jeffreys, 1883)

Figures 923–926

Chresonymy.

Scissurella umbilicata Jeffreys, 1883a: 88–89, pl. 19, fig. 1–1a (Fig. 923).
Scissurella umbilicata: Kobelt, 1888: 258–259.
Scissurella umbilicata: Paetel, 1888: 289.
Scissurella umbilicata: Pilsbry, 1890: 52, pl. 51, figs 31–32.
Scissurella (*Schizotrochus*) *umbilicatus* [*sic*]: Dautzenberg & Fischer, 1897: 487.
Scissurella (*Schizotrochus*) *umbilicata*: Dautzenberg & Fischer, 1898: 179.
Scsissurella umbilicata: Locard, 1898: 69.
Scissurella umbilicata: Locard, 1899: 107.
Scissurella umbilicata: Nobre, 1900: 135
Scissurella umbilicata: Thiele, 1912a: 4–5, pl. 1, figs 6–7 [copy figures Jeffreys, 1883a].
Scissurella umbilicata: Pallary, 1912: 70.
Scissurella (*Schizotrochus*) *umbilicata*: Dautzenberg, 1927: 214.
Scissurella umbilicata: Nobre, 1932: 242, pl. 14, figs 10–11.
Scissurella umbilicata: Nobre, 1936: 179.
Scissurella umbilicata: Nobre, 1938–1940: 469–470.
Scissurella umbilicata: Clarke, 1962: 7.
Scissurella umbilicata: Pasteur-Humbert, 1962: 123.
Scissurella (*Schizotrochus*) *umbilicata*: Nordsieck, 1968: 10, pl. 1, fig. 2.11.
Scissurella (*Anatoma*) *umbilicata*: Piani, 1980: 117.
Scissurella (*Schizotrochus*) *umbilicata*: Nordsieck, 1982: 1, pl. 1, fig. 2.11.
Anatoma umbilicata: Sabelli *et al.*, 1990: 12.
Anatoma umbilicata: Bodon *et al.*, 1995: 14.
? *Anatoma umbilicata*: Arduino *et al.*, 1995: 17, text-figs [unidentifiable illustration].
Anatoma umbilicata: Geiger, 2003: 74.
Anatoma umbilicata: Høisaeter & Geiger, 2011: 109.

Figure 923. Original illustration of *Scissurella umbilicata.*

Misidentifications.

not *Scissurella umbilicata*: Poppe & Goto, 1991: 32, 64, pl. 2, fig. 6 [is *An. aspera*].
not *Anatoma umbilicata*: Giannuzzi-Savelli *et al.*, 1994: 13, figs 93a–b [is juvenile *An. aspera*].
not *Hainella umbilicata*: Bandel, 1998: 37–38, pl. 12, fig. 8 [is *An. crispata*].
not *Anatoma umbilicata*: Adrovini & Coossignani, 1999: 32, 2 text-figs p. 33 [is *An. tenuisculpta*].

not *Scissurella* (*Anatoma*) *umbilicata*: Macedo *et al.*, 1999: 267 [size indication too large: 4.3 × 4.5 mm vs. 2.9 mm].

not *Anatoma umbilicata*: Landau *et al.*, 2003: pl. 1, fig. 4 [is *An. eximia*].

not *Anatoma umbilicata*: Ardovini & Cossignani, 2004: 19, text-fig. p. 65 [is *An. tenuisculpta*].

not *Anatoma umbilicata*: Beck *et al.*, 2006: 43, text-fig. [is *An. tenuisculpta*].

not *Anatoma umbilicata*: Portalatina, 2008: 155, fig. 2L–M [is *An. micalii*].

not *Anatoma umbilicata*: Cossignani & Ardovini, 2011, 97, text-figs [is *An. tenuisculpta*].

not *Anatoma umbilicata*: Hernández *et al.*, 2011: 62, fig. 12A–B [is *An. micalii*].

Type material. Syntypes Stations 16, 17, 17a (NHMUK 1885.11.5.1663–1666, 4). Station 16 (USNM 181638, 10). W off Portugal, 39°42'N, 9°43'W, 600–1095 fms. [= 1097–2003 m] (USNM 181639, 9). Station 17a (USNM 181640, 11), 0.1 × 0.1".

The specimen from USNM 181638 shown in Figure 924C is here designated as the lectotype with the express taxonomic purpose of taxon stabilization. That specimen corresponds well with the original illustration, and is well-preserved, showing all characters necessary for positive species identification. All other syntypes become paralectotypes.

Type locality. Porcupine Expedition 1870: Atlantic Station 16 [39°55'N 9°56'W, 994 fms = 1817 m], 17 [39°42'N 9°43'W, 740–1095 fms. = 1354–2004 m], 17a [39°39'N 9°39'W, 795 fms = 1454 m] [OD: coordinates and depth from Jeffreys (1878)].

The type locality is here restricted explicitly to W off Portugal, Station 16, 39°55'N, 9°56'W, 994 fms. [= 1817 m], the locality of the here designated lectotype.

Etymology. Most likely referring to the distinct umbilicus.

Description. Shell of moderate size (to 2.9 mm. Figs 924, 925), trochiform, biconical. Protoconch of one whorl, sculpture flocculent, weak apertural varix not connected to embryonic cap, apertural margin convex. Teleoconch I of 0.55 whorl, smooth, with very fine axial growth marks only. Teleoconch II of 2.875 whorls; suture lightly impressed; sutsel as wide as selenizone on early teleoconch II whorls, at aperture of mature specimen >2× width of selenizone. Shoulder slightly convex, superficially smooth; dense, indistinct, low axial cords towards apertural margin, crossed by approximately six barely perceptible spiral threads. Base with same axial sculpture as on shoulder; approximately 20 axials, barely perceptible fine spiral threads below selenizone, none at periphery, fine spiral threads in umbilical region. Umbilicus moderately wide, continuously sloping with base. Aperture rounded oblong, basoadumbilical region flared. Selenizone slightly above periphery; lunules indistinct; keels of moderate strength, distinctly elevated; slit open, margins somewhat converging.

Animal unknown.

Comparisons. *Anatoma bertheloti* from the Canary Islands has distinct and dense axial cords over the entire shoulder and base and a more subquadrate shell outline.

Anatoma crispata from the northeastern Atlantic has a more globular overall shell shape, stronger axial cords on shoulder and base, and a distinct funiculus in the more narrow umbilicus.

Anatoma eximia from the Mediterranean to Liberia has a much more turreted overall shell shape, a much wider sutsel at any given whorl count, and the selenizone higher on the whorl.

Anatoma janusa from the subtropical northeastern Atlantic has strong axial cords on the shoulder, strong reticulate sculpture on the base, and a more narrow umbilicus.

Anatoma micalii from the Mediterranean to Angola has much stronger axial cords on the shoulder, reticulate sculpture on the base, and a longer teleoconch I (0.75 vs. 0.5 whorl).

Figure 924. *Anatoma umbilicata.* **A–B.** Paralectotypes of *Scissurella umbilicata* here designated. **A.** W off Portugal, *Porcupine* Station 17a, 39°39'N, 9°29'W, 740 fms. [= 1353 m] (USNM 181640, 11). **B.** W off Portugal, 39°42'N, 9°43'W, 600–1095 fms. [= 1097–2003 m) (USNM 181639, 9). **C.** Lectotype of *Scissurella umbilicata* here designated. W off Portugal, , *Porcupine* Station 16, 39°55'N, 9°56'W, 994 fms. [= 1817 m] (ex USNM 181638, 10). Scale bars shell = 1 mm. Scale bars protoconch = 100 μm.

Anatoma orbiculata from West Africa has strong axial sculpture on the shoulder, an overall more lenticular shell shape, and a more narrow umbilicus.

Anatoma tenuis from the North Atlantic has no sutsel up to a shell size of at least 4.4 mm, distinct spiral and axial sculpture, and an overall more lenticular shell shape.

Figure 925. *Anatoma umbilicata*. **A.** Best of four paralectotypes of *Scissurella umbilicata* (NHMUK 1885.11.5.1663–1666). **B.** 2306 m, off Sahara (USNM 181641, 1). **C.** *Anatoma* cf. *umbilicata*. 1664 m, Cuba, 23.633°N, 82.083°W (USNM 780313, 2). Scale bars shell = 1 mm. Scale bars protoconch = 100 μm.

Anatoma tenuisculpta from the Mediterranean and Norway to Angola has a more lenticular shell shape, distinct axial and spiral cords, and a more narrow umbilicus.

Anatoma aedonia from the Caribbean to Brazil has a taller and more turreted overall shell shape, a more narrow umbilicus, and a wider sutsel after one whorl (>1.5× vs. <0.5× width of selenizone).

Anatoma austrolissa from the Indian Ocean to the Western Pacific has an overall taller and more turreted shell shape, a wider sutsel after 1 whorl (>1× vs. <0.5× width of selenizone), and a shorter teleoconch I (0.33 vs. >0.5 whorl).

Anatoma breveprima from the Western Pacific has a shorter teleoconch I (0.1 vs. >0.5 whorl), the shoulder at a flatter angle towards the coiling axis, and a more bulging base.

Anatoma hyposculpta from the Western Pacific has a shorter teleoconch I (0.25 vs. >0.5 whorl), a wider sutsel after one teleoconch II whorl (1x vs. 0.5× width of selenizone, and a shoulder that is at a lower angle towards the coiling axis of the shell.

Anatoma planapex from the Indo-Malayan Archipelago and Australia has the shoulder at a flatter angle towards the coiling axis of the shell, a longer teleoconch I (>0.8 vs. 0.5 whorl), and stronger spiral sculpture on the base.

Distribution. Western Mediterranean, tropical to temperate northeastern Atlantic (and possibly Caribbean. Fig. 926).

Specimen records.

Monaco. 1340 m, near Monaco St. 1713, 43.500°N, 7.500°E (KBIN, 2).

France. 1600–1630 m, Gulf of Lion, 42.345°N, 3.982°E (MNHN, 4). 450–800 m, NE canyon of Toulon, 42.955°N, 6.022°E (MNHN, 3). Gulf of Gascogne, 43.750°N, 1.750°W (MNHN, 1).

Italy. 200–300 m, K2, off Sardegna, 40.250°N, 10.083°E (RMNH AD 12356, 3).

Spain. 392 m, off Galicia, 44.083°N, 14.750°W (MNHN, 1).

Portugal. 2003 m (NMSZ 1961.61, 1). 1245 m, Azores, 38.500°N, 28.000°W (USNM 181642, 1). 90 m, E of Faial, Azores, 38.567°N, 28.567°W (RMNH, 2). 90 m, E of Faial, Azores, 38.583°N, 28.567°W (RMNH, 2). 1600 m, *Alice* DR 90, [140 km ESE off Flores, Azores], 39.183°N, 32.742°W (KBIN, 8). 1097–2002 m, off Lisbon, 39.700°N, 9.717°W (AMS C.24124, 20+).

West Africa. 2306 m, off Sahara (USNM 181641, 1).

Cuba. 1664 m, 23.633°N, 82.083°W (USNM 780313, 2).

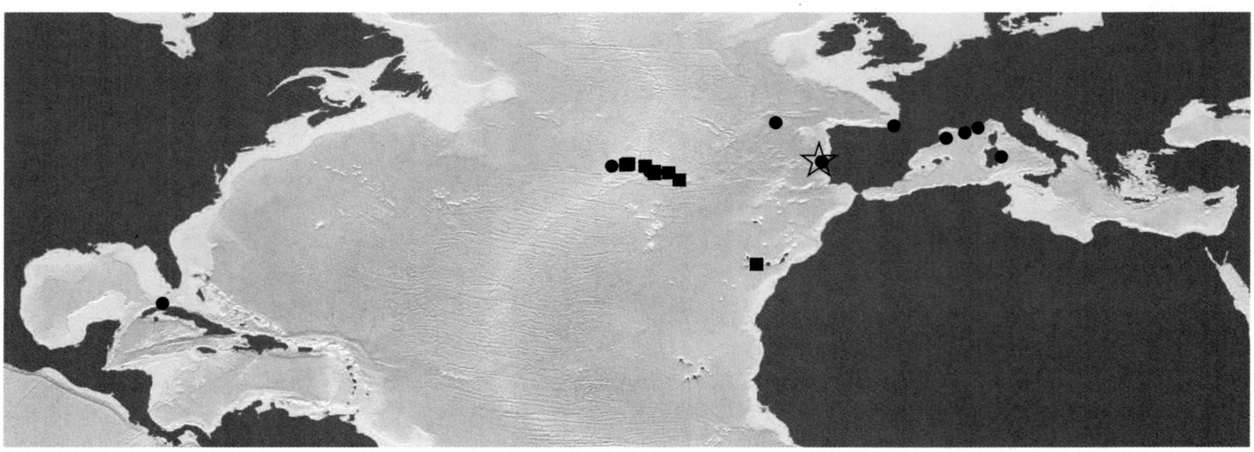

Figure 926. Distribution of *Anatoma umbilicata*. ☆ type locality. ● specimen record. ■ literature record.

Literature records.

Portugal, Azores. 1187 m, 6 mi S of San Miguel, 37.617°N, 25.346°W. 1550 m, S of Pico, 38.333°N, 28.079°W. 1250 m, between Pico and St. Jorge, 38.592°N, 28.096°W. 1600 m, SE of Corvo, 39.183°N, 29.088°W 1450 m, near Tenerife, 28.067°N, 16.825°W. 1165 m, 38.433°N, 26.513°W. 1372 m, 39.301°N, 31.200°W. 1360 m, 39.356°N, 31.100°W. 1557 m, 39.451°N, 30.917°W (Dautzenberg, 1927).

Remarks. Type specimens in NHMUK and USNM are very smooth and not conspecific with *An. crispata*, though Warén (1980) doubted its distinctness from *An. crispata.*

USNM 181639 has slightly different depth indication compared to those of Station 17 (600–1095 fms vs. 740–1095 fms). The rest of the locality data match, for which reason the lot is considered part of the original syntype series, now part of the paralectotypes.

Given the extensive confusion about the identity of true *An. umbilicata*, the references to *An. umbilicata* lacking illustrations have to be viewed with much caution. The erroneous synonymization of *An. richardi* (= *An. tenuis*) under *An. umbilicata* by Locard (1898) that was subsequently accepted (Burnay & Rolán, 1990), has led to further ambiguities in the application of *umbilicata*.

Anatoma weddelliana (Zelaya & Geiger, 2007)

Figures 927–931

Chresonymy.

Thieleella weddelliana Zelaya & Geiger, 2007: 412–415, figs 51 [map], 71–74 [shell SEM], 75–85 [radula, external anatomy].

Type material. Holotype (ZSM-Moll 20062198). Paratypes from type locality (ZSM-Moll 20062199, 2).

Type locality. 70°31.08' S, 14°34.83' W, Weddell Sea, 4384–4420 m (OD).

Etymology. The species is named after the type locality (OD).

Description. Shell large (up to 5.6 mm. Fig. 927), trochiform, laterally expanded. Protoconch of one whorl, about 225 µm in diameter, with reticulated ornamentation; apertural margin convex, no varix. Teleoconch I of 0.75–0.83 rounded whorl, sculptured with 10–12 axials cords, one spiral cord in position of selenizone forming strong angulation. Teleoconch II of up to 2.75 whorls; suture moderately impressed; sutsel less than width of selenizone at onset of selenizone, about 1.5× as wide at apertural margin. Shoulder rounded; last whorl nearly straight. Shell with cancellate sculpture; axial cords strong: 22–26 on shoulder of first whorl, 27–28 on second whorl, 31–41 on base; spiral threads weak, beginning at start of selenizone, gradually increasing in number with growth: 5–8 on first whorl, 25–30 at apertural margin. Selenizone sculptured with low axial cords, surrounded by narrow, strong keels; lunules moderately distinct; slit open, margins parallel. Umbilicus open, wide, deep; funiculus present. Aperture rounded.

Animal (Figs 928, 930) pale cream, uniform. Cephalic tentacles long, lanceolate, flat, sparsely papillate; papillae tubular, of about 110 µm long, spirally arranged, slightly expanded distally with a handful of cilia. No eyes, eye stalks. Head separated from foot by wide neck, with one pair of smooth cylindrical tentacles on each side. Three pairs of conical papillate epipodial tentacles; first epipodial tentacle divided basally, forming two tentacles. Papillae similar to that of cephalic tentacles, shorter (42–50 µm). Columellar muscle thick, elongate in outline. Pallial cavity with enlarged mantle border, deeply sinuated in medial line, one rounded papillate tentacle extending anteriorly in medial line. Two large bipectinate ctenidia; left ctenidium about 10% larger than right one. Hindgut very large, sinuous, drawing posteriorly directed loop at left side before ending in anus.

Radula (Fig. 929). Rachidian tooth trapezoidal at base, tip nearly straight, with one main central cusp, 6–10 narrower, sharper cusps on each side. Lateral teeth 1–3 with rectangular bases, innermost tooth widest; tip with prominent innermost cusp, 7–11 narrower cusps along outer margin; inner margin smooth. Lateral tooth 4 narrower than lateral tooth 3. Lateral tooth 5 claw-shaped, wider at base, with wide, sharply serrated tip. Marginal teeth numerous, long, narrow, becoming more delicate outwards, with prominent, delicately serrated tip; no food groove. Radular interlock of central field moderate.

Figure 927. *Anatoma weddelliana*. Holotype. 4384–4420 m, Weddell Sea, 70°31.08'S, 14°34.83'W (ZSM Mol 20062198). Protoconch from paratype (ZSM-Moll 20062199, 2). Scale bar shell = 1 mm. Scale bar protoconch = 200 µm. Images by Diego Zelaya. From Zelaya & Geiger (2007: modified).

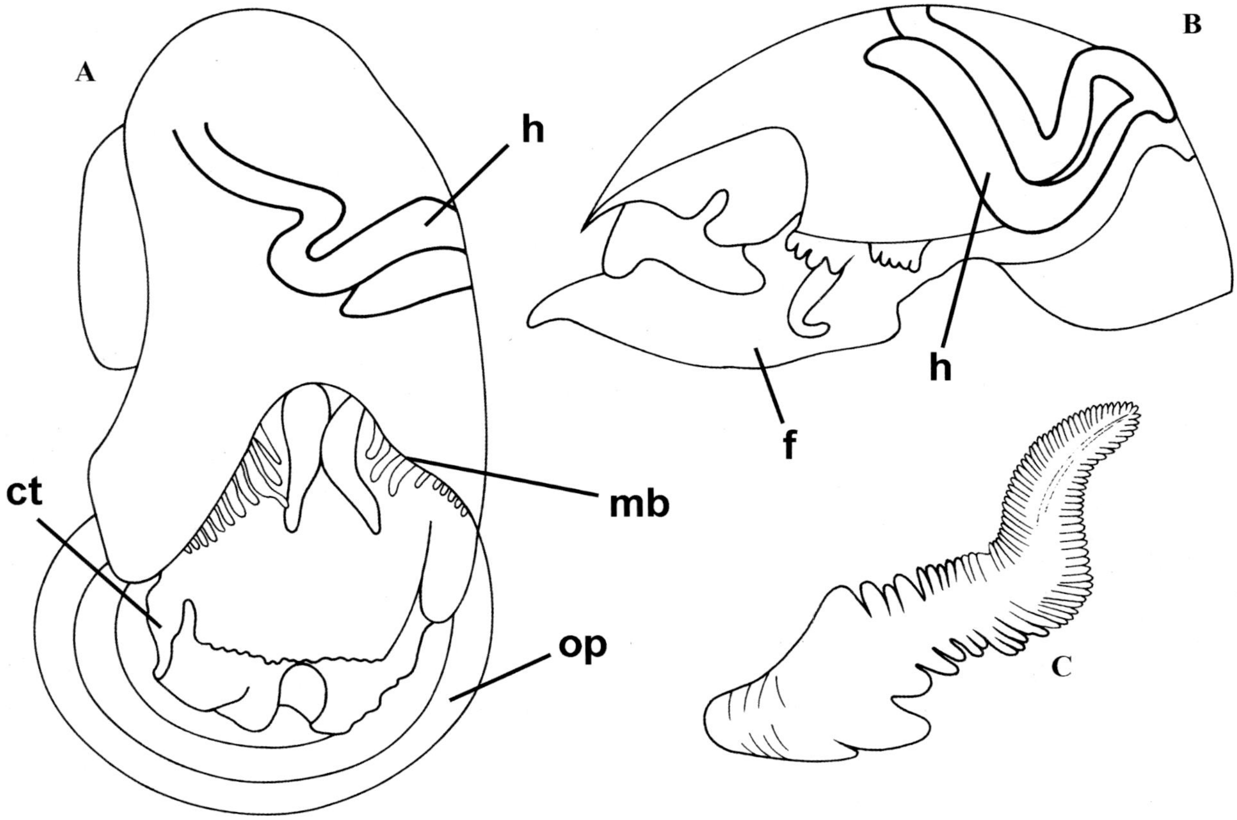

Figure 928. *Anatoma weddelliana*. External anatomy. **A–B.** Detail of hindgut (h); **A.** Frontal view. **B.** Lateral view. **C.** Detail of cephalic tentacle. ct: cephalic tentacle; f: foot; mb: mantle border; op: operculum. Drawings by Diego Zelaya; from Zelaya & Geiger (2007: modified).

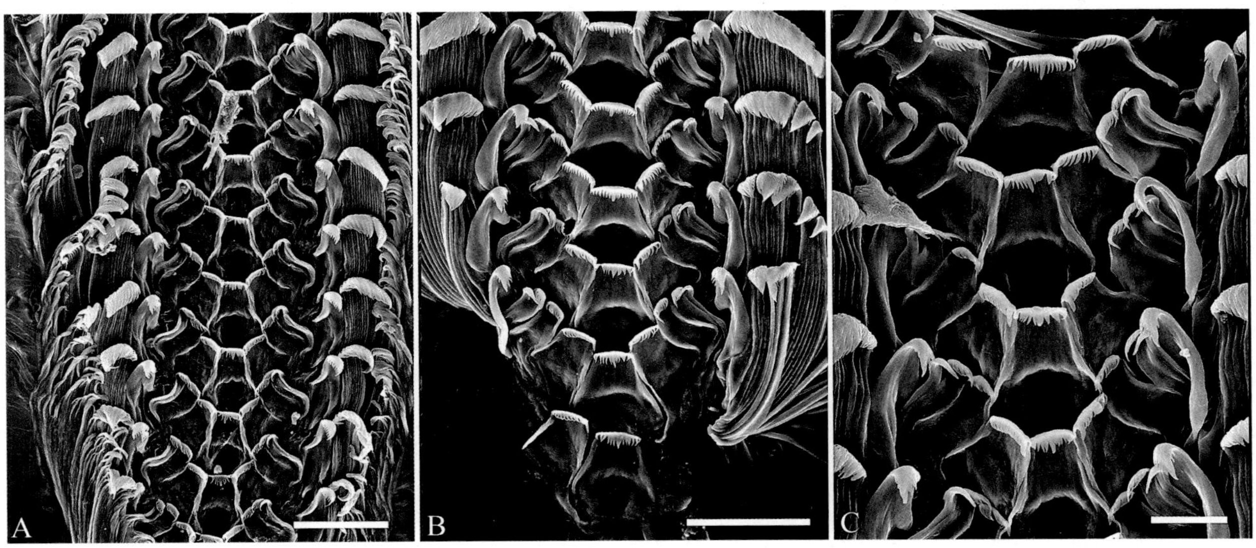

Figure 929. *Anatoma weddelliana*. Radula from paratype. 4384–4420 m, Weddell Sea, 70°31.08'S, 14°34.83'W (ZSM Mol 20062199, 2). **A.** Overview of radula. **B.** Central field and inner marginal teeth. **C.** Central field enlarged. Scale bar A–B = 50 µm. Scale bar C = 20 µm. Images by Diego Zelaya. From Zelaya & Geiger (2007: modified).

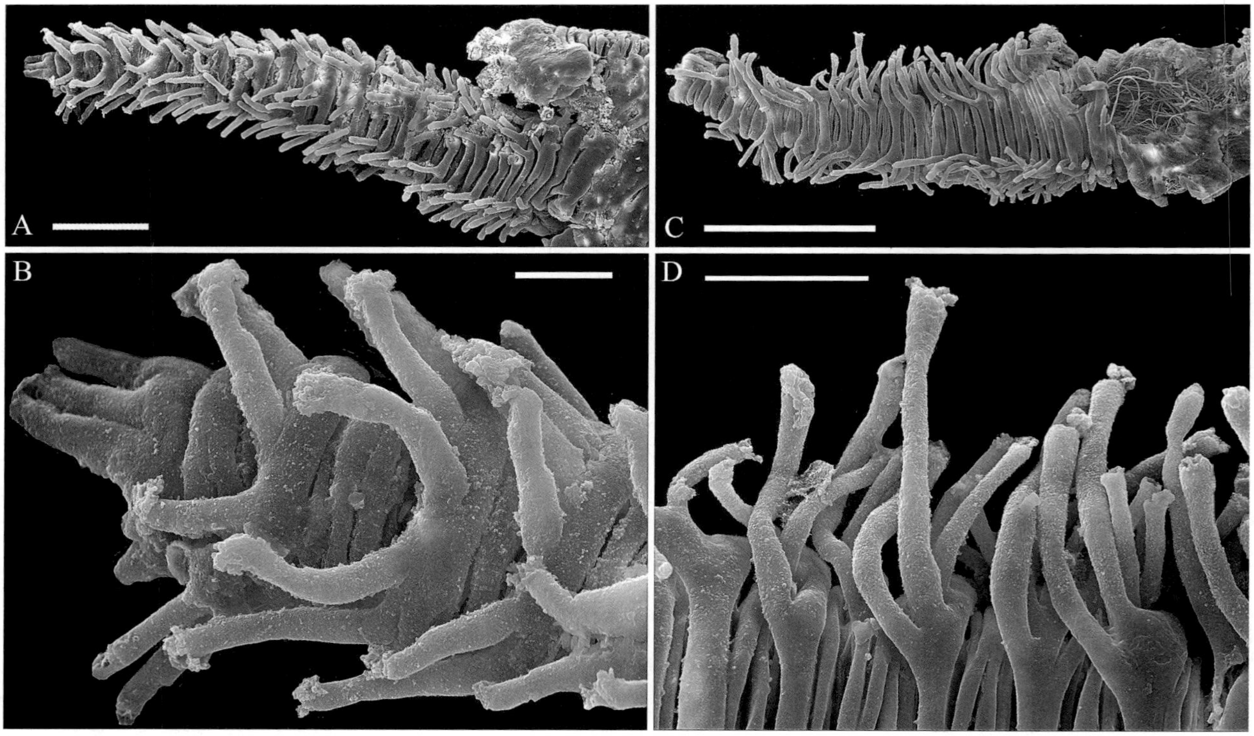

Figure 930. *Anatoma weddelliana*. **A–B.** Epipodial tentacle. **A.** Overview. Scale bar = 100 μm. **B.** Detail of papillae. Scale bar = 20 μm. **C–D.** Cephalic tentacle. **C.** Overview. Scale bar = 250 μm. **D.** Detail of papillae. Scale bar = 50 μm. Images by Diego Zelaya. From Zelaya & Geiger (2007: modified).

Comparisons. *Anatoma amoena* from Antarctica has a smaller shell, a more narrow umbilicus, a less bulging base, and a wider selenizone.

Anatoma conica from Antarctica has a smaller shell, a more narrow umbilicus, a less bulging base, a protoconch with flocculent sculpture, and more axial cords on the first teleoconch I whorl (37–39 vs. 22–26).

Anatoma euglypta from Antarctica has a biconical overall shell shape and weaker axial sculpture.

Anatoma shiraseae from Antarctica has a biconical overall shell shape, weaker axial sculpture, a prominent spiral keel on the shoulder, and one spiral keel on the base below the selenizone.

Anatoma finlayi from the Indian Ocean to the Central Pacific (and possibly Caribbean) lacks spiral lines on the base immediately below the selenizone, has fewer but stronger axial lamellae (teleoconch II whorl 2: 20 vs. 30), and inturned keels of the selenizone.

Anatoma indonesica from the Indo-Malayan Archipelago to the Western Pacific has a smooth band on the base below the selenizone, only strong axial lamellae on the shoulder, and at least twice as many axial cords on the base.

Figure 931. Distribution of *Anatoma weddelliana*. ☆ type locality.

Anatoma lamellata from the Western to the Central Pacific has wider but lower axial lamellae, a spiral cord on the protoconch, and a proportionally smaller aperture.

Anatoma lyra from the northeastern Pacific has more numerous (30 vs. 13) weaker axial cords on teleoconch I and lacks a spiral cord in the position of the selenizone on teleoconch I.

Distribution. Known only from the type locality (Fig. 931).

Remarks. The reticulate protoconch sculpture of this species is composed of small and irregular cells.

Anatoma yaroni Herbert, 1986

Figures 932–936

Chresonymy.

Anatoma yaroni Herbert, 1986: 613–617, figs 1–2 [radula], fig. 4 [protoconch], figs 15–17 [shell] (Fig. 932).

Hainella yaroni: Bandel, 1998: 39.

Anatoma yaroni: Geiger, 2003: 74.

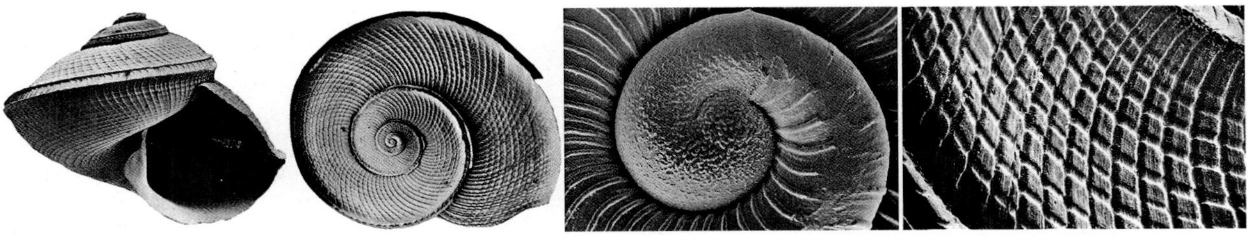

Figure 932. Original illustration of *Anatoma yaroni*. Left holotype, remainder paratypes.

Type material. Holotype (NMSA C6590/T3258. Fig. 932), 3.8 × 4.5 mm. Paratype from type locality (NMSA C9515/T3259; NMSA C9535/T3257, 8; ZMMU Lc-13974, 1: Ivanov & Sysoev, 2000). 455 m, off Qora River, Transkei, South Africa, 32.433°S, 28.683°E (NMSA C6641/T3260, 1). 440 m, off Whale Rock, Transkei, South Africa, 31.916°S, 29.217°E (NMSA C9514/T3261, 1). 410 m, off Whale Rock, Transkei, South Africa, 31.916°S, 29.217°E (NMSA C2046/T3262, 1). 380 m, off Rame Head, Transkei, South Africa, 31.868°S, 29.483°E (NMSA C9512/C3263, 3). 395 m, off Stony Point, Transkei, South Africa, 32.516°S, 28.600°E (NMSA C9511/T3264, 3). 450 m, off Kei River, Transkei, South Africa, 32.650°S, 28.367°E (NMSA C6904/T3265, 2). 510 m, off Qolora River, Transkei, South Africa, 32.616°S, 28.433°E (NM SAC7010/T3266, 2). 412 m, off Mendu Point, Transkei, South Africa, 32.266°S, 28.883°E (NMSA C4991/T3267, 1). 325 m, off Mbashe River, Transkei, South Africa, 32.250°S, 28.883°E (NMSA C9513/T3268, 1). 405 m, off Nthlonyane River, Transkei, South Africa, 32.050°S, 28.800°E (NMSA C8604/T3269, 2). 350 m, off Whale Rock, South Africa, 31.916°S, 29.217°E (NMSA

C9414/T3270, 1). 500 m, off Shixini Point, Transkei, South Africa, 32.528°S, 28.878°E (NMSA C9534/T3256, 12).

Type locality. 490 m, off Shixini Point, Transkei [South Africa] (32°31.7'S, 28°52.7'E), muddy sand, coral rubble (OD).

Etymology. Named after Isaac (Jitzack) Yaron [1934–1985: Mienis, 1992] for his study of Red Sea Scissurellidae (OD).

Description. Shell large (to 4.8 mm. Figs 932–934), biconical. Protoconch of 0.75 whorl, sculpture flocculent, no apertural varix, apertural margin rounded. Teleoconch I 0.75–0.825 whorl, approximately 25–30 strong axials, interstices smooth, indistinct spiral in position of selenizone. Shoulder with curved axials, approximately 90 on body whorl of fully grown specimen, interstices with very fine growth lines; number of spirals increasing from two at beginning of selenizone to 10–15 at apertural margin of fully grown specimen; axials and spirals of similar strength forming cancellate pattern; intersections of axials and spiral at best forming weak nodules. Base with similar cancellation as on shoulder. Umbilicus open, narrow, deep, sloping

Figure 933. *Anatoma yaroni*. 300 m, Mbotyi, Transkei, South Africa, 31.557°S, 29.844°E (DLG 1405, 1). Scale bar shell = 1 mm. Scale bar protoconch = 100 µm.

Figure 934. *Anatoma yaroni*. **A–C.** 500 m, off Shixini Point, Transkei, South Africa, 32.528°S, 28.878°E (NMSA C7071). Scale bars shell = 1 mm. Scale bars protoconch = 100 μm.

continuously from base, with weak funiculus. Selenizone at periphery; lunules distinct; keels moderately elevated, moderately strong, slightly folded adapically; slit open, margins converging.

Operculum corneous, thin, multispiral, with central nucleus.

Radula (based on Herbert, 1986. Fig. 935). Rachidian tooth broad, cutting edge parallel to rows, central cusp twice size of six cusps on either side. Lateral teeth 1–3 similar, cusps decreasing in number, increasing in size towards periphery. Lateral tooth 4 reduced, with three cusps, base with distinct lateral projections for interlock. Lateral tooth 5 moderately enlarged, V-shaped tip, central cusp enlarged, 3–5 cusps on each side. Marginal teeth with elongated, triangular tip, carrying multiple cusps; no food groove on shaft. Radular interlock of central field moderate.

External anatomy based on Herbert (1986). Animal with well developed, papillate cephalic tentacles. No cephalic lappets. Snout smooth, short, broad; small subcircular mouth. Eye black, eye-stalk short, fused to cephalic tentacle. Accessory cephalic tentacle smooth; additional smooth neck tentacle. Epipodial tentacles papillate, three on each side; anterior-most divided at base. Epipodial sense organ as large papilla at base of posterior two epipodial tentacles. One or two pallial tentacles in slit, mantle edge along slit papillate. Foot folded longitudinally, with anterior mucous gland, metapodium triangular. Gills attached close to mantle edge. Intestine with multiple coils, filled with detrital matter.

Comparisons. *Anatoma africanae* from the western Indian Ocean has a more globular overall shell shape, has at least twice as many axial cords and spiral lines on shoulder and base, and has the selenizone higher on the whorl.

Anatoma boucheti from the western Indian Ocean has a sunken apex, has a longer teleoconch I (0.9 vs. 0.75 whorl) and has the selenizone higher on the whorl.

Anatoma japonica from the Indian Ocean to the Central Pacific has stronger sculpture, and more spiral cords (at two teleoconch whorls, 10 vs. 4).

Anatoma tabulata from the western Indian Ocean has a more elevated spire, a sutsel wider than the selenizone on early whorls, and at most half as many axial cords.

Anatoma atlantica from the northwestern Atlantic lacks a spiral cord in the position of the selenizone, and has stronger sculpture, and more spiral cords (at two teleoconch whorls, 10 vs. 4).

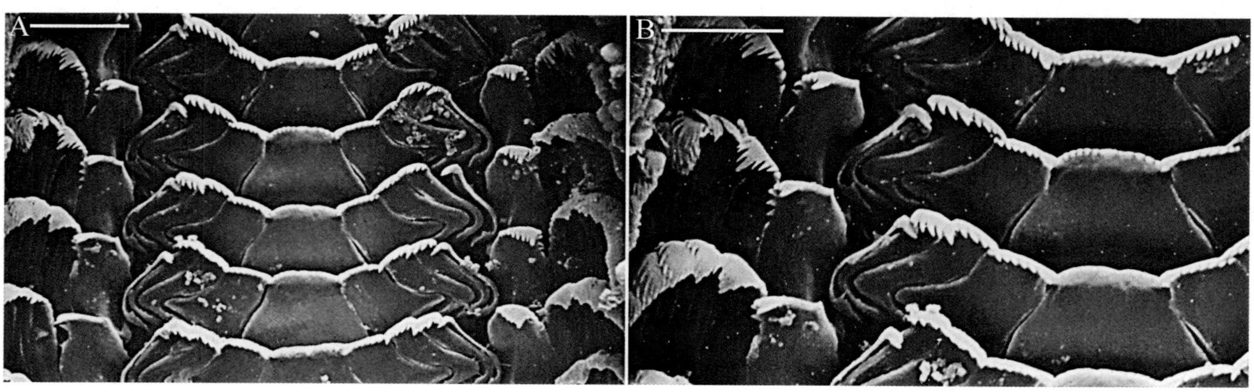

Figure 935. *Anatoma yaroni*. Radula. **A.** Full width of radula. Scale bar = 25 μm. **B.** Half width of radula. Scale bar = 20 μm. Images from Herbert (1986) with kind permission of D. Herbert. Source lot not identified.

Anatoma amydra from the Indo-Malayan Archipelago to the Central Pacific has a wider sutsel on early whorls (>0.5× vs. 0.25× width of selenizone), has at least 50% more axial cords on the base, and has a shorter teleoconch I (0.5 vs. 0.75 whorl).

Anatoma australis from eastern Australia has more than one spiral cord on teleoconch I, only every 3–4 axial cord on the base reaches the umbilicus, and the base has at most half as many spiral lines and steps.

Anatoma crispata from the temperate to boreal North Atlantic has predominantly axial sculpture and is less depressed in overall shell shape, whereas in *An. yaroni* axials and spirals are of very similar strength.

Anatoma tangaroa from northernmost New Zealand has a shorter teleoconch I (0.3 vs. 0.75 whorl), has stronger axial cords on shoulder and base, and has a more convex shoulder.

Anatoma euglypta from the Antarctic has no sutsel after 2 teleoconch II whorls, has a shorter teleoconch I (0.5 vs. 0.75 whorl), and has at least 50% more axial cords.

Anatoma finlayi from the Indian Ocean to the Central Pacific (and possibly Caribbean) has strong axial cords on shoulder and base, is more globular in overall shell shape, and has more convex shoulder and base.

Anatoma flexidentata from the Indian Ocean to the Western Pacific has a wider selenizone, a more convex shoulder, and a shorter teleoconch I (0.3 vs. 0.75 whorl).

Anatoma maxima from the Indo-Malayan Archipelago to the Central Pacific has a wider selenizone and wider keels of the selenizone.

Anatoma proxima from the Caribbean to Brazil lacks a spiral cord in the position of the selenizone on teleoconch I, has a more convex shoulder, and a wider umbilicus.

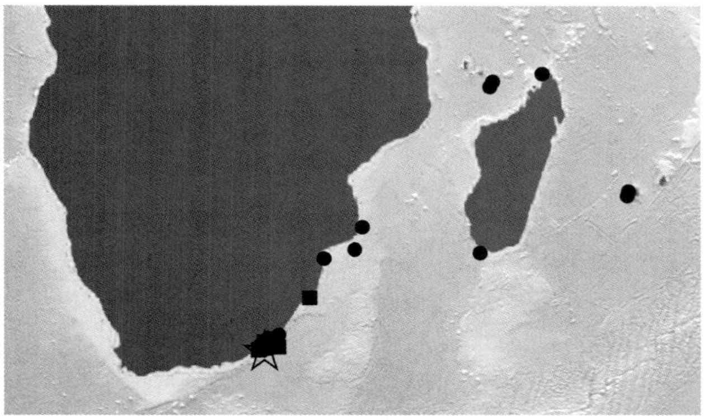

Figure 936. Distribution of *Anatoma yaroni*. ☆ type locality. ● specimen record. ■ literature record.

Anatoma regia from temperate New Zealand has a shorter teleoconch I (0.5 vs. 0.75 whorl), more spiral lines (after two teleoconch II whorls: 10–12 vs. 4), and a more inflated base.

Anatoma tenuis from the North Atlantic has no sutsel up to a size of at least 4.4 mm, more spiral lines on the shoulder (after two teleoconch II whorls: 14 vs. 4), and a shorter teleoconch I (0.5 vs. 0.75 whorl).

Anatoma tenuisculpta from the Mediterranean and adjacent Atlantic has more spiral lines on the shoulder (after two teleoconch II whorls: 10–12 vs. 4), a more convex shoulder, and a wider umbilicus.

Anatoma umbilicata from the Mediterranean and adjacent Atlantic has a more turreted shell shape, very little sculpture, and a wider umbilicus.

Distribution. Western Indian Ocean, 295–1200 m (Fig. 936).

Specimen records.

Madagascar. 238–249 m, E of Cape d'Ambre, MIRIKY DW3196, 12.133°S, 48.933°E (MNHN, 4). 203–210 m, SE Barrow Point, ATIMO VATAE CP3584, 25.466°S, 44.416°E (MNHN, 3).

Mozambique. 684–698 m, Maputo transect, Mozambique channel, MAINBAZA CP3141, 23.550°S, 35.916°E (MNHN, 1). 700–707 m, Maputo Transect, Mozambique channel, MAINBAZA CP3138, 25.216°S, 35.350°E (MNHN, 1). 112–127 m, Maputo transect, Mozambique channel, MAINBAZA CP3130, 25.883°S, 33.116°E (MNHN, 1).

France. 400–520 m, S Reef, Mayotte, 13.087°S, 45.132°E (MNHN, 1). 705 m, NE Pamanzi Island, Mayotte, 12.705°S, 45.332°E (MNHN, 1).

Reunion Island. 1150–1180 m, 21.200°S, 55.067°E (MNHN, 2). 740–780 m, 20.883°S, 55.150°E (MNHN, 2).

South Africa. 366 m, 29.317°S, 23.000°E (USNM 1003920, 2). 300 m, Mbotyi, Transkei, 31.557°S, 29.844°E (DLG 1405, 1).

Literature records.

South Africa. 390 m, off Kei River, Transkei, 32.650°S, 28.367°E (NMSA C3871). 495 m, off Kei River, Transkei, 32.650°S, 28.367°E (NMSA C6921). 410 m, off Shixini Point, Transkei, 32.528°S, 28.878°E (NMSA C6505). 490 m, off Shixini Point, Transkei, 32.528°S, 28.878°E (NMSA C9536). 500 m, off Shixini Point, Transkei, 32.528°S, 28.878°E (NMSA C7071). 395 m, off Stony Point, Transkei, 32.517°S, 28.600°E (NMSA C4345). 460 m, off Stony Point, Transkei, 32.517°S, 28.600°E (NMSA C6682). 410 m, off Qora River, 32.433°S, 28.683°E (NMSA C6837). 300 m, off Waterfall Bluff, 32.433°S, 29.833°E (NMSA C9641). 450 m, off Mendu Point, 32.267°S, 28.883°E (NMSA C6316). 475 m, off Mbashe River, Transkei, 32.250°S, 28.883°E (NMSA C9035). 297 m, off Mbashe River, Transkei, 32.250°S, 28.883°E (NMSA C9007). 322 m, off Mbashe River, Transkei, 32.250°S, 28.883°E (NMSA C9097). 375 m, off Nthlonyane River, 32.050°S, 28.800°E (NMSA C9225). 475 m, off Nthlonyane River, 32.050°S, 28.800°E (NMSA C8681). 410 m, off Whale Rock, 31.917°S, 29.217°E (NMSA C7940). 440 m, off Whale Rock, 31.917°S, 29.217°E (NMSA C8898). 420 m, off Rame Head, Transkei, 31.800°S, 29.350°E (NMSA C1879; NMSA C7999). 380 m, off Rame Head, Transkei, 31.800°S, 29.350°E (NMSA C2070). 474 m, off Sandy Point, Transkei, 28.800°S, 32.083°E (NMSA C4103). 485 m, off Sandy Point, Transkei, 28.800°S, 32.083°E (NMSA C9637) (Herbert, 1986).

Anatoma zancliformis new species

Figures 937–940

Type material. Holotype (NMNZ M.303299. Fig. 937). Paratypes from type locality (NMNZ M.174845, 39. Fig. 938A–B). 1760–1799 m, off Cape Foulwind, New Zealand, 41.34°S, 166.913°E (NMNZ M.174846, 4. Fig. 938C). 1280 m, St. 5637, S of Bouro Island, Moluccas, 3.889°N, 126.800°E (USNM 279265, 3).

Type locality. 1676 m, Bounty Trough, New Zealand, 45.872°S, 174.082°E.

Etymology. Zanklon, Greek for sickle; - formis, Latin for shaped. Referring to the sickle-shaped lateral tooth 5 in this species.

Description. Shell medium-sized (holotype 2.6 mm; to 4.7 mm. Figs 937, 938), trochiform, globular. Protoconch of 0.75 whorl, flocculent to irregular reticulate sculpture, no apertural varix, shape of apertural margin unknown. Teleoconch I of 0.33 whorl, approximately seven strong axial cords, usually no spiral in position of selenizone. Teleoconch II of up to four whorls, suture moderately impressed; no sutsel on early whorls, sutsel up to 2× width of selenizone at apertural margin of fully grown specimen. Shoulder convex, many fine axial cords, approximately 30–40 axial cords on first whorl, over 100 axial cords on last whorl of largest specimen; first spiral thread after 0.2–1.6 whorls, approximately 10 spiral threads at apertural margin of fully grown specimen; spirals running over axials. Base with weak constriction below selenizone, same density and strength of axials and spirals, approximately 30 spiral threads on last whorl of large specimen. Umbilicus moderately wide, no funiculus. Aperture rounded with moderately flaring lip; roof overhanging. Selenizone at periphery; keels moderately strong, moderately elevated; lunules distinct; slit open, margins somewhat converging.

Operculum (Fig. 939A) corneous, round, thin, multispiral, nucleus central.

Radula (Fig. 939). Rachidian tooth triangular, central cusp largest, wide, 5–6 minor cusps on each side in straight line from outline of central cusp. Lateral teeth 1–2 similar, apical cusp largest, 4–5 cusps along outer edge, 1–2 cusps along inner edge. Lateral tooth 3 with triangular tip. Lateral tooth 4 moderately enlarged with elongate tip. Lateral tooth 5 much enlarged by elongation, sickle-shaped tip with minute cusps at base of tip. Inner marginal teeth similar to lateral tooth 5, cusps at base of tip more distinct, increasing in relative size in more distal teeth. Outermost marginal teeth with thin triangular tips, apical cusp largest, approximately six smaller

Figure 937. *Anatoma zancliformis* new species. Holotype. 1676 m, Bounty Trough, New Zealand, 45.872°S, 174.082°E (NMNZ M.303299). Scale bar shell = 1 mm. Scale bar protoconch = 100 µm. Arrow indicates apertural margin of protoconch.

cusps on each side. Shafts of marginal teeth without food groove. Radular interlock of central field moderate.

Comparisons. *Anatoma australis* from eastern Australia has a longer teleoconch I (0.66 vs. 0.33 whorl) with a distinct spiral cord in position of the selenizone, a sutsel of the same width as the selenizone at the onset of the selenizone, and spiral steps on the base.

Anatoma biconica from the Indian Ocean to the Central Pacific has a longer teleoconch I (0.75 vs. 0.33 whorl) with a spiral cord in position of the selenizone, a sutsel as wide or wider as the selenizone at the onset of the selenizone, and lacks a constriction of the base below the selenizone.

Anatoma tangaroa from northernmost New Zealand has at least twice as many and stronger axial cords, a less bulging base, and more elevated keels of the selenizone.

Figure 938. *Anatoma zancliformis* new species. Paratypes. **A–B.** 1676 m, Bounty Trough, New Zealand, 45.872°S, 174.082°E (NMNZ M.174845, 39). Radula shown in Figure 939. **C.** 1760–1799 m, off Cape Foulwind, New Zealand, 41.34°S, 166.913°E (NMNZ M.174846, 4). Scale bars shell = 1 mm.

Anatoma finlayi from the Indian Ocean to the Central Pacific (and possibly Caribbean) has strong axial folds on the shoulder, stronger keels of the selenizone, and a distinct funiculus.

Anatoma flemingi from temperate New Zealand has a protoconch with reticulate sculpture, a longer teleoconch I (0.75 vs. 0.33 whorl), a sutsel barely as wide as the selenizone at the onset of the selenizone, and axial lamellae on the shoulder.

Anatoma flexidentata from the Indian Ocean to the Western Pacific has an overall more lenticular shell shape, at most half as many spiral lines on shoulder and base, a wider selenizone, and a flexoglossate radula.

Anatoma globulus from the Indo-Malayan Archipelago has a more globular overall shell shape, stronger axial cords, and the intersections of axials and spirals form minute points.

Anatoma japonica from the Indian Ocean to the Central Pacific has an overall more lenticular shell shape and denser axial and spiral cords.

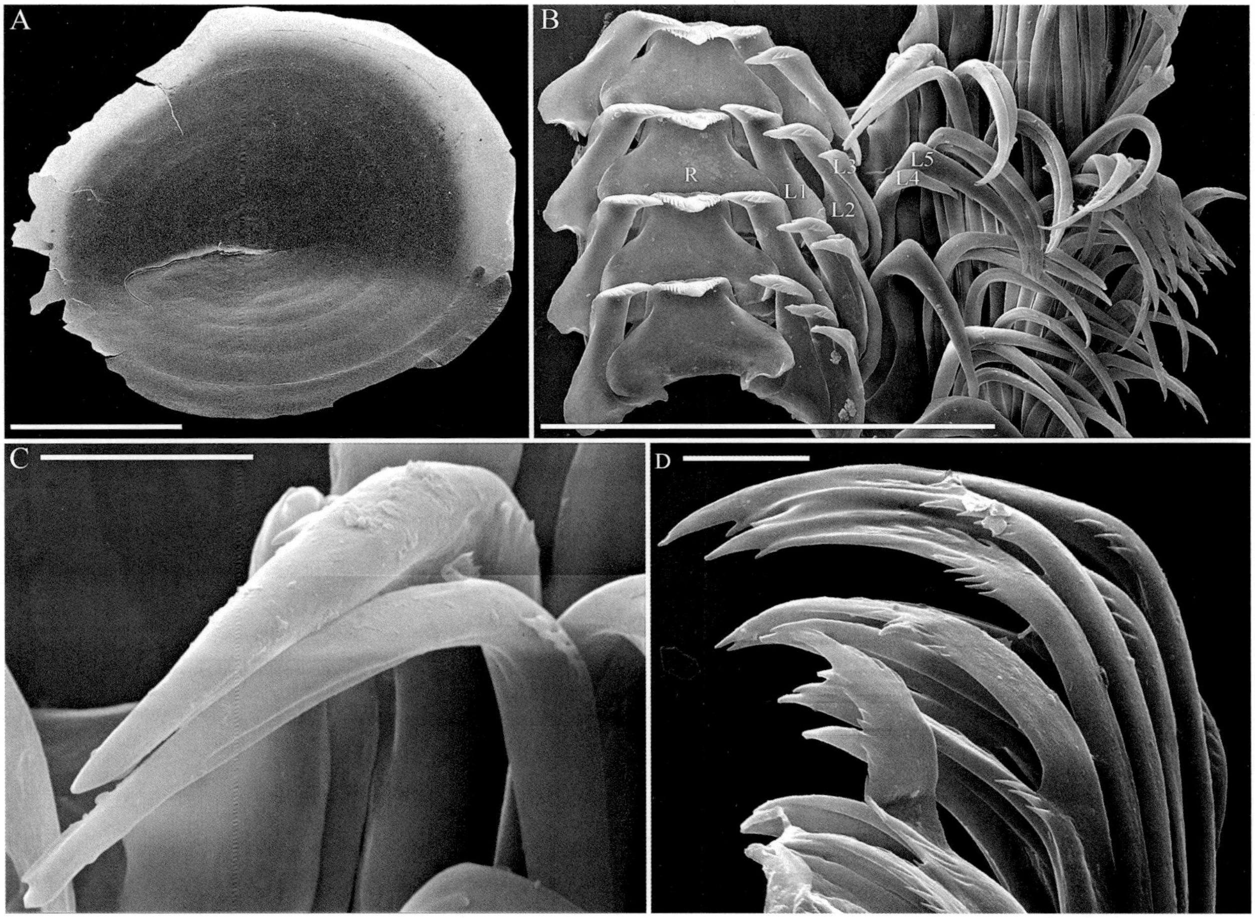

Figure 939. *Anatoma zancliformis* new species. 1676 m, Bounty Trough, New Zealand, 45.872°S, 174.082°E (NMNZ M.174845, 39). Shell shown in Figure 938A–B. **A.** Operculum. Scale bar shell = 1 mm. **B–D.** Radula. **B.** Central field and marginal teeth. Scale bar = 100 µm. **C.** Lateral tooth 5. Scale bar = 10 µm. D. Marginal teeth in side view. Scale bar = 10 µm.

Anatoma kopua from eastern Australia to New Zealand has a more turreted overall shell shape, the selenizone higher on the whorl, and a wider sutsel after two whorls (>1× vs. <0.5× width of selenizone).

Anatoma tobeyoides from southeastern Australia has very fine oblique lines on the teleoconch I and early teleoconch II and more densely spaced axial cords on shoulder and base with interstices as wide as cords (at least twice as wide in *An. zancliformis*).

Distribution. Indo-Malayan Archipelago to Western Pacific (Fig. 940).

Remarks. The radula of this species is among the more atypical of the genus. The distinctness of lateral tooth 3 from lateral teeth 1–2, the relatively large size of lateral tooth 4, the sickle-shaped lateral tooth 5 (also known from *An. australis, An. austrolissa,* and *An. flexidentata,* and to a lesser extent in *An. porcellana* and *An. sagamiana*), and the outer marginal teeth with thin and broadly triangular tip, are noteworthy.

The similar shell morphological to *An. flexidentata,* particularly the rather uncommonly short teleoconch I, and shared sickle-shaped lateral tooth 5 have to be noted. The radular differences in the central field suggest that they are distinct, which is supported by a subtle shell morphological variance: the number of spiral lines on the shoulder (10 in *An. zancliformis* vs. 3–4 in *An. flexidentata*).

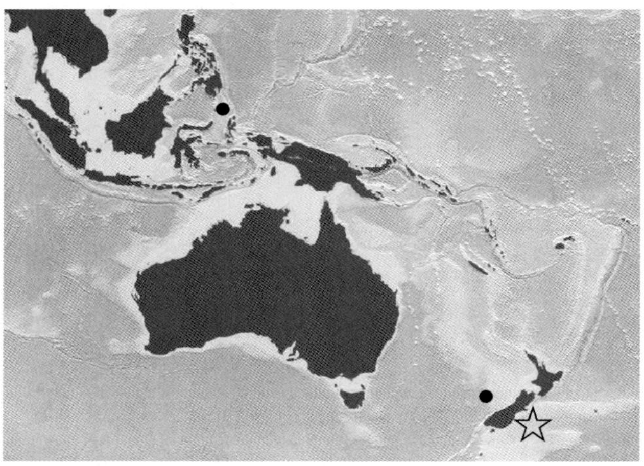

Figure 940. Distribution of *Anatoma zancliformis* new species. ☆ type locality. ● specimen record.

Anatoma gephyra Maxwell, 1992 †

Figures 941–943

Chresonymy.

Anatoma gephyra Maxwell, 1992: 82, pl. 8g, j, l (Fig. 941).

Figure 941. Original illustration of *Anatoma gephyra*.

Type material. Holotype (GNS TM 7003. Fig. 942).

Type locality. GS 9508, Tahu Member, McCulloch's Bridge (Kaiatan = late Eocene) [South Island, New Zealand, 44.800°S, 170.977°E] (OD).

Etymology. Geohyra: Greek for bridge, most likely referring to the name of the type locality.

Description. Shell of moderate size (2.9 × 2.6 mm H × W: holotype), trochiform, globular. Protoconch 0.5 (?) whorl, sculpture unknown (eroded in holotype), apertural varix not connected to embryonic cap, shape of apertural margin unknown. Teleoconch I 0.33 whorl, approximately 12 axial cords, other sculptural elements uncertain. Teleoconch II >3.125 whorls. Shoulder low sinusoidal profile with deeply impressed suture, many axial cords increasing in strength with growth, crossed by two spiral lines in early whorls (fading on last whorl). Suture just below selenizone in early whorl, at apertural margin sutsel 2× width of selenizone. Base rounded, without constriction below selenizone, continuously sloping with umbilicus; same density and strength of axial cords as on shoulder, crossed by approximately 11 spiral cord of equal strength, resulting in coarse reticulate pattern. Umbilicus closed. Selenizone above periphery; keels low, of moderate strength; lunules low, regularly spaced; condition of slit uncertain (broken aperture). Aperture rounded.

Comparisons. *Anatoma miocenica* from New Zealand has much finer axial and spiral sculpture and more numerous spiral lines on shoulder (8 vs. 2).

Anatoma aspera from the Paleogene to Recent of Europe has stronger axial cords on the shoulder and a distinctly open umbilicus.

Anatoma subaspera from the Oligocene to Miocene of Europe has a more discoidal overall shell and a distinct umbilicus.

Anatoma tongaensis from the late Eocene of Tonga has a more depressed shell (see *An. tongaensis* Remarks) and a wider sutsel.

Anatoma kopua from the temperate eastern Australia and New Zealand has an open umbilicus.

Anatoma porcellana from the Recent of the Indo-Malayan Archipelago to the Western Pacific has a longer teleoconch I (0.825 vs. 0.33 whorl) and fewer but stronger axial cords, which are crossed by more, finer spiral threads.

Figure 942. *Anatoma gephyra*. Holotype. GS 9508, Tahu Member, McCulloch's Bridge (Kaiatan = late Eocene) [South Island, New Zealand, 44.800°S, 170.977°E] (GNS TM 7003). Scale bar shell = 1 mm. Scale bar protoconch = 100 μm.

Distribution. Known only from type locality (Fig. 943).

Remarks. The species is remarkably similar to the Recent *An. porcellana* of broad Indo-Pacific distribution. The main distinction is the length of teleoconch I (0.33 vs. 0.825 whorl); a difference by 2× and >0.25 whorls usually is indicative of specimens to be distinct. Unfortunately, the protoconch sculpture of *An. gephyra* is unknown. Most paleobiologists would argue that conspecificity of Eocene and Recent species is unlikely. The original description mentioned additional axial threads on the shoulder; however, they are invisible on the illustrations.

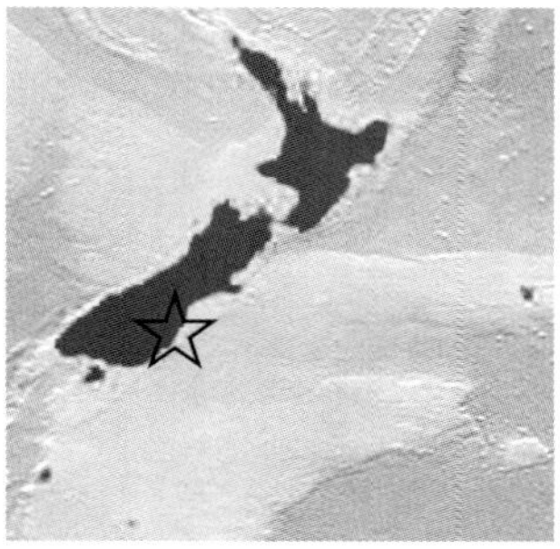

Figure 943. Distribution of *Anatoma gephyra*. ☆ type locality.

Anatoma miocenica (Laws, 1939) †

Figures 944–948

Chresonymy.

Schizotrochus miocenica [sic] Laws, 1939: 475, fig. 48 (Fig. 944).
Anatoma miocenica: Lozouet, 1986: 113.
Anatoma miocenica: Beu & Mawell, 1990: 402.

Figure 944. Original illustration of *Schizotrochus miocenicus*.

Type material. Holotype (GNS TM 1456, ex coll. Laws. Fig. 945), 1.6 × 1.7 mm. Paratypes from type locality (coll. Laws, 16: OD; GNS TM 8645–8655, 11 ex coll. Laws: Fig. 946; whereabouts unknown, 5).

Type locality. Pakaurangi Point, Kaipara, New Zealand[36.239°S, 174.293°E]. Tertiary beds (OD: inferred from title). Lower Miocene (Lozouet, 1986; specimen label).

Etymology. Most likely referring to the Miocene age of the specimen.

Description. Shell small for genus (to 1.7 mm. Figs 945–946), trochiform, inflated. Protoconch 0.75 whorl, flocculent sculpture, apertural varix not connected to embryonic cap, apertural margin somewhat sinusoidal. Teleoconch I of 0.5 whorl, approximately 15 axial cords, no spiral sculpture. Teleoconch II of 2.5 whorls, suture moderately impressed. Shoulder with many fine, sharp axial cords; first spiral line after 0.5 whorl, 7–8 spirals at peristome. Base with many fine, sharp axial cords; approximately

16 finer spiral lines. Umbilicus open, continuously sloping with base, partially covered by recurved apertural margin; presence of funiculus uncertain because of in-fill. Selenizone at periphery; lunules distinct; keels distinct, moderately strong; slit open.

Comparisons. *Anatoma gephyra* from early Eocene of New Zealand lacks spiral lines on the shoulder and has overall coarser sculpture.

Anatoma aspera from the Paleogene to Recent of Europe has stronger axial cords on the shoulder, which are markedly stronger than the axials on the base, and a wider umbilicus.

Anatoma subaspera from the Oligocene and Miocene of Europe has a more discoidal overall shell shape with the slit at the periphery.

Figure 945. *Anatoma miocenica*. Holotype. Pakaurangi Point, Kaipara, New Zealand, 36.239°S, 174.293°E (GNS TM 1456). Scale bar shell = 1 mm. Scale bar protoconch = 100 µm.

Anatoma tongaensis from the Eocene of Tonga has an overall more discoidal shell (see *An. tongaensis* Remarks) and lacks the reticulate pattern on the base.

Distribution. New Zealand (Fig. 948).

Fossil. Otaian - Altonian = early Miocene of New Zealand.

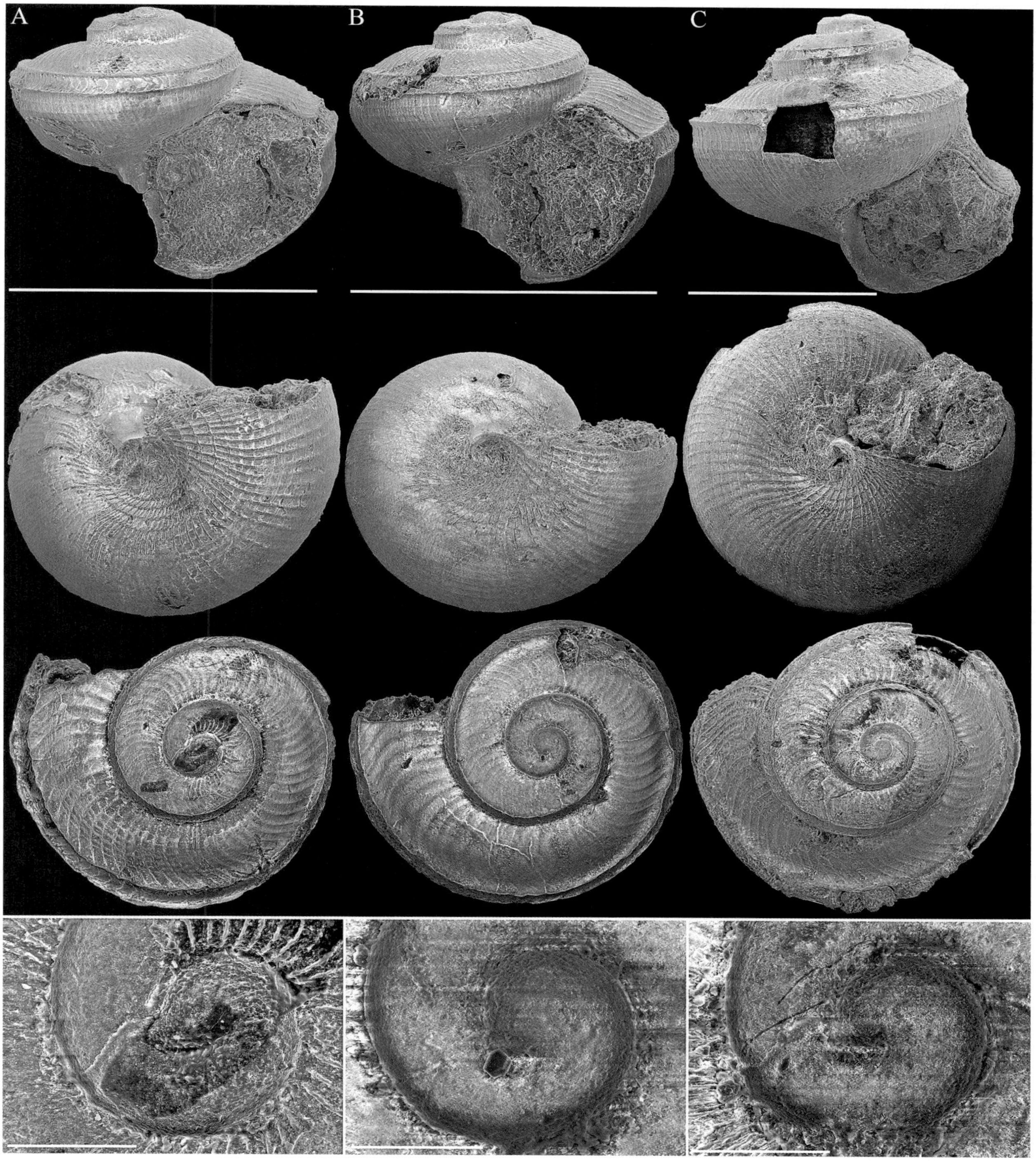

Figure 946. *Anatoma miocenica*. Paratypes. **A–C.** Pakaurangi Point, Kaipara Harbor, Northland, New Zealand, 36.239°S, 174.293°E (GNS TM 8645–8655, 11). Scale bar shells = 1 mm. Scale bars protoconchs = 100 μm.

Figure 947. *Anatoma miocenica*. Small specimen with more discoid shell shape and well-preserved protoconch. GS 12163, Moutapu Point, NE Pitt Island, Chatham Islands, Whenuataru Tuff, Waipipian-Mangapanian (middle to late Pliocene), New Zealand, 36.239°S, 174.293°E (GNS, 12). Scale bar shell = 1 mm. Scale bar protoconch = 100 μm.

Specimen records.

New Zealand. GS 12163, Moutapu Point, NE Pitt Island, Chatham Islands, Whenuataru Tuff, Waipipian-Mangapanian (middle to late Pliocene), 36.239°S, 174.293°W (GNS, 12). GS 12161, cliffs near "The Point" homestead, W of Flowerpot Bay, Pitt Island, Chatham Islands. Whenuataru Tuff, Waipipian-Mangapanian (middle to late Pliocene), 44.240°S, 176.239°W (GNS, 6: *An.* cf. *miocenica*).

Figure 948. Distribution of *Anatoma miocenica*. ☆ type locality.

Remarks. The species is indistinguishable from the Recent *An. equatoria*, particularly the more coarsely sculptured specimens.

Anatoma muelleri new species †

Figures 949–950

Type material. Holotype (MMG AnT 1).

Type locality. Site M3, Mammendorf, Germany, 52.177°N, 11.437°E, early Oligocene (Rupelian: NP22 zone: see Müller, 2011, for details).

Etymology. Named for Arnold Müller, Professor and Curator at the Geological-Paleontological Collection of the University Leipzig, Germany, who collected the material and kindly made it available for study.

Description. Shell medium-sized (2.03 mm: holotype). Protoconch of 0.75 whorl, flocculent sculpture, no apertural varix, apertural margin straight. Teleoconch I of 0.6 whorl, 12 axial cordlets, spiral cord in position of selenizone uncertain. Teleoconch II of two whorls; suture moderately impressed; sutsel as wide as selenizone at onset of

Figure 949. *Anatoma muelleri* new species. Holotype (MMG AnT 1). Mammendorf, Germany, 52.177°N, 11.437°E, early Oligocene (Rupelian: NP22 zone: see Müller, 2011, for details). Scale bar shell = 1 mm. Scale bar protoconch = 100 µm.

selenizone, narrowing to very thin line, as wide as selenizone at apertural margin. Shoulder with 32 axials in first teleoconch II whorl, increasing from cordlets to low fold; occasional supernumerary spiral lines on keels of selenizone; first spiral line after 0.1 whorl, four lines after one teleoconch II whorl, approximately six at apertural margin in centripetal ⅔ of shoulder. Base without constriction below selenizone; twice as many axial cords crossed by 22 spiral lines running over axials. Umbilicus at sharp angle to base, walls straight, smooth. Aperture likely D-shaped. Selenizone slightly above periphery; lunules indistinct; keels moderately strong, moderately elevated; condition of slit unknown

Comparisons. This is the only know species of *Anatoma* with a sharp carina at the transition between base and umbilicus. The position of the selenizone almost at the periphery of the shell instantly distinguishes it from any member of Scissurellidae.

Distribution. Only known from type locality (Fig. 950).

Remarks. The present species is remarkable as it is the only species in *Anatoma*, which has the strong angulation between base and umbilicus. The specimen is not an member of Scissurellidae due to the placement of the selenizone at the periphery. The protoconch with flocculent sculpture (common in *Anatoma* but rare in Scissurellidae) and without an apertural varix (common in *Anatoma*, but uncommon in Scissurellidae) further support the generic placement of the species.

The only Anatomidae with strong angulation between base and umbilicus is the Recent *Sasakiconcha elegantissima*, which has the distinct umbilical trough not seen in *An. muelleri*. *Sasakiconcha elegantissima* additionally has a very short teleoconch I (0.1 whorl), a greater expansion rate of the shell, and grows much larger (8 mm).

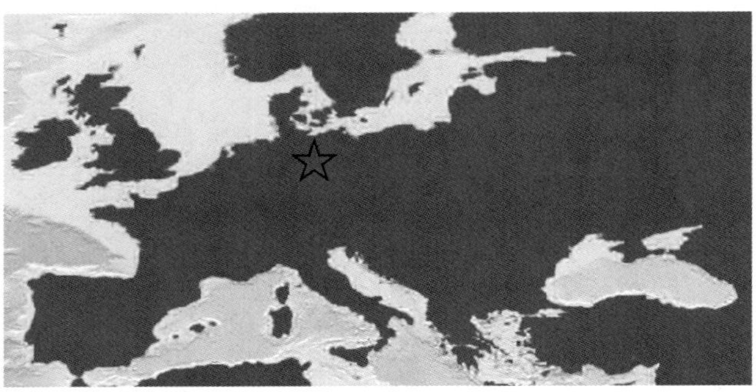

Figure 950. Distribution of *Anatoma muelleri* new species. ☆ type locality.

Anatoma philippiana (Semper, 1865) †

Figures 951–957

Type material. (Semper collection: whereabouts unknown), 3.5 × 3 mm.

Type locality. Latdorf, [Saxony, Germany, 51.800°N, 11.800°E]. Lower Oligocene (OD). Middle Eocene (Lozouet, 1986).

Etymology. [not given].

Chresonymy.

Scissurella philippiana Semper, 1865: 284, 285–286: pl. 12, fig. 2 (Fig. 951).

Scissurella philippiana: Koenen, 1867: 125–126(271–272), pl. 7(30), fig. 10.

Scissurella philippiana: Koenen, 1892: 890–891, pl. 54, figs 9a–e.

Scissurella (*Anatoma*) *philippiana*: Janssen, 1978a: 17 [*fide* Amitrov, 1996].

? *Scissurella* (*Anatoma*) *philippiana*: Kuster-Wendenburg, 1982: 84 [confused identification as both Semper (1865) and Neuffer (1979) are cited, the latter showing *Sci. depontaillieri*].

Scissurella philippiana: Gougerot & LeRenard, 1977: 5.

Praecissurella ? *philippiana*: Lozouet, 1986: 110.

Anatoma philippiana: Amitrov, 1996: 101–102, figs 1O–R.

Anatoma philippiana: Amitrov, 1998: 7, fig. 1A–F [protoconch].

Anatoma philippiana: Müller, 2011: 22, pl. 5, fig. 3 [possibly mixed species concept with *An. tenuisculpta*: size up to 5.5 mm].

Figure 951. Original illustration of *Scissurella philippiana*.

Synonyms.

+ *Scissurella* (*Anatoma*) *koeneniana* Janssen, 1978b: 142, pl. 9, fig. 6 (Fig. 952). ***Type material.*** Holotype SMF 250387a, 0.93 × 1.2 mm. ***Type locality.*** Former mine "Höllkopf" near Glimnerode, Hessisch-Lichtenau, Niederhessen, Germany [51.200°N, 9.733°E]. Marine sand of Kassel, Layer 6 of Janssen (1978), Eochatt. Upper Oligocene (Lozouet, 1986). ***Etymology.*** Named for A. V. Koenen (1837–1915) (OD) [synonymy after Amitrov, 1996].

? *Scissurella* (*Schizotrochus*) aff. *koeneniana*: Janssen, 1984: 118–119, pl. 43, figs 2a–d.

Anatoma koeniana [*sic*]: Lozouet, 1986: 113.

Figure 952. Original illustration of *Scissurella koeneniana*.

+ *Anatoma stephanensis* Bandel, 1998: 29–30, pl. 9, figs 2–4 (Fig. 953). ***Type material.*** Holotype (RMNH), 1.2 × 1 mm. ***Type locality.*** Peyrère, Aquitaine, SW France [44.670°N, 5.620°E]. Late Oligocene (OD). ***Etymology.*** Bandel adopted the suggestion in Lozouet's unpublished (1986) thesis (OD). ***Remarks.*** Bandel (1986: 29) used "*Anatoma stephanensis* Lozouet 1986, n. sp." in the heading for the species description. The species was described by Bandel (1998), the description being taken from the thesis of Lozouet (1986).

Anatoma stephanensis: Lozouet, 1986: 113, 131–132, pl. 2, fig. 6 [unavailable, unpublished].

Anatoma stephanensis: Geiger, 2003: 74.

Figure 953. Original illustration of holotype of *Anatoma stephanensis*.

+ *Anatoma rosenkrantzi* Schnetler, Lozouet & Pacaud, 2000: 88–89, pl. 22, figs 4–6, pl. 3, figs 3a–c (Fig. 954). ***Type material.*** Holotype (MGUH 25757), 1.2 × 1.5 mm. Paratype from type locality (MGUH 25760, 1; MGUH 25760, 1). ***Type locality.*** Fakse quarry, Denmark [55.250°N, 12.130°E]. Coral limestone, middle Danian, early Paleocene (OD). ***Etymology.*** Named for A. Rosenkrantz (OD).

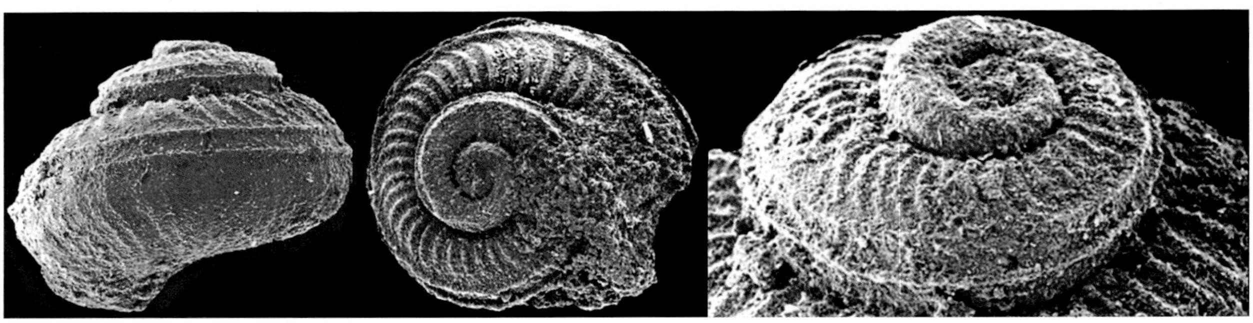

Figure 954. Original illustration of *Scissurella rosenkrantzi*. Left and right: holotype. Middle: paratype.

Misidentifications.

not *Anatoma philippiana*: Gürs, 1995: 28–29, pl. 3, fig. 2a–c [is *Sci. depontaillieri*].

Description. Shell medium-sized (to 2.9 mm) trochiform, subquadrate. Protoconch of 0.75 whorls, flocculent sculpture, no apertural varix. Teleoconch I of 0.4–0.8 whorls, spiral cord in position of selenizone, approximately 15–18 axial cords. Teleoconch II of up to 2.6 whorls; sutsel as wide as selenizone at onset of selenizone, subsequently narrowing to 0.25–0.5× width of selenizone, up to 1.5× width of selenizone at apertural margin. Shoulder with regular axial cords to low folds, approximately 33 on first teleoconch II whorl, approximately 40 on last whorl of largest specimen;

first spiral cordlet after 0.1–0.25 whorls, approximately 3 spiral cordlets after 1 teleoconch II whorl, 7–8 at apertural margin of large specimen; spirals running over axials. Base with weak constriction below selenizone followed by weak spiral keel, then continuous with umbilicus; about twice as dense axial cordlets crossed by

Figure 955. *Anatoma philippiana*. Granite quarry Rybalskij near Dnipropetrovsk, Ukraine, 48.450°N, 35.142°E. Late Eocene (Priabonian: NP19 zone: see Müller & Rosenberg, 2003, for details). **A–B.** Horizon U3. **C.** Horizon U5. Scale bars shells = 1 mm. Scale bars protoconch = 100 µm.

18–20 slightly weaker, equally spaced, spiral cordlets, forming slight thickenings at intersection with axials; axials in umbilical region as crunched folds. Umbilicus narrow, with distinct funiculus. Selenizone slightly above periphery, rather wide; keels moderately elevated, moderately strong; lunules indistinct; slit open, margins parallel.

Figure 956. *Anatoma philippiana*. Mammendorf, Germany, 52.177°N, 11.437°E, early Oligocene (Rupelian: NP22 zone: see Müller, 2011, for details). **A.** Site M7 (MMG AnT 5). **B–C.** Site M6 (MMG AnT 4). Scale bars shell = 1 mm. Scale bars protoconch = 100 μm.

Comparisons. *Anatoma aspera* from the Pleistocene to Recent of Europe has a protoconch with an apertural varix and a wider sutsel on intermediate whorls (1× vs. <0.5× width of selenizone).

Anatoma tenuisculpta from the Pleistocene to the Recent of Europe has a protoconch with an apertural varix, a more biconical overall shell shape, and at least 50% denser axial sculpture on shoulder and base.

Anatoma orbiculata from the Recent of the subtropical eastern Atlantic has a protoconch with an apertural varix and usually denser spiral lines on the shoulder.

Anatoma alternatisculpta from the Caribbean has an overall taller shell, fewer spirals on the shoulder, and a smooth protoconch with an apertural varix.

Anatoma eximia from the Pleistocene to Recent of Europe has a more quadrate, tabulated overall shell morphology, a consistently wider sutsel, and finer sculpture on shoulder and base.

Anatoma janusa from the subtropical eastern Atlantic has sharper axial cords on the shoulder, supernumerary axials towards the apical keel of the selenizone, at least twice as dense axials on the base, and spiral lines restricted to the upper part of the base.

Anatoma lamellata from Japan to the Central Pacific has a protoconch with a median spiral cord and a wider sutsel on early whorls (>0.5× vs. <0.5× width of selenizone).

Anatoma micalii from the Mediterranean to Liberia has a more tabulated overall shell shape and supernumerary axial lines on the more elevated keels of the selenizone.

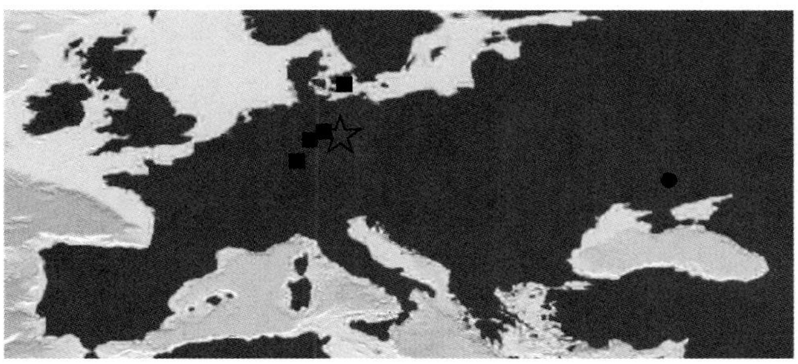

Figure 957. Distribution of *Anatoma philippiana*. ☆ type locality. ● specimen record. ■ literature record.

Distribution. Paleogene of Eurasia (Fig. 957).

Fossil. S of Kallo, East Flanders, Belgium, 51.295°N, 4.249°E (Marquet, 1984; 1998: as *An. crispata*). Early Paleocene (Danian) of Denmark, late Eocene (Priabonian) of northern Germany to late Oligocene (Chattian) of middle and northern Germany (Amitrov, 1996).

Specimen records.

Germany. Mammendorf, 52.177°N, 11.437°E, early Oligocene (Rupelian: NP22 zone: see Müller, 2011, for details. MMG AnT 4; MMG AnT 5).

Ukraine. Granite quarry Rybalskij near Dnipropetrovsk, 48.450°N, 35.142°E. Late Eocene (Priabonian: NP19 zone: see Müller & Rosenberg, 2003, for details. MMG UkT 13; MMG UkT 16; MMG UkT 17; MMG UkT 18; MMG UkT 19).

Literature records.

Germany. Former mine "Höllkopf" near Glimnerode, Hessisch-Lichtenau, Niederhessen, 51.200°N, 9.733°E (Janssen, 1978b: as *An. koeneniana*).— Latdorf, Saxony, 51.800°N, 11.800°E (Semper, 1865).—Lower Oligocene of Mammendorf, Saxony, 52.78°N, 11.436°E (Müller, 2011).—Mainz Basin, ~50°N, 8.25°E (Gürs, 1995).

Denmark. Fakse, 55.250°N, 12.130°E (Schnetler *et al.*, 2000: as *An. rosenkrantzi*).

Remarks. *Scissurella koeneniana*, *An. stephanensis*, and *Sci. rosenkrantzi* are here synonymized under *An. philippiana*. They are all European species from the Paleogene (Paleocene to Oligocene), sharing the same overall shell morphology, same arrangement of the sutsel, and where observable (*An. stephanensis*) the same protoconch attributes with the specimens figured here. There are slight differences in figured specimens in terms of teleoconch I length: late Eocene of Ukraine with 0.4–0.6 whorls, early Oligocene of Germany with 0.6–0.8 whorls. It is still within the usual range of 0.25 whorls of intraspecific variability, and there are no other characters that could separate specimens of the two populations (Figs 955, 956).

Amitrov (1996) synonymized *An. koeneniana* under *An. philippiana*, noting that the cited differences are within the intraspecific variation of the species. The types of *An. koeneniana* (Fig. 952) and *An. rosenkranzi* (Fig. 954) are indistinguishable. Apparent shape differences can be accounted for by differences in shell size. Müller (2011) also stressed the importance of comparing the same portions of the whorls or specimens of the same age/size.

Anatoma tongaensis Ladd, 1970 †

Figures 958–960

Chresonymy.

Scissurella (*Anatoma*) *tongaensis* Ladd, 1970: C2, pl. 2, figs 1–4.
Anatoma tongaensis: Lozouet, 1986: 114.

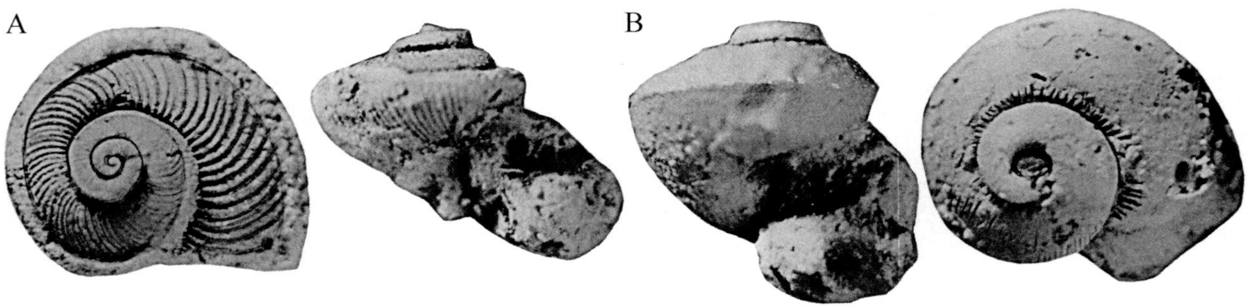

Figure 958. Original illustration of *Scissurella tongaensis*. **A.** Holotype. **B.** Paratype.

Type material. Holotype (USNM 650547), 1.6 × 2.0 mm (H × W). Paratype from type locality (USNM 650548, 1), 1.9 × 2.1 mm (H × W).

Type locality. Eua, Tonga [21.370°S, 174.930°W]. Upper Eocene (OD).

Etymology. Named for its type locality in Tonga (OD).

Description. Shell of moderate size (to 2.1 mm), trochiform, biconical. Protoconch unknown (eroded). Teleoconch I whorl count unknown; sculpture eroded. Teleoconch II of two whorls. Shoulder convex, 57 distinct axial cords on last whorl, axials crossed by approximately 12 weak, indistinct spiral lines forming hardly perceptible nodes at intersection with axials. Base with same density and strength of axials as on shoulder, approximately 12 weak, indistinct spiral lines forming hardly perceptible nodes at intersection with axials. Umbilicus open; details obscured by fill-in. Selenizone slightly above periphery, keels of moderate elevation, moderate strength. Aperture subquadrate.

Comparisons. *Anatoma gephyra* from the late Eocene of New Zealand, and *An. miocenica* from the Miocene of New Zealand, both have more a globular shell with no to a very narrow umbilicus.

Anatoma aspera from the Paleogene to Recent of Europe has distinctly stronger axial cords on the shoulder compared to the base.

Distribution. Known only from type material (Fig. 922). 100 fms. [= 183 m] depositional environment (Ladd, 1970).

Figure 959. *Anatoma tongaensis*. Eua, Tonga [21.370°S, 174.930°W]. Upper Eocene. **A.** Holotype (USNM 650547). **B.** Paratype (USNM 650548). Scale bars shells = 1 mm. Scale bar apex = 100 μm.

Remarks. The two known specimens are in moderate state of preservation (paratype is more or less a steinkern with some attached shell remnants), making detailed comparisons to other species very difficult; the pattern of the distinct axial sculpture make it distinct. The strength of the sculpture is exaggerated in the photographs accompanying the original description. The holotype is notably more depressed than the paratype, and both are of approximately the same size ruling out ontogenetic differences. Either the two specimens are not conspecific, which is not entirely unlikely; the Recent fauna in the region harbors several species, which are distinguished based on rather subtle difference in teleoconch sculpture. Alternatively, the holotype may have been compressed during diagenesis. These uncertainties can only be addressed with additional material. I treat the two specimens here as tentatively conspecific.

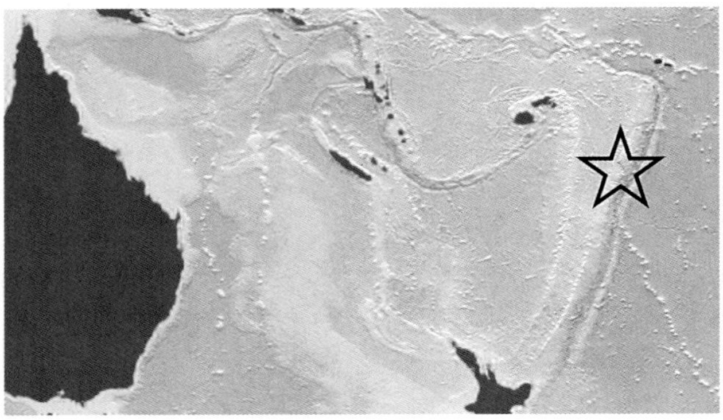

Figure 960. Distribution of *Anatoma tongaensis*. ☆ type locality.

Sasakiconcha Geiger, 2006

Chresonymy.

Sasakiconcha Geiger, 2006b: 45–46.

Type species. *Sasakiconcha elegantissima* Geiger, 2006 (OD).

Etymology. Sasaki-: honoring Takenori Sasaki of the University Museum of the University of Tokyo for his significant contributions to the systematics of basal gastropods. -concha: Latin for shell. Sasaki's shell. Gender feminine (OD).

Description. Shell depressed biconical, whorls rapidly increasing in size, no nacre. Protoconch sometimes partially covered by suture of subsequent whorl. Whorls not touching at base forming umbilical trough; no columella. Selenizone and foramen at periphery. Suture mostly at or above upper keel of selenizone of previous whorl. Umbilicus wide.

Animal unknown.

Comparisons. *Anatoma* shares the position of the slit at the periphery, but the slit remains open, and the suture is always below the lower keel of the selenizone, even in the most depressed species, *An. megascutula*.

Distribution. Western Pacific (Fig. 963). For specific localities, see individual species.

Geologic range. Recent.

Remarks. *Sasakiconcha* is tentatively placed in Anatomidae because of the peripheral position of the selenizone. That position was shown to discriminate between major scissurellid/anatomid lineages in an exploratory phylogenetic analysis of shell characters (Geiger, 2003. Fig. 21). The most likely alternative hypothesis is that *Sasakiconcha* represents yet an other independent lineage of slit-bearing, small vetigastropods.

Many species of *Anatoma* show converging margins of the slit, but in no species do they fuse and enclose a foramen. *Sasakiconcha* is not based on senile specimens of *Anatoma*, as the foramen is found over a wide range of shell sizes (4–8 mm). Not a single specimen of *Anatoma* with a closed foramen is known.

Sasakiconcha must form its foramen by a different process compared to most *Sinezona* or *Sukashitrochus*. *Sasakiconcha* closes the foramen from fairly early growth onwards, i.e., at about half-maximum size, which contrasts to most scissurellids, which close the foramen at about 80% of maximum size. It implies that with growth, the anterior margin of the foramen is resorbed, while shell material is added at the apertural margin, leaving a foramen of approximately constant proportion over a wide range of shell sizes (4–8 mm). In contrast, juveniles of *Sinezona* and *Sukashitrochus* have an open slit, which only closes once the animal is fully grown. Only *Sinezona globosa* also closes the foramen in not fully grown specimens (Geiger, 2006a). Comparison to other foramen-bearing small vetigastropod genera of the scissurellid grade (*Sutilizona, Temnocinclis, Temnozaga, Depressizona*) is not possible due to the scarcity of material.

The overgrowth of the selenizone by the shoulder of the subsequent whorl is unique among Recent Vetigastropoda, but is known from some fossil genera (Brookes-Knight *et al.*, 1960): e.g., *Agniesella* Cossmann, 1909 (Portlockiellidae), *Umbotropis* Perner, 1903 (Gosseletinidae), and *Catachisma* Branson, 1909 (Gosseletinidae). The overall shell morphology is most similar to the caenogastropod genus *Cochliolepis* Stimpson, 1858 (Vitrinellidae), though missing the selenizone indicative of a vetigastropod placement of *Sasakiconcha*. Sowerby I (1820–1834) illustrated *Scissurella elatior* on an unnumbered plate with the selenizone overgrown by the subsequent whorl

(Fig. 675). This character is not discussed as the species was introduced as a *nomen et figura*, without indication of where it was found. Sowerby II (1842, 1852) placed it as a Recent species occurring off the coast of Britain, from where mainly *An. crispata* is known. It is thus most likely that Sowerby I's (1820–1834) figure was stylized and is erroneous with respect to the overgrowth of the selenizone. The whereabouts of the type material of Sowerby's species is unknown.

Sasakiconcha elegantissima Geiger, 2006

Figures 961–963

Chresonymy.

Sasakiconcha elegantissima Geiger, 2006b: 46–49, figs 1–2.
Sasakiconcha elegantissima: Anonymous, 2007: 239, fig. 1 [copy Geiger, 2006b: fig. 1].
Sasakiconcha elegantissima: Geiger, 2008b: fig. 1.
Sasakiconcha elegantissima: Bouchet *et al.*, 2008: 33.
Sasakiconcha elegantissima: Tröndlé & Boutet, 2009: 12.

Type material. Holotype (MNHN type collection. Fig. 961). Paratypes 477–493 m, BORDAU1, DW1432, off Fiji, 17.333°S, 178.733°W (MNHN, 1. Fig. 962A). 300–302 m, BORDAU2, DW1523, Eua, Tonga, 21.300°S, 175.000°W (MNHN, 1. Fig. 962B). 391–421 m, BORDAU2, DW1537, haut-fond, Tonga, 21.683°S, 175.317°W (MNHN, 2. Fig. 962C). 700–707 m, BORDAU1, DW1485, Fiji, 19.050°S, 178.500°W (UMUT RM29213, 1; NMNZ M.273290, 1).

Type locality. 970–983 m, BORDAU2, DW1531, Eua, Tonga, 21.200°S, 174.933°W (OD).

Etymology. Elegantissima: elative of Latin adjective: extremely elegant referring to the overall aspect of the shell (OD).

Description. Shell large (to 8.5 mm: holotype. Figs 961, 962), biconical depressed, *Sinum*-shaped; thin, fragile. Protoconch 285–325 µm (mean = 311 µm, n = 4, 325 µm in holotype), 0.75 whorl, sculpture flocculent (mostly eroded in holotype), no apertural varix, apertural margin convex; sometimes partially covered by suture of subsequent whorl. Teleoconch I of <0.05 whorl. Teleoconch II of up to 1.75 rapidly expanding whorls. Shoulder slightly convex, early teleoconch II with axial cordlets only, interstices irregular with tiny pits; after 0.5 whorl, axials with approximately one dozen nodules, nodules becoming spirally elongated with strength of axials decreasing; after 1.25 whorls, sculpture transformed into approximately 50 spirally arranged, irregularly spaced, nodular cordlets; interstices with very fine axial growth marks. Base with fine, stepped, longitudinally undulating spirals, reaching just into uppermost portion of umbilical cavity, approximately 30 in 4 mm specimen, approximately 40 in 8 mm specimen. Umbilicus wide, walls steep, smooth; umbilical wall not connecting with previous whorl, forming spiral trough between umbilical wall and previous whorl produced by inner surface of shoulder. Selenizone at periphery, usually hidden by suture of subsequent whorl; slit closed anteriorly to form foramen in specimens ≥ 4 mm. In fully mature specimens (holotype), last 0.125 whorl descending. Aperture oblong D-shaped, roof overhanging.

Animal unknown.

Comparisons. The cited generic characters will readily identify *Sas. elegantissima*.

Anatoma pseudoequatoria from the Indo-Malayan Archipelago to the Central Pacific is the most similar *Anatoma* species, but has a protoconch of half the size (150 µm vs. 325 µm), a longer teleoconch I (0.125 vs. 0.05 whorl), the suture is below the lower keel of the selenizone of the previous whorl (i.e., the selenizone is not obscured), the sculpture of the shoulder has distinct axial and spiral components, the base has reticulate sculpture, and the slit is open.

Anatoma megascutula from the Indo-Malayan Archipelago and the Western Pacific has a longer teleoconch I (>0.75 vs. 0.05 whorl), has no umbilical trough, and the slit remains open.

Specimen records.

Fiji. 700–707 m, BORDAU1, DW1485, 19.050°S, 178.500°W (MNHN, 1).
Tonga. 456–510 m, BORDAU2, DW1509, Tongatapu, 21.083°S, 175.367°W (MNHN, 1).
578 m, BORDAU2, DW1585, Vava'u group, 18.550°S, 173.950°W (MNHN, 2).

Figure 961. *Sasakiconcha elegantissima*. Holotype. 970–983 m, BORDAU2, DW1531, Eua, Tonga, 21.200°S, 174.933°W (MNHN type collection). Scale bar shell = 2 mm. Scale bar protoconch = 200 µm.

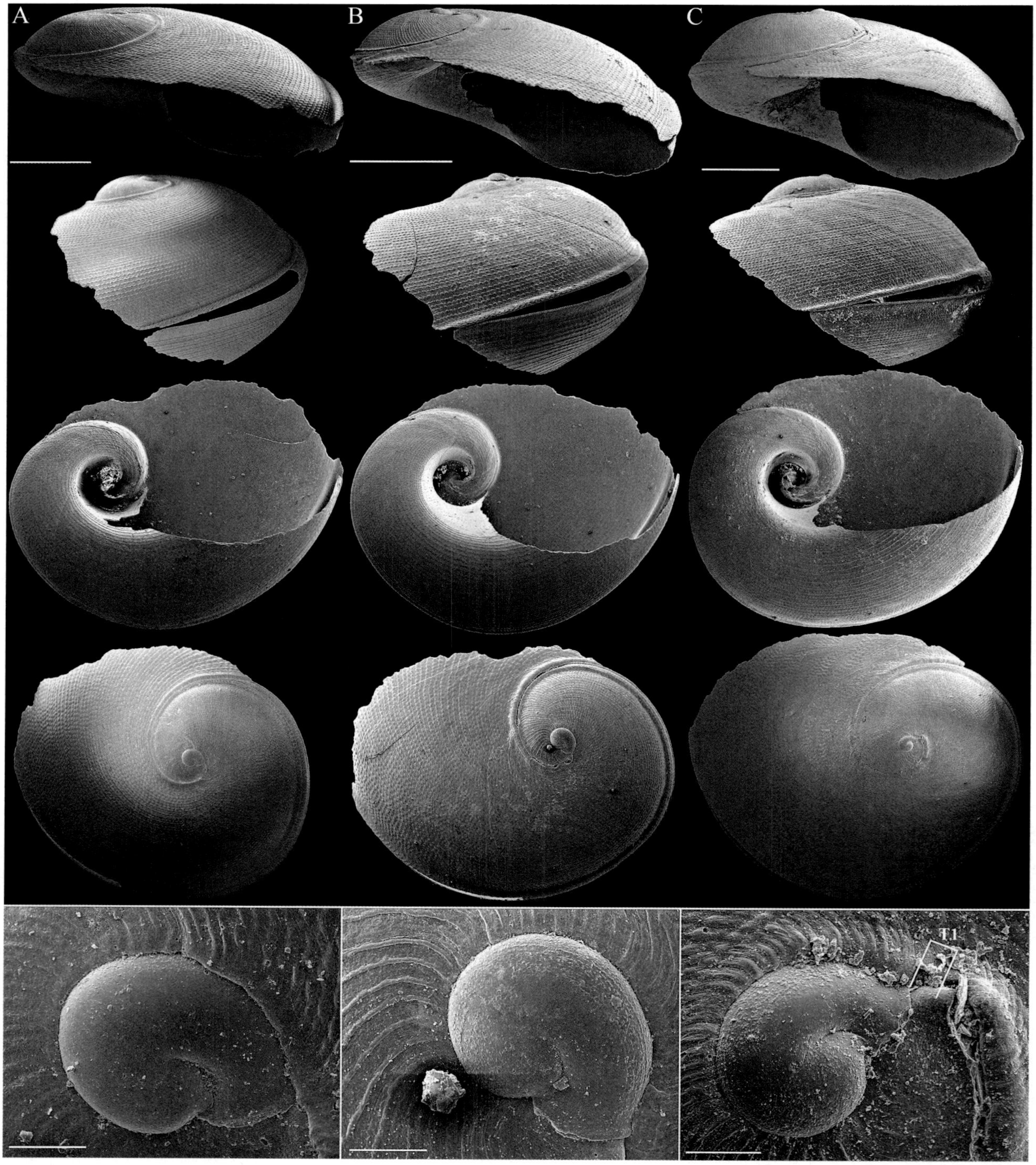

Figure 962. *Sasakiconcha elegantissima*. Paratypes. **A.** 477–493 m, off Fiji, 17.333°S, 178.733°W (MNHN, 1). **B.** 300–302 m, Eua, Tonga, 21.300°S, 175.000°W (MNHN, 1). **C.** 391–421 m, haut-fond, Tonga, 21.683°S, 175.317°W (MNHN, 2). Only in one of the smaller specimens on the left the slit is still open. In the other two, which are approximately half the size of the holotype, the slit is already closed to a foramen. Scale bars shell = 1 mm. Scale bars protoconch = 100 μm; T1: extent of teleoconch I.

French Polynesia. 600–620 m, BENTHAUS, DW1889, E of Rapa, 27.613°S, 144.262°W (MNHN, 1). 554–630 m, Mont Ari'i Moana, Tarava Seamounts, TARASOC DW3324, 19.250°S, 151.566°W (MNHN, 1). 557–569 m, Mount Ari'i Moana, Tarava Seamounts, TARASOC DW3318, 19.250°S, 151.516°W (MNHN, 1).

Distribution. Western to Central Pacific, 300–981 m [all shells only] (Fig. 963).

Remarks. *Sasakiconcha* is not a morphological endpoint in a series of undescribed *Anatoma* species. None even remotely approaches *Sasakiconcha.*

The umbilical trough is a character not known from any other scissurellid s.l. or anatomid and is to my knowledge a novel configuration in gastropods; its function is unclear. Umbilical modifications have been shown to function as brood pouches in Larocheidae (Marshall, 1993).

Sasakiconcha elegantissima is a fairly large species of anatomid, and with a diameter of 8 mm does not even qualify as a micromollusk, which usually is defined as shells smaller than 5 mm (Geiger *et al.*, 2007; Geiger & Ruthensteiner, eds., 2008). Despite its size, it is not the largest known anatomid. A specimen of *An. euglypta* from Antarctica is 10.5 mm in diameter (Zelaya & Geiger, 2007).

The protoconch of *Sas. elegantissima* is about twice as large as that of a typical scissurellid/anatomid species. The only species with a still significantly larger protoconch is *An. argentinae* (480 µm). It implies that rather large eggs are formed and that the number of eggs is relatively small, as has been observed in other scissurellids and anatomids (*Sin. rimuloides*; *An. euglypta*: Geiger, unpubl. data. Fig. 29, Color Plate 5), which is quite unlike the more familiar gonad condition in the large, broadcast spawning Vetigastropoda with millions of eggs (e.g., Poore, 1973; Ault, 1985; Bevelander, 1988; Rogers-Bennet *et al.*, 2004). Large eggs themselves do, however, not entail brooding or parental care. Eggs may also be deposited onto any available hard substrate as shown for *Sci. jucunda* by Strasoldo (1991).

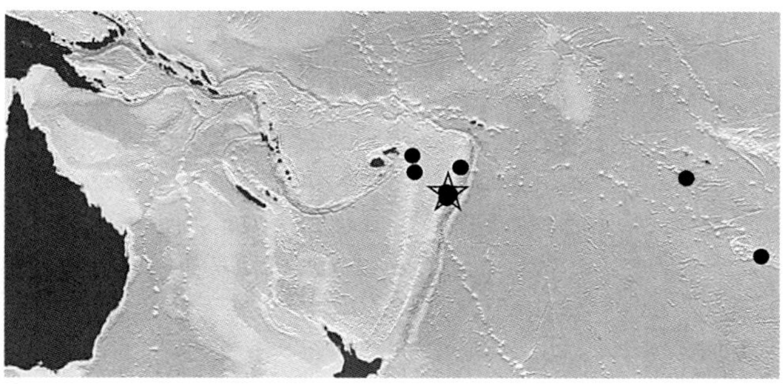

Figure 963. Distribution of *Sasakiconcha elegantissima.* ☆ type locality. ● specimen record.

The very short teleoconch I is not necessarily a corollary of the large protoconch. The selenizone of *Sas. elegantissima* forms at a size of 350–400 µm, in *An. argentinae* at 540 µm, in *Sin. globosa* with 1.5 teleoconch I whorls at 840 µm, and the slitless *La. spirata* and *Tro. lamellinodosa* grow to 1.5 mm and 1.6 mm, respectively.

Larocheidae Finlay, 1927

Chresonymy.

Larocheinae: Marshall, 1993: 285–286.

Type genus. *Larochea* Finlay, 1927 (M).

Other genera. *Larocheopsis* Marshall, 1993.
Trogloconcha Kase & Kano, 2002.

Description. Shell trochiform to patelliform due to increased expansion rate of whorl. No slit, foramen, selenizone. Anomphalous or umbilicate with profile of base and umbilicus describing continuous curve. Protoconch smooth or with flocculent sculpture, no apertural varix, aperture simple convex curve. Operculum reduced or missing. Radula R - 5 - n, lateral teeth 1–4 similar, lateral tooth 5 enlarged or similar in size to other lateral teeth, tip similar to other lateral teeth.

Distribution. Indian Ocean and Red Sea through Central Pacific. For specific localities, see individual species.

Geologic range. Oligocene to Recent.

Comparisons. All other subfamilies have a slit, foramen, or selenizone, which is missing in all Larocheidae. The only exception is the scissurellid *Coronadoa*, which can be distinguished by its protoconch with axial sculpture, and the typical scissurellid radula, with the much reduced lateral tooth 4 and a broadened lateral tooth 5.

Remarks. Finlay (1927) suggested a placement near Merriidae (Vanikoridae: Hipponicoidea). The family was placed in the vicinity of Lamellariidae due to the overall globular shell without a slit by Powell (1937).

Marshall (1993) noted as a diagnostic character of Larocheidae the non-papillate nature of the cephalic and epipodial tentacles. Baborka (2007) demonstrated the presence of papillae on both types of tentacles in *La. miranda*. The relatively large right suboptic tentacle found in *La. miranda* and *Tro. ohashii* may be a synapomorphy of Larocheidae; a penial function cannot be supported (Baborka, 2007).

Larochea Finlay 1927

Chresonymy.

Larochea Finlay 1927: 486.
Larochea: Marshall, 1993: 287.
Larochea: Bandel, 1998: 66–67.

Type species. *Larochea miranda* Finlay, 1927 (M).

Other species. *Larochea scitula* Marshall, 1993.
Larochea secunda Powell, 1937.
Larochea spirata Geiger & Marshall, 2012.

Etymology. Named after Mr. W. La Roche of Auckland, an enthusiastic collector (OD). Gender feminine.

Description. Shell small (to 1.3 mm), thin, depressed trochiform. Protoconch with granular sculpture. Teleoconch without selenizone, slit, or foramen. No umbilicus. No operculum. Young brooded in large subpallial cavity.

Distribution. New Zealand (Fig. 964). For specific localities, see individual species.

Geologic range. Recent.

Comparisons. *Larocheopsis* has a smooth protoconch, has fine pits on the early teleoconch, and lacks the columellar brood pouch.

Trogloconcha lacks the brood pouch.

Differentiation of species. Species are differentiated by overall shell shape and details of shell sculpture.

Intraspecific variability extends predominantly to the ontogenetic change in shell height, larger specimens having a proportionally taller shell.

Figure 964. Global distribution of *Larochea* species. ☆ type locality. ● specimen record. ■ literature record.

Larochea miranda Finlay, 1927

Figures 965–970

Chresonymy.

Larochea miranda Finlay, 1927: 486, pl. 24, figs 6–8 (Fig. 965).
Larochea miranda: Wenz, 1940: 884, fig. 2603.
Larochea miranda: Powell, 1979: 151, pl. 38, figs 1–2.
Larochea miranda: Marshall, 1993: 287–289, figs 1B–H [shell], 3 B–D [radula].
Larochea miranda: Spencer & Willan, 1995: 13.
Larochea miranda: Geiger, 2003: 75.
Larochea miranda: Baborka, 2007: 38–67 [histology, 3D reconstruction].

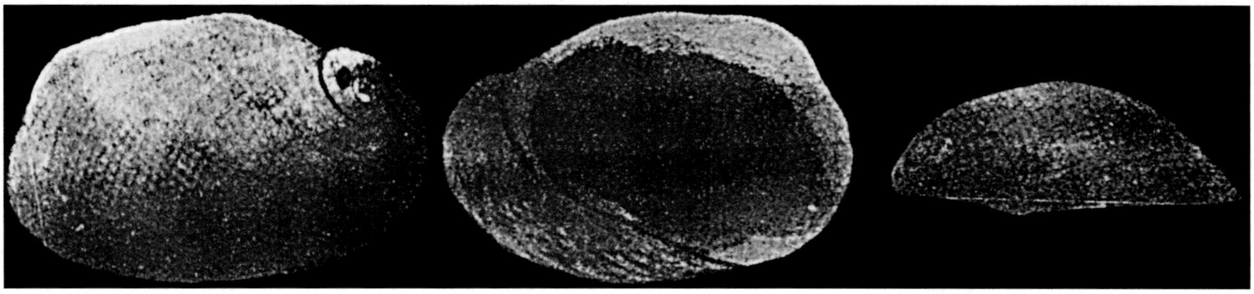

Figure 965. Original illustration of *Larochea miranda*.

Larochea miranda: Spencer *et al.*, 2009: 202.

Larochea miranda: Raven & Bracegirdle, 2010: 10, text-fig.

Misidentification.

not *Larochea miranda*: Bandel, 1998: 66–67, pl. 23, figs 4–5 [is *Tro. tesselata*].

Type material. Holotype (AIM AK 70386: 965, 966), 0.5 × 1.1 mm. Paratypes from type locality (AIM AK 72305, 17: 15 remaining but very fragile, two destroyed by Byne's/glass disease: W. Blom, pers. comm. 4/2009).

Figure 966. Holotype of *Larochea miranda*. Scale bar = 1 mm. Image courtesy Peter Quin/AIM.

Type locality. Awanui Bay, North Auckland, New Zealand [34.836°S, 173.250°E], [12 fms = 22 m] (OD).

Etymology. Miranda, Latin for wonderful, strange, singular.

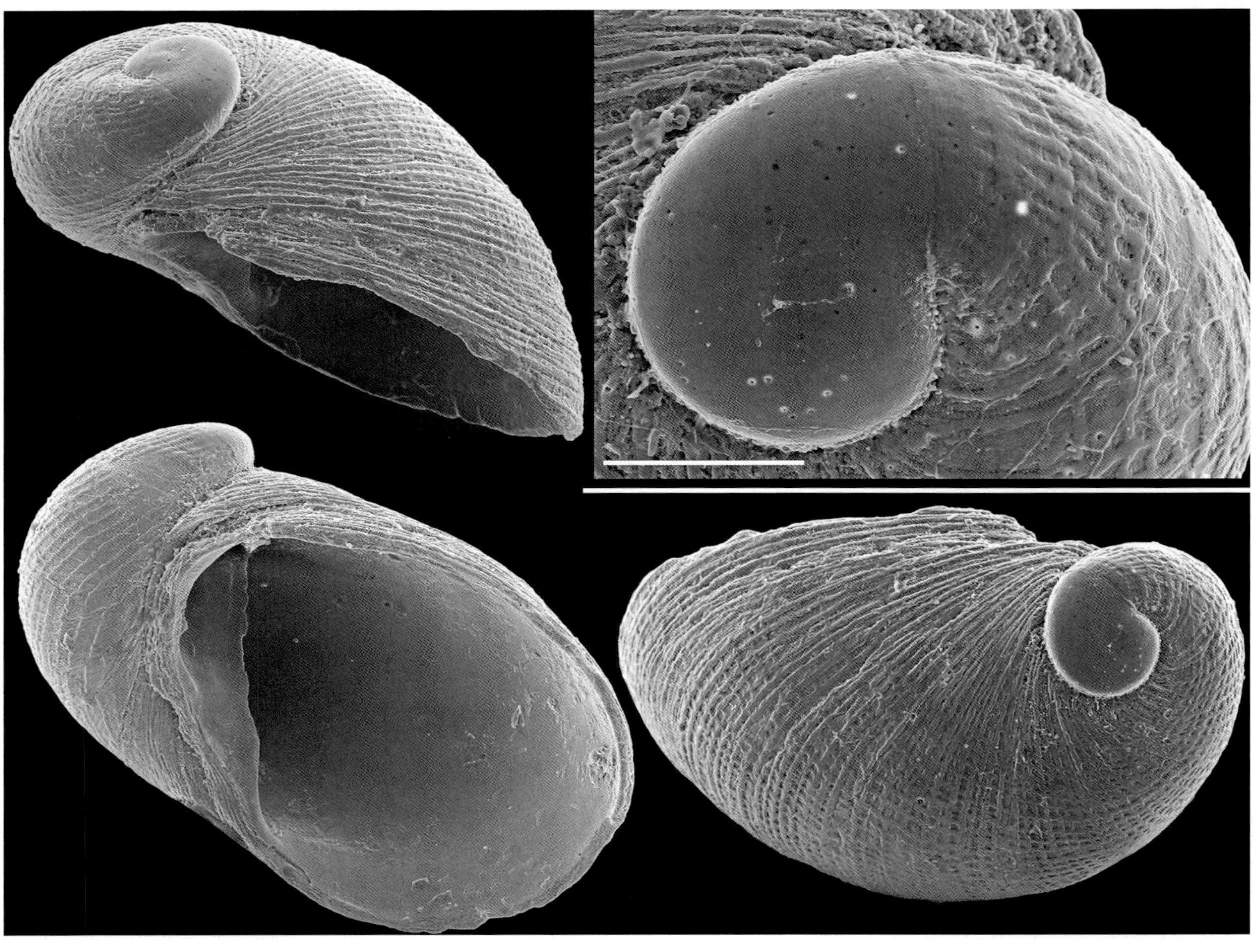

Figure 967. *Larochea miranda*. 4 m, Awanui, N Auckland, New Zealand, 50.533°S, 166.233°E (USNM 681135, 2). Scale bar shell = 1 mm. Scale bar protoconch = 100 µm.

Description. Shell small (to 1.3 mm. Figs 967, 968), depressed trochiform, thin fragile. Protoconch of 1.25 whorls, sculpture granular, no apertural varix, apertural margin convex. Teleoconch of up to 1.25 rapidly expanding whorls; sculpture of many fine and low spiral and axial cordlets producing fine reticulate pattern. No selenizone, slit, or foramen. Columella folded to form brood pouch. No operculum (Baborka, 2007).

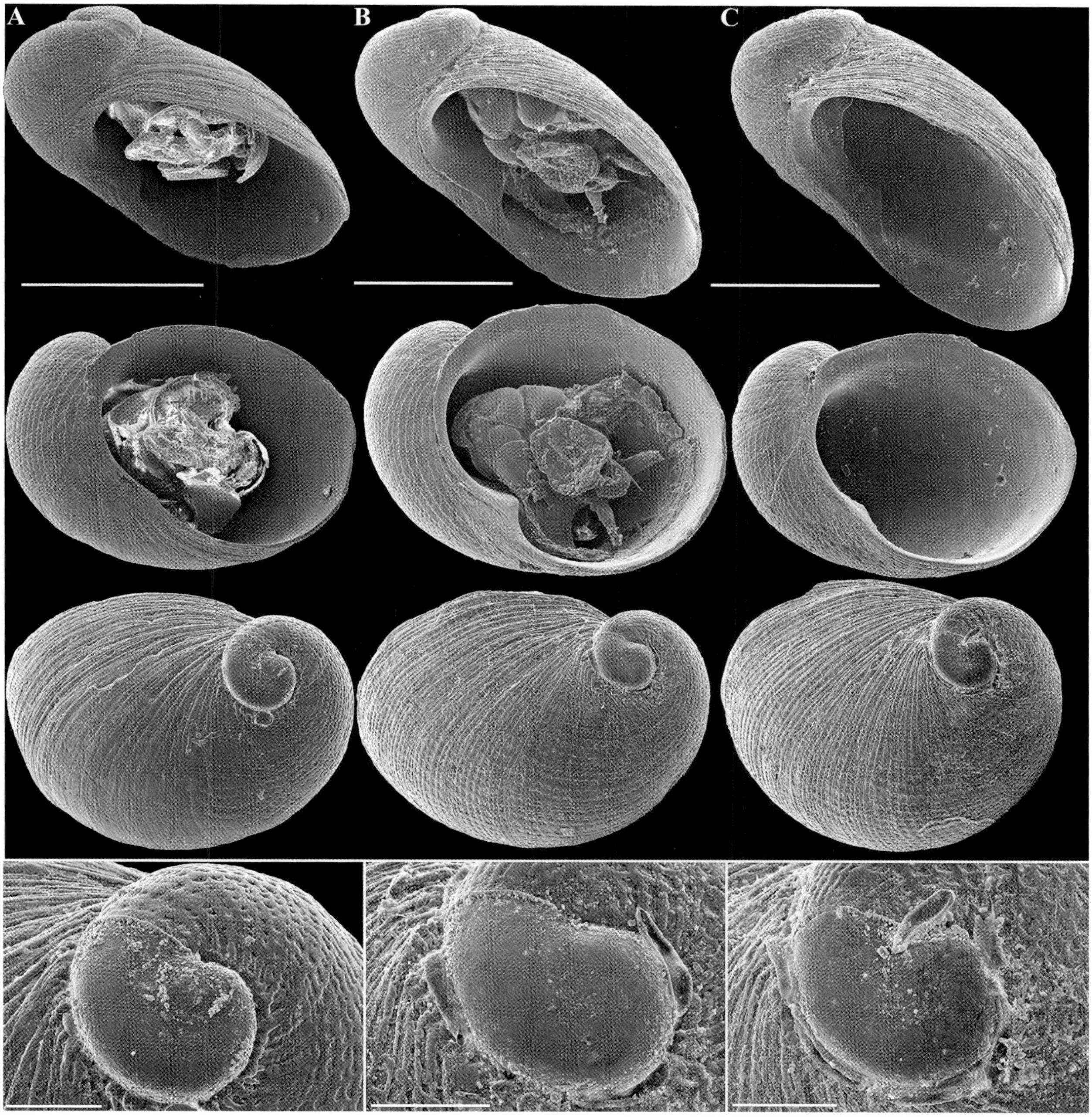

Figure 968. *Larochea miranda*. **A.** 44 m, Princess Islands, Arch Pinnacle, Three Kings Islands, New Zealand, 34.175°S, 172.050°E (NMNZ M.150125, 10). **B–C.** 25 m, Northland, Rangaunu Bay, North Island, New Zealand, 34.807°S, 173.325°E (NMNZ M.109198, 20). Scale bars shell = 0.5 mm. Scale bars protoconch = 100 µm.

Radula (Fig. 969). Rachidian tooth broadest, subtriangular with convex sides of shaft; tip with row of cusps parallel to tooth row, approximately 11 cusps of very similar size. Lateral teeth 1–4 similar to one another; shafts C-shaped, convex side towards rachidian tooth, tips asymmetrical, outer portion larger, largest cusp pointing towards rachidian tooth, 2–3 cusps on either side. Marginal teeth numerous, tips with approximately seven cusps on either side.

Anatomy (based on Baborka, 2007). Cephalic tentacles papillate in distal half, with right suboptic tentacle. Eyes present. Two shell muscles, pedal gland present in propodium. Three pairs of epipodial tentacles, some certainly papillate, ESO not recognized (insufficient preservation). Single monopectinate gill, leaflets with skeletal rods and bursicles. Hypobranchial gland and osphradium present. Heart most likely monotocardian. Intestine passing through pericardium and ventricle. Only sperm in gonad of examined specimen. Buccal cavity with paired jaws, pharynx with salivary glands, long coiled intestine. Nervous system hyparthroid with statocysts.

Comparisons. *Larochea secunda* and *La. scitula* from New Zealand are more elevated trochiform, and have coarser, more widely spaced axials and spirals.

Larocha spirata from New Zealand has an even more elevated shell shape and has distinct axial sculpture.

Distribution. Northern New Zealand. 0–622 m; living intertidal to 76 m, most from 22–76 m. Live material from clean shell substrate encrusted with red or brown algae, or rocky bottom with bryozoans, corals, gorgonians and hydroids (Marshall, 1993. Fig. 970).

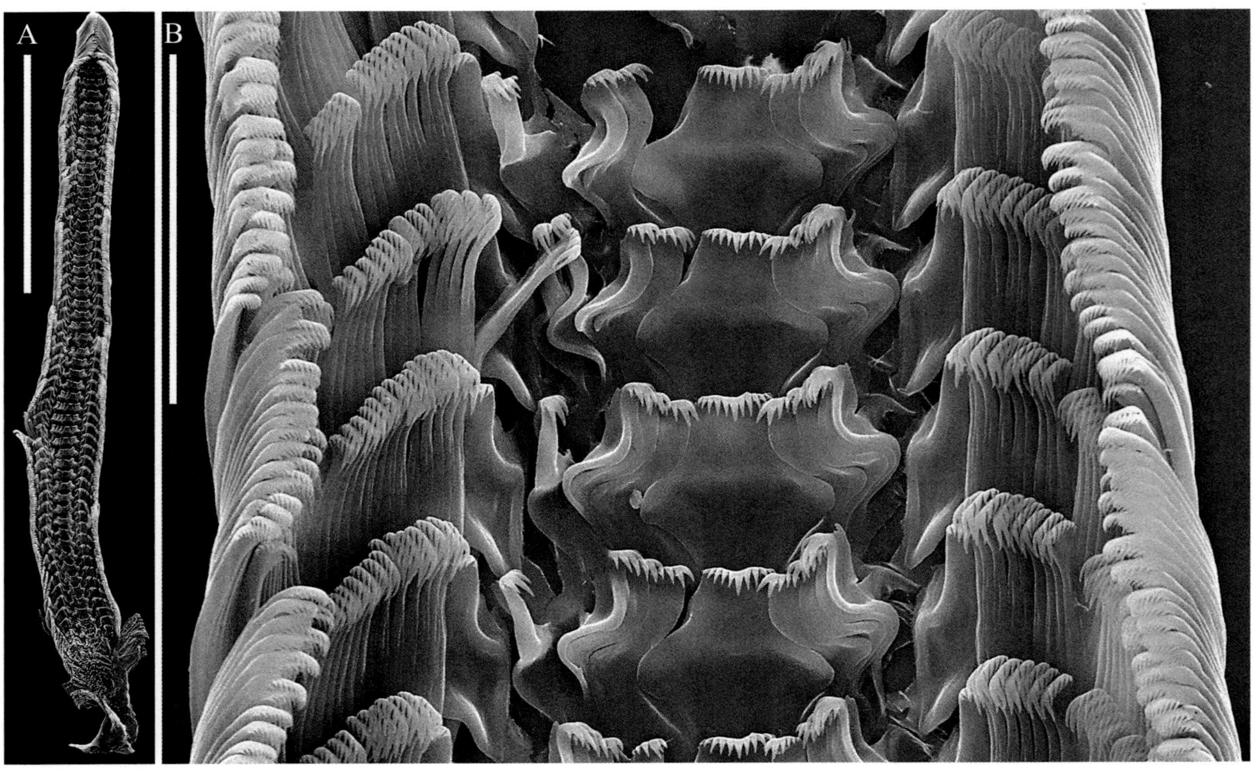

Figure 969. *Larochea miranda*. Radula. 71–76 m, East Cape, Ranfurly Bank, North Island, New Zealand, 37.553°S, 178.838°E (NMNZ M.072672, 25). **A.** Entire radula. Scale bar = 200 µm. **B.** Full width of radula. Scale bar = 20 µm.

Specimen records.

New Zealand. 44 m, Princess Islands, Arch Pinnacle, Three Kings Islands, 34.175°S, 172.050°E (NMNZ M.150125, 10). 25 m, Northland, Rangaunu Bay, North Island, 34.806°S, 173.325°E (NMNZ M.109198, 20). Doubtless Bay, 34.916°S, 173.467°E (MV F 96918, 2). 33 m, NE side Stephenson Island, 34.967°S, 173.783°E (AMS C.461511, 1). 11–18 m, Northland, Mangonui Heads, 34.983°S, 173.550°E (AMS C.405819, 2). 24 m, Whangaroa Harbour entrance, 35.100°S, 173.867°E (AMS C.61503, 3). 71–76 m, East Cape, Ranfurly Bank, North Island, 37.553°S, 178.838°E (NMNZ M.072672, 25). Waitangi, Chatham Islands, 43.950°S, 176.533°W (USNM 681136, 1). 4 m, Awanui, N Auckland, 50.533°S, 166.233°E (USNM 681135, 2). Awanui Bay, North Auckland, 35.050°S, 173.300°E (AIM AK 70386).

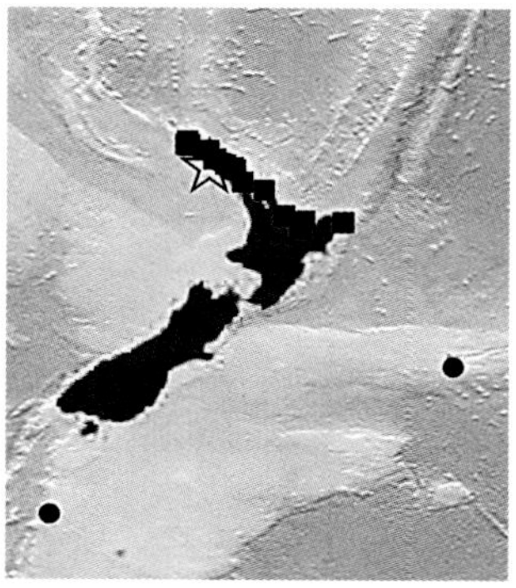

Figure 970. Distribution of *Larochea miranda.* ☆ type locality. ● specimen record. ■ literature record.

Literature records.

New Zealand. 622 m, N of Three Kings Island, 34.016°S, 172.117°E (NMNZ BS391). 53 m, reef between Great Island, and Farmer Rock, Three Kings Island, 34.150°S, 172.150°E (NMNZ 969). 44 m, off Arch Pinnacle, Prince's Rocks, Three King Island, 34.166°S, 172.050°E (NMNZ). 23 m, off N face of Hinemoa Island, Three Kings Island, 34.166°S, 172.083°E (NMNZ BS997). 252 m, off Three Kings Island, 34.166°S, 172.200°E (NMNZ BS395; NMNZ). 18–24 m, off N tip of West Island, Three Kings Island, 34.183°S, 172.033°E (NMNZ). 92 m, off Three Kings Island, 34.193°S, 172.182°E (NMNZ). 192–202 m, off Three Kings Island, 34.235°S, 172.150°E (NMNZ). Spirits Bay, 34.416°S, 172.783°E (NMNZ). 0 m, E side of North Cape, 34.416°S, 173.033°E (NMNZ). 22 m, off Rangaunu Bay, 34.806°S, 173.325°E (NMNZ). 22 m, off Rangaunu Bay, 34.826°S, 173.250°E (NMNZ). 22 m, Rangaunu Bay, Northland, 34.833°S, 173.250°E (NMNZ). 20 m, Main Channel, Whangaroa Harbour entrance, 35.050°S, 173.750°E (NMNZ). 31 m, between Black Island and Moturoa, Bay of Islands, 35.183°S, 174.083°E (NMNZ). 7 m, off Raraoa Bay Point, Bay of Islands, 35.263°S, 174.173°E (NMNZ). 6–7 m, off Castle Rock, Taurikura Bay, Whangerei Heads, 35.816°S, 174.500°E (NMNZ). 15–27 m, off Great Barrier Island, 36.166°S, 175.417°E (NMNZ). 104–109 m, off Mayor Island, 37.278°S, 176.292°E (NMNZ). 37–91 m, off SE side of Mayor Island, 37.300°S, 176.267°E (NMNZ). 29–34 m, off Mayor Island, 37.305°S, 176.278°E (NMNZ). 59–74 m, off Mayor Island, 37.315°S, 176.270°E (NMNZ). 64–69 m, off White Island, 37.508°S, 177.162°E (NMNZ; NMNZ). 71–76 m, Ranfurly Bank, East Cape, 37.553°S, 178.838°E (NMNZ). 49 m, Ranfurly Bank, East Cape, 37.596°S, 178.878°E (NMNZ). 55–59 m, Ranfurly Bank, East Cape, 37.606°S, 178.858°E (NMNZ). 0 m, Boulder Bay, Motuhora Island, 37.850°S, 176.983°E (NMNZ) (Marshall, 1993).

Remarks. Marshall (1993) noted that all live specimens larger than 0.93 mm were brooding females. In histological examination of a 0.9 mm body (without shell: Baborka, 2007) only male gametes were found in the gonad. Whether sexual dimorphism as reported by Hickman (1999) for *I. auriformis* [as *Sinezona* sp.] is found in *Larochea*, or whether the species is protandrous or hermaphroditic is unknown.

Larochea secunda Powell, 1937

Figures 971–975

Chresonymy.

Larochea secunda Powell, 1937: 207–208, pl. 50, fig. 7 (Fig. 971).
Larochea secunda: Powell, 1979: 151, pl. 38, fig. 3.
Larochea secunda: Marshall, 1993: 289–291, figs 2A–G [shell], 3E–H [radula].
Larochea secunda: Spencer & Willan, 1995: 13.
Larochea secunda: Geiger, 2003: 75.
Larochea secunda: Spencer *et al.*, 2009: 202.

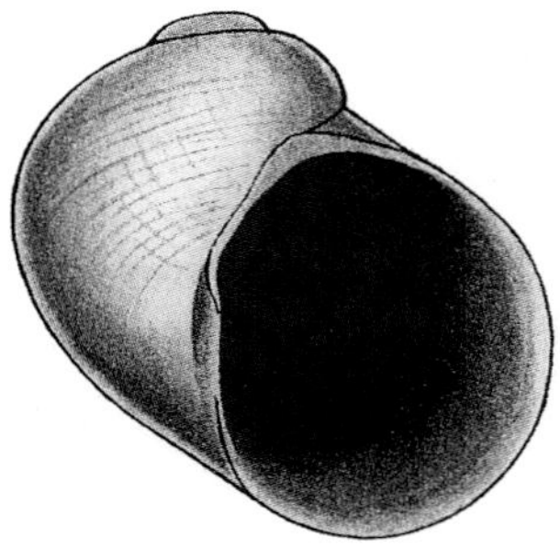

Figure 971. Original illustration of *Larochea secunda*.

Type material. Holotype (NHMUK 19621044. Fig. 974A), 0.9 × 0.9 mm. Paratype from type locality (AIM AK 72286, 1).

Type locality. 260 m, off Three Kings Islands, [New Zealand, 34.090°S, 172.080°E] *Discovery* station 933 (OD).

Etymology. Secunda, Latin for second; second species in genus (OD).

Description. Shell medium-sized (to 1.12 mm. Figs 973, 974), turbiniform, thin fragile, translucent. On upper 0.25 portion of whorl, spiral milky-white band of increasing width with growth. Protoconch of slightly less than one whorl, sculpture finely granular, no apertural varix, apertural margin convex. Teleoconch of slightly more than one whorl, with numerous soft axial cords, becoming less distinct on base, finer less distinct spiral lines turning into steps towards umbilicus. Anomphalous. Columella with deep brood pouch.

Radula (Fig. 975). Rachidian tooth triangular, sides of shaft strongly convex, tip broad V-shaped, central cusp less than twice as large as 5–6 cusps on either side. Lateral teeth 1–4 similar, tip somewhat asymmetrical, central cusp largest, 3–5 smaller cusps on either side. Marginal teeth with long shafts, no food groove, tips with approximately eight cusps. Radular interlock moderate.

Comparisons. *Larochea miranda* from New Zealand has a more depressed shell and a more coarsely granulated protoconch.

Larochea scitula from New Zealand has more distinct axials and spirals, as well as a more coarsely granulated protoconch.

Larochea spirata from New Zealand has an overall more elevated shell and has strong spirals on the protoconch.

Distribution. Off Three Kings Islands, northern New Zealand, 91–805 m, live 91–310 m, on shell/bryozoan substrate with other invertebrates (Marshall, 1993. Fig. 972).

Specimen records.

New Zealand. 260 m, off Three Kings Islands, *Discovery* Station 933, 34.222°S, 172.200°E (NHMUK, 1). 622 m, N of Three Kings Islands, 34.017°S, 172.117°E (NMNZ M.034791, 5). 192–202 m, SE of Great Island, Three Kings Islands, 34.235°S, 172.150°E (NMNZ M.150089, 1). 310 m, 11 km NW of Great Island, Three Kings

Islands, 34.108°S, 172.078°E (NMNZ M.110553, 10). 173–178 m, SE of Great Island, Three Kings Islands, 34.247°S, 172.227°E (NMNZ M.110552, 2).

Remarks. The milky white band on the upper portion of the whorl is quite pronounced, however is only visible in un-coated specimens under the light microscope.

Figure 972. Distribution of *Larochea secunda*. ☆ type locality. ● specimen record.

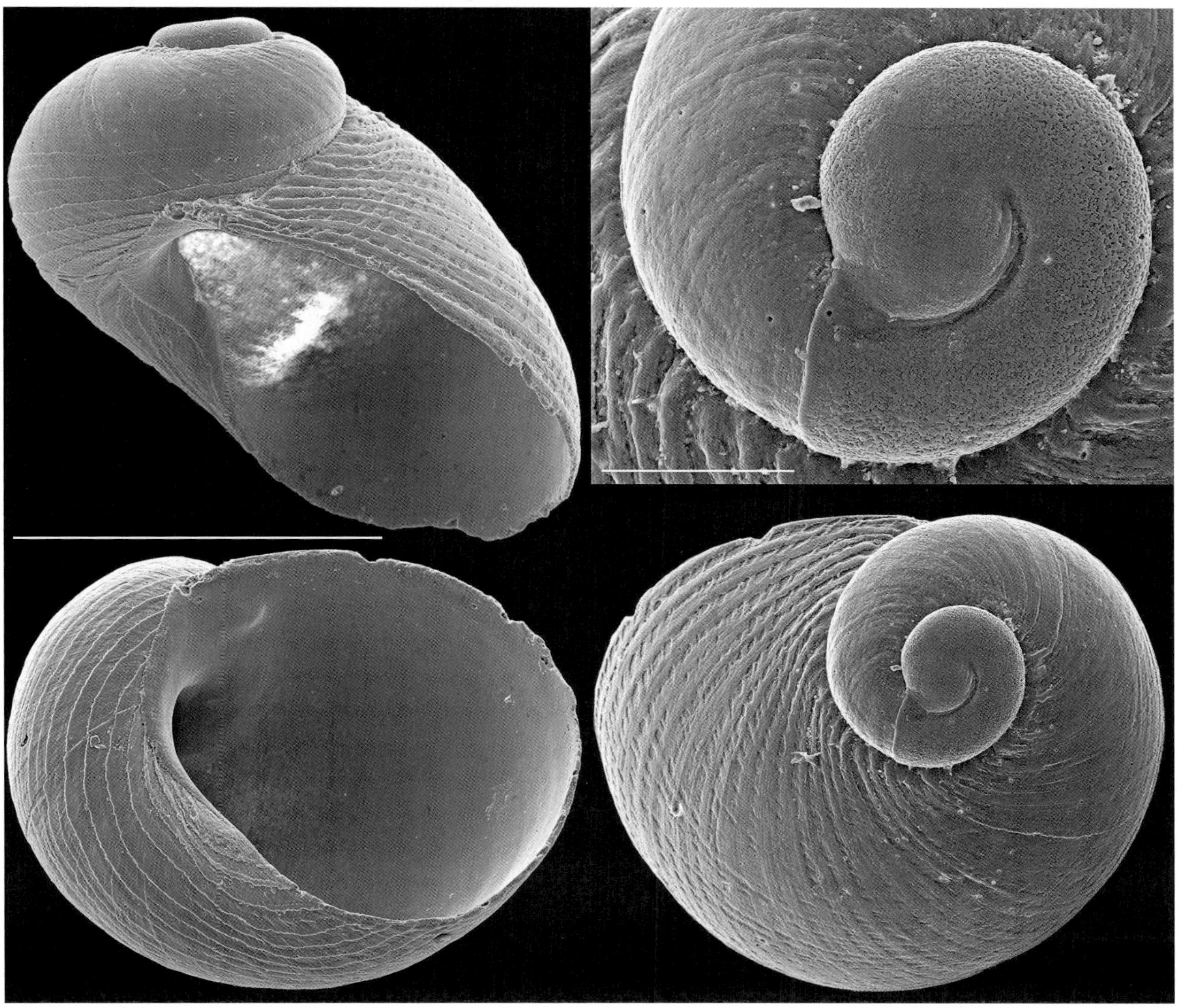

Figure 973. *Larochea secunda*. 622 m, N of Three Kings Islands, New Zealand, 34.017°S, 172.117°E (NMNZ M.034791, 5). Scale bar shell = 0.5 mm. Scale bar protoconch = 100 μm.

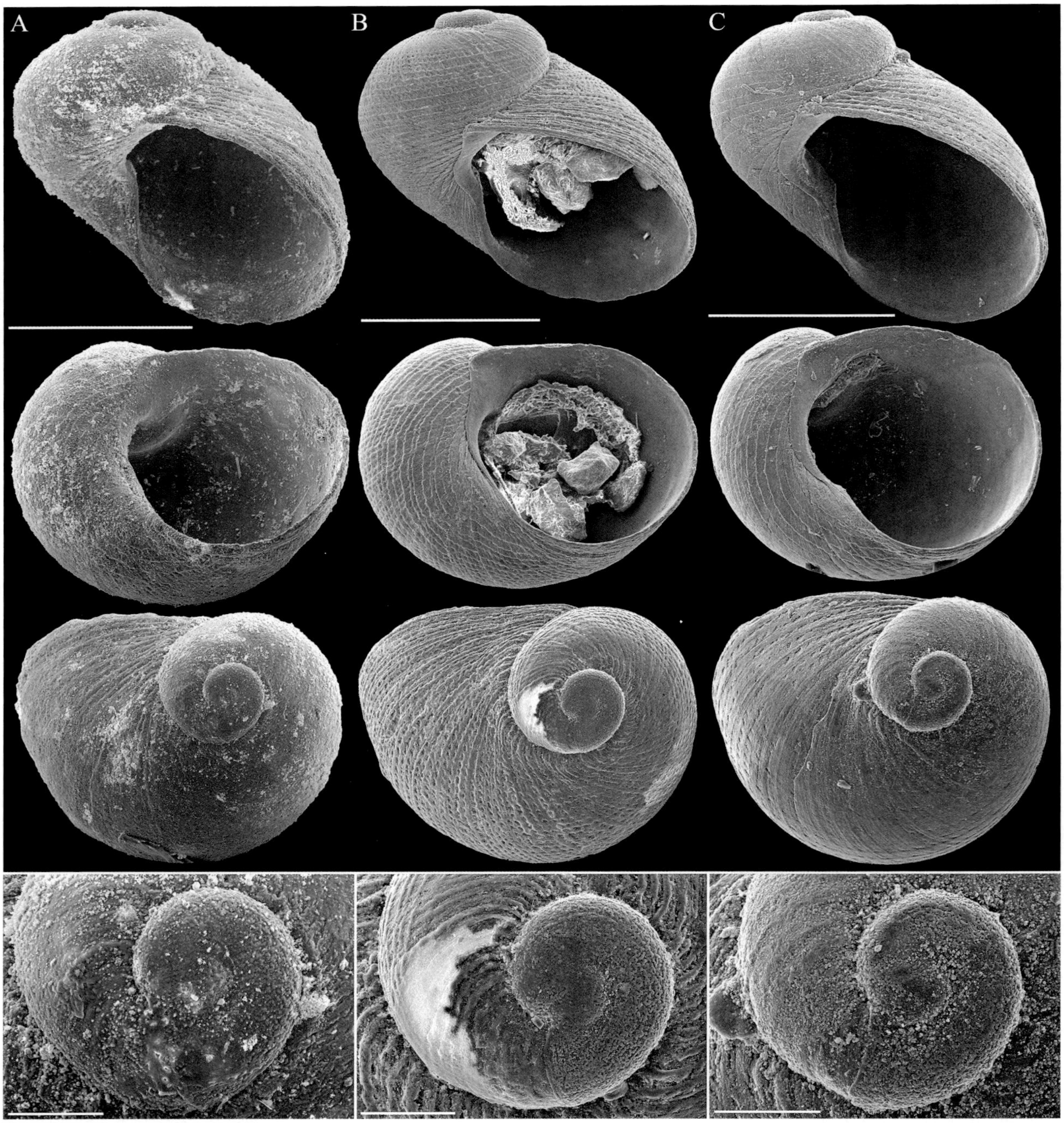

Figure 974. *Larochea secunda*. **A.** Holotype. 260 m, off Three Kings Islands, [New Zealand, 34.090°S, 172.080°E] *Discovery* station 933 (NHMUK 19621044). **B.** 173–178 m, SE of Great Island, Three Kings Islands, New Zealand, 34.247°S, 172.227°E (NMNZ M.110552, 2). **C.** 310 m, 11 km NW of Great Island, Three Kings Islands, New Zealand, 34.108°S, 172.078°E (NMNZ M.110553, 10). Scale bars shell = 0.5 mm. Scale bars protoconch = 100 μm.

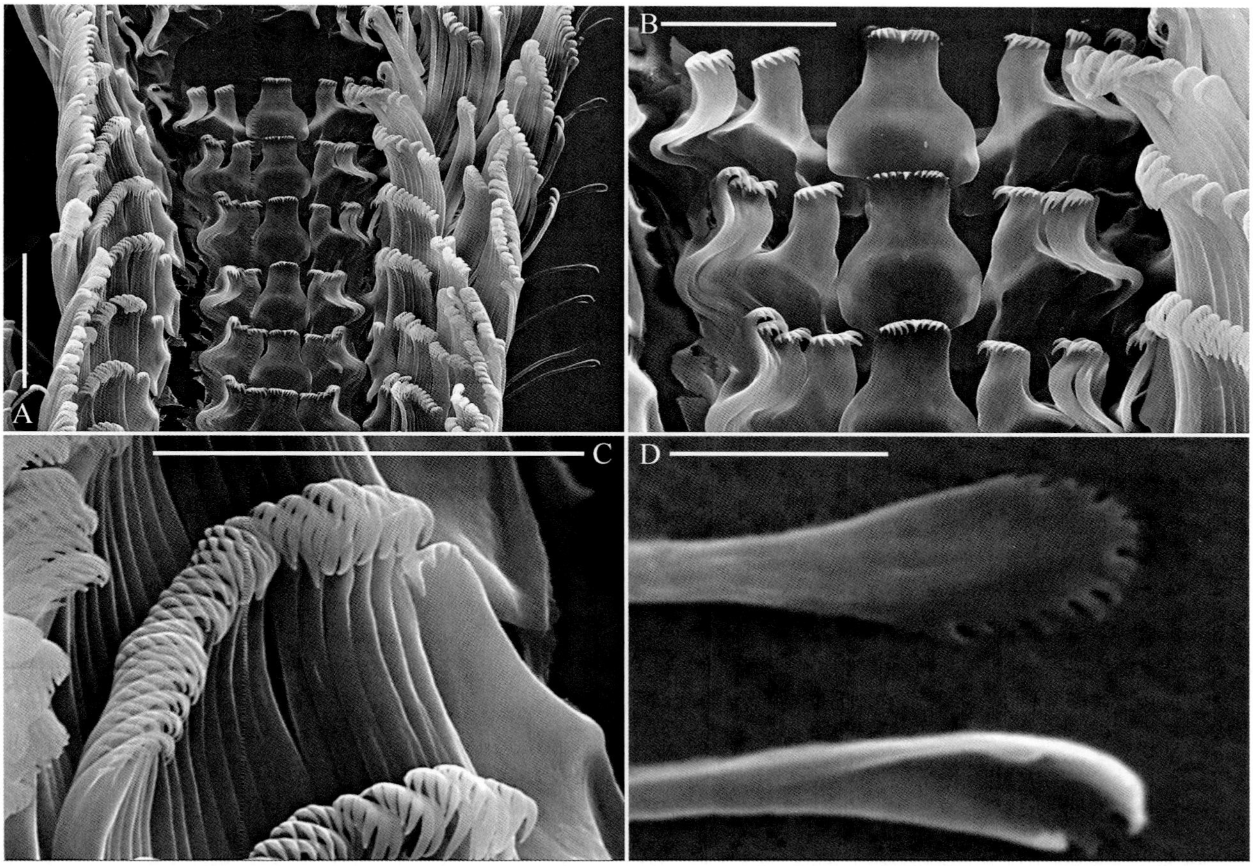

Figure 975. *Larochea secunda*. Radula. 173–178 m, SE of Great Island, Three Kings Islands, New Zealand, 34.247°S, 172.227°E (NMNZ M.110552, 2). **A.** Full width of radula. Scale bar = 20 µm. **B.** Central field enlarged. Scale bar = 10 µm. **C.** Marginal teeth. Scale bar = 10 µm. **D.** Outermost paddle-shaped marginal teeth. Scale bar = 2 µm.

Larochea scitula Marshall, 1993

Figures 976–979

Chresonymy.

Larochea scitula Marshall, 1993: 291, pl. 2, figs H–K (Fig. 976).

Larochea scitula: Geiger, 2003: 76.

Type material. Holotype (NMNZ M.260630. Fig. 976), 1.22 × 1.10 mm. Paratypes from type locality (NMNZ BS890 (0.636), 2). 113 m, Wanganella Bank summit, 32°34.4'S, 167°31'E (MNNZ BS 885 (0.631), 2). 133 m, Wanganella Bank summit, 32°39.2'S, 167°31.7'E (NMNZ BS 891 (0.637), 1). 757–660 m, Wanganella Bank slope, 37°40.8'S, 167°6.9'E (National Institute of Water and Atmospheric Research, Wellington P8. 1).

Type locality. 133 m, Wanganella Bank summit, southern Norfolk Ridge, New Zealand, 32°40'S, 167°33.6'E (OD).

Etymology. Scitula, Latin for neat (OD).

Description. Shell medium-sized (to 1.22 mm. Figs 977, 978), turbiniform, thin, fragile, translucent. Protoconch of slightly less than one whorl, sculpture flocculent, no apertural varix, apertural margin convex. Teleoconch of up to 1.75 whorls, sculpture of approximately 33 strong axial cords on last whorl, decreasing in strength from

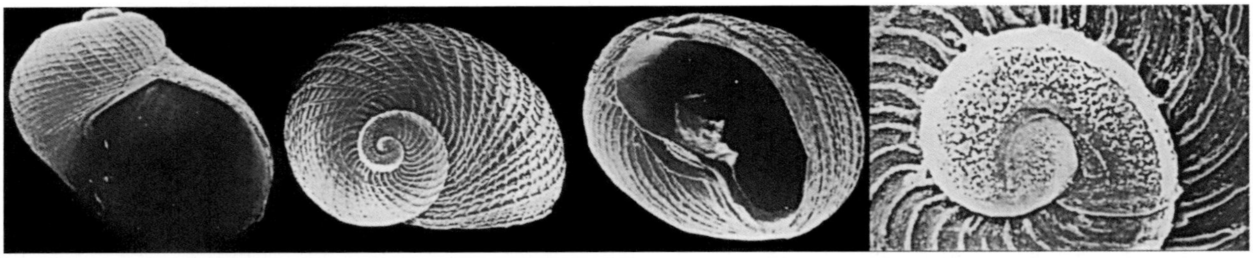

Figure 976. Original illustration of *Larochea scitula*. All of holotype except umbilical view.

Figure 977. *Larochea scitula*. 120–126 m, Wanganella Bank summit, West Norfolk Ridge, New Zealand, 32.611°S, 167.584°E (NMNZ M.171769, 20). Scale bar shell = 1 mm. Scale bar protoconch = 100 µm.

suture to base, approximately 14 weaker spiral lines visible in top apical view; axials denser towards apertural margin. No umbilicus. Columella with brood pouch.

Animal unknown.

Comparisons. *Larochea miranda* from New Zealand is flatter in overall shape, has more rapidly expanding whorls, and finer sculpture with axials and spirals of equal strength.

Larochea secunda from New Zealand has a more finely sculptured protoconch and less distinct axials and spirals.

Figure 978. *Larochea scitula*. **A–C.** 120–126 m, Wanganella Bank summit, West Norfolk Ridge, New Zealand, 32.611°S, 167.584°E (NMNZ M.171769, 20). Scale bars shell = 1 mm. Scale bars protoconch = 100 μm.

Larochea spirata from New Zealand has an overall much taller shell, and has weaker axial sculpture.

Distribution. Northernmost New Zealand (Fig. 979).

Specimen records.

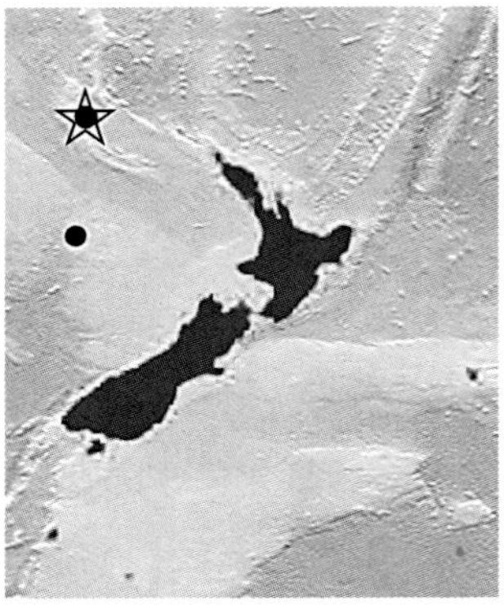

New Zealand. 120–126 m, Wanganella Bank summit, West Norfolk Ridge, 32.611°S, 167.584°E (NMNZ M.171769, 20). 133 m, Wanganella Bank summit, southern Norfolk Ridge, 32.667°S, 167.560°E (NMNZ M.260630, 1). 113 m, Wanganella Bank summit, southern Norfolk Ridge, 32.573°S, 167.517°E (NMNZ BS 885, 2). 133 m, Wanganella Bank summit, southern Norfolk Ridge, 32.653°S, 167.528°E (NMNZ BS891, 1). 660–757 m, Wanganella Bank slope, southern Norfolk Ridge, 37.680°S, 167.115°E (NMNZ NIWA P8, 1). 133 m, Wanganella Bank summit, southern Norfolk Ridge, 32.667°S, 167.560°E (NMNZ M.260630, 1).

Figure 979. Distribution of *Larochea scitula*. ☆ type locality. ● specimen record.

Remarks. The species is evidently gonochoristic as specimens of similar size with and without brood pouch have been found. This can be interpreted as sexual dimorphism (Marshall, 1993).

Larochea spirata Geiger & Marshall, 2012

Figures 980–982

Chresonymy.

Larochea spirata Geiger & Marshall, 2012: 29–31, figs 23–25.

Type material. Holotype (NMNZ M.172130, 1. Fig. 981). Paratypes from type locality (NMNZ M.172127, 36).

Type locality. 785–800 m, W of Cape Reinga, West Norfolk Ridge, New Zealand, 34.285°S, 168.430°E (OD).

Etymology. Spira, Latin for the spire, referring to the pronounced spire compared to other species in the genus (OD).

Description. Shell medium-sized (to 1.5 mm. Figs 981–982), trochiform, elevated, up to 20% taller than wide. Protoconch of 0.75 whorl, sculpture flocculent with three indistinct spiral cords, no apertural varix, apertural margin straight. Teleoconch of up to 1.8 whorls, suture moderately impressed. Early teleoconch with shoulder showing approximately 20–22 indistinct axial cords, becoming more distinct with growth; spirals about one third weaker than axials, first spiral after 0.33–0.5 whorl, approximately 25 at apertural margin; spirals running over axials forming minute elevated points. Axials becoming less distinct from suture towards base, sometimes vanishing entirely. Spirals transitioning at periphery from cords on shoulder to low steps on base of same density as on shoulder; axials barely perceptible on base. Anomphalous. Aperture oval, roof overhanging.

Animal unknown.

Comparisons. The elevated shell shape immediately distinguishes this species from any other. Additionally, *La. secunda* from New Zealand lacks the spiral sculpture on the protoconch, while *La. scitula* from New Zealand has much stronger, elevated axial cords, and *La. miranda* from New Zealand lacks axial sculpture.

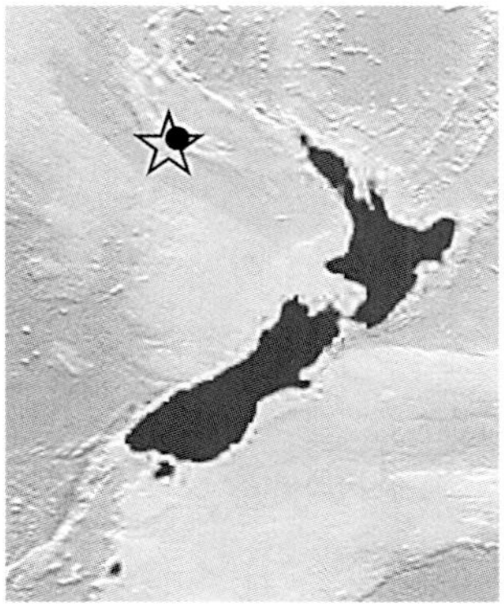

Figure 980. Distribution of *Larochea spirata*. ☆ type locality. ● specimen record.

Figure 981. *Larochea spirata*. Holotype (NMNZ M.172130). 785–800 m, W of Cape Reinga, West Norfolk Ridge, New Zealand, 34.285°S, 168.430°E Scale bar shell = 1 mm. Scale bars protoconch = 100 μm.

Distribution. Known only from Norfolk Ridge, New Zealand (Fig. 980).

Figure 982. *Larochea spirata*. **A–C.** Paratypes. 785–800 m, W of Cape Reinga, West Norfolk Ridge, New Zealand, 34.285°S, 168.430°E (NMNZ M.172127, 36). Scale bars shell A–B = 1 mm. Scale bar shell C = 0.5 mm. Scale bars protoconch = 100 μm.

Larocheopsis Marshall, 1993

Chresonymy.

Larocheopsis Marshall, 1993: 291.
Larocheopsis: Bandel, 1998: 67.

Type species. *Larocheopsis amplexa* Marshall, 1993 (OD).
Other species. *Larocheopsis macrostoma* Geiger & Marshall, 2012.
Species removed from *Larocheopsis*. *Trogloconcha marshalli* (Lozouet, 1998) †.
Etymology. *Larochea*: similar genus. -opsis, Greek for look alike. Gender feminine.
Description. Shell small (to 0.87 mm), turbiniform, thin, fragile. Protoconch smooth. Early teleoconch with fine pits. No columellar brood pouch.
Distribution. New Zealand (Fig. 983). For specific localities, see individual species.
Geologic range. Recent.
Comparisons. *Larochea* lacks the finely pitted early teleoconch and has a distinct brood pouch. *Trogloconcha* lacks the finely pitted early teleoconch.
Differentiation of species. Species are differentiated by overall shell shape, and sculpture.

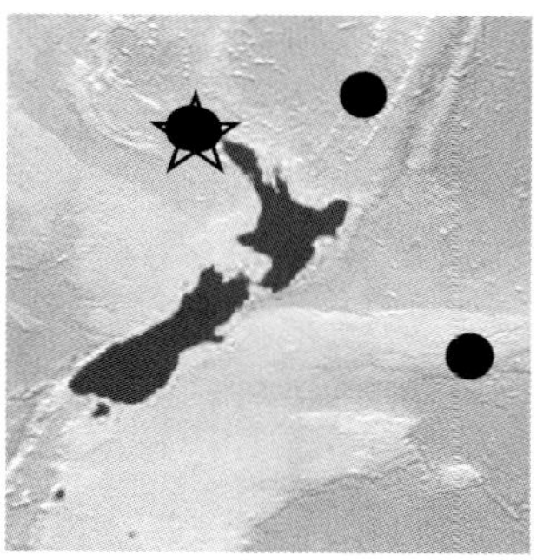

Figure 983. Global distribution of *Larocheopsis* species. ☆ type locality. ● specimen record. ■ literature record.

Larocheidae *Larocheopsis*

Larocheopsis amplexa Marshall, 1993

Figures 984–985

Chresonymy.

Larocheopsis amplexa Marshall, 1993: 291–294, figs 4A–E [shell], 4F–G [radula] (Fig. 984).
Larocheopsis amplexa: Spencer & Willan, 1995: 13.
Larocheopsis amplexa: Geiger, 2003: 75.
Larocheopsis amplexa: Spencer *et al.*, 2009: 202.

Type material. Holotype (NMNZ M.74954. Fig. 984), 0.8 × 0.75 mm. Paratypes from type locality (NMNZ M.074955, 10; AMS C.311886, 1; MNHN, 1). 246–291 m, Middlesex Bank, NW of Three Kings Island, [New Zealand,] 34°2'S, 171°44'E

Figure 984. Original illustration of *Larocheopsis amplexa*. All of holotype (NMNZ M.74954) except lower left and radula.

(NMNZ M.093882, 1). 310 m, 11 km NW of Great Island, Three Kings Islands, [New Zealand,] 34°6.5'S 172° 4.7'E, (MNHM, 1: not found in 4/2005).

Type locality. 310 m, off Three Kings Island, [New Zealand], 34°6.5'S, 172°4.7'E (OD).

Etymology. Amplexus, Latin for embrace.

Description. Shell small (to 0.87 mm. Figs 984, 986), turbiniform, thin, fragile, translucent. Protoconch of slightly less than one whorl, with distinct outpocketing at tip of embryonic cap, smooth, no apertural varix, apertural margin convex. Teleoconch of up to 1.5 whorls, early whorl with fine pits; sculpture of fine axial growth lines, approximately 20 fine spirals from periphery to base. Aperture round. Anomphalous.

Radula (Fig. 984). Rachidian tooth with tip parallel to tooth row, central cusp slightly larger, seven smaller cusps on either side. Lateral tooth 1 subtriangular, tip with 8–9 cusps. Lateral teeth 2–4 narrower, lateral tooth 2 with seven cusps, lateral teeth 3–4 with four cusps. Lateral tooth 5 slender pointed, with 3–4 cusps. Marginal teeth with long shaft, no food groove, tips finely serrated.

Figure 985. Distribution of *Larocheopsis amplexa*. ☆ type locality. ● specimen record.

Comparisons. *Larocheopsis macrostoma* from New Zealand is much wider than tall.

Distribution. Off Three Kings Islands, New Zealand, 246–310 m, on shell/bryozoan material with invertebrates (Fig. 985).

Remarks. The species seems to exhibit sexual dimorphism, with the smaller males being firmly attached to the larger females. Whether this is a protandric type of sequential hermaphrodism or a permanent gonochoristic dimorphism is unclear.

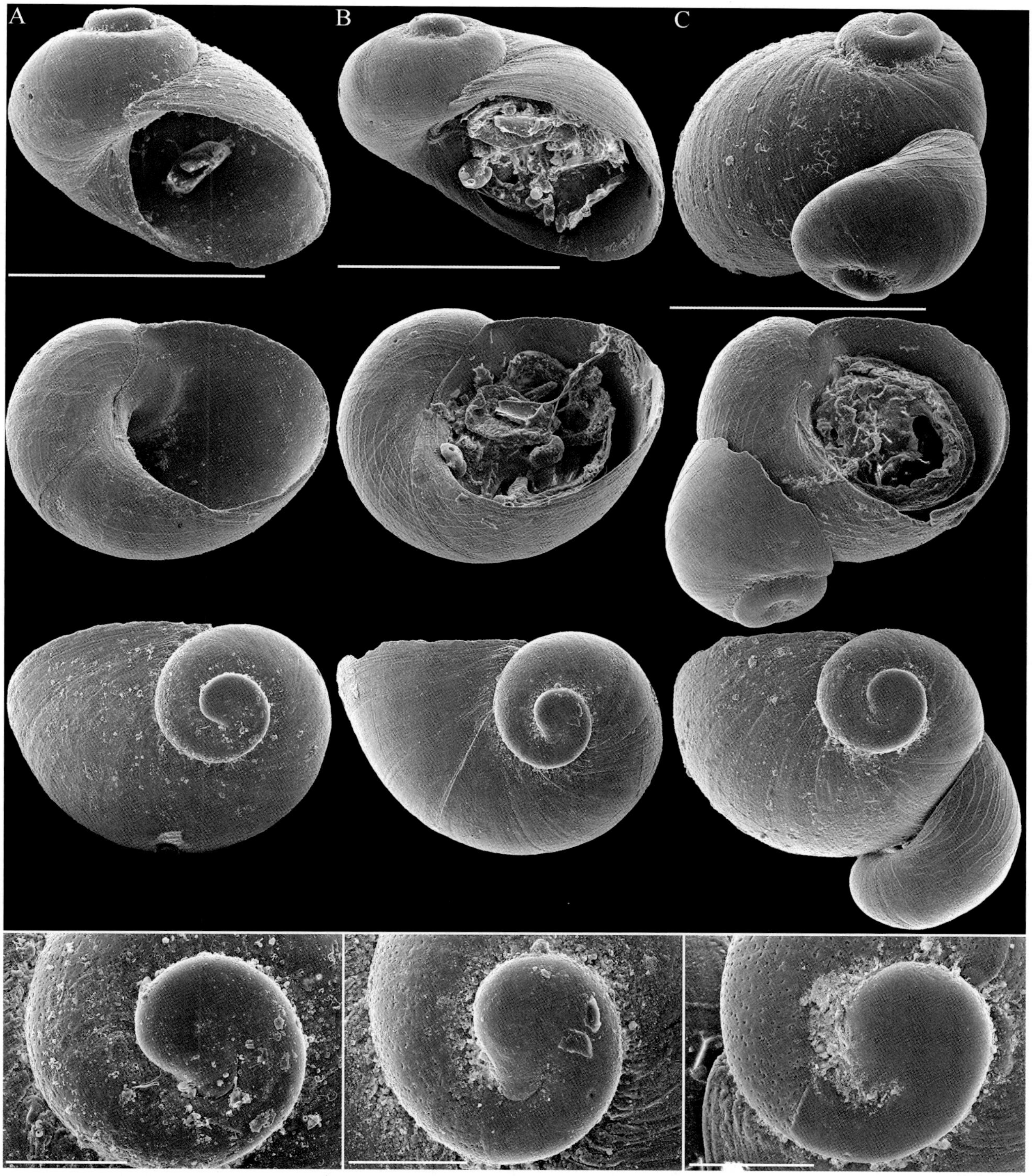

Figure 986. *Larocheopsis amplexa*. Paratypes. **A.** 246–291 m, Middlesex Bank, NW of Three Kings Islands, New Zealand, 34.033°S, 171.733°E (NMNZ M.093882, 1). **B–C.** 310 m, 11 km NW of Great Island, Three Kings Islands, New Zealand, 34.108°S, 172.078°E (NMNZ M.074955, 10). Scale bars shell = 0.5 mm. Scale bars protoconch = 100 μm.

Larocheopsis macrostoma Geiger & Marshall 2012

Figure 987–988

Chresonymy.

Larocheopsis macrostoma Geiger & Marshall, 2012: 32, fig. 26.

Type material. Holotype (NMNZ M.137435. Fig. 987).

Type locality. 538 m, seamount 130 km S of Esperance Rock, Kermadec Ridge, New Zealand, 32.472°S, 179.265°W (OD).

Etymology. Macro-, Latin prefix for large, -stoma, Latin suffix for mouth or aperture, referring to the comparatively large aperture of the species (OD).

Description. Shell medium-sized (to 1.2 mm. Fig. 987), trochiform, depressed. Protoconch of 0.75 whorl, smooth, no apertural varix, apertural margin sinusoidal. Teleoconch of 1.66 whorls, suture little impressed. Sculpture of first 1.5 whorls consisting of pits, shoulder of last portion with distinct comarginal axial cords, crossed by

Figure 987. *Larocheopsis macrostoma*. Holotype. 538 m, seamount 130 km S of Esperance Rock, Kermadec Ridge, New Zealand, 32.472°S, 179.265°W (NMNZ M.137435). Scale bar shell = 1 mm. Scale bar protoconch = 100 µm.

approximately eight irregularly-spaced, indistinct spiral threads. Base from periphery to closed umbilicus with up to 16 low, spiral steps, starting after 0.66 teleoconch whorl. Umbilicus closed covered by callus. Aperture elongated D-shaped oval, roof overhanging.

Animal unknown.

Comparisons. *Larocheopsis amplexa* from New Zealand has an much taller shell and lacks distinct axial sculpture on the shoulder near the apertural margin.

Figure 988. Distribution of *Larocheopsis macrostoma*. ☆ type locality.

Distribution. Known only from type locality (Fig. 988).

Trogloconcha Kase & Kano, 2002

Chresonymy.
Trogloconcha Kase & Kano, 2002: 26.
Trogloconcha: Geiger, 2003: 47–48.

Type species. *Trogloconcha ohashii* Kase & Kano, 2002 (OD).

Other species. *Trogloconcha christinae* Geiger, 2003.
Trogloconcha lamellinodosa new species.
Trogloconcha lozoueti Geiger, 2008.
Trogloconcha marshalli (Lozouet, 1998) †.
Trogloconcha tesselata Kase & Kano, 2002.

Etymology. Trogle, Greek for cave; -concha, Latin for shell. Referring to the cave habitat of the genus (OD). Gender feminine.

Description. Shell naticiform, globular, thin, fragile, with or without umbilicus. No selenizone, slit, or foramen. Teleoconch sculpture with axial, often with spiral sculpture. Aperture round to obliquely oval. Protoconch smooth or covered in flocculent sculpture, no apertural varix, apertural margin curved. Radula rhipidoglossate, central tooth broadest, lateral teeth approximately equal in size and shape. Operculum rudimentary, diameter 0.25 of aperture. Animal with papillate cephalic tentacles, non-papillate epipodial tentacles, without brood pouch. Gonochoristic, no sexual dimorphism.

Distribution. Indian Ocean to the Central Pacific (Fig. 989). For specific localities, see individual species.

Geologic range. Oligocene to Recent.

Comparisons. *Larochea* has a brood pouch not found in *Trogloconcha*, and lacks an operculum, which is only reduced in *Trogloconcha*.

Larocheopsis differs in the teleoconch sculpture consisting of fine pits, as opposed to distinct axials and spirals, and lacks an operculum, which is only reduced in *Trogloconcha*.

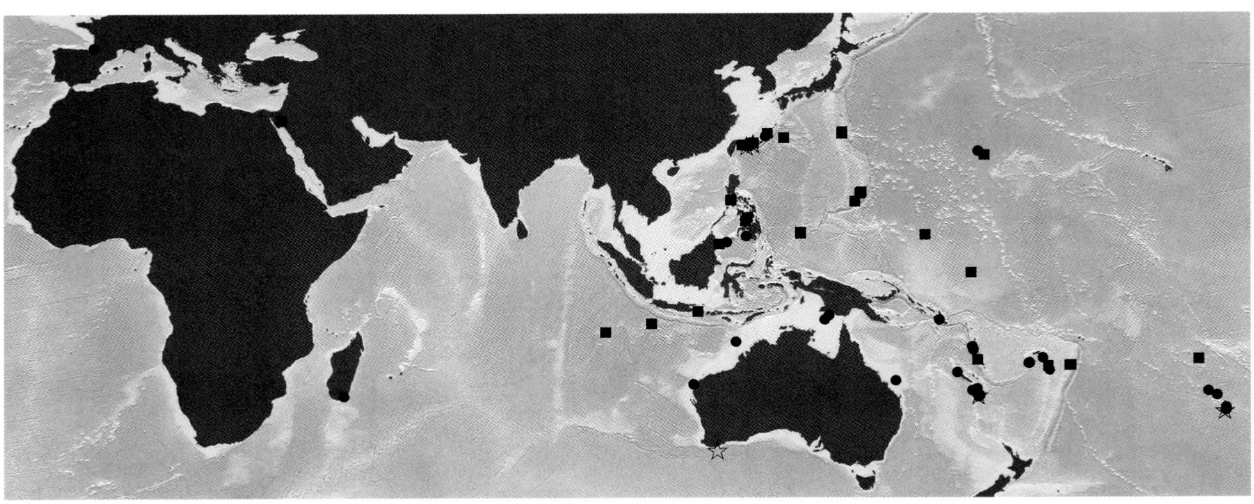

Figure 989. Global distribution of *Trogloconcha* species. ☆ type locality. ● specimen record. ■ literature record.

Differentiation of species. Species are differentiated by overall shell shape, presence or absence of an umbilicus, and details of shell sculpture.

Intraspecific variability extends predominantly to the ontogenetic change in shell height, larger specimens having a proportionally taller shell.

Remarks. *Triadostoma rachelae* Nützel, 2006 from the Late Triassic Cassian Formation of northern Italy, was tentatively classified in Vetigastropoda: Trochoidea: Stomatellidae (Nützel, 2006). At a size of 1.6 × 1.2 × 1.2 mm it would be one of the smallest members of the family, and shows uncharacteristic reticulate and honeycomb teleoconch sculpture. The potential nacre tablets are arrayed at a distinct angle to the surface of that layer, contrary to the parallel orientation seen in Vetigastropoda. It is likely that *T. rachelae* in fact can be classified in Larocheidae, and the *Triadostoma* may be synonymous with *Trogloconcha*. None of the other observable characters, (protoconch, umbilicus, overall shell shape) contradict such a placement.

Trogloconcha christinae Geiger, 2003

Figures 990–991

Chresonymy.

Trogloconcha christinae Geiger, 2003: 48–49, 80, fig. 13.

Trogloconcha christinae: Geiger, 2003: text-fig. p. 7.

Trogloconcha christinae: Geiger, 2006c: text-fig. 6.

Type material. Holotype (WAM S10886. Fig. 990A), 0.67 mm. Paratype from type locality (AMS C.402713, 1. Fig. 990B).

Type locality. 112 m, off Albany, Western Australia, Australia, 35.333°S, 118.333°E. In sand and shells, March 12 1980 (OD).

Etymology. Named for Christine Thacker of [formerly] Pasadena, California, USA, for continuing collaboration and advice in the field as well as in the laboratory (OD).

Description. Shell small (holotype 0.67 mm; paratype 0.59 mm. Fig. 990), as wide as tall (width/height ratio = 1.03, 0.96). Color off white. Protoconch 133, 141 μm, 1.125 whorls, covered entirely with finely flocculent sculpture, no apertural varix, apertural margin curved. Teleoconch of 1.25 whorls, shoulder rounded. Sculpture of only axials, approximately 75 on last whorl. Anomphalous. Aperture obliquely oval. No selenizone, slit, or foramen.

Animal unknown.

Comparisons. *Trogloconcha lamellinodosa* from the Indo-Malayan Archipelago to the Western Pacific has spiral sculpture and a narrow umbilicus.

Trogloconcha lozoueti from the Indo-Malayan Archipelago to the Central Pacific has a wider overall shell, a narrow umbilicus, and spiral sculpture.

Trogloconcha ohashii from the Indo-Malayan Archipelago tot he Central Pacific lacks spiral sculpture and has distinct umbilicus.

Trogloconcha tesselata from the Indian Ocean to the Western Pacific has spiral sculpture.

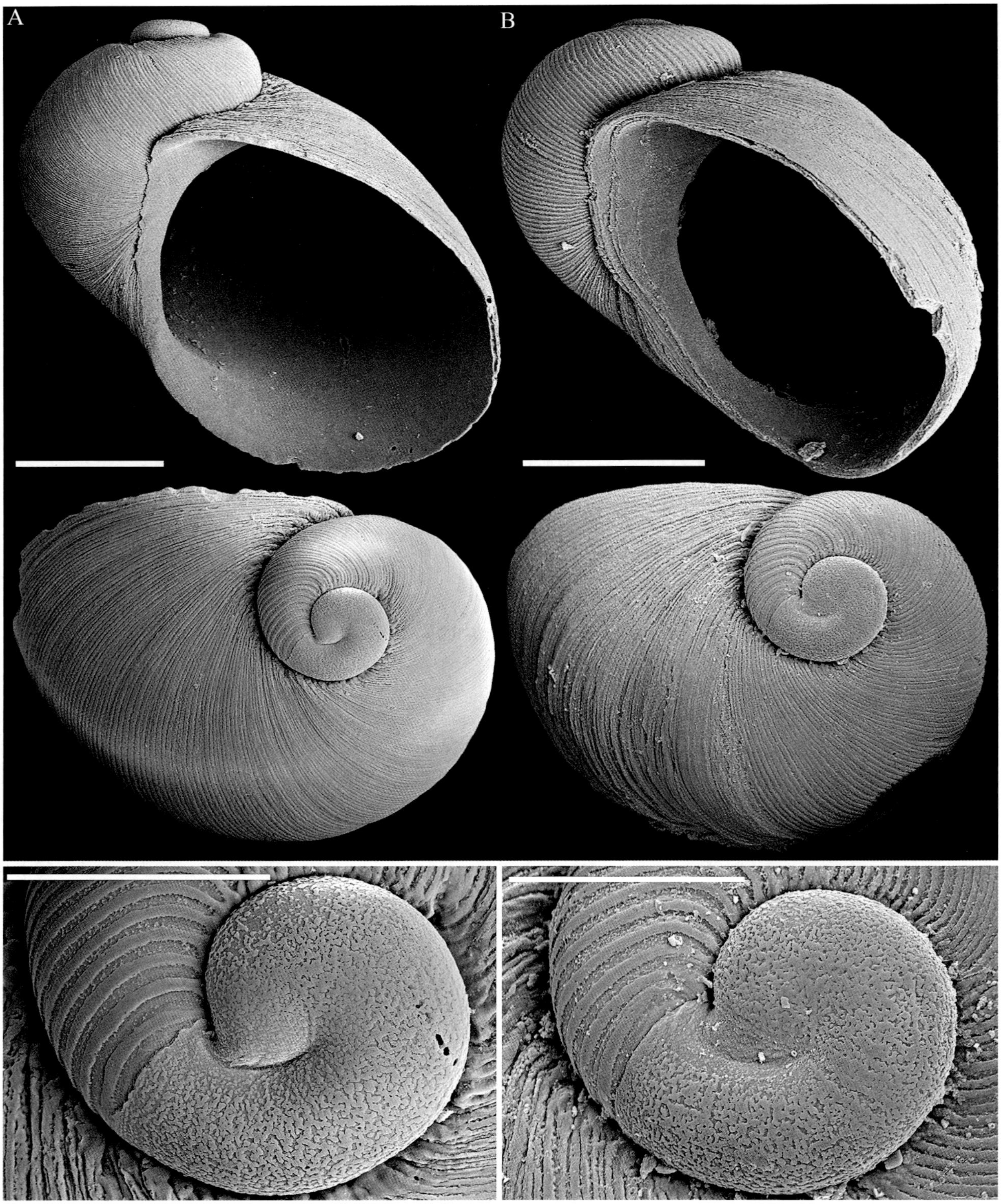

Figure 990. *Trogloconcha christinae*. 112 m, off Albany, Western Australia, Australia (35.333°S, 118.333°E). **A.** Holotype (WAM S10886). **B.** Paratype (AMS C.402713, 1). Scale bars shell = 0.2 mm. Scale bars protoconch = 100 µm.

Trogloconcha marshalli from the Oligocene of Europe is taller and has a very narrow umbilicus.

Distribution. Known only from type locality (Fig. 991)

Remarks. The type numbers were changed in proof of the original description. The original intention was to only have the two figured specimens as type material. Inadvertently, the AMS C.402713 lot was cited as having two specimens. There was a second, non-conspecific specimen in that lot (*Sci. cyprina* AMS C.461553), but it was never the intention of the original author to have this second specimen designated as a paratype.

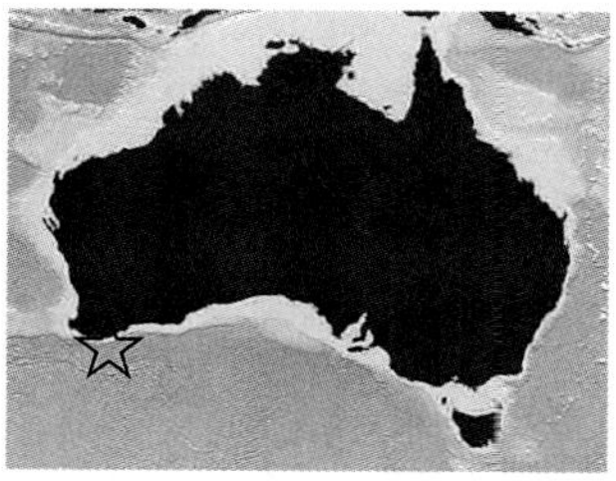

Figure 991. Distribution of *Trogloconcha christinae*. ☆ type locality.

Trogloconcha lamellinodosa new species

Figures 992–995

Type material. Holotype (MNHN 25106. Fig. 992). Paratypes from type locality (MNHN 25129, 4). 200–220 m, Ua Huka Island, Marquesas Islands, N.O. *Alis*, MUSORSTOM 9 DW1288, 8.900°S, 139.633°E (MNHN 25107, 9. Fig. 993B). 507–967 m, N.O. *Alis*, NORFOLK 1 DW1692, Norfolk Ridge, Sponge Bank, New Caledonia, 24.933°S, 168.350°E (MNHN 25108, 1. Fig. 993A). 305–330 m, Norfolk Ridge, Aztec Bank, N.O. *Alis*, SMIB 8 DW182–184, 23.308°S, 168.083°E (MNHN 25109, 11. Fig. 993C).

Type locality. 508–532 m, N.O. *Alis* SMIB 8 DW146–147, Sponge Bank, B mound, Norfolk Ridge, New Caledonia, 24.917°S, 168.367°E.

Etymology. Lamelli-, Latin for lamella referring to the lamellae on the early portion of the teleoconch; nodosa, Latin for nodulous, referring to the base of the teleoconch of larger specimens.

Description. Shell medium-sized (1.61 × 1.57 mm: holotype. Figs 992–993), trochiform. Protoconch of 0.9 whorl, smooth or with very fine flocculent sculpture, no apertural varix, apertural margin opisthocline. Teleoconch of up to 2.25 whorls, suture below periphery, moderately incised. Approximately 25–30 strong axial cords on first whorl, continuing at same spacing, increasing in number with growth. First spiral thread after 0.75 whorl, after one whorl approximately seven on shoulder, five on base; increasing in number to approximately 20 on shoulder, approximately 15 on base on fully grown specimen; spacing increasing towards umbilicus. On small specimens, spiral threads running over axial cords, on larger specimens, spirals forming small points on shoulder at intersection with axial cords, on base forming distinct nodules. Umbilicus very narrow, partially obscured by callus, walls at discrete angle to base, walls straight.

Operculum not found.

Animal. Eyes present, cephalic tentacles papillate. Based on single rehydrated specimen.

Radula (Fig. 994). Rachidian tooth trapezoidal, central cusp largest, 5–6 smaller cusps arranged in shallow arc on each side. Lateral tooth 1 broadened, with approximately six cusps on outer edge; lateral teeth 2–4 similar, tip tear-drop-shaped, apical cusp largest, 3–4 cusps on each side; lateral tooth 5 with strong triangular tip, apical cusp largest, 4–5 strong cusps along inner margin, 1–2 cusps along outer edge. Inner marginal teeth with distinct tip, apical cusp largest, 5–6 cusps along each side. Outer marginal teeth with minute tips with fine bristles. Radular interlock of central field moderately weak.

Comparisons. *Trogloconcha christinae* from southwestern Australia lacks spiral sculpture on the teleoconch.

Figure 992. *Trogloconcha lamellinodosa* new species. Holotype. 508–532 m, N.O. *Alis* SMIB 8 DW146–147, Sponge Bank, B mound, Norfolk Ridge, New Caledonia, 24.917°S, 168.367°E (MNHN 25106). Scale bar shell = 1 mm. Scale bar protoconch = 100 μm.

Figure 993. *Trogloconcha lamellinodosa* new species. Ontogenetic series from paratypes. **A.** 507–967 m, N.O. *Alis*, NORFOLK 1 DW1692, Norfolk Ridge, Sponge Bank, New Caledonia, 24.933°S, 168.350°E (MNHN 25108, 1). **B.** 200–220 m, Ua Huka Island, Marquesas Islands, N.O. *Alis*, MUSORSTOM 9 DW1288, 8.900°S, 139.633°E (MNHN 25107, 9). **C.** 305–330 m, Norfolk Ridge, Aztec Bank, N.O. *Alis*, SMIB 8 DW182–184, 23.308°S, 168.083°E (MNHN 25109, 11). Scale bars shell A–B= 1 mm. Scale bar shell C = 0.5 mm. Scale bars protoconch = 100 μm.

Trogloconcha lozoueti from the Indo-Malayan Archipelago to the Central Pacific has a protoconch with fine spiral patterns, and the umbilicus is fully closed by a callus.

Trogloconcha ohashii from the Indo-Malayan Archipelago to the Central Pacific has a distinct umbilicus.

Trogloconcha tesselata from the Indian Ocean to the Western Pacific lacks the strong axial cords on the early shell, as well as the distinct nodules on the base of larger specimens.

Distribution. Eastern Indo-Malayan Archipelago to Western Pacific (Fig. 995).

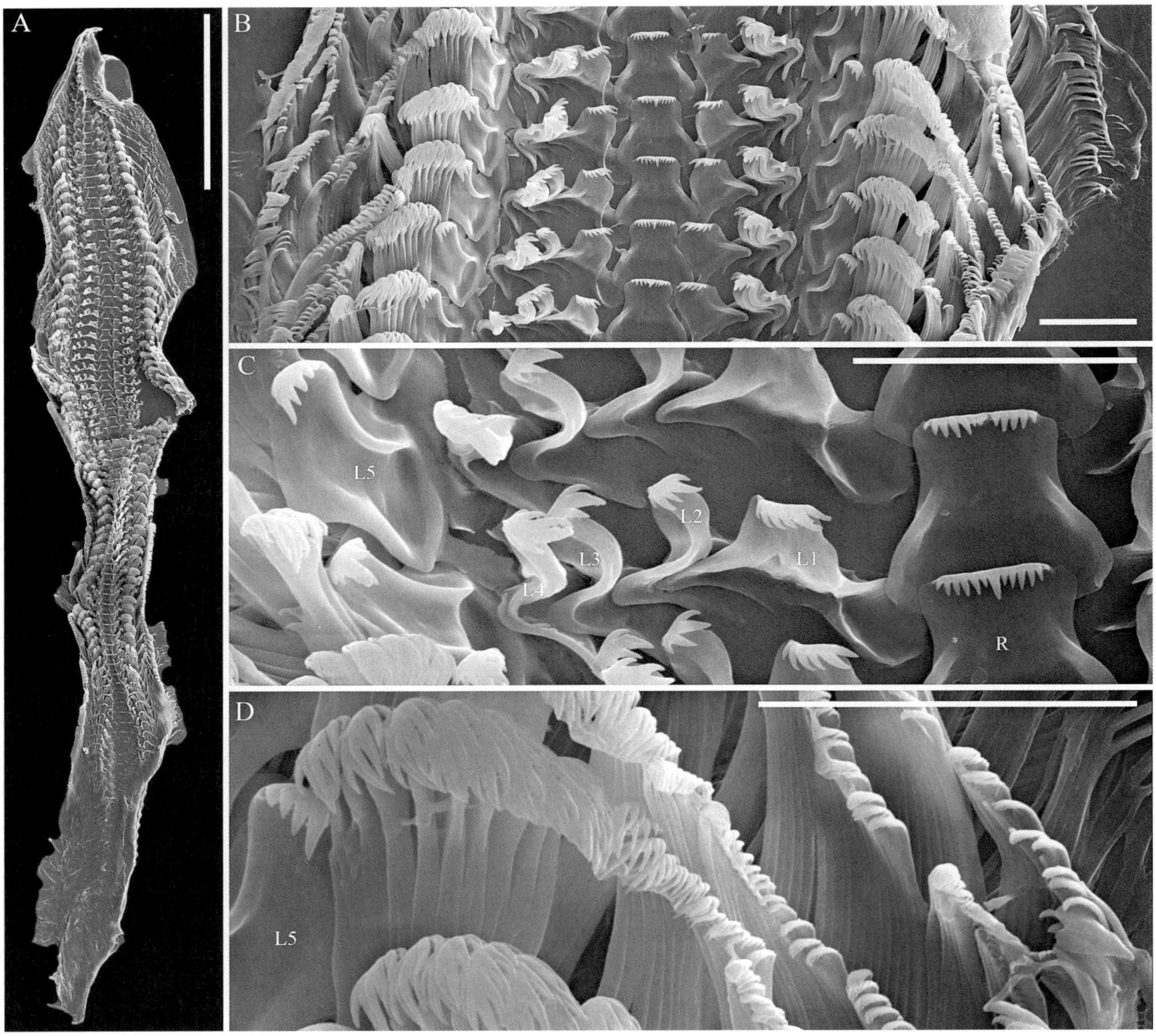

Figure 994. *Trogloconcha lamellinodosa* new species. Radula from paratype. 508–532 m, N.O. *Alis* SMIB 8 DW146–147, Sponge Bank, B mound, Norfolk Ridge, New Caledonia (MNHN 25129, 4) **A.** Entire radula. **B.** Whole width of radula. **C.** Central field enlarged, **D.** Lateral teeth. Scale bar A = 200 μm. Scale bars C–D = 20 μm.

Specimen records.

Fiji. 820–863 m, 17.300°S, 179.550°W (MNHN, 1). 820–863 m, BORDAU 1 DW1459, 17.300°S, 179.550°W (MNHN, 3; MNHN, 2). 240–319 m, BORDAU 1 DW1494, 18.916°S, 178.483°W (MNHN, 1).

Solomon Islands. 53 m, 9.795°S, 160.842°E (MNHN, 1). 396–411 m, SALOMON 1 DW1762, 8.665°S, 160.065°E (MNHN, 1). 766–1100 m, N Russel Island, SALOMON 2 DW2167, 8.913°S, 159.121°E (MNHN, 1).

Marquesas Islands. 340–352 m, Hiva Oa Island, 9.733°S, 138.850°E (MNHN, 3).

New Caledonia. 381–469 m, Norfolk Ridge, 23.367°S, 168.017°E (MNHN, 1). 600–620 m, Norfolk Ridge, Sponge Bank, 24.900°S, 168.433°E (MNHN, 1). 420–438 m, Norfolk Ridge, 23.667°S, 168.483°E (MNHN, 1). 386–390 m, Norfolk Ridge, 23.350°S, 168.033°E (MNHN, 2). 569–616 m, Norfolk Ridge, Banc Introuvable, 24.650°S, 168.633°E (MNHN, 1). 200–291 m, Norfolk Ridge, 23.283°N, 168.233°E (MNHN, 6). 280 m, Loyalty Ridge, MUSORSTOM 6 DW423, 20.433°S, 166.683°E (MNHN, 1). 538 m, between Lifou and Uvea, CALSUB 1989 PL 15, 20.616°S, 166.966°E (MNHN, 1). 400 m, south, BATHUS 2 DW729, 22.866°S, 167.200°E (MNHN, 1). 397–400 m, south, BATHUS 2 DW730, 23.050°S, 166.966°E (MNHN, 1). 405–456 m, Norfolk Ridge, NORFOLK 2 DW2132, 23.283°S, 168.233°E (MNHN, 7). 187–197 m, Crypthelia Bank, Norfolk Ridge, NORFOLK 2 DW2123, 23.300°S, 168.250°E (MNHN, 1). 601–608 m, Norfolk Ridge, BATHUS 3 DW824, 23.316°S, 168.000°E (MNHN, 1). 290–378 m, Bank N, Norfolk Ridge, NORFOLK 1 DW1652, 23.435°S, 167.838°E (MNHN, 1). 250–350 m, LAGON 40, 23.500°S, 166.233°E (MNHN, 3). 433–450 m, Stylaster Bank, SMIB 8 DW166, 23.633°S, 167.716°E (MNHN, 1). 555–565 m, Unfindable Bank, Norfolk Ridge, NORFOLK 2 DW2057, 24.666°S, 168.650°E (MNHN, 4). 635–680 m, south, BERYX 11 DW09, 24.866°S, 168.366°E (MNHN, 4). 565–600 m, south, BERYX 11, DW10, 24.883°S, 168.350°E (MNHN, 2). 518–586 m, Sponge Bank, Norfolk Ridge, NORFOLK 2 DW2087, 24.933°S, 168.366°E (MNHN, 1). 444–445 m, south, BATHUS 2 DW719, 22.800°S, 167.267°E (MNHN, 8).

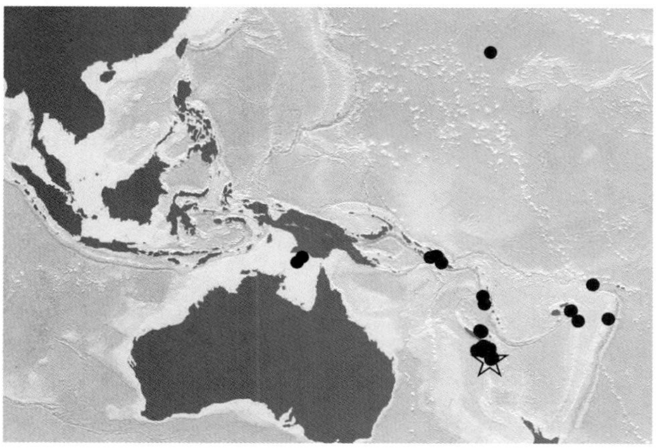

Figure 995. Distribution of *Trogloconcha lamellinodosa* new species. ☆ type locality. ● specimen record.

Vanuatu. 154–179 m, MUSORSTOM 8 DW1105, 15.050°S, 167.116°E (MNHN, 1; MNHN, 1). 360–419 m, MUSORSTOM 8 DW1065, 16.266°S, 167.350°E (MNHN, 1).

Tonga. 440–487 m, Vava'u, BORDAU 2 DW1586, 18.566°S, 173.916°W (MNHN, 1).

Wallis and Futuna. 455–515 m, Wallis Island, MUSORSTOM 7 DW523, 13.200°S, 176.266°W (MNHN, 1).

Remarks. The strong ontogenetic change in sculpture is notable. Comparison of the shoulder of small specimens with the corresponding early portion of whorls of larger specimens do not show any perceptible differences, and specimens of intermediate size demonstrate that the sculptural differences smoothly intergrade between specimens, regardless of locality of the specimens.

Trogloconcha lozoueti Geiger, 2008

Figures 996–999, Color Plate 4H

Chresonymy.

Trogloconcha lozoueti Geiger, 2008c: 196–199, figs 25–30.

Trogloconcha lozoueti: Tröndlé & Boutet, 2009: 12.

Type material. Holotype (MNHN 20825. Fig. 996). Paratypes: 15–20 m, Vavai, Rapa, Austral Islands, 27.590°S, 144.381°W (MNHN 20826, 6. Fig. 997A–B). 3–24 m, Hiri Bay, Rapa, Austral Islands, 27.622°S, 144.370°W (MNHN 20827, 20. Fig. 997C).

Type locality. 15–20 m, St. 32, Vavai, Rapa, Austral Islands, French Polynesia, 27.590°S, 144.381°W (OD).

Etymology. The name honors Pierre Lozouet of MNHN for his many accomplishments in fossil and Recent malacology, and specifically for his initiative with the Austral Islands expedition (OD).

Figure 996. *Trogloconcha lozoueti*. Holotype. 15–20 m, Vavai, Rapa, Austral Islands, French Polynesia, 27.590°S, 144.381°W (MNHN 20825). Scale bar shell = 1 mm. Scale bar protoconch = 100 µm.

Description. Shell medium-sized (to 1.17 mm. Figs 996–997), trochiform globular, with rapidly increasing whorls. Protoconch of 0.75 whorl, sculpture flocculent somewhat spirally arranged, weak apertural varix, apertural margin convex. Teleoconch of two whorls, approximately 15 axial cords on first 0.5 whorl with no spiral sculpture, interstices with fine flocculent sculpture; onset of spirals after 0.5 whorl; approximately 31 spirals, from suture to mid-base as fine spiral cords, transitioning to low spiral steps in 5–6 adumbilical spirals; approximately 70 fine axial cords on last whorl, forming distinct

Figure 997. *Trogloconcha lozoueti*. **A–B.** 15–20 m, Vavai, Rapa, Austral Islands, French Polynesia, 27.590°S, 144.381°W (MNHN 20826). **C.** 3–24 m, Hiri Bay, Rapa, Austral Islands, French Polynesia, 27.622°S, 144.370°W (MNHN 20827, 20). Scale bar shell A = 1 mm. Scale bars shell B–C = 0.5 mm. Scale bars protoconch = 100 μm.

points at intersections; overall appearance of shell finely reticulate and spiky. Suture bordered by strong irregularly lamellate thickening on shoulder. Anomphalous with weak callus in umbilical region. Aperture oval, roof overhanging.

Radula (Fig. 998). Rachidian tooth triangular, tip with central cusp largest, three cusps on each side, arranged in convex curve. Lateral teeth 1–4 similar, L-shaped, tip with apical cusp largest, 3–4 cusps along outer margin, 1–3 cusps along inner margin; lateral tooth 5 enlarged by half, apical cusp largest, 3–4 cusps along each side. Inner marginal teeth with elongated shaft central cusp largest, 3–4 cusps along inner margin, 5–6 cusps along outer margin; outer marginal teeth with spoon-shaped tip, with many fine cusps on each side. Radular interlock of central field strong.

Operculum, corneous, round, multispiral, with central nucleus, covering only approximately ⅓ of aperture.

Comparisons. *Trogloconcha lamellinodosa* from the Indo-Malayan Archipelago to the Western Pacific has a protoconch with smooth to flocculent sculpture, but lacking spiral elements and the umbilicus is not fully covered by the callus.

Trogloconcha ohashii from the Indian Ocean to the Western Pacific has fewer spiral and axial elements that also form elevated points, but has a wide umbilicus.

Trogloconcha tesselata from the Indo-Malayan Archipelago has a smooth protoconch, and lacks the fine points at the intersection of axials and spirals.

Trogloconcha christinae from Western Australia lacks spiral sculpture and has flattened-flocculent protoconch sculpture.

Distribution. Indo-Malayan Archipelago to Central Pacific (Fig. 999).

Specimen records.

Philippines. 95–128 m, Cervera Shoal, Pamilacan Island, PANGLAO 2004 T36, 9.488°N, 123.858°E (MNHN, 1). 120–139 m, Balicasag Island, PANGLAO 2004 L74–75,

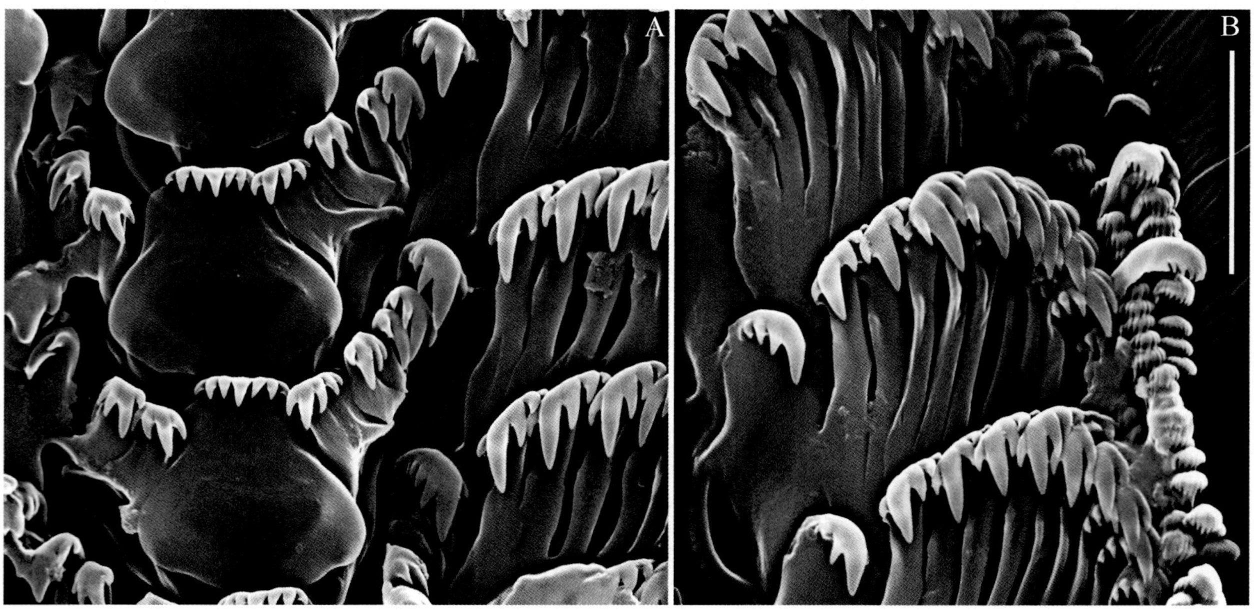

Figure 998. *Trogloconcha lozoueti.* Radula from specimen shown in Figure 997C. 3–24 m, Hiri Bay, Rapa, Austral Islands, French Polynesia, 27.622°S, 144.370°W (MNHN 20827, 20). **A.** Central field. **B.** Lateral tooth 5 and marginal teeth. Scale bars = 10 µm.

9.511°N, 123.682°E (MNHN, 1). 80–140 m, Balicasag, PANGLAO 2004 T38, 9.538°N, 123.705°E (MNHN, 1).

Vanuatu. 42–45 m, Palikulo Bay, SANTO 2006 DB77, 15.465°S, 167.245°E (MNHN, 1). 45–50 m, Aoré Island, Aimbuei Bay, SANTO 2006 FP46–49, 15.540°S, 167.211°E (MNHN, 1). 100–105 m, NW Tutuba Island, SANTO 2006 DS99, 15.541°S, 167.281°E (MNHN, 12). 155 m, Tutuba Island, SANTO 2006 EP27, 15.550°S, 167.266°E (MNHN, 1). 92 m, NW Tutuba Island, SANTO 2006 DS105, 15.550°S, 167.278°E (MNHN, 4). 125–156 m, W Tutuba Island, SANTO 2006 EP40, 15.555°S,

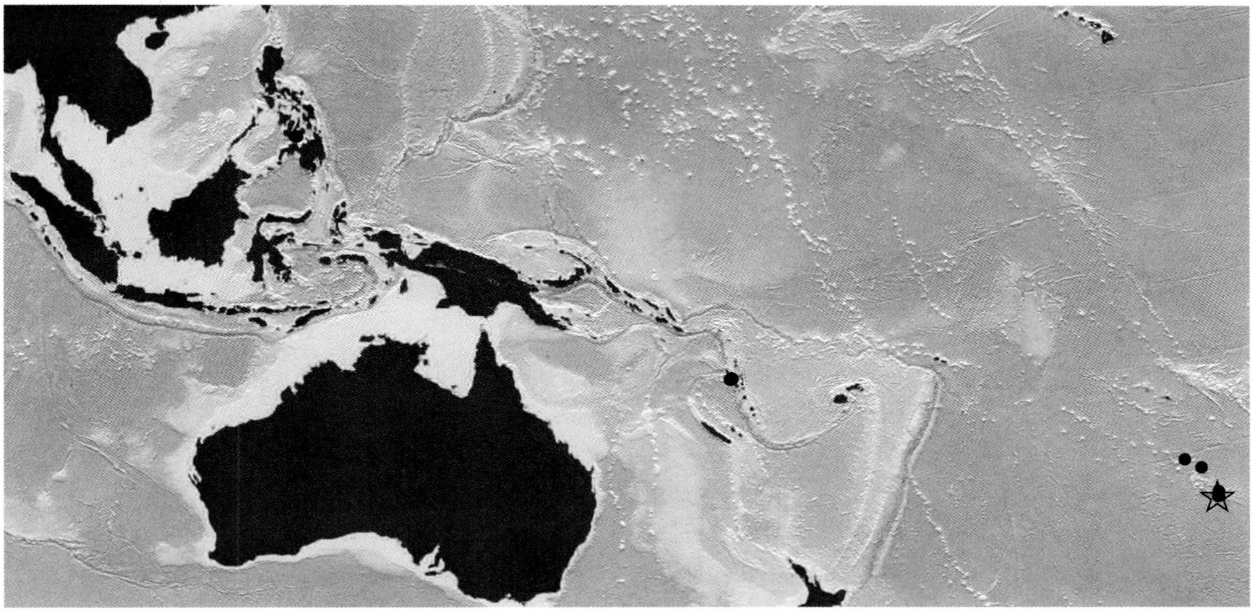

Figure 999. Distribution of *Trogloconcha lozoueti*. ☆ type locality. ● specimen record.

167.274°E (MNHN, 1). 100 m, W Tutuba Island, Aoré Island, SANTO 2006 DS93, 15.560°S, 167.275°E (MNHN, 1). 114 m, W Tutuba Island, SANTO 2006 DS96, 15.560°S, 167.275°E (MNHN, 2). 98–100 m, W Tutuba Island, SANTO 2006 DS102, 15.568°S, 167.266°E (MNHN, 3). 80 m, W Tutuba Island, SANTO 2006 DS104, 15.568°S, 167.266°E (MNHN, 8). 108–121 m, W Urélapa Island, SANTO 2006 EP24, 15.608°S, 167.021°E (MNHN, 1). 99 m, NW coast of Malo Island, SANTO 2006 EP21, 15.628°S, 167.087°E (MNHN, 3).

French Polynesia, Austral Islands. 36 m, Around Rukuaga Point, Rapa, 27.568°S, 144.368°W (MNHN, 100). 33 m, Haurei Bay, Rapa, 27.612°S, 144.318°W (MNHN, 6). 52–57 m, SE of Tauna Island, Rapa, 27.608°S, 144.295°W (MNHN, 20). 30 m, NW of Tauna Island, Rapa, 27.605°S, 144.303°W (MNHN, 3). 8 m, S of Tarakoi Island, Rapa, 27.093°S, 144.308°W (MNHN, 1). 52–57 m, SE of Tauna Island, Rapa, 27.608°S, 144.295°W (MNHN, 9). 3–24 m, Hiri Bay, Rapa, 27.622°S, 144.370°W (MNHN, 20: paratypes). 15–20 m, Vavai, Rapa, 27.590°S, 144.381°W (MNHN, 6: paratypes).

Trogloconcha ohashii Kase & Kano, 2002

Figures 1000–1002

Chresonymy.

Trogloconcha ohashii Kase & Kano, 2002: 26–31, figs 1A–C [shell], 2A–B [protoconch], 3 [external anatomy, radula] (Fig. 1000).

Trogloconcha ohashii: Geiger, 2003: 80.

Figure 1000. Original illustrations of *Trogloconcha ohashii.*

Misidentification.

Trogloconcha tesselata: Geiger, 2008b: fig. 1 [is *Tro. ohashii*].

Type material. Holotype (NSMT Mo72828). 0.81 × 1.13 mm. Paratypes [from type locality?] (NSMT Mo72829; MNHN: not found in 4/2005, DLG pers. obs.).

Type locality. "Black Hole" diving site, northwest of Shimoji Island, Miyago Group, Okinawa, [Japan] 24°49.1'N, 125°8.3'E, 35 m, totally dark inside, calcareous muddy sand (OD).

Etymology. Named after professional diver Mr. Shu-ichi Ohashi from Naha, Okinawa, Japan (OD).

Description. Shell medium-sized (to 1.34 mm. Figs 1000, 1001), naticiform inflated, thin. Periostracum thin, colorless. Protoconch of 0.75 whorl, smooth with traces of flocculent sculpture, no apertural varix, apertural margin convex. Teleoconch of up to two whorls, approximately 43 fine axial cords, approximately 30 fine spiral cordlets, at intersection forming minute points. Umbilicus open, wide, with smooth sides. Aperture oblique oval oblong, roof overhanging. Operculum small (¼ diameter of aperture), multispiral, with central nucleus, thin.

External anatomy (adapted from Kase & Kano, 2002). Cephalic tentacles with papillae, eyes on short stalks, only right suboptic tentacle present. Left ctenidium only, monopectinate, 11 leaflets. Three pairs of posterior epipodial tentacles surrounding operculum; presence/absence of papillae not known. Sexes separate.

Radula (adapted from Kase & Kano, 2002). Rachidian tooth broad triangular, margins of shaft convex, cutting surface obtuse V-shaped, central cusp approximately twice as large as three to four on either side. Lateral teeth 1–5 similar, shafts approximately as long as rachidian tooth, half as long as marginal teeth. Lateral tooth 1 with second cusp from midline of radula largest of six, broad triangular base. Lateral tooth 4 with central cusp largest of seven, shaft slender. Lateral teeth 2–3 intermediate. Lateral tooth 5 with outermost cusp largest of three, base strongly elongated. Marginal teeth with central cusp at least twice as long as additional two on either side; shafts long.

Comparisons. *Trogloconcha tesselata* from the Indian Ocean to the Western Pacific lacks the umbilicus, has denser sculpture with no distinct points being formed on the intersection between axials and spirals.

Trogloconcha christinae from the southern coast of Western Australia, lacks an umbilicus and lacks spiral sculpture.

Trogloconcha lamellinodosa from the Indo-Malayan Archipelago to the Western Pacific has a much narrower umbilicus covered by a callus.

Figure 1001. *Trogloconcha ohashii*. **A.** 20 m, E side, Sandy Islet, Scott Reef, Western Australia, Australia, 14.044°S, 121.780°E (SBMNH 369510, 3). **B.** 260–305 m, S of Viti Levu, Fiji, 18.308°S, 178.097°E (MNHN, 1). **C.** 80 m, W Tutuba Island, SANTO 2006 DS104, Vanuatu, 15.568°S, 167.267°E (MNHN, 6). Scale bars shell = 0.5 mm. Scale bars protoconch = 100 µm.

Trogloconcha marshalli from the Oligocene of France has a very narrow umbilicus and lacks the points at the intersections of axials and spirals.

Trogloconcha lozoueti from the Indo-Malayan Archipelago to the Central Pacific lacks an umbilicus.

Distribution. Eastern Indian Ocean to Central Pacific in shallow water (Fig. 1002).

Specimen records.

Philippines. 35 m, Pamilacan Island, PANGLAO 2004 B26, 9.490°N, 123.933°E (MNHN, 1). 25 m, Pamilacan Island, PANGLAO 2004 B30, 9.490°N, 123.935°E (MNHN, 2). 17 m, Balicasag Island, PANGLAO 2004 S45, 9.515°N, 123.680°E (MNHN, 2).

Australia. 20 m, E side, Sandy Islet, Scott Reef, Western Australia, 14.044°S, 121.780°E (SBMNH 369510, 3).

Vanuatu. 80 m, W Tutuba Island, SANTO 2006 DS104, 15.568°S, 167.266°E (MNHN, 6). 360–419 m, MUSORSTOM 8 DW1065, 16.266°S, 167.350°E (MNHN, 1).

Fiji. 260–305 m, S of Viti Levu, 18.308°S, 178.097°E (MNHN, 1).

Literature records.

Christmas Island. 9 m, Thunder Dome diving site, 10.500°S, 105.667°E (3) (Kase & Kano, 2002: repository unknown).

Cocos Keeling. 49 m, W of West Island, 12.180°S, 96.813°E (1) (Kase & Kano, 2002: repository unknown).

Fiji. 7 m, N of Ono Island, Great Astrolabe Reef, 18.863°S, 178.450°W (1). 5 m, W of Haano Island, Ha'apai Group, 18.718°S, 174.290°W (3). 17–18 m, Swallows Cave, SW of Falevai Island, Vava'u Group, 18.682°S, 174.048°W (15) (Kase & Kano, 2002: repository unknown).

French Polynesia. 22–30 m, Cave Arue, W of Tahiti, 17.515°S, 149.535°W (72) (Kase & Kano, 2002: repository unknown).

Guam. 10 m, Apra Point, 13.450°N, 144.617°E (1) (Kase & Kano, 2002: repository unknown).

Indonesia. 20–30 m, Menjangan Island National Park, S side, Bali, 8.117°S, 114.517°E (20) (Kase & Kano, 2002: repository unknown).

Japan. 15–20 m, N of Kohama Island, Yaeyama Islands, Okinawa, 24.358°N, 123.982°E (23). Sabachi Cave, Yonaguni Islands, Okinawa, 24.467°N, 123.000°E (2). 12–40 m, Toriike, Irabu Islet, Miyako Islands, Okinawa, 24.818°N, 125.138°E (3). 25 m, Devil's Palace, Shimoji Islet, Miyako Islands, Okinawa, 24.822°N, 125.138°E (6). 35 m, Whitch's House, Shimoji Islet, Miyako Islands, Okinawa, 24.822°N, 125.138°E (107). 15 m, Cross Hole, Irabu Islet, Miyako Islands, Okinawa, 24.860°N, 125.158°E (43). 4 m, Lunch Hole, Irabu Islet, Miyako Islands, Okinawa, 24.860°N, 125.167°E (3). 25 m, L-Arch, Irabu Islet, Miyako Islands, Okinawa, 24.862°N, 125.162°E (1). 23 m, Gon-gon-ana Cave, Minami-Daito Island, Borodino Islands Okinawa, 25.845°N, 131.031°E (2). Giant Cave, Tatejima, Ogasawara Islands, 26.689°N, 142.132°E (2). 20 m, Shodokutsu, E of Ie Islet, Okinawa, 26.715°N, 127.835°E (112). Daidokutsu, E of Ie Islet, Okinawa, 26.715°N, 127.835°E (4). Otoutopjima, Ogasawara Islands, 27°N, 142.167°E (1) (Kase & Kano, 2002: repository unknown).

Malaysia. 15 m, Turtle Cavern, Sipadan Island, Borneo, 5.080°N, 118.608°E (116) (Kase & Kano, 2002: repository unknown).

Nauru. 15–25 m, AW Aiwo, 0.542°S, 166.908°E (7) (Kase & Kano, 2002: repository unknown).

New Caledonia. 17–21 m, E of Nuu Powa, Ile des Pins, 22.527°N, 169.432°E (3) (Kase & Kano, 2002: repository unknown).

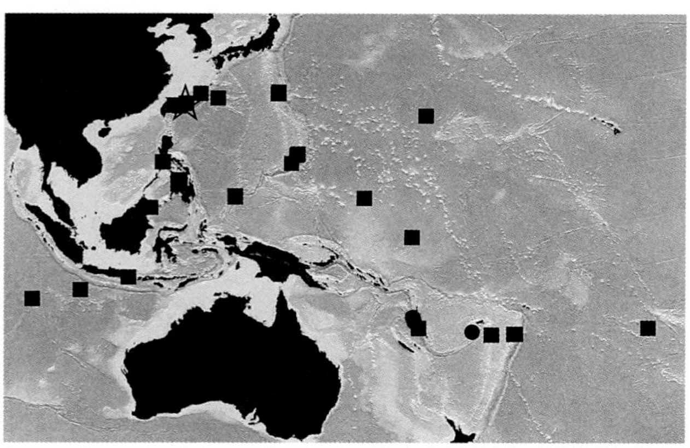

Figure 1002. Distribution of *Trogloconcha ohashii*. ☆ type locality. ● specimen record. ■ literature record.

Belau. 17 m, Virgin Hole, 7.120°N, 134.235°E (>100). 36–38 m, Blue Hole, 7.138°N, 134.222°E (55). 24–44 m, Siaes Tunnel, 7.312°N, 134.227°E (5) (Kase & Kano, 2002: repository unknown).

Philippines. 17 m, Balicasag Island, Bohol, 9.545°N, 123.678°E (86). 27 m, Marigondon Cave, S of Mactan Island, 10.263°N, 123.987°E (39). 46 m, Maptin Cave, SW of Maricaban Island, Bantagas, 13.670°N, 120.817°E (38). 40 m, in front of Vistamar Resort, Anilao, Bantagas, 13.752°N, 120.917°E (>100) (Kase & Kano, 2002: repository unknown).

Micronesia. 26 m, Tawag Point, Pohnpei, 6.927°N, 6.000°E (1). 20 m, Grotto, Saipan, 15.255°N, 145.827°E (26). 50–51 m, close to Tinian Grotto, Tinian, 15.018°N, 145.583°E (>200) (Kase & Kano, 2002: repository unknown).

Vanuatu. 15–18 m, Taj Mahal, W of Efate Island, 17.640°S, 168.145°E (6).(Kase & Kano, 2002: repository unknown).

Remarks. Kase & Kano (2002) identified Bandel's (1998) "*La. miranda*" as *Tro. ohashii*, although it rather corresponds to *Tro. tesselata*.

Trogloconcha tesselata Kase & Kano, 2002

Figures 1003–1006, Color Plate 2D

Chresonymy.

Trogloconcha tesselata Kase & Kano, 2002: 31, figs 1D–F, 2C–D (Fig. 1003).
Trogloconcha tesselata: Marshall, 2002: 179.
Trogloconcha tesselata: Geiger, 2003: 80.
Trogloconcha tesselata: Geiger & Jansen, 2004b: 64–65, figs 32d [map], 33 [shell].
Trogloconcha tesselata: Geiger, 2004: text-fig. p. 7.

Figure 1003. Original illustration of *Trogloconcha tesselata*.

Misidentifications.

not *Larochea miranda*: Bandel, 1998: 66–67, pl. 23, figs 4–5 [is *Tro. tesselata*].
not *Trogloconcha tesselata*: Geiger, 2008b: fig. 1 [is *Tro. ohashii*].

Type material. Holotype (NSMT Mo72830. Fig. 1003), 1.02 × 1.11 mm. Paratypes (NSMT Mo72831, 7) from type locality.

Type locality. North of Kohama Island, Yaeyama Group, Okinawa (24°21.5'N, 123°58.9'E), 15–20 m; crevices, coral sand (OD).

Etymology. Tesselatus, Lati for tesselate, reticulate. Named for the sculpture of crossed spirals and axials (OD).

Description. Shell small (Holotype: 1.1 × 1.02 mm, maximum 1.24 × 1.15 mm. Figs 1003–1005), as wide as high (width/height ratio 1.04–1.11). Protoconch approximately 240 µm in diameter, 0.75 whorl, mostly smooth (possibly eroded) with some traces of flocculent sculpture, no apertural varix, apertural margin curved. Teleoconch with 1.75–2 whorls. Sculpture with dense axials and spirals of equal strength producing fine reticulate pattern, approximately 70 axials on body whorl; spirals stepped on lowermost base. Anomphalous. Aperture round. No selenizone, slit, or foramen.

Animal (Color Plate 2D) with large protruding eyes below cephalic tentacles, three pairs of epipodial tentacles

Comparisons. *Trogloconcha christinae* from the southern shore of Western Australia lacks spiral sculpture.

Trogloconcha ohashii from the Indian Ocean and Western Pacific Ocean is umbilicate.

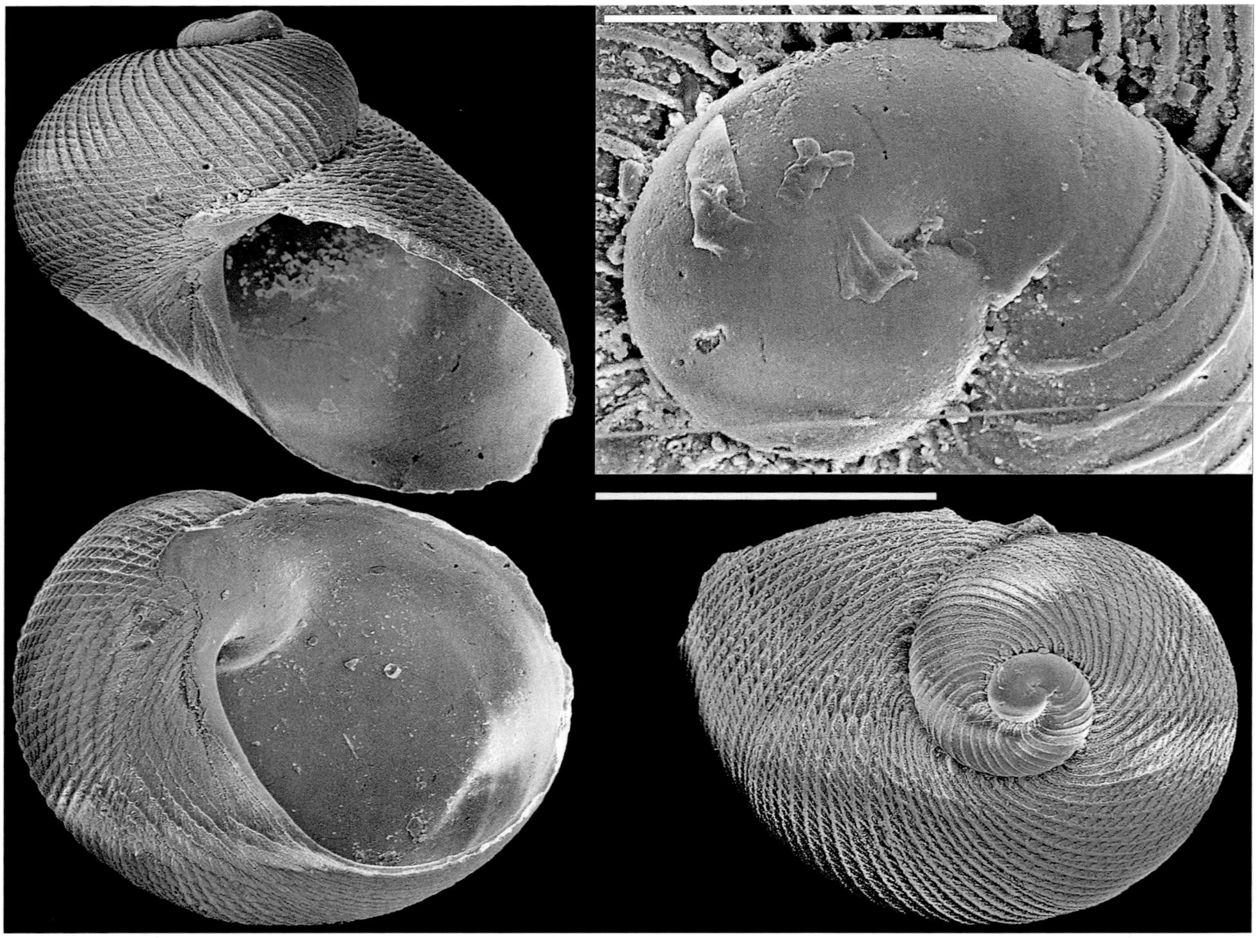

Figure 1004. *Trogloconcha tesselata*. 64 m, Great Barrier Reef, Swain Reefs, 3 km NE of West side of Bylund Gillett Cay, Queensland, Australia, 21.700°S, 152.433°E (AMS C.406375, 1). Scale bar shell = 0.5 mm. Scale bar protoconch = 100 µm.

Troglococha lamellinodosa from the Indo-Malayan Archipelago to the Central Pacific has an overall much taller shell, has spiral sculpture on the protoconch and a very narrow but open umbilicus not constricted by an umbilicus

Trogloconcha marshalli from the Oligocene of France has weaker spirals towards the very narrow umbilicus.

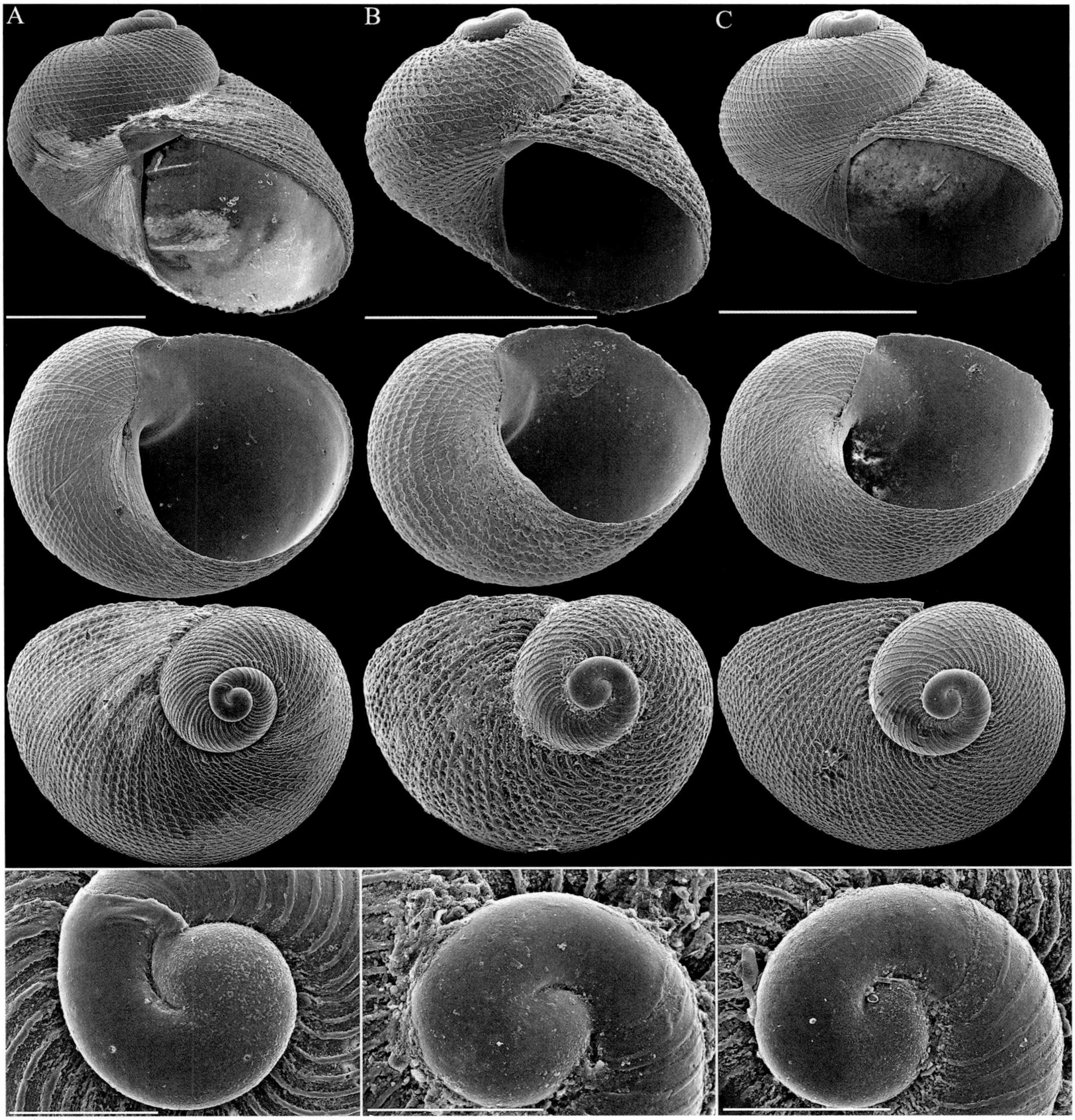

Figure 1005. *Trogloconcha tesselata*. **A.** 538 m, seamount 130 km S of Esperance Rock, Kermadec Ridge, New Zealand, 32.472°S, 179.265°E (NMNZ M.137470, 2). **B.** 190 m, off North West Cape, Western Australia, Australia, 22.386°S, 113.677°E (AMS C.462428, 5). **C.** 49 m, off Marsat Abu Samra, Israel (AMS C.462427, 5). Scale bars shell = 0.5 mm. Scale bars protoconch = 100 μm.

Trogloconcha lozoueti from the Indo-Malayan Archipelago to the Central Pacific, shows distinct points at the intersection of axial and spirals, and has a proportionally wider shell.

Distribution. Indian Ocean and Red Sea to Western Pacific, 15–73 m (Fig. 1006).

Specimens Records.

Israel. 49 m, off Marsat Abu Samra, 29.272°N, 34.745°E (AMS C.462427, 5).

Madagascar. 200–220 m, Manantenina sector, DW3523, 24.383°S, 47.517°E (MNHN, 1). 307–319 m, Manantenina sector, DW3524, 24.383°S, 47.533°E (MNHN, 1).

Japan, Okinawa. N of Kohama Island, Yaeyama Group, 24°21.5'N 123°58.9'E (NSMT Mo72830: holotype; NSMT Mo72831, 7: paratypes). 43 m, Horseshoe Beach, 26.505°N, 127.853°E (DLG 1550, 1). 13–33 m, Flab, Horseshoe Bay, 26.509°N, 127.856°E (DLG 1487, 1).

Philippines. 90–200 m, Maribohoc Bay, Bohol Island, 6.602°N, 123.750°E (MNHN, 1: complete). 507 m, St. 5576, off Mount Dromedarie, N Tawi Tawi, 5.432°N, 120.061°E (USNM ex 257870, 2). 35 m, Pamilacan Island, PANGLAO 2004 B26, 9.490°N, 123.933°E (MNHN, 1). 25 m, Pamilacan Island, PANGLAO 2004 B30, 9.490°N, 123.935°E (MNHN, 2). 17 m, Balicasag Island, PANGLAO 2004 S45, 9.515°N, 123.680°E (MNHN, 2).

Australia, Western Australia. 20 m, E side, Sandy Islet, Scott Reef, 14.044°S, 121.780°E (SBMNH 369510, 3). 190 m, off North West Cape, 22.386°S, 113.677°E (AMS C.462428, 5).

Australia, Queensland. 64–73 m, Swains Reef, 3 km NE of W side Bylund Gillettt Cay, 21.700°S, 152.433°E (AMS C.406375, 1).

Vanuatu. 80 m, W Tutuba Island, SANTO 2006 DS104, 15.568°S, 167.266°E (MNHN, 6). 360–419 m, MUSORSTOM 8 DW1065, 16.266°S, 167.350°E (MNHN, 1).

Fiji. 260–305 m, S of Viti Levu, 18.308°S, 178.097°E (MNHN, 4; MNHN, 1). 275–430 m, S of Viti Levu, 18.297°S, 177.907°E (MNHN, 1, MNHN, 4).

New Zealand. 538 m, seamount 130 km S of Esperance Rock, Kermadec Ridge, 32.472°S, 179.265°E (NMNZ M.137470, 2).

Literature records.

Australia, Queensland. Swains Reef (Bandel, 1998).

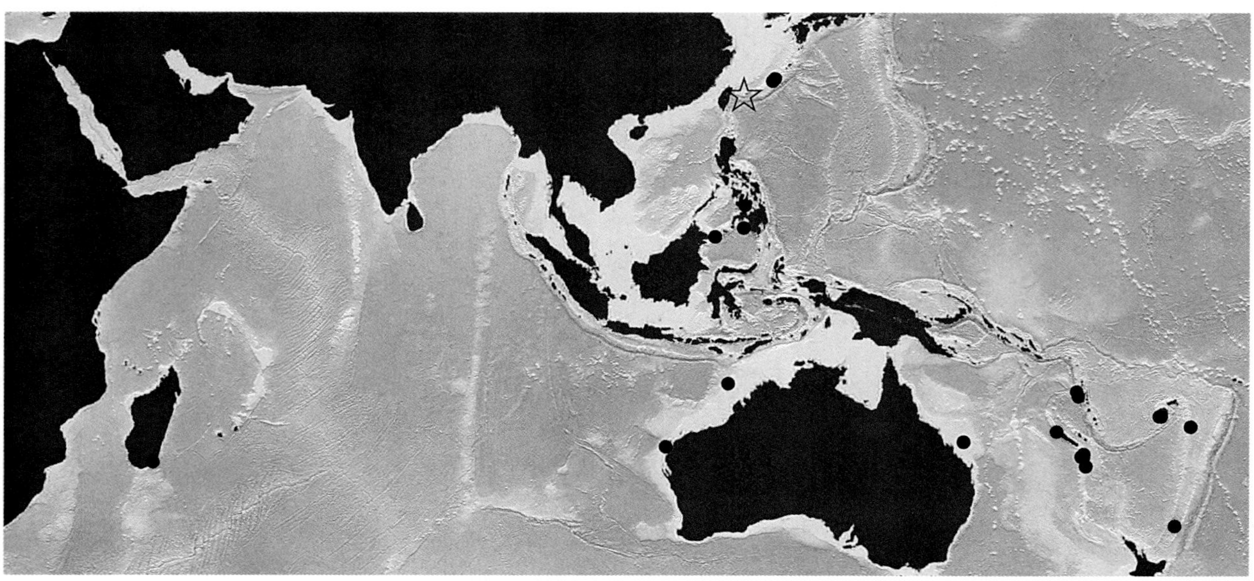

Figure 1006. Distribution of *Trogloconcha tesselata*. ☆ type locality. ● specimen record. ■ literature record.

Remarks. Kase & Kano (2002) referred the specimen illustrated by Bandel (1998: pl. 23, figs 4–5, repository unknown) as "*La. miranda*" to *Tro. ohashii*. Bandel's specimen, however, represents *Tro. tesselata* not *Tro. ohashii* (cf. Marshall, 2002). *Trogloconcha ohashii* is characterized by pustules at the intersection of the axials and the spirals, which are not evident in Bandel's illustration. *Trogloconcha ohashii* seems also to be wider than Bandel's specimen. Bandel's specimen shares with *Tro. tesselata* that the axials are stronger than the spirals, whereas in *Tro. ohashii* axials and spirals are of equal strength.

Trogloconcha marshalli (Lozouet, 1998) †

Figures 1007–1008

Chresonymy.

Larocheopsis marshalli Lozouet, 1998: 68–69, figs 5A–C.

Trogloconcha marshalli: Geiger, 2003: 80.

Type material. Holotype (MNHN. Fig. 1007). Paratypes from type locality (MNHN, 2). St.-Étienne-d'Orthe "R. Verdun," Landes, Chattian (MNHN, 5).

Type locality. Landes, Peyrehorade (Peyrère), France [44.670°N, 5.620°E], late Oligocene (Chattian), *Eulepidina* marls (OD).

Etymology. Honors Bruce Marshall of NMNZ (OD).

Description. Shell medium-sized (to 1.6 mm. Fig. 1007), naticiform inflated, thin, fragile. Protoconch of 0.75 whorl, smooth, no apertural varix, slight apertural thickening, apertural margin convex. Teleoconch of 1.75 whorls, reticulate sculpture with approximately 65 slightly stronger colabral axials on body whorl, spirals starting after approximately one whorl, approximately 35 spirals at aperture. Umbilicus very narrow, deep. Aperture round.

Distribution. Known only from type locality (Fig. 1008).

Comparisons. *Trogloconcha ohashii* from the Indo-Malayan Archipelago to the Central Pacific has a large umbilicus and pointed intersections of the axials and spirals.

Trogloconcha tesselata from the Indian Ocean to the Western Pacific has no umbilicus and more prominent spiral sculpture on the base.

Trogloconcha christinae from southern Western Australia lacks an umbilicus and spiral sculpture.

Trogloconcha lozoueti from the Austral Islands shows marked points at the intersection between axials and spirals, and has an earlier onset of spirals.

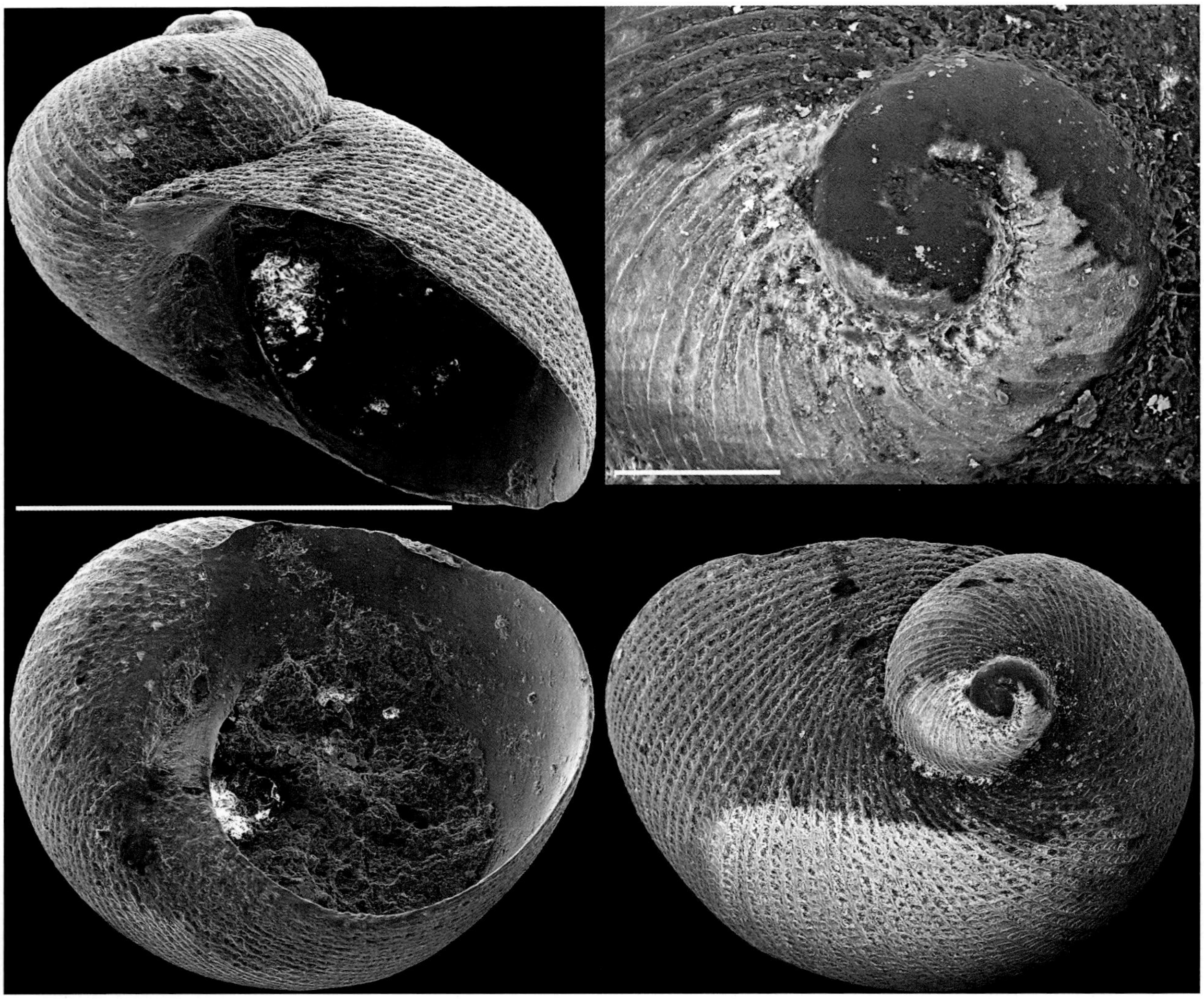

Figure 1007. Holotype of *Trogloconcha marshalli* (MNHN). Landes, Peyrehorade (Peyrère), France [44.670°N, 5.620°E]. Scale bar shell = 1 mm. Scale bar protoconch = 100 µm.

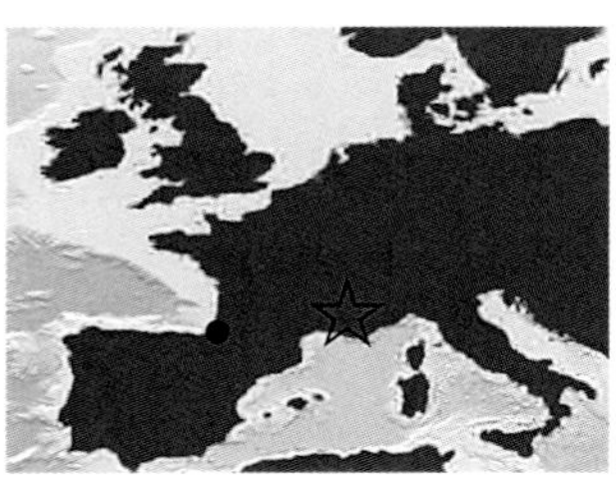

Figure 1008. Distribution of *Trogloconcha marshalli*. ☆ type locality. ● specimen record.

Depressizonidae Geiger, 2003
Vetigastropoda *incertae sedis*

Chresonymy.

Depressizoninae Geiger, 2003: 50.

Depressizonidae: Geiger, 2009: 57.

Description. Shell calyptraeiform, without nacre. Teleoconch with beaded sculpture, selenizone above periphery, slit closed to foramen. Umbilicate, no brood pouch.

Animal unknown.

Distribution. Tropical Western and Eastern Pacific (Fig. 1009). For specific localities, see individual species.

Geologic range. Recent.

Comparisons. The calyptraeiform shell distinguishes Depressizonidae from any other scissurellid family. The beaded sculpture is unknown from any other scissurellid.

Anatomidae have a trochoid shell, an open slit, and the selenizone is at the periphery of the shell.

Scissurellidae usually have a trochoid shell, and the limpet shape of *Incisura* is attained by strongly increased expansion rate of the shell, but maintains the height of the shell. In Depressizonidae, however, the axis is compressed, while the expansion rate of the shell is comparable to that of typical Scissurellidae.

In Temnocinclidae, the limpet shape is also attained by strongly increased expansion rate of the whorl, but retaining the height of the shell. All species of Temnocinclidae are endemic to the hydrothermal vent environment.

Sutilizonidae have a stronger expansion rate than Depressizonidae, and the axis is at a 45° angle to the apertural plane, maintaining more of the height of the shell. The protoconch in Sutilizonidae is pitted, whereas Depressizonidae have flocculent sculpture. All sutilizonid species are endemic to the hydrothermal vent environment.

Larocheidae lack a slit, foramen, or selenizone.

Remarks. Batten (1984) noted that the calyptraeiform shell is unknown in archaeogastropods, for which reason the placement of *Neomphalus fretterae* McLean, 1989, remained obscure. *Depressizona* also exhibits this shell morphology, hence, shows that the

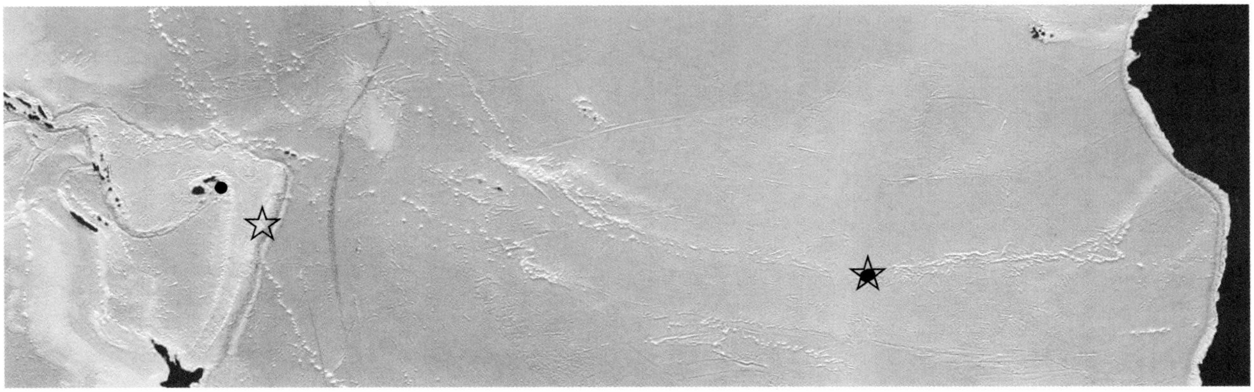

Figure 1009. Global distribution of *Depressizona* species. ☆ type locality. ● specimen record.

placement of Neomphalidae in Vetigastropoda does not pose a problem given its shell morphology.

Bouchet & Rocroi (2005) synonymized Depressizonidae under Scissurellidae without comment or justification. The overall shell shape is certainly as different from Scissurellidae as that of Temnocinclidae and Sutilizonidae. Additionally, the beaded sculpture is unknown in Scissurellidae, further supporting the distinctness of Depressizonidae as a family (Geiger, 2009).

The elevation of subfamily to family is due to the overall elevation of ranks of former scissurellid subfamilies to full families: Scissurellidae, Anatomidae, Larocheidae, Temnocinclidae and Sutilizonidae are now all treated as families (Geiger & Jansen, 2004; Geiger & Thacker, 2005; Geiger, 2008b). Depressizonidae is as distinct from any of those families, as they are from one another. It is recognized that the change in rank per se has no meaning with regard to phylogenetic relations, and is only carried out to keep the overall classification in balance (Geiger, 2009).

Depressizona Geiger, 2003

Chresonymy.

Depressizona Geiger, 2003: 50.

Type species. *Depressizona exorum* Geiger, 2003 (OD).

Other species. *Depressizona axiosculpta* Geiger, 2009.

Etymology. The name describes the overall, depressed shape of the shell (Depressi-), and noting that the foramen is closed anteriorly as in *Sinezona* (-zona). Gender feminine (OD).

Description, Comparison. As for family.

Differentiation of species. Species are differentiated by shell sculpture.

Distribution. As for family.

Remarks. *Depressizona* is assigned to its own family due to the unique way in which limpet shape has been obtained and due to the beaded sculpture on the teleoconch. Protoconch sculpture is very different from any limpet-shaped scissurellids: *Sutilizona* has a pitted protoconch and *Incisura* has strong axials. Flocculent sculpture is found in *Anatoma* with a trochiform shell and an open slit at the periphery, and in Larocheidae with a naticiform shell and no selenizone, slit, or foramen. The combination of limpet-shaped shell with a protoconch showing flocculent sculpture cannot be placed in any of the existing scissurellid genera or families.

Depressizona axiosculpta Geiger, 2009

Figure 1010 –1011

Chresonymy.

Depressizona axiosculpta Geiger, 2009: 57–59, fig. 1.

Type material. Holotype (AMS C.461464. Fig. 1011B), 1.08 × 0.80 × 0.55 mm (L × W × H).

Type locality. 260 m, SE of Tongatapu, Tonga, 21.345°S, 175.042°W (OD).

Etymology. Named because of the pronounced axial sculpture on the shell (OD).

Description. Shell medium-sized (to 1.56 mm. Fig. 1011A. Holotype = 1.08 mm. Fig. 1011B), calyptraeiform, moderately thick. Protoconch 0.75 whorl, sculpture flocculent, no apertural varix, apertural margin convex. Teleoconch I of 0.6 whorl, eight strong raised axial cords, crossed by eight spiral cordlets running over axials. Teleoconch II of approximately 1.1–1.2 whorls. Shoulder flat, axial cords low, broad, little distinct; spiral lines expanding to low spiral cords up to ⅔ as wide as axial cords; at intersection forming distinct points. Base with distinct constriction below selenizone, strong angulation at mid-base forming periphery; approximately 13–18 axial lamellae near periphery only; weak spiral lines between selenizone and periphery forming weak nodules at intersection with axials; on underside from adjacent to periphery to umbilicus series of strongly beaded spiral cords, weakly beaded spiral lines, approximately three spiral lines. Umbilicus wide, open, walls straight. Selenizone above periphery; keels strong, low; slit closed to foramen (larger due to shell damage in holotype), anteriorly closed by raphe. Aperture D-shaped, roof overhanging; peristome thickened, recurved, with fine spiral lines over thickened portion.

Animal unknown.

Comparisons. *Depressizona exorum* from Easter Island has a much thinner shell, lacks the strong axial lamellae at the periphery and none of the three known specimens have a thickened apertural margin.

Specimen records.

Fiji. 820–863 m, BORDAU 1 DW1459, 17.300°S, 179.550°W (MNHN, 1).

Distribution. Tonga and Fiji (Fig. 1010).

Remarks. Contrary to expectations voiced by Geiger (2009), a second, much better preserved specimen was discovered in newly collected material by MNHN, and is shown here in Figure 1011A.

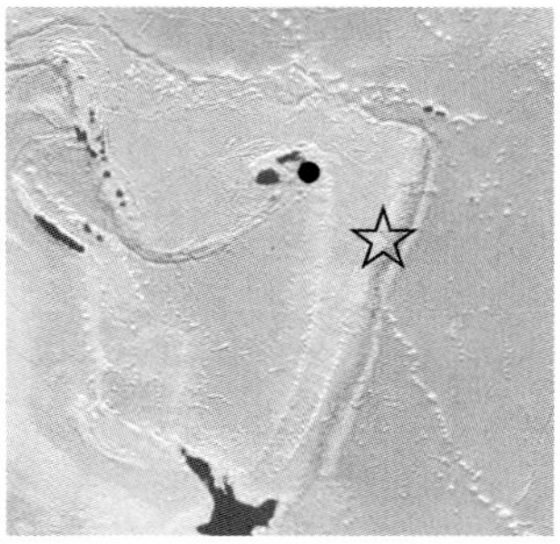

Figure 1010. Distribution of *Depressizona axiosculpta*. ☆ type locality. ● specimen record.

Figure 1011. *Depressizona axiosculpta.* **A.** 820–863 m, BORDAU 1 DW1459, Fiji, 17.300°S, 179.550°W (MNHN, 1). **B.** Holotype. 260 m, SE of Tongatapu, Tonga, 21.345°S, 175.042°W. (AMS C.461464). Scale bars shell = 1 mm. Scale bar protoconch = 100 μm.

Depressizona exorum Geiger, 2003

Figures 1012–1014

Chresonymy.

Depressizona exorum Geiger, 2003: 51–52, 75, fig. 14.

Depressizona exorum: The Ex, 2004: inside cover [copy Geiger, 2003: fig. 14].

Depressizona exorum: Hallström & Wabäck, 2004: inside cover (paratype), extras [copy Geiger, 2003: fig. 14].

Depressizona exorum: Geiger, 2004: text-fig. p. 7 [third specimen].

Depressizona exorum: Geiger, 2006c: text-fig. 6.

Depressizona exorum: Geiger, 2008b: fig. 1.

Type material. Holotype (LACM 2932), 1.17 × 0.3 mm (W × H). Paratype from type locality (LACM 2933, 1), 1.24 × 0.35 mm.

Type locality. 10–20 m, off Rosalia Point, E of Anakena, Easter Island, Chile, in sand and rubble at base of cliff, Dec. 2000. Leg. B. Raines [27.067°S, 109.333°E] (OD).

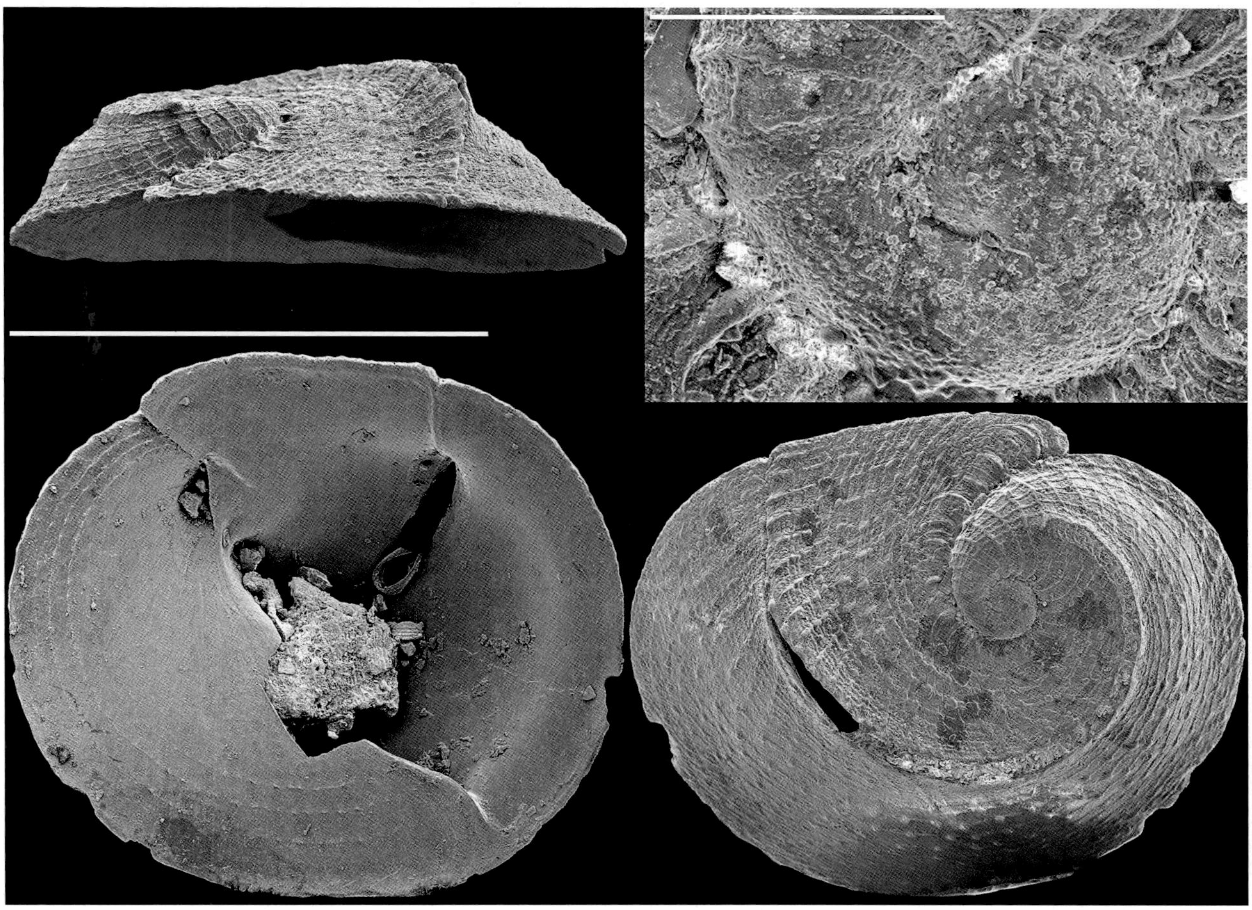

Figure 1012. *Depressizona exorum*. 10–20 m, off Rosalia Point, Anakena Beach, Easter Island, Chile, 27.067°S, 109.333°W (DLG 205, 1). Scale bar shell = 1 mm. Scale bar protoconch = 100 μm.

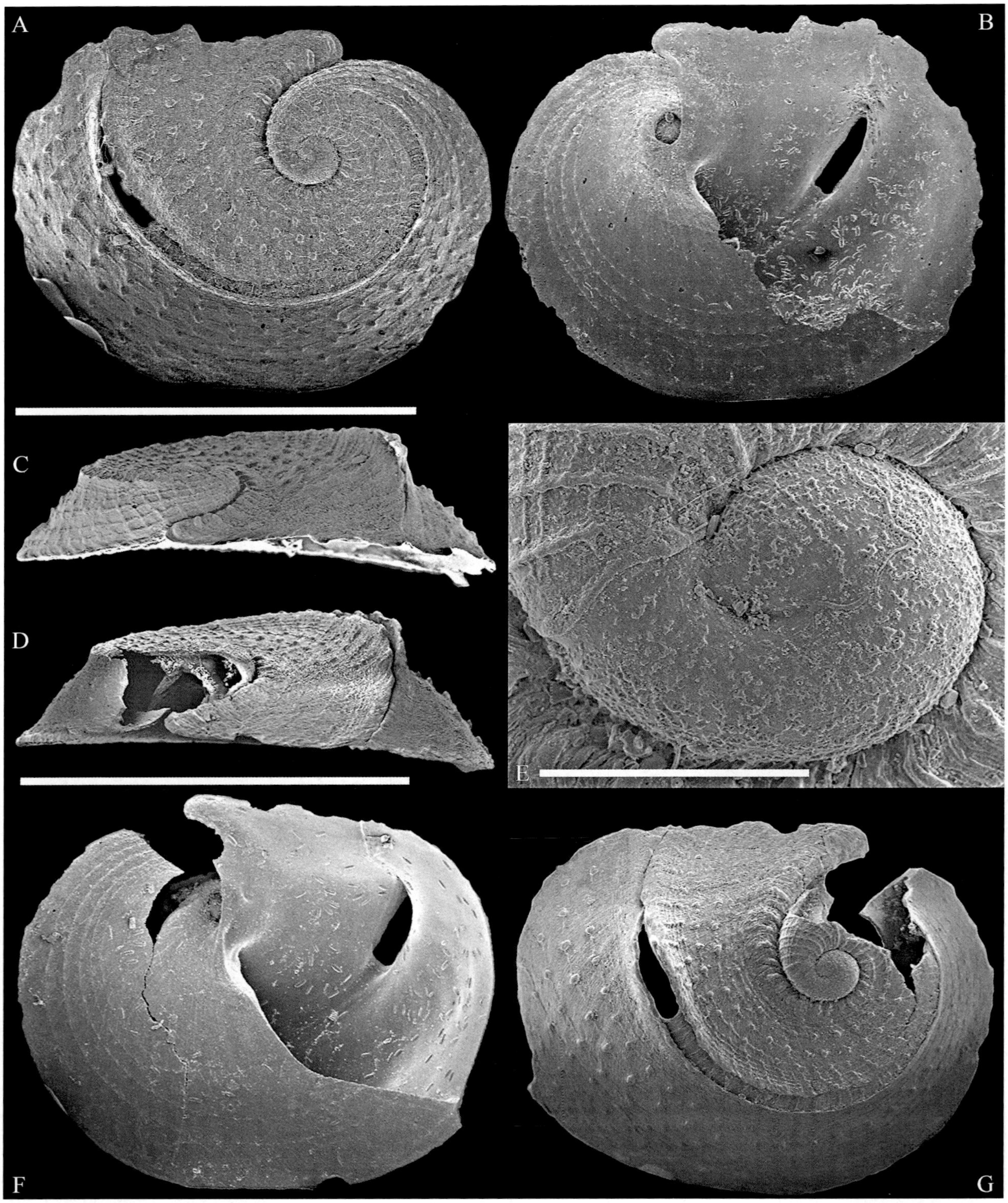

Figure 1013. *Depressizona exorum*. 10–20 m, off Rosalia Point, Anakena Beach, Easter Island, Chile, 27.067°S, 109.333°W. **A–C.** Holotype. 1.17 × 0.30 mm (LACM 2932). **D–G.** Paratype. 1.24 × 0.35 mm (LACM 2933). Scale bars shell = 1 mm. Scale bar protoconch = 100 µm.

Depressizonidae *Depressizona*

Etymology. Named for Andy, G. W. Sok, Katrin, Luc, and Terrie Ex, of Amsterdam, The Low Countries. Genitive masculine plural of surname (OD; see also Remarks).

Description. Shell medium-sized (to 1.24 mm: paratype. Figs 1012–1013), calyptraeiform, thin, fragile. Shoulder convex, angulated at selenizone, acute angle at mid-base forming periphery of shell. Shoulder and dorsal surface of base with spiral rows of beads, sometimes connected by thin spirals, on young shell with cancellate sculpture and raised intersections of axials and spirals. Ventral surface of base with spiral sculpture of elevated elongated beads. Umbilicus forming continuous curve with base. Selenizone on shoulder, after 0.5 teleoconch I whorl, selenizone of 200°, foramen teardrop-shaped closed anteriorly. Protoconch 160 µm (paratype), 0.75 whorl, with flocculent sculpture, no apertural varix, aperture sinusoidal, keels at foramen raised forming narrow elongated chimney.

Animal unknown.

Comparisons. *Depressizona axiosculpta* from Tonga and Fiji has an overall thicker shell, strong axial lamellae at the periphery, has a thickened apertural margin.

Distribution. Known only from type locality (Fig. 1014).

Specimen records.

Chile. 10–20 m, off Rosalia Point, Anakena Beach, Easter Island, 27.067°S, 109.333°W (DLG 205, 1; DLG 1042, 1).

Remarks. The persons named in the etymology were the then current members of the Dutch band *The Ex* (1979–present; see Hallström & Wabäck, 2004). The membership of the project continues to change, and all members and collaborators have certainly affected how the band has gone repeatedly off any known path. Their music has provided me with great inspiration and joy while working on this book and other projects.

The quality of the original material had been critiqued by one of the reviewers of the original description. The two illustrated specimens were the only ones known at the time; hence, they were used despite being somewhat damaged, because they were highly distinct.

Figure 1014. Distribution of *Depressizona exorum*. ☆ type locality. ● specimen record.

Depressizonidae
Depressizona

Sutilizonidae McLean, 1989
Vetigastropoda *incertae sedis*: Lepetodriloidea?

Chresonymy.

Sutilizoninae McLean, 1989: 11–12. Marshall, 1993: 286.
Sutilizonidae: Warén & Bouchet, 2001: 141.
Sutilizonidae: Bouchet & Rocroi, 2005: 168 [Lepetodriloidea].

Type genus. *Sutilizona* McLean, 1989.

Description. Shell strongly asymmetrical, limpet-shaped, coiling axis at approximately 45° angle to apertural plane. Slit closed forming foramen. Protoconch finely pitted, with finely anastomosing reticulations, no apertural varix. Operculum, thin, central nucleus, much smaller than aperture. Radula without enlarged lateral tooth 5. Lateral and marginal teeth difficult to distinguish; number of lateral teeth uncertain, at least four. Ctenidia monopectinate, left one much larger. Cephalic and epipodial tentacles without papillae. No eyes.

Distribution. Northeastern Pacific Rise and northern Mid-Atlantic Ridge at hydrothermal vents (Fig. 1015). For specific localities, see individual species.

Geologic range. Recent.

Comparisons. Temnocinclidae (*Temnocinclis*, *Temnozaga*, *Triassurella* †) have almost symmetrical shells with the coiling axis approximately parallel to the plane of the aperture. In the calyptraeiform *Depressizona* (Depressizonidae), the coiling axis is perpendicular to the apertural plane.

Remarks. Warén & Bouchet (2001) elevated the subfamily Sutilizoninae to family rank and placed it in the superfamily Lepetodriloidea, an arrangement also followed by Warén *et al.* (2006). [Note that Warén & Bouchet (2001) as well as Schwarzpaul (2002) referred to "Warén (in press)" (Zoologica Scripta), but that work does not appear to have been published.] Warén & Bouchet *in* Bouchet & Rocroi (2005: 168) synonymized Temnocinclidae under Sutilizonidae without further justification applying first reviser's principle. Their opinion is not followed here. The differences in shell geometry, and coiling mode are considered sufficient to warrant recognition of those two lineages at the family level.

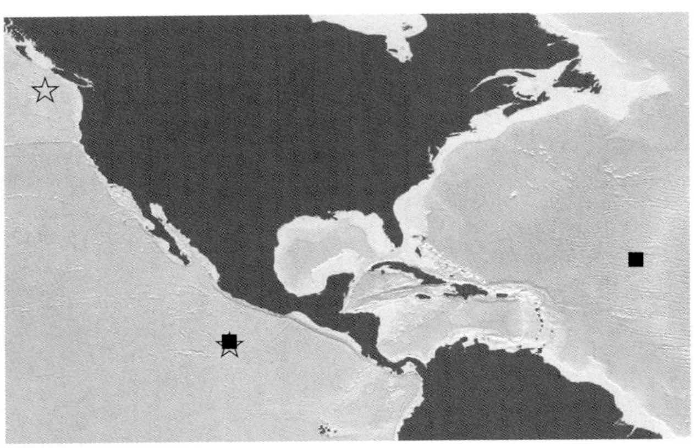

Figure 1015. Global distribution of *Sutilizona* species. ☆ type locality. ● specimen record. ■ literature record.

The systematic affinity of Sutilizonidae among Vetigastropoda is uncertain. The non-papillate nature of their cephalic and epipodial tentacles (both papillate in Scissurellidae and Anatomidae), as well as the radula with poorly differentiated lateral and marginal teeth (clearly differentiated in Scissurellidae and Anatomidae), suggest that Sutilizonidae is not closely related to Scissurellidae or Anatomidae. Lepetodrilidae share the non-papillate nature of the tentacles, but have additional head appendages (neck appendage, ventral sensory papilla) not evident in *Sutilizona*, and the radula of *Lepetodrilus* has clearly differentiated lateral and marginal teeth (see Warén & Bouchet, 2001). Whether the placement of Sutilizonidae in the larger group of Lepetodriloidea is justified or only creates a "waste basket taxon" remains to be determined.

Sutilizona McLean, 1989

Chresonymy.

Sutilizona McLean, 1989: 14–15.

Sutilizona: Bandel, 1998: 49.

Sutilizona: Warén & Bouchet, 2001: 141.

Sutilizona: Warén *et al.*, 2006: 98.

Type species. *Sutilizona theca* McLean, 1989 (OD).

Other species. *Sutilizona pterodon* Warén & Bouchet, 2001.
Sutilizona tunnicliffae Warén & Bouchet, 2001.

Etymology. Sutilis: Latin for sewn together; zona: Latin for belt, referring to the sealed selenizone in mature specimens. Gender feminine (OD).

Description, Comparison. See family.

Differentiation of species. Species are differentiated by expansion rate of the whorls and placement of the apex, open slit or foramen, and details of the sculpture.

In foramen-bearing species, small specimens most likely will have an open slit, but no material to confirm this assumption is available.

Distribution. See family.

Sutilizona pterodon Warén & Bouchet, 2001

Figures 1016–1018

Chresonymy.

Sutilizona pterodon Warén & Bouchet, 2001: 141–143, figs 15g [protoconch], 17 c, e, g, h [shell]; 18 a, f [radula].

Sutilizona pterodon: Geiger, 2003: 79.

Sutilizona pterodon: Warén *et al.*, 2006: 98, text-fig. 1.

Type material. Holotype (MNHN. Fig. 1016B), 2.0 mm. Paratypes from type locality (MNHN, 5. Fig. 1016A). Type material not found in MNHN in 4/2005 (DLG pers. obs.).

Type locality. 3490–3520 m, Mid-Atlantic Rise, Snake Pit, Ruches or Elan, *Alvin* dive 2619, 23°22.13'N, 44°57.13'W (OD).

Etymology. Pteros, Greek for wing and odous: Greek for teeth, referring to the winglike outer marginal teeth (OD).

Description. Shell medium-sized (to 2 mm: holotype. Fig. 1016), inflated, thin, fragile. Overall outline of shell rounded. Covered in thick deposits of iron and manganese oxides. Protoconch of 0.66 whorl, bulbous, almost detached, finely pitted with coarse anastomosing ridges, no apertural varix, apertural margin convex. Teleoconch I of 0.5 whorl, smooth except for fine growth lines. Teleoconch II of 0.75 whorl, with relatively weak axials and spirals of equal strength, forming pointed nodules at intersection. At apertural margin, approximately 7–8 spirals on shoulder and base. Selenizone in slightly elevated channel, slit open, keels moderately elevated, sharp. Shell adjacent to selenizone with somewhat thickened callus. Umbilicus open, deep, with smooth internal walls. Transition from base to umbilicus at angle, without keel. Operculum extremely thin, multispiral with central nucleus, folded in crawling condition, approximately ⅓ of animal length.

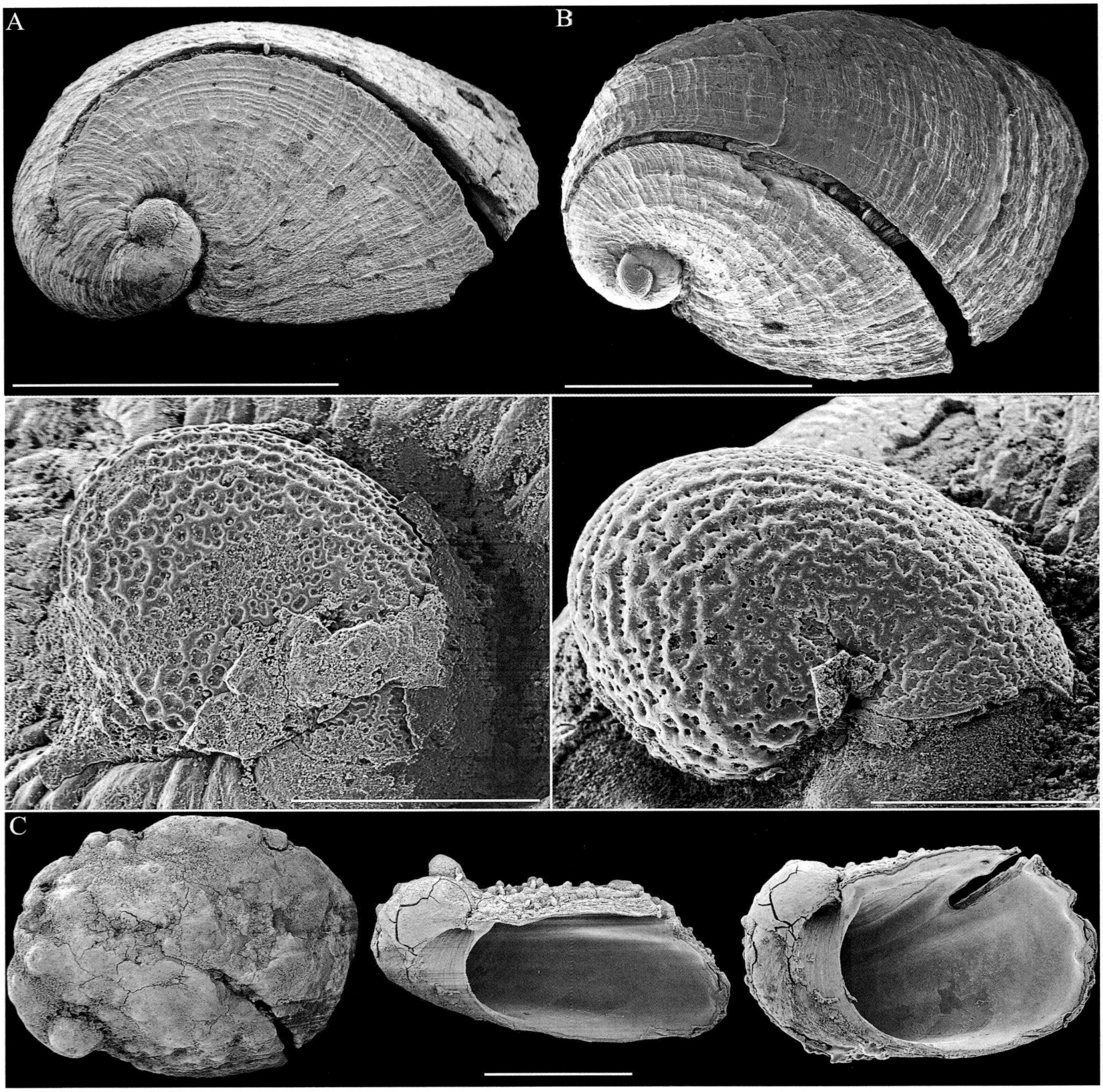

Figure 1016. *Sutilizona pterodon*. 3490–3520 m, Mid-Atlantic Rise, Snake Pit, Ruches or Elan, *Alvin* dive 2619, 23°22.13'N, 44°57.13'W. **A.** Paratype (MNHN). **B.** Holotype (MNHN). **C.** Paratype (MNHN). Scale bars shell = 1 mm. Scale bars protoconch = 100 µm. Images by and with kind permission of A. Warén.

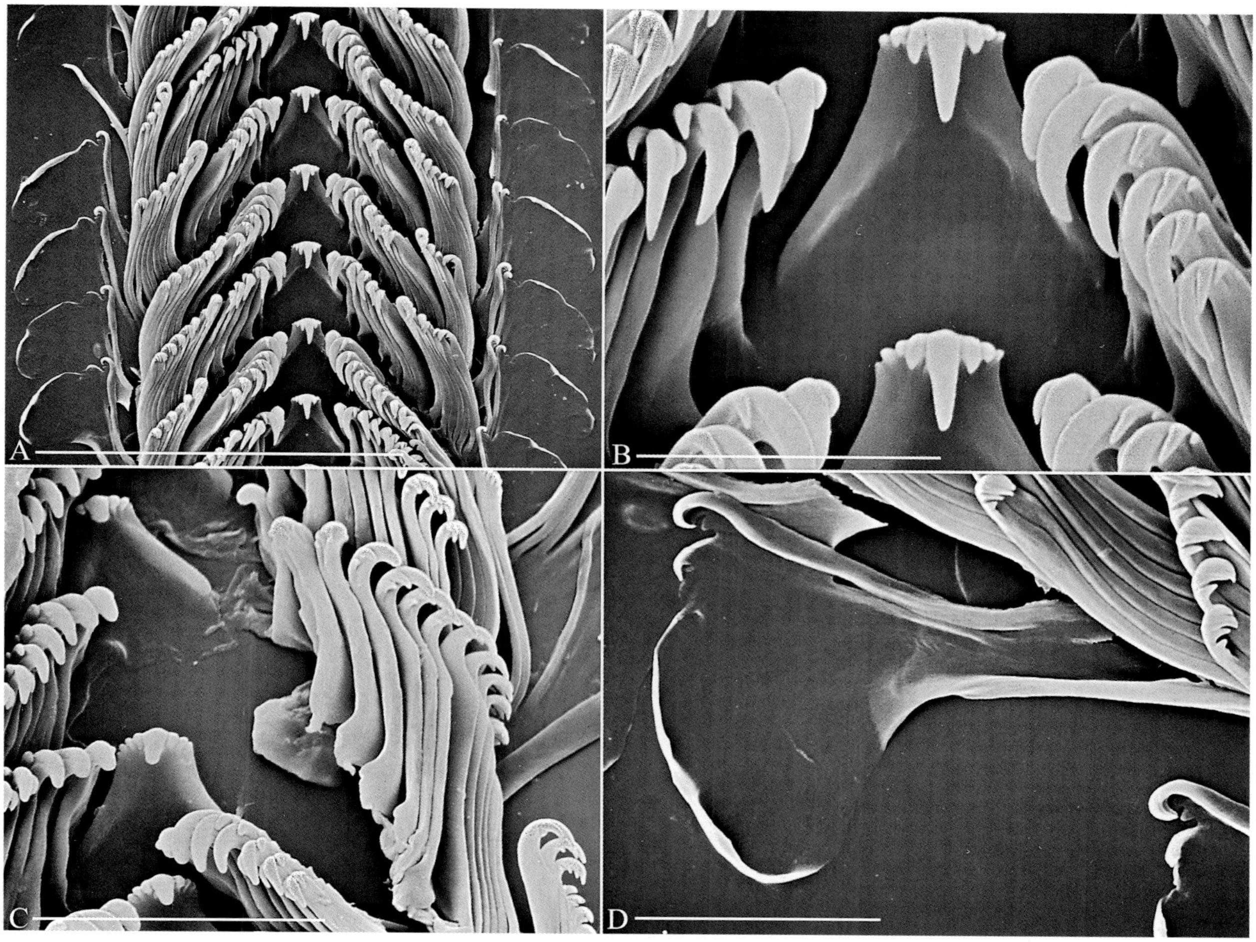

Figure 1017. *Sutilizona pterodon*. Radula. **A.** Overview of radula. Scale bar = 50 µm. **B.** Central field enlarged. Scale bar = 10 µm. **C.** Lateral and marginal teeth. Scale bar = 20 µm. **D.** Outermost marginal teeth. Scale bar = 10 µm. Images by and with kind permission of A. Warén.

External anatomy (adapted from Warén & Bouchet, 2001. Fig. 1018). Cephalic tentacles slender, no papillae. Eyeless, eye stalk posterior to cephalic tentacle. No neck lobes, no accessory cephalic tentacle. One pair of posterior epipodial tentacles, just behind the anterior margin of the operculum, without papillae. Ctenidia monopectinate, left with seven leaflets, right with five leaflets.

Radula (Fig. 1017). Ribbon symmetrical. Rachidian tooth triangular, central cusp 2.5 × as large as 3–4 on either side. Lateral and marginal teeth poorly differentiated: central cusp much enlarged, usually with two outer cusps. Outermost marginal teeth form broad fan with convex outline of shaft. Radular interlock moderate.

Comparisons. *Sutilizona pterodon* is the only species in the genus with an open slit. The rachidian tooth is more slender in *Sut. pterodon* compared to *Sut. theca*. The fan-shaped marginal teeth of *Sut. pterodon* are cited as another distinguishing character, although it is not certain that these delicate structures are indeed absent in *Sut. theca*.

Sutilizona tunnicliffae from the Northeastern Pacific Rise has a closed foramen, and more dense spiral sculpture on the shoulder.

Distribution. Known only from type locality (Fig. 1019).

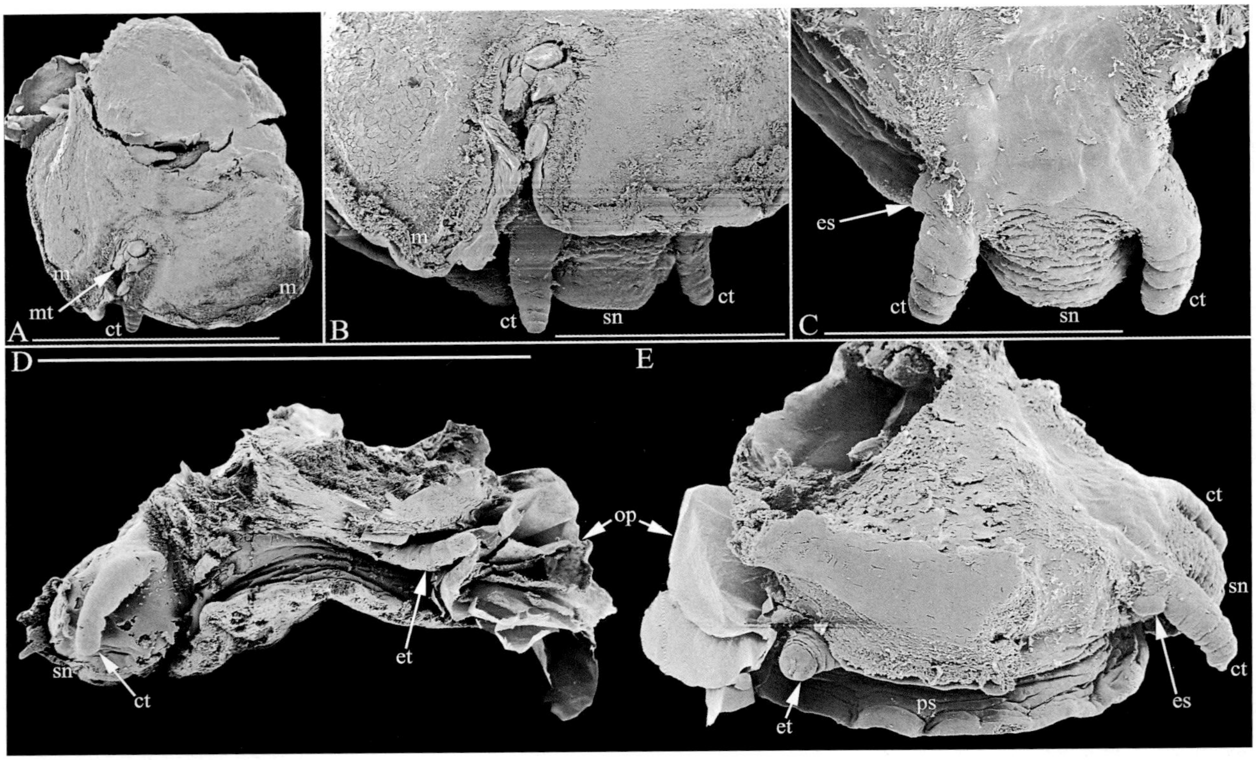

Figure 1018. *Sutilizona pterodon*. Animal. **A.** Dorsal view of whole animal with mantle. Scale bar = 1 mm. **B.** Enlargement of A showing slit in mantle with mantle tentacles. Scale bar = 0.5 mm. **C.** Same as B but with mantle removed revealing eyestalk. Scale bar = 0.5 mm. **D–E.** Lateral views of entire animals showing single epipodial tentacle and operculum. Scale bar = 1 mm. ct: cephalic tentacle. es: eye stalk. m: mantle. mt: mantle tentacle. op: operculum. ps: pedal sole. sn: snout. Images by and with kind permission of A. Warén.

Remarks. *Sutilizona pterodon* is the only species in the genus with an open slit; whether the species is therefore based on juveniles, as Warén & Bouchet (2001) hinted, is uncertain. *Incisura* also has species with open and closed slit, hence, intrageneric variability may be greater than usually assumed. Warén & Bouchet (2001) noted the pointed overall outline of the shell of *Sut. theca*, and gave it as a character to discriminate if from *Sut. pterodon*. However, their figure 17f of *Sut. theca* shows a specimen with a very rounded overall appearance.

The type locality is given as Snake Pit and the species description was based entirely on topotypical material. The figure captions to their figures 17c, e, g, h and 18a, f erroneously give as locality for the specimens Lucky Strike (A. Warén, pers. comm.).

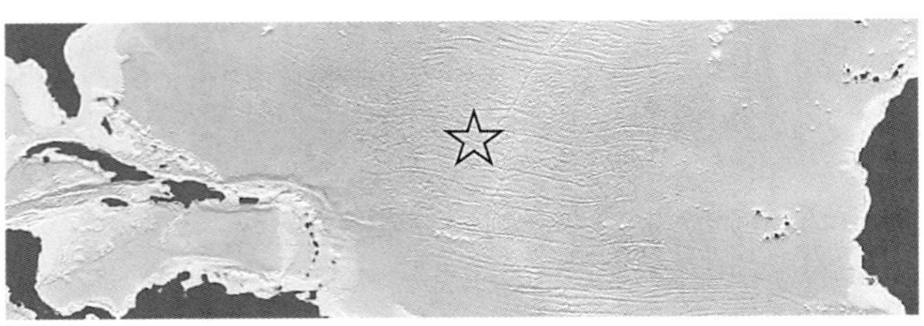

Figure 1019. Distribution of *Sutilizona pterodon*. ☆ type locality.

Sutilizona theca McLean, 1989

Figures 1020–1023

Chresonymy.

Sutilizona theca McLean, 1989: 15, figs 3A–C, 4D–F [shell, protoconch], 5E–F [radula] (Fig. 1020).

Sutilizona theca: Haszprunar, 1989: 3–5, fig. 2.

Sutilizona theca: Skoglund, 1992: 4.

Sutilizona theca: Warén & Bouchet, 2001: 141, figs 17f, i.

Sutilizona theca: Skoglund, 2002: 3.

Sutilizona theca: Geiger, 2003: 80.

Sutilizona theca: Warén *et al.*, 2006: 98.

Sutilizona theca: Adams *et al.*, 2010: 1054.

Figure 1020. Original illustration of *Sutilizona theca*.

Type material. Holotype (USNM 859967: not received 11/1991; 2/2004), 0.9 × 0.5 mm. Paratypes from type locality (LACM 2355, 2).

Type locality. 2715 m, on sulfide mounds at base of inactive chimney, East Pacific Rise near 12°N, 11°46'N, 103°47'W (OD).

Etymology. Theca, Latin for sheath, noun in apposition, referring to the coiled portion of the shell that is retained in the interior (ventral) view (OD).

Description. Shell medium-sized to large (2.4 mm; 1.3 mm in holotype. Figs 1020, 1021), strongly asymmetric, limpet-shaped, coiling axis at approximately 45° angle to apertural plane. Overall outline pointed at termination of selenizone in juveniles, rounded in large specimens. Periostracum thin. Protoconch of 0.75 whorl, bulbous, with deep pits and irregular fine meshwork; no apertural varix, apertural margin convex. Teleoconch I of 0.625 whorl, approximately a dozen axial cords to lamellae, increasing in strength with growth, interstices smooth, with very fine growth lines; on last quarter whorl, spiral band in position of selenizone forming axial scales at intersection with last five axials; some specimens with ruffled axials (Warén & Bouchet, 2001: fig. 17f). Teleoconch II of 0.375 whorl, with 11–15 axials, approximately 15 weaker spiral cords, forming at intersection scales on axials; interstices with somewhat coarser growth lines. Slit closed to form foramen. Selenizone with moderately raised, sharp keels. Inside of shell with coiled portion visible; no shelf; no umbilicus. Operculum small (0.5 mm in larger specimen).

External anatomy (adapted from McLean, 1989. Fig. 1021C). Body asymmetrical. Cephalic tentacles moderately long, without papillae. No eyes. One pair of posterior epipodial tentacles without papillae. Two monopectinate ctenidia, left ctenidium larger than right one, with seven and four leaflets, respectively. Shell muscle horseshoe-shaped, longer on left side, terminations expanded inwardly. Foot with double anterior edge, indicating opening of pedal gland.

Radula (Fig. 1022). Ribbon nearly symmetrical. Rachidian tooth triangular with central tip twice as large as 3–4 cusps on either side. Lateral and marginal teeth poorly differentiated; number of lateral teeth ambiguous. Lateral teeth similar in height to rachidian tooth, marginal teeth with long shafts, no food groove. Tips of lateral/marginal teeth almost symmetrical, central tip strongest, 2–4 smaller tip on either side. Radular interlock moderate.

Comparisons. *Sutilizona pterodon* from the Mid-North-Atlantic Rise has an open slit.

Sutilizona tunnicliffae from the Northeastern Pacific Rise has a more centrally located apex, and the spiral sculpture is more prominent than the axial sculpture.

Distribution. 2500–2700 m, 12–13°N, East Pacific Rise, off Washington State, USA (Fig. 1023).

Figure 1021. *Sutilizona theca.* **A–B.** East Pacific Rise, 13°N vent field. Scale bar = 1 mm. **C.** Animal in shell in ventral view. Scale bar = 1 mm. ct: cephalic tentacle. et: epipodial tentacle. m: mantle. sn: snout. **D.** Protoconch of B. Scale bar = 100 µm. Images by and with kind permission of A. Warén.

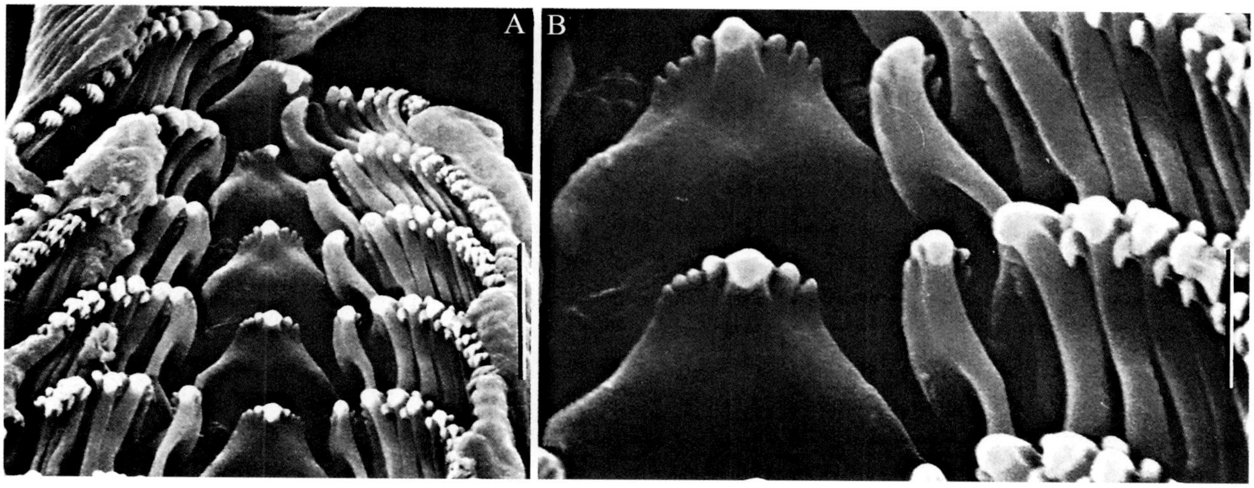

Figure 1022. *Sutilizona theca*. Radula. 2715 m, Sulfide mounds at base of inactive chimney, East Pacific Rise near 12°N, 11°46'N, 103°47'W (LACM 2355b). **A.** Full width of radula. Scale bar = 10 µm. **B.** Central field enlarged. Scale bar = 4 µm. From McLean (1989).

Literature records. East Pacific Rise at 13°N, 12.150°N, 103.942°W (5) (Warén & Bouchet, 2001).

Remarks. All type material is broken. The holotype shell is mounted for SEM and the body was sectioned; the shell of one paratype was broken when the animal was removed for sectioning; the shell of the second paratype was heavily decalcified, and the body was used for radular extraction. Warén & Bouchet (2001) noted the extensive variation in sculpture, with coarse and very finely sculptured specimens known.

Figure 1023. Distribution of *Sutilizona theca*. ☆ type locality. ■ literature record.

Sutilizona tunnicliffae Warén & Bouchet, 2001

Figure 1024–1026

Chresonymy.

Sutilizona tunnicliffae Warén & Bouchet, 2001: 143, figs 15f [protoconch], 17d [shell], 18 b, d, g [radula].

Sutilizona tunnicliffae: Geiger, 2003: 80.

Sutilizona tunnicliffae: Warén *et al.*, 2006: 98.

Type material. Holotype (Fig. 1024 A), destroyed during original investigation, 2.2 mm.

Type locality. Juan de Fuca ridge, Endavour Segment, off Washington state, USA, ROPOS #278 fissure at Main Field, 47°56.9'N, 129°6.9'W, 2202 m (OD).

Etymology. Named in honor of Verena Tunnicliffe (OD).

Description. Shell medium-size (2.2 mm. Fig. 1024), thin, fragile; covered in thin deposit of iron oxide. Periostracum thin. Protoconch of approximately 0.5 whorl, bulbous, almost detached, finely pitted, irregular anastomosing ridges, no apertural varix, apertural margin convex. Teleoconch I of 0.5 whorl, with coarse irregular growth lines, a few very fine spirals, particularly at periphery. Teleoconch II of slightly more than 0.5 whorl, with approximately 20 spiral cordlets, crossed by fine, irregular growth lines. Selenizone with moderately elevated sharp keels. Slit closed forming foramen. Umbilicus open, wide, deep. Operculum thin, transparent, with central nucleus.

Radula (Fig. 1025). Ribbon slightly asymmetrical. Rachidian tooth broad triangular, central cusp twice as large as 2–3 smaller ones on either side. Lateral and marginal teeth poorly differentiated; number of lateral teeth ambiguous. Tips of lateral/marginal teeth almost symmetric, central tip strongest, 2–4 smaller cusps on either side. Outermost marginal teeth forming triangular fan with linear outline of shaft, fine outer denticulation. Radula interlock moderate.

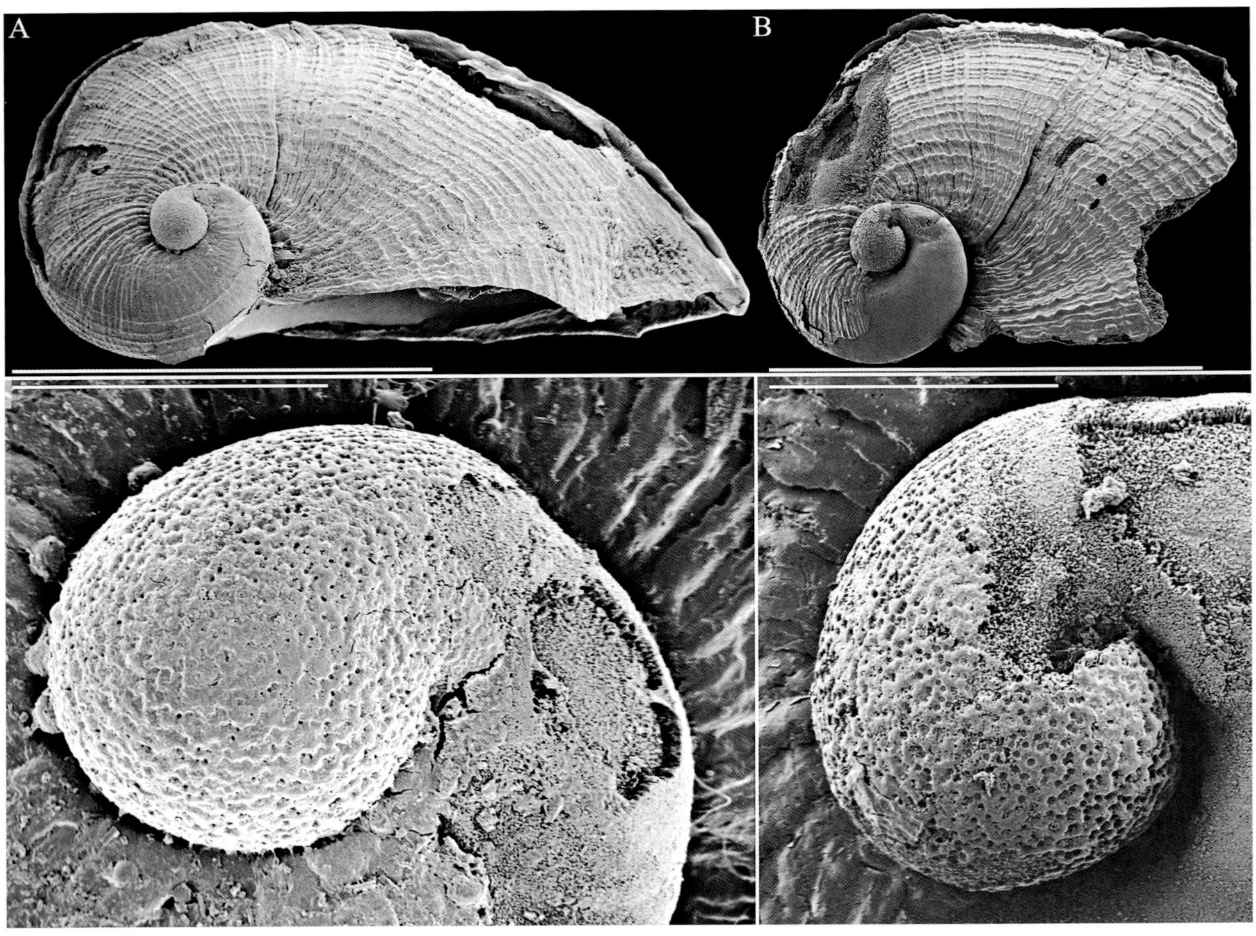

Figure 1024. *Sutilizona tunnicliffae*. 2202 m, Juan de Fuca Ridge, Endeavour Segment, off Washington state, USA, ROPOS #278 fissure at Main Field, 47°56.9'N, 129°6.9'W. **A.** Holotype. **B.** Topotypical specimen (see Remarks). Scale bars shell = 1 mm. Scale bars protoconch = 100 µm. Images by and with kind permission of A. Warén.

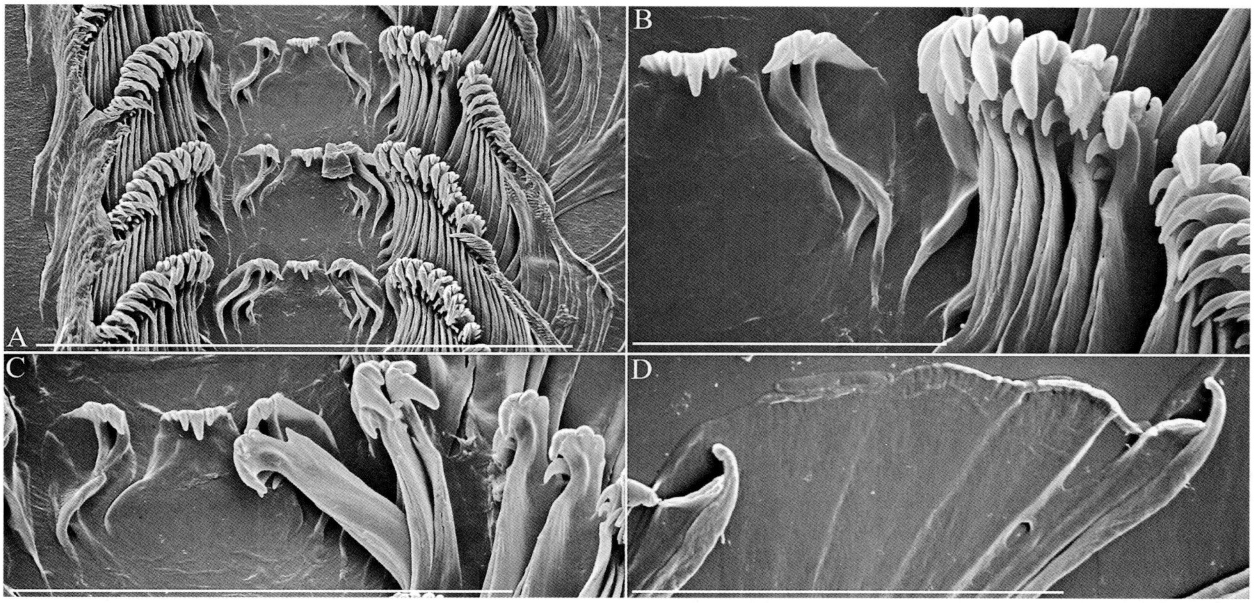

Figure 1025. *Sutilizona tunnicliffae*. Radula. **A.** Entire width of radula. Scale bar = 100 μm. **B.** Central field enlarged. Scale bar = 20 μm. **C.** Exposed shafts of lateral/inner marginal teeth. Scale bar = 50 μm. **D.** Outermost marginal teeth. Scale bar = 20 μm. Images by and with kind permission of A. Warén.

Comparisons. *Sutilizona theca* from the northeastern Pacific Rise has much stronger axial sculpture and the apex is more peripheral in location.

Sutilizona pterodon from the mid-northern Mid-Atlantic Rise has an open slit, and the shafts of the fan-shaped outer marginal teeth have a convex outline.

Distribution. Known only from type locality (Fig. 1026).

Remarks. The original description refers to a single known specimen, but the images kindly made available by A. Warén illustrate a second specimen of that species (Fig. 1024B), apparently from the same submersible dive. As this specimen was not mentioned in Warén & Bouchet (2001), it is not part of the type series (ICZN Art. 72.4.1); ICZN Art. 72.4.1.1. regarding consideration of unpublished information to determine the type series only applies to taxa established before 2000. The original description is sufficient to identify the species; accordingly, there is no need to designated a neotype.

Figure 1026. Distribution of *Sutilizona tunnicliffae*. ☆ type locality.

Sutilizonidae *Sutilizona*

Temnocinclidae McLean, 1989
Vetigastropoda *incertae sedis*: Lepetodriloidea?

Chresonymy.

Temnocinclinae McLean, 1989: 5.
Temnocinclinae: Marshall, 1993: 286.

Type genus. *Temnocinclis* McLean, 1989.

Other genera. *Temnozaga* McLean, 1989.
Triassurella Nützel & Geiger, 2006 †.

Description. Limpet-shaped shell with last whorl rapidly expanding. Usually with closed foramen. Protoconch finely reticulate, no apertural varix (unknown in *Temnocinclis*). Operculum multispiral, reduced. Radula without enlarged lateral tooth 5; distinction of lateral and marginal teeth difficult. Ctenidia two, monopectinate. Cephalic tentacles without papillae. No eyes.

Comparisons. Sutilizonidae (*Sutilizona*) have asymmetrical shells with the coiling axis at ~45° to the plane of the aperture.

In the calyptraeiform *Depressizona* (Depressizonidae), the coiling axis is perpendicular to the apertural plane.

Distribution. Northeast Pacific Rise at hydrothermal vents (Fig. 1027). For specific localities, see individual species.

Remarks. Temnocinclidae has been included with Sutilizonidae in the superfamily Lepetodriloidea by Warén & Bouchet (2001) and Warén *et al.* (2006). Warén & Bouchet *in* Bouchet & Rocroi (2005: 168) synonymized Temnocinclidae under Sutilizonidae without further justification applying first reviser's principle. Their opinion is not followed here. The differences in shell geometry, and coiling mode are considered sufficient to warrant recognition of those two lineages at the family level.

The systematic affinity of Temnocinclidae among Vetigastropoda is uncertain. The non-papillate nature of their cephalic and epipodial tentacles (both papillate in Scissurellidae and Anatomidae), as well as the radula with poorly differentiated lateral and marginal teeth (clearly differentiated in Scissurellidae and Anatomidae), suggest that Temnocinclidae is not closely related to Scissurellidae or Anatomidae. Lepetodrilidae share the non-papillate nature of the tentacles, but the radula of *Lepetodrilus* has clearly differentiated lateral and marginal teeth (see Warén & Bouchet, 2001). Whether the placement of Temnocinclidae in the larger group of Lepetodriloidea is justified or only creates a "waste basket taxon" remains to be determined.

Figure 1027. Distribution of Temnocinclidae. ☆ type locality. ● specimen record. ■ literature record.

Temnocinclis McLean, 1989

Chresonymy.

Temnocinclis McLean, 1989: 5–7.

Temnocinclis: Bandel, 1998: 49–50.

Type species. *Temnocinclis euripes* McLean, 1989 (OD).

Etymology. Temno-, Greek verb to cut; kinklis, Greek for lattice, referring to the shell sculpture. Gender feminine (OD).

Description. Shell large (to 4 mm), almost planispiral, limpet-shaped. Protoconch sculpture unknown. Last half whorl of teleoconch II rapidly expanding. Apex overhanging posterior end of shell. Operculum multispiral, reduced. No nacre.

Distribution. Northeastern Pacific Rise at hydrothermal vents (Fig. 1029). For specific localities, see individual species.

Geologic Range. Recent.

Comparisons. *Temnocinclis* has a shell with the apex overhanging the posterior end, while in *Temnozaga* the apical whorl does not reach beyond the posterior margin.

Temnocinclis euripes McLean, 1989

Figures 1028–1031

Chresonymy.

Temnocinclis euripes McLean, 1989: 7–9, figs 1A–H [shell and animal], 4A [shell], 5A–B [radula].

Temnocinclis euripes: Haszprunar, 1989: 3, fig. 1.

Temnocinclis euripes: Bouchet & Warén, 1997: 114, text-fig.

Temnocinclis euripes: Warén & Bouchet, 2001: 143.

Temnocinclis euripes: Geiger, 2003: 80.

Temnocinclis euripes: Geiger, 2006c: text-fig. 6.

Temnocinclis euripes: Nützel & Geiger, 2006: fig. 2A–C.

Temnocinclis euripes: Warén *et al.*, 2006: 99, text figs 1–2.

Temnocinclis euripes: Geiger, 2008b: fig. 1.

Type material. Holotype (USNM 859964. Fig. 1028), 3.8 mm. Paratype from type locality (LACM 2352a–b, 1: 2.7 mm; USNM 859965, 1). Southern Juan de Fuca Ridge, *Alvin* dive 1463, 44.653°N, 130.367°W (LACM 2353a, b, 2. MCZ 297060, 1. Fig. 1030B, 3.4 mm). 2213 m, Endeavour Segment, Juan de Fuca Ridge, *Alvin* dive 1447, 47.950°N, 129.100°W (LACM 2351: sectioned; USNM 859962, 1). 2208 m, Endeavour Segment, Endeavour Segment, Juan de Fuca Ridge, off Washington State, USA, *Pisces IV* dive 1452, 47.950°N, 129.100°W (USNM 859963, 1).

Type locality. 1575 m, hydrothermal vent community, Axial Seamount, Endeavour Segment, Juan de Fuca Ridge, off Washington State, USA, *Pisces IV* dive 1726, 45°59.5'N, 130°3.5'W (OD).

Etymology. Euripos, Greek for channel, referring to the slit (OD).

Description. Shell large (to 4.0 mm. Figs 1028–1030), almost planispiral, limpet-shaped, thin, transparent. Periostracum thick, brown. Protoconch 250 μm, sculpture unknown. Teleoconch I of 0.75 whorl. Teleoconch II of 0.75 whorl. First whorl of teleoconch I & II tightly coiled, last half whorl rapidly expanding. Strong axial cords intersected by weaker spiral cords, forming small nodules at intersection. Number of spiral cords increasing with growth forming uniform density of nodules. Umbilicus closed. Aperture large, peristome thin at periphery, thickened to form shelf in position of columella. Slit open, extending over 50% of shell length, converging towards anterior margin, slightly above mid-height of shell; keels distinct, but low. Muscle scars barely detectable.

Operculum (Fig. 1031E), multispiral, covering approximately ¼ of apertural area, exceedingly thin.

External anatomy (adapted from McLean, 1989). Cephalic tentacles without papillae, thick, short (contracted). No eyes. Three pairs of posterior epipodial tentacles, without papillae. Shell muscle horseshoe-shaped, terminators rounded. Ctenidia two, monopectinate, left slightly larger. Foot sole with double anterior edge marking opening of pedal gland. Operculum attached to metapodium, approximately ⅓ of shell length.

Radula (Fig. 1031). Ribbon almost symmetrical; teeth thin, weak. Rachidian tooth broad, tip with approximately 5–6 cusps on each side of enlarged central

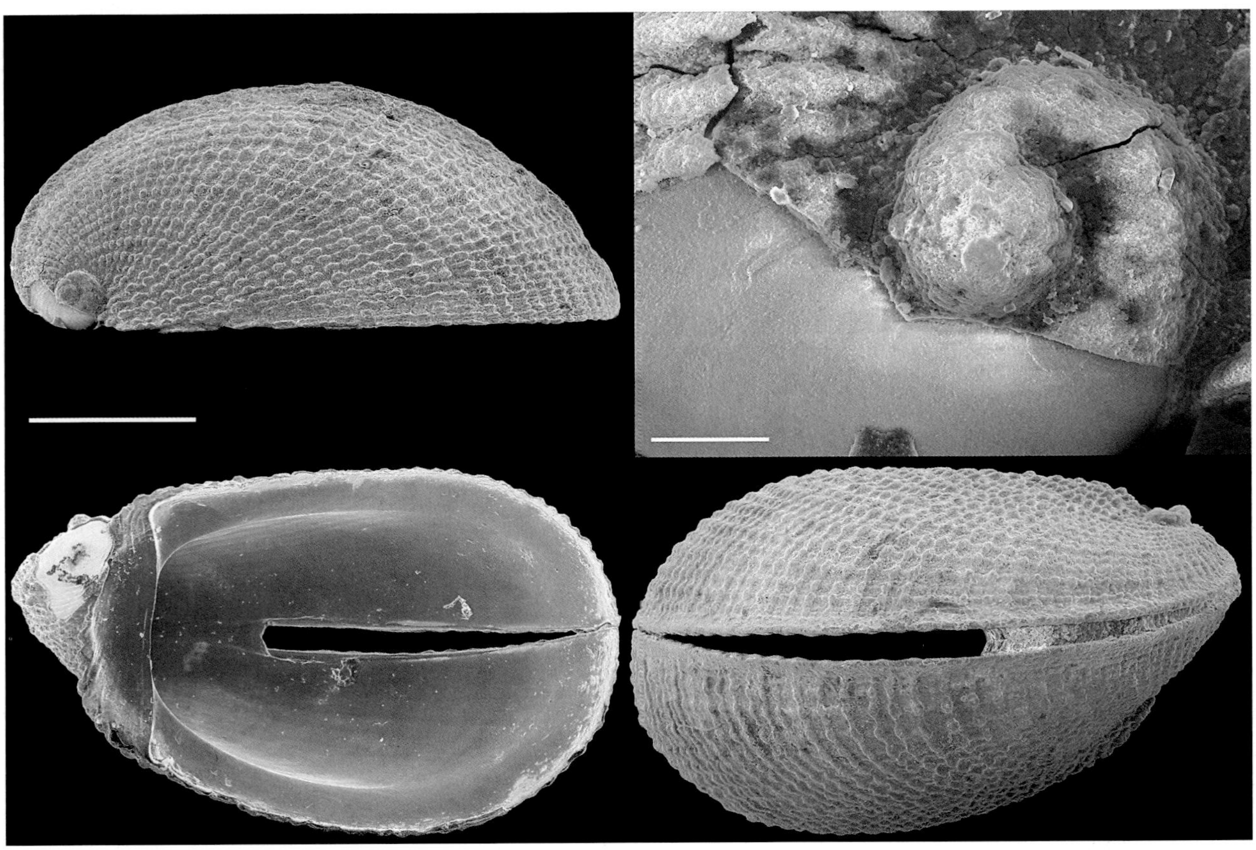

Figure 1028. *Temnocinclis euripes*. Holotype. 1575 m, Axial Seamount, Juan de Fuca Ridge, *Pisces IV* dive 1726, Washington State, USA, 45.992°N, 130.058°W (USNM 859964). Scale bar shell = 1 mm. Scale bar protoconch = 100 μm.

cusp. Distinction of lateral and marginal teeth difficult: number of lateral teeth ambiguous. Innermost lateral/marginal teeth with five cusps on tip, decreasing to three cusps with distance from rachidian tooth. Some weak radular interlock.

Distribution. Hydrothermal vent field off Western North America, Southern Canada to Oregon, 1500–2300 m (Fig. 1029).

Specimen records.

USA, Washington State. 2213 m, Endeavour Segment, Juan de Fuca Ridge, 47.950°N, 129.100°W (USNM 859962, 1). 2208 m, Endeavour Segment, Juan de Fuca Ridge, 47.950°N, 129.100°W (USNM 859963, 1). 2200 m, Clam Bed Vent Field, Endeavour Segment, Juan de Fuca Ridge, 47.963°N, 129.092°W (FMNH 279657, 2: complete). 2200 m, Clam Bed Vent Field, Endeavour Segment, Juan de Fuca Ridge, 47.963°N, 129.091°W (FMNH 279652, 1). 1575 m, Axial Seamount, Juan de Fuca Ridge, *Pisces IV* dive 1726, 45.992°N, 130.058°W (LACM 2352b, 1; USNM 859965, 1).

USA, Oregon. 2250 m, Juan de Fuca Ridge, off, 44.653°N, 130.367°W (MCZ 297060, 1: complete).

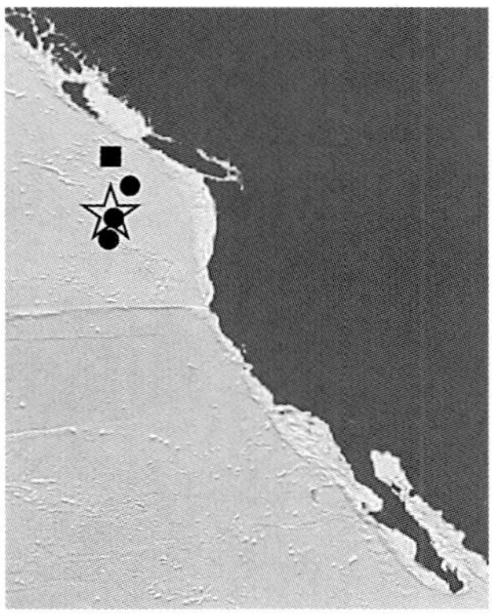

Figure 1029. Distribution of *Temnocinclis euripes*. ☆ type locality. ● specimen record. ■ literature record.

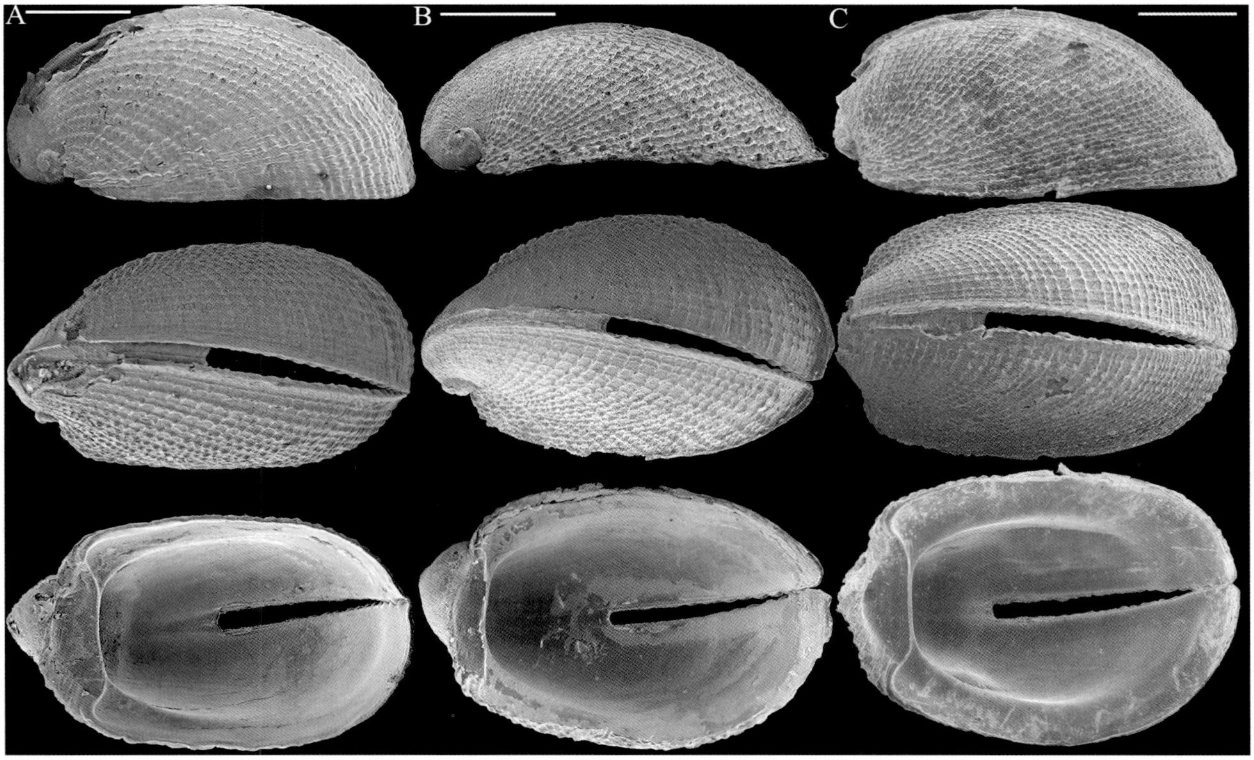

Figure 1030. *Temnocinclis euripes*. **A.** 2200 m, Clam Bed vent field, Endeavour Segment, Juan de Fuca Ridge, Washington State, USA, 47.963°N, 129.091°W (FMNH 279652, 1). Radula and operculum shown in Figure 1031. **B.** Paratype. 2250 m, Juan de Fuca Ridge, off Oregon, USA, 44.653°N, 130.367°W (MCZ 297060, 1). **C.** 2200 m, Juan de Fuca Ridge, Endeavour Segment, Clam Bed Vent Field, Washington State, USA, 47.963°N, 129.092°W (FMNH 279657, 2: complete). Scale bars = 1 mm. Protoconchs eroded.

Literature records.

Canada. Explorer Ridge. 49.750°N, 130.250°W (MNHN, 1) (Warén & Bouchet, 2001).
Eastern Pacific Rise. Megaplume South (MNHN, 3). Juan de Fuca Ridge (Warén & Bouchet, 2001).

Remarks. Paratype MCZ 267060 is annotated that specimen label and data given in original description do not match.

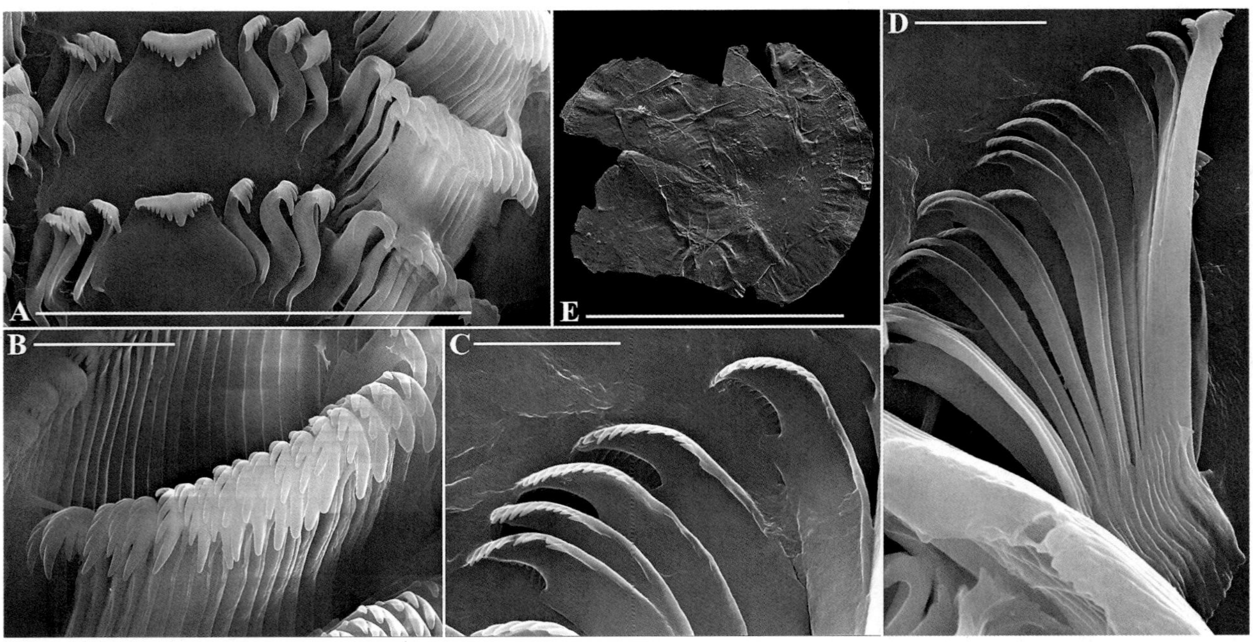

Figure 1031. *Temnocinclis euripes*. Radula and operculum from specimen shown in Figure 1030A. 2200 m, Clam Bed vent field, Endeavour Segment, Juan de Fuca Ridge, Washington, USA, 47.963°N, 129.091°W (FMNH 279652, 2: complete). **A–D.** Radula. **A.** Half-row of radula. Scale bar = 100 µm. **B.** Marginal teeth in apical view. Scale bar = 20 µm. **C.** Tips of marginal teeth in lateral view. Scale bar = 20 µm. **D.** Entire marginal teeth with shafts. Scale bar = 20 µm. **E.** Operculum. SEM in high vacuum, secondary electron detector with bias at -50V turning it into a low angle backscatter detector. Scale bar = 1 mm.

Temnozaga McLean, 1989

Chresonymy.

Temnozaga McLean, 1989: 9.

Temnozaga: Bandel, 1998: 50.

Type species. *Temnozaga parilis* McLean, 1989 (OD).

Etymology. Temno-: Greek verb to cut; zaga: part of zig-zag, referring to the deeply cut slit and the interlocking projections on the inside margin of the open slit. Gender neuter (OD).

Distribution. Northeastern Pacific Rise (Fig. 1034). For specific localities, see individual species.

Geologic range. Recent.

Description. Shell limpet-shaped, with rapidly expanding last whorl, almost symmetrical. Apex not overhanging posterior margin of shell. Protoconch smooth, with reticulated sculpture, no apertural rim, almost detached. Operculum multispiral, vestigial.

Comparisons. In *Temnocinclis*, the apex overhangs the posterior margin of the shell.

Temnozaga parilis McLean, 1989

Figures 1032–1034

Chresonymy.

Rimula (?): Turner *et al.*, 1985: figs 11a–c (juvenile shell and protoconch).

Temnozaga parilis McLean, 1989: 9–11, figs 2A–H [shell, animal], 4B–C [shell, protoconch], 5C–D [radula].

Temnozaga parilis: Haszprunar, 1989: 3.

Temnozaga parilis: Skoglund, 1992: 4.

Temnozaga parilis: Bouchet & Warén, 1997: 115, text-fig.

Temnozaga parilis: Skoglund, 2002: 3.

Temnozaga parilis: Geiger, 2003: 80.

Temnozaga parilis: Geiger, 2008b: fig. 1.

Temnozaga parilis: Warén *et al.*, 2006: 100: text-fig. 1.

Temnozaga parilis: Adams *et al.*, 2010: 1054.

Type material. Holotype (USNM 859966: body sectioned. Fig. 1032), 4.1 × 3.0 × 1.7 mm (L × H × W). Paratype from type locality (LACM 2354, 1: body only used for radula extraction), 1.0 mm

Type locality. 2616 m, East Pacific Rise at 21°N, 20°50.0'N, 109°6.0'W (OD).

Etymology. Latin for equal, referring to the symmetry of the teleoconch (OD).

Description. Shell large (to 4.0 mm. Fig. 1032), limpet-shaped with rapidly expanding body whorl, nearly symmetrical. Periostracum thick. Protoconch smooth, no apertural varix, almost entirely separate from teleoconch. Teleoconch I of 0.33–0.5 whorl, smooth except for faint growth lines. Teleoconch II of 0.75–0.625 whorl. Sculpture of approximately 30 low rounded spiral primary ribs, intersected by indistinct axial lines, at intersection raised to prominent scales. Between primary ribs 3–4 secondary and tertiary ribs. Slit open over ⅔ of shell length, margins

Temnocinclidae
Temnozaga

Figure 1032. *Temnozaga parilis*. Holotype. 2616 m, East Pacific Rise at 21°N, 20°50.0'N, 109°6.0'W (USNM 859966). Scale bar shell = 1 mm. Scale bar protoconch = 100 μm.

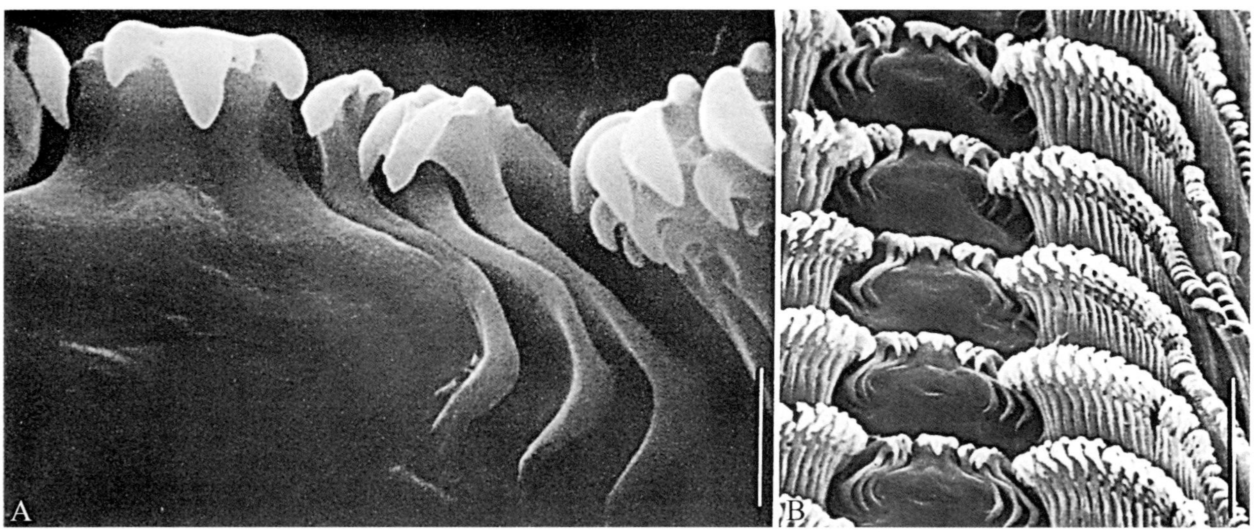

Figure 1033. *Temnozaga parilis*. Radula. 2616 m, East Pacific Rise at 21°N, 20°50.0'N, 109°6.0'W (LACM 2354, 1). **A.** Central field. Scale bar = 4 μm. **B.** Entire half row. Scale bar 20 μm. From McLean (1989).

converging towards aperture, near aperture beads on inside of slit alternating on either side forming zig-zag or zipper pattern. Inside of slit border thickened by callus; keels broad, low. Umbilicus closed. Aperture large, rounded oblong, almost in one plane, with anterior and posterior margin slightly raised; posterior margin thickened to form columella. Muscle scar barely detectable. Operculum present, multispiral (>six whorls), diameter 1.3 mm covering approximately one-quarter of apertural area.

External anatomy (adapted from McLean, 1989). Body nearly symmetrical. Cephalic tentacles short, thick, contracted, without papillae. No eyes. Four pairs of posterior epipodial tentacles, without papillae, decreasing in size posteriorly. Shell muscle horseshoe-shaped, terminators rounded. Two monopectinate ctenidia of equal size. Foot sole with double edge marking opening of pedal gland.

Radula (Fig. 1033). Ribbon slightly asymmetric. Rachidian tooth with three cusps on tip, central cusp larger, base 3× as broad as tip. Three lateral teeth, with two cusps each, inner cusp much larger. Marginal teeth numerous, decreasing in size towards margin, shafts much longer than in lateral teeth. Radular interlock moderate.

Figure 1034. Distribution of *Temnozaga parilis*. ☆ type locality. ● specimen record.

Comparisons. See genus.

Distribution. Known only from type locality (Fig. 1034).

Remarks. The protoconch of the paratype is eroded (Geiger, 2003).

Triassurella Nützel & Geiger, 2006 †

Temnocinclidae *Triassurella*

Chresonymy.

Triassurella Nützel & Geiger, 2006: 279.
Triassurella: Kiel *et al.*, 2010: 28.

Type species. *Triassurella carnica* † (OD).

Other species. *Triassurella goederti* Kiel, Campbell & Gaillard, 2010 †.

Etymology. After the Triassic period and the ending of *Scissurella* (OD).

Description. Shell with high whorl expansion rate, forming wide aperture and asymmetrical cap shape; selenizone with constant width, situated asymmetrically on right side (dorsal view) of shell; strong ornament of broad radial ribs, about 18 visible in dorsal view. Slit open or closed. Aperture large, oval.

Distribution. Triassic of Italy, Cretaceous of Washington State (Fig. 1035). For specific localities, see individual species.

Geologic range. Triassic to Cretaceous.

Comparisons. *Triassurella* resembles several extant hydrothermal vent limpets, particularly *Temnocinclis*. However, the shell of *Temnocinclis* is more elongated, has stronger concentric ornamentation, the slit is open for approximately 40% of total shell length, the inner margin of the slit shows distinct crenulations, and the columellar region contains two distinct shelves. Some *Incisura* have a similar overall morphology although they are more trochiform than *Triassurella*, and lack any strong ornament and have a much later onset of the selenizone. The monotypic genus *Zardinitrochus* (see Introduction for taxonomic-nomenclatural remarks) is more trochi-stomatelliform, with a third of the whorl extending past the posterior margin of the shell's coiling axis.

Differentiation of species. Species are differentiated by details of the sculpture.

Remarks. The systematic affinity of the genus is debatable. Nützel & Geiger (2006) compared *Triassurella* to the hydrothermal vent family Temnocinclidae. Kiel *et al.* (2010) placed *Triassurella* in the hydrothermal vent family Sutilizonidae, most likely following Warén & Bouchet *in* Bouchet & Rocroi (2005: 168) in synonymizing Temnocinclidae under Sutilizonidae applying first reviser's principle. Here the genus is tentatively assigned to Temnocinclidae based on similarities in overall shell geometry. The coiling axis and the apertural plane of Temnocinclidae as well as *Triassurella* are at an angle of 70–80° (i.e., close to at right angle), while in Sutilizonidae it is at an angle of ~45°. While *Tri. carnica* was placed in a shallow tropical lagoon paleosetting, *Tri. goederti* was found in a confirmed cold-seep biome. Given the difficulty of determining fossil hydrothermal vent environments, the

Figure 1035. Distribution of *Triassurella* species. ☆ type locality. ● specimen record.

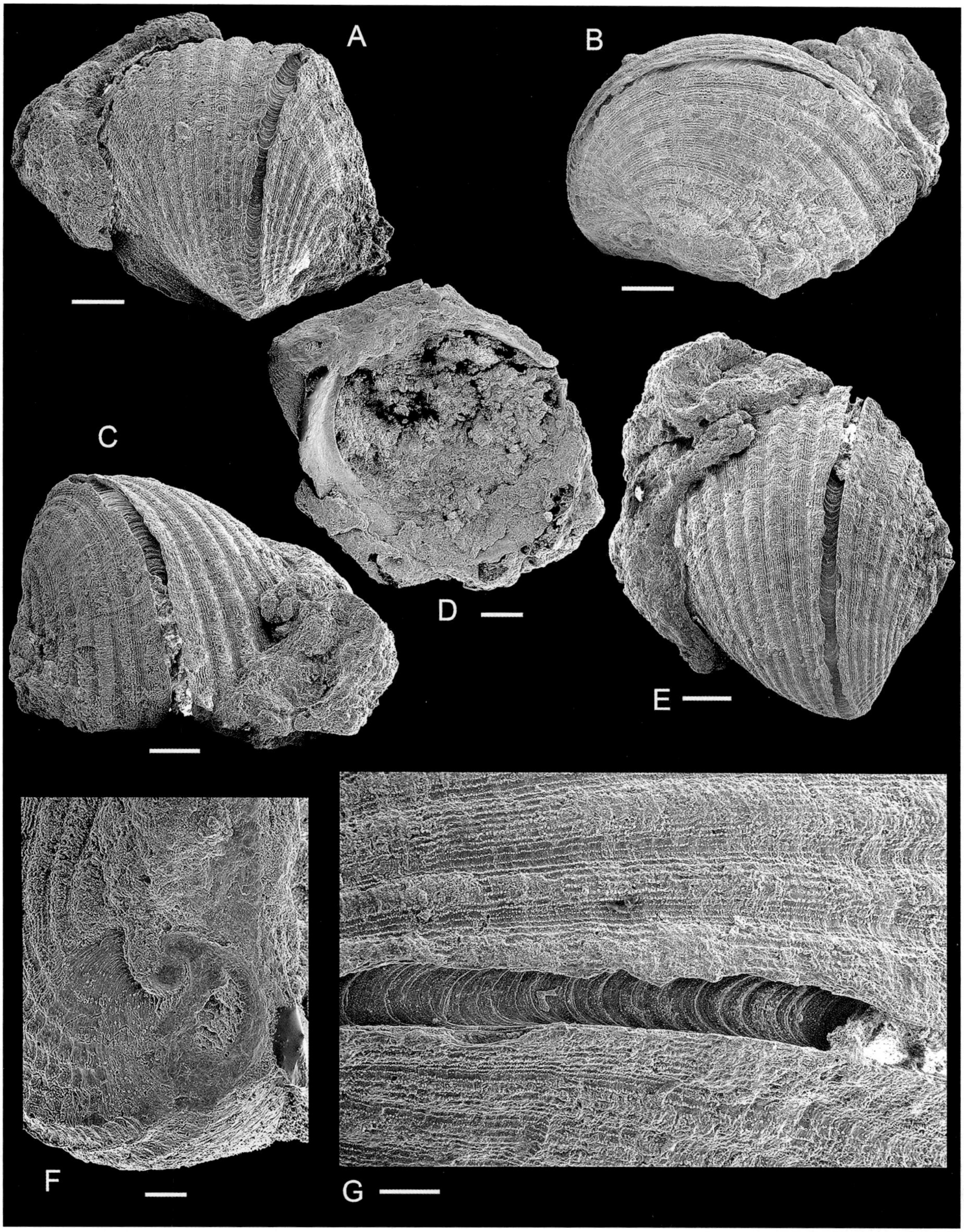

Figure 1036. *Triassurella carnica*. Holotype. Carnian St. Cassian Formation, near Misurina, 46.591°N, 12.261°E, Italy (NHMB MB.Ga.2388). **A.** Oblique view, posterior. **B.** Lateral view. **C.** Anterior view. **D.** Apertural view. **E.** Dorsal view. **F.** Protoconch with onset of teleoconch and selenizone. **G.** Detail of selenizone and teleoconch ornament. Scale bars A–E = 300 µm. Scale bars F–G = 100 µm. Images by A. Nützel, from Nützel & Geiger (2006).

indication of Kiel *et al* (2010) is taken as a convincing supplementary fact to permit placement of *Triassurella* in Temnocinclidae restricted to Recent chemosynthetic environments.

Triassurella carnica Nützel & Geiger, 2006 †

Figures 1036–1037

Chresonymy.

Triassurella carnica Nützel & Geiger, 2006: 279, fig. 1 (Fig. 1036).
Triassurella carnica: Bandel, 2009: 20.
Triassurella carnica: Kiel *et al.*, 2010: 28.

Type material. Holotype (NHMB MB.Ga.2388. Fig. 1036).

Type locality. Misurina, near Cortina d'Ampezzo, northern Italy, 46.591°N, 12.261°E. St. Cassian Formation, early Carnian, Upper Triassic, aonoides-zone (OD).

Etymology. Named after the Carnian stage (OD).

Description. Shell medium-sized (2 mm) coiled, dextral of about 1.6 whorls with a very high whorl expansion rate producing wide aperture and flat limpet shape; shell asymmetrical, trapezoidal in dorsal view with coiled apex at posterior end; protoconch plus teleoconch I approximately one whorl; teleoconch II approximately 0.6 whorl, early teleoconch II with cancellate sculpture, remainder with 18 strong broad, nodulated spiral ribs, running parallel to selenizone; additionally shell covered with micro-ornament of minute dots, arranged in fine, spiral threads; distinct selenizone situated asymmetrically on right side (dorsal view) of shell, running parallel to strong ribs; keels near selenizone forming slight crest; slit open for quarter of shell length, with parallel margins; aperture large, oval; posterior margin with thin crescent-shaped shelf.

Comparisons. *Triassurella goederti* from the Cretaceous of Washington State, USA, has finer spiral cords, and a foramen is formed.

Distribution. Known only from holotype at type locality (Fig. 1037).

Remarks. The early growth stage of the shell is too insufficiently preserved to provide a detailed account. Even the distinction between protoconch on teleoconch I, which is usually quite distinct cannot be ascertained. Interestingly enough, for the most similar genus, the Recent *Temnocinclis*, the protoconch sculpture is also still unknown. The expansion of the posterior apertural margin into a thin crescent-shaped shelf is reminiscent of the fissurellid *Zeidora*. However, *Zeidora* species are almost perfectly bilaterally symmetrical and more elongated than *Triassurella*. A crescent-shaped apertural shelf is also found in very mature specimens of the Recent *Sat. tabulata*.

Triassurella carnica resembles *Emarginula abnormis* Stoppani, 1860, from the Ladinian Esino Limestone (South Alps) as illustrated by Stoppani (1858–1860) and Kittl (1899) in having a dominant radial ornament and in the position of the selenizone, which is situated distinctly to the right of the middle of the shell. However, this species is known from a few steinkerns only so that details of the shell

Figure 1037. Distribution of *Triassurella carnica*. ☆ type locality.

ornament are unknown. As it may be, *E. abnormis* is much higher in lateral view and more symmetrical in dorsal view than *Tri. carnica*. Other early Mesozoic fissurelloid species are much higher limpet-shaped and have strong cancellate ornaments (e.g., *Emarginula muensteri* Pictet, 1855, and others, see Zardini, 1978).

Triassurella goederti Kiel, Campbell & Gaillard, 2010 †

Figures 1038–1039

Chresonymy.

Triassurella goederti Kiel *et al.*, 2010: 28–29, fig. 1 (Fig. 1038).

Type material. Holotype (USNM 538733, 2.6 mm. Fig. 1038A–C). Paratypes (USNM 538734, 1: 3.5 mm. Fig. 1038D–E; USNM 538735, 1: 1.6 mm. Fig. 1038F), .

Type locality. The Bear Creek seep site, California, USA; Crack Canyon Formation of the Great Valley Group 39°03'14" N 122°24'40" W; Lower Cretaceous, Valanginian (OD).

Etymology. For James L. Goedert, Wauna, Washington, who collected and provided the specimens (OD).

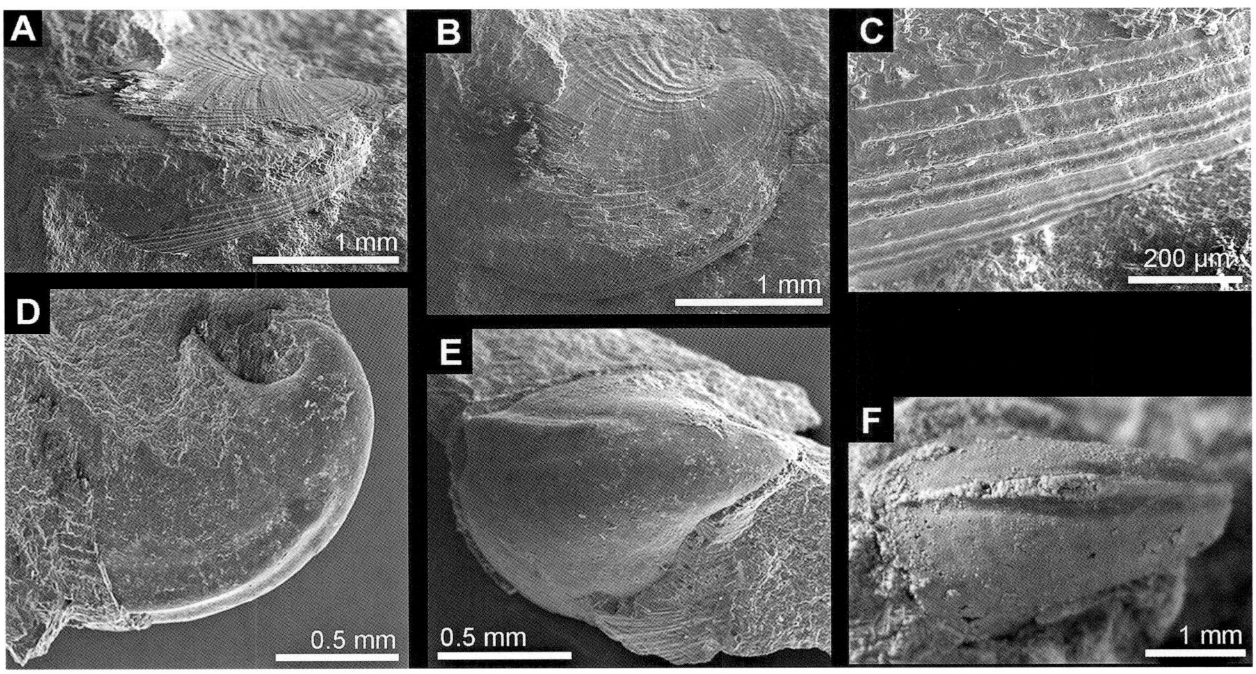

Figure 1038. Original illustration of *Triassurella goederti*. The Bear Creek seep site, California, USA; Crack Canyon Formation of the Great Valley Group 39°03'14"N 122°24'40"W; Lower Cretaceous, Valanginian **A–C.** Holotype (USNM 538733) with preserved shell material; note detail of sculpture on whorl's flank (C). **D–E.** Paratype (USNM 538735), internal mold of juvenile specimen showing onset of slit. **F.** Paratype (USNM 538734), steinkern showing the slit as it widens abruptly and then tapers towards the aperture. From Kiel *et al.* (2010) with kind permission of S. Kiel.

Description. Shell medium-sized (to 3.5 mm), shape of asymmetrical limpet. Protoconch unknown. Teleoconch of at least 0.5 whorl, rapidly expanding. Shoulder with fine spiral cords of unequal strength, stronger near selenizone and near suture, no axial sculpture. Details of selenizone unknown, foramen closed anteriorly.

Comparisons. *Triassurella carnica* from the Triassic of Italy has much broader spiral cords, and the slit remains open.

Distribution. Known from type locality only (Fig. 1039).

Remarks. Some additional specimens have been collected at the type locality in California (S. Kiel, pers. comm. 2/2012).

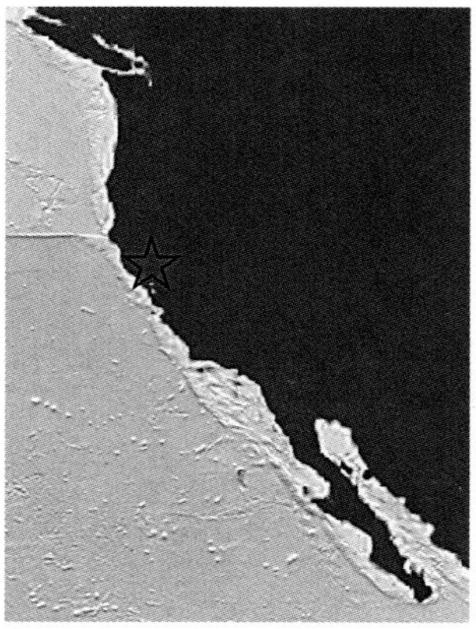

Figure 1039. Distribution of *Triassurella goederti*. ☆ type locality.

Problem specimens

Despite best efforts, some problem specimens of uncertain identity remain. They are discussed here, but neither assigned to a described species, nor described as new. To resolve the identity of those specimens will require additional material.

***Scissurella* 1.** The specimen (Fig. 1040A) is similar to *Sin. calumnior* from the Indo-Malayan Archipelago to Central Pacific. It shares the overall shell shape, the identical protoconch (strong and relatively widely spaced axial cords, an apertural varix, and a straight apertural margin), the narrow umbilicus with periumbilical cord, and the teleoconch I whorl count of one. However, it has overall denser and weaker axial sculpture, a longer selenizone (0.4 vs. 0.3 whorls), and retains an open slit with parallel margins even with the longer selenizone. It differs from *Sci. phenax* by the protoconch lacking a median spiral cord, having weaker axial sculpture, and by the more narrow umbilicus.

Specimen records.

Australia. Tumby Bay, Eyre Peninsula, South Australia, 34.383°S, 136.117°E (AMS C.378555, 2). Great Barrier Reef, Capricorn Group, Heron Island, Queensland, 23.433°S, 151.950°E (AMS C.406378, 1).

***Sinezona* 2.** This unique specimen from 30 m, off Spirits Bay, North Island, New Zealand, 34.397°S, 172.760°E (NMNZ M.174854, 1. Fig. 1040B) shows some affinities with *Sin. calumnior*. It shares the squat overall shell shape, the protoconch (strong and relatively widely spaced axial cords, an apertural varix, and a straight apertural

Figure 1040. Unidentified Scissurellidae. **A.** *Scissurella* sp. 1. Tumby Bay, Eyre Peninsula, South Australia, Australia, 34.383°S, 136.117°E (AMS C.378555, 2). Scale bar shell = 0.5 mm. **B.** *Sinezona* sp. 2. 30 m, off Spirits Bay, North Island, New Zealand, 34.397°S, 172.760°E (NMNZ M.174854, 1). Scale bar shell = 0.5 mm. **C.** Scissurellidae 3. Great Barrier Reef, Capricorn Group, Wilson Island, N side, Queensland, Australia, 23.300°S, 151.950°E (AMS C.377899, 1). Scale bar = 0.25 mm. Scale bars all protoconch = 100 μm.

Problem Specimens

margin), and the narrow umbilicus. However, it has a shorter teleoconch I (0.75 vs. 1 whorl) and overall weaker sculpture. This specimen is known in the NMNZ collection as *Sinezona* sp. Ξ.

Scissurellidae 3. This juvenile specimen without a slit from Great Barrier Reef, Capricorn Group, Wilson Island, N side, Queensland, Australia, 23.300°S, 151.950°E (AMS 377899, 1. Fig. 1040C) appears to be a scissurellid. It has an uncommon protoconch configuration with the axials alternatively reaching past the mid-point and stopping short on the adsutural slope; it is also known from *Scissurella* species1 (Fig. 1040A) from temperate South Australia, and *Sin. calumnior*; it differs from the latter from the Indo-Malayan Archipelago to the Central Pacific by the more numerous spiral lines on teleoconch I. Strong axials on the protoconch are rare among tropical scissurellids; most either show fine axials (*Sci. evaensis, Sci. lorenzi, Sci. mirifica, Sci. quadrata, Sin. wileyi*) or have smooth protoconchs or fine spirals (*Sci. cebuana, Sci. xandaros, Sin. ferriezi, Sin. globosa, Sin. macleani, Sin. plicata*). The only species with strong axials on the protoconch as well as strong axial cords to lamellae on teleoconch I of at least one whorl is *Sin. marrowi*. However, it has some weakly developed median thickening on the axials of the protoconch not evident in the specimen from Wilson Island. *Sinezona danieldrieri* has moderately strong axials, of which some may be bifurcated, but has a shorter teleoconch I (0.75 vs. 1 whorl).

Problem specimens

***Satondella* species 1.** Two broken specimens of a species with flat spire, strong axial lamellae, and a protoconch with flocculent sculpture and an apertural varix are distinct (Fig. 1041A). The overall shell shape suggests membership in the genus *Satondella*, but none of the species in the tropical Pacific has a protoconch with flocculent sculpture and an apertural varix.

Specimen records.

New Caledonia. 455 m, Southern, 22.267°S, 166.183°E (MNHN, 1).
French Polynesia. 36 m, Around Rukuaga Point, Rapa, Austral Islands, 27.568°S, 144.368°W (MNHN, 1).

***Sinezona* aff *platyspira*.** A single specimen not in best state of preservation is known (Fig. 1041B). It shows affinities to *Sin. platyspira*: same overall shell shape, same protoconch, teleoconch I of >1 whorl, axial lamellae. But it has a closed foramen already after 0.25 teleoconch II whorl, while *Sin. platyspira* has a teleoconch of 0.66 whorl. In most *Sinezona* species, the slit is open in early growth stages. Second, the axial lamellae are restricted to the upper portion of the shoulder, while in *Sin. platyspira* they reach to the umbilicus particularly in earlier ontogenetic stages. Third, the early teleoconch I has hardly any axial sculpture, while *Sin. platyspira* shows well-developed and dense axial lamellae, which are unlikely to be obliterated by wear.

Sinezona platyspira is known only from two specimens; hence, the question of intraspecific variability is open. It seems that the single *Sin.* aff. *platyspira* is not conspecific with it. Given that New Zealand is a fairly well collected and collectable region, and that the specimen is not in good state of preservation, I abstain from describing it as new.

Specimen records.

New Zealand. 38 m, Arch Pinnacle, Princes Rocks, Three Kings Islands, 34.175°S, 172.050°E (NMNZ M.174857, 1).

Figure 1041. Unidentified Scissurellidae. **A.** Cf. *Satondella* sp. 455 m, Southern, New Caledonia, 22.267°S, 166.183°E (MNHN, 1). **B.** *Sinezona* aff. *platyspira*. 38 m, Arch Pinnacle, Princes Rocks, Three Kings Islands, New Zealand, 34.175°S, 172.050°E (NMNZ M.174857, 1). Scale bars shell = 0.5 mm. Scale bars protoconch = 100 μm.

Anatoma species 1. A single *Anatoma* specimen from the Caribbean (Fig. 1042A) does not fit any of the species known from there (*An. aedonia, An. alta, An. plicatazona, An. proxima, An. rolani*). It is most similar to *An. aedonia*, but has at least twice as dense axial and spiral sculpture and lacks a spiral cord in the position of the selenizone on teleoconch I. It is even more similar to *An. eximia* from the Western Mediterranean and the tropical/subtropical eastern Atlantic, which, has only about at least 50% less dense axial and spiral sculpture, a steeper slope of the shoulder, and also has a spiral cord in the position of the selenizone on teleoconch I. Whether *An. eximia* is a species that ranges like *An. alta* from the Caribbean to the eastern Atlantic, or whether this specimen represents a yet undescribed species remains to be determined. *Anatoma exima* is only know from relatively few specimens and intraspecific variability is not fully assessed.

Anatoma species 2. Four lots from Fiji, Indonesia, Philippines, and Japan are presently considered to belong to this species. The material is not in the best state of preservation, and it can neither be ruled out that the specimens belong to more than one species, nor that they are simply variants of already described species. The most notable character is the fine dot pattern on the shoulder produced by the intersection of axial and spiral sculptural elements. Figure 1042B shows the most distinct of the available specimens. The degree of the fine dot pattern varies among the specimens; hence, those may simply be peculiar specimens of some Indo-Pacific species.

Specimen records.

Fiji. 251–257 m, Bligh Water, 16.857°S, 178.232°E (MNHN, 1).
Indonesia. 175 m, SW Salayer, 6.373°S, 120.437°E (RMNH, 5).
Japan. 401–404 m, Nago Bay, Okinawa, 26.525°N, 127.760°E (SBMNH 149456, 1).
Philippines. 100 m, Manga, Bohol Island, PANGLAO 2004 L48, 9.705°N, 123.832°E (MNHN, 1).

Figure 1042. Unidentified *Anatoma* species. **A.** *Anatoma* species 1. 165–183 m, off Lazaretto, Barbados, 13.133°N, 59.617°W (USNM 500682, 1). **B.** *Anatoma* species 2. 251–257 m, Bligh Water, Fiji, 16.857°S, 178.232°E (MNHN, 1). Scale bars shell = 1 mm. Scale bars all protoconch = 100 μm.

Literature cited

Aartsen, J. J. van & R. Kinzelbach. 1990. Marine molluscs from the Iztuzu beach near Dalyan (Mediterranean coast of Turkey). *Zoology in the Middle East* 4: 103–112.

Aartsen, J. J. van, M. P. M. G. Menkhorst & E. Gittenberg. 1984. The marine Mollusca of the Bay of Algeciras, Spain, with general notes on *Mitrella*, Marginellidae and Turridae. *Basteria* Supplement 2: 1–135.

Abbott, R. T. 1954. *American Seashells, First Edition*. Van Nostrand, Princeton, New Jersey. 541 pp.

Abbott, R. T. 1974. *American Seashells, Second Edition*. Van Nostrand, Princeton, New Jersey. 663 pp.

Absalão, R. S. 1989. Distribution patterns and zoogeography of the mollusks from the Brazilian continental Shelf. Part III. Oceanographic Expedition Espírito Santo I. *Memoria Instituto Oswalso Cruz* 84, Supplement IV: 1–6.

Adams, A. 1862. On some new species of Scissurellidae from the Seas of China and Japan. *Annals and Magazine of Natural History* Series 3, 10: 346–348.

Adams, H. & A. Adams. 1853–1858. *The Genera of Recent Mollusca Arranged According to Their Organization*, 3 Volumes. John van Voorst, London. 448 pp., 661 pp., 138 pls.

Adams, D. K., S. W. Mills, T. M. Shank & L. S. Mullineux. 2010. Expanding dispersal studies at hydrothermal vents through species identification of cryptic larval forms. *Marine Biology* 157: 1049–1062.

Adock, D. J. 1893. *A Hand List of the Aquatic Mollusca Inhabiting South Australia*. D. J. Adock, Adelaide. 14 pp. [not seen].

Aktipis, S. W. & G. Giribet. 2012. Testing relationships among the vetigastropod taxa: a molecular approach. *Journal of Molluscan Studies* 78: 12–27.

Albanesi, O., B. Berardelli, M. Consenza, P. Crovato, G. Fasulo, R. Ferro, F. Izzillo, M. Miline, E. Perna, F. Toscano & G. Spada. 1986. Malacofauna marina della Campania — Area 042. *Notiziario SIM* 4(7–8): 123–12.

Albuquerque, M., J. P. Borges & G. Callado. 2009. *Moluscos Marinhos Atlas das Ilhas Selvagens*. Direcção Regional do Ambiente, Funchal. 309 pp.

Aldea, C. & J. S. Troncoso. 2008. Systematics and distribution of shelled molluscs (Gastropoda, Bivalvia and Scaphopoda) from the South Shetland Islands to the Bellingshausen Sea, West Antarctica. *Iberus* 26: 43–117.

Allan, J. 1934. Australian Shells. Slit shells and false limpets. *The Australian Museum Magazine 1934*: 277–282.

Amitrov, O. V. 1996. Scissurellids (Gastropoda, Scissurellidae) from the Upper Eocene of Ukraine. *Ruthenica* 5: 93–104.

Amitrov, O. V. & E. A. Zhagallo. 1998. On the protoconchs of Ukrainian Eocene Scissurellidae (Gastropoda). *Ruthenica* 8: 7–11.

Angas, G. F. 1878. A list of additional species of marine Mollusca to be included in the fauna of the province of South Australia; with notes on their habitats and local distribution. *Proceedings of the Zoological Society, London* 1878: 864–871.

Anistratenko, V. v. & Y. I. Starobogatov. 1997. Which species of *Scissurella* (Gastropoda: Pleurotomariiformes) inhabits the Black Sea? *Vestnik Zoologii* 31: 75–77.

Anonymous. 1899. Summary of the fauna of Funafuti. *Memoirs of the Australian Museum* 3: 513–535.

Anonymous. 1906. Malacological section of the Royal Society of South Australia (Incorporated). Annual Report for 1905–6. *Transactions of the Royal Society of South Australia* 49: 364–367.

Anonymous. 2002. National Science Foundation: science hard. *The Onion* Issue 45-01: [http://www.theonion.com/articles/national-science-foundation-science-hard,1405/: accessed 8/2/2010].

Apellius, F. L. 1869. Le conchiglie del Mar Tirreno, parte seconda. *Bolletino Malacologico Italiano* 2: 177–206.

Appelius, F. L. 1870. Catalogo delle conchiglie fossili del Livornese, desunto dalle collezioni e manoscritti del defunto G. B. Caterini. *Bollettino Malacologico Italiano* 3: 177–297.

Aradas, A. & L. Benoit. 1870. *Conchigliologia Vivente Marina della Sicilia e delle Isole che la Circondano.* Galatola, Catania. 324 pp., 5 pls.

Ardovini, R. & T. Cossignani. 1999. *Atlante delle Conchiglie de Profundità de Mediterraneo.* L'Informatore Piceno, Ancona, Italy. 111 pp.

Ardovini, R. & T. Cossignani. 2004. *West African Seashells (including Azores, Madeira and Canary Is.).* L'Informatore Piceno, Ancona. 320 pp.

Arduino, G., B. Locatelli, F. Orlando, & G. Repetto. 1995. *Catalogo Illustrato delle Conchiglie Marine del Mediterraneo.* Associazione Amici del Museo "Frederico Eusebio", Alba, Italy. 173 pp.

Arnaud, M. 1972a. Sur une petite collection de gastéropodes prosobranches et de pélécypodes de l'île Petermann (Antarctique). *Tethys* 4: 429–436.

Arnaud, P. M. 1972b. Invertébrés marins des XIIème et XVème expéditions antarctiques françaises en Terre Adélie 8. — Gastéropodes prosobranches. *Tethys* Supplement 4: 105–134.

Arnaud, P. M. 1974. Contribution à la bionomie marine benthique des régions antarctiques et subantarctiques. *Téthys* 6: 467–653.

Arntz, W. E. & T. Brey. 2003. Expedition ANTARKTIS XIX/5 (LAMPOS) of RV "Polarstern" in 2002. *Berichte zur Polarforschung* 462: 1–120.

Audouin, V. 1828. *Explication Sommaire des Planches de Mollusques de I'Egypte et de la Syrie Publiées par J. C. Savigny. Description de I'Égypte ou Recueil des Observations et des Recherches qui ont été Faites en Égypte Pendant I'Expédition de l'Armée Française, Publié par les Ordres de sa Majesté I'Empéreur Napoléon le Grand. Histoire Naturelle, Animaux Invertébrés. 2ème Édition, Dediée au Roi.* Imprimérie impériale, Paris. l(4): 7–56.

Ault, J. S. 1985. Some quantitative aspects of reproduction and growth of the red abalone, *H. rufescens* Swainson. *Journal of the World Mariculture Society* 16: 398–425.

Ávila, S. P., R. Amen, J. M. N Azevedo, M. Cachão & F. García-Talavera. 2002. Checklist of the Pleistocene marine molluscs of Praínha and Lagoínhas (Santa Maria Island, Azores). *Açoreana* 9: 343–370.

Ávila, S. P. & J. M. N. Azevedo. 1997. Shallow-water molluscs of the Formigas Islets, Azores, collected during the "Santa Maria e Formigas 1990" Scientific Expedition. *Açoreana* 8: 323–330.

Ávila, S. P., J. M. N. Azevedo, J. M. Goncalves, J. Fontes & F. Cardigos. 1998. Checklist of the shallow-water marine molluscs of the Azores: 1 — Pico, Faial, Flores and Corvo. *Açoreana* 8: 487–523.

Ávila, S. P., F. Cardigos & R. S. Santos. 2004. D. João de Castro Bank, a shallow water hydrothermal-vent in the Azores: Checklist of the marine molluscs. *Arquipélago* 21A: 75–80.

Ávila, S. P., F. Cardigos & R. D. Santos. 2006. Comparison of the community structure of the marine molluscs of the "Banco D. João de Castro" seamount (Azores, Portugal) with that of typical inshore habitats on the Azores archipelago. *Helogoland Marine Research* 61: 43–53.

Azuma, M. 1960. A *Catalogue of the Shell-Bearing Mollusca of Okinoshima, Kashiwajima and the Adjacent Area (Tosa Province) Shikoku, Japan.* Privately published?, Japan. 7 pp., 101 pp., 5 pls, 17 pp. index.

Baluk, W. 1975. Lower Tortonian gastropods from Korytnica, Poland. Part 1. *Palaeontologica Polonica* 32: 1–186, pls 1–21.

Baborka, C. J. 2007. *Computergestützte 3D-Rekonstruktion und vergleichende Anatomie scissurellider Mikrogastropoden an* Scissurella jucunda *Smith, 1910 und* Larochea miranda *Finlay, 1927.* Diplomarbeit, Ludwig-Maximilians-Universität, Munich. 72 pp.

Bandel, K. 1982. Morphologie und Bildung der frühontogenetischen Gehäuse bei conchiferen Mollusken. *Facies* 7: 1–198.

Bandel, K. 1991. Character of a microgastropod fauna from a carbonate sand of Cebu (Philippines). *Mitteilungen des Geologisch-Paläontologischen Instituts der Universität Hamburg* 71: 441–485.

Bandel, K. 1998. Scissurellidae als Modell für die Variationsbreite einer natürlichen Einheit der Schlitzbandschnecken (Mollusca, Archaeogastropoda). *Mitteilungen des Geologisch-Paläontologischen Instituts der Universität Hamburg* 81: 1–120.

Bandel, K. 2000. Gastropods from the upper Early Jurassic/ early Middle Jurassic of Kaiwara Valley, North Canterbury, New Zealand. *Freiberger Forschungshefte* C490: 67–132.

Bandel, K. 2009. The slit bearing nacreous Archaeogastropoda of the Triassic tropical reefs in the St. Cassian Formation with evaluation of the taxonomic value of the selenizone. *Berliner paläobiologische Abhandlungen* 10: 5–47.

Barash, A. & Z. Danin. 1992. *Fauna Palaestina Mollusca I— Annotated List of Mediterranean Molluscs of Israel and Sinai*. The Israel Academy of Sciences and Humanities, Jerusalem. 405 pp., 372 pls.

Barnard, K. H. 1963. Deep-sea Mollusca from the region south of Madagascar. *Commerce & Industry Investigational Report* No. 44: 1–19.

Barnard, K. H. 1964. The work of the *S. S. Pieter Faure* in Natal waters, with special reference to the Crustacea and Mollusca; with description of new species of Mollusca from Natal. *Annals of the Natal Museum* 16: 9–29.

Barnard, K. H. 1963. Contribution to the knowledge of South African marine Mollusca Part 4. *Annals of the South African Museum* 47: 201–360.

Barrera, N. C. 2001. M. Sc. Thesis. *Micromolluscan Assemblages on the Flower Garden Banks, Northwestern Gulf of Mexico*. Texas A&M University, Corpus Christi. 90 pp.

Barrett, L. 1856. Memorandum on the animal of *Scissurella crispata*. *Annals and Magazine of Natural History* Series 2, 17: 206.

Barros, J. C. N. de, F. N. dos Santos, M. do Carmo Ferrão Santos, E. Cabral & F. D. Acioli. 2001. Redescoberta de moluscos obtidos durante a "Challenger Expedition" (1873–1876): micromoluscos de águas profunda. *Boletim Téchnico-Cientifico DEPENE* 9: 9–24.

Bartsch, P. 1903. A new *Scissurella* from Patagonia. *The Nautilus* 17: 90–91.

Bartsch, P. 1915. Report on the Turton collection of South African marine mollusks, with additional notes on other South African shells contained in the USNM. *United States National Museum Bulletin* 91: 1–267, pls 1–54.

Bartsch, P. 1946. A remarkable new genus and species of West American marine mollusks. *Journal of the Washington Academy of Sciences* 36: 447–448.

Batten, R. L. 1975. The Scissurellidae-Are they neotenously derived Fissurellidae? (Archaeogastropoda). *American Museum Novitates* 2567: 1–29.

Baxter, R. 1983. *Mollusks of Alaska*. Privately published, Bethel, Alaska. 69 pp., 17 pp.

Beck, T., T. Metzger, & A. Freiwald. 2006. *BIAS: Biodiversity Inventorial Atlas of Macrobenthic Seamount Animals*. OASIS: University of Hamburg, Hamburg Downloaded from http://www.1uni-hamburg.de/OASIS//Pages/publications/BIASpdf Nov 2007. 127 pp.

Bedulli, D. 1994. Censimento della malacofauna marina Italiana. *Notiziario SIM* 12(7–9): 76–77.

Bedulli, D. & C. Magistrali. 1995. Censimento della malacofauna marina Italiana. *Notiziario SIM* 13(4–6): 43–50.

Bedulli, D., G. Spada, B. Sabelli, V. Iacono & C. Magistrali. 1993a. Censimento della malacofauna marina Italiana II. Elenco delle specie rinventute nella regione Toscana con indicazione del biotopo. *Notiziario SIM* 11(3): 69–85.

Bedulli, D., G. Spada, B. Sabelli, V. Iacono & C. Magistrali. 1993b. Censimento della malacofauna marina Italiana III. Elenco delle specie rinventute nella regione Lazio con indicazione del biotopo. *Notiziario SIM* 11(6–7): 105–113.

Bedulli, D., G. Spada, B. Sabelli, V. Iacono & C. Magistrali. 1994. Censimento della malacofauna marina Italiana IV. Elenco delle specie rinventute nella regione Campania con indicazione del biotopo. *Notiziario SIM* 12(7–9): 79–98.

Beets, C. 1943. Beiträge zur Kenntnis der angeblich oberoligocänen Mollusken-Faund der Insel Buton, Niederländisch-Ostindien. *Leidsche Geologische Mededeelingen* 13: 256–328, pls 27–30.

Bellini, R. 1901. Contribuzione alla conoscenza della fauna dei molluschi marini dell'isola di Capri. *Bollettino della Società dei Naturalisti in Napoli* 15: 85–121.

Benoist, M. E. 1875. Description de coquilles fossils des terrains tertiaries moyens. *Procès Verbaux de la Société Linéenne de Bordeaux* 30: 67–73, pl. 1.

Berry, S. S. 1947. New Mollusca from the Pleistocene of San Pedro, California—III. *Bulletins of American Paleontology* 31: 257–272, pls 26–27.

Bertolosi, L. & S. Palazzi. 2000. Nota sulla raccolta Seguenza di Moluschi Plio-Pleistocenichi della Provincia di Messina conservata presso il Museo di Geologia e Paleontologia dell'Università di Firenze. *Bollettino Malacologico* 35: 3–44.

Bertozzi, A., F. Biagi, V. Biagi, L. Cerri, A. Gaglini, F. Mannucci, E. Miari, P. Micali, D. Poli, F. Sbaragli. 1984. Malacofauna marina della Toscana — Area 024. *Notiziario SIM* 2(5–6): 75–83.

Bertozzi, A., F. Dinetti, E. Miari, P. Micali, D. Poli, G. Terzer. 1984. Malacofauna marina della Toscana — Area 025. *Notiziario SIM* 2(9–10): 125–128.

Beu, A. G., P. A. Maxwell & R. C. Brazier. 1990. Cenozoic Mollusca of New Zealand. *New Zealand Geological Survey Paleontological Bulletin* 58: 1–518.

Biagi, V., V. Bientinesi, C. Corselli, D. Poli. 1985. Malacofauna marina della Toscana — Area 024. *Notiziario SIM* 3(3–4): 51–52.

Bieler, R. & K. J. Boss. 1989. Johannes Thiele and his contributions to Zoology. Part 1. Biography and bibliography. *Nemouria* 34: 1–30.

Bieler, R. & R. E. Petit. 2005. Catalogue of Recent and fossil taxa of the family Architectonicidae Gray, 1850 (Mollusca: Gastropoda). *Zootaxa* 1101: 1–119.

Bishop, M. J. & S. J. Bishop. 1973. A census of marine prosobranch gastropods at San Diego, California. *The Veliger* 16: 143–152.

Bizet, E. 1888. Catalogue des mollusques observés à l'état vivant dans le département de la Somme. *Mémoires de la Société Linnéenne du Nord de la France* 7: 3–206.

Bodon, M., L. Favilli, R. Giannuzzi-Savelli, F. Giovine, F. Guisti, G. Manganelli, G. Melone, M. Olivero, B. Sabelli & G. Spada. 1995. Gastropoda Prosobranchia, Heterobranchia Heterostropha. In: *Checklist Delle Species Della Fauna Italiana*. A. Minelli, S. Ruffo & S. La Posta (eds): pp. 1–60. Edizioni Claderini, Bologna.

Boettger, O. 1902. Zur Kenntnis der Fauna der mittelmiocänen Schichten von Kostej im Krassó-Szörényer Kominat. *Verhandlungen und Mitteilungen des Siebenbürgischen Vereins für Naturwissenschaften zu Hermannstadt* 51: 1–200.

Boettger, O. 1906. Zur Kenntnis der Fauna der mittelmiocänen Schichten von Kostej im Krassó-Szörényer Kominat (Gasteropoden und Anneliden) III. *Verhandlungen und Mitteilungen des Siebenbürgischen Vereins für Naturwissenschaften zu Hermannstadt* 54: I–VIII, 1–99.

Boettger, O. 1907. Zur Kenntnis der Fauna der mittelmiocänen Schichten von Kostej im Krassó-Szörényer Kominat (Gasteropoden und Anneliden) III (Fortsetzung). *Verhandlungen und Mitteilungen des Siebenbürgischen Vereins für Naturwissenschaften zu Hermannstadt* 55: 101–243.

Bofill i Poch, A. & M. de Chia. 1913. *Fauna de Catalunya: Fauna Malacologica Fascicle I Introducció*. Institut d'Estudis Catalans, Barcelona. 49 pp.

Borja, A. & E. Muxika. 2001. Update of the checklist of marine mollusks from the Basque from surveys conducted by AZTI. *Iberus* 19: 67–73.

Bosch, D. T., S. P. Dance, R. G. Moolenbeek & P. G. Oliver. 1995. *Seashells of Eastern Arabia*. Motivate Publishing, Dubai. 296 pp.

Boss, K. J., J. Rosewater & F. A. Ruhoff. 1968. The zoological taxa of William Healey Dall.

United States National Museum Bulletin 287: 1–427.
Boucher, H. du. 1884. L'atlas conchyliologique de Grateloups révisé et complété. *Bulletin de la Société Borda-Dan* 1884: 165–184.
Bouchet, P. 1983. Revue de presse (Novembre 1983). *Xenophora* 18: 5–7.
Bouchet, P. 1987. Revue de presse (Mars 1987). *Xenophora* 38: 4–6.
Bouchet, P. 1990. Turrid genera and mode of development: the use and abuse of protoconch morphology. *Malacologia* 32: 69–77.
Bouchet, P. & F. Danrigal. 1982. Napoleon's Egyptian campaign (1798–1801) and the Savigny collection of shells. *The Nautilus* 96: 9–24.
Bouchet, P. & J. P. Rocroi. 2005. Classification and nomenclator of gastropod families. *Malacologia* 47: 1–397.
Bouchet, P. & E. E. Strong. 2010. Historical name-bearing types in marine molluscs: an impediment to biodiversity studies? In: *Systema Naturae 250*. A. Polaszek (ed.): pp. 63–74. CRC Press, London.
Bouchet, P. & A. Warén. 1979. The abyssal molluscan fauna of the Norwegian sea and its relationship to other faunas. *Sarsia* 64: 211–243.
Bouchet, P. & A. Warén. 1997. [Mollusca: Lower Gastropoda]. In: *Handbook of Deep-Sea Hydrothermal vent Fauna*. D. Desbruyères & M. Segonzac (eds): pp. 109–132. IFREMER éditions, plouzané.
Bouchet, P., V. Héros, P. Lozouet & Maestrati, P. 2008. A quarter century of deep-sea malacological exploration in the South and West Pacific: Where do we stand? How far to go? In: *Tropical Deep-Sea Benthos 25. Mémoirs du Muséum National d'Histoire Naturelle* 196. V. Héros, R. H. Cowie, P. & Bouchet (eds): pp. 9–40. Muséum National d'Histoire Naturelle, Paris.
Bouchet, P., V. Héros, P. Lozouet, P. Maestrati & R. von Cosel. 2011. The marine molluscs of Santo. In: *The Natural History of Santo*. P. Bouchet, H. Le Guyader & O. Pascal (eds): pp. 421–431. IRD, Paris.
Bouchet, P., P. Lozouet, P. Maestrati & V. Héros. 2002. Assessing the magnitude of species richness in tropical marine environments: exceptionally high numbers of molluscs at a New Caledonia site. *Biological Journal of the Linnean Society* 75: 421–436.
Bourne, G. C. 1910. On the anatomy and systematic position of *Incisura* (*Scissurella*) *lytteltonensis*. *Quarterly Journal of Microscopical Sciences* 55: 1–45.
Boutan, L. 1886. Recherches sur l'anatomie et le dévelopement de la fissurelle. *Archive de Zoologie Expérimentale et Générale* 2nd Series, 3, bis, Supplement: 1–173, pls 31–44.
Branch, M. L., P. M. Arnaud, J. Cantera & D. Gianakouras. 1991. The benthic Mollusca and Brachiopoda of subantarctic Marion and Prince Edward Islands: 1) Illustrated keys to the species 2) Records of the 1982–1989 University of Cape Town Surveys. *South African Journal of Antarctic Research* 21: 45–64.
Brann, D. C. & L. S. Kent. 1960. Catalogue of the type and figured specimens in the Paleontological Research Institution. *Bulletins of American Paleontology* 40: 5–995, 1 pl.
Brokenshire, A. 2006. Recent/subfossil molluscs from Roskilde Fjord, near Frederikssund, Denmark. *Palidula* 36(1): 5–6.
Brook, F. J. & B. A. Marshall. 1998. Appendix: checklist of benthic coastal marine chitons, bivalves, gastropods and cephalopods of the northern Kermadec Islands. *Journal of the Royal Society of New Zealand* 28: 210–233.
Brookes Knight, J., L. R. Cox, A. Myra Keen, R. L. Batten, E. L. Yochelson & R. Robertson. 1960. Systematic descriptions [Archaeogastropoda]. In: *Treatise on Invertebrate Paleontology, (I) Mollusca Part 1*. R. C. Moore & C. W. Pitrat (eds): pp. I1–I351. Geological Society of America and University of Kansas Press, Lawrence, Kansas.
Brown, T. 1843. *The Elements of Fossil Conchology*. Houlston & Stoneman, London. 1–138 pls, 12 pls.
Brown, T. 1844. *Illustrations of the Conchology of Great Britain and Ireland Drawn from Nature, Second Edition*. W. H. Lizars, London. 144 pp., 59 pls.

Brøgger, W. C. 1901. Om de senglaciale og postglaciale Nivåforandringer I Kristianiafeltet (Molluskfaunan). *Norges Geologiske Undersøgelse* 31: pp. 1–731, pls 1–19.

Brusina, S. 1864. Conchiglie Dalmate inedite. *Verhandlungen der K. K. Zoologisch-Botanischen Gesellschaft in Wien* 15: 1–42.

Brusina, S. 1866. Contribuzione pella Fauna dei molluschi Dalmati. *Verhandlungen der K. K. Zoologisch-Botanischen Gesellschaft in Wien* 16: 1–134, pl. 1.

Brusina, S. 1870. Ipsia Chiereghinii Conchylia ovvero contribuzione pella Malacologia Adriatica desunta dal manoscritto descrizione de' Crostacei, de' Testacei, e de' Pesci che abianto la Lagune e Golfo Veneto, rapressantati in figure, a chiaroscuro ed a colori dall' Abate S. Chireghini illustrata da S. Brusina [not seen].

Brusina, S. 1886. Appunti ed osservazioni sull' ultimo lavoro di J. G. Jeffreys "on Mollusca procured during the *Lightning* and *Porcupine* Expeditions, 1868–1870". *Societas Historico-Naturalis Croatica* 1: 182–221.

Bruyne, R. H. de. 2003. *Geïllustreede Schelpen Enzyclopedie.* Rebo Productions, Lisse, The Netherlands. 336 pp.

Buchner, O. 1913. *Einführung in die europäische Meeresmollusken-Fauna.* K. G. Lutz Verlag, Stuttgart. 166 pp., 26 pls.

Bucquoy, E., P. Dautzenberg, & G. Dollfus. 1882–1886. *Les Mollusques Marins du Roussillon.* J. B. Baillière & Fils, Paris. 570 pp., 66 pls.

Bullock, R. C., R. D. Turner & R. A. Fralick. 1990. Species richness and diversity of algal-associated micromolluscan communities from São Miguel, Açores. *Açoreana* 1990, Supplement: 39–58.

Burch, J. Q. 1946a. *Check List of West North American Marine Mollusks.* Burch, Los Angeles. 30 pp.

Burch, J. Q. 1946b. [No title]. *Minutes of the Conchological Club of Southern California* 58: 1–27.

Burnay, L. P. & E. Rolán. 1990. The family Scissurellidae in the Cape Verde Islands. *Archiv für Molluskenkunde* 120: 31–45.

Bush, K. J. 1885. List of deep-water Mollusca dredged by the United States Fish Commission steamer Fish Hawk in 1880, 1881, and 1882 with their range in depth. *United States Commission of Fish and Fisheries Report of the Commissioner for 1883* vol. 11: 701–707.

Cachia, C., C. Mifsud & P. M. Sammut. 1991. *The Marine Shelled Mollusca of the Maltese Islands: Part One: Archaeogastropoda.* Grima Printing and Publishing Industries, Marsa, Malta. 112 pp., 6 pls.

Cachia, C., C. Mifsud & P. M. Sammut. 1993. *An Annotated Check-List of the Marine Mollusca of the Maltese Islands.* Erste Voralberger Malakologische Gesellschaft, Rankweil. 80 pp.

Cadien, D. 1980. Mollusca. *Allan Hancock Foundation Technical Report* 3: 130–195.

Camerano, L. & M. Lessona. 1895. *Compendio della Fauna Italiana.* G. B. Paravia & co., Torino. 311 pp.

Camp, D. K., W. G. Lyons & T. H. Perkins. 1998. *Checklists of Selected Shallow-Water Marine Invertebrates of Florida.* Florida Marine Research Institute Technical Reports, St. Petersburg. 239 pp.

Canefri, C. T. 1869. Indice sistematico dei molluschi testacei dei dintorni di spezia e del suo Golfo. *Atti della Società Italiana di Scienze Naturali* 12: 261–406.

Cantera, J. R. & P. M. Arnaud. 1985. Les gastéropodes prosobranches des Îles Kerguelen et Crozet (Sude de l'Océan Indien) comparaison écologique et particularités biologiques. *Comité National Français des Recherches Antarctiques* 56: 1–169.

Cantraine, F. 1840. *Malacologie Méditerranéenne et Littorale.* Hayez, Bruxelles. 173 pp., 6 pls.

Cantraine, F. 1842. Diagnoses de quelque espèces nouvelles de coquilles soit natives soit fossiles, appartenant au bassin méditerranéen. *Bulletin de l'Académie Royale des Sciences et Belle-Lettres de Bruxelles* 9: 340–349.

Capellini, G. 1860. [see Jeffreys, 1860].
Carcelles, A. R. 1950. Catalogo de los moluscos marinos de Patagonia. *Anales del Museo Naturel Huapi* 2: 41–250, pls 1–5, 1 map.
Carcelles, A. R. 1953. Catalogo de la malacofauna Antarctica Argentina. *Anales del Museu Naturel Huapi* 3: 155–250.
Carcelles, A. R. & S. I. Williamson. 1951. Catalogo de los moluscos marinos de la provincia Magellanica. *Revista del Instituto Nacional de Investigacion de las Ciencias Naturales, Ciencias Zoologica* 2: 225–383.
Cardigos, F., A. Colac, P. R. Dando, S. P. Ávila, P.-M. Sarradin, F. Tempera, P. Conceição, A. Pascoal & R. S. Santos. 2005. Shallow water hydrothermal vent field fluids and communities of the D. João de Castro Seamount (Azores). *Chemical Geology* 224: 153–168.
Carpenter, P. P. 1861. Lectures on Mollusca; or "shell-fish" and their allies. *Annual Report of the Board of Regents of the Smithsonian Institution* 15: 151–283.
Carpenter, P. P. 1864. Supplementary report on the present state of our knowledge with regard to the Mollusca of the West Coast of North America. *Report of the British Association for the Advancement of Science* for 1863: 517–686.
Carpenter, P. P. 1865. Diagnoses of new species and a new genus of mollusks from the Reigen Mazatlan collection: with an account of additional specimens presented to the British Museum. *Proceedings of the Zoological Society of London* 1865: 268–271.
Carrozza, F. 1976. Microdoride di malacologia mediterranea (Contributo secondo). *Conchiglie (Milan)* 12: 163–169.
Carus, J. V. 1889–1893. *Prodromus Faunae Mediterraneae sive Descriptio Animalium Maris Mediterranei Incolarum*. E. Schweizerbart'sche Verlagshandlung, Stuttgart. 854 pp.
Castellanos, Z. A. de. 1980. Micromoluscos poco conocidoes del sur argentino-chileeno. *Neotropica* 25: 133–140.
Castellanos, Z. A. de, S. Bartolotta & E. Rolán. 1987. Aportes a la malacofauna del talud superior del Atlantico sur. *Thalassas* 5: 57–70.
Castellanos, Z. J. A. de & N. A. Landoni. 1988. *Catalogo Descriptivo de la Malacofauna Marina Magellanica 2. Archigastropoda: Scissurellidae Fissurellidae Patellidae Acmaeidae y Lepetidae*. Provincia de Buenos Aires Comisión de Investigaciones Cientificas, Buenos Aires. 40 pp.
Cavallo, O. & G. Repetto. 1992. Conchiglie Fossili de Roero. Atlante Iconigrafico. *Memorie dell Associazione Naturalistica Piemontese (Associazione Amici del Museo "Frederio Eusebio")* 2: 1–251.
Cernohorsky, W. O. 1972. Type specimens of Recent and fossil Mollusca described by H. J. Finlay. Part I (Scissurellidae - Turbinidae). *Records of the Auckland Institute and Museum* 9: 231–247.
Cerulli-Irelli, S. 1916. Fauna Malacologica Mariana, Parte Ottava: Phasianellidae, Turbinidae, Trochidae, Cyclostrematidae, Cocculinidae, Tornidae, Scissurellidae, Fissurellidae, Tecturidae. *Paleontographia Italica* 22: 171–242, pls 20–24.
Chaster, G. W., J. C. Melvill, G. A. F. Knight & W. E. Hoyle. 1901. List of the British marine Mollusca and Brachiopoda. *Journal of Conchology* 10: 9–27.
Chemello, R., P. Micali & R. Giannuzzi-Savelli. 1984. Malacofauna marina della Sicilia — Area 141. *Notiziario SIM* 2(11–12): 169–171.
Chenu, J. C. 1847. *Leçons Élémentaires d'Histoire Naturelle*. J. J. Dubochet, Paris. 284 pp., 12 pls.
Chenu, J. C. 1859. *Manuel de Conchyliologie et de Paléontologie Conchiologique, Vol. 1*. Victor Masson, Paris. 507 pp.
Clark, W. 1855. *A History of the British Marine Testaceous Mollusca*. John van Voorst, London. 536 pp.
Clark, W. 1856. On *Scissurella crispata*. *Annals and Magazine of Natural History* Series 2, 17: 269–272.
Clarke, A. H. 1962. Annotated list and bibliography of the abyssal marine molluscs of the

Literature cited

world. *National Museum of Canada Bulletin* 181: 1–114.
Claudon, E. 1902. Faunule malacologique marine de Saint-Raphaël (Var). *La Feuille des Jeunes Naturalistes* Series 4, 381/382: 184–199.
Coan, E. V., A. R. Kabat & R. E. Petit. 2007. *2,400 Years of Malacology, 4th Edition* [http://www.malacological.org/publications/2400_malacology.php accessed Feb. 2007]. Self published. 721 pp. + 14 pp.
Coen, G. 1933. *Saggio di una Sylloge Molluscorum Adriaticorum*. Carlo Ferrari, Venezia. 186 pp., 10 pls.
Coen, G. 1937. *Nuovo Saggio di una Sylloge Molluscorum Adriaticorum*. Carlo Ferrari, Venezia. 173 pp., 10 pls.
Committee for Celebrating Dr. Okutani's Retirement from Tokyo University of Fisheries. 1996. *Bibliography on the Malacological Works of Dr. Takashi Okutani 1955–1994*. Tokyo University of Fisheries, Tokyo. 101 pp., 11 pls.
Cosentino, A. & S. Giacobbe. 2007. Aspects of epizoobiontic mollusc assemblages on *Pinna* shells. I. Composition and structure. *Cahiers de Biologie Marine* 48: 187–197.
Conti, A. 1864. *Il Monte Mario ed i suoi Fossili Subapennini*. Giovanni Cesaretti, Rome. 57 pp., 1 pl.
Conti, M. A. & S. Monari. 1991. Bivalve and gastropod fauna from the Liassic Ammonitico Rosso facies in the Bilecik area (western Pontides, Turkey). *Geologica Romana* 27: 245–301.
Cooke, A. H. 1882. On the MacAndrew collection of British shells. *Journal of Conchology* 3: 340–390.
Coppa, M. G. & P. Crovato. 1983. Osservazioni biostratigrafiche e paleoecologiche sui depositi argillosi emiliani e siciliani di Gallipoli (Puglia). *Bollettino della Società dei Naturalisti in Napoli* 92: 159–225.
Coraggio, L. 1985. Malacofauna marina del Lazio — Area 032. *Notiziario SIM* 3(9–10): 162.
Cossignani, T. & R. Ardovini. 2011. *Malacologia Mediterranea*. L'Informatore Piceno, Ancona. 536 pp.
Coraggio, L. & P. Micali. 1985. Malacofauna marina del Lazio — Area 031. *Notiziario SIM* 3(9–10): 161.
Cossignani, T., V. Cossignani, A. Di Nisio & M. Passamonti. 1992. *Atlante delle Conchiglie del Medio Adriatico*. Monstra Mondiale Malacologia, Cupra Marina, Ancona. 40 pp., 417 figs.
Cossmann, M. & A. Peyrot. 1917. Conchologie Néogénique de l'Aquitaine. *Actes de la Société Linnénne de Bordeaux* 69: 157–305.
Cossmann, M. 1879. Description de deux espèces nouvelles de Tongrien des environs d'Étampes. *Journal de Conchyliologie* Series 3, 27: 346–348, pl. 13.
Cossmann, M. 1883. Description d'espèces de terrain tertiaire des environs de Paris (suite). *Journal de Conchyliologie* Series 3, 31: 153–174, pls 5–7.
Cossmann, M. 1888. Catalog illustrées des coquilles fossiles de l'Éocène de Paris. *Annales de la Société Royale Malacologique de Belgique* 23: 3–324, pls 1–12.
Cossmann, M. 1892. Catalog illustrées des coquilles fossiles de l'Éocène des environs de Paris. *Annales de la Société Royale Malacologique de Belgique* 26: 3–168, pls 1–26.
Cossmann, M. 1902. Catalog illustrées des coquilles fossiles de l'Éocène de Paris. Appendice no 3. *Annales de la Société Royale Malacologique de Belgique* 36: 9–110, pls 2–7.
Cossmann, M. 1914. Révision des scaphopodes, gastropodes, et céphalopodes du Montien de Belgique. *Musée Royal d'Histoire Naturelle de Belgique* 6: 1–71, pls 1–4.
Cossmann, M. 1923. Mollusques éocéniques de la Loire-inférieure. *Bulletin de la Société des Sciences Naturelles de l'Ouest de la France* Series 4, 2: 109–160, pls 3–4.
Cossmann, M. & J. Lambert. 1884. Étude paléontologique et stratigraphique sur le terrain Oligocène marin aux environs d'Étampes. *Mémoires de la Société Géologique de France* Series 3, 3: 1–187, pls 1–6.
Cossmann, M. & G. Pissarro. 1901. Faune éocénique du Cotentin (mollusques). *Bulletin de*

la Société Géologique de Normandie 19: 3–295, pls 1–32.

Cossmann, M. & G. Pissarro. 1906. *Iconographie Complète des Coquilles Fossiles de l'Éocène des Environs de Paris. Tome 2. Scaphopodes, Gastropodes, Brachiopodes, Céphalopodes et Supplément.* Privately published (?), Paris. 22 pp., 65 pls.

Costa, O. G. da 1828. *Testacei e Crustacei microsc.* [not traceable, not seen].

Costa, O. G. da 1829. *Catalogo Sistematico e Ragionato de' Testacei delle due Sicilie.* Tipografia della Minerva, Napoli. 132 pp., 2 pls.

Costa, O. G. da. 1838. Ordine VIII. Gasteropodi scutibranchi. In: *Fauna del Regno die Napoli.* O. G. da Costa (ed.): pp. 1–30, pls 1–3. Azzolino E. Compagno, Napoli [for dating see Sherborn, 1937].

Costa, O. G. da 1844. Catalogo de' testacei viventi nel piccolo e grande mare di Taranto redatto sul sistema di Lamarkc. *Atti della Reale Accademia della Scienze, sezione della Società Reale Borbonica* 5(2): 14–66, pls. 1–4.

Costa, O. G. da 1861. *Microdoride Mediterranea o Descrizione de' poco ben Conoscuiti od Affatto Ignoti Viventi Minuti e Microscopici del Mediterraneo.* Vol. 1. Dell'Iride, Napoli. 80 pp., 12 pls.

Cotton, B. C. 1945. Southern Australian Gastropoda Part I. Streptoneura. *Transactions of the Royal Society South Australia* 69: 150–171, pls 12–13.

Cotton, B. C. 1959. *South Australian Mollusca, Archaeogastropoda.* Handbook of the Flora and Fauna of South Australia. W. L. Hawes, Government Printer, Adelaide. 449 pp.

Cotton, B. C. 1963. *South Australian Shells.* South Australian Museum, Adelaide. 6 pp., 14 pls.

Cotton, B. C. & F. K. Godfrey. 1933. South Australia Shells (including description of new genera and species). *South Australian Naturalist* 15: 14–24, pl. 1.

Cotton, B. C. & F. K. Godfrey. 1938a. New species of South Australian Gastropoda. *Records of the South Australian Museum* 6: 199–206, pl. 17.

Cotton, B. C. & F. K. Godfrey. 1938b. A systematic list of the Gastropoda. The marine, freshwater and land univalve Mollusca of south and central Australia. *Malacological Society of South Australia Publication* 1: 1–44.

Crofts, D. R. 1937. The development of *Haliotis tuberculata* with special reference to organogenesis during torsion. *Philosophical Transactions London* Series B 228: 219–266.

Crosse, H. 1863. Bibliographie. *Journal de Conchyliologie* 11: 101–111.

Crosse, H. 1865. Observations sur la note précédente. *Journal de Conchyliologie* 13: 237–239.

Crosse, H. 1867. Description d'un genre nouveau et plusieurs espèces inédites provenant de la Nouvelle-Calédonie. *Journal de Conchyliologie* 15: 312–321, pl. 11.

Crosse, H. 1877. Fauna malacologique des îles Kerguelen. *Journal de Conchyliologie* Series 3, 17: 5–15.

Crosse, H. 1880. Description de mollusques inédits, provenant de la Nouvelle-Calédonie et de la Nouvelle-Bretagne. *Journal de Conchyliologie* 28: 142–149, pl. 4.

Crosse, H. & P. Fischer. 1861. Sur le genre *Schismope* de M. Gwyn Jeffreys. *Journal de Conchyliologie* 9: 257–260.

Crosse, H. & P. Fischer. 1861. Observations sur le genre pleurotomaire, et description d'une deuxième espèce appartenant au même genre. *Journal de Conchyliologie* 9: 155–167, pl. 5.

Cruz, F. V. & Gándara, C. G. 2006. Updated checklist of gastropods from flat Lobos Reef, Veracruz, Mexico. *Revista UDO Agricola* 6: 128–137.

Cuomo, M. & G. Fasculo. 1983. Malacofauna Marina della Campania — Area 043. *Notiziario SIM* 1(71–2): 41–42.

Cuvier, B. 1834. *The Mollusca and Radiata.* Whittaker & Co, London. 601 pp.

Daccarett, E. Y & V. S. Bossio. 2011. *Colombian Seashells from the Caribbean Sea.* L'Informatore Piceno, Ancona. 384 pp. + 16 pp. index.

Dall, W. H. 1889a. Reports on the results of dredging, under the supervision of Alexander

Literature cited

Agassiz, in the Gulf of Mexico (1877–1878) and in the Caribbean Sea (1879–1880), by the U. S. Coast Survey Steamer "Blake," Lieut.-Commander C. D. Sigsbee, U. S. N., and Commander J. R. Barlett, U. S N., commanding. XXIX.—Report on the Mollusca. Part II—Gastropoda and Scaphopoda. *Bulletin of the Museum of Comparative Zoology* 18: 1–492, pls 10–40.

Dall, W. H. 1889b. *A Preliminary Catalogue of the Shell-Bearing Marine Molluscs and Brachiopods of the South-Eastern Coast of the United States, with Illustrations of Many of the Species*. Government Printing Office, Washington (DC). 221 pp., 74 pls.

Dall, W. H. 1905. Some new species of mollusks from California. *The Nautilus* 18: 123–125.

Dall, W. H. 1919. Descriptions of new species of Mollusca from the North Pacific Ocean in the collection of the United States National Museum. *Proceedings of the United States National Museum* 56: 293–371.

Dall, W. H. 1921. Summary of the marine shell-bearing mollusks of the northwest coast of America, from San Diego, California, to the Polar Sea, mostly contained in the collection of the United States National Museum, with illustration of hitherto unfigured species. *United States National Museum Bulletin* 112: 1–217, pls 1–22.

Dall, W. H. 1923. Additions and emendations to *United States National Museum Bulletin* No. 112. *Proceedings of the United States National Museum* 63: 1–4.

Dall, W. H. 1927a. Small shells from dredgings off the southeast coast of the United States by the United States fisheries steamer "Albatross" in 1885 and 1886. *Proceedings of the United States National Museum* 70(18): 1–134.

Dall, W. H. 1927b. Diagnoses of undescribed new species of mollusks in the collection of the United States National Museum. *Proceedings of the United States National Museum* 70(19): 1–11.

Dance, S. P. 1966. *Shell Collecting, an Illustrated History*. Faber & Faber, London. 344 pp., 35 pls.

Danielssen, D. C. 1859. Beretning om en zoologisk Reise foretagen i Sommeren 1858. *Det Kongelige Norske Videnskabers-Selskabs Skrifter i det 19de Årh* 4: 97–174.

Dautzenberg, P. 1883. Liste de coquilles du Golfe de Gabès. *Journal de Conchyliologie* Series 3, 31: 1–42.

Dautzenberg, P. 1889. Contribution à la faune malacologique des Îles Açores. *Résultats des Dragages Effectués par le Yacht l'*Hirondelle *Pendant sa Campagne Scientifique de 1887* 1: 1–112, pls 1–4.

Dautzenberg, P. 1891. Contribution à la faune malacologique du Golfe de Gascogne. *Mémoires de la Société Zoologique de France* 4: 604–619, pls 16–17.

Dautzenberg, P. 1910. Contribution à la faune malacologique de l'Afrique occidentale. *Actes de la Société Linnéenne de Bordeaux* 1910: 1–174, pls 1–4.

Dautzenberg, P. 1913. Mission Gruvel sur la Côte Occidentale d'Afrique (1909–1910). *Annales de L'Institut Océanographique* 1912 (3): 1–111, pls 1–3.

Dautzenberg, P. 1927. Mollusques provenant des campagnes scientifiques de Prince Albert 1er de Monaco dans l'Océan Atlantique et dans le Golfe de Gascogne. *Résultats des Campagnes Scientifiques du Prince de Monaco* 72: 1–400.

Dautzenberg, P. & J.-L. Bouge. 1933. Les mollusques testacés marins des établissements Français de l'Océanie. *Journal de Conchyliologie* 77: 1–469.

Dautzenberg, P. & H. Fischer. 1897. Dragages effectués par l'*Hirondelle* et par la *Princesse-Alice* 1888–1895. *Mémoires de la Société Zoologique de France* 9: 395–498, pls 15–22.

Dautzenberg, P. & H. Fischer. 1898. Dragages effectués par l'*Hirondelle* et par la *Princesse-Alice*. 1888–1896. *Mémoires de la Société Zoologique de France* 10: 139–234, pls 3–7.

Dautzenberg, P. & H. Fischer. 1912. Mollusques provenant des campagnes de l'*Hirondelle* et par la *Princesse-Alice* dans les Mers du Nord. *Résultats des Campagnes Scientifiques du Prince de Monaco* 37: 629 pp., 11 pls.

Davolos, J. & R. G. Moolenbeek. 2006. The intertidal gastropods of South Georgia. Part 1. Patellidae, Scissurellidae, Trochidae and Cerithiidae. *Gloria Maris* 44: 128–145.

de Jong, K. M. & H. E. Coomans. 1988. *Marine Gastropods from Curaçao, Aruba and Bonaire.* E. J. Brill, Leiden. 261 pp.

Dekker, H. & Z. Orlin. 2000. Check-list of Red Sea Mollusca. *Spirula* 47 Supplement: 1–46.

Delhaes, W. 1909. Beiträge zur Morphologie und Phylogenie von *Haliotis* Linné. *Zeitschrift für Induktive Abstammungs- und Vererbungslehre* 2: 1–55, pls 2–3.

Dell, R. K. 1951. A molluscan fauna from the Mernoo Bank, New Zealand. *New Zealand Journal of Science and Technology* Section B, 33: 15–18.

Dell, R. K. 1956. The archibenthal Mollusca of New Zealand. *Dominion Museum Bulletin* 18: 1–235.

Dell, R. K. 1960. Chatham Island marine Mollusca based upon the collections of the Chatham Islands Expedition 1954. *New Zealand Department of Scientific and Industrial Research Bulletin* 139: 141–157.

Dell, R. K. 1990. Antarctic Mollusca with special reference to the Fauna of the Ross Sea. *Royal Society of New Zealand Bulletin* 27: 1–311.

Demir, M. 2003. Shells of Mollusca collected from the Seas of Turkey. *Turkish Journal of Zoology* 27: 101–140.

Depontaillier, J. 1881. Description de deux nouvelles espèces fossiles. *Journal de Conchyliologie* Series 3, 29: 173–178, pl. 7.

de Rayneval, Van den Hecke & Ponzi. 1854. *Catalogue des fossiles du Monte Mario (près Rome).* 20 + 6 pp.

Deshayes, G. P. 1865. Note sur le genre *Trochotoma* et description d'une espèce nouvelle des sables de Bordeaux. *Journal de Conchyliologie* 13: 230–237, pl. 7.

Deshayes, G. P. 1864–1866. *Description des Animaux Sans Vertèbres Découverts Dans le Bassin de Paris.* J.-B. Bailliere et fils, Paris. 667 pp., 42 pls.

Dewalque, G. 1888. François-Léopold Cornet. Notice Biographique. *Annales de la Société Royale Malacologique de Belgique* 23: v–xxi.

Doneddu, M. & B. Manunza. 1992. Catalogo dei molluschi conchiferi marini della Sardegna. *Notiziario SIM* 10(1–2): 15–36.

Doneddu, M. & E. Trainito. 2005. *Conchiglie del Mediterraneo.* Il Castello, Trezzoro su Naviglio (MI). 256 pp.

Dornellas, A. P. S. & L. R. L. Simone. 2011. Annotated list of type specimens of mollusks deposited in Museu de Zoologia da Universidade de São Paulo, Brazil. *Arquivos de Zoologia* 42: 1–81.

Draper, B. 1973. Minute shells, part 3. *Tabulata* 6(2): 8–12.

Dunker, G. 1882. *Index Molluscorum Maris Japonici.* Sumptibus Theodori Fischer, Casselis Cattorum. 301 pp., 16 pls.

Dunker, W. 1861. Beschreibungen neuer Mollusken. *Malakozoologische Blätter* 8: 35–45.

Egorova, E. 1982. Biological results of the Soviet Antarctic Expeditions, 7. *Explorations of the Fauna of the Seas* 26(34): 1–143.

Faber, M. J. & R. G. Moolenbeek. 1987. On the doubtful records of *Alvania platycephala, Alvania pagodula* and *Alvania didyma*, with the description of two new rissoid species (Mollusca; Gastropoda: Rissoidae). *Beaufortia* 37: 67–71.

Faustino, L. A. 1928. *Summary of Philippines Marine and Fresh-Water Mollusks.* Bureau of Printing, Manila. 384 pp.

Fekih, M. 1975. Paléoécologie du Pliocène marin au nord de la Tunisie. *Annales des Mines et de Géologie, Tunis* 27: 1–195.

Feng, W.-M. 1996. Microgastropods from Nansha Sea area, China. In: *Studies on Marine Fauna and Flora and Biogeography of the Nansha Islands and Neighbouring Waters II.* The Multidisciplinary Oceanographic Expedition Team of Academia Sinica to Nansha Islands (eds): pp. 85–205. Ocean Press, Beijing.

Feng, W.-M. & J. A. Todd. 2007. Late Holocene microgastropods from the Yongshu Reef Lagoon of the South China Sea—Order Cocculinifornia, Vetigastropoda, Discopoda. *Acta Micropalaeontologica Sinica* 24: 28–52.

Fernald, H. T. 1939. On type nomenclature. *Annals of the Entomological Society of America* 22: 689–702.

Finet, Y. 1985. Preliminary faunal list of the marine mollusks of the Galapagos Islands. *Institut Royal des Sciences Naturelles de Belgique, Documents de Travail* 20: 1–50.

Finet, Y. 1991. The marine mollusks of the Galápagos Islands. In: *Galápagos Marine Invertebrates. Taxonomy Biogeography and Evolution in Darwin's Islands.* M. J. James (ed.): pp. 253–280. Plenum Press, New York.

Finet, Y. 1993. Marine Molluscs of the Galápagos Islands. *Vita Marina* 42: 33–39.

Finet, Y. 1994a. *Marine Mollusks of the Galápagos. Gastropods - A Monograph and Revision of the Families Haliotidae, Scissurellidae, Fissurellidae and Lottiidae. Monographs on Galápagos Mollusca no 1.* L'Informatore Piceno, Ancona. 110 pp.

Finet, Y. 1994b. *The Marine Mollusks of the Galapagos Islands: a Documented Faunal List.* Muséum d'Histoire Naturelle de Genève, Geneva. 180 pp.

Finlay, H. J. 1924. Additions to the Recent molluscan fauna of New Zealand. *Transactions and Proceedings of the New Zealand Institute* 55: 517–526, pl. 52.

Finlay, H. J. 1926. A further commentary on New Zealand molluscan systematics. *Transactions and Proceedings of the New Zealand Institute* 57: 320–485, pls 18–23.

Finlay, H. J. 1927. Additions to the Recent molluscan fauna of New Zealand No. 2. *Transactions and Proceedings of the New Zealand Institute* 57: 485–487, pls 24–25.

Finlay, H. J. 1929. The Recent Mollusca of the Chatham Islands. *Transactions and Proceedings of the New Zealand Institute* 59: 232–286, pls 38–43.

Fischer, P. & A. C. Bernardi. 1856. Description d'un pleurotomarie vivant. *Journal de Conchyliologie* 5: 160–166, pl. 5.

Fischer-Piette, E. 1950. Liste des types décrits dans le *Journal de Conchyliologie* et conservés dans la collection de ce journal. *Journal de Conchyliologie* 90: 8–23, 65–82, 149–180, 310–333, pls 1–5.

Fischer, M. P. 1882. Sur la faune malacologique abyssale de la Méditerranée. *Comptes Rendus Hebdomadaires des Séances de l'Académie des Sciences, Paris* 94: 1201–1203.

Fischer, P. 1862. Description d'une espèce nouvelle de *Scissurella. Journal de Conchyliologie* 10: 390–391.

Fischer, P. 1867. Description d'une espèce des îles Chusan. *Journal de Conchyliologie* 15: 304–305, 468, pl. 4.

Fischer, P. 1878. Essai sur la distribution géographique des brachiopodes et des mollusques du littoral océanique de la France. *Actes de la Société Linnéenne de Bordeaux* 32: 171–215.

Fischer, P.-H. 1962. Catalogue des types et exemplaires figurés, conservés dans la collection de l'école des mines de Paris (gastéropodes prosobranches) partie première: Archaeogastropoda. *Journal de Conchyliologie* 102: 98–113.

Fleming, C. A. 1948. New species and genera of marine Mollusca from the Southland Fiords. *Transactions of the Royal Society of New Zealand* 77: 72–92, pls 4–8.

Fleming, C. A. 1950. The molluscan fauna of the Fiords of western Southland. *New Zealand Oceanographic Committee Publication* 1: 20–40.

Fleming, C. A. 1951. Some Australasian Mollusca in the British Museum (Natural History). *Transactions of the Royal Society of New Zealand* 79: 126–139.

Fleming, J. 1828a. Remarks on the genus *Scissurella* of M. d'Orbigny, with a description of a Recent British species. *Memoirs of the Wernerian Natural History Society* 6: 384–387.

Fleming, J. 1828b. *History of British Animals Exhibiting the Descriptive Characters and Systematical Arrangement of the Genera and Species of Quadrupeds, Birds, Reptiles, Fishes, Mollusca and Radiata of the United Kingdom; Including the Indigenous, Extirpated, and Extinct Kinds, Together with Periodical and Occasional Visitants.* Bell & Bradflute, Edinburgh. 565 pp.

Flor, G., E. M. Llera & J. A. Ortea. 1982. Los carbonatos biogenicos de los sedimentos de las playas arenosas de Asturia y Cantabria: su origen y significado dinamico. Primera parte (a). *Caudernos del Crinas* 2: 3–77.

Folin, L. de. 1869. Chapitre XXVIII. Rade de Saint-Vincent du Cap-Vert (supplément). *Les Fonds de Mer* 1: 136–146, pls. 17–22.

Folin, L. de. 1871. Catalogue des Mollusques des vases de Syra. *Les Fonds de Mer* 1: 266–267.

Forbes, E. 1843. Report on the Mollusca and Radiata of the Aegean Sea, and on their distribution, considered as bearing on Geology. *Report of the British Association for Advance of Sciences* 13: 130–193.

Forbes, E. 1850. Report on the investigation of British marine zoology by means of the dredge. Part 1 The infra-littoral distribution of marine invertebrates on the southern, western, and northern coasts of Great Britain. *Report of the British Association for the Advancement of Science* 20: 192–264.

Forbes, E. & S. Hanley. 1853. *A History of British Mollusca and their Shells.* Vol. 2. John van Voorst, London. 557 pp., 133 pls.

Forcelli, D. O. 2000. *Moluscos Magallanicos.* Vazquez Mazzini, Buenos Aires. 200 pp.

Ford, B. J. 2011. The clarity of images from early single-lens microscopes captured on video. *Microscopy and Analysis* 25(2): 15–17.

Foster, N. R. 1981. *A Synopsis of the Marine Prosobranch Gastropod and Bivalve Mollusks in Alaskan Waters.* Institute of Marine Science, University of Alaska, Fairbanks, Alaska. 479 pp.

Franc, A. 1956. Mollusque marins. *Annales de l'Institut Océanographique* New Series 32: 19–60.

Franceschini, C. P. 1906. Elenco delle conchiglie del Golfo di Napoli e del Mediterraneo esistenti nel Museo Zoologico di Napoli. *Annuario del Museo Zoologico della Universitá di Napoli* Nuova Serie 2, 5: 1–68.

Freeman, A. N. D., B. A. Marshall, P. A. Maxwell, M. W. Walker, & D. C. Nicholls. 1997. Recent molluscan name-bearing types in the Canterbury Museum, Christchurch, New Zealand. *Records of the Canterbury Museum* 11: 17–41.

Fretter, V. & A. Grabham. 1976. The prosobranch molluscs of Britain and Denmark. Part 1—Pleurotomariacea, Fissurellacaea and Patellacea. *Journal of Molluscan Studies* Supplement 1: 1–37.

Friedberg, W. 1938. Katalog meiner Sammlung der Miozänmollusken Polens. *Mémoires de l'Académie Polonaise des Sciences et des Lettres, Classe des Sciences Mathématiques et Naturelles. Série B, Sciences Naturelles* 12: 1–94 [not seen].

Friele, H. 1873. Oversigt over de i Bergens omegn forekommende skaldakte Mollusker. *Christiania Videnskabs-Selskaps Forhandlinger* 1873: 1–24.

Friele, H. 1877. Tungebavaebingen hos de norske Rhipidoglossa. *Archiv for Mathematik og Naturvidenskab* 2: 199–217.

Friele, H. 1902. Mollusken der ersten Nordmeerfahrt des Fischereidampfers "Michael Sars" 1900 under Leitung von Herrn Dr. Johan Hjort. *Bergens Museums Aarbog* 3: 1–18, pls 1–4.

Friele, H. & J. A. Grieg. 1901. *The Norwegian North-Atlantic Expedition 1876–1878. Zoology. Mollusca III.* Christiania, Bergen. 131 pp., 1 map.

Fukuda, H. 1993. Marine Gastropoda (Mollusca) of the Ogasawara (Bonin) Islands Part 1: Archaeogastropoda and Neotaenioglossa. *Ogasawara Research* 19: 1–85.

Fukuda, H., K. Mashino, & T. Sugimura. 1992. *A Review of the Molluscan Fauna of Yamaguchi Prefecture Western Japan.* Yamaguchi Museum, Yamaguchi. 99 pp., 50 pls, xxvi.

Gaillard, J. M. 1971. Mollusca. In: *Marion and Prince Edward Islands.* E. M. van Zinderen Bakker Sr., J. M. Winterbottom & R. A. Dyer (eds): pp. 291–300. Balkema, Cape Town.

Galindo, E. S. 1977. *Index and Register of Seashells (with Cross References).* Librería de Porrúa Hermanus Y Cia, S A, Mexico City. 524 pp.

Garavelli, C. L. & N. Melone. 1971. Ritrovamenti malacologici in Mediterraneo. *Conchiglie* 7: 11–19.

García-Rios, C. I., F. J. Soto-Santiago, R. J. Colón-Riveria & J. R. Medina-Hernández. 2007.

Gasterópodos asociados al alga calcárea *Halimeda opuntia* (Udotaceae) en Puerto Rico. *International Journal of Tropical Biology* 56: 1665–1675.

García-Talavera Casañas, F. 1983. Los moluscos gasteropodos anfiatlanticos. *Universidad de la Laguna, Secretariado Publicaciorzes, Coleccion Monografias* 10: 1–352.

García, E. F. 2005. A dredging trip in the Gulf of Mexico. *Of Sea and Shore* 27: 4–8.

Garrett, A. 1872. Description of a new species of *Scissurella* form the Paumotu Islands. *Proceedings of the Zoological Society of London* 1872: 878.

Gatliff, J. H. & C. J. Gabriel. 1910. Additions to the catalogue of the marine shells of Victoria. *Proceedings of the Royal Society of Victoria* New Series, 23: 87–98.

Geiger, D. L. 2002. Australian Scissurellidae. *Australasian Shell News* 113: 1.

Geiger, D. L. 2003. Phylogenetic assessment of characters proposed for the generic classification of Recent Scissurellidae (Gastropoda: Vetigastropoda) with a description of one new genus and six new species from Easter Island and Australia. *Molluscan Research* 23: 21–83.

Geiger, D. L. 2004a. Vetigastropod evolution: summary of knowledge, insights from Scissurellidae, macroevolutionary prospects. In: *Molluscan Megadiversity: Sea, Land and Freshwater. World Congress of Malacology Perth, Western Australia 11–16 July 2004.* F. E. Wells (ed.): pp. 46–47. Western Australian Museum, Perth.

Geiger, D. L. 2004b. Made for SEM: the fascinating world of scissurellids. *American Conchologist* 32(4): 4–7.

Geiger, D. L. 2006a. Eight new species of Scissurellidae and Anatomidae (Mollusca: Gastropoda: Vetigastropoda) from around the world, with discussion of two new senior synonyms. *Zootaxa* 1128: 1–33.

Geiger, D. L. 2006b. *Sasakiconcha elegantissima* new genus and species (Gastropoda: Vetigastropoda: Anatomidae?) with disjointly coiled base. *The Nautilus* 120: 45–51.

Geiger, D. L. 2006c. Kleinste Schnecken gross geschrieben — der aktuelle Stand der Scissurelliden-Forschung. *Club Conchylia Informationen* 38 (1/2): 33–46.

Geiger, D. L. 2006d. Schaufelbaggern für Grosse: Mit dem Petite-Ponar Greiffer in Neu Britannien unterwegs. *Club Conchylia Mitteilungen* 5: 31–33.

Geiger, D. L. 2006e. Chapter 7. Applied film photography in systematic malacology. In: *The Mollusks, A Guide to their Study, Collection, and Preservation.* C. F. Sturm, T. A. Pearce & A. Valdés (eds): pp. 73–100. Universal Publishers, Inc., Boca Raton, Florida.

Geiger, D. L. 2006f. A new blind *Anatoma* species from the bathyal of the northeastern Pacific (Vetigastropoda: Anatomidae). *Molluscan Research* 28: 108–112.

Geiger, D. L. 2008a. Anatomidae & Scissurellidae. In: *Philippine Marine Mollusks, Volume 1.* G. T. Poppe (ed.): pp. 158–159. ConchBooks, Hackenheim.

Geiger, D. L. 2008b. Monographing micromolluscs: a case study on Scissurellidae s. l. (Vetigastropoda). *Zoosymposia* 1: 133–145.

Geiger, D. L. 2008c. New species of scissurellids from the Austral Islands, French Polynesia, and the Indo-Malayan Archipelago (Gastropoda: Vetigastropoda: Scissurellidae, Anatomidae, Larocheidae). *The Nautilus* 122: 185–200.

Geiger, D. L. 2009. A new species of *Depressizona* and the family rank of Depressizonidae. *Zootaxa* 2095: 57–59.

Geiger, D. L. 2011a. Family Scissurellidae. In: *Shells of the Hawaiian Islands - The Sea Shells.* Severn, M. (ed.): pp. 38–39. ConchBooks, Hackenheim.

Geiger, D. L. 2011b. Family Anatomidae. In: *Shells of the Hawaiian Islands - The Sea Shells.* Severn, M. (ed.): pp. 40–41. ConchBooks, Hackenheim.

Geiger, D. L. 2011c. *Anatoma emilioi,* a new species of Anatomidae McLean, 1989 from the Hawaiian Islands. In: *Shells of the Hawaiian Islands - The Sea Shells.* Severn, M. (ed.): pp. 534–535. ConchBooks, Hackenheim.

Geiger, D. L. & R. Fernández-Garcés. 2010. *Anatoma rolani* n. sp. and new records of rare Caribbean *Anatoma* species (Gastropoda: Vetigastropoda: Anatomidae). *Zootaxa* 2488: 65–68.

Geiger, D. L. & P. Jansen. 2004a. Revision of the Australian species of Anatomidae

Literature cited

(Gastropoda: Vetigastropoda). *Zootaxa* 435: 1–35.

Geiger, D. L. & P. Jansen. 2004b. New species of Australian Scissurellidae (Mollusca: Gastropoda: Vetigastropoda) with remarks on Australian and Indo-Malayan species. *Zootaxa* 714: 1–72.

Geiger, D. L. & B. A. Marshall. 2012. New species of Scissurellidae, Anatomidae, and Larocheidae (Mollusca: Gastropoda: Vetigastropoda) from New Zealand and beyond. *Zootaxa* 334: 1–33.

Geiger, D. L., B. A. Marshall, W. F. Ponder, T. Sasaki & A. Warén. 2007. Techniques for collecting, handling, and preparing small molluscan specimens. *Molluscan Research* 7: 1–50.

Geiger, D. L. & J. H. McLean. 2010. New species and records of Scissurellidae and Anatomidae from the Americas (Mollusca: Gastropoda: Vetigastropoda). *Zootaxa* 2356: 1–35.

Geiger, D. L., A. Nützel & T. Sasaki. 2008. Vetigastropoda. In: *Phylogeny and Evolution of the Mollusca.* W. Ponder & D. Lindberg (eds): pp. 297–330. University of California Press, Berkeley.

Geiger, D. L. & B. Owen. 2012. *Abalone: World-Wide Haliotidae.* ConchBooks, Hackenheim. viii, 361 pp.

Geiger, D. L. & B. Ruthensteiner (eds). 2008. Micromolluscs: Methodological Challenges — Exciting Results. *Zoosymposia* 1: i–vi, 1–308.

Geiger, D. L. & T. Sasaki. 2008. Four new species of Anatomidae (Mollusca: Vetigastropoda) from the Indian Ocean (Reunion, Mayotte) and Australia, with notes on a novel radular type for the family. *Zoosymposia* 1: 247–264.

Geiger, D. L. & T. Sasaki. 2009. New Scissurellidae and Anatomidae from Manazuru, Sagami Bay, and Okinawa, Japan (Mollusca: Gastropoda: Vetigastropoda). *Molluscan Research* 29: 1–16.

Geiger, D. L. & C. E. Thacker. 2005. Molecular phylogeny of Vetigastropoda reveals non-monophyletic Scissurellidae, Trochoidea, and Fissurelloidea. *Molluscan Research* 25: 47–55.

Geiger, D. L. & C. E. Thacker. 2008. Micromolluscs in molecular systematics: experiences and best practices. *Zoosymposia* 1: 39–45.

George, D. 2005. Marine Invertebrates. In: *The Emirates — A Natural History.* P. Hellyer & S. Aspinall (eds): pp. 197–222. Trident Press, London.

Geuze, G. J. 19994. *Schismope cingulata* niet zeldzaam op Cyprus. *Correspondentieblad van de Nederlandse Malacologische Vereniging* 278: 81.

Ghisotti, F., R. Giannuzzi-Savelli, E. Miari, G. Repetto, F. Pusateri. 1984. Malacofauna marina della Liguria — Area 013. *Notiziario SIM* 2(57–58): 104–105.

Giacobbe, S. 2002. Epibiontic mollusc communities on *Pinna nobilis* L. (Bivalvia, Mollusca). *Journal of Natural History* 36: 1385–1396.

Giannuzzi-Savelli, R. & P. Piani. 1990. Revisione nomenclatorale del taxa specifici Monterosatiani. *Lavori della Società Italiana di Malacologia* 23: 141–199.

Giannuzzi-Savelli, R., F. Pusateri, A. Palmeri & C. Ebreo, C. 1994. *Atlas of Mediterranean Sea Shells, Volume* 1. La Conchiglia, Rome. 125 pp.

Gignoux, M. 1913. Les formations marines Pliocènes et Quaternaires de l'Italie du sud et de al Sicilie. *Annales de l'Université de Lyon* Nouvelle Serie, 1: 1–633, pls 1–21.

Glayzer, B., D. T. Glayzer & K. R. Smythe. 1984. The marine Mollusca of Kuwait, Arabian Gulf. *Journal of Conchology* 31: 311–330.

Glibert, M. 1962. Les Archaeogastropoda fossiles du Cénozoïque étranger des collections de L'Institut Royal des Sciences Naturelles de Belgique. *Institut Royal des Sciences Naturelles de Belgique* Deuxième Serie, 68: 1–131.

Glibert, M. 1973. Révision des Gastropoda du Danien et du Montien de la Belgique I. Les Gastropoda du Calcaire de Mons. *Mémoires de l'Institut Royal des Sciences Naturelles de Belgique* 173: 3–116, pls 1–11.

Gofas, S. 2011. La distribución geográfica de los moluscos. In: *Moluscos Marinos de Andalucia*

S. Gofas, D. Moreno & C. Salas (eds). Servicio de Publicaciones e Intercambio Científico, Universidad de Málaga: pp. 23–30.

Gofas, S., J. Pinto Afonso & M. Brandào. nd [~1985]. *Coquillages et Mollusques d'Angola*. Universidade Agnostinho Neto, Palanca. 144 pp.

Goldfuss, G. A. 1841–1844. *Abbildungen und Beschreibungen der Petrefacten Deutschlands und der angränzenden Länder III*. Arnz & Co, Düsseldorf. 121 pp. [not seen].

Goldstein, J. I., D. E. Newbury, P. Echlin, D. C. Joy, A. D. Roming, Jr., C. E. Lyman, C. Fiori & E. Lifshin. 1992. *Scanning Electron Microscopy and X-Ray Microanalysis. A Text for Biologists, Material Scientists, and Geologists*. Plenum Press, New York. 820 pp.

Golikov, A. N. 1995. *Shell-bearing Gastropoda of the Arctic*. Colus, Moskow. 108 pp.

Golikov, A. N. & V. V. Gublin. 1978. Prosobranchial gastropods of the Kurile Islands: I: Orders Docoglossa — Entomostoma. In: *Fauna and Vegetation of the Shelf of the Kurile Islands*. O. G. Kussakin (ed.): pp. 159–223. Nauka, Moscow.

Golikov, A. N. & B. I. Sirenko. 1980. New species of the subclass Scutibranchia from the Sea of Japan. *Academy of Sciences of the USSR, Zoological Institute, Explorations of the Fauna of the Seas* 25: 105–108.

Golikov, A. N. & B. I. Sirenko. 2004. Class Gastropoda, subclass Cyclobranchia, subclass Scutibranchia, subclass Pectinibranchia, subclass Sinistrobranchia. In: *Fauna and Ecosystems of the Laptev Sea and Adjacent Deep Waters of the Arctic Basin. Issledovaniya Fany Morei*, 54(62), pt. 2: pp. 145–147 [not seen].

Golikov, A. N., B. I. Sirenko, V. V. Gulbin, & E. M. Chaban. 2001. Checklist of the shell-bearing gastropods of the northwestern Pacific. *Ruthenica* 11: 153–173.

Gorbunov, G. P. 1946. Bottom life of the New Siberian shallows and the central part of the Arctic Ocean. In: *Transactions of the Drifting Expedition of Glasvsevmrput on Ice-Breaking Steamer (G. Sedov) in 1937–1940. Vol. 3. Biology*. Buinitzky, V. Kh. (ed.): pp. 30–138. Izdatelstovo Glavsevmorputi, Moscow-Leningrad [not seen].

Gordillo, S., A. M. J. Coronato & J. O. Rabassa. 2005. Quaternary molluscan faunas from the island of Tierra del Fuego after the last glacial maximum. *Scientia Marina* 69, Supplement 2: 337–348.

Gordon, W. J. 1901. *Our Country's Shells and How to Know Them*. Simpkin, Marshall, Hamilton, Kent & Co Ltd, London. 152 pp., 33 pls.

Gosse, P. H. 1856. *A Manual of Marine Zoology for the British Isles, Part II*. John van Voorst, London. 239 pp.

Götting, K.-J. 1974. *Malakozoologie*. Gustav Fischer, Stuttgart. 320 pp.

Gougerot, L. & J. LeRenard. 1977. Nouvelles espèces de petits gastéropodes marins de l'Éocene des bassins de Paris, de Nantes et du Cotentin. *Bulletin d'Information des Géologues du Bassin de Paris* 14: 3–33.

Graham, A. 1971. *Synopsis of the British Fauna (New Series) No. 2*. British Prosobranchs. Academic Press, London. 112 pp.

Graham, A. 1988. *Molluscs: Prosobranch and Pyramidellid Gastropods (Second Edition)*. Linnean Society, London. 662 pp.

Granger, A. 1885. *Histoire Naturelle de la France. 6e partie Mollusques (Céphalopodes, Gastéropodes)*. Émile Deyrolle, Paris. 272 pp., 20 pls.

Grasset, J.-P. A. 1884. *Index Testaceorum Viventium*. Casablanca, Alger. 324 pp.

Grasso, S., A. Giudice, F. Mannucci, E. Miari, & R. Giannuzzi-Savelli. 1983. Malacofauna marina della Sicilia — Area 144. *Notiziario SIM* 1(7–8): 36–40.

Gray, J. E. 1847. A list of the genera of Recent Mollusca, their synonyms and types. *Proceedings of the Zoological Society of London* 15: 129–219.

Gray, J. E. 1854a. *List of the Shells of South America in the Collection of the British Museum Collected and Described by M. Alcide d'Orbigny, in the "Voyage dans l'Amérique Méridonale"*. Taylor & Francis, London. 89 pp.

Gray, J. E. 1854b. *List of Shells of the Canaries in the Collection of the British Museum Collected by MM Webb and Berthelot, Described and Figured by Pro. Alcide D'Orbigny, in the "Histoire Naturelle des Îles Canaries."*. Taylor and Francis, London. 32 pp.

Gregorio, A. de. 1889. Esame di taluni molluschi viventi e terziari del Bacino Mediterraneo. *Naturalista Siciliano* 8: 1–26, pls 1–2.

Grieg, J. A. 1896. Bidrag til kundskaben om vestlandets mollusker. *Bergens Museums Aarbog* 10: 3–32, pl. 1.

Grieg, J. A. 1913. Marine mollusker fra Indre Sogn. *Nyt Magazin for Naturviden-skaberne* 51: 27–42.

Grossu, A. V. 1986. *Gastropoda Romaniae I: Characterele generale, istoricul si biologia gasteropodelor II. Subclasa Prosobranchia si Opisthobranchia.* Ditura Litera, Bucharest. 523 pp.

Gublin, V. V. 2006. Catalogue of the shell-bearing gastropods in the Russian waters of the Sea of Japan. Part 1. *The Bulletin of the Russian Far East Malacological Society* 10: 5–28.

Gürs, K. 1995. *Revision der marinen Molluskenfauna des Unteren Meeressandes (Oligozän, Rupelium) des Mainzer Beckens.* Dissertation, Mainz. 314 pp.

Gutt, J., B. I. Sirenko, W. E. Arntz, I. S. Smirnov, & C. De Broyer. 2000. Biodiversity of the Weddell Sea: macrozoobenthic species (demersal fish included) sampled during the expedition ANT XIll13 (EASIZ I) with RV "Polarstern". *Berichte zur Polarforschung* 372: 1–103.

Habe, T. 1951. Scissurellidae in Japan. *Illustrated Catalogue of Japanese Shells* 11: 65–69.

Habe, T. 1961. *Colored Illustrations of Shells of Japan, Volume 2, First Edition.* Hoikusha, Osaka. 183 pp., 42 pp. Appendix.

Habe, T. 1963. Eight minute species of shells from Amami Islands, far south of Kyushu, Japan, including six new species. *Venus* 22: 229–237.

Habe, T. & T. Kikuchi. 1960. *Fauna and flora of the sea round the Amakusa Marine Biological Laboratory. Part 1. Mollusca.* The Amakusa Marine Biological Laboratory, Kyushu University, Tomioka. 70 pp.

Habe, T., T. Kubota, A. Kawakami & O. Masuda. 1986. Check list of the shell-bearing Mollusca of Suruga Bay, Japan. *Science Reports of the Natural History Museum of Tokai University* 1: 1–44.

Habe, T. & S. Kosuge. 1964. Superfamily Pleurotomarioidea. List of the Indo-Pacific Mollusca concerning to the Japanese Fauna 1: 1–8.

Habe, T. & O. Masuda. 1990. Catalogue of the molluscan shells donated by Mr. Hiroshi Noguchi to the Natural History Museum, Tokai University. *Science Reports of the Natural History Museum Tokai University* 4: 1–152, pls 1–4.

Habe, T. & K. Tsuchiya. 1998. Checklist of the malacofauna of Akajima Island and adjacent area, Okinawa Prefecture. *Midoriishi, Report of the Akajima Marine Science Laboratory, Establishment of Tropical Marine Ecological Research* 9: 15–25.

Hain, S. 1989. *Beiträge zur Biologie der Beschalten Mollusken (Kl. Gastropoda und Bivalvia) des Weddellmeeres, Antarktis.* Dissertation. University Bremen, Bremen. 289 pp.

Hain, S. 1990. The benthic seashells (Gastropoda and Bivalvia) of the Weddell Sea, Antarctica. *Berichte zur Polarforschung* 70: 1–181.

Haller, B. 1894. *Studien über Docoglosse und Rhipidoglosse Prosobranchier.* Wilhelm Engelmann, Leipzig. 173 pp., 12 pls.

Hallström, C & M. Wabäck. 2004. *The Ex in Beautiful Frenzy.* Cutproductions, Haarlem, The Netherlands. DVD.

Hansson, H. G. 2003. Faktablad *Anatoma crispata. ArtDatabanken* 2003-09-10. [http://www.artdata.slu.se/rodlista/Faktablad/anat_cri.PDF: accessed 12/8/2008]

Harasewych, M. G. & F. Moretzsohn. 2010. *The Book of Shells: a Life-Size Guide to Identifying and Classifying Six Hundred Seashells.* Ivy Press, Lewes, Great Britain. 656 pp.

Hardy, G. H. 1916. List of the Tenison-Woods types of Recent Mollusca in the Tasmanian Museum. *Papers and Proceedings of the Royal Society of Tasmania* 1915: 61–73.

Hartman, O. 1959. Catalogue of the polychaetous annelids of the world, part II. *Allan Hancock Foundation Publications* Occasional Paper 23: 355–628.

Hasegawa, K. 2001. Deep-sea gastropods of Tosa Bay, Japan, collected by R/V *Kotaka-*

Maru and *Tansei-Maru* during the years 1997–2000. In: *Deep-Sea Fauna and Pollutants in Tosa Bay. National Science Museum Monograph, No 20.* T. Fujita, H. Saito, & M. Takeda (eds): pp. 121–165. National Science Museum, Tokyo.

Hasegawa, K. 2004. Swimming behavior of scissurellids (Gastropoda: Scissurellidae) and its taxonomic significance. In: *Molluscan Megadiversity: Sea, Land and Freshwater. World Congress of Malacology Perth, Western Australia 11–16 July 2004.* F. E. Wells (ed.): p. 62. Western Australian Museum, Perth.

Hasegawa, K. 2005. A preliminary list of deep-sea gastropods collected from the Nansei Islands, Southwestern Japan. In: *Deep-Sea Fauna and Pollutants in Nansei Islands. National Science Museum Monographs, No. 29.* K. Hasegawa, G. Shinohara & M. Takeda (eds): pp. 137–190. National Science Museum, Tokyo.

Hasegawa, K., H. Saito, T. Kubodera & F. Xu. 2001. Marine molluscs collected from the shallow waters of Hainan Island, South China Sea, by China-Japan Joint Research in 1997. In: *Marine Fauna of the Shallow Waters around Hainan Island, South China Sea. National Science Museum Monograph, No 21.* K. Matsuura (ed.): pp. 1–43, pls 1–9. National Science Museum, Tokyo.

Haszprunar, G. 1987. The fine structure of the ctenidial sense organs (bursicles) of Vetigastropoda (Zeugobranchia, Trochoidea) and their functional and phylogenetic significance. *Journal of Molluscan Studies* 53: 46–51.

Haszprunar, G. 1988a. On the origin and evolution of major gastropod groups, with special reference to the Streptoneura. *Journal of Molluscan Studies* 54: 367–441.

Haszprunar, G. 1988b. *Sukashitrochus* sp., a scissurellid with heteropod-like locomotion (Mollusca, Archaeogastropoda). *Annalen des Naturhistorischen Museums Wien* 90/B: 367–371.

Haszprunar, G. 1989. New slit-limpets (Scissurellacea and Fissurellacea) from hydrothermal vents. Part 2 Anatomy and relationship. *Contributions in Science of the Los Angeles County Museum of Natural History* 408: 1–17.

Hayashi, I. 1983. Shell morphology of veliger larvae of juveniles of the Japanese top shell, *Turbo (Batillus) cornutus. Venus* 42: 212–216.

Hayward, P. J., G. D. Wigham, G. D. & N. Yonow. 1990. Mollusca I: Polyplacophora, Scaphopoda, and Gastropoda. In: *The Marine Fauna of the British Isles and North-West Europe, Volume 2.* P. J. Hayward & J. S. Ryland (eds): pp. 628–730. Clarendon Press, Oxford.

Healy, J. 1990. Sperm structure in the scissurellid gastropod *Sinezona* sp. (Prosobranchia, Pleurotomariidae). *Zoologica Scripta* 19: 189–193.

Hedegaard, C. 1990. *Shell Structures of the Recent Archaeogastropoda, 2 Vols.* Department of Ecology and Genetics, University of Aarhus, Denmark. 154 pp., 78 pls.

Hedegaard, C. 1997. Shell structures of the Recent Vetigastropoda. *Journal of Molluscan Studies* 63: 369–377.

Hedley, C. 1899. The Mollusca of Funafuti (supplement). *Memoirs of the Australian Museum* 3: 549–565.

Hedley, C. 1901. Studies on Australian Mollusca Part 3. *Proceedings of the Linnean Society of New South Wales* 25: 721–732, pl. 48.

Hedley, C. 1902. Studies on Australian Mollusca Part 4. *Proceedings of the Linnean Society of New South Wales* 26: 16–23, pl. 2.

Hedley, C. 1903. Mollusca Part II. Scaphopoda and Gastropoda. *Memoirs of the Australian Museum* 4: 327–402, pls 36–38.

Hedley, C. 1904. Additions to the marine molluscan fauna of New Zealand. *Records of the Australian Museum* 5: 86–97.

Hedley, C. 1905. Mollusca from one hundred and eleven fathoms, east of Cape Byron, New South Wales. *Records of the Australian Museum* 6: 41–54.

Hedley, C. 1907a. The Mollusca of Mast Head Reef, Capricorn Group, Queensland. Part II. *Proceedings of the Linnean Society of New South Wales* 32: 476–513, pls 16–21.

Hedley, C. 1907b. The results of deep-sea investigation in the Tasman Sea. I.—The expedition of H.M.C.S. "Miner." 3.—Mollusca from eighty fathoms off Narrabeen. *Records of the Australian Museum* 6: 283–304, pls 54–56.

Hedley, C. 1911. Report on the Mollusca obtained by the F.I.S. "Endeavour," chiefly off Cape Wiles, South Australia. Part I. *Zoological Results of the Fishing Experi-ments Carried out by the F. I. S. "Endavour," 1909–1910.* 1: 90–114, pls 17–20.

Hedley, C. 1916a. Mollusca. Australasian Antarctic Expedition 1911–1914. *Scientific Reports Section C.—Zoology & Botany* 4(1): 1–80, pls 1–9.

Hedley, C. 1916b. Report on Mollusca from elevated marine beds, "raised beaches," off McMurdo Sound. In: *Geology (Vol. 2) British Antarctic Expedition, 1907–1909*: pp. 85–88. Heinemann, London.

Hedley, C. 1918. A check-list of the marine fauna of New South Wales. *Journal of the Royal Society of New South Wales*, Supplement 51: M1–M120.

Hedley, C. & W. L. May. 1908. Mollusca from one hundred fathoms, seven miles east of Cape Pillar, Tasmania. *Records of the Australian Museum* 7: 108–125, pls 22–25.

Hendrickx, M. E. & R. C. Brusca. 2005. Mollusca: Pelecypoda, Polyplacophora, Gastropoda, Monoplacophora, Scaphopoda & Cephalopoda. In: *A Distrbutional Checklist of the Macrofauna of the Gulf of California, Mexico.* M. D. Hendrickx, R. C. Brusca & L. T. Findley (eds): pp. 195–310. Arizona-Sonora Desert Museum.

Herbert, D. G. 1986. A revision of the southern African Scissurellidae (Mollusca: Gastropoda: Prosobranchia). *Annals of the Natal Museum* 16: 9–29.

Herbert, D. G. 1987. Scissurellids — tiny treasures. *The Strandloper* 220: 1–2.

Hernández, J. M., E. Rolán, & F. Swinnen. 2011. Parte 3: Gastropoda: Prosobranchia. In: *Moluscos y Conchas Marinas de Canarias.* E. Rolán (ed.): pp. 54–269. ConchBooks, Hackenheim.

Herzenstein, S. 1885. Beiträge zur Kenntniss der Fauna der Murmanküste und des Weissen Meeres. I. Mollusca. *Transactions of the St.-Petersburg Society of Naturalists* 16: 635–814 [in Russian with German summary].

Hescheler, K. 1901. Über die Gattung *Pleurotomaria. Biologisches Centralblatt* 21: 569–582.

Hickman, C. S. 1980. Gastropod radulae and the assessment of form in evolutionary paleobiology. *Paleobiology* 6: 276–294.

Hickman, C. S. 1981. Selective deposit feeding by the deep-sea archaeogastropod *Bathybembix aeola. Marine Ecology - Progress Series* 6: 339–342.

Hickman, C. S. 1998. Superfamily Pleurotomarioidea. In: *Mollusca: the Southern Synthesis.* P. L. Beesley, G. J. B. Ross & A. Wells (eds): pp. 664–669. CSIRO Publishing, Melbourne.

Hickman, C. S. 1999. Sexual dimorphism and contact pairing in *Sinezona* sp. (Vetigastropoda: Scissurellidae). In: *The Seagrass Flora and Fauna of Rottnest Island, Western Australia.* D. I. Walker & F. E. Wells (eds): pp. 129–135. Western Australian Museum, Perth.

Hickman, C. S. & S. S. Porter. 2007. Nocturnal swimming, aggregation at light traps, and mass spawning of scissurellid gastropods (Mollusca: Vetigastropoda). *Invertebrate Biology* 126: 10–17.

Higo, S. 1973. A *Catalogue of Molluscan Fauna of the Japanese Islands and the Adjacent Area.* Privately published, Japan. 397 pp., 61 pp. index.

Higo, S. & Y. Goto. 1993. *A Systematic List of Molluscan Shells from the Japanese Islands and the Adjacent Area.* Tenoji, Osaka. 22 pp., 693 pp., 13 pp., 148 pp.

Higo, S., P. Callomon, P. & Y. Goto. 2001. *Catalogue and Bibliography of the Marine Shell-Bearing Mollusca of Japan. Gastropoda, Bivalvia, Polyplacophora, Scaphopoda. Type Figures.* Elle Scientific Publications, Osaka. 749 pp., 2 maps.

Hildago, J. G. 1867. Catalogue des mollusques testacés marins des côtes de l'Espagne et des Îles Baléares. *Journal de Conchyliologie* 15: 1–163, pl. 12.

Hildago, J. G. 1890. Estudios preliminares subre los moluscos terrestres y marinos de España, Portugal y las Baleares. *Memorias de la Real Acedemia de Ciencias Exactas, Fisica y Naturales* 15: 1–735.

Hildago, J. G. 1904. Catalogo de los moluscaos testáceos de las Island Filipinas, Joló y Marinanas. *Revista de la Real Academia de Ciencias Exactas, Fisicas y Naturales de Madrid* 2: 9–406.

Hildago, J. G. 1916. *Fauna Malacológica de España, Portugal y las Baleares*. Museo Nacional de Ciencias Naturales, Madrid. 752 pp.

Hoel, A. 1907. Kvaraergelogiske undersøgelser i nordre Trondhjems og Nordlands Amter. *Archiv for Mathematik og Naturvidensab* 28: 3–64.

Honma, Y. & T. Kitami. 1979. Fauna and flora in the waters adjacent to the Sado Marine Biological Station, Niigata University: Supplement 1. *Annual Report of the Sado Marine Biological Station Niigata University* 9: 27–36.

Horst, R. & M. M. Schepman. 1908. Catalogue systématique des mollusques (gastropodes prosobranches et polyplacophores). *Muséum d'Histoire Naturelle des Pays-Bas* 13: 1–572.

Høisaeter, T. 1986. An annotated check-list of marine molluscs of the Norwegian coast and adjacent waters. *Sarsia* 71: 73–145.

Høisaeter, T. 2010. The shell-bearing, benthic gastropods on the southern part of the continental slope off Norway. *Journal of Molluscan Studies* 76: 234–244.

Høisaeter, T. & D. L. Geiger. 2011. Species of *Anatoma* (Gastropoda: Anatomidae) in Norwegian and adjacent waters, with the description of two new species. *The Nautilus* 125: 89–112.

Hu, C-H. & H-J. Tao. 2000. Molluscan fossil fauna from the Leuchungchi Formation (Pliocene), eastern Chiayi District. *Molluscan Fauna of Taiwan* 5: 43–68.

Hubendick, B. & A. Warén. 1974. Småsnäckor från svenska västkusten 6. Släktena *Emarginula, Lacuna, Natica, Philbertia, Lora* m. fl. *Årstyrck Göteborgs Naturhistoriska Museum* 1974: 25–32.

Hutton, F. W. 1885. Revision of the Recent rhipidoglossate and docoglossate Mollusca of New Zealand. *Proceedings of the Linnean Society of New South Wales* 9: 354–378.

Hutton, F. W. 1893. The Pliocene Mollusca of New Zealand. In: *The Macleay Memorial Volume*. J. J. Fletcher (ed.): pp. 35–92, pls 6–9. Linnean Society of New South Wales, Sydney.

Hutton, F. W. 1904. *Index Faunae Novae Zealandiae*. Philosophical Institute of Canterbury/ Dulay & co., London. 372 pp.

Inaba, A. 1982. *Fauna and Flora of the Seto Inland Sea, Japan*. Hiroshima Shell Club, Hiroshima, Japan. 181 pp.

Inaba, A. 1983. *Fauna and Flora of the Seto Inland Sea, Second Edition I (Mollusca)*. Mukaishima Marine Biological Station, Hiroshima, Japan. 181 pp.

Iredale, T. 1908a. Notes on some New Zealand marine molluscs. *Transactions and Proceedings of the New Zealand Institute* 40: 373–387.

Iredale, T. 1908b. A preliminary list of the marine Mollusca of Banks Peninsula, New Zealand. *Transactions and Proceedings of the Royal Society of New Zealand* 40: 387–403.

Iredale, T. 1908c. List of marine molluscs collected in Otago. *Transactions and Proceedings of the New Zealand Institute* 40: 404–410.

Iredale, T. 1924. Results from Roy Bell's molluscan collections. *Proceedings of the Linnean Society of New South Wales* 49: 179–278, pls 33–36.

Iredale, T. & D. F. McMichael. 1962. A reference list of the marine Mollusca of New South Wales. *The Australian Museum, Sydney, Memoir* 11: 1–109.

Issel, A. 1870. *Malacologia del Mar Rosso*. Biblioteca Malacologica, Pisa. 387 pp., 5 pls.

Itoigawa, J. & H. Ogawa. 1973. Pleistocene molluscan fauna of the Sakishima Formation, Shima Peninsula, Central Japan. *The Science Reports of the Tohoku University 2nd Series, Geology*, Special Volume 6: 69–80.

Ivanov, D. L. & A. V. Sysoev. 2000. Types of Mollusca in the Zoological Museum of Moscow University. *Archives of the Zoological Museum of the Moscow State University* 40: 1–187.

Izawa, N. & K. Matsuoka. 1999. Catalogue of shell collection by Mr. Hiroshi Takakuwa presented to Toyohashi Museum of Natural History. *Miscellaneous Report of the Toyohashi Museum of Natural History* 7: 1–25.

Jansen, P. 1995. *Seashells of Central New South Wales.* Jansen, Townsville. 129 pp.

Jansen, P. 1998. *Sinezona pacifica* in Australia. *Australasian Shell News* 99–100: 4.

Jansen, P. 1999. The Australian Scissurellidae. *La Conchiglia* 30: 47–55, 64.

Janssen, A. W. 1984. *Mollusken uit het Mioceen van Winterswijik-Miste.* Koninklijke Nederlandse Natuurhistorische Vereniging, Amsterdam. 451 pp., 82 pls.

Janssen, R. 1978. Die Mollusken des Oberoligozäns (Chattium) im Nordsee-Becken. *Archiv für Molluskenkunde* 109: 137–227.

Jaume, M. L. & H. Sarasúa. 1943. Notas sobre moluscos marinos cubanos. *Revista de la Sociedad Malacológica "Carlos de la Torre"* 1: 52–61.

Jeffreys, C. 1887. *A Complete Catalogue of British Mollusca.* Herbert W. Marsden, Gloucester. 53 pp.

Jeffreys, J. G. 1856a. On the marine Testacea of the Piedmontese coast. *Annals and Magazine of Natural History* Series 2, 17: 155–189, pl. 2.

Jeffreys, J. G. 1856b. Further notices of Piedmontese Mollusca. *Annals and Magazine of Natural History* Series 2, 17: 271–272.

Jeffreys, J. G. 1856c. Note on the genus *Scissurella*. *Annals and Magazine of Natural History* Series 2, 17: 319–322.

Jeffreys, J. G. 1856d. On *Scissurella* and *Schismope*. *Annals and Magazine of Natural History* Series 2, 17: 470–471.

Jeffreys, J. G. 1856e. A last word on *Scissurella*. *Annals and Magazine of Natural History* Series 2, 18: 188–189.

Jeffreys, J. G. 1860. *Sui testacei marini delle Costa del Piemonte: Traduzione Con Note Ed Un Catalogo Speciale Per Il Golfo Della Spezia Del Prof. G. Capellini.* Co' Tipi del R. I. De' Sordo-Muti, Genova. 87 pp., 1 pl. [in older works credited to Capellini].

Jeffreys, J. G. 1865. *British Conchology or an Account of the Mollusca which now Inhabit the British Isles and the Surrounding Seas.* John Van Voorst, London. 393 pp., 8 pls.

Jeffreys, J. G. 1868. Last report on dredging among the Shetland Isles. *Annals and Magazine of Natural History* Series 4, 2: 298–316.

Jeffreys, J. G. 1870a. Mediterranean Mollusca. *Annals and Magazine of Natural History* Series 4, 6: 65–87.

Jeffreys, J. G. 1870b. Norwegian Mollusca. *Annals and Magazine of Natural History* Series 4, 5: 438–448.

Jeffreys, J. G. 1874. Some remarks on the Mollusca of the Mediterranean. *Report of the British Association for the Advancement of Science* 1874: 111–116.

Jeffreys, J. G. 1877. New and peculiar Mollusca of the Patellidae and other families of Gastropoda procured in the 'Valorous' Expedition. *Annals and Magazine of Natural History* Series 4, 19: 231–243.

Jeffreys, J. G. 1878. On the Mollusca procured during the 'Lightning' and 'Porcupine' Expeditions, 1868–70. (Part I.). *Proceedings of the Zoological Society, London* 1878: 393–416.

Jeffreys, J. G. 1883a. On the Mollusca procured during the 'Lightning' and 'Porcupine' Expeditions, 1868–1870 (Part VI.). *Proceedings of the Zoological Society of London* 1883: 88–115.

Jeffreys, J. G. 1883b. Mediterranean Mollusca (No. 3) and other Invertebrata. *Annals and Magazine of Natural History* Series 5, 11: 393–401.

Johnson, C. W. 1915. Fauna of New England 13. List of the Mollusca. *Boston Society of Natural History Occasional Papers* 7: 3–231.

Johnson, C. W. 1934. List of marine Mollusca of the Atlantic coast from Labrador to Texas. *Proceedings of the Boston Society of Natural History* 40: 1–204.

Jousseaume, F. 1888. Description des mollusques recueillis par M. de Dr. Faurot dans la Mer Rouge et le Golfe d'Aden. *Mémoires de la Société Zoologique de France* 1: 165–223.

Kabasakal, H. & E. Kabasakal. 2005. Prosobranch gastropods collected from the Anatolian coast of North Aegean Sea. *Club Conchylia Informationen* 37: 49–57.

Kabat, A. R. & K. J. Boss. 1989. An index catalog of publications on molluscan type specimens. *Occasional Papers on Mollusks, Museum of Comparative Zoology* 5(69): 157–336.

Kaiser, K. L. 1993. An overview of the known Recent micromolluscan marine fauna of the Islas Galápagos including microfaunal list. *The Festivus* 25: 90–109.

Kaiser, K. L. 2009. Les Mollusques. In: *Clipperton Environment et Biodiversité d'un Microcosme Océanique*. L. Charpy (ed.): pp. 217–234. Muséum National d'Histoire Naturelle, Paris.

Kaiser, K. L. & C. W. Bryce. 2001. The Recent molluscan marine fauna of Isla de Malpelo, Colombia. *The Festivus* Occasional Paper 1: 1–149.

Kano, Y. 2008. Vetigastropod phylogeny and a new concept of Seguenzioidea: independent evolution of copulatory organs in the deep-sea habitats. *Zoologica Scripta* 37: 1–21.

Kano, Y., E. Chikyu & A. Warén. 2009. Morphological, ecological and molecular characterization of the enigmatic planispiral snail genus *Adeuomphalus* (Vetigastropoda: Seguenzioidea). *Journal of Molluscan Studies* 75: 397–418.

Kantor, I. Yu. & A. V. Sysoev. 2005. *Catalogue of Molluscs of Russia and Adjacent Countries*. KMK Scientific Press Ltd, Moscow. 627 pp.

Kantor, I. Yu. & A. V. Sysoev. 2006. *Marine and Brackish Water Gastropoda of Russia and Adjacent Countries: an Illustrated Catalogue*. KMK Scientific Press Ltd, Moscow. 371 pp., 140 pls.

Kase, T. & Y. Kano. 2002. *Trogloconcha*, a new genus of larocheine Scissurellidae (Gastropoda: Vetigastropoda) from tropical Indo-Pacific submarine caves. *The Veliger* 45: 25–32.

Kawamoto, T. & Tanabe, S. 1956. *Catalogue of Molluscan Shells of Yamaguti Prefecture*. Yamaguti Prefectural Yamagui Museum, Yamaguti. 170 pp.

Kawamura, T., H. Takami, R. D. Roberts & Y. Yamashita. 2001. Radula development in abalone *Haliotis discus hannai* from larva to adult in relation to feeding transition. *Fisheries Science* 67: 596–605.

Kay, E. A. 1979. *Hawaiian Marine Shells*. Reef and Shore Fauna of Hawaii, Section 4: Mollusca. Bishop Museum Press, Honolulu. 653 pp.

Kay, E. A. & S. Johnson. 1987. Mollusca of Enewetak Atoll. In: *The Natural History of Enewetak Atoll, Volume 2 Biogeography and Systematics*. D. M. Devaney, E. S. Reese, B. L. Burch, & P. Helfrich (eds): pp. 105–146. United States Department of Energy, Oak Ridge, Tennessee.

Keen, A. M. 1937. *An Abridged Check List and Bibliography of West North American Marine Mollusca*. Stanford University Press, Stanford. 87 pp.

Keen, A. M. 1958. *Sea Shells of Tropical West America*. Stanford University Press, Stanford. 624 pp.

Keen, A. M. 1968. West American mollusks types at the British Museum (Natural History) IV. Carpenter's Mazatlan Collection. *The Veliger* 10: 389–439, pls 55–59.

Keen, A. M. & J. C. Pearson. 1952. *Illustrated Key to West North American Gastropod Genera*. Stanford University Press, Stanford. 39 pp.

Kensley, B. 1973. *Sea-Shells of Southern Africa: Gastropods*. Maskew Miller, Cape Town. 225 + 11 pp.

Kidder, J. H. 1876. Contributions to the natural history of Kerguelen Islands. *Smithsonian Miscellaneous Collections* 13(2): 1–122.

Kiel, S. 2001. *Taxonomy and Biogeography of Late Cretaceous Gastropoda*. Dissertation University of Hamburg, Hamburg. 162 pp., 37 pls.

Kiel, S. & K. Bandel. 2000. New slit bearing Archaeogastropoda from the Late Cretaceous of Spain. *Berliner Geowissenschaftliche Abhandlungen* E34: 269–277.

Kiel, S., K. A. Campbell & C. Gaillard. 2010. New and little known mollusks form the ancient chemosynthetic environments. *Zootaxa* 2390: 26–48.

Kilburn, R. N. 1996. Notes on the benthic stations sampled by the Deutsche Tiefsee-Expedition (1898–1899) in Southern African waters. *South African Journal of Marine*

Literature cited

Sciences 17: 325–328.

Kisch, B. S. 1960. Les mollusques décrits par de Folin à part les Caecidae et Cheminitzidae avec catalog. *Journal de Conchyliologie* 100: 137–162.

Kittl, E. 1891. Die Gastropoden der Schichten von St. Cassian der südalpinen Trias. *Annalen des K. K. Naturhistorischen Hofmuseums, Wien* 6: 166–262, pls 1–7.

Kittl, E. 1899. Die Gastropoden der Esinokalke, nebst einer Revision der Gastropoden der Marmolatakalke. *Annalen des Kaiserlich Königlichen Naturhistorischen Hofmuseums* 14: 1–237.

Kobelt, W. 1878. *Illustrirtes Conchylienbuch*. Bauer & Raspe, Nürnberg. 391 pp., 112 pls.

Kobelt, W. 1887. Bericht über die wissenschaftlichen Leistungen im Gebiete der Malakologie während des Jahres 1886. *Archiv für Naturgeschichte* 53: 373–448.

Kobelt, W. 1888. *Prodromus Faunae Molluscorum Testaceorum Marina Europaea Inhabitantium*. Bauer & Raspe, Nürnberg. 550 pp.

Kobelt, W. 1891. Bericht über die geographische Verbreitung die Systematik und die Biologie etc. der Mollusken. *Archiv für Naturgeschichte* 57: 325–380.

Kobelt, W. 1905. Mollusca für 1901–1904. Geographische Verbreitung, Systematik, Biologie. *Archiv für Naturgeschichte* 65: 79–264.

Koenen, A. v. 1867. Das marine Mittel-Oligocän Norddeutschlands und seine Mollusken-Fauna, erster Theil. *Palaeontographica* 16: 1–148, pls 1–30.

Koenen, A. v. 1892. *Das Norddeutsche Unter-Oligocän und seine Mollusken-Fauna, Lieferung IV*. Simon Schropp'sche Hof-Landkartenhandlung, Berlin. pp. 819–1004, pls 53–62.

Kollmann, H. A. & J. S. Peel. 1983. Paleocene gastropods from Nûgssuaq, West Greenland. *Grønlands Geologiske Undersøgelse Bulletin* 146: 1–115.

Kosuge, S. 1979. Report on the Mollusca collected from Ishikari Bay and its adjacent waters by the R. V. *Tansei-Maru* during Cruise KT-67-7 (1967). *Bulletin of the Institute of Malacology Tokyo* 1: 9–12.

Koyama, Y. 2008. Catalogue of molluscan order Vetigastropoda from Wakayama Prefecture (1). Families Scissurellidae, Pleurotomariidae, Haliotidae and Fissurellidae. *Nanki Seibutu* 50: 253–260.

Kuntz, P. 1988. À la pêche aux scissurelles en Méditerranée. *Xenophora* 44: 19.

Kuroda, T. & T. Habe. 1952. *Check List and Bibliography of the Recent Marine Mollusca of Japan*. Leo W. Strach, Tokyo. 210 pp., 1 map.

Kuroda, T. & T. Habe. 1954. On some Japanese Mollusca described by A. Adams, whose specimens are deposited in the Redpath Museum of Canada. *Venus* 18: 1–16, pls 1–2.

Kuroda, T., T. Habe & K. Oyama. 1971. *The Sea Shells of Sagami Bay*. Maruzen, Tokyo. 489 + 51 pp., 121 pls.

Kuroda, T. & T. Kinoshita. 1951. A catalogue of marine molluscan shells of Hokkaido. *Bulletin of the Hokkaido Regional Fisheries Research Laboratory Fisheries Agency* 2: 1–40.

Kurtz, I. 2001. Spanien: Cabo da Gata — ein Reisebericht. *Club Conchylia Inforationen* 33(1–3): 78–81.

Kurtz, I. 2011. Sammeln auf Mauritius. *Club Conchylia Mitteilungen* 17: 34–39.

Kuster-Wendenburg, E. 1982. Bestandesaufnahme der Gastropoden im "prä-aquitanen" Tertiär des Mainzer Beckens. *Mainzer geowissenschaftliche Mitteilungen* 10: 83–130.

Kwon, O. K., D. K. Min, J. R. Lee, J. S. Lee, J. G. Je & B. L. Choe. 2001. *Korean Mollusks with Color Illustrations*. Hanguel, Busan. 332 pp.

Ladd, H. S. 1982. Cenozoid fossil mollusks from Western Pacific Islands; gastropods (Eulimidae and Volutidae through Terebridae). *United States Geological Survey Professional Paper* 1171: 1–100, pls 1–41.

Ladd, H. S. 1970. Eocene Mollusks from Eua, Tonga. *United States Geological Survey Professional Paper* 640-C: 1–12: pls 1–5.

Lamy, E. 1910. Mission dans l'Antarctique dirigée par M. Le Dr. Charcot (1908–1910). Collections recueillies par M. le Dr. J. Liouville. Gastropodes prosobranches et

Literature cited

scaphopode. *Bulletin du Muséum National d'Histoire Naturelle, Paris* 16: 318–325.

Lamy, E. 1911. Gastropodes prosobranches, scaphopode et pélécypodes. In: *Deuxième Expédition Antartique Française (1908–1910)*. L. Joubin (ed.): pp. 1–31, pl. 1. Masson et cie, Paris.

Landau, B., R. Marquet & M. Grigis. 2003. The early Pliocene Gastropoda (Mollusca) of Estepona, southern Spain Part 1: Vetigastropoda. *Palaeontos* 3: 1–87.

Laube, G. C. 1869. Die Fauna der Schichten von St. Cassian. Ein Beitrag zur Paläontologie der Alpinen Trias. IV Abtheilung. Gastropoden II. Hälfte. *Denkschrift der Kaiserlichen Akademie der Wissenschaften, Mathematisch-Naturwissenschaftliche Classe* 30: 1–48, pls 29–35.

Laws, C. R. 1935. New Tertiary Mollusca from New Zealand—No. 3. *Transactions and Proceedings of the Royal Society of New Zealand* 65: 30–43, pls 5–7.

Laws, C. R. 1936. Waitotaran faunule of Kaawa Creek, Part 2. *Transactions and Proceedings of the Royal Society of New Zealand* 66: 99–124, pls 15–17.

Laws, C. R. 1939. The Molluscan faunule at Pakaurangi Point Kaipara—No. 1. *Transactions and Proceedings of the Royal Society of New Zealand* 69: 466–503, pls 62–67.

Laws, C. R. 1940a. The Waitotaran Faunule at Kaawa Creek—Part 3. *Transactions and Proceedings of the Royal Society of New Zealand* 69: 427–447, pls 59–61.

Laws, C. R. 1940b. Paleontological study of Nukumaruan and Waitotaran Rocks near Wanganui. *Transactions and Proceedings of the Royal Society of New Zealand* 70: 34–56, pls 5–7.

Laws, C. R. 1941. The molluscan faunule at Pakaurangi Point, Kaipara No. 2. *Transactions and Proceedings of the Royal Society of New Zealand* 71: 134–151, pls 16–19.

Laws, C. R. 1948. Further Tertiary Mollusca from Hokianga District, North Auckland. *Transactions and Proceedings of the Royal Society of New Zealand* 77: 142–150, pls 11–12.

Lazzari, G., F. Mannucci, E. Miari, & P. Micali. 1984. Malacofauna marina della Veneto — Area 122. *Notiziario SIM* 2(11–12): 167–169.

Leal, J. 1991. *Marine Prosobranch Gastropods from Oceanic Islands off Brazil*. Backhuys, Oegstgeest. 418 pp.

Lesueur, [C. A.]. 1817. Mémoire sur deux nouveaux genres de mollusques, Atlante et Atlas. *Journal de Physique, de Chimie, et d'Histoire Naturelle* 85: 390–393, pl. 2.

Liff-Grieff, P. 2005. Southern California's rich micro-fauna: a cursory look at some local intertidal marine mollusks. *Las Conchas* 37(2): 2–6.

Liff-Grieff, P. 2006. Errata. *Las Conchas* 37(3): 2.

Lindberg, D. R. & W. F. Ponder. 2001. The influence of classification on the evolutionary interpretation of structure — a re-evaluation of the evolution of the pallial cavity in gastropod molluscs. *Organisms Diversity & Evolution* 1: 273–299.

Lingwood, P. F. 1981. The dispersal of the collections of *H.M.S. Challenger*: an example of the importance of historical research in tracing a systematically important collection. In: *History in the Service of Systematics*. A. Wheeler & J. H. Price (eds): pp. 71–78. Society for the Bibliography of Natural History, London.

Linse, K. 1997. Die Verbreitung epibenthischer Mollusken im chilenischen Beagel-Kanal. *Berichte zur Polarforschung* 228: 1–131.

Linse, K. 1999. Mollusca of the Magellan region. A checklist of the species and their distribution. In: *Magellan-Antarctic: Ecosytems that Drifted Apart*. W. E. Arntz & C. Rios (eds): pp. 399–407. Scientia Marina, Barcelona 63, Supplement 1.

Linse, K. 2002. *The Shelled Magellanic Mollusca: with Special Reference to Biogeographic Relations in the Southern Ocean*. A. R. A. Gnater Verlag, Rugell, Liechtenstein. 252 pp.

Linse, K. & A. Brandt. 1998. Distribution of the epibenthic Mollusca on a transect through the Beagle Channel (Southern Chile). *Journal of the Marine Biological Association of the United Kingdom* 78: 875–889.

Llera Gonzales, E. M., J. A. Ortea Rato & A. Vizcaino Fernandez. 1983. *Fauna Marina de*

Asturias 1. Moluscos. 1. Archaeogastropoda (Prosobranchia). Centro de Investigaciones Acuáticas de Asturias, Gijón. 75 pp.

Locard, A. 1886. *Prodrome de Malacologie Française. Catalogue Général des Mollusques Vivants de France: Mollusques Marins.* Henri Georg, Lyon. 778 pp.

Locard, A. 1892. *Les Coquilles Marines des Côtes de France.* J.-B. Baillière et fils, Paris. 384 pp.

Locard, A. 1898. Mollusques Testacés. In: *Expéditions Scientifiques du* Travailleur *et du* Talisman *Pendent les Années 1880, 1881, 1882, 1883. Vol. 2.* A. Milne-Edwards (ed.): pp. 1–545, pls 1–18. Masson et Cie, Paris.

Locard, A. 1899. *Les Coquilles Marines au Large des Côtes de France.* J.-B. Baillière et fils, Paris. 198 pp.

Locard, A.& E. Caziot. 1900. Les coquilles marines des côtes de Corse, suite. *Annales de la Société Linnéenne* 47: 159–274.

Long, P. E. & J. A. Zalasiewicz. 2011. The molluscan fauna of the Coralline Crag (Pliocene, Zanclean) at Raydon Hall, Suffolk, UK: Palaeoecological significance reassessed. *Palaeogeography, Palaeoclimatology, Palaeoecology* 309: 53–72.

Lovén, S. 1846. Index Molluscorum Litora Scandinaviae Occidentalia Habitantium. *Öfversigt af Kongl Vetenskaps-Akademiens Förhandlingar* 3: 135–160, 182–204.

Lovén, S. 1846. *Index Molluscorum Litora Scandinaviae Occidentalia Habitantium.* Norstedt, Holmiae. 50 pp.

Lozouet, P. 1986. *Les Gastropods Prosobranches de l'Oligocène supérieur du Bassin de l'Adour (Systématique, Paléoenvironnements, Paléoclimatologie, Paléobiogeographie).* Diploma Thesis, École Pratique des Hautes Études, Paris. 475 pp., 34 pls.

Lozouet, P. 1998. Nouvelles espèces de gastéropodes (Mollusca: Gastropoda) de l'Oligocène et du Miocène inférieur de l'Aquitaine (sud-ouest de la France). *Cossmanniana* 5: 61–102.

Lozouet, P. 1999. Nouvelles espèces de gastéropodes (Mollusca: Gastropoda) de l'Oligocène et du Miocène inférieur d'Aquitaine (Sud-ouet de la France). Partie 2. *Cossmanniana* 6: 1–68.

Lozouet, P., J.-F. Lesport & P. Renard. 2001. Révision des Gastropoda (Mollusca) du stratotype de l'Aquitaenien (Miocène inf.): site de Saucats "Lariey", Gironde, France. *Cossmanniana* hors-série 3: 1–189, pls 1–37.

Luque, A. 1986. *Contribution al Connocimiento de los Moluscos Gasteropodos de las Costas de Malaga y Grenada.* Universidad Complutense de Madrid, Madrid. 695 pp.

Luque, A. L., D. L. Geiger & E. Rolán. 2011. A revision of the genus *Satondella* Bandel, 1998 (Gastropoda. Scissurellidae). *Molluscan Research* 31: 1–14.

Lyman, F. 1948. New price list. *Shell Notes* 2(5): 74–77.

MacAndrew, R. 1856. Report on the marine testaceous Mollusca of the north-east Atlantic and neighbouring Seas, and the physical conditions affecting their development. *Report of the British Association for the Advancement of Science* 1856: 101–158.

MacAndrew, R. 1852. Notes of the Mollusca observed during a short visit to the Canary and Madeira Islands, &c., in the months of April and May 1852. *Annals and Magazine of Natural History* Series 2, 10: 100–108.

MacAndrew, R. 1854. *On the Geographical Distribution of Testaceous Mollusca in the North Atlantic and Neighbouring Seas.* H. Greenwood, Liverpool. 51 pp.

MacAndrew, R. 1858. Dredg. coast of Norway [not traceable, not seen].

MacAndrew, R. 1861. *List of the British Marine Invertebrate Fauna.* Taylor and Francis, London. 71 pp.

Macedo, M. C. C., M. I. C. Macedo & J. P. Borges. 1999. *Seashells of Portugal.* Verbo, Lisbon. 516 pp.

Macpherson, E. 1971. *The Marine Molluscs of Arctic Canada.* National Museums of Canada, Ottawa. 149 pp.

MacPherson, J. H. & C. J. Gabriel. 1962. *Marine Molluscs of Victoria.* Melbourne University Press, Melbourne. 475 pp.

Maes, V. O. 1967. The littoral marine mollusks of Cocos-Keeling Islands (Indian Ocean). *Proceedings of the Academy of Natural Sciences Philadelphia* 119: 93–217.

Marais, J. & A. Marais. 2003. Rare shells in a bottle. *The Strandloper* 271: 12.

Marin, A. & J. Ros. 1987. Los moluscos gasteropodos del sudeste Español: adiciones. *Lavori Societa Italiana di Malacologia* 23: 201–216.

Marincovich, L. 1973. Intertidal mollusks of Iquique, Chile. *Natural History Museum of Los Angeles County Science Bulletin* 16: 1–49.

Marion, A. F. 1883a. Considerations sur les faunes profondes de la Méditerranée. *Annales Muséum d'Histoire Naturelle de Marseille, Zoologie* 1: 1–50 [not seen].

Marion, A. F. 1883b. Faune du bassin Mediterranéen. *Annales du Musée d'Histoire Naturelle de Marseille. Zoologie* 1: [not seen].

Marquet, R. 1984. A remarkable molluscan fauna from the Kattendijk Formation (Lower Pliocene) at Kallo (Oost-Vlaanderen, Belgium). *Bulletin de la Société Belge de Géologie* 93: 335–345.

Marquet, R. 1995. Pliocene gastropod faunas from Kallo (Oost-Vlaanderen, Belgium) — Part 1. Introduction and Archaeogastropoda. *Contributions to Tertiary and Quaternary Geology* 32: 53–85.

Marquet, R. 1998. *De Pliocene Gastropodenfauna van Kallo (Oost-Vlaanderen, Belgie).* Belgische Vereniging voor Paleontologie, Antwerpen. 246 pp.

Marshall, B. A. 1993. The systematic position of *Larochea* Finlay, 1927, and introduction of a new genus and two new species (Gastropoda: Scissurellidae). *Journal of Molluscan Studies* 59: 285–294.

Marshall, B. A. 2002. Some Recent scissurellids form the New Zealand region, and remarks on some scissurellid genus group names (Mollusca: Gastropoda). *Molluscan Research* 22: 165–181.

Marshall, J. T. 1895. Alterations in "British Conchology." *Journal of Conchology* 8: 24–41.

Marshall, J. T. 1897. Alterations in "British Conchology". *Journal of Conchology* 9: 61–75.

Marshall, P. 1915. Cainozoic fossils from Oamaru. *Transactions and Proceedings of the New Zealand Institute* 47: 377–387.

Marshall, P. & G. H. Uttley. 1914. Localities for fossils near Oamaru. *Transactions and Proceedings of the Royal Society of New Zealand* 46: 279–280.

Marshall, W. 1888. *Die Tiefsee und ihr Leben. Nach den neuesten Quellen gemeinfasslich dargestellt.* Ferdinand Hirt & Sohn, Leipzig. 343 pp., 4 pls.

Mastaller, M. 1979. *Beiträge zur Faunistik unf Ökologie der Mollusken und Echinodermen in den Korallenriffen bei Aquaba, Rotes Meer.* Dissertation, Ruhr University, Bochum. 344 pp. [not seen].

Matsumoto, Y. 1979. *Molluscan Shells of the Mie Prefecture, Japan.* Toba Aquarium, Toba. 179 pp., 21 pls.

Mattar, A. S. 1987. Um novo arqueogastrópode da família Scissurellidae Gray, 1847, da plataforma continental Brasileira. *Pesquisas* 20: 75–84, 1 pl.

Maxwell, P. A. 1992. Eocene Mollusca from the vicinity of McCulloch's Bridge, Waihao River, South Canterbury, New Zealand: paleoecology and systematics. *New Zealand Geological Survey Paleontological Bulletin* 65: 1–280.

May, W. L. 1908. Additions to the Tasmanian molluscan fauna. *Papers and Proceedings of the Royal Society of Tasmania* 1908: 53–59.

May, W. L. 1921. *Check-List of the Mollusca of Tasmania.* John Vail, Hobart. 99 pp., 47 pls.

May, W. L. 1923. *An Illustrated Index of Tasmanian* Shells. John Vail, Hobart. 47 pls.

May, W. L. & J. H. Macpherson. 1958. *An Illustrated Index of Tasmanian Shells.* Shea, Government Printer, Tasmania. 54 pp., 18 pp. index.

Mazziotti, C., F. Agamennone & M. Tisselli. 2008. Checklist della malacofauna delle Isola Tremiti (Medio Adriatico). *Bollettino Malacologico* 44: 71–86.

McClelland, H. n.d. [>1923]. *General Index of all Families, Genera, Species, and Varieties Described and Noted in* The Journal of Conchology *Vols. I-XVI, 1874–1922,* Proceedings of the Malacological Society of London *Vols. I-XV, 1893–1923,* The Conchologist

Literature cited

Vols. I & II, 1891–1893, The Journal of Conchology *Vols. III-XII, 1894–1905*. Birbeck & Sons, Limited, Birmingham. 165 pp.

McLean, J. H. 1966. *West American Prosobranch Gastropoda: Superfamilies Patellacea, Pleurotomariacea and Fissurellacea*. Stanford University, Stanford. 245 pp., 7 pls.

McLean, J. H. 1967. West American Scissurellidae. *The Veliger* 9: 404–410.

McLean, J. H. 1969. Marine shells of Southern California. *Science Series Los Angeles County Museum of Natural History* 11: 1–104.

McLean, J. H. 1970. New species of tropical eastern Pacific Gastropoda. *Malacological Reviews* 2: 115–130.

McLean, J. H. 1971. Archaeogastropoda. In: *Sea Shells of the Tropical West Pacific, Second Edition*. A. M. Keen (ed.): pp. 307–363. Stanford University Press, Stanford.

McLean, J. H. 1978. Marine shells of Southern California. *Natural History Museum of Los Angeles County, Science Series* 24, revised edition: 1–104.

McLean, J. H. 1984. New species of northeast Pacific archaeogastropods. *The Veliger* 26: 233–239.

McLean, J. H. 1989. New slit-limpets (Scissurellacea and Fissurellacea) from hydrothermal vents. Part 1 Systematic description and comparison based on shell and radular characters. *Contributions in Science of the Los Angeles County Museum of Natural History* 407: 1–29.

McLean, J. H. 1996. Gastropoda, the Prosobranchia. In: *Taxonomic Atlas of the Benthic Fauna of the Santa Maria Basin and Western Santa Barbara Channel, Volume 9*. P. H. Scott, J. A. Blake & A. L. Lissner (eds): pp. i–v, 1–160. Santa Barbara Museum of Natural History, Santa Barbara.

McLean, J. H. 2007. Shelled Gastropoda. In: *The Light and Smith Manual. Intertidal Invertebrates from Central California to Oregon. Fourth Edition, Completely Revised and Expanded*. J. T. Carlton (ed.): pp. 713–753. University of California Press, Berkeley.

McLean, J. H. & E. V. Coan. 1996. Marine mollusks of Rocas Alijos. In: *Rocas Alijos*. R. W. Schmieder (ed.): pp. 305–318. Kluwer Academic Publishers, Dordrecht.

Melvill, J. C. 1904. Descriptions of twenty-eight species of Gastropoda from the Persian Gulf, Gulf of Oman, and Arabian Sea, dredged by Mr. F. W. Townsend, of the Indo-European Telegraph Service, 1900–1904. *Proceedings of the Malacological Society London* 6: 159–169, pl. 10.

Melvill, J. C. & Standen. 1903. Descriptions of sixty-eight new Gastropoda from the Persian Gulf, Gulf of Oman, and North Arabian Sea, dredged by Mr. F. W. Townsend, of the Indo-European Telegraph Service, (1901–1903). *Annals and Magazine of Natural History* Series 7, 12: 289–324, pls 20–23.

Melvill, J. C. & R. Standen. 1912. The marine Mollusca of the Scottish National Antarctic Expedition. *Transactions of the Royal Society of Edinburgh* 48: 333–366, 1 pl.

Merlano, J. M. D. & M. P. Hegedus. 1994. *Moluscos del Caribe Colombiano: un Catálogo Ilustrado*. Colciencias Fundación Natura, Invemar, Santafé de Bogotá. 291 pp., 74 pls.

Mestayer, M. K. 1916. Preliminary list of Mollusca from dredgings taken off the Northern coasts of New Zealand. Transactions and *Proceedings of the New Zealand Institute* 48: 122–128, pl. 12.

Mestayer, M. K. 1919. New species of Mollusca, from various dredgings taken off the coast of New Zealand, the Snares Islands, and the Bounty Islands. *Transactions and Proceedings of the New Zealand Institute* 51: 130–135.

Miari, E. 1983. Malacofauna marina della Sardegna — Area 154. *Notiziario SIM* 1 (9–10): 17–18.

Miari, E. & P. Micali. 1984a. Malacofauna marina della Puglia — Area 072. *Notiziario SIM* 2(7–8): 108.

Miari, E. & P. Micali. 1984b. Malacofauna marina della Sardegna — Area 151. *Notiziario SIM* 2(11–12): 175.

Mienis, H. K. & R. Ben-David-Zaslow. 2004. A preliminary list of the marine molluscs of the national park and nature reserve of Akhziv-Rosh Haniqra. *Triton* 10: 13–37.

Mienis, H. K. 1992. Isaac Yaron (1934–1985). *Argamon* 8: 1–7.

Mifsud, C. 1998. The molluscan species living in Maltese *Posidonia* meadows. *La Conchiglia* 30: 37–48.

Millard, V. 2003. *Classification of Mollusca, A Classification of Worldwide Mollusca, Volume 1.* Self published, Rhine Road, Republic South Africa. 890 pp.

Min, D.-K. 2004. *Mollusks in Korea (Revised Supplementary Edition).* Min Molluscan Research Institute, Seoul, Korea. 566 pp.

Monterosato, T. A. di. 1872a. *Notizie Intorno Alle Conchiglie Mediterranee.* Ufficio Tipografico di Michele Amenta, Palermo. 61 pp.

Monterosato, T. A. di. 1872b. *Notizie Intorno alle Conchiglie Fossili di Monte Pellegrino e Ficarazzi.* Ufficio Tipografico di Michele Amenta, Palermo. 44 pp.

Monterosato, T. A. di. 1875a. Nuova rivista delle conchiglie mediterranee. *Atti dell'Accademia Palermitana di Scienze, Lettere ed Arti, Palermo* 2(5): 1–50.

Monterosato, T. A. di. 1875b. *Poche Note Sulla Conchiologia Mediterranea.* Tipografia del Giornale di Sicilia, Palermo. 15 pp.

Monterosato, T. A. di. 1877a. Notizie sulle conchiglie della rada di Civitavecchia. *Annali Museum Civico di Genova* 9: 407–428.

Monterosato, T. A. di. 1877b. Catalogo delle conchiglie fossili di Monte Pellegrino e Ficarazzi presso Palermo. *Bollettino del Reale Comitato Geologico d'Italia* 1–2: 28–42.

Monterosato, T. A. di. 1878a. Notes sur quelque coquilles draguées dans les eaux de Palerme. *Journal de Conchyliologie* 26: 143–160.

Monterosato, T. A. di. 1878b. Enumerazione e sinonimia delle conchiglie mediterranee. *Giornale Scienze Naturali ed Economiche, Palermo* 13: 61–115.

Monterosato, T. A. di. 1884. *Nomenclatura Generica e Specifica di Alcune Conchiglie Mediterranee.* Virzi, Palermo. 152 pp.

Monterosato, T. A. di. 1888. Molluschi del Porto di Palermo. *Bollettino della Società Malacologica Italiana* 14: 161–180.

Monterosato, T. A. di. 1890. Conchiglie della produndtà del mare di Palermo. *Naturalista Siciliano, Palermo* 9: 140–151.

Montfort, D. de. 1810. *Conchyliologie Systématique et Classifiction Méthodique des Coquilles.* Schoell, Paris. 676 pp.

Montouchet, P. C. 1972. Three new species of Scissurellidae (Gastropoda, Prosobranchia) from the coast of Brazil. *Boletim Instituto Oceanographico San Paulo* 21: 1–13.

Montouchet, P. C. unpublished manuscript. On the bibliography for the genera of *Scissurella.*

Mörch, O. A. L. 1857. Tillaeg Nr. 4, Fortegnelse over Grønlands Bloddnr. In: *Naturhistoriske Bidrag til En Beskrioelse af Grønland.* J. Reinhardt, J. Schiødte, O. A. F. Mörch, G. H. Lütken, F. Lange & H. Rink (eds): pp. 75–100. Louis Klein, Kjobenhaven.

Mörch, O. A. L. 1863. Revisio critica serpulidarum. Et bidrag til røromenes Naturhistorie. *Naturhistorisk Tidsskrift stiftet af Hendrik Krøyer, København* (1861–1863) Series 3, 1: 347–470, 1 pl.

Mörch, O. A. L. 1869. Catalogue des mollusques du Spitzberg. *Annales de la Société Malacologique de Belgique* 4: 7–32.

Moreno, D. 2011. Familia Scissurellidae. In: S. Gofas, D. Moreno & C. Salas (eds) *Moluscos Marinos de Andalucia.* Servicio de Publicaciones e Intercambio Científico, Universidad de Málaga: pp. 94–96.

Morgan, J. de. 1915. Observations sur la stratigraphie et la paléontologie du Falunien de la Touraine. *Bulletin de la Société Géologique de France,* 4th Series, 15: 217–241.

Morretess, F. L. de. 1949. Ensaio de catálogo dos moluscos do Brasil. *Arquivos do Museum Paranaense* 7: 5–126.

Morse, E. S. 1909. Shells new to the New England Fauna. *The Nautilus* 22: 95.

Müller, A. 2011. Der Steinbruch Mammendorf NW Magdeburg - ein Felslitoral der unteroligozänen Nordsee. *Geologica Saxoniaca* 57: 3–120.

Müller, A. & A. Rosenberg. 2003. Teleostei-Otoliths from the Mandrikovka Beds (Priabonian) of Dnepropetrovsk (Ukraine). *Paläontologische Zeitschrift* 77: 361–387.

Munier-Chalmas, E. 1862. Description d'une nouvelle *Scissurella*, suivie de la liste monographique des espèces connues de ce genre. *Journal de Conchyliologie* 10: 391–397.

Murdoch, R. 1905. Additions to the marine Mollusca of New Zealand. *Transactions and Proceedings of the New Zealand Institute* 37: 217–232, pls 7–8.

Musial, T. & G. Jakubowski. 1975. Nowe stanowiska fauny mioceneskiej w okolicy Monastryrza i Huty Lubyckiej (Poludniowe Roztocze) [New stations of Miocene faunas around Monastyrza and Huty Lubyckiej (circle of Roztocze)]. *Przeglad Geologiczny* 23: 336–338.

Neuffer, O. 1978. Zwei für den unteren Meeressand (Rupelium) des Mainzer Beckens neue Mollusken. *Mainzer Geowissenschaftliche Mitteilungen* 6: 93–97.

Nicklès, M. 1950. *Mollusques Testacés Marins de la Côte Occidentale d'Afrique*. Paul Lechevalier, Paris. 269 pp.

Nicolay, K. & M. Angioy. 1993. Mediterranean sea jewels from shell grit. *La Conchiglia* 25: 52–55.

Nobre, A. 1900. Mollusques et Brachiopodes du Portugal. *Annales de Sciencias Naturaes* 6: 79–151.

Nobre, A. 1932. *Moluscos Marinhos de Portugal*. Impressa Portuguesa, Porto. 466 pp., 80 pls.

Nobre, A. 1936. *Moluscos Marinhos de Portugal*. Companhia Editora do Minho, Barcelos. 378 pp., 6 pls.

Nobre, A. 1937. Moluscos tesáceos marinhos do arquipélago da Madeira. *Memórias e Estudos do Museu Zoológico da Universidade de Coimbra* Series 1, 98: 1–101.

Nobre, A. 1938–1940. *Moluscos Marinhos e das Águas Salobras*. Imprensa Portuguesa, Porto. 806 pp., 86 pls.

Nolt, J. M. 2008. A new species of *Scissurella* from the Azores with discussions on *Sinezona semicostata* (Burnay & Rolán, 1990 and *Sinezona cingulata* (O. G. Costa, 1861) (Gastropoda: Vetigastropoda: Scissurellidae). *Zootaxa* 1678: 51–62.

Nordsieck, F. 1968. *Die Europäischen Meeres-Gehäuseschnecken (Prosobranchia) vom Eismeer bis Kapveden und Mittelmeer*. Gustav Fischer Verlag, Stuttgart. 331 pp., 31 pls.

Nordsieck, F. 1969. *Die Europäischen Meeresmuscheln (Bivalvia) vom Eismeer bis Kapveden, Mittelmeer und Schwarzes Meer*. Gustav Fischer Verlag, Stuttgart. 256 pp.

Nordsieck, F. 1973. Abyssal molluscs from the Ionic-Sea. *La Conchiglia* 5(11–12): 4–7, pls 1–2.

Nordsieck, F. 1982. *Die Europäischen Meeresgehäuseschnecken, 2. Auflage*. Gustav Fischer, Stuttgart. 539 pp.

Nordsieck, F. & F. García-Talavera. 1979. *Moluscaos Marinos de Canarias y Madera (Gastropoda)*. Aula de Cultura de Tenerife, Madrid. 208 pp., 46 pls.

Norman, A. M. 1879. The Mollusca of the Fiords near Bergen, Norway. *Journal of Conchology* 2: 8–61.

Norman, A. M. 1893. A month on the Trondhjem Fiord. *Annals and Magazine of Natural History* Series 6, 12: 341–367, pl. 7.

Norman, A. M. 1888. *Museum Normanianum, or A Catalogue of the Invertebrata of Europe and the Arctic and North Atlantic Oceans, which are Contained in the Collection of the Rev. Canon A. M. Norman, M. A., D. C. L., F. L. S. IV. Mollusca Marina. V. Brachiopoda*. Privately Published, Durham. 30 pp.

Numanami, H. 1996. Taxonomic study of Antarctic gastropods collected by Japanese Antarctic research expeditions. *Memoirs of National Institute of Polar Research Series E (Biology and Medical Sciences)* 39: 1–244.

Numanami, H. & T. Okutani. 1990. A new and two known species of the genus *Anatoma* collected by the Icebreaker *Shirase* from Breid Bay and Günnerus Bank, Antarctica (Gastropoda: Scissurellidae). *Venus* 49: 93–106.

Numanami, H. & T. Okutani. 1992. Archaeogastropods collected from Breid Bay and Günners Bank, Antarctica by Japanese Antarctic Research Expeditions (extended abstract). *Proceedings of the NIPR Symposium on Polar Biology* 5: 171–173.

Nützel, A. 2006. A new Vetigastropod (Gastropoda Stomatellidae) from the Late Triassic Cassian Formation. *Neues Jahrbuch Geologisches und Paläontologisches Monatsheft* 2006: 634–640.

Nützel, A. & D. L. Geiger. 2006. A new scissurelloid genus and species (Mollusca, Gastropoda) from the Late Triassic Cassian Formation. *Paläontologische Zeitschrift* 80: 277–283.

Odé, H. 1986. Distribution and records of the marine Mollusca in the northwest Gulf of Mexico (A continuing monograph). *Texas Conchologist* 22: 69–73.

Odé, H. 1988. Observations on a naughty beach. *Texas Conchologist* 25: 10–25.

Odhner, N. H. 1912. Northern and Arctic invertebrates in the collection of the Swedish State Museum. V. Prosobranchia. 1. Diotocardia. *Kungliga Svenska Vetenskapsakademiens Handlingar* 48: 3–92, pls 1–7.

Odhner, N. H. 1960. Mollusca. *Reports of the Swedish Deep-Sea Expedition* 2: 367–400, pls 1–2.

Okamoto, M. 2001. *Molluscan Taxa Described by Tadeshige Habe, Volume 1. Gastropoda (excl. Heterobranchia).* The Committee for Celebrating Dr. T. Habe's Eightieth Birthday, Tokyo, Japan. 891 pp.

Okutani, T. 1964. Report on the archibenthal and abyssal gastropod Mollusca mainly collected from Sagami Bay and adjacent waters by the R. V. *Soya-Maru* [should read *Soyo-Maru*] during the years 1955–1963. *Journal of the Faculty of Science, University of Tokyo*, Section II 15: 371–447, pls 1–7.

Okutani, T. 1966. Archibenthal and abyssal Mollusca collected by the R. V. *Soyo-Maru* from Japanese waters during 1964. *Bulletin of the tokai Regional Fisheries Research Laboratory* 46: 1–32, pls 1–2.

Okutani, T. 1974. Review and new records of abyssal and hadal molluscan fauna in Japanese and adjacent waters. *Venus* 33: 23–39.

Okutani, T. 1986. *The Definitive Edition. Big Illustrated Book of Life Shells.* World Culture Col Ltd, Tokyo. 399 pp.

Okutani, T. & Hasegawa. K. 2000. Family Scissurellidae. In: *Marine Mollusks in Japan.* T. Okutani (ed.): pp. 36–37. Tokai University Press, Tokyo.

Oldroyd, I. S. 1924. Marine shells of Puget Sound and vicinity. *Publications Puget Sound Biological Station of the University of Washington* 4: 1–272.

Oldroyd, I. S. 1927. *The Marine Shells of the West Coast of North America Volume 2, Part 3.* Stanford University Press, Stanford. 339 pp., pls 73–108.

Oliveira, P. S. de, M. J. Martins-Silva, J. C. N. de Barros, I. Swoboda, D. A. Francisco, M. Schneider, J. C. L. Batista, P. de P. U. D. Aquino, G. Vianna & M. F. Coutinho. 2009. Malacofauna. In: *O Arquipélago de São Pedro e São Paulo: 10 anos de Estação Científica.* F. H. V. Hazin (ed.): pp. 174–185. SECRIM, Brasilia.

Oliver, W. R. B. 1915. The Mollusca of the Kermadec Islands. *Transactions and Proceedings of the New Zealand Institute* 1914: 509–568, pls 9–12.

Olson, D. A. 1968a. Banding patterns of *Haliotis rufescens* as indicators of botanical and animal succession. *Biological Bulletin* 134: 139–144.

Olson, D. A. 1968b. Banding patterns in *Haliotis* — II Some behavioral considerations and the effect of diet on shell coloration for *H. rufescens*, *H. sorenseni* and *H. assimilis*. *The Veliger* 11: 135–139.

Olson, O. P. 1957. Appendix 4: Fossil Mollusca. In: *The Geology of Waitaki Subdivision.* M. Gage (ed.). *New Zealand Geological Survey Bulletin* New Series 55: 1–135, 4 maps [pp. 108–127].

Olsson, A. A. & T. L. McGinty. 1958. Recent marine mollusks from the Caribbean coast of Panama with the description of some new genera and species. *Bulletins of American Paleontology* 39: 1–52, pls 1–5.

Onitsuka, T., T. Kawamura, S. Ohashi, T. Horii, & Y Watanabe. 2004. Morphological changes in the radula of abalone *Haliotis diversicolor aquatilis* from post-larva to adult. *Journal of Shellfish Research* 23: 1079–1085.

Orbigny, A. d'. 1824. Monographie d'un nouveau genre de mollusque gastéropode, de la famille des trochoides. *Mémoire de la Société d'Histoire Naturelle de Paris* 1: 340–345, pl. 23.

Orbigny, A. d'. 1835–1843. *Voyage dans l'Amérique Méridionale 5(3). Mollusques.* Levrault, Strasbourg. xliii, 758 pp., 79 pls.

Orbigny, A. d'. 1840. Mollusques echinodermes foraminifères et polypes recoltés aux îles Canaries. In: *Histoire Naturelle des Îles Canaries.* P. B. Webb & S. Berthelot (eds): pp. 1–117, pls 1–8. Bethune, Paris.

Orbigny, A. d' 1855. *Mollusques Vivants et Fossiles.* Simon Raçon et co., Paris. 605 pp.

Orcutt, C. R. 1915. *Molluscan World Volume 1.* Privately published [?], San Diego, California. 208 pp.

Ortiz-Corps, E. 1985. *An Annotated Checklist of the Recent Marine Gastropoda (Mollusca) from Puerto Rico.* University of Puerto Rico, Hunacao, Puerto Rico. 220 pp.

Óskarsson, I. 1982. *Skeldyrafá Íslands Samlokur Í sjó Saesniglar Med Skel.* Prentsmidjan Leiftur Hf, Reykjavík. 351 pp.

Otero-Schmitt, J. & J. Trigo-Trigo. 1989. Moluscos de la Ria de Muros. *Thalassas* 7: 779–790.

Oyama, K. 1973. Revision of Matajiro Yokoyama's type Mollusca from the Tertiary and Quarternary of the Kanto area. *Paleontological Society of Japan, Special Papers* 17: 1–148, pls 1–57.

Öztürk, B. & Z. Ergen. 2000. Les archéogastéropodes (Mollusca - Gastropoda) du litoral Turc de la Mer Egée. *Acta Adriatica* 41: 59–70.

Øyen, P. A. 1915. Kvartaer-studier I Trondhjemsfeltft III. *Det Kongelige Norske Videnskabers Selskabs Skrifter* 2: 1–506.

Paetel, F. 1869. *Molluscorum Systema et Catalogus.* Oscar Weiske, Dresden. 119 pp.

Paetel, F. 1883. *Catalog der Conchylien-Sammlung von Fr. Paetel. Neubearbeitung. Erste Abtheilung: Die Cephalopoden, Pteropoden und Meeres-Gastropoden.* Gebrüder Paetel, Berlin. 271 pp.

Paetel, F. 1888. *Catalog der Conchylien-Sammlung von Fr. Paetel. Vierte Neubearbeitung. Erste Abtheilung: Die Cephalopoden, Pteropoden und Meeres-Gastropoden.* Gebrüder Paetel, Berlin. 639 pp.

Palaeontological Society of Japan and Geological Survey of Japan, AIST. 2008. *The database of Japanese fossil type specimens described during the 20th century* (Web Version). http://riodb02.ibase.aist.go.jp/dform/FossilType/ [accessed 3/2008].

Palazzi, S. 1985. Su alcuni molluschi marini di Isola de Pinos (Cuba). *Notiziario Centro Italiano Studi Malacologici* 5: 39–47.

Palazzi, S. & A. Villari. 1994. Malacofaune batiali Plio-Pleistoceniche del Messinese. 1: Venetico Marina. *Naturalista Siciliana,* Series 4, 18: 73–106.

Palazzi, S. & A. Villari. 1996. Malacofaune batiali Plio-Pleistoceniche del Messinese. 2: Capo Milazzo. *Naturalista Siciliana,* Series 4, 20: 237–279.

Palazzi, S. & A. Villari. 2000. Mollusks and brachipods from the submarine caves of Taormina, Sicily. *La Conchiglia* Supplement 297: 1–56.

Pallary, M. G. 1912. Exploration scientifique du Maroc. *Archives Scientifiques du Protectorat Français* 2: 1–107.

Pallary, P. 1900. Coquilles marines du littoral du Département d'Oran. *Journal de Conchyliologie* 48: 211–422, pls 6–8.

Pallary, P. 1909. Catalogue de la faune malacologique de L'Égypte. *Mémoires de l'Institut Égyptien* 6: 1–200, pls 15–18.

Pallary, P. 1926. Explication des planches de J. C. Savigny. *Memoires de l'Institut d'Egypte* 11: vii + 1–138, pls 1–18 [not seen].

Parenzan, P. 1961. Malacologia Jonica. *Istituto Sperimentale Talassografico di Taranto* 6:

1–183.

Parenzan, P. 1970. *Carta d'identità delle Conchiglie del Mediterraneo Vol. 1 Gasteropodi.* Bios Taras, Raranto. 283 pp.

Pasteur-Humbert, C. 1962. Les mollusques marins testacés du Maroc. Catalogue non critique I. Les gastéropodes. *Travaux de l'Institute Scientifique Chérifien, Série Zoologie* 23: 1–245.

Paulmier, G. 1994. Deep water molluscs of the French Antilles. *Xenophora* 68: 3–22.

Pelseneer, P. 1899. Recherches morphologiques et phylogénétiques sur les mollusques archaïques. *Mémoires Couronnés et Mémoires des Savants Étrangers, Académie Royale des Sciences, Lettres et des Beaux-Arts de Belgique* 57: 3–112, pls 1–24.

Pelseneer, P. 1903. Mollusques (amphineurs, gastropodes et lamellibranches). *Résultats du Voyage du S. Y. Belgica, Rapports Scientifiques, Zoologie* 9: 1–85, pls 1–9.

Pelseneer, P. 1906. *A Treatise in Zoology, Part 5 Mollusca.* Adam & Charles Black, London. 355 pp.

Perretti, E. 1986. Malacofauna marina della Basilicata — Area 051. *Notiziario SIM* 4(7–8): 126.

Peñas, A., E. Rolán, A. A. Luque, J. Templado, D. Moreno, F. Rubio, C. Salas & A. Sierra. 2006. Marine Mollusca from the Alboran Islands. *Iberus* 24: 23–151.

Petit, de la Saussaye, M. 1860. Nouveau supplément au catalogue des mollusques marins qui vivent sur les côtes de France. *Journal de Conchyliologie* Series 2, 8: 243–260.

Petit, de la Saussaye, M. 1869. *Catalogue des Mollusques Testacés des Mer d'Europe.* F Savy, Paris. 312 pp.

Petit, R. E. 2006. Notes on Sowerby's *The Genera of Recent and Fossil Shells* (1821–1834). *Archives of Natural History* 33: 71–89.

Petit, R. E. 2009. George Brettingham Sowerby, I, II & III: their conchological publications and molluscan taxa. *Zootaxa* 2189: 1–218.

Petterd, W. F. 1879. New species of Tasmanian marine shells. *Journal of Conchology* 2: 102–105.

Petterd, W. F. 1884. Description of new Tasmanian shells. *Journal of Conchology* 4: 135–145.

Philippi, R. A. 1836. *Enumeratio Molluscorum Siciliae Vol. I.* Simon Schrapp and co., Berolini. 267 pp., 12 pls.

Philippi, R. A. 1844. *Enumeratio Molluscorum Siciliae Vol. II.* Sumptibus Eduardi Anton, Halis Saxonum. 303 pp., 28 pls.

Philippi, R. A. 1853. Die Gattungen *Delphinula, Scissurella* und *Globulus.* In: *Systematisches Conchylien-Cabinet.* Martini & Chemnitz (eds): pp. 1–57, pls 1–8. Bauer und Raspe, Nürnberg.

Piani, P. 1980. Catalogo dei molluschi conchiferi viventi nel Mediterraneo. *Bolletino Malacologico* 16: 113–224.

Pilsbry, H. A. 1895. *Catalogue of the Marine Mollusks of Japan.* Frederick Stearns, Detroit. 196 pp., 6 pls.

Pilsbry, H. A. 1903. *Schismope rimuloides* (Cpr.) at San Diego. *The Nautilus* 17: 84.

Pilsbry, H. A. 1890. *Manual of Conchology. Structural and Systematic. Volume 12.* Academy of Natural Sciences of Philadelphia, Philadelphia. 323 pp., 68 pls. [The penultimate plate is numbered 64, while the last bears number 68; the captions to this plate were not printed, and most text refers to it as 58, an existing plate showing Fissurellidae. The text references for *Schismope* species on this plate refer to pl. 68; therefore, 68 considered the correct plate number.]

Pisera, A. 1985. Paleocology and lithogenesis of the Middle Miocene (Badenian) algal-vermetid reefs from the Roztocze Hills, southeastern Poland. *Acta Geologica Polonica* 35: 89–155.

Poirier, H. 1954. *An up-to-date Systematic List of 3200 Seashells from Greenland to Texas, Translation, Explanation and Gender of their Names.* Poirier, [place not given]. 215 pp.

Literature cited

Ponder, W. F. & R. G. de Keyzer. 1998. Superfamily Rissooidea. In: *Mollusca: the Southern Synthesis*. P. L. Beesley, G. J. B. Ross & A. Wells (eds): pp. 745–766. CSIRO Publishing, Melbourne.

Poore, G. C. B. 1973. Ecology of New Zealand abalones, *Haliotis* species (Mollusca: Gastropoda). *Journal of Marine and Freshwater Research* 7: 67–84.

Poppe, G. T. & Y. Goto. 1991. *European Seashells, Volume I.* Christa Hemmen, Wiesbaden. 354 pp.

Poppe, G. T. & S. P. Tagaro. 2011. Addendum 1. Aclididae, Acteonidae, Anatomidae, Amathinidae & Assimineidae: pp. 546–547 in G. T. Poppe (ed.) *Philippine Marine Mollusks Vol. 4. Bivalvia Part 2, Scaphopoda, Polyplacophora, Cephalopoda & Addenda.* ConchBooks, Hackenheim.

Portalatina, M. 2008. Dati sulla malacofauna marina del Salento Ionico (Puglia). Parte 1. Archaeogastropoda. *Bolletino Malacologico* 44: 145–161.

Posset, H. J. 1898. Conspectus Faunae Groenlandicae. Brachiopoda et Mollusca. Grønlands Brachiopoder og Bløddyr. *Meddelelser om Grønland* 23: 1–298, 1 pl.

Powell, A. W. B. 1933a. The marine Mollusca of the Chatham Islands. *Records of the Auckland Institute and Museum* 1: 181–208, pls 33–36. [Sept. 25, 1933]

Powell, A. W. B. 1933b. Marine Mollusca from the Bounty Islands. *Records of the Canterbury Museum* 4: 29–39, pl. 4. [Oct. 6, 1933]

Powell, A. W. B. 1937. New species of marine Mollusca from New Zealand. *Discovery Reports* 15: 155–221, pls 45–56.

Powell, A. W. B. 1940. The marine Mollusca of the Aupourian Province, New Zealand. *Transactions and Proceedings of the Royal Society of New Zealand* 70: 205–248, pls 28–33.

Powell, A. W. B. 1941. Biological primary types in the Auckland Museum. *Records of the Auckland Institute and Museum* 2: 239–259.

Powell, A. W. B. 1951. Antarctic and subantarctic Mollusca: Pelecypoda and Gastropoda. In: *Discovery Reports* 26: pp. 47–196, pls 5–10. Cambridge University Press, Cambridge.

Powell, A. W. B. 1955. Mollusca of the southern Islands of New Zealand. *Cape Expedition Series Bulletin* 15: 1–151.

Powell, A. W. B. 1957. Mollusca of Kerguelen and Macquarie Islands. *B. A. N. Z. A. R. E. Reports* Series B, 6: 109–149.

Powell, A. W. B. 1958. Mollusca from the Victoria-Ross quadrants of Antarctica. *B. A. N. Z. A. R. E. Reports* Series B, 6: 165–215.

Powell, A. W. B. 1960. Antarctic and subantarctic Mollusca. *Records of the Auckland Institute and Museum* 5: 117–193.

Powell, A. W. B. 1979. *New Zealand Mollusca. Marine, Land and Freshwater Shells.* Collins, Auckland. 500 pp.

Priolo, O. 1948. Nuova revisione delle conchiglie marine di Sicilia. *Atti della Accademia Gioenia di Scienze Naturali in Catania* Series 6, 6: 1–26, pl. 1.

Pritchard, G. B. & J. H. Gatliff. 1902. Catalogue of the marine shells of Victoria. *Proceedings of the Royal Society Victoria* New Series, 15: 176–223.

Pritchard, G. B. & J. H. Gatliff. 1906. Catalogue of the marine shells of Victoria.—Part IX; with complete index to the whole catalogue. *Proceedings of the Royal Society Victoria* New Series, 18: 39–92.

Quinn, J. F. Jr. 1983. A revision of the Seguenziacea Verrill, 1884 (Gastropoda: Prosobranchia). I. Summary and evaluation of the superfamily. *Proceedings of the Biological Society of Washington* 96: 725–757.

Raines, B. K. 2002. Contributions to the knowledge of Easter Island Mollusca. *La Conchiglia* 34: 11–40.

Ramazzotti, D., D. Trono, A. Pacetta, M. Sanniu, M. Doneddu, F. Daga, R. Petrone, E. Perna, E. Trainito, F. Favero, C. Giannobi & M. Piera Amato. 2007. *Catalogo Annotato e Atlante Iconografico dei Mulluschi Marini del Mediterraneo Vers. 0.1.9.9 Beta.* www.malachia.it/molluschi_marini_mediterranei.htm (accessed 6/2011).

Randall, J. E. & D. W. Greenfield. 2001. A preliminary review of the Indo-Pacific gobiid fishes of the genus *Gnatholepis*. *Ichthyological Bulletin* 69: 1–17.

Raven, J. & S. Bracegirdle. 2010. *New Zealand Seashells Visual Guide*. Creatus Design, Wellington. 76 pp.

Ravn, J. P. J. 1933. Études sur les pélécypodes et gastropodes Daniens du calcaire de Faxe. *Mémoire de l'Académie Royale des Sciences et des Letteres de Danemark, Copenhague* Series 9, 5: 3–74, pls 1–7.

Redfern, C. 2001. *Bahamian Seashells*. Bahamianseashells.com, Boca Raton, Florida. 280 pp.

Rehder, H. A. 1946. Additional notes on the dates of publication of *Les Fonds de la Mer*. *Proceedings of the Malacological Society of London* 27: 74–75.

Rendall, R. 1956. Mollusca Orcadensia. *Proceedings of the Royal Society of Edinburgh Section B (Biology)* 66: 131–201.

Requien, E. 1848. *Catalogue des Coquilles de l'Île de Corse*. Seguin Ainé, Avignon. 109 pp.

Richard, G. 1985. Fauna and flora. A first compendium of French Polynesian sea-dwellers. In: *French Polynesian Coral Reefs*. B. Delesalle, R. Galzin & B. Slavat (eds): pp. 379–520. Muséum National d'Histoire Naturelle, Paris.

Richmond, M. H. 1992. *Tasmanian Sea Shells, Volume 2*. Selfpublished (?), Tasmania. 112 pp.

Rios, E. C. 1970. *Coastal Brazilian Seashells*. Fundação Cicade do Rio Grande, Museu Oceanográfico de Rio Grande, Rio Grande. 255 pp., 60 pls.

Rios, E. C. 1975. *Brazilian Marine Mollusks Iconography*. Fundação Universidade do Rio Grande, Centro de Ciencias do Mar, Museum Oceanográfico, Rio Grande. 331 pp.

Rios, E. C. 1985. *Seashells of Brazil*. Fundação Cicade do Rio Grande, Rio Grande. 328 pp., 102 pls.

Rios, E. C. 1994. *Seashells of Brazil, 2nd edition*. Fundação Cicade do Rio Grande, Rio Grande. 368 pp., 113 pls.

Rios, E. C. 2009. *Compendium of Brazilian Sea Shells*. Evangraf, Rio Grande, RS. 668 pp.

Rios, E. C., I. S. Calvo & L. J. Barcellos. 1987. Moluscos marinos de la Isla Trinidad. I. *Comunicaciones de la Sociedad Malacologica del Uruguay* 7: 57–62.

Roberts, R. D., T. Kawamura & H. Takami. 1999. Morphological changes in the radula of abalone (*Haliotis iris*) during post-larval development. *Journal of Shellfish Research* 18: 637–644.

Roberts, S. R. 1898. C. E. Beddome. *The Nautilus* 12: 80.

Robertson, R. 1969. On some molluscs collected from Southwest Ceylon during the international Indian Ocean Expedition, 1964. *Spolia Zeylanica* 31: 1–8.

Robertson, R., L. Richardson, G. M. Davis & A. E. Bogan. 1981. Catalog of the types of neontological Mollusca of the Academy of Natural Sciences of Philadelphia pt. 2 Gastropoda, Archaeogastropoda: Pleurotomariacea, Fissurellacea, Patellacea. *Tryonia* 5: 1–48.

Rochebrune, A.-T. de & J. Mabille. 1889. Mollusques. In: *Mission Scientifique du Cap Horn 1882–1883, VI*: pp. H3–H129, pls 1–8. Gauthier-Villars, Paris.

Rogers-Bennett, L., R. F. Dondanville & J. Kashiwada. 2004. Size specific fecundity of red abalone (*Haliotis rufescens*): evidence for reproductive senescence? *Journal of Shellfish Research* 23: 553–556.

Rolán, E. M. 1983. Moluscos de la Ria de Vigo I. Gasteropodos. *Thalassas* Supplement 1: 1–383 pp.

Rolán, E. M. 1991. Adiciones y correcciones taxonomicas a "Moluscos de la Ria de Vigo". *Thalassas* 9: 123–130.

Rolán, E. 1996. A new species of *Sinezona* (Gastropoda, Scissurellidae) from the Caribbean Sea. *Iberus* 14: 109–112.

Rolán, E. 2005. *Malacological Fauna from the Cape Verde Archipelago. Part 1 Pyloplacophora and Gastropoda*. ConchBooks, Hackenheim. 455 pp.

Rolán, E. & A. A. Luque. 1994. A new species of *Sinezona* (Gastropoda, Scissurellidae) from

Literature cited

the Caribbean Sea. *Iberus* 12: 1–5.

Rolán, E. & F. Rubio. 1998. The genera *Megalomphalus* and *Macromphalina* (Mollusca, Caenogastropoda, Vanikoridae) in the Caribbean area, with the description of thirteen new species. *Iberus* 16: 21–72.

Rolán, E. & P. Ryall. 1999. Checklist of the Angolan marine molluscs. *Reseñas Malacológicas* 10: 1–132.

Rosewater, J. 1975. An annotated list of the marine mollusks of Ascension Island, South Atlantic Ocean. *Smithsonian Contributions to Zoology* 189: 1–41.

Rueda, J., C. Salas & S. Gofas. 2000. A molluscan community from coastal bioclastic bottoms in the Strait of Gibraltar area. *Iberus* 18: 95–123.

Ruhoff, F. A. 1973. Bibliography and zoological taxa of Paul Bartsch. *Smithsonian Contributions to Zoology* 143: 1–166.

Ruthensteiner, B. 2008. Soft part 3D visualization by serial sectioning and computer reconstruction. *Zoosymposia* 1: 63–100.

Sabelli, B., R. Giannuzzi-Savelli, & D. Bedulli. 1990. *Annotated Check-List of Mediterranean Marine Mollusks, Volume 1.* Libreria Naturalistica Bolognese, Bologna. 348 pp.

Sacco, F. 1896. I molluschi dei terreni terziarii del Piemonte e della Liguria. Parti XIX, XX, XXI e XXII (fine dei Gasteropodi e Scafopodi). *Bollettino dei Musei di Zoologia ed Anatomia Comparata della R Università di Torino* 11: 79–98.

Salvini-Plawen, L. v. 1983. Gastropoda. In: *Fauna und Flora des Mittelmeeres. 3. Auflage.* R. Riedl (ed.): pp. 259–340. Paul Parey, Hamburg.

Salvini-Plawen, L. v. & Haszprunar, G. 1987. The Vetigastropoda and the systematics of streptoneurous Gastropoda (Mollusca). *Journal of Zoology, London* 211: 747–770.

Sandberg, L. & A. Warén. 1993. Molluscan taxa introduced by Nils Hjalmar Odhner. *Malakologische Abhandlungen Staatliches Museum für Tierkunde Dresden* 16: 117–134.

Sars, G. O. 1878. *Bidrag til Kundskaben om Noreges Arktiske fauna I. Mollusca Regionis Arctricae Norvegiae.* Christiania, Brøgger. 466, 52 pls.

Sars, G. O./M. [?] 1858. Arct. Moll. Norges. *Forhandlinger Videnskabers Selskabs* 1: [not seen].

Sars, G. O./M. [?] 1868. *Forhandlinger Videnskabers Selskabs*[not seen].

Sars, M. 1851. Beretning om en i Sommeren 1849 foretagen zoologisk reise i Lofoten og Finmaken. *Nyt Magazin for Naturvidenkaberne* 6: 121–211.

Sars, M. 1863. Geologiske og zoologiske Iagttagelser, anstillede paa en Reise i en Deel af Trondhjems Stift i Sommeren 1862. *Nyt Magazin for Naturvidenkaberne* 12: 253–272.

Sars, M. 1870. *Bidrad til Kundskab om Christianiafjordens Fauna pt 2.* Christianaia. 114 pp., 13 pls.

Sasaki, T. 1998. Comparative anatomy and phylogeny of the Recent Archaeogastropoda (Mollusca: Gastropoda). *The University Museum, University of Tokyo, Bulletin* 38: 1–224.

Sasaki, T. 2006. Molluscs collected from the intertidal zone in Manazuru, Kanagawa Prefecture, Japan. *Chiribotan* 37: 100–108.

Sasaki, T. 2008. Micromolluscs in Japan: taxonomic composition, habitats, and future topics. *Zoosymposia* 1: 147–232.

Sasaki, T., Geiger, D. L., & Okutani, T. 2010. A new species of *Anatoma* (Vetigastropoda: Anatomidae) from a hydrothermal vent field in Myojin Knoll Caldera, Izo-Ogaswara Arc, Japan. *The Veliger* 51: 63–75.

Savigny, J.-C. 1817. *Description de I'Égypte, ou Recueil des Observations et des Recherches qui ont été Faites en Égypte Pendant I'Expedition de I'Armée Française, Publiée par Ordre du Gouvernement. Histoire Naturelle. Planches, Vol. II.* Imprimérie Royale, Paris.

Scacchi, A. 1836. *Catalogus Conchyliorium Regni Neapolitani Quea Usque Adhuc Repertit.* Filiatre Sebetii, Naples, Italy. 18 pp., 1 pl.

Schepman, M. M. 1908. Prosobranchia Part I: Rhipidoglossa and Docoglossa. *Siboga-*

Expeditie 49: 1–363, 49 pls.

Schiaparelli, S., A-N. Lörz & R. Cattaneo-Vietti. 2006. Diversity and distribution of molluscs assemblages on the Victoria Land coast and the Balleny Islands, Ross Sea, Antarctica. *Antarctic Science* 18: 615–631.

Schiøtte, T. 2005. Marine, benthic molluscs common to North Greenland and the Faroes: A first investigation of a previously unrecognized, North Atlantic fauna element. *BIOFAR Proceedings* 2005: 94–108.

Schiro, G. 1986. Scissurellidae of the Mediterranean Sea. *La Conchiglia* 18: 22–23.

Schliewen, U. K. & M. Kovačić. 2008. *Didogobius amicuscaridis* spec. nov. and *D. wirtzi* spec. nov., two new species of symbiotic gobiid fish from São Tomé and Cape Verde islands. *Spixiana* 31: 247–261.

Schlotheim, E. F. 1820. *Die Petrefactenkunde auf Ihrem Jetzigen Standpunkte durch die Beschreibung Seiner Sammlung Versteinerter und Fossiler Überreste des Thier- und Pflanzenreichs der Vorwelt Erläutert.* Becker, Gotha. lxii, 437 pp.

Schnetler, K. I., P. Lozouet & J.-M. Pacaud. 2000. Revision of the gastropod family Scissurellidae from the Middle Danian (Paleocene) of Denmark. *Bulletin of the Geological Society of Denmark* 48: 79–90.

Schwarzpaul, K. 2002. *Phylogenie Hydrothermaler "Archaeogastropoden" der Tiefsee — Morphologische und Molekulare Untersuchungen.* Doctoral Dissertation. Philipps University, Marburg, Germany. 110 pp LXXXV.

Segers, W., F. Swinnen & A. Abreu. 2009a. *Satondella danieli,* a new species of Scissurellidae (Mollusca, Gastropoda, Prosobranchia) from the Archipelago of Madeira (NE Atlantic Ocean). *Bocagiana* 225: 1–6.

Segers, W., F. Swinnen & A. Abreu. 2009b. An annotated checklist of the marine molluscs from the Archipelago of Madeira and the Selvagens (NE Atlantic Ocean). *Bocagiana* 226: 1–60.

Segers, W., F. Swinnen & R. de Prins. 2009c. *Marine Molluscs of Madeira.* Snoeck Publishers, Belgium. 612 pp.

Seguenza, G. 1876. Di alcun molluschi corallig. di Messina. *ubi?* [not traceable, not seen].

Seguenza, G. 1877. La Formazioni Terziarie nella Provincia di Reggio (Calabria). *Reale Accademia Dei Lincei — Memorie Della Classe di Scienze Fisiche, Matematiche e Naturali* Series 3a, 6: 1–416, pls 1–17.

Semper, O. 1865. Additions au genre *Scissurella. Journal de Conchyliologie* Series 3, 13: 283–289, pl. 12.

Senn, D. G. 2010. *Ozeanische und Universitäre Vernetzung. Festvorlesung am 3. Juni 2010 gehalten in der Alten Univeritӓt Basel.* Vertebrate Lab Basel. 56 pp.

Sevilla-Hernández, M. L. 1998. Datos acerca de los micromolucos del arrecife isla Lobos, Veracruz, México. *Anales de la Escuela Nacional de Cincias Biológicas, México* 44: 44–56 [not seen].

Shasky, D. R. 1961. Notes on rare and little known Panamic mollusks. *The Veliger* 4: 22–24, pl. 4.

Shasky, D. R. 1989. An update on the molluscs from the Galapagos Islands as listed in the preliminary faunal list of the marine mollusks of the Galapagos Islands by Yves Finet. *Western Society of Malacologists Annual Report* 21: 8–10.

Sherborn, C. D. 1937. On the dates of publication of Costa (O. G.) and (A.) Fauna del Regno di Napoli, (1829–1886). *Journal of the Society for the Bibliography of Natural History* 1: 35–47.

Sherborn, D. & F. J. Griffin. 1934. On the dates of publication of the natural history portion of Alcide d'Orbigny's "Voyage Amérique Méridionale". *Annals and Magazine of Natural History* Series 10, 8: 130–134.

Shuto, T. 1983. New turrid taxa from Australian waters. *Memoirs of the Faculty of Science, Kyushu University, Series D, Geology* 25: 1–26, pls 1–2.

Simone, L. R. L. 2009. New gastropods from the São Pedro e São Paulo Archipelago, Brazil (Vetigastropoda and Caenogastropoda). *Strombus* 16: 11–18.

Singer, B. S. 2006. Two new marine species from the Gulf of Aqaba. *Triton* 13: 35.
Singer, B. S., H. K. Mienis & D. L. Geiger. 2000. *Clanculus korkosi* sp. nov. from the Red Sea (Gastropoda, Vetigastropoda, Trochidae). *La Conchiglia* 32(294–296): 32–38.
Skoglund, C. 1992. Additions to the Panamic Province Gastropod (Mollusca) Literature 1971–1992. *The Festivus* 24 Supplement: 1–169.
Skoglund, C. 2002. Panamic province molluscan literature addition and changes from 1971–2001: III Gastropoda. *The Festivus* 33 Supplement: i–xi, 1–286.
Smith, B. D. 2003. Prosobranch gastropods of Guam. *Micronesia* 35–36: 244–270.
Smith, E. A. 1875. Descriptions of some new shells from Kerguelen's Island. *Annals and Magazine of Natural History* Series 4, 4: 67–73.
Smith, E. A. 1879. Mollusca. *Philosophical Transactions of the Royal Society of London* 168: 167–192, pl. 9.
Smith, E. A. 1890. Report on the marine molluscan fauna of the Island of St. Helena. *Proceedings of the Zoological Society London* 1890: 247–317.
Smith, E. A. 1894. On some new species of shells from New Zealand and Australia, and remarks upon some Atlantic forms occurring in deep water of Sydney. *Proceedings of the Malacological Society of London* 1: 57–60, pl. 7.
Smith, E. A. & H. W. England. 1937. Martini & Chemnitz (Kuester Edition) Systematisches Conchylien-Cabinet, 1837–1918. *Journal of the Society for the Bibliography of Natural History* 1: 89–99.
Smith, E. A. 1910. On South African marine Mollusca with description of new species. *Annals of the Natal Museum* 2: 175–220, pls 7–8.
Smith, J. T. 1978. Primary types in the Stanford Paleontological type collection. *Bulletin of American Paleontologists* 72: 317–552.
Smith, M. 1944. *Panamic Marine Shells.* Tropical Photographic Laboratory, Winter Park, FL. 127 pp.
Smith, M. 1951. *East Coast Marine Shells, Fourth Revised Edition.* Edwards Brothers, Inc., Ann Arbor, Michigan. 314 pp.
Smith, M. 1961. *Universal Shells.* Alpine Press Inc, Asheville, NC. 254 pp.
Smith, S. M. & D. Heppell. 1991. Checklist of British Marine Mollusca. *National Museums of Scotland Information Series* 11: 1–114.
Sneli, J.-A. 1970. Archaeogastropoda from Hardangerfjorden, western Norway. *Sarsia* 42: 63–72.
Sneli, J.-A. 1974. A collection of marine Mollusca from Möre and Romsdal, northwestern Norway. *Det Kgl. Norske Videnskabers Selskap Museet, Miscellanea* 20: 1–17.
Sneli, J.-A. 1979. Prosobranchia from Sobnefjorden, western Norway. *Sarsia* 64: 259–267.
Sneli, J.-A., T. Schiøtte, K. R. Jensen. P. B. Wikander. Ø. Stokland, & J. Sørensen. 2005. The marine Mollusca of the Faroes. *Annales Societatis Scientiarum Faeroensis* Supplementum 42: 15–176.
Sohl, N. F. 1992. Upper Cretaceous gastropods (Fissurellidae, Haliotidae, Scissurellidae) from Puerto Rico and Jamaica. *Journal of Paleontology* 66: 414–434.
Solandi, A. 1789. *Testaceographiae ac Zoophytographiae Parvae et Microscopicae Tomus Primus.* Franciscus Rossi, Florence. xxx, 80 pp., 32 pls.
Soot-Ryen, T. 1924. Faunistische Untersuchungen im Ramfjorde. *Tromsø Museums Årshefter* 45: 5–106, 5 maps, tables.
Sowerby, G. B., I. 1821–1834. *The Genera of Recent and Fossil Shells, 3 Volumes.* Privately published, London. Unnumbered pages and 81 unnumbered plates.
Sowerby, G. B., I 1824. Abstract of a monograph on a new genus of gasteropodous Mollusca, named *Scissurella.* By M. Alcide Dessalines D'Orbigny. *Zoological Journal* 1: 255–258.
Sowerby, G. B., II. 1842. *Conchological Manual, Second Edition.* H. G. Bohn, London. 313 pp., 26 pls.
Sowerby, G. B., II. 1852. *Conchological Manual, Fourth Edition.* H. G. Bohn, London. 337 pp., 26 pls.

Sowerby, G. B., II 1854. *Popular British Conchology*. Lovell Reeve, London. 304 pp.

Sowerby, G. B., II. 1859. *Illustrated Index of British Shells*. Simpkin, Marshall, and Co, London. 24 pls.

Sowerby, G. B., III & Fulton. 1902. *Mollusca. A Catalogue (with Prices) of a few Species of Recent Shells*. Dryden Press, London. 38 pp.

Sowerby, G. B., III & Fulton. 1903. *Recent Mollusca. A Catalogue of Cephalopoda, Pteropoda, Gastropoda (Marine and Freshwater) and Scaphopoda*. Dryden Press, London. 58 pp.

Sparre-Schneider, J. 1886. *Underøgelser over Dyrlivet i de arktiske Fjorde* [not seen].

Spärck, R. & G. Thorson. 1933. Marine Gastropoda Prosobranchiata. *The Zoology of the Faroes* 3: 1–56 [not seen].

Spencer, H. G. & R. C. Willan. 1995. The marine fauna of New Zealand: index to the fauna: 3. Mollusca. *New Zealand Oceanographic Institute Memoir* 105: 1–125.

Spencer, H. G., B. A. Marshall & R. C. Willan. 2009. Checklist of New Zealand living Mollusca. In: *New Zealand Inventory of Biodiversity, Volume 1 Kingdom Animalia: Radiata, Lophotrochozoa, Deuterstomia*. D. P. Gordon (ed.): pp. 196–219. Canterbury University Press, Christchurch.

Sphon, G. G. 1971. Type specimens of Recent mollusks in the Los Angeles County Museum of Natural History. *Los Angeles County Museum of Natural History Contributions in Science* 213: 1–37.

Squires, R. L. & Goedert, J. L. 1996. New species of small to minute gastropods of early Eocene age from the Crescent Formation, Black Hills, Southwest Washington. *The Veliger* 39: 226–240.

Step, E. 1901. *Shell Life, an Introduction to the British Mollusca*. Frederick Warne & Co., London. 414 pp.

Stokes, D. J. 2008. *Principles and Practice of Variable Pressure/Environmental Scanning Electron Microscopy (VP-SEM)*. John Wiley, Southern Gate. 221 pp.

Strasoldo, M. 1991. *Anatomie und Ontogenie von* Scissurella jucunda *(Smith, 1890) und Anatomie von* Anatoma *sp.* Dissertation. University of Vienna, Austria, Vienna. 145 pp., 37 pls.

Strausz. L. 1955. Adatok a Várpalotai Miocén Fanuához. *Földtani Közlöny* 85: 198–210, pl. 12.

Strausz, L. 1966. *Die Miozän-Mediterranen Gastropoden Ungarns*. Akadémia Kiadó, Budapest. 692 pp., 79 pls.

Strebel, H. 1908. *Wissenschaftliche Ergebnisse der Schwedischen Südpolar-Expedition 1901–1903. Band 6, Lieferung 1. Die Gastropoden*. Asher, Berlin. 111 pp., 6 pls.

Stoppani, A. 1858–1860. *Les Pétrifications d'Ésino, Monographie des Gastéropodes Apparteneant au Dépot des pétrifications d'Ésino Paléontologie Lombarde ou Description des Fossiles de Lombardie*. J. Bernardoni, Milano. 72 pp.

Sturnay, R. 1896. Zoologische Ergebnisse. VII, Mollusken I. (Prosobanchier und Opisthobranchier: Scaphopoden: Lamellibranchier) gesammelt vom S. M. Schiff "Pola" 1890–1894. *Denkschriften der Mathematisch-Natur-wissenschaftlichen Classe der Kaiserlichen Akademie der Wissenschaften* 63: 1–36, pls 1–2.

Sturnay, R. 1905. [not seen, reference not in Yaron (1983)].

Suter, H. 1898. List of New Zealand Mollusca described in foreign publications since 1890. *Transactions of the New Zealand Institute* 31: 255–261.

Suter, H. 1902. List of the species described in F. W. Hutton's *Manual of the New Zealand Mollusca*, with the corresponding names used at the present time. *Transactions and Proceedings of the New Zealand Institute* 34: 207–224.

Suter, H. 1907. Notes on, and additions to, the New Zealand molluscan fauna. *Transactions of the New Zealand Institute* 39: 265–270.

Suter, H. 1909. The Mollusca of the Subantarctic Islands of New Zealand. In: *The Subantarctic Islands of New Zealand*. C. Chilton (ed.): pp. 6–57. John Mackay, Wellington.

Suter, H. 1910. List of Recent shells found fossil in New Zealand. *Transactions of the New Zealand Institute* 42: 8–13.

Suter, H. 1913. *A Manual of the New Zealand Mollusca, with Atlas of Quarto Plates (1915).* Government Printer, Wellington. xxiii, 1120 pp.

Suter, H. 1915. *Alphabetical Hand-List of New Zealand Tertiary Mollusca.* John Macky, Wellington. 28 pp.

Sykes, E. R. 1906. On the dates of publication of Sowerby's "Mineral Conchology" and "Genera of Recent and fossil shells". *Proceedings of the Malacological Society of London* 7: 191–194.

Szabó, J., M. A. Conti & S. Monari. 1993. Jurassic gastropods from Sicily; new data to the classification of Ataphridae (Trochoidea). *Scripta Geologica* Special Issue 2: 407–416.

Szöts, E. 1953. Mollusques Éocènes de la Hongrie I. Les mollusques Éocènes des environs de Gánt. *Geologica Hungarica, Seria Palaeontologica* 22: 1–270, pls 1–10.

Taki, I. & K. Oyama. 1954. The Pliocene and later faunas from the Kwanto Region in Japan. *Palaeontological Society of Japan Special Papers* 2: 1–68, pls 1–49.

Talmadge, R. R. 1966. Notes on the Mollusca of Prince William Sound, Alaska. *The Veliger* 9: 82–86.

Tate, R. 1868. *Appendix to the* Manual of Mollusca, *of S. P. Woodward, A. L. S.* Virtue & Co., London. 86 pp.

Tate, R. & W. L. May. 1902. A revised census of the marine Mollusca of Tasmania. *Proceedings of the Linnean Society of New South Wales* 26: 344–470, pls 23–27.

Templado, J., L. Baratech, M. Calvo, M. Villena & T. Aparicio. 1993. *Los "Ejemplares Tipo" de las Colecciones Malacologicas del Museo Nacional de Ciencias Naturales.* Museo Nacional de Ciencias Naturales, Spain. 328, 12 pls.

Ten Hove, H. A. & E. K. Kupriyanova. 2009. Taxonomy of Serpulidae (Annelida, Polychaeta): The state of affairs. *Zootaxa* 2036: 1–126.

Tenekides, N. S. 1989. *On a Collection of Shells from the Greek Seas [in Greek].* Privately published?, Athens. 187 pp.

Tenison-Woods, J. E. 1877. On some new Tasmanian marine shells. *Proceedings of the Royal Society of Tasmania* 1876: 131–159.

Tenison-Woods, J. E. 1878. Census; with brief descriptions of the marine shells of Tasmania and the adjacent islands. *Proceedings of the Royal Society of Tasmania* 1877: 26–57.

Terreni, G. 1981. *Molluschi Conchiferi del Mare Antistante La Costa Toscana.* Museo di Storia Naturale della Provincia di Livorno, Livorno. 100 pp.

Thacker, C. 2004. Phylogeny and species boundaries in the gobiid genus *Gnatholepis* (Teleostei: Perciformes). *Zoological Journal of the Linnean Society* 142: 573–582.

Thacker, C. E. & H. Grier. 2005. Unusual gonad structure in the paedomorphic teleost *Schindleria praematura* (Teleostei: Gobioidei): a comparison with other gobioid fishes. *Journal of Fish Biology* 66: 378–391.

The Ex. 2004. *Turn.* Ex Music/Touch and Go, Chicago. 2 CDs.

Thiele, J. H. 1891. *Das Gebiss der Schnecken, Second Volume, Delivery* 7. Nicolaische Verlags-Buchhandlung, Berlin. 251–334, 6 pls.

Thiele, J. H. 1895. Zur Phylogenie der Gastropoden. *Biologisches Centralblatt 15: 220–236.*

Thiele, J. H. 1912a. Scissurelliden und Fissurelliden. In: *Systematisches Conchylien-Cabinet von Martini und Chemnitz.* H. C. Kobelt & W. Küster (eds): pp. 1–34, pls 1–4. Bauer & Raspe, Nürnberg.

Thiele, J. H. 1912b. Die antarktischen Schnecken und Muscheln. In: *Deutsche Südpolar-Expedition 1901–1903.* E. von Drygalsky (ed.): pp. 185–285, pls 11–19. Georg Reimer, Berlin.

Thiele, J. H. 1925. Gastropoda der Deutschen Tiefsee-Expedition. II. Teil. In: *Wissenschaftliche Ergebnisse der Deutschen Tiefsee-Expedition auf dem Dampfer "Valdivia" 1898–1899.* C. Chun (ed.): pp. 1–348, pls 1–34. Gustav Fischer, Berlin.

Thiele, J. H. 1931. *Handbuch der Systematischen Weichtierkunde.* Gustav Fischer, Jena. 1153 pp.

Thorpe, C. 1844. *British Marine Conchology.* Edward Lumley, London. L, 7 pls, 267 pp.

Thorson, G. 1941. Marine Gastropoda Prosobranchiata. *The Zoology of Iceland* 4(60): 1–150.

Thorson, G. 1944. Marine Gastropoda Prosobranchiata. In: the Zoology of East Greenland. *Meddeleleser om Grønland* 121: 1–181 [not seen].

Tichy, G. 1970. Typen-Katalog: Verzeichnis der in der Geologisch-Paläontologischen Abteilung des Naturhistorischen Museums in Wien aufbewahrten Typen sowie der Abbildungsorginale. *Annalen des Naturhistorischen Museums Wien* 74: 607–655.

Traub, F. 1984. Weitere Paläozän-Gastropoden aus dem Helveticum des Hansberges nördlich von Salzburg 3. Fortsetzung. *Mitteilungen der Bayerischen Staatssammlung für Paläontologie und Historische Geologie* 24: 3–26.

Trew, A. 1983. *The Melvill-Tomlin Collection, Parts 19, 20 and 21, Pleurotomariacea, Fissurellacea and Patellacea.* Handlists of the Molluscan Collections in the Department of Zoology, National Museum of Wales, Cardiff. 94 pp.

Tröndlé, J. 1986. Premières données en écologie et faunistique sur la microfauna malacologique de Tahiti (Société - Polynésie française). *Haliotis* 15: 61–72.

Tröndlé, J. & M. Boutet. 2009. Inventory of marine molluscs of French Polynesia. *Atoll Research Bulletin* 570: 1–86.

Troschel, F. H. 1863. Bericht über die Leistungen in der Naturgeschichte der Mollusken während des Jahres 1862. *Archiv für Naturgeschichte* 29: 678–719.

Tryon, G. W. 1883. *Structural and Systematic Conchology: an Introduction to the Study of the Mollusca. Vol. 2.* Tryon, Philadelphia. 430 pp., 91 pls.

Tsuchida, E. & S. Hori. 1996. Marine mollusks around Mishima and Tsunoshima Islands, Japan Sea collected by the R/V *Tansei-Maru*. *Bulletin of the National Science Museum Series A (Zoology)* 22: 219–261.

Tsuchida, E., Y. Shikano, S. Hori & T. Mitoki. 1991. Study on the Mollusca of Yamaguchi Pref. — 3 Revision of the remarkable molluscan shells from Yamaguchi Pref. in Mr. T. Kawamoto's Collection. (2) Gastropoda and Pelecypoda. *Bulletin of the Yamaguchi Museum* 17: 1–40.

Tucker, J. K. 2004. Catalog of Recent and fossil turrids (Mollusca: Gastropoda). *Zootaxa* 682: 1–1295.

Tunnell, J. W., J. C. Woods, M. E. Kindinger & J. L. Kindinger. 1978. *Fauna of Shelf-Edge Submarine Banks in the Northwestern Gulf of Mexico.* Report to the U. S. Geological Survey, Office of Marine Geology in fulfillment of contract 14-08-001-G-381: 1–49, 4 tables.

Tunnell, J. W. Jr., J. Andrews, N. C. Barrera, & F. Moretzsohn. 2010. *Encyclopedia of Texas Seashells: Identification, Ecology, Distribution, and History.* Harte Research Institute for Gulf of Mexico Studies Series.

Turner, R. D., R. A. Lutz & D. Jablonski. 1985. Modes of molluscan larval development at deep-sea hydrothermal vents. *Biological Society of Washington Bulletin* 6: 167–184.

Turton, W. H. 1932. *The Marine Shells of Port Alfred S. Africa.* Oxford University Press, London. 331 pp., 70 pls.

Van der Bijl, A. N., R. G. Moolenbeek & J. Goud. 2010. *Mattheus Marinus Schepman (1847–1919) and his Contributions to Malacology.* Netherlands Malacological Society, Leiden. 200 pp.

van der Land, J. 1987. Report on the CANCAP-project for marine biological research in the Canarian-Cape Verde Region of the North Atlantic Ocean (1976–1986). Part I. List of Stations. *Zoolgische Verhandelingen, Leiden* 243: 1–94.

Vávra, N. 2002. August Emanuel Reuss — portrait of an early bryozoologist. In: *Annals of Bryozoology: Aspects of the History of Research on Mryozoans.* P. N. Wyse Jackson & M. E. Spencer Jones (eds): pp. 327–338. International Bryozoa Association, Dublin.

Vayssière, A. 1894. Étude zoologique de la *Scissurella costata*, var. *laevigata*. *Journal de Conchyliologie* 42: 19–29, pl. 2.

Vazzana, A. 1996. Malacofauna batiale del Pleistocene Inferiore del Vallone Catrica (Reggio Calabria, Italia). *Bolletino Malacologico* 31: 143–162.

Vazzana, A. 2010. La malacofauna del circalitorale di Scilla (Streto di Messina). *Bolletino Malacologico* 46: 66–75.

Vélain, C. 1876. Sur la faune malacologique des îles Saint-Paul et Amsterdam. *Comptes Rendus Hébdomadaires des Séances de l'Académie des Sciences* 83: 284–287.

Vélain, M. C. 1877. Passage de Vénus sur le soleil (9 Décembre 1874) Expédition française aux Îles Saint-Paul et Amsterdam: Zoologie. *Archives de Zoologie Expérimentale et Générale* 6: 1–144, pls 1–5.

Verco, J. C. 1910. Notes on South Australian marine Mollusca, with descriptions of new species.—Part XIII. *Transactions of the Royal Society of South Australia* 34: 115–145, pls 29–30.

Verco, J. C. 1912a. Notes on South Australian marine Mollusca, with description of new species.–Part XV. *Transactions of the Royal Society of South Australia* 36: 181–191.

Verco, J. C. 1912b. Notes on the marine shells of Western Australia, with descriptions of new species. *Transactions and Proceedings of the Royal Society of South Australia* 36: 192–205, pls 15–16.

Verrill, A. E. 1882. Catalogue of marine Mollusca added to the fauna of the New England region, during the past ten years. *Transactions of the Connecticut Academy of Arts and Sciences* 5: 447–587, pls 57–58.

Verrill, A. E. & K. J. Bush. 1900. Additions to the marine Mollusca of the Bermudas. *Transactions of the Connecticut Academy of Arts and Sciences* 10: 513–544, pls 63–67.

Vokes, H. E. & E. H. Vokes. 1983. *Distribution of Shallow-Water Marine Mollusca, Yucatan Peninsula, Mexico*. Middle American Research Institute Publication 58. Tulane University, Tulane, Oklahoma. 183 pp.

Walsby, J. & J. Morton. 1982. *Marine Molluscs. Part I Amphineura, Archaeogastropoda and Pulmonata*. Leigh Marine Laboratory, Leigh. 89 pp.

Warén, A. 1980. Marine Mollusca described by John Gwyn Jeffreys, with the location of the type material. *Conchological Society of Great Britain and Ireland*, Special Publication 1: 1–60, pls 1–8.

Warén, A. 2011. Molluscs of biogenic substrates. In: *The Natural History of Santo*. Bouchet, P., H. Le Guyader & O. Pascal (eds.): pp. 438–448. IRD, Paris.

Warén, A. & P. Bouchet, P. 1989. New gastropods from East Pacific hydrothermal vents. *Zoologica Scripta* 18: 67–102.

Warén, A. & P. Bouchet. 2001. Gastropoda and Monoplacophora from hydrothermal vents and seeps; new taxa and records. *The Veliger* 44: 116–231.

Warén, A., P. Bouchet & R. von Cosel. 2006. *Sutilizona, Temnocinclis euripes, Temnozaga parilis*. In: *Handbook of Deep-Sea Hydrothermal Vent Fauna, Second Completely Revised Edition. Densia* 18. D. Desbruières, M. Segonzac & M. Bright (eds): pp. 98–100. Land Oberöstrerreich, Linz.

Warmke, G. L. & L. R. Almodóvar. 1963. Some associations of marine mollusks and algae in Puerto Rico. *Malacologia* 1: 163–177.

Watanabe, T. 1988. A catalogue of marine Mollusca from the district of Choshi: pp. 1–96, pls 1–15. In: Watanabe, T. (ed.) *A Catalogue of the Mollusca of Chochi*.

Watson, R. B. 1897. On the marine Mollusca of Madeira; with description of thirty-five new species, and an index-list of all the known sea-dwelling species of that Island. *Journal of the Linnean Society (Zoology)* 26: 233–329, pls 19–20.

Watson, R. B. 1885–1886. Report on the Scaphopoda and Gastropoda collected by *H. M. S. Challenger* during the years 1873–1876. *Report of the Scientific Results of the Voyage of* H.M.S. Challenger. *Zoology* 15(II): i–v, 1–756, 59 pls + 3 pls.

Weinkauff, H. C. 1862. Catalogue des coquilles marines recueillies sur les côtes de l'Algérie. *Journal de Conchyliologie* 10: 301–371.

Weinkauff, H. C. 1868. *Die Conchylien des Mittelmeeres, ihre Geographische und Geologische Verbreitung. Band II Mollusca Cephala*. Theodor Fischer, Cassel. 512 pp.

Weinkauff, H. C. 1973. *Catalog der im Europäischen Faunengebiet Lebenden Meeresconchylien*. R. Voigtländer, Creuznach. i–xvi, 86 pp.

Wenz, W. 1961. Gastropoda. In: *Handbuch der Paläozoologie* 6(1). O. H. Schindewolf (ed.): pp. 1–471. Gebrüder Borntraeger, Berlin.

Wesselingh, F. P. & R. Pouwer. 2011. De fossiele schelpen van de Nederlandse kust II. Patellogastropoda en Vetigastropoda (deel 1). *Spirula* 383: 129–142.

Williams, S. T. & T. Ozawa. 2006. Molecular phylogeny suggests polyphyly of both the turban shells (family Turbinidae) and the superfamily Trochoidea (Mollusca: Vetigastropoda). *Molecular Phylogenetics and Evolution* 39: 33–51.

Williams, S. T., S. Karube & T. Ozawa. 2008. Molecular systematics of Vetigastropoda: Trochidae, Turbinidae and Trochoidea redefined. *Zoologica Scripta* 37: 483–506.

Wilson, B. 1993. *Australian Marine Shells, Volume 1.* Odyssey, Kallaroo, Western Australia. 408 pp.

Wilson, E. C. & G. L. Kennedy. 1967. Type specimens of Recent invertebrates (except Arachnida and Insecta) in the San Diego Natural History Museum. *Transactions of the San Diego Society of Natural History* 14: 237–280.

Winckworth, M. A. 1932. The British marine Mollusca. *Journal of Conchology* 19: 211–252.

Winckworth, R. 1936. *Journal de Conchyliologie*: dates of publication. *Proceedings of the Malacological Society of London* 22: 153–156.

Wood, S. V. 1842. A catalogue of shells from the Crag (concluded). *Annals and Magazine of Natural History* 9: 527–544, pl. 5.

Wood, S. V. 1848. *A Monograph of the Crag Mollusca or Descriptions of Shells from the Middle and Upper Tertiaries of the East of England. Part I. Univalves.* Paleontographical Society, London. 208 pp., 21 pls.

Woodward, S. P. 1851. *Manual of the Mollusca; or a Rudimentary Treatise of Recent and Fossil Shells.* John Weale, London. 158 pp., 12 pls, 12 pp.

Woodward, S. P. 1856a. On the evils of increasing synonyms. *Annals and Magazine of Natural History* Series 2, 18: 36–37.

Woodward, S. P. 1856b. Another note on *Scissurella*. *Annals and Magazine of Natural History* Series 2, 17: 401–402.

Woodward, S. P. 1859. On a new species of mollusk of the genus *Scissurella* d'Orb. *Proceedings of the Zoological Society of London* 27: 202–204, pl. 46.

Woodward, S. P. 1866. *Manual of the Mollusca: a Treatise of Recent and Fossil Shells, Second Edition.* Virtue Brothers, London. 519 pp., 23 pls.

Woodward, S. P. 1870. *Manuel de Conchyliologie ou Histoire Naturelle des Mollusques Vivantes et Fossiles par S. P. Woodward; augmenté d'un appendice par Ralph Tate; traduit de l'anglais sur la deuxième édition par Aloïs Humbert.* F. Savy, Paris. iii, 656 pp., 23 pls.

Yamada, M. *Biyoai.* [http://shell.kwansei.ac.jp/~shell/pic_book/data36/r003590.html accessed 2/2004].

Yaron, I. 1983. A review of the Scissurellidae (Mollusca, Gastropoda) of the Red Sea. *Annalen des Naturhistorischen Museum Wien* 84/B: 263–279.

Yokoyama, M. 1924. Mollusca from the coral-bed of Awa. *Journal of the College of Science, Tokyo Imperial University* 45: 1–62, i–iii, pls 1–5.

Yoon, S. H. & W. Kim. 2005. Phylogenetic relationships among six vetigastropod subgroups (Mollusca, Gastropoda) based on 18S rDNA sequences. *Molecules and Cells* 19: 283–288.

Yu, W. & W.-M. Feng. 1996. Holocene gastropods from the Nansha Sea area, China. In: *Quaternary Biological Groups of the Nansha Islands and the Neighbouring Waters.* K. K. H. Y. Chung (ed.): pp. 355–434. Zhongshan University Publishing House, China.

Zardini, R. 1978. *Fossili Cassiani (Trias medio-superiore).* Edizioni Ghedina, Cortina d'Ampezzo, Italy. 58 pp., 41 pls, 1 map.

Zardini, R. 1978. *Fossili Cassiani (Trias Medio-superiore) Atlante dei Gasteropodi della Formazione di S. Cassinao Raccolti Nella Regione Dolomitica Attorno a Cortina d'Ampezzo.* Ghedina, Cortina d'Ampezzo. 58 pp., 41 pls.

Zardini, R. 1980. *Fossili Cassiani (Trias medio-superiore).* Edizioni Ghedina, Cortina d'Ampezzo, Italy. 17 pp., 6 pls.

Zelaya, D. G. 2010. Gastropoda (gastropods). In: *Marine Benthic Fauna of Chilean Patagonia; Illustrated Identification Guide*. V. Häussermann & G. Försterra (eds.): pp. 461–504. Nature in Focus, Chile.

Zelaya, D. G. & D. L. Geiger. 2007. Species of Scissurellidae and Anatomidae from sub-antarctic and antarctic waters (Gastropoda: Vetigastropoda). *Malacologia* 49: 393–443.

Zelaya, D. G. & D. L. Geiger. 2008. *Scissurella clathrata* in the San Jorge Gulf (Argentina) and the identity of the enigmatic *S. dalli* Bartsch, 1903 (Gastropoda: Scissurellidae). *Zoosymposia* 1: 265–269.

Zelaya, D. G. & S. Gordillo. 2011. Gastropods from raised marine deposits along the Beagle Channel, southern Argentina: the ancestors of the living fauna. *Marine Biodiversity Records* 4(e64): 1-20.

Ziegelmeier, E. 1966. Die Schnecken (Gastropoda Prosobranchia) der deutschen Meeresgebiete und brackigen Küstengewässer. *Helgoländer Meeresuntersuchungen* 13: 1–61.

Zilch, A. 1934. Zur Fauna des Mittel-Miocäns von Kostej (Banat). *Senkenbergiana* 16: 193–302, pls 1–22.

Zuschin, M., R. Janssen & C. Baal. 2009. Gastropods and their habitats from the northern Red Sea (Egypt: Safaga) Part 1: Patellogastropoda, Vetigastropoda and Cycloneritimorpha. *Annalen des Naturhistorischen Museums Wien* 111A: 73–158.

Index

All species are listed both alphabetically by species epithet, and under the respective genus.

Symbols

A

B

Index

C

D

Index

J

K

L

M

Index

N

O

P

Q

R

S

Index

T

Index